CONVERSION FACTORS

Length

1 m = 39.37 in. = 3.281 ft
1 in. = 2.54 cm
1 km = 0.621 mi
1 mi = 5 280 ft = 1.609 km
1 light year (ly) = 9.461×10^{15} m
1 angstrom (Å) = 10^{-10} m

Mass

1 kg = 10^3 g = 6.85×10^{-2} slug
1 slug = 14.59 kg
1 u = 1.66×10^{-27} kg = 931.5 MeV/c^2

Time

1 min = 60 s
1 h = 3 600 s
1 day = 8.64×10^4 s
1 yr = 365.242 days = 3.156×10^7 s

Volume

1 L = 1 000 cm^3 = 3.531×10^{-2} ft^3
1 ft^3 = 2.832×10^{-2} m^3
1 gal = 3.786 L = 231 in.^{3}

Angle

180° = π rad
1 rad = 57.30°
1° = 60 min = 1.745×10^{-2} rad

Speed

1 km/h = 0.278 m/s = 0.621 mi/h
1 m/s = 2.237 mi/h = 3.281 ft/s
1 mi/h = 1.61 km/h = 0.447 m/s = 1.47 ft/s

Force

1 N = 0.224 8 lb = 10^5 dynes
1 lb = 4.448 N
1 dyne = 10^{-5} N = 2.248×10^{-6} lb

Work and energy

1 J = 10^7 erg = 0.738 ft · lb = 0.239 cal
1 cal = 4.186 J
1 ft · lb = 1.356 J
1 Btu = 1.054×10^3 J = 252 cal
1 J = 6.24×10^{18} eV
1 eV = 1.602×10^{-19} J
1 kWh = 3.60×10^6 J

Pressure

1 atm = 1.013×10^5 N/m^2 (or Pa) = 14.70 lb/in.^{2}
1 Pa = 1 N/m^2 = 1.45×10^{-4} lb/in.^{2}
1 lb/in.^{2} = 6.895×10^3 N/m^2

Power

1 hp = 550 ft · lb/s = 0.746 kW
1 W = 1 J/s = 0.738 ft · lb/s
1 Btu/h = 0.293 W

College Physics

ENHANCED seventh edition

Raymond A. Serway
Emeritus, James Madison University

Jerry S. Faughn
Emeritus, Eastern Kentucky University

Chris Vuille
Embry-Riddle Aeronautical University

Charles A. Bennett
The University of North Carolian, Asheville
Contributing Author, Media

Australia · Canada · Mexico · Singapore · Spain · United Kingdom · United States

Physics Acquisitions Editor: Chris Hall
Publisher: David Harris
Vice President, Editor-in-Chief, Sciences: Michelle Julet
Development Editor: Ed Dodd
Assistant Editor: Annie Mac
Editorial Assistant: Kyra Engelberg
Technology Project Manager: Sam Subity
Marketing Manager: Erik Evans, Julie Conover
Marketing Assistant: Leyla Jowza
Advertising Project Manager: Stacey Purviance
Project Manager, Editorial Production: Teri Hyde
Print/Media Buyer: Judy Inouye
Permissions Editor: Audrey Pettengill
Production Service: Progressive Publishing Alternatives
Text Designer: Patrick Devine
Photo Researcher: Jane Sanders
Copy Editor: Progressive Publishing Alternatives
Illustrator: Rolin Graphics, Progressive Information Technologies
Cover Designer: Patrick Devine
Cover Image: ML Sinibaldi/CORBIS
Cover Printer: Phoenix Color Corp., MD
Compositor: Progressive Information Technologies
Printer: R.R. Donnelley/Willard

Printed in the United States of America
1 2 3 4 5 6 7 09 08 07 06

Thomson Brooks/Cole
10 Davis Drive
Belmont, CA 94002
USA

Library of Congress Control Number: 2004113839

Student Edition: ISBN 0-495-11369-7
Instructor's Edition: ISBN 0-534-99724-4
International Student Edition: ISBN 0-534-49318-1

Asia
Thomson Learning
5 Shenton Way #01-01
UIC Building
Singapore 068808

Australia/New Zealand
Thomson Learning
102 Dodds Street
Southbank, Victoria 3006
Australia

Canada
Nelson
1120 Birchmount Road
Toronto, Ontario M1K 5G4
Canada

Europe/Middle East/Africa
Thomson Learning
High Holborn House
50/51 Bedford Row
London WC1R 4LR
United Kingdom

Latin America
Thomson Learning
Seneca, 53
Colonia Polanco
11560 Mexico D.F.
Mexico

Spain/Portugal
Paraninfo
Calle Magallanes, 25
28015 Madrid, Spain

Welcome to PhysicsNow, your fully integrated system for physics tutorials and self-assessment on the web. To get started, just follow these simple instructions.

Your first visit to PhysicsNow

1. Go to **http://www.cp7e.com** and click the **Register** button.
2. The first time you visit, you will be asked to select your school. Choose your state from the drop-down menu, then type in your school's name in the box provided and click **Search**. A list of schools with names similar to what you entered will show on the right. Find your school and click on it.
3. On the next screen, enter the access code from the card that came with your textbook in the "Content or Course Access Code" box*. Enter your email address in the next box and click **Submit.**

   * **PhysicsNow** access codes may be purchased separately. Should you need to purchase an access code, go back to **http://www.cp7e.com** and click the **Buy** button.
4. On the next screen, choose a password and click **Submit.**
5. Lastly, fill out the registration form and click **Register and Enter iLrn.** This information will only be used to contact you if there is a problem with your account.
6. You should now see the **PhysicsNow** homepage. Select a chapter and begin!

   Note: Your account information will be sent to the email address that you entered in Step 3, so be sure to enter a valid email address. You will use your email address as your username the next time you login.

Second and later visits

1. Go to **http://www.cp7e.com** and click the **Login** button.
2. Enter your user name (the email address you entered when you registered) and your password and then click **Login.**

SYSTEM REQUIREMENTS:
(Please see the System Requirements link at* www.ilrn.com *for complete list.)
PC: Windows 98 or higher, Internet Explorer 5.5 or higher
Mac: OS X or higher, Mozilla browser 1.2.1 or higher

TECHNICAL SUPPORT:
For online help, click on **Technical Support** in the upper right corner of the screen, or contact us at:

1-800-423-0563 Monday–Friday • 8:30 A.M. to 6:00 P.M. EST

tl.support@thomson.com

Turn the page to learn more about PhysicsNow and how it can help you achieve success in your course!

THOMSON
BROOKS/COLE

What do you need to learn now?

Take charge of your learning with **PhysicsNow™**, a powerful student-learning tool for physics! This interactive resource helps you gauge your unique study needs, then gives you a *Personalized Learning Plan* that will help you focus in on the concepts and problems that will most enhance your understanding. With **PhysicsNow**, you have the resources you need to take charge of your learning!

The access code card included with this new copy of ***College Physics*** is your ticket to all of the resources in **PhysicsNow.** (See the previous page for login instructions.)

Interact at every turn with the POWER and SIMPLICITY of PhysicsNow!

PhysicsNow combines Serway and Faughn's best-selling *College Physics* with carefully crafted media resources that will help you learn. This dynamic resource and the Seventh Edition of the text were developed in concert, to enhance each other and provide you with a seamless, integrated learning system.

Physics Now™

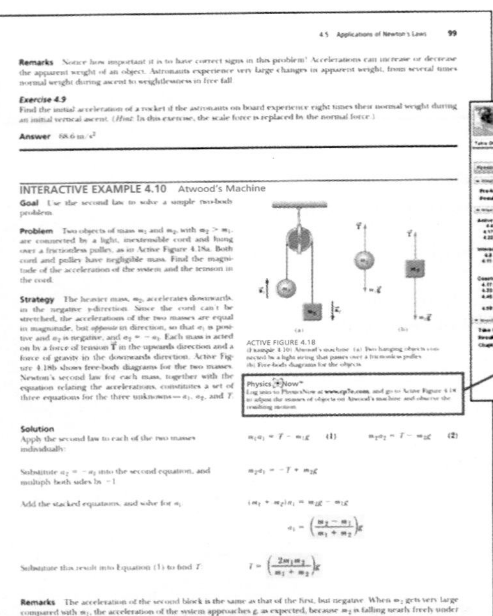

As you work through the text, you will see notes that direct you to the media-enhanced activities in PhysicsNow. This precise page-by-page integration means you'll spend less time flipping through pages or navigating websites looking for useful exercises. These multimedia exercises will make all the difference when you're studying and taking exams . . . after all, it's far easier to understand physics if it's seen in action, and PhysicsNow enables you to become a part of the action!

Begin at http://www.cp7e.com and build your own Personalized Learning Plan now!

Log into PhysicsNow at **http://www.cp7e.com** by using the free access code packaged with the text. You'll immediately notice the system's simple, browser-based format. You can build a complete *Personalized Learning Plan* for yourself by taking advantage of all three powerful components found on PhysicsNow:

- **What I Know**
- **What I Need to Learn**
- **What I've Learned**

The best way to maximize the system and optimize your time is to start by taking the *Pre-Test* ▶▶▶

PhysicsNow™ Quick Start Guide

What I Know

▲ You take a *Pre-Test* to measure your level of comprehension after reading a chapter. Each *Pre-Test* includes approximately 15 questions. The *Pre-Test* is your first step in creating your custom-tailored *Personalized Learning Plan.*

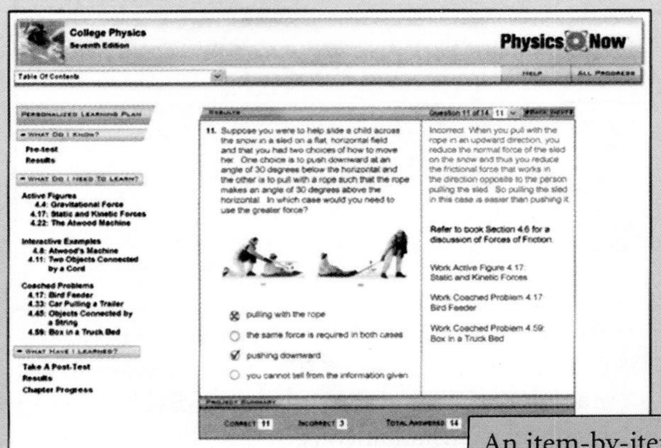

An item-by-item analysis gives you feedback on each of your answers.

▲ Once you've completed the "What I Know" *Pre-Test,* you are presented with a detailed *Personalized Learning Plan,* with text references that outline the elements you need to review in order to master the chapter's most essential concepts. This roadmap to concept mastery guides you to exercises designed to improve skills and to increase your understanding of the basic concepts.

At each stage, the *Personalized Learning Plan* refers to ***College Physics*** to reinforce the connection between text and technology as a powerful learning tool.

What I Need to Learn

Once you've completed the *Pre-Test,* you're ready to work through tutorials and exercises that will help you master the concepts that are essential to your success in the course.

ACTIVE FIGURES

A remarkable bank of more than 200 animated figures helps you visualize physics in action. Taken straight from illustrations in the text, these *Active Figures* help you master key concepts from the book. By interacting with the animations and accompanying quiz questions, you come to an even greater understanding of the concepts you need to learn from each chapter. ▼

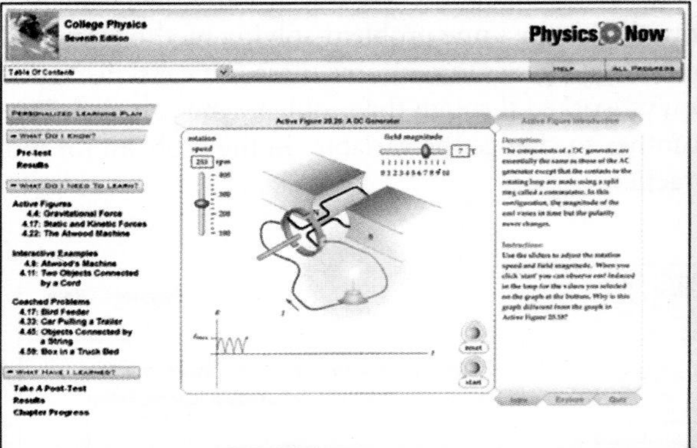

Each figure is titled so you can easily identify the concept you are seeing. The final tab features a *Quiz*. The *Explore* tab guides you through the animation so you understand what you should be seeing and learning.

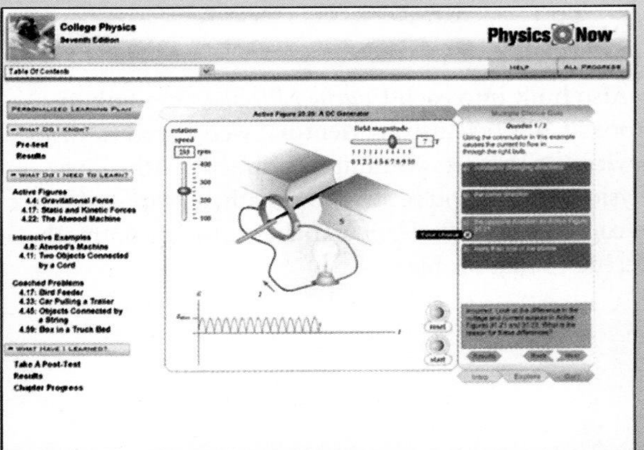

▲ The brief *Quiz* ensures that you mastered the concept played out in the animation—and gives you feedback on each response.

Continued on the next page

COACHED PROBLEMS

Engaging *Coached Problems* reinforce the lessons in the text by taking a step-by-step approach to problem-solving methodology. Each *Coached Problem* gives you the option of breaking down a problem from the text into steps with feedback to 'coach' you toward the solution. There are approximately five *Coached Problems* per chapter.

You can choose to work through the *Coached Problems* by inputting an answer directly or working in steps with the program. If you choose to work in steps, the problem is solved with the same problem-solving methodology used in ***College Physics*** to reinforce these critical skills. Once you've worked through the problem, you can click **Try Another** to change the variables in the problem for more practice.

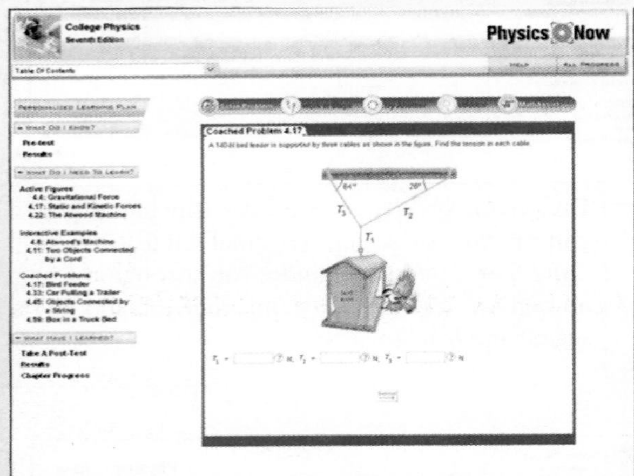

▲ Also built into each *Coached Problem* is a link to Brooks/Cole's exclusive **vMentor**™ web-based tutoring service site that lets you interact directly with a live physics tutor. If you're stuck on math, a *MathAssist* link on each *Coached Problem* launches tutorials on math specific to that problem.

INTERACTIVE EXAMPLES

You'll strengthen your problem-solving and visualization skills with *Interactive Examples.* Extending selected examples from the text, *Interactive Examples* utilize the proven and trusted problem-solving methodology presented in ***College Physics.*** These animated learning modules give you all the tools you need to solve a problem type—you're then asked to apply what you have learned to different scenarios. You will find approximately two *Interactive Examples* for each chapter of the text. ▼

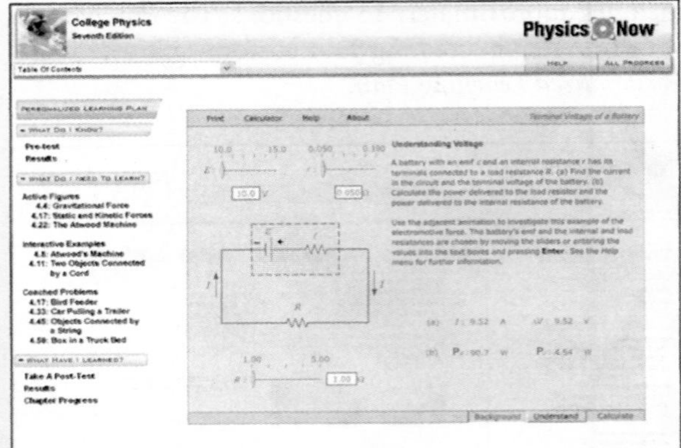

You're guided through the steps to solve the problem and then asked to input an answer in a simulation to see if your result is correct. Feedback is instantaneous.

What I've Learned

► After working through the problems highlighted in your *Personalized Learning Plan,* you move on to a *Post Test,* about 15 questions per chapter.

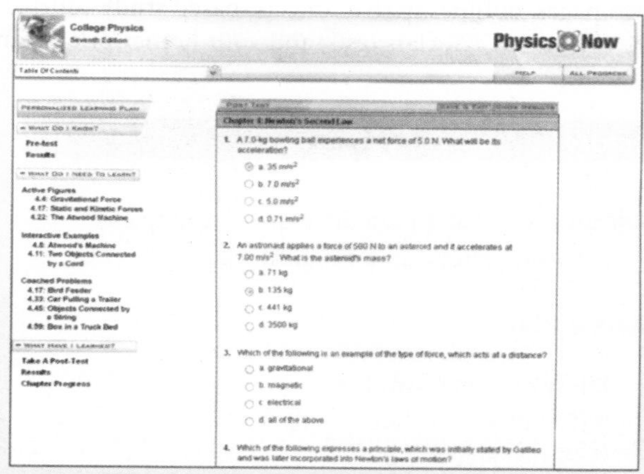

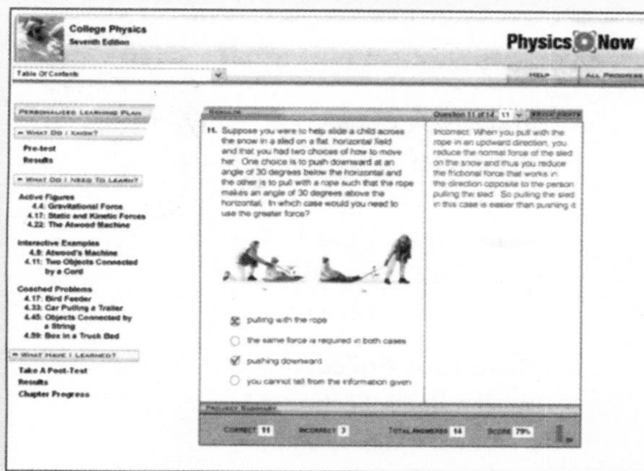

◄ Once you've completed the *Post Test,* you receive your percentage score and specific feedback on each answer. The *Post Tests* give you a new set of questions with each attempt, so you can take them over and over as you continue to build your knowledge and skills and master concepts.

Also available to help you succeed in your course

Student Solutions Manual and Study Guide

Volume I (Ch. 1–14) ISBN: 0-534-99920-4
Volume II (Ch. 15–30) ISBN: 0-534-99930-1

These manuals contain detailed solutions to approximately 12 problems per chapter. These problems are indicated in the textbook with boxed problem numbers. Each manual also features a skills section, important notes from key sections of the text, and a list of important equations and concepts.

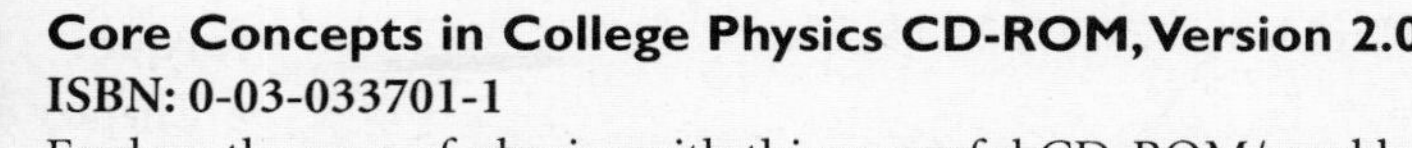

Core Concepts in College Physics CD-ROM, Version 2.0

ISBN: 0-03-033701-1

Explore the core of physics with this powerful CD-ROM/workbook program! Content screens provide in-depth coverage of abstract and often difficult principles, building connections between physical concepts and mathematics. The presentation contains more than 350 movies—both animated and live video—including laboratory demonstrations, real-world examples, graphic models, and step-by-step explanations of essential mathematics. An accompanying workbook contains practical physics problems directly related to the presentation, along with worked solutions. Package includes three discs and a workbook.

Welcome to your MCAT Test Preparation Guide

The MCAT Test Preparation Guide makes your copy of *College Physics,* **Seventh Edition,** the most comprehensive MCAT study tool and classroom resource in introductory physics. The grid, which begins below and continues on the next two pages, outlines twelve concept-based **study courses** for the physics part of your MCAT exam. Use it to prepare for the MCAT, class tests, and your homework assignments.

Vectors

Skill Objectives: To calculate distance, angles between vectors, and magnitudes.

Review Plan:

Distance and Angles:
- Chapter 1, Sections 1.7, 1.8
- Active Figure 1.6
- Chapter Problems 35, 39, 42

Using Vectors:
- Chapter 3, Sections 3.1, 3.2
- Quick Quizzes 3.1–3.3
- Examples 3.1–3.3
- Active Figure 3.3
- Chapter Problems 4, 11, 19

Force

Skill Objectives: To know and understand Newton's Laws, to calculate resultant forces, and weight.

Review Plan:

Newton's Laws:
- Chapter 4, Sections 4.1–4.4
- Quick Quizzes 4.1–4.5
- Examples 4.1–4.4
- Active Figure 4.6
- Chapter Problems 5, 7, 11

Resultant Forces:
- Chapter 4, Section 4.5
- Quick Quizzes 4.6, 4.7
- Examples 4.7, 4.9, 4.10
- Chapter Problems 17, 25, 33

Motion

Skill Objectives: To understand motion in two dimensions, to calculate speed and velocity, centripetal acceleration, and acceleration in free fall problems.

Review Plan:

Motion in 1 Dimension:
- Chapter 2, Sections 2.1–2.6
- Quick Quizzes 2.1–2.8
- Examples 2.1–2.10
- Active Figure 2.15
- Chapter Problems 1, 8, 17, 21, 29, 33, 38, 47, 51

Motion in 2 Dimensions:
- Chapter 3, Sections 3.3, 3.4
- Quick Quizzes 3.4–3.7
- Examples 3.4–3.8
- Active Figures 3.14, 3.15
- Chapter Problems 27, 33

Centripetal Acceleration:
- Chapter 7, Section 7.4
- Quick Quizzes 7.6, 7.7
- Example 7.6

Equilibrium

Skill Objectives: To calculate momentum and impulse, center of gravity, and torque.

Review Plan:

Momentum:
- Chapter 6, Sections 6.1–6.3
- Quick Quizzes 6.2–6.6
- Examples 6.1–6.4, 6.6
- Active Figures 6.7, 6.10, 6.13
- Chapter Problems 7, 16, 21

Torque:
- Chapter 8, Sections 8.1–8.4
- Examples 8.1–8.7
- Chapter Problems 5, 9

Work

Skill Objectives: To calculate friction, work, kinetic energy, potential energy, and power.

Review Plan:

Friction:
- Chapter 4, Section 4.6
- Quick Quizzes 4.8–4.10
- Active Figure 4.19

Work:
- Chapter 5, Section 5.1
- Quick Quiz 5.1
- Example 5.1
- Active Figure 5.5
- Chapter Problems 11, 17

Energy:
- Chapter 5, Sections 5.2, 5.3
- Examples 5.4, 5.5
- Quick Quizzes 5.2, 5.3

Power:
- Chapter 5, Section 5.6
- Examples 5.12, 5.13

Matter

Skill Objectives: To calculate pressure, density, specific gravity, and flow rates.

Review Plan:

Properties:
- Chapter 9, Sections 9.1–9.3
- Quick Quiz 9.1
- Examples 9.1, 9.3, 9.4
- Active Figure 9.3
- Chapter Problem 7

Pressure:
- Chapter 9, Sections 9.3–9.6
- Quick Quizzes 9.2–9.6
- Examples 9.4–9.9
- Active Figures 9.19, 9.20
- Chapter Problems 13, 23, 39

Flow rates:
- Chapter 9, Sections 9.7, 9.8
- Quick Quiz 9.7
- Examples 9.11–9.14
- Chapter Problem 42

Waves

Skill Objectives: To understand interference of waves, to calculate basic properties of waves, properties of springs, and properties of pendulums.

Review Plan:

Wave Properties:
- Chapters 13, Sections 13.1–13.4, 13.7–13.11
- Quick Quizzes 13.1–13.6
- Examples 13.1, 13.6, 13.8–13.10
- Active Figures 13.1, 13.8, 13.12, 13.13, 13.24,
- 13.26, 13.32, 13.33, 13.34, 13.35
- Chapter Problems 9, 15, 23, 29, 41, 47, 53

Pendulum:
- Chapter 13, Sections 13.5
- Quick Quizzes 13.7–13.9
- Example 13.7
- Active Figures 13.15, 13.16
- Chapter Problem 35

Sound

Skill Objectives: To understand interference of waves, to calculate properties of waves, the speed of sound, Doppler shifts, and intensity.

Review Plan:

Sound Properties:
- Chapter 14, Sections 14.1–14.4, 14.6
- Quick Quizzes 14.1, 14.2
- Examples 14.1, 14.2, 14.4, 14.5
- Active Figures 14.6, 14.11
- Chapter Problems 7, 13,25

Interference/Beats:
- Chapter 14, Sections 14.7, 14.8, 14.11
- Quick Quiz 14.7
- Examples 14.6, 14.11
- Active Figures 14.18, 14.25
- Chapter Problems 33, 37, 51

Light

Skill Objectives: To understand mirrors and lenses, to calculate the angles of reflection, to use the index of refraction, and to find focal lengths.

Review Plan:

Reflection and Refraction:
- Chapter 22, Sections 22.1–22.4
- Quick Quizzes 22.2–22.4
- Examples 22.1–22.4
- Active Figures 22.4, 22.6, 22.7
- Chapter Problems 9, 17, 21, 25

Mirrors and Lenses:
- Chapter 23, Sections 23.1–23.6
- Quick Quizzes 23.1, 23.2, 23.4–23.6
- Examples 23.7, 23.8, 23.9
- Active Figures 23.2, 23.16, 23.25
- Chapter Problems 25, 29, 33, 39

Electrostatics

Skill Objectives: To understand and calculate the electric field, the electrostatic force, and the electric potential.

Review Plan:

Coulomb's Law:
- Chapter 15, Sections 15.1–15.3
- Quick Quiz 15.2
- Examples 15.1–15.3
- Active Figure 15.6
- Chapter Problems 11

Electric Field:
- Chapter 15, Sections 15.4, 15.5
- Quick Quizzes 15.3–15.6
- Examples 15.4, 15.5
- Active Figures 15.11, 15.16
- Chapter Problems 19, 23, 27

Potential:
- Chapter 16, Sections 16.1–16.3
- Quick Quizzes 16.1–16.5
- Examples 16.1, 16.4
- Active Figure 16.7
- Chapter Problems 1, 7, 15

Circuits

Skill Objectives: To understand and calculate current, resistance, voltage, power, and energy, and to use circuit analysis.

Review Plan:

Ohm's Law:
- Chapter 17, Sections 17.1–17.4
- Quick Quizzes 17.1, 17.3
- Examples 17.1, 17.3
- Chapter Problem 15

Power and energy:
- Chapter 17, Section 17.8
- Quick Quizzes 17.6–17.8
- Example 17.6
- Active Figure 17.11
- Chapter Problem 36

Circuits:
- Chapter 18, Sections 18.2, 18.3
- Quick Quizzes 18.1–18.5
- Examples 18.1–18.3
- Active Figures 18.2, 18.6
- Chapter Problems 4, 13

Atoms

Skill Objectives: To calculate half-life, and to understand decay processes and nuclear reactions.

Review Plan:

Atoms:
- Chapter 29, Sections 29.1, 29.2

Radioactive Decay:
- Chapter 29, Sections 29.3–29.5
- Examples 29.3, 29.4, 29.7
- Active Figures 29.6, 29.7
- Chapter Problems 15, 21, 29, 35

Nuclear reactions:
- Chapter 29, Sections 29.6
- Quick Quiz 29.4
- Examples 29.8, 29.9
- Chapter Problems 41, 45

WE DEDICATE THIS BOOK

To our wives, children, grandchildren, and relatives who have provided so much love, support, and understanding through the years, and to the students for whom this book was written.

Contents Overview

Contents

Part 1: Mechanics

Chapter 21
Alternating Current Circuits and Electromagnetic Waves 693

Part 5: Light and Optics

Chapter 22
Reflection and Refraction of Light 726

Chapter 23
Mirrors and Lenses 754

Chapter 24
Wave Optics 786

Chapter 25
Optical Instruments 819

Part 6: Modern Physics

Chapter 26
Relativity 843

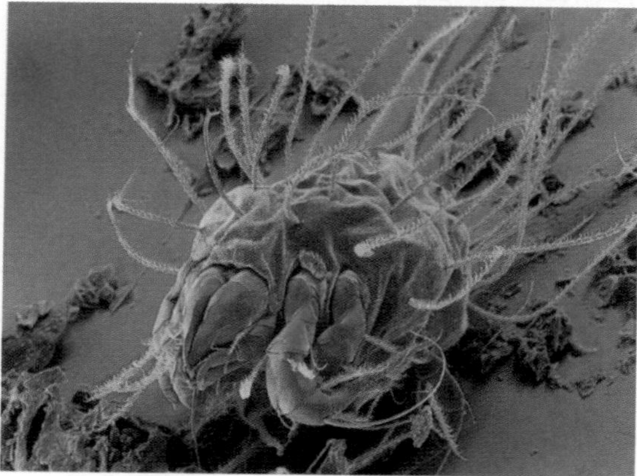

About the Authors

Raymond A. Serway received his doctorate at Illinois Institute of Technology and is Professor Emeritus at James Madison University. In 1990, he received the Madison Scholar Award at James Madison University, where he taught for 17 years. Dr. Serway began his teaching career at Clarkson University, where he conducted research and taught from 1967 to 1980. He was the recipient of the Distinguished Teaching Award at Clarkson University in 1977 and of the Alumni Achievement Award from Utica College in 1985. As Guest Scientist at the IBM Research Laboratory in Zurich, Switzerland, he worked with K. Alex Müller, 1987 Nobel Prize recipient. Dr. Serway also was a visiting scientist at Argonne National Laboratory, where he collaborated with his mentor and friend, Sam Marshall. In addition to earlier editions of this textbook, Dr. Serway is the co-author of *Physics for Scientists and Engineers,* Sixth Edition; *Principles of Physics,* Fourth Edition; and *Modern Physics,* Third Edition. He also is the author of the high-school textbook *Physics,* published by Holt, Rinehart, & Winston. In addition, Dr. Serway has published more than 40 research papers in the field of condensed matter physics and has given more than 70 presentations at professional meetings. Dr. Serway and his wife Elizabeth enjoy traveling, golfing, and spending quality time with their four children and six grandchildren.

Jerry S. Faughn earned his doctorate at the University of Mississippi. He is Professor Emeritus and former Chair of the Department of Physics and Astronomy at Eastern Kentucky University. Dr. Faughn has also written a microprocessor interfacing text for upper-division physics students. He is co-author of a non-mathematical physics text and a physical science text for general education students, and (with Dr. Serway) the high-school textbook *Physics,* published by Holt, Reinhart, & Winston. He has taught courses ranging from the lower division to the graduate level, but his primary interest is in students just beginning to learn physics. Dr. Faughn has a wide variety of hobbies, among which are reading, travel, genealogy, and old-time radio. His wife Mary Ann is an avid gardener, and he contributes to her efforts by staying out of the way. His daughter Laura is in family practice and his son David is an attorney.

Chris Vuille is an associate professor of physics at Embry-Riddle Aeronautical University, Daytona Beach, Florida, the world's premier institution for aviation higher education. He received his doctorate in physics at the University of Florida in 1989, moving to Daytona after a year at ERAU's Prescott, Arizona campus. While he has taught courses at all levels, including post-graduate, his primary interest has been the delivery of introductory physics. He has received several awards for teaching excellence, including the Senior Class Appreciation Award (three times). He conducts research in general relativity and quantum theory, and was a participant in the JOVE program, a special three-year NASA grant program during which he studied neutron stars. His work has appeared in a number of scientific journals, and he has been a featured science writer in Analog Science Fiction/Science Fact magazine. Dr. Vuille enjoys tennis, lap swimming, guitar and classical piano, and is a former chess champion of St. Petersburg and Atlanta. In his spare time he writes science fiction and goes to the beach. His wife, Dianne Kowing, is an optometrist for a local VA clinic. Teen daughter Kira Vuille-Kowing is an accomplished swimmer, violinist, and mall shopper. He has two sons, twelve-year-old Christopher, a cellist and electronic game expert, and three-year-old James, master of tinkertoys.

Charles A. Bennett received his doctorate at North Carolina State University, and is Professor of Physics at the University of North Carolina at Asheville. His research interests include quantum and physical optics, and laser applications in environmental and fusion energy research. He has collaborated with Oak Ridge National Laboratory since 1983, where he is currently an adjunct research and development associate of the Advanced Laser and Optical Technology and Development group. In addition to his work in optics, Dr. Bennett has a long record of innovation in educational technology, particularly in the integration of active media into on-line homework. He is a past director of the UNCA Center for Teaching and Learning, and has received UNCA's most prestigious recognition for scholarship: the Ruth and Leon Feldman Professorship for 1996–1997. He is an avid hiker, and enjoys spending time with his wife Karen and four children.

Preface

College Physics is written for a one-year course in introductory physics usually taken by students majoring in biology, the health professions, and other disciplines including environmental, earth, and social sciences, and technical fields such as architecture. The mathematical techniques used in this book include algebra, geometry, and trigonometry, but not calculus.

The main objectives of this introductory textbook are twofold: to provide the student with a clear and logical presentation of the basic concepts and principles of physics, and to strengthen an understanding of the concepts and principles through a broad range of interesting applications to the real world. To meet these objectives, we have emphasized sound physical arguments and problem-solving methodology. At the same time, we have attempted to motivate the student through practical examples that demonstrate the role of physics in other disciplines.

This textbook, which covers the standard topics in classical physics and 20th-century physics, is divided into six parts. Part I (Chapters 1–9) deals with Newtonian mechanics and the physics of fluids; Part II (Chapters 10–12) is concerned with heat and thermodynamics; Part III (Chapters 13–14) covers wave motion and sound; Part IV (Chapters 15–21) develops the concepts of electricity and magnetism; Part V (Chapters 22–25) treats the properties of light and the field of geometric and wave optics; and Part VI (Chapters 26–30) provides an introduction to special relativity, quantum physics, atomic physics, and nuclear physics.

CHANGES TO THE SEVENTH EDITION

A number of new features, changes, and improvements have been added to this edition. Based on comments from users of the sixth edition and reviewers' suggestions, a major effort was made to improve clarity of presentation, precision of language, and accuracy throughout. The new pedagogical features added to this edition are based on current trends in science education. The following represent the major changes in the seventh edition.

Pedagogical Changes

- **Examples** All in-text worked examples have been reconstituted for the seventh edition, and now are presented in a two-column format that better aids student learning and helps to better reinforce physical concepts. Special care has been taken in this new edition to present a range of levels within each chapter's collection of worked examples, so that students are better prepared to solve the end-of-chapter problems. The examples are set off from the text for ease of location and are given titles to describe their content. All worked examples now include the following parts:

  (a) **Goal,** which describes the physical concepts being explored within the Worked Example.
  (b) **Problem,** which presents the problem itself.
  (c) **Strategy,** which helps students analyze the problem and create a framework for working out the solution.
  (d) **Solution,** which uses a two-column format that gives the explanation for each step of the solution in the left-hand column, while giving each accompanying mathematical step in the right-hand column. This layout facilitates matching the idea with its execution, and helps students learn how to organize their work. Another benefit: students can easily use this format as a training tool, covering up the solution on the right and solving the problem using the comments on the left as a guide.
  (e) **Remarks,** which follow each Solution and highlight some of the underlying concepts and methodology used in arriving at a correct solution. In addition, the remarks are often used to put the problem into a larger, real world context.
  (f) **Exercise/Answer** By popular demand of users and reviewers, every worked example in the new edition is now followed immediately by an exercise with an answer. These exercises allow the students to reinforce their understanding by working a similar or related problem, with the answers giving them instant feedback. At the option of the instructor, the exercises can also be assigned as homework. Students who work through these exercises on a regular basis will find the end-of-chapter problems less intimidating.

- ***PhysicsNow*™** Directly links with *College Physics, Seventh Edition. PhysicsNow*™ and *College Physics, Seventh Edition* were built in concert to enhance each other and provide a seamless, integrated learning system. Throughout the text, readers will see **PhysicsNow™** icons that direct them to media-enhanced activities at the *PhysicsNow*™ Web site (**www.cp7e.com**). The precise page-by-page integration means professors and students will spend less time flipping through pages and navigating Web sites for useful exercises.

- **Active Figures** Many diagrams from the text have been animated to form Active Figures, part of the new *PhysicsNow*™ integrated Web-based learning system. There are over 120 Active Figures in the seventh edition, available at **www.cp7e.com.** By visualizing phenomena and processes that cannot be fully represented on a static page, students greatly increase their conceptual understanding. Active Figures are easily identified by their red figure legend and PhysicsNow™ icon, and the caption after the icon describes briefly the nature and contents of the animation. In addition to viewing animations of the figures, students can change variables to see the effects, conduct suggested explorations of the principles involved in the figure, and take and receive feedback on quizzes related to the figure.
- **Interactive Examples** Thirty-four of the worked examples in the seventh edition have been identified as interactive. As part of the *PhysicsNow*™ Web-based learning system, students can engage in an extension of the problem solved in the example. This often includes elements of both visualization and calculation, and may also involve prediction and intuition-building. Interactive Examples are available at **www.cp7e.com.**
- **Quick Quizzes** All of the Quick Quizzes in this edition have been cast in an objective format, including multiple choice, true-false, and ranking. Quick Quizzes provide students with opportunities to test their understanding of the physical concepts presented. The questions require students to make decisions on the basis of sound reasoning, and some of them have been written to help students overcome common misconceptions. Answers to all Quick Quiz questions are found at the end of the textbook, while answers with detailed explanations are provided in the *Instructor's Manual.* Many instructors choose to use Quick Quiz questions in a "peer instruction" teaching style.
- **Summary** The Summary is now organized by each individual chapter heading, for ease of reference.
- **Problems and Conceptual Questions** All problems have been carefully worded and have been checked for clarity and accuracy, and approximately 15% of the conceptual questions and problems are new to this edition. Solutions to approximately 20% of the end-of-chapter problems are included in the *Student Solutions Manual and Study Guide.* These problems are identified with a box around the problem number. A smaller subset of problems will be available with coached solutions as part of the *PhysicsNow*™ Web-based learning system and will be accessible to students and instructors using *College Physics.* The PhysicsNow™ icon identifies these problems.
- **Line-by-Line Revision** The text has been carefully edited to improve clarity of presentation and precision of language. We hope that the result is a book both accurate and enjoyable to read.

CONTENT CHANGES

Although the overall content and organization of the textbook is similar to that of the sixth edition, several changes were implemented.

- Vectors are now denoted in bold face with arrows over them (for example, $\vec{\mathbf{v}}$), making them easier to recognize.
- In many chapters, one or two examples have been modified so as to combine new concepts with previously studied concepts. This not only reviews prior material, but also integrates different concepts, showing how they work together to enhance our understanding of the physical world.
- In Chapter 4, on Newton's laws, we have doubled the number of examples, from seven in the sixth edition to fourteen in the seventh, to help aid student understanding.
- Chapter 5, on work and energy, has been extensively reorganized, with gravitational potential energy and spring potential energy handled in separate sections. The connection between work done by conservative forces and the definition of the potential energy of a system has been emphasized and clarified.
- The number of examples in Chapter 9, Solids and Fluids, has been increased from thirteen to nineteen.
- Chapter 12, The Laws of Thermodynamics, now covers the important concepts of degrees of freedom and the equipartition of energy, and relates them to determining molar specific heats. The different kinds of thermodynamic processes have been organized into their own subsections, and the work done during an isothermal process is now presented and used in examples. Generic processes are also briefly addressed. The methods for calculating the quantities appearing in the first law are organized into a table that makes clear the similarities and differences between the processes.
- The concepts of electric potential and electric potential energy have been clarified in Chapter 16.
- In Chapter 23, the tables for sign conventions for mirrors and lenses have been reorganized for greater clarity.
- The methods of handling interference in thin films have been organized into a table in Chapter 24, facilitating student understanding and problem-solving in that area.

- Additional detail on prescribing lenses has been added to the example problems of Chapter 25, Optical Instruments.
- In Chapter 26, the Twin Paradox and concepts of general relativity were clarified and enhanced.

TEXTBOOK FEATURES

Most instructors would agree that the textbook assigned in a course should be the student's primary guide for understanding and learning the subject matter. Furthermore, the textbook should be easily accessible and written in a style that facilitates instruction and learning. With this in mind, we have included many pedagogical features that are intended to enhance the textbook's usefulness to both students and instructors. These features are as follows:

Style To facilitate rapid comprehension, we have attempted to write the book in a style that is clear, logical, relaxed, and engaging. The somewhat informal and relaxed writing style is designed to connect better with students and enhance their reading enjoyment. New terms are carefully defined, and we have tried to avoid the use of jargon.

Previews All chapters begin with a brief preview that includes a discussion of the chapter's objectives and content.

Units The international system of units (SI) is used throughout the text. The U.S. customary system of units is used only to a limited extent in the chapters on mechanics and thermodynamics.

Marginal Notes Comments and notes appearing in the margin can be used to locate important statements, equations, and concepts in the text.

Problem-Solving Strategies A general strategy to be followed by the student is outlined at the end of Chapter 1 and provides students with a structured process for solving problems. In most chapters, more specific strategies and suggestions are included for solving the types of problems featured in both the worked examples and end-of-chapter problems. This feature, highlighted by a surrounding box, is intended to help students identify the essential steps in solving problems and increases their skills as problem solvers.

Biomedical Applications For biology and pre-med students, icons point the way to various practical and interesting applications of physical principles to biology and medicine. Where possible, an effort was made to include more problems that would be relevant to these disciplines.

Applying Physics provide students with an additional means of reviewing concepts presented in that section. Some Applying Physics examples demonstrate the connection between the concepts presented in that chapter and other scientific disciplines. These examples also serve as models for students when assigned the task of responding to the Conceptual Questions presented at the end of each chapter.

Tips are placed in the margins of the text and address common student misconceptions and situations in which students often follow unproductive paths. Approximately 40 Tips are provided in Volume 2 to help students avoid common mistakes and misunderstandings.

Important Statements and Equations Most important statements and definitions are set in **boldface** type or are highlighted with a background screen for added emphasis and ease of review. Similarly, important equations are highlighted with a tan background screen to facilitate location.

Illustrations and Tables The readability and effectiveness of the text material, worked examples, and end-of-chapter conceptual questions and problems are enhanced by the large number of figures, diagrams, photographs, and tables. Full color adds clarity to the artwork and makes illustrations as realistic as possible. Three-dimensional effects are rendered with the use of shaded and lightened areas where appropriate. Vectors are color coded, and curves in graphs are drawn in color. Color photographs have been carefully selected, and their accompanying captions have been written to serve as an added instructional tool. A complete description of the pedagogical use of color appears on the inside front cover.

Significant Figures Significant figures in both worked examples and end-of-chapter problems have been handled with care. Most numerical examples and problems are worked out to either two or three significant figures, depending on the accuracy of the data provided. Intermediate results presented in the examples are rounded to the proper number of significant figures, and only those digits are carried forward.

Conceptual Questions are provided at the end of each chapter (almost 550 questions are included in the seventh edition). The **Applying Physics** examples presented in the text serve as models for students when conceptual questions are assigned, and show how the concepts can be applied to understanding the physical world. The conceptual questions provide the student with a means of self-testing the concepts presented in the chapter. Some conceptual questions are appropriate for initiating classroom discussions. Answers to all odd-numbered conceptual questions are located in the answer section at the end of the book.

End-of-Chapter Problems An extensive set of problems is included at the end of each chapter (in all, over 2,000 problems are provided in the seventh edition). Answers to odd-numbered problems are given at the end of the book. For the convenience of both the student and instructor, about two thirds of the problems are keyed to specific sections of the chapter. The remaining problems, labeled "Additional Problems," are not keyed to specific sections. There are three levels of problems that are graded according to their difficulty. Straightforward problems are numbered in black, intermediate level problems are numbered in blue, and the most challenging problems are numbered in magenta. The icon identifies problems dealing with applications to the life sciences and medicine.

Solutions to approximately 20% of the problems in each chapter are in the *Student Solutions Manual and Study Guide*. Among these, selected problems are identified with *PhysicsNow*™ icons and have coached solutions available at **www.cp7e.com.**

Activities have been included at the end of each chapter's problem set to encourage students to engage in scientific activities outside of the classroom.

Appendices and Endpapers Several appendices are provided at the end of the textbook. Most of the appendix material represents a review of mathematical concepts and techniques used in the text, including scientific notation, algebra, geometry, trigonometry, differential calculus, and integral calculus. Reference to these appendices is made as needed throughout the text. Most of the mathematical review sections include worked examples and exercises with answers. In addition to the mathematical review, some appendices contain useful tables that supplement textual information. For easy reference, the front endpapers contain a chart explaining the use of color throughout the book and a list of frequently used conversion factors.

TEACHING OPTIONS

This book contains more than enough material for a one-year course in introductory physics. This serves two purposes. First, it gives the instructor more flexibility in choosing topics for a specific course. Second, the book becomes more useful as a resource for students. On the average, it should be possible to cover about one chapter each week for a class that meets three hours per week. Those sections, examples, and end-of-chapter problems dealing with applications of physics to life sciences are identified with the DNA icon. We offer the following suggestions for shorter courses for those instructors who choose to move at a slower pace through the year.

Option A: If you choose to place more emphasis on contemporary topics in physics, you should consider omitting all or parts of Chapter 8 (Rotational Equilibrium and Rotational Dynamics), Chapter 21 (Alternating Current Circuits and Electromagnetic Waves), and Chapter 25 (Optical Instruments).

Option B: If you choose to place more emphasis on classical physics, you could omit all or parts of Part VI of the textbook, which deals with special relativity and other topics in 20th-century physics.

The *Instructor's Manual* offers additional suggestions for specific sections and topics that may be omitted without loss of continuity if time presses.

ANCILLARIES

The ancillary package has been updated substantially and streamlined in response to suggestions from users of the sixth edition. The most essential parts of the student package are the two-volume *Student Solutions Manual and Study Guide* with a tight focus on problem-solving and the Web-based *PhysicsNow*™ learning system. Instructors will find increased support for their teaching efforts with new electronic materials.

STUDENT ANCILLARIES

Thomson • Brooks/Cole offers several items to supplement and enhance the classroom experience. These ancillaries will allow instructors to customize the textbook to their students' needs and to their own style of instruction. One or more of these ancillaries may be shrink-wrapped with the text at a reduced price:

Student Solutions Manual and Study Guide by John R. Gordon, Charles Teague, and Raymond A. Serway. Now offered in two volumes, this manual features detailed solutions to approximately 12 problems per chapter. Boxed numbers identify those problems in the textbook for which complete solutions are found in the manual. The manual also features a skills section, important notes from key sections of the text, and a list of important equations and concepts. Volume 1 contains Chapters 1–14 and Volume 2 contains Chapters 15–30.

PhysicsNow™ Students log into *PhysicsNow*™ at **www.cp7e.com** by using the free access code packaged with this text.* The *PhysicsNow*™ system is made up of three interrelated parts:

- How much do you know?
- What do you need to learn?
- What have you learned?

Students maximize their success by starting with the Pre-Test for the relevant chapter. Each Pre-Test is a mix of conceptual and numerical questions. After completing the Pre-Test, each student is presented with a detailed Learning Plan. The Learning Plan outlines elements to review in the text and Web-based media (Active Figures, Interactive Examples, and Coached Problems) in order to master the chapter's most essential concepts. After working through these materials, students move on to a multiple-choice Post-Test presenting them with questions similar to those that might appear on an exam. Results can be e-mailed to instructors.

Physics Laboratory Manual, 2nd edition by David Loyd. This manual supplements the learning of basic physical principles while introducing laboratory procedures and equipment. Each chapter of the manual includes a pre-laboratory assignment, objectives, an equipment list, the theory behind the experiment, experimental procedures, graphs, and questions. A laboratory report is provided for each experiment so the student can record data, calculations, and experimental results. Students are encouraged to apply statistical analysis to their data in order to develop their ability to judge the validity of their results.

WebTutor™ on WebCT and Blackboard WebTutor™ offers students real-time access to a full array of study tools, including PhysicsNow's Active Figures, chapter outlines, summaries, and learning objectives. New to this edition is a bank of end-of-chapter problems from each chapter of the book coded in multiple choice format and ready to assign as quizzes or homework.

The Brooks/Cole Physics Resource Center You'll find additional online quizzes, Web links, and animations at **http://physics.brookscole.com.**

INSTRUCTOR ANCILLARIES

The following ancillaries are available to qualified adopters. Please contact your local Thomson • Brooks/Cole sales representative for details.
Ancillaries offered in two volumes are split as follows: Volume 1 contains Chapters 1–14 and Volume 2 contains Chapters 15–30.

Instructor's Solutions Manual by Jerry Faughn and Charles Teague. Available in two volumes, this manual consists of complete solutions to all the problems in the text, answers to the even-numbered problems and conceptual questions, and full answers with explanations to the Quick Quizzes.

Test Bank by Ed Oberhofer. Contains approximately 1 750 multiple-choice problems and questions. Answers are provided in a separate key. It is provided in print form (in two volumes) for the instructor who does not have access to a computer, and instructors may duplicate pages for distribution to students. The questions in the ***Test Bank*** are also available in electronic format with complete answers and solutions in iLrn's Computerized Testing.

Overhead Transparency Acetates The collection of transparencies in two volumes consists of approximately 200 full-color figures and photographs from the text to enhance lectures. These transparencies feature large print for easy viewing in the classroom.

Instructor's Manual for Physics Laboratory Manual 2nd edition by David Loyd. Each chapter contains a discussion of the experiment, teaching hints, answers to selected questions from the student laboratory manual, and a post-laboratory quiz with short answers and essay questions. The author has also included a list of the suppliers of scientific equipment and a summary of the equipment needed for all the experiments in the manual.

*Free pincodes are only available with new copies of *College Physics,* 7th edition.

Personal Response System Content To support your use of personal response systems in the classroom, we've coded the questions from the Quick Quiz questions for use with the system of your choice. Contact your Thomson representative for more details.

Multimedia Manager Instructor's Resource CD This easy-to-use lecture tool allows you to quickly assemble art, photos, and multimedia with notes to create fluid lectures. The CD-ROM set (Volume 1, Chapters 1–14; Volume 2, Chapters 15–30) includes a database of animations, video clips and digital art from the text as well as PowerPoint lectures and electronic files of the Instructor's Solutions Manual and **Test Bank.** You'll also find additional content from the textbook like the Quick Quizzes and Conceptual Questions.

Physics Now™ ***PhysicsNow™*** **Course Management Tools powered by iLrn** This extension to the student tutorial environment of PhysicsNow™ allows instructors to deliver online assignments in an environment that is familiar to students. This powerful system is your gateway to managing online homework, testing, and course administration all in one shell with the proven content to make your course a success. To see a demonstration of this powerful system, contact your Thomson representative or go to **www.cp7e.com.**

Physics Now™ ***PhysicsNow™*** **Homework Management powered by iLrn** To further enhance PhysicsNow, contributing author Chuck Bennett has created ready-to-use homework assignments which are based on the in-chapter examples. These assignments feature step-by-step guides through algorithmic versions of the in-text examples paired with more challenging examples to extend your students' problem-solving abilities. The problems include hints and feedback to help students pinpoint their errors. The goal of this set of exercises is to bring students' problem solving abilities up to the level they will need to work the end-of-chapter problems.

PhysicsNow™ gives you a rich array of problem types and grading options. Its library of assignable questions includes all of the end-of-chapter problems from the text so that you can select the problems you want to include in your online homework assignments. These well-crafted problems are algorithmically generated so that you can assign the same problem with different variables for each student. A flexible grading tolerance feature allows you to specify a percentage range of correct answers so that your students are not penalized for rounding errors. You can give students the option to work an assignment multiple times and record the highest score or limit the times they are able to attempt it. In addition, you can create your own problems to complement the problems from the text. Results flow automatically to an exportable grade book so that instructors are better able to assess student understanding of the material, even prior to class or to an actual test.

Physics Now™ **iLrn Computerized Testing** Extend the student experience with iLrn into a testing or quizzing environment. The test item file from the text is included to give you a bank of well-crafted questions that you can deliver online or print out. As with the homework problems, you can use the program's friendly interface to craft your own questions to complement the Serway/Faughn questions. You have complete control over grading, deadlines, and availability and can create multiple tests based on the same material.

WebTutor™ on WebCT and Blackboard With **WebTutor**™'s text-specific, pre-formatted content and total flexibility, instructors can easily create and manage their own personal Web site. WebTutor™'s course management tool gives instructors the ability to provide virtual office hours, post syllabi, set up threaded discussions, track student progress with the quizzing material, and much more. WebTutor™ also provides robust communication tools, such as a course calendar, asynchronous discussion, real-time chat, a whiteboard, and an integrated e-mail system.

Additional Options for Online Homework

WebAssign: A Web-Based Homework System WebAssign is the most utilized homework system in physics. Designed by physicists for physicists, this system is a trusted companion to your teaching. An enhanced version of WebAssign is available for College Physics. In addition to selected end-of-chapter problems, this enhanced version includes animations with conceptual questions. We've also added tutorial problems with feedback and hints to guide students' content mastery. Take a look at this new innovation from the most trusted name in physics homework. **www.webassign.net**

LON-CAPA This Web-based course management system developed at Michigan State University includes selected end of chapter problems from *College Physics.* For more information, visit the LON-CAPA Web site at **www.lon-capa.org.**

UT Austin Homework Service Details about and a demonstration of this service are available at **http://hw.ph.utexas.edu/hw.html.**

ACKNOWLEDGEMENTS

In preparing the seventh edition of this textbook, we have been guided by the expertise of many people who have reviewed manuscript and/or provided pre-revision suggestions. We wish to acknowledge the following reviewers and express our sincere appreciation for their helpful suggestions, criticism, and encouragement.

Seventh edition reviewers:

Gary B. Adams, *Arizona State University*
Ricardo Alarcon, *Arizona State University*
Natalie Batalha, *San Jose State University*
Ken Bolland, *Ohio State University*
Kapila Clara Castoldi, *Oakland University*
Andrew Cornelius, *University of Nevada, Las Vegas*
Yesim Darici, *Florida International University*
N. John DiNardo, *Drexel University*
Steve Ellis, *University of Kentucky*
Hasan Fakhruddin, *Ball State University/ The Indiana Academy*
Emily Flynn
Lewis Ford, *Texas A & M University*
James R. Goff, *Pima Community College*
Yadin Y. Goldschmidt, *University of Pittsburgh*
Steve Hagen, *University of Florida*
Raymond Hall, *California State University, Fresno*
Patrick Hamill, *San Jose State University*
Joel Handley
Grant W. Hart, *Brigham Young University*
James E. Heath, *Austin Community College*
Grady Hendricks, *Blinn College*
Rhett Herman, *Radford University*
Aleksey Holloway, *University of Nebraska at Omaha*
Joey Huston, *Michigan State University*
Mark James, *Northern Arizona University*
Randall Jones, *Loyola College in Maryland*
Joseph Keane, *St. Thomas Aquinas College*
Dorina Kosztin, *University of Missouri-Columbia*
Martha Lietz, *Niles West High School*
Edwin Lo
Rafael Lopez-Mobilia, *University of Texas at San Antonio*
John A. Milsom, *University of Arizona*
Monty Mola, *Humboldt State University*
Charles W. Myles, *Texas Tech University*
Ed Oberhofer, *Lake Sumter Community College*
Chris Pearson, *University of Michigan-Flint*
Alexey A. Petrov, *Wayne State University*
M. Anthony Reynolds, *Embry-Riddle Aeronautical University*
Dubravka Rupnik, *Louisiana State University*
Surajit Sen, *State University of New York at Buffalo*
Marllin L. Simon, *Auburn University*
Matthew Sirocky
George Strobel, *University of Georgia*
James Wanliss, *Embry-Riddle Aeronautical University*

College Physics, seventh edition was carefully checked for accuracy by Ken Bolland (Ohio State University), Steve Ellis (University of Kentucky), Steve Hagen (University of Florida), Mark James (Northern Arizona University), Randall Jones (Loyola College in Maryland), Edwin Lo, Marllin L. Simon (Auburn University), George Strobel (University of Georgia), M. Anthony Reynolds (Embry-Riddle Aeronautical University), Emily Flynn, Joel Handley and Matthew Sirocky. Though responsibility for any remaining errors rests with us, we thank them for their dedication and vigilance.

We thank the following people for their suggestions and assistance during the preparation of earlier editions of this textbook:

Marilyn Akins, *Broome Community College;* Albert Altman, *University of Lowell;* John Anderson, *University of Pittsburgh;* Lawrence Anderson-Huang, *University of Toledo;* Subhash Antani, *Edgewood College;* Neil W. Ashcroft, *Cornell University;* Charles R. Bacon, *Ferris State University;* Dilip Balamore, *Nassau Community College;* Ralph Barnett, *Florissant Valley Community College;* Lois Barrett, *Western Washington University;* Paul D. Beale, *University of Colorado at Boulder;* Paul Bender, *Washington State University;* David H. Bennum, *University of Nevada at Reno;* Jeffery Braun, *University of Evansville;* John Brennan, *University of Central Florida;* Michael Bretz, *University of Michigan, Ann Arbor;* Michael E. Browne, *University of Idaho;* Joseph Cantazarite, *Cypress College;* Ronald W. Canterna, *University of Wyoming;* Clinton M. Case, *Western Nevada Community College;* Neal M. Cason, *University of Notre Dame;* Roger W. Clapp, *University of South Florida;* Giuseppe Colaccico, *University of South Florida;* Lattie F. Collins, *East Tennessee State University;* Lawrence B. Colman, *University of California, Davis;* Jorge Cossio, *Miami Dade Community College;* Terry T. Crow, *Mississippi State College;* Stephen D. Davis, *University of Arkansas at Little Rock;* John DeFord, *University of Utah;* Chris J. DeMarco, *Jackson Community College;* Michael Dennin, *University of California, Irvine;* N. John DiNardo, *Drexel University;* Robert J. Endorf, *University of Cincinnati;* Paul Feldker, *Florissant Valley Community College;* Leonard X. Finegold, *Drexel University;* Tom French, *Montgomery County Community College;* Albert Thomas Frommhold, Jr., *Auburn University;* Lothar Frommhold, *University of Texas at Austin;* Eric Ganz, *University of Minnesota;* Teymoor Gedayloo, *California Polytechnic State University;* Simon George, *California State University, Long Beach;* John R. Gordon, *James Madison University;* George W. Greenlees, *University of Minnesota;* Wlodzimierz Guryn, *Brookhaven National Laboratory;* James Harmon, *Oklahoma State University;* Grant W. Hart, *Brigham Young University;* Christopher Herbert, *New Jersey City University;* John Ho, *State University of New York at Buffalo;* Murshed Hossain, *Rowan University;* Robert C. Hudson, *Roanoke College;* Fred Inman, *Mankato State University;* Ronald E. Jodoin, *Rochester Institute of Technology;* Drasko Jovanovic, *Fermilab;* George W. Kattawar, *Texas A & M University;* Frank Kolp, *Trenton State University;* Joan P.S. Kowalski, *George Mason University;* Ivan Kramer, *University of Maryland, Baltimore County;* Sol Krasner, *University of Chicago;* Karl F. Kuhn, *Eastern Kentucky University;* David Lamp, *Texas Tech University;* Harvey S. Leff, *California State Polytechnic University;*

Joel Levine, *Orange Coast College;* Michael Lieber, *University of Arkansas;* James Linbald, *Saddleback Community College;* Bill Lochslet, *Pennsylvania State University;* Michael LoPresto, *Henry Ford Community College;* Bo Lou, *Ferris State University;* Jeffrey V. Mallow, *Loyola University of Chicago;* David Markowitz, *University of Connecticut;* Steven McCauley, *California State Polytechnic University, Pomona;* Joe McCauley, Jr., *University of Houston;* Ralph V. McGrew, *Broome Community College;* Bill F. Melton, *University of North Carolina at Charlotte;* H. Kent Moore, *James Madison University;* John Morack, *University of Alaska, Fairbanks;* Steven Morris, *Los Angeles Harbor College;* Carl R. Nave, *Georgia State University;* Martin Nikolo, *Saint Louis University;* Blaine Norum, *University of Virginia;* M. E. Oakes, *University of Texas at Austin;* Lewis J. Oakland, *University of Minnesota;* Ed Oberhofer, *Lake Sumter Community College;* Lewis O'Kelly, *Memphis State University;* David G. Onn, *University of Delaware;* J. Scott Payson, *Wayne State University;* T. A. K. Pillai, *University of Wisconsin, La Crosse;* Lawrence S. Pinsky, *University of Houston;* William D. Ploughe, *Ohio State University;* Patrick Polley, *Beloit College;* Brooke M. Pridmore, *Clayton State University;* Joseph Priest, *Miami University;* James Purcell, *Georgia State University;* W. Steve Quon, *Ventura College;* Michael Ram, *State University of New York at Buffalo;* Kurt Reibel, *Ohio State University;* Barry Robertson, *Queen's University;* Virginia Roundy, *California State University, Fullerton;* Larry Rowan, *University of North Carolina, Chapel Hill;* William R. Savage, *The University of Iowa;* Reinhard A. Schumacher, *Carnegie Mellon University;* John Simon, *University of Toledo;* Donald D. Snyder, *Indiana University at Southbend;* Carey E. Stronach, *Virginia State University;* Thomas W. Taylor, *Cleveland State University;* Perry A. Tompkins, *Samford University;* L. L. Van Zandt, *Purdue University;* Howard G. Voss, *Arizona State University;* Larry Weaver, *Kansas State University;* Donald H. White, *Western Oregon State College;* Bernard Whiting, *University of Florida;* George A. Williams, *The University of Utah;* Jerry H. Wilson, *Metropolitan State College;* Robert M. Wood, *University of Georgia;* Clyde A. Zaidins, *University of Colorado at Denver*

Randall Jones generously contributed many of the new end-of-chapter problems, especially those of interest to the life sciences. Edward F. Redish of the University of Maryland graciously allowed us to list some of his problems from the Activity Based Physics Project as both problems and activities. Some of the remaining activities were written by Robert J. Beichner of North Carolina State University.

We are extremely grateful to the publishing team at the Brooks/Cole Publishing Company for their expertise and outstanding work in all aspects of this project. In particular, we'd like to thank Ed Dodd, who tirelessly coordinated and directed our efforts in preparing the manuscript in its various stages, and Annie Mac, who worked behind the scenes securing text and accuracy reviews, and coordinated and transmitted all the print ancillaries. Jane Sanders, the photo researcher, did a great job finding photos of physical phenomena, Sam Subity coordinated the building of the ***PhysicsNow***™ Web site, and Rob Hugel helped translate our rough sketches into accurate, compelling art. Donna King of Progressive Publishing Alternatives managed the difficult task of keeping production moving and on schedule. Chris Hall and Teri Hyde also made numerous valuable contributions. Chris provided just the right amount of guidance and vision throughout the project. We would like to thank David Harris, a great team builder and motivator with loads of enthusiasm and an infectious sense of humor.

Finally, we are deeply indebted to our wives and children for their love, support, and long-term sacrifices.

Raymond A. Serway
St. Petersburg, Florida

Jerry S. Faughn
Richmond, Kentucky

Chris Vuille
Daytona Beach, Florida

Charles A. Bennett
Asheville, North Carolina

Applications

APPLICATION

Directs you to sections discussing applied principles of physics

BIOMEDICAL APPLICATION

Although physics is relevant to so much in our modern lives, this may not be obvious to students in an introductory course. In this seventh edition of *College Physics,* we continue a design feature begun in the previous edition. This feature makes the relevance of physics to everyday life more obvious by pointing out specific applications in the form of a marginal note. Some of these applications pertain to the life sciences and are marked with the DNA icon. The list below is not intended to be a complete listing of all the applications of the principles of physics found in this textbook. Many other applications are to be found within the text and especially in the worked examples, conceptual questions, and end-of-chapter problems.

To the Student

As a student, it's important that you understand how to use the book most effectively, and how best to go about learning physics. Scanning through the Preface will acquaint you with the various features available, both in the book and online. Awareness of your educational resources and how to use them is essential. Though physics is challenging, it can be mastered with the correct approach.

HOW TO STUDY

Students often ask how best to study physics and prepare for examinations. There's no simple answer to this question, but we'd like to offer some suggestions based on our own experiences in learning and teaching over the years.

First and foremost, maintain a positive attitude toward the subject matter. Like learning a language, physics takes time. Those who keep applying themselves on a daily basis can expect to reach understanding. Keep in mind that physics is the most fundamental of all natural sciences. Other science courses that follow will use the same physical principles, so it is important that you understand and are able to apply the various concepts and theories discussed in the text.

CONCEPTS AND PRINCIPLES

Students often try to do their homework without first studying the basic concepts. It is essential that you understand the basic concepts and principles *before* attempting to solve assigned problems. You can best accomplish this goal by carefully reading the textbook *before* you attend your lecture on the covered material. When reading the text, you should jot down those points that are not clear to you. Also be sure to make a diligent attempt at answering the questions in the Quick Quizzes as you come to them in your reading. We have worked hard to prepare questions that help you judge for yourself how well you understand the material. Pay careful attention to the many Tips throughout the text. These will help you avoid misconceptions, mistakes, and misunderstandings as well as maximize the efficiency of your time by minimizing adventures along fruitless paths. During class, take careful notes and ask questions about those ideas that are unclear to you. Keep in mind that few people are able to absorb the full meaning of scientific material after only one reading.

Be sure to take advantage of the features available in the ***PhysicsNow***™ learning system, such as the Active Figures, Interactive Examples, and Coached Problems. Your lectures and laboratory work supplement your textbook and should clarify some of the more difficult material. You should minimize rote memorization of material. Successful memorization of passages from the text, equations, and derivations does not necessarily indicate that you understand the fundamental principles.

Your understanding will be enhanced through a combination of efficient study habits, discussions with other students and with instructors, and your ability to solve the problems presented in the textbook. Ask questions whenever you feel clarification of a concept is necessary.

STUDY SCHEDULE

It is important for you to set up a regular study schedule, preferably a daily one. Make sure you read the syllabus for the course and adhere to the schedule set by your instructor. As a general rule, you should devote about two hours of study time for every hour you are in class. If you are having trouble with the course, seek the advice of the instructor or other students who have taken the course. You may find it necessary to seek further instruction from experienced students. Very often, instructors offer review sessions in addition to regular class periods. It is important that you avoid the practice of delaying study until a day or two before an exam. One hour of study a day for fourteen days is far more effective than fourteen hours the day before the exam. "Cramming" usually produces disastrous results, especially in science. Rather than undertake an all-night study session just before an exam, briefly review the basic concepts and equations and get a good night's rest. If you feel you need additional help in understanding the concepts, in preparing for exams, or in problem-solving, we suggest that you acquire a copy of the *Student Solutions Manual and Study Guide* that accompanies this textbook; this manual should be available at your college bookstore.

USE THE FEATURES

You should make full use of the various features of the text discussed in the preface. For example, marginal notes are useful for locating and describing important equations and concepts, and **boldfaced** type indicates important statements and definitions. Many useful tables are contained in the Appendices, but most tables are incorporated in the text where they are most often referenced. Appendix A is a convenient review of mathematical techniques.

Answers to all Quick Quizzes, as well as odd-numbered conceptual questions and problems, are given at the end of the textbook. Answers to selected end-of-chapter problems are provided in the *Student Solutions Manual and Study Guide.* Problem-Solving Strategies included in selected chapters throughout the text give you additional information about how you should solve problems. The Table of Contents provides an overview of the entire text, while the Index enables you to locate specific material quickly. Footnotes sometimes are used to supplement the text or to cite other references on the subject discussed.

After reading a chapter, you should be able to define any new quantities introduced in that chapter and to discuss the principles and assumptions used to arrive at certain key relations. The chapter summaries and the review sections of the *Student Solutions Manual and Study Guide* should help you in this regard. In some cases, it may be necessary for you to refer to the index of the text to locate certain topics. You should be able to correctly associate with each physical quantity the symbol used to represent that quantity and the unit in which the quantity is specified. Furthermore, you should be able to express each important relation in a concise and accurate prose statement.

PROBLEM-SOLVING

R. P. Feynman, Nobel laureate in physics, once said, "You do not know anything until you have practiced." In keeping with this statement, we strongly advise that you develop the skills necessary to solve a wide range of problems. Your ability to solve problems will be one of the main tests of your knowledge of physics; therefore, you should try to solve as many problems as possible. It is essential that you understand basic concepts and principles before attempting to solve problems. It is good practice to try to find alternate solutions to the same problem. For example, you can solve problems in mechanics using Newton's laws, but very often an alternate method that draws on energy considerations is more direct. You should not deceive yourself into thinking you understand a problem merely because you have seen it solved in class. You must be able to solve the problem and similar problems on your own. We have recast the examples in this book to help you in this regard. After studying an example, see if you can cover up the right-hand side and do it yourself, using only the written descriptions on the left as hints. Once you succeed at that, try solving the example completely on your own. Finally, solve the exercise. Once you have accomplished all this, you will have a good mastery of the problem, its concepts, and mathematical technique.

The approach to solving problems should be carefully planned. A systematic plan is especially important when a problem involves several concepts. First, read the problem several times until you are confident you understand what is being asked. Look for any key words that will help you interpret the problem and perhaps allow you to make certain assumptions. Your ability to interpret a question properly is an integral part of problem-solving. Second, you should acquire the habit of writing down the information given in a problem and those quantities that need to be found; for example, you might construct a table listing both the quantities given and the quantities to be found. This procedure is sometimes used in the worked examples of the textbook. After you have decided on the method you feel is appropriate for a given problem, proceed with your solution. Finally, check your results to see if they are reasonable and consistent with your initial understanding of the problem. General problem-solving strategies of this type are included in the text and are highlighted with a surrounding box. If you follow the steps of this procedure, you will find it easier to come up with a solution and also gain more from your efforts.

Often, students fail to recognize the limitations of certain equations or physical laws in a particular situation. It is very important that you understand and remember the assumptions underlying a particular theory or formalism. For example, certain equations in kinematics apply only to a particle moving with constant acceleration. These equations are not valid for describing motion whose acceleration is not constant, such as the motion of an object connected to a spring or the motion of an object through a fluid.

EXPERIMENTS

Physics is a science based on experimental observations. In view of this fact, we recommend that you try to supplement the text by performing various types of "hands-on" experiments, either at home or in the laboratory. For example, the common Slinky™ toy is excellent for studying traveling waves; a ball swinging on the end of a long string can be used to investigate pendulum motion; various masses attached to the end of a vertical spring or rubber band can be used to determine their elastic nature; an old pair of Polaroid sunglasses and some discarded lenses and a magnifying glass are the components of various experiments in optics; and the approximate measure of the free-fall acceleration can be determined simply by measuring with a stopwatch the time it takes for a ball to drop from a known height. The list of such experiments is endless. When physical models are not available, be imaginative and try to develop models of your own.

Some of the Activities at the end of each chapter can be performed on your own or with the help of a friend and will give you specific guidelines for acquiring "hands-on" experience.

PhysicsNow™ with Active Figures and Interactive Examples

We strongly encourage you to use the ***PhysicsNow***™ Web-based learning system that accompanies this textbook. It is far easier to understand physics if you see it in action, and these new materials will enable you to become a part of that action. ***PhysicsNow***™ media described in the Preface are accessed at the URL **www.cp7e.com**, and feature a three-step learning process consisting of a Pre-Test, a personalized learning plan, and a Post-Test.

In addition to the Coached Problems identified with icons, ***PhysicsNow***™ includes the following Active Figures and Interactive Examples:

Chapter 1 Active Figures 1.6 and 1.7

Chapter 2 Active Figures 2.2, 2.12, 2.13, and 2.15; Interactive Examples 2.5 and 2.9

Chapter 3 Active Figures 3.3, 3.14, and 3.15; Interactive Examples 3.3 and 3.7

Chapter 4 Active Figures 4.6, 4.18, and 4.19; Interactive Example 4.10

Chapter 5 Active Figures 5.5, 5.15, 5.20, and 5.29; Interactive Example 5.5

Chapter 6 Active Figure 6.7, 6.10, 6.13, and 6.15; Interactive Examples 6.3 and 6.7

Chapter 7 Active Figures 7.5, 7.17, and 7.21; Interactive Example 7.7

Chapter 8 Active Figure 8.25; Interactive Examples 8.8 and 8.11

Chapter 9 Active Figures 9.3, 9.5, 9.6, and 9.19; Interactive Examples 9.7 and 9.13

Chapter 10 Active Figures 10.10, 10.12, and 10.15

Chapter 11 Interactive Example 11.10

Chapter 12 Active Figures 12.1, 12.9, 12.12, 12.15, and 12.16

Chapter 13 Active Figures 13.1, 13.8, 13.12, 13.13, 13.15, 13.16, 13.19, 13.24, 13.26, 13.32, 13.33, 13.34, and 13.35; Interactive Example 13.10

Chapter 14 Active Figures 14.8, 14.10, 14.18, and 14.25; Interactive Examples 14.1, 14.5, and 14.7

Chapter 15 Active Figures 15.6, 15.11, 15.21, and 15.28; Interactive Example 15.2

Chapter 16 Active Figures 16.7, 16.17, and 16.19; Interactive Examples 16.2 and 16.8

Chapter 17 Active Figures 17.4 and 17.11; Interactive Example 17.4

Chapter 18 Active Figures 18.1, 18.6, 18.16, and 18.17; Interactive Examples 18.2 and 18.5

Chapter 19 Active Figures 19.2, 19.17, 19.19, 19.20, 19.23, and 19.28

Chapter 20 Active Figures 20.4, 20.13, 20.20, 20.22, 20.27, and 20.28

Chapter 21 Active Figures 21.1, 21.2, 21.4, 21.5, 21.6, 21.7, 21.8, 21.9, and 21.20

Chapter 22 Active Figures 22.4, 22.6, 22.7, 22.19, and 22.25; Interactive Examples 22.1 and 22.4

Chapter 23 Active Figures 23.2, 23.13, 23.16, and 23.25; Interactive Examples 23.2, 23.7, 23.8, and 23.9

Chapter 24 Active Figures 24.1, 24.16, 24.20, 24.21, and 24.26; Interactive Examples 24.1, 24.3, 24.6, and 24.7

Chapter 25 Active Figures 25.7, 25.8, and 25.15

Chapter 26 Active Figures 26.4, 26.8, and 26.11

Chapter 27 Active Figures 27.2, 27.3, and 27.4; Interactive Examples 27.3 and 27.6

Chapter 28 Active Figures 28.6, 28.7, 28.17, 28.18, and 28.19; Interactive Example 28.1

Chapter 29 Active Figures 29.1, 29.6, and 29.7; Interactive Example 29.3

Chapter 30 Active Figures 30.2 and 30.10

An Invitation to Physics

It is our hope that you too will find physics an exciting and enjoyable experience and that you will profit from this experience, regardless of your chosen profession. Welcome to the exciting world of physics!

To see the World in a Grain of Sand
And a Heaven in a Wild Flower,
Hold infinity in the palm of your hand
And Eternity in an hour.

William Blake, "Auguries of Innocence"

Stone/Getty Images

Thousands of years ago, people in Southern England built Stonehenge, which was used as a calendar. The position of the sun and stars relative to the stones determined seasons for planting or harvesting.

CHAPTER

1

Introduction

OUTLINE

The goal of physics is to provide an understanding of the physical world by developing theories based on experiments. A physical theory is essentially a guess, usually expressed mathematically, about how a given physical system works. The theory makes certain predictions about the physical system which can then be checked by observations and experiments. If the predictions turn out to correspond closely to what is actually observed, then the theory stands, although it remains provisional. No theory to date has given a complete description of all physical phenomena, even within a given subdiscipline of physics. Every theory is a work in progress.

The basic laws of physics involve such physical quantities as force, velocity, volume, and acceleration, all of which can be described in terms of more fundamental quantities. In mechanics, the three most fundamental quantities are **length** (L), **mass** (M), and **time** (T); all other physical quantities can be constructed from these three.

1.1 STANDARDS OF LENGTH, MASS, AND TIME

To communicate the result of a measurement of a certain physical quantity, a *unit* for the quantity must be defined. For example, if our fundamental unit of length is defined to be 1.0 meter, and someone familiar with our system of measurement reports that a wall is 2.0 meters high, we know that the height of the wall is twice the fundamental unit of length. Likewise, if our fundamental unit of mass is defined as 1.0 kilogram, and we are told that a person has a mass of 75 kilograms, then that person has a mass 75 times as great as the fundamental unit of mass.

In 1960, an international committee agreed on a standard system of units for the fundamental quantities of science, called **SI** (Système International). Its units of length, mass, and time are the meter, kilogram, and second, respectively.

PhysicsNow™ Throughout the text, the PhysicsNow icon indicates an opportunity for you to test yourself on key concepts and to explore animations and interactions on the PhysicsNow website at **www.cp7e.com.**

Length

In 1799, the legal standard of length in France became the meter, defined as one ten-millionth of the distance from the equator to the North Pole. Until 1960, the official length of the meter was the distance between two lines on a specific bar of platinum-iridium alloy stored under controlled conditions. This standard was abandoned for several reasons, the principal one being that measurements of the separation between the lines are not precise enough. In 1960, the meter was defined as 1 650 763.73 wavelengths of orange-red light emitted from a krypton-86 lamp. In October 1983, this definition was abandoned also, and **the meter was redefined as the distance traveled by light in vacuum during a time interval of 1/299 792 458 second**. This latest definition establishes the speed of light at 299 792 458 meters per second.

Definition of the meter ▶

Mass

Definition of the kilogram ▶

The SI unit of mass, the kilogram, is defined as the mass of a specific platinum-iridium alloy cylinder kept at the International Bureau of Weights and Measures at Sèvres, France (similar to that shown in Figure 1.1a). As we'll see in Chapter 4, mass is a quantity used to measure the resistance to a change in the motion of an object. It's more difficult to cause a change in the motion of an object with a large mass than an object with a small mass.

Time

Before 1960, the time standard was defined in terms of the average length of a solar day in the year 1900. (A solar day is the time between successive appearances of the Sun at the highest point it reaches in the sky each day.) The basic unit of time, the second, was defined to be $(1/60)(1/60)(1/24) = 1/86\ 400$ of the average solar day. In 1967, the second was redefined to take advantage of the high precision attainable with an atomic clock, which uses the characteristic frequency of the light emitted from the cesium-133 atom as its "reference clock." **The second is now defined as 9 192 631 700 times the period of oscillation of radiation from the cesium atom**. The newest type of cesium atomic clock is shown in Figure 1.1b.

Definition of the second ▶

(a)

(b)

Courtesy of National Institute of Standards and Technology, U.S. Dept. of Commerce

Figure 1.1 (a) The National Standard Kilogram No. 20, an accurate copy of the International Standard Kilogram kept at Sèvres, France, is housed under a double bell jar in a vault at the National Institute of Standards and Technology. (b) The nation's primary time standard is a cesium fountain atomic clock developed at the National Institute of Standards and Technology laboratories in Boulder, Colorado. This clock will neither gain nor lose a second in 20 million years.

TABLE 1.1

Approximate Values of Some Measured Lengths

	Length (m)
Distance from Earth to most remote known quasar	1×10^{26}
Distance from Earth to most remote known normal galaxies	4×10^{25}
Distance from Earth to nearest large galaxy (M31, the Andromeda galaxy)	2×10^{22}
Distance from Earth to nearest star (Proxima Centauri)	4×10^{16}
One light year	9×10^{15}
Mean orbit radius of Earth about Sun	2×10^{11}
Mean distance from Earth to Moon	4×10^{8}
Mean radius of Earth	6×10^{6}
Typical altitude of satellite orbiting Earth	2×10^{5}
Length of football field	9×10^{1}
Length of housefly	5×10^{-3}
Size of smallest dust particles	1×10^{-4}
Size of cells in most living organisms	1×10^{-5}
Diameter of hydrogen atom	1×10^{-10}
Diameter of atomic nucleus	1×10^{-14}
Diameter of proton	1×10^{-15}

Approximate Values for Length, Mass, and Time Intervals

Approximate values of some lengths, masses, and time intervals are presented in Tables 1.1, 1.2, and 1.3, respectively. Note the wide ranges of values. Study these tables to get a feel for a kilogram of mass (this book has a mass of about 2 kilograms), a time interval of 10^{10} seconds (one century is about 3×10^9 seconds), or two meters of length (the approximate height of a forward on a basketball team). Appendix A reviews the notation for powers of 10, such as the expression of the number 50 000 in the form 5×10^4.

Systems of units commonly used in physics are the Système International, in which the units of length, mass, and time are the meter (m), kilogram (kg), and second (s); the cgs, or Gaussian, system, in which the units of length, mass, and time are the centimeter (cm), gram (g), and second; and the U.S. customary system, in which the units of length, mass, and time are the foot (ft), slug, and second. SI units are almost universally accepted in science and industry, and will be used throughout the book. Limited use will be made of Gaussian and U.S. customary units.

TABLE 1.2

Approximate Values of Some Masses

	Mass (kg)
Observable Universe	1×10^{52}
Milky Way galaxy	7×10^{41}
Sun	2×10^{30}
Earth	6×10^{24}
Moon	7×10^{22}
Shark	1×10^{2}
Human	7×10^{1}
Frog	1×10^{-1}
Mosquito	1×10^{-5}
Bacterium	1×10^{-15}
Hydrogen atom	2×10^{-27}
Electron	9×10^{-31}

TABLE 1.3

Approximate Values of Some Time Intervals

	Time Interval (s)
Age of Universe	5×10^{17}
Age of Earth	1×10^{17}
Average age of college student	6×10^{8}
One year	3×10^{7}
One day (time required for one revolution of Earth about its axis)	9×10^{4}
Time between normal heartbeats	8×10^{-1}
Period[a] of audible sound waves	1×10^{-3}
Period[a] of typical radio waves	1×10^{-6}
Period[a] of vibration of atom in solid	1×10^{-13}
Period[a] of visible light waves	2×10^{-15}
Duration of nuclear collision	1×10^{-22}
Time required for light to travel across a proton	3×10^{-24}

[a]A *period* is defined as the time required for one complete vibration.

TABLE 1.4

Some Prefixes for Powers of Ten Used with "Metric" (SI and cgs) Units

Power	Prefix	Abbreviation
10^{-18}	atto-	a
10^{-15}	femto-	f
10^{-12}	pico-	p
10^{-9}	nano-	n
10^{-6}	micro-	μ
10^{-3}	milli-	m
10^{-2}	centi-	c
10^{-1}	deci-	d
10^{1}	deka-	da
10^{3}	kilo-	k
10^{6}	mega-	M
10^{9}	giga-	G
10^{12}	tera-	T
10^{15}	peta-	P
10^{18}	exa-	E

Some of the most frequently used "metric" (SI and cgs) prefixes representing powers of 10 and their abbreviations are listed in Table 1.4. For example, 10^{-3} m is equivalent to 1 millimeter (mm), and 10^3 m is 1 kilometer (km). Likewise, 1 kg is equal to 10^3 g, and 1 megavolt (MV) is 10^6 volts (V).

1.2 THE BUILDING BLOCKS OF MATTER

A 1-kg ($\approx$ 2-lb) cube of solid gold has a length of about 3.73 cm ($\approx$ 1.5 in.) on a side. Is this cube nothing but wall-to-wall gold, with no empty space? If the cube is cut in half, the two resulting pieces retain their chemical identity as solid gold. But what if the pieces of the cube are cut again and again, indefinitely? Will the smaller and smaller pieces always be the same substance, gold? Questions such as these can be traced back to early Greek philosophers. Two of them—Leucippus and Democritus—couldn't accept the idea that such cutting could go on forever. They speculated that the process ultimately would end when it produced a particle that could no longer be cut. In Greek, *atomos* means "not sliceable." From this term comes our English word *atom*, once believed to be the smallest, ultimate particle of matter, but since found to be a composite of more elementary particles.

The atom can be visualized as a miniature Solar System, with a dense, positively charged nucleus occupying the position of the Sun, with negatively charged electrons orbiting like planets. This model of the atom, first developed by the great Danish physicist Niels Bohr nearly a century ago, led to the understanding of certain properties of the simpler atoms such as hydrogen, but failed to explain many fine details of atomic structure.

Notice the size of a hydrogen atom, listed in Table 1.1, and the size of a proton—the nucleus of a hydrogen atom—one hundred thousand times smaller. If the proton were the size of a Ping Pong ball, the electron would be a tiny speck about the size of a bacterium, orbiting the proton a kilometer away! Other atoms are similarly constructed. So there is a surprising amount of empty space in ordinary matter.

After the discovery of the nucleus in the early 1900s, questions arose concerning its structure. Is the nucleus a single particle or a collection of particles? The exact composition of the nucleus hasn't been defined completely even today, but by the early 1930s a model evolved that helped us understand how the nucleus behaves. Scientists determined that two basic entities—protons and neutrons—occupy the nucleus. The *proton* is nature's fundamental carrier of positive charge (equal in magnitude but opposite in sign to the charge on the electron), and the number of protons in a nucleus determines what the element is. For instance, a nucleus containing only one proton is the nucleus of an atom of hydrogen, regardless of how many neutrons may be present. Extra neutrons correspond to different isotopes of hydrogen—deuterium and tritium—which react chemically in exactly the same way as hydrogen, but are more massive. An atom having two protons in its nucleus, similarly, is always helium, although again, differing numbers of neutrons are possible.

The existence of *neutrons* was verified conclusively in 1932. A neutron has no charge and has a mass about equal to that of a proton. One of its primary purposes is to act as a "glue" to hold the nucleus together. If neutrons were not present, the repulsive electrical force between the positively charged protons would cause the nucleus to fly apart.

The division doesn't stop here; it turns out that protons, neutrons, and a zoo of other exotic particles are now thought to be composed of six particles called **quarks** (rhymes with "forks," though some rhyme it with "sharks"). These particles have been given the names *up, down, strange, charm, bottom,* and *top.* The up, charm, and top quarks each carry a charge equal to $+\frac{2}{3}$ that of the proton, whereas the down, strange, and bottom quarks each carry a charge equal to $-\frac{1}{3}$ the proton charge. The proton consists of two up quarks and one down quark (see Fig. 1.2), giving the correct charge for the proton, $+1$. The neutron is composed of two down quarks and one up quark and has a net charge of zero.

The up and down quarks are sufficient to describe all normal matter, so the existence of the other four quarks indirectly observed in high-energy experiments,

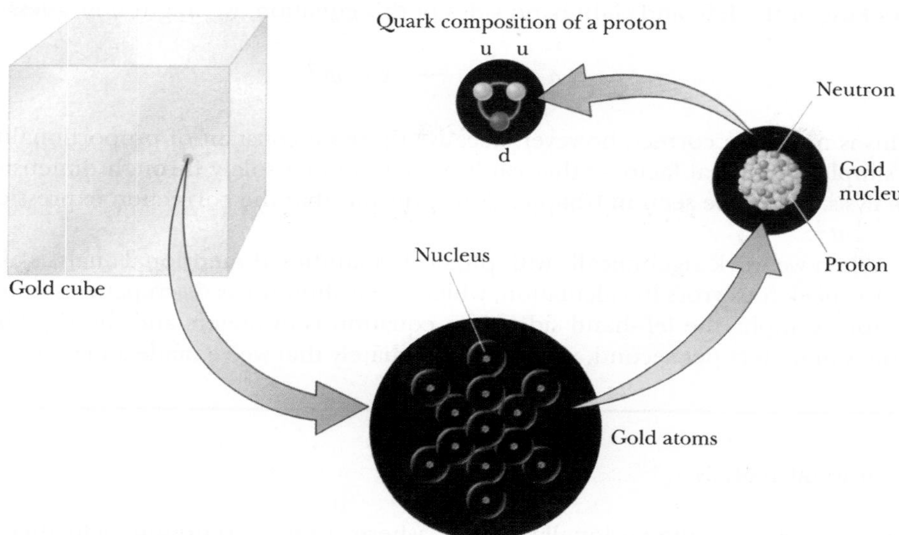

Figure 1.2 Levels of organization in matter. Ordinary matter consists of atoms, and at the center of each atom is a compact nucleus consisting of protons and neutrons. Protons and neutrons are composed of quarks. The quark composition of a proton is shown.

is something of a mystery. It's also possible that quarks themselves have internal structure. Many physicists believe that the most fundamental particles may be tiny loops of vibrating string.

1.3 DIMENSIONAL ANALYSIS

In physics, the word *dimension* denotes the physical nature of a quantity. The distance between two points, for example, can be measured in feet, meters, or furlongs, which are different ways of expressing the dimension of *length.*

The symbols that we use in this section to specify the dimensions of length, mass, and time are L, M, and T, respectively. Brackets [ ] will often be used to denote the dimensions of a physical quantity. For example, in this notation the dimensions of velocity v are written $[v] = \mathrm{L/T}$, and the dimensions of area A are $[A] = \mathrm{L}^2$. The dimensions of area, volume, velocity, and acceleration are listed in Table 1.5, along with their units in the three common systems. The dimensions of other quantities, such as force and energy, will be described later as they are introduced.

In physics, it's often necessary either to derive a mathematical expression or equation or to check its correctness. A useful procedure for doing this is called **dimensional analysis**, which makes use of the fact that **dimensions can be treated as algebraic quantities**. Such quantities can be added or subtracted only if they have the same dimensions. It follows that the terms on the opposite sides of an equation must have the same dimensions. If they don't, the equation is wrong. If they do, the equation is probably correct, except for a possible constant factor.

To illustrate this procedure, suppose we wish to derive a formula for the distance x traveled by a car in a time t if the car starts from rest and moves with constant acceleration a. The quantity x has the dimension length: $[x] = \mathrm{L}$. Time t, of course, has dimension $[t] = \mathrm{T}$. Acceleration is the change in velocity v with time. Since v has dimensions of length per unit time, or $[v] = \mathrm{L/T}$, acceleration must have dimensions $[a] = \mathrm{L/T}^2$. We organize this information in the form of an equation:

$$[a] = \frac{[v]}{[t]} = \frac{\mathrm{L/T}}{\mathrm{T}} = \frac{\mathrm{L}}{\mathrm{T}^2} = \frac{[x]}{[t]^2}$$

TABLE 1.5

Dimensions and Some Units of Area, Volume, Velocity, and Acceleration

System	Area (L^2)	Volume (L^3)	Velocity (L/T)	Acceleration ($\mathrm{L/T}^2$)
SI	m^2	m^3	m/s	$\mathrm{m/s}^2$
cgs	cm^2	cm^3	cm/s	$\mathrm{cm/s}^2$
U.S. customary	ft^2	ft^3	ft/s	$\mathrm{ft/s}^2$

Looking at the left- and right-hand sides of this equation, we might now guess that

$$a = \frac{x}{t^2} \quad \rightarrow \quad x = at^2$$

This is not quite correct, however, because there's a constant of proportionality—a simple numerical factor—that can't be determined solely through dimensional analysis. As will be seen in Chapter 2, it turns out that the correction expression is $x = \frac{1}{2}at^2$.

When we work algebraically with physical quantities, dimensional analysis allows us to check for errors in calculation, which often show up as discrepancies in units. If, for example, the left-hand side of an equation is in meters and the right-hand side is in meters per second, we know immediately that we've made an error.

EXAMPLE 1.1 Analysis of an Equation

Goal Check an equation using dimensional analysis.

Problem Show that the expression $v = v_0 + at$, is dimensionally correct, where v and v_0 represent velocities, a is acceleration, and t is a time interval.

Strategy Analyze each term, finding its dimensions, and then check to see if all the terms agree with each other.

Solution

Find dimensions for v and v_0.

$$[v] = [v_0] = \frac{\mathrm{L}}{\mathrm{T}}$$

Find the dimensions of at.

$$[at] = \frac{\mathrm{L}}{\mathrm{T}^2}(\mathrm{T}) = \frac{\mathrm{L}}{\mathrm{T}}$$

Remarks All the terms agree, so the equation is dimensionally correct.

Exercise 1.1

Determine whether the equation $x = vt^2$ is dimensionally correct. If not, provide a correct expression, up to an overall constant of proportionality.

Answer Incorrect. The expression $x = vt$ is dimensionally correct.

EXAMPLE 1.2 Find an Equation

Goal Derive an equation by using dimensional analysis.

Problem Find a relationship between a constant acceleration a, speed v, and distance r from the origin for a particle traveling in a circle.

Strategy Start with the term having the most dimensionality, a. Find its dimensions, and then rewrite those dimensions in terms of the dimensions of v and r. The dimensions of time will have to be eliminated with v, since that's the only quantity in which the dimension of time appears.

Solution

Write down the dimensions of a:

$$[a] = \frac{\mathrm{L}}{\mathrm{T}^2}$$

Solve the dimensions of speed for T:

$$[v] = \frac{\mathrm{L}}{\mathrm{T}} \quad \rightarrow \quad \mathrm{T} = \frac{\mathrm{L}}{[v]}$$

Substitute the expression for T into the equation for $[a]$:

$$[a] = \frac{\mathrm{L}}{\mathrm{T}^2} = \frac{\mathrm{L}}{(\mathrm{L}/[v])^2} = \frac{[v]^2}{\mathrm{L}}$$

Substitute L = $[r]$, and guess at the equation: $[a] = \frac{[v]^2}{[r]} \rightarrow a = \frac{v^2}{r}$

Remarks This is the correct equation for centripetal acceleration—acceleration towards the center of motion—to be discussed in Chapter 7. There isn't any need in this case, to introduce a numerical factor. Such a factor is often displayed explicitly as a constant k in front of the right-hand side—for example, $a = kv^2/r$. As it turns out, $k = 1$ gives the correct expression.

Exercise 1.2

In physics, energy E carries dimensions of mass times length squared, divided by time squared. Use dimensional analysis to derive a relationship for energy in terms of mass m and speed v, up to a constant of proportionality. Set the speed equal to c, the speed of light, and the constant of proportionality equal to 1 to get the most famous equation in physics.

Answer $E = kmv^2 \rightarrow E = mc^2$ when $k = 1$ and $v = c$.

1.4 UNCERTAINTY IN MEASUREMENT AND SIGNIFICANT FIGURES

Physics is a science in which mathematical laws are tested by experiment. No physical quantity can be determined with complete accuracy, because our senses are physically limited, even when extended with microscopes, cyclotrons, and other gadgets.

Knowing the experimental uncertainties in any measurement is very important. Without this information, little can be said about the final measurement. Using a crude scale, for example, we might find that a gold nugget has a mass of 3 kilograms. A prospective client interested in purchasing the nugget would naturally want to know about the accuracy of the measurement, to ensure paying a fair price. He wouldn't be happy to find that the measurement was good only to within a kilogram, because he might pay for three kilograms and get only two. Of course, he might get four kilograms for the price of three, but most people would be hesitant to gamble that an error would turn out in their favor.

Accuracy of measurement depends on the sensitivity of the apparatus, the skill of the person carrying out the measurement, and the number of times the measurement is repeated. There are many ways of handling uncertainties, and here we'll develop a basic and reliable method of keeping track of them in the measurement itself and in subsequent calculations.

Suppose that in a laboratory experiment we measure the area of a rectangular plate with a meter stick. Let's assume that the accuracy to which we can measure a particular dimension of the plate is ± 0.1 cm. If the length of the plate is measured to be 16.3 cm, we can claim only that it lies somewhere between 16.2 cm and 16.4 cm. In this case, we say that the measured value has three significant figures. Likewise, if the plate's width is measured to be 4.5 cm, the actual value lies between 4.4 cm and 4.6 cm. This measured value has only two significant figures. We could write the measured values as 16.3 ± 0.1 cm and 4.5 ± 0.1 cm. In general, **a significant figure is a reliably known digit** (other than a zero used to locate a decimal point).

Suppose we would like to find the area of the plate by multiplying the two measured values together. The final value can range between $(16.3 - 0.1 \text{ cm})(4.5 - 0.1 \text{ cm}) = (16.2 \text{ cm})(4.4 \text{ cm}) = 71.28 \text{ cm}^2$ and $(16.3 + 0.1 \text{ cm})(4.5 + 0.1 \text{ cm}) = (16.4 \text{ cm})(4.6 \text{ cm}) = 75.44 \text{ cm}^2$. Claiming to know anything about the hundredths place, or even the tenths place, doesn't make any sense, because it's clear we can't even be certain of the units place, whether it's the 1 in 71, the 5 in 75, or somewhere in between. The tenths and the hundredths places are clearly not significant. We have some information about the units place, so that number is significant. Multiplying the numbers at the middle of the uncertainty ranges gives (16.3 cm)(4.5 cm) = 73.35 cm^2, which is also in the middle of the area's uncertainty range. Since the hundredths and tenths are not significant, we drop them and take the answer to be 73 cm^2, with an uncertainty of ± 2 cm^2. Note that the answer has two significant figures, the same number of figures as the least accurately known quantity being multiplied, the 4.5-cm width.

There are two useful rules of thumb for determining the number of significant figures. The first, concerning multiplication and division, is as follows: **In multiplying (dividing) two or more quantities, the number of significant figures in the final product (quotient) is the same as the number of significant figures in the *least accurate* of the factors being combined, where *least accurate* means *having the lowest number of significant figures*.**

To get the final number of significant figures, it's usually necessary to do some rounding. If the last digit dropped is less than 5, simply drop the digit. If the last digit dropped is greater than or equal to 5, raise the last retained digit by one.

EXAMPLE 1.3 Installing a Carpet

Goal Apply the multiplication rule for significant figures.

Problem A carpet is to be installed in a room of length 12.71 m and width 3.46 m. Find the area of the room, retaining the proper number of significant figures.

Strategy Count the significant figures in each number. The smaller result is the number of significant figures in the answer.

Solution

Count significant figures:

12.71 m $\rightarrow$ 4 significant figures

3.46 m $\rightarrow$ 3 significant figures

Multiply the numbers, keeping only three digits:

$12.71 \text{ m} \times 3.46 \text{ m} = 43.9766 \text{ m}^2 \rightarrow 44.0 \text{ m}^2$

Remarks In reducing 43.976 6 to three significant figures, we used our rounding rule, adding 1 to the 9, which made 10 and resulted in carrying 1 to the unit's place.

Exercise 1.3

Repeat this problem, but with a room measuring 9.72 m long by 5.3 m wide.

Answer 52 m^2

TIP 1.1 Using Calculators

Calculators were designed by engineers to yield as many digits as the memory of the calculator chip permitted, so be sure to round the final answer down to the correct number of significant figures.

Zeros may or may not be significant figures. Zeros used to position the decimal point in such numbers as 0.03 and 0.007 5 are not significant (but are useful in avoiding errors). Hence, 0.03 has one significant figure, and 0.007 5 has two.

When zeros are placed after other digits in a whole number, there is a possibility of misinterpretation. For example, suppose the mass of an object is given as 1 500 g. This value is ambiguous, because we don't know whether the last two zeros are being used to locate the decimal point or whether they represent significant figures in the measurement.

Using scientific notation to indicate the number of significant figures removes this ambiguity. In this case, we express the mass as 1.5×10^3 g if there are two significant figures in the measured value, 1.50×10^3 g if there are three significant figures, and 1.500×10^3 g if there are four. Likewise, 0.000 15 is expressed in scientific notation as 1.5×10^{-4} if it has two significant figures or as 1.50×10^{-4} if it has three significant figures. The three zeros between the decimal point and the digit 1 in the number 0.000 15 are not counted as significant figures because they only locate the decimal point. In this book, **most of the numerical examples and end-of-chapter problems will yield answers having two or three significant figures**.

For addition and subtraction, it's best to focus on the number of decimal places in the quantities involved rather than on the number of significant figures. **When numbers are added (subtracted), the number of decimal places in the result should equal the smallest number of decimal places of any term in the sum (difference)**. For example, if we wish to compute 123 (zero decimal places) + 5.35 (two decimal places), the answer is 128 (zero decimal places) and not 128.35. If we compute the sum 1.000 1 (four decimal places) + 0.000 3 (four decimal places) = 1.000 4, the result has the correct number of decimal places, namely four. Observe that the rules for multiplying significant figures don't work here because the answer has five significant figures even though one of the terms in the sum, 0.000 3, has only one significant figure. Likewise, if we perform the subtraction 1.002 − 0.998 = 0.004, the result has three decimal places because each term in the subtraction has three decimal places.

TIP 1.2 No Commas in Numbers with Many Digits

In science, numbers with more than three digits are written in groups of three digits separated by spaces rather than commas; so that 10 000 is the same as the common American notation 10,000. Similarly, $\pi = 3.14159265$ is written as 3.141 592 65.

To show why this rule should hold, we return to the first example in which we added 123 and 5.35, and rewrite these numbers as 123.*xxx* and 5.35*x*. Digits written with an *x* are completely unknown and can be any digit from 0 to 9. Now we line up 123.*xxx* and 5.35*x* relative to the decimal point and perform the addition, using the rule that an unknown digit added to a known or unknown digit yields an unknown:

$$\begin{array}{r} 123.xxx \\ +\quad 5.35x \\ \hline 128.xxx \end{array}$$

The answer of 128.*xxx* means that we are justified only in keeping the number 128 because everything after the decimal point in the sum is actually unknown. The example shows that the controlling uncertainty is introduced into an addition or subtraction by the term with the smallest number of decimal places.

In performing any calculation, especially one involving a number of steps, there will always be slight discrepancies introduced by both the rounding process and the algebraic order in which steps are carried out. For example, consider 2.35 × 5.89/1.57. This computation can be performed in three different orders. First, we have 2.35 × 5.89 = 13.842, which rounds to 13.8, followed by 13.8/1.57 = 8.789 8, rounding to 8.79. Second, 5.89/1.57 = 3.751 6, which rounds to 3.75, resulting in 2.35 × 3.75 = 8.812 5, rounding to 8.81. Finally, 2.35/1.57 = 1.496 8 rounds to 1.50, and 1.50 × 5.89 = 8.835 rounds to 8.84. So three different algebraic orders, following the rules of rounding, lead to answers of 8.79, 8.81, and 8.84, respectively. Such minor discrepancies are to be expected, because the last significant digit is only one representative from a range of possible values, depending on experimental uncertainty. The discrepancies can be reduced by carrying one or more extra digits during the calculation. In our examples, however, intermediate results will be rounded off to the proper number of significant figures, and only those digits will be carried forward. In experimental work, more sophisticated techniques are used to determine the accuracy of an experimental result.

1.5 CONVERSION OF UNITS

Sometimes it's necessary to convert units from one system to another. Conversion factors between the SI and U.S. customary systems for units of length are as follows:

$$1 \text{ mile} = 1\,609 \text{ m} = 1.609 \text{ km} \qquad 1 \text{ ft} = 0.304\,8 \text{ m} = 30.48 \text{ cm}$$

$$1 \text{ m} = 39.37 \text{ in.} = 3.281 \text{ ft} \qquad 1 \text{ in.} = 0.025\,4 \text{ m} = 2.54 \text{ cm}$$

A more extensive list of conversion factors can be found on the inside front cover of this book.

Billy E. Barnes/Stock Boston

This road sign near Raleigh, North Carolina, shows distances in miles and kilometers. How accurate are the conversions?

Units can be treated as algebraic quantities that can "cancel" each other. We can make a fraction with the conversion that will cancel the units we don't want, and multiply that fraction by the quantity in question. For example, suppose we want to convert 15.0 in. to centimeters. Because 1 in. = 2.54 cm, we find that

$$15.0 \text{ in.} = 15.0 \text{ in.} \times \left(\frac{2.54 \text{ cm}}{1.00 \text{ in.}}\right) = 38.1 \text{ cm}$$

The next two examples show how to deal with problems involving more than one conversion and with powers.

EXAMPLE 1.4 Pull Over, Buddy!

Goal Convert units using several conversion factors.

Problem If a car is traveling at a speed of 28.0 m/s, is it exceeding the speed limit of 55.0 mi/h?

Strategy Meters must be converted to miles and seconds to hours, using the conversion factors listed on the inside front cover of the book. This requires two or three conversion ratios.

Solution

Convert meters to miles:

$$28.0 \text{ m/s} = \left(28.0 \frac{\text{m}}{\text{s}}\right)\left(\frac{1.00 \text{ mi}}{1\,609 \text{ m}}\right) = 1.74 \times 10^{-2} \text{ mi/s}$$

Convert seconds to hours:

$$1.74 \times 10^{-2} \text{ mi/s} = \left(1.74 \times 10^{-2} \frac{\text{mi}}{\text{s}}\right)\left(60.0 \frac{\text{s}}{\text{min}}\right)\left(60.0 \frac{\text{min}}{\text{h}}\right) = 62.6 \text{ mi/h}$$

Remarks The driver should slow down because he's exceeding the speed limit. An alternate approach is to use the single conversion relationship 1.00 m/s = 2.24 mi/h:

$$28.0 \text{ m/s} = \left(28.0 \frac{\text{m}}{\text{s}}\right)\left(\frac{2.24 \text{ mi/h}}{1.00 \text{ m/s}}\right) = 62.7 \text{ mi/h}$$

Answers to conversion problems may differ slightly, as here, due to rounding during intermediate steps.

Exercise 1.4

Convert 152 mi/h to m/s.

Answer 68.0 m/s

EXAMPLE 1.5 Press the Pedal to the Metal

Goal Convert a quantity featuring powers of a unit.

Problem The traffic light turns green, and the driver of a high-performance car slams the accelerator to the floor. The accelerometer registers 22.0 m/s^2. Convert this reading to km/min^2.

Strategy Here we need one factor to convert meters to kilometers and another two factors to convert seconds squared to minutes squared.

Solution

Insert the necessary factors:

$$\frac{22.0 \text{ m}}{1.00 \text{ s}^2}\left(\frac{1.00 \text{ km}}{1.00 \times 10^3 \text{ m}}\right)\left(\frac{60.0 \text{ s}}{1.00 \text{ min}}\right)^2 = 79.2 \frac{\text{km}}{\text{min}^2}$$

Remarks Notice that in each conversion factor the numerator equals the denominator when units are taken into account. A common error in dealing with squares is to square the units inside the parentheses while forgetting to square the numbers!

Exercise 1.5
Convert 4.50×10^3 kg/m^3 to g/cm^3.

Answer 4.50 g/cm^3

1.6 ESTIMATES AND ORDER-OF-MAGNITUDE CALCULATIONS

Getting an exact answer to a calculation may often be difficult or impossible, either for mathematical reasons or because limited information is available. In these cases, estimates can yield useful approximate answers that can determine whether a more precise calculation is necessary. Estimates also serve as a partial check if the exact calculations are actually carried out. If a large answer is expected but a small exact answer is obtained, there's an error somewhere.

For many problems, knowing the approximate value of a quantity—within a factor of 10 or so—is sufficient. This approximate value is called an **order-of-magnitude** estimate, and requires finding the power of 10 that is closest to the actual value of the quantity. For example, 75 kg $\sim 10^2$ kg, where the symbol $\sim$ means "is on the order of" or "is approximately." Increasing a quantity by three orders of magnitude means that its value increases by a factor of $10^3 = 1\,000$.

Occasionally, the process of making such estimates results in fairly crude answers, but answers ten times or more too large or small are still useful. For example, suppose you're interested in how many people have contracted a certain disease. Any estimates under ten thousand are small compared with Earth's total population, but a million or more would be alarming. So even relatively imprecise information can provide valuable guidance.

In developing these estimates, you can take considerable liberties with the numbers. For example, $\pi \sim 1$, $27 \sim 10$, and $65 \sim 100$. To get a less crude estimate, it's permissible to use slightly more accurate numbers (e.g., $\pi \sim 3$, $27 \sim 30$, $65 \sim 70$). Better accuracy can also be obtained by systematically underestimating as many numbers as you overestimate. Some quantities may be completely unknown, but it's standard to make reasonable guesses, as the examples show.

EXAMPLE 1.6 How Much Gasoline Do We Use?

Goal Develop a complex estimate.

Problem Estimate the number of gallons of gasoline used by all cars in the United States each year.

Strategy Estimate the number of people in the United States, and then estimate the number of cars per person. Multiply to get the number of cars. Guess at the number of miles per gallon obtained by a typical car and the number of miles driven per year, and from that get the number of gallons each car uses every year. Multiply by the estimated number of cars to get a final answer, the number of gallons of gas used.

Solution

The number of cars equals the number of people times the number of cars per person:

$$\text{number of cars} = (3.00 \times 10^8 \text{ people}) \times (0.5 \text{ cars/person}) \sim 10^8 \text{ cars}$$

The number of gallons used by one car in a year is the number of miles driven divided by the miles per gallon.

$$\frac{\#\text{ gal/yr}}{\text{car}} \approx \frac{\left(\dfrac{10^4 \text{ mi/yr}}{\text{car}}\right)}{10\,\dfrac{\text{mi}}{\text{gal}}} = 10^3\,\frac{\text{gal/yr}}{\text{car}}$$

Multiply these two results together to get an estimate of the number of gallons of gas used per year.

$$\#\text{ gal} \sim (10^8 \text{ cars}) \times \left(10^3\,\frac{\text{gal/yr}}{\text{car}}\right) = 10^{11} \text{ gal/yr}$$

Remarks Notice the inexact, and somewhat high, figure for the number of people in the United States, the estimate on the number of cars per person (figuring that every other person has a car of one kind or another), and the truncation of 1.5×10^8 cars to 10^8 cars. A similar estimate was used on the number of miles driven per year by a typical vehicle, and the average fuel economy, 10 mi/gal, looks low. None of this is important, because we are interested only in an order-of-magnitude answer. Few people owning a car would drive just 1 000 miles in a year, and very few would drive 100 000 miles, so 10 000 miles is a good estimate. Similarly, most cars get between 10 and 30 mi/gal, so using 10 is a reasonable estimate, while very few cars would get 100 mi/gal or 1 mi/gal.

In making estimates, it's okay to be cavalier! Feel free to take liberties ordinarily denied.

Exercise 1.6

How many new car tires are purchased in the United States each year? (Use the fact that tires wear out after about 50 000 miles.)

Answer $\sim 10^8$ tires (Individual answers may vary.)

Example 1.7 Stack One-Dollar Bills to the Moon

Goal Estimate the number of stacked objects required to reach a given height.

Problem How many one-dollar bills, stacked one on top of the other, would reach the Moon?

Strategy The distance to the Moon is about 400 000 km. Guess at the number of dollar bills in a millimeter, and multiply the distance by this number, after converting to consistent units.

Solution

We estimate that ten stacked bills form a layer of 1 mm. Convert mm to km:

$$\frac{10 \text{ bills}}{1 \cancel{\text{mm}}}\left(\frac{10^3 \cancel{\text{mm}}}{1 \cancel{\text{m}}}\right)\left(\frac{10^3 \cancel{\text{m}}}{1 \text{ km}}\right) = \frac{10^7 \text{ bills}}{1 \text{ km}}$$

Multiply this value by the approximate lunar distance:

$$\# \text{ of dollar bills} \sim (4 \times 10^5 \text{ km})\left(\frac{10^7 \text{ bills}}{1 \text{ km}}\right) = \boxed{4 \times 10^{12} \text{ bills}}$$

Remarks That's the same order of magnitude as the U.S. national debt!

Exercise 1.7

How many pieces of cardboard, typically found at the back of a bound pad of paper, would you have to stack up to match the height of the Washington monument, about 170 m tall?

Answer $\sim 10^5$ (Answers may vary.)

EXAMPLE 1.8 Number of Galaxies in the Universe

Goal Estimate a volume and a number density, and combine.

Problem Given that astronomers can see about 10 billion light years into space and that there are 14 galaxies in our local group, 2 million light years from the next local group, estimate the number of galaxies in the observable universe. (*Note*: One light year is the distance traveled by light in one year, about 9.5×10^{15} m.) (See Fig. 1.3.)

Strategy From the known information, we can estimate the number of galaxies per unit volume. The local group of 14 galaxies is contained in a sphere a million light years in radius, with the Andromeda group in a similar sphere, so there are about 10 galaxies within a volume of radius 1 million light years. Multiply that number density by the volume of the observable universe.

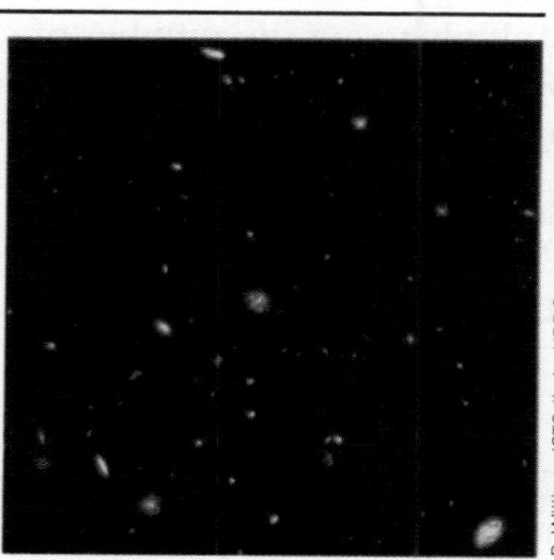

R. Williams (STScI), the HDF-S team, and NASA

Figure 1.3 In this deep-space photograph, there are few stars—just galaxies without end.

Solution

Compute the approximate volume V_{lg} of the local group of galaxies:

$$V_{lg} = \frac{4}{3}\pi r^3 \sim (10^6\ \text{ly})^3 = 10^{18}\ \text{ly}^3$$

Compute the number of galaxies per cubic light year:

$$\frac{\#\text{ of galaxies}}{\text{ly}^3} = \frac{\#\text{ of galaxies}}{V_{lg}} \sim \frac{10\text{ galaxies}}{10^{18}\ \text{ly}^3} = 10^{-17}\ \frac{\text{galaxies}}{\text{ly}^3}$$

Compute the approximate volume of the observable universe:

$$V_u = \frac{4}{3}\pi r^3 \sim (10^{10}\ \text{ly})^3 = 10^{30}\ \text{ly}^3$$

Multiply the density of galaxies by V_u:

$$\#\text{ of galaxies} \sim \left(\frac{\#\text{ of galaxies}}{\text{ly}^3}\right)V_u = \left(10^{-17}\ \frac{\text{galaxies}}{\text{ly}^3}\right)(10^{30}\ \text{ly}^3) = 10^{13}\text{ galaxies}$$

Remarks Notice the approximate nature of the computation, which uses $4\pi/3 \sim 1$ on two occasions and $14 \sim 10$ for the number of galaxies in the local group. This is completely justified: Using the actual numbers would be pointless, because the other assumptions in the problem—the size of the observable universe and the idea that the local galaxy density is representative of the density everywhere—are also very rough approximations. Further, there was nothing in the problem that required using volumes of spheres rather than volumes of cubes. Despite all these arbitrary choices, the answer still gives useful information, because it rules out a lot of reasonable possible answers. Before doing the calculation, a guess of a billion galaxies might have seemed plausible.

Exercise 1.8

Given that the nearest star is about 4 light years away and that the galaxy is roughly a disk 100 000 light years across and a thousand light years thick, estimate the number of stars in the Milky Way galaxy.

Answer $\sim 10^{12}$ stars (Estimates will vary. The actual answer is probably close to 4×10^{11} stars.)

1.7 COORDINATE SYSTEMS

Many aspects of physics deal with locations in space, which require the definition of a coordinate system. A point on a line can be located with one coordinate, a point in a plane with two coordinates, and a point in space with three.

A coordinate system used to specify locations in space consists of the following:

- A fixed reference point O, called the *origin*
- A set of specified axes, or directions, with an appropriate scale and labels on the axes
- Instructions on labeling a point in space relative to the origin and axes

One convenient and commonly used coordinate system is the **Cartesian coordinate system**, sometimes called the **rectangular coordinate system**. Such a system in two dimensions is illustrated in Figure 1.4. An arbitrary point in this system is labeled with the coordinates (x, y). For example, the point P in the figure has coordinates (5, 3). If we start at the origin O, we can reach P by moving 5 meters horizontally to the right and then 3 meters vertically upwards. In the same way, the point Q has coordinates $(-3, 4)$, which corresponds to going 3 meters horizontally to the left of the origin and 4 meters vertically upwards from there.

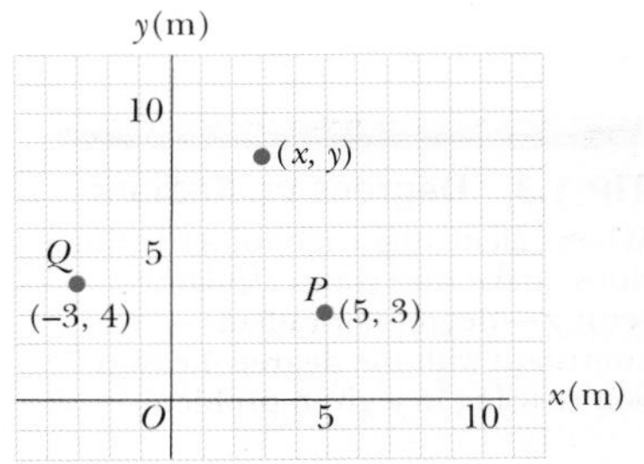

Figure 1.4 Designation of points in a two-dimensional Cartesian coordinate system. Every point is labeled with coordinates (x, y).

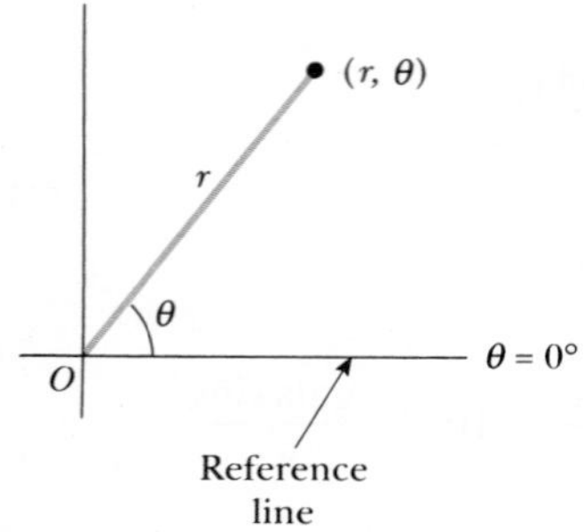

Figure 1.5 A polar coordinate system.

Positive x is usually selected as right of the origin and positive y upward from the origin, but in two dimensions this choice is largely a matter of taste. (In three dimensions, however, there are "right-handed" and "left-handed" coordinates, which lead to minus sign differences in certain operations. These will be addressed as needed.)

Sometimes it's more convenient to locate a point in space by its **plane polar coordinates** (r, θ), as in Figure 1.5. In this coordinate system, an origin O and a reference line are selected as shown. A point is then specified by the distance r from the origin to the point and by the angle θ between the reference line and a line drawn from the origin to the point. The standard reference line is usually selected to be the positive x-axis of a Cartesian coordinate system. The angle θ is considered positive when measured counterclockwise from the reference line and negative when measured clockwise. For example, if a point is specified by the polar coordinates 3 m and 60°, we locate this point by moving out 3 m from the origin at an angle of 60° above (counterclockwise from) the reference line. A point specified by polar coordinates 3 m and −60° is located 3 m out from the origin and 60° below (clockwise from) the reference line.

1.8 TRIGONOMETRY

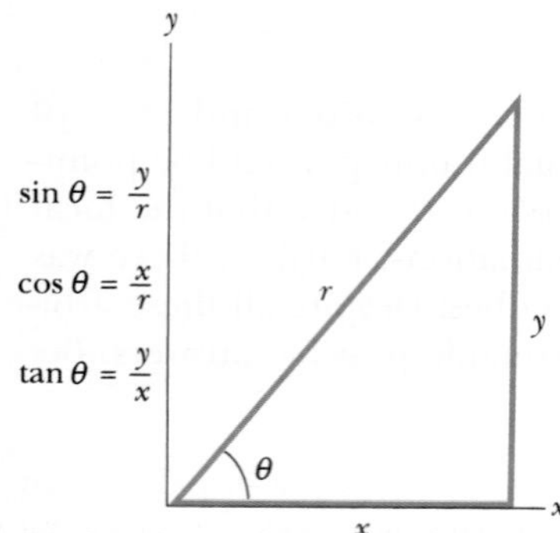

ACTIVE FIGURE 1.6
Certain trigonometric functions of a right triangle.

PhysicsNow™ Log into PhysicsNow at **www.cp7e.com**, and go to Active Figure 1.6 to move the point and see the changes to the rectangular and polar coordinates and to the sine, cosine, and tangent of angle θ.

Consider the right triangle shown in Active Figure 1.6, where side y is opposite the angle θ, side x is adjacent to the angle θ, and side r is the hypotenuse of the triangle. The basic trigonometric functions defined by such a triangle are the ratios of the lengths of the sides of the triangle. These relationships are called the sine (sin), cosine (cos), and tangent (tan) functions. In terms of θ, the basic trigonometric functions are as follows:[1]

$$\sin\theta = \frac{\text{side opposite } \theta}{\text{hypotenuse}} = \frac{y}{r}$$

$$\cos\theta = \frac{\text{side adjacent to } \theta}{\text{hypotenuse}} = \frac{x}{r} \qquad [1.1]$$

$$\tan\theta = \frac{\text{side opposite } \theta}{\text{side adjacent to } \theta} = \frac{y}{x}$$

For example, if the angle θ is equal to 30°, then the ratio of y to r is always 0.50; that is, $\sin 30° = 0.50$. Note that the sine, cosine, and tangent functions are quantities without units because each represents the ratio of two lengths.

Another important relationship, called the **Pythagorean theorem**, exists between the lengths of the sides of a right triangle:

$$r^2 = x^2 + y^2 \qquad [1.2]$$

Finally, it will often be necessary to find the values of inverse relationships. For example, suppose you know that the sine of an angle is 0.866, but you need to know the value of the angle itself. The inverse sine function may be expressed as $\sin^{-1}(0.866)$, which is a shorthand way of asking the question "What angle has a sine of 0.866?" Punching a couple of buttons on your calculator reveals that this angle is 60.0°. Try it for yourself and show that $\tan^{-1}(0.400) = 21.8°$. Be sure that your calculator is set for degrees and not radians. In addition, the inverse tangent function can return only values between −90° and +90°, so when an angle is in the second or third quadrant, it's necessary to add 180° to the answer in the calculator window.

TIP 1.3 Degrees vs. Radians
When calculating trigonometric functions, make sure your calculator setting—degrees or radians—is consistent with the degree measure you're using in a given problem.

The definitions of the trigonometric functions and the inverse trigonometric functions, as well as the Pythagorean theorem, can be applied to *any* right triangle, regardless of whether its sides correspond to x- and y-coordinates.

These results from trigonometry are useful in converting from rectangular coordinates to polar coordinates, or vice versa, as the next example shows.

[1]Many people use the mnemonic *SOHCAHTOA* to remember the basic trigonometric formulas: *S*ine = *O*pposite/*H*ypotenuse, *C*osine = *A*djacent/*H*ypotenuse, and *T*angent = *O*pposite/*A*djacent. (Thanks go to Professor Don Chodrow for pointing this out.)

EXAMPLE 1.9 Cartesian and Polar Coordinates

Goal Understand how to convert from plane rectangular coordinates to plane polar coordinates and vice versa.

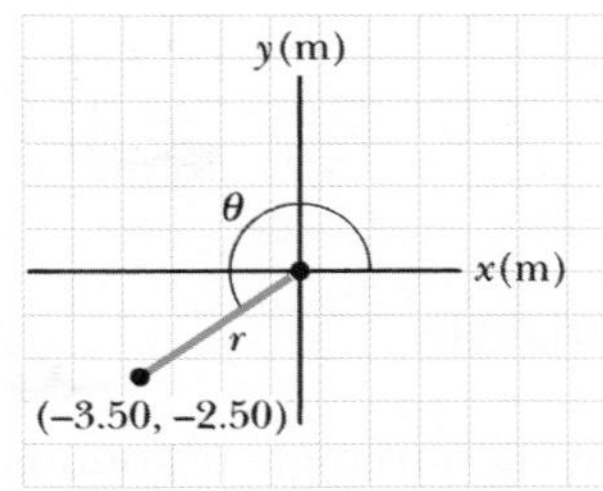

ACTIVE FIGURE 1.7 (Example 1.9) Converting from Cartesian coordinates to polar coordinates.

Physics Now™ Log into PhysicsNow at **www.cp7e.com**, and go to Active Figure 1.7 to move the point in the xy-plane and see how its Cartesian and polar coordinates change.

Problem **(a)** The Cartesian coordinates of a point in the xy-plane are $(x, y) = (-3.50, \ -2.50)$ m, as shown in Active Figure 1.7. Find the polar coordinates of this point. **(b)** Convert $(r, \theta) = (5.00 \text{ m}, 37.0°)$ to rectangular coordinates.

Strategy Apply the trigonometric functions and their inverses, together with the Pythagorean theorem.

Solution

(a) Cartesian to Polar

Take the square root of both sides of Equation 1.2 to find the radial coordinate:

$$r = \sqrt{x^2 + y^2} = \sqrt{(-3.50 \text{ m})^2 + (-2.50 \text{ m})^2} = \boxed{4.30 \text{ m}}$$

Use Equation 1.1 for the tangent function to find the angle with the inverse tangent, adding 180° because the angle is actually in third quadrant:

$$\tan\theta = \frac{y}{x} = \frac{-2.50 \text{ m}}{-3.50 \text{ m}} = 0.714$$

$$\theta = \tan^{-1}(0.714) = 35.5° + 180° = \boxed{216°}$$

(b) Polar to Cartesian

Use the trigonometric definitions, Equation 1.1.

$$x = r\cos\theta = (5.00 \text{ m}) \cos 37.0° = \boxed{3.99 \text{ m}}$$

$$y = r\sin\theta = (5.00 \text{ m}) \sin 37.0° = \boxed{3.01 \text{ m}}$$

Remarks When we take up vectors in two dimensions in Chapter 3, we will routinely use a similar process to find the direction and magnitude of a given vector from its components, or, conversely, to find the components from the vector's magnitude and direction.

Exercise 1.9

(a) Find the polar coordinates corresponding to $(x, y) = (-3.25, 1.50)$ m. **(b)** Find the Cartesian coordinates corresponding to $(r, \theta) = (4.00 \text{ m}, 53.0°)$

Answers **(a)** $(r, \theta) = (3.58 \text{ m}, 155°)$ **(b)** $(x, y) = (2.41 \text{ m}, 3.19 \text{ m})$

EXAMPLE 1.10 How High Is the Building?

Goal Apply basic results of trigonometry.

Problem A person measures the height of a building by walking out a distance of 46.0 m from its base and shining a flashlight beam toward the top. When the beam is elevated at an angle of 39.0° with respect to the horizontal, as shown in Figure 1.8, the beam just strikes the top of the building. Find the height of the building and the distance the flashlight beam has to travel before it strikes the top of the building.

Strategy Refer to the right triangle shown in the figure. We know the angle, 39.0°, and the length of the side adjacent to it. Since the height of the building is the side opposite the angle, we can use the tangent function. With the adjacent and opposite sides known, we can then find the hypotenuse with the Pythagorean theorem.

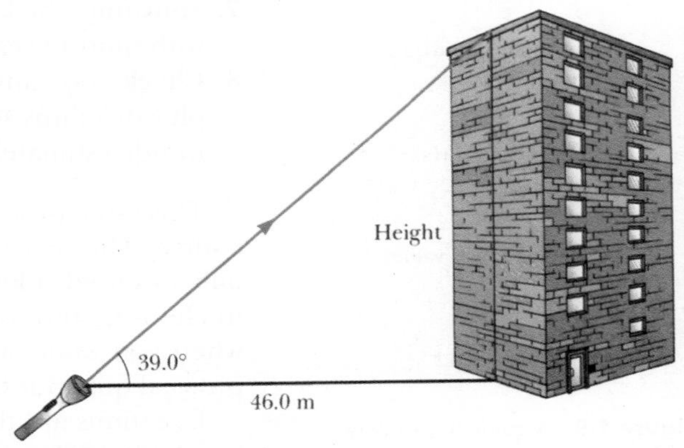

Figure 1.8 (Example 1.10)

Solution

Use the tangent of the given angle: $\tan 39.0° = \dfrac{\text{height}}{46.0 \text{ m}}$

Solve for the height: $\text{Height} = (\tan 39.0°)(46.0 \text{ m}) = (0.810)(46.0 \text{ m})$
$= 37.3 \text{ m}$

Find the hypotenuse of the triangle: $r = \sqrt{x^2 + y^2} = \sqrt{(37.3 \text{ m})^2 + (46.0 \text{ m})^2} = 59.2 \text{ m}$

Remarks In a later chapter, right-triangle trigonometry is often used when working with vectors.

Exercise 1.10

High atop a building 50.0 m tall, you spot a friend standing on a street corner. Using a protractor and dangling a plumb bob, you find that the angle between the horizontal and the direction of your friend is 25.0°. Your eyes are located 1.75 m above the top of the building. How far away from the foot of the building is your friend?

Answer 111 m

1.9 PROBLEM-SOLVING STRATEGY

Most courses in general physics require the student to learn the skills used in solving problems, and examinations usually include problems that test such skills. This brief section presents some useful suggestions that will help increase your success in solving problems. An organized approach to problem solving will also enhance your understanding of physical concepts and reduce exam stress. Throughout the book, there will be a number of sections labeled "Problem-Solving Strategy," many of them just a specializing of the list given below (and illustrated in Figure 1.9).

General Problem-Solving Strategy

1. **Read** the problem carefully at least twice. Be sure you understand the nature of the problem before proceeding further.
2. **Draw** a diagram while rereading the problem.
3. **Label** all physical quantities in the diagram, using letters that remind you what the quantity is (e.g., m for mass). Choose a coordinate system and label it.
4. **Identify** physical principles, the knowns and unknowns, and list them. Put circles around the unknowns.
5. **Equations**, the relationships between the labeled physical quantities, should be written down next. Naturally, the selected equations should be consistent with the physical principles identified in the previous step.
6. **Solve** the set of equations for the unknown quantities in terms of the known. Do this algebraically, without substituting values until the next step, except where terms are zero.
7. **Substitute** the known values, together with their units. Obtain a numerical value with units for each unknown.
8. **Check** your answer. Do the units match? Is the answer reasonable? Does the plus or minus sign make sense? Is your answer consistent with an order of magnitude estimate?

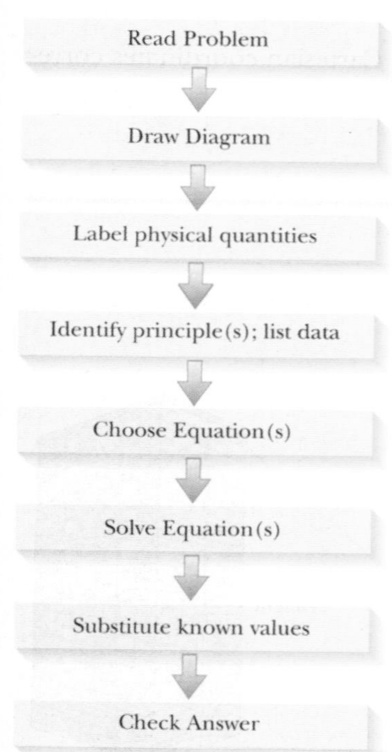

Figure 1.9 A guide to problem solving.

This same procedure, with minor variations, should be followed throughout the course. The first three steps are extremely important, because they get you mentally oriented. Identifying the proper concepts and physical principles assists you in choosing the correct equations. The equations themselves are essential, because when you understand them, you also understand the relationships between the physical quantities. This understanding comes through a lot of daily practice.

Equations are the tools of physics: To solve problems, you have to have them at hand, like a plumber and his wrenches. Know the equations, and understand what they mean and how to use them. Just as you can't have a conversation without knowing the local language, you can't solve physics problems without knowing and understanding the equations. This understanding grows as you study and apply the concepts and the equations relating them.

Carrying through the algebra for as long as possible, substituting numbers only at the end, is also important, because it helps you think in terms of the physical quantities involved, not merely the numbers that represent them. Many beginning physics students are eager to substitute, but once numbers are substituted, it's harder to understand relationships and easier to make mistakes.

The physical layout and organization of your work will make the final product more understandable and easier to follow. Although physics is a challenging discipline, your chances of success are excellent if you maintain a positive attitude and keep trying.

EXAMPLE 1.11 A Round Trip by Air

Goal Illustrate the Problem-Solving Strategy.

Problem An airplane travels 4.50×10^2 km due east and then travels an unknown distance due north. Finally, it returns to its starting point by traveling a distance of 525 km. How far did the airplane travel in the northerly direction?

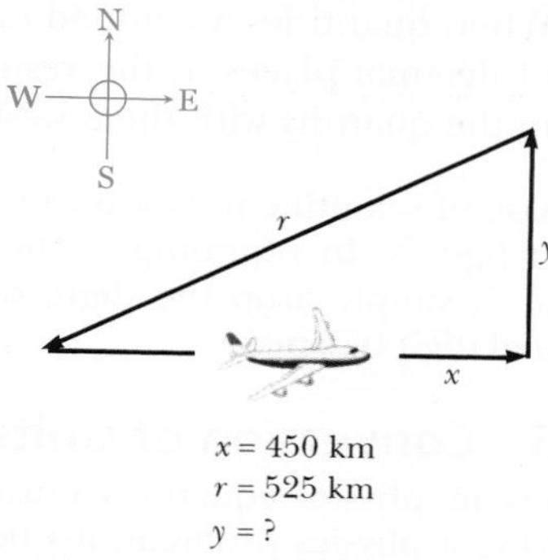

Figure 1.10 (Example 1.11)

Strategy We've finished reading the problem (step 1), and have drawn a diagram (step 2) in Figure 1.10 and labeled it (step 3). From the diagram, we recognize a right triangle and identify (step 4) the principle involved: the Pythagorean theorem. Side y is the unknown quantity, and the other sides are known.

Solution

Write the Pythagorean theorem (step 5):

$$r^2 = x^2 + y^2$$

Solve symbolically for y (step 6):

$$y^2 = r^2 - x^2 \quad \rightarrow \quad y = +\sqrt{r^2 - x^2}$$

Substitute the numbers, with units (step 7):

$$y = \sqrt{(525 \text{ km})^2 - (4.50 \times 10^2 \text{ km})^2} = \boxed{270 \text{ km}}$$

Remarks Note that the negative solution has been disregarded, because it's not physically meaningful. In checking (step 8), note that the units are correct and that an approximate answer can be obtained by using the easier quantities, 500 km and 400 km. Doing so gives an answer of 300 km, which is approximately the same as our calculated answer of 270 km.

Exercise 1.11

A plane flies 345 km due south, then turns and flies northeast 615 km, until it's due east of its starting point. If the plane now turns and heads for home, how far will it have to go?

Answer 509 km

SUMMARY

PhysicsNow™ Take a practice test by logging into PhysicsNow at **www.cp7e.com** and clicking on the Pre-Test link for this chapter.

1.1 Standards of Length, Mass, and Time

The physical quantities in the study of mechanics can be expressed in terms of three fundamental quantities: length, mass, and time, which have the SI units meters (m), kilograms (kg), and seconds (s), respectively.

1.2 The Building Blocks of Matter

Matter is made of atoms, which in turn are made up of a relatively small nucleus of protons and neutrons within a cloud of electrons. Protons and neutrons are composed of still smaller particles, called quarks.

1.3 Dimensional Analysis

Dimensional analysis can be used to check equations and to assist in deriving them. When the dimensions on both sides of the equation agree, the equation is often correct up to a numerical factor. When the dimensions don't agree, the equation must be wrong.

1.4 Uncertainty in Measurement and Significant Figures

No physical quantity can be determined with complete accuracy. The concept of significant figures affords a basic method of handling these uncertainties. A significant figure is a reliably known digit, other than a zero, used to locate the decimal point. The two rules of significant figures are as follows:

1. When multiplying or dividing using two or more quantities, the result should have the same number of significant figures as the quantity having the fewest significant figures.
2. When quantities are added or subtracted, the number of decimal places in the result should be the same as in the quantity with the fewest decimal places.

Use of scientific notation can avoid ambiguity in significant figures. In rounding, if the last digit dropped is less than 5, simply drop the digit, otherwise raise the last retained digit by one.

1.5 Conversion of Units

Units in physics equations must always be consistent. In solving a physics problem, it's best to start with consistent units, using the table of conversion factors on the inside front cover as necessary.

Converting units is a matter of multiplying the given quantity by a fraction, with one unit in the numerator and its equivalent in the other units in the denominator, arranged so the unwanted units in the given quantity are cancelled out in favor of the desired units.

1.6 Estimates and Order-of-Magnitude Calculations

Sometimes it's useful to find an approximate answer to a question, either because the math is difficult or because information is incomplete. A quick estimate can also be used to check a more detailed calculation. In an order-of-magnitude calculation, each value is replaced by the closest power of ten, which sometimes must be guessed or estimated when the value is unknown. The computation is then carried out. For quick estimates involving known values, each value can first be rounded to one significant figure.

1.7 Coordinate Systems

The Cartesian coordinate system consists of two perpendicular axes, usually called the x-axis and y-axis, with each axis labeled with all numbers from negative infinity to positive infinity. Points are located by specifying the x- and y-values. Polar coordinates consist of a radial coordinate r which is the distance from the origin, and an angular coordinate θ, which is the angular displacement from the positive x-axis.

1.8 Trigonometry

The three most basic trigonometric functions of a right triangle are the sine, cosine, and tangent, defined as follows:

$$\sin\theta = \frac{\text{side opposite } \theta}{\text{hypotenuse}} = \frac{y}{r}$$

$$\cos\theta = \frac{\text{side adjacent to } \theta}{\text{hypotenuse}} = \frac{x}{r} \qquad \textbf{[1.1]}$$

$$\tan\theta = \frac{\text{side opposite } \theta}{\text{side adjacent to } \theta} = \frac{y}{x}$$

The **Pythagorean theorem** is an important relationship between the lengths of the sides of a right triangle:

$$r^2 = x^2 + y^2 \qquad \textbf{[1.2]}$$

where r is the hypotenuse of the triangle and x and y are the other two sides.

CONCEPTUAL QUESTIONS

1. Estimate the order of magnitude of the length, in meters, of each of the following: (a) a mouse, (b) a pool cue, (c) a basketball court, (d) an elephant, (e) a city block.
2. What types of natural phenomena could serve as time standards?
3. (a) Estimate the number of times your heart beats in a month. (b) Estimate the number of human heartbeats in an average lifetime.
4. An object with a mass of 1 kg weighs approximately 2 lb. Use this information to estimate the mass of the following objects: (a) a baseball; (b) your physics textbook; (c) a pickup truck.
5. Find the order of magnitude of your age in seconds.
6. Estimate the number of atoms in 1 cm^3 of a solid. (Note that the diameter of an atom is about 10^{-10} m.)
7. The height of a horse is sometimes given in units of "hands." Why is this a poor standard of length?
8. How many of the lengths or time intervals given in Tables 1.2 and 1.3 could you verify, using only equipment found in a typical dormitory room?
9. An ancient unit of length called the *cubit* was equal to six palms, where a palm was the width of the four fingers of an open hand. Noah's ark was 300 cubits long, 50 cubits wide, and 30 cubits high. Estimate the volume of the ark in cubic meters. Also, estimate the volume of a typical home in cubic meters, and compare it with the volume of the ark.
10. Do an order-of-magnitude calculation for an everyday situation you encounter. For example, how far do you walk or drive each day?
11. If an equation is dimensionally correct, does this mean that the equation must be true? If an equation is not dimensionally correct, does this mean that the equation can't be true?

12. Figure Q1.12 is a photograph showing unit conversions on the labels of some grocery-store items. Check the accuracy of these conversions. Are the manufacturers using significant figures correctly?

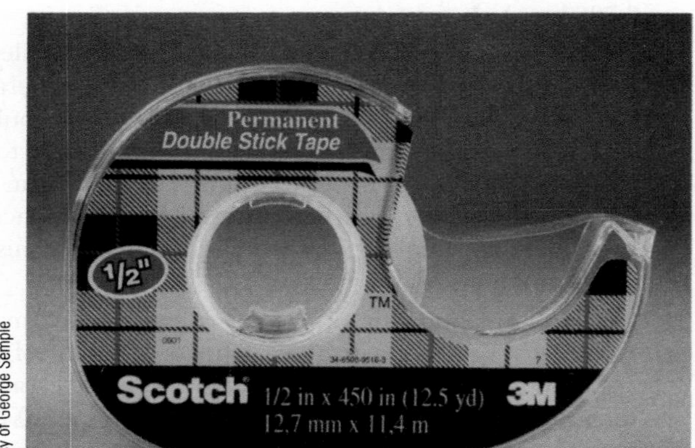

Courtesy of George Semple

PROBLEMS

Figure Q1.12

PROBLEMS

1, 2, 3 = straightforward, intermediate, challenging ☐ = full solution available in *Student Solutions Manual/Study Guide*
PhysicsNow™ = coached solution with hints available at **www.cp7e.com** [icon] = biomedical application

Section 1.3 Dimensional Analysis

1. A shape that covers an area A and has a uniform height h has a volume $V = Ah$. (a) Show that $V = Ah$ is dimensionally correct. (b) Show that the volumes of a cylinder and of a rectangular box can be written in the form $V = Ah$, identifying A in each case. (Note that A, sometimes called the "footprint" of the object, can have any shape and that the height can, in general, be replaced by the average thickness of the object.)

2. (a) Suppose that the displacement of an object is related to time according to the expression $x = Bt^2$. What are the dimensions of B? (b) A displacement is related to time as $x = A\sin(2\pi ft)$, where A and f are constants. Find the dimensions of A. (*Hint*: A trigonometric function appearing in an equation must be dimensionless.)

3. The period of a simple pendulum, defined as the time necessary for one complete oscillation, is measured in time units and is given by

$$T = 2\pi\sqrt{\frac{\ell}{g}}$$

where ℓ is the length of the pendulum and g is the acceleration due to gravity, in units of length divided by time squared. Show that this equation is dimensionally consistent. (You might want to check the formula using your keys at the end of a string and a stopwatch.)

4. Each of the following equations was given by a student during an examination:

$$\tfrac{1}{2}mv^2 = \tfrac{1}{2}mv_0^2 + \sqrt{mgh} \qquad v = v_0 + at^2 \qquad ma = v^2$$

Do a dimensional analysis of each equation and explain why the equation can't be correct.

5. Newton's law of universal gravitation is represented by

$$F = G\frac{Mm}{r^2}$$

where F is the gravitational force, M and m are masses, and r is a length. Force has the SI units $\text{kg}\cdot\text{m/s}^2$. What are the SI units of the proportionality constant G?

6. (a) One of the fundamental laws of motion states that the acceleration of an object is directly proportional to the resultant force on it and inversely proportional to its mass. If the proportionality constant is defined to have no dimensions, determine the dimensions of force. (b) The newton is the SI unit of force. According to the results for (a), how can you express a force having units of newtons by using the fundamental units of mass, length, and time?

Section 1.4 Uncertainty in Measurement and Significant Figures

7. How many significant figures are there in (a) 78.9 ± 0.2, (b) 3.788×10^9, (c) 2.46×10^{-6}, (d) 0.0032?

8. A rectangular plate has a length of (21.3 ± 0.2) cm and a width of (9.8 ± 0.1) cm. Calculate the area of the plate, including its uncertainty.

9. PhysicsNow™ Carry out the following arithmetic operations: (a) the sum of the measured values 756, 37.2, 0.83, and 2.5; (b) the product 0.0032×356.3; (c) the product $5.620 \times \pi$.

10. The speed of light is now defined to be $2.99\ 7924\ 58 \times 10^8$ m/s. Express the speed of light to (a) three significant figures, (b) five significant figures, and (c) seven significant figures.

11. A farmer measures the perimeter of a rectangular field. The length of each long side of the rectangle is found to be 38.44 m, and the length of each short side is found to be 19.5 m. What is the perimeter of the field?

12. The radius of a circle is measured to be (10.5 ± 0.2) m. Calculate (a) the area and (b) the circumference of the circle, and give the uncertainty in each value.

13. A fisherman catches two striped bass. The smaller of the two has a measured length of 93.46 cm (two decimal places, four significant figures), and the larger fish has a measured length of 135.3 cm (one decimal place, four significant figures). What is the total length of fish caught for the day?

14. (a) Using your calculator, find, in scientific notation with appropriate rounding, (a) the value of $(2.437 \times 10^4)(6.5211 \times 10^9)/(5.37 \times 10^4)$ and (b) the value of $(3.14159 \times 10^2)(27.01 \times 10^4)/(1\,234 \times 10^6)$.

Section 1.5 Conversion of Units

15. A fathom is a unit of length, usually reserved for measuring the depth of water. A fathom is approximately 6 ft in length. Take the distance from Earth to the Moon to be 250 000 miles, and use the given approximation to find the distance in fathoms.

16. Find the height or length of these natural wonders in kilometers, meters, and centimeters: (a) The longest cave system in the world is the Mammoth Cave system in Central Kentucky, with a mapped length of 348 miles. (b) In the United States, the waterfall with the greatest single drop is Ribbon Falls in California, which drops 1 612 ft. (c) At 20 320 feet, Mount McKinley in Alaska is America's highest mountain. (d) The deepest canyon in the United States is King's Canyon in California, with a depth of 8 200 ft.

17. A rectangular building lot measures 100 ft by 150 ft. Determine the area of this lot in square meters (m^2).

18. Suppose your hair grows at the rate of 1/32 inch per day. Find the rate at which it grows in nanometers per second. Since the distance between atoms in a molecule is on the order of 0.1 nm, your answer suggests how rapidly atoms are assembled in this protein synthesis.

19. Using the data in Table 1.1 and the appropriate conversion factors, find the distance to the nearest star, in feet.

20. Using the data in Table 1.3 and the appropriate conversion factors, find the age of Earth in years.

21. The speed of light is about 3.00×10^8 m/s. Convert this figure to miles per hour.

22. A house is 50.0 ft long and 26 ft wide and has 8.0-ft-high ceilings. What is the volume of the interior of the house in cubic meters and in cubic centimeters?

23. The amount of water in reservoirs is often measured in acre-ft. One acre-ft is a volume that covers an area of one acre to a depth of one foot. An acre is 43 560 ft^2. Find the volume in SI units of a reservoir containing 25.0 acre-ft of water.

24. The base of a pyramid covers an area of 13.0 acres (1 acre = 43 560 ft^2) and has a height of 481 ft (Fig. P1.24). If the volume of a pyramid is given by the expression $V = bh/3$, where b is the area of the base and h is the height, find the volume of this pyramid in cubic meters.

© Sylvain Grandadam/Photo Researchers, Inc.

Figure P1.24

25. PhysicsNow™ A quart container of ice cream is to be made in the form of a cube. What should be the length of a side, in centimeters? (Use the conversion 1 gallon = 3.786 liter.)

26. (a) Find a conversion factor to convert from miles per hour to kilometers per hour. (b) For a while, federal law mandated that the maximum highway speed would be 55 mi/h. Use the conversion factor from part (a) to find the speed in kilometers per hour. (c) The maximum highway speed has been raised to 65 mi/h in some places. In kilometers per hour, how much of an increase is this over the 55-mi/h limit?

27. One cubic centimeter (1.0 cm^3) of water has a mass of 1.0×10^{-3} kg. (a) Determine the mass of 1.0 m^3 of water. (b) Assuming that biological substances are 98% water, estimate the masses of a cell with a diameter of 1.0 μm, a human kidney, and a fly. Take a kidney to be roughly a sphere with a radius of 4.0 cm and a fly to be roughly a cylinder 4.0 mm long and 2.0 mm in diameter.

28. A billionaire offers to give you $1 billion if you can count out that sum with only $1 bills. Should you accept her offer? Assume that you can count at an average rate of one bill every second, and be sure to allow for the fact that you need about 8 hours a day for sleeping and eating.

Section 1.6 Estimates and Order-of-Magnitude Calculations

Note: In developing answers to the problems in this section, you should state your important assumptions, including the numerical values assigned to parameters used in the solution.

29. Imagine that you are the equipment manager of a professional baseball team. One of your jobs is to keep baseballs on hand for games. Balls are sometimes lost when players hit them into the stands as either home runs or foul balls. Estimate how many baseballs you have to buy per season in order to make up for such losses. Assume that your team plays an 81-game home schedule in a season.

30. A hamburger chain advertises that it has sold more than 50 billion hamburgers. Estimate how many pounds of hamburger meat must have been used by the chain and how many head of cattle were required to furnish the meat.

31. An automobile tire is rated to last for 50 000 miles. Estimate the number of revolutions the tire will make in its lifetime.

32. Grass grows densely everywhere on a quarter-acre plot of land. What is the order of magnitude of the number of blades of grass? Explain your reasoning. Note that 1 acre = 43 560 ft^2.

33. Estimate the number of Ping-Pong balls that would fit into a typical-size room (without being crushed). In your solution, state the quantities you measure or estimate and the values you take for them.

34. Soft drinks are commonly sold in aluminum containers. To an order of magnitude, how many such containers are thrown away or recycled each year by U.S. consumers? How many tons of aluminum does this represent? In your solution, state the quantities you measure or estimate and the values you take for them.

Section 1.7 Coordinate Systems

35. A point is located in a polar coordinate system by the coordinates $r = 2.5$ m and $\theta = 35°$. Find the x- and

y-coordinates of this point, assuming that the two coordinate systems have the same origin.

36. A certain corner of a room is selected as the origin of a rectangular coordinate system. If a fly is crawling on an adjacent wall at a point having coordinates (2.0, 1.0), where the units are meters, what is the distance of the fly from the corner of the room?

37. Express the location of the fly in Problem 36 in polar coordinates.

38. Two points in a rectangular coordinate system have the coordinates (5.0, 3.0) and (−3.0, 4.0), where the units are centimeters. Determine the distance between these points.

Section 1.8 Trigonometry

39. PhysicsNow™ For the triangle shown in Figure P1.39, what are (a) the length of the unknown side, (b) the tangent of θ, and (c) the sine of ϕ?

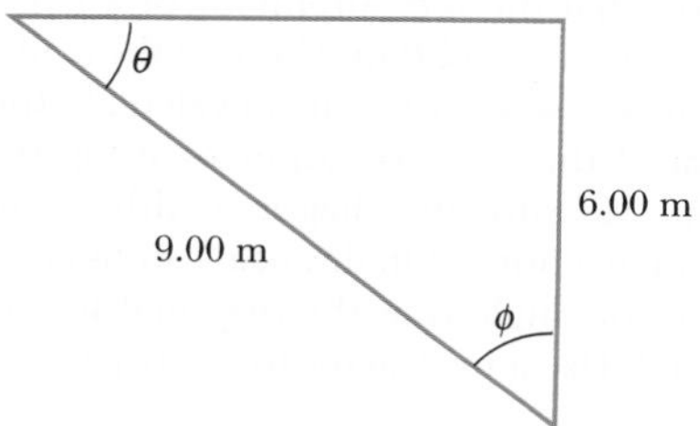

Figure P1.39

40. A ladder 9.00 m long leans against the side of a building. If the ladder is inclined at an angle of 75.0° to the horizontal, what is the horizontal distance from the bottom of the ladder to the building?

41. A high fountain of water is located at the center of a circular pool as shown in Figure P1.41. Not wishing to get his feet wet, a student walks around the pool and measures its circumference to be 15.0 m. Next, the student stands at the edge of the pool and uses a protractor to gauge the angle of elevation at the bottom of the fountain to be 55.0°. How high is the fountain?

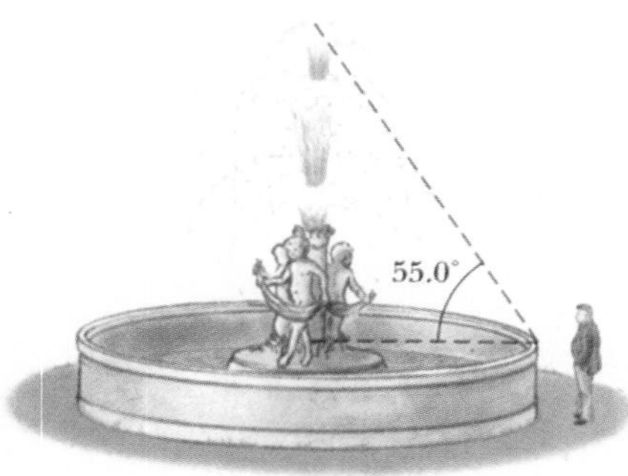

Figure P1.41

42. A right triangle has a hypotenuse of length 3.00 m, and one of its angles is 30.0°. What are the lengths of (a) the side opposite the 30.0° angle and (b) the side adjacent to the 30.0° angle?

43. In Figure P1.43, find (a) the side opposite θ, (b) the side adjacent to ϕ, (c) cos θ, (d) sin ϕ, and (e) tan ϕ.

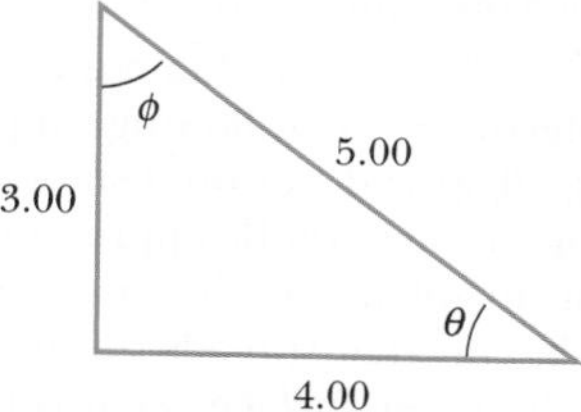

Figure P1.43

44. In a certain right triangle, the two sides that are perpendicular to each other are 5.00 m and 7.00 m long. What is the length of the third side of the triangle?

45. In Problem 44, what is the tangent of the angle for which 5.00 m is the opposite side?

46. A surveyor measures the distance across a straight river by the following method: Starting directly across from a tree on the opposite bank, he walks 100 m along the riverbank to establish a baseline. Then he sights across to the tree. The angle from his baseline to the tree is 35.0°. How wide is the river?

ADDITIONAL PROBLEMS

47. A restaurant offers pizzas in two sizes: small, with a radius of six inches; and large, with a radius of nine inches. A customer argues that if the small one sells for six dollars, the large should sell for nine dollars. Without doing any calculations, is the customer correct? Defend your answer. Calculate the area of each pizza to find out how much pie you are getting in each case. If the small one costs six dollars how much should the large cost?

48. The radius of the planet Saturn is 5.85×10^7 m, and its mass is 5.68×10^{26} kg (Fig. P1.48). (a) Find the density of Saturn (its mass divided by its volume) in grams per cubic centimeter. (The volume of a sphere is given by $(4/3)\pi r^3$.) (b) Find the area of Saturn in square feet. (The surface area of a sphere is given by $4\pi r^2$.)

Figure P1.48 A view of Saturn.

49. The displacement of an object moving under uniform acceleration is some function of time and the acceleration. Suppose we write this displacement as $s = ka^m t^n$, where k is a dimensionless constant. Show by dimensional analysis that this expression is satisfied if $m = 1$ and $n = 2$. Can the analysis give the value of k?

50. Compute the order of magnitude of the mass of (a) a bathtub filled with water and (b) a bathtub filled with pennies. In your solution, list the quantities you estimate and the value you estimate for each.

51. PhysicsNow™ You can obtain a rough estimate of the size of a molecule by the following simple experiment: Let a droplet of oil spread out on a smooth surface of water. The resulting oil slick will be approximately one molecule thick. Given an oil droplet of mass 9.00×10^{-7} kg and density 918 kg/m^3 that spreads out into a circle of radius 41.8 cm on the water surface, what is the order of magnitude of the diameter of an oil molecule?

52. In 2003, the U.S. national debt was about \$7 trillion. (a) If payments were made at the rate of \$1 000 per second, how many years would it take to pay off the debt, assuming that no interest were charged? (b) A dollar bill is about 15.5 cm long. If seven trillion dollar bills were laid end to end around the Earth's equator, how many times would they encircle the planet? Take the radius of the Earth at the equator to be 6 378 km. (*Note:* Before doing any of these calculations, try to guess at the answers. You may be very surprised.)

53. Estimate the number of piano tuners living in New York City. This question was raised by the physicist Enrico Fermi, who was well known for making order-of-magnitude calculations.

54. Sphere 1 has surface area A_1 and volume V_1, and sphere 2 has surface area A_2 and volume V_2. If the radius of sphere 2 is double the radius of sphere 1, what is the ratio of (a) the areas, A_2/A_1 and (b) the volumes, V_2/V_1?

55. (a) How many seconds are there in a year? (b) If one micrometeorite (a sphere with a diameter on the order of 10^{-6} m) struck each square meter of the Moon each second, estimate the number of years it would take to cover the Moon with micrometeorites to a depth of one meter. (*Hint:* Consider a cubic box, 1 m on a side, on the Moon, and find how long it would take to fill the box.)

ACTIVITIES

A.1. Choose a variety of objects that range in length from a few centimeters to a few meters. Try guessing the lengths in a unit appropriate to their size, and then use a meter stick supplied by your instructor to check your guesses. Keep trying until you can estimate consistently to within 20% of an object's actual length.

A.2. Choose a variety of objects that range in mass from a few grams to a few kilograms. Estimate the masses by hefting the objects, then use a balance supplied by your instructor to check your guesses. Keep trying until you can estimate consistently to within 30% of an object's actual mass.

A.3. You know that the measurements of a typical sheet of paper are 8.50 in. by 11.0 in. Convert these measurements to millimeters. Use your results to calculate the length of the diagonal of the sheet of paper by using the Pythagorean theorem. Measure the diagonal with a ruler to see how well you have done. Finally, use a suitable trig function to calculate the angle that the diagonal line makes with the horizontal. Use a protractor to verify your calculation.

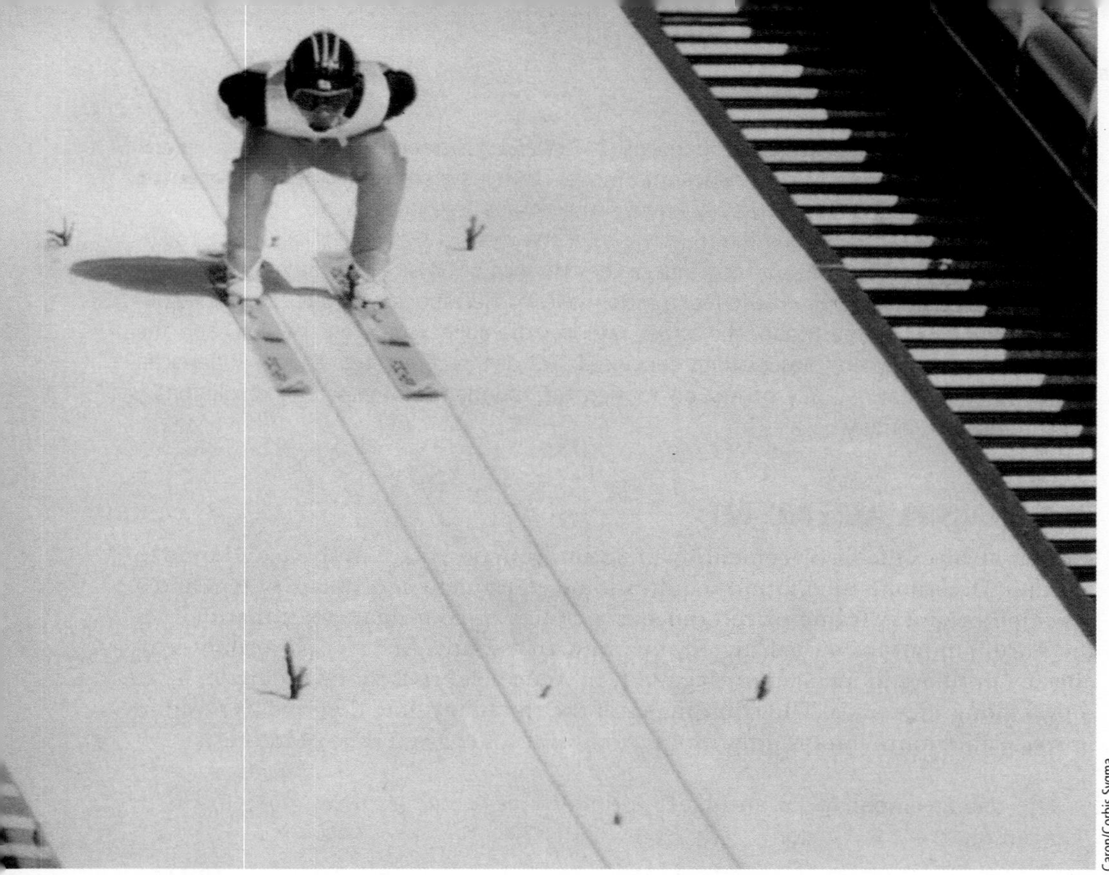

Caron/Corbis Sygma

Gravity propels a ski jumper down a straight, snow-covered slope at an acceleration that is approximately constant. The equations of kinematics, studied in this chapter, can give his position and velocity along the slope at any time.

CHAPTER

2

Motion In One Dimension

OUTLINE

Life is motion. Our muscles coordinate motion microscopically to enable us to walk and jog. Our hearts pump tirelessly for decades, moving blood through our bodies. Cell wall mechanisms move select atoms and molecules in and out of cells. From the prehistoric chase of antelopes across the savanna to the pursuit of satellites in space, mastery of motion has been critical to our survival and success as a species.

The study of motion and of physical concepts such as force and mass is called **dynamics**. The part of dynamics that describes motion without regard to its causes is called **kinematics**. In this chapter, the focus is on kinematics in one dimension: motion along a straight line. This kind of motion—and, indeed, *any* motion—involves the concepts of displacement, velocity, and acceleration. Here, we use these concepts to study the motion of objects undergoing constant acceleration. In Chapter 3 we will repeat this discussion for objects moving in two dimensions.

The first recorded evidence of the study of mechanics can be traced to the people of ancient Sumeria and Egypt, who were interested primarily in understanding the motions of heavenly bodies. The most systematic and detailed early studies of the heavens were conducted by the Greeks from about 300 B.C. to A.D. 300. Ancient scientists and laypeople regarded the Earth as the center of the Universe. This **geocentric model** was accepted by such notables as Aristotle (384–322 B.C.) and Claudius Ptolemy (about A.D. 140). Largely because of the authority of Aristotle, the geocentric model became the accepted theory of the Universe until the 17th century.

About 250 B.C., the Greek philosopher Aristarchus worked out the details of a model of the Solar System based on a spherical Earth that rotated on its axis and revolved around the Sun. He proposed that the sky appeared to turn westward because the Earth was turning eastward. This model wasn't given much consideration, because it was believed that if the Earth turned, it would set up a great wind as it moved through the air. We know now that the Earth carries the air and everything else with it as it rotates.

"Miss Dempswell, send in a frame of reference."

Figure 2.1

The Polish astronomer Nicolaus Copernicus (1473–1543) is credited with initiating the revolution that finally replaced the geocentric model. In his system, called the **heliocentric model**, Earth and the other planets revolve in circular orbits around the Sun.

This early knowledge formed the foundation for the work of Galileo Galilei (1564–1642), who stands out as the dominant facilitator of the entrance of physics into the modern era. In 1609, he became one of the first to make astronomical observations with a telescope. He observed mountains on the Moon, the larger satellites of Jupiter, spots on the Sun, and the phases of Venus. Galileo's observations convinced him of the correctness of the Copernican theory. His quantitative study of motion formed the foundation of Newton's revolutionary work in the next century.

2.1 DISPLACEMENT

Motion involves the displacement of an object from one place in space and time to another. Describing the motion requires some convenient coordinate system and a specified origin. A **frame of reference** is a choice of coordinate axes that defines the starting point for measuring any quantity, an essential first step in solving virtually any problem in mechanics (Fig. 2.1). In Active Figure 2.2a, for example, a car moves along the x-axis. The coordinates of the car at any time describe its position in space and, more importantly, its *displacement* at some given time of interest.

Definition of displacement ▶

The **displacement** Δx of an object is defined as its *change in position*, and is given by

$$\Delta x \equiv x_f - x_i \qquad [2.1]$$

where the initial position of the car is labeled x_i and the final position is x_f. (The indices i and f stand for initial and final, respectively.)

SI unit: meter (m)

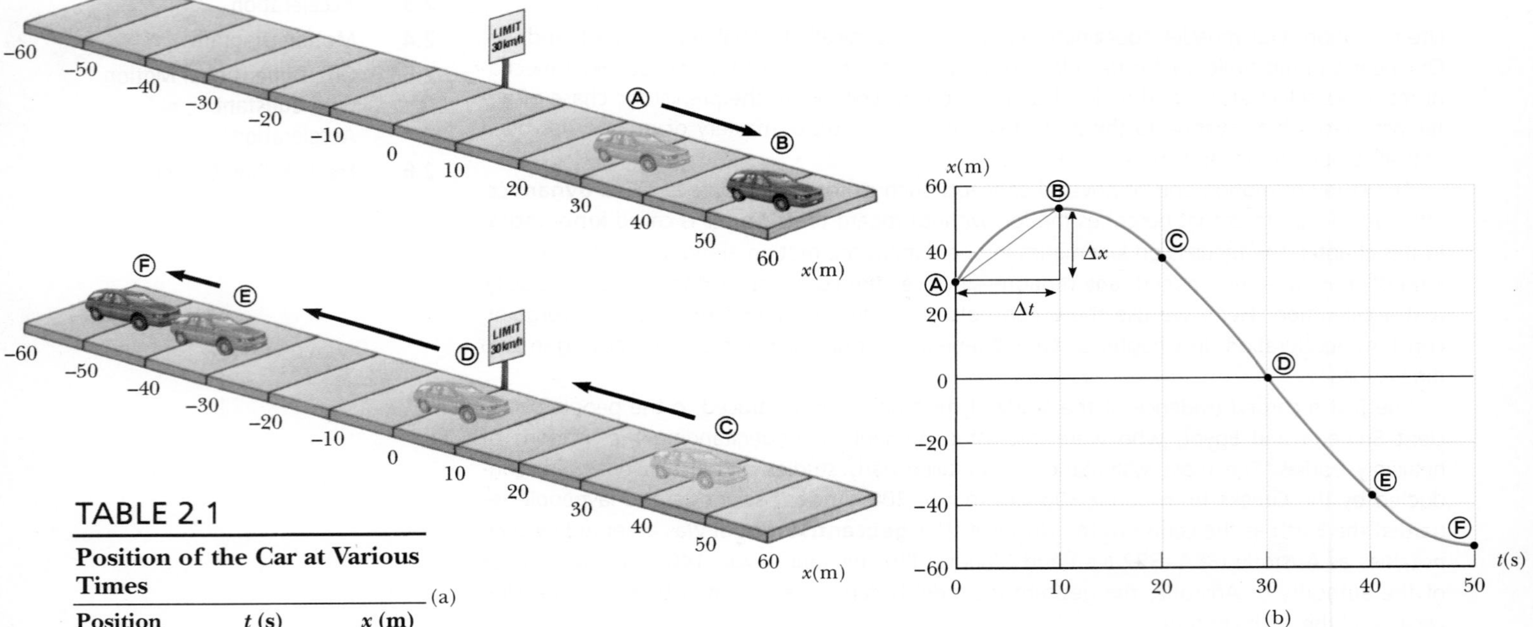

TABLE 2.1

Position of the Car at Various Times

Position	t (s)	x (m)
Ⓐ	0	30
Ⓑ	10	52
Ⓒ	20	38
Ⓓ	30	0
Ⓔ	40	−37
Ⓕ	50	−53

ACTIVE FIGURE 2.2

(a) A car moves back and forth along a straight line taken to be the x-axis. Because we are interested only in the car's translational motion, we can model it as a particle. (b) Graph of position vs. time for the motion of the "particle."

Physics Now™

Log into PhysicsNow at **www.cp7e.com**, and go to Active Figure 2.2 to move each of the six points Ⓐ through Ⓕ and observe the motion of the car pictorially and graphically as it follows a smooth path through the points.

We will use the Greek letter delta, Δ, to denote a change in any physical quantity. From the definition of displacement, we see that Δx (read "delta ex") is positive if x_f is greater than x_i and negative if x_f is less than x_i. For example, if the car moves from point Ⓐ to point Ⓑ, so that the initial position is $x_i = 30$ m and the final position is $x_f = 52$ m, the displacement is $\Delta x = x_f - x_i = 52 \text{ m} - 30 \text{ m} = +22 \text{ m}$. However, if the car moves from point Ⓒ to point Ⓕ, then the initial position is $x_i = 38$ m and the final position is $x_f = -53$ m, the displacement is $\Delta x = x_f - x_i = -53 \text{ m} - 38 \text{ m} = -91 \text{ m}$. A positive answer indicates a displacement in the positive x-direction, whereas a negative answer indicates a displacement in the negative x-direction. Active Figure 2.2b displays the graph of the car's position as a function of time.

TIP 2.1 A Displacement Isn't a Distance!

The displacement of an object is *not* the same as the distance it travels. Toss a tennis ball up and catch it. The ball travels a *distance* equal to twice the maximum height reached, but its *displacement* is zero.

Because displacement has both a magnitude (size) and a direction, it's a vector quantity, as are velocity and acceleration. In general, **a vector quantity is characterized by having both a magnitude and a direction**. By contrast, **a scalar quantity has magnitude, but no direction**. Scalar quantities such as mass and temperature are completely specified by a numeric value with appropriate units; no direction is involved.

Vector quantities will be usually denoted in boldface type with an arrow over the top of the letter. For example, $\vec{\mathbf{v}}$ represents velocity and $\vec{\mathbf{a}}$ denotes an acceleration, both vector quantities. In this chapter, however, it won't be necessary to use that notation, because in one-dimensional motion an object can only move in one of two directions, and these directions are easily specified by plus and minus signs.

TIP 2.2 Vectors Have Both a Magnitude and a Direction.

Scalars have size. Vectors, too, have size, and they also point in a direction.

2.2 VELOCITY

In day-to-day usage, the terms *speed* and *velocity* are interchangeable. In physics, however, there's a clear distinction between them: Speed is a scalar quantity, having only magnitude, while velocity is a vector, having both magnitude and direction.

Why must velocity be a vector? If you want to get to a town 70 km away in an hour's time, it's not enough to drive at a speed of 70 km/h; you must travel in the correct direction as well. This is obvious, but shows that velocity gives considerably more information than speed, as will be made more precise in the formal definitions.

◀ Definition of average speed

The **average speed** of an object over a given time interval is defined as the total distance traveled divided by the total time elapsed:

$$\text{Average speed} \equiv \frac{\text{total distance}}{\text{total time}}$$

SI unit: meter per second (m/s)

In symbols, this equation might be written $v = d/t$, with the letter v understood in context to be the average speed, and not a velocity. Because total distance and total time are always positive, the average speed will be positive, also. The definition of average speed completely ignores what may happen between the beginning and the end of the motion. For example, you might drive from Atlanta, Georgia, to St. Petersburg, Florida, a distance of about 500 miles, in 10 hours. Your average speed is 500 mi/10 h = 50 mi/h. It doesn't matter if you spent two hours in a traffic jam traveling only 5 mi/h and another hour at a rest stop. For average speed, only the total distance traveled and total elapsed time are important.

EXAMPLE 2.1 The Tortoise and The Hare

Goal Apply the concept of average speed.

Problem A turtle and a rabbit engage in a footrace over a distance of 4.00 km. The rabbit runs 0.500 km and then stops for a 90.0-min nap. Upon awakening, he remembers the race and runs twice as fast. Finishing the course in a total time of 1.75 h, the rabbit wins the race. **(a)** Calculate the average speed of the rabbit. **(b)** What was his average speed before he stopped for a nap?

Strategy Finding the overall average speed in part (a) is just a matter of dividing the total distance by the total time. Part (b) requires two equations and two unknowns, the latter turning out to be the two different average speeds: v_1 before the nap and v_2 after the nap. One equation is given in the statement of the problem ($v_2 = 2v_1$), while the other comes from the fact the rabbit ran for only fifteen minutes because he napped for ninety minutes.

Solution

(a) Find the rabbit's overall average speed.

Apply the equation for average speed:

$$\text{Average speed} \equiv \frac{\text{total distance}}{\text{total time}} = \frac{4.00 \text{ km}}{1.75 \text{ h}}$$
$$= \boxed{2.29 \text{ km/h}}$$

(b) Find the rabbit's average speed before his nap.

Sum the running times, and set the sum equal to 0.25 h:

$$t_1 + t_2 = 0.250 \text{ h}$$

Substitute $t_1 = d_1/v_1$ and $t_2 = d_2/v_2$:

$$\frac{d_1}{v_1} + \frac{d_2}{v_2} = 0.250 \text{ h} \qquad (1)$$

Equation (1) and $v_2 = 2v_1$ are the two equations needed, and d_1 and d_2 are known. Solve for v_1 by substitution:

$$\frac{d_1}{v_1} + \frac{d_2}{v_2} = \frac{0.500 \text{ km}}{v_1} + \frac{3.50 \text{ km}}{2v_1} = 0.250 \text{ h}$$
$$v_1 = \boxed{9.00 \text{ km/h}}$$

Remark As seen in this example, average speed can be calculated regardless of any variation in speed over the given time interval.

Exercise 2.1

Estimate the average speed of the Apollo spacecraft in m/s, given that the craft took five days to reach the Moon from Earth. (The Moon is 3.8×10^8 m from Earth.)

Answer ~ 900 m/s

Unlike average speed, **average velocity** is a vector quantity, having both a magnitude and a direction. Consider again the car of Figure 2.2, moving along the road (the x-axis). Let the car's position be x_i at some time t_i and x_f at a later time t_f. In the time interval $\Delta t = t_f - t_i$, the displacement of the car is $\Delta x = x_f - x_i$.

Definition of average velocity ▶

The average velocity $\bar{v}$ during a time interval Δt is the displacement Δx divided by Δt:

$$\bar{v} = \frac{\Delta x}{\Delta t} = \frac{x_f - x_i}{t_f - t_i} \qquad [2.2]$$

SI Unit: meter per second (m/s)

Unlike the average speed, which is always positive, the average velocity of an object in one dimension can be either positive or negative, depending on the sign of the displacement. (The time interval Δt is always positive.) For example, in Figure 2.2a, the average velocity of the car is positive in the upper illustration, a positive sign indicating motion to the right along the x-axis. Similarly, a negative average velocity for the car in the lower illustration of the figure indicates that it moves to the left along the x-axis.

As an example, we can use the data in Table 2.1 to find the average velocity in the time interval from point Ⓐ to point Ⓑ (assume two digits are significant):

$$\bar{v} = \frac{\Delta x}{\Delta t} = \frac{52 \text{ m} - 30 \text{ m}}{10 \text{ s} - 0 \text{ s}} = 2.2 \text{ m/s}$$

Aside from meters per second, other common units for average velocity are feet per second (ft/s) in the U.S. customary system and centimeters per second (cm/s) in the cgs system.

To further illustrate the distinction between speed and velocity, suppose we're watching a drag race from the Goodyear blimp. In one run we see a car follow the straight-line path from Ⓟ to Ⓠ shown in Figure 2.3 during the time interval Δt, and in a second run a car follows the curved path during the same interval. From the definition in Equation 2.2, the two cars had the same average velocity, because they had the same displacement $\Delta x = x_f - x_i$ during the same time interval Δt. The car taking the curved route, however, traveled a greater distance and had the higher average speed.

Figure 2.3 A drag race viewed from a blimp. One car follows the red straight-line path from Ⓟ to Ⓠ, and a second car follows the blue curved path.

Quick Quiz 2.1

Figure 2.4 shows the unusual path of a confused football player. After receiving a kickoff at his own goal, he runs downfield to within inches of a touchdown, then reverses direction and races back until he's tackled at the exact location where he first caught the ball. During this run, what is (a) the total distance he travels, (b) his displacement, and (c) his average velocity in the x-direction?

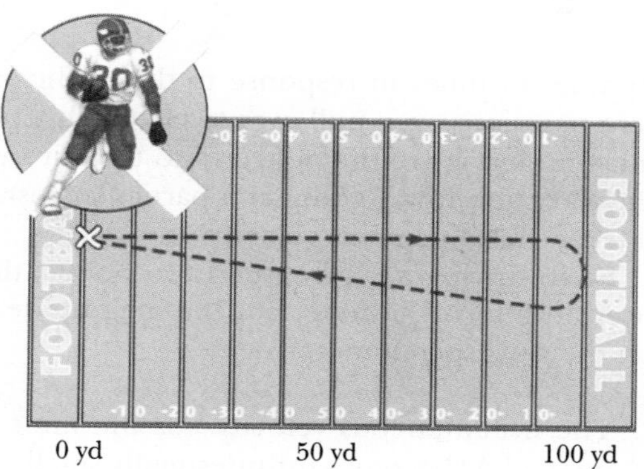

Figure 2.4 (Quick Quiz 2.1) The path followed by a confused football player.

Graphical Interpretation of Velocity

If a car moves along the x-axis from Ⓐ to Ⓑ to Ⓒ, and so forth, we can plot the positions of these points as a function of the time elapsed since the start of the motion. The result is a **position vs. time graph** like those of Figure 2.5. In Figure 2.5a, the graph is a straight line, because the car is moving at constant velocity. The same displacement Δx occurs in each time interval Δt. In this case, the average velocity is always the same and is equal to $\Delta x/\Delta t$. Figure 2.5b is a graph of the data in Table 2.1. Here, the position vs. time graph is not a straight line, because the velocity of the car is changing. Between any two points, however, we can draw a straight line just as in Figure 2.5a, and the slope of that line is the average velocity $\Delta x/\Delta t$ in that time interval. In general, **the average velocity of an object during the time interval Δt is equal to the slope of the straight line joining the initial and final points on a graph of the object's position versus time.**

From the data in Table 2.1 and the graph in Figure 2.5b, we see that the car first moves in the positive x-direction as it travels from Ⓐ to Ⓑ, reaches a position of 52 m at time $t = 10$ s, then reverses direction and heads backwards. In the first 10 s of its motion, as the car travels from Ⓐ to Ⓑ, its average velocity is 2.2 m/s, as previously calculated. In the first 40 seconds, as the car goes from Ⓐ to Ⓔ, its displacement is $\Delta x = -37 \text{ m} - (30 \text{ m}) = -67$ m. So the average velocity in this interval, which equals the slope of the blue line in Figure 2.5b from Ⓐ to Ⓔ, is $\bar{v} = \Delta x/\Delta t = (-67 \text{ m})/(40 \text{ s}) = -1.7$ m/s. In general, there will be a different average velocity between any distinct pair of points.

TIP 2.3 Slopes of Graphs

The word *slope* is often used in reference to the graphs of physical data. Regardless of the type of data, the *slope* is given by

$$\text{Slope} = \frac{\text{change in vertical axis}}{\text{change in horizontal axis}}$$

Slope carries units.

TIP 2.4 Average Velocity Versus Average Speed

Average velocity is *not* the same as average speed. If you run from $x = 0$ m to $x = 25$ m and back to your starting point in a time interval of 5 s, the average velocity is zero, while the average speed is 10 m/s.

Instantaneous Velocity

Average velocity doesn't take into account the details of what happens *during* an interval of time. On a car trip, for example, you may speed up or slow down a

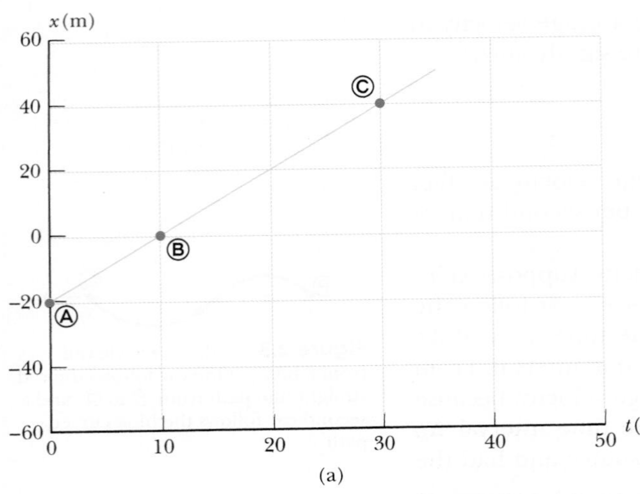

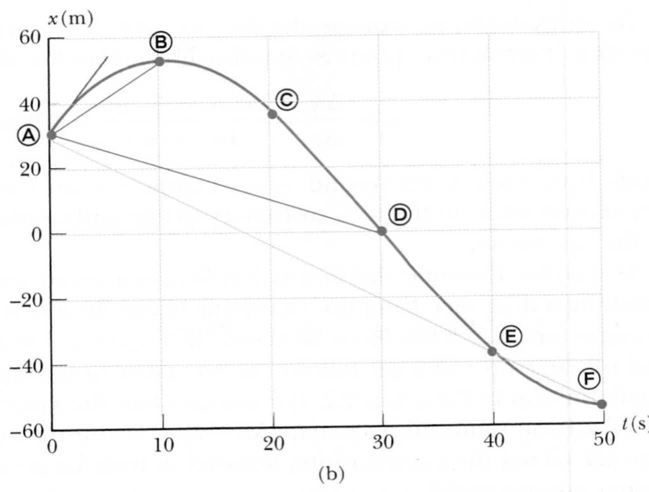

Figure 2.5 (a) Position vs. time graph for the motion of a car moving along the x-axis at constant velocity. (b) Position vs. time graph for the motion of a car with changing velocity, using the data in Table 2.1. The average velocity in the time interval Δt is the slope of the blue straight line connecting Ⓐ and Ⓓ.

number of times in response to the traffic and the condition of the road, and on rare occasions even pull over to chat with a police officer about your speed. What is most important to the police (and to your own safety) is the speed of your car and the direction it was going at a particular instant in time, which together determine the car's **instantaneous velocity**.

So in driving a car between two points, the average velocity must be computed over an interval of time, but the magnitude of instantaneous velocity can be read on the car's speedometer.

Definition of instantaneous velocity ▶

The instantaneous velocity v is the limit of the average velocity as the time interval Δt becomes infinitesimally small:

$$v \equiv \lim_{\Delta t \to 0} \frac{\Delta x}{\Delta t} \qquad [2.3]$$

SI unit: meter per second (m/s)

The notation $\lim_{\Delta t \to 0}$ means that the ratio $\Delta x/\Delta t$ is repeatedly evaluated for smaller and smaller time intervals Δt. As Δt gets extremely close to zero, the ratio $\Delta x/\Delta t$ gets closer and closer to a fixed number, which is defined as the instantaneous velocity.

To better understand the formal definition, consider data obtained on our vehicle via radar (Table 2.2). At $t = 1.00$ s, the car is at $x = 5.00$ m, and at $t = 3.00$ s, it's at $x = 52.5$ m. The average velocity computed for this interval is $\Delta x/\Delta t = (52.5 \text{ m} - 5.00 \text{ m})/(3.00 \text{ s} - 1.00 \text{ s}) = 23.8$ m/s. This result could be used as an estimate for the velocity at $t = 1.00$ s, but it wouldn't be very accurate, because the

TABLE 2.2

Positions of a Car at Specific Instants of Time

t (s)	x (m)
1.00	5.00
1.01	5.47
1.10	9.67
1.20	14.3
1.50	26.3
2.00	34.7
3.00	52.5

TABLE 2.3

Calculated Values of the Time Intervals, Displacements, and Average Velocities for the Car of Table 2.2

Time Interval (s)	Δt (s)	Δx (m)	$\bar{v}$ (m/s)
1.00 to 3.00	2.00	47.5	23.8
1.00 to 2.00	1.00	29.7	29.7
1.00 to 1.50	0.50	21.3	42.6
1.00 to 1.20	0.20	9.30	46.5
1.00 to 1.10	0.10	4.67	46.7
1.00 to 1.01	0.01	0.470	47.0

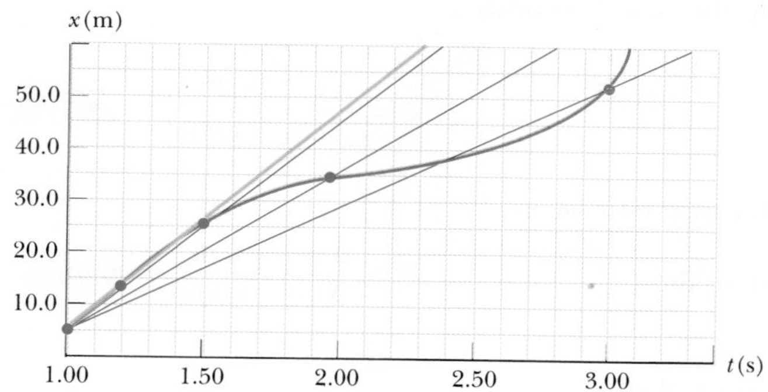

Figure 2.6 Graph representing the motion of the car from the data in Table 2.2. The slope of the blue line represents the average velocity for smaller and smaller time intervals and approaches the slope of the green tangent line.

speed changes considerably in the two-second time interval. Using the rest of the data, we can construct Table 2.3. As the time interval gets smaller, the average velocity more closely approaches the instantaneous velocity. Using the final interval of only 0.010 0 s, we find that the average velocity is $\overline{v} = \Delta x/\Delta t = 0.470 \text{ m}/0.010\,0 \text{ s} = 47.0 \text{ m/s}$. Since 0.010 0 s is a very short time interval, the actual instantaneous velocity is likely to be very close to this latter average velocity. Finally using the conversion factor on the inside front cover of the book, we see that this is 105 mi/h, a likely violation of the speed limit.

As can be seen in Figure 2.6, the chord formed by the line gradually approaches a tangent line as the time interval becomes smaller. **The slope of the line tangent to the position vs. time curve at "a given time" is defined to be the instantaneous velocity at that time.**

The instantaneous speed of an object, which is a scalar quantity, is defined as the magnitude of the instantaneous velocity. Like average speed, instantaneous speed (which we will usually call, simply, "speed") has no direction associated with it and hence carries no algebraic sign. For example, if one object has an instantaneous velocity of +15 m/s along a given line and another object has an instantaneous velocity of −15 m/s along the same line, both have an instantaneous speed of 15 m/s.

EXAMPLE 2.2 Slowly Moving Train

Goal Obtain average and instantaneous velocities from a graph.

Problem A train moves slowly along a straight portion of track according to the graph of position versus time in Figure 2.7a. Find **(a)** the average velocity for the total trip, **(b)** the average velocity during the first 4.00 s of motion, **(c)** the average velocity during the next 4.00 s of motion, **(d)** the instantaneous velocity at $t = 2.00$ s, and **(e)** the instantaneous velocity at $t = 9.00$ s.

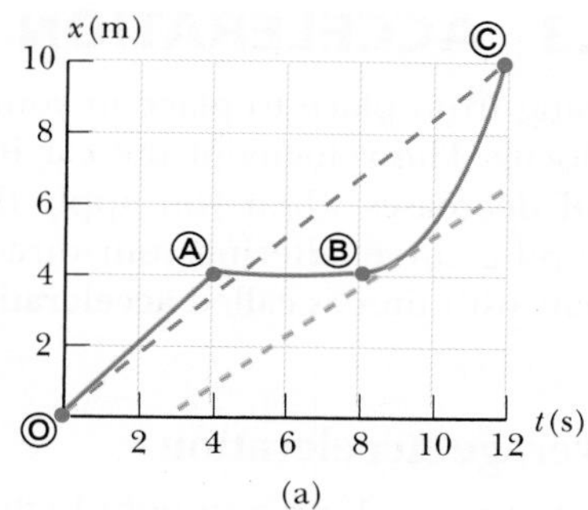

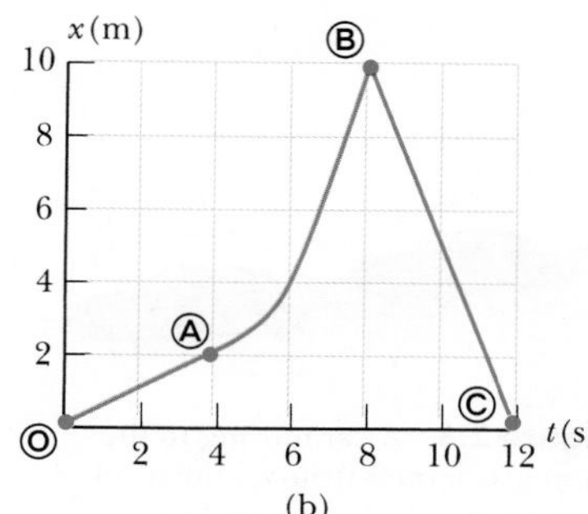

Figure 2.7 (a) (Example 2.2) (b) (Exercise 2.2).

Strategy The average velocities can be obtained by substituting the data into the definition. The instantaneous velocity at $t = 2.00$ s is the same as the average velocity at that point, because the position vs. time graph is a straight line, indicating constant velocity. Finding the instantaneous velocity when $t = 9.00$ s requires sketching a line tangent to the curve at that point and finding its slope.

Solution

(a) Find the average velocity from Ⓞ to Ⓒ.

Calculate the slope of the dashed blue line:

$$\overline{v} = \frac{\Delta x}{\Delta t} = \frac{10.0 \text{ m}}{12.0 \text{ s}} = +0.833 \text{ m/s}$$

(b) Find the average velocity during the first 4 seconds of the train's motion.

Again, find the slope:

$$\overline{v} = \frac{\Delta x}{\Delta t} = \frac{4.00 \text{ m}}{4.00 \text{ s}} = \boxed{+1.00 \text{ m/s}}$$

(c) Find the average velocity during the next four seconds.

Here, there is no change in position, so the displacement Δx is zero:

$$\overline{v} = \frac{\Delta x}{\Delta t} = \frac{0 \text{ m}}{4.00 \text{ s}} = \boxed{0 \text{ m/s}}$$

(d) Find the instantaneous velocity at $t = 2.00$ s.

This is the same as the average velocity found in **(b)**, because the graph is a straight line:

$$v = \boxed{1.00 \text{ m/s}}$$

(e) Find the instantaneous velocity at $t = 9.00$ s.

The tangent line appears to intercept the x-axis at (3.0 s, 0 m) and graze the curve at (9.0 s, 4.5 m). The instantaneous velocity at $t = 9.00$ s equals the slope of the tangent line through these points.

$$v = \frac{\Delta x}{\Delta t} = \frac{4.5 \text{ m} - 0 \text{ m}}{9.0 \text{ s} - 3.0 \text{ s}} = \boxed{0.75 \text{ m/s}}$$

Remarks From the origin to Ⓐ, the train moves at constant speed in the positive x-direction for the first 4.00 s, because the position vs. time curve is rising steadily toward positive values. From Ⓐ to Ⓑ, the train stops at $x = 4.00$ m for 4.00 s. From Ⓑ to Ⓒ, the train travels at increasing speed in the positive x-direction.

Exercise 2.2

Figure 2.7b graphs another run of the train. Find (a) the average velocity from Ⓞ to Ⓒ; (b) the average and instantaneous velocities from Ⓞ to Ⓐ; (c) the approximate instantaneous velocity at $t = 6.0$ s; and (d) the average and instantaneous velocity at $t = 9.0$ s.

Answers (a) 0 m/s (b) both are +0.5 m/s (c) 2 m/s (d) both are − 2.5 m/s

2.3 ACCELERATION

Going from place to place in your car, you rarely travel long distances at constant velocity. The velocity of the car increases when you step harder on the gas pedal and decreases when you apply the brakes. The velocity also changes when you round a curve, altering your direction of motion. The changing of an object's velocity with time is called **acceleration.**

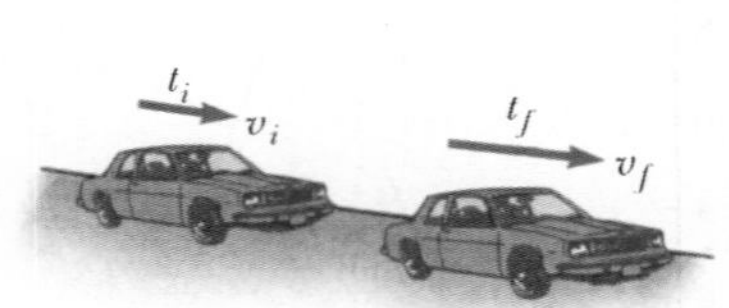

Figure 2.8 A car moving to the right accelerates from a velocity of v_i to a velocity of v_f in the time interval $\Delta t = t_f - t_i$.

Average Acceleration

A car moves along a straight highway as in Figure 2.8. At time t_i it has a velocity of v_i, and at time t_f its velocity is v_f, with $\Delta v = v_f - v_i$ and $\Delta t = t_f - t_i$.

Definition of average acceleration ▶

The average acceleration $\overline{a}$ during the time interval Δt is the change in velocity Δv divided by Δt:

$$\overline{a} \equiv \frac{\Delta v}{\Delta t} = \frac{v_f - v_i}{t_f - t_i} \quad [2.4]$$

SI unit: meter per second per second (m/s²)

For example, suppose the car shown in Figure 2.8 accelerates from an initial velocity of $v_i = +10$ m/s to a final velocity of $v_f = +20$ m/s in a time interval of 2 s.

(Both velocities are toward the right, selected as the positive direction.) These values can be inserted into Equation 2.4 to find the average acceleration:

$$\bar{a} = \frac{\Delta v}{\Delta t} = \frac{20 \text{ m/s} - 10 \text{ m/s}}{2 \text{ s}} = +5 \text{ m/s}^2$$

Acceleration is a vector quantity having dimensions of length divided by the time squared. Common units of acceleration are meters per second per second ((m/s)/s, which is usually written m/s^2) and feet per second per second (ft/s^2). An average acceleration of $+5 \text{ m/s}^2$ means that, on average, the car increases its velocity by 5 m/s every second in the positive x-direction.

For the case of motion in a straight line, the direction of the velocity of an object and the direction of its acceleration are related as follows: **When the object's velocity and acceleration are in the same direction, the speed of the object increases with time. When the object's velocity and acceleration are in opposite directions, the speed of the object decreases with time.**

To clarify this point, suppose the velocity of a car changes from -10 m/s to -20 m/s in a time interval of 2 s. The minus signs indicate that the velocities of the car are in the negative x-direction; they do *not* mean that the car is slowing down! The average acceleration of the car in this time interval is

$$\bar{a} = \frac{\Delta v}{\Delta t} = \frac{-20 \text{ m/s} - (-10 \text{ m/s})}{2 \text{ s}} = -5 \text{ m/s}^2$$

The minus sign indicates that the acceleration vector is also in the negative x-direction. Because the velocity and acceleration vectors are in the same direction, the speed of the car must increase as the car moves to the left. Positive and negative accelerations specify directions relative to chosen axes, not "speeding up" or "slowing down." The terms "speeding up" or "slowing down" refer to an increase and a decrease in speed, respectively.

TIP 2.5 Negative Acceleration

Negative acceleration doesn't necessarily mean an object is slowing down. If the acceleration is negative and the velocity is also negative, the object is speeding up!

TIP 2.6 Deceleration

The word *deceleration* means a reduction in speed, a slowing down. Some confuse it with a negative acceleration, which can speed something up. (See Tip 2.5.)

Quick Quiz 2.2

True or False? Define east as the negative direction and west as the positive direction. **(a)** If a car is traveling east, its acceleration must be eastward. **(b)** If a car is slowing down, its acceleration may be positive. **(c)** An object with constant nonzero acceleration can never stop and stay stopped.

Instantaneous Acceleration

The value of the average acceleration often differs in different time intervals, so it's useful to define the **instantaneous acceleration**, which is analogous to the instantaneous velocity discussed in Section 2.2.

◀ Definition of instantaneous acceleration

The instantaneous acceleration a is the limit of the average acceleration as the time interval Δt goes to zero:

$$a \equiv \lim_{\Delta t \to 0} \frac{\Delta v}{\Delta t} \qquad [2.5]$$

SI unit: meter per second per second (m/s^2)

Here again, the notation $\lim_{\Delta t \to 0}$ means that the ratio $\Delta v/\Delta t$ is evaluated for smaller and smaller values of Δt. The closer Δt gets to zero, the closer the ratio gets to a fixed number, which is the instantaneous acceleration.

Figure 2.9, a **velocity vs. time graph**, plots the velocity of an object against time. The graph could represent, for example, the motion of a car along a busy street. The average acceleration of the car between times t_i and t_f can be found by determining the slope of the line joining points Ⓟ and Ⓠ. If we imagine that point Ⓠ is brought closer and closer to point Ⓟ, the line comes closer and closer to becoming tangent at Ⓟ. **The instantaneous acceleration of an object at a given time**

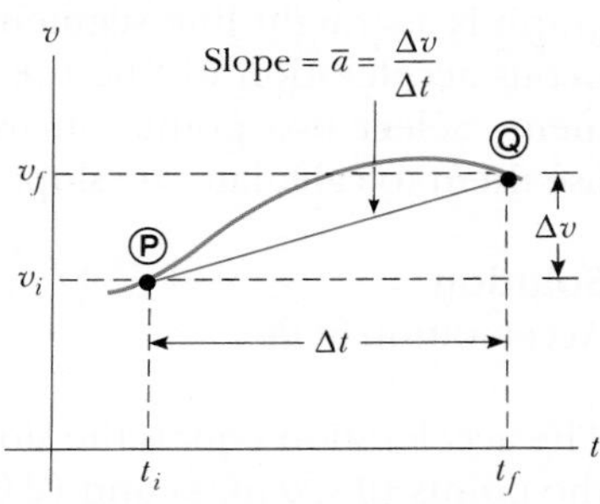

Figure 2.9 Velocity vs. time graph for an object moving in a straight line. The slope of the blue line connecting points Ⓟ and Ⓠ is defined as the average acceleration in the time interval $\Delta t = t_f - t_i$.

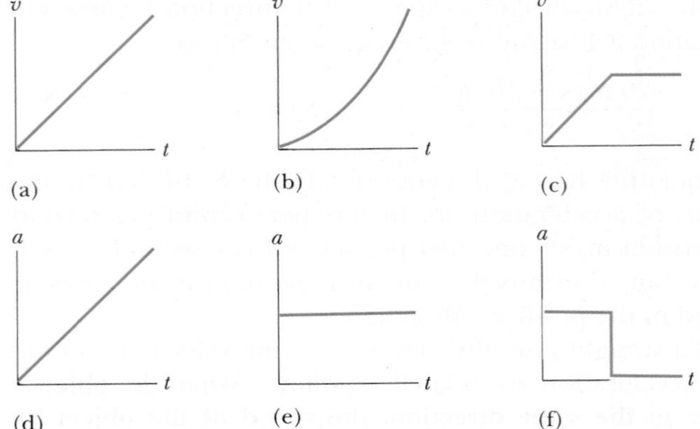

Figure 2.10 (Quick Quiz 2.3) Match each velocity vs. time graph to its corresponding acceleration vs. time graph.

equals the slope of the tangent to the velocity vs. time graph at that time. From now on, we will use the term *acceleration* to mean "instantaneous acceleration."

In the special case where the velocity vs. time graph of an object's motion is a straight line, the instantaneous acceleration of the object at any point is equal to its average acceleration. This also means that the tangent line to the graph overlaps the graph itself. In that case, the object's acceleration is said to be *uniform*, which means that it has a constant value. Constant acceleration problems are important in kinematics and will be studied extensively in this and the next chapter.

Quick Quiz 2.3

Parts (a), (b), and (c) of Figure 2.10 represent three graphs of the velocities of different objects moving in straight-line paths as functions of time. The possible accelerations of each object as functions of time are shown in parts (d), (e), and (f). Match each velocity vs. time graph with the acceleration vs. time graph that best describes the motion.

EXAMPLE 2.3 Catching a Fly Ball

Goal Apply the definition of instantaneous acceleration.

Problem A baseball player moves in a straight-line path in order to catch a fly ball hit to the outfield. His velocity as a function of time is shown in Figure 2.11a. Find his instantaneous acceleration at points Ⓐ, Ⓑ, and Ⓒ.

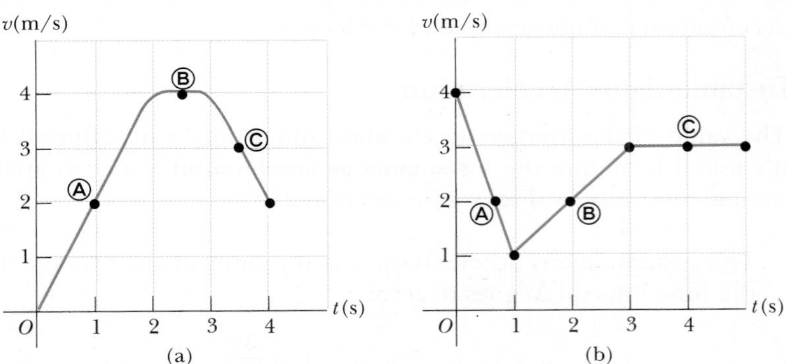

Figure 2.11 (a) (Example 2.3) (b) (Exercise 2.3)

Strategy At each point, the velocity vs. time graph is a straight line segment, so the instantaneous acceleration will be the slope of that segment. Select two points on each segment and use them to calculate the slope.

Solution

Acceleration at Ⓐ.

The acceleration equals the slope of the line connecting the points (0 s, 0 m/s) and (2.0 s, 4.0 m/s):

$$a = \frac{\Delta v}{\Delta t} = \frac{4.0 \text{ m/s} - 0}{2.0 \text{ s} - 0} = \boxed{+2.0 \text{ m/s}^2}$$

Acceleration at Ⓑ.

$\Delta v = 0$, because the segment is horizontal:

$$a = \frac{\Delta v}{\Delta t} = \frac{4.0 \text{ m/s} - 4.0 \text{ m/s}}{3.0 \text{ s} - 2.0 \text{ s}} = \boxed{0 \text{ m/s}^2}$$

Acceleration at Ⓒ.

The acceleration equals the slope of the line connecting the points (3.0 s, 4.0 m/s) and (4.0 s, 2.0 m/s):

$$a = \frac{\Delta v}{\Delta t} = \frac{2.0\ \text{m/s} - 4.0\ \text{m/s}}{4.0\ \text{s} - 3.0\ \text{s}} = -2.0\ \text{m/s}^2$$

Remarks For the first 2.0 s, the ballplayer moves in the positive x-direction (the velocity is positive) and steadily accelerates (the curve is steadily rising) to a maximum speed of 4.0 m/s. He moves for 1.0 s at a steady speed of 4.0 m/s and then slows down in the last second (the v vs. t curve is falling), still moving in the positive x-direction (v is always positive).

Exercise 2.3

Repeat the problem, using Figure 2.11b.

Answer The accelerations at Ⓐ, Ⓑ, and Ⓒ are $-3.0\ \text{m/s}^2$, $1.0\ \text{m/s}^2$, and $0\ \text{m/s}^2$, respectively.

2.4 MOTION DIAGRAMS

Velocity and acceleration are sometimes confused with each other, but they're very different concepts, as can be illustrated with the help of motion diagrams. A **motion diagram** is a representation of a moving object at successive time intervals, with velocity and acceleration vectors sketched at each position, red for velocity vectors and violet for acceleration vectors, as in Active Figure 2.12. The time intervals between adjacent positions in the motion diagram are assumed equal.

A motion diagram is analogous to images resulting from a stroboscopic photograph of a moving object. Each image is made as the strobe light flashes. Active Figure 2.12 represents three sets of strobe photographs of cars moving along a straight roadway from left to right. The time intervals between flashes of the stroboscope are equal in each diagram.

In Active Figure 2.12a, the images of the car are equally spaced: The car moves the same distance in each time interval. This means that the car moves with *constant positive velocity* and has *zero acceleration.* The red arrows are all the same length (constant velocity) and there are no violet arrows (zero acceleration).

In Active Figure 2.12b, the images of the car become farther apart as time progresses and the velocity vector increases with time, because the car's displacement between adjacent positions increases as time progresses. The car is moving with a *positive velocity* and a constant *positive acceleration.* The red arrows are successively longer in each image, and the violet arrows point to the right.

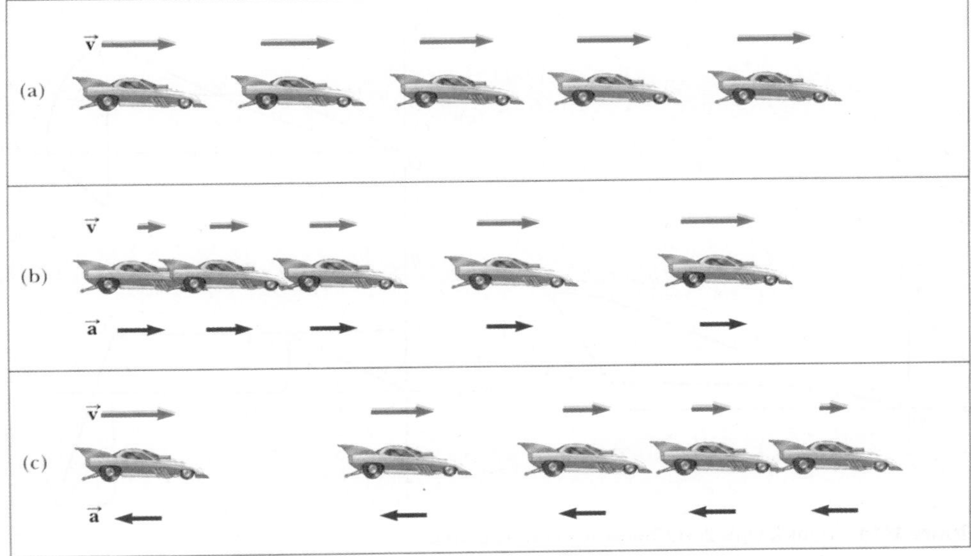

ACTIVE FIGURE 2.12
(a) Motion diagram for a car moving at constant velocity (zero acceleration). (b) Motion diagram for a car undergoing constant acceleration in the direction of its velocity. The velocity vector at each instant is indicated by a red arrow, and the constant acceleration vector by a violet arrow. (c) Motion diagram for a car undergoing constant acceleration in the direction *opposite* the velocity at each instant.

Physics Now™
Log into PhysicsNow at **www.cp7e.com**, and go to Active Figure 2.12, where you can select the constant acceleration and initial velocity of the car and observe pictorial and graphical representations of its motion.

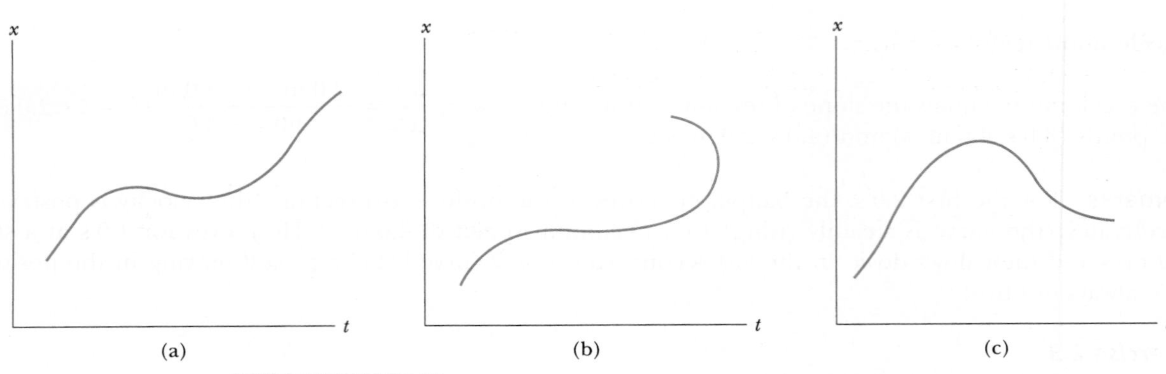

ACTIVE FIGURE 2.13
(Quick Quiz 2.4) Which position vs. time curve is impossible?

Physics Now™
Log into PhysicsNow at **www.cp7e.com**, and go to Active Figure 2.13, where you can practice matching appropriate velocity vs. time graphs and acceleration vs. time graphs.

In Active Figure 2.12c, the car slows as it moves to the right because its displacement between adjacent positions decreases with time. In this case, the car moves initially to the right with a constant negative acceleration. The velocity vector decreases in time (the red arrows get shorter) and eventually reaches zero, as would happen when the brakes are applied. Note that the acceleration and velocity vectors are *not* in the same direction. The car is moving with a *positive velocity,* but with a *negative acceleration.*

Try constructing your own diagrams for various problems involving kinematics.

Quick Quiz 2.4

The three graphs in Active Figure 2.13 represent the position vs. time for objects moving along the x-axis. Which, if any, of these graphs is not physically possible?

Quick Quiz 2.5

Figure 2.14a is a diagram of a multiflash image of an air puck moving to the right on a horizontal surface. The images sketched are separated by equal time intervals, and the first and last images show the puck at rest. (a) In Figure 2.14b, which color graph best shows the puck's position as a function of time? (b) In Figure 2.14c, which color graph best shows the puck's velocity as a function of time? (c) In Figure 2.14d, which color graph best shows the puck's acceleration as a function of time?

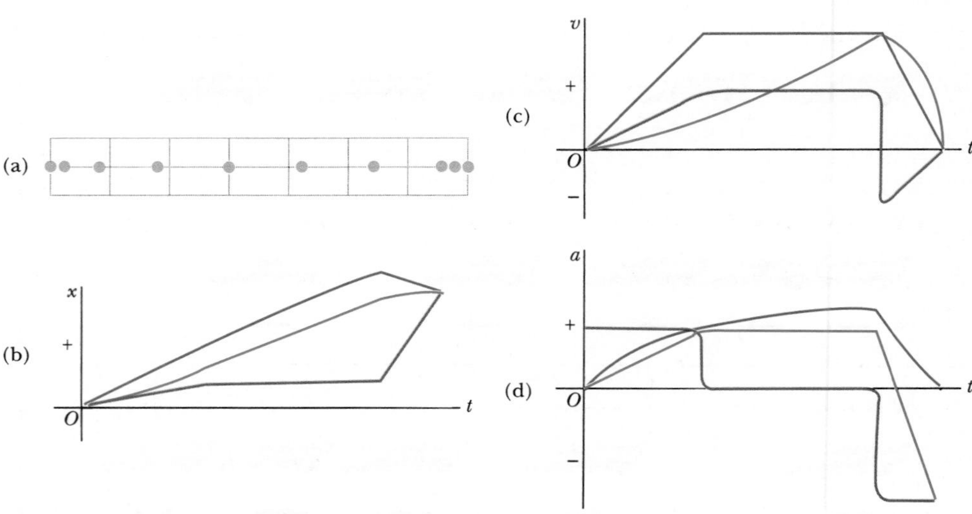

Figure 2.14 (Quick Quiz 2.5) Choose the correct graphs.

2.5 ONE-DIMENSIONAL MOTION WITH CONSTANT ACCELERATION

Many applications of mechanics involve objects moving with *constant acceleration*. This type of motion is important because it applies to numerous objects in nature, such as an object in free fall near Earth's surface (assuming that air resistance can be neglected). A graph of acceleration versus time for motion with constant acceleration is shown in Active Figure 2.15a. **When an object moves with constant acceleration, the instantaneous acceleration at any point in a time interval is equal to the value of the average acceleration over the entire time interval.** Consequently, the velocity increases or decreases at the same rate throughout the motion, and a plot of v versus t gives a straight line with either positive, zero, or negative slope.

Because the average acceleration equals the instantaneous acceleration when a is constant, we can eliminate the bar used to denote average values from our defining equation for acceleration, writing $\overline{a} = a$, so that Equation 2.4 becomes

$$a = \frac{v_f - v_i}{t_f - t_i}$$

The observer timing the motion is always at liberty to choose the initial time, so for convenience, let $t_i = 0$ and t_f be any arbitrary time t. Also, let $v_i = v_0$ (the initial velocity at $t = 0$) and $v_f = v$ (the velocity at any arbitrary time t). With this notation, we can express the acceleration as

$$a = \frac{v - v_0}{t}$$

or

$$v = v_0 + at \qquad \text{(for constant } a\text{)} \qquad [2.6]$$

Equation 2.6 states that the acceleration a steadily changes the initial velocity v_0 by an amount at. For example, if a car starts with a velocity of $+2.0$ m/s to the right and accelerates to the right with $a = +6.0 \text{ m/s}^2$, it will have a velocity of $+14$ m/s after 2.0 s have elapsed:

$$v = v_0 + at = +2.0 \text{ m/s} + (6.0 \text{ m/s}^2)(2.0 \text{ s}) = +14 \text{ m/s}$$

The graphical interpretation of v is shown in Active Figure 2.15b. The velocity varies linearly with time according to Equation 2.6, as it should for constant acceleration.

Because the velocity is increasing or decreasing *uniformly* with time, we can express the average velocity in any time interval as the arithmetic average of the initial velocity v_0 and the final velocity v:

$$\overline{v} = \frac{v_0 + v}{2} \qquad \text{(for constant } a\text{)} \qquad [2.7]$$

Remember that this expression is valid only when the acceleration is constant, in which case the velocity increases uniformly.

We can now use this result along with the defining equation for average velocity, Equation 2.2, to obtain an expression for the displacement of an object as a function of time. Again, we choose $t_i = 0$ and $t_f = t$, and for convenience, we write $\Delta x = x_f - x_i = x - x_0$. This results in

$$\Delta x = \overline{v}t = \left(\frac{v_0 + v}{2}\right)t$$

$$\Delta x = \tfrac{1}{2}(v_0 + v)t \qquad \text{(for constant } a\text{)} \qquad [2.8]$$

We can obtain another useful expression for displacement by substituting the equation for v (Eq. 2.6) into Equation 2.8:

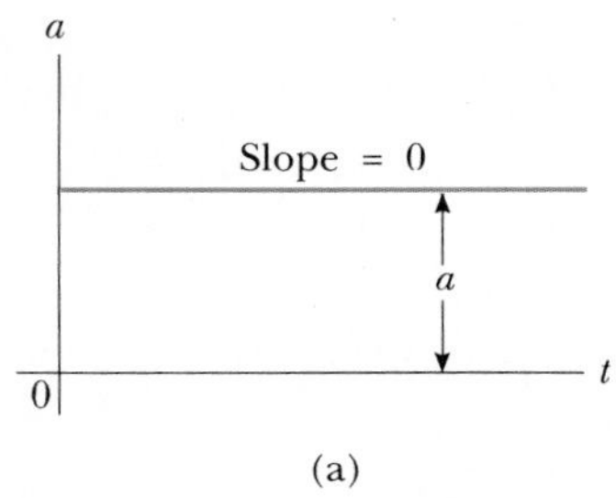

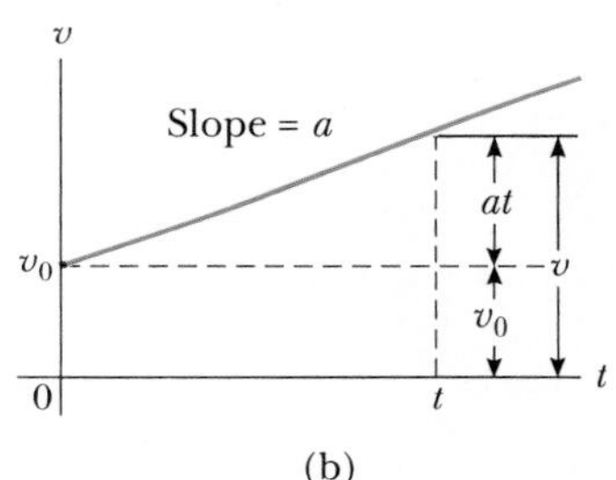

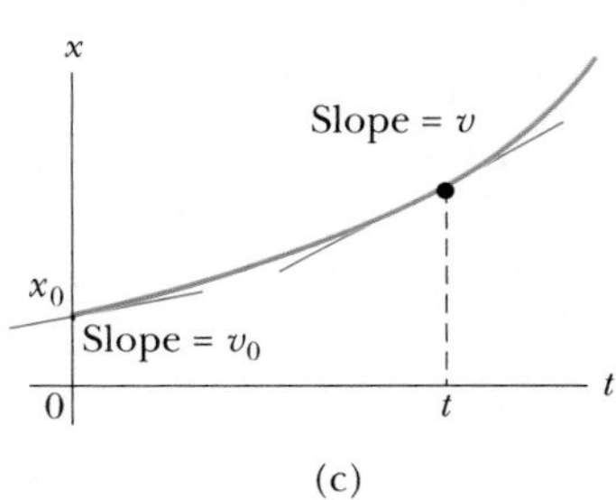

ACTIVE FIGURE 2.15
A particle moving along the x-axis with constant acceleration a.
(a) the acceleration vs. time graph,
(b) the velocity vs. time graph, and
(c) the position vs. time graph.

Physics Now™
Log into PhysicsNow at **www.cp7e.com**, and go to Active Figure 2.15, where you can adjust the constant acceleration and observe the effect on the position and velocity graphs.

TABLE 2.4

Equations for Motion in a Straight Line Under Constant Acceleration

Equation	Information Given by Equation
$v = v_0 + at$	Velocity as a function of time
$\Delta x = v_0 t + \frac{1}{2}at^2$	Displacement as a function of time
$v^2 = v_0^2 + 2a\Delta x$	Velocity as a function of displacement

Note: Motion is along the x-axis. At $t = 0$, the velocity of the particle is v_0.

$$\Delta x = \tfrac{1}{2}(v_0 + v_0 + at)\,t$$

$$\Delta x = v_0 t + \tfrac{1}{2}at^2 \qquad \text{(for constant } a\text{)} \qquad [2.9]$$

This equation can also be written in terms of the position x, since $\Delta x = x - x_0$. Active Figure 2.15c shows a plot of x versus t for Equation 2.9, which is related to the graph of velocity vs. time: The area under the curve in Active Figure 2.15b is equal to $v_0 t + \frac{1}{2}at^2$, which is equal to the displacement Δx. In fact, **the area under the graph of v versus t for any object is equal to the displacement Δx of the object.**

Finally, we can obtain an expression that doesn't contain time by solving Equation 2.6 for t and substituting into Equation 2.8, resulting in

$$\Delta x = \tfrac{1}{2}(v + v_0)\left(\frac{v - v_0}{a}\right) = \frac{v^2 - v_0^2}{2a}$$

$$v^2 = v_0^2 + 2a\Delta x \qquad \text{(for constant } a\text{)} \qquad [2.10]$$

Equations 2.6 and 2.9 together can solve any problem in one-dimensional motion with constant acceleration, but Equations 2.7, 2.8, and, especially, 2.10 are sometimes convenient. The three most useful equations—Equations 2.6, 2.9, and 2.10—are listed in Table 2.4.

The best way to gain confidence in the use of these equations is to work a number of problems. There is usually more than one way to solve a given problem, depending on which equations are selected and what quantities are given. The difference lies mainly in the algebra.

Problem-Solving Strategy Accelerated Motion

The following procedure is recommended for solving problems involving accelerated motion.

1. **Read** the problem.
2. **Draw** a diagram, choosing a coordinate system, labeling initial and final points, and indicating directions of velocities and accelerations with arrows.
3. **Label** all quantities, circling the unknowns. Convert units as needed.
4. **Equations** from Table 2.4 should be selected next. All kinematics problems in this chapter can be solved with the first two equations, and the third is often convenient.
5. **Solve** for the unknowns. Doing so often involves solving two equations for two unknowns. It's usually more convenient to substitute all known values before solving.
6. **Check** your answer, using common sense and estimates.

TIP 2.7 Pigs Don't Fly

After solving a problem, you should think about your answer and decide whether it seems reasonable. If it isn't, look for your mistake!

Most of these problems reduce to writing the kinematic equations from Table 2.4 and then substituting the correct values into the constants a, v_0, and x_0 from the given information. Doing this produces two equations—one linear and one quadratic—for two unknown quantities.

EXAMPLE 2.4 The Daytona 500

Goal Apply the basic kinematic equations.

Problem A race car starting from rest accelerates at a constant rate of 5.00 m/s^2. What is the velocity of the car after it has traveled 1.00×10^2 ft?

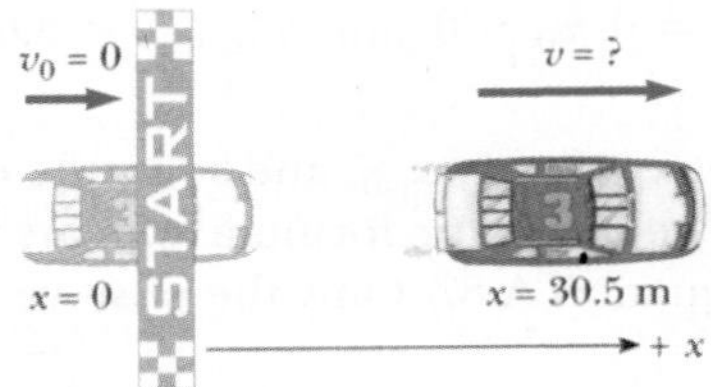

Figure 2.16 (Example 2.4)

Strategy We've read the problem, drawn the diagram in Figure 2.16, and chosen a coordinate system (steps 1 and 2). We'd like to find the velocity v after a certain known displacement Δx. The acceleration a is also known, as is the initial velocity v_0 (step 3, labeling, is complete), so the third equation in Table 2.4 looks most useful. The rest is simple substitution.

Solution

Convert units of Δx to SI, using the information in the inside front cover.

$$1.00 \times 10^2 \text{ ft} = (1.00 \times 10^2 \text{ ft})\left(\frac{1 \text{ m}}{3.28 \text{ ft}}\right) = 30.5 \text{ m}$$

Write the kinematics equation for v^2 (step 4):

$$v^2 = v_0{}^2 + 2a\,\Delta x$$

Solve for v, taking the positive square root because the car moves to the right (step 5):

$$v = \sqrt{v_0{}^2 + 2a\,\Delta x}$$

Substitute $v_0 = 0$, $a = 5.00$ m/s^2, and $\Delta x = 30.5$ m:

$$v = \sqrt{v_0{}^2 + 2a\,\Delta x} = \sqrt{(0)^2 + 2(5.00 \text{ m/s}^2)(30.5 \text{ m})}$$
$$= 17.5 \text{ m/s}$$

Remarks The answer is easy to check. An alternate technique is to use $\Delta x = v_0t + \frac{1}{2}at^2$ to find t and then use the equation $v = v_0 + at$ to find v.

Exercise 2.4

Suppose the driver in this example now slams on the brakes, stopping the car in 4.00 s. Find (a) the acceleration and (b) the distance the car travels, assuming the acceleration is constant.

Answers (a) $a = -4.38$ m/s^2 (b) $d = 35.0$ m

INTERACTIVE EXAMPLE 2.5 Car Chase

Goal Solve a problem involving two objects, one moving at constant acceleration and the other at constant velocity.

Problem A car traveling at a constant speed of 24.0 m/s passes a trooper hidden behind a billboard, as in Figure 2.17. One second after the speeding car passes the billboard, the trooper sets off in chase with a constant acceleration of 3.00 m/s^2. **(a)** How long does it take the trooper to overtake the speeding car? **(b)** How fast is the trooper going at that time?

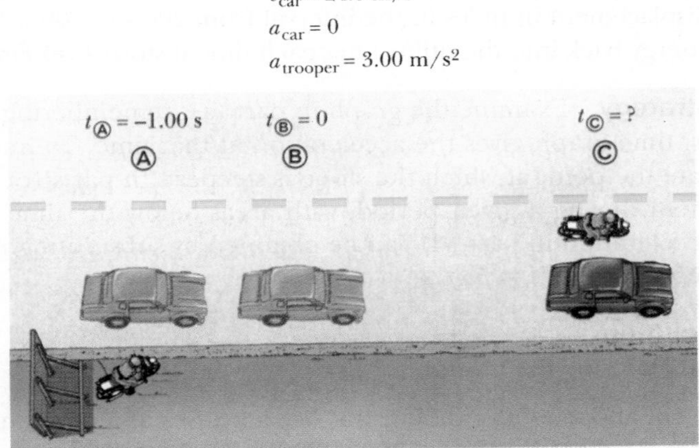

Figure 2.17 (Example 2.5) A speeding car passes a hidden trooper. When does the trooper catch up to the car?

Strategy Solving this problem involves two simultaneous kinematics equations of position, one for the police motorcycle and the other for the car. Choose $t = 0$ to correspond to the time the trooper takes up the chase, when the car is at $x_{car} = 24.0$ m because of its head start (24.0 m/s $\times$ 1.00 s). The trooper catches up with the car when their positions are the same, which suggests setting $x_{trooper} = x_{car}$ and solving for time, which can then be used to find the trooper's speed in part (b).

Solution

(a) How long does it take the trooper to overtake the car?

Write the equation for the car's displacement:

$$\Delta x_{car} = x_{car} - x_0 = v_0t + \tfrac{1}{2}a_{car}t^2$$

Take $x_0 = 24.0$ m, $v_0 = 24.0$ m/s and $a_{car} = 0$. Solve for x_{car}:

$$x_{car} = x_0 + vt = 24.0 \text{ m} + (24.0 \text{ m/s})t$$

Write the equation for the trooper's position, taking $x_0 = 0$, $v_0 = 0$, and $a_{trooper} = 3.00 \text{ m/s}^2$:

$$x_{trooper} = \tfrac{1}{2}a_{trooper}t^2 = \tfrac{1}{2}(3.00 \text{ m/s}^2)t^2 = (1.50 \text{ m/s}^2)t^2$$

Set $x_{trooper} = x_{car}$, and solve the quadratic equation. (The quadratic formula appears in Appendix A, Equation A.8.) Only the positive root is meaningful.

$$(1.50 \text{ m/s}^2)t^2 = 24.0 \text{ m} + (24.0 \text{ m/s})t$$

$$(1.50 \text{ m/s}^2)t^2 - (24.0 \text{ m/s})t - 24.0 \text{ m} = 0$$

$$t = 16.9 \text{ s}$$

(b) Find the trooper's speed at this time.

Substitute the time into the trooper's velocity equation:

$$v_{trooper} = v_0 + a_{trooper}\,t = 0 + (3.00 \text{ m/s}^2)(16.9 \text{ s})$$

$$= 50.7 \text{ m/s}$$

Remarks The trooper, traveling about twice as fast as the car, must swerve or apply his brakes strongly to avoid a collision! This problem can also be solved graphically, by plotting position versus time for each vehicle on the same graph. The intersection of the two graphs corresponds to the time and position at which the trooper overtakes the car.

Exercise 2.5

A motorist with an expired license tag is traveling at 10.0 m/s down a street, and a policeman on a motorcycle, taking another 5.00 s to finish his donut, gives chase at an acceleration of 2.00 m/s^2. Find (a) the time required to catch the car and (b) the distance the trooper travels while overtaking the motorist.

Answers (a) 13.7 s (b) 188 m

PhysicsNow™ You can study the motion of the car and the trooper for various velocities of the car by logging into PhysicsNow at **www.cp7e.com** and going to Interactive Example 2.5.

EXAMPLE 2.6 The Acela: The Porsche of American Trains

Problem The sleek high-speed electric train known as the Acela (pronounced ahh-sell-ah) is currently in service on the Washington-New York-Boston run and is shown in Figure 2.18a. The Acela consists of two power cars and six coaches and can carry 304 passengers at speeds up to 170 mi/h. In order to negotiate curves comfortably at high speeds, the train carriages tilt as much as 6° from the vertical, to prevent passengers from being pushed to the side. A velocity vs. time graph for the Acela is shown in Figure 2.18b. **(a)** Describe the motion of the Acela. **(b)** Find the peak acceleration of the Acela in miles per hour per second ((mi/h)/s) as the train speeds up from 45 mi/h to 170 mi/h. **(c)** Find the train's displacement in miles between $t = 0$ and $t = 200$ s. **(d)** Find the average acceleration of the Acela and its displacement in miles in the interval from 200 s to 300 s. (The train has regenerative braking, which means that it feeds energy back into the utility lines each time it stops!) **(e)** Find the total displacement in the interval from 0 to 400 s.

Strategy Examine the graph in part (a), remembering that the slope of the tangent line at any point of the velocity vs. time graph gives the acceleration at that time. To find the peak acceleration in part (b), study the graph and locate the point at which the slope is steepest. In parts (c)–(e), estimating the area under the curve gives the displacement during a given period, with areas below the time axis, as in part (e), subtracted from the total. The average acceleration in part (d) can be obtained by substituting numbers taken from the graph into the definition of average acceleration, $\bar{a} = \Delta v/\Delta t$.

Solution

(a) Describe the motion.

From about −50 s to 50 s, the Acela cruises at a constant velocity in the +x-direction. Then the train accelerates in the +x-direction from 50 s to 200 s, reaching a top speed of about 170 mi/h, whereupon it brakes to rest at 350 s and reverses, steadily gaining speed in the −x-direction.

(b) Find the peak acceleration.

Calculate the slope of the steepest tangent line, which connects the points (50 s, 50 mi/h) and (100 s, 150 mi/h) (the light blue line in Figure 2.18c):

$$a = \text{slope} = \frac{\Delta v}{\Delta t} = \frac{(1.5 \times 10^2 - 5.0 \times 10^1) \text{ mi/h}}{(1.0 \times 10^2 - 5.0 \times 10^1) \text{ s}}$$

$$= 2.0 \text{ (mi/h)/s}$$

(c) Find the displacement between 0 s and 200 s.

Using triangles and rectangles, approximate the area in Figure 2.18d:

$$\begin{aligned}\Delta x_{0\to 200\text{ s}} &= \text{area}_1 + \text{area}_2 + \text{area}_3 + \text{area}_4 + \text{area}_5 \\ &\approx (5.0 \times 10^1 \text{ mi/h})(5.0 \times 10^1 \text{ s}) \\ &+ (5.0 \times 10^1 \text{ mi/h})(5.0 \times 10^1 \text{ s}) \\ &+ (1.6 \times 10^2 \text{ mi/h})(1.0 \times 10^2 \text{ s}) \\ &+ \tfrac{1}{2}(5.0 \times 10^1 \text{ s})(1.0 \times 10^2 \text{ mi/h}) \\ &+ \tfrac{1}{2}(1.0 \times 10^2 \text{ s})(1.7 \times 10^2 \text{ mi/h} - 1.6 \times 10^2) \\ &= 2.4 \times 10^4 (\text{mi/h})\text{s}\end{aligned}$$

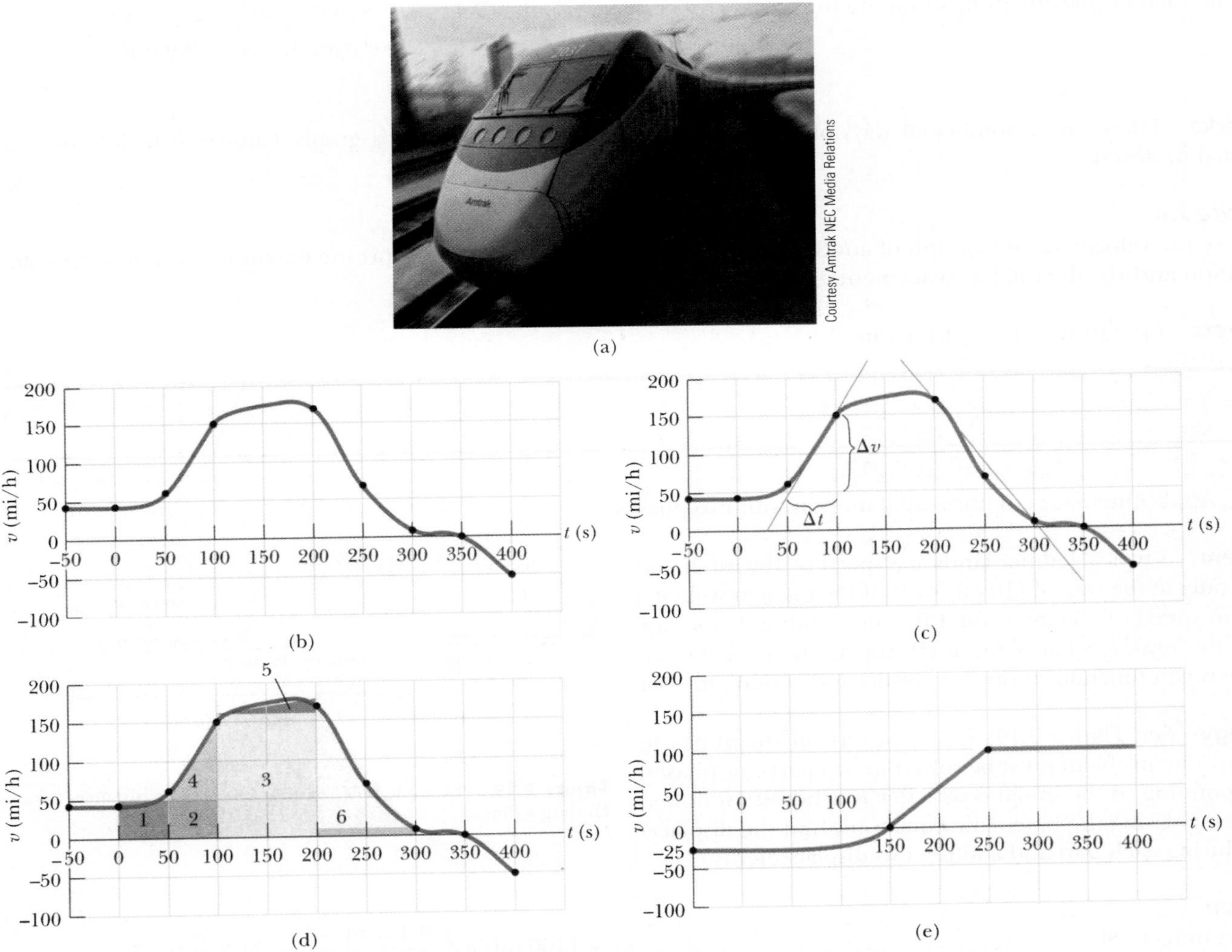

Figure 2.18 (Example 2.6) (a) The Acela, 1 250 000 lb of cold steel thundering along at 170 mi/h. (b) Velocity vs. time graph for the Acela. (c) The slope of the steepest tangent blue line gives the peak acceleration, while the slope of the green line is the average acceleration between 200 s and 300 s. (d) The area under the velocity vs. time graph in some time interval gives the displacement of the Acela in that time interval. (e) (Exercise 2.6).

Convert units to miles by converting hours to seconds:

$$\Delta x_{0\to 200\text{ s}} \approx 2.4 \times 10^4 \, \frac{\text{mi}\cdot\text{s}}{\text{h}}\left(\frac{1 \text{ h}}{3\,600 \text{ s}}\right) = \boxed{6.7 \text{ mi}}$$

(d) Find the average acceleration from 200 s to 300 s, and find the displacement.

The slope of the green line is the average acceleration from 200 s to 300 s (Fig. 2.18c):

$$\bar{a} = \text{slope} = \frac{\Delta v}{\Delta t} = \frac{(1.0 \times 10^1 - 1.7 \times 10^2) \text{ mi/h}}{1.0 \times 10^2 \text{ s}}$$

$$= \boxed{-1.6 (\text{mi/h})/\text{s}}$$

The displacement from 200 s to 300 s is equal to area$_6$, which is the area of a triangle plus the area of a very narrow rectangle beneath the triangle:

$$\Delta x_{200\to 300\text{ s}} \approx \tfrac{1}{2}(1.0\times 10^2\text{ s})(1.7\times 10^2 - 1.0\times 10^1)\text{ mi/h} + (1.0\times 10^1\text{ mi/h})(1.0\times 10^2\text{ s})$$
$$= 9.0\times 10^3(\text{mi/h})(\text{s}) = \boxed{2.5\text{ mi}}$$

(e) Find the total displacement from 0 s to 400 s.

The total displacement is the sum of all the individual displacements. We still need to calculate the displacements for the time intervals from 300 s to 350 s and from 350 s to 400 s. The latter is negative, because it's below the time axis.

$$\Delta x_{300\to 350\text{ s}} \approx \tfrac{1}{2}(5.0\times 10^1\text{ s})(1.0\times 10^1\text{ mi/h}) = 2.5\times 10^2(\text{mi/h})(\text{s})$$

$$\Delta x_{350\to 400\text{ s}} \approx \tfrac{1}{2}(5.0\times 10^1\text{ s})(-5.0\times 10^1\text{ mi/h}) = -1.3\times 10^3(\text{mi/h})(\text{s})$$

Find the total displacement by summing the parts:

$$\Delta x_{0\to 400\text{ s}} \approx (2.4\times 10^4 + 9.0\times 10^3 + 2.5\times 10^2 - 1.3\times 10^3)(\text{mi/h})(\text{s}) = \boxed{8.9\text{ mi}}$$

Remarks There are a number of ways of finding the approximate area under a graph. Choice of technique is a personal preference.

Exercise 2.6

Suppose the velocity vs. time graph of another train is given in Figure 2.18e. Find (a) the maximum instantaneous acceleration and (b) the total displacement in the interval from 0 s to 4.00×10^2 s.

Answers (a) 1.0(mi/h)/s (b) 4.7 mi

EXAMPLE 2.7 Runway Length

Goal Apply kinematics to horizontal motion with two phases.

Problem A typical jetliner lands at a speed of 160 mi/h and decelerates at the rate of (10 mi/h)/s. If the plane travels at a constant speed of 160 mi/h for 1.0 s after landing before applying the brakes, what is the total displacement of the aircraft between touchdown on the runway and coming to rest?

Strategy See Figure 2.19. First, convert all quantities to SI units. The problem must be solved in two parts, or phases, corresponding to the initial coast after touchdown, followed by braking. Using the kinematic equations, find the displacement during each part and add the two displacements.

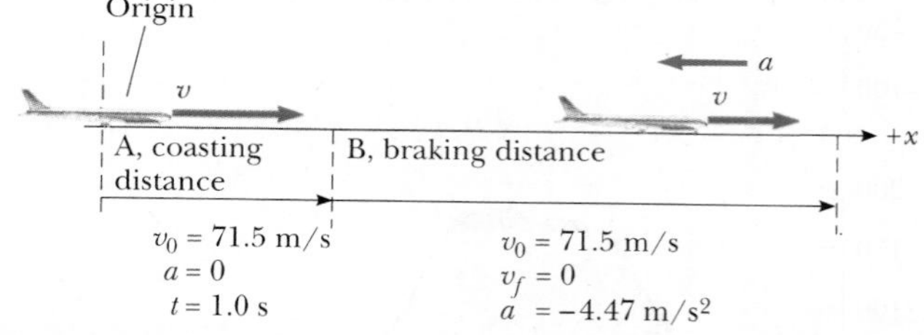

Figure 2.19 (Example 2.7) Coasting and braking distances for a landing jetliner.

Solution

Convert units to SI:

$$v_0 = (160\text{ mi/h})\left(\frac{0.447\text{ m/s}}{1.00\text{ mi/h}}\right) = 71.5\text{ m/s}$$

$$a = (-10.0\text{ (mi/h)/s})\left(\frac{0.447\text{ m/s}}{1.00\text{ mi/h}}\right) = -4.47\text{ m/s}^2$$

Taking $a = 0$, $v_0 = 71.5$ m/s, and $t = 1.00$ s, find the displacement while the plane is coasting:

$$\Delta x_{\text{coasting}} = v_0t + \tfrac{1}{2}at^2 = (71.5\text{ m/s})(1.00\text{ s}) + 0 = 71.5\text{ m}$$

Use the time-independent kinematic equation to find the displacement while the plane is braking.

$$v^2 = v_0^2 + 2a\,\Delta x_{\text{braking}}$$

Take $a = -4.47$ m/s^2 and $v_0 = 71.5$ m/s. The negative sign on a means that the plane is slowing down.

$$\Delta x_{\text{braking}} = \frac{v^2 - v_0^2}{2a} = \frac{0 - (71.5\text{ m/s})^2}{2.00(-4.47\text{ m/s}^2)} = 572\text{ m}$$

Sum the two results to find the total displacement: $\Delta x_{\text{coasting}} + \Delta x_{\text{braking}} = 72 \text{ m} + 572 \text{ m} = 644 \text{ m}$

Remarks To find the displacement while braking, we could have used the two kinematics equations involving time, namely, $\Delta x = v_0 t + \frac{1}{2}at^2$ and $v = v_0 + at$, but because we weren't interested in time, the time-independent equation was easier to use.

Exercise 2.7

A jet lands at 80.0 m/s, applying the brakes 2.00 s after landing. Find the acceleration needed to stop the jet within 5.00×10^2 m.

Answer $a = -9.41 \text{ m/s}^2$

2.6 FREELY FALLING OBJECTS

When air resistance is negligible, all objects dropped under the influence of gravity near Earth's surface fall toward Earth with the same constant acceleration. This idea may seem obvious today, but it wasn't until about 1600 that it was accepted. Prior to that time, the teachings of the great philosopher Aristotle (384–322 B.C.) had held that heavier objects fell faster than lighter ones.

According to legend, Galileo discovered the law of falling objects by observing that two different weights dropped simultaneously from the Leaning Tower of Pisa hit the ground at approximately the same time. Although it's unlikely that this particular experiment was carried out, we know that Galileo performed many systematic experiments with objects moving on inclined planes. In his experiments, he rolled balls down a slight incline and measured the distances they covered in successive time intervals. The purpose of the incline was to reduce the acceleration and enable Galileo to make accurate measurements of the intervals. (Some people refer to this experiment as "diluting gravity.") By gradually increasing the slope of the incline, he was finally able to draw mathematical conclusions about freely falling objects, because a falling ball is equivalent to a ball going down a vertical incline. Galileo's achievements in the science of mechanics paved the way for Newton in his development of the laws of motion, which we will study in Chapter 4.

North Wind Archive

GALILEO GALILEI Italian Physicist and Astronomer (1564–1642)

Galileo formulated the laws that govern the motion of objects in free fall. He also investigated the motion of an object on an inclined plane, established the concept of relative motion, invented the thermometer, and discovered that the motion of a swinging pendulum could be used to measure time intervals. After designing and constructing his own telescope, he discovered four of Jupiter's moons, found that our own Moon's surface is rough, discovered sunspots and the phases of Venus, and showed that the Milky Way consists of an enormous number of stars. Galileo publicly defended Nicolaus Copernicus's assertion that the Sun is at the center of the Universe (the heliocentric system). He published *Dialogue Concerning Two New World Systems* to support the Copernican model, a view the Church declared to be heretical. After being taken to Rome in 1633 on a charge of heresy, he was sentenced to life imprisonment and later was confined to his villa at Arcetri, near Florence, where he died in 1642.

Try the following experiment: Drop a hammer and a feather simultaneously from the same height. The hammer hits the floor first, because air drag has a greater effect on the much lighter feather. On August 2, 1971, this same experiment was conducted on the Moon by astronaut David Scott, and the hammer and feather fell with exactly the same acceleration, as expected, hitting the lunar surface at the same time. In the idealized case where air resistance is negligible, such motion is called *free fall.*

The expression *freely falling object* doesn't necessarily refer to an object dropped from rest. **A freely falling object is any object moving freely under the influence of gravity alone, regardless of its initial motion.** Objects thrown upward or downward and those released from rest are all considered freely falling.

We denote the magnitude of the **free-fall acceleration** by the symbol g. The value of g decreases with increasing altitude, and varies slightly with latitude, as well. At Earth's surface, the value of g is approximately 9.80 m/s^2. Unless stated otherwise, we will use this value for g in doing calculations. For quick estimates, use $g \approx 10 \text{ m/s}^2$.

If we neglect air resistance and assume that the free-fall acceleration doesn't vary with altitude over short vertical distances, then the motion of a freely falling object is the same as motion in one dimension under constant acceleration. This means that the kinematics equations developed in Section 2.6 can be applied. It's conventional to define "up" as the $+y$-direction and to use y as the position variable. In that case, the acceleration is $a = -g = -9.80 \text{ m/s}^2$. In Chapter 7, we study how to deal with variations in g with altitude.

Quick Quiz 2.6

A tennis player on serve tosses a ball straight up. While the ball is in free fall, does its acceleration (a) increase, (b) decrease, (c) increase and then decrease, (d) decrease and then increase, or (e) remain constant?

Quick Quiz 2.7

As the tennis ball of Quick Quiz 2.6 travels through the air, its speed (a) increases, (b) decreases, (c) decreases and then increases, (d) increases and then decreases, or (e) remains the same.

Quick Quiz 2.8

A skydiver jumps out of a hovering helicopter. A few seconds later, another skydiver jumps out, so they both fall along the same vertical line relative to the helicopter. Both sky divers fall with the same acceleration. Does the vertical distance between them (a) increase, (b) decrease, or (c) stay the same? Does the difference in their velocities (d) increase, (e) decrease, or (f) stay the same? (Assume that g is constant.)

EXAMPLE 2.8 Look Out Below!

Goal Apply the basic kinematics equations to an object falling from rest under the influence of gravity.

Problem A golf ball is released from rest at the top of a very tall building. Neglecting air resistance, calculate the position and velocity of the ball after 1.00 s, 2.00 s, and 3.00 s.

Strategy Make a simple sketch. Because the height of the building isn't given, it's convenient to choose coordinates so that $y = 0$ at the top of the building. Use the velocity and position kinematic equations, substituting known and given values.

Solution

Write the kinematics Equations 2.6 and 2.9:

$$v = at + v_0$$

$$\Delta y = y - y_0 = v_0 t + \tfrac{1}{2}at^2$$

Substitute $y_0 = 0$, $v_0 = 0$, and $a = -g = -9.80\text{ m/s}^2$ into the preceding two equations:

$$v = at = (-9.80\text{ m/s}^2)t$$

$$y = \tfrac{1}{2}(-9.80\text{ m/s}^2)t^2 = -(4.90\text{ m/s}^2)t^2$$

Substitute in the different times, and create a table.

t (s)	v (m/s)	y (m)
1.00	−9.8	−4.9
2.00	−19.6	−19.6
3.00	−29.4	−44.1

Remarks The minus signs on v mean that the velocity vectors are directed downward, while the minus signs on y indicates positions below the origin. The velocity of a falling object is directly proportional to the time, and the position is proportional to the time squared, results first proven by Galileo.

Exercise 2.8

Calculate the position and velocity of the ball after 4.00 s has elapsed.

Answer −78.4 m, −39.2 m/s

INTERACTIVE EXAMPLE 2.9
Not a Bad Throw for a Rookie!

Goal Apply the kinematic equations to a freely falling object with a nonzero initial velocity.

Problem A stone is thrown from the top of a building with an initial velocity of 20.0 m/s straight upward, at an initial height of 50.0 m above the ground. The stone just misses the edge of the roof on its way down, as shown in Figure 2.20. Determine **(a)** the time needed for the stone to reach its maximum height, **(b)** the maximum height, **(c)** the time needed for the stone to return to the height from which it was thrown and the velocity of the stone at that instant; **(d)** the time needed for the stone to reach the ground, and **(e)** the velocity and position of the stone at $t = 5.00$ s.

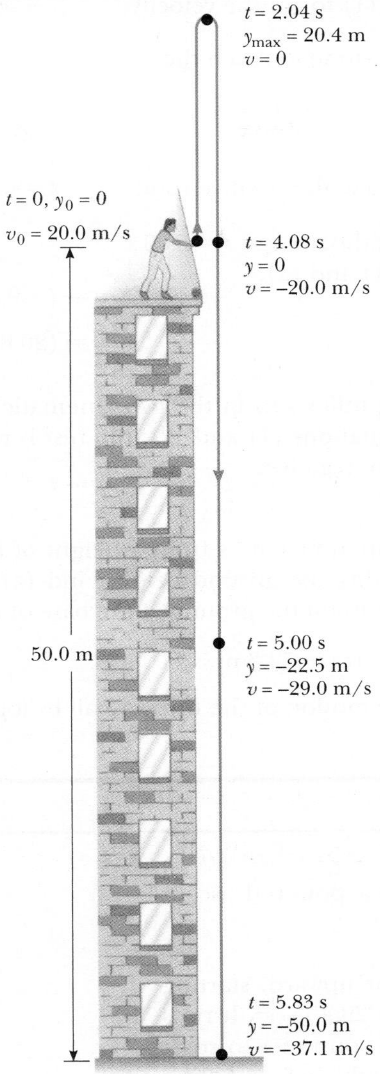

Figure 2.20 (Example 2.9) A freely falling object is thrown upward with an initial velocity of $v_0 = +20.0$ m/s. Positions and velocities are given for several times.

Strategy The diagram in Figure 2.20 establishes a coordinate system with $y_0 = 0$ at the level at which the stone is released from the thrower's hand, with y positive upward. Write the velocity and position kinematic equations for the stone, and substitute the given information. All the answers come from these two equations by using simple algebra or by just substituting the time. In part (a), for example, the stone comes to rest for an instant at its maximum height, so set $v = 0$ at this point and solve for time. Then substitute the time into the displacement equation, obtaining the maximum height.

Solution

(a) Find the time when the stone reaches its maximum height.

Write the velocity and position kinematic equations:

$$v = at + v_0$$

$$\Delta y = y - y_0 = v_0 t + \tfrac{1}{2}at^2$$

Substitute $a = -9.80 \text{ m/s}^2$, $v_0 = 20.0$ m/s, and $y_0 = 0$ into the preceding two equations:

$$v = (-9.80 \text{ m/s}^2)t + 20.0 \text{ m/s} \quad \textbf{(1)}$$

$$y = (20.0 \text{ m/s})t - (4.90 \text{ m/s}^2)t^2 \quad \textbf{(2)}$$

Substitute $v = 0$, the velocity at maximum height, into Equation (1) and solve for time:

$$0 = (-9.80 \text{ m/s}^2)t + 20.0 \text{ m/s}$$

$$t = \frac{-20.0 \text{ m/s}}{-9.80 \text{ m/s}^2} = \boxed{2.04 \text{ s}}$$

(b) Determine the stone's maximum height.

Substitute the time $t = 2.04$ s into Equation (2):

$$y_{max} = (20.0 \text{ m/s})(2.04) - (4.90 \text{ m/s}^2)(2.04)^2 = \boxed{20.4 \text{ m}}$$

(c) Find the time the stone takes to return to its initial position, and find the velocity of the stone at that time.

Set $y = 0$ in Equation (2) and solve t:

$$0 = (20.0 \text{ m/s})t - (4.90 \text{ m/s}^2)t^2$$

$$= t(20.0 \text{ m/s} - 4.90 \text{ m/s}^2\, t)$$

$$t = \boxed{4.08 \text{ s}}$$

Substitute the time into Equation (1) to get the velocity: $v = 20.0\ \text{m/s} + (-9.80\ \text{m/s}^2)(4.08\ \text{s}) = \boxed{-20.0\ \text{m/s}}$

(d) Find the time required for the stone to reach the ground.

In Equation 2, set $y = -50.0$ m:

$$-50.0\ \text{m} = (20.0\ \text{m/s})t - (4.90\ \text{m/s}^2)t^2$$

Apply the quadratic formula and take the positive root: $\boxed{t = 5.83\ \text{s}}$

(e) Find the velocity and position of the stone at $t = 5.00$ s. Substitute values into Equations (1) and (2):

$$v = (-9.80\ \text{m/s}^2)(5.00\ \text{s}) + 20.0\ \text{m/s} = \boxed{-29.0\ \text{m/s}}$$

$$y = (20.0\ \text{m/s})(5.00\ \text{s}) - (4.90\ \text{m/s}^2)(5.00\ \text{s})^2 = \boxed{-22.5\ \text{m}}$$

Remarks Notice how everything follows from the two kinematic equations. Once they are written down, and constants correctly identified as in Equations (1) and (2), the rest is relatively easy. If the stone were thrown downward, the initial velocity would have been negative.

Exercise 2.9

A projectile is launched straight up at 60.0 m/s from a height of 80.0 m, at the edge of a sheer cliff. The projectile falls, just missing the cliff and hitting the ground below. Find (a) the maximum height of the projectile above the point of firing, (b) the time it takes to hit the ground at the base of the cliff, and (c) its velocity at impact.

Answers (a) 184 m (b) 13.5 s (c) -72.3 m/s

PhysicsNow™ You can study the motion of the thrown ball by logging into PhysicsNow at **www.cp7e.com** and going to Interactive Example 2.9.

EXAMPLE 2.10 A Rocket Goes Ballistic

Goal Solve a problem involving a powered ascent followed by free fall motion.

Problem A rocket moves straight upward, starting from rest with an acceleration of $+29.4\ \text{m/s}^2$. It runs out of fuel at the end of 4.00 s and continues to coast upward, reaching a maximum height before falling back to Earth. **(a)** Find the rocket's velocity and position at the end of 4.00 s. **(b)** Find the maximum height the rocket reaches. **(c)** Find the velocity the instant before the rocket crashes on the ground.

Strategy Take $y = 0$ at the launch point and y positive upward, as in Figure 2.21. The problem consists of two phases. In phase 1, the rocket has a net *upward* acceleration of $29.4\ \text{m/s}^2$, and we can use the kinematic equations with constant acceleration a to find the height and velocity of the rocket at the end of phase 1, when the fuel is burned up. In phase 2, the rocket is in free fall and has an acceleration of $-9.80\ \text{m/s}^2$, with initial velocity and position given by the results of phase 1. Apply the kinematic equations for free fall.

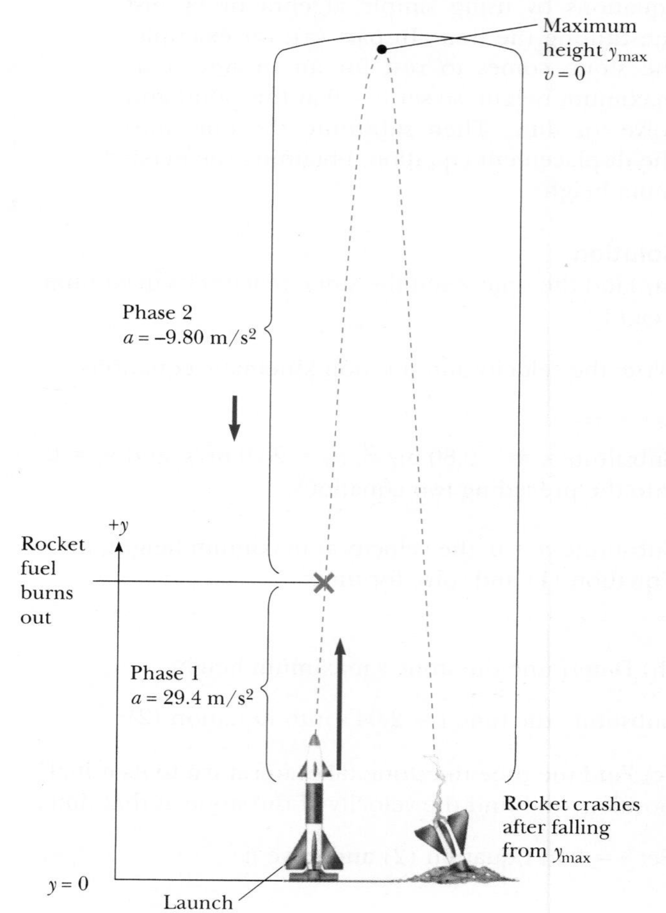

Figure 2.21 (Example 2.10) Two linked phases of motion for a rocket that is launched, uses up its fuel, and crashes.

Solution

(a) Phase 1: Find the rocket's velocity and position after 4.00 s.

Write the velocity and position kinematic equations:

$$v = v_0 + at \quad (1)$$

$$\Delta y = y - y_0 = v_0 t + \tfrac{1}{2}at^2 \quad (2)$$

Adapt these equations to phase 1, substituting $a = 29.4\ \text{m/s}^2$, $v_0 = 0$, and $y_0 = 0$:

$$v = (29.4\ \text{m/s}^2)t \quad (3)$$

$$y = \tfrac{1}{2}(29.4\ \text{m/s}^2)t^2 = (14.7\ \text{m/s}^2)t^2 \quad (4)$$

Substitute $t = 4.00$ s into Equations (3) and (4) to find the rocket's velocity v and position y at the time of burnout. These will be called v_b and y_b, respectively.

$$v_b = 118\ \text{m/s} \quad \text{and} \quad y_b = 235\ \text{m}$$

(b) Phase 2: Find the maximum height the rocket attains.

Adapt Equations (1) and (2) to phase 2, substituting $a = -9.8\ \text{m/s}^2$, $v_0 = v_b = 118\ \text{m/s}$, and $y_0 = y_b = 235$ m:

$$v = (-9.8\ \text{m/s}^2)t + 118\ \text{m/s} \quad (5)$$

$$y = 235\ \text{m} + (118\ \text{m/s})t - (4.90\ \text{m/s}^2)t^2 \quad (6)$$

Substitute $v = 0$ (the rocket's velocity at maximum height) in Equation 5 to get the time it takes the rocket to reach its maximum height:

$$0 = (-9.8\ \text{m/s}^2)t + 118\ \text{m/s} \rightarrow t = \frac{118\ \text{m/s}}{9.80\ \text{m/s}^2} = 12.0\ \text{s}$$

Substitute $t = 12.0$ s into Equation (6) to find the rocket's maximum height:

$$y_{max} = 235\ \text{m} + (118\ \text{m/s})(12.0\ \text{s}) - (4.90\ \text{m/s}^2)(12.0\ \text{s})^2$$
$$= 945\ \text{m}$$

(c) Phase 2: Find the velocity of the rocket just prior to impact.

Find the time to impact by setting $y = 0$ in Equation (6) and using the quadratic formula:

$$0 = 235\ \text{m} + (118\ \text{m/s})t - (4.90\ \text{m/s}^2)t^2$$

$$t = 25.9\ \text{s}$$

Substitute this value of t into Equation (5):

$$v = (-9.80\ \text{m/s}^2)(25.9\ \text{s}) + 118\ \text{m/s} = -136\ \text{m/s}$$

Remarks You may think that it is more natural to break this problem into three phases, with the second phase ending at the maximum height and the third phase a free fall from maximum height to the ground. Although this approach gives the correct answer, it's an unnecessary complication. Two phases are sufficient, one for each different acceleration.

Exercise 2.10

An experimental rocket designed to land upright falls freely from a height of 2.00×10^2 m, starting at rest. At a height of 80.0 m, the rocket's engines start and provide constant upward acceleration until the rocket lands. What acceleration is required if the speed on touchdown is to be zero? (Neglect air resistance.)

Answer $14.7\ \text{m/s}^2$

SUMMARY

PhysicsNow™ Take a practice test by logging into PhysicsNow at **www.cp7e.com** and clicking on the Pre-Test link for this chapter.

2.1 Displacement

The **displacement** of an object moving along the x-axis is defined as the change in position of the object,

$$\Delta x \equiv x_f - x_i \quad [2.1]$$

where x_i is the initial position of the object and x_f is its final position.

A **vector** quantity is characterized by both a magnitude and a direction. A **scalar** quantity has a magnitude only.

2.2 Velocity

The **average speed** of an object is given by

$$\text{Average speed} \equiv \frac{\text{total distance}}{\text{total time}}$$

The **average velocity** $\overline{v}$ during a time interval Δt is the displacement Δx divided by Δt.

$$\overline{v} \equiv \frac{\Delta x}{\Delta t} = \frac{x_f - x_i}{t_f - t_i} \qquad \textbf{[2.2]}$$

The average velocity is equal to the slope of the straight line joining the initial and final points on a graph of the position of the object versus time.

The slope of the line tangent to the position vs. time curve at some point is equal to the **instantaneous velocity** at that time. The **instantaneous speed** of an object is defined as the magnitude of the instantaneous velocity.

2.3 Acceleration

The **average acceleration** $\overline{a}$ of an object undergoing a change in velocity Δv during a time interval Δt is

$$\overline{a} \equiv \frac{\Delta v}{\Delta t} = \frac{v_f - v_i}{t_f - t_i} \qquad \textbf{[2.4]}$$

The **instantaneous acceleration** of an object at a certain time equals the slope of a velocity vs. time graph at that instant.

2.5 One-Dimensional Motion with Constant Acceleration

The most useful equations that describe the motion of an object moving with constant acceleration along the x axis are as follows:

$$v = v_0 + at \qquad \textbf{[2.6]}$$

$$\Delta x = v_0 t + \tfrac{1}{2}at^2 \qquad \textbf{[2.9]}$$

$$v^2 = v_0{}^2 + 2a\Delta x \qquad \textbf{[2.10]}$$

All problems can be solved with the first two equations alone, the last being convenient when time doesn't explicitly enter the problem. After the constants are properly identified, most problems reduce to one or two equations in as many unknowns.

2.6 Freely Falling Objects

An object falling in the presence of Earth's gravity exhibits a free-fall acceleration directed toward Earth's center. If air friction is neglected and if the altitude of the falling object is small compared with Earth's radius, then we can assume that the free-fall acceleration $g = 9.8\ \text{m/s}^2$ is constant over the range of motion. Equations 2.6, 2.9, and 2.10 apply, with $a = -g$.

CONCEPTUAL QUESTIONS

1. If the velocity of a particle is nonzero, can the particle's acceleration be zero? Explain.
2. If the velocity of a particle is zero, can the particle's acceleration be zero? Explain.
3. If a car is traveling eastward, can its acceleration be westward? Explain.
4. The speed of sound in air is 331 m/s. During the next thunderstorm, try to estimate your distance from a lightning bolt by measuring the time lag between the flash and the thunderclap. You can ignore the time it takes for the light flash to reach you. Why?
5. Can the equations of kinematics be used in a situation where the acceleration varies with time? Can they be used when the acceleration is zero?
6. If the average velocity of an object is zero in some time interval, what can you say about the displacement of the object during that interval?
7. A child throws a marble into the air with an initial speed v_0. Another child drops a ball at the same instant. Compare the accelerations of the two objects while they are in flight.
8. Figure Q2.8 shows strobe photographs taken of a disk moving from left to right under different conditions. The time interval between images is constant. Taking the direction to the right to be positive, describe the motion of the disk in each case. For which case is (a) the acceleration positive? (b) the acceleration negative? (c) the velocity constant?

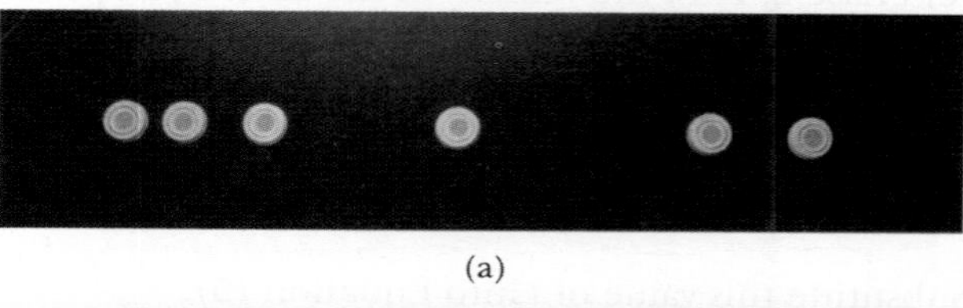

(a)

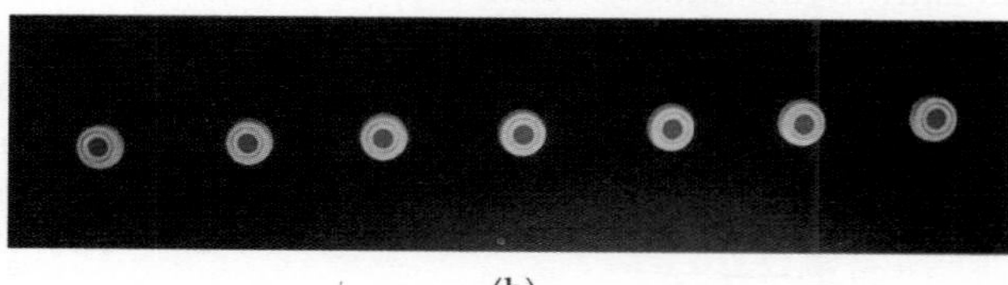

(b)

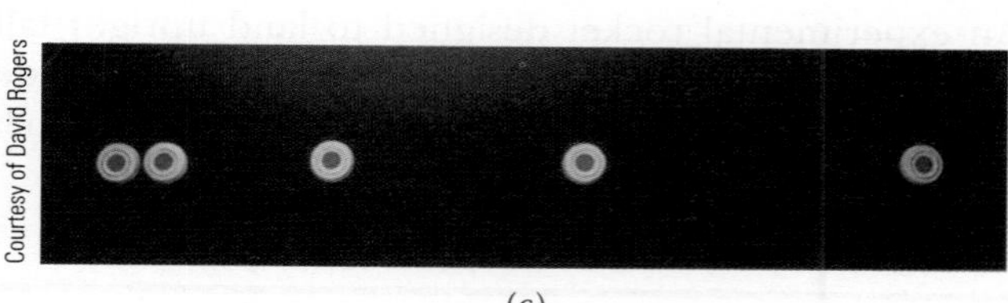

(c)

Courtesy of David Rogers

Figure Q2.8

9. Can the instantaneous velocity of an object at an instant of time ever be greater in magnitude than the average velocity over a time interval containing that instant? Can it ever be less?
10. Car A, traveling from New York to Miami, has a speed of 25 m/s. Car B, traveling from New York to Chicago, also has a speed of 25 m/s. Are their velocities equal? Explain.
11. A ball is thrown vertically upward. (a) What are its velocity and acceleration when it reaches its maximum altitude? (b) What is the acceleration of the ball just before it hits the ground?

12. A rule of thumb for driving is that a separation of one car length for each 10 mi/h of speed should be maintained between moving vehicles. Assuming a constant reaction time, discuss the relevance of this rule for (a) motion with constant velocity and (b) motion with constant acceleration.

13. Two cars are moving in the same direction in parallel lanes along a highway. At some instant, the velocity of car A exceeds the velocity of car B. Does this mean that the acceleration of A is greater than that of B? Explain.

14. Consider the following combinations of signs and values for the velocity and acceleration of a particle with respect to a one-dimensional x-axis:

	Velocity	Acceleration
a.	Positive	Positive
b.	Positive	Negative
c.	Positive	Zero
d.	Negative	Positive
e.	Negative	Negative
f.	Negative	Zero
g.	Zero	Positive
h.	Zero	Negative

Describe what the particle is doing in each case, and give a real-life example for an automobile on an east-west one-dimensional axis, with east considered the positive direction.

15. A student at the top of a building of height h throws one ball upward with a speed of v_0 and then throws a second ball downward with the same initial speed, v_0. How do the final velocities compare when the balls reach the ground?

16. A ball is thrown straight upward and moves in free fall. Choose a coordinate system with its origin at the release point of the ball and the positive direction upward. (a) What is the sign of the velocity of the ball just before the ball reaches its maximum height, just after it reaches its maximum height, and at its maximum height. (b) What is the sign of the acceleration of the ball just before the ball reaches its maximum height, just after it reaches its maximum height, and at its maximum height. (c) If the ball takes time t_1 to reach its maximum height, how long will it take to return to ground level? (d) If the ball is thrown upward with a velocity of $+v_0$, what will be the ball's velocity upon returning to ground level?

17. A pebble is dropped into a water well, and the splash is heard 16 s later, as illustrated in the cartoon strip shown in Figure Q2.17. Estimate the distance from the rim of the well to the water's surface.

18. A ball rolls in a straight line along the horizontal direction. Using motion diagrams (or multiflash photographs), describe the velocity and acceleration of the ball for each of the following situations: (a) The ball moves to the right at a constant speed. (b) The ball moves from right to left and continually slows down. (c) The ball moves from right to left and continually speeds up. (d) The ball moves to the right, first speeding up at a constant rate and then slowing down at a constant rate.

19. You drop a ball from a window on an upper floor of a building. The ball strikes the ground with speed v. You now repeat the drop, but you have a friend down on the street who throws another ball upward at speed v. Your friend throws the ball upward at exactly the same time that you drop yours from the window. At some location, the balls pass each other. Is this location *at* the halfway point between window and ground, *above* that point, or *below* that point?

Figure Q2.17

PROBLEMS

1, 2, 3 = straightforward, intermediate, challenging ☐ = full solution available in *Student Solutions Manual/Study Guide*
PhysicsNow™ = coached problem with hints available at **www.cp7e.com** = biomedical application

Section 2.1 Displacement
Section 2.2 Velocity

1. A person travels by car from one city to another with different constant speeds between pairs of cities. She drives for 30.0 min at 80.0 km/h, 12.0 min at 100 km/h, and 45.0 min at 40.0 km/h and spends 15.0 min eating lunch and buying gas. (a) Determine the average speed for the trip. (b) Determine the distance between the initial and final cities along the route.

2. (a) Sand dunes on a desert island move as sand is swept up the windward side to settle in the leeward side. Such "walking" dunes have been known to travel 20 feet in a

year and can travel as much as 100 feet per year in particularly windy times. Calculate the average speed in each case in m/s. (b) Fingernails grow at the rate of drifting continents, about 10 mm/yr. Approximately how long did it take for North America to separate from Europe, a distance of about 3 000 mi?

3. Two boats start together and race across a 60-km-wide lake and back. Boat A goes across at 60 km/h and returns at 60 km/h. Boat B goes across at 30 km/h, and its crew, realizing how far behind it is getting, returns at 90 km/h. Turnaround times are negligible, and the boat that completes the round trip first wins. (a) Which boat wins and by how much? (Or is it a tie?) (b) What is the average velocity of the winning boat?

4. The Olympic record for the marathon is 2 h, 9 min, 21 s. The marathon distance is 26 mi, 385 yd. Determine the average speed (in miles per hour) of the record.

5. A motorist drives north for 35.0 minutes at 85.0 km/h and then stops for 15.0 minutes. He then continues north, traveling 130 km in 2.00 h. (a) What is his total displacement? (b) What is his average velocity?

6. A graph of position versus time for a certain particle moving along the x-axis is shown in Figure P2.6. Find the average velocity in the time intervals from (a) 0 to 2.00 s, (b) 0 to 4.00 s, (c) 2.00 s to 4.00 s, (d) 4.00 s to 7.00 s, and (e) 0 to 8.00 s.

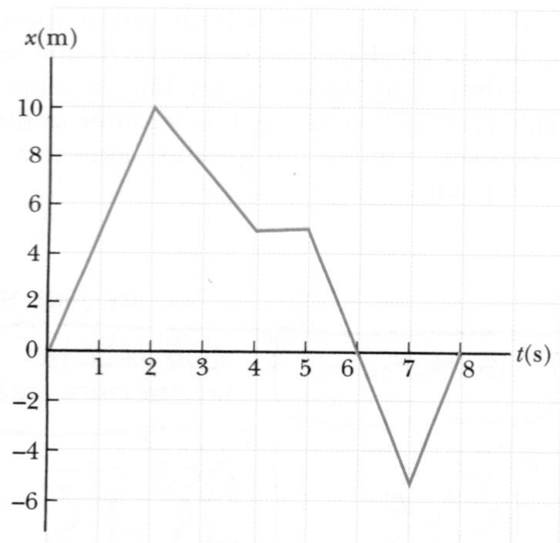

Figure P2.6 (Problems 6 and 15)

7. A tennis player moves in a straight-line path as shown in Figure P2.7. Find her average velocity in the time intervals from (a) 0 to 1.0 s, (b) 0 to 4.0 s, (c) 1.0 s to 5.0 s, and (d) 0 to 5.0 s.

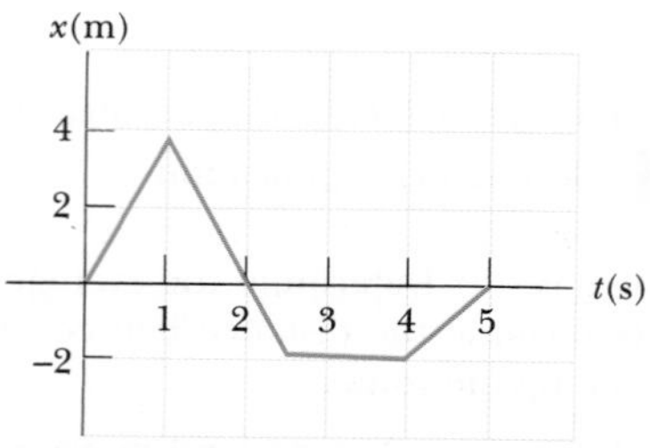

Figure P2.7 (Problems 7 and 17)

8. Two cars travel in the same direction along a straight highway, one at a constant speed of 55 mi/h and the other at 70 mi/h. (a) Assuming that they start at the same point, how much sooner does the faster car arrive at a destination 10 mi away? (b) How far must the faster car travel before it has a 15-min lead on the slower car?

9. An athlete swims the length of a 50.0-m pool in 20.0 s and makes the return trip to the starting position in 22.0 s. Determine her average velocities in (a) the first half of the swim, (b) the second half of the swim, and (c) the round trip.

10. If the average speed of an orbiting space shuttle is 19 800 mi/h, determine the time required for it to circle Earth. Make sure you consider the fact that the shuttle is orbiting about 200 mi above Earth's surface, and assume that Earth's radius is 3 963 miles.

11. A person takes a trip, driving with a constant speed of 89.5 km/h, except for a 22.0-min rest stop. If the person's average speed is 77.8 km/h, how much time is spent on the trip and how far does the person travel?

12. A tortoise can run with a speed of 0.10 m/s, and a hare can run 20 times as fast. In a race, they both start at the same time, but the hare stops to rest for 2.0 minutes. The tortoise wins by a shell (20 cm). (a) How long does the race take? (b) What is the length of the race?

13. **PhysicsNow™** In order to qualify for the finals in a racing event, a race car must achieve an average speed of 250 km/h on a track with a total length of 1 600 m. If a particular car covers the first half of the track at an average speed of 230 km/h, what minimum average speed must it have in the second half of the event in order to qualify?

14. Runner A is initially 4.0 mi west of a flagpole and is running with a constant velocity of 6.0 mi/h due east. Runner B is initially 3.0 mi east of the flagpole and is running with a constant velocity of 5.0 mi/h due west. How far are the runners from the flagpole when they meet?

15. A graph of position versus time for a certain particle moving along the x-axis is shown in Figure P2.6. Find the instantaneous velocity at the instants (a) $t = 1.00$ s, (b) $t = 3.00$ s, (c) $t = 4.50$ s, and (d) $t = 7.50$ s.

16. A race car moves such that its position fits the relationship

$$x = (5.0\text{ m/s})t + (0.75\text{ m/s}^3)t^3$$

where x is measured in meters and t in seconds. (a) Plot a graph of the car's position versus time. (b) Determine the instantaneous velocity of the car at $t = 4.0$ s, using time intervals of 0.40 s, 0.20 s, and 0.10 s. (c) Compare the average velocity during the first 4.0 s with the results of (b).

17. Find the instantaneous velocities of the tennis player of Figure P2.7 at (a) 0.50 s, (b) 2.0 s, (c) 3.0 s, and (d) 4.5 s.

Section 2.3 Acceleration

18. Secretariat ran the Kentucky Derby with times of 25.2 s, 24.0 s, 23.8 s, and 23.0 s for the quarter mile. (a) Find his average speed during each quarter-mile segment. (b) Assuming that Secretariat's instantaneous speed at the finish line was the same as his average speed during the final quarter mile, find his average acceleration for the entire race. (*Hint:* Recall that horses in the Derby start from rest.)

19. A steam catapult launches a jet aircraft from the aircraft carrier *John C. Stennis*, giving it a speed of 175 mi/h in 2.50 s. (a) Find the average acceleration of the plane. (b) Assuming that the acceleration is constant, find the distance the plane moves.

20. A car traveling in a straight line has a velocity of +5.0 m/s at some instant. After 4.0 s, its velocity is +8.0 m/s. What is the car's average acceleration during the 4.0-s time interval?

21. PhysicsNow™ A certain car is capable of accelerating at a rate of +0.60 m/s². How long does it take for this car to go from a speed of 55 mi/h to a speed of 60 mi/h?

22. The velocity vs. time graph for an object moving along a straight path is shown in Figure P2.22. (a) Find the average acceleration of the object during the time intervals 0 to 5.0 s, 5.0 s to 15 s, and 0 to 20 s. (b) Find the instantaneous acceleration at 2.0 s, 10 s, and 18 s.

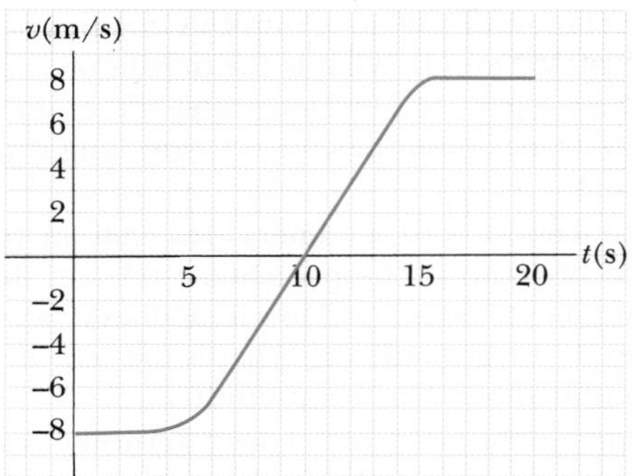

Figure P2.22

23. The engine of a model rocket accelerates the rocket vertically upward for 2.0 s as follows: At $t = 0$, the rocket's speed is zero; at $t = 1.0$ s, its speed is 5.0 m/s; and at $t = 2.0$ s, its speed is 16 m/s. Plot a velocity vs. time graph for this motion, and use the graph to determine (a) the rocket's average acceleration during the 2.0-s interval and (b) the instantaneous acceleration of the rocket at $t = 1.5$ s.

Section 2.5 One-Dimensional Motion with Constant Acceleration

24. A car traveling in a straight-line path has a velocity of +10.0 m/s at some instant. After 3.00 s, its velocity is +6.00 m/s. What is the average acceleration of the car during this time interval?

25. In 1865, Jules Verne proposed sending men to the Moon by firing a space capsule from a 220-m-long cannon with final speed of 10.97 km/s. What would have been the unrealistically large acceleration experienced by the space travelers during their launch? (A human can stand an acceleration of $15g$ for a short time.) Compare your answer with the free-fall acceleration, 9.80 m/s².

26. A truck covers 40.0 m in 8.50 s while smoothly slowing down to a final speed of 2.80 m/s. (a) Find the truck's original speed. (b) Find its acceleration.

27. A speedboat increases its speed uniformly from 20 m/s to 30 m/s in a distance of 200 m. Find (a) the magnitude of its acceleration and (b) the time it takes the boat to travel the 200-m distance.

28. Two cars are traveling along a straight line in the same direction, the lead car at 25.0 m/s and the other car at 30.0 m/s. At the moment the cars are 40.0 m apart, the lead driver applies the brakes, causing his car to have an acceleration of −2.00 m/s². (a) How long does it take for the lead car to stop? (b) Assuming that the chasing car brakes at the same time as the lead car, what must be the chasing car's minimum negative acceleration so as not to hit the lead car? (c) How long does it take for the chasing car to stop?

29. A Cessna aircraft has a lift-off speed of 120 km/h. (a) What minimum constant acceleration does the aircraft require if it is to be airborne after a takeoff run of 240 m? (b) How long does it take the aircraft to become airborne?

30. A truck on a straight road starts from rest and accelerates at 2.0 m/s² until it reaches a speed of 20 m/s. Then the truck travels for 20 s at constant speed until the brakes are applied, stopping the truck in a uniform manner in an additional 5.0 s. (a) How long is the truck in motion? (b) What is the average velocity of the truck during the motion described?

31. A drag racer starts her car from rest and accelerates at 10.0 m/s² for a distance of 400 m ($\frac{1}{4}$ mile). (a) How long did it take the race car to travel this distance? (b) What is the speed of the race car at the end of the run?

32. A jet plane lands with a speed of 100 m/s and can accelerate at a maximum rate of −5.00 m/s² as it comes to rest. (a) From the instant the plane touches the runway, what is the minimum time needed before it can come to rest? (b) Can this plane land on a small tropical island airport where the runway is 0.800 km long?

33. A driver in a car traveling at a speed of 60 mi/h sees a deer 100 m away on the road. Calculate the minimum constant acceleration that is necessary for the car to stop without hitting the deer (assuming that the deer does not move in the meantime).

34. A record of travel along a straight path is as follows:

1. Start from rest with a constant acceleration of 2.77 m/s² for 15.0 s.
2. Maintain a constant velocity for the next 2.05 min.
3. Apply a constant negative acceleration of −9.47 m/s² for 4.39 s.

(a) What was the total displacement for the trip? (b) What were the average speeds for legs 1, 2, and 3 of the trip, as well as for the complete trip?

35. A train is traveling down a straight track at 20 m/s when the engineer applies the brakes, resulting in an acceleration of −1.0 m/s² as long as the train is in motion. How far does the train move during a 40-s time interval starting at the instant the brakes are applied?

36. A car accelerates uniformly from rest to a speed of 40.0 mi/h in 12.0 s. Find (a) the distance the car travels during this time and (b) the constant acceleration of the car.

37. A car starts from rest and travels for 5.0 s with a uniform acceleration of +1.5 m/s². The driver then applies the brakes, causing a uniform acceleration of −2.0 m/s². If the brakes are applied for 3.0 s, (a) how fast is the car going at the end of the braking period, and (b) how far has the car gone?

38. A train 400 m long is moving on a straight track with a speed of 82.4 km/h. The engineer applies the brakes at a crossing, and later the last car passes the crossing with a speed of 16.4 km/h. Assuming constant acceleration, determine how long the train blocked the crossing. Disregard the width of the crossing.

39. A hockey player is standing on his skates on a frozen pond when an opposing player, moving with a uniform speed of 12 m/s, skates by with the puck. After 3.0 s, the first player makes up his mind to chase his opponent. If he accelerates uniformly at 4.0 m/s^2, (a) how long does it take him to catch his opponent, and (b) how far has he traveled in that time? (Assume that the player with the puck remains in motion at constant speed.)

40. A glider on an air track carries a flag of length ℓ through a stationary photogate that measures the time interval Δt_d during which the flag blocks a beam of infrared light passing across the gate. The ratio $v_d = \ell/\Delta t_d$ is the average velocity of the glider over this part of its motion. Suppose the glider moves with constant acceleration. (a) Argue for or against the idea that v_d is equal to the instantaneous velocity of the glider when it is halfway through the photogate in terms of distance. (b) Argue for or against the idea that v_d is equal to the instantaneous velocity of the glider when it is halfway through the photogate in terms of time.

41. In the Daytona 500 auto race, a Ford Thunderbird and a Mercedes Benz are moving side by side down a straightaway at 71.5 m/s. The driver of the Thunderbird realizes that she must make a pit stop, and she smoothly slows to a stop over a distance of 250 m. She spends 5.00 s in the pit and then accelerates out, reaching her previous speed of 71.5 m/s after a distance of 350 m. At this point, how far has the Thunderbird fallen behind the Mercedes Benz, which has continued at a constant speed?

42. A certain cable car in San Francisco can stop in 10 s when traveling at maximum speed. On one occasion, the driver sees a dog a distance d m in front of the car and slams on the brakes instantly. The car reaches the dog 8.0 s later, and the dog jumps off the track just in time. If the car travels 4.0 m beyond the position of the dog before coming to a stop, how far was the car from the dog? (*Hint*: You will need three equations.)

Section 2.6 Freely Falling Objects

43. A ball is thrown vertically upward with a speed of 25.0 m/s. (a) How high does it rise? (b) How long does it take to reach its highest point? (c) How long does the ball take to hit the ground after it reaches its highest point? (d) What is its velocity when it returns to the level from which it started?

44. It is possible to shoot an arrow at a speed as high as 100 m/s. (a) If friction is neglected, how high would an arrow launched at this speed rise if shot straight up? (b) How long would the arrow be in the air?

45. A certain freely falling object requires 1.50 s to travel the last 30.0 m before it hits the ground. From what height above the ground did it fall?

46. Traumatic brain injury such as concussion results when the head undergoes a very large acceleration. Generally, an acceleration less than 800 m/s^2 lasting for any length of time will not cause injury, whereas an acceleration greater than 1 000 m/s^2 lasting for at least 1 ms will cause injury. Suppose a small child rolls off a bed that is 0.40 m above the floor. If the floor is hardwood, the child's head is brought to rest in approximately 2.0 mm. If the floor is carpeted, this stopping distance is increased to about 1.0 cm. Calculate the magnitude and duration of the deceleration in both cases, to determine the risk of injury. Assume that the child remains horizontal during the fall to the floor. Note that a more complicated fall could result in a head velocity greater or less than the speed you calculate.

47. A small mailbag is released from a helicopter that is descending steadily at 1.50 m/s. After 2.00 s, (a) what is the speed of the mailbag, and (b) how far is it below the helicopter? (c) What are your answers to parts (a) and (b) if the helicopter is *rising* steadily at 1.50 m/s?

48. A ball thrown vertically upward is caught by the thrower after 2.00 s. Find (a) the initial velocity of the ball and (b) the maximum height the ball reaches.

49. A model rocket is launched straight upward with an initial speed of 50.0 m/s. It accelerates with a constant upward acceleration of 2.00 m/s^2 until its engines stop at an altitude of 150 m. (a) What is the maximum height reached by the rocket? (b) How long after lift-off does the rocket reach its maximum height? (c) How long is the rocket in the air?

50. A parachutist with a camera descends in free fall at a speed of 10 m/s. The parachutist releases the camera at an altitude of 50 m. (a) How long does it take the camera to reach the ground? (b) What is the velocity of the camera just before it hits the ground?

51. A student throws a set of keys vertically upward to his fraternity brother, who is in a window 4.00 m above. The brother's outstretched hand catches the keys 1.50 s later. (a) With what initial velocity were the keys thrown? (b) What was the velocity of the keys just before they were caught?

52. It has been claimed that an insect called the froghopper (*Philaenus spumarius*) is the best jumper in the animal kingdom. This insect can accelerate at 4 000 m/s^2 over a distance of 2.0 mm as it straightens its specially designed "jumping legs." (a) Assuming a uniform acceleration, what is the velocity of the insect after it has accelerated through this short distance, and how long did it take to reach that velocity? (b) How high would the insect jump if air resistance could be ignored? Note that the actual height obtained is about 0.7 m, so air resistance is important here.

ADDITIONAL PROBLEMS

53. A truck tractor pulls two trailers, one behind the other, at a constant speed of 100 km/h. It takes 0.600 s for the big rig to completely pass onto a bridge 400 m long. For what duration of time is all or part of the truck-trailer combination on the bridge?

54. A speedboat moving at 30.0 m/s approaches a no-wake buoy marker 100 m ahead. The pilot slows the boat with a constant acceleration of -3.50 m/s^2 by reducing the throttle. (a) How long does it take the boat to reach the buoy? (b) What is the velocity of the boat when it reaches the buoy?

55. A bullet is fired through a board 10.0 cm thick in such a way that the bullet's line of motion is perpendicular to the face of the board. If the initial speed of the bullet is 400 m/s and it emerges from the other side of the board with a speed of 300 m/s, find (a) the acceleration of the bullet as it passes through the board and (b) the total time the bullet is in contact with the board.

56. An indestructible bullet 2.00 cm long is fired straight through a board that is 10.0 cm thick. The bullet strikes the board with a speed of 420 m/s and emerges with a speed of 280 m/s. (a) What is the average acceleration of the bullet through the board? (b) What is the total time that the bullet is in contact with the board? (c) What thickness of board (calculated to 0.1 cm) would it take to stop the bullet, assuming that the acceleration through all boards is the same?

57. A ball is thrown upward from the ground with an initial speed of 25 m/s; at the same instant, another ball is dropped from a building 15 m high. After how long will the balls be at the same height?

58. PhysicsNow™ A ranger in a national park is driving at 35.0 mi/h when a deer jumps into the road 200 ft ahead of the vehicle. After a reaction time t, the ranger applies the brakes to produce an acceleration $a = -9.00 \text{ ft/s}^2$. What is the maximum reaction time allowed if she is to avoid hitting the deer?

59. Two students are on a balcony 19.6 m above the street. One student throws a ball vertically downward at 14.7 m/s; at the same instant, the other student throws a ball vertically upward at the same speed. The second ball just misses the balcony on the way down. (a) What is the difference in the two balls' time in the air? (b) What is the velocity of each ball as it strikes the ground? (c) How far apart are the balls 0.800 s after they are thrown?

60. The driver of a truck slams on the brakes when he sees a tree blocking the road. The truck slows down uniformly with an acceleration of -5.60 m/s^2 for 4.20 s, making skid marks 62.4 m long that end at the tree. With what speed does the truck then strike the tree?

61. A young woman named Kathy Kool buys a sports car that can accelerate at the rate of 4.90 m/s^2. She decides to test the car by drag racing with another speedster, Stan Speedy. Both start from rest, but experienced Stan leaves the starting line 1.00 s before Kathy. If Stan moves with a constant acceleration of 3.50 m/s^2 and Kathy maintains an acceleration of 4.90 m/s^2, find (a) the time it takes Kathy to overtake Stan, (b) the distance she travels before she catches him, and (c) the speeds of both cars at the instant she overtakes him.

62. A mountain climber stands at the top of a 50.0-m cliff that overhangs a calm pool of water. She throws two stones vertically downward 1.00 s apart and observes that they cause a single splash. The first stone had an initial velocity of -2.00 m/s. (a) How long after release of the first stone did the two stones hit the water? (b) What initial velocity must the second stone have had, given that they hit the water simultaneously? (c) What was the velocity of each stone at the instant it hit the water?

63. An ice sled powered by a rocket engine starts from rest on a large frozen lake and accelerates at $+40 \text{ ft/s}^2$. After some time t_1, the rocket engine is shut down and the sled moves with constant velocity v for a time t_2. If the total distance traveled by the sled is 17 500 ft and the total time is 90 s, find (a) the times t_1 and t_2 and (b) the velocity v. At the 17 500-ft mark, the sled begins to accelerate at -20 ft/s^2. (c) What is the final position of the sled when it comes to rest? (d) How long does it take to come to rest?

64. In Bosnia, the ultimate test of a young man's courage used to be to jump off a 400-year-old bridge (now destroyed) into the River Neretva, 23 m below the bridge. (a) How long did the jump last? (b) How fast was the jumper traveling upon impact with the river? (c) If the speed of sound in air is 340 m/s, how long after the jumper took off did a spectator on the bridge hear the splash?

65. A person sees a lightning bolt pass close to an airplane that is flying in the distance. The person hears thunder 5.0 s after seeing the bolt and sees the airplane overhead 10 s after hearing the thunder. The speed of sound in air is 1 100 ft/s. (a) Find the distance of the airplane from the person at the instant of the bolt. (Neglect the time it takes the light to travel from the bolt to the eye.) (b) Assuming that the plane travels with a constant speed toward the person, find the velocity of the airplane. (c) Look up the speed of light in air, and defend the approximation used in (a).

66. Another scheme to catch the roadrunner has failed! Now a safe falls from rest from the top of a 25.0-m-high cliff toward Wile E. Coyote, who is standing at the base. Wile first notices the safe after it has fallen 15.0 m. How long does he have to get out of the way?

67. PhysicsNow™ A stunt man sitting on a tree limb wishes to drop vertically onto a horse galloping under the tree. The constant speed of the horse is 10.0 m/s, and the man is initially 3.00 m above the level of the saddle. (a) What must be the horizontal distance between the saddle and the limb when the man makes his move? (b) How long is he in the air?

68. A hard rubber ball, released at chest height, falls to the pavement and bounces back to nearly the same height. When the ball is in contact with the pavement, its lower side is temporarily flattened. Before the dent in the ball pops out, suppose that its maximum depth is on the order of 1 cm. Compute an order-of-magnitude estimate for the maximum acceleration of the ball. State your assumptions, the quantities you estimate, and the values you estimate for them.

69. *Vroom—vroom!* As soon as a traffic light turns green, a car speeds up from rest to 50.0 mi/h with a constant acceleration of 9.00 mi/h · s. In the adjoining bike lane, a cyclist speeds up from rest to 20.0 mi/h with a constant acceleration of 13.0 mi/h · s. Each vehicle maintains a constant velocity after reaching its cruising speed. (a) For how long is the bicycle ahead of the car? (b) By what maximum distance does the bicycle lead the car?

70. In order to pass a physical education class at a university, a student must run 1.0 mi in 12 min. After running for 10 min, she still has 500 yd to go. If her maximum acceleration is 0.15 m/s^2, can she make it? If the answer is no, determine what acceleration she would need to be successful.

71. One swimmer in a relay race has a 0.50-s lead and is swimming at a constant speed of 4.0 m/s. He has 50 m to swim before reaching the end of the pool. A second swimmer moves in the same direction as the leader. What constant speed must the second swimmer have in order to catch up to the leader at the end of the pool?

ACTIVITIES

A.1. Estimate a few speeds in metric units, using a stopwatch or a wristwatch. For example, roll a ball across a table and estimate the number of centimeters it moves each second to find its speed. Other speeds you might try are for someone walking across the room, a jogger running, a car moving through some distance, and so forth. To see how well you did, make some actual measurements for those situations in which it is feasible to do so.

A.2. Use what you know about falling objects to measure your reaction time. Hold the index finger and thumb of your dominant hand about 2.5 cm apart, and then have your co-worker hold a ruler vertically in the space between your finger and thumb, as shown in Figure A2.2. Note the position of the ruler relative to your index finger. Your co-worker must release the ruler, and you must catch it (without moving your hand downward) as quickly as you can. The ruler (a freely falling object) falls through a distance $d = \frac{1}{2}gt^2$, where t is the reaction time and $g = 9.80\ \text{m/s}^2$. Repeat this measurement of d five times, average your results, and calculate an average value of t. Now measure your co-worker's reaction time, using the same procedure. Compare your results. For most people, the reaction time is at best about 0.2 s. As an extension to this experiment, replace the ruler with a crisp dollar bill. Hold the bill such that your thumb and index finger are just at the level of Washington's face. Unless you are anticipating the time of release, you will not be able to catch the bill when it is released, because the time required for the top to pass out of your hand is less than the typical 0.2-s reaction time.

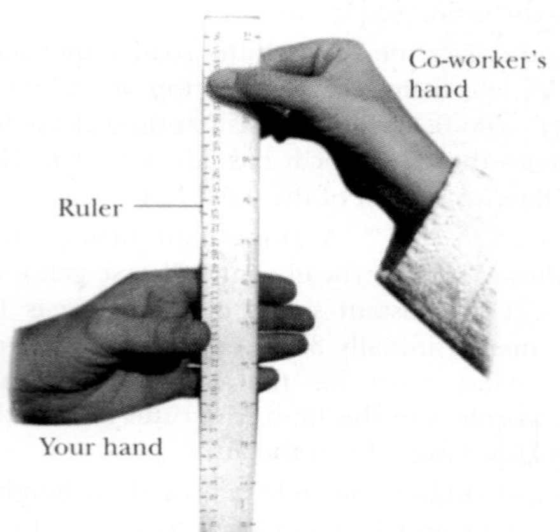

Figure A2.2

A.3. Galileo studied accelerated motion by allowing objects to roll down inclined planes so that their motion would be slow enough to make reasonable observations. Try a similar procedure. Make a mark at the top of an inclined plane as the starting point for the motion, and use a metal barrier at the end as a sound cue for stopping a stopwatch. Measure the length of the plane. Record the average time for several trials of a ball rolling down the plane at a measured angle. From the information you obtain, calculate the acceleration. Repeat the experiment for a larger angle of inclination. Do this for several trials, until you can plot a graph of acceleration versus angle. From your graph, can you guess what the acceleration would be if the inclined plane were vertical? Would the results of your experiment be different if you had used a significantly more massive ball? If you are unsure, repeat the experiment to see if there is a difference.

A.4. Perform the activities that follow to verify that all objects fall with the same acceleration. First, try dropping a coffee filter oriented horizontally and also dropping a pencil. Then repeat the experiment with the filter in a loose ball, a tight ball, and, finally, in a compacted wad. You should note that compacting the filter tends to reduce the effects of air resistance and makes the two objects fall more nearly at the same rate.

© Bettmann/Corbis

Legendary motorcycle stuntman Evel Knievel blasts off in his custom rocket-powered Harley-Davidson Skycycle in an attempt to jump the Snake River Canyon in 1974. A parachute prematurely deployed and caused the craft to fall into the canyon, just short of the other side. Knievel survived.

CHAPTER

3

Vectors and Two-Dimensional Motion

OUTLINE

In our discussion of one-dimensional motion in Chapter 2, we used the concept of vectors only to a limited extent. In our further study of motion, manipulating vector quantities will become increasingly important, so much of this chapter is devoted to vector techniques. We'll then apply these mathematical tools to two-dimensional motion, especially that of projectiles, and to the understanding of relative motion.

3.1 VECTORS AND THEIR PROPERTIES

Each of the physical quantities we will encounter in this book can be categorized as either a *vector quantity* or a *scalar quantity*. As noted in Chapter 2, a vector has both direction and magnitude (size). A scalar can be completely specified by its magnitude with appropriate units; it has no direction. An example of each kind of quantity is shown in Figure 3.1 (page 54).

As described in Chapter 2, displacement, velocity, and acceleration are vector quantities. Temperature is an example of a scalar quantity. If the temperature of an object is −5°C, that information completely specifies the temperature of the object; no direction is required. Masses, time intervals, and volumes are scalars as well. Scalar quantities can be manipulated with the rules of ordinary arithmetic. Vectors can also be added and subtracted from each other, and multiplied, but there are a number of important differences, as will be seen in the following sections.

When a vector quantity is handwritten, it is often represented with an arrow over the letter ($\vec{A}$). As mentioned in Section 2.1, a vector quantity in this book

Figure 3.1 (a) The number of grapes in this bunch ripe for picking is one example of a scalar quantity. Can you think of other examples? (b) This helpful person pointing in the right direction tells us to travel five blocks north to reach the courthouse. A vector is a physical quantity that must be specified by both magnitude and direction.

will be represented by boldface type with an arrow on top (for example, $\vec{\mathbf{A}}$). The magnitude of the vector $\vec{\mathbf{A}}$ will be represented by italic type, as A. Italic type will also be used to represent scalars.

Equality of Two Vectors. Two vectors $\vec{\mathbf{A}}$ and $\vec{\mathbf{B}}$ are equal if they have the same magnitude and the same direction. This property allows us to translate a vector parallel to itself in a diagram without affecting the vector. In fact, for most purposes, any vector can be moved parallel to itself without being affected. (See Fig. 3.2.)

Adding Vectors. When two or more vectors are added, they must all have the same units. For example, it doesn't make sense to add a velocity vector, carrying units of meters per second, to a displacement vector, carrying units of meters. Scalars obey the same rule: It would be similarly meaningless to add temperatures to volumes or masses to time intervals.

Vectors can be added geometrically or algebraically. (The latter is discussed at the end of the next section.) To add vector $\vec{\mathbf{B}}$ to vector $\vec{\mathbf{A}}$ geometrically, first draw $\vec{\mathbf{A}}$ on a piece of graph paper to some scale, such as 1 cm = 1 m, and so that its direction is specified relative a coordinate system. Then draw vector $\vec{\mathbf{B}}$ to the same scale with the tail of $\vec{\mathbf{B}}$ starting at the tip of $\vec{\mathbf{A}}$, as in Active Figure 3.3a. Vector $\vec{\mathbf{B}}$ must be drawn along the direction that makes the proper angle relative vector $\vec{\mathbf{A}}$. The **resultant vector** $\vec{\mathbf{R}} = \vec{\mathbf{A}} + \vec{\mathbf{B}}$ is the vector drawn from the tail of $\vec{\mathbf{A}}$ to the tip of $\vec{\mathbf{B}}$. This procedure is known as the **triangle method of addition**.

When two vectors are added, their sum is independent of the order of the addition: $\vec{\mathbf{A}} + \vec{\mathbf{B}} = \vec{\mathbf{B}} + \vec{\mathbf{A}}$. This relationship can be seen from the geometric construction in Active Figure 3.3b, and is called the **commutative law of addition**.

This same general approach can also be used to add more than two vectors, as is done in Figure 3.4 for four vectors. The resultant vector sum $\vec{\mathbf{R}} = \vec{\mathbf{A}} + \vec{\mathbf{B}} + \vec{\mathbf{C}} + \vec{\mathbf{D}}$ is the vector drawn from the tail of the first vector to the tip of the last. Again, the order in which the vectors are added is unimportant.

TIP 3.1 Vector Addition Versus Scalar Addition

$\vec{\mathbf{A}} + \vec{\mathbf{B}} = \vec{\mathbf{C}}$ is very different from $A + B = C$. The first is a vector sum, which must be handled graphically or with components, while the second is a simple arithmetic sum of numbers.

Negative of a Vector. The negative of the vector $\vec{\mathbf{A}}$ is defined as the vector that gives zero when added to $\vec{\mathbf{A}}$. This means that $\vec{\mathbf{A}}$ and $-\vec{\mathbf{A}}$ have the same magnitude but opposite directions.

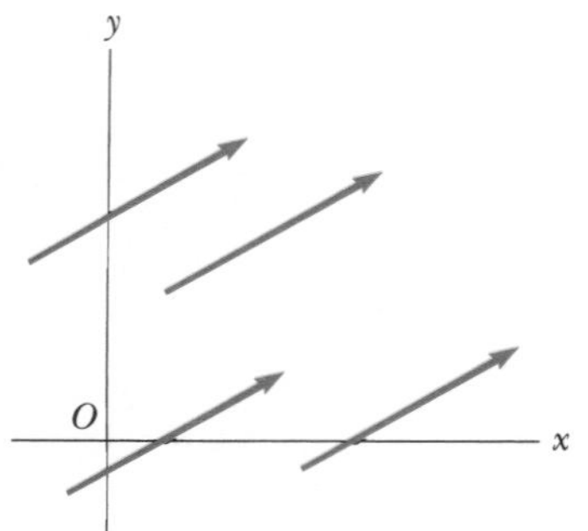

Figure 3.2 These four vectors are equal because they have equal lengths and point in the same direction.

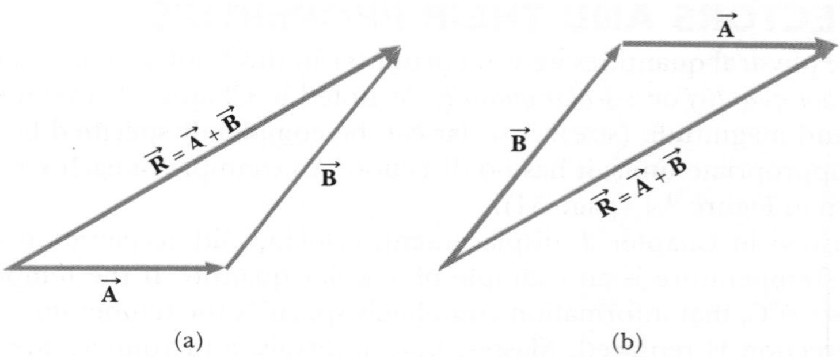

ACTIVE FIGURE 3.3
(a) When vector $\vec{\mathbf{B}}$ is added to vector $\vec{\mathbf{A}}$, the vector sum $\vec{\mathbf{R}}$ is the vector that runs from the tail of $\vec{\mathbf{A}}$ to the tip of $\vec{\mathbf{B}}$. (b) Here the resultant runs from the tail of $\vec{\mathbf{B}}$ to the tip of $\vec{\mathbf{A}}$. These constructions prove that $\vec{\mathbf{A}} + \vec{\mathbf{B}} = \vec{\mathbf{B}} + \vec{\mathbf{A}}$.

PhysicsNow™
Log into PhysicsNow at **www.cp7e.com**, and go to Active Figure 3.3 to vary $\vec{\mathbf{A}}$ and $\vec{\mathbf{B}}$ and see the effect on the resultant.

Subtracting Vectors. Vector subtraction makes use of the definition of the negative of a vector. We define the operation $\vec{\mathbf{A}} - \vec{\mathbf{B}}$ as the vector $-\vec{\mathbf{B}}$ added to the vector $\vec{\mathbf{A}}$:

$$\vec{\mathbf{A}} - \vec{\mathbf{B}} = \vec{\mathbf{A}} + (-\vec{\mathbf{B}}) \qquad [3.1]$$

Vector subtraction is really a special case of vector addition. The geometric construction for subtracting two vectors is shown in Figure 3.5.

Multiplying or Dividing a Vector by a Scalar. Multiplying or dividing a vector by a scalar gives a vector. For example, if vector $\vec{\mathbf{A}}$ is multiplied by the scalar number 3, the result, written $3\vec{\mathbf{A}}$, is a vector with a magnitude three times that of $\vec{\mathbf{A}}$ and pointing in the same direction. If we multiply vector $\vec{\mathbf{A}}$ by the scalar -3, the result is $-3\vec{\mathbf{A}}$, a vector with a magnitude three times that of $\vec{\mathbf{A}}$ and pointing in the opposite direction (because of the negative sign).

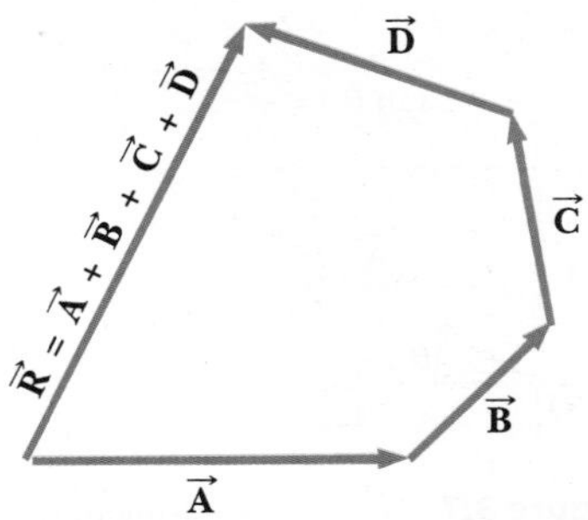

Figure 3.4 A geometric construction for summing four vectors. The resultant vector $\vec{\mathbf{R}}$ is the vector that completes the polygon.

Quick Quiz 3.1

The magnitudes of two vectors $\vec{\mathbf{A}}$ and $\vec{\mathbf{B}}$ are 12 units and 8 units, respectively. What are the largest and smallest possible values for the magnitude of the resultant vector $\vec{\mathbf{R}} = \vec{\mathbf{A}} + \vec{\mathbf{B}}$? (a) 14.4 and 4; (b) 12 and 8; (c) 20 and 4; (d) none of these.

Quick Quiz 3.2

If vector $\vec{\mathbf{B}}$ is added to vector $\vec{\mathbf{A}}$, the resultant vector $\vec{\mathbf{A}} + \vec{\mathbf{B}}$ has magnitude $A + B$ when $\vec{\mathbf{A}}$ and $\vec{\mathbf{B}}$ are (a) perpendicular to each other; (b) oriented in the same direction; (c) oriented in opposite directions; (d) none of these answers.

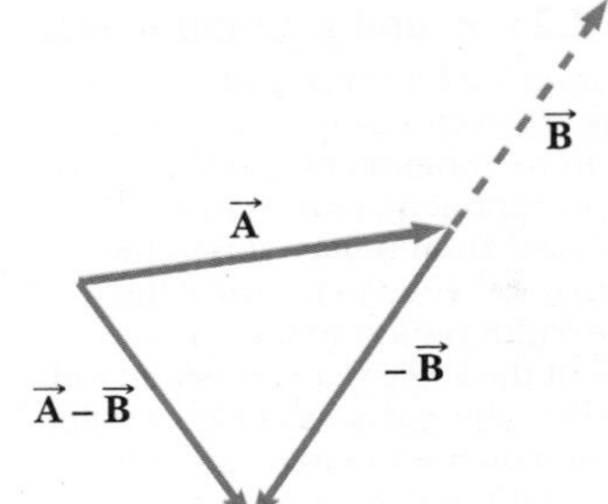

Figure 3.5 This construction shows how to subtract vector $\vec{\mathbf{B}}$ from vector $\vec{\mathbf{A}}$. The vector $-\vec{\mathbf{B}}$ has the same magnitude as the vector $\vec{\mathbf{B}}$, but points in the opposite direction.

EXAMPLE 3.1 Taking a Trip

Goal Find the sum of two vectors by using a graph.

Problem A car travels 20.0 km due north and then 35.0 km in a direction 60° west of north, as in Figure 3.6. Using a graph, find the magnitude and direction of a single vector that gives the net effect of the car's trip. This vector is called the car's *resultant displacement*.

Strategy Draw a graph, and represent the displacement vectors as arrows. Graphically locate the vector resulting from the sum of the two displacement vectors. Measure its length and angle with respect to the vertical.

Solution

Let $\vec{\mathbf{A}}$ represent the first displacement vector, 20.0 km north, $\vec{\mathbf{B}}$ the second displacement vector, extending west of north. Carefully graph the two vectors, drawing a resultant vector $\vec{\mathbf{R}}$ with its base touching the base of $\vec{\mathbf{A}}$ and extending to the tip of $\vec{\mathbf{B}}$. Measure the length of this vector, which turns out to be about 48 km. The angle β, measured with a protractor, is about 39° west of north.

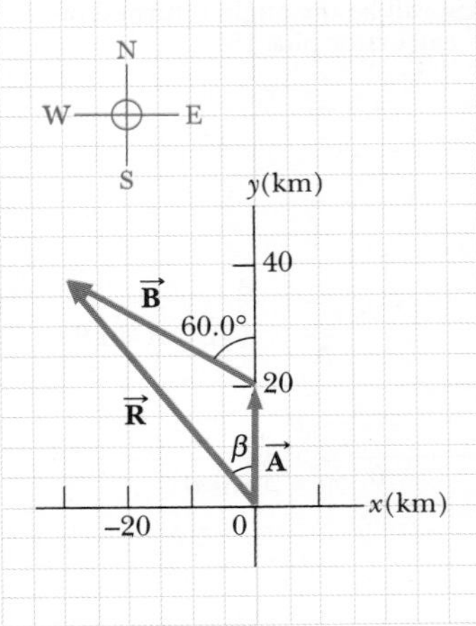

Figure 3.6 (Example 3.1) A graphical method for finding the resultant displacement vector $\vec{\mathbf{R}} = \vec{\mathbf{A}} + \vec{\mathbf{B}}$.

Remarks Notice that ordinary arithmetic doesn't work here: the correct answer of 48 km is not equal to 20.0 km + 35.0 km = 55.0 km!

Exercise 3.1

Graphically determine the magnitude and direction of the displacement if a man walks 30.0 km 45° north of east and then walks due east 20.0 km.

Answer 46 km, 27° north of east

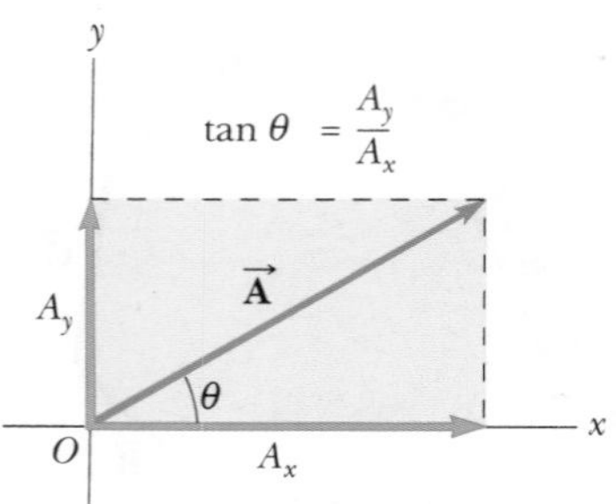

Figure 3.7 Any vector $\vec{\mathbf{A}}$ lying in the *xy*-plane can be represented by its rectangular components A_x and A_y.

3.2 COMPONENTS OF A VECTOR

One method of adding vectors makes use of the projections of a vector along the axes of a rectangular coordinate system. These projections are called **components.** Any vector can be completely described by its components.

Consider a vector $\vec{\mathbf{A}}$ in a rectangular coordinate system, as shown in Figure 3.7. $\vec{\mathbf{A}}$ can be expressed as the sum of two vectors: $\vec{\mathbf{A}}_x$, parallel to the *x*-axis; and $\vec{\mathbf{A}}_y$, parallel to the *y*-axis. Mathematically,

$$\vec{\mathbf{A}} = \vec{\mathbf{A}}_x + \vec{\mathbf{A}}_y$$

where $\vec{\mathbf{A}}_x$ and $\vec{\mathbf{A}}_y$ are the component vectors of $\vec{\mathbf{A}}$. The projection of $\vec{\mathbf{A}}$ along the *x*-axis, A_x, is called the *x*-component of $\vec{\mathbf{A}}$, and the projection of $\vec{\mathbf{A}}$ along the *y*-axis, A_y, is called the *y*-component of $\vec{\mathbf{A}}$. These components can be either positive or negative numbers with units. From the definitions of sine and cosine, we see that $\cos\theta = A_x/A$ and $\sin\theta = A_y/A$, so the components of $\vec{\mathbf{A}}$ are

$$A_x = A\cos\theta$$
$$A_y = A\sin\theta \qquad [3.2]$$

TIP 3.2 *x*- and *y*-Components

Equation 3.2 for the *x*- and *y*-components of a vector associates cosine with the *x*-component and sine with the *y*-component, as in Figure 3.8a. This association is due *solely* to the fact that we chose to measure the angle θ with respect to the positive *x*-axis. If the angle were measured with respect to the *y*-axis, as in Figure 3.8b, the components would be given by $A_x = A\sin\theta$ and $A_y = A\cos\theta$.

These components form two sides of a right triangle having a hypotenuse with magnitude A. It follows that $\vec{\mathbf{A}}$'s magnitude and direction are related to its components through the Pythagorean theorem and the definition of the tangent:

$$A = \sqrt{A_x^{\,2} + A_y^{\,2}} \qquad [3.3]$$

$$\tan\theta = \frac{A_y}{A_x} \qquad [3.4]$$

TIP 3.3 Inverse Tangents on Calculators: Right Half the Time

The inverse tangent function on calculators returns an angle between $-90°$ and $+90°$. If the vector lies in the second or third quadrant, the angle, as measured from the positive *x*-axis, will be the angle returned by your calculator plus $180°$.

To solve for the angle θ, which is measured from the positive *x*-axis by convention, we can write Equation 3.4 in the form

$$\theta = \tan^{-1}\left(\frac{A_y}{A_x}\right)$$

This formula gives the right answer only half the time! The inverse tangent function returns values only from $-90°$ to $+90°$, so the answer in your calculator window will only be correct if the vector happens to lie in first or fourth quadrant. If it lies in second or third quadrant, adding $180°$ to the number in the calculator window will always give the right answer. The angle in Equations 3.2 and 3.4 must be measured from the positive *x*-axis. Other choices of reference line are possible, but certain adjustments must then be made. (See Tip 3.2 and Figure 3.8.)

If a coordinate system other than the one shown in Figure 3.7 is chosen, the components of the vector must be modified accordingly. In many applications it's more convenient to express the components of a vector in a coordinate system having axes that are not horizontal and vertical, but are still perpendicular to each other. Suppose a vector $\vec{\mathbf{B}}$ makes an angle θ' with the x'-axis defined in Figure 3.9. The rectangular components of $\vec{\mathbf{B}}$ along the axes of the figure are given by $B_{x'} = B\cos\theta'$ and $B_{y'} = B\sin\theta'$, as in Equations 3.2. The magnitude and direction of $\vec{\mathbf{B}}$ are then obtained from expressions equivalent to Equations 3.3 and 3.4.

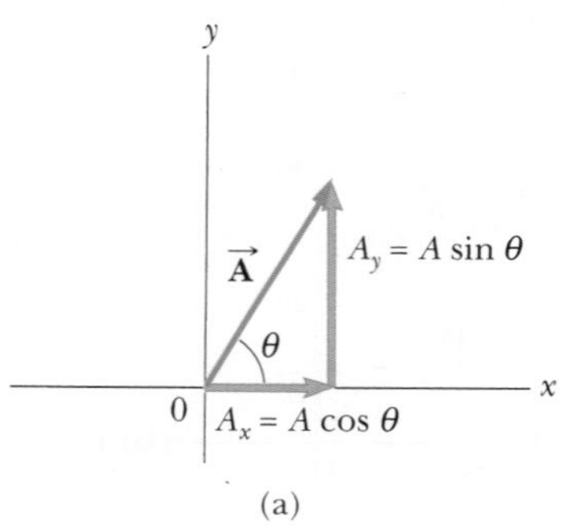

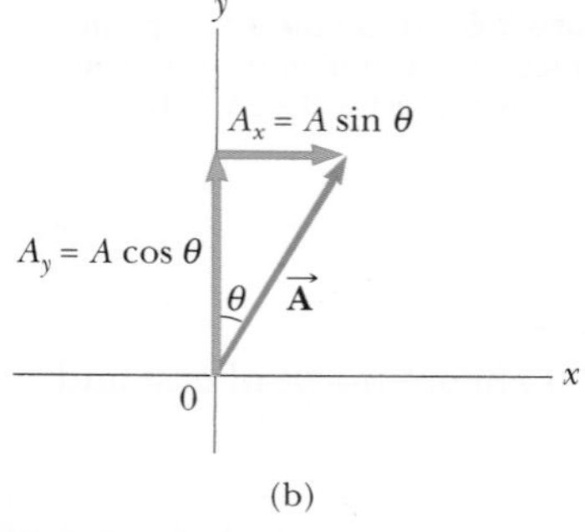

Figure 3.8 The angle θ need not always be defined from the positive *x*-axis.

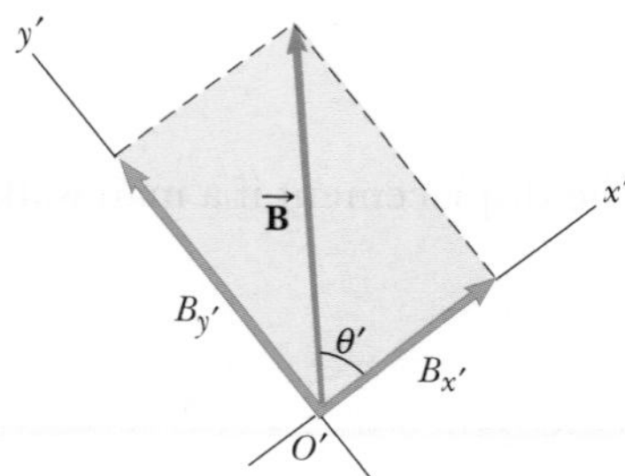

Figure 3.9 The components of vector $\vec{\mathbf{B}}$ in a tilted coordinate system.

Quick Quiz 3.3

Figure 3.10 shows two vectors lying in the xy-plane. Determine the signs of the x- and y-components of $\vec{\mathbf{A}}$, $\vec{\mathbf{B}}$, and $\vec{\mathbf{A}} + \vec{\mathbf{B}}$, and place your answers in the following table:

Vector	x-component	y-component
$\vec{\mathbf{A}}$		
$\vec{\mathbf{B}}$		
$\vec{\mathbf{A}} + \vec{\mathbf{B}}$		

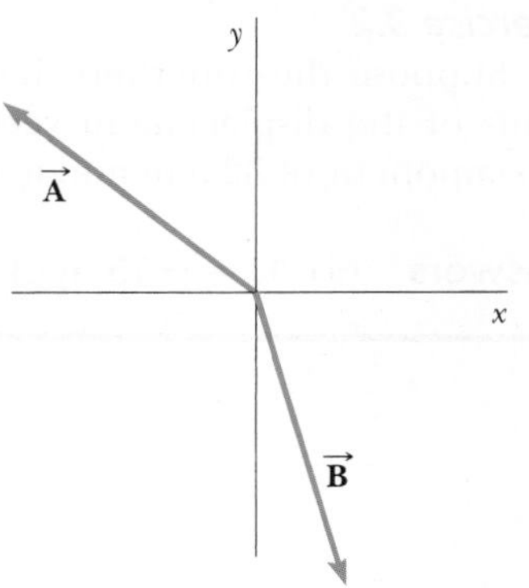

Figure 3.10 (Quick Quiz 3.3)

EXAMPLE 3.2 Help Is on the Way!

Goal Find vector components, given a magnitude and direction, and vice versa.

Problem **(a)** Find the horizontal and vertical components of the 1.00×10^2 m displacement of a superhero who flies from the top of a tall building along the path shown in Figure 3.11a. **(b)** Suppose instead the superhero leaps in the other direction along a displacement vector $\vec{\mathbf{B}}$, to the top of a flagpole where the displacement components are given by $B_x = -25.0$ m and $B_y = 10.0$ m. Find the magnitude and direction of the displacement vector.

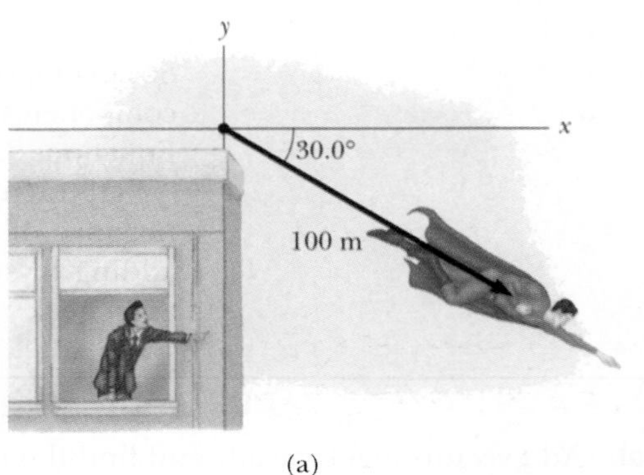

(a)

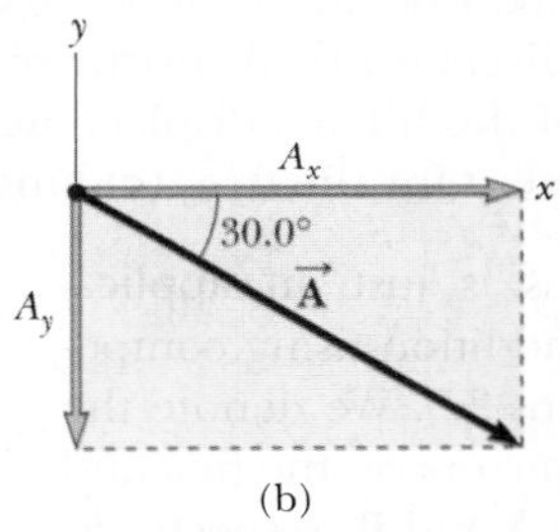

(b)

Figure 3.11 (Example 3.2)

Strategy **(a)** The triangle formed by the displacement and its components is shown in Figure 3.11b. Simple trigonometry gives the components relative the standard x-y coordinate system: $A_x = A \cos \theta$ and $A_y = A \sin \theta$ (Equations 3.2). Note that $\theta = -30.0°$, negative because it's measured clockwise from the positive x-axis. **(b)** Apply Equations 3.3 and 3.4 to find the magnitude and direction of the vector.

Solution

(a) Find the vector components of $\vec{\mathbf{A}}$ from its magnitude and direction.

Use Equations 3.2 to find the components of the displacement vector $\vec{\mathbf{A}}$:

$$A_x = A \cos \theta = (1.00 \times 10^2 \text{ m}) \cos(-30.0°) = \boxed{+86.6 \text{ m}}$$

$$A_y = A \sin \theta = (1.00 \times 10^2 \text{ m}) \sin(-30.0°) = \boxed{-50.0 \text{ m}}$$

(b) Find the magnitude and direction of the displacement vector $\vec{\mathbf{B}}$ from its components.

Compute the magnitude of $\vec{\mathbf{B}}$ from the Pythagorean theorem:

$$B = \sqrt{B_x^2 + B_y^2} = \sqrt{(-25.0 \text{ m})^2 + (10.0 \text{ m})^2} = \boxed{26.9 \text{ m}}$$

Calculate the direction of $\vec{\mathbf{B}}$ using the inverse tangent, remembering to add 180° to the answer in your calculator window, because the vector lies in the second quadrant:

$$\theta = \tan^{-1}\left(\frac{B_y}{B_x}\right) = \tan^{-1}\left(\frac{10.0}{-25.0}\right) = -21.8°$$

$$\theta = \boxed{158°}$$

Remarks In part (a), note that $\cos(-\theta) = \cos \theta$; however, $\sin(-\theta) = -\sin \theta$. The negative sign of A_y reflects the fact that displacement in the y-direction is *downward.*

Exercise 3.2

(a) Suppose the superhero had flown 150 m at a 120° angle with respect to the positive x-axis. Find the components of the displacement vector. (b) Suppose instead, the superhero had leaped with a displacement having an x-component of 32.5 m and a y-component of 24.3 m. Find the magnitude and direction of the displacement vector.

Answers (a) $A_x = -75$ m, $A_y = 130$ m (b) 40.6 m, 36.8°

Adding Vectors Algebraically

The graphical method of adding vectors is valuable in understanding how vectors can be manipulated, but most of the time vectors are added algebraically in terms of their components. Suppose $\vec{\mathbf{R}} = \vec{\mathbf{A}} + \vec{\mathbf{B}}$. Then the components of the resultant vector $\vec{\mathbf{R}}$ are given by

$$R_x = A_x + B_x \quad [3.5a]$$

$$R_y = A_y + B_y \quad [3.5b]$$

So x-components are added only to x-components, and y-components only to y-components. The magnitude and direction of $\vec{\mathbf{R}}$ can subsequently be found with Equations 3.3 and 3.4.

Subtracting two vectors works the same way, because it's a matter of adding the negative of one vector to another vector. You should make a rough sketch when adding or subtracting vectors, in order to get an approximate geometric solution as a check.

INTERACTIVE EXAMPLE 3.3 Take a Hike

Goal Add vectors algebraically and find the resultant vector.

Problem A hiker begins a trip by first walking 25.0 km southeast from her base camp. On the second day she walks 40.0 km in a direction 60.0° north of east, at which point she discovers a forest ranger's tower. **(a)** Determine the components of the hiker's displacements in the first and second days. **(b)** Determine the components of the hiker's total displacement for the trip. **(c)** Find the magnitude and direction of the displacement from base camp.

Strategy This is just an application of vector addition using components, Equations 3.5. We denote the displacement vectors on the first and second days by $\vec{\mathbf{A}}$ and $\vec{\mathbf{B}}$, respectively. Using the camp as the origin of the coordinates, we get the vectors shown in Figure 3.12a. After finding x- and y-components for each vector, we add them "componentwise." Finally, we determine the magnitude and direction of the resultant vector $\vec{\mathbf{R}}$, using the Pythagorean theorem and the inverse tangent function.

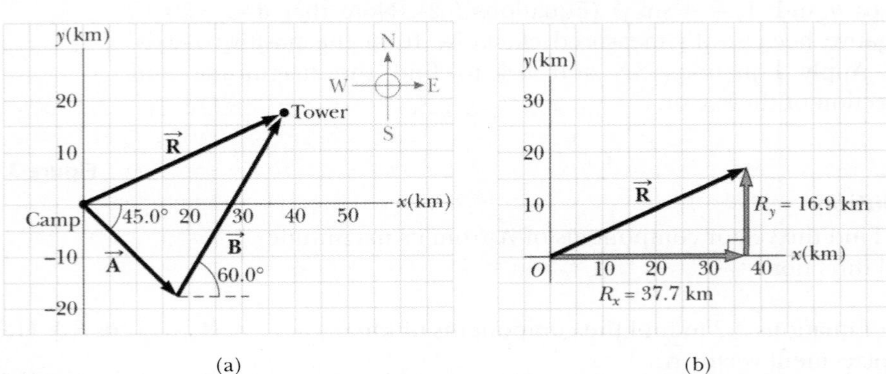

Figure 3.12 (Example 3.3) (a) Hiker's path and the resultant vector. (b) Components of the hiker's total displacement from camp.

Solution

(a) Find the components of $\vec{\mathbf{A}}$.

Use Equations 3.2 to find the components of $\vec{\mathbf{A}}$:

$$A_x = A\cos(-45.0°) = (25.0 \text{ km})(0.707) = \boxed{17.7 \text{ km}}$$

$$A_y = A\sin(-45.0°) = -(25.0 \text{ km})(0.707) = \boxed{-17.7 \text{ km}}$$

Find the components of $\vec{\mathbf{B}}$:

$$B_x = B\cos 60.0° = (40.0 \text{ km})(0.500) = \boxed{20.0 \text{ km}}$$

$$B_y = B\sin 60.0° = (40.0 \text{ km})(0.866) = \boxed{34.6 \text{ km}}$$

(b) Find the components of the resultant vector, $\vec{\mathbf{R}} = \vec{\mathbf{A}} + \vec{\mathbf{B}}$.

To find R_x, add the x-components of $\vec{\mathbf{A}}$ and $\vec{\mathbf{B}}$:

$$R_x = A_x + B_x = 17.7 \text{ km} + 20.0 \text{ km} = 37.7 \text{ km}$$

To find R_y, add the y-components of $\vec{\mathbf{A}}$ and $\vec{\mathbf{B}}$:

$$R_y = A_y + B_y = -17.7 \text{ km} + 34.6 \text{ km} = 16.9 \text{ km}$$

(c) Find the magnitude and direction of $\vec{\mathbf{R}}$.

Use the Pythagorean theorem to get the magnitude:

$$R = \sqrt{R_x^2 + R_y^2} = \sqrt{(37.7 \text{ km})^2 + (16.9 \text{ km})^2} = 41.3 \text{ km}$$

Calculate the direction of $\vec{\mathbf{R}}$ using the inverse tangent function:

$$\theta = \tan^{-1}\left(\frac{16.9 \text{ km}}{37.7 \text{ km}}\right) = 24.1°$$

Remarks Figure 3.12b shows a sketch of the components of $\vec{\mathbf{R}}$ and their directions in space. The magnitude and direction of the resultant can also be determined from such a sketch.

Exercise 3.3

A cruise ship leaving port, travels 50.0 km 45.0° north of west and then 70.0 km at a heading 30.0° north of east. Find (a) the ship's displacement vector and (b) the displacement vector's magnitude and direction.

Answer (a) $R_x = 25.3$ km, $R_y = 70.4$ km (b) 74.8 km, 70.2° north of east

PhysicsNow™ Investigate this problem further by logging into PhysicsNow at **www.cp7e.com** and going to Interactive Example 3.3.

3.3 DISPLACEMENT, VELOCITY, AND ACCELERATION IN TWO DIMENSIONS

In one-dimensional motion, as discussed in Chapter 2, the direction of a vector quantity such as a velocity or acceleration can be taken into account by specifying whether the quantity is positive or negative. The velocity of a rocket, for example, is positive if the rocket is going up and negative if it's going down. This simple solution is no longer available in two or three dimensions. Instead, we must make full use of the vector concept.

Consider an object moving through space as shown in Figure 3.13. When the object is at some point Ⓟ at time t_i, its position is described by the position vector $\vec{\mathbf{r}}_i$, drawn from the origin to Ⓟ. When the object has moved to some other point Ⓠ at time t_f, its position vector is $\vec{\mathbf{r}}_f$. From the vector diagram in Figure 3.13, the final position vector is the sum of the initial position vector and the displacement $\Delta\vec{\mathbf{r}}$: $\vec{\mathbf{r}}_f = \vec{\mathbf{r}}_i + \Delta\vec{\mathbf{r}}$. From this relationship, we obtain the following one:

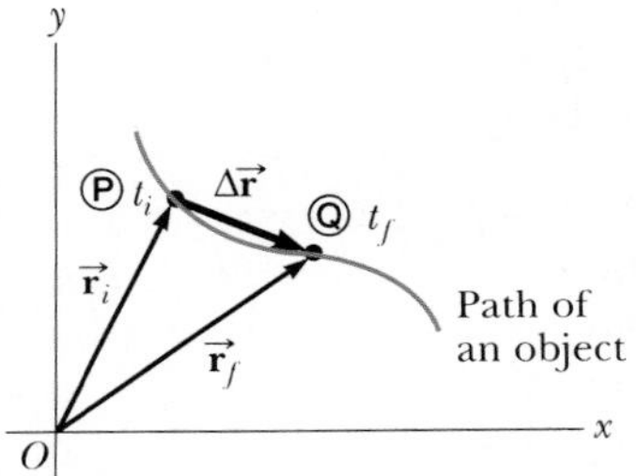

Figure 3.13 An object moving along some curved path between points Ⓟ and Ⓠ. The displacement vector $\Delta\vec{\mathbf{r}}$ is the difference in the position vectors: $\Delta\vec{\mathbf{r}} = \vec{\mathbf{r}}_f - \vec{\mathbf{r}}_i$.

An object's **displacement** is defined as the change in its position vector, or

$$\Delta\vec{\mathbf{r}} \equiv \vec{\mathbf{r}}_f - \vec{\mathbf{r}}_i \qquad [3.6]$$

◀ Displacement vector

SI unit: meter (m)

We now present several generalizations of the definitions of velocity and acceleration given in Chapter 2.

An object's **average velocity** during a time interval Δt is its displacement divided by Δt:

$$\vec{\mathbf{v}}_{\text{av}} \equiv \frac{\Delta\vec{\mathbf{r}}}{\Delta t} \qquad [3.7]$$

◀ Average velocity

SI unit: meter per second (m/s)

Because the displacement is a vector quantity and the time interval is a scalar quantity, we conclude that the average velocity is a *vector* quantity directed along $\Delta\vec{\mathbf{r}}$.

Instantaneous velocity ▶

An object's **instantaneous velocity** $\vec{\mathbf{v}}$ is the limit of its average velocity as Δt goes to zero:

$$\vec{\mathbf{v}} \equiv \lim_{\Delta t \to 0} \frac{\Delta\vec{\mathbf{r}}}{\Delta t} \quad [3.8]$$

SI unit: meter per second (m/s)

The direction of the instantaneous velocity vector is along a line that is tangent to the object's path and in the direction of its motion.

Average acceleration ▶

An object's **average acceleration** during a time interval Δt is the change in its velocity $\Delta\vec{\mathbf{v}}$ divided by Δt, or

$$\vec{\mathbf{a}}_{\text{av}} \equiv \frac{\Delta\vec{\mathbf{v}}}{\Delta t} \quad [3.9]$$

SI unit: meter per second squared (m/s²)

Instantaneous acceleration ▶

An object's **instantaneous acceleration** vector $\vec{\mathbf{a}}$ is the limit of its average acceleration vector as Δt goes to zero:

$$\vec{\mathbf{a}} \equiv \lim_{\Delta t \to 0} \frac{\Delta\vec{\mathbf{v}}}{\Delta t} \quad [3.10]$$

SI unit: meter per second squared (m/s²)

It's important to recognize that an object can accelerate in several ways. First, the magnitude of the velocity vector (the speed) may change with time. Second, the direction of the velocity vector may change with time, even though the speed is constant, as can happen along a curved path. Third, both the magnitude and the direction of the velocity vector may change at the same time.

Quick Quiz 3.4

Which of the following objects can't be accelerating? (a) An object moving with a constant speed; (b) an object moving with a constant velocity; (c) an object moving along a curve.

Quick Quiz 3.5

Consider the following controls in an automobile: gas pedal, brake, steering wheel. The controls in this list that cause an acceleration of the car are (a) all three controls, (b) the gas pedal and the brake, (c) only the brake, or (d) only the gas pedal.

3.4 MOTION IN TWO DIMENSIONS

In Chapter 2, we studied objects moving along straight-line paths, such as the x-axis. In this chapter, we look at objects that move in both the x- and y-directions simultaneously under constant acceleration. An important special case of this two-dimensional motion is called **projectile motion**.

Projectile motion ▶

Anyone who has tossed any kind of object into the air has observed projectile motion. If the effects of air resistance and the rotation of Earth are neglected, the

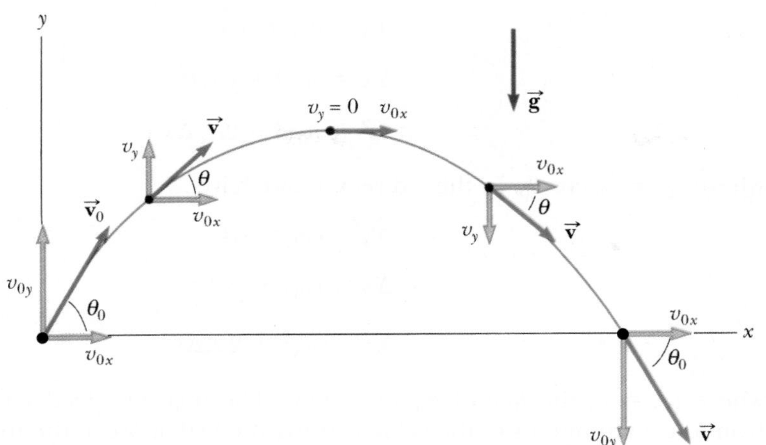

ACTIVE FIGURE 3.14
The parabolic trajectory of a particle that leaves the origin with a velocity of $\vec{v}_0$. Note that $\vec{v}$ changes with time. However, the x-component of the velocity, v_x, remains constant in time. Also, $v_y = 0$ at the peak of the trajectory, but the acceleration is always equal to the free-fall acceleration and acts vertically downward.

Physics Now™
Log into PhysicsNow at **www.cp7e.com**, and go to Active Figure 3.14, where you can change the particle's launch angle and initial speed. You can also observe the changing components of velocity along the trajectory of the projectile.

path of a projectile in Earth's gravity field is curved in the shape of a parabola, as shown in Active Figure 3.14.

The positive x-direction is horizontal and to the right, and the y-direction is vertical and positive upward. The most important experimental fact about projectile motion in two dimensions is that **the horizontal and vertical motions are completely independent of each other**. This means that motion in one direction has no effect on motion in the other direction. If a baseball is tossed in a parabolic path, as in Active Figure 3.14, the motion in the y-direction will look just like a ball tossed straight up under the influence of gravity. Active Figure 3.15 shows the effect of various initial angles; note that complementary angles give the same horizontal range.

In general, the equations of constant acceleration developed in Chapter 2 follow separately for both the x-direction and the y-direction. An important difference is that the initial velocity now has two components, not just one as in that chapter. We assume that at $t = 0$, the projectile leaves the origin with an initial velocity $\vec{v}_0$. If the velocity vector makes an angle θ_0 with the horizontal, where θ_0 is called the *projection angle*, then from the definitions of the cosine and sine functions and Active Figure 3.14, we have

$$v_{0x} = v_0 \cos \theta_0 \qquad \text{and} \qquad v_{0y} = v_0 \sin \theta_0$$

where v_{0x} is the initial velocity (at $t = 0$) in the x-direction and v_{0y} is the initial velocity in the y-direction.

Now, Equations 2.6, 2.9, and 2.10 developed in Chapter 2 for motion with constant acceleration in one dimension carry over to the two-dimensional case; there is one set of three equations for each direction, with the initial velocities modified as just discussed. In the x-direction, with a_x constant, we have

TIP 3.4 Acceleration at the Highest Point

The acceleration in the y-direction is *not* zero at the top of a projectile's trajectory. Only the y-component of the velocity is zero there. If the acceleration were zero, too, the projectile would never come down!

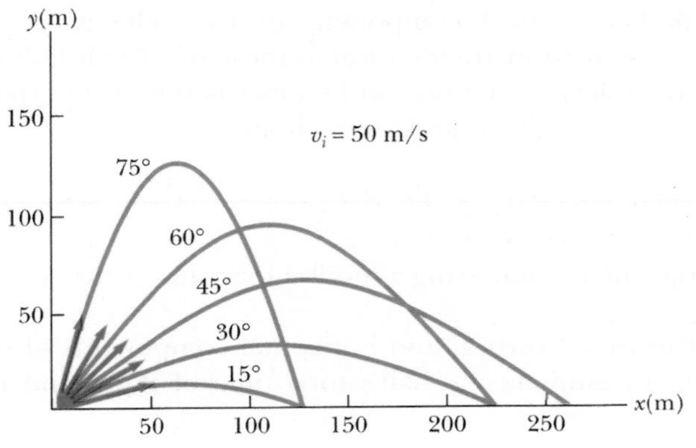

ACTIVE FIGURE 3.15
A projectile launched from the origin with an initial speed of 50 m/s at various angles of projection. Note that complementary values of the initial angle θ result in the same value of R (the range of the projectile).

Physics Now™
Log into PhysicsNow at **www.cp7e.com**, and go to Active Figure 3.15, where you can vary the projection angle to observe the effect on the trajectory and measure the flight time.

A water fountain. The individual water streams follow parabolic trajectories. The horizontal range and maximum height of a given stream of water depend on the elevation angle of that stream's initial velocity as well as its initial speed.

$$v_x = v_{0x} + a_x t \qquad [3.11a]$$

$$\Delta x = v_{0x} t + \tfrac{1}{2} a_x t^2 \qquad [3.11b]$$

$$v_x^2 = v_{0x}^2 + 2a_x \Delta x \qquad [3.11c]$$

where $v_{0x} = v_0 \cos \theta_0$. In the y-direction, we have

$$v_y = v_{0y} + a_y t \qquad [3.12a]$$

$$\Delta y = v_{0y} t + \tfrac{1}{2} a_y t^2 \qquad [3.12b]$$

$$v_y^2 = v_{0y}^2 + 2a_y \Delta y \qquad [3.12c]$$

where $v_{0y} = v_0 \sin \theta_0$ and a_y is constant. The object's speed v can be calculated from the components of the velocity using the Pythagorean theorem:

$$v = \sqrt{v_x^2 + v_y^2}$$

The angle that the velocity vector makes with the x-axis is given by

$$\theta = \tan^{-1}\left(\frac{v_y}{v_x}\right)$$

This formula for θ, as previously stated, must be used with care, because the inverse tangent function returns values only between $-90°$ and $+90°$. Adding 180° is necessary for vectors lying in the second or third quadrant.

The kinematic equations are easily adapted and simplified for projectiles close to the surface of the Earth. In that case, assuming air friction is negligible, the acceleration in the x-direction is 0 (because air resistance is neglected). **This means that $a_x = 0$, and the projectile's velocity component along the x-direction remains constant.** If the initial value of the velocity component in the x-direction is $v_{0x} = v_0 \cos \theta_0$, then this is also the value of v_x at any later time, so

$$v_x = v_{0x} = v_0 \cos \theta_0 = \text{constant} \qquad [3.13a]$$

while the horizontal displacement is simply

$$\Delta x = v_{0x} t = (v_0 \cos \theta_0) t \qquad [3.13b]$$

For the motion in the y-direction, we make the substitution $a_y = -g$ and $v_{0y} = v_0 \sin \theta_0$ in Equations 3.12, giving

$$v_y = v_0 \sin \theta_0 - gt \qquad [3.14a]$$

$$\Delta y = (v_0 \sin \theta_0) t - \tfrac{1}{2} g t^2 \qquad [3.14b]$$

$$v_y^2 = (v_0 \sin \theta_0)^2 - 2g\,\Delta y \qquad [3.14c]$$

The important facts of projectile motion can be summarized as follows:

1. Provided air resistance is negligible, the horizontal component of the velocity v_x remains constant because there is no horizontal component of acceleration.
2. The vertical component of the acceleration is equal to the free fall acceleration $-g$.
3. The vertical component of the velocity v_y and the displacement in the y-direction are identical to those of a freely falling body.
4. Projectile motion can be described as a superposition of two independent motions in the x- and y-directions.

EXAMPLE 3.4 Projectile Motion with Diagrams

Goal Approximate answers in projectile motion using a motion diagram.

Problem A ball is thrown so that its initial vertical and horizontal components of velocity are 40 m/s and 20 m/s, respectively. Use a motion diagram to estimate the ball's total time of flight and the distance it traverses before hitting the ground.

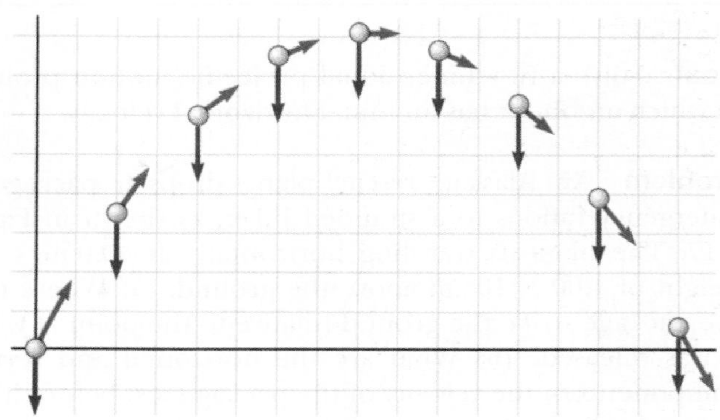
Figure 3.16 (Example 3.4) Motion diagram for a projectile.

Strategy Use the diagram, estimating the acceleration of gravity as $-10\ \mathrm{m/s^2}$. By symmetry, the ball goes up and comes back down to the ground at the same y-velocity as when it left, except with opposite sign. With this fact and the fact that the acceleration of gravity decreases the velocity in the y-direction by 10 m/s every second, we can find the total time of flight and then the horizontal range.

Solution

In the motion diagram shown in Figure 3.16, the acceleration vectors are all the same, pointing downward with magnitude of nearly $10\ \mathrm{m/s^2}$. By symmetry, we know that the ball will hit the ground at the same speed in the y-direction as when it was thrown, so the velocity in the y-direction goes from 40 m/s to $-$ 40 m/s in steps of $-$ 10 m/s every second; hence, approximately 8 seconds elapse during the motion.

The velocity vector constantly changes direction, but the horizontal velocity never changes, because the acceleration in the horizontal direction is zero. Therefore, the displacement of the ball in the x-direction is given by Equation 3.13b, $\Delta x \approx v_{0x}t = (20\ \mathrm{m/s})(8\ \mathrm{s}) = 160\ \mathrm{m}$.

Remarks This example emphasizes the independence of the x- and y-components in projectile motion problems.

Exercise 3.4

Estimate the maximum height in this same problem.

Answer 80 m

Quick Quiz 3.6

Suppose you are carrying a ball and running at constant speed, and wish to throw the ball and catch it as it comes back down. Should you (a) throw the ball at an angle of about 45° above the horizontal and maintain the same speed, (b) throw the ball straight up in the air and slow down to catch it, or (c) throw the ball straight up in the air and maintain the same speed?

Quick Quiz 3.7

As a projectile moves in its parabolic path, the velocity and acceleration vectors are perpendicular to each other (a) everywhere along the projectile's path, (b) at the peak of its path, (c) nowhere along its path, or (d) not enough information is given.

Problem-Solving Strategy Projectile Motion

1. Select a coordinate system and sketch the path of the projectile, including initial and final positions, velocities, and accelerations.
2. Resolve the initial velocity vector into x- and y-components.
3. Treat the horizontal motion and the vertical motion independently.
4. Follow the techniques for solving problems with constant velocity to analyze the horizontal motion of the projectile.
5. Follow the techniques for solving problems with constant acceleration to analyze the vertical motion of the projectile.

Remarks In fact, drag forces generally get larger with increasing speed.

Exercise 5.14
What average power must be supplied to push a 5.00-kg block from rest to 10.0 m/s in 5.00 s when the coefficient of kinetic friction between the block and surface is 0.250? Assume the acceleration is uniform.

Answer 111 W

Energy and Power in a Vertical Jump

The stationary jump consists of two parts: extension and free flight.[2] In the extension phase the person jumps up from a crouch, straightening the legs and throwing up the arms; the free-flight phase occurs when the jumper leaves the ground. Because the body is an extended object and different parts move with different speeds, we describe the motion of the jumper in terms of the position and velocity of the **center of mass (CM)**, which is the point in the body at which all the mass may be considered to be concentrated. Figure 5.27 shows the position and velocity of the CM at different stages of the jump.

Using the principle of the conservation of mechanical energy, we can find H, the maximum increase in height of the CM, in terms of the velocity v_{CM} of the CM at liftoff. Taking PE_i, the gravitational potential energy of the jumper–Earth system just as the jumper lifts off from the ground to be zero, and noting that the kinetic energy KE_f of the jumper at the peak is zero, we have

$$PE_i + KE_i = PE_f + KE_f$$

$$\tfrac{1}{2} m v_{\text{CM}}{}^2 = mgH \qquad \text{or} \qquad H = \frac{v_{\text{CM}}{}^2}{2g}$$

We can estimate v_{CM} by assuming that the acceleration of the CM is constant during the extension phase. If the depth of the crouch is h and the time for extension is Δt, we find that $v_{\text{CM}} = 2\overline{v} = 2h/\Delta t$. Measurements on a group of male college students show typical values of $h = 0.40$ m and $\Delta t = 0.25$ s, the latter value being set by the fixed speed with which muscle can contract. Substituting, we obtain

$$v_{\text{CM}} = 2(0.40 \text{ m})/(0.25 \text{ s}) = 3.2 \text{ m/s}$$

and

$$H = \frac{v_{\text{CM}}{}^2}{2g} = \frac{(3.2 \text{ m/s})^2}{2(9.80 \text{ m/s}^2)} = 0.52 \text{ m}$$

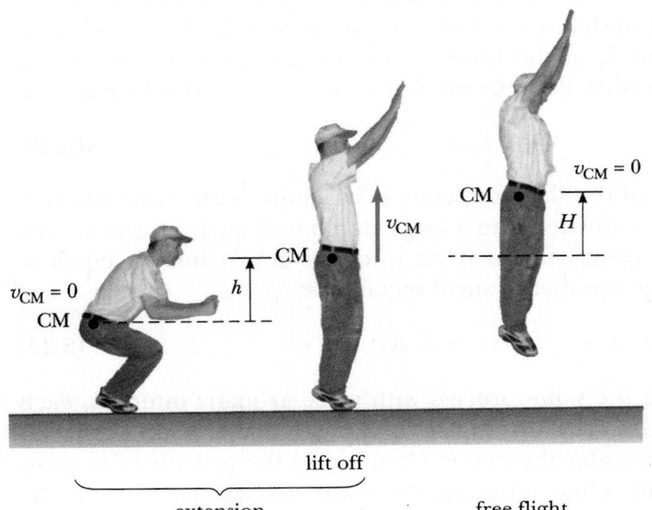

Figure 5.27 Extension and free flight in the vertical jump.

[2]For more information on this topic, see E. J. Offenbacher, *American Journal of Physics,* **38**, 829 (1969).

TABLE 5.1

Maximum Power Output from Humans over Various Periods

Power	Time
2 hp, or 1 500 W	6 s
1 hp, or 750 W	60 s
0.35 hp, or 260 W	35 min
0.2 hp, or 150 W	5 h
0.1 hp, or 75 W (safe daily level)	8 h

Measurements on this same group of students found that H was between 0.45 m and 0.61 m in all cases, confirming the basic validity of our simple calculation.

In order to relate the abstract concepts of energy, power, and efficiency to humans, it's interesting to calculate these values for the vertical jump. The kinetic energy given to the body in a jump is $KE = \frac{1}{2}mv_{\text{CM}}^2$, and for a person of mass 68 kg, the kinetic energy is

$$KE = \tfrac{1}{2}(68 \text{ kg})(3.2 \text{ m/s})^2 = 3.5 \times 10^2 \text{ J}$$

APPLICATION

Diet Versus Exercise in Weight-loss Programs

Although this may seem like a large expenditure of energy, we can make a simple calculation to show that jumping and exercise in general are not good ways to lose weight, in spite of their many health benefits. Since the muscles are at most 25% efficient at producing kinetic energy from chemical energy (muscles always produce a lot of internal energy and kinetic energy as well as work—that's why you sweat when you work out), they use up four times the 350 J (about 1 400 J) of chemical energy in one jump. This chemical energy ultimately comes from the food we eat, with energy content given in units of food calories and one food calorie equal to 4 200 J. So the total energy supplied by the body as internal energy and kinetic energy in a vertical jump is only about one-third of a food calorie! You are a lot better off not eating that piece of cheesecake than trying to work it off by jumping.

Finally, it's interesting to calculate the mechanical power that can be generated by the body in strenuous activity for brief periods. Here we find that

$$\mathcal{P} = \frac{KE}{\Delta t} = \frac{3.5 \times 10^2 \text{ J}}{0.25 \text{ s}} = 1.4 \times 10^3 \text{ W}$$

or $(1\,400 \text{ W})(1 \text{ hp}/746 \text{ W}) = 1.9$ hp. So humans can produce about 2 hp of mechanical power for periods on the order of seconds. Table 5.1 shows the maximum power outputs from humans for various periods while bicycling and rowing, activities in which it is possible to measure power output accurately.

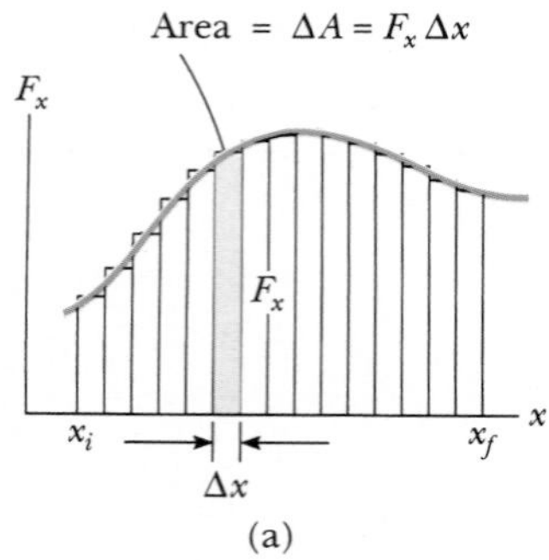

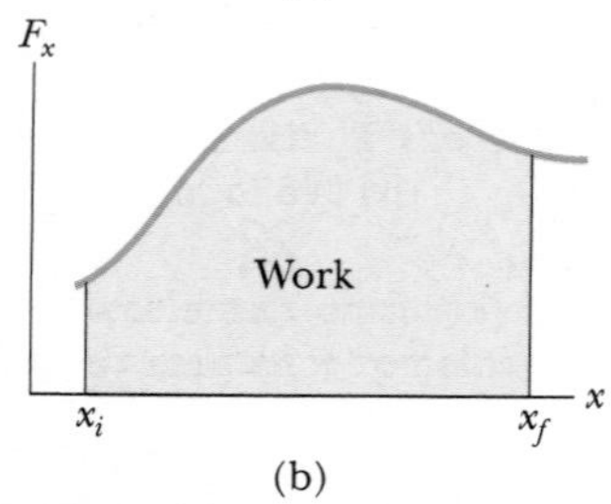

Figure 5.28 (a) The work done by the force component F_x for the small displacement Δx is $F_x \Delta x$, which equals the area of the shaded rectangle. The total work done for the displacement from x_i to x_f is approximately equal to the sum of the areas of all the rectangles. (b) The work done by the component F_x of the varying force as the particle moves from x_i to x_f is exactly equal to the area under the curve shown.

5.7 WORK DONE BY A VARYING FORCE

Suppose an object is displaced along the x-axis under the action of a force F_x that acts in the x-direction and varies with position, as shown in Figure 5.28. The object is displaced in the direction of increasing x from $x = x_i$ to $x = x_f$. In such a situation, we can't use Equation 5.1 to calculate the work done by the force because this relationship applies only when $\vec{\mathbf{F}}$ is constant in magnitude and direction. However, if we imagine that the object undergoes the *small* displacement Δx shown in Figure 5.28a, then the x-component F_x of the force is nearly constant over this interval and we can approximate the work done by the force for this small displacement as

$$W_1 \cong F_x \Delta x \qquad [5.26]$$

This quantity is just the area of the shaded rectangle in Figure 5.28a. If we imagine that the curve of F_x versus x is divided into a large number of such intervals, then the total work done for the displacement from x_i to x_f is approximately equal to the sum of the areas of a large number of small rectangles:

$$W \cong F_1 \Delta x_1 + F_2 \Delta x_2 + F_3 \Delta x_3 + \cdots \qquad [5.27]$$

Now imagine going through the same process with twice as many intervals, each half the size of the original Δx. The rectangles then have smaller widths and will better approximate the area under the curve. Continuing the process of increasing the number of intervals while allowing their size to approach zero, the number of terms in the sum increases without limit, but the value of the sum approaches a definite value equal to the area under the curve bounded by F_x and the x-axis in Figure 5.28b. In other words, **the work done by a variable force acting on an object that undergoes a displacement is equal to the area under the graph of F_x versus x.**

A common physical system in which force varies with position consists of a block on a horizontal, frictionless surface connected to a spring, as discussed in Section 5.4. When the spring is stretched or compressed a small distance x from its equilibrium position $x = 0$, it exerts a force on the block given by $F_s = -kx$, where k is the force constant of the spring.

Now let's determine the work done by an *external agent* on the block as the spring is stretched *very slowly* from $x_i = 0$ to $x_f = x_{\text{max}}$, as in Active Figure 5.29a. This work can be easily calculated by noting that at any value of the displacement, Newton's third law tells us that the applied force $\vec{\mathbf{F}}_{\text{app}}$ is equal in magnitude to the spring force $\vec{\mathbf{F}}_s$ and acts in the opposite direction, so that $F_{\text{app}} = -(-kx) = kx$. A plot of F_{app} versus x is a straight line, as shown in Active Figure 5.29b. Therefore, the work done by this applied force in stretching the spring from $x = 0$ to $x = x_{\text{max}}$ is the area under the straight line in that figure, which in this case is the area of the shaded triangle:

$$W_{F_{\text{app}}} = \tfrac{1}{2}kx_{\text{max}}^2$$

During this same time the spring has done exactly the same amount of work, but that work is negative, because the spring force points in the direction opposite the motion. The potential energy of the system is exactly equal to the work done by the applied force and is the same sign, which is why potential energy is thought of as stored work.

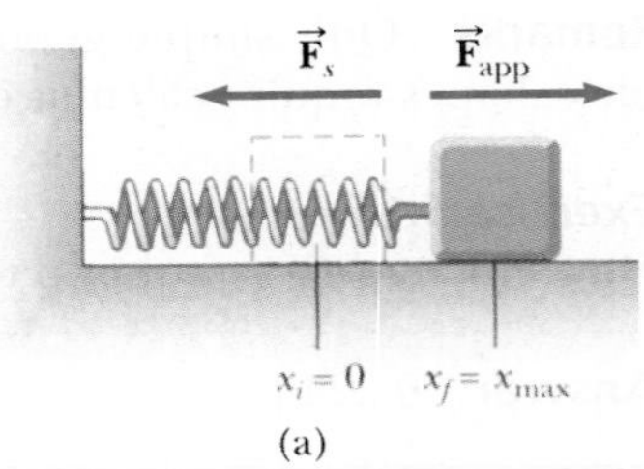

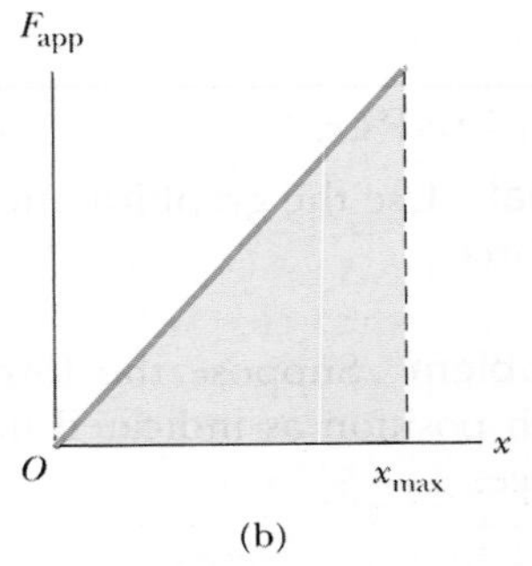

ACTIVE FIGURE 5.29
(a) A block being pulled from $x_i = 0$ to $x_f = x_{\text{max}}$ on a frictionless surface by a force $\vec{\mathbf{F}}_{\text{app}}$. If the process is carried out very slowly, the applied force is equal in magnitude and opposite in direction to the spring force at all times. (b) A graph of F_{app} versus x.

Physics Now™
Log into PhysicsNow at **www.cp7e.com**, and go to Active Figure 5.29 to observe the block's motion for various maximum displacements and spring constants.

EXAMPLE 5.15 Work Required to Stretch a Spring

Goal Apply the graphical method of finding work.

Problem One end of a horizontal spring ($k = 80.0$ N/m) is held fixed while an external force is applied to the free end, stretching it slowly from $x_Ⓐ = 0$ to $x_Ⓑ = 4.00$ cm. **(a)** Find the work done by the applied force on the spring. **(b)** Find the additional work done in stretching the spring from $x_Ⓑ = 4.00$ cm to $x_Ⓒ = 7.00$ cm.

Strategy For part (a), simply find the area of the smaller triangle, using $A = \frac{1}{2}bh$, one-half the base times the height. For part (b), the easiest way to find the additional work done from $x_Ⓑ = 4.00$ cm to $x_Ⓒ = 7.00$ cm is to find the area of the new, larger triangle and subtract the area of the smaller triangle.

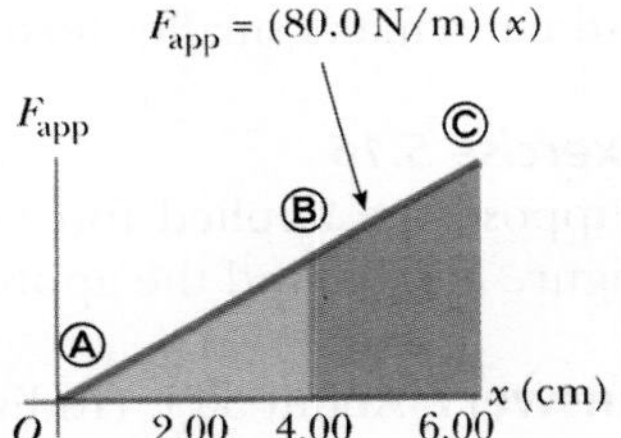

Figure 5.30 (Example 5.15) A graph of the external force required to stretch a spring that obeys Hooke's law versus the elongation of the spring.

Solution

(a) Find the work from $x_Ⓐ = 0$ cm to $x_Ⓑ = 4.00$ cm.

Compute the area of the smaller triangle:

$$W = \tfrac{1}{2}kx_B^2 = \tfrac{1}{2}(80.0 \text{ N/m})(0.040 \text{ m})^2 = 0.064\,0 \text{ J}$$

(b) Find the work from $x_Ⓑ = 4.00$ cm to $x_Ⓒ = 7.00$ cm.

Compute the area of the large triangle, and subtract the area of the smaller triangle:

$$W = \tfrac{1}{2}kx_C^2 - \tfrac{1}{2}kx_B^2$$

$$W = \tfrac{1}{2}(80.0 \text{ N/m})(0.070\,0 \text{ m})^2 - 0.064\,0 \text{ J}$$

$$= 0.196 \text{ J} - 0.064\,0 \text{ J}$$

$$= 0.132 \text{ J}$$

Remarks Only simple geometries—rectangles and triangles—can be solved exactly with this method. More complex shapes require calculus or the square-counting technique in the next worked example.

Exercise 5.15

How much work is required to stretch this same spring from $x_i = 5.00$ cm to $x_f = 9.00$ cm?

Answer 0.224 J

EXAMPLE 5.16 Estimating Work by Counting Boxes

Goal Use the graphical method and counting boxes to estimate the work done by a force.

Problem Suppose the force applied to stretch a thick piece of elastic changes with position as indicated in Figure 5.31a. Estimate the work done by the applied force.

Strategy To find the work, simply count the number of boxes underneath the curve, and multiply that number by the area of each box. The curve will pass through the middle of some boxes, in which case only an estimated fractional part should be counted.

Solution

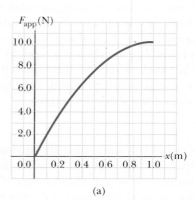

There are 62 complete or nearly complete boxes under the curve, 6 boxes that are about half under the curve, and a triangular area from $x = 0$ m to $x = 0.10$ m that is equivalent to 1 box, for a total of about 66 boxes. Since the area of each box is 0.10 J, the total work done is approximately $66 \times 0.10\text{ J} = 6.6\text{ J}$.

Remarks Mathematically, there are a number of other methods for creating such estimates, all involving adding up regions approximating the area. To get a better estimate, make smaller boxes.

Exercise 5.16

Suppose the applied force necessary to pull the drawstring on a bow is given by Figure 5.31b. Find the approximate work done by counting boxes.

Answer About 50 J. (Individual answers may vary.)

(a)

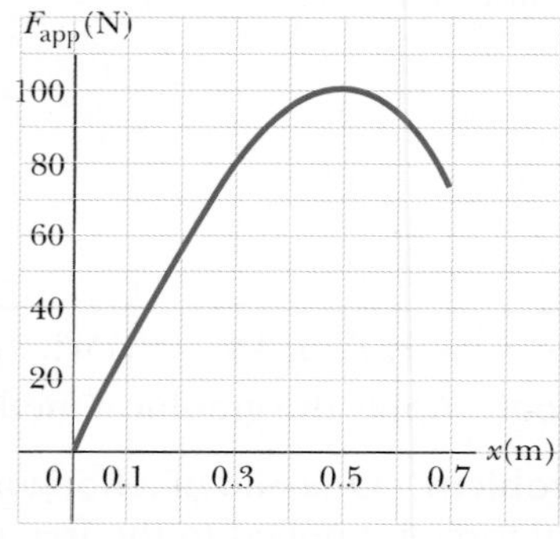

(b)

Figure 5.31 (a) (Example 5.16) (b) (Exercise 5.16)

SUMMARY

PhysicsNow™ Take a practice test by logging into PhysicsNow at **www.cp7e.com** and clicking on the Pre-Test link for this chapter.

5.1 Work

The work done on an object by a constant force is

$$W = (F\cos\theta)\Delta x \qquad [5.2]$$

where F is the magnitude of the force, Δx is the object's displacement, and θ is the angle between the direction of the force $\vec{\mathbf{F}}$ and the displacement $\Delta\vec{\mathbf{x}}$. Solving simple problems requires substituting values into this equation. More complex problems, such as those involving friction, often require using Newton's second law, $m\vec{\mathbf{a}} = \Sigma\vec{\mathbf{F}}$, to determine forces.

5.2 Kinetic Energy and the Work–Energy Theorem

The kinetic energy of a body with mass m and speed v is given by

$$KE \equiv \tfrac{1}{2}mv^2 \qquad [5.5]$$

The work–energy theorem states that the net work done on an object of mass m is equal to the change in its kinetic energy, or

$$W_{\text{net}} = KE_f - KE_i = \Delta KE \qquad [5.6]$$

Work and energy of any kind carry units of joules. Solving problems involves finding the work done by each force acting on the object and summing them up, which is W_{net}, followed by substituting known quantities into Equation 5.6, solving for the unknown quantity.

Conservative forces are special: Work done against them can be recovered—it's conserved. An example is gravity: The work done in lifting an object through a height is effectively stored in the gravity field and can be recovered in the kinetic energy of the object simply by letting it fall. Nonconservative forces, such as surface friction and drag, dissipate energy in a form that can't be readily recovered. To account for such forces, the work–energy theorem can be rewritten as

$$W_{nc} + W_c = \Delta KE \quad \textbf{[5.7]}$$

where W_{nc} is the work done by nonconservative forces and W_c is the work done by conservative forces.

5.3 Gravitational Potential Energy

The gravitational force is a conservative field. Gravitational potential energy is another way of accounting for gravitational work W_g:

$$W_g = -(PE_f - PE_i) = -(mgy_f - mgy_i) \quad \textbf{[5.11]}$$

To find the change in gravitational potential energy as an object of mass m moves between two points in a gravitational field, substitute the values of the object's y-coordinates.

The work–energy theorem can be generalized to include gravitational potential energy:

$$W_{nc} = (KE_f - KE_i) + (PE_f - PE_i) \quad \textbf{[5.12]}$$

Gravitational work and gravitational potential energy should not both appear in the work–energy theorem at the same time, only one or the other, because they're equivalent. Setting the work due to nonconservative forces to zero and substituting the expressions for *KE* and *PE*, a form of the conservation of mechanical energy with gravitation can be obtained:

$$\tfrac{1}{2}mv_i^2 + mgy_i = \tfrac{1}{2}mv_f^2 + mgy_f \quad \textbf{[5.14]}$$

To solve problems with this equation, identify two points in the system—one where information is known and the other where information is desired. Substitute and solve for the unknown quantity.

The work done by other forces, as when frictional forces are present, isn't always zero. In that case, identify two points as before, calculate the work due to all other forces, and solve for the unknown in Equation 5.12.

5.4 Spring Potential Energy

The spring force is conservative, and its potential energy is given by

$$PE_s \equiv \tfrac{1}{2}kx^2 \quad \textbf{[5.16]}$$

Spring potential energy can be put into the work–energy theorem, which then reads

$$W_{nc} = (KE_f - KE_i) + (PE_{gf} - PE_{gi}) + (PE_{sf} - PE_{si}) \quad \textbf{[5.17]}$$

When nonconservative forces are absent, $W_{nc} = 0$ and mechanical energy is conserved.

5.5 Systems and Energy Conservation

The principle of the conservation of energy states that energy can't be created or destroyed. It can be transformed, but the total energy content of any isolated system is always constant. The same is true for the universe at large. The work done by all nonconservative forces acting on a system equals the change in the total mechanical energy of the system:

$$W_{nc} = (KE_f + PE_f) - (KE_i + PE_i) = E_f - E_i \quad \textbf{[5.20–21]}$$

where *PE* represents all potential energies present.

5.6 Power

Average power is the amount of energy transferred divided by the time taken for the transfer:

$$\overline{\mathcal{P}} = \frac{W}{\Delta t} \quad \textbf{[5.22]}$$

This expression can also be written

$$\overline{\mathcal{P}} = F\overline{v} \quad \textbf{[5.23]}$$

where $\overline{v}$ is the object's average speed. The unit of power is the watt (W = J/s). To solve simple problems, substitute given quantities into one of these equations. More difficult problems usually require finding the work done on the object using the work–energy theorem or the definition of work.

CONCEPTUAL QUESTIONS

1. Consider a tug-of-war as in Figure Q5.1, in which two teams pulling on a rope are evenly matched, so that no motion takes place. Is work done on the rope? On the pullers? On the ground? Is work done on anything?
2. Discuss whether any work is being done by each of the following agents and, if so, whether the work is positive or negative: (a) a chicken scratching the ground, (b) a person studying, (c) a crane lifting a bucket of concrete, (d) the force of gravity on the bucket in part (c), (e) the leg muscles of a person in the act of sitting down.
3. If the height of a playground slide is kept constant, will the length of the slide or whether it has bumps make any difference in the final speed of children playing on it? Assume that the slide is slick enough to be considered

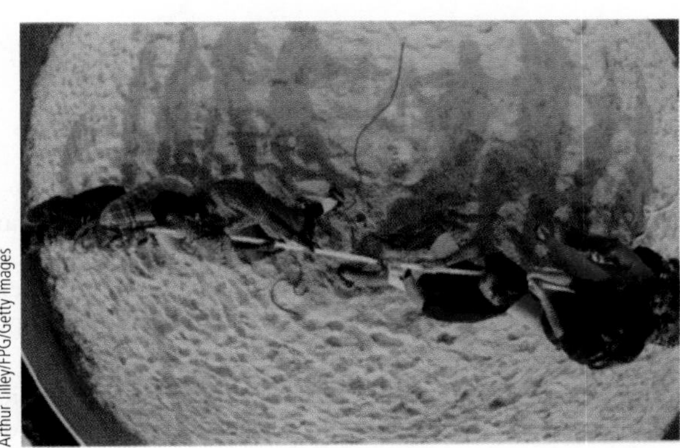

Arthur Tilley/FPG/Getty Images

Figure Q5.1

frictionless. Repeat this question, assuming that the slide is not frictionless.

4. (a) Can the kinetic energy of a system be negative? (b) Can the gravitational potential energy of a system be negative? Explain.

5. Roads going up mountains are formed into switchbacks, with the road weaving back and forth along the face of the slope such that there is only a gentle rise on any portion of the roadway. Does this configuration require any less work to be done by an automobile climbing the mountain, compared with one traveling on a roadway that is straight up the slope? Why are switchbacks used?

6. (a) If the speed of a particle is doubled, what happens to its kinetic energy? (b) If the net work done on a particle is zero, what can be said about its speed?

7. As a simple pendulum swings back and forth, the forces acting on the suspended object are the force of gravity, the tension in the supporting cord, and air resistance. (a) Which of these forces, if any, does no work on the pendulum? (b) Which of these forces does negative work at all times during the pendulum's motion? (c) Describe the work done by the force of gravity while the pendulum is swinging.

8. A bowling ball is suspended from the ceiling of a lecture hall by a strong cord. The ball is drawn away from its equilibrium position and released from rest at the tip of the demonstrator's nose, as shown in Figure Q5.8. If the demonstrator remains stationary, explain why the ball does not strike her on its return swing. Would this demonstrator be safe if the ball were given a push from its starting position at her nose?

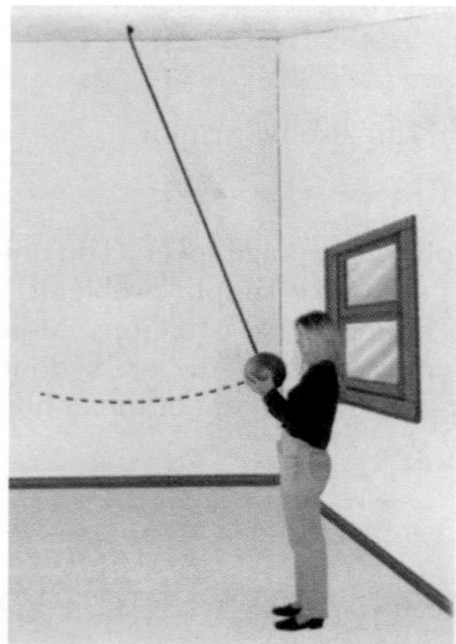

Figure Q5.8

9. An older model car accelerates from 0 to speed v in 10 seconds. A newer, more powerful sports car accelerates from 0 to $2v$ in the same time. What is the ratio of the powers expended by the two cars? Assume the energy coming from the engine appears only as kinetic energy of the cars.

10. During a stress test of the cardiovascular system, a patient walks and runs on a treadmill. (a) Is the energy expended by the patient equivalent to the energy of walking and running on the ground? Explain. (b) What effect, if any, does tilting the treadmill upward have? Discuss.

11. When a punter kicks a football, is he doing any work on the ball while the toe of his foot is in contact with it? Is he doing any work on the ball after it loses contact with his toe? Are any forces doing work on the ball while it is in flight?

12. As a sled moves across a flat, snow-covered field at constant velocity, is any work done? How does air resistance enter into the picture?

13. A weight is connected to a spring that is suspended vertically from the ceiling. If the weight is displaced downward from its equilibrium position and released, it will oscillate up and down. If air resistance is neglected, will the total mechanical energy of the system (weight plus Earth plus spring) be conserved? How many forms of potential energy are there for this situation?

14. The driver of a car slams on her brakes to avoid colliding with a deer crossing the highway. What happens to the car's kinetic energy as it comes to rest?

15. Suppose you are reshelving books in a library. You lift a book from the floor to the top shelf. The kinetic energy of the book on the floor was zero, and the kinetic energy of the book on the top shelf is zero, so there is no change in kinetic energy. Yet you did some work in lifting the book. Is the work–energy theorem violated?

16. The feet of a standing person of mass m exert a force equal to mg on the floor, and the floor exerts an equal and opposite force upwards on the feet, which we call the normal force. During the extension phase of a vertical jump (see page 145), the feet exert a force on the floor that is greater than mg, so the normal force is greater than mg. As you learned in Chapter 4, we can use this result and Newton's second law to calculate the acceleration of the jumper: $a = F_{\text{net}}/m = (n - mg)/m$.

Using energy ideas, we know that work is performed on the jumper to give him or her kinetic energy. But the normal force can't perform any work here, because the feet don't undergo any displacement. How is energy transferred to the jumper?

17. An Earth satellite is in a circular orbit at an altitude of 500 km. Explain why the work done by the gravitational force acting on the satellite is zero. Using the work–energy theorem, what can you say about the speed of the satellite?

18. In most circumstances, the normal force acting on an object and the force of static friction do no work on the object. However, the reason that the work is zero is different for the two cases. In each case, explain why the work done by the force is zero.

19. In most situations we have encountered in this chapter, frictional forces tend to reduce the kinetic energy of an object. However, frictional forces can sometimes increase an object's kinetic energy. Describe a few situations in which friction causes an increase in kinetic energy.

20. Discuss the energy transformations that occur as a pole vaulter runs at high speeds and attempts to clear a bar that is about 5 m from the ground. In your analysis, you must consider changes in the kinetic energy of the runner, the elastic potential energy of the pole as it bends, and the gravitational potential energy of the vaulter. Ignore rotational motion.

PROBLEMS

1, 2, 3 = straightforward, intermediate, challenging □ = full solution available in *Student Solutions Manual/Study Guide*
Physics Now™ = coached problem with hints available at **www.cp7e.com** = biomedical application

Section 5.1 Work

1. A weight lifter lifts a 350-N set of weights from ground level to a position over his head, a vertical distance of 2.00 m. How much work does the weight lifter do, assuming he moves the weights at constant speed?

2. If a man lifts a 20.0-kg bucket from a well and does 6.00 kJ of work, how deep is the well? Assume that the speed of the bucket remains constant as it is lifted.

3. A tugboat exerts a constant force of 5.00×10^3 N on a ship moving at constant speed through a harbor. How much work does the tugboat do on the ship if each moves a distance of 3.00 km?

4. A shopper in a supermarket pushes a cart with a force of 35 N directed at an angle of 25° downward from the horizontal. Find the work done by the shopper as she moves down a 50-m length of aisle.

5. **Physics Now™** Starting from rest, a 5.00-kg block slides 2.50 m down a rough 30.0° incline. The coefficient of kinetic friction between the block and the incline is $\mu_k = 0.436$. Determine (a) the work done by the force of gravity, (b) the work done by the friction force between block and incline, and (c) the work done by the normal force.

6. A horizontal force of 150 N is used to push a 40.0-kg packing crate a distance of 6.00 m on a rough horizontal surface. If the crate moves at constant speed, find (a) the work done by the 150-N force and (b) the coefficient of kinetic friction between the crate and surface.

7. A sledge loaded with bricks has a total mass of 18.0 kg and is pulled at constant speed by a rope inclined at 20.0° above the horizontal. The sledge moves a distance of 20.0 m on a horizontal surface. The coefficient of kinetic friction between the sledge and surface is 0.500. (a) What is the tension in the rope? (b) How much work is done by the rope on the sledge? (c) What is the mechanical energy lost due to friction?

8. A block of mass 2.50 kg is pushed 2.20 m along a frictionless horizontal table by a constant 16.0-N force directed 25.0° below the horizontal. Determine the work done by (a) the applied force, (b) the normal force exerted by the table, (c) the force of gravity, and (d) the net force on the block.

Section 5.2 Kinetic Energy and the WorK–Energy Theorem

9. A mechanic pushes a 2.50×10^3-kg car from rest to a speed of v, doing 5 000 J of work in the process. During this time, the car moves 25.0 m. Neglecting friction between car and road, find (a) v and (b) the horizontal force exerted on the car.

10. A 7.00-kg bowling ball moves at 3.00 m/s. How fast must a 2.45-g Ping-Pong ball move so that the two balls have the same kinetic energy?

11. A person doing a chin-up weighs 700 N, exclusive of the arms. During the first 25.0 cm of the lift, each arm exerts an upward force of 355 N on the torso. If the upward movement starts from rest, what is the person's velocity at that point?

12. A crate of mass 10.0 kg is pulled up a rough incline with an initial speed of 1.50 m/s. The pulling force is 100 N parallel to the incline, which makes an angle of 20.0° with the horizontal. The coefficient of kinetic friction is 0.400, and the crate is pulled 5.00 m. (a) How much work is done by gravity? (b) How much mechanical energy is lost due to friction? (c) How much work is done by the 100-N force? (d) What is the change in kinetic energy of the crate? (e) What is the speed of the crate after being pulled 5.00 m?

13. A 70-kg base runner begins his slide into second base when he is moving at a speed of 4.0 m/s. The coefficient of friction between his clothes and Earth is 0.70. He slides so that his speed is zero just as he reaches the base. (a) How much mechanical energy is lost due to friction acting on the runner? (b) How far does he slide?

14. An outfielder throws a 0.150-kg baseball at a speed of 40.0 m/s and an initial angle of 30.0°. What is the kinetic energy of the ball at the highest point of its motion?

15. A 2.0-g bullet leaves the barrel of a gun at a speed of 300 m/s. (a) Find its kinetic energy. (b) Find the average force exerted by the expanding gases on the bullet as it moves the length of the 50-cm-long barrel.

16. A 0.60-kg particle has a speed of 2.0 m/s at point A and a kinetic energy of 7.5 J at point B. What is (a) its kinetic energy at A? (b) its speed at point B? (c) the total work done on the particle as it moves from A to B?

17. A 2 000-kg car moves down a level highway under the actions of two forces: a 1 000-N forward force exerted on the drive wheels by the road and a 950-N resistive force. Use the work–energy theorem to find the speed of the car after it has moved a distance of 20 m, assuming that it starts from rest.

18. On a frozen pond, a 10-kg sled is given a kick that imparts to it an initial speed of $v_0 = 2.0$ m/s. The coefficient of kinetic friction between sled and ice is $\mu_k = 0.10$. Use the work–energy theorem to find the distance the sled moves before coming to rest.

Section 5.3 Gravitational Potential Energy

Section 5.4 Spring Potential Energy

19. Find the height from which you would have to drop a ball so that it would have a speed of 9.0 m/s just before it hits the ground.

20. A flea is able to jump about 0.5 m. It has been said that if a flea were as big as a human, it would be able to jump over a 100-story building! When an animal jumps, it converts work done in contracting muscles into gravitational potential energy (with some steps in between). The maximum force exerted by a muscle is proportional to its cross-sectional area, and the work done by the muscle is this force times the length of contraction. If we magnified a flea by a factor of 1 000, the cross section of its muscle would increase by $1\,000^2$ and the length of contraction would increase by 1 000. How high would this "superflea" be able to jump? (Don't forget that the mass of the "superflea" increases as well.)

21. An athlete on a trampoline leaps straight up into the air with an initial speed of 9.0 m/s. Find (a) the maximum height reached by the athlete relative to the trampoline and (b) the speed of the athlete when she is halfway up to her maximum height.

22. Truck suspensions often have "helper springs" that engage at high loads. One such arrangement is a leaf spring with a helper coil spring mounted on the axle, as shown in Figure P5.22. When the main leaf spring is compressed by distance y_0, the helper spring engages and then helps to support any additional load. Suppose the leaf spring constant is 5.25×10^5 N/m, the helper spring constant is 3.60×10^5 N/m, and $y_0 = 0.500$ m. (a) What is the compression of the leaf spring for a load of 5.00×10^5 N? (b) How much work is done in compressing the springs?

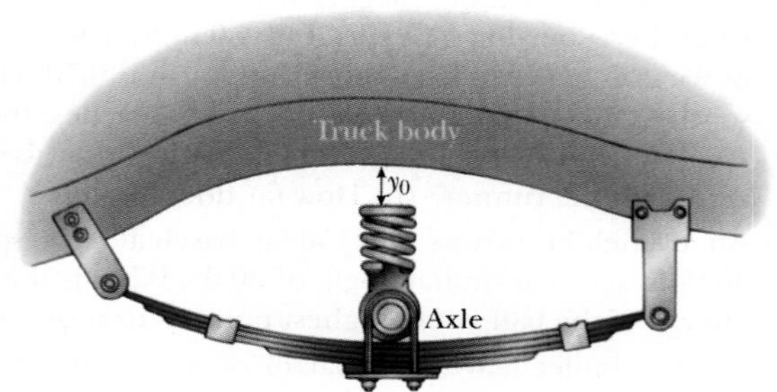

Figure P5.22

23. A daredevil on a motorcycle leaves the end of a ramp with a speed of 35.0 m/s as in Figure P5.23. If his speed is 33.0 m/s when he reaches the peak of the path, what is the maximum height that he reaches? Ignore friction and air resistance.

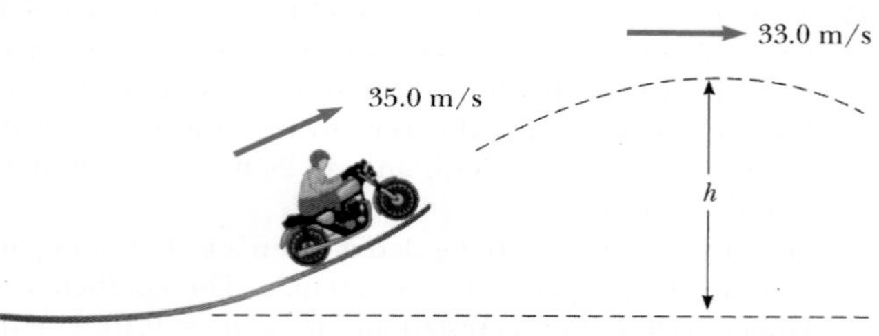

Figure P5.23

24. A softball pitcher rotates a 0.250-kg ball around a vertical circular path of radius 0.600 m before releasing it. The pitcher exerts a 30.0-N force directed parallel to the motion of the ball around the complete circular path. The speed of the ball at the top of the circle is 15.0 m/s. If the ball is released at the bottom of the circle, what is its speed upon release?

25. The chin-up is one exercise that can be used to strengthen the biceps muscle. This muscle can exert a force of approximately 800 N as it contracts a distance of 7.5 cm in a 75-kg male[3]. How much work can the biceps muscles (one in each arm) perform in a single contraction? Compare this amount of work with the energy required to lift a 75-kg person 40 cm in performing a chin-up. Do you think the biceps muscle is the only muscle involved in performing a chin-up?

Section 5.5 Systems and Energy Conservation

26. A 50-kg pole vaulter running at 10 m/s vaults over the bar. Her speed when she is above the bar is 1.0 m/s. Neglect air resistance, as well as any energy absorbed by the pole, and determine her altitude as she crosses the bar.

27. PhysicsNow™ A child and a sled with a combined mass of 50.0 kg slide down a frictionless slope. If the sled starts from rest and has a speed of 3.00 m/s at the bottom, what is the height of the hill?

28. A 0.400-kg bead slides on a curved wire, starting from rest at point Ⓐ in Figure P5.28. If the wire is frictionless, find the speed of the bead (a) at Ⓑ and (b) at Ⓒ.

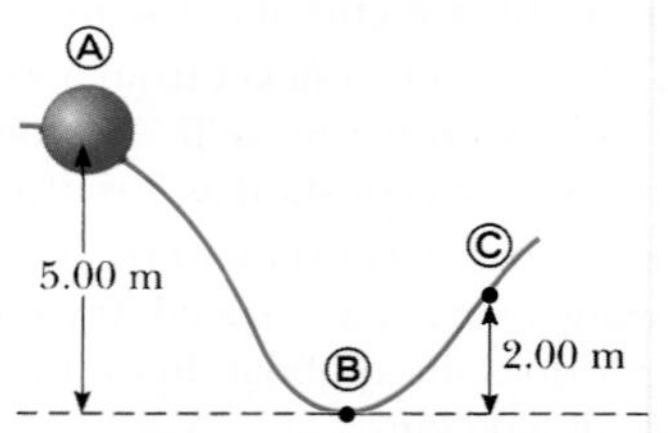

Figure P5.28 (Problems 28 and 36)

29. A 5.00-kg steel ball is dropped onto a copper plate from a height of 10.0 m. If the ball leaves a dent 3.20 mm deep in the plate, what is the average force exerted by the plate on the ball during the impact?

30. A bead of mass $m = 5.00$ kg is released from point Ⓐ and slides on the frictionless track shown in Figure P5.30. Determine (a) the bead's speed at points Ⓑ and Ⓒ and (b) the net work done by the force of gravity in moving the bead from Ⓐ to Ⓒ.

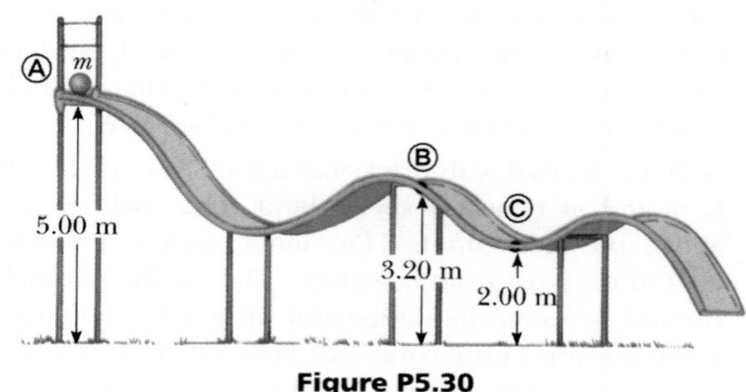

Figure P5.30

31. Tarzan swings on a 30.0-m-long vine initially inclined at an angle of 37.0° with the vertical. What is his speed at the bottom of the swing (a) if he starts from rest? (b) if he pushes off with a speed of 4.00 m/s?

32. Three objects with masses $m_1 = 5.0$ kg, $m_2 = 10$ kg, and $m_3 = 15$ kg, respectively, are attached by strings over frictionless pulleys, as indicated in Figure P5.32. The horizontal surface is frictionless and the system is released from rest. Using energy concepts, find the speed of m_3 after it moves down a distance of 4.0 m.

Figure P5.32 (Problems 32 and 89)

[3] G. P. Pappas et.al., "Nonuniform shortening in the biceps brachii during elbow flexion," *Journal of Applied Physiology* **92**, 2381, 2002.

33. The launching mechanism of a toy gun consists of a spring of unknown spring constant, as shown in Figure P5.33a. If the spring is compressed a distance of 0.120 m and the gun fired vertically as shown, the gun can launch a 20.0-g projectile from rest to a maximum height of 20.0 m above the starting point of the projectile. Neglecting all resistive forces, determine (a) the spring constant and (b) the speed of the projectile as it moves through the equilibrium position of the spring (where $x = 0$), as shown in Figure P5.33b.

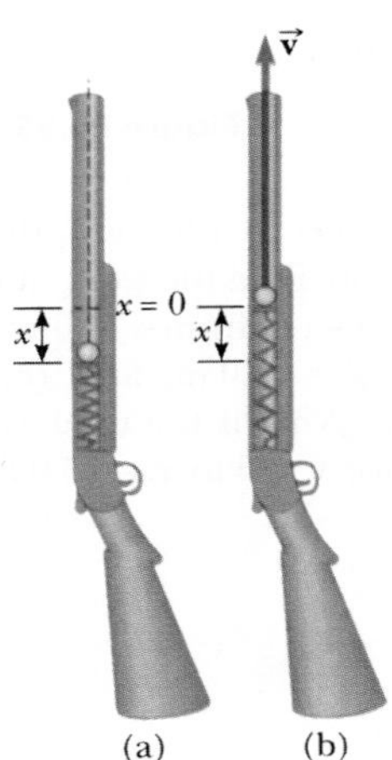

Figure P5.33

34. A projectile is launched with a speed of 40 m/s at an angle of 60° above the horizontal. Use conservation of energy to find the maximum height reached by the projectile during its flight.

35. A 0.250-kg block is placed on a light vertical spring ($k = 5.00 \times 10^3$ N/m) and pushed downwards, compressing the spring 0.100 m. After the block is released, it leaves the spring and continues to travel upwards. What height above the point of release will the block reach if air resistance is negligible?

36. The wire in Problem 28 (Fig. P5.28) is frictionless between points Ⓐ and Ⓑ and rough between Ⓑ and Ⓒ. The 0.400-kg bead starts from rest at Ⓐ. (a) Find its speed at Ⓑ. (b) If the bead comes to rest at Ⓒ, find the loss in mechanical energy as it goes from Ⓑ to Ⓒ.

37. (a) A child slides down a water slide at an amusement park from an initial height h. The slide can be considered frictionless because of the water flowing down it. Can the equation for conservation of mechanical energy be used on the child? (b) Is the mass of the child a factor in determining his speed at the bottom of the slide? (c) The child drops straight down rather than following the curved ramp of the slide. In which case will he be traveling faster at ground level? (d) If friction is present, how would the conservation-of-energy equation be modified? (e) Find the maximum speed of the child when the slide is frictionless if the initial height of the slide is 12.0 m.

38. (a) A block with a mass m is pulled along a horizontal surface for a distance x by a constant force $\vec{\mathbf{F}}$ at an angle θ with respect to the horizontal. The coefficient of kinetic friction between block and table is μ_k. Is the force exerted by friction equal to $\mu_k mg$? If not, what is the force exerted by friction? (b) How much work is done by the friction force and by $\vec{\mathbf{F}}$? (Don't forget the signs.) (c) Identify all the forces that do no work on the block. (d) Let $m = 2.00$ kg, $x = 4.00$ m, $\theta = 37.0°$, $F = 15.0$ N, and $\mu_k = 0.400$, and find the answers to parts (a) and (b).

39. **Physics Now™** A 70-kg diver steps off a 10-m tower and drops from rest straight down into the water. If he comes to rest 5.0 m beneath the surface, determine the average resistive force exerted on him by the water.

40. An airplane of mass 1.5×10^4 kg is moving at 60 m/s. The pilot then revs up the engine so that the forward thrust by the air around the propeller becomes 7.5×10^4 N. If the force exerted by air resistance on the body of the airplane has a magnitude of 4.0×10^4 N, find the speed of the airplane after it has traveled 500 m. Assume that the airplane is in level flight throughout this motion.

41. A 2.1×10^3-kg car starts from rest at the top of a 5.0-m-long driveway that is inclined at 20° with the horizontal. If an average friction force of 4.0×10^3 N impedes the motion, find the speed of the car at the bottom of the driveway.

42. A 25.0-kg child on a 2.00-m-long swing is released from rest when the ropes of the swing make an angle of 30.0° with the vertical. (a) Neglecting friction, find the child's speed at the lowest position. (b) If the actual speed of the child at the lowest position is 2.00 m/s, what is the mechanical energy lost due to friction?

43. Starting from rest, a 10.0-kg block slides 3.00 m down to the bottom of a frictionless ramp inclined 30.0° from the floor. The block then slides an additional 5.00 m along the floor before coming to a stop. Determine (a) the speed of the block at the bottom of the ramp, (b) the coefficient of kinetic friction between block and floor, and (c) the mechanical energy lost due to friction.

44. A child slides without friction from a height h along a curved water slide (Fig. P5.44). She is launched from a height $h/5$ into the pool. Determine her maximum airborne height y in terms of h and the launch angle θ.

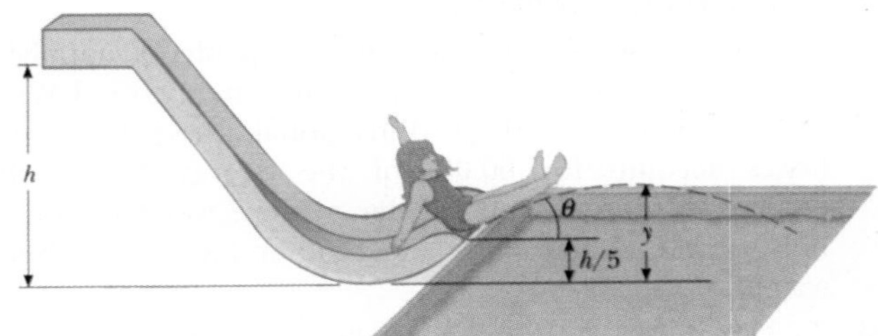

Figure P5.44

45. A skier starts from rest at the top of a hill that is inclined 10.5° with respect to the horizontal. The hillside is 200 m long, and the coefficient of friction between snow and skis is 0.075 0. At the bottom of the hill, the snow is level and the coefficient of friction is unchanged. How far does the skier glide along the horizontal portion of the snow before coming to rest?

46. In a circus performance, a monkey is strapped to a sled and both are given an initial speed of 4.0 m/s up a 20° inclined track. The combined mass of monkey and sled is 20 kg, and the coefficient of kinetic friction between sled and incline is 0.20. How far up the incline do the monkey and sled move?

47. An 80.0-kg skydiver jumps out of a balloon at an altitude of 1 000 m and opens the parachute at an altitude of

200.0 m. (a) Assuming that the total retarding force on the diver is constant at 50.0 N with the parachute closed and constant at 3 600 N with the parachute open, what is the speed of the diver when he lands on the ground? (b) Do you think the skydiver will get hurt? Explain. (c) At what height should the parachute be opened so that the final speed of the skydiver when he hits the ground is 5.00 m/s? (d) How realistic is the assumption that the total retarding force is constant? Explain.

Section 5.6 Power

48. A skier of mass 70 kg is pulled up a slope by a motor-driven cable. (a) How much work is required to pull him 60 m up a 30° slope (assumed frictionless) at a constant speed of 2.0 m/s? (b) What power must a motor have to perform this task?

49. Columnist Dave Barry poked fun at the name "The Grand Cities," adopted by Grand Forks, North Dakota, and East Grand Forks, Minnesota. Residents of the prairie towns then named a sewage pumping station for him. At the Dave Barry Lift Station No. 16, untreated sewage is raised vertically by 5.49 m in the amount of 1 890 000 liters each day. With a density of 1 050 kg/m^3, the waste enters and leaves the pump at atmospheric pressure through pipes of equal diameter. (a) Find the output power of the lift station. (b) Assume that a continuously operating electric motor with average power 5.90 kW runs the pump. Find its efficiency. In January 2002, Barry attended the outdoor dedication of the lift station and a festive potluck supper to which the residents of the different Grand Forks sewer districts brought casseroles, Jell-O® salads, and "bars" (desserts).

50. While running, a person dissipates about 0.60 J of mechanical energy per step per kilogram of body mass. If a 60-kg person develops a power of 70 W during a race, how fast is the person running? (Assume a running step is 1.5 m long.)

51. The electric motor of a model train accelerates the train from rest to 0.620 m/s in 21.0 ms. The total mass of the train is 875 g. Find the average power delivered to the train during its acceleration.

52. An electric scooter has a battery capable of supplying 120 Wh of energy. [Note that an energy of 1 Wh = (1 J/s)(3600 s) = 3600 J] If frictional forces and other losses account for 60.0% of the energy usage, what change in altitude can a rider achieve when driving in hilly terrain if the rider and scooter have a combined weight of 890 N?

53. **PhysicsNow™** A 1.50 × 10^3-kg car starts from rest and accelerates uniformly to 18.0 m/s in 12.0 s. Assume that air resistance remains constant at 400 N during this time. Find (a) the average power developed by the engine and (b) the instantaneous power output of the engine at t = 12.0 s, just before the car stops accelerating.

54. A 650-kg elevator starts from rest and moves upwards for 3.00 s with constant acceleration until it reaches its cruising speed, 1.75 m/s. (a) What is the average power of the elevator motor during this period? (b) How does this amount of power compare with its power during an upward trip with constant speed?

Section 5.7 Work Done by a Varying Force

55. The force acting on a particle varies as in Figure P5.55. Find the work done by the force as the particle moves (a) from x = 0 to x = 8.00 m, (b) from x = 8.00 m to x = 10.0 m, and (c) from x = 0 to x = 10.0 m.

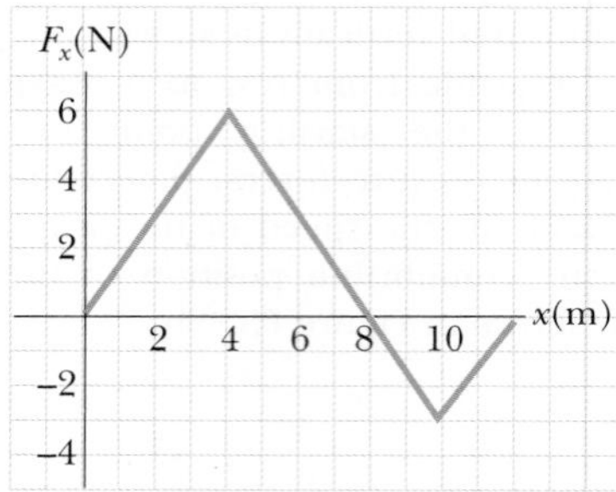

Figure P5.55

56. An object is subject to a force F_x that varies with position as in Figure P5.56. Find the work done by the force on the object as it moves (a) from x = 0 to x = 5.00 m, (b) from x = 5.00 m to x = 10.0 m, and (c) from x = 10.0 m to x = 15.0 m. (d) What is the total work done by the force over the distance x = 0 to x = 15.0 m?

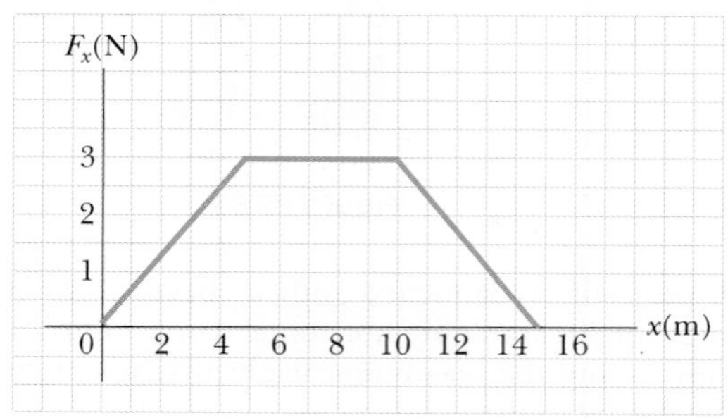

Figure P5.56

57. The force acting on an object is given by $F_x = (8x - 16)$ N, where x is in meters. (a) Make a plot of this force versus x from x = 0 to x = 3.00 m. (b) From your graph, find the net work done by the force as the object moves from x = 0 to x = 3.00 m.

ADDITIONAL PROBLEMS

58. A 2.0-m-long pendulum is released from rest when the support string is at an angle of 25° with the vertical. What is the speed of the bob at the bottom of the swing?

59. An archer pulls her bowstring back 0.400 m by exerting a force that increases uniformly from zero to 230 N. (a) What is the equivalent spring constant of the bow? (b) How much work does the archer do in pulling the bow?

61. A block of mass 12.0 kg slides from rest down a frictionless 35.0° incline and is stopped by a strong spring with k = 3.00 × 10^4 N/m. The block slides 3.00 m from the point of release to the point where it comes to rest against the spring. When the block comes to rest, how far has the spring been compressed?

61. (a) A 75-kg man steps out a window and falls (from rest) 1.0 m to a sidewalk. What is his speed just before his feet strike the pavement? (b) If the man falls with his knees and ankles locked, the only cushion for his fall is an approximately 0.50-cm give in the pads of his feet. Calculate the average force exerted on him by the ground in this situation. This average force is sufficient to cause damage to cartilage in the joints or to break bones.

62. A toy gun uses a spring to project a 5.3-g soft rubber sphere horizontally. The spring constant is 8.0 N/m, the barrel of the gun is 15 cm long, and a constant frictional force of 0.032 N exists between barrel and projectile. With what speed does the projectile leave the barrel if the spring was compressed 5.0 cm for this launch?

63. Two objects are connected by a light string passing over a light, frictionless pulley as in Figure P5.63. The 5.00-kg object is released from rest at a point 4.00 m above the floor. (a) Determine the speed of each object when the two pass each other. (b) Determine the speed of each object at the moment the 5.00-kg object hits the floor. (c) How much higher does the 3.00-kg object travel after the 5.00-kg object hits the floor?

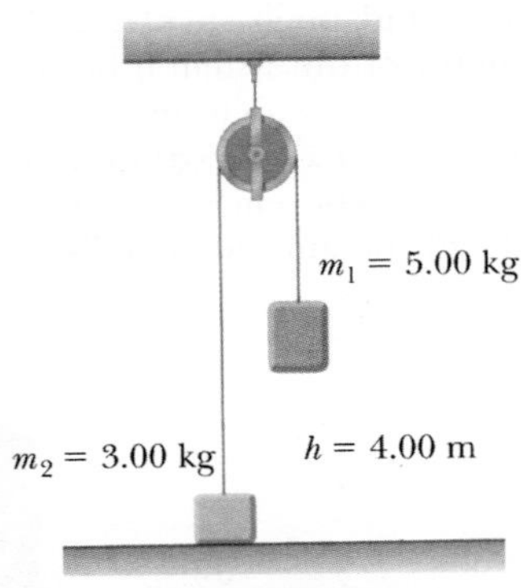

Figure P5.63

64. Two blocks, A and B (with mass 50 kg and 100 kg, respectively), are connected by a string, as shown in Figure P5.64. The pulley is frictionless and of negligible mass. The coefficient of kinetic friction between block A and the incline is $\mu_k = 0.25$. Determine the change in the kinetic energy of block A as it moves from Ⓒ to Ⓓ, a distance of 20 m up the incline if the system starts from rest.

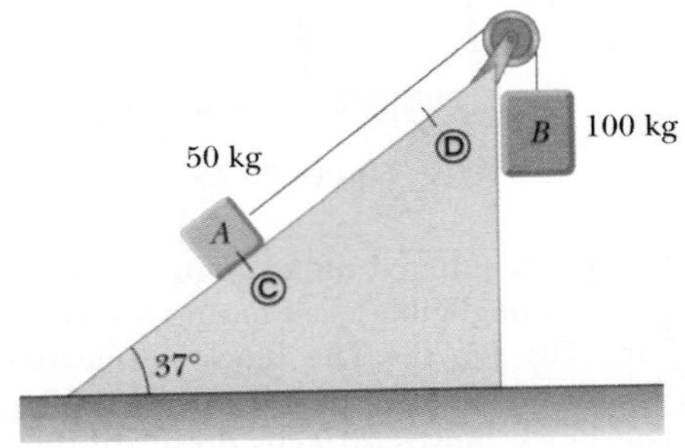

Figure P5.64

65. A 200-g particle is released from rest at point A on the inside of a smooth hemispherical bowl of radius $R = 30.0$ cm (Fig. P5.65). Calculate (a) its gravitational potential energy at A relative to B, (b) its kinetic energy at B, (c) its speed at B, (d) its potential energy at C relative to B, and (e) its kinetic energy at C.

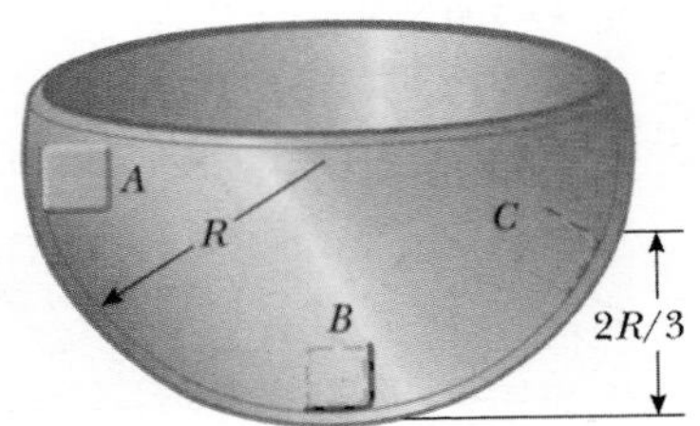

Figure P5.65

66. Energy is conventionally measured in Calories as well as in joules. One Calorie in nutrition is 1 kilocalorie, which we define in Chapter 11 as 1 kcal = 4 186 J. Metabolizing 1 gram of fat can release 9.00 kcal. A student decides to try to lose weight by exercising. She plans to run up and down the stairs in a football stadium as fast as she can and as many times as necessary. Is this in itself a practical way to lose weight? To evaluate the program, suppose she runs up a flight of 80 steps, each 0.150 m high, in 65.0 s. For simplicity, ignore the energy she uses in coming down (which is small). Assume that a typical efficiency for human muscles is 20.0%. This means that when your body converts 100 J from metabolizing fat, 20 J goes into doing mechanical work (here, climbing stairs). The remainder goes into internal energy. Assume the student's mass is 50.0 kg. (a) How many times must she run the flight of stairs to lose 1 pound of fat? (b) What is her average power output, in watts and in horsepower, as she is running up the stairs?

67. In terms of saving energy, bicycling and walking are far more efficient means of transportation than is travel by automobile. For example, when riding at 10.0 mi/h, a cyclist uses food energy at a rate of about 400 kcal/h above what he would use if he were merely sitting still. (In exercise physiology, power is often measured in kcal/h rather than in watts. Here, 1 kcal = 1 nutritionist's Calorie = 4 186 J.) Walking at 3.00 mi/h requires about 220 kcal/h. It is interesting to compare these values with the energy consumption required for travel by car. Gasoline yields about 1.30×10^8 J/gal. Find the fuel economy in equivalent miles per gallon for a person (a) walking and (b) bicycling.

68. An 80.0-N box is pulled 20.0 m up a 30° incline by an applied force of 100 N that points upwards, parallel to the incline. If the coefficient of kinetic friction between box and incline is 0.220, calculate the change in the kinetic energy of the box.

69. A ski jumper starts from rest 50.0 m above the ground on a frictionless track and flies off the track at an angle of 45.0° above the horizontal and at a height of 10.0 m above the level ground. Neglect air resistance. (a) What is her speed when she leaves the track? (b) What is the maximum altitude she attains after leaving the track? (c) Where does she land relative to the end of the track?

70. A 5.0-kg block is pushed 3.0 m up a vertical wall with constant speed by a constant force of magnitude F applied at an angle of $\theta = 30°$ with the horizontal, as shown in Figure P5.70. If the coefficient of kinetic friction between

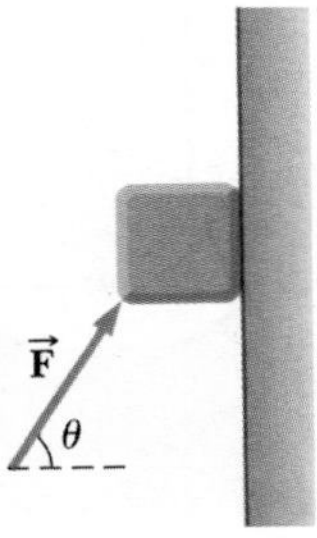

Figure P5.70

block and wall is 0.30, determine the work done by (a) $\vec{F}$, (b) the force of gravity, and (c) the normal force between block and wall. (d) By how much does the gravitational potential energy increase during the block's motion?

71. The ball launcher in a pinball machine has a spring with a force constant of 1.20 N/cm (Fig. P5.71). The surface on which the ball moves is inclined 10.0° with respect to the horizontal. If the spring is initially compressed 5.00 cm, find the launching speed of a 0.100-kg ball when the plunger is released. Friction and the mass of the plunger are negligible.

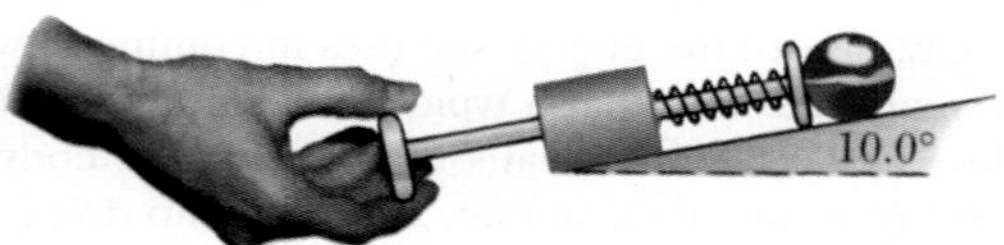

Figure P5.71

72. The masses of the javelin, discus, and shot are 0.80 kg, 2.0 kg, and 7.2 kg, respectively, and record throws in the corresponding track events are about 98 m, 74 m, and 23 m, respectively. Neglecting air resistance, (a) calculate the minimum initial kinetic energies that would produce these throws, and (b) estimate the average force exerted on each object during the throw, assuming the force acts over a distance of 2.0 m. (c) Do your results suggest that air resistance is an important factor?

73. Jane, whose mass is 50.0 kg, needs to swing across a river filled with crocodiles in order to rescue Tarzan, whose mass is 80.0 kg. However, she must swing into a *constant* horizontal wind force $\vec{F}$ on a vine that is initially at an angle of θ with the vertical. (See Fig. P5.73.) In the figure, $D = 50.0$ m, $F = 110$ N, $L = 40.0$ m, and $\theta = 50.0°$. (a) With what minimum speed must Jane begin her swing in order to just make it to the other side? (*Hint:* First determine the potential energy that can be associated with the wind force. Because the wind force is constant, use an analogy with the constant gravitational force.) (b) Once the rescue is complete, Tarzan and Jane must swing back across the river. With what minimum speed must they begin their swing?

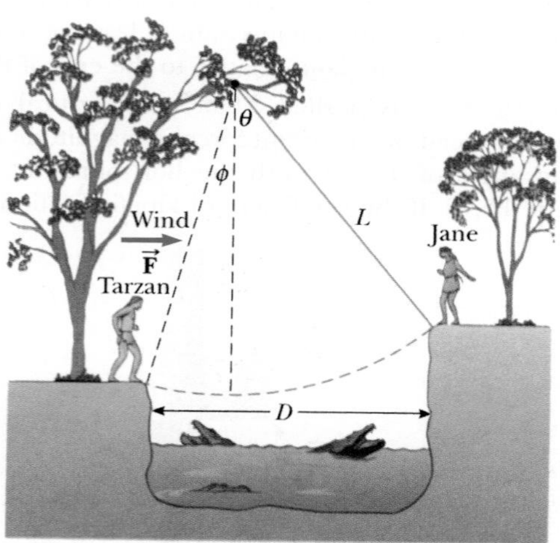

Figure P5.73

74. A hummingbird is able to hover because, as the wings move downwards, they exert a downward force on the air. Newton's third law tells us that the air exerts an equal and opposite force (upwards) on the wings. The average of this force must be equal to the weight of the bird when it hovers. If the wings move through a distance of 3.5 cm with each stroke, and the wings beat 80 times per second, determine the work performed by the wings on the air in 1 minute if the mass of the hummingbird is 3.0 grams.

75. A child's pogo stick (Fig. P5.75) stores energy in a spring ($k = 2.50 \times 10^4$ N/m). At position Ⓐ ($x_1 = -0.100$ m), the spring compression is a maximum and the child is momentarily at rest. At position Ⓑ ($x = 0$), the spring is relaxed and the child is moving upwards. At position Ⓒ, the child is again momentarily at rest at the top of the jump. Assuming that the combined mass of child and pogo stick is 25.0 kg, (a) calculate the total energy of the system if both potential energies are zero at $x = 0$, (b) determine x_2, (c) calculate the speed of the child at $x = 0$, (d) determine the value of x for which the kinetic energy of the system is a maximum, and (e) obtain the child's maximum upward speed.

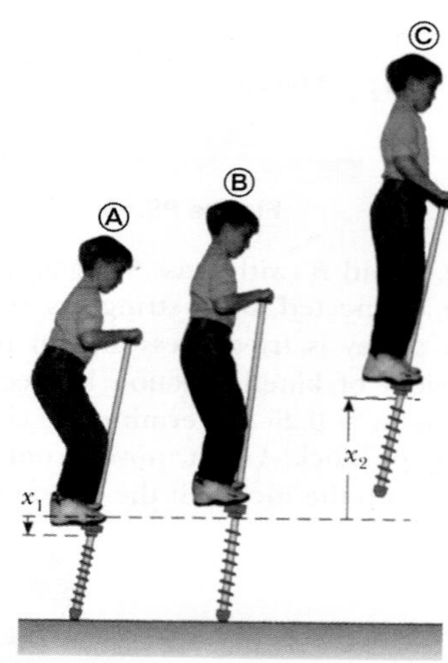

Figure P5.75

76. A 2.00-kg block situated on a rough incline is connected to a spring of negligible mass having a spring constant of 100 N/m (Fig. P5.76). The block is released from rest when the spring is unstretched, and the pulley is frictionless. The block moves 20.0 cm down the incline before coming to rest. Find the coefficient of kinetic friction between block and incline.

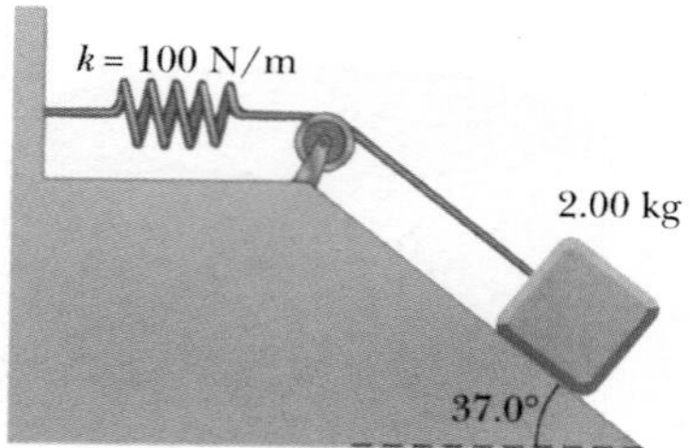

Figure P5.76

77. In the dangerous "sport" of bungee jumping, a daring student jumps from a hot-air balloon with a specially designed elastic cord attached to his waist, as shown in Figure P5.77. The unstretched length of the cord is 25.0 m, the student weighs 700 N, and the balloon is 36.0 m above the surface of a river below. Calculate the required force constant of the cord if the student is to stop safely 4.00 m above the river.

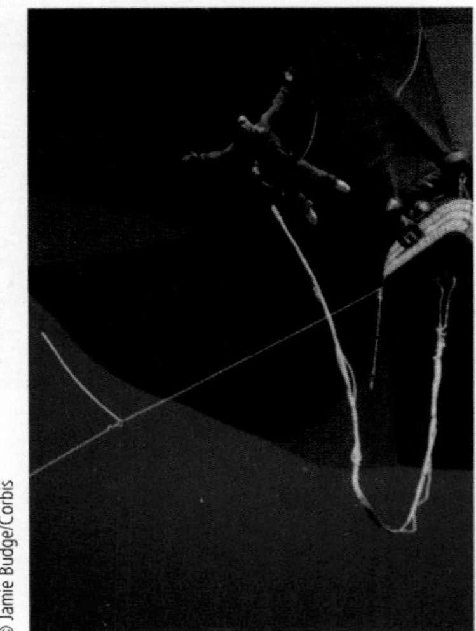

Figure P5.77 Bungee jumping. (Problems 77 and 82)

78. An object of mass m is suspended from the top of a cart by a string of length L as in Figure P5.78a. The cart and object are initially moving to the right at a constant speed v_0. The cart comes to rest after colliding and sticking to a bumper, as in Figure P5.78b, and the suspended object swings through an angle θ. (a) Show that the initial speed is $v_0 = \sqrt{2gL(1 - \cos\theta)}$. (b) If $L = 1.20$ m and $\theta = 35.0°$, find the initial speed of the cart. (*Hint:* The force exerted by the string on the object does no work on the object.)

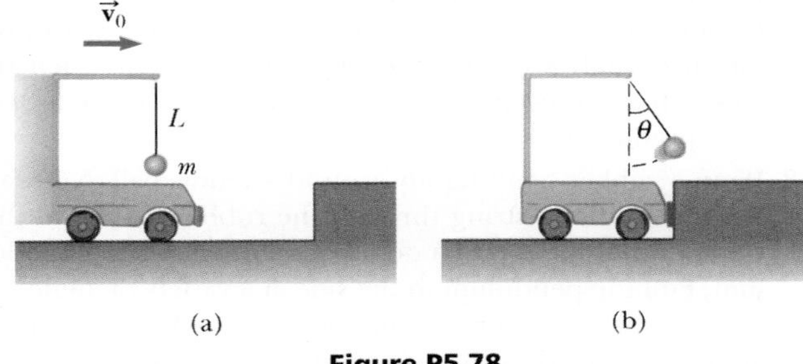

Figure P5.78

79. A truck travels uphill with constant velocity on a highway with a 7.0° slope. A 50-kg package sits on the floor of the back of the truck and does not slide, due to a static frictional force. During an interval in which the truck travels 340 m, what is the net work done on the package? What is the work done on the package by the force of gravity, the normal force, and the friction force?

80. As part of a curriculum unit on earthquakes, suppose that 375 000 British schoolchildren stand on their chairs and simultaneously jump down to the floor. Seismographers around the country see whether they can detect the resulting ground tremor. (This experiment was actually based on a suggestion by the children themselves.) (a) Find the energy released in the experiment. Model the children as having average mass 36.0 kg and as stepping from chair seats 38.0 cm above the floor. (b) Most of the energy is converted very rapidly into internal energy within the bodies of the children and the floors of the school buildings. Assume that 1% of the energy is carried away by a seismic wave. The magnitude of an earthquake on the Richter scale is given by

$$M = \frac{\log E - 4.8}{1.5}$$

where E is the seismic wave energy in joules. According to this model, what is the magnitude of the demonstration quake?

81. A loaded ore car has a mass of 950 kg and rolls on rails with negligible friction. It starts from rest and is pulled up a mine shaft by a cable connected to a winch. The shaft is inclined at 30.0° above the horizontal. The car accelerates uniformly to a speed of 2.20 m/s in 12.0 s and then continues at constant speed. (a) What power must the winch motor provide when the car is moving at constant speed? (b) What maximum power must the motor provide? (c) What total energy transfers out of the motor by work by the time the car moves off the end of the track, which is of length 1 250 m?

82. A daredevil wishes to bungee-jump from a hot-air balloon 65.0 m above a carnival midway (Fig. P5.77). He will use a piece of uniform elastic cord tied to a harness around his body to stop his fall at a point 10.0 m above the ground. Model his body as a particle and the cord as having negligible mass and a tension force described by Hooke's force law. In a preliminary test, hanging at rest from a 5.00-m length of the cord, the jumper finds that his body weight stretches it by 1.50 m. He will drop from rest at the point where the top end of a longer section of the cord is attached to the stationary balloon. (a) What length of cord should he use? (b) What maximum acceleration will he experience?

83. The system shown in Figure P5.83 consists of a light, inextensible cord, light frictionless pulleys, and blocks of equal mass. Initially, the blocks are at rest the same height above the ground. The blocks are then released. Find the speed of block A at the moment when the vertical separation of the blocks is h.

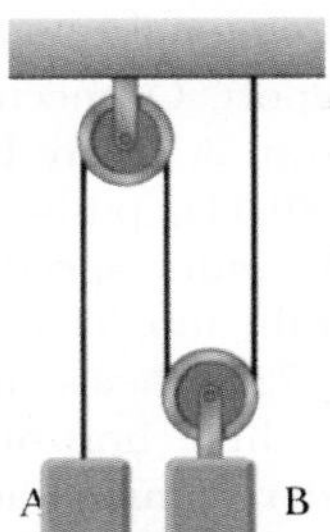

Figure P5.83

84. A cafeteria tray dispenser supports a stack of trays on a shelf that hangs from four identical spiral springs under tension, one near each corner of the shelf. Each tray has a mass of 580 g and is rectangular, 45.3 cm by 35.6 cm, and 0.450 cm thick. (a) Show that the top tray in the stack can

always be at the same height above the floor, however many trays are in the dispenser. (b) Find the spring constant each spring should have in order for the dispenser to function in this convenient way. Is any piece of data unnecessary for this determination?

85. In bicycling for aerobic exercise, a woman wants her heart rate to be between 136 and 166 beats per minute. Assume that her heart rate is directly proportional to her mechanical power output. Ignore all forces on the woman-plus-bicycle system, except for static friction forward on the drive wheel of the bicycle and an air resistance force proportional to the square of the bicycler's speed. When her speed is 22.0 km/h, her heart rate is 90.0 beats per minute. In what range should her speed be so that her heart rate will be in the range she wants?

86. In a needle biopsy, a narrow strip of tissue is extracted from a patient with a hollow needle. Rather than being pushed by hand, to ensure a clean cut the needle can be fired into the patient's body by a spring. Assume the needle has mass 5.60 g, the light spring has force constant 375 N/m, and the spring is originally compressed 8.10 cm to project the needle horizontally without friction. The tip of the needle then moves through 2.40 cm of skin and soft tissue, which exerts a resistive force of 7.60 N on it. Next, the needle cuts 3.50 cm into an organ, which exerts a backward force of 9.20 N on it. Find (a) the maximum speed of the needle and (b) the speed at which a flange on the back end of the needle runs into a stop, set to limit the penetration to 5.90 cm.

87. The power of sunlight reaching each square meter of the Earth's surface on a clear day in the tropics is close to 1 000 W. On a winter day in Manitoba, the power concentration of sunlight can be 100 W/m^2. Many human activities are described by a power-per-footprint-area on the order of 10^2 W/m^2 or less. (a) Consider, for example, a family of four paying $80 to the electric company every 30 days for 600 kWh of energy carried by electric transmission to their house, with floor area 13.0 m by 9.50 m. Compute the power-per-area measure of this energy use. (b) Consider a car 2.10 m wide and 4.90 m long traveling at 55.0 mi/h using gasoline having a "heat of combustion" of 44.0 MJ/kg with fuel economy 25.0 mi/gallon. One gallon of gasoline has a mass of 2.54 kg. Find the power-per-area measure of the car's energy use. It can be similar to that of a steel mill where rocks are melted in blast furnaces. (c) Explain why the direct use of solar energy is not practical for a conventional automobile.

88. In 1887 in Bridgeport, Connecticut, C. J. Belknap built the water slide shown in Figure P5.88. A rider on a small sled, of total mass 80.0 kg, pushed off to start at the top of the slide (point Ⓐ) with a speed of 2.50 m/s. The chute was 9.76 m high at the top, 54.3 m long, and 0.51 m wide. Along its length, 725 wheels made friction negligible. Upon leaving the chute horizontally at its bottom end (point Ⓒ), the rider skimmed across the water of Long Island Sound for as much as 50 m, "skipping along like a flat pebble," before at last coming to rest and swimming ashore, pulling his sled after him. (a) Find the speed of the sled and rider at point Ⓒ. (b) Model the force of water friction as a constant retarding force acting on a particle. Find the work done by water friction in stopping the sled and rider. (c) Find the magnitude of the force the water exerts on the sled. (d) Find the magnitude of the force the chute exerts on the sled at point Ⓑ.

Engraving from *Scientific American*, July 1888.

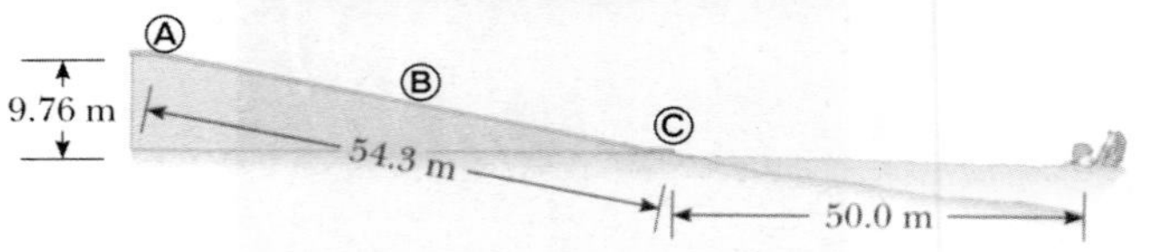

Figure P5.88

89. Three objects with masses $m_1 = 5.0$ kg, $m_2 = 10$ kg, and $m_3 = 15$ kg, respectively, are attached by strings over frictionless pulleys as indicated in Figure P5.32. The horizontal surface exerts a force of friction of 30 N on m_2. If the system is released from rest, use energy concepts to find the speed of m_3 after it moves down 4.0 m.

ACTIVITIES

A.1. Suspend a rubber band from a support and borrow some weights from your instructor to measure the rubber band's spring constant for small extensions. Calculate how much elastic potential energy is stored in the rubber band for a given extension. Use conservation of energy to predict how high a paper wad will go into the air when released from a given extension of the band. Try it to test your prediction.

A.2. Wrap a rubber band tightly around a tennis ball. Now fasten one end of a string through the rubber band and the other end to the top of a doorframe to construct a pendulum. Pull the pendulum to the side at a variety of angles to observe that the energy of the pendulum–Earth system is always conserved as the pendulum swings (almost) to the same height of its arc as the height from which it was released. (The word "almost" in the last sentence applies because some energy is lost to friction at the point of support and to air resistance. You can observe this slight loss of energy by pulling the ball to the side and letting it go from a point about half an inch from your chin. Let it go—but don't push it!—and test your belief in conservation of energy by seeing if you can avoid flinching when the ball swings back toward your chin.)

A.3. While you have your pendulum from the last activity set up, predict what will happen in the following situation and then test your guess: When a pendulum is released from a given height, it swings to the same height at the other end of its arc as you noted above. However, suppose you place a meterstick across the door opening such that the pendulum string strikes the stick about halfway up the string when it moves through the opening. How high will the ball swing in this case? Will it return to the same height as that at which it started, swing to a lower height, or swing to a greater height? Explain your answer.

A.4. Measure your pulse rate while at rest. Now slowly walk up a flight of stairs and measure your pulse rate at the top of the stairs. Repeat this activity, starting with about the same rest pulse rate at the bottom of the stairs. This time, run up the stairs. Based on your pulse rate readings, what can you conclude about the amount of work and power expended in each case? Repeat this experiment with a series of 10 push-ups.

A.5. Many fitness centers have stepper machines that enable a person to climb continuously without actually moving, because the steps move downwards as the person climbs. The work that the climber performs on the step is determined by the force exerted on the step times the distance the step moves. Since the net force on the climber is zero, the force exerted on the step must equal the climber's weight. A reasonably strenuous workout on this machine is 90 steps per minute, with each step being 8 inches (15.2 cm) high. What is the rate (in watts) at which a 130-lb (60 kg) climber does work on the stair steps? The energy actually expended by the climber is approximately five times the work done. (You may notice that a lot of heat is generated!) The machines usually report this rate in Calories/hour (1 Calorie = 1 kcal = 4186 J). Determine the rate, in Cal/h, at which energy is expended by the 130-lb climber.

Activity: If you have access to a stepper, find the ratio used by the manufacturer to determine the energy expended from the work performed. In some cases, this ratio may vary as the step speed changes. If so, generate a graph of the ratio as a function of step speed.

A small buck from the massive bull transfers a large amount of momentum to the cowboy, resulting in an involuntary dismount.

CHAPTER

6 Momentum and Collisions

OUTLINE

What happens when two automobiles collide? How does the impact affect the motion of each vehicle, and what basic physical principles determine the likelihood of serious injury? How do rockets work, and what mechanisms can be used to overcome the limitations imposed by exhaust speed? Why do we have to brace ourselves when firing small projectiles at high velocity? Finally, how can we use physics to improve our golf game?

To begin answering such questions, we introduce *momentum*. Intuitively, anyone or anything that has a lot of momentum is going to be hard to stop. In politics, the term is metaphorical. Physically, the more momentum an object has, the more force has to be applied to stop it in a given time. This concept leads to one of the most powerful principles in physics: *conservation of momentum*. Using this law, complex collision problems can be solved without knowing much about the forces involved during contact. We'll also be able to derive information about the average force delivered in an impact. With conservation of momentum, we'll have a better understanding of what choices to make when designing an automobile or a moon rocket, or when addressing a golf ball on a tee.

6.1 MOMENTUM AND IMPULSE

In physics, momentum has a precise definition. A slowly moving brontosaurus has a lot of momentum, but so does a little hot lead shot from the muzzle of a gun. We therefore expect that momentum will depend on an object's mass and velocity.

Linear momentum ►

The linear momentum $\vec{\mathbf{p}}$ of an object of mass m moving with velocity $\vec{\mathbf{v}}$ is the product of its mass and velocity:

$$\vec{\mathbf{p}} \equiv m\vec{\mathbf{v}} \qquad [6.1]$$

SI unit: kilogram-meter per second (kg · m/s)

Doubling either the mass or the velocity of an object doubles its momentum; doubling both quantities quadruples its momentum. Momentum is a vector quantity with the same direction as the object's velocity. Its components are given in two dimensions by

$$p_x = mv_x \qquad p_y = mv_y$$

where p_x is the momentum of the object in the x-direction and p_y its momentum in the y-direction.

Quick Quiz 6.1

Two objects with masses m_1 and m_2 have equal kinetic energy. How do the magnitudes of their momenta compare? (a) $p_1 < p_2$ (b) $p_1 = p_2$ (c) $p_1 > p_2$ (d) not enough information is given

Changing the momentum of an object requires the application of a force. This is, in fact, how Newton originally stated his second law of motion. Starting from the more common version of the second law, we have

$$\vec{\mathbf{F}}_{\text{net}} = m\vec{\mathbf{a}} = m\frac{\Delta\vec{\mathbf{v}}}{\Delta t} = \frac{\Delta(m\vec{\mathbf{v}})}{\Delta t} \qquad [6.2]$$

where the mass m and the forces are assumed constant. The quantity in parentheses is just the momentum, so we have the following result:

The change in an object's momentum $\Delta\vec{\mathbf{p}}$ divided by the elapsed time Δt equals the constant net force $\vec{\mathbf{F}}_{\text{net}}$ acting on the object:

$$\frac{\Delta\vec{\mathbf{p}}}{\Delta t} = \frac{\text{change in momentum}}{\text{time interval}} = \vec{\mathbf{F}}_{\text{net}} \qquad [6.3]$$

◀ Newton's second law

This equation is also valid when the forces are not constant, provided the limit is taken as Δt becomes infinitesimally small. Equation 6.3 says that if the net force on an object is zero, the object's momentum doesn't change. In other words, the linear momentum of an object is *conserved* when $\vec{\mathbf{F}}_{\text{net}} = 0$. Equation 6.3 also tells us that changing an object's momentum requires the continuous application of a force over a period of time Δt, leading to the definition of *impulse:*

If a constant force $\vec{\mathbf{F}}$ acts on an object, the **impulse $\vec{\mathbf{I}}$** delivered to the object over a time interval Δt is given by

$$\vec{\mathbf{I}} \equiv \vec{\mathbf{F}}\Delta t \qquad [6.4]$$

SI unit: kilogram meter per second (kg · m/s)

Impulse is a vector quantity with the same direction as the constant force acting on the object. When a single constant force $\vec{\mathbf{F}}$ acts on an object, Equation 6.3 can be written as

$$\vec{\mathbf{F}}\Delta t = \Delta\vec{\mathbf{p}} = m\vec{\mathbf{v}}_f - m\vec{\mathbf{v}}_i \qquad [6.5]$$

◀ Impulse–momentum theorem

This is a special case of the **impulse–momentum theorem**. Equation 6.5 shows that **the impulse of the force acting on an object equals the change in momentum of that object**. This equality is true even if the force is not constant, as long as the time interval Δt is taken to be arbitrarily small. (The proof of the general case requires concepts from calculus.)

In real-life situations, the force on an object is only rarely constant. For example, when a bat hits a baseball, the force increases sharply, reaches some maximum value, and then decreases just as rapidly. Figure 6.1(a) shows a typical graph of

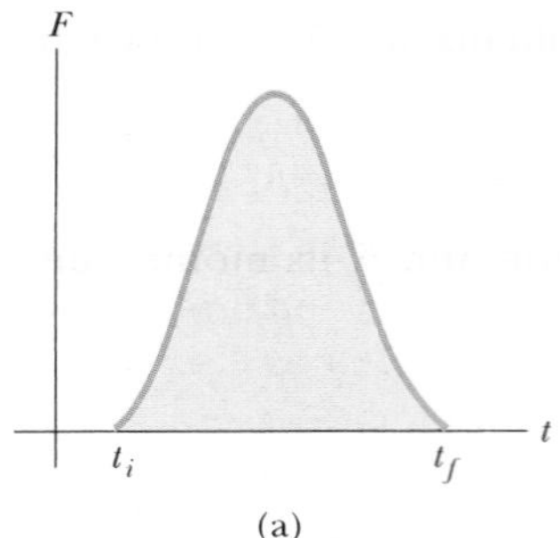

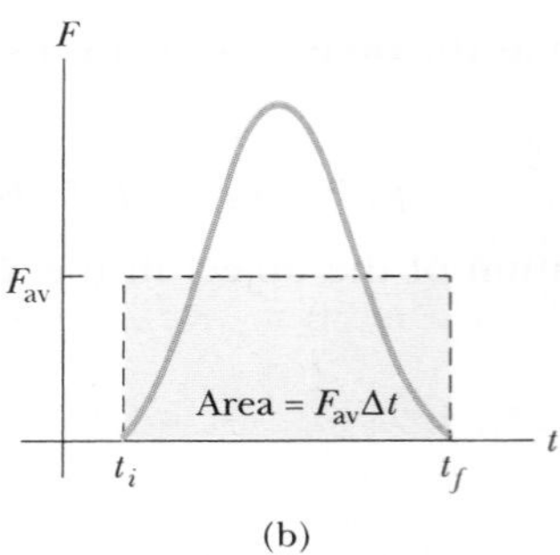

Figure 6.1 (a) A force acting on an object may vary in time. The impulse is the area under the force vs. time curve. (b) The average force (horizontal dashed line) gives the same impulse to the object in the time interval Δt as the real time-varying force described in (a).

force versus time for such incidents. The force starts out small as the bat comes in contact with the ball, rises to a maximum value when they are firmly in contact, and then drops off as the ball leaves the bat. In order to analyze this rather complex interaction, it's useful to define an **average force** $\vec{\mathbf{F}}_{av}$, shown as the dashed line in Figure 6.1b. This average force is the constant force delivering the same impulse to the object in the time interval Δt as the actual time-varying force. We can then write the impulse–momentum theorem as

$$\vec{\mathbf{F}}_{av}\Delta t = \Delta\vec{\mathbf{p}} \qquad [6.6]$$

The magnitude of the impulse delivered by a force during the time interval Δt is equal to the area under the force vs. time graph as in Figure 6.1a or, equivalently, to $F_{av}\Delta t$ as shown in Figure 6.1b. The brief collision between a bullet and an apple is illustrated in Figure 6.2.

Applying Physics 6.1 Boxing and Brain Injury

In boxing matches of the 19th century, bare fists were used. In modern boxing, fighters wear padded gloves. How do gloves protect the brain of the boxer from injury? Also, why do boxers often "roll with the punch"?

Explanation The brain is immersed in a cushioning fluid inside the skull. If the head is struck suddenly by a bare fist, the skull accelerates rapidly. The brain matches this acceleration only because of the large impulsive force exerted by the skull on the brain. This large and sudden force (large F_{av} and small Δt) can cause severe brain injury. Padded gloves extend the time Δt over which the force is applied to the head. For a given impulse $F_{av}\Delta t$, a glove results in a longer time interval than a bare fist, decreasing the average force. Because the average force is decreased, the acceleration of the skull is decreased, reducing (but not eliminating) the chance of brain injury. The same argument can be made for "rolling with the punch": If the head is held steady while being struck, the time interval over which the force is applied is relatively short and the average force is large. If the head is allowed to move in the same direction as the punch, the time interval is lengthened and the average force reduced.

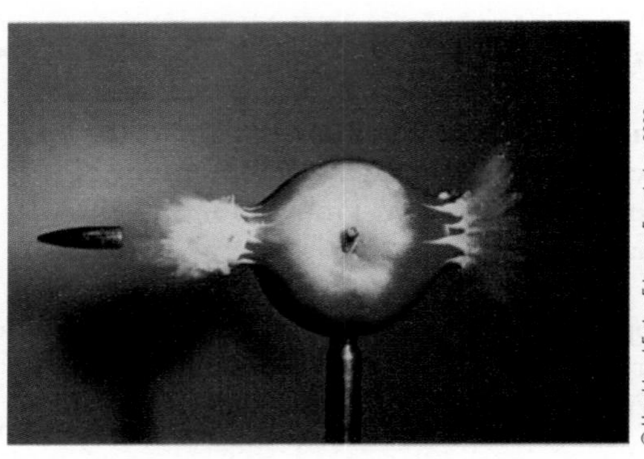

Figure 6.2 An apple being pierced by a 30-caliber bullet traveling at a supersonic speed of 900 m/s. This collision was photographed with a microflash stroboscope using an exposure time of 0.33 μs. Shortly after the photograph was taken, the apple disintegrated completely. Note that the points of entry and exit of the bullet are visually explosive.

EXAMPLE 6.1 Teeing Off

Goal Use the impulse–momentum theorem to estimate the average force exerted during an impact.

Problem A golf ball with mass 5.0×10^{-2} kg is struck with a club as in Figure 6.3. The force on the ball varies from zero when contact is made up to some maximum value (when the ball is maximally deformed) and then back to zero when the ball leaves the club, as in the graph of force vs. time in Figure 6.1. Assume that the ball leaves the club face with a velocity of $+44$ m/s. **(a)** Find the magnitude of the impulse due to the collision. **(b)** Estimate the duration of the collision and the average force acting on the ball.

© Harold and Esther Edgerton Foundation, 2002, Courtesy of Palm Press, Inc.

Figure 6.3 (Example 6.1) A golf ball being struck by a club.

Strategy In part (a), use the fact that the impulse is equal to the change in momentum. The mass and the initial and final speeds are known, so this change can be computed. In part (b), the average force is just the change in momentum computed in part (a) divided by an estimate of the duration of the collision. Guess at the distance the ball travels on the face of the club (about 2 cm, roughly the same as the radius of the ball). Divide this distance by the average velocity (half the final velocity) to get an estimate of the time of contact.

Solution

(a) Find the impulse delivered to the ball.

The problem is essentially one dimensional. Note that $v_i = 0$, and calculate the change in momentum, which equals the impulse:

$$I = \Delta p = p_f - p_i = (5.0 \times 10^{-2}\text{ kg})(44\text{ m/s}) - 0$$
$$= +2.2\text{ kg}\cdot\text{m/s}$$

(b) Estimate the duration of the collision and the average force acting on the ball.

Estimate the time interval of the collision, Δt, using the approximate displacement (radius of the ball) and its average speed (half the maximum speed):

$$\Delta t = \frac{\Delta x}{v_{\text{av}}} = \frac{2.0 \times 10^{-2}\text{ m}}{22\text{ m/s}} = 9.1 \times 10^{-4}\text{ s}$$

Estimate the average force from Equation 6.6:

$$F_{\text{av}} = \frac{\Delta p}{\Delta t} = \frac{2.2\text{ kg}\cdot\text{m/s}}{9.1 \times 10^{-4}\text{ s}} = +2.4 \times 10^{3}\text{ N}$$

Remarks This estimate shows just how large such contact forces can be. A good golfer achieves maximum momentum transfer by shifting weight from the back foot to the front foot, transmitting the body's momentum through the shaft and head of the club. This timing, involving a short movement of the hips, is more effective than a shot powered exclusively by the arms and shoulders. Following through with the swing ensures that the motion isn't slowed at the critical instant of impact.

Exercise 6.1

A 0.150-kg baseball, thrown with a speed of 40.0 m/s, is hit straight back at the pitcher with a speed of 50.0 m/s. (a) What is the impulse delivered by the bat to the baseball? (b) Find the magnitude of the average force exerted by the bat on the ball if the two are in contact for 2.00×10^{-3} s.

Answer (a) 13.5 kg·m/s (b) 6.75 kN

EXAMPLE 6.2 How Good Are the Bumpers?

Goal Find an impulse and estimate a force in a collision of a moving object with a stationary object.

Problem In a crash test, a car of mass 1.50×10^{3} kg collides with a wall and rebounds as in Figure 6.4a. The initial and final velocities of the car are $v_i = -15.0$ m/s and $v_f = 2.60$ m/s, respectively. If the collision lasts for 0.150 s, find

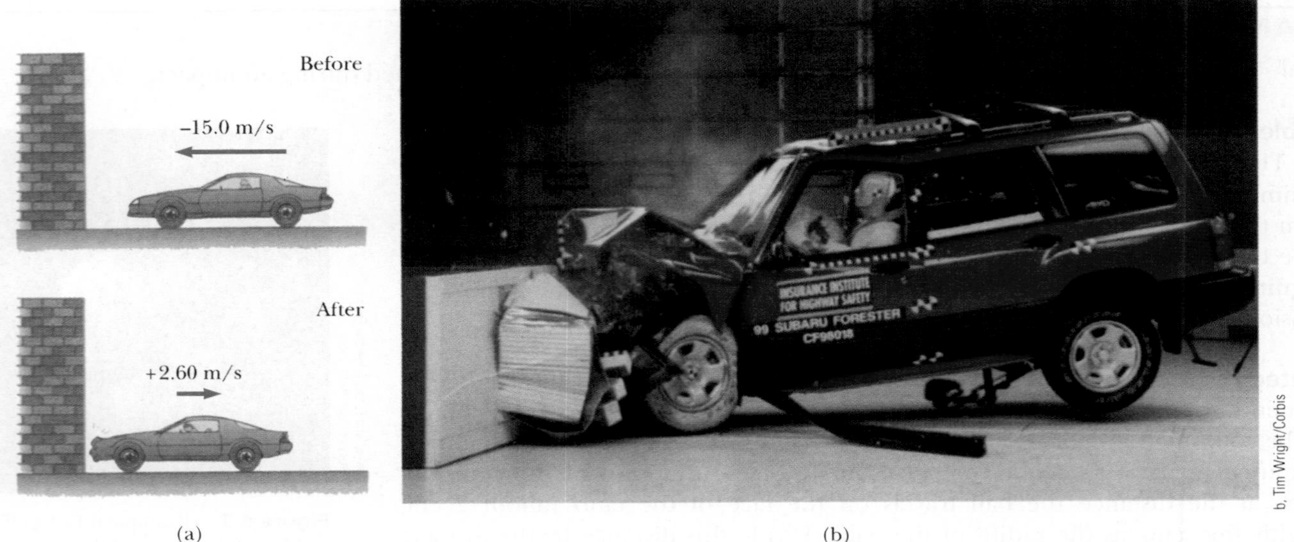

Figure 6.4 (Example 6.2) (a) This car's momentum changes as a result of its collision with the wall. (b) In a crash test (an inelastic collision), much of the car's initial kinetic energy is transformed into the energy it took to damage the vehicle.

(a) the impulse delivered to the car due to the collision and **(b)** the size and direction of the average force exerted on the car.

Strategy This problem is similar to the previous example, except that the initial and final momenta are both nonzero. Find the momenta and substitute into the impulse–momentum theorem, Equation 6.6, solving for F_{av}.

Solution

(a) Find the impulse delivered to the car.

Calculate the initial and final momenta of the car:

$$p_i = mv_i = (1.50 \times 10^3 \text{ kg})(-15.0 \text{ m/s}) = -2.25 \times 10^4 \text{ kg}\cdot\text{m/s}$$

$$p_f = mv_f = (1.50 \times 10^3 \text{ kg})(+2.60 \text{ m/s}) = +0.390 \times 10^4 \text{ kg}\cdot\text{m/s}$$

The impulse is just the difference between the final and initial momenta:

$$I = p_f - p_i = +0.390 \times 10^4 \text{ kg}\cdot\text{m/s} - (-2.25 \times 10^4 \text{ kg}\cdot\text{m/s})$$

$$I = 2.64 \times 10^4 \text{ kg}\cdot\text{m/s}$$

(b) Find the average force exerted on the car.

Apply Equation 6.6, the impulse–momentum theorem:

$$F_{\text{av}} = \frac{\Delta p}{\Delta t} = \frac{2.64 \times 10^4 \text{ kg}\cdot\text{m/s}}{0.150 \text{ s}} = +1.76 \times 10^5 \text{ N}$$

Remarks When the car doesn't rebound off the wall, the average force exerted on the car is smaller than the value just calculated. With a final momentum of zero, the car undergoes a smaller change in momentum.

Exercise 6.2

Suppose the car doesn't rebound off the wall, but the time interval of the collision remains at 0.150 s. In this case, the final velocity of the car is zero. Find the average force exerted on the car.

Answer $+1.50 \times 10^5$ N

Injury in Automobile Collisions

APPLICATION

Injury to Passengers in Car Collisions

The main injuries that occur to a person hitting the interior of a car in a crash are brain damage, bone fracture, and trauma to the skin, blood vessels, and internal organs. Here, we compare the rather imprecisely known thresholds for human injury with typical forces and accelerations experienced in a car crash.

A force of about 90 kN (20 000 lb) compressing the tibia can cause fracture. Although the breaking force varies with the bone considered, we may take this value as the threshold force for fracture. It's well known that rapid acceleration of the head, even without skull fracture, can be fatal. Estimates show that head accelerations of 150g experienced for about 4 ms or 50g for 60 ms are fatal 50% of the time. Such injuries from rapid acceleration often result in nerve damage to the spinal cord where the nerves enter the base of the brain. The threshold for damage to skin, blood vessels, and internal organs may be estimated from whole-body impact data, where the force is uniformly distributed over the entire front surface area of 0.7 m^2 to 0.9 m^2. These data show that if the collision lasts for less than about 70 ms, a person will survive if the whole-body impact pressure (force per unit area) is less than 1.9×10^5 N/m^2 (28 lb/in.^{2}). Death results in 50% of cases in which the whole-body impact pressure reaches 3.4×10^5 N/m^2 (50 lb/in.^{2}).

Armed with the data above, we can estimate the forces and accelerations in a typical car crash and see how seat belts, air bags, and padded interiors can reduce the chance of death or serious injury in a collision. Consider a typical collision involving a 75-kg passenger not wearing a seat belt, traveling at 27 m/s (60 mi/h) who comes to rest in about 0.010 s after striking an unpadded dashboard. Using $F_{av}\Delta t = mv_f - mv_i$, we find that

$$F_{av} = \frac{mv_f - mv_i}{\Delta t} = \frac{0 - (75\text{ kg})(27\text{ m/s})}{0.010\text{ s}} = -2.0 \times 10^5\text{ N}$$

and

$$a = \left|\frac{\Delta v}{\Delta t}\right| = \frac{27\text{ m/s}}{0.010\text{ s}} = 2\,700\text{ m/s}^2 = \frac{2\,700\text{ m/s}^2}{9.8\text{ m/s}^2}\,g = 280g$$

If we assume the passenger crashes into the dashboard and windshield so that the head and chest, with a combined surface area of 0.5 m^2, experience the force, we find a whole-body pressure of

$$\frac{F_{av}}{A} = \frac{2.0 \times 10^5\text{ N}}{0.5\text{ m}^2} \cong 4 \times 10^5\text{ N/m}^2$$

We see that the force, the acceleration, and the whole-body pressure all *exceed* the threshold for fatality or broken bones and that an unprotected collision at 60 mi/h is almost certainly fatal.

What can be done to reduce or eliminate the chance of dying in a car crash? The most important factor is the collision time, or the time it takes the person to come to rest. If this time can be increased by 10 to 100 times the value of 0.01 s for a hard collision, the chances of survival in a car crash are much higher, because the increase in Δt makes the contact force 10 to 100 times smaller. Seat belts restrain people so that they come to rest in about the same amount of time it takes to stop the car, typically about 0.15 s. This increases the effective collision time by an order of magnitude. Figure 6.5 shows the measured force on a car versus time for a car crash.

Air bags also increase the collision time, absorb energy from the body as they rapidly deflate, and spread the contact force over an area of the body of about 0.5 m^2, preventing penetration wounds and fractures. Air bags must deploy very rapidly (in less than 10 ms) in order to stop a human traveling at 27 m/s before he or she comes to rest against the steering column about 0.3 m away. In order to achieve this rapid deployment, accelerometers send a signal to discharge a bank of capacitors (devices that store electric charge), which then ignites an explosive, thereby filling the air bag with gas very quickly. The electrical charge for ignition is

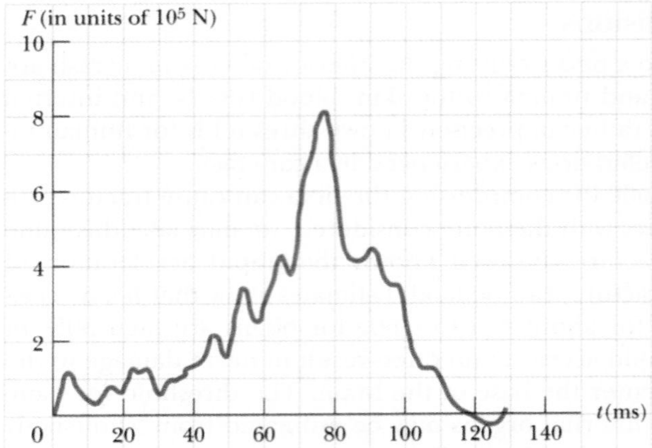

Figure 6.5 Force on a car versus time for a typical collision.

stored in capacitors to ensure that the air bag continues to operate in the event of damage to the battery or the car's electrical system in a severe collision.

The important reduction in potentially fatal forces, accelerations, and pressures to tolerable levels by the simultaneous use of seat belts and air bags is summarized as follows: If a 75-kg person traveling at 27 m/s is stopped by a seat belt in 0.15 s, the person experiences an average force of 9.8 kN, an average acceleration of 18*g*, and a whole-body pressure of $2.8 \times 10^4\ \text{N/m}^2$ for a contact area of $0.5\ \text{m}^2$. These values are about one order of magnitude less than the values estimated earlier for an unprotected person and well below the thresholds for life-threatening injuries.

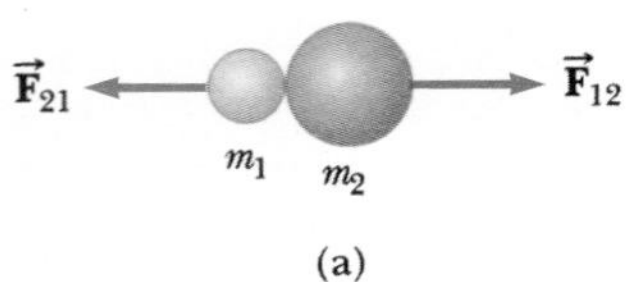

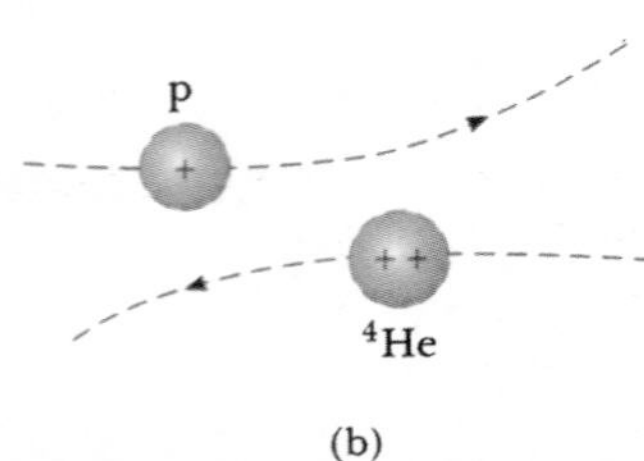

Figure 6.6 (a) A collision between two objects resulting from direct contact. (b) A collision between two charged objects (in this case, a proton and a helium nucleus).

6.2 CONSERVATION OF MOMENTUM

When a collision occurs in an isolated system, the total momentum of the system doesn't change with the passage of time. Instead, it remains constant both in magnitude and in direction. The momenta of the individual objects in the system may change, but the vector sum of *all* the momenta will not change. The total momentum, therefore, is said to be *conserved.* In this section, we will see how the laws of motion lead us to this important conservation law.

A collision may be the result of physical contact between two objects, as illustrated in Figure 6.6a. This is a common macroscopic event, as when a pair of billiard balls or a baseball and a bat strike each other. By contrast, because contact on a submicroscopic scale is hard to define accurately, the notion of *collision* must be generalized to that scale. Forces between two objects arise from the electrostatic interaction of the electrons in the surface atoms of the objects. As will be discussed in Chapter 15, electric charges are either positive or negative. Charges with the same sign repel each other, while charges with opposite sign attract each other. To understand the distinction between macroscopic and microscopic collisions, consider the collision between two positive charges, as shown in Figure 6.6b. Because the two particles in the figure are both positively charged, they repel each other. During such a microscopic collision, particles need not touch in the normal sense in order to interact and transfer momentum.

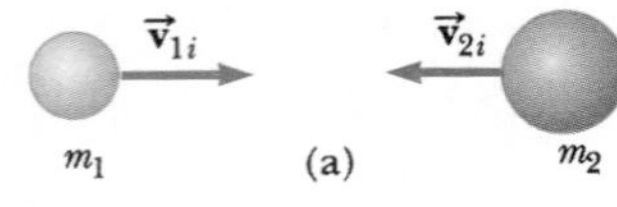

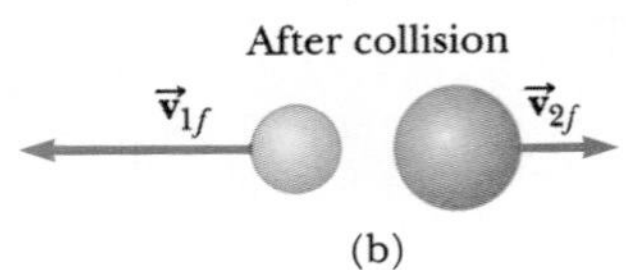

ACTIVE FIGURE 6.7
Before and after a head-on collision between two objects. The momentum of each object changes as a result of the collision, but the total momentum of the system remains constant.

Physics Now™
Log into PhysicsNow at **www.cp7e.com**, and go to Active Figure 6.7 to adjust the masses and velocities of the colliding objects and see the effect on the final velocities.

Active Figure 6.7 shows an isolated system of two particles before and after they collide. By "isolated," we mean that no external forces, such as the gravitational force or friction, act on the system. Before the collision, the velocities of the two particles are $\vec{v}_{1i}$ and $\vec{v}_{2i}$; after the collision, the velocities are $\vec{v}_{1f}$ and $\vec{v}_{2f}$. The impulse–momentum theorem applied to m_1 becomes

$$\vec{F}_{21}\Delta t = m_1\vec{v}_{1f} - m_1\vec{v}_{1i}$$

Likewise, for m_2, we have

$$\vec{F}_{12}\Delta t = m_2\vec{v}_{2f} - m_2\vec{v}_{2i}$$

where $\vec{\mathbf{F}}_{21}$ is the average force exerted by m_2 on m_1 during the collision and $\vec{\mathbf{F}}_{12}$ is the average force exerted by m_1 on m_2 during the collision, as in Figure 6.6a.

We use average values for $\vec{\mathbf{F}}_{21}$ and $\vec{\mathbf{F}}_{12}$ even though the actual forces may vary in time in a complicated way, as is the case in Figure 6.8. Newton's third law states that at all times these two forces are equal in magnitude and opposite in direction: $\vec{\mathbf{F}}_{21} = -\vec{\mathbf{F}}_{12}$. In addition, the two forces act over the same time interval. As a result, we have

$$\vec{\mathbf{F}}_{21}\Delta t = -\vec{\mathbf{F}}_{12}\Delta t$$

or

$$m_1\vec{\mathbf{v}}_{1f} - m_1\vec{\mathbf{v}}_{1i} = -(m_2\vec{\mathbf{v}}_{2f} - m_2\vec{\mathbf{v}}_{2i})$$

after substituting the expressions obtained for $\vec{\mathbf{F}}_{21}$ and $\vec{\mathbf{F}}_{12}$. This equation can be rearranged to give the following important result:

$$m_1\vec{\mathbf{v}}_{1i} + m_2\vec{\mathbf{v}}_{2i} = m_1\vec{\mathbf{v}}_{1f} + m_2\vec{\mathbf{v}}_{2f} \qquad [6.7]$$

This result is a special case of the law of **conservation of momentum** and is true of isolated systems containing any number of interacting objects.

◀ Conservation of momentum

When no net external force acts on a system, the total momentum of the system remains constant in time.

Defining the isolated system is an important feature of applying this conservation law. A cheerleader jumping upwards from rest might appear to violate conservation of momentum, because initially her momentum is zero and suddenly she's leaving the ground with velocity $\vec{\mathbf{v}}$. The flaw in this reasoning lies in the fact that the cheerleader isn't an isolated system. In jumping, she exerts a downward force on the Earth, changing its momentum. This change in the Earth's momentum isn't noticeable, however, because of the Earth's gargantuan mass compared to the cheerleader's. When we define the system to be *the cheerleader and the Earth*, momentum is conserved.

Action and reaction, together with the accompanying exchange of momentum between two objects, is responsible for the phenomenon known as *recoil*. Everyone knows that throwing a baseball while standing straight up, without bracing your feet against the Earth, is a good way to fall over backwards. This reaction, an example of recoil, also happens when you fire a gun or shoot an arrow. Conservation of momentum provides a straightforward way to calculate such effects, as the next example shows.

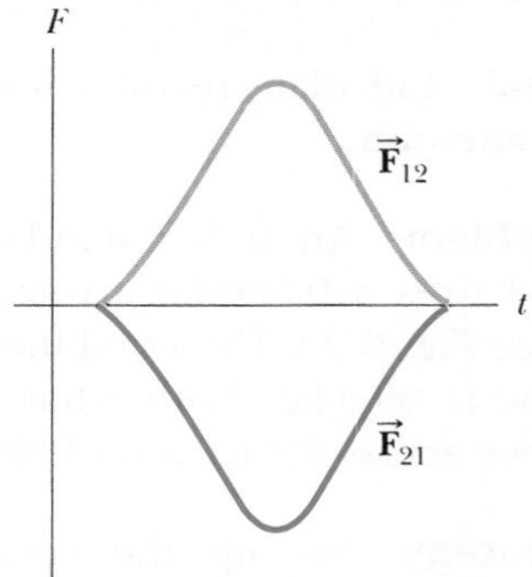

Figure 6.8 Force as a function of time for the two colliding particles in Figures 6.6(a) and 6.7. Note that $\vec{\mathbf{F}}_{21} = -\vec{\mathbf{F}}_{12}$.

TIP 6.1 Momentum Conservation Applies to a *System*!

The momentum of an isolated system is conserved, but not necessarily the momentum of one particle within that system, because other particles in the system may be interacting with it. Apply conservation of momentum to an isolated system *only*.

Courtesy of NASA

(a)

Mike Severns/Stone/Getty Images

(b)

Conservation of momentum is the principle behind these two propulsion systems. (a) The force from a nitrogen-propelled, hand-controlled device allows an astronaut to move about freely in space without restrictive tethers. (b) A squid propels itself by expelling water at a high velocity.

INTERACTIVE EXAMPLE 6.3 The Archer

Goal Calculate recoil velocity using conservation of momentum

Problem An archer stands at rest on frictionless ice and fires a 0.500-kg arrow horizontally at 50.0 m/s. (See Fig. 6.9.) The combined mass of the archer and bow is 60.0 kg. With what velocity does the archer move across the ice after firing the arrow?

Figure 6.9 (Interactive Example 6.3) An archer fires an arrow horizontally to the right. Because he is standing on frictionless ice, he will begin to slide to the left across the ice.

Strategy Set up the conservation of momentum equation in the horizontal direction and solve for the final velocity of the archer. The system of the archer (including the bow) and the arrow is *not* isolated, because the gravitational and normal forces act on it. These forces, however, are perpendicular to the motion of the system and hence do no work on it.

Solution

Write the conservation of momentum equation. Let v_{1f} be the archer's velocity and v_{2f} the arrow's velocity.

$$p_i = p_f$$

$$0 = m_1 v_{1f} + m_2 v_{2f}$$

Substitute $m_1 = 60.0$ kg, $m_2 = 0.500$ kg, and $v_{2f} = 50.0$ m/s, and solve for v_{1f}:

$$v_{1f} = -\frac{m_2}{m_1} v_{2f} = -\left(\frac{0.500 \text{ kg}}{60.0 \text{ kg}}\right)(50.0 \text{ m/s})$$

$$= \boxed{-0.417 \text{ m/s}}$$

Remarks The negative sign on v_{1f} indicates that the archer is moving opposite the direction of motion of the arrow, in accordance with Newton's third law. Because the archer is much more massive than the arrow, his acceleration and consequent velocity are much smaller than the acceleration and velocity of the arrow.

Newton's second law, $\Sigma F = ma$, can't be used in this problem because we have no information about the force on the arrow or its acceleration. An energy approach can't be used either, because we don't know how much work is done in pulling the string back or how much potential energy is stored in the bow. Conservation of momentum, however, readily solves the problem.

Exercise 6.3

A 70.0-kg man and a 55.0-kg woman on ice skates stand facing each other. If the woman pushes the man backwards so that his final speed is 1.50 m/s, at what speed does she recoil?

Answer 1.91 m/s

Physics Now™ You can change the mass of the archer and the mass and speed of the arrow by logging into PhysicsNow at **www.cp7e.com** and going to Interactive Example 6.3.

Quick Quiz 6.2

A boy standing at one end of a floating raft that is stationary relative to the shore walks to the opposite end of the raft, away from the shore. As a consequence, the raft (a) remains stationary, (b) moves away from the shore, or (c) moves toward the shore. (*Hint*: Use conservation of momentum.)

6.3 COLLISIONS

We have seen that for any type of collision, the total momentum of the system just before the collision equals the total momentum just after the collision as long as the system may be considered isolated. The total kinetic energy, on the other hand, is generally not conserved in a collision because some of the kinetic energy is converted to internal energy, sound energy, and the work needed to permanently deform the objects involved, such as cars in a car crash. **We define an inelastic collision as a collision in which momentum is conserved, but kinetic energy is not.** The collision of a rubber ball with a hard surface is inelastic, because some of the kinetic energy is lost when the ball is deformed during contact with the surface. **When two objects collide and stick together, the collision is called *perfectly inelastic*.** For example, if two pieces of putty collide, they stick together and move with some common velocity after the collision. If a meteorite collides head on with the Earth, it becomes buried in the Earth and the collision is considered perfectly inelastic. Only in very special circumstances is all the initial kinetic energy lost in a perfectly inelastic collision.

An elastic collision is defined as one in which both momentum and kinetic energy are conserved. Billiard ball collisions and the collisions of air molecules with the walls of a container at ordinary temperatures are highly elastic. Macroscopic collisions such as those between billiard balls are only approximately elastic, because some loss of kinetic energy takes place—for example, in the clicking sound when two balls strike each other. Perfectly elastic collisions do occur, however, between atomic and subatomic particles. Elastic and perfectly inelastic collisions are *limiting* cases; most actual collisions fall into a range in between them.

As a practical application, an inelastic collision is used to detect glaucoma, a disease in which the pressure inside the eye builds up and leads to blindness by damaging the cells of the retina. In this application, medical professionals use a device called a *tonometer* to measure the pressure inside the eye. This device releases a puff of air against the outer surface of the eye and measures the speed of the air after reflection from the eye. At normal pressure, the eye is slightly spongy, and the pulse is reflected at low speed. As the pressure inside the eye increases, the outer surface becomes more rigid, and the speed of the reflected pulse increases. In this way, the speed of the reflected puff of air can measure the internal pressure of the eye.

We can summarize the types of collisions as follows:

- In an elastic collision, both momentum and kinetic energy are conserved.
- In an inelastic collision, momentum is conserved but kinetic energy is not.
- In a *perfectly* inelastic collision, momentum is conserved, kinetic energy is not, and the two objects stick together after the collision, so their final velocities are the same.

◀ Elastic collision

◀ Inelastic collision

In the remainder of this section, we will treat perfectly inelastic collisions and elastic collisions in one dimension.

TIP 6.2 Momentum and Kinetic Energy in Collisions

The momentum of an isolated system is conserved in all collisions. However, the kinetic energy of an isolated system is conserved only when the collision is elastic.

TIP 6.3 Inelastic vs. Perfectly Inelastic Collisions

If the colliding particles stick together, the collision is perfectly inelastic. If they bounce off each other (and kinetic energy is not conserved), the collision is inelastic.

APPLICATION

Glaucoma Testing

Quick Quiz 6.3

A car and a large truck traveling at the same speed collide head-on and stick together. Which vehicle experiences the larger change in the magnitude of its momentum? (a) the car (b) the truck (c) the change in the magnitude of momentum is the same for both (d) impossible to determine

Before collision

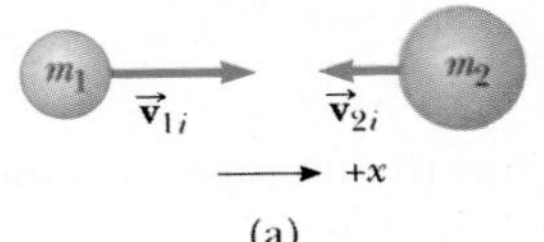

(a)

After collision

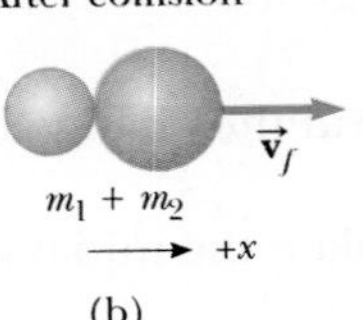

(b)

ACTIVE FIGURE 6.10
(a) Before and (b) after a perfectly inelastic head-on collision between two objects.

Physics Now™
Log into PhysicsNow at **www.cp7e.com**, and go to Active Figure 6.10 to adjust the masses and velocities of the colliding objects and see the effect on the final velocity.

Perfectly Inelastic Collisions

Consider two objects having masses m_1 and m_2 moving with known initial velocity components v_{1i} and v_{2i} along a straight line, as in Active Figure 6.10. If the two objects collide head-on, stick together, and move with a common velocity component v_f after the collision, then the collision is perfectly inelastic. Because the total momentum of the two-object isolated system before the collision equals the total momentum of the combined-object system after the collision, we can solve for the final velocity using conservation of momentum alone:

$$m_1 v_{1i} + m_2 v_{2i} = (m_1 + m_2) v_f \quad [6.8]$$

Final velocity of two objects in a one-dimensional perfectly inelastic collision ▶

$$v_f = \frac{m_1 v_{1i} + m_2 v_{2i}}{m_1 + m_2} \qquad [6.9]$$

It's important to notice that v_{1i}, v_{2i}, and v_f represent the x-components of the velocity vectors, so care is needed in entering their known values, particularly with regard to signs. For example, in Active Figure 6.10, v_{1i} would have a positive value (m_1 moving to the right), whereas v_{2i} would have a negative value (m_2 moving to the left). Once these values are entered, Equation 6.9 can be used to find the correct final velocity, as shown in Examples 6.4 and 6.5.

EXAMPLE 6.4 An SUV Versus a Compact

Goal Apply conservation of momentum to a one-dimensional inelastic collision.

Problem An SUV with mass 1.80×10^3 kg is traveling eastbound at $+15.0$ m/s, while a compact car with mass 9.00×10^2 kg is traveling westbound at -15.0 m/s. (See Fig. 6.11.) The cars collide head-on, becoming entangled. **(a)** Find the speed of the entangled cars after the collision. **(b)** Find the change in the velocity of each car. **(c)** Find the change in the kinetic energy of the system consisting of both cars.

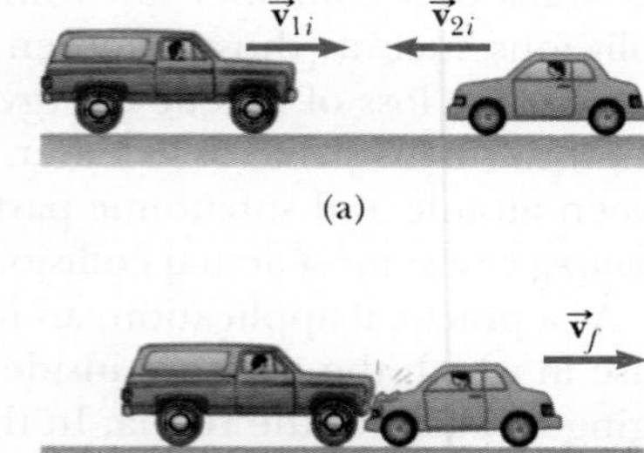

Figure 6.11 (Example 6.4)

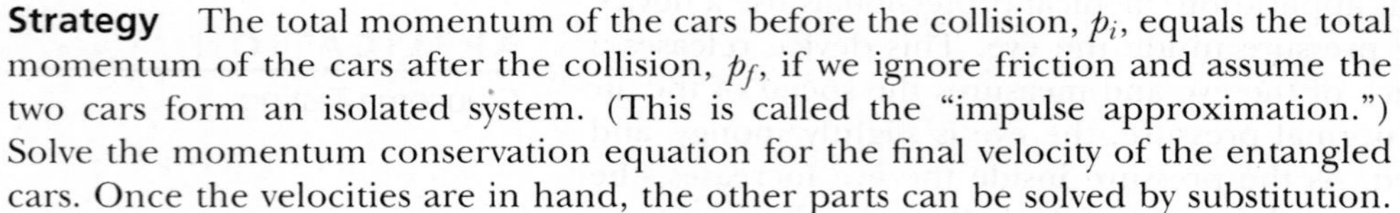

Strategy The total momentum of the cars before the collision, p_i, equals the total momentum of the cars after the collision, p_f, if we ignore friction and assume the two cars form an isolated system. (This is called the "impulse approximation.") Solve the momentum conservation equation for the final velocity of the entangled cars. Once the velocities are in hand, the other parts can be solved by substitution.

Solution

(a) Find the final speed after collision.

Let m_1 and v_{1i} represent the mass and initial velocity of the SUV, while m_2 and v_{2i} pertain to the compact. Apply conservation of momentum:

$$p_i = p_f$$

$$m_1 v_{1i} + m_2 v_{2i} = (m_1 + m_2) v_f$$

Substitute the values and solve for the final velocity, v_f:

$$(1.80 \times 10^3 \text{ kg})(15.0 \text{ m/s}) + (9.00 \times 10^2 \text{ kg})(-15.0 \text{ m/s})$$
$$= (1.80 \times 10^3 \text{ kg} + 9.00 \times 10^2 \text{ kg}) v_f$$

$$v_f = \boxed{+5.00 \text{ m/s}}$$

(b) Find the change in velocity for each car.

Change in velocity of the SUV:

$$\Delta v_1 = v_f - v_{1i} = 5.00 \text{ m/s} - 15.0 \text{ m/s} = \boxed{-10.0 \text{ m/s}}$$

Change in velocity of the compact car:

$$\Delta v_2 = v_f - v_{2i} = 5.00 \text{ m/s} - (-15.0 \text{ m/s}) = \boxed{20.0 \text{ m/s}}$$

(c) Find the change in kinetic energy of the system.

Calculate the initial kinetic energy of the system:

$$KE_i = \tfrac{1}{2} m_1 v_{1i}^2 + \tfrac{1}{2} m_2 v_{2i}^2 = \tfrac{1}{2}(1.80 \times 10^3 \text{ kg})(15.0 \text{ m/s})^2$$
$$+ \tfrac{1}{2}(9.00 \times 10^2 \text{ kg})(-15.0 \text{ m/s})^2$$
$$= 3.04 \times 10^5 \text{ J}$$

Calculate the final kinetic energy of the system and the change in kinetic energy, ΔKE.

$$KE_f = \tfrac{1}{2}(m_1 + m_2) v_f^2$$
$$= \tfrac{1}{2}(1.80 \times 10^3 \text{ kg} + 9.00 \times 10^2 \text{ kg})(5.00 \text{ m/s})^2$$
$$= 3.38 \times 10^4 \text{ J}$$

$$\Delta KE = KE_f - KE_i = \boxed{-2.70 \times 10^5 \text{ J}}$$

Remarks During the collision, the system lost almost 90% of its kinetic energy. The change in velocity of the SUV was only 10.0 m/s, compared to twice that for the compact car. This example underscores perhaps the most important safety feature of any car: its mass. Injury is caused by a change in velocity, and the more massive vehicle undergoes a smaller velocity change in a typical accident.

Exercise 6.4
Suppose the same two vehicles are both traveling eastward, the compact car leading the SUV. The driver of the compact car slams on the brakes suddenly, slowing the vehicle to 6.00 m/s. If the SUV traveling at 18.0 m/s crashes into the compact car, find (a) the speed of the system right after the collision, assuming the two vehicles become entangled, (b) the change in velocity for both vehicles, and (c) the change in kinetic energy of the system, from the instant before impact (when the compact car is traveling at 6.00 m/s) to the instant right after the collision.

Answers (a) 14.0 m/s (b) SUV: $\Delta v_1 = -4.0$ m/s Compact car: $\Delta v_2 = 8.0$ m/s (c) -4.32×10^4 J

EXAMPLE 6.5 The Ballistic Pendulum

Goal Combine the concepts of conservation of energy and conservation of momentum in inelastic collisions.

Problem The ballistic pendulum (Fig. 6.12a) is a device used to measure the speed of a fast-moving projectile such as a bullet. The bullet is fired into a large block of wood suspended from some light wires. The bullet is stopped by the block, and the entire system swings up to a height h. It is possible to obtain the initial speed of the bullet by measuring h and the two masses. As an example of the technique, assume that the mass of the bullet, m_1, is 5.00 g, the mass of the pendulum, m_2, is 1.000 kg, and h is 5.00 cm. Find the initial speed of the bullet, v_{1i}.

Strategy First, use conservation of momentum and the properties of perfectly inelastic collisions to find the initial speed of the bullet, v_{1i}, in terms of the final velocity of the block–bullet system, v_f. Second, use conservation of energy and the height reached by the pendulum to find v_f. Finally, substitute this value of v_f into the previous result to obtain the initial speed of the bullet.

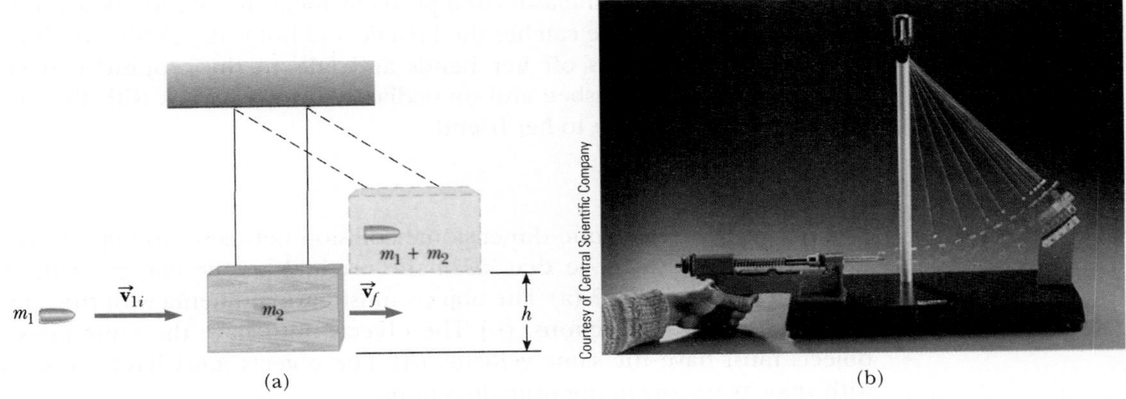

Figure 6.12 (Example 6.5) (a) Diagram of a ballistic pendulum. Note that $\vec{\mathbf{v}}_f$ is the velocity of the system just *after* the perfectly inelastic collision. (b) Multiflash photograph of a laboratory ballistic pendulum.

Solution

Use conservation of momentum, and substitute the known masses. Note that $v_{2i} = 0$ and v_f is the velocity of the system (block + bullet) just after the collision.

$$p_i = p_f$$

$$m_1 v_{1i} + m_2 v_{2i} = (m_1 + m_2) v_f$$

$$(5.00 \times 10^{-3}\ \text{kg}) v_{1i} + 0 = (1.005\ \text{kg}) v_f \qquad (1)$$

Apply conservation of energy to the block–bullet system after the collision:

$$(KE + PE)_{\text{after collision}} = (KE + PE)_{\text{top}}$$

Both the potential energy at the bottom and the kinetic energy at the top are zero. Solve for the final velocity of the block–bullet system, v_f:

$$\tfrac{1}{2}(m_1 + m_2)v_f^2 + 0 = 0 + (m_1 + m_2)gh$$

$$v_f^2 = 2gh$$

$$v_f = \sqrt{2gh} = \sqrt{2(9.80 \text{ m/s}^2)(5.00 \times 10^{-2} \text{ m})}$$

$$v_f = 0.990 \text{ m/s}$$

Finally, substitute v_f into Equation 1 to find v_{1i}, the initial speed of the bullet:

$$v_{1i} = \frac{(1.005 \text{ kg})(0.990 \text{ m/s})}{5.00 \times 10^{-3} \text{ kg}} = 199 \text{ m/s}$$

Remarks Because the impact is inelastic, it would be incorrect to equate the initial kinetic energy of the incoming bullet to the final gravitational potential energy associated with the bullet–block combination. The energy isn't conserved!

Exercise 6.5

A bullet with mass 5.00 g is fired horizontally into a 2.000-kg block attached to a horizontal spring. The spring has a constant 6.00×10^2 N/m and reaches a maximum compression of 6.00 cm. (a) Find the initial speed of the bullet–block system. (b) Find the speed of the bullet.

Answer (a) 1.04 m/s (b) 417 m/s

Quick Quiz 6.4

An object of mass m moves to the right with a speed v. It collides head-on with an object of mass $3m$ moving with speed $v/3$ in the opposite direction. If the two objects stick together, what is the speed of the combined object, of mass $4m$, after the collision?

(a) 0 (b) $v/2$ (c) v (d) $2v$

Quick Quiz 6.5

A skater is using very low friction rollerblades. A friend throws a Frisbee® at her, on the straight line along which she is coasting. Describe each of the following events as an elastic, an inelastic, or a perfectly inelastic collision between the skater and the Frisbee: (a) She catches the Frisbee and holds it. (b) She tries to catch the Frisbee, but it bounces off her hands and falls to the ground in front of her. (c) She catches the Frisbee and immediately throws it back with the same speed (relative to the ground) to her friend.

Quick Quiz 6.6

In a perfectly inelastic one-dimensional collision between two objects, what condition alone is necessary so that *all* of the original kinetic energy of the system is gone after the collision? (a) The objects must have momenta with the same magnitude but opposite directions. (b) The objects must have the same mass. (c) The objects must have the same velocity. (d) The objects must have the same speed, with velocity vectors in opposite directions.

Elastic Collisions

Now consider two objects that undergo an **elastic head-on collision** (Active Fig. 6.13). In this situation, **both the momentum and the kinetic energy of the system of two objects are conserved.** We can write these conditions as

$$m_1v_{1i} + m_2v_{2i} = m_1v_{1f} + m_2v_{2f} \quad [6.10]$$

and

$$\tfrac{1}{2}m_1v_{1i}^2 + \tfrac{1}{2}m_2v_{2i}^2 = \tfrac{1}{2}m_1v_{1f}^2 + \tfrac{1}{2}m_2v_{2f}^2 \quad [6.11]$$

where v is positive if an object moves to the right and negative if it moves to the left.

In a typical problem involving elastic collisions, there are two unknown quantities, and Equations 6.10 and 6.11 can be solved simultaneously to find them. These two equations are linear and quadratic, respectively. An alternate approach simplifies the quadratic equation to another linear equation, facilitating solution. Canceling the factor $\frac{1}{2}$ in Equation 6.11, we rewrite the equation as

$$m_1(v_{1i}^2 - v_{1f}^2) = m_2(v_{2f}^2 - v_{2i}^2)$$

Here we have moved the terms containing m_1 to one side of the equation and those containing m_2 to the other. Next, we factor both sides of the equation:

$$m_1(v_{1i} - v_{1f})(v_{1i} + v_{1f}) = m_2(v_{2f} - v_{2i})(v_{2f} + v_{2i}) \qquad [6.12]$$

Now we separate the terms containing m_1 and m_2 in the equation for the conservation of momentum (Equation 6.10) to get

$$m_1(v_{1i} - v_{1f}) = m_2(v_{2f} - v_{2i}) \qquad [6.13]$$

To obtain our final result, we divide Equation 6.12 by Equation 6.13, producing

$$v_{1i} + v_{1f} = v_{2f} + v_{2i}$$

Gathering initial and final values on opposite sides of the equation gives

$$v_{1i} - v_{2i} = -(v_{1f} - v_{2f}) \qquad [6.14]$$

This equation, in combination with Equation 6.10, will be used to solve problems dealing with perfectly elastic head-on collisions. According to Equation 6.14, the relative velocity of the two objects before the collision, $v_{1i} - v_{2i}$, equals the negative of the relative velocity of the two objects after the collision, $-(v_{1f} - v_{2f})$. To better understand the equation, imagine that you are riding along on one of the objects. As you measure the velocity of the other object from your vantage point, you will be measuring the relative velocity of the two objects. In your view of the collision, the other object comes toward you and bounces off, leaving the collision with the same speed, but in the opposite direction. This is just what Equation 6.14 states.

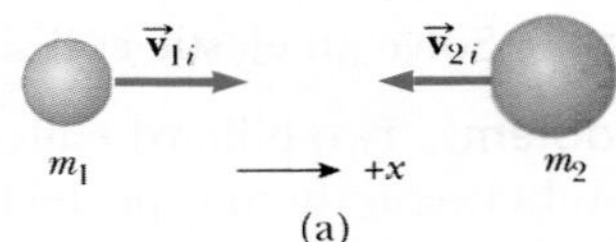

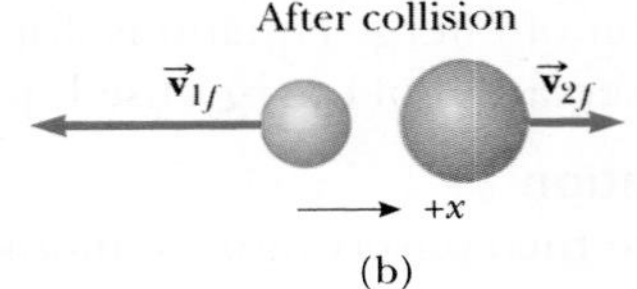

ACTIVE FIGURE 6.13
(a) Before and (b) after an elastic head-on collision between two hard spheres.

PhysicsNow™
Log into PhysicsNow at **www.cp7e.com**, and go to Active Figure 6.13 to adjust the masses and velocities of the colliding objects and see the effect on the final velocities.

Problem-Solving Strategy
One-Dimensional Collisions

The following procedure is recommended for solving one-dimensional problems involving collisions between two objects:

1. **Coordinates**. Choose a coordinate axis that lies along the direction of motion.
2. **Diagram**. Sketch the problem, representing the two objects as blocks and labeling velocity vectors and masses.
3. **Conservation of Momentum**. Write a general expression for the *total* momentum of the system of two objects *before* and *after* the collision, and equate the two, as in Equation 6.10. On the next line, fill in the known values.
4. **Conservation of Energy**. If the collision is elastic, write a general expression for the total energy before and after the collision, and equate the two quantities, as in Equation 6.11 or (preferably) Equation 6.14. Fill in the known values. (***Skip* this step if the collision is *not* perfectly elastic**.)
5. **Solve** the equations simultaneously. Equations 6.10 and 6.14 form a system of two linear equations and two unknowns. If you have forgotten Equation 6.14, use Equation 6.11 instead.

Steps 1 and 2 of the problem-solving strategy are generally carried out in the process of sketching and labeling a diagram of the problem. This is clearly the case in our next example, which makes use of Figure 6.13. Other steps are pointed out as they are applied.

EXAMPLE 6.6 Let's Play Pool

Goal Solve an elastic collision in one dimension.

Problem Two billiard balls of identical mass move toward each other as in Active Figure 6.13. Assume that the collision between them is perfectly elastic. If the initial velocities of the balls are $+30.0$ cm/s and -20.0 cm/s, what is the velocity of each ball after the collision? Assume friction and rotation are unimportant.

Strategy Solution of this problem is a matter of solving two equations, the conservation of momentum and conservation of energy equations, for two unknowns, the final velocities of the two balls. Instead of using Equation 6.11 for conservation of energy, use Equation 6.14, which is linear, hence easier to handle.

Solution

Write the conservation of momentum equation. Because $m_1 = m_2$, we can cancel the masses, then substitute $v_{1i} = +30.0$ m/s and $v_{2i} = -20.0$ cm/s (Step 3).

$$m_1v_{1i} + m_2v_{2i} = m_1v_{1f} + m_2v_{2f}$$

$$30.0 \text{ cm/s} + (-20.0 \text{ cm/s}) = v_{1f} + v_{2f}$$

$$10.0 \text{ cm/s} = v_{1f} + v_{2f} \quad \textbf{(1)}$$

Next, apply conservation of energy in the form of Equation 6.14 (Step 4):

$$v_{1i} - v_{2i} = -(v_{1f} - v_{2f})$$

$$30.0 \text{ cm/s} - (-20.0 \text{ cm/s}) = v_{2f} - v_{1f}$$

$$50.0 \text{ cm/s} = v_{2f} - v_{1f} \quad \textbf{(2)}$$

Now solve (1) and (2) simultaneously (Step 5): $v_{1f} = -20.0$ cm/s $\quad v_{2f} = +30.0$ cm/s

Remarks Notice the balls exchanged velocities—almost as if they'd passed through each other. This is always the case when two objects of equal mass undergo an elastic head-on collision.

Exercise 6.6

Find the final velocity of the two balls if the ball with initial velocity $v_{2i} = -20.0$ cm/s has a mass equal to one-half that of the ball with initial velocity $v_{1i} = +30.0$ cm/s.

Answer $v_{1f} = -3.33$ cm/s; $v_{2f} = +46.7$ cm/s

INTERACTIVE EXAMPLE 6.7 Two Blocks and a Spring

Goal Solve an elastic collision involving spring potential energy.

Problem A block of mass $m_1 = 1.60$ kg, initially moving to the right with a velocity of $+4.00$ m/s on a frictionless horizontal track, collides with a massless spring attached to a second block of mass $m_2 = 2.10$ kg moving to the left with a velocity of -2.50 m/s, as in Figure 6.14a. The spring has a spring constant of 6.00×10^2 N/m. **(a)** Determine the velocity of block 2 at the instant when block 1 is moving to the right with a velocity of $+3.00$ m/s, as in Figure 6.14b. **(b)** Find the compression of the spring.

$\vec{\mathbf{v}}_{1i} = +4.00$ m/s $\quad \vec{\mathbf{v}}_{2i} = -2.50$ m/s

(a)

$\vec{\mathbf{v}}_{1f} = +3.00$ m/s $\quad \vec{\mathbf{v}}_{2f}$

(b)

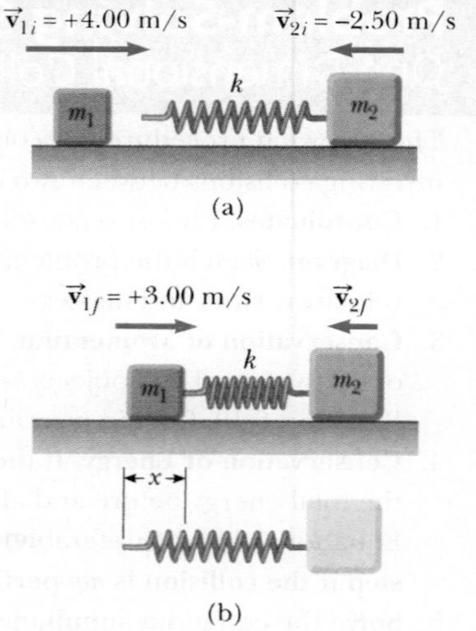

Figure 6.14 (Example 6.7)

Strategy We identify the system as the two blocks and the spring. Write down the conservation of momentum equations, and solve for the final velocity of block 2, v_{2f}. Then use conservation of energy to find the compression of the spring.

Solution

(a) Find the velocity v_{2f} when block 1 has velocity $+3.00$ m/s.

Write the conservation of momentum equation for the system and solve for v_{2f}:

$$m_1v_{1i} + m_2v_{2i} = m_1v_{1f} + m_2v_{2f}$$

$$v_{2f} = \frac{m_1v_{1i} + m_2v_{2i} - m_1v_{1f}}{m_2}$$

$$= \frac{(1.60 \text{ kg})(4.00 \text{ m/s}) + (2.10 \text{ kg})(-2.50 \text{ m/s}) - (1.60 \text{ kg})(3.00 \text{ m/s})}{2.10 \text{ kg}}$$

$$v_{2f} = -1.74 \text{ m/s}$$

(b) Find the compression of the spring.

Use energy conservation for the system, noticing that potential energy is stored in the spring when it is compressed a distance x:

$$E_i = E_f$$

$$\frac{1}{2}m_1{v_{1i}}^2 + \frac{1}{2}m_2{v_{2i}}^2 + 0 = \frac{1}{2}m_1{v_{1f}}^2 + \frac{1}{2}m_2{v_{2f}}^2 + \frac{1}{2}kx^2$$

Substitute the given values and the result of part (a) into the preceding expression, solving for x.

$$x = 0.173 \text{ m}$$

Remarks The initial velocity component of block 2 is -2.50 m/s because the block is moving to the left. The negative value for v_{2f} means that block 2 is still moving to the left at the instant under consideration.

Exercise 6.7
Find (a) the velocity of block 1 and (b) the compression of the spring at the instant that block 2 is at rest.

Answer (a) 0.719 m/s to the right (b) 0.251 m

Physics Now™ You can change the masses and speeds of the blocks and freeze the motion at the maximum compression of the spring by logging into PhysicsNow at **www.cp7e.com** and going to Interactive Example 6.7.

6.4 GLANCING COLLISIONS

In Section 6.2 we showed that the total linear momentum of a system is conserved when the system is isolated (that is, when no external forces act on the system). For a general collision of two objects in three-dimensional space, the conservation of momentum principle implies that the total momentum of the system in each direction is conserved. However, an important subset of collisions takes place in a plane. The game of billiards is a familiar example involving multiple collisions of objects moving on a two-dimensional surface. We restrict our attention to a single two-dimensional collision between two objects that takes place in a plane, and ignore any possible rotation. For such collisions, we obtain two component equations for the conservation of momentum:

$$m_1v_{1ix} + m_2v_{2ix} = m_1v_{1fx} + m_2v_{2fx}$$

$$m_1v_{1iy} + m_2v_{2iy} = m_1v_{1fy} + m_2v_{2fy}$$

We must use three subscripts in this general equation, to represent, respectively, (1) the object in question, and (2) the initial and final values of the components of velocity.

Now, consider a two-dimensional problem in which an object of mass m_1 collides with an object of mass m_2 that is initially at rest, as in Active Figure 6.15. After the collision, object 1 moves at an angle θ with respect to the horizontal, and object 2 moves at an angle ϕ with respect to the horizontal. This is called a *glancing* collision. Applying the law of conservation of momentum in component form, and noting that the initial y-component of momentum is zero, we have

x-component: $$m_1v_{1i} + 0 = m_1v_{1f}\cos\theta + m_2v_{2f}\cos\phi \qquad \textbf{[6.15]}$$

y-component: $$0 + 0 = m_1v_{1f}\sin\theta - m_2v_{2f}\sin\phi \qquad \textbf{[6.16]}$$

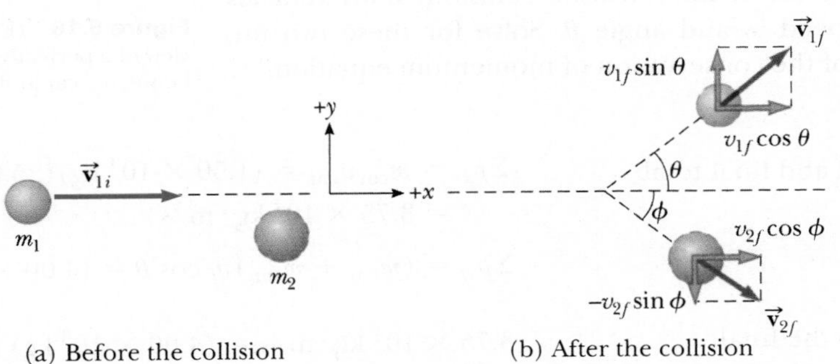

ACTIVE FIGURE 6.15
(a) Before and (b) after a glancing collision between two balls.

Physics Now™
Log into PhysicsNow at **www.cp7e.com**, and go to Active Figure 6.15 to adjust the speed and position of the blue particle, adjust the masses of both particles, and see the effects.

If the collision is elastic, we can write a third equation, for conservation of energy, in the form

$$\tfrac{1}{2}m_1v_{1i}^2 = \tfrac{1}{2}m_1v_{1f}^2 + \tfrac{1}{2}m_2v_{2f}^2 \qquad \textbf{[6.17]}$$

If we know the initial velocity v_{1i} and the masses, we are left with four unknowns (v_{1f}, v_{2f}, θ, and ϕ). Because we have only three equations, one of the four remaining quantities must be given in order to determine the motion after the collision from conservation principles alone.

If the collision is inelastic, the kinetic energy of the system is *not* conserved, and Equation 6.17 does *not* apply.

Problem-Solving Strategy
Two-Dimensional Collisions

To solve two-dimensional collisions, follow this procedure:

1. **Coordinate Axes.** Use both x- and y-coordinates. It's convenient to have either the x-axis or the y-axis coincide with the direction of one of the initial velocities.
2. **Diagram**. Sketch the problem, labeling velocity vectors and masses.
3. **Conservation of Momentum**. Write a separate conservation of momentum equation for each of the x- and y-directions. In each case, the total initial momentum in a given direction equals the total final momentum in that direction.
4. **Conservation of Energy**. If the collision is elastic, write a general expression for the total energy before and after the collision, and equate the two expressions, as in Equation 6.11. Fill in the known values. **(*Skip* this step if the collision is *not* perfectly elastic.)** The energy equation can't be simplified as in the one-dimensional case, so a quadratic expression such as Equation 6.11 or 6.17 must be used when the collision is elastic.
5. **Solve** the equations simultaneously. There are two equations for inelastic collisions and three for elastic collisions.

EXAMPLE 6.8 Collision at an Intersection

Goal Analyze a two-dimensional inelastic collision.

Problem A car with mass 1.50×10^3 kg traveling east at a speed of 25.0 m/s collides at an intersection with a 2.50×10^3-kg van traveling north at a speed of 20.0 m/s, as shown in Figure 6.16. Find the magnitude and direction of the velocity of the wreckage after the collision, assuming that the vehicles undergo a perfectly inelastic collision (that is, they stick together) and assuming that friction between the vehicles and the road can be neglected.

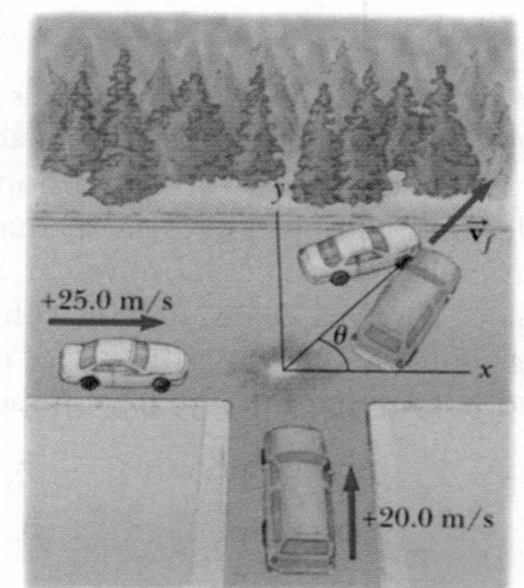

Figure 6.16 (Example 6.8) A top view of a perfectly inelastic collision between a car and a van.

Strategy Use conservation of momentum in two dimensions. (Kinetic energy is *not* conserved.) Choose coordinates as in Figure 6.16. Before the collision, the only object having momentum in the x-direction is the car, while the van carries all the momentum in the y-direction. After the totally inelastic collision, both vehicles move together at some common speed v_f and angle θ. Solve for these two unknowns, using the two components of the conservation of momentum equation.

Solution

Find the x-components of the initial and final total momenta:

$$\Sigma p_{xi} = m_{\text{car}}v_{\text{car}} = (1.50 \times 10^3 \text{ kg})(25.0 \text{ m/s})$$
$$= 3.75 \times 10^4 \text{ kg} \cdot \text{m/s}$$

$$\Sigma p_{xf} = (m_{\text{car}} + m_{\text{van}})v_f \cos\theta = (4.00 \times 10^3 \text{ kg})v_f \cos\theta$$

Set the initial x-momentum equal to the final x-momentum:

$$3.75 \times 10^4 \text{ kg} \cdot \text{m/s} = (4.00 \times 10^3 \text{ kg})\, v_f \cos\theta \qquad \textbf{(1)}$$

Find the y-components of the initial and final total momenta:

$$\Sigma p_{iy} = m_{\text{van}} v_{\text{van}} = (2.50 \times 10^3 \text{ kg})(20.0 \text{ m/s})$$
$$= 5.00 \times 10^4 \text{ kg} \cdot \text{m/s}$$
$$\Sigma p_{fy} = (m_{\text{car}} + m_{\text{van}}) v_f \sin \theta = (4.00 \times 10^3 \text{ kg}) v_f \sin \theta$$

Set the initial y-momentum equal to the final y-momentum:

$$5.00 \times 10^4 \text{ kg} \cdot \text{m/s} = (4.00 \times 10^3 \text{ kg}) v_f \sin \theta \qquad (2)$$

Divide Equation (2) by Equation (1) and solve for θ:

$$\tan \theta = \frac{5.00 \times 10^4 \text{ kg} \cdot \text{m/s}}{3.75 \times 10^4 \text{ kg} \cdot \text{m}} = 1.33$$

$$\theta = \boxed{53.1^\circ}$$

Substitute this angle back into Equation (2) to find v_f:

$$v_f = \frac{5.00 \times 10^4 \text{ kg} \cdot \text{m/s}}{(4.00 \times 10^3 \text{ kg}) \sin 53.1^\circ} = \boxed{15.6 \text{ m/s}}$$

Remark It's also possible to first find the x- and y-components v_{fx} and v_{fy} of the resultant velocity. The magnitude and direction of the resultant velocity can then be found with the Pythagorean theorem, $v_f = \sqrt{v_{fx}^2 + v_{fy}^2}$, and the inverse tangent function $\theta = \tan^{-1}(v_{fy}/v_{fx})$. Setting up this alternate approach is a simple matter of substituting $v_{fx} = v_f \cos \theta$ and $v_{fy} = v_f \sin \theta$ in Equations (1) and (2).

Exercise 6.8

A 3.00-kg object initially moving in the positive x-direction with a velocity of $+5.00$ m/s collides with and sticks to a 2.00-kg object initially moving in the negative y-direction with a velocity of -3.00 m/s. Find the final components of velocity of the composite object.

Answer $v_{fx} = 3.00$ m/s; $v_{fy} = -1.20$ m/s

6.5 ROCKET PROPULSION

When ordinary vehicles such as cars and locomotives move, the driving force of the motion is friction. In the case of the car, this driving force is exerted by the road on the car, a reaction to the force exerted by the wheels against the road. Similarly, a locomotive "pushes" against the tracks; hence, the driving force is the reaction force exerted by the tracks on the locomotive. However, a rocket moving in space has no road or tracks to push against. How can it move forward?

In fact, reaction forces also propel a rocket. (You should review Newton's third law, discussed in Chapter 4.) To illustrate this point, we model our rocket with a spherical chamber containing a combustible gas, as in Figure 6.17a. When an explosion occurs in the chamber, the hot gas expands and presses against all sides of the chamber, as indicated by the arrows. Because the sum of the forces exerted on the rocket is zero, it doesn't move. Now suppose a hole is drilled in the bottom of the chamber, as in Figure 6.17b. When the explosion occurs, the gas presses against the chamber in all directions, but can't press against anything at the hole, where it simply escapes into space. Adding the forces on the spherical chamber now results in a net force upwards. Just as in the case of cars and locomotives, this is a reaction force. A car's wheels press against the ground, and the reaction force of the ground on the car pushes it forward. The wall of the rocket's combustion chamber exerts a force on the gas expanding against it. The reaction force of the gas on the wall then pushes the rocket upward.

In a now infamous article in *The New York Times*, rocket pioneer Robert Goddard was ridiculed for thinking that rockets would work in space, where, according to the *Times*, there was nothing to push against. The *Times* retracted, rather belatedly, during the first Apollo moon landing mission in 1969. The hot gases are not pushing against anything external, but against the rocket itself—and ironically, rockets actually work *better* in a vacuum. In an atmosphere, the gases have to do work against the outside air pressure to escape the combustion chamber, slowing the exhaust velocity and reducing the reaction force.

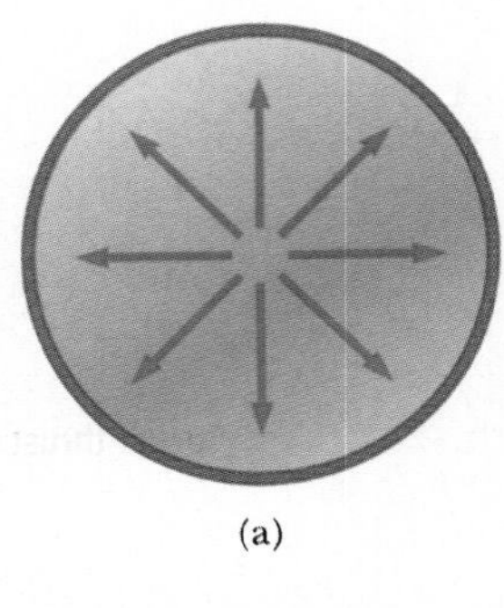

(a)

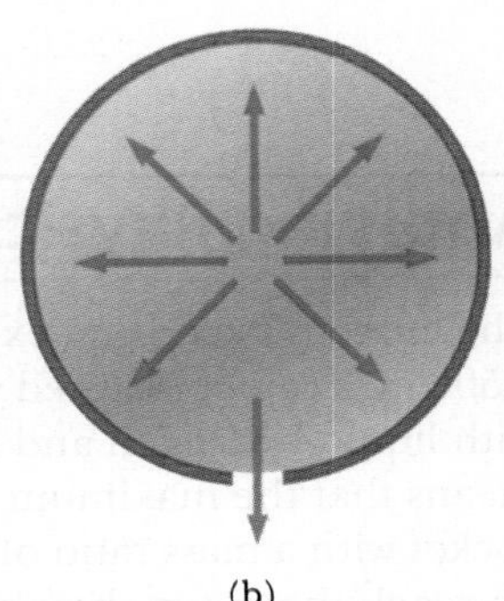

(b)

Figure 6.17 (a) A rocket reaction chamber without a nozzle has reaction forces pushing equally in all directions, so no motion results. (b) An opening at the bottom of the chamber removes the downward reaction force, resulting in a net upward reaction force.

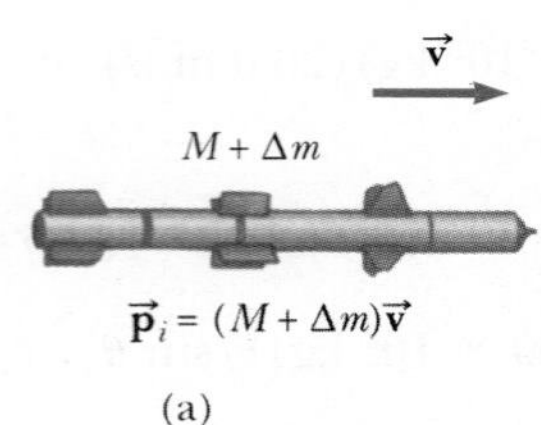

(a)

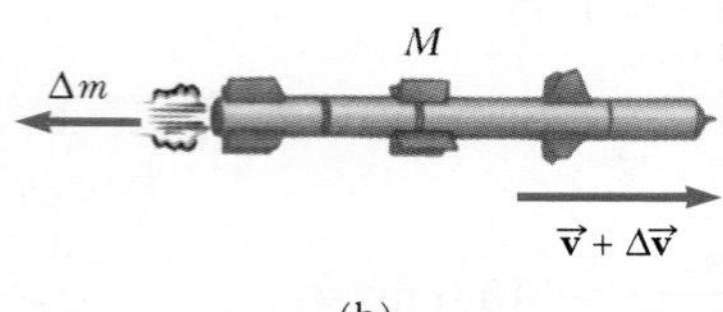

(b)

Figure 6.18 Rocket propulsion. (a) The initial mass of the rocket and fuel is $M + \Delta m$ at a time t, and the rocket's speed is v. (b) At a time $t + \Delta t$, the rocket's mass has been reduced to M, and an amount of fuel Δm has been ejected. The rocket's speed increases by an amount Δv.

At the microscopic level, this process is complicated, but it can be simplified by applying conservation of momentum to the rocket and its ejected fuel. In principle, the solution is similar to that in Example 6.3, with the archer representing the rocket and the arrows the exhaust gases.

Suppose that at some time t, the momentum of the rocket plus the fuel is $(M + \Delta m)v$, where Δm is an amount of fuel about to be burned (Fig. 6.18a). This fuel is traveling at a speed v relative to, say, the Earth, just like the rest of the rocket. During a short time interval Δt, the rocket ejects fuel of mass Δm, and the rocket's speed increases to $v + \Delta v$ (Fig. 6.18b). If the fuel is ejected with exhaust speed v_e *relative to the rocket*, the speed of the fuel relative to the Earth is $v - v_e$. Equating the total initial momentum of the system with the total final momentum, we have

$$(M + \Delta m)v = M(v + \Delta v) + \Delta m(v - v_e)$$

Simplifying this expression gives

$$M\Delta v = v_e \Delta m$$

The increase Δm in the mass of the exhaust corresponds to an equal decrease in the mass of the rocket, so that $\Delta m = -\Delta M$. Using this fact, we have

$$M\Delta v = -v_e \Delta M \qquad [6.18]$$

This result, together with the methods of calculus, can be used to obtain the following equation:

$$v_f - v_i = v_e \ln\left(\frac{M_i}{M_f}\right) \qquad [6.19]$$

where M_i is the initial mass of the rocket plus fuel and M_f is the final mass of the rocket plus its remaining fuel. This is the basic expression for rocket propulsion; it tells us that the increase in velocity is proportional to the exhaust speed v_e and to the natural logarithm of M_i/M_f. Because the maximum ratio of M_i to M_f for a single-stage rocket is about 10:1, the increase in speed can reach $v_e \ln 10 = 2.3v_e$ or about twice the exhaust speed! For best results, therefore, the exhaust speed should be as high as possible. Currently, typical rocket exhaust speeds are several kilometers per second.

The **thrust** on the rocket is defined as the force exerted on the rocket by the ejected exhaust gases. We can obtain an expression for the instantaneous thrust by dividing Equation 6.18 by Δt:

Rocket thrust ►

$$\text{Instantaneous thrust} = Ma = M\frac{\Delta v}{\Delta t} = \left|v_e \frac{\Delta M}{\Delta t}\right| \qquad [6.20]$$

The absolute value signs are used for clarity: In Equation 6.18, $-\Delta M$ is a positive quantity (as is v_e, a speed). Here we see that the thrust increases as the exhaust velocity increases and as the rate of change of mass $\Delta M/\Delta t$ (the burn rate) increases.

Applying Physics 6.2 Multistage Rockets

The current maximum exhaust speed of $v_e =$ 4 500 m/s can be realized with rocket engines fueled with liquid hydrogen and liquid oxygen. But this means that the maximum speed attainable for a given rocket with a mass ratio of 10 is $v_e \ln 10 \approx 10\ 000$ m/s. To reach the moon, however, requires a change in velocity of over 11 000 m/s. Further, this change must occur while working against gravity and atmospheric friction. How can that be managed without developing better engines?

Explanation The answer is the multistage rocket. By dropping stages, the spacecraft becomes lighter, so that fuel burned later in the mission doesn't have to accelerate mass that no longer serves any purpose. Strap-on boosters, as used by the Space Shuttle and a number of other rockets, such as the Titan 4 or Russian Proton, is a similar concept. The boosters are jettisoned after their fuel is exhausted, so the rocket is no longer burdened by their weight.

EXAMPLE 6.9 Single Stage to Orbit (SSTO)

Goal Apply the velocity and thrust equations of a rocket.

Problem A rocket has a total mass of 1.00×10^5 kg and a burnout mass of 1.00×10^4 kg, including engines, shell, and payload. The rocket blasts off from Earth and exhausts all its fuel in 4.00 min, burning the fuel at a steady rate with an exhaust velocity of $v_e = 4.50 \times 10^3$ m/s. **(a)** If air friction and gravity are neglected, what is the speed of the rocket at burnout? **(b)** What thrust does the engine develop at liftoff? **(c)** What is the initial acceleration of the rocket if gravity is not neglected? **(d)** Estimate the speed at burnout if gravity isn't neglected.

Strategy Although it sounds sophisticated, this problem is mainly a matter of substituting values into the appropriate equations. Part (a) requires substituting values into Equation 6.19 for the velocity. For part (b), divide the change in the rocket's mass by the total time, getting $\Delta M/\Delta t$, then substitute into Equation 6.20 to find the thrust. (c) Using Newton's second law, the force of gravity, and the result of (b), we can find the initial acceleration. For part (d), the acceleration of gravity is approximately constant over the few kilometers involved, so the velocity found in (b) will be reduced by roughly $\Delta v_g = -gt$. Add this loss to the result of part (a).

Solution

(a) Calculate the velocity at burnout.

Substitute $v_i = 0$, $v_e = 4.50 \times 10^3$ m/s, $M_i = 1.00 \times 10^5$ kg, and $M_f = 1.00 \times 10^4$ kg into Equation 6.19:

$$v_f = v_i + v_e \ln\left(\frac{M_i}{M_f}\right)$$

$$= 0 + (4.5 \times 10^3 \text{ m/s}) \ln\left(\frac{1.00 \times 10^5 \text{ kg}}{1.00 \times 10^4 \text{ kg}}\right)$$

$$v_f = \boxed{1.04 \times 10^4 \text{ m/s}}$$

(b) Find the thrust at liftoff.

Compute the change in the rocket's mass:

$$\Delta M = M_f - M_i = 1.00 \times 10^4 \text{ kg} - 1.00 \times 10^5 \text{ kg}$$

$$= -9.00 \times 10^4 \text{ kg}$$

Calculate the rate at which rocket mass changes by dividing the change in mass by the time (4.00 min, converted to seconds):

$$\frac{\Delta M}{\Delta t} = \frac{-9.00 \times 10^4 \text{ kg}}{2.40 \times 10^2 \text{ s}} = -3.75 \times 10^2 \text{ kg/s}$$

Substitute this rate into Equation 6.20, obtaining the thrust:

$$\text{Thrust} = \left|v_e \frac{\Delta M}{\Delta t}\right| = (4.50 \times 10^3 \text{ m/s})(3.75 \times 10^2 \text{ kg/s})$$

$$= \boxed{1.69 \times 10^6 \text{ N}}$$

(c) Find the initial acceleration.

Write Newton's second law, where T stands for thrust, and solve for the acceleration a:

$$Ma = \Sigma F = T - Mg$$

$$a = \frac{T}{M} - g = \frac{1.69 \times 10^6 \text{ N}}{1.00 \times 10^5 \text{ kg}} - 9.80 \text{ m/s}^2$$

$$= \boxed{7.10 \text{ m/s}^2}$$

(d) Estimate the speed at burnout when gravity is not neglected.

Find the approximate loss of speed due to gravity:

$$\Delta v_g = -g\Delta t = -(9.80 \text{ m/s}^2)(2.40 \times 10^2 \text{ s})$$

$$= -2.35 \times 10^3 \text{ m/s}$$

Add this loss to the result of part (b):

$$v_f = 1.04 \times 10^4 \text{ m/s} - 2.35 \times 10^3 \text{ m/s}$$

$$= \boxed{8.05 \times 10^3 \text{ m/s}}$$

Remarks Even taking gravity into account, the speed is sufficient to attain orbit. Some additional boost may be required to overcome air drag.

Exercise 6.9

A spaceship with a mass of 5.00×10^4 kg is traveling at 6.00×10^3 m/s relative a space station. What mass will the ship have after it fires its engines in order to reach a relative speed of 8.00×10^3 m/s, traveling the same direction? Assume an exhaust velocity of 4.50×10^3 m/s.

Answer 3.21×10^4 kg

SUMMARY

Physics Now™ Take a practice test by logging into PhysicsNow at **www.cp7e.com** and clicking on the Pre-Test link for this chapter.

6.1 Momentum and Impulse

The **linear momentum** $\vec{\mathbf{p}}$ of an object of mass m moving with velocity $\vec{\mathbf{v}}$ is defined as

$$\vec{\mathbf{p}} \equiv m\vec{\mathbf{v}} \qquad [6.1]$$

Momentum carries units of kg · m/s. The **impulse** $\vec{\mathbf{I}}$ of a constant force $\vec{\mathbf{F}}$ delivered to an object is equal to the product of the force and the time interval during which the force acts:

$$\vec{\mathbf{I}} \equiv \vec{\mathbf{F}}\Delta t \qquad [6.4]$$

These two concepts are unifed in the **impulse–momentum theorem**, which states that the impulse of a constant force delivered to an object is equal to the change in momentum of the object:

$$\vec{\mathbf{F}}\Delta t = \Delta\vec{\mathbf{p}} \equiv m\vec{\mathbf{v}}_f - m\vec{\mathbf{v}}_i \qquad [6.5]$$

Solving problems with this theorem often involves estimating speeds or contact times (or both), leading to an average force.

6.2 Conservation of Momentum

When no net external force acts on an isolated system, the total momentum of the system is constant. This principle is called **conservation of momentum**. In particular, if the isolated system consists of two objects undergoing a collision, the total momentum of the system is the same before and after the collision. Conservation of momentum can be written mathematically for this case as

$$m_1\vec{\mathbf{v}}_{1i} + m_2\vec{\mathbf{v}}_{2i} = m_1\vec{\mathbf{v}}_{1f} + m_2\vec{\mathbf{v}}_{2f} \qquad [6.7]$$

Collision and recoil problems typically require finding unknown velocities in one or two dimensions. Each vector component gives an equation, and the resulting equations are solved simultaneously.

6.3 Collisions

In an **inelastic collision**, the momentum of the system is conserved, but kinetic energy is not. In a **perfectly inelastic collision**, the colliding objects stick together. In an **elastic collision**, both the momentum and the kinetic energy of the system are conserved.

A one-dimensional **elastic collision** between two objects can be solved by using the conservation of momentum and conservation of energy equations:

$$m_1v_{1i} + m_2v_{2i} = m_1v_{1f} + m_2v_{2f} \qquad [6.10]$$

$$\tfrac{1}{2}m_1v_{1i}^2 + \tfrac{1}{2}m_2v_{2i}^2 = \tfrac{1}{2}m_1v_{1f}^2 + \tfrac{1}{2}m_2v_{2f}^2 \qquad [6.11]$$

The following equation, derived from Equations 6.10 and 6.11, is usually more convenient to use than the original conservation of energy equation:

$$v_{1i} - v_{2i} = -(v_{1f} - v_{2f}) \qquad [6.14]$$

These equations can be solved simultaneously for the unknown velocities. Energy is not conserved in **inelastic collisions**, so such problems must be solved with Equation 6.10 alone.

6.4 Glancing Collisions

In glancing collisions, conservation of momentum can be applied along two perpendicular directions: an x-axis and a y-axis. Problems can be solved by using the x- and y-components of Equation 6.7. Elastic two-dimensional collisions will usually require Equation 6.11 as well. (Equation 6.14 doesn't apply to two dimensions.) Generally, one of the two objects is taken to be traveling along the x-axis, undergoing a deflection at some angle θ after the collision. The final velocities and angles can be found with elementary trigonometry.

CONCEPTUAL QUESTIONS

1. A batter bunts a pitched baseball, blocking the ball without swinging. (a) Can the baseball deliver more kinetic energy to the bat and batter than the ball carries initially? (b) Can the baseball deliver more momentum to the bat and batter than the ball carries initially? Explain each of your answers.
2. America will never forget the terrorist attack on September 11, 2001. One commentator remarked that the force of the explosion at the Twin Towers of the World Trade Center was strong enough to blow glass and parts of the steel structure to small fragments. Yet the television coverage showed thousands of sheets of paper floating down, many still intact. Explain how that could be.
3. In perfectly inelastic collisions between two objects, there are events in which all of the original kinetic energy is transformed to forms other than kinetic. Give an example of such an event.
4. If two objects collide and one is initially at rest, is it possible for both to be at rest after the collision? Is it possible for only one to be at rest after the collision? Explain.

5. A ball of clay of mass m is thrown with a speed v against a brick wall. The clay sticks to the wall and stops. Is the principle of conservation of momentum violated in this example?

6. A skater is standing still on a frictionless ice rink. Her friend throws a Frisbee straight at her. In which of the following cases is the largest momentum transferred to the skater? (a) The skater catches the Frisbee and holds onto it. (b) The skater catches the Frisbee momentarily, but then drops it vertically downward. (c) The skater catches the Frisbee, holds it momentarily, and throws it back to her friend.

7. You are standing perfectly still and then you take a step forward. Before the step your momentum was zero, but afterwards you have some momentum. Is the conservation of momentum violated in this case?

8. If two particles have equal kinetic energies, are their momenta necessarily equal? Explain.

9. A more ordinary example of conservation of momentum than a rocket ship occurs in a kitchen dishwashing machine. In this device, water at high pressure is forced out of small holes on the spray arms. Use conservation of momentum to explain why the arms rotate, directing water to all the dishes.

10. If two automobiles collide, they usually do not stick together. Does this mean the collision is elastic? Explain why a head-on collision is likely to be more dangerous than other types of collisions.

11. An open box slides across a frictionless, icy surface of a frozen lake. What happens to the speed of the box as water from a rain shower collect in it, assuming that the rain falls vertically downward into the box? Explain.

12. Consider a perfectly inelastic collision between a car and a large truck. Which vehicle loses more kinetic energy as a result of the collision?

13. Your physical education teacher throws you a tennis ball at a certain velocity, and you catch it. You are now given the following choice: The teacher can throw you a medicine ball (which is much more massive than the tennis ball) with the same velocity, the same momentum, or the same kinetic energy as the tennis ball. Which option would you choose in order to make the easiest catch, and why?

14. While watching a movie about a superhero, you notice that the superhero hovers in the air and throws a piano at some bad guys while remaining stationary in the air. What's wrong with this scenario?

15. In golf, novice players are often advised to be sure to "follow through" with their swing. Why does this make the ball travel a longer distance? If a shot is taken near the green, very little follow-through is required. Why?

16. An air bag inflates when a collision occurs, protecting a passenger (the dummy in Figure Q6.16) from serious injury. Why does the air bag soften the blow? Discuss the physics involved in this dramatic photograph.

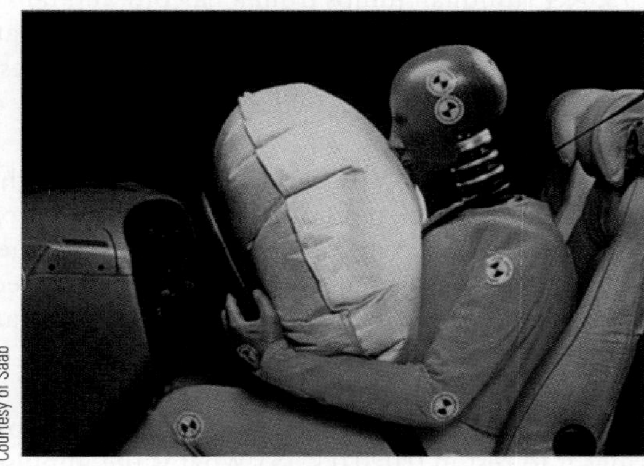
Courtesy of Saab

Figure Q6.16

17. A sharpshooter fires a rifle while standing with the butt of the gun against his shoulder. If the forward momentum of a bullet is the same as the backward momentum of the gun, why isn't it as dangerous to be hit by the gun as by the bullet?

18. A large bedsheet is held vertically by two students. A third student, who happens to be the star pitcher on the baseball team, throws a raw egg at the sheet. Explain why the egg doesn't break when it hits the sheet, regardless of its initial speed. (If you try this, make sure the pitcher hits the sheet near its center, and don't allow the egg to fall on the floor after being caught.)

PROBLEMS

1, 2, 3 = straightforward, intermediate, challenging ☐ = full solution available in *Student Solutions Manual/Study Guide*

Physics Now™ = coached problem with hints available at **www.cp7e.com** = biomedical application

Section 6.1 Momentum and Impulse

1. A ball of mass 0.150 kg is dropped from rest from a height of 1.25 m. It rebounds from the floor to reach a height of 0.960 m. What impulse was given to the ball by the floor?

2. A tennis player receives a shot with the ball (0.060 0 kg) traveling horizontally at 50.0 m/s and returns the shot with the ball traveling horizontally at 40.0 m/s in the opposite direction. (a) What is the impulse delivered to the ball by the racquet? (b) What work does the racquet do on the ball?

3. Calculate the magnitude of the linear momentum for the following cases: (a) a proton with mass 1.67×10^{-27} kg, moving with a speed of 5.00×10^{6} m/s; (b) a 15.0-g bullet moving with a speed of 300 m/s; (c) a 75.0-kg sprinter running with a speed of 10.0 m/s; (d) the Earth (mass $= 5.98 \times 10^{24}$ kg) moving with an orbital speed equal to 2.98×10^{4} m/s.

4. A 0.10-kg ball is thrown straight up into the air with an initial speed of 15 m/s. Find the momentum of the ball (a) at its maximum height and (b) halfway to its maximum height.

5. A pitcher claims he can throw a 0.145-kg baseball with as much momentum as a 3.00-g bullet moving with a speed of 1.50×10^{3} m/s. (a) What must the baseball's speed be

if the pitcher's claim is valid? (b) Which has greater kinetic energy, the ball or the bullet?

6. A stroboscopic photo of a club hitting a golf ball, such as the photo shown in Figure 6.3, was made by Harold Edgerton in 1933. The ball was initially at rest, and the club was shown to be in contact with the club for about 0.002 0 s. Also, the ball was found to end up with a speed of 2.0×10^2 ft/s. Assuming that the golf ball had a mass of 55 g, find the average force exerted by the club on the ball.

7. A professional diver performs a dive from a platform 10 m above the water surface. Estimate the order of magnitude of the average impact force she experiences in her collision with the water. State the quantities you take as data and their values.

8. A 75.0-kg stuntman jumps from a balcony and falls 25.0 m before colliding with a pile of mattresses. If the mattresses are compressed 1.00 m before he is brought to rest, what is the average force exerted by the mattresses on the stuntman?

9. A car is stopped for a traffic signal. When the light turns green, the car accelerates, increasing its speed from 0 to 5.20 m/s in 0.832 s. What are the magnitudes of the linear impulse and the average total force experienced by a 70.0-kg passenger in the car during the time the car accelerates?

10. A 0.500-kg football is thrown toward the east with a speed of 15.0 m/s. A stationary receiver catches the ball and brings it to rest in 0.020 0 s. (a) What is the impulse delivered to the ball as it's caught? (b) What is the average force exerted on the receiver?

11. The force shown in the force vs. time diagram in Figure P6.11 acts on a 1.5-kg object. Find (a) the impulse of the force, (b) the final velocity of the object if it is initially at rest, and (c) the final velocity of the object if it is initially moving along the x-axis with a velocity of -2.0 m/s.

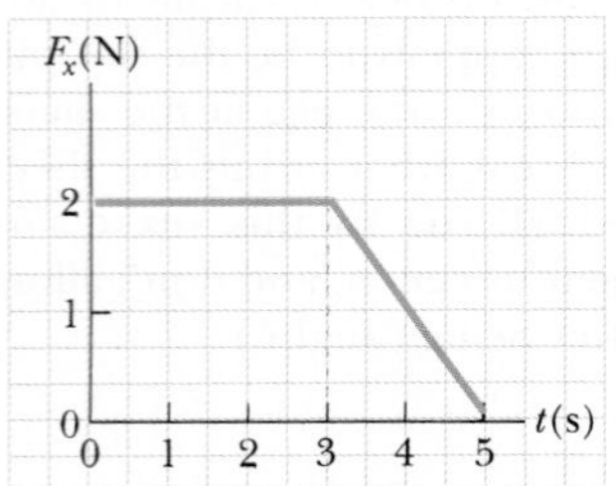

Figure P6.11

12. A force of magnitude F_x acting in the x-direction on a 2.00-kg particle varies in time as shown in Figure P6.12. Find (a) the impulse of the force, (b) the final velocity of the particle if it is initially at rest, and (c) the final velocity of the particle if it is initially moving along the x-axis with a velocity of -2.00 m/s.

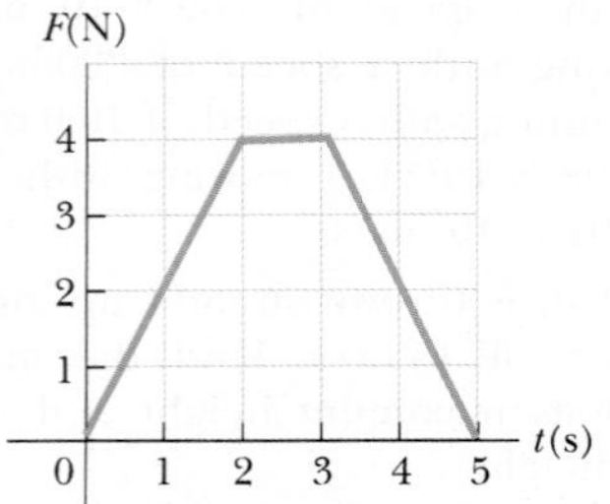

Figure P6.12

13. The forces shown in the force vs. time diagram in Figure P6.13 act on a 1.5-kg particle. Find (a) the impulse for the interval from $t = 0$ to $t = 3.0$ s and (b) the impulse for the interval from $t = 0$ to $t = 5.0$ s. (c) If the forces act on a 1.5-kg particle that is initially at rest, find the particle's speed at $t = 3.0$ s and at $t = 5.0$ s.

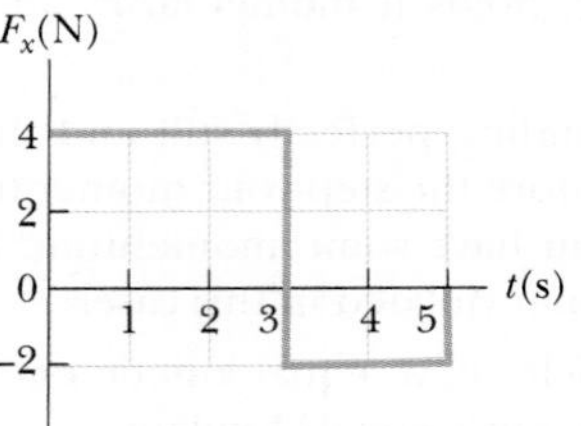

Figure P6.13

14. A 3.00-kg steel ball strikes a massive wall at 10.0 m/s at an angle of 60.0° with the plane of the wall. It bounces off the wall with the same speed and angle (Fig. P6.14). If the ball is in contact with the wall for 0.200 s, what is the average force exerted by the wall on the ball?

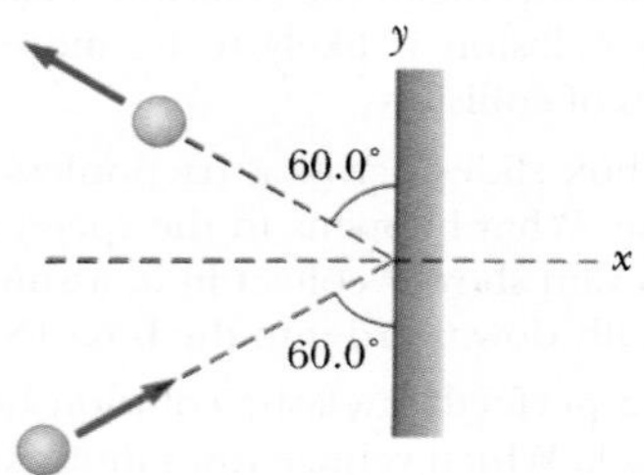

Figure P6.14

15. The front 1.20 m of a 1 400-kg car is designed as a "crumple zone" that collapses to absorb the shock of a collision. If a car traveling 25.0 m/s stops uniformly in 1.20 m, (a) how long does the collision last, (b) what is the magnitude of the average force on the car, and (c) what is the acceleration of the car? Express the acceleration as a multiple of the acceleration of gravity.

16. **PhysicsNow™** A pitcher throws a 0.15-kg baseball so that it crosses home plate horizontally with a speed of 20 m/s. The ball is hit straight back at the pitcher with a final speed of 22 m/s. (a) What is the impulse delivered to the ball? (b) Find the average force exerted by the bat on the ball if the two are in contact for 2.0×10^{-3} s.

17. A car of mass 1.6×10^3 kg is traveling east at a speed of 25 m/s along a horizontal roadway. When its brakes are applied, the car stops in 6.0 s. What is the average horizontal force exerted on the car while it is braking?

Section 6.2 Conservation of Momentum

18. A 730-N man stands in the middle of a frozen pond of radius 5.0 m. He is unable to get to the other side because of a lack of friction between his shoes and the ice. To overcome this difficulty, he throws his 1.2-kg physics textbook horizontally toward the north shore at a speed of 5.0 m/s. How long does it take him to reach the south shore?

19. High-speed stroboscopic photographs show that the head of a 200-g golf club is traveling at 55 m/s just before it strikes a 46-g golf ball at rest on a tee. After the collision, the club head travels (in the same direction) at 40 m/s. Find the speed of the golf ball just after impact.

20. A rifle with a weight of 30 N fires a 5.0-g bullet with a speed of 300 m/s. (a) Find the recoil speed of the rifle. (b) If a 700-N man holds the rifle firmly against his shoulder, find the recoil speed of the man and rifle.

21. A 45.0-kg girl is standing on a 150-kg plank. The plank, originally at rest, is free to slide on a frozen lake, which is a flat, frictionless surface. The girl begins to walk along the plank at a constant velocity of 1.50 m/s to the right relative to the plank. (a) What is her velocity relative to the surface of the ice? (b) What is the velocity of the plank relative to the surface of the ice?

22. A 65.0-kg person throws a 0.045 0-kg snowball forward with a ground speed of 30.0 m/s. A second person, with a mass of 60.0 kg, catches the snowball. Both people are on skates. The first person is initially moving forward with a speed of 2.50 m/s, and the second person is initially at rest. What are the velocities of the two people after the snowball is exchanged? Disregard friction between the skates and the ice.

23. In Section 6.2, we implied that the kinetic energy of the Earth can be ignored when considering the energy of a system consisting of the Earth and a dropped ball of mass m_b. Verify this statement by first setting up a ratio of the kinetic energy of the Earth to that of the ball as they collide. Then use conservation of momentum to show that

$$\frac{v_E}{v_b} = -\frac{m_b}{m_E} \quad \text{and} \quad \frac{KE_E}{KE_b} = \frac{m_b}{m_E}$$

Find the order of magnitude of the ratio of the kinetic energies, based on data that you specify.

24. Two ice skaters are holding hands at the center of a frozen pond when an argument ensues. Skater A shoves skater B along a horizontal direction. Identify (a) the horizontal forces acting on A and (b) those acting on B. (c) Which force is greater, the force on A or the force on B? (d) Can conservation of momentum be used for the system of A and B? Defend your answer. (e) If A has a mass of 0.900 times that of B, and B begins to move away with a speed of 2.00 m/s, find the speed of A.

Section 6.3 Collisions
Section 6.4 Glancing Collisions

25. An archer shoots an arrow toward a 300-g target that is sliding in her direction at a speed of 2.50 m/s on a smooth, slippery surface. The 22.5-g arrow is shot with a speed of 35.0 m/s and passes through the target, which is stopped by the impact. What is the speed of the arrow after passing through the target?

26. A 75.0-kg ice skater moving at 10.0 m/s crashes into a stationary skater of equal mass. After the collision, the two skaters move as a unit at 5.00 m/s. Suppose the average force a skater can experience without breaking a bone is 4 500 N. If the impact time is 0.100 s, does a bone break?

27. A railroad car of mass 2.00×10^4 kg moving at 3.00 m/s collides and couples with two coupled railroad cars, each of the same mass as the single car and moving in the same direction at 1.20 m/s. (a) What is the speed of the three coupled cars after the collision? (b) How much kinetic energy is lost in the collision?

28. A 7.0-g bullet is fired into a 1.5-kg ballistic pendulum. The bullet emerges from the block with a speed of 200 m/s, and the block rises to a maximum height of 12 cm. Find the initial speed of the bullet.

29. **PhysicsNow™** A 0.030-kg bullet is fired vertically at 200 m/s into a 0.15-kg baseball that is initially at rest. How high does the combined bullet and baseball rise after the collision, assuming the bullet embeds itself in the ball?

30. An 8.00-g bullet is fired into a 250-g block that is initially at rest at the edge of a table of height 1.00 m (Fig. P6.30). The bullet remains in the block, and after the impact the block lands 2.00 m from the bottom of the table. Determine the initial speed of the bullet.

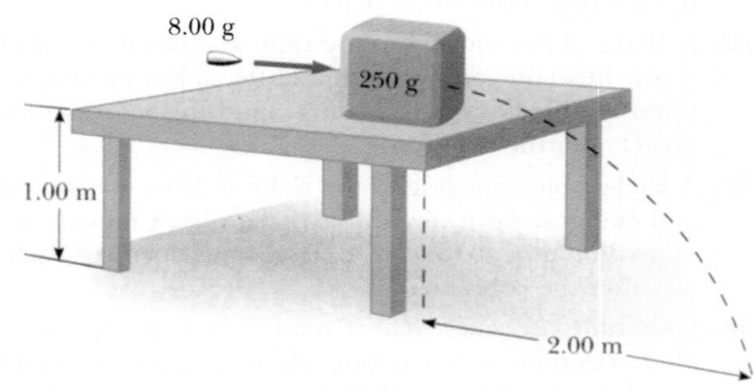

Figure P6.30

31. Gayle runs at a speed of 4.00 m/s and dives on a sled, initially at rest on the top of a frictionless, snow-covered hill. After she has descended a vertical distance of 5.00 m, her brother, who is initially at rest, hops on her back, and they continue down the hill together. What is their speed at the bottom of the hill if the total vertical drop is 15.0 m? Gayle's mass is 50.0 kg, the sled has a mass of 5.00 kg, and her brother has a mass of 30.0 kg.

32. A 1 200-kg car traveling initially with a speed of 25.0 m/s in an easterly direction crashes into the rear end of a 9 000-kg truck moving in the same direction at 20.0 m/s (Fig. P6.32). The velocity of the car right after the collision is 18.0 m/s to the east. (a) What is the velocity of the truck right after the collision? (b) How much mechanical energy is lost in the collision? Account for this loss in energy.

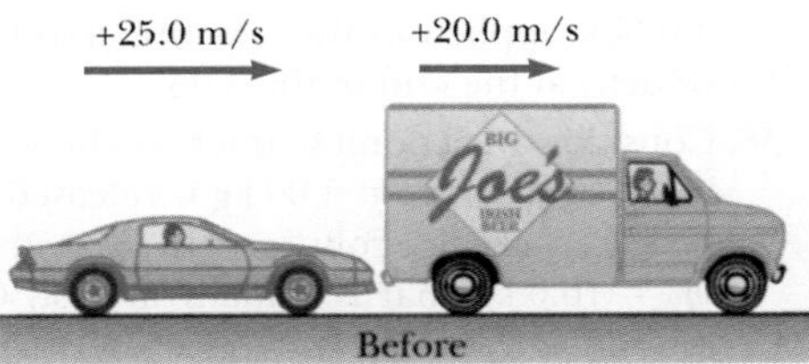

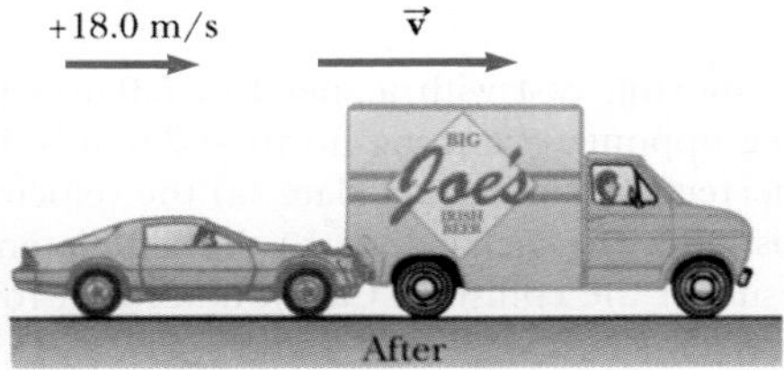

Figure P6.32

33. A 12.0-g bullet is fired horizontally into a 100-g wooden block that is initially at rest on a frictionless horizontal surface and connected to a spring having spring constant 150 N/m. The bullet becomes embedded in the block. If the bullet–block system compresses the spring by a maximum of 80.0 cm, what was the speed of the bullet at impact with the block?

34. (a) Three carts of masses 4.0 kg, 10 kg, and 3.0 kg move on a frictionless horizontal track with speeds of 5.0 m/s, 3.0 m/s, and 4.0 m/s, as shown in Figure P6.34. The carts stick together after colliding. Find the final velocity of the three carts. (b) Does your answer require that all carts collide and stick together at the same time?

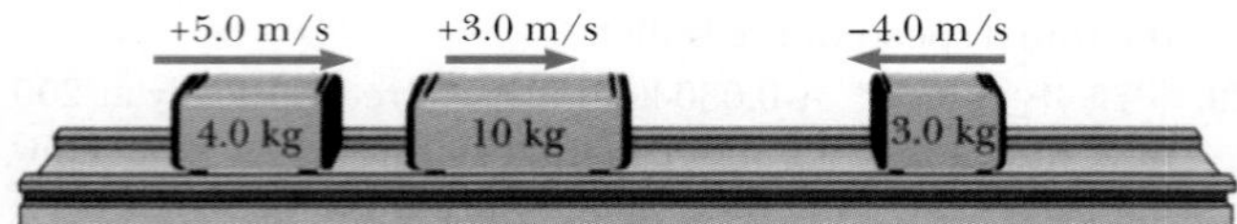

Figure P6.34

35. A 5.00-g object moving to the right at 20.0 cm/s makes an elastic head-on collision with a 10.0-g object that is initially at rest. Find (a) the velocity of each object after the collision and (b) the fraction of the initial kinetic energy transferred to the 10.0-g object.

36. A 10.0-g object moving to the right at 20.0 cm/s makes an elastic head-on collision with a 15.0-g object moving in the opposite direction at 30.0 cm/s. Find the velocity of each object after the collision.

37. A 25.0-g object moving to the right at 20.0 cm/s overtakes and collides elastically with a 10.0-g object moving in the same direction at 15.0 cm/s. Find the velocity of each object after the collision.

38. Four railroad cars, each of mass 2.50×10^4 kg, are coupled together and coasting along horizontal tracks at speed v_i toward the south. A very strong but foolish movie actor riding on the second car uncouples the front car and gives it a big push, increasing its speed to 4.00 m/s south. The remaining three cars continue moving south, now at 2.00 m/s. (a) Find the initial speed of the cars. (b) How much work did the actor do?

39. When fired from a gun into a 1.00-kg block of wood held in a vise, a 7.00-g bullet penetrates the block to a depth of 8.00 cm. The block is then placed on a frictionless, horizontal surface, and a second 7.00-g bullet is fired from the gun into the block. To what depth does the bullet penetrate the block in this case?

40. A billiard ball rolling across a table at 1.50 m/s makes a head-on elastic collision with an identical ball. Find the speed of each ball after the collision (a) when the second ball is initially at rest, (b) when the second ball is moving toward the first at a speed of 1.00 m/s, and (c) when the second ball is moving away from the first at a speed of 1.00 m/s.

41. A 90-kg fullback moving east with a speed of 5.0 m/s is tackled by a 95-kg opponent running north at 3.0 m/s. If the collision is perfectly inelastic, calculate (a) the velocity of the players just after the tackle and (b) the kinetic energy lost as a result of the collision. Can you account for the missing energy?

42. An 8.00-kg object moving east at 15.0 m/s on a frictionless horizontal surface collides with a 10.0-kg object that is initially at rest. After the collision, the 8.00-kg object moves south at 4.00 m/s. (a) What is the velocity of the 10.0-kg object after the collision? (b) What percentage of the initial kinetic energy is lost in the collision?

43. A 2 000-kg car moving east at 10.0 m/s collides with a 3 000-kg car moving north. The cars stick together and move as a unit after the collision, at an angle of 40.0° north of east and a speed of 5.22 m/s. Find the speed of the 3 000-kg car before the collision.

44. Two automobiles of equal mass approach an intersection. One vehicle is traveling with velocity 13.0 m/s toward the east, and the other is traveling north with speed v_{2i}. Neither driver sees the other. The vehicles collide in the intersection and stick together, leaving parallel skid marks at an angle of 55.0° north of east. The speed limit for both roads is 35 mi/h, and the driver of the northward-moving vehicle claims he was within the limit when the collision occurred. Is he telling the truth?

45. PhysicsNow™ A billiard ball moving at 5.00 m/s strikes a stationary ball of the same mass. After the collision, the first ball moves at 4.33 m/s at an angle of 30° with respect to the original line of motion. (a) Find the velocity (magnitude and direction) of the second ball after collision. (b) Was the collision inelastic or elastic?

ADDITIONAL PROBLEMS

46. In research in cardiology and exercise physiology, it is often important to know the mass of blood pumped by a person's heart in one stroke. This information can be obtained by means of a *ballistocardiograph.* The instrument works as follows: The subject lies on a horizontal pallet floating on a film of air. Friction on the pallet is negligible. Initially, the momentum of the system is zero. When the heart beats, it expels a mass m of blood into the aorta with speed v, and the body and platform move in the opposite direction with speed V. The speed of the blood can be determined independently (for example, by observing an ultrasound Doppler shift). Assume that the blood's speed is 50.0 cm/s in one typical trial. The mass of the subject plus the pallet is 54.0 kg. The pallet moves 6.00×10^{-5} m in 0.160 s after one heartbeat. Calculate the mass of blood that leaves the heart. Assume that the mass of blood is negligible compared with the total mass of the person. This simplified example illustrates the principle of ballistocardiography, but in practice a more sophisticated model of heart function is used.

47. A 0.50-kg object is at rest at the origin of a coordinate system. A 3.0-N force in the $+x$-direction acts on the object for 1.50 s. (a) What is the velocity at the end of this interval? (b) At the end of the interval, a constant force of 4.0 N is applied in the $-x$-direction for 3.0 s. What is the velocity at the end of the 3.0 s?

48. Consider a frictionless track as shown in Figure P6.48. A block of mass $m_1 = 5.00$ kg is released from Ⓐ. It makes a head-on elastic collision at Ⓑ with a block of mass $m_2 = 10.0$ kg that is initially at rest. Calculate the maximum height to which m_1 rises after the collision.

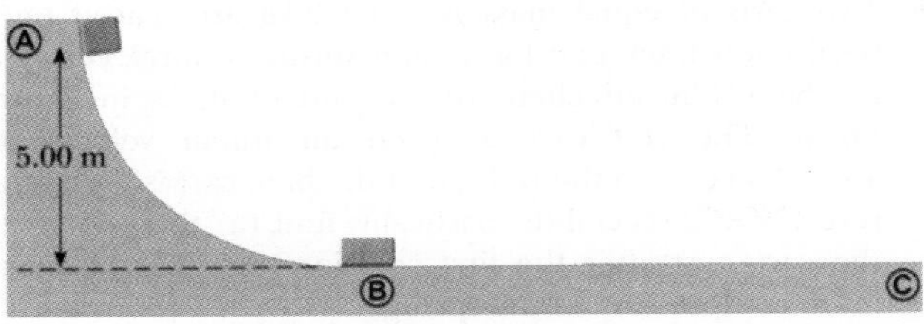

Figure P6.48

49. Most of us know intuitively that in a head-on collision between a large dump truck and a subcompact car, you are better off being in the truck than in the car. Why is this? Many people imagine that the collision force exerted on the car is much greater than that exerted on the truck. To substantiate this view, they point out that the car is crushed, whereas the truck is only dented. This idea of unequal forces, of course, is false; Newton's third law tells us that both objects are acted upon by forces of the same magnitude. The truck suffers less damage because it is made of stronger metal. But what about the two drivers? Do they experience the same forces? To answer this question, suppose that each vehicle is initially moving at 8.00 m/s and that they undergo a perfectly inelastic head-on collision. Each driver has mass 80.0 kg. Including the masses of the drivers, the total masses of the vehicles are 800 kg for the car and 4 000 kg for the truck. If the collision time is 0.120 s, what force does the seat belt exert on each driver?

50. A bullet of mass m and speed v passes completely through a pendulum bob of mass M as shown in Figure P6.50. The bullet emerges with a speed of $v/2$. The pendulum bob is suspended by a stiff rod of length ℓ and negligible mass. What is the minimum value of v such that the bob will barely swing through a complete vertical circle?

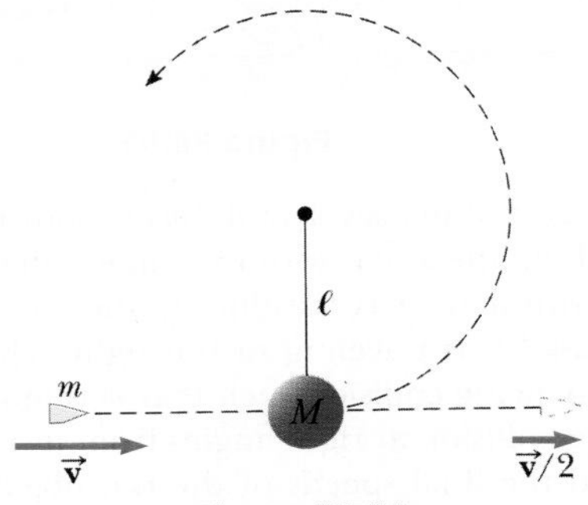

Figure P6.50

51. A 2.0-g particle moving at 8.0 m/s makes a perfectly elastic head-on collision with a resting 1.0-g object. (a) Find the speed of each particle after the collision. (b) Find the speed of each particle after the collision if the stationary particle has a mass of 10 g. (c) Find the final kinetic energy of the incident 2.0-g particle in the situations described in (a) and (b). In which case does the incident particle lose more kinetic energy?

52. A 0.400-kg green bead slides on a curved frictionless wire, starting from rest at point Ⓐ in Figure P6.52. At point Ⓑ, the bead collides elastically with a 0.600-kg blue ball at rest. Find the maximum height the blue ball rises as it moves up the wire.

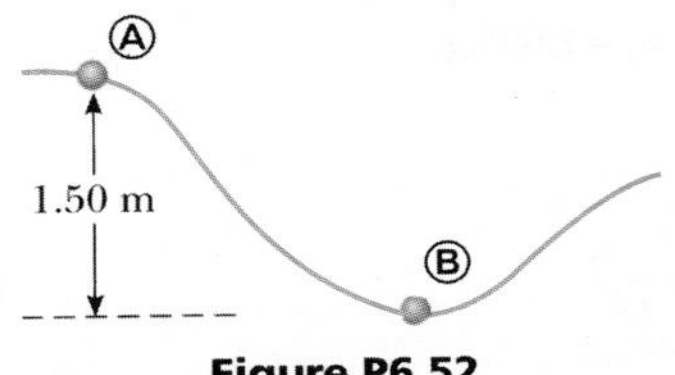

Figure P6.52

53. An 80-kg man standing erect steps off a 3.0-m-high diving platform and begins to fall from rest. The man again comes to rest 2.0 s after reaching the water. What average force did the water exert on him?

54. **PhysicsNow™** A 12.0-g bullet is fired horizontally into a 100-g wooden block initially at rest on a horizontal surface. After impact, the block slides 7.5 m before coming to rest. If the coefficient of kinetic friction between block and surface is 0.650, what was the speed of the bullet immediately before impact?

55. A 60.0-kg person running at an initial speed of 4.00 m/s jumps onto a 120-kg cart that is initially at rest (Figure P6.55). The person slides on the cart's top surface and finally comes to rest relative to the cart. The coefficient of kinetic friction between the person and the cart is 0.400. Friction between the cart and ground can be neglected. (a) Find the final speed of the person and cart relative to the ground. (b) Find the frictional force acting on the person while he is sliding across the top surface of the cart. (c) How long does the frictional force act on the person? (d) Find the change in momentum of the person and the change in momentum of the cart. (e) Determine the displacement of the person relative to the ground while he is sliding on the cart. (f) Determine the displacement of the cart relative to the ground while the person is sliding. (g) Find the change in kinetic energy of the person. (h) Find the change in kinetic energy of the cart. (i) Explain why the answers to (g) and (h) differ. (What kind of collision is this, and what accounts for the loss of mechanical energy?)

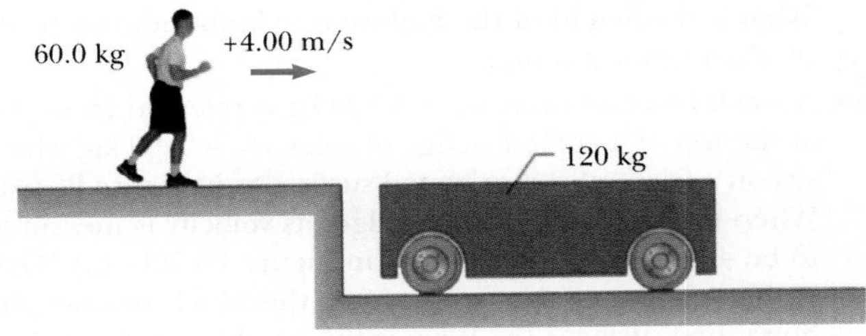

Figure P6.55

56. Two blocks of masses $m_1 = 2.00$ kg and $m_2 = 4.00$ kg are each released from rest at a height of 5.00 m on a frictionless track, as shown in Figure P6.56, and undergo an elastic head-on collision. (a) Determine the velocity of each block just before the collision. (b) Determine the velocity of each block immediately after the collision. (c) Determine the maximum heights to which m_1 and m_2 rise after the collision.

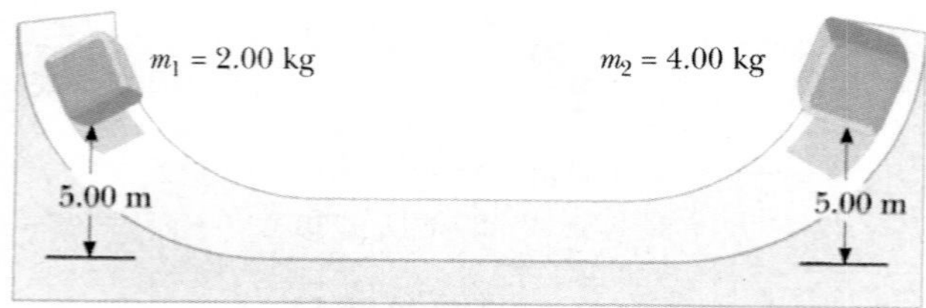

Figure P6.56

57. A 0.500-kg block is released from rest at the top of a frictionless track 2.50 m above the top of a table. It then collides elastically with a 1.00-kg object that is initially at rest on the table, as shown in Figure P6.57. (a) Determine the velocities of the two objects just after the collision. (b) How high up the track does the 0.500-kg object travel back after the collision? (c) How far away from the bottom of the table does the 1.00-kg object land, given that the table is 2.00 m high? (d) How far away from the bottom of the table does the 0.500-kg object eventually land?

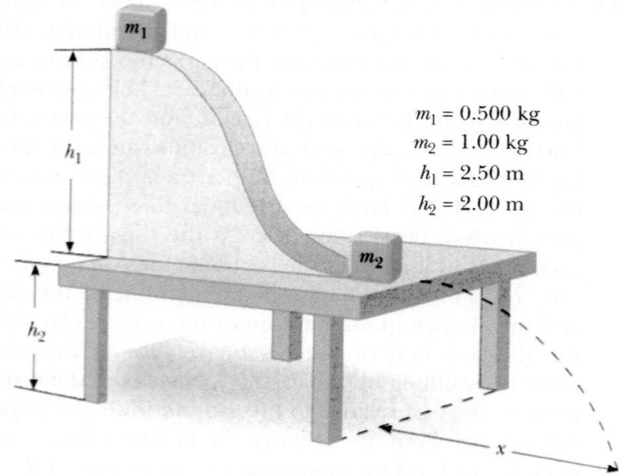

Figure P6.57

58. Tarzan, whose mass is 80.0 kg, swings from a 3.00-m vine that is horizontal when he starts. At the bottom of his arc, he picks up 60.0-kg Jane in a perfectly inelastic collision. What is the height of the highest tree limb they can reach on their upward swing?

59. A small block of mass m_1 = 0.500 kg is released from rest at the top of a curved wedge of mass m_2 = 3.00 kg, which sits on a frictionless horizontal surface as in Figure P6.59a. When the block leaves the wedge, its velocity is measured to be 4.00 m/s to the right, as in Figure P6.59b. (a) What is the velocity of the wedge after the block reaches the horizontal surface? (b) What is the height h of the wedge?

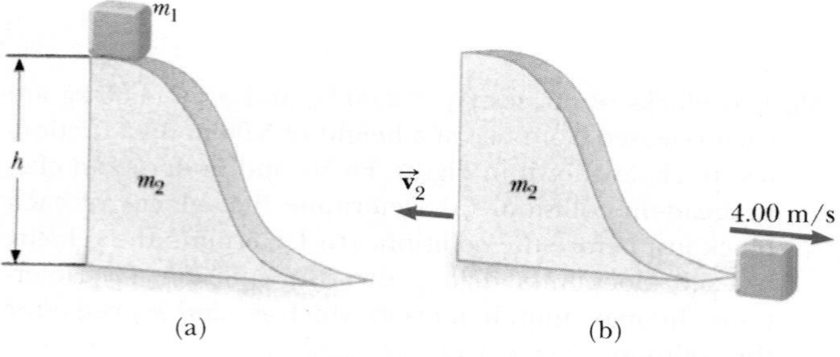

Figure P6.59

60. Two carts of equal mass m = 0.250 kg are placed on a frictionless track that has a light spring of force constant k = 50.0 N/m attached to one end of it, as in Figure P6.60. The red cart is given an initial velocity of $\vec{v}_0$ = 3.00 m/s to the right, and the blue cart is initially at rest. If the carts collide elastically, find (a) the velocity of the carts just after the first collision and (b) the maximum compression of the spring.

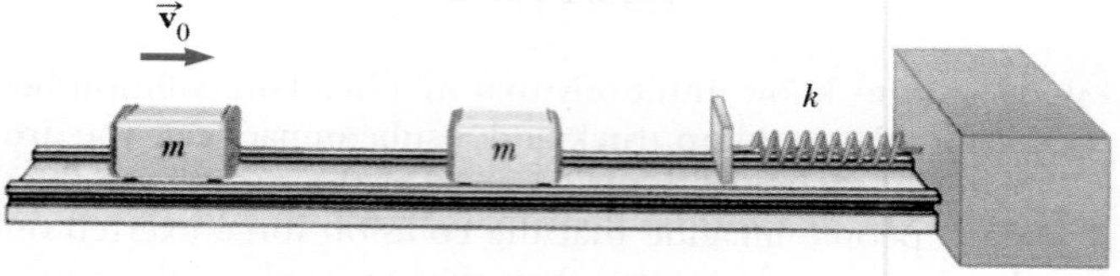

Figure P6.60

61. A cannon is rigidly attached to a carriage, which can move along horizontal rails, but is connected to a post by a large spring, initially unstretched and with force constant $k = 2.00 \times 10^4$ N/m, as in Figure P6.61. The cannon fires a 200-kg projectile at a velocity of 125 m/s directed 45.0° above the horizontal. (a) If the mass of the cannon and its carriage is 5 000 kg, find the recoil speed of the cannon. (b) Determine the maximum extension of the spring. (c) Find the maximum force the spring exerts on the carriage. (d) Consider the system consisting of the cannon, the carriage, and the shell. Is the momentum of this system conserved during the firing? Why or why not?

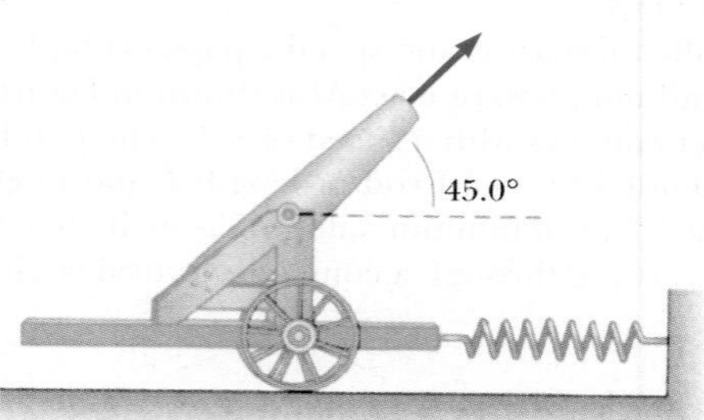

Figure P6.61

62. Two objects of masses m and $3m$ are moving toward each other along the x-axis with the same initial speed v_0. The object with mass m is traveling to the left, and the object with mass $3m$ is traveling to the right. They undergo an elastic glancing collision such that m is moving downward after the collision at right angles from its initial direction. (a) Find the final speeds of the two objects. (b) What is the angle θ at which the object with mass $3m$ is scattered?

63. A neutron in a reactor makes an elastic head-on collision with a carbon atom that is initially at rest. (The mass of the carbon nucleus is about 12 times that of the neutron.) (a) What fraction of the neutron's kinetic energy is transferred to the carbon nucleus? (b) If the neutron's initial kinetic energy is 1.6×10^{-13} J, find its final kinetic energy and the kinetic energy of the carbon nucleus after the collision.

64. A cue ball traveling at 4.00 m/s makes a glancing, elastic collision with a target ball of equal mass that is initially at rest. The cue ball is deflected so that it makes an angle of 30.0° with its original direction of travel. Find (a) the angle between the velocity vectors of the two balls after the collision and (b) the speed of each ball after the collision.

Again the vectors $\vec{\mathbf{r}}$ and $\vec{\mathbf{F}}$ lie in a plane and for our purposes the chosen point O will usually correspond to an axis of rotation perpendicular to the plane.

A second way of understanding the sin θ factor is to associate it with the magnitude r of the position vector $\vec{\mathbf{r}}$. The quantity $d = r \sin \theta$ is called the **lever arm**, which is the perpendicular distance from the axis of rotation to a line drawn along the direction of the force. This alternate interpretation is illustrated in Figure 8.3c.

It's important to remember that **the value of τ depends on the chosen axis of rotation.** Torques can be computed around any axis, regardless of whether there is some actual, physical rotation axis present. Once the point is chosen, however, it must be used consistently throughout a given problem.

Torque is a vector perpendicular to the plane determined by the position and force vectors, as illustrated in Figure 8.4. The direction can be determined by the *right-hand rule*:

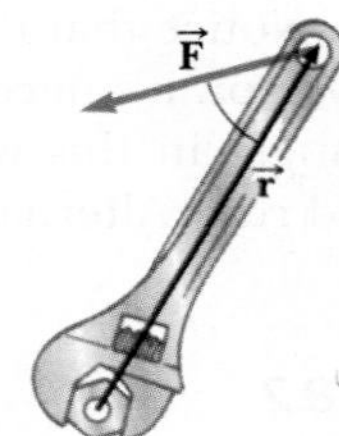

Figure 8.4 The right-hand rule: Point the fingers of your right hand along $\vec{\mathbf{r}}$ and curl them in the direction of $\vec{\mathbf{F}}$. Your thumb then points in the direction of the torque (out of the page, in this case).

1. Point the fingers of your right hand in the direction of $\vec{\mathbf{r}}$.
2. Curl your fingers toward the direction of vector $\vec{\mathbf{F}}$.
3. Your thumb then points approximately in the direction of the torque, in this case out of the page.

Problems used in this book will be confined to objects rotating around an axis perpendicular to the plane containing $\vec{\mathbf{r}}$ and $\vec{\mathbf{F}}$, so if these vectors are in the plane of the page, the torque will always point either into or out of the page, parallel to the axis of rotation. If your right thumb is pointed in the direction of a torque, your fingers curl naturally in the direction of rotation that the torque would produce on an object at rest.

EXAMPLE 8.2 The Swinging Door

Goal Apply the more general definition of torque.

Problem (a) A man applies a force of $F = 3.00 \times 10^2$ N at an angle of 60.0° to the door of Figure 8.5a, 2.00 m from the hinges. Find the torque on the door, choosing the position of the hinges as the axis of rotation. (b) Suppose a wedge is placed 1.50 m from the hinges on the other side of the door. What minimum force must the wedge exert so that the force applied in part (a) won't open the door?

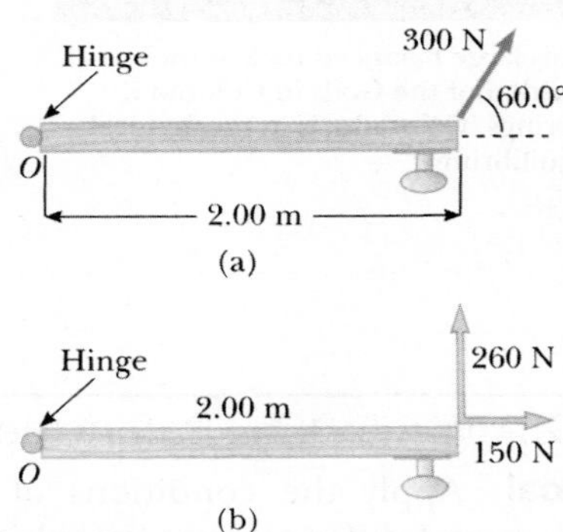

Figure 8.5 (Example 8.2a) (a) Top view of a door being pushed by a 300-N force. (b) The components of the 300-N force.

Strategy Part (a) can be solved by substitution into the general torque equation. In part (b), the hinges, the wedge, and the applied force all exert torques on the door. The door doesn't open, so the sum of these torques must be zero, a condition that can be used to find the wedge force.

Solution

(a) Compute the torque due to the applied force exerted at 60.0°.

Substitute into the general torque equation:

$$\tau_F = rF \sin \theta = (2.00 \text{ m})(3.00 \times 10^2 \text{ N}) \sin 60.0°$$
$$= (2.00 \text{ m})(2.60 \times 10^2 \text{ N}) = \boxed{5.20 \times 10^2 \text{ N}\cdot\text{m}}$$

(b) Calculate the force exerted by the wedge on the other side of the door.

Set the sum of the torques equal to zero:

$$\tau_{\text{hinge}} + \tau_{\text{wedge}} + \tau_F = 0$$

The hinge force provides no torque because it acts at the axis ($r = 0$). The wedge force acts at an angle of $-90.0°$, opposite F_y.

$$0 + F_{\text{wedge}}(1.50 \text{ m}) \sin(-90.0°) + 5.20 \times 10^2 \text{ N}\cdot\text{m} = 0$$

$$F_{\text{wedge}} = \boxed{347 \text{ N}}$$

Remark Notice that the angle from the position vector to the wedge force is $-90°$. This is because, starting at the position vector, it's necessary to go 90° clockwise (the negative angular direction) to get to the force vector. Measuring the angle in this way automatically supplies the correct sign for the torque term and is consistent with the right-hand rule. Alternately, the magnitude of the torque can be found and the correct sign chosen based on physical intuition.

Exercise 8.2

A man ties one end of a strong rope 8.00 m long to the bumper of his truck, 0.500 m from the ground, and the other end to a vertical tree trunk at a height of 3.00 m. He uses the truck to create a tension of 8.00×10^2 N in the rope. Compute the magnitude of the torque on the tree due to the tension in the rope, with the base of the tree acting as the reference point.

Answer $2.28 \times 10^3 \text{ N} \cdot \text{m}$

This large balanced rock at the Garden of the Gods in Colorado, Springs, Colorado, is in mechanical equilibrium.

8.2 TORQUE AND THE TWO CONDITIONS FOR EQUILIBRIUM

An object in mechanical equilibrium must satisfy the following two conditions:

1. **The net external force must be zero:** $\Sigma\vec{\mathbf{F}} = 0$

2. **The net external torque must be zero:** $\Sigma\vec{\boldsymbol{\tau}} = 0$

The first condition is a statement of translational equilibrium: The sum of all forces acting on the object must be zero, so the object has no translational acceleration, $\vec{\mathbf{a}} = 0$. The second condition is a statement of rotational equilibrium: The sum of all torques on the object must be zero, so the object has no angular acceleration, $\vec{\boldsymbol{\alpha}} = 0$. For an object to be in equilibrium, it must both translate and rotate at a constant rate.

Because we can choose any location for calculating torques, it's usually best to select an axis that will make at least one torque equal to zero, just to simplify the net torque equation.

EXAMPLE 8.3 Balancing Act

Goal Apply the conditions of equilibrium and illustrate the use of different axes for calculating the net torque on an object.

Problem A woman of mass $m = 55.0$ kg sits on the left end of a seesaw—a plank of length $L = 4.00$ m, pivoted in the middle as in Figure 8.6. **(a)** First compute the torques on the seesaw about an axis that passes through the pivot point. Where should a man of mass $M = 75.0$ kg sit if the system (seesaw plus man and woman) is to be balanced? **(b)** Find the normal force exerted by the pivot if the plank has a mass of $m_{\text{pl}} = 12.0$ kg. **(c)** Repeat part (b), but this time compute the torques about an axis through the left end of the plank.

Strategy In part (a), apply the second condition of equilibrium, $\Sigma\tau = 0$, computing torques around the pivot point. The mass of the plank forming the seesaw is distributed evenly on either side of the pivot point, so the torque exerted by gravity on the plank, τ_{gravity}, can be computed as if all the plank's mass is concentrated at the pivot point. Then

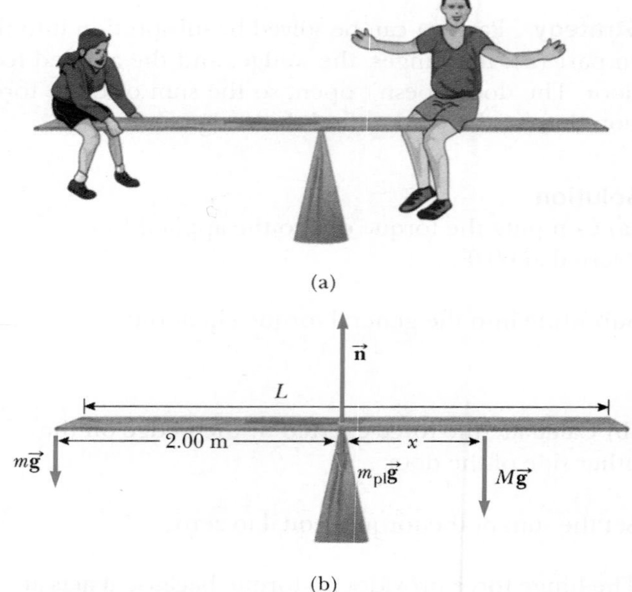

Figure 8.6 (a) (Example 8.3) Two people on a see-saw. (b) Free body diagram for the plank.

τ_{gravity} is zero, as is the torque exerted by the pivot, because their lever arms are zero. In part (b), the first condition of equilibrium, $\Sigma\vec{\mathbf{F}} = 0$, must be applied. Part (c) is a repeat of part (a) showing that choice of a different axis yields the same answer.

Solution

(a) Where should the man sit to balance the seesaw?

Apply the second condition of equilibrium to the plank by setting the sum of the torques equal to zero:

$$\tau_{\text{pivot}} + \tau_{\text{gravity}} + \tau_{\text{man}} + \tau_{\text{woman}} = 0$$

The first two torques are zero. Let x represent the man's distance from the pivot. The woman is at a distance $L/2$ from the pivot.

$$0 + 0 - Mgx + mg(L/2) = 0$$

Solve this equation for x and evaluate it:

$$x = \frac{m(L/2)}{M} = \frac{(55.0\text{ kg})(2.00\text{ m})}{75.0\text{ kg}} = \boxed{1.47\text{ m}}$$

(b) Find the normal force n exerted by the pivot on the seesaw.

Apply for first condition of equilibrium to the plank, solving the resulting equation for the unknown normal force, n:

$$-Mg - mg - m_{\text{pl}}g + n = 0$$

$$n = (M + m + m_{\text{pl}})g$$

$$= (75.0\text{ kg} + 55.0\text{ kg} + 12.0\text{ kg})(9.80\text{ m/s}^2)$$

$$n = \boxed{1.39 \times 10^3\text{ N}}$$

(c) Repeat part (a), choosing a new axis through the left end of the plank.

Compute the torques using this axis, and set their sum equal to zero. Now the pivot and gravity forces on the plank result in nonzero torques.

$$\tau_{\text{man}} + \tau_{\text{woman}} + \tau_{\text{plank}} + \tau_{\text{pivot}} = 0$$

$$-Mg(L/2 + x) + mg(0) - m_{\text{pl}}g(L/2) + n(L/2) = 0$$

Substitute all known quantities:

$$-(75.0\text{ kg})(9.80\text{ m/s}^2)(2.00\text{ m} + x) + 0$$

$$-(12.0\text{ kg})(9.80\text{ m/s}^2)(2.00\text{ m}) + n(2.00\text{ m}) = 0$$

$$-(1.47 \times 10^3\text{ N}\cdot\text{m}) - (735\text{ N})x - (235\text{ N}\cdot\text{m})$$

$$+ (2.00\text{ m})n = 0$$

Solve for x, substituting the normal force found in part (b):

$$x = \boxed{1.46\text{ m}}$$

Remarks The answers for x in parts (a) and (c) agree except for a small round-off discrepancy. This illustrates how choosing a different axis leads to the same solution.

Exercise 8.3

Suppose a 30.0-kg child sits 1.50 m to the left of center on the same seesaw. A second child sits at the end on the opposite side, and the system is balanced. (a) Find the mass of the second child. (b) Find the normal force acting at the pivot point.

Answers (a) 22.5 kg (b) 632 N

TIP 8.1 Specify Your Axis

Choose the axis of rotation, and use that axis exclusively throughout a given problem. The axis need not correspond to a physical axle or pivot point. Any convenient point will do.

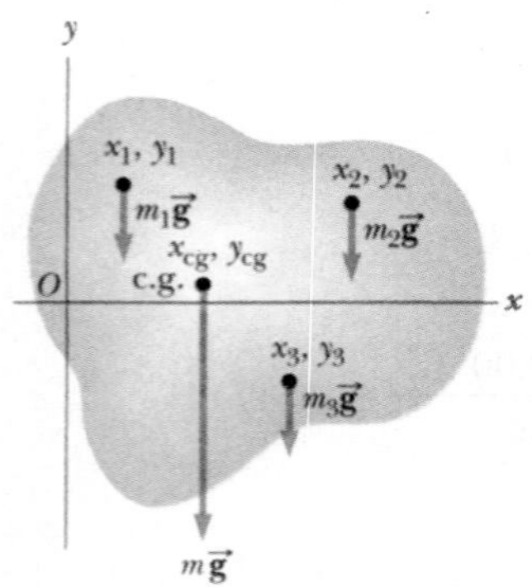

Figure 8.7 The net gravitational torque on an object is zero if computed around the center of gravity. The object will balance if supported at that point (or at any point along a vertical line above or below that point).

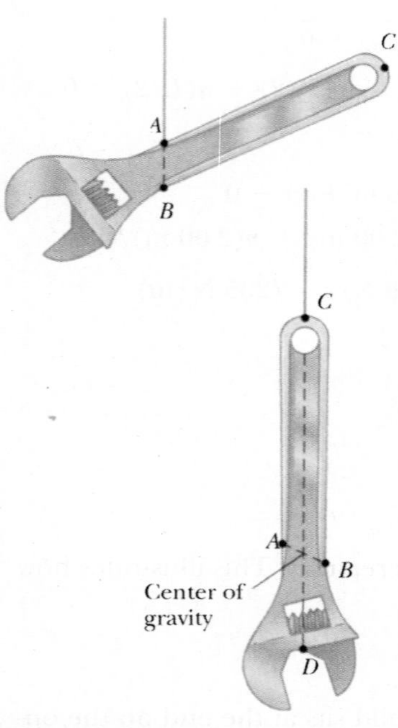

Figure 8.8 An experimental technique for determining the center of gravity of a wrench. The wrench is hung freely from two different pivots, A and C. The intersection of the two vertical lines, AB and CD, locates the center of gravity.

8.3 THE CENTER OF GRAVITY

In the example of the seesaw in the previous section, we guessed that the torque due to the force of gravity on the plank was the same as if all the plank's weight were concentrated at its center. This is a general procedure: To compute the torque on a rigid body due to the force of gravity, the body's entire weight can be thought of as concentrated at a single point. The problem then reduces to finding the location of that point. If the body is homogeneous (its mass is distributed evenly) and symmetric, it's usually possible to guess the location of that point, as in Example 8.3. Otherwise, it's necessary to calculate the point's location, as explained in this section.

Consider an object of arbitrary shape lying in the xy-plane, as in Figure 8.7. The object is divided into a large number of very small particles of weight m_1g, m_2g, m_3g, . . . having coordinates (x_1, y_1), (x_2, y_2), (x_3, y_3), ... If the object is free to rotate around the origin, each particle contributes a torque about the origin that is equal to its weight multiplied by its lever arm. For example, the torque due to the weight m_1g is m_1gx_1, and so forth.

We wish to locate the point of application of the single force of magnitude $w = F_g = Mg$ (the total weight of the object), where the effect on the rotation of the object is the same as that of the individual particles. This point is called the object's **center of gravity**. Equating the torque exerted by w at the center of gravity to the sum of the torques acting on the individual particles gives

$$(m_1g + m_2g + m_3g + \cdots)x_{cg} = m_1gx_1 + m_2gx_2 + m_3gx_3 + \cdots$$

We assume that g is the same everywhere in the object (which is true for all objects we will encounter). Then the g factors in the preceding equation cancel, resulting in

$$x_{cg} = \frac{m_1x_1 + m_2x_2 + m_3x_3 + \cdots}{m_1 + m_2 + m_3 + \cdots} = \frac{\Sigma m_ix_i}{\Sigma m_i} \qquad [8.3a]$$

where x_{cg} is the x-coordinate of the center of gravity. Similarly, the y-coordinate and z-coordinate of the center of gravity of the system can be found from

$$y_{cg} = \frac{\Sigma m_iy_i}{\Sigma m_i} \qquad [8.3b]$$

and

$$z_{cg} = \frac{\Sigma m_iz_i}{\Sigma m_i} \qquad [8.3c]$$

These three equations are identical to the equations for a similar concept called **center of mass**. The center of mass and center of gravity of an object are exactly the same when g doesn't vary significantly over the object.

It's often possible to guess the location of the center of gravity. **The center of gravity of a homogeneous, symmetric body must lie on the axis of symmetry.** For example, the center of gravity of a homogeneous rod lies midway between the ends of the rod, and the center of gravity of a homogeneous sphere or a homogeneous cube lies at the geometric center of the object. The center of gravity of an irregularly shaped object, such as a wrench, can be determined experimentally by suspending the wrench from two different arbitrary points (Fig. 8.8). The wrench is first hung from point A, and a vertical line AB (which can be established with a plumb bob) is drawn when the wrench is in equilibrium. The wrench is then hung from point C, and a second vertical line CD is drawn. The center of gravity coincides with the intersection of these two lines. In fact, if the wrench is hung freely from any point, the center of gravity always lies straight below the point of support, so the vertical line through that point must pass through the center of gravity.

Several examples in Section 8.4 involve homogeneous, symmetric objects where the centers of gravity coincide with their geometric centers. A rigid object in a uniform gravitational field can be balanced by a single force equal in magnitude to the weight of the object, as long as the force is directed upward through the object's center of gravity.

EXAMPLE 8.4 Where Is the Center of Gravity?

Goal Find the center of gravity of a system of particles.

Problem Three particles are located in a coordinate system as shown in Figure 8.9. Find the center of gravity.

Strategy The y-coordinate and z-coordinate of the center of gravity are both zero because all the particles are on the x-axis. We can find the x-coordinate of the center of gravity using Equation 8.3a.

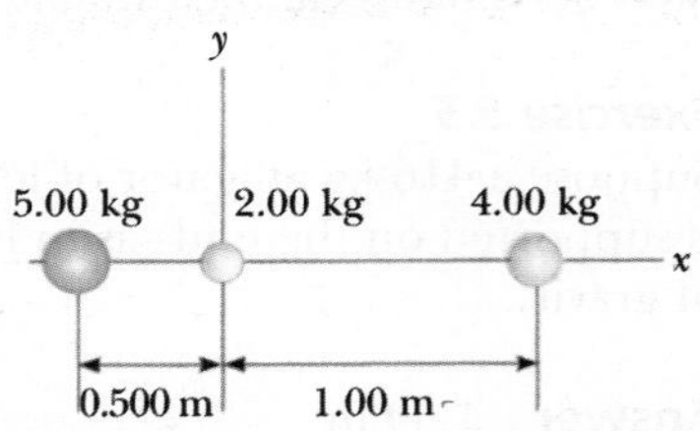

Figure 8.9 (Example 8.4) Locating the center of gravity of a system of three particles.

Solution

Apply Equation 8.3a to the system of three particles:

$$x_{cg} = \frac{\Sigma m_i x_i}{\Sigma m_i} = \frac{m_1 x_1 + m_2 x_2 + m_3 x_3}{m_1 + m_2 + m_3} \quad (1)$$

Compute the numerator of Equation (1):

$$\begin{aligned}\Sigma m_i x_i &= m_1 x_1 + m_2 x_2 + m_3 x_3 \\ &= (5.00\text{ kg})(-0.500\text{ m}) + (2.00\text{ kg})(0\text{ m}) \\ &\quad + (4.00\text{ kg})(1.00\text{ m}) \\ &= 1.50\text{ kg}\cdot\text{m}\end{aligned}$$

Substitute the denominator, $\Sigma m_i = 11.0$ kg, and the numerator into Equation (1).

$$x_{cg} = \frac{1.50\text{ kg}\cdot\text{m}}{11.0\text{ kg}} = 0.136\text{ m}$$

Exercise 8.4

If a fourth particle of mass 2.00 kg is placed at $x = 0$, $y = 0.250$ m, find the x-and y-coordinates of the center of gravity for this system of four particles.

Answer $x_{cg} = 0.115$ m; $y_{cg} = 0.0385$ m

EXAMPLE 8.5 Locating Your Lab Partner's Center of Gravity

Goal Use torque to find a center of gravity.

Problem In this example, we show how to find the location of a person's center of gravity. Suppose your lab partner has a height L of 173 cm (5 ft, 8 in) and a weight w of 715 N (160 lb). You can determine the position of his center of gravity by having him stretch out on a uniform board supported at one end by a scale, as shown in Figure 8.10. If the board's weight w_b is 49 N and the scale reading F is 3.50×10^2 N, find the distance of your lab partner's center of gravity from the left end of the board.

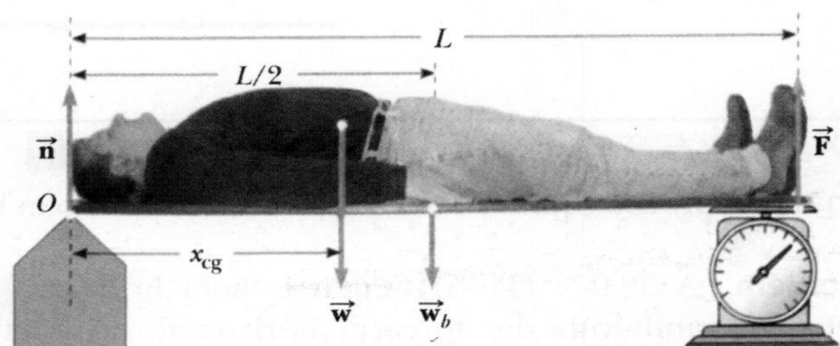

Figure 8.10 (Example 8.5) Determining your lab partner's center of gravity.

Strategy To find the position x_{cg} of the center of gravity, compute the torques using an axis through O. Set the sum of the torques equal to zero and solve for x_{cg}.

Solution

Apply the second condition of equilibrium. There is no torque due to the normal force $\vec{\mathbf{n}}$ because its moment arm is zero.

$$\Sigma \tau_O = 0$$

$$-wx_{cg} - w_b(L/2) + FL = 0$$

Solve for x_{cg} and substitute known values:

$$\begin{aligned}x_{cg} &= \frac{FL - w_b(L/2)}{w} \\ &= \frac{(350\text{ N})(173\text{ cm}) - (49\text{ N})(86.5\text{ cm})}{715\text{ N}} = 79\text{ cm}\end{aligned}$$

Remarks The given information is sufficient only to determine the x-coordinate of the center of gravity. The other two coordinates can be estimated, based on the body's symmetry.

Exercise 8.5

Suppose a 416-kg alligator of length 3.5 m is stretched out on a board of the same length weighing 65 N. If the board is supported on the ends as in Figure 8.10, and the scale reads 1 880 N, find the x-component of the alligator's center of gravity.

Answer 1.59 m

8.4 EXAMPLES OF OBJECTS IN EQUILIBRIUM

TIP 8.2 Rotary Motion Under Zero Torque

If a net torque of zero is exerted on an object, it will continue to rotate at a constant angular speed—which need not be zero. However, zero torque *does* imply that the angular acceleration is zero.

Recall from Chapter 4 that when an object is treated as a geometric point, equilibrium requires only that the net force on the object is zero. In this chapter, we have shown that for extended objects a second condition for equilibrium must also be satisfied: The net torque on the object must be zero. The following general procedure is recommended for solving problems that involve objects in equilibrium.

Problem-Solving Strategy Objects in Equilibrium

1. **Diagram the system.** Include coordinates and choose a convenient rotation axis for computing the net torque on the object.
2. **Draw a free-body diagram** of the object of interest, showing all external forces acting on it. For systems with more than one object, draw a *separate* diagram for each object. (Most problems will have a single object of interest.)
3. **Apply $\Sigma\tau_i = 0$, the second condition of equilibrium.** This condition yields a single equation for each object of interest. If the axis of rotation has been carefully chosen, the equation often has only one unknown and can be solved immediately.
4. **Apply $\Sigma F_x = 0$ and $\Sigma F_y = 0$, the first condition of equilibrium.** This yields two more equations per object of interest.
5. **Solve the system of equations.** For each object, the two conditions of equilibrium yield three equations, usually with three unknowns. Solve by substitution.

EXAMPLE 8.6 A Weighted Forearm

Goal Apply the equilibrium conditions to the human body.

Problem A 50.0-N (11-lb) weight is held in a person's hand with the forearm horizontal, as in Figure 8.11. The biceps muscle is attached 0.030 0 m from the joint, and the weight is 0.350 m from the joint. Find the upward force $\vec{F}$ exerted by the biceps on the forearm (the ulna) and the downward force $\vec{R}$ exerted by the humerus on the forearm, acting at the joint. Neglect the weight of the forearm.

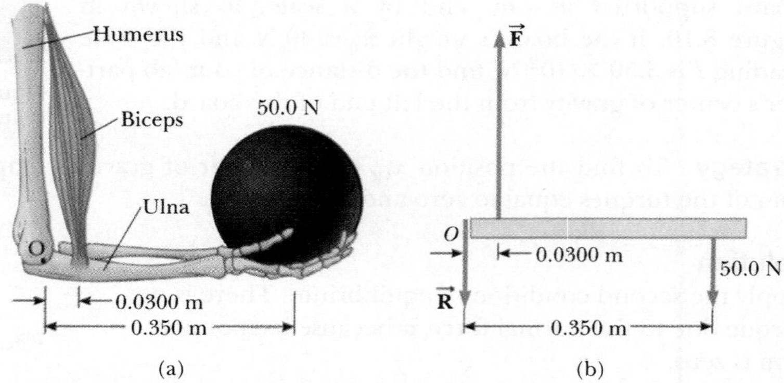

Figure 8.11 (Example 8.6) (a) A weight held with the forearm horizontal. (b) The mechanical model for the system.

Strategy The forces acting on the forearm are equivalent to those acting on a bar of length 0.350 m, as shown in Figure 8.11b. Choose the usual x- and y-coordinates as shown and the axis at O on the left end. (This completes Steps 1 and 2.) Use the conditions of equilibrium to generate equations for the unknowns, and solve.

Solution

Apply the second condition for equilibrium (step 3):

$$\Sigma\tau_i = \tau_R + \tau_F + \tau_{BB} = 0$$

$$R(0) + F(0.0300 \text{ m}) - (50.0 \text{ N})(0.350 \text{ m}) = 0$$

$$F = 583 \text{ N } (131 \text{ lb})$$

Apply the first condition for equilibrium (step 4):

$$\Sigma F_y = F - R - 50.0 \text{ N} = 0$$

$$R = F - 50.0 \text{ N} = 583 \text{ N} - 50 \text{ N} = 533 \text{ N } (120 \text{ lb})$$

Exercise 8.6

Suppose you wanted to limit the force acting on your joint to a maximum value of 8.00×10^2 N. (a) Under these circumstances, what maximum weight would you attempt to lift? (b) What force would your biceps apply while lifting this weight?

Answers (a) 75.0 N (b) 875 N

EXAMPLE 8.7 Walking a Horizontal Beam

Goal Solve an equilibrium problem with nonperpendicular torques.

Problem A uniform horizontal beam 5.00 m long and weighing 3.00×10^2 N is attached to a wall by a pin connection that allows the beam to rotate. Its far end is supported by a cable that makes an angle of 53.0° with the horizontal (Fig. 8.12a). If a person weighing 6.00×10^2 N stands 1.50 m from the wall, find the magnitude of the tension $\vec{\mathbf{T}}$ in the cable and the force $\vec{\mathbf{R}}$ exerted by the wall on the beam.

Strategy

The second condition of equilibrium, $\Sigma\tau_i = 0$, with torques computed around the pin, can be solved for the tension T in the cable. The first condition of equilibrium, $\Sigma\vec{\mathbf{F}}_i = 0$, gives two equations and two unknowns for the two components of the force exerted by the wall, R_x and R_y.

Solution

From Figure 8.12, the forces causing torques are the wall force $\vec{\mathbf{R}}$, the gravity forces on the beam and the man, w_B and w_M, and the tension force $\vec{\mathbf{T}}$. Apply the condition of rotational equilibrium:

$$\Sigma\tau_i = \tau_R + \tau_B + \tau_M + \tau_T = 0$$

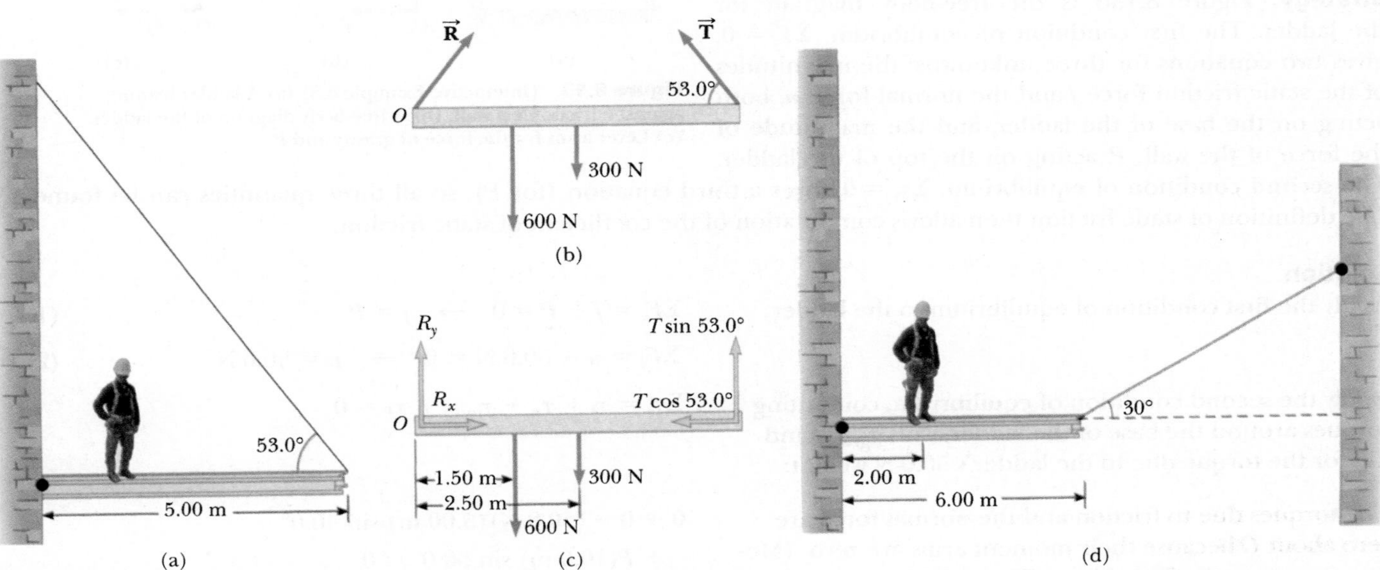

Figure 8.12 (Example 8.7) (a) A uniform beam attached to a wall and supported by a cable. (b) A free-body diagram for the beam. (c) The component form of the free-body diagram. (d) (Exercise 8.7)

Compute torques around the pin at O, so $\tau_R = 0$ (zero moment arm). The torque due to the beam's weight acts at the beam's center of gravity.

$$\Sigma\tau_i = 0 - w_B(L/2) - w_M(1.50\text{ m}) + TL\sin(53°) = 0$$

Substitute $L = 5.00$ m and the weights, solving for T:

$$-(3.00 \times 10^2\text{ N})(2.50\text{ m})$$
$$-(6.00 \times 10^2\text{ N})(1.50\text{ m})$$
$$+(T\sin 53.0°)(5.00\text{ m}) = 0$$

$$T = 413\text{ N}$$

Now apply the first condition of equilibrium to the beam:

$$\Sigma F_x = R_x - T\cos 53.0° = 0 \quad \textbf{(1)}$$
$$\Sigma F_y = R_y - w_B - w_M + T\sin 53.0° = 0 \quad \textbf{(2)}$$

Substituting the value of T found in the previous step and the weights, obtain the components of $\vec{\mathbf{R}}$:

$$R_x = 249\text{ N} \qquad R_y = 5.70 \times 10^2\text{ N}$$

Remarks Even if we selected some other axis for the torque equation, the solution would be the same. For example, if the axis were to pass through the center of gravity of the beam, the torque equation would involve both T and R_y. Together with equations (1) and (2), however, the unknowns could still be found—a good exercise.

Exercise 8.7

A person with mass 55.0 kg stands 2.00 m away from the wall on a 6.00-m beam, as shown in Figure 8.12d. The mass of the beam is 40.0 kg. Find the hinge force components and the tension in the wire.

Answers $T = 751$ N, $R_x = -6.50 \times 10^2$ N, $R_y = 556$ N

INTERACTIVE EXAMPLE 8.8 Don't Climb the Ladder

Goal Apply the two conditions of equilibrium.

Problem A uniform ladder 10.0 m long and weighing 50.0 N rests against a smooth vertical wall as in Figure 8.13a. If the ladder is just on the verge of slipping when it makes a 50.0° angle with the ground, find the coefficient of static friction between the ladder and ground.

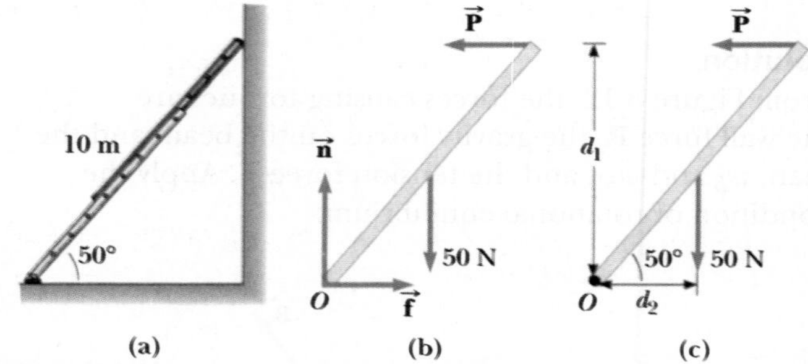

Figure 8.13 (Interactive Example 8.8) (a) A ladder leaning against a frictionless wall. (b) A free-body diagram of the ladder. (c) Lever arms for the force of gravity and $\vec{\mathbf{P}}$.

Strategy Figure 8.13b is the free-body diagram for the ladder. The first condition of equilibrium, $\Sigma\vec{\mathbf{F}}_i = 0$, gives two equations for three unknowns: the magnitudes of the static friction force f and the normal force n, both acting on the base of the ladder, and the magnitude of the force of the wall, P, acting on the top of the ladder. The second condition of equilibrium, $\Sigma\tau_i = 0$, gives a third equation (for P), so all three quantities can be found. The definition of static friction then allows computation of the coefficient of static friction.

Solution

Apply the first condition of equilibrium to the ladder:

$$\Sigma F_x = f - P = 0 \quad \rightarrow \quad f = P \quad \textbf{(1)}$$
$$\Sigma F_y = n - 50.0\text{ N} = 0 \quad \rightarrow \quad n = 50.0\text{ N} \quad \textbf{(2)}$$

Apply the second condition of equilibrium, computing torques around the base of the ladder, with τ_{grav} standing for the torque due to the ladder's 50.0-N weight:

$$\Sigma\tau_i = \tau_f + \tau_n + \tau_{\text{grav}} + \tau_P = 0$$

The torques due to friction and the normal force are zero about O because their moment arms are zero. (Moment arms can be found from Figure 8.13c.)

$$0 + 0 - (50.0\text{ N})(5.00\text{ m})\sin 40.0°$$
$$+ P(10.0\text{ m})\sin 50.0° = 0$$

$$P = 21.0\text{ N}$$

From Equation (1), we now have $f = P = 21.0$ N. The ladder is on the verge of slipping, so write an expression for the maximum force of static friction and solve for μ_s:

$$21.0 \text{ N} = f = f_{s,\text{max}} = \mu_s n = \mu_s(50.0 \text{ N})$$

$$\mu_s = \frac{21.0 \text{ N}}{50.0 \text{ N}} = 0.420$$

Remarks Note that torques were computed around an axis through the bottom of the ladder so that only $\vec{\mathbf{P}}$ and the force of gravity contributed nonzero torques. This choice of axis reduces the complexity of the torque equation, often resulting in an equation with only one unknown.

Exercise 8.8

If the coefficient of static friction is 0.360, and the same ladder makes a 60.0° angle with respect to the horizontal, how far along the length of the ladder can a 70.0-kg painter climb before the ladder begins to slip?

Answer 6.33 m

PhysicsNow™ You can adjust the angle of the ladder and watch what happens when it is released by logging into PhysicsNow at **www.cp7e.com** and going to Interactive Example 8.8.

8.5 RELATIONSHIP BETWEEN TORQUE AND ANGULAR ACCELERATION

When a rigid object is subject to a net torque, it undergoes an angular acceleration that is directly proportional to the net torque. This result, which is analogous to Newton's second law, is derived as follows.

The system shown in Figure 8.14 consists of an object of mass m connected to a very light rod of length r. The rod is pivoted at the point O, and its movement is confined to rotation on a frictionless *horizontal* table. Assume that a force F_t acts perpendicular to the rod and hence is tangent to the circular path of the object. Because there is no force to oppose this tangential force, the object undergoes a tangential acceleration a_t in accordance with Newton's second law:

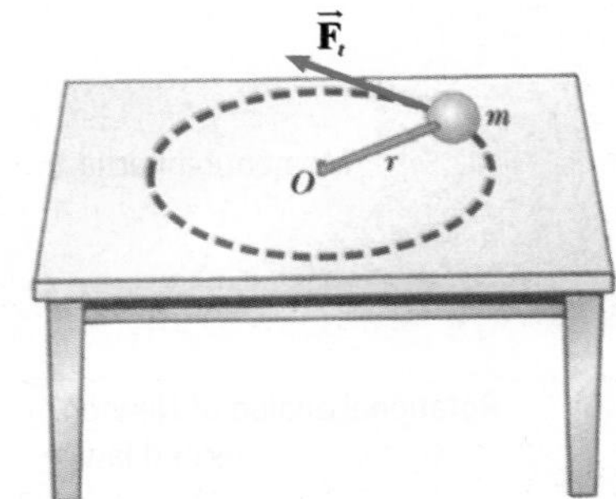

Figure 8.14 An object of mass m attached to a light rod of length r moves in a circular path on a frictionless horizontal surface while a tangential force $\vec{\mathbf{F}}_t$ acts on it.

$$F_t = ma_t$$

Multiply both sides of this equation by r:

$$F_t r = mra_t$$

Substituting the equation $a_t = r\alpha$ relating tangential and angular acceleration into the above expression gives

$$F_t r = mr^2\alpha \qquad [8.4]$$

The left side of Equation 8.4 is the torque acting on the object about its axis of rotation, so we can rewrite it as

$$\tau = mr^2\alpha \qquad [8.5]$$

Equation 8.5 shows that the torque on the object is proportional to the angular acceleration of the object, where the constant of proportionality mr^2 is called the **moment of inertia** of the object of mass m. (Because the rod is very light, its moment of inertia can be neglected.)

Quick Quiz 8.1

Using a screwdriver, you try to remove a screw from a piece of furniture, but can't get it to turn. To increase the chances of success, you should use a screwdriver that (a) is longer, (b) is shorter, (c) has a narrower handle, or (d) has a wider handle.

Torque on a Rotating Object

Consider a solid disk rotating about its axis as in Figure 8.15a. The disk consists of many particles at various distances from the axis of rotation. (See Fig. 8.15b.) The

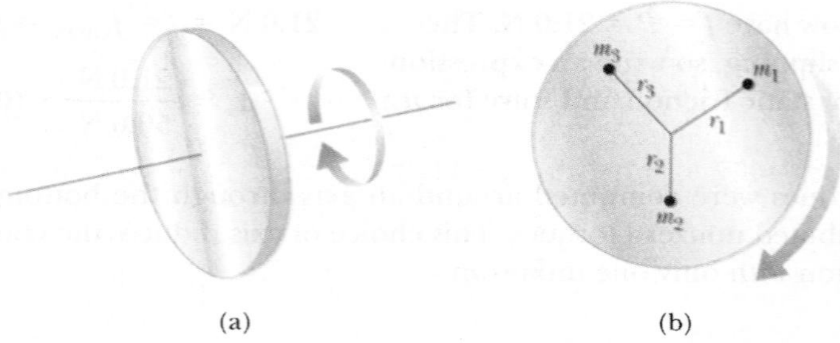

Figure 8.15 (a) A solid disk rotating about its axis. (b) The disk consists of many particles, all with the same angular acceleration.

torque on each one of these particles is given by Equation 8.5. The *net* torque on the disk is given by the sum of the individual torques on all the particles:

$$\Sigma\tau = (\Sigma mr^2)\alpha \qquad [8.6]$$

Because the disk is rigid, all of its particles have the *same* angular acceleration, so α is not involved in the sum. If the masses and distances of the particles are labeled with subscripts as in Figure 8.15b, then

$$\Sigma mr^2 = m_1r_1^2 + m_2r_2^2 + m_3r_3^2 + \cdots$$

This quantity is the moment of inertia, I, of the whole body:

Moment of inertia ▶

$$I \equiv \sum mr^2 \qquad [8.7]$$

The moment of inertia has the SI units kg · m^2. Using this result in Equation 8.6, we see that the net torque on a rigid body rotating about a fixed axis is given by

Rotational analog of Newton's second law ▶

$$\sum\tau = I\alpha \qquad [8.8]$$

Equation 8.8 says that **the angular acceleration of an extended rigid object is proportional to the net torque acting on it.** This equation is the rotational analog of Newton's second law of motion, with torque replacing force, moment of inertia replacing mass, and angular acceleration replacing linear acceleration. Although the moment of inertia of an object is related to its mass, there is an important difference between them. The mass m depends only on the quantity of matter in an object while the moment of inertia, I, depends on both the quantity of matter and its distribution (through the r^2 term in $I = \Sigma mr^2$) in the rigid object.

Quick Quiz 8.2

A constant net torque is applied to an object. Which one of the following will *not* be constant? (a) angular acceleration, (b) angular velocity, (c) moment of inertia, or (d) center of gravity.

Quick Quiz 8.3

The two rigid objects shown in Figure 8.16 have the same mass, radius, and angular speed. If the same braking torque is applied to each, which takes longer to stop? (a) A (b) B (c) more information is needed

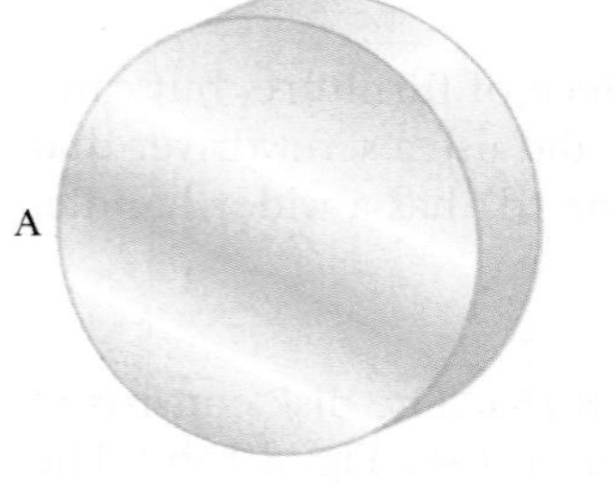

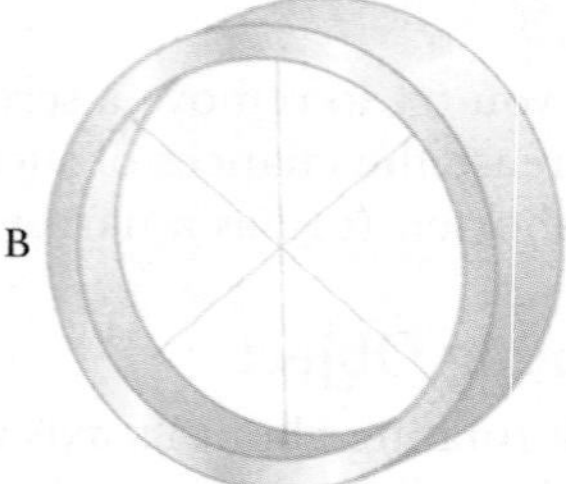

Figure 8.16 (Quick Quiz 8.3)

The gear system on a bicycle provides an easily visible example of the relationship between torque and angular acceleration. Consider first a five-speed gear system in which the drive chain can be adjusted to wrap around any of five gears attached to the back wheel (Fig. 8.17). The gears, with different radii, are concentric with the wheel hub. When the cyclist begins pedaling from rest, the chain is attached to the largest gear. Because it has the largest radius, this gear provides the largest torque to the drive wheel. A large torque is required initially, because the bicycle starts from rest. As the bicycle rolls faster, the tangential speed of the chain increases, eventually becoming too fast for the cyclist to maintain by pushing the pedals. The chain is then moved to a gear with a smaller radius, so the chain has a smaller tangential speed that the cyclist can more easily maintain. This gear doesn't provide as much torque as the first, but the cyclist needs to accelerate only to a somewhat higher speed. This process continues as the bicycle moves faster and faster and the cyclist shifts through all five gears. The fifth gear supplies the lowest torque, but now the main function of that torque is to counter the frictional torque from the rolling tires, which tends to reduce the speed of the bicycle. The small radius of the fifth gear allows the cyclist to keep up with the chain's movement by pushing the pedals.

A 15-speed bicycle has the same gear structure on the drive wheel, but has three gears on the sprocket connected to the pedals. By combining different positions of the chain on the rear gears and the sprocket gears, 15 different torques are available.

APPLICATION

Bicycle Gears

George Semple

Figure 8.17 The drive wheel and gears of a bicycle.

More on the Moment of Inertia

As we have seen, a small object (or a particle) has a moment of inertia equal to mr^2 about some axis. The moment of inertia of a *composite* object about some axis is just the sum of the moments of inertia of the object's components. For example, suppose a majorette twirls a baton as in Figure 8.18. Assume that the baton can be modeled as a very light rod of length 2ℓ with a heavy object at each end. (The rod of a real baton has a significant mass relative to its ends.) Because we are neglecting the mass of the rod, the moment of inertia of the baton about an axis through its center and perpendicular to its length is given by Equation 8.7:

$$I = \Sigma mr^2$$

Because this system consists of two objects with equal masses equidistant from the axis of rotation, $r = \ell$ for each object, and the sum is

$$I = \Sigma mr^2 = m\ell^2 + m\ell^2 = 2m\ell^2$$

If the mass of the rod were not neglected, we would have to include its moment of inertia to find the total moment of inertia of the baton.

We pointed out earlier that I is the rotational counterpart of m. However, there are some important distinctions between the two. For example, mass is an intrinsic property of an object that doesn't change, whereas **the moment of inertia of a system depends on how the mass is distributed and on the location of the axis of rotation**. Example 8.9 illustrates this point.

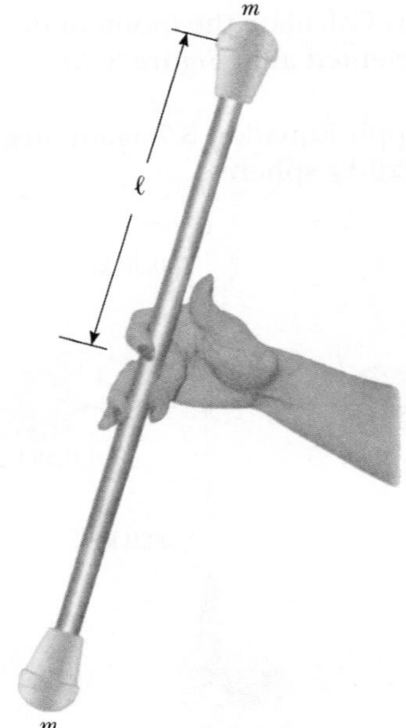

Figure 8.18 A baton of length 2ℓ and mass $2m$. (The mass of the connecting rod is neglected.) The moment of inertia about the axis through the baton's center and perpendicular to its length is $2m\ell^2$.

EXAMPLE 8.9 The Baton Twirler

Goal Calculate a moment of inertia.

Problem In an effort to be the star of the half-time show, a majorette twirls an unusual baton made up of four spheres fastened to the ends of very light rods (Fig. 8.19). Each rod is 1.0 m long. **(a)** Find the moment of inertia of the baton about an axis perpendicular to the page and passing through the point where the rods cross. **(b)** The majorette tries spinning her strange baton about the axis OO', as shown in Figure 8.20. Calculate the moment of inertia of the baton about this axis.

Strategy In Figure 8.19, all four balls contribute to the moment of inertia, whereas in Figure 8.20, with the new axis, only the two balls on the left and right contribute. Technically, the balls on the top and bottom still make a small contribution because they're not really point particles. However, their moment of inertia can be neglected, because the radius of the sphere is much smaller than the radius formed by the rods.

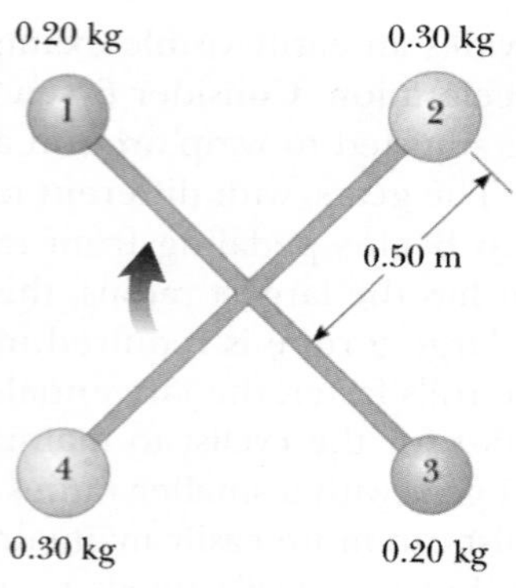

Figure 8.19 (Example 8.9a) Four objects connected to light rods rotating in the plane of the page.

Solution

(a) Calculate the moment of inertia of the baton when oriented as in Figure 8.19.

Apply Equation 8.7, neglecting the mass of the connecting rods:

$$I = \Sigma mr^2 = m_1r_1^2 + m_2r_2^2 + m_3r_3^2 + m_4r_4^2$$
$$= (0.20 \text{ kg})(0.50 \text{ m})^2 + (0.30 \text{ kg})(0.50 \text{ m})^2$$
$$+ (0.20 \text{ kg})(0.50 \text{ m})^2 + (0.30 \text{ kg})(0.50 \text{ m})^2$$
$$I = 0.25 \text{ kg}\cdot\text{m}^2$$

(b) Calculate the moment of inertia of the baton when oriented as in Figure 8.20.

Apply Equation 8.7 again, neglecting the radii of the 0.20-kg spheres.

$$I = \Sigma mr^2 = m_1r_1^2 + m_2r_2^2 + m_3r_3^2 + m_4r_4^2$$
$$= (0.20 \text{ kg})(0)^2 + (0.30 \text{ kg})(0.50 \text{ m})^2$$
$$+ (0.20 \text{ kg})(0)^2 + (0.30 \text{ kg})(0.50 \text{ m})^2$$
$$I = 0.15 \text{ kg}\cdot\text{m}^2$$

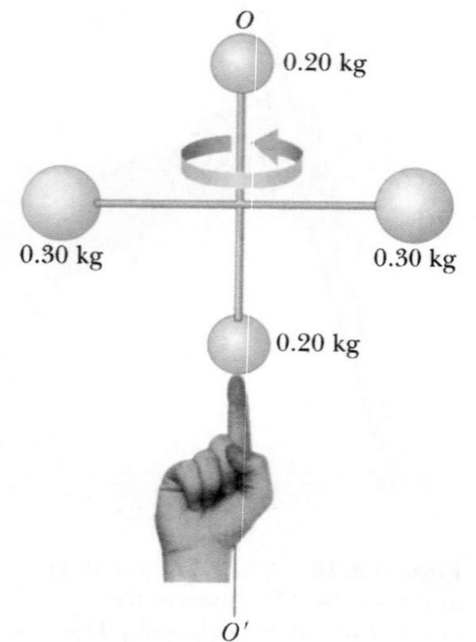

Figure 8.20 (Example 8.9b) A double baton rotating about the axis OO'.

Remarks The moment of inertia is smaller in part (b) because in this configuration the 0.20-kg spheres are essentially located on the axis of rotation.

Exercise 8.9

Yet another bizarre baton is created by taking four identical balls, each with mass 0.300 kg, and fixing them as before, except that one of the rods has a length of 1.00 m and the other has a length of 1.50 m. Calculate the moment of inertia of this baton (a) when oriented as in Figure 8.19; (b) when oriented as in Figure 8.20, with the shorter rod vertical; and (c) when oriented as in Figure 8.20, but with longer rod vertical.

Answers (a) $0.488 \text{ kg}\cdot\text{m}^2$ (b) $0.338 \text{ kg}\cdot\text{m}^2$ (c) $0.150 \text{ kg}\cdot\text{m}^2$

Calculation of Moments of Inertia for Extended Objects

The method used for calculating moments of inertia in Example 8.9 is simple when only a few small objects rotate about an axis. When the object is an extended one, such as a sphere, a cylinder, or a cone, techniques of calculus are often required, unless some simplifying symmetry is present. One such extended object amenable to a simple solution is a hoop rotating about an axis perpendicular to its plane and passing through its center, as shown in Figure 8.21. (A bicycle tire, for example, would approximately fit into this category.)

To evaluate the moment of inertia of the hoop, we can still use the equation $I = \Sigma mr^2$ and imagine that the mass of the hoop M is divided into n small segments having masses $m_1, m_2, m_3, \ldots, m_n$, as in Figure 8.21, with $M = m_1 + m_2 + m_3 + \ldots + m_n$. This approach is just an extension of the baton problem described in the preceding examples, except that now we have a large number of small masses in rotation instead of only four.

We can express the sum for I as

$$I = \Sigma mr^2 = m_1 r_1^2 + m_2 r_2^2 + m_3 r_3^2 + \cdots + m_n r_n^2$$

All of the segments around the hoop are at the *same distance* R from the axis of rotation, so we can drop the subscripts on the distances and factor out R^2 to obtain

$$I = (m_1 + m_2 + m_3 + \cdots + m_n)R^2 = MR^2 \quad [8.9]$$

This expression can be used for the moment of inertia of any ring-shaped object rotating about an axis through its center and perpendicular to its plane. Note that the result is strictly valid only if the thickness of the ring is small relative to its inner radius.

The hoop we selected as an example is unique in that we were able to find an expression for its moment of inertia by using only simple algebra. Unfortunately, for most extended objects the calculation is much more difficult because the mass elements are not all located at the same distance from the axis, so the methods of integral calculus are required. The moments of inertia for some other common shapes are given without proof in Table 8.1. You can use this table as needed to determine the moment of inertia of a body having any one of the listed shapes.

If mass elements in an object are redistributed parallel to the axis of rotation, the moment of inertia of the object doesn't change. Consequently, the expression $I = MR^2$ can be used equally well to find the axial moment of inertia of an embroidery hoop or of a long sewer pipe. Likewise, a door turning on its hinges is described by the same moment-of-inertia expression as that tabulated for a long thin rod rotating about an axis through its end.

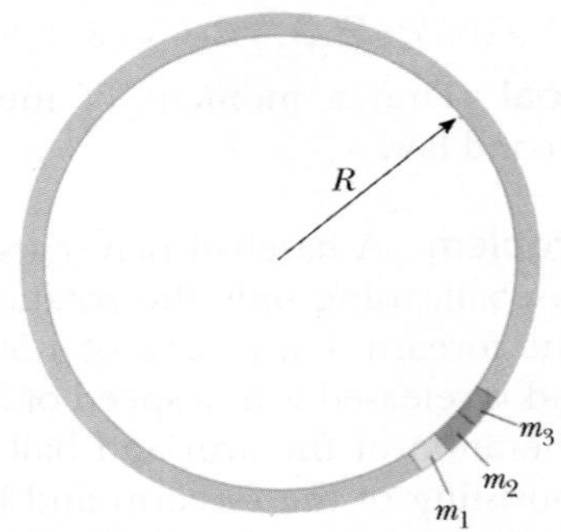

Figure 8.21 A uniform hoop can be divided into a large number of small segments that are equidistant from the center of the hoop.

TIP 8.3 No Single Moment of Inertia

Moment of inertia is analogous to mass, but there are major differences. Mass is an inherent property of an object. The moment of inertia of an object depends on the shape of the object, its mass, and the choice of rotation axis.

TABLE 8.1

Moments of Inertia for Various Rigid Objects of Uniform Composition

Hoop or thin cylindrical shell
$I = MR^2$

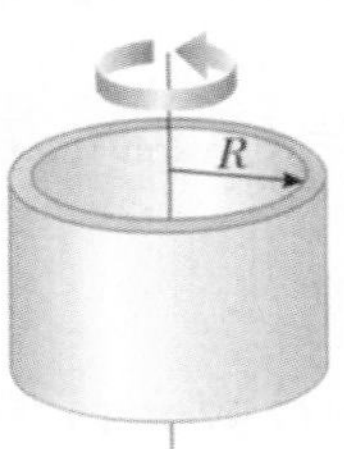

Solid sphere
$I = \frac{2}{5} MR^2$

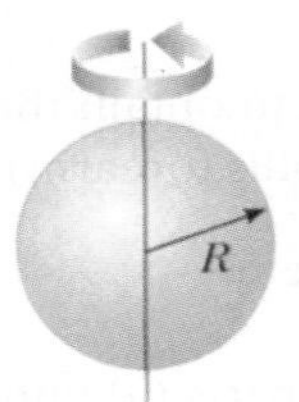

Solid cylinder or disk
$I = \frac{1}{2} MR^2$

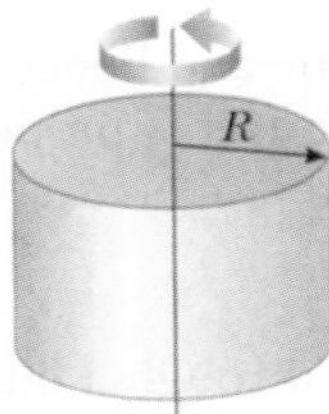

Thin spherical shell
$I = \frac{2}{3} MR^2$

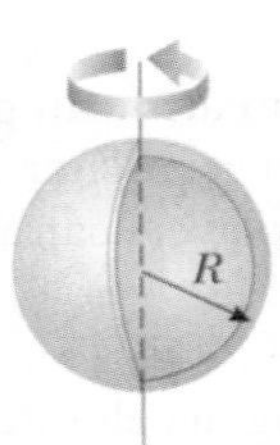

Long thin rod with rotation axis through center
$I = \frac{1}{12} ML^2$

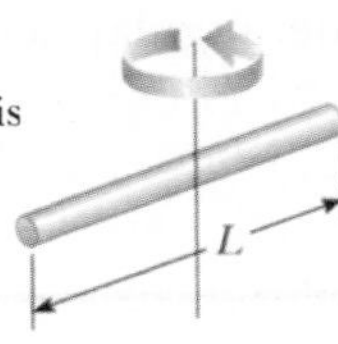

Long thin rod with rotation axis through end
$I = \frac{1}{3} ML^2$

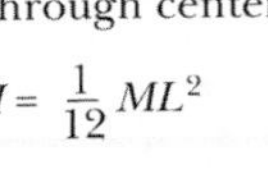

EXAMPLE 8.10 Warming Up

Goal Find a moment of inertia and apply the rotational analog of Newton's second law.

Problem A baseball player loosening up his arm before a game tosses a 0.150-kg baseball, using only the rotation of his forearm to accelerate the ball (Fig. 8.22). The forearm has a mass of 1.50 kg and a length of 0.350 m. The ball starts at rest and is released with a speed of 30.0 m/s in 0.300 s. **(a)** Find the constant angular acceleration of the arm and ball. **(b)** Calculate the moment of inertia of the system consisting of the forearm and ball. **(c)** Find the torque exerted on the system that results in the angular acceleration found in part (a).

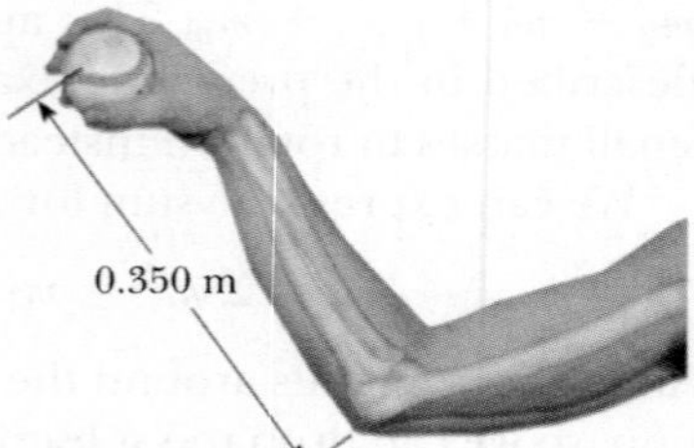

Figure 8.22 (Example 8.10) A ball being tossed by a pitcher. The forearm is used to accelerate the ball.

Strategy The angular acceleration can be found with rotational kinematic equations, while the moment of inertia of the system can be obtained by summing the separate moments of inertia of the ball and forearm. Multiplying these two results together gives the torque.

Solution

(a) Find the angular acceleration of the ball.

The angular acceleration is constant, so use the angular velocity kinematic equation with $\omega_i = 0$.

$$\omega = \omega_i + \alpha t \quad \rightarrow \quad \alpha = \frac{\omega}{t}$$

The ball accelerates along a circular arc with radius given by the length of the forearm. Solve $v = r\omega$ for ω and substitute:

$$\alpha = \frac{\omega}{t} = \frac{v}{rt} = \frac{30.0 \text{ m/s}}{(0.350 \text{ m})(0.300 \text{ s})} = \boxed{286 \text{ rad/s}^2}$$

(b) Find the moment of inertia of the system (forearm plus ball).

Find the moment of inertia of the ball about an axis that passes through the elbow, perpendicular to the arm:

$$I_{\text{ball}} = mr^2 = (0.150 \text{ kg})(0.350 \text{ m})^2 = 1.84 \times 10^{-2} \text{ kg} \cdot \text{m}^2$$

Obtain the moment of inertia of the forearm, modeled as a rod, by consulting Table 8.1:

$$I_{\text{forearm}} = \tfrac{1}{3} ML^2 = \tfrac{1}{3}(1.50 \text{ kg})(0.350 \text{ m})^2 = 6.13 \times 10^{-2} \text{ kg} \cdot \text{m}^2$$

Sum the individual moments of inertia to obtain the moment of inertia of the system (ball plus forearm):

$$I_{\text{system}} = I_{\text{ball}} + I_{\text{forearm}} = \boxed{7.97 \times 10^{-2} \text{ kg} \cdot \text{m}^2}$$

(c) Find the torque exerted on the system.

Apply Equation 8.8, using the results of parts (a) and (b):

$$\tau = I_{\text{system}}\alpha = (7.97 \times 10^{-2} \text{ kg} \cdot \text{m}^2)(286 \text{ rad/s}^2) = \boxed{22.8 \text{ N} \cdot \text{m}}$$

Remarks Notice that having a long forearm can greatly increase the torque and hence the acceleration of the ball. This is one reason it's advantageous for a pitcher to be tall—the pitching arm is proportionately longer. A similar advantage holds in tennis, where taller players can usually deliver faster serves.

Exercise 8.10

A catapult with a radial arm 4.00 m long accelerates a ball of mass 20.0 kg through a quarter circle. The ball leaves the apparatus at 45.0 m/s. If the mass of the arm is 25.0 kg and the acceleration is uniform, find (a) the angular acceleration, (b) the moment of inertia of the arm and ball, and (c) the net torque exerted on the ball and arm. *Hint:* Use the time-independent rotational kinematics equation to find the angular acceleration, rather than the angular velocity equation.

Answers (a) 40.3 rad/s^2 (b) 453 kg · m^2 (c) 1.83×10^4 N · m

INTERACTIVE EXAMPLE 8.11 The Falling Bucket

Goal Combine Newton's second law with its rotational analog.

Problem A solid, frictionless cylindrical reel of mass $M = 3.00$ kg and radius $R = 0.400$ m is used to draw water from a well (Fig. 8.23a). A bucket of mass $m = 2.00$ kg is attached to a cord that is wrapped around the cylinder. **(a)** Find the tension T in the cord and acceleration a of the bucket. **(b)** If the bucket starts from rest at the top of the well and falls for 3.00 s before hitting the water, how far does it fall?

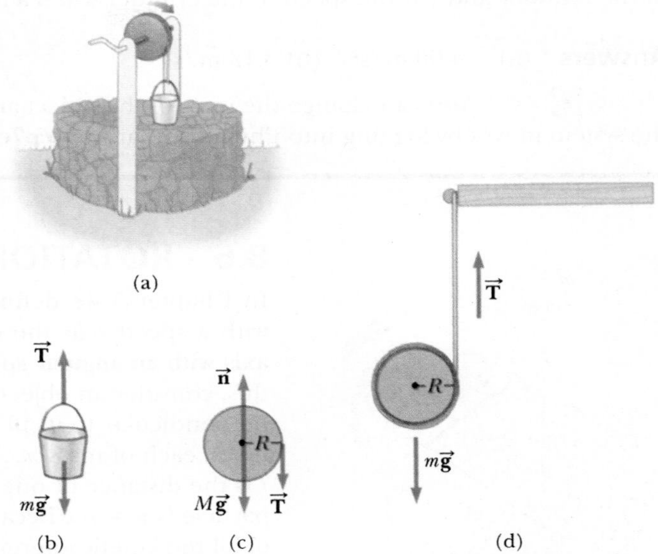

Figure 8.23 (Interactive Example 8.11) (a) A water bucket attached to a rope passing over a frictionless reel. (b) A free-body diagram for the bucket. (c) The tension produces a torque on the cylinder about its axis of rotation. (d) A falling cylinder (Exercise 8.11).

Strategy This problem involves three equations and three unknowns. The three equations are Newton's second law applied to the bucket, $ma = \Sigma F_i$; the rotational version of the second law applied to the cylinder, $I\alpha = \Sigma\tau_i$; and the relationship between linear and angular acceleration, $a = r\alpha$, which connects the dynamics of the bucket and cylinder. The three unknowns are the acceleration a of the bucket, the angular acceleration α of the cylinder, and the tension T in the rope. Assemble the terms of the three equations and solve for the three unknowns by substitution. Part (b) is a review of kinematics.

Solution

(a) Find the tension in the cord and the acceleration of the bucket.

Apply Newton's second law to the bucket in Figure 8.23b. There are two forces: the tension $\vec{\mathbf{T}}$ acting upwards and gravity $m\vec{\mathbf{g}}$ acting downwards.

$$ma = -mg + T \qquad (1)$$

Apply $\tau = I\alpha$ to the cylinder in Figure 8.23c:

$$\Sigma\tau = I\alpha = \tfrac{1}{2}MR^2\alpha \quad \text{(solid cylinder)}$$

Notice the angular acceleration is clockwise, so the torque is negative. The normal and gravity forces have zero moment arm, and don't contribute any torque.

$$-TR = \tfrac{1}{2}MR^2\alpha \qquad (2)$$

Solve for T and substitute $\alpha = a/R$ (notice that both α and a are negative):

$$T = -\tfrac{1}{2}MR\alpha = -\tfrac{1}{2}Ma \qquad (3)$$

Substitute the expression for T in Equation (3) into Equation (1), and solve for the acceleration:

$$ma = -mg - \tfrac{1}{2}Ma \quad \rightarrow \quad a = -\frac{mg}{m + \frac{1}{2}M}$$

Substitute the values for m, M, and g, getting a, then substitute a into Equation (3) to get T.

$$a = -5.60 \text{ m/s}^2 \qquad T = 8.40 \text{ N}$$

(b) Find the distance the bucket falls in 3.00 s.

Apply the displacement kinematic equation for constant acceleration, with $t = 3.00$ s and $v_0 = 0$:

$$\Delta y = v_0 t + \tfrac{1}{2}at^2 = -\tfrac{1}{2}(5.60 \text{ m/s}^2)(3.00 \text{ s})^2 = -25.2 \text{ m}$$

Remarks Proper handling of signs is very important in these problems. All such signs should be chosen initially and checked mathematically and physically. In this problem, for example, both the angular acceleration α and the acceleration a are negative, so $\alpha = a/R$ applies. If the rope had been wound the other way on the cylinder, causing counterclockwise rotation, the torque would have been positive, and the relationship would have been $\alpha = -a/R$, with the double negative making the right-hand side positive, just like the left-hand side.

Exercise 8.11

A hollow cylinder of mass 0.100 kg and radius 4.00 cm has a string wrapped several times around it, as in Figure 8.23d. If the string is attached to a rigid support and the cylinder allowed to drop from rest, find (a) the acceleration of the cylinder and (b) the speed of the cylinder when a meter of string has unwound off of it.

Answers (a) -4.90 m/s^2 (b) 3.13 m/s

PhysicsNow™ You can change the mass of the object and the mass and radius of the wheel to see the effect on how the system moves by logging into PhysicsNow at **www.cp7e.com** and going to Interactive Example 8.11.

8.6 ROTATIONAL KINETIC ENERGY

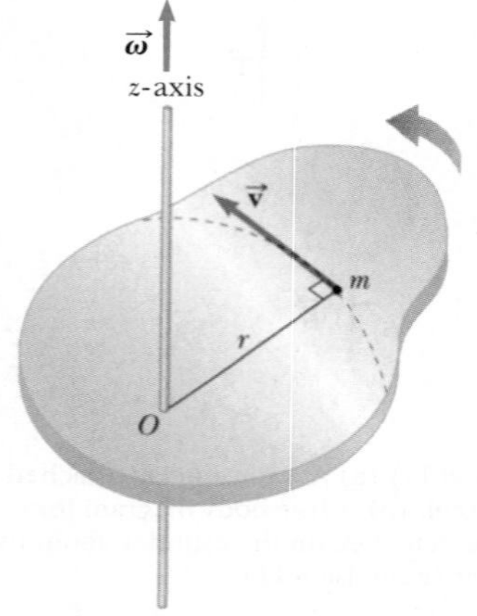

Figure 8.24 A rigid plate rotating about the z-axis with angular speed ω. The kinetic energy of a particle of mass m is $\frac{1}{2}mv^2$. The total kinetic energy of the plate is $\frac{1}{2}I\omega^2$.

In Chapter 5 we defined the kinetic energy of a particle moving through space with a speed v as the quantity $\frac{1}{2}mv^2$. Analogously, **an object rotating about some axis with an angular speed ω has rotational kinetic energy given by $\frac{1}{2}I\omega^2$**. To prove this, consider an object in the shape of a thin rigid plate rotating around some axis perpendicular to its plane, as in Figure 8.24. The plate consists of many small particles, each of mass m. All these particles rotate in circular paths around the axis. If r is the distance of one of the particles from the axis of rotation, the speed of that particle is $v = r\omega$. Because the *total* kinetic energy of the plate's rotation is the sum of all the kinetic energies associated with its particles, we have

$$KE_r = \Sigma(\tfrac{1}{2}mv^2) = \Sigma(\tfrac{1}{2}mr^2\omega^2) = \tfrac{1}{2}(\Sigma mr^2)\omega^2$$

In the last step, the ω^2 term is factored out because it's the same for every particle. Now, the quantity in parentheses on the right is the moment of inertia of the plate in the limit as the particles become vanishingly small, so

$$KE_r = \tfrac{1}{2}I\omega^2 \quad [8.10]$$

where $I = \Sigma mr^2$ is the moment of inertia of the plate.

A system such as a bowling ball rolling down a ramp is described by three types of energy: **gravitational potential energy PE_g, translational kinetic energy KE_t,** and **rotational kinetic energy KE_r**. All these forms of energy, plus the potential energies of any other conservative forces, must be included in our equation for the conservation of mechanical energy of an isolated system:

Conservation of mechanical energy ▶

$$(KE_t + KE_r + PE)_i = (KE_t + KE_r + PE)_f \quad [8.11]$$

where i and f refer to initial and final values, respectively, and PE includes the potential energies of all conservative forces in a given problem. This relation is true *only* if we ignore dissipative forces such as friction. In that case, it's necessary to resort to a generalization of the work–energy theorem:

Work–energy theorem including rotational energy ▶

$$W_{nc} = \Delta KE_t + \Delta KE_r + \Delta PE \quad [8.12]$$

Problem-Solving Strategy Energy Methods and Rotation

1. **Choose two points of interest**, one where all necessary information is known, and the other where information is desired.
2. **Identify** the conservative and nonconservative forces acting on the system being analyzed.
3. **Write the general work–energy theorem**, Equation 8.12, or Equation 8.11 if all forces are conservative.
4. **Substitute general expressions** for the terms in the equation.
5. **Use $v = r\omega$** to eliminate either ω or v from the equation.
6. **Solve** for the unknown.

EXAMPLE 8.12 A Ball Rolling Down an Incline

Goal Combine gravitational, translational, and rotational energy.

Problem A ball of mass M and radius R starts from rest at a height of 2.00 m and rolls down a 30.0° slope, as in Active Figure 8.25. What is the linear speed of the ball when it leaves the incline? Assume that the ball rolls without slipping.

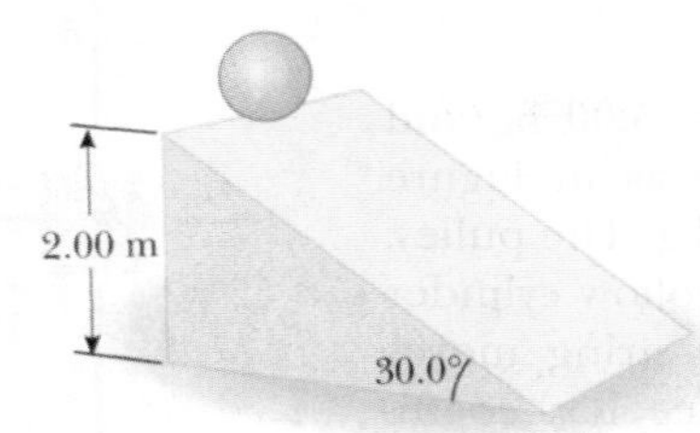

ACTIVE FIGURE 8.25
(Example 8.12) A ball starts from rest at the top of an incline and rolls to the bottom without slipping.

Physics Now™
Log into PhysicsNow at **www.cp7e.com**, and go to Active Figure 8.25 to roll several objects down a hill and see how the final speed depends on the shape of the object.

Strategy The two points of interest are the top and bottom of the incline, with the bottom acting as the zero point of gravitational potential energy. The force of static friction converts translational kinetic energy to rotational kinetic energy without dissipating any energy, so mechanical energy is conserved and Equation 8.11 can be applied.

Solution

Apply conservation of energy with $PE = PE_g$, the potential energy associated with gravity.

$$(KE_t + KE_r + PE_g)_i = (KE_t + KE_r + PE_g)_f$$

Substitute the appropriate general expressions, noting that $(KE_t)_i = (KE_r)_i = 0$ and $(PE_g)_f = 0$ (obtain the moment of inertia of a ball from Table 8.1):

$$0 + 0 + Mgh = \tfrac{1}{2}Mv^2 + \tfrac{1}{2}(\tfrac{2}{5}MR^2)\omega^2 + 0$$

The ball rolls without slipping, so $R\omega = v$, the "no-slip condition," can be applied:

$$Mgh = \tfrac{1}{2}Mv^2 + \tfrac{1}{5}Mv^2 = \tfrac{7}{10}Mv^2$$

Solve for v, noting that M cancels.

$$v = \sqrt{\frac{10gh}{7}} = \sqrt{\frac{10(9.80\ \text{m/s}^2)(2.00\ \text{m})}{7}} = 5.29\ \text{m/s}$$

Exercise 8.12

Repeat this example for a solid cylinder of the same mass and radius as the ball and released from the same height. In a race between the two objects on the incline, which one would win?

Answer $v = \sqrt{4gh/3} = 5.11$ m/s; the ball would win.

Quick Quiz 8.4

Two spheres, one hollow and one solid, are rotating with the same angular speed around an axis through their centers. Both spheres have the same mass and radius. Which sphere, if either, has the higher rotational kinetic energy? (a) The hollow sphere. (b) The solid sphere. (c) They have the same kinetic energy.

Quick Quiz 8.5

Which arrives at the bottom first, (a) a ball rolling without sliding down a certain incline A, (b) a solid cylinder rolling without sliding down incline A, or (c) a box of the same mass as the ball sliding down a frictionless incline B having the same dimensions as A? Assume that each object is released from rest at the top of its incline.

EXAMPLE 8.13 Blocks and Pulley

Goal Solve a system requiring rotation concepts and the work–energy theorem.

Problem Two blocks with masses $m_1 = 5.00$ kg and $m_2 = 7.00$ kg are attached by a string as in Figure 8.26a, over a pulley with mass $M = 2.00$ kg. The pulley, which turns on a frictionless axle, is a hollow cylinder with radius 0.050 0 m over which the string moves without slipping. The horizontal surface has coefficient of kinetic friction 0.350. Find the speed of the system when the block of mass m_2 has dropped 2.00 m.

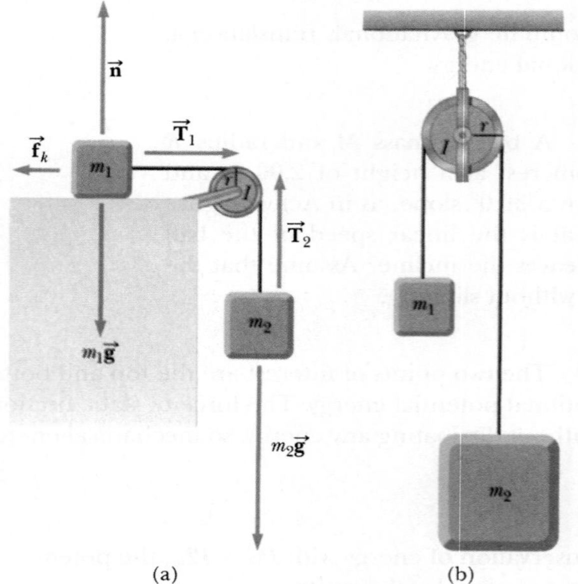

Figure 8.26 (a) (Example 8.13) (b) (Exercise 8.13) In both cases, $\vec{T}_1$ and $\vec{T}_2$ exert torques on the pulley.

Strategy This problem can be solved with the extension of the work–energy theorem, Equation 8.12. If the block of mass m_2 falls from height h to 0, then the block of mass m_1 moves the same distance, $\Delta x = h$. Apply the work–energy theorem, solve for v, and substitute. Kinetic friction the sole nonconservative force.

Solution

Apply the work–energy theorem, with $PE = PE_g$, the potential energy associated with gravity.

$$W_{nc} = \Delta KE_t + \Delta KE_r + \Delta PE_g$$

Substitute the frictional work for W_{nc}, kinetic energy changes for the two blocks, the rotational kinetic energy change for the pulley, and the potential energy change for the second block.

$$-\mu_k n \Delta x = -\mu_k(m_1 g)\Delta x = (\tfrac{1}{2}m_1v^2 - 0) + (\tfrac{1}{2}m_2v^2 - 0) + (\tfrac{1}{2}I\omega^2 - 0) + (0 - m_2gh)$$

Substitute $\Delta x = h$, and write I as $(I/r^2)r^2$:

$$-\mu_k(m_1g)h = \tfrac{1}{2}m_1v^2 + \tfrac{1}{2}m_2v^2 + \tfrac{1}{2}\left(\frac{I}{r^2}\right)r^2\omega^2 - m_2gh$$

For a hoop, $I = Mr^2$ so $(I/r^2) = M$. Substitute this quantity and $v = r\omega$:

$$-\mu_k(m_1g)h = \tfrac{1}{2}m_1v^2 + \tfrac{1}{2}m_2v^2 + \tfrac{1}{2}Mv^2 - m_2gh$$

Solve for v:

$$m_2gh - \mu_k(m_1g)h = \tfrac{1}{2}m_1v^2 + \tfrac{1}{2}m_2v^2 + \tfrac{1}{2}Mv^2 = \tfrac{1}{2}(m_1 + m_2 + M)v^2$$

$$v = \sqrt{\frac{2gh(m_2 - \mu_k m_1)}{m_1 + m_2 + M}}$$

Substitute $m_1 = 5.00$ kg, $m_2 = 7.00$ kg, $M = 2.00$ kg, $g = 9.80$ m/s^2, $h = 2.00$ m, and $\mu_k = 0.350$:

$$v = 3.83 \text{ m/s}$$

Remarks In the expression for the speed v, the mass m_1 of the first block and the mass M of the pulley all appear in the denominator, reducing the speed, as they should. In the numerator, m_2 is positive while the friction term is negative. Both assertions are reasonable, because the force of gravity on m_2 increases the speed of the system while the force of friction on m_1 slows it down. This problem can also be solved with Newton's second law together with $\tau = I\alpha$, a difficult exercise (though it can be facilitated with a system approach).

Exercise 8.13

Two blocks with masses $m_1 = 2.00$ kg and $m_2 = 9.00$ kg are attached over a pulley with mass $M = 3.00$ kg, hanging straight down as in Atwood's machine (Fig. 8.26b). The pulley is a solid cylinder with radius 0.050 0 m, and there is

some friction in the axle. The system is released from rest, and the string moves without slipping over the pulley. If the larger mass is traveling at a speed of 2.50 m/s when it has dropped 1.00 m, how much mechanical energy was lost due to friction in the pulley's axle?
[*Hint:* This exercise is slightly easier than the associated example because the friction force need not be determined.]

Answer 29.5 J

8.7 ANGULAR MOMENTUM

In Figure 8.27, an object of mass m rotates in a circular path of radius r, acted on by a net force, $\vec{\mathbf{F}}_{\text{net}}$. The resulting net torque on the object increases its angular speed from the value ω_0 to the value ω in a time interval Δt. Therefore, we can write

$$\Sigma\tau = I\alpha = I\frac{\Delta\omega}{\Delta t} = I\left(\frac{\omega - \omega_0}{\Delta t}\right) = \frac{I\omega - I\omega_0}{\Delta t}$$

If we define the product

$$L \equiv I\omega \qquad [8.13]$$

◀ Definition of angular momentum

as the **angular momentum** of the object, then we can write

$$\Sigma\tau = \frac{\text{change in angular momentum}}{\text{time interval}} = \frac{\Delta L}{\Delta t} \qquad [8.14]$$

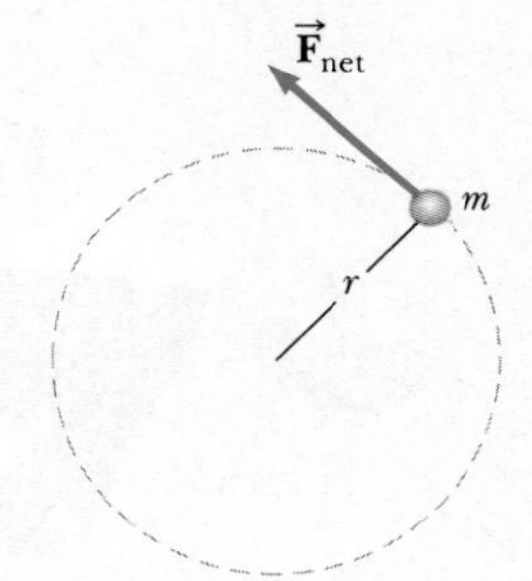

Figure 8.27 An object of mass m rotating in a circular path under the action of a constant torque.

Equation 8.14 is the rotational analog of Newton's second law in the form $F = \Delta p/\Delta t$ and states that **the net torque acting on an object is equal to the time rate of change of the object's angular momentum**. Recall that this equation also parallels the impulse–momentum theorem.

When the net external torque ($\Sigma\tau$) acting on a system is zero, Equation 8.14 gives that $\Delta L/\Delta t = 0$, which says that the time rate of change of the system's angular momentum is zero. We then have the following important result:

◀ Conservation of angular momentum

Let L_i and L_f be the angular momenta of a system at two different times, and suppose there is no net external torque, so $\Sigma\tau = 0$. Then

$$L_i = L_f \qquad [8.15]$$

and angular momentum is said to be *conserved.*

Equation 8.15 gives us a third conservation law to add to our list: **conservation of angular momentum**. We can now state that **the mechanical energy, linear momentum, and angular momentum of an isolated system all remain constant.**

If the moment of inertia of an isolated rotating system changes, the system's angular speed will change. Conservation of angular momentum then requires that

$$I_i\omega_i = I_f\omega_f \quad \text{if} \quad \Sigma\tau = 0 \qquad [8.16]$$

Note that conservation of angular momentum applies to macroscopic objects such as planets and people, as well as to atoms and molecules. There are many examples of conservation of angular momentum; one of the most dramatic is that of a figure skater spinning in the finale of her act. In Figure 8.28a, the skater has pulled her arms and legs close to her body, reducing their distance from her axis of rotation and hence also reducing her moment of inertia. By conservation of angular momentum, a reduction in her moment of inertia must increase her angular velocity. Coming out of the spin in Figure 8.28b, she needs to reduce her angular velocity, so she extends her arms and legs again, increasing her moment of inertia and thereby slowing her rotation.

APPLICATION

Figure Skating

© Benson Krista Hicks/Corbis Sygma

(a) (b)

Figure 8.28 Michelle Kwan controls her moment of inertia. (a) By pulling in her arms and legs, she reduces her moment of inertia and increases her angular velocity (rate of spin). (b) Upon landing, extending her arms and legs increases her moment of inertia and helps slow her spin.

APPLICATION

Aerial Somersaults

© Patrick Giardino/Corbis

Tightly curling her body, a diver decreases her moment of inertia, increasing her angular velocity.

Similarly, when a diver or an acrobat wishes to make several somersaults, she pulls her hands and feet close to the trunk of her body in order to rotate at a greater angular speed. In this case, the external force due to gravity acts through her center of gravity and hence exerts no torque about her axis of rotation, so the angular momentum about her center of gravity is conserved. For example, when a diver wishes to double her angular speed, she must reduce her moment of inertia to half its initial value.

An interesting astrophysical example of conservation of angular momentum occurs when a massive star, at the end of its lifetime, uses up all its fuel and collapses under the influence of gravitational forces, causing a gigantic outburst of energy called a supernova. The best-studied example of a remnant of a supernova explosion is the Crab Nebula, a chaotic, expanding mass of gas (Fig. 8.29). In a supernova, part of the star's mass is ejected into space, where it eventually condenses into new stars and planets. Most of what is left behind typically collapses into a **neutron star**—an extremely dense sphere of matter with a diameter of about 10 km, greatly reduced from the 10^6-km diameter of the original star and containing a large fraction of the star's original mass. In a neutron star, pressures become so great that atomic electrons combine with protons, becoming neutrons. As the moment of inertia of the system decreases

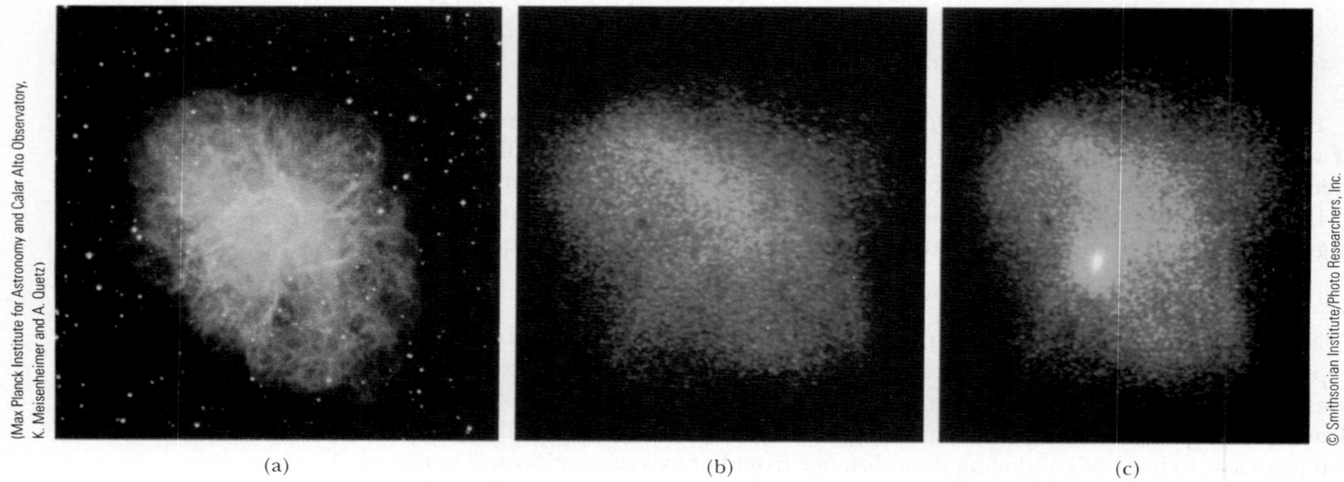

(Max Planck Institute for Astronomy and Calar Alto Observatory, K. Meisenheimer and A. Quetz)

© Smithsonian Institute/Photo Researchers, Inc.

(a) (b) (c)

Figure 8.29 (a) The Crab Nebula in the constellation Taurus. This nebula is the remnant of a supernova seen on Earth in A.D. 1054. It is located some 6 300 lightyears away and is approximately 6 lightyears in diameter, still expanding outward. A pulsar deep inside the nebula flashes 30 times every second. (b) Pulsar off. (c) Pulsar on.

during the collapse, the star's rotational speed increases. More than 700 rapidly rotating neutron stars have been identified since their first discovery in 1967, with periods of rotation ranging from a millisecond to several seconds. The neutron star is an amazing system—an object with a mass greater than the Sun, fitting comfortably within the space of a small county and rotating so fast that the tangential speed of the surface approaches a sizeable fraction of the speed of light!

APPLICATION

Rotating Neutron Stars

Quick Quiz 8.6

A horizontal disk with moment of inertia I_1 rotates with angular speed ω_1 about a vertical frictionless axle. A second horizontal disk, with moment of inertia I_2 and initially not rotating, drops onto the first. Because their surfaces are rough, the two eventually reach the same angular speed ω. The ratio ω/ω_1 is equal to (a) I_1/I_2 (b) I_2/I_1 (c) $I_1/(I_1 + I_2)$ (d) $I_2/(I_1 + I_2)$

Quick Quiz 8.7

If global warming continues, it's likely that some ice from the polar ice caps of the Earth will melt and the water will be distributed closer to the Equator. If this occurs, would the length of the day (one revolution) (a) increase, (b) decrease, or (c) remain the same?

EXAMPLE 8.14 The Spinning Stool

Goal Apply conservation of angular momentum to a simple system.

Problem A student sits on a pivoted stool while holding a pair of weights. (See Fig. 8.30.) The stool is free to rotate about a vertical axis with negligible friction. The moment of inertia of student, weights, and stool is 2.25 kg · m^2. The student is set in rotation with arms outstretched, making one complete turn every 1.26 s, arms outstretched. **(a)** What is the initial angular speed of the system? **(b)** As he rotates, he pulls the weights inward so that the new moment of inertia of the system (student, objects, and stool) becomes 1.80 kg · m^2. What is the new angular speed of the system? **(c)** Find the work done by the student on the system while pulling in the weights. (Ignore energy lost through dissipation in his muscles.)

Figure 8.30 (Example 8.14) (a) The student is given an initial angular speed while holding two weights out. (b) The angular speed increases as the student draws the weights inwards.

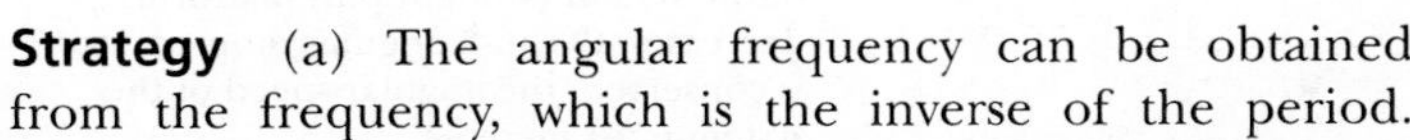

Strategy (a) The angular frequency can be obtained from the frequency, which is the inverse of the period. (b) There are no external torques acting on the system, so the new angular speed can be found with the principle of conservation of angular momentum. (c) The work done on the system during this process is the same as the system's change in rotational kinetic energy.

Solution

(a) Find the initial angular speed of the system.

Invert the period to get the frequency, and multiply by 2π:

$$\omega_i = 2\pi f = 2\pi/T = \boxed{4.99 \text{ rad/s}}$$

(b) After he pulls the weights in, what's the system's new angular speed?

Equate the initial and final angular momenta of the system:

$$L_i = L_f \quad \rightarrow \quad I_i\omega_i = I_f\omega_f \qquad (1)$$

Substitute and solve for the final angular speed ω_f:

$$(2.25\ \text{kg}\cdot\text{m}^2)(4.99\ \text{rad/s}) = (1.80\ \text{kg}\cdot\text{m}^2)\omega_f \quad (2)$$

$$\omega_f = 6.24\ \text{rad/s}$$

(c) Find the work the student does on the system.

Apply the work–energy theorem:

$$\begin{aligned} W_{\text{student}} &= \Delta K_r = \tfrac{1}{2}I_f\omega_f^2 - \tfrac{1}{2}I_i\omega_i^2 \\ &= \tfrac{1}{2}(1.80\ \text{kg}\cdot\text{m}^2)(6.24\ \text{rad/s})^2 \\ &\quad - \tfrac{1}{2}(2.25\ \text{kg}\cdot\text{m}^2)(4.99\ \text{rad/s})^2 \end{aligned}$$

$$W_{\text{student}} = 7.03\ \text{J}$$

Remarks Although the angular momentum of the system is conserved, mechanical energy is not conserved because the student does work on the system.

Exercise 8.14

A star with an initial radius of 1.0×10^8 m and period of 30.0 days collapses suddenly to a radius of 1.0×10^4 m. (a) Find the period of rotation after collapse. (b) Find the work done by gravity during the collapse if the mass of the star is 2.0×10^{30} kg. (c) What is the speed of an indestructible person standing on the equator of the collapsed star? (Neglect any relativistic or thermal effects, and assume the star is spherical before and after it collapses.)

Answers (a) 2.6×10^{-2} s (b) 2.3×10^{42} J (c) 2.4×10^6 m/s

EXAMPLE 8.15 The Merry-Go-Round

Goal Apply conservation of angular momentum while combining two moments of inertia.

Problem A merry-go-round modeled as a disk of mass $M = 1.00 \times 10^2$ kg and radius $R = 2.00$ m is rotating in a horizontal plane about a frictionless vertical axle (Fig. 8.31). **(a)** After a student with mass $m = 60.0$ kg jumps onto the merry-go-round, the system's angular speed decreases to 2.00 rad/s. If the student walks slowly from the edge toward the center, find the angular speed of the system when she reaches a point 0.500 m from the center. **(b)** Find the change in the system's rotational kinetic energy caused by her movement to the center. **(c)** Find the work done on the student as she walks to $r = 0.500$ m.

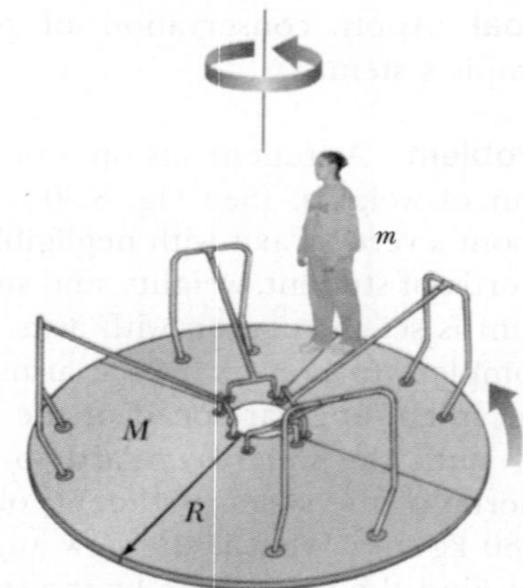

Figure 8.31 (Example 8.15) As the student walks toward the center of the rotating platform, the moment of inertia of the system (student plus platform) decreases. Because angular momentum is conserved, the angular speed of the system must increase.

Strategy This problem can be solved with conservation of angular momentum by equating the system's initial angular momentum when the student stands at the rim to the angular momentum when the student has reached $r = 0.500$ m. The key is to find the different moments of inertia.

Solution

(a) Find the angular speed when the student reaches a point 0.500 m from the center.

Calculate the moment of inertia of the disk, I_D:

$$\begin{aligned} I_D &= \tfrac{1}{2}MR^2 = \tfrac{1}{2}(1.00 \times 10^2\ \text{kg})(2.00\ \text{m})^2 \\ &= 2.00 \times 10^2\ \text{kg}\cdot\text{m}^2 \end{aligned}$$

Calculate the initial moment of inertia of the student. This is the same as the moment of inertia of a mass a distance R from the axis:

$$I_S = mR^2 = (60.0\ \text{kg})(2.00\ \text{m})^2 = 2.40 \times 10^2\ \text{kg}\cdot\text{m}^2$$

Sum the two moments of inertia and multiply by the initial angular speed to find L_i, the initial angular momentum of the system:

$$\begin{aligned} L_i &= (I_D + I_S)\omega_i \\ &= (2.00 \times 10^2\ \text{kg}\cdot\text{m}^2 + 2.40 \times 10^2\ \text{kg}\cdot\text{m}^2)(2.00\ \text{rad/s}) \\ &= 8.80 \times 10^2\ \text{kg}\cdot\text{m}^2/\text{s} \end{aligned}$$

Calculate the student's final moment of inertia, I_{Sf}, when she is 0.500 m from the center:

$$I_{Sf} = mr_f^2 = (60.0 \text{ kg})(0.50 \text{ m})^2 = 15.0 \text{ kg}\cdot\text{m}^2$$

The moment of inertia of the platform is unchanged. Add it to the student's final moment of inertia, and multiply by the unknown final angular speed to find L_f:

$$L_f = (I_D + I_{Sf})\omega_f = (2.00 \times 10^2 \text{ kg}\cdot\text{m}^2 + 15.0 \text{ kg}\cdot\text{m}^2)\omega_f$$

Equate the initial and final angular momenta and solve for the final angular speed of the system:

$$L_i = L_f$$

$$(8.80 \times 10^2 \text{ kg}\cdot\text{m}^2/\text{s}) = (2.15 \times 10^2 \text{ kg}\cdot\text{m}^2)\omega_f$$

$$\omega_f = 4.09 \text{ rad/s}$$

(b) Find the change in the rotational kinetic energy of the system.

Calculate the initial kinetic energy of the system:

$$KE_i = \tfrac{1}{2}I_i\omega_i^2 = \tfrac{1}{2}(4.40 \times 10^2 \text{ kg}\cdot\text{m}^2)(2.00 \text{ rad/s})^2 = 8.80 \times 10^2 \text{ J}$$

Calculate the final kinetic energy of the system:

$$KE_f = \tfrac{1}{2}I_f\omega_f^2 = \tfrac{1}{2}(215 \text{ kg}\cdot\text{m}^2)(4.09 \text{ rad/s})^2 = 1.80 \times 10^3 \text{ J}$$

Calculate the change in kinetic energy of the system.

$$KE_f - KE_i = 920 \text{ J}$$

(c) Find the work done on the student.

The student undergoes a change in kinetic energy that equals the work done on her. Apply the work–energy theorem:

$$W = \Delta KE_{\text{student}} = \tfrac{1}{2}I_{Sf}\omega_f^2 - \tfrac{1}{2}I_S\omega_i^2$$
$$= \tfrac{1}{2}(15.0 \text{ kg}\cdot\text{m}^2)(4.09 \text{ rad/s})^2 - \tfrac{1}{2}(2.40 \times 10^2 \text{ kg}\cdot\text{m}^2)(2.00 \text{ rad/s})^2$$

$$W = -355 \text{ J}$$

Remarks The angular momentum is unchanged by internal forces; however, the kinetic energy increases, because the student must perform positive work in order to walk toward the center of the platform.

Exercise 8.15

(a) Find the angular speed of the merry-go-round before the student jumped on, assuming the student didn't transfer any momentum or energy as she jumped on the merry-go-round. (b) By how much did the kinetic energy of the system change when the student jumped on? Notice that energy is lost in this process, as should be expected, since it is essentially a perfectly inelastic collision.

Answers (a) 4.4 rad/s (b) $KE_f - KE_i = -1.06 \times 10^3$ J.

SUMMARY

Physics Now™ Take a practice test by logging into PhysicsNow at **www.cp7e.com** and clicking on the Pre-Test link for this chapter.

8.1 Torque

Let $\vec{\mathbf{F}}$ be a force acting on an object, and let $\vec{\mathbf{r}}$ be a position vector from a chosen point O to the point of application of the force. Then the magnitude of the torque $\vec{\tau}$ of the force $\vec{\mathbf{F}}$ is given by

$$\tau = rF\sin\theta \qquad [8.2]$$

where r is the length of the position vector, F the magnitude of the force, and θ the angle between $\vec{\mathbf{F}}$ and $\vec{\mathbf{r}}$.

The quantity $d = r\sin\theta$ is called the *lever arm* of the force.

8.2 Torque and the Two Conditions for Equilibrium

An object in mechanical equilibrium must satisfy the following two conditions:

1. The net external force must be zero: $\Sigma\vec{\mathbf{F}} = 0$.
2. The net external torque must be zero: $\Sigma\vec{\tau} = 0$.

These two conditions, used in solving problems involving rotation in a plane—result in three equations and three unknowns—two from the first condition (corresponding to the x- and y-components of the force) and one from the second condition, on torques. These equations must be solved simultaneously.

8.5 Relationship Between Torque and Angular Acceleration

The **moment of inertia** of a group of particles is

$$I \equiv \Sigma mr^2 \quad [8.7]$$

If a rigid object free to rotate about a fixed axis has a net external torque $\Sigma\tau$ acting on it, then the object undergoes an angular acceleration α, where

$$\Sigma\tau = I\alpha \quad [8.8]$$

This equation is the rotational equivalent of the second law of motion.

Problems are solved by using Equation 8.8 together with Newton's second law and solving the resulting equations simultaneously. The relation $a = r\alpha$ is often key in relating the translational equations to the rotational equations.

8.6 Rotational Kinetic Energy

If a rigid object rotates about a fixed axis with angular speed ω, its **rotational kinetic energy** is

$$KE_r \equiv \tfrac{1}{2}I\omega^2 \quad [8.10]$$

where I is the moment of inertia of the object around the axis of rotation.

A system involving rotation is described by three types of energy: potential energy PE, translational kinetic energy KE_t, and rotational kinetic energy KE_r. All these forms of energy must be included in the equation for conservation of mechanical energy for an isolated system:

$$(KE_t + KE_r + PE)_i = (KE_t + KE_r + PE)_f \quad [8.11]$$

where i and f refer to initial and final values, respectively. When non-conservative forces are present, it's necessary to use a generalization of the work–energy theorem:

$$W_{nc} = \Delta KE_t + \Delta KE_r + \Delta PE \quad [8.12]$$

8.7 Angular Momentum

The **angular momentum** of a rotating object is given by

$$L \equiv I\omega \quad [8.13]$$

Angular momentum is related to torque in the following equation:

$$\Sigma\tau = \frac{\text{change in angular momentum}}{\text{time interval}} = \frac{\Delta L}{\Delta t} \quad [8.14]$$

If the net external torque acting on a system is zero, then the total angular momentum of the system is constant,

$$L_i = L_f \quad [8.15]$$

and is said to be conserved. Solving problems usually involves substituting into the expression

$$I_i\omega_i = I_f\omega_f \quad [8.16]$$

and solving for the unknown.

CONCEPTUAL QUESTIONS

1. Why can't you put your heels firmly against a wall and then bend over without falling?
2. Why does a tall athlete have an advantage over a smaller one when the two are competing in the high jump?
3. Both torque and work are products of force and distance. How are they different? Do they have the same units?
4. Is it possible to calculate the torque acting on a rigid object without specifying an origin? Is the torque independent of the location of the origin?
5. Can an object be in equilibrium when only one force acts on it? If you believe the answer is yes, give an example to support your conclusion.
6. The polar ice caps contain about 2.3×10^{19} kg of ice. This mass contributes almost nothing to the moment of inertia of the Earth because it is located at the poles, close to the Earth's axis of rotation. *Estimate* the change in the length of the day that would be expected if the polar ice caps were to melt and the water were distributed uniformly over the surface of the Earth. (Note that the moment of inertia of a thin spherical shell of radius r and mass m is $2mr^2/3$.) (Question 6 is courtesy of Edward F. Redish. For more questions of this type, see www.physics.umd.edu/perg/.)
7. In some motorcycle races, the riders drive over small hills, and the motorcycle becomes airborne for a short time. If the motorcycle racer keeps the throttle open while leaving the hill and going into the air, the motorcycle's nose tends to rise upwards. Why does this happen?
8. In the movie *Jurassic Park*, there is a scene in which some members of the visiting group are trapped in the kitchen with dinosaurs outside. The paleontologist is pressing against the center of the door, trying to keep out the dinosaurs on the other side. The botanist throws herself against the door at the edge near the hinge. A pivotal point in the film is that she cannot reach a gun on the floor because she is trying to hold the door closed. If the paleontologist is pressing at the center of the door, and the botanist is pressing at the edge about 8 cm from the hinge, estimate how far the paleontologist would have to relocate in order to have a greater effect on keeping the door closed than both of them pushing together have in their original positions. (Question 8 is courtesy of Edward F. Redish. For more questions of this type, see www.physics.umd.edu/perg/.)
9. Suppose you are designing a car for a coasting race—a race in which the cars have no engines, but simply coast downhill. Do you want large wheels or small wheels? Do you want solid, disklike wheels or hooplike wheels? Should the wheels be heavy or light?
10. If you toss a textbook into the air, rotating it each time about one of the three axes perpendicular to it, you will find that it will not rotate smoothly about one of those axes. (Try placing a strong rubber band around the book

before the toss so that it will stay closed.) The book's rotation is stable about those axes having the largest and smallest moments of inertia, but unstable about the axis of intermediate moment. Try this on your own to find the axis that has this intermediate moment of inertia.

11. Stars originate as large bodies of slowly rotating gas. Because of gravity, these clumps of gas slowly decrease in size. What happens to the angular speed of a star as it shrinks? Explain.

12. If a high jumper positions his body correctly when going over the bar, the center of gravity of the athlete may actually pass under the bar. (See Fig. Q8.12.) Explain how this is possible.

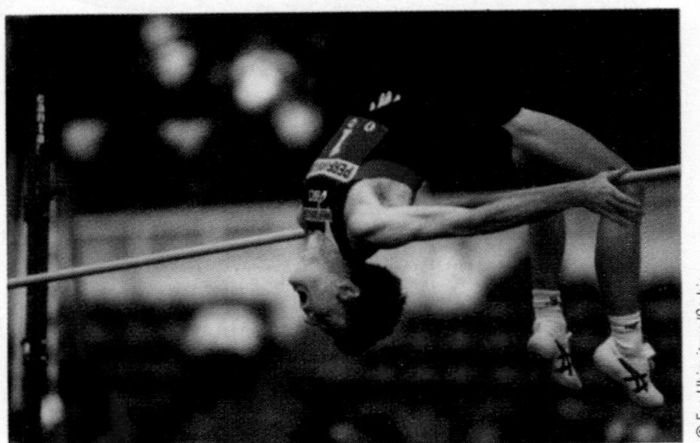

Figure Q8.12

13. In a tape recorder, the tape is pulled past the read–write heads at a constant speed by the drive mechanism. Consider the reel from which the tape is pulled: As the tape is pulled off, the radius of the roll of remaining tape decreases. How does the torque on the reel change with time? How does the angular speed of the reel change with time? If the tape mechanism is suddenly turned on so that the tape is quickly pulled with a large force, is the tape more likely to break when pulled from a nearly full reel or from a nearly empty reel?

14. (a) Give an example in which the net force acting on an object is zero, yet the net torque is nonzero. (b) Give an example in which the net torque acting on an object is zero, yet the net force is nonzero.

15. A mouse is initially at rest on a horizontal turntable mounted on a frictionless vertical axle. If the mouse begins to walk clockwise around the perimeter of the table, what happens to the turntable? Explain.

16. A cat usually lands on its feet regardless of the position from which it is dropped. A slow-motion film of a cat falling shows that the upper half of its body twists in one direction while the lower half twists in the opposite direction. (See Fig. Q8.16.) Why does this type of rotation occur?

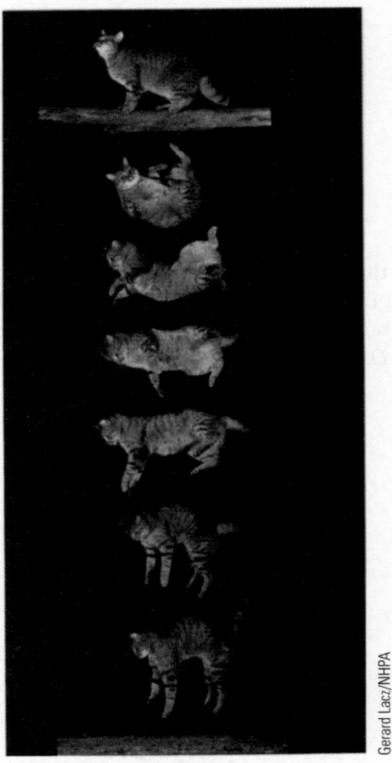

Figure Q8.16 A falling, twisting cat.

17. A ladder rests inclined against a wall. Would you feel safer climbing up the ladder if you were told that the floor was frictionless, but the wall was rough, or that the wall was frictionless, but the floor was rough? Justify your answer.

18. Two solid spheres—a large, massive sphere and a small sphere with low mass—are rolled down a hill. Which one reaches the bottom of the hill first? Next, we roll a large, low-density sphere and a small, high-density sphere, both with the same mass. Which one wins the race?

PROBLEMS

1, 2, 3 = straightforward, intermediate, challenging ☐ = full solution available in *Student Solutions Manual/Study Guide*
PhysicsNow™ = coachedproblem with hints available at **www.cp7e.com** = biomedical application

Section 8.1 Torque

1. If the torque required to loosen a nut that is holding a flat tire in place on a car has a magnitude of 40.0 N·m, what *minimum* force must be exerted by the mechanic at the end of a 30.0-cm lug wrench to accomplish the task?

2. A steel band exerts a horizontal force of 80.0 N on a tooth at point B in Figure P8.2. What is the torque on the root of the tooth about point A?

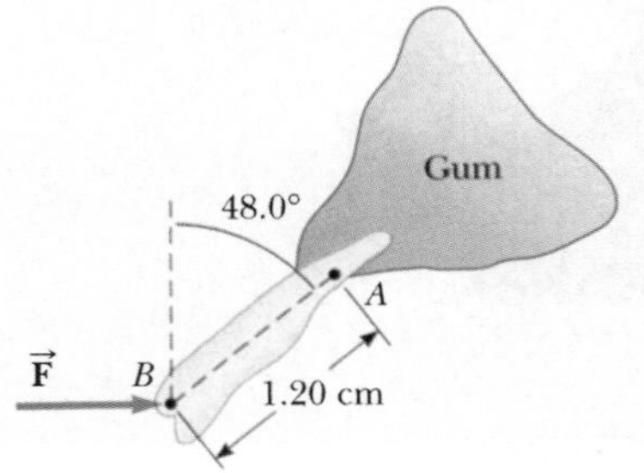

Figure P8.2

3. Calculate the net torque (magnitude and direction) on the beam in Figure P8.3 about (a) an axis through O perpendicular to the page and (b) an axis through C perpendicular to the page.

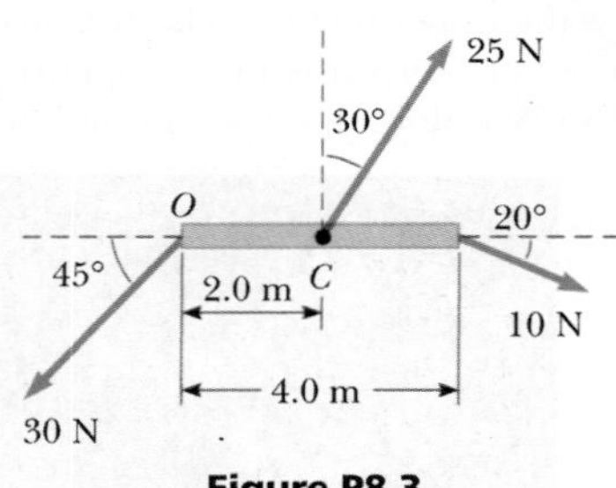

Figure P8.3

4. Write the necessary equations of equilibrium of the object shown in Figure P8.4. Take the origin of the torque equation about an axis perpendicular to the page through the point O.

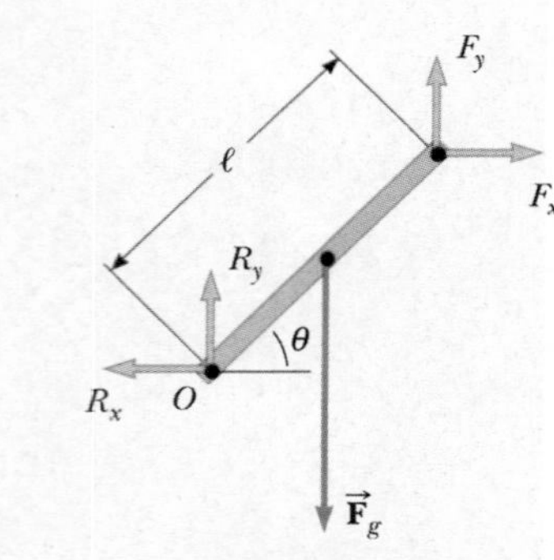

Figure P8.4

5. A simple pendulum consists of a small object of mass 3.0 kg hanging at the end of a 2.0-m-long light string that is connected to a pivot point. Calculate the magnitude of the torque (due to the force of gravity) about this pivot point when the string makes a 5.0° angle with the vertical.

6. A fishing pole is 2.00 m long and inclined to the horizontal at an angle of 20.0° (Fig. P8.6). What is the torque exerted by the fish about an axis perpendicular to the page and passing through the hand of the person holding the pole?

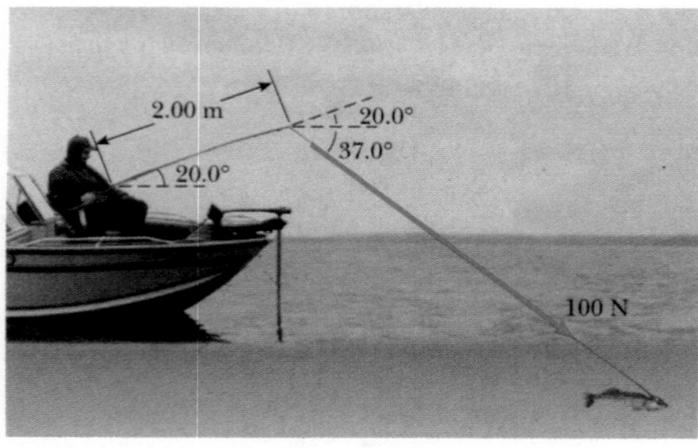

Figure P8.6

Section 8.2 Torque and the Two Conditions for Equilibrium
Section 8.3 The Center of Gravity
Section 8.4 Examples of Objects in Equilibrium

7. The arm in Figure P8.7 weighs 41.5 N. The force of gravity acting on the arm acts through point A. Determine the magnitudes of the tension force $\vec{F}_t$ in the deltoid muscle and the force $\vec{F}_s$ exerted by the shoulder on the humerus (upper-arm bone) to hold the arm in the position shown.

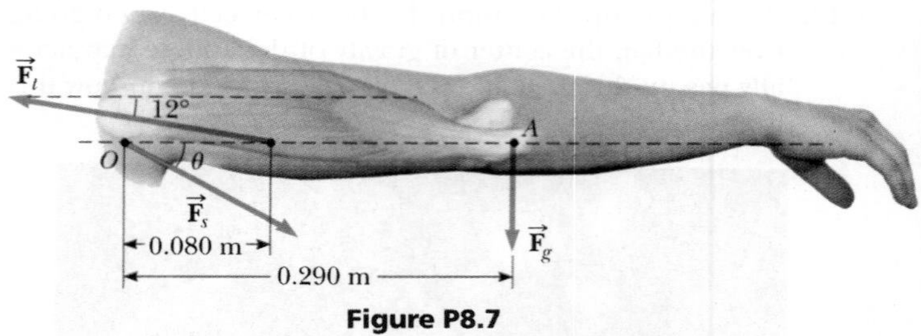

Figure P8.7

8. A water molecule consists of an oxygen atom with two hydrogen atoms bound to it as shown in Figure P8.8. The bonds are 0.100 nm in length, and the angle between the two bonds is 106°. Use the coordinate axes shown, and determine the location of the center of gravity of the molecule. Take the mass of an oxygen atom to be 16 times the mass of a hydrogen atom.

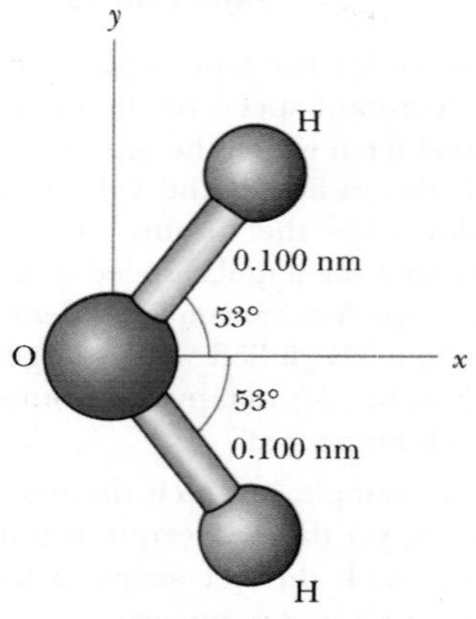

Figure P8.8

9. A cook holds a 2.00-kg carton of milk at arm's length (Fig. P8.9). What force $\vec{F}_B$ must be exerted by the biceps muscle? (Ignore the weight of the forearm.)

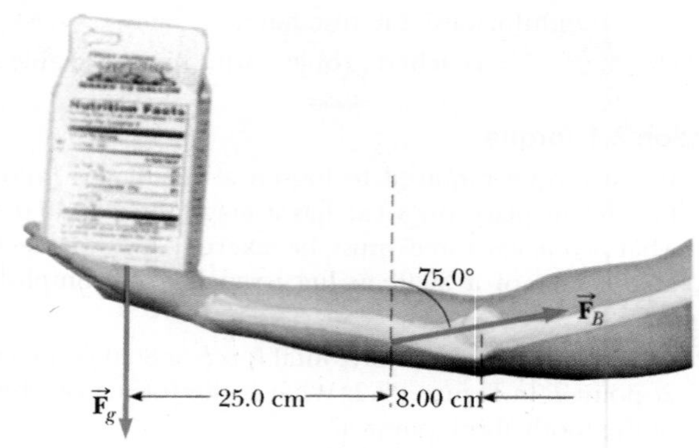

Figure P8.9

10. A meterstick is found to balance at the 49.7-cm mark when placed on a fulcrum. When a 50.0-gram mass is attached at the 10.0-cm mark, the fulcrum must be moved to the 39.2-cm mark for balance. What is the mass of the meter stick?

11. Find the x- and y-coordinates of the center of gravity of a 4.00-ft by 8.00-ft uniform sheet of plywood with the upper right quadrant removed as shown in Figure P8.11.

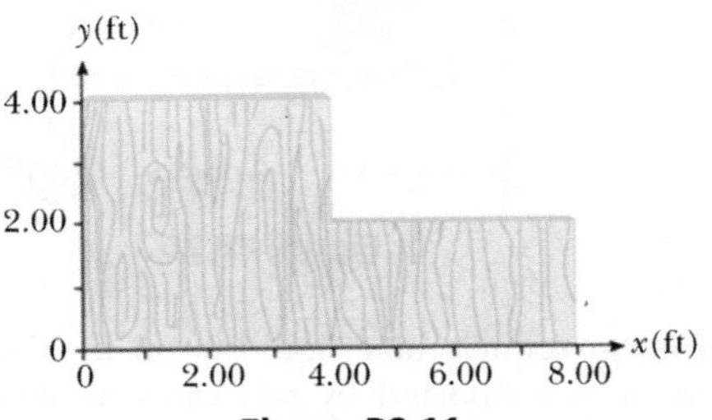

Figure P8.11

12. Consider the following mass distribution, where x- and y-coordinates are given in meters: 5.0 kg at (0.0, 0.0) m, 3.0 kg at (0.0, 4.0) m, and 4.0 kg at (3.0, 0.0) m. Where should a fourth object of 8.0 kg be placed so that the center of gravity of the four-object arrangement will be at (0.0, 0.0) m?

13. Many of the elements in horizontal-bar exercises can be modeled by representing the gymnast by four segments consisting of arms, torso (including the head), thighs, and lower legs, as shown in Figure P8.13a. Inertial parameters for a particular gymnast are as follows:

Segment	Mass (kg)	Length (m)	r_{cg} (m)	I (kg-m^2)
Arms	6.87	0.548	0.239	0.205
Torso	33.57	0.601	0.337	1.610
Thighs	14.07	0.374	0.151	0.173
Legs	7.54	—	0.227	0.164

Note that in Figure P8.13a r_{cg} is the distance to the center of gravity measured from the joint closest to the bar and the masses for the arms, thighs, and legs include both appendages. I is the moment of inertia of each segment about its center of gravity. Determine the distance from the bar to the center of gravity of the gymnast for the two positions shown in Figures P8.13b and P8.13c.

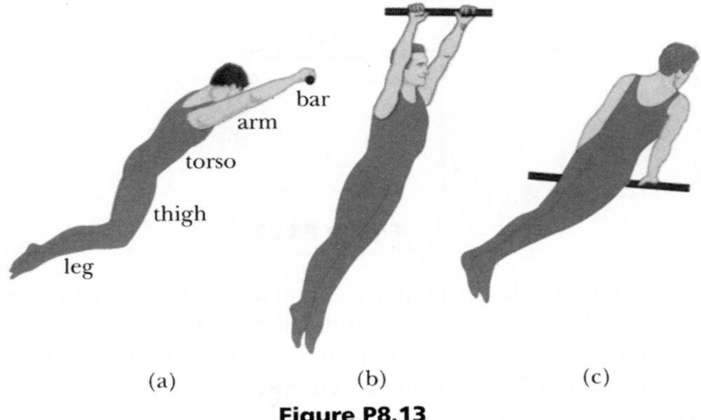

Figure P8.13

14. Using the data given in Problem 13 and the coordinate system shown in Figure P8.14b, calculate the position of the center of gravity of the gymnast shown in Figure P8.14a. Pay close attention to the definition of r_{cg} in the table.

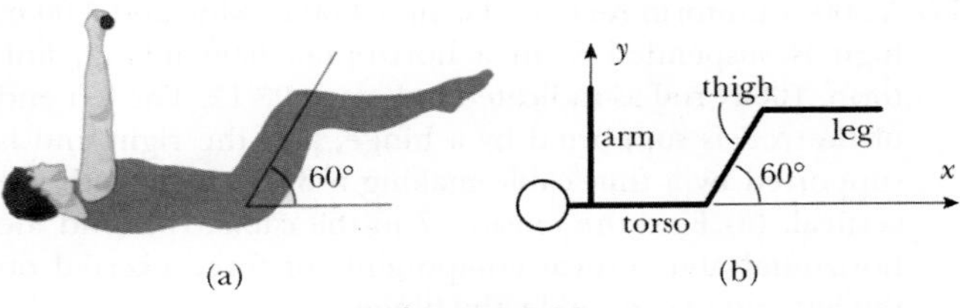

Figure P8.14

15. A person bending forward to lift a load "with his back" (Fig. P8.15a) rather than "with his knees" can be injured by large forces exerted on the muscles and vertebrae. The spine pivots mainly at the fifth lumbar vertebra, with the principal supporting force provided by the erector spinalis muscle in the back. To see the magnitude of the forces involved, and to understand why back problems are common among humans, consider the model shown in Fig. P8.15b of a person bending forward to lift a 200-N object. The spine and upper body are represented as a uniform horizontal rod of weight 350 N, pivoted at the base of the spine. The erector spinalis muscle, attached at a point two-thirds of the way up the spine, maintains the position of the back. The angle between the spine and this muscle is 12.0°. Find the tension in the back muscle and the compressional force in the spine.

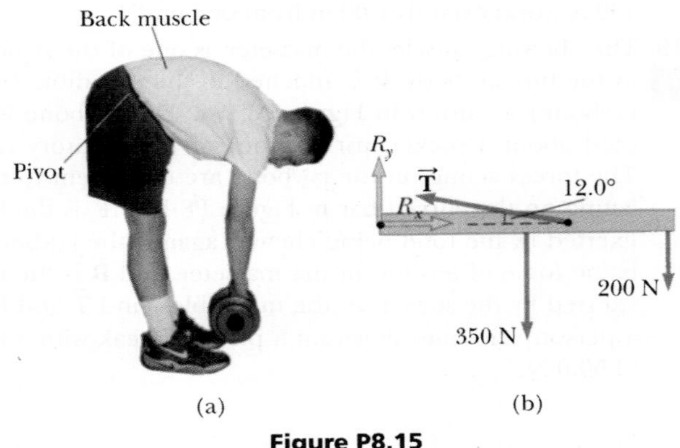

Figure P8.15

16. When a person stands on tiptoe (a strenuous position), the position of the foot is as shown in Figure P8.16a. The total gravitational force on the body, $\vec{F}_g$, is supported by the force $\vec{n}$ exerted by the floor on the toes of one foot. A mechanical model of the situation is shown in Figure P8.16b, where $\vec{T}$ is the force exerted by the Achilles tendon on the foot and $\vec{R}$ is the force exerted by the tibia on the foot. Find the values of T, R, and θ when $F_g = 700$ N.

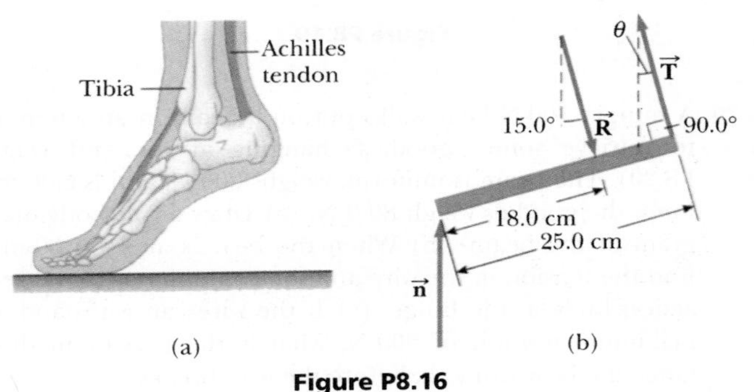

Figure P8.16

17. A 500-N uniform rectangular sign 4.00 m wide and 3.00 m high is suspended from a horizontal, 6.00-m-long, uniform, 100-N rod as indicated in Figure P8.17. The left end of the rod is supported by a hinge, and the right end is supported by a thin cable making a 30.0° angle with the vertical. (a) Find the tension T in the cable. (b) Find the horizontal and vertical components of force exerted on the left end of the rod by the hinge.

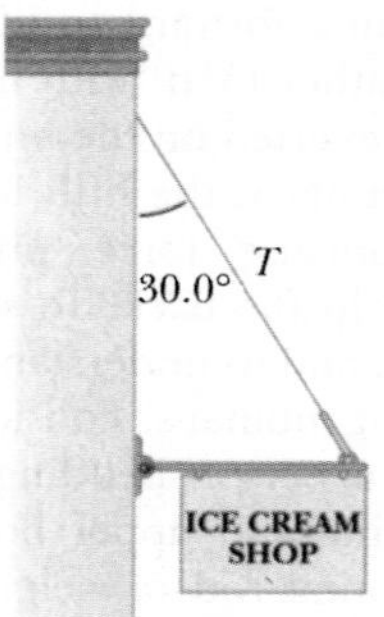

Figure P8.17

18. A window washer is standing on a scaffold supported by a vertical rope at each end. The scaffold weighs 200 N and is 3.00 m long. What is the tension in each rope when the 700-N worker stands 1.00 m from one end?

19. The chewing muscle, the masseter, is one of the strongest in the human body. It is attached to the mandible (lower jawbone) as shown in Figure P8.19a. The jawbone is pivoted about a socket just in front of the auditory canal. The forces acting on the jawbone are equivalent to those acting on the curved bar in Figure P8.19b: $\vec{\mathbf{F}}_c$ is the force exerted by the food being chewed against the jawbone, $\vec{\mathbf{T}}$ is the force of tension in the masseter, and $\vec{\mathbf{R}}$ is the force exerted by the socket on the mandible. Find $\vec{\mathbf{T}}$ and $\vec{\mathbf{R}}$ for a person who bites down on a piece of steak with a force of 50.0 N.

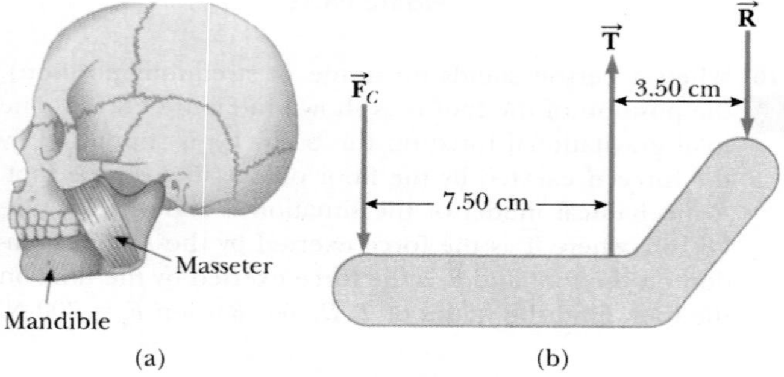

Figure P8.19

20. A hungry 700-N bear walks out on a beam in an attempt to retrieve some "goodies" hanging at the end (Fig. P8.20). The beam is uniform, weighs 200 N, and is 6.00 m long; the goodies weigh 80.0 N. (a) Draw a free-body diagram of the beam. (b) When the bear is at $x = 1.00$ m, find the tension in the wire and the components of the reaction force at the hinge. (c) If the wire can withstand a maximum tension of 900 N, what is the maximum distance the bear can walk before the wire breaks?

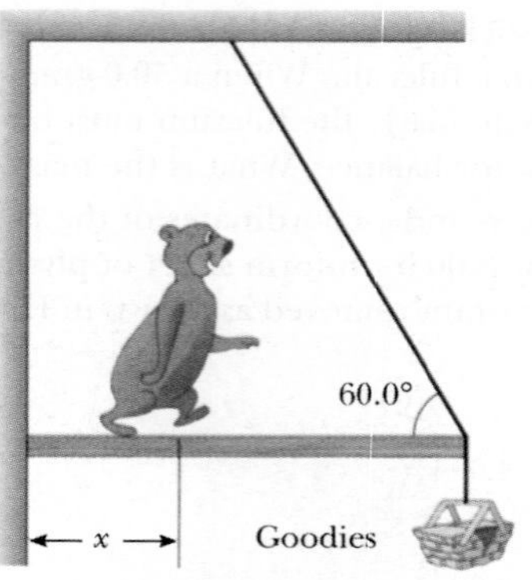

Figure P8.20

21. A uniform semicircular sign 1.00 m in diameter and of weight w is supported by two wires as shown in Figure P8.21. What is the tension in each of the wires supporting the sign?

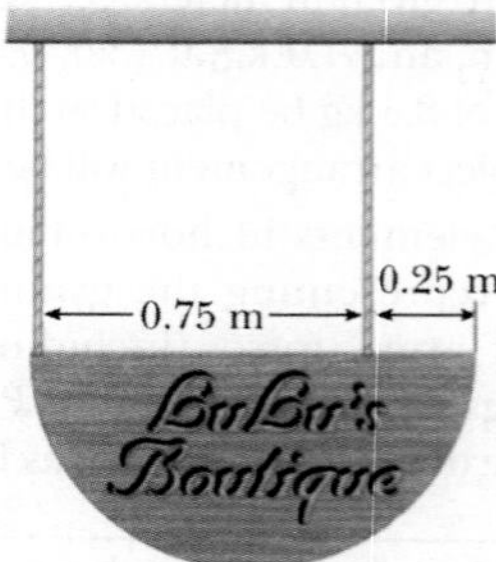

Figure P8.21

22. A 20.0-kg floodlight in a park is supported at the end of a horizontal beam of negligible mass that is hinged to a pole, as shown in Figure P8.22. A cable at an angle of 30.0° with the beam helps to support the light. Find (a) the tension in the cable and (b) the horizontal and vertical forces exerted on the beam by the pole.

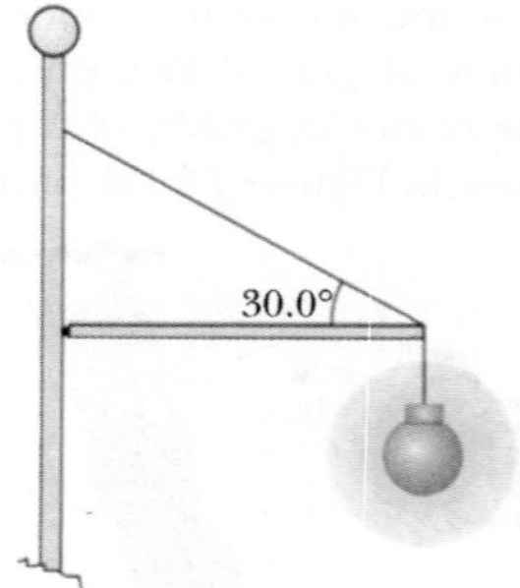

Figure P8.22

23. A uniform plank of length 2.00 m and mass 30.0 kg is supported by three ropes, as indicated by the blue vectors in Figure P8.23. Find the tension in each rope when a 700-N person is 0.500 m from the left end.

24. A 15.0-m, 500-N uniform ladder rests against a frictionless wall, making an angle of 60.0° with the horizontal. (a) Find the horizontal and vertical forces exerted on the base of the ladder by the Earth when an 800-N firefighter is 4.00 m from the bottom. (b) If the ladder is just on the

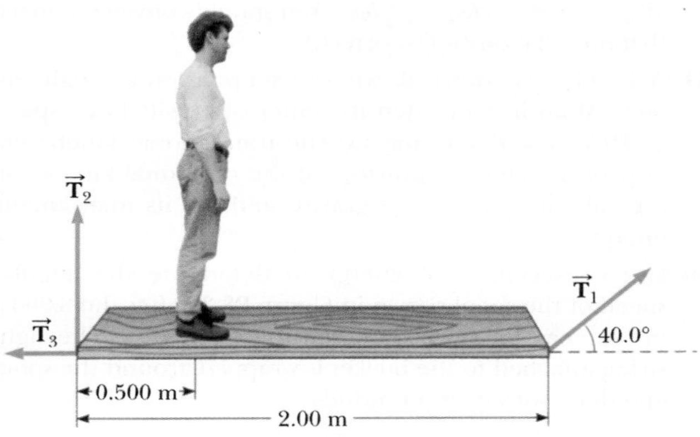

Figure P8.23

verge of slipping when the firefighter is 9.00 m up, what is the coefficient of static friction between ladder and ground?

25. **PhysicsNow™** An 8.00-m, 200-N uniform ladder rests against a smooth wall. The coefficient of static friction between the ladder and the ground is 0.600, and the ladder makes a 50.0° angle with the ground. How far up the ladder can an 800-N person climb before the ladder begins to slip?

26. A 1 200-N uniform boom is supported by a cable perpendicular to the boom as in Figure P8.26. The boom is hinged at the bottom, and a 2 000-N weight hangs from its top. Find the tension in the supporting cable and the components of the reaction force exerted on the boom by the hinge.

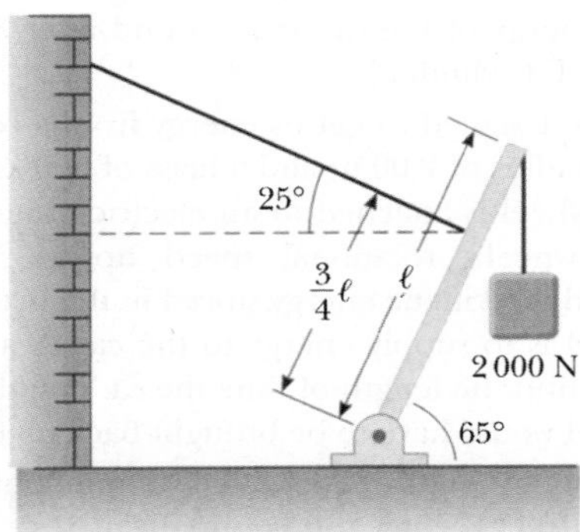

Figure P8.26

27. The large quadriceps muscle in the upper leg terminates at its lower end in a tendon attached to the upper end of the tibia (Fig. P8.27a). The forces on the lower leg when the leg is extended are modeled as in Figure P8.27b, where $\vec{T}$ is the force of tension in the tendon, $\vec{w}$ is the force of gravity acting on the lower leg, and $\vec{F}$ is the force of gravity acting on the foot. Find $\vec{T}$ when the tendon is at an angle of 25.0° with the tibia, assuming that $w = 30.0$ N, $F = 12.5$ N, and the leg is extended at an angle θ of 40.0° with the vertical. Assume that the center of gravity of the lower leg is at its center and that the tendon attaches to the lower leg at a point one-fifth of the way down the leg.

28. One end of a uniform 4.0-m-long rod of weight w is supported by a cable. The other end rests against a wall, where it is held by friction. (See Fig. P8.28.) The coefficient of static friction between the wall and the rod is

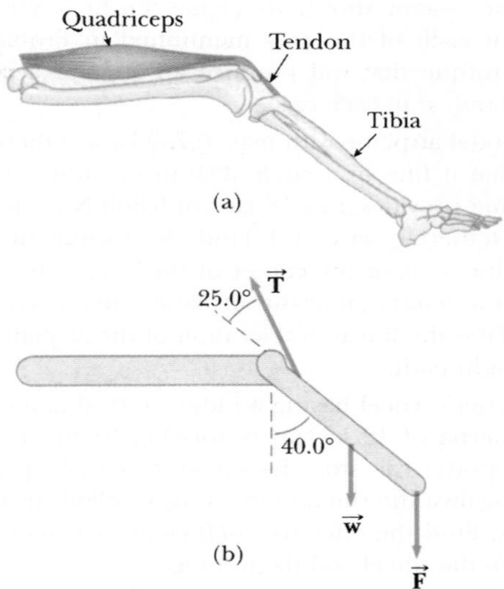

Figure P8.27

$\mu_s = 0.50$. Determine the minimum distance x from point A at which an additional weight w (the same as the weight of the rod) can be hung without causing the rod to slip at point A.

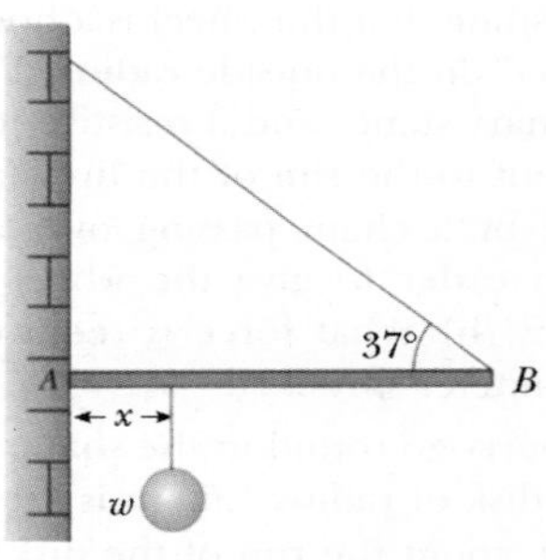

Figure P8.28

Section 8.5 Relationship Between Torque and Angular Acceleration

29. Four objects are held in position at the corners of a rectangle by light rods as shown in Figure P8.29. Find the moment of inertia of the system about (a) the x-axis, (b) the y-axis, and (c) an axis through O and perpendicular to the page.

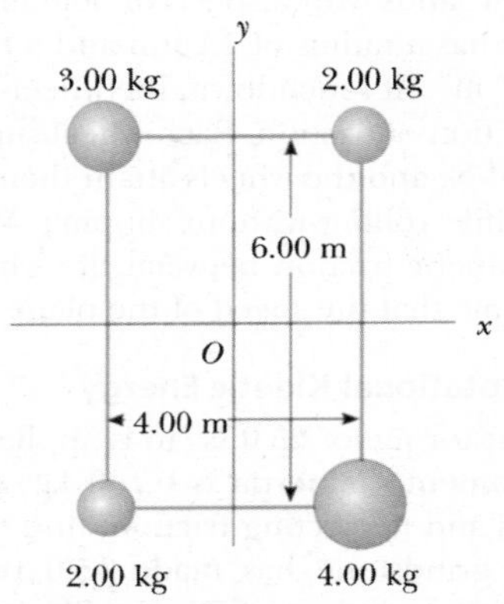

Figure P8.29 (Problems 29 and 30)

30. If the system shown in Figure P8.29 is set in rotation about each of the axes mentioned in Problem 29, find the torque that will produce an angular acceleration of 1.50 rad/s^2 in each case.

31. A model airplane with mass 0.750 kg is tethered by a wire so that it flies in a circle 30.0 m in radius. The airplane engine provides a net thrust of 0.800 N perpendicular to the tethering wire. (a) Find the torque the net thrust produces about the center of the circle. (b) Find the angular acceleration of the airplane when it is in level flight. (c) Find the linear acceleration of the airplane tangent to its flight path.

32. A potter's wheel having a radius of 0.50 m and a moment of inertia of 12 kg · m^2 is rotating freely at 50 rev/min. The potter can stop the wheel in 6.0 s by pressing a wet rag against the rim and exerting a radially inward force of 70 N. Find the effective coefficient of kinetic friction between the wheel and the wet rag.

33. A cylindrical fishing reel has a moment of inertia $I = 6.8 \times 10^{-4}$ kg · m^2 and a radius of 4.0 cm. A friction clutch in the reel exerts a restraining torque of 1.3 N · m if a fish pulls on the line. The fisherman gets a bite, and the reel begins to spin with an angular acceleration of 66 rad/s^2. (a) What is the force exerted by the fish on the line? (b) How much line unwinds in 0.50 s?

34. A bicycle wheel has a diameter of 64.0 cm and a mass of 1.80 kg. Assume that the wheel is a hoop with all the mass concentrated on the outside radius. The bicycle is placed on a stationary stand, and a resistive force of 120 N is applied tangent to the rim of the tire. (a) What force must be applied by a chain passing over a 9.00-cm-diameter sprocket in order to give the wheel an acceleration of 4.50 rad/s^2? (b) What force is required if you shift to a 5.60-cm-diameter sprocket?

35. A 150-kg merry-go-round in the shape of a uniform, solid, horizontal disk of radius 1.50 m is set in motion by wrapping a rope about the rim of the disk and pulling on the rope. What constant force must be exerted on the rope to bring the merry-go-round from rest to an angular speed of 0.500 rev/s in 2.00 s?

36. **Physics Now™** A 5.00-kg cylindrical reel with a radius of 0.600 m and a frictionless axle starts from rest and speeds up uniformly as a 3.00-kg bucket falls into a well, making a light rope unwind from the reel (Fig. P8.36). The bucket starts from rest and falls for 4.00 s. (a) What is the linear acceleration of the falling bucket? (b) How far does it drop? (c) What is the angular acceleration of the reel?

37. An airliner lands with a speed of 50.0 m/s. Each wheel of the plane has a radius of 1.25 m and a moment of inertia of 110 kg · m^2. At touchdown, the wheels begin to spin under the action of friction. Each wheel supports a weight of 1.40×10^4 N, and the wheels attain their angular speed in 0.480 s while rolling without slipping. What is the coefficient of kinetic friction between the wheels and the runway? Assume that the speed of the plane is constant.

Section 8.6 Rotational Kinetic Energy

38. A constant torque of 25.0 N · m is applied to a grindstone whose moment of inertia is 0.130 kg · m^2. Using energy principles and neglecting friction, find the angular speed after the grindstone has made 15.0 revolutions. [*Hint:* The angular equivalent of $W_{net} = F\Delta x = \frac{1}{2}mv_f^2 - \frac{1}{2}mv_i^2$ is $W_{net} = \tau\Delta\theta = \frac{1}{2}I\omega_f^2 - \frac{1}{2}I\omega_i^2$. You should convince yourself that this relationship is correct.)

39. A 10.0-kg cylinder rolls without slipping on a rough surface. At an instant when its center of gravity has a speed of 10.0 m/s, determine (a) the translational kinetic energy of its center of gravity, (b) the rotational kinetic energy about its center of gravity, and (c) its total kinetic energy.

40. Use conservation of energy to determine the angular speed of the spool shown in Figure P8.36 after the 3.00-kg bucket has fallen 4.00 m, starting from rest. The light string attached to the bucket is wrapped around the spool and does not slip as it unwinds.

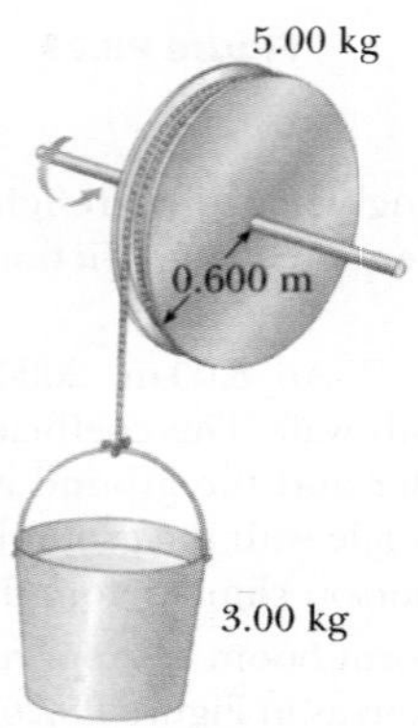

Figure P8.36 (Problems 36 and 40)

41. A horizontal 800-N merry-go-round of radius 1.50 m is started from rest by a constant horizontal force of 50.0 N applied tangentially to the merry-go-round. Find the kinetic energy of the merry-go-round after 3.00 s. (Assume it is a solid cylinder.)

42. A car is designed to get its energy from a rotating flywheel with a radius of 2.00 m and a mass of 500 kg. Before a trip, the flywheel is attached to an electric motor, which brings the flywheel's rotational speed up to 5 000 rev/min. (a) Find the kinetic energy stored in the flywheel. (b) If the flywheel is to supply energy to the car as a 10.0-hp motor would, find the length of time the car could run before the flywheel would have to be brought back up to speed.

43. The top in Figure P8.43 has a moment of inertia of 4.00×10^{-4} kg · m^2 and is initially at rest. It is free to rotate about a stationary axis AA'. A string wrapped around a peg along the axis of the top is pulled in such a manner as to maintain a constant tension of 5.57 N in the string. If the string does not slip while wound around the peg, what

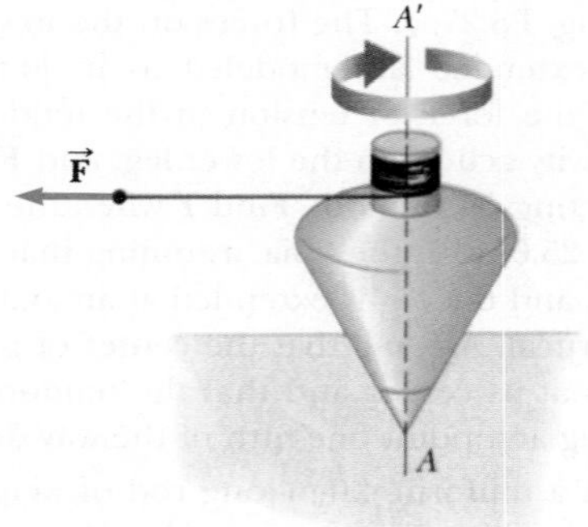

Figure P8.43

is the angular speed of the top after 80.0 cm of string has been pulled off the peg? [*Hint:* Consider the work that is done.]

44. A 240-N sphere 0.20 m in radius rolls without slipping 6.0 m down a ramp that is inclined at 37° with the horizontal. What is the angular speed of the sphere at the bottom of the slope if it starts from rest?

Section 8.7 Angular Momentum

45. A light rigid rod 1.00 m in length rotates about an axis perpendicular to its length and through its center, as shown in Figure P8.45. Two particles of masses 4.00 kg and 3.00 kg are connected to the ends of the rod. What is the angular momentum of the system if the speed of each particle is 5.00 m/s? (Neglect the rod's mass.)

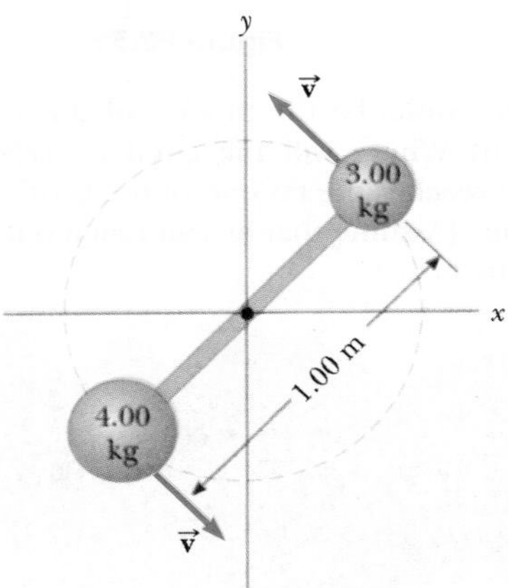

Figure P8.45

46. Halley's comet moves about the Sun in an elliptical orbit, with its closest approach to the Sun being 0.59 A.U. and its greatest distance being 35 A.U. (1 A.U. is the Earth–Sun distance). If the comet's speed at closest approach is 54 km/s, what is its speed when it is farthest from the Sun? You may neglect any change in the comet's mass and assume that its angular momentum about the Sun is conserved.

47. The system of small objects shown in Figure P8.47 is rotating at an angular speed of 2.0 rev/s. The objects are connected by light, flexible spokes that can be lengthened or shortened. What is the new angular speed if the spokes are shortened to 0.50 m? (An effect similar to that illustrated in this problem occurred in the early stages of the formation of our galaxy. As the massive cloud of dust and gas that was the source of the stars and planets contracted, an initially small angular speed increased with time.)

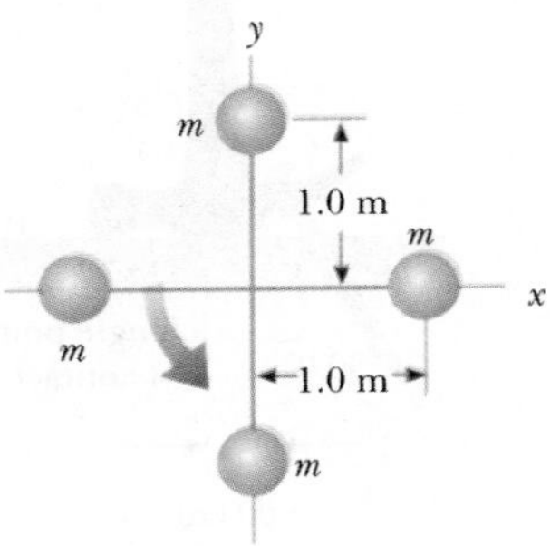

Figure P8.47

48. A playground merry-go-round of radius 2.00 m has a moment of inertia $I = 275\ \text{kg}\cdot\text{m}^2$ and is rotating about a frictionless vertical axle. As a child of mass 25.0 kg stands at a distance of 1.00 m from the axle, the system (merry-go-round and child) rotates at the rate of 14.0 rev/min. The child then proceeds to walk toward the edge of the merry-go-round. What is the angular speed of the system when the child reaches the edge?

49. A solid, horizontal cylinder of mass 10.0 kg and radius 1.00 m rotates with an angular speed of 7.00 rad/s about a fixed vertical axis through its center. A 0.250-kg piece of putty is dropped vertically onto the cylinder at a point 0.900 m from the center of rotation and sticks to the cylinder. Determine the final angular speed of the system.

50. A student sits on a rotating stool holding two 3.0-kg objects. When his arms are extended horizontally, the objects are 1.0 m from the axis of rotation and he rotates with an angular speed of 0.75 rad/s. The moment of inertia of the student plus stool is $3.0\ \text{kg}\cdot\text{m}^2$ and is assumed to be constant. The student then pulls in the objects horizontally to 0.30 m from the rotation axis. (a) Find the new angular speed of the student. (b) Find the kinetic energy of the student before and after the objects are pulled in.

51. **PhysicsNow™** The puck in Figure P8.51 has a mass of 0.120 kg. Its original distance from the center of rotation is 40.0 cm, and it moves with a speed of 80.0 cm/s. The string is pulled downward 15.0 cm through the hole in the frictionless table. Determine the work done on the puck. [*Hint:* Consider the change in kinetic energy of the puck.]

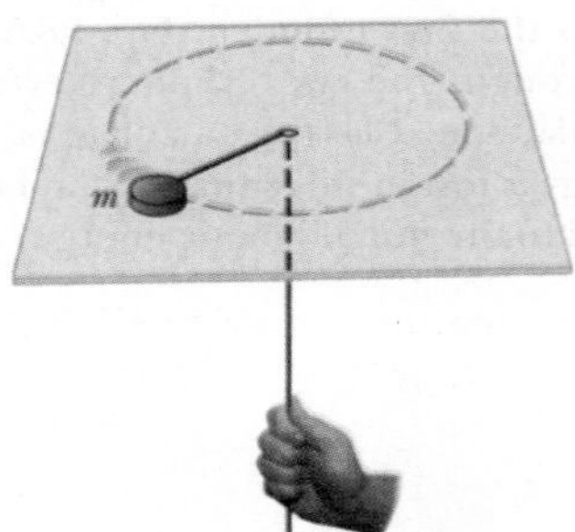

Figure P8.51

52. A merry-go-round rotates at the rate of 0.20 rev/s with an 80-kg man standing at a point 2.0 m from the axis of rotation. (a) What is the new angular speed when the man walks to a point 1.0 m from the center? Assume that the merry-go-round is a solid 25-kg cylinder of radius 2.0 m. (b) Calculate the change in kinetic energy due to the man's movement. How do you account for this change in kinetic energy?

53. A 60.0-kg woman stands at the rim of a horizontal turntable having a moment of inertia of $500\ \text{kg}\cdot\text{m}^2$ and a radius of 2.00 m. The turntable is initially at rest and is free to rotate about a frictionless, vertical axle through its center. The woman then starts walking around the rim clockwise (as viewed from above the system) at a constant speed of 1.50 m/s relative to the Earth. (a) In what direction and with what angular speed does the turntable rotate? (b) How much work does the woman do to set herself and the turntable into motion?

54. A space station shaped like a giant wheel has a radius of 100 m and a moment of inertia of 5.00×10^8 kg · m^2. A crew of 150 lives on the rim, and the station is rotating so that the crew experiences an apparent acceleration of 1g (Fig. P8.54). When 100 people move to the center of the station for a union meeting, the angular speed changes. What apparent acceleration is experienced by the managers remaining at the rim? Assume the average mass of a crew member is 65.0 kg.

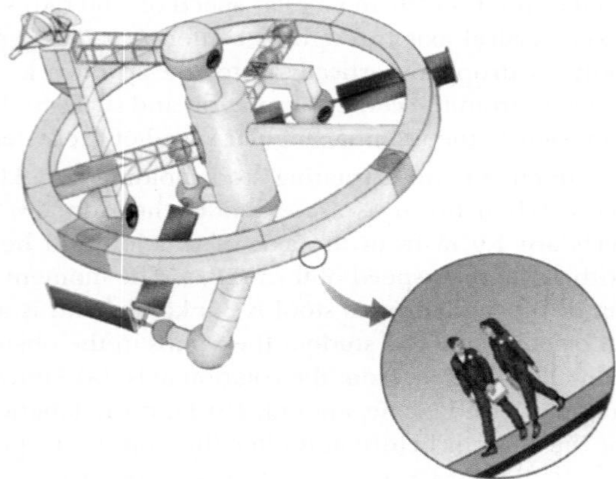

Figure P8.54

ADDITIONAL PROBLEMS

55. A cylinder with moment of inertia I_1 rotates with angular velocity ω_0 about a frictionless vertical axle. A second cylinder, with moment of inertia I_2, initially not rotating, drops onto the first cylinder (Fig. P8.55). Since the surfaces are rough, the two cylinders eventually reach the same angular speed ω. (a) Calculate ω. (b) Show that kinetic energy is lost in this situation, and calculate the ratio of the final to the initial kinetic energy.

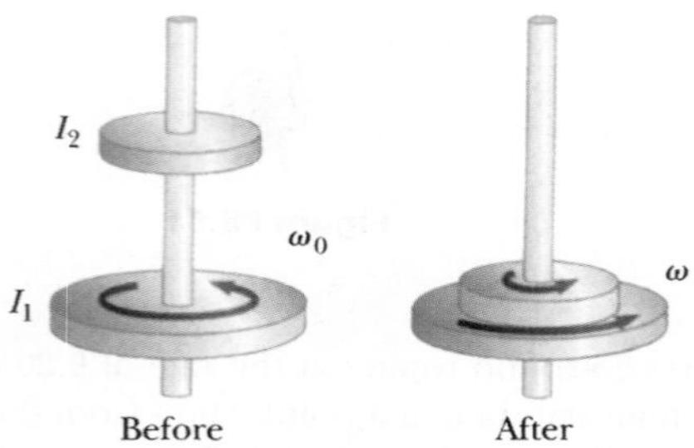

Figure P8.55

56. A new General Electric stove has a mass of 68.0 kg and the dimensions shown in Figure P8.56. The stove comes with a warning that it can tip forward if a person stands or sits on the oven door when it is open. What can you conclude about the weight of such a person? Could it be a child? List the assumptions you make in solving this problem. (The stove is supplied with a wall bracket to prevent the accident.)

57. A 40.0-kg child stands at one end of a 70.0-kg boat that is 4.00 m long (Fig. P8.57). The boat is initially 3.00 m from the pier. The child notices a turtle on a rock beyond the far end of the boat and proceeds to walk to that end to catch the turtle. (a) Neglecting friction between the boat and water, describe the motion of the system (child plus boat). (b) Where will the child be relative to the pier when he reaches the far end of the boat? (c) Will he catch the turtle? (Assume that he can reach out 1.00 m from the end of the boat.)

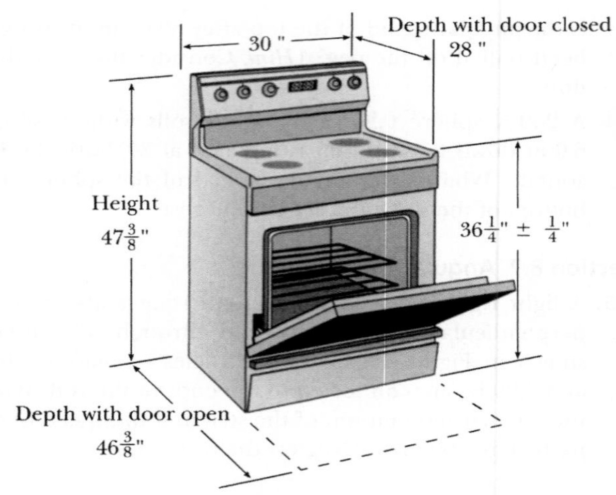

Figure P8.56

Figure P8.57

58. Figure P8.58 shows a clawhammer as it is being used to pull a nail out of a horizontal board. If a force of magnitude 150 N is exerted horizontally as shown, find (a) the

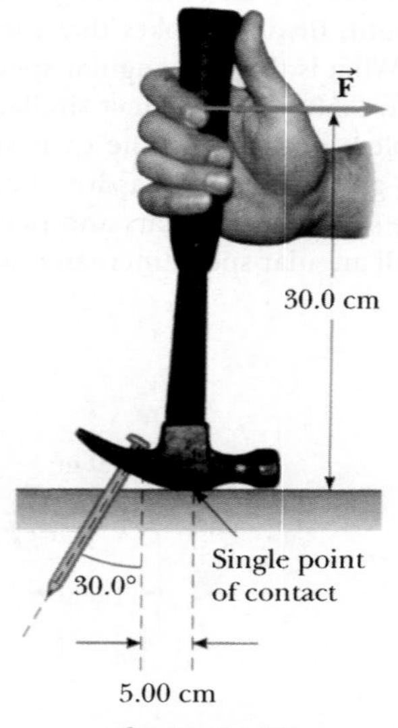

Figure P8.58

force exerted by the hammer claws on the nail and (b) the force exerted by the surface at the point of contact with the hammer head. Assume that the force the hammer exerts on the nail is parallel to the nail.

59. The pulley in Figure P8.59 has a moment of inertia of 5.0 kg · m^2 and a radius of 0.50 m. The cord supporting the masses m_1 and m_2 does not slip, and the axle is frictionless. (a) Find the acceleration of each mass when $m_1 = 2.0$ kg and $m_2 = 5.0$ kg. (b) Find the tension in the cable supporting m_1 and the tension in the cable supporting m_2. [*Note:* The two tensions are different].

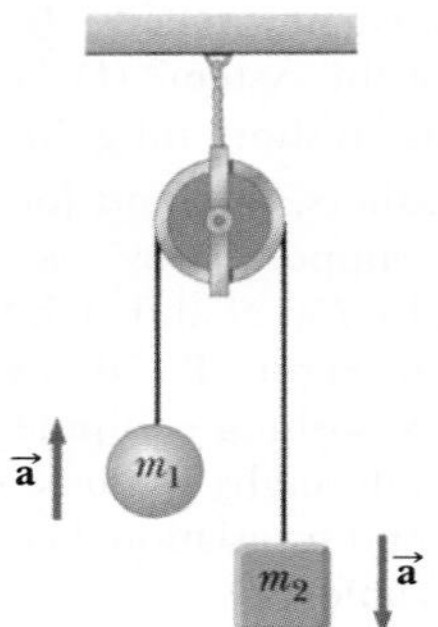

Figure P8.59

60. A 12.0-kg object is attached to a cord that is wrapped around a wheel of radius $r = 10.0$ cm (Fig. P8.60). The acceleration of the object down the frictionless incline is measured to be 2.00 m/s^2. Assuming the axle of the wheel to be frictionless, determine (a) the tension in the rope, (b) the moment of inertia of the wheel, and (c) the angular speed of the wheel 2.00 s after it begins rotating, starting from rest.

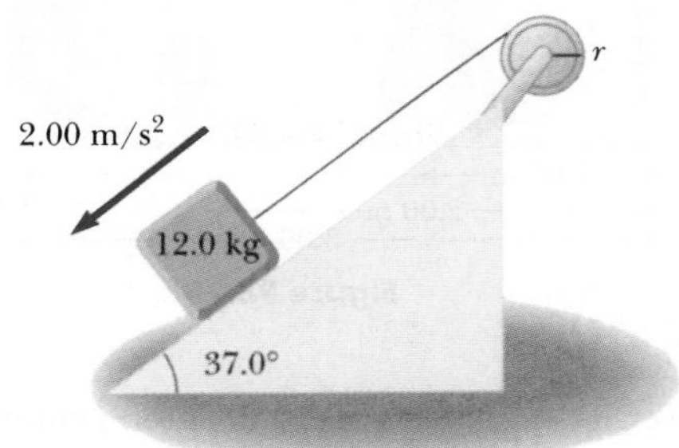

Figure P8.60

61. A uniform ladder of length L and weight w is leaning against a vertical wall. The coefficient of static friction between the ladder and the floor is the same as that between the ladder and the wall. If this coefficient of static friction is $\mu_s = 0.500$, determine the smallest angle the ladder can make with the floor without slipping.

62. A uniform 10.0-N picture frame is supported as shown in Figure P8.62. Find the tension in the cords and the magnitude of the horizontal force at P that are required to hold the frame in the position shown.

63. A solid 2.0-kg ball of radius 0.50 m starts at a height of 3.0 m above the surface of the Earth and *rolls* down a 20° slope. A solid disk and a ring start at the same time and the same height. The ring and disk each have the same mass and radius as the ball. Which of the three wins the race to the bottom if all roll without slipping?

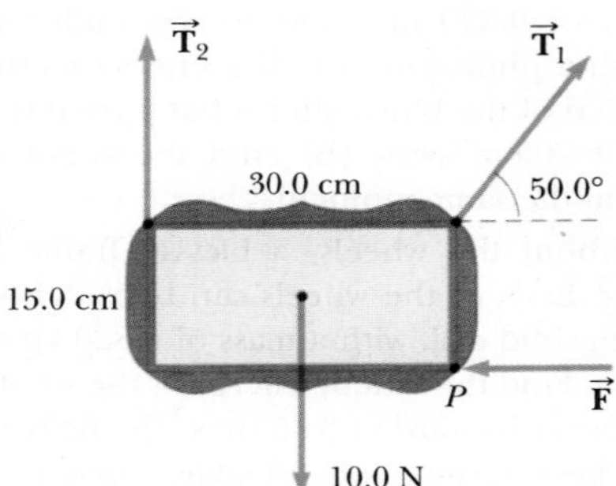

Figure P8.62

64. A common physics demonstration (Fig. P8.64) consists of a ball resting at the end of a board of length ℓ that is elevated at an angle θ with the horizontal. A light cup is attached to the board at r_c so that it will catch the ball when the support stick is suddenly removed. Show that (a) the ball will lag behind the falling board when $\theta < 35.3°$ and (b) the ball will fall into the cup when the board is supported at this limiting angle and the cup is placed at

$$r_c = \frac{2\ell}{3 \cos \theta}$$

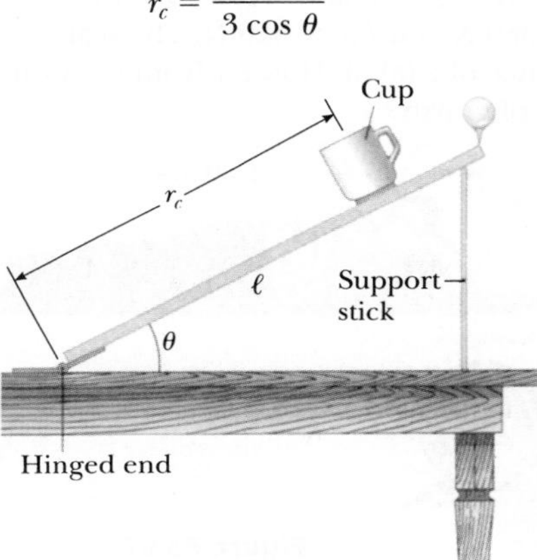

Figure P8.64

65. In Figure P8.65, the sliding block has a mass of 0.850 kg, the counterweight has a mass of 0.420 kg, and the pulley is a uniform solid cylinder with a mass of 0.350 kg and an outer radius of 0.0300 m. The coefficient of kinetic friction between the block and the horizontal surface is 0.250. The pulley turns without friction on its axle. The light cord does not stretch and does not slip on the pulley. The block has a

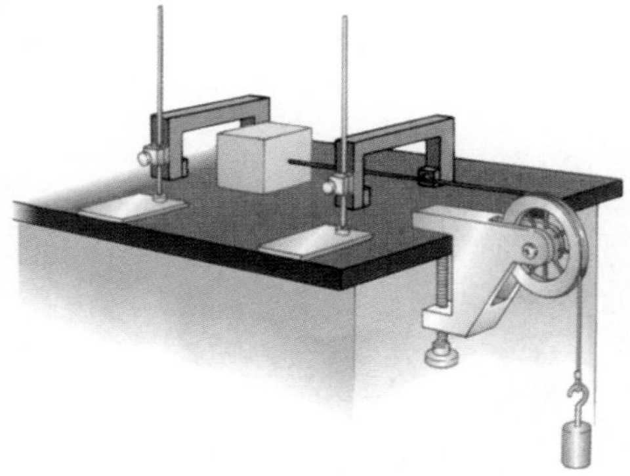

Figure P8.65

velocity of 0.820 m/s toward the pulley when it passes through a photogate. (a) Use energy methods to predict the speed of the block after it has moved to a second photogate 0.700 m away. (b) Find the angular speed of the pulley at the same moment.

66. (a) Without the wheels, a bicycle frame has a mass of 8.44 kg. Each of the wheels can be roughly modeled as a uniform solid disk with a mass of 0.820 kg and a radius of 0.343 m. Find the kinetic energy of the whole bicycle when it is moving forward at 3.35 m/s. (b) Before the invention of a wheel turning on an axle, ancient people moved heavy loads by placing rollers under them. (Modern people use rollers, too: Any hardware store will sell you a roller bearing for a lazy Susan.) A stone block of mass 844 kg moves forward at 0.335 m/s, supported by two uniform cylindrical tree trunks, each of mass 82.0 kg and radius 0.343 m. There is no slipping between the block and the rollers or between the rollers and the ground. Find the total kinetic energy of the moving objects.

67. In exercise physiology studies, it is sometimes important to determine the location of a person's center of gravity. This can be done with the arrangement shown in Figure P8.67. A light plank rests on two scales that read F_{g1} = 380 N and F_{g2} = 320 N. The scales are separated by a distance of 2.00 m. How far from the woman's feet is her center of gravity?

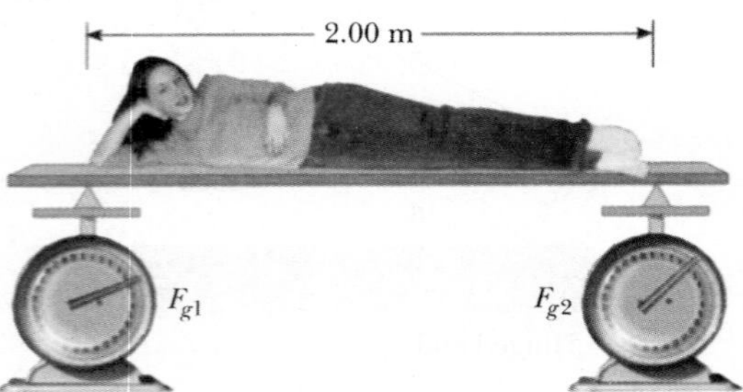

Figure P8.67

68. Physics Now™ Two astronauts (Fig. P8.68), each having a mass of 75.0 kg, are connected by a 10.0-m rope of negligible mass. They are isolated in space, moving in circles around the point halfway between them at a speed of 5.00 m/s. Treating the astronauts as particles, calculate (a) the magnitude of the angular momentum and (b) the rotational energy of the system. By pulling on the rope, the astronauts shorten the distance between them to 5.00 m. (c) What is the new angular momentum of the system? (d) What are their new speeds? (e) What is the new rotational energy of the system? (f) How much work is done by the astronauts in shortening the rope?

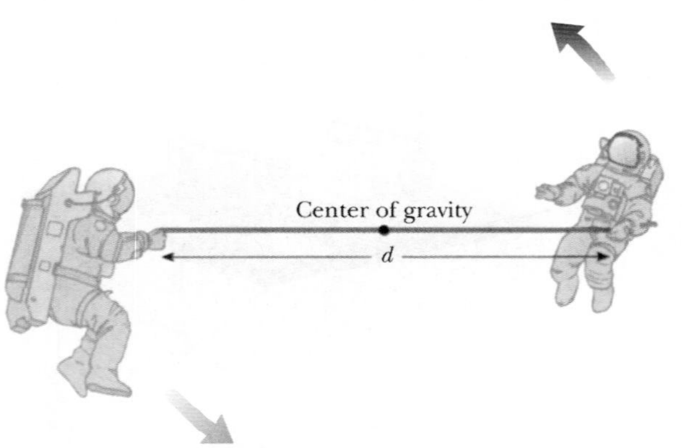

Figure P8.68 (Problems 68 and 69)

69. Two astronauts (Fig. P8.68), each having a mass M, are connected by a rope of length d having negligible mass. They are isolated in space, moving in circles around the point halfway between them at a speed v. (a) Calculate the magnitude of the angular momentum of the system by treating the astronauts as particles. (b) Calculate the rotational energy of the system. By pulling on the rope, the astronauts shorten the distance between them to $d/2$. (c) What is the new angular momentum of the system? (d) What are their new speeds? (e) What is the new rotational energy of the system? (f) How much work is done by the astronauts in shortening the rope?

70. Two window washers, Bob and Joe, are on a 3.00-m-long, 345-N scaffold supported by two cables attached to its ends. Bob weighs 750 N and stands 1.00 m from the left end, as shown in Figure P8.70. Two meters from the left end is the 500-N washing equipment. Joe is 0.500 m from the right end and weighs 1 000 N. Given that the scaffold is in rotational and translational equilibrium, what are the forces on each cable?

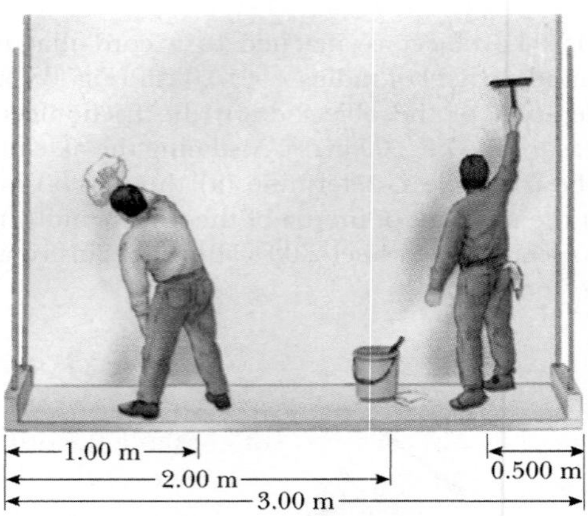

Figure P8.70

71. We have all complained that there aren't enough hours in a day. In an attempt to change that, suppose that all the people in the world lined up at the equator and started running east at 2.5 m/s relative to the surface of the Earth. By how much would the length of a day increase? (Assume that there are 5.5×10^9 people in the world with an average mass of 70 kg each and that the Earth is a solid, homogeneous sphere. In addition, you may use the result $1/(1 - x) \approx 1 + x$ for small x.)

72. In a circus performance, a large 5.0-kg hoop of radius 3.0 m rolls without slipping. If the hoop is given an angular speed of 3.0 rad/s while rolling on the horizontal ground and is then allowed to roll up a ramp inclined at 20° with the horizontal, how far along the incline does the hoop roll?

73. A uniform solid cylinder of mass M and radius R rotates on a frictionless horizontal axle (Fig. P8.73). Two objects with equal masses m hang from light cords wrapped around the cylinder. If the system is released from rest, find (a) the tension in each cord and (b) the acceleration of each object after the objects have descended a distance h.

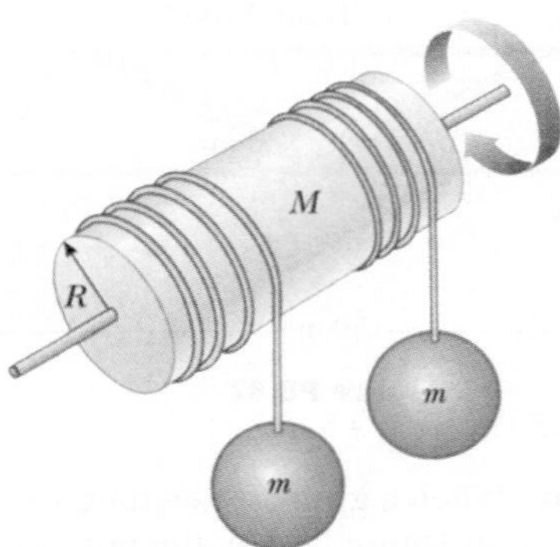

Figure P8.73

74. Figure P8.74 shows a vertical force applied tangentially to a uniform cylinder of weight w. The coefficient of static friction between the cylinder and all surfaces is 0.500. Find, in terms of w, the maximum force $\vec{\mathbf{F}}$ that can be applied without causing the cylinder to rotate. [*Hint:* When the cylinder is on the verge of slipping, both friction forces are at their maximum values. Why?]

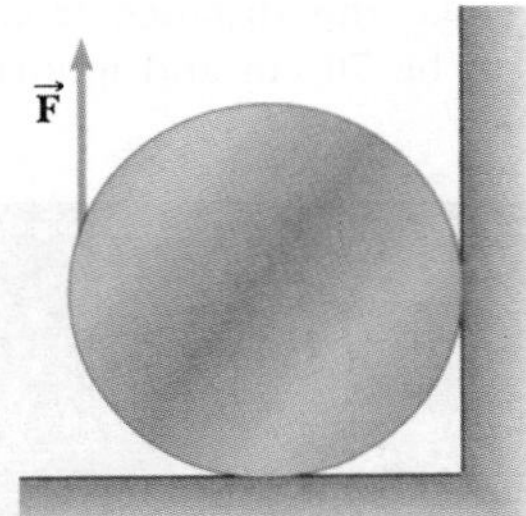

Figure P8.74

75. Due to a gravitational torque exerted by the Moon on the Earth, our planet's period of rotation slows at a rate on the order of 1 ms/century. (a) Determine the order of magnitude of Earth's angular acceleration. (b) Find the order of magnitude of the torque. (c) Find the order of magnitude of the size of the wrench an ordinary person would need to exert such a torque, as in Figure P8.75. Assume the person can brace his feet against a solid firmament.

Figure P8.75

76. A uniform pole is propped between the floor and the ceiling of a room. The height of the room is 7.80 ft, and the coefficient of static friction between the pole and the ceiling is 0.576. The coefficient of static friction between the pole and the floor is greater than that. What is the length of the longest pole that can be propped between the floor and the ceiling?

77. A *war-wolf,* or *trebuchet,* is a device used during the Middle Ages to throw rocks at castles and now sometimes used to fling pumpkins and pianos. A simple trebuchet is shown in Figure P8.77. Model it as a stiff rod of negligible mass 3.00 m long and joining particles of mass 60.0 kg and 0.120 kg at its ends. It can turn on a frictionless horizontal axle perpendicular to the rod and 14.0 cm from the particle of larger mass. The rod is released from rest in a horizontal orientation. Find the maximum speed that the object of smaller mass attains.

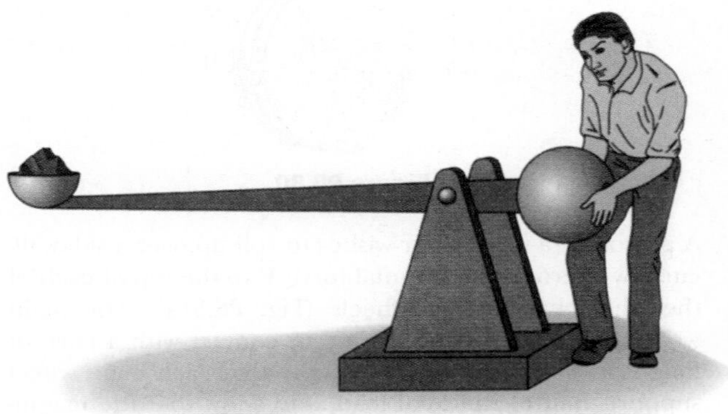

Figure P8.77

78. A painter climbs a ladder leaning against a smooth wall. At a certain height, the ladder is on the verge of slipping. (a) Explain why the force exerted by the vertical wall on the ladder is horizontal. (b) If the ladder of length L leans at an angle θ with the horizontal, what is the lever arm for this horizontal force with the axis of rotation taken at the base of the ladder? (c) If the ladder is uniform, what is the lever arm for the force of gravity acting on the ladder? (d) Let the mass of the painter be 80 kg, $L = 4.0$ m, the ladder's mass be 30 kg, $\theta = 53°$, and the coefficient of friction between ground and ladder be 0.45. Find the maximum distance the painter can climb up the ladder.

79. A 4.00-kg mass is connected by a light cord to a 3.00-kg mass on a smooth surface (Fig. P8.79). The pulley rotates about a frictionless axle and has a moment of inertia of $0.500\ \text{kg}\cdot\text{m}^2$ and a radius of 0.300 m. Assuming that the cord does not slip on the pulley, find (a) the acceleration of the two masses and (b) the tensions T_1 and T_2.

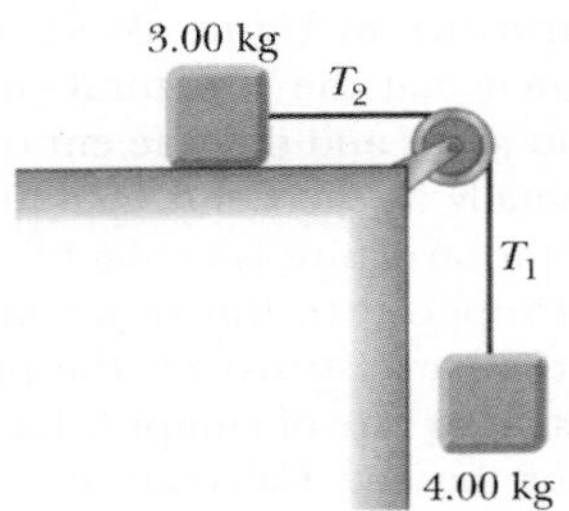

Figure P8.79

80. A string is wrapped around a uniform cylinder of mass M and radius R. The cylinder is released from rest with the string vertical and its top end tied to a fixed bar (Fig. P8.80). Show that (a) the tension in the string is one-third the weight of the cylinder, (b) the magnitude of the acceleration of the center of gravity is $2g/3$, and (c) the speed of the center of gravity is $(4gh/3)^{1/2}$ after the cylinder has descended through distance h. Verify your answer to (c) with the energy approach.

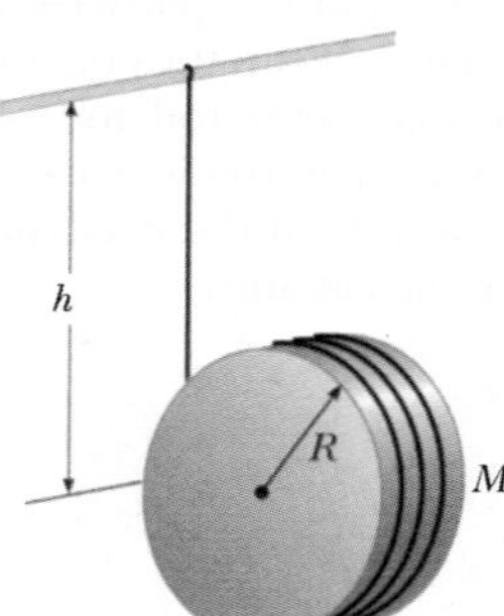

Figure P8.80

81. A person in a wheelchair wishes to roll up over a sidewalk curb by exerting a horizontal force $\vec{F}$ to the top of each of the wheelchair's main wheels (Fig. P8.81a). The main wheels have radius r and come in contact with a curb of height h (Fig. P8.81b). (a) Assume that each main wheel supports half of the total load, and show that the magnitude of the minimum force necessary to raise the wheelchair from the street is given by

$$F = \frac{mg\sqrt{2rh - h^2}}{2(2r - h)}$$

where mg is the combined weight of the wheelchair and person. (b) Estimate the value of F, taking $mg = 1\ 400$ N, $r = 30$ cm, and $h = 10$ cm.

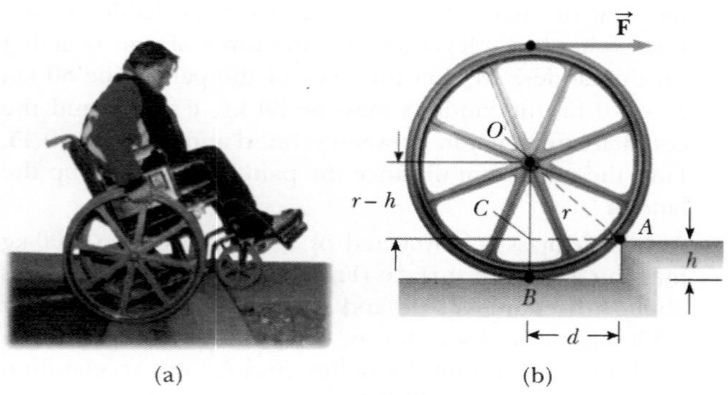

Figure P8.81

82. The truss structure in Figure P8.82 represents part of a bridge. Assume that the structural components are connected by pin joints and that the entire structure is free to slide horizontally at each end. Assume furthermore that the mass of the structure is negligible compared with the load it must support. In this situation, the force exerted by each of the bars (struts) on the pin joints is either a force of tension or one of compression and must be along the length of the bar. Calculate the force in each strut when the bridge supports a 7 200-N load at its center.

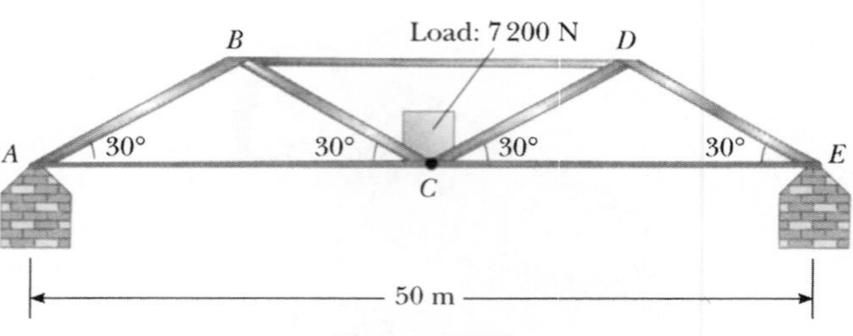

Figure P8.82

83. **The Iron Cross** When a gymnast weighing 750 N executes the iron cross as in Figure P8.83a, the primary muscles involved in supporting this position are the latissimus dorsi ("lats") and the pectoralis major ("pecs"). The rings exert an upward force on the arms and support the weight of the gymnast. The force exerted by the shoulder joint on the arm is labeled $\vec{F}_s$ while the two muscles exert a total force $\vec{F}_m$ on the arm. Estimate the magnitude of the force $\vec{F}_m$. Note that one ring supports half the weight of the gymnast, which is 375 N as indicated in Figure P8.83b. Assume that the force $\vec{F}_m$ acts at an angle of 45° below the horizontal at a distance of 4.0 cm from the shoulder joint. In your estimate, take the distance from the shoulder joint to the hand to be 70 cm and ignore the weight of the arm.

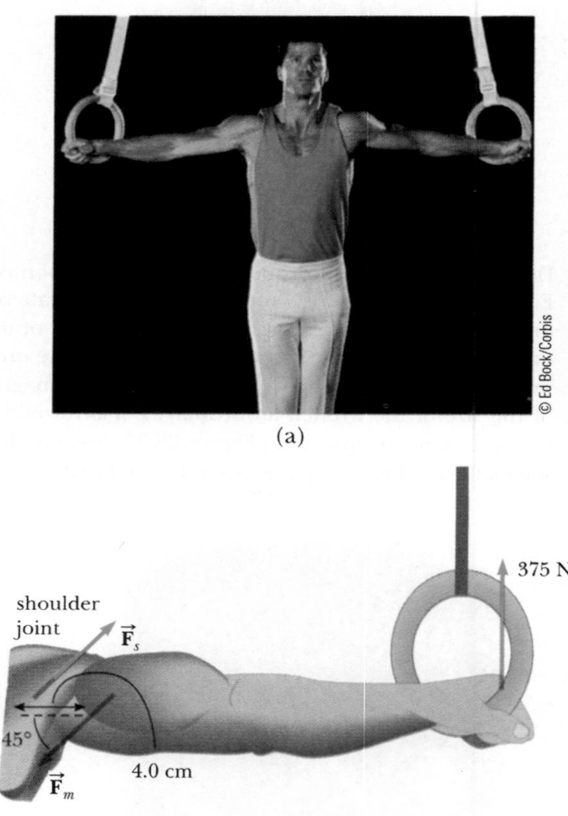

Figure P8.83

84. **Swinging on a high bar** The gymnast shown in Figure P8.84 is performing a backwards giant swing on the high bar. Starting from rest in a near-vertical orientation, he rotates around the bar in a counterclockwise direction,

keeping his body and arms straight. Friction between the bar and the gymnast's hands exerts a constant torque opposing the rotational motion. If the angular velocity of the gymnast at position 2 is measured to be 4.0 rad/s, determine his angular velocity at position 3. (Note that this maneuver is called a backwards giant swing, even though the motion of the gymnast would seem to be forwards.)

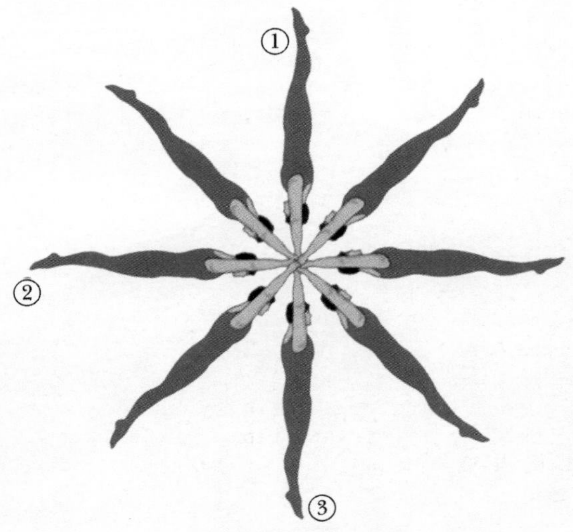

Figure P8.84

ACTIVITIES

A.1. Compare the motion of an empty soup can and a filled soup can down the same incline, such as a tilted table. If they are released from rest at the same height on the incline, which one reaches the bottom first? Repeat your observations with different kinds of soup (tomato, chicken noodle, etc.) and a can of beans. Compare their motions, and try to explain your observations. Finally, compare the motion of a filled soup can with a tennis ball, and explain your results to a friend.

A.2. Before attempting this exercise, review Example 8.14, dealing with the spinning stool. The techniques used here are similar to those used there. (a) First, make an estimate of the moment of inertia of your body. One way to do this would be to model your body as a solid cylinder and find I from $I = \frac{1}{2}MR^2$. You would have to determine your mass and estimate your average "radius" for this approach. Can you think of an alternative way to estimate I? (b) Now use the approach of Example 8.13 to measure I: Sit on a rotating stool, hold two weights (say, two books), and determine the angular speed of rotation with the books extended and after they are pulled in. The angular speed can be found by estimating the time taken for a given number of rotations. Use conservation of angular momentum to determine I. Do this five times to determine an average value for I. How well do your results for (a) and (b) compare, and if they differ greatly, what might cause the discrepancy?

A.3. This experiment demonstrates a simple way to find the center of gravity of an irregularly shaped object. Cut out an irregular shape from a piece of cardboard, and punch three to five holes around the edge of the shape. Put a pushpin through one of the holes, and tack the shape to a corkboard so that the shape can rotate freely. Now tie a weight to one end of a string, and hang the other end of the string from the pushpin. When the string stops moving, trace a line on the cardboard that follows the string. Repeat this for each of the holes in the cardboard. You will find that there is a point where all the lines intersect. This point is the center of gravity of the object.

In the Dead Sea, a lake between Jordan and Israel, the high percentage of salt dissolved in the water raises the fluid's density, dramatically increasing the buoyant force. Bathers can kick back and enjoy a good read, dispensing with the floating lounge chairs.

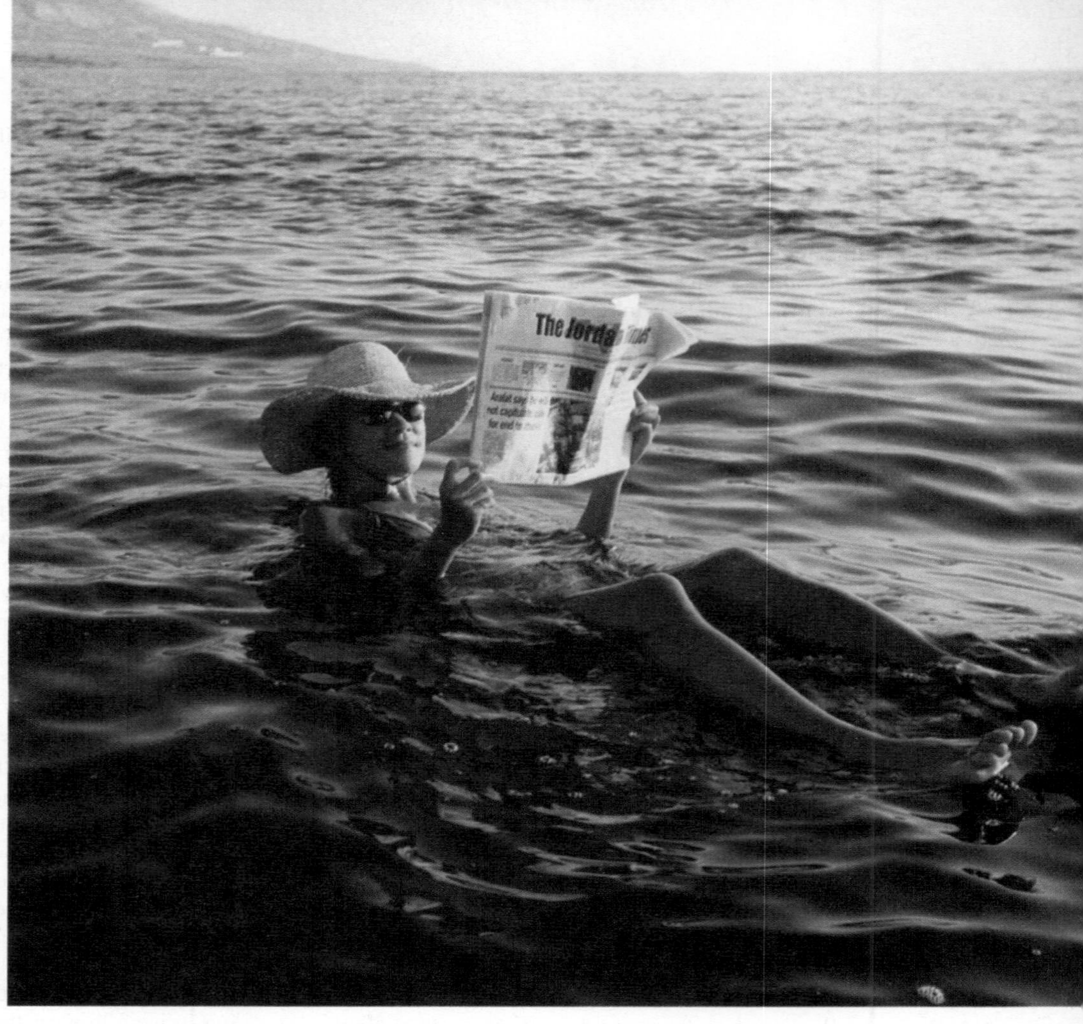

© Alison Wright/Corbis

CHAPTER

9

OUTLINE

Solids and Fluids

There are four known states of matter: solids, liquids, gases, and plasmas. In the universe at large, plasmas—systems of charged particles interacting electromagnetically—are the most common. In our environment on Earth, solids, liquids, and gases predominate.

An understanding of the fundamental properties of these different states of matter is important in all the sciences, in engineering, and in medicine. Forces put stresses on solids, and stresses can strain, deform, and break those solids, whether they are steel beams or bones. Fluids under pressure can perform work, or they can carry nutrients and essential solutes, like the blood flowing through our arteries and veins. Flowing gases cause pressure differences that can lift a massive cargo plane or the roof off a house in a hurricane. High-temperature plasmas created in fusion reactors may someday allow humankind to harness the energy source of the sun.

The study of any one of these states of matter is itself a vast discipline. Here, we'll introduce basic properties of solids and liquids, the latter including some properties of gases. In addition, we'll take a brief look at surface tension, viscosity, osmosis, and diffusion.

9.1 STATES OF MATTER

Matter is normally classified as being in one of three states: **solid**, **liquid**, or **gas**. Often this classification system is extended to include a fourth state of matter, called a **plasma**.

Everyday experience tells us that a solid has a definite volume and shape. A brick, for example, maintains its familiar shape and size day in and day out. A liquid has a definite volume but no definite shape. When you fill the tank on a lawn mower, the gasoline changes its shape from that of the original container to that of the tank on the mower, but the original volume is unchanged. A gas differs from solids and liquids in that it has neither definite volume nor definite shape. Because gas can flow, however, it shares many properties with liquids.

All matter consists of some distribution of atoms or molecules. The atoms in a solid, held together by forces that are mainly electrical, are located at specific positions with respect to one another and vibrate about those positions. At low temperatures, the vibrating motion is slight and the atoms can be considered essentially fixed. As energy is added to the material, the amplitude of the vibrations increases. A vibrating atom can be viewed as being bound in its equilibrium position by springs attached to neighboring atoms. A collection of such atoms and imaginary springs is shown in Figure 9.1. We can picture applied external forces as compressing these tiny internal springs. When the external forces are removed, the solid tends to return to its original shape and size. Consequently, a solid is said to have *elasticity*.

Crystals of natural quartz (SiO_2), one of the most common minerals on Earth. Quartz crystals are used to make special lenses and prisms and are employed in certain electronic applications.

Solids can be classified as either crystalline or amorphous. In a **crystalline solid** the atoms have an ordered structure. For example, in the sodium chloride crystal (common table salt), sodium and chlorine atoms occupy alternate corners of a cube, as in Figure 9.2a. In an **amorphous solid**, such as glass, the atoms are arranged almost randomly, as in Figure 9.2b.

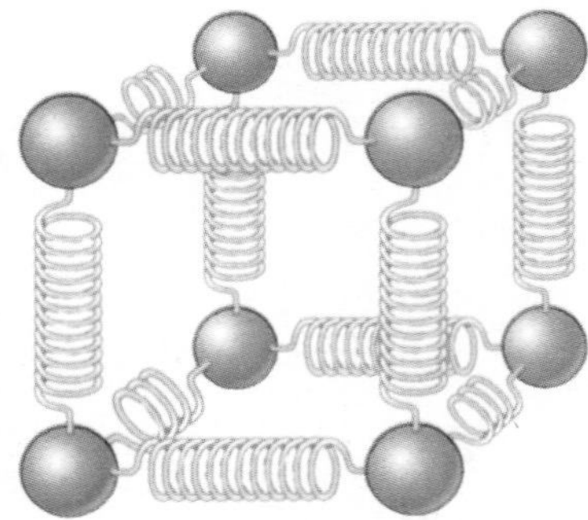

Figure 9.1 A model of a portion of a solid. The atoms (spheres) are imagined as being attached to each other by springs, which represent the elastic nature of the interatomic forces. A solid consists of trillions of segments like this, with springs connecting all of them.

For any given substance, the liquid state exists at a higher temperature than the solid state. The intermolecular forces in a liquid aren't strong enough to keep the molecules in fixed positions, and they wander through the liquid in random fashion (Fig. 9.2c). Solids and liquids both have the property that when an attempt is made to compress them, strong repulsive atomic forces act internally to resist the compression.

In the gaseous state, molecules are in constant random motion and exert only weak forces on each other. The average distance between the molecules of a gas is quite large compared with the size of the molecules. Occasionally the molecules collide with each other, but most of the time they move as nearly free, noninteracting particles. As a result, unlike solids and liquids, gases can be easily compressed. We'll say more about gases in subsequent chapters.

When a gas is heated to high temperature, many of the electrons surrounding each atom are freed from the nucleus. The resulting system is a collection of free, electrically charged particles—negatively charged electrons and positively charged ions. Such a highly ionized state of matter containing equal amounts of positive and negative charges is called a **plasma**. Unlike a neutral gas, the long-range electric and magnetic forces allow the constituents of a plasma to interact with each other. Plasmas are found inside stars and in accretion disks around black holes, for example, and are far more common than the solid, liquid, and gaseous states because there are far more stars around than any other form of celestial matter,

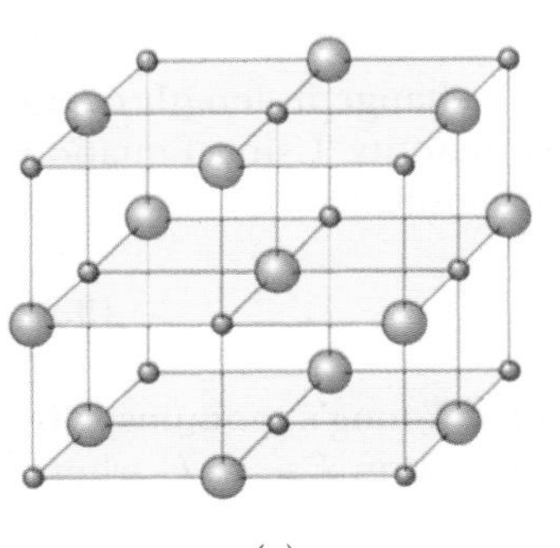

(a)

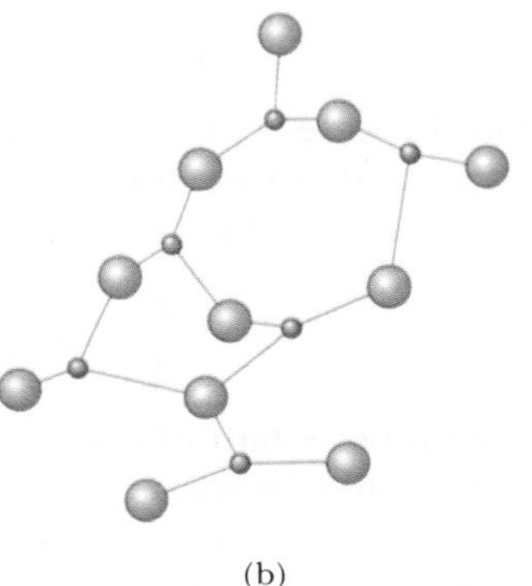

(b)

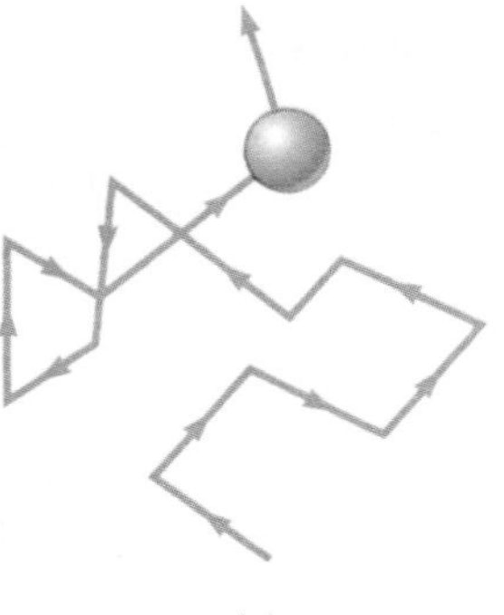

(c)

Figure 9.2 (a) The NaCl structure, with the Na^+ (gray) and Cl^- (green) ions at alternate corners of a cube. (b) In an amorphous solid, the atoms are arranged randomly. (c) Erratic motion of a molecule in a liquid.

except possibly **dark matter**. Dark matter, inferred by observations of the motion of stars around the galaxy, makes up about 90% of the matter in the universe and is of unknown composition. In this chapter, however, we ignore plasmas and dark matter and concentrate on the more familiar solid, liquid, and gaseous forms that make up the environment of our planet.

9.2 THE DEFORMATION OF SOLIDS

While a solid may be thought of as having a definite shape and volume, it's possible to change its shape and volume by applying external forces. A sufficiently large force will permanently deform or break an object, but otherwise, when the external forces are removed, the object tends to return to its original shape and size. This is called *elastic behavior.*

The elastic properties of solids are discussed in terms of stress and strain. **Stress** is the force per unit area causing a deformation; **strain** is a measure of the amount of the deformation. For sufficiently small stresses, **stress is proportional to strain,** with the constant of proportionality depending on the material being deformed and on the nature of the deformation. We call this proportionality constant the **elastic modulus:**

$$\text{stress} = \text{elastic modulus} \times \text{strain} \qquad [9.1]$$

The elastic modulus is analogous to a spring constant. It can be taken as the stiffness of a material: A material having a large elastic modulus is very stiff and difficult to deform. There are three relationships having the form of Equation 9.1, corresponding to tensile, shear, and bulk deformation, and all of them satisfy an equation similar to Hooke's law for springs:

$$F = k\Delta x \qquad [9.2]$$

where F is the applied force, k is the spring constant, and Δx is the amount by which the spring is compressed.

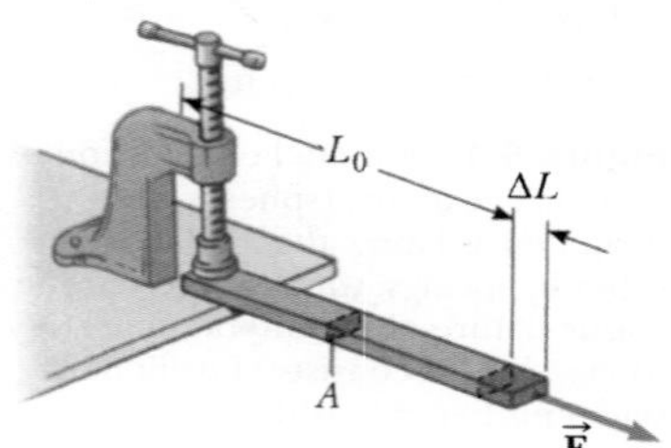

ACTIVE FIGURE 9.3
A long bar clamped at one end is stretched by the amount ΔL under the action of a force $\vec{\mathbf{F}}$.

Physics Now™
Log into PhysicsNow at **www.cp7e.com**, and go to Active Figure 9.3 to adjust the values of the applied force and Young's modulus and observe the change in length of the bar.

Young's Modulus: Elasticity in Length

Consider a long bar of cross-sectional area A and length L_0, clamped at one end (Active Fig. 9.3). When an external force $\vec{\mathbf{F}}$ is applied along the bar, perpendicular to the cross section, internal forces in the bar resist the distortion ("stretching") that $\vec{\mathbf{F}}$ tends to produce. Nevertheless, the bar attains an equilibrium in which (1) its length is greater than L_0 and (2) the external force is balanced by internal forces. Under these circumstances, the bar is said to be *stressed.* We define the **tensile stress** as the ratio of the magnitude of the external force F to the cross-sectional area A. The word "tensile" has the same root as the word "tension" and is used because the bar is under tension. The SI unit of stress is the newton per square meter (N/m^2), called the **pascal** (Pa):

The pascal ▶

$$1\ \text{Pa} \equiv 1\ \text{N/m}^2$$

The **tensile strain** in this case is defined as the ratio of the change in length ΔL to the original length L_0 and is therefore a dimensionless quantity. Using Equation 9.1, we can write an equation relating tensile stress to tensile strain:

$$\frac{F}{A} = Y\frac{\Delta L}{L_0} \qquad [9.3]$$

In this equation, Y is the constant of proportionality, called **Young's modulus.** Notice that Equation 9.3 could be solved for F and put in the form $F = k\Delta L$, where $k = YA/L_0$, making it look just like Hooke's law, Equation 9.2.

A material having a large Young's modulus is difficult to stretch or compress. This quantity is typically used to characterize a rod or wire stressed under either tension or compression. Because strain is a dimensionless quantity, Y is in pascals.

TABLE 9.1

Typical Values for the Elastic Modulus

Substance	Young's Modulus (Pa)	Shear Modulus (Pa)	Bulk Modulus (Pa)
Aluminum	7.0×10^{10}	2.5×10^{10}	7.0×10^{10}
Bone	1.8×10^{10}	8.0×10^{10}	—
Brass	9.1×10^{10}	3.5×10^{10}	6.1×10^{10}
Copper	11×10^{10}	4.2×10^{10}	14×10^{10}
Steel	20×10^{10}	8.4×10^{10}	16×10^{10}
Tungsten	35×10^{10}	14×10^{10}	20×10^{10}
Glass	$6.5\text{–}7.8 \times 10^{10}$	$2.6\text{–}3.2 \times 10^{10}$	$5.0\text{–}5.5 \times 10^{10}$
Quartz	5.6×10^{10}	2.6×10^{10}	2.7×10^{10}
Rib Cartilage	1.2×10^{7}	—	—
Rubber	0.1×10^{7}	—	—
Tendon	2×10^{7}	—	—
Water	—	—	0.21×10^{10}
Mercury	—	—	2.8×10^{10}

Typical values are given in Table 9.1. Experiments show that (1) the change in length for a fixed external force is proportional to the original length and (2) the force necessary to produce a given strain is proportional to the cross-sectional area. The value of Young's modulus for a given material depends on whether the material is stretched or compressed. A human femur, for example, is stronger under tension than compression.

It's possible to exceed the **elastic limit** of a substance by applying a sufficiently great stress (Fig. 9.4). At the elastic limit, the stress–strain curve departs from a straight line. A material subjected to a stress beyond this limit ordinarily doesn't return to its original length when the external force is removed. As the stress is increased further, it surpasses the **ultimate strength**: the greatest stress the substance can withstand without breaking. The **breaking point** for brittle materials is just beyond the ultimate strength. For ductile metals like copper and gold, after passing the point of ultimate strength, the metal thins and stretches at a lower stress level before breaking.

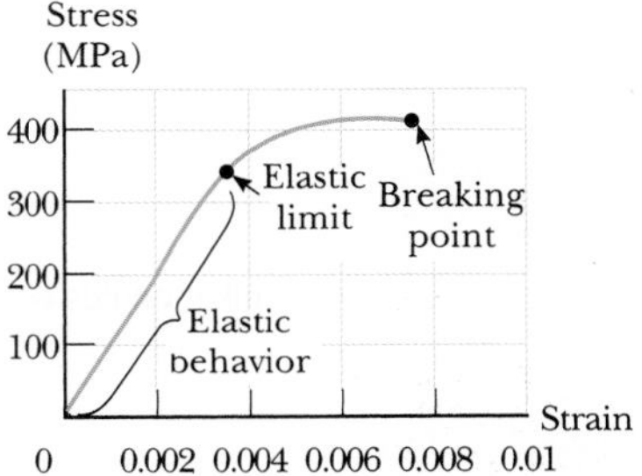

Figure 9.4 Stress-versus-strain curve for an elastic solid.

Shear Modulus: Elasticity of Shape

Another type of deformation occurs when an object is subjected to a force $\vec{\mathbf{F}}$ *parallel* to one of its faces while the opposite face is held fixed by a second force (Active Fig. 9.5a). If the object is originally a rectangular block, such a parallel force results in a shape with the cross section of a parallelogram. This kind of stress is called a **shear stress**. A book pushed sideways, as in Active Figure 9.5b, is being subjected to a shear stress. There is no change in volume with this kind of deformation. It's important to remember that in shear stress, the applied force is *parallel* to the cross-sectional area, whereas in tensile stress the force is *perpendicular* to the cross-sectional area. We define **the shear stress as F/A, the ratio of the magnitude of the**

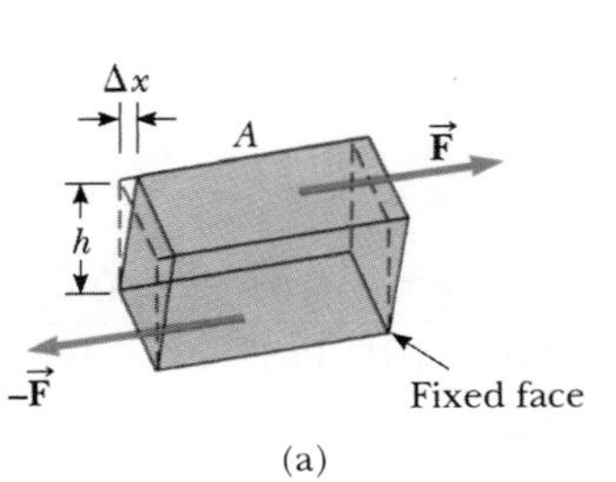

(a)

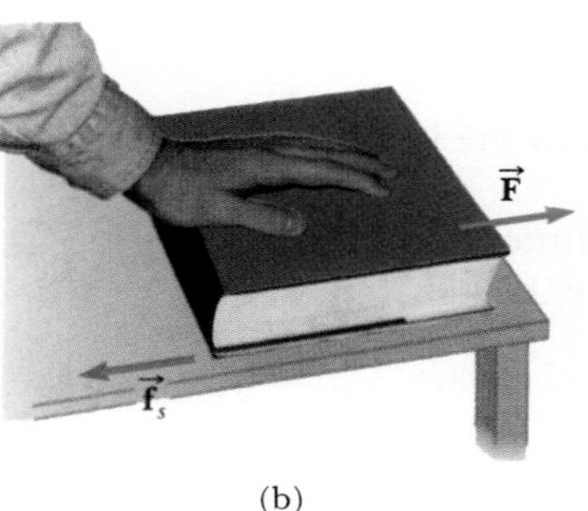

(b)

ACTIVE FIGURE 9.5
(a) A shear deformation in which a rectangular block is distorted by forces applied tangent to two of its faces. (b) A book under shear stress.

Physics Now™
Log into PhysicsNow at **www.cp7e.com**, and go to Active Figure 9.5 to adjust the values of the applied force and the shear modulus and observe the change in shape of the block in part (a).

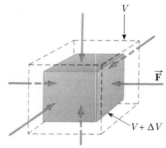

ACTIVE FIGURE 9.6
When a solid is under uniform pressure, it undergoes a change in volume, but no change in shape. This cube is compressed on all sides by forces normal to its six faces.

Physics Now™
Log into PhysicsNow at **www.cp7e.com**, and go to Active Figure 9.6 to adjust the values of the applied force and the bulk modulus and observe the change in volume of the cube.

parallel force to the area A of the face being sheared. The shear strain is the ratio $\Delta x/h$, where Δx is the horizontal distance the sheared face moves and h is the height of the object. The shear stress is related to the shear strain according to

$$\frac{F}{A} = S\frac{\Delta x}{h} \qquad [9.4]$$

where S is the **shear modulus** of the material, with units of pascals (force per unit area). Once again, notice the similarity to Hooke's law.

A material having a large shear modulus is difficult to bend. Shear moduli for some representative materials are listed in Table 9.1.

Bulk Modulus: Volume Elasticity

The bulk modulus characterizes the response of a substance to uniform squeezing. Suppose that the external forces acting on an object are all perpendicular to the surface on which the force acts and are distributed uniformly over the surface of the object (Active Fig. 9.6). This occurs when an object is immersed in a fluid. An object subject to this type of deformation undergoes a change in volume but no change in shape. **The volume stress ΔP is defined as the ratio of the magnitude of the change in the applied force ΔF to the surface area A.** (In dealing with fluids, we'll refer to the quantity F/A as the **pressure**, to be defined and discussed more formally in the next section.) The volume strain is equal to the change in volume ΔV divided by the original volume V. Again using Equation 9.1, we can relate a volume stress to a volume strain by the formula

Bulk modulus ▶

$$\Delta P = -B\frac{\Delta V}{V} \qquad [9.5]$$

A material having a large bulk modulus doesn't compress easily. Note that a negative sign is included in this defining equation so that B is always positive. An increase in pressure (positive ΔP) causes a decrease in volume (negative ΔV) and vice versa.

Table 9.1 lists bulk modulus values for some materials. If you look up such values in a different source, you may find that the reciprocal of the bulk modulus, called the **compressibility** of the material, is listed. Note from the table that both solids and liquids have bulk moduli. There is neither a Young's modulus nor shear modulus for liquids, however, because liquids simply flow when subjected to a tensile or shearing stress.

EXAMPLE 9.1 Built to Last

Goal Calculate a compression due to tensile stress, and maximum load.

Problem A vertical steel beam in a building supports a load of 6.0×10^4 N. **(a)** If the length of the beam is 4.0 m and its cross-sectional area is 8.0×10^{-3} m^2, find the distance the beam is compressed along its length. **(b)** What maximum load in newtons could the steel beam support before failing?

Strategy Equation 9.3 pertains to compressive stress and strain and can be solved for ΔL, followed by substitution of known values. For part (b), set the compressive stress equal to the ultimate strength of steel from Table 9.2. Solve for the magnitude of the force, which is the total weight the structure can support.

Solution

(a) Find the amount of compression in the beam.

Solve Equation 9.3 for ΔL and substitute, using the value of Young's modulus from Table 9.1:

$$\frac{F}{A} = Y\frac{\Delta L}{L_0}$$

$$\Delta L = \frac{FL_0}{YA} = \frac{(6.0 \times 10^4\ \text{N})(4.0\ \text{m})}{(2.0 \times 10^{11}\ \text{Pa})(8.0 \times 10^{-3}\ \text{m}^2)}$$

$$= 1.5 \times 10^{-4}\ \text{m}$$

(b) Find the maximum load that the beam can support.

Set the compressive stress equal to the ultimate compressive strength from Table 9.2, and solve for F:

$$\frac{F}{A} = \frac{F}{8.0 \times 10^{-3}\ \mathrm{m}^2} = 5.0 \times 10^8\ \mathrm{Pa}$$

$$F = \boxed{4.0 \times 10^6\ \mathrm{N}}$$

Remarks In designing load-bearing structures of any kind, it's always necessary to build in a safety factor. No one would drive a car over a bridge that had been designed to supply the minimum necessary strength to keep it from collapsing.

Exercise 9.1

A cable used to lift heavy materials like steel I-beams must be strong enough to resist breaking even under a load of 1.0×10^6 N. For safety, the cable must support twice that load. (a) What cross-sectional area should the cable have if it's to be made of steel? (b) By how much will an 8.0-m length of this cable stretch when subject to the 1.0×10^6-N load?

Answers (a) $4.0 \times 10^{-3}\ \mathrm{m}^2$ (b) 1.0×10^{-2} m

TABLE 9.2

Ultimate Strength of Materials

Material	Tensile Strength ($\mathrm{N/m^2}$)	Compressive Strength ($\mathrm{N/m^2}$)
Iron	1.7×10^8	5.5×10^8
Steel	5.0×10^8	5.0×10^8
Aluminum	2.0×10^8	2.0×10^8
Bone	1.2×10^8	1.5×10^8
Marble	—	8.0×10^7
Brick	1×10^6	3.5×10^7
Concrete	2×10^6	2×10^7

EXAMPLE 9.2 Explosive Bolts

Goal Calculate the maximum shear stress supported by a set of bolts.

Problem Until launch, rockets are generally held to the launch pad by explosive bolts. Such bolts are also used in escape hatches and to secure different stages of the rocket, external tanks, and strap-on boosters, allowing rapid release when a part needs to be separated from the rest of the vehicle. Suppose a rocket has a strap-on booster supported by eight horizontal steel bolts, each 9.00 cm in diameter and oriented horizontally. (Bolts similar to these would be used in rockets like the Titan IV, shown at right.) **(a)** What maximum load can be placed on these bolts before they are sheared off? Assume the load is shared equally by the eight bolts. The ultimate shear strength of steel is 2.50×10^8 Pa. **(b)** If the booster has a mass of 3.00×10^5 kg, calculate the shear deformation of one of the bolts if the length of the bolt between rocket and booster is 8.00 cm.

NASA

The Titan IV launch vehicle, with its two solid rocket boosters, is capable of placing large satellites in geosynchronous orbit.

Strategy (a) The total force required to shear off the bolts increases with the number of bolts, but the necessary shear stress does not. Set the ultimate shear strength equal to the shear stress and solve for the force, multiplying the answer by eight to find the total shear force that can be applied. Part (b) can be solved by substituting values into Equation 9.4, which relates shear stress to shear strain.

Solution

(a) Find the maximum load the bolts can support.

Set the shear stress for one bolt equal to its ultimate shear strength:

$$\frac{F_1}{A} = \frac{F_1}{\pi r^2} = 2.50 \times 10^8\ \mathrm{N/m^2}$$

Solve for the force needed to shear off one bolt:

$$F_1 = \pi r^2(2.50 \times 10^8 \text{ Pa})$$
$$= \pi(4.50 \times 10^{-2} \text{ m})^2(2.50 \times 10^8 \text{ Pa})$$
$$= 1.59 \times 10^6 \text{ N}$$

Multiply this result by eight to get the total shear force that the eight bolts can support:

$$F_{\text{tot}} = \boxed{1.27 \times 10^7 \text{ N}}$$

(b) Calculate the shear deformation of one bolt at half the maximum load.

The bolts must support the booster against the force of gravity. The reaction to this force is the force exerted by the booster on the bolts, equal in magnitude to the weight of the booster.

$$F_{\text{tot}} = mg = (3.00 \times 10^5 \text{ kg})(9.80 \text{ m/s}^2) = 2.94 \times 10^6 \text{ N}$$

This force is shared among eight bolts. Use Equation 9.4, taking $F = F_{\text{tot}}/8$ and solving for Δx:

$$\frac{F}{A} = S\frac{\Delta x}{h}$$

$$\Delta x = \frac{(F_{\text{tot}}/8)h}{AS} = \frac{(3.68 \times 10^5 \text{ N})(0.080 \text{ m})}{\pi(4.50 \times 10^{-2} \text{ m})^2(8.40 \times 10^{10} \text{ Pa})}$$
$$= \boxed{5.51 \times 10^{-5} \text{ m}}$$

Remarks The Titan IV launch vehicle, which is capable of putting large satellites into geosynchronous orbit, has a pair of strap-on boosters similar to this one. Notice that the bolts in this example are capable of supporting about four times as much weight as needed. Because all materials either have or develop microscopic defects, a safety factor has to be built in, so structures are designed to tolerate several times the maximum stresses they are expected to undergo.

Exercise 9.2

Calculate the diameter of a single steel horizontal bolt if it is expected to support a maximum load having a mass of 2.00×10^3 kg, but for safety reasons must be designed to support three times that load.

Answer 1.73 cm

EXAMPLE 9.3 Stressing a Lead Ball

Goal Apply the concepts of bulk stress and strain.

Problem A solid lead sphere of volume 0.50 m^3, dropped in the ocean, sinks to a depth of 2.0×10^3 m (about 1 mile), where the pressure increases by 2.0×10^7 Pa. Lead has a bulk modulus of 4.2×10^{10} Pa. What is the change in volume of the sphere?

Strategy Solve Equation 9.5 for ΔV and substitute the given quantities.

Solution

Start with the definition of bulk modulus:

$$B = -\frac{\Delta P}{\Delta V/V}$$

Solve for ΔV:

$$\Delta V = -\frac{V\Delta P}{B}$$

Substitute the known values:

$$\Delta V = -\frac{(0.50 \text{ m}^3)(2.0 \times 10^7 \text{ Pa})}{4.2 \times 10^{10} \text{ Pa}} = \boxed{-2.4 \times 10^{-4} \text{ m}^3}$$

Remarks The negative sign indicates a *decrease* in volume. The following exercise shows that even water can be compressed, though not by much, despite the depth.

Exercise 9.3

(a) By what percentage does a similar globe of water shrink at that same depth? (b) What is the ratio of the new radius to the initial radius?

Answer (a) 0.95% (b) 0.997

Arches and the Ultimate Strength of Materials

As we have seen, the ultimate strength of a material is the maximum force per unit area the material can withstand before it breaks or fractures. Such values are of great importance, particularly in the construction of buildings, bridges, and roads. Table 9.2 gives the ultimate strength of a variety of materials under both tension and compression. Note that bone and a variety of building materials (concrete, brick, and marble) are stronger under compression than under tension. The greater ability of brick and stone to resist compression is the basis of the semicircular arch, developed and used extensively by the Romans in everything from memorial arches to expansive temples and aqueduct supports.

APPLICATION

Arch Structures in Buildings

Before the development of the arch, the principal method of spanning a space was the simple post-and-beam construction (Fig. 9.7a), in which a horizontal beam is supported by two columns. This type of construction was used to build the great Greek temples. The columns of these temples were closely spaced because of the limited length of available stones and the low ultimate tensile strength of a sagging stone beam.

The semicircular arch (Fig. 9.7b) developed by the Romans was a great technological achievement in architectural design. It effectively allowed the heavy load of a wide roof span to be channeled into horizontal and vertical forces on narrow supporting columns. The stability of this arch depends on the compression between its wedge-shaped stones. The stones are forced to squeeze against each other by the uniform loading, as shown in the figure. This compression results in horizontal outward forces at the base of the arch where it starts curving away from the vertical. These forces must then be balanced by the stone walls shown on the sides of the arch. It's common to use very heavy walls (buttresses) on either side of the arch to provide horizontal stability. If the foundation of the arch should move, the compressive forces between the wedge-shaped stones may decrease to the extent that the arch collapses. The stone surfaces used in the arches constructed by the Romans were cut to make very tight joints; mortar was usually not used. The resistance to slipping between stones was provided by the compression force and the friction between the stone faces.

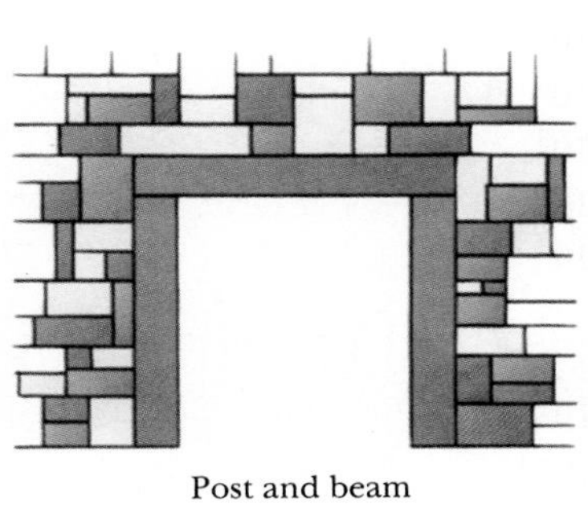

Post and beam
(a)

Semicircular arch (Roman)
(b)

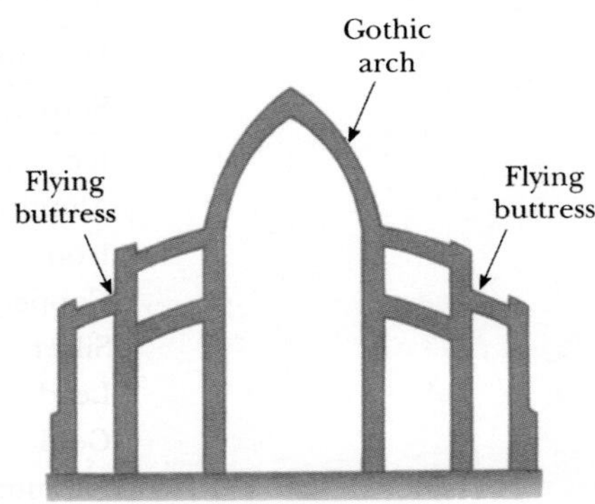

Pointed arch (Gothic)
(c)

Figure 9.7 (a) A simple post-and-beam structure. (b) The semicircular arch developed by the Romans. (c) Gothic arch with flying buttresses to provide lateral support.

Another important architectural innovation was the pointed Gothic arch, shown in Figure 9.7c. This type of structure was first used in Europe beginning in the 12th century, followed by the construction of several magnificent Gothic cathedrals in France in the 13th century. One of the most striking features of these cathedrals is their extreme height. For example, the cathedral at Chartres rises to 118 ft, and the one at Reims has a height of 137 ft. Such magnificent buildings evolved over a very short time, without the benefit of any mathematical theory of structures. However, Gothic arches required flying buttresses to prevent the spreading of the arch supported by the tall, narrow columns.

9.3 DENSITY AND PRESSURE

Equal masses of aluminum and gold have an important physical difference: The aluminum takes up over seven times as much space as the gold. While the reasons for the difference lie at the atomic and nuclear levels, a simple measure of this difference is the concept of *density.*

Density ▶

The **density** ρ of an object having uniform composition is defined as its mass M divided by its volume V:

$$\rho \equiv \frac{M}{V} \qquad [9.6]$$

SI unit: kilogram per meter cubed (kg/m³)

The most common units used for density are kilograms per cubic meter in the SI system and grams per cubic centimeter in the cgs system. Table 9.3 lists the densities of some substances. The densities of most liquids and solids vary slightly with changes in temperature and pressure; the densities of gases vary greatly with such changes. Under normal conditions, the densities of solids and liquids are about 1 000 times greater than the densities of gases. This difference implies that the average spacing between molecules in a gas under such conditions is about ten times greater than in a solid or liquid.

The **specific gravity** of a substance is the ratio of its density to the density of water at 4°C, which is 1.0×10^3 kg/m³. (The size of the kilogram was originally defined to make the density of water 1.0×10^3 kg/m³ at 4°C.) By definition, specific gravity is a dimensionless quantity. For example, if the specific gravity of a substance is 3.0, its density is $3.0(1.0 \times 10^3 \text{ kg/m}^3) = 3.0 \times 10^3 \text{ kg/m}^3$.

TABLE 9.3

Densities of Some Common Substances

Substance	ρ(kg/m³)[a]	Substance	ρ(kg/m³)[a]
Ice	0.917×10^3	Water	1.00×10^3
Aluminum	2.70×10^3	Glycerin	1.26×10^3
Iron	7.86×10^3	Ethyl alcohol	0.806×10^3
Copper	8.92×10^3	Benzene	0.879×10^3
Silver	10.5×10^3	Mercury	13.6×10^3
Lead	11.3×10^3	Air	1.29
Gold	19.3×10^3	Oxygen	1.43
Platinum	21.4×10^3	Hydrogen	8.99×10^{-2}
Uranium	18.7×10^3	Helium	1.79×10^{-1}

[a]All values are at standard atmospheric temperature and pressure (STP), defined as 0°C (273 K) and 1 atm (1.013×10^5 Pa). To convert to grams per cubic centimeter, multiply by 10^{-3}.

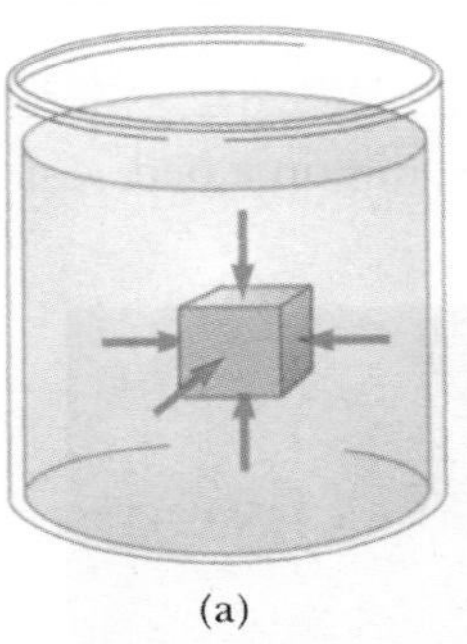

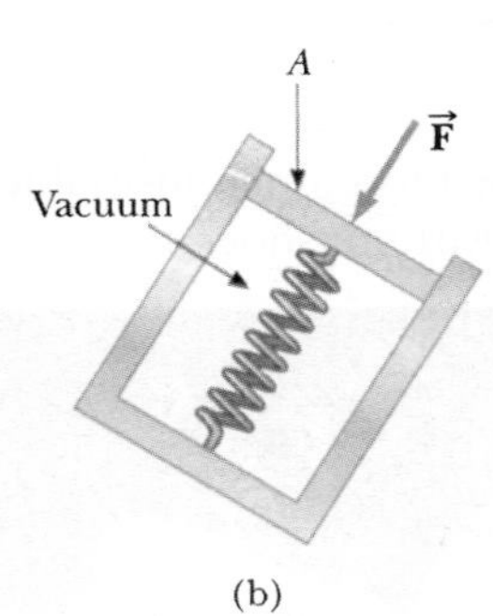

(a) (b)

FIGURE 9.8 (a) The force exerted by a fluid on a submerged object at any point is perpendicular to the surface of the object. The force exerted by the fluid on the walls of the container is perpendicular to the walls at all points and increases with depth. (b) A simple device for measuring pressure in a fluid.

Quick Quiz 9.1

Suppose you have one cubic meter of gold, two cubic meters of silver, and six cubic meters of aluminum. Rank them by mass, from smallest to largest. (a) gold, aluminum, silver (b) gold, silver, aluminum (c) aluminum, gold, silver (d) silver, aluminum, gold

Fluids don't sustain shearing stresses, so the only stress that a fluid can exert on a submerged object is one that tends to compress it, which is bulk stress. The force exerted by the fluid on the object is always perpendicular to the surfaces of the object, as shown in Figure 9.8a.

The pressure at a specific point in a fluid can be measured with the device pictured in Figure 9.8b: an evacuated cylinder enclosing a light piston connected to a spring that has been previously calibrated with known weights. As the device is submerged in a fluid, the fluid presses down on the top of the piston and compresses the spring until the inward force exerted by the fluid is balanced by the outward force exerted by the spring. Let F be the magnitude of the force on the piston and A the area of the top surface of the piston. Notice that the force that compresses the spring is spread out over the entire area, motivating our formal definition of pressure:

TIP 9.1 Force and Pressure

Equation 9.7 makes a clear distinction between force and pressure. Another important distinction is that *force is a vector* and *pressure is a scalar.* There is no direction associated with pressure, but the direction of the force associated with the pressure is perpendicular to the surface of interest.

◀ Pressure

If F is the magnitude of a force exerted perpendicular to a given surface of area A, then the pressure P is the force divided by the area:

$$P \equiv \frac{F}{A} \qquad [9.7]$$

SI unit: pascal (Pa)

Because pressure is defined as force per unit area, it has units of pascals (newtons per square meter). The English customary unit for pressure is the pound per inch squared. Atmospheric pressure at sea level is 14.7 lb/in^2, which in SI units is 1.01×10^5 Pa.

As we see from Equation 9.7, the effect of a given force depends critically on the area to which it's applied. A 700-N man can stand on a vinyl-covered floor in regular street shoes without damaging the surface, but if he wears golf shoes, the metal cleats protruding from the soles can do considerable damage to the floor. With the cleats, the same force is concentrated into a smaller area, greatly elevating the pressure in those areas, resulting in a greater likelihood of exceeding the ultimate strength of the floor material.

Snowshoes use the same principle (Fig. 9.9). The snow exerts an upward normal force on the shoes to support the person's weight. According to Newton's third law, this upward force is accompanied by a downward force exerted by the shoes on the snow. If the person is wearing snowshoes, that force is distributed over the very large area of each snowshoe, so that the pressure at any given point is relatively low and the person doesn't penetrate very deeply into the snow.

Figure 9.9 Snowshoes prevent the person from sinking into the soft snow because the force on the snow is spread over a larger area, reducing the pressure on the snow's surface.

Applying Physics 9.1 Bed of Nails Trick

After an exciting but exhausting lecture, a physics professor stretches out for a nap on a bed of nails, as in Figure 9.10, suffering no injury and only moderate discomfort. How is this possible?

Explanation If you try to support your entire weight on a single nail, the pressure on your body is your weight divided by the very small area of the end of the nail. The resulting pressure is large enough to penetrate the skin. If you distribute your weight over several hundred nails, however, as demonstrated by the professor, the pressure is considerably reduced because the area that supports your weight is the total area of all nails in contact with your body. (Why is lying on a bed of nails more comfortable than sitting on the same bed? Extend the logic to show that it would be more uncomfortable yet to stand on a bed of nails without shoes.)

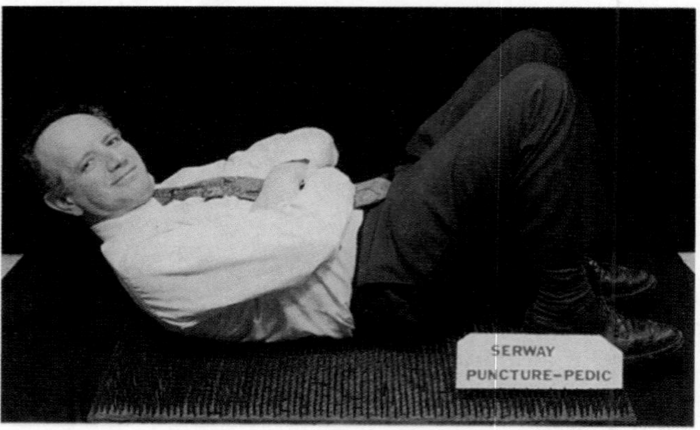

Figure 9.10 (Applying Physics 9.1) Does anyone have a pillow?

EXAMPLE 9.4 The Water Bed

Goal Calculate a density and a pressure from a weight.

Problem A water bed is 2.00 m on a side and 30.0 cm deep. **(a)** Find its weight. **(b)** Find the pressure that the water bed exerts on the floor. Assume that the entire lower surface of the bed makes contact with the floor.

Strategy Density is mass per unit volume: first, find the volume of the bed and multiply it by the density of water to get the bed's mass. Multiplying by the acceleration of gravity then gives the weight of the bed. The weight divided by the area of floor the bed rests upon gives the pressure exerted on the floor.

Solution

(a) Find the weight of the water bed.

First, find the volume of the bed:

$$V = lwh = (2.00\text{ m})(2.00\text{ m})(0.300\text{ m}) = 1.20\text{ m}^3$$

Solve the density equation for the mass and substitute, then multiply the result by g to get the weight:

$$\rho = \frac{M}{V}$$

$$M = \rho V = (1.00 \times 10^3\text{ kg/m}^3)(1.20\text{ m}^3) = 1.20 \times 10^3\text{ kg}$$

$$w = Mg = (1.20 \times 10^3\text{ kg})(9.80\text{ m/s}^2) = 1.18 \times 10^4\text{ N}$$

(b) Find the pressure that the bed exerts on the floor.

Use the cross-sectional area $A = 4.00\text{ m}^2$ and the value of w from part (a) to get the pressure:

$$P = \frac{F}{A} = \frac{w}{A} = \frac{1.18 \times 10^4\text{ N}}{4.00\text{ m}^2} = 2.95 \times 10^3\text{ Pa}$$

Remarks Notice that the answer to part (b) is far less than atmospheric pressure. Water is heavier than air for a given volume, but the air is stacked up considerably higher (100 km!). The total pressure exerted on the floor would include the pressure of the atmosphere.

Exercise 9.4

Calculate the pressure exerted by the water bed on the floor if the bed rests on its side.

Answer 1.97×10^4 Pa

9.4 VARIATION OF PRESSURE WITH DEPTH

When a fluid is at rest in a container, **all portions of the fluid must be in static equilibrium**—at rest with respect to the observer. Furthermore, **all points at the same depth must be at the same pressure**. If this were not the case, fluid would flow from the higher pressure region to the lower pressure region. For example, consider the small block of fluid shown in Figure 9.11a. If the pressure were greater on the left side of the block than on the right, $\vec{\mathbf{F}}_1$ would be greater than $\vec{\mathbf{F}}_2$, and the block would accelerate to the right and thus would not be in equilibrium.

Next, let's examine the fluid contained within the volume indicated by the darker region in Figure 9.11b. This region has cross-sectional area A and extends from position y_1 to position y_2 below the surface of the liquid. Three external forces act on this volume of fluid: the force of gravity, Mg; the upward force P_2A exerted by the liquid below it; and a downward force P_1A exerted by the fluid above it. Because the given volume of fluid is in equilibrium, these forces must add to zero, so we get

$$P_2A - P_1A - Mg = 0 \qquad [9.8]$$

From the definition of density, we have

$$M = \rho V = \rho A(y_1 - y_2) \qquad [9.9]$$

Substituting Equation 9.9 into Equation 9.8, canceling the area A, and rearranging terms, we get

$$P_2 = P_1 + \rho g(y_1 - y_2) \qquad [9.10]$$

Notice that $(y_1 - y_2)$ is positive, because $y_2 < y_1$. The force P_2A is greater than the force P_1A by exactly the weight of water between the two points. This is the same principle experienced by the person at the bottom of a pileup in football or rugby.

Atmospheric pressure is also caused by a piling up of fluid—in this case, the fluid is the gas of the atmosphere. The weight of all the air from sea level to the edge of space results in an atmospheric pressure of $P_0 = 1.013 \times 10^5$ Pa (equivalent to 14.7 lb/in.^{2}) at sea level. This result can be adapted to find the pressure P at any depth $h = (y_1 - y_2) = (0 - y_2)$ below the surface of the water:

$$P = P_0 + \rho gh \qquad [9.11]$$

According to Equation 9.11, **the pressure *P* at a depth *h* below the surface of a liquid open to the atmosphere is greater than atmospheric pressure by the amount *ρgh***. Moreover, the pressure isn't affected by the shape of the vessel, as shown in Figure 9.12.

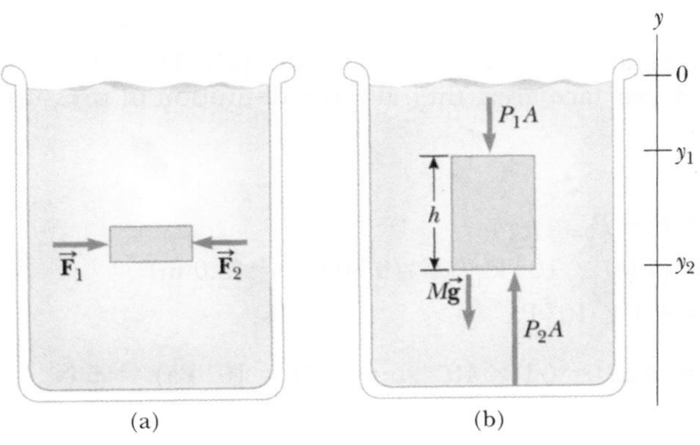

Figure 9.11 (a) If the block of fluid is to be in equilibrium, the force $\vec{\mathbf{F}}_1$ must balance the force $\vec{\mathbf{F}}_2$. (b) The net force on the volume of liquid within the darker region must be zero.

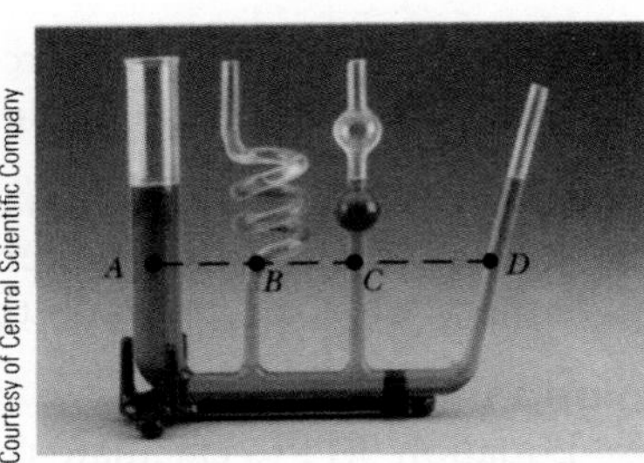

Courtesy of Central Scientific Company

Figure 9.12 This photograph illustrates the fact that the pressure in a liquid is the same at all points lying at the same elevation. For example, the pressure is the same at points *A*, *B*, *C*, and *D*. Note that the shape of the vessel does not affect the pressure.

Quick Quiz 9.2

The pressure at the bottom of a glass filled with water ($\rho = 1\ 000\ \text{kg/m}^3$) is P. The water is poured out and the glass is filled with ethyl alcohol ($\rho = 806\ \text{kg/m}^3$). The pressure at the bottom of the glass is now (a) smaller than P (b) equal to P (c) larger than P (d) indeterminate.

EXAMPLE 9.5 Oil and Water

Goal Calculate pressures created by layers of different fluids.

Problem In a huge oil tanker, salt water has flooded an oil tank to a depth of 5.00 m. On top of the water is a layer of oil 8.00 m deep, as in the cross-sectional view of the tank in Figure 9.13. The oil has a density of $0.700\ \text{g/cm}^3$. Find the pressure at the bottom of the tank. (Take $1\ 025\ \text{kg/m}^3$ as the density of salt water.)

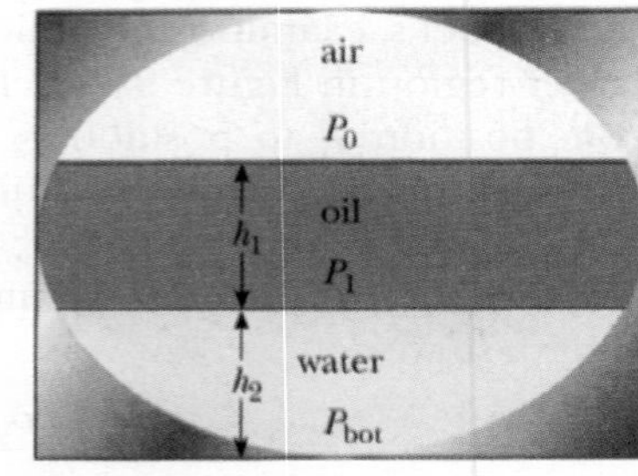

Figure 9.13 (Example 9.5)

Strategy Equation 9.11 must be used twice. First, use it to calculate the pressure P_1 at the bottom of the oil layer. Then use this pressure in place of P_0 in Equation 9.11 and calculate the pressure P_{bot} at the bottom of the water layer.

Solution

Use Equation 9.11 to calculate the pressure at the bottom of the oil layer:

$$P_1 = P_0 + \rho g h_1 \tag{1}$$
$$= 1.01 \times 10^5\ \text{Pa} + (7.00 \times 10^2\ \text{kg/m}^3)(9.80\ \text{m/s}^2)(8.00\ \text{m})$$
$$P_1 = 1.56 \times 10^5\ \text{Pa}$$

Now adapt Equation 9.11 to the new starting pressure, and use it to calculate the pressure at the bottom of the water layer:

$$P_{bot} = P_1 + \rho g h_2 \tag{2}$$
$$= 1.56 \times 10^5\ \text{Pa} + (1.025 \times 10^3\ \text{kg/m}^3)(9.80\ \text{m/s}^2)(5.00\ \text{m})$$
$$P_{bot} = 2.06 \times 10^5\ \text{Pa}$$

Remark The weight of the atmosphere results in P_0 at the surface of the oil layer. Then the weight of the oil and the weight of the water combine to create the pressure at the bottom.

Exercise 9.5

Calculate the pressure on the top lid of a chest buried under 4.00 meters of mud with density $1.75 \times 10^3\ \text{kg/m}^3$ at the bottom of a 10.0-m-deep lake.

Answer 2.68×10^5 Pa

EXAMPLE 9.6 A Pain in the Ear

Goal Calculate a pressure difference at a given depth, and estimate a force.

Problem Estimate the net force exerted on your eardrum due to the water above when you are swimming at the bottom of a pool that is 5.0 m deep.

Strategy Use Equation 9.11 to find the pressure difference across the eardrum at the given depth. The air inside the ear is generally at atmospheric pressure. Estimate the eardrum's surface area, then use the definition of pressure to get the net force exerted on the eardrum.

Solution

Use Equation 9.11 to calculate the difference between the water pressure at the depth h and the pressure inside the ear:

$$\Delta P = P - P_0 = \rho g h$$
$$= (1.00 \times 10^3\ \text{kg/m}^3)(9.80\ \text{m/s}^2)(5.0\ \text{m})$$
$$= 4.9 \times 10^4\ \text{Pa}$$

Mutliply by area A to get the net force on the eardrum associated with this pressure difference, estimating the area of the eardrum as $1\ \text{cm}^2$

$$F_{net} = A\Delta P \approx (1 \times 10^{-4}\ \text{m}^2)(4.9 \times 10^4\ \text{Pa}) \approx 5\ \text{N}$$

Remarks Because a force on the eardrum of this magnitude is uncomfortable, swimmers often "pop their ears" by swallowing or expanding their jaws while underwater, an action that pushes air from the lungs into the middle ear. Using this technique equalizes the pressure on the two sides of the eardrum and relieves the discomfort.

Exercise 9.6

An airplane takes off at sea level and climbs to a height of 425 m. Estimate the net outward force on a passenger's eardrum assuming the density of air is approximately constant at 1.3 kg/m^3 and that the inner ear pressure hasn't been equalized.

Answer 0.54 N

In view of the fact that the pressure in a fluid depends on depth and on the value of P_0, any increase in pressure at the surface must be transmitted to every point in the fluid. This was first recognized by the French scientist Blaise Pascal (1623–1662) and is called **Pascal's principle**:

> A change in pressure applied to an enclosed fluid is transmitted undiminished to every point of the fluid and to the walls of the container.

APPLICATION

Hydraulic Lifts

An important application of Pascal's principle is the hydraulic press (Fig. 9.14a). A downward force $\vec{\mathbf{F}}_1$ is applied to a small piston of area A_1. The pressure is transmitted through a fluid to a larger piston of area A_2. As the pistons move and the fluids in the left and right cylinders change their relative heights, there are slight differences in the pressures at the input and output pistons. Neglecting these small differences, the fluid pressure on each of the pistons may be taken to be the same; $P_1 = P_2$. From the definition of pressure, it then follows that $F_1/A_1 = F_2/A_2$. Therefore, the magnitude of the force $\vec{\mathbf{F}}_2$ is larger than the magnitude of $\vec{\mathbf{F}}_1$ by the factor A_2/A_1. That's why a large load, such as a car, can be moved on the large piston by a much smaller force on the smaller piston. Hydraulic brakes, car lifts, hydraulic jacks, forklifts, and other machines make use of this principle.

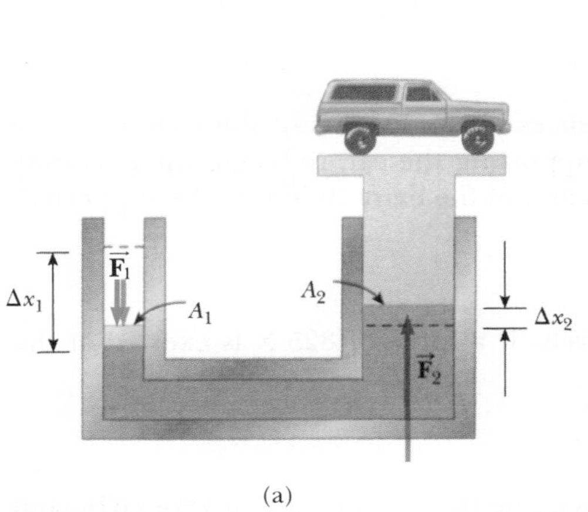

(a)

(b)

Figure 9.14 (a) Diagram of a hydraulic press (Example 9.7). Because the pressure is the same at the left and right sides, a small force $\vec{\mathbf{F}}_1$ at the left produces a much larger force $\vec{\mathbf{F}}_2$ at the right. (b) A vehicle under repair is supported by a hydraulic lift in a garage.

INTERACTIVE EXAMPLE 9.7 The Car Lift

Goal Apply Pascal's principle to a car lift, and show that the input work is the same as the output work.

Problem In a car lift used in a service station, compressed air exerts a force on a small piston of circular cross section having a radius of $r_1 = 5.00$ cm. This pressure is transmitted by an incompressible liquid to a second piston of radius $r_2 = 15.0$ cm. **(a)** What force must the compressed air exert on the small piston in order to lift a car weighing 13 300 N? Neglect the weights of the pistons. **(b)** What air pressure will produce a force of that magnitude? **(c)** Show that the work done by the input and output pistons is the same.

Strategy Substitute into Pascal's principle in part (a), while recognizing that the magnitude of the output force, F_2, must be equal to the car's weight in order to support it. Use the definition of pressure in part (b). In part (c), use $W = F\Delta x$ to find the ratio W_1/W_2, showing that it must equal 1. This requires combining Pascal's principle with the fact that the input and output pistons move through the same volume.

Solution

(a) Find the necessary force on the small piston.

Substitute known values into Pascal's principle, using $A = \pi r^2$ for the area.

$$F_1 = \left(\frac{A_1}{A_2}\right)F_2 = \frac{\pi r_1^2}{\pi r_2^2}F_2$$

$$= \frac{\pi(5.00 \times 10^{-2}\text{ m})^2}{\pi(15.0 \times 10^{-2}\text{ m})^2}(1.33 \times 10^4\text{ N})$$

$$= 1.48 \times 10^3\text{ N}$$

(b) Find the air pressure producing F_1.

Substitute into the definition of pressure:

$$P = \frac{F_1}{A_1} = \frac{1.48 \times 10^3\text{ N}}{\pi(5.00 \times 10^{-2}\text{ m})^2} = 1.88 \times 10^5\text{ Pa}$$

(c) Show that the work done by the input and output pistons is the same.

First equate the volumes, and solve for the ratio of A_2 to A_1:

$$V_1 = V_2 \rightarrow A_1\Delta x_1 = A_2\Delta x_2$$

$$\frac{A_2}{A_1} = \frac{\Delta x_1}{\Delta x_2}$$

Now use Pascal's principle to get a relationship for F_1/F_2:

$$\frac{F_1}{A_1} = \frac{F_2}{A_2} \rightarrow \frac{F_1}{F_2} = \frac{A_1}{A_2}$$

Evaluate the work ratio, substituting the preceding two results:

$$\frac{W_1}{W_2} = \frac{F_1\Delta x_1}{F_2\Delta x_2} = \left(\frac{F_1}{F_2}\right)\left(\frac{\Delta x_1}{\Delta x_2}\right) = \left(\frac{A_1}{A_2}\right)\left(\frac{A_2}{A_1}\right) = 1$$

$$W_1 = W_2$$

Remark In this problem, we didn't address the effect of possible differences in the heights of the pistons. If the column of fluid is higher in the small piston, the fluid weight assists in supporting the car, reducing the necessary applied force. If the column of fluid is higher in the large piston, both the car and the extra fluid must be supported, so additional applied force is required.

Exercise 9.7

A hydraulic lift has pistons with diameters 8.00 cm and 36.0 cm, respectively. If a force of 825 N is exerted at the input piston, what maximum mass can be lifted by the output piston?

Answer 1.70×10^3 kg

PhysicsNow™ You can adjust the weight of the truck in Figure 9.14a by logging into PhysicsNow at **www.cp7e.com** and going to Interactive Example 9.7.

Applying Physics 9.2 Building the Pyramids

A corollary to the statement that pressure in a fluid increases with depth is that water always seeks its own level. This means that if a vessel is filled with water, then regardless of the vessel's shape the surface of the water is perfectly flat and at the same height at all points. The ancient Egyptians used this fact to make the pyramids level. Devise a scheme showing how this could be done.

Explanation There are many ways it could be done, but Figure 9.15 shows the scheme used by the Egyptians. The builders cut grooves in the base of the pyramid as in (a) and partially filled the grooves with water. The height of the water was marked as in (b), and the rock was chiseled down to the mark, as in (c). Finally, the groove was filled with crushed rock and gravel, as in (d).

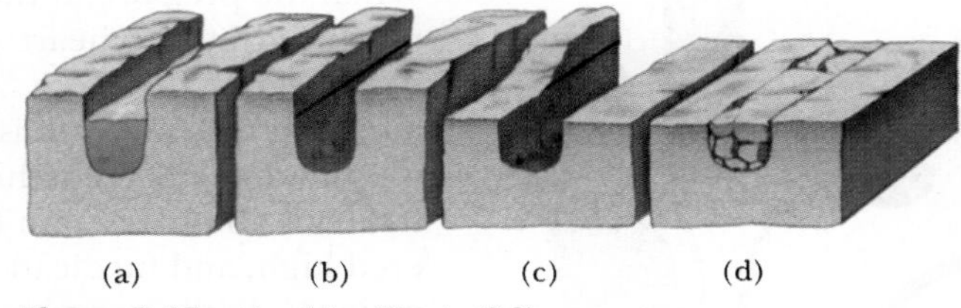

Figure 9.15 (Applying Physics 9.2)

9.5 PRESSURE MEASUREMENTS

A simple device for measuring pressure is the open-tube manometer (Fig. 9.16a). One end of a U-shaped tube containing a liquid is open to the atmosphere, and the other end is connected to a system of unknown pressure P. The pressure at point B equals $P_0 + \rho gh$, where ρ is the density of the fluid. The pressure at B, however, equals the pressure at A, which is also the unknown pressure P. We conclude that $P = P_0 + \rho gh$.

The pressure P is called the **absolute pressure**, and $P - P_0$ is called the **gauge pressure**. If P in the system is greater than atmospheric pressure, h is positive. If P is less than atmospheric pressure (a partial vacuum), h is negative, meaning that the right-hand column in Figure 9.16a is lower than the left-hand column.

Another instrument used to measure pressure is the **barometer** (Fig. 9.16b), invented by Evangelista Torricelli (1608–1647). A long tube closed at one end is filled with mercury and then inverted into a dish of mercury. The closed end of the tube is nearly a vacuum, so its pressure can be taken to be zero. It follows that $P_0 = \rho gh$, where ρ is the density of the mercury and h is the height of the mercury column. Note that the barometer measures the pressure of the atmosphere, whereas the manometer measures pressure in an enclosed fluid.

One atmosphere of pressure is defined to be the pressure equivalent of a column of mercury that is exactly 0.76 m in height at 0°C with $g = 9.806\ 65\ \text{m/s}^2$. At this temperature, mercury has a density of $13.595 \times 10^3\ \text{kg/m}^3$; therefore,

$$P_0 = \rho gh = (13.595 \times 10^3\ \text{kg/m}^3)(9.806\ 65\ \text{m/s}^2)(0.760\ 0\ \text{m})$$
$$= 1.013 \times 10^5\ \text{Pa} = 1\ \text{atm}$$

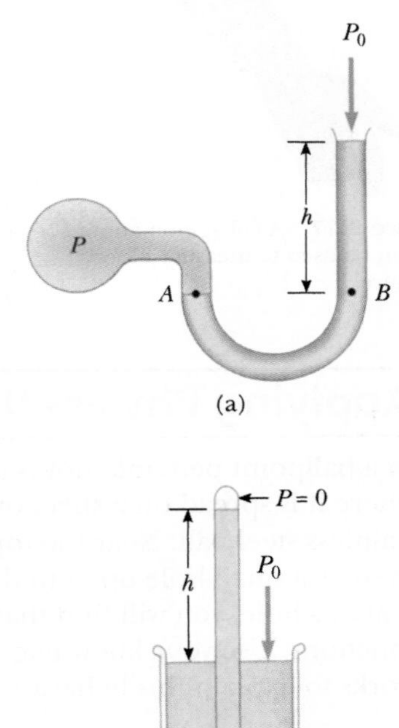

Figure 9.16 Two devices for measuring pressure: (a) an open-tube manometer and (b) a mercury barometer.

It is interesting to note that the force of the atmosphere on our bodies (assuming a body area of 2 000 in^2) is extremely large, on the order of 30 000 lb! If it were not for the fluids permeating our tissues and body cavities, our bodies would collapse. The fluids provide equal and opposite forces. In the upper atmosphere or in space, sudden decompression can lead to serious injury and death. Air retained in the lungs can damage the tiny alveolar sacs, and intestinal gas can even rupture internal organs.

Quick Quiz 9.3

Several common barometers are built using a variety of fluids. For which fluid will the column of fluid in the barometer be the highest? (Refer to Table 9.3.) (a) mercury (b) water (c) ethyl alcohol (d) benzene.

APPLICATION

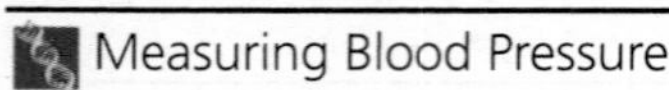
Measuring Blood Pressure

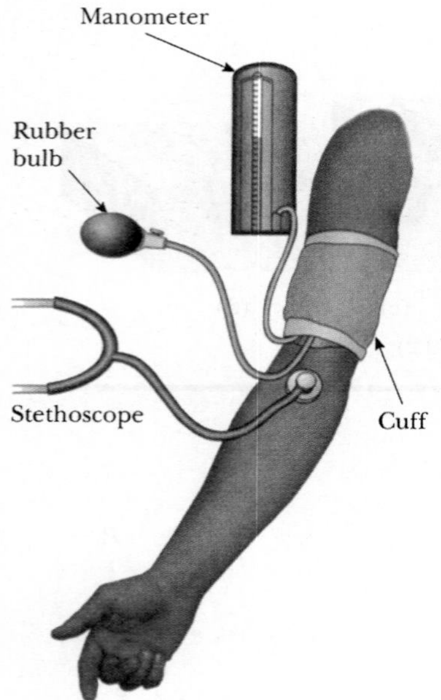

Figure 9.17 A sphygmomanometer can be used to measure blood pressure.

Blood Pressure Measurements

A specialized manometer (called a sphygmomanometer) is often used to measure blood pressure. In this application, a rubber bulb forces air into a cuff wrapped tightly around the upper arm and simultaneously into a manometer, as in Figure 9.17. The pressure in the cuff is increased until the flow of blood through the brachial artery in the arm is stopped. A valve on the bulb is then opened, and the measurer listens with a stethoscope to the artery at a point just below the cuff. When the pressure in the cuff and brachial artery is just below the maximum value produced by the heart (the systolic pressure), the artery opens momentarily on each beat of the heart. At this point, the velocity of the blood is high and turbulent, and the flow is noisy and can be heard with the stethoscope. The manometer is calibrated to read the pressure in millimeters of mercury, and the value obtained is about 120 mm for a normal heart. Values of 130 mm or above are considered high, and medication to lower the blood pressure is often prescribed for such patients. As the pressure in the cuff is lowered further, intermittent sounds are still heard until the pressure falls just below the minimum heart pressure (the diastolic pressure). At this point, continuous sounds are heard. In the normal heart, this transition occurs at about 80 mm of mercury, and values above 90 require medical intervention. Blood pressure readings are usually expressed as the ratio of the systolic pressure to the diastolic pressure, which is 120/80 for a healthy heart.

Quick Quiz 9.4

Blood pressure is normally measured with the cuff of the sphygmomanometer around the arm. Suppose that the blood pressure is measured with the cuff around the calf of the leg of a standing person. Would the reading of the blood pressure be (a) the same here as it is for the arm? (b) greater than it is for the arm? or (c) less than it is for the arm?

Applying Physics 9.3 Ballpoint Pens

In a ballpoint pen, ink moves down a tube to the tip, where it is spread on a sheet of paper by a rolling stainless steel ball. Near the top of the ink cartridge, there is a small hole open to the atmosphere. If you seal this hole, you will find that the pen no longer functions. Use your knowledge of how a barometer works to explain this behavior.

Explanation If the hole is sealed, or if it were not present, the pressure of the air above the ink would decrease as the ink gets used. Consequently, atmospheric pressure exerted against the ink at the bottom of the cartridge would prevent some of the ink from flowing out. The hole allows the pressure above the ink to remain at atmospheric pressure. Why does a ballpoint pen seem to run out of ink when you write on a vertical surface?

9.6 BUOYANT FORCES AND ARCHIMEDES'S PRINCIPLE

A fundamental principle affecting objects submerged in fluids was discovered by the Greek mathematician and natural philosopher Archimedes. **Archimedes's principle** can be stated as follows:

Archimedes's principle ▶

> Any object completely or partially submerged in a fluid is buoyed up by a force with magnitude equal to the weight of the fluid displaced by the object.

Many historians attribute the concept of buoyancy to Archimedes's "bathtub epiphany," when he noticed an apparent change in his weight upon lowering himself into a tub of water. As will be seen in Example 9.8, buoyancy yields a method of determining density.

Everyone has experienced Archimedes's principle. It's relatively easy, for example, to lift someone if you're both standing in a swimming pool, whereas lifting that same individual on dry land may be a difficult task. Water provides partial support to any object placed in it. We often say that an object placed in a fluid is buoyed up by the fluid, so we call this upward force the **buoyant force**.

The buoyant force is *not* a mysterious new force that arises in fluids. In fact, the physical cause of the buoyant force is the pressure difference between the upper and lower sides of the object, which can be shown to be equal to the weight of the displaced fluid. In Figure 9.18a, the fluid inside the indicated sphere, colored darker blue, is pressed on all sides by the surrounding fluid. Arrows indicate the forces arising from the pressure. Because pressure increases with depth, the arrows on the underside are larger than those on top. Adding them all up, the horizontal components cancel, but there is a net force upwards. This force, due to differences in pressure, is the buoyant force $\vec{\mathbf{B}}$. The sphere of water neither rises nor falls, so the vector sum of the buoyant force and the force of gravity on the sphere of fluid must be zero, and it follows that $B = Mg$, where M is the mass of the fluid.

Replacing the shaded fluid with a bowling ball of the same volume, as in Figure 9.18b, changes only the mass on which the pressure acts, so the buoyant force is the same: $B = Mg$, where M is the mass of the displaced fluid, *not* the mass of the bowling ball. The force of gravity on the heavier ball is greater than it was on the fluid, so the bowling ball sinks.

Archimedes's principle can also be obtained from Equation 9.8, relating pressure and depth, using Figure 9.11b. Horizontal forces from the pressure cancel, but in the vertical direction P_2A acts upwards on the bottom of the block of fluid and P_1A and the gravity force on the fluid, Mg, act downwards, giving

$$B = P_2A - P_1A = Mg \qquad \textbf{[9.12a]}$$

where the buoyancy force has been identified as a difference in pressure equal in magnitude to the weight of the displaced fluid. This buoyancy force remains the same regardless of the material occupying the volume in question because it's due to the *surrounding* fluid. Using the definition of density, Equation 9.12a becomes

$$B = \rho_{\text{fluid}} V_{\text{fluid}} g \qquad \textbf{[9.12b]}$$

where ρ_{fluid} is the density of the fluid and V_{fluid} is the volume of the displaced fluid. This result applies equally to all shapes, because any irregular shape can be approximated by a large number of infinitesimal cubes.

It's instructive to compare the forces on a totally submerged object with those on a floating object.

Case I: A Totally Submerged Object. When an object is *totally* submerged in a fluid of density ρ_{fluid}, the upward buoyant force acting on the object has a magnitude of $B = \rho_{\text{fluid}} V_{\text{obj}} g$, where V_{obj} is the volume of the object. If the object has density ρ_{obj}, the downward gravitational force acting on the object has a magni-

ARCHIMEDES: Greek Mathematician, Physicist, and Engineer (287–212 B.C.)

Archimedes was perhaps the greatest scientist of antiquity. He is well known for discovering the nature of the buoyant force and was a gifted inventor. According to legend, Archimedes was asked by King Hieron to determine whether the king's crown was made of pure gold or merely a gold alloy. The task was to be performed without damaging the crown. Archimedes allegedly arrived at a solution while taking a bath, noting a partial loss of weight after submerging his arms and legs in the water. As the story goes, he was so excited about his great discovery that he ran naked through the streets of Syracuse, shouting, "Eureka!" which is Greek for "I have found it."

TIP 9.2 Buoyant Force is Exerted by the Fluid

The buoyant force on an object is exerted by the fluid and is the same, regardless of the density of the object. Objects more dense than the fluid sink; objects less dense rise.

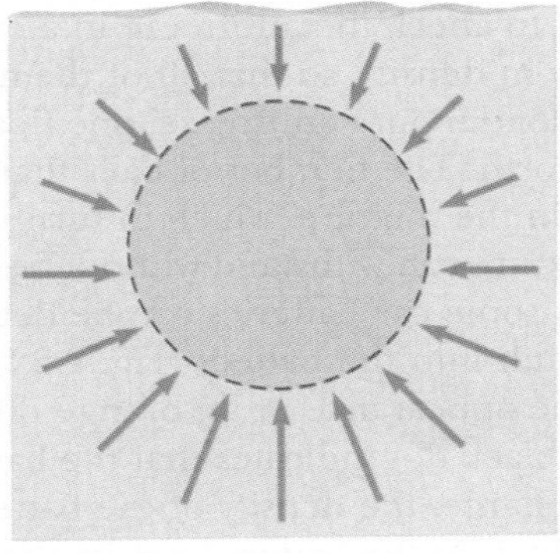
(a)

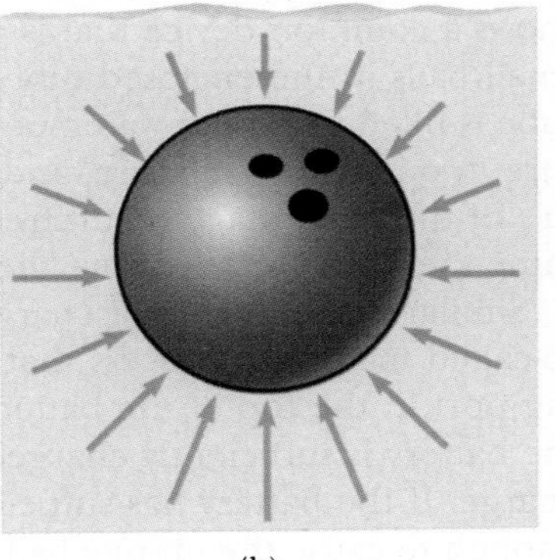
(b)

Figure 9.18 (a) The arrows indicate forces on the sphere of fluid due to pressure, larger on the underside because pressure increases with depth. The net upward force is the buoyant force. (b) The buoyant force, which is caused by the *surrounding* fluid, is the same on any object of the same volume, including this bowling ball. The magnitude of the buoyant force is equal to the weight of the displaced fluid.

Hot-air balloons. Because hot air is less dense than cold air, there is a net upward force on the balloons.

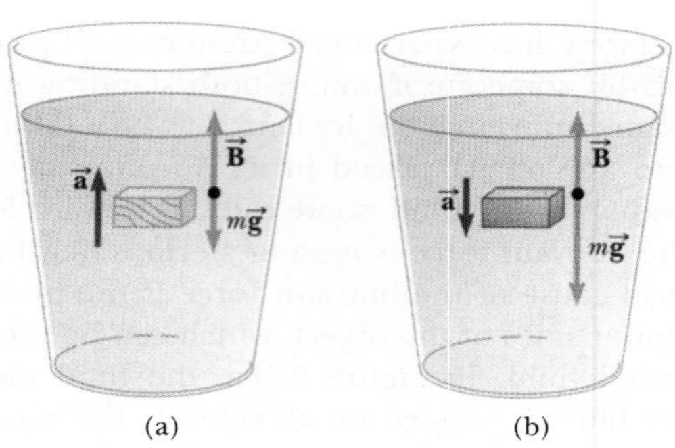

ACTIVE FIGURE 9.19
(a) A totally submerged object that is less dense than the fluid in which it is submerged is acted upon by a net upward force. (b) A totally submerged object that is denser than the fluid sinks.

Physics Now™
Log into PhysicsNow at **www.cp7e.com**, and go to Active Figure 9.19 to move the object to new positions, as well as change the density of the object, and see the results.

tude equal to $w = mg = \rho_{\text{obj}} V_{\text{obj}} g$, and the net force on it is $B - w = (\rho_{\text{fluid}} - \rho_{\text{obj}}) V_{\text{obj}} g$. Therefore, if the density of the object is *less* than the density of the fluid, as in Active Figure 9.19a, the net force exerted on the object is *positive* (upward) and the object accelerates *upward.* If the density of the object is *greater* than the density of the fluid, as in Active Figure 9.19b, the net force is *negative* and the object accelerates *downwards.*

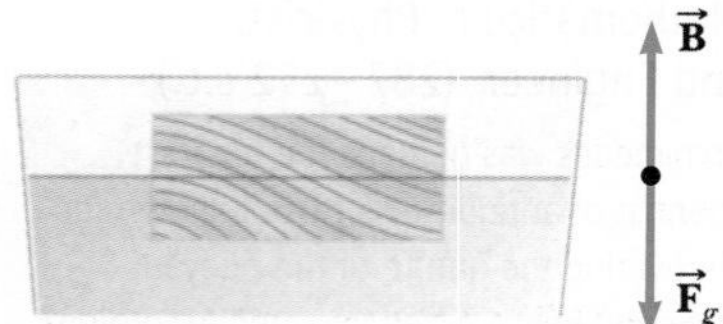

ACTIVE FIGURE 9.20
An object floating on the surface of a fluid is acted upon by two forces: the gravitational force $\vec{F}_g$ and the buoyant force $\vec{B}$. These two forces are equal in magnitude and opposite in direction.

Physics Now™
Log into PhysicsNow at **www.cp7e.com**, and go to Active Figure 9.20 to change the densities of the object and the fluid and see the results.

Case II: A Floating Object. Now consider a partially submerged object in static equilibrium floating in a fluid, as in Active Figure 9.20. In this case, the upward buoyant force is balanced by the downward force of gravity acting on the object. If V_{fluid} is the volume of the fluid displaced by the object (which corresponds to the volume of the part of the object beneath the fluid level), then the magnitude of the buoyant force is given by $B = \rho_{\text{fluid}} V_{\text{fluid}} g$. Because the weight of the object is $w = mg = \rho_{\text{obj}} V_{\text{obj}} g$, and because $w = B$, it follows that $\rho_{\text{fluid}} V_{\text{fluid}} g = \rho_{\text{obj}} V_{\text{obj}} g$, or

$$\frac{\rho_{\text{obj}}}{\rho_{\text{fluid}}} = \frac{V_{\text{fluid}}}{V_{\text{obj}}} \qquad [9.13]$$

Equation 9.13 neglects the buoyant force of the air, which is slight, because the density of air is only 1.29 kg/m^3 at sea level.

Under normal circumstances, the average density of a fish is slightly greater than the density of water, so it would sink if it didn't have a mechanism for adjusting its density. By changing the size of an internal swim bladder, fish maintain neutral buoyancy as they swim to various depths.

APPLICATION

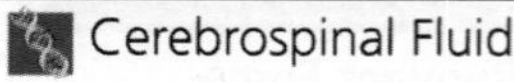
Cerebrospinal Fluid

The human brain is immersed in a fluid (the cerebrospinal fluid) of density 1 007 kg/m^3, which is slightly less than the average density of the brain, 1 040 kg/m^3. Consequently, most of the weight of the brain is supported by the buoyant force of the surrounding fluid. In some clinical procedures, a portion of this fluid must be removed for diagnostic purposes. During such procedures, the nerves and blood vessels in the brain are placed under great strain, which in turn can cause extreme discomfort and pain. Great care must be exercised with such patients until the initial volume of brain fluid has been restored by the body.

When service station attendants check the antifreeze in your car or the condition of your battery, they often use devices that apply Archimedes's principle. Figure 9.21 shows a common device that is used to check the antifreeze in a car radiator. The small balls in the enclosed tube vary in density, so that all of them float when the tube is filled with pure water, none float in pure antifreeze, one floats in a 5% mixture, two in a 10% mixture, and so forth. The number of balls that float is a measure of the percentage of antifreeze in the mixture, which in turn is used to determine the lowest temperature the mixture can withstand without freezing.

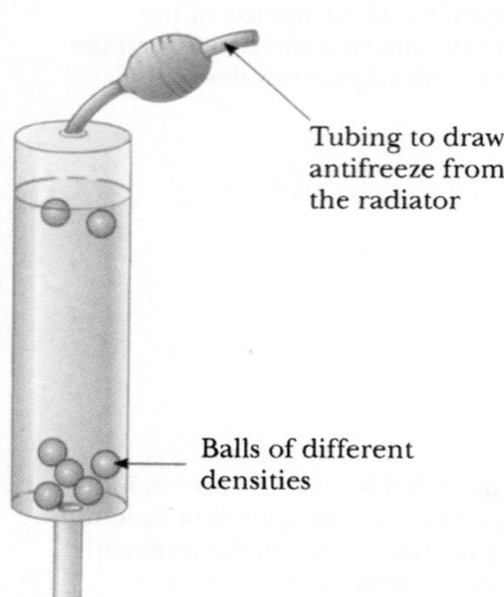

Figure 9.21 The number of balls that float in this device is a measure of the density of the antifreeze solution in a vehicle's radiator and, consequently, a measure of the temperature at which freezing will occur.

Similarly, the degree of charge in some car batteries can be determined with a so-called magic-dot process that is built into the battery (Fig. 9.22). Inside a viewing port in the top of the battery, the appearance of an orange dot indicates that the battery is sufficiently charged; a black dot indicates that the battery has lost its charge. If the battery has sufficient charge, the density of the battery fluid is high enough to cause the orange ball to float. As the battery loses its charge, the density

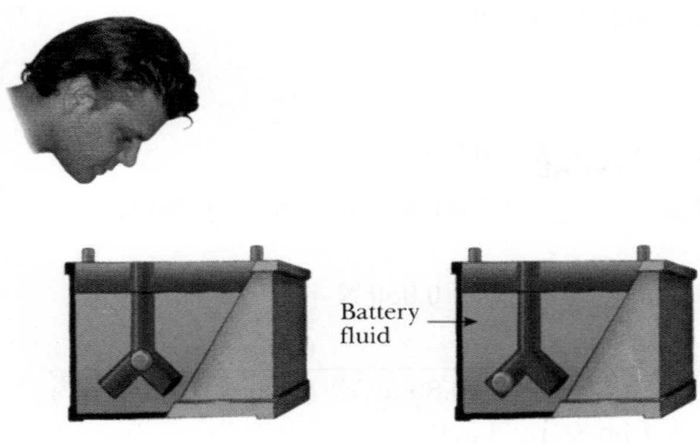

Figure 9.22 The orange ball in the plastic tube inside the battery serves as an indicator of whether the battery is (a) charged or (b) discharged. As the battery loses its charge, the density of the battery fluid decreases, and the ball sinks out of sight.

APPLICATION

Checking the Battery Charge

of the battery fluid decreases and the ball sinks beneath the surface of the fluid, leaving the dot to appear black.

Quick Quiz 9.5

Atmospheric pressure varies from day to day. The level of a floating ship on a high-pressure day is (a) higher (b) lower, or (c) no different than on a low-pressure day.

Quick Quiz 9.6

The density of lead is greater than iron, and both metals are denser than water. Is the buoyant force on a solid lead object (a) greater than, (b) equal to, or (c) less than the buoyant force acting on a solid iron object of the same dimensions?

Most of the volume of this iceberg is beneath the water. Can you determine what fraction of the total volume is under water?

EXAMPLE 9.8 A Red-Tag Special on Crowns

Goal Apply Archimedes's principle to a submerged object.

Problem A bargain hunter purchases a "gold" crown at a flea market. After she gets home, she hangs it from a scale and finds its weight to be 7.84 N (Fig. 9.23a). She then weighs the crown while it is immersed in water, as in Figure 9.23b, and now the scale reads 6.86 N. Is the crown made of pure gold?

Strategy The goal is to find the density of the crown and compare it to the density of gold. We already have the weight of the crown in air, so we can get the mass by dividing by the acceleration of gravity. If we can find the volume of the crown, we can obtain the desired density by dividing the mass by this volume.

When the crown is fully immersed, the displaced water is equal to the volume of the crown. This same volume is used in calculating the buoyant force. So our strategy is as follows: (1) Apply Newton's second law to the crown, both in the water and in the air to find the buoyant force. (2) Use the buoyant force to find the crown's volume. (3) Divide the crown's scale weight in air by the acceleration of gravity to get the mass, then by the volume to get the density.

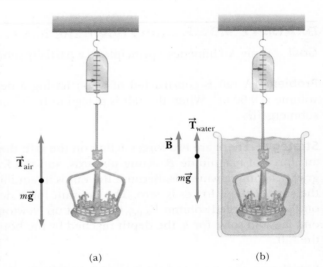

Figure 9.23 (Example 9.8) (a) When the crown is suspended in air, the scale reads $T_{\text{air}} = mg$, the crown's true weight. (b) When the crown is immersed in water, the buoyant force $\vec{\mathbf{B}}$ reduces the scale reading by the magnitude of the buoyant force, $T_{\text{water}} = mg - B$.

Solution

Apply Newton's second law to the crown when it's weighed in air. There are two forces on the crown—gravity $m\vec{\mathbf{g}}$ and $\vec{\mathbf{T}}_{\text{air}}$, the force exerted by the scale on the crown, with magnitude equal to the reading on the scale.

$$T_{\text{air}} - mg = 0 \qquad (1)$$

When the crown is immersed in water, the scale force is $\vec{\mathbf{T}}_{\text{water}}$, with magnitude equal to the scale reading, and there is an upward buoyant force $\vec{\mathbf{B}}$ and the force of gravity.

$$T_{\text{water}} - mg + B = 0 \quad \textbf{(2)}$$

Solve Equation (1) for mg, substitute into Equation (2), and solve for the buoyant force, which equals the difference in scale readings:

$$T_{\text{water}} - T_{\text{air}} + B = 0$$

$$B = T_{\text{air}} - T_{\text{water}} = 7.84 \text{ N} - 6.86 \text{ N} = 0.980 \text{ N}$$

Find the volume of the displaced water, using the fact that the magnitude of the buoyant force equals the weight of the displaced water:

$$B = \rho_{\text{water}} g V_{\text{water}} = 0.980 \text{ N}$$

$$V_{\text{water}} = \frac{0.980 \text{ N}}{g\rho_{\text{water}}} = \frac{0.980 \text{ N}}{(9.80 \text{ m/s}^2)(1.00 \times 10^3 \text{ kg/m}^3)}$$

$$= 1.00 \times 10^{-4} \text{ m}^3$$

The crown is totally submerged, so $V_{\text{crown}} = V_{\text{water}}$. From Equation (1), the mass is the crown's weight in air, T_{air}, divided by g:

$$m = \frac{T_{\text{air}}}{g} = \frac{7.84 \text{ N}}{9.80 \text{ m/s}^2} = 0.800 \text{ kg}$$

Find the density of the crown:

$$\rho_{\text{crown}} = \frac{m}{V_{\text{crown}}} = \frac{0.800 \text{ kg}}{1.00 \times 10^{-4} \text{ m}^3} = 8.00 \times 10^3 \text{ kg/m}^3$$

Remarks Because the density of gold is 19.3×10^3 kg/m^3, the crown is either hollow, made of an alloy, or both. Despite the mathematical complexity, it is certainly conceivable that this was the method that occurred to Archimedes. Conceptually, it's a matter of realizing (or guessing) that equal weights of gold and a silver–gold alloy would have different scale readings when immersed in water, because their densities and hence their volumes are different, leading to differing buoyant forces.

Exercise 9.8

The weight of a metal bracelet is measured to be 0.100 N in air and 0.092 N when immersed in water. Find its density.

Answer 1.25×10^4 kg/m^3

EXAMPLE 9.9 Floating down the River

Goal Apply Archimedes's principle to a partially submerged object.

Problem A raft is constructed of wood having a density of 6.00×10^2 kg/m^3. Its surface area is 5.70 m^2, and its volume is 0.60 m^3. When the raft is placed in fresh water as in Figure 9.24, to what depth h is the bottom of the raft submerged?

Strategy There are two forces acting on the raft: the buoyant force of magnitude B, acting upwards, and the force of gravity, acting downwards. Because the raft is in equilibrium, the sum of these forces is zero. The buoyant force depends on the submerged volume $V_{\text{water}} = Ah$. Set up Newton's second law and solve for h, the depth reached by the bottom of the raft.

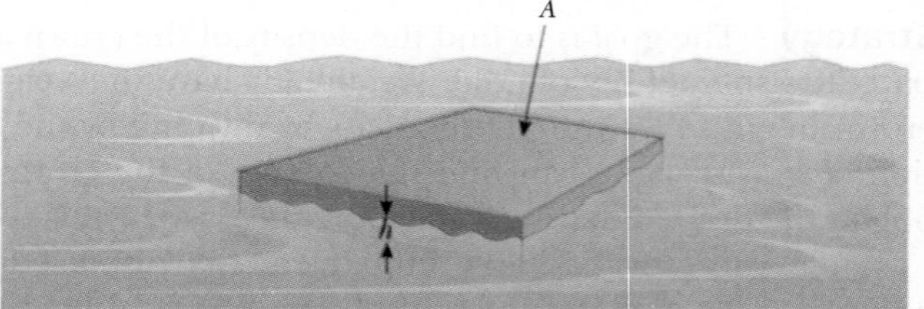

Figure 9.24 (Example 9.9) A raft partially submerged in water.

Solution

Apply Newton's second law to the raft, which is in equilibrium:

$$B - m_{\text{raft}} g = 0 \quad \rightarrow \quad B = m_{\text{raft}} g$$

The volume of the raft submerged in water is given by $V_{\text{water}} = Ah$. The magnitude of the buoyant force is equal to the weight of this displaced volume of water:

$$B = m_{\text{water}} g = (\rho_{\text{water}} V_{\text{water}}) g = (\rho_{\text{water}} Ah) g$$

Now rewrite the gravity force on the raft using the raft's density and volume:

$$m_{\text{raft}} g = (\rho_{\text{raft}} V_{\text{raft}}) g$$

Substitute these two expressions into Newton's second law, $B = m_{\text{raft}}g$, and solve for h (note that g cancels):

$$(\rho_{\text{water}}Ah)\cancel{g} = (\rho_{\text{raft}}V_{\text{raft}})\cancel{g}$$

$$h = \frac{\rho_{\text{raft}}V_{\text{raft}}}{\rho_{\text{water}}A}$$

$$= \frac{(6.00 \times 10^2 \text{ kg/m}^3)(0.600 \text{ m}^3)}{(1.00 \times 10^3 \text{ kg/m}^3)(5.70 \text{ m}^2)}$$

$$= 0.063\,2 \text{ m}$$

Remarks How low the raft rides in the water depends on the density of the raft. The same is true of the human body: Fat is less dense than muscle and bone, so those with a higher percentage of body fat float better.

Exercise 9.9

Calculate how much of an iceberg is beneath the surface of the ocean, given that the density of ice is 917 kg/m^3, and salt water has density 1 025 kg/m^3.

Answer 89.5%

EXAMPLE 9.10 Floating in Two Fluids

Goal Apply Archimedes's principle to an object floating in a fluid having two layers with different densities.

Problem A 1.00×10^3-kg cube of aluminum is placed in a tank. Water is then added to the tank until half the cube is immersed. **(a)** What is the normal force on the cube? (See Fig. 9.25a.) **(b)** Mercury is now slowly poured into the tank until the normal force on the cube goes to zero. (See Fig. 9.25b.) How deep is the layer of mercury?

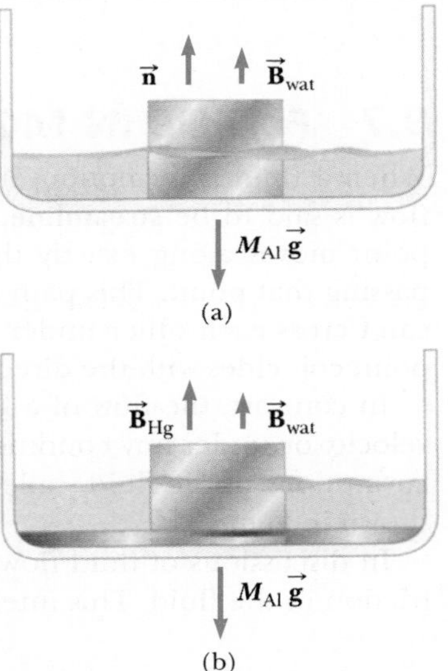

Figure 9.25 (Example 9.10)

Strategy Both parts of this problem involve applications of Newton's second law for a body in equilibrium, together with the concept of a buoyant force. In part (a), the normal, gravitational, and buoyant force of water act on the cube. In part (b), there is an additional buoyant force of mercury, while the normal force goes to zero. Using $V_{\text{Hg}} = Ah$, solve for the height of mercury, h.

Solution

(a) Find the normal force on the cube when half-immersed in water.

Calculate the volume V of the cube and the length d of one side, for future reference (both quantities will be needed for what follows):

$$V_{\text{Al}} = \frac{M_{\text{Al}}}{\rho_{\text{Al}}} = \frac{1.00 \times 10^3 \text{ kg}}{2.70 \times 10^3 \text{ kg/m}^3} = 0.370 \text{ m}^3$$

$$d = V_{\text{Al}}^{1/3} = 0.718 \text{ m}$$

Write Newton's second law for the cube, and solve for the normal force. The buoyant force is equal to the weight of the displaced water (half the volume of the cube).

$$n - M_{\text{Al}}g + B_{\text{wat}} = 0$$

$$n = M_{\text{Al}}g - B_{\text{wat}} = M_{\text{Al}}g - \rho_{\text{wat}}(V/2)g$$

$$= (1.00 \times 10^3 \text{ kg})(9.80 \text{ m/s}^2)$$

$$- (1.00 \times 10^3 \text{ kg/m}^3)(0.370 \text{ m}^3/2.00)(9.80 \text{ m/s}^2)$$

$$n = 9.80 \times 10^3 \text{ N} - 1.81 \times 10^3 \text{ N} = 7.99 \times 10^3 \text{ N}$$

(b) Calculate the level *h* of added mercury.

Apply Newton's second law to the cube: $n - M_{\text{Al}}g + B_{\text{wat}} + B_{\text{Hg}} = 0$

Set $n = 0$ and solve for the buoyant force of mercury: $B_{\text{Hg}} = (\rho_{\text{Hg}} Ah)g = M_{\text{Al}}g - B_{\text{wat}} = 7.99 \times 10^3 \text{ N}$

Solve for h, noting that $A = d^2$:

$$h = \frac{M_{\text{Al}}g - B_{\text{wat}}}{\rho_{\text{Hg}}Ag} = \frac{7.99 \times 10^3 \text{ N}}{(13.6 \times 10^3 \text{ kg/m}^3)(0.718 \text{ m})^2(9.80 \text{ m/s}^2)}$$

$$h = 0.116 \text{ m}$$

Remarks Notice that the buoyant force of mercury calculated in part (b) is the same as the normal force in part (a). This is naturally the case, because enough mercury was added to exactly cancel out the normal force. We could have used this fact to take a shortcut, simply writing $B_{\text{Hg}} = 7.99 \times 10^3$ N immediately, solving for *h*, and avoiding a second use of Newton's law. Most of the time, however, we won't be so lucky! Try calculating the normal force when the level of mercury is 4.00 cm.

Exercise 9.10

A cube of aluminum 1.00 m on a side is immersed one-third in water and two-thirds in glycerin. What is the normal force on the cube?

Answer 1.50×10^4 N

9.7 FLUIDS IN MOTION

When a fluid is in motion, its flow can be characterized in one of two ways. The flow is said to be **streamline**, or **laminar**, if every particle that passes a particular point moves along exactly the same smooth path followed by previous particles passing that point. This path is called a *streamline* (Fig. 9.26). Different streamlines can't cross each other under this steady-flow condition, and the streamline at any point coincides with the direction of the velocity of the fluid at that point.

In contrast, the flow of a fluid becomes irregular, or **turbulent**, above a certain velocity or under any conditions that can cause abrupt changes in velocity. Irregular motions of the fluid, called *eddy currents*, are characteristic in turbulent flow, as shown in Figure 9.27.

In discussions of fluid flow, the term **viscosity** is used for the degree of internal friction in the fluid. This internal friction is associated with the resistance between

Figure 9.26 An illustration of streamline flow around an automobile in a test wind tunnel. The streamlines in the airflow are made visible by smoke particles.

two adjacent layers of the fluid moving relative to each other. A fluid such as kerosene has a lower viscosity than does crude oil or molasses.

Many features of fluid motion can be understood by considering the behavior of an **ideal fluid**, which satisfies the following conditions:

1. **The fluid is nonviscous**, which means there is no internal friction force between adjacent layers.
2. **The fluid is incompressible**, which means its density is constant.
3. **The fluid motion is steady**, meaning that the velocity, density, and pressure at each point in the fluid don't change with time.
4. **The fluid moves without turbulence**. This implies that each element of the fluid has zero angular velocity about its center, so there can't be any eddy currents present in the moving fluid. A small wheel placed in the fluid would translate but not rotate.

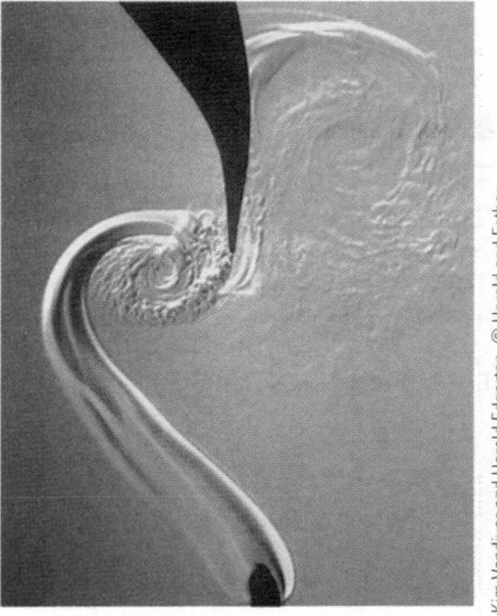

Kim Vandiver and Harold Edgerton. © Harold and Esther Edgerton Foundation, 2002, courtesy of Palm Press, Inc.

Figure 9.27 Turbulent flow: The tip of a rotating blade (the dark region at the top) forms a vortex in air that is being heated by an alcohol lamp. (The wick is at the bottom.) Note the air turbulence on both sides of the blade.

Equation of Continuity

Figure 9.28a represents a fluid flowing through a pipe of nonuniform size. The particles in the fluid move along the streamlines in steady-state flow. In a small time interval Δt, the fluid entering the bottom end of the pipe moves a distance $\Delta x_1 = v_1 \Delta t$, where v_1 is the speed of the fluid at that location. If A_1 is the cross-sectional area in this region, then the mass contained in the bottom blue region is $\Delta M_1 = \rho_1 A_1 \Delta x_1 = \rho_1 A_1 v_1 \Delta t$, where ρ_1 is the density of the fluid at A_1. Similarly, the fluid that moves out of the upper end of the pipe in the same time interval Δt has a mass of $\Delta M_2 = \rho_2 A_2 v_2 \Delta t$. However, **because mass is conserved and because the flow is steady**, the mass that flows into the bottom of the pipe through A_1 in the time Δt must equal the mass that flows out through A_2 in the same interval. Therefore, $\Delta M_1 = \Delta M_2$, or

$$\rho_1 A_1 v_1 = \rho_2 A_2 v_2 \qquad [9.14]$$

For the case of an incompressible fluid, $\rho_1 = \rho_2$ and Equation 9.14 reduces to

$$A_1 v_1 = A_2 v_2 \qquad [9.15]$$

◀ Equation of continuity

This expression is called the **equation of continuity**. From this result, we see that **the product of the cross-sectional area of the pipe and the fluid speed at that cross section is a constant**. Therefore, the speed is high where the tube is constricted and low where the tube has a larger diameter. The product Av, which has dimensions of volume per unit time, is called the **flow rate**. **The condition Av = constant is equivalent to the fact that the volume of fluid that enters one end of the tube in a given time interval equals the volume of fluid leaving the tube in the same interval, assuming that the fluid is incompressible and there are no leaks.** Figure 9.28b

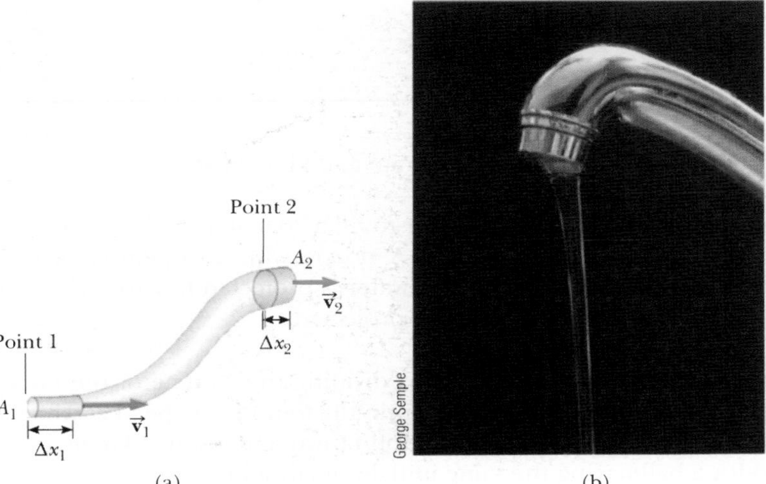

Figure 9.28 (a) A fluid moving with streamline flow through a pipe of varying cross-sectional area. The volume of fluid flowing through A_1 in a time interval Δt must equal the volume flowing through A_2 in the same time interval. Therefore, $A_1 v_1 = A_2 v_2$. (b) Water flowing slowly out of a faucet. The width of the stream narrows as the water falls and speeds up in accord with the continuity equation.

is an example of an application of the equation of continuity: As the stream of water flows continuously from a faucet, the width of the stream narrows as it falls and speeds up.

TIP 9.3 Continuity Equations

The rate of flow of fluid into a system equals the rate of flow out of the system. The incoming fluid occupies a certain volume and can enter the system only if the fluid already inside goes out, thereby making room.

There are many instances in everyday experience that involve the equation of continuity. Reducing the cross-sectional area of a garden hose by putting a thumb over the open end makes the water spray out with greater speed; hence the stream goes farther. Similar reasoning explains why smoke from a smoldering piece of wood first rises in a streamline pattern, getting thinner with height, eventually breaking up into a swirling, turbulent pattern. The smoke rises because it's less dense than air and the buoyant force of the air accelerates it upward. As the speed of the smoke stream increases, the cross-sectional area of the stream decreases, in accordance with the equation of continuity. The stream soon reaches a speed so great that streamline flow is not possible. We will study the relationship between speed of fluid flow and turbulence in a later discussion on the Reynolds number.

EXAMPLE 9.11 Niagara Falls

Goal Apply the equation of continuity.

Problem Each second, 5 525 m^3 of water flows over the 670-m-wide cliff of the Horseshoe Falls portion of Niagara Falls. The water is approximately 2 m deep as it reaches the cliff. Estimate its speed at that instant?

Strategy This is an estimate, so only one significant figure will be retained in the answer. The volume flow rate is given, and according to the equation of continuity, is a constant equal to Av. Find the cross-sectional area, substitute, and solve for the speed.

Solution

Calculate the cross-sectional area of the water as it reaches the edge of the cliff:

$$A = (670 \text{ m})(2 \text{ m}) = 1\,340 \text{ m}^2$$

Multiply this result by the speed and set it equal to the flow rate. Then solve for v.

$$Av = \text{volume flow rate}$$

$$(1340 \text{ m}^2)v = 5\,525 \text{ m}^3/\text{s} \quad \rightarrow \quad v \approx 4 \text{ m/s}$$

Exercise 9.11

The Garfield Thomas water tunnel at Pennsylvania State University has a circular cross section that constricts from a diameter of 3.6 m to the test section, which is 1.2 m in diameter. If the speed of flow is 3.0 m/s in the larger-diameter pipe, determine the speed of flow in the test section.

Answer 27 m/s

EXAMPLE 9.12 Watering a Garden

Goal Combine the equation of continuity with concepts of flow rate and kinematics.

Problem A water hose 2.50 cm in diameter is used by a gardener to fill a 30.0-liter bucket. (One liter = 1 000 cm^3.) The gardener notices that it takes 1.00 min to fill the bucket. A nozzle with an opening of cross-sectional area 0.500 cm^2 is then attached to the hose. The nozzle is held so that water is projected horizontally from a point 1.00 m above the ground. Over what horizontal distance can the water be projected?

Strategy We can find the volume flow rate through the hose by dividing the volume of the bucket by the time it takes to fill it. After finding the flow rate, apply the equation of continuity to find the speed at which the water shoots horizontally out the nozzle. The rest of the problem is an application of two-dimensional kinematics. The answer obtained is the same as would be found for a ball having the same initial velocity and height.

Solution

Calculate the volume flow rate into the bucket, and convert to m^3/s:

$$\text{volume flow rate} = \frac{30.0 \text{ L}}{1.00 \text{ min}}\left(\frac{1.00 \times 10^3 \text{ cm}^3}{1.00 \text{ L}}\right)\left(\frac{1.00 \text{ m}}{100.0 \text{ cm}}\right)^3\left(\frac{1.00 \text{ min}}{60.0 \text{ s}}\right)$$
$$= 5.00 \times 10^{-4} \text{ m}^3/\text{s}$$

Solve the equation of continuity for v_{0x}, the x-component of the initial velocity of the stream exiting the hose:

$$A_1v_1 = A_2v_2 = A_2v_{0x}$$

$$v_{0x} = \frac{A_1v_1}{A_2} = \frac{5.00 \times 10^{-4} \text{ m}^3/\text{s}}{0.500 \times 10^{-4} \text{ m}^2} = 10.0 \text{ m/s}$$

Calculate the time for the stream to fall 1.00 m, using kinematics. Initially, the stream is horizontal, so v_{0y} is zero

$$\Delta y = v_{0y}t - \tfrac{1}{2}gt^2$$

Set $v_{0y} = 0$ in the preceding equation and solve for t noting that $\Delta y = -1.00$ m:

$$t = \sqrt{\frac{-2\Delta y}{g}} = \sqrt{\frac{-2(-1.00 \text{ m})}{9.80 \text{ m/s}^2}} = 0.452 \text{ s}$$

Find the horizontal distance the stream travels:

$$x = v_{0x}t = (10.0 \text{ m/s})(0.452 \text{ s}) = 4.52 \text{ m}$$

Remark It's interesting that the motion of fluids can be treated with the same kinematics equations as individual objects.

Exercise 9.12

The nozzle is replaced with a Y-shaped fitting that splits the flow in half. Garden hoses are connected to each end of the Y, with each hose having a 0.400 cm^2 nozzle. (a) How fast does the water come out of one of the nozzles? (b) How far would one of the nozzles squirt water if both were operated simultaneously and held horizontally 1.00 m off the ground? [*Hint:* Find the volume flow rate through each 0.400-cm^2 nozzle, then follow the same steps as before.]

Answer (a) 6.25 m/s (b) 2.82 m

Bernoulli's Equation

As a fluid moves through a pipe of varying cross section and elevation, the pressure changes along the pipe. In 1738 the Swiss physicist Daniel Bernoulli (1700–1782) derived an expression that relates the pressure of a fluid to its speed and elevation. Bernoulli's equation is not a freestanding law of physics; rather, it's **a consequence of energy conservation as applied to an ideal fluid.**

In deriving Bernoulli's equation, we again assume that the fluid is incompressible, nonviscous, and flows in a nonturbulent, steady-state manner. Consider the flow through a nonuniform pipe in the time Δt, as in Figure 9.29. The force on the lower end of the fluid is P_1A_1, where P_1 is the pressure at the lower end. The work done on the lower end of the fluid by the fluid behind it is

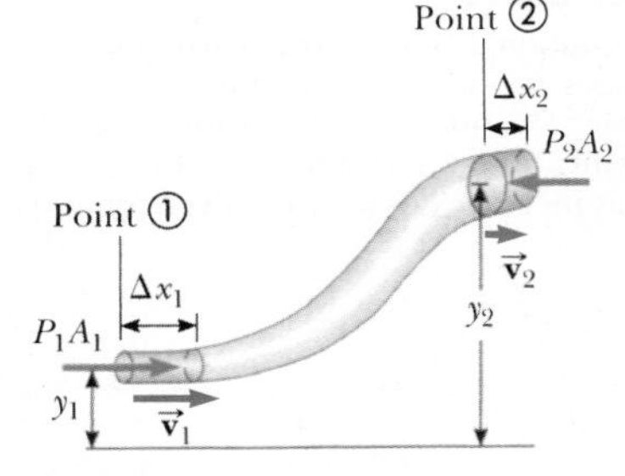

Figure 9.29 A fluid flowing through a constricted pipe with streamline flow. The fluid in the section with a length of Δx_1 moves to the section with a length of Δx_2. The volumes of fluid in the two sections are equal.

$$W_1 = F_1\,\Delta x_1 = P_1A_1\,\Delta x_1 = P_1V$$

where V is the volume of the lower blue region in the figure. In a similar manner, the work done on the fluid on the upper portion in the time Δt is

$$W_2 = -P_2A_2\Delta x_2 = -P_2V$$

The volume is the same because by the equation of continuity, the volume of fluid that passes through A_1 in the time Δt equals the volume that passes through A_2 in the same interval. The work W_2 is negative because the force on the fluid at the top is opposite its displacement. The net work done by these forces in the time Δt is

$$W_{\text{fluid}} = P_1V - P_2V$$

Part of this work goes into changing the fluid's kinetic energy, and part goes into changing the gravitational potential energy of the fluid–Earth system. If m is the

DANIEL BERNOULLI, Swiss Physicist and Mathematician (1700–1782)

In his most famous work, *Hydrodynamica*, Bernoulli showed that, as the velocity of fluid flow increases, its pressure decreases. In this same publication, Bernoulli also attempted the first explanation of the behavior of gases with changing pressure and temperature; this was the beginning of the kinetic theory of gases.

mass of the fluid passing through the pipe in the time interval Δt, then the change in kinetic energy of the volume of fluid is

$$\Delta KE = \tfrac{1}{2}mv_2^2 - \tfrac{1}{2}mv_1^2$$

The change in the gravitational potential energy is

$$\Delta PE = mgy_2 - mgy_1$$

Because the net work done by the fluid on the segment of fluid shown in Figure 9.29 changes the kinetic energy and the potential energy of the nonisolated system, we have

$$W_{\text{fluid}} = \Delta KE + \Delta PE$$

The three terms in this equation are those we have just evaluated. Substituting expressions for each of the terms gives

$$P_1V - P_2V = \tfrac{1}{2}mv_2^2 - \tfrac{1}{2}mv_1^2 + mgy_2 - mgy_1$$

If we divide each term by V and recall that $\rho = m/V$, this expression becomes

$$P_1 - P_2 = \tfrac{1}{2}\rho v_2^2 - \tfrac{1}{2}\rho v_1^2 + \rho gy_2 - \rho gy_1$$

Rearrange the terms as follows:

$$P_1 + \tfrac{1}{2}\rho v_1^2 + \rho gy_1 = P_2 + \tfrac{1}{2}\rho v_2^2 + \rho gy_2 \qquad [9.16]$$

This is **Bernoulli's equation**, often expressed as

Bernoulli's equation ▶

$$P + \tfrac{1}{2}\rho v^2 + \rho gy = \text{constant} \qquad [9.17]$$

Bernoulli's equation states that the sum of the pressure P, the kinetic energy per unit volume, $\frac{1}{2}\rho v^2$, and the potential energy per unit volume, ρgy, has the same value at all points along a streamline.

TIP 9.4 Bernoulli's Principle for Gases

Equation 9.16 isn't strictly true for gases because they aren't incompressible. The qualitative behavior is the same, however: As the speed of the gas increases, its pressure decreases.

An important consequence of Bernoulli's equation can be demonstrated by considering Figure 9.30, which shows water flowing through a horizontal constricted pipe from a region of large cross-sectional area into a region of smaller cross-sectional area. This device, called a **Venturi tube**, can be used to measure the speed of fluid flow. Because the pipe is horizontal, $y_1 = y_2$, and Equation 9.16 applied to points 1 and 2 gives

$$P_1 + \tfrac{1}{2}\rho v_1^2 = P_2 + \tfrac{1}{2}\rho v_2^2 \qquad [9.18]$$

Because the water is not backing up in the pipe, its speed v_2 in the constricted region must be greater than its speed v_1 in the region of greater diameter. From Equation 9.18, we see that P_2 must be less than P_1 because $v_2 > v_1$. This result is often expressed by the statement that **swiftly moving fluids exert less pressure than do slowly moving fluids**. This important fact enables us to understand a wide range of everyday phenomena.

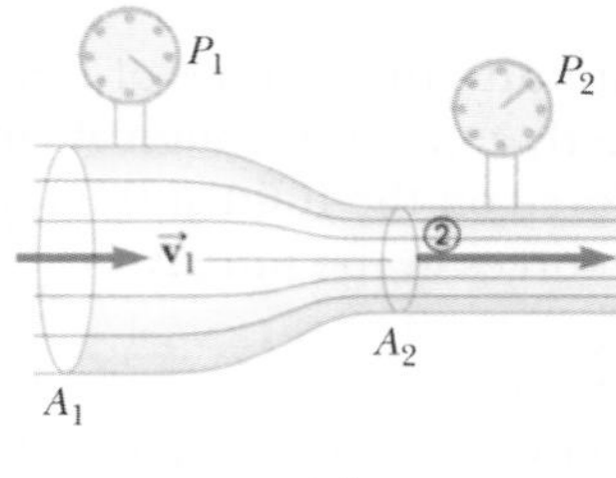

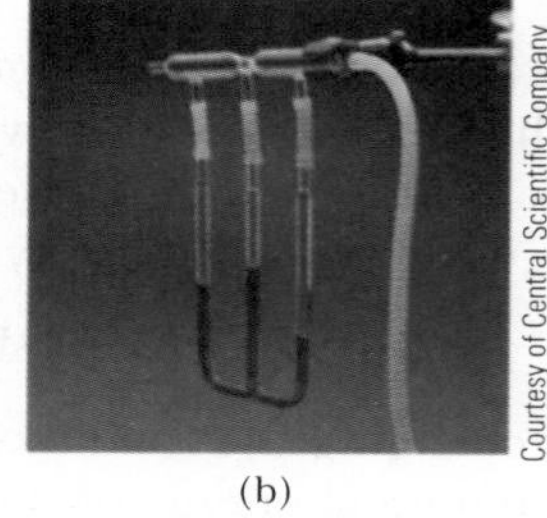

Figure 9.30 (a) The pressure P_1 is greater than the pressure P_2, because $v_1 < v_2$. This device can be used to measure the speed of fluid flow. (b) A Venturi tube, located at the top of the photograph. The higher level of fluid in the middle column shows that the pressure at the top of the column, which is in the constricted region of the Venturi tube, is lower than the pressure elsewhere in the column.

Quick Quiz 9.7

You observe two helium balloons floating next to each other at the ends of strings secured to a table. The facing surfaces of the balloons are separated by 1–2 cm. You blow through the opening between the balloons. What happens to the balloons? (a) They move toward each other; (b) they move away from each other; (c) they are unaffected.

INTERACTIVE EXAMPLE 9.13 Shoot-Out at the Old Water Tank

Goal Apply Bernoulli's equation to find the speed of a fluid.

Problem A nearsighted sheriff fires at a cattle rustler with his trusty six-shooter. Fortunately for the rustler, the bullet misses him and penetrates the town water tank, causing a leak (Fig. 9.31). **(a)** If the top of the tank is open to the atmosphere, determine the speed at which the water leaves the hole when the water level is 0.500 m above the hole. **(b)** Where does the stream hit the ground if the hole is 3.00 m above the ground?

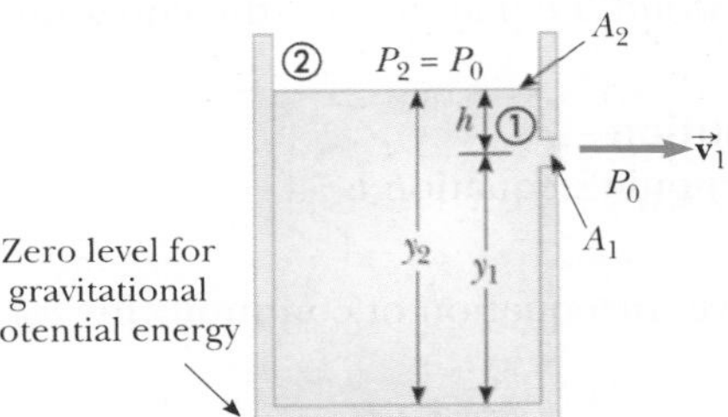

Figure 9.31 (Example 9.13) The water speed v_1 from the hole in the side of the container is given by $v_1 = \sqrt{2gh}$.

Strategy (a) Assume the tank's cross-sectional area is large compared to the hole's ($A_2 >> A_1$), so the water level drops very slowly and $v_2 \approx 0$. Apply Bernoulli's equation to points ① and ② in Figure 9.31, noting that P_1 equals atmospheric pressure P_0 at the hole and is approximately the same at the top of the water tank. Part (b) can be solved with kinematics, just as if the water were a ball thrown horizontally.

Solution

(a) Find the speed of the water leaving the hole.

Substitute $P_1 = P_2 = P_0$ and $v_2 \approx 0$ into Bernoulli's equation, and solve for v_1:

$$P_0 + \tfrac{1}{2}\rho v_1^2 + \rho g y_1 = P_0 + \rho g y_2$$

$$v_1 = \sqrt{2g(y_2 - y_1)} = \sqrt{2gh}$$

$$v_1 = \sqrt{2(9.80 \text{ m/s}^2)(0.500 \text{ m})} = 3.13 \text{ m/s}$$

(b) Find where the stream hits the ground.

Use the displacement equation to find the time of the fall, noting that the stream is initially horizontal, so $v_{0y} = 0$.

$$\Delta y = -\tfrac{1}{2}gt^2 + v_{0y}t$$

$$-3.00 \text{ m} = -(4.90 \text{ m/s}^2)t^2$$

$$t = 0.782 \text{ s}$$

Compute the horizontal distance the stream travels in this time:

$$x = v_{0x}t = (3.13 \text{ m/s})(0.782 \text{ s}) = 2.45 \text{ m}$$

Remarks As the analysis of part (a) shows, the speed of the water emerging from the hole is equal to the speed acquired by an object falling freely through the vertical distance h. This is known as **Torricelli's law**.

Exercise 9.13

Suppose, in a similar situation, the water hits the ground 4.20 m from the hole in the tank. If the hole is 2.00 m above the ground, how far above the hole is the water level?

Answer 2.21 m above the hole

Physics Now™ You can move the hole vertically and see where the water lands by logging into PhysicsNow at **www.cp7e.com** and going to Interactive Example 9.13.

Example 9.14 Fluid Flow in a Pipe

Goal Solve a problem combining Bernoulli's equation and the equation of continuity.

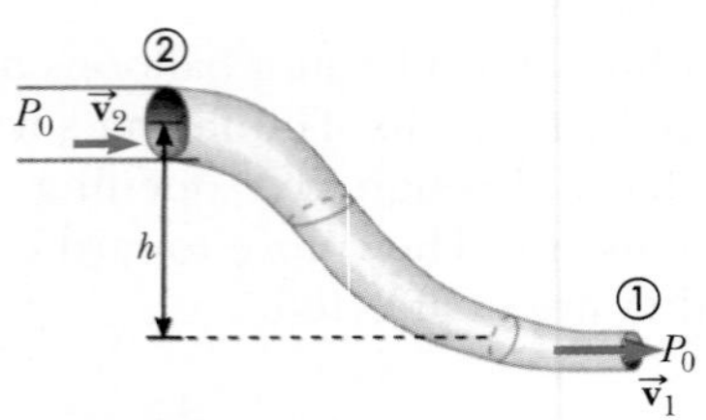

Figure 9.32 (Example 9.14)

Problem A large pipe with a cross-sectional area of 1.00 m^2 descends 5.00 m and narrows to 0.500 m^2, where it terminates in a valve (Fig. 9.32). If the pressure at point ② is atmospheric pressure, and the valve is opened wide and water allowed to flow freely, find the speed of the water leaving the pipe.

Strategy The equation of continuity, together with Bernoulli's equation, constitute two equations in two unknowns: the speeds v_1 and v_2. Eliminate v_2 from Bernoulli's equation with the equation of continuity, and solve for v_1.

Solution

Bernoulli's equation is

$$P_1 + \tfrac{1}{2}\rho v_1^2 + \rho g y_1 = P_2 + \tfrac{1}{2}\rho v_2^2 + \rho g y_2 \qquad \textbf{(1)}$$

Solve the equation of continuity for v_2:

$$A_2 v_2 = A_1 v_1$$

$$v_2 = \frac{A_1}{A_2} v_1 \qquad \textbf{(2)}$$

In Equation 1, set $P_1 = P_2 = P_0$, and substitute the expression for v_2. Then solve for v_1.

$$P_0 + \tfrac{1}{2}\rho v_1^2 + \rho g y_1 = P_0 + \tfrac{1}{2}\rho\left(\frac{A_1}{A_2} v_1\right)^2 + \rho g y_2$$

$$v_1^2\left(1 - \left(\frac{A_1}{A_2}\right)^2\right) = 2g(y_2 - y_1) = 2gh$$

$$v_1 = \frac{\sqrt{2gh}}{\sqrt{1 - (A_1/A_2)^2}}$$

Substitute the given values:

$$v_1 = 11.4 \text{ m/s}$$

Remarks This speed is slightly higher than the speed predicted by Torricelli's law, because the narrowing pipe squeezes the fluid.

Exercise 9.14

Water flowing in a horizontal pipe is at a pressure of 1.4×10^5 Pa at a point where its cross-sectional area is 1.00 m^2. When the pipe narrows to 0.400 m^2, the pressure drops to 1.16×10^5 Pa. Find the water's speed (a) in the wider pipe and (b) in the narrower pipe.

Answer (a) 3.02 m/s (b) 7.56 m/s

9.8 OTHER APPLICATIONS OF FLUID DYNAMICS

In this section we describe some common phenomena that can be explained, at least in part, by Bernoulli's equation.

In general, an object moving through a fluid is acted upon by a net upward force as the result of any effect that causes the fluid to change its direction as it flows past the object. For example, a golf ball struck with a club is given a rapid backspin, as shown in Figure 9.33. The dimples on the ball help entrain the air along the curve of the ball's surface. The figure shows a thin layer of air wrapping partway around the ball and being deflected downward as a result. Because the

Solution

Calculate the mass of liquid water:

$$m_{\text{water}} = \rho_{\text{water}} V$$
$$= (1.00 \times 10^3 \text{ kg/m}^3)(30.0 \text{ L})\frac{1.00 \text{ m}^3}{1.00 \times 10^3 \text{ L}}$$
$$= 30.0 \text{ kg}$$

Write the equation of thermal equilibrium:

$$Q_{\text{ice}} + Q_{\text{melt}} + Q_{\text{ice-water}} + Q_{\text{water}} = 0 \qquad (1)$$

Construct a comprehensive table:

Q	m (kg)	c (J/kg · °C)	L (J/kg)	T_f (°C)	T_i (°C)	Expression
Q_{ice}	6.00	2 090		0	−5.00	$m_{\text{ice}}c_{\text{ice}}(T_f - T_i)$
Q_{melt}	6.00		3.33×10^5	0	0	$m_{\text{ice}}L_f$
$Q_{\text{ice-water}}$	6.00	4 190		T	0	$m_{\text{ice}}c_{\text{wat}}(T_f - T_i)$
Q_{water}	30.0	4 190		T	20.0	$m_{\text{wat}}c_{\text{wat}}(T_f - T_i)$

Substitute all quantities in the second through sixth columns into the last column and sum (which is the evaluation of Equation 1), and solve for T:

$$6.27 \times 10^4 \text{ J} + 2.00 \times 10^6 \text{ J} + (2.51 \times 10^4 \text{ J/°C})(T - 0\text{°C}) + (1.26 \times 10^5 \text{ J/°C})(T - 20.0\text{°C}) = 0$$

$$T = 3.03\text{°C}$$

Remarks Making a table is optional. However, simple substitution errors are extremely common, and the table makes such errors less likely.

Exercise 11.6

What mass of ice at −10.0°C is needed to cool a whale's water tank, holding 1.20×10^3 m^3 of water, from 20.0°C down to a more comfortable 10.0°C?

Answer 1.27×10^5 kg

EXAMPLE 11.7 Partial Melting

Goal Understand how to handle an incomplete phase change.

Problem A 5.00-kg block of ice at 0°C is added to an insulated container partially filled with 10.0 kg of water at 15.0°C. **(a)** Find the final temperature, neglecting the heat capacity of the container. **(b)** Find the mass of the ice that was melted.

Strategy Part (a) is tricky, because the ice does not entirely melt in this example. When there is any doubt concerning whether there will be a complete phase change, some preliminary calculations are necessary. First, find the total energy required to melt the ice, Q_{melt}, and then find Q_{water}, the maximum energy that can be delivered by the water above 0°C. If the energy delivered by the water is high enough, all the ice melts. If not, there will usually be a final mixture of ice and water at 0°C, unless the ice starts at a temperature far below 0°C, in which case all the liquid water freezes.

Solution

(a) Find the equilibrium temperature.

First, compute the amount of energy necessary to completely melt the ice:

$$Q_{\text{melt}} = m_{\text{ice}}L_f = (5.00 \text{ kg})(3.33 \times 10^5 \text{ J/kg})$$
$$= 1.67 \times 10^6 \text{ J}$$

Next, calculate the maximum energy that can be lost by the initial mass of liquid water without freezing it:

$$Q_{\text{water}} = m_{\text{water}}c\Delta T$$
$$= (10.0 \text{ kg})(4\,190 \text{ J/kg}\cdot\text{°C})(0\text{°C} - 15.0\text{°C})$$
$$= -6.29 \times 10^5 \text{ J}$$

This is less than half the energy necessary to melt all the ice, so the final state of the system is a mixture of water and ice at the freezing point:

$$T = 0°\text{C}$$

(b) Compute the mass of ice melted.

Set the total available energy equal to the heat of fusion of m grams of ice, mL_f:

$$6.29 \times 10^5\ \text{J} = mL_f = m(3.33 \times 10^5\ \text{J/kg})$$

$$m = 1.89\ \text{kg}$$

Remarks If this problem is solved assuming (wrongly) that all the ice melts, a final temperature of $T = -16.5°\text{C}$ is obtained. The only way that could happen is if the system were not isolated, contrary to the statement of the problem. In the following exercise, you must also compute the thermal energy needed to warm the ice to its melting point.

Exercise 11.7

If 8.00 kg of ice at $-5.00°\text{C}$ is added to 12.0 kg of water at 20.0°C, compute the final temperature. How much ice remains, if any?

Answer $T = 0°\text{C}$, 5.23 kg

Sometimes problems involve changes in mechanical energy. During a collision, for example, some kinetic energy can be transformed to the internal energy of the colliding objects. This kind of transformation is illustrated in Example 11.8, involving a possible impact of a comet on Earth. In this example, a number of liberties will be taken in order to estimate the magnitude of the destructive power of such a catastrophic event. The specific heats depend on temperature and pressure, for example, but that will be ignored. Also, the ideal gas law doesn't apply at the temperatures and pressures attained, and the result of the collision wouldn't be superheated steam, but a plasma of charged particles. Despite all these simplifications, the example yields good order-of-magnitude results.

EXAMPLE 11.8 Armageddon!

Goal Link mechanical energy to thermal energy, phase changes, and the ideal gas law to create an estimate.

Problem A comet half a kilometer in radius consisting of ice at 273 K hits Earth at a speed of 4.00×10^4 m/s. For simplicity, assume that all the kinetic energy converts to thermal energy on impact and that all the thermal energy goes into warming the comet. **(a)** Calculate the volume and mass of the ice. **(b)** Use conservation of energy to find the final temperature of the comet material. Assume, contrary to fact, that the result is superheated steam and that the usual specific heats are valid, though in fact they depend on both temperature and pressure. **(c)** Assuming the steam retains a spherical shape and has the same initial volume as the comet, calculate the pressure of the steam using the ideal gas law. This law actually doesn't apply to a system at such high pressure and temperature, but can be used to get an estimate.

Strategy Part (a) requires the volume formula for a sphere and the definition of density. In part (b), conservation of energy can be applied. There are four processes involved: (1) melting the ice, (2) warming the ice water to the boiling point, (3) converting the boiling water to steam, and (4) warming the steam. The energy needed for these processes will be designated Q_{melt}, Q_{water}, Q_{vapor}, and Q_{steam}, respectively. These quantities plus the change in kinetic energy ΔK sum to zero because they are assumed to be internal to the system. In this case, the first three Q's can be neglected compared to the (extremely large) kinetic energy term. Solve for the unknown temperature, and substitute it into the ideal gas law in part (c).

Solution

(a) Find the volume and mass of the ice.

Apply the volume formula for a sphere:

$$V = \frac{4}{3}\pi r^3 = \frac{4}{3}(3.14)(5.00 \times 10^2\ \text{m})^3$$

$$= 5.23 \times 10^8\ \text{m}^3$$

Apply the density formula to find the mass of the ice:

$$m = \rho V = (917\ \text{kg/m}^3)(5.23 \times 10^8\ \text{m}^3)$$
$$= \boxed{4.80 \times 10^{11}\ \text{kg}}$$

(b) Find the final temperature of the cometary material.

Use conservation of energy:

$$Q_{\text{melt}} + Q_{\text{water}} + Q_{\text{vapor}} + Q_{\text{steam}} + \Delta K = 0 \quad \textbf{(1)}$$
$$mL_f + mc_{\text{water}}\,\Delta T_{\text{water}} + mL_v + mc_{\text{steam}}\,\Delta T_{\text{steam}} + \left(0 - \tfrac{1}{2}mv^2\right) = 0 \quad \textbf{(2)}$$

The first three terms are negligible compared to the kinetic energy. The steam term involves the unknown final temperature, so retain only it and the kinetic energy, canceling the mass and solving for T:

$$mc_{\text{steam}}\,(T - 373\ \text{K}) - \tfrac{1}{2}mv^2 = 0$$
$$T = \frac{\frac{1}{2}v^2}{c_{\text{steam}}} + 373\ \text{K} = \frac{\frac{1}{2}(4.00 \times 10^4\,\text{m/s})^2}{2\,010\,\text{J/kg}\cdot\text{K}} + 373\ \text{K}$$
$$T = \boxed{3.98 \times 10^5\ \text{K}}$$

(c) Estimate the pressure of the gas, using the ideal gas law.

First, compute the number of moles of steam:

$$n = (4.80 \times 10^{11}\,\text{kg})\left(\frac{1\ \text{mol}}{0.018\ \text{kg}}\right) = 2.67 \times 10^{13}\,\text{mol}$$

Solve for the pressure, using $PV = nRT$:

$$P = \frac{nRT}{V}$$
$$= \frac{(2.67 \times 10^{13}\,\text{mol})(8.31\,\text{J/mol}\cdot\text{K})(3.98 \times 10^5\,\text{K})}{5.23 \times 10^8\,\text{m}^3}$$
$$P = \boxed{1.69 \times 10^{11}\ \text{Pa}}$$

Remarks The estimated pressure is several hundred times greater than the ultimate shear stress of steel! This high-pressure region would expand rapidly, destroying everything within a very large radius. Fires would ignite across a continent-sized region, and tidal waves would wrap around the world, wiping out coastal regions everywhere. The sun would be obscured for at least a decade, and numerous species, possibly including *Homo sapiens,* would become extinct. Such extinction events are rare, but in the long run represent a significant threat to life on Earth.

Exercise 11.8

Suppose a lead bullet with mass 5.00 g and an initial temperature of 65.0°C hits a wall and completely liquefies. What minimum speed did it have before impact? (*Hint:* The minimum speed corresponds to the case where all the kinetic energy becomes internal energy of the lead and the final temperature of the lead is at its melting point. Don't neglect any terms here!)

Answer 341 m/s

11.5 ENERGY TRANSFER

For some applications it's necessary to know the rate at which energy is transferred between a system and its surroundings and the mechanisms responsible for the transfer. This is particularly important in weatherproofing buildings or in medical applications, such as human survival time when exposed to the elements.

Earlier in this chapter we defined heat as a transfer of energy between a system and its surroundings due to a temperature difference between them. In this section, we take a closer look at heat as a means of energy transfer and consider the processes of thermal conduction, convection, and radiation.

Figure 11.4 Conduction makes the metal handle of a cooking pan hot.

Thermal Conduction

The energy transfer process most closely associated with a temperature difference is called **thermal conduction** or simply **conduction**. In this process, the transfer can be viewed on an atomic scale as an exchange of kinetic energy between microscopic particles—molecules, atoms, and electrons—with less energetic particles gaining energy as they collide with more energetic particles. An inexpensive pot, as in Figure 11.4, may have a metal handle with no surrounding insulation. As the pot is warmed, the temperature of the metal handle increases, and the cook must hold it with a cloth potholder to avoid being burned.

The way the handle warms up can be understood by looking at what happens to the microscopic particles in the metal. Before the pot is placed on the stove, the particles are vibrating about their equilibrium positions. As the stove coil warms up, those particles in contact with it begin to vibrate with larger amplitudes. These particles collide with their neighbors and transfer some of their energy in the collisions. Metal atoms and electrons farther and farther from the flame gradually increase the amplitude of their vibrations, until eventually those in the metal near your hand are affected. This increased vibration represents an increase in temperature of the metal (and possibly a burned hand!).

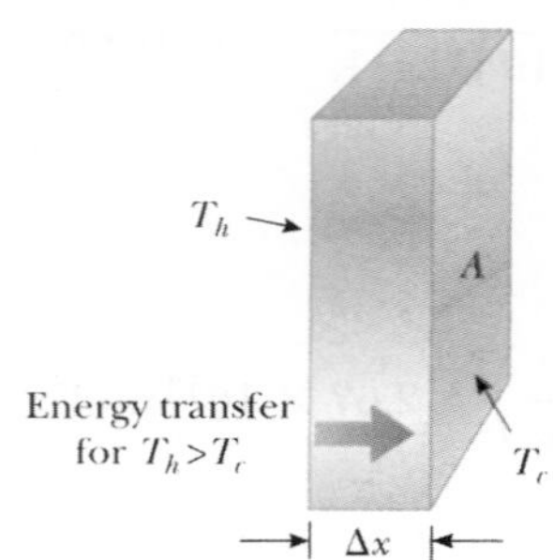

Figure 11.5 Energy transfer through a conducting slab of cross-sectional area A and thickness L. The opposite faces are at different temperatures T_c and T_h.

Although the transfer of energy through a substance can be partly explained by atomic vibrations, the rate of conduction depends on the properties of the substance. For example, it's possible to hold a piece of asbestos in a flame indefinitely. This fact implies that very little energy is conducted through the asbestos. In general, metals are good thermal conductors because they contain large numbers of electrons that are relatively free to move through the metal and can transport energy from one region to another. In a good conductor such as copper, conduction takes place via the vibration of atoms and the motion of free electrons. Materials such as asbestos, cork, paper, and fiberglass are poor thermal conductors. Gases are also poor thermal conductors because of the large distance between their molecules.

Conduction occurs only if there is a difference in temperature between two parts of the conducting medium. The temperature difference drives the flow of energy. Consider a slab of material of thickness Δx and cross-sectional area A with its opposite faces at different temperatures T_c and T_h, where $T_h > T_c$ (Fig. 11.5). The slab allows energy to transfer from the region of higher temperature to the region of lower temperature by thermal conduction. The rate of energy transfer, $\mathcal{P} = Q/\Delta t$, is proportional to the cross-sectional area of the slab and the temperature difference and is inversely proportional to the thickness of the slab:

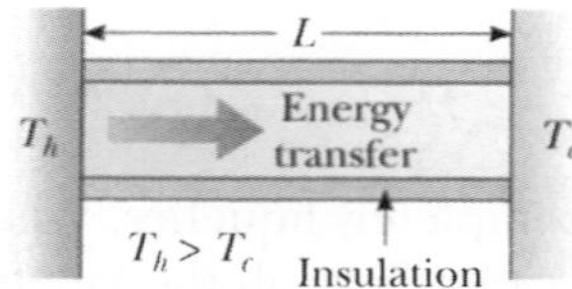

Figure 11.6 Conduction of energy through a uniform, insulated rod of length L. The opposite ends are in thermal contact with energy reservoirs at different temperatures.

$$\mathcal{P} = \frac{Q}{\Delta t} \propto A\frac{\Delta T}{\Delta x}$$

Note that $\mathcal{P}$ has units of watts when Q is in joules and Δt is in seconds.

Suppose a substance is in the shape of a long, uniform rod of length L, as in Figure 11.6. We assume the rod is insulated, so thermal energy can't escape by conduction from its surface except at the ends. One end is in thermal contact with an energy reservoir at temperature T_c and the other end is in thermal contact with a reservoir at temperature $T_h > T_c$. When a steady state is reached, the temperature at each point along the rod is constant in time. In this case, $\Delta T = T_h - T_c$ and $\Delta x = L$, so

$$\frac{\Delta T}{\Delta x} = \frac{T_h - T_c}{L}$$

The rate of energy transfer by conduction through the rod is given by

$$\mathcal{P} = kA\frac{(T_h - T_c)}{L} \qquad [11.7]$$

TIP 11.4 Blankets and Coats in Cold Weather

When you sleep under a blanket in the winter or wear a warm coat outside, the blanket or coat serves as a layer of material with low thermal conductivity in order to reduce the transfer of energy away from your body by heat. The primary insulating medium is the air trapped in small pockets within the material.

where k, a proportionality constant that depends on the material, is called the **thermal conductivity**. Substances that are good conductors have large thermal conductivities, whereas good insulators have low thermal conductivities. Table 11.3 lists the thermal conductivities for various substances.

TABLE 11.3
Thermal Conductivities

Substance	Thermal Conductivity (J/s · m · °C)
Metals (at 25°C)	
Aluminum	238
Copper	397
Gold	314
Iron	79.5
Lead	34.7
Silver	427
Gases (at 20°C)	
Air	0.023 4
Helium	0.138
Hydrogen	0.172
Nitrogen	0.023 4
Oxygen	0.023 8
Nonmetals	
Asbestos	0.25
Concrete	1.3
Glass	0.84
Ice	1.6
Rubber	0.2
Water	0.60
Wood	0.10

Quick Quiz 11.3

Will an ice cube wrapped in a wool blanket remain frozen for (a) less time, (b) the same length of time, or (c) a longer time than an identical ice cube exposed to air at room temperature?

Quick Quiz 11.4

Two rods of the same length and diameter are made from different materials. The rods are to connect two regions of different temperature so that energy will transfer through the rods by heat. They can be connected in series, as in Figure 11.7a, or in parallel, as in Figure 11.7b. In which case is the rate of energy transfer by heat larger? (a) When the rods are in series (b) When the rods are in parallel (c) The rate is the same in both cases.

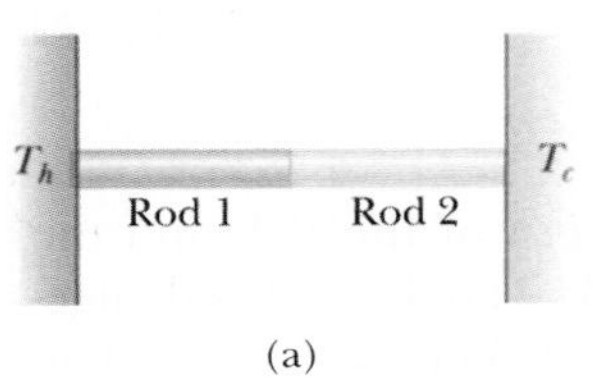

(a)

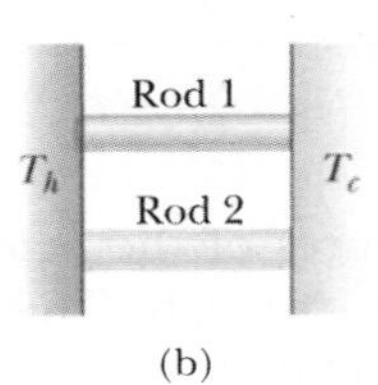

(b)

Figure 11.7 (Quick Quiz 11.4) In which case is the rate of energy transfer larger?

EXAMPLE 11.9 Energy Transfer through a Concrete Wall

Goal Apply the equation of heat conduction.

Problem Find the energy transferred in 1.00 h by conduction through a concrete wall 2.0 m high, 3.65 m long, and 0.20 m thick if one side of the wall is held at 20°C and the other side is at 5°C.

Strategy Equation 11.7 gives the rate of energy transfer by conduction in joules per second. Multiply by the time and substitute given values to find the total thermal energy transferred.

Solution

Multiply Equation 11.7 by Δt to find an expression for the total energy Q transferred through the wall:

$$Q = \mathcal{P}\,\Delta t = kA\left(\frac{T_h - T_c}{L}\right)\Delta t$$

Substitute the numerical values to obtain Q, consulting Table 11.3 for k:

$$Q = (1.3\ \text{J/s}\cdot\text{m}\cdot{}^\circ\text{C})(7.3\ \text{m}^2)\left(\frac{15^\circ\text{C}}{0.20\ \text{m}}\right)(3\,600\ \text{s})$$

$$= \boxed{2.6 \times 10^6\ \text{J}}$$

Remarks Early houses were insulated with thick masonry walls, which restrict energy loss by conduction because k is relatively low. The large thickness L also decreases energy loss by conduction, as shown by Equation 11.7. There are much better insulating materials, however, and layering is also helpful. Despite the low thermal conductivity of masonry, the amount of energy lost is still rather large—enough to raise the temperature of 600 kg of water by more than 1°C. There are better insulating materials than masonry.

Exercise 11.9
A wooden shelter has walls constructed of wooden planks 1.00 cm thick. If the exterior temperature is $-20.0°C$ and the interior is $5.00°C$, find the rate of energy loss through a wall that has dimensions 2.00 m by 2.00 m.

Answer 1.00×10^3 W

Home Insulation

To determine whether to add insulation to a ceiling or some other part of a building, the preceding discussion of conduction must be extended, for two reasons:

1. The insulating properties of materials used in buildings are usually expressed in engineering (U.S. customary) rather than SI units. Measurements stamped on a package of fiberglass insulating board will be in units such as British thermal units, feet, and degrees Fahrenheit.
2. In dealing with the insulation of a building, conduction through a compound slab must be considered, with each portion of the slab having a certain thickness and a specific thermal conductivity. A typical wall in a house consists of an array of materials, such as wood paneling, drywall, insulation, sheathing, and wood siding.

The rate of energy transfer by conduction through a compound slab is

$$\frac{Q}{\Delta t} = \frac{A(T_h - T_c)}{\sum_i L_i/k_i} \qquad [11.8]$$

where T_h and T_c are the temperatures of the *outer extremities* of the slab and the summation is over all portions of the slab. This formula can be derived algebraically, using the facts that the temperature at the interface between two insulating materials must be the same and that the rate of energy transfer through one insulator must be the same as through all the other insulators. If the slab consists of three different materials, the denominator is the sum of three terms. In engineering practice, the term L/k for a particular substance is referred to as the

TABLE 11.4
***R* Values for Some Common Building Materials**

Material	*R* value ($ft^2 \cdot °F \cdot h/Btu$)
Hardwood siding (1.0 in. thick)	0.91
Wood shingles (lapped)	0.87
Brick (4.0 in. thick)	4.00
Concrete block (filled cores)	1.93
Styrofoam (1.0 in. thick)	5.0
Fiber glass batting (3.5 in. thick)	10.90
Fiber glass batting (6.0 in. thick)	18.80
Fiber glass board (1.0 in. thick)	4.35
Cellulose fiber (1.0 in. thick)	3.70
Flat glass (0.125 in. thick)	0.89
Insulating glass (0.25-in. space)	1.54
Vertical air space (3.5 in. thick)	1.01
Stagnant layer of air	0.17
Dry wall (0.50 in. thick)	0.45
Sheathing (0.50 in. thick)	1.32

R value of the material, so Equation 11.8 reduces to

$$\frac{Q}{\Delta t} = \frac{A(T_h - T_c)}{\sum_i R_i} \qquad [11.9]$$

The R values for a few common building materials are listed in Table 11.4. Note the unit of R and the fact that the R values are defined for specific thicknesses.

Next to any vertical outside surface is a very thin, stagnant layer of air that must be considered when the total R value for a wall is computed. The thickness of this stagnant layer depends on the speed of the wind. As a result, energy loss by conduction from a house on a day when the wind is blowing is greater than energy loss on a day when the wind speed is zero. A representative R value for a stagnant air layer is given in Table 11.4.

INTERACTIVE EXAMPLE 11.10 The *R* Value of a Typical Wall

Goal Calculate the R value of a wall consisting of several layers of insulating material.

Problem Calculate the total R value for a wall constructed as shown in Figure 11.8a. Starting outside the house (to the left in the figure) and moving inward, the wall consists of 4.0 in. brick, 0.50 in. sheathing, an air space 3.5 in. thick, and 0.50 in. drywall.

Strategy Add all the R values together, remembering the stagnant air layers inside and outside the house.

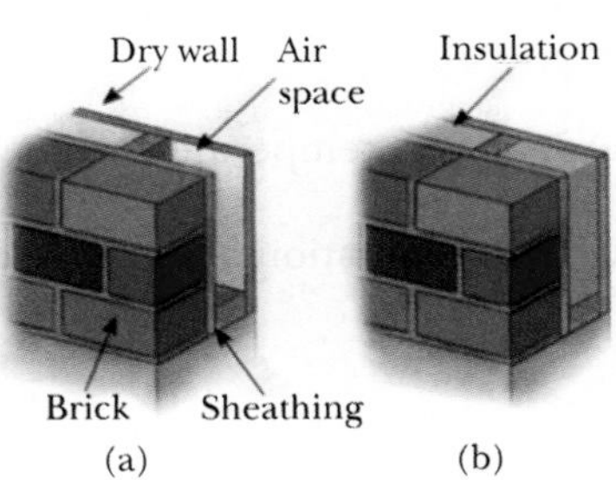

Figure 11.8 (Example 11.10) A cross-sectional view of an exterior wall containing (a) an air space and (b) insulation.

Solution

Refer to Table 11.4, and sum. All quantities are in units of $\text{ft}^2 \cdot {}^\circ\text{F} \cdot \text{h/Btu}$.

$$R_{\text{total}} = R_{\text{outside air layer}} + R_{\text{brick}} + R_{\text{sheath}} + R_{\text{air space}} + R_{\text{drywall}} + R_{\text{inside air layer}} = (0.17 + 4.00 + 1.32 + 1.01 + 0.45 + 0.17)\,\text{ft}^2 \cdot {}^\circ\text{F} \cdot \text{h/Btu}$$

$$R_{\text{total}} = 7.12\ \text{ft}^2 \cdot {}^\circ\text{F} \cdot \text{h/Btu}$$

Exercise 11.10

If a layer of fiber glass insulation 3.5 in. thick is placed inside the wall to replace the air space, as in Figure 11.8b, what is the new total R value? By what factor is the rate of energy loss reduced?

Answer $R = 17\ \text{ft}^2 \cdot {}^\circ\text{F} \cdot \text{h/Btu}$; 2.4

PhysicsNow™ Study the R values of various types of common building materials by logging into PhysicsNow at **www.cp7e.com** and going to Interactive Example 11.10.

EXAMPLE 11.11 Staying Warm in the Arctic

Goal Combine two layers of insulation.

Problem An arctic explorer builds a wooden shelter out of wooden planks that are 1.0 cm thick. To improve the insulation, he covers the shelter with a layer of ice 3.2 cm thick. **(a)** Compute the R factors for the wooden planks and the ice. **(b)** If the temperature outside the shelter is $-20.0°\text{C}$ and the temperature inside is 5.00°C, find the rate of energy loss through one of the walls, if the wall has dimensions 2.00 m by 2.00 m. **(c)** Find the temperature at the interface between the wood and the ice. Disregard stagnant air layers.

Strategy After finding the R values, substitute into Equation 11.9 to get the rate of energy transfer. To answer part (c), use Equation 11.7 for one of the layers, setting it equal to the rate found in part (b), solving for the temperature.

Solution

(a) Compute the R values using the data in Table 11.3.

Find the R value for the wooden wall:

$$R_{\text{wood}} = \frac{L_{\text{wood}}}{k_{\text{wood}}} = \frac{0.01 \text{ m}}{0.10 \text{ J/s}\cdot\text{m}\cdot{}^\circ\text{C}} = 0.10 \text{ m}^2\cdot\text{s}\cdot{}^\circ\text{C/J}$$

Find the R-value for the ice layer:

$$R_{\text{ice}} = \frac{L_{\text{ice}}}{k_{\text{ice}}} = \frac{0.032 \text{ m}}{1.6 \text{ J/s}\cdot\text{m}\cdot{}^\circ\text{C}} = 0.020 \text{ m}^2\cdot\text{s}\cdot{}^\circ\text{C/J}$$

(b) Find the rate of heat loss.

Apply Equation 11.9:

$$\mathcal{P} = \frac{Q}{\Delta t} = \frac{A(T_h - T_c)}{\sum_i R_i}$$

$$= \frac{(4.00 \text{ m}^2)(5.00^\circ\text{C} - (-20.0^\circ\text{C}))}{0.12 \text{ m}^2\cdot\text{s}\cdot{}^\circ\text{C/J}}$$

$$\mathcal{P} = 830 \text{ W}$$

(c) Find the temperature in between the ice and wood.

Apply the equation of heat conduction to the wood:

$$\frac{k_{\text{wood}} A(T_h - T_c)}{L} = \mathcal{P}$$

$$\frac{(0.10 \text{ J/s}\cdot\text{m}\cdot{}^\circ\text{C})(4.00 \text{ m}^2)(5.00^\circ\text{C} - T)}{0.010 \text{ m}} = 830 \text{ W}$$

Solve for the unknown temperature:

$$T = -16^\circ\text{C}$$

Remarks The outer side of the wooden wall and the inner surface of the ice must have the same temperature, and the rate of energy transfer through the ice must be the same as through the wooden wall. Using Equation 11.7 for ice instead of wood gives the same answer. This rate of energy transfer is only a modest improvement over the thousand-watt rate in Exercise 11.9. The choice of insulating material is important!

Exercise 11.11

Rather than use ice to cover the wooden shelter, the explorer glues pressed cork with thickness 0.500 cm to the outside of his wooden shelter. Find the new rate of energy loss through the same wall. (Note that $k_{\text{cork}} = 0.046 \text{ J/s}\cdot\text{m}\cdot{}^\circ\text{C}$.)

Answer 480 W

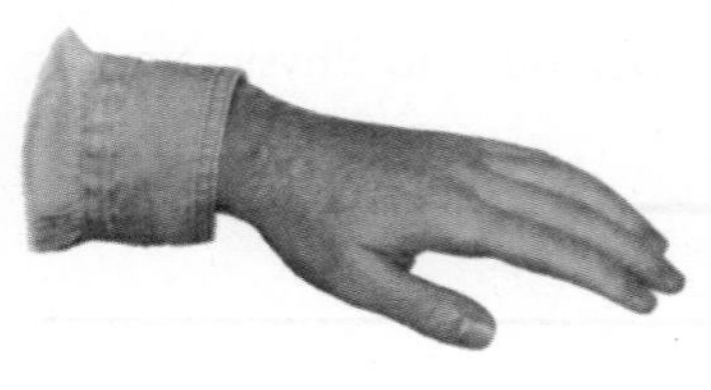

Figure 11.9 Warming a hand by convection.

Convection

When you warm your hands over an open flame, as illustrated in Figure 11.9, the air directly above the flame, being warmed, expands. As a result, the density of this air decreases and the air rises, warming your hands as it flows by. **The transfer of energy by the movement of a substance is called convection**. When the movement results from differences in density, as with air around a fire, it's referred to as *natural convection*. Airflow at a beach is an example of natural convection, as is the mixing that occurs as surface water in a lake cools and sinks. When the substance is forced to move by a fan or pump, as in some hot air and hot water heating systems, the process is called *forced convection*.

Convection currents assist in the boiling of water. In a teakettle on a hot stovetop, the lower layers of water are warmed first. The warmed water has a lower density and rises to the top, while the denser, cool water at the surface sinks to the bottom of the kettle and is warmed.

The same process occurs when a radiator raises the temperature of a room. The hot radiator warms the air in the lower regions of the room. The warm air expands and, because of its lower density, rises to the ceiling. The denser cooler air from above sinks, setting up the continuous air current pattern shown in Figure 11.10.

An automobile engine is maintained at a safe operating temperature by a combination of conduction and forced convection. Water (actually, a mixture of water and antifreeze) circulates in the interior of the engine. As the metal of the engine block increases in temperature, energy passes from the hot metal to the cooler water by thermal conduction. The water pump forces water out of the engine and into the radiator, carrying energy along with it (by forced convection). In the radiator, the hot water passes through metal pipes that are in contact with the cooler outside air, and energy passes into the air by conduction. The cooled water is then returned to the engine by the water pump to absorb more energy. The process of air being pulled past the radiator by the fan is also forced convection.

The algal blooms often seen in temperate lakes and ponds during the spring or fall are caused by convection currents in the water. To understand this process, consider Figure 11.11. During the summer, bodies of water develop temperature gradients, with an upper, warm layer of water separated from a lower, cold layer by a buffer zone called a thermocline. In the spring or fall, temperature changes in the water break down this thermocline, setting up convection currents that mix the water. The mixing process transports nutrients from the bottom to the surface. The nutrient-rich water forming at the surface can cause a rapid, temporary increase in the algae population.

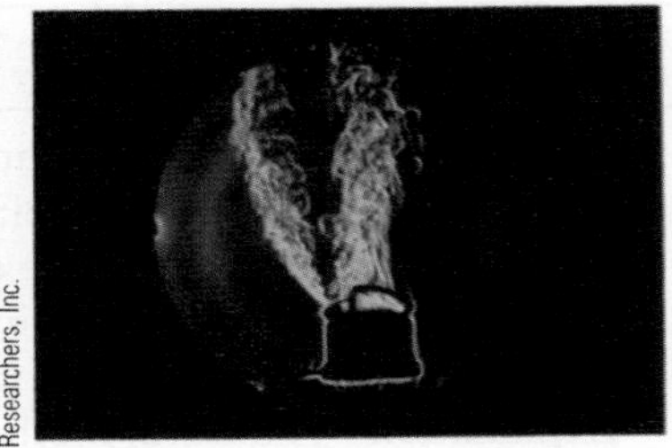

Photograph of a teakettle, showing steam and turbulent convection air currents.

APPLICATION

Cooling Automobile Engines

APPLICATION

Algal Blooms in Ponds and Lakes

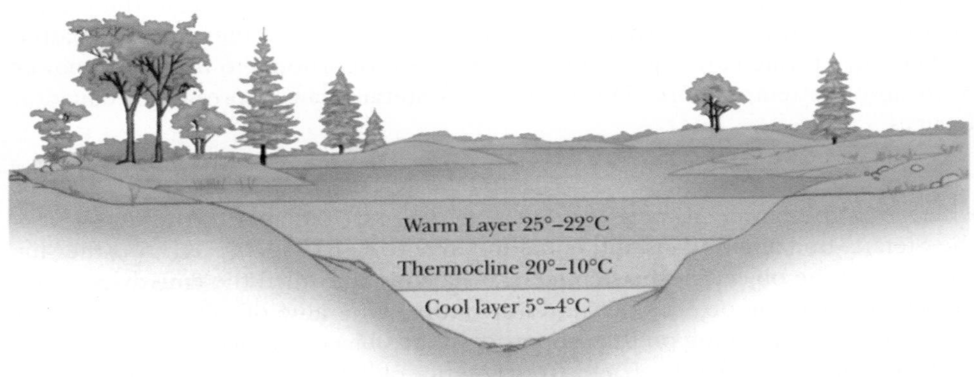

(a) Summer layering of water

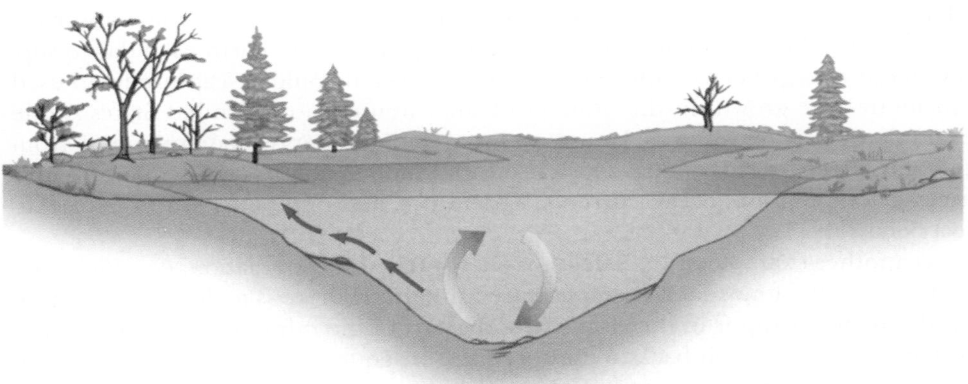

(b) Fall and spring upwelling

Figure 11.11 (a) During the summer, a warm upper layer of water is separated from a cooler lower layer by a thermocline. (b) Convection currents during the spring or fall mix the water and can cause algal blooms.

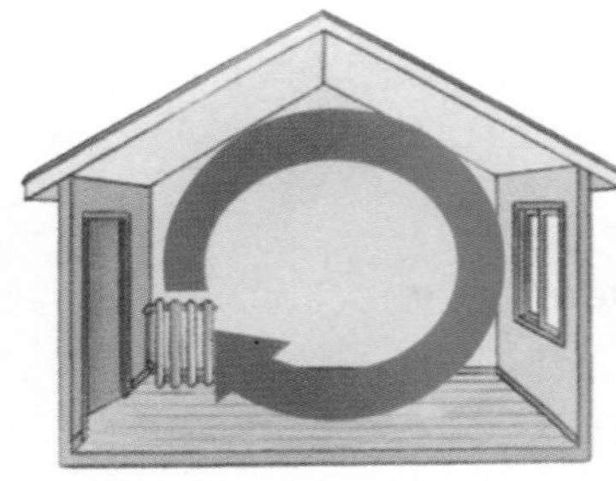

Figure 11.10 Convection currents are set up in a room warmed by a radiator.

Applying Physics 11.1 Body Temperature 

The body temperature of mammals ranges from about 35°C to 38°C, while that of birds ranges from about 40°C to 43°C. How can these narrow ranges of body temperature be maintained in cold weather?

Explanation A natural method of maintaining body temperature is via layers of fat beneath the skin. Fat protects against both conduction and convection because of its low thermal conductivity and because there are few blood vessels in fat to carry blood to the surface, where energy losses by convection can occur. Birds ruffle their feathers in cold weather in order to trap a layer of air with a low thermal conductivity between the feathers and the skin. Bristling the fur produces the same effect in fur-bearing animals.

Humans keep warm with wool sweaters and down jackets that trap the warmer air in regions close to their bodies, reducing energy loss by convection and conduction.

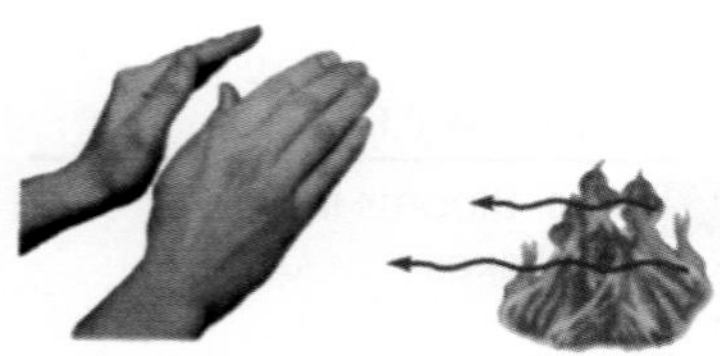

Figure 11.12 Warming hands by radiation.

Radiation

Another process of transferring energy is through **radiation**. Figure 11.12 shows how your hands can be warmed at an open flame through radiation. Because your hands aren't in physical contact with the flame and the conductivity of air is very low, conduction can't account for the energy transfer. Nor can convection be responsible for any transfer of energy, because your hands aren't above the flame in the path of convection currents. The warmth felt in your hands, therefore, must come from the transfer of energy by radiation.

All objects radiate energy continuously in the form of electromagnetic waves due to thermal vibrations of their molecules. These vibrations create the orange glow of an electric stove burner, an electric space heater, and the coils of a toaster.

The rate at which an object radiates energy is proportional to the fourth power of its absolute temperature. This is known as **Stefan's law**, expressed in equation form as

Stefan's law ▶

$$\mathcal{P} = \sigma A e T^4 \qquad [11.10]$$

where $\mathcal{P}$ is the power in watts (or joules per second) radiated by the object, σ is the Stefan–Boltzmann constant, equal to $5.669\,6 \times 10^{-8}\ \mathrm{W/m^2 \cdot K^4}$, A is the surface area of the object in square meters, e is a constant called the **emissivity** of the object, and T is the object's Kelvin temperature. The value of e can vary between zero and one, depending on the properties of the object's surface.

Approximately 1 340 J of electromagnetic radiation from the Sun passes through each square meter at the top of the Earth's atmosphere every second. This radiation is primarily visible light, accompanied by significant amounts of infrared and ultraviolet. We will study these types of radiation in detail in Chapter 21. Some of this energy is reflected back into space, and some is absorbed by the atmosphere, but enough arrives at the surface of the Earth each day to supply all our energy needs hundreds of times over—if it could be captured and used efficiently. The growth in the number of solar houses in the United States is one example of an attempt to make use of this abundant energy. Radiant energy from the Sun affects our day-to-day existence in a number of ways, influencing Earth's average temperature, ocean currents, agriculture, and rain patterns. It can also affect behavior.

As another example of the effects of energy transfer by radiation, consider what happens to the atmospheric temperature at night. If there is a cloud cover above Earth, the water vapor in the clouds absorbs part of the infrared radiation emitted by Earth and re-emits it back to the surface. Consequently, the temperature at the surface remains at moderate levels. In the absence of cloud cover, there is nothing to prevent the radiation from escaping into space, so the temperature drops more on a clear night than when it's cloudy.

As an object radiates energy at a rate given by Equation 11.10, it also absorbs radiation. If it didn't, the object would eventually radiate all its energy and its temperature would reach absolute zero. The energy an object absorbs comes from its environment, which consists of other bodies that radiate energy. If an object is at a temperature T, and its surroundings are at a temperature T_0, the net energy gained or lost each second by the object as a result of radiation is

$$\mathcal{P}_{\text{net}} = \sigma A e(T^4 - T_0^4) \qquad [11.11]$$

When an object is in equilibrium with its surroundings, it radiates and absorbs energy at the same rate, so its temperature remains constant. When an object is hotter than its surroundings, it radiates more energy than it absorbs and so cools.

An **ideal absorber** is an object that absorbs all the light radiation incident on it, including invisible infrared and ultraviolet light. Such an object is called a **black body** because a room-temperature black body would look black. Since a black body doesn't reflect radiation at any wavelength, any light coming from it is due to atomic and molecular vibrations alone. A perfect black body has emissivity $e = 1$. An ideal absorber is also an ideal radiator of energy. The Sun, for example, is nearly a perfect black body. This statement may seem contradictory, because the Sun is bright, not dark; however, the light that comes from the Sun is emitted, not reflected. Black bodies are perfect absorbers that look black at room temperature because they don't reflect any light. All black bodies, except those at absolute zero, emit light that has a characteristic spectrum, to be discussed in Chapter 27. In contrast to black bodies, an object for which $e = 0$ absorbs none of the energy incident on it, reflecting it all. Such a body is an **ideal reflector**.

APPLICATION
Light-Colored Summer Clothing

White clothing is more comfortable to wear in the summer than black clothing. Black fabric acts as a good absorber of incoming sunlight and as a good emitter of this absorbed energy. About half of the emitted energy, however, travels toward the body, causing the person wearing the garment to feel uncomfortably warm. White or light-colored clothing reflects away much of the incoming energy.

APPLICATION
Thermography

The amount of energy radiated by an object can be measured with temperature-sensitive recording equipment via a technique called **thermography**. An image of the pattern formed by varying radiation levels, called a **thermogram**, is brightest in the warmest areas. Figure 11.13 reproduces a thermogram of a house. More energy escapes in the lighter regions, such as the door and windows. The owners of this house could conserve energy and reduce their heating costs by adding insulation to the attic area and by installing thermal draperies over the windows. Thermograms have also been used to image injured or diseased tissue in medicine, since such areas are often at a different temperature than surrounding healthy tissue, though many radiologists consider thermograms inadequate as a diagnostic tool.

APPLICATION
Radiation Thermometers for Measuring Body Temperature

Figure 11.14 shows a recently developed radiation thermometer that has removed most of the risk of taking the temperature of young children or the aged with a rectal thermometer—risks such as bowel perforation or bacterial contamination. The instrument measures the intensity of the infrared radiation leaving

Daedalus Enterprises, Inc./Peter Arnold, Inc.

Figure 11.13 This thermogram of a house, made during cold weather, shows colors ranging from white and yellow (areas of greatest energy loss) to blue and purple (areas of least energy loss).

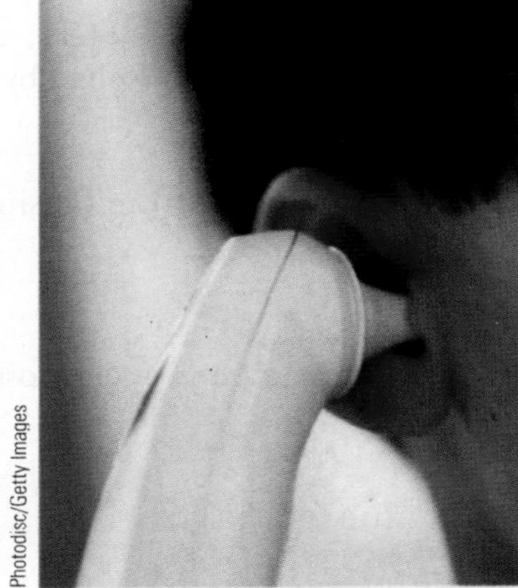
Photodisc/Getty Images

Figure 11.14 A radiation thermometer measures a patient's temperature by monitoring the intensity of infrared radiation leaving the ear.

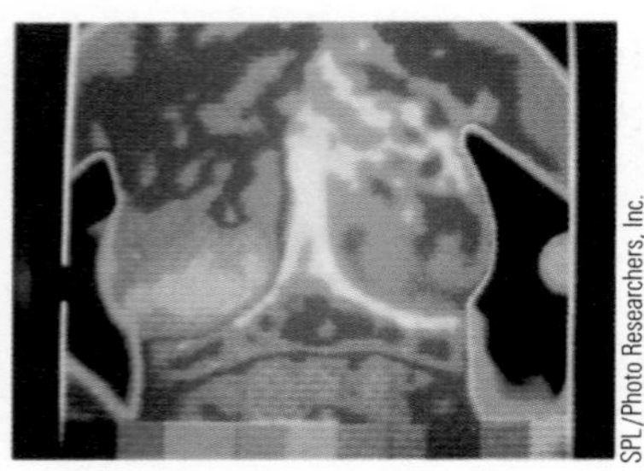

Thermogram of a woman's breasts. Her left breast is diseased (red and orange) and her right breast (blue) is healthy.

the eardrum and surrounding tissues and converts this information to a standard numerical reading. The eardrum is a particularly good location to measure body temperature because it's near the hypothalamus—the body's temperature control center.

Quick Quiz 11.5

Stars A and B have the same temperature, but star A has twice the radius of star B. (a) What is the ratio of star A's power output to star B's output due to electromagnetic radiation? The emissivity of both stars can assumed to be 1. (b) Repeat the question if the stars have the same radius, but star A has twice the absolute temperature of star B. (c) What's the ratio if star A has both twice the radius and twice the absolute temperature of star B?

Applying Physics 11.2 Thermal Radiation and Night Vision

How can thermal radiation be used to see objects in near total darkness?

Explanation There are two methods of night vision, one enhancing a combination of very faint visible light and infrared light, and another using infrared light only. The latter is valuable for creating images in absolute darkness. Because all objects above absolute zero emit thermal radiation due to the vibrations of their atoms, the infrared (invisible) light can be focused by a special lens and scanned by an array of infrared detector elements. These elements create a thermogram. The information from thousands of separate points in the field of view is converted to electrical impulses and translated by a microchip into a form suitable for display. Different temperature areas are assigned different colors, which can then be easily discerned on the display.

EXAMPLE 11.12 Polar Bear Club

Goal Apply Stefan's law.

Problem A member of the Polar Bear Club, dressed only in bathing trunks of negligible size, prepares to plunge into the Baltic Sea from the beach in St. Petersburg, Russia. The air is calm, with a temperature of 5°C. If the swimmer's surface body temperature is 25°C, compute the net rate of energy loss from his skin due to radiation. How much energy is lost in 10.0 min? Assume his emissivity is 0.900, and his surface area is 1.50 m^2.

Strategy Use Equation 11.11, the thermal radiation equation, substituting the given information. Remember to convert temperatures to Kelvin by adding 273 to each value in degrees Celsius!

Solution

Convert temperatures from Celsius to Kelvin:

$$T_{5^\circ\text{C}} = T_C + 273 = 5 + 273 = 278\text{ K}$$

$$T_{25^\circ\text{C}} = T_C + 273 = 25 + 273 = 298\text{ K}$$

Compute the net rate of energy loss, using Equation 11.11:

$$\mathcal{P}_{\text{net}} = \sigma A e(T^4 - T_0^{\,4})$$

$$= (5.67 \times 10^{-8}\text{ W/m}^2\cdot\text{K}^4)(1.50\text{ m}^2) \times (0.90)[(298\text{ K})^4 - (278\text{ K})^4]$$

$$\mathcal{P}_{\text{net}} = \boxed{146\text{ W}}$$

Multiply the preceding result by the time, 10 minutes, to get the energy lost in that time due to radiation:

$$Q = \mathcal{P}_{\text{net}} \times \Delta t = (146)(6.00 \times 10^2\text{ s}) = \boxed{8.76 \times 10^4\text{ J}}$$

Remarks Energy is also lost from the body through convection and conduction. Clothing traps layers of air next to the skin, which are warmed by radiation and conduction. In still air these warm layers are more readily retained. Even a Polar Bear Club member enjoys some benefit from the still air, better retaining a stagnant air layer next to the surface of his skin.

Exercise 11.12

Repeat the calculation when the man is standing in his bedroom, with an ambient temperature of 20.0°C. Assume his body surface temperature is 27.0°C, with emissivity of 0.900.

Answer 55.9 W, 3.35×10^4 J

The Dewar Flask

The Thermos bottle, also called a **Dewar flask** (after its inventor), is designed to minimize energy transfer by conduction, convection, and radiation. The thermos can store either cold or hot liquids for long periods. The standard vessel (Fig. 11.15) is a double-walled Pyrex glass with silvered walls. The space between the walls is evacuated to minimize energy transfer by conduction and convection. The silvered surface minimizes energy transfer by radiation because silver is a very good reflector and has very low emissivity. A further reduction in energy loss is achieved by reducing the size of the neck. Dewar flasks are commonly used to store liquid nitrogen (boiling point 77 K) and liquid oxygen (boiling point 90 K).

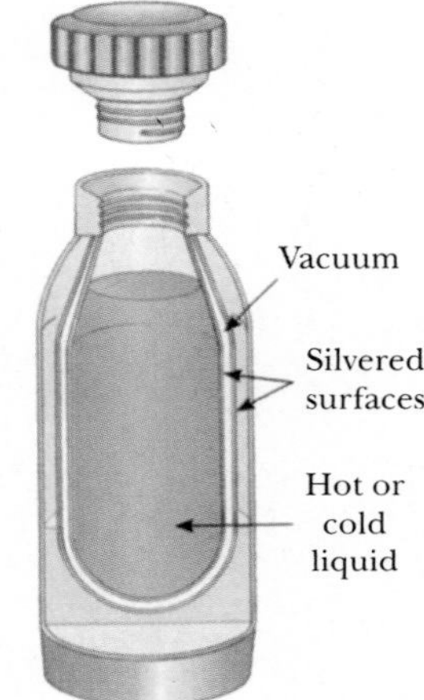

Figure 11.15 A cross-sectional view of a Thermos bottle designed to store hot or cold liquids.

To confine liquid helium (boiling point 4.2 K), which has a very low heat of vaporization, it's often necessary to use a double Dewar system in which the Dewar flask containing the liquid is surrounded by a second Dewar flask. The space between the two flasks is filled with liquid nitrogen.

Some of the principles of the Thermos bottle are used in the protection of sensitive electronic instruments in orbiting space satellites. In half of its orbit around the Earth a satellite is exposed to intense radiation from the Sun, and in the other half it lies in the Earth's cold shadow. Without protection, its interior would be subjected to tremendous extremes of temperature. The interior of the satellite is wrapped with blankets of highly reflective aluminum foil. The foil's shiny surface reflects away much of the Sun's radiation while the satellite is in the unshaded part of the orbit and helps retain interior energy while the satellite is in the Earth's shadow.

APPLICATION

Thermos Bottles

11.6 GLOBAL WARMING AND GREENHOUSE GASES

Many of the principles of energy transfer, and opposition to it, can be understood by studying the operation of a glass greenhouse. During the day, sunlight passes into the greenhouse and is absorbed by the walls, soil, plants, and so on. This absorbed visible light is subsequently reradiated as infrared radiation, causing the temperature of the interior to rise.

In addition, convection currents are inhibited in a greenhouse. As a result, warmed air can't rapidly pass over the surfaces of the greenhouse that are exposed to the outside air and thereby cause an energy loss by conduction through those surfaces. Most experts now consider this restriction to be a more important warming effect than the trapping of infrared radiation. In fact, experiments have shown that when the glass over a greenhouse is replaced by a special glass known to transmit infrared light, the temperature inside is lowered only slightly. On the basis of this evidence, the primary mechanism that raises the temperature of a greenhouse is not the trapping of infrared radiation, but the inhibition of airflow that occurs under any roof (in an attic, for example).

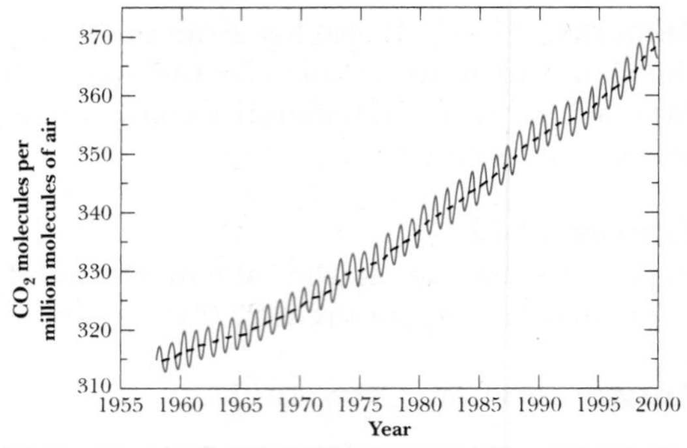

Figure 11.16 The concentration of atmospheric carbon dioxide in parts per million (ppm) of dry air as a function of time during the latter part of the 20th century. These data were recorded at Mauna Loa Observatory in Hawaii. The yearly variations (red curve) coincide with growing seasons, because vegetation absorbs carbon dioxide from the air. The steady increase (black curve) is of concern to scientists.

A phenomenon commonly known as the **greenhouse effect** can also play a major role in determining the Earth's temperature. First, note that the Earth's atmosphere is a good transmitter (and hence a poor absorber) of visible radiation and a good absorber of infrared radiation. The visible light that reaches the Earth's surface is absorbed and reradiated as infrared light, which in turn is absorbed (trapped) by the Earth's atmosphere. An extreme case is the warmest planet, Venus, which has a carbon dioxide (CO_2) atmosphere and temperatures approaching 850°F.

As fossil fuels (coal, oil, and natural gas) are burned, large amounts of carbon dioxide are released into the atmosphere, causing it to retain more energy. This is of great concern to scientists and governments throughout the world. Many scientists are convinced that the 10% increase in the amount of atmospheric carbon dioxide in the past 30 years could lead to drastic changes in world climate. The increase in concentration of atmospheric carbon dioxide in the latter part of the 20th century is shown in Figure 11.16. According to one estimate, doubling the carbon dioxide content in the atmosphere will cause temperatures to increase by 2°C. In temperate regions, such as Europe and the United States, a 2°C temperature rise would save billions of dollars per year in fuel costs. Unfortunately, it would also melt a large amount of ice from the polar ice caps, which could cause flooding and destroy many coastal areas. A 2°C rise would also increase the frequency of droughts, and consequently decrease already low crop yields in tropical and subtropical countries. Even slightly higher average temperatures might make it impossible for certain plants and animals to survive in their customary ranges.

At present, about 3.5×10^{11} tons of CO_2 are released into the atmosphere each year. Most of this gas results from human activities such as the burning of fossil fuels, the cutting of forests, and manufacturing processes. Another greenhouse gas is methane (CH_4), which is released in the digestive process of cows and other ruminants. This gas originates from that part of the animal's stomach called the *rumen,* where cellulose is digested. Termites are also major producers of this gas. Finally, greenhouse gases such as nitrous oxide (N_2O) and sulfur dioxide (SO_2) are increasing due to automobile and industrial pollution.

Whether the increasing greenhouse gases are responsible or not, there is convincing evidence that global warming is underway. The evidence comes from the melting of ice in Antarctica and the retreat of glaciers at widely scattered sites throughout the world (see Fig. 11.17). For example, satellite images of Antarctica show James Ross Island completely surrounded by water for the first time since maps were made, about 100 years ago. Previously, the island was connected to the mainland by an ice bridge. In addition, at various places across the continent, ice shelves are retreating, some at a rapid rate.

Perhaps at no place in the world are glaciers monitored with greater interest than in Switzerland. There, it is found that the Alps have lost about 50% of their glacial ice compared to 130 years ago. The retreat of glaciers on high-altitude peaks in the tropics is even more severe than in Switzerland. The Lewis glacier on Mount Kenya and the snows of Kilimanjaro are two examples. However, in certain regions of the planet where glaciers are near large bodies of water and are fed by

Tom Bean

(a)

(b)

Figure 11.17 Glacier melting in Alaska. (a) Hiker on a ridge above the Muir Glacier in Glacier Bay National Park, August 26, 1978. (b) Hiker on the same ridge, June 27, 1993.

large and frequent snows, glaciers continue to advance, so the overall picture of a catastrophic global-warming scenario may be premature. In about 50 years, however, the amount of carbon dioxide in the atmosphere is expected to be about twice what it was in the preindustrial era. Because of the possible catastrophic consequences, most scientists voice the concern that reductions in greenhouse gas emissions need to be made now.

SUMMARY

Physics Now™ Take a practice test by logging into PhysicsNow at **www.cp7e.com** and clicking on the Pre-Test link for this chapter.

11.1 Heat and Internal Energy

Internal energy is associated with a system's microscopic components. Internal energy includes the kinetic energy of translation, rotation, and vibration of molecules, as well as potential energy.

Heat is the transfer of energy across the boundary of a system resulting from a temperature difference between the system and its surroundings. The symbol Q represents the amount of energy transferred.

The **calorie** is the amount of energy necessary to raise the temperature of 1 g of water from 14.5°C to 15.5°C. The **mechanical equivalent of heat** is 4.186 J/cal.

11.2 Specific Heat

11.3 Calorimetry

The energy required to change the temperature of a substance of mass m by an amount ΔT is

$$Q = mc\,\Delta T \qquad [11.3]$$

where c is the **specific heat** of the substance. In calorimetry problems, the specific heat of a substance can be determined by placing it in water of known temperature, isolating the system, and measuring the temperature at equilibrium. The sum of all energy gains and losses for all the objects in an isolated system is given by

$$\Sigma Q_k = 0 \qquad [11.5]$$

where Q_k is the energy change in the kth object in the system. This equation can be solved for the unknown specific heat, or used to determine an equilibrium temperature.

11.4 Latent Heat and Phase Change

The energy required to change the phase of a pure substance of mass m is

$$Q = \pm\, mL \qquad [11.6]$$

where L is the **latent heat** of the substance. The latent heat of fusion, L_f, describes an energy transfer during a change from a solid phase to a liquid phase (or vice-versa), while the latent heat of vaporizaion, L_v, describes an energy transfer during a change from a liquid phase to a gaseous phase (or vice-versa). Calorimetry problems involving phase changes are handled with Equation 11.5, with latent heat terms added to the specific heat terms.

11.5 Energy Transfer

Energy can be transferred by several different processes, including work, discussed in Chapter 5, and by conduction, convection, and radiation. **Conduction** can be viewed as an exchange of kinetic energy between colliding molecules or electrons. The rate at which energy transfers by conduction through a slab of area A and thickness L is

$$\mathcal{P} = kA\frac{(T_h - T_c)}{L} \qquad [11.7]$$

where k is the **thermal conductivity** of the material making up the slab.

Energy is transferred by **convection** as a substance moves from one place to another.

All objects emit **radiation** from their surfaces in the form of electromagnetic waves at a net rate of

$$\mathcal{P}_{\text{net}} = \sigma A e(T^4 - T_0{}^4) \qquad [11.11]$$

where T is the temperature of the object and T_0 is the temperature of the surroundings. An object that is hotter than its surroundings radiates more energy than it absorbs, whereas a body that is cooler than its surroundings absorbs more energy than it radiates.

CONCEPTUAL QUESTIONS

1. Rub the palm of your hand on a metal surface for 30–45 seconds. Place the palm of your other hand on an unrubbed portion of the surface and then the rubbed portion. The rubbed portion will feel warmer. Now repeat this process on a wooden surface. Why does the temperature difference between the rubbed and unrubbed portions of the wood surface seem larger than for the metal surface?
2. Pioneers stored fruits and vegetables in underground cellars. Discuss fully this choice for a storage site.
3. In usually warm climates that experience an occasional hard freeze, fruit growers will spray the fruit trees with water, hoping that a layer of ice will form on the fruit. Why would such a layer be advantageous?
4. In winter, why did the pioneers (mentioned in Question 2) store an open barrel of water alongside their produce?
5. Cups of water for coffee or tea can be warmed with a coil that is immersed in the water and raised to a high temperature by means of electricity. Why do the instructions warn users not to operate the coils in the absence of water? Can the immersion coil be used to warm up a cup of stew?
6. The U.S. penny is now made of copper-coated zinc. Can a calorimetric experiment be devised to test for the metal content in a collection of pennies? If so, describe the procedure.
7. On a clear, cold night, why does frost tend to form on the tops, rather than the sides, of mailboxes and cars?
8. A warning sign often seen on highways just before a bridge is "Caution—Bridge Surface Freezes before Road Surface." Of the three energy transfer processes discussed in Section 11.5, which is most important in causing a bridge surface to freeze before the road surface on very cold days?
9. A tile floor may feel uncomfortably cold to your bare feet, but a carpeted floor in an adjoining room at the same temperature feels warm. Why?
10. On a very hot day, it's possible to cook an egg on the hood of a car. Would you select a black car or a white car on which to cook your egg? Why?
11. Concrete has a higher specific heat than does soil. Use this fact to explain (partially) why a city has a higher average temperature than the surrounding countryside. Would you expect evening breezes to blow from city to country or from country to city? Explain.
12. You need to pick up a very hot cooking pot in your kitchen. You have a pair of hot pads. Should you soak them in cold water or keep them dry in order to pick up the pot most comfortably?
13. In a daring demonstration, a professor dips her wetted fingers into molten lead (327°C) and withdraws them quickly without getting burned. How is this possible?
14. The air temperature above coastal areas is profoundly influenced by the large specific heat of water. One reason is that the energy released when 1 cubic meter of water cools by 1.0°C will raise the temperature of an enormously larger volume of air by 1.0°C. Estimate that volume of air. The specific heat of air is approximately 1.0 kJ/kg · °C. Take the density of air to be 1.3 kg/m^3.
15. Ethyl alcohol has about one-half the specific heat of water. Compare the temperature increases of equal masses of alcohol and water in separate beakers that are supplied with the same amount of energy.
16. Energy is added to ice, raising its temperature from −10°C to −5°C. A larger amount of energy is added to the same mass of liquid water, raising its temperature from 15°C to 20°C. From these results, we can conclude that (a) overcoming the latent heat of fusion of ice requires an input of energy (b) the latent heat of fusion of ice delivers some energy to the system (c) the specific heat of ice is less than that of water (d) the specific heat of ice is greater than that of water.
17. The specific heat of substance A is greater than the specific heat of substance B. Both A and B are at the same initial temperature when equal amounts of energy are added to them. Assuming no melting, freezing, or evaporation occurs, which of the following can be concluded about the final temperature T_A of substance A and the final temperature T_B of substance B? (a) $T_A > T_B$ (b) $T_A < T_B$ (c) $T_A = T_B$ (d) more information is needed.

PROBLEMS

1, 2, 3 = straightforward, intermediate, challenging □ = full solution available in *Student Solutions Manual/Study Guide*
PhysicsNow™ = coached problem with hints available at **www.cp7e.com** = biomedical application

Section 11.1 Heat and Internal Energy
Section 11.2 Specific Heat

1. Water at the top of Niagara Falls has a temperature of 10.0°C. If it falls a distance of 50.0 m and all of its potential energy goes into heating the water, calculate the temperature of the water at the bottom of the falls.
2. A 50.0-g piece of cadmium is at 20°C. If 400 cal of energy is transferred to the cadmium, what is its final temperature?
3. □ Lake Erie contains roughly 4.00×10^{11} m^3 of water. (a) How much energy is required to raise the temperature of that volume of water from 11.0°C to 12.0°C? (b) How many years would it take to supply this amount of energy by

using the 1 000-MW exhaust energy of an electric power plant?

4. An aluminum rod is 20.0 cm long at 20°C and has a mass of 350 g. If 10 000 J of energy is added to the rod by heat, what is the change in length of the rod?

5. How many joules of energy are required to raise the temperature of 100 g of gold from 20.0°C to 100°C?

6. As part of an exercise routine, a 50.0-kg person climbs 10.0 meters up a vertical rope. How many (food) Calories are expended in a single climb up the rope? (1 food Calorie = 10^3 calories)

7. A 75.0-kg weight watcher wishes to climb a mountain to work off the equivalent of a large piece of chocolate cake rated at 500 (food) Calories. How high must the person climb? (1 food Calorie = 10^3 calories)

8. The apparatus shown in Figure P11.8 was used by Joule to measure the mechanical equivalent of heat. Work is done on the water by a rotating paddle wheel, which is driven by two blocks falling at a constant speed. The temperature of the stirred water increases due to the friction between the water and the paddles. If the energy lost in the bearings and through the walls is neglected, then the loss in potential energy associated with the blocks equals the work done by the paddle wheel on the water. If each block has a mass of 1.50 kg and the insulated tank is filled with 200 g of water, what is the increase in temperature of the water after the blocks fall through a distance of 3.00 m?

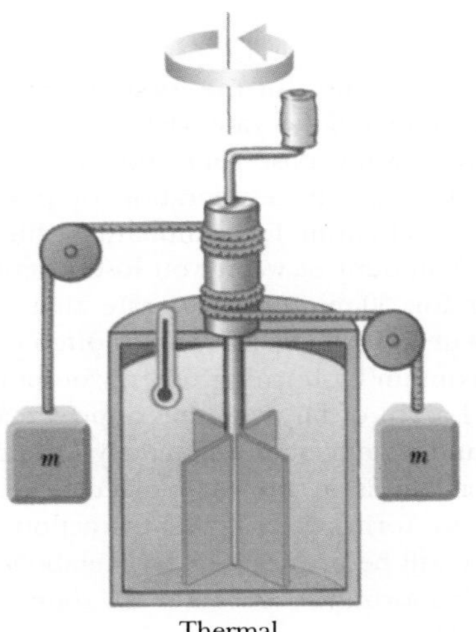

Figure P11.8 The falling weights rotate the paddles, causing the temperature of the water to increase.

9. A 5.00-g lead bullet traveling at 300 m/s is stopped by a large tree. If half the kinetic energy of the bullet is transformed into internal energy and remains with the bullet while the other half is transmitted to the tree, what is the increase in temperature of the bullet?

10. A 1.5-kg copper block is given an initial speed of 3.0 m/s on a rough horizontal surface. Because of friction, the block finally comes to rest. (a) If the block absorbs 85% of its initial kinetic energy as internal energy, calculate its increase in temperature. (b) What happens to the remaining energy?

11. A 200-g aluminum cup contains 800 g of water in thermal equilibrium with the cup at 80°C. The combination of cup and water is cooled uniformly so that the temperature decreases by 1.5°C per minute. At what rate is energy being removed? Express your answer in watts.

Section 11.3 Calorimetry

12. Lead pellets, each of mass 1.00 g, are heated to 200°C. How many pellets must be added to 500 g of water that is initially at 20.0°C to make the equilibrium temperature 25.0°C? Neglect any energy transfer to or from the container.

13. What mass of water at 25.0°C must be allowed to come to thermal equilibrium with a 3.00-kg gold bar at 100°C in order to lower the temperature of the bar to 50.0°C?

14. In a showdown on the streets of Laredo, the good guy drops a 5.0-g silver bullet at a temperature of 20°C into a 100-cm^3 cup of water at 90°C. Simultaneously, the bad guy drops a 5.0-g copper bullet at the same initial temperature into an identical cup of water. Which one ends the showdown with the coolest cup of water in the west? Neglect any energy transfer into or away from the container.

15. PhysicsNow™ An aluminum cup contains 225 g of water and a 40-g copper stirrer, all at 27°C. A 400-g sample of silver at an initial temperature of 87°C is placed in the water. The stirrer is used to stir the mixture until it reaches its final equilibrium temperature of 32°C. Calculate the mass of the aluminum cup.

16. It is desired to cool iron parts from 500°F to 100°F by dropping them into water that is initially at 75°F. Assuming that all the heat from the iron is transferred to the water and that none of the water evaporates, how many kilograms of water are needed per kilogram of iron?

17. A 100-g aluminum calorimeter contains 250 g of water. The two substances are in thermal equilibrium at 10°C. Two metallic blocks are placed in the water. One is a 50-g piece of copper at 80°C. The other sample has a mass of 70 g and is originally at a temperature of 100°C.

The entire system stabilizes at a final temperature of 20°C. Determine the specific heat of the unknown second sample.

18. When a driver brakes an automobile, the friction between the brake drums and the brake shoes converts the car's kinetic energy to thermal energy. If a 1 500-kg automobile traveling at 30 m/s comes to a halt, how much does the temperature rise in each of the four 8.0-kg iron brake drums? (The specific heat of iron is 448 J/kg · °C.)

19. A student drops two metallic objects into a 120-g steel container holding 150 g of water at 25°C. One object is a 200-g cube of copper that is initially at 85°C, and the other is a chunk of aluminum that is initially at 5.0°C. To the surprise of the student, the water reaches a final temperature of 25°C, precisely where it started. What is the mass of the aluminum chunk?

Section 11.4 Latent Heat and Phase Change

20. A 50-g ice cube at 0°C is heated until 45 g has become water at 100°C and 5.0 g has become steam at 100°C. How much energy was added to accomplish the transformation?

21. A 100-g cube of ice at 0°C is dropped into 1.0 kg of water that was originally at 80°C. What is the final temperature of the water after the ice has melted?

22. How much energy is required to change a 40-g ice cube from ice at −10°C to steam at 110°C?

23. What mass of steam that is initially at 120°C is needed to warm 350 g of water and its 300-g aluminum container from 20°C to 50°C?

24. A resting adult of average size converts chemical energy in food into internal energy at the rate of 120 W, called her *basal metabolic rate.* To stay at a constant temperature, energy must be transferred out of the body at the same rate. Several processes exhaust energy from your body. Usually the most important is thermal conduction into the air in contact with your exposed skin. If you are not wearing a hat, a convection current of warm air rises vertically from your head like a plume from a smokestack. Your body also loses energy by electromagnetic radiation, by your exhaling warm air, and by the evaporation of perspiration. Now consider still another pathway for energy loss: moisture in exhaled breath. Suppose you breathe out 22.0 breaths per minute, each with a volume of 0.600 L. Suppose also that you inhale dry air and exhale air at 37°C containing water vapor with a vapor pressure of 3.20 kPa. The vapor comes from the evaporation of liquid water in your body. Model the water vapor as an ideal gas. Assume its latent heat of evaporation at 37°C is the same as its heat of vaporization at 100°C. Calculate the rate at which you lose energy by exhaling humid air.

25. A 75-kg cross-country skier glides over snow as in Figure P11.25. The coefficient of friction between skis and snow is 0.20. Assume all the snow beneath his skis is at 0°C and that all the internal energy generated by friction is added to snow, which sticks to his skis until it melts. How far would he have to ski to melt 1.0 kg of snow?

Nathan Bilow/Leo de Wys, Inc.

Figure P11.25 A cross-country skier.

26. When you jog, most of the food energy you burn above your basal metabolic rate (BMR) ends up as internal energy that would raise your body temperature if it were not eliminated. The evaporation of perspiration is the primary mechanism for eliminating this energy. Determine the amount of water you lose to evaporation when running for 30 minutes at a rate that uses 400 kcal/h above your BMR. (That amount is often considered to be the "maximum fat-burning" energy output.) The metabolism of 1 gram of fat generates approximately 9.0 kcal of energy and produces approximately 1 gram of water. (The hydrogen atoms in the fat molecule are transferred to oxygen to form water.) What fraction of your need for water will be provided by fat metabolism? (The latent heat of vaporization of water at room temperature is 2.5×10^6 J/kg).

27. **PhysicsNow™** A 40-g block of ice is cooled to −78°C and is then added to 560 g of water in an 80-g copper calorimeter at a temperature of 25°C. Determine the final temperature of the system consisting of the ice, water, and calorimeter. (If not all the ice melts, determine how much ice is left.) Remember that the ice must first warm to 0°C, melt, and then continue warming as water. The specific heat of ice is 0.500 cal/g · °C = 2090 J/kg · °C.

28. A 60.0-kg runner expends 300 W of power while running a marathon. Assuming that 10.0% of the energy is delivered to the muscle tissue and that the excess energy is

removed from the body primarily by sweating, determine the volume of bodily fluid (assume it is water) lost per hour. (At 37.0°C, the latent heat of vaporization of water is 2.41×10^6 J/kg.)

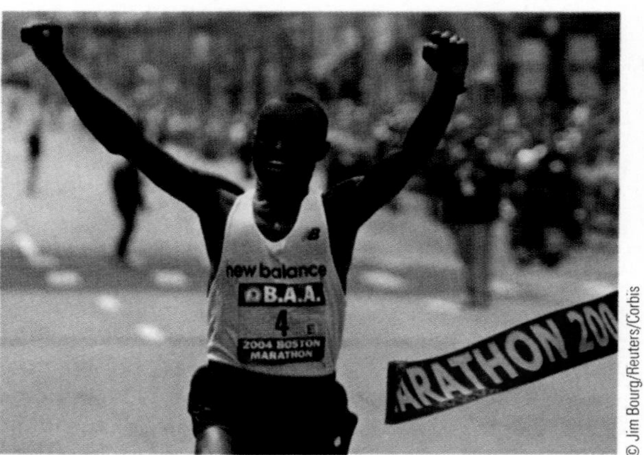

Figure P11.28 Timothy Cherigat of Kenya, winner of the Boston Marathon in 2004.

29. A high-end gas stove usually has at least one burner rated at 14 000 Btu/h. If you place a 0.25-kg aluminum pot containing 2.0 liters of water at 20°C on this burner, how long will it take to bring the water to a boil, assuming all of the heat from the burner goes into the pot? How long will it take to boil all of the water out of the pot?

30. A beaker of water sits in the sun until it reaches an equilibrium temperature of 30°C. The beaker is made of 100 g of aluminum and contains 180 g of water. In an attempt to cool this system, 100 g of ice at 0°C is added to the water. (a) Determine the final temperature of the system. If $T_f = 0$°C, determine how much ice remains. (b) Repeat your calculations for 50 g of ice.

31. Steam at 100°C is added to ice at 0°C. (a) Find the amount of ice melted and the final temperature when the mass of steam is 10 g and the mass of ice is 50 g. (b) Repeat with steam of mass 1.0 g and ice of mass 50 g.

Section 11.5 Energy Transfer

32. The average thermal conductivity of the walls (including windows) and roof of a house in Figure P11.32 is 4.8×10^{-4} kW/m · °C, and their average thickness is 21.0 cm. The house is heated with natural gas, with a heat of combustion (energy released per cubic meter of gas burned) of 9300 kcal/m^3. How many cubic meters of gas must be burned each day to maintain an inside temperature of 25.0°C if the outside temperature is 0.0°C? Disregard radiation and energy loss by heat through the ground.

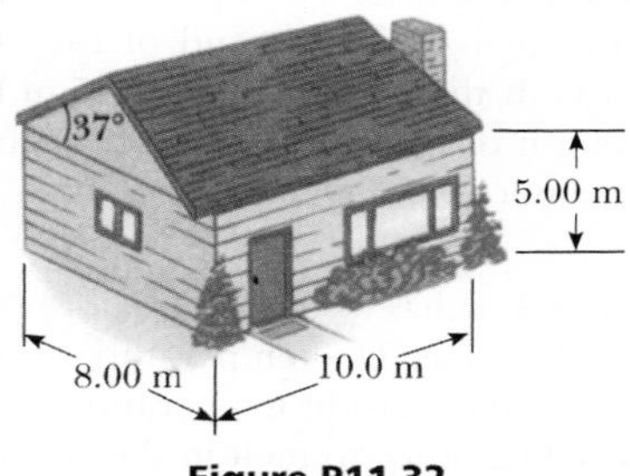

Figure P11.32

33. (a) Find the rate of energy flow through a copper block of cross-sectional area 15 cm^2 and length 8.0 cm when a temperature difference of 30°C is established across the block. Repeat the calculation, assuming that the material is (b) a block of stagnant air with the given dimensions; (c) a block of wood with the given dimensions.

34. A window has a glass surface area of 1.6×10^3 cm^2 and a thickness of 3.0 mm. (a) Find the rate of energy transfer by conduction through the window when the temperature of the inside surface of the glass is 70°F and the outside temperature is 90°F. (b) Repeat for the same inside temperature and an outside temperature of 0°F.

35. A steam pipe is covered with 1.50-cm-thick insulating material of thermal conductivity 0.200 cal/ cm · °C · s. How much energy is lost every second when the steam is at 200°C and the surrounding air is at 20.0°C? The pipe has a circumference of 800 cm and a length of 50.0 m. Neglect losses through the ends of the pipe.

36. A box with a total surface area of 1.20 m^2 and a wall thickness of 4.00 cm is made of an insulating material. A 10.0-W electric heater inside the box maintains the inside temperature at 15.0°C above the outside temperature. Find the thermal conductivity k of the insulating material.

37. Determine the R value for a wall constructed as follows: The outside of the house consists of lapped wood shingles placed over 0.50-in.-thick sheathing, over 3.0 in. of cellulose fiber, over 0.50 in. of drywall.

38. A thermopane window consists of two glass panes, each 0.50 cm thick, with a 1.0-cm-thick sealed layer of air in between. If the inside surface temperature is 23°C and the outside surface temperature is 0.0°C, determine the rate

of energy transfer through 1.0 m^2 of the window. Compare your answer with the rate of energy transfer through 1.0 m^2 of a single 1.0-cm-thick pane of glass.

39. A copper rod and an aluminum rod of equal diameter are joined end to end in good thermal contact. The temperature of the free end of the copper rod is held constant at 100°C, and that of the far end of the aluminum rod is held at 0°C. If the copper rod is 0.15 m long, what must be the length of the aluminum rod so that the temperature at the junction is 50°C?

40. A Styrofoam box has a surface area of 0.80 m^2 and a wall thickness of 2.0 cm. The temperature of the inner surface is 5.0°C, and the outside temperature is 25°C. If it takes 8.0 h for 5.0 kg of ice to melt in the container, determine the thermal conductivity of the Styrofoam.

41. PhysicsNow™ A sphere that is a perfect blackbody radiator has a radius of 0.060 m and is at 200°C in a room where the temperature is 22°C. Calculate the net rate at which the sphere radiates energy.

42. The surface temperature of the Sun is about 5 800 K. Taking the radius of the Sun to be 6.96×10^8 m, calculate the total energy radiated by the Sun each second. (Assume $e = 0.965$.)

43. A large, hot pizza 70 cm in diameter and 2.0 cm thick, at a temperature of 100°C, floats in outer space. Assume its emissivity is 0.8. What is the order of magnitude of its rate of energy loss?

44. Calculate the temperature at which a tungsten filament that has an emissivity of 0.90 and a surface area of 2.5×10^{-5} m^2 will radiate energy at the rate of 25 W in a room where the temperature is 22°C.

45. Measurements on two stars indicate that Star X has a surface temperature of 5 727°C and Star Y has a surface temperature of 11 727°C. If both stars have the same radius, what is the ratio of the luminosity (total power output) of Star Y to the luminosity of Star X? Both stars can be considered to have an emissivity of 1.0.

46. At high noon, the Sun delivers 1.00 kW to each square meter of a blacktop road. If the hot asphalt loses energy only by radiation, what is its equilibrium temperature?

ADDITIONAL PROBLEMS

47. The bottom of a copper kettle has a 10-cm radius and is 2.0 mm thick. The temperature of the outside surface is 102°C, and the water inside the kettle is boiling at 1 atm of pressure. Find the rate at which energy is being transferred through the bottom of the kettle.

48. A family comes home from a long vacation with laundry to do and showers to take. The water heater has been turned off during the vacation. If the heater has a capacity of 50.0 gallons and a 4 800-W heating element, how much time is required to raise the temperature of the water from 20.0°C to 60.0°C? Assume that the heater is well insulated and no water is withdrawn from the tank during that time.

49. Solar energy can be the primary source of winter space heating for a typical house (with floor area 130 m^2 = 1 400 ft^2) in the north central United States. If the house has very good insulation, you may model it as losing energy by heat steadily at the rate of 1 000 W during the winter, when the average exterior temperature is −5°C. The passive solar-energy collector can consist simply of large windows facing south. Sunlight shining in during the daytime is absorbed by the floor, interior walls, and objects in the house, raising their temperature to 30°C. As the sun goes down, insulating draperies or shutters are closed over the windows. During the period between 4 PM and 8 AM, the temperature of the house will drop, and a sufficiently large "thermal mass" is required to keep it from dropping too far. The thermal mass can be a large quantity of stone (with specific heat 800 J/kg · °C) in the floor and the interior walls exposed to sunlight. What mass of stone is required if the temperature is not to drop below 18°C overnight?

50. A water heater is operated by solar power. If the solar collector has an area of 6.00 m^2, and the intensity delivered by sunlight is 550 W/m^2, how long does it take to increase the temperature of 1.00 m^3 of water from 20.0°C to 60.0°C?

51. A 40-g ice cube floats in 200 g of water in a 100-g copper cup; all are at a temperature of 0°C. A piece of lead at 98°C is dropped into the cup, and the final equilibrium temperature is 12°C. What is the mass of the lead?

52. The evaporation of perspiration is the primary mechanism for cooling the human body. Estimate the amount of water you will lose when you bake in the sun on the beach for an hour. Use a value of 1 000 W/m^2 for the intensity of sunlight, and note that the energy required to evaporate a liquid at a particular temperature is approximately equal to the sum of the energy required to raise its temperature to the boiling point and the latent heat of vaporization (determined at the boiling point).

53. A 200-g block of copper at a temperature of 90°C is dropped into 400 g of water at 27°C. The water is contained in a 300-g glass container. What is the final temperature of the mixture?

54. A class of 10 students taking an exam has a power output per student of about 200 W. Assume that the initial temperature of the room is 20°C and that its dimensions are

6.0 m by 15.0 m by 3.0 m. What is the temperature of the room at the end of 1.0 h if all the energy remains in the air in the room and none is added by an outside source? The specific heat of air is 837 J/kg · °C, and its density is about 1.3×10^{-3} g/cm^3.

55. The human body must maintain its core temperature inside a rather narrow range around 37°C. Metabolic processes (notably, muscular exertion) convert chemical energy into internal energy deep in the interior. From the interior, energy must flow out to the skin or lungs, to be lost by heat to the environment. During moderate exercise, an 80-kg man can metabolize food energy at the rate of 300 kcal/h, do 60 kcal/h of mechanical work, and put out the remaining 240 kcal/h of energy by heat. Most of the energy is carried from the interior of the body out to the skin by "forced convection" (as a plumber would say): Blood is warmed in the interior and then cooled at the skin, which is a few degrees cooler than the body core. Without blood flow, living tissue is a good thermal insulator, with a thermal conductivity about 0.210 W/m · °C. Show that blood flow is essential to keeping the body cool by calculating the rate of energy conduction, in kcal/h, through the tissue layer under the skin. Assume that its area is 1.40 m^2, its thickness is 2.50 cm, and it is maintained at 37.0°C on one side and at 34.0°C on the other side.

56. **Physics Now™** An aluminum rod and an iron rod are joined end to end in good thermal contact. The two rods have equal lengths and radii. The free end of the aluminum rod is maintained at a temperature of 100°C, and the free end of the iron rod is maintained at 0°C. (a) Determine the temperature of the interface between the two rods. (b) If each rod is 15 cm long and each has a cross-sectional area of 5.0 cm^2, what quantity of energy is conducted across the combination in 30 min?

57. Water is being boiled in an open kettle that has a 0.500-cm-thick circular aluminum bottom with a radius of 12.0 cm. If the water boils away at a rate of 0.500 kg/min, what is the temperature of the lower surface of the bottom of the kettle? Assume that the top surface of the bottom of the kettle is at 100°C.

58. A 3.00-g copper penny at 25.0°C drops 50.0 m to the ground. (a) If 60.0% of the initial potential energy associated with the penny goes into increasing its internal energy, determine the final temperature of the penny. (b) Does the result depend on the mass of the coin? Explain.

59. A bar of gold (Au) is in thermal contact with a bar of silver (Ag) of the same length and area (Fig. P11.59). One end of the compound bar is maintained at 80.0°C, while the opposite end is at 30.0°C. Find the temperature at the junction when the energy flow reaches a steady state.

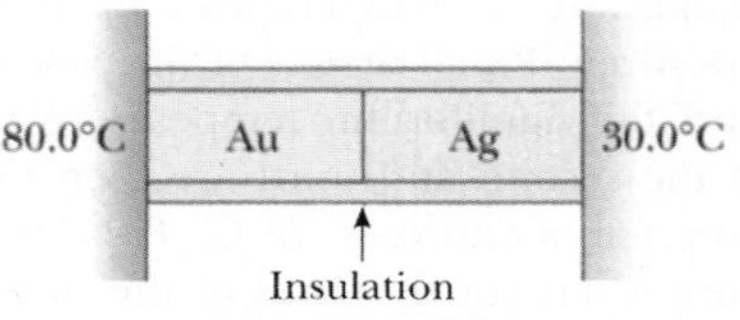

Figure P11.59

60. An iron plate is held against an iron wheel so that a sliding frictional force of 50 N acts between the two pieces of metal. The relative speed at which the two surfaces slide over each other is 40 m/s. (a) Calculate the rate at which mechanical energy is converted to internal energy. (b) The plate and the wheel have masses of 5.0 kg each, and each receives 50% of the internal energy. If the system is run as described for 10 s, and each object is then allowed to reach a uniform internal temperature, what is the resultant temperature increase?

61. An automobile has a mass of 1 500 kg, and its aluminum brakes have an overall mass of 6.0 kg. (a) Assuming that all of the internal energy transformed by friction when the car stops is deposited in the brakes, and neglecting energy transfer, how many times could the car be braked to rest starting from 25 m/s (56 mi/h) before the brakes would begin to melt? (Assume an initial temperature of 20°C.) (b) Identify some effects that are neglected in part (a), but are likely to be important in a more realistic assessment of the temperature increase of the brakes.

62. A 1.0-m-long aluminum rod of cross-sectional area 2.0 cm^2 is inserted vertically into a thermally insulated vessel containing liquid helium at 4.2 K. The rod is initially at 300 K. If half of the rod is inserted into the helium, how many liters of helium boil off in the very short time while the inserted half cools to 4.2 K? The density of liquid helium at 4.2 K is 122 kg/m^3.

63. A *flow calorimeter* is an apparatus used to measure the specific heat of a liquid. The technique is to measure the temperature difference between the input and output points of a flowing stream of the liquid while adding energy at a known rate. (a) Start with the equations $Q = mc(\Delta T)$ and $m = \rho V$, and show that the rate at which energy is added to the liquid is given by the expression $\Delta Q/\Delta t = \rho c(\Delta T)(\Delta V/\Delta t)$. (b) In a particular experiment, a liquid of density 0.72 g/cm^3 flows through the calorimeter at the rate of 3.5 cm^3/s. At steady state, a temperature difference of 5.8°C is established between the input and output points when energy is supplied at the rate of 40 J/s. What is the specific heat of the liquid?

64. Three liquids are at temperatures of 10°C, 20°C, and 30°C, respectively. Equal masses of the first two liquids are mixed, and the equilibrium temperature is 17°C. Equal masses of the second and third are then mixed, and the equilibrium temperature is 28°C. Find the equilibrium temperature when equal masses of the first and third are mixed.

65. At time $t = 0$, a vessel contains a mixture of 10 kg of water and an unknown mass of ice in equilibrium at 0°C. The temperature of the mixture is measured over a period of an hour, with the following results: During the first 50 min, the mixture remains at 0°C; from 50 min to 60 min, the temperature increases steadily from 0°C to 2°C. Neglecting the heat capacity of the vessel, determine the mass of ice that was initially placed in it. Assume a constant power input to the container.

66. A wood stove is used to heat a single room. The stove is cylindrical in shape, with a diameter of 40.0 cm and a length of 50.0 cm and operates at a temperature of 400°F. (a) If the temperature of the room is 70.0°F determine the amount of radiant energy delivered to the room by the stove each second if the emissivity is 0.920. (b) If the room is a square with walls that are 8.00 ft high and 25.0 ft wide, determine the R value needed in the walls and ceiling to maintain the inside temperature at 70.0°F if the outside temperature is 32.0°F. Note that we are ignoring any heat conveyed by the stove via convection and any energy lost through the walls (and windows!) via convection or radiation.

67. A "solar cooker" consists of a curved reflecting mirror that focuses sunlight onto the object to be heated (Fig. P11.67). The solar power per unit area reaching the Earth at the location of a 0.50-m-diameter solar cooker is 600 W/m^2. Assuming that 50% of the incident energy is converted to thermal energy, how long would it take to boil away 1.0 L of water initially at 20°C? (Neglect the specific heat of the container.)

Figure P11.67

68. For bacteriological testing of water supplies and in medical clinics, samples must routinely be incubated for 24 h at 37°C. A standard constant temperature bath with electric heating and thermostatic control is not suitable in developing nations without continuously operating electric power lines. Peace Corps volunteer and MIT engineer Amy Smith invented a low cost, low maintenance incubator to fill the need. The device consists of a foam-insulated box containing several packets of a waxy material that melts at 37.0°C, interspersed among tubes, dishes, or bottles containing the test samples and growth medium (food for bacteria). Outside the box, the waxy material is first melted by a stove or solar energy collector. Then it is put into the box to keep the test samples warm as it solidifies. The heat of fusion of the phase-change material is 205 kJ/kg. Model the insulation as a panel with surface area 0.490 m^2, thickness 9.50 cm, and conductivity 0.012 0 W/m°C. Assume the exterior temperature is 23.0°C for 12.0 h and 16.0°C for 12.0 h. (a) What mass of the waxy material is required to conduct the bacteriological test? (b) Explain why your calculation can be done without knowing the mass of the test samples or of the insulation.

69. What mass of steam initially at 130°C is needed to warm 200 g of water in a 100-g glass container from 20.0°C to 50.0°C?

ACTIVITIES

1. A plot of the decreasing temperature of a substance over time is called a cooling curve and has the same shape and basic explanation as the curve shown in Figure 11.3. You can plot such a curve by observing some water in a container in the freezer compartment of a refrigerator. Place a thermometer in the liquid, and record the reading of the thermometer every minute until about five minutes after the liquid has frozen completely. Explain your observations. A material that is a little easier to work with is naphthalene (mothballs). You can plot the cooling curve in this case without a freezer. Melt a small amount of naphthalene in a container, and plot a graph of temperature versus time as before. Again, explain your observations.

2. You have probably heard someone say that hot water freezes faster than cold water. Is this an urban legend or is it true? To test this hypothesis, fill one container with hot water, at about 200°F, and another with cooler water, at about 70°F. Place the two containers in the freezer compartment of a refrigerator, and find out for yourself. There are a number of variables that you need to attempt to control in such an experiment: (1) The two containers need to be placed at similar locations in the freezer compartment. That is, one should not be near the door while the other is in the back of the compartment. (2) The two containers should not be placed close together, or an un-

wanted exchange of energy will take place between them. (3) If the freezer is one of the old types that forms frost on its walls, the hot container should not be allowed to melt through the frost and make intimate contact with the cold walls of the freezer. Can you list any more variables that you need to control?

3. You may have heard that you can greatly reduce the baking time for potatoes in a *conventional* oven by inserting a nail through each potato. Are there any scientific reasons for believing that this hypothesis is true? Test it with a couple of similar-sized potatoes—but don't bake them in a microwave oven!

Lance Armstrong is an engine: he requires fuel and oxygen to burn it, and the result is work that drives him up the mountainside as his excess, waste energy is expelled in his evaporating sweat.

© Tim De Waele/Corbis

CHAPTER

12

OUTLINE

The Laws of Thermodynamics

According to the first law of thermodynamics, the internal energy of a system can be increased either by adding energy to the system or by doing work on it. This means the internal energy of a system, which is just the sum of the molecular kinetic and potential energies, can change as a result of two separate types of energy transfer across the boundary of the system. Although the first law imposes conservation of energy for both energy added by heat and work done on a system, it doesn't predict which of several possible energy-conserving processes actually occur in nature.

The second law of thermodynamics constrains the first law by establishing which processes allowed by the first law actually occur. For example, the second law tells us that energy never flows by heat spontaneously from a cold object to a hot object. One important application of this law is in the study of heat engines (such as the internal combustion engine) and the principles that limit their efficiency.

12.1 WORK IN THERMODYNAMIC PROCESSES

Energy can be transferred to a system by heat and by work done on the system. In most cases of interest treated here, the system is a volume of gas, which is important in understanding engines. All such systems of gas will be assumed to be in thermodynamic equilibrium, so that every part of the gas is at the same temperature and pressure. If that were not the case, the ideal gas law wouldn't apply and most of the results presented here wouldn't be valid. Consider a gas contained by a cylinder fitted with a movable piston (Active Fig. 12.1a) and in equilibrium. The gas occupies a volume V and exerts a uniform pressure P on the cylinder walls and the piston. The gas is compressed slowly enough so the system remains essentially in thermodynamic equilibrium at all times. As the piston is pushed downward by an external force F through a distance Δy, the work done on the gas is

$$W = -F\Delta y = -PA\,\Delta y$$

where we have set the magnitude F of the external force equal to PA, possible because the pressure is the same everywhere in the system (by the assumption of equilibrium). Note that if the piston is pushed downward, $\Delta y = y_f - y_i$ is negative, so we need an explicit negative sign in the expression for W to make the work positive. The change in volume of the gas is $\Delta V = A\,\Delta y$, which leads to the following definition:

The **work W done on a gas** at constant pressure is given by

$$W = -P\Delta V \qquad [12.1]$$

where P is the pressure throughout the gas and ΔV is the change in volume of the gas during the process.

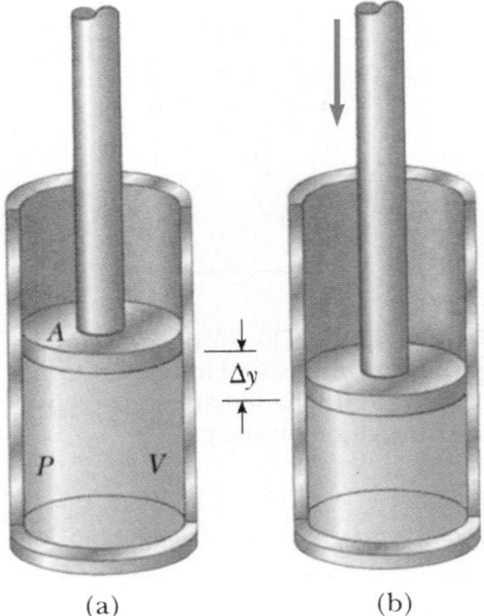

ACTIVE FIGURE 12.1
(a) A gas in a cylinder occupying a volume V at a pressure P. (b) Pushing the piston down compresses the gas.

Physics Now™
Log into PhysicsNow at **www.cp7e.com**, and go to Active Figure 12.1 to move the piston and see the resulting work done on the gas.

If the gas is compressed as in Active Figure 12.1b, ΔV is negative and the work done on the gas is positive. If the gas expands, ΔV is positive and the work done on the gas is negative. The work done by the gas on its environment, W_{env}, is simply the negative of the work done on the gas. In the absence of a change in volume, the work is zero.

EXAMPLE 12.1 Work Done by an Expanding Gas

Goal Apply the definition of work at constant pressure.

Problem In a system similar to that shown in Active Figure 12.1, the gas in the cylinder is at a pressure of 1.01×10^5 Pa and the piston has an area of $0.100\ \text{m}^2$. As energy is slowly added to the gas by heat, the piston is pushed up a distance of 4.00 cm. Calculate the work done by the expanding gas on the surroundings, W_{env}, assuming the pressure remains constant.

Strategy The work done on the environment is the negative of the work done on the gas given in Equation 12.1. Compute the change in volume and multiply by the pressure.

Solution

Find the change in volume of the gas, ΔV, which is the cross-sectional area times the displacement:

$$\Delta V = A\,\Delta y = (0.100\ \text{m}^2)(4.00 \times 10^{-2}\ \text{m}) = 4.00 \times 10^{-3}\ \text{m}^3$$

Multiply this result by the pressure, getting the work the gas does on the environment, W_{env}:

$$W_{env} = P\Delta V = (1.01 \times 10^5\ \text{Pa})(4.00 \times 10^{-3}\ \text{m}^3) = 404\ \text{J}$$

Remark The volume of the gas increases, so the work done on the environment is positive. The work done on the system during this process is $W = -404$ J. The energy required to perform positive work on the environment must come from the energy of the gas. (See the next section for more details.)

Exercise 12.1
Gas in a cylinder similar to Figure 12.1 moves a piston with area $0.20\ \text{m}^2$ as energy is slowly added to the system. If 2.00×10^3 J of work is done on the environment and the pressure of the gas in the cylinder remains constant at 1.01×10^5 Pa, find the displacement of the piston.

Answer 9.90×10^{-2} m

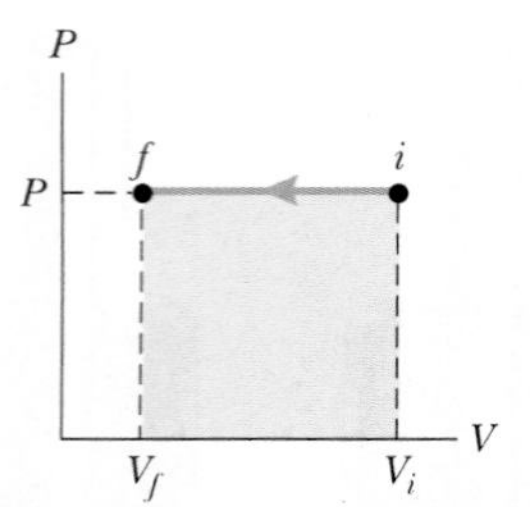

Figure 12.2 The *PV* diagram for a gas being compressed at constant pressure. The shaded area represents the work done on the gas.

Equation 12.1 can be used to calculate the work done on the system *only* when the pressure of the gas remains constant during the expansion or compression. A process in which the pressure remains constant is called an **isobaric process.** The pressure vs. volume graph, or ***PV* diagram**, of an isobaric process is shown in Figure 12.2. The curve on such a graph is called the *path* taken between the initial and final states, with the arrow indicating the direction the process is going, in this case from smaller to larger volume. The area under the graph is

$$\text{Area} = P(V_f - V_i) = P\Delta V$$

The area under the graph in a *PV* diagram is equal in magnitude to the work done on the gas.

This is true in general, whether or not the process proceeds at constant pressure. Just draw the *PV* diagram of the process, find the area underneath the graph (and above the horizontal axis), and that area will be the equal to the magnitude of the work done on the gas. If the arrow on the graph points toward larger volumes, the work done on the gas is negative. If the arrow on the graph points toward smaller volumes, the work done on the gas is positive.

Whenever negative work is done on a system, positive work is done by the system on its environment. The negative work done on the system represents a loss of energy from the system—the cost of doing positive work on the environment.

Quick Quiz 12.1

By visual inspection, order the *PV* diagrams shown in Figure 12.3 from the most negative work done on the system to the most positive work done on the system. (a) a,b,c,d (b) a,c,b,d (c) d,b,c,a (d) d,a,c,b

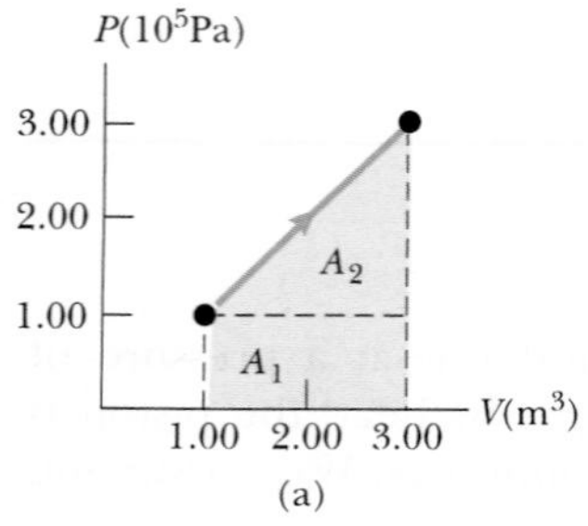

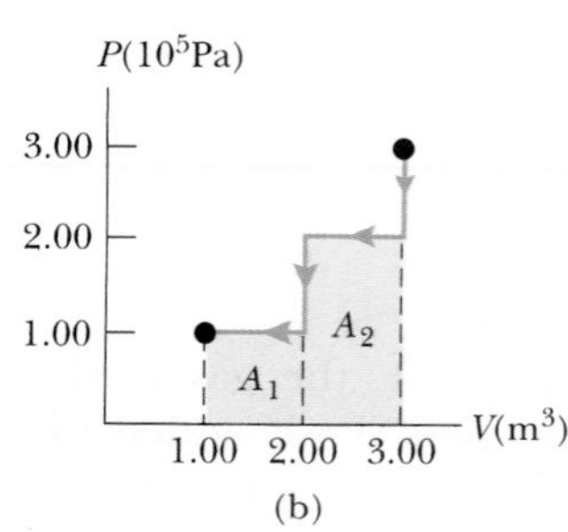

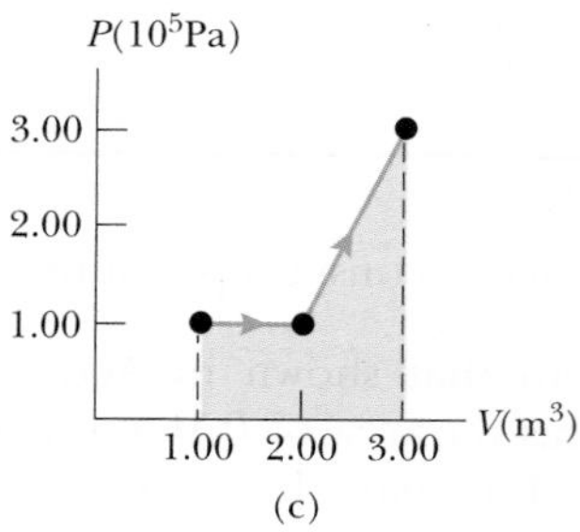

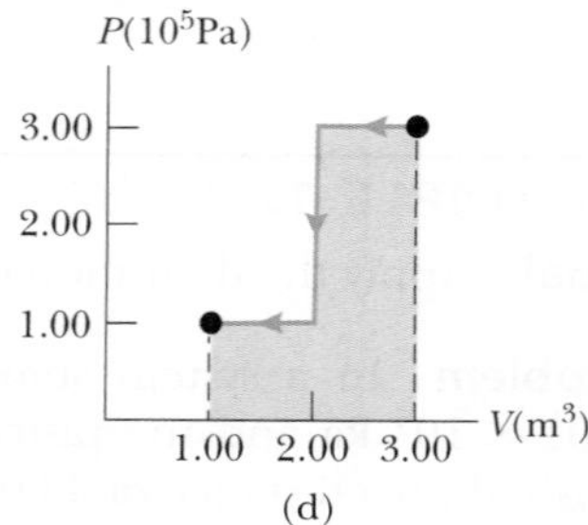

Figure 12.3 (Quick Quiz 12.1 and Example 12.2)

Notice that the graphs in Figure 12.3 all have the same endpoints, but the areas beneath the curves are different. The work done on a system depends on the path taken in the *PV* diagram.

EXAMPLE 12.2 Work and *PV* Diagrams

Goal Calculate work from a *PV* diagram.

Problem Find the numeric value of the work done on the gas in **(a)** Figure 12.3a and **(b)** Figure 12.3b.

Strategy The regions in question are composed of rectangles and triangles. Use basic geometric formulas to find the area underneath each curve. Check the direction of the arrow to determine signs.

Solution

(a) Find the work done on the gas in Figure 12.3a.

Compute the areas A_1 and A_2 in Figure 12.3a. A_1 is a rectangle and A_2 is a triangle.

$$A_1 = \text{height} \times \text{width} = (1.00 \times 10^5 \text{ Pa})(2.00 \text{ m}^3) = 2.00 \times 10^5 \text{ J}$$

$$A_2 = \text{one-half base} \times \text{height} = \tfrac{1}{2}(2.00 \text{ m}^3)(2.00 \times 10^5 \text{ Pa}) = 2.00 \times 10^5 \text{ J}$$

Sum the areas (the arrows point to increasing volume, so the work done on the gas is negative):

$$\text{Area} = A_1 + A_2 = 4.00 \times 10^5\text{ J} \quad \rightarrow$$
$$W = -4.00 \times 10^5\text{ J}$$

(b) Find the work done on the gas in Figure 12.3b.

Compute the areas of the two rectangular regions:

$$A_1 = \text{height} \times \text{width} = (1.00 \times 10^5\text{ Pa})(1.00\text{ m}^3) = 1.00 \times 10^5\text{ J}$$
$$A_2 = \text{height} \times \text{width} = (2.00 \times 10^5\text{ Pa})(1.00\text{ m}^3) = 2.00 \times 10^5\text{ J}$$

Sum the areas (the arrows point to decreasing volume, so the work done on the gas is positive):

$$\text{Area} = A_1 + A_2 = 3.00 \times 10^5\text{ J} \quad \rightarrow$$
$$W = +3.00 \times 10^5\text{ J}$$

Remarks Notice that in both cases the paths in the *PV* diagrams start and end at the same points, but the answers are different.

Exercise 12.2
Compute the work done on the system in Figures 12.3c and 12.3d.

Answers -3.00×10^5 J, $+4.00 \times 10^5$ J

12.2 THE FIRST LAW OF THERMODYNAMICS

The **first law of thermodynamics** is another energy conservation law that relates changes in internal energy—the energy associated with the position and jiggling of all the molecules of a system—to energy transfers due to heat and work. The first law is universally valid, applicable to all kinds of processes, providing a connection between the microscopic and macroscopic worlds.

There are two ways energy can be transferred between a system and its surroundings: by doing work, which requires a macroscopic displacement of an object through the application of a force; and by heat, which occurs through random molecular collisions. Both mechanisms result in a *change in internal energy*, ΔU, of the system and therefore in measurable changes in the macroscopic variables of the system, such as the pressure, temperature, and volume. This change in the internal energy can be summarized in the **first law of thermodynamics**:

◀ First law of thermodynamics

If a system undergoes a change from an initial state to a final state, where Q is the energy transferred to the system by heat and W is the work done on the system, the change in the internal energy of the system, ΔU, is given by

$$\Delta U = U_f - U_i = Q + W \qquad [12.2]$$

The quantity Q is positive when energy is transferred into the system by heat and negative when energy is transferred out of the system by heat. The quantity W is positive when work is done on the system and negative when the system does work on its environment. All quantities in the first law, Equation 12.2, must have the same energy units. Any change in the internal energy of a system—the positions and vibrations of the molecules—is due to the transfer of energy by heat or work (or both).

From Equation 12.2, we also see that the internal energy of any isolated system must remain constant, so that $\Delta U = 0$. Even when a system isn't isolated, the change in internal energy will be zero if the system goes through a cyclic process in which all the thermodynamic variables—pressure, volume, temperature, and moles of gas—return to their original values.

It's important to remember that the quantities in Equation 12.2 concern a *system*, not the effect on the system's environment through work. If the system is hot

TIP 12.1 Dual Sign Conventions

Many physics and engineering textbooks present the first law as $\Delta U = Q - W$, with a minus sign between the heat and the work. The reason is that work is defined in these treatments as the work done *by* the gas rather than *on* the gas, as in our treatment. This form of the first law represents the original interest in applying it to steam engines, where the primary concern is the work extracted from the engine.

steam expanding against a piston, for example, the system work W is *negative*, because the piston can only expand at the expense of the internal energy of the gas. The work W_{env} done by the hot steam on the *environment*—in this case, moving a piston which moves the train—is positive, but that's not the work W in Equation 12.2. This way of defining work in the first law makes it consistent with the concept of work defined in Chapter 5. There, positive work done on a system (for example, a block) increased its mechanical energy, while negative work decreased its energy. In this chapter, positive work done on a system (typically, a volume of gas) increases its internal energy, and negative work decreases that internal energy. In both the mechanical and thermal cases, the effect on the system is the same: positive work increases the system's energy, and negative work decreases the system's energy.

Some textbooks identify W as the work done by the gas on its environment. This is an equivalent formulation, but it means that W must carry a minus sign in the first law. That convention isn't consistent with previous discussions of the energy of a system, because when W is positive the system *loses* energy, whereas in Chapter 5 positive W means the system *gains* energy. For that reason, the old convention is not used in this book.

EXAMPLE 12.3 Heating a Gas

Goal Combine the first law of thermodynamics with work done during a constant pressure process.

Problem An ideal gas absorbs 5.00×10^3 J of energy while doing 2.00×10^3 J of work on the environment during a constant pressure process. **(a)** Compute the change in the internal energy of the gas. **(b)** If the internal energy now drops by 4.50×10^3 J and 7.50×10^3 J is expelled from the system, find the change in volume, assuming a constant pressure process at 1.01×10^5 Pa.

Strategy Part (a) requires substitution of the given information into the first law, Equation 12.2. Notice, however, that the given work is done on the *environment*. The negative of this amount is the work done on the *system*, representing a loss of internal energy. Part (b) is a matter of substituting the equation for work at constant pressure into the first law and solving for the change in volume.

Solution

(a) Compute the change in internal energy.

Substitute values into the first law, noting that the work done on the gas is negative:

$$\Delta U = Q + W = 5.00 \times 10^3\ \text{J} - 2.00 \times 10^3\ \text{J}$$
$$= \boxed{3.00 \times 10^3\ \text{J}}$$

(b) Find the change in volume, noting that ΔU and Q are both negative in this case.

Substitute the equation for work done at constant pressure into the first law:

$$\Delta U = Q + W = Q - P\Delta V$$
$$-4.50 \times 10^3\ \text{J} = -7.50 \times 10^3\ \text{J} - (1.01 \times 10^5\ \text{J})\Delta V$$

Solve for the change in volume, ΔV:

$$\Delta V = \boxed{-2.97 \times 10^{-2}\ \text{m}^3}$$

Remarks The change in volume is negative, so the system contracts, doing negative work on the environment, while the work W on the system is positive.

Exercise 12.3

Suppose the internal energy of an ideal gas rises by 3.00×10^3 J at a constant pressure of 1.00×10^5 Pa, while the system gains 4.20×10^3 J of energy by heat. Find the change in volume of the system.

Answer $1.20 \times 10^{-2}\ \text{m}^3$

Recall that an expression for the internal energy of an ideal gas is

$$U = \frac{3}{2}nRT \quad [12.3a]$$

This expression is valid only for a *monatomic* ideal gas, which means the particles of the gas consist of single atoms. The change in the internal energy, ΔU, for such a gas is given by

$$\Delta U = \frac{3}{2}nR\Delta T \quad [12.3b]$$

The **molar specific heat at constant volume** of a monatomic ideal gas, C_v, is defined by

$$C_v \equiv \frac{3}{2}R \quad [12.4]$$

The change in internal energy of an ideal gas can then be written

$$\Delta U = nC_v\Delta T \quad [12.5]$$

For ideal gases, this expression is always valid, even when the volume isn't constant. The value of the molar specific heat, however, depends on the gas and can vary under different conditions of temperature and pressure.

A gas with a larger molar specific heat requires more energy to realize a given temperature change. The size of the molar specific heat depends on the structure of the gas molecule and how many different ways it can store energy. A monatomic gas such as helium can store energy as motion in three different directions. A gas such as hydrogen, on the other hand, is diatomic in normal temperature ranges, and aside from moving in three directions, it can also tumble, rotating in two different directions. So hydrogen molecules can store energy in the form of translational motion, and in addition can store energy through tumbling. Further, molecules can also store energy in the vibrations of their constituent atoms. A gas composed of molecules with more ways to store energy will have a larger molar specific heat.

Each different way a gas molecule can store energy is called a *degree of freedom.* Each degree of freedom contributes $\frac{1}{2}R$ to the molar specific heat. Because an atomic ideal gas can move in three directions, it has a molar specific heat capacity $C_v = 3(\frac{1}{2}R) = \frac{3}{2}R$. A diatomic gas like molecular oxygen, O_2, can also tumble in two different directions. This adds $2 \times \frac{1}{2}R = R$ to the molar heat specific heat, so $C_v = \frac{5}{2}R$ for diatomic gases. The spinning about the long axis connecting the two atoms is generally negligible. Vibration of the atoms in a molecule can also contribute to the heat capacity. A full analysis of a given system is often complex, so in general, molar specific heats must be determined by experiment. Some representative values of C_v can be found in Table 12.1 (page 392).

There are four basic types of thermal processes, which will be studied and illustrated by their effect on an ideal gas.

Isobaric Processes

Recall from Section 12.1 that in an isobaric process the pressure remains constant as the gas expands or is compressed. An expanding gas does work on its environment, given by $W_{env} = P\Delta V$. The PV diagram of an isobaric expansion is given in Figure 12.2. As previously discussed, the magnitude of the work done on the gas is just the area under the path in its PV diagram: height times length, or $P\Delta V$. The negative of this quantity, $W = -P\Delta V$, is the energy lost by the gas because the gas does work as it expands. This is the quantity that should be substituted into the first law.

The work done by the gas on its environment must come at the expense of the change in its internal energy, ΔU. Because the change in the internal energy of an ideal gas is given by $\Delta U = nC_v\Delta T$, the temperature of an expanding gas must decrease as the internal energy decreases. Expanding volume and decreasing temperature means the pressure must also decrease, in conformity with the ideal gas law, $PV = nRT$. Consequently, the only way such a process can remain at constant

TABLE 12.1

Molar Specific Heats of Various Gases

Gas	Molar Specific Heat (J/mol · K)[a]			
	C_P	C_V	$C_P - C_V$	$\gamma = C_P/C_V$
Monatomic Gases				
He	20.8	12.5	8.33	1.67
Ar	20.8	12.5	8.33	1.67
Ne	20.8	12.7	8.12	1.64
Kr	20.8	12.3	8.49	1.69
Diatomic Gases				
H_2	28.8	20.4	8.33	1.41
N_2	29.1	20.8	8.33	1.40
O_2	29.4	21.1	8.33	1.40
CO	29.3	21.0	8.33	1.40
Cl_2	34.7	25.7	8.96	1.35
Polyatomic Gases				
CO_2	37.0	28.5	8.50	1.30
SO_2	40.4	31.4	9.00	1.29
H_2O	35.4	27.0	8.37	1.30
CH_4	35.5	27.1	8.41	1.31

[a]All values except that for water were obtained at 300 K.

pressure is if thermal energy Q is transferred into the gas by heat. Rearranging the first law, we obtain

$$Q = \Delta U - W = \Delta U + P\Delta V$$

Now we can substitute the expression in Equation 12.3b for ΔU and use the ideal gas law to substitute $P\Delta V = nR\Delta T$:

$$Q = \tfrac{3}{2}nR\Delta T + nR\Delta T = \tfrac{5}{2}nR\Delta T$$

Another way to express this transfer by heat is

$$Q = nC_p\Delta T \qquad [12.6]$$

where $C_p = \frac{5}{2}R$. For ideal gases, the molar heat capacity at constant pressure, C_p, is the sum of the molar heat capacity at constant volume, C_v, and the gas constant R:

$$C_p = C_v + R \qquad [12.7]$$

This can be seen in the fourth column of Table 12.1, where $C_p - C_v$ is calculated for a number of different gases. The difference works out to be approximately R in virtually every case.

EXAMPLE 12.4 Expanding Gas

Goal Use molar specific heats and the first law in a constant pressure process.

Problem Suppose a system of monatomic ideal gas at 2.00×10^5 Pa and an initial temperature of 293 K slowly expands at constant pressure from a volume of 1.00 L to 2.50 L. **(a)** Find the work done on the environment. **(b)** Find the change in internal energy of the gas. **(c)** Use the first law of thermodynamics to obtain the thermal energy absorbed by the gas during the process. **(d)** Use the molar heat capacity at constant pressure to find the thermal energy absorbed. **(e)** How would the answers change for a diatomic ideal gas?

Strategy This problem mainly involves substituting into the appropriate equations. Substitute into the equation for work at constant pressure to obtain the answer to part (a). In part (b), use the ideal gas law twice, to find the temperature when $V = 2.00$ L and to find the number of moles of the gas. These quantities can then be used to obtain the change in internal energy, ΔU. Part (c) can then be solved by substituting into the first law, yielding Q, the answer checked in part (d) with Equation 12.6. Repeat these steps for part (e) after increasing the molar specific heats by R because of the extra two degrees of freedom associated with a diatomic gas.

Solution

(a) Find the work done on the environment.

Apply the definition of work at constant pressure:

$$W_{\text{env}} = P\Delta V = (2.00 \times 10^5 \text{ Pa})(2.50 \times 10^{-3} \text{ m}^3 - 1.00 \times 10^{-3} \text{ m}^3)$$

$$W_{\text{env}} = \boxed{3.00 \times 10^2 \text{ J}}$$

(b) Find the change in the internal energy of the gas.

First, obtain the final temperature, using the ideal gas law, noting that $P_i = P_f$:

$$\frac{P_f V_f}{P_i V_i} = \frac{T_f}{T_i} \quad \rightarrow \quad T_f = T_i \frac{V_f}{V_i} = (293 \text{ K}) \frac{(2.50 \times 10^{-3} \text{ m}^3)}{(1.00 \times 10^{-3} \text{ m}^3)}$$

$$T_f = 733 \text{ K}$$

Again using the ideal gas law, obtain the number of moles of gas:

$$n = \frac{P_i V_i}{R T_i} = \frac{(2.00 \times 10^5 \text{ Pa})(1.00 \times 10^{-3} \text{ m}^3)}{(8.31 \text{ J/K} \cdot \text{mol})(293 \text{ K})}$$

$$= 8.21 \times 10^{-2} \text{ mol}$$

Use these results and given quantities to calculate the change in internal energy, ΔU:

$$\Delta U = nC_v\Delta T = \tfrac{3}{2}nR\Delta T$$

$$= \tfrac{3}{2}(8.21 \times 10^{-2} \text{ mol})(8.31 \text{ J/K} \cdot \text{mol})(733 \text{ K} - 293 \text{ K})$$

$$\Delta U = \boxed{4.50 \times 10^2 \text{ J}}$$

(c) Use the first law to obtain the energy transferred by heat.

Solve the first law for Q, and substitute ΔU and $W = -W_{\text{env}} = -3.00 \times 10^2$ J:

$$\Delta U = Q + W \quad \rightarrow \quad Q = \Delta U - W$$

$$Q = 4.50 \times 10^2 \text{ J} - (-3.00 \times 10^2 \text{ J}) = \boxed{7.50 \times 10^2 \text{ J}}$$

(d) Use the molar heat capacity at constant pressure to obtain Q:

Substitute values into Equation 12.6:

$$Q = nC_p\Delta T = \tfrac{5}{2}nR\Delta T$$

$$= \tfrac{5}{2}(8.21 \times 10^{-2} \text{ mol})(8.31 \text{ J/K} \cdot \text{mol})(733 \text{ K} - 293 \text{ K})$$

$$= \boxed{7.50 \times 10^2 \text{ J}}$$

(e) How would the answers change for a diatomic gas?

Obtain the new change in internal energy, ΔU, noting that $C_v = \frac{5}{2}R$ for a diatomic gas:

$$\Delta U = nC_v\Delta T = (\tfrac{3}{2} + 1)nR\Delta T$$

$$= \tfrac{5}{2}(8.21 \times 10^{-2} \text{ mol})(8.31 \text{ J/K} \cdot \text{mol})(733 \text{ K} - 293 \text{ K})$$

$$\Delta U = \boxed{7.50 \times 10^2 \text{ J}}$$

Obtain the new energy transferred by heat, Q:

$$Q = nC_p\Delta T = (\tfrac{5}{2} + 1)nR\Delta T$$

$$= \tfrac{7}{2}(8.21 \times 10^{-2} \text{ mol})(8.31 \text{ J/K} \cdot \text{mol})(733 \text{ K} - 293 \text{ K})$$

$$Q = \boxed{1.05 \times 10^3 \text{ J}}$$

Remarks Notice that problems involving diatomic gases are no harder than those with monatomic gases. It's just a matter of adjusting the molar specific heats.

Exercise 12.4

Suppose an ideal monatomic gas at an initial temperature of 475 K is compressed from 3.00 L to 2.00 L while its pressure remains constant at 1.00×10^5 Pa. Find (a) the work done on the gas, (b) the change in internal energy, and (c) the energy transferred by heat, Q.

Answers (a) 1.00×10^2 J (b) -150 J (c) -250 J

Adiabatic Processes

In an adiabatic process, no energy enters or leaves the system by heat. Such a system is insulated—thermally isolated from its environment. In general, however, the system isn't mechanically isolated, so it can still do work. A sufficiently rapid process may be considered approximately adiabatic because there isn't time for any significant transfer of energy by heat.

For adiabatic processes $Q = 0$, so the first law becomes

$$\Delta U = W$$

The work done during an adiabatic process can be calculated by finding the change in the internal energy. Alternately, the work can be computed from a PV diagram. For an ideal gas undergoing an adiabatic process, it can be shown that

$$PV^{\gamma} = \text{constant} \qquad \textbf{[12.8a]}$$

where

$$\gamma = \frac{C_p}{C_v} \qquad \textbf{[12.8b]}$$

is called the *adiabatic index* of the gas. Values of the adiabatic index for several different gases are given in Table 12.1. After computing the constant on the right-hand side of Equation 12.8a and solving for the pressure P, the area under the curve in the PV diagram can be found by counting boxes, yielding the work.

If a hot gas is allowed to expand so quickly that there is no time for energy to enter or leave the system by heat, the work done on the gas is negative and the internal energy decreases. This decrease occurs because kinetic energy is transferred from the gas molecules to the moving piston. Such an adiabatic expansion is of practical importance and is nearly realized in an internal combustion engine when a gasoline–air mixture is ignited and expands rapidly against a piston. The following example illustrates this process.

EXAMPLE 12.5 Work and an Engine Cylinder

Goal Use the first law to find the work done in an adiabatic expansion.

Problem In a car engine operating at 1.80×10^3 rev/min, the expansion of hot, high-pressure gas against a piston occurs in about 10 ms. Because energy transfer by heat typically takes a time on the order of minutes or hours, it's safe to assume that little energy leaves the hot gas during the expansion. Estimate the work done by the gas on the piston during this adiabatic expansion by assuming the engine cylinder contains 0.100 moles of an ideal monatomic gas which goes from 1.200×10^3 K to 4.00×10^2 K, typical engine temperatures, during the expansion.

Strategy Find the change in internal energy using the given temperatures. For an adiabatic process, this equals the work done on the gas, which is the negative of the work done on the environment—in this case, the piston.

Solution

Start with the first law, taking $Q = 0$.

$$W = \Delta U - Q = \Delta U - 0 = \Delta U$$

Find ΔU from the expression for the internal energy of an ideal monatomic gas.

$$\Delta U = U_f - U_i = \tfrac{3}{2} nR(T_f - T_i)$$
$$= \tfrac{3}{2}(0.100 \text{ mol})(8.31 \text{ J/mol}\cdot\text{K})(4.00 \times 10^2 \text{ K} - 1.20 \times 10^3 \text{ K})$$
$$\Delta U = -9.97 \times 10^2 \text{ J}$$

The change in internal energy equals the work done on the system, which is the negative of the work done on the piston.

$$W_{\text{piston}} = -W = -\Delta U = 9.97 \times 10^2 \text{ J}$$

Remarks The work done on the piston comes at the expense of the internal energy of the gas. In an ideal adiabatic expansion, the loss of internal energy is completely converted into useful work. In a real engine, there are always losses.

Exercise 12.5

A monatomic ideal gas with volume 0.200 L is rapidly compressed, so the process can be considered adiabatic. If the gas is initially at 1.01×10^5 Pa and 3.00×10^2 K and the final temperature is 477 K, find the work done by the gas on the environment, W_{env}.

Answer -17.9 J

EXAMPLE 12.6 An Adiabatic Expansion

Goal Use the adiabatic pressure vs. volume relation to find a change in pressure and the work done on a gas.

Problem A monatomic ideal gas at a pressure 1.01×10^5 Pa expands adiabatically from an initial volume of $1.50\ \text{m}^3$, doubling its volume. **(a)** Find the new pressure. **(b)** Sketch the *PV* diagram and estimate the work done on the gas.

Strategy There isn't enough information to solve this problem with the ideal gas law. Instead, use Equation 12.8 and the given information to find the adiabatic index and the constant *C* for the process. For part (b), sketch the *PV* diagram and count boxes to estimate the area under the graph, which gives the work.

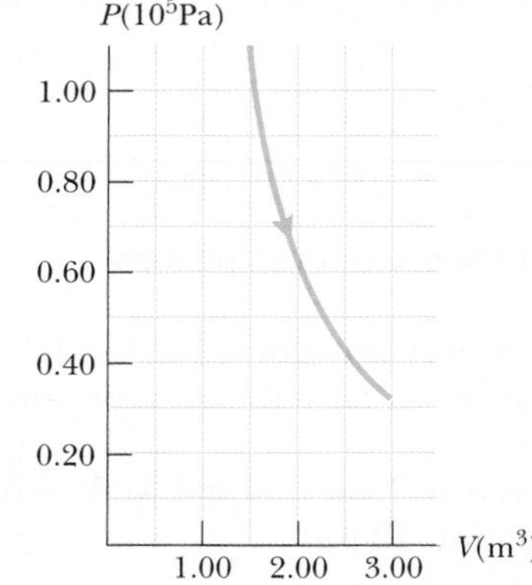

Figure 12.4 (Example 12.6) The *PV* diagram of an adiabatic expansion: the graph of $P = CV^{-\gamma}$, where *C* is a constant and $\gamma = C_p/C_v$.

Solution

(a) Find the new pressure.

First, calculate the adiabatic index:

$$\gamma = \frac{C_p}{C_v} = \frac{\frac{5}{2}R}{\frac{3}{2}R} = \frac{5}{3}$$

Use Equation 12.8a to find the constant *C*:

$$C = P_1V_1^{\gamma} = (1.01 \times 10^5\ \text{Pa})(1.50\ \text{m}^3)^{5/3}$$
$$= 1.99 \times 10^5\ \text{Pa}\cdot\text{m}^5$$

The constant *C* is fixed for the entire process and can be used to find P_2:

$$C = P_2V_2^{\gamma} = P_2(3.00\ \text{m}^3)^{5/3}$$
$$1.99 \times 10^5\ \text{Pa}\cdot\text{m}^5 = P_2\,(6.24\ \text{m}^5)$$
$$P_2 = 3.19 \times 10^4\ \text{Pa}$$

(b) Estimate the work done on the gas from a *PV* diagram.

Count the boxes between $V_1 = 1.50\ \text{m}^3$ and $V_2 = 3.00\ \text{m}^3$ in the graph of $P = (1.99 \times 10^5\ \text{Pa}\cdot\text{m}^5)\,V^{-5/3}$ in the *PV* diagram shown in Figure 12.4:

number of boxes ≈ 17

Each box has 'area' 5.00×10^3 J.

$$W \approx -17 \cdot 5.00 \times 10^3\ \text{J} = -8.5 \times 10^4\ \text{J}$$

Remarks The exact answer, obtained with calculus, is -8.43×10^4 J, so our result is a very good estimate. The answer is negative because the gas is expanding, doing positive work on the environment, thereby reducing its own internal energy.

Exercise 12.6

Repeat the preceding calculations for an ideal diatomic gas expanding adiabatically from an initial volume of $0.500\ \text{m}^3$ to a final volume of $1.25\ \text{m}^3$, starting at a pressure of $P_1 = 1.01 \times 10^5$ Pa. (You must sketch the curve to find the work.)

Answers $P_2 = 2.80 \times 10^4$ Pa, $W \approx -4 \times 10^4$ J

Isovolumetric Processes

An **isovolumetric process**, sometimes called an *isochoric* process (which is harder to remember), proceeds at constant volume, corresponding to vertical lines in a PV diagram. If the volume doesn't change, no work is done on or by the system, so $W = 0$, and the first law of thermodynamics reads

$$\Delta U = Q \quad \text{(isovolumetric process)}$$

This result tells us that **in an isovolumetric process, the change in internal energy of a system equals the energy transferred to the system by heat**. From Equation 12.5, the energy transferred by heat in constant volume processes is given by

$$Q = nC_v \Delta T \qquad [12.9]$$

EXAMPLE 12.7 An Isovolumetric Process

Goal Apply the first law to a constant-volume process.

Problem How much thermal energy must be added to 5.00 moles of monatomic ideal gas at 3.00×10^2 K and with a constant volume of 1.50 L in order to raise the temperature of the gas to 3.80×10^2 K?

Strategy The energy transferred by heat is equal to the change in the internal energy of the gas, which can be calculated by substitution into Equation 12.9.

Solution

Apply Equation 12.9, using the fact that $C_v = 3R/2$ for an ideal monatomic gas:

$$Q = \Delta U = nC_v \Delta T = \tfrac{3}{2} nR \Delta T$$

$$= \tfrac{3}{2}(5.00 \text{ mol})(8.31 \text{ J/K} \cdot \text{mol})(80.0 \text{ K})$$

$$Q = \boxed{4.99 \times 10^3 \text{ J}}$$

Remark Constant volume processes are the simplest to handle, and include such processes as heating a solid or liquid, in which the work of expansion is negligible.

Exercise 12.7

Find the change in temperature of 22.0 mol of a monatomic ideal gas if it absorbs 9 750 J at constant volume.

Answer 35.6 K

Isothermal Processes

During an isothermal process, the temperature of a system doesn't change. In an ideal gas the internal energy U depends only on the temperature, so it follows that $\Delta U = 0$ because $\Delta T = 0$. In this case, the first law of thermodynamics gives

$$W = -Q \quad \text{(isothermal process)}$$

We see that if the system is an ideal gas undergoing an isothermal process, the work done on the system is equal to the negative of the thermal energy transferred to the system. Such a process can be visualized in Figure 12.5. A cylinder filled with gas is in contact with a large energy reservoir that can exchange energy with the gas without changing its temperature. For a constant temperature ideal gas,

$$P = \frac{nRT}{V}$$

where the numerator on the right-hand side is constant. The PV diagram of a typical isothermal process is graphed in Figure 12.6, contrasted with an adiabatic process. When the process is adiabatic, the pressure falls off more rapidly.

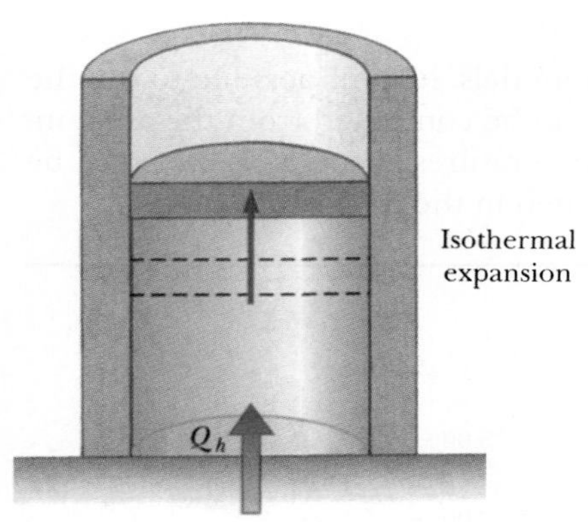

Figure 12.5 The gas in the cylinder expands isothermally while in contact with a reservoir at temperature T_h.

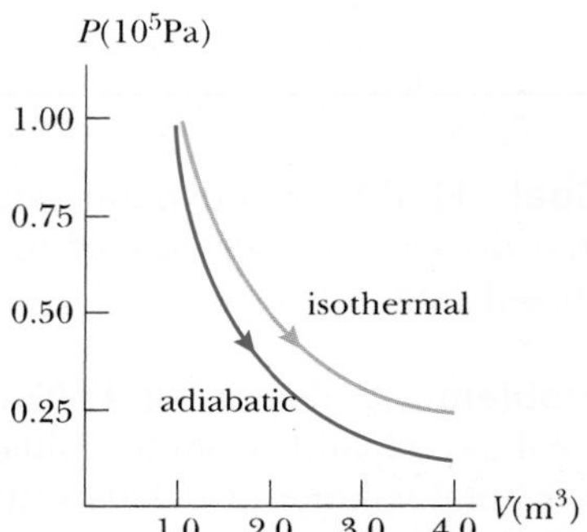

Figure 12.6 The *PV* diagram of an isothermal expansion, graph of $P = CV^{-1}$, where C is a constant. Contrasted with an adiabatic expansion, $P = C_A V^{-\gamma}$. C_A is a constant equal in magnitude to C in this case, but carrying different units.

Using methods of calculus, it can be shown that the work done on the environment during an isothermal process is given by

$$W_{\text{env}} = nRT \ln\left(\frac{V_f}{V_i}\right) \qquad [12.10]$$

The symbol "ln" in Equation 12.10 is an abbreviation for the natural logarithm, discussed in Appendix A. The work W done on the gas is just the negative of W_{env}.

EXAMPLE 12.8 An Isothermally Expanding Balloon

Goal Find the work done during an isothermal expansion.

Problem A balloon contains 5.00 moles of a monatomic ideal gas. As energy is added to the system by heat (say, by absorption from the Sun), the volume increases by 25% at a constant temperature of 27.0°C. Find the work W_{env} done by the gas in expanding the balloon, the thermal energy Q transferred to the gas, and the work W done on the gas.

Strategy Be sure to convert temperatures to kelvins. Use the equation for isothermal work to find the work done on the balloon, which is the work done on the environment. The latter is equal to the thermal energy Q transferred to the gas, and the negative of this quantity is the work done on the gas.

Solution

Substitute into Equation 12.10, finding the work done during the isothermal expansion. Note that $T = 27.0°\text{C} = 3.00 \times 10^2$ K.

$$W_{\text{env}} = nRT\ln\left(\frac{V_f}{V_i}\right)$$

$$= (5.00 \text{ mol})(8.31 \text{ J/K}\cdot\text{mol})(3.00 \times 10^2 \text{ K}) \times \ln\left(\frac{1.25V_0}{V_0}\right)$$

$$W_{\text{env}} = 2.78 \times 10^3 \text{ J}$$

$$Q = W_{\text{env}} = 2.78 \times 10^3 \text{ J}$$

The negative of this amount is the work done on the gas:

$$W = -W_{\text{env}} = -2.78 \times 10^3 \text{ J}$$

Remarks Notice the relationship between the work done on the gas, the work done on the environment, and the energy transferred. These relationships are true of all isothermal processes.

Exercise 12.8

Suppose that subsequent to this heating, 1.50×10^4 J of thermal energy is removed from the gas isothermally. Find the final volume in terms of the initial volume of the example, V_0. (*Hint:* Follow the same steps as in the example, but in reverse. Also note that the initial volume in this exercise is $1.25V_0$.)

Answer $0.375V_0$

General Case

When a process follows none of the four given models, it's still possible to use the first law to get information about it. The work can be computed from the area under the curve of the *PV* diagram, and if the temperatures at the endpoints can be found, ΔU follows from Equation 12.5, as illustrated in the following example.

EXAMPLE 12.9 A General Process

Goal Find thermodynamic quantities for a process that doesn't fall into any of the four previously discussed categories.

Problem A quantity of 4.00 moles of a monatomic ideal gas expands from an initial volume of 0.100 m^3 to a final volume of 0.300 m^3 and pressure of 2.5 × 10^5 Pa (Fig. 12.7a). Compute **(a)** the work done on the gas, **(b)** the change in internal energy of the gas, and **(c)** the thermal energy transferred to the gas.

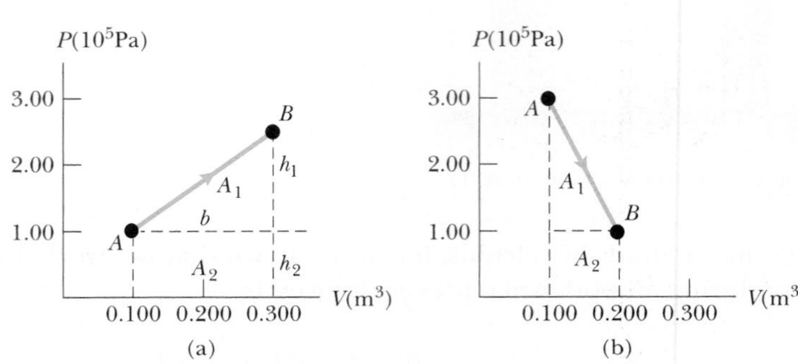

Figure 12.7 (a) (Example 12.9) (b) (Exercise 12.9)

Strategy The work done on the gas is just equal to the negative of the area under the curve in the *PV* diagram. Use the ideal gas law to get the temperature change and, subsequently, the change in internal energy. Finally, the first law gives the thermal energy transferred by heat.

Solution

(a) Find the work done on the gas by computing the area under the curve in Figure 12.7a.

Find A_1, the area of the triangle:

$$A_1 = \tfrac{1}{2}bh_1 = \tfrac{1}{2}(0.200\ \text{m}^3)(1.50 \times 10^5\ \text{Pa}) = 1.50 \times 10^4\ \text{J}$$

Find A_2, the area of the rectangle:

$$A_2 = bh_2 = (0.200\ \text{m}^3)(1.00 \times 10^5\ \text{Pa}) = 2.00 \times 10^4\ \text{J}$$

Sum the two areas (the gas is expanding, so the work done on the gas is negative and a minus sign must be supplied):

$$W = -(A_1 + A_2) = \boxed{-3.50 \times 10^4\ \text{J}}$$

(b) Find the change in the internal energy during the process.

Compute the temperature at points A and B with the ideal gas law:

$$T_A = \frac{P_A V_A}{nR} = \frac{(1.00 \times 10^5\ \text{Pa})(0.100\ \text{m}^3)}{(4.00\ \text{mol})(8.31\ \text{J/K}\cdot\text{mol})} = 301\ \text{K}$$

$$T_B = \frac{P_B V_B}{nR} = \frac{(2.50 \times 10^5\ \text{Pa})(0.300\ \text{m}^3)}{(4.00\ \text{mol})(8.31\ \text{J/K}\cdot\text{mol})} = 2.26 \times 10^3\ \text{K}$$

Compute the change in internal energy:

$$\Delta U = \tfrac{3}{2}nR\Delta T$$
$$= \tfrac{3}{2}(4.00\ \text{mol})(8.31\ \text{J/K}\cdot\text{mol})(2.26 \times 10^3\ \text{K} - 301\ \text{K})$$
$$\Delta U = \boxed{9.77 \times 10^4\ \text{J}}$$

(c) Compute Q with the first law:

$$Q = \Delta U - W = 9.77 \times 10^4\ \text{J} - (-3.50 \times 10^4\ \text{J})$$
$$= \boxed{1.33 \times 10^5\ \text{J}}$$

Remarks As long as it's possible to compute the work, cycles involving these more exotic processes can be completely analyzed. Usually, however, it's necessary to use calculus.

Exercise 12.9

Figure 12.7b represents a process involving 3.00 moles of a monatomic ideal gas expanding from 0.100 m^3 to 0.200 m^3. Find the work done on the system, the change in the internal energy of the system, and the thermal energy transferred in the process.

Answers $W = -2.00 \times 10^4$ J, $\Delta U = -1.50 \times 10^4$ J, $Q = 5.00 \times 10^3$ J

TABLE 12.2

The First Law and Thermodynamic Processes (Ideal Gases)

Process	ΔU	Q	W
Isobaric	$nC_v\Delta T$	$nC_p\Delta T$	$-P\Delta V$
Adiabatic	$nC_v\Delta T$	0	ΔU
Isovolumetric	$nC_v\Delta T$	ΔU	0
Isothermal	0	$-W$	$-nRT\ln\left(\frac{V_f}{V_i}\right)$
General	$nC_v\Delta T$	$\Delta U - W$	(PV Area)

Given all the different processes and formulae, it's easy to become confused when approaching one of these ideal gas problems, though most of the time only substitution into the correct formula is required. The essential facts and formulas are compiled in Table 12.2, both for easy reference and also to display the similarities and differences between the processes.

Quick Quiz 12.2

Identify the paths A, B, C, and D in Figure 12.8 as isobaric, isothermal, isovolumetric, or adiabatic. For path B, $Q = 0$.

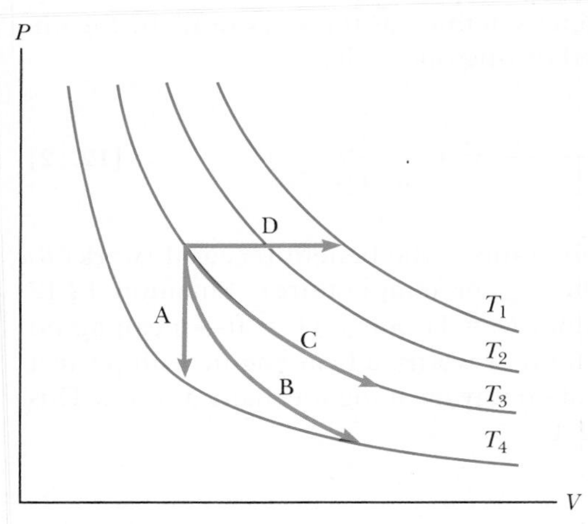

Figure 12.8 (Quick Quiz 12.2) Identify the nature of paths A, B, C, and D.

12.3 HEAT ENGINES AND THE SECOND LAW OF THERMODYNAMICS

A **heat engine** takes in energy by heat and partially converts it to other forms, such as electrical and mechanical energy. In a typical process for producing electricity in a power plant, for instance, coal or some other fuel is burned, and the resulting internal energy is used to convert water to steam. The steam is then directed at the blades of a turbine, setting it rotating. Finally, the mechanical energy associated with this rotation is used to drive an electric generator. In another heat engine—the internal combustion engine in an automobile—energy enters the engine as fuel is injected into the cylinder and combusted, and a fraction of this energy is converted to mechanical energy.

In general, a heat engine carries some working substance through a **cyclic process**[1] during which (1) energy is transferred by heat from a source at a high

◀ Cyclic process

[1]Strictly speaking, the internal combustion engine is not a heat engine according to the description of the cyclic process, because the air–fuel mixture undergoes only one cycle and is then expelled through the exhaust system.

temperature, (2) work is done by the engine, and (3) energy is expelled by the engine by heat to a source at lower temperature. As an example, consider the operation of a steam engine in which the working substance is water. The water in the engine is carried through a cycle in which it first evaporates into steam in a boiler and then expands against a piston. After the steam is condensed with cooling water, it returns to the boiler, and the process is repeated.

It's useful to draw a heat engine schematically, as in Active Figure 12.9. The engine absorbs energy Q_h from the hot reservoir, does work W_{eng}, then gives up energy Q_c to the cold reservoir. (Note that *negative* work is done *on* the engine, so that $W = -W_{\text{eng}}$.) Because the working substance goes through a cycle, always returning to its initial thermodynamic state, its initial and final internal energies are equal, so $\Delta U = 0$. From the first law of thermodynamics, therefore,

$$\Delta U = 0 = Q + W \quad \rightarrow \quad Q_{\text{net}} = -W = W_{\text{eng}}$$

The last equation shows that **the work W_{eng} done by a heat engine equals the net energy absorbed by the engine**. As we can see from Active Figure 12.9, $Q_{\text{net}} = |Q_h| - |Q_c|$. Therefore,

$$W_{\text{eng}} = |Q_h| - |Q_c| \qquad [12.11]$$

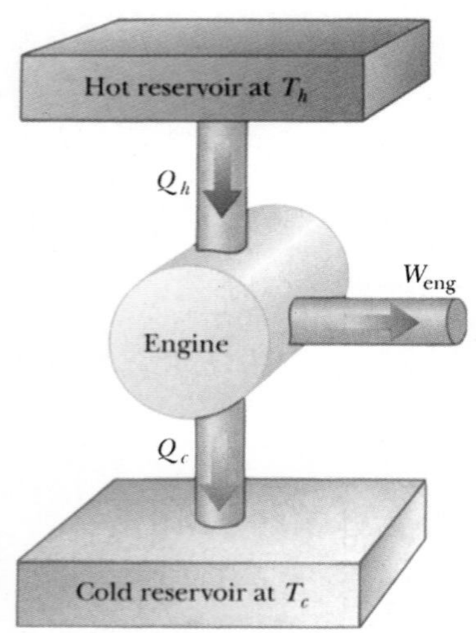

ACTIVE FIGURE 12.9
A schematic representation of a heat engine. The engine receives energy Q_h from the hot reservoir, expels energy Q_c to the cold reservoir, and does work W.

Physics Now™
Log into PhysicsNow at **www.cp7e.com** and go to Active Figure 12.9 to select the efficiency of the engine and observe the transfer of energy.

Ordinarily, a transfer of thermal energy Q can be either positive or negative, so the use of absolute value signs makes the signs of Q_h and Q_c explicit.

If the working substance is a gas, then **the work done by the engine for a cyclic process is the area enclosed by the curve representing the process on a *PV* diagram**. This area is shown for an arbitrary cyclic process in Figure 12.10.

The **thermal efficiency** e of a heat engine is defined as the work done by the engine, W_{eng}, divided by the energy absorbed during one cycle:

$$e \equiv \frac{W_{\text{eng}}}{|Q_h|} = \frac{|Q_h| - |Q_c|}{|Q_h|} = 1 - \frac{|Q_c|}{|Q_h|} \qquad [12.12]$$

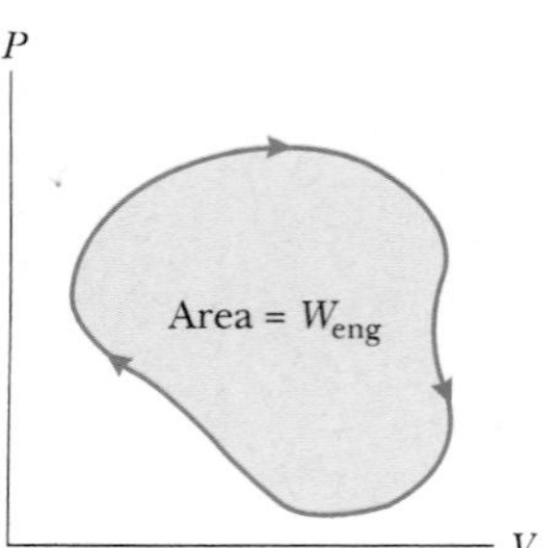

Figure 12.10 The *PV* diagram for an arbitrary cyclic process. The area enclosed by the curve equals the net work done.

We can think of thermal efficiency as the ratio of the benefit received (work) to the cost incurred (energy transfer at the higher temperature). Equation 12.12 shows that a heat engine has 100% efficiency ($e = 1$) only if $Q_c = 0$—meaning no energy is expelled to the cold reservoir. In other words, a heat engine with perfect efficiency would have to expel all the input energy by doing mechanical work. This isn't possible, as will be seen in Section 12.4.

EXAMPLE 12.10 The Efficiency of an Engine

Goal Apply the efficiency formula to a heat engine.

Problem During one cycle, an engine extracts 2.00×10^3 J of energy from a hot reservoir and transfers 1.50×10^3 J to a cold reservoir. **(a)** Find the thermal efficiency of the engine. **(b)** How much work does this engine do in one cycle? **(c)** How much power does the engine generate if it goes through four cycles in 2.50 s?

Strategy Apply Equation 12.12 to obtain the thermal efficiency, then use the first law, adapted to engines (Equation 12.11), to find the work done in one cycle. To obtain the power generated, just divide the work done in four cycles by the time it takes to run those cycles.

Solution

(a) Find the engine's thermal efficiency.

Substitute Q_c and Q_h into Equation 12.12:

$$e = 1 - \frac{|Q_c|}{|Q_h|} = 1 - \frac{1.50 \times 10^3\text{ J}}{2.00 \times 10^3\text{ J}} = \boxed{0.250, \text{ or } 25.0\%}$$

(b) How much work does this engine do in one cycle?

Apply the first law in the form of Equation 12.11 to find the work done by the engine:

$$W_{\text{eng}} = |Q_h| - |Q_c| = 2.00 \times 10^3\text{ J} - 1.50 \times 10^3\text{ J}$$
$$= \boxed{5.00 \times 10^2\text{ J}}$$

(c) Find the power output of the engine.

Multiply the answer of part (b) by four and divide by time:

$$\mathcal{P} = \frac{W}{\Delta t} = \frac{4.00 \times (5.00 \times 10^2\text{ J})}{2.50\text{ s}} = \boxed{8.00 \times 10^2\text{ W}}$$

Remark Problems like this usually reduce to solving two equations and two unknowns, as here, where the two equations are the efficiency equation and the first law and the unknowns are the efficiency and the work done by the engine.

Exercise 12.10

The energy absorbed by an engine is three times as great as the work it performs. (a) What is its thermal efficiency? (b) What fraction of the energy absorbed is expelled to the cold reservoir? (c) What is the power output of the engine if the energy input is 1 650 J each cycle and it goes through two cycles every 3 seconds?

Answer (a) 1/3 (b) 2/3 (c) 367 W

EXAMPLE 12.11 Analyzing an Engine Cycle

Goal Combine several concepts to analyze an engine cycle.

Problem A heat engine contains an ideal monatomic gas confined to a cylinder by a movable piston. The gas starts at A, where $T = 3.00 \times 10^2$ K. (See Fig. 12.11a.) $B \rightarrow C$ is an isothermal expansion. **(a)** Find the number n of moles of gas and the temperature at B. **(b)** Find ΔU, Q, and W for the isovolumetric process $A \rightarrow B$. **(c)** Repeat for the isothermal process $B \rightarrow C$. **(d)** Repeat for the isobaric process $C \rightarrow A$. **(e)** Find the net change in the internal energy for the complete cycle. **(f)** Find the thermal energy Q_h transferred into the system, the thermal energy rejected, Q_c, the thermal efficiency, and net work on the environment performed by the engine.

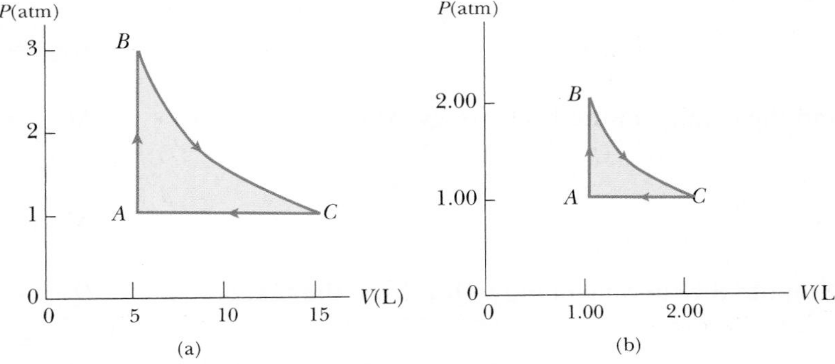

Figure 12.11 (a) (Example 12.11) (b) (Exercise 12.11)

Strategy In part (a) n, T, and V can be found from the ideal gas law, which connects the equilibrium values of P, V, and T. Once the temperature T is known at the points A, B, and C, the change in internal energy, ΔU, can be computed from the formula in Table 12.2 for each process. Q and W can be similarly computed, or deduced from the first law, using the techniques applied in the single process examples.

Solution

(a) Find n and T_B with the ideal gas law:

$$n = \frac{P_A V_A}{RT_A} = \frac{(1.00\text{ atm})(5.00\text{ L})}{(0.0821\text{ L}\cdot\text{atm/mol}\cdot\text{ K})(3.00 \times 10^2\text{ K})}$$
$$= \boxed{0.203\text{ mol}}$$

$$T_B = \frac{P_B V_B}{nR} = \frac{(3.00\text{ atm})(5.00\text{ L})}{(0.203\text{ mol})(0.0821\text{ L}\cdot\text{atm/mol}\cdot\text{K})}$$
$$= \boxed{9.00 \times 10^2\text{ K}}$$

(b) Find ΔU_{AB}, Q_{AB}, and W_{AB} for the constant volume process $A \rightarrow B$.

Compute ΔU_{AB}, noting that $C_v = \frac{3}{2}R = 12.5 \text{ J/mol} \cdot \text{K}$:

$$\Delta U_{AB} = nC_v\Delta T = (0.203 \text{ mol})(12.5 \text{ J/mol} \cdot \text{K}) \times (9.00 \times 10^2 \text{ K} - 3.00 \times 10^2 \text{ K})$$

$$\Delta U_{AB} = \boxed{1.52 \times 10^3 \text{ J}}$$

$\Delta V = 0$ for isovolumetric processes, so no work is done:

$$W_{AB} = \boxed{0}$$

We can find Q_{AB} from the first law:

$$Q_{AB} = \Delta U_{AB} = \boxed{1.52 \times 10^3 \text{ J}}$$

(c) Find ΔU_{BC}, Q_{BC}, and W_{BC} for the isothermal process $B \rightarrow C$.

This process is isothermal, so the temperature doesn't change, and the change in internal energy is zero:

$$\Delta U_{BC} = nC_v\Delta T = \boxed{0}$$

Compute the work done on the system, using the negative of Equation 12.10:

$$W_{BC} = -nRT \ln\left(\frac{V_C}{V_B}\right)$$

$$= -(0.203 \text{ mol})(8.31 \text{ J/mol} \cdot \text{K})(9.00 \times 10^2 \text{ K}) \times \ln\left(\frac{1.50 \times 10^{-2} \text{ m}^3}{5.00 \times 10^{-3} \text{ m}^3}\right)$$

$$W_{BC} = \boxed{-1.67 \times 10^3 \text{ Pa}}$$

Compute Q_{BC} from the first law:

$$0 = Q_{BC} + W_{BC} \quad \rightarrow \quad Q_{BC} = -W_{BC} = \boxed{1.67 \times 10^3 \text{ J}}$$

(d) Find ΔU_{CA}, Q_{CA}, and W_{CA} for the isobaric process $C \rightarrow A$.

Compute the work on the system, with pressure constant:

$$W_{CA} = -P\Delta V = -(1.01 \times 10^5 \text{ Pa})(5.00 \times 10^{-3} \text{ m}^3 - 1.50 \times 10^{-2} \text{ m}^3)$$

$$W_{CA} = \boxed{1.01 \times 10^3 \text{ J}}$$

Find the change in internal energy, ΔU_{CA}:

$$\Delta U_{CA} = \tfrac{3}{2}nRT = \tfrac{3}{2}(0.203 \text{ mol})(8.31 \text{ J/K} \cdot \text{mol}) \times (3.00 \times 10^2 \text{ K} - 9.00 \times 10^2 \text{ K})$$

$$\Delta U_{CA} = \boxed{-1.52 \times 10^3 \text{ J}}$$

Compute the thermal energy, Q_{CA}, from the first law:

$$Q_{CA} = \Delta U_{CA} - W_{CA} = -1.52 \times 10^3 \text{ J} - 1.01 \times 10^3 \text{ J}$$

$$= \boxed{-2.53 \times 10^3 \text{ J}}$$

(e) Find the net change in internal energy, ΔU_{net}, for the cycle:

$$\Delta U_{\text{net}} = \Delta U_{AB} + \Delta U_{BC} + \Delta U_{CA}$$

$$= 1.52 \times 10^3 \text{ J} + 0 - 1.52 \times 10^3 \text{ J} = \boxed{0}$$

(f) Find the energy input, Q_h; the energy rejected, Q_c; the thermal efficiency; and the net work performed by the engine:

Sum all the positive contributions to find Q_h:

$$Q_h = Q_{AB} + Q_{BC} = 1.52 \times 10^3 \text{ J} + 1.67 \times 10^3 \text{ J}$$

$$= \boxed{3.19 \times 10^3 \text{ J}}$$

Sum any negative contributions (in this case, there is only one):

$$Q_c = \boxed{-2.53 \times 10^3 \text{ J}}$$

Find the engine efficiency and the net work done by the engine:

$$e = 1 - \frac{|Q_c|}{|Q_h|} = 1 - \frac{2.53 \times 10^3\text{ J}}{3.19 \times 10^3\text{ J}} = \boxed{0.207}$$

$$W_{\text{eng}} = -(W_{AB} + W_{BC} + W_{CA})$$
$$= -(0 - 1.67 \times 10^3\text{ J} + 1.01 \times 10^3\text{ J})$$
$$= \boxed{6.60 \times 10^2\text{ J}}$$

Remarks Cyclic problems are rather lengthy; however, the individual steps are often short substitutions. Notice that the change in internal energy for the cycle is zero and that the net work done on the environment is identical to the net thermal energy transferred, both as they should be.

Exercise 12.11

4.05×10^{-2} mol of monatomic ideal gas goes through the process shown in Figure 12.11b. The temperature at point A is 3.00×10^2 K and is 6.00×10^2 K during the isothermal process $B \rightarrow C$. (a) Find Q, ΔU, and W for the constant volume process $A \rightarrow B$. (b) Do the same for the isothermal process $B \rightarrow C$. (c) Repeat, for the constant pressure process $C \rightarrow A$. (d) Find Q_h, Q_c, and the efficiency. (e) Find W_{eng}.

Answers (a) $Q_{AB} = \Delta U_{AB} = 151\text{ J}$, $W_{AB} = 0$ (b) $\Delta U_{BC} = 0$, $Q_{BC} = -W_{BC} = 1.40 \times 10^2\text{ J}$ (c) $Q_{CA} = -252\text{ J}$, $\Delta U_{CA} = -151\text{ J}$, $W_{CA} = 101\text{ J}$ (d) $Q_h = 291\text{ J}$, $Q_c = -252\text{ J}$, $e = 0.134$ (e) $W_{\text{eng}} = 39.0\text{ J}$

Refrigerators and Heat Pumps

Heat engines can operate in reverse. In this case, energy is injected into the engine, modeled as work W in Active Figure 12.12, resulting in energy being extracted from the cold reservoir and transferred to the hot reservoir. The system now operates as a heat pump, a common example being a refrigerator (Fig. 12.13). Energy Q_c is extracted from the interior of the refrigerator and delivered as energy Q_h to the warmer air in the kitchen. The work is done in the compressor unit of the refrigerator, compressing a refrigerant such as freon, causing its temperature to increase.

A household air conditioner is another example of a heat pump. Some homes are both heated and cooled by heat pumps. In the winter, the heat pump extracts energy Q_c from the cool outside air and delivers energy Q_h to the warmer air inside. In summer, energy Q_c is removed from the cool inside air, while energy Q_h is ejected to the warm air outside.

For a refrigerator or an air conditioner—a heat pump operating in cooling mode—work W is what you pay for, in terms of electrical energy running the compressor, while Q_c is the desired benefit. The most efficient refrigerator or air conditioner is one that removes the greatest amount of energy from the cold reservoir in exchange for the least amount of work.

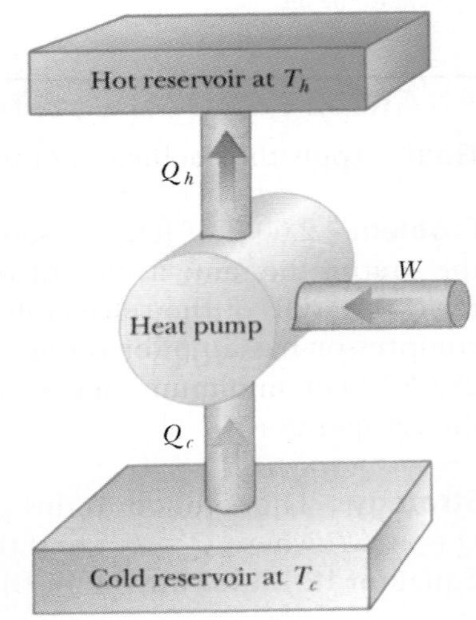

ACTIVE FIGURE 12.12
Schematic diagram of a heat pump, which takes in energy $Q_c > 0$ from a cold reservoir and expels energy $Q_h < 0$ to a hot reservoir. Work W is done *on* the heat pump. A refrigerator works the same way.

Physics Now™
Log into PhysicsNow at **www.cp7e.com**, and go to Active Figure 12.12 to select the coefficient of performance (COP) of the heat pump and observe the transfer of energy.

The coefficient of performance for a refrigerator or an air conditioner is the magnitude of the energy extracted from the cold reservoir, $|Q_c|$, divided by the work W performed by the device:

$$\text{COP(cooling mode)} = \frac{|Q_c|}{W} \qquad \textbf{[12.13]}$$

SI unit: dimensionless

The larger this ratio, the better the performance, since more energy is being removed for a given amount of work. A good refrigerator or air conditioner will have a COP of 5 or 6.

Charles D. Winters

Figure 12.13 The coils on the back of a refrigerator transfer energy by heat to the air.

A heat pump operating in heating mode warms the inside of a house in winter by extracting energy from the colder outdoor air. This may seem paradoxical, but recall that this is equivalent to a refrigerator removing energy from its interior and ejecting it into the kitchen.

The coefficient of performance of a heat pump operating in the heating mode is the magnitude of the energy rejected to the hot reservoir, $|Q_h|$, divided by the work W done by the pump:

$$\text{COP(heating mode)} = \frac{|Q_h|}{W} \qquad [12.14]$$

SI unit: dimensionless

In effect, the COP of a heat pump in the heating mode is the ratio of what you gain (energy delivered to the interior of your home) to what you give (work input). Typical values for this COP are greater than one, because $|Q_h|$ is usually greater than W.

In a groundwater heat pump, energy is extracted in the winter from water deep in the ground rather than from the outside air, while energy is delivered to that water in the summer. This strategy increases the year-round efficiency of the heating and cooling unit, because the groundwater is at a higher temperature than the air in winter and at a cooler temperature than the air in summer.

EXAMPLE 12.12 Cooling the Leftovers

Goal Apply the coefficient of performance of a refrigerator.

Problem 2.00 L of leftover soup at a temperature of 323 K is placed in a refrigerator. Assume the specific heat of the soup is the same as that of water and the density is 1.25×10^3 kg/m^3. The refrigerator cools the soup to 283 K. **(a)** If the COP of the refrigerator is 5.00, find the energy needed, in the form of work, to cool the soup. **(b)** If the compressor has a power rating of 0.250 hp, for what minimum length of time must it operate to cool the soup to 283 K? (The minimum time assumes the soup cools at the same rate that the heat pump ejects thermal energy from the refrigerator.)

Strategy The solution to this problem requires three steps. First, find the total mass m of the soup. Second, using $Q = mc\Delta T$, where $Q = Q_c$, find the energy transfer required to cool the soup. Third, substitute Q_c and the COP into Equation 12.13, solving for W. Divide the work by the power to get an estimate of the time required to cool the soup.

Solution

(a) Find the work needed to cool the soup.

Calculate the mass of the soup:

$$m = \rho V = (1.25 \times 10^3 \text{ kg/m}^3)(2.00 \times 10^{-3} \text{ m}^3) = 2.50 \text{ kg}$$

Find the energy transfer required to cool the soup:

$$\begin{aligned} Q_c = Q &= mc\Delta T \\ &= (2.50 \text{ kg})(4\,190 \text{ J/kg}\cdot\text{K})(283 \text{ K} - 323 \text{ K}) \\ &= -4.19 \times 10^5 \text{ J} \end{aligned}$$

Substitute Q_c and the COP into Equation 12.13:

$$\text{COP} = \frac{|Q_c|}{W} = \frac{4.19 \times 10^5 \text{ J}}{W} = 5.00$$

$$W = \boxed{8.38 \times 10^4 \text{ J}}$$

(b) Find the time needed to cool the food.

Convert horsepower to watts:

$$\mathcal{P} = (0.250 \text{ hp})(746 \text{ W/1 hp}) = 187 \text{ W}$$

Divide the work by the power to find the elapsed time: $\Delta t = \frac{W}{\mathcal{P}} = \frac{8.38 \times 10^4\ \text{J}}{187\ \text{W}} = 448\ \text{s}$

Remarks This example illustrates how cooling different substances requires differing amounts of work, due to differences in specific heats. The problem doesn't take into account the insulating properties of the soup container and of the soup itself, which retard the cooling process.

Exercise 12.12

(a) How much work must a heat pump with a COP of 2.50 do in order to extract 1.00 MJ of thermal energy from the outdoors (the cold reservoir)? (b) If the unit operates at 0.500 hp, how long will the process take? (Be sure to use the correct COP!)

Answers (a) 6.67×10^5 J (b) 1.79×10^3 s

The Second Law of Thermodynamics

There are limits to the efficiency of heat engines. The ideal engine would convert all input energy into useful work, but it turns out that such an engine is impossible to construct. The Kelvin–Planck formulation of the **second law of thermodynamics** can be stated as follows:

> No heat engine operating in a cycle can absorb energy from a reservoir and use it entirely for the performance of an equal amount of work.

This form of the second law means that the efficiency $e = W_{\text{eng}}/|Q_h|$ of engines must always be less than one. Some energy Q_c must always be lost to the environment. In other words, it's theoretically impossible to construct a heat engine with an efficiency of 100%.

To summarize, the first law says **we can't get a greater amount of energy out of a cyclic process than we put in**, and the second law says **we can't break even.**

LORD KELVIN, British Physicist and Mathematician (1824–1907)

Born William Thomson in Belfast, Kelvin was the first to propose the use of an absolute scale of temperature. His study of Carnot's theory led to the idea that energy cannot pass spontaneously from a colder object to a hotter object; this principle is known as the second law of thermodynamics.

Reversible and Irreversible Processes

No engine can operate with 100% efficiency, but different designs yield different efficiencies, and it turns out one design in particular delivers the maximum possible efficiency. This design is the Carnot cycle, discussed in the next subsection. Understanding it requires the concepts of reversible and irreversible processes. In a **reversible** process, every state along the path is an equilibrium state, so the system can return to its initial conditions by going along the same path in the reverse direction. A process that doesn't satisfy this requirement is **irreversible.**

Most natural processes are known to be irreversible—the reversible process is an idealization. Although real processes are always irreversible, some are *almost* reversible. If a real process occurs so slowly that the system is virtually always in equilibrium, the process can be considered reversible. Imagine compressing a gas very slowly by dropping grains of sand onto a frictionless piston, as in Figure 12.14. The temperature can be kept constant by placing the gas in thermal contact with an energy reservoir. The pressure, volume, and temperature of the gas are well defined during this isothermal compression. Each added grain of sand represents a change to a new equilibrium state. The process can be reversed by slowly removing grains of sand from the piston.

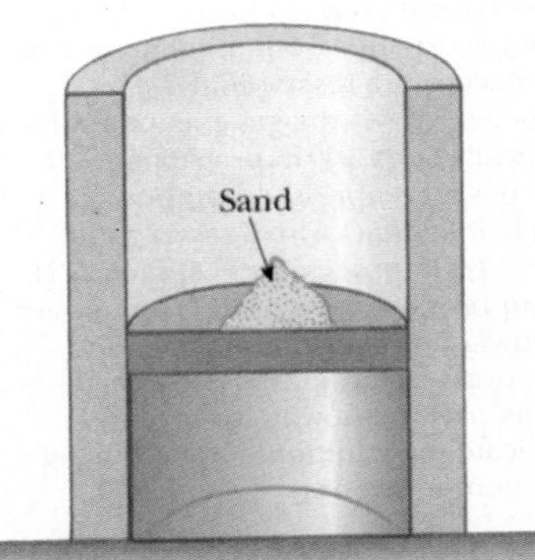

Figure 12.14 A gas in thermal contact with an energy reservoir is compressed slowly by grains of sand dropped onto a piston. The compression is isothermal and reversible.

The Carnot Engine

In 1824, in an effort to understand the efficiency of real engines, a French engineer named Sadi Carnot (1796–1832) described a theoretical engine now called a *Carnot engine* that is of great importance from both a practical and a theoretical

J.-L. Charmet/SPL/Photo Researchers, Inc.

SADI CARNOT, French Engineer (1796–1832)

Carnot is considered to be the founder of the science of thermodynamics. Some of his notes found after his death indicate that he was the first to recognize the relationship between work and heat.

viewpoint. He showed that a heat engine operating in an ideal, reversible cycle—now called a **Carnot cycle**—between two energy reservoirs is the most efficient engine possible. Such an engine establishes an upper limit on the efficiencies of all real engines. **Carnot's theorem** can be stated as follows:

> No real engine operating between two energy reservoirs can be more efficient than a Carnot engine operating between the same two reservoirs.

In a Carnot cycle, an ideal gas is contained in a cylinder with a movable piston at one end. The temperature of the gas varies between T_c and T_h. The cylinder walls and the piston are thermally nonconducting. Active Figure 12.15 shows the four stages of the Carnot cycle, and Active Figure 12.16 is the *PV* diagram for the cycle. The cycle consists of two adiabatic and two isothermal processes, all reversible:

1. The process $A \rightarrow B$ is an isothermal expansion at temperature T_h in which the gas is placed in thermal contact with a hot reservoir (a large oven, for example) at temperature T_h (Active Fig. 12.15a). During the process, the gas absorbs energy Q_h from the reservoir and does work W_{AB} in raising the piston.
2. In the process $B \rightarrow C$, the base of the cylinder is replaced by a thermally non-conducting wall and the gas expands adiabatically, so no energy enters or

$A \rightarrow B$ Isothermal expansion

Q_h

Energy reservoir at T_h

(a)

$D \rightarrow A$ Adiabatic compression

Cycle

$B \rightarrow C$ Adiabatic expansion

$Q = 0$

$Q = 0$

(d)

(b)

$C \rightarrow D$ Isothermal compression

Q_c

Energy reservoir at T_c

(c)

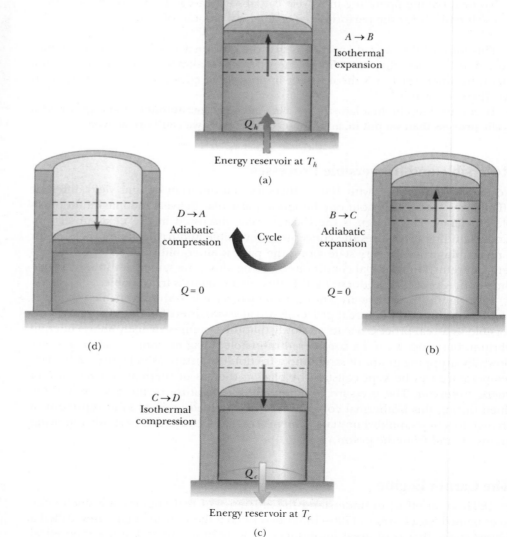

ACTIVE FIGURE 12.15
The Carnot cycle. In process $A \rightarrow B$, the gas expands isothermally while in contact with a reservoir at T_h. In process $B \rightarrow C$, the gas expands adiabatically ($Q = 0$). In process $C \rightarrow D$, the gas is compressed isothermally while in contact with a reservoir at $T_c < T_h$. In process $D \rightarrow A$, the gas is compressed adiabatically. The upward arrows on the piston indicate the removal of sand during the expansions, and the downward arrows indicate the addition of sand during the compressions.

Physics Now™
Log into PhysicsNow at **www.cp7e.com**, and go to Active Figure 12.15 to observe the motion of the piston in the Carnot cycle while you also observe the cycle on the *PV* diagram of Active Figure 12.16.

frequency $f = 1.50$ Hz. Block B rests on it, as shown in Figure P13.55, and the coefficient of static friction between the two is $\mu_s = 0.600$. What maximum amplitude of oscillation can the system have if block B is not to slip?

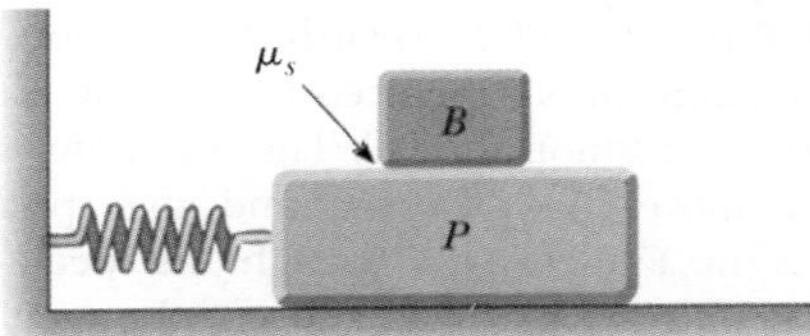

Figure P13.55

56. A 500-g block is released from rest and slides down a frictionless track that begins 2.00 m above the horizontal, as shown in Figure P13.56. At the bottom of the track, where the surface is horizontal, the block strikes and sticks to a light spring with a spring constant of 20.0 N/m. Find the maximum distance the spring is compressed.

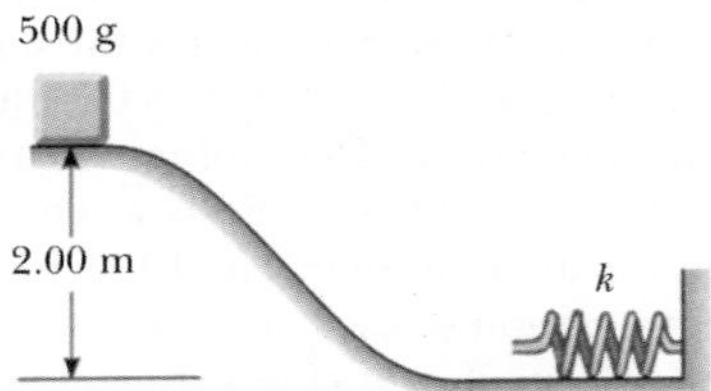

Figure P13.56

57. **Physics Now™** A 3.00-kg object is fastened to a light spring, with the intervening cord passing over a pulley (Fig. P13.57). The pulley is frictionless, and its inertia may be neglected. The object is released from rest when the spring is unstretched. If the object drops 10.0 cm before stopping, find (a) the spring constant of the spring and (b) the speed of the object when it is 5.00 cm below its starting point.

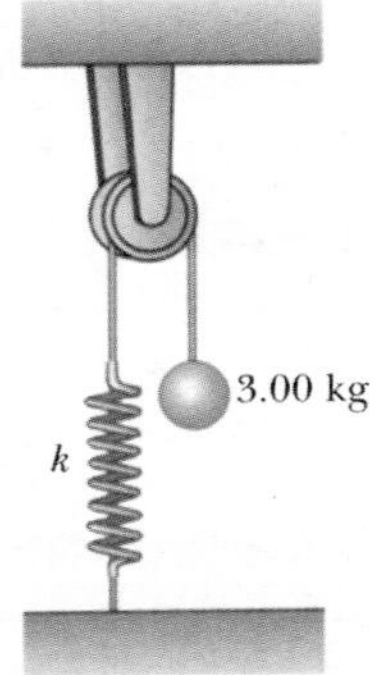

Figure P13.57

58. A 5.00-g bullet moving with an initial speed of 400 m/s is fired into and passes through a 1.00-kg block, as in Figure P13.58. The block, initially at rest on a frictionless horizontal surface, is connected to a spring with a spring constant of 900 N/m. If the block moves 5.00 cm to the right after impact, find (a) the speed at which the bullet emerges from the block and (b) the mechanical energy lost in the collision.

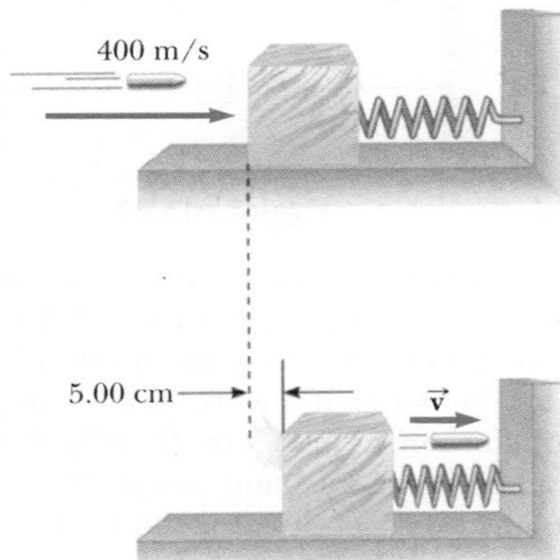

Figure P13.58

59. A 25-kg block is connected to a 30-kg block by a light string that passes over a frictionless pulley. The 30-kg block is connected to a light spring of force constant 200 N/m, as in Figure P13.59. The spring is unstretched when the system is as shown in the figure, and the incline is smooth. The 25-kg block is pulled 20 cm down the incline (so that the 30-kg block is 40 cm above the floor) and is released from rest. Find the speed of each block when the 30-kg block is 20 cm above the floor (that is, when the spring is unstretched).

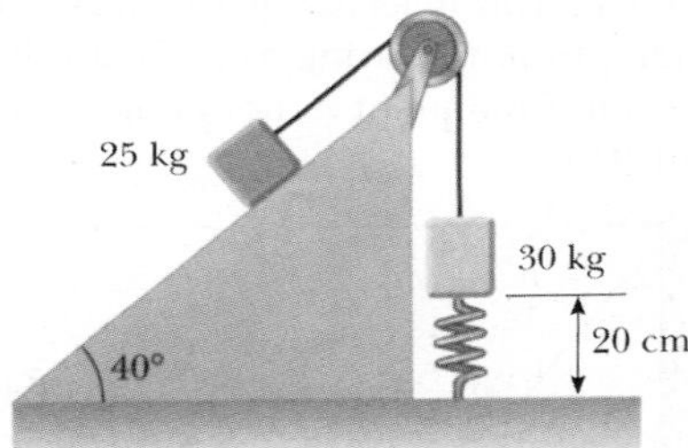

Figure P13.59

60. A spring in a toy gun has a spring constant of 9.80 N/m and can be compressed 20.0 cm beyond the equilibrium position. A 1.00-g pellet resting against the spring is propelled forward when the spring is released. (a) Find the muzzle speed of the pellet. (b) If the pellet is fired horizontally from a height of 1.00 m above the floor, what is its range?

61. A 2.00-kg block hangs without vibrating at the end of a spring ($k = 500$ N/m) that is attached to the ceiling of an elevator car. The car is rising with an upward acceleration of $g/3$ when the acceleration suddenly ceases (at $t = 0$). (a) What is the angular frequency of oscillation of the block after the acceleration ceases? (b) By what amount is the spring stretched during the time that the elevator car is accelerating? This distance will be the amplitude of the ensuing oscillation of the block.

62. An object of mass m is connected to two rubber bands of length L, each under tension F, as in Figure P13.62. The object is displaced vertically by a small distance y. Assuming the tension does not change, show that (a) the restoring force is $-(2F/L)y$ and (b) the system exhibits simple harmonic motion with an angular frequency $\omega = \sqrt{2F/mL}$.

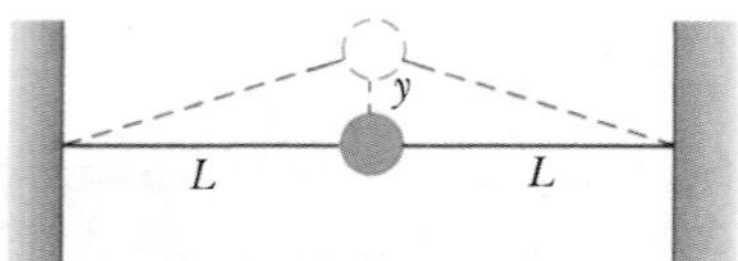

Figure P13.62

63. A light balloon filled with helium of density 0.180 kg/m^3 is tied to a light string of length $L = 3.00$ m. The string is tied to the ground, forming an "inverted" simple pendulum (Fig. P13.63a). If the balloon is displaced slightly from equilibrium, as in Figure P13.63b, show that the motion is simple harmonic, and determine the period of the motion. Take the density of air to be 1.29 kg/m^3. [*Hint:* Use an analogy with the simple pendulum discussed in the text, and see Chapter 9.]

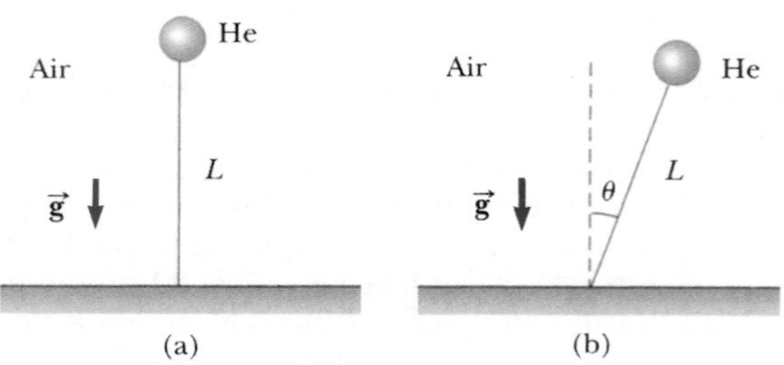

Figure P13.63

64. A light string of mass 10.0 g and length $L = 3.00$ m has its ends tied to two walls that are separated by the distance $D = 2.00$ m. Two objects, each of mass $M = 2.00$ kg, are suspended from the string as in Figure P13.64. If a wave pulse is sent from point A, how long does it take to travel to point B?

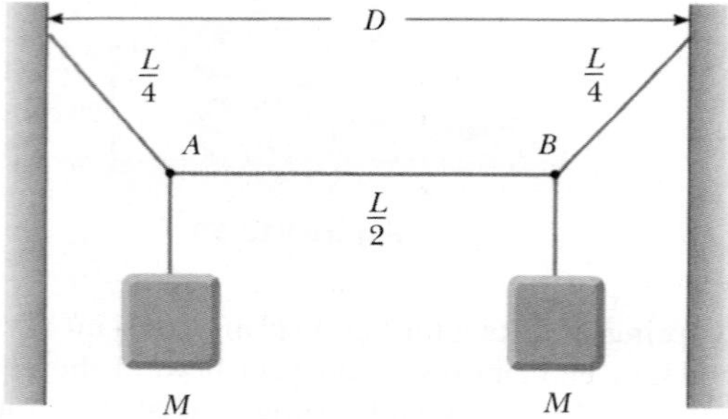

Figure P13.64

65. Assume that a hole is drilled through the center of the Earth. It can be shown that an object of mass m at a distance r from the center of the Earth is pulled toward the center only by the material in the shaded portion of Figure P13.65.

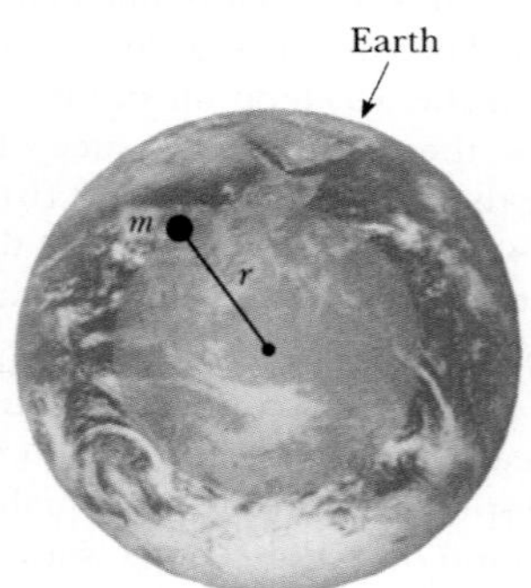

Figure P13.65

Assume Earth has a uniform density ρ. Write down Newton's law of gravitation for an object at a distance r from the center of the Earth, and show that the force on it is of the form of Hooke's law, $F = -kr$, with an effective force constant of $k = (\frac{4}{3})\pi\rho Gm$, where G is the gravitational constant.

66. A 60.0-kg firefighter slides down a pole while a constant frictional force of 300 N retards his motion. A horizontal 20.0-kg platform is supported by a spring at the bottom of the pole to cushion the fall. The firefighter starts from rest 5.00 m above the platform, and the spring constant is 2500 N/m. Find (a) the firefighter's speed just before he collides with the platform and (b) the maximum distance the spring is compressed. Assume that the frictional force acts during the entire motion. [*Hint:* The collision between the firefighter and the platform is perfectly inelastic.]

67. An object of mass $m_1 = 9.0$ kg is in equilibrium while connected to a light spring of constant $k = 100$ N/m that is fastened to a wall, as in Figure P13.67a. A second object, of mass $m_2 = 7.0$ kg, is slowly pushed up against m_1, compressing the spring by the amount $A = 0.20$ m, as shown in Figure P13.67b. The system is then released, causing both objects to start moving to the right on the frictionless surface. (a) When m_1 reaches the equilibrium point, m_2 loses contact with m_1 (Fig. P13.67c) and moves to the right with velocity $\vec{v}$. Determine the magnitude of $\vec{v}$. (b) How far apart are the objects when the spring is fully stretched for the first time (Fig. P13.67d)? [*Hint:* First determine the period of oscillation and the amplitude of the m_1–spring system after m_2 loses contact with m_1.]

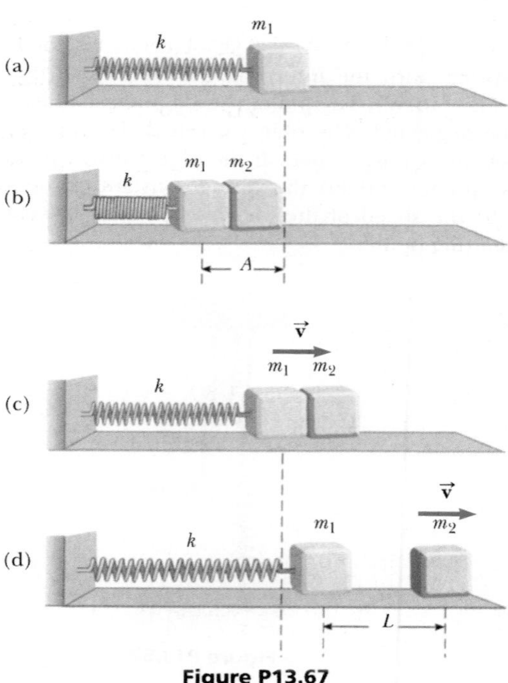

Figure P13.67

68. An 8.00-kg block travels on a rough horizontal surface and collides with a spring. The speed of the block just before the collision is 4.00 m/s. As it rebounds to the left with the spring uncompressed, the block travels at 3.00 m/s. If the coefficient of kinetic friction between the block and the surface is 0.400, determine (a) the loss in

mechanical energy due to friction while the block is in contact with the spring and (b) the maximum distance the spring is compressed.

69. Two points, A and B, on Earth are at the same longitude and 60.0° apart in latitude. An earthquake at point A sends two waves toward B. A transverse wave travels along the surface of Earth at 4.50 km/s, and a longitudinal wave travels through Earth at 7.80 km/s. (a) Which wave arrives at B first? (b) What is the time difference between the arrivals of the two waves at B? Take the radius of Earth to be 6.37×10^6 m.

70. Figure P13.70 shows a crude model of an insect wing. The mass m represents the entire mass of the wing, which pivots about the fulcrum F. The spring represents the surrounding connective tissue. Motion of the wing corresponds to vibration of the spring. Suppose the mass of the wing is 0.30 g and the effective spring constant of the tissue is 4.7×10^{-4} N/m. If the mass m moves up and down a distance of 2.0 mm from its position of equilibrium, what is the maximum speed of the outer tip of the wing?

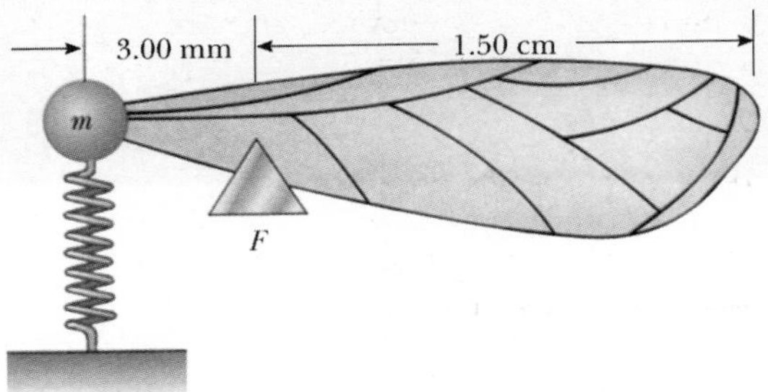

Figure P13.70

71. A 1.6-kg block on a horizontal surface is attached to a spring with a force constant of 1.0×10^3 N/m, as in Active Figure 13.1. The spring is compressed a distance of 2.0 cm, and the block is released from rest. (a) Calculate the speed of the block as it passes through the equilibrium position, $x = 0$, if the surface is frictionless. (b) Calculate the speed of the block as it passes through the equilibrium position if a constant frictional force of 4.0 N retards its motion. (c) How far does the block travel before coming to rest in part (b)?

ACTIVITIES

A.1. Construct a simple pendulum by tying a metal bolt to one end of a string and taping the other end to the top of a doorframe. Adjust the length of the pendulum to be about 1.0 m, and measure it precisely. Obtain the period by timing 25 complete oscillations, making sure the string always makes small angles with the vertical. Repeat the measurements for precisely measured pendulum lengths ranging from 0.4 m to 1.6 m in increments of 0.2 m. Plot the square of the period versus the length of the pendulum and measure the slope of the line best fitting your data points. Does your slope agree with that predicted by $T^2 = (4\pi^2/g)L$? What value of g do you obtain from your data?

While you have your pendulum in position, use a procedure similar to the preceding to verify that the period is also independent of the amplitude for small angles and that the period is also independent of the mass.

A.2. Attach one end of a rope (or a spring such as a Slinky™) to a wall, stretch it taut, and use the system to study the following aspects of wave motion:

(a) Send a pulse down the rope by striking it sharply from the side. An observer watching from the side can measure the time elapsed during 3–5 trips of the pulse from one end to the other. Dividing the total distance traveled by the elapsed time yields the wave speed. To get reliable results, take a number of readings and average them.

(b) Use the same setup as in part (a) to test whether the initial amplitude of the pulse changes the wave speed.

(c) Have two people hold opposite ends of the rope (or spring). Then, at the same instant, have both people hit the rope sharply from the side. Observe what happens when the two pulses meet. The superposition lasts only for the short time that the pulses overlap, and you must look carefully to see the effect.

(d) Using the same setup, devise a way to check whether pulses traveling toward one another pass through when they collide or reflect off each other.

(e) Tie one end of a rope to a doorknob and send a pulse down it. Observe what happens when the pulse reflects from the door. Does it return on the same side of the rope, or does it invert?

The characteristic sound of any instrument is referred to as the quality of that sound. What is it about the sound from the tube that allows us to distinguish between it and the sound from a flute?

Patrick Ward/CORBIS

CHAPTER

14

OUTLINE

Sound

Sound waves are the most important example of longitudinal waves. In this chapter we discuss the characteristics of sound waves: how they are produced, what they are, and how they travel through matter. We then investigate what happens when sound waves interfere with each other. The insights gained in this chapter will help you understand how we hear.

14.1 PRODUCING A SOUND WAVE

Whether it conveys the shrill whine of a jet engine or the soft melodies of a crooner, any sound wave has its source in a vibrating object. Musical instruments produce sounds in a variety of ways. The sound of a clarinet is produced by a vibrating reed, the sound of a drum by the vibration of the taut drumhead, the sound of a piano by vibrating strings, and the sound from a singer by vibrating vocal cords.

Sound waves are longitudinal waves traveling through a medium, such as air. In order to investigate how sound waves are produced, we focus our attention on the tuning fork, a common device for producing pure musical notes. A tuning fork consists of two metal prongs, or tines, that vibrate when struck. Their vibration disturbs the air near them, as shown in Figure 14.1. (The amplitude of vibration of the tine shown in the figure has been greatly exaggerated for clarity.) When a tine swings to the right, as in Figure 14.1a, the molecules in an element of air in front of its movement are forced closer together than normal. Such a region of high molecular density and high air pressure is called a **compression**. This compression moves away from the fork like a ripple on a pond. When the tine swings to the left, as in Figure 14.1b, the molecules in an element of air to the right of the tine spread apart, and the density and air pressure in this region are then lower than normal. Such a region of reduced density is called a **rarefaction** (pronounced "rare a fak' shun"). Molecules to the right of the rarefaction in the figure move to the left. The rarefaction itself therefore moves to the right, following the previously produced compression.

As the tuning fork continues to vibrate, a succession of compressions and rarefactions forms and spreads out from it. The resultant pattern in the air is somewhat like that pictured in Figure 14.2a. We can use a sinusoidal curve to represent a sound wave, as in Figure 14.2b. Notice that there are crests in the sinusoidal wave at the points where the sound wave has compressions and troughs where the sound wave has rarefactions. The compressions and rarefactions of the sound waves are superposed on the random thermal motion of the atoms and molecules of the air (discussed in Chapter 10), so sound waves in gases travel at about the molecular rms speed.

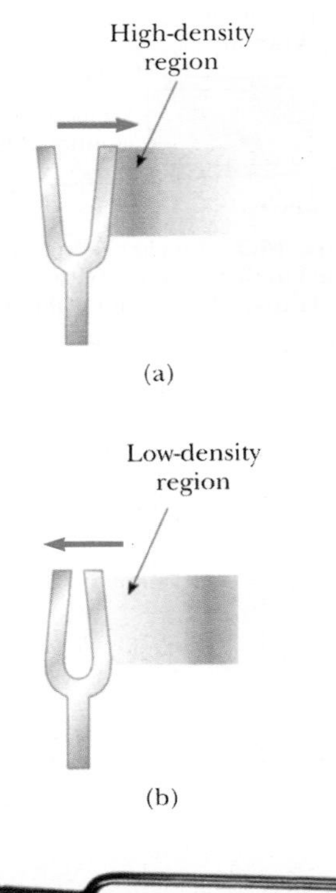

Figure 14.1 A vibrating tuning fork. (a) As the right tine of the fork moves to the right, a high-density region (compression) of air is formed in front of its movement. (b) As the right tine moves to the left, a low-density region (rarefaction) of air is formed behind it. (c) A tuning fork.

14.2 CHARACTERISTICS OF SOUND WAVES

As already noted, the general motion of elements of air near a vibrating object is back and forth between regions of compression and rarefaction. This back-and-forth motion of elements of the medium in the direction of the disturbance is characteristic of a longitudinal wave. **The motion of the elements of the medium in a longitudinal sound wave is back and forth along the direction in which the wave travels.** By contrast, **in a transverse wave, the vibrations of the elements of the medium are at right angles to the direction of travel of the wave.**

Categories of Sound Waves

Sound waves fall into three categories covering different ranges of frequencies. **Audible waves** are longitudinal waves that lie within the range of sensitivity of the human ear, approximately 20 to 20 000 Hz. **Infrasonic waves** are longitudinal waves with frequencies below the audible range. Earthquake waves are an example. **Ultrasonic waves** are longitudinal waves with frequencies above the audible range for humans and are produced by certain types of whistles. Animals such as dogs can hear the waves emitted by these whistles.

Applications of Ultrasound

APPLICATION

Medical Uses of Ultrasound

Ultrasonic waves are sound waves with frequencies greater than 20 kHz. Because of their high frequency and corresponding short wavelengths, ultrasonic waves can be used to produce images of small objects and are currently in wide use in medical applications, both as a diagnostic tool and in certain treatments. Internal organs can be examined via the images produced by the reflection and absorption of ultrasonic waves. Although ultrasonic waves are far safer than x-rays, their images don't always have as much detail. Certain organs, however, such as the liver and the spleen, are invisible to x-rays but can be imaged with ultrasonic waves.

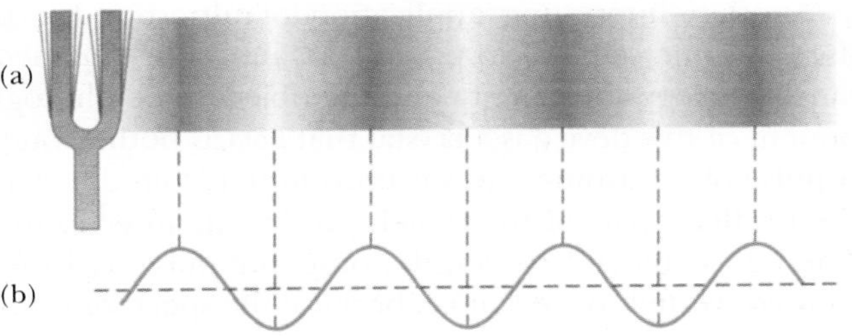

Figure 14.2 (a) As the tuning fork vibrates, a series of compressions and rarefactions moves outward, away from the fork. (b) The crests of the wave correspond to compressions, the troughs to rarefactions.

Direction of vibration

Electrical connections

Crystal

Figure 14.3 An alternating voltage applied to the faces of a piezoelectric crystal causes the crystal to vibrate.

Medical workers can measure the speed of the blood flow in the body with a device called an ultrasonic flow meter, which makes use of the Doppler effect (discussed in Section 14.6). The flow speed is found by comparing the frequency of the waves scattered by the flowing blood with the incident frequency.

Figure 14.3 illustrates the technique that produces ultrasonic waves for clinical use. Electrical contacts are made to the opposite faces of a crystal, such as quartz or strontium titanate. If an alternating voltage of high frequency is applied to these contacts, the crystal vibrates at the same frequency as the applied voltage, emitting a beam of ultrasonic waves. At one time, a technique like this was used to produce sound in nearly all headphones. This method of transforming electrical energy into mechanical energy, called the **piezoelectric effect**, is reversible: If some external source causes the crystal to vibrate, an alternating voltage is produced across it. A single crystal can therefore be used to both generate and receive ultrasonic waves.

The primary physical principle that makes ultrasound imaging possible is the fact that a sound wave is partially reflected whenever it is incident on a boundary between two materials having different densities. If a sound wave is traveling in a material of density ρ_i and strikes a material of density ρ_t, the percentage of the incident sound wave intensity reflected, PR, is given by

$$PR = \left(\frac{\rho_i - \rho_t}{\rho_i + \rho_t}\right)^2 \times 100$$

This equation assumes that the direction of the incident sound wave is perpendicular to the boundary and that the speed of sound is approximately the same in the two materials. The latter assumption holds very well for the human body because the speed of sound doesn't vary much in the organs of the body.

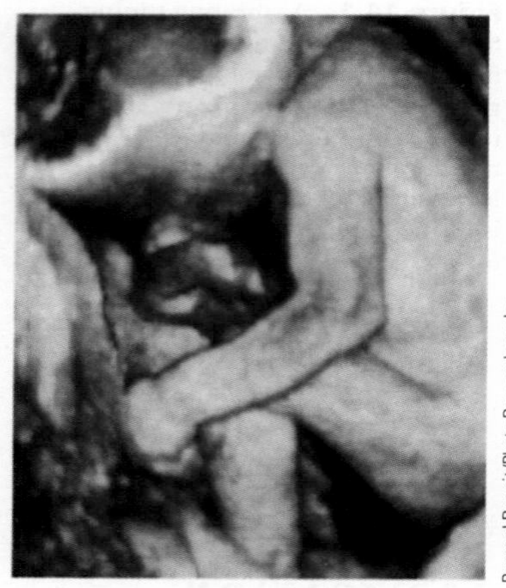

An ultrasound image of a human fetus in the womb.

Physicians commonly use ultrasonic waves to observe fetuses. This technique presents far less risk than do x-rays, which deposit more energy in cells and can produce birth defects. First the abdomen of the mother is coated with a liquid, such as mineral oil. If this were not done, most of the incident ultrasonic waves from the piezoelectric source would be reflected at the boundary between the air and the mother's skin. Mineral oil has a density similar to that of skin, and a very small fraction of the incident ultrasonic wave is reflected when $\rho_i \approx \rho_t$. The ultrasound energy is emitted in pulses rather than as a continuous wave, so the same crystal can be used as a detector as well as a transmitter. An image of the fetus is obtained by using an array of transducers placed on the abdomen. The reflected sound waves picked up by the transducers are converted to an electric signal, which is used to form an image on a fluorescent screen. Difficulties such as the likelihood of spontaneous abortion or of breech birth are easily detected with this technique. Fetal abnormalities such as spina bifida and water on the brain are also readily observed.

A relatively new medical application of ultrasonics is the *cavitron ultrasonic surgical aspirator* (CUSA). This device has made it possible to surgically remove brain tumors that were previously inoperable. The probe of the CUSA emits ultrasonic waves (at about 23 kHz) at its tip. When the tip touches a tumor, the part of the tumor near the probe is shattered and the residue can be sucked up (aspirated) through the hollow probe. Using this technique, neurosurgeons are able to remove brain tumors without causing serious damage to healthy surrounding tissue.

Ultrasound is also used to break up kidney stones that are otherwise too large to pass. Previously, invasive surgery was more often required.

APPLICATION

Ultrasonic Ranging Unit for Cameras

Another interesting application of ultrasound is the ultrasonic ranging unit used in some cameras to provide an almost instantaneous measurement of the distance between the camera and the object to be photographed. The principal component of this device is a crystal that acts as both a loudspeaker and a microphone. A pulse of ultrasonic waves is transmitted from the transducer to the object, which then reflects part of the signal, producing an echo that is detected by the device. The time interval between the outgoing pulse and the detected echo is electronically converted to a distance, because the speed of sound is a known quantity.

14.3 THE SPEED OF SOUND

The speed of a sound wave in a fluid depends on the fluid's compressibility and inertia. If the fluid has a bulk modulus B and an equilibrium density ρ, the speed of sound in it is

$$v = \sqrt{\frac{B}{\rho}} \qquad [14.1]$$

◀ Speed of sound in a fluid

Equation 14.1 also holds true for a gas. Recall from Chapter 9 that the bulk modulus is defined as the ratio of the change in pressure, ΔP, to the resulting fractional change in volume, $\Delta V/V$:

$$B \equiv -\frac{\Delta P}{\Delta V/V} \qquad [14.2]$$

B is always positive because an increase in pressure (positive ΔP) results in a decrease in volume. Hence, the ratio $\Delta P/\Delta V$ is always negative.

It's interesting to compare Equation 14.1 with Equation 13.18 for the speed of transverse waves on a string, $v = \sqrt{F/\mu}$, discussed in Chapter 13. In both cases, the wave speed depends on an elastic property of the medium (B or F) and on an inertial property of the medium (ρ or μ). In fact, the speed of all mechanical waves follows an expression of the general form

$$v = \sqrt{\frac{\text{elastic property}}{\text{inertial property}}}$$

Another example of this general form is the **speed of a longitudinal wave in a solid rod**, which is

$$v = \sqrt{\frac{Y}{\rho}} \qquad [14.3]$$

where Y is the Young's modulus of the solid (see Eqn. 9.3), and ρ is its density. This expression is valid only for a thin, solid rod.

Table 14.1 lists the speeds of sound in various media. The speed of sound is much higher in solids than in gases, because the molecules in a solid interact more strongly with each other than do molecules in a gas. Striking a long steel rail with a hammer, for example, produces two sound waves, one moving through the rail and a slower wave moving through the air. A student with an ear pressed against the rail first hears the faster sound moving through the rail, then the sound moving through air. In general, sound travels faster through solids than liquids and faster through liquids than gases, although there are exceptions.

The speed of sound also depends on the temperature of the medium. For sound traveling through air, the relationship between the speed of sound and temperature is

$$v = (331 \text{ m/s})\sqrt{\frac{T}{273 \text{ K}}} \qquad [14.4]$$

where 331 m/s is the speed of sound in air at 0°C and T is the absolute (Kelvin) temperature. Using this equation the speed of sound in air at 293 K (a typical room temperature) is approximately 343 m/s.

TABLE 14.1

Speeds of Sound in Various Media

Medium	v **(m/s)**
Gases	
Air (0°C)	331
Air (100°C)	386
Hydrogen (0°C)	1 290
Oxygen (0°C)	317
Helium (0°C)	972
Liquids at 25°C	
Water	1 490
Methyl alcohol	1 140
Sea water	1 530
Solids	
Aluminum	5 100
Copper	3 560
Iron	5 130
Lead	1 320
Vulcanized rubber	54

Quick Quiz 14.1

Which of the following actions will increase the speed of sound in air? (a) decreasing the air temperature (b) increasing the frequency of the sound (c) increasing the air temperature (d) increasing the amplitude of the sound wave (e) reducing the pressure of the air.

Applying Physics 14.1 The Sounds Heard During a Storm

How does lightning produce thunder, and what causes the extended rumble?

Explanation Assume that you're at ground level, and neglect ground reflections. When lightning strikes, a channel of ionized air carries a large electric current from a cloud to the ground. This results in a rapid temperature increase of the air in the channel as the current moves through it, causing a similarly rapid expansion of the air. The expansion is so sudden and so intense that a tremendous disturbance is produced in the air—thunder. The entire length of the channel produces the sound at essentially the same instant of time. Sound produced at the bottom of the channel reaches you first, because that's the point closest to you. Sounds from progressively higher portions of the channel reach you at later times, resulting in an extended roar. If the lightning channel were a perfectly straight line, the roar might be steady, but the zigzag shape of the path results in the rumbling variation in loudness, with different quantities of sound energy from different segments arriving at any given instant.

INTERACTIVE EXAMPLE 14.1 Sound Waves in Various Media

Goal Calculate and compare the speeds of sound in different media.

Problem **(a)** If a solid bar of aluminum 1.00 m long is struck at one end with a hammer, a longitudinal pulse propagates down the bar. Find the speed of sound in the bar, which has a Young's modulus of 7.0×10^{10} Pa and a density of 2.7×10^3 kg/m^3. **(b)** Calculate the speed of sound in ethyl alcohol, which has a density of 806 kg/m^3 and bulk modulus of 1.0×10^9 Pa. **(c)** Compute the speed of sound in air at 35.0°C.

Strategy Substitute the given values into the appropriate equations.

Solution

(a) Compute the speed of sound in an aluminum bar.

Substitute values into Equation 14.3:

$$v_{\text{Al}} = \sqrt{\frac{Y}{\rho}} = \sqrt{\frac{7.0 \times 10^{10}\ \text{Pa}}{2.7 \times 10^3\ \text{kg/m}^3}}$$

$$= 5\,100\ \text{m/s, or about } 11\,000\ \text{mi/h !}$$

(b) Compute the speed of sound in ethyl alcohol.

Substitute values into Equation 14.1:

$$v = \sqrt{\frac{B}{\rho}} = \sqrt{\frac{1.0 \times 10^9\ \text{Pa}}{806\ \text{kg/m}^3}} = 1.1 \times 10^3\ \text{m/s}$$

(c) Compute the speed of sound in air at 35.0°C.

Substitute values into Equation 14.4:

$$v = (331\ \text{m/s})\sqrt{\frac{(273\ \text{K} + 35.0\ \text{K})}{273\ \text{K}}} = 352\ \text{m/s}$$

Remark The speed of sound in aluminum is dramatically higher than in either liquid alcohol or air.

Exercise 14.1

Compute the speed of sound in the following substances at 273 K: (a) lead ($Y = 1.6 \times 10^{10}$ Pa), (b) mercury ($B = 2.8 \times 10^{10}$ Pa), and (c) air at −15.0°C.

Answers (a) 1.2×10^3 m/s (b) 1.4×10^3 m/s (c) 322 m/s

PhysicsNow™

You can compare the speeds of sound through various media by logging into PhysicsNow at **www.cp7e.com** and going to Interactive Example 14.1.

14.4 ENERGY AND INTENSITY OF SOUND WAVES

As the tines of a tuning fork move back and forth through the air, they exert a force on a layer of air and cause it to move. In other words, the tines do work on the layer of air. The fact that the fork pours sound energy into the air is one of the reasons the vibration of the fork slowly dies out. (Other factors, such as the energy lost to friction as the tines bend, are also responsible for the lessening of movement.)

The average **intensity *I*** of a wave on a given surface is defined as the rate at which energy flows through the surface, $\Delta E/\Delta t$, divided by the surface area A:

$$I \equiv \frac{1}{A}\frac{\Delta E}{\Delta t} \qquad [14.5]$$

where the direction of energy flow is perpendicular to the surface at every point.

SI unit: watt per meter squared (W/m^2)

A rate of energy transfer is power, so Equation 14.5 can be written in the alternate form

$$I \equiv \frac{\text{power}}{\text{area}} = \frac{\mathcal{P}}{A} \qquad [14.6]$$

◀ Intensity of a wave

where $\mathcal{P}$ is the sound power passing through the surface, measured in watts, and the intensity again has units of watts per square meter.

The faintest sounds the human ear can detect at a frequency of 1 000 Hz have an intensity of about 1×10^{-12} W/m^2. This intensity is called the **threshold of hearing**. The loudest sounds the ear can tolerate have an intensity of about 1 W/m^2 (the **threshold of pain**). At the threshold of hearing, the increase in pressure in the ear is approximately 3×10^{-5} Pa over normal atmospheric pressure. Because atmospheric pressure is about 1×10^5 Pa, this means the ear can detect pressure fluctuations as small as about 3 parts in 10^{10}! The maximum displacement of an air molecule at the threshold of hearing is about 1×10^{-11} m—a remarkably small number! If we compare this displacement with the diameter of a molecule (about 10^{-10} m), we see that the ear is an extremely sensitive detector of sound waves.

The loudest sounds the human ear can tolerate at 1 kHz correspond to a pressure variation of about 29 Pa away from normal atmospheric pressure, with a maximum displacement of air molecules of 1×10^{-5} m.

Intensity Level in Decibels

The loudest tolerable sounds have intensities about 1.0×10^{12} times greater than the faintest detectable sounds. The most intense sound, however, isn't perceived as being 1.0×10^{12} times louder than the faintest sound, because the sensation of loudness is approximately logarithmic in the human ear. (For a review of logarithms, see Section A.3, heading G, in Appendix A.) The relative intensity of a sound is called the **intensity level** or **decibel level**, defined by

$$\beta \equiv 10 \log\left(\frac{I}{I_0}\right) \qquad [14.7]$$

◀ Intensity level

The constant $I_0 = 1.0 \times 10^{-12}$ W/m^2 is the reference intensity, the sound intensity at the threshold of hearing—I is the intensity, and β is the corresponding intensity

TIP 14.1 Intensity Versus Intensity Level

Don't confuse intensity with intensity level. Intensity is a physical quantity with units of watts per meter squared; intensity level, or decibel level, is a convenient mathematical transformation of intensity to a logarithmic scale.

level measured in decibels (dB). (The word *decibel*, which is one-tenth of a *bel*, comes from the name of the inventor of the telephone, Alexander Graham Bell (1847–1922).

To get a feel for various decibel levels, we can substitute a few representative numbers into Equation 14.7, starting with $I = 1.0 \times 10^{-12}\ \mathrm{W/m^2}$:

$$\beta = 10 \log\left(\frac{1.0 \times 10^{-12}\ \mathrm{W/m^2}}{1.0 \times 10^{-12}\ \mathrm{W/m^2}}\right) = 10 \log(1) = 0\ \mathrm{dB}$$

From this result, we see that the lower threshold of human hearing has been chosen to be zero on the decibel scale. Progressing upward by powers of ten yields

$$\beta = 10 \log\left(\frac{1.0 \times 10^{-11}\ \mathrm{W/m^2}}{1.0 \times 10^{-12}\ \mathrm{W/m^2}}\right) = 10 \log(10) = 10\ \mathrm{dB}$$

$$\beta = 10 \log\left(\frac{1.0 \times 10^{-10}\ \mathrm{W/m^2}}{1.0 \times 10^{-12}\ \mathrm{W/m^2}}\right) = 10 \log(100) = 20\ \mathrm{dB}$$

Notice the pattern: *Multiplying* a given intensity by ten *adds* 10 db to the intensity level. This pattern holds throughout the decibel scale. For example, a 50-dB sound is 10 times as intense as a 40-dB sound, while a 60-dB sound is 100 times as intense as a 40-dB sound.

On this scale, the threshold of pain ($I = 1.0\ \mathrm{W/m^2}$) corresponds to an intensity level of $\beta = 10 \log(1/1 \times 10^{-12}) = 10 \log(10^{12}) = 120$ dB. Nearby jet airplanes can create intensity levels of 150 dB, and subways and riveting machines have levels of 90 to 100 dB. The electronically amplified sound heard at rock concerts can attain levels of up to 120 dB, the threshold of pain. Exposure to such high intensity levels can seriously damage the ear. Earplugs are recommended whenever prolonged intensity levels exceed 90 dB. Recent evidence suggests that noise pollution, which is common in most large cities and in some industrial environments, may be a contributing factor to high blood pressure, anxiety, and nervousness. Table 14.2 gives the approximate intensity levels of various sounds.

TABLE 14.2

Intensity Levels in Decibels for Different Sources

Source of Sound	β(dB)
Nearby jet airplane	150
Jackhammer, machine gun	130
Siren, rock concert	120
Subway, power mower	100
Busy traffic	80
Vacuum cleaner	70
Normal conversation	50
Mosquito buzzing	40
Whisper	30
Rustling leaves	10
Threshold of hearing	0

EXAMPLE 14.2 A Noisy Grinding Machine

Goal Working with watts and decibels.

Problem A noisy grinding machine in a factory produces a sound intensity of $1.00 \times 10^{-5}\ \mathrm{W/m^2}$. Calculate **(a)** the decibel level of this machine, and **(b)** the new intensity level when a second, identical machine is added to the factory. **(c)** A certain number of additional such machines are put into operation alongside these two. When all the machines are running at the same time the decibel level is 77.0 dB. Find the sound intensity.

Strategy Parts (a) and (b) require substituting into the decibel formula, Equation 14.7, with the intensity in part (b) twice the intensity in part (a). In part (c), the intensity level in decibels is given, and it's necessary to work backwards, using the inverse of the logarithm function, to get the intensity in watts per meter squared.

Solution

(a) Calculate the intensity level of the single grinder.

Substitute the intensity into the decibel formula:

$$\beta = 10 \log\left(\frac{1.00 \times 10^{-5}\ \mathrm{W/m^2}}{1.00 \times 10^{-12}\ \mathrm{W/m^2}}\right) = 10 \log(10^7)$$

$$= 70.0\ \mathrm{dB}$$

(b) Calculate the new intensity level when an additional machine is added.

Substitute twice the intensity of part (a) into the decibel formula:

$$\beta = 10 \log\left(\frac{2.00 \times 10^{-5}\ \mathrm{W/m^2}}{1.00 \times 10^{-12}\ \mathrm{W/m^2}}\right) = 73.0\ \mathrm{dB}$$

(c) Find the intensity corresponding to an intensity level of 77.0 dB.

Substitute 77.0 dB into the decibel formula and divide both sides by 10:

$$\beta = 77.0 \text{ dB} = 10 \log\left(\frac{I}{I_0}\right)$$

$$7.70 = \log\left(\frac{I}{10^{-12}\text{ W/m}^2}\right)$$

Make each side the exponent of 10. On the right-hand side, $10^{\log u} = u$, by definition of base ten logarithms.

$$10^{7.70} = 5.01 \times 10^7 = \frac{I}{1.00 \times 10^{-12}\text{ W/m}^2}$$

$$I = 5.01 \times 10^{-5}\text{ W/m}^2$$

Remark The answer is five times the intensity of the single grinder, so in part (c) there are five such machines operating simultaneously. Because of the logarithmic definition of intensity level, large changes in intensity correspond to small changes in intensity level.

Exercise 14.2

Suppose a manufacturing plant has an average sound intensity level of 97.0 dB created by 25 identical machines. (a) Find the total intensity created by all the machines. (b) Find the sound intensity created by one such machine. (c) What's the sound intensity level if five such machines are running?

Answers (a) $5.01 \times 10^{-3}\text{ W/m}^2$ (b) $2.00 \times 10^{-4}\text{ W/m}^2$ (c) 90.0 dB

APPLICATION

OSHA Noise-Level Regulations

Federal regulations now demand that no office or factory worker be exposed to noise levels that average more than 90 dB over an 8-h day. From a management point of view, here's the good news: one machine in the factory may produce a noise level of 70 dB, but a second machine, while doubling the total intensity, increases the noise level by only 3 dB. Because of the logarithmic nature of intensity levels, doubling the intensity doesn't double the intensity level; in fact, it alters it by a surprisingly small amount. This means that equipment can be added to the factory without appreciably altering the intensity level of the environment.

Now here's the bad news: as you remove noisy machinery, the intensity level isn't lowered appreciably. In Exercise 14.2, reducing the intensity level by 7 dB would require the removal of 20 of the 25 machines! To lower the level another 7 dB would require removing 80% of the remaining machines, in which case only one machine would remain.

14.5 SPHERICAL AND PLANE WAVES

If a small spherical object oscillates so that its radius changes periodically with time, a spherical sound wave is produced (Fig. 14.4, page 466). The wave moves outward from the source at a constant speed.

Because all points on the vibrating sphere behave in the same way, we conclude that the energy in a spherical wave propagates equally in all directions. This means that no one direction is preferred over any other. If $\mathcal{P}_{av}$ is the average power emitted by the source, then at any distance r from the source, this power must be distributed over a spherical surface of area $4\pi r^2$, assuming no absorption in the medium. (Recall that $4\pi r^2$ is the surface area of a sphere.) Hence, the **intensity** of the sound at a distance r from the source is

$$I = \frac{\text{average power}}{\text{area}} = \frac{\mathcal{P}_{av}}{A} = \frac{\mathcal{P}_{av}}{4\pi r^2} \qquad [14.8]$$

This equation shows that the intensity of a wave decreases with increasing distance from its source, as you might expect. The fact that I varies as $1/r^2$ is a result of the

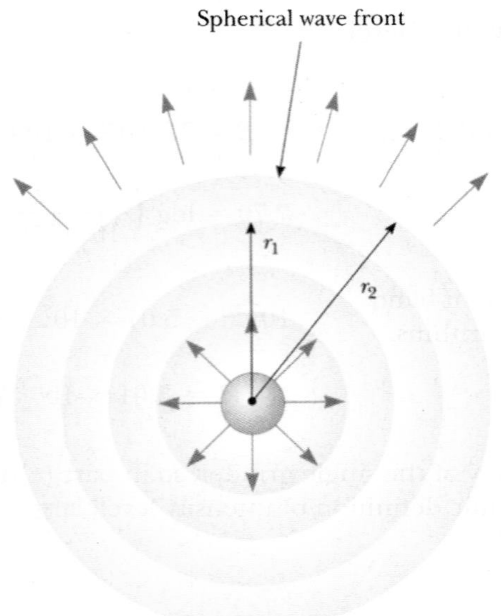

Figure 14.4 A spherical wave propagating radially outward from an oscillating sphere. The intensity of the wave varies as $1/r^2$.

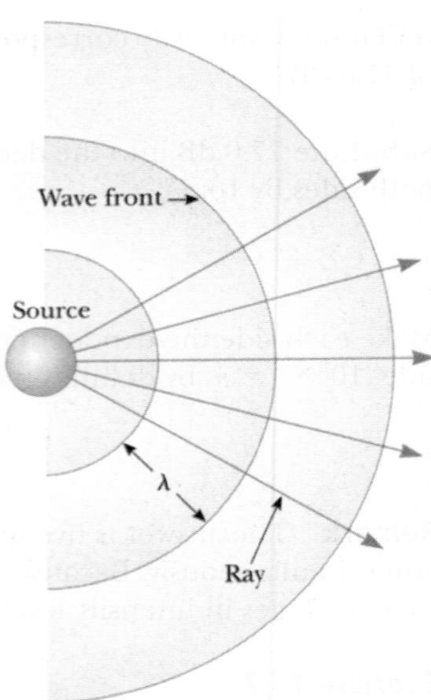

Figure 14.5 Spherical waves emitted by a point source. The circular arcs represent the spherical wave fronts concentric with the source. The rays are radial lines pointing outward from the source, perpendicular to the wavefronts.

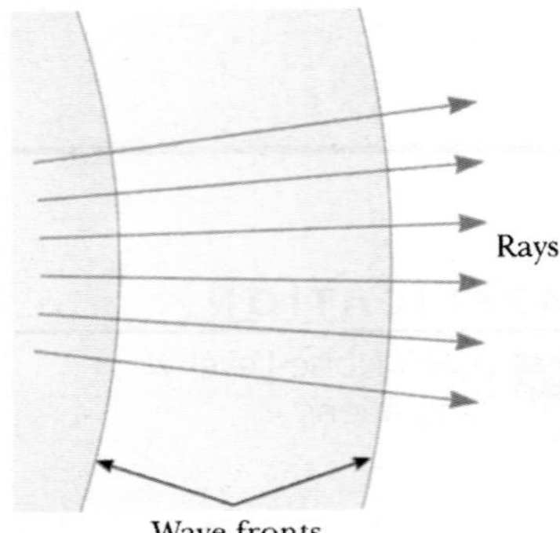

Figure 14.6 Far away from a point source, the wave fronts are nearly parallel planes and the rays are nearly parallel lines perpendicular to the planes. Hence, a small segment of a spherical wavefront is approximately a plane wave.

assumption that the small source (sometimes called a **point source**) emits a spherical wave. (In fact, light waves also obey this so-called inverse-square relationship.) Because the average power is the same through any spherical surface centered at the source, we see that the intensities at distances r_1 and r_2 (Fig. 14.4) from the center of the source are

$$I_1 = \frac{\mathcal{P}_{av}}{4\pi r_1{}^2} \qquad I_2 = \frac{\mathcal{P}_{av}}{4\pi r_2{}^2}$$

The ratio of the intensities at these two spherical surfaces is

$$\frac{I_1}{I_2} = \frac{r_2{}^2}{r_1{}^2} \qquad [14.9]$$

It's useful to represent spherical waves graphically with a series of circular arcs (lines of maximum intensity) concentric with the source representing part of a spherical surface, as in Figure 14.5. We call such an arc a **wave front.** The distance between adjacent wave fronts equals the wavelength λ. The radial lines pointing outward from the source and perpendicular to the arcs are called **rays.**

Now consider a small portion of a wave front that is at a *great* distance (relative to λ) from the source, as in Figure 14.6. In this case, the rays are nearly parallel to each other and the wave fronts are very close to being planes. At distances from the source that are great relative to the wavelength, therefore, we can approximate the wave front with parallel planes, called **plane waves.** Any small portion of a spherical wave that is far from the source can be considered a plane wave. Figure 14.7 illustrates a plane wave propagating along the x-axis. If the positive x-direction is taken to be the direction of the wave motion (or ray) in this figure, then the wave fronts are parallel to the plane containing the y- and z-axes.

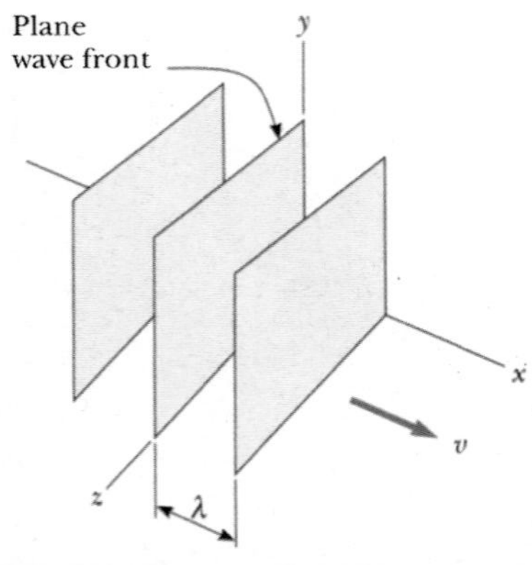

Figure 14.7 A representation of a plane wave moving in the positive x-direction with a speed v. The wavefronts are planes parallel to the yz-plane.

EXAMPLE 14.3 Intensity Variations of a Point Source

Goal Relate sound intensities and their distances from a point source.

Problem A small source emits sound waves with a power output of 80.0 W. **(a)** Find the intensity 3.00 m from the source. **(b)** At what distance would the intensity be one-fourth as much as it is at $r = 3.00$ m? **(c)** Find the distance at which the sound level is 40.0 dB.

Strategy The source is small, so the emitted waves are spherical and the intensity in part (a) can be found by substituting values into Equation 14.8. Part (b) involves solving for r in Equation 14.8 followed by substitution (though Equation 14.9 can be used instead). In part (c), convert from the sound intensity level to the intensity in W/m^2, using Equation 14.7. Then substitute into Equation 14.9 (though 14.8 could be used, instead) and solve for r_2.

Solution

(a) Find the intensity 3.00 m from the source.

Substitute $\mathcal{P}_{av} = 80.0$ W and $r = 3.00$ m into Equation 14.8:

$$I = \frac{\mathcal{P}_{av}}{4\pi r^2} = \frac{80.0 \text{ W}}{4\pi(3.00 \text{ m})^2} = \boxed{0.707 \text{ W/m}^2}$$

(b) At what distance would the intensity be one-fourth as much as it is at $r = 3.00$ m?

Take $I = (0.707 \text{ W/m}^2)/4$, and solve for r in Equation 14.8:

$$r = \left(\frac{\mathcal{P}_{av}}{4\pi I}\right)^{1/2} = \left(\frac{80.0 \text{ W}}{4\pi(0.707 \text{ W/m}^2)/4.0}\right)^{1/2} = \boxed{6.00 \text{ m}}$$

(c) Find the distance at which the sound level is 40.0 dB.

Convert the intensity level of 40.0 dB to an intensity in W/m^2 by solving Equation 14.7 for I:

$$40.0 = 10 \log\left(\frac{I}{I_0}\right) \rightarrow 4.00 = \log\left(\frac{I}{I_0}\right)$$

$$10^{4.00} = \frac{I}{I_0} \rightarrow I = 10^{4.00} I_0 = 1.00 \times 10^{-8} \text{ W/m}^2$$

Solve Equation 14.9 for $r_2{}^2$, substitute the intensity and the result of part (a), and take the square root:

$$\frac{I_1}{I_2} = \frac{r_2{}^2}{r_1{}^2} \rightarrow r_2{}^2 = r_1{}^2 \frac{I_1}{I_2}$$

$$r_2{}^2 = (3.00 \text{ m})^2 \left(\frac{0.707 \text{ W/m}^2}{1.00 \times 10^{-8} \text{ W/m}^2}\right)$$

$$r_2 = \boxed{2.52 \times 10^4 \text{ m}}$$

Remarks Once the intensity is known at one position a certain distance away from the source, it's easier to use Equation 14.9 rather than Equation 14.8 to find the intensity at any other location. This is particularly true for part (b), where, using Equation 14.9, we can see right away that doubling the distance reduces the intensity to one-quarter its previous value.

Exercise 14.3

Suppose a certain jet plane creates an intensity level of 125 dB at a distance of 5.00 m. What intensity level does it create on the ground directly underneath it when flying at an altitude of 2.00 km?

Answer 73.0 dB

14.6 THE DOPPLER EFFECT

If a car or truck is moving while its horn is blowing, the frequency of the sound you hear is higher as the vehicle approaches you and lower as it moves away from you. This is one example of the *Doppler effect,* named for the Austrian physicist Christian Doppler (1803–1853), who discovered it. The same effect is heard if you're on a motorcycle and the horn is stationary: the frequency is higher as you approach the source and lower as you move away.

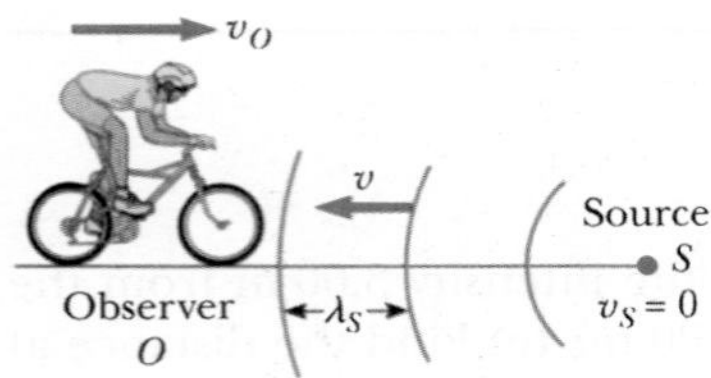

ACTIVE FIGURE 14.8
An observer moving with a speed v_O *toward* a stationary point source (S) hears a frequency f_O that is *greater* than the source frequency f_S.

Physics Now™
Log into PhysicsNow at **www.cp7e.com** and go to Active Figure 14.8 to adjust the speed of the observer.

Although the Doppler effect is most often associated with sound, it's common to all waves, including light.

In deriving the Doppler effect, we assume that the air is stationary and that all speed measurements are made relative to this stationary medium. The speed v_O is the speed of the observer, v_S is the speed of the source, and v is the speed of sound.

Case 1: The Observer Is Moving Relative to a Stationary Source

In Active Figure 14.8 an observer is moving with a speed of v_O toward the source (considered a point source), which is at rest ($v_S = 0$).

We take the frequency of the source to be f_S, the wavelength of the source to be λ_S, and the speed of sound in air to be v. If both observer and source are stationary, the observer detects f_S wave fronts per second. (That is, when $v_O = 0$ and $v_S = 0$, the observed frequency f_O equals the source frequency f_S.) When moving toward the source, the observer moves a distance of $v_O t$ in t seconds. During this interval, **the observer detects an additional number of wave fronts**. The number of extra wave fronts is equal to the distance traveled, $v_O t$, divided by the wavelength λ_S:

$$\text{Additional wave fronts detected} = \frac{v_O t}{\lambda_S}$$

Divide this equation by the time t to get the number of additional wave fronts detected *per second*, v_O/λ_S. Hence, the frequency heard by the observer is *increased* to

$$f_O = f_S + \frac{v_O}{\lambda_S}$$

Substituting $\lambda_S = v/f_S$ into this expression for f_O we obtain

$$f_O = f_S\left(\frac{v + v_O}{v}\right) \quad [14.10]$$

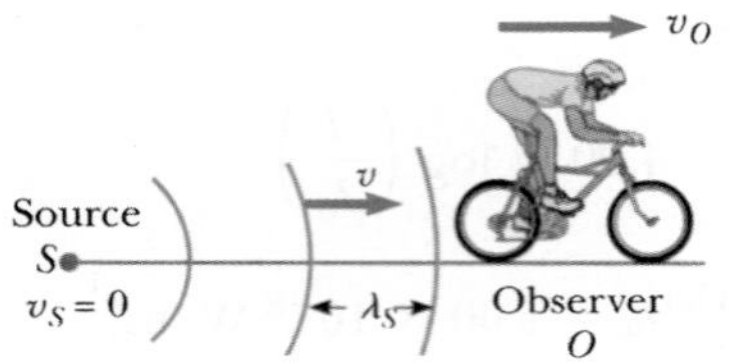

Figure 14.9 An observer moving with a speed of v_O *away* from a stationary source hears a frequency f_O that is *lower* than the source frequency f_S.

When the observer is *moving away* from a stationary source (Fig. 14.9), the observed frequency decreases. A derivation yields the same result as Equation 14.10, but with $v - v_O$ in the numerator. Therefore, when the observer is moving away from the source, substitute $-v_O$ for v_O in Equation 14.10.

Case 2: The Source Is Moving Relative to a Stationary Observer

Now consider a source moving toward an observer at rest, as in Active Figure 14.10. Here, the wave fronts passing observer A are closer together because the source is moving in the direction of the outgoing wave. As a result, the wavelength λ_O measured by observer A is shorter than the wavelength λ_S of the source at rest. During each vibration, which lasts for an interval T (the period), the source moves a distance $v_S T = v_S/f_S$ and **the wavelength is shortened by that amount**. The observed wavelength is therefore given by

$$\lambda_O = \lambda_S - \frac{v_S}{f_S}$$

Because $\lambda_S = v/f_S$, the frequency observed by A is

$$f_O = \frac{v}{\lambda_O} = \frac{v}{\lambda_S - \dfrac{v_S}{f_S}} = \frac{v}{\dfrac{v}{f_S} - \dfrac{v_S}{f_S}}$$

or

$$f_O = f_S\left(\frac{v}{v - v_S}\right) \quad [14.11]$$

As expected, **the observed frequency increases when the source is moving toward the observer**. When the source is *moving away* from an observer at rest, the minus sign in the denominator must be replaced with a plus sign, so the factor becomes $(v + v_S)$.

TIP 14.2 Doppler Effect Doesn't Depend on Distance
The sound from a source approaching at constant speed will increase in intensity, but the observed (elevated) frequency will remain unchanged. The Doppler effect doesn't depend on distance.

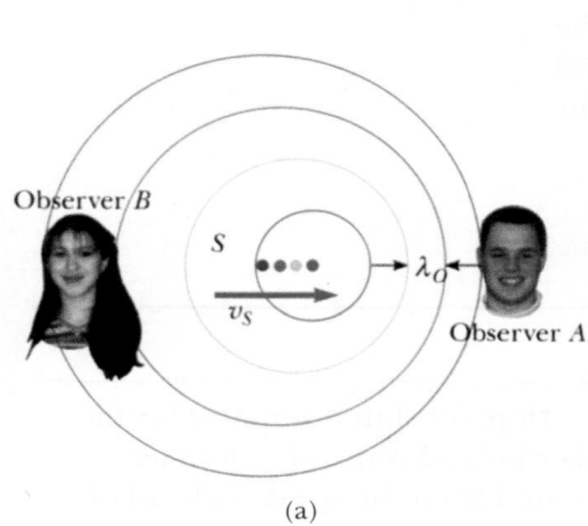

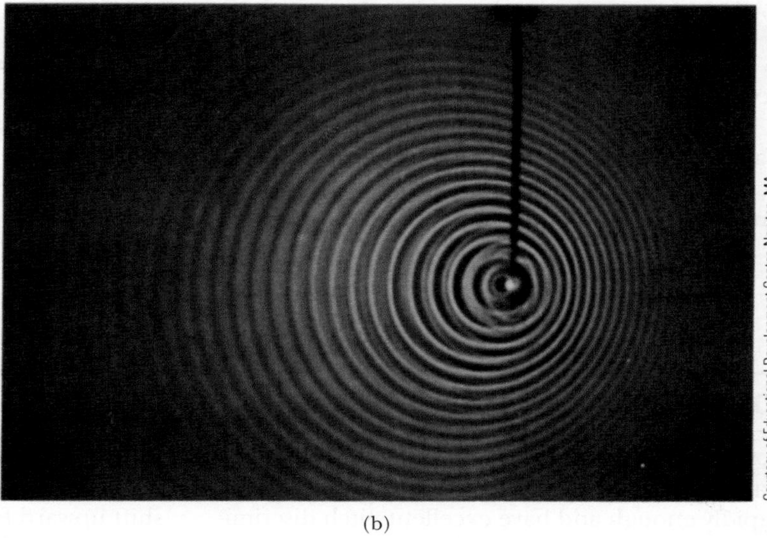

Courtesy of Educational Development Center, Newton, MA.

ACTIVE FIGURE 14.10
(a) A source S moving with speed v_S toward stationary observer A and away from stationary observer B. Observer A hears an *increased* frequency, and observer B hears a *decreased* frequency. (b) The Doppler effect in water, observed in a ripple tank. The source producing the water waves is moving to the right.

Physics Now™
Log into PhysicsNow at **www.cp7e.com**, and go to Active Figure 14.10 to adjust the speed of the source.

General Case

When both the source and the observer are in motion relative to Earth, Equations 14.10 and 14.11 can be combined to give

$$f_O = f_S \left(\frac{v + v_O}{v - v_S} \right) \qquad [14.12]$$

◀ Doppler shift equation—observer and source in motion

In this expression, the signs for the values substituted for v_O and v_S depend on the direction of the velocity. When the observer moves *toward* the source, a *positive* speed is substituted for v_O; when the observer moves *away from* the source, a *negative* speed is substituted for v_O. Similarly, a *positive* speed is substituted for v_S when the source moves *toward* the observer, a *negative* speed when the source moves away from the observer.

Choosing incorrect signs is the most common mistake made in working a Doppler effect problem. The following rules may be helpful: The word *toward* is associated with an *increase* in the observed frequency; the words *away from* are associated with a *decrease* in the observed frequency.

These two rules derive from the physical insight that when the observer is moving toward the source (or the source toward the observer), there is a smaller observed period between wave crests, hence a larger frequency, with the reverse holding—a smaller observed frequency—when the observer is moving away from the source (or the source away from the observer). Keep the physical insight in mind whenever you're in doubt about the signs in Equation 14.12: Adjust them as necessary to get the correct physical result.

The second most common mistake made in applying Equation 14.12 is to accidentally reverse numerator and denominator. Some find it helpful to remember the equation in the following form:

$$\frac{f_O}{v + v_O} = \frac{f_S}{v - v_S}$$

The advantage of this form is its symmetry: both sides are very nearly the same, with *O*'s on the left and *S*'s on the right. Forgetting which side has the plus sign and which has the minus sign is not a serious problem, as long as physical insight is used to check the answer and make adjustments as necessary.

Quick Quiz 14.2

Suppose you're on a hot air balloon ride, carrying a buzzer that emits a sound of frequency f. If you accidentally drop the buzzer over the side while the balloon is rising at constant speed, what can you conclude about the sound you hear as the buzzer falls toward the ground? (a) the frequency and intensity increase, (b) the frequency decreases and the intensity increases, (c) the frequency decreases and the intensity decreases, or (d) the frequency remains the same, but the intensity decreases.

Applying Physics 14.2 Out-of-Tune Speakers

Suppose you place your stereo speakers far apart and run past them from right to left or left to right. If you run rapidly enough and have excellent pitch discrimination, you may notice that the music playing seems to be out of tune when you're between the speakers. Why?

Explanation When you are between the speakers, you are running away from one of them and toward the other, so there is a Doppler shift downward for the sound from the speaker behind you and a Doppler shift upward for the sound from the speaker ahead of you. As a result, the sound from the two speakers will not be in tune. A calculation shows that a world-class sprinter could run fast enough to generate about a semitone difference in the sound from the two speakers.

EXAMPLE 14.4 Listen, but Don't Stand on the Track

Goal Solve a Doppler shift problem when only the source is moving.

Problem A train moving at a speed of 40.0 m/s sounds its whistle, which has a frequency of 5.00×10^2 Hz. Determine the frequency heard by a stationary observer as the train *approaches* the observer. The ambient temperature is 24.0°C.

Strategy Use Equation 14.4 to get the speed of sound at the ambient temperature, then substitute values into Equation 14.12 for the Doppler shift. Because the train approaches the observer, the observed frequency will be larger. Choose the sign of v_S to reflect this fact.

Solution

Use Equation 14.4 to calculate the speed of sound in air at $T = 24.0$°C:

$$v = (331 \text{ m/s})\sqrt{\frac{T}{273 \text{ K}}}$$

$$= (331 \text{ m/s})\sqrt{\frac{(273 + 24.0)\text{ K}}{273 \text{ K}}} = 345 \text{ m/s}$$

The observer is stationary, so $v_O = 0$. The train is moving *toward* the observer, so $v_S = +40.0$ m/s (*positive*). Substitute these values and the speed of sound into the Doppler shift equation:

$$f_O = f_S\left(\frac{v + v_O}{v - v_S}\right)$$

$$= (5.00 \times 10^2 \text{ Hz})\left(\frac{345 \text{ m/s}}{345 \text{ m/s} - 40.0 \text{ m/s}}\right)$$

$$= 566 \text{ Hz}$$

Remark If the train were going away from the observer, $v_S = -40.0$ m/s would have been chosen instead.

Exercise 14.4

Determine the frequency heard by the stationary observer as the train *recedes* from the observer.

Answer 448 Hz

INTERACTIVE EXAMPLE 14.5 The Noisy Siren

Goal Solve a Doppler shift problem when both the source and observer are moving.

Problem An ambulance travels down a highway at a speed of 75.0 mi/h, its siren emitting sound at a frequency of 4.00×10^2 Hz. What frequency is heard by a passenger in a car traveling at 55.0 mi/h in the opposite direction as the car and ambulance **(a)** *approach* each other and **(b)** pass and *move away* from each other? Take the speed of sound in air to be $v = 345$ m/s.

Strategy Aside from converting from mi/h to m/s, this problem only requires substitution into the Doppler formula, but two signs must be chosen correctly in each part. In part (a), the observer moves toward the source and the source moves toward the observer, so both v_O and v_S should be chosen to be positive. Switch signs after they pass each other.

Solution

Convert the speeds from mi/h to m/s:

$$v_S = (75.0 \text{ mi/h})\left(\frac{0.447 \text{ m/s}}{1.00 \text{ mi/h}}\right) = 33.5 \text{ m/s}$$

$$v_O = (55.0 \text{ mi/h})\left(\frac{0.447 \text{ m/s}}{1.00 \text{ mi/h}}\right) = 24.6 \text{ m/s}$$

(a) Compute the observed frequency as the ambulance and car approach each other.

Each vehicle goes toward the other, so substitute $v_O = +24.6$ m/s and $v_S = +33.5$ m/s into the Doppler shift formula:

$$f_O = f_S\left(\frac{v + v_O}{v - v_S}\right)$$

$$= (4.00 \times 10^2 \text{ Hz})\left(\frac{345 \text{ m/s} + 24.6 \text{ m/s}}{345 \text{ m/s} - 33.5 \text{ m/s}}\right) = 475 \text{ Hz}$$

(b) Compute the observed frequency as the ambulance and car recede from each other.

Each vehicle goes away from the other, so substitute $v_O = -24.6$ m/s and $v_S = -33.5$ m/s into the Doppler shift formula:

$$f_O = f_S\left(\frac{v + v_O}{v - v_S}\right)$$

$$= (4.00 \times 10^2 \text{ Hz})\left(\frac{345 \text{ m/s} + (-24.6 \text{ m/s})}{345 \text{ m/s} - (-33.5 \text{ m/s})}\right)$$

$$= 339 \text{ Hz}$$

Remarks Notice how the signs were handled: In part (b), the negative signs were required on the speeds because both observer and source were moving away from each other. Sometimes, of course, one of the speeds is negative and the other is positive.

Exercise 14.5

Repeat this problem, but assume the ambulance and car are going the same direction, with the ambulance initially behind the car. The speeds and the frequency of the siren are the same as in the example. Find the frequency heard by the observer in the car (a) before and (b) after the ambulance passes the car. [*Note:* The highway patrol subsequently gives the driver of the car a ticket for not pulling over for an emergency vehicle!]

Answers (a) 411 Hz (b) 391 Hz

PhysicsNow™

You can alter the relative speeds of two submarines and observe the Doppler-shifted frequency by logging into PhysicsNow at **www.cp7e.com** and going to Interactive Example 14.5.

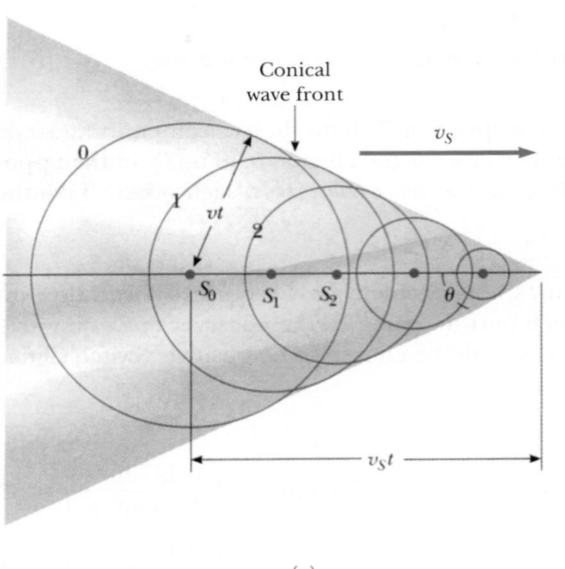

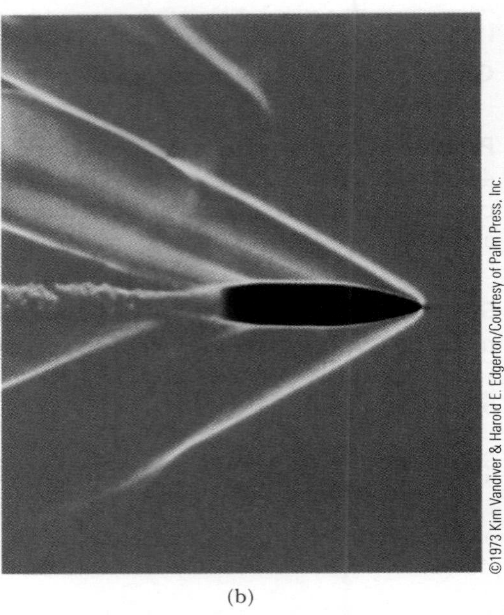

Figure 14.11 (a) A representation of a shock wave, produced when a source moves from S_0 to S_n with a speed v_s that is *greater* than the wave speed v in that medium. The envelope of the wave fronts forms a cone with half-angle of $\sin \theta = v/v_s$. (b) A stroboscopic photograph of a bullet moving at supersonic speed through the hot air above a candle.

Shock Waves

What happens when the source speed v_S *exceeds* the wave velocity v? Figure 14.11a describes this situation graphically. The circles represent spherical wave fronts emitted by the source at various times during its motion. At $t = 0$, the source is at point S_0, and at some later time t, the source is at point S_n. In the interval t, the wave front centered at S_0 reaches a radius of vt. In this same interval, the source travels to S_n, a distance of v_st. At the instant the source is at S_n, the waves just beginning to be generated at this point have wave fronts of zero radius. The line drawn from S_n to the wave front centered on S_0 is tangent to all other wave fronts generated at intermediate times. All such tangent lines lie on the surface of a cone. The angle θ between one of these tangent lines and the direction of travel is given by

$$\sin \theta = \frac{v}{v_s}$$

Figure 14.12 The V-shaped bow wave of a boat is formed because the boat travels at a speed greater than the speed of the water waves. A bow wave is analogous to a shock wave formed by an airplane traveling faster than sound.

The ratio v_s/v is called the **Mach number**. The conical wave front produced when $v_s > v$ (supersonic speeds) is known as a **shock wave**. Figure 14.11b is a photograph of a bullet traveling at supersonic speed through the hot air rising above a candle. Notice the shock waves in the vicinity of the bullet. Another interesting example of a shock wave is the V-shaped wave front produced by a boat (the bow wave) when the boat's speed exceeds the speed of the water waves (Fig. 14.12).

Jet aircraft and space shuttles traveling at supersonic speeds produce shock waves that are responsible for the loud explosion, or sonic boom, heard on the ground. A shock wave carries a great deal of energy concentrated on the surface of the cone, with correspondingly great pressure variations. Shock waves are unpleasant to hear and can damage buildings when aircraft fly supersonically at low altitudes. In fact, an airplane flying at supersonic speeds produces a double boom, because two shock waves are formed—one from the nose of the plane and one from the tail (Fig. 14.13).

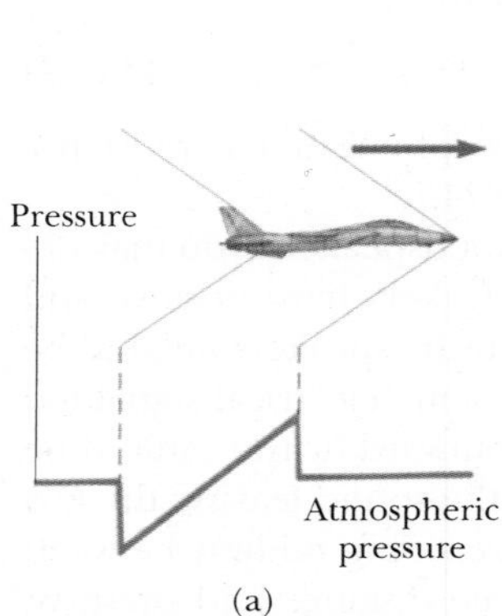

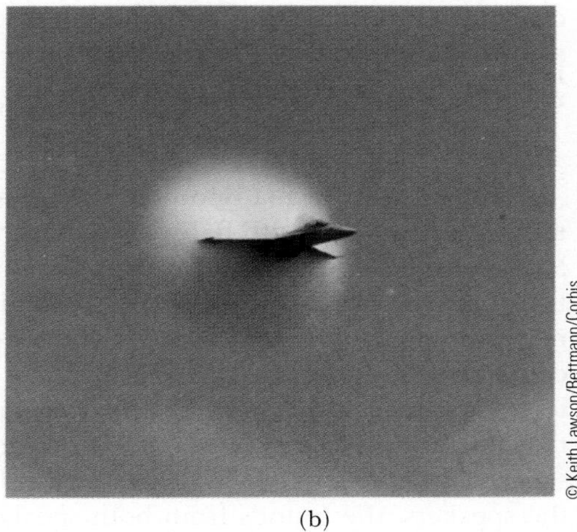

Figure 14.13 (a) The two shock waves produced by the nose and tail of a jet airplane traveling at supersonic speed. (b) A shock wave due to a jet traveling at the speed of sound is made visible as a fog of water vapor. The large pressure variation in the shock wave causes the water in the air to condense into water droplets.

Quick Quiz 14.3

As an airplane flying with constant velocity moves from a cold air mass into a warm air mass, does the Mach number (a) increase, (b) decrease, or (c) remain the same?

14.7 INTERFERENCE OF SOUND WAVES

Sound waves can be made to interfere with each other, a phenomenon that can be demonstrated with the device shown in Figure 14.14. Sound from a loudspeaker at S is sent into a tube at *P*, where there is a T-shaped junction. The sound splits and follows two separate pathways, indicated by the red arrows. Half of the sound travels upward, half downward. Finally, the two sounds merge at an opening where a listener places her ear. If the two paths r_1 and r_2 have the same length, waves that enter the junction will separate into two halves, travel the two paths, and then combine again at the ear. This reuniting of the two waves produces *constructive interference*, so the listener hears a loud sound. If the upper path is adjusted to be one full wavelength longer than the lower path, constructive interference of the two waves occurs again, and a loud sound is detected at the receiver. We have the following result: **If the path difference $r_2 - r_1$ is zero or some integer multiple of wavelengths, then constructive interference occurs and**

$$r_2 - r_1 = n\lambda \quad (n = 0, 1, 2, \ldots) \qquad [14.13]$$

◀ Condition for constructive interference

Suppose, however, that one of the path lengths, r_2, is adjusted so that the upper path is half a wavelength *longer* than the lower path r_1. In this case, an entering sound wave splits and travels the two paths as before, but now the wave along the upper path must travel a distance equivalent to half a wavelength farther than the wave traveling along the lower path. As a result, the crest of one wave meets the trough of the other when they merge at the receiver, causing the two waves to

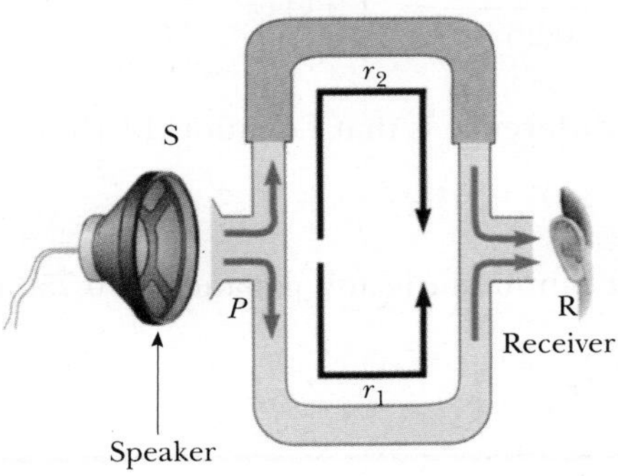

Figure 14.14 An acoustical system for demonstrating interference of sound waves. Sound from the speaker enters the tube and splits into two parts at *P*. The two waves combine at the opposite side and are detected at *R*. The upper path length is varied by the sliding section.

cancel each other. This phenomenon is called *totally destructive interference,* and no sound is detected at the receiver. In general, **if the path difference $r_2 - r_1$ is $\frac{1}{2}$, $1\frac{1}{2}$, $2\frac{1}{2}$ . . . wavelengths, destructive interference occurs** and

Condition for destructive interference ▶

$$r_2 - r_1 = (n + \tfrac{1}{2})\lambda \quad (n = 0, 1, 2, \ldots) \qquad [14.14]$$

Nature provides many other examples of interference phenomena, most notably in connection with light waves, described in Chapter 24.

APPLICATION

Connecting Your Stereo Speakers

In connecting the wires between your stereo system and loudspeakers, you may notice that the wires are usually color coded and that the speakers have positive and negative signs on the connections. The reason for this is that the speakers need to be connected with the same "polarity." If they aren't, then the same electrical signal fed to both speakers will result in one speaker cone moving outward at the same time that the other speaker cone is moving inward. In this case, the sound leaving the two speakers will be 180° out of phase with each other. If you are sitting midway between the speakers, the sounds from both speakers travel the same distance and preserve the phase difference they had when they left. In an ideal situation, for a 180° phase difference, you would get complete destructive interference and no sound! In reality, the cancellation is not complete and is much more significant for bass notes (which have a long wavelength) than for the shorter wavelength treble notes. Nevertheless, to avoid a significant reduction in the intensity of bass notes, the color-coded wires and the signs on the speaker connections should be carefully noted.

TIP 14.3 Do Waves Really Interfere?

In popular usage, to *interfere* means "to come into conflict with" or "to intervene to affect an outcome." This differs from its use in physics, where waves pass through each other and interfere, but don't affect each other in any way.

EXAMPLE 14.6 Two Speakers Driven by the Same Source

Goal Use the concept of interference to compute a frequency.

Problem Two speakers placed 3.00 m apart are driven by the same oscillator (Fig. 14.15). A listener is originally at point *O*, which is located 8.00 m from the center of the line connecting the two speakers. The listener then walks to point *P*, which is a perpendicular distance 0.350 m from *O*, before reaching the *first minimum* in sound intensity. What is the frequency of the oscillator? Take the speed of sound in air to be $v_s = 343$ m/s.

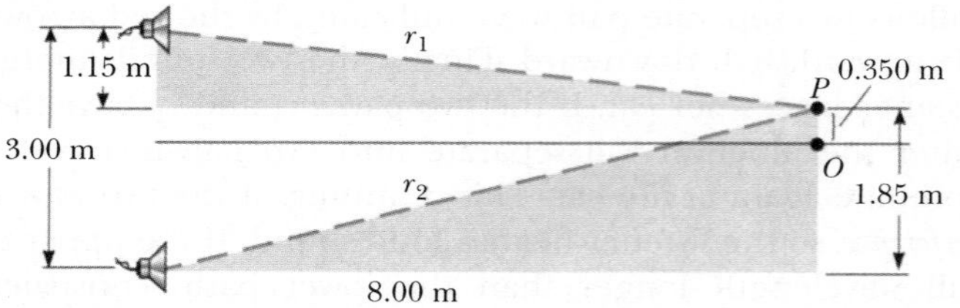

Strategy The position of the first minimum in sound intensity is given, which is a point of destructive interference. We can find the path lengths r_1 and r_2 with the Pythagorean theorem and then use Equation 14.14 for destructive interference to find the wavelength λ. Using $v = f\lambda$ then yields the frequency.

Figure 14.15 (Example 14.6) Two loudspeakers driven by the same source can produce interference.

Solution

Use the Pythagorean theorem to find the path lengths r_1 and r_2:

$$r_1 = \sqrt{(8.00 \text{ m})^2 + (1.15 \text{ m})^2} = 8.08 \text{ m}$$

$$r_2 = \sqrt{(8.00 \text{ m})^2 + (1.85 \text{ m})^2} = 8.21 \text{ m}$$

Substitute these values and $n = 0$ into Equation 14.14, solving for the wavelength:

$$r_2 - r_1 = (n + \tfrac{1}{2})\lambda$$

$$8.21 \text{ m} - 8.08 \text{ m} = 0.13 \text{ m} = \lambda/2 \quad \rightarrow \quad \lambda = 0.26 \text{ m}$$

Solve $v = \lambda f$ for the frequency f and substitute the speed of sound and the wavelength:

$$f = \frac{v}{\lambda} = \frac{343 \text{ m/s}}{0.26 \text{ m}} = \boxed{1.3 \text{ kHz}}$$

Remark For problems involving constructive interference, the only difference is that Equation 14.13, $r_2 - r_1 = n\lambda$, would be used instead of Equation 14.14.

Exercise 14.6

If the oscillator frequency is adjusted so that the location of the first minimum is at a distance of 0.750 m from *O*, what is the new frequency?

Answer 0.642 kHz

14.8 STANDING WAVES

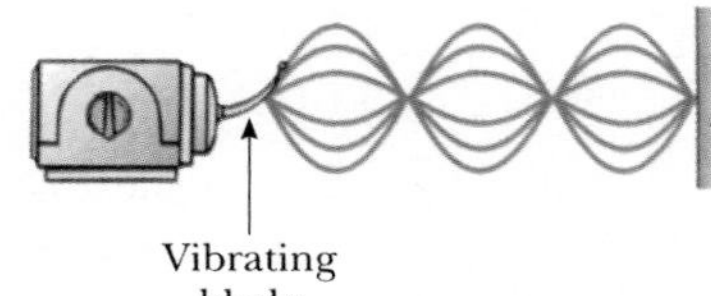

Figure 14.16 Standing waves can be set up in a stretched string by connecting one end of the string to a vibrating blade. When the blade vibrates at one of the natural frequencies of the string, large-amplitude standing waves are created.

Standing waves can be set up in a stretched string by connecting one end of the string to a stationary clamp and connecting the other end to a vibrating object, such as the end of a tuning fork, or by shaking the hand holding the string up and down at a steady rate (Fig. 14.16). Traveling waves then reflect from the ends and move in both directions on the string. The incident and reflected waves combine according to the **superposition principle**. (See Section 13.10.) If the string vibrates at exactly the right frequency, the wave appears to stand still—hence its name, **standing wave**. A **node** occurs where the two traveling waves always have the same magnitude of displacement but the opposite sign, so the net displacement is zero at that point. There is no motion in the string at the nodes, but midway between two adjacent nodes, at an **antinode**, the string vibrates with the largest amplitude.

Figure 14.17 shows snapshots of the oscillation of a standing wave during half of a cycle. The pink arrows indicate the direction of motion of different parts of the string. Notice that **all points on the string oscillate together vertically with the same frequency, but different points have different amplitudes of motion**. The points of attachment to the wall and all other stationary points on the string are called nodes, labeled N in Figure 14.17a. From the figure, observe that the distance between adjacent nodes is one-half the wavelength of the wave:

$$d_{\text{NN}} = \tfrac{1}{2}\lambda$$

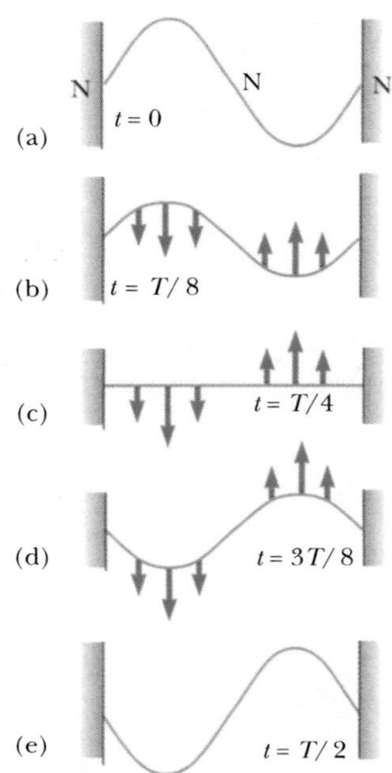

Figure 14.17 A standing-wave pattern in a stretched string, shown by snapshots of the string during one-half of a cycle. In part (a), N denotes a node.

Consider a string of length L that is fixed at both ends, as in Active Figure 14.18. For a string, we can set up standing-wave patterns at many frequencies—the more loops, the higher the frequency. Three such patterns are shown in Active Figures 14.18b, 14.18c, and 14.18d. Each has a characteristic frequency, which we will now calculate.

First, **the ends of the string must be nodes, because these points are fixed.** If the string is displaced at its midpoint and released, the vibration shown in Active Figure 14.18b can be produced, in which case the center of the string is an antinode, labeled A. Note that from end to end, the pattern is N–A–N. The distance from a node to its adjacent antinode, N–A, is always equal to a quarter wavelength, $\lambda_1/4$. There are two such segments, N–A and A–N, so $L = 2(\lambda_1/4) = \lambda_1/2$, and $\lambda_1 = 2L$. The frequency of this vibration is therefore

$$f_1 = \frac{v}{\lambda_1} = \frac{v}{2L} \qquad [14.15]$$

Recall that the speed of a wave on a string is $v = \sqrt{F/\mu}$, where F is the tension in the string and μ is its mass per unit length (Chapter 13). Substituting into Equation 14.15, we obtain

$$f_1 = \frac{1}{2L}\sqrt{\frac{F}{\mu}} \qquad [14.16]$$

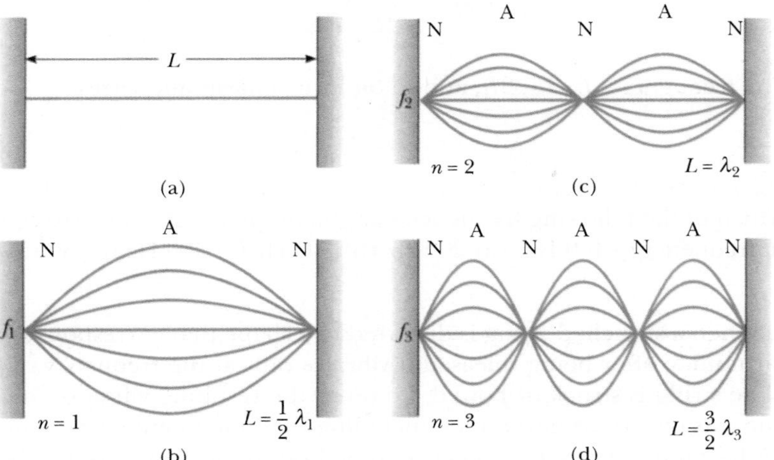

ACTIVE FIGURE 14.18
(a) Standing waves in a stretched string of length L fixed at both ends. The characteristic frequencies of vibration form a harmonic series: (b) the fundamental frequency, or first harmonic; (c) the second harmonic; and (d) the third harmonic. Note that N denotes a node, A an antinode.

Physics Now™
Log into PhysicsNow at **www.cp7e.com** and go to Active Figure 14.18 to choose the mode number and see the corresponding standing wave.

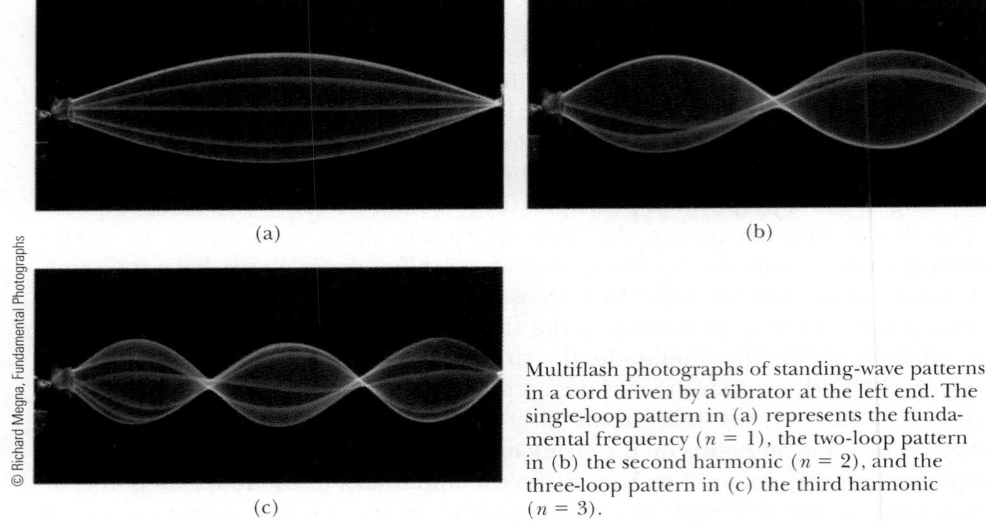

Multiflash photographs of standing-wave patterns in a cord driven by a vibrator at the left end. The single-loop pattern in (a) represents the fundamental frequency ($n = 1$), the two-loop pattern in (b) the second harmonic ($n = 2$), and the three-loop pattern in (c) the third harmonic ($n = 3$).

This lowest frequency of vibration is called the **fundamental frequency** of the vibrating string, or the **first harmonic**.

The first harmonic has nodes only at the ends—the points of attachment, with node–antinode pattern of N–A–N. The next harmonic, called the **second harmonic** (also called the **first overtone**) can be constructed by inserting an additional node–antinode segment between the endpoints. This makes the pattern N–A–N–A–N, as in Active Figure 14.18c. We count the node–antinode pairs: N–A, A–N, N–A, and A–N, four segments in all, each representing a quarter wavelength. We then have $L = 4(\lambda_2/4) = \lambda_2$, and the second harmonic (first overtone) is

$$f_2 = \frac{v}{\lambda_2} = \frac{v}{L} = 2\left(\frac{v}{2L}\right) = 2f_1$$

This frequency is equal to *twice* the fundamental frequency. The **third harmonic** (**second overtone**) is constructed similarly. Inserting one more N–A segment, we obtain Active Figure 14.18d, the pattern of nodes reading N–A–N–A–N–A–N. There are six node–antinode segments, so $L = 6(\lambda_3/4) = 3(\lambda_3/2)$, which means that $\lambda_3 = 2L/3$, giving

$$f_3 = \frac{v}{\lambda_3} = \frac{3v}{2L} = 3f_1$$

All the higher harmonics, it turns out, are positive integer multiples of the fundamental:

$$f_n = nf_1 = \frac{n}{2L}\sqrt{\frac{F}{\mu}} \qquad n = 1, 2, 3 \ldots \qquad \textbf{[14.17]}$$

The frequencies f_1, $2f_1$, $3f_1$, and so on form a **harmonic series**.

Quick Quiz 14.4

Which of the following frequencies are higher harmonics of a string with fundamental frequency of 150 Hz? (a) 200 Hz (b) 300 Hz (c) 400 Hz (d) 500 Hz (e) 600 Hz.

When a stretched string is distorted to a shape that corresponds to any one of its harmonics, after being released it vibrates only at the frequency of that harmonic. If the string is struck or bowed, however, the resulting vibration includes different amounts of various harmonics, including the fundamental frequency. Waves not in the harmonic series are quickly damped out on a string fixed at both ends. In effect, when disturbed, the string "selects" the standing-wave frequencies. As we'll see later, the presence of several harmonics on a string gives stringed instruments their characteristic sound, which enables us to distinguish one from another even when they are producing identical fundamental frequencies.

The frequency of a string on a musical instrument can be changed by varying either the tension or the length. The tension in guitar and violin strings is varied by turning pegs on the neck of the instrument. As the tension is increased, the frequency of the harmonic series increases according to Equation 14.17. Once the instrument is tuned, the musician varies the frequency by pressing the strings against the neck at a variety of positions, thereby changing the effective lengths of the vibrating portions of the strings. As the length is reduced, the frequency again increases, as follows from Equation 14.17.

APPLICATION
Tuning a Musical Instrument

Finally, Equation 14.17 shows that a string of fixed length can be made to vibrate at a lower fundamental frequency by increasing its mass per unit length. This is achieved in the bass strings of guitars and pianos by wrapping them with metal windings.

INTERACTIVE EXAMPLE 14.7 Guitar Fundamentals

Goal Apply standing-wave concepts to a stringed instrument.

Problem The high E string on a certain guitar measures 64.0 cm in length and has a fundamental frequency of 329 Hz. When a guitarist presses down so that the string is in contact with the first fret (Fig. 14.19a), the string is shortened so that it plays an F note that has a frequency of 349 Hz. **(a)** How far is the fret from the nut? **(b)** Overtones can be produced on a guitar string by gently placing the index finger in the location of a node of a higher harmonic. The string should be touched, but not depressed against a fret. (Given the width of a finger, pressing too hard will damp out higher harmonics as well.) The fundamental frequency is thereby suppressed, making it possible to hear overtones. Where on the guitar string relative to the nut should the finger be lightly placed so as to hear the second harmonic? The fourth harmonic? (This is equivalent to finding the location of the nodes in each case.)

Strategy For part (a) use Equation 14.15, corresponding to the fundamental frequency, to find the speed of waves on the string. Shortening the string by playing a higher note doesn't affect the wave speed, which depends only on the tension and linear density of the string (which are unchanged). Solve Equation 14.15 for the new length L, using the new fundamental frequency, and subtract this length from the original length to find the distance from the nut to the first fret. In part (b), remember that the distance from node to node is half a wavelength. Calculate the wavelength, divide it in two, and locate the nodes, which are integral numbers of half-wavelengths from the nut. [*Note:* The nut is a small piece of wood or ebony at the top of the fret board. The distance from the nut to the bridge (below the sound hole) is the length of the string. (See Fig. 14.19b.)]

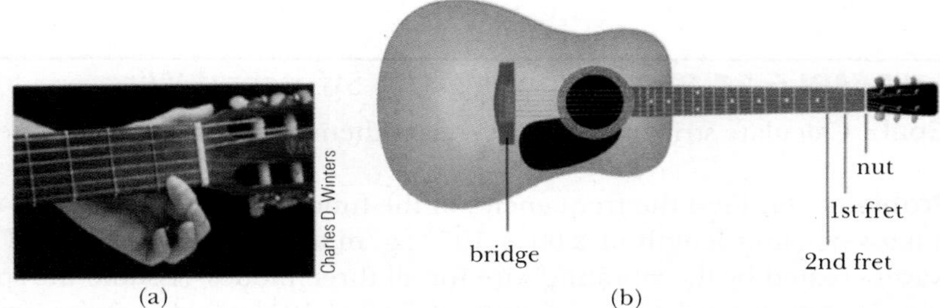

Figure 14.19 (Example 14.7) (a) Playing an F note on a guitar. (b) Some parts of a guitar.

Solution

(a) Find the distance from the nut to the first fret.

Substitute $L_0 = 0.640$ m and $f_1 = 329$ Hz into Equation 14.15, finding the wave speed on the string:

$$f_1 = \frac{v}{2L_0}$$

$$v = 2L_0 f_1 = 2(0.640 \text{ m})(329 \text{ Hz}) = 421 \text{ m/s}$$

Solve Equation 14.15 for the length L, and substitute the wave speed and the frequency of F.

$$L = \frac{v}{2f} = \frac{421 \text{ m/s}}{2(349 \text{ Hz})} = 0.603 \text{ m} = 60.3 \text{ cm}$$

Subtract this length from the original length L_0 to find the distance from the nut to the first fret:

$$\Delta x = L_0 - L = 64.0 \text{ cm} - 60.3 \text{ cm} = 3.7 \text{ cm}$$

(b) Find the locations of nodes for the second and fourth harmonics.

The second harmonic has a wavelength $\lambda_2 = L_0 = 64.0$ cm. The distance from nut to node corresponds to half a wavelength.

$$\Delta x = \tfrac{1}{2}\lambda_2 = \tfrac{1}{2}L_0 = 32.0 \text{ cm}$$

The fourth harmonic, of wavelength $\lambda_4 = \frac{1}{2}L_0 =$ 32.0 cm, has three nodes between the endpoints:

$$\Delta x = \tfrac{1}{2}\lambda_4 = \boxed{16.0 \text{ cm}}, \ \Delta x = 2(\lambda_4/2) = \boxed{32.0 \text{ cm}},$$

$$\Delta x = 3(\lambda_4/2) = \boxed{48.0 \text{ cm}}$$

Remarks Placing a finger at the position $\Delta x = 32.0$ cm damps out the fundamental and odd harmonics, but not all the higher even harmonics. The second harmonic dominates, however, because the rest of the string is free to vibrate. Placing the finger at $\Delta x = 16.0$ cm or 48.0 cm damps out the first through third harmonics, allowing the fourth harmonic to be heard.

Exercise 14.7

Pressing the E-string down on the fret board just above the second fret pinches the string firmly against the fret, giving an F sharp, which has frequency 3.70×10^2 Hz. (a) Where should the second fret be located? (b) Find two locations where you could touch the open E-string and hear the third harmonic.

Answer (a) 7.1 cm from the nut and 3.4 cm from the first fret. Note that the distance from the first to the second fret isn't the same as from the nut to the first fret. (b) 21.3 cm and 42.7 cm from the nut.

PhysicsNow™

Explore this situation by logging into PhysicsNow at **www.cp7e.com** and going to Interactive Example 14.7.

EXAMPLE 14.8 Harmonics of a Stretched Wire

Goal Calculate string harmonics, relate them to sound and combine them with tensile stress.

Problem **(a)** Find the frequencies of the fundamental, second, and third harmonics of a steel wire 1.00 m long with a mass per unit length of 2.00×10^{-3} kg/m and under a tension of 80.0 N. **(b)** Find the wavelengths of the sound waves created by the vibrating wire for all three modes. Assume the speed of sound in air is 345 m/s. **(c)** Suppose the wire is carbon steel with a density of 7.80×10^3 kg/m^3, a cross-sectional area $A = 2.56 \times 10^{-7}$ m^2, and an elastic limit of 2.80×10^8 Pa. Find the fundamental frequency if the wire is tightened to the elastic limit. Neglect any stretching of the wire (which would slightly reduce the mass per unit length).

Strategy (a) It's easiest to find the speed of waves on the wire then substitute into Equation 14.15 to find the first harmonic. The next two are multiples of the first, given by Equation 14.17. (b) The frequencies of the sound waves are the same as the frequencies of the vibrating wire, but the wavelengths are different. Use $v_s = f\lambda$, where v_s is the speed of sound in air, to find the wavelengths in air. (c) Find the force corresponding to the elastic limit, and substitute it into Equation 14.16.

Solution

(a) Find the first three harmonics at the given tension.

Use Equation 13.18 to calculate the speed of the wave on the wire:

$$v = \sqrt{\frac{F}{\mu}} = \sqrt{\frac{80.0 \text{ N}}{2.00 \times 10^{-3} \text{ kg/m}}} = 2.00 \times 10^2 \text{ m/s}$$

Find the wire's fundamental frequency from Equation 14.15:

$$f_1 = \frac{v}{2L} = \frac{2.00 \times 10^2 \text{ m/s}}{2(1.00 \text{ m})} = \boxed{1.00 \times 10^2 \text{ Hz}}$$

Find the next two harmonics by multiplication:

$$f_2 = 2f_1 = \boxed{2.00 \times 10^2 \text{ Hz}}, \ f_3 = 3f_1 = \boxed{3.00 \times 10^2 \text{ Hz}}$$

(b) Find the wavelength of the sound waves produced.

Solve $v_s = f\lambda$ for the wavelength and substitute the frequencies.

$$\lambda_1 = v_s/f_1 = (345\ \text{m/s})/(1.00 \times 10^2\ \text{Hz}) = \boxed{3.45\ \text{m}}$$

$$\lambda_2 = v_s/f_2 = (345\ \text{m/s})/(2.00 \times 10^2\ \text{Hz}) = \boxed{1.73\ \text{m}}$$

$$\lambda_3 = v_s/f_3 = (345\ \text{m/s})/(3.00 \times 10^2\ \text{Hz}) = \boxed{1.15\ \text{m}}$$

(c) Find the fundamental frequency corresponding to the elastic limit.

Calculate the tension in the wire from the elastic limit:

$$\frac{F}{A} = \text{elastic limit} \quad \rightarrow \quad F = (\text{elastic limit})A$$

$$F = (2.80 \times 10^8\ \text{Pa})(2.56 \times 10^{-7}\ \text{m}^2) = 71.7\ \text{N}$$

Substitute the values of F, μ, and L into Equation 14.16:

$$f_1 = \frac{1}{2L}\sqrt{\frac{F}{\mu}}$$

$$f_1 = \frac{1}{2(1.00\ \text{m})}\sqrt{\frac{71.7\ \text{N}}{2.00 \times 10^{-3}\ \text{kg/m}}} = \boxed{94.7\ \text{Hz}}$$

Remarks From the answer to part (c), it appears we need to choose a thicker wire or use a better grade of steel with a higher elastic limit. The frequency corresponding to the elastic limit is smaller than the fundamental!

Exercise 14.8

(a) Find the fundamental frequency and second harmonic if the tension in the wire is increased to 115 N. (Assume the wire doesn't stretch or break.) (b) Using a sound speed of 345 m/s, find the wavelengths of the sound waves produced.

Answer (a) 1.20×10^2 Hz, 2.40×10^2 Hz (b) 2.88 m, 1.44 m

14.9 FORCED VIBRATIONS AND RESONANCE

In Chapter 13 we learned that the energy of a damped oscillator decreases over time because of friction. It's possible to compensate for this energy loss by applying an external force that does positive work on the system.

For example, suppose an object–spring system having some natural frequency of vibration f_0 is pushed back and forth by a periodic force with frequency f. The system vibrates at the frequency f of the driving force. This type of motion is referred to as a **forced vibration**. Its amplitude reaches a maximum when the frequency of the driving force equals the natural frequency of the system f_0, called the **resonant frequency** of the system. Under this condition, the system is said to be in **resonance**.

In Section 14.8 we learned that a stretched string can vibrate in one or more of its natural modes. Here again, if a periodic force is applied to the string, the amplitude of vibration increases as the frequency of the applied force approaches one of the string's natural frequencies of vibration.

Resonance vibrations occur in a wide variety of circumstances. Figure 14.20 illustrates one experiment that demonstrates a resonance condition. Several pendulums of different lengths are suspended from a flexible beam. If one of them, such as A, is set in motion, the others begin to oscillate because of vibrations in the flexible beam. Pendulum C, the same length as A, oscillates with the greatest amplitude because its natural frequency matches that of pendulum A (the driving force).

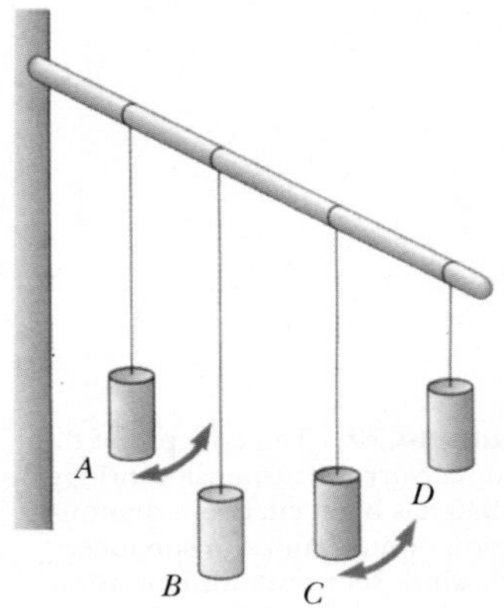

Figure 14.20 Resonance. If pendulum A is set in oscillation, only pendulum C, with a length matching that of A, will eventually oscillate with a large amplitude, or resonate. The arrows indicate motion perpendicular to the page.

Another simple example of resonance is a child being pushed on a swing, which is essentially a pendulum with a natural frequency that depends on its length. The

APPLICATION

Shattering Goblets with the Voice

swing is kept in motion by a series of appropriately timed pushes. For its amplitude to increase, the swing must be pushed each time it returns to the person's hands. This corresponds to a frequency equal to the natural frequency of the swing. If the energy put into the system per cycle of motion equals the energy lost due to friction, the amplitude remains constant.

Opera singers have been known to set crystal goblets in audible vibration with their powerful voices, as shown in Figure 14.21. This is yet another example of resonance: The sound waves emitted by the singer can set up large-amplitude vibrations in the glass. If a highly amplified sound wave has the right frequency, the amplitude of forced vibrations in the glass increases to the point where the glass becomes heavily strained and shatters.

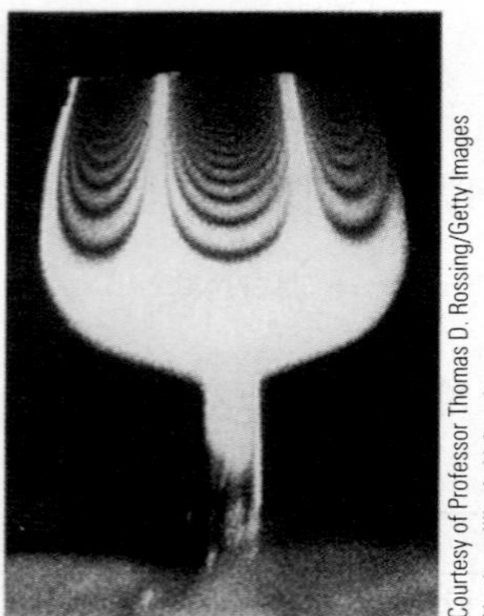

Courtesy of Professor Thomas D. Rossing/Getty Images Northern Illinois University

Ben Rose/The IMAGE Bank/Getty Images

Figure 14.21 (*Top*) Standing-wave pattern in a vibrating wineglass. The glass will shatter if the amplitude of vibration becomes too large. (*Bottom*) A wineglass shattered by the amplified sound of a human voice.

The classic example of structural resonance occurred in 1940, when the Tacoma Narrows bridge in the state of Washington was set in oscillation by the wind (Fig. 14.22). The amplitude of the oscillations increased rapidly and reached a high value until the bridge ultimately collapsed (probably because of metal fatigue). In recent years, however, a number of researchers have called this explanation into question. Gusts of wind, in general, don't provide the periodic force necessary for a sustained resonance condition, and the bridge exhibited large twisting oscillations, rather than the simple up-and-down oscillations expected of resonance.

A more recent example of destruction by structural resonance occurred during the Loma Prieta earthquake near Oakland, California, in 1989. In a mile-long section of the double-decker Nimitz Freeway, the upper deck collapsed onto the lower deck, killing several people. The collapse occurred because that particular section was built on mud fill while other parts were built on bedrock. As seismic waves pass through mud fill or other loose soil, their speed decreases and their amplitude increases. The section of the freeway that collapsed oscillated at the same frequency as other sections, but at a much larger amplitude.

14.10 STANDING WAVES IN AIR COLUMNS

Standing longitudinal waves can be set up in a tube of air, such as an organ pipe, as the result of interference between sound waves traveling in opposite directions. The relationship between the incident wave and the reflected wave depends on whether the reflecting end of the tube is open or closed. A portion of the sound wave is reflected back into the tube even at an open end. **If one end is closed, a node must exist at that end because the movement of air is restricted. If the end is open, the elements of air have complete freedom of motion, and an antinode exists.**

Figure 14.23a shows the first three modes of vibration of a pipe open at both ends. When air is directed against an edge at the left, longitudinal standing waves are formed and the pipe vibrates at its natural frequencies. Note that, from end to end, the pattern is A–N–A, the same pattern as in the vibrating string, except node and antinode have exchanged positions. As before, an antinode and its adjacent node, A–N, represent a quarter-wavelength, and there are two, A–N and N–A, so $L = 2(\lambda_1/4) = \lambda_1/2$ and $\lambda_1 = 2L$. The fundamental frequency of the pipe open at both ends is then $f_1 = v/\lambda_1 = v/2L$. The next harmonic has an addi-

United Press International Photo

Figure 14.22 The collapse of the Tacoma Narrows suspension bridge in 1940 has been cited as a demonstration of mechanical resonance. High winds set up standing waves in the bridge, causing it to oscillate at one of its natural frequencies. Once established, the resonance may have led to the bridge's collapse (although this interpretation is currently being challenged by mathematicians and physical scientists).

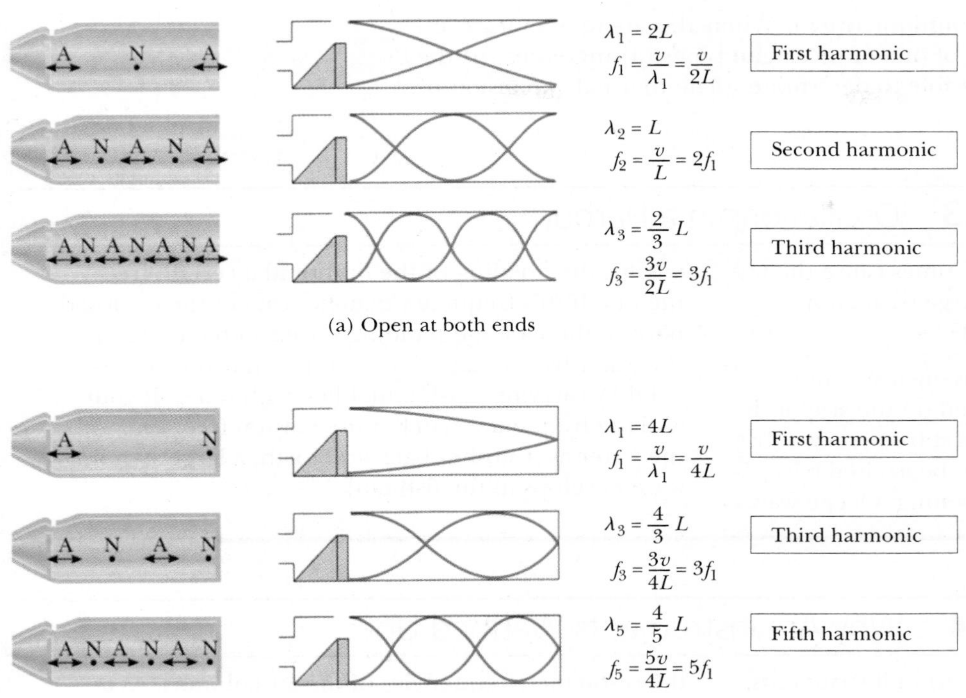

Figure 14.23 (a) Standing longitudinal waves in an organ pipe open at both ends. The natural frequencies $f_1, 2f_1, 3f_1 \ldots$ form a harmonic series. (b) Standing longitudinal waves in an organ pipe closed at one end. Only *odd* harmonics are present, and the natural frequencies are $f_1, 3f_1, 5f_1$, and so on.

TIP 14.4 Sound Waves Are Not Transverse

The standing longitudinal waves in Figure 14.23 are drawn as transverse waves only because it's difficult to draw longitudinal displacements—they're in the same direction as the wave propagation. In the figure, the vertical axis represents either pressure or horizontal displacement of the elements of the medium.

tional node and antinode between the ends, creating the pattern A–N–A–N–A. We count the pairs: A–N, N–A, A–N, and N–A, making four segments, each with length $\lambda_2/4$. We have $L = 4(\lambda_2/4) = \lambda_2$, and the second harmonic (first overtone) is $f_2 = v/\lambda_2 = v/L = 2(v/2L) = 2f_1$. All higher harmonics, it turns out, are positive integer multiples of the fundamental:

$$f_n = n\frac{v}{2L} = nf_1 \quad n = 1, 2, 3, \ldots \qquad [14.18]$$

◀ Pipe open at both ends; all harmonics are present

where v is the speed of sound in air. Notice the similarity to Equation 14.17, which also involves multiples of the fundamental.

If a pipe is open at one end and closed at the other, the open end is an antinode while the closed end is a node (Fig. 14.23b). In such a pipe, the fundamental frequency consists of a single antinode–node pair, A–N, so $L = \lambda_1/4$ and $\lambda_1 = 4L$. The fundamental harmonic for a pipe closed at one end is then $f_1 = v/\lambda_1 = v/4L$. The first overtone has another node and antinode between the open end and closed end, making the pattern A–N–A–N. There are three antinode–node segments in this pattern (A–N, N–A, and A–N), so $L = 3(\lambda_3/4)$ and $\lambda_3 = 4L/3$. The first overtone, therefore, has frequency $f_3 = v/\lambda_3 = 3v/4L = 3f_1$. Similarly, $f_5 = 5f_1$. In contrast to the pipe open at both ends, **there are no even multiples of the fundamental harmonic**. The odd harmonics for a pipe open at one end only are given by

$$f_n = n\frac{v}{4L} = nf_1 \quad n = 1, 3, 5, \ldots \qquad [14.19]$$

◀ Pipe closed at one end; only odd harmonics are present

Quick Quiz 14.5

A pipe open at both ends resonates at a fundamental frequency f_{open}. When one end is covered and the pipe is again made to resonate, the fundamental frequency is f_{closed}. Which of the following expressions describes how these two resonant frequencies compare? (a) $f_{closed} = f_{open}$, (b) $f_{closed} = \frac{3}{2}f_{open}$, (c) $f_{closed} = 2f_{open}$, (d) $f_{closed} = \frac{1}{2}f_{open}$, (e) none of these.

Quick Quiz 14.6

Balboa Park in San Diego has an outdoor organ. When the air temperature increases, the fundamental frequency of one of the organ pipes (a) increases (b) decreases (c) stays the same (d) impossible to determine. (The thermal expansion of the pipe is negligible.)

Applying Physics 14.3 Oscillations in a Harbor

Why do passing ocean waves sometimes cause the water in a harbor to undergo very large oscillations, called a *seiche* (pronounced *saysh*)?

Explanation Water in a harbor is enclosed and possesses a natural frequency based on the size of the harbor. This is similar to the natural frequency of the enclosed air in a bottle, which can be excited by blowing across the edge of the opening. Ocean waves pass by the opening of the harbor at a certain frequency. If this frequency matches that of the enclosed harbor, then a large standing wave can be set up in the water by resonance. This situation can be simulated by carrying a fish tank filled with water. If your walking frequency matches the natural frequency of the water as it sloshes back and forth, a large standing wave develops in the fish tank.

Applying Physics 14.4 Why Are Instruments Warmed Up?

Why do the strings go flat and the wind instruments go sharp during a performance if an orchestra doesn't warm up beforehand?

Explanation Without warming up, all the instruments will be at room temperature at the beginning of the concert. As the wind instruments are played, they fill with warm air from the player's exhalation. The increase in temperature of the air in the instruments causes an increase in the speed of sound, which raises the resonance frequencies of the air columns. As a result, the instruments go sharp. The strings on the stringed instruments also increase in temperature due to the friction of rubbing with the bow. This results in thermal expansion, which causes a decrease in tension in the strings. With the decrease in tension, the wave speed on the strings drops, and the fundamental frequencies decrease, so the stringed instruments go flat.

Applying Physics 14.5 How Do Bugles Work?

A bugle has no valves, keys, slides, or finger holes. How can it be used to play a song?

Explanation Songs for the bugle are limited to harmonics of the fundamental frequency, because there is no control over frequencies without valves, keys, slides, or finger holes. The player obtains different notes by changing the tension in the lips as the bugle is played, in order to excite different harmonics. The normal playing range of a bugle is among the third, fourth, fifth, and sixth harmonics of the fundamental. "Reveille," for example, is played with just the three notes G, C, and F. And "Taps" is played with these three notes and the G one octave above the lower G.

EXAMPLE 14.9 Harmonics of a Pipe

Goal Find frequencies of open and closed pipes.

Problem A pipe is 2.46 m long. **(a)** Determine the frequencies of the first three harmonics if the pipe is open at both ends. Take 343 m/s as the speed of sound in air. **(b)** How many harmonic frequencies of this pipe lie in the audible range, from 20 Hz to 20 000 Hz? **(c)** What are the three lowest possible frequencies if the pipe is closed at one end and open at the other?

Strategy Substitute into Equation 14.18 for part (a) and Equation 14.19 for part (c). All harmonics, $n = 1, 2, 3 \ldots$ are available for the pipe open at both ends, but only the harmonics with $n = 1, 3, 5, \ldots$ for the pipe closed at one end. For part (b), set the frequency in Equation 14.18 equal to 2.00×10^4 Hz.

Solution

(a) Find the frequencies if the pipe is open at both ends.

Substitute into Equation 14.18, with $n = 1$:

$$f_1 = \frac{v}{2L} = \frac{343 \text{ m/s}}{2(2.46 \text{ m})} = 69.7 \text{ Hz}$$

Multiply to find the second and third harmonics:

$$f_2 = 2f_1 = 139 \text{ Hz} \qquad f_3 = 3f_1 = 209 \text{ Hz}$$

(b) How many harmonics lie between 20 Hz and 20 000 Hz for this pipe?

Set the frequency in Equation 14.18 equal to 2.00×10^4 and solve for n:

$$f_n = n\frac{v}{2L} = n\frac{343 \text{ m/s}}{2 \cdot 2.46 \text{ m}} = 2.00 \times 10^4 \text{ Hz}$$

This works out to $n = 286.88$, which must be truncated down ($n = 287$ gives a frequency over 2.00×10^4 Hz).

$$n = 286$$

(c) Find the frequencies for the pipe closed at one end.

Apply Equation 14.19 with $n = 1$:

$$f_1 = \frac{v}{4L} = \frac{343 \text{ m/s}}{4(2.46 \text{ m})} = 34.9 \text{ Hz}$$

The next two harmonics are odd multiples of the first:

$$f_3 = 3f_1 = 105 \text{ Hz} \qquad f_5 = 5f_1 = 175 \text{ Hz}$$

Exercise 14.9

(a) What length pipe open at both ends has a fundamental frequency of 3.70×10^2 Hz? Find the first overtone. (b) If the one end of this pipe is now closed, what is the new fundamental frequency? Find the first overtone. (c) If the pipe is open at one end only, how many harmonics are possible in the normal hearing range from 20 to 20 000 Hz?

Answer (a) 0.464 m, 7.40×10^2 Hz (b) 185 Hz, 555 Hz (c) 54

EXAMPLE 14.10 Resonance in a Tube of Variable Length

Goal Understand resonance in tubes and perform elementary calculations.

Problem Figure 14.24a shows a simple apparatus for demonstrating resonance in a tube. A long tube open at both ends is partially submerged in a beaker of water, and a vibrating tuning fork of unknown frequency is placed near the top of the tube. The length of the air column, L, is adjusted by moving the tube vertically. The sound waves generated by the fork are reinforced when the length of the air column corresponds to one of the resonant frequencies of the tube. Suppose the smallest value of L for which a peak occurs in the sound intensity is 9.00 cm. **(a)** With this measurement, determine the frequency of the tuning fork. **(b)** Find the wavelength and the next two air-column lengths giving resonance. Take the speed of sound to be 345 m/s.

Strategy Once the tube is in the water, the setup is the same as a pipe closed at one end. For part (a), substitute values for v and L into Equation 14.19 with $n = 1$ and find the frequency of the tuning fork. (b) The next resonance maximum occurs when the water level is low enough to allow a second node, which is another half-wavelength in distance. The third resonance occurs when the third node is reached, requiring yet another half-wavelength of distance. The frequency in each case is the same, because it's generated by the tuning fork.

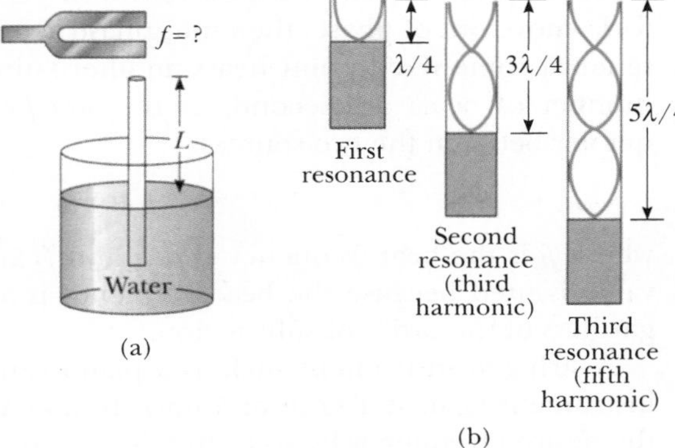

Figure 14.24 (Example 14.10) (a) Apparatus for demonstrating the resonance of sound waves in a tube closed at one end. The length L of the air column is varied by moving the tube vertically while it is partially submerged in water. (b) The first three resonances of the system.

Solution

(a) Find the frequency of the tuning fork.

Substitute $n = 1$, $v = 345$ m/s, and $L_1 = 9.00 \times 10^{-2}$ m into Equation 14.19:

$$f_1 = \frac{v}{4L_1} = \frac{345 \text{ m/s}}{4(9.00 \times 10^{-2} \text{ m})} = \boxed{958 \text{ Hz}}$$

(b) Find the wavelength and the next two water levels giving resonance.

Calculate the wavelength, using the fact that, for a tube open at one end, $\lambda = 4L$ for the fundamental.

$$\lambda = 4L_1 = 4(9.00 \times 10^{-2} \text{ m}) = \boxed{0.360 \text{ m}}$$

Add a half-wavelength of distance to L_1 to get the next resonance position:

$$L_2 = L_1 + \lambda/2 = 0.0900 \text{ m} + 0.180 \text{ m} = \boxed{0.270 \text{ m}}$$

Add another half-wavelength to L_2 to obtain the third resonance position:

$$L_3 = L_2 + \lambda/2 = 0.270 \text{ m} + 0.180 \text{ m} = \boxed{0.450 \text{ m}}$$

Remark This experimental arrangement is often used to measure the speed of sound, in which case the frequency of the tuning fork must be known in advance.

Exercise 14.10

An unknown gas is introduced into the aforementioned apparatus using the same tuning fork, and the first resonance occurs when the air column is 5.84 cm long. Find the speed of sound in the gas.

Answer 224 m/s

14.11 BEATS

The interference phenomena we have been discussing so far have involved the superposition of two or more waves with the same frequency, traveling in opposite directions. Another type of interference effect results from the superposition of two waves with slightly different frequencies. In such a situation, the waves at some fixed point are periodically in and out of phase, corresponding to an alternation in time between constructive and destructive interference. In order to understand this phenomenon, consider Active Figure 14.25. The two waves shown in Active Figure 14.25a were emitted by two tuning forks having slightly different frequencies; Active Figure 14.25b shows the superposition of these waves. At some time t_a the waves are in phase and constructive interference occurs, as demonstrated by the resultant curve in Active Figure 14.25b. At some later time, however, the vibrations of the two forks move out of step with each other. At time t_b, one fork emits a compression while the other emits a rarefaction, and destructive interference occurs, as demonstrated by the curve shown. As time passes, the vibrations of the two forks move out of phase, then into phase again, and so on. As a consequence, a listener at some fixed point hears an alternation in loudness, known as **beats**. The number of beats per second, or the *beat frequency*, equals the difference in frequency between the two sources:

$$f_b = |f_2 - f_1| \qquad [14.20]$$

where f_b is the beat frequency and f_1 and f_2 are the two frequencies. The absolute value is used because the beat frequency is a positive quantity and will occur regardless of the order of subtraction.

APPLICATION

Using Beats to Tune a Musical Instrument

A stringed instrument such as a piano can be tuned by beating a note on the instrument against a note of known frequency. The string can then be tuned to the desired frequency by adjusting the tension until no beats are heard.

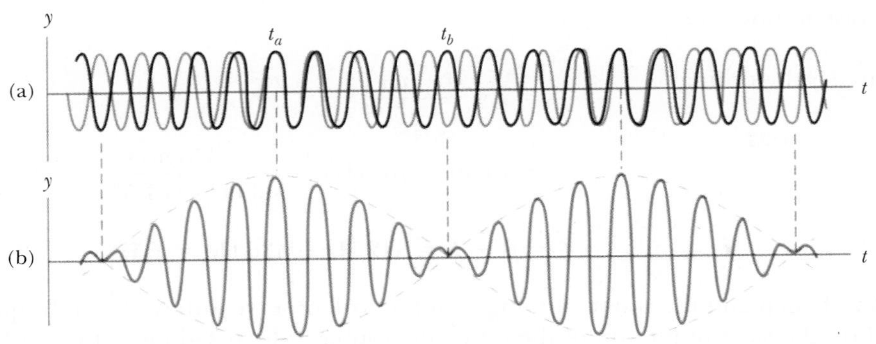

ACTIVE FIGURE 14.25
Beats are formed by the combination of two waves of slightly different frequencies traveling in the same direction. (a) The individual waves heard by an observer at a fixed point in space. (b) The combined wave has an amplitude (dashed line) that oscillates in time.

Physics Now™
Log into PhysicsNow at **www.cp7e.com**, and go to Active Figure 14.25 to choose two frequencies and see the corresponding beats.

Quick Quiz 14.7

You are tuning a guitar by comparing the sound of the string with that of a standard tuning fork. You notice a beat frequency of 5 Hz when both sounds are present. As you tighten the guitar string, the beat frequency rises steadily to 8 Hz. In order to tune the string exactly to the tuning fork, you should (a) continue to tighten the string (b) loosen the string (c) impossible to determine from the given information.

EXAMPLE 14.11 Sour Notes

Goal Apply the beat frequency concept.

Problem A certain piano string is supposed to vibrate at a frequency of 4.40×10^2 Hz. In order to check its frequency, a tuning fork known to vibrate at a frequency of 4.40×10^2 Hz is sounded at the same time the piano key is struck, and a beat frequency of 4 beats per second is heard. **(a)** Find the two possible frequencies at which the string could be vibrating. **(b)** Suppose the piano tuner runs toward the piano, holding the vibrating tuning fork while his assistant plays the note, which is at 436 Hz. At his maximum speed, the piano tuner notices the beat frequency drops from 4 Hz to 2 Hz (without going through a beat frequency of zero). How fast is he moving? Use a sound speed of 343 m/s. **(c)** While the piano tuner is running, what beat frequency is observed by the assistant? [*Note:* Assume all numbers are accurate to two decimal places, necessary for this last calculation.]

Strategy (a) The beat frequency is equal to the absolute value of the difference in frequency between the two sources of sound and occurs if the piano string is tuned either too high or too low. Solve Equation 14.20 for these two possible frequencies. (b) Moving toward the piano raises the observed piano string frequency. Solve the Doppler shift formula, Equation 14.12, for the speed of the observer. (c) The assistant observes a Doppler shift for the tuning fork. Apply Equation 14.12.

Solution

(a) Find the two possible frequencies.

Case 1: $f_2 - f_1$ is already positive, so just drop the absolute-value signs:

$$f_b = f_2 - f_1 \quad \rightarrow \quad 4 \text{ Hz} = f_2 - 4.40 \times 10^2 \text{ Hz}$$

$$f_2 = \boxed{444 \text{ Hz}}$$

Case 2: $f_2 - f_1$ is negative, so drop the absolute-value signs, but apply an overall negative sign:

$$f_b = -(f_2 - f_1) \quad \rightarrow \quad 4 \text{ Hz} = -(f_2 - 4.40 \times 10^2 \text{ Hz})$$

$$f_2 = \boxed{436 \text{ Hz}}$$

(b) Find the speed of the observer if running toward the piano results in a beat frequency of 2 Hz.

Apply the Doppler shift to the case where frequency of the piano string heard by the running observer is $f_O = 438$ Hz:

$$f_O = f_S\left(\frac{v + v_O}{v - v_S}\right)$$

$$438 \text{ Hz} = (436 \text{ Hz})\left(\frac{343 \text{ m/s} + v_O}{343 \text{ m/s}}\right)$$

$$v_O = \left(\frac{438 \text{ Hz} - 436 \text{ Hz}}{436 \text{ Hz}}\right)(343 \text{ m/s}) = \boxed{1.57 \text{ m/s}}$$

(c) What beat frequency does the assistant observe?

Apply Equation 14.12. Now the source is the tuning fork, so $f_S = 4.40 \times 10^2$ Hz.

$$f_O = f_S\left(\frac{v + v_O}{v - v_S}\right)$$

$$= (4.40 \times 10^2 \text{ Hz})\left(\frac{343 \text{ m/s}}{343 \text{ m/s} - 1.57 \text{ m/s}}\right) = 442 \text{ Hz}$$

Compute the beat frequency.

$$f_b = f_2 - f_1 = 442 \text{ Hz} - 436 \text{ Hz} = \boxed{6 \text{ Hz}}$$

Remarks The assistant on the piano bench and the tuner running with the fork observe different beat frequencies. Many physical observations depend on the state of motion of the observer, a subject discussed more fully in Chapter 26, on relativity.

Exercise 14.11

The assistant adjusts the tension in the same piano string, and a beat frequency of 2.00 Hz is heard when the note and the tuning fork are struck at the same time. (a) Find the two possible frequencies of the string. (b) Assume the actual string frequency is the higher frequency. If the piano tuner runs away from the piano at 4.00 m/s while holding the vibrating tuning fork, what beat frequency does he hear? (c) What beat frequency does the assistant on the bench hear? Use 343 m/s for the speed of sound.

Answers (a) 438 Hz, 442 Hz (b) 3 Hz (c) 7 Hz

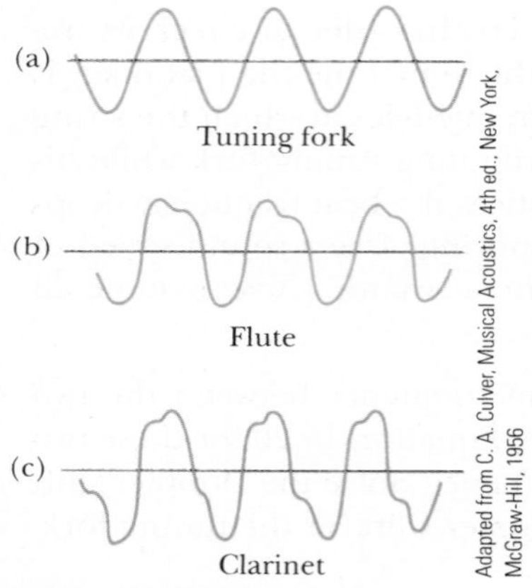

Figure 14.26 Waveforms produced by (a) a tuning fork, (b) a flute, and (c) a clarinet, all at approximately the same frequency. Pressure is plotted vertically, time horizontally.

14.12 QUALITY OF SOUND

The sound-wave patterns produced by most musical instruments are complex. Figure 14.26 shows characteristic waveforms (pressure is plotted on the vertical axis, time on the horizontal axis) produced by a tuning fork, a flute, and a clarinet, each playing the same steady note. Although each instrument has its own characteristic pattern, the figure reveals that each of the waveforms is periodic. Note that the tuning fork produces only one harmonic (the fundamental frequency), but the two instruments emit mixtures of harmonics. Figure 14.27 graphs the harmonics of the waveforms of Figure 14.26. When the note is played on the flute (Fig. 14.26b), part of the sound consists of a vibration at the fundamental frequency, an even higher intensity is contributed by the second harmonic, the fourth harmonic produces about the same intensity as the fundamental, and so on. These sounds add together according to the principle of superposition to give the complex waveform shown. The clarinet emits a certain intensity at a frequency of the first harmonic, about half as much intensity at the frequency of the second harmonic, and so forth. The resultant superposition of these frequencies produces the pattern shown in Figure 14.26c. The tuning fork (Figs. 14.26a and 14.27a) emits sound only at the frequency of the first harmonic.

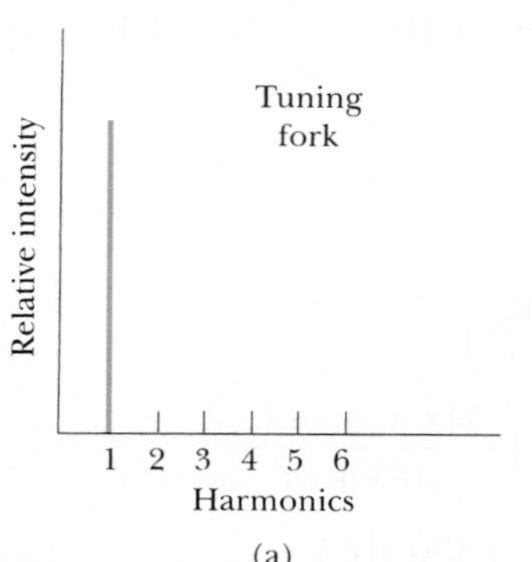

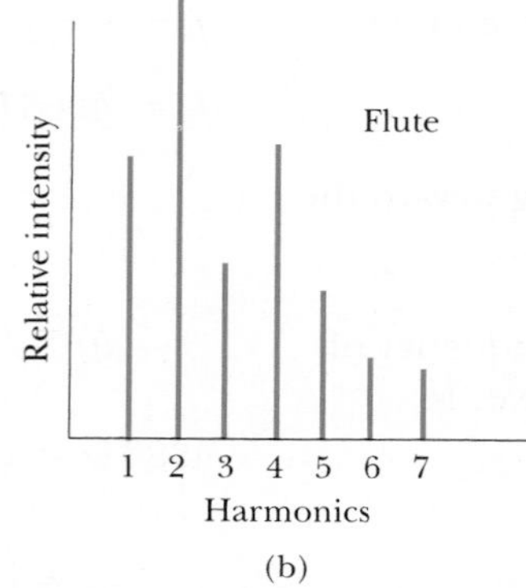

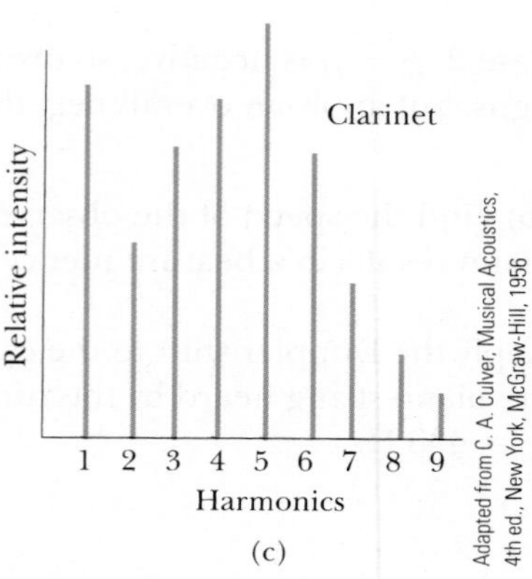

Figure 14.27 Harmonics of the waveforms in Figure 14.26. Note their variation in intensity.

In music, the characteristic sound of any instrument is referred to as the *quality*, or *timbre*, of the sound. The quality depends on the mixture of harmonics in the sound. We say that the note C on a flute differs in quality from the same C on a clarinet. Instruments such as the bugle, trumpet, violin, and tuba are rich in harmonics. A musician playing a wind instrument can emphasize one or another of these harmonics by changing the configuration of the lips, thereby playing different musical notes with the same valve openings.

TIP 14.5 Pitch is Not The Same as Frequency

Although pitch is related mostly (but not completely) to frequency, they are not the same. A phrase such as "the pitch of the sound" is incorrect, because pitch is not a physical property of the sound. Frequency is the physical measurement of the number of oscillations per second of the sound. Pitch is a psychological reaction to sound that enables a human being to place the sound on a scale from high to low or from treble to bass. Frequency is the stimulus and pitch is the response.

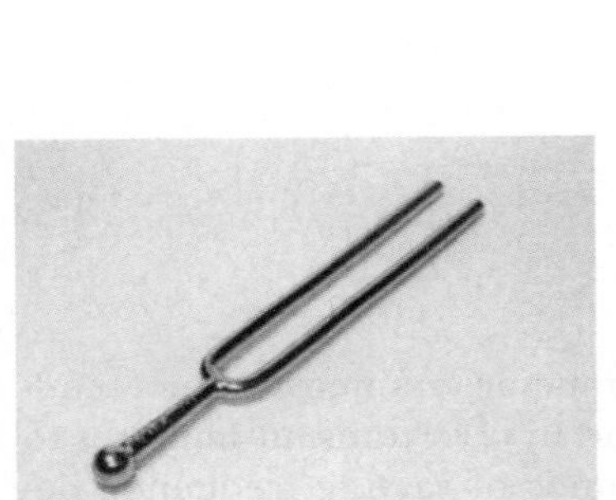

(a)

(b)

(c)

Each musical instrument has its own characteristic sound and mixture of harmonics. (See Figures 14.26 and 14.27.) Instruments shown are (a) the tuning fork, (b) the flute, and (c) the clarinet.

Applying Physics 14.6 Why Does the Professor Sound Like Donald Duck?

A professor performs a demonstration in which he breathes helium and then speaks with a comical voice. One student explains, "The velocity of sound in helium is higher than in air, so the fundamental frequency of the standing waves in the mouth is increased." Another student says, "No, the fundamental frequency is determined by the vocal folds and cannot be changed. Only the quality of the voice has changed." Which student is correct?

Explanation The second student is correct. The fundamental frequency of the complex tone from the voice is determined by the vibration of the vocal folds and is not changed by substituting a different gas in the mouth. The introduction of the helium into the mouth results in harmonics of higher frequencies being excited more than in the normal voice, but the fundamental frequency of the voice is the same—only the quality has changed. The unusual inclusion of the higher frequency harmonics results in a common description of this effect as a "high-pitched" voice, but that description is incorrect. (It is really a "quacky" timbre.)

14.13 THE EAR

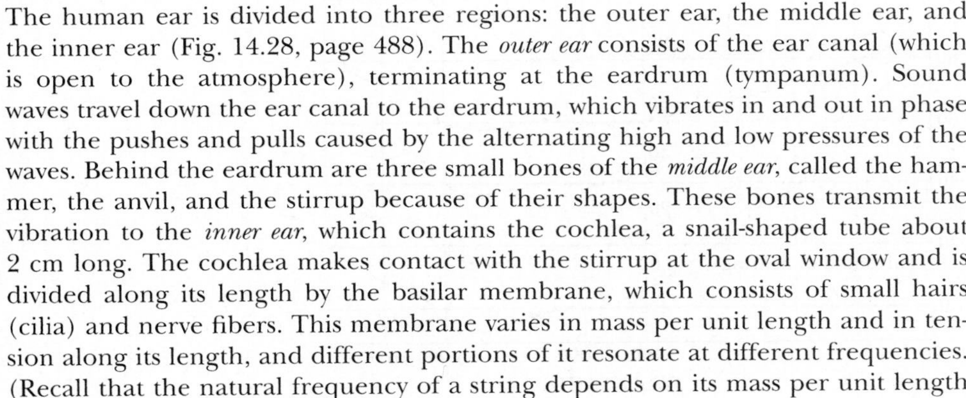

The human ear is divided into three regions: the outer ear, the middle ear, and the inner ear (Fig. 14.28, page 488). The *outer ear* consists of the ear canal (which is open to the atmosphere), terminating at the eardrum (tympanum). Sound waves travel down the ear canal to the eardrum, which vibrates in and out in phase with the pushes and pulls caused by the alternating high and low pressures of the waves. Behind the eardrum are three small bones of the *middle ear*, called the hammer, the anvil, and the stirrup because of their shapes. These bones transmit the vibration to the *inner ear*, which contains the cochlea, a snail-shaped tube about 2 cm long. The cochlea makes contact with the stirrup at the oval window and is divided along its length by the basilar membrane, which consists of small hairs (cilia) and nerve fibers. This membrane varies in mass per unit length and in tension along its length, and different portions of it resonate at different frequencies. (Recall that the natural frequency of a string depends on its mass per unit length

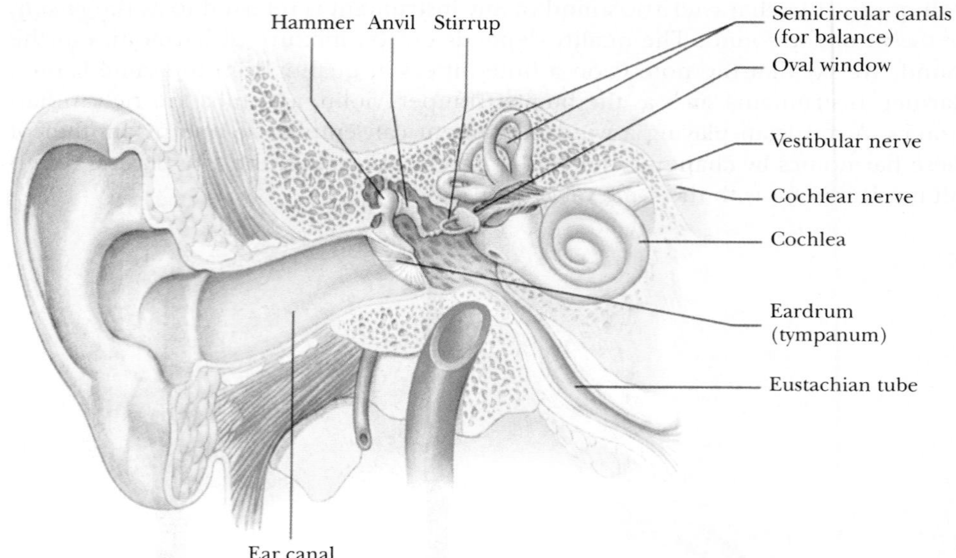

Figure 14.28 The structure of the human ear. The three tiny bones (ossicles) that connect the eardrum to the window of the cochlea act as a double-lever system to decrease the amplitude of vibration and hence increase the pressure on the fluid in the cochlea.

and on the tension in it.) Along the basilar membrane are numerous nerve endings, which sense the vibration of the membrane and in turn transmit impulses to the brain. The brain interprets the impulses as sounds of varying frequency, depending on the locations along the basilar membrane of the impulse-transmitting nerves and on the rates at which the impulses are transmitted.

Figure 14.29 shows the frequency response curves of an average human ear for sounds of equal loudness, ranging from 0 to 120 dB. To interpret this series of graphs, take the bottom curve as the threshold of hearing. Compare the intensity level on the vertical axis for the two frequencies 100 Hz and 1 000 Hz. The vertical axis shows that the 100-Hz sound must be about 38 dB greater than the 1 000-Hz sound to be at the threshold of hearing, which means that the threshold of hearing is very strongly dependent on frequency. The easiest frequencies to hear are around 3 300 Hz; those above 12 000 Hz or below about 50 Hz must be relatively intense to be heard.

Now consider the curve labeled 80. This curve uses a 1 000-Hz tone at an intensity level of 80 dB as its reference. The curve shows that a tone of frequency 100 Hz would have to be about 4 dB louder than the 80-dB, 1 000-Hz tone in order to sound as loud. Notice that the curves flatten out as the intensities levels of the sounds increase, so when sounds are loud, all frequencies can be heard equally well.

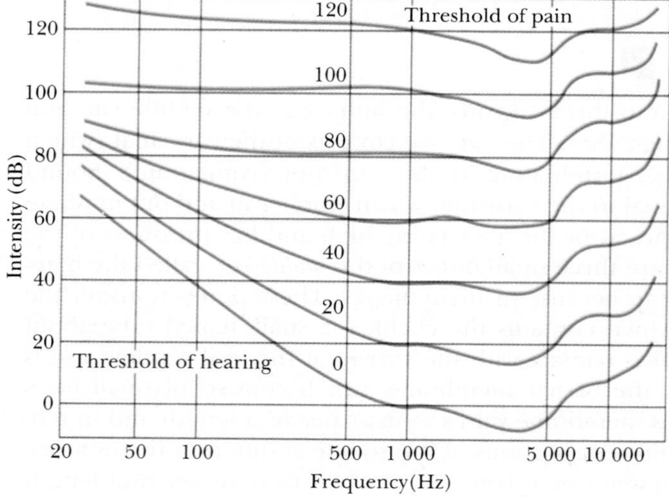

Figure 14.29 Curves of intensity level versus frequency for sounds that are perceived to be of equal loudness. Note that the ear is most sensitive at a frequency of about 3 300 Hz. The lowest curve corresponds to the threshold of hearing for only about 1% of the population.

The small bones in the middle ear represent an intricate lever system that increases the force on the oval window. The pressure is greatly magnified because the surface area of the eardrum is about 20 times that of the oval window (in analogy with a hydraulic press). The middle ear, together with the eardrum and oval window, in effect acts as a matching network between the air in the outer ear and the liquid in the inner ear. The overall energy transfer between the outer ear and the inner ear is highly efficient, with pressure amplification factors of several thousand. In other words, pressure variations in the inner ear are much greater than those in the outer ear.

The ear has its own built-in protection against loud sounds. The muscles connecting the three middle-ear bones to the walls control the volume of the sound by changing the tension on the bones as sound builds up, thus hindering their ability to transmit vibrations. In addition, the eardrum becomes stiffer as the sound intensity increases. These two events make the ear less sensitive to loud incoming sounds. There is a time delay between the onset of a loud sound and the ear's protective reaction, however, so a very sudden loud sound can still damage the ear.

The complex structure of the human ear is believed to be related to the fact that mammals evolved from seagoing creatures. In comparison, insect ears are considerably simpler in design, because insects have always been land residents. A typical insect ear consists of an eardrum exposed directly to the air on one side and to an air-filled cavity on the other side. Nerve cells communicate directly with the cavity and the brain, without the need for the complex intermediary of an inner and middle ear. This simple design allows the ear to be placed virtually anywhere on the body. For example, a grasshopper has its ears on its legs. One advantage of the simple insect ear is that the distance and orientation of the ears can be varied so that it is easier to locate sources of sound, such as other insects.

One of the most amazing medical advances in recent decades is the cochlear implant, allowing the deaf to hear. Deafness can occur when the hairlike sensors (cilia) in the cochlea break off over a lifetime or sometimes because of prolonged exposure to loud sounds. Because the cilia don't grow back, the ear loses sensitivity to certain frequencies of sound. The cochlear implant stimulates the nerves in the ear electronically to restore hearing loss that is due to damaged or absent cilia.

SUMMARY

PhysicsNow™ Take a practice test by logging into PhysicsNow at **www.cp7e.com** and clicking on the Pre-Test link for this chapter.

14.2 Characteristics of Sound Waves

Sound waves are longitudinal waves. **Audible waves** are sound waves with frequencies between 20 and 20 000 Hz. **Infrasonic waves** have frequencies below the audible range, and **ultrasonic waves** have frequencies above the audible range.

14.3 The Speed of Sound

The speed of sound in a medium of bulk modulus B and density ρ is

$$v = \sqrt{\frac{B}{\rho}} \quad [14.1]$$

The speed of sound also depends on the temperature of the medium. The relationship between temperature and the speed of sound in air is

$$v = (331 \text{ m/s}) \sqrt{\frac{T}{273 \text{ K}}} \quad [14.4]$$

where T is the absolute (Kelvin) temperature and 331 m/s is the speed of sound in air at 0°C.

14.4 Energy and Intensity of Sound Waves

The **average intensity** of sound incident on a surface is defined by

$$I \equiv \frac{\text{power}}{\text{area}} = \frac{\mathcal{P}}{A} \quad [14.6]$$

where the power $\mathcal{P}$ is the energy per unit time flowing through the surface, which has area A. The **intensity level** of a sound wave is given by

$$\beta \equiv 10 \log \left(\frac{I}{I_0}\right) \quad [14.7]$$

The constant $I_0 = 1.0 \times 10^{-12}$ W/m^2 is a reference intensity, usually taken to be at the threshold of hearing, and I is the intensity at level β, measured in **decibels** (dB).

14.5 Spherical and Plane Waves

The **intensity** of a *spherical wave* produced by a point source is proportional to the average power emitted and inversely proportional to the square of the distance from the source:

$$I = \frac{\mathcal{P}_{av}}{4\pi r^2} \quad [14.8]$$

14.6 The Doppler Effect

The change in frequency heard by an observer whenever there is relative motion between a source of sound and the observer is called the **Doppler effect**. If the observer is moving with speed v_O and the source is moving with speed v_S, the observed frequency is

$$f_O = f_S\left(\frac{v + v_O}{v - v_S}\right) \qquad [14.12]$$

where v is the speed of sound. A positive speed is substituted for v_O when the observer moves toward the source, a negative speed when the observer moves away from the source. Similarly, a positive speed is substituted for v_S when the sources moves toward the observer, a negative speed when the source moves away.

14.7 Interference of Sound Waves

When waves interfere, the resultant wave is found by adding the individual waves together point by point. When crest meets crest and trough meets trough, the waves undergo **constructive interference**, with path length difference

$$r_2 - r_1 = n\lambda \quad (n = 0, 1, 2, \ldots) \qquad [14.13]$$

When crest meets trough, **destructive interference** occurs, with path length difference

$$r_2 - r_1 = (n + \tfrac{1}{2})\lambda \quad (n = 0, 1, 2, \ldots) \qquad [14.14]$$

14.8 Standing Waves

Standing waves are formed when two waves having the same frequency, amplitude, and wavelength travel in opposite directions through a medium. The natural frequencies of vibration of a stretched string of length L, fixed at both ends, are

$$f_n = nf_1 = \frac{n}{2L}\sqrt{\frac{F}{\mu}} \quad n = 1, 2, 3, \ldots \qquad [14.17]$$

where F is the tension in the string and μ is its mass per unit length.

14.9 Forced Vibrations and Resonance

A system capable of oscillating is said to be in **resonance** with some driving force whenever the frequency of the driving force matches one of the natural frequencies of the system. When the system is resonating, it oscillates with maximum amplitude.

14.10 Standing Waves in Air Columns

Standing waves can be produced in a tube of air. If the reflecting end of the tube is *open*, all harmonics are present and the natural frequencies of vibration are

$$f_n = n\frac{v}{2L} = nf_1 \quad n = 1, 2, 3, \ldots \qquad [14.18]$$

If the tube is *closed* at the reflecting end, only the *odd* harmonics are present and the natural frequencies of vibration are

$$f_n = n\frac{v}{4L} = nf_1 \quad n = 1, 3, 5, \ldots \qquad [14.19]$$

14.11 Beats

The phenomenon of **beats** is an interference effect that occurs when two waves with slightly different frequencies combine at a fixed point in space. For sound waves, the intensity of the resultant sound changes periodically with time. The *beat frequency* is

$$f_b = |f_2 - f_1| \qquad [14.20]$$

where f_2 and f_1 are the two source frequencies.

CONCEPTUAL QUESTIONS

1. (a) You are driving down the highway in your car when a police car sounding its siren overtakes you and passes you. If its frequency at rest is f_0, is the frequency you hear while the car is catching up to you higher or lower than f_0? (b) What about the frequency you hear after the car has passed you?
2. A crude model of the human throat is that of a pipe open at both ends with a vibrating source to introduce the sound into the pipe at one end. Assuming the vibrating source produces a range of frequencies, discuss the effect of changing the pipe's length.
3. An autofocus camera sends out a pulse of sound and measures the time taken for the pulse to reach an object, reflect off of it, and return to be detected. Can the temperature affect the camera's focus?
4. To keep animals away from their cars, some people mount short, thin pipes on the fenders. The pipes give out a high-pitched wail when the cars are moving. How do they create the sound?
5. Secret agents in the movies always want to get to a secure phone with a voice scrambler. How do these devices work?
6. When a bell is rung, standing waves are set up around its circumference. What boundary conditions must be satisfied by the resonant wavelengths? How does a crack in the bell, such as in the Liberty Bell, affect the satisfying of the boundary conditions and the sound emanating from the bell?
7. How does air temperature affect the tuning of a wind instrument?
8. Explain how the distance to a lightning bolt can be determined by counting the seconds between the flash and the sound of thunder.
9. You are driving toward a cliff and you honk your horn. Is there a Doppler shift of the sound when you hear the echo? Is it like a moving source or moving observer? What if the reflection occurs not from a cliff, but from the forward edge of a huge alien spacecraft moving toward you as you drive?

10. Of the following sounds, state which is most likely to have an intensity level of 60 dB: a rock concert, the turning of a page in this text, a normal conversation, a cheering crowd at a football game, and background noise at a church?

11. Guitarists sometimes play a "harmonic" by lightly touching a string at its exact center and plucking the string. The result is a clear note one octave higher than the fundamental frequency of the string, even though the string is not pressed to the fingerboard. Why does this happen?

12. Will two separate 50-dB sounds together constitute a 100-dB sound? Explain.

13. An archer shoots an arrow from a bow. Does the string of the bow exhibit standing waves after the arrow leaves? If so, and if the bow is perfectly symmetric so that the arrow leaves from the center of the string, what harmonics are excited?

14. The radar systems used by police to detect speeders are sensitive to the Doppler shift of a pulse of radio waves. Discuss how this sensitivity can be used to measure the speed of a car.

15. As oppositely moving pulses of the same shape (one upward, one downward) on a string pass through each other, there is one instant at which the string shows no displacement from the equilibrium position at any point. Has the energy carried by the pulses disappeared at this instant of time? If not, where is it?

16. A soft drink bottle resonates as air is blown across its top. What happens to the resonant frequency as the level of fluid in the bottle decreases?

17. A blowing whistle is attached to the roof of a car that moves around a circular race track. Assuming you're standing near the outside of the track, explain the nature of the sound you hear as the whistle comes by each time.

18. Despite a reasonably steady hand, a person often spills his coffee when carrying it to his seat. Discuss resonance as a possible cause of this difficulty, and devise a means for solving the problem.

19. An airplane mechanic notices that the sound from a twin-engine aircraft varies rapidly in loudness when both engines are running. What could be causing this variation from loud to soft?

20. Why does a vibrating guitar string sound louder when placed on the instrument than it would if allowed to vibrate in the air while off the instrument?

PROBLEMS

1, 2, 3 = straightforward, intermediate, challenging ☐ = full solution available in *Student Solutions Manual/Study Guide*

Physics Now™ = coached problem with hints available at **www.cp7e.com** = biomedical application

Section 14.2 Characteristics of Sound Waves

Section 14.3 The Speed of Sound

Unless otherwise stated, use 345 m/s as the speed of sound in air.

1. Suppose that you hear a clap of thunder 16.2 s after seeing the associated lightning stroke. The speed of sound waves in air is 343 m/s and the speed of light in air is 3.00×10^8 m/s. How far are you from the lightning stroke?

2. A dolphin located in sea water at a temperature of 25°C emits a sound directed toward the bottom of the ocean 150 m below. How much time passes before it hears an echo?

3. A sound wave has a frequency of 700 Hz in air and a wavelength of 0.50 m. What is the temperature of the air?

4. The range of human hearing extends from approximately 20 Hz to 20 000 Hz. Find the wavelengths of these extremes at a temperature of 27°C.

5. A group of hikers hears an echo 3.00 s after shouting. If the temperature is 22.0°C, how far away is the mountain that reflected the sound wave?

6. A stone is dropped from rest into a well. The sound of the splash is heard exactly 2.00 s later. Find the depth of the well if the air temperature is 10.0°C.

7. You are watching a pier being constructed on the far shore of a saltwater inlet when some blasting occurs. You hear the sound in the water 4.50 s before it reaches you through the air. How wide is the inlet? [*Hint:* See Table 14.1. Assume the air temperature is 20°C.]

8. The speed of sound in a column of air is measured to be 356 m/s. What is the temperature of the air?

Section 14.4 Energy and Intensity of Sound Waves

9. The toadfish makes use of resonance in a closed tube to produce very loud sounds. The tube is its swim bladder, used as an amplifier. The sound level of this creature has been measured as high as 100 dB. (a) Calculate the intensity of the sound wave emitted. (b) What is the intensity level if three of these fish try to imitate three frogs by saying "Budweiser" at the same time.

10. The area of a typical eardrum is about 5.0×10^{-5} m^2. Calculate the sound power (the energy per second) incident on an eardrum at (a) the threshold of hearing and (b) the threshold of pain.

11. There is evidence that elephants communicate via infrasound, generating rumbling vocalizations as low as 14 hz that can travel up to 10 km. The intensity level of these sounds can reach 103 dB, measured a distance of 5.0 m from the source. Determine the intensity level of the infrasound 10 km from the source, assuming the sound energy radiates uniformly in all directions.

12. Two sounds have measured intensities of $I_1 = 100$ W/m^2 and $I_2 = 200$ W/m^2. By how many decibels is the level of sound 1 lower than that of sound 2?

13. **Physics Now™** A noisy machine in a factory produces sound with a level of 80 dB. How many identical machines could you add to the factory without exceeding the 90-dB limit?

14. A family ice show is held at an enclosed arena. The skaters perform to music playing at a level of 80.0 dB. This intensity level is too loud for your baby, who yells at 75.0 dB. (a) What total sound intensity engulfs you? (b) What is the combined sound level?

15. Calculate the sound level in decibels of a sound wave that has an intensity of 4.00 μW/m^2.

Section 14.5 Spherical and Plane Waves

16. An outside loudspeaker (considered a small source) emits sound waves with a power output of 100 W. (a) Find the intensity 10.0 m from the source. (b) Find the intensity level in decibels at that distance. (c) At what distance would you experience the sound at the threshold of pain, 120 dB?

17. A train sounds its horn as it approaches an intersection. The horn can just be heard at a level of 50 dB by an observer 10 km away. (a) What is the average power generated by the horn? (b) What intensity level of the horn's sound is observed by someone waiting at an intersection 50 m from the train? Treat the horn as a point source and neglect any absorption of sound by the air.

18. A skyrocket explodes 100 m above the ground (Fig. P14.18). Three observers are spaced 100 m apart, with the first (A) directly under the explosion. (a) What is the ratio of the sound intensity heard by observer A to that heard by observer B? (b) What is the ratio of the intensity heard by observer A to that heard by observer C?

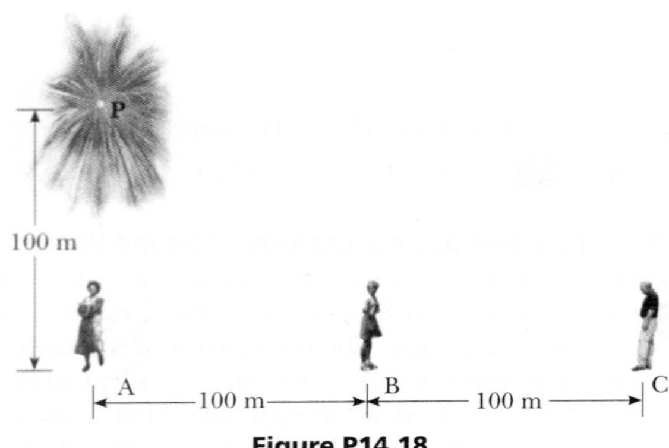

Figure P14.18

19. Show that the difference in decibel levels β_1 and β_2 of a sound source is related to the ratio of its distances r_1 and r_2 from the receivers by the formula

$$\beta_2 - \beta_1 = 20 \log \left(\frac{r_1}{r_2}\right)$$

Section 14.6 The Doppler Effect

20. An airplane traveling at half the speed of sound ($v = 172$ m/s) emits a sound of frequency 5.00 kHz. At what frequency does a stationary listener hear the sound (a) as the plane approaches? (b) after it passes?

21. A commuter train passes a passenger platform at a constant speed of 40.0 m/s. The train horn is sounded at its characteristic frequency of 320 Hz. (a) What overall change in frequency is detected by a person on the platform as the train moves from approaching to receding? (b) What wavelength is detected by a person on the platform as the train approaches?

22. At rest, a car's horn sounds the note A (440 Hz). The horn is sounded while the car is moving down the street. A bicyclist moving in the same direction with one-third the car's speed hears a frequency of 415 Hz. What is the speed of the car? Is the cyclist ahead of or behind the car?

23. Two trains on separate tracks move towards one another. Train 1 has a speed of 130 km/h, train 2 a speed of 90.0 km/h. Train 2 blows its horn, emitting a frequency of 500 Hz. What is the frequency heard by the engineer on train 1?

24. A bat flying at 5.0 m/s emits a chirp at 40 kHz. If this sound pulse is reflected by a wall, what is the frequency of the echo received by the bat?

25. **Physics Now™** An alert physics student stands beside the tracks as a train rolls slowly past. He notes that the frequency of the train whistle is 442 Hz when the train is approaching him and 441 Hz when the train is receding from him. Using these frequencies, he calculates the speed of the train. What value does he find?

26. Expectant parents are thrilled to hear their unborn baby's heartbeat, revealed by an ultrasonic motion detector. Suppose the fetus's ventricular wall moves in simple harmonic motion with amplitude 1.80 mm and frequency 115 per minute. (a) Find the maximum linear speed of the heart wall. Suppose the motion detector in contact with the mother's abdomen produces sound at precisely 2 MHz, which travels through tissue at 1.50 km/s. (b) Find the maximum frequency at which sound arrives at the wall of the baby's heart. (c) Find the maximum frequency at which reflected sound is received by the motion detector. (By electronically "listening" for echoes at a frequency different from the broadcast frequency, the motion detector can produce beeps of audible sound in synchrony with the fetal heartbeat.)

27. A tuning fork vibrating at 512 Hz falls from rest and accelerates at 9.80 m/s^2. How far below the point of release is the tuning fork when waves of frequency 485 Hz reach the release point? Take the speed of sound in air to be 340 m/s.

28. A supersonic jet traveling at Mach 3 at an altitude of 20 000 m is directly overhead at time $t = 0$, as in Figure P14.28. (a) How long will it be before the ground observer encounters the shock wave? (b) Where will the plane be when it is finally heard? (Assume an average value of 330 m/s for the speed of sound in air.)

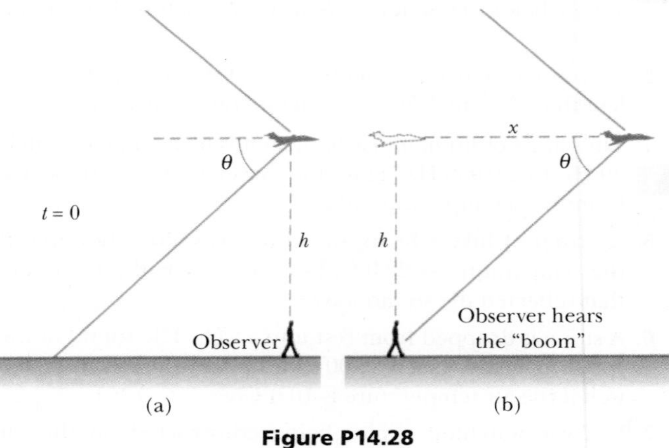

Figure P14.28

29. The now-discontinued *Concorde* flew at Mach 1.5, which meant the speed of the plane was 1.5 times the speed of sound in air. What was the angle between the direction of propagation of the shock wave and the direction of the plane's velocity?

Section 14.7 Interference of Sound Waves

30. The acoustical system shown in Figure 14.14 is driven by a speaker emitting a 400-Hz note. If *destructive* interference occurs at a particular instant, how much must the path length in the U-shaped tube be increased in order to hear (a) constructive interference and (b) destructive interference once again?

31. The ship in Figure P14.31 travels along a straight line parallel to the shore and 600 m from it. The ship's radio receives simultaneous signals of the same frequency from antennas *A* and *B*. The signals interfere constructively at point *C*, which is equidistant from *A* and *B*. The signal goes through the first minimum at point *D*. Determine the wavelength of the radio waves.

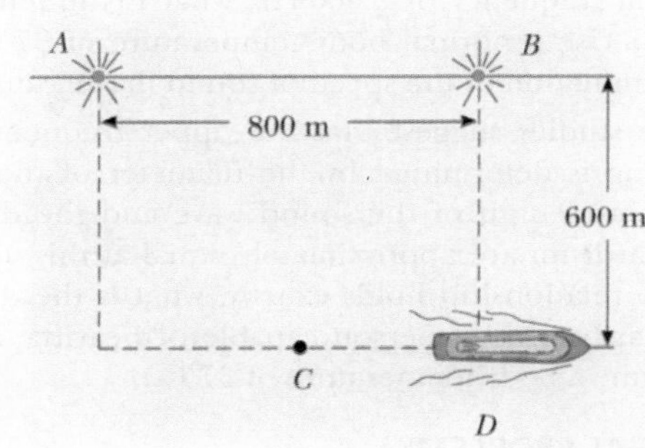

Figure P14.31

32. Two loudspeakers are placed above and below one another, as in Figure 14.15, and are driven by the same source at a frequency of 500 Hz. (a) What minimum distance should the top speaker be moved back in order to create destructive interference between the speakers? (b) If the top speaker is moved back twice the distance calculated in part (a), will there be constructive or destructive interference?

33. A pair of speakers separated by 0.700 m are driven by the same oscillator at a frequency of 690 Hz. An observer originally positioned at one of the speakers begins to walk along a line perpendicular to the line joining the speakers. (a) How far must the observer walk before reaching a relative maximum in intensity? (b) How far will the observer be from the speaker when the first relative minimum is detected in the intensity?

Section 14.8 Standing Waves

34. A steel wire in a piano has a length of 0.700 0 m and a mass of 4.300×10^{-3} kg. To what tension must this wire be stretched in order that the fundamental vibration correspond to middle C ($f_C = 261.6$ Hz on the chromatic musical scale)?

35. A stretched string fixed at each end has a mass of 40.0 g and a length of 8.00 m. The tension in the string is 49.0 N. (a) Determine the positions of the nodes and antinodes for the third harmonic. (b) What is the vibration frequency for this harmonic?

36. Resonance of sound waves can be produced within an aluminum rod by holding the rod at its midpoint and stroking it with an alcohol-saturated paper towel. In this resonance mode, the middle of the rod is a node while the ends are antinodes; no other nodes or antinodes are present. What is the frequency of the resonance if the rod is 1.00 m long?

37. Two speakers are driven by a common oscillator at 800 Hz and face each other at a distance of 1.25 m. Locate the points along a line joining the speakers where relative minima of the amplitude of the pressure would be expected. (Use $v = 343$ m/s.)

38. Two pieces of steel wire with identical cross sections have lengths of L and $2L$. The wires are each fixed at both ends and stretched so that the tension in the longer wire is four times greater than in the shorter wire. If the fundamental frequency in the shorter wire is 60 Hz, what is the frequency of the second harmonic in the longer wire?

39. A 12-kg object hangs in equilibrium from a string of total length $L = 5.0$ m and linear mass density $\mu = 0.001\,0$ kg/m. The string is wrapped around two light, frictionless pulleys that are separated by the distance $d = 2.0$ m (Fig. P14.39a). (a) Determine the tension in the string. (b) At what frequency must the string between the pulleys vibrate in order to form the standing-wave pattern shown in Figure P14.39b?

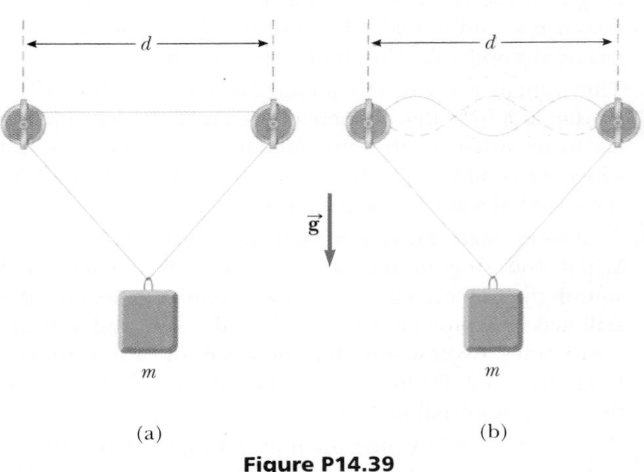

Figure P14.39

40. In the arrangement shown in Figure P14.40, an object of mass $m = 5.0$ kg hangs from a cord around a light pulley. The length of the cord between point P and the pulley is $L = 2.0$ m. (a) When the vibrator is set to a frequency of 150 Hz, a standing wave with six loops is formed. What must be the linear mass density of the cord? (b) How many loops (if any) will result if m is changed to 45 kg? (c) How many loops (if any) will result if m is changed to 10 kg?

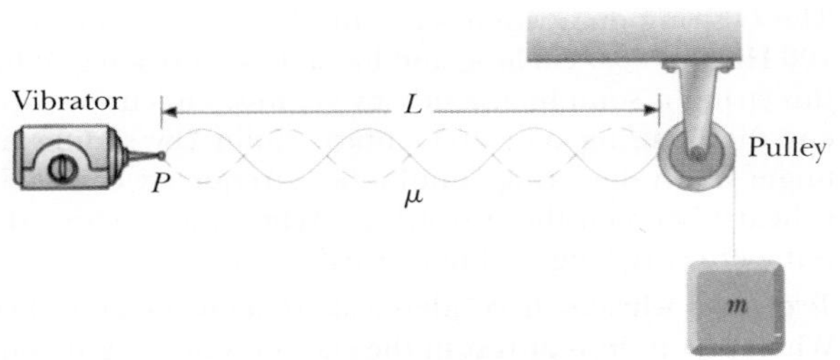

Figure P14.40

41. A 60.00-cm guitar string under a tension of 50.000 N has a mass per unit length of 0.100 00 g/cm. What is the highest resonant frequency that can be heard by a person capable of hearing frequencies up to 20 000 Hz?

Section 14.9 Forced Vibrations and Resonance

42. Standing-wave vibrations are set up in a crystal goblet with four nodes and four antinodes equally spaced around the 20.0-cm circumference of its rim. If transverse waves move around the glass at 900 m/s, an opera singer would have to produce a high harmonic with what frequency in order to shatter the glass with a resonant vibration?

Section 14.10 Standing Waves in Air Columns

43. The windpipe of a typical whooping crane is about 5.0 feet long. What is the lowest resonant frequency of this pipe, assuming that it is closed at one end? Assume a temperature of 37°C.

44. The overall length of a piccolo is 32.0 cm. The resonating air column vibrates as in a pipe that is open at both ends. (a) Find the frequency of the lowest note a piccolo can play, assuming the speed of sound in air is 340 m/s. (b) Opening holes in the side effectively shortens the length of the resonant column. If the highest note a piccolo can sound is 4 000 Hz, find the distance between adjacent antinodes for this mode of vibration.

45. The human ear canal is about 2.8 cm long. If it is regarded as a tube that is open at one end and closed at the eardrum, what is the fundamental frequency around which we would expect hearing to be most sensitive? Take the speed of sound to be 340 m/s.

46. A shower stall measures 86.0 cm × 86.0 cm × 210 cm. When you sing in the shower, which frequencies will sound the richest (because of resonance)? Assume the stall acts as a pipe closed at both ends, with nodes at opposite sides. Assume also that the voices of various singers range from 130 Hz to 2 000 Hz. Let the speed of sound in the hot shower stall be 355 m/s.

47. Physics Now™ A pipe open at both ends has a fundamental frequency of 300 Hz when the temperature is 0°C. (a) What is the length of the pipe? (b) What is the fundamental frequency at a temperature of 30°C?

48. A 2.00-m-long air column is open at both ends. The frequency of a certain harmonic is 410 Hz, and the frequency of the next-higher harmonic is 492 Hz. Determine the speed of sound in the air column.

Section 14.11 Beats

49. Two identical mandolin strings under 200 N of tension are sounding tones with frequencies of 523 Hz. The peg of one string slips slightly, and the tension in it drops to 196 N. How many beats per second are heard?

50. The G string on a violin has a fundamental frequency of 196 Hz. It is 30.0 cm long and has a mass of 0.500 g. While this string is sounding, a nearby violinist effectively shortens the G string on her identical violin (by sliding her finger down the string) until a beat frequency of 2.00 Hz is heard between the two strings. When this occurs, what is the effective length of her string?

51. Two train whistles have identical frequencies of 180 Hz. When one train is at rest in the station, sounding its whistle, a beat frequency of 2 Hz is heard from a moving train. What two possible speeds and directions can the moving train have?

52. Two pipes of equal length are each open at one end. Each has a fundamental frequency of 480 Hz at 300 K. In one pipe, the air temperature is increased to 305 K. If the two pipes are sounded together, what beat frequency results?

53. A student holds a tuning fork oscillating at 256 Hz. He walks toward a wall at a constant speed of 1.33 m/s. (a) What beat frequency does he observe between the tuning fork and its echo? (b) How fast must he walk away from the wall to observe a beat frequency of 5.00 Hz?

Section 14.13 The Ear

54. If a human ear canal can be thought of as resembling an organ pipe, closed at one end, that resonates at a fundamental frequency of 3 000 Hz, what is the length of the canal? Use a normal body temperature of 37°C for your determination of the speed of sound in the canal.

55. Some studies suggest that the upper frequency limit of hearing is determined by the diameter of the eardrum. The wavelength of the sound wave and the diameter of the eardrum are approximately equal at this upper limit. If the relationship holds exactly, what is the diameter of the eardrum of a person capable of hearing 20 000 Hz? (Assume a body temperature of 37°C.)

ADDITIONAL PROBLEMS

56. A commuter train blows its horn as it passes a passenger platform at a constant speed of 40.0 m/s. The horn sounds at a frequency of 320 Hz when the train is at rest. What is the frequency observed by a person on the platform (a) as the train approaches and (b) as the train recedes from him? (c) What wavelength does the observer find in each case?

57. A quartz watch contains a crystal oscillator in the form of a block of quartz that vibrates by contracting and expanding. Two opposite faces of the block, 7.05 mm apart, are antinodes, moving alternately towards and away from each other. The plane halfway between these two faces is a node of the vibration. The speed of sound in quartz is 3.70 km/s. Find the frequency of the vibration. An oscillating electric voltage accompanies the mechanical oscillation, so the quartz is described as *piezoelectric*. An electric circuit feeds in energy to maintain the oscillation and also counts the voltage pulses to keep time.

58. A flowerpot is knocked off a balcony 20.0 m above the sidewalk and falls toward an unsuspecting 1.75-m-tall man who is standing below. How close to the sidewalk can the flowerpot fall before it is too late for a warning shouted from the balcony to reach the man in time? Assume that the man below requires 0.300 s to respond to the warning.

59. Physics Now™ On a workday, the average decibel level of a busy street is 70 dB, with 100 cars passing a given point every minute. If the number of cars is reduced to 25 every minute on a weekend, what is the decibel level of the street?

60. A variable-length air column is placed just below a vibrating wire that is fixed at both ends. The length of the column, open at one end, is gradually increased from zero until the first position of resonance is observed at

$L = 34.0$ cm. The wire is 120 cm long and is vibrating in its third harmonic. If the speed of sound in air is 340 m/s, what is the speed of transverse waves in the wire?

61. A block with a speaker bolted to it is connected to a spring having spring constant $k = 20.0$ N/m, as shown in Figure P14.61. The total mass of the block and speaker is 5.00 kg, and the amplitude of the unit's motion is 0.500 m. If the speaker emits sound waves of frequency 440 Hz, determine the lowest and highest frequencies heard by the person to the right of the speaker.

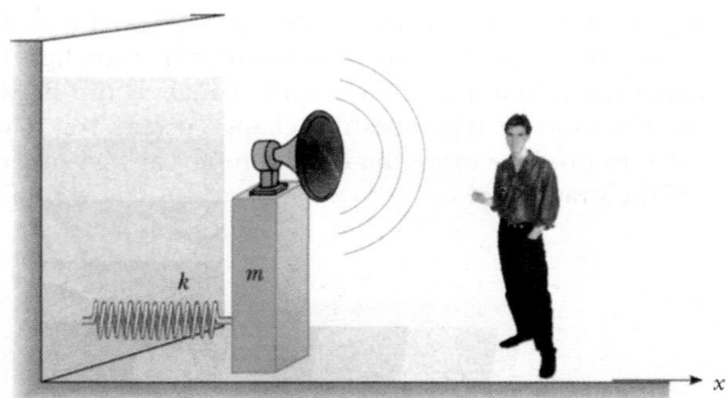

Figure P14.61

62. A flute is designed so that it plays a frequency of 261.6 Hz, middle C, when all the holes are covered and the temperature is 20.0°C. (a) Consider the flute to be a pipe open at both ends, and find its length, assuming that the middle-C frequency is the fundamental frequency. (b) A second player, nearby in a colder room, also attempts to play middle C on an identical flute. A beat frequency of 3.00 beats/s is heard. What is the temperature of the room?

63. When at rest, two trains have sirens that emit a frequency of 300 Hz. The trains travel toward one another and toward an observer stationed between them. One of the trains moves at 30.0 m/s, and the observer hears a beat frequency of 3.0 beats per second. What is the speed of the second train, which travels faster than 30.0 m/s?

64. Many artists sing very high notes in *ad lib* ornaments and cadenzas. The highest note written for a singer in a published score was F-sharp above high C, 1.480 kHz, sung by Zerbinetta in the original version of Richard Strauss's opera *Ariadne auf Naxos.* (a) Find the wavelength of this sound in air. (b) In response to complaints, Strauss later transposed the note down to F above high C, 1.397 kHz. By what increment did the wavelength change?

65. A speaker at the front of a room and an identical speaker at the rear of the room are being driven at 456 Hz by the same sound source. A student walks at a uniform rate of 1.50 m/s away from one speaker and towards the other. How many beats does the student hear per second?

66. Two identical speakers separated by 10.0 m are driven by the same oscillator with a frequency of $f = 21.5$ Hz (Fig. P14.66). Explain why a receiver at A records a minimum in sound intensity from the two speakers. (b) If the receiver is moved in the plane of the speakers, what path should it take so that the intensity remains at a minimum? That is, determine the relationship between x and y (the coordinates of the receiver) such that the receiver will record a minimum in sound intensity. Take the speed of sound to be 344 m/s.

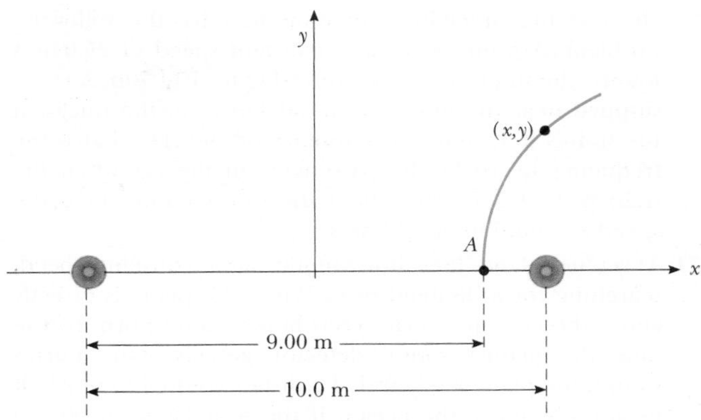

Figure P14.66

67. By proper excitation, it is possible to produce both longitudinal and transverse waves in a long metal rod. In a particular case, the rod is 150 cm long and 0.200 cm in radius and has a mass of 50.9 g. Young's modulus for the material is 6.80×10^{10} Pa. Determine the required tension in the rod so that the ratio of the speed of longitudinal waves to the speed of transverse waves is 8.

68. A student stands several meters in front of a smooth reflecting wall, holding a board on which a wire is fixed at each end. The wire, vibrating in its third harmonic, is 75.0 cm long, has a mass of 2.25 g, and is under a tension of 400 N. A second student, moving towards the wall, hears 8.30 beats per second. What is the speed of the student approaching the wall? Use 340 m/s as the speed of sound in air.

69. Two ships are moving along a line due east. The trailing vessel has a speed of 64.0 km/h relative to a land-based observation point, and the leading ship has a speed of 45.0 km/h relative to the same station. The trailing ship transmits a sonar signal at a frequency of 1 200 Hz. What frequency is monitored by the leading ship? (Use 1 520 m/s as the speed of sound in ocean water.)

70. The Doppler equation presented in the text is valid when the motion between the observer and the source occurs on a straight line, so that the source and observer are moving either directly toward or directly away from each other. If this restriction is relaxed, one must use the more general Doppler equation

$$f_O = \left[\frac{v + v_O \cos(\theta_O)}{v - v_S \cos(\theta_S)}\right] f_S$$

where θ_O and θ_S are defined in Figure P14.70a. (a) If both observer and source are moving away from each other along a straight line, show that the preceding equation yields the same result as Equation 14.12 in the text.

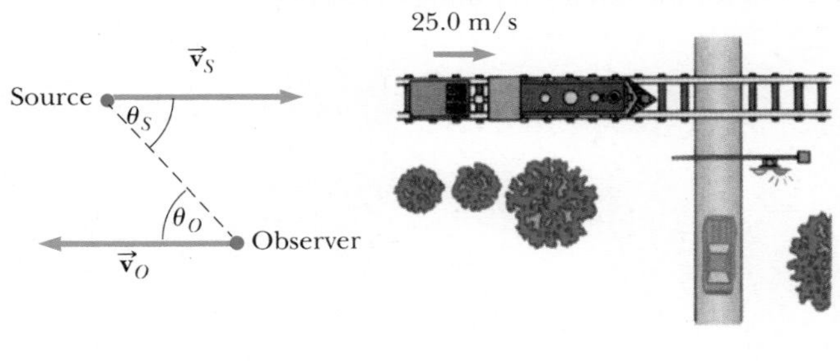

Figure P14.70

(b) Use the preceding equation to solve the following problem: A train moves at a constant speed of 25.0 m/s toward the intersection shown in Figure P14.70b. A car is stopped near the intersection, 30.0 m from the tracks. If the train's horn emits a frequency of 500 Hz, what is the frequency heard by the passengers in the car when the train is 40.0 m to the left of the intersection? Take the speed of sound to be 343 m/s.

71. A rescue plane flies horizontally at a constant speed, searching for a disabled boat. When the plane is directly above the boat, the boat's crew blows a loud horn. By the time the plane's sound detector perceives the horn's sound, the plane has traveled a distance equal to one-half its altitude above the ocean. If the sound takes 2.00 s to reach the plane, determine (a) the plane's altitude and (b) its speed.

72. In order to determine her speed, a skydiver carries a tone generator. A friend on the ground at the landing site has equipment for receiving and analyzing sound waves. While the skydiver is falling at terminal speed, her tone generator emits a steady tone of 1.80 kHz. (Assume that the air is calm, that the speed of sound is 343 m/s, independent of altitude.) (a) If her friend on the ground (directly beneath the skydiver) receives waves of frequency 2.15 kHz, what is the skydiver's speed of descent? (b) If the skydiver were also carrying sound-receiving equipment sensitive enough to detect waves reflected from the ground, what frequency of waves would she receive?

ACTIVITIES

A.1. Use an empty 1-liter soft-drink container, blow over the open end, and listen to the sound that is produced. Add some water to the container to change the height of the air column, and repeat the procedure. How does the frequency that you hear change with the height of the air column?

If you want to investigate this phenomenon in more detail, construct a musical instrument made up of several soft-drink bottles with different amounts of water in each. You can play your instrument as a wind instrument by blowing over the mouths of the bottles.

A.2. Beats can easily be heard on a guitar. When a finger is placed at the fifth fret of the second string, the note produced when the string is plucked should be identical to the note from the first string when it is played without fingering. With your finger in position on the second string, pluck the two strings simultaneously. If one of the strings is slightly out of tune, a very pronounced beat frequency will be heard. What happens to the beat frequency as the string tension is changed in small increments from too low for the intended tuning to too high?

A.3. Attach a rope to a door and shake the other end to see how many of the standing-wave patterns in Figure 14.18 you can produce. When a pattern is formed, note that the amplitude of the rope's vibration is much larger than the movement of your hand.

A.4. Snip off the corners of one end of a paper straw so that the end tapers to a point, as shown in Figure A14.4. Chew on this end to flatten it, and you have created a double-reed instrument. Put your lips around the tapered end of the straw, press them together slightly, and blow through the straw. When you hear a steady tone, slowly snip off a piece of the straw at the other end. Be careful to keep about the same amount of pressure with your lips. How does the frequency of the sound change as the straw becomes shorter? Why does this change occur? You may be able to produce more than one tone for any given length of the straw. Why?

Figure A14.4

A.5. Inflate a balloon just enough to form a small sphere. Measure its diameter. Use a marker to color in a 1-cm square on its surface. Now continue inflating the balloon until it reaches twice the original diameter. Measure the size of the square now. Note how the color of the marked area has changed. Use the information in Section 14.5 to explain these results.

"I love hearing that lonesome wail of the train whistle as the magnitude of the frequency of the wave changes due to the Doppler effect."

This nighttime view of multiple bolts of lightning was photographed in Tucson, Arizona. During a thunderstorm, a high concentration of electrical charge in a thundercloud creates a higher-than-normal electric field between the thundercloud and the negatively charged Earth's surface. This strong electric field creates an electric discharge between the charged cloud and the ground—an enormous spark. Other discharges that are observed in the sky include cloud-to -cloud discharges and the more frequent intracloud discharges.

CHAPTER

15

Electric Forces and Electric Fields

OUTLINE

Electricity is the lifeblood of technological civilization and modern society. Without it, we revert to the mid-nineteenth century: no telephones, no television, none of the household appliances that we take for granted. Modern medicine would be a fantasy, and due to the lack of sophisticated experimental equipment and fast computers—and especially the slow dissemination of information—science and technology would grow at a glacial pace.

Instead, with the discovery and harnessing of electric forces and fields, we can view arrangements of atoms, probe the inner workings of the cell, and send spacecraft beyond the limits of the solar system. All this has become possible in just the last few generations of human life, a blink of the eye compared to the million years our kind spent foraging the savannahs of Africa.

Around 700 B.C. the ancient Greeks conducted the earliest known study of electricity. It all began when someone noticed that a fossil material called amber would attract small objects after being rubbed with wool. Since then we have learned that this phenomenon is not restricted to amber and wool, but occurs (to some degree) when almost any two nonconducting substances are rubbed together.

In the current chapter we use the effect of charging by friction to begin an investigation of electric forces. We then discuss Coulomb's law, which is the fundamental law of force between any two stationary charged particles. The concept of an electric field associated with charges is introduced and its effects on other charged particles described. We end with discussions of the Van de Graaff generator and Gauss's law.

15.1 PROPERTIES OF ELECTRIC CHARGES

After running a plastic comb through your hair, you will find that the comb attracts bits of paper. The attractive force is often strong enough to suspend the

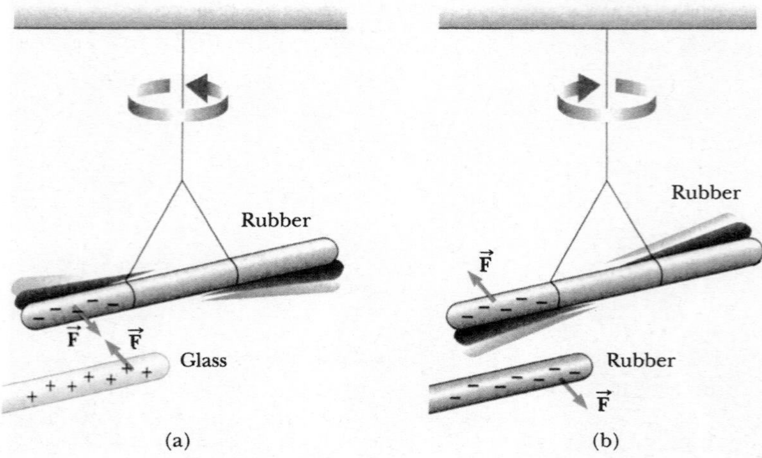

Figure 15.1 (a) A negatively charged rubber rod, suspended by a thread, is attracted to a positively charged glass rod. (b) A negatively charged rubber rod is repelled by another negatively charged rubber rod.

BENJAMIN FRANKLIN
(1706–1790)

Franklin was a printer, author, physical scientist, inventor, diplomat, and a founding father of the United States. His work on electricity in the late 1740s changed a jumbled, unrelated set of observations into a coherent science.

paper from the comb, defying the gravitational pull of the entire Earth. The same effect occurs with other rubbed materials, such as glass and hard rubber.

Another simple experiment is to rub an inflated balloon against wool (or across your hair). On a dry day, the rubbed balloon will then stick to the wall of a room, often for hours. These materials have become **electrically charged**. You can give your body an electric charge by vigorously rubbing your shoes on a wool rug or by sliding across a car seat. You can then surprise and annoy a friend or co-worker with a light touch on the arm, delivering a slight shock to both yourself and your victim. (If the co-worker is your boss, don't expect a promotion!) These experiments work best on a dry day because excessive moisture can facilitate a leaking away of the charge.

Experiments also demonstrate that there are two kinds of electric charge, which Benjamin Franklin (1706–1790) named **positive** and **negative**. Figure 15.1 illustrates the interaction of the two charges. A hard rubber (or plastic) rod that has been rubbed with fur is suspended by a piece of string. When a glass rod that has been rubbed with silk is brought near the rubber rod, the rubber rod is attracted toward the glass rod (Fig. 15.1a). If two charged rubber rods (or two charged glass rods) are brought near each other, as in Figure 15.1b, the force between them is repulsive. These observations may be explained by assuming that the rubber and glass rods have acquired different kinds of excess charge. We use the convention suggested by Franklin, where the excess electric charge on the glass rod is called positive and that on the rubber rod is called negative. On the basis of observations such as these, we conclude that **like charges repel one another and unlike charges attract one another**. Objects usually contain equal amounts of positive and negative charge—electrical forces between objects arise when those objects have net negative or positive charges.

Like charges repel
Unlike charges attract ▶

Nature's basic carriers of positive charge are protons, which, along with neutrons, are located in the nuclei of atoms. The nucleus, about 10^{-15} m in radius, is surrounded by a cloud of negatively charged electrons about ten thousand times larger in extent. An electron has the same magnitude charge as a proton, but the opposite sign. In a gram of matter there are approximately 10^{23} positively charged protons and just as many negatively charged electrons, so the net charge is zero. Because the nucleus of an atom is held firmly in place inside a solid, protons never move from one material to another. Electrons are far lighter than protons and hence more easily accelerated by forces. Furthermore, they occupy the outer regions of the atom. Consequently, objects become charged by gaining or losing electrons.

Charge transfers readily from one type of material to another. Rubbing the two materials together serves to increase the area of contact, facilitating the transfer process.

An important characteristic of charge is that **electric charge is always conserved.** Charge isn't *created* when two neutral objects are rubbed together; rather, the objects become charged because **negative charge is transferred from one object to the other.** One object gains a negative charge while the other loses an equal amount of negative charge and hence is left with a net positive charge. When a glass rod is rubbed with silk, as in Figure 15.2, electrons are transferred from the rod to the silk. As a result, the glass rod carries a net positive charge, the silk a net negative charge. Likewise, when rubber is rubbed with fur, electrons are transferred from the fur to the rubber.

◀ Charge is conserved; charge is quantized

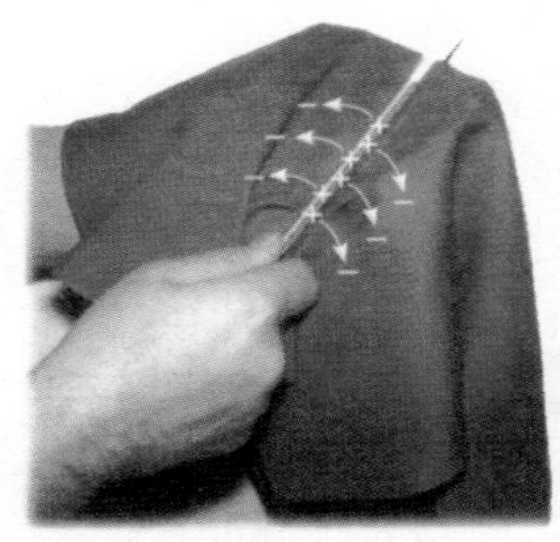

Figure 15.2 When a glass rod is rubbed with silk, electrons are transferred from the glass to the silk. Because of conservation of charge, each electron adds negative charge to the silk, and an equal positive charge is left behind on the rod. Also, because the charges are transferred in discrete bundles, the charges on the two objects are $\pm e$, $\pm 2e$, $\pm 3e$, and so on.

In 1909 Robert Millikan (1886–1953) discovered that if an object is charged, its charge is always a multiple of a fundamental unit of charge, designated by the symbol e. In modern terms, the charge is said to be **quantized**, meaning that charge occurs in discrete chunks that can't be further subdivided. An object may have a charge of $\pm e$, $\pm 2e$, $\pm 3e$, and so on, but never[1] a fractional charge of $\pm 0.5e$ or $\pm 0.22e$. Other experiments in Millikan's time showed that the electron has a charge of $-e$ and the proton has an equal and opposite charge of $+e$. Some particles, such as a neutron, have no net charge. A neutral atom (an atom with no net charge) contains as many protons as electrons. The value of e is now known to be $1.602\,19 \times 10^{-19}$ C. (The SI unit of electric charge is the **coulomb** [C].)

15.2 INSULATORS AND CONDUCTORS

Substances can be classified in terms of their ability to conduct electric charge.

> In **conductors**, electric charges move freely in response to an electric force. All other materials are called **insulators.**

Glass and rubber are insulators. When such materials are charged by rubbing, only the rubbed area becomes charged, and there is no tendency for the charge to move into other regions of the material. In contrast, materials such as copper, aluminum, and silver are good conductors. When such materials are charged in some small region, the charge readily distributes itself over the entire surface of the material. If you hold a copper rod in your hand and rub the rod with wool or fur, it will not attract a piece of paper. This might suggest that a metal can't be charged. However, if you hold the copper rod with an insulator and then rub it with wool or fur, the rod remains charged and attracts the paper. In the first case, the electric charges produced by rubbing readily move from the copper through your body and finally to ground. In the second case, the insulating handle prevents the flow of charge to ground.

Semiconductors are a third class of materials, and their electrical properties are somewhere between those of insulators and those of conductors. Silicon and germanium are well-known semiconductors that are widely used in the fabrication of a variety of electronic devices.

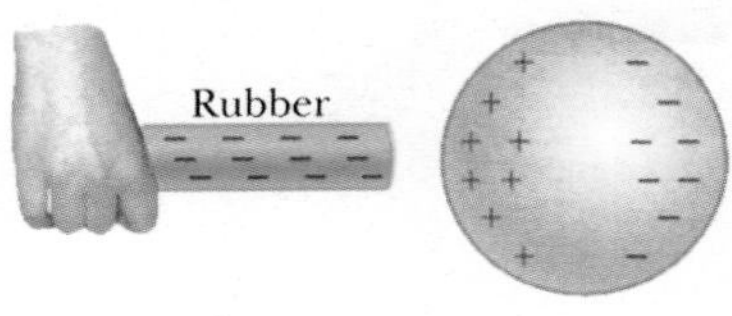

(a) Before

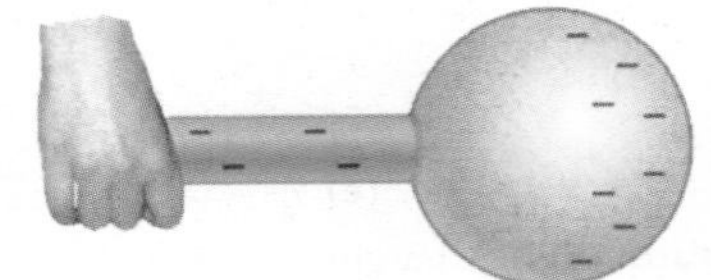

(b) Contact

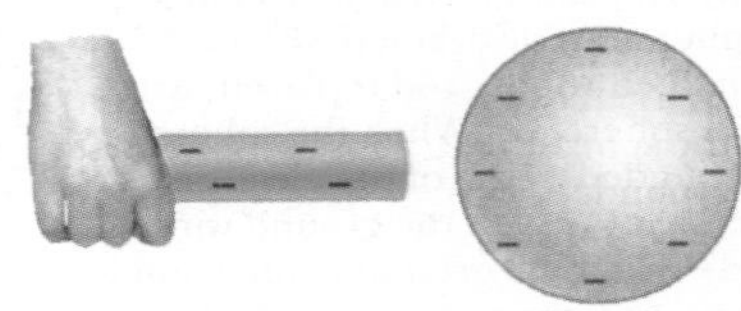

(c) After breaking contact

Figure 15.3 Charging a metallic object by conduction. (a) Just before contact, the negative rod repels the sphere's electrons, inducing a localized positive charge. (b) After contact, electrons from the rod flow onto the sphere, neutralizing the local positive charges. (c) When the rod is removed, the sphere is left with a negative charge.

Charging by Conduction

Consider a negatively charged rubber rod brought into contact with an insulated neutral conducting sphere. The excess electrons on the rod repel electrons on the sphere, creating local positive charges on the neutral sphere. On contact, some electrons on the rod are now able to move onto the sphere, as in Figure 15.3, neutralizing the positive charges. When the rod is removed, the sphere is left with a net negative charge. This process is referred to as charging by **conduction.** The object being charged in such a process (the sphere) is always left with a charge having the same sign as the object doing the charging (the rubber rod).

[1]There is strong evidence for the existence of fundamental particles called **quarks** that have charges of $\pm e/3$ or $\pm 2e/3$. The charge is *still* quantized, but in units of $\pm e/3$ rather than $\pm e$. A more complete discussion of quarks and their properties is presented in Chapter 30.

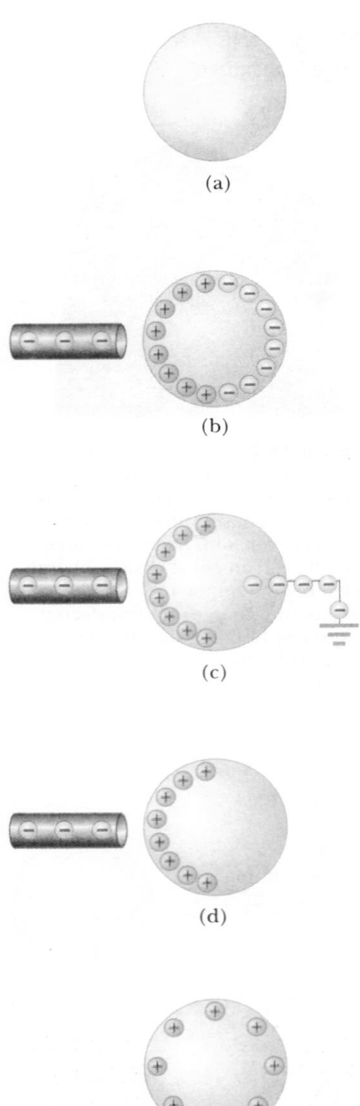

Figure 15.4 Charging a metallic object by *induction*. (a) A neutral metallic sphere with equal numbers of positive and negative charges. (b) The charge on a neutral metal sphere is redistributed when a charged rubber rod is placed near the sphere. (c) When the sphere is grounded, some of the electrons leave it through the ground wire. (d) When the ground connection is removed, the nonuniformly charged sphere is left with excess positive charge. (e) When the rubber rod is moved away, the charges on the sphere redistribute themselves until the sphere's surface becomes uniformly charged.

Charging by Induction

An object connected to a conducting wire or copper pipe buried in the Earth is said to be **grounded**. The Earth can be considered an infinite reservoir for electrons; in effect, it can accept or supply an unlimited number of electrons. With this idea in mind, we can understand the charging of a conductor by a process known as **induction**.

Consider a negatively charged rubber rod brought near a neutral (uncharged) conducting sphere that is insulated, so there is no conducting path to ground (Fig. 15.4). Initially the sphere is electrically neutral (Fig. 15.4a). When the negatively charged rod is brought close to the sphere, the repulsive force between the electrons in the rod and those in the sphere causes some electrons to move to the side of the sphere farthest away from the rod (Fig. 15.4b). The region of the sphere nearest the negatively charged rod has an excess of positive charge because of the migration of electrons away from that location. If a grounded conducting wire is then connected to the sphere, as in Figure 15.4c, some of the electrons leave the sphere and travel to ground. If the wire to ground is then removed (Fig. 15.4d), the conducting sphere is left with an excess of induced positive charge. Finally, when the rubber rod is removed from the vicinity of the sphere (Fig. 15.4e), the induced positive charge remains on the ungrounded sphere. Even though the positively charged atomic nuclei remain fixed, this excess positive charge becomes uniformly distributed over the surface of the ungrounded sphere because of the repulsive forces among the like charges and the high mobility of electrons in a metal.

In the process of inducing a charge on the sphere, the charged rubber rod doesn't lose any of its negative charge because it never comes in contact with the sphere. Furthermore, the sphere is left with a charge opposite that of the rubber rod. **Charging an object by induction requires no contact with the object inducing the charge**.

A process similar to charging by induction in conductors also takes place in insulators. In most neutral atoms or molecules, the center of positive charge coincides with the center of negative charge. However, in the presence of a charged object, these centers may separate slightly, resulting in more positive charge on one side of the molecule than on the other. This effect is known as **polarization.** The realignment of charge within individual molecules produces an induced charge on the surface of the insulator, as shown in Figure 15.5a. This explains why a balloon charged through rubbing will stick to an electrically neutral wall, or the comb you just used on your hair attracts tiny bits of neutral paper.

Quick Quiz 15.1

A suspended object A is attracted to a neutral wall. It's also attracted to a positively charged object B. Which of the following is true about object A? (a) It is uncharged. (b) It has a negative charge. (c) It has a positive charge. (d) It may be either charged or uncharged.

15.3 COULOMB'S LAW

In 1785 Charles Coulomb (1736–1806) experimentally established the fundamental law of electric force between two stationary charged particles.

An **electric force** has the following properties:

1. It is directed along a line joining the two particles and is inversely proportional to the square of the separation distance r, between them.
2. It is proportional to the product of the magnitudes of the charges, $|q_1|$ and $|q_2|$, of the two particles.
3. It is attractive if the charges are of opposite sign and repulsive if the charges have the same sign.

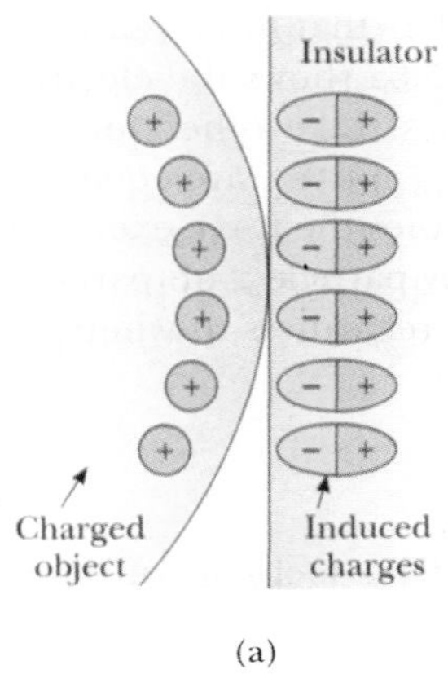

(a)

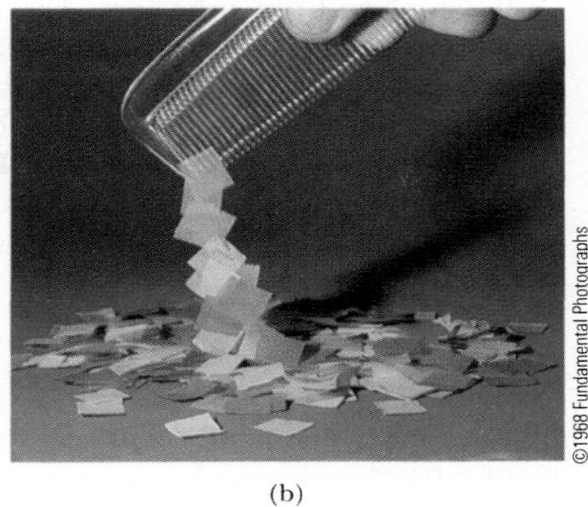

(b)

Figure 15.5 (a) The charged object on the left induces charges on the surface of an insulator. (b) A charged comb attracts bits of paper because charges are displaced in the paper.

From these observations, Coulomb proposed the following mathematical form for the electric force between two charges:

◀ Coulomb's law

The magnitude of the electric force F between charges q_1 and q_2 separated by a distance r is given by

$$F = k_e \frac{|q_1||q_2|}{r^2} \qquad [15.1]$$

where k_e is a constant called the *Coulomb constant.*

Equation 15.1, known as **Coulomb's law**, applies exactly only to point charges and to spherical distributions of charges, in which case r is the distance between the two centers of charge. Electric forces between unmoving charges are called *electrostatic* forces. Moving charges, in addition, create magnetic forces, studied in Chapter 19.

The value of the Coulomb constant in Equation 15.1 depends on the choice of units. The SI unit of charge is the **coulomb** (C). From experiment, we know that the **Coulomb constant** in SI units has the value

$$k_e = 8.9875 \times 10^9 \text{ N} \cdot \text{m}^2/\text{C}^2 \qquad [15.2]$$

This number can be rounded, depending on the accuracy of other quantities in a given problem. We'll use either two or three significant digits, as usual.

The charge on the proton has a magnitude of $e = 1.6 \times 10^{-19}$ C. Therefore, it would take $1/e = 6.3 \times 10^{18}$ protons to create a total charge of $+1.0$ C. Likewise, 6.3×10^{18} electrons would have a total charge of -1.0 C. Compare this with the number of free electrons in 1 cm^3 of copper, which is on the order of 10^{23}. Even so, 1.0 C is a very large amount of charge. In typical electrostatic experiments in which a rubber or glass rod is charged by friction, there is a net charge on the order of 10^{-6} C ($= 1\ \mu$C). Only a very small fraction of the total available charge is transferred between the rod and the rubbing material. Table 15.1 lists the charges and masses of the electron, proton, and neutron.

Photo courtesy of AIP Niels Bohr Library, E. Scott Barr Collection

CHARLES COULOMB
(1736–1806)

Coulomb's major contribution to science was in the field of electrostatics and magnetism. During his lifetime, he also investigated the strengths of materials and identified the forces that affect objects on beams, thereby contributing to the field of structural mechanics.

TABLE 15.1

Charge and Mass of the Electron, Proton, and Neutron

Particle	Charge (C)	Mass (kg)
Electron	-1.60×10^{-19}	9.11×10^{-31}
Proton	$+1.60 \times 10^{-19}$	1.67×10^{-27}
Neutron	0	1.67×10^{-27}

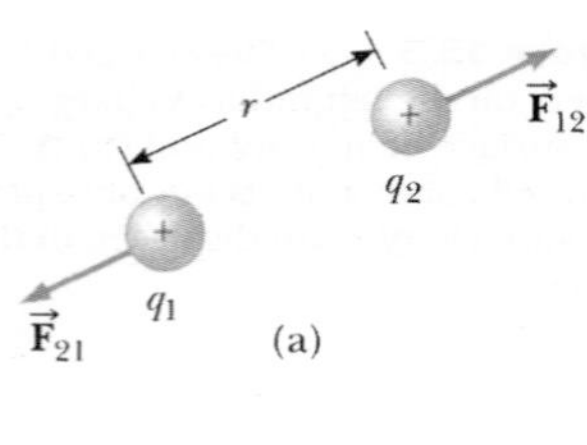

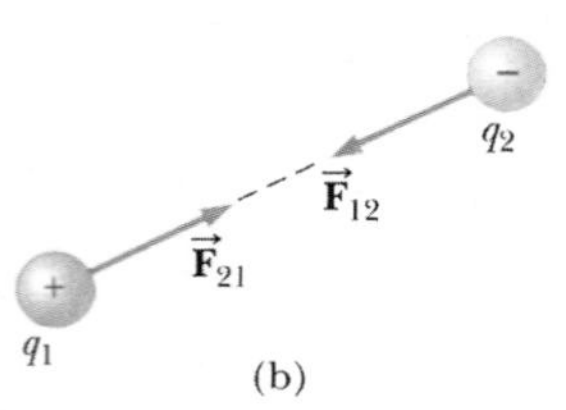

ACTIVE FIGURE 15.6
Two point charges separated by a distance r exert a force on each other given by Coulomb's law. The force on q_1 is equal in magnitude and opposite in direction to the force on q_2. (a) When the charges are of the same sign, the force is repulsive. (b) When the charges are of opposite sign, the force is attractive.

Physics Now™
Log into PhysicsNow at **www.cp7e.com** and go to Active Figure 15.6, where you can move the charges to any position in two-dimensional space and observe the electric forces acting on them.

When using Coulomb's force law, remember that force is a vector quantity and must be treated accordingly. Active Figure 15.6a shows the electric force of repulsion between two positively charged particles. Like other forces, electric forces obey Newton's third law; hence, the forces $\vec{\mathbf{F}}_{12}$ and $\vec{\mathbf{F}}_{21}$ are equal in magnitude but opposite in direction. (The notation $\vec{\mathbf{F}}_{12}$ denotes the force exerted by particle 1 on particle 2; likewise, $\vec{\mathbf{F}}_{21}$ is the force exerted by particle 2 on particle 1.) From Newton's third law, F_{12} and F_{21} are always equal regardless of whether q_1 and q_2 have the same magnitude.

Quick Quiz 15.2

Object A has a charge of $+2\ \mu$C, and object B has a charge of $+6\ \mu$C. Which statement is true?

(a) $\vec{\mathbf{F}}_{AB} = -3\vec{\mathbf{F}}_{BA}$ (b) $\vec{\mathbf{F}}_{AB} = -\vec{\mathbf{F}}_{BA}$ (c) $3\vec{\mathbf{F}}_{AB} = -\vec{\mathbf{F}}_{BA}$

The Coulomb force is similar to the gravitational force. Both act at a distance without direct contact. Both are inversely proportional to the distance squared, with the force directed along a line connecting the two bodies. The mathematical form is the same, with the masses m_1 and m_2 in Newton's law replaced by q_1 and q_2 in Coulomb's law and with Newton's constant G replaced by Coulomb's constant k_e. There are two important differences: (1) electric forces can be either attractive or repulsive, but gravitational forces are always attractive, and (2) the electric force between charged elementary particles is far stronger than the gravitational force between the same particles, as the next example shows.

EXAMPLE 15.1 Forces in a Hydrogen Atom

Goal Contrast the magnitudes of an electric force and a gravitational force.

Problem The electron and proton of a hydrogen atom are separated (on the average) by a distance of about 5.3×10^{-11} m. Find the magnitudes of the electric force and the gravitational force that each particle exerts on the other, and the ratio of the electric force F_e to the gravitational force F_g.

Strategy Solving this problem is just a matter of substituting known quantities into the two force laws and then finding the ratio.

Solution

Substitute $|q_1| = |q_2| = e$ and the distance into Coulomb's law to find the electric force:

$$F_e = k_e \frac{|e|^2}{r^2} = \left(8.99 \times 10^9 \frac{\text{N}\cdot\text{m}^2}{\text{C}^2}\right) \frac{(1.6 \times 10^{-19}\,\text{C})^2}{(5.3 \times 10^{-11}\,\text{m})^2}$$

$$= \boxed{8.2 \times 10^{-8}\,\text{N}}$$

Substitute the masses and distance into Newton's law of gravity to find the gravitational force:

$$F_g = G\frac{m_e m_p}{r^2}$$

$$= \left(6.67 \times 10^{-11} \frac{\text{N}\cdot\text{m}^2}{\text{kg}^2}\right) \frac{(9.11 \times 10^{-31}\,\text{kg})(1.67 \times 10^{-27}\,\text{kg})}{(5.3 \times 10^{-11}\,\text{m})^2}$$

$$= \boxed{3.6 \times 10^{-47}\,\text{N}}$$

Find the ratio of the two forces:

$$\frac{F_e}{F_g} = \boxed{2.27 \times 10^{39}}$$

Remarks The gravitational force between the charged constituents of the atom is negligible compared with the electric force between them. The electric force is so strong, however, that any net charge on an object quickly attracts nearby opposite charges, neutralizing the object. As a result, gravity plays a greater role in the mechanics of moving objects in everyday life.

Exercise 15.1

Find the magnitude of the electric force between two protons separated by one femtometer (10^{-15} m), approximately the distance between two protons in the nucleus of a helium atom. The answer may not appear large, but if not for the strong nuclear force, the two protons would fly apart at an initial acceleration of nearly $7 \times 10^{28}\ \mathrm{m/s^2}$!

Answers 2.30×10^2 N

The Superposition Principle

When a number of separate charges act on the charge of interest, each exerts an electric force. These electric forces can all be computed separately, one at a time, then added as vectors. This is another example of the **superposition principle**. The following example illustrates this procedure in one dimension.

INTERACTIVE EXAMPLE 15.2 May the Force Be Zero

Goal Apply Coulomb's law in one dimension.

Problem Three charges lie along the x-axis as in Figure 15.7. The positive charge $q_1 = 15\ \mu\mathrm{C}$ is at $x = 2.0$ m, and the positive charge $q_2 = 6.0\ \mu\mathrm{C}$ is at the origin. Where must a *negative* charge q_3 be placed on the x-axis so that the resultant electric force on it is zero?

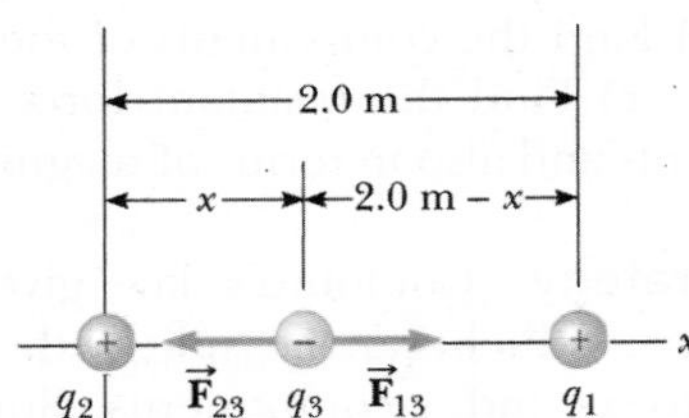

Figure 15.7 (Example 15.2) Three point charges are placed along the x-axis. The charge q_3 is negative, whereas q_1 and q_2 are positive. If the resultant force on q_3 is zero, then the force $\vec{\mathbf{F}}_{13}$ exerted by q_1 on q_3 must be equal in magnitude and opposite the force $\vec{\mathbf{F}}_{23}$ exerted by q_2 on q_3.

Strategy If q_3 is to the right or left of the other two charges, then the net force on q_3 can't be zero, because then $\vec{\mathbf{F}}_{13}$ and $\vec{\mathbf{F}}_{23}$ act in the same direction. Consequently, q_3 must lie between the two other charges. Write $\vec{\mathbf{F}}_{13}$ and $\vec{\mathbf{F}}_{23}$ in terms of the unknown coordinate position x, sum them and set them equal to zero, solving for the unknown. The solution can be obtained with the quadratic formula.

Solution

Write the x-component of $\vec{\mathbf{F}}_{13}$:

$$F_{13x} = +k_e \frac{(15 \times 10^{-6}\ \mathrm{C})|q_3|}{(2.0\ \mathrm{m} - x)^2}$$

Write the x-component of $\vec{\mathbf{F}}_{23}$:

$$F_{23x} = -k_e \frac{(6.0 \times 10^{-6}\ \mathrm{C})|q_3|}{x^2}$$

Set the sum equal to zero:

$$k_e \frac{(15 \times 10^{-6}\ \mathrm{C})|q_3|}{(2.0\ \mathrm{m} - x)^2} - k_e \frac{(6.0 \times 10^{-6}\ \mathrm{C})|q_3|}{x^2} = 0$$

Cancel k_e, 10^{-6} and q_3 from the equation, and rearrange terms (explicit significant figures and units are temporarily suspended for clarity):

$$6(2 - x)^2 = 15x^2$$

Put this equation into standard quadratic form, $ax^2 + bx + c = 0$:

$$6(4 - 4x + x^2) = 15x^2 \quad \rightarrow \quad 2(4 - 4x + x^2) = 5x^2$$

$$3x^2 + 8x - 8 = 0$$

Apply the quadratic formula:

$$x = \frac{-8 \pm \sqrt{64 - (4)(3)(-8)}}{2 \cdot 3} = \frac{-4 \pm 2\sqrt{10}}{3}$$

Only the positive root makes sense:

$$x = 77\ \mathrm{m}$$

Remarks Notice that it was necessary to use physical reasoning to choose between the two possible answers for x. This is nearly always the case when quadratic equations are involved.

Exercise 15.2

Three charges lie along the x-axis. The positive charge $q_1 = 10.0\ \mu C$ is at $x = 1.00$ m, and the *negative charge* $q_2 = -2.00\ \mu C$ is at the origin. Where must a *positive* charge q_3 be placed on the x-axis so that the resultant force on it is zero?

Answer $x = -0.809$ m

PhysicsNow™ You can predict where on the x-axis the electric force is zero for random values of of q_1 and q_2 by logging into PhysicsNow at **www.cp7e.com** and going to Interactive Example 15.2.

EXAMPLE 15.3 A Charge Triangle

Goal Apply Coulomb's law in two dimensions.

Problem Consider three point charges at the corners of a triangle, as shown in Figure 15.8, where $q_1 = 6.00 \times 10^{-9}$ C, $q_2 = -2.00 \times 10^{-9}$ C, and $q_3 = 5.00 \times 10^{-9}$ C. **(a)** Find the components of the force $\vec{F}_{23}$ exerted by q_2 on q_3. **(b)** Find the components of the force $\vec{F}_{13}$ exerted by q_1 on q_3. **(c)** Find the resultant force on q_3, in terms of components and also in terms of magnitude and direction.

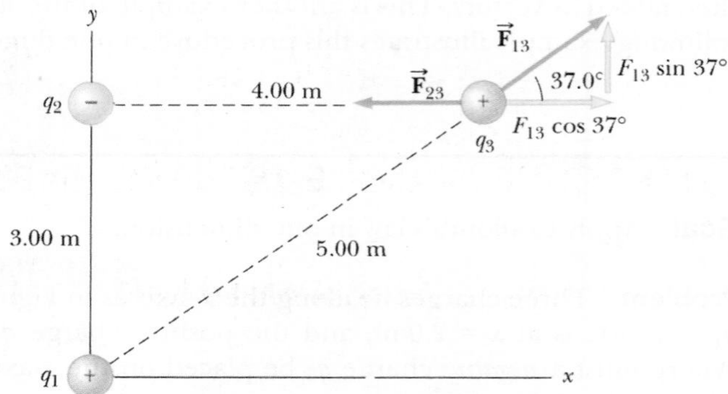

Figure 15.8 (Example 15.3) The force exerted by q_1 on q_3 is $\vec{F}_{13}$. The force exerted by q_2 on q_3 is $\vec{F}_{23}$. The *resultant force* $\vec{F}_3$ exerted on q_3 is the *vector* sum $\vec{F}_{13} + \vec{F}_{23}$.

Strategy Coulomb's law gives the magnitude of each force, which can be split with right-triangle trigonometry into x- and y-components. Sum the vectors componentwise, and then find the magnitude and direction of the resultant vector.

Solution

(a) Find the components of the force exerted by q_2 on q_3.

Find the magnitude of $\vec{F}_{23}$ with Coulomb's law:

$$F_{23} = k_e \frac{|q_2||q_3|}{r^2}$$

$$= (8.99 \times 10^9\ \text{N}\cdot\text{m}^2/\text{C}^2)\frac{(2.00 \times 10^{-9}\ \text{C})(5.00 \times 10^{-9}\ \text{C})}{(4.00\ \text{m})^2}$$

$$F_{23} = 5.62 \times 10^{-9}\ \text{N}$$

Because $\vec{F}_{23}$ is horizontal and points in the negative x-direction, the negative of the magnitude gives the x-component, and the y-component is zero:

$$F_{23x} = -5.62 \times 10^{-9}\ \text{N}$$

$$F_{23y} = 0$$

(b) Find the components of the force exerted by q_1 on q_3.

Find the magnitude of $\vec{F}_{13}$:

$$F_{13} = k_e \frac{|q_1||q_3|}{r^2}$$

$$= (8.99 \times 10^9\ \text{N}\cdot\text{m}^2/\text{C}^2)\frac{(6.00 \times 10^{-9}\ \text{C})(5.00 \times 10^{-9}\ \text{C})}{(5.00\ \text{m})^2}$$

$$F_{13} = 1.08 \times 10^{-8}\ \text{N}$$

Use the given triangle to find the components:

$$F_{13x} = F_{13} \cos\theta = (1.08 \times 10^{-8}\ \text{N})\cos(37°)$$

$$= 8.63 \times 10^{-9}\ \text{N}$$

$$F_{13y} = F_{13} \sin\theta = (1.08 \times 10^{-8}\ \text{N})\sin(37°)$$

$$= 6.50 \times 10^{-9}\ \text{N}$$

(c) Find the components of the resultant vector.

Sum the x-components to find the resultant F_x:

$$F_x = -5.62 \times 10^{-9}\text{ N} + 8.63 \times 10^{-9}\text{ N}$$
$$= 3.01 \times 10^{-9}\text{ N}$$

Sum the y-components to find the resultant F_y:

$$F_y = 0 + 6.50 \times 10^{-9}\text{ N} = 6.50 \times 10^{-9}\text{ N}$$

Find the magnitude of the resultant force on the charge q_3, using the Pythagorean theorem:

$$|\vec{\mathbf{F}}| = \sqrt{F_x{}^2 + F_y{}^2}$$
$$= \sqrt{(3.01 \times 10^{-9}\text{ N})^2 + (6.50 \times 10^{-9}\text{ N})^2}$$
$$= 7.16 \times 10^{-9}\text{ N}$$

Find the angle the force vector makes with respect to the positive x-axis:

$$\theta = \tan^{-1}\left(\frac{F_y}{F_x}\right) = \tan^{-1}\left(\frac{6.50 \times 10^{-9}\text{ N}}{3.01 \times 10^{-9}\text{ N}}\right) = 65.2°$$

Remarks The methods used here are just like those used with Newton's law of gravity in two dimensions.

Exercise 15.3

Using the same triangle, find the vector components of the electric force on q_1 and the vector's magnitude and direction.

Answers $F_x = -8.63 \times 10^{-9}$ N, $F_y = 5.50 \times 10^{-9}$ N, $F = 1.02 \times 10^{-8}$ N, $\theta = 147°$

15.4 THE ELECTRIC FIELD

The gravitational force and the electrostatic force are both capable of acting through space, producing an effect even when there isn't any physical contact between the objects involved. Field forces can be discussed in a variety of ways, but an approach developed by Michael Faraday (1791–1867) is the most practical. In this approach, an **electric field** is said to exist in the region of space around a charged object. The electric field exerts an electric force on any other charged object within the field. This differs from the Coulomb's law concept of a force exerted at a distance, in that the force is now exerted by something—the field—that is in the same location as the charged object.

Figure 15.9 shows an object with a small positive charge q_0 placed near a second object with a much larger positive charge Q.

The electric field $\vec{\mathbf{E}}$ produced by a charge Q at the location of a small "test" charge q_0 is defined as the electric force $\vec{\mathbf{F}}$ exerted by Q on q_0, divided by the test charge q_0:

$$\vec{\mathbf{E}} \equiv \frac{\vec{\mathbf{F}}}{q_0} \qquad [15.4]$$

SI Unit: newton per coulomb (N/C)

Conceptually and experimentally, the test charge q_0 is required to be very small (arbitrarily small, in fact), so it doesn't cause any significant rearrangement of the charge creating the electric field $\vec{\mathbf{E}}$. Mathematically, however, the size of the test charge makes no difference: the calculation comes out the same, regardless. In view of this, using $q_0 = 1$ C in Equation 15.4 can be convenient if not rigorous.

When a positive test charge is used, the electric field always has the same direction as the electric force on the test charge. This follows from Equation 15.4. Hence in Figure 15.9, the direction of the electric field is horizontal and to the

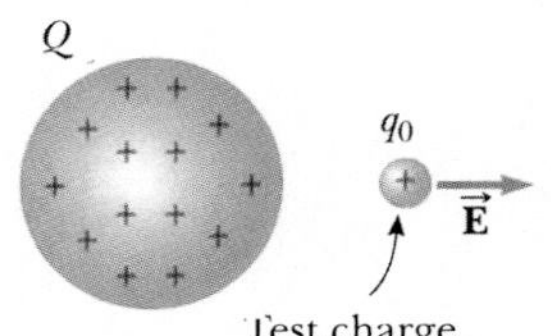

Figure 15.9 A small object with a positive charge q_0 placed near an object with a larger positive charge Q is subject to an electric field $\vec{\mathbf{E}}$ directed as shown. The magnitude of the electric field at the location of q_0 is defined as the electric force on q_0 divided by the charge q_0.

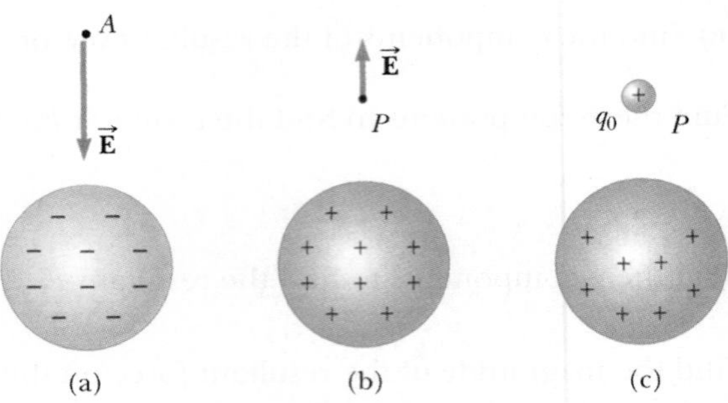

Figure 15.10 (a) The electric field at A due to the negatively charged sphere is downward, toward the negative charge. (b) The electric field at P due to the positively charged conducting sphere is upward, away from the positive charge. (c) A test charge q_0 placed at P will cause a rearrangement of charge on the sphere, unless q_0 is very small compared with the charge on the sphere.

right. The electric field at point A in Figure 15.10a is vertical and downward because at that point a positive test charge would be attracted toward the negatively charged sphere.

Once the electric field due to a given arrangement of charges is known at some point, the force on *any* particle with charge q placed at that point can be calculated from a rearrangement of Equation 15.4:

$$\vec{\mathbf{F}} = q\vec{\mathbf{E}} \tag{15.5}$$

Here q_0 has been replaced by q, which need not be a mere test charge.

As shown in Active Figure 15.11, the direction of $\vec{\mathbf{E}}$ is the direction of the force that acts on a positive test charge q_0 placed in the field. We say that **an electric field exists at a point if a test charge at that point is subject to an electric force there.**

Consider a point charge q located a distance r from a test charge q_0. According to Coulomb's law, the *magnitude* of the electric force of the charge q on the test charge is

$$F = k_e \frac{|q||q_0|}{r^2}$$

Because the magnitude of the electric field at the position of the test charge is defined as $E = F/q_0$, we see that the *magnitude* of the electric field due to the charge q at the position of q_0 is

$$E = k_e \frac{|q|}{r^2} \tag{15.6}$$

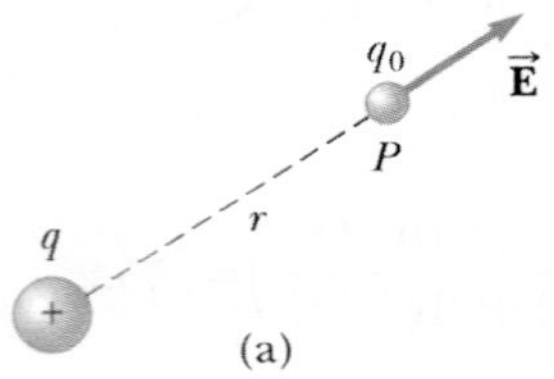

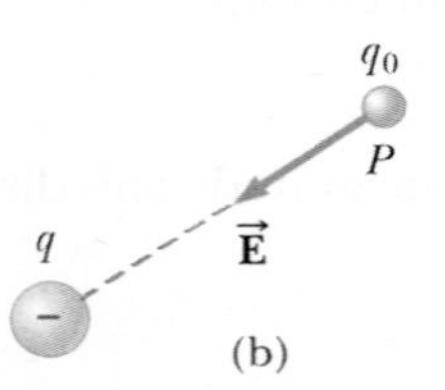

ACTIVE FIGURE 15.11
A test charge q_0 at P is a distance r from a point charge q. (a) If q is positive, the electric field at P points radially *outwards* from q. (b) If q is negative, the electric field at P points radially *inwards* toward q.

Physics Now™
Log into PhysicsNow at **www.cp7e.com** and go to Active Figure 15.11, where you can move point P to any position in two-dimensional space and observe the electric field due to q.

Equation 15.6 points out an important property of electric fields that makes them useful quantities for describing electrical phenomena. As the equation indicates, an electric field at a given point depends only on the charge q on the object setting up the field and the distance r from that object to a specific point in space. As a result, we can say that an electric field exists at point P in Active Figure 15.11 whether or not there is a test charge at P.

The principle of superposition holds when the electric field due to a group of point charges is calculated. We first use Equation 15.6 to calculate the electric field produced by each charge individually at a point and then add the electric fields together as vectors.

It's also important to exploit any symmetry of the charge distribution. For example, if equal charges are placed at $x = a$ and at $x = -a$, the electric field is zero at the origin, by symmetry. Similarly, if the x-axis has a uniform distribution of positive charge, it can be guessed by symmetry that the electric field points away from the x-axis and is zero parallel to that axis.

Quick Quiz 15.3

A test charge of $+3\ \mu C$ is at a point P where the electric field due to other charges is directed to the right and has a magnitude of 4×10^6 N/C. If the test charge is replaced with a charge of $-3\ \mu C$, the electric field at P (a) has the same magnitude as before, but changes direction, (b) increases in magnitude and changes direction, (c) remains the same, or (d) decreases in magnitude and changes direction.

Quick Quiz 15.4

A circular ring of charge of radius b has a total charge q uniformly distributed around it. The magnitude of the electric field at the center of the ring is

(a) 0 (b) $k_e q/b^2$ (c) $k_e q^2/b^2$ (d) $k_e q^2/b$ (e) none of these.

Quick Quiz 15.5

A "free" electron and a "free" proton are placed in an identical electric field. Which of the following statements are true? (a) Each particle is acted upon by the same electric force and has the same acceleration. (b) The electric force on the proton is greater in magnitude than the force on the electron, but in the opposite direction. (c) The electric force on the proton is equal in magnitude to the force on the electron, but in the opposite direction. (d) The magnitude of the acceleration of the electron is greater than that of the proton. (e) Both particles have the same acceleration.

EXAMPLE 15.4 Electrified Oil

Goal Use electric forces and fields together with Newton's second law in a one-dimensional problem.

Problem Tiny droplets of oil acquire a small negative charge while dropping through a vacuum (pressure = 0) in an experiment. An electric field of magnitude 5.92×10^4 N/C points straight down. **(a)** One particular droplet is observed to remain suspended against gravity. If the mass of the droplet is 2.93×10^{-15} kg, find the charge carried by the droplet. **(b)** Another droplet of the same mass falls 10.3 cm from rest in 0.250 s, again moving through a vacuum. Find the charge carried by the droplet.

Strategy We use Newton's second law with both gravitational and electric forces. In both parts, the electric field $\vec{\mathbf{E}}$ is pointing down, taken as the negative direction, as usual. In part (a), the acceleration is equal to zero. In part (b), the acceleration is uniform, so the kinematic equations yield the acceleration. Newton's law can then be solved for q.

Solution

(a) Find the charge on the suspended droplet.

Apply Newton's second law to the droplet in the vertical direction:

$$(1) \quad ma = \Sigma F = -mg + Eq$$

E points downward, hence is negative. Set $a = 0$ and solve for q:

$$q = \frac{mg}{E} = \frac{(2.93 \times 10^{-15}\,\text{kg})(9.80\,\text{m/s}^2)}{-5.92 \times 10^4\,\text{N/C}}$$

$$= -4.85 \times 10^{-19}\,\text{C}$$

(b) Find the charge on the falling droplet.

Use the kinematic displacement equation to find the acceleration:

$$\Delta y = \tfrac{1}{2}at^2 + v_0 t$$

Substitute $\Delta y = -0.103$ m, $t = 0.250$ s, and $v_0 = 0$:

$$-0.103\,\text{m} = \tfrac{1}{2}a(0.250\,\text{s})^2 \quad \rightarrow \quad a = -3.30\,\text{m/s}^2$$

Solve Equation 1 for q and substitute:

$$q = \frac{m(a+g)}{E}$$

$$= \frac{(2.93 \times 10^{-15}\,\text{kg})(-3.30\,\text{m/s}^2 + 9.80\,\text{m/s}^2)}{-5.92 \times 10^4\,\text{N/C}}$$

$$= -3.22 \times 10^{-19}\,\text{C}$$

Remark This example exhibits features similar to the Millikan Oil-Drop experiment discussed in Section 15.7, which determined the value of the fundamental electric charge e. Notice that in both parts of the example, the charge is very nearly a multiple of e.

Exercise 15.4
Suppose a droplet of unknown mass remains suspended against gravity when $E = -2.70 \times 10^5$ N/C. What is the minimum mass of the droplet?

Answer 4.41×10^{-15} kg

Problem-Solving Strategy
Calculating Electric Forces and Fields

The following procedure is used to calculate electric forces (the same procedure can be used to calculate an electric field, a simple matter of replacing the charge of interest, q, with a convenient test charge and dividing by the test charge at the end):

1. **Draw** a diagram of the charges in the problem.
2. **Identify** the charge of interest, q, and circle it.
3. **Convert all units** to SI, with charges in coulombs and distances in meters, so as to be consistent with the SI value of the Coulomb constant k_e.
4. **Apply Coulomb's law.** For each charge Q, find the electric force on the charge of interest, q. The magnitude of the force can be found using Coulomb's law. The vector direction of the electric force is along the line of the two charges, directed away from Q if the charges have the same sign, toward Q if the charges have the opposite sign. Find the angle θ this vector makes with the positive x-axis. The x-component of the electric force exerted by Q on q will be $F \cos \theta$, and the y-component will be $F \sin \theta$.
5. **Sum all the x-components,** getting the x-component of the resultant electric force.
6. **Sum all the y-components,** getting the y-component of the resultant electric force.
7. **Use the Pythagorean theorem and trigonometry** to find the magnitude and direction of the resultant force if desired.

EXAMPLE 15.5 Electric Field Due to Two Point Charges

Goal Use the superposition principle to calculate the electric field due to two point charges.

Problem Charge $q_1 = 7.00\ \mu$C is at the origin, and charge $q_2 = -5.00\ \mu$C is on the x-axis, 0.300 m from the origin (Fig. 15.12). **(a)** Find the magnitude and direction of the electric field at point P, which has coordinates (0, 0.400) m. **(b)** Find the force on a charge of 2.00×10^{-8} C placed at P.

Strategy Follow the problem-solving strategy, finding the electric field at point P due to each individual charge in terms of x- and y-components, then adding the components of each type to get the x- and y-components of the resultant electric field at P. The magnitude of the force in part (b) can be found by simply multiplying the magnitude of the electric field by the charge.

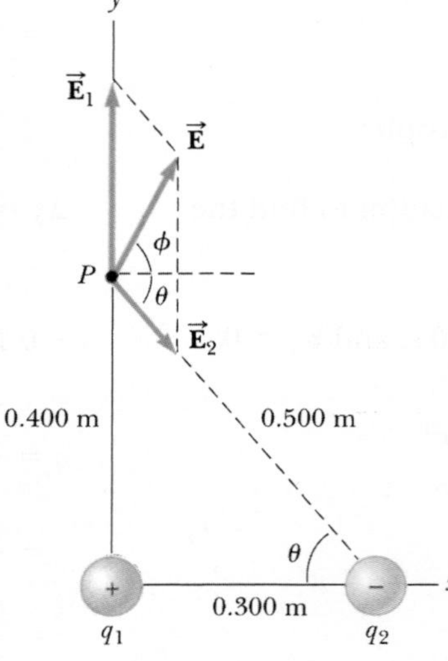

Figure 15.12 (Example 15.5) The resultant electric field $\vec{E}$ at P equals the vector sum $\vec{E}_1 + \vec{E}_2$, where $\vec{E}_1$ is the field due to the positive charge q_1 and $\vec{E}_2$ is the field due to the negative charge q_2.

Solution

(a) Calculate the electric field at *P*.

Find the magnitude of $\vec{\mathbf{E}}_1$ with Equation 15.6:

$$E_1 = k_e \frac{|q_1|}{r_1^2} = (8.99 \times 10^9 \text{ N}\cdot\text{m}^2/\text{C}^2)\frac{(7.00 \times 10^{-6}\text{ C})}{(0.400 \text{ m})^2}$$
$$= 3.93 \times 10^5 \text{ N/C}$$

The vector $\vec{\mathbf{E}}_1$ is vertical, making an angle of 90° with respect to the positive x-axis. Use this fact to find its components:

$$E_{1x} = E_1 \cos(90°) = 0$$
$$E_{1y} = E_1 \sin(90°) = 3.93 \times 10^5 \text{ N/C}$$

Next, find the magnitude of $\vec{\mathbf{E}}_2$, again with Equation 15.6:

$$E_2 = k_e \frac{|q_2|}{r_2{}^2} = (8.99 \times 10^9 \text{ N}\cdot\text{m}^2/\text{C}^2)\frac{(5.00 \times 10^{-6}\text{ C})}{(0.500 \text{ m})^2}$$
$$= 1.80 \times 10^5 \text{ N/C}$$

Obtain the x-component of $\vec{\mathbf{E}}_2$, using the triangle in Figure 15.12 to find $\cos\theta$:

$$\cos\theta = \frac{\text{adj}}{\text{hyp}} = \frac{0.300}{0.500} = 0.600$$
$$E_{2x} = E_2 \cos\theta = (1.80 \times 10^5 \text{ N/C})(0.600)$$
$$= 1.08 \times 10^5 \text{ N/C}$$

Obtain the y-component in the same way, but a minus sign has to be provided for $\sin\theta$ because this component is directed downwards:

$$\sin\theta = \frac{\text{opp}}{\text{hyp}} = \frac{0.400}{0.500} = 0.800$$
$$E_{2y} = E_2 \sin\theta = (1.80 \times 10^5 \text{ N/C})(-0.800)$$
$$= -1.44 \times 10^5 \text{ N/C}$$

Sum the x-components to get the x-component of the resultant vector:

$$E_x = E_{1x} + E_{2x} = 0 + 1.08 \times 10^5 \text{ N/C} = 1.08 \times 10^5 \text{ N/C}$$

Sum the y-components to get the y-component of the resultant vector:

$$E_y = E_{1y} + E_{2y} = 0 + 3.93 \times 10^5 \text{ N/C} - 1.44 \times 10^5 \text{ N/C}$$
$$E_y = 2.49 \times 10^5 \text{ N/C}$$

Use the Pythagorean theorem to find the magnitude of the resultant vector:

$$E = \sqrt{E_x{}^2 + E_y{}^2} = \boxed{2.71 \times 10^5 \text{ N/C}}$$

The inverse tangent function yields the direction of the resultant vector:

$$\phi = \tan^{-1}\left(\frac{E_y}{E_x}\right) = \tan^{-1}\left(\frac{2.49 \times 10^5 \text{ N/C}}{1.08 \times 10^5 \text{ N/C}}\right) = \boxed{66.6°}$$

(b) Find the force on a charge of 2.00×10^{-8} C placed at *P*.

Calculate the magnitude of the force (the direction is the same as that of $\vec{\mathbf{E}}$ because the charge is positive):

$$F = Eq = (2.71 \times 10^5 \text{ N/C})(2.00 \times 10^{-8}\text{ C})$$
$$= \boxed{5.42 \times 10^{-3}\text{ N}}$$

Remarks There were numerous steps to this problem, but each was very short. When attacking such problems, it's important to focus on one small step at a time. The solution comes not from a leap of genius, but from the assembly of a number of relatively easy parts.

Exercise 15.5

(a) Place a charge of $-7.00\ \mu$C at point *P* and find the magnitude and direction of the electric field at the location of q_2 due to q_1 and the charge at *P*. (b) Find the magnitude and direction of the force on q_2.

Answer (a) 5.84×10^5 N/C, $\phi = 20.2°$ (b) $F = 2.92$ N, $\phi = 200.°$

15.5 ELECTRIC FIELD LINES

A convenient aid for visualizing electric field patterns is to draw lines pointing in the direction of the electric field vector at any point. These lines, introduced by Michael Faraday and called **electric field lines**, are related to the electric field in any region of space in the following way:

1. The electric field vector $\vec{\mathbf{E}}$ is tangent to the electric field lines at each point.
2. The number of lines per unit area through a surface perpendicular to the lines is proportional to the strength of the electric field in a given region.

Note that $\vec{\mathbf{E}}$ is large when the field lines are close together and small when the lines are far apart.

Figure 15.13a shows some representative electric field lines for a single positive point charge. This two-dimensional drawing contains only the field lines that lie in the plane containing the point charge. The lines are actually directed radially outward from the charge in *all* directions, somewhat like the quills of an angry porcupine. Because a positive test charge placed in this field would be repelled by the charge q, the lines are directed radially away from the positive charge. The electric field lines for a single negative point charge are directed toward the charge (Fig. 15.13b), because a positive test charge is attracted by a negative charge. In either case, the lines are radial and extend all the way to infinity. Note that the lines are closer together as they get near the charge, indicating that the strength of the field is increasing. Equation 15.6 verifies that this is indeed the case.

The rules for drawing electric field lines for any charge distribution follow directly from the relationship between electric field lines and electric field vectors:

1. The lines for a group of point charges must begin on positive charges and end on negative charges. In the case of an excess of charge, some lines will begin or end infinitely far away.
2. The number of lines drawn leaving a positive charge or ending on a negative charge is proportional to the magnitude of the charge.
3. No two field lines can cross each other.

TIP 15.1 Electric Field Lines Aren't Paths of Particles

Electric field lines are *not* material objects. They are used only as a pictorial representation of the electric field at various locations. Except in special cases, they *do not* represent the path of a charged particle released in an electric field.

Figure 15.14 shows the beautifully symmetric electric field lines for two point charges of equal magnitude but opposite sign. This charge configuration is called an **electric dipole**. Note that the number of lines that begin at the positive charge must equal the number that terminate at the negative charge. At points very near either charge, the lines are nearly radial. The high density of lines between the charges indicates a strong electric field in this region.

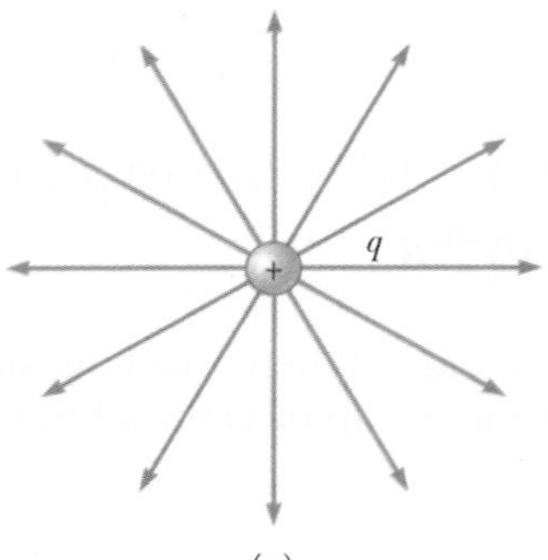

(a)

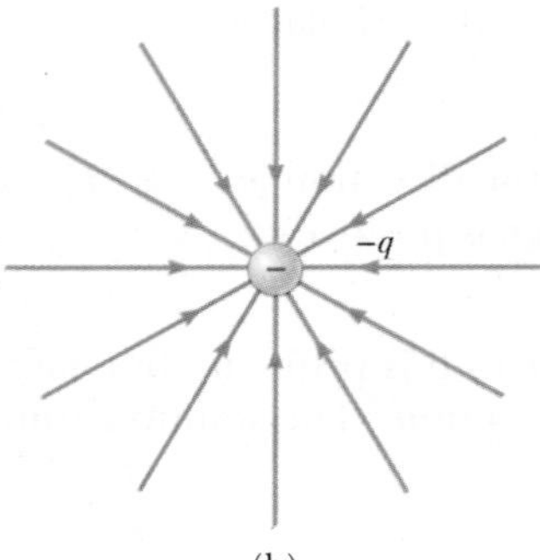

(b)

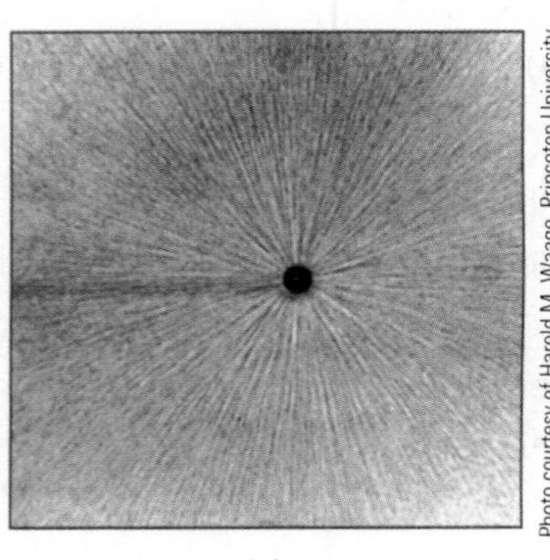

(c)

Figure 15.13 The electric field lines for a point charge. (a) For a positive point charge, the lines radiate outward. (b) For a negative point charge, the lines converge inward. Note that the figures show only those field lines which lie in the plane containing the charge. (c) The dark lines are small pieces of thread suspended in oil, which align with the electric field produced by a small charged conductor at the center.

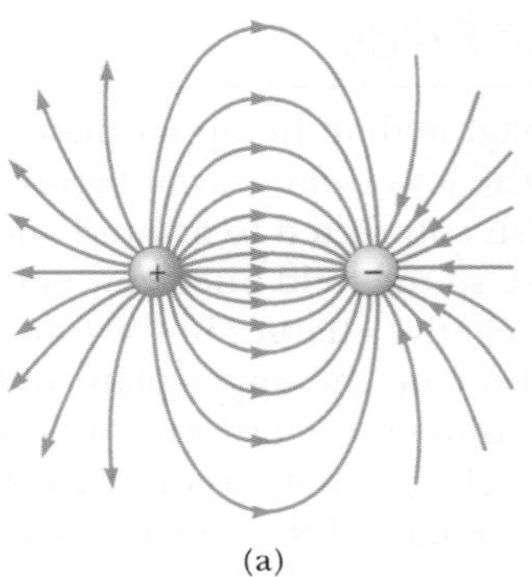

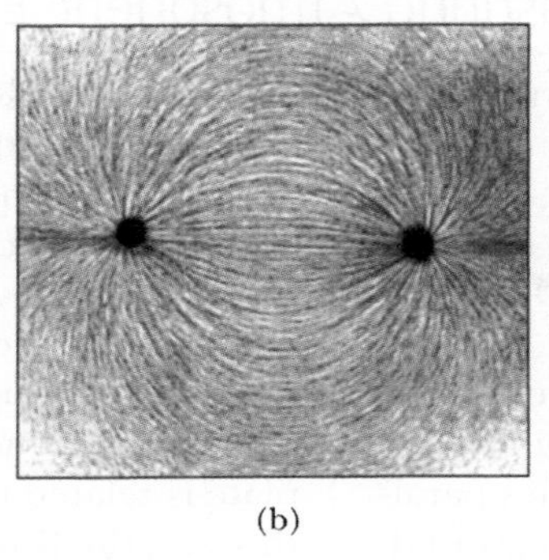

Figure 15.14 (a) The electric field lines for two equal and opposite point charges (an electric dipole). Note that the number of lines leaving the positive charge equals the number terminating at the negative charge. (b) The dark lines are small pieces of thread suspended in oil, which align with the electric field produced by two charged conductors.

Figure 15.15 shows the electric field lines in the vicinity of two equal positive point charges. Again, close to either charge the lines are nearly radial. The same number of lines emerges from each charge because the charges are equal in magnitude. At great distances from the charges, the field is approximately equal to that of a single point charge of magnitude $2q$. The bulging out of the electric field lines between the charges reflects the repulsive nature of the electric force between like charges. Also, the low density of field lines between the charges indicates a weak field in this region, unlike the dipole.

Finally, Active Figure 15.16 is a sketch of the electric field lines associated with the positive charge $+2q$ and the negative charge $-q$. In this case, the number of lines leaving charge $+2q$ is twice the number terminating on charge $-q$. Hence, only half of the lines that leave the positive charge end at the negative charge. The remaining half terminate on negative charges that we assume to be located at infinity. At great distances from the charges (great compared with the charge separation), the electric field lines are equivalent to those of a single charge $+q$.

Quick Quiz 15.6

Rank the magnitudes of the electric field at points A, B, and C in Figure 15.15, with the largest magnitude first.
(a) A, B, C (b) A, C, B (c) C, A, B (d) Can't be determined by visual inspection

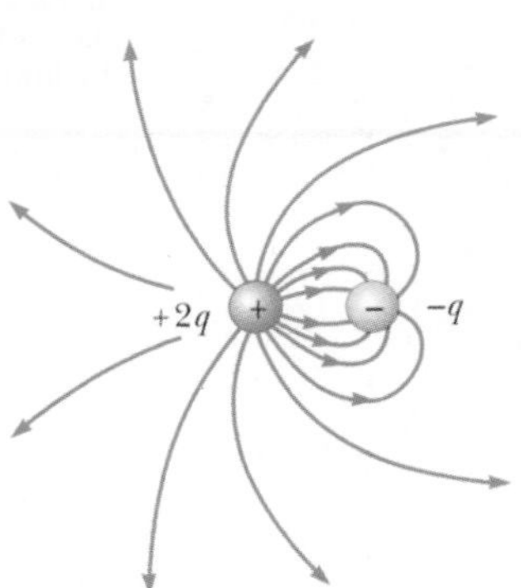

ACTIVE FIGURE 15.16
The electric field lines for a point charge of $+2q$ and a second point charge of $-q$. Note that two lines leave the charge $+2q$ for every line that terminates on $-q$.

PhysicsNow™
Log into PhysicsNow at **www.cp7e.com** and go to Active Figure 15.16, where you can choose the values and signs for the two charges and observe the resulting electric field lines.

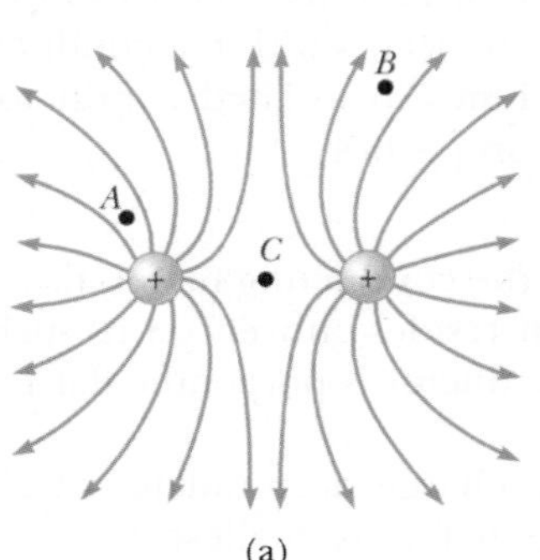

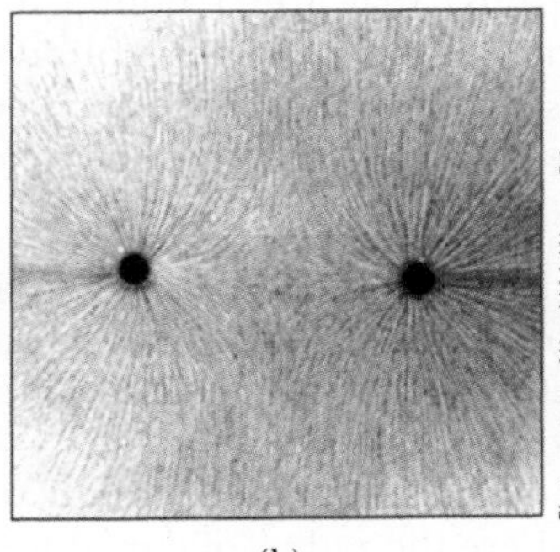

Figure 15.15 (a) The electric field lines for two positive point charges. The points A, B, and C will be discussed in Quick Quiz 15.6. (b) The dark lines are small pieces of thread suspended in oil, which align with the electric field produced by two charged conductors.

Applying Physics 15.1 Measuring Atmospheric Electric Fields

The electric field near the surface of the Earth in fair weather is about 100 N/C downward. Under a thundercloud, the electric field can be very large, on the order of 20 000 N/C. How are these electric fields measured?

Explanation A device for measuring these fields is called the *field mill.* Figure 15.17 shows the fundamental components of a field mill: two metal plates parallel to the ground. Each plate is connected to ground with a wire, with an ammeter (a low-resistance device for measuring the flow of charge, to be discussed in Section 19.6) in one path. Consider first just the lower plate. Because it's connected to ground and the ground carries a negative charge, the plate is negatively charged. The electric field lines, therefore, are directed downward, ending on the plate as in Figure 15.17a. Now imagine that the upper plate is suddenly moved over the lower plate, as in Figure 15.17b. This plate is also connected to ground and is also negatively charged, so the field lines now end on the upper plate. The negative charges in the lower plate are repelled by those on the upper plate and must pass through the ammeter, registering a flow of charge. The amount of charge that was on the lower plate is related to the strength of the electric field. In this way, the flow of charge through the ammeter can be calibrated to measure the electric field. The plates are normally designed like the blades of a fan, with the upper plate rotating so that the lower plate is alternately covered and uncovered. As a result, charges flow back and forth continually through the ammeter, and the reading can be related to the electric field strength.

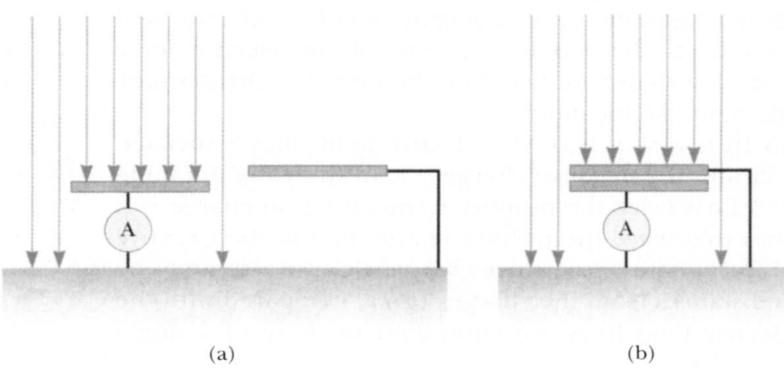

Figure 15.17 (Applying Physics 15.1) In (a), electric field lines end on negative charges on the lower plate. In (b), the second plate is moved above the lower plate. Electric field lines now end on the upper plate, and the negative charges in the lower plate are repelled through the ammeter.

15.6 CONDUCTORS IN ELECTROSTATIC EQUILIBRIUM

A good electric conductor like copper, though electrically neutral, contains charges (electrons) that aren't bound to any atom and are free to move about within the material. When no net motion of charge occurs within a conductor, the conductor is said to be in **electrostatic equilibrium**. An isolated conductor (one that is insulated from ground) has the following properties:

Properties of an isolated conductor ▶

1. The electric field is zero everywhere inside the conducting material.
2. Any excess charge on an isolated conductor resides entirely on its surface.
3. The electric field just outside a charged conductor is perpendicular to the conductor's surface.
4. On an irregularly shaped conductor, the charge accumulates at sharp points, where the radius of curvature of the surface is smallest.

The first property can be understood by examining what would happen if it were *not* true. If there were an electric field inside a conductor, the free charge

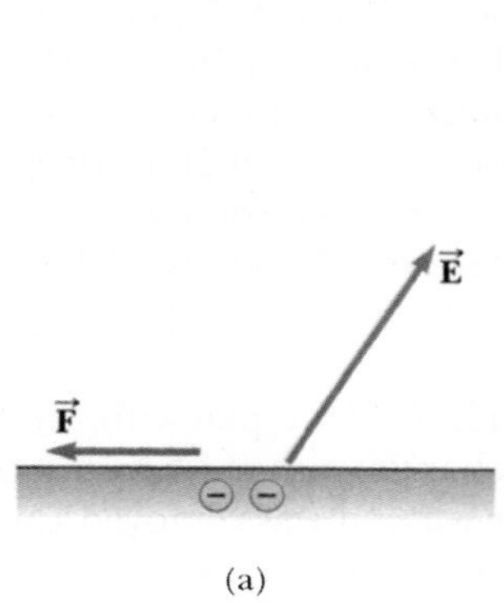

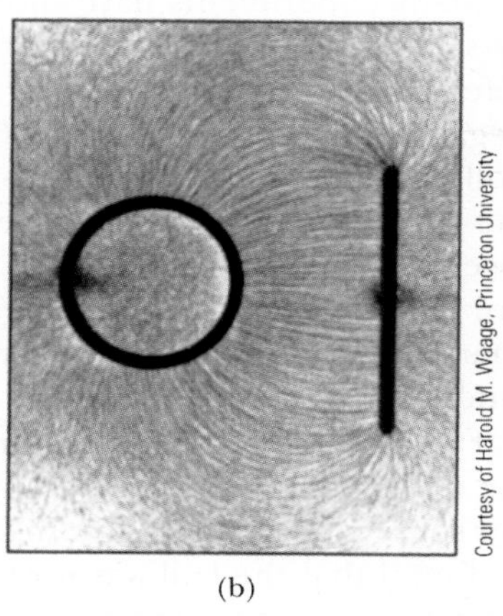

(a) (b)

Figure 15.18 (a) Negative charges at the surface of a conductor. If the electric field were at an angle to the surface, as shown, an electric force would be exerted on the charges along the surface and they would move to the left. Because the conductor is assumed to be in electrostatic equilibrium, $\vec{\mathbf{E}}$ cannot have a component along the surface and hence must be perpendicular to it. (b) The electric field pattern of a charged conducting plate near an oppositely charged conducting cylinder. Small pieces of thread suspended in oil align with the electric field lines. Note that (1) the electric field lines are perpendicular to the conductors and (2) there are no lines inside the cylinder ($\vec{\mathbf{E}} = 0$).

there would move and a flow of charge, or current, would be created. However, if there were a net movement of charge, the conductor would no longer be in electrostatic equilibrium.

Property 2 is a direct result of the $1/r^2$ repulsion between like charges described by Coulomb's law. If by some means an excess of charge is placed inside a conductor, the repulsive forces between the like charges push them as far apart as possible, causing them to quickly migrate to the surface. (We won't prove it here, but the excess charge resides on the surface due to the fact that Coulomb's law is an inverse-square law. With any other power law, an excess of charge would exist on the surface, but there would be a distribution of charge, of either the same or opposite sign, inside the conductor.)

Property 3 can be understood by again considering what would happen if it were not true. If the electric field in Figure 15.18a were not perpendicular to the surface, it would have a component along the surface, which would cause the free charges of the conductor to move (to the left in the figure). If the charges moved, however, a current would be created and the conductor would no longer be in electrostatic equilibrium. Therefore, $\vec{\mathbf{E}}$ must be perpendicular to the surface.

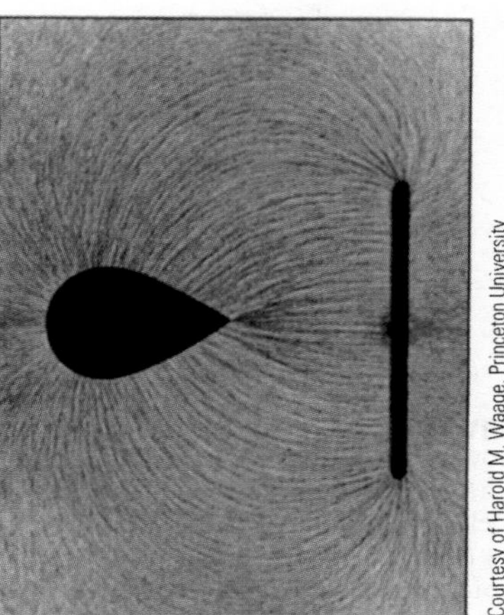

Electric field pattern of a charged conducting plate near an oppositely charged pointed conductor. Small pieces of thread suspended in oil align with the electric field lines. Note that the electric field is most intense near the pointed part of the conductor, where the radius of curvature is the smallest. Also, the lines are perpendicular to the conductors.

To see why property 4 must be true, consider Figure 15.19a, which shows a conductor that is fairly flat at one end and relatively pointed at the other. Any excess charge placed on the object moves to its surface. Figure 15.19b shows the forces between two such charges at the flatter end of the object. These forces are predominantly directed parallel to the surface, so the charges move apart until repulsive forces from other nearby charges establish an equilibrium. At the sharp end, however, the forces of repulsion between two charges are directed predominantly away from the surface, as in Figure 15.19c. As a result, there is less tendency for the charges to move apart along the surface here, and the amount of charge per unit area is greater than at the flat end. The cumulative effect of many such outward forces from nearby charges at the sharp end produces a large resultant force directed away from the surface that can be great enough to cause charges to leap from the surface into the surrounding air.

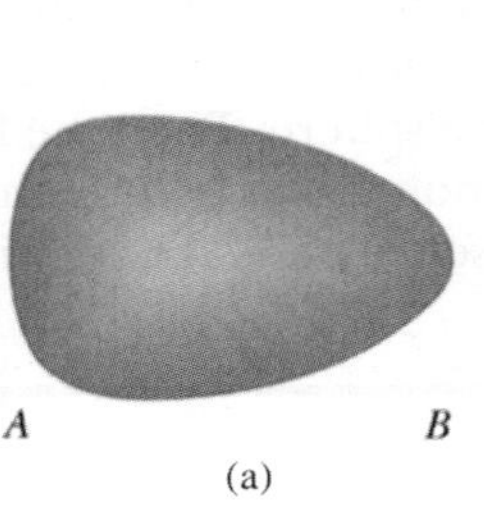

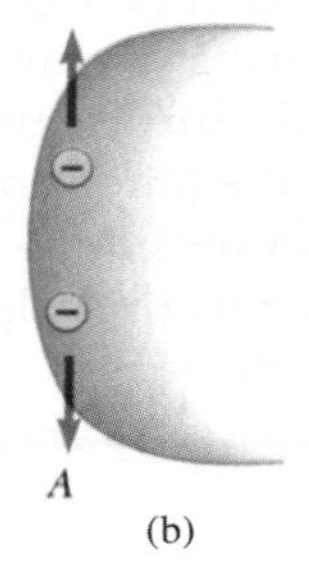

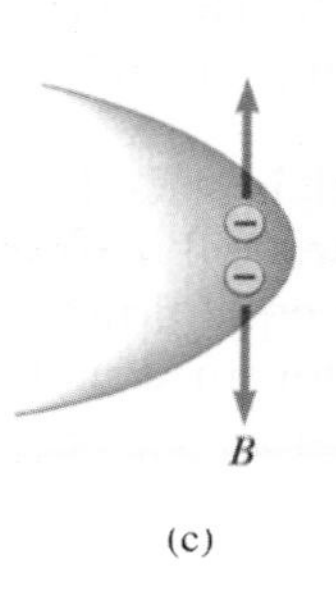

(a) (b) (c)

Figure 15.19 (a) A conductor with a flatter end A and a relatively sharp end B. Excess charge placed on this conductor resides entirely at its surface and is distributed so that (b) there is less charge per unit area on the flatter end and (c) there is a large charge per unit area on the sharper end.

Solution

(a) Calculate the resistance per unit length.

Find the cross-sectional area of the wire:

$$A = \pi r^2 = \pi(0.321 \times 10^{-3}\text{ m})^2 = 3.24 \times 10^{-7}\text{ m}^2$$

Obtain the resistivity of nichrome from Table 17.1, solve Equation 17.5 for R/l, and substitute:

$$\frac{R}{l} = \frac{\rho}{A} = \frac{1.5 \times 10^{-6}\,\Omega\cdot\text{m}}{3.24 \times 10^{-7}\text{ m}^2} = \boxed{4.6\ \Omega/\text{m}}$$

(b) Find the current in a 1.00-m segment of the wire if the potential difference across it is 10.0 V.

Substitute given values into Ohm's law:

$$I = \frac{\Delta V}{R} = \frac{10.0\text{ V}}{4.6\ \Omega} = \boxed{2.2\text{ A}}$$

(c) If the wire is melted down and recast with twice its original length, find the new resistance as a multiple of the old.

Find the new area A_N in terms of the old area A_O, using the fact the volume doesn't change and $l_N = 2l_O$:

$$V_N = V_O \quad\rightarrow\quad A_N l_N = A_O l_O \quad\rightarrow\quad A_N = A_O(l_O/l_N)$$

$$A_N = A_O(l_O/2l_O) = A_O/2$$

Substitute into Equation 17.5:

$$R_N = \frac{\rho l_N}{A_N} = \frac{\rho(2l_O)}{(A_O/2)} = 4\frac{\rho l_O}{A_O} = \boxed{4R_O}$$

Remarks From Table 17.1, the resistivity of nichrome is about 100 times that of copper, a typical good conductor. Therefore, a copper wire of the same radius would have a resistance per unit length of only 0.052 Ω/m, and a 1.00-m length of copper wire of the same radius would carry the same current (2.2 A) with an applied voltage of only 0.115 V.

Because of its resistance to oxidation, nichrome is often used for heating elements in toasters, irons, and electric heaters.

Exercise 17.4

What is the resistance of a 6.0-m length of nichrome wire that has a radius 0.321 mm? How much current does it carry when connected to a 120-V source?

Answer 28 Ω; 4.3 A

PhysicsNow™ You can explore the resistance of different materials by logging into PhysicsNow at **www.cp7e.com** and going to Interactive Example 17.4.

Quick Quiz 17.5

Suppose an electrical wire is replaced with one having every linear dimension doubled (i.e. the length and radius have twice their original values). Does the wire now have (a) more resistance than before, (b) less resistance, or (c) the same resistance?

17.6 TEMPERATURE VARIATION OF RESISTANCE

The resistivity ρ, and hence the resistance, of a conductor depends on a number of factors. One of the most important is the temperature of the metal. For most metals, resistivity increases with increasing temperature. This correlation can be understood as follows: as the temperature of the material increases, its constituent atoms vibrate with greater amplitudes. As a result, the electrons find it more difficult to get by those atoms, just as it is more difficult to weave through a crowded

In an old-fashioned carbon filament incandescent lamp, the electrical resistance is typically 10 Ω, but changes with temperature.

room when the people are in motion than when they are standing still. The increased electron scattering with increasing temperature results in increased resistivity. Technically, thermal expansion also affects resistance; however, this is a very small effect.

Over a limited temperature range, the resistivity of most metals increases linearly with increasing temperature according to the expression

$$\rho = \rho_0[1 + \alpha(T - T_0)] \qquad [17.6]$$

where ρ is the resistivity at some temperature T (in Celsius degrees), ρ_0 is the resistivity at some reference temperature T_0 (usually taken to be 20°C), and α is a parameter called the **temperature coefficient of resistivity**. Temperature coefficients for various materials are provided in Table 17.1. The interesting negative values of α for semiconductors arise because these materials possess weakly bound charge carriers that become free to move and contribute to the current as the temperature rises.

Because the resistance of a conductor with a uniform cross section is proportional to the resistivity according to Equation 17.5 ($R = \rho l/A$), the temperature variation of resistance can be written

$$R = R_0[1 + \alpha(T - T_0)] \qquad [17.7]$$

Precise temperature measurements are often made using this property, as shown by the following example.

EXAMPLE 17.5 A Platinum Resistance Thermometer

Goal Apply the temperature dependence of resistance.

Problem A resistance thermometer, which measures temperature by measuring the change in resistance of a conductor, is made of platinum and has a resistance of 50.0 Ω at 20.0°C. **(a)** When the device is immersed in a vessel containing melting indium, its resistance increases to 76.8 Ω. From this information, find the melting point of indium. **(b)** The indium is heated further until it reaches a temperature of 235°C. What is the ratio of the new current in the platinum to the current I_{mp} at the melting point?

Strategy In part (a), solve Equation 17.7 for $T - T_0$ and get α for platinum from Table 17.1, substituting known quantities. For part (b), use Ohm's law in Equation 17.7.

Solution

(a) Find the melting point of indium.

Solve Equation 17.7 for $T - T_0$:

$$T - T_0 = \frac{R - R_0}{\alpha R_0} = \frac{76.8\ \Omega - 50.0\ \Omega}{[3.92 \times 10^{-3}\ (°\text{C})^{-1}][50.0\ \Omega]}$$
$$= 137°\text{C}$$

Substitute $T_0 = 20.0°\text{C}$ and obtain the melting point of indium:

$$T = 157°\text{C}$$

(b) Find the ratio of the new current to the old when the temperature rises from 157°C to 235°C.

Write Equation 17.7, with R_0 and T_0 replaced by R_{mp} and T_{mp}, the resistance and temperature at the melting point.

$$R = R_{mp}[1 + \alpha(T - T_{mp})]$$

According to Ohm's law, $R = \Delta V/I$ and $R_{mp} = \Delta V/I_{mp}$. Substitute these expressions into Equation 17.7:

$$\frac{\Delta V}{I} = \frac{\Delta V}{I_{mp}}[1 + \alpha(T - T_{mp})]$$

Cancel the voltage differences, invert the two expressions, and then divide both sides by I_{mp}:

$$\frac{I}{I_{mp}} = \frac{1}{1 + \alpha(T - T_{mp})}$$

Substitute $T = 235°C$, $T_{mp} = 157°C$, and the value for α, obtaining the desired ratio:

$$\frac{I}{I_{mp}} = 0.766$$

Exercise 17.5

Suppose a wire made of an unknown alloy and having a temperature of 20.0°C carries a current of 0.450 A. At 52.0°C the current is 0.370 A for the same potential difference. Find the temperature coefficient of resistivity of the alloy.

Answer $6.76 \times 10^{-3}\ (°C)^{-1}$

17.7 SUPERCONDUCTORS

There is a class of metals and compounds with resistances that fall virtually to *zero* below a certain temperature T_c called the *critical temperature*. These materials are known as **superconductors**. The resistance–temperature graph for a superconductor follows that of a normal metal at temperatures above T_c (Fig. 17.9). When the temperature is at or below T_c, however, the resistance suddenly drops to zero. This phenomenon was discovered in 1911 by the Dutch physicist H. Kamerlingh Onnes as he and a graduate student worked with mercury, which is a superconductor below 4.1 K. Recent measurements have shown that the resistivities of superconductors below T_c are less than $4 \times 10^{-25}\ \Omega\cdot m$—around 10^{17} times smaller than the resistivity of copper and in practice considered to be zero.

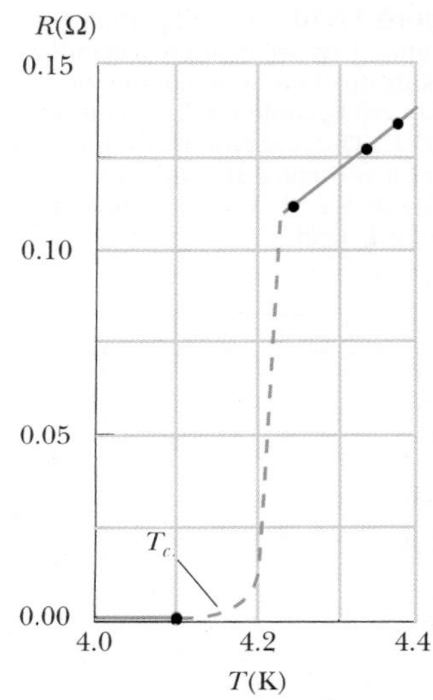

Figure 17.9 Resistance versus temperature for a sample of mercury (Hg). The graph follows that of a normal metal above the critical temperature T_c. The resistance drops to zero at the critical temperature, which is 4.2 K for mercury, and remains at zero for lower temperatures.

Today thousands of superconductors are known, including such common metals as aluminum, tin, lead, zinc, and indium. Table 17.2 lists the critical temperatures of several superconductors. The value of T_c is sensitive to chemical composition, pressure, and crystalline structure. Interestingly, copper, silver, and gold, which are excellent conductors, don't exhibit superconductivity.

One of the truly remarkable features of superconductors is the fact that once a current is set up in them, it persists *without any applied voltage* (because $R = 0$). In fact, steady currents in superconducting loops have been observed to persist for years with no apparent decay!

An important development in physics that created much excitement in the scientific community was the discovery of high-temperature copper-oxide-based superconductors. The excitement began with a 1986 publication by J. Georg Bednorz and K. Alex Müller, scientists at the IBM Zurich Research Laboratory in Switzerland, in which they reported evidence for superconductivity at a temperature near 30 K in an oxide of barium, lanthanum, and copper. Bednorz and Müller were awarded the Nobel Prize for physics in 1987 for their important discovery. The discovery was remarkable in view of the fact that the critical temperature was significantly higher than those of any previously known superconductors. Shortly thereafter a new family of compounds was investigated, and research activity in the field of superconductivity proceeded vigorously. In early 1987, groups at the University of Alabama at Huntsville and the University of Houston announced the discovery of superconductivity at about 92 K in an oxide of yttrium, barium, and copper ($YBa_2Cu_3O_7$), shown as the gray disk in Figure 17.10. Late in 1987, teams of scientists from Japan and the United States reported superconductivity at 105 K in an oxide of bismuth, strontium, calcium, and copper. More recently, scientists have reported superconductivity at temperatures as high as 150 K in an oxide containing mercury. The search for novel superconducting materials continues, with the hope of someday obtaining a room-temperature superconducting material. This research is important both for scientific reasons and for practical applications.

An important and useful application is the construction of superconducting magnets in which the magnetic field intensities are about ten times greater than those of the best normal electromagnets. Such magnets are being considered as a means of storing energy. The idea of using superconducting power lines to transmit

TABLE 17.2

Critical Temperatures for Various Superconductors

Material	T_c (K)
Zn	0.88
Al	1.19
Sn	3.72
Hg	4.15
Pb	7.18
Nb	9.46
Nb_3Sn	18.05
Nb_3Ge	23.2
$YBa_2Cu_3O_7$	90
Bi–Sr–Ca–Cu–O	105
Tl–Ba–Ca–Cu–O	125
$HgBa_2Ca_2Cu_3O_8$	134

Courtesy of IBM Research Laboratory

Figure 17.10 A small permanent magnet floats freely above a ceramic disk made of the superconductor $YBa_2Cu_3O_7$, cooled by liquid nitrogen at 77 K. The superconductor has zero electric resistance at temperatures below 92 K and expels any applied magnetic field.

power efficiently is also receiving serious consideration. Modern superconducting electronic devices consisting of two thin-film superconductors separated by a thin insulator have been constructed. Among these devices are magnetometers (magnetic-field measuring devices) and various microwave devices.

17.8 ELECTRICAL ENERGY AND POWER

If a battery is used to establish an electric current in a conductor, chemical energy stored in the battery is continuously transformed into kinetic energy of the charge carriers. This kinetic energy is quickly lost as a result of collisions between the charge carriers and fixed atoms in the conductor, causing an increase in the temperature of the conductor. In this way, the chemical energy stored in the battery is continuously transformed into thermal energy.

In order to understand the process of energy transfer in a simple circuit, consider a battery with terminals connected to a resistor (Active Fig. 17.11; remember that the positive terminal of the battery is always at the higher potential). Now imagine following a quantity of positive charge ΔQ around the circuit from point A, through the battery and resistor, and back to A. Point A is a reference point that is grounded (the ground symbol is ⏚), and its potential is taken to be zero. As the charge ΔQ moves from A to B through the battery, the electrical potential energy of the system increases by the amount $\Delta Q\,\Delta V$, and the chemical potential energy in the battery decreases by the same amount. (Recall from Chapter 16 that $\Delta PE = q\,\Delta V$.) However, as the charge moves from C to D through the resistor, it loses this electrical potential energy during collisions with atoms in the resistor. In the process, the energy is transformed to internal energy corresponding to increased vibrational motion of those atoms. Because we can ignore the very small resistance of the interconnecting wires, no energy transformation occurs for paths BC and DA. When the charge returns to point A, the net result is that some of the chemical energy in the battery has been delivered to the resistor and has caused its temperature to rise.

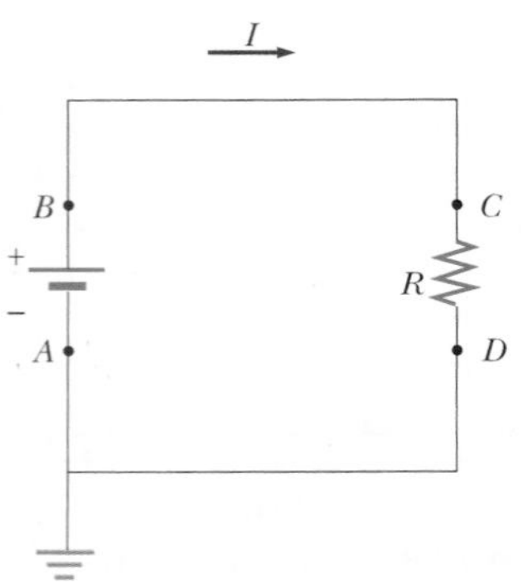

ACTIVE FIGURE 17.11
A circuit consisting of a battery and a resistance R. Positive charge flows clockwise from the positive to the negative terminal of the battery. Point A is grounded.

Physics Now™
Log into PhysicsNow at **www.cp7e.com** and go to Active Figure 17.11, where you can adjust the battery voltage and the resistance, and see the resulting current in the circuit and the power dissipated as heat by the resistor.

The charge ΔQ loses energy $\Delta Q\,\Delta V$ as it passes through the resistor. If Δt is the time it takes the charge to pass through the resistor, then the rate at which it loses electric potential energy is

$$\frac{\Delta Q}{\Delta t}\,\Delta V = I\Delta V$$

where I is the current in the resistor and ΔV is the potential difference across it. Of course, the charge regains this energy when it passes through the battery, at the expense of chemical energy in the battery. The rate at which the system loses potential energy as the charge passes through the resistor is equal to the rate at which the system gains internal energy in the resistor. Therefore, the power $\mathcal{P}$, representing the rate at which energy is delivered to the resistor, is

Power ▶

$$\mathcal{P} = I\Delta V \qquad [17.8]$$

While this result was developed by considering a battery delivering energy to a resistor, Equation 17.8 can be used to determine the power transferred from a voltage source to *any* device carrying a current I and having a potential difference ΔV between its terminals.

Using Equation 17.8 and the fact that $\Delta V = IR$ for a resistor, we can express the power delivered to the resistor in the alternate forms

Power delivered to a resistor ▶

$$\mathcal{P} = I^2R = \frac{\Delta V^2}{R} \qquad [17.9]$$

When I is in amperes, ΔV in volts, and R in ohms, the SI unit of power is the watt (introduced in Chapter 5). The power delivered to a conductor of resistance R is often referred to as an I^2R *loss.* Note that Equation 17.9 applies only to resistors and not to nonohmic devices such as lightbulbs and diodes.

Regardless of the ways in which you use electrical energy in your home, you ultimately must pay for it or risk having your power turned off. The unit of energy used by electric companies to calculate consumption, the **kilowatt-hour**, is defined in terms of the unit of power and the amount of time it's supplied. One kilowatt-hour (kWh) is the energy converted or consumed in 1 h at the constant rate of 1 kW. It has the numerical value

$$1 \text{ kWh} = (10^3 \text{ W})(3600 \text{ s}) = 3.60 \times 10^6 \text{ J} \qquad [17.10]$$

On an electric bill, the amount of electricity used in a given period is usually stated in multiples of kilowatt-hours.

TIP 17.3 Misconception About Current

Current is *not* "used up" in a resistor. Rather, some of the energy the charges have received from the voltage source is delivered to the resistor, making it hot and causing it to radiate. Also, the current doesn't slow down when going through the resistor: it's the same throughout the circuit.

Applying Physics 17.2 Lightbulb Failures

Why do lightbulbs fail so often right after they're turned on?

Explanation Once the switch is closed, the line voltage is applied across the bulb. As the voltage is applied across the cold filament when the bulb is first turned on, the resistance of the filament is low, the current is high, and a relatively large amount of power is delivered to the bulb. This current spike at the beginning of operation is the reason why lightbulbs often fail just after they are turned on. As the filament warms, its resistance rises and the current decreases. As a result, the power delivered to the bulb decreases, and the bulb is less likely to burn out.

Quick Quiz 17.6

A voltage ΔV is applied across the ends of a nichrome heater wire having a cross-sectional area A and length L. The same voltage is applied across the ends of a second heater wire having a cross-sectional area A and length $2L$. Which wire gets hotter? (a) the shorter wire, (b) the longer wire, or (c) more information is needed.

Quick Quiz 17.7

For the two resistors shown in Figure 17.12, rank the currents at points a through f from largest to smallest.

(a) $I_a = I_b > I_e = I_f > I_c = I_d$
(b) $I_a = I_b > I_c = I_d > I_e = I_f$
(c) $I_e = I_f > I_c = I_d > I_a = I_b$

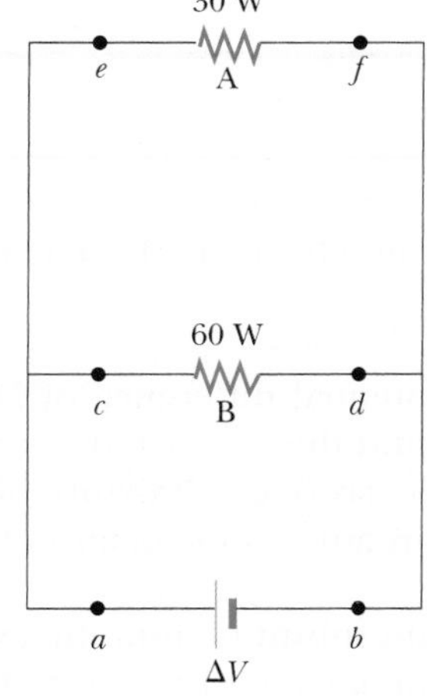

Figure 17.12 (Quick Quiz 17.7)

Quick Quiz 17.8

Two resistors, A and B, are connected in a series circuit with a battery. The resistance of A is twice that of B. Which resistor dissipates more power? (a) resistor A (b) resistor B (c) More information is needed.

Example 17.6 The Cost of Lighting Up Your Life

Goal Apply the electric power concept, and calculate the cost of power usage using kilowatt-hours.

Problem A circuit provides a maximum current of 20.0 A at an operating voltage of 1.20×10^2 V. **(a)** How many 75 W bulbs can operate with this voltage source? **(b)** At $0.120 per kilowatt-hour, how much does it cost to operate these bulbs for 8.00 h?

Strategy Find the necessary power with $\mathcal{P} = I\Delta V$, then divide by 75.0 W per bulb to get the total number of bulbs. To find the cost, convert power to kilowatts and multiply by the number of hours, then multiply by the cost per kilowatt-hour.

Solution

(a) Find the number of bulbs that can be lighted.

Substitute into Equation 17.8 to get the total power:

$$\mathcal{P}_{\text{total}} = I\Delta V = (20.0\text{ A})(1.20 \times 10^2\text{ V}) = 2.40 \times 10^3\text{ W}$$

Divide the total power by the power per bulb to get the number of bulbs.

$$\text{Number of bulbs} = \frac{\mathcal{P}_{\text{total}}}{\mathcal{P}_{\text{bulb}}} = \frac{2.40 \times 10^3\text{ W}}{75.0\text{ W}} = \boxed{32.0}$$

(b) Calculate the cost of this electricity for an 8.00-h day.

Find the energy in kilowatt-hours:

$$\text{Energy} = \mathcal{P}t = (2.40 \times 10^3\text{ W})\left(\frac{1.00\text{ kW}}{1.00 \times 10^3\text{ W}}\right)(8.00\text{ h})$$
$$= 19.2\text{ kWh}$$

Multiply by the cost per kilowatt-hour:

$$\text{Cost} = (19.2\text{ kWh})(\$0.12/\text{kWh}) = \boxed{\$2.30}$$

Remarks This amount of energy might correspond to what a small office uses in a working day, taking into account all power requirements (not just lighting). In general, resistive devices can have variable power output, depending on how the circuit is wired. Here, power outputs were specified, so such considerations were unnecessary.

Exercise 17.6

(a) How many Christmas tree lights drawing 5.00 W of power each could be run on a circuit operating at 1.20×10^2 V and providing 15.0 A of current? (b) Find the cost to operate one such string 24.0 h per day for the Christmas season (two weeks), using the rate $0.12/kWh.

Answers (a) 3.60×10^2 bulbs (b) $72.60

EXAMPLE 17.7 The Power Converted by an Electric Heater

Goal Calculate an electrical power output, and link to its effect on the environment through the first law of thermodynamics.

Problem An electric heater is operated by applying a potential difference of 50.0 V to a nichrome wire of total resistance 8.00 Ω. **(a)** Find the current carried by the wire and the power rating of the heater. **(b)** Using this heater, how long would it take to heat 2.50×10^3 moles of diatomic gas (e.g., a mixture of oxygen and nitrogen—air) from a chilly 10.0°C to 25.0°C? Take the molar specific heat at constant volume of air to be $\frac{5}{2}R$.

Strategy For part (a), find the current with Ohm's law and substitute into the expression for power. Part (b) is an isovolumetric process, so the thermal energy provided by the heater all goes into the change in internal energy, ΔU. Calculate this quantity using the first law of thermodynamics, and divide by the power to get the time.

Solution

(a) Compute the current and power output.

Apply Ohm's law to get the current:

$$I = \frac{\Delta V}{R} = \frac{50.0\text{ V}}{8.00\ \Omega} = \boxed{6.25\text{ A}}$$

Substitute into Equation 17.9 to find the power: $\mathcal{P} = I^2R = (6.25 \text{ A})^2(8.00 \ \Omega) = \boxed{313 \text{ W}}$

(b) How long does it take to heat the gas?

Calculate the thermal energy transfer from the first law. Note that $W = 0$ because the volume doesn't change.

$$\begin{aligned} Q = \Delta U &= nC_v\Delta T \\ &= (2.50 \times 10^3 \text{ mol})(\tfrac{5}{2}\cdot 8.31 \text{ J/mol}\cdot\text{K})(298 \text{ K} - 283 \text{ K}) \\ &= 7.79 \times 10^5 \text{ J} \end{aligned}$$

Divide the thermal energy by the power, to get the time: $t = \dfrac{Q}{\mathcal{P}} = \dfrac{7.79 \times 10^5 \text{ J}}{313 \text{ W}} = \boxed{2.49 \times 10^3 \text{ s}}$

Remarks The number of moles of gas given here is approximately what would be found in a bedroom. Warming the air with this space heater requires only about forty minutes. However, the calculation doesn't take into account conduction losses. Recall that a 20-cm-thick concrete wall, as calculated in Chapter 11, permitted the loss of over two megajoules an hour by conduction!

Exercise 17.7

A hot-water heater is rated at 4.50×10^3 W and operates at 2.40×10^2 V. (a) Find the resistance in the heating element, and the current. (b) How long does it take to heat 125 L of water from 20.0°C to 50.0°C, neglecting conduction and other losses?

Answers (a) 12.8 Ω, 18.8 A (b) 3.49×10^3 s

17.9 ELECTRICAL ACTIVITY IN THE HEART

Electrocardiograms

APPLICATION

Electrocardiograms

Every action involving the body's muscles is initiated by electrical activity. The voltages produced by muscular action in the heart are particularly important to physicians. Voltage pulses cause the heart to beat, and the waves of electrical excitation that sweep across the heart associated with the heartbeat are conducted through the body via the body fluids. These voltage pulses are large enough to be detected by suitable monitoring equipment attached to the skin. A sensitive voltmeter making good electrical contact with the skin by means of contacts attached with conducting paste can be used to measure heart pulses, which are typically of the order of 1 mV at the surface of the body. The voltage pulses can be recorded on an instrument called an **electrocardiograph**, and the pattern recorded by this instrument is called an **electrocardiogram** (EKG). In order to understand the information contained in an EKG pattern, it is necessary first to describe the underlying principles concerning electrical activity in the heart.

The right atrium of the heart contains a specialized set of muscle fibers called the SA (sinoatrial) node that initiates the heartbeat (Fig. 17.13). Electric impulses that originate in these fibers gradually spread from cell to cell throughout the right and left atrial muscles, causing them to contract. The pulse that passes through the muscle cells is often called a *depolarization wave* because of its effect on individual cells. If an individual muscle cell were examined in its resting state, a double-layer electric charge distribution would be found on its surface, as shown in Figure 17.14a (page 584). The impulse generated by the SA node momentarily and locally allows positive charge on the outside of the cell to flow in and neutralize the negative charge on the inside layer. This effect changes the cell's charge distribution to that shown in Figure 17.14b. Once the depolarization wave has passed through an individual heart muscle cell, the cell recovers the resting-state charge distribution (positive out, negative in) shown in Figure 17.14a in about 250 ms. When the impulse reaches the atrioventricular (AV) node (Fig. 17.13), the muscles of the atria begin to relax, and the pulse is directed to the ventricular

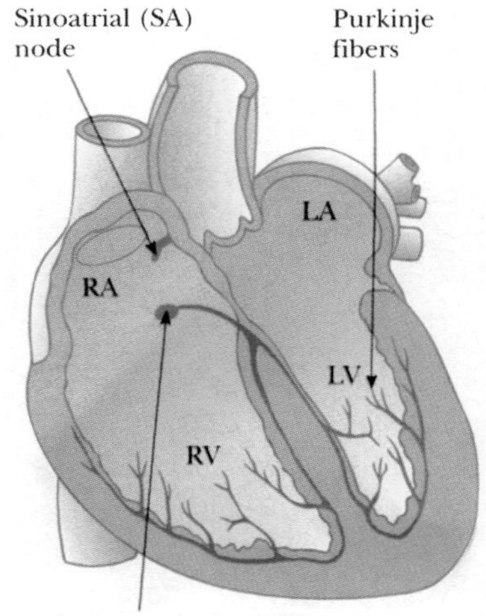

Figure 17.13 The electrical conduction system of the human heart. (RA: right atrium; LA: left atrium; RV: right ventricle; LV: left ventricle.)

Figure 17.14 (a) Charge distribution of a muscle cell in the atrium before a depolarization wave has passed through the cell. (b) Charge distribution as the wave passes.

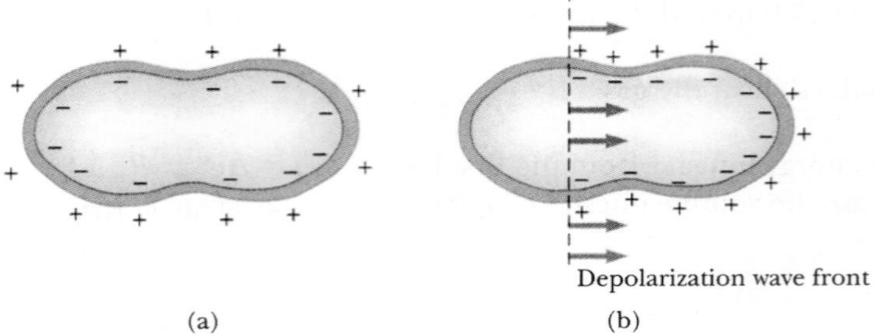

muscles by the AV node. The muscles of the ventricles contract as the depolarization wave spreads through the ventricles along a group of fibers called the *Purkinje fibers.* The ventricles then relax after the pulse has passed through. At this point, the SA node is again triggered and the cycle is repeated.

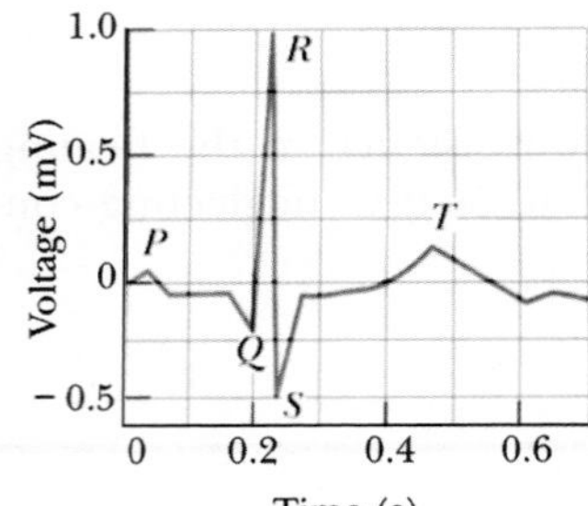

Figure 17.15 An EKG response for a normal heart.

A sketch of the electrical activity registered on an EKG for one beat of a normal heart is shown in Figure 17.15. The pulse indicated by *P* occurs just before the atria begin to contract. The *QRS* pulse occurs in the ventricles just before they contract, and the *T* pulse occurs when the cells in the ventricles begin to recover. EKGs for an abnormal heart are shown in Figure 17.16. The *QRS* portion of the pattern shown in Figure 17.16a is wider than normal, indicating that the patient may have an enlarged heart. (Why?) Figure 17.16b indicates that there is no constant relationship between the *P* pulse and the *QRS* pulse. This suggests a blockage in the electrical conduction path between the SA and AV nodes which results in the atria and ventricles beating independently and inefficient heart pumping. Finally, Figure 17.16c shows a situation in which there is no *P* pulse and an irregular spacing between the *QRS* pulses. This is symptomatic of irregular atrial contraction, which is called *fibrillation.* In this condition, the atrial and ventricular contractions are irregular.

APPLICATION

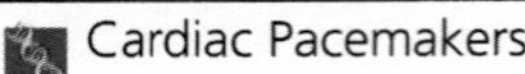

Cardiac Pacemakers

As noted previously, the sinoatrial node directs the heart to beat at the appropriate rate, usually about 72 beats per minute. However, disease or the aging process can damage the heart and slow its beating, and a medical assist may be necessary in the form of a *cardiac pacemaker* attached to the heart. This matchbox-sized electrical device implanted under the skin has a lead that is connected to the wall of the right ventricle. Pulses from this lead stimulate the heart to maintain its proper rhythm. In general, a pacemaker is designed to produce pulses at a rate of about 60 per minute, slightly slower than the normal number of beats per minute,

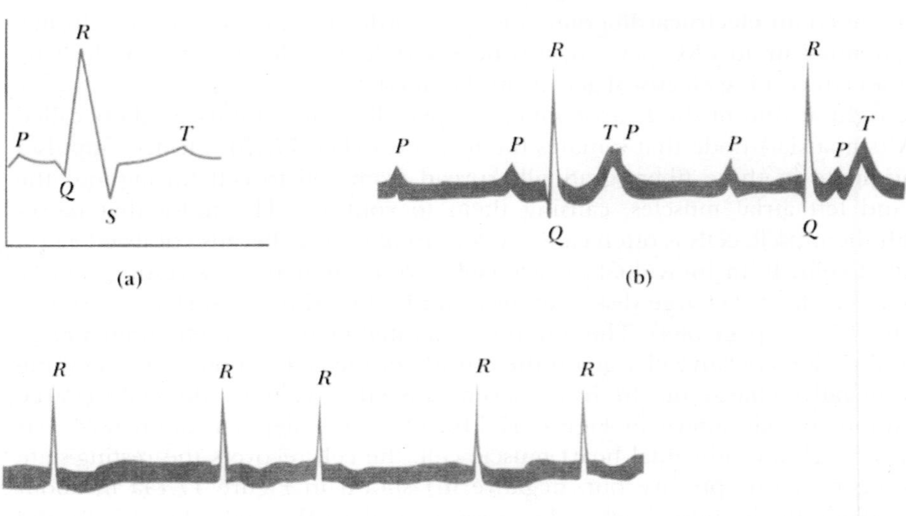

Figure 17.16 Abnormal EKGs.

but sufficient to maintain life. The circuitry basically consists of a capacitor charging up to a certain voltage from a lithium battery and then discharging. The design of the circuit is such that, if the heart is beating normally, the capacitor is not allowed to charge completely and send pulses to the heart.

An Emergency Room in Your Chest

APPLICATION

Implanted Cardioverter Defibrillators

In June 2001, an operation on Vice President Dick Cheney focused attention on the progress in treating heart problems with tiny implanted electrical devices. Aptly termed "an emergency room in your chest" by Cheney's attending physician, devices called *I*mplanted *C*ardioverter *D*efibrillators (**ICD's**) can monitor, record, and logically process heart signals and then supply different corrective signals to hearts beating too slowly, too rapidly, or irregularly. ICD's can even monitor and send signals to the atria and ventricles independently! Figure 17.17a shows a sketch of an ICD with conducting leads that are implanted in the heart. Figure 17.17b shows an actual titanium-encapsulated dual-chamber ICD.

The latest ICD's are sophisticated devices capable of a number of functions:

1. monitoring both atrial and ventricular chambers to differentiate between atrial and potentially fatal ventricular arrhythmias, which require prompt regulation;
2. storing about a half hour of heart signals that can easily be read out by a physician;
3. being easily reprogrammed with an external magnetic wand;
4. performing complicated signal analysis and comparison;
5. supplying either 0.25- to 10-V repetitive pacing signals to speed up or slow down a malfunctioning heart, or a high-voltage pulse of about 800 V to halt the potentially fatal condition of ventricular fibrillation, in which the heart quivers rapidly rather than beats (people who have experienced such a high-voltage jolt say that it feels like a kick or a bomb going off in the chest);
6. automatically adjusting the number of pacing pulses per minute to match the patient's activity.

ICD's are powered by lithium batteries and have implanted lifetimes of 4–6 years. Some basic properties of these adjustable ICD's are given in Table 17.3 (page 586). In the table, *tachycardia* means "rapid heartbeat" and *bradycardia*

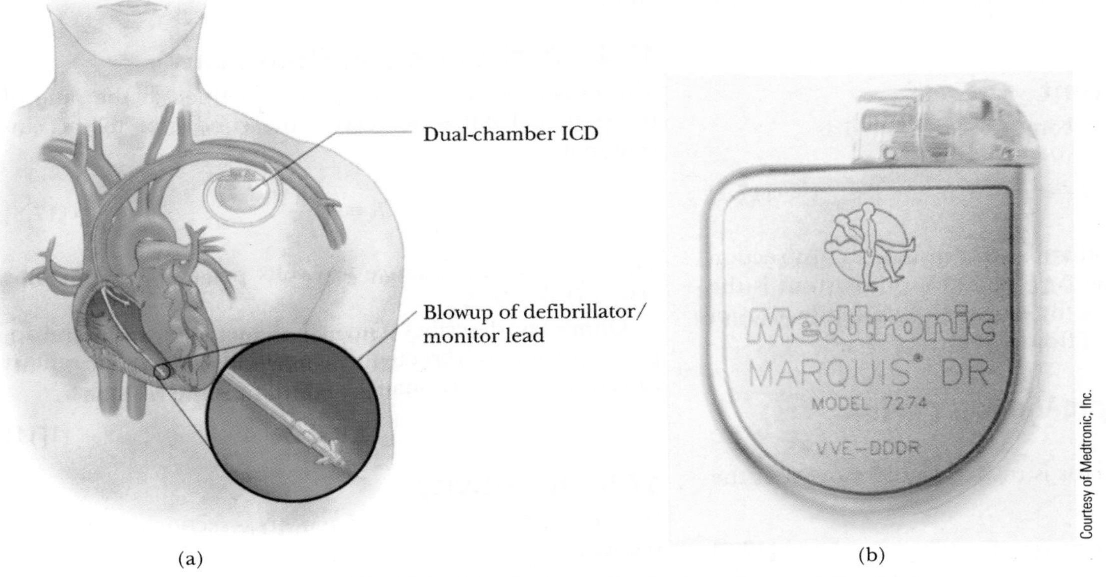

FIGURE 17.17 (a) A dual-chamber ICD with leads in the heart. One lead monitors and stimulates the right atrium, and the other monitors and stimulates the right ventricle. (b) Medtronic Dual Chamber ICD.

TABLE 17.3

Properties of Implanted Cardioverter Defibrillators[a]	
Physical Specifications	
Mass (g)	85
Size (cm)	7.3 × 6.2 × 1.3 (about five stacked silver dollars)
Antitachycardia Pacing	ICD delivers a burst of critically timed low-energy pulses
Number of Bursts	1–15
Burst Cycle Length (ms)	200–552
Number of Pulses per Burst	2–20
Pulse Amplitude (V)	7.5 or 10
Pulse Width (ms)	1.0 or 1.9
High-Voltage Defibrillation	
Pulse energy (J)	37 stored/33 delivered
Pulse Amplitude (V)	801
Bradycardia Pacing	A dual-chamber ICD can steadily deliver repetitive pulses to both the atrium and the ventricle
Base Frequency (beats/minute)	40–100
Pulse Amplitude (V)	0.25–7.5
Pulse Width (ms)	0.05, 0.1–1.5, 1.9

[a]For more information see **www.photonicd.com/specs.html**.

means "slow heartbeat." A key factor in developing tiny electrical implants that serve as defibrillators is the development of capacitors with relatively large capacitance (125 μf) and small physical size.

SUMMARY

Physics Now™ Take a practice test by logging into PhysicsNow at **www.cp7e.com** and clicking on the Pre-Test link for this chapter.

17.1 Electric Current

The **electric current** I in a conductor is defined as

$$I \equiv \frac{\Delta Q}{\Delta t} \qquad [17.1]$$

where ΔQ is the charge that passes through a cross section of the conductor in time Δt. The SI unit of current is the **ampere** (A); 1 A = 1 C/s. By convention, the direction of current is the direction of flow of positive charge.

17.2 A Microscopic View: Current and Drift Speed

The current in a conductor is related to the motion of the charge carriers by

$$I = nqv_dA \qquad [17.2]$$

where n is the number of mobile charge carriers per unit volume, q is the charge on each carrier, v_d is the drift speed of the charges, and A is the cross-sectional area of the conductor.

17.4 Resistance and Ohm's Law

The **resistance** R of a conductor is defined as the ratio of the potential difference across the conductor to the current in it:

$$R \equiv \frac{\Delta V}{I} \qquad [17.3]$$

The SI units of resistance are volts per ampere, or **ohms** (Ω); 1 Ω = 1 V/A.

Ohm's law describes many conductors, for which the applied voltage is directly proportional to the current it causes. The proportionality constant is the resistance:

$$\Delta V = IR \qquad [17.4]$$

17.5 Resistivity

If a conductor has length l and cross-sectional area A, its **resistance** is

$$R = \rho \frac{l}{A} \qquad [17.5]$$

where ρ, is an intrinsic property of the conductor called the **electrical resistivity**. The SI unit of resistivity is the **ohm-meter** ($\Omega \cdot m$).

17.6 Temperature Variation of Resistance

Over a limited temperature range, the resistivity of a conductor varies with temperature according to the expression

$$\rho = \rho_0[1 + \alpha(T - T_0)] \quad [17.6]$$

where α is the **temperature coefficient of resistivity** and ρ_0 is the resistivity at some reference temperature T_0 (usually taken to be 20°C).

The resistance of a conductor varies with temperature according to the expression

$$R = R_0[1 + \alpha(T - T_0)] \quad [17.7]$$

17.8 Electrical Energy and Power

If a potential difference ΔV is maintained across an electrical device, the **power**, or rate at which energy is supplied to the device, is

$$\mathcal{P} = I\,\Delta V \quad [17.8]$$

Because the potential difference across a resistor is $\Delta V = IR$, the **power delivered to a resistor** can be expressed as

$$\mathcal{P} = I^2R = \frac{\Delta V^2}{R} \quad [17.9]$$

A **kilowatt-hour** is the amount of energy converted or consumed in one hour by a device supplied with power at the rate of 1 kW. This is equivalent to

$$1 \text{ kWh} = 3.60 \times 10^6 \text{ J} \quad [17.10]$$

CONCEPTUAL QUESTIONS

1. Car batteries are often rated in ampere-hours. Does this unit designate the amount of current, power, energy, or charge that can be drawn from the battery?
2. We have seen that an electric field must exist inside a conductor that carries a current. How is that possible in view of the fact that in electrostatics we concluded that the electric field must be zero inside a conductor?
3. Why don't the free electrons in a metal fall to the bottom of the metal due to gravity? And charges in a conductor are supposed to reside on the surface—why don't the free electrons all go to the surface?
4. In an analogy between traffic flow and electrical current, what would correspond to the charge Q? What would correspond to the current I?
5. Newspaper articles often have statements such as "10 000 volts of electricity surged *through* the victim's body." What is wrong with this statement?
6. Two lightbulbs are each connected to a voltage of 120 V. One has a power of 25 W, the other 100 W. Which bulb has the higher resistance? Which bulb carries more current?
7. When the voltage across a certain conductor is doubled, the current is observed to triple. What can you conclude about the conductor?
8. There is an old admonition given to experimenters to "keep one hand in the pocket" when working around high voltages. Why is this warning a good idea?
9. What factors affect the resistance of a conductor?
10. Some homes have light dimmers that are operated by rotating a knob. What is being changed in the electric circuit when the knob is rotated?
11. Two wires A and B with circular cross section are made of the same metal and have equal lengths, but the resistance of wire A is three times greater than that of wire B. What is the ratio of their cross-sectional areas? How do the radii compare?
12. What single experimental requirement makes superconducting devices expensive to operate? In principle, can this limitation be overcome?
13. What could happen to the drift velocity of the electrons in a wire and to the current in the wire if the electrons could move through it freely without resistance?
14. Use the atomic theory of matter to explain why the resistance of a material should increase as its temperature increases.
15. When is more power delivered to a lightbulb, just after it is turned on and the glow of the filament is increasing or after it has been on for a few seconds and the glow is steady?

PROBLEMS

1, 2, 3 = straightforward, intermediate, challenging ☐ = full solution available in *Student Solutions Manual/Study Guide*

Physics Now™ = coached problem with hints available at **www.cp7e.com** = biomedical application

Section 17.1 Electric Current

Section 17.2 A Microscopic View: Current and Drift Speed

1. If a current of 80.0 mA exists in a metal wire, how many electrons flow past a given cross section of the wire in 10.0 min? Sketch the direction of the current and the direction of the electrons' motion.
2. A certain conductor has 7.50×10^{28} free electrons per cubic meter, a cross-sectional area of 4.00×10^{-6} m^2, and carries a current of 2.50 A. Find the drift speed of the electrons in the conductor.
3. A 1.00-V potential difference is maintained across a 10.0-Ω resistor for a period of 20.0 s. What total charge passes through the wire in this time interval?
4. In a particular television picture tube, the measured beam current is 60.0 μA. How many electrons strike the screen every second?

5. In the Bohr model of the hydrogen atom, an electron in the lowest energy state moves at a speed of 2.19×10^6 m/s in a circular path having a radius of 5.29×10^{-11} m. What is the effective current associated with this orbiting electron?

6. If 3.25×10^{-3} kg of gold is deposited on the negative electrode of an electrolytic cell in a period of 2.78 h, what is the current in the cell during that period? Assume that the gold ions carry one elementary unit of positive charge.

7. A 200-km-long high-voltage transmission line 2.0 cm in diameter carries a steady current of 1 000 A. If the conductor is copper with a free charge density of 8.5×10^{28} electrons per cubic meter, how many years does it take one electron to travel the full length of the cable?

8. Physics Now™ An aluminum wire carrying a current of 5.0 A has a cross-sectional area of 4.0×10^{-6} m^2. Find the drift speed of the electrons in the wire. The density of aluminum is 2.7 g/cm^3. (Assume that one electron is supplied by each atom.)

9. If the current carried by a conductor is doubled, what happens to (a) the charge carrier density? (b) the electron drift velocity?

Section 17.4 Resistance and Ohm's Law
Section 17.5 Resistivity

10. A lightbulb has a resistance of 240 Ω when operating at a voltage of 120 V. What is the current in the bulb?

11. A person notices a mild shock if the current along a path through the thumb and index finger exceeds 80 μA. Compare the maximum possible voltage without shock across the thumb and index finger with a dry-skin resistance of 4.0×10^5 Ω and a wet-skin resistance of 2 000 Ω.

12. Suppose that you wish to fabricate a uniform wire out of 1.00 g of copper. If the wire is to have a resistance $R = 0.500$ Ω, and if all of the copper is to be used, what will be (a) the length and (b) the diameter of the wire?

13. Calculate the diameter of a 2.0-cm length of tungsten filament in a small lightbulb if its resistance is 0.050 Ω.

14. Eighteen-gauge wire has a diameter of 1.024 mm. Calculate the resistance of 15 m of 18-gauge copper wire at 20°C.

15. A potential difference of 12 V is found to produce a current of 0.40 A in a 3.2-m length of wire with a uniform radius of 0.40 cm. What is (a) the resistance of the wire? (b) the resistivity of the wire?

16. Make an order-of-magnitude estimate of the cost of one person's routine use of a hair dryer for 1 yr. If you do not use a blow dryer yourself, observe or interview someone who does. State the quantities you estimate and their values.

17. A wire 50.0 m long and 2.00 mm in diameter is connected to a source with a potential difference of 9.11 V, and the current is found to be 36.0 A. Assume a temperature of 20°C, and, using Table 17.1, identify the metal out of which the wire is made.

18. A rectangular block of copper has sides of length 10 cm, 20 cm, and 40 cm. If the block is connected to a 6.0-V source across two of its opposite faces, what are (a) the maximum current and (b) the minimum current that the block can carry?

19. A wire of initial length L_0 and radius r_0 has a measured resistance of 1.0 Ω. The wire is drawn under tensile stress to a new uniform radius of $r = 0.25r_0$. What is the new resistance of the wire?

Section 17.6 Temperature Variation of Resistance

20. A certain lightbulb has a tungsten filament with a resistance of 19 Ω when cold and 140 Ω when hot. Assume that Equation 17.7 can be used over the large temperature range involved here, and find the temperature of the filament when it is hot. Assume an initial temperature of 20°C.

21. While taking photographs in Death Valley on a day when the temperature is 58.0°C, Bill Hiker finds that a certain voltage applied to a copper wire produces a current of 1.000 A. Bill then travels to Antarctica and applies the same voltage to the same wire. What current does he register there if the temperature is − 88.0°C? Assume that no change occurs in the wire's shape and size.

22. A metal wire has a resistance of 10.00 Ω at a temperature of 20°C. If the same wire has a resistance of 10.55 Ω at 90°C, what is the resistance of the wire when its temperature is − 20°C?

23. At 20°C, the carbon resistor in an electric circuit connected to a 5.0-V battery has a resistance of 200 Ω. What is the current in the circuit when the temperature of the carbon rises to 80°C?

24. A wire 3.00 m long and 0.450 mm^2 in cross-sectional area has a resistance of 41.0 Ω at 20°C. If its resistance increases to 41.4 Ω at 29.0°C, what is the temperature coefficient of resistivity?

25. The copper wire used in a house has a cross-sectional area of 3.00 mm^2. If 10.0 m of this wire is used to wire a circuit in the house at 20.0°C, find the resistance of the wire at temperatures of (a) 30.0°C and (b) 10.0°C.

26. A 100-cm-long copper wire of radius 0.50 cm has a potential difference across it sufficient to produce a current of 3.0 A at 20°C. (a) What is the potential difference? (b) If the temperature of the wire is increased to 200°C, what potential difference is now required to produce a current of 3.0 A?

27. Physics Now™ (a) A 34.5-m length of copper wire at 20.0°C has a radius of 0.25 mm. If a potential difference of 9.0 V is applied across the length of the wire, determine the current in the wire. (b) If the wire is heated to 30.0°C while the 9.0-V potential difference is maintained, what is the resulting current in the wire?

28. A toaster rated at 1 050 W operates on a 120-V household circuit and a 4.00-m length of nichrome wire as its heating element. The operating temperature of this element is 320°C. What is the cross-sectional area of the wire?

29. In one form of plethysmograph (a device for measuring volume), a rubber capillary tube with an inside diameter of 1.00 mm is filled with mercury at 20°C. The resistance of the mercury is measured with the aid of electrodes sealed into the ends of the tube. If 100.00 cm of the tube is wound in a spiral around a patient's upper arm, the blood flow during a heartbeat causes the arm to expand, stretching the tube to a length of 100.04 cm. From this observation, and

assuming cylindrical symmetry, you can find the change in volume of the arm, which gives an indication of blood flow. (a) Calculate the resistance of the mercury. (b) Calculate the fractional change in resistance during the heartbeat. [*Hint:* The fraction by which the cross-sectional area of the mercury thread decreases is the fraction by which the length increases, since the volume of mercury is constant.] Take $\rho_{Hg} = 9.4 \times 10^{-7}\ \Omega \cdot \text{m}$.

30. A platinum resistance thermometer has resistances of 200.0 Ω when placed in a 0°C ice bath and 253.8 Ω when immersed in a crucible containing melting potassium. What is the melting point of potassium? [*Hint:* First determine the resistance of the platinum resistance thermometer at room temperature, 20°C.]

Section 17.8 Electrical Energy and Power

31. A toaster is rated at 600 W when connected to a 120-V source. What current does the toaster carry, and what is its resistance?

32. If electrical energy costs 12 cents, or $0.12, per kilowatt-hour, how much does it cost to (a) burn a 100-W lightbulb for 24 h? (b) operate an electric oven for 5.0 h if it carries a current of 20.0 A at 220 V?

33. How many 100-W lightbulbs can you use in a 120-V circuit without tripping a 15-A circuit breaker? (The bulbs are connected in parallel, which means that the potential difference across each lightbulb is 120 V.)

34. A high-voltage transmission line with a resistance of 0.31 Ω/km carries a current of 1 000 A. The line is at a potential of 700 kV at the power station and carries the current to a city located 160 km from the station. (a) What is the power loss due to resistance in the line? (b) What fraction of the transmitted power does this loss represent?

35. The heating element of a coffeemaker operates at 120 V and carries a current of 2.00 A. Assuming that the water absorbs all of the energy converted by the resistor, calculate how long it takes to heat 0.500 kg of water from room temperature (23.0°C) to the boiling point.

36. The power supplied to a typical black-and-white television set is 90 W when the set is connected to 120 V. (a) How much electrical energy does this set consume in 1 hour? (b) A color television set draws about 2.5 A when connected to 120 V. How much time is required for it to consume the same energy as the black-and-white model consumes in 1 hour?

37. What is the required resistance of an immersion heater that will increase the temperature of 1.50 kg of water from 10.0°C to 50.0°C in 10.0 min while operating at 120 V?

38. A certain toaster has a heating element made of Nichrome resistance wire. When the toaster is first connected to a 120-V source of potential difference (and the wire is at a temperature of 20.0°C), the initial current is 1.80 A. However, the current begins to decrease as the resistive element warms up. When the toaster reaches its final operating temperature, the current has dropped to 1.53 A. (a) Find the power the toaster converts when it is at its operating temperature. (b) What is the final temperature of the heating element?

39. A copper cable is designed to carry a current of 300 A with a power loss of 2.00 W/m. What is the required radius of this cable?

40. A small motor draws a current of 1.75 A from a 120-V line. The output power of the motor is 0.20 hp. (a) At a rate of $0.060/kWh, what is the cost of operating the motor for 4.0 h? (b) What is the efficiency of the motor?

41. It has been estimated that there are 270 million plug-in electric clocks in the United States, approximately one clock for each person. The clocks convert energy at the average rate of 2.50 W. To supply this energy, how many metric tons of coal are burned per hour in coal-fired electric generating plants that are, on average, 25.0% efficient? The heat of combustion for coal is 33.0 MJ/kg.

42. The cost of electricity varies widely throughout the United States; $0.120/kWh is a typical value. At this unit price, calculate the cost of (a) leaving a 40.0-W porch light on for 2 weeks while you are on vacation, (b) making a piece of dark toast in 3.00 min with a 970-W toaster, and (c) drying a load of clothes in 40.0 min in a 5 200-W dryer.

43. How much does it cost to watch a complete 21-hour-long World Series on a 180-W television set? Assume that electricity costs $0.070/kWh.

44. A house is heated by a 24.0-kW electric furnace that uses resistance heating. The rate for electrical energy is $0.080/kWh. If the heating bill for January is $200, how long must the furnace have been running on an average January day?

45. An 11-W energy-efficient fluorescent lamp is designed to produce the same illumination as a conventional 40-W lamp. How much does the energy-efficient lamp save during 100 hours of use? Assume a cost of $0.080/kWh for electrical energy.

46. An office worker uses an immersion heater to warm 250 g of water in a light, covered, insulated cup from 20°C to 100°C in 4.00 minutes. The heater is a Nichrome resistance wire connected to a 120-V power supply. Assume that the wire is at 100°C throughout the 4.00-min time interval. Specify a diameter and a length that the wire can have. Can it be made from less than 0.5 cm^3 of Nichrome?

47. **Physics Now™** The heating coil of a hot-water heater has a resistance of 20 Ω and operates at 210 V. If electrical energy costs $0.080/kWh, what does it cost to raise the 200 kg of water in the tank from 15°C to 80°C? (See Chapter 11.)

ADDITIONAL PROBLEMS

48. One lightbulb is marked "25 W 120 V," and another "100 W 120 V"; this means that each converts its respective power when plugged into a constant 120-V potential difference. (a) Find the resistance of each bulb. (b) How long does it take for 1.00 C to pass through the dim bulb? How is this charge different upon its exit from, versus its entry into, the bulb? (c) How long does it take for 1.00 J to pass through the dim bulb? How is this energy different upon its exit from, versus its entry into, the bulb? (d) Find the cost of running the dim bulb continuously for 30.0 days if the electric company sells its product at $0.070 0 per kWh. What physical quantity *does* the electric company sell? What is its price for one SI unit of this quantity?

49. A particular wire has a resistivity of $3.0 \times 10^{-8}\ \Omega \cdot \text{m}$ and a cross-sectional area of $4.0 \times 10^{-6}\ \text{m}^2$. A length of this wire is to be used as a resistor that will develop 48 W of power when connected across a 20-V battery. What length of wire is required?

50. A steam iron draws 6.0 A from a 120-V line. (a) How many joules of internal energy are produced in 20 min? (b) How much does it cost, at \$0.080/kWh, to run the steam iron for 20 min?

51. An experiment is conducted to measure the electrical resistivity of Nichrome in the form of wires with different lengths and cross-sectional areas. For one set of measurements, a student uses 30-gauge wire, which has a cross-sectional area of $7.30 \times 10^{-8}\ \text{m}^2$. The student measures the potential difference across the wire and the current in the wire with a voltmeter and an ammeter, respectively. For each of the measurements given in the following table taken on wires of three different lengths, calculate the resistance of the wires and the corresponding value of the resistivity:

L (m)	ΔV (V)	I (A)	R (Ω)	ρ ($\Omega \cdot$m)
0.540	5.22	0.500		
1.028	5.82	0.276		
1.543	5.94	0.187		

What is the average value of the resistivity, and how does this value compare with the value given in Table 17.1?

52. Birds resting on high-voltage power lines are a common sight. The copper wire on which a bird stands is 2.2 cm in diameter and carries a current of 50 A. If the bird's feet are 4.0 cm apart, calculate the potential difference across its body.

53. You are cooking breakfast for yourself and a friend using a 1 200-W waffle iron and a 500-W coffeepot. Usually, you operate these appliances from a 110-V outlet for 0.500 h each day. (a) At 12 cents per kWh, how much do you spend to cook breakfast during a 30.0 day period? (b) You find yourself addicted to waffles and would like to upgrade to a 2 400-W waffle iron that will enable you to cook twice as many waffles during a half-hour period, but you know that the circuit breaker in your kitchen is a 20-A breaker. Can you do the upgrade?

54. The current in a conductor varies in time as shown in Figure P17.54. (a) How many coulombs of charge pass through a cross section of the conductor in the interval from $t = 0$ to $t = 5.0$ s? (b) What constant current would transport the same total charge during the 5.0-s interval as does the actual current?

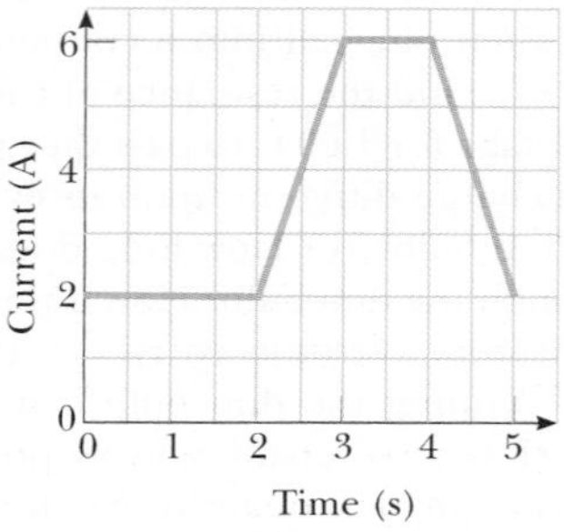

Figure P17.54

55. An electric car is designed to run off a bank of 12.0-V batteries with a total energy storage of 2.00×10^7 J. (a) If the electric motor draws 8.00 kW, what is the current delivered to the motor? (b) If the electric motor draws 8.00 kW as the car moves at a steady speed of 20.0 m/s, how far will the car travel before it is "out of juice"?

56. (a) A 115-g mass of aluminum is formed into a right circular cylinder, shaped so that its diameter equals its height. Calculate the resistance between the top and bottom faces of the cylinder at 20°C. (b) Calculate the resistance between opposite faces if the same mass of aluminum is formed into a cube.

57. A length of metal wire has a radius of 5.00×10^{-3} m and a resistance of 0.100 Ω. When the potential difference across the wire is 15.0 V, the electron drift speed is found to be 3.17×10^{-4} m/s. On the basis of these data, calculate the density of free electrons in the wire.

58. A carbon wire and a Nichrome wire are connected one after the other. If the combination has a total resistance of 10.0 kΩ at 20°C, what is the resistance of each wire at 20°C so that the resistance of the combination does not change with temperature?

59. (a) Determine the resistance of a lightbulb marked 100 W @ 120 V. (b) Assuming that the filament is tungsten and has a cross-sectional area of 0.010 mm^2, determine the length of the wire inside the bulb when the bulb is operating. (c) Why do you think the wire inside the bulb is tightly coiled? (d) If the temperature of the tungsten wire is 2 600°C when the bulb is operating, what is the length of the wire after the bulb is turned off and has cooled to 20°C? (See Chapter 10, and use 4.5×10^{-6}/°C as the coefficient of linear expansion for tungsten.)

60. In a certain stereo system, each speaker has a resistance of 4.00 Ω. The system is rated at 60.0 W in each channel. Each speaker circuit includes a fuse rated at a maximum current of 4.00 A. Is this system adequately protected against overload?

61. A resistor is constructed by forming a material of resistivity $3.5 \times 10^5\ \Omega \cdot$m into the shape of a hollow cylinder of length 4.0 cm and inner and outer radii 0.50 cm and 1.2 cm, respectively. In use, a potential difference is applied between the ends of the cylinder, producing a current parallel to the length of the cylinder. Find the resistance of the cylinder.

62. The graph in Figure P17.62a shows the current I in a diode as a function of the potential difference ΔV across the diode. Figure P17.62b shows the circuit used to make the measurements. The symbol —▶|— represents the diode. (a) Using Equation 17.3, make a table of the resistance of the diode for different values of ΔV in the range from -1.5 V to $+1.0$ V. (b) Based on your results, what amazing electrical property does a diode possess?

63. An x-ray tube used for cancer therapy operates at 4.0 MV, with a beam current of 25 mA striking the metal target. Nearly all the power in the beam is transferred to a stream of water flowing through holes drilled in the target. What rate of flow, in kilograms per second, is needed if the rise in temperature (ΔT) of the water is not to exceed 50°C?

64. A 50.0-g sample of a conducting material is all that is available. The resistivity of the material is measured to be $11 \times 10^{-8}\ \Omega \cdot$m, and the density is 7.86 g/cm^3. The material is to be shaped into a solid cylindrical wire that has a total resistance of 1.5 Ω. (a) What length of wire is required? (b) What must be the diameter of the wire?

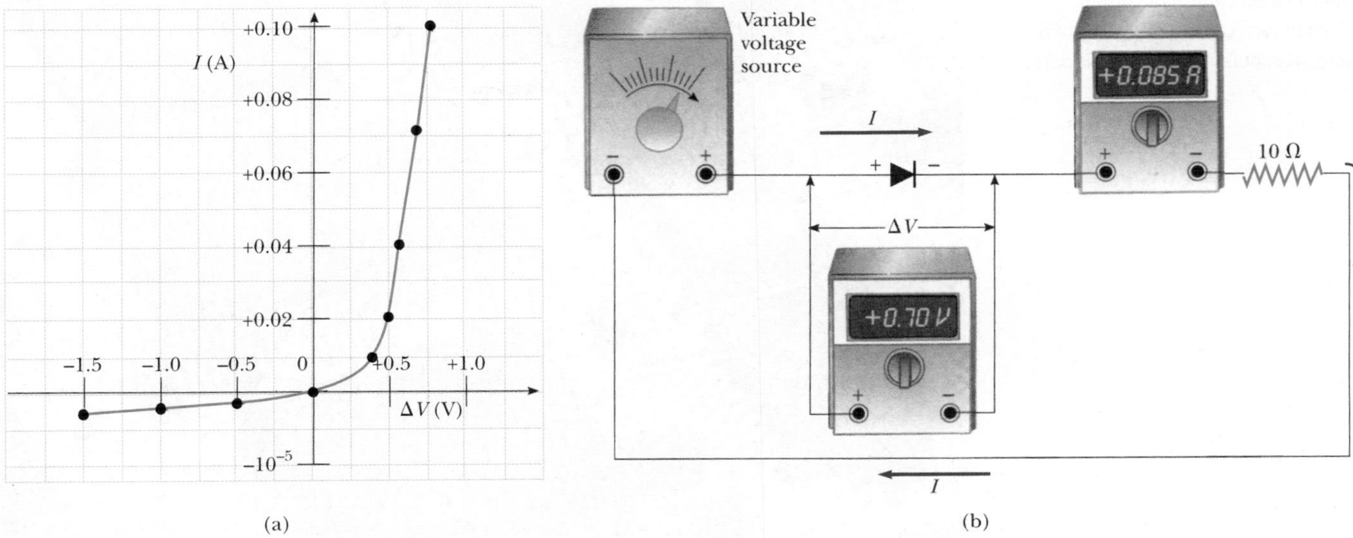

Figure P17.62

65. **PhysicsNow™** (a) A sheet of copper ($\rho = 1.7 \times 10^{-8}$ $\Omega \cdot$m) is 2.0 mm thick and has surface dimensions of 8.0 cm $\times$ 24 cm. If the long edges are joined to form a tube 24 cm in length, what is the resistance between the ends? (b) What mass of copper is required to manufacture a 1 500-m-long spool of copper cable with a total resistance of 4.5 Ω?

66. When a straight wire is heated, its resistance changes according to the equation

$$R = R_0[1 + \alpha(T - T_0)]$$

(Eq. 17.7), where α is the temperature coefficient of resistivity. (a) Show that a more precise result, which includes the fact that the length and area of a wire change when it is heated, is

$$R = \frac{R_0[1 + \alpha(T - T_0)][1 + \alpha'(T - T_0)]}{[1 + 2\alpha'(T - T_0)]}$$

where α' is the coefficient of linear expansion. (See Chapter 10.) (b) Compare the two results for a 2.00-m-long copper wire of radius 0.100 mm, starting at 20.0°C and heated to 100.0°C.

67. A man wishes to vacuum his car with a canister vacuum cleaner marked 535 W at 120 V. The car is parked far from the building, so he uses an extension cord 15.0 m long to plug the cleaner into a 120-V source. Assume that the cleaner has constant resistance. (a) If the resistance of each of the two conductors of the extension cord is 0.900 Ω, what is the actual power delivered to the cleaner? (b) If, instead, the power is to be at least 525 W, what must be the diameter of each of two identical copper conductors in the cord the young man buys? (c) Repeat part (b) if the power is to be at least 532 W. [*Suggestion*: A symbolic solution can simplify the calculations.]

ACTIVITIES

1. Connect one terminal of a D-cell battery to the base of a flashlight bulb with insulated wire, tape a second wire to the other battery terminal, and tape a third wire to the center conductor of the bulb, as in Figure A17.1. Make sure to remove about 1 cm of insulation from the ends of all wires before making the connections. Now bridge the gap between the open wires with different objects, such as a plastic pen, an aluminum can, a penny, a rubber band, and a spoon. Which objects make the bulb light up? Explain your observations.

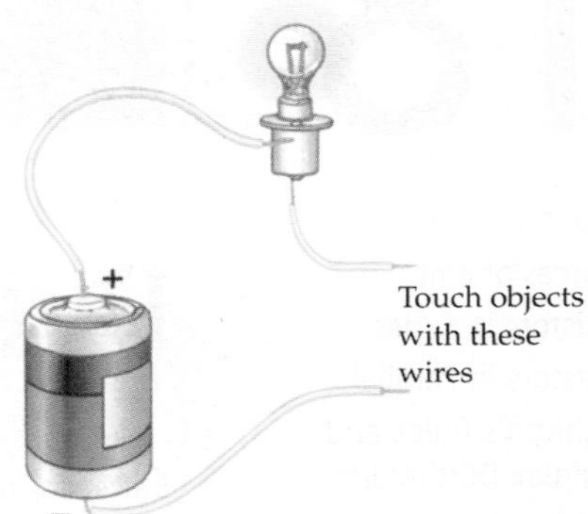

Figure A17.1

2. When the lightbulbs in your home are turned on, they are always connected across the same potential difference. Which do you believe has a filament with the highest resistance when cool, a 60-W bulb or a 100-W bulb? To check your prediction, ask your instructor to lend you a device called an ohmmeter and to instruct you in its use. A resistor must always be disconnected from a circuit when its resistance is measured with an ohmmeter.

3. Examine the labels on several appliances, such as a toaster, a television set, a lamp, a stereo system, an air conditioner, and a clock. From each label, determine the power rating of the device in watts. Check the billing statement from your electric utility company to find the cost of electrical energy per kilowatt-hour. (Prices usually range from about a nickel to 20 cents.) Calculate the cost of running each appliance for 1 h. Estimate how many hours per day each appliance is used. Then, on the basis of your daily estimate, calculate the monthly cost of using each appliance.

The complex circuits in modern electronic devices allow a highly sophisticated control of current, which in turn can be used to obtain, store, manipulate, and transmit data.

© Lester Lefkowitz/Corbis

CHAPTER

18

OUTLINE

Direct-Current Circuits

Batteries, resistors, and capacitors can be used in various combinations to construct electric circuits, which direct and control the flow of electricity and the energy it conveys. Such circuits make possible all the modern conveniences in a home—electric lights, electric stove tops and ovens, washing machines, and a host of other appliances and tools. Electric circuits are also found in our cars, in tractors that increase farming productivity, and in all types of medical equipment that saves so many lives every day.

In this chapter, we study and analyze a number of simple direct-current circuits. The analysis is simplified by the use of two rules known as Kirchhoff's rules, which follow from the principle of conservation of energy and the law of conservation of charge. Most of the circuits are assumed to be in *steady state*, which means that the currents are constant in magnitude and direction. We close the chapter with a discussion of circuits containing resistors and capacitors, in which current varies with time.

18.1 SOURCES OF EMF

A current is maintained in a closed circuit by a source of emf.[1] Among such sources are any devices (for example, batteries and generators) that increase the potential energy of the circulating charges. A source of emf can be thought of as a "charge pump" that forces electrons to move in a direction opposite the electrostatic field inside the source. The emf $\mathcal{E}$ of a source is the work done per unit charge; hence the SI unit of emf is the volt.

Consider the circuit in Active Figure 18.1a consisting of a battery connected to a resistor. We assume that the connecting wires have no resistance. If we neglect the internal resistance of the battery, the potential drop across the battery (the terminal voltage) equals the emf of the battery. Because a real battery always has some

[1]The term was originally an abbreviation for *electromotive force*, but emf is not really a force, so the long form is discouraged.

internal resistance r, however, the terminal voltage is not equal to the emf. The circuit of Active Figure 18.1a can be described schematically by the diagram in Active Figure 18.1b. The battery, represented by the dashed rectangle, consists of a source of emf $\mathcal{E}$ in series with an internal resistance r. Now imagine a positive charge moving through the battery from a to b in the figure. As the charge passes from the negative to the positive terminal of the battery, the potential of the charge increases by $\mathcal{E}$. As the charge moves through the resistance r, however, its potential decreases by the amount Ir, where I is the current in the circuit. The terminal voltage of the battery, $\Delta V = V_b - V_a$, is therefore given by

$$\Delta V = \mathcal{E} - Ir \quad [18.1]$$

From this expression, we see that **$\mathcal{E}$ is equal to the terminal voltage when the current is zero**, called the **open-circuit voltage**. By inspecting Figure 18.1b, we find that the terminal voltage ΔV must also equal the potential difference across the external resistance R, often called the **load resistance**; that is, $\Delta V = IR$. Combining this relationship with Equation 18.1, we arrive at

$$\mathcal{E} = IR + Ir \quad [18.2]$$

Solving for the current gives

$$I = \frac{\mathcal{E}}{R + r}$$

The preceding equation shows that the current in this simple circuit depends on both the resistance external to the battery and the internal resistance of the battery. If R is much greater than r, we can neglect r in our analysis (an option we usually select).

If we multiply Equation 18.2 by the current I, we get

$$I\mathcal{E} = I^2R + I^2r$$

This equation tells us that the total power output $I\mathcal{E}$ of the source of emf is converted at the rate I^2R at which energy is delivered to the load resistance, *plus* the rate I^2r at which energy is delivered to the internal resistance. Again, if $r << R$, most of the power delivered by the battery is transferred to the load resistance.

Unless otherwise stated, we will assume in our examples and end-of-chapter problems that the internal resistance of a battery in a circuit is negligible.

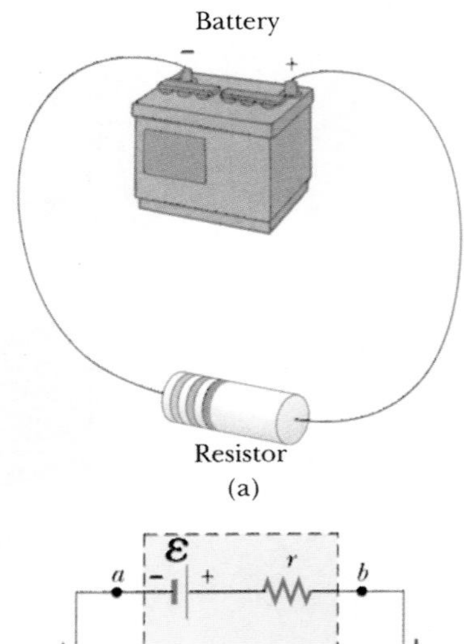

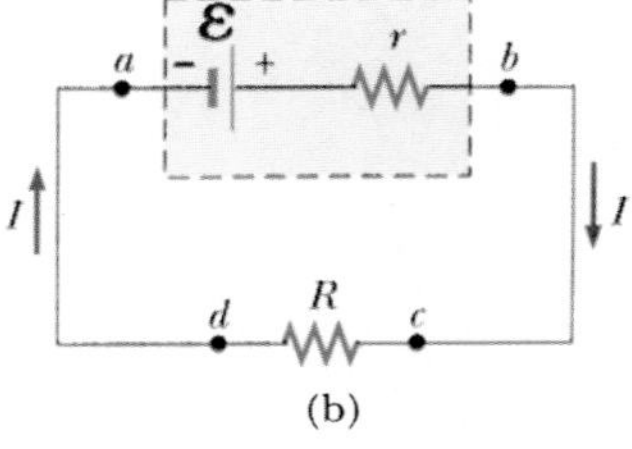

ACTIVE FIGURE 18.1
(a) A circuit consisting of a resistor connected to the terminals of a battery. (b) A circuit diagram of a source of emf $\mathcal{E}$ having internal resistance r connected to an external resistor R.

Physics Now™
Log into PhysicsNow at **www.cp7e.com** and go to Active Figure 18.1, where you can adjust the emf and the resistances r and R, and see the effect on the current in part (b).

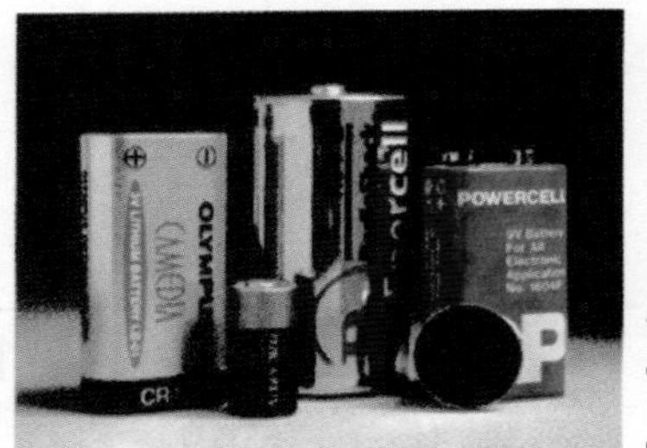

George Semple

An assortment of batteries.

18.2 RESISTORS IN SERIES

When two or more resistors are connected end to end as in Active Figure 18.2, they are said to be in *series*. The resistors could be simple devices, such as lightbulbs or heating elements. When two resistors R_1 and R_2 are connected to a battery as in Active Figure 18.2, **the current is the same in the two resistors, because any charge that flows through R_1 must also flow through R_2.** This is analogous to water flowing through a pipe with two constrictions, corresponding to R_1 and R_2. Whatever volume of water flows in one end in a given time interval must exit the opposite end.

Because the potential difference between a and b in Active Figure 18.2b equals IR_1 and the potential difference between b and c equals IR_2, the potential difference between a and c is

$$\Delta V = IR_1 + IR_2 = I(R_1 + R_2)$$

Regardless of how many resistors we have in series, the sum of the potential differences across the resistors is equal to the total potential difference across the combination. As we will show later, this is a consequence of the conservation of energy. Active Figure 18.2c shows an equivalent resistor R_{eq} that can replace the two resistors of the original circuit. The equivalent resistor has the same effect on the circuit because it results in the same current in the circuit as the two resistors. Applying Ohm's law to this equivalent resistor, we have

$$\Delta V = IR_{eq}$$

TIP 18.1 What's Constant in a Battery?
Equation 18.2 shows that the current in a circuit depends on the resistance of the battery, so a battery can't be considered a source of constant current. Even the terminal voltage of a battery given by Equation 18.1 can't be considered constant, because the internal resistance can change (due to warming, for example, during the operation of the battery). A battery is, however, a source of constant emf.

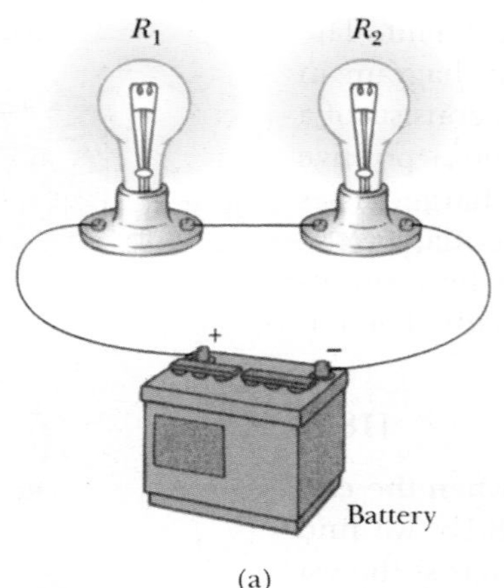

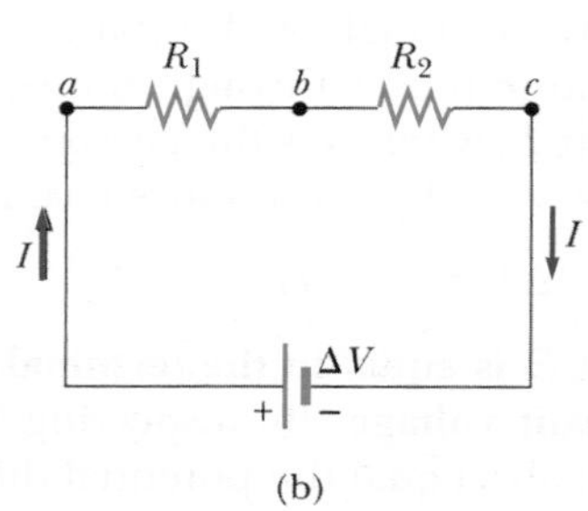

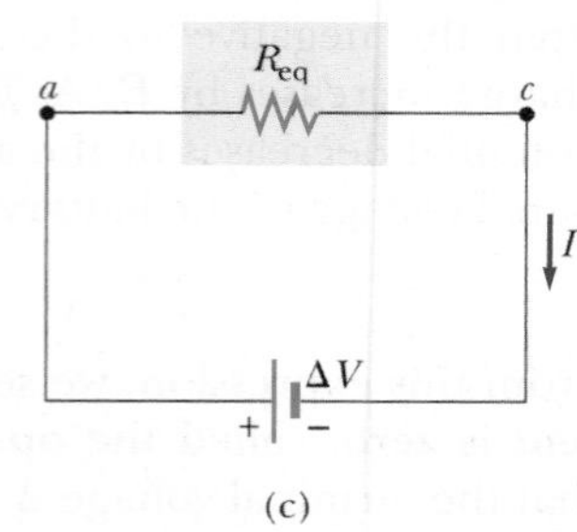

ACTIVE FIGURE 18.2
A series connection of two resistors, R_1 and R_2. The currents in the resistors are the same, and the equivalent resistance of the combination is given by $R_{eq} = R_1 + R_2$.

PhysicsNow™
Log into PhysicsNow at **www.cp7e.com** and go to Active Figure 18.2, where you can adjust the battery voltage and resistances R_1 and R_2, observing the effect on the current and voltages of the individual resistors.

Equating the preceding two expressions, we have

$$IR_{eq} = I(R_1 + R_2)$$

or

$$R_{eq} = R_1 + R_2 \quad \text{(series combination)} \qquad [18.3]$$

An extension of the preceding analysis shows that the equivalent resistance of three or more resistors connected in series is

Equivalent resistance of a series combination of resistors ▶

$$R_{eq} = R_1 + R_2 + R_3 + \cdots \qquad [18.4]$$

Therefore, **the equivalent resistance of a series combination of resistors is the algebraic sum of the individual resistances and is always greater than any individual resistance.**

Note that if the filament of one lightbulb in Active Figure 18.2 were to fail, the circuit would no longer be complete (an open-circuit condition would exist) and the second bulb would also go out.

Applying Physics 18.1 Christmas Lights in Series

A new design for Christmas tree lights allows them to be connected in series. A failed bulb in such a string would result in an open circuit, and all of the bulbs would go out. How can the bulbs be redesigned to prevent this from happening?

Explanation If the string of lights contained the usual kind of bulbs, a failed bulb would be hard to locate. Each bulb would have to be replaced with a good bulb, one by one, until the failed bulb was found. If there happened to be two or more failed bulbs in the string of lights, finding them would be a lengthy and annoying task.

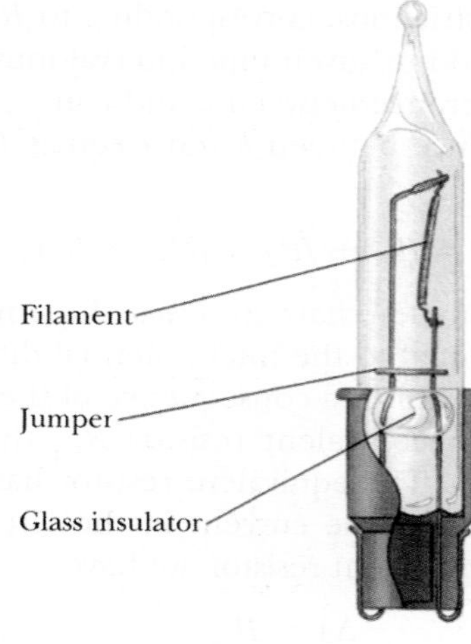

Figure 18.3 (Applying Physics 18.1) Diagram of a modern miniature holiday lightbulb, with a jumper connection to provide a current if the filament breaks.

Christmas lights use special bulbs that have an insulated loop of wire (a jumper) across the conducting supports to the bulb filaments (Fig. 18.3). If the filament breaks and the bulb fails, the bulb's resistance increases dramatically. As a result, most of the applied voltage appears across the loop of wire. This voltage causes the insulation around the loop of wire to burn, causing the metal wire to make electrical contact with the supports. This produces a conducting path through the bulb, so the other bulbs remain lit.

Quick Quiz 18.1

When a piece of wire is used to connect points b and c in Figure 18.2b, the brightness of bulb R_1 (a) increases, (b) decreases but remains lit, (c) stays the same, (d) goes out. The brightness of bulb R_2 (a) increases, (b) decreases but remains lit, (c) stays the same, (d) goes out. (Assume connecting wires have no resistance.)

Quick Quiz 18.2

In Figure 18.4a the current is measured with the ammeter at the right side of the circuit. When the switch is opened as in Figure 18.4b, the reading on the ammeter (a) increases (b) decreases (c) doesn't change.

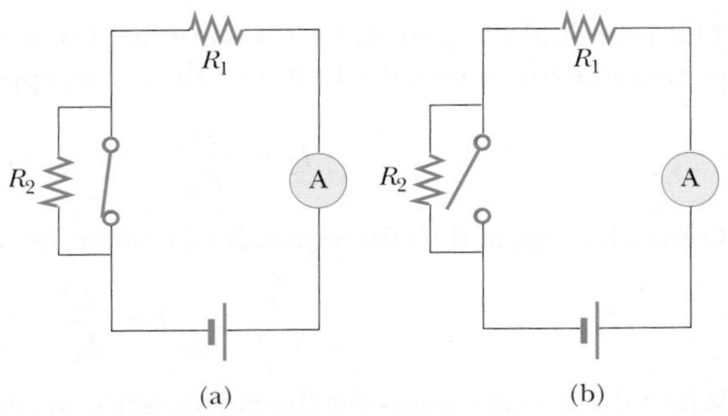

Figure 18.4 (Quick Quiz 18.2)

EXAMPLE 18.1 Four Resistors in Series

Goal Analyze several resistors connected in series.

Problem Four resistors are arranged as shown in Figure 18.5a. Find **(a)** the equivalent resistance of the circuit and **(b)** the current in the circuit if the emf of the battery is 6.0 V.

Strategy Because the resistors are connected in series, summing their resistances gives the equivalent resistance. Ohm's law can then be used to find the current.

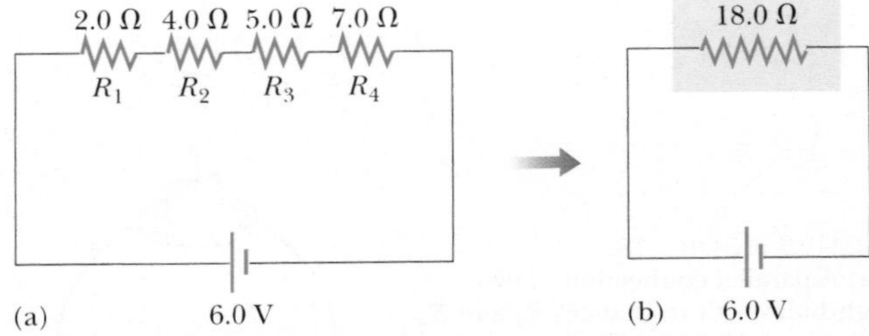

Figure 18.5 (Example 18.1) (a) Four resistors connected in series. (b) The equivalent resistance of the circuit in (a).

Solution

(a) Find the equivalent resistance of the circuit.

Apply Equation 18.4, summing the resistances:

$$R_{eq} = R_1 + R_2 + R_3 + R_4 = 2.0\ \Omega + 4.0\ \Omega + 5.0\ \Omega + 7.0\ \Omega$$
$$= 18.0\ \Omega$$

(b) Find the current in the circuit.

Apply Ohm's law to the equivalent resistor in Figure 18.5b, solving for the current:

$$I = \frac{\Delta V}{R_{eq}} = \frac{6.0\ \text{V}}{18.0\ \Omega} = \frac{1}{3}\ \text{A}$$

Exercise 18.1

Because the current in the equivalent resistor is $\frac{1}{3}$ A, this must also be the current in each resistor of the original circuit. Find the voltage drop across each resistor.

Answers $\Delta V_{2\Omega} = \frac{2}{3}$ V; $\Delta V_{4\Omega} = \frac{4}{3}$ V; $\Delta V_{5\Omega} = \frac{5}{3}$ V; $\Delta V_{7\Omega} = \frac{7}{3}$ V.

18.3 RESISTORS IN PARALLEL

Now consider two resistors connected in parallel, as in Active Figure 18.6. In this case, **the potential differences across the resistors are the same because each is connected directly across the battery terminals.** The currents are generally not the same. When charges reach point *a* (called a junction) in Active Figure 18.6b, the current splits into two parts: I_1, flowing through R_1; and I_2, flowing through R_2. If R_1 is greater than R_2, then I_1 is less than I_2. In general, more charge travels through the path with less resistance. **Because charge is conserved, the current I that enters point a must equal the total current $I_1 + I_2$ leaving that point.** Mathematically, this is written

$$I = I_1 + I_2$$

The potential drop must be the same for the two resistors and must also equal the potential drop across the battery. Ohm's law applied to each resistor yields

$$I_1 = \frac{\Delta V}{R_1} \qquad I_2 = \frac{\Delta V}{R_2}$$

Ohm's law applied to the equivalent resistor in Active Figure 18.6c gives

$$I = \frac{\Delta V}{R_{eq}}$$

When these expressions for the currents are substituted into the equation $I = I_1 + I_2$ and the ΔV's are cancelled, we obtain

$$\frac{1}{R_{eq}} = \frac{1}{R_1} + \frac{1}{R_2} \qquad \text{(parallel combination)} \qquad [18.5]$$

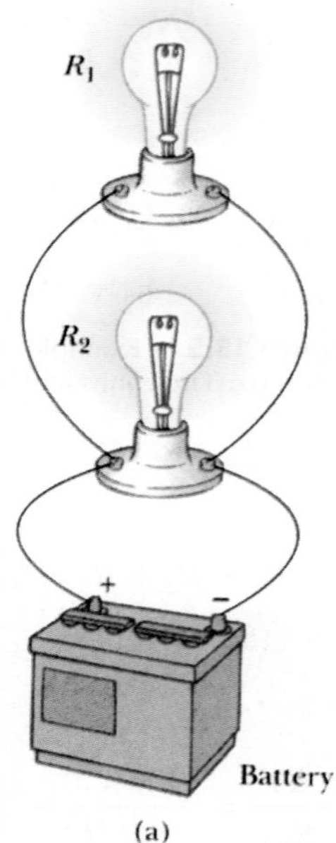

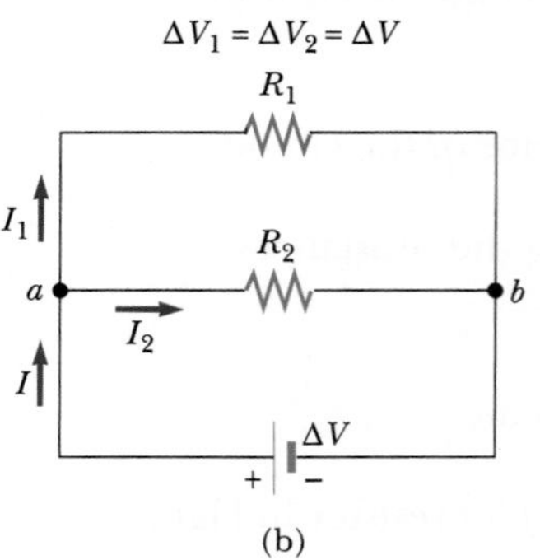

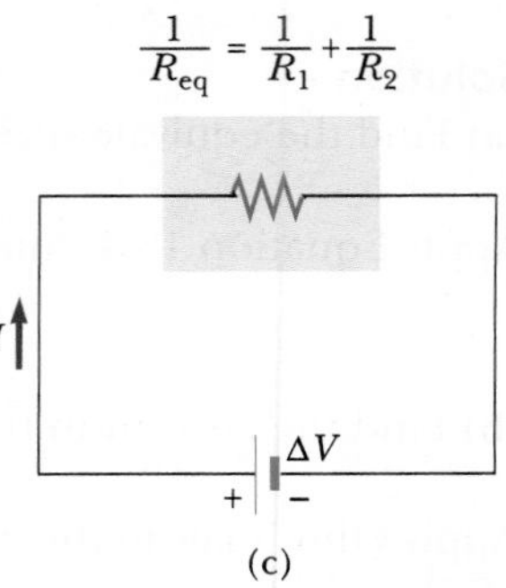

ACTIVE FIGURE 18.6
(a) A parallel connection of two lightbulbs with resistances R_1 and R_2. (b) Circuit diagram for the two-resistor circuit. The potential differences across R_1 and R_2 are the same. (c) The equivalent resistance of the combination is given by the reciprocal relationship $1/R_{eq} = 1/R_1 + 1/R_2$.

Physics Now™
Log into PhysicsNow at **www.cp7e.com**, and go to Active Figure 18.6, where you can adjust the battery voltage and resistances R_1 and R_2 and see the effect on the currents and voltages in the individual resistors.

An extension of this analysis to three or more resistors in parallel produces the following general expression for the equivalent resistance:

$$\frac{1}{R_{eq}} = \frac{1}{R_1} + \frac{1}{R_2} + \frac{1}{R_3} + \cdots \qquad [18.6]$$

◀ Equivalent resistance of a parallel combination of resistors

From this expression, we see that **the inverse of the equivalent resistance of two or more resistors connected in parallel is the sum of the inverses of the individual resistances and is always less than the smallest resistance in the group.**

INTERACTIVE EXAMPLE 18.2 Three Resistors in Parallel

Goal Analyze a circuit having resistors connected in parallel.

Problem Three resistors are connected in parallel as in Figure 18.7. A potential difference of 18 V is maintained between points a and b. **(a)** Find the current in each resistor. **(b)** Calculate the power delivered to each resistor and the total power. **(c)** Find the equivalent resistance of the circuit. **(d)** Find the total power delivered to the equivalent resistance.

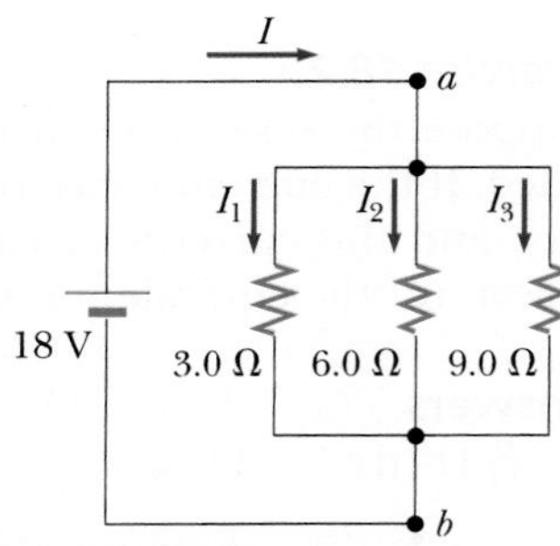

Figure 18.7 (Example 18.2) Three resistors connected in parallel. The voltage across each resistor is 18 V.

Strategy We can use Ohm's law and the fact that the voltage drops across parallel resistors are all the same to get the current in each resistor. The rest of the problem just requires substitution into the equation for power delivered to a resistor, $\mathcal{P} = I^2R$, and the reciprocal-sum law for parallel resistors.

Solution

(a) Find the current in each resistor.

Apply Ohm's law, solved for the current I delivered by the battery to find the current in each resistor:

$$I_1 = \frac{\Delta V}{R_1} = \frac{18\text{ V}}{3.0\ \Omega} = 6.0\text{ A}$$

$$I_2 = \frac{\Delta V}{R_2} = \frac{18\text{ V}}{6.0\ \Omega} = 3.0\text{ A}$$

$$I_3 = \frac{\Delta V}{R_3} = \frac{18\text{ V}}{9.0\ \Omega} = 2.0\text{ A}$$

(b) Calculate the power delivered to each resistor and the total power.

Apply $\mathcal{P} = I^2R$ to each resistor, substituting the results from part (a).

$$3\ \Omega:\quad \mathcal{P}_1 = I_1^2R_1 = (6.0\text{ A})^2(3.0\ \Omega) = 110\text{ W}$$

$$6\ \Omega:\quad \mathcal{P}_2 = I_2^2R_2 = (3.0\text{ A})^2(6.0\ \Omega) = 54\text{ W}$$

$$9\ \Omega:\quad \mathcal{P}_3 = I_3^2R_3 = (2.0\text{ A})^2(9.0\ \Omega) = 36\text{ W}$$

Sum to get the total power:

$$\mathcal{P}_{tot} = 110\text{ W} + 54\text{ W} + 36\text{ W} = 2.0 \times 10^2\text{ W}$$

(c) Find the equivalent resistance of the circuit.

Apply the reciprocal-sum rule, Equation 18.6:

$$\frac{1}{R_{eq}} = \frac{1}{R_1} + \frac{1}{R_2} + \frac{1}{R_3}$$

$$\frac{1}{R_{eq}} = \frac{1}{3.0\ \Omega} + \frac{1}{6.0\ \Omega} + \frac{1}{9.0\ \Omega} = \frac{11}{18\ \Omega}$$

$$R_{eq} = \frac{18}{11}\ \Omega = 1.6\ \Omega$$

(d) Compute the power dissipated by the equivalent resistance.

Use the alternate power equation: $\mathcal{P} = \dfrac{(\Delta V)^2}{R_{eq}} = \dfrac{(18\text{ V})^2}{(1.6\ \Omega)} = 2.0 \times 10^2\text{ W}$

Remarks There's something important to notice in part (a): the smallest 3.0 Ω resistor carries the largest current, while the other, larger resistors of 6.0 Ω and 9.0 Ω carry smaller currents. The largest current is always found in the path of least resistance. In part (b), the power could also be found with $\mathcal{P} = (\Delta V)^2/R$. Note that $\mathcal{P}_1 = 108$ W, but is rounded to 110 W because there are only two significant figures. Finally, notice that the total power dissipated in the equivalent resistor is the same as the sum of the power dissipated in the individual resistors, as it should be.

TIP 18.2 Don't Forget to Flip It!

The most common mistake in calculating the equivalent resistance for resistors in parallel is to forget to invert the answer after summing the reciprocals. Don't forget to flip it!

Exercise 18.2

Suppose the resistances in the example are 1.0 Ω, 2.0 Ω, and 3.0 Ω, respectively, and a new voltage source is provided. If the current measured in the 3.0-Ω resistor is 2.0 A, find (a) the potential difference provided by the new battery, and the currents in each of the remaining resistors, (b) the power delivered to each resistor, and the total power, (c) the equivalent resistance, and (d) the total current, and the power dissipated by the equivalent resistor.

Answers (a) $\mathcal{E} = 6.0$ V, $I_1 = 6.0$ A, $I_2 = 3.0$ A (b) $\mathcal{P}_1 = 36$ W, $\mathcal{P}_2 = 18$ W, $\mathcal{P}_3 = 12$ W, $\mathcal{P}_{tot} = 66$ W (c) $\frac{6}{11}$ Ω (d) $I = 11$ A, $\mathcal{P}_{eq} = 66$ W

PhysicsNow™ Explore different configurations of the battery and resistors by logging into PhysicsNow at **www.cp7e.com** and going to Interactive Example 18.2.

Household circuits are always wired so that the electrical devices are connected in parallel, as in Active Figure 18.6a. In this way, each device operates independently of the others, so that if one is switched off, the others remain on. For example, if one of the lightbulbs in Active Figure 18.6 were removed from its socket, the other would continue to operate. Equally important, each device operates at the same voltage. If the devices were connected in series, the voltage across any one device would depend on how many devices were in the combination and on their individual resistances.

APPLICATION

Circuit Breakers

In many household circuits, circuit breakers are used in series with other circuit elements for safety purposes. A circuit breaker is designed to switch off and open the circuit at some maximum value of the current (typically 15 A or 20 A) that depends on the nature of the circuit. If a circuit breaker were not used, excessive currents caused by operating several devices simultaneously could result in excessive wire temperatures, perhaps causing a fire. In older home construction, fuses were used in place of circuit breakers. When the current in a circuit exceeded some value, the conductor in a fuse melted and opened the circuit. The disadvantage of fuses is that they are destroyed in the process of opening the circuit, whereas circuit breakers can be reset.

Applying Physics 18.2 Lightbulb Combinations

Compare the brightness of the four identical bulbs shown in Figure 18.8. What happens if bulb A fails, so that it cannot conduct current? What if C fails? What if D fails?

Explanation Bulbs A and B are connected in series across the emf of the battery, whereas bulb C is connected by itself across the battery. This means the

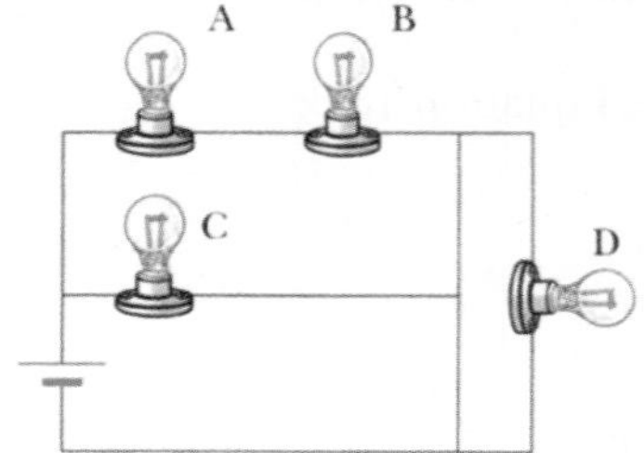

Figure 18.8 (Applying Physics 18.2)

voltage drop across C has the same magnitude as the battery emf, whereas this same emf is split between bulbs A and B. As a result, bulb C will glow more brightly than either of bulbs A and B, which will glow equally brightly. Bulb D has a wire connected across it—a short circuit—so the potential difference across bulb D is zero and it doesn't glow. If bulb A fails, B goes out, but C stays lit. If C fails, there is no effect on the other bulbs. If D fails, the event is undetectable, because D was not glowing initially.

Applying Physics 18.3 Three-Way Lightbulbs

Figure 18.9 illustrates how a three-way lightbulb is constructed to provide three levels of light intensity. The socket of the lamp is equipped with a three-way switch for selecting different light intensities. The bulb contains two filaments. Why are the filaments connected in parallel? Explain how the two filaments are used to provide the three different light intensities.

Explanation If the filaments were connected in series and one of them were to fail, there would be no current in the bulb, and the bulb would not glow, regardless of the position of the switch. However, when the filaments are connected in parallel and one of them (say, the 75-W filament) fails, the bulb will still operate in one of the switch positions because there is current in the other (100-W) filament. The three light intensities are made possible by selecting one of three values of filament resistance, using a single value of 120 V for the applied voltage. The 75-W filament offers one value of resistance, the 100-W filament offers a second value, and the third resistance is obtained by combining the two filaments in parallel. When switch S_1 is closed and switch S_2 is opened, only the 75-W filament carries current. When switch S_1 is open and switch S_2 is closed, only the 100-W filament carries current. When both switches are closed, both filaments carry current and a total illumination corresponding to 175 W is obtained.

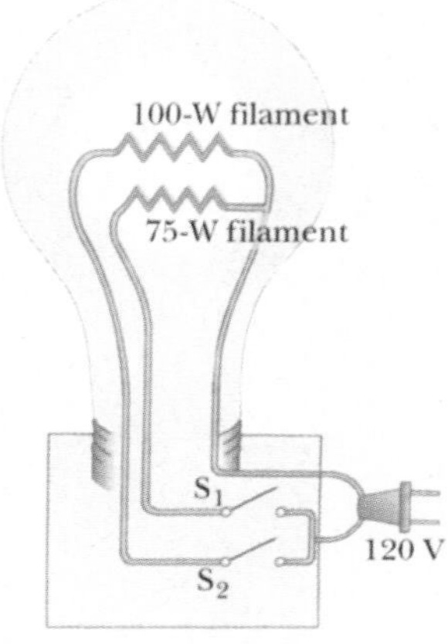

Figure 18.9 (Applying Physics 18.3)

Quick Quiz 18.3

In Figure 18.10a the current is measured with the ammeter on the right side of the circuit diagram. When the switch is closed, the reading on the ammeter (a) increases, (b) decreases, or (c) remains the same.

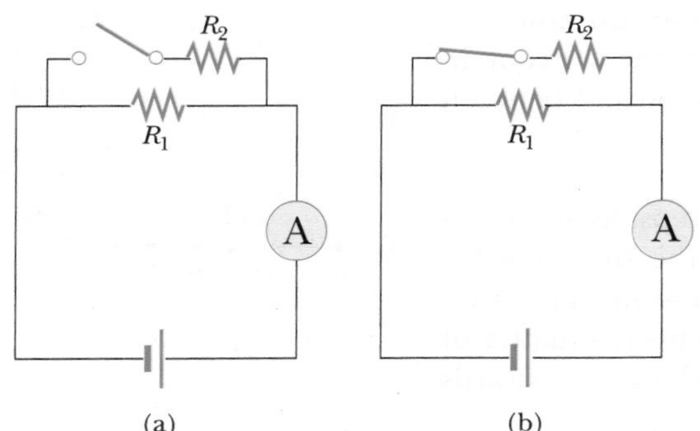

Figure 18.10 (Quick Quiz 18.3)

Quick Quiz 18.4

Suppose you have three identical lightbulbs, some wire, and a battery. You connect one lightbulb to the battery and take note of its brightness. You add a second lightbulb, connecting it in parallel with the previous bulbs, again taking note of the brightness. Repeat the process with the third bulb, connecting it in parallel with the other two. As the lightbulbs are added, what happens to (a) the brightness of the bulbs? (b) the individual currents in the bulbs? (c) the power delivered by the battery? (d) the lifetime of the battery? (Neglect the battery's internal resistance)

Quick Quiz 18.5

If the lightbulbs in Quick Quiz 18.4 are connected, one by one, in series instead of in parallel, what happens to (a) the brightness of the bulbs? (b) the individual currents in the bulbs? (c) the power delivered by the battery? (d) the lifetime of the battery? (Again, neglect the battery's internal resistance)

Problem-Solving Strategy Simplifying Circuits with Resistors

1. **Combine all resistors in series** by summing the individual resistances, and draw the new, simplified circuit diagram.
   Useful facts: $R_{eq} = R_1 + R_2 + R_3 + \cdots$
   The current in each resistor is the same.
2. **Combine all resistors in parallel** by summing the reciprocals of the resistances and then taking the reciprocal of the result. Draw the new, simplified circuit diagram.
   Useful facts: $\dfrac{1}{R_{eq}} = \dfrac{1}{R_1} + \dfrac{1}{R_2} + \dfrac{1}{R_3} + \cdots$
   The potential difference across each resistor is the same.
3. **Repeat** the first two steps as necessary, until no further combinations can be made. If there is only a single battery in the circuit, this will usually result in a single equivalent resistor in series with the battery.
4. **Use Ohm's Law**, $\Delta V = IR$, to determine the current in the equivalent resistor. Then work backwards through the diagrams, applying the useful facts listed in step 1 or step 2 to find the currents in the other resistors. (In more complex circuits, Kirchhoff's rules will be needed, as described in the next section).

EXAMPLE 18.3 Equivalent Resistance

Goal Solve a problem involving both series and parallel resistors.

Problem Four resistors are connected as shown in Figure 18.11a. **(a)** Find the equivalent resistance between points *a* and *c*. **(b)** What is the current in each resistor if a 42-V battery is connected between *a* and *c*?

Strategy Reduce the circuit in steps, as shown in Figures 18.11b and 18.11c, using the sum rule for resistors in series and the reciprocal-sum rule for resistors in parallel. Finding the currents is a matter of applying Ohm's law while working backwards through the diagrams.

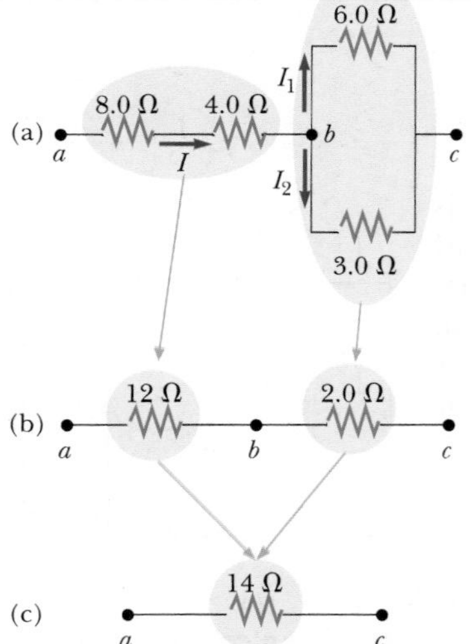

Figure 18.11 (Example 18.3) The four resistors shown in (a) can be reduced in steps to an equivalent 14-Ω resistor.

Solution

(a) Find the equivalent resistance of the circuit.

The 8.0-Ω and 4.0-Ω resistors are in series, so use the sum rule to find the equivalent resistance between *a* and *b*:

$$R_{eq} = R_1 + R_2 = 8.0\ \Omega + 4.0\ \Omega = 12\ \Omega$$

The 6.0-Ω and 3.0-Ω resistors are in parallel, so use the reciprocal-sum rule to find the equivalent resistance between *b* and *c* (don't forget to invert!):

$$\frac{1}{R_{eq}} = \frac{1}{R_1} + \frac{1}{R_2} = \frac{1}{6.0\ \Omega} + \frac{1}{3.0\ \Omega} = \frac{1}{2.0\ \Omega}$$

$$R_{eq} = 2.0\ \Omega$$

Because the filings point in the direction of $\vec{\mathbf{B}}$, we conclude that the lines of $\vec{\mathbf{B}}$ form circles about the wire. By symmetry, the magnitude of $\vec{\mathbf{B}}$ is the same everywhere on a circular path centered on the wire and lying in a plane perpendicular to the wire. By varying the current and distance from the wire, it can be experimentally determined that $\vec{\mathbf{B}}$ is proportional to the current and inversely proportional to the distance from the wire. These observations lead to a mathematical expression for the strength of the magnetic field due to the current I in a long, straight wire:

◀ Magnetic field due to a long, straight wire

$$B = \frac{\mu_0 I}{2\pi r} \qquad [19.11]$$

The proportionality constant μ_0, called the **permeability of free space**, has the value

$$\mu_0 \equiv 4\pi \times 10^{-7}\,\mathrm{T \cdot m/A} \qquad [19.12]$$

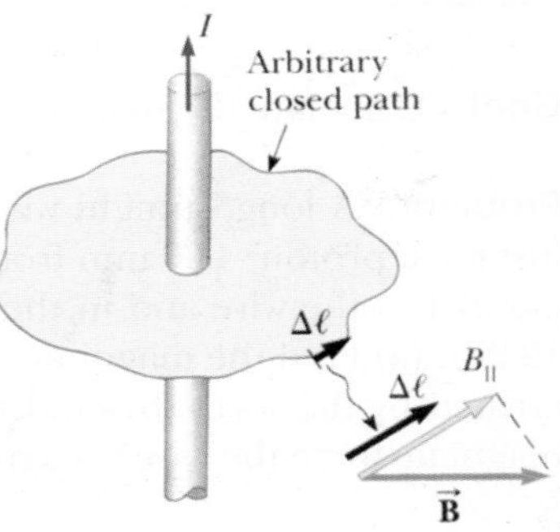

Figure 19.25 An arbitrary closed path around a current is used to calculate the magnetic field of the current by the use of Ampère's rule.

Ampère's Law and a Long, Straight Wire

Equation 19.11 enables us to calculate the magnetic field due to a long, straight wire carrying a current. A general procedure for deriving such equations was proposed by the French scientist André-Marie Ampère (1775–1836); it provides a relation between the current in an arbitrarily shaped wire and the magnetic field produced by the wire.

Consider an arbitrary closed path surrounding a current as in Figure 19.25. The path consists of many short segments, each of length $\Delta\ell$. Multiply one of these lengths by the component of the magnetic field parallel to that segment, where the product is labeled $B_{\parallel}\,\Delta\ell$. According to Ampère, the sum of all such products over the closed path is equal to μ_0 times the net current I that passes through the surface bounded by the closed path. This statement, known as **Ampère's circuital law**, can be written

$$\Sigma B_{\parallel}\,\Delta\ell = \mu_0 I \qquad [19.13]$$

where $B_{\parallel}$ is the component of $\vec{\mathbf{B}}$ parallel to the segment of length $\Delta\ell$ and $\Sigma B_{\parallel}\,\Delta\ell$ means that we take the sum over all the products $B_{\parallel}\,\Delta\ell$ around the closed path. Ampère's law is the fundamental law describing how electric currents create magnetic fields in the surrounding empty space.

We can use Ampère's circuital law to derive the magnetic field due to a long, straight wire carrying a current I. As discussed earlier, each of the magnetic field lines of this configuration forms a circle with the wire at its centers. The magnetic field is tangent to this circle at every point, and its magnitude has the same value B over the entire circumference of a circle of radius r, so that $B_{\parallel} = B$, as shown in Figure 19.26 (page 642). In calculating the sum $\Sigma B_{\parallel}\,\Delta\ell$ over the circular path, notice that $B_{\parallel}$ can be removed from the sum (because it has the same value B for each element on the circle). Equation 19.13 then gives

$$\sum B_{\parallel}\,\Delta\ell = B_{\parallel}\sum \Delta\ell = B(2\pi r) = \mu_0 I$$

Dividing both sides by the circumference $2\pi r$, we obtain

$$B = \frac{\mu_0 I}{2\pi r}$$

This is identical to Equation 19.11, which is the magnetic field due to the current I in a long, straight wire.

ANDRÉ-MARIE AMPÈRE (1775–1836)

Ampère, a Frenchman, is credited with the discovery of electromagnetism—the relationship between electric currents and magnetic fields. Ampère's genius, particularly in mathematics, became evident by the age of 12, but his personal life was filled with tragedy. His father, a wealthy city official, was guillotined during the French Revolution, and his wife died young, in 1803. Ampère died of pneumonia at the age of 61. His judgment of his life is clear from the epitaph he chose for his gravestone: *Tandem felix* (Happy at last).

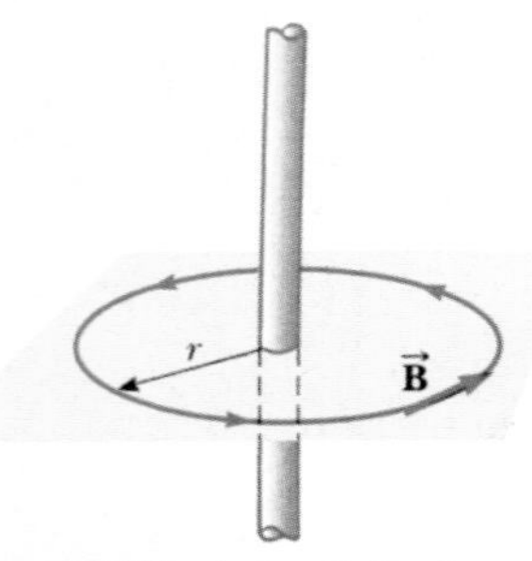

Figure 19.26 A closed circular path of radius r around a long, straight current-carrying wire is used to calculate the magnetic field set up by the wire.

Ampère's circuital law provides an elegant and simple method for calculating the magnetic fields of highly symmetric current configurations. However, it can't easily be used to calculate magnetic fields for complex current configurations that lack symmetry. In addition, Ampère's circuital law in this form is valid only when the currents and fields don't change with time.

EXAMPLE 19.7 The Magnetic Field of a Long Wire

Goal Calculate the magnetic field of a long, straight wire and the force that the field exerts on a particle.

Problem A long, straight wire carries a current of 5.00 A. At one instant, a proton, 4.00 mm from the wire, travels at 1.50×10^3 m/s parallel to the wire and in the same direction as the current (Fig. 19.27). **(a)** Find the magnitude and direction of the magnetic field created by the wire. **(b)** Find the magnitude and direction of the magnetic force the wire's magnetic field exerts on the proton.

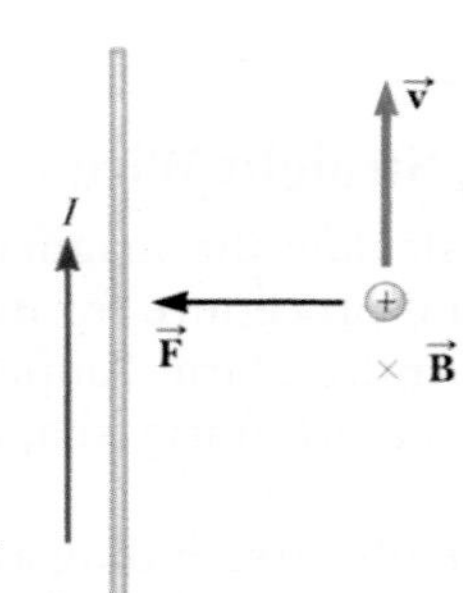

Figure 19.27 (Example 19.7) The magnetic field due to the current is into the page at the location of the proton, and the magnetic force on the proton is to the left.

Strategy First use Equation 19.11 to find the magnitude of the magnetic field at the given point. Use right-hand rule number 2 to find the direction of the field. Finally, substitute into Equation 19.1, computing the magnetic force on the proton.

Solution

(a) Find the magnitude and direction of the wire's magnetic field.

Use Equation 19.11 to calculate the magnitude of the magnetic field 4.00 mm from the wire:

$$B = \frac{\mu_0 I}{2\pi r} = \frac{(4\pi \times 10^{-7}\,\mathrm{T\cdot m/A})(5.00\,\mathrm{A})}{2\pi(4.00 \times 10^{-3}\,\mathrm{m})}$$

$$= 2.50 \times 10^{-4}\,\mathrm{T}$$

Apply right-hand rule number 2 to find the direction of the magnetic field $\vec{\mathbf{B}}$:

With the right thumb pointing in the direction of the current in Figure 19.27, the fingers curl into the page at the location of the proton. The angle θ between $\vec{\mathbf{v}}$ and $\vec{\mathbf{B}}$ is therefore 90°.

(b) Compute the magnetic force exerted by the wire on the proton.

Substitute into Equation 19.1, which gives the magnitude of the magnetic force on a charged particle:

$$F = qvB\sin\theta = (1.60 \times 10^{-19}\,\mathrm{C})(1.50 \times 10^3\,\mathrm{m/s}) \times (2.50 \times 10^{-4}\,\mathrm{T})(\sin 90°)$$

$$= 6.00 \times 10^{-20}\,\mathrm{N}$$

Find the direction of the magnetic force with right-hand rule number 1:

Point your right fingers in the direction of $\vec{\mathbf{v}}$, curling them into the page toward $\vec{\mathbf{B}}$. Your thumb points to the left, which is the direction of the magnetic force.

Remarks The location of the proton is important. On the left-hand side, the wire's magnetic field points outward, and the magnetic force on the proton is to the right.

Exercise 19.7

Find (a) the magnetic field created by the wire and (b) the magnetic force on a helium-3 nucleus located 7.50 mm to the left of the wire in Figure 19.27, traveling 2.50×10^3 m/s opposite the direction of the current. (See the data table presented in Example 19.5 on page 639).

Answers (a) 1.33×10^{-4} T (b) 1.07×10^{-19} N, directed to the left in Figure 19.27.

19.8 MAGNETIC FORCE BETWEEN TWO PARALLEL CONDUCTORS

As we have seen, a magnetic force acts on a current-carrying conductor when the conductor is placed in an external magnetic field. Because a conductor carrying a current creates a magnetic field around itself, it is easy to understand that two current-carrying wires placed close together exert magnetic forces on each other. Consider two long, straight, parallel wires separated by the distance d and carrying currents I_1 and I_2 in the same direction, as shown in Active Figure 19.28. Wire 1 is directly above wire 2. What's the magnetic force on one wire due to a magnetic field set up by the other wire?

In this calculation, we are finding the force on wire 1 due to the magnetic field of wire 2. The current I_2, sets up magnetic field $\vec{\mathbf{B}}_2$ at wire 1. The direction of $\vec{\mathbf{B}}_2$ is perpendicular to the wire, as shown in the figure. Using Equation 19.11, we find that the magnitude of this magnetic field is

$$B_2 = \frac{\mu_0 I_2}{2\pi d}$$

According to Equation 19.5, the magnitude of the magnetic force on wire 1 in the presence of field $\vec{\mathbf{B}}_2$ due to I_2 is

$$F_1 = B_2 I_1 \ell = \left(\frac{\mu_0 I_2}{2\pi d}\right) I_1 \ell = \frac{\mu_0 I_1 I_2 \ell}{2\pi d}$$

We can rewrite this relationship in terms of the force per unit length:

$$\frac{F_1}{\ell} = \frac{\mu_0 I_1 I_2}{2\pi d} \qquad [19.14]$$

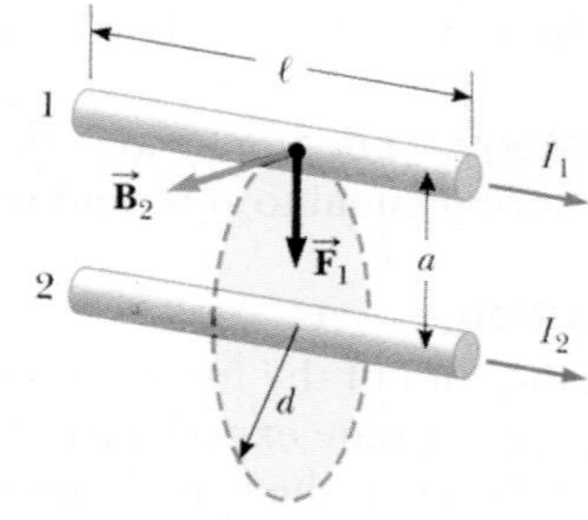

ACTIVE FIGURE 19.28
Two parallel wires, oriented vertically, carry steady currents and exert forces on each other. The field $\vec{\mathbf{B}}_2$ at wire 1 due to wire 2 produces a force on wire 1 given by $F_1 = B_2 I_1 \ell$. The force is attractive if the currents have the same direction, as shown, and repulsive if the two currents have opposite directions.

Physics Now™
Log into PhysicsNow at **www.cp7e.com** and go to Active Figure 19.28, where you can adjust the currents in the wires and the distance between them, and see the effect on the force.

The direction of $\vec{\mathbf{F}}_1$ is downward, toward wire 2, as indicated by right-hand rule number 1. This calculation is completely symmetric, which means that the force $\vec{\mathbf{F}}_2$ on wire 2 is equal to and opposite $\vec{\mathbf{F}}_1$, as expected from Newton's third law of action–reaction.

We have shown that parallel conductors carrying currents in the same direction *attract* each other. You should use the approach indicated by Figure 19.28 and the steps leading to Equation 19.14 to show that parallel conductors carrying currents in opposite directions *repel* each other.

The force between two parallel wires carrying a current is used to define the SI unit of current, the **ampere** (A), as follows:

◀ Definition of the ampere

If two long, parallel wires 1 m apart carry the same current, and the magnetic force per unit length on each wire is 2×10^{-7} N/m, then the current is defined to be 1 A.

The SI unit of charge, the **coulomb** (C), can now be defined in terms of the ampere as follows:

◀ Definition of the coulomb

If a conductor carries a steady current of 1 A, then the quantity of charge that flows through any cross section in 1 s is 1 C.

Quick Quiz 19.5

If, in Figure 19.28, $I_1 = 2$ A and $I_2 = 6$ A, which of the following is true? (a) $F_1 = 3F_2$ (b) $F_1 = F_2$ or (c) $F_1 = F_2/3$

EXAMPLE 19.8 Levitating a Wire

Goal Calculate the magnetic force of one current-carrying wire on a parallel current-carrying wire.

Problem Two wires, each having a weight per unit length of 1.00×10^{-4} N/m, are parallel with one directly above the other. Assume that the wires carry currents that are equal in magnitude and opposite in direction. The wires are 0.10 m apart, and the sum of the magnetic force and gravitational force on the upper wire is zero. Find the current in the wires. (Neglect Earth's magnetic field.)

Strategy The upper wire must be in equilibrium under the forces of magnetic repulsion and gravity. Set the sum of the forces equal to zero and solve for the unknown current, I.

Solution

Set the sum of the forces equal to zero, and substitute the appropriate expressions. Notice that the magnetic force between the wires is repulsive.

$$\vec{\mathbf{F}}_{\text{grav}} + \vec{\mathbf{F}}_{\text{mag}} = 0$$

$$-mg + \frac{\mu_0 I_1 I_2}{2\pi d}\ell = 0$$

The currents are equal, so $I_1 = I_2 = I$. Make these substitutions and solve for I^2.

$$\frac{\mu_0 I^2}{2\pi d}\ell = mg \quad \rightarrow \quad I^2 = \frac{(2\pi d)(mg/\ell)}{\mu_0}$$

Substitute given values, finding I^2, then take the square root. Notice that the weight per unit length, mg/ℓ, is given.

$$I^2 = \frac{(2\pi \cdot 0.100 \text{ m})(1.00 \times 10^{-4} \text{ N/m})}{(4\pi \times 10^{-7} \text{ T} \cdot \text{m})} = 50.0 \text{ A}^2$$

$$I = 7.07 \text{ A}$$

Remark Exercise 19.3 showed that using the Earth's magnetic field to levitate a wire required extremely large currents. Currents in wires can create much stronger magnetic fields than Earth's in regions near the wire.

Exercise 19.8

If the current in each wire is doubled, how far apart should the wires be placed if the magnitudes of the gravitational and magnetic forces on the upper wire are to be equal?

Answer 0.400 m

19.9 MAGNETIC FIELDS OF CURRENT LOOPS AND SOLENOIDS

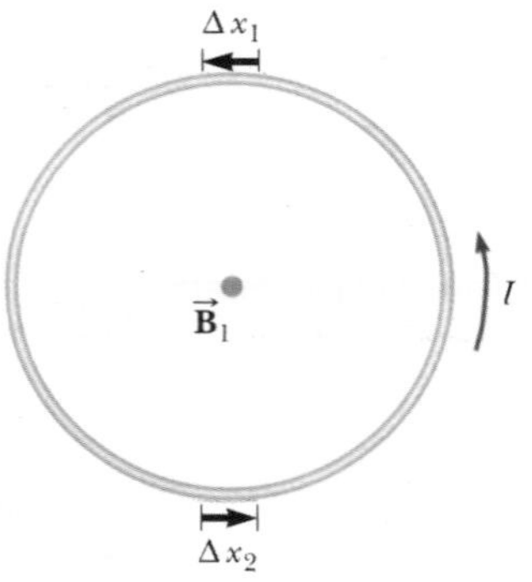

The strength of the magnetic field set up by a piece of wire carrying a current can be enhanced at a specific location if the wire is formed into a loop. You can understand this by considering the effect of several small segments of the current loop, as in Figure 19.29. The small segment at the top of the loop, labeled Δx_1, produces a magnetic field of magnitude B_1 at the loop's center, directed out of the page. The direction of $\vec{\mathbf{B}}$ can be verified using right-hand rule number 2 for a long, straight wire. Imagine holding the wire with your right hand, with your thumb pointing in the direction of the current. Your fingers then curl around in the direction of $\vec{\mathbf{B}}$.

Δx_1

$\vec{\mathbf{B}}_1$ I

Δx_2

Figure 19.29 All segments of the current loop produce a magnetic field at the center of the loop, directed *out of the page.*

A segment of length Δx_2 at the bottom of the loop also contributes to the field at the center, increasing its strength. The field produced at the center by the segment Δx_2 has the same magnitude as B_1 and is also directed out of the page. Similarly, all other such segments of the current loop contribute to the field. The net effect is a magnetic field for the current loop as pictured in Figure 19.30a.

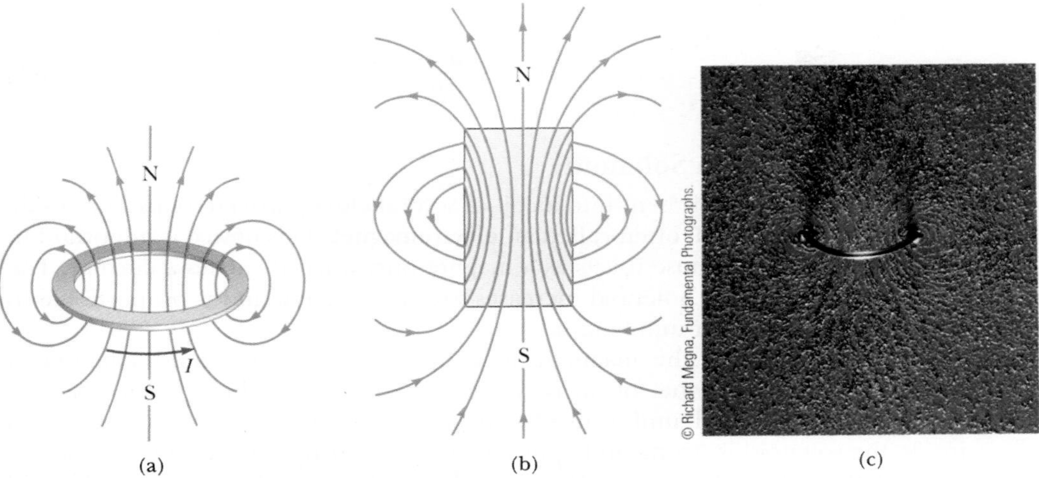

Figure 19.30 (a) Magnetic field lines for a current loop. Note that the lines resemble those of a bar magnet. (b) The magnetic field of a bar magnet is similar to that of a current loop. (c) Field lines of a current loop, displayed by iron filings.

Notice in Figure 19.30a that the magnetic field lines enter at the bottom of the current loop and exit at the top. Compare this to Figure 19.30b, illustrating the field of a bar magnet. The two fields are similar. One side of the loop acts as though it were the north pole of a magnet, and the other acts as a south pole. The similarity of these two fields will be used to discuss magnetism in matter in an upcoming section.

Applying Physics 19.4 Twisted Wires

In electrical circuits, it is often the case that insulated wires carrying currents in opposite directions are twisted together. What is the advantage of doing this?

Explanation If the wires are not twisted together, the combination of the two wires forms a current loop, which produces a relatively strong magnetic field. This magnetic field generated by the loop could be strong enough to affect adjacent circuits or components. When the wires are twisted together, their magnetic fields tend to cancel.

The magnitude of the magnetic field at the center of a circular loop carrying current I as in Figure 19.30a is given by

$$B = \frac{\mu_0 I}{2R}$$

This must be derived with calculus. However, it can be shown to be reasonable by calculating the field at the center of four long wires, each carrying current I and forming a square, as in Figure 19.31, with a circle of radius R inscribed within it. Intuitively, this arrangement should give a magnetic field at the center that is similar in magnitude to the field produced by the circular loop. The current in the circular wire is closer to the center, so that wire would have a magnetic field somewhat stronger than just the four legs of the rectangle, but the lengths of the straight wires beyond the rectangle compensate for this. Each wire contributes the same magnetic field at the exact center, so the total field is given by

$$B = 4 \times \frac{\mu_0 I}{2\pi R} = \frac{4}{\pi}\left(\frac{\mu_0 I}{2R}\right) = (1.27)\left(\frac{\mu_0 I}{2R}\right)$$

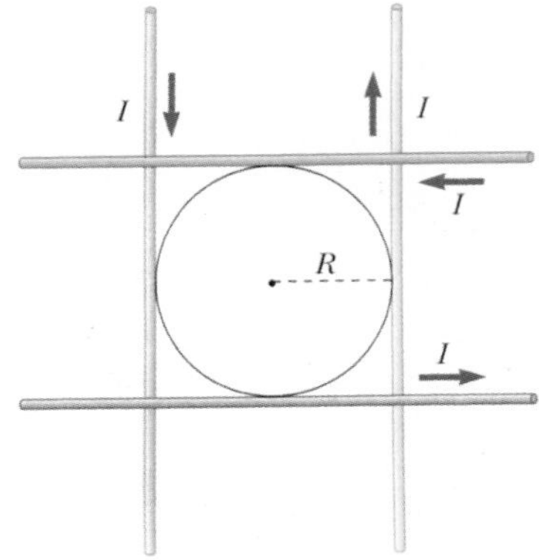

Figure 19.31 The field of a circular loop carrying current I can be approximated by the field due to four straight wires, each carrying current I.

This is *approximately* the same as the field produced by the circular loop of current.

When the coil has N loops, each carrying current I, the magnetic field at the

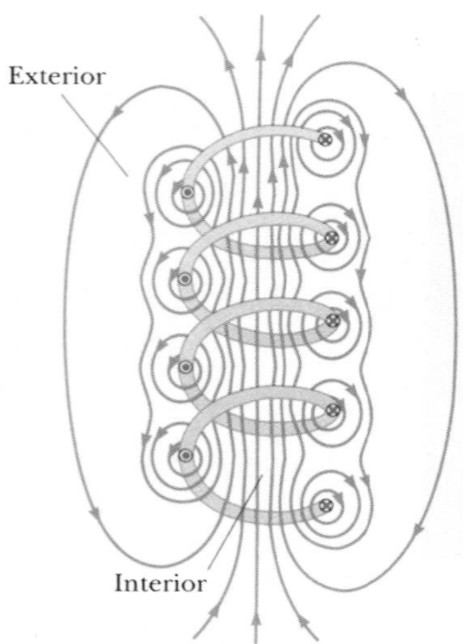

Figure 19.32 The magnetic field lines for a loosely wound solenoid.

center is given by

$$B = N\frac{\mu_0 I}{2R} \qquad [19.15]$$

Magnetic Field of a Solenoid

If a long, straight wire is bent into a coil of several closely spaced loops, the resulting device is a **solenoid**, often called an **electromagnet**. This device is important in many applications because it acts as a magnet only when it carries a current. The magnetic field inside a solenoid increases with the current and is proportional to the number of coils per unit length.

Figure 19.32 shows the magnetic field lines of a loosely wound solenoid of length ℓ and total number of turns N. Notice that the field lines inside the solenoid are nearly parallel, uniformly spaced, and close together. As a result, the field inside the solenoid is strong and approximately uniform. The exterior field at the sides of the solenoid is nonuniform, much weaker than the interior field, and *opposite in direction* to the field inside the solenoid.

If the turns are closely spaced, the field lines are as shown in Figure 19.33a, entering at one end of the solenoid and emerging at the other. This means that one end of the solenoid acts as a north pole and the other end acts as a south pole. If the length of the solenoid is much greater than its radius, the lines that leave the north end of the solenoid spread out over a wide region before returning to enter the south end. The more widely separated the field lines are, the weaker the field. This is in contrast to a much stronger field *inside* the solenoid, where the lines are close together. Also, the field inside the solenoid has a constant magnitude at all points far from its ends. As will be shown subsequently, these considerations allow the application of Ampere's law to the solenoid, giving a result of

The magnetic field inside a solenoid ▶

$$B = \mu_0 n I \qquad [19.16]$$

for the field inside the solenoid, where $n = N/\ell$ is the number of turns per unit length of the solenoid.

So-called steering magnets placed along the neck of the picture tube in a televi-

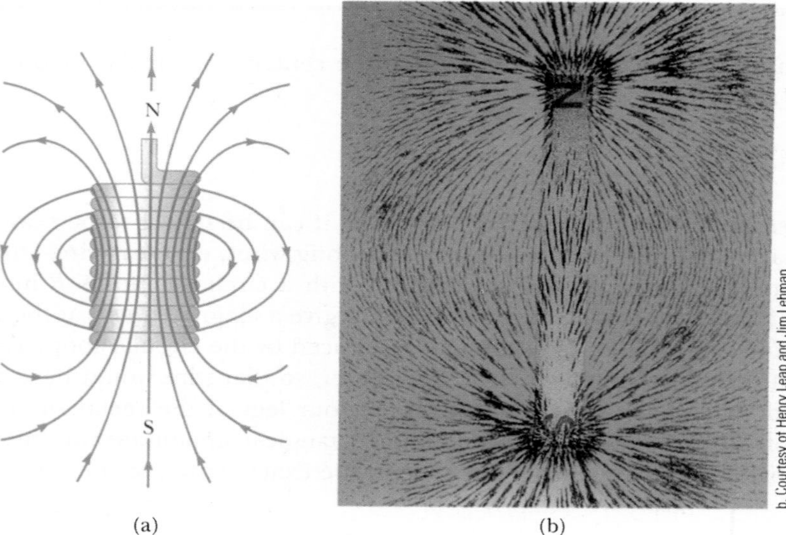

(a) (b)

Figure 19.33 (a) Magnetic field lines for a tightly wound solenoid of finite length carrying a steady current. The field inside the solenoid is nearly uniform and strong. Note that the field lines resemble those of a bar magnet, so the solenoid effectively has north and south poles. (b) The magnetic field pattern of a bar magnet, displayed by small iron filings on a sheet of paper.

EXAMPLE 19.9 The Magnetic Field inside a Solenoid

Goal Calculate the magnetic field of a solenoid from given data and the momentum of a charged particle in this field.

Problem A certain solenoid consists of 100 turns of wire and has a length of 10.0 cm. **(a)** Find the magnitude of the magnetic field inside the solenoid when it carries a current of 0.500 A. **(b)** What is the momentum of a proton orbiting inside the solenoid in a circle with a radius of 0.020 m? The axis of the solenoid is perpendicular to the plane of the orbit. **(c)** Approximately how much wire would be needed to build this solenoid? Assume the solenoid's radius is 5.00 cm.

Strategy In part (a), calculate the number of turns per meter and substitute that and given information into Equation 19.16, getting the magnitude of the magnetic field. Part (b) is an application of Newton's second law.

Solution

(a) Find the magnitude of the magnetic field inside the solenoid when it carries a current of 0.500 A.

Calculate the number of turns per unit length:

$$n = \frac{N}{\ell} = \frac{100 \text{ turns}}{0.100 \text{ m}} = 1.00 \times 10^3 \text{ turns/m}$$

Substitute n and I into Equation 19.16 to find the magnitude of the magnetic field:

$$\begin{aligned} B &= \mu_0 n I \\ &= (4\pi \times 10^{-7} \text{ T}\cdot\text{m/A})(1.00 \times 10^3 \text{ turns/m})(0.500 \text{ A}) \\ &= 6.28 \times 10^{-4} \text{ T} \end{aligned}$$

(b) Find the momentum of a proton orbiting in a circle of radius 0.020 m near the center of the solenoid.

Write Newton's second law for the proton:

$$ma = F = qvB$$

Substitute the centripetal acceleration $a = v^2/r$:

$$m\frac{v^2}{r} = qvB$$

Cancel one factor of v on both sides and multiply by r, getting the momentum mv:

$$mv = rqB = (0.020 \text{ m})(1.60 \times 10^{-19} \text{ C})(6.28 \times 10^{-4} \text{ T})$$

$$p = mv = 2.01 \times 10^{-24} \text{ kg}\cdot\text{m/s}$$

(c) Approximately how much wire would be needed to build this solenoid?

Multiply the number of turns by the circumference of one loop:

$$\begin{aligned} \text{length of wire} &\approx (\text{number of turns})(2\pi r) \\ &= (1.00 \times 10^2 \text{ turns})(2\pi \cdot 0.0500 \text{ m}) \\ &= 31.4 \text{ m} \end{aligned}$$

Remarks An electron in part (b) would have the same momentum as the proton, but a much higher speed. It would also orbit in the opposite direction. The length of wire in part (c) is only an estimate, because the wire has a certain thickness, slightly increasing the size of each loop. In addition the wire loops aren't perfect circles, because they wind slowly up along the solenoid.

Exercise 19.9

Suppose you have a 32.0-m length of copper wire. If the wire is wrapped into a solenoid 0.240 m long and having a radius of 0.0400 m, how strong is the resulting magnetic field in its center when the current is 12.0 A?

Answer 8.00×10^{-3} T

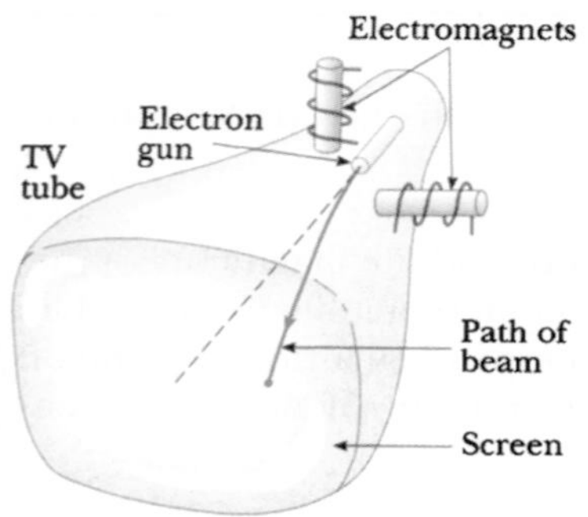

Figure 19.34 Electromagnets are used to deflect electrons to desired positions on the screen of a television tube.

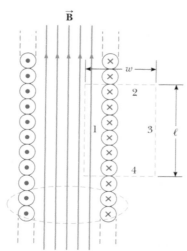

sion set, as in Figure 19.34, are used to make the electron beam move to the desired locations on the screen, tracing out the images. The rate at which the electron beam sweeps over the screen is so fast that to the eye it looks like a picture rather than a sequence of dots.

Ampère's Law Applied to a Solenoid

We can use Ampère's law to obtain the expression for the magnetic field inside a solenoid carrying a current I. A cross section taken along the length of part of our solenoid is shown in Figure 19.35. $\vec{\mathbf{B}}$ inside the solenoid is uniform and parallel to the axis, and $\vec{\mathbf{B}}$ outside is approximately zero. Consider a rectangular path of length L and width w, as shown in the figure. We can apply Ampère's law to this path by evaluating the sum of $B_{\|} \Delta \ell$ over each side of the rectangle. The contribution along side 3 is clearly zero, because $\vec{\mathbf{B}} = 0$ in this region. The contributions from sides 2 and 4 are both zero, because $\vec{\mathbf{B}}$ is perpendicular to $\Delta \ell$ along these paths. Side 1 of length L gives a contribution BL to the sum, because $\vec{\mathbf{B}}$ is uniform along this path, and parallel to $\Delta \ell$. Therefore, the sum over the closed rectangular path has the value

$$\Sigma\, B_{\|}\, \Delta \ell = BL$$

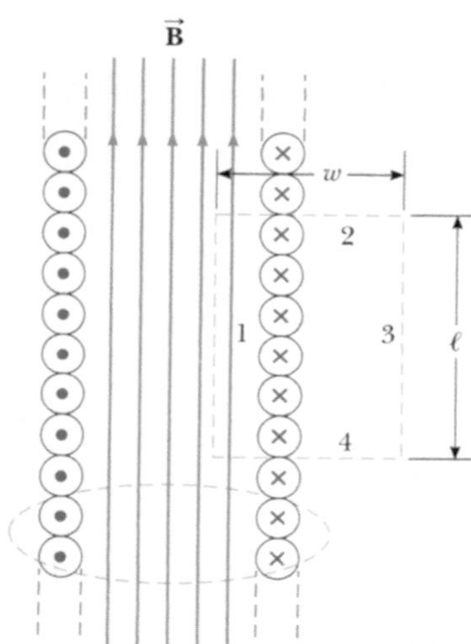

Figure 19.35 A cross-sectional view of a tightly wound solenoid. If the solenoid is long relative to its radius, we can assume that the magnetic field inside is uniform and the field outside is zero. Ampère's law applied to the blue dashed rectangular path can then be used to calculate the field inside the solenoid.

The right side of Ampère's law involves the total current that passes through the area bounded by the path chosen. In this case, the total current through the rectangular path equals the current through each turn of the solenoid, multiplied by the number of turns. If N is the number of turns in the length L, then the total current through the rectangular path equals NI. Ampère's law applied to this path therefore gives

$$\Sigma\, B_{\|}\, \Delta \ell = BL = \mu_0 N I$$

or

$$B = \mu_0 \frac{N}{L} I = \mu_0 n I$$

where $n = N/L$ is the number of turns per unit length.

19.10 MAGNETIC DOMAINS

The magnetic field produced by a current in a coil of wire gives us a hint as to what might cause certain materials to exhibit strong magnetic properties. A single coil like that in Figure 19.30a has a north pole and a south pole, but if this is true for a coil of wire, it should also be true for any current confined to a circular path. In particular, *an individual atom should act as a magnet because of the motion of the electrons about the nucleus.* Each electron, with its charge of 1.6×10^{-19} C, circles the atom once in about 10^{-16} s. If we divide the electric charge by this time interval, we see that the orbiting electron is equivalent to a current of 1.6×10^{-3} A. Such a current produces a magnetic field on the order of 20 T at the center of the circular path. From this we see that a very strong magnetic field would be produced if several of these atomic magnets could be aligned inside a material. This doesn't occur, however, because the simple model we have described is not the complete story. A thorough analysis of atomic structure shows that the magnetic field produced by one electron in an atom is often canceled by an oppositely revolving electron in the same atom. The net result is that **the magnetic effect produced by the electrons orbiting the nucleus is either zero or very small for most materials.**

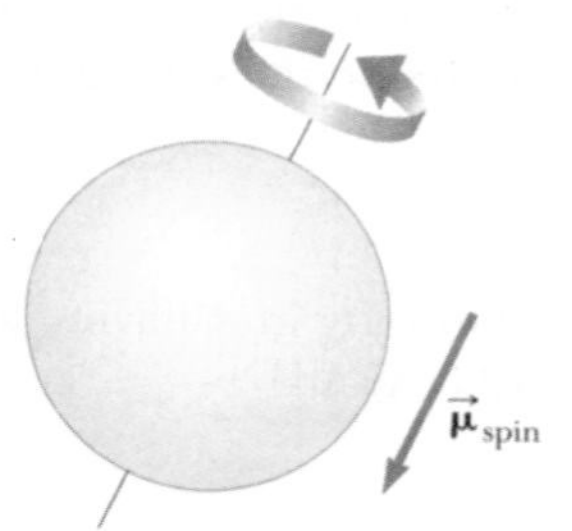

Figure 19.36 Classical model of a spinning electron.

The magnetic properties of many materials can be explained by the fact that an electron not only circles in an orbit, but also spins on its axis like a top, with spin magnetic moment as shown (Fig. 19.36). (This classical description should not be taken too literally. The property of electron *spin* can be understood only in the context of quantum mechanics, which we will not discuss here.) The spinning elec-

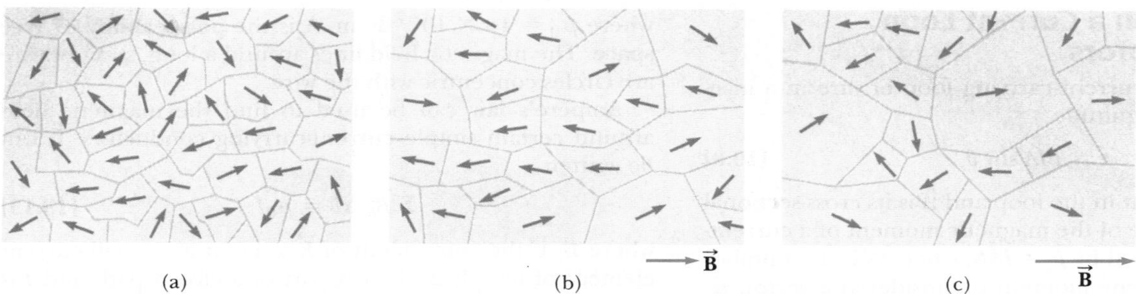

Figure 19.37 (a) Random orientation of domains in an unmagnetized substance. (b) When an external magnetic field $\vec{\mathbf{B}}$ is applied, the domains tend to align with the magnetic field. (c) As the field is made even stronger, the domains not aligned with the external field become very small.

tron represents a charge in motion that produces a magnetic field. The field due to the spinning is generally stronger than the field due to the orbital motion. In atoms containing many electrons, the electrons usually pair up with their spins opposite each other, so that their fields cancel each other. That is why most substances are not magnets. However, in certain strongly magnetic materials, such as iron, cobalt, and nickel, the magnetic fields produced by the electron spins don't cancel completely. Such materials are said to be **ferromagnetic**. In ferromagnetic materials, strong coupling occurs between neighboring atoms, forming large groups of atoms with spins that are aligned. Called **domains**, the sizes of these groups typically range from about 10^{-4} cm to 0.1 cm. In an unmagnetized substance the domains are randomly oriented, as shown in Figure 19.37a. When an external field is applied, as in Figure 19.37b, the magnetic field of each domain tends to come nearer to alignment with the external field, resulting in magnetization.

In what are called hard magnetic materials, domains remain aligned even after the external field is removed; the result is a **permanent magnet**. In soft magnetic materials, such as iron, once the external field is removed, thermal agitation produces motion of the domains and the material quickly returns to an unmagnetized state.

The alignment of domains explains why the strength of an electromagnet is increased dramatically by the insertion of an iron core into the magnet's center. The magnetic field produced by the current in the loops causes the domains to align, thus producing a large net external field. The use of iron as a core is also advantageous because it is a soft magnetic material that loses its magnetism almost instantaneously after the current in the coils is turned off.

TIP 19.4 The Electron Spins—but Doesn't!

Even though we use the word *spin*, the electron, unlike a child's top, isn't physically spinning in this sense. The electron has an intrinsic angular momentum that causes it to act *as if it were spinning*, but the concept of spin angular momentum is actually a relativistic quantum effect.

SUMMARY

PhysicsNow™ Take a practice test by logging into PhysicsNow at **www.cp7e.com** and clicking on the Pre-Test link for this chapter.

19.3 Magnetic Fields

The **magnetic force** that acts on a charge q moving with velocity $\vec{\mathbf{v}}$ in a magnetic field $\vec{\mathbf{B}}$ has magnitude

$$F = qvB \sin \theta \qquad [19.1]$$

where θ is the angle between $\vec{\mathbf{v}}$ and $\vec{\mathbf{B}}$.

To find the direction of this force, use **right-hand rule number 1**: point the fingers of your open right hand in the direction of $\vec{\mathbf{v}}$ and then curl them in the direction of $\vec{\mathbf{B}}$. Your thumb then points in the direction of the magnetic force $\vec{\mathbf{F}}$.

If the charge is *negative* rather than positive, the force is directed opposite the force given by the right-hand rule.

The SI unit of the magnetic field is the **tesla** (T), or weber per square meter (Wb/m^2). An additional commonly used unit for the magnetic field is the **gauss** (G); 1 T = 10^4 G.

19.4 Magnetic Force on a Current-Carrying Conductor

If a straight conductor of length ℓ carries current I, the magnetic force on that conductor when it is placed in a uniform external magnetic field $\vec{\mathbf{B}}$ is

$$F = BI\ell \sin \theta \qquad [19.6]$$

where θ is the angle between the direction of the current and the direction of the magnetic field.

Right-hand rule number 1 also gives the direction of the magnetic force on the conductor. In this case, however, you must point your fingers in the direction of the current rather than in the direction of $\vec{\mathbf{v}}$.

19.5 Torque on a Current Loop and Electric Motors

The torque τ on a current-carrying loop of wire in a magnetic field $\vec{\mathbf{B}}$ has magnitude

$$\tau = BIA \sin \theta \qquad [19.8]$$

where I is the current in the loop and A is its cross-sectional area. The magnitude of the magnetic moment of a current-carrying coil is defined by $\mu = IAN$, where N is the number of loops. The magnetic moment is considered a vector, $\vec{\mu}$, that is perpendicular to the plane of the loop. The angle between $\vec{\mathbf{B}}$ and $\vec{\mu}$ is θ.

19.6 Motion of a Charged Particle in a Magnetic Field

If a charged particle moves in a uniform magnetic field so that its initial velocity is perpendicular to the field, it will move in a circular path in a plane perpendicular to the magnetic field. The radius r of the circular path can be found from Newton's second law and centripetal acceleration, and is given by

$$r = \frac{mv}{qB}. \qquad [19.10]$$

where m is the mass of the particle and q is its charge.

19.7 Magnetic Field of a Long, Straight Wire and Ampère's Law

The magnetic field at distance r from a **long, straight wire** carrying current I has the magnitude

$$B = \frac{\mu_0 I}{2\pi r} \qquad [19.11]$$

where $\mu_0 = 4\pi \times 10^{-7}\ \mathrm{T \cdot m/A}$ is the **permeability of free space**. The magnetic field lines around a long, straight wire are circles concentric with the wire.

Ampère's law can be used to find the magnetic field around certain simple current-carrying conductors. It can be written

$$\Sigma B_{\parallel}\, \Delta \ell = \mu_0 I \qquad [19.13]$$

where $B_{\parallel}$ is the component of $\vec{\mathbf{B}}$ tangent to a small current element of length $\Delta \ell$ that is part of a closed path and I is the total current that penetrates the closed path.

19.8 Magnetic Force between Two Parallel Conductors

The force per unit length on each of two parallel wires separated by the distance d and carrying currents I_1 and I_2 has the magnitude

$$\frac{F}{\ell} = \frac{\mu_0 I_1 I_2}{2\pi d} \qquad [19.14]$$

The forces are attractive if the currents are in the same direction and repulsive if they are in opposite directions.

19.9 Magnetic Field of Current Loops and Solenoids

The magnetic field at the center of a coil of N circular loops of radius R, each carrying current I, is given by

$$B = N\frac{\mu_0 I}{2R} \qquad [19.15]$$

The magnetic field inside a solenoid has the magnitude

$$B = \mu_0 n I \qquad [19.16]$$

where $n = N/\ell$ is the number of turns of wire per unit length.

CONCEPTUAL QUESTIONS

1. In your home television set, a beam of electrons moves from the back of the picture tube to the screen, where it strikes a fluorescent dot that glows with a particular color when hit. The Earth's magnetic field at the location of the television set is horizontal and toward the north. In which direction(s) should the set be oriented so that the beam undergoes the largest deflection?
2. Can a constant magnetic field set a proton at rest into motion? Explain your answer.
3. A proton moving horizontally enters a region where a uniform magnetic field is directed perpendicular to the proton's velocity, as shown in Figure Q19.3. Describe the subsequent motion of the proton. How would an electron behave under the same circumstances.

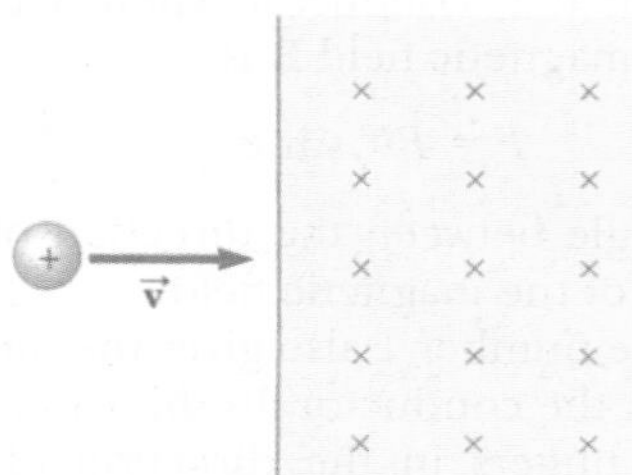

Figure Q19.3

4. No magnetic force acts upon a current-carrying conductor when it is placed in a certain manner in a uniform magnetic field. Explain.
5. How can the motion of a charged particle be used to distinguish between a magnetic field and an electric field in a certain region?
6. Which way would a compass point if you were at Earth's north magnetic pole?
7. Why does the picture on a television screen become distorted when a magnet is brought near the screen as in Figure Q19.7? [*Caution:* You should not do this at home on a color television set, because it may permanently affect the picture quality.]

Figure Q19.7

8. A magnet attracts a piece of iron. The iron can then attract another piece of iron. On the basis of domain alignment, explain what happens in each piece of iron.
9. A Hindu ruler once suggested that he be entombed in a magnetic coffin with the polarity arranged so that he could be forever suspended between heaven and Earth. Is such magnetic levitation possible? Discuss.
10. Will a nail be attracted to either pole of a magnet? Explain what is happening inside the nail when it is placed near the magnet.
11. Suppose you move along a wire at the same speed as the drift speed of the electrons in the wire. Do you now measure a magnetic field of zero?
12. Describe the change in the magnetic field in the space enclosed by a solenoid carrying a steady current I if (a) the length of the solenoid is doubled, but the number of turns remains the same, and (b) the number of turns is doubled, but the length remains the same.
13. Can you use a compass to detect the currents in wires in the walls near light switches in your home?
14. Why do charged particles from outer space, called cosmic rays, strike Earth more frequently at the poles than at the equator?
15. Two wires carry currents in opposite directions and are oriented parallel, with one above the other. The wires repel each other. Is the upper wire in a stable levitation over the lower wire? Suppose the current in one wire is reversed, so that the wires now attract. Is the lower wire hanging in a stable attraction to the upper wire?
16. How can a current loop be used to determine the presence of a magnetic field in a given region of space?
17. A hanging Slinky® toy is attached to a powerful battery and a switch. When the switch is closed so that the toy now carries current, does the Slinky® compress or expand?
18. Is it possible to orient a current loop in a uniform magnetic field such that the loop will not tend to rotate?
19. Parallel wires exert magnetic forces on each other. What about perpendicular wires? Imagine two wires oriented perpendicular to each other and almost touching. Each wire carries a current. Is there a force between the wires?
20. Is the magnetic field created by a current loop uniform? Explain.
21. The electron beam in Figure Q19.21 is projected to the right. The beam deflects downward in the presence of a magnetic field produced by a pair of current-carrying coils. (a) What is the direction of the magnetic field? (b) What would happen to the beam if the magnetic field were reversed in direction?

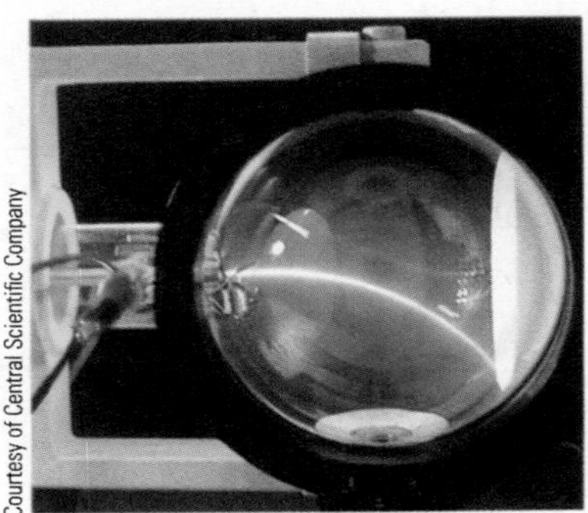

Figure Q19.21

22. Figure Q19.22 shows two permanent magnets, each having a hole through its center. Note that the upper magnet is levitated above the lower one. (a) How does this occur? (b) What purpose does the pencil serve? (c) What can you say about the poles of the magnets from this observation? (d) If the upper magnet were inverted, what do you suppose would happen?

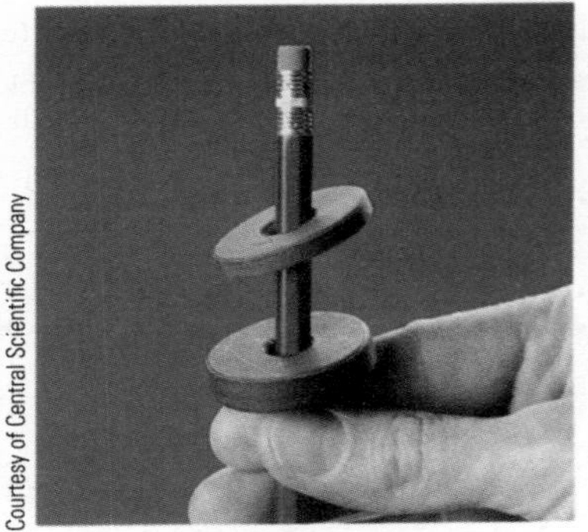

Figure Q19.22

PROBLEMS

1, 2, 3 = straightforward, intermediate, challenging ☐ = full solution available in *Student Solutions Manual/Study Guide*
Physics Now™ = coached problem with hints available at **www.cp7e.com** = biomedical application

Section 19.3 Magnetic Fields

1. An electron gun fires electrons into a magnetic field directed straight downward. Find the direction of the force exerted by the field on an electron for each of the following directions of the electron's velocity: (a) horizontal and due north; (b) horizontal and 30° west of north; (c) due north, but at 30° below the horizontal; (d) straight upward. (Remember that an electron has a negative charge.)
2. (a) Find the direction of the force on a proton (a positively charged particle) moving through the magnetic fields in Figure P19.2 (page 652), as shown. (b) Repeat part (a), assuming the moving particle is an electron.
3. Find the direction of the magnetic field acting on the positively charged particle moving in the various situations shown in Figure P19.3 (page 652) if the direction of the magnetic force acting on it is as indicated.
4. Determine the initial direction of the deflection of charged particles as they enter the magnetic fields, as shown in Figure P19.4 (page 652).
5. At the equator, near the surface of Earth, the magnetic field is approximately 50.0 μT northward, and the electric field is about 100 N/C downward in fair weather. Find the gravitational, electric, and magnetic forces on an electron with an instantaneous velocity of 6.00×10^6 m/s directed to the east in this environment.
6. The magnetic field of the Earth at a certain location is directed vertically downward and has a magnitude of 50.0 μT. A proton is moving horizontally toward the west

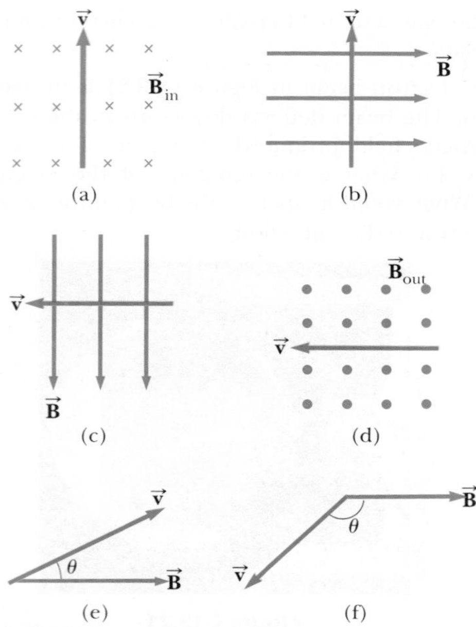

Figure P19.2 (Problems 2 and 12) For Problem 12, replace the velocity vector with a current in that direction.

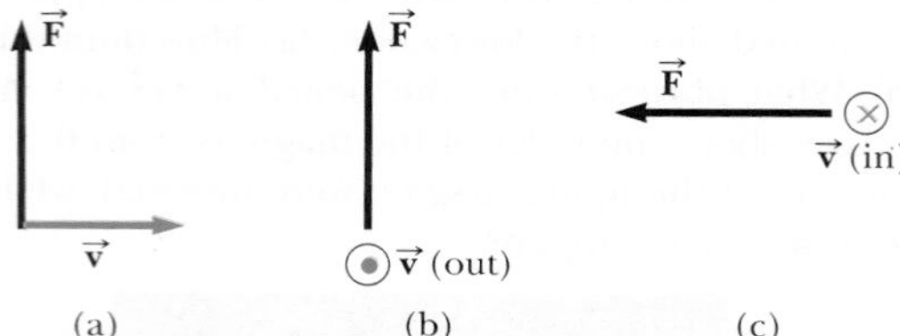

Figure P19.3 (Problems 3 and 13) For Problem 13, replace the velocity vector with a current in that direction.

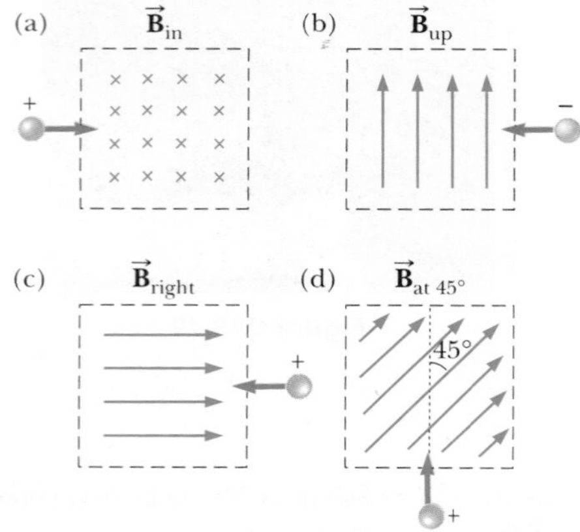

Figure P19.4

in this field with a speed of 6.20×10^6 m/s. What are the direction and magnitude of the magnetic force the field exerts on the proton?

7. **PhysicsNow™** What velocity would a proton need to circle Earth 1 000 km above the magnetic equator, where Earth's magnetic field is directed horizontally north and has a magnitude of 4.00×10^{-8} T?

8. An electron is accelerated through 2 400 V from rest and then enters a region where there is a uniform 1.70-T magnetic field. What are (a) the maximum and (b) the minimum magnitudes of the magnetic force acting on this electron?

9. A proton moves perpendicularly to a uniform magnetic field $\vec{B}$ at 1.0×10^7 m/s and exhibits an acceleration of 2.0×10^{13} m/s² in the $+x$-direction when its velocity is in the $+z$-direction. Determine the magnitude and direction of the field.

10. Sodium ions (Na^+) move at 0.851 m/s through a bloodstream in the arm of a person standing near a large magnet. The magnetic field has a strength of 0.254 T and makes an angle of 51.0° with the motion of the sodium ions. The arm contains 100 cm³ of blood with 3.00×10^{20} Na^+ ions per cubic centimeter. If no other ions were present in the arm, what would be the magnetic force on the arm?

Section 19.4 Magnetic Force on a Current-Carrying Conductor

11. A current $I = 15$ A is directed along the positive x-axis and perpendicular to a magnetic field. A magnetic force per unit length of 0.12 N/m acts on the conductor in the negative y-direction. Calculate the magnitude and direction of the magnetic field in the region through which the current passes.

12. In Figure P19.2, assume that in each case the velocity vector shown is replaced with a wire carrying a current in the direction of the velocity vector. For each case, find the direction of the magnetic force acting on the wire.

13. In Figure P19.3, assume that in each case the velocity vector shown is replaced with a wire carrying a current in the direction of the velocity vector. For each case, find the direction of the magnetic field that will produce the magnetic force shown.

14. A wire having a mass per unit length of 0.500 g/cm carries a 2.00-A current horizontally to the south. What are the direction and magnitude of the minimum magnetic field needed to lift this wire vertically upward?

15. A wire carries a current of 10.0 A in a direction that makes an angle of 30.0° with the direction of a magnetic field of strength 0.300 T. Find the magnetic force on a 5.00-m length of the wire.

16. At a certain location, Earth has a magnetic field of 0.60×10^{-4} T, pointing 75° below the horizontal in a north–south plane. A 10.0-m-long straight wire carries a 15-A current. (a) If the current is directed horizontally toward the east, what are the magnitude and direction of the magnetic force on the wire? (b) What are the magnitude and direction of the force if the current is directed vertically upward?

17. A wire with a mass of 1.00 g/cm is placed on a horizontal surface with a coefficient of friction of 0.200. The wire carries a current of 1.50 A eastward and moves horizontally to the north. What are the magnitude and the direction of the *smallest* vertical magnetic field that enables the wire to move in this fashion?

18. A conductor suspended by two flexible wires as shown in Figure P19.18 has a mass per unit length of 0.040 0 kg/m. What current must exist in the conductor for the tension in the supporting wires to be zero when the magnetic field is 3.60 T into the page? What is the required direction for the current?

19. **PhysicsNow™** An unusual message delivery system is pictured in Figure P19.19. A 15-cm length of conductor that is free to move is held in place between two thin conduc-

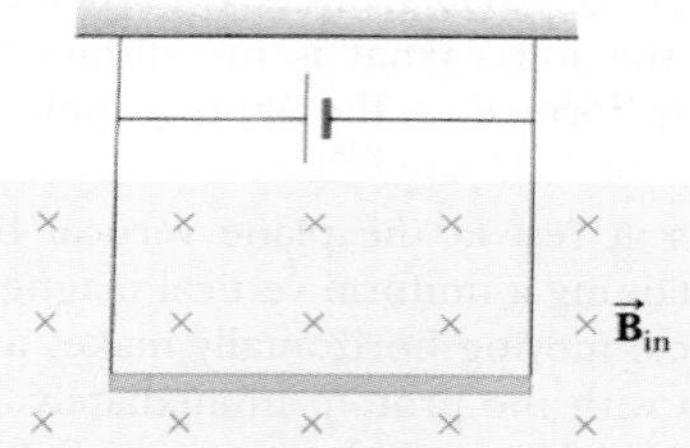

Figure P19.18

tors. When a 5.0-A current is directed as shown in the figure, the wire segment moves upward at a constant velocity. If the mass of the wire is 15 g, find the magnitude and direction of the minimum magnetic field that is required to move the wire. (The wire slides without friction on the two vertical conductors.)

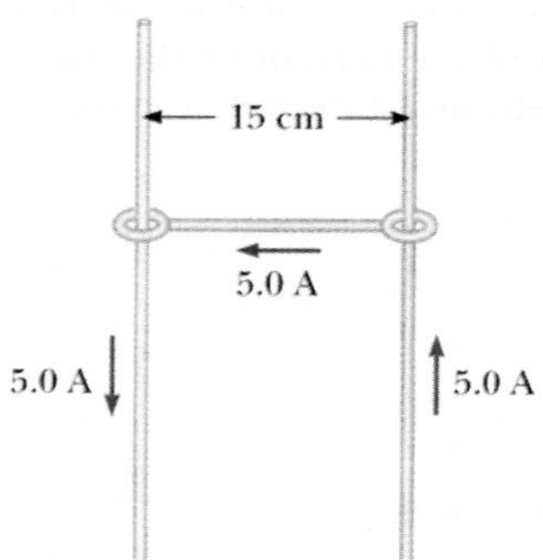

Figure P19.19

20. A wire 2.80 m in length carries a current of 5.00 A in a region where a uniform magnetic field has a magnitude of 0.390 T. Calculate the magnitude of the magnetic force on the wire, assuming the angle between the magnetic field and the current is (a) 60.0°, (b) 90.0°, (c) 120°.

21. In Figure P19.21, the cube is 40.0 cm on each edge. Four straight segments of wire—*ab*, *bc*, *cd*, and *da*—form a closed loop that carries a current $I = 5.00$ A in the direction shown. A uniform magnetic field of magnitude $B = 0.020\ 0$ T is in the positive *y*-direction. Determine the magnitude and direction of the magnetic force on each segment.

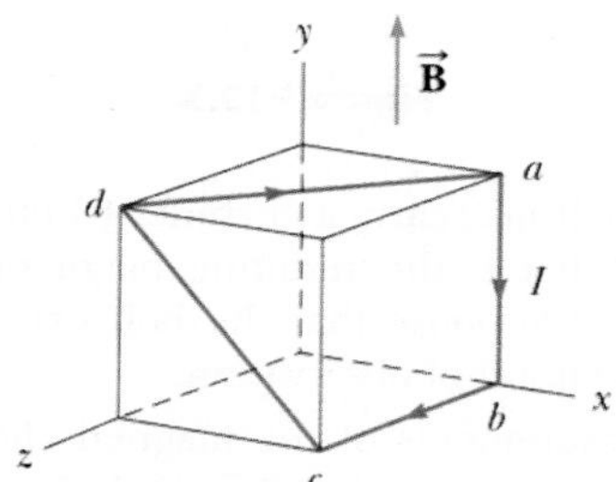

Figure P19.21

Section 19.5 Torque on a Current Loop and Electric Motors

22. A current of 17.0 mA is maintained in a single circular loop with a circumference of 2.00 m. A magnetic field of 0.800 T is directed parallel to the plane of the loop. What is the magnitude of the torque exerted by the magnetic field on the loop?

23. An eight-turn coil encloses an elliptical area having a major axis of 40.0 cm and a minor axis of 30.0 cm (Fig. P19.23). The coil lies in the plane of the page and has a 6.00-A current flowing clockwise around it. If the coil is in a uniform magnetic field of 2.00×10^{-4} T directed toward the left of the page, what is the magnitude of the torque on the coil? [*Hint*: The area of an ellipse is $A = \pi ab$, where a and b are, respectively, the semimajor and semiminor axes of the ellipse.]

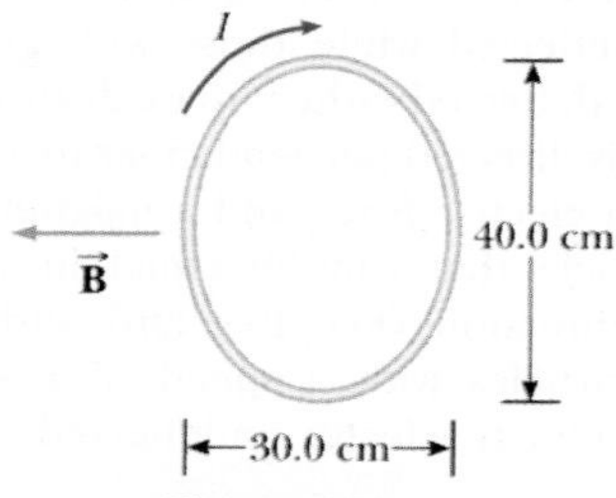

Figure P19.23

24. A rectangular loop consists of 100 closely wrapped turns and has dimensions 0.40 m by 0.30 m. The loop is hinged along the *y*-axis, and the plane of the coil makes an angle of 30.0° with the *x*-axis (Fig. P19.24). What is the magnitude of the torque exerted on the loop by a uniform magnetic field of 0.80 T directed along the *x*-axis when the current in the windings has a value of 1.2 A in the direction shown? What is the expected direction of rotation of the loop?

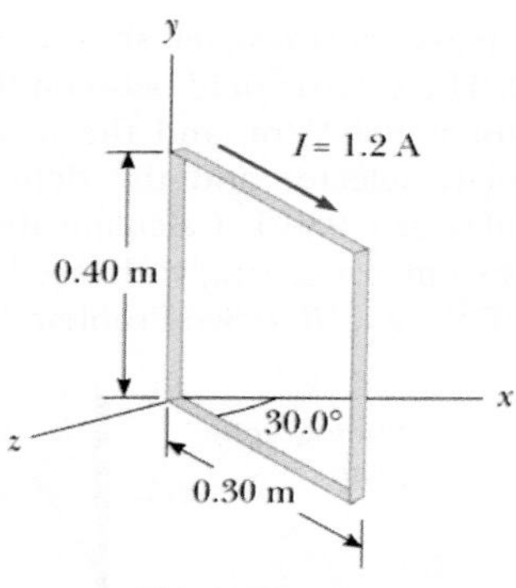

Figure P19.24

25. A long piece of wire with a mass of 0.100 kg and a total length of 4.00 m is used to make a square coil with a side of 0.100 m. The coil is hinged along a horizontal side, carries a 3.40-A current, and is placed in a vertical magnetic field with a magnitude of 0.010 0 T. (a) Determine the angle that the plane of the coil makes with the vertical when the coil is in equilibrium. (b) Find the torque acting on the coil due to the magnetic force at equilibrium.

26. A copper wire is 8.00 m long and has a cross-sectional area of $1.00 \times 10^{-4}\ m^2$. The wire forms a one-turn loop in the shape of square and is then connected to a battery that applies a potential difference of 0.100 V. If the loop is placed in a uniform magnetic field of magnitude 0.400 T, what is the maximum torque that can act on it? The resistivity of copper is $1.70 \times 10^{-8}\ \Omega \cdot m$.

Section 19.6 Motion of a Charged Particle in a Magnetic Field

27. A proton moving freely in a circular path perpendicular to a constant magnetic field takes 1.00 μs to complete one revolution. Determine the magnitude of the magnetic field

28. A cosmic-ray proton in interstellar space has an energy of 10.0 MeV and executes a circular orbit having a radius equal to that of Mercury's orbit around the Sun, which is 5.80×10^{10} m. What is the magnetic field in that region of space?

29. Figure P19.29a is a diagram of a device called a velocity selector, in which particles of a specific velocity pass through undeflected while those with greater or lesser velocities are deflected either upwards or downwards. An electric field is directed perpendicular to a magnetic field, producing an electric force and a magnetic force on the charged particle that can be equal in magnitude and opposite in direction (Fig. P19.29b) and hence cancel. Show that particles with a speed of $v = E/B$ will pass through the velocity selector undeflected.

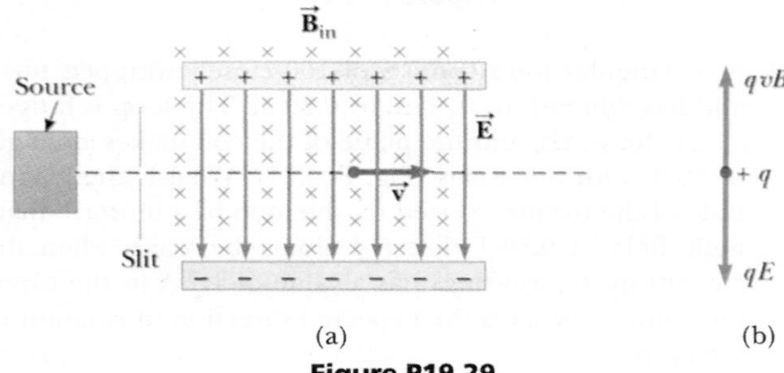

Figure P19.29

30. Consider the mass spectrometer shown schematically in Figure P19.30. The electric field between the plates of the velocity selector is 950 V/m, and the magnetic fields in both the velocity selector and the deflection chamber have magnitudes of 0.930 T. Calculate the radius of the path in the system for a singly charged ion with mass $m = 2.18 \times 10^{-26}$ kg. [*Hint:* See Problem 29.]

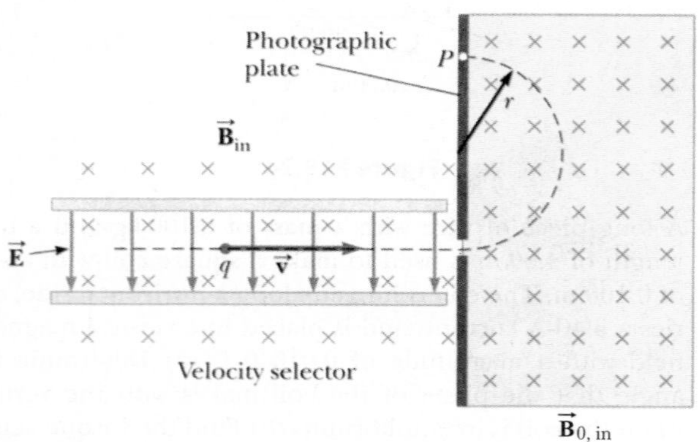

Figure P19.30 A mass spectrometer. Charged particles are first sent through a velocity selector. They then enter a region where a magnetic field $\vec{B}_0$ (directed inward) causes positive ions to move in a semicircular path and strike a photographic film at *P*.

31. A singly charged positive ion has a mass of 2.50×10^{-26} kg. After being accelerated through a potential difference of 250 V, the ion enters a magnetic field of 0.500 T, in a direction perpendicular to the field. Calculate the radius of the path of the ion in the field.

32. A mass spectrometer is used to examine the isotopes of uranium. Ions in the beam emerge from the velocity selector at a speed of 3.00×10^5 m/s and enter a uniform magnetic field of 0.600 T directed perpendicularly to the velocity of the ions. What is the distance between the impact points formed on the photographic plate by singly charged ions of ^{235}U and ^{238}U?

33. A proton is at rest at the plane vertical boundary of a region containing a uniform vertical magnetic field *B*. An alpha particle moving horizontally makes a head-on elastic collision with the proton. Immediately after the collision, both particles enter the magnetic field, moving perpendicular to the direction of the field. The radius of the proton's trajectory is *R*. Find the radius of the alpha particle's trajectory. The mass of the alpha particle is four times that of the proton, and its charge is twice that of the proton.

Section 19.7 Magnetic Field of a Long, Straight Wire and Ampère's Law

34. In each of parts (a), (b), and (c) of Figure P19.34, find the direction of the current in the wire that would produce a magnetic field directed as shown.

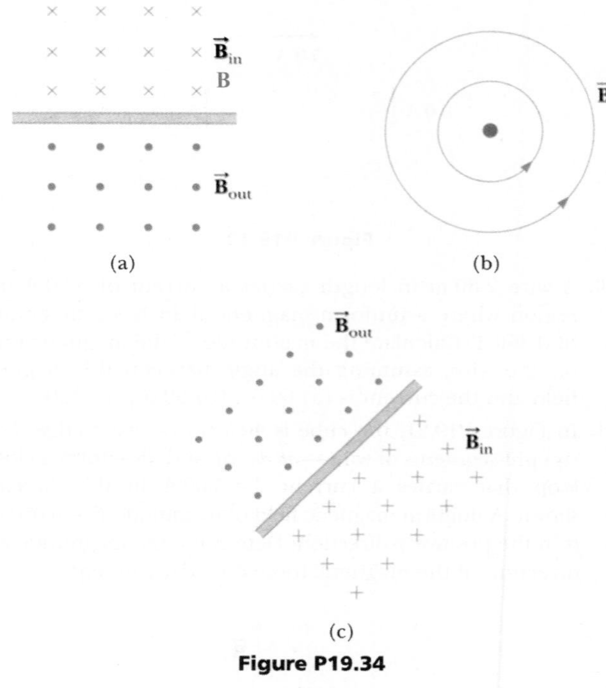

Figure P19.34

35. A lightning bolt may carry a current of 1.00×10^4 A for a short time. What is the resulting magnetic field 100 m from the bolt? Suppose that the bolt extends far above and below the point of observation.

36. In 1962, measurements of the magnetic field of a large tornado were made at the Geophysical Observatory in Tulsa, Oklahoma. If the magnitude of the tornado's field was $B = 1.50 \times 10^{-8}$ T pointing north when the tornado was 9.00 km east of the observatory, what current was carried up or down the funnel of the tornado? Model the vortex as a long, straight wire carrying a current.

37. A cardiac pacemaker can be affected by a static magnetic field as small as 1.7 mT. How close can a pacemaker wearer come to a long, straight wire carrying 20 A?

38. The two wires shown in Figure P19.38 carry currents of 5.00 A in opposite directions and are separated by 10.0 cm. Find the direction and magnitude of the net magnetic

field (a) at a point midway between the wires; (b) at point P_1, 10.0 cm to the right of the wire on the right, and (c) at point P_2, 20.0 cm to the left of the wire on the left.

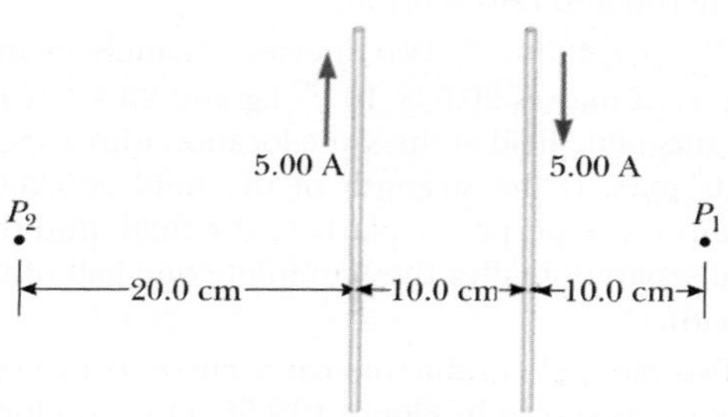

Figure P19.38

39. Four long, parallel conductors carry equal currents of $I = 5.00$ A. Figure P19.39 is an end view of the conductors. The direction of the current is into the page at points A and B (indicated by the crosses) and out of the page at C and D (indicated by the dots). Calculate the magnitude and direction of the magnetic field at point P, located at the center of the square with edge of length 0.200 m.

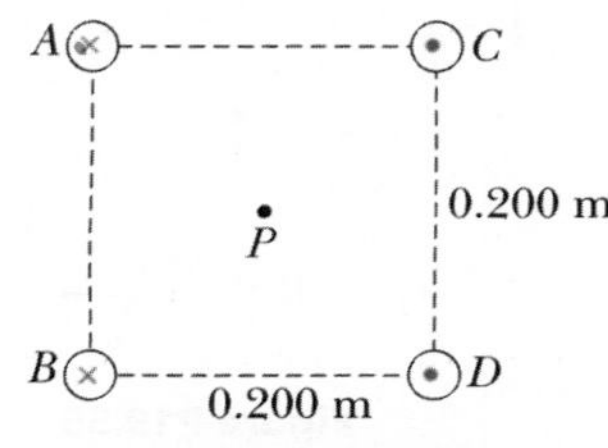

Figure P19.39

40. The two wires in Figure P19.40 carry currents of 3.00 A and 5.00 A in the direction indicated. (a) Find the direction and magnitude of the magnetic field at a point midway between the wires. (b) Find the magnitude and direction of the magnetic field at point P, located 20.0 cm above the wire carrying the 5.00-A current.

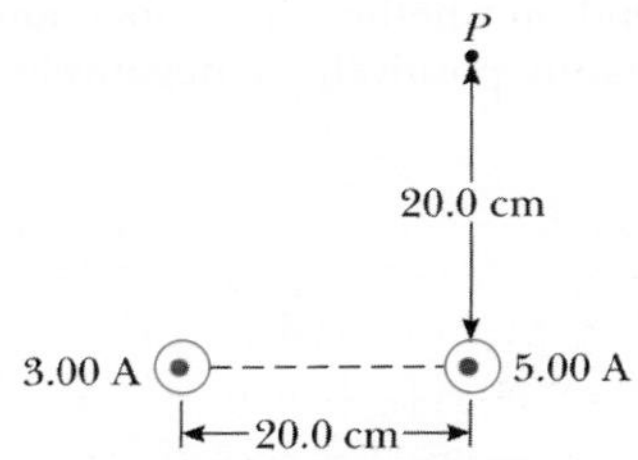

Figure P19.40

41. A wire carries a 7.00-A current along the x-axis, and another wire carries a 6.00-A current along the y-axis, as shown in Figure P19.41. What is the magnetic field at point P, located at $x = 4.00$ m, $y = 3.00$ m?

42. A long, straight wire lies on a horizontal table and carries a current of 1.20 μA. In a vacuum, a proton moves parallel to the wire (opposite the direction of the current) with a constant velocity of 2.30×10^4 m/s at a constant distance d above the wire. Determine the value of d. (You may ignore the magnetic field due to Earth.)

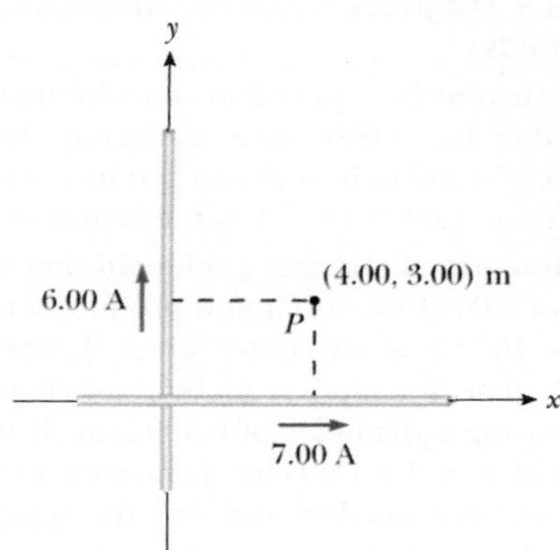

Figure P19.41

43. The magnetic field 40.0 cm away from a long, straight wire carrying current 2.00 A is 1.00 μT. (a) At what distance is it 0.100 μT? (b) At one instant, the two conductors in a long household extension cord carry equal 2.00-A currents in opposite directions. The two wires are 3.00 mm apart. Find the magnetic field 40.0 cm away from the middle of the straight cord, in the plane of the two wires. (c) At what distance is it one-tenth as large? (d) The center wire in a coaxial cable carries current 2.00 A in one direction, and the sheath around it carries current 2.00 A in the opposite direction. What magnetic field does the cable create at points outside?

Section 19.8 Magnetic Force between Two Parallel Conductors

44. Two parallel wires are 10.0 cm apart, and each carries a current of 10.0 A. (a) If the currents are in the same direction, find the force per unit length exerted on one of the wires by the other. Are the wires attracted to or repelled by each other? (b) Repeat the problem with the currents in opposite directions.

45. PhysicsNow™ A wire with a weight per unit length of 0.080 N/m is suspended directly above a second wire. The top wire carries a current of 30.0 A and the bottom wire carries a current of 60.0 A. Find the distance of separation between the wires so that the top wire will be held in place by magnetic repulsion.

46. In Figure P19.46, the current in the long, straight wire is $I_1 = 5.00$ A, and the wire lies in the plane of the rectangular loop, which carries 10.0 A. The dimensions shown are $c = 0.100$ m, $a = 0.150$ m, and $\ell = 0.450$ m. Find the magnitude and direction of the net force exerted by the magnetic field due to the straight wire on the loop.

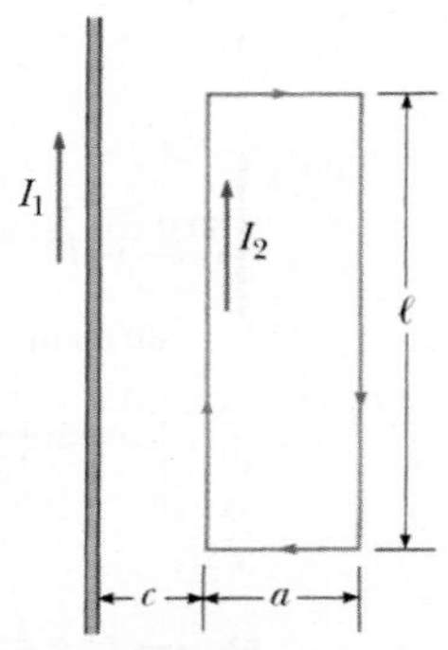

Figure P19.46

Section 19.9 Magnetic Fields of Current Loops and Solenoids

47. What current is required in the windings of a long solenoid that has 1 000 turns uniformly distributed over a length of 0.400 m in order to produce a magnetic field of magnitude 1.00×10^{-4} T at the center of the solenoid?

48. It is desired to construct a solenoid that will have a resistance of 5.00 Ω (at 20°C) and produce a magnetic field of 4.00×10^{-2} T at its center when it carries a current of 4.00 A. The solenoid is to be constructed from copper wire having a diameter of 0.500 mm. If the radius of the solenoid is to be 1.00 cm, determine (a) the number of turns of wire needed and (b) the length the solenoid should have.

49. A single-turn square loop of wire 2.00 cm on a side carries a counterclockwise current of 0.200 A. The loop is inside a solenoid, with the plane of the loop perpendicular to the magnetic field of the solenoid. The solenoid has 30 turns per centimeter and carries a counterclockwise current of 15.0 A. Find the force on each side of the loop and the torque acting on the loop.

50. An electron is moving at a speed of 1.0×10^4 m/s in a circular path of radius of 2.0 cm inside a solenoid. The magnetic field of the solenoid is perpendicular to the plane of the electron's path. Find (a) the strength of the magnetic field inside the solenoid and (b) the current in the solenoid if it has 25 turns per centimeter.

ADDITIONAL PROBLEMS

51. A circular coil consisting of a single loop of wire has a radius of 30.0 cm and carries a current of 25 A. It is placed in an external magnetic field of 0.30 T. Find the torque on the wire when the plane of the coil makes an angle of 35° with the direction of the field.

52. An electron enters a region of magnetic field of magnitude 0.010 0 T, traveling perpendicular to the linear boundary of the region. The direction of the field is perpendicular to the velocity of the electron. (a) Determine the time it takes for the electron to leave the "field-filled" region, noting that its path is a semicircle. (b) Find the kinetic energy of the electron if the radius of its semicircular path is 2.00 cm.

53. Two long, straight wires cross each other at right angles, as shown in Figure P19.53. (a) Find the direction and magnitude of the magnetic field at point P, which is in the same plane as the two wires. (b) Find the magnetic field at a point 30.0 cm above the point of intersection (30.0 cm out of the page, toward you).

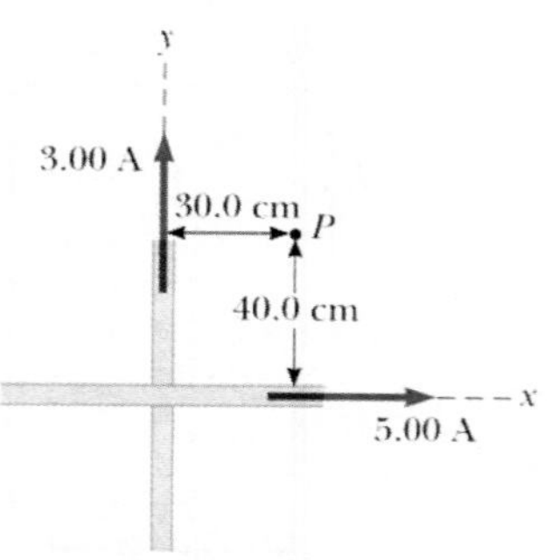

Figure P19.53

54. A 0.200-kg metal rod carrying a current of 10.0 A glides on two horizontal rails 0.500 m apart. What vertical magnetic field is required to keep the rod moving at a constant speed if the coefficient of kinetic friction between the rod and rails is 0.100?

55. PhysicsNow™ Two species of singly charged positive ions of masses 20.0×10^{-27} kg and 23.4×10^{-27} kg enter a magnetic field at the same location with a speed of 1.00×10^5 m/s. If the strength of the field is 0.200 T, and the ions move perpendicularly to the field, find their distance of separation after they complete one-half of their circular path.

56. Two parallel conductors carry currents in opposite directions, as shown in Figure P19.56. One conductor carries a current of 10.0 A. Point A is the midpoint between the wires, and point C is 5.00 cm to the right of the 10.0-A current. I is adjusted so that the magnetic field at C is zero. Find (a) the value of the current I and (b) the value of the magnetic field at A.

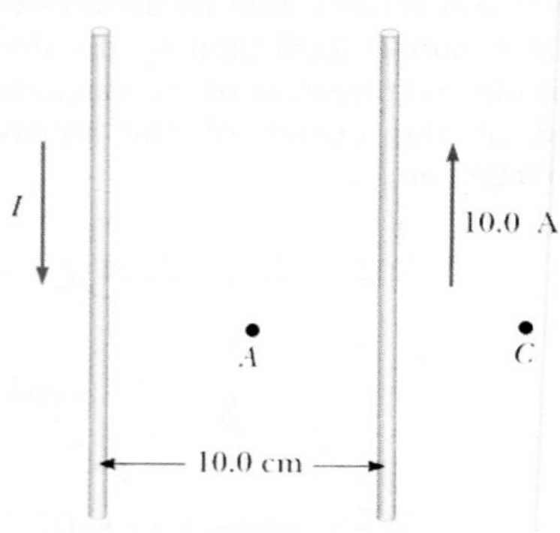

Figure P19.56

57. Using an electromagnetic flowmeter (Fig. P19.57), a heart surgeon monitors the flow rate of blood through an artery. Electrodes A and B make contact with the outer surface of the blood vessel, which has interior diameter 3.00 mm. (a) For a magnetic field magnitude of 0.040 0 T, a potential difference of 160 μV appears between the electrodes. Calculate the speed of the blood. (b) Verify that electrode A is positive, as shown. Does the sign of the emf depend on whether the mobile ions in the blood are predominantly positively or negatively charged? Explain.

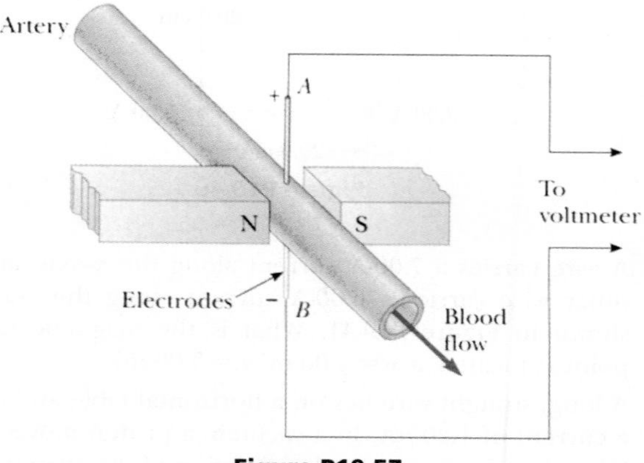

Figure P19.57

58. Two circular loops are parallel, coaxial, and almost in contact 1.00 mm apart (Fig. P19.58). Each loop is 10.0 cm in

radius. The top loop carries a clockwise current of 140 A. The bottom loop carries a counterclockwise current of 140 A. (a) Calculate the magnetic force that the bottom loop exerts on the top loop. (b) The upper loop has a mass of 0.021 0 kg. Calculate its acceleration, assuming that the only forces acting on it are the force in part (a) and its weight. [*Hint*: The distance between the loops is small in comparison to their radius of curvature, so the loops may be treated as long, straight parallel wires.]

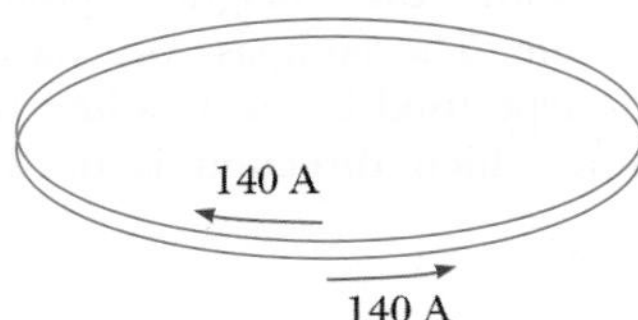

Figure P19.58

59. A 1.00-kg ball having net charge $Q = 5.00\ \mu$C is thrown out of a window horizontally at a speed $v = 20.0$ m/s. The window is at a height $h = 20.0$ m above the ground. A uniform horizontal magnetic field of magnitude $B = 0.010\,0$ T is perpendicular to the plane of the ball's trajectory. Find the magnitude of the magnetic force acting on the ball just before it hits the ground. [*Hint*: Ignore magnetic forces in finding the ball's final velocity.]

60. The idea that static magnetic fields might have a therapeutic value has been around for centuries. A currently available rare-Earth magnet that is advertised to relieve joint pain is shown in Figure P19.60. It is 1.0 mm thick and has a field strength of 5.0×10^{-2} T at the center of the flat surface. The magnetic field strength at points away from the center of the disk is inversely proportional to h^3, where h is the distance from the midplane of the disk. How far from the surface of the disk will the field strength be reduced to that of Earth (5.0×10^{-5} T)?

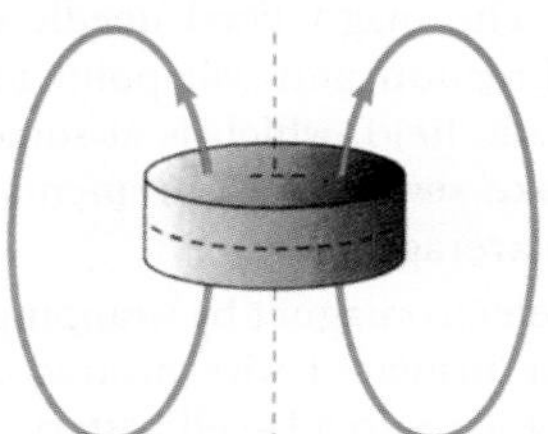

Figure P19.60

61. Two long parallel conductors carry currents $I_1 = 3.00$ A and $I_2 = 3.00$ A, both directed into the page in Figure P19.61. Determine the magnitude and direction of the resultant magnetic field at P.

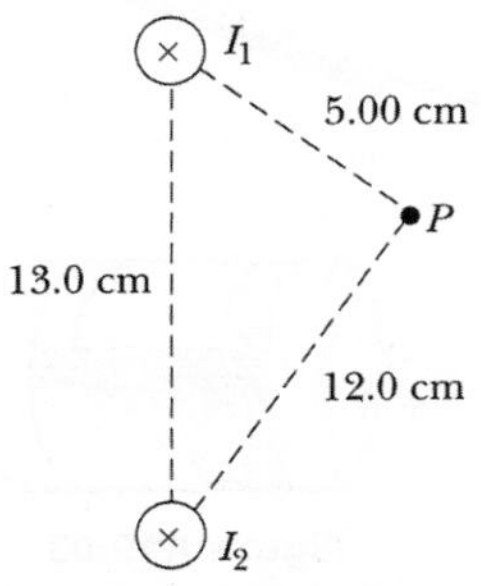

Figure P19.61

62. A uniform horizontal wire with a linear mass density of 0.50 g/m carries a 2.0-A current. It is placed in a constant magnetic field with a strength of 4.0×10^{-3} T. The field is horizontal and perpendicular to the wire. As the wire moves upward starting from rest, (a) what is its acceleration and (b) how long does it take to rise 50 cm? Neglect the magnetic field of Earth.

63. Three long parallel conductors carry currents of $I = 2.0$ A. Figure P19.63 is an end view of the conductors, with each current coming out of the page. Given that $a = 1.0$ cm, determine the magnitude and direction of the magnetic field at points A, B, and C.

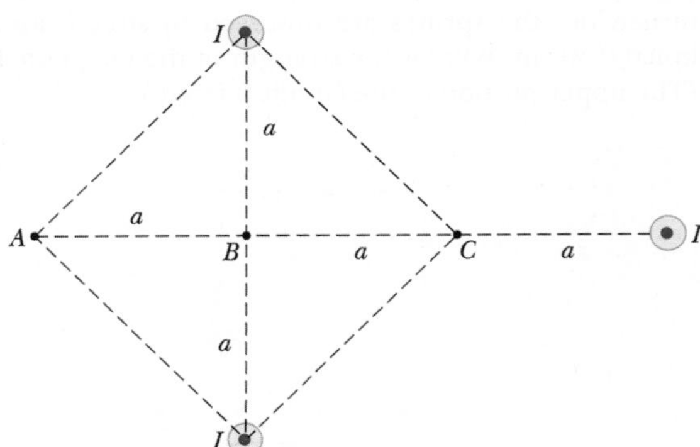

Figure P19.63

64. Two long parallel wires, each with a mass per unit length of 40 g/m, are supported in a horizontal plane by 6.0-cm-long strings, as shown in Figure P19.64. Each wire carries the same current I, causing the wires to repel each other so that the angle θ between the supporting strings is 16°. (a) Are the currents in the same or opposite directions? (b) Determine the magnitude of each current.

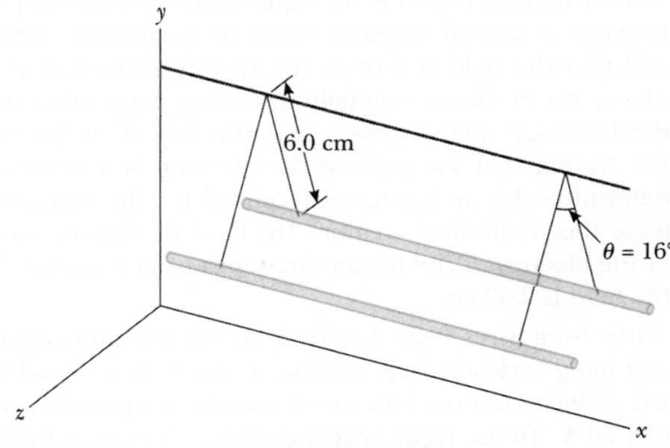

Figure P19.64

65. Protons having a kinetic energy of 5.00 MeV are moving in the positive x-direction and enter a magnetic field of 0.050 0 T in the z-direction, out of the plane of the page, and extending from $x = 0$ to $x = 1.00$ m as in Figure P19.65 (page 658). (a) Calculate the y-component of the protons' momentum as they leave the magnetic field. (b) Find the angle α between the initial velocity vector of the proton beam and the velocity vector after the beam emerges from the field. [*Hint*: Neglect relativistic effects and note that 1 eV = 1.60×10^{-19} J.]

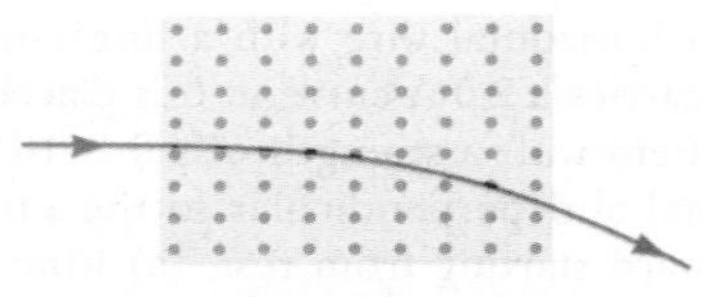

Figure P19.65

66. A straight wire of mass 10.0 g and length 5.0 cm is suspended from two identical springs that, in turn, form a closed circuit (Fig. P19.66). The springs stretch a distance of 0.50 cm under the weight of the wire. The circuit has a total resistance of 12 Ω. When a magnetic field directed out of the page (indicated by the dots in the figure is turned on, the springs are observed to stretch an additional 0.30 cm. What is the strength of the magnetic field? (The upper portion of the circuit is fixed.)

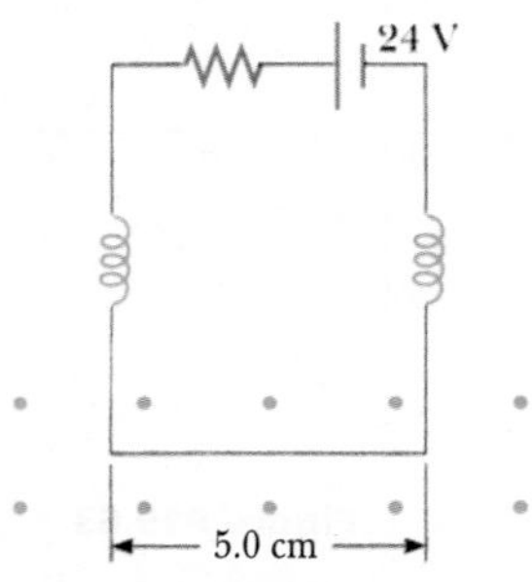

Figure P19.66

67. A solenoid 10.0 cm in diameter and 75.0 cm long is made from copper wire of diameter 0.100 cm with very thin insulation. The wire is wound onto a cardboard tube in a single layer, with adjacent turns touching each other. To produce a field of magnitude 20.0 mT at the center of the solenoid, what power must be delivered to the solenoid?

68. Assume that the region to the right of a certain vertical plane contains a vertical magnetic field of magnitude 1.00 mT and that the field is zero in the region to the left of the plane. An electron, originally traveling perpendicular to the boundary plane, passes into the region of the field. (a) Noting that the path of the electron is a semicircle, determine the time interval required for the electron to leave the "field-filled" region. (b) Find the kinetic energy of the electron if the maximum depth of penetration into the field is 2.00 cm.

69. Three long wires (wire 1, wire 2, and wire 3) are coplanar and hang vertically. The distance between wire 1 and wire 2 is 20.0 cm. On the left, wire 1 carries an upward current of 1.50 A. To the right, wire 2 carries a downward current of 4.00 A. Wire 3 is located such that when it carries a certain current, no net force acts upon any of the wires. Find (a) the position of wire 3 and (b) the magnitude and direction of the current in wire 3.

70. Two long parallel conductors separated by 10.0 cm carry currents in the same direction. The first wire carries a current $I_1 = 5.00$ A and the second carries $I_2 = 8.00$ A. (a) What is the magnitude of the magnetic field created by I_1 at the location of I_2? (b) What is the force per unit length exerted by I_1 on I_2? (c) What is the magnitude of the magnetic field created by I_2 at the location of I_1? (d) What is the force per length exerted by I_2 on I_1?

ACTIVITIES

1. For this activity, you will need a small bar magnet, a small plastic container, and a bowl of water. Tape the magnet to the bottom of the container, and float the container and magnet on the surface of the bowl as in Figure A19.1. The magnet and the container should rotate and come to equilibrium, with the magnet pointing along a north–south line. The compass you have constructed is similar to the type used by early sailing vessels. How can you determine which direction is north and which is south?

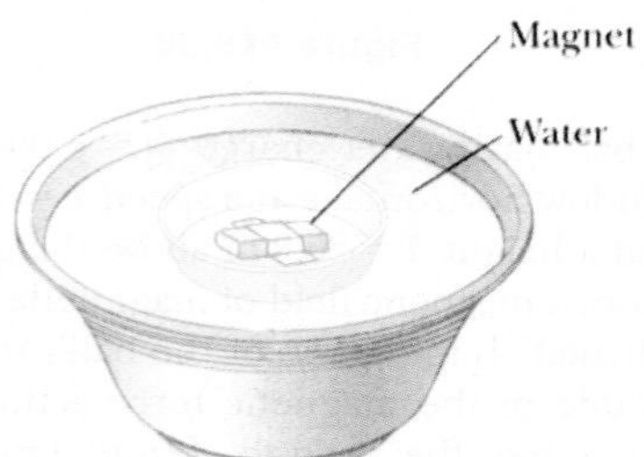

Figure A19.01

2. In the Northern Hemisphere, the direction of Earth's magnetic field becomes more and more nearly vertical the farther north one goes. To find the variation from the horizontal of the magnetic field in your locale, try the following: press an unmagnetized needle through a Ping-Pong® ball, and balance the structure between two drinking glasses that are lined up along an east–west line. Next, press a magnetized needle through the ball at right angles to the unmagnetized needle so that the needle points north. The magnetized needle can now rotate in the vertical direction and will point in the direction of Earth's magnetic field, which is at some angle below the horizontal. Take several measurements of this dip angle and obtain an average value.

3. Construct an electromagnet by wrapping about 1 meter of small-diameter insulated wire around a steel nail. Tape the ends of the wires to a D-cell battery as in Figure A19.3. How many staples or paper clips can you pick up with your electromagnet? How would you increase the magnetic field set up by the nail? Disconnect the wires from

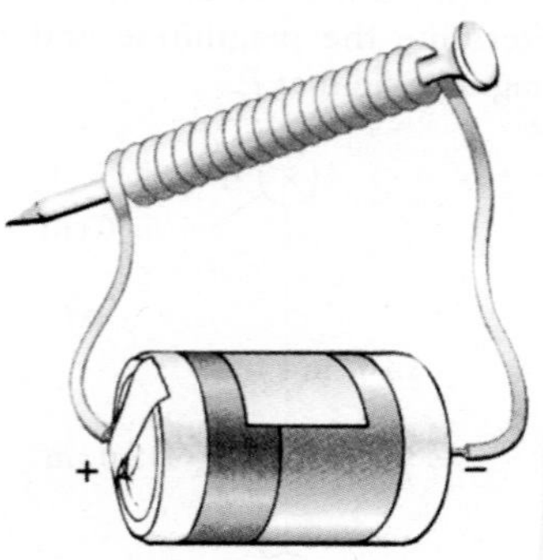

Figure A19.03

the battery and test how much magnetism is retained by the nail by seeing how many staples it can pick up. A convenient way to test the strength of a magnet is to attach a paper clip to a rubber band. Note how far the rubber band is stretched before the clip comes free of the magnet. Test your electromagnet in this way. Where is the magnetic field of the electromagnet strongest, at the ends of the nail or near its center? When you have your nail magnetized, bang it against a table or the floor and then check its magnetism. Why does the nail lose its magnetism by this procedure?

4. You can trace out the field pattern of a magnet with iron filings. Any machine shop will supply the filings, which should be soaked in a soap solution to remove grit and oil and then dried. Scatter them lightly over the surface of a paper covering the magnet, and then tap the paper gently to jar the filings into alignment. Explain why the filings form their pattern. Examine the field pattern set up in the following situations: (a) Arrange two bar magnets about 4 cm apart, aligned with opposite poles facing each other. (b) Use two bar magnets about 4 cm apart, aligned with like poles facing each other. (c) Use a horseshoe magnet.

The vibrating strings induce a voltage in pickup coils that detect and amplify the musical sounds being produced The details of how this phenomenon works are discussed in this chapter.

PhotoDisc/Getty Images

CHAPTER

20

OUTLINE

Induced Voltages and Inductance

In 1819, Hans Christian Oersted discovered that an electric current exerted a force on a magnetic compass. Although there had long been speculation that such a relationship existed, Oersted's finding was the first evidence of a link between electricity and magnetism. Because nature is often symmetric, the discovery that electric currents produce magnetic fields led scientists to suspect that magnetic fields could produce electric currents. Indeed, experiments conducted by Michael Faraday in England and independently by Joseph Henry in the United States in 1831 showed that a changing magnetic field could induce an electric current in a circuit. The results of these experiments led to a basic and important law known as Faraday's law. In this chapter we discuss Faraday's law and several practical applications, one of which is the production of electrical energy in power generation plants throughout the world.

20.1 INDUCED EMF AND MAGNETIC FLUX

An experiment first conducted by Faraday demonstrated that a current can be produced by a changing magnetic field. The apparatus shown in Figure 20.1 (page 661) consists of a coil connected to a switch and a battery. We will refer to this coil as the *primary coil* and to the corresponding circuit as the primary circuit. The coil is wrapped around an iron ring to intensify the magnetic field produced by the current in the coil. A second coil, at the right, is wrapped around the iron ring and is connected to an ammeter. This is called the *secondary coil,* and the corresponding circuit is called the secondary circuit. It's important to notice that **there is no battery in the secondary circuit.**

At first glance, you might guess that no current would ever be detected in the secondary circuit. However, when the switch in the primary circuit in Figure 20.1 is suddenly closed, something amazing happens: the ammeter measures a current

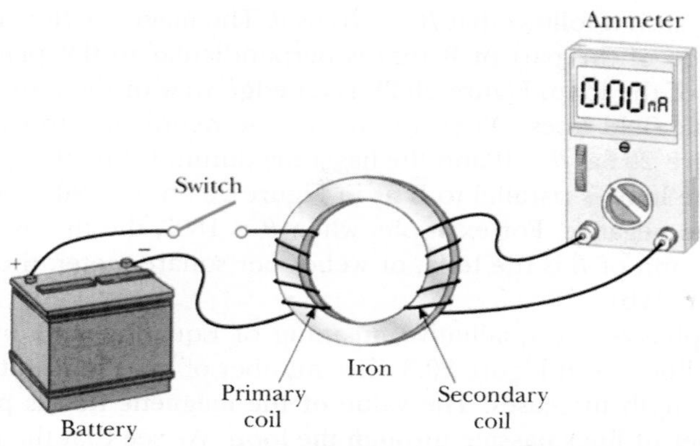

FIGURE 20.1 Faraday's experiment. When the switch in the primary circuit at the left is closed, the ammeter in the secondary circuit at the right measures a momentary current. The emf in the secondary circuit is induced by the changing magnetic field through the coil in that circuit.

MICHAEL FARADAY, British physicist and chemist (1791–1867)

Faraday is often regarded as the greatest experimental scientist of the 1800s. His many contributions to the study of electricity include the invention of the electric motor, electric generator, and transformer, as well as the discovery of electromagnetic induction and the laws of electrolysis. Greatly influenced by religion, he refused to work on military poison gas for the British government.

in the secondary circuit and then returns to zero! When the switch is opened again, the ammeter reads a current in the opposite direction and again returns to zero. Finally, whenever there is a steady current in the primary circuit, the ammeter reads zero.

From observations such as these, Faraday concluded that an electric current could be produced by a *changing* magnetic field. (A steady magnetic field doesn't produce a current, unless the coil is moving, as explained below.) The current produced in the secondary circuit occurs only for an instant while the magnetic field through the secondary coil is changing. In effect, the secondary circuit behaves as though a source of emf were connected to it for a short time. It's customary to say that **an induced emf is produced in the secondary circuit by the changing magnetic field.**

Magnetic Flux

In order to evaluate induced emfs quantitatively, we need to understand what factors affect the phenomenon. While changing magnetic fields always induce electric fields, there are also situations in which the magnetic field remains constant, yet an induced electric field is still produced. The best example of this is an electric generator: A loop of conductor rotating in a constant magnetic field creates an electric current.

The physical quantity associated with magnetism that creates an electric field is a **changing magnetic flux**. Magnetic flux is defined in the same way as electric flux (Section 15.9) and is proportional to both the strength of the magnetic field passing through the plane of a loop of wire and the area of the loop.

◀ Magnetic flux

The **magnetic flux Φ_B** through a loop of wire with area A is defined by

$$\Phi_B \equiv B_\perp A = BA \cos\theta \qquad [20.1]$$

where $B_\perp$ is the component of $\vec{\mathbf{B}}$ perpendicular to the plane of the loop, as in Figure 20.2a, and θ is the angle between $\vec{\mathbf{B}}$ and the normal (perpendicular) to the plane of the loop.

SI unit: weber (Wb)

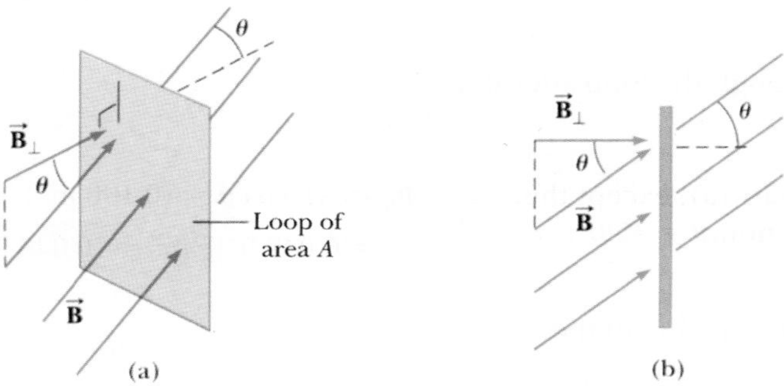

Figure 20.2 (a) A uniform magnetic field $\vec{\mathbf{B}}$ making an angle θ with a direction normal to the plane of a wire loop of area A. (b) An edge view of the loop.

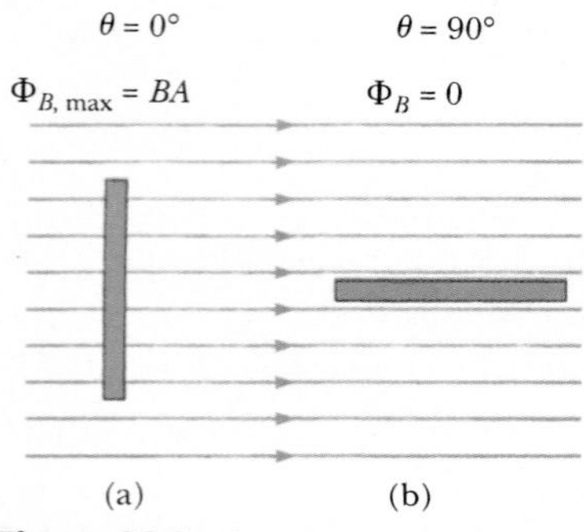

Figure 20.3 An edge view of a loop in a uniform magnetic field. (a) When the field lines are perpendicular to the plane of the loop, the magnetic flux through the loop is a maximum and equal to $\Phi_B = BA$. (b) When the field lines are parallel to the plane of the loop, the magnetic flux through the loop is zero.

From Equation 20.1, it follows that $B_\perp = B\cos\theta$. The magnetic flux, in other words, is the magnitude of the part of $\vec{\mathbf{B}}$ that is perpendicular to the plane of the loop times the area of the loop. Figure 20.2b is an edge view of the loop and the penetrating magnetic field lines. When the field is perpendicular to the plane of the loop as in Figure 20.3a, $\theta = 0$ and Φ_B has a maximum value, $\Phi_{B,\,\max} = BA$. When the plane of the loop is parallel to $\vec{\mathbf{B}}$ as in Figure 20.3b, $\theta = 90°$ and $\Phi_B = 0$. The flux can also be negative. For example, when $\theta = 180°$, the flux is equal to $-BA$. Because the SI unit of B is the tesla, or weber per square meter, the unit of flux is $\mathrm{T \cdot m^2}$, or weber (Wb).

We can emphasize the qualitative meaning of Equation 20.1 by first drawing magnetic field lines, as in Figure 20.3. The number of lines per unit area increases as the field strength increases. **The value of the magnetic flux is proportional to the total number of lines passing through the loop.** We see that the most lines pass through the loop when its plane is perpendicular to the field, as in Figure 20.3a, so the flux has its maximum value at that time. As Figure 20.3b shows, no lines pass through the loop when its plane is parallel to the field, so in that case $\Phi_B = 0$.

Applying Physics 20.1 Flux Compared

Argentina has more land area ($2.8 \times 10^6\ \mathrm{km^2}$) than Greenland ($2.2 \times 10^6\ \mathrm{km^2}$). Why is the magnetic flux of the Earth's magnetic field larger through Greenland than through Argentina?

Explanation Greenland (latitude 60° north to 80° north) is closer to a magnetic pole than Argentina (latitude 20° south to 50° south), so the magnetic field is stronger there. That in itself isn't sufficient to conclude that the magnetic flux is greater, but Greenland's proximity to a pole also means the angle magnetic field lines make with the vertical is smaller than in Argentina. As a result, more field lines penetrate the surface in Greenland, despite Argentina's slightly larger area.

EXAMPLE 20.1 Magnetic Flux

Goal Calculate magnetic flux and a change in flux.

Problem A conducting circular loop of radius 0.250 m is placed in the *xy*-plane in a uniform magnetic field of 0.360 T that points in the positive *z*-direction, the same direction as the normal to the plane. **(a)** Calculate the magnetic flux through the loop. **(b)** Suppose the loop is rotated clockwise around the *x*-axis, so the normal direction now points at a 45.0° angle with respect to the *z*-axis. Recalculate the magnetic flux through the loop. **(c)** What is the change in flux due to the rotation of the loop?

Strategy After finding the area, substitute values into the equation for magnetic flux for each part.

Solution

(a) Calculate the initial magnetic flux through the loop.

First, calculate the area of the loop:

$$A = \pi r^2 = \pi(0.250\ \mathrm{m})^2 = 0.196\ \mathrm{m^2}$$

Substitute A, B, and $\theta = 0°$ into Equation 20.1 to find the initial magnetic flux:

$$\Phi_B = AB\cos\theta = (0.196\ \mathrm{m^2})(0.360\ \mathrm{T})\cos(0°)$$
$$= 0.070\,6\ \mathrm{T\cdot m^2} = \boxed{0.070\,6\ \mathrm{Wb}}$$

(b) Calculate the magnetic flux through the loop after it has rotated 45.0° around the *x*-axis.

Make the same substitutions as in part (a), except the angle between $\vec{\mathbf{B}}$ and the normal is now $\theta = 45.0°$:

$$\Phi_B = AB\cos\theta = (0.196\ \mathrm{m^2})(0.360\ \mathrm{T})\cos(45.0°)$$
$$= 0.049\,9\ \mathrm{T\cdot m^2} = \boxed{0.049\,9\ \mathrm{Wb}}$$

(c) Find the change in the magnetic flux due to the rotation of the loop.

Subtract the result of part (a) from the result of part (b): $\Delta\Phi_B = 0.049\,9\text{ Wb} - 0.070\,6\text{ Wb} = -0.020\,7\text{ Wb}$

Remarks Notice that the rotation of the loop, not any change in the magnetic field, is responsible for the change in flux. This changing magnetic flux is essential in the functioning of electric motors and generators.

Exercise 20.1

The loop, having rotated by 45°, rotates clockwise another 30°, so the normal to the plane points at an angle of 75° with respect to the direction of the magnetic field. Find (a) the magnetic flux through the loop when $\theta = 75°$ and (b) the change in magnetic flux during the rotation from 45° to 75°.

Answers (a) 0.018 3 Wb (b) −0.031 6 Wb

20.2 FARADAY'S LAW OF INDUCTION

The usefulness of the concept of magnetic flux can be made obvious by another simple experiment that demonstrates the basic idea of electromagnetic induction. Consider a wire loop connected to an ammeter as in Active Figure 20.4. If a magnet is moved toward the loop, the ammeter reads a current in one direction, as in Active Figure 20.4a. When the magnet is held stationary, as in Active Figure 20.4b, the ammeter reads zero current. If the magnet is moved away from the loop, the ammeter reads a current in the opposite direction, as in Active Figure 20.4c. If the magnet is held stationary and the loop is moved either toward or away from the magnet, the ammeter also reads a current. From these observations, it can be concluded that **a current is set up in the circuit as long as there is relative motion between the magnet and the loop**. The same experimental results are found whether the loop moves or the magnet moves. We call such a current an **induced current**, because it is produced by an **induced emf**.

TIP 20.1 Induced Current Requires a Change in Magnetic Flux

The existence of magnetic flux through an area is not sufficient to create an induced emf. A *change* in the magnetic flux over some time interval Δt must occur for an emf to be induced.

This experiment is similar to the Faraday experiment discussed in Section 20.1. In each case, an emf is induced in a circuit when the magnetic flux through the circuit changes with time. It turns out that the instantaneous emf induced in a circuit equals the negative of the rate of change of magnetic flux with respect to time through the circuit. This is **Faraday's law of magnetic induction**.

◀ Faraday's law

If a circuit contains N tightly wound loops and the magnetic flux through each loop changes by the amount $\Delta\Phi_B$ during the interval Δt, the average emf induced in the circuit during time Δt is

$$\mathcal{E} = -N\frac{\Delta\Phi_B}{\Delta t} \quad [20.2]$$

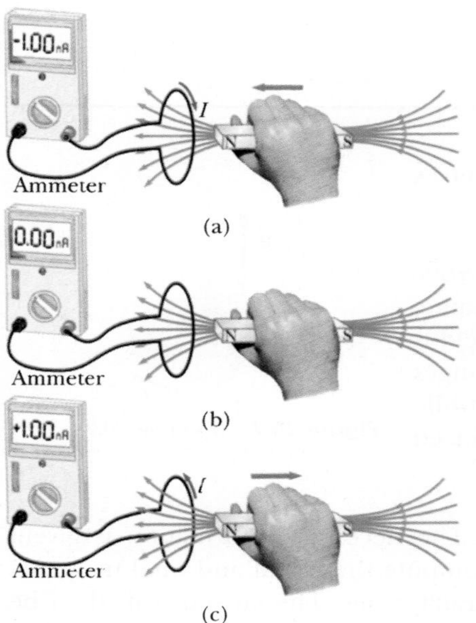

ACTIVE FIGURE 20.4

(a) When a magnet is moved toward a wire loop connected to an ammeter, the ammeter reads a current as shown, indicating that a current I is induced in the loop. (b) When the magnet is held stationary, no current is induced in the loop, even when the magnet is inside the loop. (c) When the magnet is moved away from the loop, the ammeter reads a current in the opposite direction, indicating an induced current going opposite the direction of the current in part (a).

PhysicsNow™

Log into PhysicsNow at **www.cp7e.com** and go to Active Figure 20.4, where you can move the magnet and observe the current in the ammeter.

Because $\Phi_B = BA \cos \theta$, a change of any of the factors B, A, or θ with time produces an emf. We explore the effect of a change in each of these factors in the following sections. The minus sign in Equation 20.2 is included to indicate the polarity of the induced emf. This polarity simply determines which of two different directions current will flow in a loop, a direction given by **Lenz's law**:

Lenz's law ▶

> The current caused by the induced emf travels in the direction that creates a magnetic field with flux opposing the change in the original flux through the circuit.

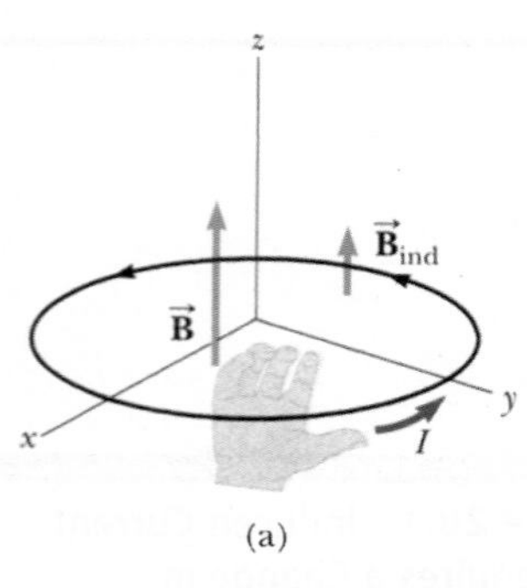

(a)

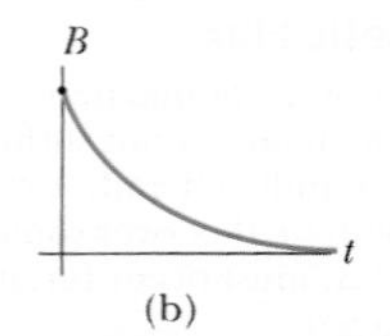

(b)

Figure 20.5 (a) The magnetic field $\vec{\mathbf{B}}$ becomes smaller with time, reducing the flux, so current is induced in a direction that creates an induced magnetic field $\vec{\mathbf{B}}_{\text{ind}}$ opposing the change in magnetic flux. (b) Graph of the magnitude of the magnetic field as a function of time.

Lenz's law says that if the magnetic flux through a loop is becoming more positive, say, then the induced emf creates a current and associated magnetic field that produces negative magnetic flux. Some mistakenly think this "counter magnetic field" created by the induced current, called $\vec{\mathbf{B}}_{\text{ind}}$ ("ind" for induced) will always point in a direction opposite the applied magnetic field $\vec{\mathbf{B}}$, but this is only true half the time! Figure 20.5 shows a field penetrating a loop. The graph in Figure 20.5b shows that the magnitude of the magnetic field $\vec{\mathbf{B}}$ shrinks with time. This means the flux of $\vec{\mathbf{B}}$ is shrinking with time, so the induced field $\vec{\mathbf{B}}_{\text{ind}}$ will actually be in the same direction as $\vec{\mathbf{B}}$. In effect, $\vec{\mathbf{B}}_{\text{ind}}$ "shores up" the field $\vec{\mathbf{B}}$, slowing the loss of flux through the loop.

The direction of the current in Figure 20.5 can be determined by right-hand rule number 2: Point your right thumb in the direction that will cause the fingers on your right hand to curl in the direction of the induced field $\vec{\mathbf{B}}_{\text{ind}}$. In this case, that direction is counterclockwise: with the right thumb pointed in the direction of the current, your fingers curl down outside the loop and around and ***up through the inside of the loop***. Remember, inside the loop is where it's important for the induced magnetic field to be pointing up.

Quick Quiz 20.1

Figure 20.6 is a graph of the magnitude B versus time for a magnetic field that passes through a fixed loop and is oriented perpendicular to the plane of the loop. Rank the magnitudes of the emf generated in the loop at the three instants indicated, from largest to smallest.

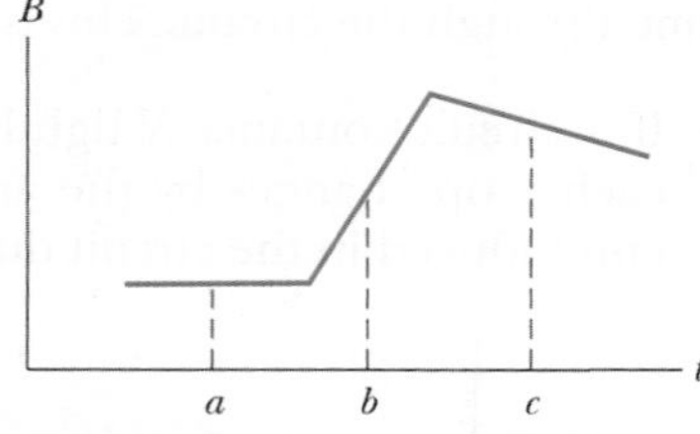

Figure 20.6 (Quick Quiz 20.1)

EXAMPLE 20.2 Faraday and Lenz to the Rescue

Goal Calculate an induced emf and current with Faraday's law, and apply Lenz's law, when the magnetic field changes with time.

Problem A coil with 25 turns of wire is wrapped on a frame with a square cross-section 1.80 cm on a side. Each turn has the same area, equal to that of the frame, and the total resistance of the coil is 0.350 Ω. An applied uniform magnetic field is perpendicular to the plane of the coil, as in Figure 20.7. **(a)** If the field changes uniformly from 0.00 T to 0.500 T in 0.800 s, find the induced emf in the coil while the field is changing. Find **(b)** the magnitude and **(c)** the direction of the induced current in the coil while the field is changing.

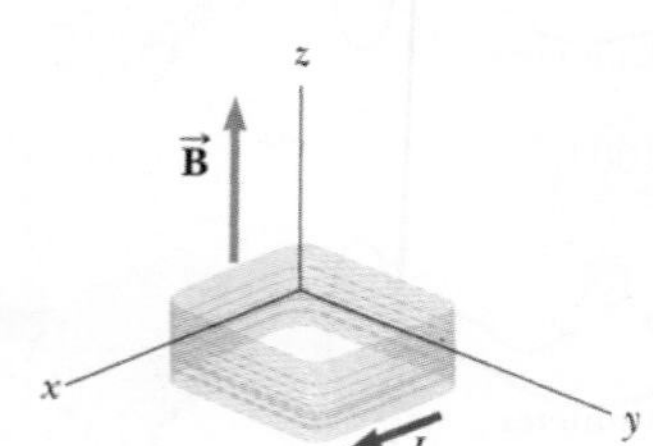

Figure 20.7 (Example 20.2)

Strategy Part (a) requires substituting into Faraday's law, Equation 20.2. The necessary information is given, except for $\Delta\Phi_B$, the change in the magnetic flux during the elapsed time. Compute the initial and final magnetic fluxes with Equation 20.1, find the difference, and assemble all terms in Faraday's law. The current can then be found with Ohm's law, and its direction with Lenz's law.

Solution

(a) Find the induced emf in the coil.

To compute the flux, the area of the coil is needed:

$$A = L^2 = (0.018\,0\text{ m})^2 = 3.24 \times 10^{-4}\text{ m}^2$$

The magnetic flux $\Phi_{B,i}$ through the coil at $t = 0$ is zero because $B = 0$. Calculate the flux at $t = 0.800$ s:

$$\Phi_{B,f} = BA\cos\theta = (0.500\text{ T})(3.24 \times 10^{-4}\text{ m}^2)\cos(0°)$$
$$= 1.62 \times 10^{-4}\text{ Wb}$$

Compute the change in the magnetic flux through the cross-section of the coil over the 0.800-s interval:

$$\Delta\Phi_B = \Phi_{B,f} - \Phi_{B,i} = 1.62 \times 10^{-4}\text{ Wb}$$

Substitute into Faraday's law of induction to find the induced emf in the coil:

$$\mathcal{E} = -N\frac{\Delta\Phi_B}{\Delta t} = (25\text{ turns})\left(\frac{1.62 \times 10^{-4}\text{ Wb}}{0.800\text{ s}}\right)$$
$$= \boxed{-5.06 \times 10^{-3}\text{ V}}$$

(b) Find the magnitude of the induced current in the coil.

Substitute the voltage difference and the resistance into Ohm's law:

$$I = \frac{\Delta V}{R} = \frac{5.06 \times 10^{-3}\text{ V}}{0.350\ \Omega} = \boxed{1.45 \times 10^{-2}\text{ A}}$$

(c) Find the direction of the induced current in the coil.

The magnetic field is increasing up through the loop, in the same direction as the normal to the plane; hence the flux is positive and increasing, also. A downward-pointing induced magnetic field will create negative flux, opposing the change. If you point your right thumb in the clockwise direction along the loop, your fingers curl down through the loop—the correct direction for the counter magnetic field.

Remark Lenz's law can best be handled by sketching a diagram, first.

Exercise 20.2

Suppose the magnetic field changes uniformly from 0.500 T to 0.200 T in the next 0.600 s. Compute (a) the induced emf in the coil and (b) the magnitude and direction of the induced current.

Answers (a) 4.05×10^{-3} V (b) 1.16×10^{-2} A, counterclockwise

The ground fault interrupter (GFI) is an interesting safety device that protects people against electric shock when they touch appliances and power tools. Its operation makes use of Faraday's law. Figure 20.8 shows the essential parts of a ground fault interrupter. Wire 1 leads from the wall outlet to the appliance to be protected, and wire 2 leads from the appliance back to the wall outlet. An iron ring surrounds the two wires to confine the magnetic field set up by each wire. A sensing coil, which can activate a circuit breaker when changes in magnetic flux occur, is wrapped around part of the iron ring. Because the currents in the wires are in opposite directions, the net magnetic field through the sensing coil due to the currents is zero. However, if a short circuit occurs in the appliance so that there is no returning current, the net magnetic field through the sensing coil is no longer zero. This can happen if, for example, one of the wires loses its insulation, providing a path through you to ground if you happen to be touching the appliance and are grounded as in Figure 18.23a. Because the current is alternating, the magnetic flux through the sensing coil changes with time, producing an induced voltage in the coil. This induced voltage is used to trigger a circuit breaker, stopping the

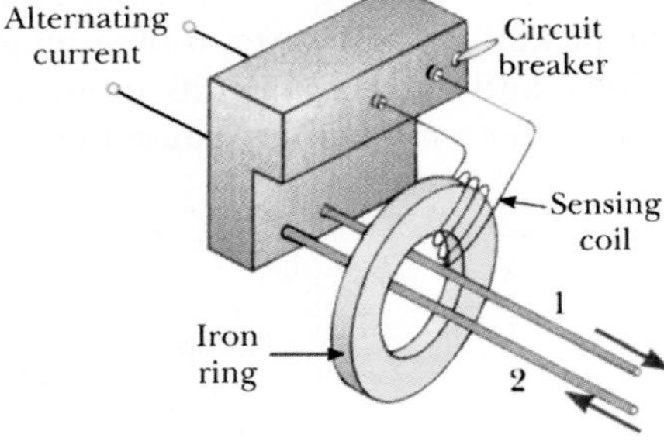

Figure 20.8 Essential components of a ground fault interrupter (contents of the gray box in Fig. 20.9a). In newer homes, such devices are built directly into wall outlets. The purpose of the sensing coil and circuit breaker is to cut off the current before damage is done.

APPLICATION

Ground Fault Interrupters

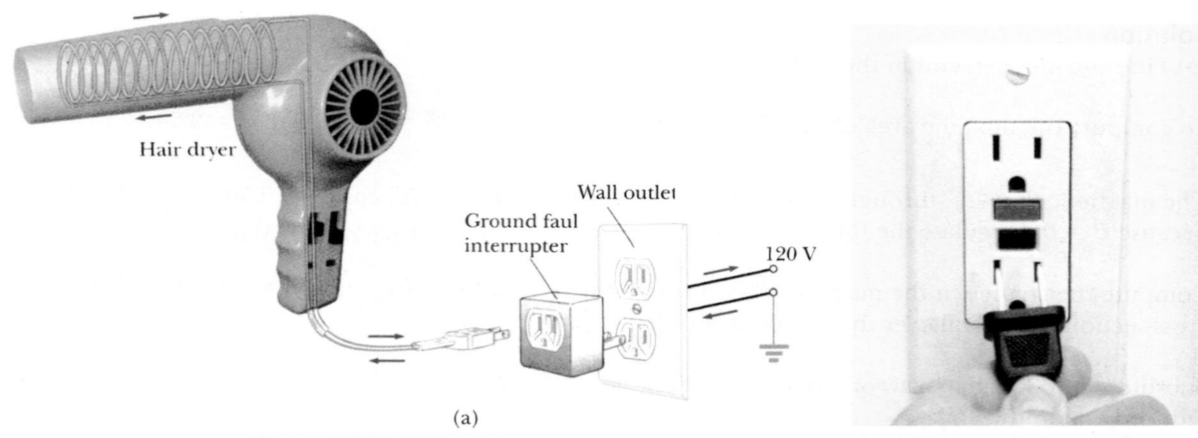

b, George Semple

Figure 20.9 (a) This hair dryer has been plugged into a ground fault interrupter that is in turn plugged into an unprotected wall outlet. (b) You likely have seen this kind of ground fault interrupter in a hotel bathroom, where hair dryers and electric shavers are often used by people just out of the shower or who might touch a water pipe, providing a ready path to ground in the event of a short circuit.

current quickly in about a millisecond before it reaches a level that might be harmful to the person using the appliance. A ground fault interrupter provides faster and more complete protection than even the case-ground-and-circuit-breaker combination shown in Figure 18.23b. For this reason, ground fault interrupters are commonly found in bathrooms, where electricity poses a hazard to people. (See Fig. 20.9.)

APPLICATION
Electric Guitar Pickups

Another interesting application of Faraday's law is the production of sound in an electric guitar. A vibrating string induces an emf in a coil (Fig. 20.10). The pickup coil is placed near the vibrating guitar string, which is made of a metal that can be magnetized. The permanent magnet inside the coil magnetizes the portion of the string nearest the coil. When the guitar string vibrates at some frequency, its magnetized segment produces a changing magnetic flux through the pickup coil. The changing flux induces a voltage in the coil; the voltage is fed to an amplifier. The output of the amplifier is sent to the loudspeakers, producing the sound waves that we hear.

APPLICATION
Apnea Monitors

Sudden infant death syndrome (SIDS) is a devastating affliction in which a baby suddenly stops breathing during sleep without an apparent cause. One type of monitoring device, called an apnea monitor, is sometimes used to alert caregivers of the cessation of breathing. The device uses induced currents, as shown in Figure 20.11. A coil of wire attached to one side of the chest carries an alternating current. The varying magnetic flux produced by this current passes through a pickup coil attached to the opposite side of the chest. Expansion and contraction of the chest caused by breathing or movement changes the strength of the voltage induced in the pickup coil. However, if breathing stops, the pattern of the induced voltage stabilizes, and external circuits monitoring the voltage sound an alarm to the caregivers after a momentary pause to ensure that a problem actually does exist.

Pickup coil
Magnet
N
S
N
S
Magnetized portion of string
To amplifier
Guitar string
(a)

b, Charles D. Winters

(b)

Figure 20.10 (a) In an electric guitar, a vibrating string induces a voltage in the pickup coil. (b) Several pickups allow the vibration to be detected from different portions of the string.

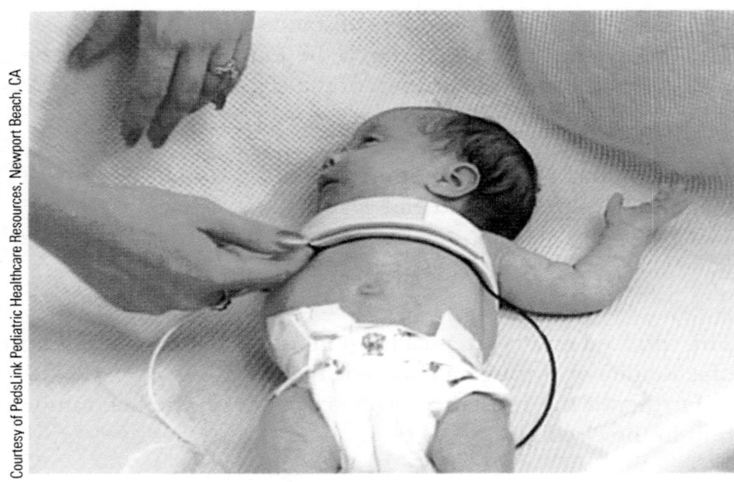

Figure 20.11 This infant is wearing a monitor designed to alert caregivers if breathing stops. Note the two wires attached to opposite sides of the chest.

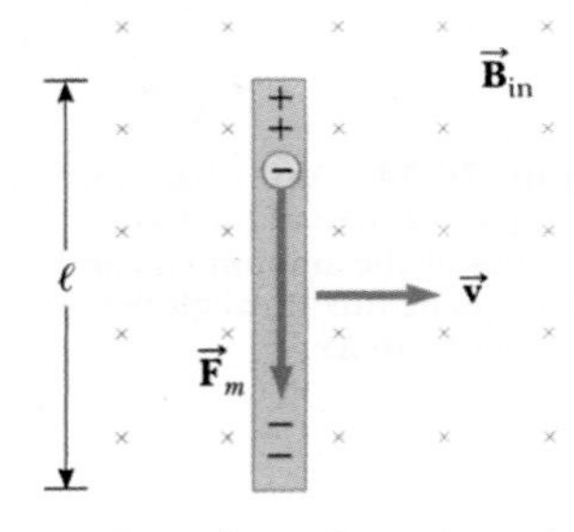

Figure 20.12 A straight conductor of length ℓ moving with velocity $\vec{\mathbf{v}}$ through a uniform magnetic field $\vec{\mathbf{B}}$ directed perpendicular to $\vec{\mathbf{v}}$. The vector $\vec{\mathbf{F}}_m$ is the magnetic force on an electron in the conductor. An emf of $B\ell v$ is induced between the ends of the bar.

20.3 MOTIONAL emf

In Section 20.2, we considered emfs induced in a circuit when the magnetic field changes with time. In this section we describe a particular application of Faraday's law in which a so-called **motional emf** is produced. This is the emf induced in a conductor moving through a magnetic field.

First consider a straight conductor of length ℓ moving with constant velocity through a uniform magnetic field directed into the paper, as in Figure 20.12. For simplicity, we assume that the conductor moves in a direction perpendicular to the field. A magnetic force of magnitude $F_m = qvB$, directed downward, acts on the electrons in the conductor. Because of this magnetic force, the free electrons move to the lower end of the conductor and accumulate there, leaving a net positive charge at the upper end. As a result of this charge separation, an electric field is produced in the conductor. The charge at the ends builds up until the downward magnetic force qvB is balanced by the upward electric force qE. At this point, charge stops flowing and the condition for equilibrium requires that

$$qE = qvB \quad \text{or} \quad E = vB$$

Because the electric field is uniform, the field produced in the conductor is related to the potential difference across the ends by $\Delta V = E\ell$, giving

$$\Delta V = E\ell = B\ell v \qquad [20.3]$$

Because there is an excess of positive charge at the upper end of the conductor and an excess of negative charge at the lower end, the upper end is at a higher potential than the lower end. There is a potential difference across a conductor as long as it moves through a field. If the motion is reversed, the polarity of the potential difference is also reversed.

A more interesting situation occurs if the moving conductor is part of a closed conducting path. This situation is particularly useful for illustrating how a changing loop area induces a current in a closed circuit described by Faraday's law. Consider a circuit consisting of a conducting bar of length ℓ, sliding along two fixed parallel conducting rails, as in Active Figure 20.13a. For simplicity, assume that the moving bar has zero resistance and that the stationary part of the circuit has constant resistance R. A uniform and constant magnetic field $\vec{\mathbf{B}}$ is applied perpendicular to the plane of the circuit. As the bar is pulled to the right with velocity $\vec{\mathbf{v}}$ under the influence of an applied force $\vec{\mathbf{F}}_{\text{app}}$, a magnetic force along the length of the bar acts on the free charges in the bar. This force in turn sets up an induced current because the charges are free to move in a closed conducting path. In this case, the changing magnetic flux through the loop and the corresponding induced emf across the moving bar arise from the *change in area of the loop* as the bar moves through the magnetic field.

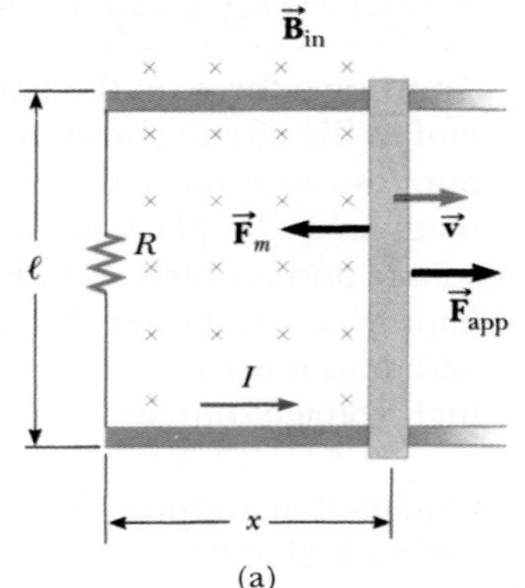

(a)

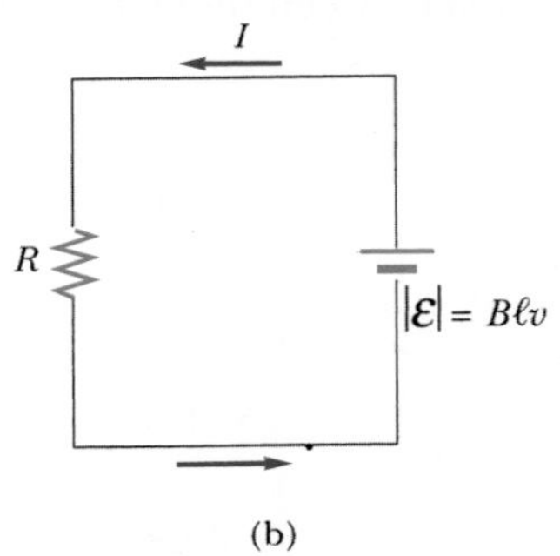

(b)

ACTIVE FIGURE 20.13
(a) A conducting bar sliding with velocity $\vec{\mathbf{v}}$ along two conducting rails under the action of an applied force $\vec{\mathbf{F}}_{\text{app}}$. The magnetic force $\vec{\mathbf{F}}_m$ opposes the motion, and a counterclockwise current is induced in the loop. (b) The equivalent circuit of that in (a).

Physics Now™
Log into PhysicsNow at **www.cp7e.com** and go to Active Figure 20.13, where you can adjust the applied force, the magnetic field, and the resistance, and observe the effects on the motion of the bar.

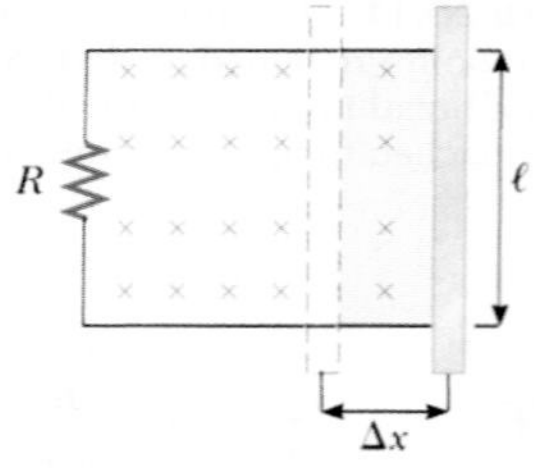

Figure 20.14 As the bar moves to the right, the area of the loop increases by the amount $\ell\Delta x$ and the magnetic flux through the loop increases by $B\ell\Delta x$.

Assume that the bar moves a distance Δx in time Δt, as shown in Figure 20.14. The increase in flux $\Delta\Phi_B$ through the loop in that time is the amount of flux that now passes through the portion of the circuit that has area $\ell\Delta x$:

$$\Delta\Phi_B = BA = B\ell\,\Delta x$$

Using Faraday's law and noting that there is one loop ($N = 1$), we find that the magnitude of the induced emf is

$$|\mathcal{E}| = \frac{\Delta\Phi_B}{\Delta t} = B\ell\frac{\Delta x}{\Delta t} = B\ell v \qquad [20.4]$$

This induced emf is often called a **motional emf** because it arises from the motion of a conductor through a magnetic field.

Further, if the resistance of the circuit is R, the magnitude of the induced current in the circuit is

$$I = \frac{|\mathcal{E}|}{R} = \frac{B\ell v}{R} \qquad [20.5]$$

Active Figure 20.13b shows the equivalent circuit diagram for this example.

Applying Physics 20.2 Space Catapult

Applying a force on the bar will result in an induced emf in the circuit shown in Active Figure 20.13. Suppose we remove the external magnetic field in the diagram and replace the resistor with a high-voltage source and a switch, as in Figure 20.15. What will happen when the switch is closed? Will the bar move, and does it matter which way we connect the high-voltage source?

Explanation Suppose the source is capable of establishing high current. Then the two horizontal conducting rods will create a strong magnetic field in the area between them, directed into the page. (The movable bar also creates a magnetic field, but this field can't exert force on the bar itself.) Because the moving bar carries a downward current, a magnetic force is exerted on the bar, directed to the right. Hence, the bar accelerates along the rails away from the power supply. If the polarity of the power were reversed, the magnetic field would be out of the page, the current in the bar would be upward, and the force on the bar would still be directed to the right. The $BI\ell$ force exerted by a magnetic field according to Equation 19.6 causes the bar to accelerate away from the voltage source. Studies have shown that it's possible to launch payloads into space with this technology. (This is the working principle of a rail gun.) Very large accelerations can be obtained with currently available technology, with payloads being accelerated to a speed of several kilometers per second in a fraction of a second. This is a larger acceleration than humans can tolerate.

Rail guns have been proposed as propulsion systems for moving asteroids into more useful orbits. The material of the asteroid could be mined and launched off the surface by a rail gun, which would act like a rocket engine, modifying the velocity and hence the orbit of the asteroid. Some asteroids contain trillions of dollars worth of valuable metals.

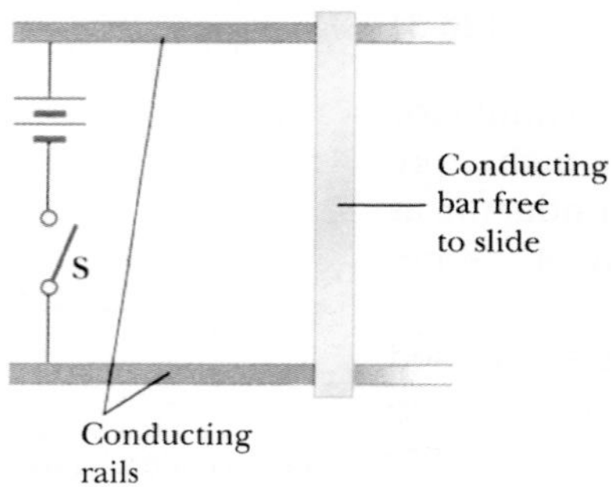

Figure 20.15 (Applying Physics 20.2)

Quick Quiz 20.2

A horizontal metal bar oriented east–west drops straight down in a location where the Earth's magnetic field is due north. As a result, an emf develops between the ends. Which end is positively charged? (a) the east end (b) the west end (c) neither end carries a charge

Quick Quiz 20.3

You intend to move a rectangular loop of wire into a region of uniform magnetic field at a given speed so as to induce an emf in the loop. The plane of the loop must remain perpendicular to the magnetic field lines. In which orientation should you hold the loop while you move it into the region with the magnetic field in order to generate the largest emf? (a) With the long dimension of the loop parallel to the velocity vector, (b) with the short dimension of the loop parallel to the velocity vector, or (c) either way—the emf is the same regardless of orientation.

EXAMPLE 20.3 The Electrified Airplane Wing

Goal Find the emf induced by motion through a magnetic field.

Problem An airplane with a wingspan of 30.0 m flies due north at a location where the downward component of the Earth's magnetic field is 0.600×10^{-4} T. There is also a component pointing due north which has a magnitude of 0.470×10^{-4} T. **(a)** Find the difference in potential between the wingtips when the speed of the plane is 2.50×10^2 m/s. **(b)** Which wingtip is positive?

Strategy Because the plane is flying north, the northern component of the magnetic field won't have any effect on the induced emf. The induced emf across the wing is caused solely by the downward component of the Earth's magnetic field. Substitute the given quantities into Equation 20.4. Use right-hand rule number 1 to find the direction positive charges would be propelled by the magnetic force.

Solution

(a) Calculate the difference in potential across the wingtips.

Write the motional emf equation and substitute the given quantities:

$$\mathcal{E} = B\ell v = (0.600 \times 10^{-4}\ \text{T})(30.0\ \text{m})(2.50 \times 10^2\ \text{m/s})$$

$$= \boxed{0.450\ \text{V}}$$

(b) Which wingtip is positive?

Apply right hand rule number 1:

Point your right fingers north, in the direction of the velocity, curl them down, in the direction of the magnetic field. Your thumb points west.

Remark An induced emf such as this can cause problems on an aircraft.

Exercise 20.3

Suppose the magnetic field in a given region of space is parallel to the Earth's surface, points north, and has magnitude 1.80×10^{-4} T. A metal cable attached to a space station stretches radially outwards 2.50 km. (a) Estimate the potential difference that develops between the ends of the cable if it's traveling eastward around Earth at 7.70×10^3 m/s. (b) Which end of the cable is positive, the lower end or the upper end?

Answer (a) 3.47×10^3 V (b) The upper end is positive.

EXAMPLE 20.4 Where Is the Energy Source?

Goal Use motional emf to find an induced emf and a current.

Problem **(a)** The sliding bar in Figure 20.13a has a length of 0.500 m and moves at 2.00 m/s in a magnetic field of magnitude 0.250 T. Using the concept of motional emf, find the induced voltage in the moving rod. **(b)** If the resistance in the circuit is 0.500 Ω, find the current in the circuit and the power delivered to the resistor. (*Note*: The current, in this case, goes counterclockwise around the loop.) **(c)** Calculate the magnetic force on the bar. **(d)** Use the concepts of work and power to calculate the applied force.

Strategy For part (a), substitute into Equation 20.4 for the motional emf. Once the emf is found, substitution into Ohm's law gives the current. In part (c), use Equation 19.6 for the magnetic force on a current-carrying conductor. In part (d), use the fact that the power dissipated by the resistor multiplied by the elapsed time must equal the work done by the applied force.

Solution

(a) Find the induced emf with the concept of motional emf.

Substitute into Equation 20.4 to find the induced emf:

$$\mathcal{E} = B\ell v = (0.250\ \text{T})(0.500\ \text{m})(2.00\ \text{m/s}) = \boxed{0.250\ \text{V}}$$

(b) Find the induced current in the circuit and the power dissipated by the resistor.

Substitute the emf and the resistance into Ohm's law to find the induced current:

$$I = \frac{\mathcal{E}}{R} = \frac{0.250\ \text{V}}{0.500\ \Omega} = \boxed{0.500\ \text{A}}$$

Substitute I and $\mathcal{E} = 0.250$ V into Equation 17.8 to find the power dissipated by the 0.500-Ω resistor:

$$\mathcal{P} = I\Delta V = (0.500\ \text{A})(0.250\ \text{V}) = \boxed{0.125\ \text{W}}$$

(c) Calculate the magnitude and direction of the magnetic force on the bar.

Substitute values for I, B, and ℓ into Equation 19.6 (with $\sin\theta = \sin(90°) = 1$) to find the magnitude of the force:

$$F_m = IB\ell = (0.500\ \text{A})(0.250\ \text{T})(0.500\ \text{m}) = \boxed{6.25 \times 10^{-2}\ \text{N}}$$

Apply right hand rule number 2 to find the direction of the force:

Point the fingers of your right hand in the direction of the positive current, then curl them in the direction of the magnetic field. Your thumb points in the negative x-direction.

(d) Find the value of F_{app}, the applied force.

Set the work done by the applied force equal to the dissipated power times the elapsed time:

$$W_{\text{app}} = F_{\text{app}}d = \mathcal{P}\Delta t$$

Solve for F_{app} and substitute $d = v\Delta t$:

$$F_{\text{app}} = \frac{\mathcal{P}\Delta t}{d} = \frac{\mathcal{P}\Delta t}{v\Delta t} = \frac{\mathcal{P}}{v} = \frac{0.125\ \text{W}}{2.00\ \text{m/s}} = \boxed{6.25 \times 10^{-2}\ \text{N}}$$

Remarks Part (d) could be solved by using Newton's second law for an object in equilibrium: two forces act horizontally on the bar, and the acceleration of the bar is zero, so the forces must be equal in magnitude and opposite in direction. Notice the agreement between the answers for F_m and F_{app}, despite the very different concepts used.

Exercise 20.4

Suppose the current suddenly increases to 1.25 A in the same direction as before, due to an increase in speed of the bar. Find (a) the emf induced in the rod, (b) the new speed of the rod.

Answers (a) 0.625 V (b) 5.00 m/s

20.4 LENZ'S LAW REVISITED (The Minus Sign in Faraday's Law)

To reach a better understanding of Lenz's law, consider the example of a bar moving to the right on two parallel rails in the presence of a uniform magnetic field directed into the paper (Fig. 20.16a). As the bar moves to the right, the magnetic flux through the circuit increases with time because the area of the loop increases. Lenz's law says that the induced current must be in a direction such that the flux *it* produces opposes the change in the external magnetic flux. Because the flux due to the external field is increasing *into* the paper, the induced current, to oppose the change, must produce a flux *out* of the paper. Hence, the induced current must be counterclockwise when the bar moves to the right. (Use right-hand rule number 2 from Chapter 19 to verify this direction.) On the other hand, if the bar is moving to the left, as in Figure 20.16b, the magnetic flux through the loop decreases with time. Because the flux is into the paper, the induced current has to be clockwise to produce its own flux into the paper (which opposes the decrease in the external flux). In either case, the induced current tends to maintain the original flux through the circuit.

Now we examine this situation from the viewpoint of energy conservation. Suppose that the bar is given a slight push to the right. In the preceding analysis, we found that this motion led to a counterclockwise current in the loop. Let's see what would happen if we assume that the current is clockwise, opposite the direction required by Lenz's law. For a clockwise current I, the direction of the magnetic force $BI\ell$ on the sliding bar is to the right. This force accelerates the rod and increases its velocity. This, in turn, causes the area of the loop to increase more rapidly, thereby increasing the induced current, which increases the force, which increases the current, and so forth. In effect, the system acquires energy with zero input energy. This is inconsistent with all experience and with the law of conservation of energy, so we're forced to conclude that the current must be counterclockwise.

Consider another situation. A bar magnet is moved to the right toward a stationary loop of wire, as in Figure 20.17a. As the magnet moves, the magnetic flux through the loop increases with time. To counteract this rise in flux, the induced current produces a flux to the left, as in Figure 20.17b; hence, the induced current is in the direction shown. Note that the magnetic field lines associated with the induced current oppose the motion of the magnet. The left face of the current loop is therefore a north pole and the right face is a south pole.

On the other hand, if the magnet were moving to the left, as in Figure 20.17c, its flux through the loop, which is toward the right, would decrease with time. Under these circumstances, the induced current in the loop would be in a direction to set up a field directed from left to right through the loop, in an effort to maintain a constant number of flux lines. Hence, the induced current in the loop would be as shown in Figure 20.17d. In this case, the left face of the loop would be a south pole and the right face would be a north pole.

As another example, consider a coil of wire placed near an electromagnet, as in Figure 20.18a (page 672). We wish to find the direction of the induced current in the coil at various times: at the instant the switch is closed, after the switch has been closed for several seconds, and when the switch is opened.

When the switch is closed, the situation changes from a condition in which no lines of flux pass through the coil to one in which lines of flux pass through in the

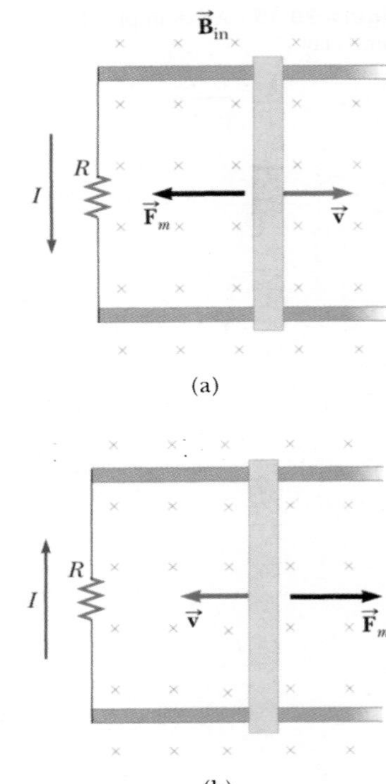

Figure 20.16 (a) As the conducting bar slides on the two fixed conducting rails, the magnetic flux through the loop increases with time. By Lenz's law, the induced current must be *counterclockwise* so as to produce a counteracting flux *out of the paper*. (b) When the bar moves to the left, the induced current must be *clockwise*. Why?

TIP 20.2 There are Two Magnetic Fields to Consider

When applying Lenz's law, there are *two* magnetic fields to consider. The first is the external changing magnetic field that induces the current in a conducting loop. The second is the magnetic field produced by the induced current in the loop.

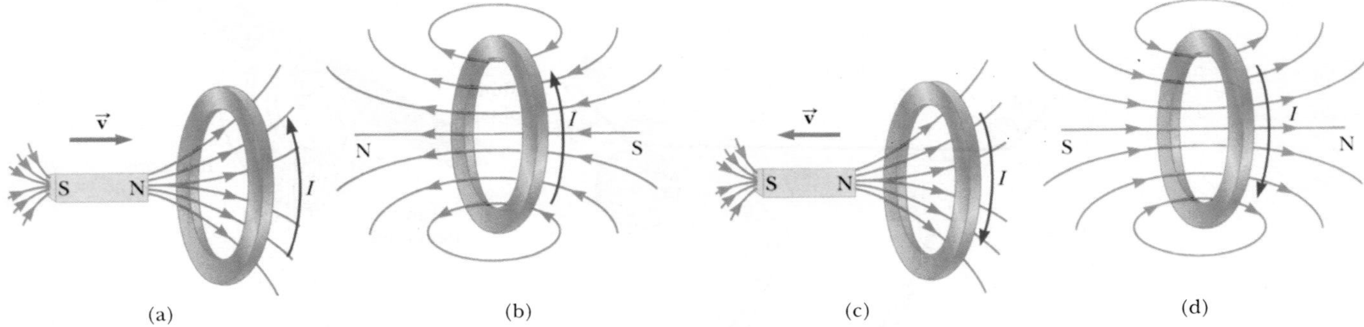

Figure 20.17 (a) When the magnet is moved toward the stationary conducting loop, a current is induced in the direction shown. (b) This induced current produces its own flux to the left to counteract the increasing external flux to the right. (c) When the magnet is moved away from the stationary conducting loop, a current is induced in the direction shown. (d) This induced current produces its own flux to the right to counteract the decreasing external flux to the right.

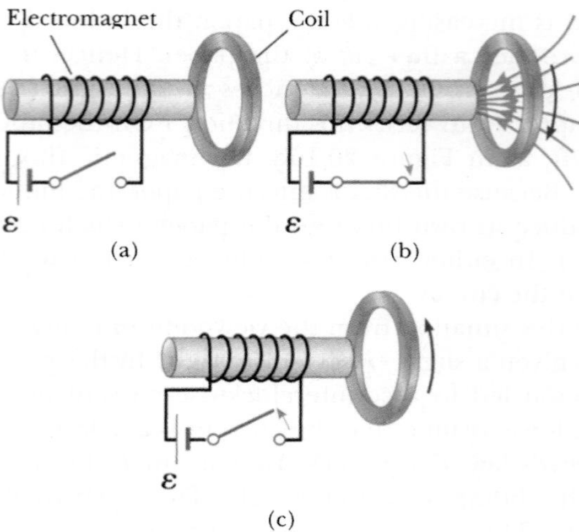

Figure 20.18 An example of Lenz's law.

direction shown in Figure 20.18b. To counteract this change in the number of lines, the coil must set up a field from left to right in the figure. This requires a current directed as shown in Figure 20.18b.

After the switch has been closed for several seconds, there is no change in the number of lines through the loop; hence, the induced current is zero.

Opening the switch causes the magnetic field to change from a condition in which flux lines thread through the coil from right to left to a condition of zero flux. The induced current must then be as shown in Figure 20.18c, so as to set up its own field from right to left.

Quick Quiz 20.4

A bar magnet is falling through a loop of wire with constant velocity with the north pole entering first. Viewed from the same side of the loop as the magnet, as the north pole approaches the loop, what is the direction of the induced current? (a) clockwise (b) zero (c) counterclockwise (d) along the length of the magnet

Tape Recorders

APPLICATION
Magnetic Tape Recorders

One common practical use of induced currents and emfs is associated with the tape recorder. Many different types of tape recorders are made, but the basic principles are the same for all. A magnetic tape moves past a recording and playback head, as in Figure 20.19a. The tape is a plastic ribbon coated with iron oxide or

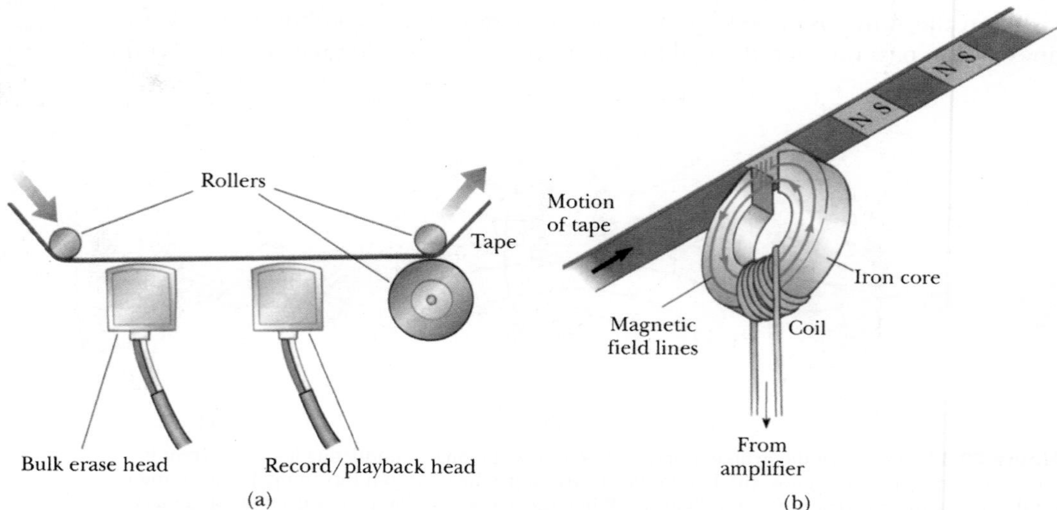

Figure 20.19 (a) Major parts of a magnetic tape recorder. If a new recording is to be made, the bulk erase head wipes the tape clean of signals before recording. (b) The fringing magnetic field magnetizes the tape during recording.

chromium oxide. The hard drives in computers work on the same principle, but use a coated disk instead of tape, allowing for faster access.

The recording process uses the fact that a current in an electromagnet produces a magnetic field. Figure 20.19b illustrates the steps in the process. A sound wave sent into a microphone is transformed into an electric current, amplified, and allowed to pass through a wire coiled around a doughnut-shaped piece of iron, which functions as the recording head. The iron ring and the wire constitute an electromagnet, in which the lines of the magnetic field are contained completely inside the iron except at the point where a slot is cut in the ring. Here the magnetic field fringes out of the iron and magnetizes the small pieces of iron oxide embedded in the tape. As the tape moves past the slot, it becomes magnetized in a pattern that reproduces both the frequency and the intensity of the sound signal entering the microphone.

To reconstruct the sound signal, the previously magnetized tape is allowed to pass through a recorder head operating in the playback mode. A second wire-wound doughnut-shaped piece of iron with a slot in it passes close to the tape, so that the varying magnetic fields on the tape produce changing field lines through the wire coil. The changing flux induces a current in the coil which corresponds to the current in the recording head that originally produced the tape. This changing electric current can be amplified and used to drive a speaker. Playback is thus an example of induction of a current by a moving magnet.

20.5 GENERATORS

APPLICATION

Alternating-Current Generators

Generators and motors are important practical devices that operate on the principle of electromagnetic induction. First, consider the **alternating-current** (AC) **generator**, a device that converts mechanical energy to electrical energy. In its simplest form, the AC generator consists of a wire loop rotated in a magnetic field by some external means (Active Fig. 20.20a). In commercial power plants, the energy required to rotate the loop can be derived from a variety of sources. For example, in a hydroelectric plant, falling water directed against the blades of a turbine produces the rotary motion; in a coal-fired plant, heat produced by burning coal is used to convert water to steam, and this steam is directed against the turbine blades. As the loop rotates, the magnetic flux through it changes with time, inducing an emf and a current in an external circuit. The ends of the loop are connected to slip rings that rotate with the loop.

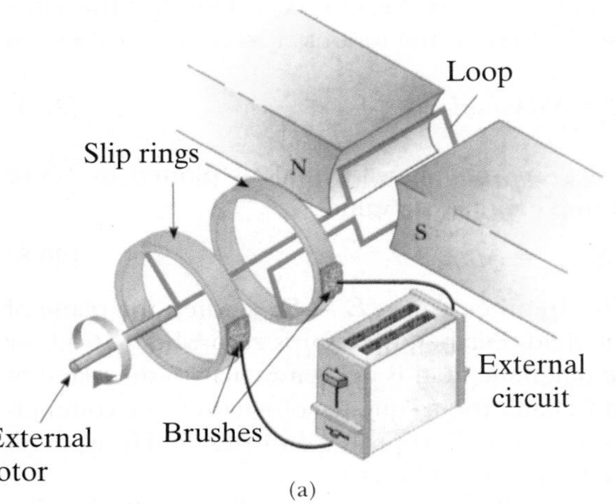

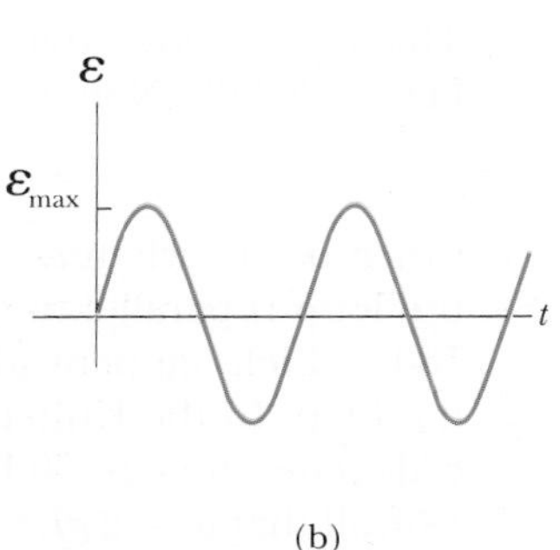

ACTIVE FIGURE 20.20

(a) A schematic diagram of an AC generator. An emf is induced in a coil, which rotates by some external means in a magnetic field. (b) A plot of the alternating emf induced in the loop versus time.

Physics Now™

Log into PhysicsNow at **www.cp7e.com** and go to Active Figure 20.20, where you can adjust the speed of rotation and the strength of the field to see the effects on the generated emf.

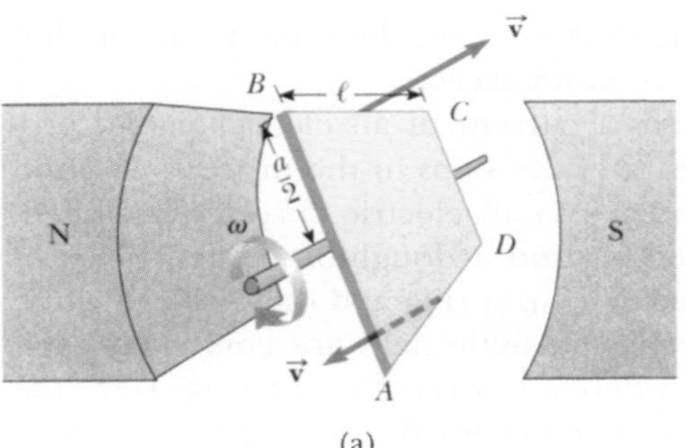

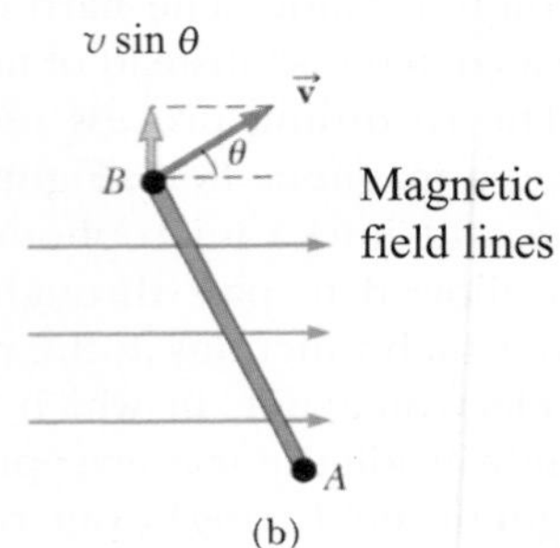

Figure 20.21 (a) A loop rotating at a constant angular velocity in an external magnetic field. The emf induced in the loop varies sinusoidally with time. (b) An edge view of the rotating loop.

Connections to the external circuit are made by stationary brushes in contact with the slip rings.

We can derive an expression for the emf generated in the rotating loop by making use of the equation for motional emf, $\mathcal{E} = B\ell v$. Figure 20.21a shows a loop of wire rotating clockwise in a uniform magnetic field directed to the right. The magnetic force (qvB) on the charges in wires AB and CD is not along the lengths of the wires. (The force on the electrons in these wires is perpendicular to the wires.) Hence, an emf is generated only in wires BC and AD. At any instant, wire BC has velocity $\vec{\mathbf{v}}$ at an angle θ with the magnetic field, as shown in Figure 20.21b. (Note that the component of velocity parallel to the field has no effect on the charges in the wire, whereas the component of velocity perpendicular to the field produces a magnetic force on the charges that moves electrons from C to B.) The emf generated in wire BC equals $B\ell v_\perp$, where ℓ is the length of the wire and $v_\perp$ is the component of velocity perpendicular to the field. An emf of $B\ell v_\perp$ is also generated in wire DA, and the sense of this emf is the same as that in wire BC. Because $v_\perp = v \sin \theta$, the total induced emf is

$$\mathcal{E} = 2B\ell v_\perp = 2B\ell v \sin \theta \qquad [20.6]$$

Turbines turn electric generators at a hydroelectric power plant.

If the loop rotates with a constant angular speed ω, we can use the relation $\theta = \omega t$ in Equation 20.6. Furthermore, because every point on the wires BC and DA rotates in a circle about the axis of rotation with the same angular speed ω, we have $v = r\omega = (a/2)\omega$, where a is the length of sides AB and CD. Equation 20.6 therefore reduces to

$$\mathcal{E} = 2B\ell\left(\frac{a}{2}\right)\omega \sin \omega t = B\ell a\omega \sin \omega t$$

If a coil has N turns, the emf is N times as large because each loop has the same emf induced in it. Further, because the area of the loop is $A = \ell a$, the total emf is

$$\mathcal{E} = NBA\omega \sin \omega t \qquad [20.7]$$

This result shows that the emf varies sinusoidally with time, as plotted in Active Figure 20.20b. Note that the maximum emf has the value

$$\mathcal{E}_{\text{max}} = NBA\omega \qquad [20.8]$$

which occurs when $\omega t = 90°$ or $270°$. In other words, $\mathcal{E} = \mathcal{E}_{\text{max}}$ when the plane of the loop is parallel to the magnetic field. Further, the emf is zero when $\omega t = 0$ or $180°$, which happens whenever the magnetic field is perpendicular to the plane of the loop. In the United States and Canada the frequency of rotation for commercial generators is 60 Hz, whereas in some European countries 50 Hz is used. (Recall that $\omega = 2\pi f$, where f is the frequency in hertz.)

APPLICATION

Direct Current Generators

The **direct current** (DC) **generator** is illustrated in Active Figure 20.22a. The components are essentially the same as those of the AC generator, except that the contacts to the rotating loop are made by a split ring, or commutator. In this design, the output voltage always has the same polarity and the current is a pulsating direct current, as in Active Figure 20.22b. This can be understood by noting

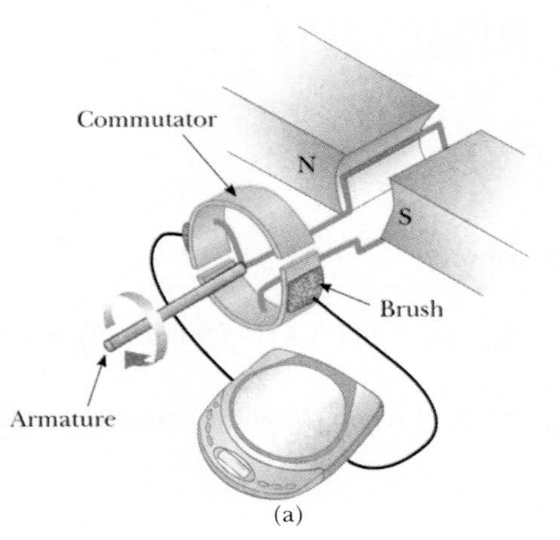

(a)

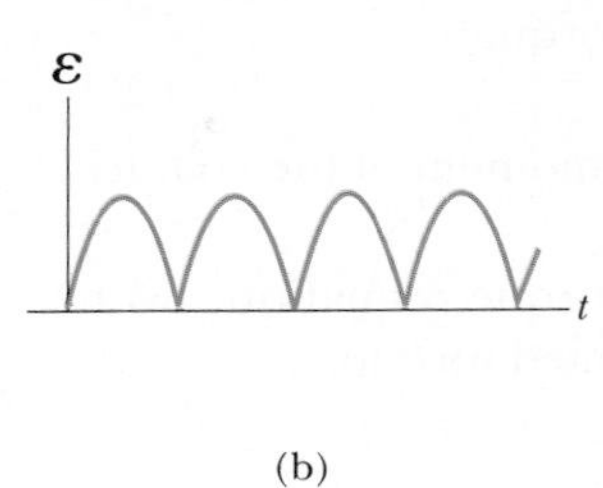

(b)

ACTIVE FIGURE 20.22
(a) A schematic diagram of a DC generator. (b) The emf fluctuates in magnitude, but always has the same polarity.

Physics Now™
Log into PhysicsNow at **www.cp7e.com** and go to Active Figure 20.22, where you can adjust the speed of rotation and the strength of the field, observing the effects on the generated emf.

that the contacts to the split ring reverse their roles every half cycle. At the same time, the polarity of the induced emf reverses. Hence, the polarity of the split ring remains the same.

A pulsating DC current is not suitable for most applications. To produce a steady DC current, commercial DC generators use many loops and commutators distributed around the axis of rotation so that the sinusoidal pulses from the loops overlap in phase. When these pulses are superimposed, the DC output is almost free of fluctuations.

EXAMPLE 20.5 Emf Induced in an AC Generator

Goal Understand physical aspects of an AC generator.

Problem An AC generator consists of eight turns of wire, each having area $A = 0.090\,0\ \text{m}^2$, with a total resistance of 12.0 Ω. The loop rotates in a magnetic field of 0.500 T at a constant frequency of 60.0 Hz. **(a)** Find the maximum induced emf. **(b)** What is the maximum induced current? **(c)** Determine the induced emf and current as functions of time. **(d)** What maximum torque must be applied to keep the coil turning?

Strategy From the given frequency, calculate the angular frequency ω and substitute it, together with given quantities, into Equation 20.8. As functions of time, the emf and current have the form $A \sin \omega t$, where A is the maximum emf or current, respectively. For part (d), calculate the magnetic torque on the coil when the current is at a maximum. (See Chapter 19.) The applied torque must do work against this magnetic torque to keep the coil turning.

Solution

(a) Find the maximum induced emf.

First, calculate the angular frequency of the rotational motion:

$$\omega = 2\pi f = 2\pi(60.0\ \text{Hz}) = 377\ \text{rad/s}.$$

Substitute the values for N, A, B, and ω into Equation 20.8, obtaining the maximum induced emf:

$$\mathcal{E}_{\text{max}} = NAB\omega = 8(0.090\,0\ \text{m}^2)(0.500\ \text{T})(377\ \text{rad/s})$$
$$= \boxed{136\ \text{V}}$$

(b) What is the maximum induced current?

Substitute the maximum induced emf $\mathcal{E}_{\text{max}}$ and the resistance R into Ohm's law to find the maximum induced current:

$$I_{\text{max}} = \frac{\mathcal{E}_{\text{max}}}{R} = \frac{136\ \text{V}}{12.0\ \Omega} = \boxed{11.3\ \text{A}}$$

(c) Determine the induced emf and the current as functions of time.

Substitute $\mathcal{E}_{\text{max}}$ and ω into Equation 20.7 to obtain the variation of $\mathcal{E}$ with time t in seconds:

$$\mathcal{E} = \mathcal{E}_{\text{max}} \sin \omega t = \boxed{(136\ \text{V}) \sin 377t}$$

The time variation of the current looks just like this, except with the maximum current out in front:

$$I = \boxed{(11.3 \text{ A}) \sin 377t}$$

(d) Calculate the maximum applied torque necessary to keep the coil turning.

Write the equation for magnetic torque:

$$\tau = \mu B \sin \theta$$

Calculate the maximum magnetic moment of the coil, μ:

$$\mu = I_{max}AN = (11.3 \text{ A})(0.090 \text{ m}^2)(8) = 8.14 \text{ A}\cdot\text{m}^2$$

Substitute into the magnetic torque equation, with $\theta = 90°$ to find the maximum applied torque:

$$\tau_{max} = (8.14 \text{ A}\cdot\text{m}^2)(0.500 \text{ T}) \sin 90° = \boxed{4.07 \text{ N}\cdot\text{m}}$$

Remarks The number of loops, N, can't be arbitrary, because there must be a force strong enough to turn the coil.

Exercise 20.5

An AC generator is to have a maximum output of 301 V. Each coil has an area of 0.100 m^2 and a resistance of 16.0 Ω and rotates in a magnetic field of 0.600 T with a frequency of 40.0 Hz. (a) How many turns of wire should the coil have to produce the desired emf? (b) Find the maximum current induced in the coil. (c) Determine the induced emf as a function of time.

Answers (a) 20 turns (b) 18.8 A (c) $\mathcal{E} = (301 \text{ V}) \sin 251t$

Motors and Back emf

APPLICATION

Motors

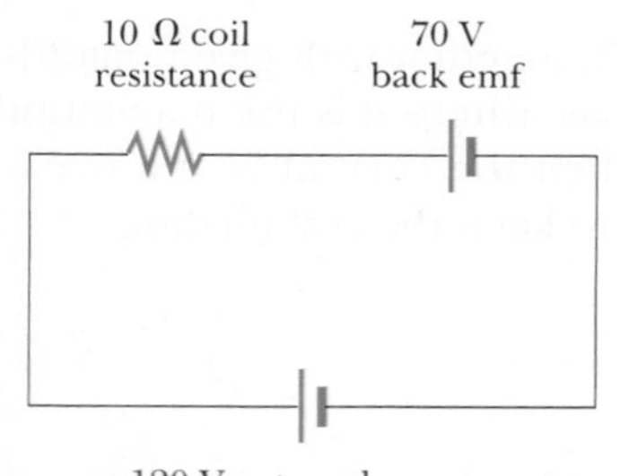

Figure 20.23 A motor can be represented as a resistance plus a back emf.

Motors are devices that convert electrical energy to mechanical energy. Essentially, **a motor is a generator run in reverse**: instead of a current being generated by a rotating loop, a current is supplied to the loop by a source of emf, and the magnetic torque on the current-carrying loop causes it to rotate.

A motor can perform useful mechanical work when a shaft connected to its rotating coil is attached to some external device. As the coil in the motor rotates, however, the changing magnetic flux through it induces an emf which acts to reduce the current in the coil. If it *increased* the current, Lenz's law would be violated. The phrase **back emf** is used for an emf that tends to reduce the applied current. The back emf increases in magnitude as the rotational speed of the coil increases. We can picture this state of affairs as the equivalent circuit in Figure 20.23. For illustrative purposes, assume that the external power source supplying current in the coil of the motor has a voltage of 120 V, that the coil has a resistance of 10 Ω, and that the back emf induced in the coil at this instant is 70 V. The voltage available to supply current equals the difference between the applied voltage and the back emf, 50 V in this case. The current is always reduced by the back emf.

When a motor is turned on, there is no back emf initially, and the current is very large because it's limited only by the resistance of the coil. As the coil begins to rotate, the induced back emf opposes the applied voltage and the current in the coil is reduced. If the mechanical load increases, the motor slows down, which decreases the back emf. This reduction in the back emf increases the current in the coil and therefore also increases the power needed from the external voltage source. As a result, the power requirements for starting a motor and for running it under heavy loads are greater than those for running the motor under average loads. If the motor is allowed to run under no mechanical load, the back emf reduces the current to a value just large enough to balance energy losses by heat and friction.

EXAMPLE 20.6 Induced Current in a Motor

Goal Apply the concept of a back emf in calculating the induced current in a motor.

Problem A motor has coils with a resistance of 10.0 Ω and is supplied by a voltage of $\Delta V = 1.20 \times 10^2$ V. When the motor is running at its maximum speed, the back emf is 70.0 V. Find the current in the coils **(a)** when the motor is first turned on and **(b)** when the motor has reached its maximum rotation rate.

Strategy For each part, find the net voltage, which is the applied voltage minus the induced emf. Divide the net voltage by the resistance to get the current.

Solution

(a) Find the initial current, when the motor is first turned on.

If the coil isn't rotating, the back emf is zero and the current has its maximum value. Calculate the difference between the emf and the initial back emf and divide by the resistance R, obtaining the initial current:

$$I = \frac{\mathcal{E} - \mathcal{E}_{\text{back}}}{R} = \frac{1.20 \times 10^2\text{ V} - 0}{10.0\ \Omega} = 12.0\text{ A}$$

(b) Find the current when the motor is rotating at its maximum rate.

Repeat the calculation, using the maximum value of the back emf:

$$I = \frac{\mathcal{E} - \mathcal{E}_{\text{back}}}{R} = \frac{1.20 \times 10^2\text{ V} - 70.0\text{ V}}{10.0\ \Omega} = \frac{50.0\ V}{10.0\ \Omega}$$

$$= 5.00\text{ A}$$

Remark The phenomenon of back emf is one way in which the rotation rate of electric motors is limited.

Exercise 20.6

If the current in the motor is 8.00 A at some instant, what is the back emf at that time?

Answer 40.0 V

20.6 SELF-INDUCTANCE

Consider a circuit consisting of a switch, a resistor, and a source of emf, as in Figure 20.24. When the switch is closed, the current doesn't immediately change from zero to its maximum value, $\mathcal{E}/R$. The law of electromagnetic induction—Faraday's law—prevents this. What happens instead is the following: as the current increases with time, the magnetic flux through the loop due to this current also increases. The increasing flux induces an emf in the circuit that opposes the change in magnetic flux. By Lenz's law, the induced emf is in the direction indicated by the dashed battery in the figure. The net potential difference across the resistor is the emf of the battery minus the opposing induced emf. As the magnitude of the current increases, the *rate* of increase lessens and hence the induced emf decreases. This opposing emf results in a gradual increase in the current. For the same reason, when the switch is opened, the current doesn't immediately fall to zero. This effect is called **self-induction** because the changing flux through the circuit arises from the circuit itself. The emf that is set up in the circuit is called a **self-induced emf**.

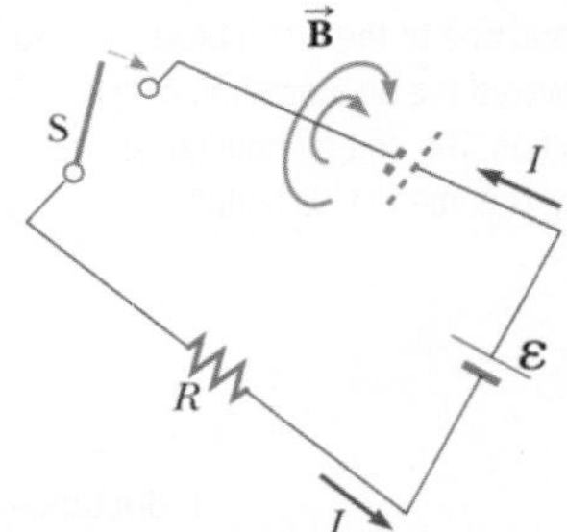

Figure 20.24 After the switch in the circuit is closed, the current produces its own magnetic flux through the loop. As the current increases towards its equilibrium value, the flux changes in time and induces an emf in the loop. The battery drawn with dashed lines is a symbol for the self-induced emf.

As a second example of self-inductance, consider Figure 20.25 (page 678), which shows a coil wound on a cylindrical iron core. (A practical device would have several hundred turns.) Assume that the current changes with time. When the current is in the direction shown, a magnetic field is set up inside the coil, directed from right to left. As a result, some lines of magnetic flux pass through

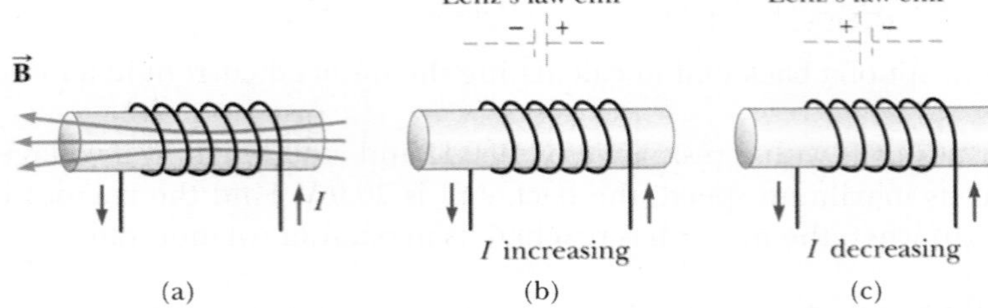

Figure 20.25 (a) A current in the coil produces a magnetic field directed to the left. (b) If the current increases, the coil acts as a source of emf directed as shown by the dashed battery. (c) The induced emf in the coil changes its polarity if the current decreases.

the cross-sectional area of the coil. As the current changes with time, the flux through the coil changes and induces an emf in the coil. Lenz's law shows that this induced emf has a direction so as to oppose the change in the current. If the current is increasing, the induced emf is as pictured in Figure 20.25b, and if the current is decreasing, the induced emf is as shown in Figure 20.25c.

JOSEPH HENRY, American physicist (1797–1878)

Henry became the first director of the Smithsonian Institution and first president of the Academy of Natural Science. He was the first to produce an electric current with a magnetic field, but he failed to publish his results as early as Faraday because of his heavy teaching duties at the Albany Academy in New York State. He improved the design of the electromagnet and constructed one of the first motors. He also discovered the phenomenon of self-induction. The unit of inductance, the henry, is named in his honor.

To evaluate self-inductance quantitatively, first note that, according to Faraday's law, the induced emf is given by Equation 20.2:

$$\mathcal{E} = -N\frac{\Delta\Phi_B}{\Delta t}$$

The magnetic flux is proportional to the magnetic field, which is proportional to the current in the coil. Therefore, **the self-induced emf must be proportional to the rate of change of the current with time**, or

$$\mathcal{E} \equiv -L\frac{\Delta I}{\Delta t} \qquad [20.9]$$

where L is a proportionality constant called the **inductance** of the device. The negative sign indicates that a changing current induces an emf in opposition to the change. This means that if the current is increasing (ΔI positive), the induced emf is negative, indicating opposition to the increase in current. Likewise, if the current is decreasing (ΔI negative), the sign of the induced emf is positive, indicating that the emf is acting to oppose the decrease.

The inductance of a coil depends on the cross-sectional area of the coil and other quantities, all of which can be grouped under the general heading of geometric factors. The SI unit of inductance is the **henry** (H), which, from Equation 20.9, is equal to 1 volt-second per ampere:

$$1\ \text{H} = 1\ \text{V}\cdot\text{s/A}$$

In the process of calculating self-inductance, it is often convenient to equate Equations 20.2 and 20.9 to find an expression for L:

$$N\frac{\Delta\Phi_B}{\Delta t} = L\frac{\Delta I}{\Delta t}$$

Inductance ▶

$$L = N\frac{\Delta\Phi_B}{\Delta I} = \frac{N\Phi_B}{I} \qquad [20.10]$$

Applying Physics 20.3 Making Sparks Fly

In some circuits, a spark occurs between the poles of a switch when the switch is opened. Why isn't there a spark when the switch for this circuit is closed?

Explanation According to Lenz's law, the direction of induced emfs is such that the induced magnetic field opposes change in the original magnetic flux. When the switch is opened, the sudden drop in the magnetic field in the circuit induces an emf in a direction that opposes change in the original current. This induced emf can cause a spark as the current bridges the air gap between the poles of the switch. The spark doesn't occur when the switch is closed, because the original current is zero and the induced emf opposes any change in that current.

In general, determining the inductance of a given current element can be challenging. Finding an expression for the inductance of a common solenoid, however, is straightforward. Let the solenoid have N turns and length ℓ. Assume that ℓ is large compared with the radius and that the core of the solenoid is air. We take the interior magnetic field to be uniform and given by Equation 19.16,

$$B = \mu_0 nI = \mu_0 \frac{N}{\ell} I$$

where $n = N/\ell$ is the number of turns per unit length. The magnetic flux through each turn is therefore

$$\Phi_B = BA = \mu_0 \frac{N}{\ell} AI$$

where A is the cross-sectional area of the solenoid. From this expression and Equation 20.10, we find that

$$L = \frac{N\Phi_B}{I} = \frac{\mu_0 N^2 A}{\ell} \quad \textbf{[20.11a]}$$

This shows that L depends on the geometric factors ℓ and A and on μ_0 and is proportional to the square of the number of turns. Because $N = n\ell$, we can also express the result in the form

$$L = \mu_0 \frac{(n\ell)^2}{\ell} A = \mu_0 n^2 A\ell = \mu_0 n^2 V \quad \textbf{[20.11b]}$$

where $V = A\ell$ is the volume of the solenoid.

EXAMPLE 20.7 Inductance, Self-Induced emf, and Solenoids

Goal Calculate the inductance and self-induced emf of a solenoid.

Problem (a) Calculate the inductance of a solenoid containing 300 turns if the length of the solenoid is 25.0 cm and its cross-sectional area is $4.00 \times 10^{-4}\ \text{m}^2$. **(b)** Calculate the self-induced emf in the solenoid described in (a) if the current in the solenoid decreases at the rate of 50.0 A/s.

Strategy Substituting given quantities into Equation 20.11a gives the inductance L. For part (b), substitute the result of part (a) and $\Delta I/\Delta t = -50.0$ A/s into Equation 20.9 to get the self-induced emf.

Solution

(a) Calculate the inductance of the solenoid.

Substitute the number N of turns, the area A, and the length ℓ into Equation 20.11a to find the inductance:

$$L = \frac{\mu_0 N^2 A}{\ell}$$

$$= (4\pi \times 10^{-7}\ \text{T}\cdot\text{m/A}) \frac{(300)^2 (4.00 \times 10^{-4}\ \text{m}^2)}{25.0 \times 10^{-2}\ \text{m}}$$

$$= 1.81 \times 10^{-4}\ \text{T}\cdot\text{m}^2/\text{A} = \boxed{0.181\ \text{mH}}$$

(b) Calculate the self-induced emf in the solenoid.

Substitute L and $\Delta I/\Delta t = -50.0$ A/s into Equation 20.9, finding the self-induced emf:

$$\mathcal{E} = -L\frac{\Delta I}{\Delta t} = -(1.81 \times 10^{-4}\ \text{H})(-50.0\ \text{A/s})$$

$$= \boxed{9.05\ \text{mV}}$$

Remark Notice that $\Delta I/\Delta t$ is negative because the current is decreasing with time.

Exercise 20.7

A solenoid is to have an inductance of 0.285 mH, a cross-sectional area of $6.00 \times 10^{-4}\ m^2$, and a length of 36.0 cm. (a) How many turns per unit length should it have? (b) If the self-induced emf is -12.5 mV at a given time, at what rate is the current changing at that instant?

Answers (a) 1 025 turns/m (b) 43.9 A/s

20.7 *RL* CIRCUITS

A circuit element that has a large inductance, such as a closely wrapped coil of many turns, is called an **inductor**. The circuit symbol for an inductor is ─ꝏ─. We will always assume that the self-inductance of the remainder of the circuit is negligible compared with that of the inductor in the circuit.

To gain some insight into the effect of an inductor in a circuit, consider the two circuits in Figure 20.26. Figure 20.26a shows a resistor connected to the terminals of a battery. For this circuit, Kirchhoff's loop rule is $\mathcal{E} - IR = 0$. The voltage drop across the resistor is

$$\Delta V_R = -IR \qquad [20.12]$$

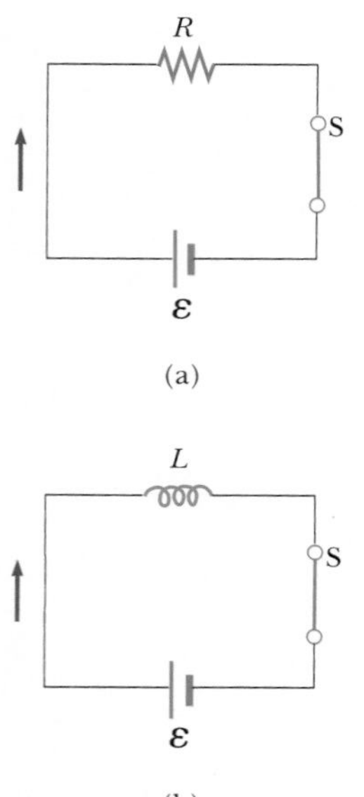

Figure 20.26 A comparison of the effect of a resistor with that of an inductor in a simple circuit.

In this case, **we interpret resistance as a measure of opposition to the current**. Now consider the circuit in Figure 20.26b, consisting of an inductor connected to the terminals of a battery. At the instant the switch in this circuit is closed, because $IR = 0$, the emf of the battery equals the back emf generated in the coil. Hence, we have

$$\mathcal{E}_L = -L\frac{\Delta I}{\Delta t} \qquad [20.13]$$

From this expression, **we can interpret L as a measure of opposition to the rate of change of current.**

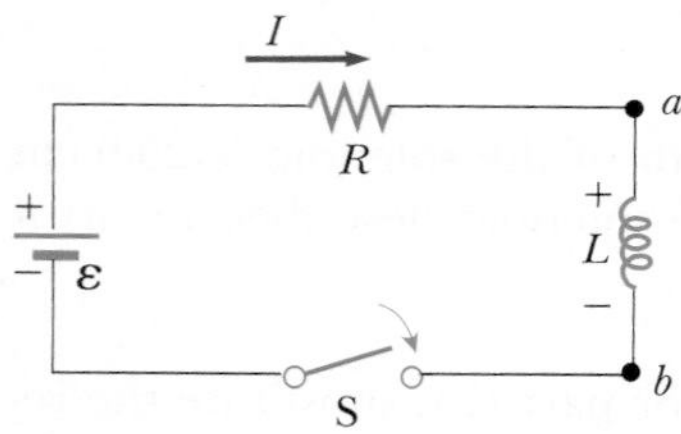

ACTIVE FIGURE 20.27
A series *RL* circuit. As the current increases towards its maximum value, the inductor produces an emf that opposes the increasing current.

Physics Now™
Log into PhysicsNow at **www.cp7e.com** and go to Active Figure 20.27, where you can adjust the values of R and L and observe the effect on current. A graphical display as in Active Figure 20.28 is available.

Active Figure 20.27 shows a circuit consisting of a resistor, an inductor, and a battery. Suppose the switch is closed at $t = 0$. The current begins to increase, but the inductor produces an emf that opposes the increasing current. As a result, the current can't change from zero to its maximum value of $\mathcal{E}/R$ instantaneously. Equation 20.13 shows that the induced emf is a maximum when the current is changing most rapidly, which occurs when the switch is first closed. As the current approaches its steady-state value, the back emf of the coil falls off because the current is changing more slowly. Finally, when the current reaches its steady-state value, the rate of change is zero and the back emf is also zero. Active Figure 20.28 plots current in the circuit as a function of time.[1] This plot is similar to that of the charge on a capacitor as a function of time, discussed in Chapter 18. In that case, we found it convenient to introduce a quantity called the *time constant of the circuit*, which told us something about the time required for the capacitor to approach its steady-state charge. In the same way, time constants are defined for circuits containing resistors and inductors. The **time constant** τ for an *RL* circuit is the time required for the current in the circuit to reach 63.2% of its final value $\mathcal{E}/R$; the time constant of an *RL* circuit is given by

Time constant for an *RL* circuit ▶

$$\tau = \frac{L}{R} \qquad [20.14]$$

[1]The equation for the current in the circuit as a function of time may be obtained from calculus and is

$$I = \frac{\mathcal{E}}{R}(1 - e^{-Rt/L})$$

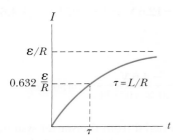

ACTIVE FIGURE 20.28
A plot of current versus time for the *RL* circuit shown in Figure 20.27. The switch is closed at $t = 0$, and the current increases towards its maximum value $\mathcal{E}/R$. The time constant τ is the time it takes the current to reach 63.2% of its maximum value.

Physics Now™
Log into PhysicsNow at **www.cp7e.com** and go to Active Figure 20.27, where you can observe this graph develop after the switch in Active Figure 20.27 is closed.

Quick Quiz 20.5

The switch in the circuit shown in Figure 20.29 is closed and the lightbulb glows steadily. The inductor is a simple air-core solenoid. An iron rod is inserted into the interior of the solenoid, increasing the magnitude of the magnetic field in the solenoid. As the rod is inserted, the brightness of the lightbulb (a) increases, (b) decreases, or (c) remains the same.

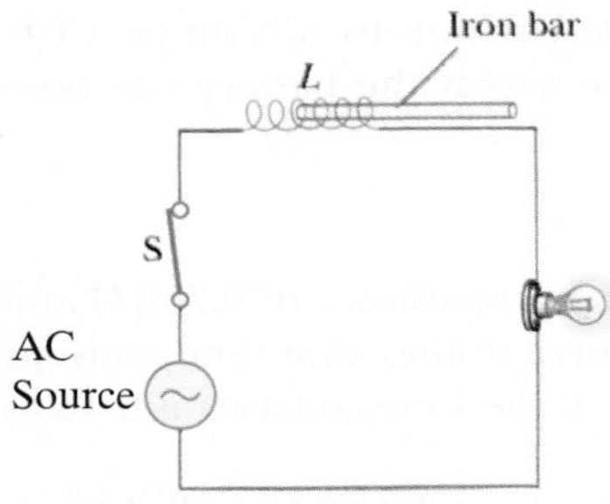

Figure 20.29 (Quick Quiz 20.5)

EXAMPLE 20.8 An *RL* Circuit

Goal Calculate a time constant and relate it to current in an *RL* circuit.

Problem A 12.6-V battery is in a circuit with a 30.0-mH inductor and a 0.150-Ω resistor, as in Active Figure 20.27. The switch is closed at $t = 0$. **(a)** Find the time constant of the circuit. **(b)** Find the current after one time constant has elapsed. **(c)** Find the voltage drops across the resistance when $t = 0$ and $t =$ one time constant. **(d)** What's the rate of change of the current after one time constant?

Solution

(a) What's the time constant of the circuit?

Substitute the inductance L and resistance R into Equation 20.14, finding the time constant:

$$\tau = \frac{L}{R} = \frac{30.0 \times 10^{-3}\ \text{H}}{0.150\ \Omega} = \boxed{0.200\ \text{s}}$$

(b) Find the current after one time constant has elapsed.

First, use Ohm's law to compute the final value of the current after many time constants have elapsed:

$$I_{\text{max}} = \frac{\mathcal{E}}{R} = \frac{12.6\ \text{V}}{0.150\ \Omega} = 84.0\ \text{A}$$

After one time constant, the current rises to 63.2% of its final value:

$$I_{1\tau} = (0.632)I_{\text{max}} = (0.632)(84.0\ \text{A}) = \boxed{53.1\ \text{A}}$$

(c) Find the voltage drops across the resistance when $t = 0$ and $t =$ one time constant.

Initially, the current in the circuit is zero, so, from Ohm's law, the voltage across the resistor is zero:

$$\Delta V_R = IR$$

$$\Delta V_R\,(t = 0\ \text{s}) = (0\ \text{A})(0.150\ \Omega) = \boxed{0}$$

Next, using Ohm's law, find the magnitude of the voltage drop across the resistor after one time constant:

$$\Delta V_R\,(t = 0.200\ \text{s}) = (53.1\ \text{A})(0.150\ \Omega) = \boxed{7.97\ \text{V}}$$

(d) What's the rate of change of the current after one time constant?

Using Kirchhoff's voltage rule, calculate the voltage drop across the inductor at that time:

$$\mathcal{E} + \Delta V_R + \Delta V_L = 0$$

Solve for ΔV_L: $\Delta V_L = -\mathcal{E} - \Delta V_R = -12.6\text{ V} - (-7.97\text{ V}) = -4.6\text{ V}$

Now solve Equation 20.13 for $\Delta I/\Delta t$ and substitute: $\Delta V_L = -L\dfrac{\Delta I}{\Delta t}$

$$\frac{\Delta I}{\Delta t} = -\frac{\Delta V_L}{L} = -\frac{-4.6\text{ V}}{30.0 \times 10^{-3}\text{ H}} = \boxed{150\text{ A/s}}$$

Remarks The values used in this problem were taken from actual components salvaged from the starter system of a car. Because the current in such an RL circuit is initially zero, inductors are sometimes referred to as "chokes," since they temporarily choke off the current. In solving part (d), we traversed the circuit in the direction of positive current, so the voltage difference across the battery was positive and the differences across the resistor and inductor were negative.

Exercise 20.8
A 12.6-V battery is in series with a resistance of 0.350 Ω and an inductor. (a) After a long time, what is the current in the circuit? (b) What is the current after one time constant? (c) What's the voltage drop across the inductor at this time? (d) Find the inductance if the time constant is 0.130 s.

Answers (a) 36.0 A (b) 22.8 A (c) 4.62 V (d) 4.55×10^{-2} H

20.8 ENERGY STORED IN A MAGNETIC FIELD

The emf induced by an inductor prevents a battery from establishing an instantaneous current in a circuit. The battery has to do work to produce a current. We can think of this needed work as energy stored in the inductor in its magnetic field. In a manner similar to that used in Section 16.9 to find the energy stored in a capacitor, we find that the energy stored by an inductor is

Energy stored in an inductor ▶

$$PE_L = \tfrac{1}{2}LI^2 \qquad [20.15]$$

Note that the result is similar in form to the expression for the energy stored in a charged capacitor (Equation 16.18):

Energy stored in a capacitor ▶

$$PE_C = \tfrac{1}{2}C(\Delta V)^2$$

EXAMPLE 20.9 Magnetic Energy

Goal Relate the storage of magnetic energy to currents in an RL circuit.

Problem A 12.0-V battery is connected in series to a 25.0-Ω resistor and a 5.00-H inductor. **(a)** Find the maximum current in the circuit. **(b)** Find the energy stored in the inductor at this time. **(c)** How much energy is stored in the inductor when the current is changing at a rate of 1.50 A/s?

Strategy In part (a), Ohm's law and Kirchhoff's voltage rule yield the maximum current, because the voltage across the inductor is zero when the current is maximal. Substituting the current into Equation 20.15 gives the energy stored in the inductor. In part (c), the given rate of change of the current can be used to calculate the voltage drop across the inductor at the specified time. Kirchhoff's voltage rule and Ohm's law then give the current I at that time, which can be used to find the energy stored in the inductor.

Solution
(a) Find the maximum current in the circuit.

Apply Kirchhoff's voltage rule to the circuit: $\Delta V_{\text{batt}} + \Delta V_R + \Delta V_L = 0$

$$\mathcal{E} - IR - L\frac{\Delta I}{\Delta t} = 0$$

When the maximum current is reached, $\Delta I/\Delta t$ is zero, so the voltage drop across the inductor is zero. Solve for the maximum current I_{max}:

$$I_{\text{max}} = \frac{\mathcal{E}}{R} = \frac{12.0 \text{ V}}{25.0 \ \Omega} = \boxed{0.480 \text{ A}}$$

(b) Find the energy stored in the inductor at this time.

Substitute known values into Equation 20.15:

$$PE_L = \tfrac{1}{2}LI_{\text{max}}^2 = \tfrac{1}{2}(5.00 \text{ H})(0.480 \text{ A})^2 = \boxed{0.576 \text{ J}}$$

(c) Find the energy in the inductor when the current changes at a rate of 1.50 A/s.

Apply Kirchhoff's voltage rule to the circuit, once again:

$$\mathcal{E} - IR - L\frac{\Delta I}{\Delta t} = 0$$

Solve this equation for the current I and substitute:

$$I = \frac{1}{R}\left(\mathcal{E} - L\frac{\Delta I}{\Delta t}\right)$$

$$= \frac{1}{25.0 \ \Omega}[12.0 \text{ V} - (5.00 \text{ H})(1.50 \text{ A/s})] = 0.180 \text{ A}$$

Finally, substitute the value for the current into Equation 20.15, finding the energy stored in the inductor:

$$PE_L = \tfrac{1}{2}LI^2 = \tfrac{1}{2}(5.00 \text{ H})(0.180 \text{ A})^2 = \boxed{0.081\,0 \text{ J}}$$

Remark Notice how important it is to combine concepts from previous chapters. Here, Ohm's law and Kirchhoff's loop rule were essential to the solution of the problem.

Exercise 20.9
For the same circuit, find the energy stored in the inductor when the rate of change of the current is 1.00 A/s.

Answer 0.196 J

SUMMARY

PhysicsNow™ Take a practice test by logging into PhysicsNow at **www.cp7e.com** and clicking on the Pre-Test link for this chapter.

20.1 Induced emf and Magnetic Flux

The magnetic flux Φ_B through a closed loop is defined as

$$\Phi_B \equiv BA\cos\theta \qquad \textbf{[20.1]}$$

where B is the strength of the uniform magnetic field, A is the cross-sectional area of the loop, and θ is the angle between $\vec{\mathbf{B}}$ and a direction perpendicular to the plane of the loop.

20.2 Faraday's Law of Induction

Faraday's law of induction states that the instantaneous emf induced in a circuit equals the negative of the rate of change of magnetic flux through the circuit,

$$\mathcal{E} = -N\frac{\Delta\Phi_B}{\Delta t} \qquad \textbf{[20.2]}$$

where N is the number of loops in the circuit. The magnetic flux Φ_B can change with time whenever the magnetic field $\vec{\mathbf{B}}$, the area A, or the angle θ changes with time.

Lenz's law states that the current from the induced emf creates a magnetic field with flux opposing the *change* in magnetic flux through a circuit.

20.3 Motional emf

If a conducting bar of length ℓ moves through a magnetic field with a speed v so that $\vec{\mathbf{B}}$ is perpendicular to the bar, then the emf induced in the bar, often called a **motional emf**, is

$$|\mathcal{E}| = B\ell v \qquad \textbf{[20.4]}$$

20.5 Generators

When a coil of wire with N turns, each of area A, rotates with constant angular speed ω in a uniform magnetic field $\vec{\mathbf{B}}$, the emf induced in the coil is

$$\mathcal{E} = NAB\omega\sin\omega t \qquad \textbf{[20.7]}$$

Such generators naturally produce alternating current (AC), which changes direction with frequency $\omega/2\pi$. The AC current can be transformed to direct current.

20.7 *RL* Circuits

When the current in a coil changes with time, an emf is induced in the coil according to Faraday's law. This

self-induced emf is defined by the expression

$$\mathcal{E} \equiv -L\frac{\Delta I}{\Delta t} \qquad [20.9]$$

where L is the inductance of the coil. The SI unit for inductance is the henry (H); $1\ \text{H} = 1\ \text{V}\cdot\text{s/A}$.

The **inductance** of a coil can be found from the expression

$$L = \frac{N\Phi_B}{I} \qquad [20.10]$$

where N is the number of turns on the coil, I is the current in the coil, and Φ_B is the magnetic flux through the coil produced by that current. For a solenoid, the inductance is given by

$$L = \frac{\mu_0 N^2 A}{\ell} \qquad [20.11]$$

If a resistor and inductor are connected in series to a battery and a switch is closed at $t = 0$, the current in the circuit doesn't rise instantly to its maximum value. After one **time constant** $\tau = L/R$, the current in the circuit is 63.2% of its final value $\mathcal{E}/R$. As the current approaches its final, maximum value, the voltage drop across the inductor approaches zero.

20.8 Energy Stored in a Magnetic Field

The **energy stored** in the magnetic field of an inductor carrying current I is

$$PE_L = \tfrac{1}{2}LI^2 \qquad [20.15]$$

As the current in an RL circuit approaches its maximum value, the stored energy also approaches a maximum value.

CONCEPTUAL QUESTIONS

1. A circular loop is located in a uniform and constant magnetic field. Describe how an emf can be induced in the loop in this situation.
2. Does dropping a magnet down a copper tube produce a current in the tube? Explain.
3. A spacecraft orbiting the Earth has a coil of wire in it. An astronaut measures a small current in the coil, although there is no battery connected to it and there are no magnets in the spacecraft. What is causing the current?
4. A loop of wire is placed in a uniform magnetic field. For what orientation of the loop is the magnetic flux a maximum? For what orientation is the flux zero?
5. As the conducting bar in Figure Q20.5 moves to the right, an electric field directed downward is set up. If the bar were moving to the left, explain why the electric field would be upward.

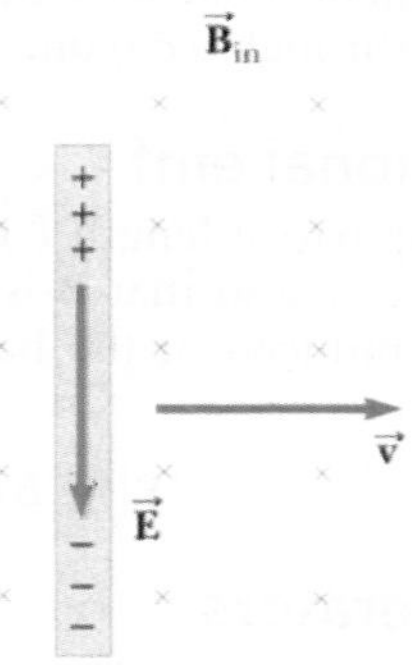

Figure Q20.5 (Conceptual Questions 5 and 6)

6. As the bar in Figure Q20.5 moves perpendicular to the field, is an external force required to keep it moving with constant speed?
7. Wearing a metal bracelet in a region of strong magnetic field could be hazardous. Discuss this statement.
8. How is electrical energy produced in dams? (That is, how is the energy of motion of the water converted to AC electricity?)
9. Eddy currents are induced currents set up in a piece of metal when it moves through a nonuniform magnetic field. For example, consider the flat metal plate swinging at the end of a bar as a pendulum, as shown in Figure Q20.9. At position 1, the pendulum is moving from a region where there is no magnetic field into a region where the field $\vec{B}_{in}$ is directed into the paper. Show that at position 1 the direction of the eddy current is counterclockwise. Also, at position 2 the pendulum is moving out of the field into a region of zero field. Show that the direction of the eddy current is clockwise in this case. Use right-hand rule number 2 to show that these eddy currents lead to a magnetic force on the plate directed as shown in the figure. Because the induced eddy current always produces a retarding force when the plate enters or leaves the field, the swinging plate quickly comes to rest.

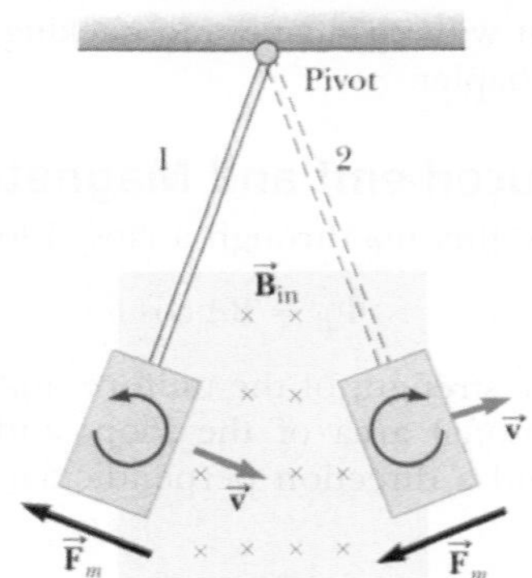

Figure Q20.9

10. Suppose you would like to steal power for your home from the electric company by placing a loop of wire near a transmission cable in order to induce an emf in the loop (Don't do this; it's illegal.) Should you locate the loop so that the transmission cable passes through your loop or simply place your loop near the transmission cable? Does the orientation of the loop matter?
11. A piece of aluminum is dropped vertically downward between the poles of an electromagnet. Does the magnetic

field affect the velocity of the aluminum? [*Hint*: See Conceptual Question 9.]

12. A bar magnet is dropped toward a conducting ring lying on the floor. As the magnet falls toward the ring, does it move as a freely falling object?

13. If the current in an inductor is doubled, by what factor does the stored energy change?

14. Is it possible to induce a constant emf for an infinite amount of time?

15. Why is the induced emf that appears in an inductor called a back (counter) emf?

16. A magneto is used to cause the spark in a spark plug in many lawn mowers today. A magneto consists of a permanent magnet mounted on a flywheel so that it spins past a fixed coil. Explain how this arrangement generates a large enough potential difference to cause the spark.

17. A ramp runs from the bed of a truck down to the level ground. The ramp holds two parallel conducting rails connected at its base. A metal bar slides on the rails without friction. A magnet supplies an external magnetic field directed toward the ground. It is found that the bar slides down the ramp at a constant speed. (a) What is the direction of the induced current in the bar as viewed from above? (b) What can you conclude about the forces exerted on the bar?

18. A bar magnet is held above the center of a wire loop in a horizontal plane, as shown in Figure Q20.18. The south end of the magnet is toward the loop. The magnet is dropped. Find the direction of the current in the resistor as viewed from above (a) while the magnet is falling toward the loop and (b) after the magnet has passed through the loop and moved away from it.

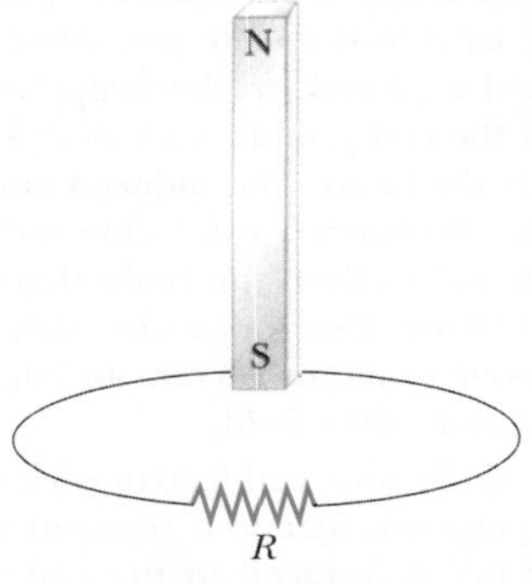

Figure Q20.18

19. What is the direction of the current induced in the resistor when the current in the long, straight wire in Figure Q20.19 decreases rapidly to zero?

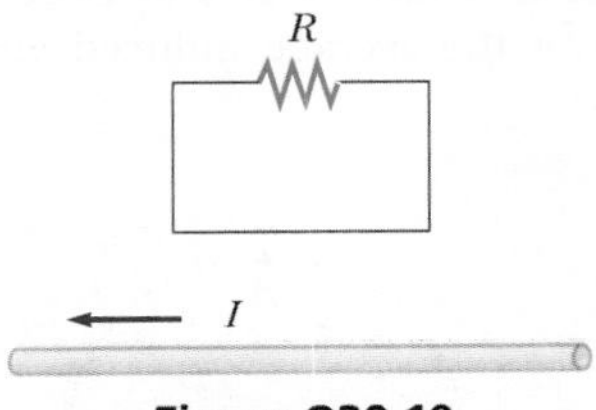

Figure Q20.19

PROBLEMS

1, 2, 3 = straightforward, intermediate challenging ☐ = full solution available in *Student Solutions Manual/Study Guide*

PhysicsNow™ = coached problem with hints available at **www.cp7e.com** = biomedical application

Section 20.1 Induced emf and Magnetic Flux

1. A magnetic field of strength 0.30 T is directed perpendicular to a plane circular loop of wire of radius 25 cm. Find the magnetic flux through the area enclosed by this loop.

2. Find the flux of the Earth's magnetic field of magnitude 5.00×10^{-5} T through a square loop of area 20.0 cm^2 (a) when the field is perpendicular to the plane of the loop, (b) when the field makes a 30.0° angle with the normal to the plane of the loop, and (c) when the field makes a 90.0° angle with the normal to the plane.

3. A square loop 2.00 m on a side is placed in a magnetic field of magnitude 0.300 T. If the field makes an angle of 50.0° with the normal to the plane of the loop, find the magnetic flux through the loop.

4. A long, straight wire carrying a current of 2.00 A is placed along the axis of a cylinder of radius 0.500 m and a length of 3.00 m. Determine the total magnetic flux through the cylinder.

5. A long, straight wire lies in the plane of a circular coil with a radius of 0.010 m. The wire carries a current of 2.0 A and is placed along a diameter of the coil. (a) What is the net flux through the coil? (b) If the wire passes through the center of the coil and is perpendicular to the plane of the coil, find the net flux through the coil.

6. A solenoid 4.00 cm in diameter and 20.0 cm long has 250 turns and carries a current of 15.0 A. Calculate the magnetic flux through the circular cross-sectional area of the solenoid.

7. A cube of edge length ℓ = 2.5 cm is positioned as shown in Figure P20.7. There is a uniform magnetic field throughout the region with components $B_x = +5.0$ T, $B_y = +4.0$ T, and $B_z = +3.0$ T. (a) Calculate the flux through the shaded face of the cube. (b) What is the total flux

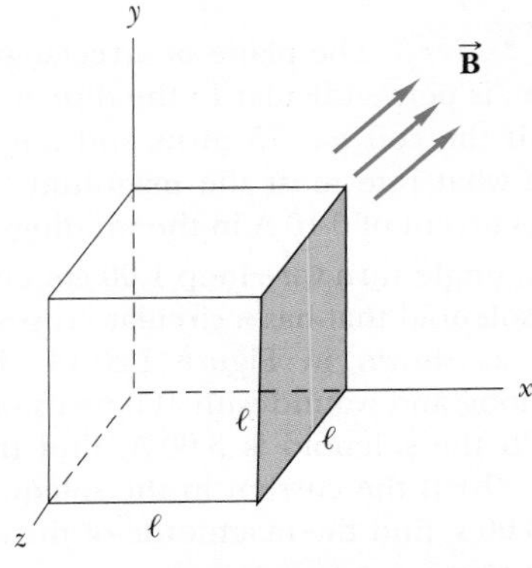

Figure P20.7

emerging from the volume enclosed by the cube (i.e., the total flux through all six faces)?

Section 20.2 Faraday's Law of Induction

8. Transcranial magnetic stimulation (TMS) is a noninvasive technique used to stimulate regions of the human brain. A small coil is placed on the scalp, and a brief burst of current in the coil produces a rapidly changing magnetic field inside the brain. The induced emf can be sufficient to stimulate neuronal activity. One such device generates a magnetic field within the brain that rises from zero to 1.5 T in 120 ms. Determine the induced emf within a circle of tissue of radius 1.6 mm and that is perpendicular to the direction of the field.

9. A square, single-turn coil 0.20 m on a side is placed with its plane perpendicular to a constant magnetic field. An emf of 18 mV is induced in the coil winding when the area of the coil decreases at the rate of 0.10 m^2/s. What is the magnitude of the magnetic field?

10. The flexible loop in Figure P20.10 has a radius of 12 cm and is in a magnetic field of strength 0.15 T. The loop is grasped at points A and B and stretched until its area is nearly zero. If it takes 0.20 s to close the loop, find the magnitude of the average induced emf in it during this time.

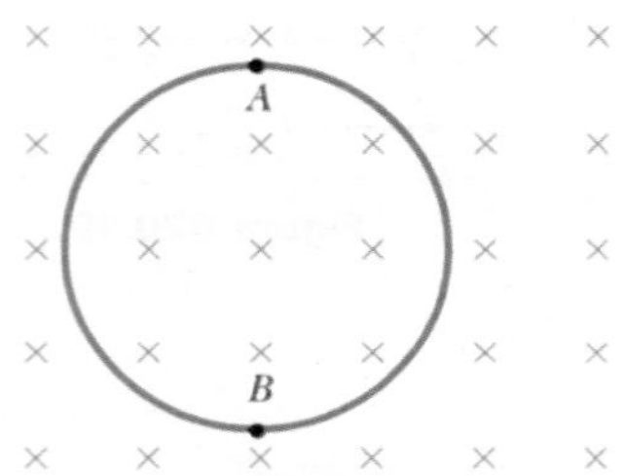

Figure P20.10

11. A wire loop of radius 0.30 m lies so that an external magnetic field of magnitude 0.30 T is perpendicular to the loop. The field reverses its direction, and its magnitude changes to 0.20 T in 1.5 s. Find the magnitude of the average induced emf in the loop during this time.

12. A 500-turn circular-loop coil 15.0 cm in diameter is initially aligned so that its axis is parallel to the Earth's magnetic field. In 2.77 ms, the coil is flipped so that its axis is perpendicular to the Earth's magnetic field. If an average voltage of 0.166 V is thereby induced in the coil, what is the value of the Earth's magnetic field at that location?

13. Physics Now™ The plane of a rectangular coil, 5.0 cm by 8.0 cm, is perpendicular to the direction of a magnetic field $\vec{\mathbf{B}}$. If the coil has 75 turns and a total resistance of 8.0 Ω, at what rate must the magnitude of $\vec{\mathbf{B}}$ change to induce a current of 0.10 A in the windings of the coil?

14. A square, single-turn wire loop 1.00 cm on a side is placed inside a solenoid that has a circular cross section of radius 3.00 cm, as shown in Figure P20.14. The solenoid is 20.0 cm long and wound with 100 turns of wire. (a) If the current in the solenoid is 3.00 A, find the flux through the loop. (b) If the current in the solenoid is reduced to zero in 3.00 s, find the magnitude of the average induced emf in the loop.

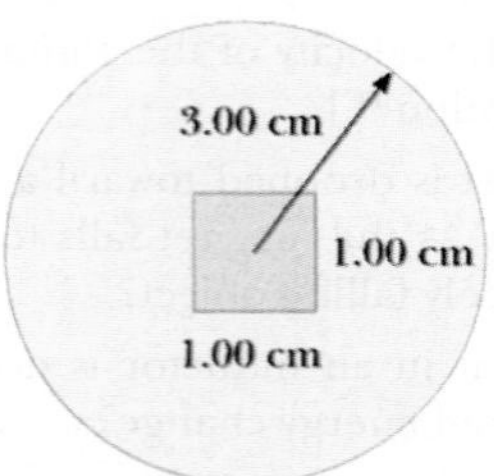

Figure P20.14

15. A 300-turn solenoid with a length of 20 cm and a radius of 1.5 cm carries a current of 2.0 A. A second coil of four turns is wrapped tightly about this solenoid so that it can be considered to have the same radius as the solenoid. Find (a) the change in the magnetic flux through the coil and (b) the magnitude of the average induced emf in the coil when the current in the solenoid increases to 5.0 A in a period of 0.90 s.

16. A circular coil enclosing an area of 100 cm^2 is made of 200 turns of copper wire. The wire making up the coil has resistance of 5.0 Ω, and the ends of the wire are connected to form a closed circuit. Initially, a 1.1-T uniform magnetic field points perpendicularly upward through the plane of the coil. The direction of the field then reverses so that the final magnetic field has a magnitude of 1.1 T and points downward through the coil. If the time required for the field to reverse directions is 0.10 s, what average current flows through the coil during that time?

17. To monitor the breathing of a hospital patient, a thin belt is girded around the patient's chest as in Figure P20.17. The belt is a 200-turn coil. When the patient inhales, the area encircled by the coil increases by 39.0 cm^2. The magnitude of the Earth's magnetic field is 50.0 μT and makes an angle of 28.0° with the plane of the coil. Assuming a patient takes 1.80 s to inhale, find the magnitude of the average induced emf in the coil during that time.

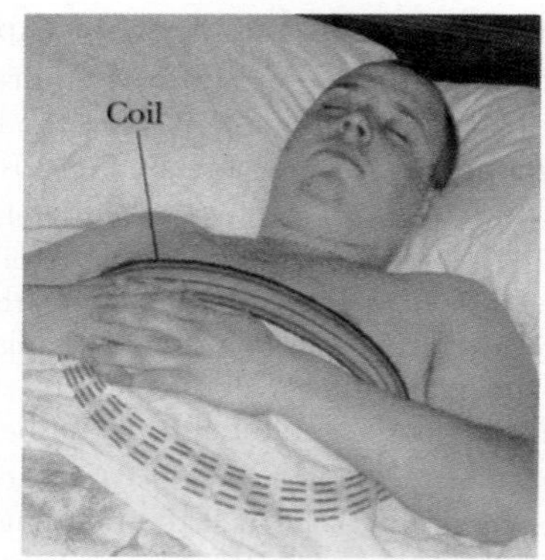

Figure P20.17

Section 20.3 Motional emf

18. Consider the arrangement shown in Figure P20.18. Assume that $R = 6.00\ \Omega$ and $\ell = 1.20$ m, and that a uniform 2.50-T magnetic field is directed *into* the page. At what speed should the bar be moved to produce a current of 0.500 A in the resistor?

19. A Boeing 747 jet with a wingspan of 60.0 m is flying horizontally at a speed of 300 m/s over Phoenix, Arizona, at a location where the Earth's magnetic field is 50.0 μT at

58.0° below the horizontal. What voltage is generated between the wingtips?

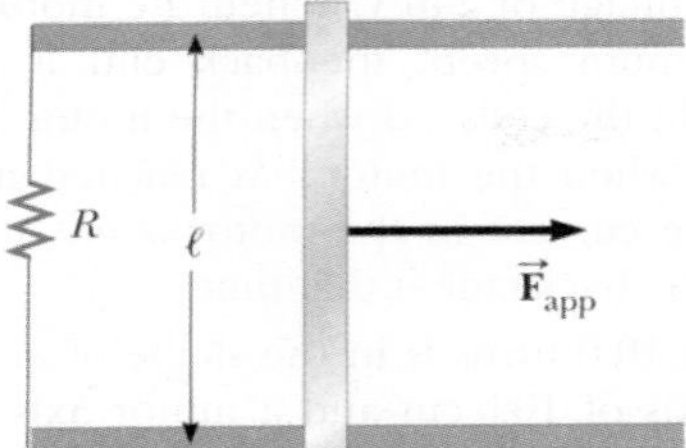

Figure P20.18 (Problems 18 and 57)

20. A 12.0-m-long steel beam is accidentally dropped by a construction crane from a height of 9.00 m. The horizontal component of the Earth's magnetic field over the region is 18.0 μT. What is the induced emf in the beam just before impact with the Earth? Assume the long dimension of the beam remains in a horizontal plane, oriented perpendicular to the horizontal component of the Earth's magnetic field.

21. An automobile has a vertical radio antenna 1.20 m long. The automobile travels at 65.0 km/h on a horizontal road where the Earth's magnetic field is 50.0 μT, directed toward the north and downwards at an angle of 65.0° below the horizontal. (a) Specify the direction the automobile should move in order to generate the maximum motional emf in the antenna, with the top of the antenna positive relative to the bottom. (b) Calculate the magnitude of this induced emf.

22. A helicopter has blades of length 3.0 m, rotating at 2.0 rev/s about a central hub. If the vertical component of Earth's magnetic field is 5.0×10^{-5} T, what is the emf induced between the blade tip and the central hub?

Section 20.4 Lenz's Law Revisited (the Minus Signin Faraday's Law)

23. A bar magnet is positioned near a coil of wire as shown in Figure P20.23. What is the direction of the current in the resistor when the magnet is moved (a) to the left? (b) to the right?

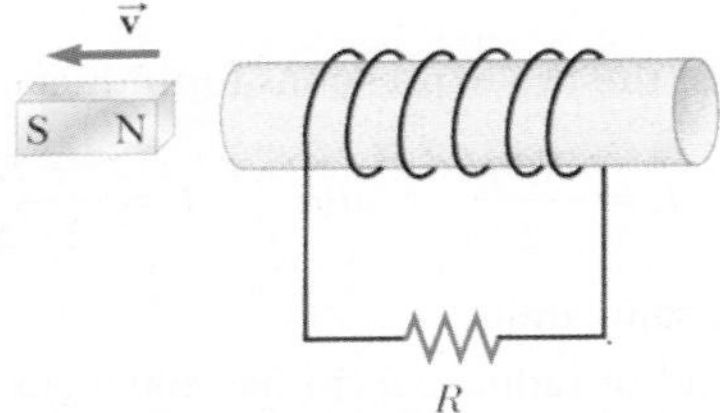

Figure P20.23

24. A conducting rectangular loop of mass M, resistance R, and dimensions w by ℓ falls from rest into a magnetic field $\vec{B}$ as shown in Figure P20.24. During the time interval before the top edge of the loop reaches the field, the loop approaches a terminal speed v_T. (a) Show that

$$v_T = \frac{MgR}{B^2w^2}$$

(b) Why is v_T proportional to R? (c) Why is it inversely proportional to B^2?

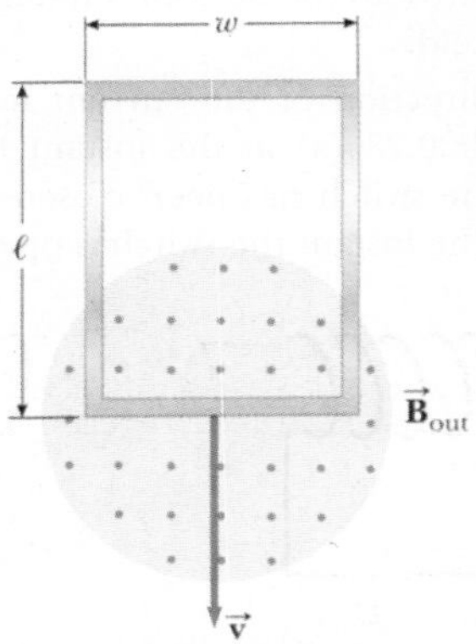

Figure P20.24

25. A rectangular coil with resistance R has N turns, each of length ℓ and width w as shown in Figure P20.25. The coil moves into a uniform magnetic field $\vec{B}$ with constant velocity $\vec{v}$. What are the magnitude and direction of the total magnetic force on the coil (a) as it enters the magnetic field, (b) as it moves within the field, and (c) as it leaves the field?

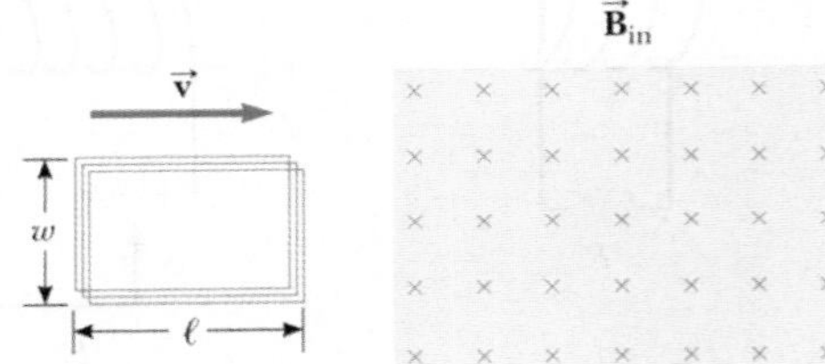

Figure P20.25

26. In Figure P20.26, what is the direction of the current induced in the resistor at the instant the switch is closed?

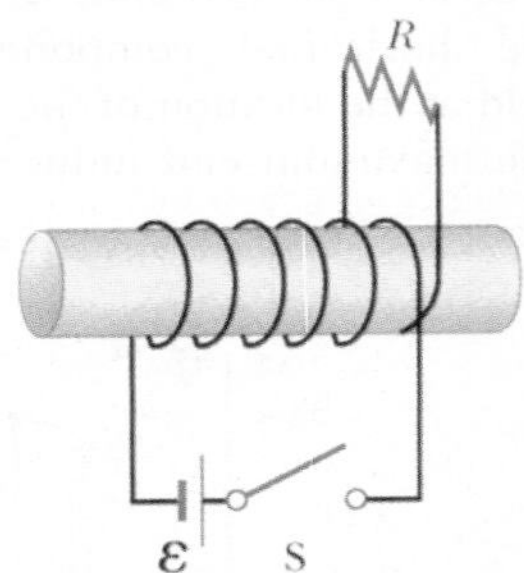

Figure P20.26

27. A copper bar is moved to the right while its axis is maintained in a direction perpendicular to a magnetic field, as shown in Figure P20.27. If the top of the bar becomes

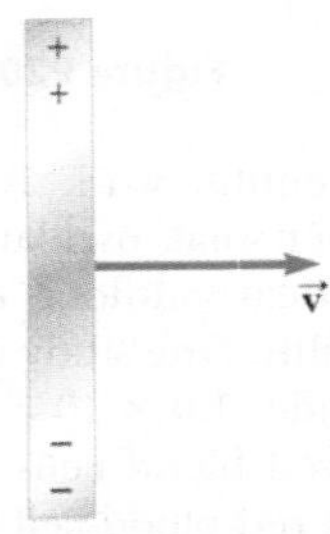

Figure P20.27

positive relative to the bottom, what is the direction of the magnetic field?

28. Find the direction of the current in the resistor shown in Figure P20.28 (a) at the instant the switch is closed, (b) after the switch has been closed for several minutes, and (c) at the instant the switch is opened.

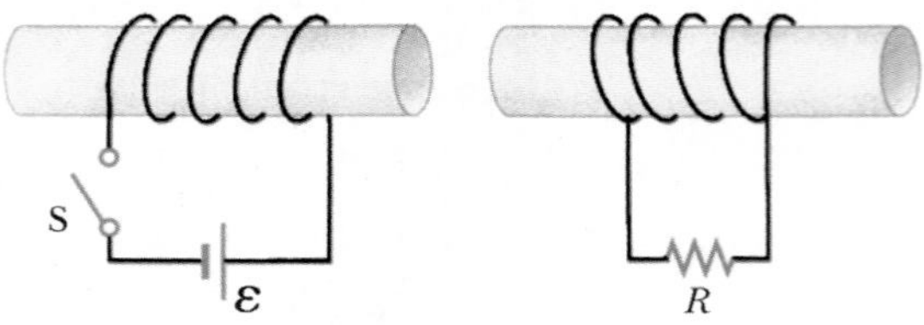

Figure P20.28

29. Find the direction of the current in the resistor R shown in Figure P20.29 after each of the following steps (taken in the order given): (a) The switch is closed. (b) The variable resistance in series with the battery is decreased. (c) The circuit containing resistor R is moved to the left. (d) The switch is opened.

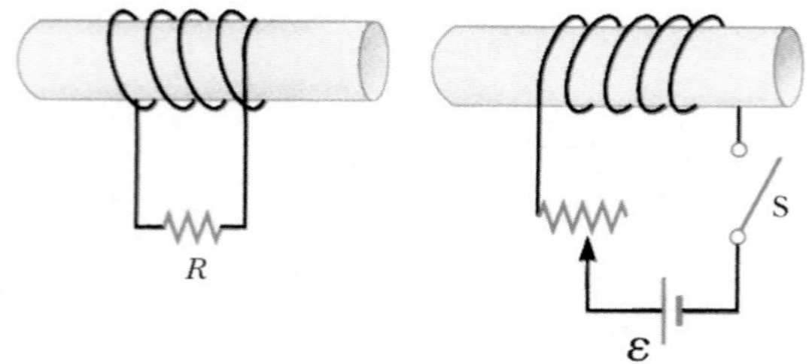

Figure P20.29

Section 20.5 Generators

30. A 100-turn square wire coil of area 0.040 m^2 rotates about a vertical axis at 1 500 rev/min, as indicated in Figure P20.30. The horizontal component of the Earth's magnetic field at the location of the loop is 2.0×10^{-5} T. Calculate the maximum emf induced in the coil by the Earth's field.

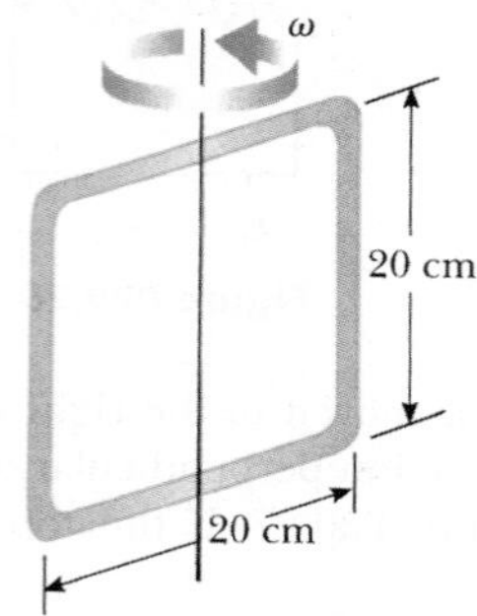

Figure P20.30

31. Considerable scientific work is currently underway to determine whether weak oscillating magnetic fields such as those found near outdoor electric power lines can effect human health. One study indicated that a magnetic field of magnitude 1.0×10^{-3} T, oscillating at 60 Hz, might stimulate red blood cells to become cancerous. If the diameter of a red blood cell is 8.0 μm, determine the maximum emf that can be generated around the perimeter of the cell.

32. A motor has coils with a resistance of 30 Ω and operates from a voltage of 240 V. When the motor is operating at its maximum speed, the back emf is 145 V. Find the current in the coils (a) when the motor is first turned on and (b) when the motor has reached maximum speed. (c) If the current in the motor is 6.0 A at some instant, what is the back emf at that time?

33. A coil of 10.0 turns is in the shape of an ellipse having a major axis of 10.0 cm and a minor axis of 4.00 cm. The coil rotates at 100 rpm in a region in which the magnitude of the Earth's magnetic field is 55 μT. What is the maximum voltage induced in the coil if the axis of rotation of the coil is along its major axis and is aligned (a) perpendicular to the Earth's magnetic field and (b) parallel to the Earth's magnetic field? (Note that the area of an ellipse is given by $A = \pi ab$, where a is the length of the *semi*major axis and b is the length of the *semi*minor axis.)

34. A flat coil enclosing an area of 0.10 m^2 is rotating at 60 rev/s, with its axis of rotation perpendicular to a 0.20-T magnetic field. (a) If there are 1 000 turns on the coil, what is the maximum voltage induced in the coil? (b) When the maximum induced voltage occurs, what is the orientation of the coil with respect to the magnetic field?

35. **PhysicsNow™** In a model AC generator, a 500-turn rectangular coil 8.0 cm by 20 cm rotates at 120 rev/min in a uniform magnetic field of 0.60 T. (a) What is the maximum emf induced in the coil? (b) What is the instantaneous value of the emf in the coil at $t = (\pi/32)$ s? Assume that the emf is zero at $t = 0$. (c) What is the smallest value of t for which the emf will have its maximum value?

Section 20.6 Self-Inductance

36. A coiled telephone cord forms a spiral with 70.0 turns, a diameter of 1.30 cm, and an unstretched length of 60.0 cm. Determine the self-inductance of one conductor in the unstretched cord.

37. A coil has an inductance of 3.0 mH, and the current in it changes from 0.20 A to 1.5 A in 0.20 s. Find the magnitude of the average induced emf in the coil during this period.

38. Show that the two expressions for inductance given by

$$L = \frac{N\Phi_B}{I} \quad \text{and} \quad L = \frac{-\mathcal{E}}{\Delta I/\Delta t}$$

have the same units.

39. A solenoid of radius 2.5 cm has 400 turns and a length of 20 cm. Find (a) its inductance and (b) the rate at which current must change through it to produce an emf of 75 mV.

40. An emf of 24.0 mV is induced in a 500-turn coil when the current is changing at a rate of 10.0 A/s. What is the magnetic flux through each turn of the coil at an instant when the current is 4.00 A?

Section 20.7 *RL* Circuits

41. Show that the SI units for the inductive time constant $\tau = L/R$ are seconds.

42. An RL circuit with $L = 3.00$ H and an RC circuit with $C = 3.00\ \mu$F have the same time constant. If the two circuits have the same resistance R, (a) what is the value of R and (b) what is this common time constant?

43. A 6.0-V battery is connected in series with a resistor and an inductor. The series circuit has a time constant of 600 μs, and the maximum current is 300 mA. What is the value of the inductance?

44. A 25-mH inductor, an 8.0-Ω resistor, and a 6.0-V battery are connected in series. The switch is closed at $t = 0$. Find the voltage drop across the resistor (a) at $t = 0$ and (b) after one time constant has passed. Also, find the voltage drop across the inductor (c) at $t = 0$ and (d) after one time constant has elapsed.

45. PhysicsNow™ Calculate the resistance in an RL circuit in which $L = 2.50$ H and the current increases to 90.0% of its final value in 3.00 s.

46. Consider the circuit shown in Figure P20.46. Take $\mathcal{E} = 6.00$ V, $L = 8.00$ mH, and $R = 4.00\ \Omega$. (a) What is the inductive time constant of the circuit? (b) Calculate the current in the circuit 250 μs after the switch is closed. (c) What is the value of the final steady-state current? (d) How long does it take the current to reach 80.0% of its maximum value?

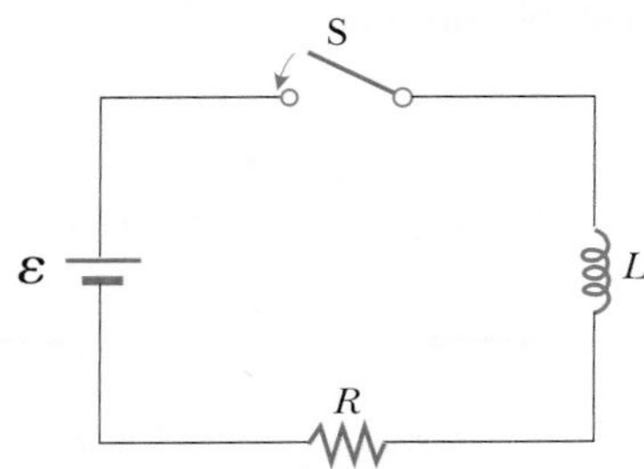

Figure P20.46

20.8 Energy Stored in a Magnetic Field

47. How much energy is stored in a 70.0-mH inductor at an instant when the current is 2.00 A?

48. A 300-turn solenoid has a radius of 5.00 cm and a length of 20.0 cm. Find (a) the inductance of the solenoid and (b) the energy stored in the solenoid when the current in its windings is 0.500 A.

49. A 24-V battery is connected in series with a resistor and an inductor, with $R = 8.0\ \Omega$ and $L = 4.0$ H, respectively. Find the energy stored in the inductor (a) when the current reaches its maximum value and (b) one time constant after the switch is closed.

ADDITIONAL PROBLEMS

50. What is the time constant for (a) the circuit shown in Figure P20.50a and (b) the circuit shown in Figure P20.50b?

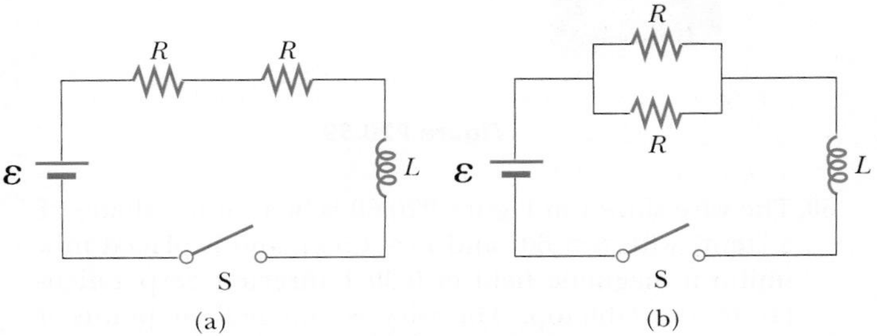

Figure P20.50

51. In Figure P20.51, the bar magnet is being moved toward the loop. Is $(V_a - V_b)$ positive, negative, or zero during this motion? Explain.

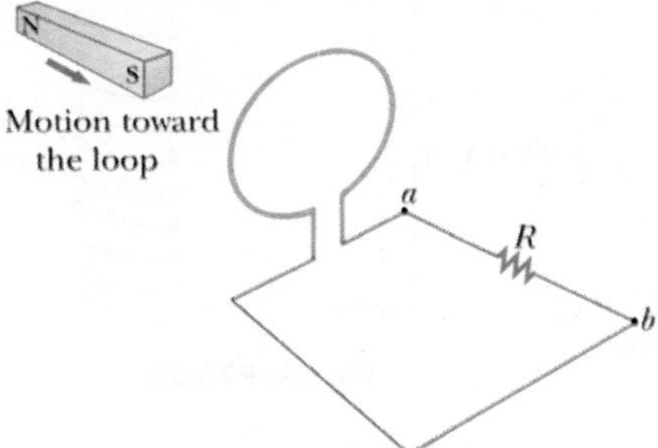

Figure P20.51

52. Your physics teacher asks you to help her set up a demonstration of Faraday's law for the class. The apparatus consists of a strong permanent magnet that has a field of 0.10 T, a small 10-turn coil of radius 2.0 cm cemented on a wood frame with a handle, some flexible connecting wires, and an ammeter, as in Figure P20.52. The idea is to pull the coil out of the center of the magnetic field as quickly as possible and read the average current registered on the meter. The combined resistance of the coil, leads, and meter is 2.0 Ω, and you must flip the coil out of the field in about 0.20 s. The ammeter you must use has a full-scale sensitivity of 1 000 μA. Will this meter be sensitive enough to show the induced current clearly?

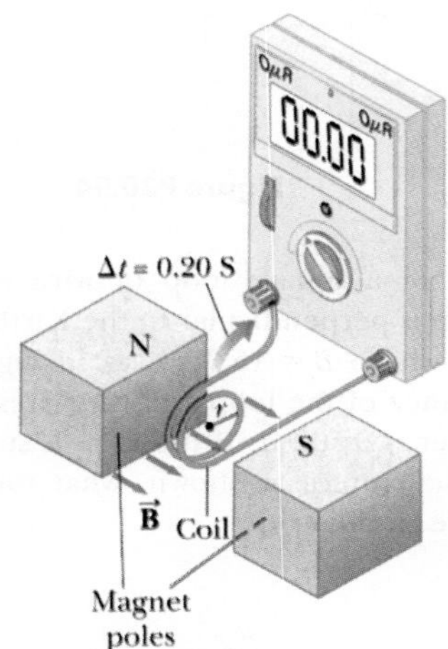

Figure P20.52

53. PhysicsNow™ An 820-turn wire coil of resistance 24.0 Ω is placed on top of a 12 500-turn, 7.00-cm-long solenoid, as in Figure P20.53 (page 690). Both coil and solenoid have cross-sectional areas of $1.00 \times 10^{-4}\ m^2$. (a) How long does it take the solenoid current to reach 0.632 times its maximum value? (b) Determine the average back emf caused by the self-inductance of the solenoid during this interval. The magnetic field produced by the solenoid at the location of the coil is one-half as strong as the field at the center of the solenoid. (c) Determine the average rate of change in magnetic flux through each turn of the coil during the stated interval. (d) Find the magnitude of the average induced current in the coil.

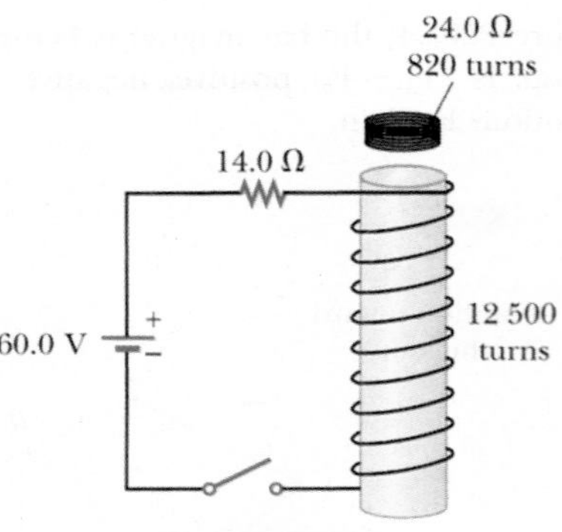

Figure P20.53

54. Figure P20.54 is a graph of induced emf versus time for a coil of N turns rotating with angular speed ω in a uniform magnetic field directed perpendicular to the axis of rotation of the coil. Copy this sketch (increasing the scale), and, on the same set of axes, show the graph of emf versus t when (a) the number of turns in the coil is doubled, (b) the angular speed is doubled, and (c) the angular speed is doubled while the number of turns in the coil is halved.

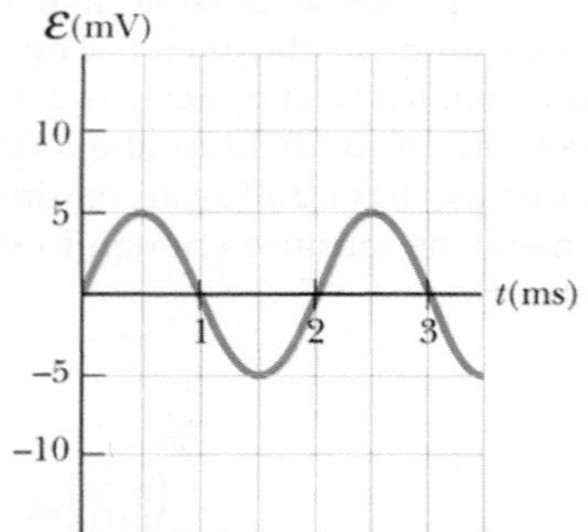

Figure P20.54

55. The plane of a square loop of wire with edge length $a = 0.200$ m is perpendicular to the Earth's magnetic field at a point where $B = 15.0\ \mu$T, as in Figure P20.55. The total resistance of the loop and the wires connecting it to the ammeter is 0.500 Ω. If the loop is suddenly collapsed by horizontal forces as shown, what total charge passes through the ammeter?

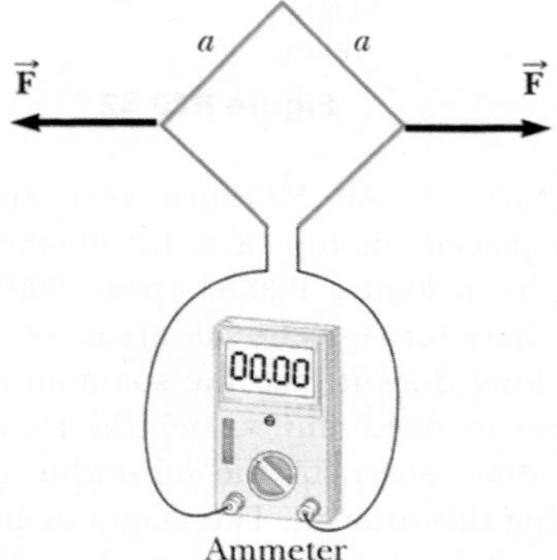

Figure P20.55

56. A novel method of storing electrical energy has been proposed. A huge underground superconducting coil 1.00 km in diameter would be fabricated. It would carry a maximum current of 50.0 kA through each winding of a 150-turn Nb_3Sn solenoid. (a) If the inductance of this huge coil were 50.0 H, what is the total stored energy? (b) What is the compressive force per meter acting between two adjacent windings 0.250 m apart? [*Hint:* Because the radius of the coil is so large, the magnetic field created by one winding and acting on an adjacent turn can be considered to be that of a long, straight wire.]

57. A conducting rod of length ℓ moves on two horizontal frictionless rails, as in Figure P20.18. A constant force of magnitude 1.00 N moves the bar at a uniform speed of 2.00 m/s through a magnetic field $\vec{B}$ that is directed into the page. (a) What is the current in an 8.00-Ω resistor R? (b) What is the rate of energy dissipation in the resistor? (c) What is the mechanical power delivered by the constant force?

58. The square loop in Figure P20.58 is made of wires with a total series resistance of 10.0 Ω. It is placed in a uniform 0.100-T magnetic field directed perpendicular into the plane of the paper. The loop, which is hinged at each corner, is pulled as shown until the separation between points A and B is 3.00 m. If this process takes 0.100 s, what is the average current generated in the loop? What is the direction of the current?

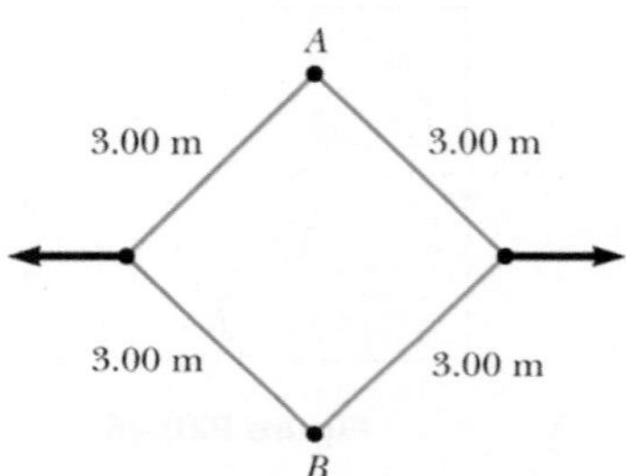

Figure P20.58

59. The bolt of lightning depicted in Figure P20.59 passes 200 m from a 100-turn coil oriented as shown. If the current in the lightning bolt falls from 6.02×10^6 A to zero in 10.5 μs, what is the average voltage induced in the coil? Assume that the distance to the center of the coil determines the average magnetic field at the coil's position. Treat the lightning bolt as a long, vertical wire.

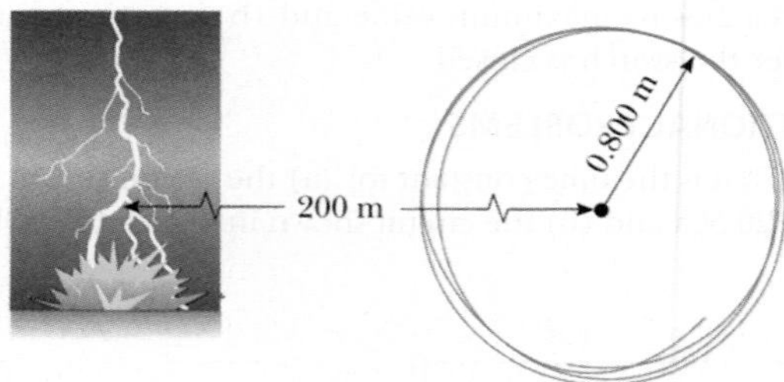

Figure P20.59

60. The wire shown in Figure P20.60 is bent in the shape of a "tent" with $\theta = 60°$ and $L = 1.5$ m, and is placed in a uniform magnetic field of 0.30 T directed perpendicular to the tabletop. The wire is "hinged" at points a and b. If the tent is flattened out on the table in 0.10 s,

what is the average induced emf in the wire during this time?

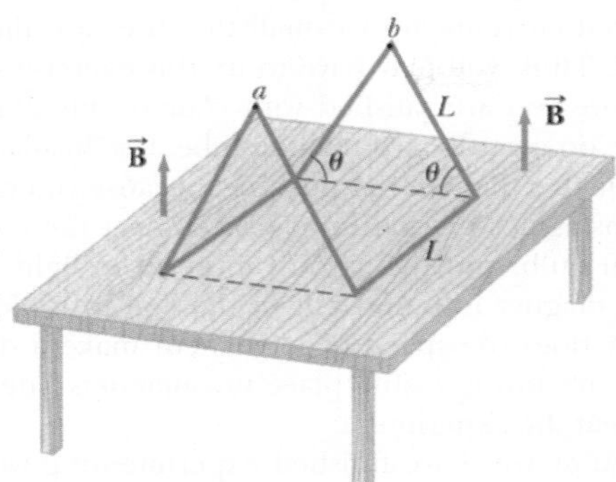

Figure P20.60

61. The magnetic field shown in Figure P20.61 has a uniform magnitude of 25.0 mT directed into the paper. The initial diameter of the kink is 2.00 cm. (a) The wire is quickly pulled taut, and the kink shrinks to a diameter of zero in 50.0 ms. Determine the average voltage induced between endpoints A and B. Include the polarity. (b) Suppose the kink is undisturbed, but the magnetic field increases to 100 mT in 4.00×10^{-3} s. Determine the average voltage across terminals A and B, including polarity, during this period.

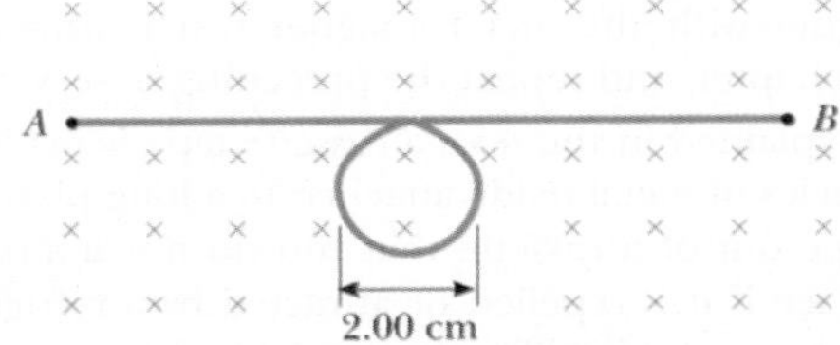

Figure P20.61

62. An aluminum ring of radius 5.00 cm and resistance $3.00 \times 10^{-4}\ \Omega$ is placed around the top of a long air-core solenoid with 1 000 turns per meter and a smaller radius of 3.00 cm, as in Figure P20.62. If the current in the solenoid is increasing at a constant rate of 270 A/s, what is the induced current in the ring? Assume that the magnetic field produced by the solenoid over the area at the end of the solenoid is one-half as strong as the field at the center of the solenoid. Assume also that the solenoid produces a negligible field outside its cross-sectional area.

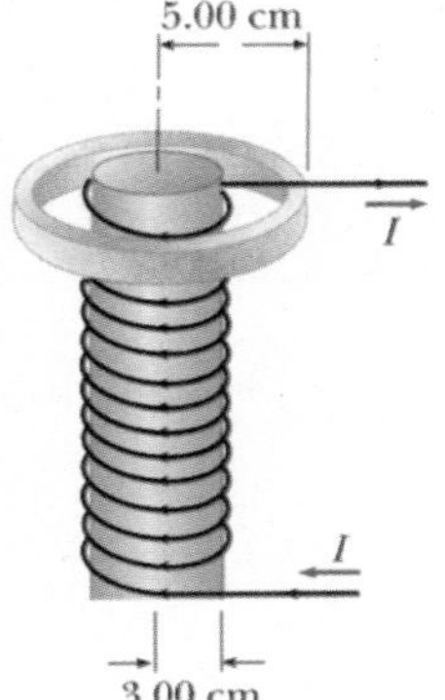

Figure P20.62

63. In Figure P20.63, the rolling axle, 1.50 m long, is pushed along horizontal rails at a constant speed v = 3.00 m/s. A resistor R = 0.400 Ω is connected to the rails at points a and b, directly opposite each other. (The wheels make good electrical contact with the rails, so the axle, rails, and R form a closed-loop circuit. The only significant resistance in the circuit is R.) A uniform magnetic field B = 0.800 T is directed vertically downwards. (a) Find the induced current I in the resistor. (b) What horizontal force $\vec{\mathbf{F}}$ is required to keep the axle rolling at constant speed? (c) Which end of the resistor, a or b, is at the higher electric potential? (d) After the axle rolls past the resistor, does the current in R reverse direction?

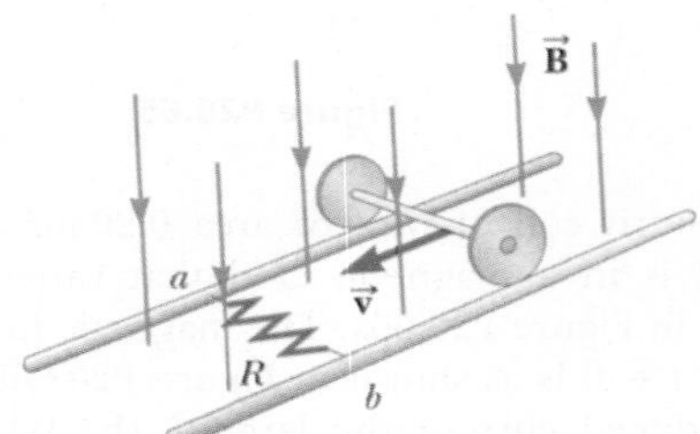

Figure P20.63

64. In 1832, Faraday proposed that the apparatus shown in Figure P20.64 could be used to generate electric current from the water flowing in the Thames River.[2] Two conducting plates of length a and width b are placed facing one another on opposite sides of the river, a distance w apart and immersed entirely. The flow velocity of the river is $\vec{\mathbf{v}}$, and the vertical component of Earth's magnetic field is B. Show that the current in the load resistor R is

$$I = \frac{abvB}{\rho + abR/w}$$

where ρ is the resistivity of the water. (b) Calculate the short-circuit current (R = 0) if a = 100 m, b = 5.00 m, v = 3.00 m/s, B = 50.0 μT, and ρ = 100 Ω · m.

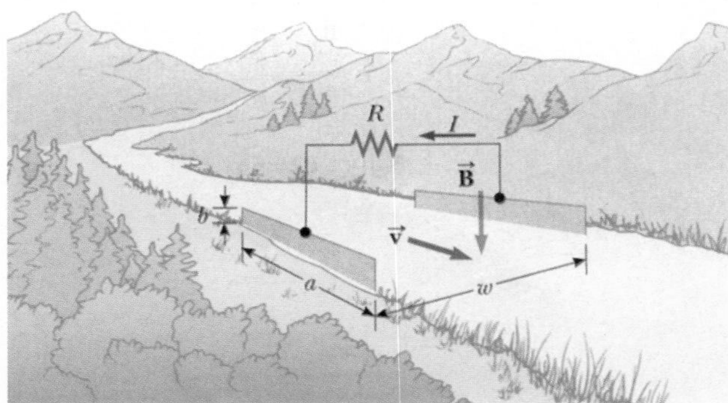

Figure P20.64

65. A horizontal wire is free to slide on the vertical rails of a conducting frame, as in Figure P20.65 (page 692). The wire has mass m and length ℓ, and the resistance of the circuit is R. If a uniform magnetic field is directed perpendicular to

[2]The idea for this problem and Figure P20.64 is from Oleg D. Jefimenko, *Electricity and Magnetism: An Introduction to the Theory of Electric and Magnetic Fields* (Star City, WV, Electret Scientific Co., 1989).

the frame, what is the terminal speed of the wire as it falls under the force of gravity? (Neglect friction.)

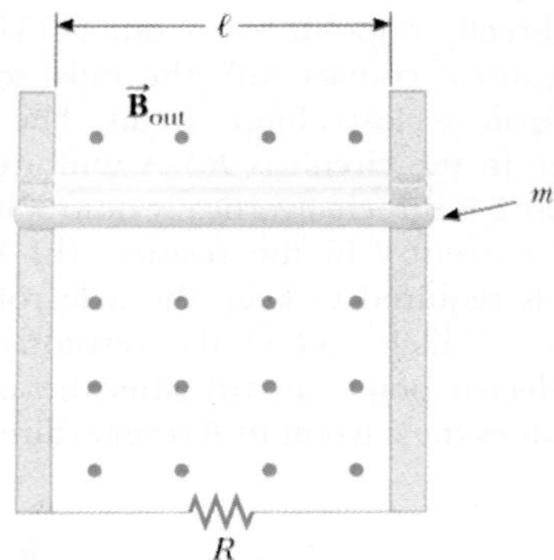

Figure P20.65

66. A one-turn coil of wire of area 0.20 m^2 and resistance 0.25 Ω is in a magnetic field that varies with time as shown in Figure P20.66a. The magnetic flux through the coil at $t = 0$ is as shown in Figure P20.66b. (a) When is the induced current the largest? (b) When is it zero? (c) Is the induced current always in the same direction? (d) Find the direction (clockwise or counterclockwise) and magnitude of the current between times $t = 0$ and $t = 2.0$ s, between $t = 2.0$ s and $t = 4.0$ s, and between $t = 4.0$ s and $t = 6.0$ s.

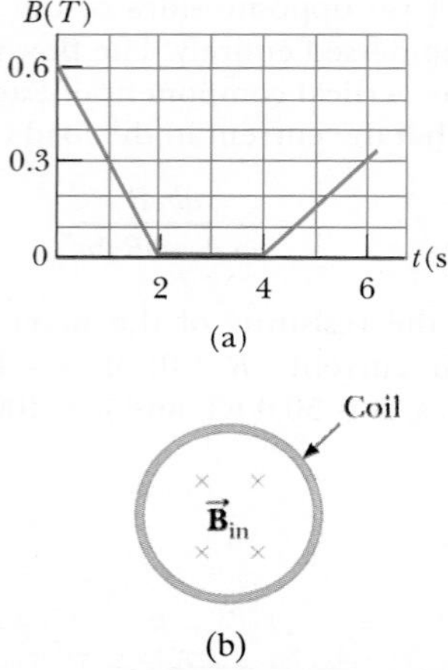

Figure P20.66

ACTIVITIES

1. Experimenting with induced currents is not easy. For small magnets and small coils of wire, the resulting induced currents are so small that they are difficult to detect. Thus, you may have to try this exercise several times before you are satisfied with your results. Wind a coil of wire on a cardboard mailing tube. Use insulated wire with as small a diameter as possible, because you need as many turns as possible on the coil. Connect the coil to a flashlight bulb, and see if you can get it to light by moving a bar magnet into and out of the coil in rapid succession. Why does the speed of movement make a difference? If you are unsuccessful, place two magnets side by side and repeat the experiment.

After you have finished experimenting with the bulb, ask your instructor to let you use a galvanometer as a current detector. These devices are capable of measuring very small currents, and they have the added advantage of detecting the direction of the current in a circuit.

Use your equipment to observe or test the following: (a) Does the magnitude of the induced current depend on the speed of movement of the magnet? (b) Can you induce a current by holding the magnet still and moving the coil over it? (c) Does the direction of the current depend on whether the magnet is pushed in or pulled out of the coil? (d) Does the direction of the current depend on whether the inserted pole of the magnet is the north pole or the south pole? (e) Can you predict the direction of the current by using Lenz's law? (f) Replace your bar magnet with the electromagnet you constructed in the last chapter, and repeat the preceding observations.

2. As explained in the text, a cassette tape is made up of tiny particles of metal oxide attached to a long plastic strip. Pull a tape out of a cassette that you do not mind destroying, and see if it is repelled or attracted by a refrigerator magnet. Also, try this with an expendable floppy computer disk.

3. This experiment takes steady hands, a dime, and a strong magnet. After verifying that a dime is not attracted to the magnet, carefully balance the coin on its edge. (This will not work with other coins, because they require too much force to topple them.) Hold one pole of the magnet within a millimeter of the face of the dime, but do not make contact with it. Now very rapidly pull the magnet straight back away from the coin. Which way does the dime tip? Does the coin fall the same way most of the time? Explain what is going on in terms of Lenz's law.

Arecibo, a large radio telescope in Puerto Rico, gathers electromagnetic radiation in the form of radio waves. These long wavelengths pass through obscuring dust clouds, allowing astronomers to create images of the core region of the Milky Way Galaxy, which can't be observed in the visible spectrum.

CHAPTER

21

Alternating Current Circuits and Electromagnetic Waves

OUTLINE

Every time we turn on a television set, a stereo system, or any of a multitude of other electric appliances, we call on alternating currents (AC) to provide the power to operate them. We begin our study of AC circuits by examining the characteristics of a circuit containing a source of emf and one other circuit element: a resistor, a capacitor, or an inductor. Then we examine what happens when these elements are connected in combination with each other. Our discussion is limited to simple series configurations of the three kinds of elements.

We conclude this chapter with a discussion of **electromagnetic waves**, which are composed of fluctuating electric and magnetic fields. Electromagnetic waves in the form of visible light enable us to view the world around us; infrared waves warm our environment; radio-frequency waves carry our television and radio programs, as well as information about processes in the core of our galaxy. X-rays allow us to perceive structures hidden inside our bodies, and study properties of distant, collapsed stars. Light is key to our understanding of the universe.

21.1 RESISTORS IN AN AC CIRCUIT

An AC circuit consists of combinations of circuit elements and an AC generator or an AC source, which provides the alternating current. We have seen that the output of an AC generator is sinusoidal and varies with time according to

$$\Delta v = \Delta V_{\max} \sin 2\pi f t \qquad [21.1]$$

where Δv is the instantaneous voltage, $\Delta V_{\max}$ is the maximum voltage of the AC generator, and f is the frequency at which the voltage changes, measured in hertz (Hz). (Compare Equations 20.7 and 20.8 with Equation 21.1.) We first consider a simple

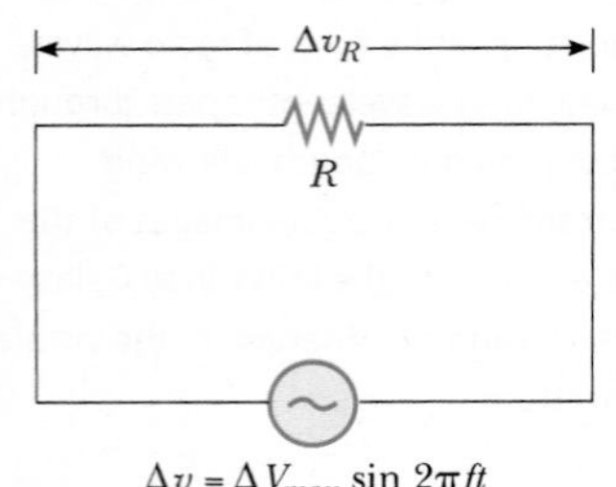

ACTIVE FIGURE 21.1
A series circuit consisting of a resistor R connected to an AC generator, designated by the symbol

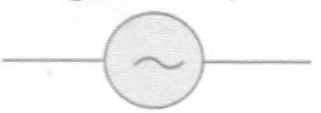

Physics Now™
Log into PhysicsNow at **www.cp7e.com** and go to Active Figure 21.1, where you can adjust the resistance, the frequency, and the maximum voltage of the circuit shown. The results can be studied with the graph and phasor diagram in Active Figure 21.2.

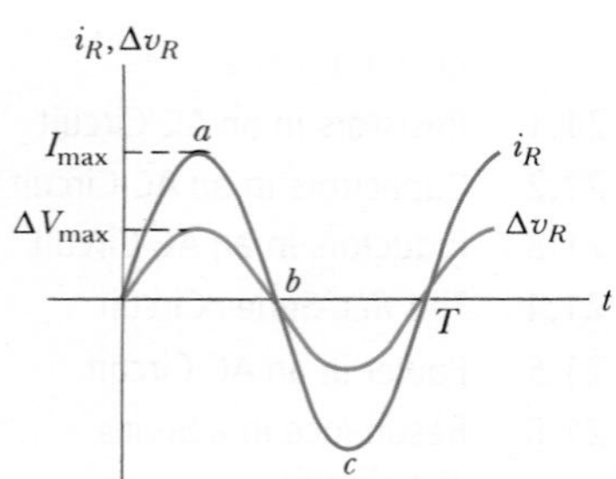

ACTIVE FIGURE 21.2
A plot of current and voltage across a resistor versus time.

Physics Now™
Log into PhysicsNow at **www.cp7e.com** and go to Active Figure 21.1, where you can adjust the resistance, the frequency, and the maximum voltage of the circuit in Active Figure 21.1. The results can be studied with the graph and phasor diagram in Active Figure 21.2.

circuit consisting of a resistor and an AC source (designated by the symbol ——(~)——), as in Active Figure 21.1. The current and the voltage across the resistor are shown in Active Figure 21.2.

To explain the concept of alternating current, we begin by discussing the current-versus-time curve in Active Figure 21.2. At point a on the curve, the current has a maximum value in one direction, arbitrarily called the positive direction. Between points a and b, the current is decreasing in magnitude but is still in the positive direction. At point b, the current is momentarily zero; it then begins to increase in the opposite (negative) direction between points b and c. At point c, the current has reached its maximum value in the negative direction.

The current and voltage are in step with each other because they vary identically with time. **Because the current and the voltage reach their maximum values at the same time, they are said to be in phase**. Notice that **the average value of the current over one cycle is zero**. This is because the current is maintained in one direction (the positive direction) for the same amount of time and at the same magnitude as it is in the opposite direction (the negative direction). However, the direction of the current has no effect on the behavior of the resistor in the circuit: the collisions between electrons and the fixed atoms of the resistor result in an increase in the resistor's temperature regardless of the direction of the current.

We can quantify this discussion by recalling that the rate at which electrical energy is dissipated in a resistor, the power $\mathcal{P}$, is

$$\mathcal{P} = i^2 R$$

where i is the *instantaneous* current in the resistor. Because the heating effect of a current is proportional to the *square* of the current, it makes no difference whether the sign associated with the current is positive or negative. However, the heating effect produced by an alternating current with a maximum value of I_{max} *is not the same* as that produced by a direct current of the same value. The reason is that the alternating current has this maximum value for only an instant of time during a cycle. The important quantity in an AC circuit is a special kind of average value of current, called the **rms current**—the direct current that dissipates the same amount of energy in a resistor that is dissipated by the actual alternating current. To find the rms current, we first square the current, Then find its average value, and finally take the square root of this average value. Hence, the rms current is the square *root* of the average (*mean*) of the *square* of the current. Because i^2 varies as $\sin^2 2\pi ft$, the average value of i^2 is $\frac{1}{2} I_{max}^2$(Fig. 21.3b).[1] Therefore, the rms current I_{rms} is related to the maximum value of the alternating current I_{max} by

$$I_{rms} = \frac{I_{max}}{\sqrt{2}} = 0.707 I_{max} \qquad [21.2]$$

This equation says that an alternating current with a maximum value of 3 A produces the same heating effect in a resistor as a direct current of $(3/\sqrt{2})$ A. We can therefore say that the average power dissipated in a resistor that carries alternating current I is

$$\mathcal{P}_{av} = I_{rms}^2 R.$$

[1]The fact that $(i^2)_{av} = I_{max}^2/2$ can be shown as follows: The current in the circuit varies with time according to the expression $i = I_{max} \sin 2\pi ft$, so $i^2 = I_{max}^2 \sin^2 2\pi ft$. Therefore, we can find the average value of i^2 by calculating the average value of $\sin^2 2\pi ft$. Note that a graph of $\cos^2 2\pi ft$ versus time is identical to a graph of $\sin^2 2\pi ft$ versus time, except that the points are shifted on the time axis. Thus, the time average of $\sin^2 2\pi ft$ is equal to the time average of $\cos^2 2\pi ft$, taken over one or more cycles. That is,

$$(\sin^2 2\pi ft)_{av} = (\cos^2 2\pi ft)_{av}$$

With this fact and the trigonometric identity $\sin^2 \theta + \cos^2 \theta = 1$, we get

$$(\sin^2 2\pi ft)_{av} + (\cos^2 2\pi ft)_{av} = 2(\sin^2 2\pi ft)_{av} = 1$$

$$(\sin^2 2\pi ft)_{av} = \tfrac{1}{2}$$

When this result is substituted into the expression $i^2 = I_{max}^2 \sin^2 2\pi ft$, we get $(i^2)_{av} = I_{rms}^2 = I_{max}^2/2$, or $I_{rms} = I_{max}/\sqrt{2}$, where I_{rms} is the rms current.

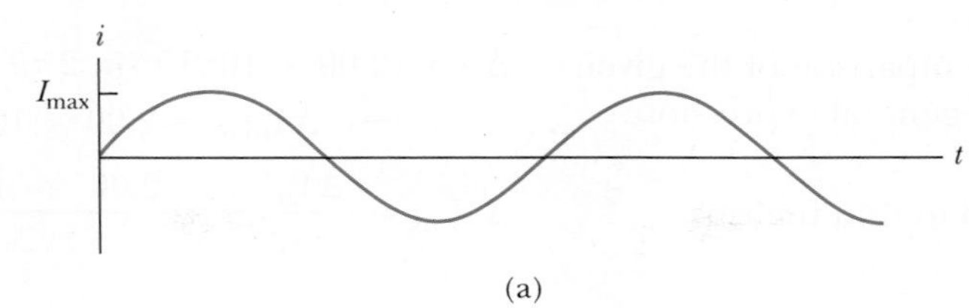

(a)

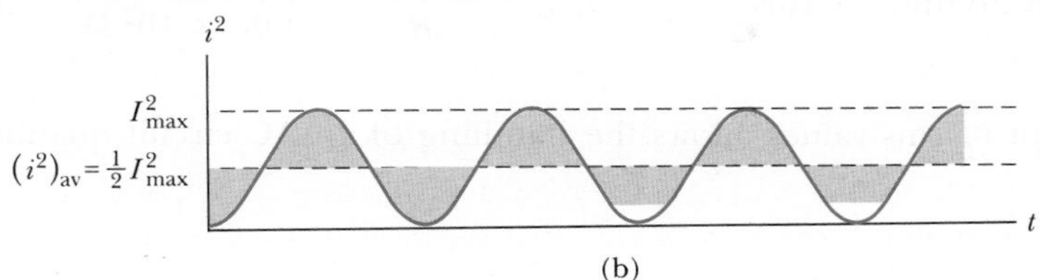

(b)

Figure 21.3 (a) Plot of the current in a resistor as a function of time. (b) Plot of the square of the current in a resistor as a function of time. Notice that the gray shaded regions *under* the curve and *above* the dashed line for $I^2_{max}/2$ have the same area as the gray shaded regions *above* the curve and *below* the dashed line for $I^2_{max}/2$. Thus, the average value of I^2 is $I^2_{max}/2$.

Alternating voltages are also best discussed in terms of rms voltages, with a relationship identical to the preceding one,

$$\Delta V_{rms} = \frac{\Delta V_{max}}{\sqrt{2}} = 0.707\,\Delta V_{max} \qquad [21.3]$$

◀ rms voltage

where ΔV_{rms} is the rms voltage and ΔV_{max} is the maximum value of the alternating voltage.

When we speak of measuring an AC voltage of 120 V from an electric outlet, we really mean an rms voltage of 120 V. A quick calculation using Equation 21.3 shows that such an AC voltage actually has a peak value of about 170 V. In this chapter we use rms values when discussing alternating currents and voltages. One reason is that AC ammeters and voltmeters are designed to read rms values. Further, if we use rms values, many of the equations for alternating current will have the same form as those used in the study of direct-current (DC) circuits. Table 21.1 summarizes the notations used throughout the chapter.

TABLE 21.1

Notation Used in This Chapter

	Voltage	Current
Instantaneous value	Δv	i
Maximum value	ΔV_{max}	I_{max}
rms value	ΔV_{rms}	I_{rms}

Consider the series circuit in Figure 21.1, consisting of a resistor connected to an AC generator. A resistor impedes the current in an AC circuit, just as it does in a DC circuit. Ohm's law is therefore valid for an AC circuit, and we have

$$\Delta V_{R,rms} = I_{rms}R \qquad [21.4a]$$

The rms voltage across a resistor is equal to the rms current in the circuit times the resistance. This equation is also true if maximum values of current and voltage are used:

$$\Delta V_{R,max} = I_{max}R \qquad [21.4b]$$

Quick Quiz 21.1

Which of the following statements can be true for a resistor connected in a simple series circuit to an operating AC generator? (a) $\mathcal{P}_{av} = 0$ and $i_{av} = 0$ (b) $\mathcal{P}_{av} = 0$ and $i_{av} > 0$ (c) $\mathcal{P}_{av} > 0$ and $i_{av} = 0$ (d) $\mathcal{P}_{av} > 0$ and $i_{av} > 0$

EXAMPLE 21.1 What Is the rms Current?

Goal Perform basic AC circuit calculations for a purely resistive circuit.

Problem An AC voltage source has an output of $\Delta v = (2.00 \times 10^2 \text{ V}) \sin 2\pi ft$. This source is connected to a $1.00 \times 10^2\text{-}\Omega$ resistor as in Figure 21.1. Find the rms voltage and rms current in the resistor.

Strategy Compare the expression for the voltage output just given with the general form, $\Delta v = \Delta V_{max} \sin 2\pi ft$, finding the maximum voltage. Substitute this result into the expression for the rms voltage.

Solution

Obtain the maximum voltage by comparison of the given expression for the output with the general expression:

$$\Delta v = (2.00 \times 10^2 \text{ V}) \sin 2\pi ft \qquad \Delta v = \Delta V_{\text{max}} \sin 2\pi ft$$

$$\rightarrow \Delta V_{\text{max}} = 2.00 \times 10^2 \text{ V}$$

Next, substitute into Equation 21.3 to find the rms voltage of the source:

$$\Delta V_{\text{rms}} = \frac{\Delta V_{\text{max}}}{\sqrt{2}} = \frac{2.00 \times 10^2 \text{ V}}{\sqrt{2}} = \boxed{141 \text{ V}}$$

Substitute this result into Ohm's law to find the rms current:

$$I_{\text{rms}} = \frac{\Delta V_{\text{rms}}}{R} = \frac{141 \text{ V}}{1.00 \times 10^2 \, \Omega} = \boxed{1.41 \text{ A}}$$

Remarks Notice how the concept of rms values allows the handling of an AC circuit quantitatively in much the same way as a DC circuit.

Exercise 21.1

Find the maximum current in the circuit and the average power delivered to the circuit.

Answer 2.00 A; 2.00×10^2 W

21.2 CAPACITORS IN AN AC CIRCUIT

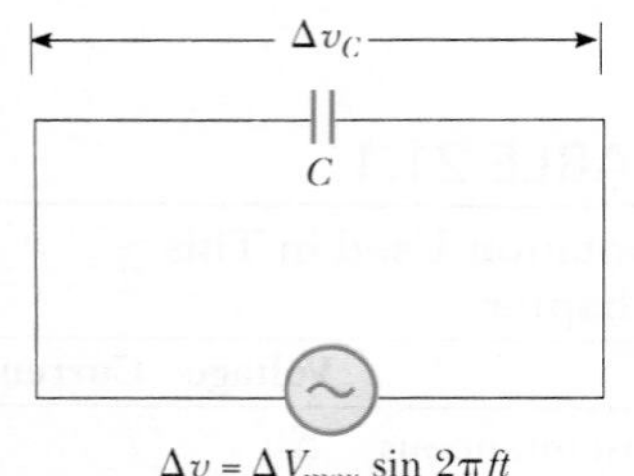

Figure 21.4 A series circuit consisting of a capacitor C connected to an AC generator.

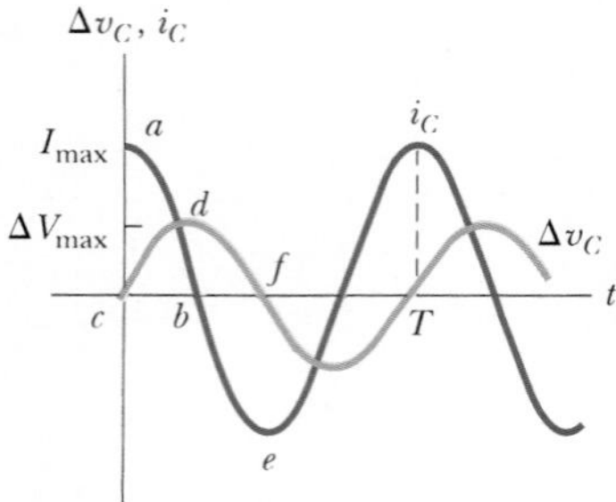

Figure 21.5 Plots of current and voltage across a capacitor versus time in an AC circuit. The voltage lags the current by 90°.

To understand the effect of a capacitor on the behavior of a circuit containing an AC voltage source, we first review what happens when a capacitor is placed in a circuit containing a DC source, such as a battery. When the switch is closed in a series circuit containing a battery, a resistor, and a capacitor, the initial charge on the plates of the capacitor is zero. The motion of charge through the circuit is therefore relatively free, and there is a large current in the circuit. As more charge accumulates on the capacitor, the voltage across it increases, opposing the current. After some time interval, which depends on the time constant RC, the current approaches zero. Consequently, a capacitor in a DC circuit limits or impedes the current so that it approaches zero after a brief time.

Now consider the simple series circuit in Figure 21.4, consisting of a capacitor connected to an AC generator. We sketch curves of current versus time and voltage versus time, and then attempt to make the graphs seem reasonable. The curves are shown in Figure 21.5. First, note that the segment of the current curve from a to b indicates that the current starts out at a rather large value. This can be understood by recognizing that there is no charge on the capacitor at $t = 0$; as a consequence, there is nothing in the circuit except the resistance of the wires to hinder the flow of charge at this instant. However, the current decreases as the voltage across the capacitor increases from c to d on the voltage curve. When the voltage is at point d, the current reverses and begins to increase in the opposite direction (from b to e on the current curve). During this time, the voltage across the capacitor decreases from d to f because the plates are now losing the charge they accumulated earlier. The remainder of the cycle for both voltage and current is a repeat of what happened during the first half of the cycle. The current reaches a maximum value in the opposite direction at point e on the current curve and then decreases as the voltage across the capacitor builds up.

In a purely resistive circuit, the current and voltage are always in step with each other. This isn't the case when a capacitor is in the circuit. In Figure 21.5, when an alternating voltage is applied across a capacitor, the voltage reaches its maximum value one-quarter of a cycle after the current reaches its maximum value. We say that **the voltage across a capacitor always lags the current by 90°**.

The voltage across a capacitor lags the current by 90° ▶

The impeding effect of a capacitor on the current in an AC circuit is expressed in terms of a factor called the **capacitive reactance** X_C, defined as

Capacitive reactance ▶

$$X_C \equiv \frac{1}{2\pi fC} \qquad [21.5]$$

When C is in farads and f is in hertz, the unit of X_C is the ohm. Notice that $2\pi f = \omega$, the angular frequency.

From Equation 21.5, as the frequency f of the voltage source increases, the capacitive reactance X_C (the impeding effect of the capacitor) decreases, so the current increases. At high frequency, there is less time available to charge the capacitor, so less charge and voltage accumulate on the capacitor, which translates into less opposition to the flow of charge and, consequently, a higher current. The analogy between capacitive reactance and resistance means that we can write an equation of the same form as Ohm's law to describe AC circuits containing capacitors. This equation relates the rms voltage and rms current in the circuit to the capacitive reactance:

$$\Delta V_{C,\mathrm{rms}} = I_{\mathrm{rms}} X_C \qquad [21.6]$$

EXAMPLE 21.2 A Purely Capactive AC Circuit

Goal Perform basic AC circuit calculations for a capacitive circuit.

Problem An 8.00-μF capacitor is connected to the terminals of an AC generator with an rms voltage of 1.50×10^2 V and a frequency of 60.0 Hz. Find the capacitive reactance and the rms current in the circuit.

Strategy Substitute values into Equations 21.5 and 21.6.

Solution

Substitute the values of f and C into Equation 21.5:

$$X_C = \frac{1}{2\pi f C} = \frac{1}{2\pi\,(60.0\text{ Hz})\,(8.00 \times 10^{-6}\text{ F})} = 332\ \Omega$$

Solve Equation 21.6 for the current, and substitute X_C and the rms voltage to find the rms current:

$$I_{\mathrm{rms}} = \frac{\Delta V_{C,\mathrm{rms}}}{X_C} = \frac{1.50 \times 10^2\text{ V}}{332\ \Omega} = 0.452\text{ A}$$

Remark Again, notice how similar the technique is to that of analyzing a DC circuit with a resistor.

Exercise 21.2

If the frequency is doubled, what happens to the capacitive reactance and the rms current?

Answer X_C is halved, and I_{rms} is doubled.

21.3 INDUCTORS IN AN AC CIRCUIT

Now consider an AC circuit consisting only of an inductor connected to the terminals of an AC source, as in Active Figure 21.6. (In any real circuit, there is some resistance in the wire forming the inductive coil, but we ignore this for now.) The changing current output of the generator produces a back emf that impedes the current in the circuit. The magnitude of this back emf is

$$\Delta v_L = L\frac{\Delta I}{\Delta t} \qquad [21.7]$$

The effective resistance of the coil in an AC circuit is measured by a quantity called the **inductive reactance**, X_L:

$$X_L \equiv 2\pi f L \qquad [21.8]$$

When f is in hertz and L is in henries, the unit of X_L is the ohm. The inductive reactance *increases* with increasing frequency and increasing inductance. Contrast these facts with capacitors, where increasing frequency or capacitance *decreases* the capacitive reactance.

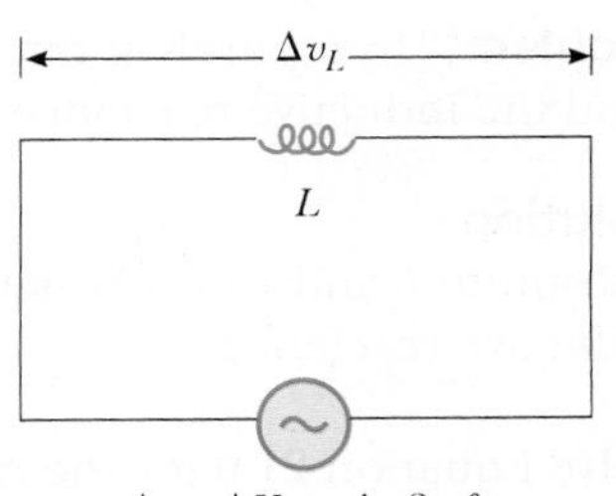

ACTIVE FIGURE 21.6
A series circuit consisting of an inductor L connected to an AC generator.

Physics Now™
Log into PhysicsNow at **www.cp7e.com** and go to Active Figure 21.6, where you can adjust the inductance, the frequency, and the maximum voltage. The results can be studied with the graph and phasor diagram in Active Figure 21.7.

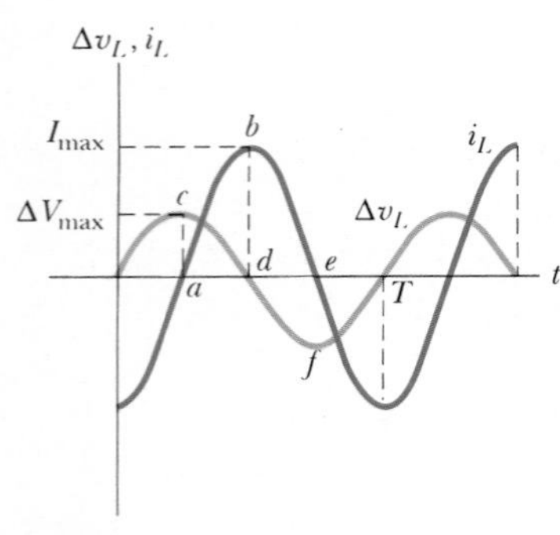

ACTIVE FIGURE 21.7
Plots of current and voltage across an inductor versus time in an AC circuit. The voltage leads the current by 90°.

Physics Now™
Log into PhysicsNow at **www.cp7e.com** and go to Active Figure 21.6, where you can adjust the inductance, the frequency, and the maximum voltage. The results can be studied with the graph and phasor diagram in Active Figure 21.7.

To understand the meaning of inductive reactance, compare Equation 21.8 with Equation 21.7. First, note from Equation 21.8 that the inductive reactance depends on the inductance L. This is reasonable, because the back emf (Eq. 21.7) is large for large values of L. Second, note that the inductive reactance depends on the frequency f. This, too, is reasonable, because the back emf depends on $\Delta I/\Delta t$, a quantity that is large when the current changes rapidly, as it would for high frequencies.

With inductive reactance defined in this way, we can write an equation of the same form as Ohm's law for the voltage across the coil or inductor:

$$\Delta V_{L,\text{rms}} = I_{\text{rms}} X_L \qquad [21.9]$$

where $\Delta V_{L,\text{rms}}$ is the rms voltage across the coil and I_{rms} is the rms current in the coil.

Active Figure 21.7 shows the instantaneous voltage and instantaneous current across the coil as functions of time. When a sinusoidal voltage is applied across an inductor, the voltage reaches its maximum value one-quarter of an oscillation period before the current reaches its maximum value. In this situation, we say that **the voltage across an inductor always leads the current by 90°**.

To see why there is a phase relationship between voltage and current, we examine a few points on the curves of Active Figure 21.7. At point a on the current curve, the current is beginning to increase in the positive direction. At this instant, the rate of change of current, $\Delta I/\Delta t$ (the slope of the current curve), is at a maximum, and we see from Equation 21.7 that the voltage across the inductor is consequently also at a maximum. As the current rises between points a and b on the curve, $\Delta I/\Delta t$ gradually decreases until it reaches zero at point b. As a result, the voltage across the inductor is decreasing during this same time interval, as the segment between c and d on the voltage curve indicates. Immediately after point b, the current begins to decrease, although it still has the same direction it had during the previous quarter cycle. As the current decreases to zero (from b to e on the curve), a voltage is again induced in the coil (from d to f), but the polarity of this voltage is opposite the polarity of the voltage induced between c and d. This occurs because back emfs always oppose the change in the current.

We could continue to examine other segments of the curves, but no new information would be gained because the current and voltage variations are repetitive.

EXAMPLE 21.3 A Purely Inductive AC Circuit

Goal Perform basic AC circuit calculations for an inductive circuit.

Problem In a purely inductive AC circuit (see Active Fig. 21.6), $L = 25.0$ mH and the rms voltage is 1.50×10^2 V. Find the inductive reactance and rms current in the circuit if the frequency is 60.0 Hz.

Solution

Substitute L and f into Equation 21.8 to get the inductive reactance:

$$X_L = 2\pi f L = 2\pi(60.0\ \text{s}^{-1})(25.0 \times 10^{-3}\ \text{H}) = 9.42\ \Omega$$

Solve Equation 21.9 for the rms current and substitute:

$$I_{\text{rms}} = \frac{\Delta V_{L,\text{rms}}}{X_L} = \frac{1.50 \times 10^2\ \text{V}}{9.42\ \Omega} = 15.9\ \text{A}$$

Remark The analogy with DC circuits is even closer than in the capacitive case, because in the inductive equivalent of Ohm's law, the voltage across an inductor is *proportional* to the inductance L, just as the voltage across a resistor is proportional to R in Ohm's law.

Exercise 21.3

Calculate the inductive reactance and rms current in a similar circuit if the frequency is again 60.0 Hz, but the rms voltage is 85.0 V and the inductance is 47.0 mH.

Answers $X_L = 17.7\ \Omega$; $I = 4.80$ A

21.4 THE *RLC* SERIES CIRCUIT

In the foregoing sections, we examined the effects of an inductor, a capacitor, and a resistor when they are connected separately across an AC voltage source. We now consider what happens when these devices are combined.

Active Figure 21.8 shows a circuit containing a resistor, an inductor, and a capacitor connected in series across an AC source that supplies a total voltage Δv at some instant. The current in the circuit is the same at all points in the circuit at any instant and varies sinusoidally with time, as indicated in Active Figure 21.9a. This fact can be expressed mathematically as

$$i = I_{max} \sin 2\pi ft$$

Earlier, we learned that the voltage across each element may or may not be in phase with the current. The instantaneous voltages across the three elements, shown in Active Figure 21.9, have the following phase relations to the instantaneous current:

1. The instantaneous voltage Δv_R across the resistor is *in phase* with the instantaneous current. (See Active Fig. 21.9b.)
2. The instantaneous voltage Δv_L across the inductor *leads* the current by 90°. (See Active Fig. 21.9c.)
3. The instantaneous voltage Δv_C across the capacitor *lags* the current by 90°. (See Active Fig. 21.9d.)

The net instantaneous voltage Δv supplied by the AC source equals the sum of the instantaneous voltages across the separate elements: $\Delta v = \Delta v_R + \Delta v_C + \Delta v_L$. This doesn't mean, however, that the voltages measured with an AC voltmeter across R, C, and L sum to the measured source voltage! In fact, the measured voltages *don't* sum to the measured source voltage, because the voltages across R, C, and L all have different phases.

To account for the different phases of the voltage drops, we use a technique involving vectors. We represent the voltage across each element with a rotating vector, as in Figure 21.10. The rotating vectors are referred to as **phasors**, and the diagram is called a **phasor diagram**. This particular diagram represents the circuit voltage given by the expression $\Delta v = \Delta V_{max} \sin(2\pi ft + \phi)$, where ΔV_{max} is the maximum voltage (the magnitude or length of the rotating vector or phasor) and ϕ is the angle between the phasor and the + x-axis when $t = 0$. The phasor can be viewed as a vector of magnitude ΔV_{max} rotating at a constant frequency f so that its projection along the y-axis is the instantaneous voltage in the circuit. Because ϕ is the phase angle between the voltage and current in the circuit, the phasor for the current (not shown in Fig. 21.10) lies along the positive x-axis when $t = 0$ and is expressed by the relation $i = I_{max} \sin(2\pi ft)$.

The phasor diagrams in Figure 21.11 (page 700) are useful for analyzing the *series RLC* circuit. Voltages in phase with the current are represented by vectors along the positive x-axis, and voltages out of phase with the current lie along other directions. ΔV_R is horizontal and to the right because it's in phase with the current. Likewise, ΔV_L is represented by a phasor along the positive y-axis because it leads the current by 90°. Finally, ΔV_C is along the negative y-axis because it lags the current[2] by 90°. If the phasors are added as vector quantities in order to account for the different phases of the voltages across R, L, and C, Figure 21.11a shows that the only x-component for the voltages is ΔV_R and the net y-component is $\Delta V_L - \Delta V_C$. We now add the phasors vectorially to find the phasor ΔV_{max} (Fig. 21.11b), which represents the maximum voltage. The right triangle in Figure 21.11b gives the following equations for the maximum voltage and the phase angle ϕ between the maximum voltage and the current:

[2]A mnemonic to help you remember the phase relationships in *RLC* circuits is "*ELI* the *ICE* man." *E* represents the voltage $\mathcal{E}$, *I* the current, *L* the inductance, and *C* the capacitance. Thus, the name *ELI* means that, in an inductive circuit, the voltage $\mathcal{E}$ leads the current *I*. In a capacitive circuit, *ICE* means that the current leads the voltage.

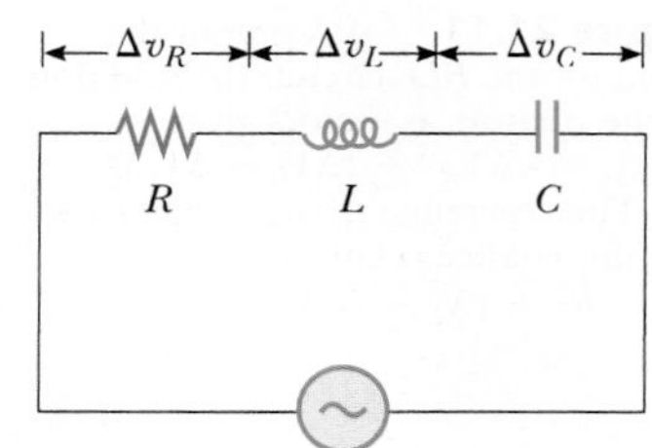

ACTIVE FIGURE 21.8
A series circuit consisting of a resistor, an inductor, and a capacitor connected to an AC generator.

Physics Now™
Log into PhysicsNow at **www.cp7e.com** and go to Active Figure 21.8, where you can adjust the resistance, the inductance, and the capacitance. The results can be studied with the graph in Active Figure 21.9 and the phasor diagram in Figure 21.10.

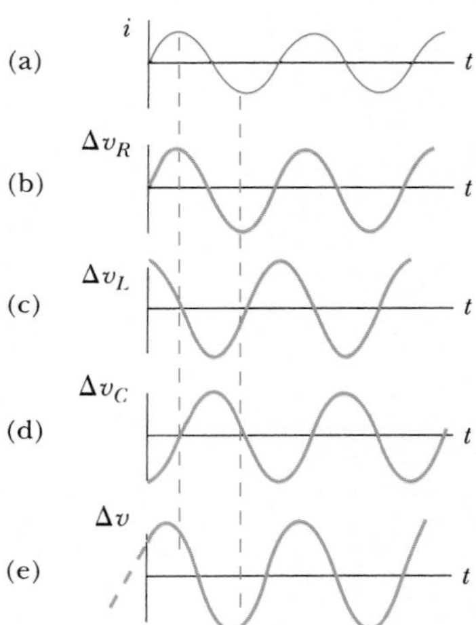

ACTIVE FIGURE 21.9
Phase relations in the series *RLC* circuit shown in Figure 21.8.

Physics Now™
Log into PhysicsNow at **www.cp7e.com** and go to Active Figure 21.8, where you can adjust the resistance, the inductance, and the capacitance. The results can be studied with the graph in this figure and the phasor diagram in Figure 21.10.

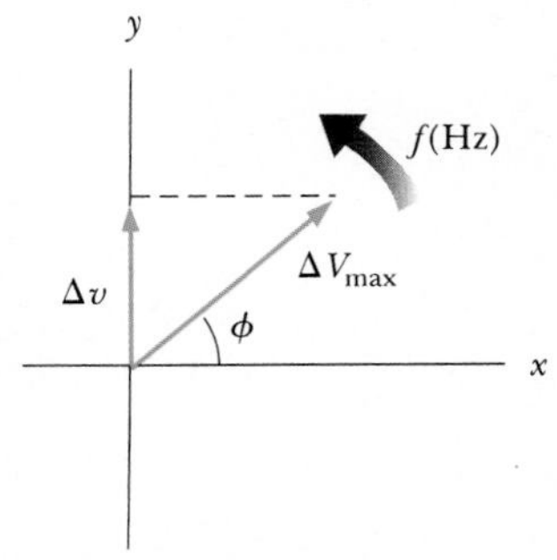

Figure 21.10 A phasor diagram for the voltage in an AC circuit, where ϕ is the phase angle between the voltage and the current and Δv is the instantaneous voltage.

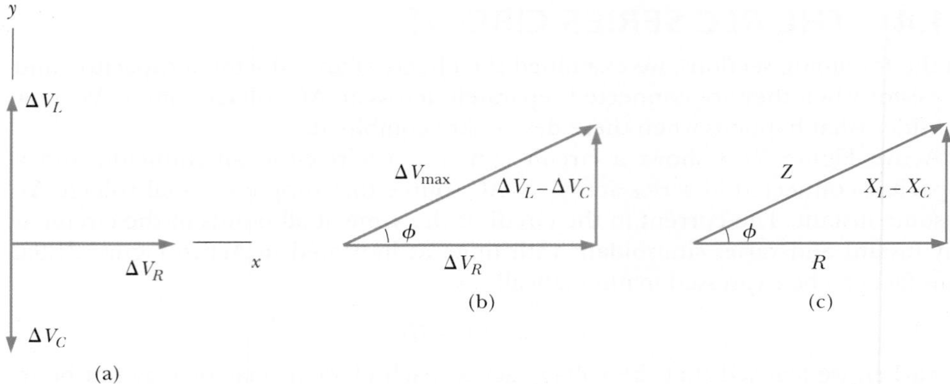

Figure 21.11 (a) A phasor diagram for the *RLC* circuit. (b) Addition of the phasors as vectors gives $\Delta V_{max} = \sqrt{\Delta V_R^2 + (\Delta V_L - \Delta V_C)^2}$. (c) The reactance triangle that gives the impedance relation $Z = \sqrt{R^2 + (X_L - X_C)^2}$.

$$\Delta V_{max} = \sqrt{\Delta V_R^2 + (\Delta V_L - \Delta V_C)^2} \quad [21.10]$$

$$\tan \phi = \frac{\Delta V_L - \Delta V_C}{\Delta V_R} \quad [21.11]$$

In these equations, all voltages are maximum values. Although we choose to use maximum voltages in our analysis, the preceding equations apply equally well to rms voltages, because the two quantities are related to each other by the same factor for all circuit elements. The result for the maximum voltage ΔV_{max} given by Equation 21.10 reinforces the fact that **the voltages across the resistor, capacitor, and inductor are not in phase, so one cannot simply add them to get the voltage across the combination of element, or the source voltage**.

Quick Quiz 21.2

For the circuit of Figure 21.8, is the instantaneous voltage of the source equal to (a) the sum of the maximum voltages across the elements, (b) the sum of the instantaneous voltages across the elements, or (c) the sum of the rms voltages across the elements?

We can write Equation 21.10 in the form of Ohm's law, using the relations $\Delta V_R = I_{max}R$, $\Delta V_L = I_{max}X_L$, and $\Delta V_C = I_{max}X_C$, where I_{max} is the maximum current in the circuit:

$$\Delta V_{max} = I_{max}\sqrt{R^2 + (X_L - X_C)^2} \quad [21.12]$$

It's convenient to define a parameter called the **impedance** Z of the circuit as

Impedance ▶

$$Z \equiv \sqrt{R^2 + (X_L - X_C)^2} \quad [21.13]$$

so that Equation 21.12 becomes

$$\Delta V_{max} = I_{max} Z \quad [21.14]$$

Equation 21.14 is in the form of Ohm's law, $\Delta V = IR$, with R replaced by the impedance in ohms. Indeed, Equation 21.14 can be regarded as a generalized form of Ohm's law applied to a series AC circuit. Both the impedance and, therefore, the current in an AC circuit depend on the resistance, the inductance, the capacitance, *and* the frequency (because the reactances are frequency dependent).

It's useful to represent the impedance Z with a vector diagram such as the one depicted in Figure 21.11c. A right triangle is constructed with right side $X_L - X_C$, base R, and hypotenuse Z. Applying the Pythagorean theorem to this triangle, we see that

$$Z = \sqrt{R^2 + (X_L - X_C)^2}$$

which is Equation 21.13. Furthermore, we see from the vector diagram in Figure 21.11c that the phase angle ϕ between the current and the voltage obeys the

TABLE 21.2

Impedance Values and Phase Angles for Various Combinations of Circuit Elements[a]

Circuit Elements	Impedance Z	Phase Angle ϕ
R	R	0°
C	X_C	–90°
L	X_L	+90°
R C	$\sqrt{R^2 + X_C^2}$	Negative, between –90° and 0°
R L	$\sqrt{R^2 + X_L^2}$	Positive, between 0° and 90°
R L C	$\sqrt{R^2 + (X_L - X_C)^2}$	Negative if $X_C > X_L$ Positive if $X_C < X_L$

[a] In each case, an AC voltage (not shown) is applied across the combination of elements (that is, across the dots).

relationship

$$\tan \phi = \frac{X_L - X_C}{R} \qquad [21.15]$$

◀ Phase angle ϕ

The physical significance of the phase angle will become apparent in Section 21.5.

Table 21.2 provides impedance values and phase angles for some series circuits containing different combinations of circuit elements.

Parallel alternating current circuits are also useful in everyday applications. We won't discuss them here, however, because their analysis is beyond the scope of this book.

NIKOLA TESLA (1856–1943)

Tesla was born in Croatia, but spent most of his professional life as an inventor in the United States. He was a key figure in the development of alternating-current electricity, high-voltage transformers, and the transport of electrical power via AC transmission lines. Tesla's viewpoint was at odds with the ideas of Edison, who committed himself to the use of direct current in power transmission. Tesla's AC approach won out.

Quick Quiz 21.3

The switch in the circuit shown in Figure 21.12 is closed and the lightbulb glows steadily. The inductor is a simple air-core solenoid. As an iron rod is being inserted into the interior of the solenoid, the brightness of the lightbulb (a) increases, (b) decreases, or (c) remains the same.

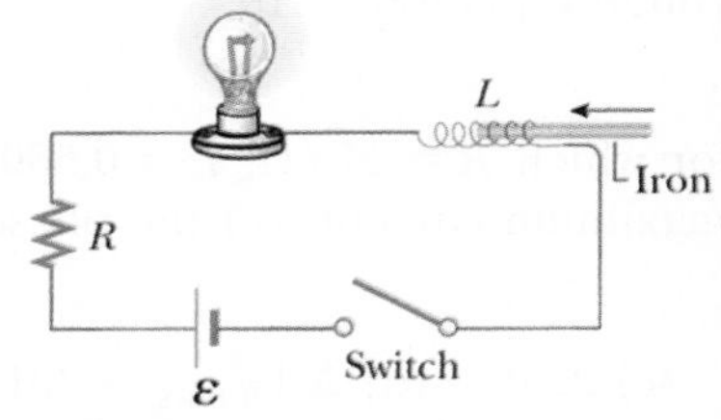

Figure 21.12 (Quick Quiz 21.3)

Problem-Solving Strategy Alternating Current

The following procedure is recommended for solving alternating-current problems:

1. Calculate as many of the unknown quantities, such as X_L and X_C, as possible.
2. Apply the equation $\Delta V_{\max} = I_{\max} Z$ to the portion of the circuit of interest. For example, if you want to know the voltage drop across the combination of an inductor and a resistor, the equation for the voltage drop reduces to $\Delta V_{\max} = I_{\max}\sqrt{R^2 + X_L^2}$.

EXAMPLE 21.4 An *RLC* Circuit

Goal Analyze a series *RLC* AC circuit and find the phase angle.

Problem A series *RLC* AC circuit has resistance $R = 2.50 \times 10^2\ \Omega$, inductance $L = 0.600$ H, capacitance $C = 3.50\ \mu$F, frequency $f = 60.0$ Hz, and maximum voltage $\Delta V_{\max} = 1.50 \times 10^2$ V. Find **(a)** the impedance, **(b)** the maximum current in the circuit, **(c)** the phase angle, and **(d)** the maximum voltages across the elements.

Strategy Calculate the inductive and capacitive reactances, then substitute them and given quantities into the appropriate equations.

Solution

(a) Find the impedance of the circuit.

First, calculate the inductive and capacitive reactances:

$$X_L = 2\pi fL = 226\ \Omega \qquad X_C = 1/2\pi fC = 758\ \Omega$$

Substitute these results and the resistance R into Equation 21.13 to obtain the impedance of the circuit:

$$Z = \sqrt{R^2 + (X_L - X_C)^2}$$
$$= \sqrt{(2.50 \times 10^2\ \Omega)^2 + (226\ \Omega - 758\ \Omega)^2} = \boxed{588\ \Omega}$$

(b) Find the maximum current.

Use Equation 21.12, the equivalent of Ohm's law, to find the maximum current:

$$I_{\text{max}} = \frac{\Delta V_{\text{max}}}{Z} = \frac{1.50 \times 10^2\ \text{V}}{588\ \Omega} = \boxed{0.255\ \text{A}}$$

(c) Find the phase angle.

Calculate the phase angle between the current and the voltage with Equation 21.15:

$$\phi = \tan^{-1}\frac{X_L - X_C}{R} = \tan^{-1}\left(\frac{226\ \Omega - 758\ \Omega}{2.50 \times 10^2\ \Omega}\right) = \boxed{-64.8^\circ}$$

(d) Find the maximum voltages across the elements.

Substitute into the "Ohm's law" expressions for each individual type of current element:

$$\Delta V_{R,\,\text{max}} = I_{\text{max}}R = (0.255\ \text{A})(2.50 \times 10^2\ \Omega) = \boxed{63.8\ \text{V}}$$

$$\Delta V_{L,\,\text{max}} = I_{\text{max}}X_L = (0.255\ \text{A})(2.26 \times 10^2\ \Omega) = \boxed{57.6\ \text{V}}$$

$$\Delta V_{C,\,\text{max}} = I_{\text{max}}X_C = (0.255\ \text{A})(7.58 \times 10^2\ \Omega) = \boxed{193\ \text{V}}$$

Remarks Because the circuit is more capacitive than inductive ($X_C > X_L$), ϕ is negative. A negative phase angle means that the current leads the applied voltage. Notice also that the sum of the maximum voltages across the elements is $\Delta V_R + \Delta V_L + \Delta V_C = 314$ V, which is much greater than the maximum voltage of the generator, 150 V. As we saw in Quick Quiz 21.2, the sum of the maximum voltages is a meaningless quantity because when alternating voltages are added, *both their amplitudes and their phases* must be taken into account. We know that the maximum voltages across the various elements occur at different times, so it doesn't make sense to add all the maximum values. The correct way to "add" the voltages is through Equation 21.10.

Exercise 21.4

Analyze a series RLC AC circuit for which $R = 175\ \Omega$, $L = 0.500$ H, $C = 22.5\ \mu$F, $f = 60.0$ Hz, and $\Delta V_{\text{max}} = 325$ V. Find (a) the impedance, (b) the maximum current, (c) the phase angle, and (d) the maximum voltages across the elements.

Answers (a) 189 Ω (b) 1.72 A (c) 22.0° (d) $\Delta V_{R,\text{max}} = 301$ V, $\Delta V_{L,\text{max}} = 324$ V, $\Delta V_{C,\text{max}} = 203$ V

21.5 POWER IN AN AC CIRCUIT

No power losses are associated with pure capacitors and pure inductors in an AC circuit. A pure capacitor, by definition, has no resistance or inductance, while a pure inductor has no resistance or capacitance. (These are idealizations: in a real capacitor, for example, inductive effects could become important at high frequencies.) We begin by analyzing the power dissipated in an AC circuit that contains only a generator and a capacitor.

When the current increases in one direction in an AC circuit, charge accumulates on the capacitor and a voltage drop appears across it. When the voltage reaches its maximum value, the energy stored in the capacitor is

$$PE_C = \tfrac{1}{2}C(\Delta V_{\text{max}})^2$$

However, this energy storage is only momentary: When the current reverses direction, the charge leaves the capacitor plates and returns to the voltage source. During one-half of each cycle the capacitor is being charged, and during the other half the charge is being returned to the voltage source. Therefore, the average power supplied by the source is zero. In other words, **no power losses occur in a capacitor in an AC circuit.**

Similarly, the source must do work against the back emf of an inductor that is carrying a current. When the current reaches its maximum value, the energy stored in the inductor is a maximum and is given by

$$PE_L = \tfrac{1}{2}LI_{\text{max}}^2$$

When the current begins to decrease in the circuit, this stored energy is returned to the source as the inductor attempts to maintain the current in the circuit. The average power delivered to a resistor in an *RLC* circuit is

$$\mathcal{P}_{\text{av}} = I_{\text{rms}}^2 R \qquad [21.16]$$

The average power delivered by the generator is converted to internal energy in the resistor. No power loss occurs in an ideal capacitor or inductor.

An alternate equation for the average power loss in an AC circuit can be found by substituting (from Ohm's law) $R = \Delta V_R/I_{\text{rms}}$ into Equation 21.16:

$$\mathcal{P}_{\text{av}} = I_{\text{rms}}\Delta V_R$$

It's convenient to refer to a voltage triangle that shows the relationship among ΔV_{rms}, ΔV_R, and $\Delta V_L - \Delta V_C$, such as Figure 21.11b. (Remember that Fig. 21.11 applies to *both* maximum and rms voltages.) From this figure, we see that the voltage drop across a resistor can be written in terms of the voltage of the source, ΔV_{rms}:

$$\Delta V_R = \Delta V_{\text{rms}} \cos \phi$$

Hence, the average power delivered by a generator in an AC circuit is

$$\mathcal{P}_{\text{av}} = I_{\text{rms}}\,\Delta V_{\text{rms}} \cos \phi \qquad [21.17]$$

◀ Average power

where the quantity cos ϕ is called the **power factor**.

Equation 21.17 shows that the power delivered by an AC source to any circuit depends on the phase difference between the source voltage and the resulting current. This fact has many interesting applications. For example, factories often use devices such as large motors in machines, generators, and transformers that have a large inductive load due to all the windings. To deliver greater power to such devices without using excessively high voltages, factory technicians introduce capacitance in the circuits to shift the phase.

APPLICATION

Shifting Phase to Deliver More Power

EXAMPLE 21.5 Average Power in an *RLC* Series Circuit

Goal Understand power in *RLC* series circuits.

Problem Calculate the average power delivered to the series *RLC* circuit described in Example 21.4.

Strategy After finding the rms current and rms voltage with Equations 21.2 and 21.3, substitute into Equation 21.17, using the phase angle found in Example 21.4.

Solution

First, use Equations 21.2 and 21.3 to calculate the rms current and rms voltage:

$$I_{\text{rms}} = \frac{I_{\text{max}}}{\sqrt{2}} = \frac{0.255\text{ A}}{\sqrt{2}} = 0.180\text{ A}$$

$$\Delta V_{\text{rms}} = \frac{\Delta V_{\text{max}}}{\sqrt{2}} = \frac{1.50 \times 10^2\text{ V}}{\sqrt{2}} = 106\text{ V}$$

Substitute these results and the phase angle $\phi = -64.8°$ into Equation 21.17 to find the average power:

$$\mathcal{P}_{\text{av}} = I_{\text{rms}}\Delta V_{\text{rms}} \cos \phi = (0.180\text{ A})(106\text{ V}) \cos(-64.8°)$$
$$= 8.12\text{ W}$$

Remark The same result can be obtained from Equation 21.16, $\mathcal{P}_{av} = I_{rms}^2 R$.

Exercise 21.5
Repeat this problem, using the system described in Exercise 21.4.

Answer 259 W

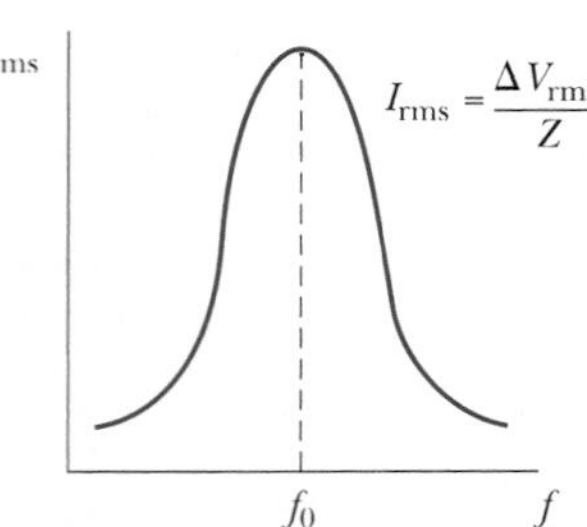

Figure 21.13 A plot of current amplitude in a series *RLC* circuit versus frequency of the generator voltage. Note that the current reaches its maximum value at the resonance frequency f_0.

21.6 RESONANCE IN A SERIES *RLC* CIRCUIT

In general, the rms current in a series *RLC* circuit can be written

$$I_{rms} = \frac{\Delta V_{rms}}{Z} = \frac{\Delta V_{rms}}{\sqrt{R^2 + (X_L - X_C)^2}} \quad [21.18]$$

From this equation, we see that if the frequency is varied, the current has its *maximum* value when the impedance has its *minimum* value. This occurs when $X_L = X_C$. In such a circumstance, the impedance of the circuit reduces to $Z = R$. The frequency f_0 at which this happens is called the **resonance frequency** of the circuit. To find f_0, we set $X_L = X_C$, which gives, from Equations 21.5 and 21.8,

$$2\pi f_0 L = \frac{1}{2\pi f_0 C}$$

Resonance frequency ▶

$$f_0 = \frac{1}{2\pi\sqrt{LC}} \quad [21.19]$$

Figure 21.13 is a plot of current as a function of frequency for a circuit containing a fixed value for both the capacitance and the inductance. From Equation 21.18, it must be concluded that the current would become infinite at resonance when $R = 0$. Although Equation 21.18 predicts this result, real circuits always have some resistance, which limits the value of the current.

APPLICATION
Tuning Your Radio

The tuning circuit of a radio is an important application of a series resonance circuit. The radio is tuned to a particular station (which transmits a specific radio-frequency signal) by varying a capacitor, which changes the resonance frequency of the tuning circuit. When this resonance frequency matches that of the incoming radio wave, the current in the tuning circuit increases.

Applying Physics 21.1 Metal Detectors in Airports

When you walk through the doorway of an airport metal detector, as the person in Figure 21.14 is doing, you are really walking through a coil of many turns. How might the metal detector work?

Explanation The metal detector is essentially a resonant circuit. The portal you step through is an inductor (a large loop of conducting wire) that is part of the circuit. The frequency of the circuit is tuned to the resonant frequency of the circuit when there is no metal in the inductor. When you walk through with metal in your pocket, you change the effective inductance of the resonance circuit, resulting in a change in the current in the circuit. This change in current is detected, and an electronic circuit causes a sound to be emitted as an alarm.

Figure 21.14 (Applying Physics 21.1) An airport metal detector.

Ryan Williams/International Stock Photography

EXAMPLE 21.6 A Circuit in Resonance

Goal Understand resonance frequency and its relation to inductance, capacitance, and the rms current.

Problem Consider a series *RLC* circuit for which $R = 1.50 \times 10^2\ \Omega$, $L = 20.0$ mH, $\Delta V_{\text{rms}} = 20.0$ V, and $f = 796\ \text{s}^{-1}$. **(a)** Determine the value of the capacitance for which the rms current is a maximum. **(b)** Find the maximum rms current in the circuit.

Strategy The current is a maximum at the resonance frequency f_0, which should be set equal to the driving frequency, $796\ \text{s}^{-1}$. The resulting equation can be solved for *C*. For part (b), substitute into Equation 21.18 to get the maximum rms current.

Solution

(a) Find the capacitance giving the maximum current in the circuit (the resonance condition).

Solve the resonance frequency for the capacitance:

$$f_0 = \frac{1}{2\pi\sqrt{LC}} \rightarrow \sqrt{LC} = \frac{1}{2\pi f_0} \rightarrow LC = \frac{1}{4\pi^2 f_0^2}$$

$$C = \frac{1}{4\pi^2 f_0^2 L}$$

Insert the given values, substituting the source frequency for the resonance frequency, f_o:

$$C = \frac{1}{4\pi^2(796\ \text{Hz})^2(20.0 \times 10^{-3}\ \text{H})} = 2.00 \times 10^{-6}\ \text{F}$$

(b) Find the maximum rms current in the circuit.

The capacitive and inductive reactances are equal, so $Z = R = 1.50 \times 10^2\ \Omega$. Substitute into Equation 21.18 to find the rms current:

$$I_{\text{rms}} = \frac{\Delta V_{\text{rms}}}{Z} = \frac{20.0\ \text{V}}{1.50 \times 10^2\ \Omega} = 0.133\ \text{A}$$

Remark Because the impedance *Z* is in the denominator of Equation 21.18, the maximum current will always occur when $X_L = X_C$, since that yields the minimum value of *Z*.

Exercise 21.6

Consider a series *RLC* circuit for which $R = 1.20 \times 10^2\ \Omega$, $C = 3.10 \times 10^{-5}$ F, $\Delta V_{\text{rms}} = 35.0$ V, and $f = 60.0\ \text{s}^{-1}$. (a) Determine the value of the inductance for which the rms current is a maximum. (b) Find the maximum rms current in the circuit.

Answers (a) 0.227 H (b) 0.292 A

21.7 THE TRANSFORMER

It's often necessary to change a small AC voltage to a larger one or vice versa. Such changes are effected with a device called a transformer.

In its simplest form, the **AC transformer** consists of two coils of wire wound around a core of soft iron, as shown in Figure 21.15. The coil on the left, which is connected to the input AC voltage source and has N_1 turns, is called the primary winding, or the *primary*. The coil on the right, which is connected to a resistor *R* and consists of N_2 turns, is the *secondary*. The purpose of the common iron core is to increase the magnetic flux and to provide a medium in which nearly all the flux through one coil passes through the other.

When an input AC voltage ΔV_1 is applied to the primary, the induced voltage across it is given by

$$\Delta V_1 = -N_1 \frac{\Delta \Phi_B}{\Delta t} \qquad [21.20]$$

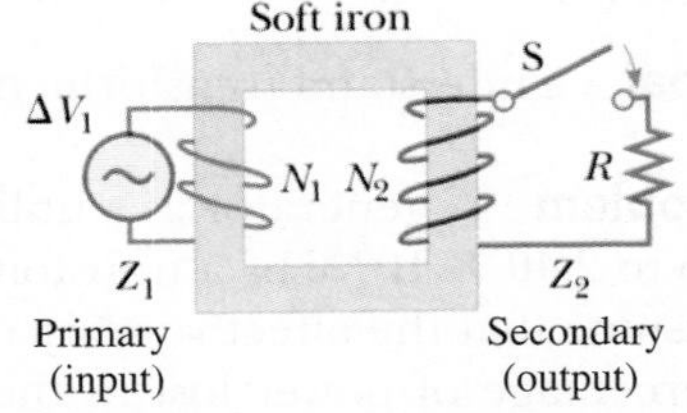

Figure 21.15 An ideal transformer consists of two coils wound on the same soft iron core. An AC voltage ΔV_1 is applied to the primary coil, and the output voltage ΔV_2 is observed across the load resistance *R* after the switch is closed.

where Φ_B is the magnetic flux through each turn. If we assume that no flux leaks from the iron core, then the flux through each turn of the primary equals the flux through each turn of the secondary. Hence, the voltage across the secondary coil is

$$\Delta V_2 = -N_2 \frac{\Delta \Phi_B}{\Delta t} \qquad [21.21]$$

The term $\Delta\Phi_B/\Delta t$ is common to Equations 21.20 and 21.21 and can be algebraically eliminated, giving

$$\Delta V_2 = \frac{N_2}{N_1}\Delta V_1 \qquad [21.22]$$

When N_2 is greater than N_1, ΔV_2 exceeds ΔV_1 and the transformer is referred to as a *step-up transformer.* When N_2 is less than N_1, making ΔV_2 less than ΔV_1, we have a *step-down transformer.*

By Faraday's law, a voltage is generated across the secondary only when there is a *change* in the number of flux lines passing through the secondary. The input current in the primary must therefore change with time, which is what happens when an alternating current is used. When the input at the primary is a direct current, however, a voltage output occurs at the secondary only at the instant a switch in the primary circuit is opened or closed. Once the current in the primary reaches a steady value, the output voltage at the secondary is zero.

It may seem that a transformer is a device in which it is possible to get something for nothing. For example, a step-up transformer can change an input voltage from, say, 10 V to 100 V. This means that each coulomb of charge leaving the secondary has 100 J of energy, whereas each coulomb of charge entering the primary has only 10 J of energy. That is not the case, however, because **the power input to the primary equals the power output at the secondary**:

In an ideal transformer, the input power equals the output power. ▶

$$I_1 \Delta V_1 = I_2 \Delta V_2 \qquad [21.23]$$

While the *voltage* at the secondary may be, say, ten times greater than the voltage at the primary, the *current* in the secondary will be smaller than the primary's current by a factor of ten. Equation 21.23 assumes an **ideal transformer**, in which there are no power losses between the primary and the secondary. Real transformers typically have power efficiencies ranging from 90% to 99%. Power losses occur because of such factors as eddy currents induced in the iron core of the transformer, which dissipate energy in the form of I^2R losses.

APPLICATION

Long-Distance Electric Power Transmission

When electric power is transmitted over large distances, it's economical to use a high voltage and a low current because the power lost via resistive heating in the transmission lines varies as I^2R. This means that if a utility company can reduce the current by a factor of ten, for example, the power loss is reduced by a factor of one hundred. In practice, the voltage is stepped up to around 230 000 V at the generating station, then stepped down to around 20 000 V at a distribution station, and finally stepped down to 120 V at the customer's utility pole.

EXAMPLE 21.7 Distributing Power to a City

Goal Understand transformers and their role in reducing power loss.

Problem A generator at a utility company produces 1.00×10^2 A of current at 4.00×10^3 V. The voltage is stepped up to 2.40×10^5 V by a transformer before being sent on a high-voltage transmission line across a rural area to a city. Assume that the effective resistance of the power line is 30.0 Ω and that the transformers are ideal. **(a)** Determine the percentage of power lost in the transmission line. **(b)** What percentage of the original power would be lost in the transmission line if the voltage were not stepped up?

Strategy Solving this problem is just a matter of substitution into the equation for transformers and the equation for power loss. To obtain the fraction of power lost, it's also necessary to compute the power output of the generator—the current times the potential difference created by the generator.

Solution

(a) Determine the percentage of power lost in the line.

Substitute into Equation 21.23 to find the current in the transmission line:

$$I_2 = \frac{I_1 \Delta V_1}{\Delta V_2} = \frac{(1.00 \times 10^2 \text{ A})(4.00 \times 10^3 \text{ V})}{2.40 \times 10^5 \text{ V}} = 1.67 \text{ A}$$

Now use Equation 21.16 to find the power lost in the transmission line:

$$(1) \quad \mathcal{P}_{\text{lost}} = I_2{}^2 R = (1.67 \text{ A})^2(30.0 \ \Omega) = 83.7 \text{ W}$$

Calculate the power output of the generator:

$$\mathcal{P} = I_1 \Delta V_1 = (1.00 \times 10^2 \text{ A})(4.00 \times 10^3 \text{ V}) = 4.00 \times 10^5 \text{ W}$$

Finally, divide $\mathcal{P}_{\text{lost}}$ by the power output and multiply by 100 to find the percentage of power lost:

$$\% \text{ power lost} = \left(\frac{83.7 \text{ W}}{4.00 \times 10^5 \text{ W}}\right) \times 100 = 0.020\,9\%$$

(b) What percentage of the original power would be lost in the transmission line if the voltage were not stepped up?

Replace the stepped-up current in equation (1) by the original current of 1.00×10^2 A.

$$\mathcal{P}_{\text{lost}} = I^2 R = (1.00 \times 10^2 \text{ A})^2(30.0 \ \Omega) = 3.00 \times 10^5 \text{ W}$$

Calculate the percentage loss, as before:

$$\% \text{ power lost} = \left(\frac{3.00 \times 10^5 \text{ W}}{4.00 \times 10^5 \text{ W}}\right) \times 100 = 75\%$$

Remarks This example illustrates the advantage of high-voltage transmission lines. At the city, a transformer at a substation steps the voltage back down to about 4 000 V, and this voltage is maintained across utility lines throughout the city. When the power is to be used at a home or business, a transformer on a utility pole near the establishment reduces the voltage to 240 V or 120 V.

Exercise 21.7

Suppose the same generator has the voltage stepped up to only 7.50×10^4 V and the resistance of the line is 85.0 Ω. Find the percentage of power lost in this case.

Answer 0.604%

George Semple

This cylindrical step-down transformer drops the voltage from 4 000 V to 220 V for delivery to a group of residences.

21.8 MAXWELL'S PREDICTIONS

During the early stages of their study and development, electric and magnetic phenomena were thought to be unrelated. In 1865, however, James Clerk Maxwell (1831–1879) provided a mathematical theory that showed a close relationship between all electric and magnetic phenomena. In addition to unifying the formerly separate fields of electricity and magnetism, his brilliant theory predicted that electric and magnetic fields can move through space as waves. The theory he developed is based on the following four pieces of information:

1. Electric field lines originate on positive charges and terminate on negative charges.
2. Magnetic field lines always form closed loops—they don't begin or end anywhere.
3. A varying magnetic field induces an emf and hence an electric field. This is a statement of Faraday's law (Chapter 20).
4. Magnetic fields are generated by moving charges (or currents), as summarized in Ampère's law (Chapter 19).

JAMES CLERK MAXWELL,
Scottish Theoretical Physicist (1831–1879)

Maxwell developed the electromagnetic theory of light, the kinetic theory of gases, and explained the nature of Saturn's rings and color vision. Maxwell's successful interpretation of the electromagnetic field resulted in the equations that bear his name. Formidable mathematical ability combined with great insight enabled him to lead the way in the study of electromagnetism and kinetic theory.

The first statement is a consequence of the nature of the electrostatic force between charged particles, given by Coulomb's law. It embodies the fact that **free charges (electric monopoles) exist in nature**.

The second statement—that magnetic fields form continuous loops—is exemplified by the magnetic field lines around a long, straight wire, which are closed circles, and the magnetic field lines of a bar magnet, which form closed loops. It says, in contrast to the first statement, that **free magnetic charges (magnetic monopoles) don't exist in nature**.

The third statement is equivalent to Faraday's law of induction, and the fourth is equivalent to Ampère's law.

In one of the greatest theoretical developments of the 19th century, Maxwell used these four statements within a corresponding mathematical framework to prove that electric and magnetic fields play symmetric roles in nature. It was already known from experiments that a changing magnetic field produced an electric field according to Faraday's law. Maxwell believed that nature was symmetric, and he therefore hypothesized that a changing electric field should produce a magnetic field. This hypothesis could not be proven experimentally at the time it was developed, because the magnetic fields generated by changing electric fields are generally very weak and therefore difficult to detect.

To justify his hypothesis, Maxwell searched for other phenomena that might be explained by it. He turned his attention to the motion of rapidly oscillating (accelerating) charges, such as those in a conducting rod connected to an alternating voltage. Such charges are accelerated and, according to Maxwell's predictions, generate changing electric and magnetic fields. The changing fields cause electromagnetic disturbances that travel through space as waves, similar to the spreading water waves created by a pebble thrown into a pool. The waves sent out by the oscillating charges are fluctuating electric and magnetic fields, so they are called *electromagnetic waves.* From Faraday's law and from Maxwell's own generalization of Ampère's law, Maxwell calculated the speed of the waves to be equal to the speed of light, $c = 3 \times 10^8$ m/s. He concluded that visible light and other electromagnetic waves consist of fluctuating electric and magnetic fields traveling through empty space, with each varying field inducing the other! This was truly one of the greatest discoveries of science, on a par with Newton's discovery of the laws of motion. Like Newton's laws, it had a profound influence on later scientific developments.

HEINRICH RUDOLF HERTZ,
German Physicist (1857–1894)

Hertz made his most important discovery of radio waves in 1887. After finding that the speed of a radio wave was the same as that of light, Hertz showed that radio waves, like light waves, could be reflected, refracted, and diffracted. Hertz died of blood poisoning at the age of 36. During his short life, he made many contributions to science. The hertz, equal to one complete vibration or cycle per second, is named after him.

21.9 HERTZ'S CONFIRMATION OF MAXWELL'S PREDICTIONS

In 1887, after Maxwell's death, Heinrich Hertz (1857–1894) was the first to generate and detect electromagnetic waves in a laboratory setting, using *LC* circuits. In such a circuit, a charged capacitor is connected to an inductor, as in Figure 21.16. When the switch is closed, oscillations occur in the current in the circuit and in the charge on the capacitor. If the resistance of the circuit is neglected, no energy is dissipated and the oscillations continue.

In the following analysis, we neglect the resistance in the circuit. We assume that the capacitor has an initial charge of $Q_{\max}$ and that the switch is closed at $t = 0$. When the capacitor is fully charged, the total energy in the circuit is stored in the electric field of the capacitor and is equal to $Q_{\max}^2/2C$. At this time, the current is zero, so no energy is stored in the inductor. As the capacitor begins to discharge, the energy stored in its electric field decreases. At the same time, the current increases and energy equal to $LI^2/2$ is now stored in the magnetic field of the inductor. Thus, energy is transferred from the electric field of the capacitor to the magnetic field of the inductor. When the capacitor is fully discharged, it stores no energy. At this time, the current reaches its maximum value and all of the energy is stored in the inductor. The process then repeats in the reverse direction. The energy continues to transfer between the inductor and the capacitor, corresponding to oscillations in the current and charge.

As we saw in Section 21.6, the frequency of oscillation of an *LC* circuit is called the *resonance frequency* of the circuit and is given by

$$f_0 = \frac{1}{2\pi\sqrt{LC}}$$

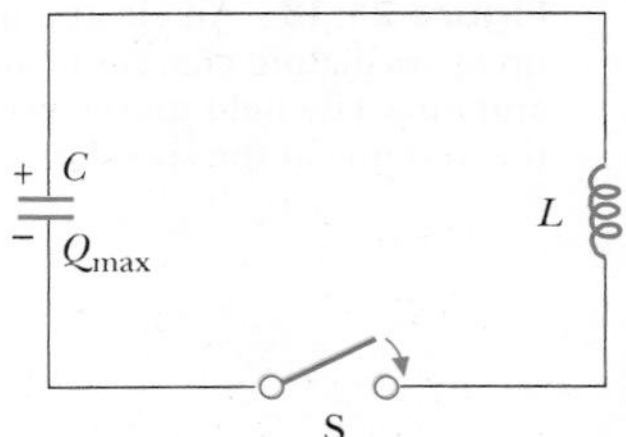

Figure 21.16 A simple *LC* circuit. The capacitor has an initial charge of Q_{max} and the switch is closed at $t = 0$.

The circuit Hertz used in his investigations of electromagnetic waves is similar to that just discussed and is shown schematically in Figure 21.17. An induction coil (a large coil of wire) is connected to two metal spheres with a narrow gap between them to form a capacitor. Oscillations are initiated in the circuit by short voltage pulses sent via the coil to the spheres, charging one positive, the other negative. Because *L* and *C* are quite small in this circuit, the frequency of oscillation is quite high, $f \approx$ 100 MHz. This circuit is called a transmitter because it produces electromagnetic waves.

Several meters from the transmitter circuit, Hertz placed a second circuit, the receiver, which consisted of a single loop of wire connected to two spheres. It had its own effective inductance, capacitance, and natural frequency of oscillation. Hertz found that energy was being sent from the transmitter to the receiver when the resonance frequency of the receiver was adjusted to match that of the transmitter. The energy transfer was detected when the voltage across the spheres in the receiver circuit became high enough to produce ionization in the air, which caused sparks to appear in the air gap separating the spheres. Hertz's experiment is analogous to the mechanical phenomenon in which a tuning fork picks up the vibrations from another, identical tuning fork.

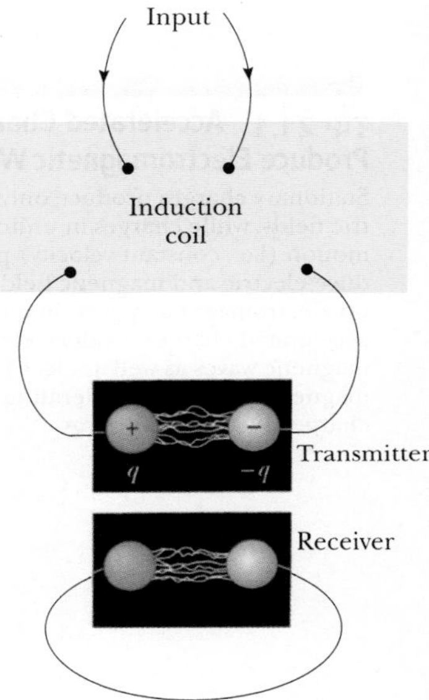

Figure 21.17 A schematic diagram of Hertz's apparatus for generating and detecting electromagnetic waves. The transmitter consists of two spherical electrodes connected to an induction coil, which provides short voltage surges to the spheres, setting up oscillations in the discharge. The receiver is a nearby single loop of wire containing a second spark gap.

Hertz hypothesized that the energy transferred from the transmitter to the receiver is carried in the form of waves, now recognized as electromagnetic waves. In a series of experiments, he also showed that the radiation generated by the transmitter exhibits wave properties: interference, diffraction, reflection, refraction, and polarization. As you will see shortly, all of these properties are exhibited by light. It became evident that Hertz's electromagnetic waves had the same known properties of light waves and differed only in frequency and wavelength. Hertz effectively confirmed Maxwell's theory by showing that Maxwell's mysterious electromagnetic waves existed and had all the properties of light waves.

Perhaps the most convincing experiment Hertz performed was the measurement of the speed of waves from the transmitter, accomplished as follows: waves of known frequency from the transmitter were reflected from a metal sheet so that an interference pattern was set up, much like the standing-wave pattern on a stretched string. As we learned in our discussion of standing waves, the distance between nodes is $\lambda/2$, so Hertz was able to determine the wavelength λ. Using the relationship $v = \lambda f$, he found that v was close to 3×10^8 m/s, the known speed of visible light. Hertz's experiments thus provided the first evidence in support of Maxwell's theory.

21.10 PRODUCTIONOF ELECTROMAGNETIC WAVES BY AN ANTENNA

In the previous section, we found that the energy stored in an *LC* circuit is continually transferred between the electric field of the capacitor and the magnetic field of the inductor. However, this energy transfer continues for prolonged periods of time only when the changes occur slowly. If the current alternates rapidly, the circuit loses some of its energy in the form of electromagnetic waves. In fact, electromagnetic waves are radiated by *any* circuit carrying an alternating current. The fundamental mechanism responsible for this radiation is the acceleration of a charged particle. **Whenever a charged particle accelerates it radiates energy.**

APPLICATION
Radio-Wave Transmission

An alternating voltage applied to the wires of an antenna forces electric charges in the antenna to oscillate. This is a common technique for accelerating charged particles and is the source of the radio waves emitted by the broadcast antenna of a radio station.

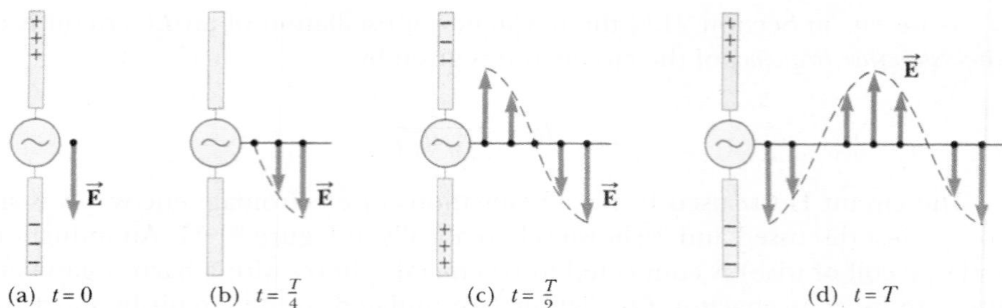

Figure 21.18 An electric field set up by oscillating charges in an antenna. The field moves away from the antenna at the speed of light.

TIP 21.1 Accelerated Charges Produce Electromagnetic Waves

Stationary charges produce only electric fields, while charges in uniform motion (i.e., constant velocity) produce electric and magnetic fields, but no electromagnetic waves. In contrast, accelerated charges produce electromagnetic waves as well as electric and magnetic fields. An accelerating charge also radiates energy.

Figure 21.18 illustrates the production of an electromagnetic wave by oscillating electric charges in an antenna. Two metal rods are connected to an AC source, which causes charges to oscillate between the rods. The output voltage of the generator is sinusoidal. At $t = 0$, the upper rod is given a maximum positive charge and the bottom rod an equal negative charge, as in Figure 21.18a. The electric field near the antenna at this instant is also shown in the figure. As the charges oscillate, the rods become less charged, the field near the rods decreases in strength, and the downward-directed maximum electric field produced at $t = 0$ moves away from the rod. When the charges are neutralized, as in Figure 21.18b, the electric field has dropped to zero, after an interval equal to one-quarter of the period of oscillation. Continuing in this fashion, the upper rod soon obtains a maximum negative charge and the lower rod becomes positive, as in Figure 21.18c, resulting in an electric field directed upward. This occurs after an interval equal to one-half the period of oscillation. The oscillations continue as indicated in Figure 21.18d. Note that the electric field near the antenna oscillates in phase with the charge distribution: the field points down when the upper rod is positive and up when the upper rod is negative. Further, the magnitude of the field at any instant depends on the amount of charge on the rods at that instant.

As the charges continue to oscillate (and accelerate) between the rods, the electric field set up by the charges moves away from the antenna in all directions at the speed of light. Figure 21.18 shows the electric field pattern on one side of the antenna at certain times during the oscillation cycle. As you can see, one cycle of charge oscillation produces one full wavelength in the electric field pattern.

Because the oscillating charges create a current in the rods, a magnetic field is also generated when the current in the rods is upward, as shown in Figure 21.19. The magnetic field lines circle the antenna (recall right-hand rule number 2) and are perpendicular to the electric field at all points. As the current changes with time, the magnetic field lines spread out from the antenna. At great distances from the antenna, the strengths of the electric and magnetic fields become very weak. At these distances, however, it is necessary to take into account the facts that (1) a changing magnetic field produces an electric field and (2) a changing electric field produces a magnetic field, as predicted by Maxwell. These induced electric and magnetic fields are in phase: at any point, the two fields reach their maximum values at the same instant. This synchrony is illustrated at one instant of time in Active Figure 21.20. Note that (1) the $\vec{\mathbf{E}}$ and $\vec{\mathbf{B}}$ fields are perpendicular to each other, and (2) both fields are perpendicular to the direction of motion of the wave. This second property is characteristic of transverse waves. Hence, we see that **an electromagnetic wave is a transverse wave.**

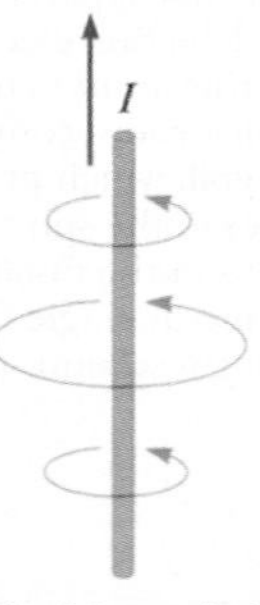

Figure 21.19 Magnetic field lines around an antenna carrying a changing current.

21.11 PROPERTIES OF ELECTROMAGNETIC WAVES

We have seen that Maxwell's detailed analysis predicted the existence and properties of electromagnetic waves. In this section we summarize what we know about electromagnetic waves thus far and consider some additional properties. In our discussion here and in future sections, we will often make reference to a type of wave called a **plane wave**. A plane electromagnetic wave is a wave traveling from a very distant source. Active Figure 21.20 pictures such a wave at a given instant of time. In

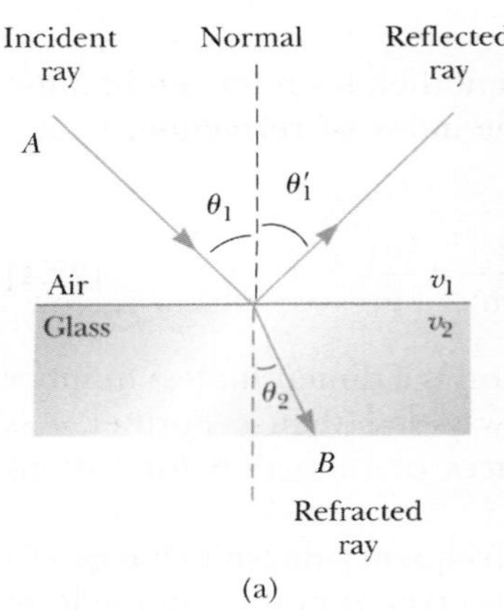

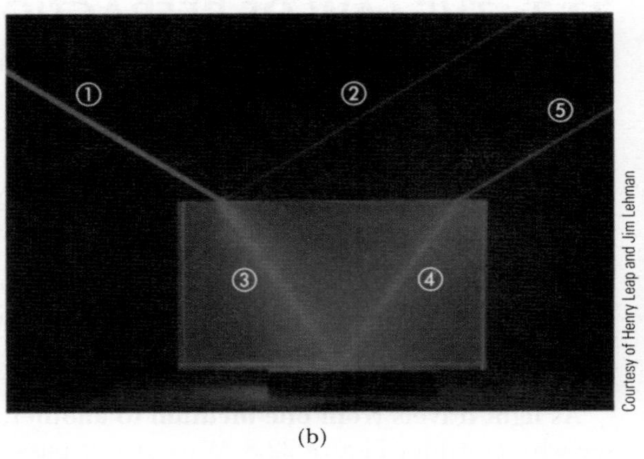

ACTIVE FIGURE 22.6
(a) A ray obliquely incident on an air–glass interface. The refracted ray is bent toward the normal because $v_2 < v_1$. (b) Light incident on the Lucite block bends both when it enters the block and when it leaves the block.

Physics Now™
Log into PhysicsNow at **www.cp7e.com** and go to Active Figure 22.6, where you can vary the incident angle and see the effect on the reflected and refracted rays.

in the same plane. The **angle of refraction**, θ_2, in Active Figure 22.6a depends on the properties of the two media and on the angle of incidence, through the relationship

$$\frac{\sin\theta_2}{\sin\theta_1} = \frac{v_2}{v_1} = \text{constant} \qquad [22.3]$$

where v_1 is the speed of light in medium 1 and v_2 is the speed of light in medium 2. Note that the angle of refraction is also measured with respect to the normal. In Section 22.7 we will derive the laws of reflection and refraction using Huygens' principle.

Experiment shows that **the path of a light ray through a refracting surface is reversible**. For example, the ray in Active Figure 22.6a travels from point A to point B. If the ray originated at B, it would follow the same path to reach point A, but the reflected ray would be in the glass.

Quick Quiz 22.2

If beam 1 is the incoming beam in Active Figure 22.6b, which of the other four beams are due to reflection? Which are due to refraction?

When light moves from a material in which its speed is high to a material in which its speed is lower, the angle of refraction θ_2 is less than the angle of incidence. The refracted ray therefore bends toward the normal, as shown in Active Figure 22.7a. If the ray moves from a material in which it travels slowly to a material in which it travels more rapidly, θ_2 is greater than θ_1, so the ray bends away from the normal, as shown in Active Figure 22.7b.

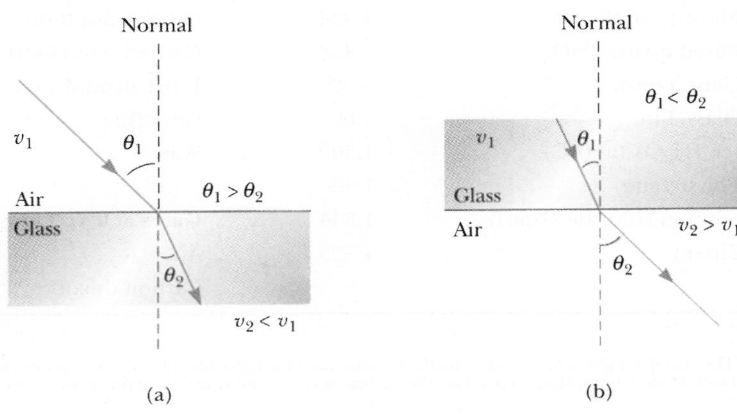

ACTIVE FIGURE 22.7
(a) When the light beam moves from air into glass, its path is bent toward the normal. (b) When the beam moves from glass into air, its path is bent away from the normal.

Physics Now™
Log into PhysicsNow at **www.cp7e.com** and go to Active Figure 22.7. Light passes through three layers of material. You can vary the incident angle and see the effect on the refracted rays for a variety of values of the index of refraction (Table 22.1, page 732) of the three materials.

22.3 THE LAW OF REFRACTION

When light passes from one transparent medium to another, it's refracted because the speed of light is different in the two media[1]. The **index of refraction**, n, of a medium is defined as the ratio c/v;

Index of refraction ▶

$$n \equiv \frac{\text{speed of light in vacuum}}{\text{speed of light in a medium}} = \frac{c}{v} \qquad [22.4]$$

From this definition, we see that the index of refraction is a dimensionless number that is greater than or equal to one because v is always less than c. Further, n is equal to one for vacuum. Table 22.1 lists the indices of refraction for various substances.

TIP 22.1 An Inverse Relationship

The index of refraction is *inversely* proportional to the wave speed. Therefore, as the wave speed v decreases, the index of refraction n *increases.*

As light travels from one medium to another, its frequency doesn't change. To see why, consider Figure 22.8. Wave fronts pass an observer at point A in medium 1 with a certain frequency and are incident on the boundary between medium 1 and medium 2. The frequency at which the wave fronts pass an observer at point B in medium 2 must equal the frequency at which they arrive at point A. If this were not the case, the wave fronts would either pile up at the boundary or be destroyed or created at the boundary. Because neither of these events occurs, the frequency must remain the same as a light ray passes from one medium into another.

TIP 22.2 The Frequency Remains the Same

The *frequency* of a wave does *not* change as the wave passes from one medium to another. Both the wave speed and the wavelength *do* change, but the frequency remains the same.

Therefore, because the relation $v = f\lambda$ must be valid in both media, and because $f_1 = f_2 = f$, we see that

$$v_1 = f\lambda_1 \qquad \text{and} \qquad v_2 = f\lambda_2$$

Because $v_1 \neq v_2$, it follows that $\lambda_1 \neq \lambda_2$. A relationship between the index of refraction and the wavelength can be obtained by dividing these two equations and making use of the definition of the index of refraction given by Equation 22.4:

$$\frac{\lambda_1}{\lambda_2} = \frac{v_1}{v_2} = \frac{c/n_1}{c/n_2} = \frac{n_2}{n_1} \qquad [22.5]$$

which gives

$$\lambda_1 n_1 = \lambda_2 n_2 \qquad [22.6]$$

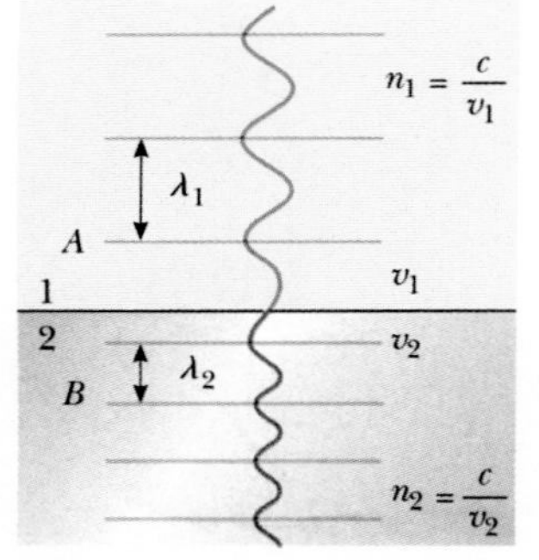

Figure 22.8 As the wave moves from medium 1 to medium 2, its wavelength changes, but its frequency remains constant.

TABLE 22.1

Indices of Refraction for Various Substances, Measured with Light of Vacuum Wavelength λ_0 = 589 mn

Substance	Index of Refraction	Substance	Index of Refraction
Solids at 20°C		**Liquids at 20°C**	
Diamond (C)	2.419	Benzene	1.501
Fluorite (CaF_2)	1.434	Carbon disulfide	1.628
Fused quartz (SiO_2)	1.458	Carbon tetrachloride	1.461
Glass, crown	1.52	Ethyl alcohol	1.361
Glass, flint	1.66	Glycerine	1.473
Ice (H_2O) (at 0°C)	1.309	Water	1.333
Polystyrene	1.49		
Sodium chloride (NaCl)	1.544	**Gases at 0°C, 1 atm**	
Zircon	1.923	Air	1.000 293
		Carbon dioxide	1.000 45

[1]The speed of light varies between media because the time lags caused by the absorption and reemission of light as it travels from atom to atom depend on the particular electronic structure of the atoms constituting each material.

Let medium 1 be the vacuum, so that $n_1 = 1$. It follows from Equation 22.6 that the index of refraction of any medium can be expressed as the ratio

$$n = \frac{\lambda_0}{\lambda_n} \quad [22.7]$$

where λ_0 is the wavelength of light in vacuum and λ_n is the wavelength in the medium having index of refraction n. Figure 22.9 is a schematic representation of this reduction in wavelength when light passes from a vacuum into a transparent medium.

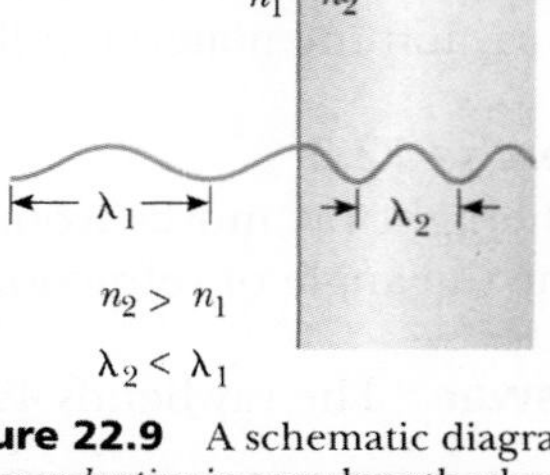

Figure 22.9 A schematic diagram of the *reduction* in wavelength when light travels from a medium with a low index of refraction to one with a higher index of refraction.

We are now in a position to express Equation 22.3 in an alternate form. If we substitute Equation 22.5 into Equation 22.3, we get

$$n_1 \sin\theta_1 = n_2 \sin\theta_2 \quad [22.8]$$

◀ Snell's law of refraction

The experimental discovery of this relationship is usually credited to Willebord Snell (1591–1627) and is therefore known as **Snell's law of refraction**.

Quick Quiz 22.3

A material has an index of refraction that increases continuously from top to bottom. Of the three paths shown in Figure 22.10, which path will a light ray follow as it passes through the material?

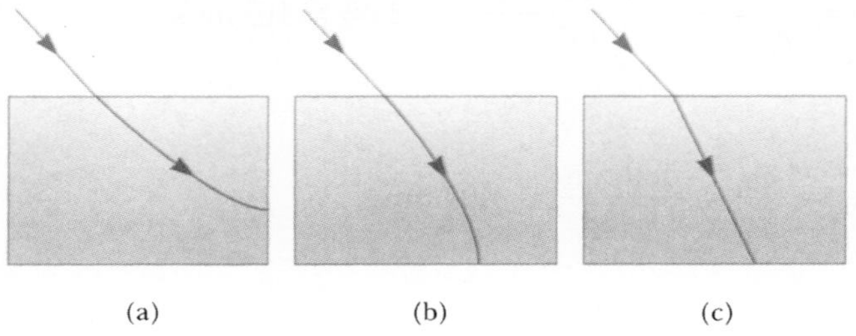

Figure 22.10 (Quick Quiz 22.3)

Quick Quiz 22.4

As light travels from a vacuum ($n = 1$) to a medium such as glass ($n > 1$), which of the following properties remains the same? (a) wavelength, (b) wave speed, or (c) frequency?

EXAMPLE 22.2 Angle of Refraction for Glass

Goal Apply Snell's law to a slab of glass.

Problem A light ray of wavelength 589 nm (produced by a sodium lamp) traveling through air is incident on a smooth, flat slab of crown glass at an angle of 30.0° to the normal, as sketched in Figure 22.11. Find the angle of refraction, θ_2.

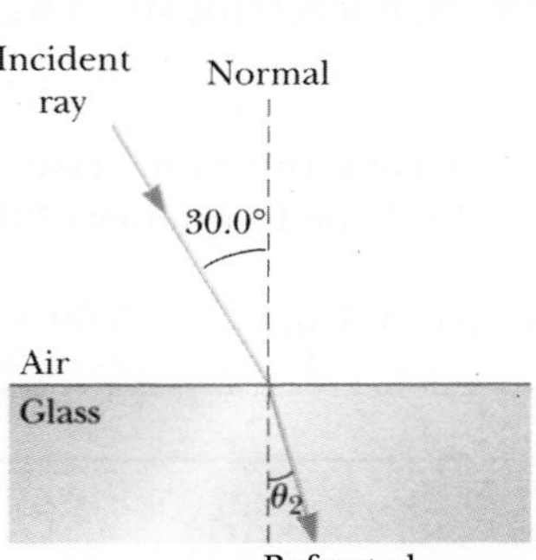

Figure 22.11 (Example 22.2) Refraction of light by glass.

Strategy Substitute quantities into Snell's law and solve for the unknown angle of refraction, θ_2.

Solution

Solve Snell's law (Eq. 22.8) for $\sin\theta_2$:

$$(1) \quad \sin\theta_2 = \frac{n_1}{n_2}\sin\theta_1$$

From Table 22.1, find $n_1 = 1.00$ for air and $n_2 = 1.52$ for crown glass. Substitute these values into (1) and take the inverse sine of both sides:

$$\sin\theta_2 = \left(\frac{1.00}{1.52}\right)(\sin 30.0°) = 0.329$$

$$\theta_2 = \sin^{-1}(0.329) = 19.2°$$

Remarks Notice the light ray bends toward the normal when it enters a material of a higher index of refraction. If the ray left the material following the same path in reverse, it would bend away from the normal.

Exercise 22.2

If the light ray moves from inside the glass toward the glass–air interface at an angle of 30.0° to the normal, determine the angle of refraction.

Answer The ray bends 49.5° *away* from the normal, as expected.

EXAMPLE 22.3 Light in Fused Quartz

Goal Use the index of refraction to determine the effect of a medium on light's speed and wavelength.

Problem Light of wavelength 589 nm in vacuum passes through a piece of fused quartz of index of refraction $n = 1.458$. **(a)** Find the speed of light in fused quartz. **(b)** What is the wavelength of this light in fused quartz? **(c)** What is the frequency of the light in fused quartz?

Strategy Substitute values into Equations 22.4 and 22.7.

Solution

(a) Find the speed of light in fused quartz.

Obtain the speed from Equation 22.4:

$$v = \frac{c}{n} = \frac{3.00 \times 10^8 \text{ m/s}}{1.458} = \boxed{2.06 \times 10^8 \text{ m/s}}$$

(b) What is the wavelength of this light in fused quartz?

Use Eq. 22.7 to calculate the wavelength:

$$\lambda_n = \frac{\lambda_0}{n} = \frac{589 \text{ nm}}{1.458} = \boxed{404 \text{ nm}}$$

(c) What is the frequency of the light in fused quartz?

The frequency in quartz is the same as in vacuum. Solve $c = f\lambda$ for the frequency:

$$f = \frac{c}{\lambda} = \frac{3.00 \times 10^8 \text{ m/s}}{589 \times 10^{-9} \text{ m}} = \boxed{5.09 \times 10^{14} \text{ Hz}}$$

Remarks It's interesting to note that the speed of light in vacuum, 3.00×10^8 m/s, is an upper limit for the speed of material objects. In our treatment of relativity in Chapter 26, we will find that this upper limit is consistent with experimental observations. However, it's possible for a particle moving in a medium to have a speed that exceeds the speed of light in that medium. For example, it's theoretically possible for a particle to travel through fused quartz at a speed greater than 2.06×10^8 m/s, but it must still have a speed less than 3.00×10^8 m/s.

Exercise 22.3

Light with wavelength 589 nm passes through crystalline sodium chloride. Find (a) the speed of light in this medium, (b) the wavelength, and (c) the frequency of the light.

Answer (a) 1.94×10^8 m/s (b) 381 nm (c) 5.09×10^{14} Hz

INTERACTIVE EXAMPLE 22.4 Light Passing through a Slab

Goal Apply Snell's law when a ray passes into and out of another medium.

Problem A light beam traveling through a transparent medium of index of refraction n_1 passes through a thick transparent slab with parallel faces and index of refraction n_2 (Fig. 22.12). Show that the emerging beam is parallel to the incident beam.

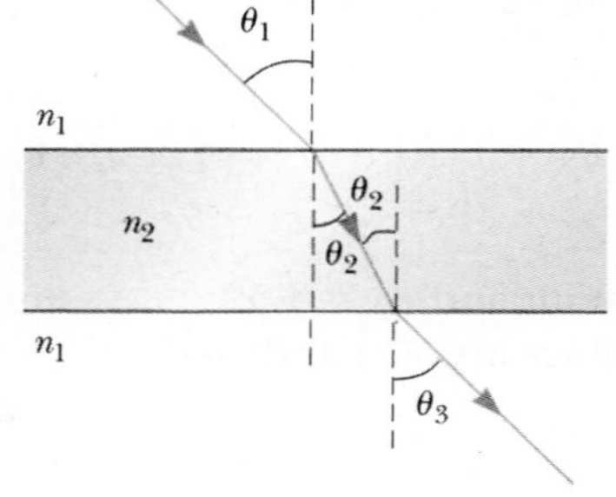

Figure 22.12 (Example 22.4) When light passes through a flat slab of material, the emerging beam is parallel to the incident beam, and therefore $\theta_1 = \theta_3$.

Strategy Apply Snell's law twice, once at the upper surface and once at the lower surface. The two equations will be related because the angle of refraction at the upper surface equals the angle of incidence at the lower surface. The ray passing through the slab makes equal angles with the normal at the entry and exit points. This procedure will enable us to compare angles θ_1 and θ_3.

Solution

Apply Snell's law to the upper surface:

$$(1) \qquad \sin\theta_2 = \frac{n_1}{n_2}\sin\theta_1$$

Apply Snell's law to the lower surface:

$$(2) \qquad \sin\theta_3 = \frac{n_2}{n_1}\sin\theta_2$$

Substitute Equation 1 into Equation 2:

$$\sin\theta_3 = \frac{n_2}{n_1}\left(\frac{n_1}{n_2}\sin\theta_1\right) = \sin\theta_1$$

Take the inverse sine of both sides, noting that the angles are positive and less than 90°:

$$\theta_3 = \theta_1$$

Remarks The preceding result proves that the slab doesn't alter the direction of the beam. It does, however, produce a lateral displacement of the beam, as shown in Figure 22.12.

Exercise 22.4

Suppose the ray, in air with $n = 1.00$, enters a slab with $n = 2.50$ at a 45.0° angle with respect to the normal, then exits the bottom of the slab into water, with $n = 1.33$. At what angle to the normal does the ray leave the slab?

Answer 32.1°

Physics Now™ Explore refraction through slabs of various thickness by logging into PhysicsNow at **www.cp7e.com** and going to Interactive Example 22.4.

EXAMPLE 22.5 Refraction of Laser Light in a Digital Video Disk (DVD)

Goal Apply Snell's law together with geometric constraints.

Problem A DVD is a video recording consisting of a spiral track about 1.0 μm wide with digital information. (See Fig. 22.13a.) The digital information consists of a series of pits that are "read" by a laser beam sharply focused on a track in the information layer. If the width a of the beam at the information layer must equal 1.0 μm to distinguish individual tracks, and the width w of the beam as it enters the plastic is 0.7000 mm, find the angle θ_1 at which the conical beam should enter the plastic. (See Fig. 22.13b.) Assume the plastic has a thickness $t = 1.20$ mm and an index of refraction $n = 1.55$. Note that this system is relatively immune to small dust particles degrading the video

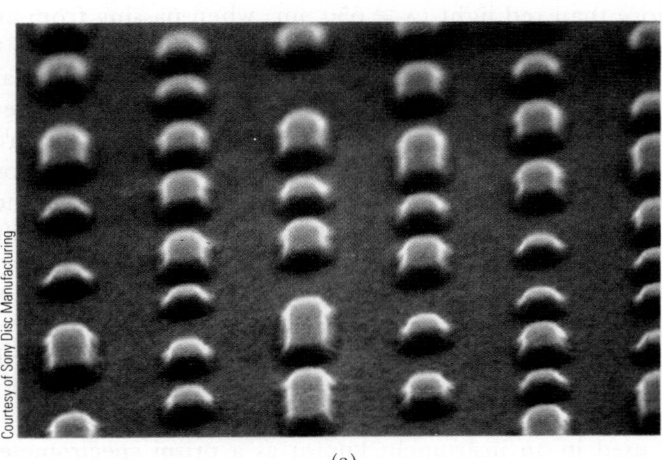

(a)

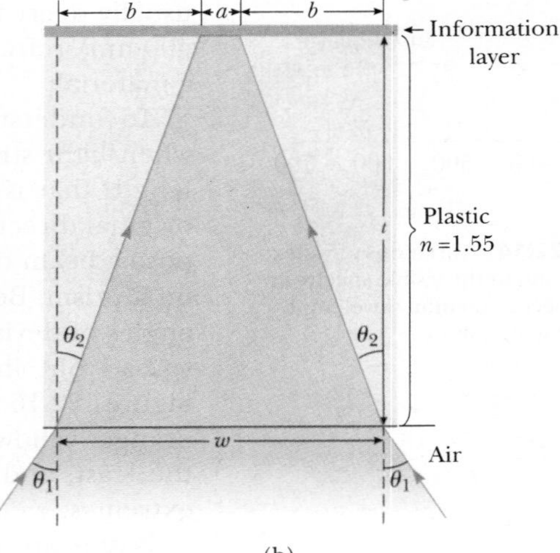

(b)

Figure 22.13 (Example 22.5) A micrograph of a DVD surface showing tracks and pits along each track. (b) Cross section of a cone-shaped laser beam used to read a DVD.

quality, because particles would have to be as large as 0.700 mm to obscure the beam at the point where it enters the plastic.

Strategy Use right-triangle trigonometry to determine the angle θ_2, and then apply Snell's law to obtain the angle θ_1.

Solution

From the top and bottom of Figure 22.13b, obtain an equation relating w, b, and a:

$$w = 2b + a$$

Solve this equation for b and substitute given values:

$$b = \frac{w - a}{2} = \frac{700.0 \times 10^{-6}\,\text{m} - 1.0 \times 10^{-6}\,\text{m}}{2} = 349.5\,\mu\text{m}$$

Now use the tangent function to find θ_2:

$$\tan\theta_2 = \frac{b}{t} = \frac{349.5\,\mu\text{m}}{1.20 \times 10^3\,\mu\text{m}} \quad\rightarrow\quad \theta_2 = 16.2^\circ$$

Finally, use Snell's law to find θ_1:

$$n_1 \sin\theta_1 = n_2 \sin\theta_2$$

$$\sin\theta_1 = \frac{n_2 \sin\theta_2}{n_1} = \frac{1.55 \sin 16.2^\circ}{1.00} = 0.433$$

$$\theta_1 = \sin^{-1}(0.433) = \boxed{25.7^\circ}$$

Remarks Despite its apparent complexity, the problem isn't that different from Example 22.2.

Exercise 22.5

Suppose you wish to redesign the system to decrease the initial width of the beam from 0.700 0 mm to 0.600 0 mm, but leave the incident angle θ_1 and all other parameters the same as before, except the index of refraction for the plastic material (n_2) and the angle θ_2. What index of refraction should the plastic have?

Answer 1.79

22.4 DISPERSION AND PRISMS

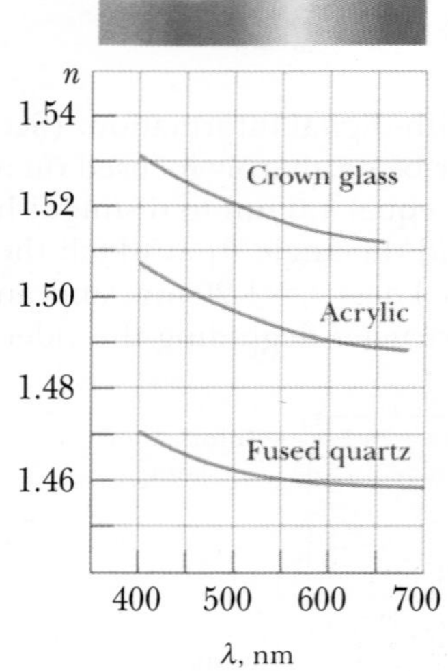

Figure 22.14 Variations of index of refraction in the visible spectrum with respect to vacuum wavelength for three materials.

In Table 22.1, we presented values for the index of refraction of various materials. If we make careful measurements, however, we find that the index of refraction in anything but vacuum depends on the wavelength of light. The dependence of the index of refraction on wavelength is called **dispersion**. Figure 22.14 is a graphical representation of this variation in the index of refraction with wavelength. Because n is a function of wavelength, Snell's law indicates that **the angle of refraction made when light enters a material depends on the wavelength of the light**. As seen in the figure, the index of refraction for a material usually decreases with increasing wavelength. This means that violet light ($\lambda \cong 400$ nm) refracts more than red light ($\lambda \cong 650$ nm) when passing from air into a material.

To understand the effects of dispersion on light, consider what happens when light strikes a prism, as in Figure 22.15a. A ray of light of a single wavelength that is incident on the prism from the left emerges bent away from its original direction of travel by an angle δ, called the **angle of deviation**. Now suppose a beam of white light (a combination of all visible wavelengths) is incident on a prism. Because of dispersion, the different colors refract through different angles of deviation, and the rays that emerge from the second face of the prism spread out in a series of colors known as a visible **spectrum**, as shown in Figure 22.16. These colors, in order of decreasing wavelength, are red, orange, yellow, green, blue, and violet. Violet light deviates the most, red light the least, and the remaining colors in the visible spectrum fall between these extremes.

Prisms are often used in an instrument known as a **prism spectrometer**, the essential elements of which are shown in Figure 22.17a (page 738). This instrument

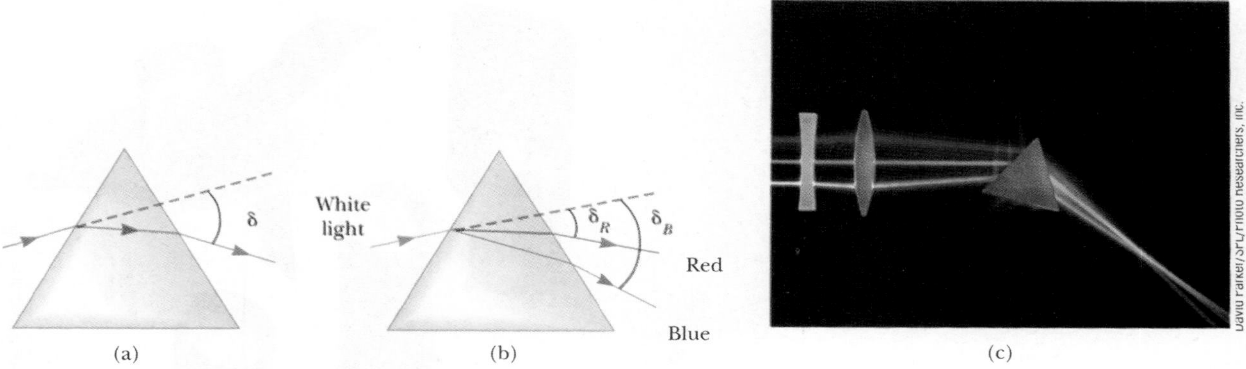

Figure 22.15 (a) A prism refracts a light ray and deviates the light through the angle δ. (b) When light is incident on a prism, the blue light is bent more than the red. (c) Light of different colors passes through a prism and two lenses. Note that as the light passes through the prism, different wavelengths are refracted at different angles.

is commonly used to study the wavelengths emitted by a light source, such as a sodium vapor lamp. Light from the source is sent through a narrow, adjustable slit and lens to produce a parallel, or collimated, beam. The light then passes through the prism and is dispersed into a spectrum. The refracted light is observed through a telescope. The experimenter sees different colored images of the slit through the eyepiece of the telescope. The telescope can be moved or the prism can be rotated in order to view the various wavelengths, which have different angles of deviation. Figure 22.17b (page 738) shows one type of prism spectrometer used in undergraduate laboratories.

APPLICATION
Identifying Gases with a Spectrometer

All hot, low-pressure gases emit their own characteristic spectra. Thus, one use of a prism spectrometer is to identify gases. For example, sodium emits only two wavelengths in the visible spectrum: two closely spaced yellow lines. (The bright linelike images of the slit seen in a spectroscope are called *spectral lines.*) A gas emitting these, and only these, colors can thus be identified as sodium. Likewise, mercury vapor has its own characteristic spectrum, consisting of four prominent wavelengths—orange, green, blue, and violet lines—along with some wavelengths of lower intensity. The particular wavelengths emitted by a gas serve as "fingerprints" of that gas. Spectral analysis, which is the measurement of the wavelengths emitted or absorbed by a substance, is a powerful general tool in many scientific areas. As examples, chemists and biologists use infrared spectroscopy to identify molecules, astronomers use visible-light spectroscopy to identify elements on distant stars, and geologists use spectral analysis to identify minerals.

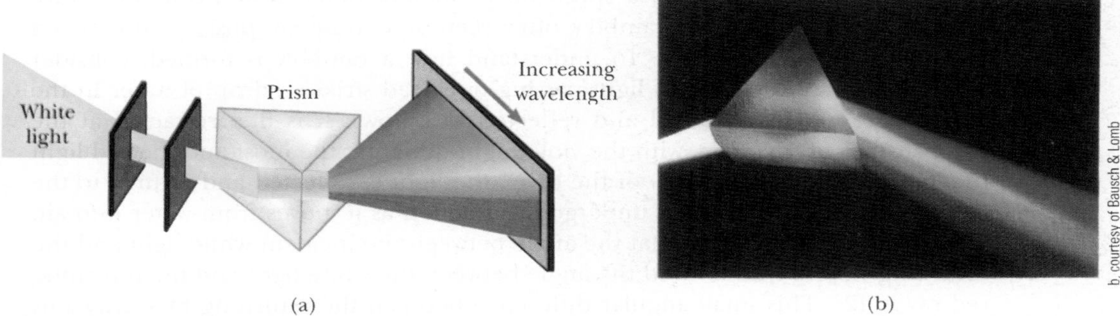

Figure 22.16 (a) Dispersion of white light by a prism. Since n varies with wavelength, the prism disperses the white light into its various spectral components. (b) Different colors of light that pass through a prism are refracted at different angles because the index of refraction of the glass depends on wavelength. Violet light bends the most, red light the least.

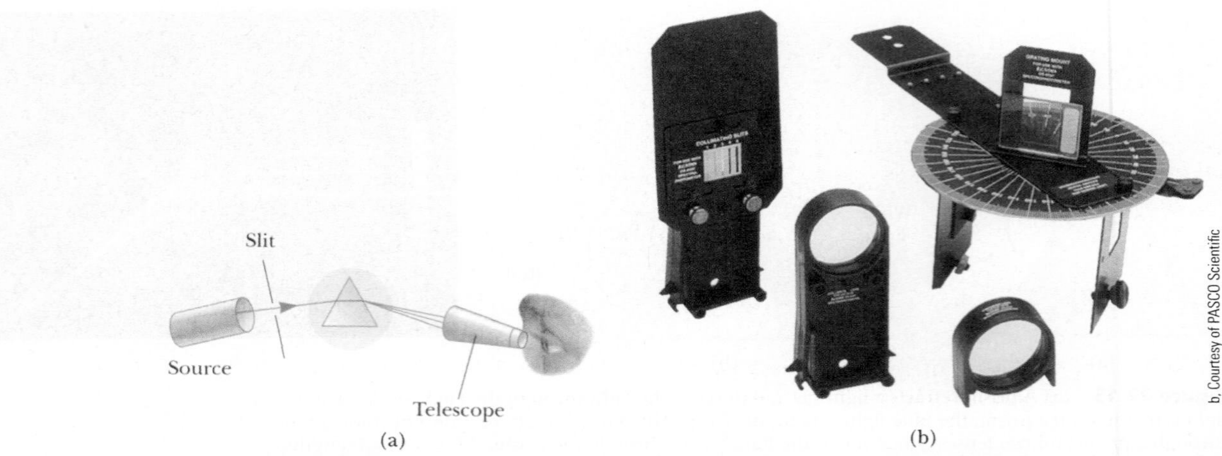

Figure 22.17 (a) A diagram of a prism spectrometer. The colors in the spectrum are viewed through a telescope. (b) A prism spectrometer with interchangeable components.

Applying Physics 22.3 Dispersion

When a beam of light enters a glass prism, which has nonparallel sides, the rainbow of color exiting the prism is a testimonial to the dispersion occurring in the glass. Suppose a beam of light enters a slab of material with parallel sides. When the beam exits the other side, traveling in the same direction as the original beam, is there any evidence of dispersion?

Explanation Due to dispersion, light at the violet end of the spectrum exhibits a larger angle of refraction on entering the glass than light at the red end. All colors of light return to their original direction of propagation as they refract back out into the air. As a result, the outgoing beam is white. But the net shift in the position of the violet light along the edge of the slab is larger than the shift of the red light, so one edge of the outgoing beam has a bluish tinge to it (it appears blue rather than violet, because the eye is not very sensitive to violet light), whereas the other edge has a reddish tinge. This effect is indicated in Figure 22.18. The colored edges of the outgoing beam of white light are evidence of dispersion.

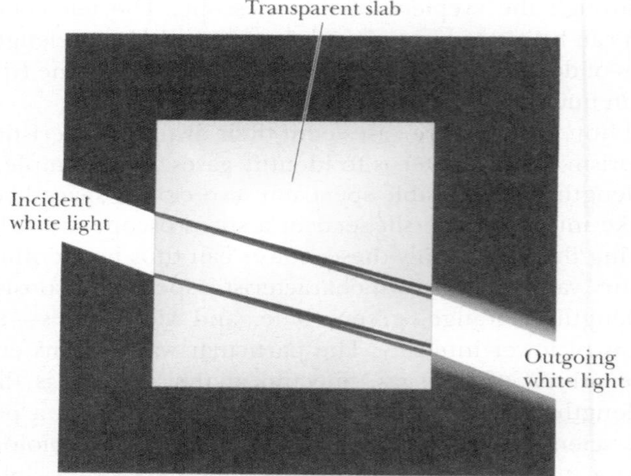

Figure 22.18 (Applying Physics 22.3)

22.5 THE RAINBOW

The dispersion of light into a spectrum is demonstrated most vividly in nature through the formation of a rainbow, often seen by an observer positioned between the Sun and a rain shower. To understand how a rainbow is formed, consider Active Figure 22.19. A ray of light passing overhead strikes a drop of water in the atmosphere and is refracted and reflected as follows: It is first refracted at the front surface of the drop, with the violet light deviating the most and the red light the least. At the back surface of the drop, the light is reflected and returns to the front surface, where it again undergoes refraction as it moves from water into air. The rays leave the drop so that the angle between the incident white light and the returning violet ray is 40° and the angle between the white light and the returning red ray is 42°. This small angular difference between the returning rays causes us to see the bow as explained in the next paragraph.

Now consider an observer viewing a rainbow, as in Figure 22.20a. If a raindrop high in the sky is being observed, the red light returning from the drop can reach the observer because it is deviated the most, but the violet light passes over the observer because it is deviated the least. Hence, the observer sees this drop as being red. Similarly, a drop lower in the sky would direct violet light toward the observer and appear to be violet. (The red light from this drop would strike the ground and not be seen.) The remaining colors of the spectrum would reach the observer from raindrops lying between these two extreme positions. Figure 22.20b shows a beautiful rainbow and a secondary rainbow with its colors reversed.

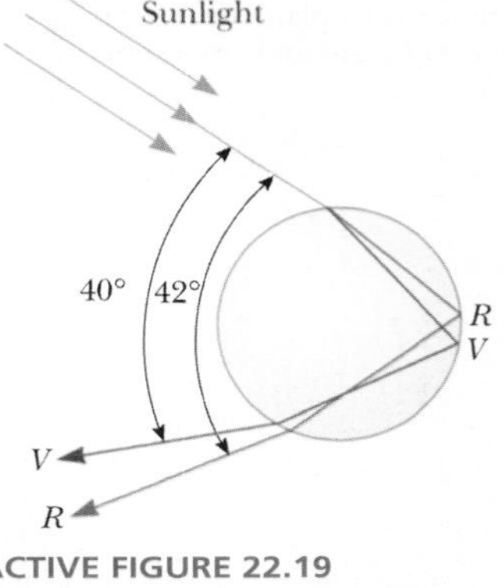

ACTIVE FIGURE 22.19
Refraction of sunlight by a spherical raindrop.

Physics Now™
Log into PhysicsNow at **www.cp7e.com** and go to Active Figure 22.19, where you can vary the point at which the sunlight enters the raindrop and verify that the angles shown are maximum angles.

22.6 HUYGENS' PRINCIPLE

The laws of reflection and refraction can be deduced using a geometric method proposed by Huygens in 1678. Huygens assumed that light is a form of wave motion rather than a stream of particles. He had no knowledge of the nature of light or of its electromagnetic character. Nevertheless, his simplified wave model is adequate for understanding many practical aspects of the propagation of light.

Huygens' principle is a geometric construction for determining at some instant the position of a new wave front from knowledge of the wave front that preceded it. (A wave front is a surface passing through those points of a wave which have the same phase and amplitude. For instance, a wave front could be a surface passing

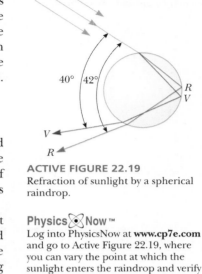

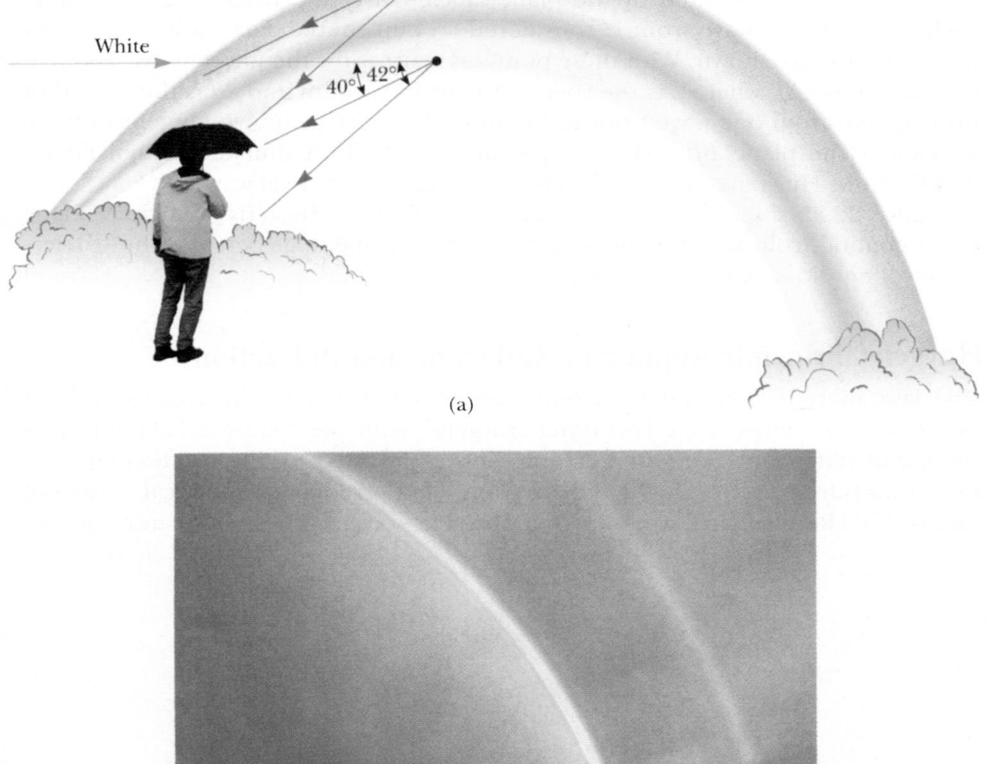

(b)

Mark D. Phillips/Photo Researchers, Inc.

Figure 22.20 (a) The formation of a rainbow. (b) This photograph of a rainbow shows a distinct secondary rainbow with the colors reversed.

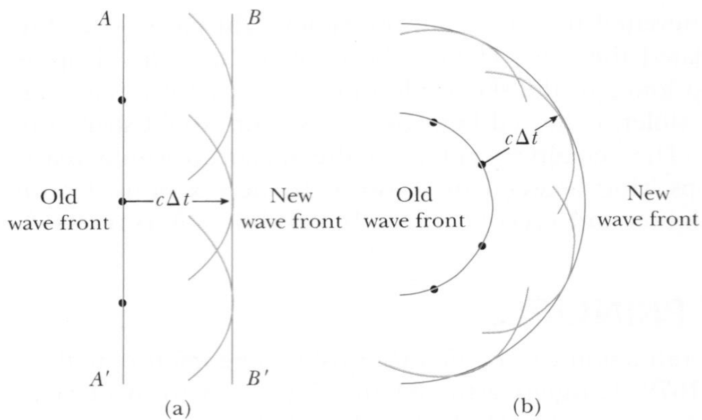

Figure 22.21 Huygens' constructions for (a) a plane wave propagating to the right and (b) a spherical wave.

through the crests of waves.) In Huygens' construction, **all points on a given wave front are taken as point sources for the production of spherical secondary waves, called wavelets, which propagate in the forward direction with speeds characteristic of waves in that medium. After some time has elapsed, the new position of the wave front is the surface tangent to the wavelets.**

Huygens' principle ►

Figure 22.21 illustrates two simple examples of Huygens' construction. First, consider a plane wave moving through free space, as in Figure 22.21a. At $t = 0$, the wave front is indicated by the plane labeled AA'. In Huygens' construction, each point on this wave front is considered a point source. For clarity, only a few points on AA' are shown. With these points as sources for the wavelets, we draw circles of radius $c\Delta t$, where c is the speed of light in vacuum and Δt is the period of propagation from one wave front to the next. The surface drawn tangent to these wavelets is the plane BB', which is parallel to AA'. In a similar manner, Figure 22.21b shows Huygens' construction for an outgoing spherical wave.

Figure 22.22 shows a convincing demonstration of Huygens' principle. Plane waves coming from far off shore emerge from the openings between the barriers as two-dimensional circular waves propagating outward.

Huygens' Principle Applied to Reflection and Refraction

The laws of reflection and refraction were stated earlier in the chapter without proof. We now derive these laws using Huygens' principle. Figure 22.23a illustrates the law of reflection. The line AA' represents a wave front of the incident light. As ray 3 travels from A' to C, ray 1 reflects from A and produces a spherical wavelet of radius AD. (Recall that the radius of a Huygens wavelet is $v\Delta t$.) Because the two

Figure 22.22 This photograph of the beach at Tel Aviv, Israel, shows Huygens wavelets radiating from each opening between breakwalls. Note how the beach has been shaped by the wave action.

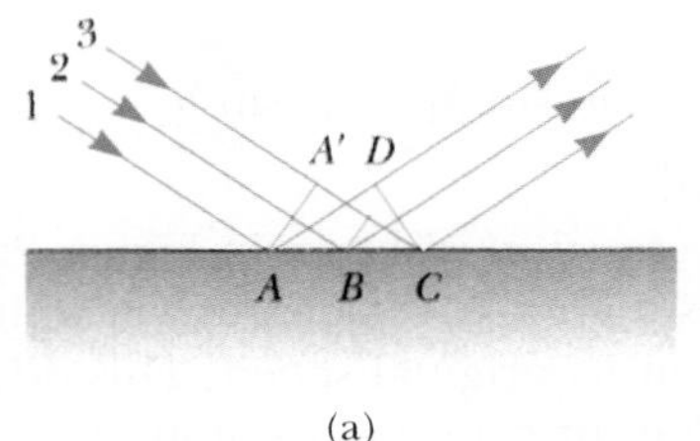

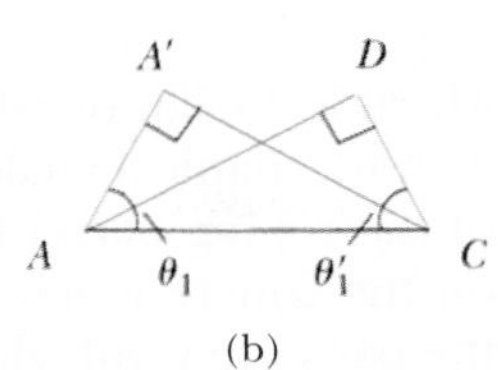

Figure 22.23 (a) Huygens' construction for proving the law of reflection. (b) Triangle ADC is congruent to triangle $AA'C$.

wavelets having radii $A'C$ and AD are in the same medium, they have the same speed v, so $AD = A'C$. Meanwhile, the spherical wavelet centered at B has spread only half as far as the one centered at A, because ray 2 strikes the surface later than ray 1.

From Huygens' principle, we find that the reflected wave front is CD, a line tangent to all the outgoing spherical wavelets. The remainder of our analysis depends on geometry, as summarized in Figure 22.23b. Note that the right triangles ADC and $AA'C$ are congruent because they have the same hypotenuse, AC, and because $AD = A'C$. From the figure, we have

$$\sin\theta_1 = \frac{A'C}{AC} \quad \text{and} \quad \sin\theta_1' = \frac{AD}{AC}$$

The right-hand sides are equal, so $\sin\,\theta = \sin\,\theta_1'$, and it follows that $\theta_1 = \theta_1'$, which is the law of reflection.

Huygens' principle and Figure 22.24a can be used to derive Snell's law of refraction. In the time interval Δt, ray 1 moves from A to B and ray 2 moves from A' to C. The radius of the outgoing spherical wavelet centered at A is equal to $v_2\,\Delta t$. The distance $A'C$ is equal to $v_1\,\Delta t$. Geometric considerations show that angle $A'AC$ equals θ_1 and angle ACB equals θ_2. From triangles $AA'C$ and ACB, we find that

$$\sin\theta_1 = \frac{v_1\,\Delta t}{AC} \quad \text{and} \quad \sin\theta_2 = \frac{v_2\,\Delta t}{AC}$$

If we divide the first equation by the second, we get

$$\frac{\sin\theta_1}{\sin\theta_2} = \frac{v_1}{v_2}$$

But from Equation 22.4 we know that $v_1 = c/n_1$ and $v_2 = c/n_2$. Therefore,

$$\frac{\sin\theta_1}{\sin\theta_2} = \frac{c/n_1}{c/n_2} = \frac{n_2}{n_1}$$

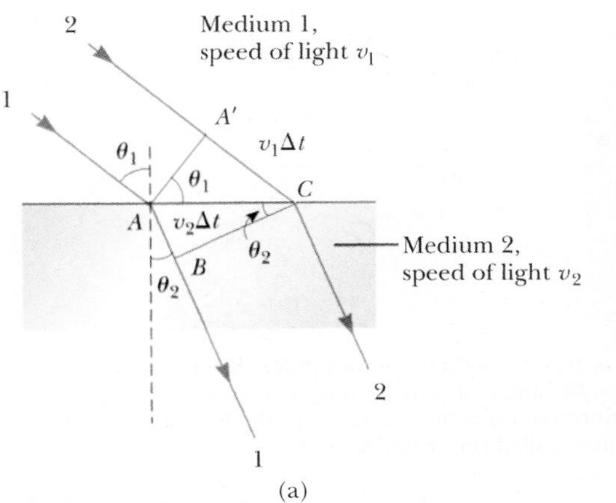

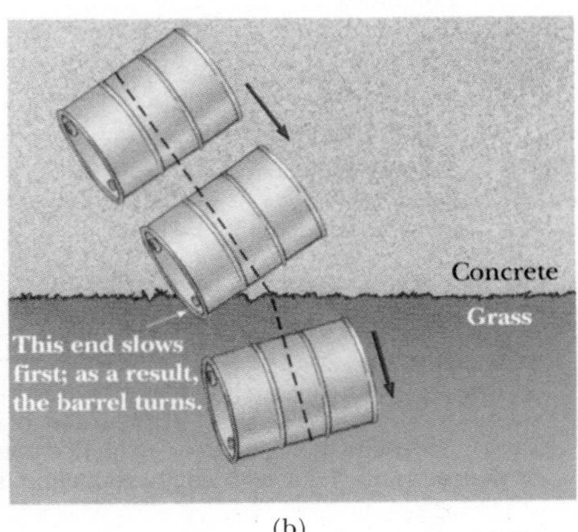

Figure 22.24 (a) Huygens' construction for proving the law of refraction. (b) Overhead view of a barrel rolling from concrete onto grass.

and it follows that

$$n_1 \sin \theta_1 = n_2 \sin \theta_2$$

which is the law of refraction.

A mechanical analog of refraction is shown in Figure 22.24b. When the left end of the rolling barrel reaches the grass, it slows down, while the right end remains on the concrete and moves at its original speed. This difference in speeds causes the barrel to pivot, changing its direction of its motion.

Courtesy of Henry Leap and Jim Lehman

This photograph shows nonparallel light rays entering a glass prism. The bottom two rays undergo total internal reflection at the longest side of the prism. The top three rays are refracted at the longest side as they leave the prism.

22.7 TOTAL INTERNAL REFLECTION

An interesting effect called *total internal reflection* can occur when light encounters the boundary between a medium with a *higher* index of refraction and one with a *lower* index of refraction. Consider a light beam traveling in medium 1 and meeting the boundary between medium 1 and medium 2, where n_1 is greater than n_2 (Active Fig. 22.25). Possible directions of the beam are indicated by rays 1 through 5. Note that the refracted rays are bent away from the normal because n_1 is greater than n_2. At some particular angle of incidence θ_c, called the **critical angle**, the refracted light ray moves parallel to the boundary, so that $\theta_2 = 90°$ (Active Fig. 22.25b). *For angles of incidence greater than* θ_c, the beam is entirely reflected at the boundary, as is ray 5 in Active Figure 22.25a. This ray is reflected as though it had struck a perfectly reflecting surface. It and all rays like it obey the law of reflection: the angle of incidence equals the angle of reflection.

We can use Snell's law to find the critical angle. When $\theta_1 = \theta_c$, $\theta_2 = 90°$, Snell's law (Eq. 22.8) gives

$$n_1 \sin \theta_c = n_2 \sin 90° = n_2$$

$$\sin\theta_c = \frac{n_2}{n_1} \quad \text{for} \quad n_1 > n_2 \qquad [22.9]$$

Equation 22.9 can be used only when n_1 is greater than n_2, because **total internal reflection occurs only when light attempts to move from a medium of higher index of refraction to a medium of lower index of refraction**. If n_1 were less than n_2, Equation 22.9 would give $\sin \theta_c > 1$, which is an absurd result because the sine of an angle can never be greater than one.

When medium 2 is air, the critical angle is small for substances with large indices of refraction, such as diamond, where $n = 2.42$ and $\theta_c = 24.0°$. By comparison, for crown glass, $n = 1.52$ and $\theta_c = 41.0°$. This property, combined with proper faceting, causes a diamond to sparkle brilliantly.

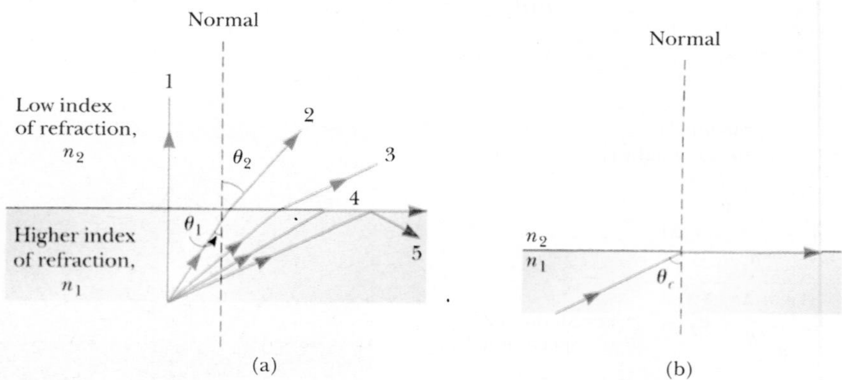

ACTIVE FIGURE 22.25
(a) Rays from a medium with index of refraction n_1 travel to a medium with index of refraction n_2, where $n_1 > n_2$. As the angle of incidence increases, the angle of refraction θ_2 increases until θ_2 is 90° (ray 4). For even larger angles of incidence, total internal reflection occurs (ray 5). (b) The angle of incidence producing a 90° angle of refraction is often called the *critical angle* θ_c.

Physics Now™
Log into PhysicsNow at **www.cp7e.com** and go to Active Figure 22.25, where you can vary the incident angle and see how the refracted ray undergoes total internal reflection when the incident angle exceeds the critical angle.

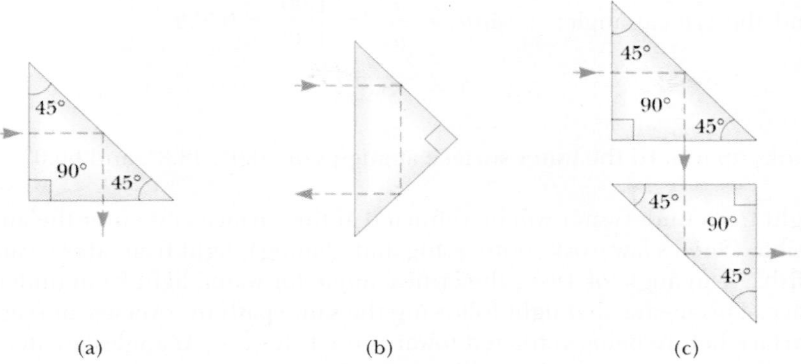

Figure 22.26 Internal reflection in a prism. (a) The ray is deviated by 90°. (b) The direction of the ray is reversed. (c) Two prisms used as a periscope.

APPLICATION

Submarine Periscopes

A prism and the phenomenon of total internal reflection can alter the direction of travel of a light beam. Figure 22.26 illustrates two such possibilities. In one case the light beam is deflected by 90.0° (Fig. 22.26a), and in the second case the path of the beam is reversed (Fig. 22.26b). A common application of total internal reflection is a submarine periscope. In this device, two prisms are arranged as in Figure 22.26c, so that an incident beam of light follows the path shown and the user can "see around corners."

Applying Physics 22.4 Total Internal Reflection and Dispersion

A beam of white light is incident on the curved edge of a semicircular piece of glass, as shown in Figure 22.27. The light enters the curved surface along the normal, so it shows no refraction. It encounters the straight side of the glass at the center of curvature of the curved side and refracts into the air. The incoming beam is moved clockwise (so that the angle θ increases) such that the beam always enters along the normal to the curved side and encounters the straight side at the center of curvature of the curved side. Why does the refracted beam become redder as it approaches a direction parallel to the straight side?

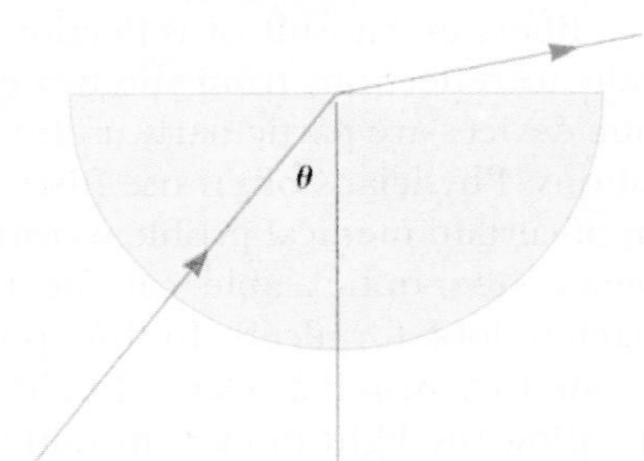

Figure 22.27 (Applying Physics 22.4)

Explanation When the outgoing beam approaches the direction parallel to the straight side, the incident angle is approaching the critical angle for total internal reflection. Dispersion occurs as the light passes out of the glass. The index of refraction for light at the violet end of the visible spectrum is larger than at the red end. As a result, as the outgoing beam approaches the straight side, the violet light undergoes total internal reflection, followed by the other colors. The red light is the last to undergo total internal reflection, so just before the outgoing light disappears, it's composed of light from the red end of the visible spectrum.

EXAMPLE 22.6 A View from the Fish's Eye

Goal Apply the concept of total internal reflection. **(a)** Find the critical angle for a water–air boundary if the index of refraction of water is 1.33. **(b)** Use the result of part (a) to predict what a fish will see (Fig. 22.28) if it looks up toward the water surface at angles of 40.0°, 48.8°, and 60.0°.

Strategy After finding the critical angle by substitution, use the fact that the path of a light ray is reversible: at a given angle, wherever a light beam can go is also where a beam of light can come from, along the same path.

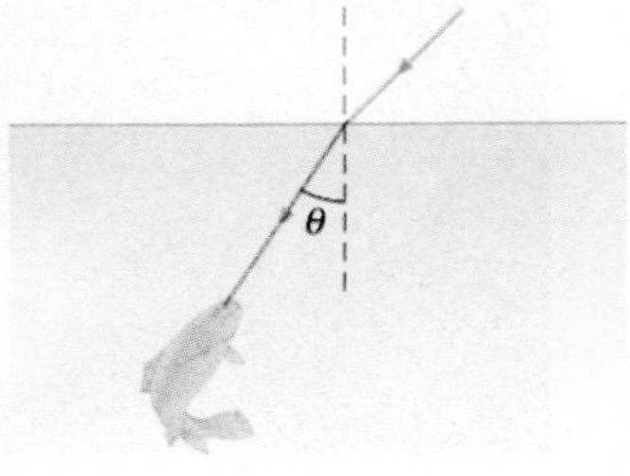

Figure 22.28 (Example 22.6) A fish looks upward toward the water's surface.

Solution

(a) Find the critical angle for a water–air boundary.

Substitute into Equation 22.9 to find the critical angle:

$$\sin\theta_c = \frac{n_2}{n_1} = \frac{1.00}{1.33} = 0.752$$

$$\theta_c = 48.8^\circ$$

(b) Predict what a fish will see if it looks up toward the water surface at angles of 40.0°, 48.8°, and 60.0°

At an angle of 40.0°, a beam of light from underwater will be refracted at the surface and enter the air above. Because the path of a light ray is reversible (Snell's law works both going and coming), light from above can follow the same path and be perceived by the fish. At an angle of 48.8°, the critical angle for water, light from underwater is bent so that it travels along the surface. This means that light following the same path in reverse can reach the fish only by skimming along the water surface before being refracted towards the fish's eye. At angles greater than the critical angle of 48.8°, a beam of light shot toward the surface will be completely reflected down toward the bottom of the pool. Reversing the path, the fish sees a reflection of some object on the bottom.

Exercise 22.6

Suppose a layer of oil with $n = 1.50$ coats the surface of the water. What is the critical angle for total internal reflection for light traveling in the oil layer and encountering the oil-water boundary?

Answer 62.7°

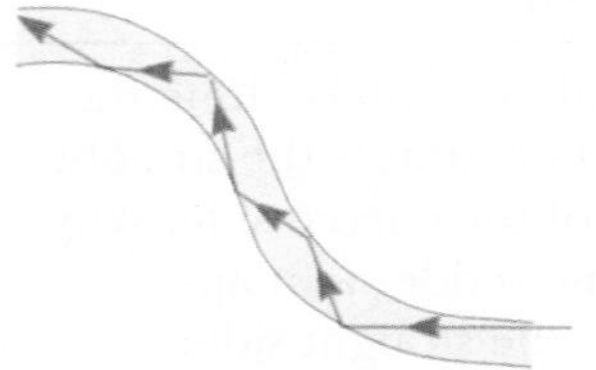

Figure 22.29 Light travels in a curved transparent rod by multiple internal reflections.

APPLICATION

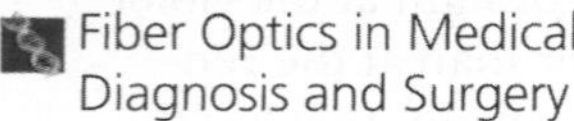

Fiber Optics in Medical Diagnosis and Surgery

Fiber Optics

Another interesting application of total internal reflection is the use of solid glass or transparent plastic rods to "pipe" light from one place to another. As indicated in Figure 22.29, light is confined to traveling within the rods, even around gentle curves, as a result of successive internal reflections. Such a light pipe can be quite flexible if thin fibers are used rather than thick rods. If a bundle of parallel fibers is used to construct an optical transmission line, images can be transferred from one point to another.

Very little light intensity is lost in these fibers as a result of reflections on the sides. Any loss of intensity is due essentially to reflections from the two ends and absorption by the fiber material. Fiber-optic devices are particularly useful for viewing images produced at inaccessible locations. Physicians often use fiber-optic cables to aid in the diagnosis and correction of certain medical problems without the intrusion of major surgery. For example, a fiber-optic cable can be threaded through the esophagus and into the stomach to look for ulcers. In this application, the cable consists of two fiber-optic lines: one to transmit a beam of light into the stomach for illumination and the other to allow the light to be transmitted out of the stomach. The resulting image can, in some cases, be viewed directly by the physician, but more often is displayed on a television monitor or captured on film.

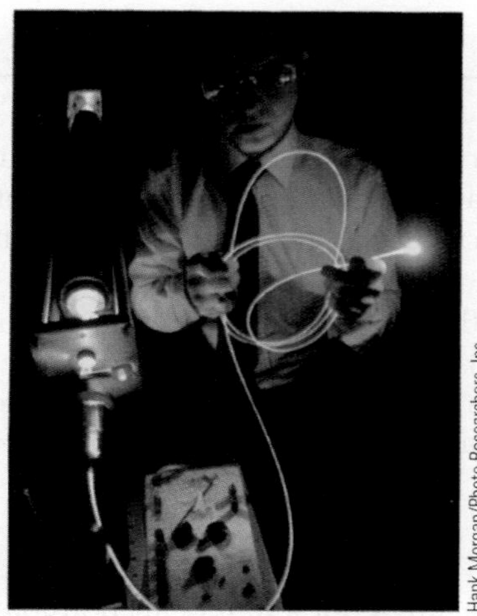

(*Left*) Strands of glass optical fibers are used to carry voice, video, and data signals in telecommunication networks. (*Right*) A bundle of optical fibers is illuminated by a laser.

In a similar way, fiber-optic cables can be used to examine the colon or to help physicians perform surgery without the need for large incisions.

The field of fiber optics has revolutionized the entire communications industry. Billions of kilometers of optical fiber have been installed in the United States to carry high-speed internet traffic, radio and television signals, and telephone calls. The fibers can carry much higher volumes of telephone calls and other forms of communication than electrical wires because of the higher frequency of the infrared light used to carry the information on optical fibers. Optical fibers are also preferable to copper wires because they are insulators and don't pick up stray electric and magnetic fields or electronic "noise."

APPLICATION

Fiber Optics in Telecommunications

Applying Physics 22.5 Design of an Optical Fiber

An optical fiber consists of a transparent core surrounded by cladding, which is a material with a lower index of refraction than the core (Fig. 22.30). A cone of angles, called the acceptance cone, is at the entrance to the fiber. Incoming light at angles within this cone will be transmitted through the fiber, whereas light entering the core from angles outside the cone will not be transmitted. The figure shows a light ray entering the fiber just within the acceptance cone and undergoing total internal reflection at the interface between the core and the cladding. If it is technologically difficult to produce light so that it enters the fiber from a small range of angles, how could you adjust the indices of refraction of the core and cladding to increase the size of the acceptance cone—would you design the indices to be farther apart or closer together?

Explanation The acceptance cone would become larger if the critical angle (θ_c in the figure) could be made smaller. This can be done by making the index of refraction of the cladding material smaller, so that the indices of refraction of the core and cladding material would be farther apart.

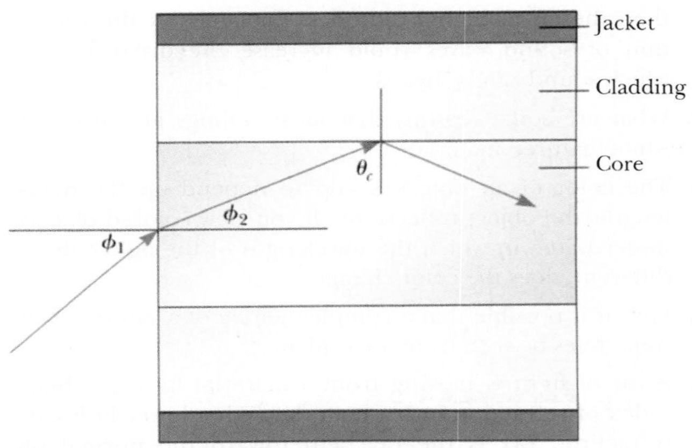

Figure 22.30 (Applying Physics 22.5)

SUMMARY

Physics Now™ Take a practice test by logging into PhysicsNow at **www.cp7e.com** and clicking on the Pre-Test link for this chapter.

22.1 The Nature of Light

Light has a dual nature. In some experiments it acts like a wave, in others like a particle, called a photon by Einstein. The energy of a photon is proportional to its frequency,

$$E = hf \qquad [22.1]$$

where $h = 6.63 \times 10^{-34}\ \text{J}\cdot\text{s}$ is *Planck's constant.*

22.2 Reflection and Refraction

In the reflection of light off a flat, smooth surface, the angle of incidence, θ_1, with respect to a line perpendicular to the surface is equal to the angle of reflection, θ_1':

$$\theta_1' = \theta_1 \qquad [22.2]$$

Light that passes into a transparent medium is bent at the boundary and is said to be *refracted.* The angle of refraction is the angle the ray makes with respect to a line perpendicular to the surface after it has entered the new medium.

22.3 The Law of Refraction

The **index of refraction** of a material, n, is defined as

$$n \equiv \frac{c}{v} \qquad [22.4]$$

where c is the speed of light in a vacuum and v is the speed of light in the material. The index of refraction of a material is also

$$n = \frac{\lambda_0}{\lambda_n} \qquad [22.7]$$

where λ_0 is the wavelength of the light in vacuum and λ_n is its wavelength in the material.

The **law of refraction**, or **Snell's law**, states that

$$n_1 \sin\theta_1 = n_2 \sin\theta_2 \qquad [22.8]$$

where n_1 and n_2 are the indices of refraction in the two media. The incident ray, the reflected ray, the refracted ray, and the normal to the surface all lie in the same plane.

Remarks This calculation depends on the angle θ being small, because the small-angle approximation was implicitly used. The measurement of the position of the bright fringes yields the wavelength of light, which in turn is a signature of atomic processes, as will be discussed in the chapters on modern physics. This kind of measurement, therefore, helped open the world of the atom.

Exercise 24.1
Suppose the same experiment is run with a different light source. If the first-order maximum is found at 1.85 cm from the centerline, what is the wavelength of the light?

Answer 463 nm

PhysicsNow™ Investigate the double-slit interference pattern by logging into PhysicsNow at **www.cp7e.com** and going to Interactive Example 24.1.

24.3 CHANGE OF PHASE DUE TO REFLECTION

Young's method of producing two coherent light sources involves illuminating a pair of slits with a single source. Another simple, yet ingenious, arrangement for producing an interference pattern with a single light source is known as *Lloyd's mirror*. A point source of light is placed at point S, close to a mirror, as illustrated in Figure 24.5. Light waves can reach the viewing point P either by the direct path SP or by the path involving reflection from the mirror. The reflected ray can be treated as a ray originating at the source S' behind the mirror. Source S', which is the image of S, can be considered a virtual source.

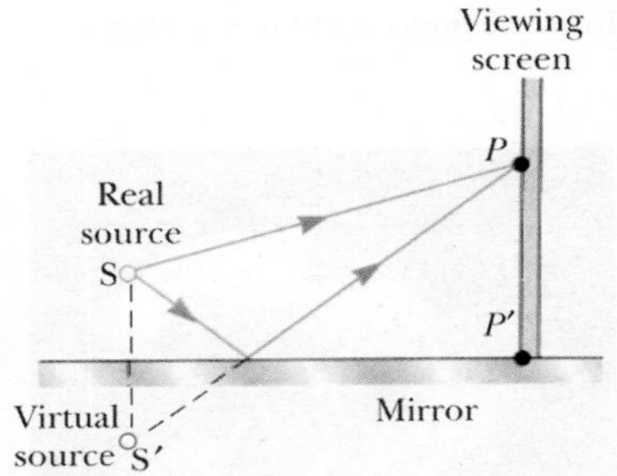

Figure 24.5 Lloyd's mirror. An interference pattern is produced on a screen at P as a result of the combination of the direct ray (blue) and the reflected ray (brown). The reflected ray undergoes a phase change of 180°.

At points far from the source, an interference pattern due to waves from S and S' is observed, just as for two real coherent sources. However, the positions of the dark and bright fringes are *reversed* relative to the pattern obtained from two real coherent sources (Young's experiment). This is because the coherent sources S and S' differ in phase by 180°, a phase change produced by reflection.

To illustrate the point further, consider P', the point where the mirror intersects the screen. This point is equidistant from S and S'. If path difference alone were responsible for the phase difference, a bright fringe would be observed at P' (because the path difference is zero for this point), corresponding to the central fringe of the two-slit interference pattern. Instead, we observe a *dark* fringe at P, from which we conclude that a 180° phase change must be produced by reflection from the mirror. In general, **an electromagnetic wave undergoes a phase change of 180° upon reflection from a medium that has an index of refraction higher than the one in which the wave was traveling**.

An analogy can be drawn between reflected light waves and the reflections of a transverse wave on a stretched string when the wave meets a boundary, as in Figure 24.6. The reflected pulse on a string undergoes a phase change of 180°

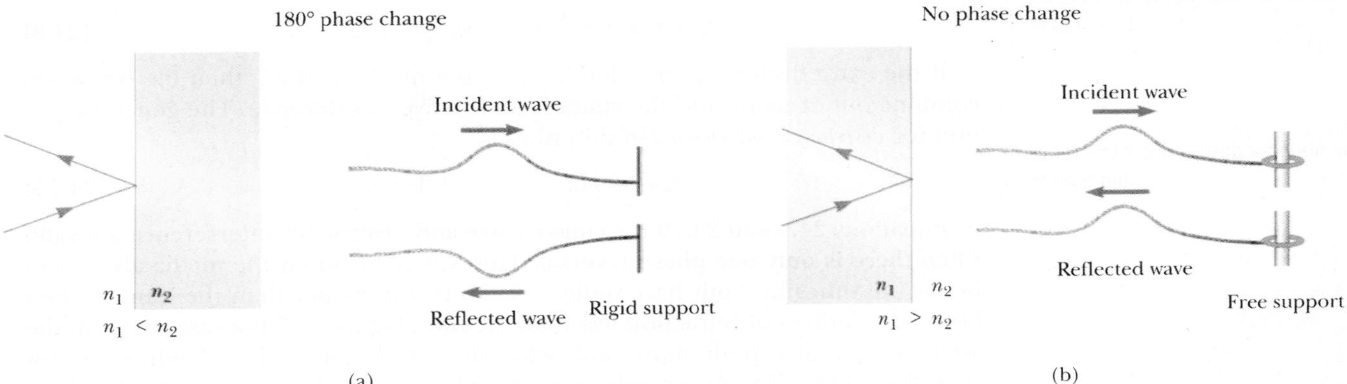

Figure 24.6 (a) A ray reflecting from a medium of higher refractive index undergoes a 180° phase change. The right side shows the analogy with a reflected pulse on a string. (b) A ray reflecting from a medium of lower refractive index undergoes no phase change.

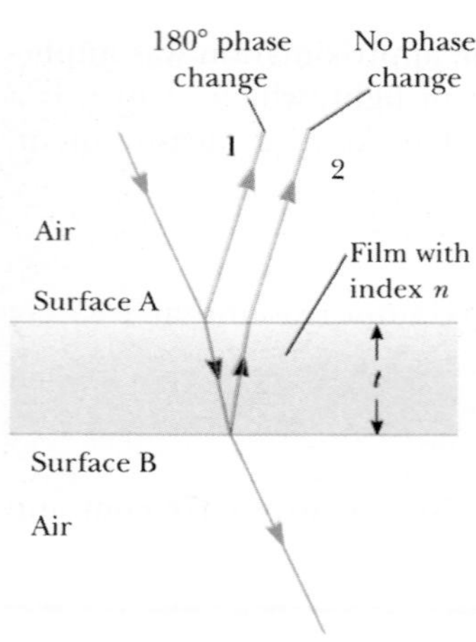

Figure 24.7 Interference observed in light reflected from a thin film is due to a combination of rays reflected from the upper and lower surfaces.

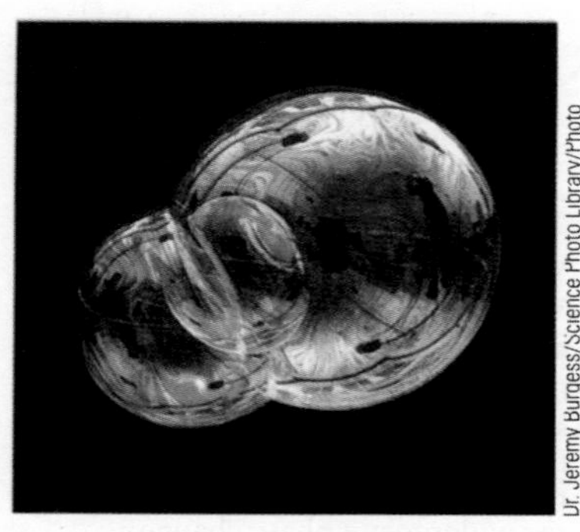

Dr. Jeremy Burgess/Science Photo Library/Photo Researchers, Inc.

Soap bubbles on water. The colors are due to interference between light rays reflected from the front and back of the thin film of soap making up the bubble. The color depends on the thickness of the film, ranging from black where the film is at its thinnest to magenta where it is thickest.

when it is reflected from the boundary of a denser string or from a rigid barrier and undergoes no phase change when it is reflected from the boundary of a less dense string. Similarly, an electromagnetic wave undergoes a 180° phase change when reflected from the boundary of a medium with index of refraction higher than the one in which it has been traveling. There is no phase change when the wave is reflected from a boundary leading to a medium of lower index of refraction. The transmitted wave that crosses the boundary also undergoes no phase change.

24.4 INTERFERENCE IN THIN FILMS

Interference effects are commonly observed in thin films, such as the thin surface of a soap bubble or thin layers of oil on water. The varied colors observed when incoherent white light is incident on such films result from the interference of waves reflected from the two surfaces of the film.

Consider a film of uniform thickness t and index of refraction n, as in Figure 24.7. Assume that the light rays traveling in air are nearly normal to the two surfaces of the film. To determine whether the reflected rays interfere constructively or destructively, we first note the following facts:

1. An electromagnetic wave traveling from a medium of index of refraction n_1 toward a medium of index of refraction n_2 undergoes a 180° phase change on reflection when $n_2 > n_1$. There is no phase change in the reflected wave if $n_2 < n_1$.
2. The wavelength of light λ_n in a medium with index of refraction n is

$$\lambda_n = \frac{\lambda}{n} \qquad [24.7]$$

where λ is the wavelength of light in vacuum.

We apply these rules to the film of Figure 24.7. According to the first rule, ray 1, which is reflected from the upper surface A, undergoes a phase change of 180° with respect to the incident wave. Ray 2, which is reflected from the lower surface B, undergoes no phase change with respect to the incident wave. Therefore, ray 1 is 180° out of phase with respect to ray 2, which is equivalent to a path difference of $\lambda_n/2$. However, we must also consider the fact that ray 2 travels an extra distance of $2t$ before the waves recombine in the air above the surface. For example, if $2t = \lambda_n/2$, then rays 1 and 2 recombine in phase, and constructive interference results. In general, the condition for *constructive interference* in thin films is

$$2t = (m + \tfrac{1}{2})\lambda_n \qquad m = 0, 1, 2, \ldots \qquad [24.8]$$

This condition takes into account two factors: (1) the difference in path length for the two rays (the term $m\lambda_n$) and (2) the 180° phase change upon reflection (the term $\lambda_n/2$). Because $\lambda_n = \lambda/n$, we can write Equation 24.8 in the form

Condition for constructive interference (thin film) ▶

$$2nt = (m + \tfrac{1}{2})\lambda \qquad m = 0, 1, 2, \ldots \qquad [24.9]$$

If the extra distance $2t$ traveled by ray 2 is a multiple of λ_n, then the two waves combine out of phase and the result is destructive interference. The general equation for *destructive interference* in thin films is

Condition for destructive interference (thin film) ▶

$$2nt = m\lambda \qquad m = 0, 1, 2, \ldots \qquad [24.10]$$

Equations 24.9 and 24.10 for constructive and destructive interference are valid when there is only one phase reversal. This will occur when the media above and below the thin film both have indices of refraction greater than the film or when both have indices of refraction less than the film. Figure 24.7 is a case in point: the air ($n = 1$) that is both above and below the film has an index of refraction less than that of the film. As a result, there is a phase reversal on reflection off the top layer of the film, but not the bottom, and Equations 24.9 and 24.10 apply. **If the film is placed between two different media, one of lower refractive index than the**

film and one of higher refractive index, Equations 24.9 and 24.10 are reversed: Equation 24.9 is used for destructive interference, and Equation 24.10 for obstructive interference. In this case, either there is a phase change of 180° for both ray 1 reflecting from surface A and ray 2 reflecting from surface B, as in Figure 24.9 of Example 24.3 (page 794), or there is no phase change for either ray, which would be the case if the incident ray came from underneath the film. Hence, the net change in relative phase due to the reflections is *zero*.

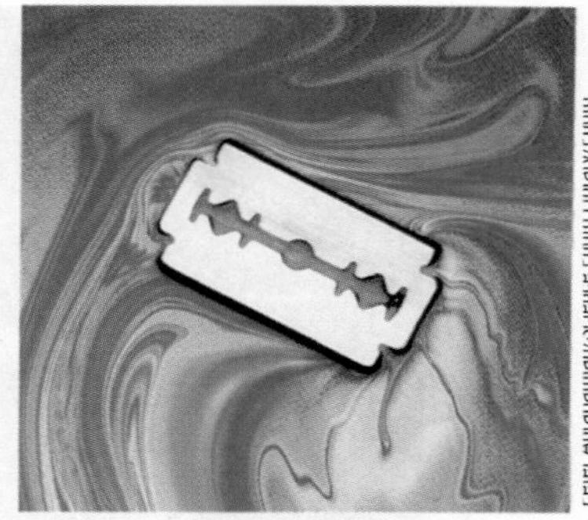

Peter Aprahamian/Science Photo Library/Photo Researchers, Inc.

A thin film of oil on water displays interference, evidenced by the pattern of colors when white light is incident on the film. Variations in the film's thickness produce the intersecting color pattern. The razor blade gives you an idea of the size of the colored bands.

Quick Quiz 24.2

Suppose Young's experiment is carried out in air, and then, in a second experiment, the apparatus is immersed in water. In what way does the distance between bright fringes change? (a) They move further apart. (b) They move closer together. (c) There is no change.

TIP 24.2 The Two Tricks of Thin Films

Be sure to include *both* effects—path length and phase change—when you analyze an interference pattern from a thin film.

Newton's Rings

Another method for observing interference in light waves is to place a planoconvex lens on top of a flat glass surface, as in Figure 24.8a. With this arrangement, the air film between the glass surfaces varies in thickness from zero at the point of contact to some value t at P. If the radius of curvature R of the lens is much greater than the distance r, and if the system is viewed from above light of wavelength λ, a pattern of light and dark rings is observed (Fig. 24.8b). These circular fringes, discovered by Newton, are called **Newton's rings**. The interference is due to the combination of ray 1, reflected from the plate, with ray 2, reflected from the lower surface of the lens. Ray 1 undergoes a phase change of 180° on reflection, because it is reflected from a boundary leading into a medium of higher refractive index, whereas ray 2 undergoes no phase change, because it is reflected from a medium of lower refractive index. Hence, the conditions for constructive and destructive interference are given by Equations 24.9 and 24.10, respectively, with $n = 1$ because the "film" is air. The contact point at O is dark, as seen in Figure 24.8b, because there is no path difference and the total phase change is due only to the 180° phase change upon reflection. Using the geometry shown in Figure 24.8a, we can obtain expressions for the radii of the bright and dark bands in terms of the radius of curvature R, and vacuum wavelength λ. For example, the dark rings have radii of $r \approx \sqrt{m\lambda R/n}$.

One of the important uses of Newton's rings is in the testing of optical lenses. A circular pattern like that in Figure 24.8b is achieved only when the lens is ground to a perfectly spherical curvature. Variations from such symmetry might produce a

APPLICATION

Checking for Imperfections in Optical Lenses

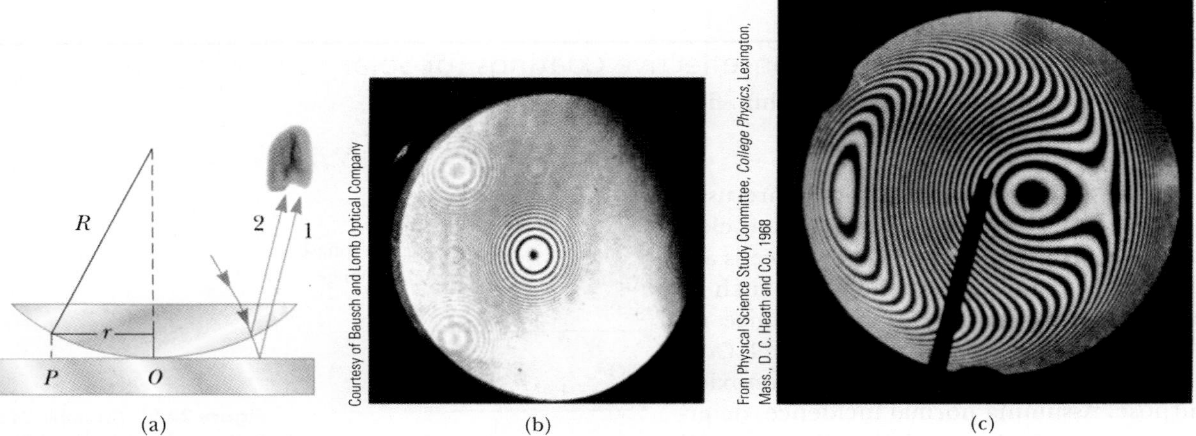

Figure 24.8 (a) The combination of rays reflected from the glass plate and the curved surface of the lens gives rise to an interference pattern known as Newton's rings. (b) A photograph of Newton's rings. (c) This asymmetric interference pattern indicates imperfections in the lens.

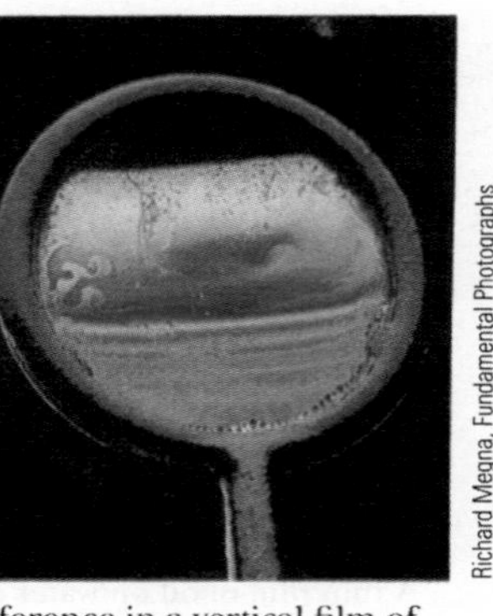

Interference in a vertical film of variable thickness. The top of the film appears darkest where the film is thinnest.

pattern like that in Figure 24.8c. These variations give an indication of how the lens must be reground and repolished to remove imperfections.

Problem-Solving Strategy Thin-Film Interference

The following steps are recommended in addressing thin-film interference problems:

1. Identify the thin film causing the interference, and the indices of refraction in the film and in the media on either side of it.
2. Determine the number of phase reversals: zero, one, or two.
3. Consult the following table, which contains Equations 24.9 and 24.10, and select the correct column for the problem in question:

Equation ($m = 0, 1, \ldots$)	1 phase reversal	0 or 2 phase reversals
$2nt = (m + \frac{1}{2})\lambda$ **[24.9]**	constructive	destructive
$2nt = m\lambda$ **[24.10]**	destructive	constructive

4. Substitute values in the appropriate equations, as selected in the previous step.

EXAMPLE 24.2 Interference in a Soap-Film

Goal Calculate interference effects in a thin film when there is one phase reversal.

Problem Calculate the minimum thickness of a soap-bubble film ($n = 1.33$) that will result in constructive interference in the reflected light if the film is illuminated by light with wavelength 602 nm in free space.

Strategy There is only one inversion, so the condition for constructive interference is $2nt = (m + \frac{1}{2})\lambda$. The minimum film thickness for constructive interference corresponds to $m = 0$ in this equation.

Solution

Solve $2nt = \lambda/2$ for the thickness t, and substitute:

$$t = \frac{\lambda}{4n} = \frac{602 \text{ nm}}{4(1.33)} = 113 \text{ nm}$$

Remark The swirling colors in a soap bubble are due to the fact that the thickness of the soap layer varies from one place to another.

Exercise 24.2

What other film thicknesses will produce constructive interference?

Answer 339 nm, 566 nm, 792 nm, and so on

INTERACTIVE EXAMPLE 24.3 Nonreflective Coatings for Solar Cells and Optical Lenses

Goal Calculate interference effects in a thin film when there are two inversions.

Problem Semiconductors such as silicon are used to fabricate solar cells—devices that generate electric energy when exposed to sunlight. Solar cells are often coated with a transparent thin film, such as silicon monoxide (SiO; $n = 1.45$) to minimize reflective losses (Fig. 24.9). A silicon solar cell ($n = 3.50$) is coated with a thin film of silicon monoxide for this purpose. Assuming normal incidence, determine the minimum thickness of the film that will produce the least reflection at a wavelength of 552 nm.

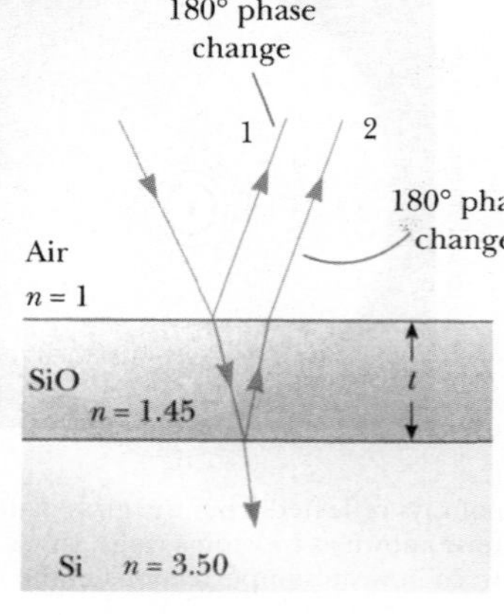

Figure 24.9 (Example 24.3) Reflective losses from a silicon solar cell are minimized by coating it with a thin film of silicon monoxide (SiO).

Strategy Reflection is least when rays 1 and 2 in Figure 24.9 meet the condition for destructive interference. Note that *both* rays undergo 180° phase changes on reflection. The condition for a reflection *minimum* is therefore $2nt = \lambda/2$.

Solution

Solve $2nt = \lambda/2$ for t, the required thickness:

$$t = \frac{\lambda}{4n} = \frac{552 \text{ nm}}{4(1.45)} = \boxed{95.2 \text{ nm}}$$

Remarks Typically, such coatings reduce the reflective loss from 30% (with no coating) to 10% (with a coating), thereby increasing the cell's efficiency because more light is available to create charge carriers in the cell. In reality, the coating is never perfectly nonreflecting, because the required thickness is wavelength dependent and the incident light covers a wide range of wavelengths.

Exercise 24.3

Glass lenses used in cameras and other optical instruments are usually coated with one or more transparent thin films, such as magnesium fluoride (MgF_2), to reduce or eliminate unwanted reflection. Carl Zeiss developed this method; his first coating was 1.00×10^2 nm thick. Using $n = 1.38$ for MgF_2, what visible wavelength would be eliminated by destructive interference in the reflected light?

Answer 552 nm

PhysicsNow™ Investigate the interference for various film properties by logging into PhysicsNow at **www.cp7e.com** and going to Interactive Example 24.3.

EXAMPLE 24.4 Interference in a Wedge-Shaoed Film

Goal Calculate interference effects when the film has variable thickness.

Problem A pair of glass slides 10.0 cm long and with $n = 1.52$ are separated on one end by a hair, forming a triangular wedge of air as illustrated in Figure 24.10. When coherent light from a helium–neon laser with wavelength 633 nm is incident on the film from above, 15.0 dark fringes per centimeter are observed. How thick is the hair?

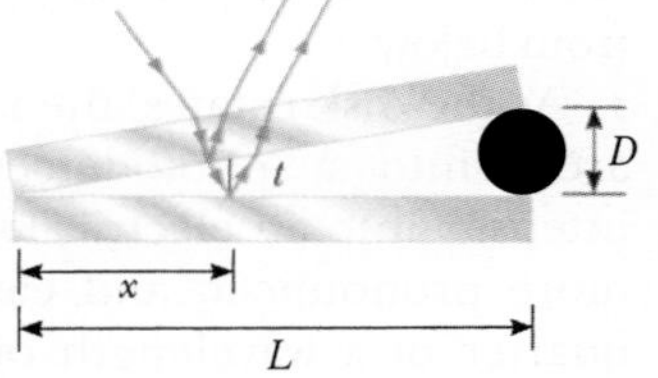

Figure 24.10 (Example 24.4) Interference bands in reflected light can be observed by illuminating a wedge-shaped film with monochromatic light. The dark areas correspond to positions of destructive interference.

Strategy The interference pattern is created by the thin film of air having variable thickness. The pattern is a series of alternating bright and dark parallel bands. A dark band corresponds to destructive interference, and there is one phase reversal, so $2nt = m\lambda$ should be used. We can also use the similar triangles in Figure 24.10 to obtain the relation $t/x = D/L$. We can find the thickness for any m, and if the position x can also be found, this last equation gives the diameter of the hair, D.

Solution

Solve the destructive-interference equation for the thickness of the film, t, with $n = 1$ for air:

$$t = \frac{m\lambda}{2}$$

If d is the distance from one dark band to the next, then the x-coordinate of the mth band is a multiple of d:

$$x = md$$

By dimensional analysis, d is just the inverse of the number of bands per centimeter.

$$d = \left(15.0 \frac{\text{bands}}{\text{cm}}\right)^{-1} = 6.67 \times 10^{-2} \frac{\text{cm}}{\text{band}}$$

Now use similar triangles, and substitute all the information:

$$\frac{t}{x} = \frac{m\lambda/2}{md} = \frac{\lambda}{2d} = \frac{D}{L}$$

Solve for D and substitute given values:

$$D = \frac{\lambda L}{2d} = \frac{(633 \times 10^{-9} \text{ m})(0.100 \text{ m})}{2(6.67 \times 10^{-4} \text{ m})} = \boxed{4.75 \times 10^{-5} \text{ m}}$$

Remarks Some may be concerned about interference caused by light bouncing off the top and bottom of, say, the upper glass slide. It's unlikely, however, that the thickness of the slide will be half an integer multiple of the wavelength of the helium-neon laser (for some very large value of m). In addition, in contrast to the air wedge, the thickness of the glass doesn't vary.

Exercise 24.4
The air wedge is replaced with water, with $n = 1.33$. Find the distance between dark bands when the helium–neon laser light hits the glass slides.

Answer 5.01×10^{-4} m

24.5 USING INTERFERENCE TO READ CD'S AND DVD'S

APPLICATION
The Physics of CD's and DVD's

Compact disks (CD's) and digital video disks (DVD's) have revolutionized the computer and entertainment industries by providing fast access; high-density storage of text, graphics, and movies; and high-quality sound recordings. The data on these disks are stored digitally as a series of zeros and ones, and these zeros and ones are read by laser light reflected from the disk. Strong reflections (constructive interference) from the disk are chosen to represent zeros and weak reflections (destructive interference) represent ones.

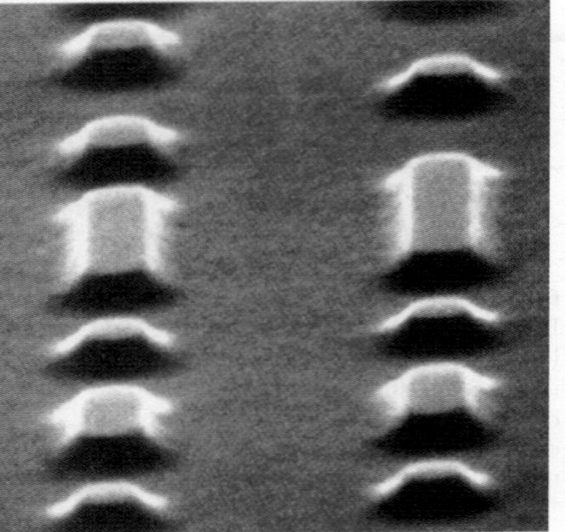
Courtesy of Sony Disc Manufacturing

Figure 24.11 A photomicrograph of adjacent tracks on a compact disc (CD). The information encoded in these pits and smooth areas is read by a laser beam.

To see in more detail how thin-film interference plays a crucial role in reading CD's, consider Figure 24.11. This shows a photomicrograph of several CD tracks which consist of a sequence of pits (when viewed from the top or label side of the disk) of varying length formed in a reflecting-metal information layer. A cross-sectional view of a CD as shown in Figure 24.12 reveals that the pits appear as bumps to the laser beam, which shines on the metallic layer through a clear plastic coating from below.

As the disk rotates, the laser beam reflects off the sequence of bumps and lower areas into a photodetector, which converts the fluctuating reflected light intensity into an electrical string of zeros and ones. To make the light fluctuations more pronounced and easier to detect, the pit depth t is made equal to one-quarter of a wavelength of the laser light in the plastic. When the beam hits a rising or falling bump edge, part of the beam reflects from the top of the bump and part from the lower adjacent area, ensuring destructive interference and very low intensity when the reflected beams combine at the detector. Bump edges are read as ones, and flat bump tops and intervening flat plains are read as zeros.

In Example 24.5 the pit depth for a standard CD, using an infrared laser of wavelength 780 nm, is calculated. DVDs use shorter wavelength lasers of 635 nm,

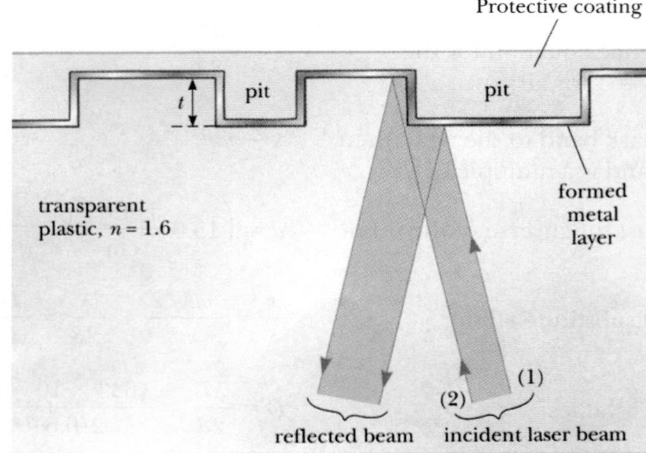

Figure 24.12 Cross section of a CD showing metallic pits of depth t and a laser beam detecting the edge of a pit.

and the track separation, pit depth, and minimum pit length are all smaller. This allows a DVD to store about 30 times more information than a CD.

EXAMPLE 24.5 Pit Depth in a CD

Goal Apply interference principles to a CD.

Problem Find the pit depth in a CD that has a plastic transparent layer with index of refraction of 1.60 and is designed for use in a CD player using a laser with a wavelength of 7.80×10^2 nm in air.

Strategy (See Fig. 24.12.) Rays (1) and (2) both reflect from the metal layer which acts like a mirror, so there is no phase difference due to reflection between those rays. There is, however, the usual phase difference caused by the extra distance $2t$ traveled by ray (2). The wavelength is λ/n, where n is the index of refraction in the substance.

Solution

Use the appropriate condition for destructive interference in a thin film:

$$2t = \frac{\lambda}{2n}$$

Solve for the thickness t and substitute:

$$t = \frac{\lambda}{4n} = \frac{7.80 \times 10^2 \text{ nm}}{(4)(1.60)} = 1.22 \times 10^2 \text{ nm}$$

Remarks Different CD systems have different tolerances for scratches. Anything that changes the reflective properties of the disk can affect the readability of the disk.

Exercise 24.5

Repeat the example for a laser with wavelength 635 nm.

Answer 99.2 nm

24.6 DIFFRACTION

Suppose a light beam is incident on two slits, as in Young's double-slit experiment. If the light truly traveled in straight-line paths after passing through the slits, as in Figure 24.13a, the waves wouldn't overlap and no interference pattern would be seen. Instead, Huygens's principle requires that the waves spread out from the slits, as shown in Figure 24.13b. In other words, the light bends from a straight-line path and enters the region that would otherwise be shadowed. This spreading out of light from its initial line of travel is called **diffraction**.

In general, diffraction occurs when waves pass through small openings, around obstacles, or by sharp edges. For example, when a single narrow slit is placed between a distant light source (or a laser beam) and a screen, the light produces a diffraction pattern like that in Figure 24.14. The pattern consists of a broad, intense central band flanked by a series of narrower, less intense secondary bands (called **secondary maxima**) and a series of dark bands, or **minima**. This phenomenon

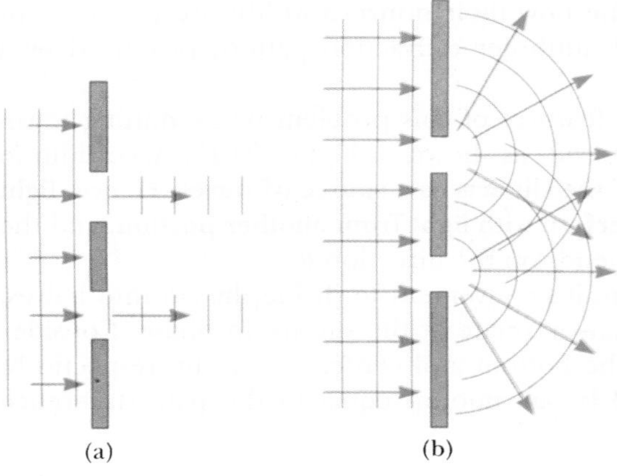

Figure 24.13 (a) If light did not spread out after passing through the slits, no interference would occur. (b) The light from the two slits overlaps as it spreads out, filling the expected shadowed regions with light and producing interference fringes.

Figure 24.14 The diffraction pattern that appears on a screen when light passes through a narrow vertical slit. The pattern consists of a broad central band and a series of less intense and narrower side bands.

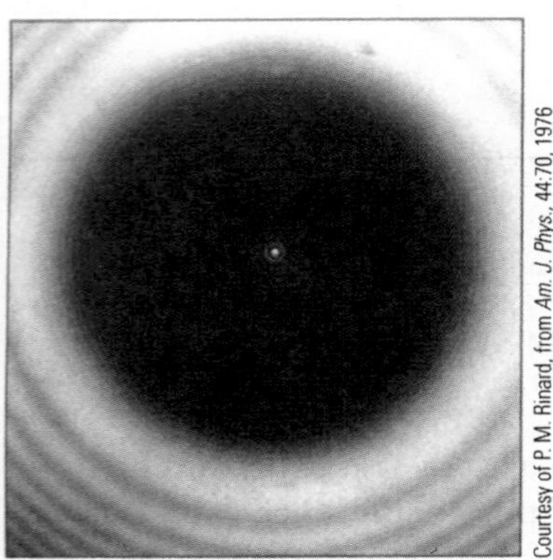

Courtesy of P. M. Rinard, from Am. J. Phys., 44:70, 1976

Figure 24.15 The diffraction pattern of a penny placed midway between the screen and the source.

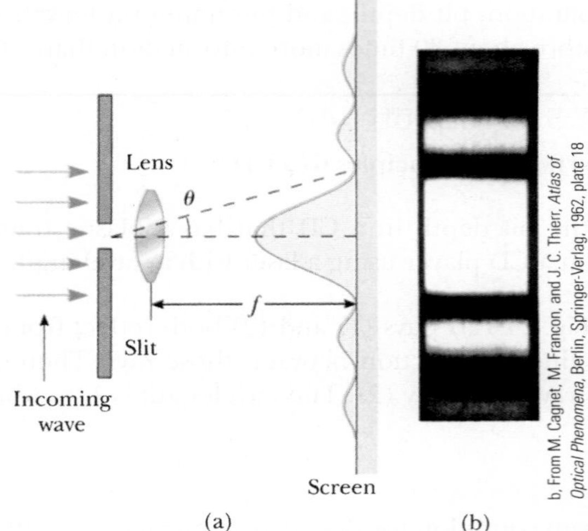

ACTIVE FIGURE 24.16
(a) The Fraunhofer diffraction pattern of a single slit. The parallel rays are brought into focus on the screen with a converging lens. The pattern consists of a central bright region flanked by much weaker maxima. (This drawing is not to scale.) (b) A photograph of a single-slit Fraunhofer diffraction pattern.

Physics Now™
Log into PhysicsNow at **www.cp7e.com** and go to Active Figure 24.16, where you can adjust the slit width and the wavelength of the light, observing the effect on the diffraction pattern.

can't be explained within the framework of geometric optics, which says that light rays traveling in straight lines should cast a sharp image of the slit on the screen.

Figure 24.15 shows the diffraction pattern and shadow of a penny. The pattern consists of the shadow, a bright spot at its center, and a series of bright and dark circular bands of light near the edge of the shadow. The bright spot at the center (called the *Fresnel bright spot*) is explained by Augustin Fresnel's wave theory of light, which predicts constructive interference at this point for certain locations of the penny. From the viewpoint of geometric optics, there shouldn't be any bright spot: the center of the pattern would be completely screened by the penny.

One type of diffraction, called **Fraunhofer diffraction**, occurs when the rays leave the diffracting object in parallel directions. Fraunhofer diffraction can be achieved experimentally either by placing the observing screen far from the slit or by using a converging lens to focus the parallel rays on a nearby screen, as in Active Figure 24.16a. A bright fringe is observed along the axis at $\theta = 0$, with alternating dark and bright fringes on each side of the central bright fringe. Active Figure 24.16b is a photograph of a single-slit Fraunhofer diffraction pattern.

24.7 SINGLE-SLIT DIFFRACTION

Until now we have assumed that slits have negligible width, acting as line sources of light. In this section we determine how their nonzero widths are the basis for understanding the nature of the Fraunhofer diffraction pattern produced by a single slit.

We can deduce some important features of this problem by examining waves coming from various portions of the slit, as shown in Figure 24.17. According to Huygens' principle, **each portion of the slit acts as a source of waves. Hence, light from one portion of the slit can interfere with light from another portion**, and the resultant intensity on the screen depends on the direction θ.

To analyze the diffraction pattern, it's convenient to divide the slit into halves, as in Figure 24.17. All the waves that originate at the slit are in phase. Consider waves 1 and 3, which originate at the bottom and center of the slit, respectively. Wave 1 travels farther than wave 3 by an amount equal to the path difference

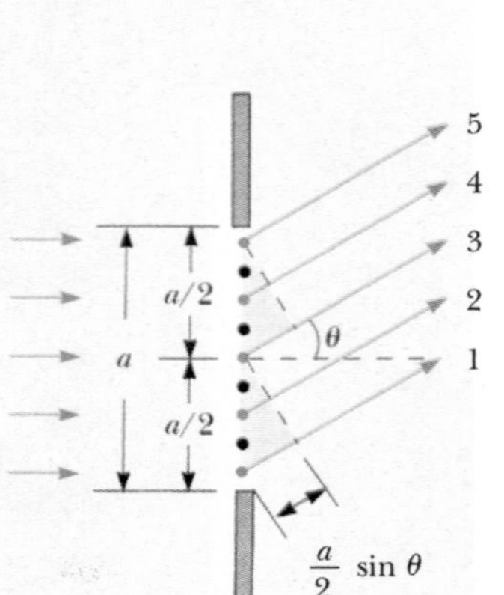

Figure 24.17 Diffraction of light by a narrow slit of width a. Each portion of the slit acts as a point source of waves. The path difference between rays 1 and 3 or between rays 2 and 4 is equal to $(a/2)\sin\theta$. (This drawing is not to scale, and the rays are assumed to converge at a distant point.)

$(a/2)\sin\theta$, where a is the width of the slit. Similarly, the path difference between waves 3 and 5 is $(a/2)\sin\theta$. If this path difference is exactly half of a wavelength (corresponding to a phase difference of 180°), the two waves cancel each other and destructive interference results. This is true, in fact, for any two waves that originate at points separated by half the slit width, because the phase difference between two such points is 180°. Therefore, waves from the upper half of the slit interfere *destructively* with waves from the lower half of the slit when

$$\frac{a}{2}\sin\theta = \frac{\lambda}{2}$$

or when

$$\sin\theta = \frac{\lambda}{a}$$

If we divide the slit into four parts rather than two and use similar reasoning, we find that the screen is also dark when

$$\sin\theta = \frac{2\lambda}{a}$$

Continuing in this way, we can divide the slit into six parts and show that darkness occurs on the screen when

$$\sin\theta = \frac{3\lambda}{a}$$

Therefore, the general condition for **destructive interference** for a single slit of width a is

$$\sin\theta_{\text{dark}} = m\frac{\lambda}{a} \qquad m = \pm 1, \pm 2, \pm 3, \ldots \qquad [24.11]$$

◀ Condition for destructive interference (single slit)

TIP 24.3 The Same, But Different

Although Equations 24.2 and 24.11 have the same form, they have different meanings. Equation 24.2 describes the *bright* regions in a two-slit interference pattern, while Equation 24.11 describes the *dark* regions in a single-slit interference pattern.

Equation 24.11 gives the values of θ for which the diffraction pattern has zero intensity, where a dark fringe forms. However, the equation tells us nothing about the variation in intensity along the screen. The general features of the intensity distribution along the screen are shown in Figure 24.18. A broad central bright fringe is flanked by much weaker bright fringes alternating with dark fringes. The various dark fringes (points of zero intensity) occur at the values of θ that satisfy Equation 24.11. The points of constructive interference lie approximately halfway between the dark fringes. Note that the central bright fringe is twice as wide as the weaker maxima having $m > 1$.

Quick Quiz 24.3

In a single-slit diffraction experiment, as the width of the slit is made smaller, the width of the central maximum of the diffraction pattern (a) becomes smaller, (b) becomes larger, or (c) remains the same.

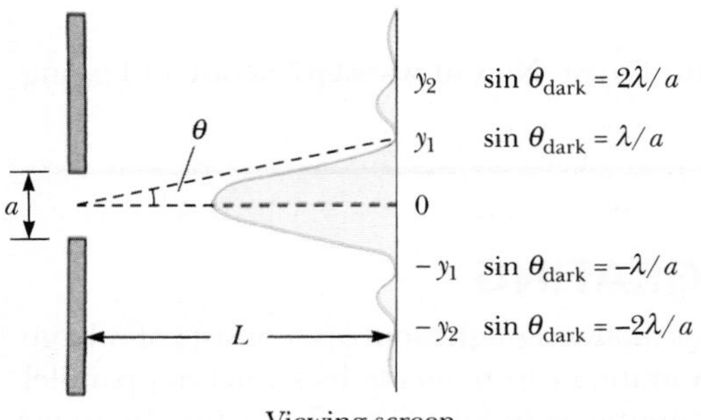

Figure 24.18 Positions of the minima for the Fraunhofer diffraction pattern of a single slit of width a. (This drawing is not to scale.)

Applying Physics 24.3 Diffraction of Sound Waves

If a classroom door is open even just a small amount, you can hear sounds coming from the hallway. Yet you can't see what is going on in the hallway. How can this difference be explained?

Explanation The space between the slightly open door and the wall is acting as a single slit for waves. Sound waves have wavelengths larger than the width of the slit, so sound is effectively diffracted by the opening, and the central maximum spreads throughout the room. Light wavelengths are much smaller than the slit width, so there is virtually no diffraction for the light. You must have a direct line of sight to detect the light waves.

INTERACTIVE EXAMPLE 24.6 A Single-Slit Experiment

Goal Find the positions of the dark fringes in single-slit diffraction.

Problem Light of wavelength 5.80×10^2 nm is incident on a slit of width 0.300 mm. The observing screen is placed 2.00 m from the slit. Find the positions of the first dark fringes and the width of the central bright fringe.

Strategy This problem requires substitution into Equation 24.11 to find the sines of the angles of the first dark fringes. The positions can then be found with the tangent function, since for small angles $\sin\theta \approx \tan\theta$. The extent of the central maximum is defined by these two dark fringes.

Solution

The first dark fringes that flank the central bright fringe correspond to $m = \pm 1$ in Equation 24.11:

$$\sin\theta = \pm\frac{\lambda}{a} = \pm\frac{5.80 \times 10^{-7}\text{ m}}{0.300 \times 10^{-3}\text{ m}} = \pm 1.93 \times 10^{-3}$$

Use the triangle in Figure 24.18 to relate the position of the fringe to the tangent function:

$$\tan\theta = \frac{y_1}{L}$$

Because θ is very small, we can use the approximation $\sin\theta \approx \tan\theta$ and then solve for y_1:

$$\sin\theta \approx \tan\theta \approx \frac{y_1}{L}$$

$$y_1 \approx L\sin\theta = \pm L\frac{\lambda}{a} = \boxed{\pm 3.86 \times 10^{-3}\text{ m}}$$

Compute the distance between the positive and negative first-order maxima, which is the width w of the central maximum:

$$w = +3.86 \times 10^{-3}\text{ m} - (-3.86 \times 10^{-3}\text{ m}) = \boxed{7.72 \times 10^{-3}\text{ m}}$$

Remarks Note that this value of w is much greater than the width of the slit. However, as the width of the slit is *increased*, the diffraction pattern *narrows*, corresponding to smaller values of θ. In fact, for large values of a, the maxima and minima are so closely spaced that the only observable pattern is a large central bright area resembling the geometric image of the slit. Because the width of the geometric image increases as the slit width increases, the narrowest image occurs when the geometric and diffraction widths are equal.

Exercise 24.6

Determine the width of the first-order bright fringe.

Answer 3.86 mm

Physics Now™ Investigate the single-slit diffraction pattern by logging into PhysicsNow at **www.cp7e.com** and going to Interactive Example 24.6.

24.8 THE DIFFRACTION GRATING

The diffraction grating, a useful device for analyzing light sources, consists of a large number of equally spaced parallel slits. A grating can be made by scratching parallel lines on a glass plate with a precision machining technique. The clear panes between

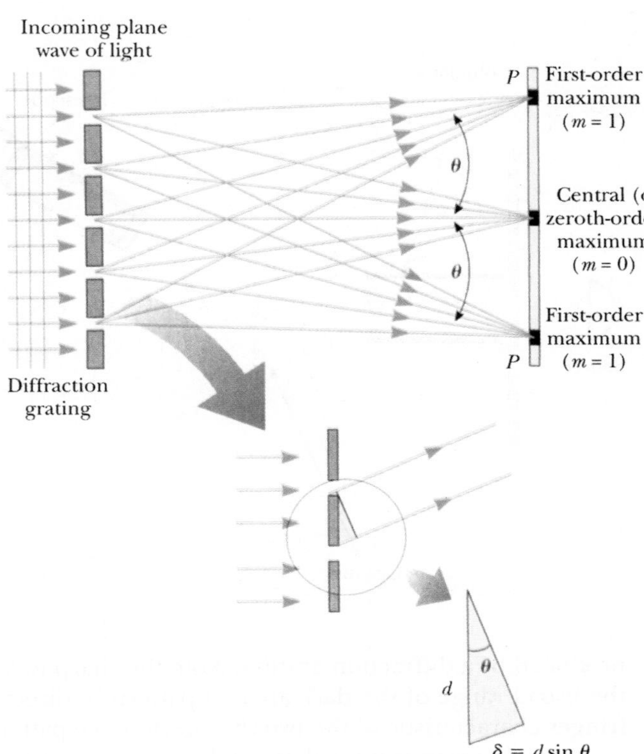

Figure 24.19 A side view of a diffraction grating. The slit separation is d, and the path difference between adjacent slits is $d \sin \theta$.

scratches act like slits. A typical grating contains several thousand lines per centimeter. For example, a grating ruled with 5 000 lines/cm has a slit spacing d equal to the reciprocal of that number; hence, $d = (1/5\ 000)\ \text{cm} = 2 \times 10^{-4}\ \text{cm}$.

Figure 24.19 is a schematic diagram of a section of a plane diffraction grating. A plane wave is incident from the left, normal to the plane of the grating. The intensity of the pattern on the screen is the result of the combined effects of interference and diffraction. Each slit causes diffraction, and the diffracted beams in turn interfere with one another to produce the pattern. Moreover, each slit acts as a source of waves, and all waves start in phase at the slits. For some arbitrary direction θ measured from the horizontal, however, the waves must travel *different* path lengths before reaching a particular point P on the screen. From Figure 24.19, note that the path difference between waves from any two adjacent slits is $d \sin \theta$. If this path difference equals one wavelength or some integral multiple of a wavelength, waves from all slits will be in phase at P and a bright line will be observed at that point. Therefore, the condition for **maxima** in the interference pattern at the angle θ is

$$d \sin \theta_{\text{bright}} = m\lambda \qquad m = 0, 1, 2, \ldots \qquad [24.12]$$

◀ Condition for maxima in the interference pattern of a diffraction grating

Light emerging from a slit at an angle other than that for a maximum interferes nearly completely destructively with light from some other slit on the grating. All such pairs will result in little or no transmission in that direction, as illustrated in Active Figure 24.20.

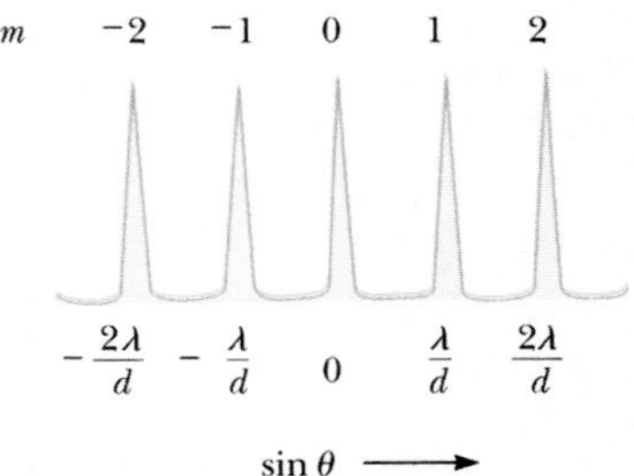

ACTIVE FIGURE 24.20
Intensity versus sin θ for the diffraction grating. The zeroth-, first-, and second-order principal maxima are shown.

Physics Now™
Log into PhysicsNow at **www.cp7e.com** and go to Active Figure 24.20, where you can choose the number of slits to be illuminated and observe the effect on the diffraction pattern.

Equation 24.12 can be used to calculate the wavelength from the grating spacing and the angle of deviation, θ. The integer m is the **order number** of the diffraction pattern. If the incident radiation contains several wavelengths, each wavelength deviates through a specific angle, which can be found from Equation 24.12. All wavelengths are focused at $\theta = 0$, corresponding to $m = 0$. This is called the *zeroth-order maximum*. The *first-order maximum*, corresponding to $m = 1$, is observed at an angle that satisfies the relationship $\sin \theta = \lambda/d$; the *second-order maximum*, corresponding to $m = 2$, is observed at a larger angle θ, and so on. Active Figure 24.20 is a sketch of the intensity distribution for some of the orders

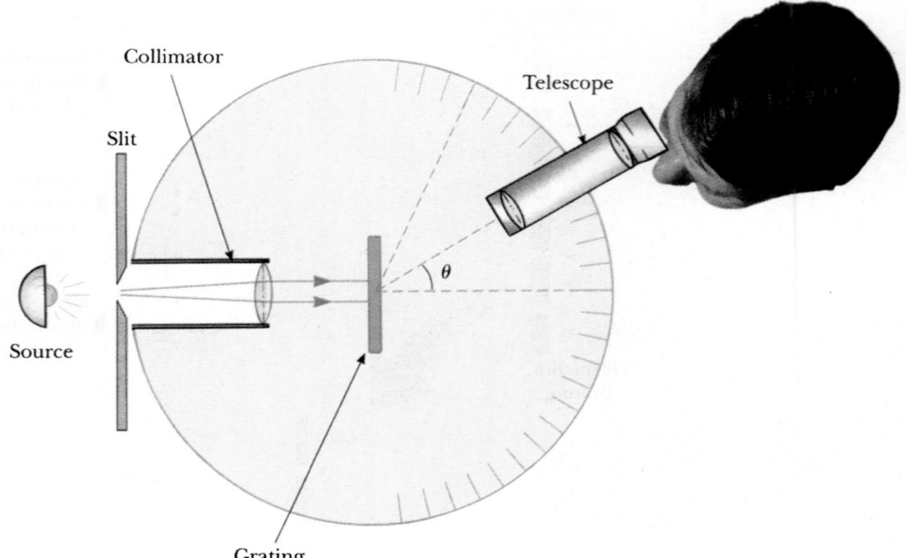

ACTIVE FIGURE 24.21
A diagram of a diffraction grating spectrometer. The collimated beam incident on the grating is diffracted into the various orders at the angles θ that satisfy the equation $d \sin \theta = m\lambda$, where $m = 0, 1, 2, \ldots$

Physics Now™
Log into PhysicsNow at **www.cp7e.com** and go to Active Figure 24.21, where you can use the spectrometer and understand how spectra are measured.

produced by a diffraction grating. Note the sharpness of the principal maxima and the broad range of the dark areas, a pattern in direct contrast to the broad bright fringes characteristic of the two-slit interference pattern.

A simple arrangement that can be used to measure the angles in a diffraction pattern is shown in Active Figure 24.21. This is a form of diffraction-grating spectrometer. The light to be analyzed passes through a slit and is formed into a parallel beam by a lens. The light then strikes the grating at a 90° angle. The diffracted light leaves the grating at angles that satisfy Equation 24.12. A telescope is used to view the image of the slit. The wavelength can be determined by measuring the angles at which the images of the slit appear for the various orders.

Quick Quiz 24.4

If laser light is reflected from a phonograph record or a compact disc, a diffraction pattern appears. The pattern arises because both devices contain parallel tracks of information that act as a reflection diffraction grating. Which device, record or compact disc, results in diffraction maxima that are farther apart?

Applying Physics 24.4 Prism vs. Grating

When white light enters through an opening in an opaque box and exits through an opening on the other side of the box, a spectrum of colors appears on the wall. From this observation, how would you be able to determine whether the box contains a prism or a diffraction grating?

Explanation The determination could be made by noticing the order of the colors in the spectrum relative to the direction of the original beam of white light. For a prism, in which the separation of light is a result of dispersion, the violet light will be refracted more than the red light. Hence, the order of the spectrum from a prism will be from red, closest to the original direction, to violet. For a diffraction grating, the angle of diffraction increases with wavelength, so the spectrum from the diffraction grating will have colors in the order from violet, closest to the original direction, to red. Furthermore, the diffraction grating will produce *two* first-order spectra on either side of the grating, while the prism will produce only a single spectrum.

Applying Physics 24.5 Rainbows from a Compact Disc

White light reflected from the surface of a compact disc has a multicolored appearance, as shown in Figure 24.22. The observation depends on the orientation of the disc relative to the eye and the position of the light source. Explain how all this works.

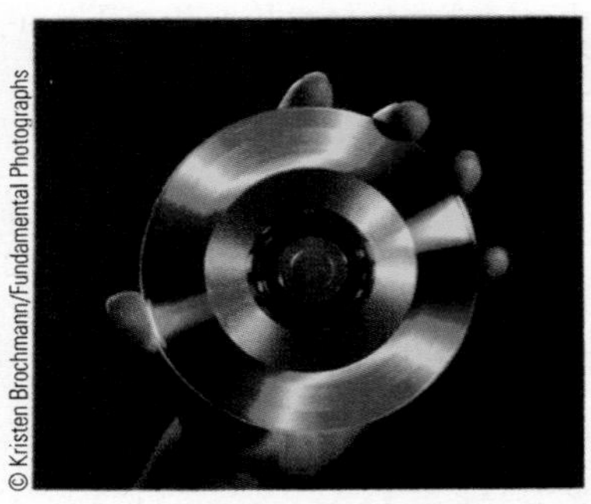

Figure 24.22 (Applying Physics 24.5) Compact discs act as diffraction gratings when observed under white light.

Explanation The surface of a compact disc has a spiral-shaped track (with a spacing of approximately 1 μm) that acts as a reflection grating. The light scattered by these closely spaced parallel tracks interferes constructively in certain directions that depend on both the wavelength and the direction of the incident light. Any one section of the disc serves as a diffraction grating for white light, sending beams of constructive interference for different colors in different directions. The different colors you see when viewing one section of the disc change as the light source, the disc, or you move to change the angles of incidence or diffraction.

Use of a Diffraction Grating in CD Tracking

APPLICATION

Tracking Information on a CD

If a CD player is to reproduce sound faithfully, the laser beam must follow the spiral track of information perfectly. Sometimes the laser beam can drift off track, however, and without a feedback procedure to let the player know this is happening, the fidelity of the music can be greatly reduced.

Figure 24.23 shows how a diffraction grating is used in a three-beam method to keep the beam on track. The central maximum of the diffraction pattern reads the information on the CD track, and the two first-order maxima steer the beam. The grating is designed so that the first-order maxima fall on the smooth surfaces on either side of the information track. Both of these reflected beams have their own detectors, and because both beams are reflected from smooth surfaces, they should have the same strong intensity when they are detected. If the central beam wanders off the track, however, one of the steering beams will begin to strike bumps on the information track and the amount of light reflected will decrease. This information is then used by electronic circuits to drive the main beam back to its desired location.

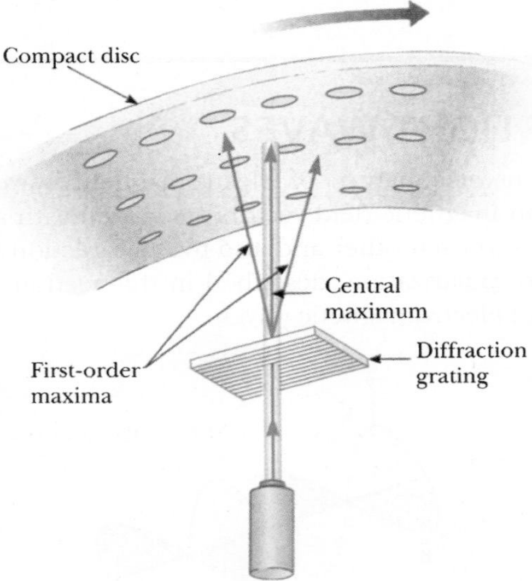

Figure 24.23 The laser beam in a CD player is able to follow the spiral track by using three beams produced with a diffraction grating.

INTERACTIVE EXAMPLE 24.7 A Diffraction Grating

Goal Calculate different-order principal maxima for a diffraction grating.

Problem Monochromatic light from a helium–neon laser ($\lambda = 632.8$ nm) is incident normally on a diffraction grating containing 6.00×10^3 lines/cm. Find the angles at which one would observe the first-order maximum, the second-order maximum, and so forth.

Strategy Find the slit separation by inverting the number of lines per centimeter, then substitute values into Equation 24.12.

Solution

Invert the number of lines per centimeter to obtain the slit separation:

$$d = \frac{1}{6.00 \times 10^3\,\text{cm}^{-1}} = 1.67 \times 10^{-4}\,\text{cm} = 1.67 \times 10^3\,\text{nm}$$

Substitute $m = 1$ into Equation 24.12 to find the sine of the angle corresponding to the first-order maximum:

$$\sin\theta_1 = \frac{\lambda}{d} = \frac{632.8\text{ nm}}{1.67 \times 10^3\text{ nm}} = 0.379$$

Take the inverse sine of the preceding result to find θ_1:

$$\theta_1 = \sin^{-1} 0.379 = 22.3^\circ$$

Repeat the calculation for $m = 2$:

$$\sin\theta_2 = \frac{2\lambda}{d} = \frac{2(632.8\text{ nm})}{1.67 \times 10^3\text{ nm}} = 0.758$$

$$\theta_2 = 49.3^\circ$$

Repeat the calculation for $m = 3$:

$$\sin\theta_3 = \frac{3\lambda}{d} = \frac{3(632.8\text{ nm})}{1.67 \times 10^3\text{ nm}} = 1.14$$

Because $\sin\theta$ can't exceed one, there is no solution for θ_3.

Remarks The foregoing calculation shows that there can only be a finite number of principal maxima. In this case, only zeroth-, first-, and second-order maxima would be observed.

Exercise 24.7

Suppose light with wavelength 7.80×10^2 nm is used instead and the diffraction grating has 3.30×10^3 lines per centimeter. Find the angles of all the principal maxima.

Answers 0°, 14.9°, 31.0°, 50.6°

PhysicsNow™ Investigate the diffraction pattern from a diffraction grating by logging into PhysicsNow at **www.cp7e.com** and going to Interactive Example 24.7.

24.9 POLARIZATION OF LIGHT WAVES

In Chapter 21, we described the transverse nature of electromagnetic waves. Figure 24.24 shows that the electric and magnetic field vectors associated with an electromagnetic wave are at right angles to each other and also to the direction of wave propagation. The phenomenon of polarization, described in this section, is firm evidence of the transverse nature of electromagnetic waves.

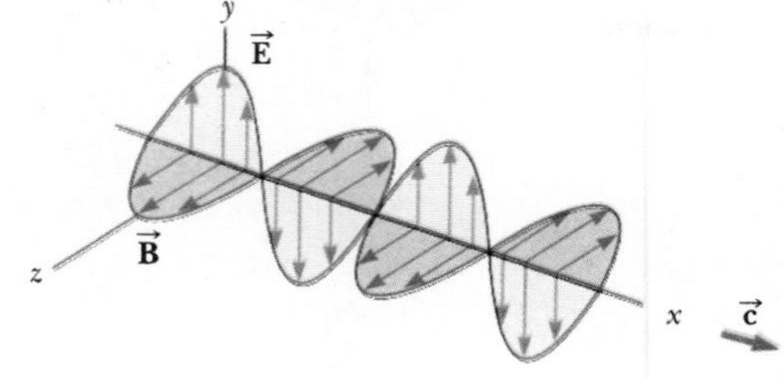

Figure 24.24 A schematic diagram of a polarized electromagnetic wave propagating in the x-direction. The electric field vector $\vec{E}$ vibrates in the xy-plane, while the magnetic field vector $\vec{B}$ vibrates in the xz-plane.

An ordinary beam of light consists of a large number of electromagnetic waves emitted by the atoms or molecules of the light source. The vibrating charges associated with the atoms act as tiny antennas. Each atom produces a wave with its own orientation of $\vec{\mathbf{E}}$, as in Figure 24.24, corresponding to the direction of atomic vibration. However, because all directions of vibration are possible, the resultant electromagnetic wave is a superposition of waves produced by the individual atomic sources. The result is an **unpolarized** light wave, represented schematically in Figure 24.25a. The direction of wave propagation shown in the figure is perpendicular to the page. Note that *all* directions of the electric field vector are equally probable and lie in a plane (such as the plane of this page) perpendicular to the direction of propagation.

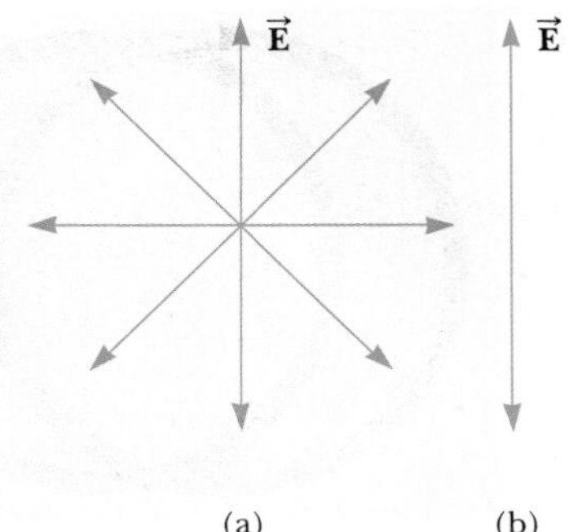

Figure 24.25 (a) An unpolarized light beam viewed along the direction of propagation (perpendicular to the page). The transverse electric field vector can vibrate in any direction with equal probability. (b) A linearly polarized light beam with the electric field vector vibrating in the vertical direction.

A wave is said to be **linearly polarized** if the resultant electric field $\vec{\mathbf{E}}$ vibrates in the same direction *at all times* at a particular point, as in Figure 24.25b. (Sometimes such a wave is described as *plane polarized* or simply *polarized*.) The wave in Figure 24.24 is an example of a wave that is linearly polarized in the y-direction. As the wave propagates in the x-direction, $\vec{\mathbf{E}}$ is always in the y-direction. The plane formed by $\vec{\mathbf{E}}$ and the direction of propagation is called the *plane of polarization* of the wave. In Figure 24.24, the plane of polarization is the xy-plane.

It's possible to obtain a linearly polarized beam from an unpolarized beam by removing all waves from the beam except those with electric field vectors that oscillate in a single plane. We now discuss three processes for doing this: (1) selective absorption, (2) reflection, and (3) scattering.

Polarization by Selective Absorption

The most common technique for polarizing light is to use a material that transmits waves having electric field vectors that vibrate in a plane parallel to a certain direction and absorbs those waves with electric field vectors vibrating in directions perpendicular to that direction.

In 1932, E. H. Land discovered a material, which he called **Polaroid**, that polarizes light through selective absorption by oriented molecules. This material is fabricated in thin sheets of long-chain hydrocarbons, which are stretched during manufacture so that the molecules align. After a sheet is dipped into a solution containing iodine, the molecules become good electrical conductors. However, conduction takes place primarily along the hydrocarbon chains, because the valence electrons of the molecules can move easily only along those chains. (Recall that valence electrons are "free" electrons that can move easily through the conductor.) As a result, the molecules readily *absorb* light having an electric field vector parallel to their lengths and *transmit* light with an electric field vector perpendicular to their lengths. It's common to refer to the direction perpendicular to the molecular chains as the **transmission axis**. In an ideal polarizer, all light with $\vec{\mathbf{E}}$ parallel to the transmission axis is transmitted and all light with $\vec{\mathbf{E}}$ perpendicular to the transmission axis is absorbed.

Polarizing material reduces the intensity of light passing through it. In Active Figure 24.26, an unpolarized light beam is incident on the first polarizing sheet, called the **polarizer**; the transmission axis is as indicated. The light that passes

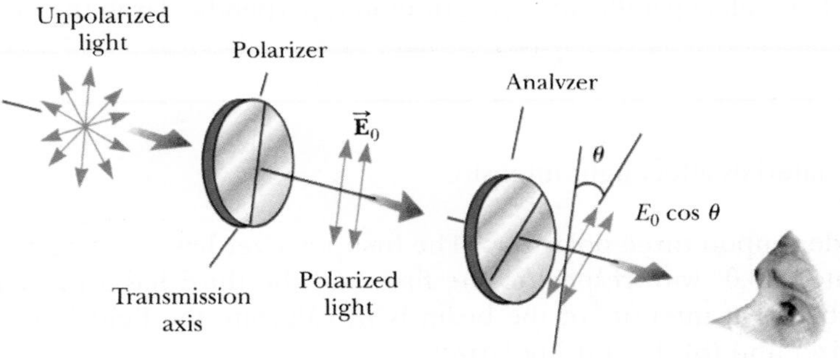

ACTIVE FIGURE 24.26
Two polarizing sheets whose transmission axes make an angle θ with each other. Only a fraction of the polarized light incident on the analyzer is transmitted.

Physics Now™
Log into PhysicsNow at **www.cp7e.com** and go to Active Figure 24.26, where you can rotate the second polarizer and see the effect on the transmitted light.

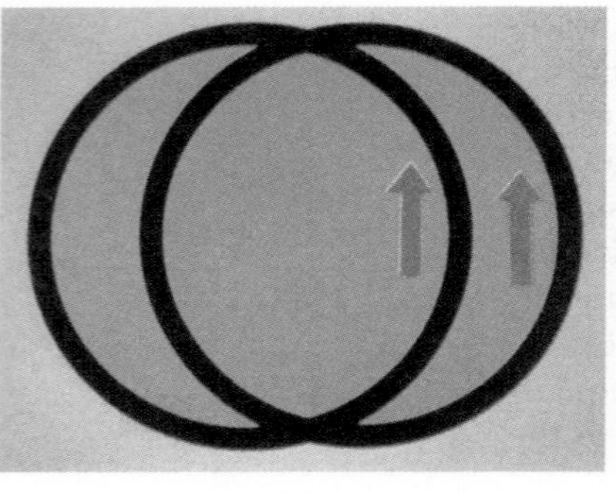
(a)

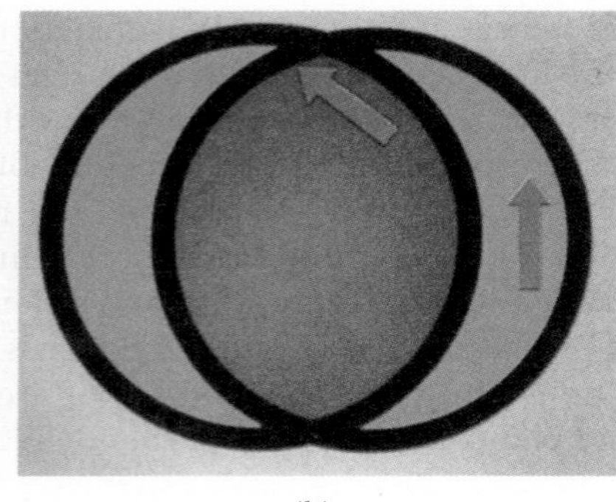
(b)

(c)

Figure 24.27 The intensity of light transmitted through two polarizers depends on the relative orientations of their transmission axes. (a) The transmitted light has *maximum* intensity when the transmission axes are *aligned* with each other. (b) The transmitted light intensity diminishes when the transmission axes are at an angle of 45° with each other. (c) The transmitted light intensity is a *minimum* when the transmission axes are at *right angles* to each other.

through this sheet is polarized vertically, and the transmitted electric field vector is $\vec{\mathbf{E}}_0$. A second polarizing sheet, called the **analyzer**, intercepts this beam with its transmission axis at an angle of θ to the axis of the polarizer. The component of $\vec{\mathbf{E}}_0$ that is perpendicular to the axis of the analyzer is completely absorbed. The component of $\vec{\mathbf{E}}_0$ that is parallel to the analyzer axis, $E_0 \cos\theta$, is allowed to pass through the analyzer. Because the intensity of the transmitted beam varies as the *square* of its amplitude E, we conclude that the intensity of the (polarized) beam transmitted through the analyzer varies as

Malus's law ▶

$$I = I_0 \cos^2\theta \qquad [24.13]$$

where I_0 is the intensity of the polarized wave incident on the analyzer. This expression, known as **Malus's law**, applies to any two polarizing materials having transmission axes at an angle of θ to each other. Note from Equation 24.13 that the transmitted intensity is a maximum when the transmission axes are parallel ($\theta = 0$ or $180°$) and is zero (complete absorption by the analyzer) when the transmission axes are perpendicular to each other. This variation in transmitted intensity through a pair of polarizing sheets is illustrated in Figure 24.27.

When unpolarized light of intensity I_0 is sent through a single ideal polarizer, the transmitted linearly polarized light has intensity $I_0/2$. This fact follows from Malus's law, because the average value of $\cos^2\theta$ is one-half.

Applying Physics 24.6 Polarizing Microwaves

A polarizer for microwaves can be made as a grid of parallel metal wires about a centimeter apart. Is the electric field vector for microwaves transmitted through this polarizer parallel or perpendicular to the metal wires?

Explanation Electric field vectors parallel to the metal wires cause electrons in the metal to oscillate parallel to the wires. Thus, the energy from the waves with these electric field vectors is transferred to the metal by accelerating the electrons and is eventually transformed to internal energy through the resistance of the metal. Waves with electric field vectors perpendicular to the metal wires are not able to accelerate electrons and pass through the wires. Consequently, the electric field polarization is perpendicular to the metal wires.

EXAMPLE 24.8 Polarizer

Goal Understand how polarizing materials affect light intensity.

Problem Unpolarized light is incident upon three polarizers. The first polarizer has a vertical transmission axis, the second has a transmission axis rotated 30.0° with respect to the first, and the third has a transmission axis rotated 75.0° relative to the first. If the initial light intensity of the beam is I_b, calculate the light intensity after the beam passes through **(a)** the second polarizer and **(b)** the third polarizer.

Strategy After the beam passes through the first polarizer, it is polarized and its intensity is cut in half. Malus's law can then be applied to the second and third polarizers. The angle used in Malus's law must be relative to the immediately preceding transmission axis.

Solution

(a) Calculate the intensity of the beam after it passes through the second polarizer:

The incident intensity is $I_b/2$. Apply Malus's law to the second polarizer:

$$I_2 = I_0 \cos^2 \theta = \frac{I_b}{2} \cos^2 (30.0^\circ) = \frac{I_b}{2}\left(\frac{\sqrt{3}}{2}\right)^2 = \frac{3}{8} I_b$$

(b) Calculate the intensity of the beam after it passes through the third polarizer.

The incident intensity is now $3I_b/8$. Apply Malus's law to the third polarizer:

$$I_3 = I_2 \cos^2 \theta = \frac{3}{8} I_b \cos^2 (45.0^\circ) = \frac{3}{8} I_b \left(\frac{\sqrt{2}}{2}\right)^2 = \frac{3}{16} I_b$$

Remarks Notice that the angle used in part (b) was not 75.0°, but 75.0° − 30.0° = 45.0°. The angle is always with respect to the previous polarizer's transmission axis, because the polarizing material physically determines what direction the transmitted electric fields can have.

Exercise 24.8

The polarizers are rotated, so that the second polarizer has a transmission axis of 40.0° with respect to the first polarizer and the third polarizer has an angle of 90.0° with respect to the first. If I_b is the intensity of the original unpolarized light, what is the intensity of the beam after it passes through (a) the second polarizer, and (b) the third polarizer? (c) What is the final transmitted intensity if the second polarizer is removed?

Answers (a) $0.293 I_b$ (b) $0.121 I_b$ (c) 0

Polarization by Reflection

When an unpolarized light beam is reflected from a surface, the reflected light is completely polarized, partially polarized, or unpolarized, depending on the angle of incidence. If the angle of incidence is either 0° or 90° (a normal or grazing angle), the reflected beam is unpolarized. For angles of incidence between 0° and 90°, however, the reflected light is polarized to some extent. For one particular angle of incidence the reflected beam is completely polarized.

Suppose an unpolarized light beam is incident on a surface, as in Figure 24.28a (page 808). The beam can be described by two electric field components, one parallel to the surface (represented by dots) and the other perpendicular to the first component and to the direction of propagation (represented by brown arrows). It is found that the parallel component reflects more strongly than the other components, and this results in a partially polarized beam. In addition, the refracted beam is also partially polarized.

Now suppose that the angle of incidence, θ_1, is varied until the angle between the reflected and refracted beams is 90° (Fig. 24.28b). At this particular angle of incidence, called the **polarizing angle** θ_p, the reflected beam is completely polarized, with its electric field vector parallel to the surface, while the refracted beam is partially polarized.

An expression relating the polarizing angle to the index of refraction of the reflecting surface can be obtained by the use of Figure 24.28b. From this figure we see that at the polarizing angle, $\theta_p + 90^\circ + \theta_2 = 180^\circ$, so that $\theta_2 = 90^\circ - \theta_p$. Using Snell's law and taking $n_1 = n_{\text{air}} = 1.00$ and $n_2 = n$ yields

$$n = \frac{\sin \theta_1}{\sin \theta_2} = \frac{\sin \theta_p}{\sin \theta_2}$$

Because $\sin \theta_2 = \sin(90^\circ - \theta_p) = \cos \theta_p$, the expression for n can be written

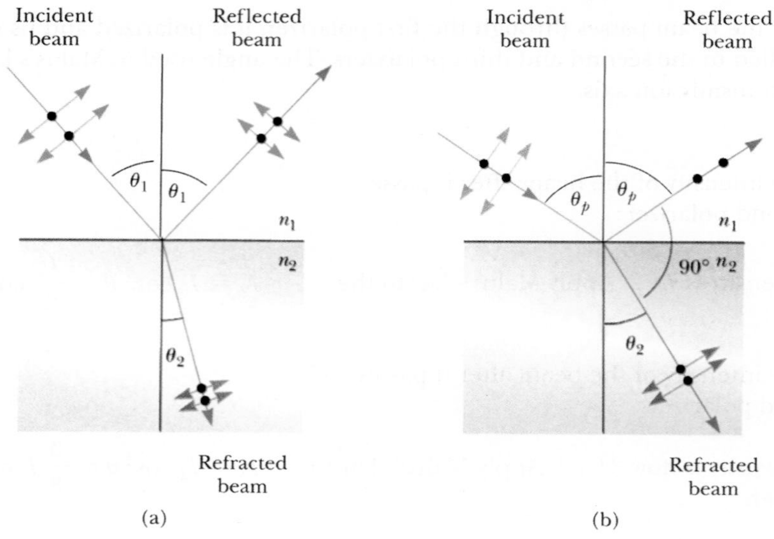

Figure 24.28 (a) When unpolarized light is incident on a reflecting surface, the reflected and refracted beams are partially polarized. (b) The reflected beam is completely polarized when the angle of incidence equals the polarizing angle θ_p, satisfying the equation $n = \tan \theta_p$.

Brewster's law ▶

$$n = \frac{\sin \theta_p}{\cos \theta_p} = \tan \theta_p \qquad [24.14]$$

Equation 24.14 is called **Brewster's law**, and the polarizing angle θ_p is sometimes called **Brewster's angle** after its discoverer, Sir David Brewster (1781–1868). For example, Brewster's angle for crown glass (where $n = 1.52$) has the value $\theta_p = \tan^{-1}(1.52) = 56.7°$. Because n varies with wavelength for a given substance, Brewster's angle is also a function of wavelength.

APPLICATION
Polaroid Sunglasses

Polarization by reflection is a common phenomenon. Sunlight reflected from water, glass, or snow is partially polarized. If the surface is horizontal, the electric field vector of the reflected light has a strong horizontal component. Sunglasses made of polarizing material reduce the glare, which *is* the reflected light. The transmission axes of the lenses are oriented vertically to absorb the strong horizontal component of the reflected light. Because the reflected light is mostly polarized, most of the glare can be eliminated without removing most of the normal light.

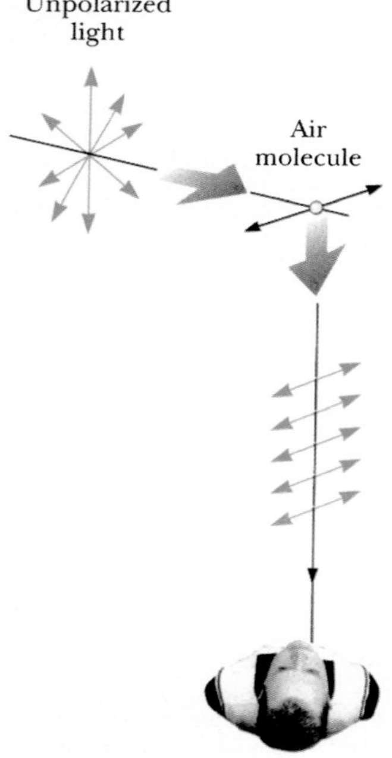

Figure 24.29 The scattering of unpolarized sunlight by air molecules. The light observed at right angles is linearly polarized because the vibrating molecule has a horizontal component of vibration.

Polarization by Scattering

When light is incident on a system of particles, such as a gas, the electrons in the medium can absorb and reradiate part of the light. The absorption and reradiation of light by the medium, called **scattering**, is what causes sunlight reaching an observer on Earth from straight overhead to be polarized. You can observe this effect by looking directly up through a pair of sunglasses made of polarizing glass. Less light passes through at certain orientations of the lenses than at others.

Figure 24.29 illustrates how the sunlight becomes polarized. The left side of the figure shows an incident unpolarized beam of sunlight on the verge of striking an air molecule. When the beam strikes the air molecule, it sets the electrons of the molecule into vibration. These vibrating charges act like those in an antenna except that they vibrate in a complicated pattern. The horizontal part of the electric field vector in the incident wave causes the charges to vibrate horizontally, and the vertical part of the vector simultaneously causes them to vibrate vertically. A horizontally polarized wave is emitted by the electrons as a result of their horizontal motion, and a vertically polarized wave is emitted parallel to the Earth as a result of their vertical motion.

Scientists have found that bees and homing pigeons use the polarization of sunlight as a navigational aid.

Optical Activity

Many important practical applications of polarized light involve the use of certain materials that display the property of **optical activity**. A substance is said to be optically active if it rotates the plane of polarization of transmitted light. Suppose

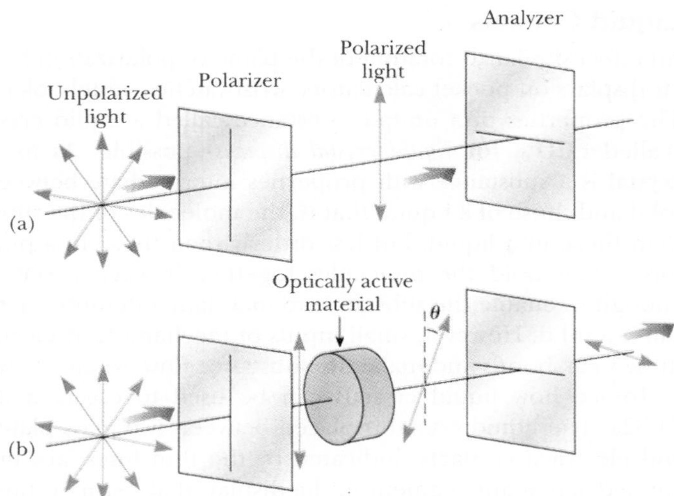

Figure 24.30 (a) When crossed polarizers are used, none of the polarized light can pass through the analyzer. (b) An optically active material rotates the direction of polarization through the angle θ, enabling some of the polarized light to pass through the analyzer.

unpolarized light is incident on a polarizer from the left, as in Figure 24.30a. The transmitted light is polarized vertically, as shown. If this light is then incident on an analyzer with its axis perpendicular to that of the polarizer, no light emerges from it. If an optically active material is placed between the polarizer and analyzer, as in Figure 24.30b, the material causes the direction of the polarized beam to rotate through the angle θ. As a result, some light is able to pass through the analyzer. The angle through which the light is rotated by the material can be found by rotating the polarizer until the light is again extinguished. It is found that the angle of rotation depends on the length of the sample and, if the substance is in solution, on the concentration. One optically active material is a solution of common sugar, dextrose. A standard method for determining the concentration of a sugar solution is to measure the rotation produced by a fixed length of the solution.

APPLICATION

Finding the Concentrations of Solutions by Means of Their Optical Activity

Optical activity occurs in a material because of an asymmetry in the shape of its constituent molecules. For example, some proteins are optically active because of their spiral shapes. Other materials, such as glass and plastic, become optically active when placed under stress. If polarized light is passed through an unstressed piece of plastic and then through an analyzer with an axis perpendicular to that of the polarizer, none of the polarized light is transmitted. If the plastic is placed under stress, however, the regions of greatest stress produce the largest angles of rotation of polarized light, and a series of light and dark bands are observed in the transmitted light. Engineers often use this property in the design of structures ranging from bridges to small tools. A plastic model is built and analyzed under different load conditions to determine positions of potential weakness and failure under stress. If the design is poor, patterns of light and dark bands will indicate the points of greatest weakness, and the design can be corrected at an early stage. Figure 24.31 shows examples of stress patterns in plastic.

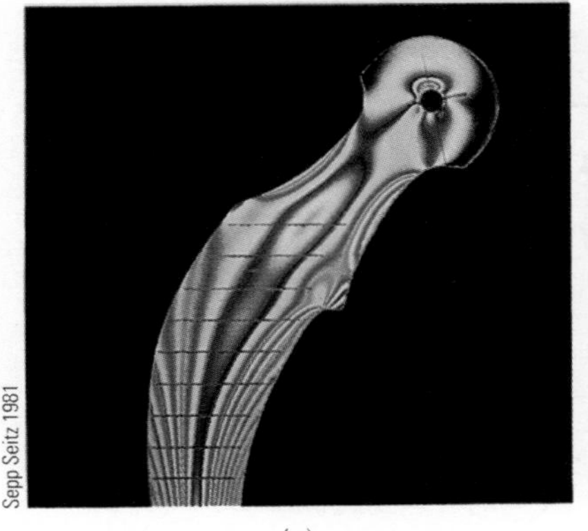

Sepp Seitz 1981

(a)

Peter Aprahamian/Science Photo Library/Photo Researchers, Inc.

(b)

Figure 24.31 (a) Strain distribution in a plastic model of a replacement hip used in a medical research laboratory. The pattern is produced when the model is placed between a polarizer and an analyzer oriented perpendicular to each other. (b) A plastic model of an arch structure under load conditions observed between perpendicular polarizers. Such patterns are useful in the optimum design of architectural components.

APPLICATION

Liquid Crystal Displays (LCD's)

Liquid Crystals

An effect similar to rotation of the plane of polarization is used to create the familiar displays on pocket calculators, wristwatches, notebook computers, and so forth. The properties of a unique substance called a liquid crystal make these displays (called LCD's, for *liquid crystal displays*) possible. As its name implies, a **liquid crystal** is a substance with properties intermediate between those of a crystalline solid and those of a liquid; that is, the molecules of the substance are more orderly than those in a liquid, but less orderly than those in a pure crystalline solid. The forces that hold the molecules together in such a state are just barely strong enough to enable the substance to maintain a definite shape, so it is reasonable to call it a solid. However, small inputs of mechanical or electrical energy can disrupt these weak bonds and make the substance flow, rotate, or twist.

To see how liquid crystals can be used to create a display, consider Figure 24.32a. The liquid crystal is placed between two glass plates in the pattern shown, and electrical contacts, indicated by the thin lines, are made. When a voltage is applied across any segment in the display, that segment turns dark. In this fashion, any number between 0 and 9 can be formed by the pattern, depending on the voltages applied to the seven segments.

To see why a segment can be changed from dark to light by the application of a voltage, consider Figure 24.32b, which shows the basic construction of a portion of

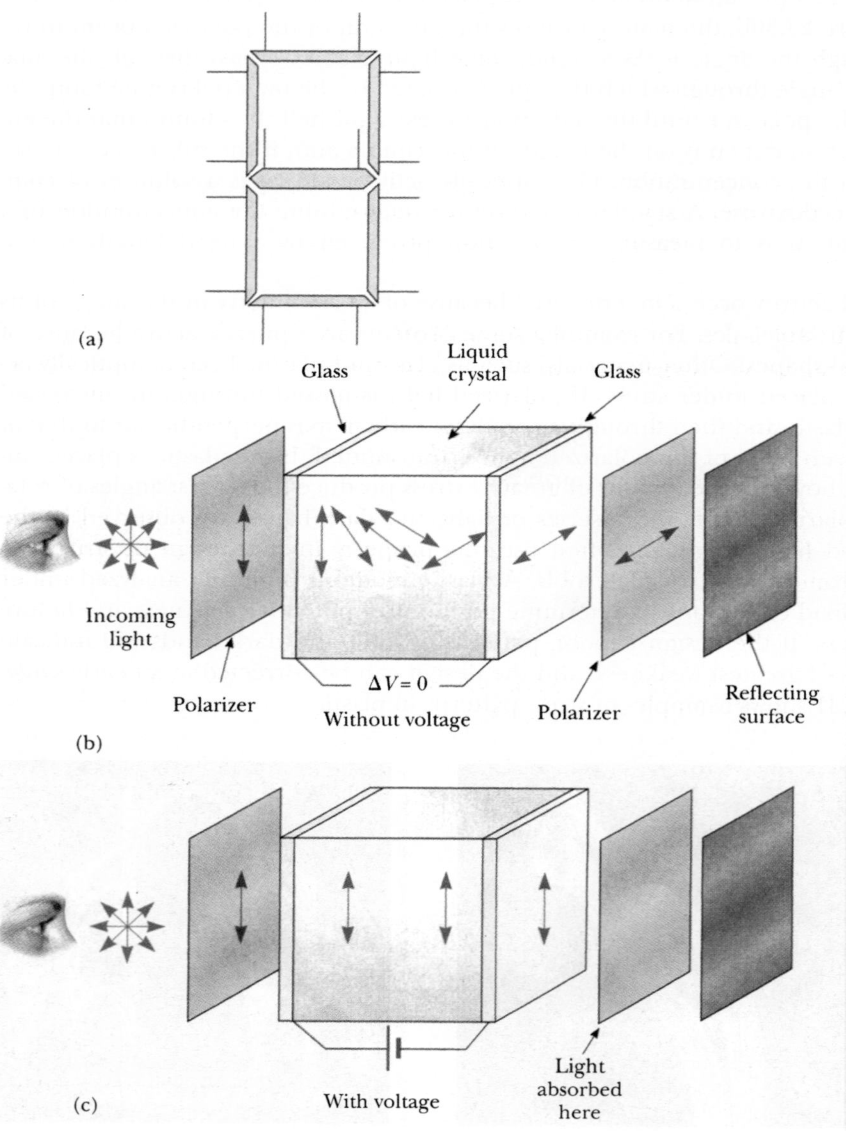

Figure 24.32 (a) The light-segment pattern of a liquid crystal display. (b) Rotation of a polarized light beam by a liquid crystal when the applied voltage is zero. (c) Molecules of the liquid crystal align with the electric field when a voltage is applied.

the display. The liquid crystal is placed between two glass substrates that are packaged between two pieces of Polaroid material with their transmission axes perpendicular. A reflecting surface is placed behind one of the pieces of Polaroid. First consider what happens when light falls on this package and no voltages are applied to the liquid crystal, as shown in Figure 24.32b. Incoming light is polarized by the polarizer on the left and then falls on the liquid crystal. As the light passes through the crystal, its plane of polarization is rotated by 90°, allowing it to pass through the polarizer on the right. It reflects from the reflecting surface and retraces its path through the crystal. Thus, an observer to the left of the crystal sees the segment as being bright. When a voltage is applied as in Figure 24.32c, the molecules of the liquid crystal don't rotate the plane of polarization of the light. In this case, the light is absorbed by the polarizer on the right and none is reflected back to the observer to the left of the crystal. As a result, the observer sees this segment as black. Changing the applied voltage to the crystal in a precise pattern at precise times can make the pattern tick off the seconds on a watch, display a letter on a computer display, and so forth.

SUMMARY

Physics Now™ Take a practice test by logging into PhysicsNow at **www.cp7e.com** and clicking on the Pre-Test link for this chapter.

24.1 Conditions for Interference

Interference occurs when two or more light waves overlap at a given point. A sustained interference pattern is observed if (1) the sources are coherent (that is, they maintain a constant phase relationship with one another), (2) the sources have identical wavelengths, and (3) the superposition principle is applicable.

24.2 Young's Double-Slit Experiment

In **Young's double-slit experiment**, two slits separated by distance d are illuminated by a single-wavelength light source. An interference pattern consisting of bright and dark fringes is observed on a screen a distance L from the slits. The condition for **bright fringes** (constructive interference) is

$$d \sin \theta_{\text{bright}} = m\lambda \qquad m = 0, \pm 1, \pm 2, \ldots \qquad [24.2]$$

The number m is called the **order number** of the fringe. The condition for **dark fringes** (destructive interference) is

$$d \sin \theta_{\text{dark}} = (m + \tfrac{1}{2})\lambda \qquad m = 0, \pm 1, \pm 2, \ldots \qquad [24.3]$$

The position y_m of the bright fringes on the screen can be determined by using the relation $\sin \theta \approx \tan \theta = y_m/L$, which is true for small angles. This can be substituted into Equations 24.2 and 24.3, yielding the location of the bright fringes:

$$y_{\text{bright}} = \frac{\lambda L}{d} m \qquad m = 0, \pm 1, \pm 2, \ldots \qquad [24.5]$$

A similar expression can be derived for the dark fringes. This equation can be used either to locate the maxima or to determine the wavelength of light by measuring y_m.

24.3 Change of Phase Due to Reflection &
24.4 Interference in Thin Films

An electromagnetic wave undergoes a phase change of 180° on reflection from a medium with an index of refraction higher than that of the medium in which the wave is traveling. There is no change when the wave, traveling in a medium with higher index of refraction, reflects from a medium with a lower index of refraction.

The wavelength λ_n of light in a medium with index of refraction n is

$$\lambda_n = \frac{\lambda}{n} \qquad [24.7]$$

where λ is the wavelength of the light in free space. Light encountering a thin film of thickness t will reflect off the top and bottom of the film, each ray undergoing a possible phase change as described above. The two rays recombine, and bright and dark fringes will be observed, with the conditions of interference given by the following table:

Equation ($m = 0, 1, \ldots$)	1 phase reversal	0 or 2 phase reversals
$2nt = (m + \frac{1}{2})\lambda$ **[24.9]**	constructive	destructive
$2nt = m\lambda$ **[24.10]**	destructive	constructive

24.6 Diffraction &
24.7 Single-Slit Diffraction

Diffraction occurs when waves pass through small openings, around obstacles, or by sharp edges. The **diffraction pattern** produced by a single slit on a distant screen consists of a central bright maximum flanked by less bright fringes alternating with dark regions. The angles θ at which the diffraction pattern has zero intensity (regions of destructive interference) are described by

$$\sin \theta_{\text{dark}} = m \frac{\lambda}{a} \qquad m = \pm 1, \pm 2, \pm 3, \ldots \qquad [24.11]$$

where a is the width of the slit and λ is the wavelength of the light incident on the slit.

24.8 The Diffraction Grating

A **diffraction grating** consists of many equally spaced, identical slits. The condition for **maximum intensity** in the interference pattern of a diffraction grating is

$$d \sin \theta_{\text{bright}} = m\lambda \qquad m = 0, 1, 2, \ldots \qquad [24.12]$$

where d is the spacing between adjacent slits and m is the order number of the diffraction pattern. A diffraction grating can be made by putting a large number of evenly spaced scratches on a glass slide. The number of such lines per centimeter is the inverse of the spacing d.

24.9 Polarization of Light Waves

Unpolarized light can be polarized by selective absorption, reflection, or scattering. A material can polarize light if it transmits waves having electric field vectors that vibrate in a plane parallel to a certain direction and absorbs waves with electric field vectors vibrating in directions perpendicular to that direction. When unpolarized light pass through a polarizing sheet, its intensity is reduced by half, and the light becomes polarized. When this light passes through a second polarizing sheet with transmission axis at an angle of θ with respect to the transmission axis of the first sheet, the transmitted intensity is given by

$$I = I_0 \cos^2 \theta \qquad [24.13]$$

where I_0 is the intensity of the light after passing through the first polarizing sheet.

In general, light reflected from an amorphous material, such as glass, is partially polarized. Reflected light is completely polarized, with its electric field parallel to the surface, when the angle of incidence produces a 90° angle between the reflected and refracted beams. This angle of incidence, called the **polarizing angle** θ_p, satisfies **Brewster's law**, given by

$$n = \tan \theta_p \qquad [24.14]$$

where n is the index of refraction of the reflecting medium.

CONCEPTUAL QUESTIONS

1. Your automobile has two headlights. What sort of interference pattern do you expect to see from them? Why?
2. Holding your hand at arm's length, you can readily block sunlight from your eyes. Why can you not block sound from your ears this way?
3. Consider a dark fringe in an interference pattern, at which almost no light energy is arriving. Light from both slits is arriving at this point, but the waves cancel. Where does the energy go?
4. If Young's double-slit experiment were performed under water, how would the observed interference pattern be affected?
5. In a laboratory accident, you spill two liquids onto water, neither of which mixes with the water. They both form thin films on the water surface. As the films spread and become very thin, you notice that one film becomes bright and the other black in reflected light. Why might this be?
6. If white light is used in Young's double-slit experiment, rather than monochromatic light, how does the interference pattern change?
7. In our discussion of thin-film interference, we looked at light *reflecting* from a thin film. Consider one light ray, the direct ray, that transmits through the film without reflecting. Then consider a second ray, the reflected ray, that transmits through the first surface, reflects back to the second, reflects again from the first, and then transmits out into the air, parallel to the direct ray. For normal incidence, how thick must the film be, in terms of the wavelength of the light, for the outgoing rays to interfere destructively? Is it the same thickness as for reflected destructive interference?
8. What is the necessary condition on the difference in path length between two waves that interfere (a) constructively and (b) destructively? Assume that the wave sources are coherent.
9. A lens with outer radius of curvature R and index of refraction n rests on a flat glass plate, and the combination is illuminated from white light from above. Is there a dark spot or a light spot at the center of the lens? What does it mean if the observed rings are noncircular?
10. Often, fingerprints left on a piece of glass such as a windowpane show colored spectra like that from a diffraction grating. Why?
11. In everyday experience, why are radio waves polarized, while light is not?
12. Suppose reflected white light is used to observe a thin, transparent coating on glass as the coating material is gradually deposited by evaporation in a vacuum. Describe some color changes that might occur during the process of building up the thickness of the coating.
13. Would it be possible to place a nonreflective coating on an airplane to cancel radar waves of wavelength 3 cm?
14. Certain sunglasses use a polarizing material to reduce the intensity of light reflected from shiny surfaces, such as water or the hood of a car. What orientation of the transmission axis should the material have to be most effective?
15. Why is it so much easier to perform interference experiments with a laser than with an ordinary light source?
16. A simple way of observing an interference pattern is to look at a distant light source through a stretched handkerchief or an open umbrella. Explain how this works.
17. When you receive a chest x-ray at a hospital, the x-rays pass through a series of parallel ribs in your chest. Do the ribs act as a diffraction grating for x-rays?
18. Can a sound wave be polarized? Explain.
19. Astronomers often observe occulations, in which a star passes behind another object, such as the Moon. During an occultation, the intensity of light from the star doesn't suddenly drop to zero as the star passes behind the edge of the Moon. Instead, the intensity fluctuates for a short time before dropping to zero. Why should this happen?
20. In one experiment, light from a laser passes through a double slit and forms an interference pattern on a distant screen. The experiment is repeated after increasing the slit separation by 50%. In which experiment is the distance from the central maximum to the next maximum the greatest?

21. Light in air that is reflected from a water surface is found to be completely polarized at an angle θ. If the light is instead reflected from a glass coffee table, will the new angle for complete polarization be larger or smaller?

22. In one experiment, blue light passes through a diffraction grating and forms an interference pattern on a screen. In a second experiment, red light passes through the same diffraction grating and forms another interference pattern. How do the separations between bright lines in the two experiments compare with each other?

PROBLEMS

1, 2, 3 = straightforward, intermediate, challenging   □ = full solution available in *Student Solutions Manual/Study Guide*
Physics Now™ = coached solution with hints available at **www.cp7e.com**   = biomedical application

Section 24.2 Young's Double-Slit Experiment

1. A laser beam (λ = 632.8 nm) is incident on two slits 0.200 mm apart. How far apart are the bright interference fringes on a screen 5.00 m away from the double slits?

2. In a Young's double-slit experiment, a set of parallel slits with a separation of 0.100 mm is illuminated by light having a wavelength of 589 nm, and the interference pattern is observed on a screen 4.00 m from the slits. (a) What is the difference in path lengths from each of the slits to the location of a third-order bright fringe on the screen? (b) What is the difference in path lengths from the two slits to the location of the third dark fringe on the screen, away from the center of the pattern?

3. A pair of narrow, parallel slits separated by 0.250 mm is illuminated by the green component from a mercury vapor lamp (λ = 546.1 nm). The interference pattern is observed on a screen 1.20 m from the plane of the parallel slits. Calculate the distance (a) from the central maximum to the first bright region on either side of the central maximum and (b) between the first and second dark bands in the interference pattern.

4. Light of wavelength 460 nm falls on two slits spaced 0.300 mm apart. What is the required distance from the slit to a screen if the spacing between the first and second dark fringes is to be 4.00 mm?

5. In a location where the speed of sound is 354 m/s, a 2 000-Hz sound wave impinges on two slits 30.0 cm apart. (a) At what angle is the first maximum located? (b) If the sound wave is replaced by 3.00-cm microwaves, what slit separation gives the same angle for the first maximum? (c) If the slit separation is 1.00 μm, what frequency of light gives the same first maximum angle?

6. White light spans the wavelength range between about 400 nm and 700 nm. If white light passes through two slits 0.30 mm apart and falls on a screen 1.5 m from the slits, find the distance between the first-order violet and the first-order red fringes.

7. Two radio antennas separated by 300 m, as shown in Figure P24.7, simultaneously transmit identical signals of the same wavelength. A radio in a car traveling due north receives the signals. (a) If the car is at the position of the second maximum, what is the wavelength of the signals? (b) How much farther must the car travel to encounter the next minimum in reception? [*Hint:* Determine the path difference between the two signals at the two locations of the car.]

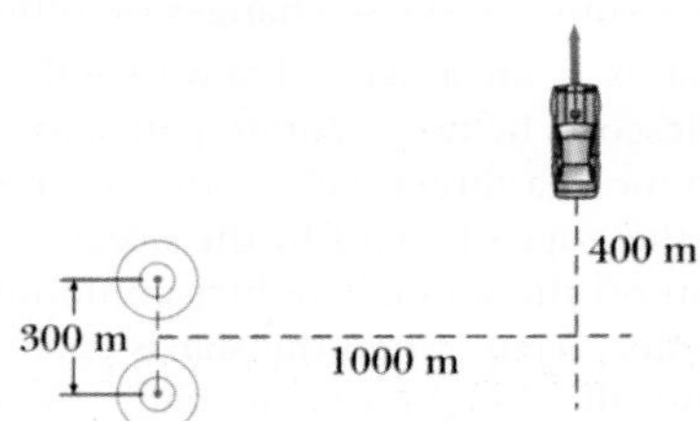

Figure P24.7

8. If the distance between two slits is 0.050 mm and the distance to a screen is 2.50 m, find the spacing between the first- and second-order bright fringes for yellow light of 600-nm wavelength.

9. Waves from a radio station have a wavelength of 300 m. They travel by two paths to a home receiver 20.0 km from the transmitter. One path is a direct path, and the second is by reflection from a mountain directly behind the home receiver. What is the minimum distance from the mountain to the receiver that produces destructive interference at the receiver? (Assume that no phase change occurs on reflection from the mountain.)

10. A pair of slits, separated by 0.150 mm, is illuminated by light having a wavelength of λ = 643 nm. An interference pattern is observed on a screen 140 cm from the slits. Consider a point on the screen located at y = 1.80 cm from the central maximum of this pattern. (a) What is the path difference δ for the two slits at the location y? (b) Express this path difference in terms of the wavelength. (c) Will the interference correspond to a maximum, a minimum, or an intermediate condition?

11. A riverside warehouse has two open doors, as in Figure P24.11. Its interior is lined with a sound-absorbing material. A boat on the river sounds its horn. To person A, the

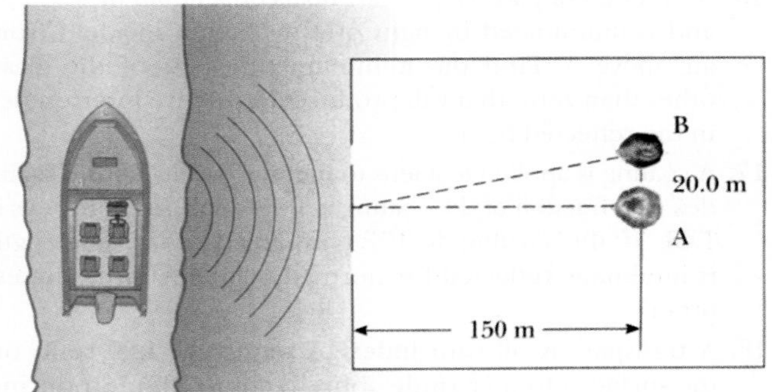

Figure P24.11

sound is loud and clear. To person B, the sound is barely audible. The principal wavelength of the sound waves is 3.00 m. Assuming person B is at the position of the first minimum, determine the distance between the doors, center to center.

12. Physics Now™ The waves from a radio station can reach a home receiver by two different paths. One is a straight-line path from the transmitter to the home, a distance of 30.0 km. The second path is by reflection from a storm cloud. Assume that this reflection takes place at a point midway between receiver and transmitter. If the wavelength broadcast by the radio station is 400 m, find the minimum height of the storm cloud that will produce destructive interference between the direct and reflected beams. (Assume no phase changes on reflection.)

13. Radio waves from a star, of wavelength 250 m, reach a radio telescope by two separate paths, as shown in Figure P24.13. One is a direct path to the receiver, which is situated on the edge of a cliff by the ocean. The second is by reflection off the water. The first minimum of destructive interference occurs when the star is 25.0° above the horizon. Find the height of the cliff. (Assume no phase change on reflection.)

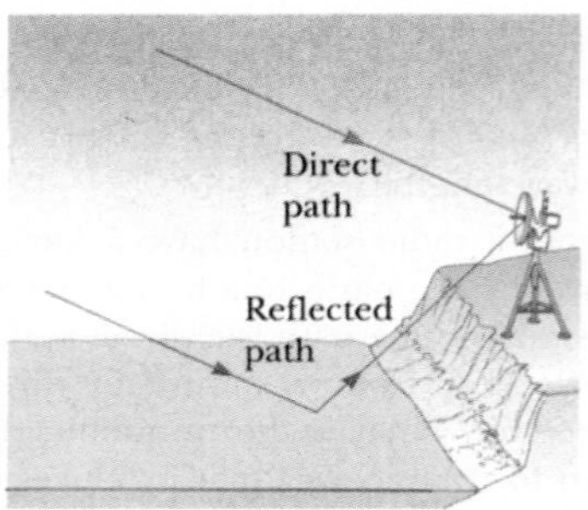

Figure P24.13

Section 24.3 Change of Phase Due to Reflection
Section 24.4 Interference in Thin Films

14. Determine the minimum thickness of a soap film ($n =$ 1.330) that will result in constructive interference of (a) the red H_α line ($\lambda = 656.3$ nm); (b) the blue H_γ line ($\lambda = 434.0$ nm).

15. Suppose the film shown in Figure 24.7 has an index of refraction of 1.36 and is surrounded by air on both sides. Find the minimum thickness that will produce constructive interference in the reflected light when the film is illuminated by light of wavelength 500 nm.

16. A thin film of glass ($n = 1.50$) floats on a liquid of $n = 1.35$ and is illuminated by light of $\lambda = 580$ nm incident from air above it. Find the minimum thickness of the glass, other than zero, that will produce destructive interference in the reflected light.

17. A coating is applied to a lens to minimize reflections. The index of refraction of the coating is 1.55, and that of the lens is 1.48. If the coating is 177.4 nm thick, what wavelength is minimally reflected for normal incidence in the lowest order?

18. A transparent oil with index of refraction 1.29 spills on the surface of water (index of refraction 1.33), producing a maximum of reflection with normally incident orange light (wavelength 600 nm in air). Assuming the maximum occurs in the first order, determine the thickness of the oil slick.

19. A possible means for making an airplane invisible to radar is to coat the plane with an antireflective polymer. If radar waves have a wavelength of 3.00 cm and the index of refraction of the polymer is $n = 1.50$, how thick would you make the coating?

20. A beam of light of wavelength 580 nm passes through two closely spaced glass plates, as shown in Figure P24.20. For what minimum non-zero value of the plate separation d will the transmitted light be bright? This arrangement is often used to measure the wavelength of light and is called a Fabry–Perot interferometer.

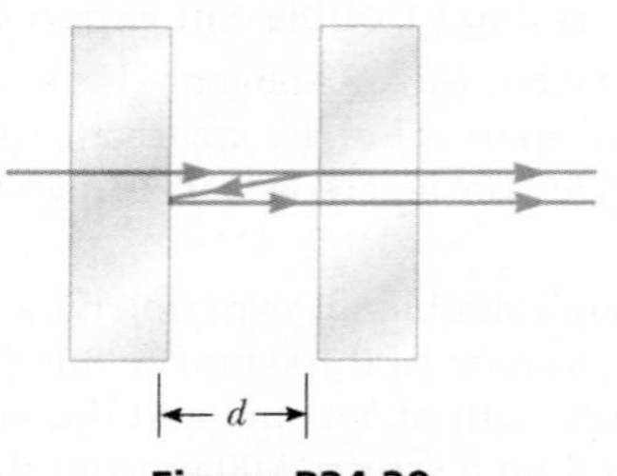

Figure P24.20

21. Astronomers observe the chromosphere of the sun with a filter that passes the red hydrogen spectral line of wavelength 656.3 nm, called the H_α line. The filter consists of a transparent dielectric of thickness d held between two partially aluminized glass plates. The filter is kept at a constant temperature. (a) Find the minimum value of d that will produce maximum transmission of perpendicular H_α light if the dielectric has an index of refraction of 1.378. (b) If the temperature of the filter increases above the normal value increasing its thickness, what happens to the transmitted wavelength? (c) The dielectric will also pass what near-visible wavelength? One of the glass plates is colored red to absorb this light.

22. Two rectangular optically flat plates ($n = 1.52$) are in contact along one end and are separated along the other end by a 2.00-μm-thick spacer (Fig. P24.22). The top plate is illuminated by monochromatic light of wavelength 546.1 nm. Calculate the number of dark parallel bands crossing the top plate (including the dark band at zero thickness along the edge of contact between the plates).

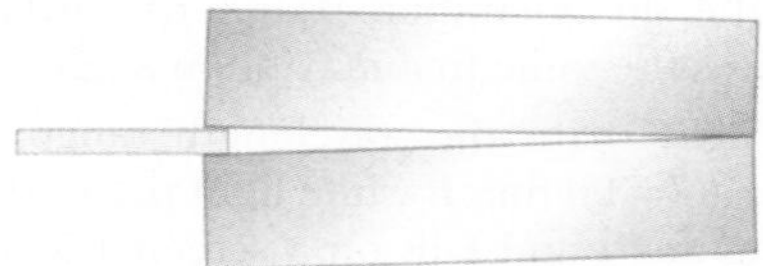
Figure P24.22 (Problems 22 and 23)

23. An air wedge is formed between two glass plates separated at one edge by a very fine wire, as in Figure P24.22. When the wedge is illuminated from above by 600-nm light, 30 dark fringes are observed. Calculate the radius of the wire.

24. A planoconvex lens with radius of curvature $R = 3.0$ m is in contact with a flat plate of glass. A light source and the observer's eye are both close to the normal, as shown in

Figure 24.8a. The radius of the 50th bright Newton's ring is found to be 9.8 mm. What is the wavelength of the light produced by the source?

25. A planoconvex lens rests with its curved side on a flat glass surface and is illuminated from above by light of wavelength 500 nm. (See Fig. 24.8.) A dark spot is observed at the center, surrounded by 19 concentric dark rings (with bright rings in between). How much thicker is the air wedge at the position of the 19th dark ring than at the center?

26. Nonreflective coatings on camera lenses reduce the loss of light at the surfaces of multilens systems and prevent internal reflections that might mar the image. Find the minimum thickness of a layer of magnesium fluoride ($n = 1.38$) on flint glass ($n = 1.66$) that will cause destructive interference of reflected light of wavelength 550 nm near the middle of the visible spectrum.

27. **Physics Now™** A thin film of MgF_2 ($n = 1.38$) with thickness 1.00×10^{-5} cm is used to coat a camera lens. Are any wavelengths in the visible spectrum intensified in the reflected light?

28. A flat piece of glass is supported horizontally above the flat end of a 10.0-cm-long metal rod that has its lower end rigidly fixed. The thin film of air between the rod and the glass is observed to be bright when illuminated by light of wavelength 500 nm. As the temperature is slowly increased by 25.0°C, the film changes from bright to dark and back to bright 200 times. What is the coefficient of linear expansion of the metal?

Section 24.7 Single-Slit Diffraction

29. Helium–neon laser light ($\lambda = 632.8$ nm) is sent through a 0.300-mm-wide single slit. What is the width of the central maximum on a screen 1.00 m from the slit?

30. Light of wavelength 600 nm falls on a 0.40-mm-wide slit and forms a diffraction pattern on a screen 1.5 m away. (a) Find the position of the first dark band on each side of the central maximum. (b) Find the width of the central maximum.

31. Light of wavelength 587.5 nm illuminates a slit of width 0.75 mm. (a) At what distance from the slit should a screen be placed if the first minimum in the diffraction pattern is to be 0.85 mm from the central maximum? (b) Calculate the width of the central maximum.

32. Microwaves of wavelength 5.00 cm enter a long, narrow window in a building that is otherwise essentially opaque to the incoming waves. If the window is 36.0 cm wide, what is the distance from the central maximum to the first-order minimum along a wall 6.50 m from the window?

33. A slit of width 0.50 mm is illuminated with light of wavelength 500 nm, and a screen is placed 120 cm in front of the slit. Find the widths of the first and second maxima on each side of the central maximum.

34. A screen is placed 50.0 cm from a single slit, which is illuminated with light of wavelength 680 nm. If the distance between the first and third minima in the diffraction pattern is 3.00 mm, what is the width of the slit?

Section 24.8 The Diffraction Grating

35. Three discrete spectral lines occur at angles of 10.1°, 13.7°, and 14.8°, respectively, in the first-order spectrum of a diffraction-grating spectrometer. (a) If the grating has 3 660 slits/cm, what are the wavelengths of the light? (b) At what angles are these lines found in the second-order spectra?

36. Intense white light is incident on a diffraction grating that has 600 lines/mm. (a) What is the highest order in which the complete visible spectrum can be seen with this grating? (b) What is the angular separation between the violet edge (400 nm) and the red edge (700 nm) of the first-order spectrum produced by the grating?

37. The hydrogen spectrum has a red line at 656 nm and a violet line at 434 nm. What angular separation between these two spectral lines obtained with a diffraction grating that has 4 500 lines/cm?

38. A grating with 1 500 slits per centimeter is illuminated with light of wavelength 500 nm. (a) What is the highest-order number that can be observed with this grating? (b) Repeat for a grating of 15 000 slits per centimeter.

39. A light source emits two major spectral lines: an orange line of wavelength 610 nm and a blue-green line of wavelength 480 nm. If the spectrum is resolved by a diffraction grating having 5 000 lines/cm and viewed on a screen 2.00 m from the grating, what is the distance (in centimeters) between the two spectral lines in the second-order spectrum?

40. White light is spread out into its spectral components by a diffraction grating. If the grating has 2 000 lines per centimeter, at what angle does red light of wavelength 640 nm appear in the first-order spectrum?

41. Sunlight is incident on a diffraction grating that has 2 750 lines/cm. The second-order spectrum over the visible range (400–700 nm) is to be limited to 1.75 cm along a screen that is a distance L from the grating. What is the required value of L?

42. Light containing two different wavelengths passes through a diffraction grating with 1 200 slits/cm. On a screen 15.0 cm from the grating, the third-order maximum of the shorter wavelength falls midway between the central maximum and the first side maximum for the longer wavelength. If the neighboring maxima of the longer wavelength are 8.44 mm apart on the screen, what are the wavelengths in the light? [*Hint:* Use the small-angle approximation.]

43. **Physics Now™** A beam of 541-nm light is incident on a diffraction grating that has 400 lines/mm. (a) Determine the angle of the second-order ray. (b) If the entire apparatus is immersed in water, determine the new second-order angle of diffraction. (c) Show that the two diffracted rays of parts (a) and (b) are related through the law of refraction.

44. Light from a helium–neon laser ($\lambda = 632.8$ nm) is incident on a single slit. What is the maximum width for which no diffraction minima are observed? [*Hint:* Values of $\sin\theta > 1$ are not possible.]

Section 24.9 Polarization of Light Waves

45. The angle of incidence of a light beam in air onto a reflecting surface is continuously variable. The reflected ray is found to be completely polarized when the angle of incidence is 48.0°. (a) What is the index of refraction of the reflecting material? (b) If some of the incident light

(at an angle of 48.0°) passes into the material below the surface, what is the angle of refraction?

46. Unpolarized light passes through two polaroid sheets. The axis of the first is vertical, and that of the second is at 30.0° to the vertical. What fraction of the initial light is transmitted?

47. The index of refraction of a glass plate is 1.52. What is the Brewster's angle when the plate is (a) in air? (b) in water? (See Problem 51.)

48. At what angle above the horizon is the Sun if light from it is completely polarized upon reflection from water?

49. A light beam is incident on heavy flint glass ($n = 1.65$) at the polarizing angle. Calculate the angle of refraction for the transmitted ray.

50. The critical angle for total internal reflection for sapphire surrounded by air is 34.4°. Calculate the Brewster angle for sapphire if the light is incident from the air.

51. Equation 24.14 assumes that the incident light is in air. If the light is incident from a medium of index n_1 onto a medium of index n_2, follow the procedure used to derive Equation 24.14 to show that $\tan \theta_p = n_2/n_1$.

52. Plane-polarized light is incident on a single polarizing disk, with the direction of E_0 parallel to the direction of the transmission axis. Through what angle should the disk be rotated so that the intensity in the transmitted beam is reduced by a factor of (a) 3.00, (b) 5.00, (c) 10.0?

53. Three polarizing plates whose planes are parallel are centered on a common axis. The directions of the transmission axes relative to the common vertical direction are shown in Figure P24.53. A linearly polarized beam of light with plane of polarization parallel to the vertical reference direction is incident from the left onto the first disk with intensity $I_i = 10.0$ units (arbitrary). Calculate the transmitted intensity I_f when $\theta_1 = 20.0°$, $\theta_2 = 40.0°$, and $\theta_3 = 60.0°$. [*Hint:* Make repeated use of Malus's law.]

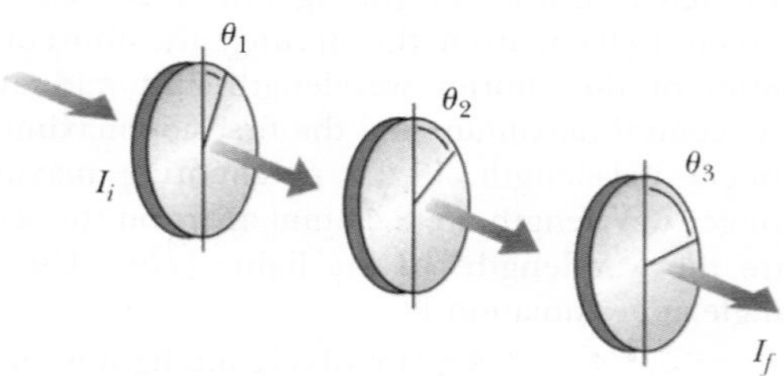

Figure P24.53 (Problems 53 and 62)

54. **Physics Now™** Light of intensity I_0 and polarized parallel to the transmission axis of a polarizer, is incident on an analyzer. (a) If the transmission axis of the analyzer makes an angle of 45° with the axis of the polarizer, what is the intensity of the transmitted light? (b) What should the angle between the transmission axes be to make $I/I_0 = 1/3$?

55. Light with a wavelength in vacuum of 546.1 nm falls perpendicularly on a biological specimen that is 1.000 μm thick. The light splits into two beams polarized at right angles, for which the indices of refraction are 1.320 and 1.333, respectively. (a) Calculate the wavelength of each component of the light while it is traversing the specimen. (b) Calculate the phase difference between the two beams when they emerge from the specimen.

ADDITIONAL PROBLEMS

56. A beam containing light of wavelengths λ_1 and λ_2 is incident on a set of parallel slits. In the interference pattern, the fourth bright line of the λ_1 light occurs at the same position as the fifth bright line of the λ_2 light. If λ_1 is known to be 540 nm, what is the value of λ_2?

57. Light of wavelength 546 nm (the intense green line from a mercury source) produces a Young's interference pattern in which the second minimum from the central maximum is along a direction that makes an angle of 18.0 min of arc with the axis through the central maximum. What is the distance between the parallel slits?

58. The two speakers are placed 35.0 cm apart. A single oscillator makes the speakers vibrate in phase at a frequency of 2.00 kHz. At what angles, measured from the perpendicular bisector of the line joining the speakers, would a distant observer hear maximum sound intensity? Minimum sound intensity? (Take the speed of sound to be 340 m/s.)

59. Interference effects are produced at point P on a screen as a result of direct rays from a 500-nm source and reflected rays off a mirror, as in Figure P24.59. If the source is 100 m to the left of the screen and 1.00 cm above the mirror, find the distance y (in millimeters) to the first dark band above the mirror.

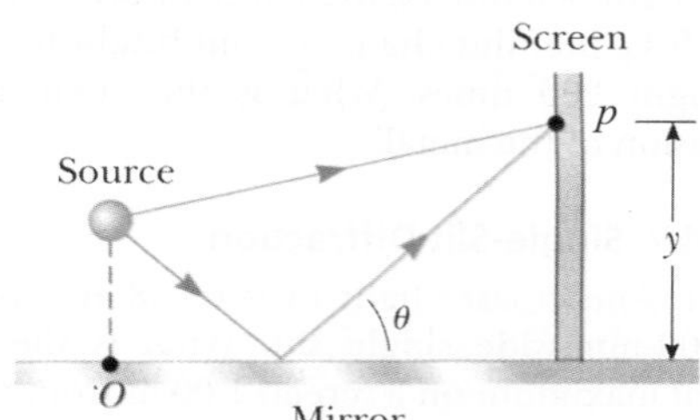

Figure P24.59

60. Many cells are transparent and colorless. Structures of great interest in biology and medicine can be practically invisible to ordinary microscopy. An *interference microscope* reveals a difference in refractive index as a shift in interference fringes, to indicate the size and shape of cell structures. The idea is exemplified in the following problem: An air wedge is formed between two glass plates in contact along one edge and slightly separated at the opposite edge. When the plates are illuminated with monochromatic light from above, the reflected light has 85 dark fringes. Calculate the number of dark fringes that appear if water ($n = 1.33$) replaces the air between the plates.

61. A thin layer of oil ($n = 1.25$) is floating on water. How thick is the oil in the region that strongly reflects green light ($\lambda = 525$ nm)?

62. Three polarizers, centered on a common axis and with their planes parallel to each other, have transmission axes oriented at angles of θ_1, θ_2, and θ_3 from the vertical, as shown in Figure P24.53. Light of intensity I_i, polarized with its plane of polarization oriented vertically, is incident from the left onto the first polarizer. What is the ratio I_f/I_i of the final transmitted intensity to the incident intensity if (a) $\theta_1 = 45°$, $\theta_2 = 90°$, and $\theta_3 = 0°$? (b) $\theta_1 = 0°$, $\theta_2 = 45°$, and $\theta_3 = 90°$?

63. Figure P24.63 shows a radio-wave transmitter and a receiver, both $h = 50.0$ m above the ground and $d = 600$ m apart. The receiver can receive signals directly from the transmitter, and indirectly, from signals that bounce off the ground. If the ground is level between the transmitter and receiver and a $\lambda/2$ phase shift occurs upon reflection, determine the longest wavelengths that interfere (a) constructively and (b) destructively.

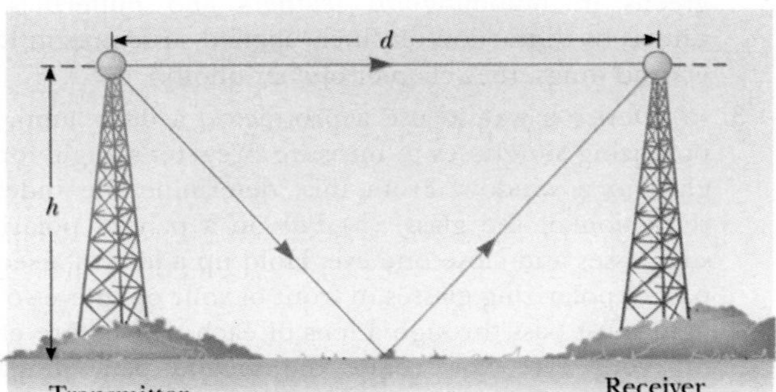

Figure P24.63

64. A planoconvex lens (flat on one side, convex on the other) with index of refraction n rests with its curved side (radius of curvature R) on a flat glass surface of the same index of refraction with a film of index n_{film} between them. The lens is illuminated from above by light of wavelength λ. Show that the dark Newton rings which appear have radii of

$$r \approx \sqrt{m\lambda R/n_{\text{film}}}$$

where m is an integer.

65. The transmitting antenna on a submarine is 5.00 m above the water when the ship surfaces. The captain wishes to transmit a message to a receiver on a 90.0-m-tall cliff at the ocean shore. If the signal is to be completely polarized by reflection off the ocean surface, how far must the ship be from the shore?

66. (a) If light is incident at an angle θ from a medium of index n_1 on a medium of index n_2 so that the angle between the reflected ray and refracted ray is β, show that

$$\tan \theta = \frac{n_2 \sin \beta}{n_1 - n_2 \cos \beta}$$

Hint: Use the trigonometric identity:

$$\sin(A + B) = \sin A \cos B + \cos A \sin B$$

(b) Show that the foregoing equation for $\tan \theta$ reduces to Brewster's law when $\beta = 90°$, $n_1 = 1$, and $n_2 = n$.

67. A diffraction pattern is produced on a screen 140 cm from a single slit, using monochromatic light of wavelength 500 nm, The distance from the center of the central maximum to the first-order maximum is 3.00 mm. Calculate the slit width. [*Hint:* Assume that the first-order maximum is halfway between the first- and second-order minima.]

68. A glass plate ($n = 1.61$) is covered with a thin, uniform layer of oil ($n = 1.20$). A light beam of variable wavelength is normally incident from air onto the oil surface. Observation of the reflected beam shows destructive interference at 500 nm and constructive interference at 750 nm. From this information, calculate the thickness of the oil film.

69. The condition for constructive interference by reflection from a thin film in air, as developed in Section 24.4, assumes nearly normal incidence. (a) Show that for large angles of incidence, the condition for constructive interference of light reflecting from a thin film of thickness t, with index of refraction n, and surrounded by air may be written as

$$2nt \cos \theta_2 = \left(m + \frac{1}{2}\right)\lambda$$

where θ_2 is the angle of refraction. (b) Calculate the minimum thickness for constructive interference if sodium light ($\lambda = 590$ nm) is incident at an angle of 30.0° on a film with an index of refraction of 1.38.

70. Figure P24.70 illustrates the formation of an interference pattern by the Lloyd's mirror method. Light from source S reaches the screen via two different pathways. One is a direct path, and the second is by reflection from a horizontal mirror. The effect is as if light from two different sources S and S' had interfered as in the Young's double-slit arrangement. Assume that the actual source S and the virtual source S' are in a plane 25 cm to the left of the mirror and the screen is a distance $L = 120$ cm to the right of that plane. Source S is a distance $h = 2.5$ mm above the top surface of the mirror, and the light is monochromatic with $\lambda = 620$ nm. Determine the distance of the first bright fringe above the surface of the mirror.

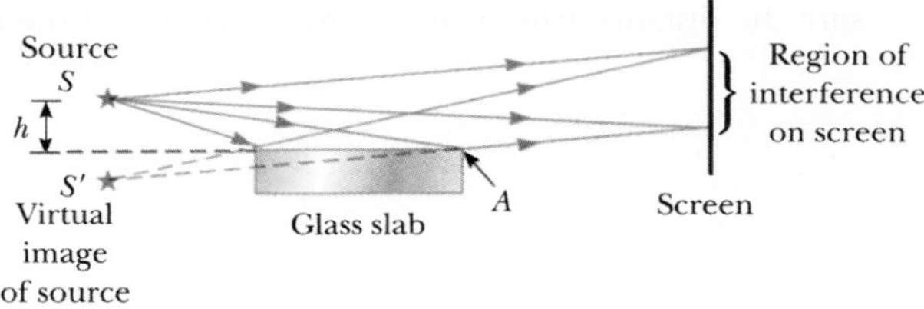

Figure P24.70

71. A piece of transparent material having an index of refraction n is cut into the shape of a wedge as shown in Figure P24.71. The angle of the wedge is small. Monochromatic light of wavelength λ is normally incident from above, and viewed from above. Let h represent the height of the wedge and ℓ its width. Show that bright fringes occur at the positions $x = \lambda\ell(m + \frac{1}{2})/2hn$ and dark fringes occur at the positions $x = \lambda\ell m/2hn$, where $m = 0, 1, 2, \ldots$ and x is measured as shown.

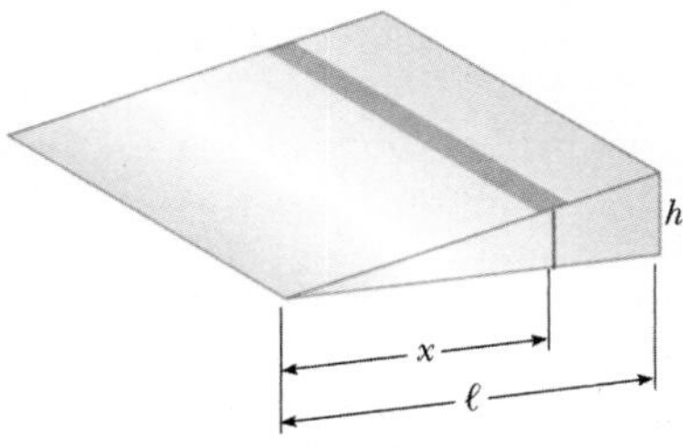

Figure P24.71

ACTIVITIES

1. Place a clear dish or plate on a black surface, such as a sheet of black construction paper. Now add a thin layer of

water to the glass, and place a few drops of kerosene or light machine oil on the water. Darken the room and shine a flashlight from an angle, as in Figure A24.1. Note the interference pattern of various colors you observe under the white light. How does the pattern change if you cover the flashlight with a sheet of red, blue, or green cellophane, which acts as a filter?

As an extension of the preceding experiment, observe the colors appearing to swirl on the surface of a soap bubble. What color do you see just before a bubble bursts?

Figure A24.1

2. Stand a couple of meters from a lightbulb. Facing away from the light, hold a compact disc about 10 cm from your eye and tilt it until the reflection of the bulb is located in the hole at the disc's center. You should see spectra radiating out from the center, with violet on the inside and red on the outside. Now move the disc away from your eye until the violet band is at the outer edge. Carefully measure the distance from your eye to the center of the disc, and also determine the radius of the disc. Use this information to find the angle θ to the first-order maximum for violet light. Now use the relationship $d \sin \theta = m\lambda$ to determine the spacing between the grooves of the disc. The industry standard is 1.6 μm. How close did you come? While you are observing the spectrum from a CD, note that the color of the light from a given point changes with the viewing angle. Explain this effect in terms of changes in the path length. It is of interest that the blues and blue-greens in hummingbird feathers and butterflies are caused by diffraction off finely aligned structures in feathers and wings. (See chapter opener photo.)

3. (a) Devise a way to use a protractor, a desk lamp, and polarizing sunglasses to measure Brewster's angle for the glass in a window. From this, determine the index of refraction of the glass. (b) Put on a pair of polarizing sunglasses and close one eye. Hold up a lens of a second pair of polarizing glasses in front of your open eye so that light must pass through a lens of each pair before entering your eye. Now rotate the second pair of glasses around. You will note that the light reaching your eye is considerably reduced at some orientations and will pass freely at others. (c) On a sunny day, rotate your polarizing sunglasses in front of your eye and observe how light reflects from a window or the surface of water. Note the change in the amount of light entering your eye for various orientations of the glasses. (d) For a final observation concerning polarized light, rotate a pair of polarizing sunglasses while looking at various areas of the sky. From what direction do you find the light to be most highly polarized?

© Denis Scott/CORBIS

The Hubble Space Telescope does its viewing above the atmosphere and doesn't suffer from the atmospheric blurring, caused by air turbulence, that plagues ground-based telescopes. Despite this advantage, it does have limitations due to diffraction effects. In this chapter, we show how the wave nature of light limits the ability of any optical system to distinguish between closely spaced objects.

CHAPTER

25

Optical Instruments

OUTLINE

We use devices made from lenses, mirrors, or other optical components every time we put on a pair of eyeglasses or contact lenses, take a photograph, look at the sky through a telescope, and so on. In this chapter we examine how these and other optical instruments work. For the most part, our analyses will involve the laws of reflection and refraction and the procedures of geometric optics. To explain certain phenomena, however, we must use the wave nature of light.

25.1 THE CAMERA

The single-lens photographic **camera** is a simple optical instrument having the features shown in Figure 25.1 (page 820). It consists of an opaque box, a converging lens that produces a real image, and a film behind the lens to receive the image. Focusing is accomplished by varying the distance between lens and film—with an adjustable bellows in antique cameras and with some other mechanical arrangements in contemporary models. For proper focusing, which leads to sharp images, the lens-to-film distance depends on the object distance as well as on the focal length of the lens. The shutter, located behind the lens, is a mechanical device that is opened for selected time intervals. With this arrangement, moving objects can be photographed by using short exposure times, dark scenes (with low light levels) by using long exposure times. If this adjustment were not available, it would be impossible to take stop-action photographs. A rapidly moving vehicle, for example, could move far enough while the shutter was open to produce a blurred image. Another major cause of blurred images is movement of the *camera* while the shutter is open. To prevent such movement, you should mount the camera on a tripod or use short exposure times. Typical shutter speeds (that is, exposure times) are 1/30, 1/60, 1/125, and 1/250 s. Stationary objects are often shot with a shutter speed of 1/60 s.

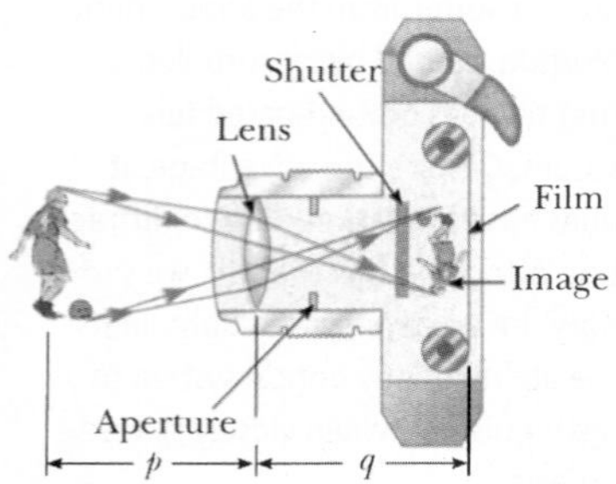

Figure 25.1 A cross-sectional view of a simple camera.

Most cameras also have an aperture of adjustable diameter to further control the intensity of the light reaching the film. When an aperture of small diameter is used, only light from the central portion of the lens reaches the film, so spherical aberration is reduced.

The intensity I of the light reaching the film is proportional to the area of the lens. Because this area in turn is proportional the square of the lens diameter D, the intensity is also proportional to D^2. Light intensity is a measure of the rate at which energy is received by the film per unit area of the image. Because the area of the image is proportional to q^2 in Figure 25.1, and $q \approx f$ (when $p >> f$, so that p can be approximated as infinite), we conclude that the intensity is also proportional to $1/f^2$, so that $I \propto D^2/f^2$. The brightness of the image formed on the film depends on the light intensity, so we see that it ultimately depends on both the focal length f and diameter D of the lens. The ratio f/D is called the ***f*-number** (or focal ratio) of a lens:

$$f\text{-number} \equiv \frac{f}{D} \qquad \textbf{[25.1]}$$

The f-number is often given as a description of the lens "speed." A lens with a low f-number is a "fast" lens. Extremely fast lenses, which have an f-number as low as approximately 1.2, are expensive because of the difficulty of keeping aberrations acceptably small with light rays passing through a large area of the lens. Camera lenses are often marked with a range of f-numbers, such as 1.4, 2, 2.8, 4, 5.6, 8, 11, ... Any one of these settings can be selected by adjusting the aperture, which changes the value of D. Increasing the setting from one f-number to the next-higher value (for example, from 2.8 to 4) decreases the area of the aperture by a factor of two. The lowest f-number setting on a camera corresponds to a wide open aperture, and the use of the maximum possible lens area.

Simple cameras usually have a fixed focal length and fixed aperture size, with an f-number of about 11. This high value for the f-number allows for a large **depth of field**. This means that objects at a wide range of distances from the lens form reasonably sharp images on the film. In other words, the camera doesn't have to be focused. Most cameras with variable f-numbers adjust them automatically.

25.2 THE EYE

Like a camera, a normal eye focuses light and produces a sharp image. However, the mechanisms by which the eye controls the amount of light admitted and adjusts to produce correctly focused images are far more complex, intricate, and effective than those in even the most sophisticated camera. In all respects, the eye is a physiological wonder.

Figure 25.2a shows the essential parts of the eye. Light entering the eye passes through a transparent structure called the *cornea,* behind which are a clear liquid (the *aqueous humor*), a variable aperture (the *pupil,* which is an opening in the *iris*), and the *crystalline lens.* Most of the refraction occurs at the outer surface of the eye, at which the cornea is covered with a film of tears. Relatively little refraction occurs in the crystalline lens, because the aqueous humor in contact with the lens has an average index of refraction close to that of the lens. The iris, which is the colored portion of the eye, is a muscular diaphragm that controls pupil size. The iris regulates the amount of light entering the eye by dilating the pupil in low-light conditions and contracting the pupil under conditions of bright light. The f-number range of the eye is from about 2.8 to 16.

The cornea–lens system focuses light onto the back surface of the eye—the *retina*—which consists of millions of sensitive receptors called *rods* and *cones.* When stimulated by light, these structures send impulses to the brain via the optic nerve, converting them into our conscious view of the world. The process by which the brain performs this conversion is not well understood and is the subject of much speculation and research. Unlike film in a camera, the rods and cones chemically

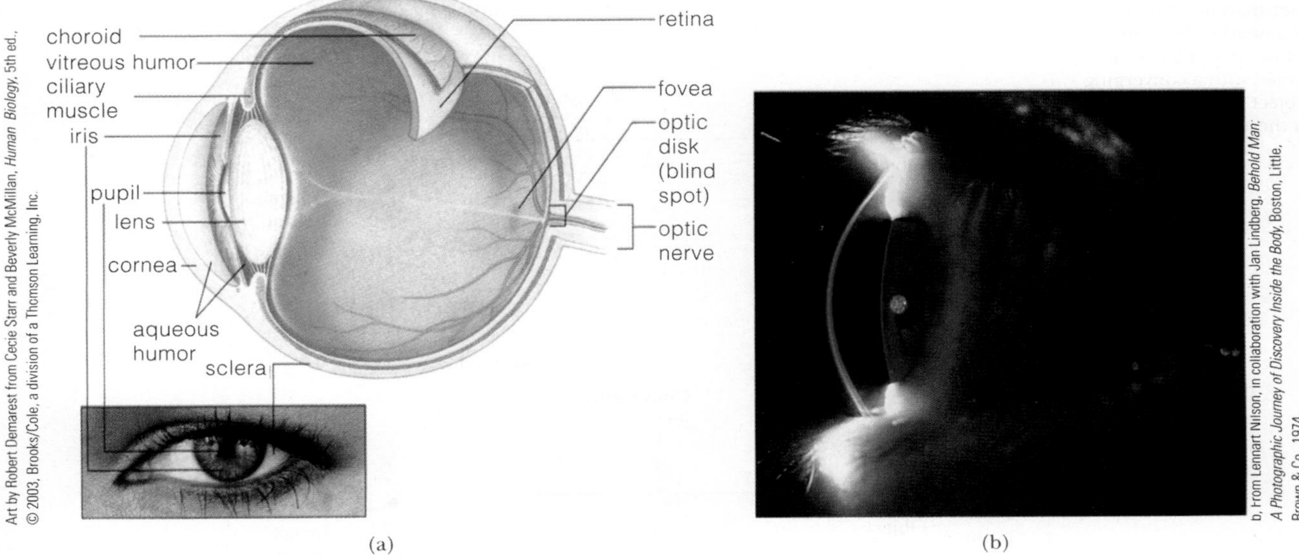

Figure 25.2 (a) Essential parts of the eye. Can you correlate the essential parts of the eye with those of the simple camera in Figure 25.1? (b) Close-up photograph of the human cornea.

adjust their sensitivity according to the prevailing light conditions. This adjustment, which takes about 15 minutes, is responsible for the experience of "getting used to the dark" in such places as movie theaters. Iris aperture control, which takes less than one second, helps protect the retina from overload in the adjustment process.

The eye focuses on an object by varying the shape of the pliable crystalline lens through an amazing process called **accommodation**. An important component in accommodation is the *ciliary muscle*, which is situated in a circle around the rim of the lens. Thin filaments, called *zonules*, run from this muscle to the edge of the lens. When the eye is focused on a distant object, the ciliary muscle is relaxed, tightening the zonules that attach the ciliary muscle to the edge of the lens. The force of the zonules causes the lens to flatten, increasing its focal length. For an object distance of infinity, the focal length of the eye is equal to the fixed distance between lens and retina, about 1.7 cm. The eye focuses on nearby objects by tensing the ciliary muscle, which relaxes the zonules. This action allows the lens to bulge a bit and its focal length decreases, resulting in the image being focused on the retina. All these lens adjustments take place so swiftly that we are not even aware of the change. In this respect, even the finest electronic camera is a toy compared with the eye.

There is a limit to accommodation because objects that are very close to the eye produce blurred images. The **near point** is the closest distance for which the lens can accommodate to focus light on the retina. This distance usually increases with age and has an average value of 25 cm. Typically, at age 10 the near point of the eye is about 18 cm. This increases to about 25 cm at age 20, 50 cm at age 40, and 500 cm or greater at age 60. The **far point** of the eye represents the farthest distance for which the lens of the relaxed eye can focus light on the retina. A person with normal vision is able to see very distant objects, such as the Moon, and so has a far point at infinity.

Conditions of the Eye

When the eye suffers a mismatch between the focusing power of the lens–cornea system and the length of the eye so that light rays reach the retina before they converge to form an image, as in Figure 25.3a, the condition is known as **farsightedness** (or *hyperopia*). A farsighted person can usually see faraway objects clearly but

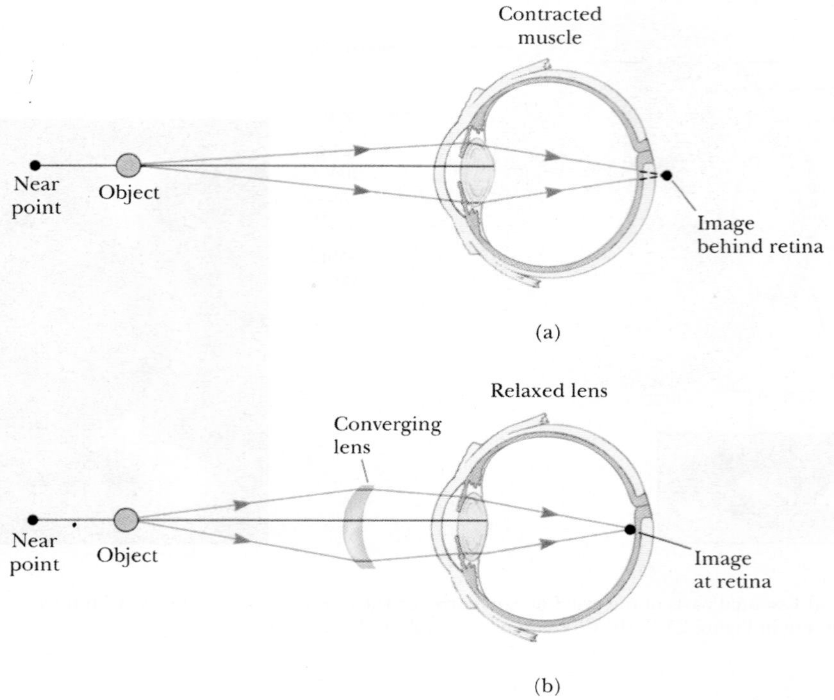

Figure 25.3 (a) A farsighted eye is slightly shorter than normal; hence, the image of a nearby object focuses *behind* the retina. (b) The condition can be corrected with a converging lens. (The object is assumed to be very small in these figures.)

not nearby objects. Although the near point of a normal eye is approximately 25 cm, the near point of a farsighted person is much farther than that. The eye of a farsighted person tries to focus by accommodation, by shortening its focal length. This works for distant objects, but because the focal length of the farsighted eye is longer than normal, the light from nearby objects can't be brought to a sharp focus before it reaches the retina, causing a blurred image. The condition can be corrected by placing a converging lens in front of the eye, as in Figure 25.3b. The lens refracts the incoming rays more toward the principal axis before entering the eye, allowing them to converge and focus on the retina.

Nearsightedness (or *myopia*) is another mismatch condition in which a person is able to focus on nearby objects, but not faraway objects. In the case of *axial myopia*, nearsightedness is caused by the lens being too far from the retina. It is also possible to have *refractive myopia*, in which the lens–cornea system is too powerful for the normal length of the eye. The far point of the nearsighted eye is not at infinity and may be less than a meter. The maximum focal length of the nearsighted eye is insufficient to produce a sharp image on the retina, and rays from a distant object converge to a focus in front of the retina. They then continue past that point, diverging before they finally reach the retina and produce a blurred image (Fig. 25.4a).

APPLICATION

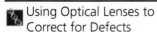

Using Optical Lenses to Correct for Defects

Nearsightedness can be corrected with a diverging lens, as shown in Figure 25.4b. The lens refracts the rays away from the principal axis before they enter the eye, allowing them to focus on the retina.

Beginning with middle age, most people lose some of their accommodation ability as the ciliary muscle weakens and the lens hardens. Unlike farsightedness, which is a mismatch of focusing power and eye length, **presbyopia** (literally, "old-age vision") is due to a reduction in accommodation ability. The cornea and lens aren't able to bring nearby objects into focus on the retina. The symptoms are the same as with farsightedness, and the condition can be corrected with converging lenses.

In the eye defect known as **astigmatism**, light from a point source produces a line image on the retina. This condition arises when either the cornea or the lens (or both) are not perfectly symmetric. Astigmatism can be corrected with lenses having different curvatures in two mutually perpendicular directions.

Optometrists and ophthalmologists usually prescribe lenses measured in **diopters**:

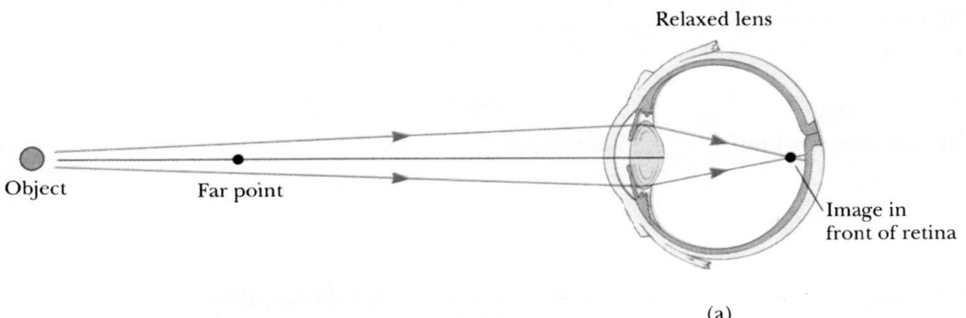

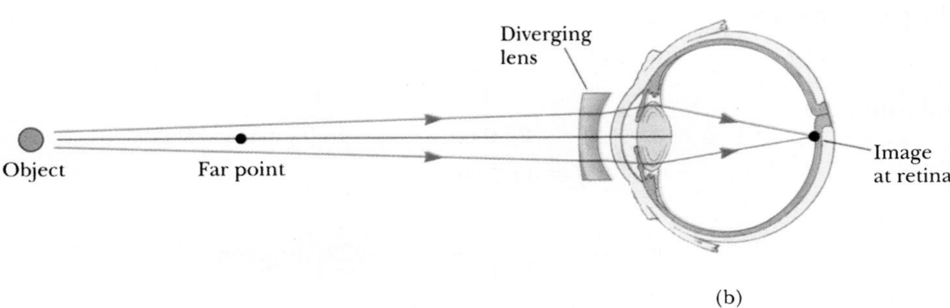

Figure 25.4 (a) A nearsighted eye is slightly longer than normal; hence, the image of a distant object focuses *in front* of the retina. (b) The condition can be corrected with a diverging lens. (The object is assumed to be very small in these figures.)

The **power** $\mathcal{P}$ of a lens in diopters equals the inverse of the focal length in meters: $\mathcal{P} = 1/f$.

For example, a converging lens with a focal length of + 20 cm has a power of + 5.0 diopters, and a diverging lens with a focal length of − 40 cm has a power of − 2.5 diopters. (Although the symbol is the same as for mechanical power, there is no relationship between the two concepts.)

The position of the lens relative the eye causes differences in power, but this usually amounts to less than a quarter diopter, which isn't noticeable to most patients. As a result, practicing optometrists deal in increments of a quarter diopter. Neglecting the eye–lens distance is equivalent to doing the calculation for a contact lens, which rests directly on the eye.

EXAMPLE 25.1 Prescribing a Corrective Lens for a Farsighted Patient 

Goal Apply geometric optics to correct farsightedness.

Problem The near point of a patient's eye is 50.0 cm. **(a)** What focal length must a corrective lens have to enable the eye to see clearly an object 25.0 cm away? Neglect the eye–lens distance. **(b)** What is the power of this lens? **(c)** Repeat the problem, taking into account the fact that, for typical eyeglasses, the corrective lens is 2.00 cm in front of the eye.

Strategy This problem requires substitution into the thin-lens equation (Eq. 23.11) and then using the definition of lens power in terms of diopters. The object is at 25.0 cm, but the lens must form an image at the patient's near point, 50.0 cm, the closest point at which the patient's eye can see clearly. In part (c), 2.00 cm must be subtracted from both the object distance and the image distance to account for the position of the lens.

Solution

(a) Find the focal length of the corrective lens, neglecting its distance from the eye.

Apply the thin-lens equation:

$$\frac{1}{p} + \frac{1}{q} = \frac{1}{f}$$

Substitute $p = 25.0$ cm and $q = -50.0$ cm (the latter is negative because the image must be virtual) on the same side of the lens as the object:

$$\frac{1}{25.0\text{ cm}} + \frac{1}{-50.0\text{ cm}} = \frac{1}{f}$$

Solve for f. The focal length is positive, corresponding to a converging lens.

$$f = \boxed{50.0\text{ cm}}$$

(b) What is the power of this lens?

The power is the reciprocal of the focal length in meters:

$$\mathcal{P} = \frac{1}{f} = \frac{1}{0.500\text{ m}} = \boxed{+2.00\text{ diopters}}$$

(c) Repeat the problem, noting that the corrective lens is actually 2.00 cm in front of the eye.

Substitute the corrected values of p and q into the thin-lens equation:

$$\frac{1}{p} + \frac{1}{q} = \frac{1}{23.0\text{ cm}} + \frac{1}{(-48.0\text{ cm})} = \frac{1}{f}$$

$$f = 44.2\text{ cm}$$

Compute the power:

$$\mathcal{P} = \frac{1}{f} = \frac{1}{0.442\text{ m}} = \boxed{+2.26\text{ diopters}}$$

Remarks Notice that the calculation in part (c), which doesn't neglect the eye–lens distance, results in a difference of 0.26 diopter.

Exercise 25.1

Suppose a lens is placed in a device that determines its power as 2.75 diopters. Find (a) the focal length of the lens and (b) the minimum distance at which a patient will be able to focus on an object if the patient's near point is 60.0 cm. Neglect the eye–lens distance.

Answers (a) 36.4 cm (b) 22.7 cm

EXAMPLE 25.2 A Corrective Lens for Nearsightedness 

Goal Apply geometric optics to correct nearsightedness.

Problem A particular nearsighted patient can't see objects clearly when they are beyond 25 cm (the far point of the eye). **(a)** What focal length should the prescribed contact lens have to correct this problem? **(b)** Find the power of the lens, in diopters. Neglect the distance between the eye and the corrective lens.

Strategy The purpose of the lens in this instance is to take objects at infinity and create an image of them at the patient's far point. Apply the thin-lens equation.

Solution

(a) Find the focal length of the corrective lens.

Apply the thin-lens equation for an object at infinity and image at 25.0 cm:

$$\frac{1}{p} + \frac{1}{q} = \frac{1}{\infty} + \frac{1}{(-25.0\text{ cm})} = \frac{1}{f}$$

$$f = \boxed{-25.0\text{ cm}}$$

(b) Find the power of the lens in diopters:

$$\mathcal{P} = \frac{1}{f} = \frac{1}{-0.250\text{ m}} = -4.00\text{ diopters}$$

Remarks The focal length is negative, consistent with a diverging lens. Notice that the power is also negative and has the same numeric value as the sum on the left side of the thin-lens equation.

Exercise 25.2
(a) What power lens would you prescribe for a patient with a far point of 35.0 cm? Neglect the eye–lens distance. (b) Repeat, assuming an eye-corrective lens distance of 2.00 cm.

Answer (a) − 2.86 diopters (b) − 3.03 diopters

Applying Physics 25.1 Vision of the Invisible Man

A classic science fiction story, *The Invisible Man* by H.G. Wells, tells of a person who becomes invisible by changing the index of refraction of his body to that of air. Students who know how the eye works have criticized this story; they claim the invisible man would be unable to see. On the basis of your knowledge of the eye, would he be able to see ?

Explanation He wouldn't be able to see. In order for the eye to see an object, incoming light must be refracted at the cornea and lens to form an image on the retina. If the cornea and lens have the same index of refraction as air, refraction can't occur, and an image wouldn't be formed.

Quick Quiz 25.1

Two campers wish to start a fire during the day. One camper is nearsighted and one is farsighted. Whose glasses should be used to focus the Sun's rays onto some paper to start the fire? (a) either camper (b) the nearsighted camper (c) the farsighted camper.

25.3 THE SIMPLE MAGNIFIER

The **simple magnifier** is one of the most basic of all optical instruments because it consists only of a single converging lens. As the name implies, this device is used to increase the apparent size of an object. Suppose an object is viewed at some distance p from the eye, as in Figure 25.5. Clearly, the size of the image formed at the retina depends on the angle θ subtended by the object at the eye. As the object moves closer to the eye, θ increases and a larger image is observed. However, a normal eye can't focus on an object closer than about 25 cm, the near point (Fig. 25.6a, page 826). (Try it!) Therefore, θ is a maximum at the near point.

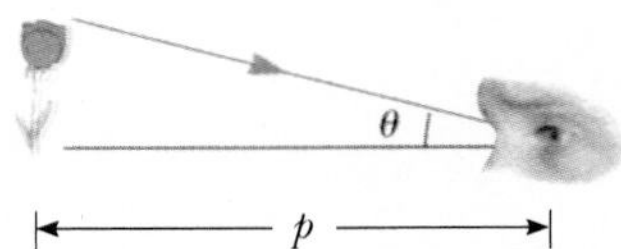

Figure 25.5 The size of the image formed on the retina depends on the angle θ subtended at the eye.

To further increase the apparent angular size of an object, a converging lens can be placed in front of the eye with the object positioned at point O, just inside the focal point of the lens, as in Figure 25.6b. At this location, the lens forms a virtual, upright, and enlarged image, as shown. The lens allows the object to be viewed closer to the eye than is possible without the lens. We define the **angular magnification** m as the ratio of the angle subtended by a small object when the lens is in use (angle θ in Fig. 25.6b) to the angle subtended by the object placed at the near point with no lens in use (angle θ_0 in Fig. 25.6a):

◀ Angular magnification with the object at the near point

$$m \equiv \frac{\theta}{\theta_0} \qquad [25.2]$$

For the case where the lens is held close to the eye, the angular magnification is a maximum when the image formed by the lens is at the near point of the eye, which corresponds to $q = -25$ cm (see Fig. 25.6b). The object distance corresponding to this image distance can be calculated from the thin-lens equation:

$$\frac{1}{p} + \frac{1}{-25\text{ cm}} = \frac{1}{f} \qquad [25.3]$$

$$p = \frac{25f}{25 + f}$$

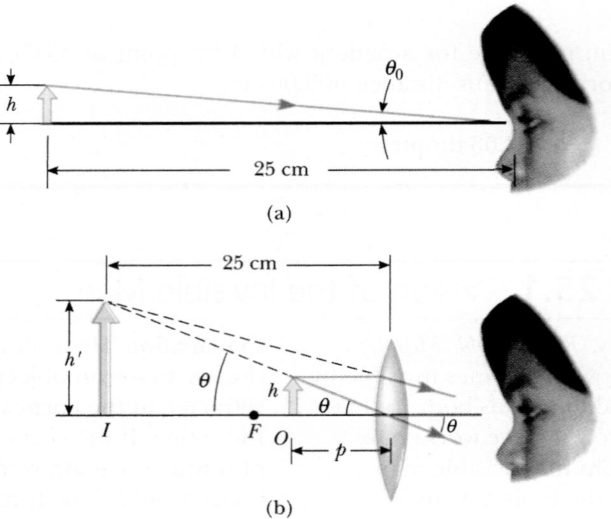

Figure 25.6 (a) An object placed at the near point ($p = 25$ cm) subtends an angle of $\theta_0 \approx h/25$ at the eye. (b) An object placed near the focal point of a converging lens produces a magnified image, which subtends an angle of $\theta \approx h'/25$ at the eye. Note that, in this situation, $q = -25$ cm.

Here, f is the focal length of the magnifier in centimeters. From Figures 25.6a and 25.6b, the small-angle approximation gives

$$\tan \theta_0 \approx \theta_0 \approx \frac{h}{25} \quad \text{and} \quad \tan \theta \approx \theta \approx \frac{h}{p} \qquad [25.4]$$

Equation 25.2 therefore becomes

$$m_{\text{max}} = \frac{\theta}{\theta_0} = \frac{h/p}{h/25} = \frac{25}{p} = \frac{25}{25f/(25+f)}$$

so that

$$m_{\text{max}} = 1 + \frac{25 \text{ cm}}{f} \qquad [25.5]$$

The maximum angular magnification given by Equation 25.5 is the ratio of the angular size seen with the lens to the angular size seen without the lens, with the object at the near point of the eye. Although the normal eye can focus on an image formed anywhere between the near point and infinity, it's most relaxed when the image is at infinity (Sec. 25.2). For the image formed by the magnifying lens to appear at infinity, the object must be placed at the focal point of the lens, so that $p = f$. In this case, Equation 25.4 becomes

$$\theta_0 \approx \frac{h}{25} \quad \text{and} \quad \theta \approx \frac{h}{f}$$

and the angular magnification is

$$m = \frac{\theta}{\theta_0} = \frac{25 \text{ cm}}{f} \qquad [25.6]$$

With a single lens, it's possible to achieve angular magnifications up to about 4 without serious aberrations. Magnifications up to about 20 can be achieved by using one or two additional lenses to correct for aberrations.

EXAMPLE 25.3 Magnification of a Lens

Goal Compute magnifications of a lens when the image is at the near point and when it's at infinity.

Problem **(a)** What is the maximum angular magnification of a lens with a focal length of 10.0 cm? **(b)** What is the angular magnification of this lens when the eye is relaxed? Assume an eye–lens distance of zero.

Strategy The maximum angular magnification occurs when the image formed by the lens is at the near point of the eye. Under these circumstances, Equation 25.5 gives us the maximum angular magnification. In part (b), the eye is relaxed only if the image is at infinity, so Equation 25.6 applies.

Solution

(a) Find the maximum angular magnification of the lens.

Substitute into Equation 25.5:

$$m_{\text{max}} = 1 + \frac{25 \text{ cm}}{f} = 1 + \frac{25 \text{ cm}}{10.0 \text{ cm}} = \boxed{3.5}$$

(b) Find the magnification of the lens when the eye is relaxed.

When the eye is relaxed, the image is at infinity, so substitute into Equation 25.6:

$$m = \frac{25 \text{ cm}}{f} = \frac{25 \text{ cm}}{10.0 \text{ cm}} = \boxed{2.5}$$

Exercise 25.3

What focal length would be necessary if the lens were to have a maximum angular magnification of 4.00?

Answer 8.3 cm

25.4 THE COMPOUND MICROSCOPE

A simple magnifier provides only limited assistance with inspection of the minute details of an object. Greater magnification can be achieved by combining two lenses in a device called a compound microscope, a schematic diagram of which is shown in Active Figure 25.7a. The instrument consists of two lenses: an objective with a very short focal length f_o (where $f_o < 1$ cm), and an ocular lens, or eyepiece, with a focal length f_e of a few centimeters. The two lenses are separated by distance L that is much greater than either f_o or f_e.

The basic approach used to analyze the image formation properties of a microscope is that of two lenses in a row: the image formed by the first becomes the object for the second. The object O placed just outside the focal length of the objective forms a real, inverted image at I_1 that is at or just inside the focal point of the eyepiece. This image is much enlarged. (For clarity, the enlargement of I_1 is not shown in Active Fig. 25.7a.) The eyepiece, which serves as a simple magnifier, uses the image at I_1 as its object and produces an image at I_2. The image seen by the eye at I_2 is virtual, inverted, and very much enlarged.

The lateral magnification M_1 of the first image is $-q_1/p_1$. Note that q_1 is approximately equal to L, because the object is placed close to the focal point of the

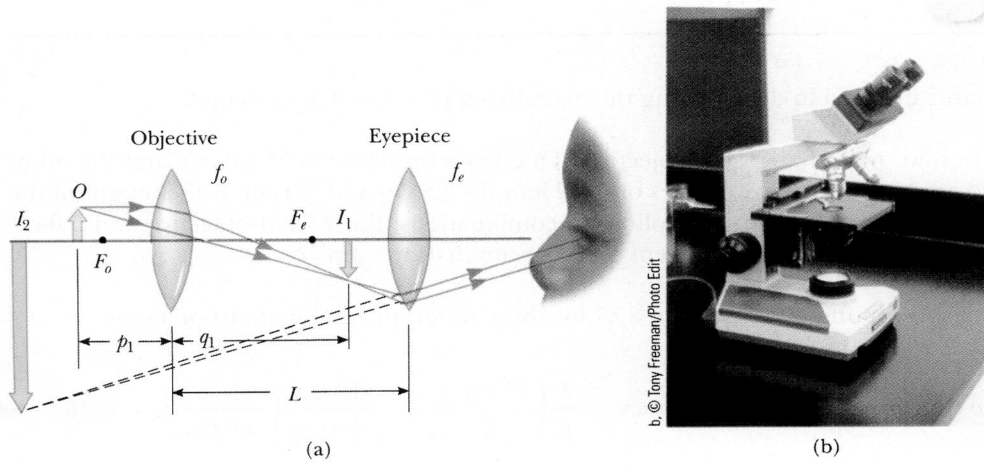

ACTIVE FIGURE 25.7
(a) A diagram of a compound microscope, which consists of an objective and an eyepiece, or ocular lens. (b) A compound microscope. The three-objective turret allows the user to switch to several different powers of magnification. Combinations of eyepieces with different focal lengths and different objectives can produce a wide range of magnifications.

Physics Now™
Log into PhysicsNow at **www.cp7e.com** and go to Active Figure 25.7, where you can adjust the focal lengths of the objective and eyepiece lenses, and see the effect on the final image.

objective lens, which ensures that the image formed will be far from the objective lens. Further, because the object is very close to the focal point of the objective lens, $p_1 \approx f_o$. Therefore, the lateral magnification of the objective is

$$M_1 = -\frac{q_1}{p_1} \approx -\frac{L}{f_o}$$

From Equation 25.6, the angular magnification of the eyepiece for an object (corresponding to the image at I_1) placed at the focal point is found to be

$$m_e = \frac{25 \text{ cm}}{f_e}$$

The overall magnification of the compound microscope is defined as the product of the lateral and angular magnifications:

Magnification of a microscope ▶

$$m = M_1 m_e = -\frac{L}{f_o}\left(\frac{25 \text{ cm}}{f_e}\right) \qquad [25.7]$$

The negative sign indicates that the image is inverted with respect to the object.

The microscope has extended our vision into the previously unknown realm of incredibly small objects, and the capabilities of this instrument have increased steadily with improved techniques in precision grinding of lenses. A natural question is whether there is any limit to how powerful a microscope could be. For example, could a microscope be made powerful enough to allow us to see an atom? The answer to this question is no, as long as visible light is used to illuminate the object. In order to be seen, the object under a microscope must be at least as large as a wavelength of light. An atom is many times smaller than the wavelength of visible light, so its mysteries must be probed via other techniques.

The wavelength dependence of the "seeing" ability of a wave can be illustrated by water waves set up in a bathtub in the following way: Imagine that you vibrate your hand in the water until waves with a wavelength of about 6 in. are moving along the surface. If you fix a small object, such as a toothpick, in the path of the waves, you will find that the waves are not appreciably disturbed by the toothpick, but continue along their path. Now suppose you fix a larger object, such as a toy sailboat, in the path of the waves. In this case, the waves are considerably disturbed by the object. The toothpick was much smaller than the wavelength of the waves, and as a result, the waves didn't "see" it. The toy sailboat, however, is about the same size as the wavelength of the waves and hence creates a disturbance. Light waves behave in this same general way. The ability of an optical microscope to view an object depends on the size of the object relative to the wavelength of the light used to observe it. Hence, it will never be possible to observe atoms or molecules with such a microscope, because their dimensions are so small ($\approx$0.1 nm) relative to the wavelength of the light ($\approx$500 nm).

EXAMPLE 25.4 Microscope Magnifications

Goal Understand the critical factors involved in determining the magnifying power of a microscope.

Problem A certain microscope has two interchangeable objectives. One has a focal length of 2.0 cm, and the other has a focal length of 0.20 cm. Also available are two eyepieces of focal lengths 2.5 cm and 5.0 cm. If the length of the microscope is 18 cm, compute the magnifications for the following combinations: the 2.0-cm objective and 5.0-cm eyepiece; the 2.0-cm objective and 2.5-cm eyepiece; the 0.20-cm objective and 5.0-cm eyepiece.

Strategy The solution consists of substituting into Equation 25.7 for three different combinations of lenses.

Solution

Apply Equation 25.7 and combine the 2.0-cm objective with the 5.0-cm eyepiece:

$$m = -\frac{L}{f_o}\left(\frac{25 \text{ cm}}{f_e}\right) = -\frac{18 \text{ cm}}{2.0 \text{ cm}}\left(\frac{25 \text{ cm}}{5.0 \text{ cm}}\right) = \boxed{-45}$$

Combine the 2.0-cm objective with the 2.5-cm eyepiece: $m = -\frac{18\text{ cm}}{2.0\text{ cm}}\left(\frac{25\text{ cm}}{2.5\text{ cm}}\right) = \boxed{-9.0 \times 10^1}$

Combine the 0.20-cm objective with the 5.0-cm eyepiece: $m = -\frac{18\text{ cm}}{0.20\text{ cm}}\left(\frac{25\text{ cm}}{5.0\text{ cm}}\right) = \boxed{-450}$

Remarks Much higher magnifications can be achieved, but the resolution starts to fall, resulting in fuzzy images that don't convey any details. (See Section 25.6 for further discussion of this point.)

Exercise 25.4
Combine the 0.20-cm objective with the 2.5-cm eyepiece.

Answer 9.0×10^2

25.5 THE TELESCOPE

There are two fundamentally different types of telescope, both designed to help us view distant objects such as the planets in our Solar System. These two types are (1) the **refracting telescope**, which uses a combination of lenses to form an image, and (2) the **reflecting telescope**, which uses a curved mirror and a lens to form an image. Once again, we will be able to analyze the telescope by considering it to be a system of two optical elements in a row. As before, the basic technique followed is that the image formed by the first element becomes the object for the second.

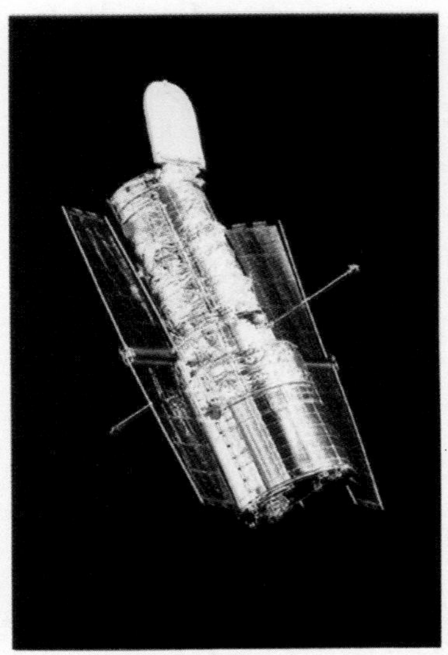

The Hubble Space Telescope enables us to see both further into space and further back in time than ever before.

In the refracting telescope, two lenses are arranged so that the objective forms a real, inverted image of the distant object very near the focal point of the eyepiece (Active Fig. 25.8a, page 830). Further, the image at I_1 is formed at the focal point of the objective because the object is essentially at infinity. Hence, the two lenses are separated by the distance $f_o + f_e$, which corresponds to the length of the telescope's tube. Finally, at I_2, the eyepiece forms an enlarged, inverted image of the image at I_1.

The angular magnification of the telescope is given by θ/θ_o, where θ_o is the angle subtended by the object at the objective and θ is the angle subtended by the final image. From the triangles in Active Figure 25.8a, and for small angles, we have

$$\theta \approx \frac{h'}{f_e} \quad \text{and} \quad \theta_o \approx \frac{h'}{f_o}$$

Therefore, the angular magnification of the telescope can be expressed as

$$m = \frac{\theta}{\theta_o} = \frac{h'/f_e}{h'/f_o} = \frac{f_o}{f_e} \qquad [25.8]$$

◀ Angular magnification of a telescope

This equation says that the angular magnification of a telescope equals the ratio of the objective focal length to the eyepiece focal length. Here again, the angular magnification is the ratio of the angular size seen with the telescope to the angular size seen with the unaided eye.

In some applications—for instance, the observation of relatively nearby objects such as the Sun, the Moon, or planets—angular magnification is important. Stars, however, are so far away that they always appear as small points of light regardless of how much angular magnification is used. The large research telescopes used to study very distant objects must have great diameters to gather as much light as possible. It's difficult and expensive to manufacture such large lenses for refracting telescopes. In addition, the heaviness of large lenses leads to sagging, which is another source of aberration.

These problems can be partially overcome by replacing the objective lens with a reflecting, concave mirror, usually having a parabolic shape so as to avoid spherical aberration. Figure 25.9 (page 830) shows the design of a typical reflecting telescope. Incoming light rays pass down the barrel of the telescope and are reflected

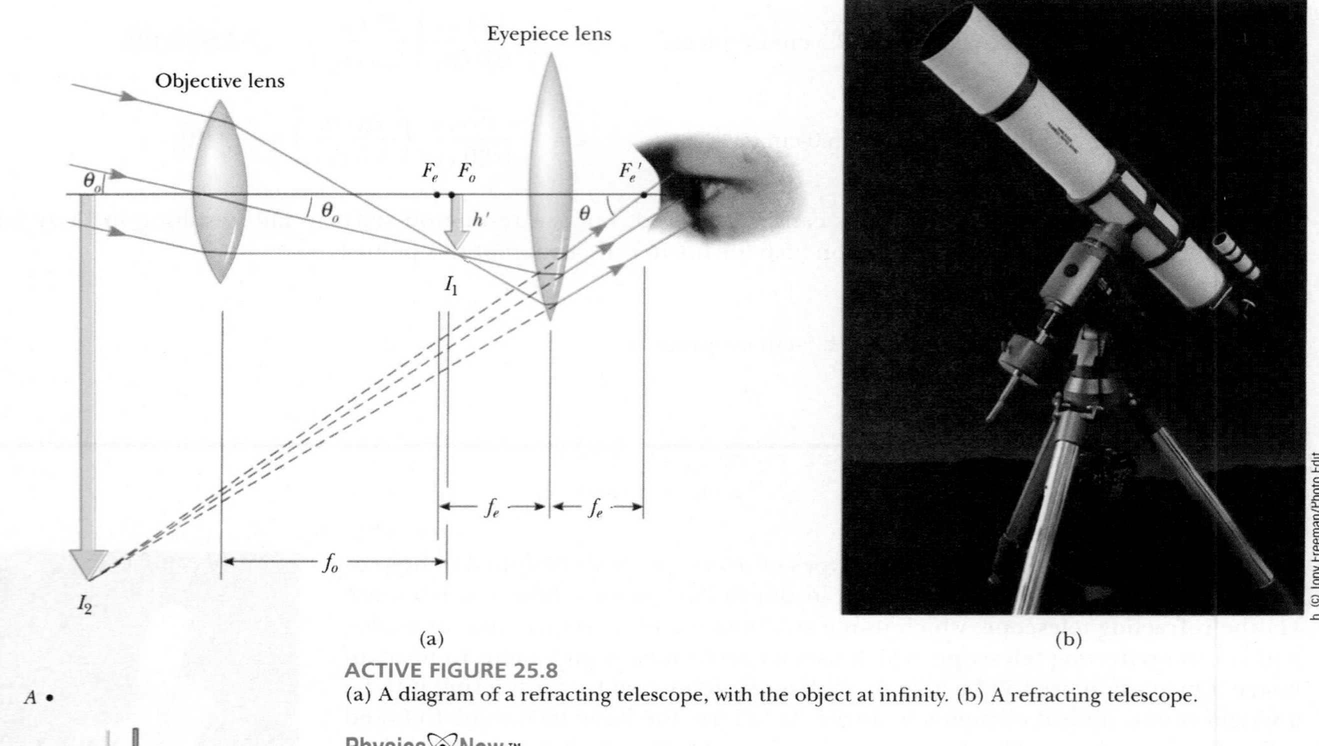

ACTIVE FIGURE 25.8
(a) A diagram of a refracting telescope, with the object at infinity. (b) A refracting telescope.

Physics Now™
Log into PhysicsNow at **www.cp7e.com** and go to Active Figure 25.8, where you can adjust the focal lengths of the objective and eyepiece lenses, and observe the effect on the final image.

A •
M
Eyepiece
Parabolic mirror

Figure 25.9 A reflecting telescope with a Newtonian focus.

by a parabolic mirror at the base. These rays converge toward point A in the figure, where an image would be formed on a photographic plate or another detector. However, before this image is formed, a small flat mirror at M reflects the light toward an opening in the side of the tube that passes into an eyepiece. This design is said to have a *Newtonian focus,* after its developer. Note that in the reflecting telescope the light never passes through glass (except for the small eyepiece). As a result, problems associated with chromatic aberration are virtually eliminated.

The largest optical telescopes in the world are the two 10-m-diameter Keck reflectors on Mauna Kea in Hawaii. The largest single-mirrored reflecting telescope in the United States is the 5-m-diameter instrument on Mount Palomar in California. (See Fig. 25.10.) In contrast, the largest refracting telescope in the world, at the Yerkes Observatory in Williams Bay, Wisconsin, has a diameter of only 1 m.

Courtesy of Palomar Observatory/California Institute of Technology

Figure 25.10 The Hale telescope at Mount Palomar Observatory. Just before taking the elevator up to the prime-focus cage, a first-time observer is always told, "Good viewing! And, if you should fall, try to miss the mirror."

EXAMPLE 25.5 Hubble Power

Goal Understand magnification in telescopes.

Problem The Hubble telescope is 13.2 m long, but has a secondary mirror that increases its effective focal length to 57.8 m. (See Fig. 25.11.) The telescope doesn't have an eyepiece, because various instruments, not a human eye, record the collected light. However, it can produce images several thousand times larger than they would appear with the unaided human eye. What focal length eyepiece used with the Hubble mirror system would produce a magnification of 8.00×10^3?

Strategy Equation 25.8 for telescope magnification can be solved for the eyepiece focal length. The equation for finding the angular magnification of a reflector is the same as that for a refractor.

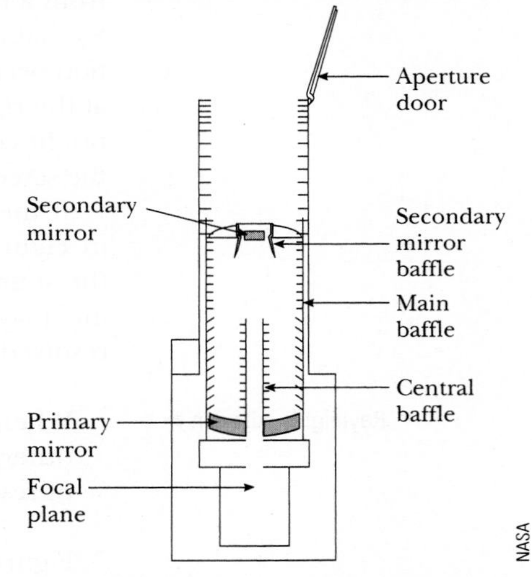

Figure 25.11 A schematic of the Hubble telescope.

Solution

Solve for f_e in Equation 25.8 and substitute values:

$$m = \frac{f_o}{f_e} \quad \rightarrow \quad f_e = \frac{f_o}{m} = \frac{57.8 \text{ m}}{8.00 \times 10^3} = 7.23 \times 10^{-3} \text{ m}$$

Remarks The result of this magnification is an image with "good" resolution. However, the light-gathering power of a telescope largely determines the resolution of the image, and is far more important than magnification. A high-resolution image can always be magnified so its details can be examined. Such details are often blurred when a low-resolution image is magnified.

Exercise 25.5

The Hale telescope on Mt. Palomar has a focal length of 16.8 m. Find the magnification of the telescope in conjunction with an eyepiece having a focal length of 5.00 mm.

Answer 3.36×10^3

25.6 RESOLUTION OF SINGLE-SLIT AND CIRCULAR APERTURES

The ability of an optical system such as the eye, a microscope, or a telescope to distinguish between closely spaced objects is limited because of the wave nature of light. To understand this difficulty, consider Figure 25.12, which shows two light sources far

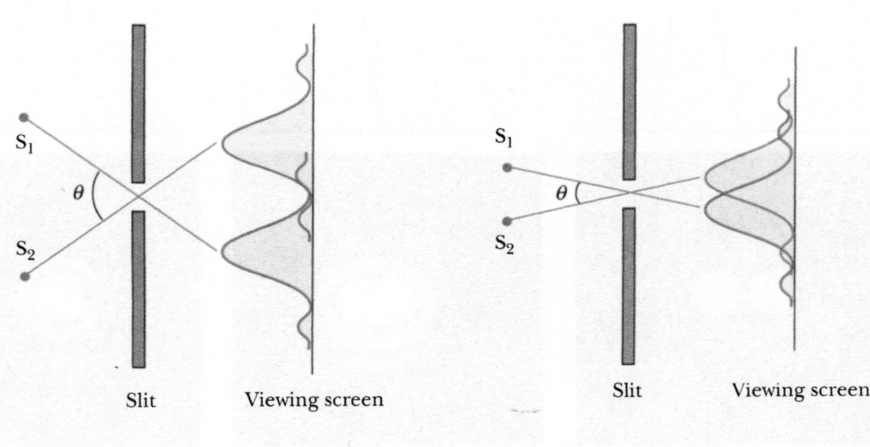

Figure 25.12 Two point sources far from a narrow slit each produce a diffraction pattern. (a) The angle subtended by the sources at the aperture is large enough so that the diffraction patterns are distinguishable. (b) The angle subtended by the sources is so small that the diffraction patterns are not distinguishable. (Note that the angles are greatly exaggerated. The drawing is not to scale.)

from a narrow slit of width a. The sources can be taken as two point sources S_1 and S_2 that are *not* coherent. For example, they could be two distant stars. If no diffraction occurred, two distinct bright spots (or images) would be observed on the screen at the right in the figure. However, because of diffraction, each source is imaged as a bright central region flanked by weaker bright and dark rings. What is observed on the screen is the sum of two diffraction patterns, one from S_1 and the other from S_2.

If the two sources are separated so that their central maxima don't overlap, as in Figure 25.12a, their images can be distinguished and are said to be *resolved*. If the sources are close together, however, as in Figure 25.12b, the two central maxima may overlap and the images are *not resolved*. To decide whether two images are resolved, the following condition is often applied to their diffraction patterns:

Rayleigh's criterion ▶

When the central maximum of one image falls on the first minimum of another image, the images are said to be just resolved. This limiting condition of resolution is known as **Rayleigh's criterion.**

Figure 25.13 shows diffraction patterns in three situations. The images are just resolved when their angular separation satisfies Rayleigh's criterion (Fig. 25.13a). As the objects are brought closer together, their images are barely resolved (Fig. 25.13b). Finally, when the sources are very close to each other, their images are not resolved (Fig. 25.13c).

From Rayleigh's criterion, we can determine the minimum angular separation θ_{min} subtended by the source at the slit so that the images will be just resolved. In Chapter 24 we found that the first minimum in a single-slit diffraction pattern occurs at the angle that satisfies the relationship

$$\sin\theta = \frac{\lambda}{a}$$

where a is the width of the slit. According to Rayleigh's criterion, this expression gives the smallest angular separation for which the two images can be resolved. Because $\lambda \ll a$ in most situations, $\sin\theta$ is small and we can use the approximation $\sin\theta \approx \theta$. Therefore, the limiting angle of resolution for a slit of width a is

Limiting angle for a slit ▶

$$\theta_{\text{min}} \approx \frac{\lambda}{a} \quad [25.9]$$

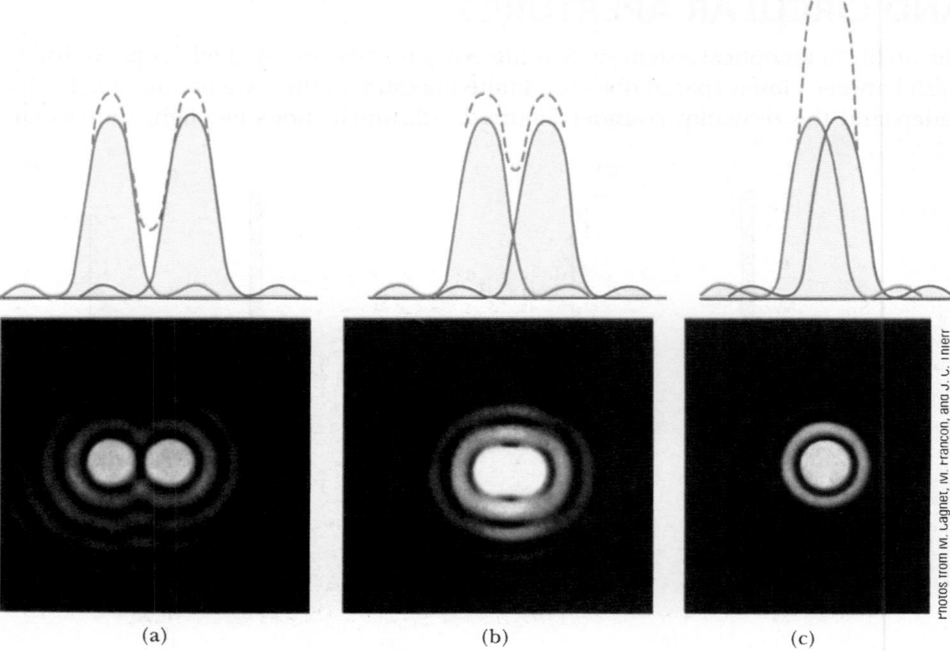

Figure 25.13 The diffraction patterns of two point sources (solid curves) and the resultant pattern (dashed curve) for three angular separations of the sources. (a) The sources are separated such that their patterns are just resolved. (b) The sources are closer together, and their patterns are barely resolved. (c) The sources are so close together that their patterns are not resolved.

where θ_{min} is in radians. Hence, the angle subtended by the two sources at the slit must be *greater* than λ/a if the images are to be resolved.

Many optical systems use circular apertures rather than slits. The diffraction pattern of a circular aperture (Fig. 25.14) consists of a central circular bright region surrounded by progressively fainter rings. Analysis shows that the limiting angle of resolution of the circular aperture is

$$\theta_{min} = 1.22\frac{\lambda}{D} \qquad [25.10]$$

where D is the diameter of the aperture. Note that Equation 25.10 is similar to Equation 25.9, except for the factor 1.22, which arises from a complex mathematical analysis of diffraction from a circular aperture.

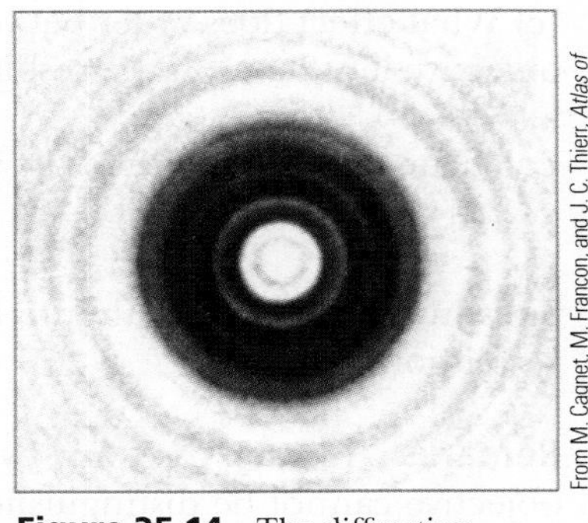

From M. Cagnet, M. Francon, and J. C. Thierr, *Atlas of Optical Phenomena*, Berlin, Springer-Verlag, 1962, plate 34

Figure 25.14 The diffraction pattern of a circular aperture consists of a central bright disk surrounded by concentric bright and dark rings.

Quick Quiz 25.2

Suppose you are observing a binary star with a telescope and are having difficulty resolving the two stars. Which color filter will better help resolve the stars? (a) blue (b) red (c) neither—colored filters have no affect on resolution

Applying Physics 25.2 Cat's Eyes

Cats' eyes have vertical pupils in dim light. Which would cats be most successful at resolving at night, headlights on a distant car or vertically separated running lights on a distant boat's mast having the same separation as the car's headlights?

Explanation The effective slit width in the vertical direction of the cat's eye is larger than that in the horizontal direction. Thus, it has more resolving power for lights separated in the vertical direction and would be more effective at resolving the mast lights on the boat.

EXAMPLE 25.6 Resolution of a Microscope

Goal Study limitations on the resolution of a microscope.

Problem Sodium light of wavelength 589 nm is used to view an object under a microscope. The aperture of the objective has a diameter of 0.90 cm. **(a)** Find the limiting angle of resolution for this microscope. **(b)** Using visible light of any wavelength you desire, find the maximum limit of resolution for this microscope. **(c)** Water of index of refraction 1.33 now fills the space between the object and the objective. What effect would this have on the resolving power of the microscope, using 589 nm light?

Strategy Parts (a) and (b) require substitution into Equation 25.10. Because the wavelength appears in the numerator, violet light, with the shortest visible wavelength, gives the maximum resolution. In part (c), the only difference is that the wavelength changes to λ/n, where n is the index of refraction of water.

Solution

(a) Find the limiting angle of resolution for this microscope.

Substitute into Equation 25.10 to obtain the limiting angle of resolution:

$$\theta_{min} = 1.22\frac{\lambda}{D} = 1.22\left(\frac{589 \times 10^{-9}\text{ m}}{0.90 \times 10^{-2}\text{ m}}\right)$$
$$= 8.0 \times 10^{-5}\text{ rad}$$

(b) Calculate the microscope's maximum limit of resolution.

To obtain the maximum resolution, substitute the shortest visible wavelength available—violet light, of wavelength 4.0×10^2 nm:

$$\theta_{min} = 1.22\frac{\lambda}{D} = 1.22\left(\frac{4.0 \times 10^{-7}\text{ m}}{0.90 \times 10^{-2}\text{ m}}\right)$$
$$= 5.4 \times 10^{-5}\text{ rad}$$

(c) What effect does water between the object and the objective lens have on the resolution, with 589-nm light?

Calculate the wavelength of the sodium light in the water:

$$\lambda_w = \frac{\lambda_a}{n} = \frac{589 \text{ nm}}{1.33} = 443 \text{ nm}$$

Substitute this wavelength into Equation 25.10 to get the resolution:

$$\theta_{\min} = 1.22\left(\frac{443 \times 10^{-9} \text{ m}}{0.90 \times 10^{-2} \text{ m}}\right) = 6.0 \times 10^{-5} \text{ rad}$$

Remarks In each case, any two points on the object subtending an angle of less than the limiting angle $\theta_{\min}$ at the objective cannot be distinguished in the image. Consequently, it may be possible to see a cell, but then be unable to clearly see smaller structures within the cell. Obtaining an increase in resolution is the motivation behind placing a drop of oil on the slide for certain objective lenses.

Exercise 25.6
Suppose oil with $n = 1.50$ fills the space between the object and the objective for this microscope. Calculate the limiting angle $\theta_{\min}$ for sodium light of wavelength 589 nm in air.

Answer 5.3×10^{-5} rad

EXAMPLE 25.7 Resolving Craters on the Moon

Goal Calculate the resolution of a telescope.

Problem The Hubble Space Telescope has an aperture of diameter 2.40 m. **(a)** What is its limiting angle of resolution at a wavelength of 6.00×10^2 nm? **(b)** What's the smallest crater it could resolve on the Moon? (The Moon is 3.84×10^8 m from Earth.)

Strategy After substituting into Equation 25.10 to find the limiting angle, use $s = r\theta$ to compute the minimum size of crater that can be resolved.

Solution
(a) What is the limiting angle of resolution at a wavelength of 6.00×10^2 nm?

Substitute $D = 2.40$ m and $\lambda = 6.00 \times 10^{-7}$ m into Equation 25.10:

$$\theta_{\min} = 1.22\frac{\lambda}{D} = 1.22\left(\frac{6.00 \times 10^{-7} \text{ m}}{2.40 \text{ m}}\right)$$

$$= 3.05 \times 10^{-7} \text{ rad}$$

(b) What's the smallest lunar crater the Hubble Space Telescope can resolve?

The two opposite sides of the crater must subtend the minimum angle. Use the arc length formula:

$$s = r\theta = (3.84 \times 10^8 \text{ m})(3.05 \times 10^{-7} \text{ rad}) = 117 \text{ m}$$

Remarks The distance is so great and the angle so small that using the arc length of a circle is justified—the circular arc is very nearly a straight line. The Hubble Space Telescope has produced several gigabytes of data every day for over 20 years.

Exercise 25.7
The Hale telescope on Mount Palomar has a diameter of 5.08 m (200 in.). (a) Find the limiting angle of resolution for a wavelength of 6.00×10^2 nm. (b) Calculate the smallest crater diameter the telescope can resolve on the Moon. (c) The answers appear better than what the Hubble can achieve. Why are the answers misleading?

Answers (a) 1.44×10^{-7} rad (b) 55.3 m (c) While the numbers are better than Hubble's, the Hale telescope must contend with the effects of atmospheric turbulence, so the smaller space-based telescope actually obtains far better results.

It's interesting to compare the resolution of the Hale telescope with that of a large radio telescope, such as the system at Arecibo, Puerto Rico, which has a diameter of 1 000 ft (305 m). This telescope detects radio waves at a wavelength of 0.75 m. The corresponding minimum angle of resolution can be calculated as 3.0×10^{-3} rad (10 min 19 s of arc), which is more than 10 000 times larger than the calculated minimum angle for the Hale telescope.

With such relatively poor resolution, why is Arecibo considered a valuable astronomical instrument? Unlike its optical counterparts, Arecibo can see through clouds of dust. The center of our Milky Way galaxy is obscured by such dust clouds, which absorb and scatter visible light. Radio waves easily penetrate the clouds, so radio telescopes allow direct observations of the galactic core.

Resolving Power of the Diffraction Grating

The diffraction grating studied in Chapter 24 is most useful for making accurate wavelength measurements. Like the prism, it can be used to disperse a spectrum into its components. Of the two devices, the grating is better suited to distinguishing between two closely spaced wavelengths. We say that the grating spectrometer has a higher *resolution* than the prism spectrometer. If λ_1 and λ_2 are two nearly equal wavelengths between which the spectrometer can just barely distinguish, the **resolving power** of the grating is defined as

$$R \equiv \frac{\lambda}{\lambda_2 - \lambda_1} = \frac{\lambda}{\Delta\lambda} \qquad [25.11]$$

where $\lambda \approx \lambda_1 \approx \lambda_2$ and $\Delta\lambda = \lambda_2 - \lambda_1$. From this equation, it's clear that a grating with a high resolving power can distinguish small differences in wavelength. Further, if N lines of the grating are illuminated, it can be shown that the resolving power in the mth-order diffraction is given by

$$R = Nm \qquad [25.12]$$

◀ Resolving power of a grating

So the resolving power R increases with the order number m and is large for a grating with a great number of illuminated slits. Note that for $m = 0$, $R = 0$, which signifies that *all wavelengths are indistinguishable* for the zeroth-order maximum. (All wavelengths fall at the same point on the screen.) However, consider the second-order diffraction pattern of a grating that has 5 000 rulings illuminated by the light source. The resolving power of such a grating in second order is $R = 5\,000 \times 2 = 10\,000$. Therefore, the *minimum* wavelength separation between two spectral lines that can be just resolved, assuming a mean wavelength of 600 nm, is calculated from Equation 25.12 to be $\Delta\lambda = \lambda/R = 6 \times 10^{-2}$ nm. For the third-order principal maximum, $R = 15\,000$ and $\Delta\lambda = 4 \times 10^{-2}$ nm, and so on.

EXAMPLE 25.8 Light from Sodium Atoms

Goal Find the necessary resolving power to distinguish spectral lines.

Problem Two bright lines in the spectrum of sodium have wavelengths of 589.00 nm and 589.59 nm, respectively. **(a)** What must the resolving power of a grating be in order to distinguish these wavelengths? **(b)** To resolve these lines in the second-order spectrum, how many lines of the grating must be illuminated?

Strategy This problem requires little more than substituting into Equations 25.11 and 25.12.

Solution

(a) What must the resolving power of a grating be in order to distinguish the given wavelengths?

Substitute into Equation 25.11 to find R:

$$R = \frac{\lambda}{\Delta\lambda} = \frac{589.00 \text{ nm}}{589.59 \text{ nm} - 589.00 \text{ nm}} = \frac{589 \text{ nm}}{0.59 \text{ nm}}$$

$$= \boxed{1.0 \times 10^3}$$

(b) To resolve these lines in the second-order spectrum, how many lines of the grating must be illuminated?

Solve Equation 25.12 for N and substitute:

$$N = \frac{R}{m} = \frac{1.0 \times 10^3}{2} = 5.0 \times 10^2 \text{ lines}$$

Remarks The ability to resolve spectral lines is particularly important in experimental atomic physics.

Exercise 25.8

When the lines of a spectrum are examined at high resolution, each line is actually found to be two closely spaced lines called a doublet, due to a phenomenon called electron spin. An example is the doublet in the hydrogen spectrum having wavelengths of 656.272 nm and 656.285 nm. (a) What must be the resolving power of a grating in order to distinguish these wavelengths? (b) How many lines of the grating must be illuminated to resolve these lines in the third-order spectrum?

Answer (a) 5.0×10^4 (b) 1.7×10^4 lines

25.7 THE MICHELSON INTERFEROMETER

The Michelson interferometer is an optical instrument having great scientific importance. Invented by the American physicist A. A. Michelson (1852–1931), it is an ingenious device that splits a light beam into two parts and then recombines them to form an interference pattern. The interferometer is used to make accurate length measurements.

Active Figure 25.15 is a schematic diagram of an interferometer. A beam of light provided by a monochromatic source is split into two rays by a partially silvered mirror M inclined at an angle of 45° relative to the incident light beam. One ray is reflected vertically upward to mirror M_1, and the other ray is transmitted horizontally through mirror M to mirror M_2. Hence, the two rays travel separate paths, L_1 and L_2. After reflecting from mirrors M_1 and M_2, the two rays eventually recombine to produce an interference pattern, which can be viewed through a telescope. The glass plate P, equal in thickness to mirror M, is placed in the path of the horizontal ray to ensure that the two rays travel the same distance through glass.

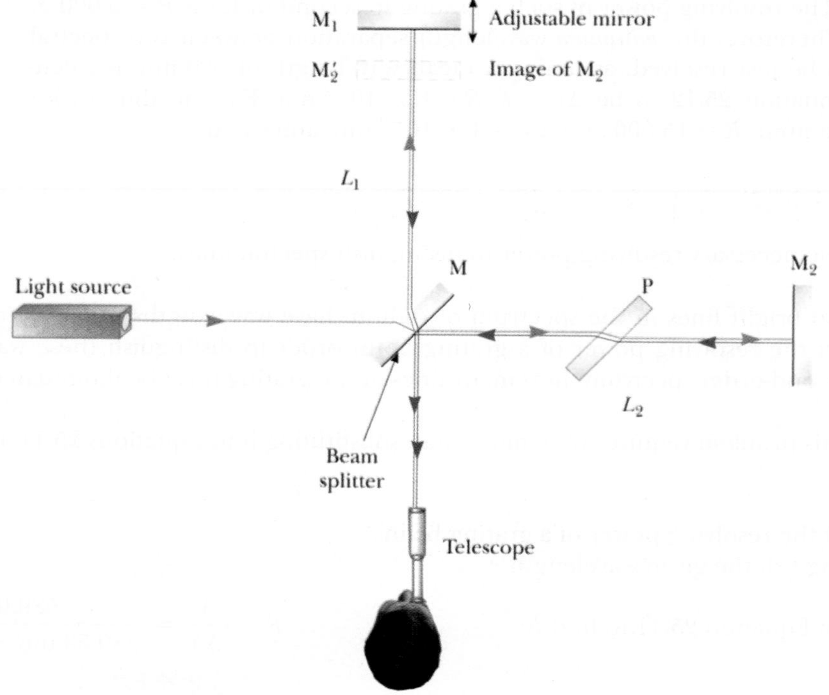

ACTIVE FIGURE 25.15
A diagram of the Michelson interferometer. A single beam is split into two rays by the half-silvered mirror M. The path difference between the two rays is varied with the adjustable mirror M_1.

Physics Now™
Log into PhysicsNow at **www.cp7e.com** and go to Active Figure 25.15, where you can move one of the mirrors, observing the effect on the interference pattern, and use the interferometer to measure the speed of light.

The interference pattern for the two rays is determined by the difference in their path lengths. When the two rays are viewed as shown, the image of M_2 is at M_2', parallel to M_1. Hence, the space between M_2' and M_1 forms the equivalent of a parallel air film. The effective thickness of the air film is varied by using a finely threaded screw to move mirror M_1 in the direction indicated by the arrows in Active Figure 25.15. If one of the mirrors is tipped slightly with respect to the other, the thin film between the two is wedge shaped, and an interference pattern consisting of parallel fringes is set up, as described in Example 24.4. Now suppose we focus on one of the dark lines with the crosshairs of a telescope. As the mirror M_1 is moved to lengthen the path L_1, the thickness of the wedge increases. When the thickness increases by $\lambda/4$, the destructive interference that initially produced the dark fringe has changed to constructive interference, and we now observe a bright fringe at the location of the crosshairs. The term *fringe shift* is used to describe the change in a fringe from dark to light or light to dark. Thus, successive light and dark fringes are formed each time M_1 is moved a distance of $\lambda/4$. The wavelength of light can be measured by counting the number of fringe shifts for a measured displacement of M_1. Conversely, if the wavelength is accurately known (as with a laser beam), the mirror displacement can be determined to within a fraction of the wavelength. Because the interferometer can measure displacements precisely, it is often used to make highly accurate measurements of the dimensions of mechanical components.

If the mirrors are perfectly aligned, rather than tipped with respect to one another, the path difference differs slightly for different angles of view. This arrangement results in an interference pattern that resembles Newton's rings. The pattern can be used in a fashion similar to that for tipped mirrors. An observer pays attention to the center spot in the interference pattern. For example, suppose the spot is initially dark, indicating that destructive interference is occurring. If M_1 is now moved a distance of $\lambda/4$, this central spot changes to a light region, corresponding to a fringe shift.

SUMMARY

PhysicsNow™ Take a practice test by logging into PhysicsNow at **www.cp7e.com** and clicking on the Pre-Test link for this chapter.

25.1 The Camera

The light-concentrating power of a lens of focal length f and diameter D is determined by the ***f*-number**, defined as

$$f\text{-number} \equiv \frac{f}{D} \qquad [25.1]$$

The smaller the f-number of a lens, the brighter is the image formed.

25.2 The Eye

Hyperopia (farsightedness) is a defect of the eye that occurs either when the eyeball is too short or when the ciliary muscle cannot change the shape of the lens enough to form a properly focused image. **Myopia** (nearsightedness) occurs either when the eye is longer than normal or when the maximum focal length of the lens is insufficient to produce a clearly focused image on the retina.

The **power** of a lens in **diopters** is the inverse of the focal length in meters.

25.3 The Simple Magnifier

The **angular magnification of a lens** is defined as

$$m \equiv \frac{\theta}{\theta_0} \qquad [25.2]$$

where θ is the angle subtended by an object at the eye with a lens in use and θ_0 is the angle subtended by the object when it is placed at the near point of the eye and no lens is used. The **maximum angular magnification of a lens** is

$$m_{\text{max}} = 1 + \frac{25\text{ cm}}{f} \qquad [25.5]$$

When the eye is relaxed, the angular magnification is

$$m = \frac{25\text{ cm}}{f} \qquad [25.6]$$

25.4 The Compound Microscope

The overall **magnification of a compound microscope** of length L is the product of the magnification produced by the objective, of focal length f_o, and the magnification produced by the eyepiece, of focal length f_e:

$$M = -\frac{L}{f_o}\left(\frac{25\text{ cm}}{f_e}\right) \quad [25.7]$$

25.5 The Telescope

The **angular magnification of a telescope** is

$$m = \frac{f_o}{f_e} \quad [25.8]$$

where f_o is the focal length of the objective and f_e is the focal length of the eyepiece.

25.6 Resolution of Single-Slit and Circular Apertures

Two images are said to be **just resolved** when the central maximum of the diffraction pattern for one image falls on the first minimum of the other image. This limiting condition of resolution is known as **Rayleigh's criterion**. The limiting angle of resolution for a **slit** of width a is

$$\theta_{\min} \approx \frac{\lambda}{a} \quad [25.9]$$

The limiting angle of resolution of a **circular aperture** is

$$\theta_{\min} = 1.22\frac{\lambda}{D} \quad [25.10]$$

where D is the diameter of the aperture.

If λ_1 and λ_2 are two nearly equal wavelengths between which a grating spectrometer can just barely distinguish, the **resolving power** R of the grating is defined as

$$R \equiv \frac{\lambda}{\lambda_2 - \lambda_1} = \frac{\lambda}{\Delta\lambda} \quad [25.11]$$

where $\lambda \approx \lambda_1 \approx \lambda_2$ and $\Delta\lambda = \lambda_2 - \lambda_1$. The **resolving power** of a diffraction grating in the mth order is

$$R = Nm \quad [25.12]$$

where N is the number of illuminated rulings on the grating.

CONCEPTUAL QUESTIONS

1. A lens is used to examine an object across a room. Is the lens probably being used as a simple magnifier?
2. Why is it difficult or impossible to focus a microscope on an object across a room?
3. The optic nerve and the brain invert the image formed on the retina. Why don't we see everything upside down?
4. If you want to examine the fine detail of an object with a magnifying glass with a power of +20.0 diopters, where should the object be placed in order to observe a magnified image of the object?
5. Suppose you are observing the interference pattern formed by a Michelson interferometer in a laboratory and a joking colleague holds a lit match in the light path of one arm of the interferometer. Will this have an effect on the interference pattern?
6. Compare and contrast the eye and a camera. What parts of the camera correspond to the iris, the retina, and the cornea of the eye?
7. Large telescopes are usually reflecting rather than refracting. List some reasons for this choice.
8. If you want to use a converging lens to set fire to a piece of paper, why should the light source be farther from the lens than its focal point?
9. Explain why it is theoretically impossible to see an object as small as an atom regardless of the quality of the light microscope being used.
10. Which is most important in the use of a camera photoflash unit, the intensity of the light (the energy per unit area per unit time) or the product of the intensity and the time of the flash, assuming the time is less than the shutter speed?
11. A patient has a near point of 1.25 m. Is she nearsighted or farsighted? Should the corrective lens be converging or diverging?
12. A lens with a certain power is used as a simple magnifier. If the power of the lens is doubled, does the angular magnification increase or decrease?

PROBLEMS

1, 2, 3 = straightforward, intermediate, challenging □ = full solution available in *Student Solutions Manual/Study Guide*
PhysicsNow™ = coached problem with hints available at **www.cp7e.com** = biomedical application

Section 25.1 The Camera

1. A camera used by a professional photographer to shoot portraits has a focal length of 25.0 cm. The photographer takes a portrait of a person 1.50 m in front of the camera. Where is the image formed, and what is the lateral magnification?
2. The lens of a certain 35 mm camera (35 mm is the width of the film strip) has a focal length of 55 mm and a speed (an f-number) of $f/1.8$. Determine the diameter of the lens.
3. A photographic image of a building is 0.092 0 m high. The image was made with a lens with a focal length of 52.0 mm. If the lens was 100 m from the building when the photograph was made, determine the height of the building.
4. The full Moon is photographed using a camera with a 120-mm-focal-length lens. Determine the diameter of the Moon's image on the film. [*Note:* The radius of the Moon is 1.74×10^6 m, and the distance from the Earth to the Moon is 3.84×10^8 m.]
5. A camera is being used with the correct exposure at $f/4$ and a shutter speed of 1/32 s. In order to "stop" a fast-moving subject, the shutter speed is changed to 1/256 s. Find the new f-stop that should be used to maintain

satisfactory exposure, assuming no change in lighting conditions.

6. **Physics Now™** (a) Use conceptual arguments to show that the intensity of light (energy per unit area per unit time) reaching the film in a camera is proportional to the square of the reciprocal of the f-number, as

$$I \propto \frac{1}{(f/D)^2}$$

(b) The correct exposure time for a camera set to $f/1.8$ is (1/500) s. Calculate the correct exposure time if the f-number is changed to $f/4$ under the same lighting conditions.

7. A certain type of film requires an exposure time of 0.010 s with an $f/11$ lens setting. Another type of film requires twice the light energy to produce the same level of exposure. What f-stop does the second type of film need with the 0.010-s exposure time?

8. Assume that the camera in Figure 25.1 has a fixed focal length of 65.0 mm and is adjusted to properly focus the image of a distant object. How far and in what direction must the lens be moved to focus the image of an object that is 2.00 m away?

Section 25.2 The Eye

9. A retired bank president can easily read the fine print of the financial page when the newspaper is held no closer than arm's length, 60.0 cm from the eye. What should be the focal length of an eyeglass lens that will allow her to read at the more comfortable distance of 24.0 cm?

10. A person has far points 84.4 cm from the right eye and 122 cm from the left eye. Write a prescription for the powers of the corrective lenses.

11. The accommodation limits for Nearsighted Nick's eyes are 18.0 cm and 80.0 cm. When he wears his glasses, he is able to see faraway objects clearly. At what minimum distance is he able to see objects clearly?

12. The near point of an eye is 100 cm. A corrective lens is to be used to allow this eye to clearly focus on objects 25.0 in front of it. (a) What should be the focal length of the lens? (b) What is the power of the needed corrective lens?

13. An individual is nearsighted; his near point is 13.0 cm and his far point is 50.0 cm. (a) What lens power is needed to correct his nearsightedness? (b) When the lenses are in use, what is this person's near point?

14. A certain child's near point is 10.0 cm; her far point (with eyes relaxed) is 125 cm. Each eye lens is 2.00 cm from the retina. (a) Between what limits, measured in diopters, does the power of this lens–cornea combination vary? (b) Calculate the power of the eyeglass lens the child should use for relaxed distance vision. Is the lens converging or diverging?

15. An artificial lens is implanted in a person's eye to replace a diseased lens. The distance between the artificial lens and the retina is 2.80 cm. In the absence of the lens, an image of a distant object (formed by refraction at the cornea) falls 2.53 cm behind the retina. The lens is designed to put the image of the distant object on the retina. What is the power of the implanted lens? [*Hint:* Consider the image formed by the cornea to be a virtual object.]

16. A person is to be fitted with bifocals. She can see clearly when the object is between 30 cm and 1.5 m from the eye. (a) The upper portions of the bifocals (Fig. P25.16) should be designed to enable her to see distant objects clearly. What power should they have? (b) The lower portions of the bifocals should enable her to see objects comfortably at 25 cm. What power should they have?

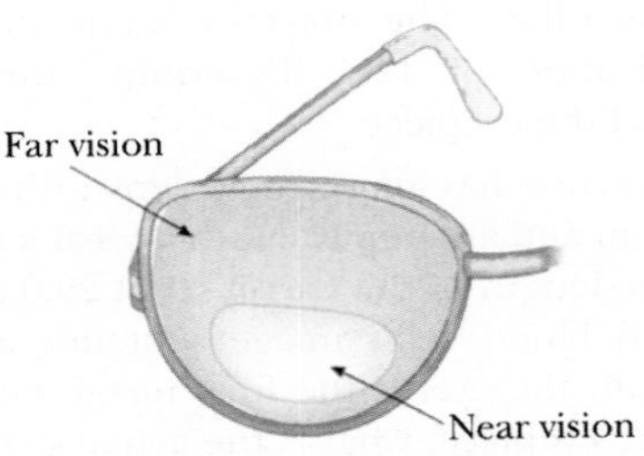

Figure P25.16

Section 25.3 The Simple Magnifier

17. A stamp collector uses a lens with 7.5-cm focal length as a simple magnifier. The virtual image is produced at the normal near point (25 cm). (a) How far from the lens should the stamp be placed? (b) What is the expected angular magnification?

18. A lens having a focal length of 25 cm is used as a simple magnifier. (a) What is the angular magnification obtained when the image is formed at the normal near point ($q = -25$ cm)? (b) What is the angular magnification produced by this lens when the eye is relaxed?

19. A biology student uses a simple magnifier to examine the structural features of the wing of an insect. The wing is held 3.50 cm in front of the lens, and the image is formed 25.0 cm from the eye. (a) What is the focal length of the lens? (b) What angular magnification is achieved?

20. A lens that has a focal length of 5.00 cm is used as a magnifying glass. (a) To obtain maximum magnification, where should the object be placed? (b) What is the magnification?

21. A leaf of length h is positioned 71.0 cm in front of a converging lens with a focal length of 39.0 cm. An observer views the image of the leaf from a position 1.26 m behind the lens, as shown in Figure P25.21. (a) What is the magnitude of the lateral magnification (the ratio of the image size to the object size) produced by the lens? (b) What angular magnification is achieved by viewing the image of the leaf rather than viewing the leaf directly?

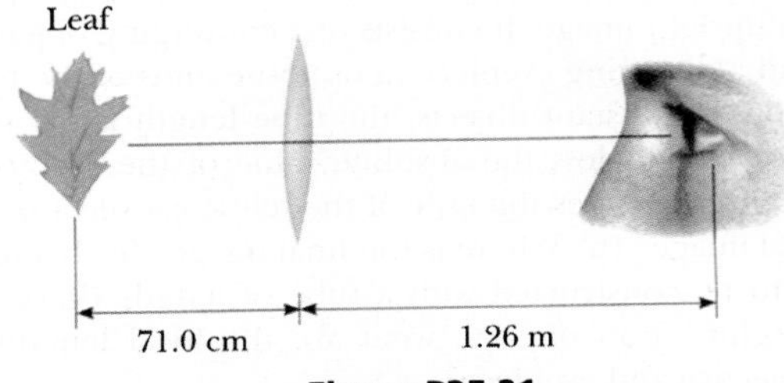

Figure P25.21

Section 25.4 The Compound Microscope & Section 25.5 The Telescope

22. The objective lens in a microscope with a 20.0-cm-long tube has a magnification of 50.0, and the eyepiece has a magnification of 20.0. What are the focal lengths of (a) the objective and (b) the eyepiece? (c) What is the overall magnification of the microscope?

23. The desired overall magnification of a compound microscope is 140×. The objective alone produces a lateral magnification of 12×. Determine the required focal length of the eyepiece.

24. A microscope has an objective lens with a focal length of 16.22 mm and an eyepiece with a focal length of 9.50 mm. With the length of the barrel set at 29.0 cm, the diameter of a red blood cell's image subtends an angle of 1.43 mrad with the eye. If the final image distance is 29.0 cm from the eyepiece, what is the actual diameter of the red blood cell?

25. **Physics Now™** The length of a microscope tube is 15.0 cm. The focal length of the objective is 1.00 cm, and the focal length of the eyepiece is 2.50 cm. What is the magnification of the microscope, assuming it is adjusted so that the eye is relaxed? [*Hint:* To solve this question go back to basics and use the thin lens equation.]

26. A certain telescope has an objective of focal length 1 500 cm. If the Moon is used as an object, a 1.0-cm-long image formed by the objective corresponds to what distance, in miles, on the Moon? Assume 3.8×10^8 m for the Earth–Moon distance.

27. The lenses of an astronomical telescope are 92 cm apart when adjusted for viewing a distant object with minimum eyestrain. The angular magnification produced by the telescope is 45. Compute the focal length of each lens.

28. An elderly sailor is shipwrecked on a desert island, but manages to save his eyeglasses. The lens for one eye has a power of +1.20 diopters, and the other lens has a power of +9.00 diopters. (a) What is the magnifying power of the telescope he can construct with these lenses? (b) How far apart are the lenses when the telescope is adjusted for minimum eyestrain?

29. Astronomers often take photographs with the objective lens or mirror of a telescope alone, without an eyepiece. (a) Show that the image size h' for a telescope used in this manner is given by $h' = fh/(f - p)$, where h is the object size, f is the objective focal length, and p is the object distance. (b) Simplify the expression in part (a) if the object distance is much greater than the objective focal length. (c) The "wingspan" of the International Space Station is 108.6 m, the overall width of its solar panel configuration. When it is orbiting at an altitude of 407 km, find the width of the image formed by a telescope objective of focal length 4.00 m.

30. Galileo devised a simple terrestrial telescope that produces an upright image. It consists of a converging objective lens and a diverging eyepiece at opposite ends of the telescope tube. For distant objects, the tube length is the objective focal length less the absolute value of the eyepiece focal length. (a) Does the user of the telescope see a real or virtual image? (b) Where is the final image? (c) If a telescope is to be constructed with a tube of length 10.0 cm and a magnification of 3.00, what are the focal lengths of the objective and eyepiece?

31. A person decides to use an old pair of eyeglasses to make some optical instruments. He knows that the near point in his left eye is 50.0 cm and the near point in his right eye is 100 cm. (a) What is the maximum angular magnification he can produce in a telescope? (b) If he places the lenses 10.0 cm apart, what is the maximum overall magnification he can produce in a microscope? (Go back to basics and use the thin-lens equation to solve part (b).)

Section 25.6 Resolution of Single-Slit and Circular Apertures

32. If the distance from the Earth to the Moon is 3.8×10^8 m, what diameter would be required for a telescope objective to resolve a Moon crater 300 m in diameter? Assume a wavelength of 500 nm.

33. A converging lens with a diameter of 30.0 cm forms an image of a satellite passing overhead. The satellite has two green lights (wavelength 500 nm) spaced 1.00 m apart. If the lights can just be resolved according to the Rayleigh criterion, what is the altitude of the satellite?

34. The pupil of a cat's eye narrows to a vertical slit of width 0.500 mm in daylight. What is the angular resolution for a pair of horizontally separated mice? (Use 500-nm light in your calculation.)

35. To increase the resolving power of a microscope, the object and the objective are immersed in oil ($n = 1.5$). If the limiting angle of resolution without the oil is 0.60 μrad, what is the limiting angle of resolution with the oil? [*Hint:* The oil changes the wavelength of the light.]

36. (a) Calculate the limiting angle of resolution for the eye, assuming a pupil diameter of 2.00 mm, a wavelength of 500 nm *in air*, and an index of refraction for the eye of 1.33. (b) What is the maximum distance from the eye at which two points separated by 1.00 cm could be resolved?

37. Two stars in a binary system are 8.0 lightyears away from the observer and can just be resolved by a 20-in. telescope equipped with a filter that allows only light of wavelength 500 nm to pass. What is the distance between the two stars?

38. A spy satellite circles the Earth at an altitude of 200 km and carries out surveillance with a special high-resolution telescopic camera having a lens diameter of 35 cm. If the angular resolution of this camera is limited by diffraction, estimate the separation of two small objects on the Earth's surface that are just resolved in yellow-green light ($\lambda =$ 550 nm).

39. Suppose a 5.00-m-diameter telescope were constructed on the Moon, where the absence of atmospheric distortion would permit excellent viewing. If observations were made using 500-nm light, what minimum separation between two objects could just be resolved on Mars at closest approach (when Mars is 8.0×10^7 km from the Moon)?

40. The H_α line in hydrogen has a wavelength of 656.20 nm. This line differs in wavelength from the corresponding spectral line in deuterium (the heavy stable isotope of hydrogen) by 0.18 nm. (a) Determine the minimum number of lines a grating must have to resolve these two wavelengths in the first order. (b) Repeat part (a) for the second order.

41. A 15.0-cm-long grating has 6 000 slits per centimeter. Can two lines of wavelengths 600.000 nm and 600.003 nm be separated with this grating? Explain.

Section 25.7 The Michelson Interferometer

42. Light of wavelength 550 nm is used to calibrate a Michelson interferometer. With the use of a micrometer screw, the platform on which one mirror is mounted is moved 0.180 mm. How many fringe shifts are counted?

43. An interferometer is used to measure the length of a bacterium. The wavelength of the light used is 650 nm. As one arm of the interferometer is moved from one end of the cell to the other, 310 fringe shifts are counted. How long is the bacterium?

44. Mirror M_1 in Figure 25.15 is displaced a distance ΔL. During this displacement, 250 fringe shifts are counted. The light being used has a wavelength of 632.8 nm. Calculate the displacement ΔL.

45. A thin sheet of transparent material has an index of refraction of 1.40 and is 15.0 μm thick. When it is inserted in the light path along one arm of an interferometer, how many fringe shifts occur in the pattern? Assume that the wavelength (in a vacuum) of the light used is 600 nm. [*Hint:* The wavelength will change within the material.]

46. The Michelson interferometer can be used to measure the index of refraction of a gas by placing an evacuated transparent tube in the light path along one arm of the device. Fringe shifts occur as the gas is slowly added to the tube. Assume that 600-nm light is used, that the tube is 5.00 cm long, and that 160 fringe shifts occur as the pressure of the gas in the tube increases to atmospheric pressure. What is the index of refraction of the gas? [*Hint:* The fringe shifts occur because the wavelength of the light changes inside the gas-filled tube.]

47. **PhysicsNow™** The light path in one arm of a Michelson interferometer includes a transparent cell that is 5.00 cm long. How many fringe shifts would be observed if all the air were evacuated from the cell? The wavelength of the light source is 590 nm and the refractive index of air is 1.000 29. (See the hint in Problem 46.)

ADDITIONAL PROBLEMS

48. A person with a nearsighted eye has near and far points of 16 cm and 25 cm, respectively. (a) Assuming a lens is placed 2.0 cm from the eye, what power must the lens have to correct this condition? (b) Suppose that contact lenses placed directly on the cornea are used to correct the person's eye. What is the power of the lens required in this case, and what is the new near point? [*Hint*: The contact lens and the eyeglass lens require slightly different powers because they are at different distances from the eye.]

49. The near point of an eye is 75.0 cm. (a) What should be the power of a corrective lens prescribed to enable the eye to see an object clearly at 25.0 cm? (b) If, using the corrective lens, the person can see an object clearly at 26.0 cm, but not at 25.0 cm, by how many diopters did the lens grinder miss the prescription?

50. If a typical eyeball is 2.00 cm long and has a pupil opening that can range from about 2.00 mm to 6.00 mm, what are (a) the focal length of the eye when it is focused on objects 1.00 m away, (b) the smallest f-number of the eye when it is focused on objects 1.00 m away, and (c) the largest f-number of the eye when it is focused on objects 1.00 m away?

51. **PhysicsNow™** A cataract-impaired lens in an eye may be surgically removed and replaced by a manufactured lens. The focal length required for the new lens is determined by the lens-to-retina distance, which is measured by a sonarlike device, and by the requirement that the implant provide for correct distance vision. (a) If the distance from lens to retina is 22.4 mm, calculate the power of the implanted lens in diopters. (b) Since there is no accommodation and the implant allows for correct distance vision, a corrective lens for close work or reading must be used. Assume a reading distance of 33.0 cm, and calculate the power of the lens in the reading glasses.

52. Estimate the minimum angle subtended at the eye of a hawk flying at an altitude of 50 m necessary to recognize a mouse on the ground.

53. The wavelengths of the sodium spectrum are $\lambda_1 =$ 589.00 nm and $\lambda_2 =$ 589.59 nm. Determine the minimum number of lines in a grating that will allow resolution of the sodium spectrum in (a) the first order and (b) the third order.

54. The text discusses the astronomical telescope. Another type is the Galilean telescope, in which an objective lens gathers light (Fig. P25.54) and tends to form an image at point *A*. An eyepiece consisting of a diverging lens intercepts the light before it comes to a focus and forms a virtual image at point *B*. When adjusted for minimum eyestrain, *B* is an infinite distance in front of the lens and parallel rays emerge from the lens, as in Figure P25.54b. An opera glass, which is a Galilean telescope, is used to view a 30.0-cm-tall singer's head that is 40.0 m from the objective lens. The focal length of the objective is +8.00 cm, and that of the eyepiece is −2.00 cm. The telescope is adjusted so parallel rays enter the eye. Compute (a) the size of the real image that would have been formed by the objective, (b) the virtual object distance for the diverging lens, (c) the distance between the lenses, and (d) the overall angular magnification.

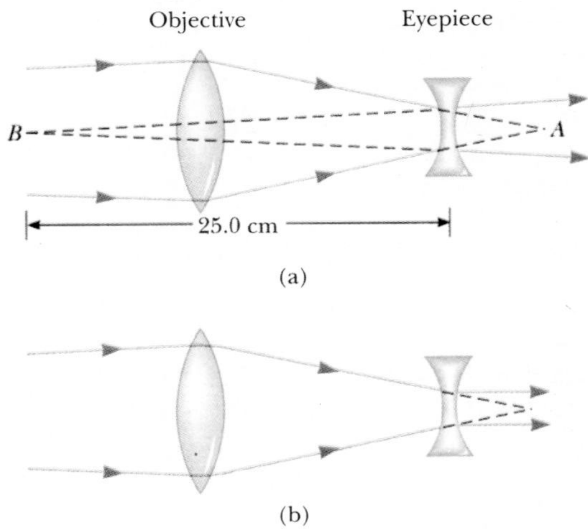

Figure P25.54

55. A laboratory (astronomical) telescope is used to view a scale that is 300 cm from the objective, which has a focal length of 20.0 cm; the eyepiece has a focal length of 2.00 cm. Calculate the angular magnification when the telescope is adjusted for minimum eyestrain. [*Note*: The object is not at infinity, so the simple expression $m = f_o/f_e$ is not sufficiently accurate for this problem. Also, assume small angles, so that $\tan \theta \approx \theta$.)

56. If the aqueous humor of the eye has an index of refraction of 1.34 and the distance from the vertex of the cornea to the retina is 2.00 cm, what is the radius of curvature of the cornea for which distant objects will be focused on the retina? (For simplicity, assume that all refraction occurs in the aqueous humor.)

57. A boy scout starts a fire by using a lens from his eyeglasses to focus sunlight on kindling 5.0 cm from the lens. The boy scout has a near point of 15 cm. When the lens is used as a simple magnifier, (a) what is the maximum magnification that can be achieved, and (b) what is the magnification when the eye is relaxed? [*Caution*: The equations derived in the text for a simple magnifier assume a "normal" eye.]

ACTIVITIES

1. (a) Move this book toward your face until the letters just begin to blur. The distance from the book to your eye is your near point. (b) On a sheet of paper, make a dot near the center. Then place an *x* about 3 inches to the left of the dot and another *x* about 3 inches to the right of the dot. With one eye shut and while looking at the dot, move the paper slowly toward your eye. You will notice that a certain distance from your eye, one of the **x**'s will disappear. This is the location of the blind spot of your eye—the point where the optic nerve enters the eye. (c) Stand before a mirror in a darkened room for a few minutes. Then turn on a light in the room and observe your pupils in the mirror as they change size. Such adaptation to the dark also takes place at the rods and cones as they chemically adjust their sensitivity. This adjustment takes 15–30 minutes, as you may have noted whenever you entered a darkened movie theater. Iris aperture control takes less than a second and helps protect the retina from overload.

2. On a sunny day, hold a magnifying glass above a nonflammable surface, such as a sidewalk, so the image of the Sun forms a round spot of light on the surface. Note where the spot formed by the lens is most distinct, or smallest. Use a ruler to measure the distance between the glass and the image. The distance is equal to the focal length of the lens.

3. Hold a pair of prescription glasses about 12 cm from your eye, and look at different objects through the lenses. Try this with different types of glasses, such as those for farsightedness and nearsightedness, and describe what effects the differences have on the image you see. If you have bifocals, how do the images produced by the top and bottom portions of the bifocal lens compare?

4. If you have never experimented with a 35-mm camera with adjustable *f*-numbers and shutter speeds, use up a couple of rolls of film to see what happens. Take several shots of the same object with different settings for these two variables. (You should record your *f*-numbers and shutter speeds for each photograph.) Explain any differences you see in the final images in terms of the settings used.

Courtesy of the Archives, California Institute of Technology

Albert Einstein revolutionized modern physics. He explained the random movements of pollen grains, which proved the existence of atoms, and the photoelectric effect, which showed that light was a particle as well as a wave. His theory of special relativity made clear the foundations of space and time, and his theory of gravitation— general relativity— is the most accurate theory in physics today. He was also deeply concerned with the social impact of scientific discovery.

CHAPTER

26

Relativity

OUTLINE

Most of our everyday experiences and observations have to do with objects that move at speeds much less than the speed of light. Newtonian mechanics was formulated to describe the motion of such objects, and its formalism is quite successful in describing a wide range of phenomena that occur at low speeds. It fails, however, when applied to particles having speeds approaching that of light.

This chapter introduces Einstein's theory of special relativity and includes a section on general relativity. The concepts of special relativity often violate our common sense. Moving clocks run slow, and the length of a moving meter stick is contracted. Nonetheless, the theory has been rigorously tested, correctly predicting the results of experiments involving speeds near the speed of light. The theory is verified daily in particle accelerators around the world.

26.1 INTRODUCTION

Experimentally, the predictions of Newtonian theory can be tested at high speeds by accelerating electrons or other charged particles through a large electric potential difference. For example, it's possible to accelerate an electron to a speed of $0.99c$ (where c is the speed of light) by using a potential difference of several million volts. According to Newtonian mechanics, if the potential difference is increased by a factor of 4, the electron's kinetic energy is four times greater and its speed should double to $1.98c$. However, experiments show that the speed of the electron—as well as the speed of any other particle that has mass—always remains *less* than the speed of light, regardless of the size of the accelerating voltage.

The existence of a universal speed limit has far-reaching consequences. It means that the usual concepts of force, momentum, and energy no longer apply for rapidly moving objects. Less obvious consequences include the fact that observers moving at different speeds will measure different time intervals and displacements between the same two events. Newtonian mechanics was contradicted by experimental observations, so it was necessary to replace it with another theory.

2. Only certain electron orbits are stable. These are orbits in which the hydrogen atom doesn't emit energy in the form of electromagnetic radiation. Hence, the total energy of the atom remains constant, and classical mechanics can be used to describe the electron's motion.
3. Radiation is emitted by the hydrogen atom when the electron "jumps" from a more energetic initial state to a less energetic state. The "jump" can't be visualized or treated classically. In particular, the frequency f of the radiation emitted in the jump is related to the change in the atom's energy and is *independent of the frequency of the electron's orbital motion.* The frequency of the emitted radiation is given by

$$E_i - E_f = hf \qquad [28.3]$$

where E_i is the energy of the initial state, E_f is the energy of the final state, h is Planck's constant, and $E_i > E_f$.
4. The size of the allowed electron orbits is determined by a condition imposed on the electron's orbital angular momentum: the allowed orbits are those for which the electron's orbital angular momentum about the nucleus is an integral multiple of $\hbar$ (pronounced "h bar"), where $\hbar = h/2\pi$:

$$m_e vr = n\hbar \quad n = 1, 2, 3, \ldots \qquad [28.4]$$

NIELS BOHR, Danish Physicist (1885–1962)

Bohr was an active participant in the early development of quantum mechanics and provided much of its philosophical framework. During the 1920s and 1930s, he headed the Institute for Advanced Studies in Copenhagen. The institute was a magnet for many of the world's best physicists and provided a forum for the exchange of ideas. When Bohr visited the United States in 1939 to attend a scientific conference, he brought news that the fission of uranium had been observed by Hahn and Strassman in Berlin. The results were the foundations of the atomic bomb developed in the United States during World War II. Bohr was awarded the 1922 Nobel Prize for his investigation of the structure of atoms and of the radiation emanating from them.

With these four assumptions, we can calculate the allowed energies and emission wavelengths of the hydrogen atom. We use the model pictured in Figure 28.5, in which the electron travels in a circular orbit of radius r with an orbital speed v. The electrical potential energy of the atom is

$$PE = k_e\frac{q_1q_2}{r} = k_e\frac{(-e)(e)}{r} = -k_e\frac{e^2}{r}$$

where k_e is the Coulomb constant. Assuming the nucleus is at rest, the total energy E of the atom is the sum of the kinetic and potential energy:

$$E = KE + PE = \tfrac{1}{2}m_e v^2 - k_e\frac{e^2}{r} \qquad [28.5]$$

We apply Newton's second law to the electron. We know that the electric force of attraction on the electron, $k_e e^2/r^2$, must equal $m_e a_r$, where $a_r = v^2/r$ is the centripetal acceleration of the electron. Thus,

$$k_e\frac{e^2}{r^2} = m_e\frac{v^2}{r} \qquad [28.6]$$

From this equation, we see that the kinetic energy of the electron is

$$\tfrac{1}{2}m_e v^2 = \frac{k_e e^2}{2r} \qquad [28.7]$$

We can combine this result with Equation 28.5 and express the energy of the atom as

$$E = -\frac{k_e e^2}{2r} \qquad [28.8]$$

◀ Energy of the hydrogen atom

where the negative value of the energy indicates that the electron is bound to the proton.

An expression for r is obtained by solving Equations 28.4 and 28.6 for v and equating the results:

$$v^2 = \frac{n^2\hbar}{m_e^{\,2}r^2} = \frac{k_e e^2}{m_e r}$$

$$r_n = \frac{n^2\hbar^2}{m_e k_e e^2} \quad n = 1, 2, 3, \ldots \qquad [28.9]$$

◀ The radii of the Bohr orbits are quantized

This equation is based on the assumption that the **electron can exist only in certain allowed orbits determined by the integer *n*.**

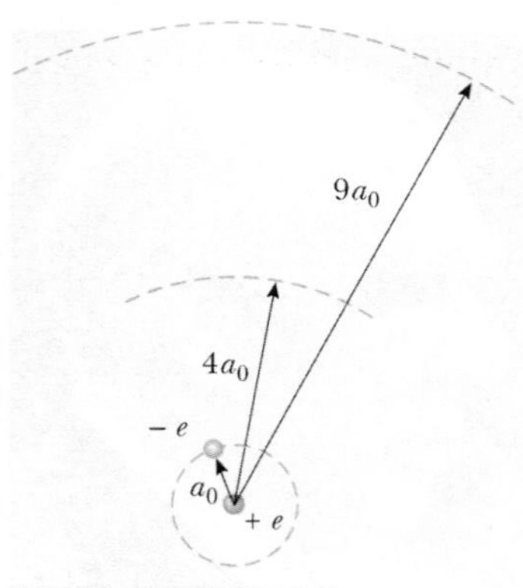

ACTIVE FIGURE 28.6
The first three circular orbits predicted by the Bohr model of the hydrogen atom.

Physics Now™
Log into PhysicsNow at **www.cp7e.com** and go to Active Figure 28.6, where you can choose the initial and final states of the hydrogen atom and observe the transition.

The orbit with the smallest radius, called the **Bohr radius**, a_0, corresponds to $n = 1$ and has the value

$$a_0 = \frac{\hbar^2}{mk_e e^2} = 0.052\,9 \text{ nm} \qquad [28.10]$$

A general expression for the radius of any orbit in the hydrogen atom is obtained by substituting Equation 28.10 into Equation 28.9:

$$r_n = n^2 a_0 = n^2(0.052\,9 \text{ nm}) \qquad [28.11]$$

The first three Bohr orbits for hydrogen are shown in Active Figure 28.6.

Equation 28.9 may be substituted into Equation 28.8 to give the following expression for the energies of the quantum states:

$$E_n = -\frac{m_e k_e^2 e^4}{2\hbar^2}\left(\frac{1}{n^2}\right) \quad n = 1, 2, 3, \ldots \qquad [28.12]$$

If we insert numerical values into Equation 28.12, we obtain

$$E_n = -\frac{13.6}{n^2} \text{ eV} \qquad [28.13]$$

The lowest energy state, or **ground state**, corresponds to $n = 1$ and has an energy $E_1 = -m_e k_e^2 e^4/2\hbar^2 = -13.6$ eV. The next state, corresponding to $n = 2$, has an energy $E_2 = E_1/4 = -3.40$ eV, and so on. An energy level diagram showing the energies of these stationary states and the corresponding quantum numbers is given in Active Figure 28.7. The uppermost level shown, corresponding to $E = 0$ and $n \rightarrow \infty$, represents the state for which the electron is completely removed from the atom. In this state, the electron's *KE* and *PE* are both zero, which means that the electron is at rest infinitely far away from the proton. The minimum energy required to ionize the atom—that is, to completely remove the electron—is called the **ionization energy**. The ionization energy for hydrogen is 13.6 eV.

Equations 28.3 and 28.12 and the third Bohr postulate show that if the electron jumps from one orbit with quantum number n_i to a second orbit with quantum number, n_f, it emits a photon of frequency f given by

$$f = \frac{E_i - E_f}{h} = \frac{m_e k_e^2 e^4}{4\pi\hbar^3}\left(\frac{1}{n_f^2} - \frac{1}{n_i^2}\right) \qquad [28.14]$$

where $n_f < n_i$.

Finally, to compare this result with the empirical formulas for the various spectral series, we use Equation 28.14 and the fact that for light, $\lambda f = c$, to get

$$\frac{1}{\lambda} = \frac{f}{c} = \frac{m_e k_e^2 e^4}{4\pi c\hbar^3}\left(\frac{1}{n_f^2} - \frac{1}{n_i^2}\right) \qquad [28.15]$$

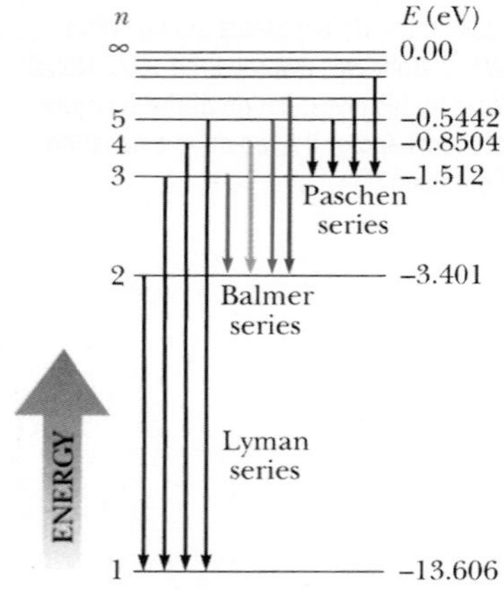

ACTIVE FIGURE 28.7
An energy level diagram for hydrogen. Quantum numbers are given on the left and energies (in electron volts) are given on the right. Vertical arrows represent the four lowest-energy transitions for each of the spectral series shown. The colored arrows for the Balmer series indicate that this series results in visible light.

Physics Now™
Log into PhysicsNow at **www.cp7e.com** and go to Active Figure 28.6, where you can choose the initial and final states of the hydrogen atom and observe the transition.

A comparison of this result with Equation 28.1 gives the following expression for the Rydberg constant:

$$R_{\text{H}} = \frac{m_e k_e^2 e^4}{4\pi c\hbar^3} \qquad [28.16]$$

If we insert the known values of m_e, k_e, e, c, and $\hbar$ into this expression, the resulting theoretical value for R_{H} is found to be in excellent agreement with the value determined experimentally for the Rydberg constant. When Bohr demonstrated this agreement, it was recognized as a major accomplishment of his theory.

In order to compare Equation 28.15 with spectroscopic data, it is convenient to express it in the form

$$\frac{1}{\lambda} = R_{\text{H}}\left(\frac{1}{n_f^2} - \frac{1}{n_i^2}\right) \qquad [28.17]$$

We can use this expression to evaluate the wavelengths for the various series in the hydrogen spectrum. For example, in the Balmer series, $n_f = 2$ and $n_i = 3, 4, 5, \ldots$ (Eq. 28.1). For the Lyman series, we take $n_f = 1$ and $n_i = 2, 3, 4, \ldots$. The energy level diagram for hydrogen shown in Active Figure 28.7 indicates the origin of the spectral lines described previously. The transitions between levels are represented by vertical arrows. Note that whenever a transition occurs between a state designated by n_i to one designated by n_f (where $n_i > n_f$), a photon with a frequency $(E_i - E_f)/h$ is emitted. This can be interpreted as follows: the lines in the visible part of the hydrogen spectrum arise when the electron jumps from the third, fourth, or even higher orbit to the second orbit. Likewise, the lines of the Lyman series (in the ultraviolet) arise when the electron jumps from the second, third, or even higher orbit to the innermost ($n_f = 1$) orbit. Hence, the Bohr theory successfully predicts the wavelengths of all the observed spectral lines of hydrogen.

TIP 28.1 Energy Depends On *n* Only for Hydrogen

According to Equation 28.13, the energy depends only on the quantum number n. Note that this is only true for the hydrogen atom. For more complicated atoms, the energy levels depend primarily on n, but also on other quantum numbers.

INTERACTIVE EXAMPLE 28.1 The Balmer Series for Hydrogen

Goal Calculate the wavelength, frequency, and energy of a photon emitted during an electron transition in an atom.

Problem The Balmer series for the hydrogen atom corresponds to electronic transitions that terminate in the state with quantum number $n = 2$, as shown in Figure 28.8. **(a)** Find the longest-wavelength photon emitted in the Balmer series and determine its frequency and energy. **(b)** Find the shortest-wavelength photon emitted in the same series.

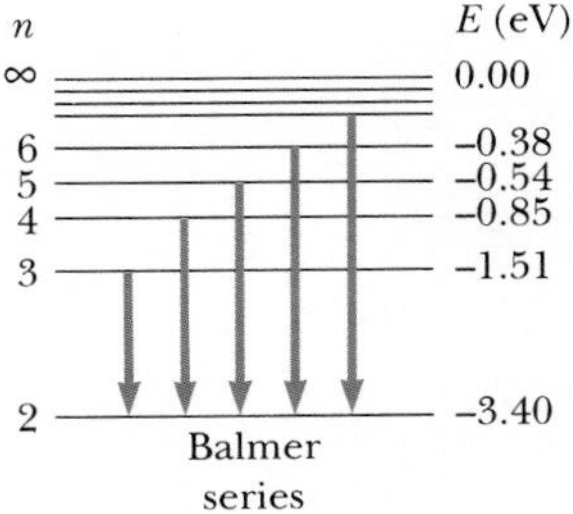

Figure 28.8 (Example 28.1) Transitions responsible for the Balmer series for the hydrogen atom. All transitions terminate at the $n = 2$ level.

Strategy This is a matter of substituting values into Equation 28.17. The frequency can then be obtained from $c = f\lambda$ and the energy from $E = hf$. The longest wavelength photon corresponds to the one that is emitted when the electron jumps from the $n_i = 3$ state to the $n_f = 2$ state. The shortest wavelength photon corresponds to the one that is emitted when the electron jumps from $n_i = \infty$ to the state $n_f = 2$.

Solution

(a) Find the longest wavelength photon emitted in the Balmer series, and determine its frequency and energy.

Substitute into Equation 28.17, with $n_i = 3$ and $n_f = 2$:

$$\frac{1}{\lambda} = R_H\left(\frac{1}{n_f^2} - \frac{1}{n_i^2}\right) = R_H\left(\frac{1}{2^2} - \frac{1}{3^2}\right) = \frac{5R_H}{36}$$

Take the reciprocal and substitute, finding the wavelength:

$$\lambda = \frac{36}{5R_H} = \frac{36}{5(1.097 \times 10^7 \text{ m}^{-1})} = 6.563 \times 10^{-7} \text{ m}$$
$$= \boxed{656.3 \text{ nm}}$$

Now use $c = f\lambda$ to obtain the frequency:

$$f = \frac{c}{\lambda} = \frac{2.998 \times 10^8 \text{ m/s}}{6.563 \times 10^{-7} \text{ m}} = \boxed{4.568 \times 10^{14} \text{ Hz}}$$

Calculate the photon's energy by substituting into Equation 27.5:

$$E = hf = (6.626 \times 10^{-34} \text{ J}\cdot\text{s})(4.568 \times 10^{14} \text{ Hz})$$
$$= 3.027 \times 10^{-19} \text{ J} = \boxed{1.892 \text{ eV}}$$

(b) Find the shortest wavelength photon emitted in the Balmer series.

Substitute into Equation 28.17, with $n_i = \infty$ and $n_f = 2$.

$$\frac{1}{\lambda} = R_H\left(\frac{1}{n_f^2} - \frac{1}{n_i^2}\right) = R_H\left(\frac{1}{2^2} - \frac{1}{\infty}\right) = \frac{R_H}{4}$$

Take the reciprocal and substitute, finding the wavelength:

$$\lambda = \frac{4}{R_H} = \frac{4}{(1.097 \times 10^7 \text{ m}^{-1})} = 3.646 \times 10^{-7} \text{ m}$$
$$= \boxed{364.6 \text{ nm}}$$

Remarks The first wavelength is in the red region of the visible spectrum. We could also obtain the energy of the photon by using Equation 28.3 in the form $hf = E_3 - E_2$, where E_2 and E_3 are the energy levels of the hydrogen atom, calculated from Equation 28.13. Note that this is the lowest energy photon in the Balmer series, because it involves the smallest energy change. The second photon, the most energetic, is in the ultraviolet region.

Exercise 28.1

(a) Calculate the energy of the shortest wavelength photon emitted in the Balmer series for hydrogen. (b) Calculate the wavelength of a transition from $n = 4$ to $n = 2$.

Answers (a) 3.40 eV (b) 486 nm

PhysicsNow™ Investigate transitions between various states by logging into PhysicsNow at **www.cp7e.com** and going to Interactive Example 28.1.

Bohr's Correspondence Principle

In our study of relativity in Chapter 26, we found that Newtonian mechanics cannot be used to describe phenomena that occur at speeds approaching the speed of light. Newtonian mechanics is a special case of relativistic mechanics and applies only when v is much smaller than c. Similarly, **quantum mechanics is in agreement with classical physics when the energy differences between quantized levels are very small**. This principle, first set forth by Bohr, is called the **correspondence principle**.

For example, consider the hydrogen atom with $n > 10\ 000$. For such large values of n, the energy differences between adjacent levels approach zero and the levels are nearly continuous, as Equation 28.13 shows. As a consequence, the classical model is reasonably accurate in describing the system for large values of n. According to the classical model, the frequency of the light emitted by the atom is equal to the frequency of revolution of the electron in its orbit about the nucleus. Calculations show that for $n > 10\ 000$, this frequency is different from that predicted by quantum mechanics by less than 0.015%.

28.4 MODIFICATION OF THE BOHR THEORY

The Bohr theory of the hydrogen atom was a tremendous success in certain areas because it explained several features of the hydrogen spectrum that had previously defied explanation. It accounted for the Balmer series and other series; it predicted a value for the Rydberg constant that is in excellent agreement with the experimental value; it gave an expression for the radius of the atom; and it predicted the energy levels of hydrogen. Although these successes were important to scientists, it is perhaps even more significant that the Bohr theory gave us a model of what the atom looks like and how it behaves. Once a basic model is constructed, refinements and modifications can be made to enlarge on the concept and to explain finer details.

The analysis used in the Bohr theory is also successful when applied to *hydrogen-like* atoms. An atom is said to be hydrogen-like when it contains only one electron. Examples are singly ionized helium, doubly ionized lithium, triply ionized beryllium, and so forth. The results of the Bohr theory for hydrogen can be extended to hydrogen-like atoms by substituting Ze^2 for e^2 in the hydrogen equations, where Z is the atomic number of the element. For example, Equations 28.12 and 28.15 become

$$E_n = -\frac{m_e k_e^2 Z^2 e^4}{2\hbar^2}\left(\frac{1}{n^2}\right) \qquad n = 1, 2, 3, \ldots \qquad [28.18]$$

and

$$\frac{1}{\lambda} = \frac{m_e k_e^2 Z^2 e^4}{4\pi c\hbar^3}\left(\frac{1}{n_f^2} - \frac{1}{n_i^2}\right) \qquad [28.19]$$

Although many attempts were made to extend the Bohr theory to more complex, multi-electron atoms, the results were unsuccessful. Even today, only approximate methods are available for treating multi-electron atoms.

Quick Quiz 28.1

Consider a hydrogen atom and a singly-ionized helium atom. Which atom has the lower ground state energy? (a) hydrogen (b) helium (c) the ground state energy is the same for both

Quick Quiz 28.2

Consider once again a singly-ionized helium atom. Suppose the remaining electron jumps from a higher to a lower energy level, resulting in the emission of photon, which we'll call photon-He. An electron in a hydrogen atom then jumps between the same two levels, resulting in an emitted photon-H. Which photon has the shorter wavelength? (a) photon-He (b) photon-H (c) The wavelengths are the same.

EXAMPLE 28.2 Singly Ionized Helium

Goal Apply the modified Bohr theory to a hydrogen-like atom.

Problem Singly ionized helium, He^+, a hydrogen-like system, has one electron in the $1s$ orbit when the atom is in its ground state. Find **(a)** the energy of the system in the ground state in electron volts, and **(b)** the radius of the ground-state orbit.

Strategy Part (a) requires substitution into the modified Bohr model, Equation 28.18. In part (b), modify Equation 28.9 for the radius of the Bohr orbits by replacing e^2 by Ze^2, where Z is the number of protons in the nucleus.

Solution

(a) Find the energy of the system in the ground state.

Write Equation 28.18 for the energies of a hydrogen-like system:

$$E_n = -\frac{m_e k_e^2 Z^2 e^4}{2\hbar^2}\left(\frac{1}{n^2}\right)$$

Substitute the constants and convert to electron volts:

$$E_n = -\frac{Z^2(13.6)}{n^2}\text{ eV}$$

Substitute $Z = 2$ (the atomic number of helium) and $n = 1$ to obtain the ground state energy:

$$E_1 = -4(13.6)\text{ eV} = -54.4\text{ eV}$$

(b) Find the radius of the ground state.

Generalize Equation 28.9 to a hydrogen-like atom by substituting Ze^2 for e^2:

$$r_n = \frac{n^2\hbar^2}{m_e k_e Z e^2} = \frac{n^2}{Z}(a_0) = \frac{n^2}{Z}(0.052\,9\text{ nm})$$

For our case, $n = 1$ and $Z = 2$:

$$r_1 = 0.026\,5\text{ nm}$$

Remarks Notice that for higher Z the energy of a hydrogen-like atom is lower, which means that the electron is more tightly bound than in hydrogen. This results in a smaller atom, as seen in part (b).

Exercise 28.2

Repeat the problem for the first excited state of doubly-ionized lithium ($Z = 3$, $n = 2$).

Answers (a) $E_2 = -30.6$ eV (b) $r_2 = 0.070\,5$ nm

TABLE 28.1

Shell and Subshell Notation

n	Shell Symbol	ℓ	Subshell Symbol
1	K	0	s
2	L	1	p
3	M	2	d
4	N	3	f
5	O	4	g
6	P	5	h
. . .		. . .	

Within a few months following the publication of Bohr's paper, Arnold Sommerfeld (1868–1951) extended the Bohr model to include elliptical orbits. We examine his model briefly because much of the nomenclature used in this treatment is still in use today. Bohr's concept of quantization of angular momentum led to the **principal quantum number** n, which determines the energy of the allowed states of hydrogen. Sommerfeld's theory retained n, but also introduced a new quantum number ℓ called the **orbital quantum number**, where the value of ℓ ranges from 0 to $n - 1$ in integer steps. According to this model, an electron in any one of the allowed energy states of a hydrogen atom may move in any one of a number of orbits corresponding to different ℓ values. For each value of n, there are n possible orbits corresponding to different ℓ values. Because $n = 1$ and $\ell = 0$ for the first energy level (ground state), there is only one possible orbit for this state. The second energy level, with $n = 2$, has two possible orbits, corresponding to $\ell = 0$ and $\ell = 1$. The third energy level, with $n = 3$, has three possible orbits, corresponding to $\ell = 0$, $\ell = 1$, and $\ell = 2$.

For historical reasons, **all states with the same principal quantum number n are said to form a shell**. Shells are identified by the letters K, L, M, . . . , which designate the states for which $n = 1, 2, 3, \ldots$. Likewise, **the states with given values of n and ℓ are said to form a subshell**. The letters *s, p, d, f, g*, . . . are used to designate the states for which $\ell = 0, 1, 2, 3, 4, \ldots$. These notations are summarized in Table 28.1.

States that violate the restriction $0 \leq \ell \leq n - 1$, for a given value of n, can't exist. A $2d$ state, for instance, would have $n = 2$ and $\ell = 2$, but can't exist because the highest allowed value of ℓ is $n - 1$, or 1 in this case. For $n = 2$, $2s$ and $2p$ are allowed subshells, but $2d$, $2f$, . . . are not. For $n = 3$, the allowed states are $3s$, $3p$, and $3d$.

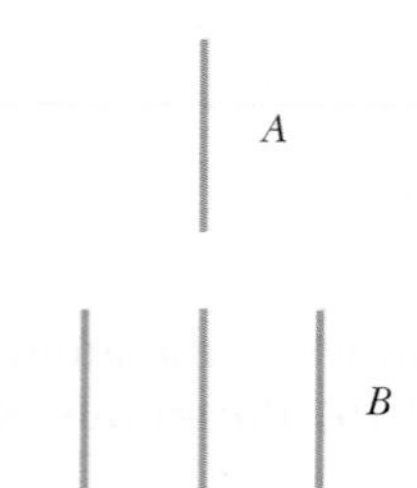

Figure 28.9 A single line (A) can split into three separate lines (B) in a magnetic field.

Another modification of the Bohr theory arose when it was discovered that the spectral lines of a gas are split into several closely spaced lines when the gas is placed in a strong magnetic field. (This is called the *Zeeman effect*, after its discoverer.) Figure 28.9 shows a single spectral line being split into three closely spaced lines. This indicates that the energy of an electron is slightly modified when the atom is immersed in a magnetic field. In order to explain this observation, a new quantum number, m_ℓ, called the **orbital magnetic quantum number**, was introduced. The theory is in accord with experimental results when m_ℓ is restricted to values ranging from $-\ell$ to $+\ell$ in integer steps. For a given value of ℓ, there are $2\ell + 1$ possible values of m_ℓ.

Finally, very high resolution spectrometers revealed that spectral lines of gases are in fact two very closely spaced lines even in the absence of an external magnetic field. This splitting was referred to as **fine structure**. In 1925 Samuel Goudsmit and George Uhlenbeck introduced the idea of an electron spinning about its own axis to explain the origin of fine structure. The results of their work introduced yet another quantum number, m_s, called the **spin magnetic quantum number**.

For each electron there are two spin states. A subshell corresponding to a given factor of ℓ can contain no more than $2(2\ell + 1)$ electrons. This number comes from the fact that electrons in a subshell must have unique pairs of the quantum numbers (m_ℓ, m_s). There are $2\ell + 1$ different magnetic quantum numbers m_ℓ, and two different spin quantum numbers m_s, making $2(2\ell + 1)$ unique pairs (m_ℓ, m_s). For example, the p subshell ($\ell = 1$) is filled when it contains $2(2 \cdot 1 + 1) = 6$ electrons. This fact can be extended to include all four quantum numbers, as will be important to us later when we discuss the *Pauli exclusion principle.*

All these quantum numbers (addressed in more detail in upcoming sections) were postulated to account for the observed spectra of elements. Only later were comprehensive mathematical theories developed that naturally yielded the same answers as these empirical models.

28.5 DE BROGLIE WAVES AND THE HYDROGEN ATOM

One of the postulates made by Bohr in his theory of the hydrogen atom was that the angular momentum of the electron is quantized in units of $\hbar$, or

$$m_e vr = n\hbar$$

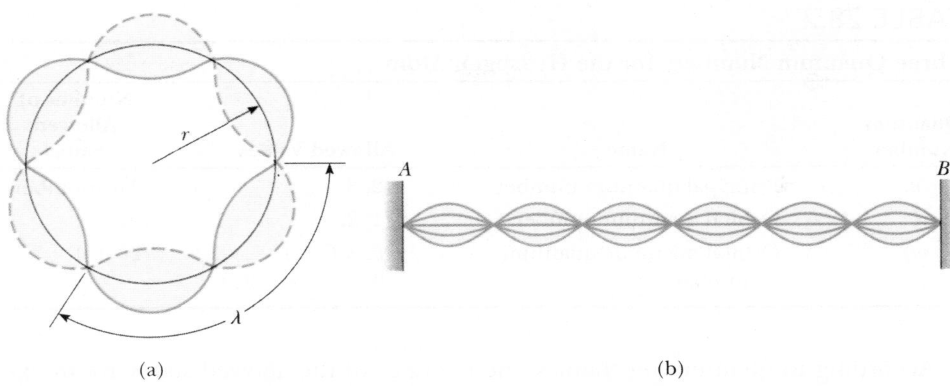

Figure 28.10 (a) Standing-wave pattern for an electron wave in a stable orbit of hydrogen. There are three full wavelengths in this orbit. (b) Standing-wave pattern for a vibrating stretched string fixed at its ends. This pattern also has three full wavelengths.

For more than a decade following Bohr's publication, no one was able to explain why the angular momentum of the electron was restricted to these discrete values. Finally, de Broglie gave a direct physical way of interpreting this condition. He assumed that an electron orbit would be stable (allowed) only if it contained an integral number of electron wavelengths. Figure 28.10a demonstrates this point when three complete wavelengths are contained in one circumference of the orbit. Similar patterns can be drawn for orbits containing one wavelength, two wavelengths, four wavelengths, five wavelengths, and so forth. These waves are analogous to standing waves on a string, discussed in Chapter 14. There, we found that strings have preferred (resonant) frequencies of vibration. Figure 28.10b shows a standing-wave pattern containing three wavelengths for a string fixed at each end. Now imagine that the vibrating string is removed from its supports at A and B and bent into a circular shape that brings those points together. The end result is a pattern such as the one shown in Figure 28.10a.

In general, the condition for a de Broglie standing wave in an electron orbit is that the circumference must contain an integral number of electron wavelengths. We can express this condition as

$$2\pi r = n\lambda \qquad n = 1, 2, 3, \ldots$$

Because the de Broglie wavelength of an electron is $\lambda = h/m_e v$, we can write the preceding equation as $2\pi r = nh/m_e v$, or

$$m_e v r = n\hbar$$

This is the same as the quantization of angular momentum condition imposed by Bohr in his original theory of hydrogen.

The electron orbit shown in Figure 28.10a contains three complete wavelengths and corresponds to the case in which the principal quantum number $n = 3$. The orbit with one complete wavelength in its circumference corresponds to the first Bohr orbit, $n = 1$; the orbit with two complete wavelengths corresponds to the second Bohr orbit, $n = 2$; and so forth.

By applying the wave theory of matter to electrons in atoms, de Broglie was able to explain the appearance of integers in the Bohr theory as a natural consequence of standing-wave patterns. This was the first convincing argument that the wave nature of matter was at the heart of the behavior of atomic systems. Although the analysis provided by de Broglie was a promising first step, gigantic strides were made subsequently with the development of Schrödinger's wave equation and its application to atomic systems.

28.6 QUANTUM MECHANICS AND THE HYDROGEN ATOM

One of the first great achievements of quantum mechanics was the solution of the wave equation for the hydrogen atom. The details of the solution are far beyond the level of this course, but we'll describe its properties and implications for atomic structure.

TABLE 28.2

Three Quantum Numbers for the Hydrogen Atom

Quantum Number	Name	Allowed Values	Number of Allowed States
n	Principal quantum number	$1, 2, 3, \ldots$	Any number
ℓ	Orbital quantum number	$0, 1, 2, \ldots, n-1$	n
m_ℓ	Orbital magnetic quantum number	$-\ell, -\ell+1, \ldots, 0, \ldots, \ell-1, \ell$	$2\ell+1$

According to quantum mechanics, the energies of the allowed states are in exact agreement with the values obtained by the Bohr theory (Eq. 28.12) when the allowed energies depend only on the principal quantum number n.

In addition to the principal quantum number, two other quantum numbers emerged from the solution of the wave equation: ℓ and m_ℓ. The quantum number ℓ is called the **orbital quantum number**, and m_ℓ is called the **orbital magnetic quantum number**. As pointed out in Section 28.4, these quantum numbers had already appeared in empirical modifications made to the Bohr theory. The significance of quantum mechanics is that those numbers and the restrictions placed on their values arose directly from mathematics and not from any ad hoc assumptions to make the theory consistent with experimental observation. Because we will need to make use of the various quantum numbers in the sections that follow, the allowed ranges of their values are repeated:

The value of n can range from 1 to ∞ in integer steps.
The value of ℓ can range from 0 to $n - 1$ in integer steps.
The value of m_ℓ can range from $-\ell$ to ℓ in integer steps.

From these rules, it can be seen that for a given value of n, there are n possible values of ℓ, while for a given value of ℓ there are $2\ell + 1$ possible values of m_ℓ. For example, if $n = 1$, there is only 1 value of ℓ, $\ell = 0$. Because $2\ell + 1 = 2 \cdot 0 + 1 = 1$, there is only one value of m_ℓ, which is $m_\ell = 0$. If $n = 2$, the value of ℓ may be 0 or 1; if $\ell = 0$, then $m_\ell = 0$, but if $\ell = 1$, then m_ℓ may be 1, 0, or -1. Table 28.2 summarizes the rules for determining the allowed values of ℓ and m_ℓ for a given value of n.

States that violate the rules given in Table 28.2 cannot exist. For instance, one state that cannot exist is the $2d$ state, which would have $n = 2$ and $\ell = 2$. This state is not allowed because the highest allowed value of ℓ is $n - 1$, or 1 in this case. Thus, for $n = 2$, $2s$ and $2p$ are allowed states, but $2d$, $2f$, . . . are not. For $n = 3$, the allowed states are $3s$, $3p$, and $3d$.

In general, for a given value of n there are n^2 states with distinct pairs of values of ℓ and m_ℓ.

Quick Quiz 28.3

When the principal quantum number is $n = 5$, how many different values of (a) ℓ and (b) m_ℓ are possible? (c) How many states have distinct pairs of values of ℓ and m_ℓ?

EXAMPLE 28.3 The $n = 2$ Level of Hydrogen

Goal Count states and determine energy based on atomic energy level.

Problem (a) Determine the number of states with a unique set of values for ℓ and m_ℓ in the hydrogen atom for $n = 2$. (b) Calculate the energies of these states.

Strategy This is a matter of counting, following the quantum rules for n, ℓ, and m_ℓ. "Unique" means that no other quantum state has the same pair of numbers for ℓ and m_ℓ. The energies are all the same because all states have the same principal quantum number, $n = 2$.

Solution

(a) Determine the number of states with a unique set of values for ℓ and m_ℓ in the hydrogen atom for $n = 2$.

Determine the different possible values of ℓ for $n = 2$: $0 \le \ell \le n - 1$, so, for $n = 2$, $0 \le \ell \le 1$ and $\ell = 0$ or 1

Find the different possible values of m_ℓ for $\ell = 0$: $-\ell \le m_\ell \le \ell$, so $-0 \le m_\ell \le 0$ implies $m_\ell = 0$

List the distinct pairs of (ℓ, m_ℓ) for $\ell = 0$: There is only one: $(\ell, m_\ell) = (0, 0)$.

Find the different possible values of m_ℓ for $\ell = 1$: $-\ell \le m_\ell \le \ell$, so $-1 \le m_\ell \le 1$ implies $m_\ell = -1, 0,$ or 1

List the distinct pairs of (ℓ, m_ℓ) for $\ell = 1$: There are three: $(\ell, m_\ell) = (1, -1), (1, 0),$ and $(1, 1)$.

Sum the results for $\ell = 0$ and $\ell = 1$: Number of states $= 1 + 3 = \boxed{4}$

(b) Calculate the energies of these states.

The common energy of all of the states can be found with Equation 28.13:

$$E_n = -\frac{13.6 \text{ eV}}{n^2} \rightarrow E_2 = -\frac{13.6 \text{ eV}}{2^2} = \boxed{-3.40 \text{ eV}}$$

Remarks While these states normally have the same energy, applying a magnetic field will result in their taking slightly different energies centered around the energy corresponding to $n = 2$. As seen in the next section, there are in fact twice as many states, corresponding to a new quantum number called *spin*.

Exercise 28.3

(a) Determine the number of states with a unique pair of values for ℓ and m_ℓ in the $n = 3$ level of hydrogen. (b) Determine the energies of those states.

Answers (a) 9 (b) $E_3 = -1.51$ eV

28.7 THE SPIN MAGNETIC QUANTUM NUMBER

As we'll see in this section, there actually are *eight* states corresponding to $n = 2$ for hydrogen, not four as given in Example 28.3. This happens because another quantum number, m_s, the **spin magnetic quantum number**, has to be introduced to explain the splitting of each level into two.

The need for this new quantum number first came about because of an unusual feature in the spectra of certain gases, such as sodium vapor. Close examination of one of the prominent lines of sodium shows that it is, in fact, two very closely spaced lines. The wavelengths of these lines occur in the yellow region of the spectrum, at 589.0 nm and 589.6 nm. In 1925, when this doublet was first noticed, atomic theory couldn't explain it. To resolve the dilemma, Samuel Goudsmit and George Uhlenbeck, following a suggestion by the Austrian physicist Wolfgang Pauli, proposed the introduction of a fourth quantum number to describe atomic energy levels, called the *spin quantum number*.

In order to describe the spin quantum number, it's convenient (but technically incorrect) to think of the electron as spinning on its axis as it orbits the nucleus, just as the Earth spins on its axis as it orbits the Sun. Strangely, there are only two ways in which the electron can spin as it orbits the nucleus, as shown in Figure 28.11. If the direction of spin is as shown in Figure 28.11a, the electron is said to have "spin up." If the direction of spin is reversed, as in Figure 28.11b, the electron is said to have "spin down." The energy of the electron is slightly different for the two spin directions, and this energy difference accounts for the sodium doublet. The quantum numbers associated with electron spin are $m_s = \frac{1}{2}$ for the spin-up state and $m_s = -\frac{1}{2}$ for the spin-down state. As we'll see in Example 28.4, this new quantum number doubles the number of allowed states specified by the quantum numbers n, ℓ, and m_ℓ.

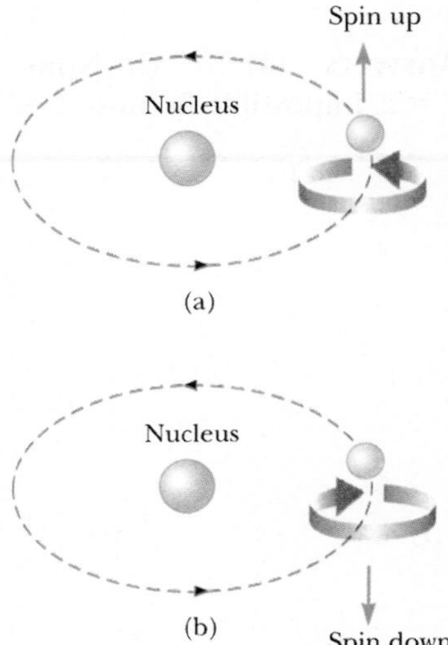

Figure 28.11 As an electron moves in its orbit about the nucleus, its spin can be either (a) up or (b) down.

TIP 28.2 The Electron Isn't Really Spinning

The electron is *not* physically spinning. Electron spin is a purely quantum effect that gives the electron an angular momentum *as if* it were physically spinning.

Any classical description of electron spin is incorrect because quantum mechanics tells us that since the electron can't be located precisely in space, it cannot be considered to be a spinning solid object, as pictured in Figure 28.11. In spite of this conceptual difficulty, all experimental evidence supports the fact that an electron does have some intrinsic property that can be described by the spin magnetic quantum number.

The spin quantum number didn't come from the original formulation of quantum mechanics by Schrodinger (and independently, by Heisenberg). The English mathematical physicist P. A. M. Dirac developed a relativistic quantum theory in which spin appears naturally.

EXAMPLE 28.4 The Quantum Numbers for the *2p* Subshell

Goal List the distinct quantum states of a subshell by their quantum numbers, including spin.

Problem List the unique sets of quantum numbers for electrons in the $2p$ subshell.

Strategy This is again a matter following the quantum rules for n, ℓ, and m_ℓ, and now m_s as well. The $2p$ subshell has $n = 2$ (that's the "2" in $2p$) and $\ell = 1$ (that's from the p in $2p$).

Solution

Because $\ell = 1$, the magnetic quantum number can have the values -1, 0, 1, and the spin quantum number is always $+\frac{1}{2}$ or $-\frac{1}{2}$. Consequently, there are $3 \times 2 = 6$ possible sets of quantum numbers with $n = 2$ and $\ell = 1$, listed in the table at right.

n	ℓ	m_ℓ	m_s
2	1	-1	$-\frac{1}{2}$
2	1	-1	$\frac{1}{2}$
2	1	0	$-\frac{1}{2}$
2	1	0	$\frac{1}{2}$
2	1	1	$-\frac{1}{2}$
2	1	1	$\frac{1}{2}$

Remark Remember that these quantum states are not just abstractions; they have real physical consequences, such as which electronic transitions can be made within an atom and, consequently, which wavelengths of radiation can be observed.

Exercise 28.4

(a) How many different sets of quantum numbers are there in the $3d$ subshell? (b) How many sets of quantum numbers are there in a $2d$ subshell?

Answers (a) 10 (b) None. A $2d$ subshell doesn't exist because that would imply a quantum state with $n = 2$ and $\ell = 2$, impossible because $\ell \leq n - 1$.

28.8 ELECTRON CLOUDS

The solution of the wave equation, discussed in Section 27.7, yields a wave function Ψ that depends on the quantum numbers n, ℓ, and m_ℓ. We assume that we have found such a wave function Ψ and see what it may tell us about the hydrogen atom. Let $n = 1$ for the principal quantum number, which corresponds to the lowest energy state for hydrogen. For $n = 1$, the restrictions placed on the remaining quantum numbers are that $\ell = 0$ and $m_\ell = 0$.

The quantity Ψ^2 has great physical significance. If p is a point and V_p a very small volume containing that point, then $\Psi^2 V_p$ is approximately the probability of finding the electron inside the volume V_p. Figure 28.12 gives the probability per unit length of finding the electron at various distances from the nucleus in the $1s$ state of hydrogen. Some useful and surprising information can be extracted from this curve. First, the curve peaks at a value of $r = 0.052\,9$ nm, the Bohr radius for the first ($n = 1$) electron orbit in hydrogen. This means that there is a maximum probability of finding the electron in a small interval centered at that distance from the nucleus. However, as the curve indicates, there is also a probability of finding the electron in a small interval centered at any other distance from the nucleus. In other words, the electron is not confined to a particular orbital distance from the nucleus, as assumed in the Bohr model. The electron may be found at various distances from the nucleus, but **the probability of finding it at a distance corresponding to the Bohr radius is a maximum**. Quantum mechanics also predicts that the wave function for the hydrogen atom in the ground state is spherically symmetric; hence the electron can be found in a spherical region surrounding the nucleus. This is in contrast to the Bohr theory, which confines the position of the electron to points in a plane. The quantum mechanical result is often interpreted by viewing the electron as a cloud surrounding the nucleus. An attempt at picturing this cloud-like behavior is shown in Figure 28.13. The densest regions of the cloud represent those locations where the electron is most likely to be found.

If a similar analysis is carried out for the $n = 2$, $\ell = 0$, state of hydrogen, a peak of the probability curve is found at $4a_0$. Likewise, for the $n = 3$, $\ell = 0$ state, the curve peaks at $9a_0$. Thus, quantum mechanics predicts a most probable electron distance to the nucleus that is in agreement with the location predicted by the Bohr theory.

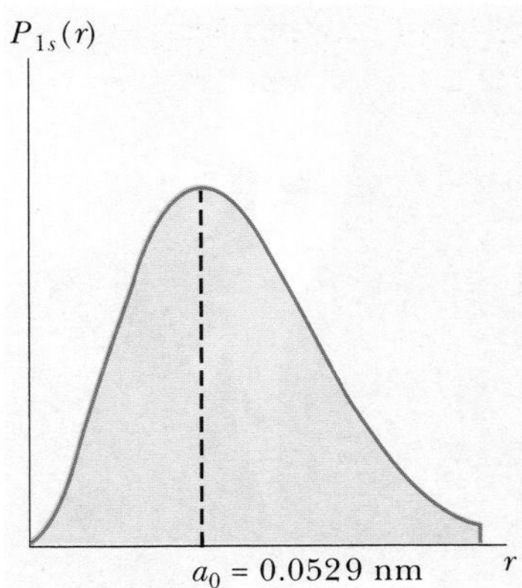

Figure 28.12 The probability per unit length of finding the electron versus distance from the nucleus for the hydrogen atom in the 1s (ground) state. Note that the graph has its maximum value when r equals the first Bohr radius, a_0.

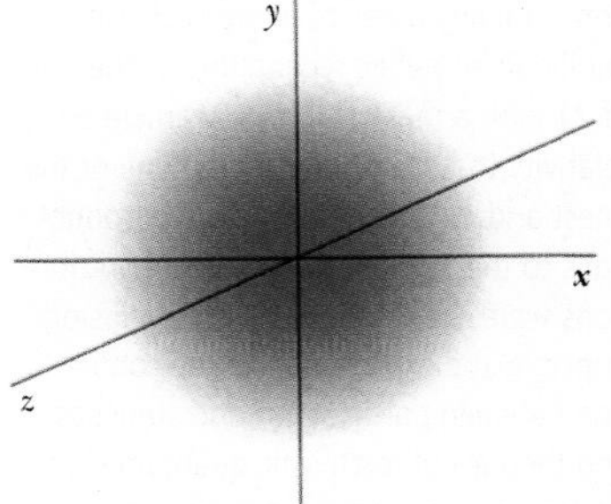

Figure 28.13 The spherical electron cloud for the hydrogen atom in its 1s state.

28.9 THE EXCLUSION PRINCIPLE AND THE PERIODIC TABLE

Earlier, we found that the state of an electron in an atom is specified by four quantum numbers: n, ℓ, m_ℓ, and m_s. For example, an electron in the ground state of hydrogen could have quantum numbers of $n = 1$, $\ell = 0$, $m_\ell = 0$, and $m_s = \frac{1}{2}$. As it turns out, the state of an electron in any other atom may also be specified by this same set of quantum numbers. In fact, these four quantum numbers can be used to describe all the electronic states of an atom, regardless of the number of electrons in its structure.

How many electrons in an atom can have a particular set of quantum numbers? This important question was answered by Pauli in 1925 in a powerful statement known as the **Pauli exclusion principle**:

◀ The Pauli exclusion principle

> No two electrons in an atom can ever have the same set of values for the set of quantum numbers n, ℓ, m_ℓ, and m_s.

The Pauli exclusion principle explains the electronic structure of complex atoms as a succession of filled levels with different quantum numbers increasing in energy, where the outermost electrons are primarily responsible for the chemical properties of the element. If this principle weren't valid, every electron would end up in the lowest energy state of the atom and the chemical behavior of the elements would be grossly different. Nature as we know it would not exist—and *we* would not exist to wonder about it!

As a general rule, the order that electrons fill an atom's subshell is as follows: once one subshell is filled, the next electron goes into the vacant subshell that is lowest in energy. If the atom were not in the lowest energy state available to it, it would radiate energy until it reached that state. A subshell is filled when it contains $2(2\ell + 1)$ electrons. This rule is based on the analysis of quantum numbers to be described later. Following the rule, shells and subshells can contain numbers of electrons according to the pattern given in Table 28.3.

TIP 28.3 The Exclusion Principle is More General

The exclusion principle stated here is a limited form of the more general exclusion principle, which states that no two *fermions* (particles with spin 1/2, 3/2, . . .) can be in the same quantum state.

The exclusion principle can be illustrated by an examination of the electronic arrangement in a few of the lighter atoms.

Hydrogen has only one electron, which, in its ground state, can be described by either of two sets of quantum numbers: 1, 0, 0, $\frac{1}{2}$ or 1, 0, 0, $-\frac{1}{2}$. The electronic configuration of this atom is often designated as $1s^1$. The notation $1s$ refers to a

CERN/Courtesy of AIP Emilio Segre Visual Archives

WOLFGANG PAULI (1900–1958)

An extremely talented Austrian theoretical physicist who made important contributions in many areas of modern physics, Pauli gained public recognition at the age of 21 with a masterful review article on relativity that is still considered one of the finest and most comprehensive introductions to the subject. Other major contributions were the discovery of the exclusion principle, the explanation of the connection between particle spin and statistics, and theories of relativistic quantum electrodynamics, the neutrino hypothesis, and the hypothesis of nuclear spin.

TABLE 28.3

Number of Electrons in Filled Subshells and Shells

Shell	Subshell	Number of Electrons in Filled Subshell	Number of Electrons in Filled Shell
K ($n = 1$)	$s(\ell = 0)$	2	2
L ($n = 2$)	$s(\ell = 0)$	2	8
	$p(\ell = 1)$	6	
M ($n = 3$)	$s(\ell = 0)$	2	18
	$p(\ell = 1)$	6	
	$d(\ell = 2)$	10	
N ($n = 4$)	$s(\ell = 0)$	2	32
	$p(\ell = 1)$	6	
	$d(\ell = 2)$	10	
	$f(\ell = 3)$	14	

state for which $n = 1$ and $\ell = 0$, and the superscript indicates that one electron is present in this level.

Neutral *helium* has two electrons. In the ground state, the quantum numbers for these two electrons are 1, 0, 0, $\frac{1}{2}$ and 1, 0, 0, $-\frac{1}{2}$. No other possible combinations of quantum numbers exist for this level, and we say that the K shell is filled. The helium electronic configuration is designated as $1s^2$.

Neutral *lithium* has three electrons. In the ground state, two of these are in the $1s$ subshell and the third is in the $2s$ subshell, because the latter is lower in energy than the $2p$ subshell. Hence, the electronic configuration for lithium is $1s^22s^1$.

A list of electronic ground-state configurations for a number of atoms is provided in Table 28.4. In 1871 Dmitri Mendeleev (1834–1907), a Russian chemist, arranged the elements known at that time into a table according to their atomic masses and chemical similarities. The first table Mendeleev proposed contained many blank spaces, and he boldly stated that the gaps were there only because those elements had not yet been discovered. By noting the column in which these missing elements should be located, he was able to make rough predictions about their chemical properties. Within 20 years of this announcement, the elements were indeed discovered.

The elements in our current version of the periodic table are still arranged so that all those in a vertical column have similar chemical properties. For example, consider the elements in the last column: He (helium), Ne (neon), Ar (argon), Kr (krypton), Xe (xenon), and Rn (radon). The outstanding characteristic of these elements is that they don't normally take part in chemical reactions, joining with other atoms to form molecules, and are therefore classified as inert. Because of this "aloofness," they are referred to as the *noble gases*. We can partially understand their behavior by looking at the electronic configurations shown in Table 28.4, page 919. The element helium has the electronic configuration $1s^2$. In other words, one shell is filled. The electrons in this filled shell are considerably separated in energy from the next available level, the $2s$ level.

The electronic configuration for neon is $1s^22s^22p^6$. Again, the outer shell is filled and there is a large difference in energy between the $2p$ level and the $3s$ level. Argon has the configuration $1s^22s^22p^63s^23p^6$. Here, the $3p$ subshell is filled and there is a wide gap in energy between the $3p$ subshell and the $3d$ subshell. Through all the noble gases, the pattern remains the same: a noble gas is formed when either a shell or a subshell is filled, and there is a large gap in energy before the next possible level is encountered.

The elements in the first column of the periodic table are called the *alkali metals* and are highly active chemically. Referring to Table 28.4, we can understand why these elements interact so strongly with other elements. All of these alkali

TABLE 28.4

Electronic Configurations of Some Elements

Z	Symbol	Ground-State Configuration	Ionization Energy (eV)	Z	Symbol	Ground-State Configuration	Ionization Energy (eV)
1	H	$1s^1$	13.595	19	K	[Ar] $4s^1$	4.339
2	He	$1s^2$	24.581	20	Ca	$4s^2$	6.111
				21	Sc	$3d4s^2$	6.54
3	Li	[He] $2s^1$	5.390	22	Ti	$3d^24s^2$	6.83
4	Be	$2s^2$	9.320	23	V	$3d^34s^2$	6.74
5	B	$2s^22p^1$	8.296	24	Cr	$3d^54s^1$	6.76
6	C	$2s^22p^2$	11.256	25	Mn	$3d^54s^2$	7.432
7	N	$2s^22p^3$	14.545	26	Fe	$3d^64s^2$	7.87
8	O	$2s^22p^4$	13.614	27	Co	$3d^74s^2$	7.86
9	F	$2s^22p^5$	17.418	28	Ni	$3d^84s^2$	7.633
10	Ne	$2s^22p^6$	21.559	29	Cu	$3d^{10}4s^1$	7.724
				30	Zn	$3d^{10}4s^2$	9.391
11	Na	[Ne] $3s^1$	5.138	31	Ga	$3d^{10}4s^24p^1$	6.00
12	Mg	$3s^2$	7.644	32	Ge	$3d^{10}4s^24p^2$	7.88
13	Al	$3s^23p^1$	5.984	33	As	$3d^{10}4s^24p^3$	9.81
14	Si	$3s^23p^2$	8.149	34	Se	$3d^{10}4s^24p^4$	9.75
15	P	$3s^23p^3$	10.484	35	Br	$3d^{10}4s^24p^5$	11.84
16	S	$3s^23p^4$	10.357	36	Kr	$3d^{10}4s^24p^6$	13.996
17	Cl	$3s^23p^5$	13.01				
18	Ar	$3s^23p^6$	15.755				

Note: The bracket notation is used as a shorthand method to avoid repetition in indicating inner-shell electrons. Thus, [He] represents $1s^2$, [Ne] represents $1s^22s^22p^6$, [Ar] represents $1s^22s^22p^63s^23p^6$, and so on.

metals have a single outer electron in an s subshell. This electron is shielded from the nucleus by all the electrons in the inner shells. Consequently, it's only loosely bound to the atom and can readily be accepted by other atoms that bind it more tightly to form molecules.

The elements in the seventh column of the periodic table are called the *halogens* and are also highly active chemically. All these elements are lacking one electron in a subshell, so they readily accept electrons from other atoms to form molecules.

Quick Quiz 28.4

Krypton (atomic number 36) has how many electrons in its next to outer shell ($n = 3$)?
(a) 2 (b) 4 (c) 8 (d) 18

Applying Physics 28.3 The Periodic Table

Scanning from left to right across one row of the periodic table, the effective size of the atoms first decreases and then increases. What would cause this behavior?

Explanation Starting on the left side of the periodic table and moving toward the middle, the nuclear charge is increasing. As a result, there is an increasing Coulomb attraction between the nucleus and the electrons, and the electrons are pulled into an average position that is closer to the nucleus. From the middle of the row to the right side, the increasing number of electrons being placed in proximity to each other results in a mutual repulsion that increases the average distance from the nucleus and causes the atomic size to grow.

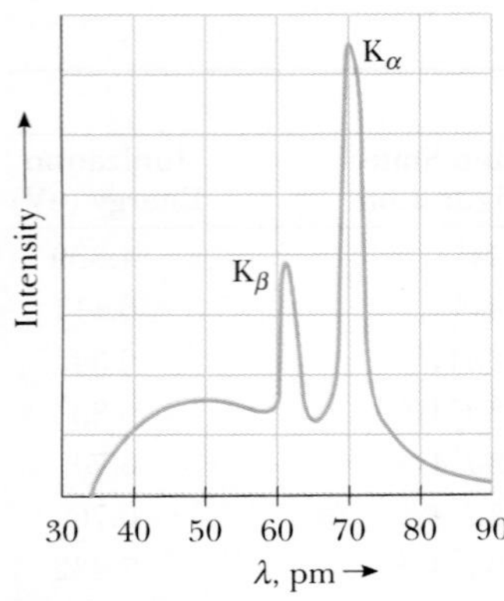

Figure 28.14 The x-ray spectrum of a metal target consists of a broad continuous spectrum (*bremstrahlung*) plus a number of sharp lines that are due to *characteristic x-rays.* The data shown were obtained when 35-keV electrons bombarded a molybdenum target. Note that 1 pm = 10^{-12} m = 0.001 nm.

28.10 CHARACTERISTIC X-RAYS

X-rays are emitted when a metal target is bombarded with high-energy electrons. The x-ray spectrum typically consists of a broad continuous band and a series of intense sharp lines that are dependent on the type of metal used for the target, as shown in Figure 28.14. These discrete lines, called **characteristic x-rays**, were discovered in 1908, but their origin remained unexplained until the details of atomic structure were developed.

The first step in the production of characteristic x-rays occurs when a bombarding electron collides with an electron in an inner shell of a target atom with sufficient energy to remove the electron from the atom. The vacancy created in the shell is filled when an electron in a higher level drops down into the lower energy level containing the vacancy. The time it takes for this to happen is very short, less than 10^{-9} s. The transition is accompanied by the emission of a photon with energy equaling the difference in energy between the two levels. Typically, the energy of such transitions is greater than 1 000 eV, and the emitted x-ray photons have wavelengths in the range of 0.01 nm to 1 nm.

We assume that the incoming electron has dislodged an atomic electron from the innermost shell, the K shell. If the vacancy is filled by an electron dropping from the next higher shell, the L shell, the photon emitted in the process is referred to as the K_α line on the curve of Figure 28.14. If the vacancy is filled by an electron dropping from the M shell, the line produced is called the K_β line.

Other characteristic x-ray lines are formed when electrons drop from upper levels to vacancies other than those in the K shell. For example, L lines are produced when vacancies in the L shell are filled by electrons dropping from higher shells. An L_α line is produced as an electron drops from the M shell to the L shell, and an L_β line is produced by a transition from the N shell to the L shell.

We can estimate the energy of the emitted x-rays as follows: consider two electrons in the K shell of an atom whose atomic number is Z. Each electron partially shields the other from the charge of the nucleus, Ze, so each is subject to an effective nuclear charge $Z_{\text{eff}} = (Z - 1)e$. We can now use a modified form of Equation 28.18 to estimate the energy of either electron in the K shell (with $n = 1$). We have

$$E_{\text{K}} = -m_e Z_{\text{eff}}^2 \frac{k_e^2 e^4}{2\hbar^2} = -Z_{\text{eff}}^2 E_0$$

where E_0 is the ground-state energy. Substituting $Z_{\text{eff}} = Z - 1$ gives

$$E_{\text{K}} = -(Z-1)^2(13.6\text{ eV}) \qquad [28.20]$$

As Example 28.5 will show, we can estimate the energy of an electron in an L or an M shell in a similar fashion. Taking the energy difference between these two levels, we can then calculate the energy and wavelength of the emitted photon.

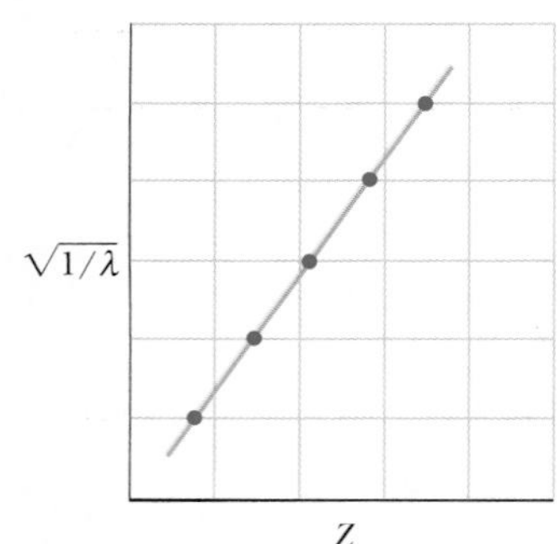

Figure 28.15 A Moseley plot of $\sqrt{1/\lambda}$ versus Z, where λ is the wavelength of the K_α x-ray line of the element of atomic number Z.

In 1914, Henry G. J. Moseley plotted the Z values for a number of elements against $\sqrt{1/\lambda}$, where λ is the wavelength of the K_α line for each element. He found that such a plot produced a straight line, as in Figure 28.15. This is consistent with our rough calculations of the energy levels based on Equation 28.20. From his plot, Moseley was able to determine the Z values of other elements, providing a periodic chart in excellent agreement with the known chemical properties of the elements.

EXAMPLE 28.5 Characteristic X-Rays

Goal Calculate the energy and wavelength of characteristic x-rays.

Problem Estimate the energy and wavelength of the characteristic x-ray emitted from a tungsten target when an electron drops from an M shell ($n = 3$ state) to a vacancy in the K shell ($n = 1$ state).

Strategy Develop two estimates, one for the electron in the K shell ($n = 1$) and one for the electron in the M shell ($n = 3$). For the K-shell estimate, we can use Equation 28.20. For the M shell, we need a new equation. There is one electron in the K shell (because one is missing) and 8 in the L shell, making 9 electrons shielding the nuclear charge. This means $Z_{\text{eff}} = 74 - 9$ and $E_{\text{M}} = -Z_{\text{eff}}^2 E_3$, where E_3 is the energy of the $n = 3$ level in hydrogen. The difference $E_{\text{M}} - E_{\text{K}}$ is the energy of the photon.

Solution

Use Equation 28.20 to estimate the energy of an electron in the K shell of tungsten, atomic number $Z = 74$:

$$E_{\text{K}} = -(74 - 1)^2(13.6\text{ eV}) = -72\,500\text{ eV}$$

Estimate the energy of an electron in the M shell in the same way:

$$E_{\text{M}} = -Z_{\text{eff}}^2 E_3 = -(Z - 9)^2 \frac{E_0}{3^2} = -(74 - 9)^2 \frac{(13.6\text{ eV})}{9}$$

$$= -6\,380\text{ eV}$$

Calculate the difference in energy between the M and K shells:

$$E_{\text{M}} - E_{\text{K}} = -6\,380\text{ eV} - (-72\,500\text{ eV}) = \boxed{66\,100\text{ eV}}$$

Find the wavelength of the emitted x-ray:

$$\Delta E = hf = h\frac{c}{\lambda} \quad \rightarrow \quad \lambda = \frac{hc}{\Delta E}$$

$$\lambda = \frac{(6.63 \times 10^{-34}\text{ J}\cdot\text{s})(3.00 \times 10^8\text{ m/s})}{(6.61 \times 10^4\text{ eV})(1.60 \times 10^{-19}\text{ J/eV})}$$

$$= 1.88 \times 10^{-11}\text{ m} = \boxed{0.018\,8\text{ nm}}$$

Exercise 28.5

Repeat the problem for a $2p$ electron transiting from the L shell to the K shell. (For technical reasons, the L shell electron must have $\ell = 1$, so a single $1s$ electron and two $2s$ electrons shield the nucleus.)

Answer (a) 5.54×10^4 eV (b) 0.022 4 nm

28.11 ATOMIC TRANSITIONS

We have seen that an atom will emit radiation only at certain frequencies that correspond to the energy separation between the various allowed states. Consider an atom with many allowed energy states, labeled $E_1, E_2, E_3, \ldots$, as in Figure 28.16. When light is incident on the atom, only those photons whose energy hf matches the energy separation ΔE between two levels can be absorbed by the atom. A schematic diagram representing this **stimulated absorption process** is shown in Active Figure 28.17. At ordinary temperatures, most of the atoms in a sample are in the ground state. If a vessel containing many atoms of a gas is illuminated with a light beam containing all possible photon frequencies (that is, a continuous spectrum), only those photons of energies $E_2 - E_1$, $E_3 - E_1$, $E_4 - E_1$, and so on, can be absorbed. As a result of this absorption, some atoms are raised to various allowed higher energy levels, called **excited states.**

Once an atom is in an excited state, there is a constant probability that it will jump back to a lower level by emitting a photon, as shown in Active Figure 28.18 (page 922).

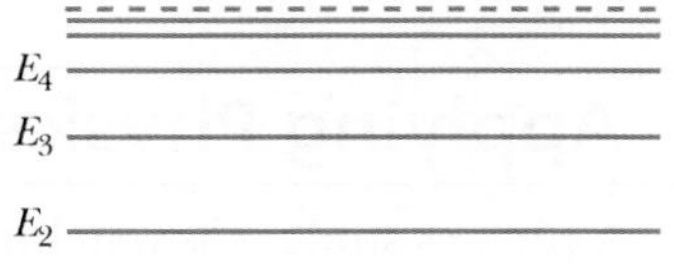

Figure 28.16 Energy level diagram of an atom with various allowed states. The lowest energy state, E_1, is the ground state. All others are excited states.

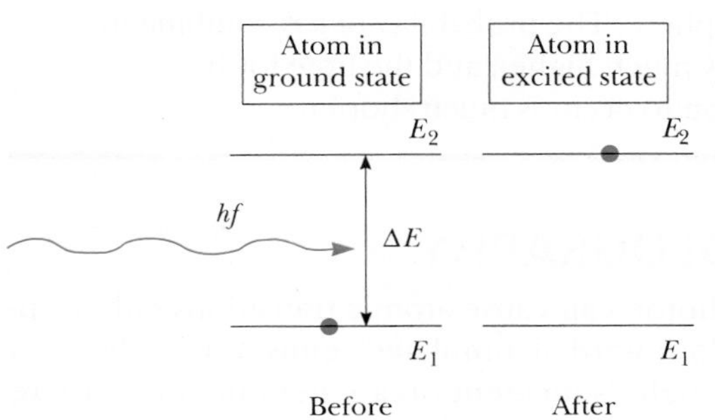

ACTIVE FIGURE 28.17
Diagram representing the process of *stimulated absorption* of a photon by an atom. The blue dot represents an electron. The electron is transferred from the ground state to the excited state when the atom absorbs a photon of energy $hf = E_2 - E_1$.

Physics Now™
Log into PhysicsNow at **www.cp7e.com** and go to Active Figure 28.17 to observe stimulated absorption.

Atom in excited state
Atom in ground state
E_2
E_2
$hf = \Delta E$
ΔE
E_1
E_1
Before
After

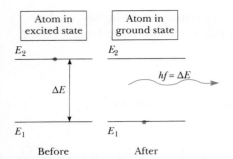

ACTIVE FIGURE 28.18
Diagram representing the process of *spontaneous emission* of a photon by an atom that is initially in the excited state E_2. When the electron falls to the ground state, the atom emits a photon of energy $hf = E_2 - E_1$.

Physics Now™
Log into PhysicsNow at **www.cp7e.com** and go to Active Figure 28.17 to observe spontaneous emission.

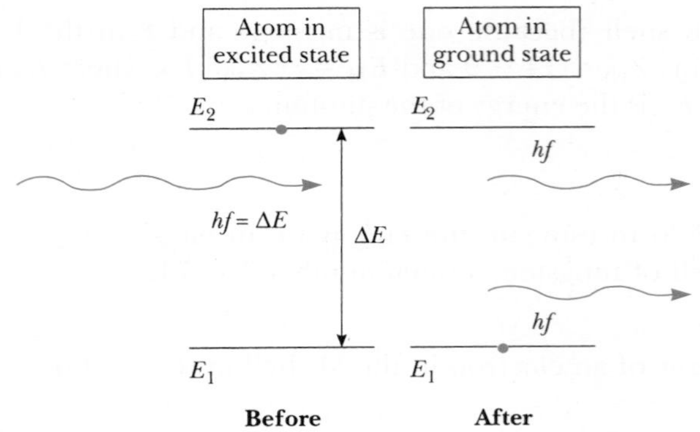

ACTIVE FIGURE 28.19
Diagram representing the process of *stimulated emission* of a photon by an incoming photon of energy hf. Initially, the atom is in the excited state. The incoming photon stimulates the atom to emit a second photon of energy $hf = E_2 - E_1$.

Physics Now™
Log into PhysicsNow at **www.cp7e.com** and go to Active Figure 28.17 to observe stimulated emission.

This process is known as **spontaneous emission**. Typically, an atom will remain in an excited state for only about 10^{-8} s.

A third process that is important in lasers, **stimulated emission**, was predicted by Einstein in 1917. Suppose an atom is in the excited state E_2, as in Active Figure 28.19, and a photon with energy $hf = E_2 - E_1$ is incident on it. The incoming photon increases the probability that the excited atom will return to the ground state and thereby emit a second photon having the same energy hf. Note that two identical photons result from stimulated emission: the incident photon and the emitted photon. *The emitted photon is exactly in phase with the incident photon.* These photons can stimulate other atoms to emit photons in a chain of similar processes. The many photons produced in this fashion are the source of the intense, coherent (in-phase) light in a laser.

Applying Physics 28.4 Streaking Meteoroids

A physics student is watching a meteor shower in the early morning hours. She notices that the streaks of light from the meteoroids entering the very high regions of the atmosphere last for as long as 2 or 3 seconds before fading. She also notices a lightning storm off in the distance. The streaks of light from the lightning fade away almost immediately after the flash, certainly in much less than 1 second. Both lightning and meteors cause the air to turn into a plasma because of the very high temperatures generated. The light is given off when the stripped electrons in the plasma recombine with the ionized atoms. Why would the light last longer for meteors than for lightning?

Explanation To answer this question, we examine the phrase "the streaks of light from the meteoroids entering the very high regions of the atmosphere." In the very high regions of the atmosphere, the pressure is very low, so the density is also very low and the atoms of the gas are relatively far apart. Low density means that after the air is ionized by the passing meteoroid, the probability of freed electrons finding an ionized atom with which to recombine is relatively low. As a result, the recombination process occurs over a relatively long time, measured in seconds. Lightning, however, occurs in the lower regions of the atmosphere (the troposphere), where the pressure and density are relatively high. After the ionization by the lightning flash, the electrons and ionized atoms are much closer together than in the upper atmosphere. The probability of a recombination is accordingly much higher, and the time for the recombination to occur is much shorter.

28.12 LASERS AND HOLOGRAPHY

We have described how an incident photon can cause atomic transitions either upward (stimulated absorption) or downward (stimulated emission). The two processes are equally probable. When light is incident on a system of atoms, there is usually a net absorption of energy, because when the system is in thermal equilibrium, there are many more atoms in the ground state than in excited states. However, if the situation can be inverted so that there are more atoms in an excited state than in the ground state, a net emission of photons can result. Such a condition is called **population inversion**. This is the fundamental principle involved in the operation of a laser, an acronym for *l*ight *a*mplification by *s*timulated *e*mission of *r*adiation. The amplification corresponds to a buildup of photons in the system as the result of a chain reaction of events. The following three conditions must be satisfied in order to achieve laser action:

1. The system must be in a state of population inversion (that is, more atoms in an excited state than in the ground state).
2. The excited state of the system must be a *metastable state*, which means its lifetime must be long compared with the otherwise usually short lifetimes of excited states. When that is the case, stimulated emission will occur before spontaneous emission.
3. The emitted photons must be confined within the system long enough to allow them to stimulate further emission from other excited atoms. This is achieved by the use of reflecting mirrors at the ends of the system. One end is totally reflecting, and the other is slightly transparent to allow the laser beam to escape.

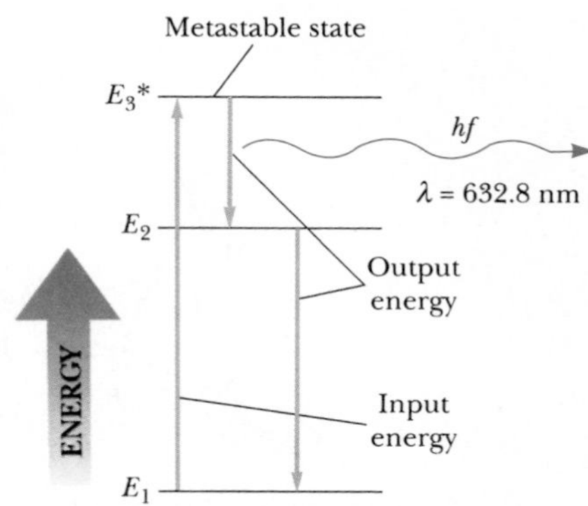

Figure 28.20 Energy-level diagram for the neon atom in a helium–neon laser. The atom emits 632.8-nm photons through stimulated emission in the transition $E_3^* \rightarrow E_2$. This is the source of coherent light in the laser.

One device that exhibits stimulated emission of radiation is the helium–neon gas laser. Figure 28.20 is an energy-level diagram for the neon atom in this system. The mixture of helium and neon is confined to a glass tube sealed at the ends by mirrors. A high voltage applied to the tube causes electrons to sweep through it, colliding with the atoms of the gas and raising them into excited states. Neon atoms are excited to state E_3* through this process and also as a result of collisions with excited helium atoms. When a neon atom makes a transition to state E_2, it stimulates emission by neighboring excited atoms. This results in the production of coherent light at a wavelength of 632.8 nm. Figure 28.21 summarizes the steps in the production of a laser beam.

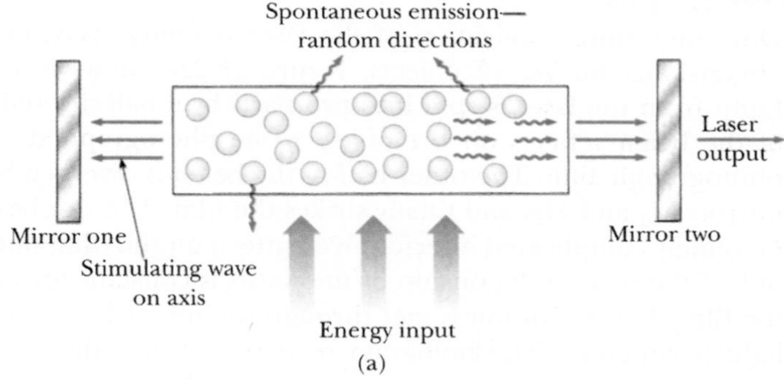

(a)

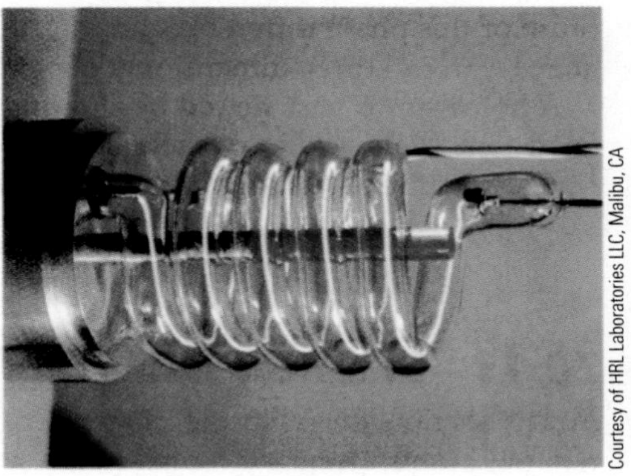

(b)

Figure 28.21 (a) Steps in the production of a laser beam. The tube contains atoms, which represent the active medium. An external source of energy (optical, electrical, etc.) is needed to "pump" the atoms to excited energy states. The parallel end mirrors provide the feedback of the stimulating wave. (b) Photograph of the first ruby laser, showing the flash lamp surrounding the ruby rod.

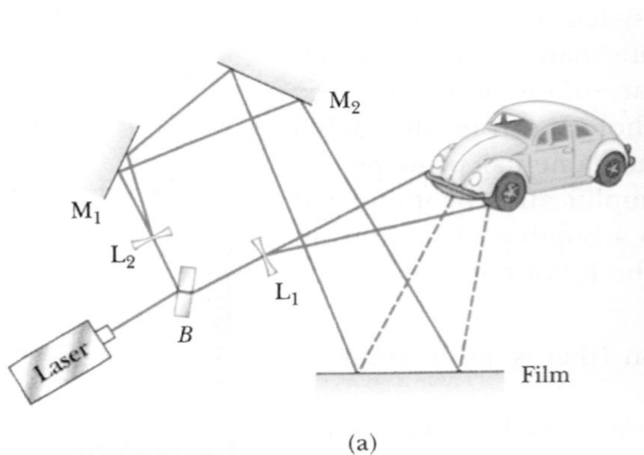

(a)

(b)

Courtesy of Central Scientific Company

Figure 28.22 (a) Experimental arrangement for producing a hologram. (b) Photograph of a hologram made with a cylindrical film. Note the detail of the Volkswagen image.

APPLICATION
Laser Technology

Since the development of the first laser in 1960, laser technology has exhibited tremendous growth. Lasers that cover wavelengths in the infrared, visible, and ultraviolet regions of the spectrum are now available. Applications include the surgical "welding" of detached retinas, "lasik" surgery, precision surveying and length measurement, a potential source for inducing nuclear fusion reactions, precision cutting of metals and other materials, and telephone communication along optical fibers. These and other applications are possible because of the unique characteristics of laser light. In addition to being highly monochromatic and coherent, laser light is also highly directional and can be sharply focused to produce regions of extremely intense light energy.

Philippe Plailly/Photo Researchers, Inc.

Scientist checking the performance of an experimental laser-cutting device mounted on a robot arm. The laser is being used to cut through a metal plate.

APPLICATION
Holography

Holography

One interesting application of the laser is holography: the production of three-dimensional images of objects. Figure 28.22a shows how a hologram is made. Light from the laser is split into two parts by a half-silvered mirror at *B*. One part of the beam reflects off the object to be photographed and strikes an ordinary photographic film. The other half of the beam is diverged by lens L_2, reflects from mirrors M_1 and M_2, and finally strikes the film. The two beams overlap to form an extremely complicated interference pattern on the film, one that can be produced only if the phase relationship of the waves is constant throughout the exposure of the film. This condition is met through the use of light from a laser, because such light is coherent. The hologram records not only the intensity of the light scattered from the object (as in a conventional photograph), but also the phase difference between the reference beam and the beam scattered from the object. Because of this phase difference, an interference pattern is formed that produces an image with full three-dimensional perspective.

A hologram is best viewed by allowing coherent light to pass through the developed film while you look back along the direction from which the beam comes. Figure 28.22b is a photograph of a hologram made using a cylindrical film.

28.13 ENERGY BANDS IN SOLIDS

In this section we trace the changes that occur in the discrete energy levels of isolated atoms when the atoms group together and form a solid. We find that in solids, the discrete levels of isolated atoms broaden into allowed energy bands

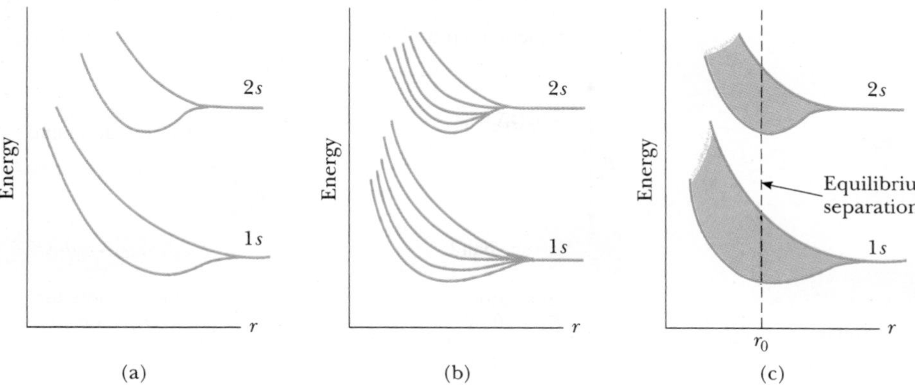

Figure 28.23 (a) Splitting of the 1s and 2s states when two atoms are brought together. (b) Splitting of the 1s and 2s states when five atoms are brought close together. (c) Formation of energy bands when a large number of sodium atoms are assembled to form a solid.

separated by forbidden gaps. The separation and electron population of the highest bands determines whether a given solid is a conductor, an insulator, or a semiconductor.

Consider two identical atoms, initially widely separated, that are brought closer and closer together. If two identical atoms are very far apart, they do not interact, and their electronic energy levels can be considered to be those of isolated atoms. Hence, the energy levels are exactly the same. As the atoms come close together, they essentially become one quantum system, and the Pauli exclusion principle demands that the electrons be in different quantum states for this single system. The exclusion principle manifests itself as a changing or splitting of electron energy levels that were identical in the widely separated atoms, as shown in Figure 28.23a. Figure 28.23b shows that with 5 atoms, each energy level in the isolated atom splits into five different, more closely spaced levels.

If we extend this argument to the large number of atoms found in solids (on the order of 10^{23} atoms/cm^3), we obtain a large number of levels so closely spaced that they may be regarded as a continuous **band** of energy levels, as in Figure 28.23c. An electron can have any energy within an allowed energy band, but cannot have an energy in the **band gap**, or the region between allowed bands. Note that the band gap energy E_g is indicated in Figure 28.23c. In practice we are only interested in the band structure of a solid at some equilibrium separation of its atoms r_0, and so we remove the distance scale on the x-axis and simply plot the allowed energy bands of a solid as a series of horizontal bands, as shown in Figure 28.24 for sodium.

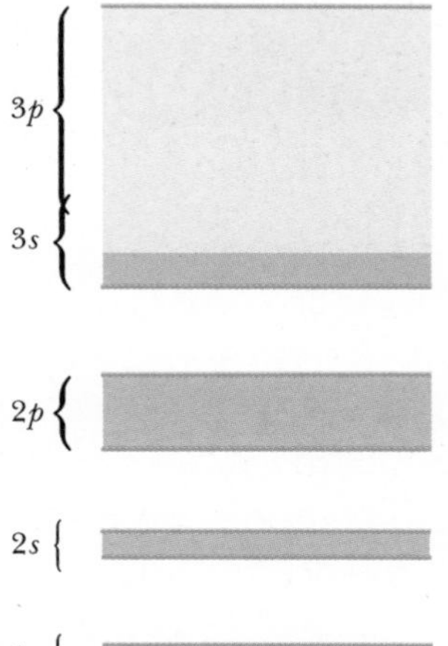

Figure 28.24 Energy bands of sodium. Note the energy gaps (white regions) between the allowed bands; electrons can't occupy states that lie in these forbidden gaps. Blue represents energy bands occupied by the sodium electrons when the atom is in its ground state. Gold represents energy bands that are empty. Note that the 3s and 3p levels broaden so much that they overlap.

Conductors and Insulators

Figure 28.24 shows that the band structure of a particular solid is quite complicated with individual atomic levels broadening by varying amounts and some levels (3s and 3p) broadening so much that they overlap. Nevertheless, it is possible to gain a qualitative understanding of whether a solid is a conductor, an insulator, or a semiconductor by considering only the structure of the upper or upper two energy bands and whether they are occupied by electrons.

Deciding whether an energy band is empty (unoccupied by electrons), partially filled, or full is carried out in basically the same way as for the energy-level population of atoms: we distribute the total number of electrons from the lowest energy levels up in a way consistent with the exclusion principle. While we omit the details of this process here, one important case is that shown in Figure 28.25a (page 926), where the highest-energy occupied band is only partially full. The other important case, where the highest occupied band is completely full, is shown in Figure 28.25b. Notice that this figure also shows that the highest filled band is called the **valence band** and the next higher empty band is called the **conduction band**. The energy band gap, which varies with the solid, is also indicated as the energy difference E_g between the top of the valence band and the bottom of the conduction band.

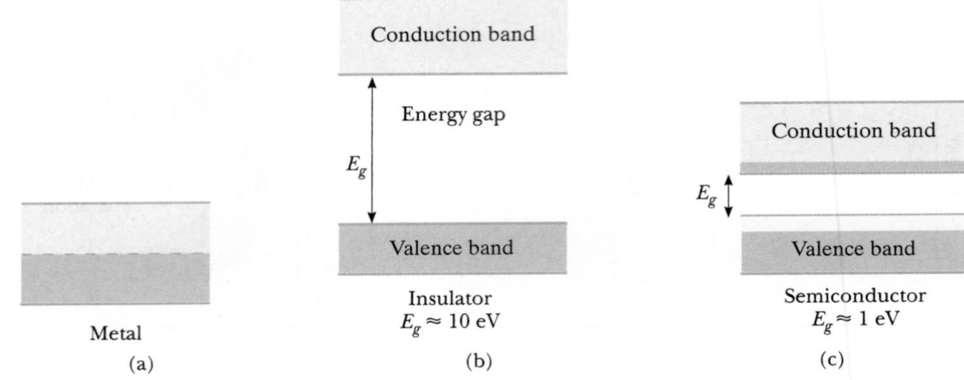

Figure 28.25 (a) Half-filled band of a metal, an electrical conductor. (b) An electrical insulator at $T = 0$ K has a filled valence band and an empty conduction band. (c) Band structure of a semiconductor at ordinary temperatures ($T \approx 300$ K). The energy gap is much smaller than in an insulator, and many electrons occupy states in the conduction band.

With these ideas and definitions we are now in a position to understand what determines, quantum mechanically, whether a solid will be a conductor or an insulator. When a modest voltage is applied to a good conductor, the electrons accelerate and gain energy. In quantum terms, electron energies increase *if there are higher unoccupied energy levels for electrons to jump to.* For example, electrons near the top of the partially filled band in sodium need to gain very little energy from the applied voltage to reach one of the nearby, closely spaced, empty states. Thus, it is easy for a small voltage to kick electrons into higher energy states, and charge flows easily in sodium, an excellent conductor.

Now consider the case of a material in which the highest occupied band is completely full of electrons and there is a band gap separating this filled valence band from the vacant conduction band, as in Figure 28.25b. A typical case might be diamond (carbon), in which the band gap is about 10 eV. When a voltage is applied, electrons can't easily gain energy, because there are no vacant energy states nearby to which electrons can make transitions. Because the only empty band is the conduction band, an electron must gain an amount of energy at least equal to the band gap in order for it to move through the solid. This large amount of energy can't be supplied by a modest applied voltage, so no charge flows and diamond is a good insulator. In summary then, a conductor has a highest-energy occupied band which is *partially filled,* and in an insulator, has a highest-energy occupied band which is *completely filled* with a large energy gap between the valence and conduction bands.

Semiconductors

To this point, we have completely ignored the influence of temperature on the electronic populations of energy bands. Recalling that the average thermal energy of a particle at temperature T is $3k_BT/2$, we find that an electron at room temperature has an average energy of about 0.04 eV. Because this energy is about 100 times smaller than the band gap in a typical insulator, very few electrons would have enough random thermal energy to jump the energy gap in an insulator and contribute to conduction. However things are different for a semiconductor. As we see in Figure 28.25c, a **semiconductor** is a material with a small band gap of about 1 eV whose conductivity results from appreciable thermal excitation of electrons across the gap into the conduction band at room temperature. The most commonly used semiconductors are silicon and gallium arsenide, with band gaps of 1.14 eV and 1.43 eV, respectively, at 300 K. As you might expect, the resistivity of semiconductors usually decreases with increasing temperature, because k_BT becomes a larger fraction of the band gap energy.

It is interesting that the electrons in the conduction band of a semiconductor don't carry the entire current when a voltage is applied, as Figure 28.26 shows. (It might be said that conduction electrons do not constitute the "whole" story.) The missing electrons in the valence band, shown as a narrow white band in the

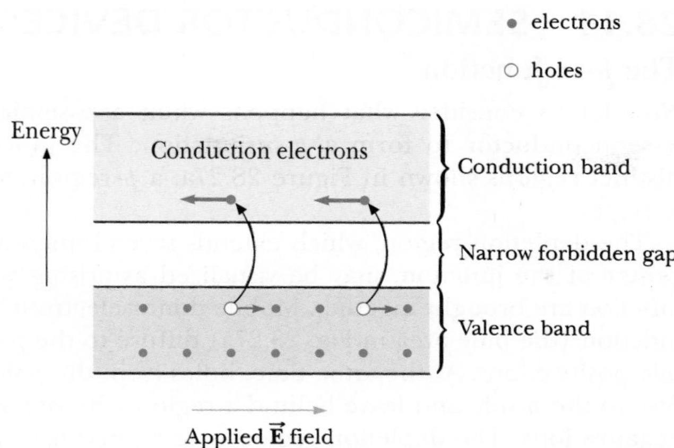

Figure 28.26 Movement of charges (holes and electrons) in a semiconductor. The electrons move in the direction opposite the direction of the external electric field, and the holes move in the direction of the field.

figure, provide a few empty states called **holes** for valence band electrons to fill; so some electrons in the valence band can gain energy and move towards a positive electrode and thus also carry the current. Since the valence band electrons that fill holes leave behind other holes, it is equally valid and more common to view the conduction process in the valence band as a flow of positive holes towards the negative electrode applied to a semiconductor. Thus, a pure semiconductor, such as silicon, can be viewed in a symmetric way: silicon has equal numbers of mobile electrons in the conduction band and holes in the valence band. Furthermore, when an external voltage is applied to the semiconductor, electrons move toward the positive electrode and holes move toward the negative electrode. In the next section we will look at the concepts of an electron and a hole in a simpler, more graphic way as the presence or absence of an outer-shell electron at a particular location in a crystal lattice.

When small amounts of impurities are added to a semiconductor such as silicon (about one impurity atom per 10^7 silicon atoms), both the band structure of the semiconductor and its resistivity are modified. The process of adding impurities, called **doping**, is important in making devices having well-defined regions of different resistivity. For example, when an atom containing five outer-shell electrons, such as arsenic, is added to a semiconductor such as silicon, four of the arsenic electrons form shared bonds with atoms of the semiconductor and one is left over. This extra electron is nearly free of its parent atom and has an energy level that lies in the energy gap, just below the conduction band. Such a pentavalent atom in effect donates an electron to the structure and hence is referred to as a **donor atom**. Because the spacing between the energy level of the electron of the donor atom and the bottom of the conduction band is very small (typically, about 0.05 eV), only a small amount of thermal energy is needed to cause this electron to move into the conduction band. (Recall that the average thermal energy of an electron at room temperature is $3k_BT/2 \approx 0.04$ eV). Semiconductors doped with donor atoms are called ***n*-type semiconductors**, because the charge carriers are electrons, the charge of which is *n*egative.

If a semiconductor is doped with atoms containing three outer-shell electrons, such as aluminum, the three electrons form shared bonds with neighboring semiconductor atoms, leaving an electron deficiency—a hole—where the fourth bond would be if an impurity-atom electron was available to form it. The energy level of this hole lies in the energy gap, just above the valence band. An electron from the valence band has enough energy at room temperature to fill that impurity level, leaving behind a hole in the valence band. Because a trivalent atom, in effect, accepts an electron from the valence band, such impurities are referred to as **acceptor atoms**. A semiconductor doped with acceptor impurities is known as a ***p*-type semiconductor**, because the majority of charge carriers are *p*ositively charged holes.

28.14 SEMICONDUCTOR DEVICES

The $p-n$ Junction

Now let us consider what happens when a p-semiconductor is joined to an n-semiconductor to form a $p-n$ junction. The junction consists of the three distinct regions shown in Figure 28.27a: a p-region, a depletion region, and an n-region.

The depletion region, which extends several micrometers to either side of the center of the junction, may be visualized as arising when the two halves of the junction are brought together. Mobile donor electrons from the n side nearest the junction (the blue area in Fig. 28.27a) diffuse to the p side, leaving behind immobile positive ions. At the same time, holes from the p side nearest the junction diffuse to the n side and leave behind a region (the red area in Fig. 28.27a) of fixed negative ions. The depletion region is so named because it is depleted of mobile charge carriers.

The depletion region contains an internal electric field (arising from the charges of the fixed ions) on the order of 10^4 to 10^6 V/cm. This field sweeps mobile charge out of the depletion region and keeps it truly depleted. This internal electric field creates an internal potential difference ΔV_0 that prevents further diffusion of holes and electrons across the junction and thereby ensures zero current in the junction when no external potential difference is applied.

Perhaps the most notable feature of the $p-n$ junction is its ability to pass current in only one direction. Such *diode* action is easiest to understand in terms of the potential-difference graph shown in Figure 28.27c. If an external voltage ΔV is applied to the junction such that the p side is connected to the positive terminal of a voltage source as in Figure 28.27a, the internal potential difference ΔV_0 across the junction is decreased, resulting in a current that increases exponentially with increasing forward voltage, or *forward bias*. In *reverse bias* (where the n side of the junction is connected to the positive terminal of a voltage source), the internal potential difference ΔV_0 *increases* with increasing reverse bias. This results in a very small reverse current that quickly reaches a saturation value I_0. The current–voltage relationship for an ideal diode is

$$I = I_0(e^{q\Delta V/k_BT} - 1) \qquad [28.21]$$

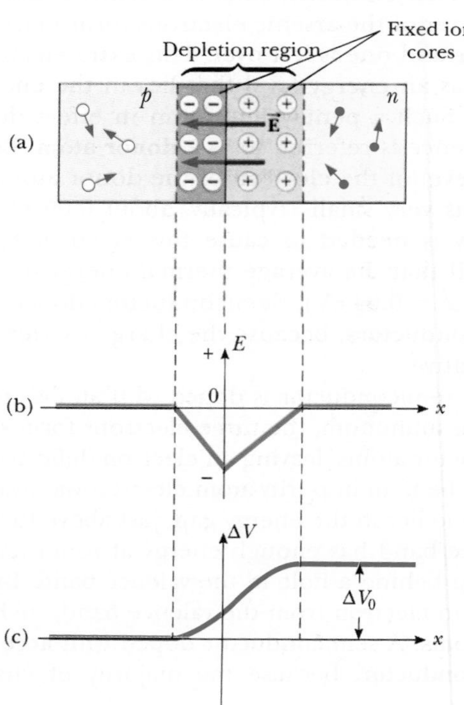

Figure 28.27 (a) Physical arrangement of a $p-n$ junction. (b) Internal electric field versus x for the $p-n$ junction. (c) Internal electric potential ΔV versus x for the $p-n$ junction. ΔV_0 represents the potential difference across the junction in the absence of an applied electric field.

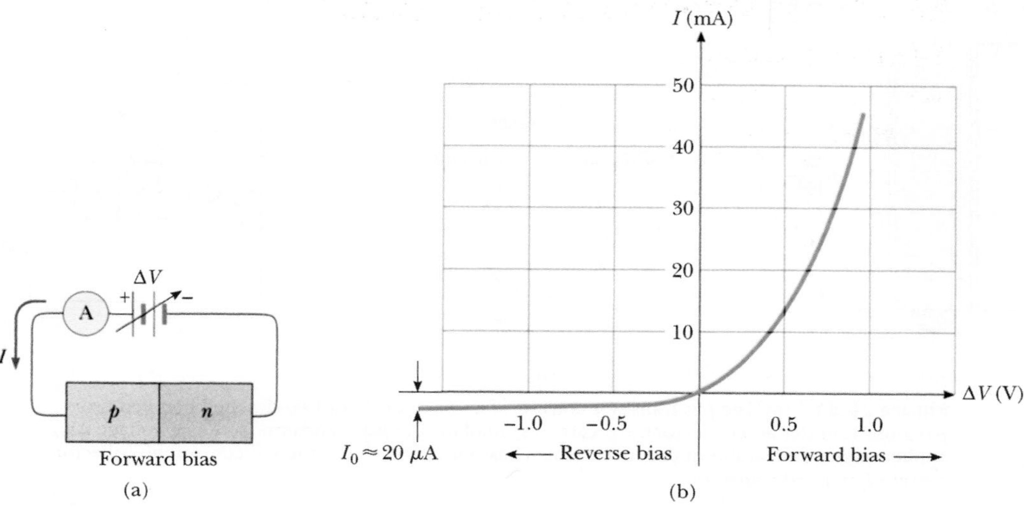

Figure 28.28 (a) Schematic of a p–n junction under forward bias. (b) The characteristic curve for a real p–n junction.

where q is the electron charge, k_B is Boltzmann's constant, and T is the temperature in kelvins. Figure 28.28 shows an I–ΔV plot characteristic of a real p–n junction, along with a schematic of such a device under forward bias.

The most common use of the semiconductor diode is as a rectifier, a device that changes 120-V AC voltage supplied by the power company to, say the 12-V DC voltage needed by your music keyboard. We can understand how a diode rectifies a current by considering Figure 28.29a, which shows a diode connected in series with a resistor and an AC source. Because appreciable current can pass through the diode in just one direction, the alternating current in the resistor is reduced to the form shown in Figure 28.29b. The diode is said to act as a **half-wave rectifier**, because there is current in the circuit during only half of each cycle.

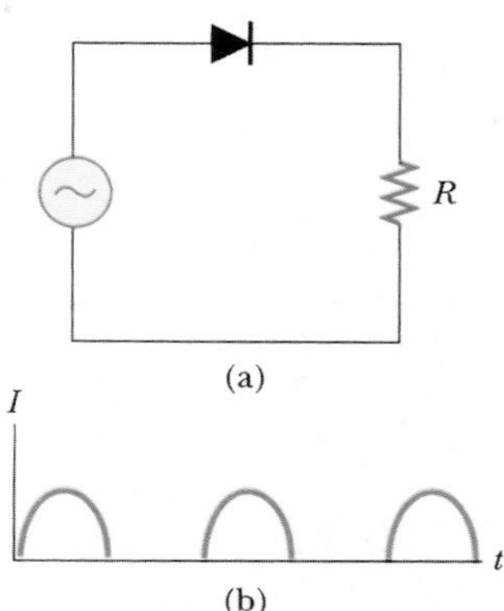

Figure 28.29 (a) A diode in series with a resistor allows current to pass in only one direction. (b) The current versus time for the circuit in (a).

Figure 28.30a shows a circuit that lowers the AC voltage to 12 V with a step-down transformer and then rectifies both halves of the 12-V AC. Such a rectifier is called a **full-wave rectifier** and when combined with a step-down transformer is the most common DC power supply around the home today. A capacitor added in parallel with the load will yield an even steadier DC voltage.

The Junction Transistor

The invention of the transistor by John Bardeen (1908–1991), Walter Brattain (1902–1987), and William Shockley (1910–1989) in 1948 totally revolutionized the world of electronics. For this work, these three men shared a Nobel prize in 1956. By 1960, the transistor had replaced the vacuum tube in many electronic applications. The advent of the transistor created a multitrillion-dollar industry that produced such popular devices as pocket radios, handheld calculators, computers, television receivers, and electronic games. In this section we explain how a transistor acts as an amplifier to boost the tiny voltages and currents generated in a microphone to the ear-splitting levels required to drive a speaker.

One simple form of the transistor, called the **junction transistor**, consists of a semiconducting material in which a very narrow n region is sandwiched between two p regions. This configuration is called a ***pnp* transistor**. Another configuration is the ***npn* transistor**, which consists of a p region sandwiched between two n regions. Because the operation of the two transistors is essentially the same, we describe only the *pnp* transistor. The structure of the *pnp* transistor, together with its circuit symbol, is shown in Figure 28.31 (page 930). The outer regions are called the **emitter** and **collector**, and the narrow central region is called the **base**. The configuration contains two junctions: the emitter–base interface and the collector–base interface.

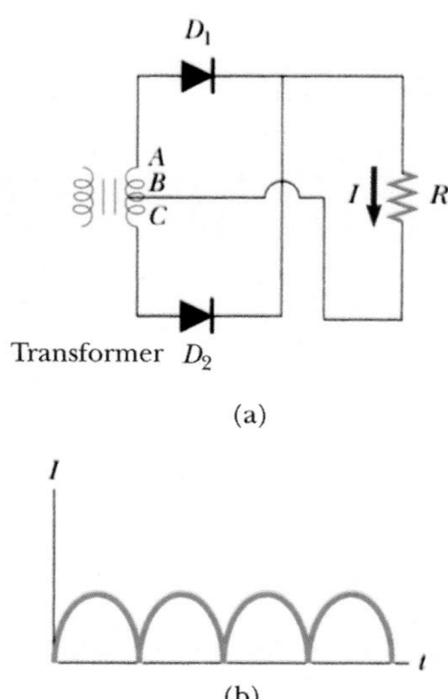

Figure 28.30 (a) A full-wave rectifier circuit. (b) The current versus time in the resistor R.

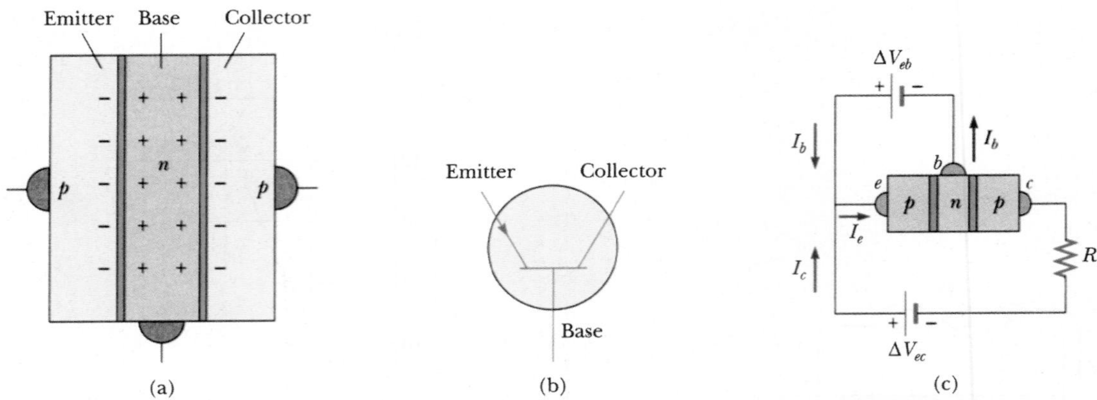

Figure 28.31 (a) The *pnp* transistor consists of an *n* region (base) sandwiched between two *p* regions (emitter and collector). (b) Circuit symbol for the *pnp* transistor. (c) A bias voltage ΔV_{eb} applied to the base as shown produces a small base current I_b that is used to control the collector current I_c in a *pnp* transistor.

Suppose a voltage is applied to the transistor so that the emitter is at a higher electric potential than the collector. (This is accomplished with the battery labeled ΔV_{ec} in Figure 28.31c.) If we think of the transistor as two diodes back to back, we see that the emitter–base junction is forward biased and the base–collector junction is reverse biased. The emitter is heavily doped relative to the base, and as a result, nearly all the current consists of holes moving across the emitter–base junction. Most of these holes do not recombine with electrons in the base because it is very narrow. Instead they are accelerated across the reverse-biased base–collector junction, producing the emitter current I_e in Figure 28.31c.

Although only a small percentage of holes recombine in the base, those that do limit the emitter current to a small value because positive charge carriers accumulating in the base prevent holes from flowing in. In order not to limit the emitter current, some of the positive charge on the base must be drawn off; this is accomplished by connecting the base to the battery labeled ΔV_{eb} in Figure 28.31c. Those positive charges that are not swept across the base–collector junction leave the base through this added pathway. **This base current I_b is very small, but a small change in it can significantly change the collector current I_c.** If the transistor is properly biased, the collector (output) current is directly proportional to the base (input) current and the transistor acts as a current amplifier. This condition may be written

$$I_c = \beta I_b$$

where β, the *current gain* factor, is typically in the range from 10 to 100. Thus, the transistor may be used to amplify a small signal. The small voltage to be amplified is placed in series with the battery V_{eb}. The input signal produces a small variation in the base current, resulting in a large change in the collector current and hence a large change in the voltage across the output resistor.

The Integrated Circuit

Invented independently by Jack Kilby (b. 1923) at Texas Instruments in late 1958 and by Robert Noyce at Fairchild Camera and Instrument in early 1959, the integrated circuit has been justly called "the most remarkable technology ever to hit mankind." Kilby's first device is shown in Figure 28.32a. Integrated circuits have indeed started a "second industrial revolution" and are found at the heart of computers, watches, cameras, automobiles, aircraft, robots, space vehicles, and all sorts of communication and switching networks.

In simplest terms, an **integrated circuit** is a collection of interconnected transistors, diodes, resistors, and capacitors fabricated on a single piece of silicon known as a chip. State-of-the-art chips easily contain several million components in

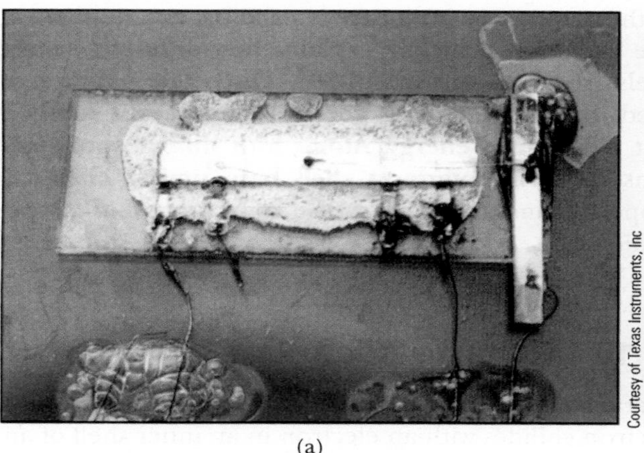

(a)

(b)

Figure 28.32 (a) Jack Kilby's first integrated circuit was tested on September 12, 1958. (b) Integrated circuits continue to shrink in size and price while simultaneously growing in capability.

a 1-cm^2 area, with the number of components per square inch having doubled every year since the integrated circuit was invented.

Integrated circuits were invented partly to solve the interconnection problem spawned by the transistor. In the era of vacuum tubes, power and size considerations of individual components set significant limits on the number of components that could be interconnected in a given circuit. With the advent of the tiny, low-power, highly reliable transistor, design limits on the number of components disappeared and were replaced by the problem of wiring together hundreds of thousands of components. The magnitude of this problem can be appreciated when we consider that second-generation computers (consisting of discrete transistors rather than integrated circuits) contained several hundred thousand components requiring more than a million hand-soldered joints to be made and tested.

In addition to solving the interconnection problem, integrated circuits possess the advantages of miniaturization and fast response, two attributes critical for high-speed computers. The fast response results from the miniaturization and close packing of components, because the response time of a circuit depends on the time it takes for electrical signals traveling at about the speed of light to pass from one component to another. This time is clearly reduced by packing components closely.

SUMMARY

Physics Now™ Take a practice test by logging into PhysicsNow at **www.cp7e.com** and clicking on the Pre-Test link for this chapter.

28.3 The Bohr Theory of Hydrogen &
28.4 Modification of the Bohr Theory

The **Bohr model** of the atom is successful in describing the spectra of atomic hydrogen and hydrogenlike ions. One of the basic assumptions of the model is that the electron can exist only in certain orbits such that its angular momentum mvr is an integral multiple of $\hbar$, where $\hbar$ is Planck's constant divided by 2π. Assuming circular orbits and a Coulomb force of attraction between electron and proton, the energies of the quantum states for hydrogen are

$$E_n = -\frac{m_e k_e^2 e^4}{2\hbar^2}\left(\frac{1}{n^2}\right) \qquad n = 1, 2, 3, \ldots \qquad \textbf{[28.12]}$$

where k_e is the Coulomb constant, e is the charge on the electron, and n is an integer called a **quantum number**.

If the electron in the hydrogen atom jumps from an orbit having quantum number n_i to an orbit having quantum number n_f, it emits a photon of frequency f, given by

$$f = \frac{m_e k_e^2 e^4}{4\pi \hbar^3}\left(\frac{1}{n_f^2} - \frac{1}{n_i^2}\right) \qquad [28.14]$$

Bohr's **correspondence principle** states that quantum mechanics is in agreement with classical physics when the quantum numbers for a system are very large.

The Bohr theory can be generalized to hydrogen-like atoms, such as singly ionized helium or doubly ionized lithium. This modification consists of replacing e^2 by Ze^2 wherever it occurs.

28.6 Quantum Mechanics and the Hydrogen Atom &

28.7 The Spin Magnetic Quantum Number

One of the many successes of quantum mechanics is that the quantum numbers n, ℓ, and m_ℓ associated with atomic structure arise directly from the mathematics of the theory. The quantum number n is called the **principal quantum number**, ℓ is the **orbital quantum number**, and m_ℓ is the **orbital magnetic quantum number**. These quantum numbers can take only certain values: $1 \leq n < \infty$ in integer steps, $0 \leq \ell \leq n - 1$, and $-\ell \leq m_\ell \leq \ell$. In addition, a fourth quantum number, called the **spin magnetic quantum number** m_s, is needed to explain a fine doubling of lines in atomic spectra, with $m_s = \pm\frac{1}{2}$.

28.9 The Exclusion Principle and the Periodic Table

An understanding of the periodic table of the elements became possible when Pauli formulated the **exclusion principle**, which states that no two electrons in an atom in the same atom can have the same values for the set of quantum numbers n, ℓ, m_ℓ, and m_s. A particular set of these quantum numbers is called a quantum state. The exclusion principle explains how different energy levels in atoms are populated. Once one subshell is filled, the next electron goes into the vacant subshell that is lowest in energy. Atoms with similar configurations in their outermost shell have similar chemical properties and are found in the same column of the periodic table.

28.10 Characteristic X-Rays

Characteristic x-rays are produced when a bombarding electron collides with an electron in an inner shell of an atom with sufficient energy to remove the electron from the atom. The vacancy is filled when an electron from a higher level drops down into the level containing the vacancy, emitting a photon in the x-ray part of the spectrum in the process.

28.11 Atomic Transitions &

28.12 Lasers and Holography

When an atom is irradiated by light of all different wavelengths, it will only absorb only wavelengths equal to the difference in energy of two of its energy levels. This phenomenon, called **stimulated absorption**, places an atom's electrons into **excited states**. Atoms in an excited state have a probability of returning to a lower level of excitation by **spontaneous emission**. The wavelengths that can be emitted are the same as the wavelengths that can be absorbed. If an atom is in an excited state and a photon with energy $hf = E_2 - E_1$ is incident on it, the probability of emission of a second photon of this energy is greatly enhanced. The emitted photon is exactly in phase with the incident photon. This process is called **stimulated emission**. The emitted and original photon can then stimulate more emission, creating an amplifying effect.

Lasers are monochromatic, coherent light sources that work on the principle of **stimulated emission** of radiation from a system of atoms.

CONCEPTUAL QUESTIONS

1. In the hydrogen atom, the quantum number n can increase without limit. Because of this, does the frequency of possible spectral lines from hydrogen also increase without limit?

2. Does the light emitted by a neon sign constitute a continuous spectrum or only a few colors? Defend your answer.

3. In an x-ray tube, if the energy with which the electrons strike the metal target is increased, the wavelengths of the characteristic x-rays do not change. Why not?

4. Must an atom first be ionized before it can emit light? Discuss.

5. Is it possible for a spectrum from an x-ray tube to show the continuous spectrum of x-rays without the presence of the characteristic x-rays?

6. Suppose that the electron in the hydrogen atom obeyed classical mechanics rather than quantum mechanics. Why should such a hypothetical atom emit a continuous spectrum rather than the observed line spectrum?

7. When a hologram is produced, the system (including light source, object, beam splitter, and so on) must be held motionless within a quarter of the light's wavelength. Why?

8. If matter has a wave nature, why is it not observable in our daily experience?

9. Discuss some consequences of the exclusion principle.

10. Can the electron in the ground state of hydrogen absorb a photon of energy less than 13.6 eV? Can it absorb a photon of energy greater than 13.6 eV? Explain.

11. Why do lithium, potassium, and sodium exhibit similar chemical properties?

12. List some ways in which quantum mechanics altered our view of the atom pictured by the Bohr theory.

13. It is easy to understand how two electrons (one with spin up, one with spin down) can fill the $1s$ shell for a helium atom. How is it possible that eight more electrons can fit into the $2s$, $2p$ level to complete the $1s2s^22p^6$ shell for a neon atom?

14. The ionization energies for Li, Na, K, Rb, and Cs are 5.390, 5.138, 4.339, 4.176, and 3.893 eV, respectively. Explain why these values are to be expected in terms of the atomic structures.

15. Why is stimulated emission so important in the operation of a laser?

16. The Bohr theory of the hydrogen atom is based upon several assumptions. Discuss these assumptions and their significance. Do any of them contradict classical physics?

17. Explain why, in the Bohr model, the total energy of the hydrogen atom is negative.

18. Consider the quantum numbers n, ℓ, m_ℓ, and m_s. (a) Which of these are integers and which are fractional? (b) Which are always positive and which can be negative? (c) If $n = 2$, what is the largest value of ℓ? (d) If $\ell = 1$, what are the possible values of m_ℓ?

19. Photon A is emitted when an electron in a hydrogen atom drops from the $n = 3$ level to the $n = 2$ level. Photon B is emitted when an electron in a hydrogen atom drops from the $n = 4$ level to the $n = 2$ level. (a) In which case is the wavelength of the emitted photon greater? (b) In which case is the energy of the emitted photon greater?

PROBLEMS

1, 2, 3 = straightforward, intermediate, challenging   □ = full solution available in *Student Solutions Manual/Study Guide*
Physics Now™ = coached problem with hints available at **www.cp7e.com**   = biomedical application

Section 28.1 Early Models of the Atom

Section 28.2 Atomic Spectra

1. Use Equation 28.1 to calculate the wavelength of the first three lines in the Balmer series for hydrogen.

2. Show that the wavelengths for the Balmer series satisfy the equation

$$\lambda = \frac{364.5n^2}{n^2 - 4}\,\text{nm} \qquad \text{where } n = 3, 4, 5, \ldots$$

3. The "size" of the *atom* in Rutherford's model is about 1.0×10^{-10} m. (a) Determine the attractive electrostatic force between an electron and a proton separated by this distance. (b) Determine (in eV) the electrostatic potential energy of the atom.

4. The "size" of the *nucleus* in Rutherford's model of the atom is about 1.0 fm $= 1.0 \times 10^{-15}$ m. (a) Determine the repulsive electrostatic force between two protons separated by this distance. (b) Determine (in MeV) the electrostatic potential energy of the pair of protons.

5. Physics Now™ The "size" of the atom in Rutherford's model is about 1.0×10^{-10} m. (a) Determine the speed of an electron moving about the proton using the attractive electrostatic force between an electron and a proton separated by this distance. (b) Does this speed suggest that Einsteinian relativity must be considered in studying the atom? (c) Compute the de Broglie wavelength of the electron as it moves about the proton. (d) Does this wavelength suggest that wave effects, such as diffraction and interference, must be considered in studying the atom?

6. In a Rutherford scattering experiment, an α-particle (charge $= +2e$) heads directly toward a gold nucleus (charge $= +79e$). The α-particle had a kinetic energy of 5.0 MeV when very far ($r \rightarrow \infty$) from the nucleus. Assuming the gold nucleus to be fixed in space, determine the distance of closest approach. [*Hint*: Use conservation of energy with $PE = k_e q_1 q_2 / r$.]

Section 28.3 The Bohr Theory of Hydrogen

7. A hydrogen atom is in its first excited state ($n = 2$). Using the Bohr theory of the atom, calculate (a) the radius of the orbit, (b) the linear momentum of the electron, (c) the angular momentum of the electron, (d) the kinetic energy, (e) the potential energy, and (f) the total energy.

8. For a hydrogen atom in its ground state, use the Bohr model to compute (a) the orbital speed of the electron, (b) the kinetic energy of the electron, and (c) the electrical potential energy of the atom.

9. Show that the speed of the electron in the nth Bohr orbit in hydrogen is given by

$$v_n = \frac{k_e e^2}{n\hbar}$$

10. A photon is emitted as a hydrogen atom undergoes a transition from the $n = 6$ state to the $n = 2$ state. Calculate (a) the energy, (b) the wavelength, and (c) the frequency of the emitted photon.

11. A hydrogen atom emits a photon of wavelength 656 nm. From what energy orbit to what lower energy orbit did the electron jump?

12. Following are four possible transitions for a hydrogen atom

I. $n_i = 2$; $n_f = 5$ II. $n_i = 5$; $n_f = 3$

III. $n_i = 7$; $n_f = 4$ IV. $n_i = 4$; $n_f = 7$

(a) Which transition will emit the shortest-wavelength photon? (b) For which transition will the atom gain the most energy? (c) For which transition(s) does the atom lose energy?

13. What is the energy of a photon that, when absorbed by a hydrogen atom, could cause (a) an electronic transition from the $n = 3$ state to the $n = 5$ state and (b) an electronic transition from the $n = 5$ state to the $n = 7$ state?

14. A hydrogen atom initially in its ground state ($n = 1$) absorbs a photon and ends up in the state for which $n = 3$. (a) What is the energy of the absorbed photon? (b) If the atom eventually returns to the ground state, what photon energies could the atom emit?

15. Determine both the longest and the shortest wavelengths in (a) the Lyman series ($n_f = 1$) and (b) the Paschen series ($n_f = 3$) of hydrogen.

16. Show that the speed of the electron in the first (ground-state) Bohr orbit of the hydrogen atom may be expressed as

$$v = (1/137)c.$$

17. A monochromatic beam of light is absorbed by a collection of ground-state hydrogen atoms in such a way that six different wavelengths are observed when the hydrogen relaxes back to the ground state. What is the wavelength of the incident beam?

18. A particle of charge q and mass m, moving with a constant speed v, perpendicular to a constant magnetic field, B, follows a circular path. If in this case the angular momentum about the center of this circle is quantized so that $mvr = 2n\hbar$, show that the allowed radii for the particle are

$$r_n = \sqrt{\frac{2\,n\hbar}{qB}}$$

where $n = 1, 2, 3, \ldots$

19. (a) If an electron makes a transition from the $n = 4$ Bohr orbit to the $n = 2$ orbit, determine the wavelength of the photon created in the process. (b) Assuming that the atom was initially at rest, determine the recoil speed of the hydrogen atom when this photon is emitted.

20. Consider a large number of hydrogen atoms, with electrons all initially in the $n = 4$ state. (a) How many different wavelengths would be observed in the emission spectrum of these atoms? (b) What is the longest wavelength that could be observed? To which series does it belong?

21. Analyze the Earth–Sun system by following the Bohr model, where the gravitational force between Earth (mass m) and Sun (mass M) replaces the Coulomb force between the electron and proton (so that $F = GMm/r^2$ and $PE = -GMm/r$). Show that (a) the total energy of the Earth in an orbit of radius r is given by (a) $E = -GMm/2r$, (b) the radius of the nth orbit is given by $r_n = r_0 n^2$, where $r_0 = \hbar^2/GMm^2 = 2.32 \times 10^{-138}$ m, and (c) the energy of the nth orbit is given by $E_n = -E_0/n^2$, where $E_0 = G^2M^2m^3/2\hbar^2 = 1.71 \times 10^{182}$ J. (d) Using the Earth–Sun orbit radius of $r = 1.49 \times 10^{11}$ m, determine the value of the quan-

tum number n. (e) Should you expect to observe quantum effects in the Earth–Sun system?

22. An electron is in the nth Bohr orbit of the hydrogen atom. (a) Show that the period of the electron is $T = t_o n^3$, and determine the numerical value of t_o. (b) On the average, an electron remains in the $n = 2$ orbit for about 10 μs before it jumps down to the $n = 1$ (ground-state) orbit. How many revolutions does the electron make before it jumps to the ground state? (c) If one revolution of the electron is defined as an "electron year" (analogous to an Earth year being one revolution of the Earth around the Sun), does the electron in the $n = 2$ orbit "live" very long? Explain. (d) How does the above calculation support the "electron cloud" concept?

23. **Physics Now™** Consider a hydrogen atom. (a) Calculate the frequency f of the $n = 2 \rightarrow n = 1$ transition, and compare it with the frequency f_{orb} of the electron orbital motion in the $n = 2$ state. (b) Make the same calculation for the $n = 10\,000 \rightarrow n = 9\,999$ transition. Comment on the results.

24. Two hydrogen atoms collide head-on and end up with zero kinetic energy. Each then emits a 121.6-nm photon ($n = 2$ to $n = 1$ transition). At what speed were the atoms moving before the collision?

25. Two hydrogen atoms, both initially in the ground state, undergo a head-on collision. If both atoms are to be excited to the $n = 2$ level in this collision, what is the minimum speed each atom can have before the collision?

26. (a) Calculate the angular momentum of the Moon due to its orbital motion about the Earth. In your calculation, use 3.84×10^8 m as the average Earth–Moon distance and 2.36×10^6 s as the period of the Moon in its orbit. (b) If the angular momentum of the moon obeys Bohr's quantization rule ($L = n\hbar$), determine the value of the quantum number n. (c) By what fraction would the Earth–Moon radius have to be increased to increase the quantum number by 1?

Section 28.4 Modification of the Bohr Theory

Section 28.5 De Broglie Waves and the Hydrogen Atom

27. (a) Find the energy of the electron in the ground state of doubly ionized lithium, which has an atomic number $Z = 3$. (b) Find the radius of its ground-state orbit.

28. (a) Construct an energy level diagram for the He^+ ion, for which $Z = 2$. (b) What is the ionization energy for He^+?

29. The orbital radii of a hydrogen-like atom is given by the equation

$$r = \frac{n^2 \hbar^2}{Z m_e k_e e^2}.$$

What is the radius of the first Bohr orbit in (a) He^+, (b) Li^{2+}, and (c) Be^{3+}?

30. (a) Substitute numerical values into Equation 28.19 to find a value for the Rydberg constant for singly ionized helium, He^+. (b) Use the result of part (a) to find the wavelength associated with a transition from the $n = 2$ state to the $n = 1$ state of He^+. (c) Identify the region of the electromagnetic spectrum associated with this transition.

31. Determine the wavelength of an electron in the third excited orbit of the hydrogen atom, with $n = 4$.

32. Using the concept of standing waves, de Broglie was able to derive Bohr's stationary orbit postulate. He assumed that a confined electron could exist only in states where its de Broglie waves form standing-wave patterns, as in Figure 28.10a. Consider a particle confined in a box of length L to be equivalent to a string of length L and fixed at both ends. Apply de Broglie's concept to show that (a) the linear momentum of this particle is quantized with $p = mv = nh/2L$ and (b) the allowed states correspond to particle energies of $E_n = n^2 E_0$, where $E_0 = h^2/(8mL^2)$.

Section 28.6 Quantum Mechanics and the Hydrogen Atom

Section 28.7 The Spin Magnetic Quantum Number

33. List the possible sets of quantum numbers for electrons in the $3p$ subshell.

34. When the principal quantum number is $n = 4$, how many different values of (a) ℓ and (b) m_ℓ are possible?

35. The ρ-meson has a charge of $-e$, a spin quantum number of 1, and a mass 1 507 times that of the electron. If the electrons in atoms were replaced by ρ-mesons, list the possible sets of quantum numbers for ρ-mesons in the $3d$ subshell.

Section 28.9 The Exclusion Principle and the Periodic Table

36. (a) Write out the electronic configuration of the ground state for oxygen ($Z = 8$). (b) Write out the values for the set of quantum numbers n, ℓ, m_ℓ, and m_s for each of the electrons in oxygen.

37. Two electrons in the same atom have $n = 3$ and $\ell = 1$. (a) List the quantum numbers for the possible states of the atom. (b) How many states would be possible if the exclusion principle did not apply to the atom?

38. How many different sets of quantum numbers are possible for an electron for which (a) $n = 1$, (b) $n = 2$, (c) $n = 3$, (d) $n = 4$, and (e) $n = 5$? Check your results to show that they agree with the general rule that the number of different sets of quantum numbers is equal to $2n^2$.

39. Physics Now™ Zirconium ($Z = 40$) has two electrons in an incomplete d subshell. (a) What are the values of n and ℓ for each electron? (b) What are all possible values of m_ℓ and m_s? (c) What is the electron configuration in the ground state of zirconium?

Section 28.10 Characteristic X-Rays

40. The K-shell ionization energy of copper is 8 979 eV. The L-shell ionization energy is 951 eV. Determine the wavelength of the K_α emission line of copper. What must the minimum voltage be on an x-ray tube with a copper target in order to see the K_α line?

41. The K_α x-ray is emitted when an electron undergoes a transition from the L shell ($n = 2$) to the K shell ($n = 1$). Use the method illustrated in Example 28.5 to calculate the wavelength of the K_α x-ray from a nickel target ($Z = 28$).

42. When an electron drops from the M shell ($n = 3$) to a vacancy in the K shell ($n = 1$), the measured wavelength of the emitted x-ray is found to be 0.101 nm. Identify the element.

43. The K series of the discrete spectrum of tungsten contains wavelengths of 0.018 5 nm, 0.020 9 nm, and 0.021 5 nm. The K-shell ionization energy is 69.5 keV. Determine the ionization energies of the L, M, and N shells. Sketch the transitions that produce the above wavelengths.

ADDITIONAL PROBLEMS

44. In a hydrogen atom, what is the principle quantum number of the electron orbit with a radius closest to 1.0 μm?

45. (a) How much energy is required to cause an electron in hydrogen to move from the $n = 1$ state to the $n = 2$ state? (b) If the electrons gain this energy by collision between hydrogen atoms in a high-temperature gas, find the minimum temperature of the heated hydrogen gas. The thermal energy of the heated atoms is given by $3k_B T/2$, where k_B is the Boltzmann constant.

46. A pulsed ruby laser emits light at 694.3 nm. For a 14.0-ps pulse containing 3.00 J of energy, find (a) the physical length of the pulse as it travels through space and (b) the number of photons in it. (c) If the beam has a circular cross section 0.600 cm in diameter, find the number of photons per cubic millimeter.

47. The Lyman series for a (new?) one-electron atom is observed in a distant galaxy. The wavelengths of the first four lines and the short-wavelength limit of this Lyman series are given by the energy-level diagram in Figure P28.47. Based on this information, calculate (a) the energies of the ground state and first four excited states for this one-electron atom and (b) the longest-wavelength (alpha) lines and the short-wavelength series limit in the Balmer series for this atom.

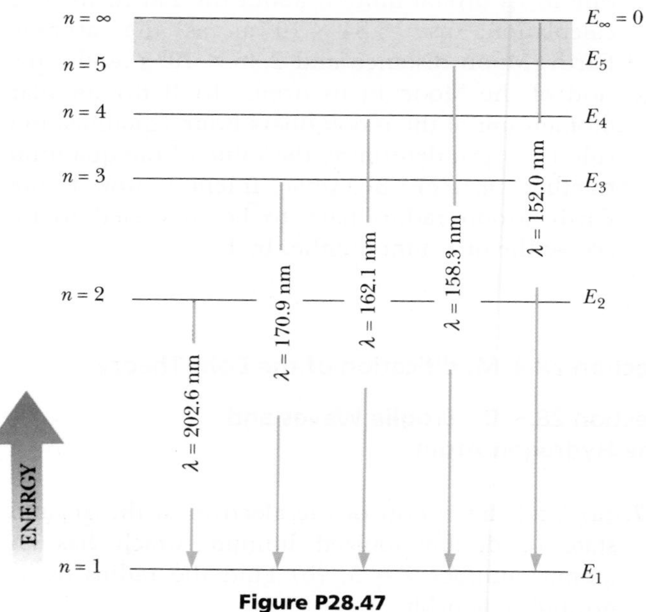

Figure P28.47

48. A dimensionless number that often appears in atomic physics is the fine-structure constant $\alpha = k_e e^2/\hbar c$, where k_e is the Coulomb constant. (a) Obtain a numerical value for $1/\alpha$. (b) In terms of α, what is the ratio of the Bohr radius a_0 to the Compton wavelength $\lambda_C = h/m_e c$? (d) In terms of α, what is the ratio of the reciprocal of the Rydberg constant $1/R_H$ to the Bohr radius?

49. Mercury's ionization energy is 10.39 eV. The three longest wavelengths of the absorption spectrum of mercury are 253.7 nm, 185.0 nm, and 158.5 nm. (a) Construct an energy-level diagram for mercury. (b) Indicate all emission lines that can occur when an electron is raised to the third level above the ground state. (c) Disregarding recoil of the mercury atom, determine the minimum speed an electron must have in order to make an inelastic collision with a mercury atom in its ground state.

50. Suppose the ionization energy of an atom is 4.100 eV. In this same atom, we observe emission lines that have wavelengths of 310.0 nm, 400.0 nm, and 1 378 nm. Use this information to construct the energy-level diagram with the least number of levels. Assume the higher energy levels are closer together.

51. A laser used in eye surgery emits a 3.00-mJ pulse in 1.00 ns, focused to a spot 30.0 μm in diameter on the retina. (a) Find (in SI units) the power per unit area at the retina. (This quantity is called the *irradiance.*) (b) What energy is delivered per pulse to an area of molecular size—say, a circular area 0.600 nm in diameter.

52. An electron has a de Broglie wavelength equal to the diameter of a hydrogen atom in its ground state. (a) What is the kinetic energy of the electron? (b) How does this energy compare with the ground-state energy of the hydrogen atom?

53. Use Bohr's model of the hydrogen atom to show that, when the atom makes a transition from the state n to the state $n - 1$, the frequency of the emitted light is given by

$$f = \frac{2\pi^2 m k_e^{\,2} e^4}{h^3}\left(\frac{2n-1}{(n-1)^2 n^2}\right)$$

54. Calculate the classical frequency for the light emitted by an atom. To do so, note that the frequency of revolution is $v/2\pi r$, where r is the Bohr radius. Show that as n approaches infinity in the equation of the preceding problem, the expression given there varies as $1/n^3$ and reduces to the classical frequency. (This is an example of the correspondence principle, which requires that the classical and quantum models agree for large values of n.)

55. **Physics Now™** A pi meson (π^-) of charge $-e$ and mass 273 times greater than that of the electron is captured by a helium nucleus ($Z = +2$) as shown in Figure P28.55. (a) Draw an energy-level diagram (in units of eV) for this "Bohr-type" atom up to the first six energy levels. (b) When the π-meson makes a transition between two orbits, a photon is emitted that Compton scatters off a free electron initially at rest, producing a scattered photon of wavelength $\lambda' = 0.089\,929\,3$ nm at an angle of $\theta = 42.68°$, as shown on the right-hand side of Figure P28.55. Between which two orbits did the π-meson make a transition?

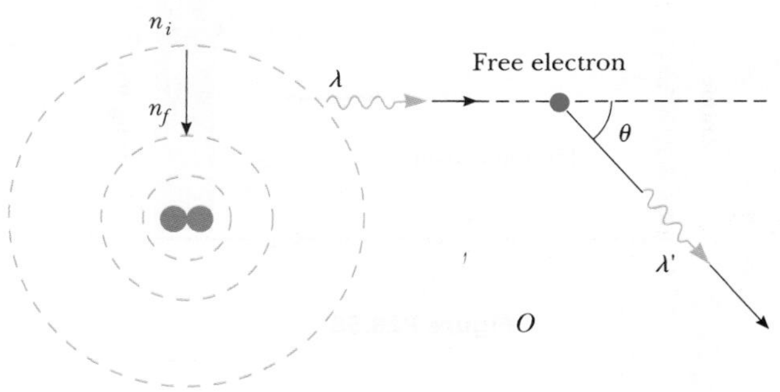

Figure P28.55

56. When a muon with charge $-e$ is captured by a proton, the resulting bound system forms a "muonic atom," which is the same as hydrogen, except with a muon (of mass 207 times the mass of an electron) replacing the electron. For this "muonic atom," determine (a) the Bohr radius and (b) the three lowest energy levels.

57. In this problem, you will estimate the classical lifetime of the hydrogen atom. An accelerating charge loses electromagnetic energy at a rate given by $\mathcal{P} = -2k_e q^2 a^2/(3c^3)$, where k_e is the Coulomb constant, q is the charge of the particle, a is its acceleration, and c is the speed of light in a vacuum. Assume that the electron is one Bohr radius (0.052 9 nm) from the center of the hydrogen atom. (a) Determine its acceleration. (b) Show that $\mathcal{P}$ has units of energy per unit time and determine the rate of energy loss. (c) Calculate the kinetic energy of the electron and determine how long it will take for all of this energy to be converted into electromagnetic waves, assuming that the rate calculated in part (b) remains constant throughout the electron's motion.

58. An electron in a hydrogen atom jumps from some initial Bohr orbit n_i to some final Bohr orbit n_f, as in

Figure P28.58. (a) If the photon emitted in the process is capable of ejecting a photoelectron from tungsten (work function = 4.58 eV), determine n_f. (b) If a minimum stopping potential of V_0 = 7.51 volts is required to prevent the photoelectron from hitting the anode, determine the value of n_i.

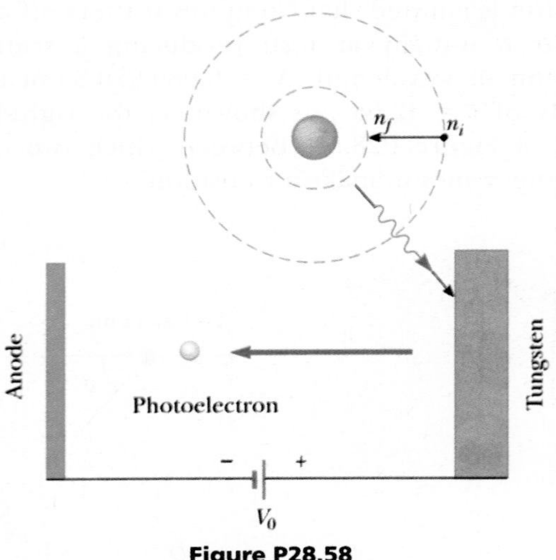

Figure P28.58

ACTIVITIES

1. With your partner not looking, use modeling clay to build one or more mounds on top of a table. Place a piece of cardboard over your mound(s), and assign your partner the task of determining the size, shape, and number of mounds without looking. He is to do this by rolling marbles at the unseen mounds and observing how they emerge. This experiment models the Rutherford scattering experiment.

2. Your instructor can probably lend you a small plastic diffraction grating to enable you to examine the spectrum of different light sources. You can use these gratings to examine a source by holding the grating very close to your eye and noting the spectrum produced by glancing out of the corner of your eye while looking at a light source. You should look at light sources such as sodium vapor lights and mercury vapor lights used in many parking lots, neon lights used in many signs, black lights, ordinary incandescent light bulbs, and so forth.

Courtesy of Public Service Electric and Gas Company

Aerial view of a nuclear power plant that generates electrical power. Energy is generated in such plants from the process of nuclear fission, in which a heavy nucleus such as ^{235}U splits into smaller particles having a large amount of kinetic energy. This surplus energy can be used to heat water into high pressure steam and drive a turbine.

CHAPTER 29

Nuclear Physics

OUTLINE

In 1896, the year that marks the birth of nuclear physics, Henri Becquerel (1852–1908) discovered radioactivity in uranium compounds. A great deal of activity followed this discovery as researchers attempted to understand and characterize the radiation that we now know to be emitted by radioactive nuclei. Pioneering work by Rutherford showed that the radiation was of three types, which he called *alpha, beta,* and *gamma rays.* These types are classified according to the nature of their electric charge and their ability to penetrate matter. Later experiments showed that alpha rays are helium nuclei, beta rays are electrons, and gamma rays are high-energy photons.

In 1911 Rutherford and his students Geiger and Marsden performed a number of important scattering experiments involving alpha particles. These experiments established the idea that the nucleus of an atom can be regarded as essentially a point mass and point charge and that most of the atomic mass is contained in the nucleus. Further, such studies demonstrated a wholly new type of force: the *nuclear force*, which is predominant at distances of less than about 10^{-14} m and drops quickly to zero at greater distances.

Other milestones in the development of nuclear physics include

- the first observations of nuclear reactions by Rutherford and coworkers in 1919, in which naturally occurring α particles bombarded nitrogen nuclei to produce oxygen,
- the first use of artificially accelerated protons to produce nuclear reactions, by Cockcroft and Walton in 1932,
- the discovery of the neutron by Chadwick in 1932,
- the discovery of artificial radioactivity by Joliot and Irene Curie in 1933,
- the discovery of nuclear fission by Hahn, Strassman, Meitner, and Frisch in 1938, and
- the development of the first controlled fission reactor by Fermi and his collaborators in 1942.

In this chapter we discuss the properties and structure of the atomic nucleus. We start by describing the basic properties of nuclei and follow with a discussion of the phenomenon of radioactivity. Finally, we explore nuclear reactions and the various processes by which nuclei decay.

ERNEST RUTHERFORD,
New Zealand Physicist
(1871–1937)

Rutherford was awarded the Nobel Prize in 1908 for discovering that atoms can be broken apart by alpha rays and for studying radioactivity. "On consideration, I realized that this scattering backward must be the result of a single collision, and when I made calculations I saw that it was impossible to get anything of that order of magnitude unless you took a system in which the greater part of the mass of the atom was concentrated in a minute nucleus. It was then that I had the idea of an atom with a minute massive center carrying a charge."

29.1 SOME PROPERTIES OF NUCLEI

All nuclei are composed of two types of particles: protons and neutrons. The only exception is the ordinary hydrogen nucleus, which is a single proton. In describing some of the properties of nuclei, such as their charge, mass, and radius, we make use of the following quantities:

- the **atomic number** Z, which equals the number of protons in the nucleus,
- the **neutron number** N, which equals the number of neutrons in the nucleus,
- the **mass number** A, which equals the number of nucleons in the nucleus (*nucleon* is a generic term used to refer to either a proton or a neutron).

The symbol we use to represent nuclei is ${}^{A}_{Z}\mathrm{X}$, where X represents the chemical symbol for the element. For example, ${}^{27}_{13}\mathrm{Al}$ has the mass number 27 and the atomic number 13; therefore, it contains 13 protons and 14 neutrons. When no confusion is likely to arise, we often omit the subscript Z, because the chemical symbol can always be used to determine Z.

The nuclei of all atoms of a particular element must contain the same number of protons, but they may contain different numbers of neutrons. Nuclei that are related in this way are called **isotopes. The isotopes of an element have the same Z value, but different N and A values.** The natural abundances of isotopes can differ substantially. For example, ${}^{11}_{6}\mathrm{C}$, ${}^{12}_{6}\mathrm{C}$, ${}^{13}_{6}\mathrm{C}$, and ${}^{14}_{6}\mathrm{C}$ are four isotopes of carbon. The natural abundance of the ${}^{12}_{6}\mathrm{C}$ isotope is about 98.9%, whereas that of the ${}^{13}_{6}\mathrm{C}$ isotope is only about 1.1%. Some isotopes don't occur naturally, but can be produced in the laboratory through nuclear reactions. Even the simplest element, hydrogen, has isotopes: ${}^{1}_{1}\mathrm{H}$, hydrogen; ${}^{2}_{1}\mathrm{H}$, deuterium; and ${}^{3}_{1}\mathrm{H}$, tritium.

Charge and Mass

The proton carries a single positive charge $+e = 1.602\,177\,33 \times 10^{-19}$ C, the electron carries a single negative charge $-e$, and the neutron is electrically neutral. Because the neutron has no charge, it's difficult to detect. The proton is about 1 836 times as massive as the electron, and the masses of the proton and the neutron are almost equal (Table 29.1).

Definition of the unified mass unit u ▶

For atomic masses, it is convenient to define the **unified mass unit** u in such a way that the mass of one atom of the isotope ${}^{12}\mathrm{C}$ is exactly 12 u, where 1 u $= 1.660\,559 \times 10^{-27}$ kg. The proton and neutron each have a mass of about 1 u, and the electron has a mass that is only a small fraction of an atomic mass unit.

Because the rest energy of a particle is given by $E_R = mc^2$, it is often convenient to express the particle's mass in terms of its energy equivalent. For one atomic mass unit, we have an energy equivalent of

$$E_R = mc^2 = (1.660\,559 \times 10^{-27}\ \mathrm{kg})(2.997\,92 \times 10^8\ \mathrm{m/s})^2$$
$$= 1.492\,431 \times 10^{-10}\ \mathrm{J} = 931.494\ \mathrm{MeV}$$

In calculations, nuclear physicists often express *mass* in terms of the unit MeV/c^2, where

$$1\ \mathrm{u} = 931.494\ \mathrm{MeV}/c^2$$

TIP 29.1 Mass Number is not the Atomic Mass
Don't confuse the mass number A with the atomic mass. Mass number is an integer that specifies an isotope and has no units—it's simply equal to the number of nucleons. Atomic mass is an average of the masses of the isotopes of a given element and has units of u.

TABLE 29.1

Masses of the Proton, Neutron, and Electron in Various Units

	Mass		
Particle	**kg**	**u**	**MeV/c^2**
Proton	1.6726×10^{-27}	1.007 276	938.28
Neutron	1.6750×10^{-27}	1.008 665	939.57
Electron	9.109×10^{-31}	5.486×10^{-4}	0.511

The Size of Nuclei

The size and structure of nuclei were first investigated in the scattering experiments of Rutherford, discussed in Section 28.1. Using the principle of conservation of energy, Rutherford found an expression for how close an alpha particle moving directly toward the nucleus can come to the nucleus before being turned around by Coulomb repulsion.

In such a head-on collision, the kinetic energy of the incoming alpha particle must be converted completely to electrical potential energy when the particle stops at the point of closest approach and turns around (Active Fig. 29.1). If we equate the initial kinetic energy of the alpha particle to the maximum electrical potential energy of the system (alpha particle plus target nucleus), we have

$$\frac{1}{2}mv^2 = k_e\frac{q_1q_2}{r} = k_e\frac{(2e)(Ze)}{d}$$

where d is the distance of closest approach. Solving for d, we get

$$d = \frac{4k_eZe^2}{mv^2}$$

From this expression, Rutherford found that alpha particles approached to within 3.2×10^{-14} m of a nucleus when the foil was made of gold. Thus, the radius of the gold nucleus must be less than this value. For silver atoms, the distance of closest approach was 2×10^{-14} m. From these results, Rutherford concluded that the positive charge in an atom is concentrated in a small sphere, which he called the nucleus, with radius no greater than about 10^{-14} m. Because such small lengths are common in nuclear physics, a convenient unit of length is the *femtometer* (fm), sometimes called the **fermi** and defined as

$$1 \text{ fm} \equiv 10^{-15} \text{ m}$$

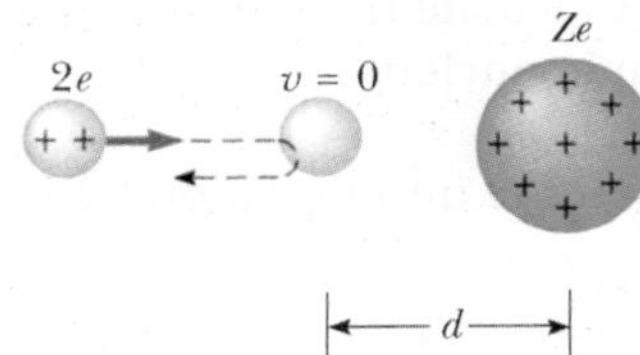

ACTIVE FIGURE 29.1
An alpha particle on a head-on collision course with a nucleus of charge Ze. Because of the Coulomb repulsion between the like charges, the alpha particle will stop instantaneously at a distance d from the nucleus, called the distance of closest approach.

Physics Now™
Log into PhysicsNow at **www.cp7e.com** and go to Active Figure 29.1, where you can adjust the atomic number of the target nucleus and the kinetic energy of the alpha particle. Then observe the approach of the alpha particle toward the nucleus.

Since the time of Rutherford's scattering experiments, a multitude of other experiments have shown that most nuclei are approximately spherical and have an average radius given by

$$r = r_0A^{1/3} \qquad [29.1]$$

where A is the total number of nucleons and r_0 is a constant equal to 1.2×10^{-15} m. Because the volume of a sphere is proportional to the cube of its radius, it follows from Equation 29.1 that the volume of a nucleus (assumed to be spherical) is directly proportional to A, the total number of nucleons. This relationship then suggests that **all nuclei have nearly the same density**. Nucleons combine to form a nucleus *as though* they were tightly packed spheres (Fig. 29.2).

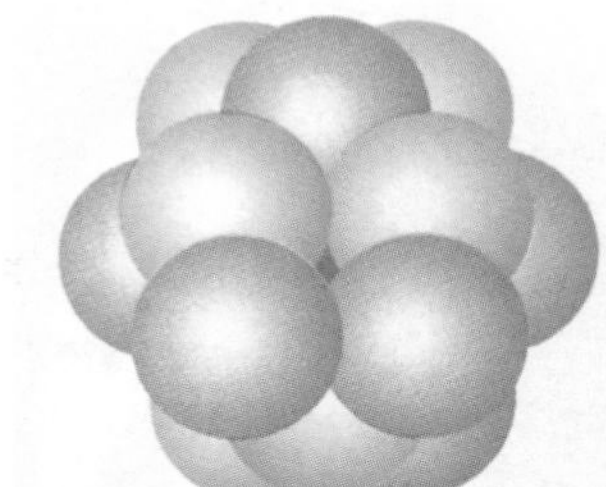

Figure 29.2 A nucleus can be visualized as a cluster of tightly packed spheres, each of which is a nucleon.

EXAMPLE 29.1 Sizing a Neutron Star

Goal Apply the concepts of nuclear size.

Problem One of the end stages of stellar life is a neutron star, where matter collapses and electrons combine with protons to form neutrons. Some liken neutron stars to a single gigantic nucleus. **(a)** Approximately how many nucleons are in a neutron star with a mass of 3.00×10^{30} kg? (This is the mass number of the star.) **(b)** Calculate the radius of the star, treating it as a giant nucleus. **(c)** Calculate the density of the star, assuming the mass is distributed uniformly.

Strategy The effective mass number of the neutron star can be found by dividing the star mass in kg by the mass of a neutron. Equation 29.1 then gives an estimate of the radius of the star, which together with the mass determines the density.

Solution

(a) Find the approximate number of nucleons in the star.

Divide the star's mass by the mass of a neutron to find A:

$$A = \left(\frac{3.00 \times 10^{30} \text{ kg}}{1.675 \times 10^{-27} \text{ kg}}\right) = 1.79 \times 10^{57}$$

(b) Calculate the radius of the star, treating it as a giant atomic nucleus.

Substitute into Equation 29.1:

$$r = r_0 A^{1/3} = (1.2 \times 10^{-15}\ \text{m})(1.79 \times 10^{57})^{1/3}$$
$$= 1.46 \times 10^4\ \text{m}$$

(c) Calculate the density of the star, assuming that its mass is distributed uniformly.

Substitute values into the equation for density and assume the star is a uniform sphere:

$$\rho = \frac{m}{V} = \frac{m}{\frac{4}{3}\pi r^3} = \frac{3.00 \times 10^{30}\ \text{kg}}{\frac{4}{3}\pi(1.46 \times 10^4\ \text{m})^3}$$
$$= 2.30 \times 10^{17}\ \text{kg/m}^3$$

Remarks This density is typical of atomic nuclei as well as of neutron stars. A ball of neutron star matter having a radius of only 1 meter would have a powerful gravity field: it could attract objects a kilometer away at an acceleration of over 50 m/s^2!

Exercise 29.1

Estimate the radius of a uranium-235 nucleus.

Answer 7.41×10^{-15} m

Courtesy of Louise Barker/AIP Niels Bohr Library

MARIA GOEPPERT-MAYER,
German Physicist (1906–1972)

Goeppert-Mayer was born and educated in Germany. She is best known for her development of the shell model of the nucleus, published in 1950. A similar model was simultaneously developed by Hans Jensen, a German scientist. Maria Goeppert-Mayer and Hans Jensen were awarded the Nobel Prize in physics in 1963 for their extraordinary work in understanding the structure of the nucleus.

Nuclear Stability

Given that the nucleus consists of a closely packed collection of protons and neutrons, you might be surprised that it can even exist. The very large repulsive electrostatic forces between protons should cause the nucleus to fly apart. However, nuclei are stable because of the presence of another, short-range (about 2 fm) force: the **nuclear force**, an attractive force that acts between all nuclear particles. The protons attract each other via the nuclear force, and at the same time they repel each other through the Coulomb force. The attractive nuclear force also acts between pairs of neutrons and between neutrons and protons.

The nuclear attractive force is stronger than the Coulomb repulsive force within the nucleus (at short ranges). If this were not the case, stable nuclei would not exist. Moreover, the strong nuclear force is nearly independent of charge. In other words, the nuclear forces associated with proton–proton, proton–neutron, and neutron–neutron interactions are approximately the same, apart from the additional repulsive Coulomb force for the proton–proton interaction.

There are about 260 stable nuclei; hundreds of others have been observed, but are unstable. A plot of N versus Z for a number of stable nuclei is given in Figure 29.3. Note that light nuclei are most stable if they contain equal numbers of protons and neutrons, so that $N = Z$, but heavy nuclei are more stable if $N > Z$. This difference can be partially understood by recognizing that as the number of protons increases, the strength of the Coulomb force increases, which tends to break the nucleus apart. As a result, more neutrons are needed to keep the nucleus stable, because neutrons are affected only by the attractive nuclear forces. In effect, the additional neutrons "dilute" the nuclear charge density. Eventually, when $Z = 83$, the repulsive forces between protons cannot be compensated for by the addition of neutrons. Elements that contain more than 83 protons don't have stable nuclei, but decay or disintegrate into other particles in various amounts of time. The masses and some other properties of selected isotopes are provided in Appendix B.

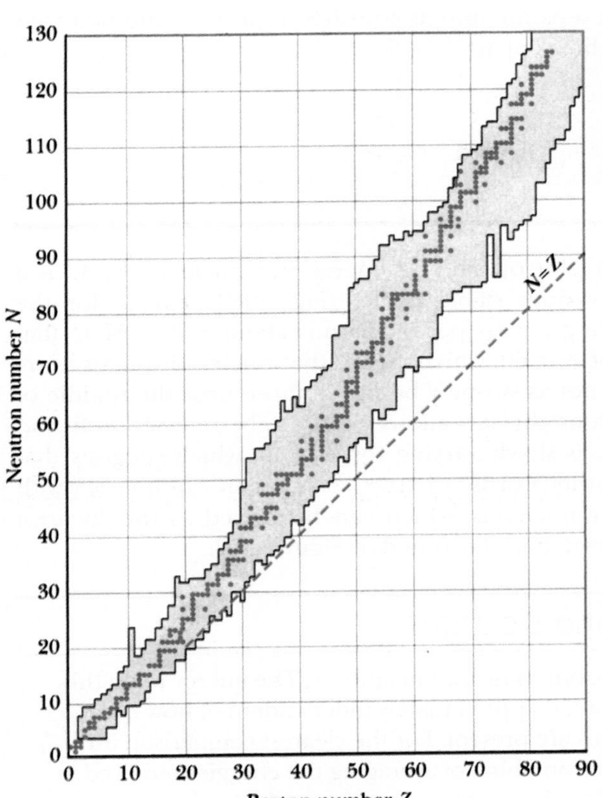

Figure 29.3 A plot of the neutron number N versus the proton number Z for the stable nuclei (solid points). The dashed straight line corresponds to the condition $N = Z$. They are centered on the so-called line of stability. The shaded area shows radioactive (unstable) nuclei.

29.2 BINDING ENERGY

The total mass of a nucleus is always less than the sum of the masses of its nucleons. Also, because mass is another manifestation of energy, **the total energy of the bound system (the nucleus) is less than the combined energy of the separated nucleons.** This difference in energy is called the **binding energy** of the nucleus and can be thought of as the energy that must be added to a nucleus to break it apart into its separated neutrons and protons.

EXAMPLE 29.2 The Binding Energy of the Deuteron

Goal Calculate the binding energy of a nucleus.

Problem The nucleus of the deuterium atom, called the deuteron, consists of a proton and a neutron. Calculate the deuteron's binding energy in MeV, given that its atomic mass—that is, *the mass of a deuterium nucleus plus an electron*—is 2.014 102 u.

Strategy Calculate the sum of the masses of the individual particles and subtract the mass of the combined particle. The masses of the neutral atoms can be used instead of the nuclei because the electron masses cancel. Use the values from Table 29.4 or Table B of the appendix. The mass of an atom given in Appendix B includes the mass of Z electrons, where Z is the atom's atomic number.

Solution

To find the binding energy, first sum the masses of the hydrogen atom and neutron and subtract the mass of the deuteron:

$$\begin{aligned}\Delta m &= (m_p + m_n) - m_d \\ &= (1.007\,825\text{ u} + 1.008\,665\text{ u}) - 2.014\,102\text{ u} \\ &= 0.002\,388\text{ u}\end{aligned}$$

Convert this mass difference to its equivalent in MeV:

$$E_b = (0.002\,388\text{ u})\frac{931.5\text{ MeV}}{1\text{ u}} = \boxed{2.224\text{ MeV}}$$

Remarks This result tells us that to separate a deuteron into a proton and a neutron, it's necessary to add 2.224 MeV of energy to the deuteron to overcome the attractive nuclear force between the proton and the neutron. One way of supplying the deuteron with this energy is by bombarding it with energetic particles.

If the binding energy of a nucleus were zero, the nucleus would separate into its constituent protons and neutrons without the addition of any energy; that is, it would spontaneously break apart.

Exercise 29.2

Calculate the binding energy of ${}^3_2\text{He}$.

Answer 7.718 MeV

It's interesting to examine a plot of binding energy per nucleon, E_b/A, as a function of mass number for various stable nuclei (Fig. 29.4). Except for the lighter nuclei, the average binding energy per nucleon is about 8 MeV. Note that the curve peaks in the vicinity of $A = 60$, which means that nuclei with mass numbers greater or less than 60 are not as strongly bound as those near the middle of the periodic table. As we'll see later, this fact allows energy to be released in fission and fusion reactions. The curve is slowly varying for $A > 40$, which suggests that the nuclear force saturates. In other words, a particular nucleon can interact with only a limited number of other nucleons, which can be viewed as the "nearest neighbors" in the close-packed structure illustrated in Figure 29.2.

Applying Physics 29.1 Binding Nucleons and Electrons

Figure 29.4 shows a graph of the amount of energy required to remove a nucleon from the nucleus. The figure indicates that an approximately constant amount of energy is necessary to remove a nucleon above $A = 40$, whereas we saw in Chapter 28 that widely varying amounts of energy are required to remove an electron from the atom. What accounts for this difference?

Explanation In the case of Figure 29.4, the approximately constant value of the nuclear binding energy is a result of the short-range nature of the nuclear force. A given nucleon interacts only with its few nearest neighbors, rather than with all of the nucleons in the nucleus. Thus, no matter how many nucleons are present in the nucleus, pulling any one nucleon out involves separating it only from its nearest neighbors. The energy to do this, therefore, is approximately independent of how many nucleons are present. For the clearest comparison with the electron, think of averaging the energies required to strip all of the electrons out of a particular atom, from the outermost valence electron to the innermost K-shell electron. This average increases steeply with increasing atomic number. The electrical force binding the electrons to the nucleus in an atom is a long-range force. An electron in an atom interacts with all the protons in the nucleus. When the nuclear charge increases, there is a stronger attraction between the nucleus and the electrons. Therefore, as the nuclear charge increases, more energy is necessary to remove an average electron.

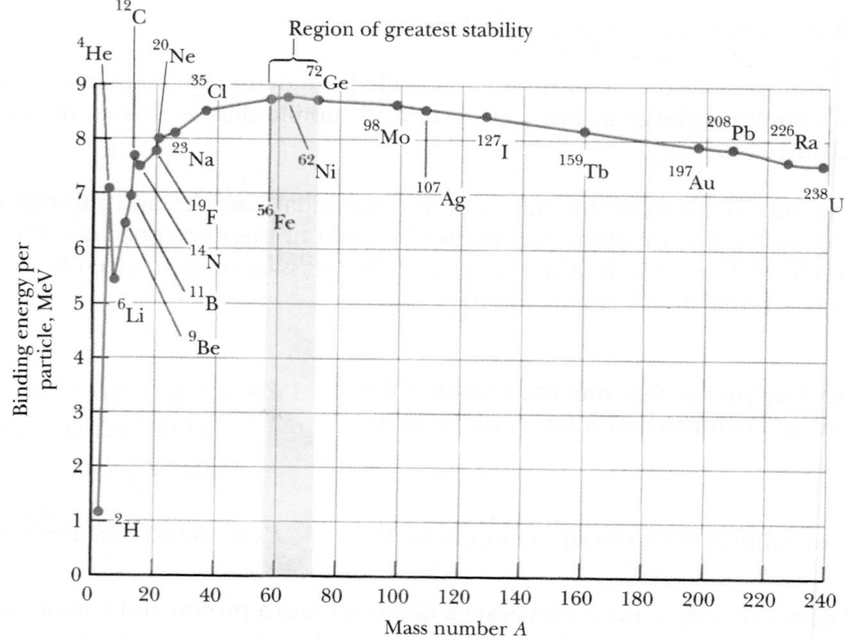

Figure 29.4 Binding energy per nucleon versus the mass number A for nuclei that are along the line of stability shown in Figure 29.3. Some representative nuclei appear as blue dots with labels. (Nuclei to the right of ${}^{208}\text{Pb}$ are unstable. The curve represents the binding energy for the most stable isotopes.)

29.3 RADIOACTIVITY

In 1896, Becquerel accidentally discovered that uranium salt crystals emit an invisible radiation that can darken a photographic plate even if the plate is covered to exclude light. After several such observations under controlled conditions, he concluded that the radiation emitted by the crystals was of a new type, one requiring no external stimulation. This spontaneous emission of radiation was soon called **radioactivity**. Subsequent experiments by other scientists showed that other substances were also radioactive.

The most significant investigations of this type were conducted by Marie and Pierre Curie. After several years of careful and laborious chemical separation processes on tons of pitchblende, a radioactive ore, the Curies reported the discovery of two previously unknown elements, both of which were radioactive. These were named polonium and radium. Subsequent experiments, including Rutherford's famous work on alpha-particle scattering, suggested that radioactivity was the result of the decay, or disintegration, of unstable nuclei.

MARIE CURIE, Polish Scientist (1867–1934)

In 1903 Marie Curie shared the Nobel Prize in physics with her husband, Pierre, and with Becquerel for their studies of radioactive substances. In 1911 she was awarded a second Nobel Prize in chemistry for the discovery of radium and polonium. Marie Curie died of leukemia caused by years of exposure to radioactive substances. "I persist in believing that the ideas that then guided us are the only ones which can lead to the true social progress. We cannot hope to build a better world without improving the individual. Toward this end, each of us must work toward his own highest development, accepting at the same time his share of responsibility in the general life of humanity."

Three types of radiation can be emitted by a radioactive substance: alpha (α) particles, in which the emitted particles are ${}^4_2\text{He}$ nuclei; beta (β) particles, in which the emitted particles are either electrons or positrons; and gamma (γ) rays, in which the emitted "rays" are high-energy photons. A **positron** is a particle similar to the electron in all respects, except that it has a charge of $+e$. (The positron is said to be the **antiparticle** of the electron.) The symbol e^- is used to designate an electron, and e^+ designates a positron.

It's possible to distinguish these three forms of radiation by using the scheme described in Figure 29.5. The radiation from a radioactive sample is directed into a region with a magnetic field, and the beam splits into three components, two bending in opposite directions and the third not changing direction. From this simple observation it can be concluded that the radiation of the undeflected beam (the gamma ray) carries no charge, the component deflected upward contains positively charged particles (alpha particles), and the component deflected downward contains negatively charged particles (e^-). If the beam includes a positron (e^+), it is deflected upward.

The three types of radiation have quite different penetrating powers. Alpha particles barely penetrate a sheet of paper, beta particles can penetrate a few millimeters of aluminum, and gamma rays can penetrate several centimeters of lead.

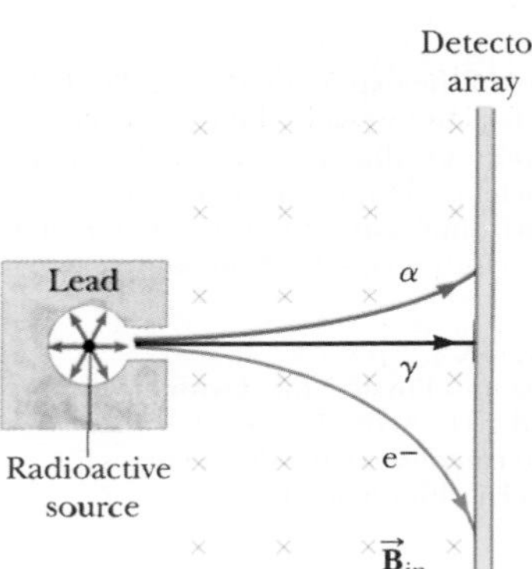

Figure 29.5 The radiation from a radioactive source, such as radium, can be separated into three components using a magnetic field to deflect the charged particles. The detector array at the right records the events. The gamma ray isn't deflected by the magnetic field.

The Decay Constant and Half-Life

Observation has shown that if a radioactive sample contains N radioactive nuclei at some instant, then the number of nuclei, ΔN, that decay in a small time interval Δt is proportional to N; mathematically,

$$\frac{\Delta N}{\Delta t} \propto N$$

or

$$\Delta N = -\lambda N \Delta t \qquad [29.2]$$

where λ is a constant called the **decay constant**. The negative sign signifies that N decreases with time; that is, ΔN is negative. The value of λ for any isotope determines the rate at which that isotope will decay. **The decay rate, or activity *R*, of a sample is defined as the number of decays per second**. From Equation 29.2, we see that the decay rate is

$$R = \left|\frac{\Delta N}{\Delta t}\right| = \lambda N \qquad [29.3]$$

◀ Decay rate

Isotopes with a large λ value decay rapidly; those with small λ decay slowly.

The hands and numbers of this luminous watch contain minute amounts of radium salt. The radioactive decay of radium causes the phosphors to glow in the dark.

A general decay curve for a radioactive sample is shown in Active Figure 29.6. It can be shown from Equation 29.2 (using calculus) that the number of nuclei present varies with time according to the equation

$$N = N_0 e^{-\lambda t} \qquad [29.4a]$$

where N is the number of radioactive nuclei present at time t, N_0 is the number present at time $t = 0$, and $e = 2.718 \ldots$ is Euler's constant. Processes that obey Equation 29.4a are sometimes said to undergo **exponential decay**.[1]

Another parameter that is useful for characterizing radioactive decay is the **half-life $T_{1/2}$. The half-life of a radioactive substance is the time it takes for half of a given number of radioactive nuclei to decay**. Using the concept of half-life, it can be shown that Equation 29.4a can also be written as

$$N = N_0 \left(\frac{1}{2}\right)^n \qquad [29.4b]$$

where n is the number of half-lives. The number n can take any non-negative value and need not be an integer. From the definition, it follows that n is related to time t and the half-life $T_{1/2}$ by

$$n = \frac{t}{T_{1/2}} \qquad [29.4c]$$

Setting $N = N_0/2$ and $t = T_{1/2}$ in Equation 29.4a gives

$$\frac{N_0}{2} = N_0 e^{-\lambda T_{1/2}}$$

Writing this in the form $e^{\lambda T_{1/2}} = 2$ and taking the natural logarithm of both sides, we get

$$T_{1/2} = \frac{\ln 2}{\lambda} = \frac{0.693}{\lambda} \qquad [29.5]$$

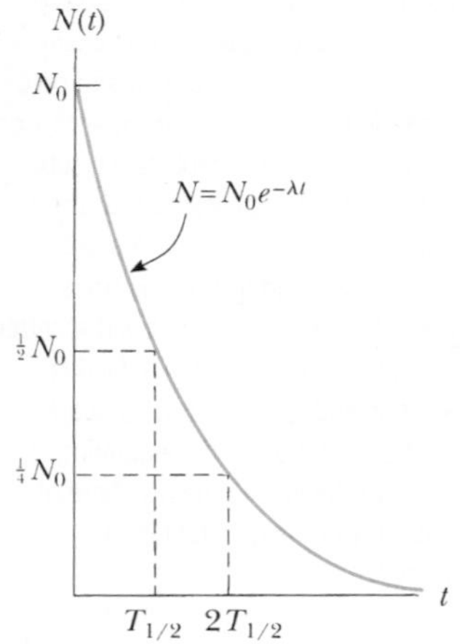

ACTIVE FIGURE 29.6
Plot of the exponential decay law for radioactive nuclei. The vertical axis represents the number of radioactive nuclei present at any time t, and the horizontal axis is time. The parameter $T_{1/2}$ is the half-life of the sample.

Physics Now™
Log into PhysicsNow at **www.cp7e.com** and go to Active Figure 29.6, where you can observe the decay curves for nuclei with varying half-lives.

This is a convenient expression relating the half-life to the decay constant. Note that after an elapsed time of one half-life, $N_0/2$ radioactive nuclei remain (by definition); after two half-lives, half of these will have decayed and $N_0/4$ radioactive nuclei will be left; after three half-lives, $N_0/8$ will be left; and so on.

The unit of activity R is the **curie** (Ci), defined as

$$1 \text{ Ci} \equiv 3.7 \times 10^{10} \text{ decays/s} \qquad [29.6]$$

This unit was selected as the original activity unit because it is the approximate activity of 1 g of radium. The SI unit of activity is the **becquerel** (Bq):

$$1 \text{ Bq} = 1 \text{ decay/s} \qquad [29.7]$$

Therefore, 1 Ci $= 3.7 \times 10^{10}$ Bq. The most commonly used units of activity are the millicurie (10^{-3} Ci) and the microcurie (10^{-6} Ci).

TIP 29.2 Two Half-Lives Don't Make a Whole-Life

A half-life is the time it takes for hal. of a given number of nuclei to decay. During a second half-life, half the remaining nuclei decay, so in two half-lives, three-quarters of the original material has decayed, not all of it.

Quick Quiz 29.1

What fraction of a radioactive sample has decayed after two half-lives have elapsed? (a) 1/4 (b) 1/2 (c) 3/4 (d) not enough information to say

Quick Quiz 29.2

Suppose the decay constant of radioactive substance A is twice the decay constant of radioactive substance B. If substance B has a half life of 4 hr, what's the half life of substance A? (a) 8 hr (b) 4 hr (c) 2 hr (d) not enough information to say

[1] Other examples of exponential decay were discussed in Chapter 18 in connection with RC circuits and in Chapter 20 in connection with RL circuits.

INTERACTIVE EXAMPLE 29.3 The Activity of Radium

Goal Calculate the activity of a radioactive substance at different times.

Problem The half-life of the radioactive nucleus $^{226}_{88}\text{Ra}$ is 1.6×10^3 yr. If a sample initially contains 3.00×10^{16} such nuclei, determine **(a)** the initial activity in curies, **(b)** the number of radium nuclei remaining after 4.8×10^3 yr, and **(c)** the activity at this later time.

Strategy For parts (a) and (c), find the decay constant and multiply it by the number of nuclei. Part (b) requires multiplying the initial number of nuclei by one-half for every elapsed half-life. (Essentially, this is an application of Equation 29.4b.)

Solution

(a) Determine the initial activity in curies.

Convert the half-life to seconds:

$$T_{1/2} = (1.6 \times 10^3\ \text{yr})(3.156 \times 10^7\ \text{s/yr}) = 5.0 \times 10^{10}\ \text{s}$$

Substitute this value into Equation 29.5 to get the decay constant:

$$\lambda = \frac{0.693}{T_{1/2}} = \frac{0.693}{5.0 \times 10^{10}\ \text{s}} = 1.4 \times 10^{-11}\ \text{s}^{-1}$$

Calculate the activity of the sample at $t = 0$, using $R_0 = \lambda N_0$, where R_0 is the decay rate at $t = 0$ and N_0 is the number of radioactive nuclei present at $t = 0$:

$$R_0 = \lambda N_0 = (1.4 \times 10^{-11}\ \text{s}^{-1})(3.0 \times 10^{16}\ \text{nuclei}) = 4.2 \times 10^5\ \text{decays/s}$$

Convert to curies to obtain the activity at $t = 0$, using the fact that $1\ \text{Ci} = 3.7 \times 10^{10}$ decays/s:

$$R_0 = (4.2 \times 10^5\ \text{decays/s})\left(\frac{1\ \text{Ci}}{3.7 \times 10^{10}\ \text{decays/s}}\right)$$

$$= 1.1 \times 10^{-5}\ \text{Ci} = \boxed{11\ \mu\text{Ci}}$$

(b) How many radium nuclei remain after 4.8×10^3 years?

Calculate the number of half-lives, n:

$$n = \frac{4.8 \times 10^3\ \text{yr}}{1.6 \times 10^3\ \text{yr/half-life}} = 3.0\ \text{half-lives}$$

Multiply the initial number of nuclei by the number of factors of one-half:

$$N = N_0\left(\frac{1}{2}\right)^n \qquad (1)$$

Substitute $N_0 = 3.0 \times 10^{16}$ and $n = 3.0$:

$$N = (3.0 \times 10^{16}\ \text{nuclei})\left(\frac{1}{2}\right)^{3.0} = \boxed{3.8 \times 10^{15}\ \text{nuclei}}$$

(c) Calculate the activity after 4.8×10^3 yr.

Multiply the number of remaining nuclei by the decay constant to find the activity R:

$$R = \lambda N = (1.4 \times 10^{-11}\ \text{s}^{-1})(3.8 \times 10^{15}\ \text{nuclei})$$

$$= 5.3 \times 10^4\ \text{decays/s}$$

$$= \boxed{1.4\ \mu\text{Ci}}$$

Remarks The activity is reduced by half every half-life, which is naturally the case because activity is proportional to the number of remaining nuclei. The precise number of nuclei at any time is never truly exact, because particles decay according to a probability. The larger the sample, however, the more accurate are the predictions from Equation 29.4.

Exercise 29.3

Find (a) the number of remaining radium nuclei after 3.2×10^3 yr and (b) the activity at this time.

Answer (a) 7.5×10^{15} nuclei (b) $2.8\ \mu$Ci

PhysicsNow™ Practice evaluating the parameters for the radioactive decay of various isotopes of radium by logging into PhysicsNow at **www.cp7e.com** and going to Interactive Example 29.3.

EXAMPLE 29.4 Radon Gas

Goal Calculate the number of nuclei after an arbitrary time and the time required for a given number of nuclei to decay.

Problem Radon $^{222}_{86}\text{Rn}$ is a radioactive gas that can be trapped in the basements of homes, and its presence in high concentrations is a known health hazard. Radon has a half-life of 3.83 days. A gas sample contains 4.00×10^8 radon atoms initially. **(a)** How many atoms will remain after 14.0 days have passed if no more radon leaks in? **(b)** What is the activity of the radon sample after 14.0 days? **(c)** How long before 99% of the sample has decayed?

Strategy The activity can be found by substitution into Equation 29.5, as before. Equation 29.4a (or Eq. 29.4b) must be used to find the number of particles remaining after 14.0 days. To obtain the time asked for in part (c), Equation 29.4a must be solved for time.

Solution

(a) How many atoms will remain after 14.0 days have passed?

Determine the decay constant from Equation 29.5:

$$\lambda = \frac{0.693}{T_{1/2}} = \frac{0.693}{3.83 \text{ days}} = 0.181 \text{ day}^{-1}$$

Now use Equation 29.4a, taking $N_0 = 4.0 \times 10^8$, and the value of λ just found to obtain the number N remaining after 14 days:

$$N = N_0 e^{-\lambda t} = (4.00 \times 10^8 \text{ atoms}) e^{-(0.181 \text{ day}^{-1})(14.0 \text{ days})}$$

$$= 3.17 \times 10^7 \text{ atoms}$$

(b) What is the activity of the radon sample after 14.0 days?

Express the decay constant in units of s^{-1}:

$$\lambda = (0.181 \text{ day}^{-1})\left(\frac{1 \text{ day}}{8.64 \times 10^4 \text{ s}}\right) = 2.09 \times 10^{-6} \text{ s}^{-1}$$

From Equation 29.3 and this value of λ, compute the activity R:

$$R = \lambda N = (2.09 \times 10^{-6} \text{ s}^{-1})(3.17 \times 10^7 \text{ atoms})$$

$$= 66.3 \text{ decays/s} = 66.3 \text{ Bq}$$

(c) How much time must pass before 99% of the sample has decayed?

Solve Equation 29.4a for t, using natural logarithms:

$$\ln(N) = \ln(N_0 e^{-\lambda t}) = \ln(N_0) + \ln(e^{-\lambda t})$$

$$\ln(N) - \ln(N_0) = \ln(e^{-\lambda t}) = -\lambda t$$

$$t = \frac{\ln(N_0) - \ln(N)}{\lambda} = \frac{\ln(N_0/N)}{\lambda}$$

Substitute values:

$$t = \frac{\ln(N_0/0.01\, N_0)}{2.09 \times 10^{-6} \text{ s}^{-1}} = 2.20 \times 10^6 \text{ s} = 25.5 \text{ days}$$

Remarks This kind of calculation is useful in determining how long you would have to wait for radioactivity at a given location to fall to safe levels.

Exercise 29.4

(a) Find the activity of the radon sample after 12.0 days have elapsed. (b) How long would it take for 85.0% of the sample to decay?

Answer (a) 95.3 Bq (b) $9.08 \times 10^5 \text{ s} = 10.5$ days

29.4 THE DECAY PROCESSES

As stated in the previous section, radioactive nuclei decay spontaneously via alpha, beta, and gamma decay. As we'll see in this section, these processes are very different from each other.

Alpha Decay

If a nucleus emits an alpha particle (^4_2He), it loses two protons and two neutrons. Therefore, the neutron number N of a single nucleus decreases by 2, Z decreases by 2, and A decreases by 4. The decay can be written symbolically as

$$^A_Z\text{X} \rightarrow {}^{A-4}_{Z-2}\text{Y} + {}^4_2\text{He} \tag{29.8}$$

where X is called the **parent nucleus** and Y is known as the **daughter nucleus**. As examples, ^{238}U and ^{226}Ra are both alpha emitters and decay according to the schemes

$$^{238}_{92}\text{U} \rightarrow {}^{234}_{90}\text{Th} + {}^4_2\text{He} \tag{29.9}$$

and

$$^{226}_{88}\text{Ra} \rightarrow {}^{222}_{86}\text{Rn} + {}^4_2\text{He} \tag{29.10}$$

The half-life for ^{238}U decay is 4.47×10^9 years, and the half-life for ^{226}Ra decay is 1.60×10^3 years. In both cases, note that the A of the daughter nucleus is four less than that of the parent nucleus, while Z is reduced by two. The differences are accounted for in the emitted alpha particle (the ^4He nucleus).

The decay of ^{226}Ra is shown in Active Figure 29.7. When one element changes into another, as happens in alpha decay, the process is called **spontaneous decay** or transmutation. As a general rule, (1) the sum of the mass numbers A must be the same on both sides of the equation, and (2) the sum of the atomic numbers Z must be the same on both sides of the equation.

In order for alpha emission to occur, the mass of the parent must be greater than the combined mass of the daughter and the alpha particle. In the decay process, this excess mass is converted into energy of other forms and appears in the form of kinetic energy in the daughter nucleus and the alpha particle. Most of the kinetic energy is carried away by the alpha particle because it is much less massive than the daughter nucleus. This can be understood by first noting that a particle's kinetic energy and momentum p are related as follows:

$$KE = \frac{p^2}{2m}$$

Because momentum is conserved, the two particles emitted in the decay of a nucleus at rest must have equal, but oppositely directed, momenta. As a result, the lighter particle, with the smaller mass in the denominator, has more kinetic energy than the more massive particle.

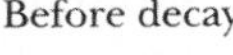

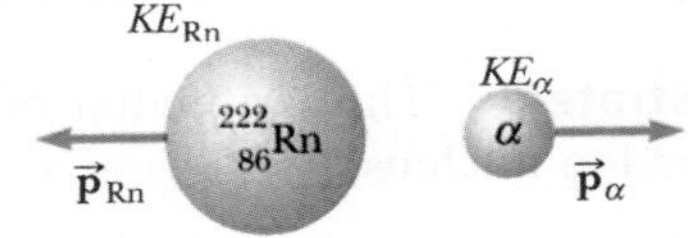

ACTIVE FIGURE 29.7
The alpha decay of radium-226. The radium nucleus is initially at rest. After the decay, the radon nucleus has kinetic energy KE_{Rn} and momentum $\vec{\mathbf{p}}_{\text{Rn}}$, and the alpha particle has kinetic energy KE_α and momentum $\vec{\mathbf{p}}_\alpha$.

Physics Now™
Log into PhysicsNow at **www.cp7e.com** and go to Active Figure 29.7, where you can observe the decay of radium-226.

Quick Quiz 29.3

If a nucleus such as ^{226}Ra that is initially at rest undergoes alpha decay, which of the following statements is true? (a) The alpha particle has more kinetic energy than the daughter nucleus. (b) The daughter nucleus has more kinetic energy than the alpha particle. (c) The daughter nucleus and the alpha particle have the same kinetic energy.

Applying Physics 29.2 Energy and Half-life

In comparing alpha decay energies from a number of radioactive nuclides, why is it found that the half-life of the decay goes down as the energy of the decay goes up?

Explanation It should seem reasonable that the higher the energy of the alpha particle, the more likely it is to escape the confines of the nucleus. The higher probability of escape translates to a faster rate of decay, which appears as a shorter half-life.

EXAMPLE 29.5 Decaying Radium

Goal Calculate the energy released during an alpha decay.

Problem We showed that the $^{226}_{88}$Ra nucleus undergoes alpha decay to $^{222}_{86}$Rn (Eq. 29.10). Calculate the amount of energy liberated in this decay. Take the mass of $^{226}_{88}$Ra to be 226.025 402 u, that of $^{222}_{86}$Rn to be 222.017 571 u, and that of $^{4}_{2}$He to be 4.002 602 u, as found in Appendix B.

Strategy This is a matter of subtracting the neutral masses of the daughter particles from the original mass of the radon nucleus.

Solution

Compute the sum of the mass of the daughter particle, m_d, and the mass of the alpha particle, m_α:

$$m_d + m_\alpha = 222.017\,571 \text{ u} + 4.002\,602 \text{ u} = 226.020\,173 \text{ u}$$

Compute the loss of mass, Δm, during the decay by subtracting the previous result from M_p, the mass of the original particle:

$$\Delta m = M_p - (m_d + m_\alpha) = 226.025\,402 \text{ u} - 226.020\,173 \text{ u} = 0.005\,229 \text{ u}$$

Convert the loss of mass Δm to its equivalent energy in MeV:

$$E = (0.005\,229 \text{ u})(931.494 \text{ MeV/u}) = 4.871 \text{ MeV}$$

Remark The potential barrier is typically higher than this value of the energy, but quantum tunneling permits the event to occur, anyway.

Exercise 29.5

Calculate the energy released when $^{8}_{4}$Be splits into two alpha particles. Beryllium-8 has an atomic mass of 8.005 305 u.

Answer 0.094 1 MeV

Beta Decay

When a radioactive nucleus undergoes beta decay, the daughter nucleus has the same number of nucleons as the parent nucleus, but the atomic number is changed by 1:

$$^{A}_{Z}\text{X} \rightarrow {}^{A}_{Z+1}\text{Y} + \text{e}^- \qquad [29.11]$$

$$^{A}_{Z}\text{X} \rightarrow {}^{A}_{Z-1}\text{Y} + \text{e}^+ \qquad [29.12]$$

Again, note that the nucleon number and total charge are both conserved in these decays. However, as we will see shortly, these processes are not described completely by these expressions. A typical beta decay event is

$$^{14}_{6}\text{C} \rightarrow {}^{14}_{7}\text{N} + \text{e}^- \qquad [29.13]$$

The emission of electrons from a *nucleus* is surprising, because, in all our previous discussions, we stated that the nucleus is composed of protons and neutrons only. This apparent discrepancy can be explained by noting that the emitted electron is created in the nucleus by a process in which a neutron is transformed into a proton. This process can be represented by the equation

$$^{1}_{0}\text{n} \rightarrow {}^{1}_{1}\text{p} + \text{e}^- \qquad [29.14]$$

Consider the energy of the system of Equation 29.13 before and after decay. As with alpha decay, energy must be conserved in beta decay. The next example illustrates how to calculate the amount of energy released in the beta decay of $^{14}_{6}$C.

EXAMPLE 29.6 The Beta Decay of Carbon-14

Goal Calculate the energy released in a beta decay.

Problem Find the energy liberated in the beta decay of $^{14}_{6}$C to $^{14}_{7}$N, as represented by Equation 29.13. That equation refers to nuclei, while Appendix B gives the masses of neutral atoms. Adding six electrons to both sides of Equation 29.13 yields

$$^{14}_{6}\text{C atom} \rightarrow {}^{14}_{7}\text{N atom}$$

Strategy As in preceding problems, finding the released energy involves computing the difference in mass between the resultant particle(s) and the initial particle(s) and converting to MeV.

Solution

Obtain the masses of ${}^{14}_{6}\text{C}$ and ${}^{14}_{7}\text{N}$ from Appendix B and compute the difference between them:

$$\Delta m = m_C - m_N = 14.003\,242\text{ u} - 14.003\,074\text{ u} = 0.000\,168\text{ u}$$

Convert the mass difference to MeV:

$$E = (0.000\,168\text{ u})(931.494\text{ MeV/u}) = 0.156\text{ MeV}$$

Remarks The calculated energy is generally more than the energy observed in this process. The discrepancy led to a crisis in physics, because it appeared that energy wasn't conserved. As discussed below, this crisis was resolved by the discovery that another particle was also produced in the reaction.

Exercise 29.6

Calculate the maximum energy liberated in the beta decay of radioactive potassium to calcium: ${}^{40}_{19}\text{K} \rightarrow {}^{40}_{20}\text{Ca}$.

Answer 1.31 MeV

From Example 29.6, we see that the energy released in the beta decay of ^{14}C is approximately 0.16 MeV. As with alpha decay, we expect the electron to carry away virtually all of this energy as kinetic energy because, apparently, it is the lightest particle produced in the decay. As Figure 29.8 shows, however, only a small number of electrons have this maximum kinetic energy, represented as KE_{max} on the graph; most of the electrons emitted have kinetic energies lower than that predicted value. If the daughter nucleus and the electron aren't carrying away this liberated energy, then where has the energy gone? As an additional complication, further analysis of beta decay shows that the principles of conservation of both angular momentum and linear momentum appear to have been violated!

In 1930 Pauli proposed that a third particle must be present to carry away the "missing" energy and to conserve momentum. Later, Enrico Fermi developed a complete theory of beta decay and named this particle the **neutrino** ("little neutral one") because it had to be electrically neutral and have little or no mass. Although it eluded detection for many years, the neutrino (ν) was finally detected experimentally in 1956. The neutrino has the following properties:

◀ Properties of the neutrino

- Zero electric charge
- A mass much smaller than that of the electron, but probably not zero. (Recent experiments suggest that the neutrino definitely has mass, but the value is uncertain—perhaps less than 1 eV/c^2.)
- A spin of $\frac{1}{2}$
- Very weak interaction with matter, making it difficult to detect

With the introduction of the neutrino, we can now represent the beta decay process of Equation 29.13 in its correct form:

$$ {}^{14}_{6}\text{C} \rightarrow {}^{14}_{7}\text{N} + \text{e}^- + \bar{\nu} \qquad [29.15]$$

The bar in the symbol $\bar{\nu}$ indicates an **antineutrino**. To explain what an antineutrino is, we first consider the following decay:

$$ {}^{12}_{7}\text{N} \rightarrow {}^{12}_{6}\text{C} + \text{e}^+ + \nu \qquad [29.16]$$

TIP 29.3 Mass Number of the Electron

Another notation that is sometimes used for an electron is ${}^{\ 0}_{-1}\text{e}$. This notation does not imply that the electron has zero rest energy. The mass of the electron is much smaller than that of the lightest nucleon, so we can approximate it as zero when we study nuclear decays and reactions.

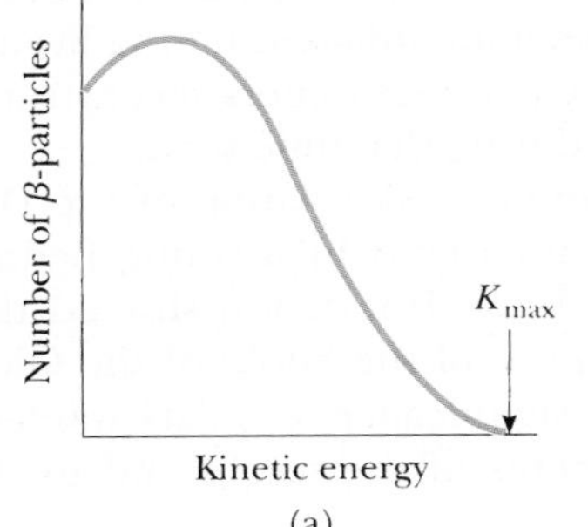

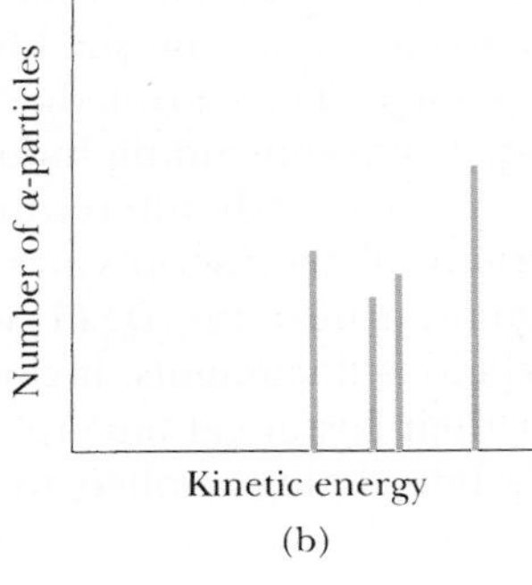

Figure 29.8 (a) Distribution of beta particle energies in a typical beta decay. All energies are observed up to a maximum value. (b) In contrast, the energies of alpha particles from an alpha decay are discrete.

Here, we see that when ^{12}N decays into ^{12}C, a particle is produced which is identical to the electron except that it has a positive charge of $+e$. This particle is called a **positron**. Because it is like the electron in all respects except charge, the positron is said to be the **antiparticle** of the electron. We will discuss antiparticles further in Chapter 30; for now, it suffices to say that, **in beta decay, an electron and an antineutrino are emitted or a positron and a neutrino are emitted.**

Unlike beta decay, which results in a daughter particle with a variety of possible kinetic energies, alpha decays come in discrete amounts, as seen in Figure 29.8b. This is because the two daughter particles have momenta with equal magnitude and opposite direction and are each composed of a fixed number of nucleons.

National Accelerator Laboratory

ENRICO FERMI, Italian Physicist (1901–1954)

Fermi was awarded the Nobel Prize in 1938 for producing the transuranic elements by neutron irradiation and for his discovery of nuclear reactions bought about by slow neutrons. He made many other outstanding contributions to physics, including his theory of beta decay, the free-electron theory of metals, and the development of the world's first fission reactor in 1942. Fermi was truly a gifted theoretical and experimental physicist. He was also well known for his ability to present physics in a clear and exciting manner. "Whatever Nature has in store for mankind, unpleasant as it may be, men must accept, for ignorance is never better than knowledge."

Gamma Decay

Very often a nucleus that undergoes radioactive decay is left in an excited energy state. The nucleus can then undergo a second decay to a lower energy state—perhaps even to the ground state—by emitting one or more high-energy photons. The process is similar to the emission of light by an atom. An atom emits radiation to release some extra energy when an electron "jumps" from a state of high energy to a state of lower energy. Likewise, the nucleus uses essentially the same method to release any extra energy it may have following a decay or some other nuclear event. In nuclear de-excitation, the "jumps" that release energy are made by protons or neutrons in the nucleus as they move from a higher energy level to a lower level. The photons emitted in the process are called **gamma rays**, which have very high energy relative to the energy of visible light.

A nucleus may reach an excited state as the result of a violent collision with another particle. However, it's more common for a nucleus to be in an excited state as a result of alpha or beta decay. The following sequence of events typifies the gamma decay processes:

$$^{12}_{5}\mathrm{B} \rightarrow {}^{12}_{6}\mathrm{C}^* + e^- + \bar{\nu} \qquad [29.17]$$

$$^{12}_{6}\mathrm{C}^* \rightarrow {}^{12}_{6}\mathrm{C} + \gamma \qquad [29.18]$$

Equation 29.17 represents a beta decay in which ^{12}B decays to $^{12}C^*$, where the asterisk indicates that the carbon nucleus is left in an excited state following the decay. The excited carbon nucleus then decays to the ground state by emitting a gamma ray, as indicated by Equation 29.18. Note that gamma emission doesn't result in any change in either Z or A.

Practical Uses of Radioactivity

Carbon Dating The beta decay of ^{14}C given by Equation 29.15 is commonly used to date organic samples. Cosmic rays (high-energy particles from outer space) in the upper atmosphere cause nuclear reactions that create ^{14}C from ^{14}N. In fact, the ratio of ^{14}C to ^{12}C (by numbers of nuclei) in the carbon dioxide molecules of our atmosphere has a constant value of about 1.3×10^{-12}, as determined by measuring carbon ratios in tree rings. All living organisms have the same ratio of ^{14}C to ^{12}C because they continuously exchange carbon dioxide with their surroundings. When an organism dies, however, it no longer absorbs ^{14}C from the atmosphere, so the ratio of ^{14}C to ^{12}C decreases as the result of the beta decay of ^{14}C. It's therefore possible to determine the age of a material by measuring its activity per unit mass as a result of the decay of ^{14}C. Through carbon dating, samples of wood, charcoal, bone, and shell have been identified as having lived from 1 000 to 25 000 years ago. This knowledge has helped researchers reconstruct the history of living organism—including human—during that time span.

APPLICATION

Carbon Dating of the Dead Sea Scrolls

A particularly interesting example is the dating of the Dead Sea Scrolls. This group of manuscripts was first discovered by a young Bedouin boy in a cave at Qumran near the Dead Sea in 1947. Translation showed the manuscripts to be religious documents, including most of the books of the Old Testament. Because of their historical and religious significance, scholars wanted to know their age. Carbon dating applied to fragments of the scrolls and to the material in which they were wrapped established that they were about 1950 years old. The scrolls are now stored at the Israel museum in Jerusalem.

Smoke Detectors Smoke detectors are frequently used in homes and industry for fire protection. Most of the common ones are the ionization-type that use radioactive materials. (See Fig. 29.9.) A smoke detector consists of an ionization chamber, a sensitive current detector, and an alarm. A weak radioactive source ionizes the air in the chamber of the detector, which creates charged particles. A voltage is maintained between the plates inside the chamber, setting up a small but detectable current in the external circuit. As long as the current is maintained, the alarm is deactivated. However, if smoke drifts into the chamber, the ions become attached to the smoke particles. These heavier particles do not drift as readily as do the lighter ions, which causes a decrease in the detector current. The external circuit senses this decrease in current and sets off the alarm.

APPLICATION
Smoke Detectors

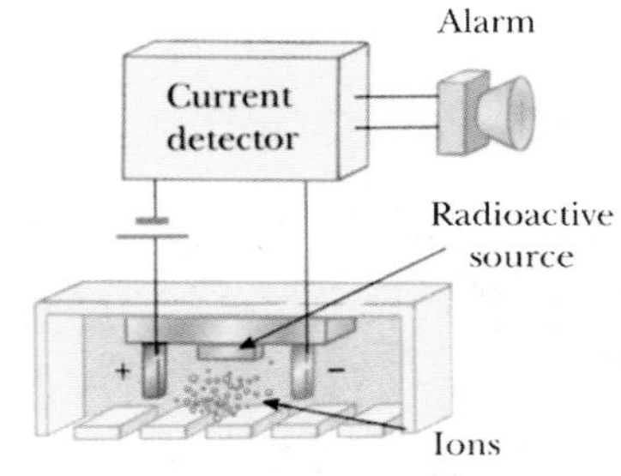

Figure 29.9 An ionization-type smoke detector. Smoke entering the chamber reduces the detected current, causing the alarm to sound.

Radon Detection Radioactivity can also affect our daily lives in harmful ways. Soon after the discovery of radium by the Curies, it was found that the air in contact with radium compounds becomes radioactive. It was then shown that this radioactivity came from the radium itself, and the product was therefore called "radium emanation." Rutherford and Soddy succeeded in condensing this "emanation," confirming that it was a real substance: the inert, gaseous element now called **radon** (Rn). Later, it was discovered that the air in uranium mines is radioactive because of the presence of radon gas. The mines must therefore be well ventilated to help protect the miners. Finally, the fear of radon pollution has moved from uranium mines into our own homes. (See Example 29.4.) Because certain types of rock, soil, brick, and concrete contain small quantities of radium, some of the resulting radon gas finds its way into our homes and other buildings. The most serious problems arise from leakage of radon from the ground into the structure. One practical remedy is to exhaust the air through a pipe just above the underlying soil or gravel directly to the outdoors by means of a small fan or blower.

APPLICATION
Radon Pollution

Applying Physics 29.3 Radioactive Dating of the Iceman

In 1991, a German tourist discovered the well-preserved remains of a man trapped in a glacier in the Italian Alps. (See Fig. 29.10.) Radioactive dating of a sample of bone from this hunter–gatherer, dubbed the "Iceman," revealed an age of 5 300 years. Why did scientists date the sample using the isotope ^{14}C, rather than ^{11}C, a beta emitter with a half-life of 20.4 min?

Explanation ^{14}C has a long half-life of 5 730 years, so the fraction of ^{14}C nuclei remaining after one half-life is high enough to accurately measure changes in the sample's activity. The ^{11}C isotope, which has a very short half-life, is not useful, because its activity decreases to a vanishingly small value over the age of the sample, making it impossible to detect.

If a sample to be dated is not very old—say, about 50 years—then you should select the isotope of some other element with half-life comparable to the age of the sample. For example, if the sample contained hydrogen, you could measure the activity of ^{3}H (tritium), a beta emitter of half-life 12.3 years. As a general rule, the expected age of the sample should be long enough to measure a change in activity, but not so long that its activity can't be detected.

Hanny Paul/Gamma Liaison

Figure 29.10 (Applying Physics 29.3) The body of an ancient man (dubbed the Iceman) was exposed by a melting glacier in the Alps.

To use radioactive dating techniques, we need to recast some of the equations already introduced. We start by multiplying both sides of Equation 29.4 by λ:

$$\lambda N = \lambda N_0 e^{-\lambda t}$$

From Equation 29.3, we have $\lambda N = R$ and $\lambda N_0 = R_0$. Substitute these expressions into the above equation and divide through by R_0:

$$\frac{R}{R_0} = e^{-\lambda t}$$

R is the present activity and R_0 was the activity when the object in question was part of a living organism. We can solve for time by taking the natural logarithm of both sides of the foregoing equation:

$$\ln\left(\frac{R}{R_0}\right) = \ln(e^{-\lambda t}) = -\lambda t$$

$$t = -\frac{\ln\left(\frac{R}{R_0}\right)}{\lambda} \qquad [29.19]$$

EXAMPLE 29.7 Should We Report This to Homicide?

Goal Apply the technique of carbon-14 dating.

Problem A 50.0-g sample of carbon is taken from the pelvis bone of a skeleton and is found to have a carbon-14 decay rate of 200.0 decays/min. It is known that carbon from a living organism has a decay rate of 15.0 decays/min · g and that ^{14}C has a half-life of 5 730 yr $= 3.01 \times 10^9$ min. Find the age of the skeleton.

Strategy Calculate the original activity and the decay constant, and then substitute those numbers and the current activity into Equation 29.19.

Solution

Calculate the original activity R_0 from the decay rate and the mass of the sample:

$$R_0 = \left(15.0 \frac{\text{decays}}{\text{min} \cdot \text{g}}\right)(50.0 \text{ g}) = 7.50 \times 10^2 \frac{\text{decays}}{\text{min}}$$

Find the decay constant from Equation 29.5:

$$\lambda = \frac{0.693}{T_{1/2}} = \frac{0.693}{3.01 \times 10^9 \text{ min}} = 2.30 \times 10^{-10} \text{ min}^{-1}$$

R is given, so now we substitute all values into Equation 29.19 to find the age of the skeleton:

$$t = -\frac{\ln\left(\frac{R}{R_0}\right)}{\lambda} = -\frac{\ln\left(\frac{200.0 \text{ decays/min}}{7.50 \times 10^2 \text{ decays/min}}\right)}{2.30 \times 10^{-10} \text{ min}^{-1}}$$

$$= \frac{1.32}{2.30 \times 10^{-10} \text{ min}^{-1}}$$

$$= 5.74 \times 10^9 \text{ min} = 1.09 \times 10^4 \text{ yr}$$

Remark For much longer periods, other radioactive substances with longer half-lives must be used to develop estimates.

Exercise 29.7

A sample of carbon of mass 7.60 g taken from an animal jawbone has an activity of 4.00 decays/min. How old is the jawbone?

Answer 2.77×10^4 yr

Carbon-14 and the Shroud of Turin

APPLICATION

Carbon-14 Dating of the Shroud of Turin

Since the Middle Ages, many people have marveled at a 14-foot-long, yellowing piece of linen found in Turin, Italy, purported to be the burial shroud of Jesus Christ (Fig. 29.11). The cloth bears a remarkable, full-size likeness of a crucified body, with wounds on the head that could have been caused by a crown of thorns and another wound in the side that could have been the cause of death. Skepticism over the authenticity of the shroud has existed since its first public showing in 1354; in fact, a French bishop declared it to be a fraud at the time. Because of its controversial nature, religious bodies have taken a neutral stance on its authenticity.

In 1978 the bishop of Turin allowed the cloth to be subjected to scientific analysis, but notably missing from these tests was carbon-14 dating. The reason for this omission was that, at the time, carbon-dating techniques required a piece of cloth about the size of a handkerchief. In 1988 the process had been refined to the point that pieces as small as one square inch were sufficient, and at that time permission was granted to allow the dating to proceed. Three labs were selected for the testing, and each was given four pieces of material. One of these was a piece of the shroud, and the other three pieces were control pieces similar in appearance to the shroud.

The testing procedure consisted of burning the cloth to produce carbon dioxide, which was then converted chemically to graphite. The graphite sample was subjected to carbon-14 analysis, and in the end all three labs agreed amazingly well on the age of the shroud. The average of their results gave a date for the cloth of A.D. 1 320 ± 60 years, with an assurance that the cloth could not be older than A.D. 1 200. Carbon-14 dating has thus unraveled the most important mystery concerning the shroud, but others remain. For example, investigators have not yet been able to explain how the image was imprinted.

Santi Visali/The IMAGE Bank

Figure 29.11 The Shroud of Turin as it appears in a photographic negative image.

29.5 NATURAL RADIOACTIVITY

Radioactive nuclei are generally classified into two groups: (1) unstable nuclei found in nature, which give rise to what is called **natural radioactivity**, and (2) nuclei produced in the laboratory through nuclear reactions, which exhibit **artificial radioactivity**.

Three series of naturally occurring radioactive nuclei exist (Table 29.2). Each starts with a specific long-lived radioactive isotope with half-life exceeding that of any of its descendants. The fourth series in Table 29.2 begins with ^{237}Np, a transuranic element (an element having an atomic number greater than that of uranium) not found in nature. This element has a half-life of "only" 2.14×10^6 yr.

The two uranium series are somewhat more complex than the ^{232}Th series (Fig. 29.12). Also, there are several other naturally occurring radioactive isotopes, such as ^{14}C and ^{40}K, that are not part of either decay series.

Natural radioactivity constantly supplies our environment with radioactive elements that would otherwise have disappeared long ago. For example, because the Solar System is about 5×10^9 years old, the supply of ^{226}Ra (with a half-life of only 1 600 yr) would have been depleted by radioactive decay long ago were it not for the decay series that starts with ^{238}U, with a half-life of 4.47×10^9 yr.

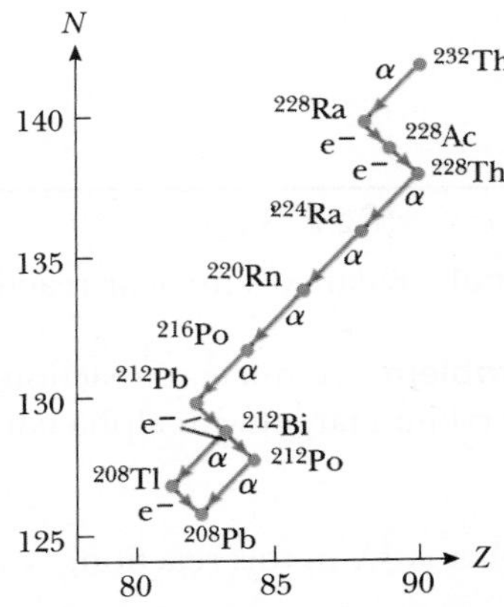

Figure 29.12 Decay series beginning with ^{232}Th.

29.6 NUCLEAR REACTIONS

It is possible to change the structure of nuclei by bombarding them with energetic particles. Such changes are called **nuclear reactions**. Rutherford was the first to observe nuclear reactions, using naturally occurring radioactive sources for the

TABLE 29.2

The Four Radioactive Series

Series	Starting Isotope	Half-life (years)	Stable End Product
Uranium	$^{238}_{92}U$	4.47×10^9	$^{206}_{82}Pb$
Actinium	$^{235}_{92}U$	7.04×10^8	$^{207}_{82}Pb$
Thorium	$^{232}_{90}Th$	1.41×10^{10}	$^{208}_{82}Pb$
Neptunium	$^{237}_{93}Np$	2.14×10^6	$^{209}_{83}Bi$

bombarding particles. He found that protons were released when alpha particles were allowed to collide with nitrogen atoms. The process can be represented symbolically as

$$ {}^{4}_{2}\text{He} + {}^{14}_{7}\text{N} \rightarrow \text{X} + {}^{1}_{1}\text{H} \qquad [29.20] $$

This equation says that an alpha particle (${}^{4}_{2}\text{He}$) strikes a nitrogen nucleus and produces an unknown product nucleus (X) and a proton (${}^{1}_{1}\text{H}$). Balancing atomic numbers and mass numbers, as we did for radioactive decay, enables us to conclude that the unknown is characterized as ${}^{17}_{8}\text{X}$. Because the element with atomic number 8 is oxygen, we see that the reaction is

$$ {}^{4}_{2}\text{He} + {}^{14}_{7}\text{N} \rightarrow {}^{17}_{8}\text{O} + {}^{1}_{1}\text{H} \qquad [29.21] $$

This nuclear reaction starts with two stable isotopes—helium and nitrogen—and produces two different stable isotopes—hydrogen and oxygen.

Since the time of Rutherford, thousands of nuclear reactions have been observed, particularly following the development of charged-particle accelerators in the 1930s. With today's advanced technology in particle accelerators and particle detectors, it is possible to achieve particle energies of at least 1 000 GeV = 1 TeV. These high-energy particles are used to create new particles whose properties are helping to solve the mysteries of the nucleus (and indeed, of the Universe itself).

Quick Quiz 29.4

Which of the following are possible reactions?

(a) ${}^{1}_{0}\text{n} + {}^{235}_{92}\text{U} \rightarrow {}^{140}_{54}\text{Xe} + {}^{94}_{38}\text{Sr} + 2({}^{1}_{0}\text{n})$

(b) ${}^{1}_{0}\text{n} + {}^{235}_{92}\text{U} \rightarrow {}^{132}_{50}\text{Sn} + {}^{101}_{42}\text{Mo} + 3({}^{1}_{0}\text{n})$

(c) ${}^{1}_{0}\text{n} + {}^{239}_{94}\text{Pu} \rightarrow {}^{127}_{53}\text{I} + {}^{93}_{41}\text{Nb} + 3({}^{1}_{0}\text{n})$

EXAMPLE 29.8 The Discovery of the Neutron

Goal Balance a nuclear reaction to determine an unknown decay product.

Problem A nuclear reaction of significant note occurred in 1932 when Chadwick, in England, bombarded a beryllium target with alpha particles. Analysis of the experiment indicated that the following reaction occurred:

$$ {}^{4}_{2}\text{He} + {}^{9}_{4}\text{Be} \rightarrow {}^{12}_{6}\text{C} + {}^{A}_{Z}\text{X} $$

What is ${}^{A}_{Z}X$ in this reaction?

Strategy Balancing mass numbers and atomic numbers yields the answer.

Solution

Write an equation relating the atomic masses on either side: $4 + 9 = 12 + A \rightarrow A = 1$

Write an equation relating the atomic numbers: $2 + 4 = 6 + Z \rightarrow Z = 0$

Identify the particle: ${}^{A}_{Z}X = {}^{1}_{0}\text{n}$ (a neutron)

Remarks This experiment was the first to provide positive proof of the existence of neutrons.

Exercise 29.8

Identify the unknown particle in this reaction:

$$ {}^{4}_{2}\text{He} + {}^{14}_{7}\text{N} \rightarrow {}^{17}_{8}\text{O} + {}^{A}_{Z}\text{X} $$

Answer ${}^{A}_{Z}\text{X} = {}^{1}_{1}\text{H}$ (a neutral hydrogen atom)

EXAMPLE 29.9 Synthetic Elements

Goal Construct equations for a series of radioactive decays.

Problem **(a)** A beam of neutrons is directed at a target of $^{238}_{92}\text{U}$. The reaction products are a gamma ray and another isotope. What is the isotope? **(b)** The isotope $^{239}_{92}\text{U}$ is radioactive and undergoes beta decay. Write the equation symbolizing this decay and identify the resulting isotope.

Strategy Balance the mass numbers and atomic numbers on both sides of the equations.

Solution

(a) Identify the isotope produced by the reaction of a neutron with a target of $^{238}_{92}\text{U}$, with production of a gamma ray.

Write an equation for the reaction in terms of the unknown isotope:

$$^{1}_{0}\text{n} + {}^{238}_{92}\text{U} \rightarrow {}^{A}_{Z}\text{X} + \gamma$$

Write and solve equations for the atomic mass and atomic number:

$$A = 1 + 238 = 239; \quad Z = 0 + 92 = 92$$

Identify the isotope:

$$^{A}_{Z}\text{X} = {}^{239}_{92}\text{U}$$

(b) Write the equation for the beta decay of $^{239}_{92}\text{U}$, identifying the resulting isotope.

Write an equation for the decay of $^{239}_{92}\text{U}$ by beta emission in terms of the unknown isotope:

$$^{239}_{92}\text{U} \rightarrow {}^{A}_{Z}\text{Y} + e^{-} + \bar{\nu}$$

Write and solve equations for the atomic mass and charge conservation (the electron counts as -1 on the right):

$$A = 239; \quad 92 = Z - 1 \rightarrow Z = 93$$

Identify the isotope:

$$^{A}_{Z}\text{Y} = {}^{239}_{93}\text{Np (neptunium)}$$

Remarks The interesting feature of these reactions is the fact that uranium is the element with the greatest number of protons (92) which exists in nature in any appreciable amount. The reactions in parts (a) and (b) do occur occasionally in nature; hence, minute traces of neptunium and plutonium are present. In 1940, however, researchers bombarded uranium with neutrons to produce plutonium and neptunium. These two elements were the first elements made in the laboratory. Since then, the list of synthetic elements has been extended to include those up to atomic number 112. Recently, elements 113 and 115 have been observed, but as of this writing, their existence has not yet been confirmed.

Exercise 29.9

The isotope $^{238}_{93}\text{U}$ is also radioactive and decays by beta emission. What is the end product?

Answer $^{239}_{94}\text{Pu}$

Q Values

We have just examined some nuclear reactions for which mass numbers and atomic numbers must be balanced in the equations. We will now consider the energy involved in these reactions, because energy is another important quantity that must be conserved.

We illustrate this procedure by analyzing the following nuclear reaction:

$$^{2}_{1}\text{H} + {}^{14}_{7}\text{N} \rightarrow {}^{12}_{6}\text{C} + {}^{4}_{2}\text{He} \qquad [29.22]$$

The total mass on the left side of the equation is the sum of the mass of $^{2}_{1}\text{H}$ (2.014 102 u) and the mass of $^{14}_{7}\text{N}$ (14.003 074 u), which equals 16.017 176 u. Similarly, the mass on the right side of the equation is the sum of the mass of $^{12}_{6}\text{C}$ (12.000 000 u) plus the mass of $^{4}_{2}\text{He}$ (4.002 602 u), for a total of 16.002 602 u. Thus, the total mass before the reaction is greater than the total mass after the reaction. The mass difference in the reaction is equal to 16.017 176 u − 16.002 602 u = 0.014 574 u. This "lost" mass is converted to the kinetic energy of the nuclei present after the reaction. In energy units, 0.014 574 u is equivalent to 13.576 MeV of kinetic energy carried away by the carbon and helium nuclei.

The energy required to balance the equation is called the Q value of the reaction. In Equation 29.22, the Q value is 13.576 MeV. Nuclear reactions in which there is a release of energy—that is, positive Q values—are said to be **exothermic reactions.**

The energy balance sheet isn't complete, however: We must also consider the kinetic energy of the incident particle before the collision. As an example, assume that the deuteron in Equation 29.22 has a kinetic energy of 5 MeV. Adding this to our Q value, we find that the carbon and helium nuclei have a total kinetic energy of 18.576 MeV following the reaction.

Now consider the reaction

$$^{4}_{2}\text{He} + {^{14}_{7}\text{N}} \rightarrow {^{17}_{8}\text{O}} + {^{1}_{1}\text{H}} \tag{29.23}$$

Before the reaction, the total mass is the sum of the masses of the alpha particle and the nitrogen nucleus: 4.002 602 u + 14.003 074 u = 18.005 676 u. After the reaction, the total mass is the sum of the masses of the oxygen nucleus and the proton: 16.999 133 u + 1.007 825 u = 18.006 958 u. In this case, the total mass after the reaction is *greater* than the total mass before the reaction. The mass deficit is 0.001 282 u, equivalent to an energy deficit of 1.194 MeV. This deficit is expressed by the negative Q value of the reaction, −1.194 MeV. Reactions with negative Q values are called **endothermic reactions.** Such reactions won't take place unless the incoming particle has at least enough kinetic energy to overcome the energy deficit.

At first it might appear that the reaction in Equation 29.23 can take place if the incoming alpha particle has a kinetic energy of 1.194 MeV. In practice, however, the alpha particle must have more energy than this. If it has an energy of only 1.194 MeV, energy is conserved but careful analysis shows that momentum isn't. This can be understood by recognizing that the incoming alpha particle has some momentum before the reaction. However, if its kinetic energy is only 1.194 MeV, the products (oxygen and a proton) would be created with zero kinetic energy and thus zero momentum. It can be shown that in order to conserve both energy and momentum, the incoming particle must have a minimum kinetic energy given by

$$KE_{\text{min}} = \left(1 + \frac{m}{M}\right)|Q| \tag{29.24}$$

where m is the mass of the incident particle, M is the mass of the target, and the absolute value of the Q value is used. For the reaction given by Equation 29.23, we find that

$$KE_{\text{min}} = \left(1 + \frac{4.002\ 602}{14.003\ 074}\right)|-1.194\ \text{MeV}| = 1.535\ \text{MeV}$$

This minimum value of the kinetic energy of the incoming particle is called the **threshold energy**. The nuclear reaction shown in Equation 29.23 won't occur if the incoming alpha particle has a kinetic energy of less than 1.535 MeV, but can occur if its kinetic energy is equal to or greater than 1.535 MeV.

Quick Quiz 29.5

If the Q value of an endothermic reaction is −2.17 MeV, then the minimum kinetic energy needed in the reactant nuclei if the reaction is to occur must be (a) equal to 2.17 MeV, (b) greater than 2.17 MeV, (c) less than 2.17 MeV, or (d) exactly half of 2.17 MeV.

29.7 MEDICAL APPLICATIONS OF RADIATION

Radiation Damage in Matter

Radiation absorbed by matter can cause severe damage. The degree and kind of damage depend on several factors, including the type and energy of the radiation and the properties of the absorbing material. Radiation damage in biological organisms is due primarily to ionization effects in cells. The normal function of a cell may be disrupted when highly reactive ions or radicals are formed as the result of ionizing radiation. For example, hydrogen and hydroxyl radicals produced from water molecules can induce chemical reactions that may break bonds in proteins and other vital molecules. Large acute doses of radiation are especially dangerous because damage to a great number of molecules in a cell may cause the cell to die. Also, cells that do survive the radiation may become defective, which can lead to cancer.

In biological systems, it is common to separate radiation damage into two categories: somatic damage and genetic damage. **Somatic damage** is radiation damage to any cells except the reproductive cells. Such damage can lead to cancer at high radiation levels or seriously alter the characteristics of specific organisms. **Genetic damage** affects only reproductive cells. Damage to the genes in reproductive cells can lead to defective offspring. Clearly, we must be concerned about the effect of diagnostic treatments, such as x-rays and other forms of exposure to radiation.

Several units are used to quantify radiation exposure and dose. The **roentgen** (R) is defined as **that amount of ionizing radiation which will produce 2.08×10^9 ion pairs in 1 cm^3 of air under standard conditions**. Equivalently, the roentgen is **that amount of radiation which deposits 8.76×10^{-3} J of energy into 1 kg of air**.

For most applications, the roentgen has been replaced by the **rad** (an acronym for *r*adiation *a*bsorbed *d*ose), defined as follows: **One rad is that amount of radiation which deposits 10^{-2} J of energy into 1 kg of absorbing material.**

Although the rad is a perfectly good physical unit, it's not the best unit for measuring the degree of biological damage produced by radiation, because the degree of damage depends not only on the dose, but also on the *type* of radiation. For example, a given dose of alpha particles causes about 10 times more biological damage than an equal dose of x-rays. The **RBE** (*r*elative *b*iological *e*ffectiveness) factor is defined as **the number of rads of x-radiation or gamma radiation that produces the same biological damage as 1 rad of the radiation being used**. The RBE factors for different types of radiation are given in Table 29.3. Note that the values are only approximate because they vary with particle energy and the form of damage.

Finally, the **rem** (*r*oentgen *e*quivalent in *m*an) is defined as the product of the dose in rads and the RBE factor:

$$\text{Dose in rem} = \text{dose in rads} \times \text{RBE}$$

According to this definition, 1 rem of any two kinds of radiation will produce the same amount of biological damage. From Table 29.3, we see that a dose of 1 rad of fast neutrons represents an effective dose of 10 rem and that 1 rad of x-radiation is equivalent to a dose of 1 rem.

TABLE 29.3

RBE Factors for Several Types of Radiation

Radiation	RBE Factor
X-rays and gamma rays	1.0
Beta particles	1.0–1.7
Alpha particles	10–20
Slow neutrons	4–5
Fast neutrons and protons	10
Heavy ions	20

APPLICATION

 Occupational Radiation Exposure Limits

Low-level radiation from natural sources, such as cosmic rays and radioactive rocks and soil, delivers a dose of about 0.13 rem/year per person. The upper limit of radiation dose recommended by the U.S. government (apart from background radiation and exposure related to medical procedures) is 0.5 rem/year. Many occupations involve higher levels of radiation exposure, and for individuals in these occupations, an upper limit of 5 rem/year has been set for whole-body exposure. Higher upper limits are permissible for certain parts of the body, such as the hands and forearms. An acute whole-body dose of 400 to 500 rem results in a mortality rate of about 50%. The most dangerous form of exposure is ingestion or inhalation of radioactive isotopes, especially those elements the body retains and concentrates, such as ^{90}Sr. In some cases, a dose of 1000 rem can result from ingesting 1 mCi of radioactive material.

APPLICATION

 Irradiation of Food and Medical Equipment

Sterilizing objects by exposing them to radiation has been going on for at least 25 years, but in recent years the methods used have become safer to use and more economical. Most bacteria, worms, and insects are easily destroyed by exposure to gamma radiation from radioactive cobalt. There is no intake of radioactive nuclei by an organism in such sterilizing processes, as there is in the use of radioactive tracers. The process is highly effective in destroying Trichinella worms in pork, salmonella bacteria in chickens, insect eggs in wheat, and surface bacteria on fruits and vegetables that can lead to rapid spoilage. Recently, the procedure has been expanded to include the sterilization of medical equipment while in its protective covering. Surgical gloves, sponges, sutures, and so forth are irradiated while packaged. Also, bone, cartilage, and skin used for grafting is often irradiated to reduce the chance of infection.

Tracing

APPLICATION

 Radioactive Tracers in Medicine

Radioactive particles can be used to trace chemicals participating in various reactions. One of the most valuable uses of radioactive tracers is in medicine. For example, ^{131}I is an artificially produced isotope of iodine. (The natural, nonradioactive isotope is ^{127}I.) Iodine, a necessary nutrient for our bodies, is obtained largely through the intake of seafood and iodized salt. The thyroid gland plays a major role in the distribution of iodine throughout the body. In order to evaluate the performance of the thyroid, the patient drinks a small amount of radioactive sodium iodide. Two hours later, the amount of iodine in the thyroid gland is determined by measuring the radiation intensity in the neck area.

A medical application of the use of radioactive tracers occurring in emergency situations is that of locating a hemorrhage inside the body. Often the location of the site cannot easily be determined, but radioactive chromium can identify the location with a high degree of precision. Chromium is taken up by red blood cells and carried uniformly throughout the body. However, the blood will be dumped at a hemorrhage site, and the radioactivity of that region will increase markedly.

APPLICATION

Radioactive Tracers in Agricultural Research

The tracer technique is also useful in agricultural research. Suppose the best method of fertilizing a plant is to be determined. A certain material in the fertilizer, such as nitrogen, can be tagged with one of its radioactive isotopes. The fertilizer is then sprayed onto one group of plants, sprinkled on the ground for a second group, and raked into the soil for a third. A Geiger counter is then used to track the nitrogen through the three types of plants.

Tracing techniques are as wide ranging as human ingenuity can devise. Present applications range from checking the absorption of fluorine by teeth to checking contamination of food-processing equipment by cleansers to monitoring deterioration inside an automobile engine. In the last case, a radioactive material is used in the manufacture of the pistons, and the oil is checked for radioactivity to determine the amount of wear on the pistons.

Computed Axial Tomography (CAT) Scans

The normal x-ray of a human body has two primary disadvantages when used as a source of clinical diagnosis. First, it is difficult to distinguish between various types of tissue in the body because they all have similar x-ray absorption properties.

only choices (b) and (c). The buoyant force depends on the amount of fluid displaced, and this force must exactly balance the gravitational force, or weight. Because the object is only three-quarters submerged, its weight must be only three-quarters that of a similar volume of water. It follows that the density of the object must be three-quarters that of water, which is answer **(c)**.

Quantitative Solution

Neglecting the buoyancy of air, two forces act on the object: the buoyant force F_B, which is equal to the weight of displaced fluid, and the gravitational force:

$$\Sigma \vec{\mathbf{F}} = \vec{\mathbf{F}}_B + \vec{\mathbf{F}}_{\text{grav}} = 0$$

Substitute the expressions for the buoyant and gravitational forces:

$$F_B - mg = 0 \qquad (1)$$

By definition, the mass of the object is given by $m = \rho_{\text{obj}} V$, where ρ_{obj} is the object's density and V is its volume. The buoyant force is the weight of the displaced water: $F_B = \rho_{\text{water}} V_{\text{sub}} g$, where V_{sub} is the volume of water displaced. Substitute these expressions into Equation (1) and solve for ρ_{obj}:

$$\rho_{\text{water}} V_{\text{sub}} g - \rho_{\text{obj}} V g = 0 \qquad (2)$$

Solve Equation (2) for ρ_{obj}, obtaining answer **(c)**:

$$\rho_{\text{obj}} = \frac{V_{\text{sub}}}{V} \rho_{\text{water}} = (0.750)(1.00 \times 10^3 \text{ kg/m}^3)$$

$$= 7.50 \times 10^2 \text{ kg/m}^3$$

EXAMPLE 4

At point 1, water flows smoothly at speed v_1 through a horizontal pipe with radius r_1. The pipe then narrows to half that radius at point 2. What can be said of the speed v_1 of the water at point 1 compared to its speed v_2 at point 2? (a) $v_1 = v_2$ (b) $v_1 > v_2$ (c) $v_1 < v_2$

Conceptual Solution

The volume flow rate is proportional to the velocity and cross-sectional area of the pipe. A larger radius at point 1 means a larger cross-sectional area. Because water is essentially incompressible, the flow rate must be the same in both sections of pipe, so the larger cross-section at point 1 results in a smaller fluid velocity, and the answer is **(c)**.

Quantitative Solution

Apply the equation of continuity for an incompressible fluid:

$$A_1 v_1 = A_2 v_2 \qquad (1)$$

Substitute an expression for the area on each side of Equation (1):

$$\pi r_1^2 v_1 = \pi r_2^2 v_2 \qquad (2)$$

Solve Equation (2) for v_1, and substitute $r_2 = \frac{1}{2} r_1$, obtaining answer **(c)**:

$$v_1 = \frac{r_2^2}{r_1^2} v_2 = \frac{r_2^2}{(2r_2)^2} v_2 = \frac{1}{4} v_2$$

MULTIPLE CHOICE PROBLEMS

1. A diver is swimming 10.0 m below the surface of the water in a reservoir. There is no current, the air has a pressure of 1.00 atmosphere, and the density of the water is 1.00×10^3 kilograms per cubic meter. What is the pressure as measured by the diver? (a) 1.10 atm (b) 1.99×10^5 Pa (c) 11.0 atm (d) 1.01×10^5 Pa
2. The aorta of a 70.0-kg man has a cross-sectional area of 3.00 cm^2 and carries blood with a speed of 30.0 cm/s. What is the average volume flow rate? (a) 10.0 cm/s (b) 33.0 cm^3/s (c) 10.0 cm^2/s (d) 90.0 cm^3/s

3. At 20.0°C the density of water is 1.00 g/cm^3. What is the density of a body that has a weight of 0.980 N in air but registers an apparent weight of only 0.245 N on a spring scale when fully immersed in water? (a) 0.245 g/cm^3 (b) 0.735 g/cm^3 (c) 1.33 g/cm^3 (d) 4.00 g/cm^3

4. Two insoluble bodies, A and B, appear to lose the same amount of weight when submerged in alcohol. Which statement is most applicable? (a) Both bodies have the same mass in air. (b) Both bodies have the same volume. (c) Both bodies have the same density. (d) Both bodies have the same weight in air.

5. The bottom of each foot of an 80-kg man has an area of about 400 cm^2. What is the effect of his wearing snowshoes with an area of about 0.400 m^2? (a) The pressure exerted on the snow becomes 10 times as great. (b) The pressure exerted on the snow becomes 1/10 as great. (c) The pressure exerted on the snow remains the same. (d) The force exerted on the snow is 1/10 as great.

6. In a hydraulic lift, the surface of the input piston is 10 cm^2 and that of the output piston is 3 000 cm^2. What is the work done if a 100 N force applied to the input piston raises the output piston by 2.0 m? (a) 20 kJ (b) 30 kJ (c) 40 kJ (d) 60 kJ

7. The Young's modulus for steel is 2.0×10^{11} N/m^2. What is the stress on a steel rod that is 100 centimeters long and 20 millimeters in diameter when it is stretched by a force of 6.3×10^3 N? (a) 2.01×10^7 N/m^2 (b) 12.6×10^{12} N/m^2 (c) 3.15×10^8 N/m^2 (d) 4.0×10^{11} N/m^2

Answers

1. (b). The fluid is at rest (no currents) so this is a hydrostatic pressure calculation. In SI units 1 atm = 1.01×10^5 Pa. Choice (d) can be eliminated because the pressure below the water must be greater than the pressure at the surface.

$$P_{\text{diver}} = P_{\text{atm}} + \rho g h = (1.01 \times 10^5\text{ Pa}) + (1.00 \times 10^3\text{ kg/m}^3)(9.80\text{ m/s}^2)(10.0\text{ m})$$
$$= 1.99 \times 10^5\text{ Pa}$$

2. (d). The answer follows from the definition:

$$\text{volume flow rate} = vA = (30\text{ cm/s})(3\text{ cm}^2) = 90\text{ cm}^3\text{/s}$$

3. (c). This problem is an application of Archimedes' Principle. The apparent loss of weight of a submerged body equals the weight of the fluid displaced. The weight w of the displaced water is $w = 0.980\text{ N} - 0.245\text{ N} = 0.735\text{ N}$. Using the definition of density, $\rho_{\text{water}} g = w/V$, so the volume occupied by this weight of water is $V = 0.735\text{ N}/\rho_{\text{water}} g$. The volume of the body must equal the volume of the water displaced. The density of the body is: $\rho = m/V = (0.980\text{ N}/0.735\text{ N})\rho_{\text{water}} = 1.33\text{ g/cm}^3$

4. (b). According to Archimedes' Principle, the apparent weight lost is equal to the weight of the displaced fluid; therefore, both must have the same volume because the volume of the fluid equals the volume of the body.

5. (b). The force exerted by the man is his weight and it is assumed to be constant. This eliminates choice (d). For a constant force, pressure and area are inversely proportional. The area of the snowshoes is ten times the area of the foot so that the pressure associated with the snowshoes is the inverse of 10 or 1/10 the pressure exerted by the foot.

6. (d). By Pascal's principle, the pressure at the input and output pistons is the same, so

$$\frac{F_{\text{in}}}{A_{\text{in}}} = \frac{F_{\text{out}}}{A_{\text{out}}} \quad \rightarrow \quad F_{\text{out}} = \frac{A_{\text{out}}}{A_{\text{in}}} F_{\text{in}}$$
$$= \left(\frac{3\,000\text{ cm}^2}{10\text{ cm}^2}\right) 100\text{ N} = 3 \times 10^4\text{ N}$$

Work is a force times a displacement:

$$W = F\Delta s = (3 \times 10^4\text{ N})(2\text{ m}) = 6 \times 10^4\text{ J} = 60\text{ kJ}$$

7. (a). Stress is force per unit area, so neither Young's modulus nor the length of the rod are needed to solve the problem.

$$\text{Stress} = F/A = (6.30 \times 10^3\text{ N})/\pi(1 \times 10^{-2}\text{ m})^2 = 2.01 \times 10^7\text{ N/m}^2$$

WAVES

EXAMPLE 1

A transverse wave travels at speed v_1 and has twice the frequency and one-quarter the wavelength of a second transverse wave. (These could be waves on two different strings, for example.) How does the speed v_1 of the first wave compare to the speed v_2 of the second wave? (a) $v_1 = v_2$ (b) $v_1 = 2v_2$ (c) $v_1 = \frac{1}{2}v_2$.

Solution

The speed of a wave is proportional to both the frequency and wavelength:

$$v = f\lambda \qquad (1)$$

Make a ratio of the v_2 to v_1 using Equation 1, and substitute $f_1 = 2f_2$ and $\lambda_1 = \lambda_2/4$:

$$\frac{v_1}{v_2} = \frac{f_1\lambda_1}{f_2\lambda_2} = \frac{(2f_2)(\lambda_2/4)}{f_2\lambda_2} = \frac{2}{4} = \frac{1}{2}$$

Solve for v_1, obtaining answer **(c)**:

$$v_1 = \frac{1}{2}v_2$$

Example 2

A block of mass m oscillates at the end of a horizontal spring on a frictionless surface. At maximum extension, an identical block drops onto the top of the first block and sticks to it. How does the new period T_{new} compare to the original period, T_0? (a) $T_{\text{new}} = \sqrt{2}T_0$ (b) $T_{\text{new}} = T_0$ (c) $T_{\text{new}} = 2T_0$ (d) $T_{\text{new}} = \frac{1}{2}T_0$

Conceptual Solution

An increased mass would increase the inertia of the system without augmenting the mechanical energy, so a mass of $2m$ should move more slowly than a mass of m. This in turn would lengthen the period, eliminating choices (b) and (d). The period is proportional to the square root of the mass, so doubling the mass increases the period by a factor of the square root of two, which is answer **(a)**.

Quantitative Solution

Apply the equation for the period of a mass-spring system:

$$T = 2\pi\sqrt{\frac{m}{k}} \qquad (1)$$

Using Equation (1), make a ratio of the new period T_{new} to the old period, T_0, canceling common terms and substituting $m_{\text{new}} = 2m_0$:

$$\frac{T_{\text{new}}}{T_0} = \frac{2\pi\sqrt{\dfrac{m_{\text{new}}}{k}}}{2\pi\sqrt{\dfrac{m_0}{k}}} = \frac{\sqrt{m_{\text{new}}}}{\sqrt{m_0}} = \sqrt{\frac{2m_0}{m_0}} = \sqrt{2} \qquad (2)$$

Solving Equation (2) for the new period T_{new} yields answer (a):

$$T_{\text{new}} = \sqrt{2}\,T_0 \text{ [answer (a)]}$$

EXAMPLE 3

A simple pendulum swings back and forth 36 times in 68 s. What is its length, if the local acceleration of gravity is 9.80 m/s^2? (a) 0.067 7 m (b) 0.356 m (c) 1.25 m (d) 0.887 m

Solution

First, find the period T of the motion:

$$T = \frac{68\text{ s}}{36} = 1.89\text{ s}$$

Write the equation for the period of a pendulum and solve it for the length L, obtaining answer (d):

$$T = 2\pi\sqrt{\frac{L}{g}} \quad \rightarrow \quad L = \frac{gT^2}{4\pi^2} = \frac{(9.80\text{ m/s}^2)(1.89\text{ s})^2}{4\pi^2}$$

$$= 0.887\text{ m} \text{ [answer (d)]}$$

MULTIPLE CHOICE PROBLEMS

1. A simple pendulum has a period of 4.63 s at a place on the Earth where the acceleration of gravity is 9.82 m/s^2. At a different location, the period increases to 4.64 seconds. What is the value of g at this second point? (a) 9.78 m/s^2 (b) 9.82 m/s^2 (c) 9.86 m/s^2 (d) Cannot be determined without knowing the length of the pendulum.
2. What is the wavelength of a transverse wave having a speed of 15 m/s and a frequency of 5.0 Hz? (a) 3.0 m (b) 10 m (c) 20 m (d) 45 m
3. What is the optimum difference in phase for maximum destructive interference between two waves of the same frequency? (a) 360° (b) 270° (c) 180° (d) 90°
4. Standing waves can be formed if coincident waves have (a) the same direction of propagation. (b) the same frequency. (c) different amplitudes. (d) different wavelengths.
5. A simple pendulum with a length L has a period of 2 s. In order for the pendulum to have a period of 4 s, we must (a) halve the length. (b) quarter the length. (c) double the length. (d) quadruple the length.
6. If a simple pendulum 12 m long has a frequency of 0.25 Hz, what will be the period of a second pendulum at the same location if its length is 3.0 m? (a) 2.0 s (b) 3.0 s (c) 4.0 s (d) 6.0 s
7. A pendulum clock runs too slowly (i.e., is losing time). Which of the following adjustments could rectify the problem? (a) The weight of the bob should be decreased so it can move faster. (b) The length of the wire holding the bob should be shortened. (c) The amplitude of the swing should be reduced so the path covered is shorter. (d) None of the above.
8. A 20.0-kg object placed on a frictionless floor is attached to a wall by a spring. A 5.00-N force horizontally displaces the object 1.00 m from its equilibrium position. What is the period of oscillation of the object? (a) 2.00 s (b) 6.08 s (c) 12.6 s (d) 16.4 s

Answers

1. **(a).** The answer can be determined without doing a numerical solution. Rearrange $T = 2\pi(L/g)^{1/2}$ to $Tg^{1/2}$ = constant. The period is inversely related to the square root of the acceleration due to gravity. Because T has increased, $g^{1/2}$ and hence g must decrease. Choice (a) is the only value of g that is less than the original 9.82 m/s^2. Quantitatively:

$$g_2 = (T_1^2 g_1/T_2^2) = (4.63\text{ s})^2(9.82\text{ m/s}^2)/(4.64\text{ s})^2 = 9.78\text{ m/s}^2$$

2. **(a).** Wavelength is velocity divided by frequency. The formula does not depend upon the type of wave involved.

$$\lambda = v/f = (15\text{ m/s})/5.0\text{ s}^{-1} = 3.0\text{ m}$$

3. **(c).** Two waves are completely out of phase when their antinodes coincide so that each crest on one wave coincides with a trough on the other. This occurs when the waves differ in phase by 180°.
4. **(b).** In standing waves, the nodes are stationary. This can be accomplished when two waves with the same frequency travel in opposite directions.
5. **(d).** In a pendulum, the period and the square root of the length are directly proportional:

$$T = 2\pi(L/g)^{1/2} \text{ so that } T/L^{1/2} = \text{constant}$$

To double the period, you must double the square root of the length. To double the square root of the length, you must quadruple the length:

$$(4L)^{1/2} = 4^{1/2}\,L^{1/2} = 2L^{1/2}$$

6. **(a).** The frequency is the reciprocal of the period: $f = 1/T$, so the first pendulum has period T_1 = 4.0 s. For pendulums, the period and the square root of the length are directly proportional, so the ratio of the two periods is:

$$T_2/T_1 = (L_2/L_1)^{1/2} = (12/3)^{1/2} = (4.0)^{1/2} = 2.0$$

It follows that T_2 = 2.0 s.

7. **(b).** The period of a pendulum is directly related to the square root of the length of the cord holding the bob. It is independent of the mass and amplitude.
8. **(c).** The force constant is $k = F/\Delta x$ = 5.0 N/1.0 m = 5.0 N·m^{-1}. The period is

$$T = 2\pi(m/k)^{1/2} = 2\pi(20.0\text{ kg}/5.00\text{ N}\cdot\text{m}^{-1})^{1/2} = 12.6\text{ s}$$

SOUND

EXAMPLE 1

When a given sound intensity doubles, by how much does the sound intensity level (or decibel level) increase?

Solution

Write the expression for a difference in decibel level, $\Delta\beta$:

$$\Delta\beta = \beta_2 - \beta_1 = 10 \log\left(\frac{I_2}{I_0}\right) - 10 \log\left(\frac{I_1}{I_0}\right)$$

Factor and apply the logarithm rule $\log a - \log b = \log(a/b) = \log(a \cdot b^{-1})$:

$$\Delta\beta = 10\left(\log\left(\frac{I_2}{I_0}\right) - \log\left(\frac{I_1}{I_0}\right)\right) = 10 \log\left(\frac{I_2}{I_0} \cdot \frac{I_0}{I_1}\right)$$

$$= 10 \log\left(\frac{I_2}{I_1}\right)$$

Substitute $I_2 = 2I_1$ and evaluate the expression:

$$\Delta\beta = 10 \log\left(\frac{2I_1}{I_1}\right) = 10 \log(2) = \boxed{3.01 \text{ dB}}$$

EXAMPLE 2

A sound wave passes from air into water. What can be said of its wavelength in water as compared to air? (a) The wavelengths are the same. (b) The wavelength in air is greater than in water. (c) The wavelength in air is less than in water.

Conceptual Solution

The frequency of the sound doesn't change in going from one type of media to the next because it's caused by periodic variations of pressure in time. The wavelength must change, however, because the speed of sound changes and during a single period the sound wave will travel a different distance, which corresponds to a single wavelength. The speed of sound in water is greater than in air, so in a single period the sound wave will travel a greater distance. Hence the wavelength of a given sound wave is greater in water than in air, which is **(c)**. This result can be made quantitative by using $v = f\lambda$, where v is the wave speed, f the frequency, and λ the wavelength.

EXAMPLE 3

A sound wave emitted from a sonar device reflects off a submarine that is traveling away from the sonar source. How does the frequency of the reflected wave, f_R, compare to the frequency of the source, f_S? (a) $f_R = f_S$ (b) $f_R < f_S$ (c) $f_R > f_S$

Solution

The reflected wave will have a lower frequency [answer **(b)**]. To see this, consider pressure maximums impinging sequentially on the submarine. The first pressure maximum hits the submarine and reflects, but the second reaches the submarine when it has moved further away. The distance between consecutive maximums in the reflected waves is thereby increased; hence the wavelength also increases. Because $v = f\lambda$ and the speed isn't affected, the frequency must decrease.

MULTIPLE CHOICE PROBLEMS

1. The foghorn of a ship echoes off an iceberg in the distance. If the echo is heard 5.00 seconds after the horn is sounded, and the air temperature is -15.0°C, how far away is the iceberg? (a) 224 m (b) 805 m (c) 827 m (d) 930 m
2. What is the sound level of a wave with an intensity of 10^{-3} W/m^2? (a) 30 dB (b) 60 dB (c) 90 dB (d) 120 dB
3. At 0°C, approximately how long does it take sound to travel 5.00 km through air? (a) 15 s (b) 30 s (c) 45 s (d) 60 s
4. If the speed of a transverse wave of a violin string is 12.0 m/s and the frequency played is 4.00 Hz, what is the wavelength of the sound in air? (Use 343 m/s for the speed of sound.) (a) 48.0 m (b) 12.0 m (c) 3.00 m (d) 85.8 m

5. If two identical sound waves interact in phase, the resulting wave will have a (a) shorter period. (b) larger amplitude. (c) higher frequency. (d) greater velocity.

6. What is the speed of a longitudinal sound wave in a steel rod if Young's modulus for steel is 2.0×10^{11} N/m^2 and the density of steel is 8.0×10^3 kg/m^3? (a) 4.0×10^{-8} m/s (b) 5.0×10^3 m/s (c) 25×10^6 m/s (d) 2.5×10^9m/s

7. If two frequencies emitted from two sources are 48 and 54 Hz, how many beats per second are heard? (a) 3 (b) 6 (c) 9 (d) 12

8. The frequency registered by a detector is higher than the frequency emitted by the source. Which of the statements below **must** be true? (a) The source must be moving away from the detector. (b) The source must be moving toward the detector. (c) The distance between the source and the detector must be decreasing. (d) The detector must be moving away from the source.

Answers

1. **(b).** The normal speed of sound at 0°C in air is 331 m/s. The speed of sound in air at various temperatures is given by

$$v = (331 \text{ m/s})\sqrt{\frac{T}{273\ K}} = (331 \text{ m/s})\sqrt{\frac{273 \text{ K} - 15.0 \text{ K}}{273 \text{ K}}}$$
$$= 322 \text{ m/s}$$

Calculate the distance traveled in half of five seconds:

$$d = vt = (322 \text{ m/s})(2.50 \text{ s}) = 805 \text{ m}.$$

2. **(c).** Substitute: $\beta = 10 \log I/I_0 = 10 \log (1 \times 10^{-3}\text{ W/m}^2/10^{-12}\text{ W/m}^2) = 90$ dB.

3. **(a).** At 0°C the speed of sound in air is 331 m/s, so $t = (5.00 \times 10^3\text{ m})/(331\text{ m/s}) \sim 15$ s

4. **(d).** The frequency is the same for the string as for the sound wave the string produces in the surrounding air, but the wavelength differs. The speed of sound is equal to the product of the frequency and wavelength, so $\lambda = v/f = (343\text{ m/s})/4.00\text{ s}^{-1} = 85.8$ m.

5. **(b).** Two waves are in phase if their crests and troughs coincide. The amplitude of the resulting wave is the algebraic sum of the amplitudes of the two waves being superposed at that point, so the amplitude is doubled.

6. **(b).** The solution requires substituting into an expression for the velocity of sound through a rod of solid material having Young's modulus Y:

$$v = (Y/\rho)^{1/2} = (2.0 \times 10^{11}\text{ kg}\cdot\text{m/s}^2)/(8.0 \times 10^3\text{ kg/m}^3)^{1/2}$$
$$= 25 \times 10^6\text{ m}^2/\text{s}^2)^{1/2} = 5.0 \times 10^3\text{ m/s}$$

7. **(b).** $f_{\text{beat}} = f_1 - f_2 = 54 - 48 = 6$ beats

8. **(c).** Movement of a sound source relative an observer is the Doppler effect. Because the frequency is shifted to a higher value, the source and the detector must be getting closer together—effectively shortening the wavelength from the observer's point of view. This observation eliminates choices (a) and (d). Choice (b) can be eliminated because it isn't necessarily true: the same effect kind of effect occurs when the source is held steady and the detector moves towards it.

LIGHT

EXAMPLE 1

Which substance will have smaller critical angle for total internal reflection, glass with an index of refraction of 1.50, or diamond with an index of refraction of 2.42? (a) glass (b) diamond (c) The critical angles are the same for both.

Conceptual Solution

A larger index of refraction indicates a slower speed of light inside the material, which in turn means a larger refraction angle—a larger bending towards the normal on entering the material from air, and a larger bending away from the normal on passing from the material into air. (Recall that a normal line is perpendicular to the surface of the material.) For total internal reflection to occur, a refraction angle of 90° must be possible, which occurs whenever the refracting medium has a lower index of refraction than the incident medium. Diamond bends light more than glass, so the incident ray can be closer to the normal and still be bent enough so the angle of refraction results in total internal reflection. Closer to the normal means a smaller critical angle, so the answer is **(b)**.

Quantitative Solution

Write Snell's law for the diamond-air interface: $n_D \sin \theta_D = n_A \sin \theta_A$

Compute the critical angle for diamond, using $n_A = 1.00$ for air and $\theta_A = 90°$, together with $n_D = 2.42$:

$$2.42 \sin \theta_D = 1.00 \quad \rightarrow \quad \theta_D = \sin^{-1}\left(\frac{1.00}{2.42}\right) = 24.4°$$

Repeat the calculation for glass:

$$1.50 \sin \theta_G = 1.00 \quad \rightarrow \quad \theta_G = \sin^{-1}\left(\frac{1.00}{1.50}\right) = 41.8°$$

The calculation explicitly shows that diamond has a smaller critical angle than glass (answer **b**).

EXAMPLE 2

A patient's near point is 85.0 cm. What focal length prescription lens will allow the patient to see objects clearly that are at a distance of 25.0 cm from the eye? Neglect the eye-lens distance.

Solution

Use the thin lens equation:

$$\frac{1}{f} = \frac{1}{p} + \frac{1}{q}$$

An object at distance $p = 25.0$ cm must form an image at the patient's near point of 85.0 cm. The image must be virtual, so $q = -85.0$ cm:

$$\frac{1}{f} = \frac{1}{25.0 \text{ cm}} + \frac{1}{-85.0 \text{ cm}} = 2.82 \times 10^{-2} \text{ cm}^{-1}$$

$$f = \boxed{35.4 \text{ cm}}$$

EXAMPLE 3

Two photons traveling in vacuum have different wavelengths. Which of the following statements is true? (a) The photon with a smaller wavelength has greater energy. (b) The photon with greater wavelength has greater energy. (c) The energy of all photons is the same. (d) The photon with the greater wavelength travels at a lesser speed.

Solution

Answer (d) can be eliminated immediately, because the speed of light in vacuum is the same for all wavelengths of light. The energy of a photon, or particle of light, is given by $E = hf$, and consequently is proportional to the frequency f, which in turn is inversely proportional to the wavelength because $f = c/\lambda$, where c is the speed of light. A smaller wavelength photon has a greater frequency, and therefore a greater energy, so answer **(a)** is true.

MULTIPLE CHOICE PROBLEMS

1. Glass has an index of refraction of 1.50. What is the frequency of light that has a wavelength of 5.00×10^2 nm in glass? (a) 1.00 Hz (b) 2.25 Hz (c) 4.00×10^{14} Hz (d) 9.00×10^{16} Hz
2. Water has an index of refraction of 1.33. If a plane mirror is submerged in water, what can be said of the angle of reflection θ if light strikes the mirror with an angle of incidence of 30°? (a) $\theta < 30°$ (b) $\theta = 30°$ (c) $30° < \theta$ (d) No light is reflected because 30° is the critical angle for water.

3. The index of refraction for water is 1.33 and that for glass is 1.50. A light ray strikes the water-glass boundary with an incident angle of 30.0° on the water side. Which of the following is the refraction angle in the glass? (a) 26.3° (b) 34.7° (c) 30.0° (d) 60.0°

4. Light is incident on a prism at an angle of 90° relative to its surface. The index of refraction of the prism material is 1.50. Which of the following statements is most accurate about the angle of refraction θ? (a) $0° < \theta < 45°$ (b) $45° < \theta < 90°$ (c) $\theta = 0°$ (d) $90° < \theta$

5. White light incident on an air-glass interface is split into a spectrum within the glass. Which color light has the greatest angle of refraction? (a) red light (b) violet light (c) yellow light (d) The angle is the same for all wavelengths.

6. A real object is placed 10.0 cm from a converging lens that has a focal length of 6.00 cm. Which statement is most accurate? (a) The image is real, upright, and enlarged. (b) The image is real, inverted, and enlarged. (c) The image is real, upright, and reduced. (d) The image is real, inverted, and reduced.

7. What is the focal length of a lens that forms a virtual image 30.0 cm from the lens when a real object is placed 15.0 cm from the lens? (a) 10.0 cm (b) 15.0 cm (c) 30.0 cm (d) 45.0 cm

8. What is the magnification of a lens that forms an image 20.0 cm to its right when a real object is placed 10.0 cm to its left? (a) 0.500 (b) 1.00 (c) 1.50 (d) −2.00

9. The human eye can respond to light with a total energy of as little as 10^{-18} J. If red light has a wavelength of 600 nm, what is the minimum number of red light photons that the eye can perceive? (a) 1 (b) 2 (c) 3 (d) 5

10. Which phenomenon occurs for transverse waves but not for longitudinal waves? (a) reflection (b) refraction (c) diffraction (d) polarization

Answers

1. (c). The velocity of light in glass can be found from the definition of refractive index, $n = c/v$; wavelength and frequency are related to velocity by the general wave relation, $v = f\lambda$. Therefore:

$$f = v/\lambda = c/n\lambda = (3.00 \times 10^8 \text{ m/s})/(1.50)(5.00 \times 10^{-7} \text{ m}) = 4.00 \times 10^{14} \text{ Hz}$$

2. (b). The law of reflection is independent of the medium involved. The angle of reflection is always equal to the angle of incidence.

3. (a). Snell's law is given by $n_1 \sin\theta_1 = n_2 \sin\theta_2$, hence: $1.33 \sin 30.0° = 1.50 \sin\theta_2$. From this expression, we see that $\sin\theta_2 < \sin 30.0°$, so θ_2 must be less than 30.0°. Therefore, (a) is the only reasonable choice.

4. (c). Snell's law is $n_1 \sin\theta_1 = n_2 \sin\theta_2$. Because the incident rays are normal to the prism surface, $\theta_1 = 0°$, and $\sin 0° = 0$. Consequently, $0 = n_2 \sin\theta_2 \rightarrow \theta_2 = 0°$

5. (b). The greater the frequency of light, the greater its energy and the faster its speed through any material medium. From Snell's Law the velocity and $\sin\theta$ with respect to the normal are inversely proportional. Of the choices, violet light has the highest frequency and therefore, the highest velocity, and the greatest angle of refraction.

6. (b). By the thin lens equation, $1/f = 1/p + 1/q$, hence with $p = 10.0$ cm and $f = 6.00$ cm, substitution results in $q = 15.0$ cm. The image distance is positive, hence the image is real. The magnification is given by $M = -q/p = -15.0 \text{ cm}/10.0 = -1.50$. Because M is negative, the image is inverted, whereas $|M| > 1$ means the image is enlarged.

7. (c). The distance is given by the thin lens equation, $1/f = 1/p + 1/q$. Because the image formed is virtual, the sign of q is negative. Therefore:

$$\frac{1}{f} = \frac{1}{p} + \frac{1}{q} = \frac{1}{15.0 \text{ cm}} + \frac{1}{-30.0 \text{ cm}} = \frac{1}{30.0 \text{ cm}} \quad \rightarrow \quad f = 30.0 \text{ cm}$$

8. (d). Magnification is given by $M = -q/p = -20.0 \text{ cm}/10.0 \text{ cm} = -2.00$. The sign of q is positive because the image is real. The negative value of M means the image is inverted.

9. (c). $E = hc/\lambda = (6.63 \times 10^{-34}\,\text{J s})(3.00 \times 10^{8}\,\text{m/s})/(6.00 \times 10^{-7}\,\text{m}) = 3.31 \times 10^{-19}\,\text{J}$. This is the energy of each red photon. The number of such photons needed to produce a total of 10^{-18} J of energy is:

$$(10^{-18}\,\text{J})/(3.31 \times 10^{-19}\,\text{J/photon}) \sim 3 \text{ photons}$$

10. (d). Polarization can only occur with transverse waves because the motion must be perpendicular to the direction of propagation.

ELECTROSTATICS

EXAMPLE 1

Two protons, each of charge $q_p = e$, exert an electric force of magnitude F_{p-p} on each other when they are a distance r_p apart. A pair of alpha particles, each of charge $q_\alpha = 2e$, exert an electric force $F_{\alpha-\alpha} = \frac{1}{4}F_{p-p}$ on each other. What is the distance between the alpha particles, r_α, in terms of the distance between the protons, r_p? (a) $r_\alpha = 4r_p$ (b) $r_\alpha = 2r_p$ (c) $r_\alpha = r_p$ (d) More information is needed.

Solution

Use Coulomb's law to find an expression for the force between the protons:

$$F_{p-p} = \frac{k_e q_p q_p}{r_p^2} = \frac{k_e e^2}{r_p^2} \qquad (1)$$

Use Coulomb's law to find and expression for the force between the alpha particles:

$$F_{\alpha-\alpha} = \frac{k_e q_\alpha q_\alpha}{r_\alpha^2} = \frac{k_e (2e)^2}{r_\alpha^2} = \frac{4k_e e^2}{r_\alpha^2} \qquad (2)$$

Divide Equation (2) by Equation (1) and cancel common terms:

$$\frac{F_{p-p}}{F_{\alpha-\alpha}} = \frac{\dfrac{k_e e^2}{r_p^2}}{\dfrac{4k_e e^2}{r_\alpha^2}} = \frac{r_\alpha^2}{4r_p^2} \qquad (3)$$

Now substitute $F_{\alpha-\alpha} = \frac{1}{4}F_{p-p}$ and solve for r_α^2:

$$\frac{r_\alpha^2}{4r_p^2} = \frac{F_{p-p}}{F_{\alpha-\alpha}} = \frac{F_{p-p}}{\frac{1}{4}F_{p-p}} = 4 \quad \to \quad r_\alpha^2 = 16r_p^2$$

Take square roots, obtaining r_α in terms of r_p:

$$r_\alpha = 4r_p$$

The distance between the alpha particles is four times the distance between the protons, which is answer **(a)**.

EXAMPLE 2

Sphere A has twice the radius of a second, very distant sphere B. Let the electric potential at infinity be taken as zero. If the electric potential at the surface of sphere A is the same as at the surface of sphere B, what can be said of the charge Q_A on sphere A as compared to charge Q_B on B? (a) $Q_A = 2Q_B$ (b) $Q_A = Q_B$ (c) $Q_A = Q_B/2$

Conceptual Solution

By Gauss's law, a spherical distribution of charge creates an electric field outside the sphere as if all the charge were concentrated as a point charge at the center of the sphere. The electric potential due to a point charge is proportional to the charge Q and inversely proportional to the distance from that charge. Twice the radius reduces the electric potential at the surface of A by a factor of one-half. The charge of sphere A must be twice that of B so that the electric potentials will be the same for both spheres. Hence the answer is **(a)**.

Quantitative Solution

Write the equation for the electric potential of a point charge q:

$$V = \frac{kQ}{r} \qquad (1)$$

Make a ratio of Equation (1) for charge A and charge B, respectively:

$$\frac{V_A}{V_B} = \frac{\dfrac{kQ_A}{r_A}}{\dfrac{kQ_B}{r_B}} = \frac{Q_A r_B}{Q_B r_A} \qquad \textbf{(2)}$$

Substitute $V_A = V_B$ and $r_A = 2r_B$ into Equation (2) and solve for Q_A, again obtaining [anwer **(a)**]:

$$1 = \frac{Q_A r_B}{Q_B(2r_B)} \quad \rightarrow \quad Q_A = 2Q_B \text{ [answer (a)]}$$

EXAMPLE 3

How much work is required to bring a proton with charge 1.6×10^{-19} C and an alpha particle with charge 3.2×10^{-19} C from rest at a great distance (effectively infinity) to rest positions a distance of 1.00×10^{-15} m away from each other?

Solution

Use the work-energy theorem:

$$W = \Delta KE + \Delta PE = KE_f - KE_i + PE_f - PE_i$$

The velocities are zero both initially and finally, so the kinetic energies are zero. Substitute values into the potential energy and find the necessary work to assemble the configuration:

$$W = 0 - 0 + \frac{k_e q_p q_\alpha}{r} - 0$$

$$= \frac{(9.00 \times 10^9 \text{ kg}\cdot\text{m}^3/\text{C}^2\cdot\text{s}^2) \times (1.60 \times 10^{-19}\text{ C})(3.20 \times 10^{-19}\text{ C})}{1.00 \times 10^{-15}\text{ m}}$$

$$= 4.61 \times 10^{-13}\text{ J}$$

EXAMPLE 4

A fixed constant electric field E accelerates a proton from rest through a displacement Δs. A fully-ionized lithium atom with three times the charge of the proton accelerates through the same constant electric field and displacement. Which of the following is true of the kinetic energies of the particles? (a) The kinetic energies of the two particles are the same. (b) The kinetic energy of the proton is larger. (c) The kinetic energy of the lithium ion is larger.

Solution

The work done by the electric field on a particle of charge q is given by $W = F\Delta s = qE\Delta s$. The electric field E and displacement Δs are the same for both particles so the field does three times as much work on the lithium ion. The Work-Energy Theorem for this physical context is $W = \Delta KE$, so the lithium ion's kinetic energy is three times that of the proton, and the answer is **(c)**.

MULTIPLE CHOICE PROBLEMS

1. What is the potential difference between point A and point B if 10.0 J of work is required to move a charge of 4.00 C from one point to the other? (a) 0.400 V (b) 2.50 V (c) 14.0 V (d) 40.0 V
2. How much work would have to be done by a nonconservative force in moving an electron through a positive potential difference of 2.0×10^6 V? Assume the electron is at rest both initially and at its final position. (a) 3.2×10^{-13} J (b) -8.0×10^{-26} J (c) 1.25×10^{25} J (d) -3.2×10^{-13} J

3. Two electrically neutral materials are rubbed together. One acquires a net positive charge. The other must have (a) lost electrons. (b) gained electrons. (c) lost protons. (d) gained protons.

4. What is the magnitude of the charge on a body that has an excess of 20 electrons? (a) 3.2×10^{-18} C (b) 1.6×10^{-18} C (c) 3.2×10^{-19} C (d) 2.4×10^{-19} C

5. Two point charges, A and B, with charges of 2.00×10^{-4} C and -4.00×10^{-4} C, respectively, are separated by a distance of 6.00 m. What is the magnitude of the electrostatic force exerted on charge A? (a) 2.20×10^{-9} N (b) 1.30 N (c) 20.0 N (d) 36.0 N

6. Two point charges, A and B, are separated by 10.0 m. If the distance between them is reduced to 5.00 m, the force exerted on each (a) decreases to one-half its original value. (b) increases to twice its original value. (c) decreases to one-quarter of its original value. (d) increases to four times its original value.

7. Sphere A with a net charge of $+3.0 \times 10^{-3}$ C is touched to a second sphere B, which has a net charge of -9.0×10^{-3} C. The two spheres, which are exactly the same size and composition, are then separated. The net charge on sphere A is now: (a) $+3.0 \times 10^{-3}$ C (b) -3.0×10^{-3} C (c) -6.0×10^{-3} C (d) -9.0×10^{-3} C

8. If the charge on a particle in an electric field is reduced to half its original value, the force exerted on the particle by the field is (a) doubled. (b) halved. (c) quadrupled. (d) unchanged.

9. In the figure below, points A, B, and C are at various distances from a given point charge.

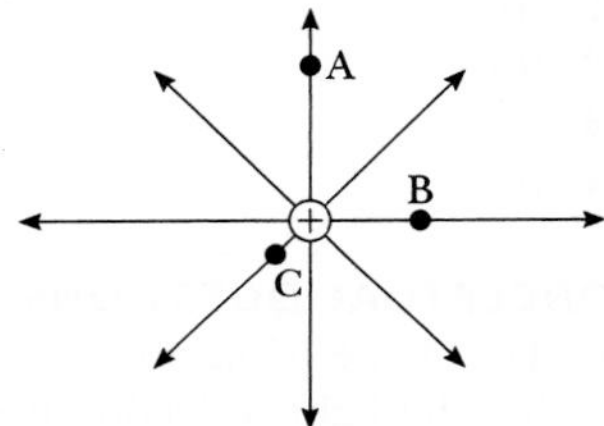

Which statement is most accurate? The electric field strength is (a) greatest at point A. (b) greatest at point B. (c) greatest at point C. (d) the same at all three points.

10. The electrostatic force between two point charges is F. If the charge of one point charge is doubled and that of the other charge is quadrupled, the force becomes which of the following? (a) $F/2$ (b) $2F$ (c) $4F$ (d) $8F$

Answers

1. **(b).** Potential difference between two points in an electric field is the work per unit charge required to move a charge between the two points:

$$\Delta V = W/q = 10.0\,\text{J}/4.00\,\text{C} = 2.50\,\text{V}$$

2. **(d).** If the only effect on the particle is a change of position, negative work must be done, because otherwise negative charges gain kinetic energy on moving through a positive potential difference.

$$W = \Delta KE + q\Delta V = 0 + (-1.6 \times 10^{-19}\,\text{C})\,(2.0 \times 10^{6}\,\text{V})$$
$$= -3.2 \times 10^{-13}\,\text{J}$$

3. **(b).** Protons are fixed in the nucleus and cannot be transferred by friction. Electrons can be transferred by friction. Therefore, net charges are due to the transfer of electrons between two bodies. Conservation of charge means that if there is a net positive charge, one body must have lost electrons and the other body must have gained the electrons.

4. **(a).** The elementary charge $e = 1.6 \times 10^{-19}$ C, so the total charge of 20 electrons has a magnitude of $20(1.6 \times 10^{-19}\,\text{C}) = 3.2 \times 10^{-18}$ C.

5. **(c).** Apply Coulomb's law: $F = k(q_A q_B)/r^2$

$$F = (9.00 \times 10^{9}\,\text{N m}^2/\text{C}^2)\,(2.00 \times 10^{-4}\,\text{C})(-4.00 \times 10^{-4}\,\text{C})/(6.00\,\text{m})^2 = -20.0\,\text{N},$$

so the magnitude is 20.0 N.

3. (a), (e); (a), (e)
4. (a) False (b) False (c) True (d) False
5. (a)

CONCEPTUAL QUESTIONS

1. An ellipsoid. The dimension in the direction of motion would be measured to be less than *D*.
3. This scenario is not possible with light. Light waves are described by the principles of special relativity. As you detect the light wave ahead of you and moving away from you (which would be a pretty good trick—think about it!), its velocity relative to you is *c*. Thus, you will not be able to catch up to the light wave.
5. No. The principle of relativity implies that nothing can travel faster than the speed of light in a *vacuum*, which is 3.00×10^8 m/s.
7. The light from the quasar moves at 3.00×10^8 m/s. The speed of light is independent of the motion of the source or the observer.
9. For a wonderful fictional exploration of this question, get a "Mr. Tompkins" book by George Gamow. All of the relativity effects would be obvious in our lives. Time dilation and length contraction would both occur. Driving home in a hurry, you would push on the gas pedal not to increase your speed very much, but to make the blocks shorter. Big Doppler shifts in wave frequencies would make red lights look green as you approached and make car horns and radios useless. High-speed transportation would be both very expensive, requiring huge fuel purchases, as well as dangerous, since a speeding car could knock down a building. When you got home, hungry for lunch, you would find that you had missed dinner; there would be a five-day delay in transit when you watch a live TV program originating in Australia. Finally, we would not be able to see the Milky Way, since the fireball of the Big Bang would surround us at the distance of Rigel or Deneb.
11. A photon transports energy. The relativistic equivalence of mass and energy means that is enough to give it momentum.
13. Your assignment: measure the length of a rod as it slides past you. Mark the position of its front end on the floor and have an assistant mark the position of the back end. Then measure the distance between the two marks. This distance will represent the length of the rod only if the two marks were made simultaneously in your frame of reference.

PROBLEMS

1. (a) $t_{OB} = 1.67 \times 10^3$ s, $t_{OA} = 2.04 \times 10^3$ s
   (b) $t_{BO} = 2.50 \times 10^3$ s, $t_{AO} = 2.04 \times 10^3$ s
   (c) $\Delta t = 90$ s
3. 5.0 s
5. (a) 20 m (b) 19 m (c) $0.31c$
7. (a) 1.3×10^{-7} s (b) 38 m (c) 7.6 m
9. (a) 2.2 μs (b) 0.65 km
11. $0.950c$
13. Yes, with 19 m to spare
15. (a) 39.2 μs (b) Accurate to one digit
17. 3.3×10^5 m/s
19. $0.285c$
21. $0.54c$ to the right
23. $0.357c$
25. $0.998c$ toward the right
27. (a) 54 min (b) 52 min
29. $c(\sqrt{3}/2)$
31. $0.786c$
33. 18.4 g/cm^3
35. 1.98 MeV
37. 2.27×10^{23} Hz, 1.32 fm for each photon
39. (a) 3.10×10^5 m/s (b) $0.758c$
41. 1.42 MeV/c
43. (a) $0.80c$ (b) 7.5×10^3 s (c) 1.4×10^{12} m, $0.38c$
45. $0.37c$ in the + x-direction
47. (a) $v/c = 1 - 1.12 \times 10^{-10}$ (b) 6.00×10^{27} J
    (c) \$$2.17 \times 10^{20}$
49. $0.80c$
51. (a) $0.946c$ (b) 0.160 ly (c) 0.114 yr (d) 7.50×10^{22} J
53. (a) 7.0 μs (b) 1.1×10^4 muons
59. 5.45 yr; Goslo is older.

Chapter 27

QUICK QUIZZES

1. (b)
2. (c)
3. (c)
4. (b)

CONCEPTUAL QUESTIONS

1. The shape of an object is normally determined by observing the light reflecting from its surface. In a kiln, the object will be very hot and will be glowing red. The emitted radiation is far stronger than the reflected radiation, and the thermal radiation emitted is only slightly dependent on the material from which the object is made. Unlike reflected light, the emitted light comes from all surfaces with equal intensity, so contrast is lost and the shape of the object is harder to discern.
3. The "blackness" of a blackbody refers to its ideal property of absorbing all radiation incident on it. If an observed room temperature object in everyday life absorbs all radiation, we describe it as (visibly) black. The black appearance, however, is due to the fact that our eyes are sensitive only to visible light. If we could detect infrared light with our eyes, we would see the object emitting radiation. If the temperature of the blackbody is raised, Wien's law tells us that the emitted radiation will move into the visible range of the spectrum. Thus, the blackbody could appear as red, white, or blue, depending on its temperature.
5. All objects do radiate energy, but at room temperature this energy is primarily in the infrared region of the electromagnetic spectrum, which our eyes cannot detect. (Pit vipers have sensory organs that *are* sensitive to infrared radiation; thus, they can seek out their warm-blooded prey in what we would consider absolute darkness.
7. Most metals have cutoff frequencies corresponding to photons in or near the visible range of the electromagnetic spectrum. AM radio wave photons have far too little energy to eject electrons from the metal.
9. We can picture higher frequency light as a stream of photons of higher energy. In a collision, one photon can give all of its energy to a single electron. The kinetic energy of such an electron is measured by the stopping potential.

than R_3, again because of a denominator larger than one, and so R_P is less than R_{combo}. The purely parallel combination therefore yields the least resistance, and the answer is **(c)**.

EXAMPLE 2

Two resistors are to dissipate as much energy as possible when a fixed voltage difference is placed across them. Should they be installed in parallel or in series?

Conceptual Solution

The power dissipated by a resistor is proportional to the voltage difference squared and inversely proportional to the resistance. The smallest possible resistance will therefore result in the largest power output. Consequently, the two resistors should be placed in parallel, which results in the lowest combined resistance.

EXAMPLE 3

A certain off-the-shelf resistor of resistivity ρ has resistance R. Suppose it is commercially desirable to design a new resistor that has one-third the length and one-fourth the cross-sectional area of the existing resistor, but have the same overall resistance. What should the resistivity ρ_n of the new resistor be, in terms of the resistivity ρ of the original device?

Solution

Write an expression for the resistance R_n of the new resistor, in terms of its resistivity, cross-sectional area A, and its length L:

$$R_n = \frac{\rho_n L_n}{A_n} \quad (1)$$

Divide Equation (1) by the same expression for the original resistor:

$$\frac{R_n}{R} = \frac{\dfrac{\rho_n L_n}{A_n}}{\dfrac{\rho L}{A}} = \frac{\rho_n L_n A}{\rho L A_n} \quad (2)$$

Substitute $A_n = A/4$, $R_n = R$, and $L_n = L/3$ into Equation (2):

$$1 = \frac{\rho_n (L/3) A}{\rho L (A/4)} = \frac{4\rho_n}{3\rho} \quad (3)$$

Solve Equation (3) for the new resistivity, ρ_n:

$$\rho_n = \frac{3}{4}\rho$$

MULTIPLE CHOICE PROBLEMS

1. Three resistors of resistance 1.0 Ω, 2.0 Ω, and 3.0 Ω, respectively, are in series. If a potential difference of 12 V is put across the combination, what is the resulting current in the circuit? (a) 0.50 A (b) 2.0 A (c) 6.0 A (d) 12 A

2. If the length of a conducting wire with resistance R is doubled, the resistance of the longer wire will be (a) $R/4$ (b) $R/2$ (c) $2R$ (d) $4R$

3. If a 2.0 Ω resistor and a 6.0 Ω resistor are connected in parallel, what is their combined resistance? (a) 1.5 Ω (b) 4.0 Ω (c) 8.0 Ω (d) 12 Ω

4. If all the components of an electric circuit are connected in series, which of the following physical quantities must be the same at all points in the circuit? (a) voltage (b) current (c) resistance (d) power

5. The current in a conductor is 3.0 A when it is connected across a 6.0-V battery. How much power is delivered? (a) 0.50 W (b) 2.0 W (c) 9.0 W (d) 18 W

6. A 12-Ω resistor is connected across a 6.0-V dc source. How much energy is delivered to the resistor in half an hour? (a) 1.5×10^{-3} kWh (b) 2.0×10^{-3} kWh (c) 3.0×10^{-3} kWh (d) 12×10^{-3} kWh

7. A battery with an emf of 6.20 V carries a current of 20.0 A. If the internal resistance of the battery is 0.01 Ω, what is the terminal voltage? (a) 1.24 V (b) 6.00 V (c) 6.40 V (d) 31.0 V
8. Devices A and B are connected in parallel to a battery. If the resistance R_A of device A is four times as great as the resistance R_B of device B, then what is true of I_A and I_B, the currents in devices A and B? (a) $I_A = 2I_B$ (b) $I_A = I_B/2$ (c) $I_A = 4I_B$ (d) $I_A = I_B/4$
9. A wire has resistance R. What is the resistance of a wire of the same substance that has the same length but twice the cross-sectional area? (a) $2R$ (b) $R/2$ (c) $4R$ (d) $R/4$
10. What must be the reading in the ammeter A for the circuit section shown?

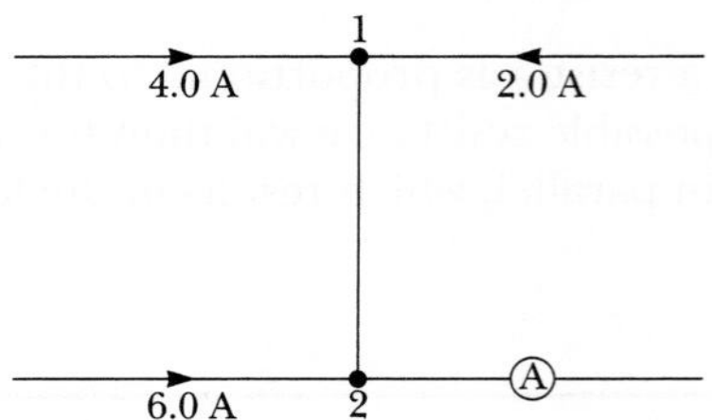

(a) 0 A (b) 6.0 A (c) 8.0 A (d) 12 A
11. What is the current in a wire if 5.00 C of charge passes through the wire in 0.500 s? (a) 1.00 A (b) 2.50 A (c) 5.00 A (d) 10.0 A

Answers

1. **(b).** The three resistors are connected in series so $R_{tot} = \Sigma R = 6.0\ \Omega$. From Ohm's law, $V_{tot}/R_{tot} = I_{tot} = 12\ V/6.0\ \Omega = 2.0\ A$.
2. **(c).** Resistance is directly proportional to length and inversely proportional to the cross-sectional area of the conductor. If the area remains constant, then doubling the length will double the resistance.
3. **(a).** The resistors are in parallel:

$$R_{eq} = R_1R_2/(R_1 + R_2) = 12\ \Omega^2/8.0\ \Omega = 1.5\ \Omega$$

In fact, the equivalent resistance of parallel resistors is always less than the smallest resistance in the combination. This means that answers (b), (c), and (d) could have been eliminated immediately.

4. **(b).** A series circuit has only one path for current so it must be the same at all points in the circuit.
5. **(d).** Power $= I\Delta V = 18$ W
6. **(a).** The energy used by a load is the product of the power it uses per unit time and the length of time it is operated:

$$E = W = \mathcal{P}\,\Delta t = (\Delta V)^2\Delta t/R = (6.0\ V)^2\,(0.50\ h)/12\ \Omega$$
$$= 1.5\ Wh = 1.5 \times 10^{-3}\ kWh$$

Note the conversion from watt-hours to kilowatt-hours.

7. **(b).** The product Ir is the potential drop occurring within the battery: $Ir = (20.0\ A)(0.01\ \Omega) = 0.20\ V$. Because the battery is producing current and not being recharged, the terminal voltage will be less than the emf by the internal potential drop. This eliminates choices (c) and (d) immediately: $V = E - Ir = 6.20\ V - 0.20\ V = 6.00\ V$
8. **(d).** More current will follow the circuit branch with less resistance. Use Kirchoff's loop law around the loop consisting of the 2 parallel resistances and the fact that $R_A = 4R_B$:

$$\sum \Delta V_i = 0 \quad \rightarrow \quad I_AR_A - I_BR_B = 0 \quad \rightarrow$$

$$\frac{I_A}{I_B} = \frac{R_B}{R_A} = \frac{R_B}{4R_B} = \frac{1}{4}$$

Hence one-fourth as much current goes through resistor A.

9. **(b).** Resistance is inversely proportional to the area; doubling the area reduces the resistance by half.
10. **(d).** Kirchhoff's current rule says that the current entering a junction must equal the current leaving the junction. The current entering junction 1 is 6.0 A which subsequently enters junction 2 from above, so the total current entering junction 2 is 12 A and the current then leaving 2 and entering the ammeter is 12 A.
11. **(d).** By definition, the current is the amount of charge that passes a point in the circuit in a given time: $I = Q/\Delta t = 5.00\text{ C}/0.500\text{ s} = 10.0\text{ A}$

ATOMS

EXAMPLE 1

Tritium is an isotope of hydrogen with a half-life of $t_{1/2} = 12.33$ yr. How long would it take for 1.60×10^2 g of tritium to decay to 20.0 g? (a) 6.17 y (b) 26.7 y (c) 37.0 y (d) 74.0 y

Solution

Calculate the number n of half-lives required. The equation can be solved by inspection (logarithms are ordinarily required):

$$\left(\frac{1}{2}\right)^n = \frac{20.0\text{ g}}{1.60 \times 10^2\text{ g}} = \frac{1}{8.00} \quad \rightarrow \quad n = 3$$

Multiply the number of half-lives by the length of the half-life to find the necessary time interval in question, verifying that the answer is **(c):**

$$\Delta t = nt_{1/2} = 3\ (12.33\text{ yr}) = \boxed{37.0\text{ yr}}$$

EXAMPLE 2

What is the mass number of a carbon atom having 6 protons and 8 neutrons? (a) 6 (b) 8 (c) 14

Solution

Don't confuse mass number with the atomic number, Z, which is the number of protons in the nucleus. The mass number A is the number of nucleons in the nucleus. To calculate the mass number, just add the number of protons and neutrons together: $6 + 8 = 14$, which is answer **(c).**

EXAMPLE 3

Which of the following particles is given by ${}^{A}_{Z}X$ in the following reaction? (a) ${}^{7}_{3}\text{Li}$ (b) ${}^{7}_{4}\text{Be}$ (c) ${}^{14}_{7}\text{N}$ (d) ${}^{10}_{6}\text{C}$

$${}^{1}_{1}\text{H} + {}^{A}_{Z}\text{X} \rightarrow {}^{4}_{2}\text{He} + {}^{4}_{2}\text{He}$$

Solution

Equate the sum of the mass numbers on both sides of the reaction:

$$1 + A = 4 + 4 = 8 \quad \rightarrow \quad A = 7$$

By charge conservation, the number of protons must also be the same on both sides:

$$1 + Z = 2 + 2 = 4 \quad \rightarrow \quad Z = 3$$

Based on these two results, the correct answer is **(a),** $\boxed{{}^{7}_{3}\text{Li}}$.

MULTIPLE CHOICE PROBLEMS

1. In the nuclear reaction below, what particle does X represent?

$$^{22}_{11}\text{Na} \rightarrow {}^{22}_{10}\text{Ne} + \text{X} + \nu_e$$

(a) an α-particle. (b) a β-particle. (c) a positron. (d) a γ-photon.

2. What is the atomic number of the daughter nuclide in the following reaction?

$$^{30}_{15}\text{P} \rightarrow {}^{A}_{Z}\text{Si} + e^{+} + \nu_e$$

(a) 14 (b) 16 (c) 30 (d) 31

3. If $^{13}_{7}$N has a half-life of about 10.0 min, how long will it take for 20 g of the isotope to decay to 2.5 g? (a) 5 min (b) 10 min (c) 20 min (d) 30 min

4. A certain radionuclide decays by emitting an α-particle. What is the difference between the atomic numbers of the parent and the daughter nuclides? (a) 1 (b) 2 (c) 4 (d) 6

5. What is the difference in mass number between the parent and daughter nuclides after a β-decay process? (a) -1 (b) 0 (c) 1 (d) 2

6. A nitrogen atom has 7 protons and 6 neutrons. What is its mass number? (a) 1 (b) 6 (c) 7 (d) 13

7. Which one of the following is an isotope of $^{182}_{63}$X? (a) $^{182}_{62}$X(b) $^{182}_{64}$X (c) $^{180}_{63}$X (d) $^{180}_{62}$X

8. In the following nuclear equation, what is X?

$$^{14}_{7}\text{N} + {}^{4}_{2}\text{He} \rightarrow {}^{17}_{8}\text{O} + \text{X}$$

(a) a proton (b) a positron (c) a β-particle (d) an α-particle

9. What is the number of neutrons in $^{140}_{54}$Xe? (a) 54 (b) 86 (c) 140 (d) 194

10. Radon-222 has a half-life of about 4 days. If a sample of ^{222}Rn gas in a container is initially doubled, the half-life will be (a) halved. (b) doubled. (c) quartered. (d) unchanged.

11. A radionuclide decays completely. In the reaction flask, the only gas that is found is helium, which was not present when the flask was sealed. The decay process was probably (a) β-decay. (b) α-decay. (c) γ-decay. (d) positron emission.

12. What is the half-life of a radionuclide if 1/16 of its initial mass is present after 2 h? (a) 15 min (b) 30 min (c) 45 min (d) 60 min

13. The half-life of $^{22}_{11}$Na is 2.6 y. If X grams of this sodium isotope are initially present, how much is left after 13 y? (a) $X/32$ (b) $X/13$ (c) $X/8$ (d) $X/5$

Answers

1. (c). Notice that the mass number is unchanged, whereas the atomic number has been reduced by one. This implies that a proton has changed into a neutron. The emitting particle must have a charge equal to a proton but an atomic mass number of zero. The positron is the only choice that has both of these attributes.

2. (a). The atomic number is Z. Conservation of charge means: $15 = Z + 1$. Therefore, $Z = 14$.

3. (d). First find the number of half lives, then multiply by the value of the half-life to get the elapsed time:

$$\frac{2.5\text{ g}}{20\text{ g}} = \frac{1}{8} = \left(\frac{1}{2}\right)^3 \rightarrow n = 3$$

$$\Delta t = nt_{1/2} = 3(10\text{ min}) = 30\text{ min}$$

4. (b). An α-particle is a $^{4}_{2}$He helium nucleus. In α-decay, two protons are effectively removed.

5. (b). β-decay emits a high-energy electron, $^{0}_{-1}$e. In the process, a neutron decays into a proton plus the emitted electron and an antineutrino. The number of nucleons remains unchanged, with the proton replacing the neutron in the sum of nucleons.

6. (d). The mass number is the sum of neutrons and pro-tons $= 6 + 7 = 13$ nucleons.

7. (c). Isotopes of an element have the same atomic number but different numbers of neutrons, so their mass numbers are different. The atomic number of X is 63.

8. (a). The mass numbers must be the same on both sides of the reaction. If A is the mass number of X, then $14 + 4 = 17 + A$, so $A = 1$. As for atomic number, $7 + 2 = 8 + Z$. Therefore, $Z = 1$. This describes a proton, $^{1}_{1}$H.

9. **(b).** The number of neutrons is the mass number minus the atomic number: $A - Z = 140 - 54 = 86$.

10. **(d).** The half-life is a constant that depends on the identity of the nuclide, not on the amount of nuclide present.

11. **(b).** The α-particle is a helium nucleus. Each α-particle then acquires two electrons to form a neutral helium atom.

12. **(b).** In every half-life, the mass decreases to half its previous value: $1/16 = 1/2^4$. It takes four half-lives to decay down to $1/16$ the original mass. Each must be 30 min long because the entire process takes two hours.

13. **(a).** In 13 y, there will be 5 half-lives of 2.6 y each $(5 \times 2.6 = 13)$. The isotope decreases to $1/2^5 = 1/32$ of its original amount.

Answers to Quick Quizzes, Odd-Numbered Conceptual Questions and Problems

Chapter 1

CONCEPTUAL QUESTIONS

1. (a) ~ 0.1 m (b) ~ 1 m (c) Between 10 m and 100 m (d) ~ 10 m (e) ~ 100 m
3. (a) $\sim 10^6$ beats (b) $\sim 10^9$ beats
5. $\sim 10^9$ s
7. The length of a hand varies from person to person, so it isn't a useful standard of length.
9. The ark has an approximate volume of 10^4 m^3, whereas a home has an approximate volume of 10^3 m^3.
11. A dimensionally correct equation isn't necessarily true. For example, the equation 2 dogs = 5 dogs is dimensionally correct, but isn't true. However, if an equation is not dimensionally correct, it cannot be correct.

PROBLEMS

1. (b) $A_{\text{cylinder}} = \pi R^2$, $A_{\text{rectangular plate}} = \text{length} \times \text{width}$
5. $\text{m}^3/(\text{kg} \cdot \text{s}^2)$
7. (a) three significant figures (b) four significant figures (c) three significant figures (d) two significant figures
9. (a) 797 (b) 1.1 (c) 17.66
11. 115.9 m
13. 228.8 cm
15. 2×10^8 fathoms
17. 1.39×10^3 m^2
19. 1×10^{17} ft
21. 6.71×10^8 mi/h
23. 3.08×10^4 m^3
25. 9.82 cm
27. (a) 1.0×10^3 kg (b) $m_{\text{cell}} = 5.2 \times 10^{-16}$ kg, $m_{\text{kidney}} = 0.27$ kg, $m_{\text{fly}} = 1.3 \times 10^{-5}$ kg
29. 1 800 balls ($\sim 10^3$ balls) (assumes 81 games per season, nine innings per game, an average of ten hitters per inning, and one ball lost for every four hitters)
31. $\sim 10^7$ rev
33. $\sim 10^6$ balls
35. (2.0 m, 1.4 m)
37. $r = 2.2$ m, $\theta = 27°$
39. (a) 6.71 m (b) 0.894 (c) 0.746
41. 3.41 m
43. (a) 3.00 (b) 3.00 (c) 0.800 (d) 0.800 (e) 1.33
45. 5.00/7.00; the angle itself is 35.5°
47. The customer is incorrect. The cost of the pizza should be proportional to the area of the pizza, so the large one should cost more than the small one by a factor of $9^2/6^2 = 81/36 = 2.25$. Therefore, if the small one costs \$6, the large one should cost \$13.50.
49. The value of k, a dimensionless constant, cannot be found by dimensional analysis.
51. $\sim 10^{-9}$ m
53. $\sim 10^3$ tuners (assumes one tuner per 10 000 residents and a population of 7.5 million)
55. (a) 3.16×10^7 s (b) Between 10^{10} yr and 10^{11} yr

Chapter 2

QUICK QUIZZES

1. (a) 200 yd (b) 0 (c) 0
2. (a) False (b) True (c) True
3. The velocity vs. time graph (a) has a constant slope, indicating a constant acceleration, which is represented by the acceleration vs. time graph (e).

   Graph (b) represents an object with increasing speed, but as time progresses, the lines drawn tangent to the curve have increasing slopes. Since the acceleration is equal to the slope of the tangent line, the acceleration must be increasing, and the acceleration vs. time graph that best indicates this behavior is (d).

   Graph (c) depicts an object which first has a velocity that increases at a constant rate, which means that the object's acceleration is constant. The velocity then stops changing, which means that the acceleration of the object is zero. This behavior is best matched by graph (f).
4. (b)
5. (a) blue graph (b) red graph (c) green graph
6. (e)
7. (c)
8. (a) and (f)

CONCEPTUAL QUESTIONS

1. Yes. If the velocity of the particle is nonzero, the particle is in motion. If the acceleration is zero, the velocity of the particle is unchanging or is constant.
3. Yes. If this occurs, the acceleration of the car is opposite to the direction of motion, and the car will be slowing down.
5. No. They can be used only when the acceleration is constant. Yes. Zero is a constant.
7. Once the objects leave the hand, both are in free fall, and both exhibit the same downward acceleration equal in magnitude to the free-fall acceleration g.
9. Yes. Yes.
11. (a) At the maximum height, the ball is momentarily at rest. (That is, it has zero velocity.) The acceleration remains constant, with magnitude equal to the free-fall acceleration g and directed downward. Thus, even though the velocity is momentarily zero, it continues to change, and the ball will begin to gain speed in the downward direction. (b) The acceleration of the ball remains constant in magnitude and direction throughout the ball's free flight, from the instant it leaves the hand until the instant just before it strikes the ground. The acceleration is directed downward and has a magnitude equal to the free-fall acceleration g.
13. No. If the velocity of A is greater than that of B, car A is moving faster than B at this instant. However, car B may be picking up speed (that is, accelerating) at a greater

rate than A. The driver of B may have stepped very hard on the gas pedal in the recent past, but car B has not yet reached the speed of A.

15. They are the same! You can see this for yourself by solving the kinematic equations for the two cases. The ball that was thrown upward will take longer to reach the ground, but when it arrives, it has the same velocity as the ball that was thrown downward.
17. Ignoring air resistance, in 16 s the pebble would fall a distance $\frac{1}{2}gt^2 = \frac{1}{2}(9.80 \text{ m/s}^2)(16.0 \text{ s})^2 = 1\ 250 \text{ m} = 1.25 \text{ km}$. Air resistance is an important force after the first few seconds, when the pebble has gained high speed. Also, part of the 16-s interval must be occupied by the sound returning up the well. Thus, the depth of the well is less than 1.25 km, but on the order of 10^3 m.
19. Above. Your ball has zero initial speed and a smaller average speed during the time of flight to the passing point.

PROBLEMS

1. (a) 52.9 km/h (b) 90.0 km
3. (a) Boat A wins by 60 km (b) 0
5. (a) 180 km (b) 63.4 km/h
7. (a) 4.0 m/s (b) $-$ 0.50 m/s (c) $-$ 1.0 m/s (d) 0
9. (a) 2.50 m/s (b) $-$ 2.27 m/s (c) 0
11. 2.80 h, 218 km
13. 274 km/h
15. (a) 5.00 m/s (b) $-$ 2.50 m/s (c) 0 (d) 5.00 m/s
17. (a) 4.0 m/s (b) $-$ 4.0 m/s (c) 0 (d) 2.0 m/s
19. (a) 70.0 mi/h $\cdot$ s = 31.3 m/s^2 = 3.19g (b) 321 ft = 97.8 m
21. 3.7 s
23. (a) 8.0 m/s^2 (b) about 12 m/s^2
25. 2.74×10^5 m/s^2, or 2.79×10^4 times g
27. (a) 1.3 m/s^2 (b) 8.0 s
29. (a) 2.32 m/s^2 (b) 14.4 s
31. (a) 8.94 s (b) 89.4 m/s
33. $-$ 3.6 m/s^2
35. 200 m
37. (a) 1.5 m/s (b) 32 m
39. (a) 8.2 s (b) 1.3×10^2 m
41. 958 m
43. (a) 31.9 m (b) 2.55 s (c) 2.55 s (d) $-$ 25.0 m/s
45. 38.2 m
47. (a) 21.1 m/s (b) 19.6 m (c) 18.1 m/s; 19.6 m
49. (a) 308 m (b) 8.51 s (c) 16.4 s
51. (a) 10.0 m/s upward (b) 4.68 m/s downward
53. 15.0 s
55. (a) $-$ 3.50×10^5 m/s^2 (b) 2.86×10^{-4} s
57. 0.60 s
59. (a) 3.00 s (b) $-$ 24.5 m/s, $-$24.5 m/s (c) 23.5 m
61. (a) 5.46 s (b) 73.0 m (c) $v_{\text{Kathy}} = 26.7$ m/s, $v_{\text{Stan}} = 22.6$ m/s
63. (a) $t_1 = 5.0$ s, $t_2 = 85$ s (b) 200 ft/s (c) 18 500 ft from starting point (d) 10 s after starting to slow down (total trip time = 100 s)
65. (a) 5.5×10^3 ft (b) 3.7×10^2 ft/s (c) The plane would travel only 0.002 ft in the time it takes the light from the bolt to reach the eye.
67. (a) 7.82 m (b) 0.782 s
69. (a) 3.45 s (b) 10.0 ft
71. 4.2 m/s

Chapter 3

QUICK QUIZZES

1. (c)
2. (b)
3.

Vector	x-component	y-component
$\vec{A}$	$-$	+
$\vec{B}$	+	$-$
$\vec{A} + \vec{B}$	$-$	$-$

4. (b)
5. (a)
6. (c)
7. (b)

CONCEPTUAL QUESTIONS

1. The components of $\vec{A}$ will both be negative when $\vec{A}$ lies in the third quadrant. The components of $\vec{A}$ will have opposite signs when $\vec{A}$ lies in either the second quadrant or the fourth quadrant.
3. Assuming that the ship sails in a straight line, the wrench will strike the deck at the base of the mast. The wrench has the same horizontal velocity as the ship when it is released, and it will maintain that component of velocity as it falls. Thus, the wrench is always above the same spot on the deck and will strike that spot when it hits the deck. To an observer on the ship, the wrench appears to fall straight downward. To a stationary observer on shore, the wrench is seen to follow a parabolic trajectory.
5. Let v_{0x} and v_{0y} be the initial velocity components for the projectile fired on Earth. Because the projectile fired on the Moon is given the same initial velocity, its initial velocity components have the same values v_{0x} and v_{0y} as those for the projectile on Earth. From $v_y^2 = v_{0y}{}^2 - 2g\Delta y$, the maximum altitude reached by a projectile is found to be $H = v_{0y}{}^2/2g$. Because the free-fall acceleration g is smaller on the Moon than on Earth, the maximum altitude will be greater for the projectile fired on the Moon. If they start from the same elevation, the projectile on the Moon will have a longer time of flight and therefore a greater range. The Apollo astronauts made several long golf drives on the Moon.
7. A vector has both magnitude and direction and can be added only to other vectors representing the same physical quantity. A scalar has only a magnitude and can be added only to other scalars representing the same physical quantity.
9. (a) At the top of the projectile's flight, its velocity is horizontal and its acceleration is downward. This is the only point at which the velocity and acceleration vectors are perpendicular. (b) If the projectile is thrown straight up or down, then the velocity and acceleration will be parallel throughout the motion. For any other kind of projectile motion, the velocity and acceleration vectors are never parallel.
11. (a) The acceleration is zero, since both the magnitude and direction of the velocity remain constant. (b) The particle has an acceleration, since the direction of $\vec{v}$ changes.
13. The spacecraft will follow a parabolic path equivalent to that of a projectile thrown off a cliff with a horizontal velocity. As regards the projectile, gravity provides an

acceleration that is always perpendicular to the initial velocity, resulting in a parabolic path. As regards the spacecraft, the initial velocity plays the role of the horizontal velocity of the projectile, and the leaking gas plays the role that gravity plays in the case of the projectile. If the orientation of the spacecraft were to change in response to the gas leak (which is by far the more likely result), then the acceleration would change direction and the motion could become very complicated.

15. For angles $\theta < 45°$, the projectile thrown at angle θ will be in the air for a shorter interval. For the smaller angle, the vertical component of the initial velocity is smaller than that for the larger angle. Thus, the projectile thrown at the smaller angle will not go as high into the air and will spend less time in the air before landing.

17. Yes. The projectile is a freely falling body, because gravity is the only force acting on it. The vertical acceleration will be the local free fall acceleration g; the horizontal acceleration will be zero.

19. If a projectile is to have zero speed at the top of its trajectory, it must be thrown straight upward with zero horizontal velocity. To give a projectile a nonzero speed at the top of its trajectory, the projectile should be thrown at any angle other than 90° to the horizontal.

PROBLEMS

1. Approximately 421 ft at 3° below the horizontal
3. Approximately 15 m at 58° S of E
5. Approximately 310 km at 57° S of W
7. (a) Approximately 5.0 units at $-53°$ (b) Approximately 5.0 units at $+53°$
9. 8.07 m at 42.0° S of E
11. (a) 5.00 blocks at 53.1° N of E (b) 13.0 blocks
13. 47.2 units at 122° from the positive x-axis
15. 157 km
17. 245 km at 21.4° W of N
19. 196 cm at 14.7° below the x-axis
21. (a) (10.0 m, 16.0 m)
23. 6.13 m
25. 12 m/s
27. 48.6 m/s
29. 25 m
31. (a) 32.5 m from the base of the cliff (b) 1.78 s
33. (a) 52.0 m/s horizontally (b) 212 m
35. 390 mi/h at 7.37° N of E
37. 1.88 m
39. 249 ft upstream
41. 18.0 s
43. 196 cm at 14.7° below the positive x-axis
45. (a) 57.7 km/h at 60.0° west of vertical (b) 28.9 km/h downward
47. 68.6 km/h
49. (a) 1.52×10^3 m (b) 36.1 s (c) 4.05×10^3 m
51. $R_{\text{moon}} = 18$ m, $R_{\text{Mars}} = 7.9$ m
53. (a) +, + (b) −, + (c) (i) must be in either the first or second quadrant
55. (a) 42 m/s (b) 3.8 s (c) $v_x = 34$ m/s, $v_y = -13$ m/s; $v = 37$ m/s
57. 7.5 m/s in the direction the ball was thrown
61. $R/2$
63. 7.5 min
65. 10.8 m
67. (b) $y = Ax^2$ with $A = \dfrac{g}{(2v_i^2)}$ where v_i is the muzzle velocity (c) 14.5 m/s
69. 227 paces at 165° from the positive x-axis
71. (a) 20.0° above the horizontal (b) 3.05 s
73. (a) 23 m/s (b) 360 m horizontally from the base of cliff

Chapter 4

QUICK QUIZZES

1. (a) True (b) False
2. (a) True (b) True (c) False
3. False
4. (a) (The value of g is smaller on the Moon than on Earth, so, for equal *weights,* the Moon *mass* must be greater.)
5. (c); (d)
6. (c)
7. (c)
8. (b)
9. (b)
10. (b) By exerting an upward force component on the sled, you reduce the normal force on the ground and so reduce the force of kinetic friction.

CONCEPTUAL QUESTIONS

1. (a) Two external forces act on the ball. (i) One is a downward gravitational force exerted by Earth. (ii) The second force on the ball is an upward normal force exerted by the hand. The reactions to these forces are (i) an upward gravitational force exerted by the ball on Earth and (ii) a downward force exerted by the ball on the hand. (b) After the ball leaves the hand, the only external force acting on the ball is the gravitational force exerted by Earth. The reaction is an upward gravitational force exerted by the ball on Earth.

3. No. If the car moves with constant acceleration, then a net force in the direction of the acceleration must be acting on it.

5. The coefficient of static friction is larger than that of kinetic friction. To start the box moving, you must counterbalance the maximum static friction force. This force exceeds the kinetic friction force that you must counterbalance to maintain the constant velocity of the box once it starts moving.

7. The inertia of the suitcase would keep it moving forward as the bus stops. There would be no tendency for the suitcase to be thrown backward toward the passenger. The case should be dismissed.

9. The force causing an automobile to move is the friction between the tires and the roadway as the automobile attempts to push the roadway backward. The force driving a propeller airplane forward is the reaction force exerted by the air on the propeller as the rotating propeller pushes the air backward (the action). In a rowboat, the rower pushes the water backward with the oars (the action). The water pushes forward on the oars and hence the boat (the reaction).

11. When the bus starts moving, Claudette's mass is accelerated by the force exerted by the back of the seat on her body. Clark is standing, however, and the only force acting on him is the friction between his shoes and the floor of the bus. Thus, when the bus starts moving, his feet accelerate forward, but the rest of his body experiences almost no accelerating force (only that due to his being attached to his accelerating feet!). As a consequence, his body tends to stay almost at rest, according to Newton's first law, relative to the ground. Relative to Claudette, however, he is moving toward her and falls into her lap. Both performers won Academy Awards.

13. The tension in the rope is the maximum force that occurs in *both* directions. In this case, then, since both are pulling with a force of magnitude 200 N, the tension is 200 N. If the rope does not move, then the force on each athlete must equal zero. Therefore, each athlete exerts 200 N against the ground.
15. (a) As the man takes the step, the action is the force his foot exerts on Earth; the reaction is the force exerted by Earth on his foot. (b) Here, the action is the force exerted by the snowball on the girl's back; the reaction is the force exerted by the girl's back on the snowball. (c) This action is the force exerted by the glove on the ball; the reaction is the force exerted by the ball on the glove. (d) This action is the force exerted by the air molecules on the window; the reaction is the force exerted by the window on the air molecules. In each case, we could equally well interchange the terms "action" and "reaction."
17. If the brakes lock, the car will travel farther than it would travel if the wheels continued to roll, because the coefficient of kinetic friction is less than that of static friction. Hence, the force of kinetic friction is less than the maximum force of static friction.
19. (a) The crate accelerates because of a friction force exerted by the floor of the truck on the crate. (b) If the driver slams on the brakes, the crate's inertia tends to keep the crate moving forward.

PROBLEMS

1. (a) 12 N (b) 3.0 m/s^2
3. 2×10^4 N
5. 3.71 N, 58.7 N, 2.27 kg
7. 9.6 N
9. 1.4×10^3 N
11. (a) 0.200 m/s^2 (b) 10.0 m (c) 2.00 m/s
13. 1.1×10^4 N
15. 600 N in vertical cable, 997 N in inclined cable, 796 N in horizontal cable
17. 150 N in vertical cable, 75 N in right-side cable, 130 N in left-side cable
19. 236 N (upper rope), 118 N (lower rope)
21. 613 N
23. 64 N
25. (a) 7.50×10^3 N backward (b) 50.0 m
27. (a) 7.0 m/s^2 horizontal and to the right (b) 21 N (c) 14 N horizontal and to the right
29. 7.90 m/s
31. (a) 14.2 m/s (b) 588 m
33. (a) 2.15×10^3 N forward (b) 645 N forward (c) 645 N rearward (d) 1.02×10^4 N at 74.1° below horizontal and rearward
35. $\mu_s = 0.38$, $\mu_k = 0.31$
37. (a) 0.256 (b) 0.509 m/s^2
39. (a) 14.7 m (b) Neither mass is necessary
41. 32.1 N
43. (a) 33 m/s (b) No. The object will speed up to 33 m/s from any lower speed and will slow down to 33 m/s from any higher speed.
45. $\mu_k = 0.287$
47. (a) 1.78 m/s^2 (b) 0.368 (c) 9.37 N (d) 2.67 m/s
49. 3.30 m/s^2
51. (a) 0.404 (b) 45.8 lb
53. (b) 2-kg block: 5.7 m/s^2 to left; 3-kg block: 5.7 m/s^2 to right; 10-kg block: 5.7 m/s^2 downward (c) 17 N, 41 N
55. (a) 84.9 N upward (b) 84.9 N downward
57. 50 m
59. (a) friction between box and truck (b) 2.94 m/s^2
61. (a) 2.22 m (b) 8.74 m/s down the incline
63. (a) 0.232 m/s^2 (b) 9.68 N
65. (a) 1.7 m/s^2, 17 N (b) 0.69 m/s^2, 17 N
67. 100 N, 204 N
69. (a) $T_1 = 78.0$ N, $T_2 = 35.9$ N (b) 0.656
71. (a) 30.7° (b) 0.843 N
73. 5.5×10^2 N
75. (a) 7.1×10^2 N (b) 8.1×10^2 N (c) 7.1×10^2 N (d) 6.5×10^2 N
77. 72.0 N
79. (a) 0.408 m/s^2 upward (b) 83.3 N
81. (b) 514 N, 558 N, 325 N

Chapter 5

QUICK QUIZZES

1. (C)
2. (d)
3. (c)
4. (c)

CONCEPTUAL QUESTIONS

1. Since no motion is taking place, the rope undergoes no displacement and no work is done on it. For the same reason, no work is being done on the pullers or the ground. Work is being done only within the bodies of the pullers. For example, the heart of each puller is applying forces on the blood to move blood through the body.
3. When the slide is frictionless, changing the length or shape of the slide will not make any difference in the final speed of the child, as long as the difference in the heights of the upper and lower ends of the slide is kept constant. If friction must be considered, the path length along which the friction force does negative work will be greater when the slide is made longer or given bumps. Thus, the child will arrive at the lower end with less kinetic energy (and hence less speed).
5. If we ignore any effects due to rolling friction on the tires of the car, we find that the same amount of work would be done in driving up the switchback and in driving straight up the mountain, since the weight of the car is moved upwards against gravity by the same vertical distance in each case. If we include friction, there is more work done in driving the switchback, since the distance over which the friction force acts is much longer. So why do we use switchbacks? The answer lies in the force required, not the work. The force required from the engine to follow a gentle rise is much smaller than that required to drive straight up the hill. To negotiate roadways running straight uphill, engines would have to be redesigned to enable them to apply much larger forces. (It is for much the same reason that ramps are designed to move heavy objects into trucks, as opposed to lifting the objects vertically.)

7. (a) The tension in the supporting cord does no work, because the motion of the pendulum is always perpendicular to the cord and therefore to the tension force. (b) The air resistance does negative work at all times, because the air resistance is always acting in a direction opposite that of the motion. (c) The force of gravity always acts downwards; therefore, the work done by gravity is positive on the downswing and negative on the upswing.
9. Because the periods are the same for both cars, we need only to compare the work done. Since the sports car is moving twice as fast as the older car at the end of the time interval, it has four times the kinetic energy. Thus, according to the work–energy theorem, four times as much work was done, and the engine must have expended four times the power.
11. During the time that the toe is in contact with the ball, the work done by the toe on the ball is given by

$$W_{\text{toe}} = \tfrac{1}{2} m_{\text{ball}} v^2 - 0 = \tfrac{1}{2} m_{\text{ball}} v^2$$

where v is the speed of the ball as it leaves the toe. After the ball loses contact with the toe, only the gravitational force and the retarding force due to air resistance continue to do work on the ball throughout its flight.
13. Yes, the total mechanical energy of the system is conserved because the only forces acting are conservative: the force of gravity and the spring force. There are two forms of potential energy in this case: gravitational potential energy and elastic potential energy stored in the spring.
15. Let's assume you lift the book slowly. In this case, there are two forces on the book that are almost equal in magnitude: the lifting force and the force of gravity. Thus, the positive work done by you and the negative work done by gravity cancel. There is no net work performed and no net change in the kinetic energy, so the work–energy theorem is satisfied.
17. As the satellite moves in a circular orbit about the Earth, its displacement during any small time interval is perpendicular to the gravitational force, which always acts toward the center of the Earth. Therefore, the work done by the gravitational force during any displacement is zero. (Recall that the work done by a force is defined to be $F\Delta x \cos\theta$, where θ is the angle between the force and the displacement. In this case, the angle is 90°, so the work done is zero.) Since the work–energy theorem says that the net work done on an object during any displacement is equal to the change in its kinetic energy, and the work done in this case is zero, the change in the satellite's kinetic energy is zero: hence, its speed remains constant.
19. If a crate is located on the bed of a truck, and the truck accelerates, the friction force exerted on the crate causes it to undergo the same acceleration as the truck, assuming that the crate doesn't slip. Another example is a car that accelerates because of the frictional forces between the road surface and its tires. This force is in the direction of the motion of the car and produces an increase in the car's kinetic energy.

PROBLEMS

1. 700 J
3. 15.0 MJ
5. (a) 61.3 J (b) −46.3 J (c) 0
7. (a) 79.4 N (b) 1.49 kJ (c) −1.49 kJ
9. (a) 2.00 m/s (b) 200 N
11. 0.265 m/s
13. (a) -5.6×10^2 J (b) 1.2 m
15. (a) 90 J (b) 1.8×10^2 N
17. 1.0 m/s
19. 4.1 m
21. (a) 4.1 m (b) 6.4 m/s
23. $h = 6.94$ m
25. $W_{\text{biceps}} = 120$ J, $W_{\text{chin-up}} = 290$ J, additional muscles must be involved
27. 0.459 m
29. 1.53×10^5 N upwards
31. (a) 10.9 m/s (b) 11.6 m/s
33. (a) 544 N/m (b) 19.7 m/s
35. 10.2 m
37. (a) Yes. There are no nonconservative forces acting on the child, so the total mechanical energy is conserved. (b) No. In the expression for conservation of mechanical energy, the mass of the child is included in every term and therefore cancels out. (c) The answer is the same in each case. (d) The expression would have to be modified to include the work done by the force of friction. (e) 15.3 m/s.
39. 2.1 kN
41. 3.8 m/s
43. (a) 5.42 m/s (b) 0.300 (c) 147 J
45. 289 m
47. (a) 24.5 m/s (b) Yes (c) 206 m (d) Unrealistic; the actual retarding force will vary with speed.
49. (a) 1.24 kW (b) 20.9%
51. 8.01 W
53. (a) 2.38×10^4 W = 32.0 hp (b) 4.77×10^4 W = 63.9 hp
55. (a) 24.0 J (b) −3.00 J (c) 21.0 J
57. (a) The graph is a straight line passing through the points (0 m, −16 N), (2 m, 0 N), and (3 m, 8 N). (b) −12.0 J
59. (a) 575 N/m (b) 46.0 J
61. (a) 4.4 m/s (b) 1.5×10^5 N
63. (a) 3.13 m/s (b) 4.43 m/s (c) 1.00 m
65. (a) 0.588 J (b) 0.588 J (c) 2.42 m/s (d) $PE_C = 0.392$ J (e) $KE_C = 0.196$ J
67. (a) 423 mi/gal (b) 776 mi/gal
69. (a) 28.0 m/s (b) 30.0 m (c) 89.0 m beyond the end of the track
71. 1.68 m/s
73. (a) 6.15 m/s (b) 9.87 m/s
75. (a) 101 J (b) 0.410 m (c) 2.84 m/s (d) −9.80 mm (e) 2.85 m/s
77. 914 N/m
79. $W_{\text{net}} = 0$, $W_{\text{grav}} = -2.0 \times 10^4$ J, $W_{\text{normal}} = 0$, $W_{\text{friction}} = 2.0 \times 10^4$ J
81. (a) 10.2 kW (b) 10.6 kW (c) 5.82×10^6 J
83. $v = (8gh/15)^{1/2}$
85. between 25.2 km/h and 27.0 km/h
87. (a) 6.75 W/m^2 (b) 6.64 kW/m^2 (c) A powerful automobile running on sunlight would have to carry on its roof a solar panel that was huge compared with the size of the car.
89. 4.3 m/s

Chapter 6

QUICK QUIZZES

1. (d)
2. (c)
3. (c)

4. (a)
5. (a) Perfectly inelastic (b) Inelastic (c) Inelastic
6. (a)

CONCEPTUAL QUESTIONS

1. (a) No. It cannot carry more kinetic energy than it possesses. That would violate the law of energy conservation. (b) Yes. By bouncing from the object it strikes, it can deliver more momentum in a collision than it possesses in its flight.

3. If all the kinetic energy disappears, there must be no motion of either of the objects after the collision. If neither is moving, the final momentum of the system is zero, and the initial momentum of the system must also have been zero. A situation in which this could be true would be the head-on collision of two objects having momenta of equal magnitude but opposite direction.

5. Initially, the clay has momentum directed toward the wall. When it collides and sticks to the wall, neither the clay nor the wall appears to have any momentum. Thus, it is tempting to (wrongfully) conclude that momentum is not conserved. However, the "lost" momentum is actually imparted to the wall and Earth, causing both to move. Because of Earth's enormous mass, its recoil speed is too small to detect.

7. Before the step the momentum was zero, so afterward the net momentum must also be zero. Obviously, you have some momentum, so something must have momentum in the opposite direction. That something is Earth, the enormous mass of which ensures that its recoil speed will be too small to detect, but if you want to make Earth move, it is as simple as taking a step.

9. As the water is forced out of the holes in the arm, the arm imparts a horizontal impulse to the water. The water then exerts an equal and opposite impulse on the spray arm, causing the spray arm to rotate in the direction opposite that of the spray.

11. Its speed decreases as its mass increases. There are no external horizontal forces acting on the box, so its momentum cannot change as it moves along the horizontal surface. As the box slowly fills with water, its mass increases with time. Because the product mv must be a constant, and because m is increasing, the speed of the box must decrease.

13. It will be easiest to catch the medicine ball when its speed (and kinetic energy) is lowest. The first option—throwing the medicine ball at the same velocity—will be the most difficult, because the speed will not be reduced at all. The second option, throwing the medicine ball with the same momentum, will reduce the velocity by the ratio of the masses. Since $m_t v_t = m_m v_m$, it follows that

$$v_m = v_t\left(\frac{m_t}{m_m}\right)$$

The third option, throwing the medicine ball with the same kinetic energy, will also reduce the velocity, but only by the square root of the ratio of the masses. Since

$$\frac{1}{2}m_t v_t^2 = \frac{1}{2}m_m v_m^2$$

it follows that

$$v_m = v_t\sqrt{\frac{m_t}{m_m}}$$

Thus, the slowest—and easiest—throw will be made when the momentum is held constant. If you wish to check this answer, try substituting in values of $v_t = 1$ m/s, $m_t = 1$ kg, and $m_m = 100$ kg. Then the same-momentum throw will be caught at 1 cm/s, while the same-energy throw will be caught at 10 cm/s.

15. The follow-through keeps the club in contact with the ball as long as possible, maximizing the impulse. Thus, the ball accrues a larger change in momentum than without the follow-through, and it leaves the club with a higher velocity and travels farther. With a short shot to the green, the primary factor is control, not distance. Hence, there is little or no follow-through, allowing the golfer to have a better feel for how hard he or she is striking the ball.

17. It is the product mv that is the same for both the bullet and the gun. The bullet has a large velocity and a small mass, while the gun has a small velocity and a large mass. Furthermore, the bullet carries much more kinetic energy than the gun.

PROBLEMS

1. 1.39 N · s up
3. (a) 8.35×10^{-21} kg · m/s (b) 4.50 kg · m/s (c) 750 kg · m/s (d) 1.78×10^{29} kg · m/s
5. (a) 31.0 m/s (b) the bullet, 3.38×10^3 J versus 69.7 J
7. $\sim 10^3$ N upward
9. 364 kg · m/s forward, 438 N forward
11. (a) 8.0 N · s (b) 5.3 m/s (c) 3.3 m/s
13. (a) 12 N · s (b) 8.0 N · s (c) 8.0 m/s, 5.3 m/s
15. (a) 9.60×10^{-2} s (b) 3.65×10^5 N (c) 26.6g
17. 6.7×10^3 N toward the west
19. 65 m/s
21. (a) 1.15 m/s (b) 0.346 m/s directed opposite to girl's motion
23. $KE_E/KE_b \sim 10^{-25}$
25. 1.67 m/s
27. (a) 1.80 m/s (b) 2.16×10^4 J
29. 57 m
31. 15.6 m/s
33. 273 m/s
35. (a) -6.67 cm/s, 13.3 cm/s (b) 0.889
37. 17.1 cm/s (25.0-g object), 22.1 cm/s (10.0-g object)
39. 7.94 cm
41. (a) 2.9 m/s at 32° N of E (b) 7.9×10^2 J converted into internal energy
43. 5.59 m/s north
45. (a) 2.50 m/s at $-60°$ (b) elastic collision
47. (a) 9.0 m/s (b) -15 m/s
49. 1.78×10^3 N on truck driver, 8.89×10^3 N on car driver
51. (a) 8/3 m/s (incident particle), 32/3 m/s (target particle) (b) $-16/3$ m/s (incident particle), 8/3 m/s (target particle) (c) 7.1×10^{-2} J in case (a), and 2.8×10^{-3} J in case (b). The incident particle loses more kinetic energy in case (a), in which the target mass is 1.0 g.
53. 1.1×10^3 N (upward)
55. (a) 1.33 m/s (b) 235 N (c) 0.681 s (d) -160 N · s, 160 N · s (e) 1.82 m (f) 0.454 m (g) -427 J (h) 107 J (i) Equal friction forces act through different distances on the person and the cart to do different amounts of work on them. This is a perfectly inelastic collision in which the total work on both person and cart together is -320 J, which becomes $+320$ J of internal energy.

57. (a) -2.33 m/s, 4.67 m/s (b) 0.277 m (c) 2.98 m (d) 1.49 m
59. (a) -0.667 m/s (b) 0.952 m
61. (a) 3.54 m/s (b) 1.77 m (c) 3.54×10^4 N (d) No, the normal force exerted by the rail contributes upward momentum to the system.
63. (a) 0.28 or 28% (b) 1.1×10^{-13} J for the neutron, 4.5×10^{-14} J for carbon
65. $\frac{2v_0^2}{9\mu_k g} - \frac{4d}{9}$
67. (a) No. After colliding, the cars, moving as a unit, would travel northeast, so they couldn't damage property on the southeast corner. (b) x-component: 16.3 km/h, y-component: 9.17 km/h, angle: the final velocity of the car is 18.7 km/h at 29.4° north of east, consistent with part (a).
69. (a) 4.85 m/s (b) 8.41 m
71. 0.312 N to the right
73. (a) 300 m/s, (b) 3.75 m/s, (c) 1.20 m

Chapter 7

QUICK QUIZZES

1. (c)
2. (b)
3. (b)
4. (b)
5. (a)
6. 1. (e) 2. (a) 3. (b)
7. (c)
8. (b), (c)
9. (e)
10. (d)

CONCEPTUAL QUESTIONS

1. (a) The head will tend to lean toward the right shoulder (that is, toward the outside of the curve). (b) When there is no strap, tension in the neck muscles must produce the centripetal acceleration. (c) With a strap, the tension in the strap performs this function, allowing the neck muscles to remain relaxed.
3. An object can move in a circle even if the total force on it is not perpendicular to its velocity, but then its speed will change. Resolve the total force into an inward radial component and a perpendicular tangential component. If the tangential force acts in the forward direction, the object will speed up, and if the tangential force acts backward, the object will slow down.
5. The speedometer will be inaccurate. The speedometer measures the number of tire revolutions per second, so its readings will be too low.
7. The car cannot round a turn at constant velocity, because "constant velocity" means the direction of the velocity is not changing. The statement is correct if the word "velocity" is replaced by the word "speed."
9. The gravitational force exerted on the moon by the planet produces a centripetal acceleration. Mathematically,

$$m_{\text{moon}}\left(\frac{v_t^2}{r}\right) = \frac{GMm_{\text{moon}}}{r^2}$$

or the mass of the planet is $M = \frac{(rv_t^2)}{G}$. Both r and v_t can be determined by observing the motion of the moon, and the mass of the planet is then easily computed.
11. Consider an individual standing against the inside wall of the cylinder with her head pointed toward the axis of the cylinder. As the cylinder rotates, the person tends to move in a straight-line path tangent to the circular path followed by the cylinder wall. As a result, the person presses against the wall, and the normal force exerted on her provides the radial force required to keep her moving in a circular path. If the rotational speed is adjusted such that this normal force is equal in magnitude to her weight on Earth, she will not be able to distinguish between the artificial gravity of the colony and ordinary gravity.
13. The tendency of the water is to move in a straight-line path tangent to the circular path followed by the container. As a result, at the top of the circular path, the water presses against the bottom of the pail, and the normal force exerted by the pail on the water provides the radial force required to keep the water moving in its circular path.
15. Any object that moves such that the *direction* of its velocity changes has an acceleration. A car moving in a circular path will always have a centripetal acceleration.
17. (b) decrease. The free-fall acceleration near the surface of the Earth is given by $g = GM_E/R_E^2$, therefore, doubling both the Earth's mass and its radius would reduce the value of g by a factor of 2.

PROBLEMS

1. (a) 3.2×10^8 rad (b) 5.0×10^7 rev
3. 1.99×10^{-7} rad/s, 0.986°/day
5. (a) 821 rad/s^2 (b) 4.21×10^3 rad
7. 3.2 rad
9. Main rotor: 179 m/s $= 0.522v_{\text{sound}}$
Tail rotor: 221 m/s $= 0.644v_{\text{sound}}$
11. $\sim 10^{-2}$ cm
13. 13.7 rad/s^2
15. (a) 3.37×10^{-2} m/s^2 downward (b) 0
17. (a) 0.35 m/s^2 (b) 1.0 m/s (c) 0.35 m/s^2, 0.94 m/s^2, 1.0 m/s^2 at 20° forward with respect to the direction of a_c
19. (a) 1.10 kN (b) 2.04 times her weight
21. 22.6 m/s
23. (a) 18.0 m/s^2 (b) 900 N (c) 1.84; this large coefficient is unrealistic, and she will not be able to stay on the merry-go-round.
25. (a) 9.8 N (b) 9.8 N (c) 6.3 m/s
27. (a) 1.58 m/s^2 (b) 455 N upward (c) 329 N upward (d) 397 N directed inward and 80.8° above horizontal
29. 321 N toward Earth
31. 1.1×10^{-10} N at 72° above the $+x$-axis
33. (a) 2.50×10^{-5} N toward the 500-kg object (b) Between the two objects and 0.245 m from the 500-kg object
35. (a) 9.58×10^6 m (b) 5.57 h
37. 1.90×10^{27} kg
39. (a) 1.48 h (b) 7.79×10^3 m/s (c) 6.43×10^9 J
41. (a) 9.40 rev/s (b) 44.1 rev/s^2; $a_r = 2\,590$ m/s^2; $a_t = 206$ m/s^2 (c) $F_r = 514$ N; $F_t = 40.7$ N
43. (a) 2.51 m/s (b) 7.90 m/s^2 (c) 4.00 m/s
45. (a) 7.76×10^3 m/s (b) 89.3 min

47. (a) $n = m\left(g - \frac{v^2}{r}\right)$ (b) 17.1 m/s

49. (a) $F_{g,\text{ true}} = F_{g,\text{ apparent}} + mR_E\omega^2$
(b) 732 N (equator), 735 N (either pole)

51. 0.131

55. 11.8 km/s

57. 0.75 m

59. (a) $v_0 = \sqrt{g\left(R - \frac{2h}{3}\right)}$ (b) $h' = \frac{R}{2} + \frac{2h}{3}$

61. (a) 15.3 km (b) 1.66×10^{16} kg (c) 1.13×10^4 s

63. 0.835 rev/s = 50.1 rev/min

65. (a) 10.6 kN (b) 14.1 m/s

67. (a) 106 N (b) 0.396

69. (a) 0.605 m/s (b) 17.3 rad/s (c) 5.82 m/s (d) The crank length is unnecessary.

71. (a) 2.0×10^{12} m/s^2 (b) 2.4×10^{11} N (c) 1.4×10^{12} J

Chapter 8

QUICK QUIZZES

1. (d)
2. (b)
3. (b)
4. (a)
5. (c)
6. (c)
7. (a)

CONCEPTUAL QUESTIONS

1. In order for you to remain in equilibrium, your center of gravity must always be over your point of support, the feet. If your heels are against a wall, your center of gravity cannot remain above your feet when you bend forward, so you lose your balance.

3. There are two major differences between torque and work. The primary difference is that the displacement in the expression for work is directed along the force, while the important distance in the torque expression is perpendicular to the force. The second difference involves whether there is motion. In the case of work, work is done only if the force succeeds in causing a displacement of the point of application of the force. By contrast, a force applied at a perpendicular distance from a rotation axis results in a torque regardless of whether there is motion.

As far as units are concerned, the mathematical expressions for both work and torque are in units that are the product of newtons and meters, but this product is called a joule in the case of work and remains a newton-meter in the case of torque.

5. No. For an object to be in equilibrium, the net external force acting on it must be zero. This is not possible when only one force acts on the object, unless that force should have zero magnitude. In that case, there is really no force acting on the object.

7. As the motorcycle leaves the ground, the friction between the tire and the ground suddenly disappears. If the motorcycle driver keeps the throttle open while leaving the ground, the rear tire will increase its angular speed and, hence, its angular momentum. The airborne motorcycle is now an isolated system, and its angular momentum must be conserved. The increase in angular momentum of the tire directed, say, clockwise must be compensated for by an increase in angular momentum of the entire motorcycle counterclockwise. This rotation results in the nose of the motorcycle rising and the tail dropping.

9. In general, you want the rotational kinetic energy of the system to be as small a fraction of the total energy as possible. That is, you want translation, not rotation. You want the wheels to have as little moment of inertia as possible, so that they present the lowest resistance to changes in rotational motion. Disklike wheels would have lower moments of inertia than hooplike wheels, so disks are preferable. The lower the mass of the wheels, the less is the moment of inertia, so light wheels are preferable. The smaller the radius of the wheels, the less is the moment of inertia, so smaller wheels are preferable—within limits: You want the wheels to be large enough to be able to travel relatively smoothly over irregularities in the road.

11. The angular momentum of the gas cloud is conserved. Thus, the product $I\omega$ remains constant. As the cloud shrinks in size, its moment of inertia decreases, so its angular speed ω must increase.

13. We can assume fairly accurately that the driving motor will run at a constant angular speed and at a constant torque. Therefore, as the radius of the take-up reel increases, the tension in the tape will decrease, in accordance with the equation.

$$T = \tau_{\text{const}}/R_{\text{take-up}} \quad \textbf{(1)}$$

As the radius of the source reel decreases, given a decreasing tension, the torque in the source reel will decrease even faster, as the following equation shows:

$$\tau_{\text{source}} = TR_{\text{source}} = \tau_{\text{const}}R_{\text{source}}/R_{\text{take-up}} \quad \textbf{(2)}$$

This torque will be partly absorbed by friction in the feed heads (which we assume to be small); some will be absorbed by friction in the source reel. Another small amount of the torque will be absorbed by the increasing angular speed of the source reel. However, in the case of a sudden jerk on the tape, the changing angular speed of the source reel becomes important. If the source reel is full, then the moment of inertia will be large and the tension in the tape will be large. If the source reel is nearly empty, then the angular acceleration will be large instead. Thus, the tape will be more likely to break when the source reel is nearly full. One sees the same effect in the case of paper towels: It is easier to snap a towel free when the roll is new than when it is nearly empty.

15. The initial angular momentum of the system (mouse plus turntable) is zero. As the mouse begins to walk clockwise, its angular momentum increases, so the turntable must rotate in the counterclockwise direction with an angular momentum whose magnitude equals that of the mouse. This conclusion follows from the fact that the final angular momentum of the system must equal the initial angular momentum (zero).

17. When a ladder leans against a wall, both the wall and the floor exert forces of friction on the ladder. If the floor is perfectly smooth, it can exert no frictional force in the horizontal direction to counterbalance the wall's normal force. Therefore, a ladder on a smooth floor cannot stand in equilibrium. However, a smooth wall can still exert a normal force to hold the ladder in equilibrium against horizontal motion. The counterclockwise torque of this force prevents rotation about the foot of the ladder. So you should choose a rough floor.

PROBLEMS

1. 133 N
3. (a) 30 N · m (counterclockwise)
(b) 36 N · m (counterclockwise)
5. 5.1 N · m
7. $F_t = 724$ N, $F_s = 716$ N
9. 312 N
11. $x_{cg} = 3.33$ ft, $y_{cg} = 1.67$ ft
13. 1.01 m in Figure P8.13b; 0.015 m towards the head in Figure P8.13c.
15. $T = 2.71$ kN, $R_x = 2.65$ kN
17. (a) 443 N, (b) 222 N (to the right), 216 N (upwards)
19. $R = 107$ N, $T = 157$ N
21. $T_{\text{left wire}} = \frac{1}{3}w$, $T_{\text{right wire}} = \frac{2}{3}w$
23. $T_1 = 501$ N, $T_2 = 672$ N, $T_3 = 384$ N
25. 6.15 m
27. 209 N
29. (a) 99.0 kg · m^2 (b) 44.0 kg · m^2 (c) 143 kg · m^2
31. (a) 24.0 N · m (b) 0.035 6 rad/s^2 (c) 1.07 m/s^2
33. (a) 34 N (b) 33 cm
35. 177 N
37. 0.524
39. (a) 500 J (b) 250 J (c) 750 J
41. 276 J
43. 149 rad/s
45. 17.5 J · s counterclockwise
47. 8.0 rev/s
49. 6.73 rad/s
51. 5.99×10^{-2} J
53. (a) 0.360 rad/s counterclockwise (b) 99.9 J
55. (a) $\omega = \left(\frac{I_1}{I_1 + I_2}\right)\omega_0$ (b) $\frac{KE_f}{KE_i} = \frac{I_1}{I_1 + I_2} < 1$
57. (a) As the child walks to the right end of the boat, the boat moves left (towards the pier). (b) The boat moves 1.45 m closer to the pier, so the child will be 5.55 m from the pier. (c) No. He will be short of reaching the turtle by 0.45 m.
59. (a) 1.1 m/s^2 (b) $T_1 = 22$ N $T_2 = 44$ N
61. 36.9°
63. $a_{\text{sphere}} = \frac{g\sin\theta}{1.4}$, $a_{\text{disk}} = \frac{g\sin\theta}{1.5}$, $a_{\text{ring}} = \frac{g\sin\theta}{2.0}$
Thus, the sphere wins and the ring comes in last.
65. (a) 1.63 m/s (b) 54.2 rad/s
67. 1.09 m
69. (a) Mvd (b) Mv^2 (c) Mvd (d) $2v$ (e) $4Mv^2$
(f) $3Mv^2$
71. 7.5×10^{-11} s
73. (a) $T = \frac{Mmg}{M + 4m}$ (b) $a_t = \frac{4mg}{M + 4m}$
75. (a) $\sim 10^{-22}$ s^{-2}, (b) $\sim 10^{16}$ N · m, (c) $\sim 10^{13}$ m
77. 24.5 m/s
79. (a) 3.12 m/s^2 (b) $T_1 = 26.7$ N, $T_2 = 9.37$ N
81. (b) 3×10^2 N
83. $6w = 4.5$ kN

Chapter 9

QUICK QUIZZES

1. (c)
2. (a)
3. (c)
4. (b)
5. (c)
6. (b)
7. (a)

CONCEPTUAL QUESTIONS

1. The density of air is lower in the mile-high city of Denver than it is at lower altitudes, so the effect of air drag is less in Denver than it would be in a city such as New York. The reduced air drag means a well-hit ball will go farther, benefiting home-run hitters. On the other hand, curve ball pitchers prefer to throw at lower altitudes where the higher density air produces greater deflecting forces on a spinning ball.

3. She exerts enough pressure on the floor to dent or puncture the floor covering. The large pressure is caused by the fact that her weight is distributed over the very small cross-sectional area of her high heels. If you are the homeowner, you might want to suggest that she remove her high heels and put on some slippers.

5. If you think of the grain stored in the silo as a fluid, the pressure the grain exerts on the walls of the silo increases with increasing depth, just as water pressure in a lake increases with increasing depth. Thus, the spacing between bands is made smaller at the lower portions to counterbalance the larger outward forces on the walls in these regions.

7. The syringe doesn't really draw the blood from the vein. By moving the plunger in the syringe back, the nurse lowers the pressure inside the syringe below the pressure inside the vein. Then the blood pressure within the body forces blood out into the syringe, which "accepts" the blood.

9. In the ocean, the ship floats due to the buoyant force from *salt water*, which is denser than fresh water. As the ship is pulled up the river, the buoyant force from the fresh water in the river is not sufficient to support the weight of the ship, and it sinks.

11. The balance will not be in equilibrium: The side with the lead will be lower. Despite the fact that the weights on both sides of the balance are the same, the Styrofoam, due to its larger volume, will experience a larger buoyant force from the surrounding air. Thus, the net force of the weight and the buoyant force is larger in the downward direction for the lead than for the Styrofoam.

13. The two cans displace the same volume of water and hence are acted upon by buoyant forces of equal magnitude. The total weight of the can of diet cola must be less than this buoyant force, whereas the total weight of the can of regular cola is greater than the buoyant force. This is possible even though the two containers are identical and contain the same volume of liquid. Because of the difference in the quantities and densities of the sweeteners used, the volume V of the diet mixture will have less mass than an equal volume of the regular mixture.

15. As the truck passes, the air between your car and the truck is squeezed into the channel between the two vehicles and moves at a higher speed than when your car is in the open. According to Bernoulli's principle, this high-speed air has a lower pressure than the air on the outer side of your car. The difference in pressure exerts a net force on your car toward the truck.
17. Opening the windows results in a smaller pressure difference between the exterior and interior of the house and, therefore, less tendency for severe damage to the structure due to the Bernoulli effect.

PROBLEMS

1. 1.3 mm
3. 1.8×10^6 Pa
5. 3.5×10^8 Pa
7. 4.4 mm
9. 0.024 mm
11. 1.9 cm
13. (a) 1.88×10^5 Pa (b) 2.65×10^5 Pa
15. 1.4 atm
17. 3.4×10^2 m
19. 0.133 m
21. 271 kN horizontally toward the cellar
23. 1.05×10^5 Pa
25. 2.1 N · m
27. 9.41 kN
29. (b) 5.9 km
31. 2.67×10^3 kg
33. 5.57 N
35. (a) 1.46×10^{-2} m^3 (b) 2.10×10^3 kg/m^3
37. 1.07 m/s^2
39. 17.3 N (upper scale), 31.7 N (lower scale)
41. (a) 80 g/s (b) 0.27 mm/s
43. 12.6 m/s
45. 9.00 cm
47. 1.47 cm
49. (a) 28.0 m/s (b) 28.0 m/s (c) 2.11 MPa
51. (b) For any y less than $y_{max} = P_0/\rho g$
53. 8.3×10^{-2} N/m
55. 5.6×10^{-2} N/m
57. 8.6 N
59. 2.1 MPa
61. 2.8 μm
63. 0.41 mm
65. $RN = 4.3 \times 10^3$; turbulent flow
67. 1.8×10^{-3} kg/m^3
69. 1.4×10^{-5} N · s/m^2
71. (a) The buoyant forces are the same because the two blocks displace equal amounts of water. (b) The spring scale reads largest value for the iron block. (c) $B = 2.0 \times 10^3$ N for both blocks, $T_{iron} = 13 \times 10^3$ N, $T_{aluminum} = 3.3 \times 10^3$ N.
73. 2.5×10^7 capillaries
75. (a) 1.57 kPa, 1.55×10^{-2} atm,.11.8 mm of Hg (b) The fluid level in the tap should rise. (c) Blockage of flow of the cerebrospinal fluid
77. 4.14×10^3 m^3
79. 833 kg/m^3
81. 2.25 m above the level of point B
83. 6.4 m
85. 1.3 cm
87. 17.0 cm above the floor
89. 532 cm^3

Chapter 10

QUICK QUIZZES

1. (c)
2. (b)
3. (c)
4. (c)
5. (a)
6. (b)

CONCEPTUAL QUESTIONS

1. An ordinary glass dish will usually break because of stresses that build up as the glass expands when heated. The expansion coefficient for Pyrex glass is much lower than that of ordinary glass. Thus, the Pyrex dish will expand much less than the dish of ordinary glass and does not normally develop sufficient stress to cause breakage.
3. The accurate answer is that it doesn't matter! Temperatures on the Kelvin and Celsius scales differ by only 273 degrees. This difference is insignificant for temperatures on the order of 10^7 degrees. If we imagine that the temperature is given in kelvins and ignore any problems with significant figures, then the Celsius temperature is $1.499\,972\,7 \times 10^7$°C.
5. Mercury must have the larger coefficient of expansion. As the temperature of a thermometer rises, both the mercury and the glass expand. If they both had the same coefficient of linear expansion, the mercury and the cavity in the glass would expand by the same amount, and there would be no apparent movement of the end of the mercury column relative to the calibration scale on the glass. If the glass expanded more than the mercury, the reading would go down as the temperature went up! Now that we have argued this conceptually, we can look in a table and find that the coefficient for mercury is about 20 times as large as that for glass, so that the expansion of the glass can sometimes be ignored.
7. We can think of each bacterium as being a small bag of liquid containing bubbles of gas at a very high pressure. The ideal gas law indicates that if the bacterium is raised rapidly to the surface, then its volume must increase dramatically. In fact, the increase in volume is sufficient to rupture the bacterium.
9. Velocity is a vector quantity, so direction must be considered. If there are the same number of particles moving to the right along the x-direction as there are to the left along the $-x$-direction, the x-components of the velocity of all the molecules will sum to zero.
11. The bags of chips contain a sealed sample of air. When the bags are taken up the mountain, the external atmospheric pressure on them is reduced. As a result, the difference between the pressure of the air inside the bags and the reduced pressure outside results in a net force pushing the plastic of the bag outward.
13. Additional water vaporizes into the bubble, so that the number of moles n increases.

PROBLEMS

1. (a) -460°C (b) 37.0°C (c) -280°C
3. (a) -423°C, 20.28 K (b) 68°F, 293 K
9. (a) 1 337 K, 2 933 K (b) 1 596°C, 1 596 K
11. 31 cm
13. 55.0°C
15. (a) -179°C (attainable) (b) -376°C (below 0 K, unattainable)

17. 2.171 cm
19. 2.7×10^2 N
21. 1.1 L (0.29 gal)
23. 0.548 gal
25. (a) increases (b) 1.603 cm
27. (a) 627°C (b) 927°C
29. (a) 2.5×10^{19} molecules (b) 4.1×10^{-21} mol
31. 287°C
33. 7.1 m
35. 16.0 cm^3
37. 6.21×10^{-21} J
39. 6.64×10^{-27} kg
41. (a) 8.76×10^{-21} J
(b) $v_{\text{rms, He}} = 1.62$ km/s, $v_{\text{rms, Ar}} = 514$ m/s
43. 16 N
45. 0.663 mm at 78.2° below the horizontal
47. 3.55 L
49. 6.57 MPa
51. The expansion of the mercury is almost 20 times that of the flask (assuming Pyrex glass).
53. shorter, by 0.061 mm
55. 2.74 m
57. 1.61 MPa
59. 0.417 L
61. (a) $\theta = \dfrac{(\alpha_2 - \alpha_1)L_0(\Delta T)}{\Delta r}$
(c) The bar bends in the opposite direction.
63. 0.53 kg

Chapter 11

QUICK QUIZZES

1. (a) Water, glass, iron. (b) Iron, glass, water.
2. (b) The slopes are proportional to the reciprocal of the specific heat, so a larger specific heat results in a smaller slope, meaning more energy is required to achieve a given temperature change.
3. (c)
4. (b)
5. (a) 4 (b) 16 (c) 64

CONCEPTUAL QUESTIONS

1. When you rub the surface, you increase the temperature of the rubbed region. With the metal surface, some of this energy is transferred away from the rubbed site by conduction. Consequently, the temperature in the rubbed area is not as high for the metal as it is for the wood, and it feels relatively cooler than the wood.
3. The fruit loses energy into the air by radiation and convection from its surface. Before ice crystals can form inside the fruit to rupture cell walls, all of the liquid water on the skin will have to freeze. The resulting time delay may prevent damage within the fruit throughout a frosty night. Further, a surface film of ice provides some insulation to slow subsequent energy loss by conduction from within the fruit.
5. The operation of an immersion coil depends on the convection of water to maintain a safe temperature. As the water near a coil warms up, the warmed water floats to the top due to Archimedes' principle. The temperature of the coil cannot go higher than the boiling temperature of water, 100°C. If the coil is operated in air, convection is reduced, and the upper limit of 100°C is removed. As a result, the coil can become hot enough to be damaged. If the coil is used in an attempt to warm a thick liquid like stew, convection cannot occur fast enough to carry energy away from the coil, so that it again may become hot enough to be damaged.
7. One of the ways that objects transfer energy is by radiation. The top of the mailbox is oriented toward the clear sky. Radiation emitted by the top of the mailbox goes upward and into space. There is little radiation coming down from space to the top of the mailbox. Radiation leaving the sides of the mailbox is absorbed by the environment. Radiation from the environment (tree, houses, cars, etc.), however, can enter the sides of the mailbox, keeping them warmer than the top. As a result, the top is the coldest portion and frost forms there first.
9. Tile is a better conductor of energy than carpet, so the tile conducts energy away from your feet more rapidly than does the carpeted floor.
11. The large amount of energy stored in the concrete during the day as the sun falls on it is released at night, resulting in an overall higher average temperature in the city than in the countryside. The heated air in a city rises as it's displaced by cooler air moving in from the countryside, so evening breezes tend to blow from country to city.
13. The fingers are wetted to create a layer of steam between the fingers and the molten lead. The steam acts as an insulator and prevents serious burns. This demonstration is dangerous: we don't recommend it.
15. The increase in the temperature of the ethyl alcohol will be about twice that of the water.
17. (d)

PROBLEMS

1. 10.1°C
3. (a) 1.67×10^{18} J (b) 53.1 yr
5. 1.03×10^3 J
7. 2.85 km
9. 176°C
11. 88 W
13. 185 g
15. 80 g
17. 1.8×10^3 J/kg · °C
19. 0.26 kg
21. 65°C
23. 21 g
25. 2.3 km
27. 16°C
29. $t_{\text{boil}} = 2.8$ min, $t_{\text{evaporate}} = 18$ min
31. (a) all ice melts, $T_f = 40$°C (b) 8.0 g melts, $T_f = 0$°C
33. (a) 0.22 kW (b) 13 mW (c) 56 mW
35. 402 MW
37. 14 ft^2 · °F · h/Btu
39. 9.0 cm
41. 0.11 kW
43. $\sim 10^3$ W
45. 16:1
47. 12 kW
49. 6.0×10^3 kg
51. 2.3 kg
53. 29°C
55. 30.3 kcal/h
57. 109°C
59. 51.2°C

61. (a) 7 stops (b) Assumes that no energy is lost to the surroundings and that all internal energy generated stays with the brakes.
63. (b) 2.7×10^3 J/kg · °C
65. 1.4 kg
67. 12 h
69. 10.9 g

Chapter 12

QUICK QUIZZES

1. (b)
2. A is isovolumetric, B is adiabatic, C is isothermal, D is isobaric.
3. (c)
4. (b)
5. The number 7 is the most probable outcome. The numbers 2 and 12 are the least probable outcomes.

CONCEPTUAL QUESTIONS

1. First, the efficiency of the automobile engine cannot exceed the Carnot efficiency: it is limited by the temperature of the burning fuel and the temperature of the environment into which the exhaust is dumped. Second, the engine block cannot be allowed to exceed a certain temperature. Third, any practical engine has friction, incomplete burning of fuel, and limits set by timing and energy transfer by heat.

3. If there is no change in internal energy, then, according to the first law of thermodynamics, the heat is equal to the negative of the work done on the gas (and thus equal to the work done *by* the gas). Thus, $Q = -W = W_{\text{by gas}}$.

5. The energy that is leaving the body by work and heat is replaced by means of biological processes that transform chemical energy in the food that the individual ate into internal energy. Thus, the temperature of the body can be maintained.

7. The statement shows a misunderstanding of the concept of heat. Heat is energy in the process of being transferred, not a form of energy that is held or contained. If you wish to speak of energy that is contained, you speak of **internal energy**, not heat. Correct statements would be (1) "Given any two objects in thermal contact, the one with the higher temperature will transfer energy by heat to the other" and (2) "Given any two objects of equal mass, the one with the higher product of absolute temperature and specific heat contains more internal energy."

9. Although no energy is transferred into or out of the system by heat, work is done on the system as the result of the agitation. Consequently, both the temperature and the internal energy of the coffee increase.

11. Practically speaking, it isn't possible to create a heat engine that creates no thermal pollution, because there must be both a hot heat source (energy reservoir) and a cold heat sink (low-temperature energy reservoir). The heat engine will warm the cold heat sink and will cool down the heat source. If either of those two events is undesirable, then there will be thermal pollution.

Under some circumstances, the thermal pollution would be negligible. For example, suppose a satellite in space were to run a heat pump between its sunny side and its dark side. The satellite would intercept some of the energy that gathered on one side and would "dump" it to the dark side. Since neither of those effects would be particularly undesirable, it could be said that such a heat pump produced no thermal pollution.

13. The rest of the Universe must have an entropy change of +8.0 J/K or more.

15. The first law is a statement of conservation of energy that says that we cannot devise a cyclic process that produces more energy than we put into it. If the cyclic process takes in energy by heat and puts out work, we call the device a heat engine. In addition to the first law's limitation, the second law says that, during the operation of a heat engine, some energy must be ejected to the environment by heat. As a result, it is theoretically impossible to construct a heat engine that will work with 100% efficiency.

17. From the point of view of energy principles, the molecules strike the piston and move it through a distance, so the molecules do work on the piston. This work represents a transfer of energy out of the gas. As a result, the internal energy of the gas drops. Because the temperature is related to internal energy, the temperature of the gas drops.

PROBLEMS

3. 1.1×10^4 J
5. (a) -810 J (b) -507 J (c) -203 J
7. (a) -6.1×10^5 J (b) 4.6×10^5 J
9. (a) 1.09×10^3 K (b) -6.81 kJ
11. (a) $Q < 0$, $W = 0$, $\Delta U < 0$ (b) $Q > 0$, $W = 0$, $\Delta U > 0$
13. (a) 567 J (b) 167 J
15. (a) -88.5 J (b) 722 J
17. (a) -180 J (b) $+188$ J
19. (a) -9.12×10^{-3} J (b) -333 J
21. (a) 0.95 J (b) 3.2×10^5 J (c) 3.2×10^5 J
23. 0.489 (or 48.9%)
25. (a) 0.333 (or 33.3%) (b) 2/3
27. (a) 0.672 (or 67.2%) (b) 58.8 kW
29. (a) 0.294 (or 29.4%) (b) 500 J (c) 1.67 kW
31. 1/3
33. 0.49°C
35. (a) -1.2 kJ/K (b) 1.2 kJ/K
37. 57 J/K
39. 3.27 J/K
41. (a)

End Result	Possible Tosses	Total Number of Same Result
All H	HHHH	1
1T, 3H	HHHT, HHTH, HTHH, THHH	4
2T, 2H	HHTT, HTHT, THHT, HTTH, THTH, TTHH	6
3T, 1H	TTTH, TTHT, THTT, HTTT	4
All T	TTTT	1

Most probable result = 2H and 2T.

(b) all H or all T (c) 2H and 2T

43. The maximum efficiency possible with these reservoirs = 50%; the claim is false.
45. (a) $-|Q_h|/T_h$ (b) $|Q_c|/T_c$ (c) $|Q_h|/T_h - |Q_c|/T_c$ (d) 0
47. 2.8°C
49. 18°C
51. (a) 12.2 kJ (b) 4.05 kJ (c) 8.15 kJ

53. (a) 26 J (b) 9.0×10^5 J (c) 9.0×10^5 J
55. (a) 2.49 kJ (b) 1.50 kJ (c) −990 J
57. (a) 0.554 (or 55.4%) (b) The Carnot efficiency is 0.749 (or 74.9%)
59. (a) 2.6×10^3 metric tons/day (b) \$$7.7 \times 10^6$/yr (c) 4.1×10^4 kg/s
61. (a) 10 J (b) No work is done; $\Delta U = 42$ J
63. 0.146; 486 kcal

Chapter 13

QUICK QUIZZES

1. (d)
2. (c)
3. (b)
4. (a)
5. (c)
6. (d)
7. (c), (b)
8. (a)
9. (b)

CONCEPTUAL QUESTIONS

1. It will increase. The speed of the wave varies inversely with the mass per unit length of the rope, so it is higher in the lighter rope. The frequency is unchanged as the wave passes from one rope to the other one. The wavelength is $\lambda = v/f$, so with the frequency constant and the speed increasing, the wavelength must increase.

3. No. Because the total energy is $E = \frac{1}{2}kA^2$, changing the mass of the object while keeping A constant has no effect on the total energy. When the object is at a displacement x from equilibrium, the potential energy is $\frac{1}{2}kx^2$, independent of the mass, and the kinetic energy is $KE = E - \frac{1}{2}kx^2$, also independent of the mass.

5. When the spring with two objects on opposite ends is set into oscillation in space, the coil at the exact center of the spring does not move. Thus, we can imagine clamping the center coil in place without affecting the motion. If we do this, we have two separate oscillating systems, one on each side of the clamp. The half-spring on each side of the clamp has twice the spring constant of the full spring, as shown by the following argument: The force exerted by a spring is proportional to the separation of the coils as the spring is extended. Imagine that we extend a spring by a given distance and measure the distance between coils. We then cut the spring in half. If one of the half-springs is now extended by the same distance, the coils will be twice as far apart as they were in the complete spring. Thus, it takes twice as much force to stretch the half-spring, from which we conclude that the half-spring has a spring constant which is twice that of the complete spring. Hence, our clamped system of objects on two half-springs will vibrate with a frequency that is higher than f by a factor of the square root of two.

7. The bouncing ball is not an example of simple harmonic motion. The ball does not follow a sinusoidal function for its position as a function of time. The daily movement of a student is also not simple harmonic motion, since the student stays at a fixed location—school—for a long time. If this motion were sinusoidal, the student would move more and more slowly as she approached her desk, and as soon as she sat down at the desk, she would start to move back toward home again.

9. We assume that the buoyant force acting on the sphere is negligible in comparison to its weight, even when the sphere is empty. We also assume that the bob is small compared with the pendulum length. Then, the frequency of the pendulum is $f = 1/T = (1/2\pi)\sqrt{g/L}$, which is independent of mass. Thus, the frequency will not change as the water leaks out.

11. As the temperature increases, the length of the pendulum will increase due to thermal expansion, and with a greater length, the period of the pendulum increases. Thus, it takes longer to execute each swing, so that each second according to the clock will take longer than an actual second. Consequently, the clock will run slow.

13. A pulse in a long line of people is longitudinal, since the movement of people is parallel to the direction of propagation of the pulse. The speed is determined by the reaction time of the people and the speed with which they can move once a space opens up. There is also a psychological factor, in that people will not want to fill a space that opens up in front of them too quickly, so as not to intimidate the person in front of them. The "wave" at a stadium is transverse, since the fans stand up vertically as the wave sweeps past them horizontally. The speed of this pulse depends on the limits of the fans' abilities to rise and sit rapidly and on psychological factors associated with the anticipation of seeing the pulse approach the observer's location.

15. A wave on a massless string would have an infinite speed of propagation, because the linear mass density of the string is zero.

17. The kinetic energy is proportional to the square of the speed, and the potential energy is proportional to the square of the displacement. Therefore, both must be positive quantities.

19. From $v = \sqrt{F/\mu}$, we see that increasing the tension by a factor of four doubles the wave speed.

PROBLEMS

1. (a) 24 N toward the equilibrium position (b) 60 m/s^2 toward the equilibrium position.
3. (b) 1.81 s (c) No, the force is not of the form of Hooke's law.
5. 0.242 kg
7. (a) 60 J (b) 49 m/s
9. 2.94×10^3 N/m
11. (a) $PE = E/4$ $KE = 3E/4$ (b) $x = A/\sqrt{2}$
13. 0.478 m
15. (a) 0.28 m/s (b) 0.26 m/s (c) 0.26 m/s (d) 3.5 cm
17. 39.2 N
19. (a) You observe uniform circular motion projected on a plane perpendicular to the motion. (b) 0.628 s
21. The horizontal displacement is described by $x(t) = A\cos\omega t$, where A is the distance from the center of the wheel to the crankpin.
23. 0.63 s
25. (a) 1.0 s (b) 0.28 m/s (c) 0.25 m/s
27. (a) 11.0 N toward the left (b) 0.881 oscillations
29. $v = \pm\,\omega A\sin\omega t$, $a = -\,\omega^2 A\cos\omega t$
31. 105 complete oscillations
33. (a) slow (b) 9:47 A.M.
35. (a) $L_{\text{Earth}} = 25$ cm, $L_{\text{Mars}} = 9.4$ cm, (b) $m_{\text{Earth}} = m_{\text{Mars}} = 0.25$ kg

37. (a) 9.00 cm (b) 20.0 cm (c) 40.0 ms (d) 5.00 m/s
39. (a) 11.4 ns (b) 3.41 m
41. 31.9 cm
43. 80.0 N
45. (a) 30.0 N (b) 25.8 m/s
47. 28.5 m/s
49. (a) 0.051 kg/m (b) 20 m/s
51. 13.5 N
53. (a) Constructive interference gives $A = 0.50$ m
(b) Destructive interference gives $A = 0.10$ m
55. 6.62 cm
57. (a) 588 N/m (b) 0.700 m/s
59. 1.1 m/s
61. (a) 15.8 rad/s (b) 5.23 cm
63. $F_{\text{tangential}} = -[(\rho_{\text{air}} - \rho_{\text{He}})Vg/L]s$, $T = 1.40$ s
67. (a) 0.50 m/s (b) 8.6 cm
69. (a) the longitudinal wave (b) 11.1 min
71. (a) 0.50 m/s (b) 0.39 m/s (c) 5.0 cm

Chapter 14

QUICK QUIZZES

1. (c)
2. (c)
3. (b)
4. (b), (e)
5. (d)
6. (a)
7. (b)

CONCEPTUAL QUESTIONS

1. (a) higher (b) lower

3. The camera is designed to operate at an assumed speed of sound of 345 m/s, the speed of sound at a room temperature of 23°C. If the temperature should decrease to, say, 0°C, the speed of sound will also decrease, and the camera will respond to the fact that it takes longer for the sound to make its round trip. Thus, it will operate as if the object is farther away than it really is.

It is of interest to note that bats use echo sounding like this to locate insects or to avoid obstacles in front of them, something that they must do because of poor eyesight and the high speeds at which they fly. Likewise, blue whales use this technique to help them avoid objects in their path. The need here is obvious, because a typical whale has a mass of 10^5 kg and travels at a relatively fast speed of 20 mi/h, so it takes a long time for it to stop its motion or to change direction.

5. Sophisticated electronic devices break the frequency range of about 60 to 4 000 Hz used in telephone conversations into several frequency bands and then mix them in a predetermined pattern so that they become unintelligible. The descrambler, of course, moves the bands back into their proper order.

7. A rise in temperature will increase the dimensions of the wind instrument much less than it increases the speed of sound in the enclosed air. This effect will raise the resonant frequencies that are produced by the instrument, which will go sharp as the temperature increases and go flat as the temperature decreases.

9. The echo is Doppler shifted, and the shift is like both a moving source and a moving observer. The sound that leaves your horn in the forward direction is Doppler shifted to a higher frequency, because it is coming from a moving source. As the sound reflects back and comes towards you, you are a moving observer, so there is a second Doppler shift to an even higher frequency. If the sound reflects from the spacecraft coming towards you, there is a different moving-source shift to an even higher frequency. The reflecting surface of the spacecraft acts as a moving source.

11. The center of the string is a node for the second harmonic, as well as for every even-numbered harmonic. By placing the finger at the center and plucking, the guitarist is eliminating any harmonic which does not have a node at that point—that is, all the odd harmonics. The even harmonics can vibrate relatively freely with the finger at the center because they exhibit no displacement at that point. The result is a sound with a mixture of frequencies that are integer multiples of the second harmonic, which is one octave higher than the fundamental.

13. The bowstring is pulled away from equilibrium and released, in a manner similar to the way a guitar string is pulled and released when it is plucked. Thus, standing waves will be excited in the bowstring. If the arrow leaves from the exact center of the string, then a series of odd harmonics will be excited. Even harmonics will not be excited, because they have a node at the point where the string exhibits its maximum displacement.

15. At the instant at which there is no displacement of the string, the string is still moving. Thus, the energy is present at that instant entirely as kinetic energy of the string.

17. As the whistle is approaching you, the sound is Doppler shifted to a frequency higher than the natural frequency of the whistle. You will momentarily hear the natural frequency just as the whistle comes parallel with you and is in the act of passing. As soon as the whistle is past, you will hear sound that is Doppler shifted to a frequency lower than the natural frequency of the whistle. As the frequency of the sound steps down from one constant value to a second one, the intensity of the sound varies continuously. The loudness increases smoothly to a maximum and then decreases.

19. The two engines are running at slightly different frequencies, thus producing a beat frequency between them.

PROBLEMS

1. 5.56 km
3. 32°C
5. 516 m
7. 1.99 km
9. (a) 1.00×10^{-2} W/m^2 (b) 105 dB
11. 37 dB
13. 9 additional machines
15. 66.0 dB
17. (a) 1.3×10^2 W (b) 96 dB
21. (a) 75.2-Hz drop (b) 0.953 m
23. 595 Hz
25. 0.391 m/s
27. 19.3 m
29. 48°
31. 800 m
33. (a) 0.240 m (b) 0.855 m
35. (a) Nodes at 0, 2.67 m, 5.33 m, and 8.00 m; antinodes at 1.33 m, 4.00 m, and 6.67 m (b) 18.6 Hz

37. At 0.0891 m, 0.303 m, 0.518 m, 0.732 m, 0.947 m, and 1.16 m from one speaker.
39. (a) 79 N (b) 2.1×10^2 Hz
41. 19.976 kHz
43. 58 Hz
45. 3.0 kHz
47. (a) 0.552 m (b) 317 Hz
49. 5.26 beats/s
51. 3.79 m/s toward the station, 3.88 m/s away from the station
53. (a) 1.98 beats/s (b) 3.40 m/s
55. 1.76 cm
57. 262 kHz
59. 64 dB
61. 439 Hz and 441 Hz
63. 32.9 m/s
65. 3.97 beats/s
67. 1.34×10^4 N
69. 1 204 Hz
71. (a) 617 m (b) 154 m/s

Chapter 15

QUICK QUIZZES

1. (b)
2. (b)
3. (c)
4. (a)
5. (c) and (d)
6. (a)
7. (c)
8. (b)
9. (d)
10. (b) and (d)

CONCEPTUAL QUESTIONS

1. Electrons have been removed from the object.

3. The configuration shown is inherently unstable. The negative charges repel each other. If there is any slight rotation of one of the rods, the repulsion can result in further rotation away from this configuration. There are three conceivable final configurations shown below. Configuration (a) is stable: If the positive upper ends are pushed towards each other, their mutual repulsion will move the system back to the original configuration. Configuration (b) is an equilibrium configuration, but it is unstable: If the lower ends are moved towards each other, their mutual attraction will be larger than that of the upper ends, and the configuration will shift to (c), another possible stable configuration.

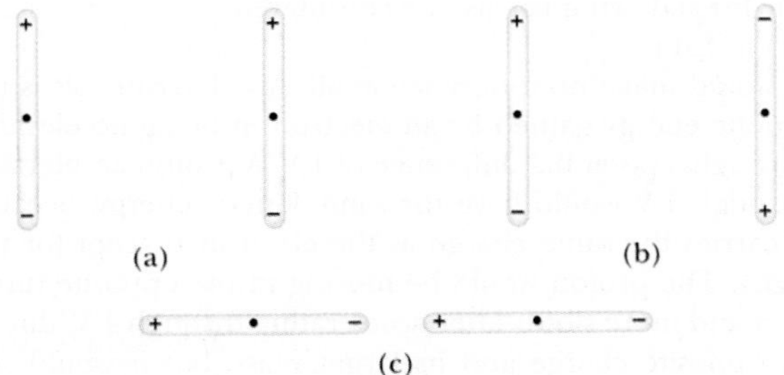

Figure Q15.3

5. Move an object A with a net positive charge so it is near, but not touching, a neutral metallic object B that is insulated from the ground. The presence of A will polarize B, causing an excess negative charge to exist on the side nearest A and an excess positive charge of equal magnitude to exist on the side farthest from A. While A is still near B, touch B with your hand. Additional electrons will then flow from ground, through your body and onto B. With A continuing to be near but not in contact with B, remove your hand from B, thus trapping the excess electrons on B. When A is now removed, B is left with excess electrons, or a net negative charge. By means of mutual repulsion, this negative charge will now spread uniformly over the entire surface of B.

7. An object's mass decreases very slightly (immeasurably) when it is given a positive charge, because it loses electrons. When the object is given a negative charge, its mass increases slightly because it gains electrons.

9. Electric field lines start on positive charges and end on negative charges. Thus, if the fair-weather field is directed into the ground, the ground must have a negative charge.

11. The two charged plates create a region with a uniform electric field between them, directed from the positive toward the negative plate. Once the ball is disturbed so as to touch one plate (say, the negative one), some negative charge will be transferred to the ball and it will be acted upon by an electric force that will accelerate it to the positive plate. Once the ball touches the positive plate, it will release its negative charge, acquire a positive charge, and accelerate back to the negative plate. The ball will continue to move back and forth between the plates until it has transferred all their net charge, thereby making both plates neutral.

13. The electric shielding effect of conductors depends on the fact that there are two kinds of charge: positive and negative. As a result, charges can move within the conductor so that the combination of positive and negative charges establishes an electric field that exactly cancels the external field within the conductor and any cavities inside the conductor. There is only one type of gravitation charge, however, because there is no negative mass. As a result, gravitational shielding is not possible.

15. The electric field patterns of each of these three configurations do not have sufficient symmetry to make the calculations practical. Gauss's law is useful only for calculating the electric fields of highly symmetric charge distributions, such as uniformly charged spheres, cylinders, and sheets.

17. No, the wall is not positively charged. The balloon induces a charge of opposite sign in the wall, causing the balloon and the wall to be attracted to each other. The balloon eventually falls because its charge slowly diminishes as it leaks to ground. Some of the balloon's charge could also be lost due to positive ions in the surrounding atmosphere, which would tend to neutralize the negative charges on the balloon.

19. When the comb is nearby, charges separate on the paper, and the paper is attracted to the comb. After contact, charges from the comb are transferred to the paper, so that it has the same type of charge as the comb. The paper is thus repelled.

21. The attraction between the ball and the object could be an attraction of unlike charges, or it could be an attraction between a charged object and a neutral object as a result of polarization of the molecules of the neutral object. Two additional experiments could help us determine whether the object is charged. First, a known neutral ball could be brought near the object, and if there is an attraction, the object is negatively charged. Another possibility is to bring a known negatively charged ball near the object. In that case, if there is a repulsion, then the object is negatively charged. If there is an attraction, then the object is neutral.

PROBLEMS

1. 1.1×10^{-8} N (attractive)
3. 91 N (repulsion)
5. (a) 36.8 N (b) 5.54×10^{27} m/s^2
7. 5.12×10^5 N
9. (a) 2.2×10^{-5} N (attraction) (b) 9.0×10^{-7} N (repulsion)
11. 1.38×10^{-5} N at 77.5° below the negative x-axis
13. 0.872 N at 30.0° below the positive x-axis
15. 7.2 nC
17. 1.5×10^{-3} C
19. 7.20×10^5 N/C (downward)
21. 1.2×10^4 N/C
23. (a) 6.12×10^{10} m/s^2 (b) 19.6 μs (c) 11.8 m (d) 1.20×10^{-15} J
25. zero
27. 1.8 m to the left of the -2.5-μC charge
33. (a) 0 (b) 5 μC inside, -5 μC outside (c) 0 inside, -5 μC outside (d) 0 inside, -5 μC outside
35. 1.3×10^{-3} C
37. (a) 4.8×10^{-15} N (b) 2.9×10^{12} m/s^2
39. (a) 858 N·m^2/C (b) 0 (c) 657 N·m^2/C
41. 4.1×10^6 N/C
43. (a) 0 (b) $k_e q/r^2$ outward
47. 57.5 N
49. 24 N/C in the positive x-direction
51. (a) $E = 2k_e qb\,(a^2 + b^2)^{-3/2}$ in the positive x-direction (b) $E = k_e Qb(a^2 + b^2)^{-3/2}$ in the positive x-direction
53. (a) 0 (b) 7.99×10^7 N/C (outward) (c) 0 (d) 7.34×10^6 N/C (outward)
55. 3.55×10^5 N·m^2/C
57. 4.4×10^5 N/C
59. (a) 10.9 nC (b) 5.44×10^{-3} N
61. $\sim 10^{-7}$ C
63. (a) 1.00×10^3 N/C (b) 3.37×10^{-8} s (c) accelerate at 1.76×10^{14} m/s^2 in the direction opposite that of the electric field

Chapter 16

QUICK QUIZZES

1. (b)
2. (b), (d)
3. (d)
4. (c)
5. (a)
6. (c)
7. (a) C decreases. (b) Q stays the same. (c) E stays the same. (d) ΔV increases. (e) The energy stored increases.
8. (a) C increases. (b) Q increases. (c) E stays the same. (d) ΔV remains the same. (e) The energy stored increases.
9. (a)

CONCEPTUAL QUESTIONS

1. (a) The proton moves in a straight line with constant acceleration in the direction of the electric field. (b) As its velocity increases, its kinetic energy increases and the electric potential energy associated with the proton decreases.

3. The work done in pulling the capacitor plates farther apart is transferred into additional electric energy stored in the capacitor. The charge is constant and the capacitance decreases, but the potential difference between the plates increases, which results in an increase in the stored electric energy.

5. If the power line makes electrical contact with the metal of the car, it will raise the potential of the car to 20 kV. It will also raise the potential of your body to 20 kV, because you are in contact with the car. In itself, this is not a problem. If you step out of the car, however, your body at 20 kV will make contact with the ground, which is at zero volts. As a result, a current will pass through your body and you will likely be injured. Thus, it is best to stay in the car until help arrives.

7. If two points on a conducting object were at different potentials, then free charges in the object would move and we would not have static conditions, in contradiction to the initial assumption. (Free positive charges would migrate from locations of higher to locations of lower potential. Free electrons would rapidly move from locations of lower to locations of higher potential.) All of the charges would continue to move until the potential became equal everywhere in the conductor.

9. The capacitor often remains charged long after the voltage source is disconnected. This residual charge can be lethal. The capacitor can be safely handled after discharging the plates by short-circuiting the device with a conductor, such as a screwdriver with an insulating handle.

11. Field lines represent the direction of the electric force on a positive test charge. If electric field lines were to cross, then, at the point of crossing, there would be an ambiguity regarding the direction of the force on the test charge, because there would be two possible forces there. Thus, electric field lines cannot cross. It is possible for equipotential surfaces to cross. (However, equipotential surfaces at different potentials cannot intersect.) For example, suppose two identical positive charges are at diagonally opposite corners of a square and two negative charges of equal magnitude are at the other two corners. Then the planes perpendicular to the sides of the square at their midpoints are equipotential surfaces. These two planes cross each other at the line perpendicular to the square at its center.

13. You should use a dielectric-filled capacitor whose dielectric constant is very large. Further, you should make the dielectric as thin as possible, keeping in mind that dielectric breakdown must also be considered.

15. (a) ii (b) i

17. It would make no difference at all. An electron volt is the kinetic energy gained by an electron in being accelerated through a potential difference of 1 V. A proton accelerated through 1 V would have the same kinetic energy, because it carries the same charge as the electron (except for the sign). The proton would be moving in the opposite direction and more slowly after accelerating through 1 V, due to its opposite charge and its larger mass, but it would still gain 1 electron volt, or 1 proton volt, of kinetic energy.

PROBLEMS

1. (a) 6.40×10^{-19} J (b) -6.40×10^{-19} J (c) -4.00 V
3. 1.4×10^{-20} J
5. 1.7×10^{6} N/C
7. (a) 1.13×10^{5} N/C (b) 1.80×10^{-14} N (c) 4.38×10^{-17} J
9. (a) 0.500 m (b) 0.250 m
11. (a) 1.44×10^{-7} V (b) -7.19×10^{-8} V
13. (a) 2.67×10^{6} V (b) 2.13×10^{6} V
15. (a) 103 V (b) -3.85×10^{-7} J; positive work must be done to separate the charges.
17. -11.0 kV
19. 2.74×10^{-14} m
21. 0.719 m, 1.44 m, 2.88 m. No. The equipotentials are not uniformly spaced. Instead, the radius of an equipotenial is inversely proportional to the potential.
23. (a) 1.1×10^{-8} F (b) 27 C
25. (a) 11.1 kV/m toward the negative plate (b) 3.74 pF (c) 74.7 pC and -74.7 pC
27. (a) 90.4 V (b) 9.04×10^{4} V/m
29. (a) 13.3 μC on each (b) 20.0 μC, 40.0 μC
31. (a) 2.00 μF (b) $Q_3 = 24.0\ \mu$C, $Q_4 = 16.0\ \mu$C, $Q_2 = 8.00\ \mu$C, $(\Delta V)_2 = (\Delta V)_4 = 4.00$ V, $(\Delta V)_3 = 8.00$ V
33. (a) 5.96 μF (b) $Q_{20} = 89.5\ \mu$C, $Q_6 = 63.2\ \mu$C, $Q_3 = Q_{15} = 26.3\ \mu$C
35. $Q_1 = 16.0\ \mu$C, $Q_5 = 80.0\ \mu$C, $Q_8 = 64.0\ \mu$C, $Q_4 = 32.0\ \mu$C
37. (a) $Q_{25} = 1.25$ mC, $Q_{40} = 2.00$ mC (b) $Q'_{25} = 288\ \mu$C, $Q'_{40} = 462\ \mu$C, $\Delta V = 11.5$ V
39. $Q'_1 = 3.33\ \mu$C, $Q'_2 = 6.67\ \mu$C
41. 83.6 μC
43. 2.55×10^{-11} J
45. 3.2×10^{10} J
47. $\kappa = 4.0$
49. (a) 8.13 nF (b) 2.40 kV
51. (a) volume 9.09×10^{-16} m^3, area 4.54×10^{-10} m^2 (b) 2.01×10^{-13} F (c) 2.01×10^{-14} C, 1.26×10^{5} electronic charges
55. 4.29 μF
57. 6.25 μF
59. 4.47 kV
61. 0.75 mC on C_1, 0.25 mC on C_2
65. 50 N

Chapter 17

QUICK QUIZZES

1. (d)
2. (b)
3. (c), (d)
4. (b)
5. (b)
6. (a)
7. (b)
8. (a)

CONCEPTUAL QUESTIONS

1. Charge. Because an ampere is a unit of current (1 A = 1 C/s) and an hour is a unit of time (1 h = 3 600 s), then 1 A · h = 3 600 C.
3. The gravitational force pulling the electron to the bottom of a piece of metal is much smaller than the electrical repulsion pushing the electrons apart. Thus, free electrons stay distributed throughout the metal. The concept of charges residing on the surface of a metal is true for a metal with an excess charge. The number of free electrons in an electrically neutral piece of metal is the same as the number of positive ions—the metal has zero net charge.
5. A voltage is not something that "surges through" a completed circuit. A voltage is a potential difference that is applied across a device or a circuit. It would be more correct to say "1 ampere of electricity surged through the victim's body." Although this amount of current would have disastrous results on the human body, a value of 1 (ampere) doesn't sound as exciting for a newspaper article as 10 000 (volts). Another possibility is to write "10 000 volts of electricity were applied across the victim's body," which still doesn't sound quite as exciting.
7. We would conclude that the conductor is nonohmic.
9. The shape, dimensions, and the resistivity affect the resistance of a conductor. Because temperature and impurities affect the conductor's resistivity, these factors also affect resistance.
11. The radius of wire B is the square root of three times the radius of wire A. Therefore the cross-sectional area of B three times larger than that of A.
13. The drift velocity might increase steadily as time goes on, because collisions between electrons and atoms in the wire would be essentially nonexistent and the conduction electrons would move with constant acceleration. The current would rise steadily without bound also, because I is proportional to the drift velocity.
15. Once the switch is closed, the line voltage is applied across the bulb. As the voltage is applied across the cold filament when it is first turned on, the resistance of the filament is low, the current is high, and a relatively large amount of power is delivered to the bulb. As the filament warms, its resistance rises and the current decreases. As a result, the power delivered to the bulb decreases. The large current spike at the beginning of the bulb's operation is the reason that lightbulbs often fail just after they are turned on.

PROBLEMS

1. 3.00×10^{20} electrons move past in the direction opposite to the current.
3. 2.00 C
5. 1.05 mA
7. 27 yr
9. (a) n is unaffected (b) v_d is doubled
11. 32 V is 200 times larger than 0.16 V
13. 0.17 mm
15. (a) 30 Ω (b) 4.7×10^{-4} Ω · m
17. silver ($\rho = 1.59 \times 10^{-8}$ Ω · m)
19. 256 Ω
21. 1.98 A
23. 26 mA
25. (a) 5.89×10^{-2} Ω (b) 5.45×10^{-2} Ω
27. (a) 3.0 A (b) 2.9 A
29. (a) 1.2 Ω (b) 8.0×10^{-4} (a 0.080% increase)
31. 5.00 A, 24.0 Ω
33. 18 bulbs
35. 11.2 min
37. 34.4 Ω
39. 1.6 cm
41. 295 metric tons/h

43. 26 cents
45. 23 cents
47. $1.2
49. 1.1 km
51. $1.47 \times 10^{-6}\ \Omega \cdot$ m; differs by 2.0% from value in Table 17.1
53. (a) $3.06 (b) No. The circuit must be able to handle at least 26 A.
55. (a) 667 A (b) 50.0 km
57. $3.77 \times 10^{28}/\text{m}^3$
59. (a) 144 Ω (b) 26 m (c) To fit the required length into a small space. (d) 25 m
61. 37 MΩ
63. 0.48 kg/s
65. (a) $2.6 \times 10^{-5}\ \Omega$ (b) 76 kg
67. (a) 470 W (b) 1.60 mm or more (c) 2.93 mm or more

Chapter 18

QUICK QUIZZES

1. (a), (d)
2. (b)
3. (a)
4. *Parallel*: (a) unchanged (b) unchanged (c) increase (d) decrease
5. *Series*: (a) decrease (b) decrease (c) decrease (d) increase
6. (c)

CONCEPTUAL QUESTIONS

1. No. When a battery serves as a source and supplies current to a circuit, the conventional current flows through the battery from the negative terminal to the positive one. However, when a source having a larger emf than the battery is used to charge the battery, the conventional current is forced to flow through the battery from the positive terminal to the negative one.

3. The total amount of energy delivered by the battery will be less than *W*. Recall that a battery can be considered an ideal, resistanceless battery in series with the internal resistance. When the battery is being charged, the energy delivered to it includes the energy necessary to charge the ideal battery, plus the energy that goes into raising the temperature of the battery due to I^2r heating in the internal resistance. This latter energy is not available during discharge of the battery, when part of the reduced available energy again transforms into internal energy in the internal resistance, further reducing the available energy below *W*.

5. The starter in the automobile draws a relatively large current from the battery. This large current causes a significant voltage drop across the internal resistance of the battery. As a result, the terminal voltage of the battery is reduced, and the headlights dim accordingly.

7. An electrical appliance has a given resistance. Thus, when it is attached to a power source with a known potential difference, a definite current will be drawn, and the device can therefore be labeled with both the voltage and the current. Batteries, however, can be applied to a number of devices. Each device will have a different resistance, so the current will vary with the device. As a result, only the voltage of the battery can be specified.

9. Connecting batteries in parallel does not increase the emf. A high-current device connected to two batteries in parallel can draw currents from both batteries. Thus, connecting the batteries in parallel increases the possible current output and, therefore, the possible power output.

11. The lightbulb will glow for a very short while as the capacitor is being charged. Once the capacitor is almost totally charged, the current in the circuit will be nearly zero and the bulb will not glow.

13. The bird is resting on a wire of fixed potential. In order to be electrocuted, a large potential difference is required between the bird's feet. The potential difference between the bird's feet is too small to harm the bird.

15. The junction rule is a statement of conservation of charge. It says that the amount of charge that enters a junction in some time interval must equal the charge that leaves the junction in that time interval. The loop rule is a statement of conservation of energy. It says that the increases and decreases in potential around a closed loop in a circuit must add to zero.

17. A few of the factors involved are as follows: the conductivity of the string (is it wet or dry?); how well you are insulated from ground (are you wearing thick rubber- or leather-soled shoes?); the magnitude of the potential difference between you and the kite; and the type and condition of the soil under your feet.

19. She will not be electrocuted if she holds onto only one high-voltage wire, because she is not completing a circuit. There is no potential difference across her body as long as she clings to only one wire. However, she should release the wire immediately once it breaks, because she will become part of a closed circuit when she reaches the ground or comes into contact with another object.

21. (a) The intensity of each lamp increases because lamp C is short circuited and there is current (which increases) only in lamps A and B. (b) The intensity of lamp C goes to zero because the current in this branch goes to zero. (c) The current in the circuit increases because the total resistance decreases from $3R$ (with the switch open) to $2R$ (after the switch is closed). (d) The voltage drop across lamps A and B increases, while the voltage drop across lamp C becomes zero. (e) The power dissipated increases from $\mathcal{E}^2/3R$ (with the switch open) to $\mathcal{E}^2/2R$ (after the switch is closed).

23. The statement is false. The current in each bulb is the same, because they are connected in series. The bulb that glows brightest has the larger resistance and hence dissipates more power

PROBLEMS

1. 4.92 Ω
3. 73.8 W. Your circuit diagram will consist of two 0.800-Ω resistors in series with the 192-Ω resistance of the bulb.
5. (a) 17.1 Ω (b) 1.99 A for 4.00 Ω and 9.00 Ω, 1.17 A for 7.00 Ω, 0.818 A for 10.0 Ω
7. $2.5R$
9. (a) 0.227 A (b) 5.68 V
11. 55 Ω
13. 0.43 A
15. (a) Connect two 50-Ω resistors in parallel, and then connect this combination in series with a 20-Ω resistor.
(b) Connect two 50-Ω resistors in parallel, connect two 20-Ω resistors in parallel, and then connect these two combinations in series with each other.

17. 0.846 A downwards in the 8.00-Ω resistor; 0.462 A downwards in the middle branch; 1.31 A upwards in the right-hand branch
19. (a) 3.00 mA (b) − 19.0 V (c) 4.50 V
21. 10.7 V
23. (a) 0.385 mA, 3.08 mA, 2.69 mA
(b) 69.2 V, with c at the higher potential
25. $I_1 = 3.5$ A, $I_2 = 2.5$ A, $I_3 = 1.0$ A
27. $I_{30} = 0.353$ A, $I_5 = 0.118$ A, $I_{20} = 0.471$ A
29. $\Delta V_2 = 3.05$ V, $\Delta V_3 = 4.57$ V, $\Delta V_4 = 7.38$ V, $\Delta V_5 = 1.62$ V
31. (a) 12 s (b) 1.2×10^{-4} C
33. 1.3×10^{-4} C
35. 0.982 s
37. (a) heater, 10.8 A; toaster, 8.33 A; grill, 12.5 A
(b) $I_{\text{total}} = 31.6$ A, so a 30-A breaker is insufficient.
39. (a) 6.25 A (b) 750 W
41. (a) 1.2×10^{-9} C, 7.3×10^{9} K^+ ions. Not large, only $1e/290\ \text{A}^2$
(b) 1.7×10^{-9} C, 1.0×10^{10} Na^+ ions (c) 0.83 μA
(d) 7.5×10^{-12} J
43. 11 nW
45. 7.5 Ω
47. (a) 15 Ω
(b) $I_1 = 1.0$ A, $I_2 = I_3 = 0.50$ A, $I_4 = 0.30$ A, and $I_5 = 0.20$ A
(c) $(\Delta V)_{ac} = 6.0$ V, $(\Delta V)_{ce} = 1.2$ V, $(\Delta V)_{ed} = (\Delta V)_{fd} = 1.8$ V, $(\Delta V)_{cd} = 3.0$ V, $(\Delta V)_{db} = 6.0$ V
(d) $\mathcal{P}_{ac} = 6.0$ W, $\mathcal{P}_{ce} = 0.60$ W, $\mathcal{P}_{ed} = 0.54$ W, $\mathcal{P}_{fd} = 0.36$ W, $\mathcal{P}_{cd} = 1.5$ W, $\mathcal{P}_{db} = 6.0$ W
49. (a) 12.4 V (b) 9.65 V
51. $I_1 = 0$, $I_2 = I_3 = 0.50$ A,
53. 112 V, 0.200 Ω
55. (a) $R_x = R_2 - \frac{1}{4}R_1$
(b) $R_x = 2.8\ \Omega$ (inadequate grounding)
59. $\mathcal{P} = \dfrac{(144\ \text{V}^2)R}{(R + 10.0\ \Omega)^2}$

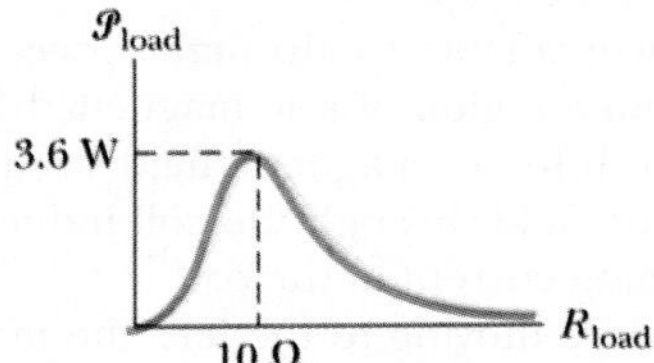

61. (a) 5.68 V (b) 0.227 A
63. 0.395 A; 1.50 V

Chapter 19

QUICK QUIZZES

1. (b)
2. (c)
3. (c)
4. (a)
5. (b)

CONCEPTUAL QUESTIONS

1. The set should be oriented such that the beam is moving either toward the east or toward the west.
3. The proton moves in a circular path upwards on the page. After completing half a circle, it exits the field and moves in a straight-line path back in the direction from whence it came. An electron will behave similarly, but the direction of traversal of the circle is downward, and the radius of the circular path is smaller.
5. The magnetic force on a moving charged particle is always perpendicular to the particle's direction of motion. There is no magnetic force on the charge when it moves parallel to the direction of the magnetic field. However, the force on a charged particle moving in an electric field is never zero and is always parallel to the direction of the field. Therefore, by projecting the charged particle in different directions, it is possible to determine the nature of the field.
7. The magnetic field produces a magnetic force on the electrons moving toward the screen that produce the image. This magnetic force deflects the electrons to regions on the screen other than the ones to which they are supposed to go. The result is a distorted image.
9. Such levitation could never occur. At the North Pole, where Earth's magnetic field is directed downward, toward the equivalent of a buried south pole, a coffin would be repelled if its south magnetic pole were directed downward. However, equilibrium would be only transitory, as any slight disturbance would upset the balance between the magnetic force and the gravitational force.
11. If you were moving along with the electrons, you would measure a zero current for the electrons, so they would not produce a magnetic field according to your observations. However, the fixed positive charges in the metal would now be moving backwards relative to you, creating a current equivalent to the forward motion of the electrons when you were stationary. Thus, you would measure the same magnetic field as when you were stationary, but it would be due to the positive charges presumed to be moving from your point of view.
13. A compass does not detect currents in wires near light switches, for two reasons. The first is that, because the cable to the light switch contains two wires, one carrying current to the switch and the other carrying it away from the switch, the net magnetic field would be very small and would fall off rapidly with increasing distance. The second reason is that the current is alternating at 60 Hz. As a result, the magnetic field is oscillating at 60 Hz also. This frequency would be too fast for the compass to follow, so the effect on the compass reading would average to zero.
15. The levitating wire is stable with respect to vertical motion: If it is displaced upward, the repulsive force weakens, and the wire drops back down. By contrast, if it drops lower, the repulsive force increases, and it moves back up. The wire is not stable, however, with respect to lateral movement: If it moves away from the vertical position directly over the lower wire, the repulsive force will have a sideways component that will push the wire away.

In the case of the attracting wires, the hanging wire is not stable with respect to vertical movement. If it rises, the attractive force increases, and the wire moves even closer to the upper wire. If the hanging wire falls, the attractive force weakens, and the wire falls farther. If the wire moves to the right, it moves farther from the upper wire and the attractive force decreases. Although there is a restoring force component pulling it back to the left, the vertical force component is not strong enough to hold the wire up, and it falls.
17. Each coil of the Slinky® will become a magnet, because a coil acts as a current loop. The sense of rotation of the

current is the same in all coils, so each coil becomes a magnet with the same orientation of poles. Thus, all of the coils attract, and the Slinky® will compress.

19. There is no net force on the wires, but there is a torque. To understand this distinction, imagine a fixed vertical wire and a free horizontal wire (see the figure below). The vertical wire carries an upward current and creates a magnetic field that circles the vertical wire, itself. To the right, the magnetic field of the vertical wire points into the page, while on the left side it points out of the page, as indicated. Each segment of the horizontal wire (of length ℓ) carries current that interacts with the magnetic field according to the equation $F = BI\ell \sin \theta$. Apply the right-hand rule on the right side: point the fingers of your right hand in the direction of the horizontal current and curl them into the page in the direction of the magnetic field. Your thumb points downward, the direction of the force on the right side of the wire. Repeating the process on the left side gives a force upward on the left side of the wire. The two forces are equal in magnitude and opposite in direction, so the net force is zero, but they create a net torque around the point where the wires cross.

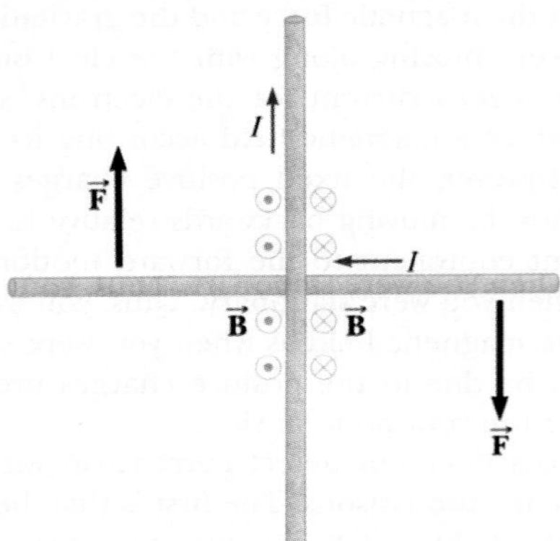

21. (a) The field is into the page. (b) The beam would deflect upwards.

PROBLEMS

1. (a) horizontal and due east (b) horizontal and 30° N of E (c) horizontal and due east (d) zero force
3. (a) into the page (b) toward the right (c) toward the bottom of the page
5. $F_g = 8.93 \times 10^{-30}$ N (downward), $F_e = 1.60 \times 10^{-17}$ N (upward), $F_m = 4.80 \times 10^{-17}$ N (downward)
7. 2.83×10^7 m/s west
9. 0.021 T in the $-y$-direction
11. 8.0×10^{-3} T in the $+z$-direction
13. (a) into the page (b) toward the right (c) toward the bottom of the page
15. 7.50 N
17. 0.131 T (downward)
19. 0.20 T directed out of the page
21. *ab*: 0, *bc*: 0.040 0 N in $-x$-direction, *cd*: 0.040 0 N in the $-z$-direction *da*: 0.056 6 N parallel to the *xz*-plane and at 45° to both the $+x$- and the $+z$-directions
23. 9.05×10^{-4} N·m, tending to make the left-hand side of the loop move toward you and the right-hand side move away.
25. (a) 3.97° (b) 3.39×10^{-3} N·m
27. 6.56×10^{-2} T
31. 1.77 cm
33. $r = 3R/4$
35. 20.0 μT
37. 2.4 mm
39. 20.0 μT toward bottom of page
41. 0.167 μT out of the page
43. (a) 4.00 m (b) 7.50 nT (c) 1.26 m (d) zero
45. 4.5 mm
47. 31.8 mA
49. 2.26×10^{-4} N away from the center, zero torque
51. 1.7 N·m
53. (a) 0.500 μT out of the page (b) 3.89 μT parallel to *xy*-plane and at 59.0° clockwise from $+x$-direction
55. 2.13 cm
57. (a) 1.33 m/s (b) the sign of the emf is independent of the charge
59. 1.41×10^{-6} N
61. 13.0 μT toward the bottom of the page
63. 53 μT toward the bottom of the page, 20 μT toward the bottom of the page, and 0
65. (a) -8.00×10^{-21} kg·m/s (b) 8.90°
67. 1.29 kW
69. (a) 12.0 cm to the left of wire 1 (b) 2.40 A, downward

Chapter 20

QUICK QUIZZES

1. *b, c, a*
2. (a)
3. (b)
4. (c)
5. (b)

CONCEPTUAL QUESTIONS

1. According to Faraday's law, an emf is induced in a wire loop if the magnetic flux through the loop changes with time. In this situation, an emf can be induced either by rotating the loop around an arbitrary axis or by changing the shape of the loop.

3. As the spacecraft moves through space, it is apparently moving from a region of one magnetic field strength to a region of a different magnetic field strength. The changing magnetic field through the coil induces an emf and a corresponding current in the coil.

5. If the bar were moving to the left, the magnetic force on the negative charges in the bar would be upward, causing an accumulation of negative charge on the top and positive charges at the bottom. Hence, the electric field in the bar would be upward, as well.

7. If, for any reason, the magnetic field should change rapidly, a large emf could be induced in the bracelet. If the bracelet were not a continuous band, this emf would cause high-voltage arcs to occur at any gap in the band. If the bracelet were a continuous band, the induced emf would produce a large induced current and result in resistance heating of the bracelet.

11. As the aluminum plate moves into the field, eddy currents are induced in the metal by the changing magnetic field at the plate. The magnetic field of the electromagnet interacts with this current, producing a retarding force on the plate that slows it down. In a similar fashion, as the plate leaves the magnetic field, a current is induced, and once again there is an upward force to slow the plate.

13. The energy stored in an inductor carrying a current I is equal to $PE_L = (1/2)LI^2$. Therefore, doubling the current will quadruple the energy stored in the inductor.
15. If an external battery is acting to increase the current in the inductor, an emf is induced in a direction to oppose the increase of current. Likewise, if we attempt to reduce the current in the inductor, the emf that is set up tends to support the current. Thus, the induced emf always acts to oppose the change occurring in the circuit, or it acts in the "back" direction to the change.
17. (a) clockwise (b) The net force exerted on the bar must be zero because it moves at constant speed. The component of the gravitational force down the incline is balanced by a component of the magnetic force up the incline.
19. from left to right

PROBLEMS

1. $5.9 \times 10^{-2}\ \text{T}\cdot\text{m}^2$
3. $7.71 \times 10^{-1}\ \text{T}\cdot\text{m}^2$
5. (a) $\Phi_{B,\text{net}} = 0$ (b) 0
7. (a) $3.1 \times 10^{-3}\ \text{T}\cdot\text{m}^2$ (b) $\Phi_{B,\text{net}} = 0$
9. 0.18 T
11. 94 mV
13. 2.7 T/s
15. (a) $4.0 \times 10^{-6}\ \text{T}\cdot\text{m}^2$ (b) 18 μV
17. 10.2 μV
19. 0.763 V
21. (a) toward the east (b) 4.58×10^{-4} V
23. (a) from left to right (b) from right to left
25. (a) $F = N^2B^2w^2v/R$ to the left (b) 0 (c) $F = N^2B^2w^2v/R$ to the left
27. into the page
29. (a) from right to left (b) from right to left (c) from left to right (d) from left to right
31. 1.9×10^{-11} V
33. (a) 18.1 μV (b) 0
35. (a) 60 V (b) 57 V (c) 0.13 s
37. 20 mV
39. (a) 2.0 mH (b) 38 A/s
43. 12 mH
45. 1.92 Ω
47. 0.140 J
49. (a) 18 J (b) 7.2 J
51. negative ($V_a < V_b$)
53. (a) 20.0 ms (b) 37.9 V (c) 1.52 mV (d) 51.8 mA
55. 1.20 μC
57. (a) 0.500 A (b) 2.00 W (c) 2.00 W
59. 115 kV
61. (a) 0.157 mV (end B is positive) (b) 5.89 mV (end A is positive)
63. (a) 9.00 A (b) 10.8 N (c) b is at the higher potential (d) No
65. $v_t = \dfrac{mgR}{B^2\ell^2}$

Chapter 21

QUICK QUIZZES

1. (c)
2. (b)
3. (b)
4. (b), (c)
5. (b), (d)

CONCEPTUAL QUESTIONS

1. For best reception, the length of the antenna should be parallel to the orientation of the oscillating electric field. Because of atmospheric variations and reflections of the wave before it arrives at your location, the orientation of this field may be in different directions for different stations.
3. The primary coil of the transformer is an inductor. When an AC voltage is applied, the back emf due to the inductance will limit the current in the coil. If DC voltage is applied, there is no back emf, and the current can rise to a higher value. It is possible that this increased current will deliver so much energy to the resistance in the coil that its temperature rises to the point at which insulation on the wire can burn.
5. An antenna that is a conducting line responds to the electric field of the electromagnetic wave—the oscillating electric field causes an electric force on electrons in the wire along its length. The movement of electrons along the wire is detected as a current by the radio and is amplified. Thus, a line antenna must have the same orientation as the broadcast antenna. A loop antenna responds to the magnetic field in the radio wave. The varying magnetic field induces a varying current in the loop (by Faraday's law), and this signal is amplified. The loop should be in the vertical plane containing the line of sight to the broadcast antenna, so the magnetic field lines go through the area of the loop.
7. The flashing of the light according to Morse code is a drastic amplitude modulation—the amplitude is changing from a maximum to zero. In this sense, it is similar to the on-and-off binary code used in computers and compact disks. The carrier frequency is that of the light, on the order of 10^{14} HZ. The frequency of the signal depends on the skill of the signal operator, but it is on the order of a single hertz, as the light is flashed on and off. The broadcasting antenna for this modulated signal is the filament of the lightbulb in the signal source. The receiving antenna is the eye.
9. The sail should be as reflective as possible, so that the maximum momentum is transferred to the sail from the reflection of sunlight.
11. Suppose the extraterrestrial looks around your kitchen. Lightbulbs and the toaster glow brightly in the infrared. Somewhat fainter are the back of the refrigerator and the back of the television set, while the television screen is dark. The pipes under the sink show the same weak glow as the walls, until you turn on the faucets. Then the pipe on the right gets darker and that on the left develops a gleam that quickly runs up along its length. The food on the plates shines, as does human skin, the same color for all races. Clothing is dark as a rule, but your seat and the chair seat glow alike after you stand up. Your face appears lit from within, like a jack-o'-lantern; your nostrils and the openings of your ear canals are bright; brighter still are the pupils of your eyes.
13. Radio waves move at the speed of light. They can travel around the curved surface of the Earth, bouncing between the ground and the ionosphere, which has an altitude that is small compared with the radius of the Earth. The distance across the lower 48 states is approximately 5 000 km, requiring a travel time that is equal to $(5 \times 10^6\ \text{m})/(3 \times 10^8\ \text{m/s}) \sim 10^{-2}$ s. Likewise, radio waves take only 0.07 s to travel halfway around the Earth. In

other words, a speech can be heard on the other side of the world (in the form of radio waves) before it is heard at the back of the room (in the form of sound waves).

15. No. The wire will emit electromagnetic waves only if the current varies in time. The radiation is the result of accelerating charges, which can occur only when the current is not constant.

17. The resonance frequency is determined by the inductance and the capacitance in the circuit. If both L and C are doubled, the resonance frequency is reduced by a factor of two.

19. It is far more economical to transmit power at a high voltage than at a low voltage because the I^2R loss on the transmission line is significantly lower at high voltage. Transmitting power at high voltage permits the use of step-down transformers to make "low" voltages and high currents available to the end user.

21. No. A voltage is induced in the secondary coil only if the flux through the core changes with time.

PROBLEMS

1. (a) 141 V (b) 20.0 A (c) 28.3 A (d) 2.00 kW
3. 70.7 V, 2.95 A
5. 6.76 W
9. 4.0×10^2 Hz
11. 17 μF
15. 3.14 A
17. $0.450 \text{ T} \cdot \text{m}^2$
19. (a) 0.361 A (b) 18.1 V (c) 23.9 V (d) $-53.0°$
21. (a) 1.4 kΩ (b) 0.10 A (c) 51° (d) voltage leads current
23. (a) 89.6 V (b) 108 V
25. 1.88 V
27. (a) 103 V (b) 150 V (c) 127 V (d) 23.6 V
29. (a) 208 Ω (b) 40.0 Ω (c) 0.541 H
31. (a) $1.8 \times 10^2\ \Omega$ (b) 0.71 H
33. 2.29 μH
35. $C_{min} = 4.9$ nF, $C_{max} = 51$ nF
37. 0.242 J
39. 0.18% is lost
41. (a) 1.1×10^3 kW (b) 3.1×10^2 A (c) 8.3×10^3 A
43. 1 000 km; there will always be better use for tax money.
45. $f_{red} = 4.55 \times 10^{14}$ Hz, $f_{IR} = 3.19 \times 10^{14}$ Hz, $E_{max,f}/E_{max,i} = 0.57$
47. 2.94×10^8 m/s
49. $E_{max} = 1.01 \times 10^3$ V/m, $B_{max} = 3.35 \times 10^{-6}$ T
51. (a) 188 m to 556 m (b) 2.78 m to 3.4 m
53. 5.2×10^{13} Hz, 5.8 μm
55. $4.299\,999\,84 \times 10^{14}$ Hz; -1.6×10^7 Hz (the frequency decreases)
57. 99.6 mH
59. 1.7 cents
61. (a) resistor and inductor (b) $R = 10\ \Omega$, $L = 30$ mH
63. (a) 6.7×10^{-16} T (b) 5.3×10^{-17} W/m^2 (c) 1.7×10^{-14} W
65. (a) 0.536 N (b) 8.93×10^{-5} m/s^2 (c) 33.9 days
67. 4.47×10^{-9} J

Chapter 22

QUICK QUIZZES

1. (a)
2. Beams 2 and 4 are reflected; beams 3 and 5 are refracted.
3. (b)
4. (c)

CONCEPTUAL QUESTIONS

1. Sound radiated upward at an acute angle with the horizontal is bent back toward Earth by refraction. This means that the sound can reach the listener by this path as well as by a direct path. Thus, the sound is louder.

3. The color will not change, for two reasons. First, despite the popular statement that color depends on wavelength, it actually depends on the *frequency* of the light, which does not change under water. Second, when the light enters the eye, it travels through the fluid within. Thus, even if color did depend on wavelength, the important wavelength is that of the light in the ocular fluid, which does not depend on the medium through which the light traveled to reach the eye.

5. (a) Away from the normal (b) increases (c) remains the same

7. No, the information in the catalog is incorrect. The index of refraction is given by $n = c/v$, where c is the speed of light in a vacuum and v is the speed of light in the material. Because light travels faster in a vacuum than in any other material, it is impossible for the index of refraction of any material to have a value less than 1.

9. There is no dependence of the angle of reflection on wavelength, because the light does not enter deeply into the material during reflection—it reflects from the surface.

11. On the one hand, a ball covered with mirrors sparkles by reflecting light from its surface. On the other hand, a faceted diamond lets in light at the top, reflects it by total internal reflection in the bottom half, and sends the light out through the top again. Because of its high index of refraction, the critical angle for diamond in air for total internal reflection, namely $\theta_c = \sin^{-1}(n_{air}/n_{diamond})$, is small. Thus, light rays enter through a large area and exit through a very small area with a much higher intensity. When a diamond is immersed in carbon disulfide, the critical angle is increased to $\theta_c = \sin^{-1}(n_{carbon\ disulfide}/n_{diamond})$. As a result, the light is emitted from the diamond over a larger area and appears less intense.

13. The index of refraction of water is 1.333, quite different from that of air, which has an index of refraction of about 1. The boundary between the air and water is therefore easy to detect, because of the differing diffraction effects above and below the boundary. (Try looking at a glass half full of water.) The index of refraction of liquid helium, however, happens to be much closer to that of air. Consequently, the defractive differences above and below the helium-air boundary are harder to see.

15. The diamond acts like a prism, dispersing the light into its spectral components. Different colors are observed as a consequence of the manner in which the index of refraction varies with the wavelength.

17. Light travels through a vacuum at a speed of 3×10^8 m/s. Thus, an image we see from a distant star or galaxy must have been generated some time ago. For example, the star Altair is 16 lightyears away; if we look at an image of Altair today, we know only what Altair looked like 16 years ago. This may not initially seem significant; however, astronomers who look at other galaxies can get an idea of what galaxies looked like when they were much younger. Thus, it does make sense to speak of "looking backward in time."

PROBLEMS

1. 3.00×10^8 m/s
3. 114 rad/s for a maximum intensity of returning light
5. (b) 3.000×10^8 m/s
7. 19.5° above the horizontal
9. (a) 1.52 (b) 417 nm (c) 4.74×10^{14} Hz (d) 1.98×10^8 m/s
11. (a) 584 nm (b) 1.12
13. 111°
15. (a) 1.559×10^8 m/s (b) 329.1 nm (c) 4.738×10^{14} Hz
17. five times from the right-hand mirror and six times from the left
19. 0.388 cm
21. $\theta = 30.4°$, $\theta' = 22.3°$
23. 6.39 ns
25. $\theta = \tan^{-1}(n_g)$
27. 3.39 m
29. $\theta_{red} = 48.22°$, $\theta_{blue} = 47.79°$
31. (a) $\theta_{1i} = 30°$, $\theta_{1r} = 19°$, $\theta_{2i} = 41°$, $\theta_{2r} = 77°$ (b) First surface: $\theta_{reflection} = 30°$; second surface: $\theta_{reflection} = 41°$
33. (a) 31.3° (b) 44.2° (c) 49.8°
35. (a) 33.4° (b) 53.4°
37. (a) 40.8° (b) 60.6°
39. 1.000 08
41. (a) 10.7° (b) air (c) Sound falling on the wall from most directions is 100% reflected.
43. 27.5°
45. 22.0°
47. (a) 53.1° (b) $\geq 38.7°$
49. (a) 38.5° (b) ≥ 1.44
53. 24.7°
55. 1.93
59. $\theta = \sin^{-1}\left(\sqrt{n^2 - 1}\,\sin\phi - \cos\phi\right)$
61. (a) 1.20 (b) 3.40 ns

Chapter 23

QUICK QUIZZES

1. At C.
2. (c)
3. (a) False (b) False (c) True
4. (b)
5. An infinite number
6. (a) False (b) True (c) False

CONCEPTUAL QUESTIONS

1. You will not be able to focus your eyes on both the picture and your image at the same time. To focus on the picture, you must adjust your eyes so that an object several centimeters away (the picture) is in focus. Thus, you are focusing on the mirror surface. But, your image in the mirror is as far behind the mirror as you are in front of it. Thus, you must focus your eyes beyond the mirror, twice as far away as the picture to bring the image into focus.
3. A single flat mirror forms a virtual image of an object due to two factors. First, the light rays from the object are necessarily diverging from the object, and second, the lack of curvature of the flat mirror cannot convert diverging rays to converging rays. If another optical element is first used to cause light rays to converge, then the flat mirror can be placed in the region in which the converging rays are present, and it will change the direction of the rays so that the real image is formed at a different location. For example, if a real image is formed by a convex lens, and the flat mirror is placed between the lens and the image position, the image formed by the mirror will be real.
5. The ultrasonic range finder sends out a sound wave and measures the time for the echo to return. Using this information, the camera calculates the distance to the subject and sets the camera lens. When the camera is facing a mirror, the ultrasonic signal reflects from the mirror surface and the camera adjusts its focus so that the mirror surface is at the correct focusing distance from the camera. But your image in the mirror is twice this distance from the camera, so it is blurry.
7. Light rays diverge from the position of a virtual image just as they do from an actual object. Thus, a virtual image can be as easily photographed as any object can. Of course, the camera would have to be placed near the axis of the lens or mirror in order to intercept the light rays.
9. We consider the two trees to be two separate objects. The far tree is an object that is farther from the lens than the near tree. Thus, the image of the far tree will be closer to the lens than the image of the near tree. The screen must be moved closer to the lens to put the far tree in focus.
11. If a converging lens is placed in a liquid having an index of refraction larger than that of the lens material, the direction of refractions at the lens surfaces will be reversed, and the lens will diverge light. A mirror depends only on reflection which is independent of the surrounding material, so a converging mirror will be converging in any liquid.
13. This is a possible scenario. When light crosses a boundary between air and ice, it will refract in the same manner as it does when crossing a boundary of the same shape between air and glass. Thus, a converging lens may be made from ice as well as glass. However, ice is such a strong absorber of infrared radiation that it is unlikely you will be able to start a fire with a small ice lens.
15. The focal length for a mirror is determined by the law of reflection from the mirror surface. The law of reflection is independent of the material of which the mirror is made and of the surrounding medium. Thus, the focal length depends only on the radius of curvature and not on the material. The focal length of a lens depends on the indices of refraction of the lens material and surrounding medium. Thus, the focal length of a lens depends on the lens material.
17. (a) all signs are positive (b) f and p are positive, q is negative
19. (c) the image becomes fuzzy and disappears

PROBLEMS

1. on the order of 10^{-9} s younger
3. 10.0 ft, 30.0 ft, 40.0 ft
5. 0.268 m behind the mirror; virtual, upright, and diminished; $M = 0.026\,8$
7. (a) 13.3 cm in front of mirror, real, inverted, $M = -0.333$ (b) 20.0 cm in front of mirror, real, inverted, $M = -1.00$ (c) No image is formed. Parallel rays leave the mirror.
9. Behind the worshipper, 3.33 m from the deepest point in the niche.
11. 5.00 cm

13. 1.0 m
15. 8.05 cm
17. − 20.0 cm
19. (a) concave with focal length $f = 0.83$ m
(b) Object must be 1.0 m in front of the mirror.
21. 38.2 cm below the upper surface of the ice
23. 3.8 mm
25. $n = 2.00$
27. 20.0 cm
29. (a) 40.0 cm beyond the lens, real, inverted, $M = -1.00$
(b) No image is formed. Parallel rays leave the lens.
(c) 20.0 cm in front of the lens, virtual, upright, $M = +2.00$
31. (a) 13.3 cm in front of the lens, virtual, upright, $M = +1/3$
(b) 10.0 cm in front of the lens, virtual, upright, $M = +1/2$
(c) 6.67 cm in front of the lens, virtual, upright, $M = +2/3$
33. (a) either 9.63 cm or 3.27 cm (b) 2.10 cm
35. (a) 39.0 mm (b) 39.5 mm
37. at distance $2|f|$ in front of lens
39. 40.0 cm
41. 30.0 cm to the left of the second lens, $M = -3.00$
43. 7.47 cm in front of the second lens; 1.07 cm; virtual, upright
45. from 0.224 m to 18.2 m
47. real image, 5.71 cm in front of the mirror
49. 38.6°
51. 160 cm to the left of the lens, inverted, $M = -0.800$
53. $q = 10.7$ cm
55. 32.0 cm to the right of the second surface (real image)
57. (a) 20.0 cm to the right of the second lens; $M = -6.00$
(b) inverted
(c) 6.67 cm to the right of the second lens; $M = -2.00$; inverted
59. (a) 1.99
(b) 10.0 cm to the left of the lens
(c) inverted
61. (a) 5.45 m to the left of the lens
(b) 8.24 m to the left of the lens
(c) 17.1 m to the left of the lens
(d) by surrounding the lens with a medium having a refractive index greater than that of the lens material.
63. (a) 263 cm (b) 79.0 cm

Chapter 24

QUICK QUIZZES

1. (c)
2. (b)
3. (b)
4. The compact disc

CONCEPTUAL QUESTIONS

1. You will *not* see an interference pattern from the automobile headlights, for two reasons. The first is that the headlights are not coherent sources and are therefore incapable of producing sustained interference. Also, the headlights are so far apart in comparison to the wavelengths emitted that, even if they were made into coherent sources, the interference maxima and minima would be too closely spaced to be observable.
3. The result of the double slit is to redistribute the energy arriving at the screen. Although there is no energy at the location of a dark fringe, there is four times as much energy at the location of a bright fringe as there would be with only a single narrow slit. The total amount of energy arriving at the screen is twice as much as with a single slit, as it must be according to the law of conservation of energy.
5. One of the materials has a higher index of refraction than water, and the other has a lower index. The material with the higher index will appear black as it approaches zero thickness. There will be a 180° phase change for the light reflected from the upper surface, but no such phase change for the light reflected from the lower surface, because the index of refraction for water on the other side is lower than that of the film. Thus, the two reflections will be out of phase and will interfere destructively. The material with index of refraction lower than water will have a phase change for the light reflected from both the upper and the lower surface, so that the reflections from the zero-thickness film will be back in phase and the film will appear bright.
7. For incidence normal to the film, the extra path length followed by the reflected ray is twice the thickness of the film. For destructive interference, this must be a distance of half a wavelength of the light in the material of the film. For a film in air, no 180° phase change will occur in these reflections, so the thickness of the film must be one-quarter wavelength, which is the same as the condition for constructive interference of reflected light. This means that the transmitted light is a minimum when the reflected light is a maximum, and vice versa.
9. Since the light reflecting at the lower surface of the film undergoes a 180° phase change, while light reflecting from the upper surface of the film does not undergo such a change, the central spot (where the film has near zero thickness) will be dark. If the observed rings are not circular, the curved surface of the lens does not have a true spherical shape.
11. For regional communication at the Earth's surface, radio waves are typically broadcast from currents oscillating in tall vertical towers. These waves have vertical planes of polarization. Light originates from the vibrations of atoms or electronic transitions within atoms, which represent oscillations in all possible directions. Thus, light generally is not polarized.
13. Yes. In order to do this, first measure the radar reflectivity of the metal of your airplane. Then choose a light, durable material that has approximately half the radar reflectivity of the metal in your plane. Measure its index of refraction, and place onto the metal a coating equal in thickness to one-quarter of 3 cm, divided by that index. Sell the plane quick, and then you can sell the supposed enemy new radars operating at 1.5 cm, which the coated metal will reflect with extra-high efficiency.
15. If you wish to perform an interference experiment, you need monochromatic coherent light. To obtain it, you must first pass light from an ordinary source through a prism or diffraction grating to disperse different colors into different directions. Using a single narrow slit, select a single color and make that light diffract to cover both slits for a Young's experiment. The procedure is much simpler with a laser because its output is already monochromatic and coherent.
17. Strictly speaking, the ribs do act as a diffraction grating, but the separation distance of the ribs is so much larger than the wavelength of the x-rays that there are no observable effects.
19. As the edge of the Moon cuts across the light from the star, edge diffraction effects occur. Thus, as the edge of

the Moon moves relative to the star, the observed light from the star proceeds through a series of maxima and minima.

21. Larger. From Brewster's law, $n = \tan \theta_p$, we see that the angle increases as n increases.

PROBLEMS

1. 1.58 cm
3. (a) 2.6 mm (b) 2.62 mm
5. (a) 36.2° (b) 5.08 cm (c) 5.08×10^{14} Hz
7. (a) 55.7 m (b) 124 m
9. 75.0 m
11. 11.3 m
13. 148 m
15. 91.9 nm
17. 550 nm
19. 0.500 cm
21. (a) 238 nm (b) λ will increase (c) 328 nm
23. 4.35 μm
25. 4.75 μm
27. No, the wavelengths intensified are 276 nm, 138 nm, 92.0 nm, . . .
29. 4.22 mm
31. (a) 1.1 m (b) 1.7 mm
33. 1.20 mm, 1.20 mm
35. (a) 479 nm, 647 nm, 698 nm (b) 20.5°, 28.3°, 30.7°
37. 5.91° in first order; 13.2° in second order; and 26.5° in third order
39. 44.5 cm
41. 9.13 cm
43. (a) 25.6° (b) 19.0°
45. (a) 1.11 (b) 42.0°
47. (a) 56.7° (b) 48.8°
49. 31.2°
53. 6.89 units
55. (a) 413.7 nm, 409.7 nm (b) 8.6°
57. 0.156 mm
59. 2.50 mm
61. Any positive integral multiple of 210 nm
63. (a) 16.6 m (b) 8.28 m
65. 127 m
67. 0.350 mm
69. 115 nm

Chapter 25

QUICK QUIZZES

1. (c)
2. (a)

CONCEPTUAL QUESTIONS

1. The observer is *not* using the lens as a simple magnifier. For a lens to be used as a simple magnifier, the object distance must be less than the focal length of the lens. Also, a simple magnifier produces a virtual image at the normal near point of the eye, or at an image distance of about $q = -25$ cm. With a large object distance and a relatively short image distance, the magnitude of the magnification by the lens would be considerably less than one. Most likely, the lens in this example is part of a lens combination being used as a telescope.
3. The image formed on the retina by the lens and cornea is already inverted.
5. There will be an effect on the interference pattern—it will be distorted. The high temperature of the flame will change the index of refraction of air for the arm of the interferometer in which the match is held. As the index of refraction varies randomly, the wavelength of the light in that region will also vary randomly. As a result, the effective difference in length between the two arms will fluctuate, resulting in a wildly varying interference pattern.
7. Large lenses are difficult to manufacture and machine with accuracy. Also, their large weight leads to sagging, which produces a distorted image. In reflecting telescopes, light does not pass through glass; hence, problems associated with chromatic aberrations are eliminated. Large-diameter reflecting telescopes are also technically easier to construct. Some designs use a rotating pool of mercury as the reflecting surface.
9. In order for someone to see an object through a microscope, the wavelength of the light in the microscope must be smaller than the size of the object. An atom is much smaller than the wavelength of light in the visible spectrum, so an atom can never be seen with the use of visible light.
11. farsighted; converging

PROBLEMS

1. 30.0 cm beyond the lens, $M = -1/5$
3. 177 m
5. $f/1.4$
7. $f/8.0$
9. 40.0 cm
11. 23.2 cm
13. (a) -2.00 diopters (b) 17.6 cm
15. $+17.0$ diopters
17. (a) 5.8 cm (b) $m = 4.3$
19. (a) 4.07 cm (b) $m = +7.14$
21. (a) $|M| = 1.22$ (b) $\theta/\theta_0 = 6.08$
23. 2.1 cm
25. $m = -115$
27. $f_o = 90$ cm, $f_e = 2.0$ cm
29. (b) $-fh/p$ (c) -1.07 mm
31. (a) $m = 1.50$ (b) $m = 1.90$
33. 492 km
35. 0.40 μrad
37. 9.1×10^7 km
39. 9.8 km
41. No. A resolving power of 2.0×10^5 is needed, and that available is only 1.8×10^5.
43. 50.4 μm
45. 40
47. 98 fringe shifts
49. (a) $+2.67$ diopters (b) 0.16 diopter too low
51. (a) $+44.6$ diopters (b) 3.03 diopters
53. (a) 1.0×10^3 lines (b) 3.3×10^2 lines
55. $m = 10.7$
57. (a) $m = 4.0$ (b) $m = 3.0$

Chapter 26

QUICK QUIZZES

1. (a)
2. No. From your perspective you're at rest with respect to the cabin, so you will measure yourself as having your normal length, and will require a normal-sized cabin.

3. (a), (e); (a), (e)
4. (a) False (b) False (c) True (d) False
5. (a)

CONCEPTUAL QUESTIONS

1. An ellipsoid. The dimension in the direction of motion would be measured to be less than *D*.
3. This scenario is not possible with light. Light waves are described by the principles of special relativity. As you detect the light wave ahead of you and moving away from you (which would be a pretty good trick—think about it!), its velocity relative to you is *c*. Thus, you will not be able to catch up to the light wave.
5. No. The principle of relativity implies that nothing can travel faster than the speed of light in a *vacuum*, which is 3.00×10^8 m/s.
7. The light from the quasar moves at 3.00×10^8 m/s. The speed of light is independent of the motion of the source or the observer.
9. For a wonderful fictional exploration of this question, get a "Mr. Tompkins" book by George Gamow. All of the relativity effects would be obvious in our lives. Time dilation and length contraction would both occur. Driving home in a hurry, you would push on the gas pedal not to increase your speed very much, but to make the blocks shorter. Big Doppler shifts in wave frequencies would make red lights look green as you approached and make car horns and radios useless. High-speed transportation would be both very expensive, requiring huge fuel purchases, as well as dangerous, since a speeding car could knock down a building. When you got home, hungry for lunch, you would find that you had missed dinner; there would be a five-day delay in transit when you watch a live TV program originating in Australia. Finally, we would not be able to see the Milky Way, since the fireball of the Big Bang would surround us at the distance of Rigel or Deneb.
11. A photon transports energy. The relativistic equivalence of mass and energy means that is enough to give it momentum.
13. Your assignment: measure the length of a rod as it slides past you. Mark the position of its front end on the floor and have an assistant mark the position of the back end. Then measure the distance between the two marks. This distance will represent the length of the rod only if the two marks were made simultaneously in your frame of reference.

PROBLEMS

1. (a) $t_{OB} = 1.67 \times 10^3$ s, $t_{OA} = 2.04 \times 10^3$ s (b) $t_{BO} = 2.50 \times 10^3$ s, $t_{AO} = 2.04 \times 10^3$ s (c) $\Delta t = 90$ s
3. 5.0 s
5. (a) 20 m (b) 19 m (c) 0.31*c*
7. (a) 1.3×10^{-7} s (b) 38 m (c) 7.6 m
9. (a) 2.2 μs (b) 0.65 km
11. 0.950*c*
13. Yes, with 19 m to spare
15. (a) 39.2 μs (b) Accurate to one digit
17. 3.3×10^5 m/s
19. 0.285*c*
21. 0.54*c* to the right
23. 0.357*c*
25. 0.998*c* toward the right
27. (a) 54 min (b) 52 min
29. $c(\sqrt{3}/2)$
31. 0.786*c*
33. 18.4 g/cm^3
35. 1.98 MeV
37. 2.27×10^{23} Hz, 1.32 fm for each photon
39. (a) 3.10×10^5 m/s (b) 0.758*c*
41. 1.42 MeV/*c*
43. (a) 0.80*c* (b) 7.5×10^3 s (c) 1.4×10^{12} m, 0.38*c*
45. 0.37*c* in the + *x*-direction
47. (a) $v/c = 1 - 1.12 \times 10^{-10}$ (b) 6.00×10^{27} J (c) \$$2.17 \times 10^{20}$
49. 0.80*c*
51. (a) 0.946*c* (b) 0.160 ly (c) 0.114 yr (d) 7.50×10^{22} J
53. (a) 7.0 μs (b) 1.1×10^4 muons
59. 5.45 yr; Goslo is older.

Chapter 27

QUICK QUIZZES

1. (b)
2. (c)
3. (c)
4. (b)

CONCEPTUAL QUESTIONS

1. The shape of an object is normally determined by observing the light reflecting from its surface. In a kiln, the object will be very hot and will be glowing red. The emitted radiation is far stronger than the reflected radiation, and the thermal radiation emitted is only slightly dependent on the material from which the object is made. Unlike reflected light, the emitted light comes from all surfaces with equal intensity, so contrast is lost and the shape of the object is harder to discern.
3. The "blackness" of a blackbody refers to its ideal property of absorbing all radiation incident on it. If an observed room temperature object in everyday life absorbs all radiation, we describe it as (visibly) black. The black appearance, however, is due to the fact that our eyes are sensitive only to visible light. If we could detect infrared light with our eyes, we would see the object emitting radiation. If the temperature of the blackbody is raised, Wien's law tells us that the emitted radiation will move into the visible range of the spectrum. Thus, the blackbody could appear as red, white, or blue, depending on its temperature.
5. All objects do radiate energy, but at room temperature this energy is primarily in the infrared region of the electromagnetic spectrum, which our eyes cannot detect. (Pit vipers have sensory organs that *are* sensitive to infrared radiation; thus, they can seek out their warm-blooded prey in what we would consider absolute darkness.
7. Most metals have cutoff frequencies corresponding to photons in or near the visible range of the electromagnetic spectrum. AM radio wave photons have far too little energy to eject electrons from the metal.
9. We can picture higher frequency light as a stream of photons of higher energy. In a collision, one photon can give all of its energy to a single electron. The kinetic energy of such an electron is measured by the stopping potential.

The reverse voltage (stopping voltage) required to stop the current is proportional to the frequency of the incoming light. More intense light consists of more photons striking a unit area each second, but atoms are so small that one emitted electron never gets a "kick" from more than one photon. Increasing the intensity of the light will generally increase the size of the current, but will not change the energy of the individual electrons that are ejected. Thus, the stopping potential remains constant.

11. Wave theory predicts that the photoelectric effect should occur at any frequency, provided that the light intensity is high enough. However, as seen in photoelectric experiments, the light must have sufficiently high frequency for the effect to occur.
13. (a) Electrons are emitted only if the photon frequency is greater than the cutoff frequency.
15. No. Suppose that the incident light frequency at which you first observed the photoelectric effect is above the cutoff frequency of the first metal, but less than the cutoff frequency of the second metal. In that case, the photoelectric effect would not be observed at all in the second metal.
17. The frequency of the scattered photon must decrease, because some of its energy is transferred to the electron.

PROBLEMS

1. (a) $\approx$ 3 000 K (b) $\approx$ 20 000 K
3. 500 nm
5. (a) 2.49×10^{-5} eV (b) 2.49 eV (c) 249 eV
7. 2.27×10^{30} photons/s
9. (a) 2.3×10^{31} (b) $\Delta E/E = 4.3 \times 10^{-32}$
11. (a) 2.24 eV (b) 555 nm (c) 5.41×10^{14} Hz
13. 234 nm
15. 148 days, incompatible with observation
17. 4.8×10^{14} Hz, 2.0 eV
19. 1.2×10^{2} V and 1.2×10^{7} V, respectively
21. 41.4 kV
23. 0.078 nm
25. 0.281 nm
27. 1.78 eV, 9.47×10^{-28} kg·m/s
29. 70°
31. 1.18×10^{-23} kg·m/s, 478 eV
33. (a) 1.2 eV (b) 6.5×10^{5} m/s
35. (a) 1.46 km/s (b) 7.28×10^{-11} m
37. (a) $\approx 10^{2}$ MeV (b) No. With kinetic energy much larger than the magnitude of the negative potential energy, the electron would immediately escape.
39. 3.58×10^{-13} m
41. (a) 15 keV (b) 1.2×10^{2} keV
43. $\sim 10^{6}$ m/s
45. 116 m/s
47. $\approx$ 5 200 K; clearly, a firefly is not at that temperature, so this cannot be blackbody radiation.
49. 18.2°
51. 1.36 eV
53. 2.00 eV
55. (a) $0.022\,0c$ (b) $0.999\,2c$
57. (b) 3.72 km/s
59. (b) 5.19×10^{-16} m
61. (a) 0.263 kg (b) 1.81 W (c) $-0.015\,3$°C/s $= -0.919$°C/min (d) 9.89 μm (e) 2.01×10^{-20} J (f) 8.98×10^{19} photon/s

Chapter 28

QUICK QUIZZES

1. (b)
2. (a)
3. (a) 5 (b) 9 (c) 25
4. (d)

CONCEPTUAL QUESTIONS

1. If the energy of the hydrogen atom were proportional to n (or any power of n), then the energy would become infinite as n grew to infinity. But the energy of the atom is inversely proportional to n^2. Thus, as n grows to infinity, the energy of the atom approaches a value that is above the ground state by a finite amount, namely, the ionization energy 13.6 eV. As the electron falls from one bound state to another, its energy loss is always less than the ionization energy. The energy and frequency of any emitted photon are finite.
3. The characteristic x-rays originate from transitions within the atoms of the target, such as an electron from the L shell making a transition to a vacancy in the K shell. The vacancy is caused when an accelerated electron in the x-ray tube supplies energy to the K shell electron to eject it from the atom. If the energy of the bombarding electrons were to be increased, the K shell electron will be ejected from the atom with more remaining kinetic energy. But the energy difference between the K and L shell has not changed, so the emitted x-ray has exactly the same wavelength.
5. A continuous spectrum without characteristic x-rays is possible. At a low accelerating potential difference for the electron, the electron may not have enough energy to eject an electron from a target atom. As a result, there will be no characteristic x-rays. The change in speed of the electron as it enters the target will result in the continuous spectrum.
7. The hologram is an interference pattern between light scattered from the object and the reference beam. If anything moves by a distance comparable to the wavelength of the light (or more), the pattern will wash out. The effect is just like making the slits vibrate in Young's experiment, to make the interference fringes vibrate wildly so that a photograph of the screen displays only the average intensity everywhere.
9. If the Pauli exclusion principle were not valid, the elements and their chemical behavior would be grossly different, because every electron would end up in the lowest energy level of the atom. All matter would therefore be nearly alike in its chemistry and composition, since the shell structures of each element would be identical. Most materials would have a much higher density, and the spectra of atoms and molecules would be very simple, resulting in the existence of less color in the world.
11. The three elements have similar electronic configurations, with filled inner shells plus a single electron in an s orbital. Since atoms typically interact through their unfilled outer shells, and since the outer shells of these atoms are similar, the chemical interactions of the three atoms are also similar.
13. Each of the eight electrons must have at least one quantum number different from each of the others. They can differ (in m_s) by being spin-up or spin-down. They can differ (in ℓ) in angular momentum and in the general shape of the wave function. Those electrons with $\ell = 1$ can differ (in m_ℓ) in orientation of angular momentum.

15. Stimulated emission is the reason laser light is coherent and tends to travel in a well-defined parallel beam. When a photon passing by an excited atom stimulates that atom to emit a photon, the emitted photon is in phase with the original photon and travels in the same direction. As this process is repeated many times, an intense, parallel beam of coherent light is produced. Without stimulated emission, the excited atoms would return to the ground state by emitting photons at random times and in random directions. The resulting light would not have the useful properties of laser light.
17. The atom is a bound system. The atomic electron does not have enough kinetic energy to escape from its electrical attraction to the nucleus. The electrical potential energy of the atom is negative and is greater than the kinetic energy, so the total energy of the atom is negative.
19. (a) The wavelength of photon A is greater. (b) The energy of photon B is greater.

PROBLEMS

1. 656 nm, 486 nm, and 434 nm
3. (a) 2.3×10^{-8} N (b) -14 eV
5. (a) 1.6×10^6 m/s (b) No, $v/c = 5.3 \times 10^{-3} << 1$ (c) 0.46 nm (d) Yes. The wavelength is roughly the same size as the atom.
7. (a) 0.212 nm (b) 9.95×10^{-25} kg · m/s (c) 2.11×10^{-34} J · s (d) 3.40 eV (e) -6.80 eV (f) -3.40 eV
11. $E = -1.51$ eV ($n = 3$) to $E = -3.40$ eV ($n = 2$)
13. (a) 0.967 eV (b) 0.266 eV
15. (a) 122 nm, 91.1 nm (b) 1.87×10^3 nm, 820 nm
17. 97.2 nm
19. (a) 488 nm (b) 0.814 m/s
21. (d) $n = 2.53 \times 10^{74}$ (e) No. At such large quantum numbers, the allowed energies are essentially continuous.
23. (a) 2.47×10^{14} Hz, $f_{orb} = 8.23 \times 10^{14}$ Hz (b) 6.59×10^3 Hz, $f_{orb} = 6.59 \times 10^3$ Hz. For large n, classical theory and quantum theory approach each other in their results.
25. 4.42×10^4 m/s
27. (a) -122 eV (b) 1.76×10^{-11} m
29. (a) 0.026 5 nm (b) 0.017 6 nm (c) 0.013 2 nm
31. 1.33 nm
33. $n = 3, \ell = 1, m_\ell = +1, m_s = \pm 1/2$; $n = 3, \ell = 1, m_\ell = 0, m_s = \pm 1/2$; $n = 3, \ell = 1, m_\ell = -1, m_s = \pm 1/2$
35. Fifteen possible states, as summarized in the following table:

n	3	3	3	3	3	3	3	3	3	3	3	3	3	3	3
ℓ	2	2	2	2	2	2	2	2	2	2	2	2	2	2	2
m_ℓ	+2	+2	+2	+1	+1	+1	0	0	0	−1	−1	−1	−2	−2	−2
m_s	+1	0	−1	+1	0	−1	+1	0	−1	+1	0	−1	+1	0	−1

37. (a) 30 possible states (b) 36
39. (a) $n = 4$ and $\ell = 2$ (b) $m_\ell = (0, \pm 1, \pm 2)$, $m_s = \pm 1/2$ (c) $1s^2 2s^2 2p^6 3s^2 3p^6 3d^{10} 4s^2 4p^6 4d^2 5s^2 = [\text{Kr}]\ 4d^2 5s^2$
41. 0.160 nm
43. L shell: 11.7 keV; M shell: 10.0 keV; N shell: 2.30 keV
45. (a) 10.2 eV (b) 7.88×10^4 K
47. (a) -8.18 eV, -2.04 eV, -0.904 eV, -0.510 eV, -0.325 eV (b) 1.09×10^3 nm and 609 nm
49. The four lowest energies are -10.39 eV, -5.502 eV, -3.687 eV, and -2.567 eV (b) The wavelengths of the emission lines are 158.5 nm, 185.0 nm, 253.7 nm, 422.5 nm, 683.2 nm, and 1 107 nm (c) 1.31×10^6 m/s
51. (a) 4.24×10^{15} W/m^2 (b) 1.20×10^{-12} J
55. (a) $E_n = (-1.49 \times 10^4 \text{ eV})/n^2$ (b) $n = 4 \rightarrow n = 1$
57. (a) 9.03×10^{22} m/s^2 (b) -4.63×10^{-8} W (c) $\sim 10^{-11}$ s

Chapter 29

QUICK QUIZZES

1. (c)
2. (c)
3. (a)
4. (a) and (b)
5. (b)

CONCEPTUAL QUESTIONS

1. Isotopes of a given element correspond to nuclei with different numbers of neutrons. This will result in a variety of different physical properties for the nuclei, including the obvious one of mass. The chemical behavior, however, is governed by the element's electrons. All isotopes of a given element have the same number of electrons and, therefore, the same chemical behavior.
3. An alpha particle contains two protons and two neutrons. Because a hydrogen nucleus contains only one proton, it cannot emit an alpha particle.
5. In alpha decay, there are only two final particles: the alpha particle and the daughter nucleus. There are also two conservation principles: of energy and of momentum. As a result, the alpha particle must be ejected with a discrete energy to satisfy both conservation principles. However, beta decay is a three-particle decay: the beta particle, the neutrino (or antineutron), and the daughter nucleus. As a result, the energy and momentum can be shared in a variety of ways among the three particles while still satisfying the two conservation principles. This allows a continuous range of energies for the beta particle.
7. The larger rest energy of the neutron means that a free proton in space will not spontaneously decay into a neutron and a positron. When the proton is in the nucleus, however, the important question is that of the total rest energy of the nucleus. If it is energetically favorable for the nucleus to have one less proton and one more neutron, then the decay process will occur to achieve this lower energy.
9. Carbon dating cannot generally be used to estimate the age of a stone, because the stone was not alive to take up carbon from the environment. Only the ages of artifacts that were once alive can be estimated with carbon dating.
11. The protons, although held together by the nuclear force, are repelled by the electrostatic force. If enough protons were placed together in a nucleus, the electrostatic force would overcome the nuclear force, which is based on the number of particles, and cause the nucleus to fission.

    The addition of neutrons prevents such fission. The neutron does not increase the electrical force, being electrically neutral, but does contribute to the nuclear force.
13. The photon and the neutrino are similar in that both particles have zero charge and very little mass. (The photon

has zero mass, but recent evidence suggests that certain kinds of neutrinos have a very small mass.) Both must travel at the speed of light and are capable of transferring both energy and momentum. They differ in that the photon has spin (intrinsic angular momentum) $\hbar$ and is involved in electromagnetic interactions, while the neutrino has spin $\hbar/2$, and is closely related to beta decays.

15. Since the two samples are of the same radioactive nuclide, they have the same half-life; the 2:1 difference in activity is due to a 2:1 difference in the mass of each sample. After 5 half lives, each will have decreased in mass by a power of $2^5 = 32$. However, since this simply means that the mass of each is 32 times smaller, the ratio of the masses will still be (2/32) : (1/32), or 2:1. Therefore, the ratio of their activities will *always* be 2:1.

PROBLEMS

1. $A = 2$, $r = 1.5$ fm; $A = 60$, $r = 4.7$ fm; $A = 197$, $r = 7.0$ fm; $A = 239$, $r = 7.4$ fm
3. 1.8×10^2 m
5. (a) 27.6 N (b) 4.16×10^{27} m/s^2 (c) 1.73 MeV
7. (a) 1.9×10^7 m/s (b) 7.1 MeV
9. 8.66 MeV/nucleon for $^{93}_{41}$Nb, 7.92 MeV/nucleon for $^{197}_{79}$Au
11. 3.54 MeV
13. 0.210 MeV/nucleon greater for $^{23}_{11}$Na, attributable to less proton repulsion
15. 0.46 Ci
17. (a) 9.98×10^{-7} s^{-1} (b) 1.9×10^{10} nuclei
19. 1.0 h
21. 4.31×10^3 yr
23. (a) 5.58×10^{-2} h^{-1}, 12.4 h (b) 2.39×10^{13} nuclei (c) 1.9 mCi
25. $^{208}_{81}$Tl, $^{95}_{37}$Rb, $^{144}_{60}$Nd
27. $^{40}_{20}$Ca, $^{94}_{42}$Mo, $^{4}_{2}$He
29. e$^+$ decay, $^{56}_{27}\text{Co} \rightarrow {}^{56}_{26}\text{Fe} + e^+ + \nu$
31. (a) cannot occur spontaneously (b) can occur spontaneously
33. 18.6 keV
35. 4.22×10^3 yr
37. (a) $^{30}_{15}$P (b) -2.64 MeV
39. (a) $^{21}_{10}$Ne (b) $^{144}_{54}$Xe (c) X = e$^+$, X$'$ = ν
41. (a) $^{13}_{6}$C (b) $^{10}_{5}$B
43. (a) $^{197}_{79}\text{Au} + n \rightarrow {}^{198}_{80}\text{Hg} + e^- + \bar{\nu}$ (b) 7.88 MeV
45. (a) $^{1}_{0}$n (b) Fluoride mass = 18.000 953 u
47. 18.8 J
49. 24 d
51. (a) 8.97×10^{11} electrons (b) 0.100 J (c) 100 rad
53. 46.5 d
55. $Q = 3.27$ MeV > 0, no threshold energy required
57. (a) 2.52×10^{24} (b) 2.29×10^{12} Bq (c) 1.07×10^6 yr
59. (a) 4.0×10^9 yr (b) It could be no older. The rock could be younger if some ^{87}Sr were initially present.
61. 54 μCi
63. 2.3×10^2 yr
65. 4.4×10^{-8} kg/h

Chapter 30

QUICK QUIZZES

1. (c)
2. (a)
3. (b)
4. (d)

CONCEPTUAL QUESTIONS

1. The experiment described is a nice analogy to the Rutherford scattering experiment. In the Rutherford experiment, alpha particles were scattered from atoms and the scattering was consistent with a small structure in the atom containing the positive charge.

3. The largest charge quark is $2e/3$, so a combination of only two particles, a quark and an antiquark forming a meson, could not have an electric charge of $+2e$. Only particles containing three quarks, each with a charge of $2e/3$, can combine to produce a total charge of $2e$.

5. Until about 700 000 years after the Big Bang, the temperature of the Universe was high enough for any atoms that formed to be ionized by ambient radiation. Once the average radiation energy dropped below the hydrogen ionization energy of 13.6 eV, hydrogen atoms could form and remain as neutral atoms for relatively long period of time.

7. In the quark model, all hadrons are composed of smaller units called quarks. Quarks have a fractional electric charge and a baryon number of $\frac{1}{3}$. There are six flavors of quarks: up (u), down (d), strange (s), charmed (c), top (t), and bottom (b). All baryons contain three quarks, and all mesons contain one quark and one antiquark. Section 30.12 has a more detailed discussion of the quark model.

9. Baryons and mesons are hadrons, interacting primarily through the strong force. They are not elementary particles, being composed of either three quarks (baryons) or a quark and an antiquark (mesons). Baryons have a nonzero baryon number with a spin of either $\frac{1}{2}$ or $\frac{3}{2}$. Mesons have a baryon number of zero and a spin of either 0 or 1.

11. All stable particles other than protons and neutrons have baryon number zero. Since the baryon number must be conserved, and the final states of the kaon decay contain no protons or neutrons, the baryon number of all kaons must be *zero*.

13. Yes, but the strong interaction predominates.

15. Unless the particles have enough kinetic energy to produce a baryon–antibaryon pair, the answer is *no*. Antibaryons have a baryon number of -1, baryons have a baryon number of $+1$, and mesons have a baryon number of 0. If such an interaction were to occur and produce a baryon, the baryon number would not be conserved.

17. Baryons and antibaryons contain three quarks, while mesons and antimesons contain two quarks. Quarks have a spin of 1/2; thus, three quarks in a baryon can only combine to form a net spin that is half-integral. Likewise, two quarks in a meson can only combine to form a net spin of 0 or 1.

19. For the first decay, the half-life is characteristic of the strong interaction, so the ρ^0 must have $S = 0$, and strangeness is conserved. The second decay must occur via the weak interaction.

PROBLEMS

1. 1.1×10^{16} fissions
3. 126 MeV
5. (a) 16.2 kg (b) 117 g
7. 2.9×10^3 km ($\approx$1 800 miles)
9. 1.01 g

11. (a) ${}^{8}_{4}\mathrm{Be}$ (b) ${}^{12}_{6}\mathrm{C}$ (c) 7.27 MeV
13. 3.07×10^{22} events/yr
15. (a) 3.44×10^{30} J (b) 1.56×10^{8} yr
17. (a) 4.53×10^{23} Hz (b) 0.622 fm
19. $\sim 10^{-23}$ s
21. $\sim 10^{-18}$ m
23. (a) conservation of electron-lepton number and conservation of muon-lepton number (b) conservation of charge (c) conservation of baryon number (d) conservation of baryon number (e) conservation of charge
25. $\overline{\nu}_\mu$
27. (a) $\overline{\nu}_\mu$ (b) ν_μ (c) $\overline{\nu}_e$ (d) ν_e (e) ν_μ (f) ν_μ and $\overline{\nu}_e$
29. (a) not allowed; violates conservation of baryon number (b) strong interaction (c) weak interaction (d) weak interaction (e) electromagnetic interaction
31. (a) not conserved (b) conserved (c) conserved (d) not conserved (e) not conserved (f) not conserved
33. (a) charge, baryon number, L_e, L_τ (b) charge, baryon number, L_e, L_μ, L_τ (c) charge, L_e, L_μ, L_τ, strangeness number (d) charge, baryon number, L_e, L_μ, L_τ, strangeness number (e) charge, baryon number, L_e, L_μ, L_τ, strangeness number (f) charge, baryon number, L_e, L_μ, L_τ, strangeness number
35. 3.34×10^{26} electrons, 9.36×10^{26} up quarks, 8.70×10^{26} down quarks
37. (a) Σ^+ (b) π^- (c) K^0 (d) Ξ^-
39.

Reaction	At Quark Level	Net Quarks (before and after)
$\pi^- + \mathrm{p} \rightarrow \mathrm{K}^0 + \Lambda^0$	$\overline{\mathrm{u}}\mathrm{d} + \mathrm{uud} \rightarrow \mathrm{d}\overline{\mathrm{s}} + \mathrm{uds}$	1 up, 2 down, 0 strange
$\pi^+ + \mathrm{p} \rightarrow \mathrm{K}^+ + \Sigma^+$	$\mathrm{u}\overline{\mathrm{d}} + \mathrm{uud} \rightarrow \mathrm{u}\overline{\mathrm{s}} + \mathrm{uus}$	3 up, 0 down, 0 strange
$\mathrm{K}^- + \mathrm{p} \rightarrow \mathrm{K}^+ + \mathrm{K}^0 + \Omega^-$	$\overline{\mathrm{u}}\mathrm{s} + \mathrm{uud} \rightarrow \mathrm{u}\overline{\mathrm{s}} + \mathrm{d}\overline{\mathrm{s}} + \mathrm{sss}$	1 up, 1 down, 1 strange

(d) The mystery particle is a Λ^0 or a Σ^0.
41. a neutron, udd
43. 70.45 MeV
45. 18.8 MeV
47. (a) electron-lepton and muon-lepton numbers not conserved (b) electron-lepton number not conserved (c) charge not conserved (d) baryon and electron-lepton numbers not conserved (e) strangeness violated by 2 units
49. (a) 2×10^{24} nuclei (b) ≈ 0.6 kg
51. (a) 1 baryon before and zero baryons after decay. Baryon number is not conserved. (b) 469 MeV, 469 MeV/c (c) $0.999\,999\,4c$
53. (b) 12 days
55. 26 collisions

Credits

Photographs

This page constitutes an extension of the copyright page. We have made every effort to trace the ownership of all copyrighted material and to secure permission from copyright holders. In the event of any question arising as to the use of any material, we will be pleased to make the necessary corrections in future printings. Thanks are due to the following authors, publishers, and agents for permission to use the material indicated.

Chapter 1. **1:** Stone/Getty Images **2:** bottom left, Courtesy of National Institute of Standards and Technology, U.S. Dept. of Commerce; bottom right, Courtesy of Institute of Standards and Technology, U.S. Dept. of Commerce **10:** Billy E. Barnes/Stock Boston **12:** R. Williams (STScl), the HDF-S team, and NASA **19:** top left, Courtesy of George Semple; top right, Courtesy of George Semple **20:** © Sylvain Grandadam/Photo Researchers, Inc. **21:** NASA

Chapter 2. **23:** Caron/Corbis Sygma **39:** Courtesy Amtrak NEC Media Relations **41:** North Wind Archive

Chapter 3. **53:** © Bettmann/Corbis **54:** top, Mack Henley/Visuals Unlimited; bottom, George Semple **62:** HIRB/Index Stock **65:** Mike Powell/Allsport/Getty Images **75:** top right, Joseph Kayne/Dembinsky Photo Associates; bottom left, AP/Wide World Photos

Chapter 4. **81:** top left, © Royalty-Free/Corbis; bottom right, © Lorenzo Ciniglio/Corbis **83:** Roger Viollet, Mill Valley, CA, University Science Books, 1982 **85:** Giraudon/Art Resource **88:** NASA **90:** Jim Gillmoure/corbisstockmarket.com **107:** Guy Sauvage/Photo Researchers, Inc.

Chapter 5. **118:** NASA **121:** top, © Chris Collins/Corbis; bottom, © McCrone Photo/Custom Medical Stock Photo **132:** Wet'n Wild Orlando **141:** top right, Jan Hinsch/Science Photo Library/Photo Researchers, Inc.; bottom right, Gareth Williams, Minor Planet Center **149:** Arthur Tilley/FPG/Getty Images **157:** © Jamie Budge/Corbis **158:** Engraving from *Scientific American,* July 1888.

Chapter 6. **160:** © Reuters/Corbis **162:** © Harold and Esther Edgerton Foundation, 2002, courtesy of Palm Press, Inc. **163:** © Harold and Esther Edgerton Foundation, 2002, Courtesy of Palm Press, Inc. **164:** top right, Tim Wright/Corbis **167:** bottom left, Courtesy of NASA; bottom right, Mike Severns/Stone/Getty Images **171:** Courtesy of Central Scientific Company **181:** Courtesy of Saab **188:** bottom left, Courtesy of CENCO

Chapter 7. **189:** Courtesy NASA **208:** top right, Courtesy of PASCO Scientific

Chapter 8. **226:** © Kevin Fleming/Corbis **230:** David Serway **239:** George Semple **248:** top left, © Benson Krista Hicks/Corbis Sygma; top right, © Benson Krista Hicks/Corbis Sygma; middle left, © Patrick Giardino/Corbis; bottom left, Max Planck Institute for Astronomy and Calar Alto Observatory, K. Meisenheimer and A. Quetz; bottom right, © Smithsonian Institute/Photo Researchers, Inc. **253:** top left, © Eye Ubiquitous/Corbis; top right, Gerard Lacz/NHPA **264:** © Ed Bock/Corbis

Chapter 9. **266:** © Alison Wright/Corbis **267:** Charles D. Winters **271:** NASA **275:** © Royalty-Free/Corbis **276:** Raymond A. Serway **277:** Courtesy of Scientific Company **279:** bottom right, David Frazier **284:** top left, © Royalty-Free/Corbis **285:** top right, © Royalty-Free/Corbis **288:** Andy Sacks/Stone/Getty Images **289:** top right, Kim Vandiver and Harold Edgerton. © Harold and Esther Edgerton Foundation, 2002, courtesy of Palm Press, Inc.; bottom right, George Semple **292:** top left, Corbis-Bettmann; bottom right, Courtesy of Central Scientific Company **299:** Herman Eisenbeiss/Photo Researchers, Inc. **300:** Charles D. Winters **310:** bottom left, Henry Leap and Jim Lehman; bottom right, Pamela Zilly/The Image Bank/Getty Images **318:** The Granger Collection

Chapter 10. **321:** © Jim Sugar/Corbis **328:** top left, George Semple; bottom left, George Semple **329:** AP/Wide World Photos **330:** George Semple **340:** © R. Folwell/Science Photo Library/Photo Researchers, Inc.

Chapter 11. **352:** Steve Bly/Getty Images **353:** By kind permission of the President and Council of the Royal Society **366:** George Semple **371:** Gary Settles/Science Source/Photo Researchers Inc. **373:** bottom left, Daedalus Enterprises, Inc./Peter Arnold, Inc.; bottom right, Photodisc/Getty Images **374:** SPL/Photo Researchers, Inc. **377:** top left, Tom Bean; top right, Tom Bean **380:** Nathan Bilow/Leo de Wys, Inc. **381:** top left, © Jim Bourg/Teuters/Corbis

Chapter 12. **386:** © Tim De Waele/Corbis **404:** Charles D. Winters **405:** J-L. Charmet/SPL/Photo Researchers, Inc. **406:** J-L. Charmet/SPL/Photo Researchers, Inc. **409:** AIP Niels Bohr Library, Lade Collection **413:** top, George Semple; bottom, George Semple **414:** © BSIP/Laurent Science Source/Photo Researchers, Inc.

Chapter 13. **424:** © Rick Doyle/Corbis **429:** Eric Lars Baleke/Black Star **433:** Link/Visuals Unlimited **434:** Telegraph Colour Library/FPG International **448:** Martin

Dohm/SPL/Photo Researchers **451:** Gregg Adams/Stone Getty Images

Chapter 14. **458:** Patrick Ward/CORBIS **460:** Bernard Benoit/Photo Researchers, Inc. **469:** top right, Courtesy Educational Development Center, Newton, Mass. **472:** top right, © 1973 Kim Vandiver & Harold E. Edgerton/Courtesy of Palm Press, Inc. **473:** © 1994, Comstock **476:** all images, © Richard Megna, Fundamental Photographs, NYC **477:** Charles D. Winters **480:** left, top, Courtesy of Professor Thomas D. Rossing, Northern Illinois University; left, bottom, Ben Rose/The IMAGE Bank Getty Images; bottom left, United Press International Photo; bottom right, United Press International Photo **487:** Royalty Free/Corbis **496:** Sidney Harris

Chapter 15. **497:** © Keith Kent/Science Photo Library/Photo Researchers, Inc. **498:** © American Philosophical Society/AIP **501:** top, © 1968 Fundamental Photographs; bottom, Photo courtesy of AIP Niels Bohr Library, E. Scott Barr Collection **510:** Photo courtesy of Harold M. Waage, Princeton University **511:** top, Photo courtesy of Harold M. Waage, Princeton University; bottom, Photo courtesy of Harold M. Waage, Princeton University **513:** top, Photo courtesy of Harold M. Waage, Princeton University; middle right, Photo courtesy of Harold M. Waage, Princeton University

Chapter 16. **531:** Courtesy of Resonance Research Corporation **543:** bottom middle, Riei O'Harra/Black Star/PNI; bottom right, Greig Cranna/Stock, Boston/PNI **546:** Photo courtesy of Harold M. Waage, Princeton University **557:** © Loren Winters/Visuals Unlimited **558:** a. Paul Silverman/Fundamental Photographs; b. George Semple

Chapter 17. **568:** Telegraph Colour Library/FPG/Getty Images **572:** Michael Dalton, Fundamental Photographs **574:** © Bettmann/CORBIS **575:** Courtesy of Henry Leap and Jim Lehman **578:** Courtesy of Central Scientific Company **580:** Courtesy of IBM Research Laboratory **585:** Courtesy of Medtronic, Inc.

Chapter 18. **592:** © Lester Lefkowitz/Corbis **593:** George Semple **602:** AIP ESVA, W.F. Meggers Collection **610:** George Semple **613:** Juergen Berger, Max-Planck Institute/Science Photo Library/Photo Researchers, Inc. **616:** middle left, Superstock; middle right, Courtesy of Henry Leap and Jim Lehman

Chapter 19. **624:** Johnny Johnson/Stone/Getty **625:** Courtesy of Central Scientific Company **626:** a–c. Henry Leap and Jim Lehman **631:** Courtesy of Henry Leap and Jim Lehman **637:** Courtesy of American Honda Motor Co., Inc. **640:** middle left, North Wind Picture Archives; bottom, © Richard Megna, Fundamental Photographs **641:** Leonard de Selva/CORBIS **645:** © Richard Megna, Fundamental Photographs **646:** Courtesy of Henry Leap and Jim Lehman **650:** © Loren Winters/Visuals Unlimited **651:** top right, Courtesy of Central Scientific Company; middle right, Courtesy of Central Scientific Company

Chapter 20. **660:** PhotoDisc/Getty Images **661:** By kind permission of the President and Council of the Royal Society **666:** Charles D. Winters **667:** Courtesy of PedsLink Pediatric Healthcare Resources, Newport Beach, CA **674:** Luis Castaneda/The Image Bank/Getty Images **678:** North Wind Picture Archives **681:** Courtesy of Saab **688:** bottom left, Courtesy of CENCO

Chapter 21. **693:** © Bettmann/Corbis **701:** Bettmann/Corbis **704:** Ryan Williams/International Stock Photography **707:** George Semple **708:** top, North Wind Photo Archives; bottom, Bettmann/Corbis **713:** John Neal/Photo Researchers, Inc. **715:** Ron Chapple/Getty Images **717:** a. NASA/CXC/SAO; b. Palomar Observatory; c. WM Keck Observatory; d, VLA/NRAO **720:** George Semple

Chapter 22. **726:** Peter Bowater/Photo Researchers, Inc. **727:** Courtesy of Rijksmuseum voor de Geschiedenis der Natuurwetenschappen. Courtesy AIP Niels Bohr Library **728:** bottom left and right, Photographs Courtesy of Henry Leap and Jim Lehman **729:** top left and right, Photos by Charles D. Winters **731:** top right, Courtesy of Henry Leap and Jim Lehman **735:** bottom left, Courtesy of Sony Disc Manufacturing **737:** top right, David Parker/SPL/Photo Researchers, Inc.; bottom right, Courtesy of Bausch & Lomb **738:** Courtesy of PASCO Scientific **739:** Mark D. Phillips/Photo Researchers, Inc. **740:** Courtesy of Sabina Zigman/Benjamin Cardozo High School **742:** Courtesy of Henry Leap and Jim Lehman **744:** bottom left, Dennis O'Clair/Tony Stone Images/Getty Images; bottom right, Hank Morgan/Photo Researchers, Inc.

Chapter 23. **754:** Don Hammond/CORBIS **757:** Courtesy of Henry Leap and Jim Lehman **759:** Courtesy of Jim Lehman, James Madison University **761:** top right, middle right, bottom right, Photos courtesy of David Rogers **762:** © Junebug Clark 1988/Photo Researchers, Inc. **768:** John M. Dunay IV, Fundamental Photographs, NYC **770:** top left, Photos Courtesy of Henry Leap and Jim Lehman **778:** a–c. Photos by Norman Goldberg **780:** middle right, Richard Megna/Fundamental Photographs, NYC **784:** Photo © Michael Levin/Opti-Gone Associates

Chapter 24. **786:** RO-MA/Index Stock Imagery **787:** M. Cagnet, M. Francon, J. C. Thierr **788:** Richard Megna/Fundamental Photographs, NYC **789:** John S. Shelton **792:** Dr. Jeremy Burgess/Science Photo Library/Photo Researchers, Inc. **793:** top right, Peter Aprahamian/Science Photo Library/Photo Researchers, Inc.; bottom middle, Courtesy of Bausch & Lomb Optical Company; bottom right, From Physical Science Study Committee, *College Physics,* Lexington Mass., D. C. Heath and Co., 1968 **794:** Richard Megna, Fundamental Photographs **796:** Courtesy of Sony Disc Manufacturing **798:** top left, Courtesy of P. M. Rinard, from *Am. J. Phys.,* 44:70, 1976; top right, From M. Cagnet, M. Francon, and J. C. Thierr, *Atlas of Optical Phenomena,* Berlin, Springer-Verlag, 1962, plate 18 **803:** © Kristen Brochmann/Fundamental Photographs **806:** a–c. Photos courtesy of Henry Leap **809:** bottom left, Sepp Seitz 1981; bottom right, Peter Aprahamian/Science Photo Library/Photo Researchers, Inc.

Chapter 25. **819:** © Denis Scott/CORBIS **821:** top right, From Lennart Nilson, in collaboration with Jan Lindberg, *Behold Man: A Photographic Journey of Discovery Inside the Body,* Boston, Little, Brown & Co., 1974 **827:** bottom right, © Tony Freeman/Photo Edit **829:** NASA **830:** top right, © Tony Freeman/Photo Edit; bottom, Courtesy of Palomar Observatory/California Institute of Technology

To Danielle, my second daughter, who demonstrates that two successive innovations can be introduced successfully

and

In memory of my parents, Jesse and Florence. *Su meile mano žmonai Marijai ir mūsu sūnui Mariui.*

Contents

PART III DESIGN PROCESS

Preface

Our approach in writing this text on new products is primarily managerial. Each section concentrates on a step in the strategy, opportunity identification, design, testing, or implementation stages of new product development. We define the managerial problems and discuss their integration in the development process. Real world examples are given to illustrate the problems and to implement the solutions. Advanced techniques of analysis are given, but our emphasis is on how actually to use these techniques in the managerial environment rather than on their technical details. Drawing on more than 15 years of experience with more than 100 products and services, and on in-depth academic research into state-of-the-art techniques, we describe how to apply these techniques and to identify pitfalls in their use.

Throughout the text we supply references to some of the most significant research in the field. Thus, the reader can gain additional detail on specific topics, learn the antecedents of each approach, and assess the validity of the recommendations. Where appropriate, we have provided technical appendices which efficiently summarize and clarify any ambiguities in verbal descriptions. Some readers may find it easier to understand and apply the techniques if they can grasp the mathematics which formalize the analysis. Other readers will find that once they grasp the fundamentals, the formal mathematics helps them implement the techniques in the business environment, and still others want to improve the techniques and undertake research to extend the state-of-the-art. Although we recommend that you study the appendices, they are not essential for a managerial understanding of this text.

Our goal is to draw together the issues, the descriptive material, and the new techniques of product development into a comprehensive book. In this way we feel you will have a single source which: (1) illustrates the product policy issues, (2) compares the modeling techniques, (3) helps you to identify market segments, to understand preferences, and to predict sales, (4) enables you to conduct pretest and test markets, and (5) allows you to guide new products to maturity.

Outline

The text is divided into six parts organized around the chronological decision steps in new product development: Innovation Strategy, Opportunity Identification, Designing New Products, Testing and Improving Products, Product Introduction and Profit Management, and Implementing the New Product Development Process. Part I discusses how to recognize cues initiating the new product process, identifies the types of new product strategies, discusses the role of R&D, and gives an overview of the development process. Part II covers the procedures for identifying market opportunities and generating new ideas. Part III provides techniques to gauge how consumers perceive products and to measure consumers' preferences so that you can design and engineer a better product. Since even the most carefully designed product must be tested, Part IV discusses methods for testing the new product's physical characteristics and its advertising. It tells you how to run and interpret a test market, and it also presents the new techniques of pretest market simulation which provide much of the information gained from a full test market, but at a fraction of the cost. It closes with techniques to analyze the tests in order to make the decision whether or not to "go national." Part V considers how to introduce a new product and to manage its profitable transition to a mature product. Finally, Part VI discusses issues of implementing the development process by customizing it to your industry and size, and by organizing the effort. The book closes with a managerial review of product development strategy.

Prerequisites

As prerequisites to reading this book we assume that you have at least a basic understanding of the fundamental principles of management and marketing such as would be taught in introductory business courses. We also presume that you have some familiarity with algebra and market research. Building on this, we hope to provide you with an understanding of the tools you will need to manage the development and marketing of new products.

Products and Services, Consumer and Industrial, Private and Public

While much of the initial development of techniques has taken place for frequently purchased consumer products such as deodorants and food products, our experience has been that these techniques can and have been applied to consumer durables such as televisions and clothes dryers, industrial products such as solar air conditioners and office copiers, and to services such as hopsitals and transportation. Furthermore, these techniques have been used for innovation in both the private and the public sectors. Since it would be cumbersome throughout the book to refer continually to produce and service, consumer and industrial, durable and nondurable, private and public, profit and nonprofit, we use the generic terms "product" and "organization" to describe the innovation and the innovating organization. When we speak of goals, we often refer to "profit," although in many cases the "profit" may be a nonmonetary goal such as membership, usage, or simply public service. Finally, we use the masculine pronouns, "he," "him," "his," to refer to both managers and consumers, but this is for semantic simplicity only.

As you might expect, no one technique can be applied without modification to the diverse set of new product problems, and few techniques are equally applicable to all "products." Thus, whenever a technique must be modified, we indicate the necessary modifications and whenever a technique is differentially applicable, we indicate the necessary caveats.

Throughout the book you will find a number of real world examples. We have tried to vary these between products and services, consumer and industrial, and private and public. But unless otherwise indicated, the techniques are applicable to the full range of "products" and "organizations."

Acknowledgments

Many people have contributed to and influenced the content of this text. Academic contributions by our colleagues at Massachusetts Institute of Technology (John D.C. Little, Alvin Silk, Gary Lilien, Manu Kalwani, Eric Von Hippel, Tom Allen), at Northwestern University (Philip Kotler, Andris Zoltners, Ken Wisniewski, Pat Lyon) and at other universities (Patricia Simmie, Scott Neslin, Allen Shocker, Steve Shugan, Robert Olsen, Len Lodish, and the reviewers—David Luck and C. Merle Crawford) were very important in our thinking and writing. Management Decision Systems, Inc. management (Rick Karash, Bob Klein, Gerry Katz, Phil Johnson, Walt Lankau, Jim Findley, Jay Wurts, Ed Wolkenmuth, Dinyar Chavda, Jan P. Willem Bol, and Kathy Moore) and Jacques Blanchard (Novaction, Inc.) supplied development and application support critical to this work. In particular we wish to thank Steve Shugan for developing the many thought provoking end-of-chapter questions.

Practicing managers (Tom Hatch, Cal Hodock, Kurt Kilty, Dawson Farber, Chuck Allen, Bob Bartz, Ed Sellers) helped us to understand decision needs in new products and how to meet them.

The students in our new product courses (1978, 1979 at MIT and Northwestern) provided many comments and suggestions that substantially improved the text.

Finally, great typing and editorial support was provided by Dianne Carpenter Smith, Sabra Van Cleef, Shirely Converse, and Peggy M. Thorpe. Thanks!

G. L. U. and **J. R. H.**
aboard *Chickadee II*
Pocasset Harbor, MA

Acknowledgments

Figure 2.1 on page 22 is reprinted by permission of the *Harvard Business Review* where it appeared as Exhibit I from "Strategies for Diversification" by H. Igor Ansoff (September-October 1957), copyright © 1957 by the President and Fellows of Harvard College, all rights reserved; Figure 2.3 on page 27 is reprinted from "Innovation in Industry and the Diffusion of Technology," by J. M. Utterback, *Science,* Vol. 183, pp. 620–26, 15 February 1974, copyright © 1974 by the American Association for the Advancement of Science; Table 3.1 on page 47 is reprinted by permission of E. Mansfield and J. Rapoport, "The Costs of Industrial Product Innovations," *Management Science,* Vol. 21#12, August 1975, copyright © 1975 by the Institute of Management Sciences; Table 3.4 on pp. 50–51 is reprinted from "Time Lag in New Product Development" by L. Adler in *Journal of Marketing,* Vol. 30, Jan. 1966, pp. 17–21, published by the American Marketing Association; Table 3.7 on page 53 is reprinted from "Organizational and Strategic Factors Associated with Probabilities of Success in Industrial R&D," by E. Mansfield and S. Wagner in *Journal of Business,* April 1975, by permission of The University of Chicago Press; Figure 5.2 on page 82 is reprinted by permission of the *Harvard Business Review,* Vol. 52, No. 5 (September-October 1974) p. 112, "Limits of the Learning Curve" by William J. Abernathy and Kenneth Wayne, copyright © 1974 by the President and Fellows of Harvard College, all rights reserved; Table 5.4 on page 97 is reprinted from "Life Cycle Concept in Marketing Research," by Wells and Gubar, in *Journal of Marketing Research,* Vol. 3, (November 1966), p. 362, published by the American Marketing Association; Figure 5.10 on page 104 is reprinted by permission of Frank M. Bass, "A New Product Growth Model for Consumer Durables," *Management Science,* Vol. 15#5, January 1969, copyright © 1969 by the Institute of Management Sciences.

Table 6.1 on page 124 and Figure 6.1 on page 125 are reprinted from "Successful Industrial Products from Customer Ideas," by E. Von Hippel in *Journal of Marketing,* Vol. 42, No. 1 (January 1978) pp. 39–49, published by the American Marketing Association; Figures 6.3 on page 132 and 6.4 on page 133 are reprinted from *Managing the Flow of Technology* by Thomas J. Allen by permission of The MIT Press, Cambridge, Massachusetts; Figures 6.5 and 6.6 on page 134 are reprinted from "Microelectronics," *Scientific American,* Vol. 237, No. 3 (September 1977), copyright © 1977 by Scientific American, Inc., all rights reserved; Figure 6.7 on page 136 is reprinted by permission of S. Basu and R. G. Schroeder, "Incorporating Judgments in Sales Forecasts: Application of the Delphia Method at American Hoist & Derrick," *Interfaces,* Vol. 7#3, May 1977, copyright © 1977 by The Institute of Management Sciences; the dialogue on pages 143–45 is from "Shortage of Psychiatrists," *The Practice of Creativity,* pp. 145–47, by George M. Prince, copyright © 1970 by George M. Prince and reprinted by permission of Harper & Row, Publishers, Inc.; Figure 9.13 on page 215 is reprinted from "Alternative Perceptual Models; Reproducibility, Validity, and Data Integrity," by P. Simmie in the *Proceedings of the 1978 AMA Education Conference,* 1978, published by the American Marketing Association.

Figure 10.9 on page 253 is reprinted by permission of the *Harvard Business Review* where it appeared as Exhibit III from "New Way to Measure Consumers' Judgments," by Paul E. Green and Yoram Wind (July-August 1975), copyright © 1975 by the President and Fellows of Harvard College, all rights reserved; Figure 11.4 on page 284 is reprinted from D. G. Morrison, "Purchase Intentions and Purchase Behavior," *Journal of Marketing,* Vol. 43, No. 2, Spring 1979, pp. 65–74, published by the American Marketing Association; Table 12.3 on page 340 is reprinted by permission of Glen L. Urban, "A New Product Analysis and Decision Model," *Management Science,* Vol. 14#8, April 1968, copyright © 1968 by The Institute of Management Sciences; Figure 14.1 on page 389 and Table 14.2 on page 390 are reprinted from H. Claycamp and L. E. Liddy, "Prediction of New Product Performance: An Analytical Approach," *Journal of Marketing Research,* Vol. 6, No. 3 (November 1969) pp. 414–420, published by The American Marketing Association; Figure 14.3 on page 398, Table 14.3 on page 399, and Table 14.4 on page 403 are reprinted from A. Silk and G. L. Urban, "Pre-test Market Evaluation of New Packaged Goods: A Model and Measurement Methodology," *Journal of Marketing Research,* Vol. 15, No. 2 (May 1978), pp. 171–191, published by The American Marketing Association.

Figure 16.9 on page 483 is reprinted from "The Relationship Between Diffusion Rates, Experience Curves, and Demand Elasticities for Consumer Durable Technological Innovations," by Frank M. Bass in the *Journal of Business,* 1980, by permission of The University of Chicago Press; Figure 17.1 on page 498 is from David B. Montgomery and Glen L. Urban, *Management Science in Marketing,* p. 18, copyright © 1969, reprinted by permission of Prentice-Hall, Inc.; Figures 18.1 on page 533 and 18.3 on page 545 are reprinted from *Managing the Flow of Technology* by Thomas J. Allen by permission of The MIT Press, Cambridge, Massachusetts.

Chapter 1

Introduction to New Products

New products are crucial to successful growth and increased profits in many organizations. Consider the following examples:

> In 1910 the dominant mode of urban transportation was the streetcar railway. It provided fast, reliable, inexpensive transportation for our nation of cities. Its growth seemed assured. Population was expanding, our cities were becoming more interdependent, and there would always be a need for urban movement. But when Henry Ford provided a more flexible and personal transit option at a slightly higher cost, consumers purchased his new product, and the automotive industry was born.
>
> In the 1950s International Business Machines recognized the need for rapid, accurate processing of business related information and captured most of the market. But in 1957 Digital Equipment Corporation (DEC) developed a rugged, specialized, low cost computer to meet the needs of a new segment of this market. This innovation resulted in the multi-billion dollar mini-computer market. DEC now has a dominant share of the sales in this rapidly growing market and has triggered entries by Data General, Prime, IBM, and others.
>
> Around 1970 S. C. Johnson and Sons recognized that the women's hair care market was big, growing, and fit their marketing and research and development skills. Based on careful marketing research they identified oiliness as a major consumer problem and introduced Agree Creme Rinse to "stop the greasies." Launched in 1977, Agree Creme Rinse has taken a large share of the market.
>
> Until 1970 Jergens dominated the hand lotion market, but Vaseline Intensive Care Lotion was introduced by Chesebrough-Pond's and took a significant share of the market based on an appeal of healing hands.

Until recently the rule of thumb in small appliances was a $50 price limit, but Cuisinart identified a need for grating, grinding, slicing, chopping, blending, pureeing, mincing, and mixing, and introduced a premium food processor at a premium price. Their recent entry revolutionized pricing in that category and they have successfully maintained much of their premium positioning despite imitation and price competition by other firms.

These anecdotes are representative of the tremendous rewards for firms who recognize consumer needs and successfully introduce new products. But how does a firm create successful products? What strategy should be used? What factors should be considered? What managerial techniques are available to aid in creating products, forecasting sales, and maximizing profits. This book gives you a managerial perspective and an operating ability to use advanced techniques to design, test, and implement successful new products and services.

INNOVATION IS IMPORTANT—BUT RISKY

Not only is the development of new products rewarding, but in many cases it is necessary to maintain a healthy organization. Basic marketing theory states that a product undergoes a product life cycle of introduction, growth, maturity, and finally decline. In the maturity or decline phase it is imperative that an organization take an active role to: (1) expand the product line and thus extend the life cycle, (2) re-align the product to make it superior, or (3) develop a new product in another category to maintain revenue. If new products are not developed, sales and profits decline as competition increases, technology and markets change, or innovation by other firms make the original product obsolete.

While there are considerable rewards to successful innovation and considerable pressures to innovate, the introduction of new products is risky. Henry Ford led the way in developing the auto market, but Ford Motor Company later introduced the Edsel and lost over $100 million. More recently General Motors abandoned its Wankel Rotary engine although over 100 million dollars had been invested in the project. Innovation in the minicomputer market was successful, but many firms such as Bomar have failed in the market for hand-held calculators and terminated operations. Technologically, DuPont's CORFAM substitute for leather resulted in hundreds of millions of dollars in losses. General Mills lost millions of dollars in the introduction of a line of snacks called "Bugles, Daisies and Butterflies." Chesebrough-Pond's introduced a successful hand lotion, but Gillette lost millions on a facial cleaning lotion called "Happy Face."

Although good statistics are not available, failure rates are substantial. Booz, Allen and Hamilton (1971) in a study of 366 new products in 54 prominent companies found one-third of products introduced in the market were not successful (10 percent outright failures and 23 percent doubtful). In some areas, such as consumer products, the failure rate is higher, where 50 to 60 percent fail in national introduction or major test markets (A. C. Nielsen, Inc., 1971, 1979; Silk and Urban, 1978).

The resources devoted to new product development include R&D, engineering, and test marketing. These costs are incurred for products before they are introduced. Since all products are not successfully developed and tested, substantial funds are spent on products that never reach the market. Booz, Allen and Hamilton found that less than 50 percent of the products in their study that began development were carried on to the testing phase. Furthermore, less than one-half of these tests resulted in commercialization of the product. This means the successful product must not only return its unique development cost, but contribute to the costs of products which received attention but were not introduced. Booz, Allen and Hamilton report in their study that 70 percent of the resources spent on new products are allocated to products that are not successful in the market. These experiences indicate there are substantial risks in new product development. The return on investment in new products will be attractive only if risks can be minimized and profits maximized. We describe these risks more fully in Chapter 3. Procedures to measure, manage, and reduce these risks are studied in subsequent chapters.

If new products are so risky, why do organizations spend considerable time and energy on new product development? To address this question we look at the forces that encourage organizations to begin new product development.

INITIATING FACTORS

A good manager is continually aware of the marketing system and the macro-environment that impacts on his organization. The manager learns to recognize factors in the environment which initiate a need for a new product and to recognize what strategies are appropriate. Among these initiating factors are financial goals, sales growth, competitive position, product life cycle, technology, regulation, material costs, inventions, and customer requests.

Financial Goals

The perceived inability to achieve financial goals of profit and earnings per share can force initiation of new product development. American Telephone and Telegraph faced these pressures when earnings stabilized as costs increased. Government rulings opened some segments of the market to competition while the demand became saturated with an average of more than one phone per household. This caused the initiation of the search for new services to meet the desired growth in dividends. Services such as mass calling from one number to many numbers as a substitute for direct mail advertising and a "letter a day" service that connects two parties at a time of slack phone network utilization are examples of new products that may renew growth and establish increased profits.

After a significant drop in stock value as a result of little growth in earnings, Kodak's president stated that "our emphasis will be on the marketplace" (*Business Week,* June 20, 1977). Similarly, after a stabilization in profit, RCA's president saw himself "reordering RCA's goals—identifying markets, segmenting them, directing RCA's R&D toward products that serve them . . . " (*Business Week,* July 4, 1977).

New product activity is intimately linked to financial planning. Financial forces can trigger development after profits have fallen, but forecasted decreases in profits are often equally compelling. The need for sound financial growth is one of the most important forces to impel new product development.

Sales Growth

Growth in sales is an important goal for many corporations; in many cases it is absolutely necessary if profits are to be increased. For example, one pharmaceutical firm identified the over-the-counter drug market as a source of sales growth. Since the firm had some brands in related drug markets, the president of the company decided that the firm should also be in the pain reliever market. This was the prime initiating factor in the firm's development of a new acetaminophen-based pain reliever.

While sales growth is a continuing force for innovation, the emphasis has shifted to profitability as the prime concern. For example, Gillette terminated its efforts to introduce pocket calculators in 1974 and digital watches in 1976 because, although they represented sales growth, they did not meet the required profitability standards. Similarly, the paper industry has shifted emphasis from tonnage to profits, and the communication industry from technology to profits. While the 1960s and 1970s were times for sales maximization, the 1980s will be times of profit maximization and risk minimization to assure a steady growth in return on investment and earnings per share.

Competitive Position

The standing of an organization relative to its competitors is a strong motivational force. For example, before 1974 there was no accepted relative ranking of university business schools. Then ratings by deans and academics appeared. This was of interest to students and faculty alike. In at least one case, these standings initiated an effort to understand the market for new master's degree programs: The Sloan School of Management at MIT undertook a systematic effort to innovate its educational program leading to a master's degree.

In some industries such as autos, market share changes of one percent are viewed as important. In one consumer product company the Nielsen bimonthly market share for a brand changed 0.1 share point. The president urgently called the market research department to find out if this was a "real" change. If it was, he was prepared to take immediate action.

Another method in which competitive pressure is felt is when a competitor enters a market first. For example, General Foods' entry of Maxim was the first freeze-dried instant coffee. This put Nestlé under considerable pressure. Nestlé responded with Taster's Choice freeze-dried instant coffee and soon dominated Maxim in the market.

In product markets competitive share is a good continuing competitive measure, but in almost all organizations the presentation of any measure of an unfavorable competitive position provides a strong incentive for change. The Russian Sputnik was the competitive indicator that spurred the U.S. to develop the space program that beat the Russians to the moon.

Product Life Cycle

Marketing theory suggests that products follow a sales pattern over time that can be divided into introduction, growth, maturity, and decline. As the product moves from maturity to decline, profits fall. To regain profit, the organization should direct effort toward rejuvenating the life cycle or at replacing the declining product with a new, more profitable product. For example, Alka Seltzer unit sales had decreased each year for five years and it appeared to be entering the decline phase of its life cycle. In response to this factor, Miles Laboratories, Inc. significantly increased its new product efforts and introduced Morning Star "Breakfast Strips," made from soybeans, and ALKA 2, a chewable antacid.

The life cycle is important in product categories as well as individual products. After years of steady growth in the 1950s and 1960s, the food industry reached a mature state and a sales plateau. In 1975 industry sales actually declined. General Mills anticipated this trend in the 1960s and developed a strategy of growth by diversifying into other industries—General Mills became the largest company in the game and toy business after it acquired the Lionel and Parker Brothers brands.

A decline in sales might not be permanent. Often the product can be rejuvenated or directed at a new market. For example, the motion picture industry declined in the 1950s and 1960s, but began a new life cycle in the 1970s. Furthermore, this pattern of one cycle followed by a recycle was found to be the predominant sales pattern in a study of 258 ethical drugs (Cox, 1967).

It is important to understand these sales patterns since, if the product is about to decline, active new product work is necessary. Cox found that it was common practice to increase promotional expenditures sharply when an ethical drug product reaches the end of the maturity phase. This promotion almost invariably increases the sales of the product, resulting in the recycle. This rejuvenation phase of the life cycle is a managerial responsibility which has led to many "new" and "improved" products (Levitt, 1965).

However, the product life cycle concept alone is difficult to apply in practice. It is difficult to identify when a product will leave the mature stage and begin its decline. In a study of 140 drugs, cigarette, and personal-use products, Polli and Cook (1969) found only 44 sales patterns consistent with the product life cycle model. They caution it is incorrect to infer that a long period of sales stability means the product has reached its maximum sales and that several periods of decline after stable sales do not mean necessarily that sales will continue to decline. Although the life cycle from maturity to decline is difficult to identify for brands of products, Polli and Cook did find the life cycle for product forms (e.g., regular vs. filter cigarettes) was strong. Dhalla and Yuspeh (1976) present further criticism of the life cycle concept. They caution it should not be used blindly to harvest products. Rather managers should try to understand the reasons for sales decline and take the appropriate actions to "save" the product before considering innovations.

Despite the difficulties in applying the product life cycle in all situations, the concept is important because it directs our attention to monitoring the sales growth or decline and initiating a search for reasons to explain this growth or decline. The compelling natural analogy of inevitable decline is a real force in placing a high priority on product development. As in humans where doctors attempt to prolong useful life, managers attempt to lengthen the maturity phase. However, managers have one advantage over doctors, since they can give a product a "rebirth" through innovation or repositioning.

Technology

One of the factors accounting for the decline of products and the shortening of life cycles is the rapid change in technology. For example, microminiaturization of computer circuits led to the demise of the mechanical adding machine. It is interesting that the leading firms in electronic calculators are not the firms that previously produced mechanical calculators. These firms apparently did not correctly forecast the changing technology and develop an active strategy to deal with it. One exception was Burroughs. Both Burroughs and National Cash Register Co. (NCR) have roots that predate the turn of the century, but Burroughs moved to computers in the 1950s. In the 1970s its profits grew steadily and exceeded NCR's (Standard and Poor's Industry Survey 1979). In the late 1970s, NCR began repositioning itself as a computer company.

Computer memories themselves have undergone rapid technical change. As storage systems changed from moving head disks to cartridge disks, and to floppy disks, the price of minicomputers dropped by more than a factor of six and the advent of the bubble memory is expected to drop prices still further. These decreasing prices have opened up many new markets such as TV Games, digital taxi meters, electronic surveying instruments, post office scales, and home computers.

Technological change puts extreme pressures on companies to innovate or decline. For those who can successfully create products based on new technology, the rewards can be high (e.g., Digital Equipment Company). In many industries, the race to be first with a new technology is important. RCA and Phillips are competing to be the first to introduce "videodiscs" that would allow home movies to be played over a TV set by using a disk on a turntable. These manufacturers see this market as possessing half a billion dollars a year sales potential in the 1980s. Although this may be optimistic, the rewards in technological and market innovation can be large. IBM and Xerox have been significant examples of technology creating new industries. A good manager follows these technological changes and puts them to profitable use by matching them to the proper market.

Invention

The invention of the Polaroid instant camera is a dramatic example of the potential of a new product. More recently, Polaroid has introduced a sound-based distance adjusting system. A study of the new technology-based enterprises surrounding Boston indicated that 160 new companies were formed by past employees of MIT's Instrument Laboratories, Lincoln Laboratories, or engineering departments (Roberts and Wainer, 1968).

While some entrepreneurs start their own companies, others approach larger companies to develop and market their ideas. New inventions, like any new product, are subject to the high risk of failure and should be carefully evaluated before any major investment is made. Even if no decision is made on an idea, the evaluation process triggers consideration of other new product opportunities and requires the firm to examine its overall goals and strategy.

Regulation

The government is becoming increasingly involved in regulating business. In many cases this causes firms to consider producing new products. For example, on May 1, 1976, prices of financial brokerage services were decontrolled. This forced major brokerage houses to consider developing new services to preempt severe price competition. Paine Webber subsequently introduced a new expanded service plan. The regulatory power of government pervades competitive practices, advertising, product safety, labeling, labor practices, and drug safety to name just a few areas. Auto companies have had to reduce pollutants and increase gasoline mileage in their autos. These regulations led GM to reduce the size of all their cars. One of the first examples was the 1977 Chevrolet Impala which was many pounds lighter and several inches shorter than the 1976 model. Since by 1985 autos must average 27.5 miles per gallon, auto manufacturers will have to aggressively pursue product development to meet government regulations. By 1985, General Motors will likely stop producing the V8 engine which has been the company's chief power plant for 25 years.

Material Costs and Availability

As raw material costs and availability change, products must be revised or dropped. The increase in gasoline prices and foreign competition were tremendous forces on Detroit to develop small cars. Producers of clam chowder find supply decreasing due to pollution and "red tide" while demand is growing. Prices have more than tripled since 1975. Therefore, the use of squid as a substitute is being considered. In 1976–77, coffee prices more than doubled. This led General Foods to introduce a new brand called "Mellow Roast" blended from instant coffee and roasted grain.

In a world of increasing shortages and supply variability, the forecasting of supply prices and the development of new products to exploit the structural shifts in raw material price will be important in many organizations.

Demographic and Life Style Changes

Just as the supply of materials changes, so does the market. The postwar baby boom brought about rapid growth in baby products, then the "youth" culture, then overflowing colleges, and then a very tight housing market. But as rapidly as the growth came, it disappeared as the demographics of the U.S. population continued to shift. Baby-food companies have diversified and babies are no longer "Gerber's only business." The population continues to change. As the average age increases some industries (e.g., pharmaceutical producers) will benefit, while others (e.g., tobacco producers, record makers, and soft drink manufacturers) may suffer. Firms such as Coca-Cola Co. which generate most of their revenues in soft drinks may have to diversify as has PepsiCo, Inc., which is marketing enough other products so that beverages account for less than half of its total sales.

Life style also generates consumption changes. More divorces and zero population growth have led to smaller families resulting in an increased need for condominiums, small washer/dryers, and other small family products. Health consciousness has led to increases in tennis and jogging equipment, and to lower cholesterol or high-fiber content foods.

In a world of mass communications and increased education, we can expect future shifts in consumption. Development of new products to exploit these shifts could lead to either success or failure for many organizations.

Customer Requests

Another source of new product stimulus is a customer request to produce a specific product that the customer has designed. Von Hippel (1976) found that in the market for scientific instruments (gas chromatography and spectometry), 80 percent of the major innovations in performance were the result of users who had a need to fill and built a prototype of what they needed. The manufacturer then produced and sold the new instrument. Similar results were found for process manufacturing innovation (manufacture of silicon semiconductor products, electronic subassembly, and so on) where in 67 percent of the cases the user recognized the need, built a prototype, and used it prior to commercial production by a manufacturer (von Hippel, 1977).

In both these cases the value of the product was sufficient enough for the consumer to develop it himself and then ask a manufacturer to produce some for the customers' needs. The manufacturer then commercialized and distributed the product. While it is common to think of the manufacturer as the innovator, in several industries the locus of innovation is not centered around the manufacturer of the final product.

Suppliers can also be a force in innovation. Alcoa designed an aluminum trailer and promoted it to manufacturers. Bakelite Company was prepared to supply vinyl, but in the 1940s it could not interest manufacturers such as Armstrong Cork. It had to develop a vinyl floor tile with a small company called Delaware Products in 1946. In consumer products, the frisbee craze was born when it was recognized that Yale students were not returning the pie tins from the Frisbie Pie Co. at Bridgeport, Conn. Rather, they were forsaking the five-cent deposit to throw them about, crying "Frisbie!"

Suppliers of materials are innovators in many industries, but even in industries where producers are the dominant force of innovation, consumer requests can spur new product development.

Future of Initiating Forces

This section has outlined factors that lead organizations to initiate product innovation. Financial goals and sales growth are internally generated pressures, while competition, life cycle, technology, inventions, regulation, and material costs are external pressures. Demographic/life style changes and customer requests are specific market stimuli that come to the company as opportunities. All of these factors act to some degree in all organizations whether they be public or private or whether they provide consumer or industrial goods or services. To be effective in developing new products it is important to act in response to these factors, but to be more effective the manager should anticipate these factors and thus avoid a hurried short-run response.

In the future this preemptive strategy will become even more crucial since:

- the cost of capital will be high,
- competitive organizations will make significant commitments to internal growth via new products development,
- more organizations will be searching into areas outside current operations,
- markets will become increasingly saturated with many product alternatives,
- consumer life styles will continue to change,
- consumers will become more sophisticated buyers,
- increasing rates of technological change will shorten life cycles of products,
- environmental constraints from government, consumers, and labor will increase,
- and more shortages of resources will occur and prices of materials critical to products will fluctuate.

The presence of the factors outlined in this chapter exert a force toward new product development. Organizations will develop explicit developmental strategies to deal with these forces and to manage innovation. The task in product development is to find a major new product so that the potential rewards will be large, but at the same time lowering the risk of failure to acceptable levels. This is a difficult task. A few examples will illustrate the magnitude of the problem.

SOME EXAMPLES OF DEVELOPMENT PROBLEMS

Problem 1. Figure 1.1 shows thirteen brands selected from more than thirty pain relievers which compete for the over 300 million dollars in sales in this market and spend over 100 million dollars on advertising and promotion per year. How would you introduce a new brand in this category?

Problem 2. Squid (Figure 1.2) are plentiful off our Atlantic and Pacific coasts and are an excellent protein source. Although consumption is over twenty pounds per capita per year in Japan, American consumers have negative attitudes toward squid. How would you help the American fishing industry develop a market for squid?

Problem 3. It is likely you now are or were enrolled in a business or management school. Suppose that your school wanted to develop new programs that fulfill the needs of students, faculty, and recruiters. You want to

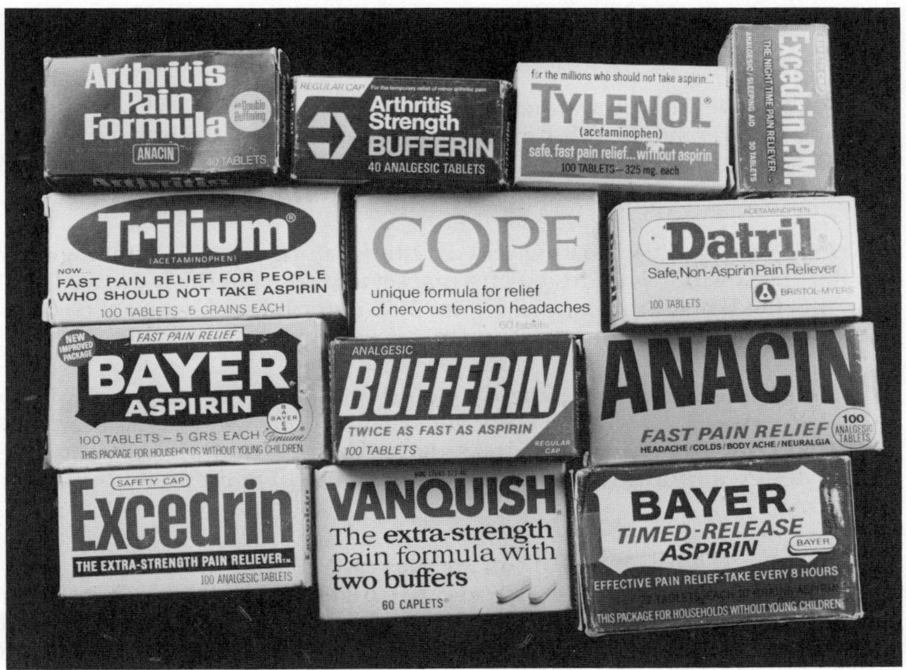

Figure 1.1. *Aspirin Brands*

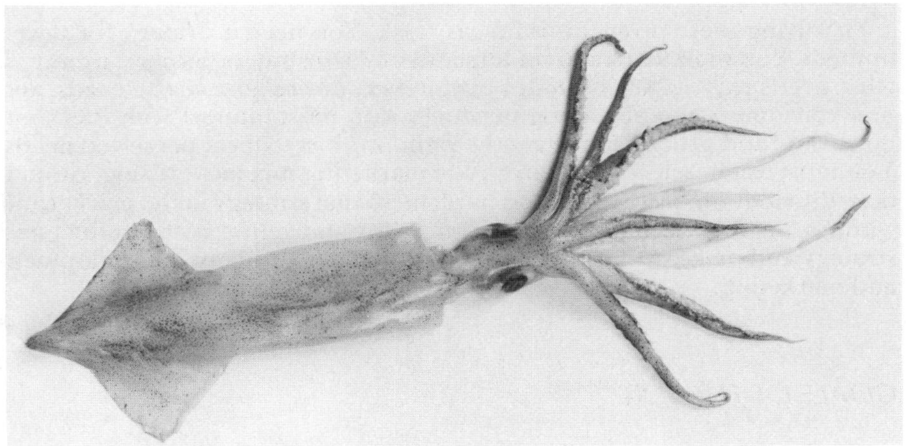

Figure 1.2. *Squid*

increase your competitive standing relative to other schools. What would you do?

Problem 4. If you wanted to innovate in the health field with a comprehensive, prepaid, health service plan, what services would you include, how would you compete with the existing medical services, and how would you forecast acceptance and financial implications?

Problem 5. Solar heating and air conditioning is now technologically feasible. How would you design and market a product in the industrial construction field? (Figure 1.3a and Figure 1.3b describe the layouts of conventional and solar air conditioning units.)

Problem 6. The Bell System's picturephone has been technologically feasible for over a decade, but it has yet to achieve any widespread use. How would you discover the reasons for its lack of success and use that information to design a product that has a greater chance of widespread consumer acceptance?

Problem 7. Imagine that you are the City Manager of a suburban community of 80,000 people. Your streets are congested with automobiles, your bus system cannot attract riders, and special groups such as the elderly and the handicapped find that they have limited mobility. Your transportation budget is limited so you are restricted in your spending on advertising, promotion, or system improvement. How do you decide how to re-allocate your budget so that you best fulfill the needs of your constituents?

Solving these problems is no easy task. You need a strategy for development. You must learn and understand how consumers perceive products, what needs exist, what consumer preferences are relative to the needs, and how consumers choose among products. You must interact with R&D, engineering, and production to get the right product to meet perceived needs. You must test it, select the appropriate marketing mix (advertising, promotion, distribution, price, etc.), and implement that strategy in the market and within your organization. You must monitor and refine both product and strategy and make profit sufficient to justify the investment in development and marketing.

GOALS OF THE TEXT

Our goal is to help you learn how to develop and market new consumer and industrial products and services in both the private and public sectors. Whether you are the vice president of growth and development, a marketing manager, a consultant to government, an R&D engineer, a hospital ad-

Conventional Absorption A/C Systems: ABSAIR

ABSAIR consists of an absorption chiller, a boiler, piping, pumps and control equipment.

In order to provide cooling, an absorption chiller utilizes a refrigerant (e.g. water) and an absorbent (e.g. lithium bromide) in conjunction with an evaporator, absorber, generator, and condenser as diagrammed below. In the evaporator, the refrigerant, in a vacuum, is vaporized by a sprayer. As it evaporates, the refrigerant absorbs heat from the water that is used to cool the building. The refrigerant vapor is then absorbed by the solution in the absorber. The resulting solution is heated in the generator to drive off the refrigerant. At the condenser, the refrigerant vapor condenses and rejects heat to the environment. The refrigerant then returns to the evaporator to start the cycle again.

The boiler uses oil, natural gas, or electricity for power. The system can also be driven by commercially produced steam. The absorption chiller is then independent of the heating system unless both use a common boiler.

The absorption chiller may be located on the roof or within a mechanical space in the building. Maintenance costs for ABSAIR are approximately 20% lower than those of compressor a/c systems. It is less efficient than compression a/c systems in that to cool a given load, ABSAIR requires between 15–20% more energy. The absorption chiller has a longer expected life than the other a/c systems, so ABSAIR's economic life is around 25 years. In addition, as it has almost no moving parts, an absorption chiller is relatively quiet and vibration free.

The other elements of ABSAIR: fans, pumps, piping, and control equipment are generally the same as those in compressor a/c systems.

The initial investment cost of ABSAIR may be significantly more than for compressor systems. The difference however tends to reduce as the installation size increases.

ABSAIR appeals to companies who need a large amount of a/c or who use steam for other industrial processes and want to make additional use of that steam. ABSAIR has been typically used by hospitals and pharmaceutical plants where abundant steam exists. Currently, there are between 3,000 and 5,000 of those industrial a/c systems in use.

DESCRIPTION OF AN ABSORPTION CHILLER

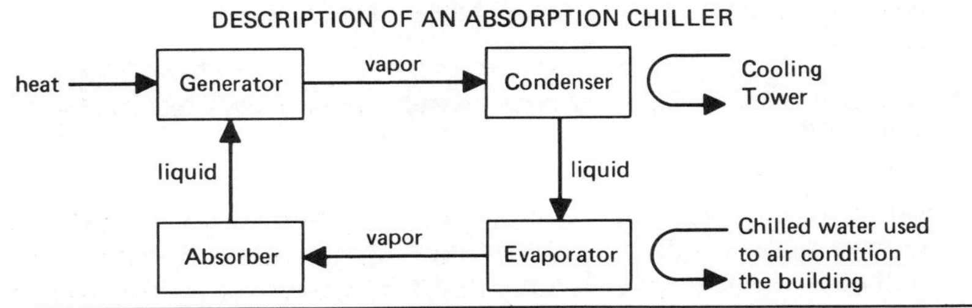

Figure 1.3a. *Solar Air-Conditioning Concepts (from a study by Choffray and Lilien, 1978)*

Solar Absorption a/c System: SOLABS

SOLABS consists of a standard absorption chiller as used in ABSAIR and a hot water solar collector which replaces the boiler in a standard absorption a/c system. As it uses solar energy as a power source, SOLABS is less sensitive to fuel shortages and power fluctuations than other industrial a/c systems.

The solar collector used by SOLABS is a flat type that is located on the roof of the building. In some cases, collectors can even replace the roof. Collectors come in panels of various standard sizes that are attached to one another by normal plumbing connections. Two water storage tanks are also part of SOLABS and are generally buried in the ground. One of these tanks is for chilled water, to meet the immediate demands of the absorption system. The other one is for hot water, to meet a/c needs during periods of little sunshine or alternatively to provide heating during these same periods. When the system is used exclusively for a/c, water storage capacity need not be large as more solar energy is available when cooling is most needed. A small backup heating and cooling system can be used to make up for prolonged periods of low sunshine.

Solar energy alone can provide 40%–60% of all building a/c requirements, significantly reducing energy costs. In addition, warm water produced by the solar collector can be used for manufacturing or domestic water needs. In colder climates, this system can provide 30%–40% of heating requirements.

The initial cost of SOLABS is at least 50% higher than for non-solar systems, depending on the site of the installation. The operating cost of SOLABS, however, is considerably lower than for other systems due to a reduction of at least 40% in a/c energy consumption (depending on the geographical location). Maintenance costs for SOLABS are similar to those for ABSAIR.

SOLABS produces no pollution. As it requires a minimum of moving parts, SOLABS is also very quiet and vibration free.

The solar a/c concept is not new. Several well-known manufacturers produce components and one such system was in operation at the University of Florida as early as 1960. Currently, there is a new school in Atlanta, Georgia that is air-conditioned by SOLABS and there are several projects to install similar a/c systems in different parts of the U.S.

Figure 1.3b. *Solar Air-Conditioning Concepts (from a study by Choffray and Lilien, 1978)*

ministrator, an entrepreneur, a market researcher, or a young executive in training, you may be called upon to make new product decisions. You may have to decide: what market to enter and how to enter it, how to physically design a product, or what image to establish for the product. Your task may be to set corporate strategy and make the decision on how to price, sell, and service the new product.

After completing this book we hope you will have the managerial perspective and tools to make these decisions and successfully manage the new product development process. We present the strategic and organizational issues of innovation management along with technical methodologies that reduce risk and enhance success.

We do not expect you to know all the technical details of new product development. We do expect you to be able to:

1. understand the need for a disciplined process of development,
2. follow the basic steps of opportunity identification design, testing, and implementation,
3. know what questions to ask and which questions can be answered,
4. know how to read and interpret new product market research and management science analyses,
5. enhance your own creativity with research information and idea generating techniques,
6. design a product and put together its marketing mix,
7. test the product and forecast its sales through pretest or test marketing,
8. monitor and refine the product and mix as it flows from test market to national introduction.

We expect that your proficiency in these tasks will enhance your career by helping you and your organization develop successful new products while at the same time reducing the risk and the consequences of failure.

REVIEW QUESTIONS

1.1 What is a new product? Why are new products important?

1.2 Why should firms innovate given the risky nature of innovation?

1.3 Booz, Allen and Hamilton report in a study that 70 percent of the resources spent on new products are allocated to products that are not successful in the market. How might you explain such an allocation of resources?

1.4 When should a new product be classified as a success and when should it be classified as a failure?

1.5 What initiating factors are strong in the automobile market? In the computer market? In the health care market?

1.6 Name some new products that have appeared on the market in the last year. For those that failed, give some reasons why the product was weak. For those that were successful, give some reasons why consumers bought the product.

1.7 Select two of the seven development problems raised in the text. How would you address these problems?

PART I

INNOVATION STRATEGY

Chapter 2

New Product Strategy

The potential rewards from developing successful new products are high, but the risks are correspondingly high. While some large organizations may be able to survive by trying one product after another in the market until success is achieved, most organizations cannot afford these costs. Even large organizations are now more profit conscious and concerned about the costs of development. We must approach new product development with an effective managerial strategy that is likely to achieve success, but at the same time minimize risk. We begin with a discussion of potential new product strategies.

STRATEGIC RESPONSE

New product development strategy should reflect the overall policy and strategy of the organization. This overall strategy formulation may have begun with an audit of the company's capabilities and its environment. What are we good at? Where are we vulnerable? Why have we succeeded in the past? Where are our major products in their life cycles? What is the forecast for material costs and availability? What technological changes can we expect? What actions will the government take that will affect us? What can we expect from our competitors and what are their strengths and weaknesses? What consumption changes can we exploit? These questions lead us to face the underlying issues: What business are we in and what are our goals?

The goals should reflect the nature of the organization, and they should be realistic. For example, a large private firm may set a target of growth of 10 percent increase in earnings per share and a 20 percent increase in sales, but set a constraint that no new product is worth considering unless it can bring at least $4 million per year in revenues representing at least 4 percent of a market. Alternatively, a small growing firm may want a 50 percent growth which can be achieved through a volume of less than $1 million per year representing less than 1 percent of a market. In a public service such as mass transportation, goals may be to reverse the decline in ridership over five years and reduce the deficit by 5 percent per annum. The important point is to set the goals and if possible quantify them to provide a measure for achievement and to help direct the organization.

After the goals and measures of performance are established, the organization must consider alternative strategies. One of the basic strategic decisions is whether to be *reactive* or *proactive*. A reactive product strategy is based on dealing with the initiating pressures as they occur while a proactive strategy would explicitly allocate resources to preempt undesirable future events and achieve goals. For example, a reactive view of the competition is to wait until the competition introduces a product and copy it if it is successful, while a proactive strategy would be based on preempting competition by being first on the market with a product competitors would find difficult to match or improve.

Each strategy is appropriate under certain conditions. A successful organization recognizes when each is appropriate and responds accordingly. We begin with examples of each strategy and then indicate when each is appropriate. Table 2.1 identifies several reactive and proactive strategies.

Table 2.1. New Product Strategies

Reactive Strategies	*Proactive Strategies*
Defensive	Research and Development
Imitative	Marketing
Second but Better	Entrepreneurial
Responsive	Acquisition

Reactive Strategies

The first reactive strategy, "defensive," guards against competitive new products after they have been successful by making changes in existing products. For example "Zenith" had a less modern production facility, but defended its existing business against new integrated circuit color TV sets by positioning itself as "hand crafted." This defensive strategy was not successful in the long run, but it did blunt the impact of the competitors' new products.

Another method of dealing with competition, "imitative," quickly copies a new product before its maker is assured of being successful. This imitator or "me too" strategy is common practice in the fashion and design industries for clothes, furniture, and small appliances. For example, once Cuisinart demonstrated that a market existed for expensive food processors, many of the major appliance companies followed with products that imitated Cuisinart.

A more sophisticated strategy to react to competition is the "second but better" strategy. In this case the firm waits until the competitor's product is revealed and then not only copies it, but improves on it. The objective here is to be flexible and efficient so as to produce a product that will be superior to the competition without incurring the heavy market developmental expense for the product.

The final reactive strategy is termed "responsive," which means purposively reacting to consumers' requests. In scientific instruments this would mean facilitating the information flow from users and manufacturing the user's prototype if a market exists. This would imply emphasis on applications engineering and manufacturing. In the case of the aerospace industry, a responsive strategy would be to submit bids for specific governmental requests for proposals. Responsive strategies are also used by manufacturers in a chain of distribution in which some other channel member is dominant. For example, teflon cookware was developed in response to customer requests which in turn were encouraged by the material supplier, Du Pont.

Proactive Strategies

A second class of strategies is proactive. In this case the organization initiates change. An aerospace company executing this strategy would not wait for a request for proposal from the government, but would estimate government needs and do preemptive R&D and product development. It would then take its work to the government and suggest a request for proposal be written around this need. There is some evidence that many companies practice this strategy (Urban, 1964). Some claim the only condition under which they would bid in a request for proposal is if they had done substantial research before the request. We have had similar experience in obtaining outside funding for academic research, i.e., that we are in a much better position to propose on a grant on the basis of internally funded pilot research. Many successful universities and consulting firms follow this proactive strategy.

The proactive strategy of an organization may be based on "Research & Development" effort to develop technically superior products. Some companies have been notably successful. IBM and Xerox represent the potential of technological innovation. In 1977 alone IBM spent over $1 billion on R&D while Xerox spent over $250 million. The figures represent 5 to 6 percent of sales for each company. (*Business Week*, July 3, 1978.)

Another approach to development is through the notion that someone must buy a product if it is to be successful and therefore success can be found by considering the consumer first. The "marketing" strategy is based on finding consumer needs and then building a product to fill them. Procter and Gamble, General Foods, McDonald's, and most consumer product companies utilize this philosophy.

One of the most proactive forms of product development is "entrepreneurial." In this mode a special person—an entrepreneur—has an idea and makes it happen by building venture enthusiasm and generating resources. Some large companies have tried to utilize this strategy. At 3 M (Minnesota Manufacturing and Mining) a separate new venture division has been established where entrepreneurs can take a leave from their regular job to work on their ventures.

A final proactive strategy is "acquisition." In this case other firms are purchased with products new to the acquiring firm and perhaps the market.

REACTIVE VERSUS PROACTIVE STRATEGIES

In order to select the appropriate strategy we must understand the situations that affect this decision. Thus we must look at the growth opportunities, the probable protection for innovation, the scale of the market, the strength of the competition, and the organization's position in the production/distribution system.

Growth Opportunities

We normally think of new product activities as an organization introducing a new product, such as a mini-cassette home recorder, to a new market. But this is only one possible strategy for growth. Let us examine the issues of reactive versus proactive strategies within the traditional framework of alternative growth opportunities (see Figure 2.1).

One opportunity is growth through existing products and markets. This strategy is one of market penetration and is characterized as developing a high market share in existing markets with the existing products (Cell 1).

	Existing Products	New Products
Existing Markets	1. Market Penetration	3. Product Development
New Markets	2. Market Development	4. Diversification

Figure 2.l. *Opportunities Matrix (Ansoff, 1957)*

This growth strategy is not based on innovation in products as much as in selling and promotion. For example, Royal Crown Cola directed its attention to its leading brands—RC Cola and Diet Rite—and devoted less time and energy to new products.

In many of today's markets, saturation occurs so frequently that firms are increasingly looking toward new markets. Cell 2 represents the strategy of taking existing products and entering new markets. For example, export to international markets can represent growth opportunities. McDonald's and Coca-Cola have effectively exploited this strategy, but Radio Shack was less successful in three years of foreign operations partly because it did not fully adapt its strategy to the special needs of the European consumer.

The usual new product development strategy is to attack existing markets with new products (Cell 3). This strategy is consistent with the notion of "building on our strength" and expanding in areas of our skill and knowledge in distribution and production. Many of the examples in this book will fall in this category.

Some companies may choose to diversify into new markets with new products (Cell 4). For example in 1969, Church and Dwight, Inc. who market "Arm and Hammer" began to diversify into the market for laundry detergents and deodorants. Sales increased from $22 million to $99 million in 1977 (Moody's OTC Industrial Manual 1978). Although diversification can be successful, a major firm moved from the aerospace market to transit vehicles and suffered. In 1976, this firm lost almost $40 million on a contract to supply railcars to the new Washington Metro.

The choice of market opportunity is an important decision that affects the strategic response. If existing products and markets are to be the primary growth vehicles (Cell 1), the organization is best at distribution, and growth rate aspirations are not high, the reactive product strategy may be most successful. In this case product development would be used to defend the existing products by reacting to competitive and environmental pressures.

However, if the organization wants growth or has a policy of innovating, and has skill in R&D and marketing, a proactive strategy would have potential to help meet its overall organization. Proactive strategies based on R&D or marketing lead to new products and new markets.

Protection for Innovation

Another major factor in selecting between reactive and proactive strategies is the amount of protection a new product can obtain. If the product can be patented, the innovating organization can be more assured that its developmental investment will be returned. Although patents are becoming more difficult to defend, Polaroid's patents have stood up well and have helped preserve its profits. Protection may be granted by the market when the first firm introduces a good product and achieves a predominant position. For example, although Burger King and Burger Chef and others have copied McDonald's food franchising operation, McDonald's is still the biggest chain, is very profitable, and continues to grow. Alternatively, in product categories such as small appliances, a first-in product can be quickly copied and the innovator has only a short period of competitive advantage. For example, six months after the first electric knife was introduced, over ten brands were on the market. Thus, firms that can achieve good protection should be proactive while those that cannot may be better off in a reactive mode.

Scale of Market

In addition to protection, market size and margin are important. In large markets with economies of scale in production, distribution, or marketing, first-in may establish market dominance and give the firm an unassailable position, especially if the firm continues to innovate. On the other hand, in small markets it is more difficult to be proactive since the resources for development cannot be returned easily. For example, special production machines and instruments may represent such small sales potential that the best strategy is to wait and be responsive to consumer requests. Similarly, in the case of teflon, each cookware manufacturer faced such a small market that the best strategy was to wait until the material supplier invested in promoting the innovation.

Competition

The competitive environment should be considered. It may make a reactive strategy of imitation feasible. If the time necessary to copy is short, there are few entry costs, the innovation is not protected by patents, and the organization can quickly achieve economies of scale, this may be appropriate. The relative size of the competitors is also important. A small firm may be particularly vulnerable to competitive reaction and thus must be very preemptive in its innovation plans. Similarly a large firm may be proactive to protect its lead. For example in appliances, although imitation is common, General Electric allocates substantial resources to design new appliances.

In many public service organizations the reactive strategy is no strategy. For example, in mass transportation there were no new products for many years and therefore little to copy. It was not until the Department of Transportation of the Federal government took on the responsibility, that there were any active new service developments. This is also true in health care where the Department of Health, Education, and Welfare has developed programs to support the new services through the introduction of Health Maintenance Organizations which provide prepaid comprehensive health services.

Position in Production/Distribution System

The best strategy for development can depend upon the organization's position and power within the production and distribution system. In some situations one firm in the chain of distribution may be proactive, with the others reacting to that firm's innovation. As noted above, the government has taken on the role of funding R&D and developing products in many public service industries. In many industrial markets the supplier of the materials or even the final user may develop the product. In consumer industries the producer is the usual innovator, but retailers like Sears will often specify innovative products and then have other firms produce them.

Whether or not a firm is proactive depends upon the stance of other firms in the distribution channel and on its relative power within that channel. Some firms actually gain power as well as profits by innovation. For example, Haines Corp. was simply another apparel producer until it introduced L'eggs, a distinctively packaged panty hose, through innovative distribution in supermarkets and drugstores. It is now a dominant force in the $1.2-billion women's hosiery market.

Synthesis and Recommendations

All organizations will not choose to or be required to bear the responsibility of developing new products. In particular, reactive strategies may be best for organizations that:

- require concentration on existing products or markets,
- can achieve little protection for innovation,
- are in markets too small to recover development costs,
- are in danger of being overwhelmed by competitive imitation, or
- are in distribution chains dominated by another innovator.

For organizations in those situations, innovation may be too large a risk.

Alternatively many organizations are in good positions to innovate. Organizations with the following characteristics should follow proactive strategies:

- overall policy of growth,
- willingness to enter new products and markets,
- capability of achieving patent or market protection,
- ability to enter high volume or high margin markets,
- resources and time necessary to develop new products,
- competition unable to rapidly enter with a "second but better" strategy,
- reasonable power in the distribution channel.

Such organizations can achieve success and reduce risk through proactive strategies.

This book addresses the issues of implementing a proactive strategy. We take the view of the innovating organization in the process. For example, this may be the manufacturer of consumer goods, a local bank, a supplier of plastics, an inventor, a government department, a user of scientific instruments, or a regional transit authority. Although these organizations are different, we describe the underlying methodologies which can be applied to any of them in creating successful new products. Even if an organization chooses a reactive strategy, it can use some of these methodologies to better understand the consumer and the market forces and thus respond quickly and more effectively to competition.

MARKETING VERSUS RESEARCH AND DEVELOPMENT

Basically a proactive strategy means taking an active role in the development of new products and markets. This active role can come by concentrating on technology (R&D), on the consumer (marketing), or both. Let us first look at the effectiveness of an R&D strategy.

The first observation is that firms spend considerable funds in R&D. In 1976 a survey by *Business Week* (June 27, 1977) reported $16.2 billion was spent in company-sponsored research in addition to $23.6 billion spent by the federal government, of which over 9 billion dollars went to private contractors. The total rate of R&D spending has been stable from 1966 to 1978, in constant dollars, but industrial spending has increased somewhat and government spending has decreased. For example, industry spent over $18 billion in 1977 (*Business Week*, July 3, 1978) and $20.5 billion in 1978 (*Business Week,* July 2, 1979). While R&D expenditures as a percentage of sales vary across industry (6.0 percent of sales in computers vs. 0.5 percent in food and beverages, and tobacco), Mansfield, et al. (1971) found that in the two industries they studied (chemical and petroleum) this rate does not vary significantly by the size of the firm.

Further investigation reveals that of this large amount, the major portion is not spent on pure research, but on specific product development. Gerstenfeld, in a 1969 survey of the R&D directions of 170 large industrial companies, found over 75 percent of the R&D expenditures in these firms was allocated to the development of new products. (Table 2.2 shows the percentage by industry.) Similarly, in a study of the chemical, drug, petroleum, and electronic industries, Mansfield, et al. (1971) found that 70 percent of R&D expenditures was on product development.

Furthermore, industries' expectations are that investments in R&D have a relatively short payback period. When McGraw-Hill surveyed companies to find their expected payback period for R&D projects, it found 90 percent of the companies expected to return their investment in R&D in five years or less (Mansfield, et al., 1971). Initial econometric analysis by Mansfield (1968a) indicates the rate of return to be high.

To evaluate the effectiveness of this spending, numerous researchers

Table 2.2. R&D Expenditures for New Product Development in Various Industries[a]

Industry	*Companies in Sample*	*Percentage of R&D*
Electrical Equip.	28	79%
Chemicals & Pharmaceuticals	34	82
Instruments	16	88
Machinery & Computers	19	68
Aircraft	6	84
Foods	7	100

[a]*Gerstenfeld (1970)*

have traced the source of technological innovations that were successful in the marketplace. Over 2,000 products spanning 100 industries and several countries have been studied* to determine the relative role of marketing and R&D. The methodologies varied, but the study by Myers and Marquis (1969) is typical. They found that of 567 successful new products in the computer, railroad, and housing industries, 60 percent originated from market needs and demand or consumer studies. This finding has been confirmed in other studies where 60 to 80 percent of successful products have been in response to market demands and needs (see Table 2.3).

Table 2.3. Product Innovations Resulting from Market Needs and Technological Opportunities[a]

Type of Innovation (Sample Size)	*Market or Product Needs*	*Technical Opportunities*
British Firms (137)	73%	27%
Winners' Industrial Research Award (108)	69	31
Weapon Systems (710)	61	34
British Innovators (84)	66	34
Computers, Railways, Housing (439)	78	22
Materials (10)	90	10
Instruments (32)	75	25
Other (303)	77	23

[a]*Adapted from Utterback (1974), p. 622*

Not only does consumer need account for 60 to 80 percent of the successful technological innovations, but even more importantly the innovations that come from the consumer tend to result in products with greater

*Arthur D. Little, 1959; Carter and Williams, 1957; Enos, 1962; Hamburg, 1963; Jewkes, et al., 1970; Langrish, 1971; Mansfield, 1968a; Miller, 1971; Myers and Marquis, 1969; Mueller, 1962; Illinois Institute of Technology, 1968; Robertson, et al., 1972; Tannenbaum, et al., 1966; Utterback, 1971.

sales. For example, in the chemical industry Meadows (1968) found that improved sales are hard to obtain, but that improvements are more likely for marketing or customer-originated ideas (see Table 2.4).

Table 2.4. Commercial Outcome for Chemical Laboratories[a]

Source of Idea	*Increase in Sales Caused by Innovation*			
	None	*Small*	*Medium*	*Large*
Projects Laboratory	66%	17%	17%	0%
Marketing	58	14	14	14
Customer	33	33	13	20

[a]*Meadows (1968)*

These studies imply that, at least for the technological industries, the primary source of successful innovation is consumer need. Thus the current emphasis of R&D on product development appears to be justified and the technological industries would do well to study the consumer as a source of ideas. This does not imply that pure research should be discontinued. It is entirely possible that greater R&D spending on pure research could result in major innovations such as with the transistor, bubble memory, or xerography. These basic research expenditures should be balanced with developmental spending. The studies cited here suggest that consumer need and demand is a prime source of successful products at the applied research phase.

We choose to illustrate this consumer dominance in the technological industries because these industries have been well studied and because they are the industries where one would expect a priori technological breakthroughs to be the dominant force. In consumer product/service industries such as personal care products, automobiles, health care, or financial services, we can anticipate this consumer dominance to be even more pronounced. For example, recently Polaroid lost over $100 million on its instant movie product—Polavision. Although a major technological innovation, it did not meet the needs of consumers. In our personal experience commercial success is highly dependent on a consumer-oriented philosophy.

Thus while a proactive strategy must include research and development, it must also have the strong marketing component that is critical to the successful development of new products.

COMPREHENSIVE STRATEGY

Of course, marketing and R&D are not mutually exclusive alternatives. In fact, a good new product strategy requires an effective integration of marketing and R&D. Marketing can help identify and assess consumer needs and R&D and engineering can develop the products to meet these needs. Both fields of knowledge are important in defining an effective proactive strategy. Experts in marketing and R&D both agree with this conclusion.

Professor Donald Marquis of MIT's R&D management group says:

> Recognition of demand is a more frequent factor in successful innovation than recognition of technical potential. It seems to me, therefore, that management ought to concentrate on any and all ways of analyzing such demands and needs. For example, more effective communication should be established among specialists in sales, marketing, production, and R&D to see that such opportunities are not overlooked (Marquis, 1969).

R&D researcher Edwin Mansfield of the Wharton Business School says:

> R&D isn't worth anything alone, it has got to be coupled with the market. The innovative firms are not necessarily the ones that produce the best technological output, but the ones that know what is marketable (*Business Week*, June 8, 1976).

This book proposes procedures to measure market demand and help integrate R&D and marketing in the design of new products. We describe methodologies to implement a comprehensive strategy that reflects the importance of the market in successful innovation, the significance of R&D, and the role of creativity in successful new product development. The output should be a superior physical product that will meet definite needs in the market.

In this book attention is on proactive development strategies. But since even organizations with an aggressive proactive new product development strategy may find the need for quick reaction when competitors are first on the market, we return in Chapter 20 to re-examine reactive strategies in light of the innovation methodologies described throughout this book.

The following chapter presents a decision process to operationalize a comprehensive new product design and marketing methodology.

REVIEW QUESTIONS

2.1 What are the roles of marketing and R&D in the new product development process? How should responsibility and authority be delegated between these two areas?

2.2 Which major corporations appear to be following a proactive new product strategy? Which corporations appear to be following a reactive strategy? Critique these strategies.

2.3 Should R&D focus on basic research or specific applied projects?

2.4 Mr. Hardy Cell, a manager with Widgets, Inc., has a view of the consumer that is not totally unique. Mr. Cell says, "Consumers are fickle. They don't know what they want! New products are necessary only because consumers quickly become bored with old products. Consumers should be treated as if they were children. They must be told what they should buy! Advertise anything and it will sell. If your product sales are falling, tell people the product is new and improved. That's all you need to do."

a. Are Mr. Cell's remarks consistent with the empirical evidence presented in the chapter?
b. Why might some managers support Mr. Cell's position?
c. Could Mr. Cell's remarks be interpreted as representing a proactive new product strategy? A reactive strategy?

2.5 Discuss the advantages and disadvantages of each of the reactive and the proactive strategies.

2.6 Consider two firms competing in the same market. Would it ever be reasonable for one to adopt a proactive strategy while another adopts a reactive strategy? If not, why not? If so, under what conditions?

Chapter 3

A Proactive New Product Development Process

Two critical issues in implementing a proactive strategy are the control of risks and the encouragement of creativity. Management is responsible for performance and must minimize risks of major loss. This can be carried to extremes if a "No" decision is always made because there will be no market failure, but also no new products, sales, or profits. The balance between risk and reward is a continuing one. The tradeoffs of risk and return are difficult because successful products are in large part the results of creative effort in R&D and Marketing and there is a danger that risk minimization will also minimize creativity. Management must develop an organization and decision structure that will allow innovation to flourish and to create an atmosphere of entrepreneurship so that profitable growth can be achieved through new products while at the same time reducing the risk inherent in any new venture.

Developing a disciplined and creative atmosphere is not an easy task. Organizations are not basically creative. They spend most of the corporate energy in maintaining established businesses and even in the new product development area they often spend too much time on routine operational aspects rather than concentrating on developing the idea to its fullest creative potential. The dominance of the operational thinking of an organization requires that management institute specific processes and systems for new product development to manage creativity and foster innovation.

The long-run survival of the organization's growth and profitability is dependent upon effective management of the creative and risk aspects of these processes and systems. Successful organizations manage the future, others are managed by the present and overwhelmed by the future.

In this chapter we consider how a proactive strategy can be converted into a sequential decision process. Then, so that you can better assess the allocation of resources to new product development, we discuss the relative time, cost, risk, and expected benefit from the process. The remainder of this book discusses, in detail, each of the decision steps.

A SEQUENTIAL DECISION PROCESS

Given that the organization wants to practice a proactive strategy and is considering the future environment for new products, we recommend a five-step decision process:

1. Opportunity Identification
2. Design
3. Testing
4. Introduction
5. Profit Management

Figure 3.1 describes these phases. As the arrows indicate this is, in reality, more than a sequential process. It can be more complex, with iterations through each step and interactions among the steps. We consider each phase in sequence for simplicity of exposition.

Opportunity identification is the definition of the best market to enter and the generation of ideas that could be the basis for entry. For example, in the squid example of Chapter 1, opportunity identification might identify the frozen prepared food (grocery) market as high potential and might generate a variety of new concepts describing frozen squid dinners. If an attractive opportunity is identified, the design phase is initiated and if not, further effort is made to find ideas and markets.

Design includes converting the ideas into a physical and psychological entity through engineering, advertising, and marketing. For example, the frozen squid concepts are refined so that they are likely to fulfill a consumer need or desire. Then an actual product (frozen dinner) and an advertising and promotion campaign are designed. In the design phase, these products and strategies are evaluated and refined based on consumer measurements until they are ready for final testing. In some cases, when the product is evaluated as less than superior, a "no" decision will be made, and effort will be directed toward other markets and designs.

If a good design is found, *testing* begins. This testing ends in a test market, but emerging techniques suggest the desirability of more pretest market analysis of the product and advertising. For example, the squid product is taste-tested, the advertising is audience-tested, and the combination is tested in a simulated store environment. Only if these pretests are successful will the product and strategy be subjected to a large-scale test market.

If final testing is successful, the product is launched. *Introduction* is the difficult task of "making the product happen" in the market. This includes

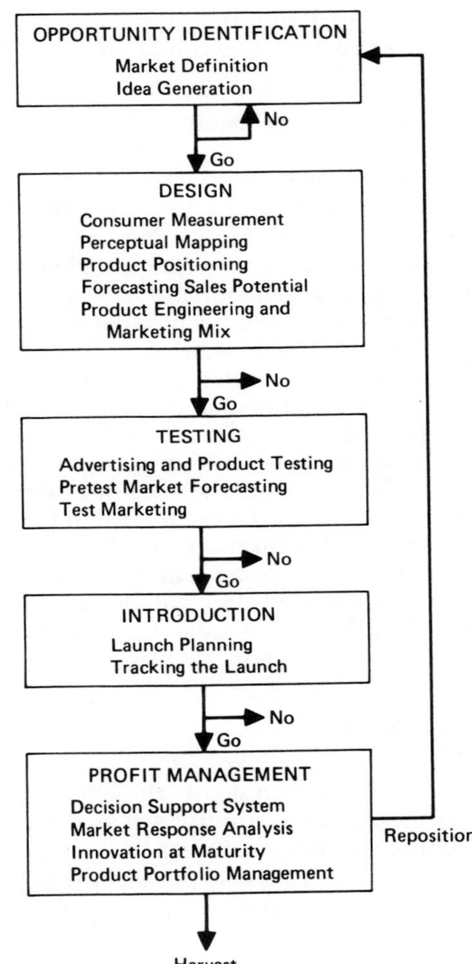

Figure 3.1. *New Product and Service Development Process*

physical implementation, as well as strategic monitoring of the national or regional launch.

If the launch is successful, the product becomes established and the process of *profit management* begins. This is critical since these profits are the reward for the risk taking and creative effort undertaken earlier in the process.

A well-designed product and marketing strategy then continues the reward phase of its life cycle, but with periodic strategy modifications to maintain maximum profitability. The organization monitors this product to identify these strategic actions and to identify if and when it enters the decline stage of its life cycle. At that point, a decision is made to harvest the product, revitalize the product, or reposition it for new markets. We now consider each of these phases in more depth.

Opportunity Identification

Basic R&D may indicate technological opportunities, but the lessons outlined earlier in this book indicate these must be integrated with market demands. At the opportunity identification phase, effort is made to find markets that are growing, profitable, and vulnerable. This requires capability to forecast demand and in some cases technology. Opportunity identification includes approaches that describe a market in terms of its structures and its component segments. This step identifies opportunities that match the strengths and capabilities of the innovative organization.

With an understanding of the market and technological potential, the next task is to generate creative ideas to tap this potential. By understanding idea sources and creative group processes, the organization can generate ideas that integrate the specific engineering, R&D, and marketing inputs. The ideas are in the form of high potential concepts which may ultimately become successful products. The output should be a large number of new ideas that are substantially different from existing products. They may not all represent ideas that will be marketed, but they should be different and new. The design phase develops the best of the ideas into feasible products and the associated marketing strategy.

Design

In the design phase, these new ideas are evaluated and refined to produce a product with physical and psychological attributes which indicate a high probability of success in the market. R&D specifications and development engineering take place in this phase.

The first lesson in the design phase is that it is necessary. Perhaps the most common mistake that students and managers make is to become too quickly over-zealous about a particular new product idea. Everyone has his own favorite ideas about what is needed. But today's markets are becoming more and more competitive, the risks of failure are greater, and the consequences more costly. Some organizations still find it deceptively easy to bring one product after another into test market until finally a "winner" is found. But emerging design techniques can identify the failures at a much lower cost to the firm while increasing the ultimate profit from the successes.

These design techniques identify and exploit needs within the overall market identified in the market definition phase of the process. The design effort begins with the newly generated ideas, selects the ideas with the greatest potential, and refines them to fulfill market needs. A key concept here is that preliminary ideas, no matter how good, change and evolve as the result of iterative cycles of evaluation and refinement.

This idea of iterative process may seem abstract, so let us take an example. Figure 3.2 gives two concept descriptions, a seafood entry made from squid and a squid chowder concept. These concept descriptions were presented to consumers for their reaction. The results of consumers' measurement were evaluated and the concepts refined to emphasize nutrition. In Figure 3.3 a picture was included in the description of Calamarios. Next, the physical product was developed and advertising was created and tested. Figure 3.4 shows the label for the Sclam Chowder can. Product and advertising were subjected to further consumer measurement and analysis until the product and its marketing mix were ready for pretest or test market. (Chapter 13 describes the squid testing.) The actual number of iterations in the development process steps would depend on the product, but each iteration provides better information, refines the product concept, and moves it closer to a marketable product.

The detailed design steps may be different for services or for industrial goods, but they also pass through iterative stages which come closer and

CALAMARIOS

CALAMARIOS* are a new and different seafood product made from tender, boneless, North Atlantic squid. The smooth white body (mantle) of the squid is thoroughly cleaned, cut into thin, bite-sized rings, then frozen to seal in their flavor. To cook CALAMARIOS, simply remove them from the package and boil them for only eight minutes. They are then ready to be used in a variety of recipes.

For example, CALAMARIOS can be combined with noodles, cheese, tomatoes, and onions to make "Baked CALAMARIO Cacciatore." Or, CALAMARIOS can be marinated in olive oil, lemon juice, mint, and garlic and served as a tasty squid salad. CALAMARIOS also are the prime ingredient for "Calamary en Casserole" and "Squid Italienne." You may simply want to steam CALAMARIOS, lightly season them with garlic, and serve dipped in melted butter. This dish brings out the fine flavor of squid. A complete CALAMARIOS recipe book will be available free of charge at your supermarket.

CALAMARIOS are both nutritious and economical. Squid, like other seafoods, is an excellent source of protein. CALAMARIOS can be found at your supermarket priced at $1.10 per pound. Each pound you buy is completely cleaned and waste-free.

Because of their convenient versatility, ample nutrition, and competitive price, we hope you will want to make CALAMARIOS a regular item on your shopping list.

*CALAMARIO is the Italian word for squid.

Figure 3.2a. *Squid Concept Alternative I*

SCLAM CHOWDER

SCLAM CHOWDER is a delicious new seafood soup made from choice New England clams and tasty, young, boneless North Atlantic squid. Small pieces of clam are combined with bite-sized strips of squid and boiled in salted water until they are soft and tender. Sautéed onions, carrots, and celery are then added together with thick, wholesome cream, a dash of white pepper, and sprinkling of fresh parsley. The entire mixture is then cooked to perfection, bringing out a fine, natural taste that will make this chowder a favorite in your household.

SCLAM CHOWDER is available canned in your supermarket. To prepare, simply combine SCLAM CHOWDER with 1½ cups of milk in a saucepan, and bring to a boil over a hot stove. After the chowder has reached a boil, simmer for 5 minutes and then serve. One can makes 2–3 servings of this hearty, robust seafood treat. Considering its ample nutrition and delicious taste, SCLAM CHOWDER is quite a bargain at 39¢ per can.

Both clams and squid are high in protein, so high in fact that SCLAM CHOWDER makes a healthy meal in itself, perfect for lunches as well as with dinner. Instead of adding milk, some will want to add ⅓ cup of sour cream, and use liquid chowder as an exquisite sauce to be served on rice, topped with grated Parmesan cheese.

However you choose to serve it, you are to find SCLAM CHOWDER a tasty, nutritious, and economical seafood dish.

Figure 3.2b. *Squid Concept Alternative II*

closer to what will be presented in an actual test. For example, a banking service may pass from concept description to brochures and testimonials to pilot services for selected customers. In industrial products, the iteration may be from concept to prototype to pilot production output with sales support materials.

Chapters 7 to 12 describe a number of techniques for understanding consumer perceptions, preferences, and choice decisions for a new product. However, since the new product design process is so iterative, no single analysis technique or new product model can provide all the information required for each step in the design process. Rather, the approach described in this book is a family of successively more accurate techniques aimed at providing the correct kind of information at the correct time in the decision process. For example, at the concept stage, a manager wants to know which concept has potential and how he can proceed to develop it. At this stage, only a rough GO/NO GO projection of ultimate revenue and profit is re-

Figure 3.3. *The Calamarios Concept*

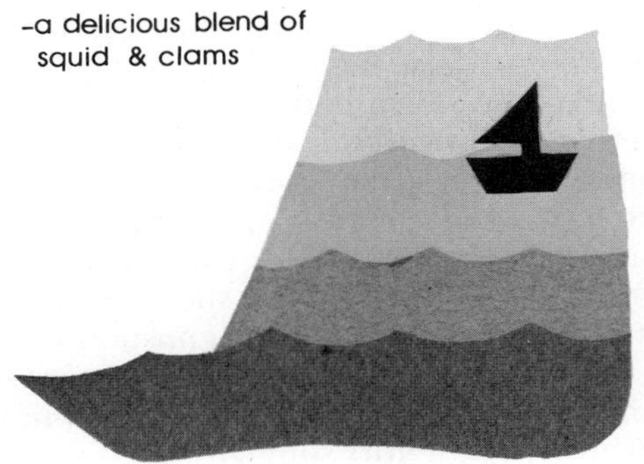

Sclam Chowder
-a delicious blend of
squid & clams
Ingredients: Squid, Clams, Milk, Water, Potatoes, Onion, Seasonings.

Figure 3.4. *Rough Design for Sclam Chowder Label*

quired. Later the manager will want to test and get predictions of costs and revenues with substantially increased accuracy.

Testing

A carefully designed product has great potential, but its success is never assured. Thus, traditionally a product passes from design into test market. Test marketing is perhaps the best-known step in new product development. It is a scaled-down version of a national introduction and is aimed at preventing a national failure. But contrary to popular opinion, it is not simply a test of whether or not the new product will be a success. Although the GO/NO GO decision is a crucial output of test market, a manager needs more information. He wants to carefully monitor consumer response as well as the company's production and distribution systems. A careful test market can evaluate every building block of the marketing strategy. It can identify improvements in advertising, promotion, pricing, distribution, and even in the product itself. A well-structured market test leads to a profit maximizing strategy for national roll out and accurate projections of national sales.

For example, a test market for sclam chowder (Figure 3.4) might be a local introduction in two cities, say Hartford, Connecticut, and Eugene, Oregon. One result might be a GO decision based on a projected national share of 10 to 12 percent. But equally important might be diagnostic information that indicates product and strategy improvements which could increase projected share to 14 to 16 percent and increase projected profits by $2 million. Among these improvements might be: (1) a slight repositioning in advertising to emphasize smoothness, (2) more aggressive promotions to retailers to obtain special displays in retail outlets, and (3) increased distribution of price-off coupons.

A full test is ultimately necessary, but test markets are expensive ($500,000 to $1.5 million), delay national introduction, and tip off competitors to a high potential idea. Fortunately, recent research has produced the analytic models and experimental designs that now make possible a "pretest" market. Pretest market is not a scaled-down, mini-test market, but rather an integrated series of careful measures and analyses in a laboratory environment in a test market city. After careful development and experimentation, such techniques have proven successful in correctly projecting national share and in identifying key diagnostic information to improve strategies. These techniques do not replace test market but rather serve as a precursor to test market. They accurately identify "winners" (GO) and "losers" (NO GO), but also they identify some products which must be improved before test market evaluation. Based on the proven success of these pretest market models, we advocate utilizing pretest procedures for evaluating the product, before proceeding to test market and national introduction, or, if necessary, dropping the product.

The need for pretest market research is especially evident in industries where test marketing is not possible. For example, in automobiles or industrial equipment, it may be difficult to test market. Premarket information may be the only information available to enhance success and eliminate the risk of national failure. In services, simulated tests may identify major needs for improvement and prevent costly pilot test programs.

Introduction

Once a product has been successfully tested, it is ready to be introduced nationally. If the firm anticipates rapid competitive entry, it will want to introduce the product quickly and establish a firm position in the market. But if the firm feels that it has a significant lead on its competitors, or if it does not have the capital to support national introduction, or if there is still some risk involved in the projected consumer response, it may introduce the product on a market-by-market basis.

For example, it is practically impossible to patent the works of an electric can opener. Once the idea is proven, many firms will enter rapidly and capture sales from the innovating firm. In this case, the innovating firm should enter rapidly throughout the target market so that it can establish a strong defensible position. On the other hand, a cold medicine may not be patentable, but its formulation of ingredients can be protected, and its image and distribution network take time to imitate. If a firm views the cold remedy product as high risk or it does not have the capital for national introduction, it may begin its national campaign by a regional introduction west of the Rockies and then "roll out" to the rest of the country.

During this introduction, whether it be rapid entry or roll out, the firm must monitor and manage the marketing strategy. Even the most carefully designed and tested product can run into trouble in a national introduction. Variations in consumer tastes, unanticipated competitive reaction, troublesome channels of distribution, or even national crises like the fuel shortage can all act to undermine the success of a national introduction. Thus, in national introduction, we use techniques to monitor the relevant aspects of the introduction so that the organization can quickly identify and react to any problems or opportunities that occur. These techniques fine-tune a product and marketing strategy (advertising, promotion, sales effort, price, distribution strategies, etc.) to ensure that the new product establishes itself as a productive component of the organization's product line.

Profit Management

After years of effort and millions of dollars of expenditure, the product is now successfully launched into the market. The profit rewards for this effort now must be returned to justify the risk and investment of developing the new product. Maximizing profit requires an effective decision support system so that marketing and production variables can be set correctly. Precise calibration of market response through statistical analysis, experimentation, and management science models can help managers increase profits for the product. Price, advertising, sales effort, and promotion strategies require changes to improve profitability as the product moves through the mature phase of the life cycle.

At the end of the mature phase, either the product must be repositioned through product innovation, or managed through its decline phase, to harvest its remaining profit potential. If it is to be rejuvenated, the new product development process is repeated to find the best target market and design to revitalize the product's life cycle.

TEXTBOOK PROCESS VERSUS REALITY

We have defined a structured approach to the new product development process. Some organizations have analogous processes written down on paper, but our experience has shown that many of these processes have not been followed. Here are a few stylized modes of operation:

> *"Who's got a new idea today?"* In spite of the structured process on paper, many organizations operate on this totally spontaneous and undisciplined approach. This process is not characterized by an organized search, but rather somebody, sometimes top management, comes up with an idea. The idea is implemented with a minimum of testing and evaluation.
>
> *"Here comes the guy in a white coat."* This is characterized by a firm with an extremely strong Research and Development department, or in an industry which is technologically oriented. The problem with this approach is that the concept can have very little meaning to the consumer in spite of the technical brilliance of the idea. Recall from the last chapter that 60 to 80 percent of successful technical innovations in a large number of fields have been in response to market needs and demands rather than in response to new scientific or technological advances.
>
> *"Me too."* Although the organization possesses an aggressive development policy, the organization has very few ideas and therefore is forced to copy competitors' new products and follows them into the marketplace. The problem is the copying organization enters with a parity product which at best produces marginal profits.
>
> *"Let's run it up the flag pole and see who salutes it."* A systematic generation of large numbers of ideas which are not well thought out or well screened prior to heavy marketing investments.

Here are three specific and real examples of the abuse of the new product development process.

A leading food producer had developed in its test kitchen "Pizza Spins" (frozen pizza four inches in diameter that you cook in the toaster). With the push from an aggressive brand manager it was determined to go national without test market since you only need "1 percent of the snack market."

However, the question should not have been how to get one percent, but how to get even one customer to try and repeat purchase. The trial appeal was limited and frequency of purchase low so that after following the brand manager's advice, the firm ended up with an inventory equal to 60 years of sales at the initial sales level. The testing stage should not have been bypassed.

A leading academic institution was responsible for developing a new transportation system which used minivans to pick up people at their homes after they phoned in a request and take them to their destination. Most of the attention focused on the computer scheduling of the buses and the operational process. It was only after several millions of dollars were spent that it was recognized that consumer response was not understood. How the product and its benefits were perceived and how decisions were made were not known. The service was not successful. It had not been carefully designed from the consumers' point of view as a superior transportation alternative that deserved consumer patronage.

In the test market of a scrubbing pad made of plastic and foam with cleanser, the new product brand manager went to the test cities and personally installed large special displays in all stores. Then he "tested" two ads at once and thereby doubled the advertising pressure. Next, he conducted his own "research" study by personally standing next to the store shelf displaying his brand and then asking people if they had heard of his product. He returned to the office to report 80 percent awareness and 30 percent market share in the first two weeks. These actions destroyed the projectability of the test market.

These examples may sound extreme, but they happened in "sophisticated" organizations with clearly defined processes. The enthusiasm and personal career or organizational interests of individuals can destroy even the best processes. The new product director of a major firm once presented the charts shown in Figure 3.5. The process on the left is similar to the one proposed in this book, but the one on the right is how he saw the process actually working in his organization. Although this was meant to be humorous, he said that it was all too real!

The lessons from these examples are that the process is important, but to make it real requires managerial discipline and control. Enthusiasm must be maintained, but discipline enforced. In the case of the cleaning pads, the brand manager was not fired. He was commended on his energy, but strongly presented with the need for accurate forecasts. In the end the product failed due to low repeat rates and increased competitive reaction from the existing brands.

Successful new product development requires creative input and analytical discipline. The process we propose maximizes creative input in the stages of idea generation, design refinement, revision after testing, and profit management. Analytic discipline is used to minimize risks in the phases of design evaluation, pretesting, test market forecasting, controlling the national launch, and transition to maturity. The purpose of this book is to give you the tools and perspective to make real an effective and efficient process of new product development.

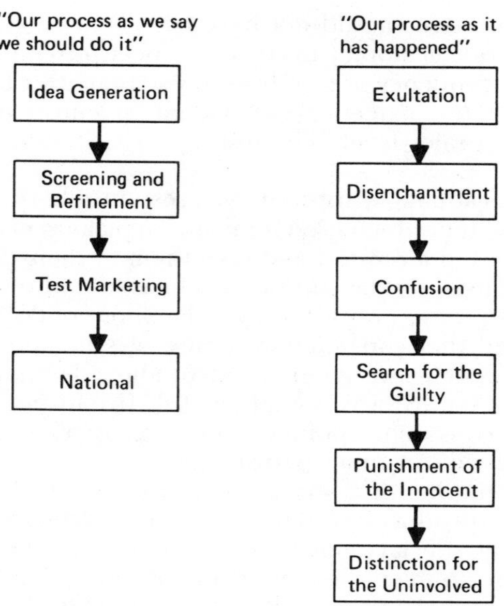

Figure 3.5. *New Product Development Processes — Plan vs. Reality*

WHY PRODUCTS FAIL, AND HOW TO AVOID SUCH FAILURES

Very few systematic studies have been carried out to diagnose why products have failed. This may be due in part to a "don't look back attitude" in organizations, but also it is difficult to untangle what happened and pinpoint the causes of failure. A number of reasons why products fail have been abstracted from research studies (Angelus, 1969; Booz, Allen and Hamilton, 1971; Briscoe, 1973; Cooper, 1975; Crawford, 1977; Davidson, 1976; Rothwell, et al., 1974) and the authors' observations. We discuss each reason for failure and indicate how the new product development process can reduce the risk of the failure or reduce the cost of failures.

Market Too Small

In some cases a good new product is developed that attains a large market share, but it fails since the target market is not large enough to generate sales and profit sufficient to earn satisfactory profits. To avoid this problem, the proposed process (see Figure 3.1) has a specific step in the opportunity identification phase for defining the market and its component parts. With the tools of market definition, potential can be roughly estimated and efforts directed only at high potential market opportunities.

Poor Match for Company

Enthusiasm for an idea may lead a company into a product that does not match its unique skills. For example, many industrial companies fail with great consumer product ideas which lack reasonable overlap with their present distribution system.

In our process we try to be sure the market opportunity matches the company's strategic plan before development is begun. In defining opportunities, strategic planning questions such as "What business are we in?" help in identifying markets and generating ideas that fit the unique capabilities of an organization.

Not New/Not Different

Many firms suffer from a lack of good ideas and therefore never offer really new or different products. We describe some specific creative procedures to generate new and different ideas. These opportunities are assessed in the design evaluation step to determine if consumers perceive them as new and different; later the idea is refined to be sure it is unique.

No Real Benefit

In the design evaluation step, specific research is undertaken to determine if consumers perceive real benefits from product concepts and actual use of the physical products. Davidson (1976) studied 100 new grocery products in Britain and found 74 percent of the successful products offered better performance and only 20 percent of the failures did so. If there is no real benefit, a "NO GO" decision should be made. If the product is only at parity with existing products, the chances of success are very small. We feel a product should be both physically and perceptually better than existing products. Although a parity performance may be marketed, we feel this is a dangerous strategy and do not recommend it.

Poor Positioning/Misunderstanding of Consumer Needs

Positioning is the identification of a set of psychological need attributes and the description of the level of each attribute for a new product. For example, in auto batteries the attributes may be power, long life, ease of maintenance, and price. Sears correctly positioned the "Diehard" as meeting needs for a powerful and long-lasting battery.

RCA failed in computers. In our judgment one factor was because they were poorly positioned. They offered computers technically equivalent to IBM, but at a lower price. This was a poor position against IBM's claims of "solving problems" and usefulness to management in "what if" situations.

RCA did not understand effectively the consumers' needs and position itself accordingly. In hand calculators Hewlett Packard maintained premium prices (about 25 percent) for several years by positioning their product as "uncompromising quality" rather than lowering price and adding computational features. Many other calculator companies tried the latter course and were soon eliminated from the market. Several studies in industrial markets show that a major contributor to failure is poor understanding of consumer needs and market (Cooper, 1975; Briscoe, 1973). In many industrial products poor pricing is a cause of failure. This means the price/benefits positioning was not correct.

We direct much attention to the positioning issue as we discuss new product design. One of the major efforts in successful new product design is to define a good psychological positioning and a set of physical features to back it up.

Little Support from Channel of Distribution

A product must be physically distributed, sold, and serviced. In many cases these functions are carried out by middlemen who have their own decision rules in accepting and promoting new products. In the pretest market phase it is essential to ascertain the response from the trade. This may mean specific research with them such as a simulated placement of the product. In later chapters (10 and 11) we describe how research was used to understand the channel for business school graduates, and the specific decision criteria recruiters have in seeking graduates for their companies.

Forecasting Error

One of the leading causes of failure is due to overestimation of sales. In this book specific management science models are described in each phase of the development process (design, pretest, and test) to forecast consumer acceptance. These tools reduce forecasting errors and provide estimates which are sufficiently accurate for the decision being considered. For example, estimates made in the design phase may be plus or minus 5 to 10 percentage points of market share. But, in most cases, these estimates are sufficient for the GO/NO GO decision on whether to proceed to pretest. Similarly, pretest estimates may be within two percentage points and test market estimates may be within one market share percentage point after adjusting for differences between test conditions and national markets. In each case, the forecasting methods are selected based on the information that is available and on the magnitude of the investment decision.

Competitive Response

Competitors probably will copy your successful products. We feel the best defense is to do very good design work so as to preempt the competitors.

One should try to make the product so good that any later competitive entry can be little more than a parity product. Then the first brand will earn its just reward. We also describe procedures for monitoring the new product launch. These emphasize quick diagnosis and response to competitive actions in order to minimize their impact.

Changes in Consumers' Tastes

Perhaps one of the most difficult problems stems from changes in consumer preference. One of the reasons for the Edsel failure was a shift in consumer taste from big cars to small cars. The initial research to support the Edsel was not bad, but Ford took over three years to reach the market and, during this time, consumers' preferences shifted. More recently, in response to changes in consumer preference and regulation toward high gas mileage, GM reduced the size of all their 1977 and 1978 cars. Although it was the big cars that sold the best in 1976 and 1977, in the subsequent period sales for full-sized (nonluxury) cars dropped.

The dynamics of consumer tastes requires a continued monitoring process so that products can be redesigned, repositioned, dropped, or delayed. In cases of long lead time, the design research to measure preferences should be continued up to introduction. Such research enables accurate assessment and response to consumer dynamics.

Change in Environmental Constraints

New regulations, technology, and material supplies can cause failure. As indicated in Chapter 1, GM aborted the rotary engine in 1977 after new pollution and mileage requirements made it infeasible. Again in the development process, monitoring and adaptive control are critical.

Insufficient Return on Investment

More products fail in profit terms than in sales terms. The product is not a disaster, but sales, margins, and costs are not good enough to meet the original financial objectives. For example, product costs may be higher than expected and prices lower. In the process proposed here, profit is protected by entering big markets with major new ideas that are effectively designed and tested so that there is a substantial margin of safety to allow for unexpected events. The profit and return on investment earned is the final test of the complete process.

Organizational Problems

Many good products fail because of poor organization. The organizational pathologies are many. For example, vested interests in R&D and marketing may prevent effective progress on a good product. Likewise, conflicts between the new products group and the sales organization may kill a good product. Poor relations between marketing research and marketing may cause important negative product information to be ignored (Crawford, 1977; Steffire, 1979). The issues of communication are serious and must be explicitly addressed in organizing for implementation of a new product development process.

Briscoe (1973) studied two new industrial products in depth and found a key problem was a lack of specific objectives for the new product in terms of the company. Rothwell et al. (1974) found in the industrial markets that the absence of higher-level management responsible for new product development contributed to failure. Organizations do need a "champion" for the new product to keep its managerial momentum high. Without clear responsibility, the best designed and tested product may fail due to poor execution of the introduction plans. Good managerial planning and control are essential in industrial, consumer, and service industries.

We attempt to detail some of the issues of organizational structure and planning in Chapter 18 where we discuss organizing the new product effort.

Summary of Reasons for New Product Failures

This section has briefly reviewed twelve of the most common reasons for new product failures. To achieve success, these pitfalls must be avoided. We feel that this requires a disciplined development process such as that in Figure 3.1. But even with such a process, new products continue to be a risky business. In fact, the profit granted by markets is a reward for risk taking. To better understand what resources are at risk in new product development, we now turn our attention to estimating the costs, time, and risks involved in new product development.

COST, TIME, RISK, AND EXPECTED BENEFIT IN NEW PRODUCT DEVELOPMENT

Once an organization selects a proactive strategy, it must allocate its resources (time, money, personnel) to projects and to stages of the new product development process. Given the inherent uncertainty in any creative effort, we do not advocate an allocation of fixed budgets to each phase of the development process, but rather, we advocate a flexible strategy in which specific goals or benchmarks are established. These goals must then be achieved before the product advances to the next phase of the process. Throughout the text we give examples of such goals and develop tests to determine when these goals are achieved. Nonetheless, an organization needs general guidelines for help in planning the allocation of resources to various phases of the development process. This section gives some historic perspectives and estimates the cost, time, risk, and expected benefit involved in the various phases of new product development. Whenever possible, we draw on experiences from a variety of industries to help you assess the specific guidelines relevant to your decision environment.

Cost

Organizations have limited resources and therefore need to manage their investment in the new product development process. To manage this investment, they must know how much each phase costs and what benefit can be expected from the investment. We look first at the cost and then return later in this section to consider the expected benefits. The first cost characteristic of new product investment is that you must invest in a number of bad ideas in order to realize one success. For example, if only one in two products which are taken to test market is successful, then on average you must take two products to test market in order to realize one success. Thus, the total investment necessary to achieve that one success will be greater, on average, than the amount invested directly in the successful product. The goal of management is then to maximize the return on this total investment.

Let us begin by first considering the costs of passing a single product through each of the various phases of development. Table 3.1 shows the costs for industrial development based on seventeen chemical products (Mansfield and Rapoport, 1975). Variances are large, but Table 3.1 does indicate the substantial investments that are involved. On average, $2.2 million are spent on a product successful at each phase of the process. Furthermore, the major part (57 percent) of this investment occurs in introduction—*after* the product is developed.

Table 3.1. Cost of Industrial Product Innovation in the Chemical Industry[a]

Phase	*Average Investment ($000's)*		*Percent of Total Development and Introd. Investment*		*Estimated Range (plus or minus one standard deviation)*
Applied Research	$ 380		17%		(0–770)
Specifications	290		13		(0–680)
Prototype/Pilot Plant	290		13		(110–470)
Total Development Cost		$ 960		43	
Tooling and Manufacturing Facilities	$ 930		41		(270–1580)
Manufacturing Start-Up	180		8		(0–360)
Marketing Start-Up	160		7		(0–430)
Total Development and Introduction Investment		$2,230		100	

[a]*Adapted from Mansfield and Rapoport (1975), p. 1382*

Table 3.2 shows our estimated costs for consumer products. Variances are again wide, but we have tried to represent the typical consumer product costs. Most of the resources for a typical consumer product are allocated to testing (16.5 percent) and introduction (78.7 percent). Although design is important, the costs are not high relative to the total costs of developing a new product. In industrial products (Table 3.1), 43 percent is spent for total development, while in consumer products only 21.3 percent is for development.

Table 3.2. Estimated Costs for New Consumer Product Introduced Nationally in the U.S.

Phase	*Estimated Investment ($000's)*		*Percent of Total Development and Introduction Investment*		*Estimated Range (plus or minus one standard deviation)*
Opportunity Identification	$ 100		1.6%		(0–250)
Design	200		3.1		(100–300)
Testing					
Pretest Market	50		0.8		(25–75)
Test Market	1,000		15.7		(500–1,500)
Total Development Cost		$1,350		21.3	
Introduction					
Advertising, Promotion, Distribution	5,000		78.7		(1,000–9,000)
Total Development and Introduction Investment		$6,350		100.0	

Although there is not much data on the costs of new services, they are probably more similar to consumer products than to industrial products. We estimate that they require fewer funds allocated to design and more to in-the-field pilot testing programs. It is not uncommon in most government programs to allocate over a million dollars to pilot testing a new service.

Table 3.3 compares consumer and industrial product costs from Tables 3.1 and 3.2. In this chart, $50,000 of Mansfield and Rapoport's applied research cost of $380,000 is classified as an "opportunity identification" expense. The remaining "applied research" and "specification" costs is called "design" and rounded off to $620,000. The prototype pilot plant is termed "testing" and can be compared to the "test market expenditure" in consumer products. Introduction includes advertising for the consumer product and manufacturing set up for the industrial product.

Certainly products can be developed for less than these costs, but between two-and six-million dollars is a rough estimate for the total cost incurred by a new product that is successful at each phase of the process.

Of course we must apply these estimates with caution as the variance is quite high and investments vary considerably depending upon the product category. For example, in a study of pharmaceutical firms, Mansfield, et al.

Table 3.3. **Estimated Typical Costs of Developing Major New Products**

	Costs for Consumer Products ($000's)		Costs for Industrial Products ($000's)	
Opportunity Identification	$ 100		$ 50	
Design	200		620	
Testing	1,050		290	
Total Development		$1,350		$ 960
Introduction	5,000		1,270	
Total Investment		$6,350		$2,230

(1971, p. 67) found an average cost of $215,000 ($\sigma$ = 332,000) for product formulation and testing—about 20 percent of the costs shown in Table 3.1. In a study of railroads, construction, and computers, Marquis (1969) found development costs of $100,000 for over two-thirds of the innovations. As rough guidelines to modify these estimates, we cite the results of a statistical analysis by Mansfield and Rapoport (1975) of the chemical industry which found that investments in development tended to be larger if: (1) the size of the new product's expected revenue was large, (2) the firm was larger, and (3) the firm had little previous technological experience in the area.

Overall, it is clear that new product development requires a major commitment of resources and that most funds are at risk in the final testing and introduction phases. Managerially, this means: (1) that the time when many creative ideas are to be encouraged is early in design of the product when less investment is at risk, and (2) that it is important to eliminate failures early before they lead to a major loss in investment.

The wide variances in cost suggest that it is difficult to forecast exact costs, but that these costs can be managed with a careful monitoring and evaluation of the process to identify major risks and prevent the expenditure of good money on bad ideas.

Time

The timing of investment can be almost as important as the magnitude of the investment. Too long a development process can result in lost opportunities, while too short a process can ignore key issues and result in failure. Unfortunately, the time required to develop a product is difficult to estimate in advance. It depends on creative breakthroughs and getting the product and marketing strategy right before continuing in the process.

We can get some estimate of the timing by historically examining the time from the first identification of an idea to product introduction. Table 3.4 shows that although some products like the Polaroid Land Cameras and Wisk laundry detergent moved to the market in one or two years, many products like penicillin, Bendix washer/dryers, and Maxim coffee took ten years or more. Some products were delayed waiting for physical product

Table 3.4. Elapsed Time From Idea to Test Market or Product Introduction for Representative Consumer and Industrial Products[a]

Product	*Idea Born*	*Test Market*	*Full-Scale Launch*	*Elapsed Time*
Birds Eye Frozen Foods	1908	1923	—[b]	15 years
Ban (roll-on deodorant)	1948	1954	1955	6 years
Calm (powder deodorant)	1959	—	1964	5 years
Chlorodent (toothpaste)	1930s	1951	1952	11-21 years
Citroid (cold compound)	1954–55	—	1956	1–2 years
Coldene (cold remedy, liquid)	1954	1955	1956	1 year
Crest (fluoride toothpaste)	1945	1955	1956	10 years
Decaf (instant coffee)	1947	1953	—	6 years
Flav-R-Straws	1953	1956	1957	3 years
Gerber (baby foods)	1927	1928	—	1 year
Hills Bros. (instant coffee)	1934	1956	—	22 years
Johnson (liquid shoe polish, inc. applicator)	1957	1960	1961	3 years
Lustre Creme (liquid shampoo)	1950	—	1958	8 years
Marlboro (filter cigarettes)	1953	1955	—	2 years
Maxim (freeze-dried instant coffee)	1954	1964	—	10 years
Minute Maid (frozen orange juice)	1944	1946	—	2 years
Minute Rice	1931	—	1949	18 years
Purina Dog Chow	1951	1955	1957	4 years
Red Kettle (dry soup)	1943	—	1962	19 years
Stripe (toothpaste)	1952	1957	1958	6 years
Wisk (liquid laundry detergent)	1955	1956	—	1 year
Bendix (washer/dryer)	before 1941	—	1953	12+ years

Table 3.4. (Continued)

Product	*Idea Born*	*Test Market*	*Full-Scale Launch*	*Elapsed Time*
Eversharp (ball-point pen)	1958	1958	1959	8 months
Fairchild (8mm sound projector)	1961	1963	—	2 years
Floron (floor tile)	1947–48	—	1953	5–6 years
G.E. (electric toothbrush)	1958–59	1961	1962	3–4 years
Polaroid Land Camera	1945–46	1947–48	—	2 years
Polaroid Color Camera	1948	1963	1963	15+ years
Scripto Felt Tip Pen	1959	1961	1961	2 years
Sinclair (gasoline, oil)	1952	—	1953	6 months
Smith Corona (portable electric typewriter)	1952	1957	—	5 years
Talon (zippers)	1883	1913	1918	30 years
Television	1884	1939	1946–47	55 years
Isothalic (chemical compound for house paints)	1951	1957–58	—	6–7 years
Krilium (soil conditioner)	1939	1952	—	12½ years
Page Master (pocket paging system)	1955	—	1957	2 years
Penicillin	1928	1943	—	15 years
Transistors	1940	1955–56	—	15–16 years
Xerox	1935	1950	—	15 years

[a] *Adapted from Adler (1966), pp. 18–21*
[b] *Dash means exact data was not given*

improvement while others waited until a market could be found. Some of these delays were the result of careful, purposive development of the idea while others were simply organizational time lags.

As with cost, it is instructive to examine where in the design process these delays occur. Table 3.5 reports the average time utilized in each phase for a sample of the chemical innovations represented in Table 3.1. Mansfield, et al. (1971) found that the allocations to categories varied across industries but overall, the majority of time was allocated to development (applied research, specification, and prototype development). The average total time in chemicals, machinery, and electronics was 51 months and the average time in the pharmaceutical industry was 60 months.

Table 3.5. Average Percentage of Total Elapsed Time in Various Stages of Innovation Activities

Stage	*Average Percentage of Total Elapsed Time*	*Range on % (Plus or Minus One Standard Deviation)*
Applied Research	62.0%	34–90
Specification	34.6	7–63
Prototype/Pilot Plant	35.0	13–57
Tool and Manufacturing	21.9	4–40
Manufacturing Start-up	7.8	2–14
(Overlap in phases[b])	59.1	10–108

[a]*Mansfield, et al. (1971), p. 120*
[b]*Overlap is the sum of the months of overlap between phases divided by the total time*

It is interesting to note that many organizations reduced total time by overlapping the various phases. This total time can be further reduced by applying more funding. Mansfield, et al. (p. 140) found that a 1-percent reduction in time requires anywhere from a 0.5 percent to 1.6 percent increase in cost, depending upon whether or not the product is already rushed.

Our experience in the consumer package goods market is expressed in Table 3.6 for the case where no major R&D breakthrough is required. Two-and-one-half years is a reasonable estimate of the average time, if a product is successful at each phase. While "me too" or minor variations of products may be rushed to market in a few months, 18 months is a very fast schedule for a significant new product. If substantial R&D work is required, at least one year could be added to the estimate.

Table 3.6. Estimated Time Required for Development of New Consumer Products

	Average Time Span (Months)	*Range Between Plus and Minus One Standard Deviation*
Opportunity Identification	5	4–8
Design	6	2–15
Testing		
Pretest Market	3	2–5
Test Market	9	6–12
Introduction Setup	4	2–6
Total Time	27	

In summary, the usual time for developing, testing, and introducing new products is probably longer than most people would think. For a typical industrial production, five years is a reasonable estimate of time for design, testing, and setup of manufacturing. In the case of consumer products, the time is about two to two-and-a-half years, if the product is successful at each phase and if no major R&D work is required. Given this major commitment of time, it is crucial that organizations have a number of ideas in various stages of development.

Organizations must manage this process by ensuring that sufficient time is allocated to each phase but that abnormal delays do not occur for nonproductive reasons. The best way to do this is to be receptive to sources of ideas and to encourage their careful development and evaluation.

Risk

Not only does it typically take several years and between two- and six-million dollars to launch a major new product, but there are also substantial risks involved. While the emerging new product design and testing techniques can greatly reduce the risk in new product development, it is useful to look at historically reported failure rates to help understand the magnitude of risk.

The risks of industrial products have been studied by Mansfield and Wagner (1975). Table 3.7 indicates conditional probabilities of success at each phase and the total probability of success. As with cost and time, there is considerable variation in these probabilities. For example, while the average probability for technical completion is .57, Mansfield and Wagner (1975) found it as low as .32 for drug innovations and as high as .73 for electronics innovations.

Table 3.7. Probabilities of Technical Completion, Commercialization, and Economic Success of New Industrial Products[a]

	Probability of Technical Completion		*Probability of Commercialization Given Technical Completion*		*Probability of Economic Success Given Commercialization*		*Overall Probability of Success*
Average	0.57	×	0.65	×	0.74	=	0.27

[a]*Mansfield and Wagner (1975), p. 181*

In analyzing the reasons for lack of technical competition, Mansfield, et al. found in an intensive study of three labs that 62 percent of the technical projects that were terminated were stopped because of poor commercial prospects rather than technical problems. Gerstenfeld (1970) similarly found that 52 percent of these R&D projects not successfully completed were failures for non-technical reasons.

In consumer products, failure rates have not been studied as systematically as in industrial products. Most of the data is for frequently purchased consumer products. In 1961, 1971 and again in 1977, the A. C. Nielsen Company reported the "success ratio" of new brands (health and beauty aids, household, and grocery products) that had been test-marketed through their facilities (Nielsen Marketing Service, 1971 and 1979). The 1961 study included 103 new brands, the 1971 study covered 204 items, and the 1977 study covered 228 items. "Success" was defined by the "manufacturer's judgment of each brand's performance in test"—namely, whether or not the brand was launched nationally. Brands withdrawn from test markets or not introduced nationally were considered "failures." By these criteria, the most recent data suggest that about one-third of the new products test marketed were launched nationally (54.4 percent in 1961, 46.6 percent in 1971, and 35.5 percent in 1977).

Similarly, Stanton's 1967 study of 28 major consumer grocery and drug product companies found that in 46 percent of the 54 specific test market experiences covered by the study, test market sales "fell short of management expectations." In contrast, Buzzell and Nourse (1967) observed in their study of the food industry that only 32 percent of 84 "distinctly new food products" developed in the 1954–1964 period were discontinued after test marketing. This somewhat lower failure rate is probably related to the special character of the sample of products studied which were "substantially different in form, ingredients, or processing methods from other products previously marketed by a *given* company" (Buzzell and Nourse, p. 96). At the individual firm level, 10-year test market success rates of 46 percent have been reported for General Foods in the U.S. (Rothwell, et al., 1974, p. 50) and 60 percent in the U.K. (Cadbury, 1975, p. 98).

Thus, failure rates ranging from 45 to 65 percent roughly bracket the publicly reported record of test market experience in the packaged goods field for failure in test market. Although there is considerable discussion of the failure rate in national markets, few empirical estimates exist. We estimate that with a well-run test market the chances of national market failure are in the 10 to 20 percent range. Finally, based on our experience, we estimate the probability of a successful design to be somewhere between 0.40 and 0.60. Table 3.8 shows the median of these estimates for consumer products. Based on the numbers, the probability of market success, given an identified opportunity, is 0.19 (0.50×0.45×0.85). This compares to 0.27 (0.57×0.65×0.74) for the case of industrial products.

Table 3.8. Estimated Probabilities of Success for Design, Testing, and Introduction of New Consumer Products

Probability of Successful Design		*Probability of Successful Test Market Given Design*		*Probability of Market Success Given Successful Test Market*		*Overall Probability of Success*
0.50	×	0.45	×	0.85	=	0.19

These figures indicate that new product development is indeed risky at all phases of development. This risk, coupled with the tremendous investment in time and money, implies that for the continued health of an organization this process must be carefully managed. We now put cost, time, and risk together to produce estimates of the expected benefit from investments in various stages of development.

Expected Cost and Time

From the analysis of cost, it is clear that the largest investment occurs in testing and in introduction. Thus, a failure in a national market is much more costly than a failure in test market or a failure in the design phase. On the other hand, the highest risk (lowest probability of success) occurs early in the design phase. Furthermore, the design and testing phases delay national introduction and could give competitors a chance to "beat you to the market." The questions naturally arise: What is the true cost of the development process? and what benefit, if any, is to be expected from the various phases of the development process? To examine these questions, we must look closely at both the cost of each phase and the probability of success in each phase. Assume that the firm has decided to undertake a program of new product development.

Suppose the new product team is under pressure from top management to produce a success every twelve months, no matter the cost. Suppose further they have many potential ideas from previous analyses. Then they may decide that to save time they will select enough ideas to be reasonably sure of a success. After thinking about this for a day or so, they realize that not only will this strategy speed things up, but it will "save" the investment in design and testing. The expected cost of this strategy is shown in Table 3.9.

Table 3.9. Expected Benefit If Organization Uses Test Methods

	"Save Development Costs (Design and Testing)"		
	Average Cost ($000's)	*Probability of Success*	*Expected Cost ($000's)*
Introduction	$5,000	0.19	$26,316[a]
		"Save Design Costs"	
	Average Cost ($000's)	*Probability of Success*	*Expected Cost ($000's)*
Test Market	$1,000	0.225	$ 5,228[b]
Introduction	5,000	0.85	5,882[c]
Total Expected Investment			$11,110

Net Benefit of Test Marketing = $26,316 − $11,110 = $15,206,000

[a]*$5,000/0.19 = 26,316;* [b]*1,000/(0.225 × 0.85) = 5,228;* [c]*5,000/0.85 = 5,882*

On average, they must bring 5.26 products to market to achieve one success. [Average attempts = 1/(probability of success), thus 5.26 = 1/(0.19).] At a cost of $5 million per try, this strategy has an expected cost of $26.3 million for one success.

From the above calculations, the new product team realizes that it is quite expensive to skip test market. Clearly, it is better to take your chances at test market when less investment is at risk. Perhaps management will be satisfied with a success in test market after twelve months and a success in the marketplace after eighteen months. Based on this strategy, they may decide to keep the test market phase, but skip the design process. These computations are shown in Table 3.9. Note that the probability of success in test market without the design phase is at most 22.5 percent (0.45 × 0.50). This is actually high since the design phase is normally used to refine the product strategy. Actual probabilities without design are more likely in the range 10 to 20 percent.

On average, over four products must be brought to test market to achieve one successful test market and 1.18 test market successes are needed for one successful product. [4.44 = 1/(0.225), 1.18 = 1/(0.85)] Together, this means that on average 5.24 products must be brought to test market resulting in 1.18 products to national introduction to achieve one national success. [5.24 = (4.44) × (1.18) = 1/(0.225 × 0.85)] As shown in Table 3.9, the total expected cost of this strategy is $11.1 million dollars or an expected savings of $15.2 million for using test markets. If the refinements due to test markets are considered in the above calculations, the savings are even greater. Test market investments pay off.

Continuing, we can calculate the expected cost if the opportunity identification and the design phase are used. Table 3.10 shows that for one new product success, the expected cost of design, testing, and introduction is $9.6 million or a net expected benefit of $1.5 million for using the design phase. In this calculation we assume that the $100,000 spent on opportunity identification will lead to a market definition suitable to the company and a set of ideas worthy of design effort. Thus we do not use an expected cost computation for opportunity identification. While not as dramatic as the benefits

Table 3.10. Expected Benefit If Organization Invests in Design

	Average Cost ($000's)	*Probability of Success*	*Expected Cost ($000's)*	
Opportunity Identification	$ 100	—	$ 100	
Design	200	0.50	1,046[a]	
Test Market	1,000	0.45	2,614[b]	
Total Development				3,760
National Introductions	5,000	0.85	5,882	
Total Expected Investment				$9,642
Net Benefit from Front-End Investment = $1,468,000[c]				

[a] *200/(0.50 × 0.45 × 0.85) = 1,046;*
[b] *1,000/(0.45 × 0.85) = 2,614;*
[c] *11,110 − 9,642 = 1,468*

from test market, this 13 percent savings from front-end spending is a major reduction in expected cost.

Finally, we see from Table 3.11 that this investment can be reduced still further by the use of pretest market which identifies failures at a cost much below test market. Investment early in the process pays off.

Table 3.11. Expected Benefit If Organization Invests in Pretest Market

	Average Cost ($000's)	Probability of Success	Expected Cost ($000's)	
Opportunity Identification	$ 100	—	$ 100	
Design	200	0.50	980	
Testing				
Pretest Market	50	0.60[a]	123	
Test Market	1,000	0.80	1,471	
Total Development Cost				2,674
National Introduction	5,000	0.85	5,882	
Total Expected Investment				$8,556
Net Benefit from Pretest Market = $1,086,000[b]				

[a] *Probabilities are based on actual pretest experience*
[b] *9,642 − 8,556 = 1,086*

These lessons of early evaluation and refinement of new products are important to new product managers. Two hundred thousand dollars, or even fifty thousand dollars, appears to be a major investment, especially since it means time delays and a seeming increase in development cost for the one successful product. It is not uncommon for managers to fall into this trap of managerial myopia. The real benefits (as computed in Tables 3.9 through 3.11) come from the decrease in cost resulting from decreased spending on failures. What is not shown in Tables 3.9 through 3.11 is that the benefits are even greater because each phase in the design process refines the product enhancing its probability of success in the next phase and often resulting in increased profitability once success is achieved.

The above computations are for consumer products. Similar calculations can be made for industrial products. For example, Table 3.12 shows that the total expected investment for a typical industrial product is about $4.6 million. Computations for the expected benefit of industrial product development can be made in an analogous way to consumer products, but should include enhanced success probabilities and the fact that some development must occur for any launch.

We have shown explicitly the calculations for the expected cost savings due to front-end investment. The same type of calculation can be done for the expected reduction (delay) in time due to front-end investment. For example, the addition of pretest market analysis to the process saves about 6 months in addition to the cost savings (see Table 3.13). The phenomenon is the same as that for expected cost. Front-end investment is time efficient because we eliminate losers rapidly and enhance the probability of success for the winners.

Of course the costs, time, and risk all have large variances, and the

Table 3.12. Expected Costs for Major Successful New Industrial Products

	Average Cost ($000's)	*Probability of Success*	*Expected Cost ($000's)*	
Opportunity Identification	$ 50	—	$ 50	
Design	620	0.57	2,261	
Testing	290	0.65	603	
Total Development				2,914
Introduction[a]	1,270	0.74	1,716	
Total Expected Investment				$4,630

[a]*Assumes no salvage of plant and equipment*

Table 3.13. Expected Time of Development of New, Frequently Purchased Consumer Product

	Expected Time of Development Without Pretest Market Analysis			
	Average Time (months)	*Probability of Success*	*Expected Time (months)*	
Opportunity Identification	5	—	5	
Design	6	0.50	31	
Testing				
Test Market	9	0.45	24	
Total Development				60
Introduction Setup	4	0.85	5	
Total Expected Time				65

	Expected Time of Development With Pretest Market Analysis			
	Average Time (months)	*Probability of Success*	*Expected Time (months)*	
Opportunity Identification	5	—	5	
Design	6	0.50	29	
Testing				
Pretest Market	3	0.60	7	
Test Market	9	0.80	13	
Total Development				54
Introduction Setup	4	0.85	5	
Total Expected Time				59

expected investments and expected benefits vary by industry and by product category. Each new product team must make its own decisions and set its own guidelines for each phase in the new product development process. Larger firms can afford the front-end costs and, in most cases, will realize benefits from investments in design and testing. Entrepreneurs and some small firms who cannot deal with averages or large firms who choose to rush the process will have to accept higher risks and higher expected costs. The new common sense is that it is better to take risks when less is at stake, and it is better to iteratively evaluate and refine ideas so that the best possible strategy is obtained before national introduction.

Planning a New Product Development Program

In planning development, managers should be sure to budget enough time and funds to carefully implement the complete process. On average, in the case of a typical frequently purchased consumer product, 2.7-million dollars and over five years is required for development. This is more than twice the development cost and time of a product that would be successful at each phase in the process. Budgeting based on the costs of such a theoretic successful product will be inadequate, unlikely to be successful, and create managerial pressures due to unrealistic time and performance expectations.

Managers should consider the industry and company they represent, define a set of new product development steps, establish their costs, duration, and probabilities of successful completion, and use the expected time and cost figures to budget for the new product development effort. Then the organization will have the capability to correctly conduct each phase in the development process and achieve the best balance between the risk and rewards of new product development.

SUMMARY

This chapter has introduced a proactive new product development process based on a sequential decision process of: opportunity identification, design, testing, introduction, and profit management. This development process is designed both to encourage creativity in organizations and to reduce risk. It suggests emphasis on the early phases and on evaluation and refinement through a development program. The emphasis is to avoid expensive failures, yet develop a product that is different from existing products. This process cannot eliminate risk, but it can manage that risk and lead to savings in expected time and cost. Its continued use should lead to more and better new products.

In the next chapter we present the experiences of one new product director in applying the new product process in the real world. You may find it insightful to compare his experience to the statistical results described in this chapter. Then we begin our formal study of the development process with the phase of opportunity identification and proceed sequentially through each stage in the new product development process.

REVIEW QUESTIONS

3.1 Outline and briefly describe the five basic steps in the sequential decision process for a new product.

3.2 Develop a new squid concept alternative not given in the chapter.

3.3 What is the difference between a test market and a pretest market?

3.4 The Weave Gotit-2 Company has a simple new product strategy. They wait until a competitor launches a successful new product and then they copy it. In this way, they invest nothing in research, nothing in developing new markets and nothing in product testing. The company has a steady stream of new products which, in many cases, perform as well as the competitive products.
 a. Why might a company adopt such a philosophy?
 b. Do you believe a company should follow this strategy? Why?
 c. What would be the long-run outlook for this company?

3.5 Can a new product be profitable and still be classified as unsuccessful? If so, how?

3.6 Consider some recent new product failures. Speculate why these new products were unsuccessful.

3.7 How can the cost of transforming a specific new product idea into a successful new product be estimated?

3.8 Why does the average total time in developing a new product vary from industry to industry?

3.9 Explain intuitively why upfront research is often so valuable.

3.10 In the text we computed the expected benefit of each stage in the new product development process. What is not shown in Tables 3.9 to 3.11 is that the benefits would be greater if we accounted for the fact that each stage enhances the product's ultimate success. Suppose that the following probabilities reflect such strategic improvements. Compute the expected benefit of each stage. What other phenomena affect the calculation of expected benefit?

	Average Cost	*Probability of Success*	*Expected Cost*
A. Introduction	5,000	.19	?
B. Test Market	1,000	.225	?
Introduction	5,000	.90	?
Total			?
C. Opportunity Identification	100	—	100
Design	200	.50	?
Test Market	1,000	.55	?
Introduction	5,000	.90	?
Total			?

Chapter 4

New Product Development — A Manager's Perspective*

EDITORIAL NOTE: In the first three chapters of this book we have described issues underlying a proactive new product development process. In practice each top manager must customize a particular process that meets the company's goals, matches the characteristics of the industry, and is compatible with the company's organizational structure. A specific sequence of developmental activities must be defined to best reflect the company's needs. In this chapter a practicing manager, Mr. Thomas Hatch, describes his experiences in building a new product development process.

Mr. Thomas Hatch has been active in new products for fifteen years in the consumer products field at Gillette, Miles Laboratories, and Mennen. In this chapter, he portrays one approach to the complexity and reality of managing product development. We feel this case history will make the remainder of this book more meaningful and relevant. Although Mr. Hatch's experience has been predominantly in consumer packaged goods, most of his comments apply equally well to consumer durables, industrial products, and services. Throughout the text, and specifically in Chapter 19, we return to the issues of implementing a new product development process for large and small organizations in consumer, industrial, and service industries.

The methods Mr. Hatch has utilized reflect the state of the art of management practice in today's leading consumer goods companies. While most companies utilize a formal new product process, fewer use management science methods to support decisions. There is a trend toward improving the new product development process through the use of analytic methods. This book is designed to equip managers to utilize such advanced methods in product development and successfully compete in such more sophisticated environments. The specific analytic techniques and concepts Mr. Hatch refers to are covered in greater detail in later chapters.

*By Thomas Hatch, Vice President of Marketing, Mennen Company.

INTRODUCTION TO PRODUCT DEVELOPMENT

I was the Vice President of Growth and Development of the Consumer Products Division of Miles Laboratories for a period of five years, and during that time developed and implemented a new product development process that integrated traditional qualitative market research techniques with quantitative management science models. The New Product Development Process implemented during this time represented a significant breakthrough in the development of new products for Miles Laboratories. The results from this integration were dramatic. This approach used management science models to significantly reduce the failure rate and priorize the work by putting scarce resources, time, money, effort, and creativity into those projects which offer the most potential. Simply stated, the use of this new product development process greatly improved the productivity in new product work.

There has been concern that management science models are too academic and are not useful in the real business world. My experience indicates management science models are effective tools for management. I have used them and they work. Currently they are not widely used, but I believe that it is the wave of the future. I believe that all successful businesses that manage to generate growth and profitability at acceptable levels in the intensely competitive markets of the 1980s will be those firms that have successfully integrated these advanced management tools into their operations.

Choice of a Strategy

The corporations in which I have worked have had a requirement for growth. Each year top management had to demonstrate growth in terms of sales and profitability to its stockholders. This is a criterion established by the financial community for all corporations. Growth is a sign of the health of the organization and the ability to generate successful new businesses is a test of validity of the management. Wall Street watches this aspect of corporate development very closely. It is the sign of a particularly well managed company. Thus, the development of growth has become a basic tenet of business.

As a result of this requirement of growth, management has been forced to look at various methods for growth. One is by internal new product development and one is by external growth through acquisitions. At one point in the 1960s a tremendous wave of acquisition took place because many firms had been badly burned in new product development. Acquisition looked like a low risk path to success. However, acquisition as an ongoing method for growth has its limitations. As more and more companies become acquired, it becomes difficult to find suitable companies to purchase at attractive prices. Another significant problem is the effective integration of the acquired company into the parent. This does not mean that acquisition as a method for generating growth is unattractive, but I feel internal development is the most effective method for maintaining consistent corporate growth. It may be reasonable to have an active acquisition program, but businessmen should not count on it alone to create the kind of ongoing growth that is necessary to fulfill objectives set down in the typical five-year long range plan of a corporation.

If a major element of growth must come from internal development, it can only come from two areas, established brands or new products. Some established brands have shown consistent vitality, have lasted long periods of time, withstood incredible competitive onslaughts, and grown significantly throughout the entire process. However, as major packaged goods markets mature in the future, it will become harder and harder to achieve growth from existing business because of increased levels of competition and a dramatic increase in the number of competitive new product introductions. As a result, I feel that a firm looking for consistent growth in profit must be active in new product development. Therefore, the strategy I recommend is to proactively pursue internal new product growth and development.

Dealing with Risk

As markets have become intensively competitive and are proliferated with many new products, the failure rate of new products has soared. This risk is not just an abstract concept; it has become a painful reality to many firms. My first two assignments in new products while working at Gillette were on Happy Face, a facial washing cream, and Lectro-Set hair spray. Both products proved to be major failures in the national marketplace, costing the firm many millions of dollars in each instance. Gillette is a very successful packaged good marketer, but its strategy for product development has had both success and failure. Gillette spends millions of dollars each year in an attempt to find a success. Ten to twenty projects are generated and then screened to find a winner. One year the success was Cricket Lighters, another year Trac II razor blades and another year Atra razors. Gillette has learned over the years to successfully manage this type of new product program in the fast moving and highly competitive markets in which they compete. My experience indicates new product development can be done most effectively by utilizing a structured and disciplined new product development process consisting of an integration of traditional qualitative marketing research tools and the new quantitative management science models.

Organizational Perspective

The uncertainty associated with any new venture makes most top managers uncomfortable. They realize they must have new products for growth, but they also have seen the tremendous damage inflicted on a corporation by a major loss. A new product plunges the corporation into the unknown and uncertain future and represents substantial change for the organization. That element creates high levels of discomfort for the organization and its employees. I have seen very few organizations that have managed to internalize the ability to cope with change and create a positive environment surrounding the whole area of new product development.

Many corporations have created a risk aversion orientation to new product work. This is an inappropriate stance if they are to be successful in creating innovations. I feel the best way to deal with the cautiousness of top management and the uninnovative propensities of most organizations is through a separate growth and development group that reports directly to the chief executive officer. This group should have a clear charter to innovate and produce major growth. With this charter, the established product business can proceed without the anxiety of new product risk, and the corporation can make the necessary commitment of resources to new product development. At Miles, the Vice President of Marketing ran the Product Management group and the Vice President of Growth and Development ran the New Products group. Both men reported organizationally to the President. This gives new products the structural advantage inside the organization necessary to combat the problems centering around priority assessments and work load assignment. The new products program has the stamp of approval of the company President and this gives these projects the leverage they need inside the organization.

In order to encourage creativity and teamwork, I feel venture teams should be used in the growth and development organization. The objective of the venture team is to break individuals out of their specific functional areas (R&D, marketing, advertising, and production) and put them into a close knit problem-solving group with a specific assignment. This creates an entrepreneurial spirit, improves the creative output, creates personal identification with the project and, in my experience, develops outstanding results. In order to be successful with new products in the future, we have to alter the traditional structure of organizations and implement a well-defined, structured, and highly sophisticated new product development department.

An important advantage of a new product department structured in this way is that it causes the company to allocate appropriate resources to growth and development. As Peter Drucker so clearly pointed out in his book, *Management Tasks, Responsibilities, and Practices* (1974), 90 percent of the revenue of a corporation is created by 10 percent of the transactions, and the other 90 percent of the transactions only account for 10 percent of the revenue. Corporations often foster the "busy work" syndrome and this becomes a very difficult climate for new products because in any setting with limited time, people, and capital resources, there is a work overload. As these corporations have grown from one product companies to multi-division, multi-national conglomerates, many have become unwieldy bureaucratic organizations that are extremely difficult to manage and maintain with any reasonable agreement on priorities and work assignments.

Usually the easiest, the safest, and the most ordinary projects are the ones that get worked on first. An example of this may be helpful. At Gillette we were working on an innovative new skin care concept with a group in the lab and work was not proceeding very well. A small project of revising the color of an existing skin care product kept sidetracking the new product work. In the larger perspective, making Deep Magic pinker should be much lower on the priority scale than the development of a breakthrough new product. Only through a formal new product commitment and an organizational structure devoted to new products with a disciplined new product development system can the correct priority be given to growth and development and the future.

Innovation Philosophy

In my view new product development is the most important mission the firm has. We must create a mental set that treats existing business as if it is deteriorating, whether that is true or not, and as a result, generate replacement products to create new sales volume.

It is important at the outset to make it understood that in order to do this with a new product, that product must be perceived as "better" than the product currently being used by the consumer. This product improvement must be on a critical dimension that is perceived as important by the consumer. Major product improvements must be created. It is not possible to take away the business that P & G has with Crest simply by offering the consumer another anti-cavity toothpaste. In these highly competitive markets it is not possible to enter the market with any hope of being successful with a me-too product. When Miles Laboratories launched Alka-2 into the antacid market, it was the 38th chewable antacid launched into the market. The world was not waiting for another chewable antacid. In that setting you really have to have something new, different and important to say to the consumer if you expect the product to sell. In product development you are changing consumer purchase habits and to do that the product must have a major consumer advantage over existing brands. The organization must develop a superior product, one that represents a real innovation.

In the next section I will outline the new product process implemented at Miles Laboratories that reflects a proactive strategy, my organizational perspective, and an innovation philosophy.

IMPLEMENTING A PRODUCT DEVELOPMENT SYSTEM

Figure 4.1 shows an overview of the new product development system. The diagram clearly represents the various work steps that occur. After defining the overall area of business development, effort is directed at concept

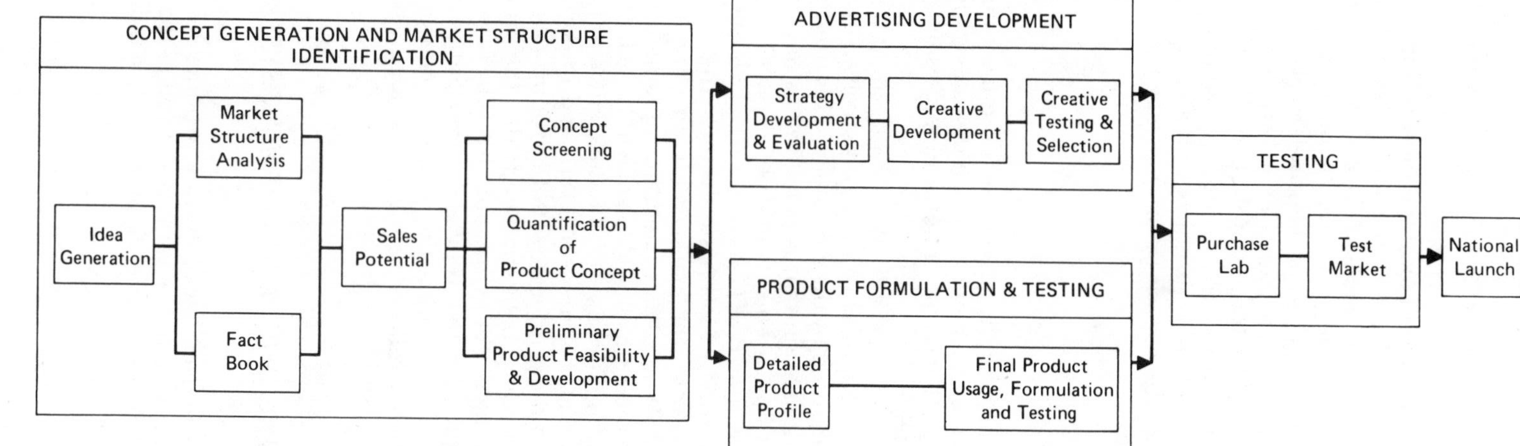

Figure 4.1. *Growth and Development New Product Development Process*

generation and market structure identification. If the idea passes the GO/NO GO screening criteria, advertising is developed and the product formulation specified and tested. If these results meet standards, pretest market and test marketing activities are undertaken. If successful, the product is launched. The system is an integration of traditional qualitative research which has been used in the package goods field for the last fifteen to twenty years and the management science computer models which have been developed in the last five years.

Concept Generation and Market Structure Identification

The first step in this new product development program centers around concept generation and market identification. In this part of the system the work falls under four headings: (1) idea generation, (2) structure identification, (3) sales potential, and (4) concept screening.

Idea Generation.* Idea generation is a creative process and requires concerted effort. In my experience the most fruitful approach is to hold meetings with the various departments inside the company: the lab, the marketing group, the sales department, and others that have expertise within the organization. Then meetings are scheduled with the advertising agency to get their input. The purpose of these meetings is to update the new products group on current technical developments, competitive activities, and consumer trends that impact on the market with the objective of developing a list of new product ideas that represent significant new product breakthroughs.

Consumer ideas are obtained from focus group interview sessions which determine the consumers' perceptions of the brands, the advertising, and the way the products are used. We are looking for competitive vulnerabilities and consumer trends in the category. How do consumers see the category changing relative to the way they use the product now, and how they anticipate they will use it in the future? The result of this idea generation step is: (1) the consumers' view of the market in qualitative terms, (2) an understanding of the technical development potential, and (3) a list of possible new product ideas in untested, raw idea form.

***Market Structure Analysis.*†** The next step in the new product process is to implement a market structure analysis (see Figure 4.1). Market structure analysis delineates the consumers' perception of the market by building a map outlining the critical consumer dimensions, placing existing brands on the map, and indicating new product opportunities. This information is derived from interviews with 300 to 400 consumers.

For example, the market structure analysis of the analgesics market indicated that the critical consumer dimensions were "effectiveness" and

*See Chapter 6 for more details.

†See Chapters 5 and 9 for more details.

"gentleness." The placement of the major brands as points on the map indicated that there was a significant marketing opportunity for a product positioned for "effective pain relief without aspirin side effects."‡

This market structure analysis or mapping technique is extremely helpful because it gets the thinking of the new products group on the right track in terms of the consumer's view of the market. It clearly isolates the critical consumer dimensions and how the consumer perceives the existing brands relative to an ideal product or products. This offers the new product group focus for further technical and marketing development work. It indicates where the marketing opportunity in the market exists, and gives a rough estimate of how big the opportunity is for a well positioned product.

In the analgesic project there was tremendous synergy between the focus group research work being done on a number of analgesic new product ideas and the market structure analysis. The following example illustrates this synergy. While working on the analgesic market we had developed four new product ideas which appeared to have some vitality and were running these ideas through consumer focus group interviews to get the consumers' reaction to the ideas. It was apparent that we did not have a clear understanding of the market because the consumer was not responding to the various new product ideas as we had anticipated. After we ran the market structure analysis, the map of that market clearly pointed out an unrecognized dimension was emerging. This gave us insight into why the consumer was not responding to these new product concepts the way we felt they should respond. The new product group was approaching the market with an old view and were not aware of the new "gentleness" dimension. The market structure pointed out this error. Clearly, without the mapping analysis we would have gone down the wrong track in our developmental effort.

In order to supplement the mapping a fact book or business review document is pulled together outlining all of the knowledge about the particular category. At this time much of the syndicated market data should be purchased for inclusion in this document so that the new products group is dealing with accurate and up-to-date market size, share, and segmentation data.

Sales Potential.* Now we estimate the potential of a new product entry into the market structure. This is implemented by another management science computer model, in this instance a trial and repeat model† based on executive judgment, since most of the data required as inputs is not available at this early stage in the process. The purpose of running such a model is two-fold: (1) to establish a rough estimate of the size of the business potential, e.g., $4,000,000 vs. $40,000,000 and (2) to establish a base case for use of this model in continually monitoring the sales forecast of the business and its advances throughout the development process. We replace executive judgment with empirical data as it becomes available. This has the effect of

‡See Chapter 9 for a map of the analgesics category.

*Forecasting of sales potential is covered in Chapter 11.

†See Chapters 11, 14, and 15 for greater detail on this model.

continually refining the market share and dollar volume estimates of the size of the business.

***Concept Screening.*‡** At this stage in the process we have been through an idea generation step, market structure analysis, and a trial and repeat model. We have developed an up-to-date view of the competitive environment from the consumer's view, and how the consumer viewed the new product concept relative to the market structure. We next implement a number of focus group interviews for concept screening and refinement. The primary objective of these groups is to develop the right wording and logical flow of the selling proposition for the new product concept.

As a validation of the concept potential, we quantify the concept statement in its most complete form through market research. We take the concept statement which was refined in focus group interviews and add to that a picture of the product and a brand name. This is exposed to 300 or 400 consumers to determine intent-to-purchase levels, primary points of difference over competition and major/minor advantages over the product currently used. This is the most stringent test of the idea. A concept that passes this test moves on to advertising development.

Advertising Development

At this point we begin advertising development* and product formulation work. As a result of the earlier work, all of the advertising and technical development has a high degree of focus. It is targeted at the opportunity pointed out in the idea generation stage of the new product development program. This section discusses advertising and the next section product formulation.

The main advertising objective is to create copy that clearly outlines the product's point of difference to the consumer. Advertising development is initiated by asking the account team and creatives to observe the concept focus groups so they can get a first hand look at real consumers and listen to them talk about the category. Then, three documents are written: (1) a strategy statement, (2) a consumer profile, and (3) an end benefit psychology statement.

The objective of the strategy statement is to get the selling message of the product down to a simple straightforward statement outlining the primary advantages of this product. This core selling proposition should be a simple declarative statement. This claim is positioned against a specific set of symptoms within the defined target market (see Figure 4.2). The core selling proposition for Alka-2, a new product introduced by Miles Labs, is "offer the heartburn sufferer fast relief." I believe that if you cannot get the selling message down into one or two simple declarative statements, you do not understand what you are trying to sell. One of the most important elements in successful advertising is simplicity and clarity. In the example, the primary product feature to obtain a point of difference over competition was

‡See Chapter 12 for more details.
*See Chapter 12 for more details.

CREATIVE STRATEGY STATEMENT

Market Target

Current users of chewable antacid tablets, primarily the O-T-C brands, Tums, Rolaids, and DiGel.

Symptom Targets

Heartburn is the symptom target for three major reasons:

1. Heartburn as a symptom has not been preempted by another brand, Tums, Rolaids, or DiGel.
2. The major thrust of the product's difference and the creative strategy is speed of relief, the primary concern of heartburn sufferers.
3. Effervescent type products are not used in significant quantities in the treatment of heartburn, thus cannibalization of Alka-Seltzer will be minimized.

Core Selling Proposition

Gives the heartburn sufferer high speed relief.

Strategy Definition

The only antacid tablet offering superior palatability and superior efficacy; i.e., speed of relief.

1. Support the efficacy claim, explaining that it disintegrates rapidly to get to work in your stomach fast.
2. Support the palatability by referring to its cool, creamy taste.

Subordinate to the texture support palatability by referring to cool and creamy taste.

Product Features

The product is a chewable antacid tablet with a difference. It is formulated to disintegrate fast; i.e., "it's built to fall apart," therefore, it goes to work fast.

This superiority offers the consumer two distinct benefits to help relieve his heartburn.

1. Superior efficacy; i.e., it disintegrates fast to get to work on your heartburn fast.
2. Superior palatability; i.e., a smooth, finer-textured tablet that "chews fast."

Figure 4.2. *Example Strategy Statement*

that it was "built to fall apart, so that it chews fast and works fast." The consumer is being offered a speed of relief promise.

In addition to the strategy statement, a consumer profile is developed. It is not the typical market research demographic description based on income, age, etc. It is rather a psychological profile of the user. In the chewable antacid example the heavy user of a chewable antacid is: Archie Bunker, blue collar, a sports fan, 45 plus in age, living in his own home, and his two children are not living at home. He is discontent with his life because the world has changed so fast around him that he feels that he is not fully in control of his existence and is somewhat frustrated by it. This psychographic profile is more useful to us in developing advertising than market research demographics because it shows us how to make advertising relevant to that type of an individual.

The third statement is a psychological description of how the consumer views the end benefit of the new product and how it relates to the problem that the product aims to correct. In the proprietary drug field we call it the "symptom psychology." In this case, "How does the consumer view heartburn?" This is not a medical or technical treatise on the symptom or end benefit, but rather the consumer's view of the problem and how it affects him or her (see Figure 4.3).

These three statements are the basis of the creative platform, and no work is begun until we have agreement from the advertising agency that this is the direction in which the work will proceed. Then the agency develops storyboards for review. At this point there are usually a number of different directions that can be pursued. To be successful you must try a number of different approaches and storyboards. In fact, the more things you try, the higher the odds are of creating breakthrough advertising.*

The next step in the development of advertising is the review of storyboards. Essentially the storyboards should be relevant to the consumers' perception of the category, pointing out the product's main point of difference from the competitive products. In today's highly segmented and saturated markets it is absolutely essential to clearly tell and demonstrate your product's point of difference. The consumer is buying these products for specific reasons. Many have established buying habits that have become ingrained over the years. You must demonstrate that the new product you are offering is better than the one they are using and you must create a reappraisal of the category choices so they will be willing to try your new product.

After the best four or five storyboards have been selected, they should be tested with consumers. I recommend making and testing commercials in anamatic form since costs of testing multiple finished ads is prohibitive. My experience is that anamatic tests are valid in identifying the best campaign. The two major approaches to testing commercials are on-air testing and in-theater testing.† I prefer in-theater testing because it is more diagnostic than

*See Chapter 13 for a further discussion of the tradeoffs in selecting an appropriate number of alternative advertising copies.

†For more details on these and other advertising testing techniques, see Chapter 13.

Heartburn: The Symptom

There is no general consensus in the medical profession as to what heartburn is. One major theory states that heartburn is pain in the esophagus caused by food, drink, tobacco, or tension/anxiety. The other major theory considers all of this a myth and attributes heartburn to distention of the esophagus or reverse peristalsis. Whatever the cause, heartburn is the third most common functional gastrointestinal disorder.

More important, however, is how the sufferer of gastric discomfort defines and treats heartburn. To him, heartburn becomes a word signifying the degree of gastric discomfort he is experiencing. Mild discomfort is usually called acid indigestion, sour stomach, upset stomach, or a gassy feeling—nor matter what the cause is or where it is occurring! If his discomfort is more severe, he tends to call it heartburn and attaches a more serious connotation to the symptom and the need for fast, effective relief.

The man who often calls his gastric upset heartburn is an overly nervous, tense, anxious person who aggravates most of his stomach problems (whether caused by food, emotion, or something else) with the addition of worry or tension. Since he is a tense person, he becomes anxious to rid himself of the heartburn fast. Since he is more often an active person, he has less time to suffer from a strong case of stomach upset and tension.

It, therefore, seems immaterial whether the man is suffering from true heartburn or a very upset stomach with accompanying gas pressure. If it is intense, he treats it as heartburn. If it is mild he treats it as acid indigestion.

Figure 4.3. *Example Symptom Profile*

on-air testing and there are many more uncontrollable variables with on-air testing. The diagnostic nature of in-theater testing allows for greater insight into the development of the advertising and as a result a higher level of problem solving. With in-theater testing I generally establish a decision rule criterion of being in the top 10 percent of the range of tested commercials. This criterion is established ahead of time so that everyone knows what constitutes a success. We continue development of advertising until we meet this goal.

Product Formulation and Testing

While all this work is going on, the lab is developing the product prototype dictated by the consumer as outlined in the early stages of the new product development process. This is a critical point of difference of this system over other systems commonly in use. The market structure analysis at the beginning of the new product development process clearly outlined the critical consumer dimensions of the category and the "ideal" product. Thus the consumer has indicated where the marketing opportunity is in the category and we are directing the development of the product and the advertising at that opportunity. The laboratory is given the assignment through a detailed product profile, and the advertising agency is given the direction from the research that has been done to date. This has the effect of targeting the work and making it much more productive and creative. Two examples of the targeting are the analgesic market where we learned that the ideal product was a product that offered effective pain relief without aspirin side effects and in the liquid antacid market the ideal product is a product which is extremely effective, i.e., concentrated, but also has an improved taste. At this point product development and advertising development are focused on opportunities in the market.

The first step in formulating a product to meet this target is a detailed product profile that outlines all the elements of the product and the product's point of difference, so that the lab can then formulate a product prototype for limited consumer use testing.

A detailed product feasibility document is written and circulated to all staff groups inside the company for their input (legal, manufacturing, consumer affairs, medical, quality assurance, etc.). At this point a number of conflicting company goals may surface. The purpose of this step is to air all points of view on this particular project and resolve any resulting problems so we can move on with the project.

Resulting from the product development work is a prototype product developed and technically tested by the lab. It is then put through a standard consumer usage test.* The concept statement used earlier is presented to the consumer with the product so that we are sure the product's benefits in use match up with the way the product will be presented to the consumer in the advertising. This research represents a rigorous test since it compares the product to the concept and to the consumer's regular brand. This is an important GO/NO GO point in the new product process.

Information from ad and product testing is put into the trial and repeat model to replace some of the prior executive judgment to generate a more accurate estimate of the product's sales potential. If the product is "GO" the lab is directed to scale up the formulation for production, establishing manufacturing standards, and produce the product for a small scale test market.

At this point in time of the new product process we have: (1) a concept statement that has been developed by market structure analysis and has been refined by focus group interviews, (2) developed advertising which is strong enough to launch the new product, and (3) a physical product that consumers perceive as a successful match with the claims made for it. If the sales forecast meets our "GO" standards, we are ready for testing.

Testing

In a traditional setting many firms would now begin a test market. However the cost of test marketing is so high in terms of dollars and time

*For more details, see Chapter 13.

that failures must be minimized. They can by the use of a purchase lab. Test marketing is not conducted until after a successful purchase lab test.

Purchase Laboratories. Purchase labs* are probably the most significant new marketing research tool in the last twenty years. The purchase lab generates the ultimate estimate of the magnitude of the new product sales using completed packaging, brand name, product, and advertising. It is the best assessment of the product program as it has been developed up to now.

The purchase lab is run in a shopping center. Users of a category are intercepted by interviewers and they are asked if they would be willing to participate in a research study. They are questioned about purchase patterns, brands purchased, reasons for purchase, attribute ratings for their brand, and other questions about the category. The consumers are shown the advertising for the new product along with competitive advertising for the category and taken to a shelf that is set up like a grocery store shelf to shop. They are given seed money and purchase or don't purchase and then check out paying for the product with the seed money. The product is taken home and used. The consumers are interviewed again to determine the product's performance. All of the data is put into the management science computer model and the model generates an estimate of share of market at various levels of distribution and advertising awareness.

I have run 15 to 20 purchase labs, and feel that they are very effective tools in estimating a new product's share of market and sales volume potential. In my experience of validating such tools, the results of the purchase lab were never more than 2 share points off the actual share of market achievement of the product in test market in any case.

Test Marketing. Assuming we have cleared the purchase lab hurdle, we are now ready for test market.† I prefer smaller controlled-store test marketing, that is markets of the size of Green Bay, Wisconsin; Bakersfield, California; and Rockford, Illinois; rather than the more traditional St. Louis, Cincinnati, and Denver markets used as "projectable" test markets. First, I don't believe that the projectable test markets achieve a substantially higher degree of accuracy than small city test markets. Secondly, I believe that a test market should be research intensive. To experiment in markets the size of Cincinnati, St. Louis, and Denver would be too costly. I prefer the smaller, controlled test markets where funds are allocated to generate in-depth research so that we understand the consumer dynamics that take place behind the share of market shifts resulting from the new brand entering into the market. The types of research that I prefer are extensive tracking studies of awareness, attitude, and purchase, as well as diary panels, store audits, and consumer usage interviews. Ultimately we spend more money in research, but that is offset by lower spending in media and distribution.

During test marketing the trial and repeat model is being used to de-

*For more details, see Chapter 14.

†For more details, see Chapter 15.

velop the most effective marketing plan for the money available. The model allows the new product group to develop the optimum marketing plan which will generate the greatest impact in the market place. It allows the marketer to diagnose his marketing plan to determine which elements of the marketing mix are working most effectively. We used this model for the test market of Alka-2 and made over 35 different computer runs to find the most effective plan. The model predicted share of market month by month for the brand prior to the start of test market. The share of market chart showed the brand performance for the first 10 months. At the end of the test market, the actual share achievement never varied more than 1.5 share points from the estimate by the model. With that kind of accuracy these models become invaluable to the new product operation.

After a successful test market, you are now ready for national launch* with a product that has a very high chance of national success.

RESULTS—EVALUATING THE SYSTEM

To sum up, the need for new products to continue the growth curve for corporations persists and, as markets mature, the number of products competing for the consumer's dollar reaches saturation level. This increases the cost of market entry and the frequency of new product failures, reduces the volume of successful new products, and shortens life cycles. These four elements have made the "New Products Game" far more complicated than it was. As the failure rate figures from past experience indicate, success for a new product is not easy. And the markets of the eighties will even be harder to crack. The system that I have presented in this chapter has helped me with these problems. There are three major elements in the success of the system I have described: (1) It improves the success/failure rate of new products; (2) it priorizes the projects based on success; and (3) it puts scarce resources (time, money, creativity, and people) behind those projects that have the most potential. The management of these three elements in new products is a most difficult task in the normal business setting, but by using this system it becomes much more easily handled.

The success/failure ratio based on my five years' experience with this system is that ten major projects resulted in three test markets and two successful national launches of multi-million dollar products. Based on this, the system performs well by comparison with the common success/failure ratios of the package goods industry. The system represents a substantial increase in productivity levels for new product development resulting from the integration of traditional qualitative marketing research tools and the management science quantitative computer models.

The system priorizes projects. A project continues to move through the system only if it has achieved an agreed-upon decision rule criteria that is

*See Chapter 16 for more details on full-scale launch.

clearly spelled out at the important GO/NO GO decision points. A project is not automatically moved from idea generation to test market. If a project fails to achieve the agreed-upon decision rule criteria, it is recycled through the step again, or dropped because the potential the project represents is not large enough. The advantage of this cannot be stressed too much. It puts the effort of the new products' operations behind those projects that offer the highest opportunity for success. The projects are being priorized by the consumer who is the individual who will ultimately be called on to put out money for the product. In my opinion this is the wisest approach. The system moves more of the decision making to the consumer; this is essential if you are to be successful. No executive or new products team can expect to second guess the consumer with a high enough success rate to warrant using approaches that do not integrate management models with consumer input and managerial judgment.

A number of positive management benefits result from using this system: (1) makes bright managers brighter through higher levels of market and consumer information, (2) channels "the impossible to manage" creative types, (3) encourages excitement about a project while maintaining disciplined rationality, (4) helps the new product group manage the difficult sales forecasting task associated with new products, and (5) unifies the organization's energies for new product development. In my mind, these management science computer models will be widely utilized. They will become the established "state of the art" in the next ten years.

If they work, why aren't these models being used more extensively in business and in new product development work? I don't have the answer to that. When I started using these models I must say that I was skeptical. I am a very pragmatic manager, but also experimental. If a tool works, then I use it. I have been able to validate these models and they have performed so well that I wouldn't consider working without them.

What does this mean for you? I believe you should take these tools and implement them in your firm and industry. Although this work was done in the package goods field I believe it has validity in industrial goods and service industries. The firms in these industries are just beginning to experience the kinds of market saturation we now see in the package goods field.

REVIEW QUESTIONS

4.1 What is the role of new product development in a company's "growth strategy"?

4.2 What is an "innovation philosophy" and why should it be adopted?

4.3 Discuss the implications of Mr. Hatch's organizational perspective.

4.4 Why are the GO/NO GO decisions important? Why does it take discipline to make a NO decision?

4.5 Discuss Mr. Hatch's new product development process (Figure 4.1).

4.6 If you were in the company as Mr. Hatch describes it, would you adopt the same or a different organizational perspective and new product process? Why?

PART II

OPPORTUNITY IDENTIFICATION

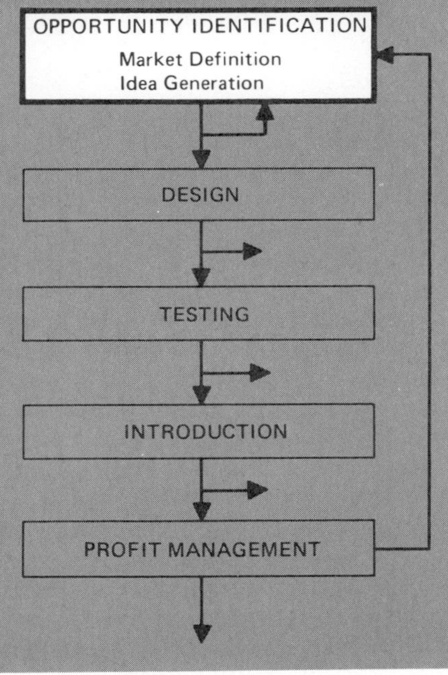

Chapter 5

Market Definition and Entry Strategy

Kodak enters the market for copy machines. Haines Corporation, after their success in introducing L'eggs, decides to enter the market for cosmetics. Warner Lambert Company introduces a new type of bifocal lens to get a share of the growing market for the "progressive power lens." Safeway introduces nongrocery items. S. C. Johnson introduces a new hair conditioner, and Union Carbide enters the markets for celery seedlings and medical test equipment.

Some of these new products will be successful, while others may be dropped after a period of financial loss. But in each case, the organization involved decided that that particular product category had profit potential. Success will depend upon whether that was a correct decision and on how that decision was implemented.

After an organization has adopted a proactive approach to new product development, the first step in implementing it is to identify an area of opportunity. This effort can be divided into two steps. In the first, markets are defined and opportunities within them are assessed. Then specific ideas are generated to tap the potential of these markets. When this opportunity identification phase is complete, design work begins based on evaluation and refinement of the idea as a physical and psychological entity.

This chapter describes market definition. We begin with a discussion of managerial criteria to evaluate alternative markets and then give a simple procedure to combine these criteria and their measures into an overall evaluation of the market. This weighting procedure gives management a method by which to screen a large number of potential markets by eliminating undesirable markets and identifying a few high potential markets.

Based on the results of this screening process, management then selects those high potential markets for detailed investigation. This detailed investigation specifies market boundaries and target consumers so that the new product can be directed at the market most likely to yield high profits. Finally, we use managerial criteria and analytic investigations of market boundaries and target consumers to set priorities and to select the best market or markets to enter.

DESIRABLE CHARACTERISTICS OF MARKETS

Success is likely in markets that have high sales potential, can be easily entered and penetrated, require small investments for large rewards, and are low risk. These general criteria are shown in Table 5.1 along with some of the specific measures that can represent them.

Table 5.1. Desirable Characteristics of Markets

General Characteristic	*Measure*
Potential	Size of Market Sales Growth Rate
Penetration	Vulnerability of Competitors
Scale	Share of Market Cumulative Sales Volume
Input	Investment in Dollars and Technology
Reward	Profits
Risk	Stability Probability of Losses

Potential

Market potential is measured by the size of the market in dollar sales and the growth rate of the market. However, growth in potential is a key to identifying a new opportunity. Although it may be possible to find a successful product opportunity in a stable market, sales will have to be taken directly from competitors. This is more difficult than getting a share of the growth in a market. A growing market is also one where prices and margins are higher and, therefore, more desirable.

Penetration

Although a market is growing, it may not be a good opportunity unless it can be penetrated. Some vulnerability to product improvement should be evident. For example, although the large computer systems market had a very high growth rate, IBM showed little vulnerability. In contrast, in calculators, Frieden was very vulnerable to exploitation by the new electronic technology.

In some cases, vulnerability is so high that even a stable category can be attractive. Jergen's dominated the hand lotion market, but were vulnerable to Chesebrough-Pond's Vaseline Intensive Care Lotion. "Sleepy" markets in which sales are stable and innovation has been absent can represent an attractive opportunity if they are penetrated effectively.

Scale

Potential and penetration are critical to achieving a large-scale operation both in terms of market share and sales volume. Market share is important because of its relationship to profitability. Large share in a market gives a firm relative strength and dominance in that market and hence control over strategy in the market. With this strength comes a flexibility of action that can lead to increased profitability. For example, Buzzell, Gale, and Sultan (1975) analyzed 600 businesses based on data collected in the PIMS project. They found that rate of return on investment was positively correlated to market share. On average, they found that a difference of 10 percentage points in market share was accompanied by a difference of about 5 points in pretax return on investment (p. 97). Businesses with market shares over 36 percent earned a rate of return on investment three times those with less than 7 percent market share (Schoeffler, et al., 1974, p. 141).

Cumulative sales volume is important due to the pressure of economies of scale in production in many industries. These economies can be described by the "experience curve." This curve indicates that for many manufacturing industries the unit cost of producing and distributing a product declines at a constant rate for each doubling of the cumulative sales by the firm (Boston Consulting Group, 1970).

Figure 5.1 shows an experience curve for an industrial chemical (polyvinyl chloride). As volume increased, price dropped. In the early phases this reduction was small, but after large volumes were achieved rapid cost reductions occurred. As volume doubled from 5 billion to 10 billion pounds, price dropped approximately 50 percent. When such experience curves exist, scale of operation is critical to product success.

Figure 5.2 shows another experience curve; this one for Ford automobiles (Abernathy and Wayne, 1974). After introduction of the Model T, costs dropped 15 percent for every doubling of sales from 1908 to 1925. This was associated with standardization and innovation in the production process, vertical integration, labor specialization, and better bargaining power over input material costs. This type of curve is common in most manufacturing based industries.

Care should be taken in planning for the cost reduction that the experience curve offers. Ford's history demonstrates one danger. The Model T enjoyed rapid cost reduction, but through greater performance and better

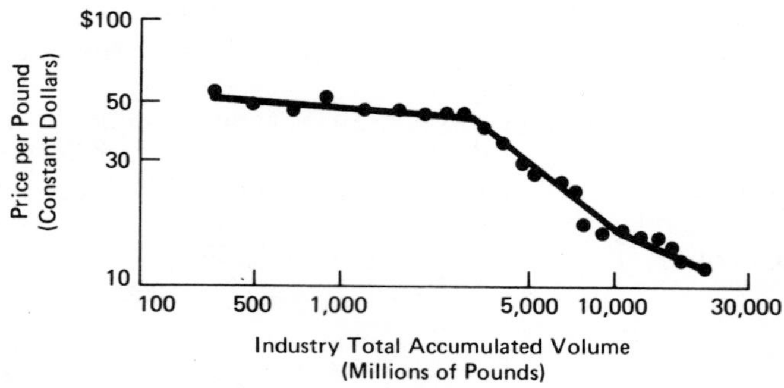

Figure 5.1. *Polyvinyl Chloride Price Curve, 1946–1968 (The Boston Consulting Group, 1970, p. 85)*

capability, GM took away Ford's market share. The new competitive environment was reflected in the 1950–1972 period where costs (list price and price per pound) increased as designs were improved and models underwent annual changes. Although Ford redesigned its car, the company never regained its initial position. The experience curve existed, but consumer preferences for comfort, power, and luxury dominated price as a determinant of market share. The experience curve is an important phenomenon, but it is not an automatic one. It results from aggressive company actions to take advantage of the potential economies of scale and must be considered along with consumer choice criteria in an overall strategy to achieve market share and profits. If the experience curve exists, high market share in a large market is a good way to move to lower unit costs. Strategically, this is important since it would imply concentrating resources in a few markets to increase volume and decrease costs. The highest priority would be towards a share in a growing market of high potential. The lowest priority would be markets in which the share potential is low, sales are stable, and total volume is low.

As pointed out above, the experience curve is most important in manufactured products where economies in scale in the production facility exist. The curve also exists in industries where scale encourages technological and production innovation. A visable example is hand-held calculators where demand and technology reduced costs by a factor of 10 in several years. Another example is in photovoltaic cells which generate electricity from the sun. Currently, they are perhaps five or more times more costly than oil. Research is underway to see if government incentives and buying can increase the scale of the industry so that costs can be reduced substantially.

In services, it is not clear that an experience curve is present. For example, in health care few cost economies exist for very large hospitals. Certainly there is a minimum scale for efficient operation, but in most service industries, expanded scale probably will not demonstrate the significant experience curves found in manufacturing.

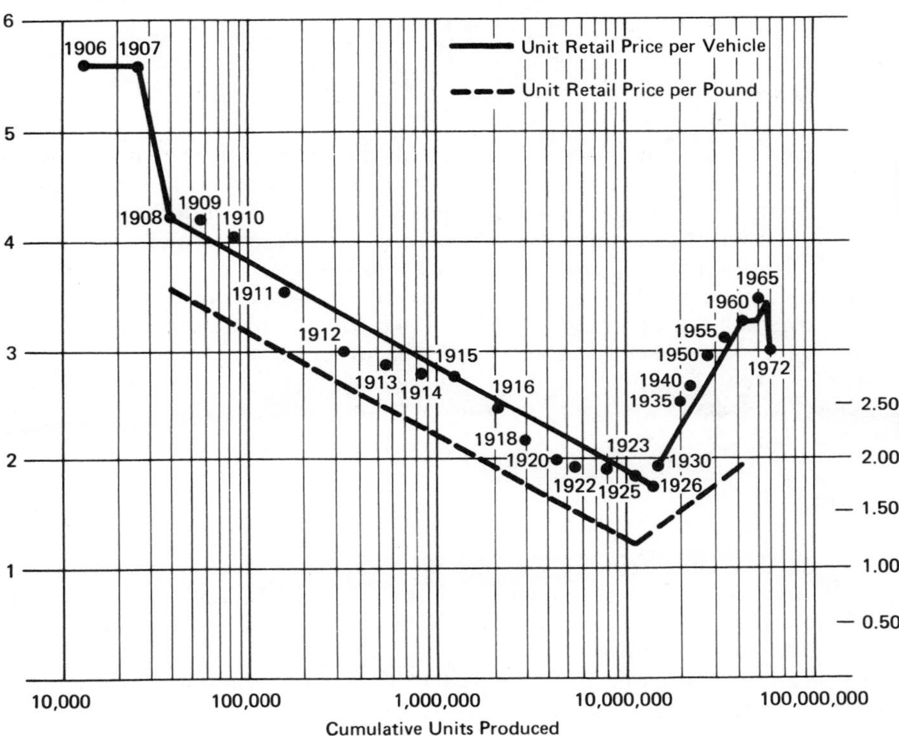

Figure 5.2. *The Ford Experience Curve in 1958 Constant Dollars (Abernathy and Wayne, 1974, p. 112)*

Input

The more investment required by a market for entry and penetration, the less attractive it is. For a given level of sales volume, a larger financial investment makes a market potentially less profitable. Furthermore, the stakes are higher and a failure is more devastating to an organization. Besides direct financial investment, an organization must consider the allocation of its other scarce resources such as managerial talent, capital equipment, and laboratory resources which represent an indirect financial investment.

On the other hand, large market entry costs act as barriers to discourage competitive entry and as a result could lead to long-term profitability. Those who pay the entry costs may find a more stable price environment in their market. Organizations with competitive strength such as excellent financial resources or good channels of distribution or a line of complementary goods or advantageous geographic location, may be well advised to consider markets with high entry costs.

Reward

Large scale is not enough to select a market. Similarly large investments in themselves are not the basis for elimination of a market. For example, a large organization may see a market requiring 500 million dollars of investment as attractive if it can return 20 percent on that investment. Similarly, an organization may reject a market with a low investment requirement if it cannot achieve a good return on that investment—especially if the administrative overhead of entering a market makes the profit insufficient. For example, one major package goods firm will not enter a market unless gross volume is greater than $10 million and payback less than three years. Thus, profitability must be considered rather than sales volume or investment alone.

Risk

A final consideration of market selection is risk. Markets characterized by instability of demand or uncertainty of supply can be very risky. In stable markets with a strong dominant firm, new products risk competitive retaliation. In markets vulnerable to rapid competitive entry or subject to changes in government regulation, risk will be high and it may be difficult to maintain a profitable product.

An organization must examine a market for characteristics of risk. Judgmental estimates or historic observations of the probability of failure and the variance in payback can be used to evaluate the risk of a given market.

MARKET PROFILE ANALYSIS

The six criteria discussed above are used to evaluate the desirability of a particular market. It is rare that a market will dominate all others in all six criteria, so an organization must have some procedure by which to select markets based on these conflicting objectives.

There are numerous sophisticated techniques based on formal decision analysis (Raiffa, 1968; Keeney and Raiffa, 1976), but these sophisticated techniques are time consuming and require reasonably accurate estimates of the various measures of market desirability. Such techniques are appropriate later in new product development when more precise estimates are feasible. At this stage we need a procedure that allows us to rapidly screen a large number of potential markets to obtain a reduced set for further analysis. This smaller set is then subjected to closer scrutiny and more detailed investigation.

The technique appropriate to the data availability and magnitude of alternative markets that must be evaluated is "market profile analysis." The technique is applied in three steps: (1) enumerate and weigh the market selection criteria for your organization, (2) rate each market on each criterion, (3) calculate the overall weighted sum of the ratings for each market, (4) evaluate the ratings to identify the market with the best overall appeal. The markets with the highest ratings would reflect the best opportunities for continued investigation. Those with very low ratings are eliminated.

Enumerate and Weigh Criteria

The first step in a market profile analysis is to generate the criteria that are important in a given organization. This is done by managerial discussion and interaction to specify their important factors in the words of the organization's management. This list becomes a mental checklist for market evaluation. The generation of a consensus on this list is an important activity since it integrates divergent points of view and begins to build communication links between key management interests in the company.

Table 5.2 lists a set of factors for a typical organization. In this list, they are grouped according to the six market characteristics cited above and by criteria that reflect the match between the market and the organization's abilities. If the organization does not have a required capability, it must be willing and able to acquire it before a market can be viewed as attractive.

There are many possible enumerations of factors. It is our view that although many lists can be found (e.g. O'Meara, 1961), each organization should go through the exercise of developing its own or customizing existing lists of factors.

Weightings

Weighing the factors in terms of their importance is a necessary step to converting the checklist into a market profile analysis. First, the major headings are assigned a relative weight and then the heading weight is divided among the component parts. For example, management might decide that 60 percent of the weight should be on market characteristics and 40 percent on the matching to organizational capabilities. Further, they may see the six market characteristics as equally important or 10 percent weight. This is continued until each factor has a weight established for it. For convenience, the sum of the weights for each factor should equal 100 percent. This process of weighing criteria is a valuable exercise for the organization in terms of structuring and setting market priorities.

Occasionally a factor is so important that it must be satisfied before a market can be considered. Such "elimination" criteria are considered by

Table 5.2. Some Factors to Be Considered in Market Profile Analysis

Market Characteristics

Sales Potential	*Input*
Size of Market	Investment Required
Growth Rate of Sales	Raw Material Availability
Length of Life Cycle	Technological Advancement Necessary
Penetration	*Reward*
Cost of Entry	Margin Size
Time to Become Established	Competitiveness of Pricing Structure
Vulnerability of Competition	Return on Investment
Potential for Product Advantage for Users	
Scale	*Risk*
Potential for Significant Market Share	Stability
Likelihood of Competitive Entry	Probability of Competitive Retaliation
Significance of Experience Curve	Chances of Failure
	Patent Protection
	Rate of Technological Change
	Possibility of Adverse Regulation

Match to Organization's Capabilities

Have Financial Resources Required
Match to Physical Distribution System
Match to Marketing Capabilities
Utilize Existing Sales Force
Can Handle Technology, R&D Experience, and Know-How
Probability of Technical Completion
Ability to Service Product
Compatibility to Other Products
Management Skills and Experience for This Market
Past Work Done in This Market
Overlap with Current Material Supply Channels

eliminating all markets that do not meet the criteria and then evaluating the remaining markets with the remaining factors and weights. An example may be an industrial company that, by policy, will not enter a frequently purchased consumer market under any conditions. In this case, all frequently purchased consumer markets are removed from the list of potential markets and the others are considered.

This judgmental weighting can be improved by technical methods to assure factors are independent and to statistically derive weights (Freimer and Simon, 1967). Our experience has been that most of the gain from weighing is due to discussion among managers that directs attention to various markets and assure a systematic first look at them. We do not recommend statistical approaches at this early phase of market consideration. In later phases where more and better data are available we suggest the use of several analytical methods.

Ratings

Next, each alternative market opportunity is rated on each factor. The rating can be done with a five-point scale relative to an average existing market for the firm (see Figure 5.3). Here a factor such as size of market is rated relative to the size of the firm's existing market. If the new market is equal to existing markets, it is rated zero, and if it is much bigger, it is rated +2. Each factor is similarly rated after management clearly articulates what is meant specifically by "better," "much better," "average," "worse," and "much worse" for each factor (see O'Meara, 1961, for an alternate set of specific scales). Each market under consideration is rated on each scale.

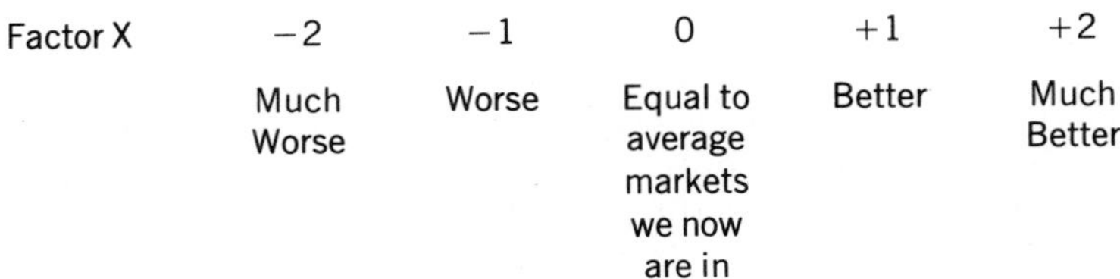

Figure 5.3. *Rating Scale for Factors*

Overall Evaluation

The next task is to combine the ratings, identify the market with the best overall appeal, and eliminate undesirable markets. This is done judgmentally by examining the profile of ratings on each factor for alternative markets. Discussion in a meeting of heads of production, marketing, finance, and R&D is very useful in generating a consensus on the best set of markets.

In many cases, this process is aided by calculating an average overall score for each market by multiplying the rating on each factor by that factor's importance and adding these (weight × rates) for each factor. The markets with the highest scores are most desirable. Those with a low score are eliminated and resources are not directed at them. If the management group confirms this, the highest scoring markets are considered further and priorities are set among the opportunities that are acceptable.

Overall evaluation scores are an aid to managerial decisions rather than a substitute for such decisions. The decisions as to weights and ratings are important because they help quantify managerial judgments, but often managers become concerned about this technique because they feel that the market priorities are very sensitive to the selection of weights. One approach is to try a few alternative weightings and/or rating schemes to determine how sensitive the priorities are to the chosen procedure. In most cases, management will find that over a variety of weightings and rating schemes some markets are always rejected while some markets always pass the screen. Management can then reexamine the markets sensitive to the selection of evaluation procedures.

If a formal rating and weighting system is used, the firm should keep records of individual judgments. These judgments are then compared to actual results in markets where products are entered. This comparison will indicate biases in the ratings of certain individuals or by the representatives of specific functional areas. For example, it may be found that marketing consistently overestimates its ability to capture a large market share. While these biases could be corrected by revising the ratings, the most important purpose of the comparison is to enable the individual evaluators to make more accurate future comparisons. Efforts should be made at each phase of the new product process to systematically learn by comparison of judgments and reality.

The output of the overall evaluation is a set of high potential markets worth further investigation plus a preliminary set of priorities to guide that investigation. We now iterate the process by narrowing the scope and carefully defining the target markets and target consumers for the top priority potential markets.

METHODS FOR MARKET DEFINITION

The previous section has described criteria and an evaluation procedure for *a* market. But what is *a* market? Is our "market" for autos, or for small autos, or for small domestic autos, or for small domestic two-door autos, or for small domestic two-door four-cylinder autos? The definition of the market is critical in specifying the size of a new product opportunity. If a new turbine powered car is going to compete only within the market for small domestic four-cylinder cars, then a 30 percent share may not be large enough to justify technical development. If, however, it would compete within the market for all small autos, a 30 percent share would represent a significant opportunity.

It is desirable to have a large share of a big market, but it is possible to build a business based on a small overall share if it in fact reflects a high penetration in some product segment. Rarely is it profitable to have a product with a low share in all segments. (See Hamermesh, Anderson, and Harris, 1978.)

Another aspect of market definition relates to the new product's relationship to existing products of the firm. If the new product will be in the same market as an existing product, cannibalization may result. The organization would like to have a product line that spans the total set of opportunities, but has little overlap and self-competition. For example, Mobil Oil found its product line had grown to over 1500 lubricants. This was considered too broad a product line and was cut by approximately one-half in the mid-nineteen seventies.

In proactive new product design, management must be very purposive in defining the boundaries of the market they are entering. We first look at the traditional product approach and then examine approaches based on consumer behavior and perceived similarity among products. A key managerial question is, "What basis should we use to define our markets?"

Traditional Approaches

Markets traditionally have been described by a generic title and then broken down by physical properties. For example, traditionally auto manufacturers have divided the market into sub-compact, compact, mid-size, full size, and luxury autos. Another definition would be based on the manufacturing company (General Motors, Ford, Chrysler, and American Motors), or foreign versus domestic brands.

In some markets, channels of distribution are used to define markets. For example, in industrial products, products sold directly by the manufacturer may be viewed separately from the products sold through independent distributors. In frequently purchased consumer products, national brands versus private labels are often used to differentiate products.

In administration of anti-trust laws, market definition is critical. For example, Procter and Gamble was ordered to divest of Clorox brand bleach. Part of the argument of obstruction of competition was based on a market definition that excluded dry bleaches from the Clorox market. The market share of Clorox was higher if only liquid bleach was considered in the relevant market.

While these traditional approaches have been useful in corporate and legal definitions, they tend to emphasize a product-oriented approach to marketing rather than a consumer-oriented approach. Economists and marketing analysts argue that market definitions should depend more on how consumers themselves view the market.

Cross Elasticity

Economists have proposed cross elasticities between products as the measure to determine if two products are in the same market. The cross elasticity of price between product "A" and product "B" is the proportional shift in sales of "A" due to a shift in the price of "B." For example, if a 10 percent reduction in the price of "B" causes a 5 percent reduction in the sales of "A," the products are substitutes. This would occur if the decrease in the price of "B" caused consumers to switch from using "A" to using "B." Markets could be defined theoretically based on identifying a set of products that were mutually substitutable.

This is a good concept, but it is difficult to implement since measures of sales and price may contain error and the elasticity may not be stable. It is difficult enough to estimate the direct price elasticity from time series data, let alone a cross elasticity. Urban (1969) estimated cross elasticities in instant coffee based on cross-sectional data from individual stores. In this case, some significant results appeared. The implications were that regular, decaffeinated, and freeze-dried coffee were substitutes across brands, but complementary (negative cross elasticity) within manufacturers' brands. However, these results were not statistically clear enough to recommend cross elasticities as the best method of market definition. Automated scanning of Uniform Product Codes (UPC) may provide a cleaner data source and make cross elasticity analysis more attractive in the future.

Homogeneous Uses

Rather than statistically estimating cross elasticities, some marketing analysts have worked with direct consumer judgments of product substitutability. For example, in the manufacture of a small electric motor and drive system, aluminum, steel, and plastic gears may be possible substitutes. In this case, they are competing for the same market. In autos, aluminum, steel, and plastic are competing in many component part markets. Timing chains for auto ignition can be made of plastic or steel. In considering recreation, TV would compete with movies as well as spectator sports such as football games if consumers judged them as substitutes for the use of recreation time.

The degree of substitution in uses can be analytically studied (Belk, 1975). Stefflre (1972) has proposed that consumers should generate the possible uses for a set of products and then specify all possible products that could be appropriate for each use. The resulting matrix of uses by products may be statistically analyzed to group similar products and use combinations. In one study, Stefflre analyzed 52 proprietary medicines for 52 medical conditions to define "markets" by clusters of products used to treat similar conditions. Other researchers have extended the measurement and statistical methods of this approach of defining markets by use (Day, Shocker, and Srivastava, 1979).

Hierarchical Market Definition

Cross elasticities depend upon econometric estimation of the results of price changes and homogeneous use is based on direct consumer judgments. A third approach based on product substitutability is hierarchical market definition. For example, Figure 5.4 depicts a hypothetical hierarchy to describe the market for light industrial structures. This hierarchy breaks the market by material, roof structure, and span. If this were the true representation of the market, consumers would first decide among material, then structure, then span. Products would be more substitutable if they were in the same branch of the hierarchical tree.

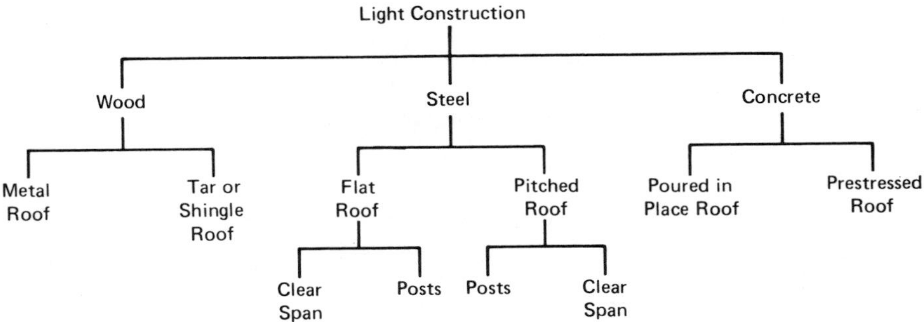

Figure 5.4. *Hypothetical Hierarchical Description of Light Industry Construction*

One method defining a hierarchy is to observe the sequence of issues considered and decisions made by consumers (Bettman, 1971). In the building case, we would determine if buyers first chose between wood, concrete, or steel and then selected within a material type. If, in these interviews, we find that since some buyers use an architect, we would have to understand the architect's decision sequence in designing the building. This kind of research can yield an in-depth understanding of the consumer choice problem. The resulting hierarchical decision network is important information. Unfortunately the decision sequence approach is very difficult to execute in the field with current techniques of measurement and analysis. Individuals can be assessed, but procedures for aggregation of total markets are just being developed (Bettman, 1974).

An interesting approach to defining hierarchies for frequently purchased goods has been developed by Butler and the Hendry Corporation (see Appendix 5.1, and Kalwani and Morrison, 1977). Butler derives a hierarchy based on consumers switching between brands of frequently purchased consumer products. In the Butler approach, physical product attributes are the branching points. For example, in deodorants, the first hierarchy might be form—aerosol vs. roll-ons vs. sticks—and the second hierarchy might be brand—Right Guard, Arrid, Ban, Mennen, etc. Figure 5.5 describes, in a simplified form, two possible hierarchies for this market. If the first hierarchy best described the consumer decision process, Mennen would be wise to consider adding a roll-on variety since it has no brand in that market. However, if the market is defined by the second hierarchy, a roll-on variety of Mennen would take much of its sales from the existing Mennen products. Right Guard saturates the market by either definition.

The criterion used to define the trees implies that switching should be high within a branch and low between branches. This switching is measured

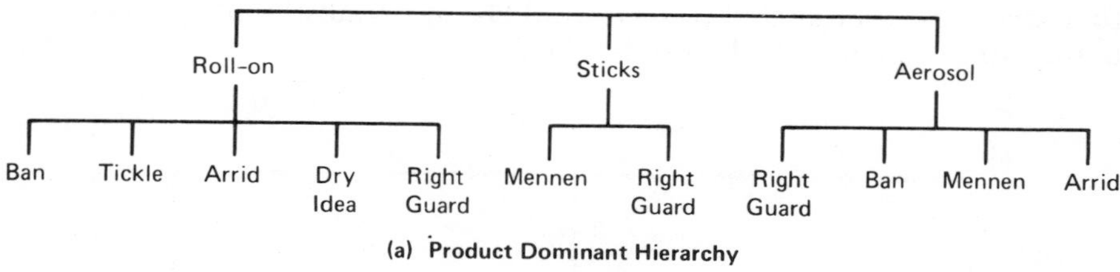

(a) Product Dominant Hierarchy

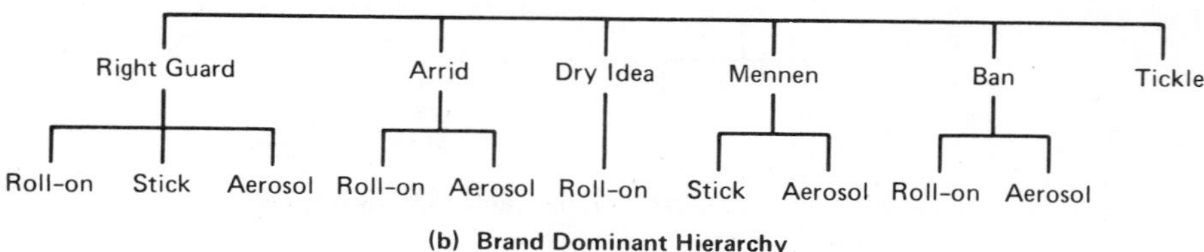

(b) Brand Dominant Hierarchy

Figure 5.5. *Two Alternative Hierarchies for Selected Deodorant Brands*

from consumer diaries of purchases or by the consumer identification of the last brand purchased and the brand purchased before that.

The technical basis for the Butler method of deriving hierarchies involves a mathematical derivation of consumer behavior based on the particular hierarchy, the observed market shares, switching data, and some rather strict behavioral assumptions about the structure of the market. (See Appendix 5.1 for technical derivations and critique.) The approach is a creative use of a hierarchical market representation and a mathematical model to examine the boundaries of a potential market for frequently purchased brands.

Perceived Similarity

The above approaches are based on observed or reported switching behavior among products. Another approach that has been used extensively in marketing is based on how consumers perceive the relative similarity among products.

This approach is based on an important technique called perceptual mapping. A perceptual map is made up of a designation of some number of dimensions and the position of each product in these dimensions. Figure 5.6 shows a perceptual map for vacation sites. Two dimensions are used to map the market and points are used to designate the position of the sites. Points that are close to each other on this map are most similar and those far from each other are most dissimilar.

Products that are clustered close to each other on the map may be grouped as a submarket. For example in Figure 5.6, Montreal, London,

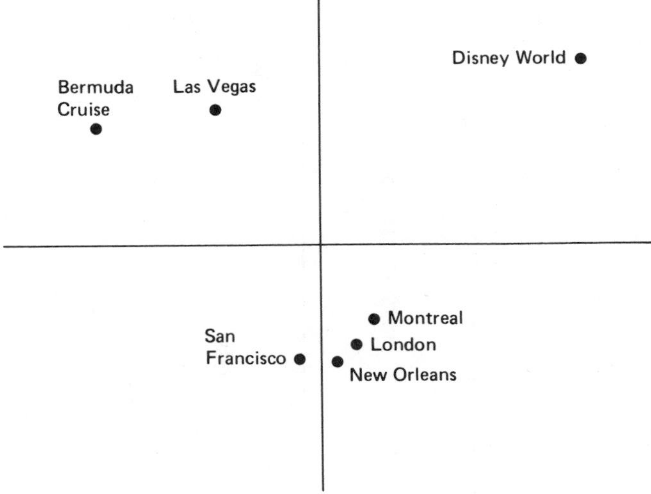

Figure 5.6. *Perceptual Map of Vacation Sites (adapted from Green, Rao, and DeSarbo, 1978, p. 189)*

New Orleans, and San Francisco form a cluster and probably compete heavily with each other for tourist business. Las Vegas and Bermuda represent a separate submarket, and Disney World could be identified as a third product segment.

Several techniques and measurement approaches are available to derive perceptual maps and name the dimensions. We consider these in detail in Chapter 9. For market definition you should recognize that perceptual maps can be readily developed from consumer measures and that they represent a basis for defining competitive sets of products.

An Emerging View

Suitable consumer-oriented analytic methods are available to define markets. They should be used rather than the traditional non-consumer-based definitions. In selecting the method for market definition, all of the analytic procedures (perceptual mapping, hierarchical trees, cross elasticities, and common usage patterns) have attractive features. Currently most applications choose one method as a basis for defining markets and identifying market entry opportunities.

Much research currently is being conducted on approaches to the market definition problem. We expect improved procedures in the near future. One emerging state of the art model integrates several of these methods. (See Appendix 5.2, and Urban, Johnson, and Brudnick, 1979). This approach is hierarchical, but branches are defined not only by physical product attribute but also by usage occasion and users. Perceptions are modeled by a

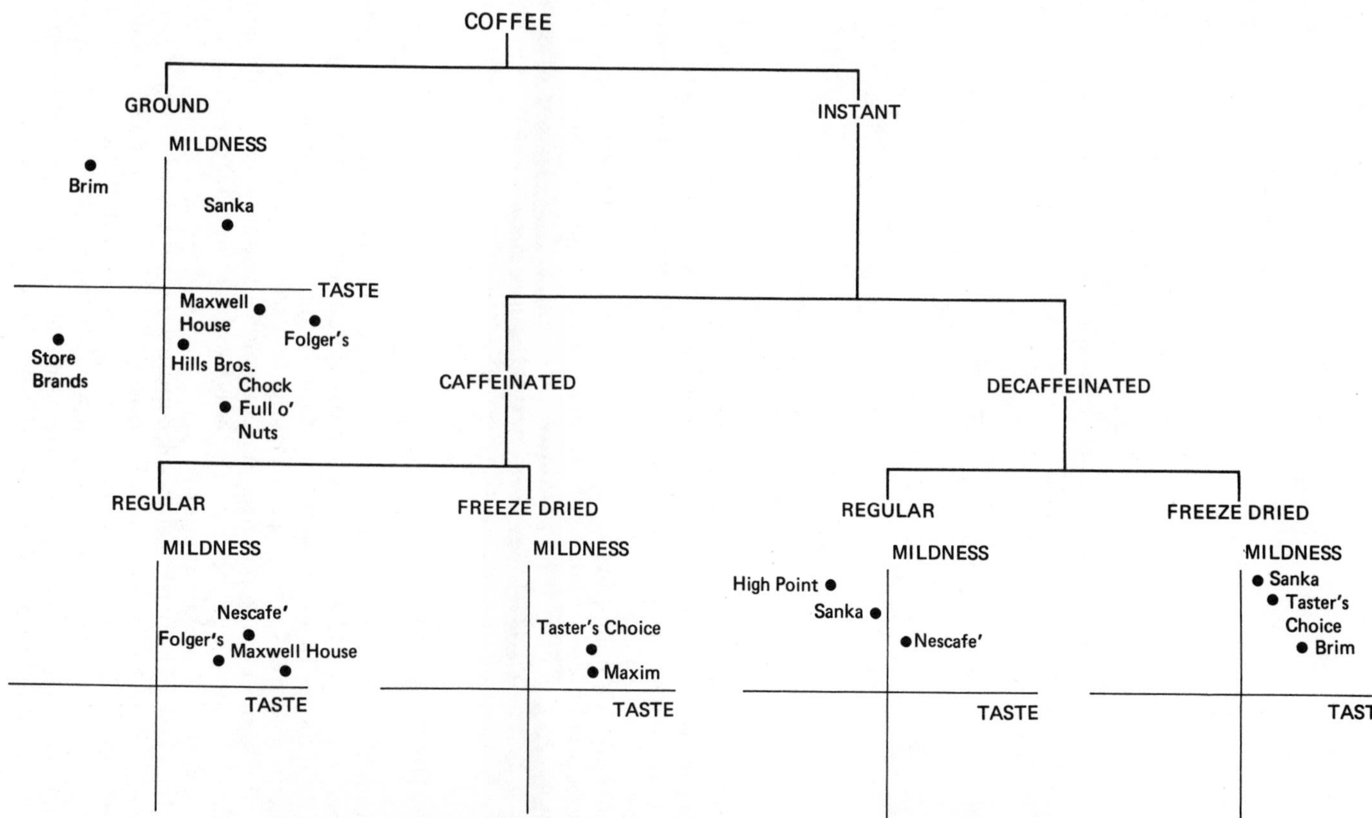

Figure 5.7. *Hierarchical Definition of Coffee Market (Urban, Johnson, and Brudnick, 1979)*

map for each branch at the bottom of the tree. Figure 5.7 shows a hierarchy for coffee which first divides coffee into ground and instant. Instant coffee is divided into caffeinated and decaffeinated brands and then each of these into freeze dried and regular varieties. For each of the five branches a perceptual map is used to describe the relative positions of brands on the dimensions of "taste" (fresh, full-bodied, rich) and "mildness" (not bitter, does not upset stomach). Spaces in these maps where no brands exist are possible entry opportunities.

The hierarchical tree implies that at a specified level more switching should occur within branches than between branches (e.g., more within decaffeinated or caffeinated than between them). This concept is operationalized by observing consumers to estimate individual probabilities of choosing products. Statistical tests then select the hierarchical tree that best represents consumer substitution behavior. (See Appendix 5.2 for the technical details.) The trees are tested by comparing predicted switching with actual substitution observed in a simulated buying situation. In the coffee case the substitution was made by observing actual buying in a laboratory store in which the consumers were forced to switch because the researchers removed their most favored brand from the shelf.

The method described above is attractive since it integrates perceptual similarity and hierarchical relationships by analytic methodology. The manager is provided with a hierarchical definition of the market plus a visual representation of the relative competitive positions of brands within that market. This approach is one of several methods that are likely to emerge in the near future to help managers more effectively define market boundaries.

We now turn to an initial identification of the target consumers. The final section of this chapter combines the analytic approaches of market definition and consumer targeting to select the best product/market combination.

TARGET GROUP SELECTION THROUGH MARKET SEGMENTATION

Identification of the products that define a market is an important part of the decision on what market to enter. An organization also must identify the consumer who will be the primary target group for the new product. This identification is important to determine the size of the potential market, to select channels of distribution that may be needed, estimate a potential pricing structure, identify initial product positioning opportunities, and guide market research.

The managerial concept underlying market segmentation is the identification of a group of consumers who have needs or responses that are different from other consumers. A product intended for the total market may have only marginal success if consumers have different preferences.

The practice of market segmentation can spread rapidly in an industry. As one competitor segments the market and successfully offers special products for them, the previous mass market suppliers find their shares decreasing and are forced to develop products that can compete effectively in these market segments. Thus in identifying a segmentation strategy, an organization must weigh consumer needs and desires with the current and potential segmentation strategies of the other firms in the product category. The best segmentation strategy will be one that identifies a group of consumers likely to purchase a new product at a price and quantity sufficient to satisfy profit goals and one that is defendable against rapid competitive responses.

Market segments can be defined in many ways (see Frank, Massy, and Wind, 1972, for an extensive discussion). Table 5.3 gives various criteria that can be used to define segments and some measures that can be used to represent them.

Table 5.3. Criteria and Measures for Market Segmentation

Criteria	*Measures*
Demographics & Socio-economic	Age, income, sex, race, marital status, family size, geographics, education, occupation, life cycle
Attitudes	Personality, life style, product perceptions
Usage Rate	Heavy/light; buying patterns
Preference/Choice	Elasticity to price, importance of product attributes, purchaser vs. user

Demographics

The most common method for segmentation is by demographic or socio-economic variables. For example, the position of a family in their life cycle described by the age, marital status, and age of children may be used to identify opportunities. Table 5.4 summarizes a set of links of life cycle to buying behavior. Since the demographic composition of the population is changing to a more uniform distribution of age, the suppliers of products used in the earlier stages of the life cycle may find decreasing markets while products sold to those in the later phases of the life cycle may represent new market opportunities. These trends are coupled with social trends such as the increased divorce rate and more single parent households.

Attitudes

Attitudes may also be used to differentiate market groups. Personality traits differ across consumers and reflect a basis of segmentation. However, these traits are difficult to accurately measure and empirical results have been mixed (Frank, Massy, and Wind, 1972, pp. 50–58). Work on personality has been extended by the use of "life style" segmentation. Life style can be defined by the person's activities, interests, opinions, and values. Typical measures may be fashion-consciousness, child orientation, interest in active sports, opinion leadership, or credit utilization. This approach tends to judgmentally describe and group consumers on the basis of a selected set of measures. This method of segmentation is most appropriate for products and services which use a high psychological appeal.

Table 5.4. An Overview of the Life Cycle and Buying Behavior[a]

Stages in Life Cycle	*Behavioral Characteristics*
Bachelor Stage: Young, Single People Not Living at Home	Few financial burdens. Fashion opinion leaders. Recreation-oriented. Buy basic kitchen equipment, basic furniture, cars, equipment for the mating game, vacations.
Newly Married Couples: Young, No Children	Better off financially than they will be in near future. Highest purchase rate and highest average purchase of durables. Buy cars, refrigerators, stoves, sensible and durable furniture, vacations.
Full Nest I: Youngest Child Under Six	Home purchasing at peak. Liquid assets low. Dissatisfied with financial position and amount of money saved. Interested in new products. Buy washers, dryers, TV, baby food, chest rubs and cough medicines, vitamins, dolls, wagons, sleds, skates.
Full Nest II: Youngest Child Six or Over Six	Financial position better. Some wives work. Less influenced by advertising. Buy larger-sized packages, multiple-unit deals. Buy many foods, cleaning materials, bicycles, music lessons, pianos.
Full Nest III: Older Couples with Dependent Children	Financial position still better. More wives work. Some children get jobs. Hard to influence with advertising. High average purchase of durables. Buy new, more tasteful furniture, auto travel, non-necessary appliances, boats, dental services, magazines.
Empty Nest I: Older Couples, No Children Living with Them, Head in Labor Force	Home ownership at peak. Most satisfied with financial position and money saved. Interested in travel, recreation, self-education. Make gifts and contributions. Not interested in new products. Buy vacations, luxuries, home improvements.
Empty Nest II: Older Married Couples, No Children Living at Home, Head Retired	Drastic cut in income. Keep home. Buy medical appliances, medical-care products that aid health, sleep, and digestion.
Solitary Survivor, in Labor Force	Income still good, but likely to sell home.
Solitary Survivor, Retired	Same medical and product needs as other retired group. Drastic cut in income. Special need for attention, affection, and security.

[a]*Adapted from Wells and Gubar (1966), p. 362*

Usage Rate

Segmentation by usage rate has been based on differentiating heavy users of a product from light users based on their profiles of demographic, socio-economic, and attitude measures. The notion is to separate heavy users so that products can be better targeted and communicated to them. For example, the "lite" beers were originally targeted towards beer drinkers who have more than one. Similarly, high and low brand loyalty could be used as a basis for grouping.

Preference/Choice

This criterion is the most relevant to managerial strategy in new product design, but also the most difficult to implement. The criterion is based directly on how consumers respond to the new product. Segments are identified based on the characteristics of the product that they will prefer and ultimately use. For example, we might like to know which buyers of office copiers have a high price elasticity. These consumers may represent a market opportunity for a low cost machine. Another response may be the feature of reliability. Some consumers may value reliability more than speed while others value speed more than reliability. This type of segmentation was called "benefit segmentation" by Haley (1968).

In the initial phase of market definition, benefit segmentation is based on a combination of managerial judgment and relationships to other segmentation criteria. This working identification is then refined in the design phase (Chapter 10) with more sophisticated procedures based on models and measures of consumer preference.

Analytic Methods for Market Segmentation

The primary method of segmentation is to use analytic clustering techniques to identify groups of consumers that are homogeneous with respect to the criteria selected to represent similarity. (See Appendix 5.3 for the details of one representative clustering technique.) In these statistical approaches, groups of similar consumers are defined by their proximity in terms of the measured traits. The attempt is to specify groups such that consumers are highly similar within each group, yet different across groups. In almost all methods, considerable judgment is added to define the best set of market segments. These groups are then examined for unique opportunities with respect to new product development. In many cases, this coupling of analytic method and managerial judgment has led to opportunity definitions that are new and creative.

One of the problems in segmentation research has been validating that the segments are indeed different not only with respect to attitude or usage variables, but with respect to purchase behavior and response to new products. It is appropriate to use target segmentation to define potential opportunities, but one must later validate these potentials before making large commitments of resources.

Market Definition and Target Group Segmentation

Market definition through the hierarchical descriptions of *products* and target group segmentation through the identification of *consumers* represent the two needed elements of a product and market strategy. These synergistic approaches must be coupled to select the best managerial strategy.

For example, suppose the first level of the hierarchical product structure for dog food is: (1) canned—all meat, (2) canned—mixed, (3) dry, and (4) semi-moist forms of product, while the target group segmentation is those who: (a) feel their dog is a member of the family, (b) have a dog for work purposes (guarding or hunting), and (c) have a dog, but really do not want it (reluctant owners). Table 5.5 shows a hypothetical estimate of the relationship of product form market definition to psychological segmentation. An X indicates the product form that is used most often by the segment. In this example, the all meat and semi-moist markets are primarily for those who love their family dog. There may be an opportunity for a new product for this segment positioned as "TV-Dinners" for your dog.

Table 5.5. Dog Food Usage Example

Product Form	*Psychological Consumer Segments*		
	Family Dog	*Work Dog*	*Reluctant Owners*
Canned/All Meat	X		
Canned/Mix			X
Dry		X	
Semi-Moist	X		

There may not always be an unambiguous relationship between the hierarchical product analysis and the target group segmentation, but at least the consumer characteristics of those at each level of a hierarchy should be defined and the product type used should be tabulated for each psychological segment.

The output of the market definition and consumer segmentation effort is: (1) a definition of the market in terms of the type of product, (2) an initial identification of the target consumers' demographic, attitude, usage rate, and preference/choice profiles, and (3) a tabulation of the relationships between product type and consumer segment. This information is then integrated to set priorities and initial strategy.

MARKET SELECTION

Market profile analysis is used to screen the potential market opportunities from a large number (say 20 to 30) down to a relatively few (say 3 to 4).

Analysis of the remaining candidate markets then identifies the market boundaries and target consumers. This information must now be used to explicitly set priorities and to select the one or two best markets to analyze in the design process.

This section begins by examining potential product portfolio managerial strategies in terms of growth rate and market share.

Managerial Strategies

An important strategic concept is that when market priorities are set, the new product must be part of the firm's complete product portfolio. The status of the set of current products must be understood before resources are allocated to a new product in a given market. It is important to recognize at this point that an addition to the firm's portfolio can be either a single product or a new product line. For example, a new product for Campbell's Soups was the line of "Soup for One." The initial introduction included more than one flavor (Golden Chicken & Noodles, Old World Vegetable, etc.), and was later extended with additional flavors.

One approach to assessing the product portfolio is to array the firm's offerings in terms of market growth rate and dominance (Day, 1977). Dominance may be expressed as share divided by the leading brand share. Figure 5.8 describes the characteristics of products in each of four cells. The Boston Consulting Group (1970) has called these cells: STARS, COWS, QUESTIONABLES, and DOGS. STARS have high growth rates, market shares, and profit plus a bright, long-term potential. COWS are ready for "milking." These high share, low growth markets usually require only sustaining investment levels and generate high cash flows and profits. QUESTIONABLES have high potential by virtue of their association with high growth rate markets, but show little market share and profit. DOGS have low profit due to small market shares in stagnant markets and drain cash reserves.

Besides indicating that new product development should be considered as part of a product portfolio, this structure implies that most new products should be aimed at high growth markets. Although a low growth market could produce a large share and hence profit, it is more difficult to enter those markets. Sales often come from competitors who try desperately to defend their share. If an organization already has a product in that category, share may come from cannibalization of the existing product's share. Furthermore, buying habits and distribution channels are reasonably stable and may be difficult to change.

Thus the best path to success is to build a share to a high level in a growth market to become a STAR and then to mature into a COW so that the rewards of innovation can be harvested. QUESTIONABLE product lines can be helped by new products that improve the existing product line by leading it to a position of higher dominance. DOGS probably should be dropped and their resources applied to new products that may become STARS.

Relative Market Share of Products

Growth Rate of Market in Which Product Competes	Low	High
High	QUESTIONABLES Medium to Low Profit	STARS High Profit
Low	DOGS Low Profit	COWS Very High Profit

Figure 5.8. *Product Portfolio Assessment*

The concept of a STAR implies the need of a high share or volume so the implication is clear. Do not enter a market unless you can earn a substantial market position. The amount of share or volume needed for a STAR varies by market, but each organization must set its target volume and select only those markets where it can obtain the necessary share or volume. If the market is vulnerable, the design process is used to develop the best positioning (psychological and physical characteristics) and marketing strategy to achieve a superior product. Before we commit the resources for the design activity, we must determine if the market is growing and vulnerable and if a good share will result in sufficient profit to justify the investment.

Market Growth Models

In order to implement the product portfolio assessment, the growth rates for each market category are forecast and the position in the product life cycle assessed. Various government agencies, trade associations, and syndicated data services provide information on total market sales. Standard Industrial Classification (SIC) data is commonly used in industrial products. In consumer goods, store audit suppliers (e.g., Nielsen) and providers of warehouse withdrawals (e.g., SAMI) can be utilized. In durable industries, numerous specialized data services exist to supply data to predict market growth rates. Many of these services are listed in Chapter 8.

Bass (1969) has developed a model that is useful in predicting the future sales of a product class based on the past sales history of the products. For example, Table 5.6 shows the actual and predicted peak sales volume and timing for 21 markets that range across industrial, consumer, durable, and services. The actual times to reach peak sales are long (median is seven years). The peaks predicted by the model match the actual data and there is reasonable agreement between the actual and predicted sales magnitudes. Statistical measures also support the model's descriptive adequacy. (See Appendix 5.4 for a more technical discussion of this model.)

The model is based on the assumption that the probability rate of initial purchase at a given time is a linear function of the total number of previous buyers. Initially, the probability rate would depend upon innovators, but would increase over time as the number of buyers [$Y(t)$] increase. Mathematically, this model implies that the rate of purchases is equal to an initial value [$p(0)$] due to innovators plus a term [$(q/m)\,Y(t)$] that reflects the impact of the

Table 5.6. Predicted Versus Actual Time and Magnitude of Sales Peak[a]

Product/Technology	*Predicted Time of Peak (no. of years)*	*Actual Time of Peak (no. of years)*	*Predicted Magnitude of Peak (Units)*	*Actual Magnitude of Peak (Units)*
Boat Trailers	9.8	10	205,240	206,000
Color TV, Retail	6.0	8	5,733,400	5,490,000
Color TV, Manuf.	5.8	7	6,637,800	5,981,000
Holiday Inns	10.9	11	131.6	141
Howard Johnson Mot.	9.0	11	38.6	48
Howard John., Hol. Inn, & Ramada Inn	9.8	11	202.6	216
McDonald's '55–'65	6.1	6	119.7	113
Continuous Bleach Range	3.2	4	16.7	18
Rapid Bleach Process	4.1	4	7.2	7
Conversion, 70 percent H_2O_2 Delivery System	3.3	4	48.5	50
Hybrid Corn	3.1	4	24.5	23
Home Freezers	11.6	13	1,200,000	1,200,000
Black & White TV	7.8	7	7,500,000	7,800,000
Water Softeners	8.9	9	500,000	500,000
Room Air Conditioners	8.6	7	1,800,000	1,700,000
Clothes Dryers	8.1	7	1,500,000	1,500,000
Power Lawnmowers	10.3	11	4,000,000	4,200,000
Electric Bed Coverings	14.9	14	4,800,000	4,500,000
Automatic Coffee Makers	9.0	10	4,800,000	4,900,000
Steam Irons	6.8	7	5,500,000	5,900,000
Record Players	4.8	5	3,800,000	3,700,000

[a]*Adapted from Bass (1969), p. 221 and Nevers (1972), p. 88*

word of mouth influence process. In equation form this model is given by:

$$P(t) = p(0) + \left(\frac{q}{m}\right)Y(t) \tag{5.1}$$

where $P(t)$ = probability of purchase given that no purchase has been made,

$p(0)$ = initial probability of trial,

$Y(t)$ = total number of people who have ever bought, and

m = the number of potential buyers of the product

q = a diffusion rate parameter to be estimated

This model presumes there is a social interaction process operating that exerts more pressure for adoption as more people adopt. The sales in each period are the number of people who have not yet purchased $[m - Y(t)]$ times the probability of their purchase.

(5.2) $S(t) = [m - Y(t)]\,P(t)$

$S(t)$ = unit sales in period t.

Substituting equation (5.1) in (5.2) and collecting terms gives

(5.3) $$S(t) = p(0)m + [q - p(0)]Y(t) - \left(\frac{q}{m}\right)Y(t)^2$$

Interpreting equation (5.3) we see that the market sales will first grow due to the effect of innovators $[p(0)m]$ and favorable word of mouth $[(q - p(0))Y(t)]$ and then decline $\left[-\left(\frac{q}{m}\right)Y(t)^2\right]$. The decline results because as more people purchase a product in the market, fewer people are left to purchase the product in the future.

The shape of this function is similar to the life cycle. Figure 5.9 compares the actual history of clothes dryer sales to the sales levels predicted with equation (5.3). In durable products, the model predicts the life of initial sales. The model does not represent replacement after the product wears out. The model helps establish the initial growth rates for a category and hence the attractiveness to new product entry.

It is important to recognize that the use of this model in market definition is for the growth rate of a *market* not a specific *product*. We introduce

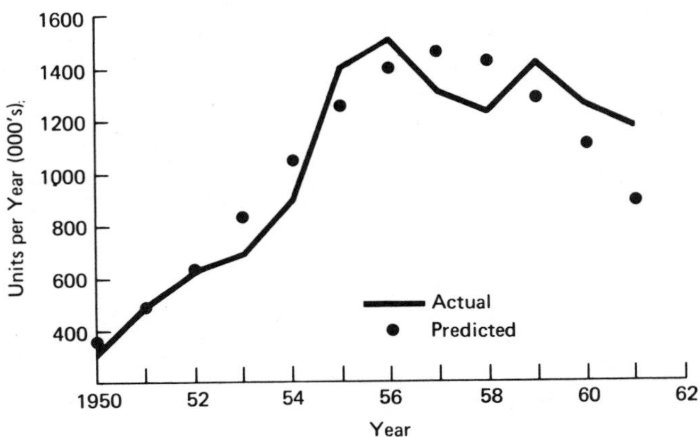

Figure 5.9. *Actual and Predicted Sales for Clothes Dryers (Bass, 1969, p. 224)*

models in Chapter 11 that forecast demand for the new product. In Chapter 12 we introduce models that are more sensitive to marketing strategies such as advertising, promotion, price, and distribution. Although other time series methods (Box and Jenkins, 1970; Nelson, 1973) could be used to forecast sales of a product class, the model in equation (5.3) is attractive since its shape is similar to the life cycle of a product and it has a behavioral foundation in the theory of diffusion of innovation.

The model was used in 1966 to forecast the sales of color TVs. Figure 5.10 compares the predictions for the period 1966–1970 with actual sales. Although the match is not exact, it is important to recognize that forecasts by RCA, Zenith, and Philco-Ford were much more optimistic and needed to be pared down in 1967 to match the market response. To forecast sales after 1970, replacement buying would have to be included. Sales grew in the 1970s due to replacement of old color sets and new sales of small color TVs.

Although the example in Figure 5.9 was reasonably accurate, it was based on very few sample points. Experience with the model shows that it is very sensitive to the variability of the initial data. Thus it is extremely important to check the parameters ($p(0)$, q, and equations 5.1 and 5.2) against historical norms and managerial judgment. Most important is the intuitive test of the plausibility of m, the market size. This sensitivity also suggests that quarterly or monthly data be obtained if possible and that the model be estimated on as large a data sample as possible. Care should be taken to add repeat sales and consider other factors which might change the structure of the diffusion of the innovation. We have presented one diffusion model to forecast growth. Others should be considered before a final sales forecasting approach is selected (see Hurter and Rubenstein, 1978; Dodson and Muller, 1978; Mahajan and Muller, 1979).

Setting Priorities

Priorities should reflect the concepts of high sales, growth rate, and high market share. This can be done by using quantitative analysis to update

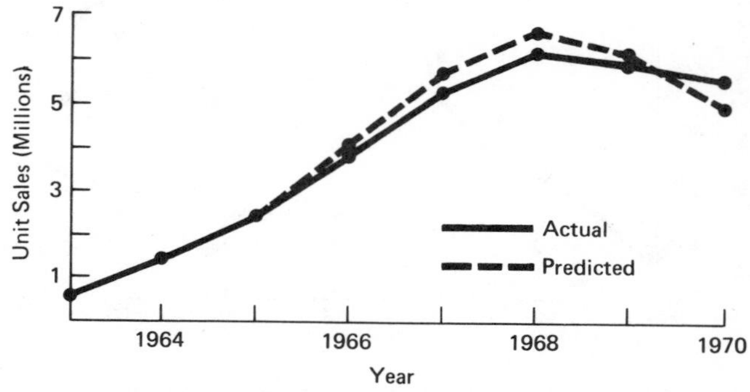

Figure 5.10. *Projected vs. Actual Sales of Color TVs (adapted from Bass, 1969, p. 225; and Standard and Poor's, 1979)*

estimates of the general factors of potential, penetration, scale, input, reward, and risk, which were described earlier in this chapter (review Table 5.1).

Each product segment in the market can be quantified by its size, growth rate, and competitive vulnerability. This potential can be compared to the investment required for entry and the profit margins in each part of the market. For example, in the coffee market defined earlier (Figure 5.7), it was found that ground and regular instant coffee types were the largest markets. They were judged to be vulnerable, since the perceptual maps showed gaps that could be exploited by a mild (not bitter) and good tasting (fresh, full bodied, rich) coffee. The estimated share potential could produce over 15 percent return on the investment of 10 to 20 million dollars required to develop and market a new brand. Only these parts of the coffee market were judged to have opportunities for new product development (see Appendix 5.2 for more details of the analysis). These opportunities would be different for each company in the market. They depend on the product portfolio each firm now offers.

To illustrate this, look at some of the firms in the coffee market and analyze their coverage and duplication in terms of the hierarchical market definition. Table 5.7 shows the offerings of General Foods, Nestlé, Procter and Gamble, and Hills Brothers. General Foods has good coverage, but there appears to be a potential duplication between Brim and Sanka. General Foods may be wise to consider further research to see if Brim and Sanka compete. If this is true, dropping Brim and investing those resources in Sanka or other new products would be appropriate. The opportunities for General Foods to introduce a new brand may not be attractive since a new ground or instant regular caffeinated coffee would cannibalize their existing brands. The new product effort at General Foods might be better directed toward creating a new market branch with a major product innovation, rather than adding brands to the existing product segments. Nestlé appears to have an opportunity to add a ground coffee. Procter and Gamble is tapping the two major opportunities and should not add other instant types unless they have an innovation feature that assures a high probability of becoming the dominant brand in those markets.

Other companies which have the necessary $20 million for investment and required marketing, production, and distribution skills could consider the opportunity for a new ground coffee. This evaluation should be weighed against the other market opportunities that pass the firm's initial market profile analysis. For example, the potential of a ground or instant caffeinated coffee might be evaluated relative to frozen dinners and chocolate snacks. The return on investment (ROI) and the factors of the market profile analysis would be used to prioritize those market opportunities and to select the most attractive market for new product development effort.

The coffee example represents a frequently purchased consumer product, but similar methods can be applied to consumer durables, industrial products, and services. An example of hierarchical definition of an industrial market was shown in Figure 5.4 for building construction.

Table 5.7. Coverage and Duplication of Products of Selected Manufacturers

Manufacturer	Ground	Caffeinated Instant		Decaffeinated Instant	
		Regular	Freeze Dried	Regular	Freeze Dried
General Foods	Maxwell House Brim Sanka Yuban	Maxwell House Yuban	Maxim	Sanka	Brim Sanka
Nestlé		Nescafé	Taster's Choice	Nescafé	Taster's Choice
Procter & Gamble	Folger's	Folger's			
Hills Bros.	Hills Bros.	Hills Bros.			

Consumer durables or services could be similarly described. The measure used to define these hierarchical trees is buyers' preferences for alternatives which can be obtained at a reasonable cost by market research surveys.

Although almost any market could be entered by a major innovation, firms should prioritize their effort and aim at those markets with the most attractive risk/return profiles. It is easier and less risky to develop a product to fill a need in an established market than to create a new market, but the rewards may be less. A firm would be unwise to only consider clearly established opportunities. Some of the biggest new product successes have been in areas where no previous product existed. For example, Xerox created the plain paper product segment in copiers. Digital Equipment created the mini-computer product segment in data processing. Even in the coffee example, Maxim created the freeze dried branch in the hierarchical tree used to describe the market.

We recommend that if a clear opportunity exists, you seriously consider exploiting it. However, also devote resources to markets that are subject to a revolutionary approach based on creating new product segments or markets. If the potential return is judged to be high enough to compensate for the high risk, consider these markets for entry. Even if you take the revolutionary approach, it is important to understand the market you will innovate. Do a market definition study to get a base line for the development effort. If your approach is not revolutionary, be sure to understand the existing market before you enter it or you may find after several failures, it is not a market you should be attempting to enter in the first place.

Profits estimated by projections of sales volume, penetration, and unit margin must be compared to the investment. Chapter 3 presented the expected costs of developing a successful consumer or industrial product. The ratio of the first five years' profit to the investment or the calculated rate of return on investment can be used to specify the most attractive market opportunities. Priority for new product development should be given to products with large sales potential and good risk/return profiles. A portfolio of high risk/high return and low risk/low return product/markets should be designated. Figure 5.11 shows a hypothetical tradeoff of risk and return and selection of a portfolio of market entry opportunities. The tradeoff line could be drawn judgmentally or analytically specified (see Chapter 15 for analytical procedures).

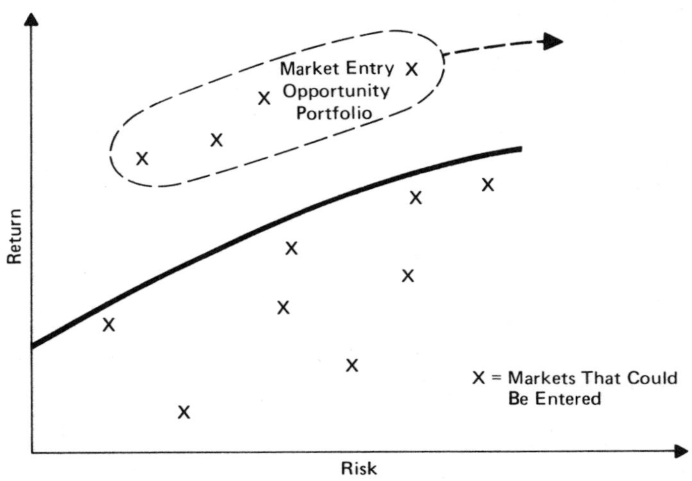

Figure 5.11. *Hypothetical Risk Return Tradeoff*

SUMMARY

It is crucial to new product success that a product category is selected that can reward the innovating organization. Market definition is that process which examines a breadth of markets to determine those that are most attractive. This process, summarized in Figure 5.12, directs resources at the markets of greatest potential while minimizing resources invested in high risk and low yield markets.

The first step is market profile analysis in which all markets are evaluated on the criteria of potential, penetration, scale, input, reward, risk, and match to the organization's capabilities. Criteria weights are established and

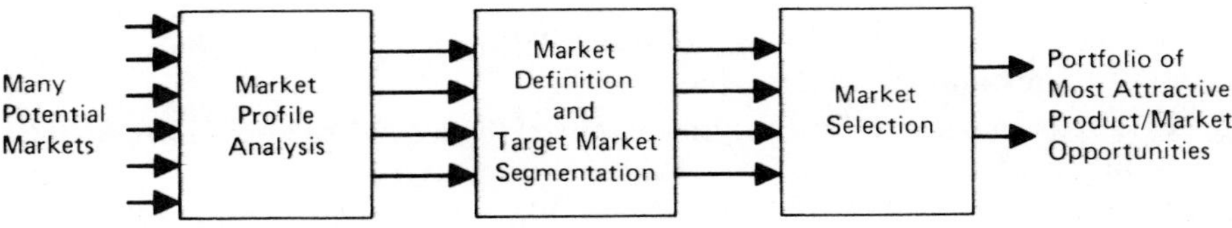

Figure 5.12. *Summary of Market Definition*

an overall evaluation is made. The result of this screening process is to eliminate poor markets and concentrate on a few highly attractive markets.

Next, these few markets are analyzed to define the boundaries of the market and to identify the target consumers. These analyses are synthesized to more exactly define the target markets and the vulnerability to entry.

Finally, this information is used along with an assessment of growth rates, profitability, return on investment, and risk to set priorities for further investigation. This market selection analysis yields a portfolio of market opportunities that are highly attractive for the organization's new product development effort.

REVIEW QUESTIONS

5.1 Can a product compete in more than one market? How should entry strategies be altered if this situation can occur?

5.2 Why should managers start new product development projects by studying consumers rather than engineering designs?

5.3 Why might businesses with large market shares be able to exert more control over their markets than businesses with smaller market shares? Is it possible a large market share might be an effect rather than a cause?

5.4 Suppose a very strong "experience curve" effect is observed in a particular market. Under what conditions would this effect encourage market entry? Hinder market entry?

5.5 How might a company measure the potential reward from entering a particular market?

5.6 What criteria might a small law firm use when deciding which legal services to offer?

5.7 Determining the importance weights of different market selection criteria is a critical activity. In what way might the selection of these weights help promote communication among managers responsible for new product development?

5.8 When rating new markets how might managers be given the incentive to provide unbiased ratings?

5.9 For a company adopting a proactive new product strategy the traditional methods for market definition might not be the best approach. Why?

5.10 Compare the different methods of defining a market. Discuss the advantages and disadvantages of each.

5.11 Why is market definition important? How could proper market definition affect the actions of a multiproduct company?

5.12 Perk-a-kup Coffee Company is thinking of launching a new brand of coffee which tastes like orange juice. Mr. Caf, the company's chief marketing executive, believes a great potential exists for this new product based on the size of the beverage market. How could proper use of market segmentation help focus on this brand's ultimate market?

5.13 What are the criteria for market segmentation listed in the text? What are the advantages and disadvantages of employing each criterion?

5.14 Consider the classification of products by market growth rate and market share as discussed in the chapter. Boston Consulting Group calls these products STARS, CASH COWS, QUESTIONABLES and DOGS. Should a company have any DOGS?

5.15 Policy Insurance Company is considering the development of several new financial policies. Their chief statistician, Alan Gorithm, estimated the parameters of Bass's diffusion model for two potential markets. The results follow:

Market 1: $S(t) = 2.2484 + .476\,Y(t) - .0048728\,Y(t)^2$

Market 2: $S(t) = 2.0500 + .011\,Y(t) - .0001024\,Y(t)^2$

where $S(t)$ = sales (1000s) in period t

$Y(t)$ = total people who have ever bought by period t

What insight do these parameters provide concerning the desirability of entering each of the markets?

Chapter 5 Appendices

5.1 The Hendry Partitioning Approach*

Key Relationship

The Hendry partitioning approach provides an understanding of direct versus indirect competition of a product. In the Hendry model, two alternatives are assumed to be in direct competition if the switching to (and between) them from any other alternative is in direct proportion to their shares. Let X and Y be dry bleach brands, with brand X having twice the share of brand Y. Furthermore, suppose that switchers from brand Z, whether liquid or dry bleach purchasers, are twice as likely to choose X as to choose Y. Then, brands X and Y are said to be in direct competition. If, on the other hand, X is a dry bleach and Y is a liquid bleach and they are not in direct competition, the switchers are no longer apt to switch to them in proportion to their shares. Paraphrasing Butler (1976): *Product alternatives are in direct competition if the switching to (and between) them is in proportion to their shares.*

Kalwani and Morrison (1977) show that assuming a zero order choice process along with the above switching proportional to share definitional relationship, the switching between alternative brands in direct competition $P(i,j)$ on two consecutive purchase occasions is given by

(5.1A-1) $$P(ij) = K_w S_i S_j$$

Where K_w is the switching constant for the set of alternatives in direct competition and S_i and S_j are shares of brands i and j. The switching constant, K_w, can be shown to be a ratio of the actual switching to the switching under homogeneity in consumer purchase probabilities. It takes the value zero when the buyers are completely loyal and always buy their favorite brand. It

*By Manohar U. Kalwani, Assistant Professor of Management, Massachusetts Institute of Technology

is equal to one, when consumers are homogeneous and each buys Brand i with probability S_i ($i = 1, 2, \ldots, g$). Thus, K_w is a measure of the degree to which consumers are not precommitted to one brand or another and is a property of the choice category as a whole. It should be noted that if the probability density function of consumers in a choice category with g alternatives is given by the Dirichlet-distribution, then their interswitching between items i and j is given by equation (5.1A-1) (Kalwani, 1979).

Procedure

The partitioning method is an iterative trial-and-error procedure. A hypothetical market structure based on "expert judgment" is set up and "theoretical" switching levels within and between product categories are computed. Empirical switching levels are then compared with the theoretical ones to determine the goodness of fit. Revisions in the hypothetical structure are surmised by noting where the theoretical switching levels exceed the empirical levels and vice versa. After one or more iterative attempts, a partitioning structure is identified which provides a reasonably good fit to the empirical data.

The empirical switching levels are obtained either from panel (or survey) data by comparing purchases on the previous choice occasion with those on the occasion prior to that. The theoretical switching levels require knowledge of the theoretical switching constant and market shares. In the model, the entropy concept is used to derive an expression for the theoretical switching constant. Its value is a function of only the brand shares within a product category

(5.1A-2) $$K_w = \frac{\displaystyle\sum_{i=1}^{g} \frac{S_i^2 \ln(1/S_i)}{1 + S_i \ln(1/S_i)}}{\displaystyle\sum_{i=l}^{g} S_i(1 - S_i)}$$

where S_i is share of brand i.

For illustration, suppose that in the bleach market the set of dry bleach brands compete directly with one another and thus form a product category. Furthermore, assume that there are only three dry bleach brands, A, B, and C, whose shares based on the purchases on the last choice occasion are 0.5, 0.3, and 0.2, respectively. Table 5.1A-1. displays the computation of the value of the theoretical switching constant. This value of K_w (0.4145) is used to compute the theoretical switching levels according to equation (5.1A-1). The theoretical repeat purchase proportions are easily obtained by subtracting the switching levels from the share data. Figure 5.1A-1 displays the theoretical switching and repeat purchase levels for brands A, B, and C. If the observed switching levels are close to the theoretical ones, then brands A, B, and C are in direct competition as assumed.

Table 5.1A-1. Computation of the Theoretical Switching

Brand	*Number of Buyers*	*Brand Share, S_i*	$\frac{S_i^2 \ln(1/S_i)}{1 + S_i \ln(1/S_i)}$	$S_i(1 - S_i)$
A	500	0.5	0.1287	0.25
B	300	0.3	0.0796	0.21
C	200	0.2	0.0487	0.16
	1,000	1.0	0.2570	0.62

$$K_w = \frac{0.2570}{0.6200} = 0.4145$$

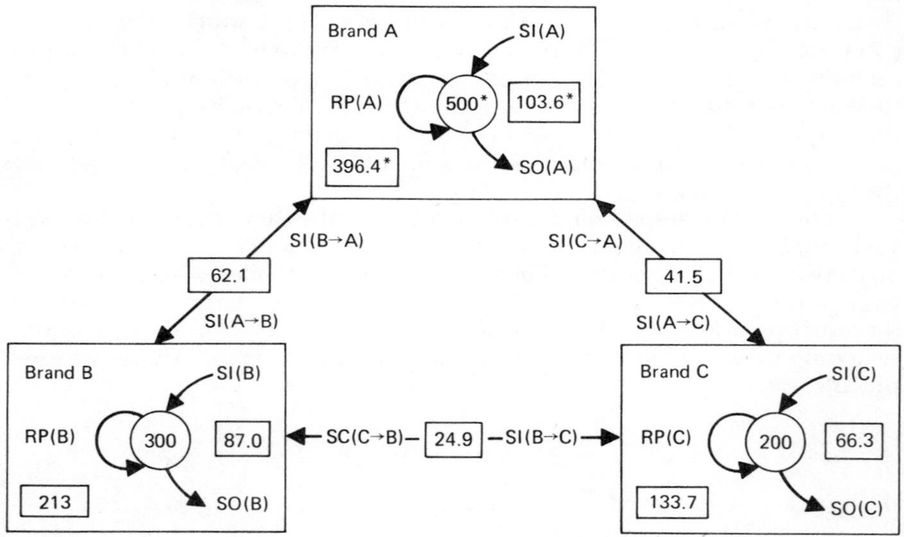

SI (·) = Observed number of consumers switching into a brand.
SO(·) = Observed number of consumers switching out of a brand.
RP(·) = Observed number of consumers repeating purchase of a brand.

Figure 5.1A-1. *Computation of Theoretical Switching Levels* [Theoretical switching levels are displayed in small rectangular boxes. For example, of a total of 500 buyers who chose Brand A on the last choice occasion, 103.6 were expected to switch to Brand A and another 396.4 (= 500 − 103.6) were expected to repeat buy it. Here 103.6 = $NK_wS_A(1 - S_A)$ = 1000(.4145)(.5)(.5) and SI(A) = SI(B→A) + SI(C→A).]

Market Partitioning Structures

The use of Hendry partitioning approach leads to the identification of two forms of partitioning structures that are nested or mixed-mode (Rubinson and Bass, 1978). In mixed-mode partitioning structures, two product characteristics—say, brand label and type (or form)—simultaneously form the primary partitioning level. The theoretical switching constants for switching between brand labels and types are obtained by finding their respective shares within the total market.

In the nested structures type-primary or brand-primary partitioning is sequential. For instance, in a type-primary structure, the theoretical switching constant for switching between types is obtained by finding the type shares within the total market. Then, the theoretical switching constants for switching between brands are calculated separately for each product type. In other words, at the secondary partitioning level, product types act as separate markets.

Concluding Comments

Knowledge of the true partitioning structure, or the market environment, is of great importance to marketers when designing marketing strategies. Type-primary structures call for promotional and new product introduction strategies different from brand-primary structures. The Hendry Corporation deserves credit for introducing the notion of partitioning and for delineating its significance in the formulation of effective marketing strategies.

As stated earlier, in Hendry model, the relationship that switching is proportional to share is used to define items in direct competition. It should be stressed, however, that this is not a theoretical but an empirical relationship which may or may not hold in a given situation. The Hendry people based on empirical analyses of consumer panel data in many different product categories contend that it most often does and use it extensively in identifying partitioning structures Kalwani (1979).

Presumably, the switching patterns implied by the switching is proportional to share relationship (in conjunction with the zero-order process assumption) can be used to identify items in direct competition. The problem, however, is that while the above switching patterns always hold for items in direct competition, they may also, although rarely, hold for items that are not in direct competition.

This suggests the need for theoretical criteria which can be used to obtain expected switching levels and thus to help verify a partitioning hypothesis. As outlined earlier, in the Hendry model, the theoretical switching constant given in equation 5.1A-2 is used to compute the expected switching levels. Kalwani questions the derivation of the theoretical switching constant by examining two of its major assumptions. He concludes that improved assumptions are needed since the theoretical switching constant plays such a central role in the Hendry partitioning approach. Future research addressing these and other related statistical issues is likely to yield the required theoretical and practical insights for market partitioning.

5.2 Prodegy Model*

Prodegy (*Prod*uct Strat*egy*) is a model and measurement system designed to guide the development of a market entry strategy. It uses a hierarchical tree to define groups of products that compete with each other as separate submarkets. In each submarket, a perceptual map is produced to allow assessment of market vulnerability and new product positioning opportunities. This market definition analysis is combined with financial and market data to determine risk/return tradeoffs and assess the appropriateness of entering the market.

Branching Criteria

Product entities are assigned to branches based on the product type, form, or brand name. No product can be assigned to more than one branch. After products are assigned, people are placed in the branch where their first preference resides. For example in the coffee case shown in Figure 5.7, the products were assigned to ground or instant branches. Then people with a first preference for a ground coffee product were assigned to the ground branch and, similarly, those with first preference for instant, to the instant branch.

The criteria for branching in the tree is based on substitution and is shown in Figure 5.2A-1 for a two-branch tree.

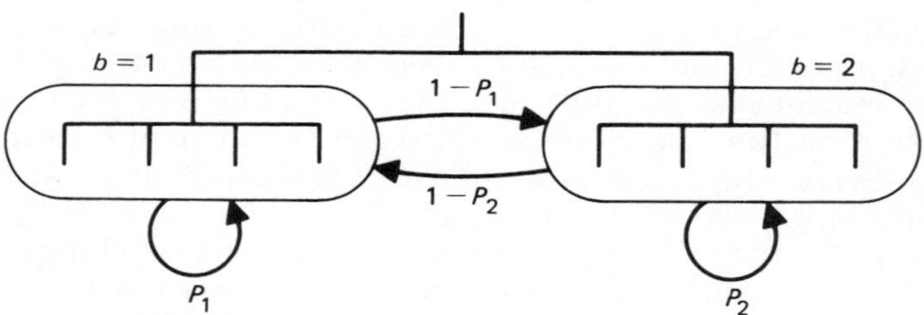

Figure 5.2A-1. *Forced Switching Probabilities*

P_b is applied only to those individuals whose first preference product is in branch b and is the average probability of buying a product in branch b if each consumer's first preference is not available. One minus P_b is the probability of consumers in branch b switching to a product in another branch if their first preference is not available. In a good branching P_b will be high.

The probabilities are estimated based on observed choices and relative preferences for products. The Logit choice model is used. This model is described in Chapter 11 and Appendix 11.1. For our purposes here it is

*This paper summarizes some of the essential aspects of the model; see Urban, Johnson and Brudnick (1979) for a complete discussion.

sufficient to indicate that after measuring preference for products and observing the last product chosen, we can calculate each individual consumer's probability of buying each product in his evoked set.

The probability of buying in branch b given that the first preference brand is unavailable is:

(5.2A-1)
$$P_b = \sum_{i \in I_b} \sum_{\substack{j \in e_i \\ j \neq J_i \\ j \in b}} \frac{P'_{ij}}{N_b}$$

where $P'_{ij} = \dfrac{P_{ij}}{1 - P^{\Delta}_{ij}}$

P_{ij} = the probability that individual i purchases product j
P^{Δ}_{ij} = the probability that individual i purchases product j if it is that individual's first preference product
P'_{ij} = the probability that i purchases product j if the first preference product is not available
e_i = set of products that individual i evokes as acceptable
J_i = individual i's first preference product
N_b = number of individuals who have their first preference product in branch b
I_b = set of individuals whose first preference is in branch b

For all consumers who have their first preference product in branch b ($i \epsilon I_b$), the probabilities of purchase (P'_{ij}) are added for all the products the consumer evokes in that branch ($j \epsilon b$), except the first preference product ($j \neq J_i$). These probabilities are summed for all individuals in the branch ($i \epsilon I_b$) and divided by the total number of individuals in the branch (N_b).

Branching Procedure

First, candidate hierarchical branchings are required to meet two necessary criterion. The first necessary criteria is that the probability of purchasing again in the branch (P_b) is greater than the probability resulting from a random assignment of products to the branches ($\hat{P}_b$). $\hat{P}_b = 1/B_1$ for the first level when B_1 is the number of first level branches. For example in the coffee case, $\hat{P}_b = ½$ for the two top level branches ($b = 1, 2$) in Table 5.2A-1. At the second level of branching the probabilities would be divided by the number of branches below each first level. The instant coffee branch is divided into four parts ($b = 5, 6, 7$, and 8) and the random probability is 0.125 (.5 ÷ 4). If N_b is sufficiently large and the P'_{ij} are not uniformly skewed, P_b will be normally distributed. The first necessary condition can be stated in statistical terms as $P_b > \hat{P}_b$ at the 10 percent level of significance.

A second necessary condition is that the switching to any other branch is not greater than the random probability of buying in that branch at the 10 percent level of significance.

If various hierarchical tree alternatives meet these necessary criteria, the one with the highest overall average P_b across the ending branches is selected. If we assume the P_b's are independent across branches, then their average, $\overline{P}_b$, is normally distributed. If we then assume that the $\overline{P}_b$'s obtained from different trees are effectively independent, the significance of the differences across hierarchies can be calculated to see if one tree is significantly better than another. The overall average, $\overline{P}_b$, also can be compared to the average random probability to measure significance.

The best branching alternative for coffee is shown in Table 5.2A-1 along with the associated probabilities. This was determined by testing first level branches and then proceeding to lower levels. For example, Table 5.2A-2 gives the first level test. Branching by product type (caffeinated vs. decaffeinated) is not acceptable since the decaffeinated branch fails the necessary condition that the predicted probability be significantly greater than the random probability. Of those people who had a first preference for a decaffeinated product, only 44 percent would buy another decaffeinated coffee if their first preference were not available. Fifty-six percent would switch to a caffeinated product. The use of brands for branching at the first level similarly fails. Ground vs. Instant passes all the tests. Applying this procedure at each level for all alternatives led to Table 5.2A-1 as the best hierarchical definition of the market.

Predictive Validity

The best fitting hierarchy is tested by comparing forced brand switching purchases made in a laboratory store with those estimated in the hierarchical model. Consumers shop in the store, but find their first preference brand "out of stock." The laboratory purchases of coffee indicated 60 percent of those with first preference for ground coffee bought another ground product when their first preference was removed. The predicted value (see Table 5.2A-1) was 60.3 percent. The overall percentage buying again in their first preference branch for the best hierarchy was 36.7 and can be compared to the predicted value of 37.6 percent. With this confirming predictive test, the confidence in the hierarchy increased. The predicted probabilities were updated by the laboratory results through Bayesian procedures; applying the branching procedure to the updated probabilities also resulted in the hierarchy shown in Table 5.2A-1.

Branching by Use

Usage can be utilized in describing a hierarchy by assigning brands to branches defined by usage situations. For example in the household cleaner market, products such as Spic and Span can be assigned to a use branch defined as "cleaning floors" and products such as Fantastic can be assigned

Table 5.2A-1. Best Coffee Hierarchy

Type of Coffee		P_b	P_b	N_b
Ground	(b = 1)	0.603	0.5	97
Instant	(b = 2)			
Caffeinated	(b = 3)			
Regular	(b = 5)	0.376	0.125	68
Freeze Dried	(b = 6)	0.217	0.125	45
Decaffeinated	(b = 4)			
Regular	(b = 7)	0.304	0.125	43
Freeze Dried	(b = 8)	0.239	0.125	42
Overall Average		0.376	0.248	295

Table 5.2A-2. First Level Branching

Forms, Types, and Brands of Coffee	P_b	$\hat{P}_b$	N_b
Alternative One: Product Form			
Ground	0.603	0.5	97
Instant	0.764	0.5	198
Overall Average	0.711	0.5	295
Alternative Two: Product Type			
Caffeinated	0.735	0.5	195
Decaffeinated	0.440[a]	0.5	100
Overall Average	0.635	0.5	295
Alternative Three: Major Brand			
Maxwell House	0.088[a]	0.143	99
Taster's Choice	0.042[a]	0.143	49
Sanka	0.073[a]	0.143	49
Brim	0.082[a]	0.143	19
Folger's	0.035[a]	0.143	15
Nescafé	0.050[a]	0.143	17
Overall Average	0.070[a]	0.143	248

[a]*Not significantly greater than random possibility at the 10 percent level.*

to a use branch defined as "metal, tile and chrome cleaning." Each product must be uniquely assigned to some branch defined by a mutually exclusive set of usage situations.

The branchings can be tested by applying the same criteria described above except that: (1) instead of assigning people to branches based on their first preference, only those who evoke the use described by the branch would be included in that branch and (2) the most preferred product in the usage branch rather than the overall first preference is utilized to calculate the forced switching probability. In this case:

(5.2A-2)
$$P_b = \sum_{i \epsilon I_{bu}} \sum_{u \epsilon U_i} \sum_{\substack{j \epsilon e_{iu} \\ j \neq J^{\Delta}_{iu} \\ j \epsilon b}} \frac{P''_{iju}}{\sum_{u \epsilon b} N_{bu}}$$

where P''_{iju} = Probability of purchase of product j for use u by individual i when first preference product for that use is not available

N_{bu} = Number of individuals to evoke a use u that is contained in branch b

U_i = Set of uses evoked by individual i

J^{Δ}_{iu} = Individual i's most preferred product for use u that is contained in branch b

I_{bu} = The set of individuals who evoke a use u that is contained in branch b

e_{iu} = The set of products that individual i evokes as acceptable for use u

Branchings based on product forms, type, or brand can be statistically compared to use branches. When usage situations are to be considered, measures of preference and last brand used must be obtained for each use the respondent evokes. In the coffee case, the evoked uses were grouped into two situations: coffee served to guests and all other uses. Products were assigned to the two usage branches and the statistical tests applied. This branching did not meet the necessary criteria — the overall probability was not significantly greater than the random probability.

Branching by Users

Branches also can be defined by segments of people after products are uniquely assigned to user groups. For example, Fruit Loops could be assigned to "children" as one user branch and 100 percent Natural Cereal to "adults" as another user group. People would be placed to branches, products assigned to branches, and forced switching probabilities calculated analogously to equation (5.2A-2).

These branchings based on users can be statistically compared to other trees based on product type, form, brand, or use. Trees based on combinations of these factors can be statistically compared to identify the best descriptor.

Entry Opportunities

After the best hierarchical tree is selected from alternatives based on combinations of a product form, type, brand, use and user, perceptual maps are generated by a factor analysis of product ratings (see Chapter 9 and Appendix 9.1 for the analytical procedures). By examining the importance of the dimensions and the positioning of brands, gaps in the maps can be identified. When a gap exists in an area of the map where preference is high, the market is vulnerable to entry (see Chapter 10 for analytic procedures).

The expected share that could be obtained by a typical new brand is estimated by examining the historical market share achievement of each successive new product to enter in the market and extrapolating this trend to estimate the expected share for the next brand to enter. This expected value is modified to reflect the vulnerabilities of the branch by multiplying the value by the ratio of the predicted share of a well positioned brand (in the gap) to the predicted share of the first product to enter (see Chapter 11 and Urban, Johnson, and Brudnick, 1979, for details).

The forecasted share is used along with a prediction of sales growth in each branch, estimated margins, expected development cost and capital requirements to calculate the return on investment (ROI) of each branch. The branch with the best ROI/RISK tradeoff is a candidate for entry. If no opportunity is present, the firm should look at other categories of products or seek to revolutionize the market by adding a new branch to the hierarchy.

Concluding Comment

Prodegy is an attractive approach since: (1) it uses individual probabilities to operationalize its explicit switching criteria, (2) statistical tests of the adequacy of a hierarchical tree and discrimination between trees are available, (3) a predictive validity test based on laboratory shopping is utilized, (4) product, use, and user segmentation can be combined and tested in a hierarchical market definition, and (5) perceptual maps and financial data are integrated to identify market entry strategies.

The model is an example of the advancing state of the art in the field of market definition. In the future we can expect the availability of various powerful models to help managers identify market opportunities. There will be a need to analytically and empirically compare them to identify the best approach.

5.3 Clustering Techniques*

The analytic problem in cluster analysis is to determine those objects, (e.g., consumers), that are most similar. The criteria are usually: (1) all objects within a group should be highly similar and (2) that objects in different groups should be highly dissimilar. Given a set of characteristics and some measure of fit, cluster analysis determines the number of groups and membership in those groups to best satisfy the above criteria. We will review one representative technique, the Howard-Harris clustering program.

Mathematically, given a set of individuals, S, and a set of clustering

*The description of the Howard-Harris clustering program is adapted from Green and Rao (1972).

variables, $Y = \{Y_1, Y_2, \ldots, Y_m\}$, the object is to partition S into disjoint segments $(S_1, S_2, \ldots, S_p)$. Let i index the individuals and let $\mathbf{y}_i = (y_{il}, y_{i2}, \ldots, y_{im})$ be the values Y takes on for individual i. Define the dissimilarity, d_{il}^2, between individuals i and l, as the square of the Euclidean distance between $\mathbf{y}_i$ and $\mathbf{y}_l$ — that is,

$$d_{il}^2 = |\mathbf{y}_i - \mathbf{y}_l|^2 = \sum_{k=1}^{m} (y_{ik} - y_{lk})^2$$

The total variance, V_T, of Y can be divided into a between-group variance, V_B, and a within-group variance, V_W, such that $V_B + V_W = V_T$. Where:

$$V_T = (\tfrac{1}{2}\, n) \sum_{i=1}^{m} \sum_{l=1}^{m} |\mathbf{y}_i - \mathbf{y}_l|^2$$

$$V_{S_j} = (\tfrac{1}{2}\, n_j) \sum_{i \epsilon S_j} \sum_{l \epsilon S_j} |\mathbf{y}_i - \mathbf{y}_l|^2$$

$$V_W = \sum_{j=1}^{p} V_{S_j}$$

where n_j = number of consumers in S_j

n = total number of consumers

Ideally, for a given p, a cluster program would find the partition to minimize V_W. At present, it is only feasible to find a heuristic solution that is locally optimal. The program starts with a $(p - 1)$-fold partition, splits one segment, S_j, on the basis of maximum within-group variance, and shifts points until a minimum V_W is found. The solution is locally optimal in the sense that shifting any single point would increase V_W.

In practice, two-fold, three-fold, etc. groupings are derived up to a prespecified limit. These groupings are then examined for interpretability and "significant" reduction in V_W. Often, a data matrix is randomly generated, cluster analyzed, and compared to V_W obtained from the consumer data.

5.4 Market Growth Model*

From the text we have that:

(5.4A-1) $$S(t) = pm + [q - p]Y(t) - \frac{q}{m} Y(t)^2$$

Where we have denoted p (0) by p for notational

But $Y(t)$ is just cumulative sales, thus

*This appendix is adapted from an article by Bass (1969).

$$Y(t) = \int_{o}^{t} S(\tau) d\tau \quad and \quad S(t) = \frac{dY}{dt}$$

and this equation can be rewritten as a differential equation. Solving this equation with boundary condition $Y(0) = 0$ yields:

$$S(t) = \frac{m(p+q)^2}{p} \left[\frac{e^{-(p+q)t}}{((q/p)e^{-(p+q)t} + 1)^2} \right]$$

To obtain the maximum sales we differentiate $S(t)$ and set the derivative to zero. If an interior solution exists, i.e., if $q > p$, then the maximum sales are given by

$$S(t^*) = \frac{m(p+q)^2}{4q}$$

$$\text{at time } t^* = \frac{\log (q/p)}{p+q}$$

Furthermore, since we are concerned with first purchase only, $f(t) = (1/m)$ $S(t)$ can be treated as a probability density function and the expected time to purchase, $E(t)$ can be calculated to be

$$E(t) = \frac{1}{q} \log \left(\frac{p+q}{q} \right)$$

The discrete analogy is such that the same maximums occur.

To test the model and to predict with the model when more than three data points are available, regression estimates are obtained from time series sales data restricted to periods where repeat purchase is not an important factor. When Bass applied this model to 11 durables including refrigerators, freezers, television, water softeners, air conditioners, clothes dryers, power lawnmowers, electric blankets, coffee makers, steam irons, and record players, he got very good R^2 values (0.828 to 0.976) and 90% of the regression coefficients were significant at the 0.05 level. It should be pointed out that these results were based on fitting the model to data for the full time period. In actual situations where the model is used to predict future sales, estimates may not be as accurate, but should provide useful managerial guidelines.

Other growth models are being developed (e.g., Lawton and Lawton, 1979) and we can expect further refinements in the future to include repeat purchasing.

Chapter 6

Idea Generation

New products are, by their very nature, innovations. These innovations result from creative insight and free thinking. Such creativity is crucial to success. This chapter explicitly deals with techniques for the enhancement and management of creativity in opportunity identification. After the priority markets have been defined and target consumers selected, the organization must generate ideas that tap the potential of these markets. These ideas are not the final product, that will be iteratively determined throughout the development process, but these ideas do form a basis for further investigation and a set of starting points for innovation.

While it may be possible to develop ideas in a vacuum, it is more productive to couple creativity with information from various sources such as marketing, R&D, and engineering. We begin with a discussion of how to tap various idea sources and then describe several methods to use this information for idea generation. The emphasis is to create major innovations and new product concepts that may expand existing markets or develop new markets. The final section discusses idea management by setting guidelines to screen initial ideas and it discusses how to select the appropriate number of ideas to advance to the design phase of the development process. The output of idea generation is a set of exciting product concepts. The design effort then evaluates their market potential, refines the product positioning, and converts them into reality.

IDEA SOURCES

New product activities start from a number of alternative initiating forces such as market needs, technological development, improvements in engineering and production, inventions and patents, and competitors' actions. These forces also act as sources of ideas. To be effective the organization should not look only at the initiation source, but at all potential idea sources. For example, a competitor may introduce a new product and thus force innovation by your organization. While there is pressure to respond with a "me too" or second but better strategy, more effective ideas could come from examination of market needs or from a recent technological development. To tap these idea sources we must first understand their potential.

Market Needs and User Solutions

One of the bases for a proactive strategy is a consumer-oriented philosophy of seeking to understand consumer needs and desires. Chapter 2 demonstrated that 60 to 80 percent of the successful technologically based products have their idea source in the recognition of market needs and demands and that the financial return from market based products tends to be higher. Since new products achieve final success through sales and profit, the consideration of market needs seems to be the most obvious source of ideas. However, many organizations do not allocate their idea generation resources to this area in proportion to its potential returns.

The power of the market in generating product opportunities has been already identified in several industrial product classes. Table 6.1 summarizes the role of the consumer of industrial products in generating successful ideas in several industries. These studies illustrate the major role of consumers in product innovation. Not only do consumers represent a source of potential needs, but they often provide solutions to these needs. In some cases a user may have solved a problem that your organization is only beginning to address with a major R&D project. There is a growing body of literature to suggest that users are active in the innovation process, not just recipients of new products (Rogers, 1978). For example, Figure 6.1 gives the level of solution content observed by a variety of researchers across a number of industries.

In consumer products, the consumer need and user solutions are equally likely to be an important source of ideas. Although it is less likely that a consumer will come to a manufacturer with a complete idea, they may have developed solutions that are a valuable basis for a new product.

Here are some examples. (1) The "banana seat" and motorcycle style "high bars" on children's bikes were first developed by kids themselves. Later manufacturers adopted the "chopper" style which is a major portion of the market today. (2) Women's use of eggs along with their shampoos, to give more body to the hair, was the solution adopted in the protein shampoo market. (3) Army corpsmen found their regular issue ammunition carriers, which fit on the belt at stomach level, were difficult to open, raised the body when crawling, and carried only half the required ammunition. Some users solved the problem by adopting a pouch, originally designed for transporting small land mines, into an ammunition pouch. A design based on this user solution is now being proposed for adoption by the military.

Table 6.1. Frequency with which Manufacturers Initiated Work on an Industrial Innovation in Response to a Customer Request[a]

Study	*Nature of Innovations and Sample Selection Criteria*	*No.*	*Data Available Regarding Presence of Customer Requests*
A. Studies of Industrial Products			
Meadows	All projects initiated during a two-year period in "Chem Lab B" — Lab of a chemical company with $100–300 million in annual sales in "industrial intermediates."	29	9 of 17 (53%) commercially successful product ideas were from customers.
Peplow	All "creative" projects carried out during a six-year period by an R&D group concerned with plant process, equipment and technique innovations.	94	30 of 48 (62%) successfully implemented projects were initiated in response to direct customer request.
von Hippel	Semiconductor and electronic sub-assembly manufacturing equipment: first of type used in commercial production (n = 7); major improvements (n = 22); minor improvements (n = 20).	49	Source of initiative for manufacture of equipment developed by users (n = 29) examined. Source clearly identified as customer request in 21% of cases. In 46% of cases frequent customer-manufacturer interaction made source of initiative unclear.
Berger	All engineering polymers developed in U.S. after 1955 with > 10 million pounds produced in 1975.	5	No project-initiating request from customers found.
Boyden	Chemical additives for plastics: all plasticizers and UV stabilizers developed post-W.W. II for use with four major polymers.	16	No project-initiating request from customers found.
Utterback	All scientific instrument innovations manufactured by Mass. firms which won "IR-100 Awards," 1963-1968 (n = 15); sample of other instruments produced by same firms (n = 17).	32	75% initiated in response to "need input." When need input originated outside product manufacturer (57%), source was "most often" customer.
Robinson et al.	Sample of standard and non-standard industrial products purchased by three firms.	NA	Customers recognize need, define functional requirements and specific goods and services needed *before contacting suppliers.*
B. Studies of Research-Engineering Interaction			
Isenson *(Project Hindsight)*	R&D accomplishments judged key to successful development of 20 weapons systems.	710	85% initiated in response to description of problem by application-engineering group.
Materials Advisory Board	Materials innovations "believed to be the result of research-engineering interaction."	10	In "almost all" cases the individual with a well-defined need initiated the communications with the basic researchers.

[a]*von Hippel (1978), p. 38*

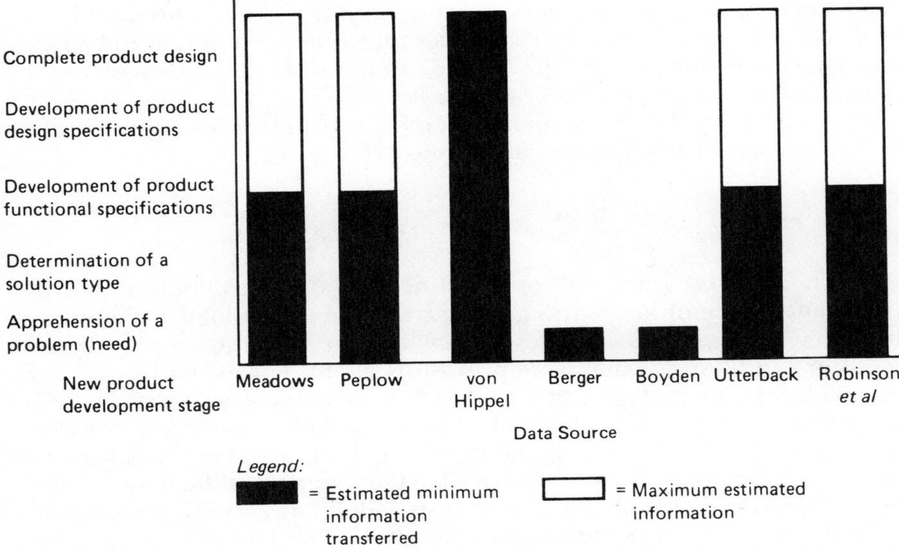

Figure 6.1. *New Product Development Data Supplied by Customer to Manufacturer (adapted from von Hippel, 1978, p. 43)*

Consumer product firms generally have not been enthusiastic about the receipt of user ideas since these suggestions present legal liabilities for the organization (Crawford, 1975). For example, if an unsolicited idea is not properly handled, a subsequent product developed independently in the organization may be claimed by the suggestor. Some firms find so many legal difficulties that they reject all consumer new product suggestions.

For example, Gillette rejects ideas unless they are patented. This means rejecting almost all unsolicited consumer input on new products.

> The inherently unpatentable ideas we refer to include suggestions relating, for example, to the advertising or sale of our products, to proposed new uses for present products, to the addition of products to our present lines, or to changes in our methods of doing business . . .
>
> We recognize that this policy does have some disadvantages for Gillette. For one thing, it does not allow us to consider a large number of ideas received from the general public, the people who use our products and on whose good will we depend for our successful operations. We regret this result, but at the same time hope the persons submitting such ideas will understand why we believe the legal risks involved outweigh the benefits which might be obtained from our considering such unprotected suggestions (Gillette, 1972).

Rejecting consumer ideas through policies like this may lead to overlooking many good ideas and profit opportunities. Although legal problems are present in accepting ideas, they may be effectively addressed by the use of a legal waiver and careful process to handle the receipt of and a response to suggestions. But in many cases these procedures are not utilized. Crawford found in a survey of 166 companies that 12.6 percent rejected all outside consumer suggestions, 67.5 used legally dangerous procedures of evaluation, and 19.9 percent of companies considered the ideas by legally sound procedures. We recommend accepting consumer ideas, but only after the organization has set up a legally sound procedure.

Technology

While the recognition of consumer needs and user solutions is the most important source of ideas, recognizing trends in technology is also important. New technologies can present new opportunities to meet consumer needs or can fill needs that were previously latent. For example, Bell Labs was exploring technologies that could allow pictures as well as voice to be transmitted over the telephone. The result was "television." A public demonstration was given in 1927. Although Bell Laboratory's technology has not yet led to a commercially successful video telephone, it is the basic technology for today's TV entertainment industry. Bell Labs is now working on a lightwave transmission system using glass fibers. This may have potential for many products other than telephones. Keeping aware of technological change and opportunities represents a valuable source of ideas for new products.

Engineering and Production

Engineering and production are often neglected as a source of innovation as firms use R&D and marketing to find new product ideas. For example, in the studies by Myers and Marquis (1969), 20 percent of the successful new technological products had their idea source in production.

In project HINDSIGHT, Isenson (1969) studied 20 military weapons systems and found that in 85 percent of the major technological developments, application engineering groups originally described the problem. The problem solving skills of engineers applied to the needs of the market produces an important source of new ideas.

Inventions and Patents

While internal R&D, engineering production, and marketing are valuable sources of ideas, external sources should not be overlooked. Contacting inventors and searching for patents may present new ideas for consideration. External consulting companies may have a portfolio of ideas that can be reviewed or suppliers of materials may have developed ideas for use of their raw materials in new final products. Sometimes acquisitions of small companies represent an opportunity for a new product. These acquisitions should be considered with ideas generated from internal and other external sources.

Competitors and Other Firms

The reason for competitors' success and knowledge about their developmental strategies are important inputs to idea generation. Even if a firm is a proactive leader in an industry, it must be ready to defend against competitors by preempting such innovation or improving on competitive firms' new products.

Noncompetitive firms may also be the source of new product ideas. Often acquisitions are possible and represent an attractive way to internalize the new product ideas of another firm.

Management and Employees

The creative potential of an organization is high. Managers and employees who are not directly involved in the new products effort may have valuable ideas and insights. This internal source of innovation can augment the creative efforts of the development team.

METHODS OF GENERATING IDEAS

The previous section listed some of the potential idea sources. To achieve success, an organization must be prepared to fully utilize these idea sources. Methods to generate such ideas can be as simple as setting up information channels that are sensitive to idea sources or as sophisticated as using creative group methods. Figure 6.2 shows some methods of idea generation that can be used to tap the areas of potential and lead to product concepts or prototype ideas.

Some creative ideas come directly from the environment and all an organization must do is to be sensitive to the ideas sources and conduct a direct search of opportunities. Other ideas require exploratory consumer studies or technology forecasting. Still others come by combining needs and technology through consumer engineering. Finally some ideas come from

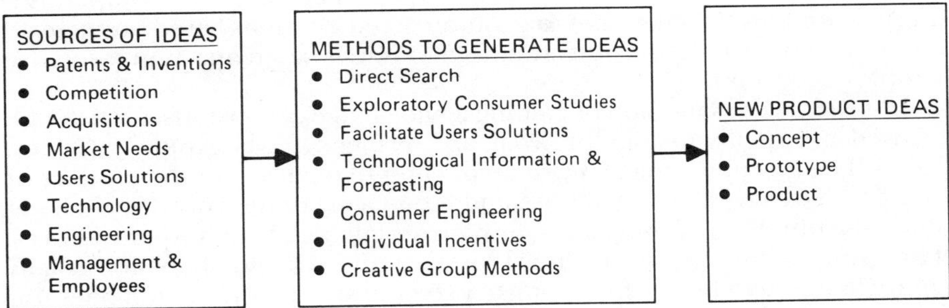

Figure 6.2. *Idea Sources and Methods for Generation*

individual effort or from creative group methods which utilize group dynamics to encourage imagination. Although an organization could rely on only one method, the best approach is to use several idea generation methods, covering all the sources of potential ideas.

The goal of idea generation is a large number of very different ideas. The more ideas generated in this step, the more likely one or two of those will pass the screening tests in the idea management phase and progress toward the design phase of the development process.

Direct Search

To effectively tap external idea sources, it is useful to allocate people to do basic information collection. For example, an organization might hire lawyers or assign a company person to search for patents that may be of interest to the firm in penetrating the target market. Competitive activity can be monitored by a feedback system which reports a competitor's sales practices, distribution, and new products. For example, some consumer products firms "read" a competitor's new product test markets. In some industrial companies, special employees travel to all the relevant trade shows to learn as much as possible about the competitors' new products. In others, the sales force reports competitive activity and new product needs.

Search efforts can be made to investigate acquisitions and thereby obtain product experience and development skills. Even such simple actions as a systematic analysis of complaints and warranty cards may identify problems that reflect a new product opportunity. There are many obvious activities such as those outlined above which should be conducted in a particular organization's target market.

Exploratory Consumer Studies

Since market needs reflect the highest potential source of ideas, the closer and earlier the new product developers get to the consumer the better. Observing how people buy and use the product can be done initially by casual observation or introspection on one's own behavior. Although this is useful, care must be taken not to generate fixed opinions based on these ad hoc observations since the observations may not be based on a representative sample of consumers.

A widely used method of gaining a knowledge of consumers is through focused group discussions or what are termed "focus groups" (Calder, 1977). The group is usually made up of eight to ten users of the product. It may be housewives talking about hand lotion, corporate controllers talking about the purchase of computers, or patients talking about hospital services. The group is brought to a central location and the members are usually compensated by $10 to $25 for one to two hours of their time. A moderator conducts the discussion. The session is taped and it is not uncommon to have a one-way mirror so that company personnel can observe the discussion.

Discussants are informed of these observations or recordings and almost always find them acceptable. The process of running a good group discussion requires experience; many professional services are available to conduct such groups. However, it is possible to run your own group. Usually the discussion begins with each person making comments on how they use or when they last bought the product under discussion. Likes and dislikes can be enumerated. In some sessions concepts of new products are presented and evaluated. In these discussions it is important for the leader to maintain control and make sure everyone has a chance to speak. At the same time, interaction should be encouraged, since group dynamics can help consumers verbalize their latent feelings. The purpose of focus groups is to learn consumers' opinions, semantic structure, usage patterns, attitudes, and buying processes. This exploratory work generates many insights, but even if three or four groups are run, the results can only be suggestive. The work is qualitative not quantitative. Insight and hypotheses are generated, not final conclusions.

A great advantage of a focus group is that it allows an early contact with users and provides an in-depth initial feeling for the products now in the target market. Consider a group discussion on hand lotion. Before the session, ten women were casually waiting with their hands on the table or in view. Within five seconds after the moderator announced that the discussion was about hand lotion, all hands had been placed under the table or out of sight and the women's faces became tense. Discussion revealed considerable guilt and shame about the condition of their hands. Further exploration revealed anger that housewives are expected to have soft hands, but must do abusive work such as washing dishes in detergents and cleaning floors while their husbands sit in offices with perfectly soft hands. Although this discussion did not directly generate a specific new product, it produced much insight into the product needs; eventually it led to a concept called "Hand Guard" which was designed to protect hands during dish washing.

While focus groups are most commonly used on consumer products, they can and have been used in developing services (e.g., health, transportation, and education) and industrial products (computer information systems). In each case the focus groups led to improved insight into consumers' needs and desires.

For example, a focus group on high efficiency electric motors conducted for the Department of Energy (*Market Facts,* 1978) indicated that the common practice is to evaluate motors on the basis of first cost rather than cost over the product's in-use life and that efficiency is not now a purchase criteria. This strongly indicates a marketing problem that the product will face. The group went on to suggest ways to educate the field through special programs, regulation, and incentives for manufacturers.

We recommend focus groups in all new product development processes. Even when a company feels they know their market, the development team will learn a great deal more at a low cost from observing focus group discussions.

Other individual exploratory consumer research is possible. In fact, the research conducted in the efforts to define the market (Chapter 5) can be useful. For example, if the perceptual dimensions of the hand lotion market are effectiveness (healing red, rough hands) and application (not greasy, soaks in), positioning gaps may become clear when existing brands are placed on the maps. A gap in an area on the map where no products now exist and where consumer preference can be expected is a good opportunity. (See chapters 9 and 10.) The hierarchical definition of markets may suggest ideas. For example, a regular instant coffee was indicated as a high potential area in the example in Chapter 5. These general ideas could be expressed as concepts and explored by personal interviews with a small sample of consumers. These would be "in-depth" interviews aimed at getting unstructured response to the ideas and insights that would be the basis of improvement.

Another kind of in-depth interview is oriented at finding problems consumers have and identifying opportunities for products to solve them. Other approaches concentrate on the benefits the consumer perceives the product as delivering and identifying situations where the consumer wants to receive more benefit (Myers, 1976).

Although there is no general rule about which method of exploratory consumer study is best, it is clear that early consumer interaction should be achieved. Whether one is interested in deodorants, machine tools, travel services, office systems, or banking, it is recommended that early contact be made with consumers so that the problem of understanding needs and the usage process can begin. This understanding is vital for successful idea generation.

Facilitate User Solutions

As indicated earlier, users not only have needs but may possess solutions. Rather than wait for the users to bring solutions to us, we could expend effort to find and facilitate this problem-solving process. For example, we could survey users to find out what special problem they have solved and inspect their prototypes. We could supply parts to innovative users at low cost to allow them to create solutions more easily.

Another approach is to set up a special program for custom one-of-a-kind work. This would put the company in a position close to the user need and solution process. A Boston electronics firm specializing in test equipment for integrated circuits decided to cut its one-of-a-kind production facility because it was not profitable. This decision was reversed when it was determined that most of the firms' products had developed from projects initiated as a one-of-a-kind machine. The user solution requirements had been generalized to a wide market.

Innovation can also occur through product modification by users. Xerox's first machines were not used for copying as intended, but rather to make offset masters. This user adaption was important in carrying the firm through its early period until the machine's speed and reliability could be improved enough to meet the market office copying requirements. Flexibility of product design and tracking user adaptation is an important method for generating ideas for overall product improvement.

Users may have solved your development problem. It is useful to examine your organization's past history and see where successful ideas have come from. If users have been active contributors, you should devote effort to cultivating them and maximizing their input to the idea generation process.

Exploiting Technology

Consumer input is crucial to success in the marketplace, but, as argued in Chapter 2, new product strategies should be comprehensive strategies based on both consumer input *and* research and development. An organization should encourage communication between marketing and the R&D laboratory and should allocate effort to understanding and monitoring the process of technological transfer. We first address these issues in R&D management which deal with information and then suggest methods of technology forecasting.

Information Flow. If technological ideas are to be generated, the most recent and relevant technical information must be in the hands of the project group charged with developing a new product within the specified market. This means that information channels within the R&D laboratory and between technical areas of capability within the company must be functioning. Also, technical material from outside the firm must find its way to the developers.

External sources of technology are important. Mansfield et al. (1971, p. 178) found that for pharmaceuticals 54 percent of the major innovations between 1935 and 1962 were based on discoveries made outside the firm. In development of numerically controlled machine tools, for example, manufacturing firms have been heavily dependent upon technology that was not present in their own firms.

While the external sources are important, information within the firm should not be neglected. Marquis (1969) found that in 41 percent of the innovations he studied, the key innovative information resulted from the person's own training and experience. This implies there is a considerable pool of talent for generating ideas if channels of communication are developed.

A key phenomenon in R&D communication has been identified by Allen (1970). He found that only a few individuals were connected to outside information sources and that these individuals acted as "gatekeepers" for their colleagues. They obtained, screened, and transmitted information. Figure 6.3 shows a typical information flow network. New information brought by individual one is transmitted to three other gatekeepers, and reaches its eventual users through their contact with the gatekeeper network. The gatekeepers read journals more extensively and have more per-

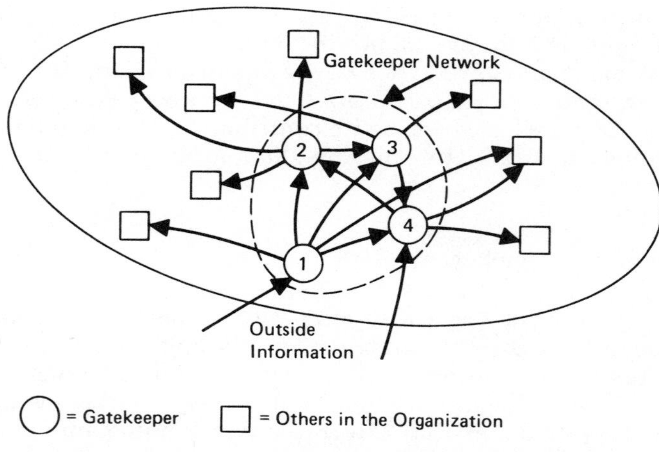

Figure 6.3. *The Functioning of the Gatekeeper Network (Allen, 1977, p. 162)*

sonal contacts outside the organization (Allen, 1977, pp. 146–147). They have a reputation for technical competence; others rely on personal contact with gatekeepers for information. To use this information source effectively, gatekeepers should be identified, rewarded, and supported. This facilitates their exposure to technological information and does not obstruct the interpersonal network used to diffuse the information in the lab.

The personal nature of information exchange can be increased by transfers of personnel within the company's divisions. In a large chemical lab, it was found that transferring persons provided an effective communication link back to their old divisions for one to one-and-a-half years (Kanno, 1968). Of course this use of transfers for information flow must be balanced against any potential detrimental effects of the transfer on the original project output.

In addition to the use of the gatekeeper network and personal contacts, physical layout is important for improving communication. Figure 6.4 shows the probability of communication between R&D lab personnel as a function of the distance between them. This is based on Allen's study of communication contacts in seven research labs (aerospace, universities, chemical, computer, and agriculture fields). The amazing fact about this curve is how fast and far it drops. After a separation of 30 feet, the probability of communication in a week is less than one-third of its maximum. It is approximately 0.05 at 65 feet. If people are on different floors or in different geographic locations, the chances of communication are even smaller. These findings imply that physical layout is important if communication is to be improved. People should have offices as close as possible, and meetings should be arranged to bring people into personal contact. Experimentation has shown

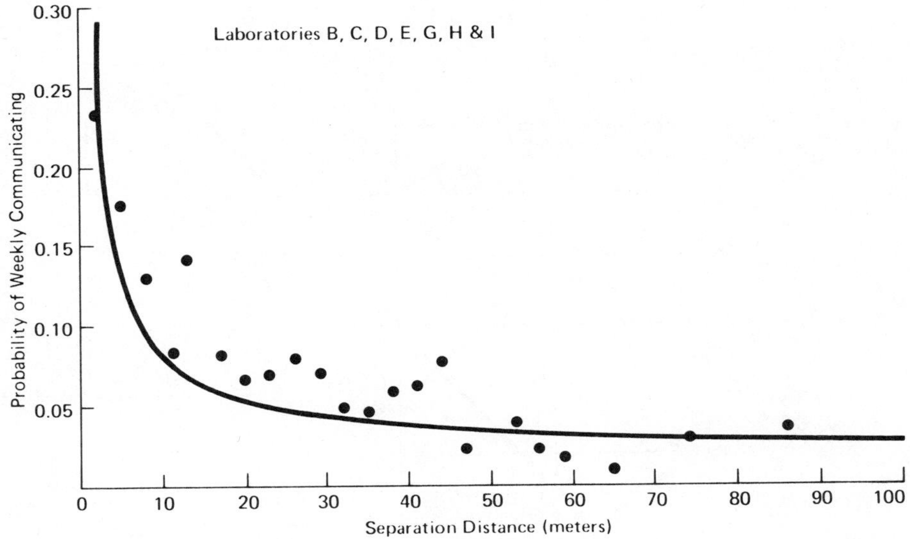

Figure 6.4. *Probability of Communication in a Week as a Function of the Distance Separating Pairs of People (Allen, 1977, p. 239)*

communication can be increased by changing architecture to reduce distances (Allen, 1977, pp. 248–263).

Technological Forecasting. Action to improve communication flows is important to exploring the potential of technological ideas. Another method of locating areas of potential is through technological forecasting. The most common approaches to technological forecasting are based on trend extrapolation and on expert judgments.

Figure 6.5 shows the trend of the number of components per integrated circuit. Clearly any firm operating in this industry must plan for products based on this trend. It is interesting to note that Gordon E. Moore, then at Fairchild Conductor, noted the trend in 1964 and predicted that it would continue (Noyce, 1977, p. 67).

Figure 6.6 shows a decline from 50 cents per bit to 2 cents per bit as the capacity of integrated circuits has increased. The primary reason for this reduction is the increasing complexity of successive circuits, but a secondary reason is that less complex circuits also continue to decline in cost. Hardware and memory costs of computers are approaching almost zero. This would indicate a major opportunity to develop computer software that utilizes the lower-cost hardware.

Examining trend data for costs and productivity can be useful in identifying ideas. Various statistical methods are available to project these trends. Most are curve fitting procedures based on extrapolation (Ayres, 1969, pp. 94–140). More elaborate dynamic models that include the effects of R&D

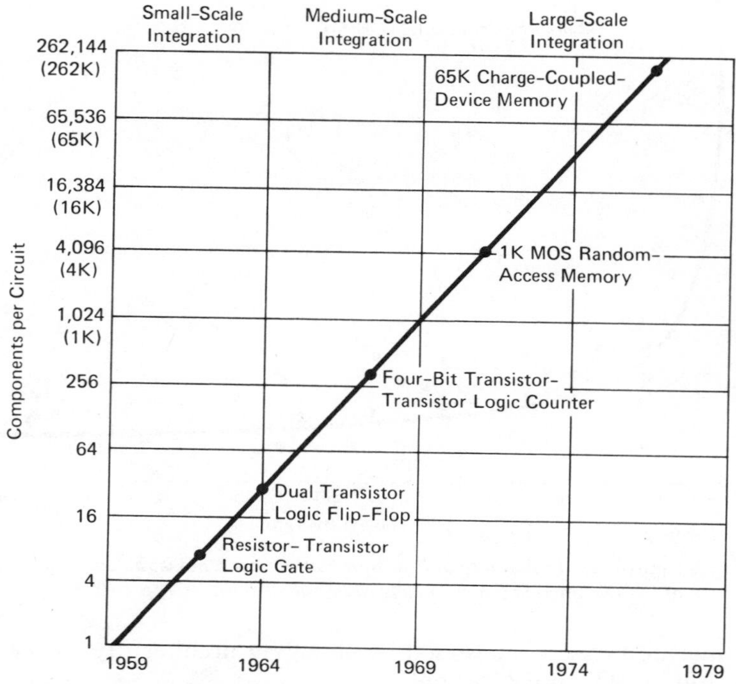

Figure 6.5. *The Trend in the Number of Components per Integrated Circuit (Noyce 1977, p. 67)*

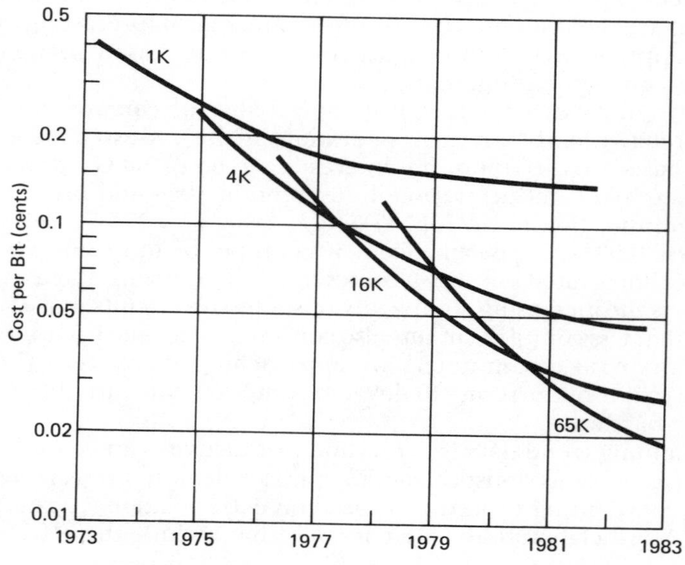

Figure 6.6. *Cost per Bit Trend (Noyce, 1977, p. 69)*

resource allocation on technological progress and specialized models for military purposes have been developed (Roberts, 1969; Sigford and Parvin, 1965).

Often it is useful to examine and extrapolate the costs of two or more related technologies (Quinn, 1967). For example, a large British chemical producer found a way to shatter used tires after they had been frozen by liquid nitrogen and then salvage the steel belting. This process initially was economically unfeasible, but the firm monitored the trend. After ten years, the price of steel had risen and the price of nitrogen had decreased so that this process became practical and could be introduced.

Technological forecasting requires careful projection and monitoring of trends. Even if the curve projections are wrong, they indicate opportunities which can be monitored to determine the appropriate time for a technologically based innovation. There is often ample warning before a technology achieves economic impact. Effort should be directed at diagnosing technological change (see Bright, 1970; and Utterback and Brown, 1972).

While trend extrapolation is an important approach to technological forecasting, the prediction of some future events are better handled by other means. For example, will nuclear power become widely available to Third World Countries by 1990? The answer to this question affects the types of products that will be feasible in these countries. Another example would be the possibility of manufacturing drugs in space. If this is practical, a whole range of ideas may be generated and regarded as feasible.

Forecasting long range and discrete technological events is often done by the use of expert judgment. Summaries of potential technological developments are generated and then expert opinions on the likelihood of them occurring are collected by "Delphi forecasting." In the Delphi approach, each expert in the group anonymously judges the likelihood and then opinion feedback and other data is reported to the group before new estimates are made. The process is repeated until a group agreement is obtained for the estimates. Figure 6.7 shows a flowchart for Delphi forecasting. About 15 to 20 experts are used as a panel. They provide inputs of a questionnaire after relevant outside data is provided. After the first estimates, statistical results of the forecasts are fed back to the experts. Usually after three to five rounds, these estimates converge (Dalkey and Helmen, 1963).

Some good results have been obtained by this method. For example, in an application to the construction industry, Delphi forecasting predicted sales within 3.3 percent for each of two years. This was better than previous errors of 20 percent and other forecasts generated by regressions and exponential smoothing.

There are some limitations to the method. In experiments with Delphi forecasting where the true answers were known (statistical almanac), convergence was towards the initial group median rather than the true value (Dalkey, 1969). Few tests of comparative forecasting accuracy have been reported. Delphi forecasting must be treated with caution, but expert opinion and re-estimation can help clarify the issues in predicting technological change.

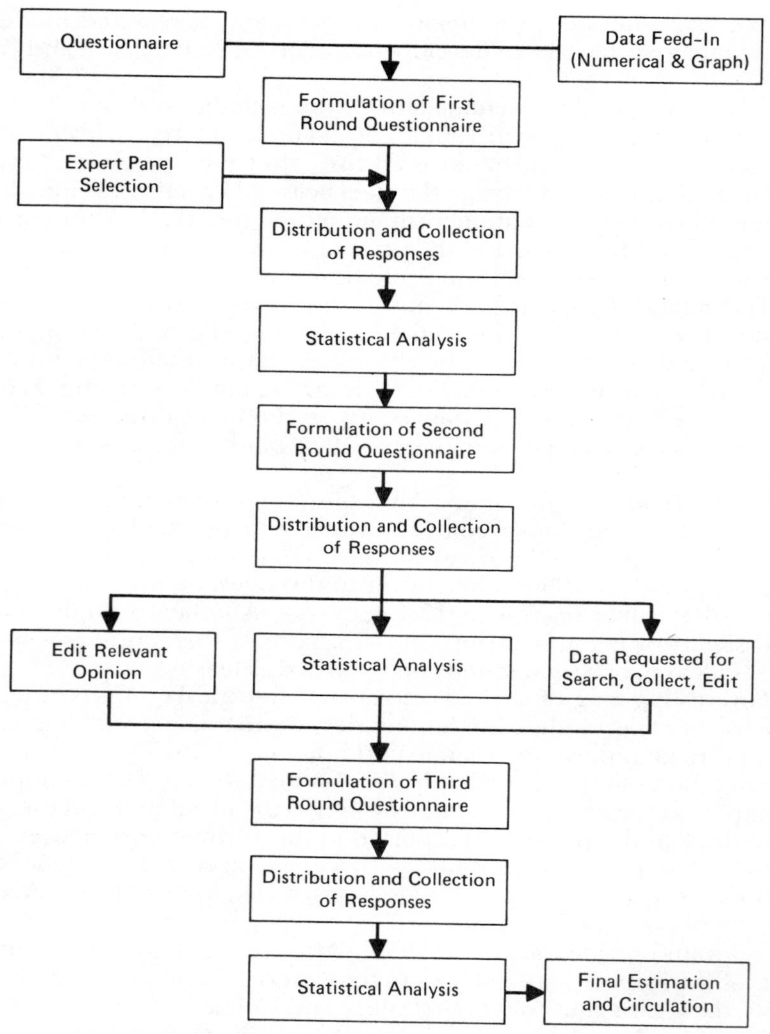

Figure 6.7. *Flowchart of Delphi Forecasting (Basu and Schroeder, 1977, p. 26)*

Several more elaborate methods such as "cross impact analysis" (Dalby, 1971) have been developed to integrate subjective judgments of social, political, and technical changes with market needs to evaluate specific ideas or indicate areas of technological opportunity.

Both technological trend extrapolation and the Delphi methods are subject to error, but it is important to allocate effort systematically to understand the movements in technology and their implications for new product idea generation. Even if an organization expends effort only to monitor rather than predict technological trends, this information proves valuable in the new product development process.

Although the above examples of technological forecasting are primarily for industrial products, they are also applicable to consumer products where such breakthroughs as sugar substitutes, fluoridated toothpastes, teflon coated cooking utensils, universal product codes, automated tellers, medical technology, and computer-aided brokerage are important.

Consumer Engineering

Focus groups and related methods identify consumer needs and desires. Technology forecasting identifies new capabilities. To be successful an organization must match the consumer needs to technological capabilities, even if this matching process requires engineering breakthroughs. For example, Figure 6.8 shows a hypothetical matrix of engineering technologies and markets for watches. The X's indicate when a given technology might be most relevant to a consumer segment.

ENGINEERING OPPORTUNITIES	MARKET OPPORTUNITIES				
	Gift	Status	Jewelry	Women	Men
Light Emitting Diodes	X				X
Liquid Crystal Diodes	X	X		X	
Thin Case		X	X		
Three Modules			X		
5 Components				X	X
Self-Charging Battery					X
Instrument Appearance	X	X	X		
Calculator plus Watch	X				X

Figure 6.8. *Matrix of Engineering Technology Versus Market Segment*

The consideration of the rows suggest ideas for digital watches based on light emitting diodes (L.E.D.) or liquid crystal diodes (L.C.D.) and watches varying in weight, thickness, color, and shape. Most firms have looked at the market this way. This is not a very imaginative use of the technology. How ordinary to put the electronics in the same round case as a mere substitution for hands. Looking at markets (columns) rather than application of specific technologies gives a different and creative perspective. We call this "consumer engineering," because it uses engineering to meet needs in particular segments. For example, Figure 6.9 shows a woman's watch to fill a jewelry positioning. Here engineering is used in a unique way to tap the

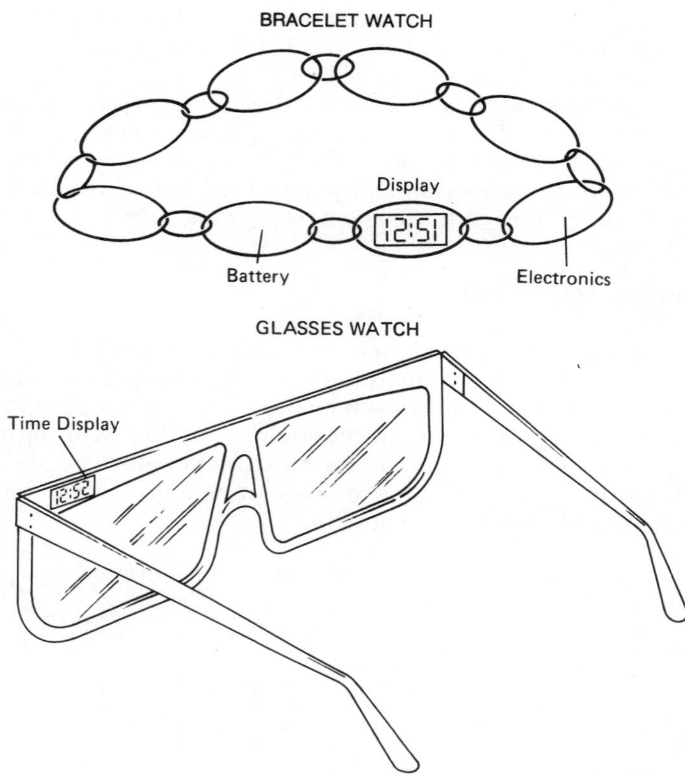

Figure 6.9 *Potential New Product Ideas Based on Consumer Engineering*

capability of the electronics to put components in separate units. Another watch could use this electronic capability to have the display mounted above a pair of glasses and be activated by touching the bow behind the ear. The use of engineering to meet specific consumer needs can be the source of many new ideas.

Another example of consumer engineering is the design of a tennis shoe for clay surfaces based on consumer needs for adhesion and safety. Since skis must "track" well and hold on ice, engineering developments to increase torsional resistance could meet market needs. The use of engineering and technical development skills to match specific market needs can yield ideas with the potential for significant innovation.

Individual Incentives

The success of idea generation depends not only on using effective methods to search idea sources, but on the creativity demonstrated in the generation process. As shown in Figure 6.2, some ideas come directly from the idea sources while others require more structured search such as technological forecasting or exploratory consumer surveys. Turning the source information into an idea for a new product may be straightforward, or it may come from the spontaneous creativity of an individual within the organization. The organization can set up reward structures to encourage ideas. One of the oldest is the suggestion box, backed by monetary rewards. Another effective method is to designate idea generation as a responsibility for certain individuals and give clear organizational recognition to these people.

Creative Group Methods

Since basic new ideas are so crucial to new product success, many organizations have set up formal creative group methods to synthesize the information on new product potentials. Creative group methods are not magic, they cannot guarantee solutions to impossible problems, but they do encourage a fertile climate for creativity by removing inhibitions and unproductive structures. They force the organization to think beyond obvious solutions to those holding more potential. These techniques assume that each individual has a wealth of knowledge and is by nature capable of creativity. The task is to encourage individuals to draw upon their personal knowledge, no matter how irrelevant it may appear, and apply this knowledge to develop creative solutions, no matter how impossible to implement they may seem. For example, the next creative breakthrough might be triggered by something you know from a recent movie or book, or some common knowledge about the weather, or specific knowledge about opera librettos, football strategies, sailing, sewing, modern art, chemistry, mythology, auto races, physics, biology, or even management. While at first this may seem absurd, experience has shown that "unrealistic" ideas can form the beginning of productive solutions.

In this section we discuss creative group methods of idea generation. We briefly outline several methods and then concentrate on some basic concepts that enhance the success of creative groups. While formal methods may not be appropriate for every new product problem, we suggest that many of the ideas behind creative groups be carefully considered and incorporated in the new product development process.

One of the first group methods of creating ideas was "brainstorming" (Arnold, 1962, pp. 251–268). In this approach a group tries to generate a large number of diverse ideas. No criticism is allowed and group members are encouraged to improve on other people's ideas. It is hoped that through this method, a wide variance of ideas will result, and some will be really new. This basic meeting format has been modified by adding structure in "attribute listing" methods. Attributes are listed for the existing products and then efforts are made to adapt, modify, magnify, minify, substitute, rearrange, reverse, or combine them. For example, among the attributes of the Oreo cookie is the filling. This was magnified to produce the highly successful double thick Oreo. In "forced" relationship techniques (Osborn, 1963), existing items are put together. For example, a new car might be both luxurious and economical.

A more highly structured creative method is represented by "morphological analysis" (Ayres, 1969). There are five steps:

1. explicitly formulate the problem,
2. identify parameters,
3. list all possible combinations of parameters,
4. examine feasibility of all alternatives,
5. select the best alternatives.

For example, in the 1930s Zwicky (Arnold, 1962, pp. 251–268) used this method on jet engines. He identified six parameters (e.g., type of ignition—self-ignition or externally ignited; and state of propellant—gas, liquid, solid). Across the values, 576 combinations could be formed. His examination of these combinations led to several radical new inventions.

Many (over 10) creative methods have been proposed. These existing methods range from unstructured methods such as meditation to structured approaches such as content analysis of advertising and functional analysis of products (Lanitis, 1970).

There are some common elements in the existing techniques that seem to work well:

1. establish openness and participation,
2. encourage many and diverse ideas,
3. build on each other's ideas,
4. orient towards problems,
5. use a leader to guide discussion.

One method, utilizing group sessions, has been developed by both Gordon (1961) and Prince (1972). These sessions are based on four simple, but powerful concepts.

The first concept is to *listen.* Unstructured meetings can often become power plays, with each participant trying to express his pet solution. As a consequence as someone starts to talk, many people will not listen with full attention, but rather will begin formulating their response. A useful method to encourage listening is for a moderator to write down each person's statement on a flip chart. This reinforces the listening and assures that no ideas are lost. Since all ideas are written down, one participant does not need to interrupt another to be sure his idea is not lost. The moderator must control the discussion so that all members can express their views and so that the discussion considers an idea completely without unproductive shifting among ideas.

The second concept is that *most ideas have some good elements.* Since most ideas have some good points, the group must build on these good points while overcoming any bad points. This is done by first identifying the good points and then the concerns the idea generates. Then explicit effort is expended in overcoming concerns and making the idea acceptable. One subprocess called "itemized response" identifies the good elements, leads to positive thinking, and rewards the idea originator. For example, a self-thinking typewriter may have been infeasible in 1960, but it was a good idea which could prove useful once its technical problems were solved. Auto-correcting typewriters and typewriters with memory have been successful innovations in office equipment in the 1970s.

The third concept is *a common understanding of a specific problem.* Frequently a group will simultaneously work on what appears to be the same problem, but is in reality several different problems. For example, one person may interpret a goal of cleaner air as the problem of preventing pollution while another sees the problem as one of cleaning already polluted air while still another views it as shifting pollution to another geographic area where it can be contained. While each may satisfy the general goal and the final solution may combine all three, the group process will be difficult if each is simultaneously working on his separately defined problem. Some group meetings use a client to define the specific problem. For example, the client might be the Vice President of Growth and Development who has been told by the President of the company to establish a position in the office copier market. The group then works with the leader and the client to achieve a working definition and common understanding that is used to address the problem of creating a new product idea.

The final concept is that of *a specific group leader.* In unstructured groups there is a tendency to jockey for leadership. The leader should be the facilitator and scribe. Since the leader does not have a personal interest in the problem, he can concentrate on encouraging effective group functioning.

These four concepts can be used to improve any group process. Prince's (1972) formal process makes good use of these concepts (see Figure 6.10). After defining the problem and understanding its background, goals and wishes are generated. Usually the client will be asked to select one goal or problem to work on first. Then ideas are generated to meet the selected goal. For example, the goal might be to develop an office copier which reuses paper or "to make copies without paper."

Simple techniques such as each participant listing one "real" and one "fantasy" idea can be used. More formal techniques based on the use of mental excursions, personal involvement, and metaphors can also be employed. Table 6.2 lists a number of specific idea-enhancing techniques. See Prince (1972) for more details. After a large number of ideas are generated, the client selects one for an itemized response. If the idea is made acceptable, it is called a "possible solution" and specific actions are outlined to implement it.

A usual session runs for three days, during which the basic flow is repeated many times. The group is usually comprised of personnel from the innovating organization, but consumer groups have also been used to create product ideas.

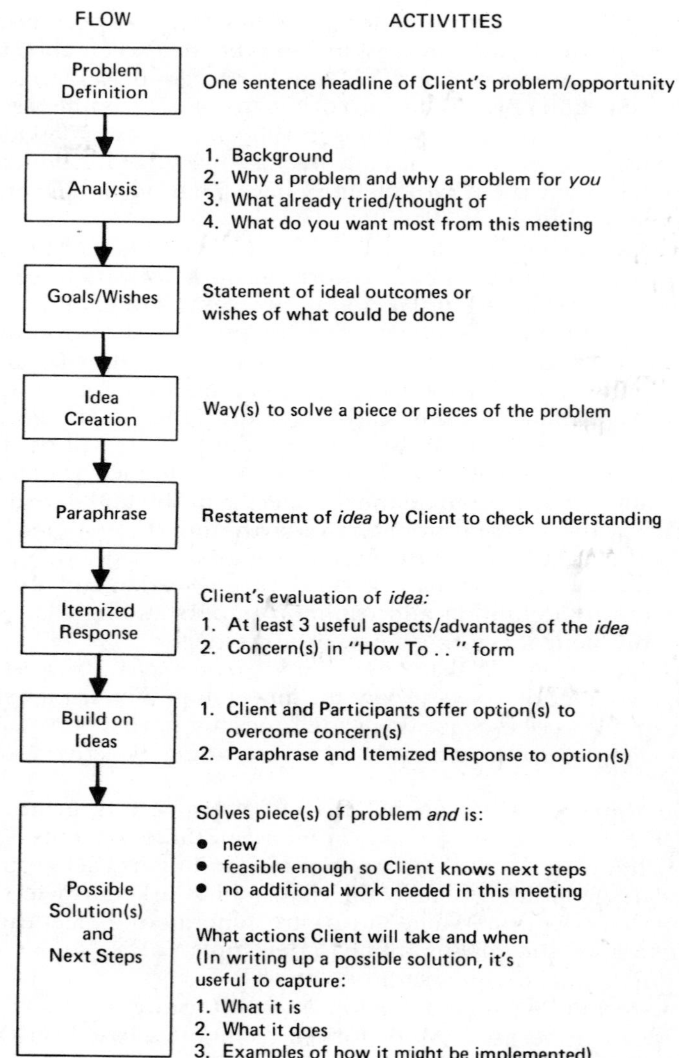
FLOW
ACTIVITIES
Problem Definition
One sentence headline of Client's problem/opportunity
Analysis
1. Background
2. Why a problem and why a problem for *you*
3. What already tried/thought of
4. What do you want most from this meeting
Goals/Wishes
Statement of ideal outcomes or wishes of what could be done
Idea Creation
Way(s) to solve a piece or pieces of the problem
Paraphrase
Restatement of *idea* by Client to check understanding
Itemized Response
Client's evaluation of *idea:*
1. At least 3 useful aspects/advantages of the *idea*
2. Concern(s) in "How To . . " form
Build on Ideas
1. Client and Participants offer option(s) to overcome concern(s)
2. Paraphrase and Itemized Response to option(s)
Possible Solution(s) and Next Steps
Solves piece(s) of problem *and* is:
• new
• feasible enough so Client knows next steps
• no additional work needed in this meeting
What actions Client will take and when
(In writing up a possible solution, it's useful to capture:
1. What it is
2. What it does
3. Examples of how it might be implemented)

Figure 6.10. *Flow of a Synetics Meeting (reproduced with permission of Synetics Inc., Cambridge. MA; Synetics is a registered trademark of Synetics Inc.)*

One remarkable feature of a creative group session is its ability to integrate marketing, R&D, engineering, and production points of view. We have found it useful to frame a problem in some sessions based on market research studies. For example, considering the market definition findings of Chapter 5, the organization may define an opportunity as a regular instant

Table 6.2. Some Formal Techniques from Synectics

Technique	*Description*
Personal Analogy	Participant puts himself in the place of a physical object, e.g., a tuning fork, and gives a first person description of what it feels like to be that object.
Book Title	Participant gives a two-word phrase that captures the essence and the paradox involved in a particular thing or set of feelings, e.g., familiar surprise.
Example Excursion	Group discusses a topic seemingly unrelated to the basic problem in order to trigger thoughts and/or "take a vacation" from the problem.
Force Fit — Get Fired	Participant thinks of an idea to force together two or more components of an idea. In the get-fired technique, the idea is to be so wild that his boss will fire him.

coffee with improved mildness and aroma. The problem would then be to feasibly create such a product. In other sessions a technical issue may be the problem. For example, with digital watches the problem may be "how to use the electronic technology to create a watch that tells time and reads blood pressure." Combining a diverse group of people in a meeting and integrating basic technical and marketing research can facilitate creation of major new ideas for innovative products that meet market needs.

These concepts of a creative group may seem abstract and hard to implement. Below is an edited transcript from an actual session that was used to develop a new service (Prince, 1972, pp. 145–147). Examine this transcript for the four concepts outlined above. You may wish to imagine yourself as either the client or a participant.

> *Problem as Given:* How can we provide more diagnostic service with no more doctors?
>
> *Analysis:* A large Boston hospital offers a 24-hour diagnostic service for emotionally disturbed people. This clinic examines patients and prescribes a course of therapy. Patient load is increasing 50 percent a year but the number of doctors remains the same. How can this service be continued with no more doctors or money?
>
> *Problems as Understood:* (1) How can one doctor be in three places at once? (2) How can patients be spaced throughout 24 hours?
>
> Leader: Let's take number 1. In the world of nature, can you give me an example of being in three places at once?
>
> Liz: Perfume.
>
> Leader: Yes?
>
> Liz: If you wear a distinctive perfume and go from one room to another you remain present in each room.

Leader: I think I see—you leave a trace or representation of yourself in each room?

Liz: Yes.

Dick: A fisherman's nets.

Leader: Go ahead.

Dick: I was thinking of a Japanese fisherman for some reason. He leaves one net in one place, another in another, and then collects them with the fish.

Leader: His nets act just as if he were there and catch the fish?

Dick: Yes.

Leader: Let's examine this fisherman's nets.

Dick: Japanese fishermen have to be efficient because they depend on a large catch to support their population. (Note: Dick is more interested in the fish than the nets and his remarks lead the team down an unexpected path. The leader, noting their interest, happily goes with them.)

Morris: (Expertly) Some fish are considered great delicacies in Japan.

Peter: Yes, what is that poisonous fish that is so popular?

Morris: Poisonous?

Dick: I forget the name, but it has one poisonous part or a spot that has to be removed.

Liz: Who removes it? The person who eats it? I'd be nervous.

Dick: I am not sure, but it seems to me you remove it when you are eating it. In the article I saw it said that it is not unusual for people to get poisoned.

Peter: I would want to know all about that poison spot so I could protect myself.

Liz: You are right. I wouldn't trust the fisherman or the chef. I'd want to remove it myself.

Leader: OK, let's take these ideas about the fish . . . How can we use them to put one doctor in three places at once?
(Long silence.)

Liz: There is something about do-it-yourself . . . get that poison out yourself.

Leader: Yes . . . this idea of protecting yourself? Can we help . . .

Peter: I don't know if it makes sense, but could the doctor use a patient as an assistant?

Morris: (Expertly) We sure have plenty of patients and they have time to help—some have to wait for hours, which is another problem.

Dick: Could they help each other? Some kind of do-it-yourself group therapy while they are waiting to see the doctor?

Dick: Yes. (To Morris, the expert): Could you?

Morris: I like the part about the patients getting some benefit while they wait—even if some just listened it would probably be reassuring. But, I am a little concerned about their working without supervision.

Liz: Could a nurse work with them, and perhaps get the history of the next one to see the doctor or something?

Peter: Or could the patient just finished take the history of the next one due to see the doctor?

Morris: If we push this thought to the end, we have a floating group-therapy session where everyone takes everyone else's history. We would want a doctor there.

This was the viewpoint. After experiments in the clinic the concept has evolved into a free-form meeting in the waiting room. The doctor presides. He and the group concentrate on helping one patient plan his own course of therapy. The doctor keeps an eye out for patients who are disturbed by the openness. Anyone who prefers can have a private interview. Most prefer the group treatment and find this waiting-room experience rewarding.

As the example indicates, an idea has its roots in many places—in this case in Dick's interest in the Japanese fishermen, in Peter's casual knowledge of Japanese delicacies, and in Liz's concern about trusting an important task to a chef or fisherman. Not all sessions will go this well and even this example is an edited transcript, but the openness, participation, teamwork, encouragement, and focus of creative group methods can lead to potential new product ideas that might not be recognized otherwise.

There have been some studies that indicate individual creative effort is superior to groups in terms of the quantity of ideas generated (see Lewis, Sadosky, and Connolly, 1975; and Bouchard and Hare, 1970). These studies are based on brainstorming and not the more advanced methods which stress the quality rather than the quantity of ideas. This area deserves more research, but one should not underrate individual effort as a component in the creative process.

In our experience, creative group methods have provided a useful tool for an organization to synthesize the diverse information obtained from direct search, exploratory consumer studies, technology forecasting, and consumer engineering, and convert this information into potential new product ideas. Whether group methods or individual efforts are used, an organization should expend energy to tap its idea sources and generate a sizable number of exciting yet diverse ideas.

IDEA MANAGEMENT

If an organization's idea generation efforts are successful, many exciting ideas are generated. Some may be the key ideas for a new product, but most of the remaining ideas will not have sufficient potential for further investigation. If each idea were advanced to the design phase costs would be prohibitive, thus we return to a managerial process to screen the ideas. Since the entire new product development process allows iteration, we select a small set of ideas based on the limited amount of information that is available for idea screening. This small set of ideas is advanced to the design phase to be analyzed in detail. Enough ideas are advanced so that it is likely that at least one is developed into an actual new product strategy. If none are successful, an organization returns to idea generation and iterates the process.

The two key managerial concepts in screening ideas are: (1) the selection process and (2) how many ideas to advance to the design phase.

Idea Selection

In setting up an idea selection process, an organization should consider information availability, the position of idea selection in the development process, and issues of intra-organizational conflict. Idea selection comes early in the development process—it is not a final strategy selection. Detailed information is not normally available and accurate estimation of financial outcomes is not always feasible. Finally there are usually a number of human aspects reflected in the varying goals of R&D, marketing, production, distribution, and top management. For these reasons, most organizations should choose a relatively simple idea selection process tailored to their own unique needs. While more complex processes may be appropriate after the design or testing phases, few organizations use such processes early in idea selection. Rather they choose a process that requires inputs that are feasible to obtain, that is appropriate for early screening, and that is flexible enough to allow the judgmental resolution of conflicting interests.

One simple approach to finding the best ideas is to apply the same procedure to each idea that was proposed for screening markets. In Chapter 5 we described "market profile analysis." The analogy for ideas is called "product profile analysis." When applied to products the scales may be more refined and some new scales added. For example, specific scales on the probability of commercial success or the probability of successful technical development could be added. Individual cost scales such as development or production costs are useful. These scales and the ratings of the ideas by managers is a good way to be sure all aspects of each idea have been considered. This checklist function is important since, while idea generation is creative outburst, this must now be tempered by a disciplined look at the feasibility and reward potential of each idea.

Product profile analysis can often be coupled with project selection indices. R&D management has specified and tested such procedures to select projects. Examining these methods and experience yields insights that are useful in not only R&D project selection, but in prioritizing ideas. The most basic procedure is to divide the expected return by the development cost:

$$I = \frac{T \times C \times P}{D} \qquad \textbf{(6.1)}$$

where I = index of attractiveness

T = probability of successful technical development

$C =$ probability of commercial success given that it is technically successful

$P =$ profit if successful

$D =$ cost of development

If an idea is already feasible, the probability of technical success (T) will be high and the cost of development low (D). For the same profitability (P) and chance of commercial success (C), this project would be preferred to an idea that requires a technological breakthrough (low T and high D). This simple formula is one way of trading off the different risks, return, and levels of knowledge of various ideas. After the index I is calculated for each project, they are ranked and the projects with the highest I are funded.

Many more complex models of project selection have been developed (Roberts, 1974; Pessemier, 1966; Baker, 1974; Baker et al., 1976). Baker and Pound (1964) list over 80 methods based on various operations research techniques. However, few of these advanced techniques are used; Baker and Pound found only 6 of 35 firms they surveyed used them. Even the basic model in equation (6.1) is used with caution due to the difficulty in accurately estimating costs and probabilities. Meadows (1968) studied five firms (three chemical labs, one electronics lab, and one equipment manufacturer) and found that the correlation of actual to estimated cost was only 0.5. The ratio of actual to estimated cost varied from 4.25 to 0.96 over the labs. Marshall and Meckling (1962) found a ratio of 1.78 in ethical drugs and 2.11 in proprietary drugs, while Norris (1971) found the ratio varied from 1.5 to 0.97 for industrial products in England. It is apparently difficult to accurately estimate the cost of an R&D project. In addition to the uncertainty of estimation, Mansfield et al. (1971, p. 213) feel that inaccuracies also stem from "deliberate underestimations" used to marshal support for a project. In studying the discrepancies, Meadows found the ratio of actual to estimated costs to be greater for project failures than for successes. One explanation may be the tendency not to give up on a project and to continue to allocate funds to it even when success sums unlikely. While costs are difficult to estimate, probabilities are even more difficult. Meadows found almost no correlation between the estimated probabilities of technical or commercial success and the observed fractions of success. Furthermore Rubenstein and Schroder (1977) found that personal, organizational, and situational variables can have major impacts on estimated probabilities. For example, project originators and those with implementation responsibility tended to give more optimistic estimates than the average while those with a "knowledge gap" about technical feasibility tended to give more pessimistic estimates.

These difficulties in cost and probability estimates suggest simplicity in the screening process because a more complex process would tend to hide the inaccuracies in these estimates. Furthermore those potential inaccuracies suggest that any estimates be treated with extreme caution and that steps be taken to minimize organizational bias on the costs and probabilities.

However, these cautions do not diminish the need for some formal process to screen ideas. Such a process allows a critical look at the many aspects of the alternative ideas as they relate to new product development. Furthermore it seems to foster dialogue among the disparate interest groups involved. It seems reasonable to consider costs and probabilities, but not to use a rigid formula to select ideas. We recommend the use of these factors as part of a product profile analysis. The profile analysis should serve as a checklist and a guide to managerial discussion and decision making. The primary goal of such a formal screening process should be to eliminate poor ideas and select for further consideration what appear to be the best ideas.

Number of Ideas

How many ideas should be identified for design work? It is reasonable to assume that there are "good" ideas and "bad" ideas, with their distribution as shown in Figure 6.11.

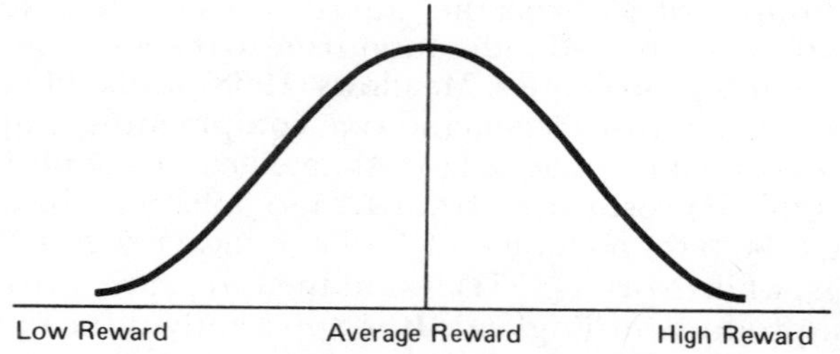

Figure 6.11. *Distribution of Potential Rewards for Ideas*

If we consider generating an idea as making a random draw from the normal distribution of Figure 6.11 and we generate only one idea each time we develop a product, we will get an average expected reward. If we could generate two independent ideas and select the "best" one, the expected value would be substantially greater. As more ideas are generated for a development opportunity, the expected reward increases. The overall gain depends on how many ideas are sampled, the variance in the distribution of ideas, the reliability and validity of our methods of finding the "best" idea, and the costs of generating ideas. (See Marschak, Glennan, and Summers, 1967, pp. 13–48; and Chapter 13 for more detail.)

In almost all cases, the rewards of generating several ideas are greater than the costs. We strongly suggest generating alternative ideas and not becoming committed to one idea alone. A common pitfall in new products is selecting the first idea and allocating large amounts of resources to it without considering alternatives that may be better.

The first idea may not reflect a major innovation in the market. It is important to allocate substantial creative attention to developing products that revolutionize markets. These may revise the structure of the markets by adding new dimensions to product performance or by creating new market segments. In order to find these major innovations the product market must be stimulated by major new idea concepts or prototypes. Consumer perceptions must be "stretched" to determine the potential of new dimensions and technologically challenged by ideas that may meet new major market needs. The examples of watches shown in Figure 6.9 stretch the market and technology. Watches that report pulse rate and blood pressure would represent new health dimensions for the product. The examples in Figure 1.3 describe the new product idea by concept statements so that consumers can evaluate the ideas. We term these revolutionary ideas "stretchers." They represent new market and technical options. Many of the stretchers will not find market acceptance, but they may lead to understanding of major new opportunities in the market.

Both the statistical considerations and the strategic notion of stretchers suggest the need to consider alternatives. The organization should use their screening process to select three, four, or more high-potential stretching ideas. These ideas should be innovative and try to cover the spectrum of potential in the selected market.

SUMMARY

The creation of major innovations and product ideas completes the opportunity identification phase of the new product development process.

This opportunity identification begins with market definition. A large number of markets are identified with characteristics compatible with the organization. These are screened via market profile analysis to identify a high potential set for further analysis. More detailed information is gathered and the competitive structure of each market is hierarchically modeled. Target groups are specified through market segmentation. Finally these alternatives are carefully examined to select a portfolio of "markets" of attractive risk/return characteristics and to select the best initial, competitive definition for that market.

Given the market definition and target group, the organization then develops ideas for products to take advantage of the identified opportunities. A large number of ideas are generated through direct search, exploratory consumer studies, facilitating user innovation, technological forecasting, consumer engineering, and creative group methods. Emphasis is on breadth and creativity since alternatives enhance the chances for success and the greatest success often comes from the most innovative ideas. Finally, a relatively simple screening process is used to select a few good ideas from those generated. These three or four innovative ideas represented by stretcher concepts or product prototypes, as well as the market definition and target group segmentation, are then the initial inputs to the design phase. Since the market was chosen because of its strategic growth and profit potential and the ideas were created to meet specific needs in this market, the opportunity defined by the set of ideas should be significant.

The next step in the new product development process is to physically and psychologically design the product to exploit the market opportunity. By management's use of measurement and models to understand consumer response and R&D and engineering skills, a final design is created. This design is then tested and, if successful, introduced.

REVIEW QUESTIONS

6.1 Why is idea generation important?

6.2 Investigate two recent and possibly significant technological advances. For each, generate three new product ideas for three different consumer needs.

6.3 Select three of the methods of generating new product ideas that were discussed in the chapter. For each method,
   a. Discuss the role of communication among individuals involved.
   b. Discuss the costs of employing the method.
   c. State which industries would find the method most valuable and why.
   d. Describe the generation of a hypothetical product when employing the method.

6.4 What are the advantages and disadvantages of creative groups?

6.5 The following table illustrates managerial judgments on the probability of successful technical development of an idea (T), the probability of commercial success given that it is technically successful (C), the profit if successful (P) and the estimated cost of development (D).

Idea	*T*	*C*	*P*	*D*
A	.7 to .9	.2 to .4	1.1 to 1.3	.1 to .9
B	.1 to .3	.7	2.2 to 2.5	1.2 to 1.9
C	.8	.4	1.3 to 1.4	.3 to .4
D	.4 to .5	.5 to .9	.1 to 2.2	.1 to .2
E	.2	.7	3.3 to 3.4	3.0
F	.6	.6	1.3 to 1.4	.9

   a. Compute the index of attractiveness for each idea.
   b. For each idea, discuss what information the index captures and what information the index ignores.
   c. What product ideas does the index dismiss as inferior? Would you dismiss those ideas?
   d. Where should managerial effort be directed when attempting idea selection?

6.6 Suppose a single individual working alone would generate ideas of similar quality. Would it be more valuable for a single individual to generate 10 separate ideas or for 10 individuals to each generate one idea?

6.7 Suppose 10 individuals working alone would come up with 10 very different ideas. However, 10 individuals working together would generate 10 similar ideas but of better average quality. Should the individuals work apart or together?

6.8 How might a hospital supply firm encourage users to innovate? How might the hospital supply firm set up an information network to discover and use these innovations?

6.9 Develop two stretcher concepts for urban transportation service.

6.10 Develop two stretcher concepts for a warehouse heating system.

6.11 What are the advantages and disadvantages of focus groups? How do focus groups provide new product ideas?

6.12 Use consumer engineering to develop ideas for a new home information system.

6.13 Do a direct search of idea sources for a new automobile accessory.

6.14 What is the role of gatekeepers in idea generation? How can this role be encouraged?

PART III

DESIGN PROCESS

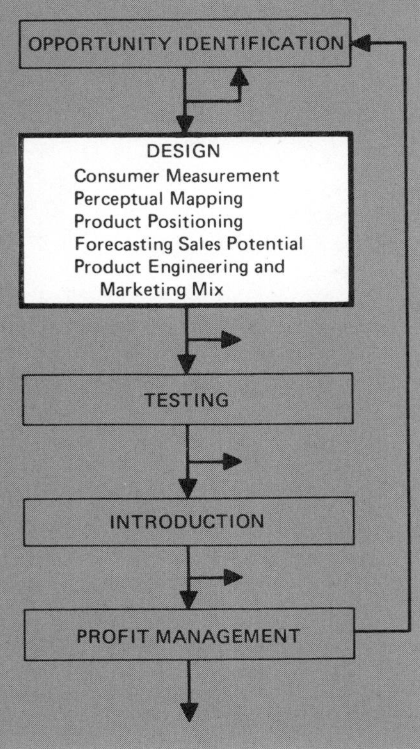

Chapter 7

Introduction to the Design Process

THE DESIGN PROBLEM

After identification of a priority market and the generation of a set of initial ideas, the next task is to "design" the product. We consider "design" as the designation of the key benefits the product is to provide, the psychological positioning of these benefits versus competitive products, and the fulfillment of the product promises by physical features. We identify the key benefits in a statement called the Core Benefit Proposition (CBP). The Core Benefit Proposition must be clear and concise, striking immediately to the essential characteristics of the strategy for the product. It should form the keystone upon which all elements of the marketing strategy are built. Here are some examples:

Sears' Die Hard Battery. Longer lasting battery with more power output.

Tylenol Pain Reliever. Effective but will not upset stomach.

The Executive Sciences Program. A flexible two-year educational experience that stresses depth of skill in specialized areas of management to facilitate a rapid rise to key positions in private and public organizations.

Personalized Transportation Service. A premium service designed to give door-to-door service on demand. It uses small, comfortable buses to give fast, prompt service when and where you need it, but at fares lower than taxis.

"Sclam" Chowder. A nutritious and economical chowder made from choice New England clams and North Atlantic squid.

Narrow Band Video Telephone. Attaches quickly and easily to an ordinary telephone. Improves the effectiveness of scientific and technical communication by transmitting high-grade, clear, black-and-white, still pictures.

New Liquid Antacid. Concentrated for more effectiveness per spoonful and homogenized to be easier to swallow.

Personal Care Hospital. A full-service hospital committed to the personalization of health care. Provides high quality primary care to the patient, increased accessibility of the staff, and a friendly, "first-name" atmosphere.

Solar Air Conditioning. Saves energy and provides reliable cooling.

Each CBP is short and to the point, stressing the key features or appeals that are special to the new product or market. In many cases the target market, e.g., scientific and technical communication, is explicit in the CBP. It identifies a selected set of benefits for the product strategy. It should specify exactly what we are offering to the consumers, what they will get from it, and how this is important and possibly unique. It is not simply an advertising appeal, but rather a basic description of the overall strategy in terms of consumer benefit. It is more than a description of the physical product because it specifies the benefits the consumer derives from the product. Consumers buy products for the benefits they deliver; good physical product features are necessary but may not be sufficient to deliver these benefits. For example, a good automobile engine is necessary to deliver power, but in 1978 General Motors found that a good Chevrolet engine in an Oldsmobile body was not sufficient to deliver Oldsmobile's image of power in consumers' minds.

The CBP forces management to come to an agreement on the basic benefits and serves to structure further work on modification of the CBP and/or strategy to achieve the CBP. The lab develops a physical product to fit the CBP, the advertising agency develops copy to stress all or part of the CBP, and top management can readily assess the new product. The CBP differs from the traditional Unique Selling Proposition (USP) found in advertising texts in that it includes all important features, not just the unique appeals to be used in selling and advertising materials. The USP can be derived from the CBP when the organization formulates the marketing mix to support the CBP.

Although a CBP is simple, it is not necessarily simple to attain. The CBP is the output of a new product design process and results from careful consideration of consumer response to product features and psychological appeals. Marketing must establish how to communicate the CBP and R&D must adapt, adopt, or invent technology to create a product that fulfills the CBP. For example, Sears' Die Hard Battery actually was a physically superior product and it was advertised as such. On the other hand, it was not sufficient to advertise Amtrak as a superior railway service without first ensuring that the promised level of service could be delivered.

To explain the design problem more clearly, let us consider an example based on designing new educational services. Many of you are currently enrolled or have been enrolled in an undergraduate or graduate business school. Pause for a moment and reflect on how you chose your school. There are over 100 business schools in the country, each with 20 to 100 faculty teaching dozens of courses with different texts. Did you find out everything there was to know about each school, plot this information on a huge chart, establish criteria, and then make a selection? Chances are that your choice process, identifiable only in retrospect, was some other, perhaps psychologically more complex process. Had we arbitrarily modeled your behavior as an exhaustive search, big chart process, we would have missed the true process and perhaps missed our chance to understand your needs and desires.

It is more likely you became aware of certain schools by brochure, by reputation, by proximity, or by advice from faculty, friends, and advisors. As you obtained more information and opinions, you began to form perceptions of those schools. Perhaps you felt that Harvard was more prestigious than MIT, but MIT could train you better for your chosen specialty. Perhaps Northwestern had friendly, cooperative students, or Stanford had the most interesting program. Perhaps you felt one school could help you get exciting, rewarding jobs, but another had a better geographic location. Based on your personal tastes and your career goals, you considered the characteristics of the schools and began to form preferences among schools. For example, school X was your first choice, school Y your second. Finally, around April or May you made your choice. Some of you came to your first choice, but others found that the first choice was not available. Perhaps you were not accepted or you could not afford the tuition or your spouse could not find employment—some external event or a change of mind caused you to switch from your preferred school to a second or even a third preference. Now think back to your high school or college class; not everyone had the same tastes and goals that you had. Others felt differently about which career was appropriate or about the importance of a school's prestige or of friendly students or of program flexibility. But some segments of the population felt about the same as you did and they formed their preferences, although perhaps not their choices, the same way that you did.

Now that you have considered how you made *your* decision, shift your perspective to that of the dean of a management school. What new programs should be offered to attract more quality students? One possibility could be a 12-month program, stressing general management skills or a specialized program as indicated in the earlier examples. In either case, a Core Benefit Proposition must be formulated and supported by specific program features. The features to be used and the psychological positioning versus other business schools represents the product design. This design must reflect the student's process in choosing a school as well as the school's goals and its ability to actually produce the new program.

A formal model of the consumer response based on the process of awareness, perception, preference, availability, and choice will form the underlying structure of our design effort. Not all consumer decisions are made in this sequential order. Many are interactive complex decisions and many are influenced by habit, by impulse, or by peer pressure, but we have found that the major part of consumer response can be better understood and accurately predicted if it is represented by this simple abstraction of consumer behavior. Furthermore, the many complexities of consumer behavior can be better understood once the basics of the consumer response are spelled out.

The next section describes the activities that surround the consumer response model and the process of designing the product's CBP and its fulfillment through physical features, psychological positioning, and the marketing mix of price, advertising, and distribution.

THE DESIGN PROCESS

The design process can be viewed as being made up of a managerial and a consumer component. Figure 7.1 gives a macro description of the two parallel subprocesses of the design process. The managerial subprocess represents a categorization of the types of managerial decisions made in new product development. The consumer response subprocess represents a categorization of the steps analysts proceed through as they study the market to help managers design new products. This figure contains the basic concepts of new product design models. (For more discussion, see a review by Shocker and Srinivasan, 1979.)

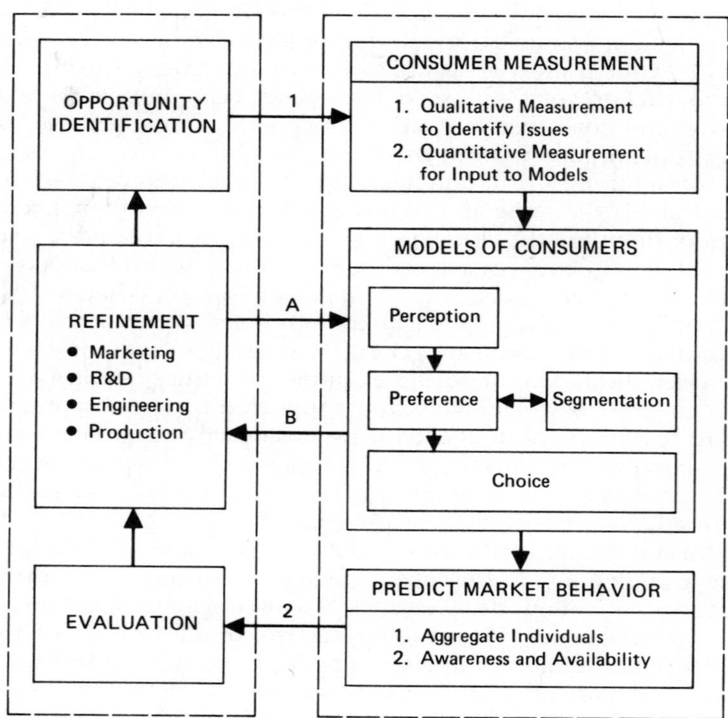

Figure 7.1. *The New Product Design Process*

Opportunity Identification

The managerial subprocess begins with Opportunity Identification. As detailed in Chapter 5, market definition is the first step in a proactive design strategy. Management examines the opportunities available, searches for new opportunities, and from these opportunities selects the market that has the greatest potential to achieve managerial goals. In market definition, management selects the basic product and the target segment of the population. The analysis, based on the techniques of Chapter 5, is necessarily only an initial description using macro-information and category potential as inputs to a provisional decision. As the product is developed based on the design process, this market definition is continually modified and refined until a final strategy is ready for formal testing. For example, consider the problem that was introduced in Chapter 1.

Seafood (lobsters, cod, sole, clams) is becoming scarce while squid, a high protein source, is plentiful off the coasts of the U.S. Presently there is little demand for squid in the American market, although many foreign markets such as Europe and Japan find squid a delicacy. In market definition it must be decided whether to promote squid in the gourmet delicacy market, the fresh produce market, the prepared frozen food market, or the canned food (chowder) market. Furthermore, is the target consumer the affluent gourmet or the bargain hunter (squid is about 69¢ per pound retail)? Is it the average household or the special nationalities (squid is considered an aphrodisiac by some)? By examining the size of the markets, the feasible price in the market, the technical capabilities (does the average consumer want to clean a squid?), and the extent that other seafood is used, one of these markets is selected.

Once management has selected the target market and target consumers, the idea generation effort is undertaken and a set of "stretcher" concepts is created for initial CBP's. These definitions of market opportunities are presented to consumers to learn their responses. Based on their responses, management evaluates and refines the concepts and through engineering turns them into physical and psychological realities.

Consumer Measurement

Early in the design process the emphasis should be on gaining an understanding of the consumer, thus the consumer response investigation begins with qualitative consumer measurement. The qualitative measurement puts management in touch with the market by providing insight on what motivates consumers, how consumers see the market, which products consumers consider to be complements and which to be substitutes, how consumers make their purchases, how consumers use the products, and so on. For example, four focus groups might be run to determine how consumers make seafood selections. Samples need not be random, questions need not be quantifiable, market share predictions need not be made. Rather, the qualitative measurement serves to focus more quantifiable measurements and models. The qualitative research raises questions, suggests answers, and directs investigation, but it does not and cannot provide the same understanding and predictions of consumer response with the level of confidence and risk reduction as can the quantifiable measurements and models. Next, quantitative measurement is undertaken to provide specific and representative inputs to models of consumer behavior. The measurement for these models is usually a series of mail, telephone, or personal interview surveys which are carefully designed to provide quantitative measures of consumer attitudes and preferences. For example, we might follow up the focus groups with 10-page questionnaires mailed to 500 randomly selected consumers asking them to evaluate existing seafood products and three squid "stretcher concepts."

Models of Consumers

Based on the consumer measures, the next step is to develop models of the consumers (see Figure 7.1). The models, focused around the awareness, perception, preference, segmentation, availability and choice factors described earlier, help management diagnose the market by providing a specific representation of each component in the consumer response process. They identify the design features and product characteristics that make the greatest impact on consumer response and direct the design process to the product or service strategy that is most likely to succeed in the marketplace. For example, airline consumers prefer 747's to 707's, punctual arrivals to late arrivals, friendly stewardesses to aloof stewardesses, movies to magazines, no stops to one stop, no crowds to crowds, and more frequent service (Green and Wind, 1975). But if we are limited in the amount of money we are willing to invest or if we are looking only for cost effective improvements, which of the above will have the greatest impact? Or should we even limit ourselves to those improvements? Maybe the greatest impact comes from having a coach lounge, or news/shorts instead of movies, or free wine, or special entertainment in the waiting area, or bargain fares, or courtesy ground transportation, or more carry-on luggage compartments, or unlimited snacks. All are improvements—but which are the key improvements?

To identify the leverage points, the models of the consumers are separated into submodels. *Perception* identifies the key dimensions that are most relevant to the consumer. For example, the perception models may tell the dean of a management school exactly what dimensions, e.g., realistic approach, good job placement, comfortable environment, describe the basic evaluation process used by potential students. Besides identifying the dimensions, perception tells the dean how potential students view his school relative to other schools along each relevant dimension. This is "psychological positioning" of the existing school. Next the dean can start on improvements relative to these dimensions and he can position a new program in the market. *Preference* identifies how consumers use the perceived dimensions to evaluate products. For example, should the school emphasize the techniques used by the majority of industries or should it emphasize new, improved techniques which, although used by only 10 percent of industry, are the most successful and likely to be the norm five years from now? Do students prefer a highly flexible program, or do they prefer a more directed, fairly rigid program? Preference answers these questions and, together with perception, helps the dean select the best "positioning" of the new program relative to competitive programs. Preference also helps the dean select what benefits to include in the CBP.

Segmentation determines whether the best strategy is to have one product for all consumers or whether to have a multiplicity of products, each directed at a specific group of consumers. Again, consider the management schools. Segmentation would investigate whether the school should offer one common program or whether it should have: (1) a "generalist" program for basic strategic management and (2) a "specialist" program for in-depth study into a specific functional area such as management science in marketing. Finally, *choice* determines what external events must be controlled to ensure that those consumers who prefer a product actually purchase or use it. For example, how does housing availability or employment for spouse affect the students' decisions? Taken together, the four component models help management understand the consumer in such a way as to make the necessary strategic decisions in new product design.

Prediction of Market Behavior

Understanding is important, but it is not enough. If management is to commit funds, perhaps millions of dollars to a new product, they need predictions of consumer response. Thus, the models of consumers combine to predict market behavior. The idea behind this last step in the consumer response models is deceivingly simple. Consumer measurement models develop measures of how each consumer (in a representative sample) perceives each existing product in the market. Management selects the characteristics for a new product and establishes an advertising and promotional strategy. The perception model predicts how the new product will be perceived by each consumer, the preference model predicts how consumers will compare the new product to the existing products, and the choice model predicts the probability that any given consumer will actually purchase the new product. Market prediction is then an aggregation of individual predictions. For example, suppose a business school dean decides on a new program. The models of consumers will predict: (1) how each potential student in the sample will perceive the prestige, the quality, the employment potential, and so on, of the new program; (2) how many will select it as their first choice, their second choice, etc.; and (3) for each potential student, the probability that he will actually come if accepted. If everyone that was aware of the program was also accepted and there were not external effects, the expected enrollment would simply be a sum of the individual predictions modified only by awareness. But since everyone is not accepted and there are external effects, the prediction model corrects for these availability effects as well as awareness when it sums the predictions.

Evaluation

The predicted market behavior forms the basis for evaluation. In evaluation, management weights this prediction as well as production costs, political constraints, technology constraints, material availability, firm image, complementarity with product line, and other aspects of new product introduction to arrive at a GO/ON/NO GO decision. "GO" means that the product concept has high promise and should go on to the next phase of development. "NO GO" means that the product is unlikely to return its investment even if it is vastly improved. It would be appropriate to abort this category and search elsewhere. Finally, "ON" means that the evaluated product has potential and that with some modification it is likely to be a successful new product. The organization should continue with the design process and look for an improved CBP and concept.

Refinement

If the evaluation is "ON," management proceeds to refinement. In refinement, the product is improved based on the diagnostic information from the models of the consumer. For example, if flexibility is important to students and a business school finds itself with a rigid program, it may drop some requirements and add electives, but only if the faculty agrees that quality education (another perceived attribute) is not sacrificed. If the average graduating student is offered exciting jobs with high salaries, but potential students perceive the school as not providing good job opportunities, the school may embark on a brochure campaign or mobilize its alumni to better communicate its true position. If the improvements are minor, the models of individual choice can predict consumer response (see arrows A and B in Figure 7.1); but if the improvements are major, management may decide to restart the design process by developing some representation of the improved new product and re-entering consumer measurement. This would restart the design process on its second interaction by returning to the consumer measurement commitment (see arrows 1 and 2 in Figure 7.1). The outcome of these iterations is an attractive new product concept.

Fulfillment of the CBP

After refining the product CBP concept to the "GO" level, the next step is to create the actual product and see that it fulfills the Core Benefit Proposition. Since a parity product seldom leads to success, R&D should try to fulfill the CBP with a "parity-plus" or a "breakthrough" physical product. In the management school case, the actual product may be preceded by a pilot program of ten students to see if a new intensive program can deliver the "depth" of knowledge and "flexibility" that the CBP promised. It is sometimes easier to evaluate physical products to see if they deliver their promised benefits. A new solar air conditioning system must be "reliable" if so promised in the CBP. Engineering, production, and marketing must work together to produce a physical product with the required engineering features to fulfill this promise. Users should be presented with the physical product to see how they perceive its performance relative to its promises. In frequently purchased consumer products, the lack of fulfillment of the CBP results in low rates of re-purchase; in durable and industrial products, dissatisfaction, bad-mouthing the product, poor repeat sales, and low long-run sales will result. After an iteration through the design process with a prototype product the GO/ON/NO GO decision is faced again. If the CBP is viable and is fulfilled by the product, a GO decision will be appropriate if the organization's goals are achieved. If an ON decision is reached, efforts would be made to improve the concept and product performance. A NO GO decision causes management to return to the opportunity identification stage to find a new market and a new idea.

The output of this iterative process is the specification of the best CBP and a product that fulfills the promises through physical and psychological features. The last step in the design process is to specify the initial levels of the marketing mix variables of price, advertising, and distribution for the product. The final design is the specification of the:

- Target market and target group of consumers,
- Core Benefits Proposition (CBP),
- Positioning of the product versus its competition,
- Physical characteristics of the product to fulfill the CBP, and
- Initial price, advertising, and distribution strategies.

After engineering, R&D, production, and marketing have produced this product and advertising, management leaves the design phase of new product development and enters the testing phase.

KEY POINTS OF THE DESIGN PROCESS

The design process in Figure 7.1 is a logical way to proceed. Although it may be obvious to some, it is not so obvious to many in the heat of new product development. Under pressure to produce a new product, some organizations stop with qualitative research, while others begin immediately with quantitative techniques. Some scale perceptions but never worry about preferences, while others try to measure tradeoffs among attributes without ever ensuring that they have the right attributes. Some concentrate on diagnostics for refinement, but make no predictions for evaluation, while others predict, but never diagnose. These mistakes can be avoided by following the process given in Figure 7.1.

Before describing the design process in detail, we would like to emphasize several important aspects of the design activity.

1. A New Product Is Both a Physical Product and a Psychological Positioning. It is essential that a product perform from an engineering point of view. However, design is not just specifying physical product characteristics. In toothpaste, the right amount of fluoride in a toothpaste may be important, but it is not enough. Psychological attributes are important too. The toothpaste needs a perceptual positioning in the space of taste and tooth decay prevention. This is created by its advertising package and by promotion, as well as by the product itself. Public water fluoridation programs have not been very successful. They relied only on the physical attribute—fluoride and its link to decay prevention. Procter & Gamble's Crest toothpaste positioned fluoride as the substantiation of the tooth decay prevention claim, but within the image that using Crest was being a good parent. Their "Look Mom, No Cavities" advertising strategy showed peer group approval for the parents who had their children use Crest.

Similarly in transit systems, the right tradeoff between cost and service is important, but these aspects must be communicated and an image created, schedules must be available, and the bus must serve the right routes at the right time of day. On the other hand, an image of fast, reliable service is difficult to achieve if the engineering is neglected and the transit system fails to deliver the promised benefits.

The design process considers both the physical and the psychological aspects of the product in the Core Benefit Proposition. This statement of the benefits to the consumer indicates the importance of understanding how consumers perceive benefits and of assuring physical features to substantiate these perceptions.

2. The Design Process Is Iterative. Design is not accomplished in one step. Evaluation, refinement, and learning take place sequentially. In the market definition phase, management is concerned with the potential of a category, they do not need detailed predictions based on a fully-developed new product strategy. Early in the process, the design team concentrates on positioning, i.e., selecting the basic appeal and image of the product. Later they concentrate on the specific physical and psychological features that will achieve this positioning, and finally they select the full marketing strategy. Thus, as the product proceeds through the process it is continually refined, but each step concentrates on particular parts of the overall strategy. Trying to simultaneously select position, specify features, and plan marketing strategy can be tremendously expensive and also confusing. Instead, at each stage of the design process, management concentrates on the component decisions and iteratively puts together the full product design.

3. Both Prediction and Understanding Are Necessary. Suppose you have a new automatic teller machine for your bank. You want to know not only how many people will use it, but what features of the machine will help you attract customers from the bank across the street. Similarly, Amtrak may know that faster service will attract more riders, but they also need to know how many riders and what level of improvement is the most cost effective. To select the best strategy, management needs predictions of how each of several potential strategies will perform, but to search out and identify strategies they need to understand what the consumer response will be.

4. The Level of Analysis Should Be Appropriate to the Strategic Decision. It is always possible to spend more time and funds on analysis. Extremely detailed models could be built or simple aggregate models could be used. The tradeoff on the required level of detail is dependent on the decision to be made. The "best" model is not necessarily the most detailed. In the sequential process the "best" model is one that most efficiently supplies the required input at the specific point in the decision process. High accuracy may not be necessary early in the concept evaluation phase, but it becomes critical in the testing process. At the concept phase, a simple model is efficient if it accurately identifies a good concept and how to improve it. At the test market phase, a more detailed set of measures and models are required to achieve a high degree of forecasting accuracy.

The models we present reflect managerial needs. No claims are made that all consumers behave exactly as specified by the models, but the models do abstract the main strategic components of behavior and the forecasts of response for groups of consumers are accurate enough for decision making.

5. The Design Process Blends Managerial Judgment with Qualitative and Quantitative Techniques. Each aspect, managerial, qualitative, and quantitative, is important. No one aspect can stand alone, but each is essential. Judgment is important, but can be led astray. Qualitative techniques uncover effects, but are subject to many interpretations. Quantitative techniques are "exact," but if they attack the wrong problem, they can be exactly wrong. Blending these aspects is an art, but once learned, it is a powerful art. The modern manager must be able to effectively combine quantitative and qualitative methods and creative thinking.

These five points are commonsense considerations that can help you avoid analysis myopia by keeping the design process in perspective and by using a balanced mix of techniques to solve the basic problem: designing a successful new product.

SUMMARY

The new product design process in Figure 7.1 is proactive. It stresses going to the consumer, understanding the consumer, and developing a new product based on consumer needs. This design process is an iterative development of the complete new product held together with the Core Benefit

Proposition. It draws on managerial judgment to design, refine, and select the strategy that best fulfills the organization's goals.

We turn now to the specific components of the design process. Chapter 8 reviews consumer measurement and gives specific insight on the consumer measurement appropriate for new products. Chapter 9 covers perceptual mapping techniques, and Chapter 10 shows how to select product positioning through preference analysis and benefit segmentation. Chapter 11 then evaluates purchase or use potential through predictions of choice, awareness, and availability. Finally, Chapter 12 completes new product design by assembling the components of the marketing mix. Throughout chapters 9 to 12 we illustrate the managerial use of the analysis with extensive examples based on a consumer product (analgesics), an industrial product (communications equipment), and a service (transportation). Other smaller examples are cited and we end with a case discussion of a new laundry detergent.

If you keep Figure 7.1 and the five key points in mind as you proceed through chapters 8 to 12, you will be able to put each technique in perspective of its use for managerial strategy and gain the ability to use these proven state-of-the-art techniques to reduce risk and enhance success in the design of new products.

REVIEW QUESTIONS

7.1 After an organization develops a new idea which appears to have potential, they should develop a core benefit proposition (CBP) for the idea. Why is a CBP necessary? What should a well-constructed CBP contain?

7.2 Consider three new products introduced in the last year. What are their core benefit propositions?

7.3 Discuss how the consumer response process in Figure 7.1 facilitates the development of a core benefit proposition. What information does the consumer response process provide that helps management refine new product ideas?

7.4 A GO/ON/NO GO decision is made in evaluating the new product idea. What costs and benefits must be considered when making this decision? Be sure to include opportunity costs.

7.5 Discuss the relationship between the physical characteristics of a product and the positioning of a product.

Chapter 8

Consumer Measurement — A Review

The success of a proactive design strategy depends on determining a product strategy that will be attractive to consumers. Throughout this book a number of managerial and analytic techniques are introduced to design and position products in such a way as to attract consumers, but these techniques are only as good as the data on which they are based. Thus the consumer measurement which produces this data is critical to the success of new product strategy. Careful, exacting market research can lead to creative insights into consumer wants and needs and result in superior products. On the other hand, careless research will cause the manager to miss opportunities and in many cases lead to errors in product design.

This chapter reviews some of the considerations in consumer measurement that are particularly important in new product design. It is not a substitute for a course in market research, rather it is meant to briefly highlight the relevant topics and provide specific examples relevant to new products. For more detail on market research in general, readers are referred to Green and Tull, 1978; Payne, 1951; Oppenheim, 1966; Churchill, 1976; Boyd, Westfall and Stasch, 1977, and Lehmann, 1979.

CONSUMER MEASUREMENT PROCESS

It is important to remember that in developing new products, market research is conducted to support decision making. Although many interesting behavioral issues might be addressed in a consumer survey, the managerial usefulness will be judged by its ability to improve decisions. In our case, the study will support models that lead to improved new product designs and higher chances for new product success.

Figure 8.1 shows a measurement process that begins with decision requirements and proceeds to model estimation. The decision requirements are key considerations in deciding the methodology for the study. Should a questionnaire be used or can a different method yield the information? Should qualitative or quantitative measures be used? Is an experimental variation of decision variables necessary? These methodological issues must reflect the needed accuracy of the information, the costs of this information, and its value in improving decisions. One explicit cost tradeoff is in sampling. What sample size is required? Should the sample be random or is a purposive sample more appropriate? If the appropriate method is a survey, the questionnaire must next be formulated and tested. Many pitfalls await the untrained person in writing a questionnaire. Inaccuracies due to ambiguity or to unwillingness or inability to respond must be minimized. We will see in Chapter 9 that part of most new product questionnaires are attitude scales. The selection and implementation of the scales is extremely important and must be done carefully.

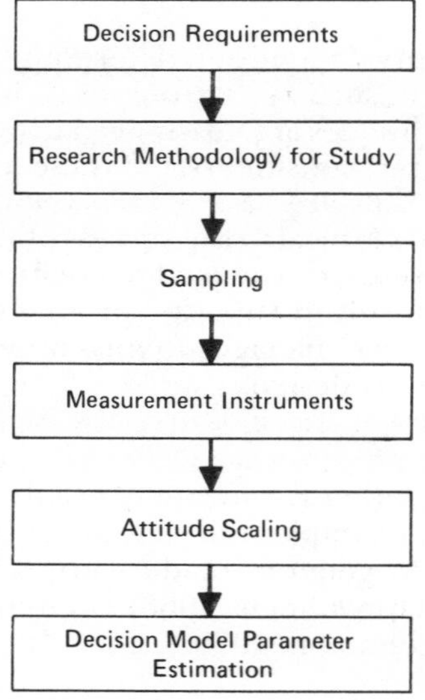

Figure 8.1. *Consumer Measurement Process*

In new products, several dimensions of attitude are usually present; the process of attaching numerical values to products along these dimensions is called "multidimensional scaling." These numerical values are especially important when models are being used to analyze decision alternatives. For example in the design process, multidimensional scaling is critical to modeling consumer perception and diagnosing consumer response to new products.

In this chapter we review selected aspects in each phase of the measurement process. The issues presented are those most critical to successful design of new products. The remaining chapters present additional measurement issues and demonstrate the link of measurement to models and to managerial decision making.

RESEARCH METHODS

Although many of the analysis and decision steps are based on quantitative measurement, an organization should not neglect archival and qualitative research. Early in the new product design process they are important. For example, focus groups are more effective for idea generation than questionnaires. However, as the product development process proceeds, the quantitative measures become essential. But even then, quantitative measurement should not be used alone. Usually, the analysis process begins with qualitative techniques and an archival search to raise issues, to identify attribute scales, to provide consumer semantics, and to help direct the quantitative measurement to the areas that appear to be most productive. This section will briefly review archival, qualitative, and quantitative methodologies.

Archival Search

A proactive strategy by definition searches for *new* ideas, but this does not mean that existing data sources are to be ignored. Secondary data, i.e., data collected for reasons other than new product development, can provide useful information which may direct the search to more productive areas or save the expense of gathering data that is available from other sources. This data might come from internal sources, U.S. government statistics, special data services, or other external sources.

Internal Sources. Some internal sources may relate specifically to the product category under investigation. If the organization already has a product in the category or if we are considering a major change to an existing product, then sales records, advertising records, complaint/compliment files, or warranty records may provide useful clues to opportunities. For example, Chapter 3 used the example of scientific instrument firms responding to consumers who build their own equipment. Occasionally market research will have been done to monitor a category or innovate in a related category, or product managers with experience in the category can provide insights worth further investigation.

Government Statistics. U.S. Census figures or other government statistics can be used to monitor demographic changes or for an estimate of category potential. For example, stackable washer/dryers are an innovation resulting from the recognition through governmental data of the number of apartment and condominium dwellers. Furthermore, the demographics can be used to check the representativeness of a consumer survey. Another government source worth investigating is the *Statistical Abstract of the United States*—this secondary source can be used to search for primary sources of government data.

Special Data Services. Some examples of special data services are: Dun and Bradstreet's Market Identifiers, Nielsen's Retail Index, SAMI Warehouse Sales, the Nielsen Television Index, the Starch Advertisement Readership Service, the Simmons Media/Marketing Services, TRENDEX, the MRCA Consumer Panel, the National Family Opinion Panel, the Consumer Mail Panel, and Predicasts. While all of these are available at some cost to the user, they may be cost-effective for specific categories and for putting together components of the marketing strategy. Finally, there are published sources of archival data such as Moody's Manuals, the Rand McNally Commercial Atlas and Marketing Guide, and the World Almanac and Book of Facts. For further information, see Churchill (1976), which contains detailed descriptions of the special data sources plus short descriptions of thirty-seven published sources and a bibliography of additional published guides to archival data.

Archival data should be used with caution since it is collected for other purposes and may not exactly fit the category under investigation. But it is quickly obtained, may be less expensive, can lead to later efficiencies, and provides an external check to any qualitative or quantitative measurement. Although archival data may not meet all the managerial requirements, it is an important source to examine before primary measurement is undertaken.

Qualitative Measurement

New product development seeks to discover what the consumer views as important in a product category. Qualitative measurement is based on in-depth probes into the consumer's viewpoint. Its purpose is not to identify the best strategy, or to project demand, or even to select the most important product features. Instead, qualitative measurement raises issues by exploring the consumers' basic needs and desires. The qualitative researcher enters the process with an open mind, seeking to learn by simply listening to the consumer.

A primary method of qualitative research is focus group interviews. We presented this method in connection with idea generation in Chapter 6. Focus groups are also useful in the design process. For example, in focus groups for "Shared Ride Auto Transit"—a form of organized hitchhiking—the U.S. Department of Transportation (1977) confirmed its intuition about the importance of safety, dependability, reliability, flexibility, and personal freedom in consumers' reactions to various strategies. But they also uncovered some surprises. Major objections to "Shared Ride" were based on not wanting to be obligated or indebted to other people. Riders felt better paying, but drivers felt very uncomfortable in accepting payment. A number of consumer misunderstandings were uncovered, e.g., most consumers felt they paid higher insurance for carpooling when in fact they paid lower premiums. When Shared Ride was described as "community carpooling," there was better initial reaction and greater willingness to experiment. When Shared Ride was called "organized hitchhiking," however, consumers' attitudes immediately turned against the system.

An important function of qualitative research in the design process is to generate a list of attributes of products within the category. For example, Table 8.1 contains a list of twenty constructs which describe management education. This list forms the basis of the identification of consumer perceptions (Chapter 9). It is important that the list be as exhaustive as possible, covering a wide array of physical and psychological characteristics, if they are at all relevant to the category. Ideas come from the transcripts of the focus groups, individual in-depth interviews, previous studies, and prior managerial beliefs. The emphasis is on breadth, even if redundancy is inherent, because later analysis will identify the underlying structure. The fo-

Table 8.1. Twenty Constructs Which Describe Management Education

1. With a degree from this school, I'd be joining a world-wide "fraternity," an effective alumni network.
2. The program emphasizes training in skills that can be used in advising decision makers on what course of action to take.
3. The content of the program would be highly relevant to the decisions I expect to make in a management position in the next five years.
4. I would not be exposed to the newest, most advanced management techniques.
5. I'd be learning a common-sense approach to problem solving.
6. The school provides excellent teaching.
7. The school is oriented toward those who want to take personal responsibility for making major decisions.
8. The program is very theoretical, with limited "real-world" focus.
9. There would not be a high level of contact with faculty members.
10. The school has a "high-prestige" reputation.
11. Financial aid in the form of loans, grants, or assistantships is widely and easily available.
12. I would have a great deal of flexibility to structure a program to suit my own needs and interests.
13. The majority of faculty are academicians and have never been practicing managers.
14. The students are not very likely to be cooperative and friendly.
15. With a degree from this school, the most exciting job opportunities will be open to me upon graduation.
16. I could not go wrong with a degree from this program.
17. I would be able to command a very high relative starting salary with a degree from this school.
18. The program is very quantitative.
19. The school has an excellent location.
20. The school would emphasize how management is now conducted, not how it should be conducted.

cus group transcripts are used to word each statement in the semantics that the consumers use.

Qualitative research does not provide final answers. It is a useful search technique which helps insure that the quantitative measurements address the issues that are relevant to the design process.

Quantitative Measurement

This measurement provides the input to the analytic techniques used throughout the remaining chapters to identify which specific strategies make the greatest impact on improving the success of a new product design. Specific perceptions and preferences are measured and preliminary estimates of purchase actions are made.

First, archival and qualitative research are used to understand issues and semantics and then a quantitative approach is used to measure attitudes and consumer response. As the product design proceeds through evaluation and refinement, estimates must become more accurate and emphasis is almost exclusively on quantitative measurement.

Since we are concerned with new products, the product concept or prototype can be an experimental treatment in the research. Usually alternatives are presented and quantitative measures are taken. For example several "Stretching" concepts may have been generated and will be evaluated. But, in some cases an explicit experimental design is used to estimate response to product attributes or features.

Since quantitative measurement is often specific to the analyses used in the decision process, we defer some of the details to those sections that cover the specific analysis techniques. We turn now to sampling, and then to questionnaire formulation and attitude scaling.

SAMPLING

After the research method is decided on, sample sizes and sampling methods must be determined. In the qualitative phase small samples are taken. Two or three groups of ten people may be sufficient if issue identification and semantics are the primary concern. Later on, quantitative surveys will require larger sample sizes. It is uncommon to collect fewer than one hundred respondents; in cases where segmentation is important there should be 100 respondents in each major segment. The samples to support design of new products will be substantial, perhaps as large as 200–500. In fact, recent work by Srinivasan (1977) and Einhorn and Hogarth (1975) has shown that naive models which provide little diagnostic information may be indistinguishable from more sophisticated models when samples are small.

Although there are theoretical methods of defining the best sample size

(Schlaifer, 1969; Green and Tull, 1978; Allaire, 1975) based on the value of information, these are very difficult to implement in the complex sequential new product development process. Classical methods of considering the standard deviation of the resulting estimate are useful, but vastly over-simplify the problem, since information is collected in many variables and often analyzed by multivariate scaling methods for which sampling properties have not been fully determined. It appears that judgment, norms, and experience are the best we can do at this time. Perhaps future research will provide more powerful and practical methods of determining the best sample size for new product design.

In general, random samples are best if they can be obtained at a reasonable cost. They tend to be most representative and are less prone to sampling errors. But non-response can be a problem. If, as is usual, 30 to 40 percent of the random sample returns a mail questionnaire, the statistical analysis may be inappropriate unless it can be established that the nonrespondents are a random subgroup of the total sample. This requires that demographic and other characteristics be compared to external sources such as census figures, and corrections made if necessary.

Alternatively, purposive samples can be used if they are done carefully and steps are taken to statistically correct any biases introduced by the sampling (see Appendix 11.3). For example, the use of a shopping mall as a site for market research is becoming common. A respondent is stopped while walking in the mall and, by means of a few questions, qualified as having the desired target group characteristics. Quotas are established for age, sex, demographic, and product use groups. Although these samples are not random, response rates are high (over 60 percent), and their representativeness can be established through statistical corrections. These mall interviews are attractive if personal interviews and product or advertising exposure is essential. Nonetheless, great care is required in the selection of criteria for screening the respondents and careful statistical analysis. They should be scrutinized for any potential biases and these must be corrected.

Purposive samples can be attractive due to their reduced cost, but lower cost should be weighed against potential biases. A good rule of thumb is that purposive samples are more appropriate early in the design process when the primary goal is exploratory analysis. Later in the design process when forecasting becomes important, the representativeness of the sample is crucial and randomness becomes more important.

Do not confuse purposive samples with convenience samples. In purposive samples, explicit selection and screening processes are established, potential biases identified, and steps taken to correct for those biases. In convenience samples such as man-in-the-street, church groups, office friends, and university students biases can be hidden. Interpretation of any analyses done on such groups can lead to incorrect strategies for new product development. Thus, in general, we recommend using a random sample with checks for non-response, but consideration should be given to the benfits and costs of purposive surveys with statistical corrections.

When questionnaires are to be used in a sample, a choice must be made between personal, mail, and telephone interviews. In some cases this is not difficult since personal interviews are the only feasible solution. For example, if a food product is to be tasted, personal contact is required. In other cases, the format (mail, telephone, personal interview) depends on the information that is being collected. For example, for complex perception and preference questions personal interviews are best, but we have found the mail format represents an acceptable tradeoff when budgets are limited. The specifics of the questionnaire, the group to be interviewed, and the research budget must all be considered before a choice is made.

Questionnaire Design

A questionnaire must be carefully constructed to collect the appropriate information with a minimum of bias. Questionnaire design is an art. In this section we cover some representative considerations in questionnaire design and indicate some of the pitfalls to avoid.

Respondents must be motivated to participate. Convincing the respondents of the importance of the work is desirable. For example, a study of health services began with an explanation that the survey would measure health needs so that improved health services could be developed. Stressing the importance of each individual respondent's input is useful. In cases where cooperation may be low, personal interviews with personable interviewers can be effective. Increasingly, respondents are given compensation for their cooperation. For example, in a study of investors a hand held calculator was given each respondent. In a study of graduate students, each respondent was given a six-pack of imported beer. In consumer products a coupon worth two dollars may be given. The use of an incentive helps to increase the response rate, thereby reducing the damage of non-response bias, but non-financial motivations should not be overlooked.

Proper motivation can be used to reduce non-response bias. But a good response rate does not necessarily mean good measurement. Improperly worded questions can be misleading to the respondent, to the analyst, and to the manager. For example:

"Do you like the taste of calamari?"

In this case the respondent may not know that calamari is the Italian word for squid. The question would underestimate consumer appreciation.

"Do you agree that the Massachusetts Health Foundation provides excellent care?"

This one-sided question, although good for public relations, tends to bias the response toward agree. A better question would give both sides of the argument, i.e.:

"In your experience, does the Massachusetts Health Foundation provide excellent care or does it provide poor care?"

Alternatively, one of the attitude scaling techniques, which are described below, could be used. The pitfalls of question wording are numerous and can only be avoided through great care and careful testing of each question or set of questions. Both Payne (1951) and Oppenheim (1966) provide examples of many of these problems and how to overcome them. Payne provides an extensive list of problem words to avoid and a checklist of 100 points to consider before each questionnaire is taken to the field. Developing a good questionnaire can be costly and time consuming, but the rewards are great through more usable information and a better understanding of the consumer.

A good questionnaire requires careful planning, which should begin long before questions are actually written. In preparing a questionnaire it is important to recall the managerial questions that need to be answered, the analytic techniques to answer those questions, and the specific information that the techniques need. First, the major sections of the questionnaire should be specified. This block layout allows the new product design team to: (1) make rational tradeoffs about the length and the necessity of various sections, (2) ensure that all needed information is collected, (3) eliminate unnecessary redundancy, (4) construct a smooth flow of response throughout the questionnaire, and (5) check ordering requirements (e.g., preference valuation should come after attribute scales to avoid what are known as "halo effects"—Beckwith and Lehmann, 1976). For example, Table 8.2 is the block design for a questionnaire used by the City of Evanston, Il. to identify new transportation services. (See Hauser, Tybout, and Koppelman, 1979.) Note that each section has a specific purpose, and each section after the warm-up questions is specific to an analytic model. (The analytic methods are those described in subsequent chapters.)

After the block design is complete, focus groups and previous experience provide the input to word and reword the questions so that they accurately gather the information needed for the analysis. Small samples are used to pretest the questionnaire. Respondents are reinterviewed about their responses to be sure they were thinking of the same issues as the researcher when answering the questions. After the pretest and possibly preanalysis, the questionnaire is revised and implemented with a representative sample of the target population.

As we have said, questionnaire design is an art. Each reader should try to develop this skill by designing a questionnaire and having a colleague critique it. Examine it in light of the pitfalls described below. Everyone who uses market research should actually administer a questionnaire to some respondents at least once in order to understand the issues of non-response, ambiguity, and inaccuracy.

Some Pitfalls and How to Avoid Them

There is no simple formula for producing a good questionnaire. Instead we identify some of the common mistakes we have seen in question-

Table 8.2. Block Design for a Questionnaire to Investigate Consumer Views on Transportation Innovation

Block Description	*Purpose*	*Analysis Technique**
1. Cover letter	Generate interest, motivate respondents	Motivation (8)
2. Warm-up questions	Gain rapport with consumer	Motivation (8)
3. Recent trip scenario sampling/Transportation choice for scenario	Representative sample of scenarios/Potential stratification by scenario	Benefit segmentation (10)
4. Instructions for rating transportation alternatives	Aid to response/Consider recent trip/Express *opinions*	Factor analysis (9)
5. Attribute scales for bus, walk, and car	Input to models of perception	Factor analysis (9)
6. Preference among available modes/ Frequency of use of all modes	Input to models of preference and choice behavior	Preference regression (10)/Logit analysis (11)
7. Self-reported travel characteristics	Input to models of choice/ Link to consumer attitudes	Logit analysis (11)
8. Stretching concepts for new modes/Attribute scales for stretching concepts	Enhance validity for changes in transportation system	Used with 4 and 5 above
9. Preference for available modes including stretching concepts	Estimate more complete models/Internal validation of preference models	Preference regression (10)
10. Conditional intent	Indicates behavior changes based upon changes in system characteristics	Intent model (11)
11. Importance ratings for the attribute scales	Alternative measure of preference	Importance model (10)
12. Transportation facts such as location of home, location of nearest bus stop, etc.	Enables measurement or estimation of benefits delivery	Purchase model (11)
13. Demographics	Personal descriptions for use in segmentation	Benefit segmentation (10)
14. Open-ended comments	Qualitative input for completeness	Qualitative input (8)

**Numbers in parentheses indicate chapter in which technique is described.*

naires designed by our students and by professional market researchers. Recognition of these pitfalls should help you avoid many of them. In addition, we give guidelines that we try to follow in developing our own questionnaires. Some of these guidelines may seem costly, such as a major pretest and preanalysis, but considering the stakes in new product development, we feel this cost is justified. Among the major pitfalls we have seen are use of the wrong semantics, low motivation, poor initial questions, difficult questions to answer, products unfamiliar to the respondents, no pretest, no preanalysis, poor sampling, and under-budgeted marketing research.

Wrong Semantics. Since you are trying to measure consumers' responses, the questions must be phrased in the language which consumers use. Rather than using the jargon of engineering, medicine, business, etc., use the language of the consumer. If a consumer talks about the time it takes to walk from his bus stop to the office, you do not ask, "What is the egress portion of your work trip via the bus mode?" If a consumer says his deodorant helps him stay "dry," you do not ask about "perspiration prevention." Rather than "minimization of propensity for cardiovascular infarction," you talk about "preventing heart attacks." A good way to discover the right semantics is through focus groups. Careful pretesting is necessary to see that respondents understand all words.

It is easy to write a biased question; difficult to write an unbiased one. While it is easy to ask, "Is the mayor doing a good job?," it is preferable to ask, "Is the mayor doing a good job, or a poor job?" A useful exercise is to ask yourself the reporter's five questions: "Who? Why? When? Where? and How?" It should be clear to the consumer which of these questions is being asked. If your question has more than one meaning, the consumer may well choose the wrong meaning. Consider Payne's example of the five-word question, "Why did you say that?" This simple question can have five different meanings, depending upon which word the consumer emphasizes in reading the question.

WHY did you say that?
Why DID you say that?
Why did YOU say that?
Why did you SAY that?
Why did you say THAT?

Without a pretest in which consumers read aloud the question, the semantics problem might well go unnoticed.

Low Motivation. It is important to the analysis that the consumer is involved with the questions and gives them the proper thought. Since poor responses are often indistinguishable from thoughtful responses, poor motivation brings ambiguity and inaccuracy. Specific questions must seem real and relevant; the questionnaire must flow smoothly and be interesting.

The credibility of the research firm and the anonymity (if appropriate) of the survey must be established rapidly. It is useful to tell the respondent what the study is about, why his inputs are crucial, how his name was selected, and how the results will be used.

Poor Initial Questions. It is tempting to start immediately with the most important questions, but it is better to begin with some simple warmup questions which will give the respondent confidence, allow the respondent to voice personal opinions, and lead smoothly into the more difficult or substantive questions of the survey. The warmup questions bring the respondents' frame of reference to the topic of the questionnaire.

Difficult Questions to Answer. How many questionnaires have you completed that require a magnifying glass to read, or have so many circles and arrows that you never know what question to answer, or have computer precoding right in the margin so that you feel like a punch card? All these characteristics have some benefits, but they must be used carefully and with much forethought. Reducing the print and eliminating white space gets more questions on a page and thus reduces the number of pages, but it may also reduce the response rate. Branching in a questionnaire directs the right questions to the right people, but it can get out of hand. "Office use only" computerese in the margin saves coding time, but the small savings in coding hardly seems worth the loss in respondent rapport.

Products Unfamiliar to the Respondents. There are over fifteen brands of deodorant on the market. Of these, how many could you seriously evaluate? In other words, how many: (1) have you used, (2) have on hand at home, (3) would you seriously consider using, or (4) would you definitely not use? Chances are that fewer than five brands of deodorant would pass any of the four criteria. In fact, in a study of over 200 consumers, Silk and Urban (1978) found that the average "evoked set" size was about three deodorants when "evoked set" was defined by the above criteria. If you ask consumers to evaluate products on the 15 to 30 attribute scales and you want these answers to be relevant, you must limit your questions to the evoked set. Alternatively, you can expand the evoked set by giving the consumer some detailed description, say a concept statement, of the product that is being evaluated. Ignoring evoked set limitations can cause erroneous interpretations and lead to inappropriate strategies. Note that for new product research all four criteria are important: "have used" to get experience, "have on hand at home" to get purchase and possibly consumption, "seriously consider" to get at dimensions the consumer considers important in that category, and "definitely not use" to uncover the bad features of a product in the category. This definition of the evoked set (due to Allaire, 1973) differs from the traditional definition (Howard and Sheth, 1969) by the inclusion of the fourth criterion. Inclusion of the fourth criterion can provide much additional information. Together these criteria ensure the evoked set gives the information that is necessary for proactive new product design.

No Pretest. No matter how much experience you have or how carefully you examine and re-examine questions, you can still misword or misdirect a question or set of questions. Perhaps the most important element in questionnaire design is pretest. No matter how carefully you design a questionnaire, it is probable that you will make at least a few mistakes. Pretest is in a sense a carefully monitored mini-study where you give your questionnaire to consumers in your target population and have them try to answer it.

A sample size of ten is common. After a question is answered, you can ask them what they thought it asked. You can try different forms of the question. You can watch for careless response or no response. You can ask a number of similar questions and check the internal consistency, or you can check the responses against your prior beliefs (Campbell and Fiske, 1959). You can do whatever is necessary to ensure that what you think is being asked is actually what is being answered. This step takes time and may delay a study, but without it you can never be confident that your measurement is reasonably unbiased.

No Preanalysis. Just as important as pretest is preanalysis where you try the analytic methods on the pretest data. Preanalysis ensures that you have included the right questions to address the issues. For example, in one questionnaire we measured frequency of purchase with a category for "5 or 6" purchases, but discovered through pretest data analysis that it was important to the analysis to distinguish between 5 times per month and 6 times per month. In another study we measured all the psychological characteristics of shopping centers, but had to construct an accessibility measure since it was not asked in the questionnaire. Preanalysis is not relevant to decision making since the sample is small, but it does help identify omitted data questions. If the results are interpretable within the small sample size limits, then you are ready for the full study; if not, then another iteration of the questionnaire may be necessary.

Poor Sampling. It is easy to hand out your questionnaire to anyone who will take it. You can increase your response rate if you have the respondent who does not want to answer your questionnaire pass it to someone who will. In transportation you have a captive audience if you do your survey on board a bus. An inexpensive way to reach consumers who buy color television is to have them fill out a questionnaire after they buy one of your television sets. You can poll your colleagues or your office staff, or you can intercept the man on the street. You can do all this, but don't treat the sample as random or representative of your target population. Although you can sometimes correct purposive sampling such as on-board surveys or point-of-purchase surveys, most of the models in this book are predicated on a random or stratified sample. Departure from randomness is sometimes done under extreme cost pressure, but methods of assuring representativeness must then be carefully considered.

Under-Budgeted Market Research. Not allowing enough funds for survey research can result in insufficient sample sizes or can result in mail surveys when personal interviews are recommended. Costs of field work vary by such factors as the incidence of a target group in a population, complexity of questions, callbacks, and screening, but for new product design personal interviews (45 minutes) usually will cost in the range of \$40 to \$70 per completed interview. Mail surveys cost less at \$10 to \$20 per completed response. Shopping mall intercepts are about \$20 to \$40 per response. A typical design study with personal interviews of 200 respondents should be budgeted at about $10,000 for the field work *alone.* Budgets for questionnaire formulation, analysis, report writing, and oral presentation bring the total for an average study to over $30,000. While this may seem high, it should be weighed against the benefits. For example, Chapter 3 estimated the average benefits due to front-end design as at least $1.5 million.

On the other hand, not all organizations can afford $30,000 for survey research. Often "in-kind" support can be substituted for cash outlays. For example, many hospitals use trained volunteers for telephone interviewing, mailing, and questionnaire coding. This "in-kind" support makes efficient use of otherwise underutilized resources. But organizations should carefully consider all cost-saving strategies and use them only when sufficient accuracy can be assured. A poorly done survey can often be worse strategically than no survey at all. If funds are extremely limited, they are better directed at qualitative research rather than casual quantitative research.

Summary. The pitfalls we have indicated should be avoided, but avoiding them does not guarantee a good questionnaire. We have found their avoidance a prerequisite for any questionnaires that we have used in new product design, but this short summary cannot replace a more complete discussion such as found in books on market research and/or questionnaire design. Before preparing a questionnaire, study what others have done, gain some firsthand experience, or use consultants with proven skill in questionnaire construction.

ATTITUDE SCALING

Much of the information collected in quantitative measurement is based on psychological scales. Such scales are used primarily to measure the attributes which are based on the constructs identified in the qualitative measurement, but they are also used to measure similarities between products, tradeoffs among attributes, preference among products, intent to purchase, propensity to innovate, and demographics. We now summarize some of the techniques that are particularly useful in measuring attitudes toward new products. The scaling of these measures and their use is discussed more completely in subsequent chapters.

Likert Scale. The most common form of scaling is to give the respondent a strongly worded statement about an attribute of a product and have him react to that statement on a five-point or seven-point agree/disagree scale. This is called a Likert type scale (see Figure 8.2a). The advantages of a Likert type scale are that it measures intensity of feeling about the statement, it is easy to administer, and consumers can respond to it easily. Its main disadvantage is that it measures attributes on an ordinal rather than on an internal scale. For example, in the scales in Figure 8.2a going from "neither

(a) Likert

I can get medical service and advice easily any time of the day and night.

Strongly Agree	Agree	Neither Agree nor Disagree	Disagree	Strongly Disagree
______	______	______	______	______
a	b	c	d	e

(b) Semantic Differential

Gentle to Natural Fabrics

Harsh on Natural Fabrics

(c) Graphical (marked)

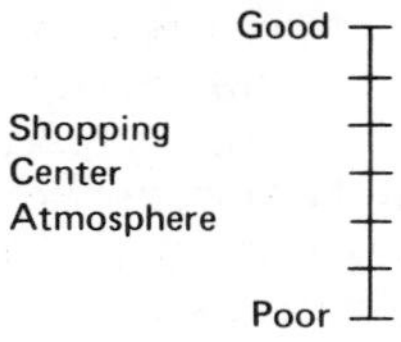

(d) Graphical (unmarked)

Low Prestige — High Prestige

Reputation

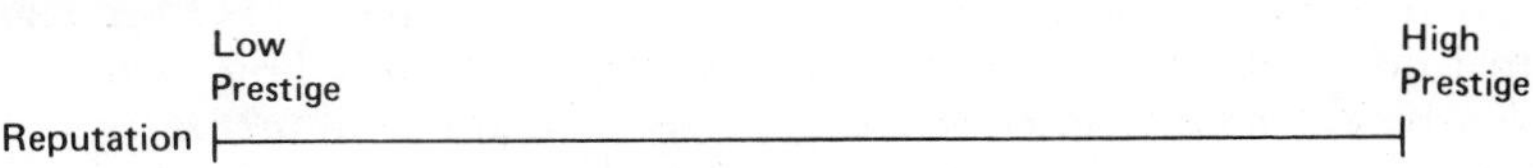

(e) Itemized

Personalness (warm, friendly, personal approach, doctors not assistants, no red tape or bureaucratic hassle).

Extremely Poor	Very Poor	Poor	Satisfactory	Good	Very Good	Excellent
______	______	______	______	______	______	______
1	2	3	4	5	6	7

(f) Pairs

Allocate 100 points between the two auto brands to reflect your preference.

V W Rabbit	or	Ford Fiesta
______		______

Figure 8.2. *Different Types of Attribute Scales*

agree nor disagree" to "agree" means the health service has higher quality, but we cannot infer that going from "neither agree nor disagree" to "strongly agree" means twice the improvement in quality. Nonetheless, in empirical experience the scales have proven rather robust with respect to the interval assumption; some manipulations (such as factor analysis and regression) can be performed if the analyst proceeds with caution.

Semantic Differential. In marketing, the most common form of a semantic differential scale is to give the respondent bipolar adjectives or phrases and have the respondent express his feelings about the product by checking a category to indicate how close his feelings are to one or the other of the phrases (see Figure 8.2b). Like the Likert scales, semantic differential scales measure intensity of feeling and are easy to administer or respond to. They can often be used interchangeably with Likert scales, but in some cases it is difficult, if not impossible, to generate the necessary bipolar adjectives or phrases.

Graphical. Rather than using categories, you can have the respondent react to a statement or to bipolar adjectives by indicating his strength of feeling as a position on a line. The line can be marked as in Figure 8.2c, or unmarked as in Figure 8.2d. If marked, the divisions can be few, as shown, or many as in a "thermometer scale." Most applications require that words be associated with at least the endpoints of the scale, although some applications, particularly those with only a few divisions, have been used with no anchor points. As with the Likert and semantic differential, these scales measure intensity of feeling and are easy to administer. Often the respondent finds them more difficult to answer, and unmarked or finely marked scales are difficult to encode for analysis. The interval properties of some of these scales may be slightly stronger than Likert or semantic differential, but not nearly as strong as might be inferred from the seemingly close relationship to the interval properties of a line. We cannot be assured that consumers react to graphical scales in a linear fashion.

Itemized. The itemized scale presents the respondent with an attribute description from which to select a category indicating his belief about how the product rates on that attribute (see Figure 8.2e). These scale questions also measure intensity of feeling and are easy to administer and answer. They do not allow the fine distinctions of graphical scales but are generally more reliable. They have only ordinal properties, but careful selection and testing of categories can get them closer to interval-like scales than either Likert, semantic differential, or graphical measures. Unfortunately, preparing the questions is difficult and tedious, and they require more space on the questionnaire than other scales.

Pairs. Paired comparisons can yield data that can be scaled to produce interval or ratio measures. The example in Figure 8f is called a constant sum comparison. Consumers are asked to allocate a fixed number of points between the two products to indicate their relative intensity of preference. If sufficient pairs are evaluated, a least squares procedure (Torgenson, 1958) or a linear programming procedure (Srinivasan and Shocker, 1973) can be used to develop an overall preference scale. For more details, see Hauser and Shugan (1980) or Huber and Sheluga (1977).

Other Scales. There are other scales used in psychology that have strong underlying theory, provide good tests of reliability and validity, and are similar in appearance to those we have discussed. Despite these properties, these scales are rarely used in studies which require consumers to rate many products on as many as 15 to 30 scales. They are not used because to achieve these properties each scale requires a major empirical effort. For example, the classical Thurstone scales require roughly 100 individuals to sort 100 to 200 items into 10 piles. Use of Thurstone scales can result in an exorbitant amount of money and effort spent on scaling rather than analysis or strategy. Guttman's scalogram analysis requires 5 to 7 statements and responses for every product on each attribute. Use of Guttman scales can transform a reasonable questionnaire into an exorbitantly long questionnaire.

Scaling. The process of turning scale measures into numerical values is called scaling. Chapters 9 and 10 consider analytic methods of scaling the attitude measures. These multidimensional scaling methods are extremely useful in defining product opportunities and consumer preferences.

SUMMARY

Consumer measurement provides key input to the design of new products. Careful consumer measurement enhances creativity, uncovers greater opportunities, leads to more accurate forecasts, and supports improved new product decision making. The guidelines presented in this chapter should be carefully adapted to the special needs of each particular study. Probably the best advice in consumer measurement is to proceed carefully and cautiously, continually critique the questionnaire, and pretest it on consumers. Seek advice when necessary and avoid making assumptions about how consumers will respond.

Good consumer measurement is difficult, but when done properly it is an extremely powerful tool for the design of new products. We provide more examples of measurement and scaling as we consider models of consumer perception, preference, and choice, and their interaction within the managerial evaluation and refinement process of new product design.

REVIEW QUESTIONS

8.1 What is the ultimate objective of consumer measurement?

8.2 Discuss the relationship between qualitative and quantitative measurement.

8.3 Generate a list of attributes for electric cars. For financial services. How would you scale these attributes?

8.4 What factors must be considered when selecting the sample size for a consumer study?

8.5 In consumer sampling, how is nonresponse a problem? How can nonresponse be minimized?

8.6 How can the manager measure and control the quality of information provided by a consumer with a questionnaire?

8.7 Why is preanalysis important when doing a consumer questionnaire study?

8.8 A company considering the marketing of a new financial service uses a *Fortune* magazine mailing list for sampling perspective buyers. Could this procedure distort the conclusions which will be reached by the study? If so, how?

8.9 Suppose you were attempting to position a new type of resume service for your fellow students. Construct a questionnaire which attempts to measure the needed information.

8.10 You are introducing a new line of premium wines and want to measure how many consumers will try your product. Write a biased question that will overestimate demand. Write a biased question that will underestimate demand. Write a confusing question. Write a question that will accurately estimate demand.

Chapter 9

Perceptual Mapping: Consumers' Perceptions of New and Existing Products

You will recall that the design is made up of a target group specification, core benefit proposition (CBP), a product positioning versus the competition, physical characteristics to fulfill the CBP, and a price, advertising, and distribution strategy. In this chapter we address how a manager can identify the perceptual component of the CBP. Succeeding chapters in Part III discuss how to position the product relative to competition, forecast sales potential, select the physical characteristics to realize that position, and orchestrate the marketing mix based on the CBP.

PERCEPTUAL POSITIONING

On any day when you turn on the TV, read the newspaper, or pick up a magazine, you will find many claims. Television tells us that SURE deodorant "goes on dry, keeps you dry," that AVON's Emprise perfume is "dazzling, rich, elegant, and gorgeous," that LIFESAVERS "has a great flavor, has been around for years, and is a part of living," that NORTHERN toilet tissue is "strong and soft—stroft," that KRAFT French dressing is "smooth, spicy, all-American, and has 100 percent natural flavors," that IVORY soap gives you that "clean, fresh feeling of clean," that GLAD bags are "strong and thick, with lots of room." Turning off the television set, you open the *Chicago Tribune* to learn that O'CONNOR & GOLDBERG's shoes are "young, sophisticated, citified," that WATER TOWER PLACE gives you the "ultimate in individuality and elegance," and that KORVETTE's L.E.D. digital watch is "good-looking, accurate, and efficient."

The Wall Street Journal's ads point out SAVIN copies cost less than XEROX, that 3 M's Secretary Two can do "eight things your current copier probably cannot do." The *New York Times'* ads say "Fly the friendly skies of UNITED" and "AMERICAN is the line for professionals." Having digested the local and national news, you may open *Newsweek* or *Time* to find that ZENITH gives you the "best picture and fewest repairs," you ponder why VOLVO, "with a reputation for being so safe, goes so fast," you discover that POLAROID's SX-70 film gives you "more beautiful, super clear pictures," your mouth waters for the "rare taste, as you like it" of J&B RARE SCOTCH, and you discover that the Chevy CITATION is "compact, but big inside, versatile, but sporty too, and quick, but also very smooth." Finally, you open a trade journal for information on mining equipment to find that General Motors' TEREX haulers provide "teamwork, a full load in the right number of passes," that CATERPILLAR offers "a unique system that goes that all important extra mile to provide you with services and programs not generally offered by other material handling equipment dealers," that American-Standard WABCO makes the "largest production mining truck in the world," that EIMCO provides "better productivity with less investment," and that INTERNATIONAL HARVESTER is "flexible enough and cares enough to make a machine that fills your special need."

These advertising claims reflect the careful efforts of these organizations not only to portray the benefits they offer consumers (CBP), but to differentiate themselves from their competition. This differentiation is achieved by uniquely positioning their products. For example, SURE positions itself as "drier than other deodorants." Other deodorants position themselves as being "more effective" or "longer lasting." In office copiers, 3M is positioned around "versatility" while others are positioned for "lower costs."

These positioning issues are critical for a new product. Not only must a product have a good CBP, but it must also be positioned differently from the competition. For example, although the CBP of dryness in a deodorant is good, a new product would be unlikely to succeed with this CBP alone if it is positioned in the same place as an existing product. A proven way for a new product to fail is by not surpassing existing products. In many markets, there is little room for a parity product.

In computers, IBM has developed a positioning based on using computers to solve management problems. RCA tried to compete with IBM with hardware similar to IBM's but at a lower price. This was unsuccessful and RCA's positioning was poor against IBM. IBM had a much better preception for delivering useful results and managers did not weigh cost as heavily as problem solving in buying a computer. CBP's and their positioning against existing products must be carefully understood when designing a new product.

Good positioning requires that we understand the dimensions used by consumers to perceive our new product and how existing products are placed on these dimensions. We must know the number of dimensions, the names of those dimensions, where the competition is positioned, and where there are gaps for a new product to fill. Then we must know how to put together product and promotion to ensure that consumers perceive the new product as it was planned. How to identify the perceptual dimensions used by consumers and the product positioning opportunities in a market is the subject of this chapter. Our approach to product positioning is through perceptual maps. We provide some comprehensive examples of perceptual maps and discuss analytic methods for deriving these maps. The chapter closes with a review of the managerial use of perceptual maps.

PERCEPTUAL MAPS

Some Examples

Perceptual maps represent the positions of products on a set of evaluative dimensions. Figure 9.1 shows a perceptual map for pain relievers. The two dimensions are "effectiveness" and "gentleness." "Effectiveness" reflects attributes of strong, fast, long-lasting relief and the ability to make headache pain go away fast. "Gentleness" represents perceptions that the product would not upset one's stomach or cause heartburn. Excedrin is positioned as the most effective relative to the other brands, while Tylenol is the most gentle. There appears to be a positioning opportunity for a CBP of gentleness and effectiveness. If a product could be created that would have the

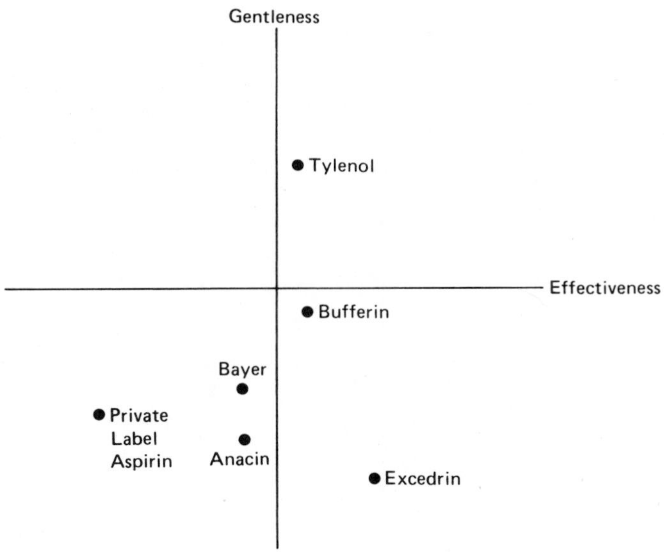

Figure 9.1. *Perceptual Map of Pain Relievers*

effectiveness of Excedrin and the gentleness of Tylenol, a unique positioning could be achieved.

Perceptual mapping is not restricted to frequently purchased consumer goods. Services and industrial products can be represented in this way also. Figure 9.2 shows consumer perceptions of transportation services. Quickness and convenience reflect the ability of a mode of travel to provide on-time service that gets consumers quickly to their destinations with no long wait, is available when needed, and allows consumers to come and go as they wish. Ease of travel includes correct temperature, no problems in bad weather, little effort needed, not tiring, and easy to carry packages or travel with children. Psychological comfort includes attributes such as relaxing, no worry about assault or injury, and not made uncomfortable or annoyed by others.

If you were a transit manager or community planner trying to increase utilization of public transportation, Figure 9.2 suggests that you would need to drastically improve consumers' perceptions of public transportation with respect to both ease of travel and quickness and convenience. You might try to modify the existing bus system or introduce a new type of service that is quicker, more convenient, and easier to use.

Product positioning for industrial products is illustrated in a study by Choffray and Lilien (1978); the new products evaluated were solar-powered systems. Table 9.1 shows the product attributes associated with each of two underlying perceptual dimensions for industrial air-conditioning systems, by corporate engineers. The first factor is "risk," and the second factor is

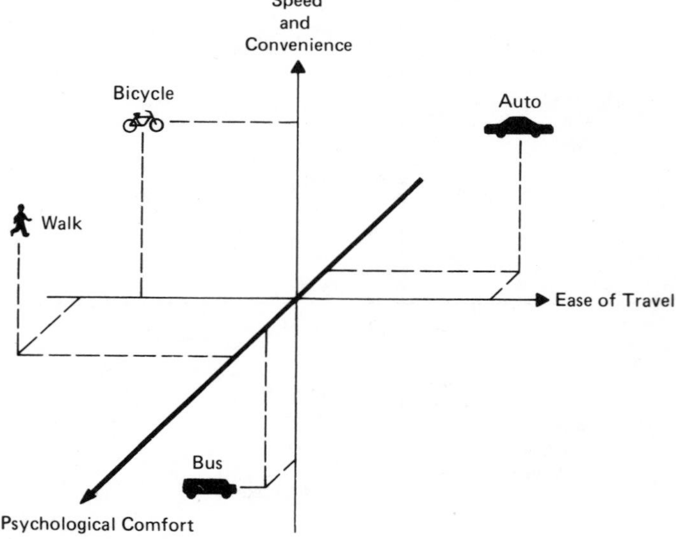

Figure 9.2. *Perceptual Map of Transportation Services (adapted from Hauser and Wisniewski, 1979)*

"benefits" of the system. In industrial buying, more than one person is usually involved. Choffray and Lilien found different participants in the decision had different numbers of dimensions, and these dimensions were made up of different compositions of the product attributes. Table 9.2 shows the

Table 9.1. Product Attributes Associated with Underlying Perceptual Dimensions of Air-Conditioning Systems[a]

	Dimension A, "Risk"	*Dimension B, "Benefits"*
Corporate Engineers	Reliability Field Tested First Cost Noise Level	Reduced Pollution Energy Savings Protection Modernness

[a]*Choffray and Lilien, (1978), p. 220*

Table 9.2. Perceptual Dimensions of Air-Conditioning Systems by Participants in the Buying Decision[a]

	Dimension A	*Dimension B*	*Dimension C*
Plant Managers	*"Benefits"* Energy Savings/ Protection Modernness Low Operating Cost	*"Risk"* Reliability/ Field Tested Modularity Noise Level	
Production Engineers	*"Benefits"* Energy Savings/ Protection Low Operating Cost Modernness Reduced Pollution	*"Noise"* Noise Level Modularity	*"Risk"* Complexity Field Tested/ Reliability
Top Managers	*"Benefits"* Energy Savings/ Protection Low Operating Cost Modernness Protection against Power Failure	*"Risk"* Reliability/ Field Tested Initial Cost Complexity	*"Noise"* Noise Level
HVAC Consultants	*"Benefits"* Energy Savings/ Protection Modernness Reduced Pollution	*"Risk"* Reliability/ Field Tested	*"Cost"* Initial Cost Noise Level

[a]*Adapted from Choffray and Lilien (1978), pp. 220–221.*

results for engineers, managers, and consultants. Generally, all perceived "benefits" and "risk," but "noise" and "cost" perceptions differed and the composition of these factors varied. In the industrial case, a good product positioning must be achieved with each member of the buying team.

Work on aircraft generators suggests that industrial customers evaluate these products along dimensions of merit such as light, inexpensive, reliable, and as small as possible. Although the dimensions in industrial products may be more engineering oriented, perceptions on such factors as reliability, service, and quality are important too, and perceptual positioning should be considered on these psychological dimensions as well as the engineering propensities.

How Many Dimensions?

In consumer and industrial markets an important question is, "How many dimensions should be on a map?" Should we use all the attributes upon which a product can be rated? Figure 9.3 shows the average consumer ratings on twenty-five scales representing attributes of a communication system.

This "snake plot," so-called because a line connecting the ratings "snakes" down the page, allows you to visually interpret the perceived positions of all existing products. This map indicates how scientists and managers at Los Alamos Scientific Laboratory in New Mexico view telephone and personal visits with respect to three new product concepts in communications. Careful examination of this map indicates the detailed strengths and weaknesses of each new product compared to existing alternatives. However, are people really using twenty-five independent dimensions to evaluate these alternatives? An analysis of data that produced Figure 9.3 indicates that two independent dimensions can adequately describe the perceptions of these communications options. The correlations between twenty-five basic attributes and each of the two dimensions are shown in Figures 9.4a and 9.4b. Upon examination, it is clear that dimension one is "effectiveness" and dimension two is "ease of use." Figure 9.5 shows how the existing products are positioned in this space. Compare Figures 9.3 and 9.5. The snake plot presented more detail, but tended to obscure the basic structure. In Figure 9.5 the market is clear. There is definitely a gap for a technology that is easier to use than personal visit, but more effective than the telephone. Note that neither narrow band video telephone or closed-circuit TV fill that gap. If the manager could develop a product in the starred (*) position, the new product might capture a significant share of the communications market.

The use of perceptual maps is important to managerial decision making. The snake plot (Figure 9.3) is difficult to interpret, while the two-dimension map (Figure 9.5) is simple, clear, and meaningful. To develop creative strategies, managers and other product development people must be able to internalize and visualize the market along a few dimensions. There is some

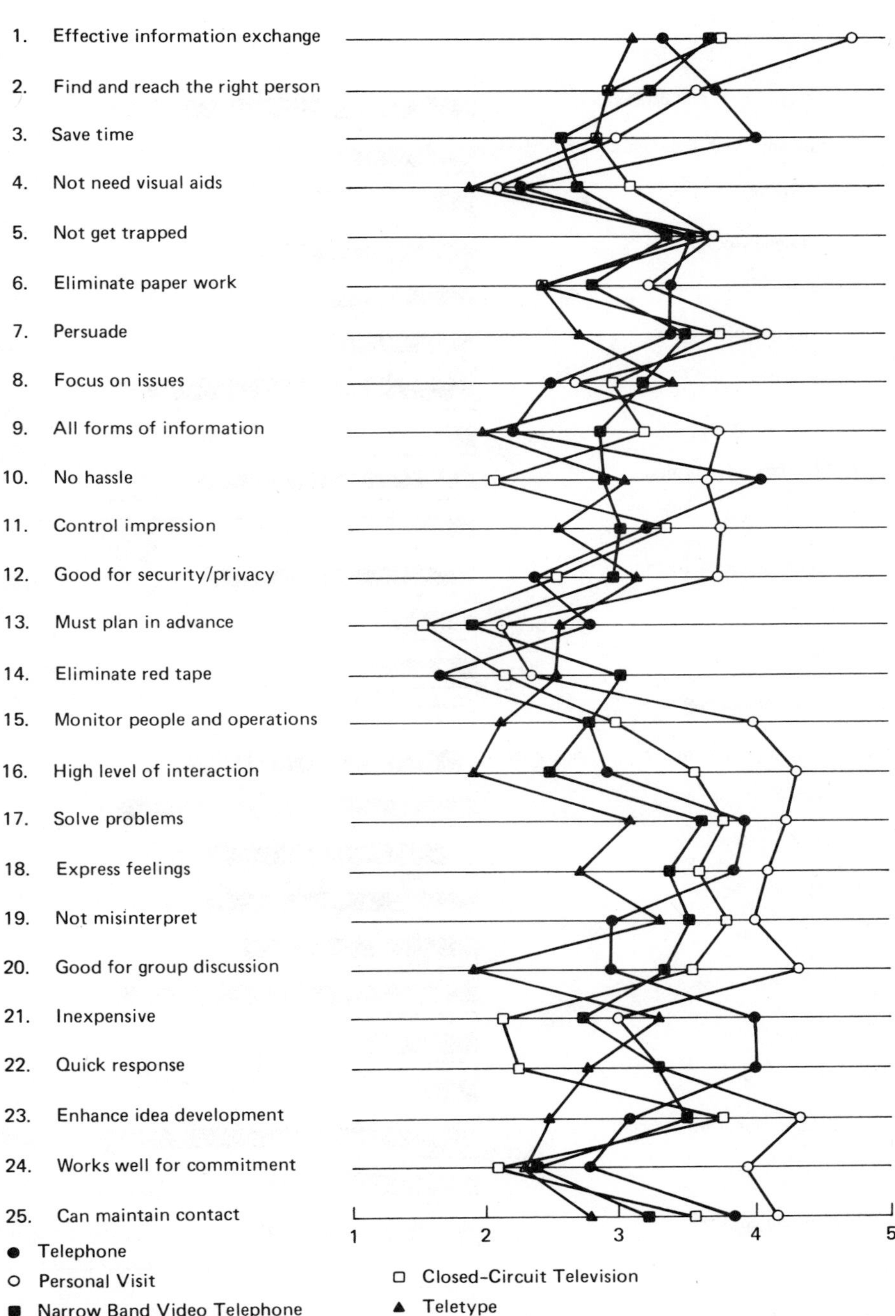

Figure 9.3. *"Snake Plot" of Scientists' and Managers' Perceptions of New Communications Options*

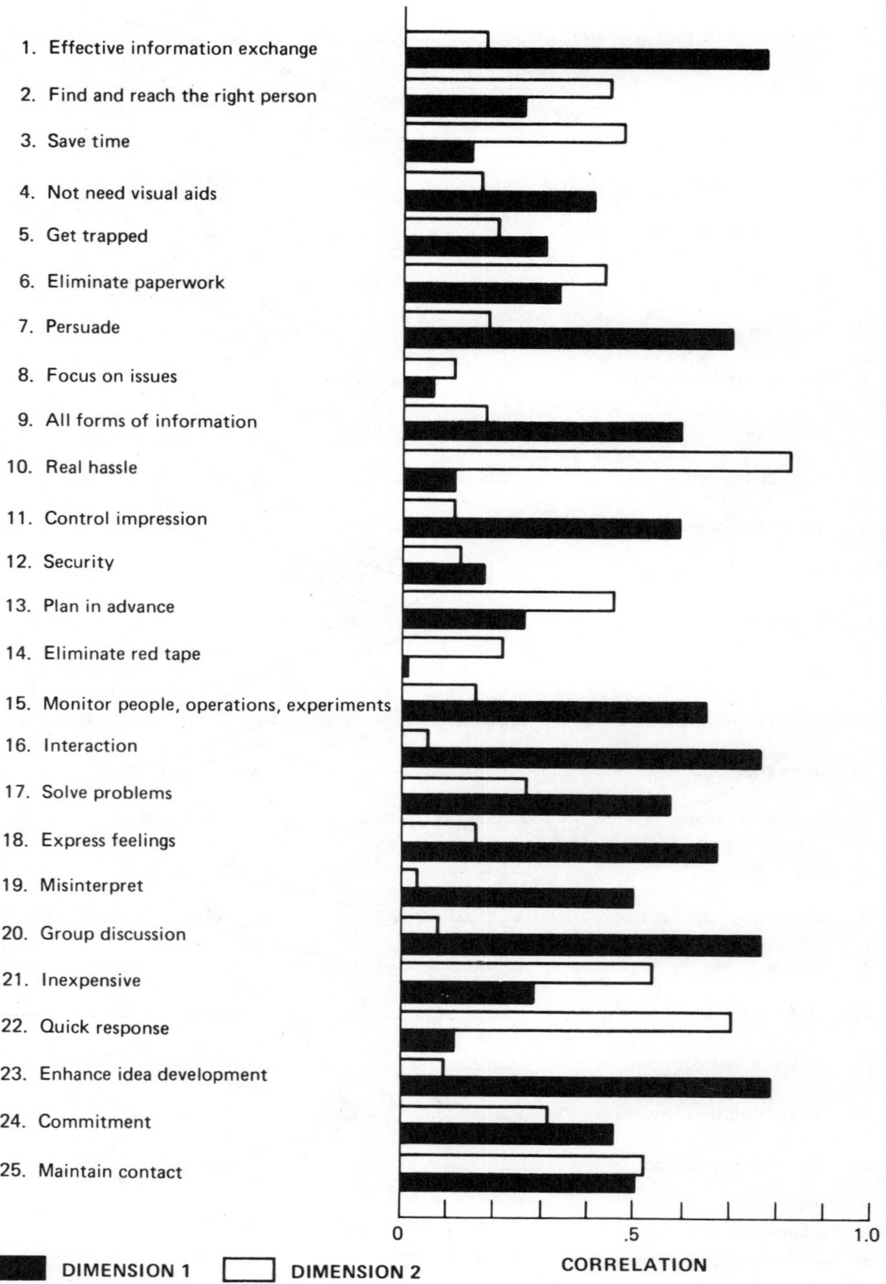

Figure 9.4. *Correlations between Communication Attributes and Perceptual Dimensions*

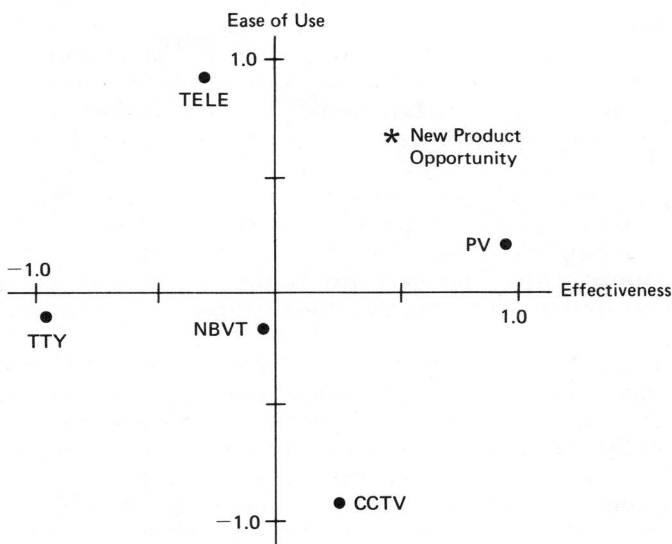

Figure 9.5. *Perceptual Map for Communications Options (Hauser and Shugan, 1980)*

evidence that consumers tend to simplify judgments by reducing dimensionality so to prevent cognitive strain and information overload. Finally, we will see in preference analysis that intercorrelations which almost always exist among attributes, make it impossible to statistically assign relative importance to the 25 attributes. Thus, to enhance creativity, to understand the market, to accurately model consumers, and to use the preference models, we must carefully analyze the basic attributes and reduce them to a set of underlying factors. This reduced set of dimensions and the positions of products upon them is the perceptual model we shall utilize in this text. The analytic techniques allow a determination of whether one, two, three, four, or more underlying dimensions are most appropriate to model consumer perceptions.

How Do Features Relate to Perception?

The examples shown above have utilized maps based on composite perceptions of products. It is common, however, to have new features in a product. For example, digital watches may have features that could be number of functions (hours, minutes, day, month, etc.), shape of face (round or rectangular), and material of strap (leather, metal, plastic). Automobiles have features such as air conditioning, thick carpets, vinyl top, or special decorative striping. These features are important in establishing perceptual positioning. Having ten functions in a watch will substantiate a perception of "computer-like." Thick carpets and a vinyl roof in an automobile will help substantiate a perception of luxury. We view features as essential to substantiating a perceptual position. IBM tries to convey the image that breakdowns will not occur in its computers. But even if a breakdown does occur, the basic reliability appeal provides a backstop for the vice president who took responsibility for the purchase decision.

Although we could ignore perceptions and concentrate only on features, real opportunities might then be lost. Hewlett-Packard concentrated on positioning their calculator as one of "uncompromising quality" and used features to support this. The market rewarded this strategy with premium prices relative to those who merely added computational features to their calculators.

Both psychological positioning and features are part of a good new product design. Early in the design process it is critical that managers visualize and internalize the consumers' perceptual process, that creativity be enhanced and constructively channeled, that the potential design not be limited by existing product characteristics, and that the analysis models represent the inputs of consumers. Thus, in most product categories, early investigation should concentrate on psychological characteristics to uncover new dimensions and to better understand the consumer. Later, as the product progresses through the design process, attention will shift to physical attributes to attain a physical product that meets the planned positioning. Finally, just before roll out, attention will return to psychological characteristics for generation of advertising and promotional strategies. Whether emphasis is on psychological or physical positioning depends on the product category. In consumer products, such as deodorants, antacids, hand-care lotion, and beer or in services such as health, transportation, banking, and education we have found that much of the design process focuses on psychological dimensions. In industrial products such as electronics, fuseboxes, and chemicals it is likely that more of the design process will focus on physical dimensions. Even when the majority of the design process emphasizes physical dimensions, managers need to know which specific dimensions to focus on.

Since the most important dimensions are those the consumer uses in selecting a product, a manager is well-advised to begin with consumers' psychological dimensions to find the structure of cognition and then use that understanding to direct the product design to produce the physical characteristics that will position the product effectively on the psychological dimensions.

Managerial Requirements

A new product manager needs to develop a core benefit proposition (CBP) and correctly position it against competitive products. To do this the manager needs to identify:

- the number of dimensions which best describe the market,
- the names of those dimensions,
- the positioning of existing products along these dimensions,
- the physical product characteristics to substantiate the perceptual positioning,
- where consumers prefer a product to be on the dimensions.

The manager needs the number, names, and existing positions to understand the market. He needs the link to physical characteristics so he can control his own position, and he needs the preferences to select the best positioning. In particular, the manager must move his product along a dimension that is not yet exploited in the existing market, or if no new dimensions can be found, the manager must identify a positioning which attracts consumers without undermining his company's sales on related products.

In this chapter we examine methods to derive perceptual maps that represent the position of existing products. We then extend these maps to represent the competitive position of a new product and its core benefit proposition. The maps allow identification of gaps in the market. The prioritizing of these gaps is addressed in Chapter 10 when preference models tell us how much relative reward can be earned by filling each gap. The position must be unique, but more importantly, it must lead to high sales and profit levels.

ANALYTICAL METHODS USED TO PRODUCE PERCEPTUAL MAPS

The examples of perceptual maps presented in the previous section were based on statistical techniques. The technique used for the telecommunications map was factor analysis. This method reduced the consumers' ratings on the 25 attribute scales to identify the structure underlying the attribute ratings. This method is dependent on prior specification of scales and it cannot identify any dimensions that do not have at least one attribute to represent them. Another method is available which does not depend upon specific scales. This procedure, called "non-metric multidimensional scaling," is based on judgments of the overall similarity of products. Instead of rating *each* product on a large number of explicit scales (see Figure 9.6a), *pairs* of brands are evaluated in terms of how similar or dissimilar the consumer views them to be (see Figure 9.6b).

There are advantages and disadvantages to both approaches (see Table 9.3). Attribute ratings directly elicit perceptions relative to psychological scales. Consumers' evaluation of products relative to these scales represents their evaluation of products in the marketplace. On the negative side, such scales can only measure the attributes that are represented by a scale. Thus

	Strongly Disagree	*Disagree*	*Neither Agree Nor Disagree*	*Agree*	*Strongly Agree*
I can convey all forms of technical information by telephone.	[ ]	[ ]	[ ]	[ ]	[ ]

(a) Attribute Rating

	Very Similar 1	2	3	4	*Very Different* 5
Telephone and narrow band video telephone	[ ]	[ ]	[ ]	[ ]	[ ]
Personal visit and narrow band video telephone	[ ]	[ ]	[ ]	[ ]	[ ]

(b) Similarity Judgement

Figure 9.6. *Alternative Consumer Measures to Product Perceptual Maps*

perceptual mapping techniques that are based on attribute scales are strongly dependent on the analyst's ability to use focus groups and to use previous experience to develop a more or less complete set of scales.

Similarity measurement has an advantage over attribute measurement because it gathers information independently of the attribute set. Consumer statements on which products are most similar help define the market and indicate which products are in direct competition and which are not. On the negative side, similarities alone cannot name the psychological dimensions and the manager must depend to some extent on attribute ratings or his personal knowledge of the market. Furthermore, minor problems result from the fact that similarity is left undefined and consumers' reaction to the measurement can vary. Finally, the computer analysis is expensive and technical problems prevent similarity maps from being used in markets where there are fewer than eight products in the average consumer's evoked set.

Both measures and the corresponding analytic techniques are useful in new product design. Techniques based on attributes are used to identify structure once a good set of attributes is identified. They provide accurate, interpretable measures and predict quite well. Techniques based on similarity measures are most useful to search for structure when nonverbal aspects of perception are present. They provide maps which are complementary to attribute-based techniques. No one technique is suited to every situation.

Table 9.3. Analytic Techniques to Produce Perceptual Map

Technique	*Consumer Measure*	*Advantages*	*Disadvantages*
Factor Analysis	Attribute Ratings	Easy, Inexpensive, Accurate Based on Assumption on How Consumers React to Attribute Scales	Requires a Complete Set of Attribute Scales
Similarity Maps	Dissimilarity Judgments	Does not Depend on Attribute Set Provides and Alternative View of Perceptual Map	Special Programs Required Statistical Limitations, if Few Products

Rather it is important that the manager recognizes when to use each technique in order to determine the structure of the market and produce the perceptual map that leads to the creative insight that will produce a successful new product.

The remainder of the chapter discusses analytic techniques that use either attribute measures or similarity measures to produce perceptual maps. In each section we give the basic conceptual ideas behind the techniques and illustrate their use with examples taken from actual new product analyses. We present the basic mathematical equations that underlie these techniques. For those who would like a more in-depth treatment of the mathematics, we have included technical appendices to introduce the statistical analysis and to serve as a guide to further study of these techniques. As we present the alternative methods, the advantages and disadvantages cited in Table 9.3 are discussed along with issues of when each technique is appropriate and how to use it in the design process. We begin with an attribute-based technique—factor analysis, that we have found to be most productive early in the design process. An alternative technique—discriminant analysis, is described in Pessemier (1977).

Since all readers may not want to follow all the mathematics, here and in subsequent chapters, we attempt to make clear non-mathematically the general methodology and its appropriateness for new product analysis. Technical details are contained in the appendices.

Factor Analysis of Attribute Ratings

Factor analysis attempts to find a number of dimensions that can represent the information in a large set of attribute ratings. This is done by examining the correlation between the ratings. If two scales are perfectly correlated, then we only need one of them. If two scales are highly correlated, we can presume they are attempting to measure the same underlying attitude. Factor analysis looks for correlations between rating scales and estimates their correlations to underlying dimensions. These correlations are called factor loadings. We describe the model that links factor loadings and ratings and consider the activities of naming dimensions, determining the number of dimensions, producing the map, and placing a new product on the map.

Underlying Model of Ratings. Consider Figures 9.4 and 9.5. Suppose that consumers really do evaluate communications options based on their "effectiveness" and their "ease of use." What would we expect to happen when consumers rate the communications options on the set of basic scales? Consider the attribute "convey all forms of technical information." When a single consumer rates a product on this attribute, we expect a "common" effect due to his underlying belief about the overall "effectiveness" of the product, plus a "unique" effect due to his knowledge about the attribute in question, plus some "residual" effect due to measurement errors. When we measure many consumers on all 25 scales, we expect these effects to surface as components that explain the variation in the way consumers rate products (see Figure 9.7).

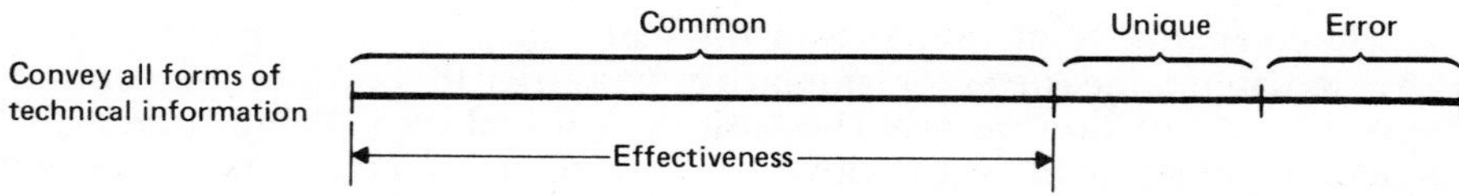

Figure 9.7. *Components of a Consumer's Response to an Attribute Scale*

Consider an individual consumer, labeled by the subscript i, a specific product (e.g., a telephone), labeled by the subscript j, and an attribute (e.g., "convey all forms of technical information"), labeled by the subscript l. Then we expect this consumer's rating (y_{ijl}) of telephone on "convey all forms of technical information" is due to some fraction (f_l) of his overall belief (x_{ij}) about a telephone's effectiveness plus some measure of unique value (u_{ijl}) on "convey all forms of technical information" and some error (ϵ_{ijl}). Mathematically:

$$\textbf{(9.1)} \qquad y_{ijl} = f_l x_{ij} + u_{ijl} + \epsilon_{ijl}$$

For example, a consumer's rating ($y_{ijl} = 4$) of telephone on "convey all forms of technical information" might be made of a component ($f_l x_{ij} = 3.5$) due to his overall evaluation of the effectiveness of telephone, a component ($u_{ijl} = 0.7$) due uniquely to the measurement scale, and a component ($\epsilon_{ijl} = -0.2$) due to error.

Actually, on some attributes we can expect common effects due to both "effectiveness" and "ease of use." Thus, if we subscript the dimensions (effectiveness and ease of use) by k and write equation (9.1) in summation notation, we get:

$$\textbf{(9.2)} \qquad y_{ijl} = \sum_{k=l}^{2} f_{lk} x_{ijk} + u_{ijl} + \epsilon_{ijl}$$

where y_{ijl} = rating of individual i for product j on attribute scale l.

f_{lk} = factor loading of scale l and underlying dimension k.

x_{ijk} = position of product j on underlying dimension k by individual i. This is termed the "factor score."

u_{ijl} = unique contribution of individual i's rating of product j on scale l.

ϵ_{ijl} = error term.

Clearly, equation (9.2) can be extended to 3, 4, or more dimensions.

Suppose now we could somehow determine the f_{lk}'s. These are called the factor loadings. In other words, suppose we knew ratings on "convey all forms of information"

$$= (0.65)\,(\text{factor } 1) + (-0.18)\,(\text{factor } 2) + \text{unique} + \text{error}.$$

Then, by examining a table of f_{lk}'s to determine which attributes load heavily on which factors, we could name the factors. This was done in Figure 9.4, which was simply a plot of the factor loadings (the f_{lk}'s).

Equation (9.2) specifies the model for what is termed "common" factor analysis. Another type is called "principal components" factor analysis. The model is the same as equation (9.2) except the u_{ijl} term (unique scale component) is omitted. In principal component analysis, all the variation in the ratings is assigned to the underlying dimensions. In both the common and principal components modeling, the analytic method estimates the factor loadings (f_{lk}).

Naming the Dimensions. Factor analysis determines the factor loadings. For example, Table 9.4 gives the factor loadings relating the two factors to the 25 attributes that describe communications options. These loadings produced the bar graphs that were shown Figure 9.4. By underlining the heavy loadings, those with correlation 0.40 or better, and examining the relationships among the heavy loaders, it is possible to name these dimensions "effectiveness" and "ease of use."

To obtain these correlations, a data matrix must be factor analyzed.

Table 9.4. Loadings Used to Name the Dimensions of the Perceptual Map for Factor Analysis[a]

Attributes	Effectiveness	Ease of Use
1. Effective Information Exchange (−)	−0.77	−0.17
2. Find and Reach Right Person	0.25	0.43
3. Save Time	0.17	0.47
4. Not Need Visual Aids	0.39	−0.16
5. Get Trapped (−)	−0.33	−0.20
6. Eliminate Paperwork	0.31	0.43
7. Persuade (−)	−0.70	−0.20
8. Focus on Issues	−0.04	−0.07
9. All Forms of Information	0.65	−0.18
10. Real Hassle (−)	−0.11	−0.83
11. Control Impression	0.56	0.07
12. Security	0.18	0.11
13. Plan in Advance (−)	0.23	−0.44
14. Eliminate Red Tape	−0.00	−0.21
15. Monitor People, Operations, Experiments	0.65	0.15
16. Interaction	0.78	0.05
17. Solve Problems (−)	−0.55	−0.27
18. Express Feelings	0.66	0.17
19. Misinterpret (−)	−0.49	0.00
20. Group Discussion	0.75	0.05
21. Inexpensive	−0.27	0.52
22. Quick Response	0.07	0.71
23. Enhance Idea Development	0.77	0.09
24. Commitment	0.44	0.32
25. Maintain Contact	0.50	0.52

[a] *(−) indicates question was worded so that a high attribute rating would mean a poor evaluation.*

This data matrix, represented in Table 9.5, is the matrix of basic attribute ratings for all individuals on each product in their evoked set. That is, if there are N individuals with an average of $\bar{J}$ products in their evoked set, and if there are L attributes, the data matrix which is factor analyzed is the $N\bar{J} \times L$ matrix of attribute ratings. In words, the basic attributes are factor analyzed across individuals and products. The entries in the matrix are the row scale ratings (e.g., assigned values one to five in Figure 9.6a) for a given individual and product on a five-category attribute scale. For example, if you agreed that with telephone you could "convey all forms of technical information" then you, individual 1, would enter a "4" for telephone on that attribute. (A more accurate procedure is to then standardize these values by taking out the mean and variance for each individual across attributes and products. This procedure reduces individual biases such as using extreme ends or limited ranges on the five-point agree/disagree scale.)

Intuitively, factor analysis searches for a common set of factors that can explain the variation in the ratings that consumers assign to their perceptions of the products relative to the attribute scales. For a given number of factors, say K factors, the computer algorithms determine the factor loadings for the K factors which explain as much as possible of this variation. The matrix is then "rotated" to find a simple structure and improve the ability to

Table 9.5. Schematic Representation of Data Matrix for Factor Analysis.

		Attribute 1	*Attribute 2*	· · ·	*Attribute 3*
Individual 1:	Product 1	y_{111}	y_{112}	· · ·	y_{11L}
	Product 2	y_{121}	y_{122}	· · ·	y_{12L}
	· · ·	· · ·			
	Product J	y_{1J1}	y_{1J2}	· · ·	y_{1JL}
Individual 2:	Product 1	y_{211}	y_{212}	· · ·	y_{21L}
	Product 2	y_{221}	y_{222}	· · ·	y_{22L}
	· · ·	· ·			
	Product J	y_{2J1}	y_{2J2}	· · ·	y_{2JL}
· · · · ·					
Individual N:	Product 1	y_{N11}	y_{N12}	· · ·	y_{N1L}
	Product 2	y_{N21}	y_{N22}	· · ·	y_{N2L}
	· · ·	· · ·			
	Product J	y_{NJ1}	y_{NJ2}	· · ·	y_{NJL}

identify the factors. Appendix 9.1 gives the mathematical details of factor analysis with this description. One of several standard statistical packages (SPSS, BIOMED, or PSTAT) can be used to perform factor analysis.

Determining the Number of Dimensions. You can name the dimensions by examining the factor loadings, but how do you determine the right number of dimensions? Statistically, the way factor analysis works is that it examines the matrix of correlations among the basic attributes and extracts factors one by one in order of variance explained. By examining the variance explained by each successive factor, an analyst can make judgments about when to cut off the factor analysis. There are two classical rules of thumb for determining the number of factors: the "scree" test and the "eigenvalue" rule.

The scree test is illustrated in Figure 9.8. Here we have plotted the incremental variance explained by each factor vs. the order in which that factor was extracted. Percentage variance explained is a standard output of most statistical packages for factor analysis. Factors are retained up to the point where incremental contribution levels off. The scree test takes its name from geology, where the "scree" is the pile of rock that accumulates at the bottom of a cliff. Some researchers prefer to plot cumulative variance explained rather than incremental variance explained, and apply related cutoff rules.

The eigenvalue rule is similar to the scree rule, except the cutoff is when

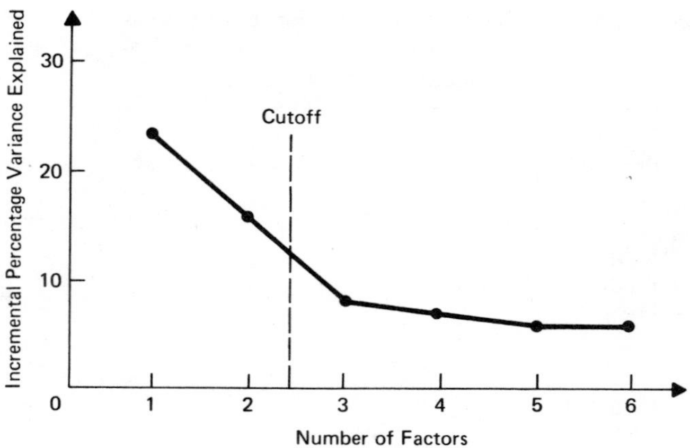

Figure 9.8. *Scree Test*

the fraction of variance explained by the next successive factor drops below ($1/L$,) where L is the number of attributes. Statistically, this point is reached when a number called the "eigenvalue," which is associated with each factor, drops below 1.0 (eigenvalues are standard outputs of the statistical packages). Recently Montanelli and Humphreys (1976) have developed a more sophisticated method of determining the appropriate number of factors which compares actual eigenvalues to eigenvalues that would be obtained from random data.

In practice, either the eigenvalue rule or scree test is used to select the appropriate number of factors, but this is not an automated procedure. The selection is often a mix of statistical rules and commonsense judgment. There is often a tradeoff between variance explained and interpretation of the structure of the factor loadings. Thus, it is good practice to interpret not only the analysis obtained with the rules of thumb, but also analyses that have one more, or one fewer factors. The final selection of structure (i.e., number of factors) should be based on the often conflicting goals of variance explained, ease of interpreting structure, usefulness to the creative team, and linkage to actionable managerial decisions.

Producing the Perceptual Map. So far we have shown how to determine the number of dimensions and name them. The next step is to produce a perceptual map by estimating the positionings of each existing product. Mathematically, we have to determine the factor scores (x_{ijk}'s) of equation (9.2) where x_{ijk} is the individual i's perception of product j on dimension k. For example, x_{111} might be the first individual's ($i = 1$) perception of telephone ($j = 1$) on effectiveness ($k = 1$).

A product's position, $\bar{x}_{jk}$, along any dimension is then obtained by averaging the factor scores (once obtained) across consumers. For example, in Table 9.6 the position of telephone on effectiveness would be obtained by

Table 9.6. Example of Data Matrix with Estimated Factor Scores (see Fig. 9.5 for abbreviations)

		Attribute Scales				Factor Scores on Dimensions	
	Effec. Info Exch. (l = 1)	*Find and Reach (l = 2)*	*Save Time (l = 3)*		*Maintain Contact (l = 25)*	*Effec- tiveness (k = 1)*	*Ease of Use (k = 2)*
Individual 1:							
Telephone ($j = 1$)	4	2	4		1	(−0.3)	0.9
Personal visit ($j = 2$)	1	3	2		3	1.0	0.2
NBVT ($j = 3$)	2	2	2	…	2	0.0	−0.2
TTY ($j = 4$)	3	2	2		1	−1.0	−0.1
CCTV ($j = 5$)	3	1	1		1	0.2	−0.9
Individual 2:							
Telephone ($j = 1$)	3	1	5		1	(−0.4)	0.8
Personal Visit ($j = 2$)	2	2	3		2	1.1	0.1
NBVT ($j = 3$)	3	1	3	…	1	0.1	−0.3
TTY ($j = 4$)	4	1	3		1	−0.9	−0.2
CCTV ($j = 5$)	2	1	2		1	0.3	−1.0
		· · ·					
Individual N:							
Telephone ($j = 1$)	5	3	3		2	(−0.2)	1.0
Personal Visit ($j = 2$)	2	4	1		4	0.9	0.3
NBVT ($j = 3$)	5	3	1	…	3	−0.1	−0.1
TTY ($j = 4$)	5	3	1		2	−1.1	0.0
CCTV ($j = 5$)	4	2	1		2	0.1	−0.8

taking the average over the factor scores for the effectiveness of telephone (circled numbers). Of course, if you wish a map for a segment of the population rather than the whole population, these averages can be taken on a segment-by-segment basis. The perceived positions of the products in Figures 9.1, 9.2, and 9.5 were computed in this way.

If *principal components factor analysis* is used, the calculation of x_{ijk} is straightforward since the ratings (y_{ijl}) and factor loadings (f_{lk}) are known and equation (9.2) is directly solved for x_{ijk}.

When *common factor analysis* is used, the task is more complex since the unique scale component is present. A regression-like procedure is used to estimate the factor scores (see Appendix 9.1). One disadvantage of this method is that the factor scores and therefore map positions are subject to greater error variance when the total variance explained is "not large," i.e., less than 80 to 90 percent. This is less true in principal components since the factor scores are exactly evaluated. Common factor analysis has a more attractive underlying model, but the factor scores from the principal components methods are more precise. We find it useful to do both common and principal components factor analyses. If they agree, as they usually do, in the number of dimensions and interpretations, use the principal component factor scores to draw the perceptual map; if they do not agree, use the common factor analysis results. For both common factor analysis and principal components the estimated factor scores are standard outputs of most statistical packages.

Another useful part of factor analysis is the description of the factor scores as a linear function of the original ratings. Parameters of this equation are called the "factor score coefficients." In principal component factor analysis, the factor score coefficients are directly calculated, but in common factor analysis, they are estimated by the regression-like procedure used to compute the factor scores. The implied multiple linear regression is:

(9.3) $$x_{ijk} = \sum_{l=1}^{L} b_{kl} y'_{ijl} + \text{error}$$

where x_{ijk} = factor scores; position of product j on underlying dimension k by individual i.

y'_{ijl} = standardized rating of individual i for product j on attribute scale l.

b_{kl} = factor score coefficient for scale l and underlying dimension k.

Since the x_{ijk}'s are unknown, a standard regression cannot be run. Fortunately, estimates, $\hat{b}_{kl}$, of the b_{kl}'s can be computed based on the correlation matrix and are standard outputs of most statistical packages for both common factor analysis and principal components.

To obtain the estimates of effectiveness and ease of use in Table 9.6, standardize the attribute ratings in the first 25 columns, (l = 1 to 25) of Table 9.6 to get the y_{ijl}'s and use the factor score coefficients ($\hat{b}_{kl}$'s) from Table 9.7. Consider individual 1. Suppose that, when standardized, the ratings for telephone become $y_{111} = 1.5, y_{112} = 0.8, y_{113} = 0.9, \ldots, y_{1125} = -1.3$. These numbers replace the first 25 columns of the first row of Table 9.6. Then the estimated effectiveness for telephone is given by

$$x_{111} = (-0.10)(1.5) + (-0.03)(0.8) + (-0.05)(0.7) + \cdots + (0.07)(-1.3) = -0.3$$

where the b_{1l}'s come from the first column of Table 9.7. The factor score estimates with principal components are obtained from the factor score coefficients in an analogous way.

We have defined three outputs—factor loadings, factor scores, and factor score coefficients. The loadings are the correlation of the scales to the underlying dimensions (Table 9.4). The factor scores (Table 9.6) when averaged across individuals specify the coordinates for a product on the per-

Table 9.7. Factor Score Coefficients Used to Estimate Product Positions in Perceptual Map

Attribute	Effectiveness	Ease of Use
1. Effective Information Exchange (−)	−0.10	−0.04
2. Find and Reach the Right Person	−0.03	0.11
3. Save Time	−0.05	0.26
4. Not Need Visual Aids	−0.06	0.04
5. Get Trapped (−)	−0.07	0.06
6. Eliminate Paperwork	−0.02	0.18
7. Persuade (−)	−0.13	0.01
8. Focus on Issues	0.01	−0.00
9. All Forms of Information	−0.02	−0.05
10. Real Hassle (−)	0.07	−0.29
11. Control Impression	0.03	−0.04
12. Security	—0.00	−0.02
13. Plan in Advance (–)	0.04	0.00
14. Eliminate Red Tape	0.00	0.01
15. Monitor People, Operations, Experiments	0.05	0.01
16. Interaction	0.25	—0.07
17. Solve Problems (–)	—0.02	—0.09
18. Express Feelings	0.13	—0.05
19. Misinterpret (–)	0.00	0.05
20. Group Discussion	0.20	—0.08
21. Inexpensive	—0.04	0.09
22. Quick Response	—0.04	0.20
23. Enhance Idea Development	0.22	—0.05
24. Commitment	0.06	0.05
25. Maintain Contact	0.07	0.18

[a] *(–) indicates question was negatively worded.*

ceptual map. The factor score coefficients (Table 9.7) link the standardized input attribute scales to the factor scores.

The managerial output of the factor analysis is a perceptual map made up of a specified number of labeled dimensions and the coordinates of each existing product on these dimensions.

Position of a New Product. If you have measured consumers' reactions to a concept statement or to an actual sample of a new product, estimation of that new product's position follows the same technique as for existing products. The factor analysis directly supplies the factor scores for each individual and each dimension for the product. When averaged across all individuals, the new brand is positioned on the map.

A manager can then examine if it is the desirable position. For example, Figure 9.9 shows the positions of two new transportation services based on ratings of one-page descriptions of these services.

Both concepts were designed to improve the quickness, convenience, and ease of travel of public transit. Budget Taxi Plan (BTP), is a privately operated system which provides service similar to that provided by taxis, but at a lower price. The only change is that if you request the budget plan, the driver may pick up and drop off other passengers on the way to your destination. Personalized Premium Service (PPS), is a publicly operated version of the same system except that service is provided with minibuses instead of taxicabs. The perceptual map in Figure 9.9 indicates that the new systems are perceived as significantly better than the existing bus service on quickness, convenience, and ease of travel but somewhat poorer on psychological comfort. Furthermore, there are some differences between perceptions of the private and public systems. Figure 9.9 suggests that if the actual systems can provide the level of service described in the one-page concepts, then they will improve consumers' perceptions of public transportation and likely lead to greater ridership. The managerial decision to further develop the system will depend upon consumer preferences (Chapter 10), the ridership forecast (Chapter 11), and the costs of providing service.

Another method of placing a new product on the map is by estimating the average attribute scale ratings of the new product and using the factor score coefficients to estimate the coordinates

$$\bar{x}_{jk} = \sum_{l=1}^{L} \hat{b}_{kl}\bar{y}_{jl} \tag{9.4}$$

where $\bar{x}_{jk}$ = coordinates of product j on perceptual map dimension k.

$\hat{b}_{kl}$ = estimated factor score coefficient for scale l and dimension k.

$\bar{y}_{jl}$ = average standardized scale ratings for product j on scale l.

For example, you might consider a new transportation service that was an improvement over PPS. To estimate its position on the perceptual map you would first estimate how consumers would perceive it, on average, on the attributes. You would then use equation (9.4) to compute the average perceptual positions and plot them on the map in Figure 9.9. The use of equation (9.4) is similar to that of equation (9.3) except you use average rather than individual scale ratings.

Earlier, we advocated "stretcher concepts" such as the BTP and PPS services which expand the range of measurement and hence the perceptual and later preference models. In prediction, stretchers provide good reference points so that one can estimate how a new product will differ from the stretchers. For example, in Figure 9.5 the new telecommunications service product may be an improved version of Narrow Band Video Telephone. The product team may achieve this by making it more available (easier to use) and by providing better resolution in transmission (more effective). The product team can then estimate the new position by one of three ways:

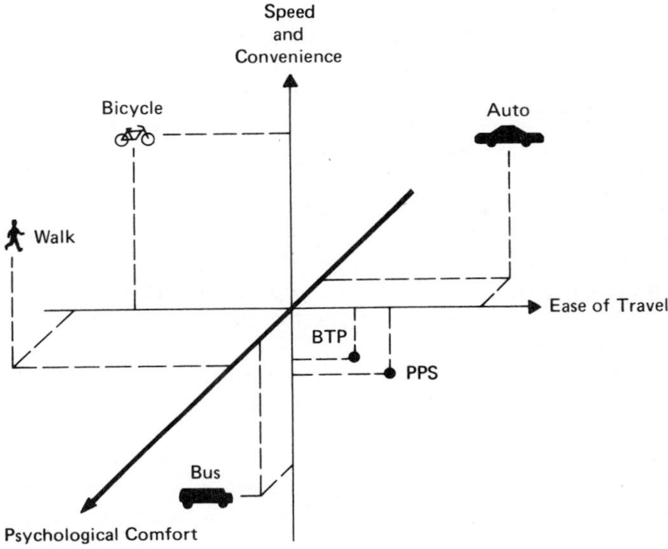

Figure 9.9. *New Service Positionings (adapted from Hauser and Wisniewski, 1979)*

- create a new concept and remeasure consumers.
- judgmentally estimate the resulting change in attribute ratings ($\bar{y}_{jl}$) by comparing it to a stretcher and use equation (9.4) to compute the change in factor scores ($\bar{x}_{jk}$'s).
- judgmentally estimate the resulting change in factor scores (x_{jk}'s) and simply move the position on the map.

The first procedure will give the best estimate of the new position, but it is costly and time consuming. If used carefully, the second and third procedures can provide excellent estimates that can give good "ballpark" indications of where the new product will be positioned. Note that the second procedure provides a direct link between measured attributes and the factor scores. It couples judgment with analytic procedure to provide estimates that are often less prone to error than the third procedure, which requires direct judgments on the factor scores.

New product managers should use the perceptual maps, the factor loadings, and the factor score coefficients to choose a positioning and then to select the attributes and features that will give them that positioning.

Example: Health Maintenance Organization. In 1975, the MIT Health Department was considering conversion to a health maintenance organization (HMO) that would provide most medical care for a fixed annual fee. The managerial problem was to design the HMO in such a way as to attract enough consumers to support the fixed costs and the operating costs of the plan. Perceptual analysis was performed to examine the new product concept relative to existing care, to a similar pilot HMO program that had been in operation for one year, and to the major competition—Harvard Community Health Plan—which in the fall of 1975 had the potential to be offered as a dual-care option to the MIT community. That is, due to federal regulations, all MIT employees would have the option to join the Harvard plan instead of the existing Blue Cross/Blue Shield plan.

Questionnaires were sent out to 1,000 members of the community. Of the 447 that were returned, 367 were prospective members of the pilot HMO and 80 were actual members of the pilot program. Figure 9.10 gives the snake plot for sixteen attribute scales, the factor loadings, and the resulting perceptual map.

By examining the high loadings on each dimension, they were labeled: (1) quality, (2) personalness, (3) value, and (4) convenience. Quality correlated with trust, preventive care, availability of good doctors, and hospitals. Personalness reflected a friendly atmosphere with privacy and no bureaucratic hassle. Value was not just price, but rather paying the right amount for services. Convenience reflected location, waiting time, and hours of operation.

Based on the perceptual map of Figure 9.10 it is clear that average perceptions of existing care are superior to the new HMO on all dimensions except convenience. Thus, it is unlikely that the concept as stated could draw a large number of consumers from existing care. Now compare the perceptions of the concept description to those of the pilot program. The pilot plan exceeds the concept in all dimensions, even though they are essentially the same product. Although this could be self-selection or post-purchase rationalization, it is more likely that this is a case of a good product where few people perceive it as such until using it. The managerial implications of this finding are that if the HMO is to be successful it would have to develop an aggressive campaign to communicate actual plan performance to perspective members. Perhaps testimonials and "live" media advertising (e.g., slide shows) would improve the image.

Next, note that relative to the major competition, Harvard Community Health Plan, the MIT concept (or pilot) is perceived better on all dimensions except quality. At this point in our analysis we have no indication of the relative importance of quality and do not know whether this difference can exert a major influence on consumers' response. Finding these importances is the next step in the analysis, which will be covered in Chapter 10. As you might expect, quality turns out to be a key variable and the gap should be closed, if possible. By examining the factor loadings, we see that attributes 3, 7, 8, 9, 12, 14, and 16 load on quality and, by examining the snake plot, we see that the lower rating for MIT in quality is based almost entirely on a low score for MIT on hospital quality. Thus, if the plan is to be successful, the design team should identify some method to move on the quality of hospitals—perhaps by switching affiliation or by emphasizing the best hospital of those that are now affiliated. In the HMO case both strategies were possible.

1. I would be able to get medical service and advice easily any time of the day and night.
2. I would have to wait a long time to get service.
3. I could trust that I am getting really good medical care.
4. The health services would be inconveniently located and would be difficult to get to.
5. I would be paying too much for my required medical services.
6. I would get a friendly, warm and personal approach to my medical problem.
7. The plan would help me prevent medical problems before they occurred.
8. I could easily find a good doctor.
9. The service would use modern, up-to-date treatment methods.
10. No one has access to my medical record except medical personnel.
11. There would not be a high continuing interest in my health care.
12. The services would use the best possible hospitals.
13. Too much work would be done by nurses and assistants rather than doctors.
14. It would be an organized and complete medical service for me and my family.
15. There would not be much red tape and bureaucratic hassle.
16. Highly competent doctors and specialists would be available to serve me.

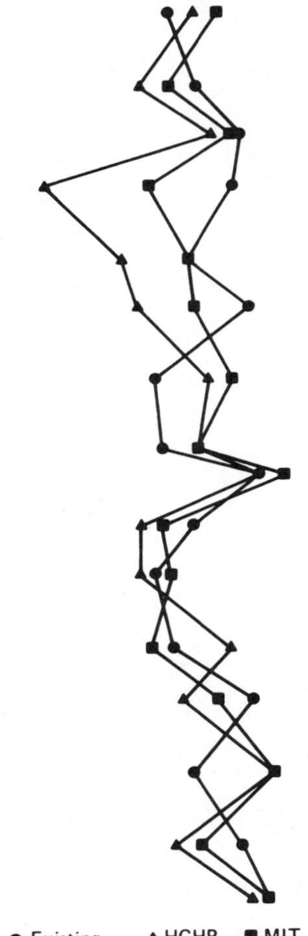

(a) Average Ratings

Figure 9.10. *Perceptual Analysis for Health Care Delivery (Hauser and Urban, 1977)*

ATTRIBUTE SCALE	QUALITY	PERSONAL	VALUE	CONVENIENCE
1. Day & Night Care	0.37244	0.07363	−0.31379	0.63939
2. Waiting Time	−0.22082	0.26204	0.15514	−0.64370
3. Trust-Good Care	0.72125	−0.21826	−0.09556	0.24703
4. Location	0.01144	0.24706	−0.12544	−0.72454
5. Price/Value	0.03066	0.12810	0.72884	−0.08461
6. Friendly/Personal	0.40986	−0.51317	−0.12265	0.16768
7. Preventive Care	0.55403	−0.14187	−0.44353	−0.01653
8. Easily Find Good M.D.	0.64412	−0.15036	−0.21491	0.27113
9. Modern Treatment	0.72288	−0.13441	−0.15906	0.08018
10. Access to Records	0.43412	−0.49053	0.18749	−0.06982
11. Continuity of Care	0.20491	0.47900	0.47727	0.04725
12. Associated Hospitals	0.68006	−0.08256	0.10854	0.00555
13. Use of Paramedicals	−0.05303	0.67083	0.12299	0.16722
14. Organized/Complete	0.47725	0.01627	−0.52893	0.14316
15. Hassle/Red Tape	−0.13031	0.69824	0.11830	−0.27903
16. Competent M.D.s	0.73953	−0.19335	−0.13971	0.18691
Eigenvalues	5.34	1.4	1.1	1.02
Cumulative Variance	0.33	0.42	0.49	0.55

(b) Factor Loadings

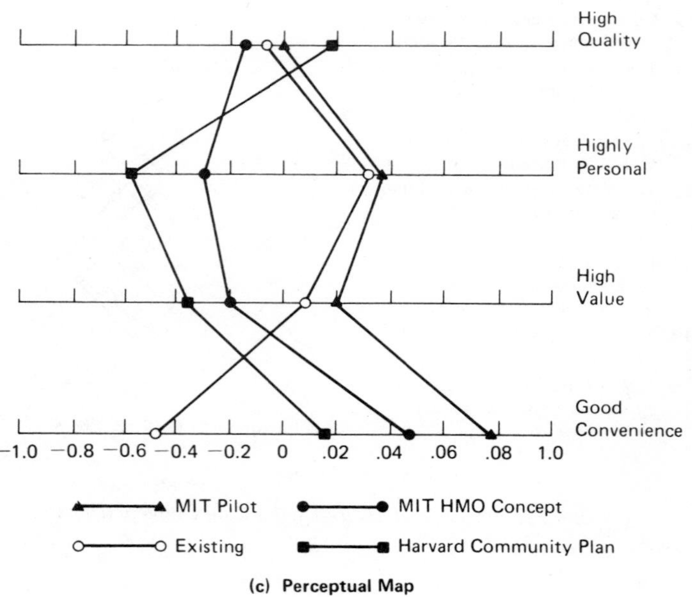

(c) Perceptual Map

Figure 9.10. *(continued)*

Thus, the perceptual map indicated two key strategies: (1) a more aggressive communication campaign and (2) a shift in hospital affiliation. Judgmentally, these improvements were expected to move the perceptual position of the MIT concept one-half of the distance from the concept to actual pilot (or the Harvard performance) on quality, personalness, and value. Based on the preference analysis of Chapter 10 and the subsequent calculations of purchase potential in Chapter 11, these changes increased the expected enrollment from 3,622 families to 5,405 families—an increase of almost 50 percent. Actual experience after making these changes confirmed these enrollment projections.

This example illustrates the power of perceptual mapping. Note the synergy between managerial judgment and analytic technique. The major changes were based on the ability of the design team to interpret the analytics. The analytics alone would not have been sufficient without the intuitive grasp of how to use the perceptual map. On the other hand, without the analytics, intuition would not have been guided to the leverage points—those managerial changes which make the greatest impact.

Pitfalls of Factor Analysis. Factor analysis is a powerful technique for perceptual analysis, but there are many dangers in its use. We list a few of the most common.

If you have a hard-to-understand scale for an attribute, most people will rate all products on the middle of that scale. If you have a number of these bad scales, they will show up as a factor. This should be avoided with careful pretesting of the scales.

If there is one or more scales that is very strong and should stand by itself, it will force mixed loadings on factors. You can recognize this by examining the "uniqueness" or "communality" of each attribute in common factor analysis. If an attribute has high mixed loadings and high uniqueness (low communality), it may be a factor by itself. If this happens, use it as a separate perceptual dimension and factor analyze only the remaining scales. Preference analysis (Chapter 10) will determine whether this is actually an important perceptual dimension.

It is tempting to use the attribute with the highest factor loadings as a surrogate for the factor score. This procedure has no sound analytic basis and ignores the information that is present in the other attributes. Most importantly it reduces control options available to the new product design team to improve positioning.

If the concept statements do not span the set of feasible alternatives, the factor score coefficients may not provide accurate estimates for alternatives outside the range of the concept statements. Thus, it is better to begin with strong stretcher concepts and interpolate into a feasible product than to have what seems to be the limit of feasible concepts and later find you must extrapolate a new product outside the range of the data.

Discussion and Summary. The above examples for analgesics, communications, health care, and transportation show the versatility of factor analysis. Factor analysis also has been used successfully in other frequently purchased products (deodorants, antacids, hand-care lotions, etc.), in industrial air conditioners, financial services, and a myriad of other product categories. To use factor analysis in another category simply:

- Obtain the structure of perceptions by factor analyzing consumers' ratings of product attributes across consumers and products. Interpret and name the factors by examining the factor loadings.
- Select the appropriate number of dimensions by judgment, based on simple structure, interpretability, and either the "scree" rule, the "eigenvalue" rule.
- Estimate the positions of existing products, test products, and concepts by the use of factor scores, or factor score coefficients and the attribute scales.
- Obtain new product positions by proactively examining the perceptual map and selecting the strategy to produce the chosen position. Calculate this position by judgment or by using the factor score coefficients.

The advantage of factor analysis is that it works well and predicts well. It is inexpensive to use (an average single run on the computer costs under $10) and relatively easy to implement with standard statistical computer packages. The behavioral assumption of how consumers react to measurement scales (Figure 9.7) is appealing; empirically it leads to results with strong face validity. The major disadvantage is that it depends on a well-specified set of fundamental attributes and can only identify structure *within* a set of attributes, not beyond them.

Before considering mapping techniques that do not depend upon rating scales, we should point out that factor analysis is not the only method of converting ratings into a perceptual map. Other researchers have successfully used *discriminant analysis* for automobiles (Pessemier, 1977), political candidates (Johnson, 1970), and other products. Managerially, this technique is used and interpreted in the same manner as factor analysis, but it is based on a different behavioral assumption and uses a different analytic technique. Intuitively, the behavioral assumption is based on selecting the dimensions to maximumly discriminate among existing products. While at first glance discriminant analysis is appealing, we have found that it concentrates on variation between products and subsequently ignores variation among consumers. Factor analysis examines both. In two comparative empirical tests to date, factor analysis was found to be more complete and to predict better than discriminant analysis (Hauser and Koppelman, 1979, Simmie, 1978).

Non-Metric Scaling of Similarity Judgments

In the last section we outlined a method of deriving a perceptual map from ratings of product attributes. In this section we describe a different set of analytical procedures that are designed to derive perceptual maps from judgments of the similarity of products. We introduce this notion with a simple example based on your perception of management schools.

Consider the business schools of Harvard, Stanford, MIT, and Northwestern. Which two are most similar? Say you judge Harvard and Stanford most similar to each other. Which two are most dissimilar? Say you judge Harvard and MIT most different. Now consider which pair is second most similar, say Stanford and Northwestern. Second most dissimilar—Stanford/MIT. Now consider the other pairs, Harvard/Northwestern and MIT/Northwestern. Let us say you think the two schools Harvard and Northwestern are more similar than MIT and Northwestern. You have now described the rank order of the pairs of schools, as shown in Table 9.8. These are your similarity judgments.

Table 9.8. Hypothetical Rank Order of Business Schools

Pair	*Rank*
Harvard/Stanford	1
Stanford/Northwestern	2
Harvard/Northwestern	3
Northwestern/MIT	4
MIT/Stanford	5
Harvard/MIT	6

Now let us derive a map of your perceptions. Consider two coordinates as shown in Figure 9.11. Place a point on the figure for each school. Try to put schools close to each other that you judged as similar. Try to put far apart any schools you judged as dissimilar. Adjust the points so that the distance is smallest between Harvard and Stanford, the next smallest between Stanford and Northwestern, then Harvard and Northwestern, and so on; the distance between Harvard and MIT should be the largest. Figure 9.12 shows one map that meets these requirements. There are many maps that could be generated by these rules, but you have in Figure 9.12 one perceptual map based on a judgmental multi-dimensional scaling of similarity data.

In practice, the maps are based on more data and more stimuli and are less ambiguous. For example, Figure 9.13 shows an actual map derived from a random sample of 100 entering students at Northwestern. In this case, we might name the dimensions "realism/theoretical" and "outcomes" based on the location of the schools in the two dimensional space. Remember when interpreting this map that Figure 9.13 represents the subjective evaluations of students who have selected, but who have not yet experienced Northwestern.

We now consider the input data and the analytic techniques used to produce similarity maps.

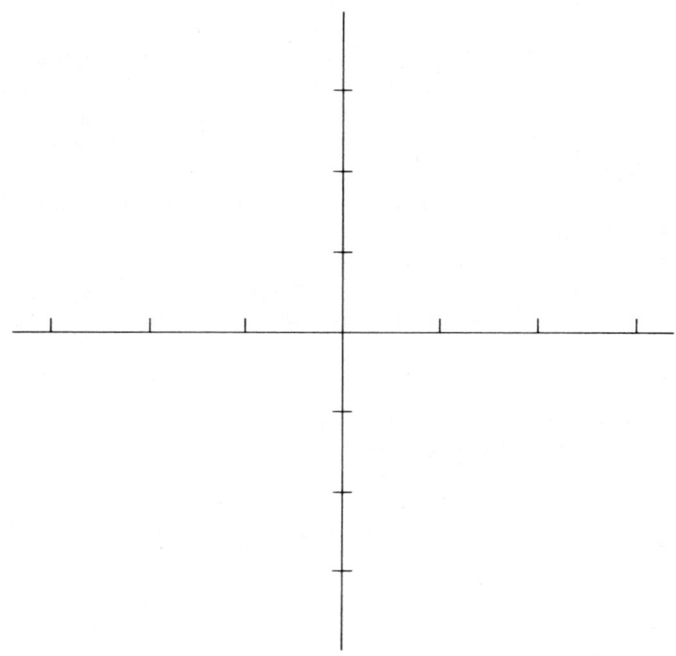

Figure 9.11. *Coordinates for Map*

Similarity Inputs. The input similarities commonly are obtained from:

- judgments of the similarity of a pair of products on scale (e.g., Figure 9.6b), or
- rank ordering pairs with respect to their similarity (e.g., Table 9.8)

These measures are tabulated in a matrix which has rows and columns designated by the products to be mapped and matrix entries which represent the similarities between products (see Table 9.9). This is called a "proximity matrix." For a market segment, each cell in the matrix contains the average

Table 9.9. Perceptual Distance (Measured in Dissimilarity)

	Personal Visit	Telephone	NBVT	CCTV	TTY
Personal Visit	—				
Telephone	10.5	—			
NBVT	7.3	8.2	—		
CCTV	8.5	13.8	5.8	—	
TTY	13.4	8.4	6.7	11.1	—

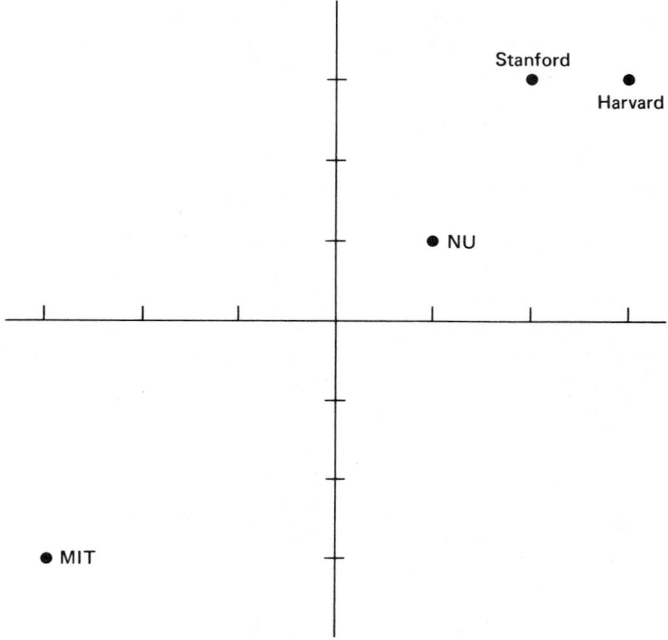

Figure 9.12. *One Map of Hypothetical Similarity Judgments*

of the scaled similarity judgments or the average of the rank orders for the appropriate pairs of products. In some cases the proximity matrix is generated by input in which the consumer rank orders all brands from most similar to most dissimilar to the brand in row one. These rank order values, when averaged across all individuals, provide the input for the first row of the proximity matrix. Next, all brands are rank ordered with respect to the brand in row two. The procedure of using each brand in turn as a reference, and rank ordering the others with respect to it is called the "anchor point" method and can be used to generate the full proximity matrix.

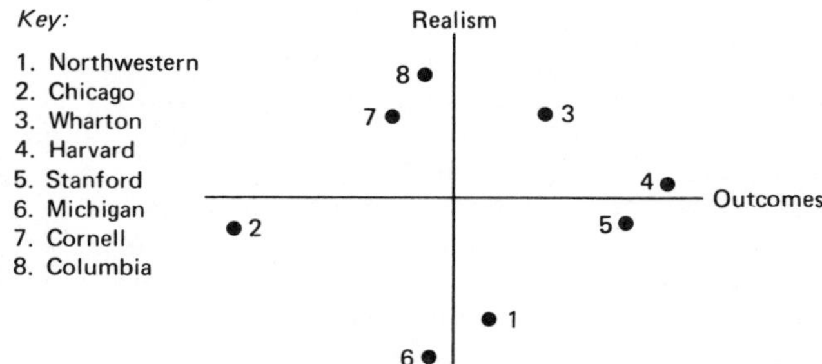

Figure 9.13. *A Similarity Map of Management Schools (adapted from Simmie, 1978)*

Producing the Perceptual Maps. In Figure 9.12 the perceptual map was produced by judgmentally shifting the points in the map to best represent the interproduct similarity. Similarity scaling algorithms automate this procedure. Given an initial starting configuration, the distance between points is measured in the usual way (Euclidean distance). A statistical measure called "stress" (see Appendix 9.2) is then applied to the map to determine how well the points reproduce the relative similarities in the proximity matrix. A large value for stress means a poor fit; a low value for stress means the interpoint distances are close to the similarities in the proximity matrix. The computer algorithm then shifts the points by a small amount in the direction that best reduces the stress. This continues until the stress has been reduced as much as is possible by the small movements. The resulting product positions for each product, j, on each dimension, k, produce the perceptual map. For example, Figure 9.14 shows a map derived by Green and Carmone (1970).

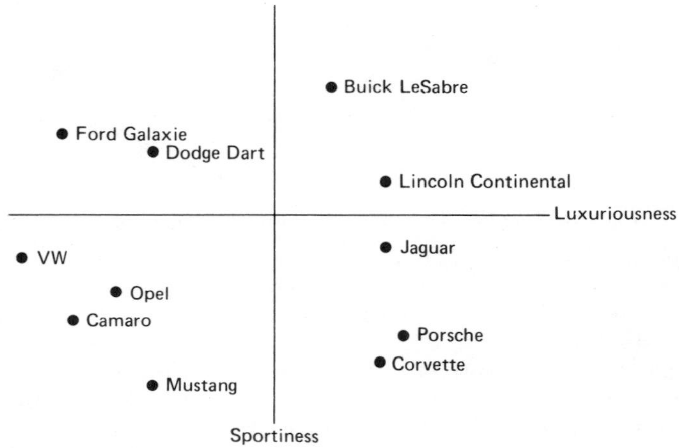

Figure 9.14. *Map of Consumer Perceptions of Autos (adapted from Green and Carmone, 1970, p. 60)*

A word of caution is in order because there are a minimum number of products needed to determine a perceptual map. Klahr (1969) shows by simulation that at least eight are needed for a good two-dimensional map. Green and Wind (1973) suggest that the number of dimensions, K, should be less than one-third the number of products, J. Thus, in certain product categories such as health care, transportation, and financial services, there may be too few stimuli for a map. Furthermore, even in categories such as deodorants, the evoked set for any given consumer may be too small for this technique. Thus, the manager is urged to proceed carefully when interpreting solutions with too few stimuli. One approach is to expand the evoked set with concept descriptions to get sufficient stimuli. Another is to use the factor analysis method of mapping described earlier.

Naming the Dimensions. Suppose we have determined the two-dimensional map in Figure 9.14. By examining the relative positions of the autos, one may name the dimensions. For example, the horizontal dimension seems to distinguish the Lincoln Continental from the Ford Galaxie, the Jaguar from the Volkswagen, and the Porsche from the Camaro. This dimension was named "luxuriousness." Similarly, the vertical dimension distinguishes the Galaxie from the Camaro, the Dart from the Mustang, and the Continental from the Jaguar. This dimension was named "sportiness." Similarity maps augment the manager's knowledge of a product category by combining the structure of the map with the manager's judgment. Used in this way, similarity maps provide an alternative measure of perception to complement factor analysis.

If a manager does not want to rely solely on judgment to name the dimensions and if attribute scale ratings are available, a technique called PROFIT (Carroll and Chang, 1964; Miller, Shepard, and Chang, 1964) can be used to name the dimensions by estimating the association between the similarity dimensions and the attribute ratings. The technique is a regression procedure with the dimension coordinates as the explanatory variables and the attribute scales as the dependent variables. For the metric version, the regression equation is:

$$\text{(9.5)} \qquad \bar{y}_{jl} = \sum_{k=1}^{K} D_{lk}\tilde{x}_{jk} + \text{error}$$

where D_{lk} is the "directional cosine" between attribute l and dimension k, $\bar{y}_{jl}$ is the average of the attributes across consumers, and $\tilde{x}_{jk}$ are the map coordinates of the product. The values of D_{lk} help indicate if the attributes are associated with a dimension. The concept is similar to that of factor loadings in the previous section. In fact, if ratings had been collected on a complete set of attributes, the products could be directly mapped by factor analysis and compared to the similarity scaling solution. For example, Figure 9.15 gives a two-dimensional PROFIT solution for a perceptual map of shopping locations serving the northern suburbs of Chicago. The dimensions were named "variety" and "quality vs. price," based on the underlined directional cosines.

Determining the Number of Dimensions. The perceptual maps in Figures 9.14 and 9.15 were produced by selecting the product positions such that the interpoint distances most nearly match the observed similarities shown in the "proximity matrix." The algorithms use the fit measure called stress to get the distances as close as possible and to judge the accuracy of a map. (See Appendix 9.2.) As we increase the number of dimensions, K, we increase the number of degrees of freedom and achieve a better fit of distance to dissimilarity. Thus, the stress value, S_K, decreases as the dimensions increase. When S_K is plotted against K, the resulting marginal changes in stress can be examined in the same spirit as in factor analysis, except that we are examining the stress value rather than the increase in incremental vari-

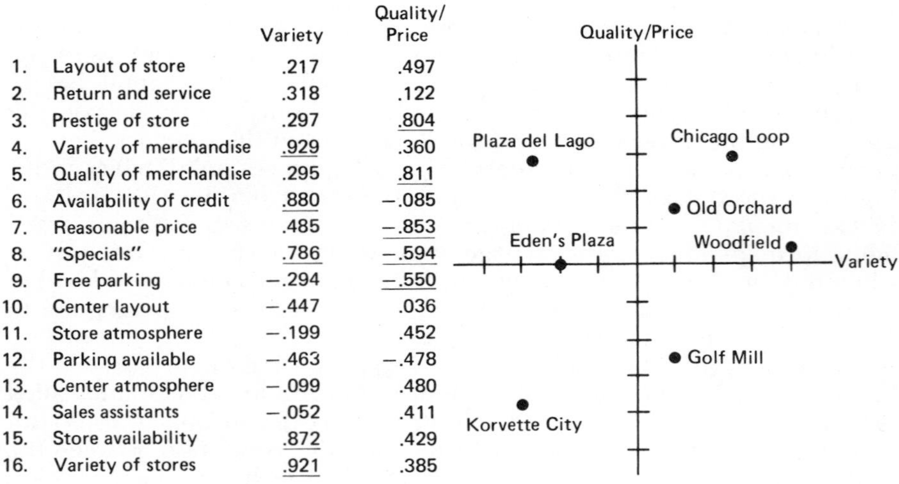

		Variety	Quality/ Price
1.	Layout of store	.217	.497
2.	Return and service	.318	.122
3.	Prestige of store	.297	.804
4.	Variety of merchandise	.929	.360
5.	Quality of merchandise	.295	.811
6.	Availability of credit	.880	−.085
7.	Reasonable price	.485	−.853
8.	"Specials"	.786	−.594
9.	Free parking	−.294	−.550
10.	Center layout	−.447	.036
11.	Store atmosphere	−.199	.452
12.	Parking available	−.463	−.478
13.	Center atmosphere	−.099	.480
14.	Sales assistants	−.052	.411
15.	Store availability	.872	.429
16.	Variety of stores	.921	.385

Figure 9.15. *Perceptual Map of Shopping Locations (Koppelman and Hauser, 1979)*

ance explained. In other words, increase the number of dimensions as long as significant reductions in stress can be achieved.

As before, the statistical rule is an indication to be weighed against the managerial criteria of interpretability of the map and its usefulness to the creative team. We suggest that you interpret both more and fewer dimensions than the statistical rule suggests and select the solution that is most useful to the design of a new product.

Individual Differences. Factor analysis produced measures of perceptions for each and every consumer. The above similarity maps are average maps. For the analysis of preferences (Chapter 10), we need estimates of each consumer's perceptions of the products in his evoked set. One method to achieve these estimates is to produce a separate map based on each consumer's similarities. Unfortunately, the computer analysis is expensive for similarity analysis and, as a result, individual level analysis would be prohibitively expensive for many reasonable sample sizes.

An alternative procedure, due to Carroll and Chang (1970) is INDSCAL. INDSCAL assumes that there is a common map, $\tilde{x}_{jk}$, as above, but the individual perceptions, $\tilde{x}_{ijk}$, result from a stretching or a contracting of the map. Thus, the distance measure and corresponding stress measure are modified by differentially weighting the axes. (See Appendix 9.2 for equations.) The mathematics of INDSCAL differ from stress-based algorithms, but should produce similar estimates of average perceptions ($\tilde{x}_{jk}$) and, in addition, estimate the individual perceptual coordinates that result from the differential stretching of dimensions.

Position of New Products. New products can be positioned in a similarity map by

- measuring new similarities for the new product, or
- judgmentally shifting the product directly in the perceptual map.

The pros and cons of each technique are similar to those discussed for factor analysis and depend on the product category and the stage of the design process.

Discussion and Summary. Similarity scaling has been applied to many product categories. For example, it has been used to map perceptions of autos (Green and Carmone, 1970), beer (Urban, 1975; Allaire, 1973), menu planning (Green and Wind, 1973), shopping locations (Koppelman and Hauser, 1979), and retail outlets (Singson, 1975). To use similarity mapping in a category:

- Have consumers evaluate existing products and/or concepts in the category according to their relative similarity and form an average similarity matrix for the group or for each consumer.
- Use one of the algorithms (MDSCAL 5, TORSCA 8, PARAMAP) to produce an average map in 2, 3, . . . dimensions.
- Use the statistical rule, based on stress and managerial judgment, to select the appropriate number of dimensions
- Name the dimensions based on the relative position of the stimuli in the perceptual map or use PROFIT to produce a matrix of directional cosines, D_{lk}, and interpret these to name the dimensions based on the attributes that load heavily.
- If desired, use INDSCAL to get individual level estimates of product perceptions.
- Obtain the new product positions by examining the perceptual map and selecting the strategy to produce the chosen position. Calculate this position by judgment or remeasurement.

The advantage of similarity maps is that they are based on measures that complement rather than depend upon the attribute measures. Thus, they provide an alternative measure of perception that can be used very early in design process to identify the relevant attributes or later to supplement the insight gained from perceptual maps developed from factor analysis.

One major disadvantage of similarity scaling is that it is difficult to use. The computer programs are less available and often require much data handling. FORTRAN programs are needed to use the full battery for analysis and the programs are quite expensive. When first introduced, there was a major concern among researchers that similarities be treated as non-metric (rank order only) data. Recent results (Green, 1975) have shown that the metric mapping methods yield similar results to non-metric scaling and are sufficient for most applications. Metric algorithms are often more available and certainly less expensive. In some cases, similarity maps provide predictions that are less accurate than those developed by factor analysis (Hauser and Koppelman, 1979; Simmie, 1978). This is probably due to the restrictive assumptions that are implicit when individual level perceptions are scaled. Nonetheless, they do provide an alternative view of the product category that can prove quite useful in new product design. For a further discussion of the advantages and disadvantages of similarity scaling, see Stefflre (1978).

Summary Comments on Analytical Mapping Methods

This section presented two analytic techniques to produce perceptual maps. Attribute-based methods like factor analysis are used to determine the basic perceptual structure by identifying the underlying cognitive structure in the attributes. These methods are easy and inexpensive to use, provide excellent managerial insight into the perceptual structure of the product category, and provide accurate measures of the perceptual dimensions which can be used in preference analysis and in forecasting the purchase potential of the new product. But to work well, factor analysis techniques depend on a well-specified and complete set of basic attribute scales.

Similarity maps provide alternative representations of the perceptual dimensions that do not require a prior specification of attributes. They are useful to gain further insight into the category. Since they are more expensive, more difficult to use, and may not predict as well as factor analysis, we feel they should be used as an augmentation to the perceptual analysis or as an exploratory method to ensure that the set of attribute scales are indeed completely and accurately specified.

We have tried to provide a description of each technique at a level that is sufficient for a basic understanding of the concept and at a level sufficient for the reader to use these techniques in new product design. Most readers will find that the mathematics become clearer as they try to actually implement the models. Others will find that they want more detail and want to see the equations behind the verbal descriptions. These readers are encouraged to read Appendices 9.1 and 9.2.

MANAGERIAL USE OF MAPS

Perceptual maps help managers understand a product category and recognize opportunities by providing a succinct representation of how consumers view and evaluate products in that category. The consumer intelligence achieved through this representation enables the new product team and the R&D team to work together and focus their creativity to achieve the desired position in the market. This focus on high potential positioning directs the design process and thus makes it more efficient and effective.

The emphasis in perceptual mapping is on psychological positioning, not physical characteristics. It is important to understand this distinction. Consumers make judgments on their own perceived reality. Thus, early in the design process it is extremely important to identify an opportunity, select a target positioning, and direct the product development accordingly. This does not mean that the physical characteristics of a product are not important. Physical characteristics (objective reality) and psychological appeals (e.g., advertising) act as cues upon which consumers base their perceptions (Brunswik, 1952). Thus, once a perceptual positioning is identified, the new product team can direct its effort toward achieving that positioning.

For example, in the transportation example it was found that the two new services, BTP and PPS, were positioned much better than the existing bus service. This is an encouraging result, but the product itself must back up this positioning. When implemented, BTP or PPS must be carefully engineered to provide the service promised in the concept statement. Furthermore, advertising and promotion must be used to communicate the core benefit proposition to consumers.

The use of concepts at the design phase is important since they test the core benefit proposition and the product's ability to fill a positioning opportunity. It also causes managers to consider how they can substantiate the positioning through physical product features and marketing tactics. The use of concepts is also very important when no products now exist in a product category. In this case you can use stretching concepts to fill out the new category. For example, in the health maintenance organization (HMO) case, no existing HMO was available to the target group. We created the category in the consumer's mind with the three new HMO concepts which were compared to existing care.

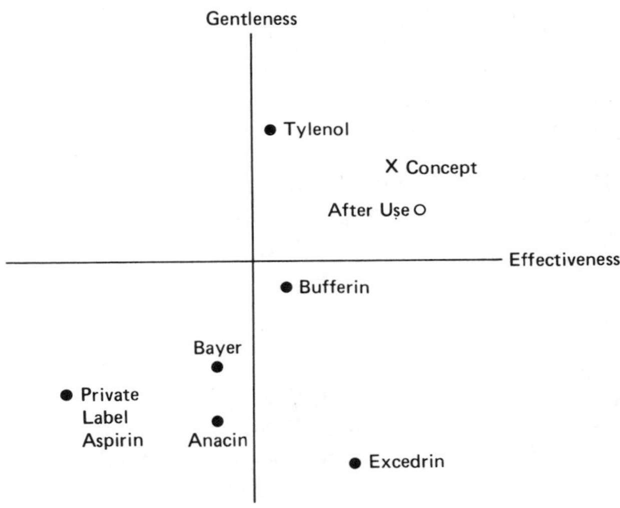

Figure 9.16. *Perception of New Pain Reliever After Home Use*

Maps are also useful to compare perceptions of a product concept to those of the actual product that is developed based on that concept. By collecting ratings after consumer use of the actual product, the fulfillment of the positioning can be measured. For example, Figure 9.16 shows the positioning of a new pain relief product after four weeks of use. Note the good match between the perceptions of the product concept and the perception of the actual physical product. When the concept and use perceptions do not match, either the product should be redesigned or the concept aligned to correspond with the actual product.

This chapter has discussed methods of mapping and how they can be used to identify gaps in markets, check positioning of concepts, and determine if physical products fulfill their core benefit promises. However, one critical managerial question has not been answered: "Which gap should we try to fill?" The next chapter shows how preference analysis is used to designate the best gap to fill.

REVIEW QUESTIONS

9.1 How would you determine the number of dimensions in a multidimensional perceptual mapping?

9.2 You are about to graduate with your MBA degree (or equivalent) and venture into the ominous job market. In this market, you view yourself as a new brand which needs proper positioning. Being an aggressive, hard-working and enterprising MBA, you decide your excellent grades and outstanding personality are not enough. A proper perceptual positioning is required at your interview. From some library research, you discover the following key attributes employers look for in potential employees:

A = Preparedness (0 − 10, 10 best)
B = Group Skills (0 − 10, 10 best)
C = Technical Skills (0 − 10, 10 best)
D = Ability to Handle Pressure (0 − 10, 10 best)

Your perceptions of yourself and your competitors follow,

	A	B	C	D
Brand 1	7	3	2	3
Brand 2	7	5	9	4
Brand 3	2	5	8	9
Yourself	6	2	8	6

a. How would you compete with Brand 1?
b. How would you compete with Brand 2?
c. How would your interview strategy depend on the behavioral model the interviewer uses to select candidates?

9.3 How do the inputs for factor analysis and nonmetric scaling differ?

9.4 Perceptual mapping techniques summarize the information contained in the perceptual data. Given a particular perceptual map derived from perceptual data, how do you know how much information was captured by the map?

9.5 When confronted with the details of factor analysis, a line manager states: "Sure analytical methods are used to generate perceptual maps and these maps are very useful. But my job is to make decisions. I must apply the perceptual maps to come up with real actions. It's not important for me to know a lot of theoretical mathematics. Academics worry about how the computer makes the map. I only worry about how to apply it."

Discuss the correctness of the manager's statements and their implications.

9.6 When perceptual data is gathered for only several specific stretcher concepts, how can slight variations on these concepts be positioned?

9.7 Draw a judgmental perceptual map for the toothpaste market. For banks in your community. Heavy duty industrial compressors.

9.8 Discuss the relationship between market definition and perceptual mapping.

9.9 Would a manager ever use both factor analysis and nonmetric scaling? Why?

9.10 How might a health maintenance organization be positioned in your community?

Chapter 9 Appendices

9.1 Factor Analysis*

The text covered the intuition behind factor analysis. The mathematics (equations 9.2 and 9.3) gave the fundamental structure in algebraic equations. In this appendix we will derive some of the estimates, but to do so we must switch to matrix notation. Let $Y = \{y_{ijl}\}$ be the $(N\bar{J} \times L)$ standardized data matrix from Table 9.5; let $F = \{f_{lk}\}$ be the $(L \times K)$ matrix of factor loadings; let $X = \{x_{ijk}\}$ be the $(N\bar{J} \times K)$ matrix of factor scores; and let $U = \{u_{ijl} + \epsilon_{ijl}\}$ be the $(N\bar{J} \times L)$ matrix of unique components of the scales. Remember that both the y_{ijl}'s and the x_{ijk}'s are standardized, thus $(1/N) \sum_i \sum_j y_{ijl} = 0$ and $(1/N) \sum_i \sum_j x_{ijk} = 0$ and the corresponding variances are equal to 1.0.

First, we will derive the correlation matrix, R. The elements of R are

$$\text{cor}\,(y_l, y_m) = \left(\frac{1}{N}\right) \sum_i \sum_j (y_{ijl} \cdot y_{ijm})$$

since the y_{ijl}'s and y_{ijm}'s are standardized. Thus, in matrix notation $R = (1/N)\, Y'Y$ where the (′) indicates transpose. Note that the standardization makes the elements of the principal diagonal of R equal to 1.0.

Fundamental Theorem. Now we can write equation (9.2) in matrix notation. This gives:

(9.1A-1) $Y = XF' + U$

*Much of the derivation is based on Rummel (1970).

Substituting equation (9.1A-1) in the formula for R gives:

(9.1A-2) $$R = \left(\frac{1}{N}\right) (XF' + U)' (XF' + U)$$

Multiplying the matrix equation out gives:

(9.1A-3) $$R = \left(\frac{1}{N}\right) (FX'XF' + U'XF' + FX'U + U'U)$$

We want the factor scores uncorrelated, thus the correlation matrix, R_x, is set equal to the identity matrix, I. But by the same reasoning as above, $R_x = (1/N)X'X = I$. Thus, $X'X = NI$ in equation (9.1A-3). Furthermore, the common components, XF', are uncorrelated with the unique components, U. Thus, $FX'U = U'XF' = \varnothing$, where $\varnothing$ is a matrix of zeros. Finally, U is a matrix of unique components, thus $(1/N)U'U = U^2$ is an $(L \times L)$ diagonal matrix. That is, the off diagonal elements are uncorrelated and hence zero, while the diagonal elements represent the unique variances of each component. Thus, rewriting equation (9.1A-3) with these substitutions gives:

(9.1A-4) $$R = FF' + U^2$$

Actually, we are more interested in the communalities, h_l^2, of each scale (that is the common variances) than the unique components. But since the y_{ijl}'s are standardized, $h_l^2 = 1 - u_l^2$ where u_l^2 is the unique component of the variance of scale l. Thus, the diagonal matrix of communalities, H^2, is given by $H^2 = I - U^2$. Making this final substitution in equation (9.1A-4) and rearranging terms gives the fundamental theorem of factor analysis:

(9.1A-5) $$FF' = R - I + H^2$$

Interpreting equation (9.1A-5), we can see that the factor loadings matrix "factors" the common matrix, $M = R - I + H^2$, which is obtained by replacing the principal diagonal, I, by the scale communalities, H^2.

One Solution. To determine F, we must first know H^2. We will cover that later, so let us for a moment assume that H^2 is known. Then the problem is to determine F such that it factors M. M is symmetric, and assume that M is positive semi-definite. (Note $Y'Y$ is positive semi-definite, as is any matrix of the form $A'A$.) Thus, its eigenvalues are non-negative, real numbers and its eigenvectors are orthogonal. (The eigenvectors of a symmetric matrix are orthogonal; the eigenvalues of a positive semi-definite matrix are non-negative and real.)

The eigenvalues, λ_l, of a matrix, M, are the roots of the characteristic equation obtained by setting the determinant of $|M - \lambda_l I|$ equal to zero. The eigenvector, e_l, corresponding to an eigenvalue, λ_l, is the vector that solves the equation $Me_l = \lambda_l e_l$.

Let Λ be the $(L \times L)$ diagonal matrix of eigenvalues, λ_l, and let $E = [e_1, e_2, ..., e_L]$ be the $(L \times L)$ matrix of L column eigenvectors. Note that because M is symmetric, E is orthogonal. We will select the eigenvectors such that they are orthonormal. Since the eigenvectors are orthogonal, $E' = E^{-1}$, and M can be factored by the eigenvector matrix, that is, $M = E\Lambda E'$. Since Λ is a diagonal matrix of non-negative numbers, we can write $\Lambda = \Lambda^{1/2}\,\Lambda^{1/2}$. Substituting these definitions and results in equation 9.1A-5 gives:

(9.1A-6) $$FF' = (E\Lambda^{1/2})\,(\Lambda^{1/2}E')$$

which gives us one potential solution to the fundamental equation (9.1A-5), except that F is $(L \times K)$ and E is $(L \times L)$.

In the text we assumed that the factor analysis identified a common structure that used fewer factors than there were attribute scales. That is, the number of factors, K, was less than the number of attribute scales, L. This means that $R - I + H^2$ is not full rank. Its rank is K, thus there are $L - K$ eigenvalues that are equal to zero. Thus, if we order the eigenvalues in order from largest to smallest, the last $L - K$ eigenvalues and hence the last $L - K$ columns of $E\Lambda^{1/2}$ will be zero. If we neglect these zero columns, then $F = E\Lambda^{1/2}$ is $L \times K$.

Thus, if we know the communalities, H^2, we can derive the "nonrotated" solution of factor analysis as the first K normalized eigenvectors (scaled by the eigenvalues) of the correlation matrix which is modified by replacing the principal diagonal by the communalities.

Determining the Communalities. The factor loadings, f_{lk}'s, are multiplicative factors for the factor scores, x_{ijk}'s, when given the attribute scales. Thus, f_{lk}^2 is the portion of the variance of the lth scale explained by the kth factor. Remember the factors are uncorrelated. Thus, the communality, h_l^2, is simply the sum of the squared factor loadings. That is:

(9.1A-7) $$h_l^2 = \sum_k f_{lk}^2$$

Thus, if we know the factor loadings, F, we can determine the communalities, H^2. But if we know the communalities, we can determine the factor loadings. This circular problem is known as the "problem of communalities." Basically, the empirical solution in most computer packages is to estimate the communalities and then iterate from equation (9.1A-5) through (9.1A-7) until successive communality estimates converge on a stable value.

Percent of Variance. In the text we quoted two statistical tests to select the number of factors: (1) the scree rule on percent of variance explained and (2) the eigenvalue greater than one rule. The total variance, V_T, of L standardized variables is L. The variance explained by a factor, V_k, is then obtained by summing the squared factor loadings for that factor: $V_k =$

$\sum_l f_{lk}^2$. But $F'F = \Lambda$, thus $\sum_l f_{lk}^2 = \lambda_k$. Then the percent of variance, P_k, explained by the kth factor is given by:

$$\text{(9.1A-8)} \qquad P_k = \frac{\sum_l f_{lk}^2}{V_T} = \frac{\lambda_k}{L}$$

Thus, we see how the eigenvalues are related to percent variance. The "eigenvalue less than 1.0" rule comes from the intuition that a factor is of little use if it explains less than what a single attribute scale would explain. (There are also numerous mathematical derivations of this cutoff.)

Finally, some programs report percent of common variance. The common variance, V_C, is the percent of variance explainable by the factors. From the definition of communalities, we get $V_C = \sum_l h_l^2$. From equation (9.1A-8) and from equation (9.1A-6), we can then show that $\sum_l h_l^2 = \sum_k \lambda_k$, thus the sum of eigenvalues for the retained factors gives the common variance.

Rotation. Equation (9.1A-6) gives one solution to the fundamental equation. Actually, there are an infinite number of solutions. What we want to do is to select the one solution that gives the best managerial insight. For example, in Figure (9.1A-1(a) the first two factor loadings, f_{l1} and f_{l2}, are plotted for each attribute, y_l. Managerially, Figure 9.1A-1(a) is difficult to interpret, but if we rotate the factor dimensions as in Figure 9.1A-1(b), then it is clear that the first dimension correlates with attributes 1, 2, 3, 9, and 10 while the second dimension correlates with attributes 4, 5, 6, 7 and 8. Remembering analytic geometry, if we rotate through an angle θ, then

$$f_{l1}^* = f_{l1} \cos\theta + f_{l2} \sin\theta \quad \text{and} \quad f_{l2}^* = -f_{l1} \sin\theta + f_{l2} \cos\theta$$

Generalizing to K dimensions, we see that rotation is just a similarity transform on F. That is, $F^* = FT$ where T is a $(K \times K)$ linear transformation and $TT' = I$. Note that $(F^*)(F^*)' = FTT'F'$, thus F^* is also a solution to equation (9.1A-5) and hence a possible factor solution. All that remains is to select the transformation, T, to maximize interpretability.

Managerially, we want the rotated matrix, F^*, to be in a form of simple structure, that is, we want each attribute to load on only one or a few factors. The criteria generally used to achieve this structure is called "varimax rotation," which attempts to maximize the variance of the columns of the squared factor matrix since the highest variance is attained when each attribute is correlated with only one factor. The criteria for varimax rotation is given by:

$$\text{(9.1A-9)} \qquad \text{Variance} = (L) \sum_k \sum_l \left(\frac{f_{lk}}{h_l}\right)^4 - \sum_k \left(\sum_l \frac{f_{lk}^2}{h_l^2}\right)^2$$

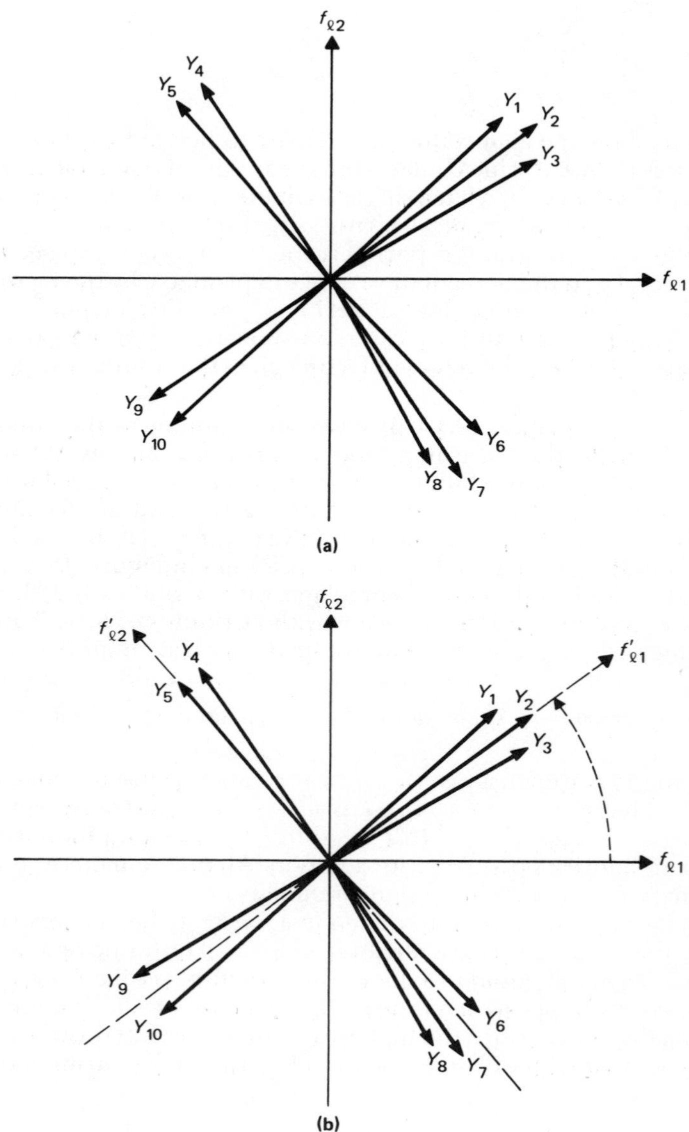

Figure 9.1A-1. *Rotation of Common Factors*

Estimating Factor Scores. From the text we want to estimate the factor scores, $\hat{X}$, from the attributes, Y, by a regression-like equation, $\hat{X} = Y\hat{B}$, where $\hat{B}$ is the estimates of the factor score coefficients (equation 9.3). At first glance, we would like to get $\hat{B}$ by inverting equation (9.1A-1), but remember that F' is not full rank. Therefore, we must estimate $\hat{B}$.

To obtain these estimates, multiply both sides of $\hat{X} = Y\hat{B}$ by Y' and divide by N. This gives:

(9.1A-10) $$\frac{1}{N}Y'\hat{X} = \frac{1}{N}Y'Y\hat{B}$$

But $(1/N)Y'Y$ is just the correlation matrix, R, and $(1/N)Y'\hat{X}$ is just an estimate of the correlations of the attributes and the factors, in other words, F. Thus, equation (9.1A-10) can be rewritten as $F = R\hat{B}$. Since any empirical R is usually non-singular, we get $\hat{B} = R^{-1}F$ and thus:

(9.1A-11) $$\hat{X} = YR^{-1}F$$

gives us the regression estimates of the factor scores.

While quite easy to obtain, one must use equation (9.1A-11) with caution. Note that the communalities to determine F are estimates. Further, although R is nonsingular, it usually has a near-zero determinant. Thus, there can be large errors in the estimates, $\hat{X}$. In fact, Guttman (1955) shows that in many cases the estimates, $\hat{X}$, are not unique and alternative estimates can be highly uncorrelated. Intuitive tests for the accuracy of equation (9.1A-11) are: (1) do the factors explain significant variance and (2) is the determinant of R sufficiently large? If not, then you may wish to use composite estimates—selecting a group of variables to represent each factor, or work with principal components for factor score estimates. Note that with the principal components technique you use common factor analysis to search for explanation, then use principal components—if it gives similar interpretations—to obtain $\hat{X}$.

Principal Components. Common factor analysis is closest to the behavioral hypothesis (Figure 9.7) and is the best technique to search for underlying structure. But occasionally the factor score estimates from common factor analysis are unstable. In this case, we turn to principal components. In principal components the analysis is essentially the same, except the unique components, U^2, are assumed to be negligible, i.e., $H^2 \rightarrow I$. Thus, the fundamental theorem is $FF' = R$ and the factors are the rotated eigenvectors/eigenvalues of the correlation matrix (i.e., the "principal components" of that matrix). In this case, if the rank of F is L (i.e., $K = L$), then F is invertable. Thus, the factor score estimates are given by $\hat{X} = Y(F')^{-1}$. If the rank of F is less than L, then you extract only the first K non-trivial eigenvectors. In this case, $YF = \hat{X}F'F$, but $F'F = \Lambda^{1/2}E'E\Lambda^{1/2}$ is just the diagonal matrix of eigenvalues ($E'E = I$), so $\hat{X} = YF\Lambda^{-1}$.

9.2 Similarity Scaling: MDSCAL 5, PROFIT, and INDSCAL*

The text gave the intuition behind the various algorithms for similarity scaling. MDSCAL 5 and related algorithms are used to produce the perceptual map based on judged similarity among the products or services in a consumer's evoked set. PROFIT was then used to name these dimensions by determining directional cosines which related the similarity dimensions to measured attributes. Finally, INDSCAL provided an alternative to MDSCAL 5 in which the relative magnitudes of the similarity dimensions were individually scaled.

The purpose of this appendix is to expand upon the basic equations given in the text to provide the reader with a more complete understanding of the algorithms. Our emphasis is on the algorithms, not on the computer code used to implement these algorithms.

MDSCAL 5

Notationally, let $\tilde{x}_{jk}$ be the measure of product j's position on the kth dimension as produced by the algorithm. Let Δ_{jm} be the judged dissimilarity between product j and product m.

The object of MDSCAL 5 is then to represent the products in a K-dimensional space such that the Euclidean distance between pairs of products, δ_{jl}, best reproduces the rank order of the judged dissimilarities, Δ_{jm}, among products. Euclidean distance is measured by

$$\textbf{(9.2A-1)} \qquad \delta_{jm} = \left[\sum_{k=1}^{K} (\tilde{x}_{jk} - \tilde{x}_{mk})^2 \right]^{1/2}$$

The goodness-of-fit criteria is called "stress" and is given by one of two formulas:

$$\textbf{(9.2A-2)} \qquad \text{Stress (formula 1)} = \left[\frac{\sum_{j=1}^{J} \sum_{m=1}^{J} (\delta_{jm} - \hat{\delta}_{jm})^2}{\sum_{j=1}^{J} \sum_{m=1}^{J} \delta_{jm}^2} \right]^{1/2}$$

$$\textbf{(9.2A-3)} \qquad \text{Stress (formula 2)} = \left[\frac{\sum_{j=1}^{J} \sum_{m=1}^{J} (\delta_{jm} - \hat{\delta}_{jm})^2}{\sum_{j=1}^{J} \sum_{m=1}^{J} (\delta_{jm} - \bar{\delta})^2} \right]^{1/2}$$

*This appendix is adapted from appendices by Green and Rao (1972), and Green and Wind (1973).

Where $\bar{\delta}$ is the average distance and the $\hat{\delta}_{jm}$'s are a set of numerical values chosen to be as close to the δ_{jm} as possible but subject the constraint that the $\hat{\delta}_{jm}$'s are monotonically related to the Δ_{jm}'s. The $\hat{\delta}_{jm}$'s are outputs of a monotonic regression procedure (Kruskal, 1964). In equations (9.2A-2) and 9.2A-3), the summations are shown as being over all product pairs. MDSCAL 5 also accepts alternative data such as the lower triangular matrices in Table 9.9. In which case the summations would all be limited to the input data. Stress formula 2 is the default option in MDSCAL 5.

Stress, S, is minimized via the non-linear programming method of steepest descent. In steepest descent, the matrix of similarity dimensions, $\tilde{X} = \{\tilde{x}_{jk}\}$, is successively modified along the direction which would cause the most rapid decrease in stress. This direction is determined by the gradient matrix of first partial deviations. That is, $G = \{-\partial S/\partial \tilde{x}_{jk}\}$. The modification in $\tilde{X}$ involves a movement of magnitude α along this direction. This parameter α is the step size and determines how rapidly the algorithm converges. If α is too small, little gain will be made on each iteration. But a gradient is a local condition, so too large an α will violate the applicability of the gradient and cause the algorithm to oscillate. The selection of α is described in Kruskal's 1964 paper. Mathematically, the algorithm is:

Given $\tilde{X}_i$, the ith interation of $\tilde{X}$

1. Compute the gradient, G.
2. Select the step size, α_i.
3. Compute the "improved matrix," $\tilde{X}_{i+1} = \tilde{X}_i + \alpha_i G_i$.

Note that steepest descent leads to a local, not necessarily a global, minima and is thus sensitive to the starting configuration, $\tilde{X}_0$. In practice, $\tilde{X}_0$ is often chosen randomly and a number of alternative starting configurations are attempted to investigate the problem of local minima. Alternatively, the user may specify an a priori judgmental configuration as a starting solution.

These are the basic ideas behind MDSCAL 5. The actual algorithm includes options to use any Minkowski p metric as well as the Euclidean metric and has options to handle missing data, ties in dissimilarity, weighting of dissimilarity, alternative monotonic regressions, splitting of input data, and iteration controls. For more details, see Green and Rao (1972), Green and Wind (1973), and Kruskal (1964).

PROFIT

Let $\tilde{x}_{jk}$ be product positions. Let y_{ijl} be individual i's ratings of product j on attribute l. For a common space, we use $\tilde{x}_{jk}$ and $\bar{y}_{jl}$ which is the average of y_{ijl} over all individuals i, but for INDSCAL, we can use $\tilde{x}_{ijk} = \lambda_{ik}^{1/2}\, \tilde{x}_{jk}$ and y_{ijl}. For simplicity, we will treat the common space procedure.

The object of PROFIT is to determine directional cosines, D_{lk}, relating the kth similarity dimension to the lth attribute. There are two procedures, a linear and a non-linear regression.

In the linear case, we wish to minimize the sum of squared distances between the actual property vector, $\bar{y}_{jl}$, and its estimate, $\hat{y}_{jl}$, as computed from the projections of the similarity dimensions. For the lth attribute, this becomes:

$$\textbf{(9.2A-4)} \qquad \bar{y}_{jl} = \sum_{k=1}^{K} D_{lk}\tilde{x}_{jk} + \text{error}$$

Let $\tilde{X} = \{\tilde{x}_{jk}\}$ and let Y_l be the (row) vector $\{\bar{y}_{jl}\}$, then the directional cosine (row) vector, $D_l = \{D_{lk}\}$, is given by

$$\textbf{(9.2A-5)} \qquad D_l' = (\tilde{X}'\tilde{X})^{-1}\tilde{X}'Y_l'$$

Computing the directional cosines for all the attributes, $l = 1$ to L, then gives the matrix, $D = \{D_{lk}\}$ as:

$$\textbf{(9.2A-6)} \qquad D' = (\tilde{X}'\tilde{X})^{-1}\tilde{X}'Y'$$

In the non-linear case, the least squares criterion, $\sum_j(\bar{y}_{jl} - \hat{y}_{jl})^2$, where $\hat{y}_{jl} = \sum_k D_{lk}\tilde{x}_{jk}$, is replaced by a K (kappa) criterion which is a generalization of von Neumann's η^2 measure. This criterion is given by:

$$\textbf{(9.2A-7)} \qquad K = \left(\frac{1}{S^2}\right) \sum_{j \neq m}\sum w_{jm}(\hat{y}_{jl} - \hat{y}_{ml})^2$$

where

$$S^2 = \left(\frac{1}{n}\right) \sum_{j=l}^{J} (\hat{y}_{jl} - \bar{\hat{y}}_{jl})^2$$

is the sample variance of $\hat{y}_{jl}$ and w_{jm} is a weighting factor that is a monotonically decreasing function of the absolute difference between the attribute ratings of products j and m, i.e., $|y_{jl} - y_{ml}|$. Usually, $w_{jm} = 1/[(\bar{y}_{jl} - \bar{y}_{ml})^2 + \alpha]$ where α is a small constant.

The solution given by Green and Wind (1973) is that the directional cosines are given by the eigenvector corresponding to the smallest non-zero eigenvalue of $\tilde{X}'A\tilde{X}$ where the elements of $A = \{a_{jm}\}$ are defined by $a_{jm} = -w_{jm}$ for $j \neq m$ and $a_{jj} = \sum_{j \neq m} w_{jm}$.

For details of the algorithms, see Green and Wind (1973), Green and Rao (1972), and Carroll and Chang (1964).

INDSCAL

As in the text, let $\tilde{X} = \{\tilde{x}_{jk}\}$ be the common coordinates of the products. Let $\tilde{x}_{ijk} = \lambda_{ik}^{1/2}\tilde{x}_{jk}$ be the INDSCAL estimate of individual i's perception of the

jth product along the kth dimension. In other words, the $\lambda_{ik}^{1/2}$'s serve to stretch the common space to fit each individual. Then the modified Euclidean distance, δ_{jm}, between two products, j and m, is given by:

$$\textbf{(9.2A-8)} \qquad \delta_{ijm} = \left[\sum_{k=1}^{K} \lambda_{ik} (\tilde{x}_{jk} - \tilde{x}_{mk})^2 \right]^{1/2}$$

INDSCAL then assumes that the dissimilarity measure, Δ_{ijm}, provided by individual i, is linearly related to the modified Euclidean distance; that is,

$$\textbf{(9.2A-9)} \qquad \delta_{ijm} = a_i + b_i \Delta_{ijm}$$

where $b_i > 0$.

The first step is to obtain appropriate values for a_i and b_i. Since we are concerned with the relative scale of the map set $b_i = 1.0$. The additive constant, a_i, is then the smallest constant that guarantees that the triangle equality is satisfied for all triples of products. This value, a_i^0, is given by

$$\textbf{(9.2A-10)} \qquad a_i^0 = \max_{j,m,n} (\Delta_{ijm} - \Delta_{ijn} - \Delta_{inm}).$$

where the maximum is taken over all possible triples, j, m, and n, of products. Any constant larger than a_i^0 would suffice and there are alternative procedures (Torgerson, 1958), but this standard procedure is both conceptually and numerically the simplest.

Next, the estimated distances, $\hat{\delta}_{ijm} = a_i^0 + \Delta_{ijm}$, are transformed to the scalar product form based on the scalar product equality for exact squared Euclidean distances:

$$\textbf{(9.2A-11)} \qquad \sum_k \tilde{x}_{ijk} \tilde{x}_{imk} = -\frac{1}{2J}\left(J\delta_{ijm}^2 - \sum_j \delta_{ijm}^2 - \sum_m \delta_{ijm}^2 + \frac{1}{J}\sum_j \sum_m \delta_{ijm}^2 \right)$$

(For the algebraic derivation of this result, expand

$$\delta_{ijm}^2 = \sum_k (\tilde{x}_{ijk} - \tilde{x}_{imk})^2$$

under the assumption that $\sum_j \tilde{x}_{ijk} = 0$ for all k. Recognize and group terms. See Green and Wind, 1973, p. 322.)

For estimated distances, let the righthand side of equation (9.2A-11) be $\hat{b}_{ijm}$ which is obtained by double-centering the matrix whose entries are $\{-\frac{1}{2}\, \hat{\delta}_{ijm}^2\}$. For the INDSCAL model, the lefthand side of equation (9.2A-11) is given by:

$$\textbf{(9.2A-12)} \qquad \hat{b}_{ijm} = \sum_{k=1}^{K} \lambda_{ik} \tilde{x}_{jk} \tilde{x}_{mk}$$

Thus, the scalar product $\tilde{b}_{ijm}$, as computed by the individual weights and common space coordinates, is an estimate of the scalar product, $\hat{b}_{ijm}$, which is also computed based on the estimated distances. Thus,

$$\hat{b}_{ijm} = \sum_{k=1}^{K} \lambda_{ik} \tilde{x}_{jk} \tilde{x}_{jm} + \text{error} \tag{9.2A-13}$$

and the elements λ_{ik}, $\tilde{x}_{jk}$ can be computed by least squares estimates of the trilinear form in equation (9.2A-13). This is achieved by a three-way canonical decomposition of the $\hat{b}_{ijm}$ matrix (Carroll and Chang, 1970).

For details of the algorithms, see Green and Wind (1973), Green and Rao (1972), and Carroll and Chang (1970).

Chapter 10

Product Positioning: Preference Analysis and Benefit Segmentation

Perceptual maps define the dimensions, the positions of existing bounds, and gaps where brands do not now exist. But maps do not tell us the best point to position a new product. To further develop the core benefit proposition (CBP), we use consumer preferences to identify the best positioning.

THE ROLE OF PREFERENCE IN PRODUCT POSITIONING

The perceptual map in Figure 9.1 identified new product opportunities for pain relievers as gaps in the very gentle/highly effective and very gentle/low effective quadrants. The new product manager must make a decision on where to place his core benefit proposition. The right choice could lead to a highly successful new product while the wrong choice could lead to a financial disaster. Analysis of consumer preferences gives the manager the information that is necessary to make this choice. For example, suppose that analysis of consumer preferences indicates positive value for the attributes of gentleness and effectiveness, with relatively higher importance placed on gentleness by consumers. These indications would lead to the development of a new product that stressed gentleness but performed well on effectiveness. But if the analysis indicated that different groups of consumers valued the dimensions differently, it would have been appropriate to segment the target groups in terms of the benefits they perceived and perhaps introduce a separate new product for each group. For example, if one group valued effectiveness five times more than gentleness and another valued gentleness five times more than effectiveness, two segments would be defined.

This segmentation of target groups, called "benefit segmentation" (Haley, 1968), is based on consumers' importances of the perceptual dimensions or product features. This segmentation approach is somewhat different from the methods discussed in Chapter 5 to define macro target groups. Benefit segmentation occurs while the target market is being analyzed and is a segmentation within that market to determine if more than one product should be offered to best meet the consumer potential.

In order to make these concepts more clear, consider some examples.

1. New long-distance communications methods are perceived relative to efficacy and ease-of-use, but where in Figure 9.5 should a new product be positioned to best attract consumers?
2. Consumers evaluate health care delivery systems with respect to quality, personalness, convenience, and value. But is the best opportunity to emphasize quality (at a premium price), or value (with adequate quality)? Both positionings may represent gaps in the market, and each positioning emphasizes one of the important dimensions, but a health care manager needs to know which of the services will best achieve his objectives.
3. Both variety of shops and parking are potentially important in the consumer selection of a shopping center. But in a given land area, more parking space means fewer stores and hence less variety. How can the shopping center manager choose between these competing objectives?
4. There is a real need for a high-speed, low-cost, easy-access, safe, and reliable transportation system. Any such system would achieve a high level of ridership. Realistically, how can a transit authority, such as San Francisco's BART, select a best system within technology, operating, labor, political, and legal constraints? If a choice must be made between high-speed and comfort, how should the transit authority decide?

If for any pair of feasible and equal cost product positions we could identify the one position that will attract the most consumers, we would select that positioning. If for any potential product position we could estimate the consumer demand, we would compute potential profit and select the best positioning. Furthermore, if we could understand the consumer choice process, we would know how to direct design efforts to find that best positioning. Preference analysis addresses these problems.

This chapter deals with techniques to select the best positioning of a product, the best physical features to achieve that positioning, and which one product or set of products is best to cover the market. First, we consider the need for understanding preferences. Then, various methods of measuring and analyzing preferences are presented, and procedures for benefit segmentation are discussed. The chapter closes with managerial issues in the use of the preference analysis and segmentation techniques.

PROACTIVE PRODUCT POSITIONING

If we have a map such as the one for pain relievers shown in Figure 9.1, how do we select the best positioning for a new product? One obvious method is to experiment with several concepts by having consumers evaluate them. Concept descriptions can be written for each potential position and tested. Although this technique will often work, it is not the most effective. This shotgun approach requires that a test be run for each new set of concepts, so that even if each individual test is relatively inexpensive, the entire process could be costly in both time and dollars. But more critically, this trial and error technique does not generate any managerial diagnostics to direct the development process. A concept may fail, be discarded, and management will never know whether an opportunity was missed because a small change could have made the concept a success. After twenty concept tests, a successful concept may not be found, and experiments would have to be begun again. Finally, even if a successful concept is found, we do not know if it is the best concept or how it might be improved.

An alternative to the trial and error technique is to direct the search via a preference analysis which measures and models consumer response to alternative product positions. By measuring consumers' "preferences" with respect to products represented by either perceptual dimensions or physical attributes and by representing the consumers' "preference process" with a mathematical model, a manager can predict the potential success of any new product positioning. A manager can look at the structure and parameters of the preference model to better understand the consumer and to proactively design a product concept to satisfy consumer needs and desires.

By using the preference model as an evaluative tool, a manager can predict how many consumers will prefer each positioning and thus select among a number of alternative product positions. Thus, although the initial effort and investment is larger to do preference analysis than to do one concept test, the investment leads to an overall process that is both more effective in identifying high potential ideas and more efficient. The cost is lower since, on average, fewer concepts need to be directly tested with consumers.

The preference models also support benefit segmentation. By knowing what dimensions or attributes consumers most prefer, a manager can look for the relevant differences among groups of consumers. These groups then represent homogeneous segments of the population to which particular products can be directed. For example, the models might indicate that one group of health care consumers, say students, prefers adequate care for a reasonable price while another group, say professional staff, prefers high quality care at a premium price. Then the health care organization may do well to offer two plans — a bargain plan and a premium plan — and direct all promotion and communication at the appropriate segments.

We begin our study of preference analysis by presenting each of the models in enough detail for the reader to use that model for his planning process. Technical issues are addressed in the appendices.

ANALYTIC PREFERENCE MODELS AND ESTIMATION METHODS

The analytic method used to estimate consumer preference depends on the measures that can be taken and the structure of the model to be measured. We present several models that are useful to the new product process and critique their strengths and weaknesses. Each of these models is powerful when used correctly and there is a large amount of literature adapting each to a myriad of purposes. Once you are comfortable with the basic models, you may wish to adapt and combine the models to meet your special needs.

Expectancy Value Models

Perhaps the easiest technique to measure consumer preferences is to use a linear model. That is, the preference (p_{ij}) that individual i has for product j is given by:

(10.1) $$p_{ij} = w_{i1}y_{ij1} + w_{i2}y_{ij2} + \cdots + w_{il}y_{ijl} + \cdots + w_{iL}y_{ijL}$$

where w_{il} = the importance that individual i places on attribute l

y_{ijl} = individual i's perception of product j relative to attribute l. L is the number of attributes.

For example, y_{111} might be the first consumer's ($i = 1$) rating of telephone ($j = 1$) on "conveys all forms of information" ($l = 1$). The importance weight, w_{i1}, would then measure how important "conveys all forms of information" is to that consumer. Expectancy value models ask consumers to directly specify the importance weights and product perceptions. This model has been used extensively and remains one of the less expensive techniques to obtain a simple model of consumer preference. (See Wilkie and Pessemier, 1973, for a review of alternative measures and empirical studies.) This model can be used early in the design process to get an initial indication of attribute importances.

Methodology. For example, suppose that you are the city manager of a suburban community. You have a fixed transportation budget and want to consider improvements in your community's transportation service. Because your budget is fixed, you are limited in what you can do, so you want to know what aspects of transportation services are most important to consumers. One method is to assume the linear form and directly measure importances relative to a list of attributes identified as relevant to the consumer.

One such measurement is shown in Figure 10.1. The attributes are the first five attributes, out of a total of twenty-five, that were measured by the City Manager of Evanston, Illinois. Note that each attribute has an indicated directionality ("will get me places on time" rather than "on time performance"). Furthermore, the five-point scale begins with "of no importance" rather than "unimportant."

More detailed measures can be collected by paired comparison of scales in terms of their importance. Measures also can be collected in terms of "goodness" or "badness" of an attribute rather than their importance in overall satisfaction. In all the measures, care must be taken to avoid observing a spurious "halo" effect. This can occur if a consumer rates his most preferred product high on all scales and least preferred product low on all scales rather than accurately assessing each attribute independently (Beck-

We would like to know how *important* the following transportation characteristics are to you, when you select the means of travel to downtown Evanston. Please be sure to tell us *how important they are to you* and not how available they are.

HOW IMPORTANT IS HAVING A MEANS OF TRANSPORTATION?	*Of No Importance*	*Moderately Important*	*Important*	*Very Important*	*Extremely Important*
1. Which will always get me places I want to go on time.	[ ]	[ ]	[ ]	[ ]	[ ]
2. Which will *not* require me to schedule trips in advance.	[ ]	[ ]	[ ]	[ ]	[ ]
3. Which will allow me to relax while traveling.	[ ]	[ ]	[ ]	[ ]	[ ]
4. In which I will not be too hot or too cold during the trip.	[ ]	[ ]	[ ]	[ ]	[ ]
5. Which will *not* cause me to worry about being mugged or assaulted.	[ ]	[ ]	[ ]	[ ]	[ ]

Figure 10.1. *Examples of Preference Measures for Expectancy Values. (This figure represents the first 5 out of 25 scales.)*

with and Lehmann, 1975, 1976; Johannson, Maclachlan and Yalch, 1976). These biases can be minimized by: (1) using objective attributes, (2) being sure all attributes are represented, (3) asking respondents only to rate products they evoke as relevant to their purchase, or (4) with post hoc statistical analyses to estimate and remove the bias (Beckwith and Kubilius, 1978).

Further extensions involve additional measures of social attitudes and a willingness to comply with the expectations of others. While these additions, called social normative beliefs, personal normative beliefs, and extraneous events (Fishbein, 1972; Wicker, 1971), add predictive power to the model, we have not yet found them managerially useful to identify product positioning.

Managerial Diagnostics. In new product design, the expectancy value model is used to get an early indication of the key product attributes. This is done by comparing importances of attributes with maps of how well existing products and concepts do with respect to those attributes.

For example, Figure 10.2 gives a plot of consumers' evaluation of bus, car, and walk, as well as two concepts. The first is Personalized Premium Service (PPS)—a publicly operated shared-ride, mini-bus system. Consumers request service by telephone and are picked up at home. The second is a Budget Taxi Plan (BTP)—a privately operated shared-ride taxi system for a reduced fare. The average importances are shown in the right margin next to each attribute scale. For example, "can come and go as I wish," "available when needed," and "no long waits," are among the most important attributes. Based on this, we might investigate an increase in the frequency of bus service which would improve the position of buses on these attribute scales. We would make initial predictions with the expectancy value model to determine if such changes appeared to be worth future investigation.

Prediction of Consumer Response. The managerial diagnostics are based on averages within a consumer segment. We can also use the model, but at the level of the individual consumer, to predict consumer response for new products or changes in existing products.

Suppose that the City Manager wants to increase the frequency of bus service. As a rough estimation procedure, he might postulate that this would increase each consumer's average rating on "can come and go as I wish," "no long waits," and "available when needed" by 50 percent of the distance to PPS. Thus, if y_{ibl} is consumer i's rating of bus on scale l before the change and y_{ipl} is i's rating of PPS, then after the change,

$$y'_{ibl} = y_{ibl} + 0.50(y_{ipl} - y_{ibl})$$

for the attributes improved by the strategy ($l = 6, 12, 20$) while $y'_{ibl} = y_{ibl}$ for all other attributes ($l \neq 6, 12, 20$). We would estimate consumer i's preference with equation (10.1). For example, if consumer i's new ratings of bus

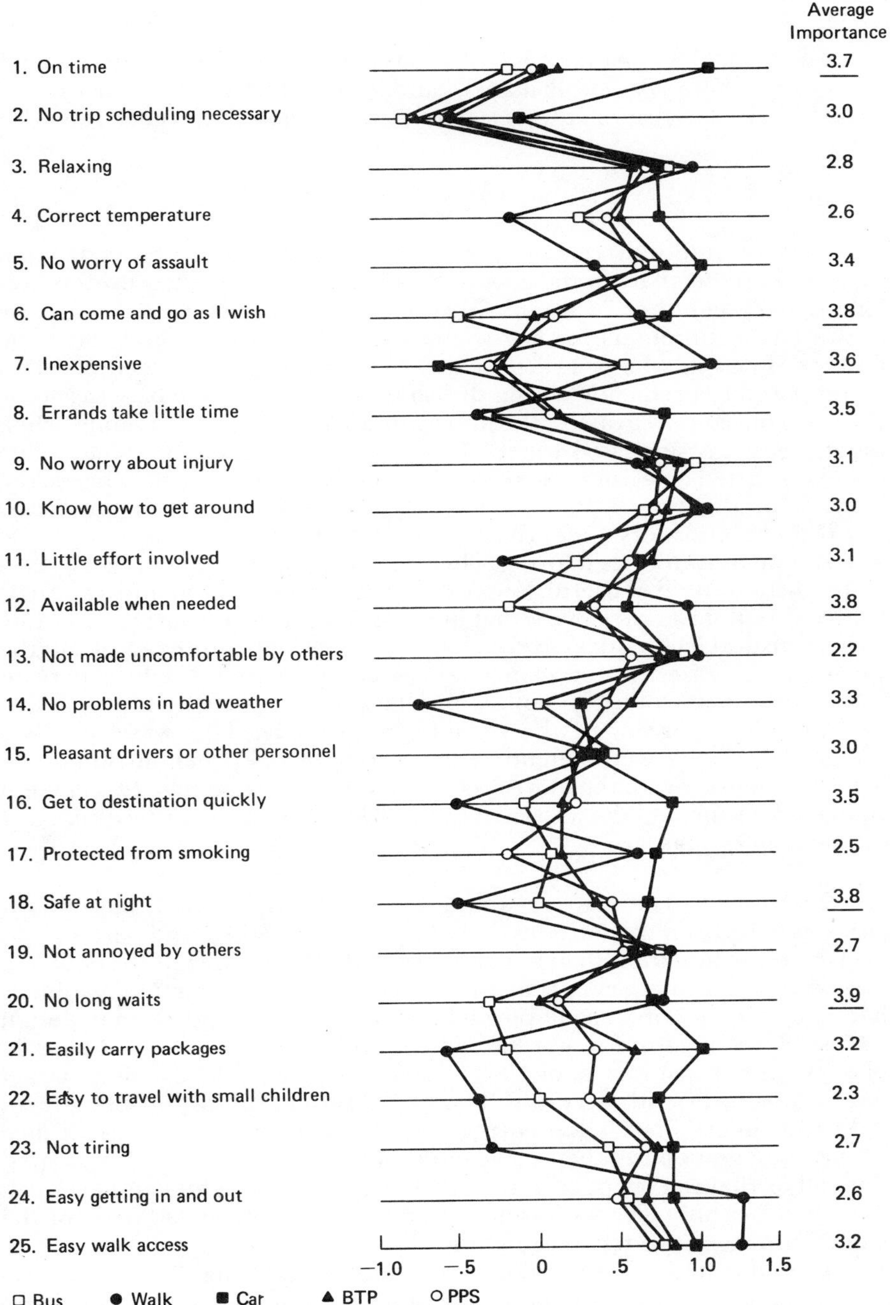

Figure 10.2. *Average Ratings on Transportation Service Attributes (adapted from Hauser, Tybout and Koppelman, 1979)*

were $y_{ib1} = -0.2$ for "on time," $y_{ib2} = -0.7$ for "no trip scheduling necessary," . . . , $y_{ib25} = 1.3$ for "easy walk access," and if his importance were $w_{i1} = 3.7$, $w_{i2} = 3.0, \ldots, w_{i25} = 3.2$, then we would compute i's preference value for bus by

$$p'_{ib} = (3.7)(-0.2) + (3.0)(-0.7) + \cdots + (3.2)(1.3).$$

Since car and walk remain unchanged, p_{ic} and p_{iw} remain unchanged. We would predict that i would choose the mode with the highest preference. That is, if $p'_{ib} > p_{ic} > p_{iw}$, i would ride the bus. To predict market share, we do this (via a computer) for every consumer in the sample and count the number of people who will ride the bus given the change. This would then be compared to predictions using the same technique on the base case (y_{ibl}, y_{icl}, y_{iwl} as measured) to assess the impact of the strategy. For example, when this strategy was simulated for the City of Evanston, the expectancy value model predicted a net increase of 28.3 percent in performance. That is, the base case prediction was 16.5 percent bus riders while the model predicted 21.2 percent bus riders under the improved strategy.

A note of caution is in order when the expectancy value model is used for prediction. It gives useful, low-cost first indications of the impact of new products, but our experience is that in many cases it is not nearly as accurate as the statistical techniques covered later in this section. For example, in the transportation case, the expectancy value was able to predict only 72 percent of the consumers correctly while a modification of preference regression predicted 80.8 percent of the consumers correctly. This was true even though the models were extended with social and personal attributes. In fact, later analyses with the more accurate model indicated that 28.3 percent change overestimated the impact of the strategy. A more realistic estimate is a 19 percent change.

Discussion. The above example was for transportation, but expectancy value has been used in a myriad of product categories and perhaps has received more attention than any other model of consumer preference. To use it in another category, write importance questions for a set of attributes that you have identified as important to the consumer and to your design process. Then measure importances in a random sample of consumers to serve as your "consumer pool." Test the model against the existing market to see if it is sufficiently accurate. The model should predict 40–60 percent of the consumers' first choices correctly and estimate market shares within a few percentage points. Then you can use it to explore potential strategies. Remember that the changes in the attributes are judgmental and, as such, the model can only give indications of potential, not exact estimates of the sales of improved product positioning.

The advantage of the expectancy value model is that it is easy to use and inexpensive. It gives an indication of what consumers feel is important on an attribute-by-attribute basis. Its disadvantage is that it is appropriate only for the attributes themselves. An expectancy value model for the perceptual dimensions derived from the attributes would require another round of surveys and there is no indication that the model could do as well as statistical methods which do not require a followup survey. Finally, since it is limited to the linear model, it is only appropriate in early design or in those categories where the consumer choice process is relatively simple, e.g., frequently purchased products. Nonetheless, the model has its managerial use, if used carefully, recognizing its limitations.

Preference Regression Methods

The expectancy value model uses importance weights "self-stated" by the consumer. There is no reason to believe that the weights the consumer gives in any way fit our mathematical model. They are useful because they indicate what the consumer believes to be important, but we cannot be assured that they capture the range and scale that we are using to mathematically represent the attributes or perceptual dimensions. One way to capture these effects is to statistically estimate the model that "best" reproduces observed preference. Such a model can produce the relative importances of the perceptual dimensions that are necessary for the new product team to develop the core benefit proposition and choose a target position in the perceptual space. Preference regression is a statistical technique which produces reasonably accurate "average" importances of the dimensions of the perceptual map.

For example, consider the pain reliever case. From the perceptual map we have already identified that effectiveness and gentleness are the relevant perceptual dimensions. Where is the best position? Is it better to concentrate on effectiveness or on gentleness? To answer this, a new product team needs to know the "average" relative weights of the dimensions for each relevant consumer segment.

Methodology. Preference regression uses a random sample of consumers to "fit" weights to best predict observed preferences. The first step is to measure preferences in the same survey as the basic attributes are measured. Figure 10.3 gives an example of this measurement for rank order preference. Let r_{ij} be individual i's rank of product j. In Figure 10.3, $r_{i,\text{Excedrin}}$ would equal 1.0 and $r_{i,\text{Tylenol}}$ would equal 5.0. We now use a linear model similar to equation (10.1) to represent preference. There are two differences. First, we require that the importance weights be the same for all consumers. Second, we use the perceptual dimensions (e.g., effectiveness and gentleness) rather than the attribute ratings to measure consumer perceptions. Then if x_{ijk} is i's perception of product j along dimension k, the preference model is given as follows:

$$p_{ij} = \sum_{k=1}^{K} w_k x_{ijk} \tag{10.2}$$

We are interested in your preferences for alternative pain relievers. Below are listed a number of products. Place a "1" next to the product you prefer. Place a "2" next to your next most preferred product. Place a "3" next to your third preference, a "4" next to your fourth preference, and a "5" next to the product you least prefer. Be sure to rate all the listed products.

3 Anacin
4 Bayer
2 Bufferin
1 Excedrin
5 Tylenol

Figure 10.3. *Example of Rank Order Preference Measurement for Preference Regression: r_{ij} Values*

For example if the importance weight for effectiveness were 0.35 and the importance weight for gentleness were 0.65, then preference is given by

$$p_{ij} = (0.35) * \text{effectiveness} + (0.65) * \text{gentleness}.$$

Ideally, we would want to select $w_1, w_2, \ldots, w_K$ such that p_{ij} is the largest of i's preference values whenever $r_{ij} = 1.0$. In the early 1970s, a number of techniques were developed to estimate the weights that do this with the least errors. These techniques include monanova (Green and Wind, 1973), monotonic regression (Johnson, 1975), and maximum score (Manski, 1975). Recent empirical and simulation results (Green, 1975; Hauser and Urban, 1977; Cattin and Wittink, 1976; Carmone, Green, and Jain, 1978) have shown that simple regression is sufficiently robust so that in most cases it can be used in place of the more expensive, more difficult to use, less available monotonic techniques.

To use preference regression, it is useful to rescale the rank orders so a high number represents first preference and a low number reflects the least preferred product. A convenient way to do this is to subtract the original ranking from the number of products ($r'_{ij} = J - r_{ij}$). In the case of pain relievers, five alternatives are present so the rank would be subtracted from 5 to yield, for the case in Figure 10.3, 4 for first preference Excedrin, 3 for Bufferin, 2 for Anacin, 1 for Bayer, and 0 for Tylenol. Then run an ordinary regression (or monotonic procedure if you have it) with r'_{ij} as the dependent variable and the x_{ijk}'s as the explanatory variables. The resulting estimates (β weights) are the importance weights. The constant does not affect the importances and can be ignored. Because there are usually insufficient degrees of freedom at the individual level, this regression is run across individuals and products. That is, for the pain reliever case, the input data matrix would be similar to that in Table 10.1; the outputs would be the β-weights shown in parentheses at the bottom of the table. If the evoked set varies as it does in Table 10.1 we still assign r'_{ij} to the evoked products.

One method of running the regression is to use a program called PREFMAP (Carroll and Chang, 1967; Carroll, 1972). It can do either a monotonic or metric regression of attributes versus preferences for individuals or on the basis of the average ratings and preferences of the total sample. If each individual has rated many products (say eight or more), individual analysis is useful, but in cases where respondents rate only an evoked set of alternatives, individual analysis would not be appropriate. In the latter cases, a regression across stimuli and individuals (see Table 10.1) would be the best procedure. PREFMAP on average ratings and preferences unnecessarily sacrifices degrees of freedom and cannot be as accurate as an equivalent regression across consumers.

In both conventional regression and PREFMAP more complex models can be estimated by the addition of nonlinear and interaction terms. PREFMAP does this by allowing four phases which process linear (phase IV), nonlinear (phases III and II), and interaction models (phase I). PREFMAP has a useful procedure which transforms the coefficients of the nonlinear regressions into co-ordinates of "ideal points." These points are chosen so that the squared distance from the ideal point to the products on the map best recovers the original preference judgments. These "ideal" points are useful in communicating with management, but keep in mind they are hypothetical points specified to represent the relative importance of attributes. In most cases, PREFMAP and regression procedures statistically cannot justify more complex models and utilize the linear model to describe preference. In this case the "ideal" is a vector whose slope is the ratio of the attribute importances. Figure 10.4 shows the "ideal" vector for the pain reliever case. The slope of this vector is the ratio of the importance weights. In Figure 10.4 the slope is (0.65)/(0.35) which is the ratio of the importance weights in Table 10.1.

Managerial Diagnostics. The ideal vector in Figure 10.4 indicates that gentleness is more important than effectiveness for pain relievers, but effectiveness has some importance (35 percent) and cannot be neglected. Thus, subject to cost considerations, the manager should try to position as far out along the ideal vector as possible. The ideal vector (or ideal point) indicates the gap that has highest priority. When evaluating alternative positionings we choose the product concept that fills the indicated gap. For example, new product 1 (*) nicely fills the gap in Figure 10.4 while new product 2 (*o*) does not fill the gap as well. Used in this way, the relative importance weights indicate which dimensions to stress in the core benefit proposition.

Prediction of Response to Product Revision. Prediction with preference regression is similar to that for expectancy values, except:(1) the manager judgmentally modifies the perceptions directly or indirectly by

Table 10.1. Input Data Matrix for Preference Regression (Estimated Importance Weights Are Output)

			Preference (r'_{ij})	*Effectiveness* (x_{ij1})	*Gentleness* (x_{ij2})
Individual 1:	Anacin	($j = 1$)	2	−0.31	−0.50
	Bayer	($j = 2$)	1	−0.40	−0.51
	Bufferin	($j = 3$)	3	0.09	−0.09
	Excedrin	($j = 4$)	4	0.39	−0.11
	Tylenol	($j = 5$)	0	−0.99	0.10
Individual 2:	Anacin	($j = 1$)	3	−0.41	−0.40
	Bufferin	($j = 3$)	4	0.19	0.12
	Excedrin	($j = 4$)	2	0.21	−0.93
		.			
		.			
		.			
Individual N:	Bayer	($j = 2$)	1	−1.1	−0.2
	Bufferin	($j = 3$)	3	0.11	0.9
	Excedrin	($j = 4$)	2	0.10	−0.3
	Tylenol	($j = 5$)	4	0.11	1.3
Estimated Importance Weights:				(0.35)	(0.65)

modifying the attributes, and (2) equation (10.2) or its PREFMAP equivalent is used for prediction. For example, if a new product was positioned at the starred (*) position in Figure 10.4 we would estimate each consumer's perception of that new product (perhaps through a modified concept statement) and use equation (10.2) to compute each consumer's preference values for the new product. These would be compared to the preference values for existing products and each consumer would be assigned to his

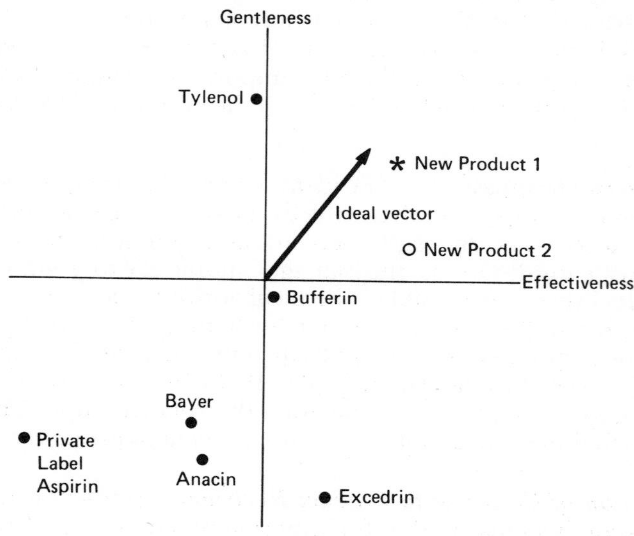

Figure 10.4. *"Ideal" Vector for Pain Relievers*

most preferred product. Suppose individual 2 in Table 10.1 perceived the new product as 0.40 on effectiveness and 0.20 on gentleness. His preference value for the new product would be $p_{2,\text{new}} = (0.35)(0.40) + (0.65)(0.20) = 0.27$. Comparing this to the values for Anacin ($p_{21} = -0.40$), Bufferin ($p_{23} = 0.14$), and Excedrin ($p_{24} = -0.53$), we would predict that he would prefer the new product. To get an overall prediction we would do the calculations (via computer) for each individual and count up the proportion of individuals preferring the new product.

Discussion. The above example is for pain relievers. This same technique has been used successfully for services, consumer durables, industrial products, and other frequently purchased consumer products.

To use it for a product category, simply measure and transform rank orders, compute the product positions on the map, and form a data matrix similar to Table 10.1. Regression is run across products and individuals with the transformed ranks as the dependent variable and the factor scores or similarity dimensions as the explanatory variables. The estimated β-weights are then the importance weights. Since R^2 from the regression is not an appropriate measure for rank order data, preference regression is tested on its ability to predict observed consumer preference. Thus, use equation (10.2) to compute preference values for the existing products and (via computer) count how many consumers rank first the product with the highest preference value. Models should predict 40–80 percent of the consumers correctly. You can also compare market shares by counting up the number of consumers that the model predicts will prefer each product. This should be within a few percentage points of the observed preference shares. You can then use the model to explore the potential product positionings and product strategies. Remember that the changes in the positionings are based on judgment and as such the model can give reasonable indications, but not exact figures.

One advantage of preference regression is that it is relatively easy to use and, because most computer installations have canned programs for regression, the tools for the technique are readily available. A more important advantage is that it can be used to estimate the importance of perceptual dimensions without a followup survey. This makes it an extremely important tool for early design when the new product team is setting the product positioning. In fact, the initial core benefit proposition is often generated based on emphasizing those dimensions that preference regression identifies as most important. For prediction, it is more accurate than expectancy values, but not as accurate in estimating final shares as models based on data collected later in the design process when the final product is actually tested with consumers.

One disadvantage of preference regression is that it is usually based on the linear model and it is not the best model to measure the nonlinear effects of decreasing returns or risk aversion. Furthermore, it should not be used on the basic attributes because they are often so highly intercorrelated that the preference regression coefficients are unstable. In fact, preference regressions on basic attributes rarely predict significantly better and often predict worse than preference regressions on reduced dimensions. For example, in predicting preference among shopping centers, preference regression on factor scores correctly predicted preference 47–51 percent of the time while preference regression on basic attributes only predicted preference 40–41 percent of the time (Hauser and Koppelman, 1979).

A final disadvantage is that the estimated weights are "average" weights. While these are useful in early design, they present some problems in benefit segmentation, especially when importances of some of the consumer population are opposite to others. Techniques to overcome this problem are discussed in the section on benefit segmentation.

Conjoint Analysis—Selection of Features

Once the perceptual positioning is identified, the new product team must select product characteristics to achieve that position. For example, a target position for a new deodorant might be that it "goes on dry, keeps you dry." To achieve this position we must determine what physical product characteristics (chemical ingredients, scent, powder, stick, roll-on, or spray, etc.) best achieve "goes on dry, keeps you dry" and, further, what other characteristics (size of can, price, package, etc.) are most preferred. One conceptually simple method to do this would be to try all combinations and see which one is most preferred. This is basically what conjoint analysis does, except that only a fraction of the feasible products need be considered and a preference model is measured to help managers understand why consumers prefer the combination they do.

Conjoint analysis is used in the intermediate design process to study the linkage of features to preference, and linkage of features to perception. Figure 10.5 shows the relationships of features, perceptions, and preference. Expectancy value models and preference regression are used to link perceptions to preference and conjoint analysis is used to study the linkage of features to perceptions and preference. The method has been used successfully in consumer durables; in frequently purchased consumer products; in industrial products such as electric generators, solar air conditioners, and computers; and in major consumer services such as education, airline services, and health care.

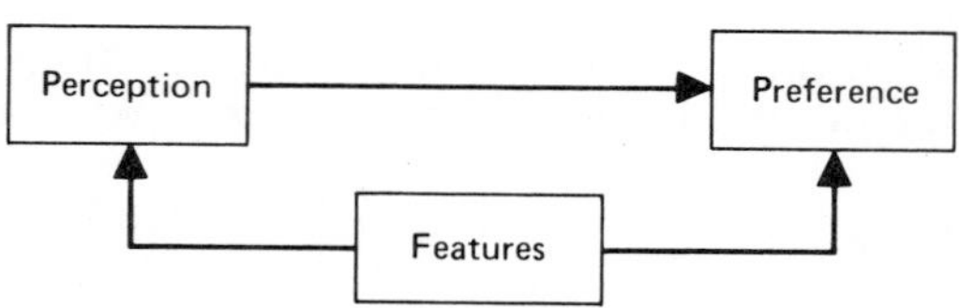

Figure 10.5. *Features, Perception, and Preference*

Methodology. In Chapter 9 we used factor analysis to identify the perceptual dimensions that consumers use to evaluate communications products. In that case, we found that the potential stretcher concept, NBVT (Narrow Band Video Telephone), did not fill the perceptual gap with respect to the dimensions of effectiveness and ease of use. Subsequent analysis with preference regression revealed importance weights of 0.57 for effectiveness and 0.43 for ease of use, thus we want to improve NBVT so that it moves out along the ideal vector to fill the gap between telephone and personal visit (see Figure 10.6).

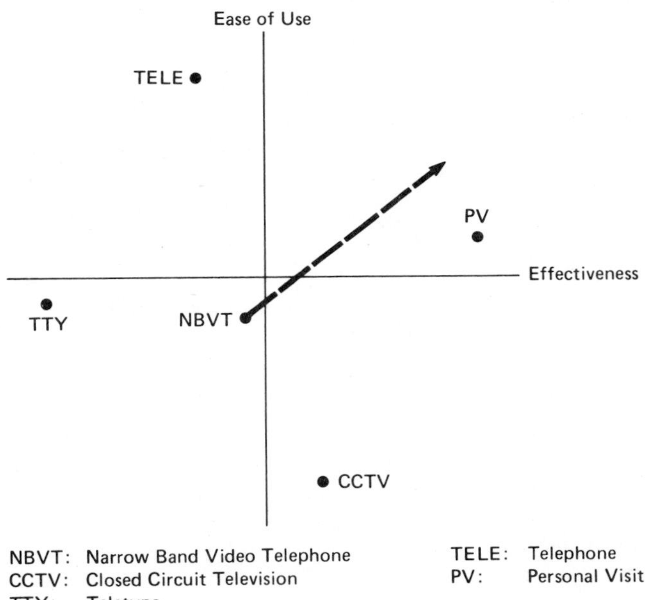

Figure 10.6. *Needed Perceptual Improvement*

The problem remains to set the features of the new product to achieve this needed improvement. For example, should we increase the resolution of the NBVT picture, decrease the transmission time, increase its accessibility to users, or make hard copy available? Table 10.2 lists these features. Note that there are $2 \times 2 \times 2 \times 3 = 24$ possible products, each of which can be described by a particular combination of characteristics as is illustrated in Figure 10.7. Notationally let us number these products $m = 1, 2, 3, \ldots, 24$. One conjoint task is to give each consumer twenty-four $3'' \times 5''$ cards, each of which describes a possible communications product, and ask each consumer to rank these cards in order of preference. Notationally, let R_{im} be the rank that individual i gives to product m. ($R_{im} = 1$ to 24.)

Table 10.2. Potential Product Features in Conjoint Design

	Resolution *(k = 1)*	*Accessibility* *(k = 2)*	*Hard Copy* *(k = 3)*	*Transmission Time* *(k = 4)*
Level 1 ($l = 1$)	Equal to Home TV	30 Minutes' Notice	None Available	30 Seconds
Level 2 ($l = 2$)	Four Times Home TV	Every Office Has One	Hard Copy Available	20 Seconds
Level 3 ($l = 3$)				10 Seconds

Conjoint analysis is then a mathematical technique to summarize the ranking information in a form that is useful to the manager. Since there is no guarantee of a linear relation between features, we use a nonlinear model. Since the features are only measured at discrete points, we use a piecewise linear function to approximate the decreasing (or increasing) returns "utility" function. Transforming the ranks, as was done in preference regression (for example, $R'_{im} = 24 - R_{im}$), we can, in theory, use a nonlinear regression to find the "utility" functions that best recover the consumer rankings. That is, set:

$$R'_{im} = u_{i1}\,(\text{resolution}) + u_{i2}\,(\text{accessibility}) + u_{i3}\,(\text{hard copy}) + u_{i4}\,(\text{transmission time}) \tag{10.3}$$

For example, u_{i1} (resolution = home TV) represents the contribution to the consumer of having the resolution of NBTV equal to that of his home television. Fortunately, we can use a "trick" to transform equation (10.3) into a linear regression. For nonlinear estimation, see Pekelman and Sen (1979).

To do the transformation, we introduce the concept of an indicator variable. This variable, d_{mkl}, is basically a switch which tells us whether product m has feature k, at level l. Formally, let

$$d_{mkl} = \begin{cases} 1 \text{ if product } m \text{ has feature } k \text{ at level } l \\ 0 \text{ otherwise} \end{cases}$$

For example, suppose Figure 10.7 is product 1 ($m = 1$) and suppose "Every office has one" is the second level ($l = 2$) of feature 2 ($k = 2$). Review Table 10.2. Then $d_{122} = 1.0$ and $d_{121} = 0.0$. Similarly for Figure 10.7, $d_{111} = d_{132} = d_{143} = 1$ and the rest of the d_{1kl}'s are equal to zero. Furthermore, let λ_{ikl} be the "utility" level that is appropriate to represent individual i's valuation of having the kth attribute at the lth level. For example, in Figure 10.8 the utility levels for accessibility are given by $\lambda_{i21} = 0.00$ and $\lambda_{i22} = 0.42$. We can now write the transformation:

$$u_{ik}(\text{feature } k \text{ for product } m) = \sum_{l=1}^{L} \lambda_{ikl}\, d_{mkl} \tag{10.4}$$

where L is the number of levels possible for feature k. Because d_{mkl} equals 1.0

NARROW BAND VIDEO TELEPHONE

Resolution: Equal to home TV

Accessibility: Every office has One

Hard Copy: Available

Transmission Time: 10 seconds

Figure 10.7. *An Example Product: One of the 24 Possible Combinations of Product Features.*

for only one level, l, this equation simply sets $u_{ik}(\cdot)$ equal to the utility that corresponds to the level of feature k that product m has. Expanding this transformation to equation (10.3) gives the conjoint estimation equation:

$$R'_{im} = \sum_{k=1}^{K} \sum_{l=1}^{L} \lambda_{ikl} d_{mkl} + \text{error} \tag{10.5}$$

where the summation is over the features ($k = 1, 2, \ldots, K$). Equation (10.5) is a linear model and the λ_{ikl}'s (called "part-worths") can be estimated with monotonic analysis of variance (Kruskal, 1965), regression, or linear programming (Srinivasan and Shocker, 1973). Figure 10.8 gives the part-worths for the potential features of NBVT. The most important feature is accessibility—more important than a resolution increase from that of a home TV to four times that of a home TV. Thus management should focus on accessibility rather than resolution.

Figure 10.8 represents the linkage of features to preference. The relationship of resolution, accessibility, hard copy, and transmission time to the perception of effectiveness and ease of use can also be studied with conjoint analysis. Combinations of features are rank ordered in terms of the consumers' perceptions of effectiveness (then ease of use) that would result from each profile. For example, when this was done for NBVT, hard copy had the greatest impact on effectiveness while accessibility had the greatest impact on ease of use. For other applications linking features to perceptions, see Neslin (1978) and Green and DeSarbo (1978). For a complete review of conjoint analysis see Green and Srinivasan (1978).

Managerial Diagnostics. Conjoint analysis is best used to select physical features for a product. The "utility" functions indicate how sensitive consumer perceptions and preference are to changes in product features. By examining the graphs of the utility functions, a manager can gain insight into which features to select for a new product. In general, the best features will be those that give the greatest gains in preference at the lowest cost. For example, Figure 10.9 gives the utility functions that Green and Wind (1975)

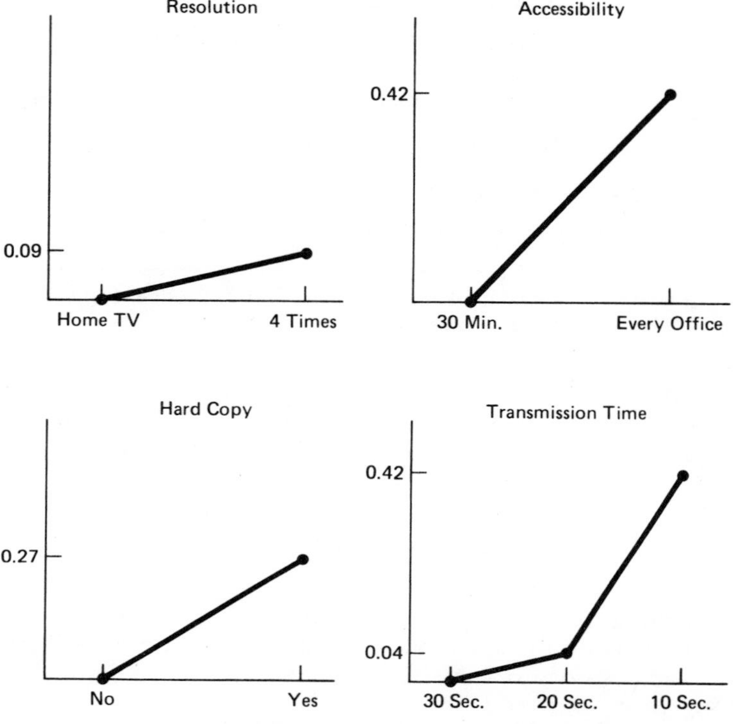

Figure 10.8. *Utility Functions for Product Features (adapted from Hauser and Simmie, 1979)*

estimated for airline service to Paris. In this study, the firm was interested in whether it was cost-effective to undertake extensive replacement of aircraft (B-747 instead of B-707). From Figure 10.9 it is clear that money is better spent on other improvements such as incentives to flight attendants, improved scheduling, and better entertainment. Similar interpretations apply when conjoint analysis is used to select features to substantiate perceptual positioning.

Prediction of New Feature Combinations. Once a consumer's preference function is known, prediction with conjoint analysis is not unlike that for preference regression and expectancy values. One first observes (in the case of physical features) the levels of the attributes for all existing products and computes the preference values by substituting these levels in equation (10.5) where the left side of equation (10.5) gives the preference (R'_{im}) for existing products. A potential new product feature profile is then selected and preference for it ($R'_{i,\text{new}}$) is computed for each consumer. Assuming each consumer will choose the product with the highest preference value, we can then count up the number of consumers who will choose the new

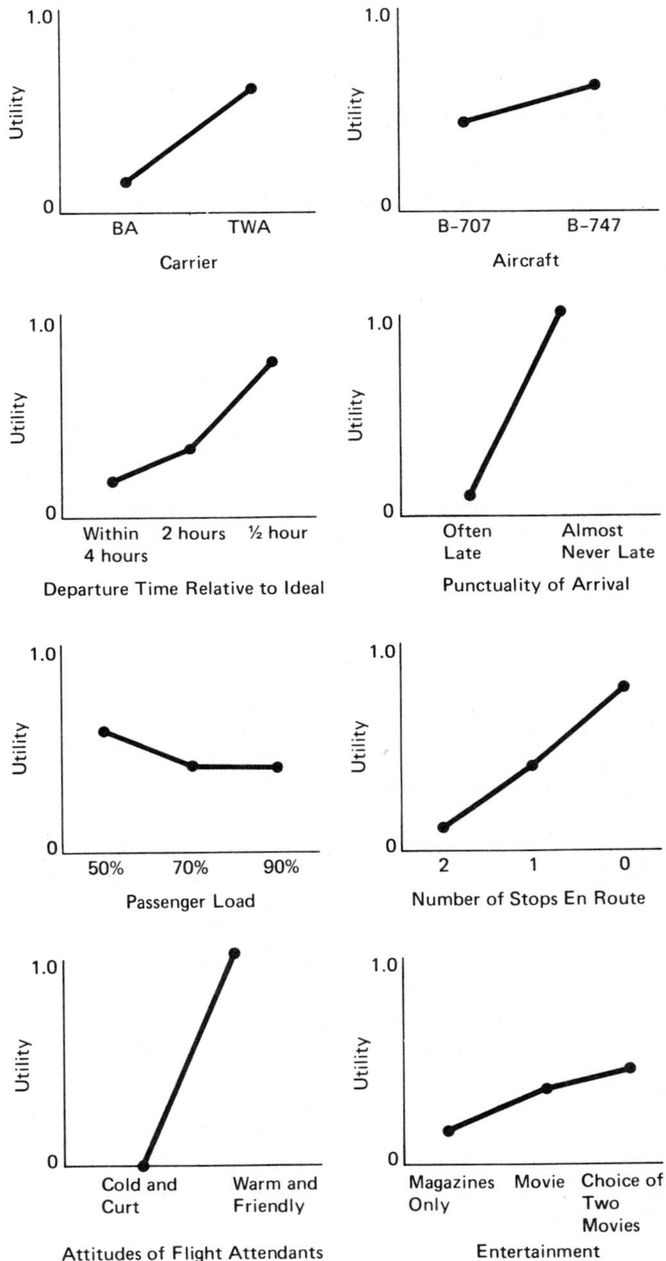

Figure 10.9. *Utility Functions for Features of Airline Service to Paris (Green and Wind, 1975)*

product. For example, in a study of carpooling preference, Market Facts (1976) used a form of conjoint analysis to test various regulatory and operating strategies designed to encourage carpooling. They found that a parking surcharge in the Chicago Loop would only raise carpooling by 0.1 of 1 percent whereas a policy such as exclusive lanes that would decrease riding time for carpools and increase riding time for single occupancy vehicles would raise carpooling by 4.9 percent. The latter policy would save over 140,000 gallons of fuel daily.

Discussion. The above examples are for telecommunication, airline service, and carpooling. These examples are but a few of the hundreds of product categories that are analyzed each year with conjoint analysis (Green and Srinivasan, 1978). Conjoint analysis is most appropriate when features need to be set for new products.

To use conjoint analysis in another category, select the features that need to be tested. These features are the result both of the perceptual analysis and decisions by the new product team in determining feasible features. Then prototypes can be made, pictures drawn, concepts written, or cards made up to represent at least part of the feasible combinations of product features. Consumers are asked to rank these products (sometimes known as pseudo-products), and equation (10.5) is used to estimate the "utility" functions. The results are then used to select features and predict consumer response.

The advantages of conjoint analysis are that it can deal with physical attributes, it can measure the more complex nonlinear compensatory preference model, and it can measure preferences for each individual consumer. One disadvantage is that it is limited to a relatively few variables because the number of potential products grows very fast as the number of features increases. For example, with 3 levels each and a full factorial design, 3 features would mean 27 products, 4 features would mean 81 products, and 5 features would mean 243 products—i.e., 3^K pseudo-products for K features. These numbers can be reduced somewhat by using fractional factorial designs (Green and Wind, 1975; Cochran and Cox, 1957), pairwise tradeoffs (Johnson, 1974), across individual designs, or intensity measures (Hauser and Shugan, 1980), but nonetheless the measurement becomes extremely tedious for a large number of features and levels. (These advanced techniques are discussed in Appendix 10.1). Another disadvantage of conjoint analysis is that it requires a special survey—usually a personal interview survey due to the complex consumer task.

Conjoint analysis is most appropriate after preference regression/perceptual analysis has been used to identify an initial core benefit positioning and critical product features. The added expense of the interviews can be justified when management feels confident in a category and is willing to probe further.

This completes the discussion of basic conjoint analysis. But because conjoint analysis is so important, a number of modifications have been made to: (1) ease the consumer task, (2) expand the capabilities to more variables,

(3) strengthen accuracy of measurement, (4) improve the estimation techniques, and (5) automate the interviews via computer. These topics are described in Appendix 10.1.

Direct Assessment of Consumer Utility

Conjoint analysis is one technique to measure the nonlinear preference models. Although continuous functions can be measured (Pekelman and Sen, 1979), the standard procedure requires that the "utility" functions be characterized by discrete values that can be separately estimated. This procedure often captures the decreasing returns aspects of the utility functions. In some applications, managers need to know about consumers' risk preferences. For example, risk is an important aspect of health care. A patient who is ill usually elects treatment and the consequences of that treatment could drastically affect the rest of his life. Preferences for the type of treatment will not only reflect the potential outcomes, but the likelihood of each outcome's occurrence. Similarly, in the purchase of durables such as refrigerators, washing machines, or cars, the consumer makes a major purchase once every 5–10 years and must live with that purchase for the interim, so risk of poor results can affect purchase. Industrial products (e.g., computers, airplanes), government programs (e.g., welfare reform, defense systems) and some consumer products (e.g., sun tan oil) also may exhibit risk aversion. To measure risk aversion, we need to know a little more about the shape of the utility function than conjoint analysis specifies. Utility theory provides the means to model and measure the risk aspects underlying the shape of the utility function. Since the technique is relatively new to new product design and since the mathematics can become complex, we refer the reader to Appendix 10.2 for an in-depth discussion of direct utility assessment.

Summary of Preference Analysis Methods

The use of the technical properties of the three basic methods of preference analysis are summarized in Table 10.3. As indicated in the introduction to this section, these methods are readily adaptable, and the properties in Table 10.3 may change somewhat depending upon use. What we have tried to do is summarize the methods as they are used in new product design. Each of these models plays an effective role in new product design. If cost is a constraint, expectancy value models offer an inexpensive way to get a rough idea of the linear effects of product attributes in formulating preferences. In defining the positioning of a new product for the core benefit proposition, preference regression excels in analyzing the importance of psychological dimensions used to define perceptual maps. Conjoint analysis is best when physical features of products are the focus of the design problem. Feature links to preference or perceptual dimensions help in selecting features to substantiate product positioning.

Table 10.3. Summary of the Preference Analysis Methods as They Are Used in New Product Design

Properties	*Expectancy Value*	*Preference Regression*	*Conjoint Analysis*
Underlying Theory	Psychology	Statistics	Mathematical Psychology
Functional Form	Linear	Linear and Non-Linear	Additive
Level of Aggregation	Individual	Group	Individual
Stimuli Presented to Respondent	Attribute Scales	Actual Alternatives or Concepts	Profiles of Attributes
Measures Taken	Attribute Importances	Attribute Ratings and Preference	Rank Order Preference
Estimation Method	Direct Consumer Input	Regression	Monotonic Analysis of Variance or Linear Programming
Use in New Product Design	Early Indications	Core Benefit Proposition	Selection of Product Features

The three techniques described here are a powerful set of methods for analyzing consumer preferences. A successful new product team knows how and when to use each of these methods. However, they are not rigid techniques and they can be done on perceived dimensions as well as features. Constant sum-paired comparisons can be used in conjoint analysis rather than rank ordering to compare profiles of features (see Appendix 10.1). Preference regression could be applied at the individual level if many (say 8) concepts were rated by consumers. The part-worths of conjoint analysis could be measured by fewer profiles (say 8) and then estimated at the group level by applying monotonic analysis of variance across individuals as well as profiles. The number of mixtures of measurement and statistical methods is large and the best method depends on the particular new product, the funds available for research, and the cognitive abilities of consumers. We have tried to outline the fundamentals of each approach. If you study the technical appendices, you will have the basic knowledge necessary to customize a preference analysis to investigate your particular new product decision.

This completes our exposition of the analytic methods of preference measurement. We now illustrate preference regression and conjoint analysis with a case.

Example: Management Education

Today's academic environment is highly competitive. Universities are continually trying to upgrade the curriculums to offer high quality programs. One of the target groups for universities is potential students. (Other target groups are faculty, alumni, current students, recruiters, government, and industry.) In 1975 the MIT Sloan School of Management undertook a study to improve the programs that they offered to graduate management students.

The first step was qualitative research, focus groups, to understand the market and to generate a set of attributes that potential students use to evaluate alternative programs. The attributes, such as "excellent teaching," are listed in Table 8.1. Questionnaires were then sent to 300 potential students who had indicated an interest in pursuing a career in management. Of these, 40 percent were returned.

Factor analysis was used to define and label three dimensions. The first dimension reflected the "outcomes" of an education program such as exciting job opportunities, high salary, membership in a worldwide alumni fraternity, prestige, and the perception that one "cannot go wrong with a degree from this program." The second dimension reflected a common-sense approach to problem-solving, real-world orientation, and relevancy of program content. This dimension was named "realism." The third dimension represented the "process" of education and was correlated to perceptions of friendly and cooperative students, high faculty contact, flexibility, and availability of financial aid. As shown in Figure 10.10a, Wharton and Harvard are high in the "outcomes" dimension, while the MIT Sloan School

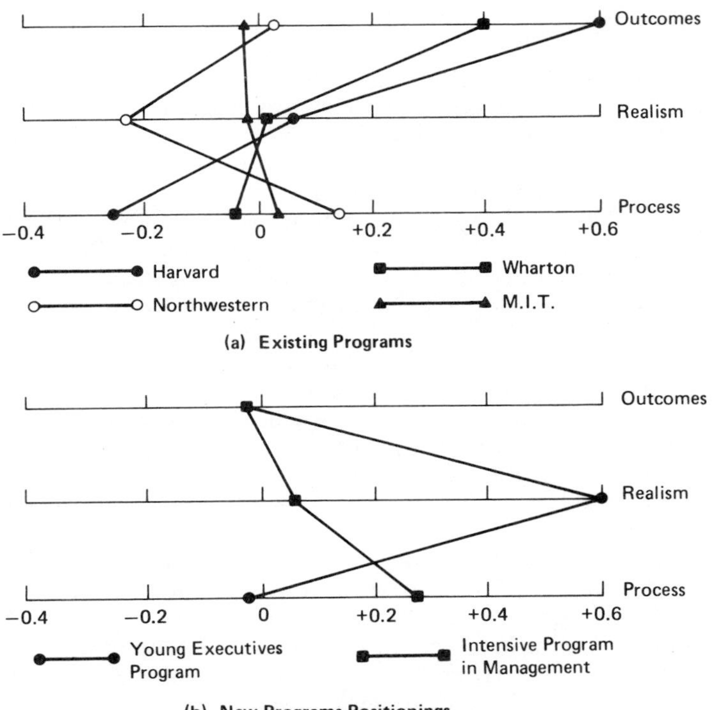

Figure 10.10. *Perceptual Maps for Management Education*

program and Northwestern are low. MIT's program was equal to others in "realism." On the "process" dimension Harvard was characterized by highly competitive students, while Sloan and Northwestern were viewed as having a friendly, cooperative student atmosphere.

In the same questionnaire potential students evaluated two new programs based on segmenting the market for a master's degree by providing one program for those interested in breadth along with a strong background in management disciplines, and a second program which allowed the student to gain in-depth competence in a particular area such as finance, economic analysis, or market research. Both were designed to facilitate career success, but by different paths. The graduate of the program stressing breadth would rise through a range of functional areas, while the graduate of the in-depth program would tend to rise through one particular functional area to become part of the top management team.

The first program, called the "Young Executives Program" (YEP), was characterized by: (1) intensive 12-month duration, (2) high emphasis on group problem solving and communication skills, (3) training in basic disciplines of economics, organization theory, quantitative methods, and information systems, (4) a two-week trip to visit leaders in business and government in New York and Washington, D.C., (5) a computer game exercise to serve as a thesis, and (6) broad exposure to law, culture, and the environment. The core benefit position was that this program provides the student with a variety of practical skills and builds confidence and competence for pursuit of a career leading to an influential management position.

The second program, called the "Intensive Program in Management" (IPM), was characterized by: (1) in-depth study through "learning cells" where three to six students work intensively with one faculty member in his area of concentration, (2) a summer job internship between the first and second years of the program, and (3) rigorous treatment of the underlying disciplines. Through in-depth study of the underlying disciplines and a specific area along with realistic experience, the student would be prepared to rapidly advance to the top of a functional area in an organization and be influential in policy and strategy formulation. Comparison of these two programs in Figure 10.10b with the positioning of existing programs shown in Figure 10.10a shows they are differentially positioned with respect to realism and process, but unchanged from the regular MIT program in terms of outcomes (jobs, salary, and prestige).

The segmented programs appear to fill a gap. To prioritize the gaps, preferences were analyzed with preference regression to reveal the relative importance of outcomes (0.65), realism (0.2), and process (0.15). Using these importances the impact of new programs was evaluated. Simulation of consumer preference indicated that the addition of the two programs improved the preference share for MIT from 11 percent to 37 percent with 87 percent of the respondents preferring the new concepts to the existing program. Thus the new concepts had potential which was investigated further.

Based on Figure 10.10b, the differential advantage of YEP is realism. To substantiate the realism positioning, a followup conjoint analysis was performed to link program features (two-week trip, computer game, intensive group sessions) to realism. As Figure 10.11 indicates, the trip was most important in establishing the realism position, followed in importance by the computer game. Other analyses indicated that a summer internship was a key feature for the IPM program, followed by flexibility in selecting a program.

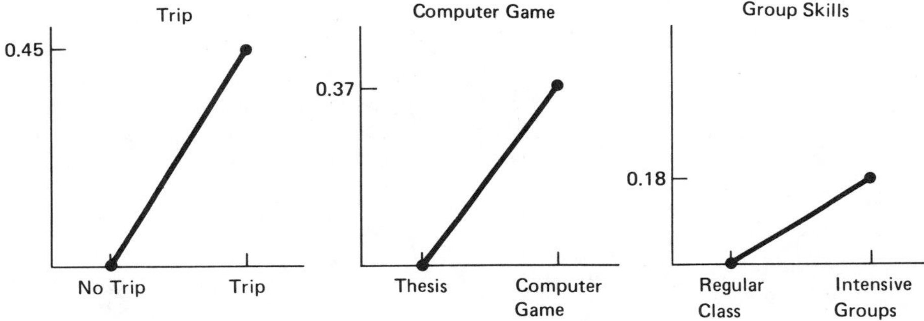

Figure 10.11. *Conjoint Analysis to Substantiate Realism for the Young Executive Program (adapted from Urban and Neslin, 1976)*

As a result of these analyses MIT moved to: (1) position a version of the YEP called the Accelerated Masters Program (AMP), and (2) improve the Sloan Masters Program (SMP). The brochure was revised to describe the new programs. Since MIT graduates commanded high salaries and since outcomes was the most important dimension, the brochure emphasized career advantages and starting salaries. At the same time the personalness of the process was better portrayed with pictures and testimonials. Realism was substantiated for the AMP with a computer game and a trip to visit business leaders. Summer job placement activities for the SMP was increased by 200 to 300 percent. In addition, students now have the option of a structured thesis within groups. Because conjoint analysis of preferences indicated that students would prefer the features even at a premium tuition, MIT now charges a premium fee to cover the cost of the added features.

The results have been significant. Applications are now up approximately 250 percent, acceptance rates have increased, and, in the judgment of the admissions committee, the average quality of the students has increased. Experimentation continues on increased specialization such as that suggested by the IPM concept.

BENEFIT SEGMENTATION

Importance weights such as those determined in preference regression are extremely useful to the new product manager because they succinctly

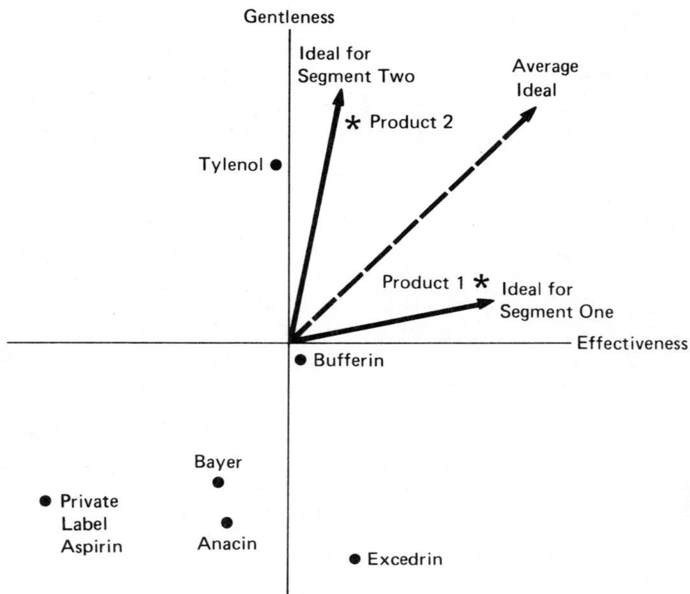

Figure 10.12. *Benefit Segmentation of Pain Relievers*

summarize consumer response and, with judicious use, they lead to improved design. For example, in the pain reliever example (Figure 10.4), the ideal direction suggested a new pain reliever that was both gentle and effective. But, if preferences are not homogeneous, average preference weights could lead to the wrong decision. For example, if two distinct segments are present as shown in Figure 10.12, a product that places equal emphasis on both gentleness and effectiveness may not be optimally positioned. A better strategy would be to introduce two products, each targeted at the appropriate segment. Product 1 would emphasize effective yet gentle while product 2 would be gentle yet effective. Another example is provided by considering the non-smoking feature of a transit system. If the number of smokers exactly balances the number of non-smokers, then the average preference weight for "smoking allowed" might be zero and a manager might conclude that smoking restrictions will have no effect on the demand for a commuter rail service. In fact, just the opposite is true for a major Chicago commuter railroad where both smokers and non-smokers favor segregation by means of smoking and non-smoking cars.

Approach

To avoid the fallacy of averages and to further direct the design of new products, we turn to benefit segmentation. Benefit segmentation is simply grouping consumers by their preferences as measured either by importance weights (in the linear case) or utility functions (in the nonlinear case). Consumers with homogeneous preference measures are a "benefit segment" and potential strategies might include different products for different segments or different communication and pricing plans.

In the cases where importance weights or their equivalent are measured at the level of the individual consumer, benefit segmentation is conceptually easy. Consumers are simply grouped by their measured importance weights. In the case of average importance weights, as in preference regression, the problem is more complex. We want to group consumers by importance weights, but we need to know the groups to estimate those weights. We begin with the case for individual measures and then turn to the case where a two-step procedure of search and test is required.

Clustering Individual Measures

The basic idea behind clustering individual importances can be seen in Figure 10.13. Suppose that a linear compensatory model has been measured for the two dimensions of pain relievers: effectiveness and gentleness. Cluster analysis then tries to group consumers who have similar relative importance weights. In Figure 10.13, cluster 1 appears to place a high premium on gentleness, while cluster 2 places a high premium on effectiveness. In this case, a manufacturer may wish to introduce two products—one that gives highly effective relief but is not necessarily gentle, and one that is very gentle but less effective. The manager may also wish to know about the demographic or psycho-social makeup of each segment so that advertising and promotion can be directed to the relevant consumers. One simple method might be to compute averages and standard deviations of the demographic

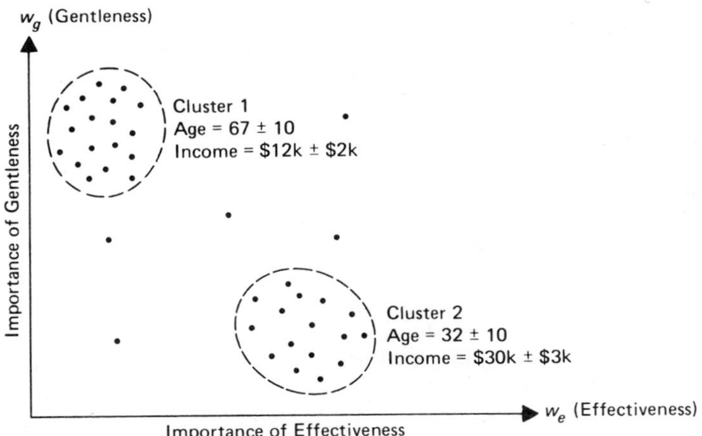

Figure 10.13. *Hypothetical Cluster Analysis to Identify Benefit Segments for Pain Relievers*

or psycho-social variables for each cluster and compare differences. The hypothetical example in Figure 10.13 would indicate that the high income young put a premium on effectiveness, while the low-income old are more concerned with gentleness. The manager would then orient the advertising and promotion accordingly.

The formal technique for cluster analysis is the same method that was used in Chapter 5 and discussed in Appendix 5.3. Dissimilarity between consumers is measured as the computed distance between the points (individuals) in importance-weight space. Individual consumers are then grouped to minimize the variation (variance) within clusters while maximizing the variation between clusters. The formal technique for finding demographic or psychosocial descriptors is discriminant analysis. The variables with the largest discriminant coefficients are the best discriminating variables among clusters.

The example in Figure 10.13 is for two variables in a linear compensatory model. The same basic technique is used with more than two variables and in the nonlinear models. Among the alternative variables that could be used in the clustering are: self-stated importance weights (w_{il}) in the expectancy value models and imputed part-worths (λ_{ikl}) in conjoint analysis.

The managerial significance of the clusters depends on the variables used. For importance weights or part-worths, the clusters indicate what dimensions or features to stress in the different products.

Search and Test

In preference regression, individual importance weights are usually not estimated. Thus, simple cluster analysis cannot be used and benefit segmentation must be performed by a two-step process. First, candidate segments are generated either by prior beliefs or by heuristic searches for preference indicators; then these segments are tested to make sure that: (1) the preference parameters are significantly different between segments and (2) the segmented model can predict preferences significantly better than the unsegmented model.

Search proceeds indirectly by examining those variables that indicate, but do not directly measure importance weights. The statistical techniques vary, but the common goal is to identify groups of consumers that are likely to have different importance weights. For example, cluster analysis can be used to group consumers by: (1) stated rank order (Pessemier, 1977), or (2) perceptual measures, e.g., factor scores, for their first preference product. Alternatively, discriminant analysis (Johnson, 1970) or automatic interaction detection (Sonquist, Baker, and Morgan, 1973) can be used on univariate measures such as: (1) intent to purchase the new product, or (2) last product purchased. For example, we might group consumers by their stated intent (definitely, probably, might, probably not, definitely not) to join a new health care plan, or in another case, we might group consumers by the last deodorant that they purchased. In either case, we would look for demographics or psycho-social variables to "explain" the clusters.

The next step is to group consumers who are homogeneous with respect to the identified demographic or psycho-social variables. A preference regression is then run within each group.

For example, in the HMO case, Greer and Suuberg used both cluster analysis on perceptions and automatic interaction detection (AID) on intent to purchase. Throughout the search there was a common but weak indication that pattern of existing care—MIT vs. private—was a possible variable for segmentation. Table 10.4 reveals that the importance weights are different, but not too different. To test this segmentation, we use a statistical test that determines whether the regressions are significantly different.

Table 10.4. Preference Regressions Within Each Segment [a]

		Quality	*Personalness*	*Value*	*Convenience*	R^2
Overall	($N = 210$)	6.2	3.9	5.7	3.3	0.27
Segmentation:						
Private	($N_1 = 88$)	6.1	4.6	4.9	4.5	0.28
MIT	($N_2 = 109$)	6.9	3.6	6.5	2.8	0.29
Mixed	($N_3 = 12$)	2.6*	5.5*	4.2*	−1.0*	0.26

[a]*All regressions are significant at the 0.01 level. All but * coefficients are significant at the 0.05 level*

The formal test for benefit segmentation is an F-test (Fisher, 1970; Johnston, 1972; Chow, 1960). Details are described in the references but the basic idea is simple. The group of segmentated regressions can be thought of as a single regression applied to the entire population, but with more variables than the unsegmented regression. The F-test then determines whether the added variables can significantly improve prediction.*

Based on Table 10.4, we can compute $F = 0.50$ with (10, 195) degrees of freedom. This indicates that pattern of care segmentation is not significant in the HMO case and should not be used for managerial strategy. In the same case, management felt that a segmentation strategy of faculty/student/staff would be effective. In that case, $F = 0.77$ with (10, 195) degrees of freedom indicating that it too was not significant.

We choose this case because it illustrates the importance of testing segments. Cluster analysis, AID, and discriminant analysis are all powerful techniques if used properly. Part of this use is a requirement that segmentation strategies based on these techniques be tested. In this way, a manager can feel confident that he has truly identified significant differences within the population.

*Specifically, the F-statistic is computed by:

$$F = \frac{(SS_g - \sum_n SS_n)/K(S-1)}{\sum_n SS_n/(N - SK)}$$

where there are K variables in the preference regression, S segments, and N observations. SS_g is the sum of squared residuals for the total group and SS_n is the sum of squared residuals for the nth segment. The statistic has $K(S-1)$ and $N - SK$ degrees of freedom.

Discussion

Since consumers are rarely homogeneous, often the most effective strategy is a two or more product strategy. For example, in the education case, MIT's Sloan School segmented the market with two programs—an intensive specialization and a generalist program.

The use of benefit segmentation is usually begun after initial preference has been collected, but it is appropriate for all phases of design. To do benefit segmentation with expectancy value or conjoint analysis simply cluster on the preference parameters and test for significant differences. For preference regression, use the search and test procedure. Examine these differences for managerial implications and, if appropriate, continue through the design stage with a multi-product or a segment-directed strategy. Remember to always test hypotheses about segmentation to ensure that the multi-product strategy is indeed the strategy with the greatest potential.

MANAGERIAL USE OF PREFERENCE MODELS

The perceptual mapping techniques identified the underlying dimensions that differentiate consumer perceptions of products and the position of existing products on the dimensions. Preference analysis identifies which gap has the most potential. This is an important input in the process of creating a core benefit proposition and differentiating it from competition.

Figure 10.14 identifies three phases where preference analysis can be useful. The first use is the understanding of existing perceptions and preferences through maps and importances. The outcome is an identification of areas of highest opportunity. Next, concept statements are created to position in these areas. This is a creative process, and the techniques outlined in Chapter 6 are usually employed here.

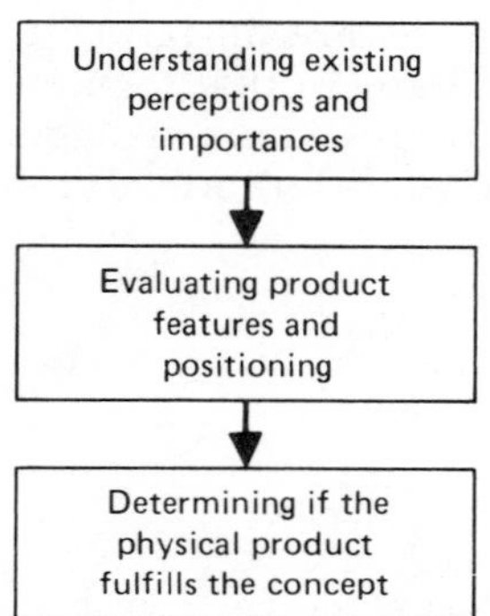

Figure 10.14. *Preference Analysis in the Design Process*

In this basic phase of understanding preferences, benefit segmentation should be conducted and the opportunity for two or more products to meet the specific needs of segments should be addressed.

After concepts have been written, they should be evaluated to see if they achieve their desired positioning. Perceptual mapping allows a check for this and preference analysis gives a confirming estimate of importances. Conjoint analysis can be used here to see what physical features are most important in concept preference and in establishing the products' psychological positioning.

Once we have performed a conjoint analysis linking features to perceptions we can place the alternative products (as defined by features) in perceptual space. Figure 10.15 shows three such products. Based on preference regression we establish the preference share for each positioning. Using the techniques of the next chapter we transform the preference share into a demand forecast and, contingent on price, change this to a revenue estimate. We then compare the expected revenue to the production cost of the required features to associate a profit with each position.

Although work to analytically define an "optimal" position (Zufryden, 1979; Albers and Brockhoff, 1977; Shocker and Srinivasan, 1974; Gavish, Horshy, and Srikanth, 1979) shows great potential, we feel that a single optimal position should not be identified at this point in the design process. There are measurement errors in the maps and the perceptual dimensions can be affected by advertising and promotion as well as feature selection. We feel preference analysis gives directional insight to the creative team which can significantly improve the product positioning, but it does not automate design.

For example, in Figure 10.15, we have labeled each point with its preference share and a rough estimate of profit. The manager then selects the target position and resulting features. The manager may wish to select product *A* over product *B* even though product *B* has a higher profit. He may do this to establish a strong initial position and preempt a "second, but better" competitive positioning.

Next, the actual physical product is designed. Here the physical features are used to achieve engineering performance. These engineering performance attributes must also be perceived and valued by consumers. A check of the evaluation of the product after in-home use by mapping and preference analysis completes the third phase of design. The result should be a uniquely positioned core benefit proposition backed up by physical features that substantiate its perceptions and performance.

While perception and preference analysis of existing markets may yield opportunities, in some cases these markets may be saturated, and major innovation in the structure of the market may be needed for success. Perhaps a new dimension should be added to the market. In the 1970s Ford added the "quietness" attribute to its cars. The use of "stretching" concepts can be valuable in determining if a new idea establishes a new dimension or merely changes the position of a new product in the existing space.

In some situations a new product will be creating a market and there

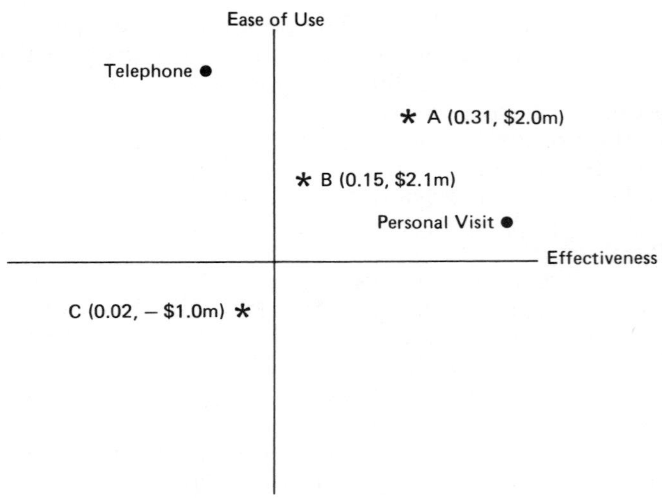

Figure 10.15. *Alternative Positions for New Telecommunications Product (numbers in parentheses indicate preference share and profit in millions of dollars)*

will be no existing competitors. For example, a new financial brokerage service which notified investors automatically by phone when their stocks' prices move outside the individuals prespecified range may not fit in the existing competitive structure. In cases such as this, multiple but different concepts around the core benefit proposition are used to learn how consumers perceive alternatives and value their features.

The techniques of preference analysis are flexible and valuable in creating new products. Understanding existing preferences, searching for ways to add new dimensions, or creating new markets are opportunities that should not be overlooked. Another opportunity is in attempting to change the importances of dimensions so a new product will be viewed as attractive. For example, increasing consumer importances for energy conservation in appliances through advertising may make low power consumption for a freezer a successful core benefit proposition. Exploiting the range of opportunities is a creative process requiring input from R&D, engineering, and marketing. Preference analysis is important for successful integration of these resources to specify a good product design.

Preference and perceptual analysis help to create a best core benefit proposition positioning, but before we can decide if we should continue developing the product we need to know if it will have a good chance of meeting our financial goals. This requires that we consider the consumer choice process. The next chapter considers the translation of preference to choice and the decision of whether the new product should move from the design phase to the testing phase of the new product development process.

REVIEW QUESTIONS

10.1 There are several approaches to selecting the best positioning for a new product. One approach might construct a concept description for each possible new product position and have consumers evaluate the various descriptions. The description with the best rating could then be chosen. A second approach would be to build a preference model which describes the entire perceptual space and then use the model to predict the best positioning. What are the advantages and disadvantages of each approach? Which approach is superior?

10.2 Preference scores for four individuals over five products were computed and are shown in the following table.

p_{ij}	$i = 1$	$i = 2$	$i = 3$	$i = 4$
$j = 1$	4.1	2.1	3.2	4.0
$j = 2$	4.3	3.8	3.0	2.6
$j = 3$	2.1	1.9	4.1	4.8
$j = 4$	4.2	2.3	3.9	4.4
$j = 5$	1.8	1.6	3.7	3.0

Suppose this sample of four individuals is perfectly representative of the general population's preferences. What information does this data provide for managerial decision-making? Can we do more than make market share predictions?

10.3 When using the results of a preference regression, should a product manager care what statistical procedure was used in determining the estimators? Specifically, should the manager care whether a monotonic procedure was used or an ordinary regression was run?

10.4 Summarize the various techniques for measuring and predicting consumer preferences. What are the advantages and disadvantages of each? When would each be used?

10.5 Benefit segmentation is useful when the firm wishes to direct several products toward several different segments of the market. What are some other possible uses and benefits of benefit segmentation?

10.6 Consider the basic conjoint analysis equations (equations 10.3 through 10.5).

a. The text states that the utility functions are nonlinear. What does this statement imply about consumer preferences? What does it imply about the result of managerial manipulation of the product features?

b. Although the utility functions of the product features may be nonlinear, the equation represents a linear independence between the product features. That is, the function is an additive combination of the individual utilities. What does this implicit assumption imply about consumer preferences? Could this assumption be misleading under some special situations?

10.7 What is the managerial significance of the relationship between product features, product perceptions, and preferences? How is conjoint analysis used to provide the manager with the information necessary to select the product features?

10.8 In Figure 10.15 product B has a higher expected profit than product A, yet the text suggests that product A might be the better entry strategy. Why? What are the implications of this decision?

10.9 An ideal vector is a visual representation to represent consumer preferences. But consumer preferences may vary. Under what conditions is an ideal vector a reasonable representation of such a market? Under what conditions is a set of segmented ideal vectors appropriate? When are neither valid?

10.10 How might preference regression be adapted to the selection of product features? How might conjoint analysis be adapted to determine an ideal vector?

Chapter 10 Appendices

10.1 Advanced Topics in Conjoint Analysis

Fractional Factorials. One difficulty of conjoint analysis is that the consumer must rank order a large number of stimuli (real or pseudo-products) when there are four or more features or when the number of levels for each feature is large. For example, consider a factorial experiment for deodorants. Suppose that there are three features—package, scent, and price—each at four levels. A full factorial, $4 \times 4 \times 4$, would require the consumer to rank order sixty-four deodorants—not an easy task for the consumer. An alternative is to use fractional factorial designs (Green and Wind, 1975). In fractional designs, a select subset of the potential products are given to the consumer in such a way that the conjoint model can still be measured. A common design is latin squares (Cochran and Cox, 1957) which is appropriate when there are no interaction terms.

A latin squares design is illustrated in Table 10.1A-1. Note that each price level appears once in each row (package level) and once in each column (scent). This is a "main effects" model in experimental design and dramatically reduces the number of stimuli required. Other designs can be used when only certain interactions among features are to be measured (see Green and Devita, 1975). For a review of more advanced designs, see Cochran and Cox (1957) and Green and Wind (1973).

Tradeoff Analysis. An alternative technique used to reduce the consumer task in large factorial designs, called tradeoff analysis, asks consumers to evaluate pairs of features independently of the other features (Johnson, 1974). The consumer is given a tradeoff matrix (see Table 10.1A-2) and asked to rank pairs of features. In the deodorant example, each consumer would be required to fill out three such matrices representing each possible

Table 10.1A-1 Latin Squares Fractional Factorial Design for Conjoint Analysis [a]

Package \ Scent	*Unscented*	*Herbal Scent*	*Spicy Scent*	*Mint Scent*
Powder	$ .75	$1.00	$1.25	$1.50
Stick	1.50	.75	1.00	1.25
Roll-on	1.25	1.50	.75	1.00
Pump Spray	1.00	1.25	1.50	.75

[a] *Consumer is asked to rank these sixteen products, usually by rank ordering 3″ × 5″ cards*

pair of features. That is, scent vs. price, package vs. price, and scent vs. package. This design has the advantage that the consumer task is relatively easy and can be used in a self-administered survey. In practice, the design tends to cause problems because consumers often simplify the task by selecting the more important feature and going row by row, column by column, or diagonal by diagonal. The carpooling example mentioned earlier was measured with this type of design.

Table 10.1A-2 Trade-off Matrix for Pairwise Tradeoffs in Conjoint Analysis [a]

	Unscented	*Herbal Scent*	*Spicy Scent*	*Mint Scent*
$ .75				
$1.00				
$1.25				
$1.50				

[a] *Consumer is asked to fill in the boxes with rank orders to indicate preference*

Intensity Measures. Originally, conjoint analysis was based on monotonic regression. Recently, many researchers have shown that metric regression can be used and the ranks treated as more than ordinal data. The natural extension of these findings is to try stronger measures that have interval or ratio properties. One of these measures is constant sum paired comparisons (CSPC). With CSPC, the consumer is given potential products two at a time and asked to allocate a fixed sum of "chips" among the products according to his preference for those products. Hauser and Shugan (1980) have shown that preference theory can be extended to this or any other form of intensity measurement. They show that a number of theories (and potential estimation techniques) are consistent with these measures including ordinal, interval, ratio, and stochastic. Further, they developed statistical tests to empirically determine how each consumer responds to the measurement task. Since they have found that many consumers in their sample respond via an intensity theory, and since ratio theory predicts best, we review it here.

If properties similar to preferential independence (Appendix 10.2) hold, then the appropriate preference model for the ratio theory is the multiplicative form:

$$\textbf{(10.1A-1)} \qquad p_{ij} = u_{il}(x_{ij1}) \cdot u_{i2}(x_{ij2}) \cdot \cdots \cdot u_{ik}(x_{ijk})$$

Suppose that a consumer is given a pair of products, m and n, and suppose he allocates a_{mn} "chips" to product m and a_{nm} "chips" to product n. Then $a_{mn}/a_{nm} = p_{im}/p_{in}$. Using a dummy variable transformation similar to equation (10.4) and taking logarithms, yields the following estimation equation:

$$\textbf{(10.1A-2)} \qquad \log\left(\frac{a_{mn}}{a_{nm}}\right) = \sum_k \sum_l (d_{mkl} - d_{nkl}) \log(\lambda_{kl}) + \text{error}$$

The consumer is given a fractional factorial of possible product pairs, equation (10.1A-2) is formed for every pair, and regression or linear programming is used to estimate the log λ_{kl}'s.

For example, CSPC measurement was used to measure consumer preference relative to the effectiveness and ease of use of the communications options. Each dimension was discretized to four levels and ten pseudo-products were chosen. From these, sixteen CSPC questions were developed. The results, shown in Figure 10.1A-1, indicate that effectiveness is more important and definitely shows decreasing returns, while the plots for ease of use indicate a thresholding (elimination phase) phenomena. Estimation was based on a linear programming formulation.

Because more information is obtained per question, the ratio theory out-performs standard conjoint analysis (ordinal theory) for a given number of product pairs. At present, it is an open question whether CSPC is more

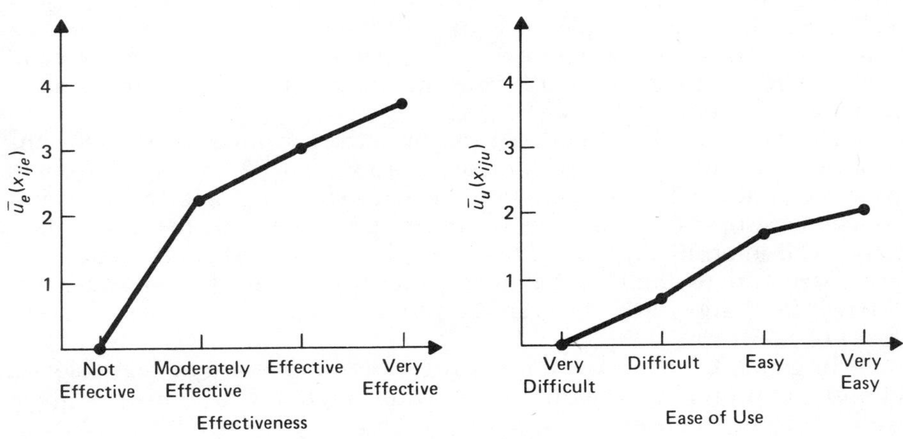

Figure 10.1A-1. *Average Utility Functions for Communications Options*

cost effective than a latin squares factorial design. Since the CSPC-based theory has certain theoretical advantages, it deserves further testing.

Other techniques based on intensity measures include dollar metric (Pessemier, 1977) and graded pairs (Huber and Sheluga, 1977; Neslin, 1978). Similar theories, estimations, and comments apply to these methods of scaling.

Linear Programming Estimation. Regression, both monotonic and ordinary least squares, is only one of the methods used to estimate part-worths in conjoint analysis. Linear programming (or more generally, integer and nonlinear programming) can also be used and, under certain circumstances, has definite advantages. Probably the most widely used technique of this type is a method called LINMAP which was developed by Srinivasan and Shocker (1973). The two major advantages of this technique are that it uses a more robust error structure and that constraints can be added to ensure certain properties of the utility function.

The error structure used in LINMAP and related techniques is absolute error rather than least squares error. This formulation has the advantage that "mistakes" or outlying data are not as severely penalized as in least squares. For example, suppose $r_{i1} = 1$ and $p_{i1} = 2$, and suppose $r_{i2} = 2$ and $p_{i2} = 4$. In regression, the latter "error" would be penalized four times as much as the former "error." In LINMAP, it would be penalized only twice as much. This property makes the estimation less sensitive to one or two aberrant answers by the consumer.

The constraints in linear programming estimation can be used to ensure that the utility functions are monotonic (e.g., in the case of price). Thus, knowledge that the analyst has can be readily incorporated into the estimation and the consumer need not be asked these questions. For example, the consumer need not be asked whether price + \$1.00 is preferred to price + \$1.25. For another example, in Figure 10.1A-1 the utility functions were constrained to be monotonically increasing in effectiveness and in ease of use. This made it possible to ask fewer questions and still estimate an accurate utility function.

In Srinivasan and Shocker's procedure the consumer is asked to rank order the potential products and from this a rank order among all pairs of products is derived. Then based on the error structure, a linearized utility function is estimated to best recover the rank order among the pairs. Alternatively, if intensity measures are collected, one can formulate an absolute error structure to minimize the discrepancy between measured preference differences (ratios) and differences (ratios) based on a linearized utility function.

In general, we can use linear programming in place of regression in almost any preference measurement. To illustrate the basic idea, suppose we have some intensity measure of preference, R'_{im}, for each product or pseudo-product, m. Then by discretizing an additive utility function as in equation (10.5), we can estimate the part-worths, λ_{ikl}, as follows:

$$
\begin{aligned}
&\text{minimize} && \sum_{m=1}^{M} |R'_{im} - P_{im}| && \\
&\text{s.t.}\quad P_{im} && = \sum_{k=1}^{K} \sum_{l=1}^{L} \lambda_{ikl}\, d_{mkl} && \text{(for all } m) \\
&\text{and}\quad \lambda_{i,k,l+1} && \geq \lambda_{i,k,l} && \text{(for all } k, \text{ for } l = 1, 2, \ldots, L-1) \\
&\text{and}\quad \lambda_{ikl} && \geq 0 && \text{(for all } k \text{ and } l)
\end{aligned}
$$

where the λ_{ikl}'s are variables in the linear program. This is then reformulated as a true LP by the standard method of adding slack variables to measure the absolute difference between R'_{im} and P_{im}. Note that the first constraint defines the linear form and the second constraint ensures a monotonic utility function. The final constraint simply defines the scale for the utility function and has no definite behavioral meaning.

Linear programming preference measurement is gaining in popularity because of its flexibility. It has been used in health care, communications services, razor blades, and many other product categories. For details of some of these applications, see Braun and Srinivasan (1975), Parker and Srinivasan (1976), Pekelman and Sen (1979), Shocker and Srinivasan (1979), and Hauser and Shugan (1980).

Interactive Computer Interviewing. In a further attempt to ease the task on the consumer and in an attempt to get overnight results to conjoint estimation, a number of interactive computer interviewing systems have been developed. In a typical system the computer prints out a pairwise comparison and requests the consumer to either choose among the pair of products or allocate chips. Most systems then automatically estimate the consumers' utility functions. Results to date have been favorable, and this form of interviewing is gaining in popularity. Among the systems that are now available are CISTS (Myers, 1976), P.A.R.I.S. (Shugan and Hauser, 1977), and a program by Market Facts (Johnson, 1975).

10.2 Risk Aversion Modeling and Measurement

Risk aversion can be measured with a technique known as von Neumann-Morgenstern utility theory (von Neumann and Morgenstern, 1947; Keeney and Raiffa, 1976). The marketing use of this theory (Hauser, 1978; Hauser and Urban, 1977, 1979) requires that the consumer be asked to make preference judgments with respect to lotteries (see Figure 10.2A-1). To use this measurement, we develop a "utility" function that satisfies the property that the consumer will choose the alternative with the highest "expected" utility.

The basic concept behind the measurement is quite simple. For example, suppose a consumer is asked to set the yellow area of the probability

Instruction to Consumer:

Imagine you can only choose between two health plans, plan 1 and plan 2. In both plans personalness, convenience, and value are good (rated 5). You are familiar with plan 1 and know that quality is satisfactory plus (rated 4). You are not sure of the quality of plan 2. If you choose plan 2, then the wheel is spun and the quality you will experience for the entire year depends on the outcome of the wheel. If it comes up yellow, the quality is very good (rated 6) and if it comes up blue the quality is just adequate (rated 2). Graphically this is stated:

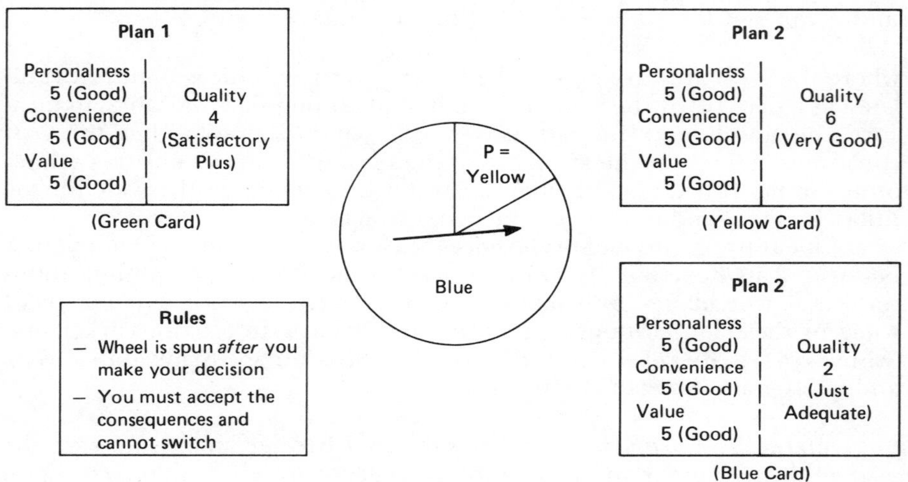

Instruction to Consumer:

At what setting of the odds (size of the yellow area) would you be indifferent between plan 1 and plan 2? (Respondent is given wheel and adjusts it until size of yellow area is appropriate. He is challenged by being given the choice with his setting. If he prefers one plan or the other, the interview iterates the question until a true indifference setting is determined.)

Figure 10.2A-1. *Schematic of Lottery Question for Direct Assessment*

wheel in Figure 10.2A-1 in such a way that he is "indifferent" between the two plans. Note that at $p = 0.999$ the consumer will prefer Plan 2, while at $p = 0.001$ he will prefer Plan 1. The consumer's task is to continually narrow this range until he neither prefers one nor the other plan. If the utility function is to measure this effect, then the utility of quality for the certain plan, Plan 1, must be equal to the "expected" utility of quality for the uncertain plan, Plan 2. That is, if the consumer sets the area of the wheel equal to 0.80 (80 percent yellow) then:

$$\textbf{(10.2A-1)} \qquad u_{iq}(x_{i1q} = 4) = (0.80)u_{iq}(x_{i2q} = 6) + (0.20)u_{iq}(x_{i2q} = 2)$$

here $u_{iq}(x_{i1q} = 4)$ represents the utility of quality for individual i when the quality of a health plan is at level 4 (satisfactory plus). Thus, if we knew $u_{iq}(x_{i2q} = 6)$ and $u_{iq}(x_{i2q} = 2)$, we could solve for $u_{iq}(x_{i1q} = 4)$. We would then fit a smooth curve through these points. For example, suppose we knew

$$u_{iq}(x_{i2q} = 2) = 0 \quad \text{and} \quad u_{iq}(x_{i2q} = 6) = 1,$$

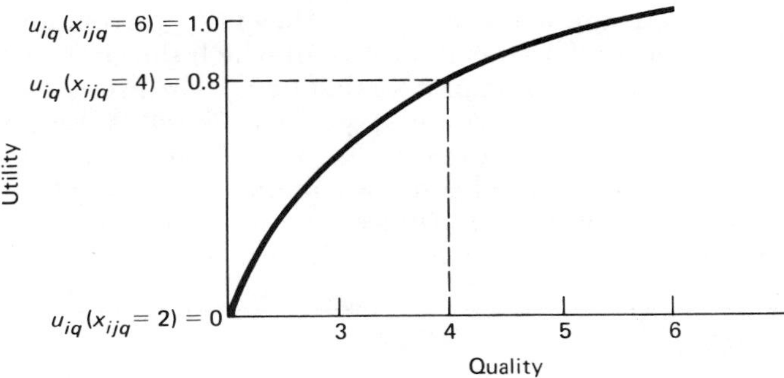

Figure 10.2A-2. *Modeling Risk Aversion*

then $u_{iq}(x_{i2q} = 4) = 0.80$ from equation (10.2A-1). This is illustrated in Figure 10.2A-2. Alternatively, we could choose a mathematical function to represent the curve and fit it to the points.

Fortunately, there is axiomatic theory (Keeney and Raiffa, 1976) which tells us what family of curves are appropriate under certain behavioral assumptions. One curve, which is extremely useful for marketing, is the constantly risk averse curve. This curve is extremely flexible and provides a good approximation to most utility functions. But more importantly for the marketing manager, the constantly risk averse curve gives an explicit measure of risk aversion that can be used to gauge consumers' feelings toward risk. This form is given by

(10.2A-2) $$u(x) = a - be^{-rx}$$

where x is the measure of the perceived dimension or product feature, r is the "risk aversion coefficient," and a, b are set by scaling conventions. Usually $u(x_{\min}) = 0.0$ and $u(x_{\max}) = 1.0$ for the minimum and maximum allowable values $x_{\min}$, $x_{\max}$, of x (see Figure 10.2A-2 for an example). Basically, the larger r is the more risk averse is the consumer. Figure 10.2A-2 shows a risk averse shape of utility curve. If $r \to 0$, the utility form becomes linear. If $r < 0$, the consumer prefers to take a risk.

Basically the constant risk aversion equation means that a consumer's marginal feelings about risk do not change as the level of an attribute increases. The interpretability of this function makes it ideal for marketing, but other forms are possible. For example, some useful functions are

$$u(x) = \log(x + b) \quad \text{and} \quad u(x) = (x + b)^c \qquad (\text{for } c < 1 \text{ but } c \neq 0)$$

which are both decreasingly risk-averse for $x \geq -b$. It is interesting to note that the quadratic utility function, $u(x) = (x + b)^2$ which is used in PREFMAP and other applications has the counterintuitive property of increasing risk aversion. (For more details, see Keeney and Raiffa, 1976.)

The lotteries give us one way to measure the utility functions. To measure the non-linear preference model, we need to measure the relative weights, w_{ik}, as well as the risk aversion coefficient (r in equation 10.2A-2). To do this, we use a similar consumer task in which the consumer is to set the level of one dimension on Plan A so that he is indifferent (would just as soon have either) between Plan A and a specified Plan B (see Figure 10.2A-3). The preference (p_{iA} and p_{iB}) of both Plan A and Plan B are written down and set equal to one another. This set of equations is then solved algebraically to give "estimates" of the importance weights. A numerical example for quality, personalness, and value is given in Hauser and Urban (1979).

Plan A	*Plan B*
Waiting time 20 minutes Price $10	Waiting time 30 minutes Price ___

Instructions: Set the price for health plan B so that you are indifferent between plans A and B.

Figure 10.2A-3 *Schematic of Tradeoff Question*

The theory can also be used to specify how the utility functions interact to produce preference. This theory and the behavioral assumptions underlying it are reviewed in Keeney and Raiffa (1976). One important interactive form for marketing is the quasi-additive form:

$$\textbf{(10.2A-3)} \quad p_{ij} = \sum_k w_{ik}u_{ik}(x_{ijk}) + \sum_k \sum_{l>k} W_i w_{ik} w_{il} u_{ik}(x_{ijk}) u_{il}(x_{ijl}) + \cdots + W_i^{K-1} w_{il} \cdots w_{iK} u_{il}(x_{ijl}) \cdots u_{ik}(x_{ijK})$$

The first terms are the direct (main) effect of the variables. The next terms reflect the pairwise interaction of the variables, while the last terms are the higher order interactions. There is one importance weight (w_{ik}) for each variable and one interaction coefficient W_i, which can be measured with only one additional consumer question. This is illustrated in Hauser and Urban (1979). While interactions can be measured with conjoint analysis (Green and Devita, 1975), they require a more complete factorial design. Furthermore, the special form in equation (10.2A-3) would require nonlinear regression.

The interaction coefficient, W_i, has a number of important managerial interpretations. If W_i is negative, then the dimensions are substitutes. This means the consumer would prefer a guarantee that at least one dimension of the product had a good level (even if it meant the other dimension had a bad level) to taking a chance that either both dimensions were good or both bad (see Figure 10.2A-4). If $W_i = 0$, there is no interaction, thus Lottery A is equivalent to Lottery B in Figure 10.2A-4. If $W_i > 0$, the dimensions are complementary and Lottery B is preferred to Lottery A. A useful interpretation of this phenomena due to Richard (1975) is to think of W_i as a measure

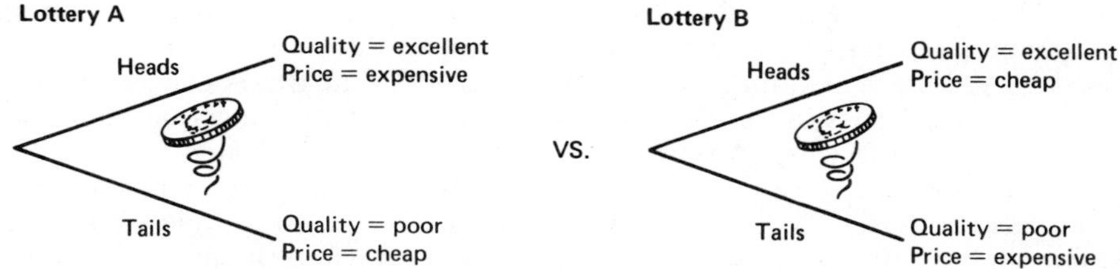

Figure 10.2A-4. *Direct Measure of Interactions Among Product Features*

of multi-attributed risk aversion. If all the attributes of a new product are uncertain, then the more negative W_i is, the less the consumer is willing to take a chance and switch to that product. Thus, if W_i is significantly negative, a new product team should consider strategies to ensure the consumer that product claims will be fulfilled.

For example, utility functions based on equations (10.2A-2) and (10.2A-3) were measured in the HMO case for quality, personalness, convenience, and value. Risk was important and the attributes were found to be strong substitutes. (The median risk aversion for quality, r_q, was 0.693 and the median W_i was -0.992. For the other three dimensions, $r_v = 0.424$, $r_p = 0.332$, and $r_c = 0.310$.) Managerially, this means that the health care manager must emphasize that the quality of care is uniform and that the patient need not fear that "mistakes" will be made in the delivery of care. Furthermore, assurances should be given to the consumer with respect to the other dimensions, but they need not be stressed as much as those for quality. The fact that W_i is so strongly negative means that the consumer is extremely concerned that everything does not go wrong at once.

Direct assessment is best used when risk phenomena or interactions are potentially important. It provides explicit measures of risk that indicate to what extent the new product manager should strive to maintain consistent levels of product features and to what extent this guarantee must be communicated to consumers. Furthermore, it provides explicit measures of interaction that indicate what multi-attribute guarantees need be made in product positioning. An example when risk was found to be important was in the provision of computer controlled bus service in Rochester, New York (Wilson, Weissberg, and Hauser, 1976). In that case, it was important to guarantee that waiting and traveling time would be reasonable. Since there was high variation in both waiting and travel time in the existing system, computer algorithms were modified to incorporate risk averse preference functions. After lottery and tradeoff judgments have been collected, the parameters of the preference equation can be calculated directly. This model can then be used to simulate the preference for a redesigned product. For example, a plan with outstanding quality but a reduction in convenience could be compared to one with good quality and good convenience, and to one with adequate quality and no wait. The richness of the prediction is reflected in the fact that the preference depends upon the level of the variables, their interactions, and risk aversion.

The advantages of direct assessment are that it can explicitly model important aspects of consumer behavior and can measure preferences for each individual consumer. The disadvantage is that the consumer task is complex. Direct assessment requires a follow-up personal interview and, because of the lotteries, is most appropriate for well-educated consumers. Thus, it is most appropriate in later design phases after preference regression has been used to define an initial product positioning. Relative to conjoint analysis, direct utility assessment is best used when the manager needs explicit measures of risk and interactions. The added expense of interviewing is appropriate only if these effects can strongly influence the outcomes of managerial decisions.

Chapter 11

Forecasting Sales Potential

The preceding chapters have discussed the perception and preference phases of developing a product's core benefit proposition. They provided procedures and techniques to find the "best" product design. But is the "best" product good enough to warrant the cost and time involved in final development and testing? To answer this question we must evaluate the purchase potential of the new product.

ROLE OF PURCHASE POTENTIAL IN THE DESIGN PROCESS

Need for Measures of Purchase Potential

To evaluate a new product we must first consider the goals of the organization developing that product. In private firms this is usually stated in terms of sales and profit. Profit can be a complex construct which may vary on a number of dimensions, including: (1) time span such as short term or long term; (2) profit measures such as net present value, return on investment, or internal rate of return; (3) scope such as contribution by product or interactions with product line; or (4) considerations such as pretax or post-tax. However, no matter how complex the profit decision is, the crucial input that marketing must provide is an estimate of the sales, i.e., consumer choice behavior, that result from the new product strategy.

In public services, goals can become even more complex. A regional transportation agency may be concerned with increased energy efficiency and less pollution, and may measure success through a reduction in vehicle miles traveled (VMT). But VMT is a result of the trips people make and what mode they use to make those trips. Thus, again, marketing's input is consumer choice behavior. In health care delivery it is enrollment, in hospitals it is admissions by illness, in universities it is applications and placement, and in energy policy it is efficiency of energy use.

Preceding chapters concentrated on designing a product or service that consumers would prefer. But this is not sufficient since the final success of a new product is based on consumer purchase behavior, not only on preference. A new toothpaste may be preferred, but it is successful only if enough consumers buy it, giving the firm reasonable return on its investment. A new material may be better than existing materials, but it is successful only if enough manufacturers use it in production so that the supplier earns a fair return on raw materials, processing, and development. A new transportation service may be viewed as better by consumers, but it will be successful only if it serves trips that would otherwise be left unserved or if the service results in decreased energy use and pollution. Thus, the manager responsible for a new product wants to know the projected demand for the new product.

To predict demand, we must estimate how preference translates into choice. For example, we must estimate how many consumers are likely to buy their first preference product. Furthermore, we must project the influence of other, nonproduct marketing mix variables. For example, based on the planned advertising and promotion, we must estimate the percentage of consumers who will become aware of the product and, based on the planned distribution strategy, we must estimate the percentage of consumers who will find that the new product is available.

In this chapter we examine models of purchase potential and sales formation. These models translate preference and intent measures into approximate sales levels and allow adjustment for awareness, distribution, and purchase dynamics.

Importance of Forecasts at Various Stages of the Design Process

The importance of forecasting actual purchase behaviors increases as a product proceeds through the development process. Early in the design phase the product is just an idea, a positioning, or at most a concept description. The investment has been relatively small and there are still many competing potential ideas. The manager wants a rough estimate—an indication of whether the product will be a "bomb," a moderate success, or a spectacular winner. Furthermore, because it is early in the design phase, estimates can be "ballpark" estimates because the idea, positioning, or concept has not yet been developed into a product. Thus, early in the design phase, we combine the results of the perception and preference analysis with rules of thumb and managerial intuition to come up early with estimates of consumer response.

As the product progresses through the design phase, it becomes more refined and more like the final product that will be taken to test market. As a result, the evaluation of purchase potential becomes more refined and the estimates become more exact. The investment in the product has grown and there are relatively fewer ideas or concepts that have passed the early screens. Thus, not only are more sophisticated techniques possible, but they are required to fine tune the marketing strategy and to make the difficult, but crucial GO/NO GO decision. These sophisticated techniques are based on a probabilistic model of the consumer that links preference to choice, with corrections for awareness and availability.

In this chapter we present both the simple rule-of-thumb methods of translating preference into purchase estimates and the more formal model of the consumers' probability of purchase. This model, called the "logit model," is valuable in the final design and evaluation. Furthermore, it will prove even more valuable in the testing place of the new product when very accurate sales forecasts are necessary to justify the large production and marketing investments that occur during full-scale launch.

MODELS OF PURCHASE POTENTIAL

Several indicators of purchase potential can be obtained from the consumer. The simplest is a direct question which asks the consumer either his intent to purchase or his probability of purchase. These measures must be treated with caution but can be translated to a rough measure of purchase potential. An alternative procedure is to use historical relationships among preference and choice to transform the preference measures developed in Chapter 10 into rough predictions of purchase probabilities. These methods are used early in the design process. Later in the design process, actual purchase can be observed for some forms of the new product. When this is possible, the more complex logit model can be used to translate preference measures into purchase predictions. We now discuss each of these models and indicate how to use them in the design process.

Intent Translation

Figure 11.1 is an intent scale administered for the communications case described in previous chapters. Consumers are simply asked to make a subjective estimate of their likelihood of using the new communications device. From past experiments and from experience in the product category, a manager can then translate consumer response to these scales into estimates of probability. For example, for frequently purchased consumer brands Gruber (1970) cross tabulated self-stated probabilities of purchase with categorical intent judgments. He found that 75.5 percent of the "definites," 31.4 percent of the "probables," and 26.8 percnt of the "mights" actually chose their preferred products. Based on our experience, we feel that if the product is well-positioned and an aggressive marketing strategy is planned,

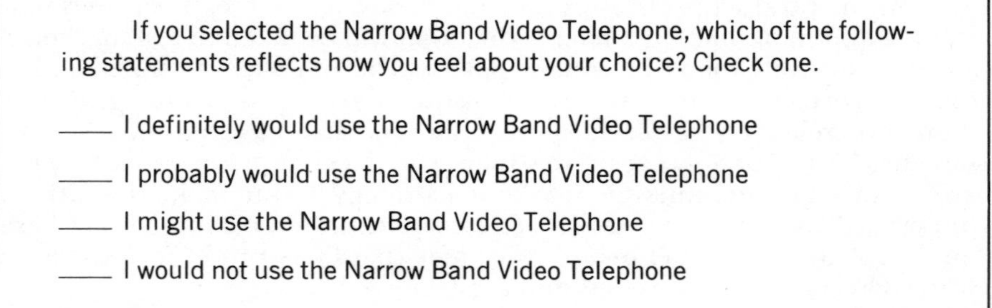
If you selected the Narrow Band Video Telephone, which of the following statements reflects how you feel about your choice? Check one.

____ I definitely would use the Narrow Band Video Telephone

____ I probably would use the Narrow Band Video Telephone

____ I might use the Narrow Band Video Telephone

____ I would not use the Narrow Band Video Telephone

Figure 11.1 *Intent Scale—Communications Example*

a reasonable but conservative estimate would be 90 percent of the "definites," 40 percent of the "probables," and 10 percent of the "mights." In each industry studies of past products or managerial judgment must be used to derive the coefficients to be used to translate the levels of intent into purchase, but they are usually in this range. Applying these numbers to the communications case yielded an overall estimate that 13.1 percent of the sample population would use Narrow Band Video Telephone (NBVT) if they were aware of it and the equipment were available. [That is, no one would definitely use NBVT, but 23.7 percent probably would use NBVT and 36.1 percent might use NBVT. Thus $0.131 = (0.90)(0.00) + (0.40)(0.237) + (0.10)(0.361)$.]

If we wanted to find the effect of design changes, the intent question would have to be asked for each design. Figure 11.2 provides an example from the telecommunication study. Only a few design alternatives can be tested in this way since the data collection burden soon becomes overwhelming.

Some market researchers prefer probability statements to intent statements because probability statements provide the respondent with more categories and because the categories are more exactly defined. For example, Juster (1966) recommends a scale similar to that in Figure 11.3.

The probability scale is used in the same way the intent scale is used. Each category is an estimate of some probability of purchase. For example, Juster compared actual purchases projected from six months data to intentions to purchase over "the next twelve months." As Table 11.1 indicates, the stated probabilities are monotonically related to the actual purchase probabilities, but the stated probabilities are not equal to the actual probabilities. Morrison (1979) suggests an underlying behavioral process model based on true intentions and extraneous events. Simply stated, Morrison's model suggests that the stated intentions, as measured by Figure 11.3, are linearly related to the observed probability of purchase. For example, Figure 11.4, which is based on the data for automobiles in Table 11.1, suggests such a linear approximation.

To use the probability scale, first establish the translation from the scale

How likely would you be to choose Narrow Band Video Telephone?

	Definitely not Choose	*Probably not Choose*	*Might Choose*	*Probably Would Choose*	*Definitely Would Choose*
If it were exactly as described:	[ ]	[ ]	[ ]	[ ]	[ ]
If every office had one:	[ ]	[ ]	[ ]	[ ]	[ ]
If hard-copy were available:	[ ]	[ ]	[ ]	[ ]	[ ]
If transmission time were improved from 30 seconds to 10 seconds:	[ ]	[ ]	[ ]	[ ]	[ ]
If resolution were improved to 4 times that of a home TV:	[ ]	[ ]	[ ]	[ ]	[ ]

Figure 11.2. *Conditional Intent Scales—Communications Example*

Taking everything into account, what are the prospects that you will adopt Narrow Band Video Telephone for your daily communications?

Certain, practically certain (99 in 100). ____

Almost sure (9 in 10). ____

Very probable (8 in 10). ____

Probable (7 in 10). ____

Good possibility (6 in 10). ____

Fairly good possibility (5 in 10). ____

Fair possibility (4 in 10). ____

Some possibility (3 in 10). ____

Slight possibility (2 in 10). ____

Very slight possibility (1 in 10). ____

No chance, almost no chance (1 in 100). ____

Figure 11.3. *Probability Scale—Communications Example*

Table 11.1. Relationship Between Scaled Probabilities and Observed Purchase Behavior[a]

Probability Scale Value	*Automobiles*	*Appliances*
1 in 100	0.07	0.017
1, 2, or 3 in 10	0.19	0.053
4, 5, or 6 in 10	0.41	0.111
7, 8, or 9 in 10	0.48	0.184
99 in 100	0.53	0.105

[a]*Adapted from Juster (1966)*

values to purchase probabilities based on past data or judgment. This can be done with a table such as Table 11.1 or a graph such as Figure 11.4. Then use your model to translate stated consumer intentions to estimated probabilities for each consumer or group of consumers. Use the average estimated probability as an estimate of market potential.

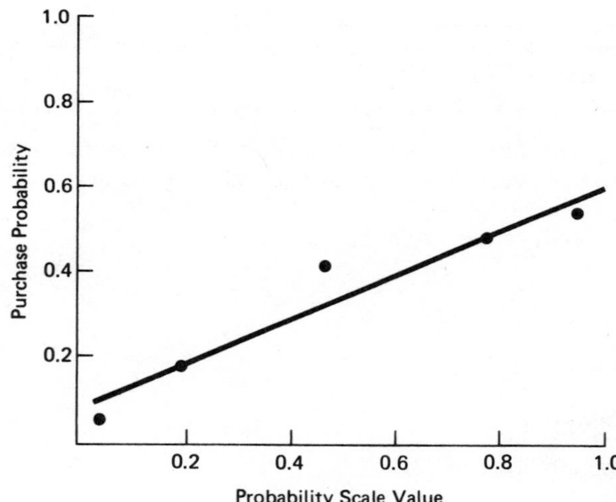

Figure 11.4. *Representative Linear Relationship Between Scaled Probabilities and Observed Purchase Probabilities (Morrison, 1979)*

A final consideration is that of nonresponse bias. If purchase intentions are not related to the likelihood that a consumer completes the survey, then the estimate of market potential is not biased by nonresponse. If, on the other hand, consumers who are more (less) likely to purchase are also more (less) likely to return the survey then the estimate of market share is biased. When such bias is suspected, you should take steps to encourage a high response rate or use an advanced statistical technique to reduce this bias. For example, Dolde, Staelin, and Yao (1979) show that one can estimate and correct for the relationship between intentions and response rates if one has observations on intentions and response rates across a variety of products. Details on this statistical technique are contained in their paper.

Preference Rank Order Transformation

Direct measures of intent are useful in evaluation of concept statements, but much of the analysis of Chapters 9 and 10 was directed at uncovering strategic refinements in positioning. In Chapter 10, expectancy values, preference regression, or conjoint analysis was used to predict preferences for the refined product. If predictions are to be made at the purchase level for each of the many design combinations, a final link between preference and choice is needed. One simple approach is to assume that everyone would choose their first preference. Alternatively, we can acknowledge that the preference measures, p_{ij}, are good approximations, but not perfect. Thus, we can expect that only some percentage, t_1, of the consumers will select their first preference. But if everyone does not select their first preference, some will select their second or even third preference product. The *preference rank order transformation* simply assigns probabilities according to whether the new product is ranked first, second, third, etc. The values reflect the probability of an individual buying his first choice product (t_1), second choice (t_2), third choice (t_3), etc. Since these probabilities vary somewhat by product category, the reader should base his estimates on past experience, reported results, and personal judgment about the product category.

For example, Silk and Urban (1978) found that for deodorants, $t_1 = 0.83$, $t_2 = 0.15$, and $t_3 = 0.02$. (t_4, t_5, etc. were negligible.) For transportation, Hauser, Tybout, and Koppelman (1979) found that $t_1 = 0.76$, $t_2 = 0.16$, and $t_3 = 0.08$.

Suppose that these numbers are reasonable for the communications case. In fact, for simplicity, take $t_1 = 0.80$ and $t_2 = 0.20$. Based on these proportions, we estimated 11.8 percent of the target population would adopt if they were aware of Narrow Band Video Telephone (NBVT) and it were available to them. [9.2 percent ranked NBVT first and 22.4 percent ranked NBVT second. Thus $0.118 = (0.80)(0.092) + (0.20)(0.224)$.] Comparing this to the 13.1 percent estimated through the intent measures, we see that although the techniques are not exact, they do give reasonably consistent estimates.

An advantage of the rank order transformation is that it can be used to estimate purchase or used for refinements in positioning without remeasuring intent to purchase for concept descriptions. Simply use the techniques of Chapter 10 to obtain estimates of the preferences (p_{ij}'s) and then convert these to choice via the ranked probabilities t_1, t_2, and t_3.

Both the intent translation and the rank order transformation are most useful early in product design when the new product is still a concept statement or a mathematical position on a perceptual map. Both models are approximations and both rest to some extent upon managerial judgment. But they do give reasonably consistent "ballpark" estimates of purchase and are best used to make early evaluations to isolate the concepts or positionings with the greatest potential.

To use these models in a product category, simply measure intent or use the ranked probabilities to transform preference to choice. In fact, some managers use both models in parallel to get alternative estimates of potential. If the estimates are similar, they have more faith in the numbers; if they diverge, then it is useful to reexamine both models and compare assumptions and measures until you understand the discrepancies and know what must be done if they are to converge.

The ranked probability model uses only the ordinal properties of the preference measures. We turn now to a more sophisticated model that bases predictions on the magnitude of the preference measures as well as their rank orders.

Logit Analysis: An Analytic Technique to Estimate Purchase Probabilities

As the design phase progresses, the measures of preference must become more refined because the risks are higher. There is a need for more accurate estimates of purchase.

For example, suppose we used preference regression to develop a preference model for the analgesics example. The first column of Table 11.2 shows the preferences from the model for four typical consumers. The second column shows the preference rank order transformation. Compare individual 1 to individual 2. From the preference values, intuitively, we would expect that individual 1 would be almost as likely to purchase Tylenol as the New Product while individual 2 would be much more likely to purchase the New Product. (Compare relative preference values.) The rank order transformation ignores the intensity information contained in the preference values and uses only the fact that one is larger than the other. For example, the rank order transformation does not distinguish between consumers 1, 2, 3 and 4.

Logit analysis is an analytic technique that uses intensity information contained in preferences values to produce potentially more accurate estimates of purchase probabilities. For example, a logit model produced the estimates in the third column of Table 11.2.

Logit Model Structure. The basic idea behind logit analysis is a mathematical function that translates the preference values into purchase probabilities based on a theory of consumer behavior developed by McFadden (1970). To present the logit model, we introduce a new notation. Let L_{ij} be an estimate of the probability that consumer i will choose product j, let p_{ij} be an observed preference measure such as was measured in Chapter 10, and

Table 11.2. **Comparison of Rank Order Transformation to Logit Predictions for Four Typical Analgesics Consumers**

	Preference Value	Predicted Probability Rank Order Transformation	Logit
Individual 1			
Bufferin	1.3	—	0.08
Excedrin	2.1	—	0.09
Tylenol	9.8	0.20	0.41
New Product	9.9	0.80	0.42
Individual 2			
Anacin	1.4	—	0.11
Private label	2.0	—	0.13
Tylenol	2.3	0.20	0.14
New Product	9.9	0.80	0.62
Individual 3			
Bayer	1.2	—	0.10
Bufferin	2.2	—	0.12
Tylenol	5.0	0.20	0.21
New Product	9.9	0.80	0.57
Individual 4			
Anacin	5.9	—	0.33
Tylenol	6.0	0.20	0.33
New Product	6.1	0.80	0.34
Share of New Product		0.80	0.49

let e_{ij} be an error term that represents the uncertainty in the preference measure. Then we hypothesize that there is some true choice indicator, p_{ij}^T, such that:

$$\textbf{(11.1)} \quad p_{ij}^T = p_{ij} + e_{ij}$$

and that consumer i will actually choose the product with the highest value of the choice indicator, p_{ij}^T.

Let the new product be indexed by $j = 1$. Then, because p_{i1}^T is the choice indicator for the new product, the probability that individual i will choose the new product is simply equal to the probability that p_{i1}^T is larger than p_{ij}^T for all the other products which are indexed by $j = 2, 3, \ldots$. We do not know the p_{ij}^T's with certainty because of measurement error and model error. This uncertainty is modeled by the error term in equation (11.1). But we do know a lot about p_{ij}^T since we know p_{ij} from our preference analysis. For example, based on column 1 in Table 11.2 we expect that individual 1 will probably choose either Tylenol or the new product and that individual 2 will probably choose the new product.

We incorporate this information by using equation (11.1) to compute the probability that p_{i1}^T is greater than p_{ij}^T for $j = 2, 3, 4$. Mathematically, this is given by:

$$\textbf{(11.2)} \qquad L_{i1} = \text{Prob}\,\{p_{i1} + e_{i1} \geq p_{ij} + e_{ij} \quad \text{for } j = 2, 3, \ldots\}$$

If we knew the values of the p_{ij}'s and if we knew the distribution of the e_{ij}'s, we could use probability theory to calculate L_{i1}. One reasonable assumption (error terms are independent and characterized by a special distribution) yields a particularly simple algebraic equation for L_{i1} called the "logit model":

$$\textbf{(11.3)} \qquad L_{i1} = \frac{\exp\,(\beta p_{i1})}{\sum_{j=1}^{J} \exp\,(\beta p_{ij})}$$

where $\exp\,(\beta p_{i1})$ is the number e ($e = 2.71828\ldots$) raised to the βp_{i1}th power and β is a statistical parameter to be estimated. (Appendix 11.1 provides a formal derivation for interested readers.)

For example, suppose that $\beta = .2$, then the probability that individual 1 in Table 11.2 chooses the new product is given by

$$p_{11} = \frac{\exp[(.2)(9.9)]}{\{\exp[(.2)(1.3)] + \exp[(.2)(2.1)] + \exp[(.2)(9.8)] + \exp[(.2)(9.9)]\}}$$
$$= .42$$

Thus once we know the value of β we can use equation (11.3) to estimate more accurate purchase probabilities. To estimate β we use statistical analysis.

Estimation of Parameters. To estimate β, we need a data sample. In that sample we must measure preferences (p_{ij}'s) by one of the techniques of Chapter 10 and observe or measure consumer choice behavior. Among the measures of the latter are: (1) brand chosen in a simulated purchase environment, (2) reported last brand chosen, or (3) reported frequency of purchase of all brands. The first measure is appropriate when an actual product is ready for testing. The other measures are used in early design or in product categories such as services, industrial products, or consumer durables where the inter-purchase interval makes simulated purchase infeasible. Of course other choice measures can be used if they are available.

Based on observed choice or frequency and based on the measured explanatory variables, we then use maximum likelihood techniques to estimate β. (See Appendix 11.1) The resulting estimate has good statistical properties. It approaches the true value as the sample sizes increase and has lower variance than values from other estimators. The estimates are normally distributed and empirical experience indicates that the estimate is quite good even in relatively small samples ($n \geq 100$). Fortunately, a number of computer programs (Berkman, et al., 1976; McFadden and Wills, 1975) are available to perform logit analyses, and in the future, we can expect that logit will be available in most standard statistical packages. (See Appendix 11.2 for methods to test the logit model.)

For example, in Chapter 10, we used preference models to project the impact on preference of a change in transportation service in Evanston, Illinois. Further analysis with a logit model based on last mode chosen produced an estimate of $\beta = 3.35$ when preferences were measured by a form of preference regression. This model was then used to predict the number of consumers who would ride the improved bus service for a number of alternative strategies as shown in Table 11.3. This table was produced by judgmentally predicting the impact of the strategies on perceptions (factor scores), then using the preference regression model to predict preference values for each individual in the sample, and finally using the logit model with $\beta = 3.35$ to change the preference values to probabilities. The usage probabilities for bus were then averaged across all consumers in the sample to predict the market share of bus. The numbers in Table 11.3 are the ranges of forecast increase in the market share for bus.

Table 11.3. Forecasts of Strategy Impacts on Bus Usage[a]

Modification	*Forecast Increase (Percent)*	*Modification*	*Forecast Increase (Percent)*
1. Information Campaign to Increase Knowledge of the Bus System	0.3 to 7.6	7. Constrain Auto Availability	3.8 to 20.8
2. Add Bus Stop Signs with Information	0.2 to 5.8	8. Reduce Perceived Auto Availability	1.4 to 3.6
3. Increase Bus Frequency by 1 Bus per Hour	2.8 to 19.3	9. Increase Perceived Bus Availability	0.9 to 6.1
4. Add Bus Shelters	2.1 to 4.1	10. Increase Perceived Bus Reliability	2.5 to 5.7
5. Improve Bus Safety	1.1 to 2.2	11. Make Bus Environment More Pleasant	0 to 1.5
6. Make the Bus More Relaxing	0.7 to 2.2		

[a]*Hauser, Tybout, and Koppelman (1979)*

Managerial Use of the Model. Intent translation and ranked probability models give estimates of demand that are used early in the design process. Logit analysis is used to refine these estimates as the product progresses through the design process. Since the logit model is based on the magnitudes or intensities of preference through a model of consumer choice behavior, it has the potential to provide more accurate estimates of demand. This accuracy is important in making a commitment to complete design development and initiation of the testing phase of development.

To use logit analysis (equation 11.3) in another product category, the analyst must first determine the parameter β. This is done by collecting preference and choice information for a sample of consumers and using standard computer packages to estimate β. To use the model to predict demand for a new idea, concept, or product the analyst must then:

1. use the preference model (expectancy values, preference regression, conjoint analysis) to determine how preferences change as the result of design changes in the new product;
2. use the logit model (equation 11.3) to estimate the probabilities, L_{i1} or L_{ij}, that each consumer will actually try (or use) the new product if it is available and he is aware of it.

These estimates of demand for the new product then serve as input to the full evaluation process to select the product and the marketing strategy that are best for further analysis or for advancement to test market.

The advantages of logit analysis are that it is based on a realistic model of consumer behavior that acknowledges and measures error and tries to explain as much of behavior as is feasible. It is relatively easy to use, provides reasonably good estimates of demand, and provides explicit statistical measures that can be used to judge the usefulness, accuracy, and significance of the model. The disadvantages are that the statistical logit estimation packages are not yet available everywhere and that equation (11.3) assumes the same preference to choice process for all consumers. Furthermore, there are a number of behavioral assumptions underlying the derivation of the model, but work is underway to relax some of the assumptions (see Appendix 11.3). To date, logit analysis is the best practical means to link preference to choice for the estimation of demand. It provides estimates of purchase potential that are sufficiently accurate for the managerial GO/NO GO decisions in the design phase. Logit analysis can also be used for the analysis of preference and for benefit segmentation. See Appendix 11.4.

Summary of Purchase Potential Models

This completes the basic discussion of purchase potential models. Together, the intent translation, preference rank order transformation, and the logit model give the new product manager a set of techniques to estimate the purchase potential of new product ideas.

The intent or probability scales give a direct measure of the consumer's beliefs about whether he will actually choose the new product. While not exact, the scales provide a good indication of behavior that is sensitive to effects that may not otherwise be captured in our models. The disadvantage of direct scales is that they are limited to testing specific concepts or a relatively few changes to those concepts.

The preference rank order transformation is more sensitive to a wide range of strategies because it is based on the multiattributed preference models of Chapter 10. Its disadvantage is that it may miss extraneous events, not modeled in the preferences. Thus, we advocate the combined use of both the direct measures—intent, and the indirect measures—preference to probability.

The logit model provides a more sophisticated and potentially more accurate model that is sensitive to the intensity of the measured preferences. It is used later in the design process than the rank order transformation and can update or confirm predictions made by earlier analysis.

We feel that it is sound practice to use these multiple techniques to get the best estimates of purchase potential. When used appropriately, these combined models provide accurate estimates that are relevant to managerial design decisions. We now turn to a set of analyses that modify purchase potential to provide estimates of sales formation.

MODELS OF SALES FORMATION

The models of purchase potential predict the sales that would occur if consumers were aware of the new product and it were available to them. Remember that the preference and intent measures were collected from respondents after exposure to a new product or a new product concept. In actual introduction, not all consumers will be aware of the product nor can an organization always achieve 100 percent availability of a new product. In fact, awareness and availability are often managerial control options that are dependent upon advertising and distribution strategy. Thus, in order to accurately forecast sales as opposed to potential, we must modify the probability of purchase estimates by the probabilities that the consumers are aware of the product and it is available to them. Figure 11.5 presents a simplified model of sales formation.

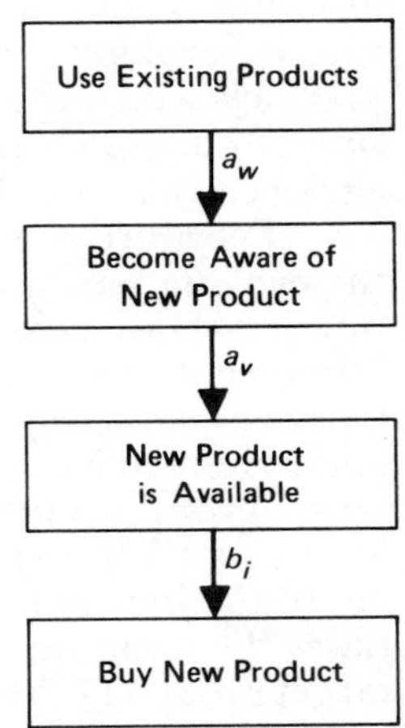

Figure 11.5. *Simplified Model for the Evaluation of Sales Formation*

The model says that before a consumer can buy a product, he must *become aware* of it (a_w) and it must be *available* (a_v). For example, suppose that there is a 0.8 chance that consumer i will become aware of a new analgesic and a 0.9 chance that his regular store will carry it. If the best model predicts that there is a 0.5 chance he will buy and continue to buy it if he is aware of it and it is available, then the expected probabilities of actual buying will be (0.8)(0.9)(0.5) = 0.36 for that consumer. If there are 100,000 consumers just like consumer i, then the expected purchase potential is 36,000 unit sales per period. Mathematically, this equation is given by:

$$P = \sum_{i=1}^{N} a_w \cdot a_v \cdot b_i \tag{11.4}$$

where P is the total purchase (for durable goods) or purchase rate (for products subject to repeat purchase) and the summation is over all consumers i. Purchase rates (b_i) can vary by individual, but unless the analysis is done by segment, all models in this chapter will compute average awareness (a_w) and availability (a_v).

Awareness and Distribution Adjustments

Take out a piece of paper and try to write down all the deodorants you can think of—this is unaided recall. Now suppose you were asked if you have heard of Right Guard, Sure, Ban, Secret, Old Spice, Mennon, Arrid, Arm & Hammer, Soft & Dri, Brut, Mitchum, Safe Day, Tickle, Dial, Dry Idea, and Calm. If you have, that is aided recall. But how many of those deodorants do you know sufficiently well that you can describe them with respect to 15–25 attribute scales that might be applicable to deodorants? For new product awareness, we want to know how many consumers will become aware of the product at a level that is sufficient so that they will have the information to seriously consider the new product. Thus, the concept we are concerned with is "evoking." This was introduced in Chapter 8 and defined as those products which the consumer has used, has on hand, would seriously consider, or definitely not consider. Thus, a_w is formally defined as the percentage of consumers who will *evoke* the new product.

There are two ways to estimate a_w. The first is to judgmentally estimate the evoking percentage. This was done for the Health Maintenance Organization (HMO) case that was introduced in Chapter 9. It was estimated that 70 percent of the target group could be made aware of the HMO by the planned marketing effort. (That is, they would be exposed to the brochure and have read it carefully.) An intent model forecast that 23.3 percent of the target group would enroll if they were made aware of the HMO plan at a level equivalent to the survey. We assumed the HMO would be available to all, so the new enrollment forecast (P) was the purchase probability (0.23) times the awareness (0.70) times the availability (1.0) times the size of the target group (17,200). The enrollment forecast was 2,800 patients.

It is often difficult to estimate a_w directly. An alternative method is to judgmentally estimate unaided recall awareness, a_u, and use it to estimate a_w. The latter quality is often easier for advertising managers to estimate, and can be readily linked to an advertising budget. These two measures, a_u and a_w, are highly related. For example, Silk and Urban (1978) estimated this linkage from a_u to a_w for deodorants with least squares regression. To estimate a_w for the design phase, we would estimate a regression equation linking a_u to a_w for our product category. Then the level of unaided brand awareness that the introductory marketing campaign is expected to obtain would be substituted in the equation to estimate the evoking proportion. If necessary, a number of estimates can be made at varying levels of marketing expenditure.

The measure of availability, a_v, is simply a measure of the percentage of consumers who could purchase the new product if they wanted to. In consumer products, a_v is judgmentally based on the percentage of retail outlets (adjusted for volume) that will carry the product in the target area. In industrial products, it may be the percentage of buyers that are within a feasible delivery area. In health it might be eligible consumers. In transportation it may be more complex, e.g., what percentage of consumers are within ¼ mile of the bus line or how many can afford an automobile. But for many categories, past experience or managerial judgment is sufficient for the design phase. The estimated availability (a_v), the awareness (a_w), and the conditioned probability of purchase (b_i) obtained from the purchase potential models are then used in equation (11.4) to predict the sales for a new product.

Dynamics

Sales potential of a new product will not occur immediately. For products that depend upon repeated purchases, we must consider the trial and repeat processes that lead to the dynamic growth of sales. For a major new product such as a home appliance, we must consider the phenomenon known as the "diffusion of an innovation."

Trial and Repeat. The sales of new products that rely on repeated purchases can be modeled by two components. The first is trial and can be represented by the simplified model that was shown in Figure 11.5. The second is repeat purchasing which applies only to those consumers who have previously tried the new product.

For example, in the HMO case, sales were made up of new enrollment plus re-enrollment from the 1,000 patients who were in the pilot HMO program for a year. Table 11.4 shows the calculation of total enrollment for the first year of full operation. The re-enrollment rate was high (0.95) since respondents evaluated the plan as excellent and had strong preferences for it over other existing health services.

Table 11.4. Calculation of Purchase Potential of New HMO

	Number Not Now in Pilot Program		Enrollment, if Aware		Estimated Awareness		Estimated New Enrollment
New Enrollment	17,200	×	0.233	×	0.70	=	2,800

	Existing Subscribers		Estimated Repeat Rate		Estimated to Remain at MIT		Estimated Repeat Enrollment
Re-enrollment	1,067	×	0.95	×	0.863	=	874

Total Enrollment = 2,800 + 874 = 3,674

The actual re-enrollment prediction was lowered somewhat since some percentage of the target group of students, faculty, and staff (.137) would graduate or leave the area for other reasons.

In frequently purchased products the projection of sales is more difficult since multiple purchases are made. Management is interested in the long-run market share. In this use, equation (11.4) is modified by providing estimates of both trial (t_i) and repeat (r_i) rather than a composite estimate (b_i). Again we make estimates for each individual i and sum up the probabilities across consumers. Mathematically this is given by:

$$\textbf{(11.5)} \qquad P = \sum_{i=1}^{N} a_w \, a_v \, t_i r_i$$

where t_i = trial probability for consumer, given awareness and availability

r_i = long-run share of purchases per period for new brand by consumer i, given that consumer i tried the product

a_w = awareness of new brand

a_v = availability of new brand

Trial probabilities are obtained from models based on consumer response after being exposed to concepts, while repeat shares are based on measures taken after home use of the new product. Note that this model is most relevant late in the design process when an actual product is available to give to consumers for in-home use. Earlier in the design process we replace r_i by some rough estimate of target repeat rate, $\bar{r}$. For example, early in the design process we may estimate a target repeat rate of $\bar{r} = 0.50$ to serve as a goal, but subject to modification.

Once one or more candidate products are selected for design evaluation, the value of P can be more easily calculated from the aggregate equation. Here we make population estimates of trial and repeat rather than summing individual estimates:

(11.6) $P = N_p a_w a_v T R$

where T = ultimate proportion of the target group who would try the product conditioned on awareness and availability, and

R = long-run share of purchases of the new product among those consumers who try the product

N_p = number of consumer purchases in category per period.

Urban (1975) shows that R can be approximated by a simple switching process as shown in Figure 11.6. In this case $R = R_{E1}/(1 + R_{E1} - R_{11})$ where R_{E1} is the proportion of consumers switching to the new product and R_{11} is the proportion of consumers who repeat purchase the new product. For example, if 40 percent of the consumers are observed to switch from existing products to the new product (R_{E1}), and if 50 percent of the consumers are observed to repeat purchase the new product after trying it (R_{11}), then the long run share among triers (R) is given by $R = (0.40)/(1 + 0.40 - 0.50) = 0.444$. Note that the long-run share (R) is not necessarily equal to the transient observed repeat rate of those who have tried the new product (R_{11}). The correction takes account of those consumers who are not 100 percent loyal to the new product.

To illustrate equation (11.6) suppose we achieved 40 percent trial (T), 80 percent awareness (a_w), 90 percent availability (a_v), and there were 1 million consumers in the target group (N_p). Equation (11.6) predicts that the long-run sales rate (P) of the new product would be

$$P = (1{,}000{,}000)\ (0.80)\ (0.90)\ (0.40)\ (0.44) = 126{,}720 \text{ purchases.}$$

Normally, these proportions are measured by giving consumers, first, an opportunity to purchase the new product and then, second, an opportunity after in-house use to repeat the purchase of the product. This model is an approximation to equation (11.6) because it confounds transient behavior (the market has not yet reached steady state) and aggregation error (the product of the averages of two numbers is not equal to the average of the product of two numbers). Despite the potential errors, we have found that these theoretical effects do not cause major empirical bias. Experience sug-

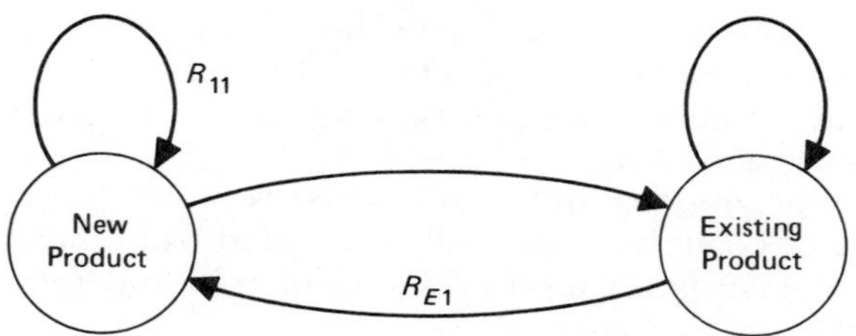

Figure 11.6. *Two-state Switching Model to Derive "Repeat" Proportion (R = representation of consumer switching among products)*

gests the model can accurately forecast long-run share. Nonetheless, it is an approximation and should be used with caution.

Diffusion of Innovation. Another type of dynamic process, diffusion of innovation, occurs when "innovators" (consumers with a high propensity to try new things) buy first and then, through an influence process, encourage others to adopt. This phenomenon is difficult to handle in new product sales forecasting since in the short run only the responses from innovators will be measured. Models based only on short-run measurements seriously underestimate the long-run adoption level. This is particularly true in consumer durables or industrial products.

The approach to modifying short-run estimates must rely on analogies to previous products. For example, Figure 11.7 describes the average acceptance for new products based on the adoption pattern of previous products. This judgmental process can be formalized by the use of a diffusion model such as the one developed by Bass and discussed in Chapter 5. We will consider such a model and measurement system in Chapter 14. The key concept in the design process is to realize that responses to new product

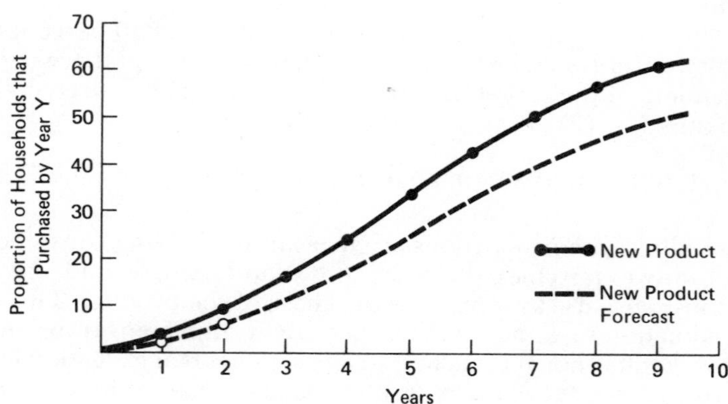

Figure 11.7. *Hypothetical Diffusion Curve for Major Kitchen Appliances*

concepts and physical products must be modified in cases where a diffusion of innovation process takes place.

Multiperson Decision Making

Not all purchase decisions are made by independent consumers. Often more than one person is involved in the purchase decision. For example, the choice of a family car or a home may be the result of some deliberation or bargaining process among the various members of the family. Multiperson decision making occurs most often in the purchase of industrial products. Engineering, purchasing, consultants, and top management may be involved in large procurements for items such as computers, production machines, and buildings.

In those cases the role of each participant must be understood. One approach to understanding roles is through a "decision matrix" which describes the role of each decision participant in each phase of the decision process. Figure 11.8 shows the decision matrix used by Choffray and Lilien in their study of solar air conditioning. The participants are divided into internal and external influences and the decision process is represented by five steps: (1) evaluation of needs, (2) budget approval, (3) search for alternatives, (4) equipment evaluation, and (5) final selection. After each participant's role is understood, his perceptions and preferences must be integrated to predict passage of the new product through each phase. Although research is underway to model this decision process, currently managers must rely on judgment to translate the multiple participants' preferences into purchase.

In some public systems, multiple groups are important to the success of a new program. While they may not interact as a formal decision-making body, their responses are important.

For example, in designing a new management program leading to the Master's degree, reactions of prospective students are important, but so are responses from recruiters who will hire the graduates of such a program. Remember that the important dimension for the student target group in the management education case was "outcomes." A good positioning of the new programs for the recruiters will lead to long-run substantiation of the "outcomes" dimension for students. In MIT's efforts to redesign its master's program, recruiters were interviewed and their perceptions and preferences modeled. Based on responses from twenty-five recruiters, a perceptual map was drawn. Four dimensions were identified by factor analysis of the recruiters' ratings of MIT, Harvard, and the new concepts (Young Executive Program and the Intensive Program in Management). The first dimension was correlated to scales that reflected graduates having real-world knowledge, action-orientation, practicality, relevant skills, the ability to choose a career, and contact with business people. It was named "job preparedness." The second reflected "group skills" and was correlated with the graduates' sensitivity to people, leadership abilities, generalist aptitude, and interpersonal

	Decision Participants / Decision phases	*Evaluation of a/c needs, specification of system requirements*	*Preliminary a/c budget approval*	*Search for alternatives, preparation of a bid list*	*Equipment and manufacturer evaluation**	*Equipment and manufacturer selection*
	Product and Maintenance Engineers	%	%	%	%	%
	Plant or Factory Manager	%	%	%	%	%
COMPANY PERSONNEL	Financial controller or accountant	%	%	%	%	%
	Procurement or purchasing department	%	%	%	%	%
	Top Management	%	%	%	%	%
	HVAC/Engineering firm	%	%	%	%	%
EXTERNAL PERSONNEL	Architects and building contractor	%	%	%	%	%
	A/C equipment manufacturers	%	%	%	%	%
	COLUMN TOTAL	100%	100%	100%	100%	100%

Figure 11.8. *Sample Decision Matrix: Industrial Cooling Study (Adapted from Choffray and Lilien.*

skills. The scales loading heavily on the third dimension involved graduates having sophisticated skills, familiarity with analytical techniques, research orientation, contact with faculty, and creativity in attacking problems. We called this the "technical skills" dimension. The final dimension represented the degree of competitive "pressure" the program placed upon the student.

The average standardized factor scores are shown in Figure 11.9. The existing program at Sloan was rated lower than Harvard on job preparedness and group skills, but higher on technical skills. Recruiters saw Harvard as a high pressure environment, while MIT was low. The recruiters rated the new programs favorably. Both the YEP and IPM programs exceeded Harvard on job preparedness. Measuring recruiter preferences for schools was difficult since preferences depend on the particular job to be filled and the individual student being considered. Therefore, in order to get an overall evaluation measure, individual ratings were weighted by the recruiter's self-designed importance for each scale (expectancy value model). Based on the weighted ratings, 63% of the recruiters rated one or both of the new concepts superior to Harvard and the existing Sloan programs. A weighted average of the importances and factor loadings indicated each of the four overall dimensions to be equally valued.

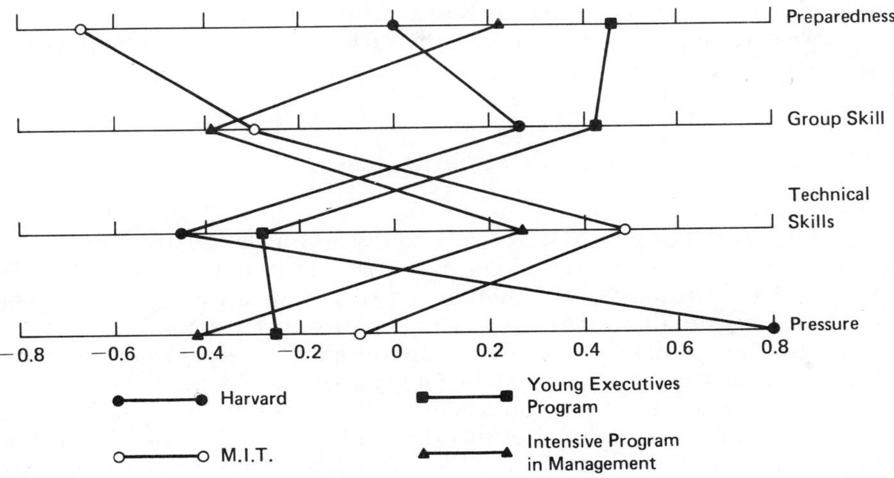

Figure 11.9. *Recruiter Perceptual Map*

In examining student and recruiter responses, there was agreement in a positive response to the new programs. Comparing the students' perceptual maps to those of recruiters revealed some common interests. The students and recruiters were both interested in career success. The student perceptions of realism and relevancy were also a component of the preparedness dimension of recruiters. Recruiters seemed particularly concerned with group skills as a positive attribute of graduates. Although both recruiters and potential students perceived Harvard high and MIT low on pressure, recruiters had a preference for more pressure, while students preferred a friendly and cooperative atmosphere.

Summary of Sales Formation Models

This completes the basic discussion of sales formation models. The basic concept is that estimates of purchase potential must be modified because of complexities that occur in new product introduction. The first modification is for awareness and availability. Purchase usually does not occur unless consumers know about the new product and can obtain it if they want, thus equation (11.4) estimates sales by conditioning purchase on estimates of awareness (a_w) and availability (a_v). The second modification is for dynamic effects. Long-run sales are not obtained immediately, thus equations (11.5) or (11.6) modify sales to account for trial/repeat phenomena and considerations such as in Figure 11.7 modify sales to account for the diffusion of innovation process. Finally, in industrial products, some consumer durables, and some public services, we must explicitly consider multiperson decision making. For more detail on sales forecasting methods see the appendices and a review article by Rao and Cox (1978).

We turn now to a review of how the various models of purchase potential and sales formation are used in the design process.

MANAGERIAL USE OF PURCHASE MODELS

The models of purchase potential and sales formation described in this chapter are designed to serve two purposes: (1) to determine if the new product has sufficient sales potential to proceed with the full design and testing effort, and (2) to aid in improving the product design.

The sales models proposed here are designed to give approximate results to enable the manager to make GO/NO GO decisions in the advancement of ideas, concepts, or products through the design process. Dynamics is used to refine the forecast after the core benefit proposition has been identified and the physical product has been formulated to substantiate the positioning. All of the models require some judgmental estimates and some averaging of individual effects, but they have proven adequate for the screening of new product designs.

The second purpose of purchase models is to aid in improving the new product design by simulating the effect on sales of improving features or perceptions of the product. For example, the HMO forecast used the preference rank order transformation to give a rough, early design estimate of demand. The result was 2,800 consumers (see Table 11.4). But we know from the case in Chapter 9 that there was an improved design based on new hospitals and more personal communication. Simulating that change through the preference model and rank order transformation gives a trial probability of 31 percent, which produces a demand estimate of 3,700 consumers (100 percent available, 70 percent aware, 17,200 consumers). But part of the improved strategy was an aggressive marketing campaign to make consumers aware of the personalness of the HMO through strong testimonials. It is quite likely that this campaign will increase awareness as well as trial. Suppose it increases awareness to 85 percent. Then the demand estimate would be 4,530 consumers who would sign up in the first year. This new enrollment, plus the returning enrollment of 875, brings the total enrollment to 5,405. These enrollment increases will justify revising the HMO design. Subsequently, the new plan was successfully implemented based on the new design.

Another example was a new aerosol hand lotion. Based on concept statements, the hand lotion was perceived as good on performance and application dimensions, but after use, it was perceived as only average on performance. In fact, after use, the product with the concept description was significantly less preferred than the product with no concept claims (i.e., "blind test"). The advertising claims of performance were not fulfilled by the product and this appeared to produce an adverse effect. This initial strategy generated trial and repeat probabilities that implied a 5 percent market share. Two alternate positions were simulated. The first represented a strategy of aligning the product claims and performance, which resulted in a 6 percent market share. The second strategy was to both improve the product physically so that it would perform better and to align the advertising to this improved performance. This strategy implied an 8 percent market share. These simulations led to design changes in the product advertising and additional R&D effort in an attempt to improve the perceived product performance. Ultimately, the product was unable to substantiate performance superiority, the project was dropped, and resources were allocated to other new product development efforts.

In general, we have found that the complete design models (perception, preference, and choice) generate their most substantial rewards in improving the design by generating creative insights through enhanced understanding of consumer response and the display of results of simulation runs of the models which allow evaluations of the costs and benefits of proceeding with a project.

SUMMARY

Success in new products is intimately linked to the ability of a new product to attract and retain consumers. Whether the goals are profit, market share, or increased utilization, the key marketing input is an estimate of how many consumers will purchase or use the new product. Previous chapters presented techniques to design a product that would be preferred by consumers. This chapter presents techniques to select the best strategy and to ensure that the chosen strategy will provide a sufficient return on investment.

The primary input to demand estimation is the preference measures developed in Chapter 10. These measures are used in the purchase potential and sales formation models to get estimates of how many consumers actually buy or use the product. On this estimate rests the potential success or failure of a product positioning. The remainder of the marketing mix is necessary to realize that potential.

We discussed how to modify these estimates for advertising (awareness) and distribution (availability), but a complete design needs detailed marketing inputs such as advertising copy, pricing schedules, sales manuals, and final engineering specifications. In the next chapter we combine the information in Chapters 8, 9, 10, and 11 to detail the complete product strategy and assemble the marketing mix and physical product to support that strategy.

REVIEW QUESTIONS

11.1 Why is sales forecasting important to the design of new products? Discuss the role of sales forecasting in the design process (Figure 7.1).

11.2 What is a preference rank order transformation? What are its uses and its accuracy?

11.3 Intent to purchase measures seem to provide the manager with all the necessary information required to make new product decisions. If the product is described in full detail including price, physical features, customer benefits, packaging, etc. and the consumer questioned states his intentions to buy, why should the manager then seek to transform this rank order preference to choice probabilities?

11.4 What are the advantages of the Logit model? The Logit model provides the most accurate demand estimates when compared to other preference models and measures discussed in the text. Given the predominance of the Logit model why employ any other technique?

11.5 What diagnostic information is produced by each of the models of sales formation? How is this information used to improve the new product?

11.6 What are the advantages of making probabilistic forecasts? Why not simply assume consumers will choose the product that preference regression or conjoint analysis says they should prefer?

11.7 Give an example of multiperson decision making in the industrial sector. How would you use the inputs from the multiple decision participants to arrive at a sales forecast?

11.8 How can Bass's model be used to forecast the dynamics of the diffusion process?

11.9 Why is it important to consider awareness and availability when forecasting sales? How do you estimate awareness and availability?

11.10 Voras Chemical Supply Company is launching a new brand of exotic dinners designed for picnics where insect pests pose a potential problem. They estimate product awareness to be approximately 99 percent because the product has received an unexpectedly high press coverage. One supermarket chain which sells 37 percent of picnic dinners in the target market will carry the product. The following probabilities of purchase given awareness and availability are given.

Percent of Population	*Purchase Probability*
50	0 – will not even buy adjacent products on shelf
40	0.2 – good souvenir
10	0.9 – definite collector's item

If the target market consists of one-half-million families each buying one dinner per purchase occasion, what will be the expected number of consumers trying the product? Do you expect that this product will be successful in the long run? Why or why not?

11.11 In problem 9.2 you considered your perceptual positioning in the job market. Reconsider your positioning in light of the recruiter perceptual map in Figure 11.9.

Chapter 11 Appendices

11.1 Derivation and Estimation of the Multinomial Logit Model*

Derivation. From the text we know that:

$$L_{i1} = \text{Prob}\,\{p_{i1} + e_{i1} \geq p_{ij} + e_{ij} \quad \text{for all } j = 2 \text{ to } J\}$$

Which is equivalent to:

$$L_{i1} = \text{Prob}\,\{e_{ij} \leq p_{i1} - p_{ij} + e_{i1} \quad \text{for all } j = 2 \text{ to } J\}$$

Now suppose that the e_{ij} for all j are independently identically distributed (i.i.d.) with the Weibull extreme value distribution, that is:

$$\text{Prob}\,\{e_{ij} \leq \epsilon_0\} = \exp\,(-\exp\,(-\beta\epsilon_0))$$

Since the e_{ij} are independent, the joint cumulative distribution for the e_{ij}'s for $j = 1$ to J can be decomposed into the product of J univariable cumulative distribution functions. Thus:

$$\text{Prob}\,\{e_{i1} \leq \epsilon_1, e_{i2} \leq \epsilon_2, \ldots, e_{iJ} \leq \epsilon_J\} = \prod_{j=1}^{J} \exp\,(-\exp\,(-\beta\epsilon_j))$$

Recognizing that for a given e_{i1}, the above equation is the joint cumulative distribution of e_{ij} for $j = 2$ to J at values $p_{i1} - p_{ij} + e_{i1}$, integrating out e_{i1} then gives:

*This derivation is based on McFadden (1970).

$$L_{i1} = \int_{e_1=-\infty}^{\infty} \beta \exp(-\beta\epsilon_1) \exp(-\exp(-\beta\epsilon_1)) \prod_{j=2}^{J} \exp(-\exp\beta(-p_{i1} + p_{ij} - \epsilon_1))d\epsilon_1$$

$$= \int_{\epsilon_1=-\infty}^{\infty} \beta \exp(-\beta\epsilon_1) \exp\{-\exp(-\beta\epsilon_1) \cdot \sum_{j=1}^{J} \exp\beta(p_{ij} - p_{i1})\}d\epsilon_1$$

Recognizing $\sum_j \exp\beta(p_{ij} - p_{i1})$ as a constant in the integration and making the substitution $x = \exp(-\beta\epsilon_1)$ yields the result that

$$L_{i1} = \frac{1}{\sum_j \exp\beta(p_{ij} - p_{i1})}$$

Multiplying numerator and denominator by $\exp(\beta p_{i1})$ yields the logit model.

Estimation. The logit model is a probability of choice model. Thus, for a sample population we would like to select that value of β that maximizes the probability that the particular observations occurred. Thus, let $L_i(\beta)$ be the probability assigned to the product that i was observed to select. Define $\delta_{ij} = 1$ if i selected j and $\delta_{ij} = 0$ otherwise. Define $L_{ij}(\beta)$ according to equation (11.3), except that β is an as yet unknown parameter.
Then:

$$L_i(\beta) = \prod_{j=1}^{J} [L_{ij}(\beta)]^{\delta_{ij}}$$

and the joint probability that the population behaved as they were observed to behave is

$$P_p = \prod_{i=1}^{N} \prod_{j=1}^{J} [L_{ij}(\beta)]^{\delta_{ij}}$$

The goal is then to maximize P_p by appropriate selection of β. Empirically, if we maximize $\log P_p$, then we maximize P_p. Therefore, the standard optimization that selects β is maximize $L(\beta)$

$$\text{subject to } L(\beta) = \sum_{i=1}^{N} \sum_{j=1}^{J} \delta_{ij} \log [L_{ij}(\beta)]$$

McFadden (1970) shows that under quite general conditions the maximum exists and is unique. Most computer packages use some modification of the Newton-Raphson method to obtain maximum likelihood estimates of β.

We have provided an intuitive description of maximum likelihood for the case when only β is to be estimated. The same intuitive holds and can be extended to the revealed preference model (Appendix 11.4) where

$$p_{ij} = \left(\frac{1}{\beta}\right) \sum_k w_k x_{ijk}$$

In that case, substitute $L_{ij}(w_1, w_2 \ldots, w_K)$ to obtain estimates for the weights w_1 through w_K. Most computer packages are written explicitly for this problem.

As stated in the text, the maximum likelihood estimates are asymptotically efficient and normally distributed. Furthermore, if the null hypothesis can be formulated as $L(w_1^0, w_2^0, \ldots, w_K^0)$ where $w_1^0, w_2^0, \ldots, w_K^0$ is restricted to some subset of the $w_1, w_2, \ldots, w_K$ space, then $2[L(w_1, w_2, \ldots, w_K) - L(w_1^0, w_2^0, \ldots, w_K^0)]$ is chi-squared distributed with degrees of freedom equal to $K - D$ where D is the degrees of freedom in the restriction. Useful null hypotheses are equally likely ($w_1 = w_2 = \cdots = w_K = 0$) and market share proportional (choice-specific variables only when they are included in the full model). The significance of each estimate is determined through a t-test based on the standard errors of the estimates. (Most packages automatically provide standard errors.)

For rigorous discussions of maximum likelihood estimation, see Mood and Greybill (1963); and for a development of the properties of the logit estimates, see McFadden (1970). For applications in marketing see Punj and Staelin (1978), and Gensch and Recker (1979).

11.2 Statistical Tests for Probability of Purchase Models

Like all models that are used in design, purchase potential models must be accurate. Thus, we introduce a statistical test for probability of purchase models. Statistical tests are based on comparing observations to predictions. Thus, let f_{ij} be the number of times consumer i actually chooses product j. Note that while f_{ij} may be based on only one observation per consumer, this presents no conceptual problem with probabilistic models. Furthermore, all tests must be compared to some null hypothesis, i.e., some naive model that uses little or no information. Let L_{ij}^0 be the probabilities associated with the null hypothesis. For example, if the null hypothesis is that all products are equally likely, then L_{ij}^0 is equal to one divided by the number of products. If the null hypothesis is that preferences do not matter and consumers choose in proportion to market shares, then $L_{ij}^0 = ms_j$, where ms_j is the observed market share of product j.

Given these definitions, we can then compute a statistic, U_0^2, that measures the amount of uncertainty that is explained by the probability model relative to the null model. This measure can be used to assess the goodness-of-fit of the model to the observed data. It can also be used to compare the model to naive models such as equally likely probability or probability proportional to market share or to simpler models. For the logit model tested against the equally likely naive model, U_0^2, is known as the likelihood ratio index and is available as an output on most logit packages. If U_0^2 is extremely high, the analyst has confidence in the numerical prediction and can base detailed marketing strategy on the output of the model. If U_0^2 is low (less than 10–20 percent), then the estimation of demand is a weak estimate of the relative effects of alternative strategies. A low U_0^2 often occurs very early in the design process and indicates that further analysis is necessary before a final decision can be made with confidence.

For example, the logit model that produced the estimates in Table 11.3 explained 50.4 percent of the uncertainty.* The preference model correctly predicted 80 percent of the preferences, but only 76 percent of the population chose their first preference. The choice model correctly predicted 78 percent of the choices. As shown in Figure 11.2A-1, this explanatory power can be decomposed into a portion that can be explained by knowing the market shares (29.1 percent) and an incremental portion due to the logit model (21.3 percent). Figures such as 11.2A-1 are useful because they graphically indicate to the new product team how well the model explains behavior. It is interesting to note that the rank order transformation could only explain 34.6 percent of the uncertainty.

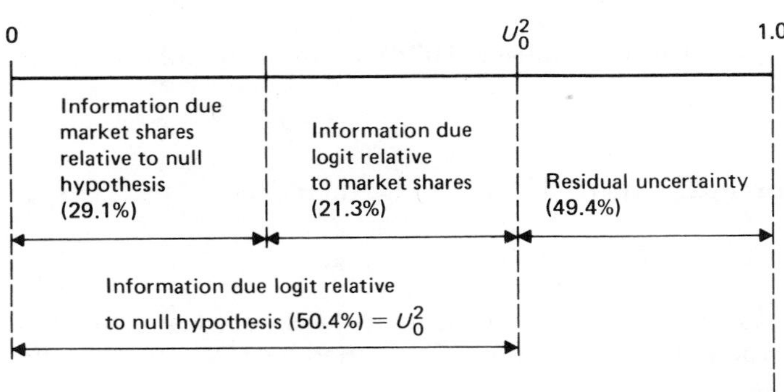

Figure 11.2A-1. *A Relative Measure for Testing the Logit Model*

The measure of uncertainty (U_0^2) is based on information theory. We provide the equations and certain theoretical results. First, it can be shown that the total uncertainty of a probabilistic system is measured by the entropy, H_0, of that system where

$$H_0 = -\sum_j L_j \log L_j$$

*The actual model included availability constraints for automobile that varied by the educational level of the consumer.

and where L_j is the prior probability for product j. Let L_{ij}^m be the estimated probability that i chooses j as given by model m. Note that $L_{ij}^m = L_{ij}$ for the logit model, $L_{ij}^0 = 1/J$ for equally likely, and $L_{ij}^1 = \text{ms}_j$ for the market share proportional model. Then the information, $I(m, n)$ that model m provides relative to model n is:

$$I(m, n) = \left(\frac{1}{N}\right) \sum_i \sum_j \delta_{ij} \log\left(\frac{L_{ij}^m}{L_{ij}^n}\right)$$

where $\delta_{ij} = 1$ if i chooses j in the sample, $\delta_{ij} = 0$ otherwise. $I(m, n)$ is actually an observation from a normal distribution with mean $EI(m, n)$ and variance $V(m, n)$ given as:

$$EI(m, n) = \left(\frac{1}{N}\right) \sum_i \sum_j L_{ij}^m \log\left(\frac{L_{ij}^m}{L_{ij}^n}\right)$$

$$V(m, n) = \left(\frac{1}{N}\right) \sum_i \left\{ \sum_j L_{ij}^m \left[\log\left(\frac{L_{ij}^m}{L_{ij}^n}\right)\right]^2 - \left[\sum_j L_{ij}^m \log\left(\frac{L_{ij}^m}{L_{ij}^n}\right)\right]^2 \right\}$$

Finally, we can derive the residual entropy, H_m, as:

$$H_m = -\sum_i \sum_j L_{ij}^m \log L_{ij}^m$$

Given these derivations, Hauser (1978) provides the following theorems:

Theorem 1: $I(\text{perfect knowledge}, n) = H_n$

Theorem 2: If $(1/N) \sum_i L_{ij} = \text{ms}_j$, then $EI(m, n) = H_n - H_m$

Theorem 3: $I(m, n) = I(m, q) + I(q, n)$

These derivations and theorems provide the theory for testing the model. If $U^2(m, n)$ is the percent uncertainty explained by m relative to n then:

$$U^2(m, n) = \frac{I(m, n)}{H_n}$$

where the statistic, U_0^2, is just $U^2(m, 0)$. Furthermore, based on theorem 3, the relative information is additive and hence Figure 11.2A-1 indeed represents relative values.

Most logit packages output a statistic called the "likelihood ratio index," which is also known as the "pseudo-R^2 measure." Hauser shows that this index is numerically equal to U_0^2. Given this measure, theorems 1, 2, and 3 can often be used to compute a number of relative measures by recognizing that if everyone has the same choice set, then

$$H_0 = \log J \qquad H_1 = -\sum_j \text{ms}_j \log \text{ms}_j$$

$$I(1, 0) = H_0 - H_1 \qquad I(m, 1) = I(m, 0) - I(1, 0), \quad \text{etc.}$$

If the choice set varies, define J_i = number of alternatives in i's choice set and then $p_{ij}^2 = (1/J_i)$ if j is in i's choice set, $p_{ij}^2 = 0$ otherwise. If N_c equals the number of consumers with c objects in their choice set, then

$$H_2 = \left(\frac{1}{N}\right) \sum_{c=1}^{J} \left(\frac{N_c}{c}\right) \log (c)$$

and $I(2, 1)$, $I(2, 0)$, $I(m, 2)$ can be easily calculated. Thus, Figure 11.2A-1 can be expanded for a multitude of successively stronger null hypotheses. Note that if $I(m, n) < 0$, then model m is counterproductive relative to model n.

If repetitive choice is allowed, then $L_{ij} = f_{ij}$ rather than $L_{ij} = \delta_{ij}$ is perfect information where f_{ij} is the frequency with which i chooses j. Define this new model as L_{ij}^f. Then for repetitive choice compute $I(m, n)$ by summing over choice occasions as well as individuals and products. It is easy to show that the maximum $I(m, n)$ is $I(f, n)$. Thus, for frequencies replace H_n by $I(f, n)$ when computing $U^2(m, n)$.

The percent uncertainty tests the usefulness of the probability model. The chi-squared test of Appendix 11.1 tests the significance of the logit model and the t-test checks each individual estimate. One final test of accuracy tests whether the logit model can accurately reproduce the observed choice. To perform this test, use the cumulative normal distribution to determine whether $I(m, n)$ is a reasonable observation from a normal distribution with mean $EI(m, n)$ and variance $V(m, n)$.

11.3 Logit Analysis: Advanced Topics

Logit analysis is but one model in a class of probabilistic choice models. We cover it in detail because: (1) it is the most practical of the choice models for use in new product design, (2) it has enjoyed widespread use in management science and social science, and (3) once the reader is familiar with basic logit analysis, he can more readily understand the advanced topics and alternative models. This section will present some of these topics on an intuitive level. For rigorous derivations and axiomatic theory, we suggest Ben-Akiva (1973); Hauser (1978); Koppelman (1976); Krantz, Luce, Suppes, and Tversky (1971); Luce (1959); Luce (1977); Luce and Suppes (1965); McFadden (1970); Manski and Lerman (1977); and Tversky (1972).

We begin with an alternative interpretation of the model (Luce's axiom) and a discussion of a key behavioral assumption underlying the derivation (independence of irrelevant alternatives). We then turn to extensions of the model for use in new product design, i.e., choice specific constants, preference inertia, choice-based sampling, and a rank order probability model. We close with a short review of alternative probabilistic models.

Luce's Axiom (Independence of Irrelevant Alternatives). One assumption in the derivation of the logit model was that the error term, e_{ij}, was independent and identically distributed across products in the consumer's evoked set. Intuitively, this means that the "unobserved" effects do not depend upon which particular products are in the evoked set. Choice probabilities depend only upon the preference values, p_{ij}, for each product in the evoked set. As a result, the logit model is in a form that can be derived from a fundamental axiom on how choice probabilities change as the evoked set changes. Stated simply, this axiom says that the ratio of the choice probabilities of any two products do not change no matter what the evoked set is as long as those two products are in the evoked set. Mathematically, if

$$P[a \mid \{a, b\}]$$

is the probability that product a is chosen when the evoked set is only a and b and if $P[a \mid C]$ is the probability that product a is chosen when the evoked set is C. (For example, C = products $a, b, c, d,$ and e.)
Then

$$\textbf{(11.3A-1)} \qquad \frac{P[a \mid \{a, b\}]}{P[b \mid \{a, b\}]} = \frac{P[a \mid C]}{P[b \mid C]}$$

where $P[b \mid \{a, b\}]$ and $P[b \mid C]$ are defined in an analogous way, $a, b, \epsilon\, C$, and the denominators are non-zero . (The obvious results hold when the denominators are zero.) We will discuss the implications of that axiom in a moment.

Define $U(a, b) = P[a \mid \{a, b\}]/P[b \mid \{a, b\}]$, then if $U(a, b) = U(a)/U(b)$ and $j = 1, 2, \ldots, J$ indexes the products in C, it is easy to show (Luce, 1959) that

$$\textbf{(11.3A-2)} \qquad P[a \mid C] = \frac{U(a)}{\sum_{j=1}^{J} U(j)}$$

Finally, if $U(j) = \exp(\beta p_{ij})$, then equation (11.3A-1) gives the logit model of equation (11.3).

It is important to note that equation (11.3A-1) is based on a behavioral axiom which determines how the choice probabilities behave if the evoked set changes. It does not explicitly state how preferences are linked to choice. To use Luce's axiom, one must provide further assumptions which dictate explicitly or implicitly the function $U(j)$. The logit model is one particularly useful model which implicitly derives this function. Alternatively, one can hypothesize other functions and be consistent with the axioms. For example, Pessemier (1977) uses Steven's Power Law to define $U(j) = \lambda_i p_{ij}^{\beta}$ and derives an alternative formulation. Note that both the logit formulation and Pessemier's formulation use a "fitting" parameter β which matches the preference scales to the particular applications. This parameter is important because preference measures are unique only to scale transformations and use of equation (11.3A-2) without any "fitting" parameters can lead to counterintuitive results. For example, if $U(j) = p_{ij}$, then adding a large positive constant to all preference values will leave the relative preferences unchanged but will make $P[a \mid C] = 1/J$ for all products. Clearly, all models are potentially subject to similar errors, but "fitting" parameters reduce these effects.

Luce's axiom leads to an alternative derivation of the logit model and to useful alternative probability of choice models. But what does it imply about choice behavior? Let us consider an example. Suppose that there are two doctors, Dr. Jones and Dr. Smith, and one HMO in a small town in Maine. Suppose that the choice probabilities for consumer i are $p_{ij} = 0.3$ for Dr. Jones, $p_{is} = 0.3$ for Dr. Smith, and $p_{ih} = 0.4$ for the HMO. Suppose Dr. Jones runs off to Lithuania. We would expect that p_{is} would increase proportionally more than p_{ih}. For example, suppose the true choice process was to first choose between private care and group care and then select the specific party to deliver that care. In this case, Dr. Jones' departure would lead to $p_{is} = 0.6$ and $p_{ih} = 0.4$. Luce's axiom assumes that the change in evoked set causes proportional shifts in probabilities, i.e.,

$$p_{is} = \frac{0.3}{(0.3 + 0.4)} = 0.43 \quad \text{and} \quad P_{ih} = \frac{0.4}{(0.3 + 0.4)} = 0.57.$$

Similarly, if we introduce a new freeze-dried coffee, we would expect to draw proportionally more shares from the freeze-dried coffees than from regular coffees; if we introduce an express bus system, we expect to draw proportionally more shares from regular buses than from the auto; and if we introduce a new subcompact car, we expect to draw proportionally more shares from other subcompacts than from large luxury cars.

Thus, Luce's axiom cautions the analyst to be careful in defining the choice set so that the detrimental effect of the axiom can be minimized. The market identification analyses in Chapter 5 can be used to search for hierarchies in the choice process and appropriately define the market. The perceptual techniques of Chapter 9 can be used to search for the relevant perceptual variables. A comprehensive set of perceptual variables more completely defines p_{ij}, minimizes the unobserved variables, and reduces the magnitude of e_{ij}. These precautions cannot ensure that the model is immune to the evoked set definition (independence of irrelevant alternatives), but they can minimize counterintuitive predictions.

One model that explicitly addresses evoked set effects is the *multinomial probit model.* This model uses a normal distribution for the error term and allows the covariances between error terms to be non-zero. (That is, cov $(e_{ij}, e_{il}) \neq 0$.) The problem is analytically complex, but recent developments have led to practical computer programs that use the multinomial probit model (Albright, Lerman, and Manski, 1977). For a discussion of the binomial probit model, see McFadden (1976).

Choice-Specific Constants. Ideally, the perceptual analysis will identify all the relevant dimensions that influence choice. But in some product categories, there may be some residual "image" characteristics that cannot be measured. For example, the strong image of IBM or Xerox may be difficult to explain with perceptual dimensions alone. Alternatively, the analyst may know that he has underspecified the dimensions due to cost or data limitations. To overcome these limitations, one can add choice-specific constants to the logit model. In other words, preference is redefined as:

(11.3A-3) $$\tilde{p}_{ij} = p_{ij} + u_j$$

where $\tilde{p}_{ij}$ is the "improved" measure of preference and u_j is a constant added to the preference of product j. Empirically, this constant is entered as a dummy variable ($d_j = 1$ for product j, $d_j = 0$ otherwise) in the logit estimation.

The advantage of this formulation is that it adds to the predictability of a model and provides an indication of the magnitude of missing effects. The disadvantage for new products is that one must judgmentally estimate u_1 for the new product. Thus, this formulation is most useful when changes to existing products are being evaluated rather than entirely new products.

One side effect of choice specific constants is that if a full set of constants is included, one for each alternative, then the logit model will always predict market shares exactly on the estimation data sample. (To prove this, use Lagrange multipliers and set up the maximum likelihood conditions.) Actually, only $J - 1$ constants are included. The Jth constant would be redundant because the probabilities must sum to 1.0. Thus, these constants are actually relative effects. Choice-specific constants have the advantage of potentially more accurate predictions for existing products but can be misleading for new products.

Preference Inertia. In general, choice-specific constants capture some product-specific effects. One particular product-specific effect, called preference inertia, has been investigated by Neslin (1976). Neslin found that consumers hesitate to select new products or "stretchers" even though the characteristics of the new products are perceived to be superior to existing products. In other words, there is a tendency by consumers to remain with existing products which have known characteristics rather than to switch to new products with uncertain characteristics. This effect is most pronounced in major purchases such as consumer durables, large industrial products, or major consumer services such as health care. In his study, Neslin introduced an inertia factor* in the form of a choice-specific constant that was the same for all "stretchers." This inertia factor alone significantly increased the preference model's ability to "explain" preference (55 percent of the first preferences were correctly predicted with the inertia factor, 46 percent without). Neslin then extended this concept to make the inertia factor dependent

*Neslin used preference regression rather than logit analysis, but the concept is the same.

upon the characteristics of the individual consumer. For example, Neslin found that relatively unhealthy people with less healthy families, who are happier with current health care, currently paying less out-of-pocket for health insurance, have a slightly higher consumption of doctor visits, and who are older, married, and female are less likely to try innovative health care delivery plans. This area of research is promising and could lead to further insights on the diffusion of innovations.

Choice-based Sampling. Often random samples are very expensive. For example, suppose we are designing a new refrigerator. Suppose we draw a random sample from the target population. Chances are that less than 10 percent of that sample is in the market for a new refrigerator. An alternative technique is to sample only those consumers interested in buying a refrigerator. This strategy is more efficient but still requires a random sample telephone screen. A third method would be to determine from store records or warranty records those consumers who bought a refrigerator in the last six months. This sampling strategy is called "choice-based sampling" because the sample is stratified by last product chosen.

The random sample minimizes sampling error; the screened sample still can have some sampling error; but the choice-based sample (CBS) is clearly biased because we are sampling on the variable we are trying to explain. For example, suppose that 90 percent of the consumers in a screened random sample actually buy Brand A, but only 50 percent of the consumers in the CBS buy Brand A. The CBS would then be biased toward consumers who bought Brand B, and any preference analysis would lead to biased importance weights. Thus, a CBS is often less expensive than a random sample, but leads to biased estimates if left uncorrected. Fortunately, there are at least three ways in which sampling errors can be corrected in a CBS.

First, if a CBS is collected randomly for each product in such a way that the number of sample consumers for each product is in proportion to the number of consumers who would have chosen that product in a random sample, then the CBS is statistically equivalent to a random sample and leads to unbiased estimates.

Second, if the proportion of consumers who would have chosen each product had a random sample been collected, is known, then a CBS is more or less equivalent to a stratified sample and each observation in the likelihood function can be weighted in such a way as to correct the bias. Lerman, Manski, and Atherton (1976) show that the correct weight for consumers who chose product j is ms_j/cbs_j where ms_j is the "true" market share from a random sample and cbs_j is the number of consumers sampled based on having chosen product j. Many logit packages have now been modified to accept these weights.

The third CBS correction is conceptually more complex but leads to tremendous practical savings in data collection. Suppose that ms_j is unknown, then how do we determine ms_j/cbs_j? While the proof is complex, the result is simple: if a full set of choice-specific constants are included in the logit model, then those constants alone are biased. In other words, if you include choice-specific constants, then the other logit parameters, β or the w_k's, are "unbiased" even in an uncorrected CBS. Intuitively, sampling bias is an unobserved error that leads to mean product-specific bias in the error term. This mean effect then becomes part of the product-specific constant. Note that an additive constant in the preference function is a multiplicative constant in exp $(\beta p_{ij} + \beta u_j) = \gamma_j \exp(\beta p_{ij})$ which ultimately "weights" each observation in the likelihood function. (See Manski and Lerman, 1977, for a full proof.) This means that the market researcher can use CBS to save sampling costs and still get valid measures of relative preference. For example, if there are seven refrigerator manufacturers on the market, but the researcher can only get samples from the warranty lists of four of those manufacturers, then he can still get "reasonable" estimates of β or of the w_k's in the revealed preference model—as long as the four manufacturers are "reasonably" representative of the market.

Rank Order Logit Model. Preference logit uses only first preference information. Is it possible to use the full rank order preference information in a logit model? Luce and Suppes (1965) show that because the logit model satisfies equation (11.3A-1) it is possible to create $J - 1$ observations for each consumer such that the resulting model gives the correct likelihood function. The $J - 1$ observations are created as follows. Observation 1 is first preference from the full evoked set. Observation 2 is second preference but from the evoked set with the first preference alternative deleted. Observation 3 is third preference from the evoked set with both the first and second preference alternatives deleted. Observations 4, 5, ..., $J - 1$ are formed in an analogous way. This procedure works because of the evoked set independence properties of Luce's axiom.

Elimination-by-Aspects. The logit model is based on a linear preference model. An alternative stochastic choice model is based on an elimination phase of the consumer model. In elimination-by-aspects, Tversky (1972) assumes that consumers select products by successively eliminating products from the evoked set on an attribute-by-attribute basis but that the order in which attributes are selected is stochastic. Based on this assumption, Tversky derives an alternative model that is not limited by Luce's axiom. While complex, these models are potentially applicable to the new product process.

Other Topics. This section has indicated some of the advanced topics that are relevant to the new product development process. Other extensions have been studied in other areas. For example, Ben-Akiva (1973) discusses how the properties of the model vary when assumptions are made about whether the choice process has hierarchies or is simultaneous; Koppelman (1976) discusses approximations to equation (11.4) when the sample is large and little is known about specific consumers; Green and Carmone (1977) use a balanced incomplete block design to improve measurement efficiency; and Krishnan (1977) introduces minimum perceivable differences.

For a state-of-the-art review of probability models, see McFadden (1976); and for an axiomatic development of the measurement for these models, see Krantz, Luce, Suppes, and Tversky (1971).

11.4 Other Uses of Logit in the Design Process

The logit model can be used in other phases of the design process. At the preference level it can be used to estimate importances either from rank order preference or as revealed by choice. Within this framework it can be used for benefit segmentation. Since some of these analyses are potentially more accurate, readers should be sure to include these logit procedures in their "tool kit" of methods to use at the preference analysis phase of new product design.

Importances by Revealed Preference. Logit analysis can be used to implicitly estimate the parameters of the linear compensatory model. Simply define

$$p_{ij} = \left(\frac{1}{\beta}\right) \sum_{k=1}^{K} w_k x_{ijk}$$

where w_k equals the importance of attribute k and x_{ijk} equals respondent's evaluation of product j with respect to attribute k. Now substitute this definition in equation (11.3). This yields:

$$\textbf{(11.4A-1)} \qquad L_{i1} = \frac{\exp\left(\sum_{k=1}^{K} w_k x_{i1k}\right)}{\sum_{j=1}^{J} \exp\left(\sum_{k=1}^{K} w_k x_{ijk}\right)}$$

Similarly, define equations for L_{i2}, L_{i3}, etc. The same theory applies and the same computer packages can now be used to estimate the "revealed" preference weights, $w_1, w_2, w_3, \ldots, w_K$. These weights are called revealed weights because statistically they are "revealed" by choice behavior rather than resulting from analysis of measured preferences. In fact, this preference model is the predominant preference model used in the analysis of transportation systems. See, for example, Ben-Akiva (1973), Charles River Associates (1972), Koppelman (1976), and McFadden (1975).

The interpretations of the importance weights, w_k's, are much the same as in preference regression. But the manager or analyst should be cautioned that non-preference effects such as awareness, availability, and trial/repeat do influence choice probabilities. These effects should be measured, estimated, or controlled, and incorporated explicitly or implicitly before equation (11.4A-1) is used to "reveal" importance weights. Revealed preference is an extremely useful technique if used correctly, but blind use of revealed preference can lead to interpretations that may not be valid.

One further use of revealed preference is worth mentioning. Some researchers estimate both a direct preference model (e.g., preference regression) and a revealed preference model and compare the relative importance weights. By examining the difference, a manager can better understand the extraneous events that influence choice above and beyond preference.

Preference Logit. Preference regression was one model for the direct statistical estimation of importance weights based on rank order preference. An alternative procedure is to use logit analysis, but with first preference as the dependent variable rather than actual choice. In other words, define $\mathcal{L}_{i1}$ as the probability that individual i ranks product 1 as first preference and replace L_{i1} with $\mathcal{L}_{i1}$ in equation (11.3). This model, called "preference logit," is used and interpreted in the same way as preference regression. In fact, in the HMO, shopping center, and Evanston transportation cases the interpretations were quite similar. Preference logit may provide slightly more accurate predictions than preference regression, and we recommend it if: (1) the sample size is greater than 100 consumers and, (2) the analyst has a logit package readily available.

Each method looks at the preference data differently. The preference logit utilizes only the first preference, while the preference regression uses the full rank order. If the preference logit and preference regression yield similar results, confidence is gained.

Benefit Segmentation. Revealed preference logit or preference logit can be used as alternatives to preference regression to estimate importance weights in a linear compensatory model. Thus, a similar search and testing procedure can be used to develop benefit segments. The formal statistical test is a chi-squared test. Simply estimate a logit model in each potential segment. Let χ_g^2 be the chi-squared statistic for the common model and let χ_n^2 be the chi-squared statistic for the model for segment n. Then, if the segments do not overlap and if the same variables and choice alternatives are used in all models, the following chi-squared statistic can be used to test whether the segmented models are significantly different:

$$\textbf{(11.4A-2)} \qquad \chi^2_{\text{seg}} = \left(\sum_{n=1}^{S} \chi_n^2 \right) - \chi_g^2$$

If there are S segments and K variables, χ^2_{seg} has $(S - 1)K$ degrees of freedom. The intuitive idea behind equation (11.4A-2) is that the segmented model can be thought of as a single model with $S \cdot K$ segment-specific variables.

Chapter 12

Completing the Design: Product Engineering and Marketing Mix

Before a new product is advanced to testing, we must complete the design. A complete design is characterized by:

- the target product market and target group of consumers;
- Core Benefit Proposition (CBP);
- positioning of the product versus its competition;
- physical characteristics of the product to fulfill the CBP;
- the initial price, advertising, and distribution strategy.

Determination of the target market was supported by the analytic procedures of market definition and consumer segmentation (Chapter 5). With this background and a designation of the target market, ideas were generated to serve as the basis of determining the Core Benefit Proposition (Chapter 6). Next, market research was used to identify and measure consumers' perceptions of the attributes that differentiate products (Chapter 9). Existing products plus the concept descriptions of the new product ideas became stimuli on the perceptional map. The relative positions determined the competitive positioning portion of the CBP.

The best positioning was determined by measuring preferences and estimating importances (Chapter 10). This analysis prioritized alternative CBP positionings. After the best concept opportunity was identified, attention turned to the physical product to substantiate the concept, and importances of physical features were estimated by conjoint analysis. After an initial prototype product was produced, consumers were given an opportunity to try the product, then perceptual mapping and preference analysis were used again, this time to assure that the product fulfilled the promise of the CBP.

With good preference response to the new product CBP positioning and benefits delivery, attention was directed at making preliminary sales forecasts to determine whether or not the new product should progress to the testing phase of the development process.

Before proceeding to the testing phase, the design must be completed. One more iteration must take place and the CBP must be implemented by final engineering of the physical product and creation of the advertising copy. Although a prototype physical product may have been tested in the analysis of preferences, formidable tasks remain. The physical product must be revised based on the results of product use and the engineering must be carried out to support production of the product. Furthermore, the organization must address the creative task of converting the CBP into finished advertising and/or sales promotion materials. Once these tasks are complete, the marketing strategy must be set by assembling the appropriate mix of advertising, pricing, couponing, sampling, distribution strategy, etc., to complement the Core Benefit Proposition.

In this chapter we first discuss the issues of implementing the CBP through the final physical product and advertising copy. This calls for integration of R&D, engineering, and marketing skills to create a complete design that effectively communicates the CBP and substantiates it by reliable product performance. With the product and communication task complete, attention is focused next on setting the initial price, distribution strategy, and budget for advertising, promotion, and selling. These decisions will then be refined at the testing phase of development when test market data become available to estimate sales response. Finally, we illustrate the integrated design process with a case study on the design of a new laundry detergent.

IMPLEMENTING THE CORE BENEFIT PROPOSITION

Creating the Physical Product

A good psychological positioning for the CBP of a new product is important, but the actual product that fulfills the CBP will determine the long-term success of the innovation. Careful attention must be directed toward the physical attributes that deliver the benefits promised in the CBP. At this point in the process the comprehensive strategy (Chapter 2) used to integrate R&D and marketing becomes crucial. The diagnostic information generated by marketing research is coupled with the initial R&D research (from idea generation) to engineer the final product. Basic research may have made the product development feasible; it is now an engineering task to use that capability, solve specific problems, and produce the product that delivers the benefits indicated by the CBP.

The success of this effort not only depends upon the quality of the organization's engineering personnel, but on their accuracy in specifying the design criteria. In the new product case, the CBP development and evaluation research should allow very clear engineering requirements to be specified. For example, the CBP in the analgesics example was partially characterized by "gentleness" (not upset stomach). The physical product had to substantiate the fact that it would not upset your stomach. However, the CBP required that a "gentle" product should not sacrifice "effectiveness." One engineering approach was to use more aspirin (7 grains rather than 5) in the tablet and to add an antacid. Another was based on utilizing the ingredients in the most popular pain reliever in England, acetaminophen. It is not an acid as is aspirin and therefore acetaminophen would be a physical substantiation of the claim that the new product would not upset one's stomach. A third approach would use a combination of these ingredients. Laboratory tests indicated that the most likely approach was an acetaminophen-based product. The product home-use tests were done with an English product that was given a new label. Recall that the home use did fulfill the CBP. Therefore, other new formulations could be tested against this one to find the best formulation. Production engineering was required to assure uniform tablet texture so the tablet would dissolve quickly and, finally, tablet size and shape had to be specified.

Engineering Design. The engineering task is to take the CBP output requirements and design the best physical product to meet them. In studying the engineering design process, Allen (1966, 1977) has found that it is important to consider alternative engineering approaches. Although many approaches could be considered, Allen found that better-rated engineering solutions were produced by groups that generated relatively few approaches during the course of the project and used a strategy of trading approaches off on a two-at-a-time basis rather than generating many alternatives and considering them simultaneously (Allen, 1966, p. 76). These results are based on studies of parallel projects. For example, Figure 12.1 shows the track of activities for two labs in designing a communication subsystem. In each lab, personnel made weekly estimates of the probability of finally adopting a specific engineering solution to the problem.

The two labs worked independently and were matched for research purposes. Notice that both labs followed different paths to solution and arrived at an answer at different times. It is also evident that the most likely approach shifted considerably over the design effort. These shifts were due to the results of various engineering studies. For example, technical approach β fell from consideration in week 12 (point A-3) because a wind tunnel study showed a wind torque much higher than predicted. Approach α was favored until the cost analysis (point A-5) of α showed it as less desirable. Approach β was selected after Weather Bureau data suggested an antenna placement that allowed a 20 percent reduction in the wind loading specification. Similar activities took place in lab B, but approach β was selected twelve weeks earlier. A second design problem showed the same shifting of the favored technical approach, but in this case, existing approaches were not acceptable and new approaches were generated in both labs.

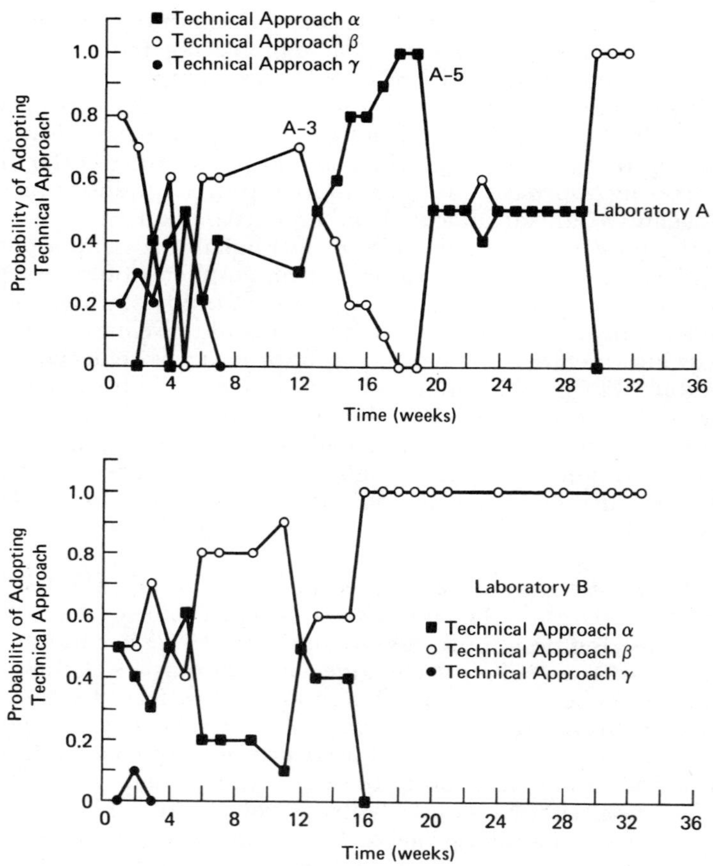

Figure 12.1. *Design of Antenna Radiation Subsystem (Allen, 1966, 1977)*

In analyzing records such as those shown, Allen found that the longer an idea was held in favor, the longer it took for it to decay from a high to low probability. There was substantial inertia in the design process, even though considerable fluctuation was observed. Based on his research Allen has proposed a model for R&D problem solving (see Figure 12.2) based on a sequential process of information gathering and analysis.

Table 12.1 shows the relative information usage from two pairs of projects at the development phase of R&D work. In the early functions, information is greatest from outside sources (customers, 36 percent; vendors, 12 percent) and in later functions, internal analysis is relied upon more heavily.

In our new product development process, the consumer research on the concept and product use fill this earlier need and help specify the "critical dimensions" and "limits of acceptability." The function of expanding alternatives is supported by consumer creative idea inputs and evaluation of

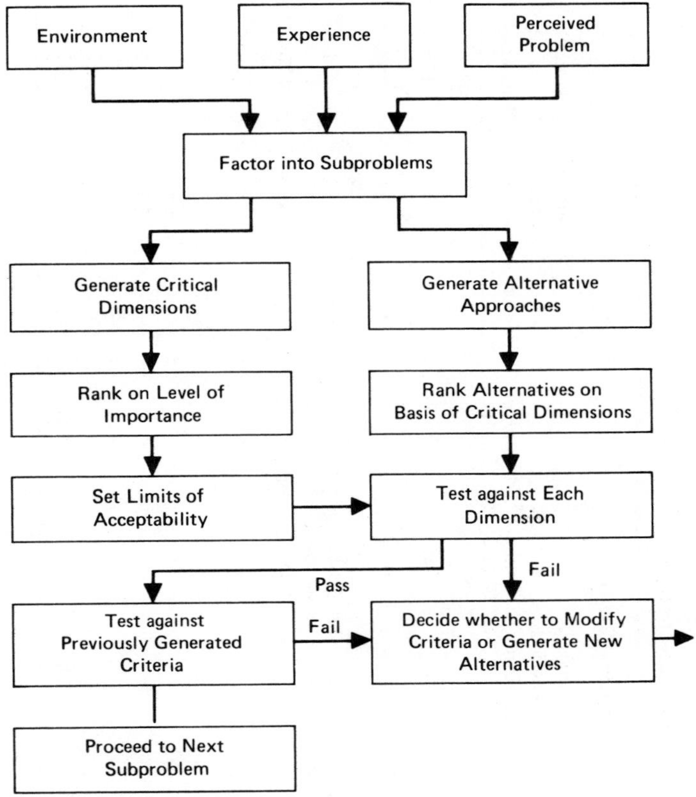

Figure 12.2. *Working Model of R&D Problem-solving Process (adapted from Allen, 1966)*

"stretchers." There is also interaction between R&D and marketing as the dimensions and limits are modified because of engineering difficulties. Here again consumer tradeoffs are important in establishing new criteria. Thus, there is an intimate relationship between good engineering design and good marketing strategy. Marketing and engineering must work together to create the physical and psychological attributes that will characterize a successful CBP.

"Engineering" Services. Engineering-like functions must be carried out for new services. For example, in defining the new Sloan Master's program the following service-delivery features had to be developed:

1. trip to visit businessmen,
2. summer job internship, and
3. close faculty-student interaction.

Table 12.1 Information Sources Used Related to Function Served in Engineering Design[a]

	Information Source							
Function	*Literature*	*Technical Staff*	*Sources Outside Laboratory*	*Vendor*	*Customer*	*Memory*	*Analysis and Experi- mentation*	*Number of Messages*
1. Generate Alternative Approach	2%	10%	2%	12%	36%	12%	26%	106
2. Reject Alternative Approach	0	12	2	13	2	10	62	68
3. Generate Critical Dimensions	3	17	4	10	9	4	54	71
4. Set Limits of Acceptability	4	13	5	19	4	5	49	75
5. Change Limits of Acceptability	0	25	17	8	25	0	25	12
6. Decide Whether to Modify Criteria or Generate New Alternatives	0	0	9	0	9	9	73	11
7. Test Alternative	9	7	4	10	3	2	65	106
8. Expand Alternative	8	0	0	33	17	17	25	12
Total	4%	11%	4%	13%	12%	7%	50%	461

[a] *Allen (1966)*

The design process was similar to Figure 12.2. Alternative specifications were examined and evaluated. The trip could be to Europe or to New York. It could be an in-depth visit to one company or an overall view of many companies. The design selected was a three-day, in-depth visit to a company. The first experimental trip was to Xerox, and post-trip evaluation indicated it met the program criteria. Similarly, an experimental program was established in the management science area to examine the issues of faculty/student interaction and summer job placement. A special seminar and pairing of faculty and students for joint research was implemented. These were successful and naturally led to successful summer placement through personal contacts of the faculty and the placement office. Although these are not the usual features developed by an R&D laboratory, the engineering design of service plans is just as critical as in products. The service plan must deliver the benefits of the CBP or it will fail.

Developing Advertising and Sales Copy

As well as delivering the CBP in a product, it must be effectively communicated through advertising and selling efforts. Before consumer products can be tested, advertising copy for television, radio and/or print media must be created. For industrial products, the salesman presentation and sales aids must be designed. In this section we briefly address the issues of creating advertising copy and sales promotion materials. Readers are referred to Aaker and Myers (1975) for a more complete discussion of advertising management.

Advertising Copy. Figure 12.3 shows a concept board for a new aspirin. It describes the CBP, but can be refined for more effective communication of the CBP. Figure 12.4 shows a storyboard that would be the basis of a television commercial. Here the CBP is refined and shown as a puzzle. The ad attempts to establish that strength, speed, and gentleness have been combined in this new product.

Figure 12.5 shows a print ad for a 3M copier—The Secretary II. It stresses "versatility" by showing eight things your copier probably cannot do. Although speed, quality, and cost of copiers are also important aspects of the CBP, this ad copy stresses versatility as the key differential selling point. An ad may be more effective if it concentrates on one unique aspect of the CBP rather than briefly describing all the attributes. The 3M ad is different than the Savin ad shown in Figure 12.6. The Savin ad does not stress versatility, although it is physically a similar machine to 3M, but rather makes a head-to-head competitive comparison to Xerox's 3100 copier in terms of "productivity." Which of these ads is best? That is a question we address in the testing phase of development, but here we consider how such ads are created and who creates ads.

In creating ads, advertising strategists concern themselves with product advantages, differentiating characteristics, and the consumers' buying

ATTACK—A New Kind of Pain Reliever

Pain relievers are one of the most widely used home medications and the main ingredient of those products is aspirin. That's because aspirin works so effectively to relieve headaches, muscular pain, fever due to cold, and "flu."

But clinical studies have recently discovered that aspirin and the extra-strength pain relievers also have side effects—and not just among people with sensitive stomachs. As many as 70% of the people who frequently take pain relievers can suffer from upset stomach, heartburn, gastric irritation, thinning of the blood, hidden stomach bleeding, and even certain allergic reactions.

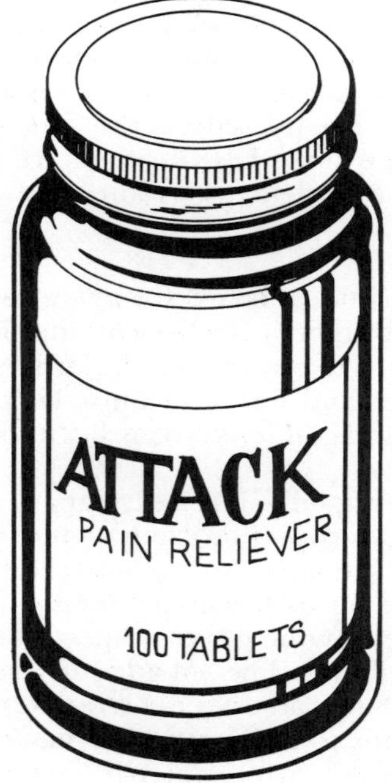

Now a new pain reliever called ATTACK is available in two strengths, regular and extra-strength, whichever is appropriate.

ATTACK has all the effectiveness but it has none of the side effects. ATTACK gives you fast, effective, extra-strength relief, but it won't irritate or upset your stomach. That's because the pain reliever in ATTACK is acetaminophen—a pain reliever which has long been known for its effectiveness and absence of side effects.

Next time you need fast, effective pain relief, without side effects, try ATTACK.

Figure 12.3. *Concept Board for "Attack"*

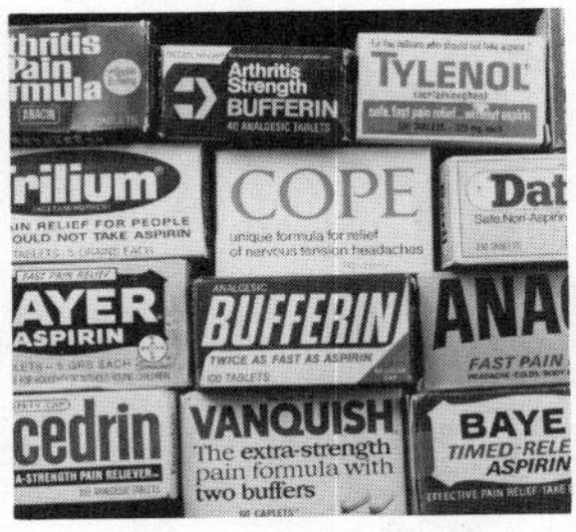

1. (Anncr VO) You wouldn't bring out a new pain relief tablet ... against all this competition if you really didn't have something.

2. Introducing ATTACK ... ATTACK has Acetaminophin.

3. ATTACK is ...

4. ... strong ...

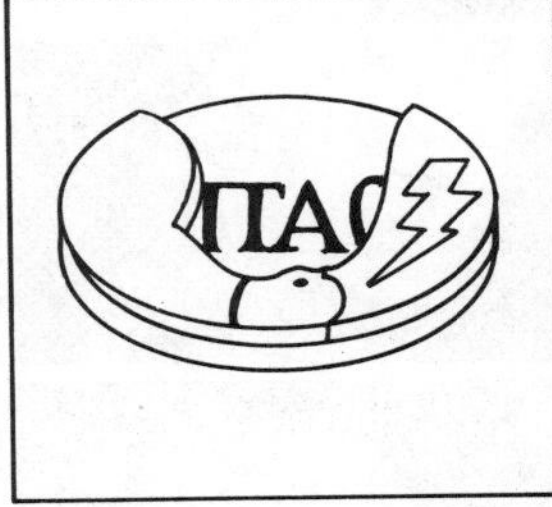

5. ... FAST and an important advantage ...

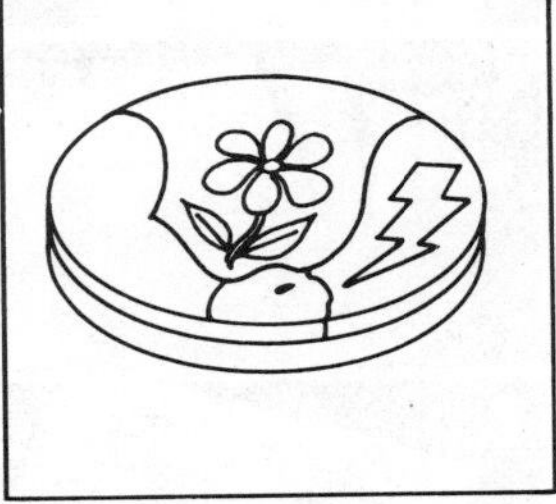

6. Gentle -- ATTACK does not cause stomach upset -- no heartburn -- no indigestion.

7. Fast and strong for effective pain relief without stomach upset.

8. Get ATTACK.

9. We think we really have something better.

Figure 12.4. *"Attack" Story Board*

New 3M plain paper copier does eight things most copiers can't.

1. High-quality plain paper copies. Even on your own letterhead. **2.** Copies on label stock for your monthly mailings. **3.** Full-size 10⅛" x 14" copies of computer print-outs. **4.** Copies on paper offset masters for in-house printing. **5.** Crisp, clear copies of halftones from books and magazines. **6.** Copies on ledger stock or other thick papers. **7.** Paper-saving two-sided copies. **8.** Copies for overhead projection. **9.** Copies on colored paper

Introducing "Secretary" II

See how the versatile 3M "Secretary" II makes clear, sharp copies on practically any paper in your supply cabinet. And gives you high-quality plain paper copies without costing you high speed copier prices.

For a no-obligation demonstration, call your local 3M Business Products Center.

Electronic Business Equipment
1500 Grand Ave., Kansas City, Mo. 64108
(816) 221-6400

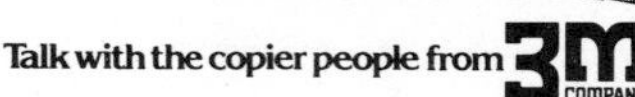

Talk with the copier people from 3M
COMPANY

Figure 12.5. *"Versatility" Selling Point*

It Takes Two Xerox 3100's To Equal The Productivity Of One Savin 780!

The people who manufacture copiers try to dazzle you with the speeds of their machines. Claims of up to 60 copies per minute, from the same original, are not unusual. But, a copier's ability to do the same thing over and over is no measure of its productivity. Most people make just a few copies from several different originals at a time. And when you add up the speed at which most copiers accomplish that, a lot of their glitter is dulled because of their lackluster performance.

Take Two Xerox 3100's For Example
Like the Savin 780, the Xerox 3100 has a repetitive copy speed of 20 copies per minute. But that's where the similarity ends. The 3100 will make copies of only 5 different originals per minute. The reason for this poor productivity is the time wasted with each original: opening and closing covers, positioning each original, pushing the print button, and waiting for the print cycle to complete so you can start the whole process over again with the next original. The result is that it takes at least two Xerox 3100's to equal the productivity of just one Savin 780! In those cases where you're making a copy from many originals it could take three or even four 3100's to equal the productivity of one Savin 780!

Take One Savin 780 For Productivity
The Savin 780 plain paper copier will not only make 20 copies per minute, it will copy 20 different originals per minute! Our machine has a document feed that automatically positions and transports each original through the copier, delivering clean, clear, needle sharp copies every time. There are no buttons to keep pushing, no covers to keep opening and closing—no shuffling of papers. When you do multiple copies, the advantages continue to add up. Need 3 copies each from 3 different originals—the Savin 780 will accomplish that twice as fast as the Xerox 3100.

Productivity Is Not The Only Benefit In The Savin 780
Paper handling devices have been available, as an option, for many years. With the Savin 780 it's standard! In addition, for more money most copiers allow you to add on a sorter attachment. With the Savin 780, for most copying needs, you don't need one. Our machine returns the originals and copies to you, collated. To copy books, periodicals, and other oversize originals, you simply lift the document feed and push the print button.

Contact us for a demonstration, today. We'll do it right in your office alongside your present copier. The odds are better than 2 to 1 you'll come out way ahead with the Savin 780. Get in touch with your nearest Savin representative or fill out the coupon below.

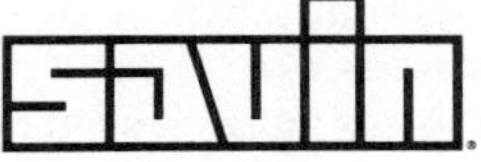

SAVIN
BUSINESS
MACHINES
CORPORATION

®Savin, Savin logotype and Savin 780 are registered trademarks of Savin Business Machines Corporation.
®Xerox and Xerox 3100 are registered trademarks of Xerox Corporation.

Savin, Valhalla, N.Y. 10595 AM78
☐ Please provide additional information about the Savin 780 plain paper copier.

Name/Title__________
Firm__________
Address__________
Telephone__________
City______ State______ Zip______

savin®

Figure 12.6. *"Productivity" Selling Point*

motives to generate a message that will "sell" the product. They use this information to generate a unique selling proposition (USP) to make a consumer *a*ware, generate *i*nterest in the new product, motivate a *d*esire to purchase, and hopefully cause the consumer to *a*ct. This is called the AIDA model (Strong, 1925). Others like it, such as hierarchy of effects (Lavidge and Steiner, 1961), which identifies a hierarchy of steps (unaware, aware, knowledge, liking, preference, conviction, and purchase), are used to identify and address the multiple strategies of advertising necessary to affect purchase. Good advertising addresses each phase in the model in order to achieve the final outcome, purchase.

Generation of the message is a creative process (Mandell, 1974; Wright, Warner, Winter, and Ziegler, 1977), but this creativity can be enhanced and focused with information from the design process. Focus groups and design analyses provide information about the buying motives for the product category, and consumer semantics for the advertising copy. The product positioning strategy directs the advertising positioning and helps define the unique aspects of the core selling proposition. The perceptual mapping tells the advertiser what dimensions to stress and what product characteristics and psychological appeals make up these dimensions.

Interest can be generated because the product fills previously unfulfilled needs. But this must be augmented with the advertiser's skill in creating interest in the advertisement per se. If the CBP is good, then the advertiser must generate sufficient belief in the appeals to cause trial, but one must be careful not to overstate product features. Otherwise, the consumer might be disappointed because the product did not live up to the advertising's claims. For example, Urban (1975) shows that a positioning that was too strong relative to blind test caused enough reaction in consumers that after using the product they perceived the product as much worse than it actually was. In his example, Urban found that the lower level of repeat purchase more than offset any gains in trial. Thus, while an advertiser might enhance a positioning with a good message, he must proceed with caution because stronger appeals are not necessarily better. The position attainable by the physical product is a good starting point.

The advertising copy (words, pictures, headlines, border, trademarks, slogans, etc., in print media; words, music, sound effects, illustrative material, action, camera cues, etc., in radio and TV) must be set to attain this message. The important consideration in the message is that it is not independent of the product strategy, but rather enhances that strategy by using the CBP and the knowledge gained in the design process.

Who actually creates ads? In most cases, advertising agencies carry out this function; the new product developer delegates much of the creative responsibility. The producer at least must select the agency and approve the final copy. Some firms take a much more active role by participating in the copy development, while others expect the agency to deliver an effective ad and do not care how it is developed.

Selecting an advertising agency is not an easy task. The sales presentations that agencies give are almost all uniformly good and usually overwhelming. While the agency's "track record" is important, we would suggest the following additional considerations in choosing an agency to participate in the new product process.

We recommend an agency which accepts the notion that they are part of the development team. Your organization has carefully identified a market opportunity and developed a CBP to exploit that market. The most successful advertising strategy will use this information and will be well integrated with the total new product strategy. While most agencies will use this information to enhance their creative output, some agencies hold the attitude that: "If you want a winner, turn the project over to us and we will give you the best ad this market has ever seen." We feel there is a real danger that the underlying R&D and marketing development strategy may be lost if the agency has this attitude. With participation by the agency and the sponsoring organization, the CBP strategy is more likely to be effectively implemented in the ad.

The specific people to be assigned to the project should have a record of successful creative output, but they also should understand marketing strategy and acknowledge the concept that their copy will be tested. Some creative copy writers believe that testing should not be done. Although copy testing is subject to limitations, we argue in the next chapter that it is a critical part of new product testing.

A final desirable attitude in an agency is the willingness to create very different alternatives. Some agencies develop one ad and give the client an accept or reject alternative. It is better to be given a set of good alternatives that reflect different creative execution and use testing and judgment to select the best one (Gross, 1972). The more creative alternatives available, the more likely it is that an outstanding ad will be found.

Sales Presentations. It is natural to think of copy development as applying to advertising, but it also is important in managing a sales force. In industrial products advertising budgets are small (Lilien and Little, 1976) and salesmen are relied upon to deliver the CBP message. The salesman identifies prospects, obtains appointments, presents his message (Core Benefit Propositions), overcomes objections, and attempts to close the sale (see Figure 12.7). An important design activity is to lay out the selling presentation. This usually is a 10 to 15 minute sales message which presents the CBP. Most companies do not spend enough time designing this message and training salesmen to deliver it at a professional level. As much care should be taken in designing the salesman's selling proposition as in designing advertising copy.

As well as preparing the layout and content of the CBP message, sales aids and brochures must be designed. These sales-support devices, brochures, and the message should be coordinated with the advertising to present the buyer with a unified CBP positioning.

Packaging, Product Name, and Point-of-Purchase Display. In consumer products, direct personal selling is not as important as in industrial

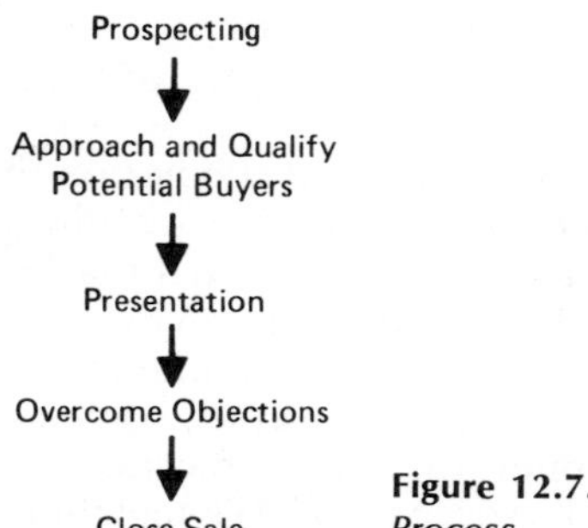

Figure 12.7. *The Personal Selling Process*

products, but packaging and the display at point of purchase are critical. L'eggs pantyhose used innovative packaging and point-of-purchase displays to create a major business in low-priced pantyhose distributed through mass merchandisers.

As with other communication, the package and display should support the CBP. If a new hair conditioner was positioned as "bright" and "uniquely you," it might be packaged in a bright yellow box with a mirrored surface on top. The package is important in enabling ease of use. Although Contact's claim is based on "time release capsules," it is packaged to allow easy dispensation of the tablets and in a package with a facing that is large enough to allow restatement of the CBP message.

A final component of the product is its name. As with advertising copy this is a creative input that communicates the CBP. For example, a product with sophistication as part of the CBP should have a sophisticated name; a product with strength as part of the CBP should have a name that connotes strength. The product name is important. A number of alternatives should be generated and tested for effectiveness and consistency with the CBP. If the new product is a potential product line extension, then its CBP should be consistent with the CBP of the product line. In some cases the product line already has a well established CBP which can be used to communicate the CBP of the new product.

In each specific product there are variables that can be designed to communicate the CBP. Advertising copy, product name, selling messages, package, and point-of-purchase displays are the most common. A good marketer is careful to consider each aspect and coordinate it with the underlying new product positioning strategy.

INITIAL DETERMINATION OF THE MARKETING MIX

The final set of decisions that must be made before a product is ready to be tested is the specification of the price levels, advertising budget, and distribution variables. Ultimately, we would like to set these variables at the best levels. We begin by reviewing attempts to determine an "optimal" marketing mix and argue that in the case of new products an adaptive strategy must be used. Simple models and judgments are used to set mix variables, then updated through premarket testing procedures, and finalized in test markets based on field experiments and more complex models.

The following sections condense some of the ideas from conventional wisdom on the details of each element of the mix. Since we are assuming the reader has a basic understanding of marketing, we do not review elementary marketing concepts, but rather stress new product decisions and how the CBP relates to the setting of the marketing mix variables.

"Optimal" Marketing Mix Models

Conceptually, it is easy to select an optimal marketing mix. Theoretical formulations go back as far as 1954 when Dorfman and Steiner derived a theory of optimal advertising and optimal product quality. Since then this formulation has been extended to include lagged effects of advertising, the effects of competition, and dynamic considerations. Conceptually, both demand and cost are functions of the marketing mix, i.e.:

(12.1) $$P = f(A, S, D, p, R)$$

(12.2) $$C = g(P)$$

where P represents volume sold and C average production cost and A, S, D, p, R represent advertising (A), selling (S), distribution (D), price (p), promotion (R) (for example, sampling) expenditures. Functions to be determined are $f(\cdot)$ and $g(\cdot)$. Profit (π) is given by:

(12.3) $$\pi = (p - C)P - A - D - S - R$$

Maximum profit is determined by maximizing π as given in equation (12.3). Many attempts have been made to structure the functional forms. To extend the model, researchers have made equation (12.1) a function of competition and have made both equations (12.1) and (12.2) functions of time. See Montgomery and Urban (1969), Kotler (1971), Fitzroy (1976), or Little (1977) for a review. The development has continued. With few exceptions, the general theory is difficult to apply in practice. Estimating equations (12.1) and (12.2) is not a trivial task even in the simplest case. One of the simplest forms of equation (12.1) is the constant elasticity equation (Kotler, 1964).

(12.4) $$P = kA^aS^sD^dp^eR^r$$

where a, s, d, e, and r are the elasticities and k is a constant. Kotler measured these quantities for tape recorders and used the model to determine "optimal" strategy. This model is relatively simple; it does not include competition, it assumes no major interactions among elements of the mix, and it does not model carry over effects of changes in variables from one period to another. Even so, the model is difficult to estimate for a new product.

Many models have been proposed to overcome the limitations of Kotler's model. Some extend the elasticity type model. Some models address the estimation problem by the use of econometric equations. (See Parsons and Schultz for a summary, 1976.) These models have good estimation properties, but require work before they capture the marketing richness necessary for new products. A "decision calculus" approach has been proposed by Little to balance the richness of the model with estimation requirements and management needs (Little, 1970). His BRANDAID model (Little, 1975) for setting the marketing mix in frequently purchased consumer good markets is discussed in Chapter 17. However, it is not appropriate here since we are facing the problem of a new product and not an established product. This limitation is also true of the other mix models. Although they are helpful once a product is established, they are of limited value for a new product since they do not model the dynamics of the new product acceptance process. Two exceptions are SPRINTER (Urban, 1970) and NEWS (Wachsley, Pringle, and Brody, 1972), which are test-market models, but which could be used based on judgmental inputs before testing. They are discussed in Chapter 15.

Many of the lessons learned from mature products apply to new products, but with some important differences. The product has not yet been on the market and, as a result, much of the information used to set the mix for mature products is not available for new products. On the other hand, the analyses of Chapters 1 through 11 provide up-to-date, in-depth knowledge of the consumers, the competition, and the product's CBP. Thus, the task is to set the mix at this stage by making maximum use of the information already available from the design process, but to remember that this is an initial selection, subject to revision, based on experimental data collected in the testing phase of development. Before test market, we have very little field experience with the new product.

Thus, while "optimal" marketing mix models are conceptually appealing and are potentially powerful tools for mature products, sufficient information is rarely available to allow optimization for new products. However, there is information available that can be used to set an "initial" marketing mix. Our approach is to utilize the simple models developed in Chapter 11 to judgmentally set initial levels of the marketing mix variables. These are certainly not "optimal" since they are improved during testing efforts, but we believe they provide a "good" starting point for improvement.

Advertising

One simple choice model proposed in Chapter 11 (equation 11.4) was

$$\textbf{(12.5)} \qquad P = \sum_{i=1}^{N} a_w a_v b_i = a_w a_v \left(\sum_{i=1}^{N} b_i \right)$$

where P is purchase volume, a_w is the percent of the target group who are aware, a_v is proportion of availability, and $\sum_{i=1}^{N} b_i$ is the number of people who will purchase the new product (this is per period if product is subject to repeat buying) if they are aware of it and it is available to them. Equations (11.5) and (11.6) can also be put in this form to make the effects of a_w and a_v explicit.

In setting the initial advertising level, we assume that the advertising copy effectively communicates the CBP. Specifically, we assume that advertising awareness will produce a purchase probability (b_i, equation 12.5) equivalent to that produced in the concept and product development research. Our attention here is on the advertising budget. Awareness is created by purchasing advertising media insertions and by these being seen and comprehended by consumers to produce attitude changes and purchases.

The budget can be set by estimating the relationship of the budget to the awareness level. One approach is to set an awareness goal and determine how much money should be budgeted to meet it. For example, if 70 percent of the target group should see at least one ad (i.e., "reach"), and on average two ads should be seen by each person across the target group (i.e., "frequency"), media schedulers can estimate the cost and schedule to meet these goals.

But this approach leaves open the question "why not 90 percent reach and frequency of three rather than 70 percent reach and frequency of two?" To answer this, expenditure must be linked to awareness. For example, Figure 12.8 shows a hypothetical graph of awareness as a function of dollar expenditure. Initially, big gains occur, but then diminishing returns set in.

We use equations (12.3) and (12.5) to calculate profit for alternative advertising expenditures. For example, suppose that net contribution per product ($p - c$) is \$1 per unit sold. Suppose that distribution costs (D) are one million dollars per year and there are no selling or promotion costs. Suppose that there is 100 percent availability ($a_v = 1$) and that our purchase model predicts that 20 million units will be sold per year if everyone were aware of the new product and it were available ($\sum_{i=1}^{N} b_i = 20$ million per year). Then purchase volume rate, P, is given by $a_w \times$ (20 million per year) and profit per year, π, is given by $\pi = P \times (\$1) - A -$ \$1 million. Table 12.2 shows explicit calculations for the response function in Figure 12.8. Of the expenditures tested, \$7.5 million per year appears to be the best expenditure for advertising. While it is tempting to "optimize" this procedure, one must understand that Figure 12.8 is based on judgment and discussions with the advertising agency. Uncertainty and unknown errors make optimization infeasible at this stage of the process. Optimization is more appropriate later in the development process when more information is available.

At the design phase, past norms on expenditures, "reach" and "frequency" targets, or judgmental response functions are the basis for setting initial advertising levels. Levels of advertising needed to support the diffusion of innovation after the first year are usually lower than the initial levels. The dynamic models described in Chapter 11 indicate that interpersonal communication will lower the need for advertising after introduction. In the

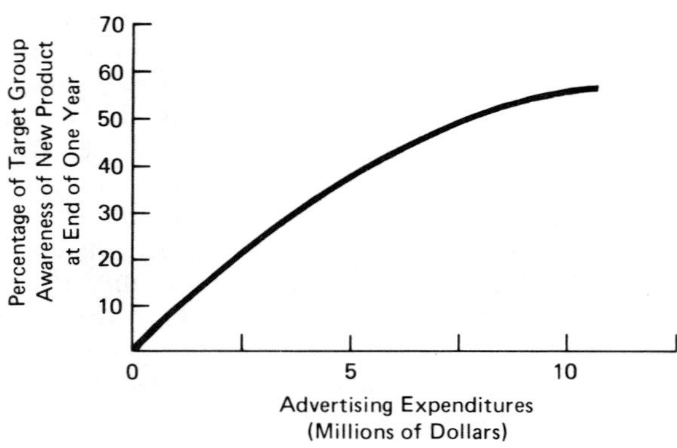

Figure 12.8. *Typical Advertising Response Function*

Table 12.2. Sample Calculations to Establish Initial Advertising Expenditures

Advertising Expenditure Rate, A (millions of dollars)	*Awareness, a_w (percentage)*	*Purchase Volume Rate, P (millions of units)*	*Profit Rate, π (millions of dollars)*
1.0	10	2.0	0.0
2.5	20	4.0	0.5
5.0	35	7.0	1.0
7.5	50	10.0	1.5
10.0	55	11.0	0.0

industrial case, which is characterized by multiple decision participants, advertising is heavily concentrated at introduction and salesmen usually carry the continuing support obligation. In frequently purchased consumer products, advertising continues well after introduction since repeat purchase decisions must be supported and competitive expenditures met.

As well as advertising budgeting, media is considered in completing the design. The message copy determined *what* to say, the media plan now determines *how* to channel it most effectively to the target group. The media plan is set to maximize results subject to advertising budget constraints. For example, a brokerage service will emphasize *The Wall Street Journal* and *Forbes,* while a household service might emphasize *Ladies' Home Journal* and daytime television. The advertiser chooses the media to achieve the best exposure, reach, and frequency levels within the target market.

Those readers interested in the details of media selection are referred to Montgomery and Urban (1969), Kotler (1971), and Fitzroy (1976). They review the mathematical programming formulations (Day, 1963; Engel and Warshaw, 1964; Bass and Lonsdale, 1966; and Wilson, 1963), and also more comprehensive models such as the one developed by Little and Lodish (1966) called MEDIAC. This model incorporates non-linearities of consumer response, lagged effects, seasonality, and other effects to select the allocation to various media, options within media, and insertions within each option by time period. These models are often made practical with an on-line computer program for interaction with media planners and can be used to consider the best media schedule for a new product.

The use of such a model is to determine which types of media to use and how to best use them. For example, the model can investigate whether television is practical for the new product. If it is, testing should include television ads. If only print media are to be used, the testing will be restricted to these media. As the product continues on to test market, a full media schedule is needed and models such as MEDIAC are particularly useful.

Selling

While advertising plays a role in industrial products, sales representatives carry most of the communication responsibility. The budget for selling is expressed in the allocation of sales time to the new product.

Each sales call exposes the potential customer to the core benefit proposition and is analogous to creating awareness levels by advertising. The number of customers and the rate at which the calls are made must be determined. Figure 12.9 shows two rates of sales coverage. Curve I shows a rapid and almost complete coverage, while curve II shows a gradual growth in coverage.

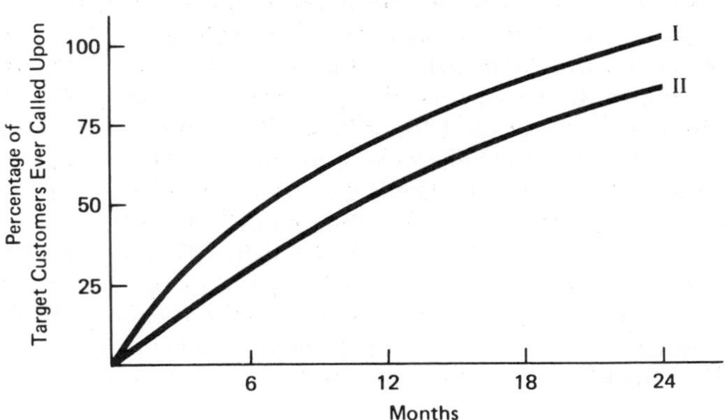

Figure 12.9. *Rates of Sales Call Coverage*

The results of sales calls can be estimated by an equation similar to that for advertising (equation 12.5). Sales calls make the consumer aware of the new product and in many cases make it available to the consumer.

In some cases there is a cumulative effect of sales calls and the probability that a customer purchases the new product grows as more sales calls are made to that customer. For example, Figure 12.10 shows one possible relationship between the probability of ordering the new product and the

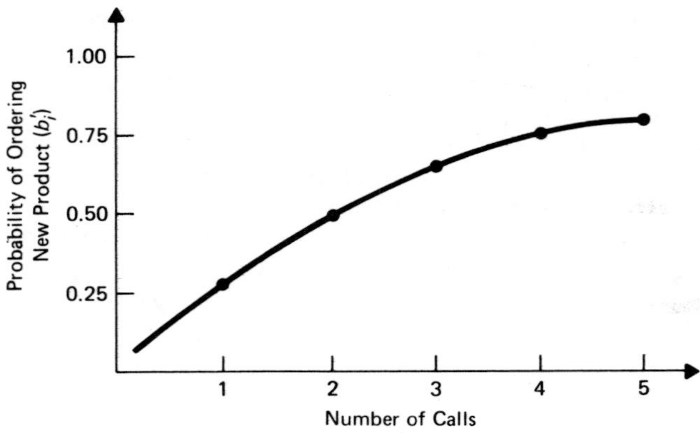

Figure 12.10. *Cumulative Effect of Sales Calls*

number of calls made on the customer. If customers vary in their susceptibility to sales calls, then separate calculations are made for each type of customer.

The *net* effect of the sales effort is more difficult to measure since the existing products may be affected. Usually total sales time is fixed and allocations of selling time to a new product reduce the time available to other products. If the effects of reduced effort on other products can be estimated, the net sales and profit effects of sales allocations can be calculated.

Montgomery, Silk, and Zaragoza (1971) developed such a model called DETAILER for the pharmaceutical industry. In ethical drugs, doctors are called upon and two, three, or four drug presentations are made to the doctor during the visit. This is called "detailing." If a new product is to be added to the detailing plan, some other existing product must be dropped. Models such as DETAILER can systematically examine the sales gain of detailing a new product and the losses of not detailing an old product.

Another model called CALLPLAN (Lodish, 1971) has been developed to analyze call policies in repetitive buying situations. Here tradeoffs are between gaining new customers and serving old customers. Both of these models are useful in setting sales allocations for a new product if estimates can be made of the effects of sales calls on new product sales.

Distribution

Availability is primarily affected by the distribution system, e.g., distributors, wholesalers, jobbers, and retailers. Like the other elements of the mix, good distribution does not ensure success, but it is necessary for success. For example, even with a good positioning, a manufacturer of a new antacid found sales down due to poor distribution. A major portion of the market buys antacids at the special displays near supermarket checkout areas. Without good coverage of these displays, the firm could not be successful.

But availability is not the only function of a distribution channel, information is also provided. By providing information, channels also play a role in communicating the CBP. For example, many audio component (Hi-Fi) stores have listening rooms where consumers can listen to components within a full audio system, or stores may supply "loaners" where consumers can try components in their home system.

In industrial products, selling can be done by the firm's sales force; but if the firm does not have its own sales force for a new market, it may utilize middlemen to carry out this function. Finally, distribution channels can perform needed service by fulfilling warranties and providing diagnostic and corrective maintenance when needed. Consider, for example, automobile distributors and small appliance dealers. This servicing function may be a crucial part of the CBP if reliability is emphasized or if the delivery of product benefits is significantly better when the product is used properly.

Channels of distribution vary widely and are usually institutionalized. Consumer goods usually have a long chain of distribution from manufacturer to distributor to wholesaler to retailer. Industrial products may use middlemen or direct distribution.

In new products the channel alternatives may not be great due to inertia in the distribution institutions and practices. However, in many cases, the new product needs more support than the channel usually supplies an average product. If the new product team wants the channel members to provide extra availability, information exchange, demand creation, and service they must pay for it or provide incentives to encourage it.

Among the incentives offered to distributors are cooperative advertising and advertising allowances. The manufacturer can give selling assistance, supply promotion material, or supply direct advertising materials. The manufacturer may supply store traffic stimulators such as free offers and product demonstrations, or the manufacturer may sponsor clinics and special events. Point-of-purchase promotions such as signs, window displays, wall displays (clocks, thermometers), display cards, and merchandise racks (e.g., razor blades) reward the retailer and serve to aid awareness and induce trial. The product itself is an incentive to the channel member. Proactively designed products are products that consumers will buy. As a result, the manufacturer can emphasize a "pull" strategy in which the consumers demand the product and "pull" it through the channel.

It is possible to support the channels' selling function by the use of missionary salesmen. These salesmen represent the company, but orders are placed with middlemen. For example, L'eggs representatives set up displays and stock them directly, but sales are by the local grocer. As discussed above, pharmaceutical companies detail products with their salesmen, but prescriptions are filled by local druggists. Another way to add more selling emphasis in the channel is by increasing the margins to the trade. It is common in consumer products to give middlemen (retail food stores, auto dealers) higher margins in exchange for special displays or to encourage extra selling push.

For new products, manufacturers must make both macro and micro distribution decisions. *Macro decisions* involve a selection of the channel itself. For example, should a manufacturer of a swimming pool chemical reach consumers through:

1. conventional retail outlets such as hardware stores and drug stores,
2. specialized swimming pool supply and equipment retailers,
3. swimming pool service companies,
4. mass retailers such as supermarkets, department stores, and discount houses,
5. direct mail supply companies, or
6. a combination of these alternatives (Bursk and Greyser, 1968).

Micro decisions involve the allocation within the channel to elements such as advertising assistance, point-of-purchase displays, and economic incentives.

A key determinant in the macro decision is the CBP which identifies the consumers who are the target market. For example, a low-priced camera would be better sold in department stores or discount houses rather than in professional photography shops; whereas, a luxury telephone attachment may be priced out of a discount store distribution strategy. Avon products are better sold by neighborhood representatives who can give personal service and advice. Tupperware is sold at "parties" where their superior seal can be demonstrated. Finally, part of the macro decision is based on economic considerations balancing the gains in availability (a_v), awareness (a_w), purchase (b_i) against the cost incurred by using the channel.

The micro decisions are the result of detailed negotiations, which are highly dependent on the relative power, conflict, and regulation involved for each potential channel (Stern and El-Ansary, 1977). At this stage, it is important for the new product team to set margin and special incentives to obtain the desired availability, information, demand creation, and service effort. These functions are filled at a cost determined by the detailed mix of channel costs and incentives and is entered directly as a cost in the profit calculation. For this expense, the new product team gets primarily availability (a_v), but may receive increased awareness (a_w), and possibly trial and repeat. The micro decision, like many other mix decisions, is based on the profit equations and is guided by the Core Benefit Proposition.

Pricing

There are several methods of selecting a price for a new product. A common approach is to mark up costs by some amount. Another is to calculate a price which will result in a break-even of profits for a conservative sales estimate. A third is to set prices at a level slightly below competitive products.

These are simple methods to implement, but may underprice or overprice a product. The best approach would be to understand the sales effects of changing price. In theory, the price of the new product is set by the general profit equation—equation (12.3). A higher price means more profit per product sold and appears explicitly in the profit equation, while a lower price means more volume, but lower profit per unit. In the design process, some measures of price response can be obtained. For example, price may be a variable in preference regression or conjoint analysis. In this case, the estimated preference for the new product depends on the product's characteristics and on price. To simulate the effect of alternative prices, the new product team uses the preference regression or conjoint model to predict preference values for different prices and uses either the rank order preference model or the logit model to transform these preferences to choice probabilities. Equation (12.5) transforms these to purchase potential, and the profit equation is used to calculate profit for each time period.

Experiments also provide information. Ryans (1974) studied the price response to electric blenders while developing perceptual maps and preference functions. He divided his sample into parts and exposed them to three different price levels for a new blender ($24.99, $29.99, $34.99).

After an initial estimate of price response, the new product team can get a reasonable range for a good price for the new product. This is done by finding the price that maximizes long-run profit. This is not necessarily the short-run optimization that results from applying the profit equation for each year without considering the effects of one year's sales on another year's profits. For example, the firm may wish to sacrifice short-run profit for longer-term goals such as penetration of a market (bigger share) and market dominance (first in with a strong position). These goals can be turned into flexibility of action and market control that can lead to later profits. Alternatively, if the firm fears rapid entry by me-too products, they may wish to skim the market with a high price.

Price becomes especially important when there are production cost reductions due to increased volume. For example, Chapter 5 introduced the concept of an experience curve where product costs declined as a function of cumulative sales volumes. When experience effects exist, the organization may wish to consider lower initial prices to achieve early volume and thus gain a cost advantage relative to competitors. Similarly, when diffusion of innovation phenomena such as strong word of mouth exist in the market, the organization may wish to lower the initial price to achieve initial penetration because initial penetration means more consumers have purchased the product and hence there are more consumers to communicate the product's features to potential purchasers.

For example, Table 12.3 shows four different pricing policies for a line of new chemical products. In this case the organization had a three- to four-year lead on its competitors. Because of diffusion effects, the organization expected gains due to penetration pricing. As Table 12.3 indicates, managerial decisions such as plant size and sales effort are dependent on volumes created by the pricing strategy and must be factored into the decision. In this case the firm could expect over $6 million in discounted incremental profit if it did not employ a penetration pricing strategy, and $10.8 million if it did. The variation in optimal price is due to lower prices to utilize expanded plant capacity and take advantage of lower costs in years four to seven. With a larger plant, prices would be further lowered and profits increased marginally. If the sales force could be expanded, more sales effort would be allocated to the new products, prices would be higher, and profits increased. (For details, see Urban, 1968. For other examples of dynamic pricing, see Robinson and Lakhani, 1975; and Dolan and Jeuland, 1979.)

Table 12.3. Example of Penetration Pricing for a Line of New Chemical Products[a]

	Reference Program	*Revised Program*	*Larger Plant*	*Larger Sales Force*
Discounted Incremental Profit	6,000,000	10,833,000	11,561,000	12,219,000
Investment	8,000,000	8,000,000	8,300,000	8,400,000
Advertising Level per Year	10,000	10,000	10,000	10,000
Percent of Sales Effort per Year	1.0	1.0	1.0	1.3
Price in Year				
1	350	250	250	250
2	350	250	250	250
3	350	250	250	250
4	250	180	170	190
5	250	200	160	200
6	250	210	170	210
7	250	170	170	180
8	250	180	180	180
9	250	180	180	190
10	250	190	190	200

[a]*Urban (1968).*

This example shows how pricing is related to other elements in the industrial mix. In consumer markets this is also true. A high price and greater advertising may bring more profit than a low price and no advertising. Price must fit with the other elements of the mix and the product line.

As in advertising, there is no general rule that applies in all categories. Again, the best strategy seems to be to start with the CBP and select a price that will support the CBP and match the planned marketing mix. When possible, this decision is supported by experimentation or analytical models. When experience or diffusion effects are present, consider penetration pricing. While this analysis does not provide an optimal price, it specifies an initial price level which can be refined in pretest market and test market analyses based on better data and more complex models that include diffusion of innovation and competitive entry. (In Chapter 16 we consider analytic pricing models that utilize this better data.)

Promotion

In many new products, special promotions are used to encourage initial acceptance. For example, in industrial products, these promotions may be special introductory prices or they may be special deals in which purchase of the new product results in lower prices for existing products or includes free service for a specified period of time.

Early in the introduction of a new consumer product, there is a desire to induce trial to obtain more rapid market penetration. Various strategies are used to entice consumers to try the new product in a way that does not lower the overall probability of repeat purchases. While there are a number of strategies including premiums (such as towels in laundry detergents), contests (such as drawings and puzzles), and combination offers (such as Bic pens and lighters), by far the most widely used strategies are the economic incentives that effectively lower the initial purchase price of a new product. These strategies (sampling, couponing, and price-off) can be used alone or in combination.

Sampling. When Sure deodorant was introduced, Procter & Gamble gave out approximately 30 million free two-ounce samples. Evelyn Wood Reading Dynamics gives free mini-lessons. Many industrial firms supply free samples of small parts like screws and transistors. The goal of these strategies is to encourage trial in the hope that if consumers try the product, they will buy more. In this way sampling substitutes for awareness and first trial. Suppose that a percentage of the target consumers are sampled. From this group initial awareness and initial availability is assured. Thus a_w becomes 1.0 and initially a_v becomes 1.0. However, repeat purchases will be affected by market availability. Though sampling trial is increased, the repeat rate may be lowered. Some people may try the sample even though they would never use the product at full price. Others may use the product because it is "free," but will not seriously consider its features (Scott, 1976). For both of these reasons, the repeat rate may be lower. While it is important to realize that the repeat rate from sampling may be altered, in our experience, setting $R_s \cong R$ provides a good first approximation. This approximation is then evaluated in the testing phase of development.

The economic effect of sampling can be considered separately considering consumers sampled and those not sampled. Expanding equation (12.5):

$$P = (1 - s) N_p a_w a_v T R + s N_p U_s R_s a_v \qquad \textbf{(12.6)}$$

where s = percent of target group sampled

N_p = number of consumer purchases in the product category per period

a_w = awareness

a_v = availability

T = ultimate proportion of target group who try the new product

R = long-run share of purchases for new product among those who have ever made a trial purchase

R_s = long-run share of purchases for new product among those who received and used a sample

U_s = proportion of samples sent that are used by member of target group

The first term represents those not sampled and the second term represents those who were. This simple equation assumes those who receive a sample, but do not use it, will not subsequently purchase the product. If this is not the case, the definition of U_s must be modified to include other forms of trial. This equation can be used to simulate the effects of sampling on sales and, after deducting the sampling costs, changes in profit can be calculated. A numerical example of equation (12.6) is given in the case following this section.

Couponing. When Pringles-Extra (rippled potato chips made from dried potatoes) was introduced, many consumers received coupons worth 20¢ off the purchase price of a package of Pringles-Extra. Amusement parks give out half-price ride tickets, and State Fairs give out half-price admission tickets. These strategies, like sampling strategies, try to produce awareness, induce trial, and hopefully not undermine repeat sales. At this stage of strategy development, couponing is handled in a similar manner to sampling. Volume is computed by separating those who receive a coupon from others and calculating their sales based on a higher trial and potentially altered repeat rate. The equation is conceptually similar to equation (12.6) except s, U_s, and R_s are replaced by c, U_c, and R_c suitably defined for couponing.

Price-Off. A third incentive is to lower the price that consumers pay. In some cases this strategy will be less effective than couponing because it requires that consumers notice the price differential in the store. In other cases, it may reach consumers who could not be reached by couponing. The effects are similar to that for sampling and couponing. For those who are presented with the price off, trial rates would be higher but repeat rates potentially lower. Again the equation is conceptually similar to equation (12.6).

Some new product teams may wish to combine sampling, couponing, and price-off strategies. In calculating the net effect, care must be exercised to avoid double counting in the volume equation. This can be done by dividing the target group into parts which receive one, two, or more specific promotions and estimate a trial (T) and repeat share (R) for each relevant combination.

Summary of Marketing Mix Determination

The models discussed above provide the new product team with usable techniques to set each of the marketing mix variables. Each model requires some direct market measurement or managerial judgment. Furthermore, each can be readily extended as the new product team becomes more sophisticated or as the stakes become higher. The important point is that these simple models summarize the essential elements of the managerial decisions and provide initial strategies which can be refined in market testing.

In using these models, the new product team should recognize that all the elements of the marketing mix are highly interrelated. Price cannot be set independently of advertising, nor advertising set independently of promotion strategy. These models need an integrating strategy to combine the variables.

For example, one strategy is an aggressive introduction through high advertising and selling effort along with a "low" price. This implies a large investment in the marketing strategy with a goal of long-run payback. An alternative strategy is a low-keyed introduction with moderate advertising, moderate sales effort, and "higher" prices. This strategy tends to pay for itself as it goes but holds sales volume at a low rate and, if the market is large enough, makes competitive entry attractive.

The choice of these or other strategies depends upon the strengths and goals of the firm. For example, large firms such as Procter & Gamble, General Mills, or General Motors, which have investment capital and marketing expertise, may opt for an aggressive strategy; whereas, a smaller firm may, by necessity, be limited to a less aggressive strategy. Alternatively, a firm may wish to move with caution, may wish to maintain a low-key image, or may be subject to regulation.

In practice, aggressive strategies are becoming more common. Such strategies are especially important if the proactive development strategy has resulted in a differential advantage based on the recognition of consumer needs. In this case, the organization can exploit the advantage by rapidly developing the market and establishing a strong market position before competitors can imitate the innovation.

The choice of basic strategy is a difficult managerial decision. The characteristics of the market opportunity (Chapter 5) and consumer needs (Chapters 8, 9, and 10) provide key input for the decision. The basic strategy must be consistent with the CBP.

We now review the design process with an illustrative case.

AN INTEGRATED EXAMPLE OF THE DESIGN PROCESS

In the last six chapters we have sequentially considered the perception, preference, and choice aspects of new product design. In practice, these elements are intertwined with each other and with the activities of physical product engineering, advertising creation, and marketing mix determination. In order to review the design phase of new product development, we present an integrated example. This short example cannot provide all the richness of the design process, but it does emphasize the key lessons. The market chosen is laundry detergents, and the data in each stage are modified or created for clear illustration of the principles of the design process. These data are not to be construed as representative of the laundry detergent market, but are representative of the design issues typically faced in practice. Readers are encouraged to concentrate on the lessons in new product design rather than the specifics of laundry detergents.

Our case concerns "Consumer Laboratories, Inc." (CLI). Suppose CLI is a major package goods manufacturer with previous successes in facial soap, shampoos, toothpaste, and antacids. CLI uses a proactive strategy based on being first in a market with a major innovation based on both an improved physical product and superior psychological positioning. R&D and marketing have worked closely in designing successful products in the past. At the time of this case, they were investigating a number of potential markets, one of which was the laundry cleaning market.

Market Definition

As a first step, CLI performed a market profile analysis based on archival search. They determined that the market was large (over $1 billion annually) and showed a stable growth of 10 percent per year. The entry cost was high (about $20 million), but a moderate share of 2 percent could return this investment in a few years. The $20 million annual sales implied by a 2 percent share exceeded CLI's minimum sales volume criteria of $10 million, and return on investment at this level was viewed as good. CLI had experience in related markets and already had a good distribution system for household laundry products. The market was highly competitive, but a recent technological breakthrough in CLI's research and development laboratory (tight-packed, homogenized particles) promised a competitive edge in the industry. The market showed high risk because four of the last five new products in the industry were failures, but appeared vulnerable to a superior product with good positioning. The current number two product was introduced in the last three years and was successfully based on a superior cleaning appeal. Finally, the lab felt they could modify the product, if necessary, to make it safer or stronger, depending upon the needed positioning. Based on these results, CLI considered the household laundry cleaning market worth further investigation.

Household floor cleaners and hair color markets also were considered, but they were not deemed as desirable in terms of vulnerability, profitability, and match to the company's R&D and marketing capabilities. (The hair color market was CLI's second choice and was identified as the market to consider once the design of a new laundry cleaning product was underway.) CLI's policy was to investigate major new markets each year with a goal of having one successful new product each year. They used a development program that resulted in a new product three to four years after initiation at an average cost of $2 to $3 million. At any given time, four or more new products would be in different phases of design, testing, and introduction.

CLI does not believe in generating ideas until the strategic market opportunity is known. CLI thus began its design effort by performing a hierarchical market definition study to select a specific target market. They wanted to be sure they selected a segment of the market where the greatest opportunity existed. CLI conducted a market research study based on simulated purchases in a mini-store to determine switching probabilities among existing brands. Based on these probabilities, the best hierarchical definition is shown in Figure 12.11. The first branching was additives versus detergents. These were further divided into liquid versus powder for detergents, and bleach versus softener versus pre-soaks for additives. Powder detergents were subdivided into hot water and cold water washing products in this analysis. CLI selected the hot water powder market because it had greater potential. The expected share was highest in this market and a typical new product share was 2.4 percent of the total market or 8 percent of the powdered, hot water detergent market. The management felt no product was positioned in this market as strong yet safe-for-delicate fabrics, and thus an opportunity existed. Demographically, this market contained a typical cross section of households. It represented the biggest part of the detergent market, and sales were growing at 9 percent per year.

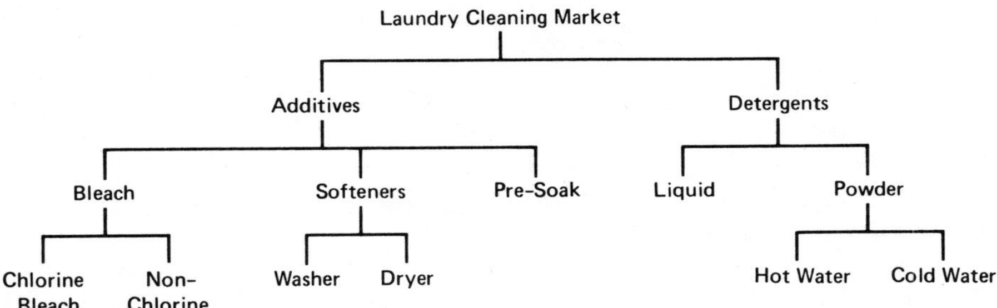

Figure 12.11. *Hypothetical Hierarchical Market Definition for Home Laundry Cleaning Products*

Idea Generation

With the target market clearly defined, creative groups were conducted to find innovative product concepts. The groups consisted of marketing, R&D, marketing research, engineering, and production personnel from within the company. Two advertising agency people (the account executive and the creative director), a retailer, and two consumers were included in the group.

In support of the idea generation effort, focus groups were conducted and perceptual maps of the existing products were developed from market research studies. Focus groups in Boston, Chicago, and New York suggested that "strong" detergents really "get the dirt out," but can "harm synthetic clothes." Complaints were raised about hard-to-handle boxes that often get spilled. An important output of the groups was a condensed list of 21 attributes in the consumers' semantics such as "get out dirt," "good for greasy oil," "won't harm synthetics," "safe for lingerie," and "economical."

Based on the focus groups, a questionnaire was developed to measure each laundry detergent in a consumer's evoked set on semantic differential scales which were constructed to measure the 21 attributes (see Figure 12.12 for typical rating questions). The questionnaire was administered to 185 consumers. Table 12.4 reports some of the results of the evoked set measurement indicating that the average number of considered products is small. Hence a sizable introduction campaign may be needed to get consumers to consider the new product.

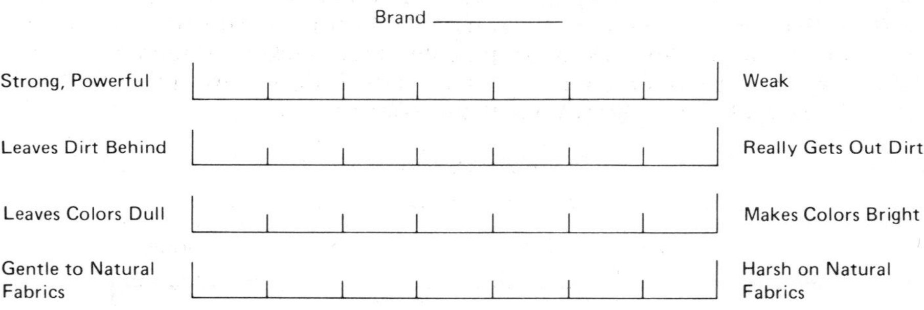

Figure 12.12. *Typical Semantic Differential Scales to Measure Attributes of Laundry Detergents*

The perceptual dimensions and market structure were identified with factor analysis of consumer ratings of the evoked products. Based on the scree and eigenvalue rules (left side of Table 12.5) backed by interpretability, two factors were selected. They were labeled "efficacy" and "mildness" based on the factor loadings in the right side of Table 12.5. Factor scores were then computed for each product and their averages were plotted in Figure 12.13.

Note that the perceptual map indicates a definite gap in this market in the upper-right quadrant. To evaluate this positioning opportunity, CLI ran a preference regression to determine the relative importances of the two dimensions. They were 0.538 for efficacy and 0.462 for mildness, indicating that movement along both dimensions is important, but with a slight emphasis on efficacy. (See vector in Figure 12.13.)

With the background information from the focus group and the perceptual and preference maps, the creative group set to work. One effort was to create a product concept that would position well on efficacy and mildness. R&D felt they could combine the ingredients used in softened additives within a powder detergent. The new "tightly packed" particles technology

Table 12.4. Hypothetical Evoked Set Results for Laundry Powders

Size of Evoked Set		*Makeup of Evoked Sets*		
Number of Brands in Evoked Set	*Number of People*	*Brand*	*Number of People*	*Percent*
1	10	1. Cheer	111	60.0
2	28	2. Tide	129	69.7
3	43	3. Bold	104	56.2
4	49	4. Oxydol	79	42.7
5	32	5. Fab	42	22.7
6	12	6. Duz	63	34.1
7	5	7. All (Powder)	40	22.0
8+	6	8. Ajax	39	21.0
		9. Others	50	NA

Table 12.5. Hypothetical Factor Analysis for Laundry Detergents

Selection of Number of Factors			*Naming Factors*		
Number of Factors	*Eigen-value*	*Percentage Variance*	*Attribute*	*Efficacy*	*Mildness*
1	5.41	0.258	Strong, Powerful	0.70	0.23
2	3.48	0.166	Gets Out Dirt	0.70	0.19
3	0.85	0.040	Makes Colors Bright	0.73	0.26
4	0.83	0.040	Gentle to Natural Fabrics	0.19	0.72
5	0.80	0.038	Inexpensive	0.09	0.20
6	0.75	0.036	Removes Grass Stains	0.64	0.26
7	0.72	0.034	Won't Harm Colors	0.15	0.77
8	0.68	0.032	Good for Greasy Oil	0.69	0.14
			Easy to Use	0.57	0.18
			Pleasant Fragrance	0.70	0.20
			Won't Harm Synthetics	0.28	0.73
			Acceptable Color	0.54	0.19
			Convenient Package	0.47	0.13
			Gets Whites Really Clean	0.72	0.27
			Good Form (Liquid vs. Powder)	0.44	0.31
			Good for the Environment	0.23	0.48
			Removes Collar Soil	0.52	0.29
			Reliable Manufacturer	0.14	0.42
			Removes Stubborn Stains	0.64	0.36
			Safe for Lingerie	0.16	0.73
			Economical	0.22	0.33

would hold the softening ingredients in a matrix of soap and softeners. Marketing felt they could effectively communicate such a product in ads which portrayed both softness and brighteners. A concept statement was created to represent the Core Benefit Proposition of "a very effective, yet mild laundry powder." A second concept was created to stretch current perceptions. Very round and hard particles were to be used to produce a "pourable pow-

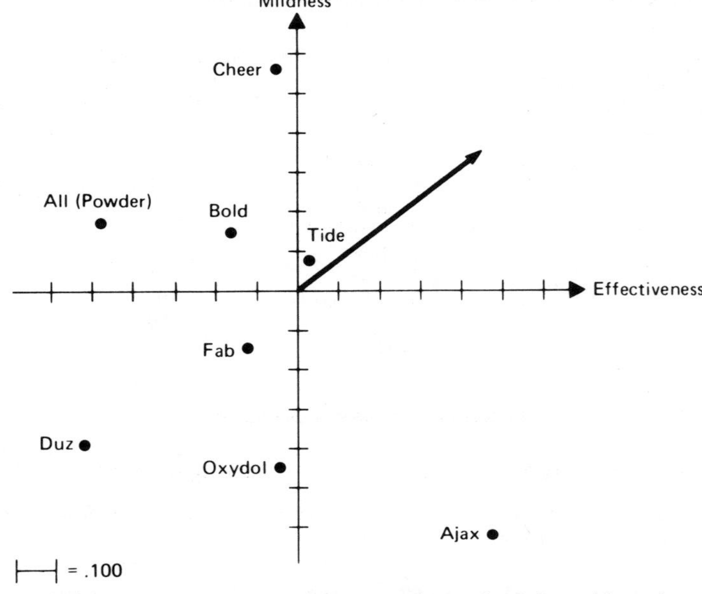

Figure 12.13. *Hypothetical Perceptual Map for Selected Laundry Powders*

der." It would be packaged like a liquid detergent, but would be powder. The idea was to give the convenience of a liquid to those who are powder users. Other concepts were suggested, but these two were carried to the design evaluation phase.

Concept Evaluation and Refinement

A new survey was conducted to see how the new concepts were positioned. Figure 12.14 shows the results. CLI-1 was the new "gentle and effective" product and did fall in the gap as desired. The pourable powder was perceived as effective, but was not in the best position. Intent to try was good for CLI-1 (40 percent definite, 20 percent probable) and low for CLI-2 (10 percent definite, 20 percent probable). Preference data indicated a 42 percent share of preference for CLI-1 if all consumers were aware and 100 percent distribution were attained. Definitely, an encouraging result!

Predictions of preference for the other brands indicated that CLI would draw significant share from Cheer and thus could expect a major competitive reaction from Cheer's manufacturer—a dominant package goods firm. In order to preempt a counter product like "new stronger Cheer," CLI decided to try to increase the positioning on efficacy without a loss in mildness (see Figure 12.14). They felt a stronger claim and improved product for Cheer could not move it very far on the efficacy dimension. If

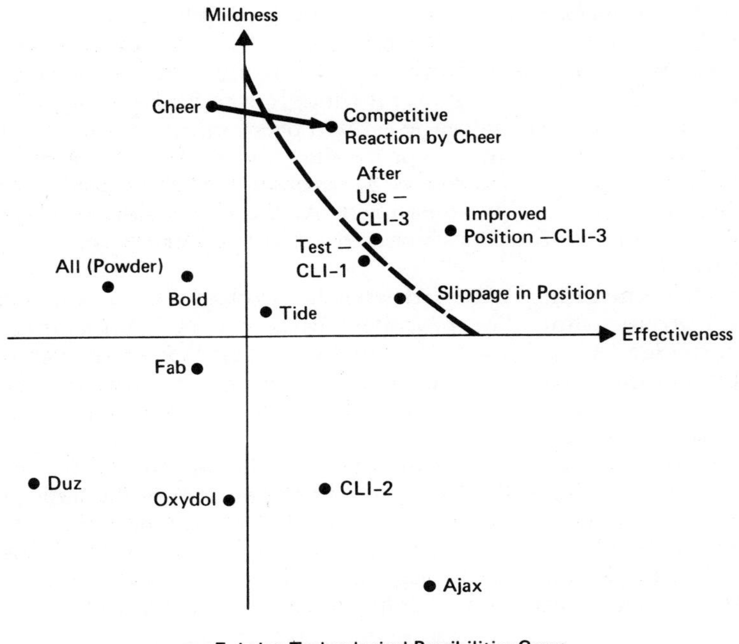

Figure 12.14. *Hypothetical Perceptual Map of Alternative Scenarios*

CLI-3 were able to obtain the improved positioning, it would be less vulnerable to defensive competitive actions. If CLI-3 could fulfill its claims by performance, a major opportunity would be present. Attention then turned to the physical product.

Product Fulfillment

R&D set out to create the new product. They felt they could attain the positioning relative to efficacy, but the laboratory was uncertain as to the mildness. The engineering tradeoffs between mildness and efficacy which existed are shown by the dashed line in Figure 12.14. Existing technologies would allow improvement in efficacy, but only with a loss in mildness. R&D felt, however, that they could advance the state of the art and shift the technology curve upward and to the right if they had six months and $50,000. The new improved position is shown in Figure 12.14.

Marketing felt that stronger claims for efficacy may reduce perceived mildness. (See "Slippage in Position" in Figure 12.14.) They felt that consumers would be skeptical of a product that claimed to be both effective and mild.

The preferences for the improved or slippage products were simulated (Figure 12.14) to assess the relative impacts. The weaker position would yield a 34 percent share of preference—still strong—and the stronger position would yield a 45 percent share. Although even with slippage the product is not bad, efforts to achieve the refined positioning were worthwhile.

After six months, the prototype product was ready to be tested by home placement. Ratings were obtained from consumers after use and preference and repurchase intents were measured. At the conclusion of their home interview, a conjoint study was done to provide information for product improvement.

Based on the ratings obtained after the product was used at home, the improved product slipped on perceived efficacy while it fulfilled the mildness claims. (See "After Use—CLI-3" in Figure 12.14) Even though lab tests showed it met the technical effectiveness requirements, consumers did not perceive this. Fortunately, the perceived efficacy could be improved with the results of the conjoint analysis.

In the conjoint analysis, consumers were shown physical formulations that varied by the color of the basic powder, the texture of the basic powder, and whether colored particles were added. The dependent variable was perceived efficacy. In addition, conjoint analyses were run for mildness to ensure that there would not be slippage along that dimension if features were added to improve efficacy. The results are shown in Table 12.6.

Table 12.6. Conjoint Analysis to Select Physical Features

		Efficacy			
Color	*Part Worth*	*Texture*	*Part Worth*	*Particles*	*Part Worth*
Green	0.5	Coarse	0.15	Added Blue Particles	0.8
Blue	0.5	Fine	0.05		
White	0.7			One Color	0.2

		Mildness			
Color	*Part Worth*	*Texture*	*Part Worth*	*Particles*	*Part Worth*
Green	0.3	Coarse	0.1	Added Blue Particles	0.05
Blue	0.4	Fine	0.9		
White	0.8			One Color	0.10

The conjoint analyses indicated that a coarse, white powder with added blue particles would best enhance perceived efficacy, but the coarseness of the powder would cause slippage on mildness. Furthermore, a coarse texture adds little to efficacy, but a fine texture greatly enhances mildness. The white color also enhances mildness, while the added particles have only a small negative impact on mildness.

Based on the conjoint analyses, a very fine, white powder with added blue particles was selected as the mix most consistent with a "very effective, yet mild" CBP. This analysis indicated that with the new color and particles, the perceived efficacy could be raised to the level measured in the lab. The "improved positioning" shown in Figure 12.14 could be achieved. It was also found in the post-use interview that a special handle and resealable spout would further improve the product.

With the assurance that the concept was good and the physical product fulfilled it, attention turned to estimating the purchase potential.

In concept and product tests, trial and repeat were directly measured for the purchase equation to estimate share (purchase potential rate, P, divided by market purchase rate N_p):

$$\textbf{(12.7)} \qquad S = \frac{P}{N_p} = a_w a_v T R$$

where a_w = awareness
a_v = availability
T = ultimate proportion of target group who will try new product
R = market share among those who have tried
S = long-run market share potential

A logit model was estimated based on the test product so that the alternative positionings could be evaluated in terms of trial and repeat probabilities.

Table 12.7 gives the results for the test product—note that the observed trial (41 percent) was close to the predicted preference share (45 percent). The predicted share of 8.3 percent of powdered, hot water detergent market ($300 million total sales) would yield a sales volume of $25 million for the new brand—enough for CLI to take the product to test market. Financial projections revealed a good rate of return on the investment of $20 million for development and introduction.

Table 12.7. Evaluation of Purchase Potential of Test Market

Trial if Aware and Available, T	41
Expected Long-Run Awareness, a_w	75
Expected Distribution, a_v	80
$a_v \cdot a_w \cdot T$ = 24.6% Cumulative Trial	
Repeat, R_{11}	70.8
Return for Competitive Brands, R_{E1}	14.8
R = 33.6 percent cumulative repeat	
Share Potential = 8.3 percent (by equation 12.7, $S = a_v \cdot a_w \cdot T \cdot R$)	

Although the opportunity was exciting, there were risks. How low would the share be if Cheer retaliated? What gains will occur if further product engineering could achieve an improved perceptual positioning which would hold up under home usage tests? The logit model was used to evaluate these scenarios, i.e., to estimate T and the components of R in equation (12.7) (see Table 12.8). Perceived improvement in efficacy would increase share to 9.8 percent. Competitive reaction by Cheer would seriously erode the share to 5.8 percent. But if the "after use" positioning were improved, share would be predicted to be 8.4 percent even with competitive retaliation. These simulations indicated an exciting opportunity with a manageable competitive situation.

Table 12.8. Simulation of Alternative Scenarios

Scenarios	*Cumulative Trial*	*Cumulative Repeat*	*Purchase Potential*
1. Improved Perceived Efficacy	24.6%	40%	9.8%
2. Competitive Reaction by Cheer	23.2	25	5.8
3. Competitive Reaction with Improved Perceived Efficacy	24.0	35	8.4

Marketing Mix

Before proceeding to test, the product advertising and promotion strategies were specified. Table 12.9 shows CLI's view of how the marketing mix elements affect sales and market share. Different levels of these variables were simulated by modifications of equation (12.6) to find a good initial specification of the marketing mix.

Table 12.9. CLI's View of Primary Effects of Mix Elements on Purchase Potential

Advertising	*Pricing*	*Couponing*	*Sampling*	*Distribution*
Causes Awareness, a_w			Forced Awareness, a_w	Primary Effect is Availability, a_v
Communicates CBP to Effect Trial, T	Strong Effect on Trial, T	Price Discount to Obtain Trial, T	Higher Chance of Trial, T	Some Effects on Trial, T
Supports Repeat and Frequency of Use, N_p	Affects Repeat, R	May Reduce Repeat, R	May Reduce Repeat, R	Affects Repeat, R

For example, purchase potential for scenario 3 was based on advertising expenditures of $7 million to produce the 75 percent awareness estimated in Table 12.7. If a_w could be increased to 90 percent, the share would increase to 10.1 percent, giving a net volume increase of $5.1 million [i.e., $(0.101 - 0.084) \times$ ($300 million)]. Thus, if CLI could attain the additional awareness and production for less than \$5.4 million, it would be worth the investment. CLI's advertising agency felt the additional awareness would cost \$3 million. Based on this figure and an analysis of production costs, the higher level of advertising was recommended.

Sampling levels were evaluated. For example, suppose we sample 40 percent of the population and 87.5 percent of those who receive samples, try the sample. Suppose that R is reduced from 35 percent (R) to 30 percent for those sampled. Then the long-run share is 14.4 percent, as computed in Table 12.10. (Note that we use scenario 3 and the increased advertising.) CLI must then weigh this increased revenue [\$12.9 million = (0.144 − 0.101) × (\$300 million)] against (1) the added costs of producing and distributing the samples and (2) the added production costs. If the 24 million households were to receive the sample at a cost of 30¢ per sample, it would be profitable to sample if the added production costs were less than \$5.7 million. That is, it would be profitable to sample if the increased revenue from sampling (\$12.9 million) exceeds sampling (\$7.2 million) and production costs.

Table 12.10. Computation of the Effects of Sampling

Sampled Consumers (s = .40)		*Unsampled Consumers (1 − s = .60)*	
		Expected Distribution, a_v	0.80
Try the Sample, U_s	0.875	Long-Run Awareness, a_w	0.90
Repeat after Sample, R_s	0.30	Expected Trial, T	0.40
Expected Distribution, a_v	0.80	Repeat after Trial, R	0.35
$sU_s R_s a_v = 0.084$		$(1 - s)a_v a_w TR = .060$	
		Long-Run Share = 14.4%	

The outcome of a series of such simulations coupled with marketing judgment was the marketing mix of \$10 million for advertising in the first year, \$7.2 million for samples, and \$5 million in trade promotion.

The brand name was selected to represent the CBP. CLI rejected "Whisper" because "Whisper" implied only mildness and may not fit with the efficacy appeals. Similarly, "Maxi-clean" missed the mildness appeal. Names like Riptide, Cleanall, Scrubsuds, Ebbtide, Softsilk, and Safewash all miss one part of the CBP. On the other hand, Gentle Power, Soft Strength, and Satinscrub all use the CBP. "Satinbright" was selected for the initial testing.

The advertising strategy was implemented in three copy themes. One stressed the personal recommendation of the product by a neighbor, another used a "scientific" demonstration to show how Satinbright brightened the colors of delicate fabrics but did not harm the fabrics; and the third used graphics to illustrate how the "tight-packing" technology combines "cleaning" and "protection" particles. These three themes were to be considered as alternatives in the subsequent testing of the product.

The outcome of the design process was:

- a target market definition of the hot water, powder detergents;
- a Core Benefit Proposition of an "effective but mild laundry detergent";
- a positioning of more effective than Cheer and milder than Ajax;
- a physical product of a very fine, white powder containing softeners and blue particles in a package with a handle and resealable spout;
- a high investment in advertising, sampling, and promotion strategy ($22.2 million);
- three advertising themes and a brand name for further testing.

SUMMARY

The CLI case has reviewed the design phase of development in the context of a frequently purchased consumer product. In durable consumer products, industrial goods, and services, the issues are similar. Based on a defined target market and creative ideas, the Core Benefit Proposition is specified and positioned against competition. The physical product characteristics necessary to fulfill the CBP are developed and the price, advertising, and distribution strategy are determined. In industrial products, more emphasis is on physical performance and personal selling than in consumer products where psychological attributes and advertising are critical to success. But industrial buyers are influenced by a supplier's reputation (perceived quality), perceptions of delivery reliability and service, as well as the personal effectiveness of the salesman and product performance. Services are delivered through people and more difficult to control, but must be oriented around a CBP and positioned against competitive service options.

In all new products, care should be taken to understand the relationship between perception, physical features, preference, and choice in the new product adoption process. Then the design should be completed by careful product engineering, creation of advertising and sales copy, and the initial specification of the advertising, selling, price, promotion, and distribution levels for the new product.

In Chapters 7 through 12, we have presented the basic ideas behind the design methodology coupled with the analytic tools necessary for the manager or analyst to actually perform the analyses. While some of the details are complex—factor analysis, preference regression, segmentation tests, logit, and the purchase model—we feel that the overall flow of analysis, as detailed in Figure 7.1 and reproduced in Figure 12.15, is quite straightforward.

At this point, you should understand the role that each component plays in the design process. You should now be comfortable with selecting a set of analyses appropriate to a particular product category and know how to use the output of these analyses to guide a proactive new product design.

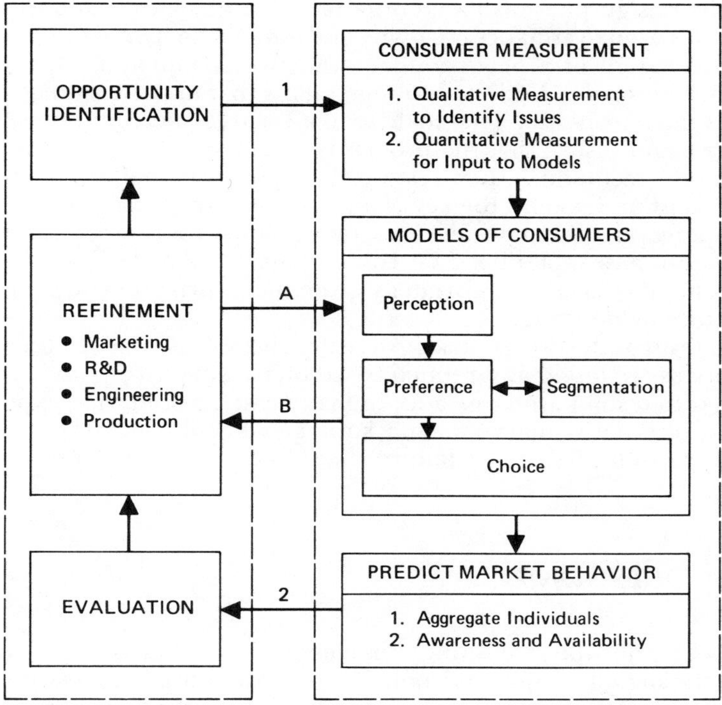

Figure 12.15. *Basic Flow of the New Product Design Process*

Complete familiarity with the details of each analytic method will require experience under fire as part of a new product design team.

To use this design process, we suggest that you:

- design the new product CBP, positioning, and marketing mix, not just a physical product;
- iterate the analysis to continually refine this strategy;
- search for understanding of consumer response to guide creativity and predictive ability to evaluate strategies;
- use analytic techniques to support strategic decisions;
- blend managerial judgment and creativity with the qualitative and quantitative techniques.

With the completion of the design, the testing phase can begin. In testing, more emphasis is placed on sales forecasting and on assessing the profitability and return on investment. In testing efforts, more complex models are used and more extensive measures are collected to evaluate and refine the marketing mix variables.

The design phase is based on a blend of qualitative and quantitative techniques used to enhance creativity and guide managerial judgment in the proactive design of a new product strategy. This analysis taps consumer needs and desires to find the leverage points that make the greatest impact on consumer response. The qualitative measures and common sense raise the issues and insure that the process is receptive to consumer input. The quantitative techniques then resolve these issues by helping management identify and understand the key elements of the response process. By measuring perceptions and preferences, by searching for segments, and by predicting purchase potential, the components of the design process help management select the best strategy given the information potentially available before field testing.

In testing, the design strategy is extensively evaluated. In these efforts, alternatives are tested and refined based on consumer response. The funds allocated to testing are large and, in many cases, test market experimentation with alternative marketing mix strategies is conducted. If the design has been done well, the risk of failure should be low and the profit potential high.

REVIEW QUESTIONS

12.1 Consider the marketing mix elements of advertising or sales copy, packaging, name, and point-of-purchase display. How is the diagnostic information generated in the new product design process used to select an appropriate marketing mix?

12.2 What is the inter-relationship between advertising (or selling), distribution, pricing, sampling, couponing, and price-off promotion for the new product? How do each of these affect awareness, availability, trial, and repeat?

12.3 How is the core benefit proposition used in setting the marketing mix?

12.4 What is the purpose of an advertiser's unique selling proposition?

12.5 Why should an advertiser bother with positioning and not just merely state the product is outstanding on every possible dimension?

12.6 The marketing department of a hypothetical manufacturing company after extensive research developed a CBP which they believed had enormous potential. The CBP was:

> A extra long lasting light bulb ideal for situations requiring extensive usage or involving difficult replacement.

The engineering group worked long and hard to come up with a better light bulb. After exhaustive research and the construction of numerous prototypes, the engineering group developed a prototype they felt marketing people would definitely appreciate. This particular prototype not only had a lifetime twice that of average bulbs but had several other remarkable features. With a novel use of gold leaf, the engineers would be able to quadruple brightness while only doubling production costs. Finally, by altering the base of the bulb and adding a special chemical to the shell, the bulb could be dropped from a height of 10 feet without breaking.

The marketing people were overwhelmed by the efforts of engineering. Not only did the bulb meet the CBP but it seemed to be the perfect light bulb.

What is your reaction to this chain of events?

12.7 What assumptions are inherent in equation (12.6)? How would you relax these assumptions? Discuss why equation (12.6) is likely to give forecasts which are sufficiently accurate for this stage of the product development process.

12.8 Discuss the basic flow of the design process (Figure 12.15) in light of the material presented in chapters 7 through 12.

12.9 Develop another name for Satinbright (the laundry detergent example). Use the CBP to develop a package and an advertising campaign for this detergent.

12.10 For a new product the following table depicts the classifications of the target market assuming 100 percent availability, 100 percent awareness and no sampling.

Number of Consumers	*Who would try product*	*Who would not try product without samples*
Who will repeat	400,000	100,000
Who will not repeat	200,000	50,000

a. Assume sampling is random, all samples are used, 100 percent availability and 100 percent awareness. How many repeat purchases will be generated per 100 units employed in sampling?

b. Now assume awareness without sampling is only 40 percent. How many repeat purchases would now be generated per 100 units employed in sampling?

c. Again assume 100 percent awareness but only 60 percent of the samples are used. Assume usage is independent of trial probability. How many repeat purchases will be generated per 100 units employed in sampling?

d. Suppose sampling was not random but could be directed at different segments of the population. Who should receive free samples? How would the effectiveness of the sample change?

PART IV

TESTING AND IMPROVING NEW PRODUCTS

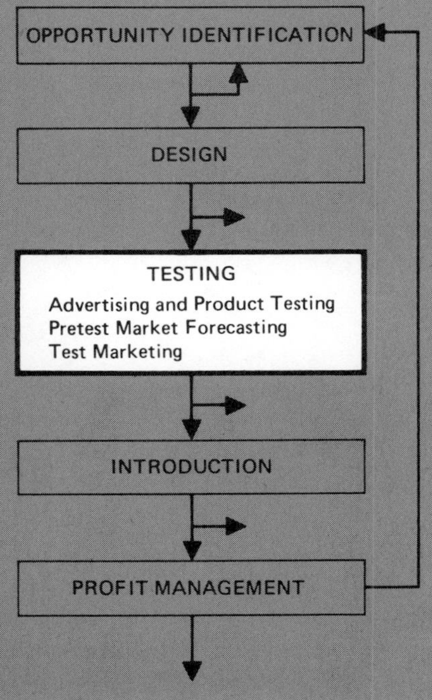

Chapter 13

Advertising and Product Testing

Your organization has invested time and money in developing a new product from the initial idea to an actual product and marketing strategy. A year or more has been devoted to this product, forecasts look good, and there is mounting organizational pressure to go to a full-scale launch. This pressure should be resisted. The design phase was based on models that are reasonable, accurate representations of the market response, but not of reality. Things can go wrong and often do.

For example, in 1978, R. J. Reynolds Tobacco Co. nationally introduced "REAL," a low-tar cigarette, and spent $40 million in advertising and promotion. They achieved a low share of the market—about half their forecast. The market was growing (the low-tar segment had come from zero to 25 percent of unit sales) and Reynolds felt they had a good strategy, but it was not a success. They had bypassed test marketing, which was a serious mistake.

A better managerial policy is to test the overall strategy and its components when the consequences of failure are small. Many of the same lessons can be learned for $100,000 or even $20,000 rather than for $40 million. Furthermore, the product can be refined during the test. In Chapter 2 we introduced a reactive strategy called "second but better." A firm following this strategy waits for another organization to create a category and then enters with a superior product. If you are first in, you want to prevent other organizations from using second but better strategies. This means that when you go national, you want to go national with the best positioning you can possibly achieve. You do this by refining the product through testing.

We can think of a new product in terms of its expected benefit and its risk. All new products have some risk. We can think of a "decision frontier" as the minimum expected benefit that is necessary for a given level of risk, or conversely as the maximum allowable risk for a given expected benefit. This decision frontier is shown conceptually in Figure 13.1. The points represent alternative product and marketing combinations. In this case all of the potential products are below the decision frontier. Suppose we select the starred (*) product for attention. The strategy of testing is to experiment, then improve the product so that it passes the decision frontier. For example, an advertising test might eliminate some uncertainty and identify better copy, thus moving the product along the dotted line. As shown, the risk/benefit position may still be unacceptable. The next step is pretest market (dashed line), and finally test market (solid line). The goal of the testing strategy is to cross the decision frontier in minimum expected cost. To do this, we must be able to reduce risk and increase expected benefit.

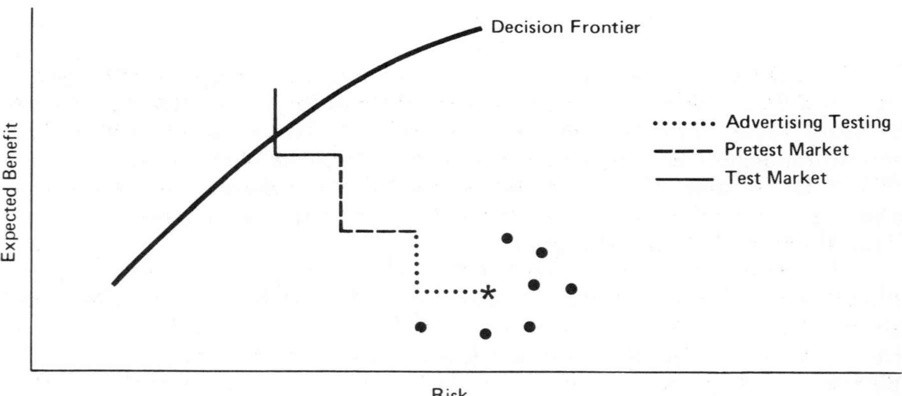

Figure 13.1. *One Testing Strategy and the Decision Frontier*

Reduction of Risk

Risk is reduced by sequentially testing first the components of the product and then the integrated product. The key concept is to delay large testing expenditures (and corresponding risk) until component risks can be minimized. In the Reynolds case, a relatively small expenditure for a pretest market laboratory test may have dramatically reduced the chances of failure in national launch. Similarly, product tests can detect flaws that could cause national failure. Testing can cause delays and give competitors a time advantage, so care must be taken to avoid unnecessary delays. In rare cases, it may be best to skip a testing stage, but only if the time pressure can justify the tremendous risks of doing so.

Uncertainty is always present. Predictions become less uncertain as our tests more closely approximate reality, but costs increase as we approach reality. The testing strategy balances the uncertainty and the cost by careful use of the sequential strategy. After the components are tested, the integrated product is tested, but not yet in a full test market. Rather, laboratory simulations, special consumer panels, and/or statistical models are used. These pretest market tests eliminate many failures at low cost without revealing the product to competitors, but they are still not full reality. Thus the final stage of testing is a test market. If the preliminary steps are used, the integrated product should have a high probability of succeeding in test market. The test market is still important because even with a good pretest, unexpected results may occur.

Maximization of Expected Benefit

One goal of testing is to reduce risk and another goal is to identify the product and marketing variables that maximize the firm's profit at a given level of risk. One way to achieve this is to test all combinations, but trying all combinations is often infeasible and inefficient. The testing strategy should maximize profit by devoting effort to finding ways to improve a product's expected benefit to the organization. In most product tests, results are lower than the desired levels. The testing procedure should produce actionable diagnostics to improve the product. In many cases entrepreneurial spirit is necessary to overcome problems and convert the product into a success based on the diagnostics. Advertising and product testing provide information on the consumer response process, as well as final buying indications. In pretest market analysis, sales forecasting is the main goal, but perceptual ratings may be collected to improve the final advertising and product to deliver the CBP positioning strategy. Test marketing produces an accurate projection of national sales, but it also produces experimental information on price, advertising, and promotion levels which are critical to improving profit.

Components of the Testing Strategy

The procedure to achieve the testing goal of reducing risk and maximizing profit is shown in Figure 13.2. First, the components such as advertising and the physical product are tested and improved. Next, the integrated product is tested in a pretest market experiment. Failures are eliminated, or sent back to the design phase, and the product and marketing mix is improved. The key idea of pretest market is that it be low-cost and not take an exorbitant amount of time, but yield sufficient accuracy and diagnostics for a GO/NO GO decision for test market. Finally, the product is put to the test of reality in a test market. If the product succeeds in test market,

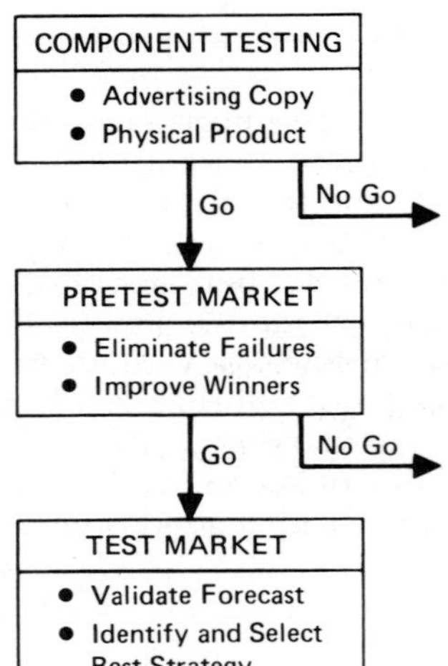

Figure 13.2. *Components of the Testing Strategy*

it is introduced to the full market. If not, it is dropped or cycled back to the design phase.

Throughout Chapters 13, 14, and 15 we discuss the details of each stage, indicating how to perform the tests, how to interpret the test results, and how to use the tests to reduce risk and maximize benefit. We consider the tradeoffs among cost, time, and benefit to the organization and indicate what testing methods are appropriate for various industries. In Chapter 15, we discuss how to select among the product and marketing combinations beyond the decision frontier shown in Figure 13.1.

ADVERTISING TESTING

Advertising is an important component of the new product. Good copy creates awareness and communicates the CBP. Advertising testing allows selection of the best ad from the available alternatives, assesses if this ad is sufficient for the new product introduction, and generates diagnostic information to improve the ad.

Why test? Managers who have worked closely with the new product have a good feel for the consumer. Because each organization has its "experts" in advertising copy and testing techniques can make errors, some firms rely on managerial judgment. Unfortunately, pure judgment has not proven to be much better than random selection. For example, in a study of twenty-four print ads (American Newspaper Publishing Association, 1969) where market results were known, managers' judgments had almost no correlation with the market results ($\rho = 0.06$). The twenty-four ads were rated by eighty-three middle management personnel (brand managers, agency account executives, creative agency personnel, and research personnel).

Only the account executives had a significant positive correlation ($\rho = 0.10$). A negative correlation ($\rho = -0.13$) was observed for agency creative personnel. Although testing techniques are not perfect, they can usually outperform pure judgment.

The question then is which technique to select, and how many ads to test. We begin by examining the criteria for evaluating advertising copy.

Criteria for Evaluating Advertising Copy

Consumer Response Hierarchy. Advertising is effective if it leads to more profitable sales, but to achieve sales, advertising must also achieve a series of intermediate goals (as shown in Figure 13.3). This consumer response hierarchy is important to the evaluation of advertising because it indicates more precisely what is effective about the advertising and what must be improved. In order to use the hierarchy, we must be able to measure the component stages. Figure 13.3 suggests measures for each stage. Some firms use these intermediate measures as the final evaluation of advertising. This is incorrect. These measures only provide clues to effectiveness. Effectiveness is difficult to evaluate without knowing advertising's effect on sales. We advocate using measures for each component, including purchase behavior.

Exposure is the prerequisite for any response and is measured by indicators of whether a consumer is in the audience for a particular media insertion. However, before any response occurs, a consumer must direct

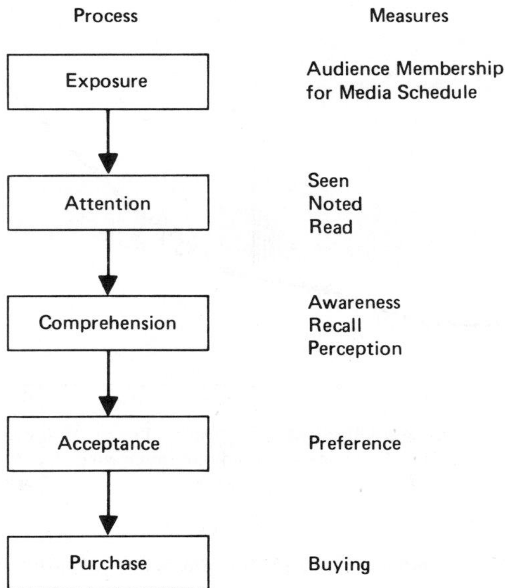

Figure 13.3. *A Model for the Response to Advertising Copy*

some attention to the exposure. Attention is measured by indicators of whether the consumer will see and/or listen to the ad, and perhaps note its content or read it. Comprehension, awareness, and understanding, are the next levels of response. Awareness can be measured by aided or unaided recall of product name and/or the unique selling proposition. It may be measured immediately, the day after ad exposure, or later. Understanding can be measured by perceptions of the product that the ad evokes. After comprehension comes acceptance which reflects the translation of comprehension into preference. Preference can be measured by first choice, a rank order of preference, intent to buy, or purchase in a simulated buying situation. The end result is buying which can be measured in market tests or in realistic purchase environments.

Likability, Believability, and Meaningfulness. In addition to measuring the effects within the hierarchy, it is useful to take measures that will determine why these effects occur. Likability, believability, and meaningfulness are three measures found to be correlated with responses in the hierarchy.

Likability improves acceptance. But some researchers believe that ads which are disliked can also produce good results (Wolfe, 1949; Britt, 1955; Schwerin, 1955). For example, probably we are all familiar with the abrasive "ring around the collar" appeal. Silk (1974) found that an unpleasant "hard sell" radio ad produced better response (brand awareness, recall, prefer-

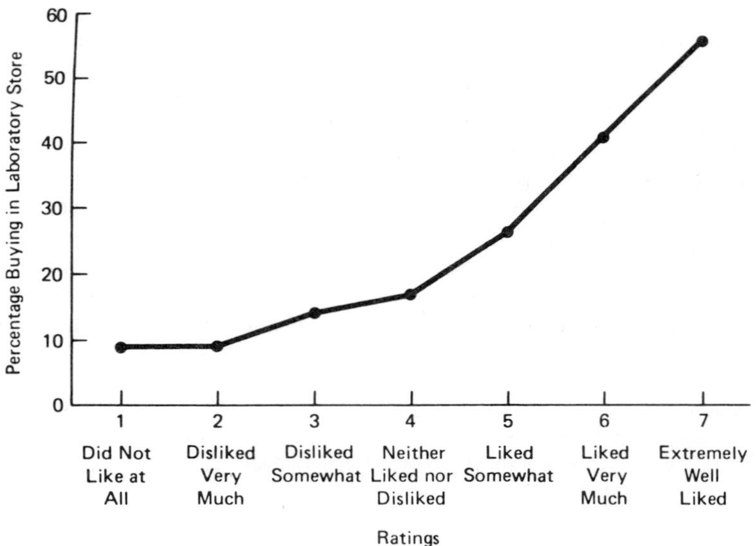

Figure 13.4. *Average Trial of 40 New Brands vs. Likability After Ad Exposure*

ence) than a pleasant "soft sell" ad. But this difference disappeared after repetition over time.

The effectiveness of low likability must be carefully evaluated. Ads that are disliked may get more attention, but may score much lower on acceptance once attention is gained. For example, Figure 13.4 shows the correlation of trial in a simulated retail store with ratings of likability. This was based on data from forty new, frequently purchased products. Three hundred women were shown ads and interviewed in each of the forty studies. Figure 13.4 suggests strongly that if attention can be achieved for a likable ad, then likability is a desirable feature of that ad. The fraction of consumers that try the product is much higher for likability ratings of 6 or 7, than those below 5.

Believability and meaningfulness also are correlated with trial once attention is gained. For example, Figure 13.5a indicates that high ratings of believability correlate with high trial. But intermediate values (4 and 5) are associated with some response. This could reflect the concept of "curious disbelief" (Malone, 1962), which implies if the consumer is uncertain as to whether to believe claims, they may try the product to find out if they are true. Meaningfulness (personal relevancy) of an ad reflects how meaningful the ad copy is to the respondent's own situation. Figure 13.5b shows high meaningfulness also correlates with higher trial.

Figures 13.4 and 13.5 imply that given attention, good ads will get more response if they are liked, believed, and perceived as personally relevant.

Methods Used to Test Advertising Copy

Various organizations may place different emphasis on the criteria for evaluating advertising copy, but each organization should select the method that is best for its needs and resources. Many methods are available; some major methods are shown in Table 13.1. These approaches differ with respect to methods, measures, completeness, accuracy, and cost.

Table 13.1. Selected Copy Testing Approaches

Approach	*Measure in Hierarchy*	*Primary Measure*
On-Air Testing	Comprehension	Day after Recall
Theater Testing	Comprehension and Acceptance	Immediate Recall, Choice of Prize
Trailer Testing	Acceptance	Coupon Redemption in Store
Simulated Buying	Comprehension and Acceptance	Buying in "Retail Store"
In-Market Test	Purchase	Panel Records of Buying

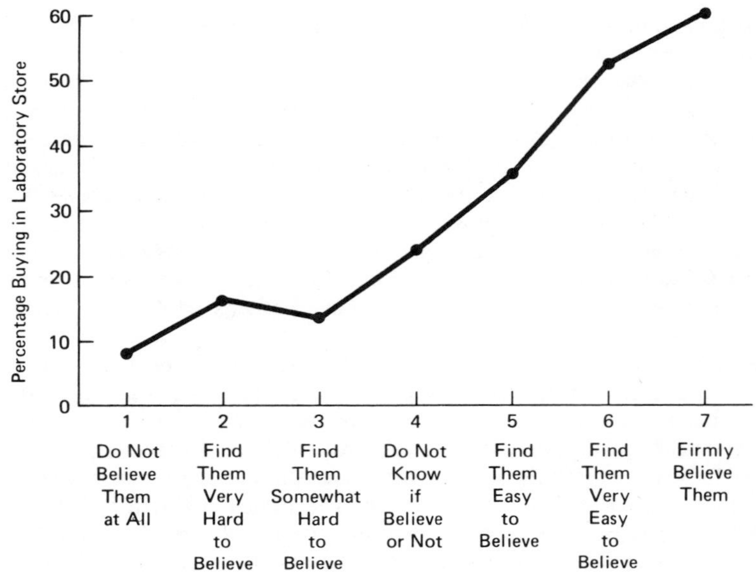

(a) Believability Ratings

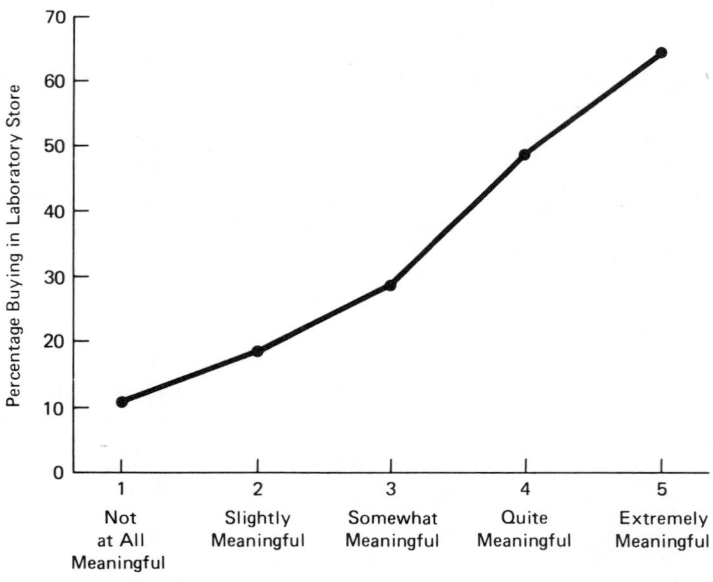

(b) Meaningfulness Ratings

Figure 13.5. *Average Trial of 40 New Brands vs. Believability and Meaningfulness After Ad Exposure*

A widely used approach is on-air testing. A test ad is used on a local TV station, and after 24 hours, a random sample of the target population is surveyed by telephone to determine whether people saw the ad, can recall the key claims, and can correctly associate the brand name with the ad. The advantages of on-air testing are that the exposure is natural and that measures of comprehension can be collected. Recall, memorability, and name association are important and necessary characteristics. However, they are incomplete measures of the purchase response since they do not consider acceptance.

One approach to measuring acceptance is through theater testing. A representative audience is exposed to new TV shows, along with commercials that include the new product ad. Response to the new product is the percentage of people who would select the new product rather than other products for a prize after seeing the programs and ads. In the theater setting, measures of interest, believability, and recall can be obtained. In some procedures, respondents turn a dial to reflect their interest in the ad, but the acceptance measure is the major output. This is usually compared to past "norms" for similar products.

When the exposure setting is changed by showing ads in a trailer outside a store, purchasing of the new product may be observed in an actual store by the measuring of how many redeem a new product coupon given the respondent after exposure. Another methodological variant is exposure in a laboratory setting followed by a purchase opportunity in a simulated retail store.

Actual purchase cannot truly be measured in either theaters, trailers, or simulated store, but each method attempts to make the choice as similar as possible to the real buying situation. The only way to measure actual purchase is through in-market tests. Some services are based on placing the ad on the air and observe purchase behavior through consumer panels and store audit data.

Selecting a Method

To select the appropriate method, an organization should weigh the reliability and validity of a technique against the cost and the suitability of the technique for the new product under consideration.

Reliability. A testing method is reliable if repeated tests of a given ad give similar results. Suppose that more than one ad is tested. Then the variation among the (true) scores for each ad should be greater than the variation within scores for a given ad. Let σ_b^2 be the variance in the true scores between ads and let σ_e^2 be the variance of the error in measuring a given score. The index of reliability, R, is given by:

(13.1) $$R = \left[\frac{\sigma_b^2}{(\sigma_b^2 + \sigma_e^2)}\right]^{1/2}$$

Reliabilities in the range of .8 can be expected in a well-designed test (Bultez, Derbaix, and Silk, 1976).

Most agencies report high reliability, but these must be viewed with caution. The majority of studies estimate R by the correlation between successive tests of the same ad. We call this measure r. The pair may be at two points in time or across two test sites at the same time. Correlation may be based on many test/retests for the same product or across various products. If the true scores are independent of the errors and the errors are independent across tests, then r equals R (Silk, 1977). However, r can be artificially inflated when these conditions are not met. For example, Clancy and Ostlund (1976) report an overall correlation of 0.67 for 106 pairs of on-air tests, but this dropped to 0.29 after adjusting for non-independence effects.

Better reliabilities can be achieved for theater, trailer, simulated store, and in-market tests than for on-air testing. This is due to a lack of control of the surrounding television program content, variations in media placement, and lack of effort to match audiences in on-air testing. Furthermore, reliabilities may decrease at higher levels in the hierarchy. For example, in a study of forced advertising exposure tests, Clancy and Ostlund (1976) report reliabilities of 0.96 for immediate recall, 0.84 for post-purchase interest ratings, and 0.64 for the change in buying interest ratings before and after ad exposure.

Thus when selecting a testing method, the new product manager should look for high reliability, but must make certain that the reliability is not inflated, and that the reliability is appropriate for the criteria being used to evaluate the copy.

Validity. A testing method is valid if its test scores correlate highly with actual market results (sales or profit). Unfortunately, much less is known about validity than reliability. Various suppliers make claims of high validity, but few of these claims are documented by published research reports since validity measurement requires careful market experimentation. The new product manager should seek high validity, but the validity claims should not be accepted unless they can be documented.

Fortunately, validity is usually higher for new product trial than for changes in the share of existing products. For example, Bloom, Jay, and Twyman (1977) found correlations of 0.9 between sales volume and test scores. Kahn and Light (1974) found that "high" scores increased sales 25 percent, while "low" scores increased sales 9 percent for the four ads they evaluated. On the other hand, Heller (1971) reports a lower correlation of 0.31 between on-air recall and actual coupon redemption.

In general, little is known about validity, but it is probably lower than reliability. With careful tests, values of .6 to .8 are reasonable for new product copy testing.

Expected Revenue Gain. The purpose of copy testing is to identify and to refine the advertising copy which will result in the most sales, revenue, or profit. To accurately identify the best ad, the test should be both reliable and valid. Some people argue that since reliability and validity are not perfect, ads should not be tested. Fortunately, expected revenue gain can be achieved with imperfect tests. To measure expected revenue gain, we use a model developed by Gross (1972).

Intuitively, we would expect the revenue gain to increase with the reliability (R) and the validity (ρ). Furthermore, we would expect that the revenue gain would be greater if there is greater variation between ads. This variation is measured by σ_E, which is the standard deviation of market profitability resulting from alternative ads. Finally, the revenue gain depends on the number of ads that are tested. Gross shows that the expected gain in revenue, E_n, for testing n ads is given by:

(13.2) $$E_n = e_n \sigma_E \rho R$$

where e_n is a value that depends only on the number of ads being tested. (See Appendix 13.1; e_n is the expected value of the largest of n observations sampled from a standardized normal distribution.)

To use equation (13.2) to evaluate a testing procedure, we must estimate σ_E and determine ρ and R. We then look up e_n in Appendix 13.1 and compute E_n for various values of n.

For example, suppose that σ_E equals \$2 million. This would mean that for an average profit of \$4 million, a profit of \$6 million or more could be earned for the top 16 percent of ads, while a profit of \$2 million or less would be earned for the bottom 16 percent of the ads. Each new product manager must estimate σ_E for his products. If the validity, ρ, were 0.7 and the reliability, R, were 0.8, the gain from testing two ads would be \$631,700. (For two ads, e_n is 0.564. Then $E_n = 0.564 \times \$2{,}000{,}000 \times 0.7 \times 0.8$.) Even if validity were 0.5 and reliability were 0.7, the expected gain would be \$394,800. These expected gains are weighed against the costs of testing n ads.

Reliability varies by method and sample size. If the sample is increased, reliability and cost increase. A tradeoff between reliability and cost can be made (see Dalal and Srinivasan, 1977). Each commercial testing service has made a choice. Bultez, Derbaix, and Silk (1976) have estimated the approximate costs, reliability, and validity of various available methods. The costs and ranges of predictability ($\rho \times R$) are shown in Figure 13.6. As the cost goes up, the predictive ability raises. These are reasonable values to use in calculating expected gains.

Recommendations. Each manager must evaluate the expected benefits and costs for the new product under consideration, but it is usually worth the cost to test alternative ads. Bultez, Derbaix, and Silk estimate that for most consumer products the best combination would cost about \$20,000 to \$30,000 with a predictability in the range of about 0.7. This procedure does not yet exist, but one would speculate that it would combine on-air and forced exposure methods (theater or trailer) with coupon redemption or buying in a simulated store. Such procedures are now under development by several suppliers of copy testing services. Until these new services are

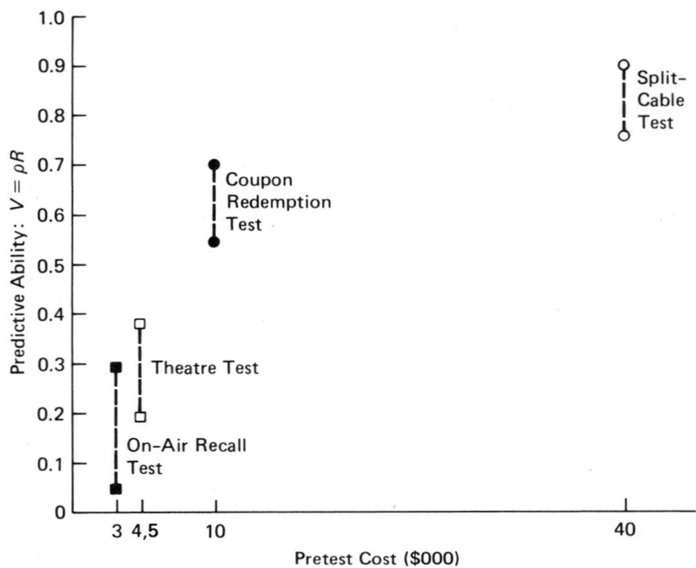

Figure 13.6. *Cost vs. Predictive Ability for Copy Testing Methods (adapted by Silk from Bultez, Derbaix, and Silk, 1976)*

available, we recommend an on-air *and* theater or an on-air *and* coupon redemption test. Split-cable testing is difficult for a new product since it requires distribution in the test city and substantial costs which would be warranted only as a final selection method during test market analysis.

The future for copy testing looks bright. With improved procedures, there will be an even more compelling argument to test two or three independent alternative ads.

Testing Personal Selling Presentations

While testing of advertisements is common in consumer products, few organizations test their personal selling presentations. In industrial products, these presentations are as important as advertising in consumer products. Alternative presentations could be made to stress particular benefit claims, use various demonstrations, or be based on direct competitive comparisons. These approaches are likely to vary in terms of sales success. If there is a variance, the advertising testing model discussed above is relevant (equation 13.2) for computing testing payoffs.

Such tests can be done by showing video-taped sales presentations to a sample of buyers and making before/after recommendations. Such a laboratory testing procedure could also test industrial advertising in conjunction with the sales presentation. By exposing one subsample to ads and sales presentations and one to only the sales presentation, data would indicate the effect of advertising in improving sales productivity.

Another approach to testing sales messages is through a panel of firms who are experimentally exposed to alternative sales strategies. Longitudinal sales records are then analyzed to determine the impact.

Research is underway in both these approaches. Although the testing is more complex due to multiple participants in the industrial buying process and infrequent purchases, we expect reasonably valid and reliable advertising copy and selling presentation testing methods to be more available in the near future.

PHYSICAL PRODUCT TESTING

A well-designed product will fulfill the core benefit proposition. Earlier, in the design phase, we used perceptual mapping to identify the CBP positioning for the product concept and a prototype. Conjoint analysis was suggested to select features that consumers felt would substantiate the CBP. Now we must determine if the product performs as planned when it is at the production level and if its formulation can be improved. Physical product testing determines whether the product will deliver the CBP and generates diagnostic information to improve the product and/or reduce costs. For example, in 1978, General Foods introduced "Mellow Roast," a blend of instant coffee and roasted grains. If the grain content is increased costs decrease, but more grain may result in a less desirable taste. What should the proportions be? Similar issues must be addressed in many consumer products where active ingredients affect cost and performance.

Product testing is also important in consumer durables, services, and industrial products. A new automobile may have a CBP based on comfort and efficiency with adequate power. How low can the horsepower be before it is viewed as "underpowered"? How much padding and carpeting should be added to the interior to provide comfort? In office copiers speed and reliability are important, but increasing speed may also increase jamming under actual office conditions. What tradeoff between speed and reliability should be made? Personal service is important in hospitals. If doctors and nurses greet their patients on a first name basis, will this enhance personalness without undermining perceived quality?

Physical products can be tested in many ways. The new product manager should try to achieve a test that indicates how the product will perform under actual consumer use situations. Inadequate tests can lead to costly mistakes in test market or full-scale launch. Some of the most costly errors have resulted in large-scale auto recalls and aircraft grounding (e.g., DC-10 aircraft, 1979). The following list indicates a variety of other things that have gone wrong in products.*

*This list is adapted from Klompmaker, Hughes, and Haley (1976); and Vinson and Heany (1977).

– Because packages would not stack, the scouring pads fell off the shelf.
– A dog food package discolored on store shelves.
– In cold weather, baby food separated into a clear liquid and a sludge.
– In hot weather, cigarettes in a new package dried out.
– A pet food produced diarrhea in the animals.
– The electric cord in a powered microscope produced shocks.
– The gas cap on a tractor disengaged due to swelling of tank filler spout in use.
– Mounting bolts on bicycle baby carrier cracked due to stress concentration.
– Sharp edges on vent slots in inner door of dishwasher caused injury.
– When it was combined with a price reduction, a product change in a liquid detergent was thought by consumers to be dilution with water.
– Excessive settling in a box of tissues caused the box to be one-third empty at purchase.

There are several approaches for product testing to prevent such problems. Laboratory tests, expert evaluation, and consumer tests can be utilized. To select the approach or combination of approaches for a particular product, the new product manager must assess the strengths and weaknesses of each approach.

Laboratory Tests

Laboratory tests can effectively answer many product performance questions. The efficiency of an automobile engine is measured on a test stand. Alternative designs for engine components such as carburetors and ignition systems are evaluated in the laboratory and on test tracks. A new copy machine is tested at various speeds to determine the relationship between speed and rates of jamming. Betty Crocker kitchens are well-equipped to test alternative cake mixes. Engineering testing practice is well-developed and provides valuable insights for product tests.

A disadvantage of laboratory tests is that they may not be completely representative of product use. Few consumers achieve the miles per gallon determined by EPA tests. Copiers may jam more often in the office than in the lab. Automated machine tools may be less reliable "on the line" than laboratory fatigue tests might indicate. Consumers may not be as careful at home with a new cake mix recipe as researchers are in the test kitchens. We suggest that "in-use" tests complement laboratory tests.

Another disadvantage is the tendency of engineers to think in engineering terms. Unfortunately, the consumer rarely does. Engineering measures for a new transit system include travel time, wait time, fare, egress and access time, but not service, quality, safety, convenience, comfort, and privacy. For a new automobile, perceived efficiency may be closely related to engineering tests, but perceived comfort may not have any good engineering surrogates. Comfort may be partially related to the depth of foam in the seats and leg room, but it is also related to the overall "lushness," "quietness," and "softness" of the interior. In many products the CBP is stated in perceptual terms because consumers buy based on perceptions. In such cases, laboratory tests should be enhanced by consumer tests.

Expert Evaluation

One way of evaluating perceptual aspects of a product is by "expert" judgment. For example, the "comfort" of new auto interiors may be judged by a panel of styling engineers and marketing executives. Expert tasters are often used for new foods. They may evaluate the "flavor" and "mildness" of different blends of coffee. Some taste tests determine whether a new product fulfills its positioning claims. In others, alternative recipes are evaluated to identify the best or lowest cost combinations. For example, if no difference is detected in flavor and mildness across alternatives, the lower cost formulations could be used.

The advantage of expert evaluation is its relatively low cost. Experts can be trained to carry out intricate comparisons. For example, many paired comparisons may be made to find the effect of each ingredient on taste perceptions. However, these methods rely on the assumption that "experts" accurately reflect consumers' perceptions. If consumers are influenced by their individual past experience or psychological attitudes, the "experts" may not completely represent the buyers reaction to the product.

Consumer Tests

To complement laboratory tests and expert evaluation, we turn to the final judge—the consumer. Products are tested under conditions close to actual use. Consumer perceptions provide key inputs to the evaluation of product performance.

Tests of physical characteristics alone are not sufficient. There is evidence that consumers are influenced by more than the physical characteristics of a product. Allison and Uhl (1964) found that when consumers tasted and rated labeled beers, they rated the brand of beer they drank most often significantly higher than other beers. But when they tasted and rated the same beers without labels, there were no significant differences between the ratings of brands. The label and its psychological associations affected the taste evaluation. In a study involving turkeys, Marquardt, Makens, and Larzelere (1965) found consumers would pay a premium price for one brand over another. Tucker (1964) observed that when confronted with a choice of loaves of bread differing only with respect to letter labels, consumers formed loyalties toward specific labels. McConnell (1968) studied the effect of price on the perceived quality of beer. In this study all of the beer was physically the same and only the price level was changed. The taste ratings (undrinkable, poor, fair, good, and very pleasant) were significantly related to price, with the higher-priced beer receiving better ratings. These studies indicate that although physical ingredients are important, past experience and perceptions may affect consumer evaluations.

Psycho-social cues also have an impact on consumer evaluations of durable products. The same physical car may be evaluated differently if it is manufactured by Volkswagen rather than Ford. Consumer perceptions of the reliability of a copying machine can depend upon whether the manufacturer is Xerox or Savin. Consumer reaction to an organized ride sharing system depends upon whether it is called "community carpooling" or "organized hitchhiking."

Even in industrial products these cues can be important. An air-conditioning system must not only be reliable, but appear and sound reliable. The right face-plate for an electronic component can help communicate that it is "state-of-the-art."

There are three commonly used procedures for consumer tests: single product evaluation, blind comparisons, and experimental variations. In each procedure it is important that consumer attitudes and experience be understood in the evaluation process.

Single Product Evaluation. The simplest approach to consumer testing is to ask consumers to evaluate the new product to see if it is "good." Evaluative ratings are usually on a five- or seven-point overall scale of liking (called the "hedonic scale"). Additional ratings may be collected on various attribute scales to evaluate whether consumers like a new product on a specific dimension of the CBP. A sample of consumers may be asked to try a cookie mix at home. A copier may be placed in a sample of offices and perceptions of reliability measured. Prospective auto buyers may be asked to rate the comfort of a new interior. Plant engineers and maintenance engineers may be asked to evaluate the reliability of an air conditioner.

Single product evaluation is useful for uncovering flaws, but because there is no reference value, it is often difficult to interpret the results. For example, how good is a "good" rating? Is "very reliable" rather than "extremely reliable" the best rating that can be achieved? To overcome this problem, organizations have turned to procedures where the comparison is more explicit.

Blind Tests. In blind tests the new product is compared to existing products. The manufacturer's identification is suppressed to measure the physical response without the confounding of brand attitudes. If superior mildness is part of the CBP for a new coffee, then the blend should be perceived as milder than existing blends without branding. In many cases brand name or positioning affects the outcomes, so blind testing is replaced by testing with label concept statements or other visuals. Blind tests provide useful information, but in some products the manufacturer's image supports the CBP and should be included in the tests.

Experimental Variations. Alternative product formulations are often under consideration, thus tests are often extended by asking consumers to evaluate these alternative formulations. Consumers may be asked to evaluate a few specific blends of coffee or they may be asked to evaluate an experimental design of alternative formulations. In the latter case, functions are estimated linking the physical features to perceptions and preference. For example, Moskowitz (1972) estimated the perceptions of sweetness in a beverage as a function of the sugar content.

In a study of paper towels, consumers were given a rack of towels which contained three different towels. They were asked to use them sequentially. Across the sample, the towels were varied in weight, adhesive content, and plastic reinforcement. Consumers rated the towels on overall preference, "softness," "absorbency," and "durability." Based on these results, a best combination of weight, adhesive, and reinforcement was selected to fulfill the CBP of soft and absorbent while maintaining a competitive cost and acceptable strength.

These procedures, which are well developed for frequently purchased products, are useful for consumer durables and industrial products. A prospective automobile buyer can rate several interiors which are systematically varied in terms of the cost and quality of vinyl materials (plain plastic to simulated leather). An office might use several copiers for one-month periods. Plant engineers can be exposed to several prototype air conditioners.

Summary of Product Testing Procedures

A new product is unlikely to succeed if the product is inferior or if consumers perceive it as such. But the optimal managerial strategy is not always to produce the best product without regard to cost. To make decisions with respect to ingredients and engineering design, the new product manager must understand how these ingredients and designs affect product performance and consumer perceptions, as well as profit. Product tests provide the manager with the information with which to make these decisions. Laboratory tests uncover flaws and provide insight on engineering measures, and expert evaluation provides an inexpensive first view of consumer perceptions.

Exposing consumers to alternative formulations allows the most complete measurement of physical and psychological product features. It should be used to supplement engineering laboratory and expert panel evaluations when perceptual attributes are present in the new product and when one cannot be sure "experts" accurately represent buyers and users.

There is no ideal method that applies to every product. The new product manager must understand the strengths, weaknesses, and complementarities of each procedure and choose the combination of tests that provides sufficient information to select the best product formulation. We now illustrate a product test for one consumer product.

Case: Taste Tests for Squid Chowder*

Conventional seafood supplies such as cod or clams are being depleted and, as a result, prices have risen. For example, the reduced supply of clams coupled with increased popularity has resulted in a tripling of the price of clam chowder. One solution might be to develop new food products which utilize the more plentiful species of fish. One such species, squid, is abundant and nutritious. It is widely accepted in the Orient and the Mediterranean countries, but is not popular in the United States.

Earlier we suggested potential new products to exploit this opportunity. One of these was a chowder based on squid and clams. (You may wish to reexamine Figures 3.2 to 3.4.)

The design process was used to develop a CBP; now the physical product and labeling must be carefully developed. Taste tests were undertaken to determine whether consumers would accept squid and whether they would like the taste. FDA regulations require that "squid" be prominent on the label. Tests were required to determine whether the psychological effect of the name "squid" would make the product unacceptable to consumers.

Study Design. Two hundred consumers were recruited at a shopping mall in a suburb of Boston. The respondents were required to be the household meal planner and to have served fish at home as a main meal item in the last month. Age quotas were used in an attempt to get a representative sample of consumers. Qualified respondents were taken to a room in the mall where they filled out a questionnaire and tasted squid chowder and clam chowder. They were paid $2.00 as compensation for approximately forty-five minutes of their time.

To address the managerial questions, the study was designed to test the effect of squid concentration in the chowder and the effect of squid identification on the label. Clam chowder was used as a reference.

Each consumer tasted the clam chowder and two squid chowders. The first squid chowder, called "Fisherman's Chowder," deemphasized the presence of squid. It was described as a "delicious blend of seafood" with ingredients of "clams, squid, milk, water, potatoes, onion, and seasonings." The second squid chowder, "Sclam Chowder," emphasized the presence of squid. It was described as a "delicious blend of squid and clams" with ingredients of "squid, clams, milk, water, potatoes, onion, and seasonings." The total weight of the clams and/or squid was constant in all chowders as were other ingredients. The respondent was presented with a bowl of chowder and a concept board describing the product. For example, the clam chowder concept board said in large letters "Clam Chowder" and smaller letters "a delicious clam soup." On the bottom, the ingredients were listed in small print as "clams, milk, water, potatoes, onion, and seasonings."

The chowders also varied in terms of the physical concentration of squid. For one-half of the respondents "Fisherman's Chowder" was a 90

*This section is adapted from Neslin and Urban (1977).

percent squid product. For the other half it was a 10 percent squid product. Similarly, for "Sclam Chowder" one-half of the respondents were given a 90 percent squid product and half a 10 percent squid product. The order of presentation was rotated and concentration randomized within the "Fisherman's" and "Sclam" chowder labels.

After each chowder was tasted, a seven-point hedonic scale and a five-point intent-to-buy scale were administered. After tasting the three chowders alone, paired comparisons were administered based on a procedure developed by Scheffe (1952). For each pair, the respondent first identified the preferred item and then specified the degree of preference (slightly better, better, much better). Paired comparisons were made in terms of overall taste, appearance, flavor, texture, and aroma.

Experimental Results. The outcome of this experiment is shown in Table 13.2. The overall preference mean of −0.09 is low compared to the scaled value of zero for pure clam chowder. The values for high salience and concentration are lower than those for low salience and concentration, respectively. Although the marginal totals indicate a negative effect for high squid identification and high squid concentration, these effects were not statistically significant [F (2,402) was 0.38 and the t for differences in salience was −0.33 and for concentration −0.79]. Overall, there was some penalty for the squid name and concentration, but it was small and not significant.

A more detailed analysis of the data indicated that the significance was much higher for some groups. For those who had previously tasted squid, there was a larger negative effect for high concentration, but no salience effect. For those who had not previously tasted squid, there was no significant penalty for squid concentration or salience.

Table 13.2. Overall Experimental Treatment Means

		Squid Concentration		
		Low (10%)	*High (90%)*	*Total*
Salience of Squid Identification	Low (Fisherman's)	0.00	−0.11	−0.06
	High (Sclam)	−0.01	−0.24	−0.13
	Total	−0.01	−0.18	−0.09

The experiment indicated small negative effects for increasing the concentration of squid in the chowder and labeling the chowder as "Sclam" rather than "Fisherman's Chowder." The greatest penalty was for the high concentration chowder by those who had previously tasted squid.

Managerial Diagnostics. Since some consumers recognize the taste of squid and dislike it, repeat rates for the chowder may be suppressed. One way to increase potential repeat purchase would be to change the recipe by lowering the proportion of squid to say 50 percent. However, such change would reduce the cost advantage of a squid chowder. To search for alternative strategies, we examined the ratings of the squid chowders on flavor, texture, and appearance.

Contrary to expectations, ratings on texture and appearance were better for the squid chowder than the clam chowder. The squid mantels were finely chopped to give a pleasing appearance and the fresh squid carefully cooked to maintain tenderness. The recommended alternative was to increase the clam flavor perception by the use of more clam juice while retaining a high level of squid concentration. Since the overall ratings were most heavily correlated to flavor, improvements in flavor without the loss of texture and appearance should improve the overall taste perceptions and the repeat purchase intents. Further taste testing of revised recipes would be appropriate.

The sensitivity to concentration and flavor also suggests that the maintenance of quality would be important if a squid product were marketed. Most of the previous trial was in restaurants where recipes and quality vary. In a canned chowder, consistency and good quality should be much easier to maintain.

The salience of squid identification had only a small effect on consumers' evaluation. "Sclam" chowder did not produce statistically significantly lower taste evaluations than the same chowder when labeled as "Fisherman's" chowder. It does not appear to be worth a legal battle over the details of labeling. "Sclam" chowder, "a blend of squid and clams" is an honest straightforward identification, and acceptable to consumers.

The experimental data suggests that relative to clam chowder, a squid chowder does not do badly. In the survey, the overall intent to buy the squid chowder after tasting was almost equal to clam chowder. There was a 27.8 percent definite and 29.8 percent probable intend for squid chowder versus 28.9 percent definite and 32.0 percent probable for clam chowder.

If trial could be obtained by price promotion and advertising to influence prior attitudes, a respectable share of the chowder market could be obtained. A reformulation could further increase the repeat rate and long run sales. If the supply price of squid remains substantially below clams, a long-range price advantage would further stimulate trial and repeat purchases. The issue of taste and price tradeoffs could be examined further by pretest market procedures.

This case illustrates the use of taste testing in formulating a new product. Our experience indicates paired comparisons are the best way to discriminate among alternative products. Large samples should be collected so that subgroups can be analyzed to find heterogeneity in response and so the power of the statistical tests will be high. If a taste test is to be done, be sure it is carefully designed, and executed, and analyzed. Then, the test can determine if the current formulation is preferred to existing products and how the formulation can be improved.

SUMMARY

Testing strategies reduce risk and maximize expected benefit. This chapter has considered procedures to test advertising copy and physical product characteristics. These testing activities are important in assuring that the best advertising has been created and that the product with its manufacturing specifications fulfills requirements and consumer expectations. As well as determining if the product and advertising are good, the procedures improve the product and advertising formulations to reduce costs and better meet consumer preferences.

In the next chapter we present procedures which test the product, advertising, price, promotion, and distribution aspects as a unified entity.

REVIEW QUESTIONS

13.1 How does a testing strategy reduce risk and increase expected benefit?

13.2 What are the advantages and disadvantages of a sequential testing strategy?

13.3 Why is it necessary to have a formal procedure for advertising copy testing? Why not have management merely choose that copy which best conveys the CBP?

13.4 What is "on-air testing" and how is it used?

13.5 Some advertisements score low on likability yet achieve excellent advertisement and product awareness. Is this enough for a successful new product? Think of an abrasive advertisement. Could you make this advertisement less abrasive yet achieve attention and communicate its unique selling proposition?

13.6 What is the relationship between reliability and validity of advertising testing? Why are both important?

13.7 Consider the expected revenue gain for copy tests as described by the Gross model (equation 13.2). Under which of the following conditions would more testing be advocated? Under which conditions would less testing be advocated? Why?

a. The reliability of the tests are improved.
b. The firm is generating some very novel and experimental forms of copy.
c. The ads generated not only differ in style but actually position the product differently.
d. One creative specialist is generating the ads rather than a staff of creative people working independently.

13.8 What are the advantages and disadvantages of laboratory tests, expert evaluation, and consumer tests for physical product testing?

13.9 How might you test a new electric turbine?

13.10 What different problems occur in testing consumer durable products as compared to consumer frequently purchased products?

Chapter 13 Appendix

13.1 Expected Benefit from Testing "n" Advertising Campaigns*

In the text we presented an equation, $E_n = e_n \sigma_E \rho R$, which gives the expected benefit from testing n alternative advertising campaigns. This appendix derives that equation. For more details, see Gross (1972).

A number of alternative advertising campaigns are under consideration. In order to choose one of the alternatives, a pretest is used, the results of which are a score for each of the alternative campaigns. The decision rule is to choose the campaign with the highest score. To describe the test, we use the one-way classification model of analysis of variance.

(13.1A-1) $$O_{ij} = \mu_o + T_j + \epsilon_{ij}$$

where O_{ij} = the observed score on the ith replication of the pretest on the jth alternative.

μ_o = the mean pretest score of campaigns which would be created by the process used to create the campaigns.

T_j = the true deviation from the mean score associated with alternative j.

ϵ_{ij} = the deviation from the true score of alternative j introduced by random error in the testing process.

$\theta_j = T_j + \mu_o$ = true test score.

*This appendix has been reproduced with notation changes from an article by Gross (1972).

The variate T is assumed to be normal with zero mean and variance σ_T^2. The variate ϵ is assumed to be normal with zero mean and variance σ_ϵ^2. Hence, the variate O is normal with mean μ_o and variance σ_o^2, where

(13.1A-2) $$\sigma_o^2 = \sigma_T^2 + \sigma_\epsilon^2$$

We have assumed that, associated with each alternative j, are its true profitability E_j, and its true score θ_j. Let us further assume that the variates E and θ, associated with the process of creating alternatives, are distributed as the bivariate normal distribution with parameters $\mu_E(=0), \mu_o, \sigma_E^2, \sigma_T^2$, and ρ, where ρ is the correlation coefficient between E and θ.

Given that n alternatives are screened in order to select the best one, we are interested in the highest observed score, $O_{\max}(n)$, as a function of n. (See Gumbel, 1958, p. 43.)

(13.1A-3) $$E[O_{max}(n)] = \mu_o + e_n\sigma_o$$

where e_n is defined in the expected value of the largest of n observations from a standardized normal distribution. (See Table 13.1A-1.)

Using equation (13.1A-2), equation (13.1A-3) may be rewritten:

(13.1A-4) $$E[O_{max}(n)] = \mu_o + e_n(\sigma_T^2 + \sigma_\epsilon^2)^{1/2}$$

Now it is necessary to derive the expected relative profitability of the alternative chosen by the screening process, taking into account the random error in the screening process, and the correlation between the true scores and the relative profitabilities of the alternatives. Let us call this $E[E_{\max}(n, \sigma_\epsilon, \rho)]$. It has already been deduced that θ is normally distributed with mean μ_o and variance σ_ϵ^2. We may also express the conditional distribution of O, given θ, as normal with mean θ and variance σ_T^2. Hence, for any value of O, we may compute via Bayes' theorem the conditional distribution $f(\theta \mid O)$, and hence $E[\theta \mid O]$. Schlaiffer has solved this problem. (See Pratt, Raiffa, and Schlaiffer, 1965, ch. 16, pp. 6–14.)

(13.1A-5) $$E[\theta \mid O] = \frac{(O\sigma_T^2 + \mu_o\sigma_\epsilon^2)}{\sigma_o^2}$$

Since $E[\theta \mid O]$ is linear in O, we may say that

(13.1A-6) $$E[\theta_{max}(n, \sigma_\epsilon)] = E\{\theta \mid E[O_{max}(n)]\}$$

Now, inserting equation (13.1A-4) for O in equation (13.1A-5), we find, after reduction, that

(13.1A-7) $$E[\theta_{max}(n, \sigma_\epsilon)] = \mu_o + e_n\sigma_T\left(1 - \frac{\sigma_\epsilon^2}{\sigma_o^2}\right)^{1/2}$$

Using the definition of reliability, equation (13.1A-7) may be expressed as

(13.1A-8) $E[\theta_{max}(n, \sigma_\epsilon)] = \mu_o + e_n \sigma_T R$

Now, to find $E[E_{\max}(n, \sigma_\epsilon \rho)]$ we shall use the expected value of E given ϵ. The conditional expectation of one variate, given the other for a bivariate normal distribution, is (see Pratt, Raiffa, and Schlaiffer, 1965, ch. 22, p. 13)

(13.1A-9) $E[E \mid \theta] = \mu_E + \dfrac{\rho \sigma_E (\theta - \mu_o)}{\sigma_T}$

Since $E[E \mid \theta]$ is linear in θ, we may write

(13.1A-10) $E[E_{max}(n, \sigma_\epsilon, \rho)] = E[E \mid E[\theta_{max}(n, \sigma_\epsilon)]]$

Substituting equation (13.1A-8) into equation (13.1A-10), reducing, and recalling that $\mu_E = 0$, we obtain

(13.1A-11) $E[E_{max}(r, \sigma_\epsilon, \rho)] = e_n \sigma_E \rho R$

In other words, $E_n = e_n \sigma_E \rho R$.

Table 13.1A-1 Expected Value (e_n) for Various Numbers of Advertising Campaigns, n[a]

n	e_n	Δe_n	n	e_n	Δe_n
1	0.000	0.564	11	1.586	0.043
2	0.564	0.282	12	1.629	0.039
3	0.846	0.183	13	1.668	0.035
4	1.029	0.134	14	1.703	0.033
5	1.163	0.104	15	1.736	0.030
6	1.267	0.085	16	1.766	0.028
7	1.352	0.071	17	1.794	0.026
8	1.423	0.061	18	1.820	0.025
9	1.484	0.054	19	1.845	–
10	1.538	0.048			

[a] $\Delta e_n = e_{n+1} - e_n$, which is the potential gain in considering one more version

Chapter 14

Pretest Market Forecasting

After the components of the new product are individually tested, we turn to pretest market to test the comprehensive new product strategy. Until the early 1970s, the only way to test the full new product strategy was to go to test market. But test markets are expensive ($1 to 1.5 million), take considerable time (nine months to two years), can prematurely alert your competitors, and the likelihood of failure in test market is greater than 50 percent. To overcome the problems of time, cost, and risk, many organizations are now using pretest markets. Pretest markets do not substitute for test markets, but they provide a lower cost, faster, and more discreet method to identify winners, eliminate losers, and provide diagnostic information for product improvement. In this way, pretest markets are a key step in reaching the decision frontier at minimum cost (review Figure 13.1).

In this chapter we describe alternative approaches and discuss how they can be used to improve the new product strategy. We illustrate their use with several minicase studies. Pretest market models are used extensively for frequently purchased consumer products, but the same methods can be valuable for examining the potential of consumer durables, industrial products, and services. We close the chapter by suggesting extensions for these categories.

CRITERIA FOR PRETEST MARKET ANALYSIS

A managerial decision to use a pretest market is justified if sufficiently accurate predictions can be achieved for an investment in time and money that is substantially below the cost for a test market. We suggest some guidelines and indicate what can be achieved with commercially available services. Each organization should assess its own costs and expected benefits before purchasing such a service or developing its own "in-house" pretest market procedure.

Accuracy

Since the final product environment may change before full-scale launch and measurement error will be present, pretest markets cannot predict perfectly, but they should be sufficiently accurate for a GO/NO GO test market decision. This means that a pretest market should reject poor products with a high probability of being correct. Also, good products should have a high probability of being correctly identified. A reasonable criteria to achieve these managerial goals is that predictions are within 25 percent of long-run sales 75 percent of the time. For most consumer brands, this would imply the predicted share should be about two points above the minimum share required for a GO decision in order to assure a 75 percent chance of success in test. Such accuracy can be obtained.

Managerial Diagnostics

Many new products (50–60 percent) will be identified as failures. One could merely drop these products, but usually there are opportunities to improve the physical product, advertising copy, or marketing mix. A pretest market model and measurement system should provide actionable diagnostics on why the product succeeded or failed, how it could be improved, and what the share implications are of such improvements. When a product is rated as "good" after the pretest analysis, diagnostics can generate information to improve the product further. Such improvement is important since the better the product is, the less likely it is to be vulnerable to competitive entries.

Time and Cost

One gains little if the pretest market takes as long as a real test market. The pretest market should provide results fast enough for managerial action. The use of a pretest market should not significantly delay full-scale launch. Two to three months is an attainable goal. Finally, the cost of a pretest market should be well below the expected gains. (The calculations in Chapter 3 suggested that the expected gain for an average product was one million dollars.) From $50,000 to $75,000 is an attainable cost for pretest market analysis.

ALTERNATIVE APPROACHES

The input to a pretest market analysis is the physical product, advertising copy, packaging, price, and the advertising and promotion budget. The output is a forecast of sales and diagnostics. To fulfill the above criteria, the pretest market must be based on careful measurement and models of consumer response. Table 14.1 lists a number of alternative approaches. Each approach has its relative strengths and weaknesses, and you may wish to combine two or more approaches to enhance accuracy. At least one available model uses such a convergent approach by combining trial/repeat and attitude change models. A final method of early analysis is to use small markets to measure results. We do not consider this a true pretest market approach since it is significantly more expensive (greater than $100,000), takes as long as a test market to attain results (six to nine months), and requires production volumes to meet sales in these markets.

Table 14.1. Alternative Modeling Approaches to Pretest Market Analysis

Judgment and Past Product Experience
Trial/Repeat Measurement
Stochastic Models
Home Delivery
Laboratory Measurement
Attitude Change Models
Convergent Approach

Judgment and Past Product Experience

We learn by experience. One reasonable approach to pretest market forecasting is to examine past experience to determine what measurable characteristics of a new product determines its success. For example, Figure 14.1 suggests one set of critical factors that affects advertising recall, initial purchase, and repeat purchase. These particular factors are used in a model developed by Claycamp and Liddy (1969). To develop such a model, one gathers these measures (both the critical factors and consumer response) and uses regression to estimate the relationships between the critical factors and consumer response. For the new product, the critical factors are measured and the regression equation is used to forecast consumer response.

In many cases, it will be infeasible or too expensive to directly measure all of the critical factors for the new product. In such cases one can substitute managerial judgment or develop a panel of experts. This makes the measurement feasible, but can introduce random error and/or bias depending upon the panel. For example, Table 14.2 suggests one way to obtain measures for the critical factors in Figure 14.1. Note that the measurement may be different for the new product.

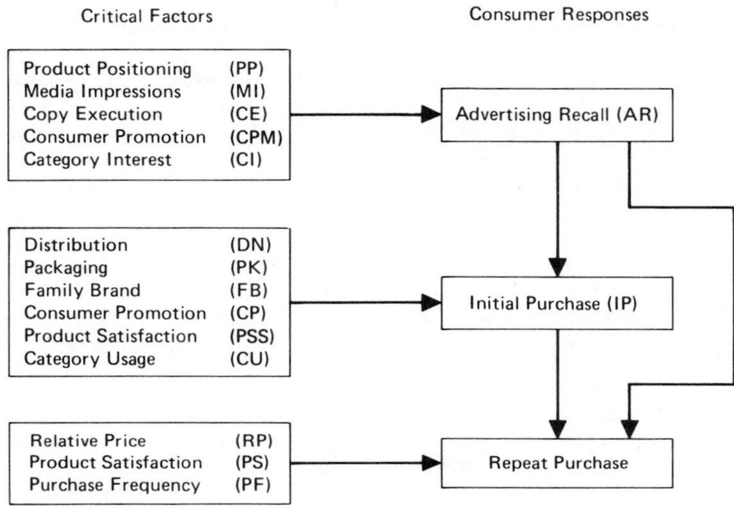

Figure 14.1. *Model Based on Past Product Experience (adapted from Claycamp and Liddy, 1969, p. 415)*

The advantage of this approach is that once the regression equation is estimated, the model can produce predictions rapidly and at a low cost. It makes use of previous experience; since the media and distribution plans appear in the model, alternative plans can be tried and the best one chosen. The approach also has a number of disadvantages. Besides its sensitivity to expert judgment, the model is extremely sensitive to the past products used to estimate the model. Since the model has no underlying theory of the consumer beyond a causal diagram, the regressions may not be appropriate for categories that are significantly different from those used to estimate the model. Finally, since few direct measures are taken from the consumer, the model may miss important consumer concerns.

Figure 14.1 and Table 14.2 are based on the model developed by Claycamp and Liddy (1969). Their model is based on data obtained from 58 new product introductions that covered 32 different types of package goods. The regressions were run on 35 of those introductions, 50 percent of which were foods. The other 23 were saved for predictive testing. Two regression equations were estimated, one for advertising recall and one for initial purchase. No results are reported for repeat purchase. The regressions explained over 70 percent of the variation in the data and were significant at the 1 percent level. The largest effect on advertising recall were produced by product positioning (PP), copy and advertising ($\sqrt{MI*CE}$), consumer promotion (CP*) and category interest (CI), in that order. Trial was affected most by packaging (PK), family branding (FB), and advertising recall (AR). When the model was tested on 23 new products, 15 of 23 fell within 10 percentage points of observed advertising recall, and 20 of 23 fell within 10

Table 14.2. One Way to Measure the Critical Factors for the Model in Figure 14.1[a]

Variable	*Measure*	*Source for Past Products*	*Source for New Products*
PP	Judged Product Positioning	Expert Panel	Expert Panel
MI	Average Number of Media Impressions/Household	Past Data	Media Plan
CE	Judged Quality of Advertising Copy Execution	Expert Panel	Expert Panel
CP*	Coverage of Consumer Promotion Containing Advertising Messages Adjusted for Type of Promotion	Past Data	Expert Panel and Plan
CI	Index of Consumer Interest in the Product Category	Expert Panel	Expert Panel
DN	Retail Distribution, Adjusted for Shelf Space and Special Displays	Past Data	Distribution Plan
PK	Judged Distinctiveness of Package	Expert Panel	Expert Panel
FB	Known or Family Brand Name	Past Data	Plan
CP	Coverage of Consumer Promotions Adjusted for Type and Value of Offer	Past Data	Expert Panel and Plan
PS*	Index of Consumer Satisfaction with New Product Samples	Past Data	Product Test Data
CU	Percent of Households Using Products in the Category	Historical Data	Historical Data
AR	Percent of Housewives Able to Accurately Recall Advertising Claims at the End of 13 Weeks	Test Market Data	Predicted by Model
IP	Percent of housewives Making One or More Purchases of the Product During the First 13 Weeks	Test Market Data	Predicted by Model

[a]*Adapted from Claycamp and Liddy (1969), p. 416, Table 2*

percentage points of the observed trial. The correlations of actual and predicted were 0.56 for recall and 0.95 for trial.

This approach is interesting since it indicates studying past products is useful. But the levels of correlation indicate that past relationships may not be sufficient. The model is limited because no experience with the repeat purchase sector has been reported. Without a valid repurchase model, the new product share cannot be predicted. The pioneering Claycamp and Liddy model has led to recent research to develop more measures and stronger models to predict market share.

Trial/Repeat Measurement

Long-run sales are based on both trial and repeat. Accurate forecasts of long-run sales can be made if the pretest analysis can predict the percentage of consumers who will try the product (cumulative trial) and the percentage of those who will become repeat users (cumulative repeat). A series of models have been developed which present the new product to consumers in a reasonably realistic setting and take direct consumer measures which are used to forecast cumulative trial and repeat purchases. The advantage of this approach is that it is based on direct observation of consumer response to the new product. The disadvantage is that errors are introduced because the direct measures may not be representative of what would happen in a test market or a full-scale launch.

Stochastic Models. In this approach, regressions based on previous purchasing experience are used to identify how the direct measures relate to trial and repeat. A number of alternative equations are possible. We illustrate the approach by describing parts of one stochastic model developed by Eskin and Malec (1976).

Their equation for cumulative trial is derived from past product trial rates. The cumulative trial after one year (α_1), depends on product class penetration (PCP, percent of households buying at least one item in the product class during one year), the total consumer direct promotional expenditures (SPN) for the new product, and the distribution (DIS, weighted percent of stores stocking the new product). Trial is then given by:

(14.1) $$\alpha_1 = a(\text{PCP})^{b_1}(\text{SPN})^{b_2}(\text{DIS})^{b_3}$$

where a, b_1, b_2, and b_3 are parameters. After taking logs of both sides of this equation, a linear regression is used to estimate b_1, b_2, and b_3 based on past experience from past new products. This equation is similar to Claycamp and Liddy, but it does not include judged variables like the product position, copy effectiveness, or packaging. However, both fit trial well. The model would be made more useful for diagnostic purposes if more trial variables were included and it was strengthened by direct measures of propensity to try.

Equation (14.1) gives cumulative trial, but it is important to predict the growth in trial. Figure 14.2 shows the most common form of growth. The stochastic models predict this growth.

The equation used by Eskin and Malec for this curve is given as follows:

(14.2) $$R_t(0) = \alpha_1(1 - \gamma_1^t)$$

where $R(0)$ = the percentage of consumers who have tried the new product by time t

α_1 = the cumulative trial given by equation (14.1)

γ_1 = a parameter to be estimated ($o \leq \gamma_1 \leq 1$).

As t gets larger, γ_1^t decreases to zero, $1 - \gamma_1^t$ approaches one, and $R_t(0)$ approaches the cumulative trial (α_1). Equation (14.2) gives a curve of cumulative trial such as that shown in Figure 14.2.

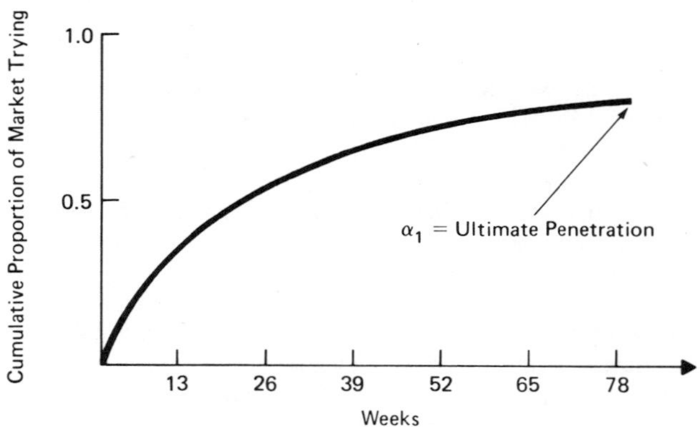

Figure 14.2. *Trial Growth*

Eskin and Malec model first repeat in a similar way except that: (1) the cumulative repeat (α_2) is estimated from a direct measure of intent to buy after usage and (2) the decay parameter (γ_2) is based on a hypothesis of consumer response and is a function of the frequency of purchase. Thus first repeat, $R_t(1)$, is given by:

$$\textbf{(14.3)} \qquad R_t(1) = \alpha_2(1 - \gamma_2^t)$$

Similar equations are used for the second repeat purchase, and the third, etc. Simply replace α_2 by α_{J+1} to get the equation for the Jth repeat purchase, $R_t(J)$. The cumulative Jth repeat, α_{J+1}, can be determined from direct observation or from a regression equation relating it to the cumulative values for the first, second, third, etc., repeats.

To use this model for a new product, penetration (PCP) is observed for the category, promotion (SPN) and distribution (DIS) are obtained from the introduction plan and these values substituted in equation (14.1) to give α_1 and in equation (14.2) to give trial over time. Intent to buy is measured for those who have tried the product and α_2 determined. Equation (14.3) predicts first repeat and similar equations give second, third, etc., repeat. Total sales comes from summing trial and the various repeat classes. (See Eskin and Malec, 1976, for details, and Silk and Kalwani, 1978, for extensions.)

Stochastic models are attractive because of their explicit modeling of trial and repeat dynamics. Using past purchasing data from new product introductions in the model helps achieve good forecasting accuracy.

Home Delivery Measures. Another approach to forecasting with a trial/repeat model is based on direct measures of trial and repeat in a sample of households served by a home delivery service. The measurement of trial and repeat is done in a panel (1,000 households) which is visited each week by a salesman. Based on information in a color catalog published each month and a biweekly promotion sheet describing price promotion, the consumer orders from a wide range of frequently purchased consumer products. The products are delivered on the same day and purchase records are computerized.

When a new product is to be tested, an ad is placed in the monthly catalog and trial and repeat purchase is observed. From this data a share of market for the product could be forecast by multiplying cumulative trial times the share of purchases the new brand receives from those who have tried it (Parfitt and Collins, 1968).

The advantage of this approach is that it is based on a home delivery panel that approximates the actual product launch. The direct measures of trial are obtained based on exposure to real advertising, and realistic repeat purchasing environments enhance forecasting accuracy. One disadvantage of this approach is that the panel must be run long enough to get good estimates, probably longer than other pretest market approaches.

This panel approach has been used both to analyze existing product promotion (Charlton and Pymont, 1975; Charlton, Ehrenberg, and Pymont, 1972) and to forecast new product sales (Pymont, Reay, and Standen, 1976).

Laboratory Measurement. The success of home delivery measures depends upon the ability of the home delivery service to closely approximate the actual purchase environment. An alternative approach is to use a central location laboratory to approximate the trial purchase environment. Consumers are recruited, exposed to ads (television or print), and given the opportunity to buy in a simulated retail store. After buying in the simulated store, the consumer takes the product home and repeat measures are taken with a call-back interview. The basic idea of laboratory measurement is to force exposure to the product and provide a realistic purchase choice environment. The success of the laboratory measurement depends on the ability of the model to either minimize the bias of such a laboratory simulation or develop procedures to correct for the bias in making the forecast.

The advantages of laboratory measurement are that results can be obtained rapidly and at a relatively low cost. The measures of trial and repeat are based on direct consumer response in a realistic purchase environment. Such measures have the potential for producing highly accurate forecasts. The disadvantage of such measurement is that the laboratory abstracts reality. The simulated store is not an actual store, measures are taken shortly after exposure to advertising, and the measurement may influence consumer behavior. While each of these problems can be overcome, they represent potential systematic biases that must be carefully considered.

One laboratory measurement model is based on a procedure development by Yankelovich, Skelly, and White. They adjust the observed trial and repeat rates based on judgment derived from past experience. For example, the observed trial is reduced by 25 percent due to "inflation" in the laboratory. Another adjustment is a "clout" factor that varies from 0.25 to 0.75 depending on introductory spending. Predictions of market share are made by multiplying trial and repeat by the frequency of purchase where frequency is adjusted based on a judged "frequency factor" which reflects departures of new products from known frequency of purchase patterns. The Yankelovich et al. approach is interesting because it blends direct observation with managerial judgment, but the predictions are very dependent upon the judgmental input. The model could be improved if it were based on a more well-developed theory and if it used statistical analysis to augment the adjustment factors.

Another laboratory measurement model is the trial/repeat component of a model developed by Silk and Urban (1978). They use the Parfitt and Collins (1968) formulation to estimate market share for the new brand by multiplying the ultimate cumulative trial by the share of purchases from those who have tried (review equation 11.6).

Trial comes about in one of two ways: direct trial or receipt and use of free samples. The direct trial, given ad exposure and availability, is the proportion of respondents who purchased the new brand in the laboratory on their simulated shopping trip. The amount of awareness from ads depends on the spending level management plans to utilize and the extent of availability depends upon how much sales force and promotional activity will be directed at the retail trade. These parameters can be obtained from judgment or from regression equations. (See Silk and Urban, 1978, or review Chapter 11.) Direct trial is the product of trial in the lab times awareness and distribution. The amount of trial by sampling depends on the number of samples sent and their use. If the probability of a consumer's direct trial is independent of the probability of receipt and use of a sample, the total trial is the sum of both sources of trial less their overlap (review equation 12.6).

The long-run share for those who have tried is modeled as a stochastic model, a first order, two-state Markov process (review Chapter 11, Figure 11.6). Two probabilities must be estimated—the probability of repeat purchase of the new brand, and the probability of purchase of the new brand if another brand had been last purchased.

Estimates of these probabilities are derived from measurements obtained in the post-usage survey. The proportion of respondents who make a mail order repurchase of the new brand when given the opportunity to do so is taken as an estimate of new product repeat. Based on those who do *not* repurchase the new brand in this situation, the probability of returning to the new product after purchase of another brand is estimated.

The advantage of the Silk and Urban formulation is that it is based on theoretical models of consumer response and most consumer response estimates come from direct measurement. The disadvantage is that this component of their model is still a laboratory simulation and thus subject to all the criticisms discussed above.

Summary of Trial/Repeat Measurement. Trial and repeat purchase models are the most logical way to represent consumer response to a new frequently purchased brand. Direct measures of trial and repeat are important inputs to forecasting and many of the models discussed above are based firmly in consumer theory. Judgments need to be applied to any model, but direct measures and explicit models promote consistency, methodological rigor, and forecasting accuracy.

Attitude Change Models

In the design phase, forecasts of purchase potential were made based on estimates of consumer preferences for the new product. The advantage of this approach is that preference is more directly predicted and intervening effects such as awareness and availability can be directly incorporated in the model. Furthermore, product characteristics or consumer perceptions can be readily incorporated in the model through the preference analysis techniques discussed in Chapter 10.

The attitude-based pretest market analysis models use the basic approach of estimating behavior from consumer preferences. Consumer attitudes (preference or beliefs about product attributes) are first measured for existing products. The consumer is then given the new product and, after use, attitudes are measured for the new product.

The advantage of this approach is that the indirect attitude measures may avoid some of the laboratory effects inherent in the direct trial and repeat measures. For example, attitudes toward existing products are often measured prior to laboratory exposure and are thus more representative of the attitudes of the consumer population. This is as opposed to the direct measures of trial which are highly dependent upon the closeness with which the laboratory approximates the real world. The disadvantage of attitude measures is that they are not direct measures. Predictions depend upon the accuracy and completeness of the model used to estimate behavior from the measured attitudes.

One attitude model, called COMP, has been developed by Burger (1972). The measurement to support this model is done by an initial interview to measure attitudes and use of existing products along with a laboratory exposure to ads and a simulated store. A call-back interview measures attitudes after use of the new brand.

Burger bases his attitude measures on the expectancy value preference model described in Chapter 10. That is, he forms a linear preference score from consumer attribute ratings and stated importances of those attributes. Burger's attitude measure is then the relative preference score, U_{ij}, given by:

$$\textbf{(14.4)} \qquad U_{ij} = \frac{\sum_k w_{ik} y_{ijk}}{\sum_l \left(\sum_k w_{ik} y_{ilk} \right)}$$

where w_{ik} is the importance consumer i places on attribute k and y_{ijk} is consumer i's rating of product j on attribute scale k. This attitude measure is then used as an explanatory variable in a purchase model.

An alternative attitude model is the second component of the pretest market forecasting model developed by Silk and Urban (1978). They augment their first component with measures of preference. Their preference measure is developed from a constant sum paired comparison task in which consumers allocate a fixed number of "chips" among each pair of products in their evoked set. A scaling technique developed by Torgerson (1958) transforms the constant sum measures into ratio-scaled preferences, p_{ij}, indicating consumer i's preference for product j. A variation of the logit model (review Chapter 11) is then used to estimate behavior from the preferences. That is:

$$\textbf{(14.5)} \qquad L_{ij} = \frac{(p_{ij})^{\beta}}{\sum_{l=1}^{m_i} (p_{il})^{\beta}}$$

where L_{ij} is the estimate of the probability that consumer i will purchase product j and β is a parameter to be estimated. The sum in the denominator is over all products in consumer i's evoked set (as defined in Chapter 8). L_{ij} is equal to zero if j is not in i's evoked set. The constant, β, is estimated by a maximum likelihood analysis of individual last purchases and preferences.

To forecast the purchase probability, L_{ib}, for the new product (brand b), preference measures for the new and existing products are obtained after the consumer has experienced a period of trial usage of the new product. The new product is assumed to be in i's evoked set and equation (14.5) is used to forecast L_{ib} for the new product.

The purchase probability is forecast assuming the new product will be in the evoked set. In order to calculate an expected market share for the new brand, Silk and Urban take into account that the new brand will not necessarily become an element of the relevant set of brands for all consumers when it does in fact become available in the market. To do this, they obtain estimates of the percent of consumers who will evoke the new product (E_b). (This procedure is described in Appendix 14.3.) The market share, M_b, of the new product is then given by:

$$\textbf{(14.6)} \qquad M_b = E_b \sum_i \frac{L_{ib}}{N}$$

A similar equation to equation (14.6) is used to forecast the new shares for established brands (see Appendix 14.2). These estimates are important to the new product manager since they indicate which established brands are likely to retaliate because of the share of the market they are likely to lose.

In comparing the Burger with the Silk and Urban models, we see that both systems base their estimates on some form of preference measures and both correct for awareness and availability. The predictions in both cases are the probabilities of purchase. The models differ in the specific equations and estimation procedures used. In particular, models should use stronger preference measures (constant sum), behavioral-based statistical techniques (logit), and limit their measurements to the evoked set (as recommended in Chapter 8).

We feel that it is a useful exercise for the new product manager and his market research staff to examine and compare alternative systems and to select the system that he is most comfortable with. We turn now to describe a technique that incorporates both direct behavior and attitude measures.

Convergent Measures and Models

Judgment, trial/repeat, and attitude models each have their strengths and their weaknesses. An emerging view on pretest market analyses is to use more than one method in parallel and compare the results. For example, one might use the Eskin and Malec model to develop estimates based on trial and repeat measures and compare this to predictions obtained from regressions on previous product experiences (say the Claycamp and Liddy model). If the models agree, then the product manager has more faith in the predictions. If they disagree, then by comparing and reconciling results, any biases in measurement or structural problems in models can be identified and corrected.

The advantage of such a convergent approach is potentially greater accuracy and more confidence in the resulting forecasts. Furthermore, a combination of approaches can give more comprehensive indication of how to improve the new product. The disadvantage of a convergent approach is the slightly greater cost. Costs do not double, however, because inputs for more than one model can often be obtained in the same set of consumer measures.

We return to the Silk and Urban model, called ASSESSOR, to illustrate the specific measures and analyses used in such a convergent approach based on a trial/repeat and attitude model.

ASSESSOR is designed to aid management in evaluating new products once a positioning strategy has been developed and executed to the point where the product, packaging, and advertising copy are available and an introductory marketing plan (price, promotion, and advertising) has been formulated. Given these inputs, the system is specifically intended to:

1. Predict the new brand's equilibrium of long-run market share.
2. Estimate the sources of the new brand's share—"cannibalization" of the firm's existing brand(s) and "draw" from competitors' brands.
3. Produce actionable diagnostic information for product improvement and develop advertising copy and other creative materials.
4. Permit low cost screening of selected elements of alternative marketing plans (advertising copy, price, and package design).

Figure 14.3 shows the overall structure of the system developed to meet these requirements. The critical task of predicting the brand's market share

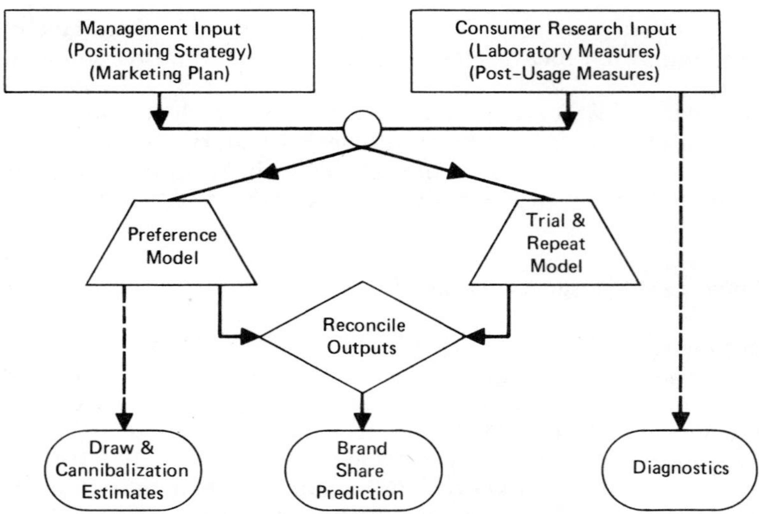

Figure 14.3. *Structure of ASSESSOR System (Silk and Urban, 1978, p. 173)*

is approached through the trial/repeat and attitude models described earlier. Convergent results strengthen confidence in the prediction while divergent outcomes signal the need for further analyses to identify sources of discrepancies and to provide bases for reconciliation. The measurement inputs required for both models are obtained from a research design involving laboratory and usage tests. The key outputs are a market share prediction plus diagnostic information which can be used to make a decision as to the brand's future.

Research Design and Measurement. The measurement inputs required to develop the desired diagnostic information and predictions for ASSESSOR are obtained from a research design structured to parallel the basic stages of the process of consumer response to a new product. Table 14.3 outlines the essential features of the design and identifies the main types of data collected at each step. To simulate the awareness-trial stages of the response process, a laboratory-based experimental procedure is employed wherein a sample of consumers are exposed to advertising for the new product and a small set of the principal competing products already established in the market. Following this, the consumers enter a simulated shopping facility where they have the opportunity to purchase quantities of the new and/or established products. The ability of the new product to attract repeat purchases is assessed by one or more waves of follow-up interviews with the same respondents conducted after sufficient time has passed for them to have used or consumed a significant quantity of the new product at home.

Table 14.3. ASSESSOR Research Design and Measurement[a]

Design	*Procedure*	*Measurement*
O_1	Respondent Screening and Recruitment (personal interview)	Criteria for Target Group Identification (e.g., product class usage)
O_2	Premeasurement for Established Brands (self-administered questionnaire)	Composition of "Relevant Set" of Established Brands, Attribute Weights and Ratings, and Preferences
X_1	Exposure to Advertising for Established Brands *and* New Brand	
$[O_3]$	Measurement of Reactions to the Advertising Materials (self-administered questionnaire)	Optional, e.g., Likability and Believability Ratings of Advertising Materials
X_2	Simulated Shopping Trip and Exposure to Display of New and Established Brands	
O_4	Purchase Opportunity (choice recorded by research personnel)	Brand(s) Purchased
X_3	Home Use/Consumption of New Brand	
O_5	Post-Usage Measurement (telephone interview)	New Brand Usage Rate, Satisfaction Ratings, and Repeat Purchase Propensity; Attribute Ratings and Preferences for "Relevant Set" of Established Brands Plus the New Brand

O = Measurement
X = Advertising or product exposure
[a]*Silk and Urban (1978), p. 174, Table 1*

The laboratory phase of the research is executed in a facility located in the immediate vicinity of a shopping center. "Intercept" interviews (O_1) are conducted with shoppers to screen and recruit a sample number of consumers representative of the target market for the new product. Field work is done at several different locations chosen to attain the heterogeneity and quotas desired in the final sample. Studies completed to date have typically employed samples of approximately 300 persons.

Upon arriving at the laboratory facility location, respondents are asked to complete a self-administered questionnaire that constitutes the before measurement (O_2). Individually, respondents then proceed to a separate area where they are shown a set of advertising materials (X_1) for the new brand plus the leading established brands. Ordinarily, respondents are exposed to 5–6 commercials, one per brand, and the order in which they are presented is rotated for different groups to avoid any systematic position effects. Measurement of reactions to the advertising materials (O_3) occurs next if such information is desired for diagnostic purposes.

The final stage of the laboratory experiment takes place in a simulated retail store where participants have the opportunity to make a purchase. When first approached, they are told that they will be given a fixed amount of compensation for their time—typically about two dollars, but always more than the sum needed to make a purchase. In the lab they are informed that they may use the money to purchase any brand or combination of brands in the product category they choose, with any unexpended cash to be kept by them. They then move to an area where quantities of the full set of competing brands, including the new one, are displayed and available for inspection (X_2). Each brand is priced at a level equal to the average price at which it is being regularly sold in mass retail outlets in the local market area. The brand (or brands) selected by each participant is (are) recorded by one of the research personnel (O_4) at the checkout counter. Although respondents are free to forego buying anything and retain the full two dollar sum, most do make a purchase. To illustrate, the proportion of participants making a purchase observed in two separate studies of deodorants and antacids were 74 percent and 64 percent, respectively. Those who do not purchase the new brand are given a quantity of it free after all buying transactions have been completed. This procedure parallels the common practice of affecting trial usage through the distribution of free samples. A record is maintained for each respondent as to whether he "purchased" or was given the new brand so as to be able to assess whether responses on the post-usage survey are differentially affected by trial purchase vs. free sampling.

The post-usage survey (O_5) is administered by telephone after sufficient time has passed for usage experience to have developed. The specific length of the pre-post measurement interval is determined by the estimated average usage rate for the new product. Respondents are offered an opportunity to make a repurchase of the new brand (to be delivered by mail) and respond to essentially the same set of perception and preference measurements that were utilized in the before or pre-measurement step (O_2), except that they now rate the new brand as well as established ones.

The details of the measurement instruments utilized in this design are discussed in Appendix 14.1.

Model Structure. As shown in Figure 14.3, ASSESSOR uses both the trial/repeat and an attitude model described earlier. The basic input to estimate equation (14.5) for the preference model is obtained from measurement O_2. The measurements for prediction are obtained from O_5. The input for the trial probability are obtained from O_4, and the repeat measures are obtained from the repurchase opportunity O_5. The models are estimated and their outputs are compared.

Convergence. The expression for market share developed from the individual preference-purchase probability model is structurally similar to that defined in terms of trial and repeat purchase levels. In the former case, market share is the product of the relevant set proportion, (E_b), and the average conditional probability of purchasing the new brand ($\sum_i L_{ib}/N$). In the latter case, market share is the product of the cumulative trial proportion and the share which repeat purchases of the new brand represent of subsequent buying by previous triers.

The sub-models and measures used to arrive at estimates of these conceptually similar quantities are quite distinct. Whereas the trial and repeat proportions are based upon essentially direct observations of these quantities obtained under controlled conditions, the relevant set proportion and the average conditional purchase probability are estimated indirectly from other measures.

Finding that the two models do yield outputs that are in close agreement can serve to strengthen confidence in the prediction. On the other hand, divergent forecasts trigger a search for, and evaluation of possible sources of error or bias that might account for the discrepancy. The first step is to compare the relevant set proportion and trial estimates. Lack of agreement here could imply that the assumptions concerning awareness and retail availability are not compatible with those made implicitly or explicitly in estimating the relevant set proportion. After reconciling the trial and relevant set estimates, attention is focused on the values of the conditional purchase probability and the repeat rate. In the end, some judgment may have to be exercised in order to reconcile differences that arise, but that process is facilitated by careful consideration of the structural comparability of the two models.

Predictions and Marketing Plans. Prediction of a new brand's market share reflects the estimated parameters and the plans for the marketing program to be employed in the future test market or launch. Frequently at this pretest market stage, management is interested in evaluating some variations in the introductory marketing mix for the new brand. The trial/repeat model can be used to advantage in performing some rough simulations of the effects of certain kinds of marketing mix modifications. Some of the changes or alternatives management may wish to consider can be approximated by judgmentally altering parameter levels. For example, increasing the level of advertising spending could be represented by raising the awareness probability and therefore the estimated trial. Differences in sampling programs could be estimated by changing the number of samples or their probability of usage. Other types of changes, such as in advertising copy or price, that affect the conditional first purchase probability can be measured by expanding the research design shown in Table 14.3 to observe the differential effects on trial purchases due to alternative price or copy treatments.

After examining the impact of strategic changes in this manner, profitability measures can then be calculated for the market share estimates. Based on these inputs and the forecasted share, management can then decide whether or not to proceed to test market the new brand.

Summary of Alternative Approaches

This section has covered a number of alternative approaches. We have separated the approaches by giving techniques rather than specific models. By understanding the basic approach behind various commercially available models, you can better assess each model and select the model that is most appropriate for your use. Alternatively, if your new product development program is sufficiently large, you can build upon these basic ideas and customize a pretest market analysis to best fit the needs of your organization.

We recommend that whatever system you select or develop, you build a convergent system. This philosophy is emerging from cumulative practical experience. Many of the original models described above are adopting positive features from other models to produce convergent systems. For example, COMP now is using trial and repeat data as well as attitude change data (Burger, Lavidge, and Gundee, 1978). ASSESSOR now is incorporating dynamic phenomena, such as that represented in Eskin and Malec's model, and repeated purchase opportunities with delivery to the home. New services are being developed commercially and it seems clear that technically strong models and measurement systems will be widely available to forecast sales of new packaged goods.

ACCURACY OF PRETEST MARKET FORECASTING

In this section we review the evidence of the accuracy of pretest models as well as suggest guidelines for using pretest forecasts. The available evidence on the predictive accuracy of pretest models is not large and a number of issues make the available evidence difficult to judge. There are commercial services who say "they have never missed" or are "97 percent correct." Many provide lists of "predicted" and "actual" share. However, we must be careful in interpreting them (Tauber, 1977). For example, Eskin and Malec report "forecasted" and actual results for their model. The authors clearly state that the model forecast is based not on estimates of repeat (α_2), but the actual value from the test market. Despite this warning, a casual reader may not realize this severely restricts implications of validity from this particular data.

In some cases, the "predicted" is not the original prediction, but one adjusted for "differences" between test market and pretest market. For example, after the test, it may have been discovered that advertising spending was less than planned and a new competitor entered the market. Although it is logical to make some adjustments, looking for reasons to make predicted and actual agree is dangerous. This may occur unintentionally if one only looks at differences and then tries to find reasons why they do not agree. If changes are allowed, they should be pursued with equal vigor in cases where actual and predicted agree as well as when they disagree. Then revision may result in higher or lower differences between predicted and actual. Another bias can occur if the product is tested concurrently with test market rather than prior to test market. It takes a high level of discipline not to be influenced by the concurrent experience. If a service claims an almost perfect record of forecasting, you should examine whether their analysis is biased, either explicitly or implicitly.

Other difficulties occur because "actual" share is itself subject to measurement error and because "actual" long-run share may not stabilize in a 9–

12 month test market. With all these cautions, we find that many claim success, but not all have used rigorous predictive testing procedures.

One set of data that has been carefully examined for bias is the predictive results from ASSESSOR (Silk and Urban, 1978, with new data made available by Urban). These are cases where new products that have been subjected to both pretest and test market evaluations; they provide a basis for a partial assessment of the accuracy of ASSESSOR.

Of the approximately 120 new package goods studied to date, 60 percent failed to meet the established pretest standards for a positive test market decision. These standards were expressed as a required minimum long run share objective. A number of these failures were improved and retested so that eventually 60 percent of the products went to test market. Fourteen test markets were still in process, and data was available for 25 completed new

Table 14.4. Predicted and Observed Market Shares for One Pretest Market Model[a]

Product Description	*Initial*	*Adjusted*	*Actual*	*Deviation (Initial-Actual)*	*Deviation (Adjusted-Actual)*
Deodorant	13.3	11.0	10.4	2.9	0.6
Antacid	9.6	10.0	10.5	−0.9	−0.5
Shampoo	3.0	3.0	3.2	−0.2	−0.2
Shampoo	1.8	1.8	1.9	−0.1	−0.1
Cleaner	12.0	12.0	12.5	−0.5	−0.5
Pet Food	17.0	21.0	22.0	−5.0	−1.0
Analgesic	3.0	3.0	2.0	1.0	1.0
Cereal	8.0	4.3	4.2	3.8	0.1
Cereal	6.0	5.0	4.4	1.6	0.6
Shampoo	15.6	15.6	15.6	0.0	0.0
Juice Drink	4.9	4.9	5.0	−0.1	−0.1
Frozen Food	2.0	2.0	2.2	−0.2	−0.2
Cereal	9.0	7.9	7.2	1.8	0.7
Detergent	8.5	8.5	8.0	0.5	0.5
Cleaner	8.4	5.5	6.3	2.1	−0.8
Shampoo	0.8	2.3	2.5	−1.7	−0.2
Shampoo	7.1	7.9	7.6	−0.5	0.3
Dog Food	2.9	2.9	2.7	0.2	0.2
Cleaner	16.5	14.7	12.9	3.6	1.8
Shampoo	1.1	0.6	0.6	0.5	0.0
Frozen Food	2.6	2.0	2.2	0.4	−0.2
Lotion	27.1	27.1	28.7	−1.6	−1.6
Food	5.6	5.0	1.5	4.1	3.5
Shampoo	5.2	2.8	1.6	3.6	1.2
Average	7.9	7.5	7.3	0.6	0.2
Average Absolute Deviation	—	—	—	1.5	0.6
Standard Deviation of Differences	—	—	—	2.0	1.0

[a]*Based on Silk and Urban (1978) and new data supplied by Urban*

product test markets. The final test market shares are known with the accuracy of the test market measures. Table 14.4 presents a summary of results for these cases. The products are listed in the chronological order in which they were studied. The studies for the first three products were performed while their test markets were in progress. These three applications occurred when the system was first developed and were conducted in this manner at the request of firms which were seeking information that would enable them to make an early evaluation of the system's predictive capability. The remainder of the studies were done before test market. The average brand had a test market share of 7.3 percent. The initial forecast average was 7.9 percent and reflects a slight upward bias (0.6 share point) in the prediction. This bias was less (0.2 share point) after adjustments to reflect the lack of execution of planned test market activities and the actual awareness achieved in test market.

The standard deviation of this difference was 2.0 share points with a positive bias in predicted versus actual shares. This experience can be used in the representation of the distribution of forecast errors (see Figure 14.4).

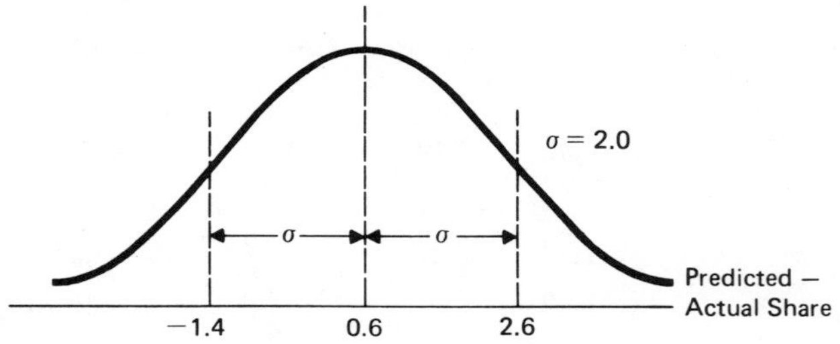

Figure 14.4. *Distribution of Errors*

The best way to interpret this predictive accuracy is to consider how the test market decision should be made. If the pretest share forecast is much higher than the minimum required share, a GO decision is appropriate. If the predicted share is much less than the required share, a NO decision is appropriate. The greater the amount by which the forecast exceeds the required share, the higher is the probability of success in test market.

For example, suppose that the minimum required share were 6.0 percent and the model predicted 9.1 percent. Then, accounting for about a 0.6 percent bias in the model, the predicted share minus the required share is 1.25 standard deviations above the average model bias. (Number of standard deviations = (9.1 − 6.0 − 0.6)/2.0 = 1.25.) We use a cumulative normal table to translate the number of standard deviations to a probability of achieving the minimum required share. For example, 1.25 standard deviations translates to a 90 percent probability of achieving or exceeding the target. Table 14.5 provides the results of these computations for a range of values for the amount by which predicted share exceeds the minimum required share.

Table 14.5. Probability of Test Market Success

Amount (Share Points) by Which Predicted Exceeds Required Share	*Probability of Success in Test Market*
−1.0	21%
−0.5	29
0	38
0.5	48
0.6	50
1.0	58
1.5	67
2.0	76
2.5	83
3.0	89
3.5	93
4.0	96

Now a difficult question is raised. How much risk should be taken in test market? If we require a 95 percent chance of success, we will need almost a 4 share point margin of forecast over actual. Few products will meet this margin so although we will have few failures, we will also have few test markets. In the final analysis each management must implicitly or explicitly make this tradeoff. Some analytical procedures are available to support the tradeoff analysis (Freimer and Simon, 1967; Raiffa and Schlaifer, 1961) which can be helpful in structuring this decision. In actual experience considering the products ASSESSOR has suggested should go to test market, 20 percent have failed in test. This implies that on average management was requiring a two and one-half share point market of safety of predicted over required share and looking for an 80 percent of chance of success in test market.

The accuracy of forecasts is higher after adjustment (the standard deviation of adjusted and actual share is 1.0 share point, rather than 2.0 for initial versus actual), so if management was sure they could execute the test as planned, the risks may be less than indicated in Table 14.5. The accuracy also varies according to the type of product. If the new product fits well into an established category the accuracy will be higher. If the product is revolutionary and creates a whole new category, variances will be higher. If the brand is going to have a small share (less than 2 percent), larger samples would be needed to achieve the indicated accuracies.

Pretest market forecasts are not perfectly valid, but pretest market analysis can be used by management to effectively control the risk of test market failure.

MANAGERIAL USE OF PRETEST MARKET ANALYSES— MINI-CASES

Pretest market analysis can provide accurate forecasts, but the new product manager is also interested in how to utilize the pretest market analyses to improve the product's performance. In this section we describe four mini-cases drawn from the over 120 applications of ASSESSOR. These cases illustrate many of the managerial actions that can result from careful pretest market analyses.

Case 1: Laundry Product

The organization introducing this product had several leading brands in the detergent and fabric softener market. It had developed a new fabric conditioner and was poised for a national launch with a share objective of 8 percent. The organization was so confident of success that test market was to be replaced by a "distribution check." This was a plan to introduce the product for three months in one test city, with advertising, to enable it to achieve the desired retail distribution and self-facings. An ASSESSOR pretest market analysis was done to get an early reading on the product and forecast sales.

The results were surprising to the brand manager. Trial was very low in the simulated store. The product was not viewed as new and the ads scored poorly on likability and believability. The "new" product was confused with the firm's existing product. The share was forecast at 2 percent. Faced with these results, the company aborted the national launch. The test market was maintained for nine months and the actual share of 1.8 percent was close to the prediction.

In this case, the use of a pretest market model prevented a product failure and the $5 million dollar loss such a failure would have entailed. The case emphasizes the danger of becoming overly optimistic and excited about a new product. A pretest market analysis should always be carried out before test market to reduce the possibility of failure.

Case 2: Household Product

This product used a new applicator to more effectively clean and condition wood surfaces. The frequency of purchase in this category was low (two or three times a year), but concept tests showed the product appealed to many people.

A pretest analysis indicated high share potential based on good trial and repeat response. However, because of the low frequency of purchase, almost all early sales would be trial, and long-run share would depend upon repeat sales. The model indicated that shares of 25 percent to 30 percent could occur in months three to six, but the long-run potential would be 18 percent.

The share dynamics were critical and the prediction was important in production planning and developing a financial plan for the product. The product was subsequently test-marketed. A share of 20 percent was observed after 16 months in the test cities. In this case, the pretest market confirmed that the product was in the "GO" state and generated a forecast to plan production and marketing. Original plans had been laid around an expectation of a revolution in the category and a 30 percent share. The pretest showed this was unlikely and plans were revised to achieve target profitability at a share objective of 18 percent. The test market was then used to find the advertising, couponing, and promotion levels to maximize profits.

Case 3: Deodorant

A new aerosol deodorant was being introduced into a saturated market based on a claim of "goes on dry." The primary goal of the pretest analysis was to forecast share and examine the effects of sampling. The claim of dryness was not effectively portrayed by the ad. In-store trial was low. However, consumer experience with the product was good with respect to the dryness dimension. The share with advertising alone was 5 percent, but if 40 percent of the households were sampled, the model predicted a 10 percent share.

The product was test-marketed and introduced nationally with heavy sampling and achieved a share very close to the predicted 10 percent in both test market and in national introduction. In this case, the pretest market analysis was used to forecast and predict the effect of introductory marketing strategies.

Case 4: OTC Drug

This organization had developed a new over-the-counter pain reliever. The decision to introduce was complex since several other firms were known to be considering introducing similar products. Two studies were run. One with the old market products and one with the set of competitive new products. The new products were represented by ads and packaged products for the competitors' products based on how they were expected to appear.

The results indicated that the organization's new product could get 8 percent of the "old" market, but in the "new" market would achieve only 3 percent share. The new entries by competitors were viewed as equal to or better than the organization's product. The pretest analysis suggested a NO GO decision.

In this case, the test market was undertaken in spite of the poor prediction. Momentum for the product was so high at the top management level that they felt they would take the risk of failure in test rather than miss the possible opportunity to enter the category. After twelve months and $1.5 million, the test market share achievement was 2 percent. The project was terminated.

It seems clear from their experience that raw determination can be costly. Organizations should develop a disciplined, new product development process and act upon pretest market models after they have been institutionalized and validated in the organization. In the above case, this was only the second application of the pretest system by the organization and management was not yet sure of its validity.

These cases demonstrate the impact of pretest market analysis. Test market failure can be reduced, better introductory market strategies (e.g., advertising, sampling) can be identified, competitive environments can be understood, and improved financial and production plans can be made.

Although pretest forecasts show a high degree of accuracy, products with large marketing and/or production investments should continue to be test marketed. However the test market should be oriented toward finding improvements in the marketing strategy as well as determining if the product can attain an adequate market share.

PRETEST MARKET ANALYSES FOR OTHER TYPES OF MARKETS

The emphasis in previous sections has been on the application of pretest market analysis to frequently purchased consumer products. The techniques have been carefully developed and have a proven record of success. Many firms in frequently purchased products are using such techniques, but the state of the art is not as advanced in other industries. We predict that in the future pretest market models and measurement procedures will be developed to forecast the sales of consumer durables, industrial products, and services. In this section, we indicate some of the issues in building such systems and their prospects for success.

Durable Consumer Products

Durable products such as appliances, automobiles, and televisions have some similarities to frequently purchased products. The consumers are the same people and the basic behavioral response is similar. However, durables differ from frequently purchased products in a number of significant ways. Although some replacement sales may take place, success is through initial sales, not through repeat sales as in frequently purchased products. The price of durables is higher and penetration of that target group is slower. In consumer products, sales may peak in the third to sixth month; in durables, it may be in the third to sixth year.

These differences affect the way in which forecasts are made. Initial measures of durable purchasing obtained in a laboratory understate sales potential that will be observed after the process of diffusion of innovation has taken place. Innovators try first and information is passed to the later adopters through word of mouth and advertising. Initial purchasing will be low relative to later levels of sales.

While the special characteristics of consumer durables may make forecasting more difficult, research is underway to use laboratory measures to forecast sales for higher priced products. It is based on comparing responses to established products that are analogous to the new product. For example, a new low-cost ($30) home burglar alarm system may be considered analogous to a smoke alarm. A new electronic thermometer may be considered analogous to a water pick. The laboratory response to the new product is compared to the existing products to predict sales.

Purchase records are examined for the analogous product and consumers are classified based on these records as early adopters (bought in first year), middle majority (bought in second or third year), and later adopters. The respondents in each category are exposed to advertising and point-of-purchase displays in a simulated store. Any initial purchases are observed. For those who do not buy, a lottery is offered, and the winners are asked to select the prize they would receive if they won where the analogous products, the new product, and cash are offered as prizes. If prices are different, the product, plus the cash necessary to match the price of the most expensive product, is offered.

The initial measures provide estimates of early sales. See the points marked "X" in Figure 14.5. Preference is measured for those consumers who receive and use the product and a logit model (equation 14.5) is used to estimate long-run sales from the preference judgments. See the points marked "0" in Figure 14.5. Estimates of competitive entry and rate of price decline are then estimated based on an experience curve.

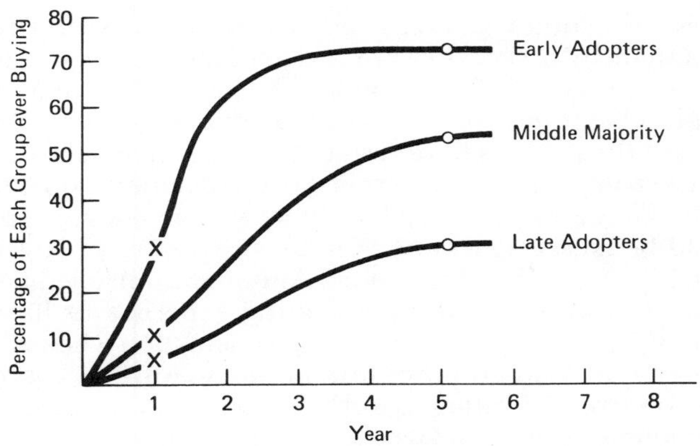

Figure 14.5. *Durable Consumer Product Pretest Market Model*

This approach works best when close analogies can be identified. When revolutionary new products are introduced, such as a new high technology product, more complex models would be needed.

We feel that the potential for a durable product pretest market forecasting tool is high. Research by the authors and others should lead to validated systems in the next few years.

Industrial Products

Some industrial markets, such as industrial cleaning supplies and photographic film, are similar to consumer markets. In these "buy and rebuy" markets, a trial/repeat model similar to that for consumer products is a viable approach.

Other industrial markets are more difficult and less well-understood than consumer markets. In these markets, the purchase decision is made by more than one person, the number of potential buyers is small, and the purchase price is large. The salesperson's job is critical in transmitting information and the selling process is complex and subject to variation between individual salesmen.

Little pretest market research has been reported for industrial products. Forecasts are generated through judgment and the results of concept and product tests or laboratory testing of sales messages. Although pretest market models have not been developed, there is a strong need for them since they could reduce the failure rate of industrial products and lower the average cost of successful innovation. We expect this area of analysis to become more important to researchers. Viable models should be available in the next 5–10 years.

Services

Services are another market where collection of pretest research has not been common. Most services are test marketed in "pilot programs" or "demonstration projects." For example, the concept of a computer controlled mini-bus has been tested in more than one city. But results have been mixed; some of the projects have resulted in expensive failures. The use of a pretest market analysis for such a transportation innovation is important if it reduces the failure rate at the pilot test level. The key concepts of such a system would be measurement of trial after exposure to advertising, and repeat ridership after use of the system. Although controlled testing of this type has not been widely used, some natural experiments have occurred. The blizzard of 1978 in Boston forced experimentation when all roads were closed for over a week. Many office workers got to work by boarding special buses in the suburbs and riding to work on roads that were not crowded by cars. Many commuters were amazed to find they arrived sooner than if they had driven on a normal day, could relax during the ride, had no parking problems, and paid a lower total cost. This experience has led some communities to introduce such a service on a regular basis. This indicates the potential of controlled exposure to new services. If the repeat rate is good, and if trial could be measured in a forced exposure laboratory setting, a basis for premarket forecasting would be available.

Currently, work is underway to develop such a system for transportation services. At present, the planned analysis is based on extensive prior measures and analogies to other communities. The model is designed to provide initial forecasts that can be updated through early demonstration project results. While this model is still under development, we expect that it, and others like it, will be available in the next 5–10 years.

SUMMARY

Pretest market models provide a low cost and rapid method to test the combined product, advertising, price, promotion, and distribution plan. The analyses are sufficiently accurate to identify most winners and eliminate most losers. Furthermore, they provide an effective way to control the risks of failure and supply actionable managerial diagnostics to improve the product.

Pretest market models are an accepted practice for most frequently purchased products, relatively new for consumer durables and services, and not yet developed for industrial products. Since pretest markets are extremely valuable in reducing risk and increasing the expected benefit of new products, we expect that models will soon be available for products other than frequently purchased consumer goods.

Pretest markets provide accurate forecasts and some diagnostic information, but they do not substitute for market testing. Bad products can be eliminated, but in most cases management will still want to refine the potentially successful products in a test market and conduct a final validation. Such test market analyses are the subject of the next chapter.

REVIEW QUESTIONS

14.1 What are the advantages and disadvantages of pretest market forecasting?

14.2 Answer the following questions for the Eskin and Malec model.

   a. What does the model assume about the relationship between product class penetration, distribution and cumulative trial of the new product after one year?

   b. Consider the people who will try the product within one year. When will one-half of these people have tried the product?

   c. How many people will try the new product within six months?

d. Suppose that one-half of all those who will first repeat purchase repeat purchase within $t = \frac{1}{2}$ week. Given this observation, what is the average time between purchases in the product category for this new product?

14.3 What are the relative advantages and disadvantages of Burger's COMP? Of Silk and Urban's ASSESSOR?

14.4 Develop a convergent model using the concepts of: (1) judgment and past history, and (2) stochastic models.

14.5 Answer the following questions for the Silk and Urban model.

a. Suppose that for a particular new product the major source of consumer awareness is store point-of-purchase displays. If these displays are the only source of awareness, what would be the effect on product trial? How would you alter the model to account for this special situation?

b. Suppose a company using the model carelessly misdefines the market. The company accidentally includes a product which, although popular, does not compete in the new product's product category. What will happen when estimates are made for purchase probabilities?

c. In the model, the final stage of the laboratory experiment takes place in a simulated retail store where participants have the opportunity to make a purchase. Those who do not purchase the new brand are given a quantity of it free after all buying transactions have been completed. This procedure parallels the common practice affecting trial usage through the distribution of free samples. How is the procedure similar to distributing free samples and how is it different? Under what conditions would this procedure duplicate the effect of random distribution of free samples by mail?

14.6 How does product sampling affect repeat rates?

14.7 How do pretest markets for durable consumer products, industrial products and services differ from frequently purchased consumer products? What special problems are involved when dealing with these categories?

14.8 Why should a manager accept less than a 100 percent probability of ultimate success when using pretest markets to screen new products?

14.9 What diagnostic information should a manager expect from pretest market analyses? How is this information used to improve the new product and its marketing mix?

14.10 How would you use pretest market analysis to test the effect of a price change and/or a price-off promotion?

Chapter 14 Appendices

14.1 ASSESSOR Measurement Instruments*

Although Table 14.3 identifies the key measures obtained at various points in the design, certain non-standard features of the methods employed deserve some additional discussion. Allaire (1973) has shown that measurement of perception and preference structures can be distorted by including unfamiliar stimuli in the set of alternatives judged. Following his methodological recommendation, we ask each respondent to provide perception and preference ratings only for those brands that comprise his "relevant set" of alternatives—i.e., that subset of available brands which are familiar to the respondent regardless of whether they are judged favorably or unfavorably as choice alternatives. (Review Chapter 8 for a detailed discussion of relevant or evoked set.) Respondents' idiosyncratic relevant sets are revealed by a series of unaided recall questions which identify brands previously purchased or used plus any others considered to be satisfactory or unsatisfactory alternatives.

The size of a typical respondent's relevant set is small relative to the total number of brands available in the market. Data presented by Urban (1975) for seven different categories of packaged goods show that the median relevant set size generally observed is about three brands. Campbell (1969) and Rao (1970) have reported evoked set sizes of approximately the same magnitude for some additional product classes. The smallness of evoked or relevant set sizes is consistent with evidence available as to the number of different brands of packaged goods actually purchased by households. Massy, Frank, and Lodahl (1968, pp. 22–24) reported some relevant statistics for a sub-sample of U.S. households in the J. Walter Thompson

*This appendix is based on an article by Silk and Urban (1978).

panel. During a one-year period, the mean number of different brands purchased per household was 3.3 for regular coffee, 2.6 for tea, and 3.0 for beer. The ranges observed in this quantity for these three product categories were: 1–12, 1–8, and 1–11, respectively. Wierenga (1974, Chapter 6) has investigated some related phenomena using purchase diary data from a panel of 2,000 Dutch households. He found that although a total of 29 different brands accounted for 85 percent of the total volume of margarine purchased, the mean number of brands purchased per household over a 2-year period was only 4.26. The comparable figures for beer and an unidentified food product were 8 and 14 brands available, respectively, with 2.57 and 2.88 being the average number of brands purchased per household in these two product categories.

After identifying a respondent's relevant set of brands, attribute importance ratings are obtained. Beliefs/perceptions about the extent to which each brand in a respondent's relevant set offers these attributes are also elicited by means of bipolar satisfaction scales. These two types of data are important components of the diagnostic information provided by the system.

A constant sum, paired comparison procedure is used to assess brand preferences. Several variants of the constant sum approach have been utilized in marketing research studies and some evidence bearing on the reliability and validity of such measures has been reported. Axelrod (1968) employed a constant sum technique as a rating scale device by asking respondents to allocate "11 cards" among a predetermined set of brands so as to indicate the likelihood of their buying each brand. An individual's preference score for a particular brand was simply the number of cards allocated to it. In a complex, multi-stage study, a number of different awareness and preference measures were compared with respect to their "sensitivity" (ability to detect an effect of advertising exposure in a before-after with control group design), "stability" (aggregate agreement between equivalent samples), and "predictive power" (ability to predict purchases at t_2 from measure obtained at t_1). Based on the results obtained, Axelrod recommended use of the constant sum scale to elicit attitude ratings for brands mentioned by consumers in response to an unaided brand awareness question.

Haley (1970) has reported the results from another comparative study of several attitudinal measures which included a combined paired comparison, constant sum procedure. For all possible pairs of brands, respondents were instructed to divide "10 points" between any two brands so as to reflect their preferences. An individual's preference score was obtained for each brand by summing the points assigned to that brand over all the relevant pairwise comparisons. Relative to the other measures investigated, Haley reported that this method proved superior in its ability to discriminate among brands. As well, it yielded scores whose distribution appeared to be approximately normal.

The findings reported by Axelrod and Haley suggested use of the constant sum technique as a desirable procedure for eliciting preference judgments from consumers. However, in both these studies as well as in other marketing research applications, the methods used to estimate scale values for brands from constant sum input data have been of an ad hoc variety. In psychophysical measurement where it was first used by Torgerson (1958, pp. 105–107), constant sum comparative judgments are the basis of an explicit scaling model for which formal estimation methods have been developed. Under the assumption that the subjects can provide ratio judgments of paired comparisons between stimuli, Torgerson (1958, pp. 108–112) devised a least-squares method for estimating ratio scale values. Hauser and Shugan (1980) provide statistical procedures to test the ratio hypothesis and related hypotheses. They suggest regression or linear programming estimations depending upon the results of the statistical tests. It is this form of constant sum, paired comparison scaling that has been employed in this work to measure a respondent's preferences for his relevant set of brands.

The measures of attribute importance weights, brand belief or attribute ratings, and preferences obtained in the before measurement (O_2) are repeated again in the post-usage survey (O_5) but with the new brand added to each respondent's "relevant set" of alternatives. Finally, respondents are given an opportunity to make a mail order repurchase of the new product.

14.2 Cannibalization and Draw Estimates with ASSESSOR*

The task of predicting how the new brand will affect the shares of existing brands requires estimation of the expected market share when equilibruim is re-established after the launching of the new brand. Under the new steady-state conditions, the market will consist of two sub-populations, distinguishable by the presence or absence of the new brand in their relevant sets. The sizes of these two groups, relative to the total target market, will be E_b and $1 - E_b$, respectively. The addition of the new brand to respondents' relevant sets is controlled experimentally. The impact of its inclusion is measured in the post-usage survey by the preferences for the established brands *after* having been exposed to the new brands. It is reasonable to assume that consumers whose relevant set does not include the new brand will continue to purchase established brands in the same manner they did prior to its entry. This is measured by established brand preferences *before* exposure to the new brand. Silk and Urban (1978) further assume that (a) the probability of the new brand being included in a consumer's relevant set is independent of relevant set size and composition or the structure of preferences for established brands, and (b) inclusion of the new brand in a consumer's relevant set does not affect the number or identity of established brands it contains.

*This appendix is based on an article by Silk and Urban (1978).

The above assumptions, plus the models in the text, imply that the probability, L_{ij}, that consumer i will purchase brand j is given by:

(14.2A-1) $$L_{ij} = \frac{(p_{ij})^\beta}{\left[(p_{ib})^\beta + \sum_{l=1}^{m_i} (p_{il})^\beta\right]}$$ (if the new product is in the evoked set)

(14.2A-2) $$L_{ij} = \frac{(p_{ij})^\beta}{\sum_{l=1}^{m_i} (p_{il})^\beta}$$ (if the new product is not in the evoked set)

where p_{ij} is the measured preference for product j. Let $\epsilon(i)$ be the set of consumers from the sample who evoke the new product and let $\bar{\epsilon}(i)$ be the set of consumers who do not evoke the new product. Let N and $\bar{N}$ be the size of the respective sets. The market share predictions, M_j' and M_j'', for the consumers who do and do not, respectively, evoke j are given by:

(14.2A-3) $$M_j' = \sum_{\epsilon(i)} \frac{L_{ij}}{N}$$

(14.2A-4) $$M_j'' = \sum_{\bar{\epsilon}(i)} \frac{L_{ij}}{\bar{N}}$$

Finally, since the evoking proportion, E_b, is a controllable quantity, the market share prediction, M_j, for product j is given by:

(14.2A-5) $$M_j = E_b M_j' + (1 - E_b) M_j''$$

14.3 Estimation of the Evoked Set Size for the New Product*

Almost all the brands comprising consumers' relevant sets are those with which they report having had some usage experience. Furthermore, there tends to be a strong and stable concurrent relationship across brands between aggregate levels of brand awareness and usage (Bird and Ehrenberg, 1966). This suggests the existence of similar relationships between relevant set and awareness proportions. Cross-sectional regressions of relevant set proportions, E_j, on unaided brand awareness, B_j, and advertising awareness, A_j, levels were performed for eighteen established brands of deodorants using measures of these variables obtained in the ASSESSOR premeasurement questionnaire (O_2 in Table 14.3). Since the observations were propor-

*This appendix is based on an article by Silk and Urban (1978).

tions which varied considerably in magnitude, an arcsin transformation was applied as a means of stabilizing the error variance and thereby obtaining efficient estimates from ordinary least-squares regressions. The following results were obtained:

(14.3A-1) $$\text{Arcsin}\, E_j = -0.599 + 0.901\ \text{Arcsin}\, B_j + e_j,\quad R^2 = 0.972,\ \text{S.E.E.} = 2.39$$

(14.3A-2) $$\text{Arcsin}\, E_j = 3.91 + 1.066\ \text{Arcsin}\, A_j + e_j,\quad R^2 = 0.894,\ \text{S.E.E.} = 4.61.$$

As expected, both brand and advertising awareness appear to co-vary with the relevant set measure. However, the values of the coefficient of determination (R^2) and the standard error of estimate (S.E.E.) indicate that the brand awareness regression provided a better fit of the data than did the estimated advertising awareness equation. Transforming the estimated values of the arcsin of E_j from the above regression back to proportions and comparing them to their corresponding observed values, we find the average residual for the brand awareness regression to be 0.021 while that for the advertising awareness regression is 0.041.

To estimate the expected relevant set proportion for the new brand, E_b, we simply apply the level of unaided brand awareness, B_b, which the introductory marketing program is expected to achieve to the above brand awareness equation. The level of brand awareness predicted for the new product is largely a judgmental estimate since it depends upon the nature and magnitude of marketing effort that will be applied to support the introduction of the new brand.

Chapter 15

Test Marketing

A test market is a major investment to acquire market information. Before an organization commits the resources necessary for test market it should carefully consider the costs and the rewards. If the decision is to test, then the organization should plan to achieve the greatest possible return on their investment. This chapter first considers the advantages and disadvantages of test markets and suggests one approach to making the decision of whether or not to undertake a test market. Next, we outline the basic testing approaches and review the behavioral theory necessary to structure and analyze a test market. We provide the analytic tools, indicate how they can be used to increase profitability, and illustrate this with a case. Finally, we review the procedures on which to base the GO/ON/NO GO decision.

DECIDING WHETHER OR NOT TO TEST MARKET

Advantages

The most compelling reason for test marketing is risk reduction. Most managers prefer to lose one million dollars in test market rather than ten million dollars in national failure. But the risk is not just monetary. A national failure endangers channel relationships, lowers the morale of the sales force, and reduces the confidence of investors.

A test market not only lowers these risks, it also identifies ways to improve profit. A carefully structured test market can identify how to improve advertising copy and placement, promotion, and price. The production facilities and channel relationships are put to the acid test. Things can go wrong, but the test market allows them to be corrected before full-scale launch.

Disadvantages

Risk reduction and strategic improvement are achieved for a price which may not always be justified. One million dollars is typical for packaged goods in a one-city test market and some firms spend $1.5 million or more. In other organizations a test market costs almost as much as a full-scale launch. For example, auto manufacturers do not test market new models. The costs of making enough cars to sell in a test market is great since the dies and production line setup represent an extremely high cost. In some industrial markets, similar costs occur. Making small amounts of a new chemical may require building a pilot plant and millions of dollars of investment. In these cases a test market may be infeasible so, the organization must rely on some extensive pretest market analysis (Chapter 14) and careful adaptive control of the launch (Chapter 16). Although losses can be minimized, these industries must endure higher risks of failure.

A test market also takes valuable time, typically 9 to 12 months, and can destroy your competitive advantage. In fact, it is not uncommon for a competitor to monitor your test market or run a pretest market analysis on your product while you are in test market. During this time, competitors can catch up and perhaps go national at the same time as the initiating firm.

Competitors can also disrupt test markets. If the competitor believes the product will be successful, he may reduce prices and increase advertising and promotion to make a good product look bad. For example, in one test market for a new shampoo, the competitor with the leading established brand tripled its advertising and sent a coupon worth $1.00 off on a $1.29 tube to each household in the test city. Alternatively, the competitor may believe your new product is poor. In this case, the competitor may reduce advertising and promotion and even buy quantities of the new product by the case in order to encourage a national launch of a potential failure.

Considering these risks in test market and considering the proven accuracy of pretest markets, it is not surprising that some organizations are questioning a testing strategy that includes test markets (Cadbury, 1975).

Making the Decision

The test market decision cannot be totally abstracted from profit considerations. The new product manager must consider seasonal timing, the financial state of the organization, the enthusiasm of top management, the relative channel strength, and a myriad of other factors, but formal decision analysis (Raiffa and Schlaifer, 1961; Schlaifer, 1969; Magee, 1964) does provide important guidelines. The technique can be used for specific situations to clarify relative risks. Even if not used formally, it does indicate the general tradeoffs.

In decision analysis, the decisions and outcomes are described by a sequential tree such as Figure 15.1a. The square (□) branches represent decisions under the control of the decision maker and the circle (○) branches represent potential outcomes of that decision. We begin at the right and compute the expected value of each potential outcome. Probabilities are given in parentheses below each branch. The branches off the right-most square are compared and a decision made. The actual analysis is shown in Figure 15.1b. The outcomes are in millions of dollars and would represent the best estimates based on earlier testing and analysis. Probabilities come from prior experience and earlier analysis. In this example a 50 percent chance of "good" test market results and 50 percent of "bad" test results are shown. If a "good" result is obtained a 90 percent chance of national success is shown and if "bad," a 90 percent chance of national failure. If no test is done, a 50 percent chance of national failure is indicated. For this illustration, we assume the success is worth \$10 million if not test marketed and \$10.5 million after improvement by test market; a failure loses \$5 million; and a test market costs \$1 million.

Examine Figure 15.1b. If we do not test then a full-scale launch returns \$2.5 million in expected profit [$0.5 \times 10 + 0.5 \times (-5)$]. The arrow at the bottom of Figure 15.1b indicates a launch without test is superior to no launch at all. If we test, it is more complicated because the GO/NO GO decision is based on the outcome of the test. A "good" result means that the product has a 90 percent chance of success, while a bad result means a 10 percent chance. Thus, we "GO" with a good test, and we "NO GO" with a bad test. This is indicated by the arrows at the top and middle of the figure. We don't know the outcome of the test before we spend the \$1 million, but we can compute probabilities, which in this case are 50–50. (See Appendix 15.1 to compute these probabilities.) Thus, 50 percent of the time we get to make a decision with a \$7.95 million expected outcome [$0.9 \times 9.5 + 0.1 \times (-6)$], while 50 percent of the time the best we can achieve is a \$1 million loss. Using the decision analysis rules, we compute the expected value of the test market and conditional launch branch as \$3.475 million [$0.5 \times 7.95 + 0.5 \times (-1)$]. As illustrated by the arrow in the left of the figure, the best decision is to test yielding \$975,000 more in expected profit [$3.475 - 2.50$].

The real world will never be as clear cut as in Figure 15.1, but this simple example does illustrate the key concepts. As our prior probability of success increases, the value of the test market decreases. For example, change success with no test to 0.80 in Figure 15.1. As the cost of testing increases, the value of testing decreases. For example, change the cost from \$1 million to \$2 million. As the reliability of the test market increases the value of testing increases. For example, change the probabilities to 95–5 with good results and 5–95 with bad results. As the value added from diagnostic information increases, the value of testing increases. For example, change the post-testing success from \$10.5 million to \$12 million.

When pretest market analysis is very accurate, the value of test market-

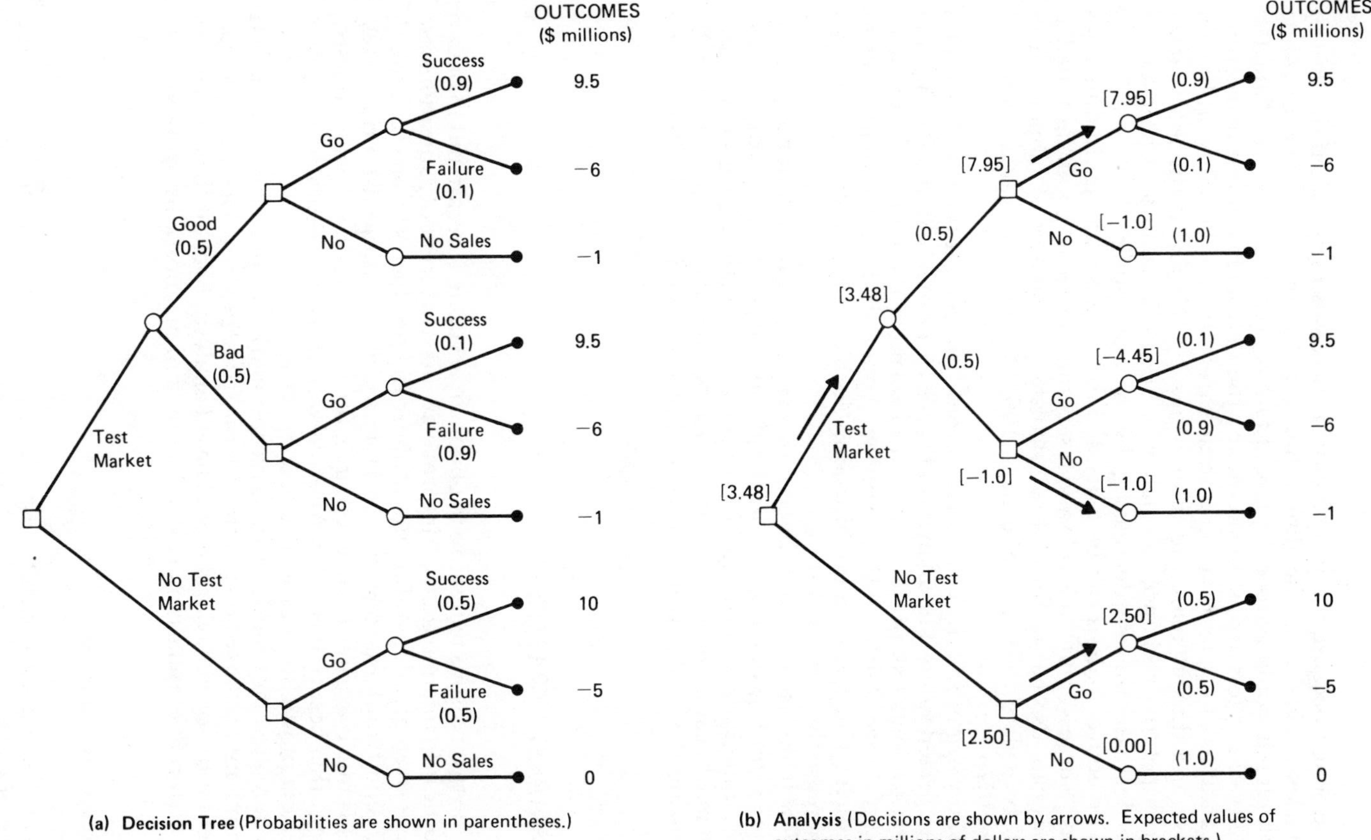

(a) **Decision Tree** (Probabilities are shown in parentheses.)

(b) **Analysis** (Decisions are shown by arrows. Expected values of outcomes in millions of dollars are shown in brackets.)

Figure 15.1. *Decision Analysis for the Decision Whether or Not to Test Market* ing is lessened because the probability of success is high. When the cost of test market is high, as in the auto industry, a test market is not worthwhile even though it would reduce the risk of failure. If competitors retaliate, the reliability of the test drops thus reducing the desirability of test marketing. If the analysis of the test market is done carefully, the reliability and profit improve and thus increase the value of testing. Finally, if the test market gives competitors a chance to catch up, the profit value of a post-testing success decreases thereby decreasing the value of testing.

The above analysis can be extended to include many more outcomes (not just "success" and "failure") and test results (not just "good" and "bad"). Analysis can also be performed to calculate the amount of money a firm should be willing to invest. (See Bass, 1963; Green and Frank, 1966; and Allaire, 1975.)

Many organizations will want to avoid risk, even if they must accept a lower expected value. The analysis in Figure 15.1 can be extended to include an organization's risk preference by transforming the monetary outcomes to "utility." (See Raiffa, 1968, or review Appendix 10.2 for techniques to assess such a risk averse utility function.) Such transformations tend to further increase the value of testing.

Few firms use the formal decision analysis approach, but the analysis and the example given in Figure 15.1 indicate the points to consider in making the strategic decision of whether or not to test. As the figure shows, careful test market analysis will increase the expected value of the test market by maximizing the profit from a successful product, and will also reduce the risks.

TEST MARKET STRATEGIES

The two reasons to test are to obtain an accurate forecast and to obtain diagnostic information that will improve profits. Different organizations use different strategies depending upon where the product is located relative to the decision frontier (see Figure 13.1). For example, if the risk prior to test market is much greater than acceptable, the organization will want to select a strategy that best reduces risk. Alternatively, if risk is acceptable, but profit is low, a diagnostic strategy will be best.

There are many models to analyze test markets, but each can be viewed as an implementation of one of three basic strategies: replicate national, experimentation, or behavioral model based analysis. Each of these strategies places different emphasis on risk reduction and diagnostic information.

Replicate National

The classical approach to test marketing is to attempt to replicate the national environment in two or three "representative" cities and duplicate the national launch plans in them. These cities have typical demographic profiles and are of moderate size, such as Peoria, Illinois, and Syracuse, New York.

The planned level of national advertising and promotion is scaled down to the test city level and the product is introduced into the distribution channel. Sales effort is set by specifying the number of calls to be made on each class of customer. Media spending is expressed in spending per capita and multiplied by the test city population. The standard scaling measure is "gross rating points" (GRPs) which is defined as the percentage of people reached by the campaign times the number of exposures per person in a specified time period (e.g., 100 GRPs could mean a reach of 50 percent and an average of two exposures per person reached).

As the size of the cities increase, the costs go up. Although some organizations have used very small cities in an attempt to reduce test marketing costs, such mini-test market approaches severely reduce the reliability of forecasts.

The classical approach emphasizes forecasting national sales. The most common measure for consumer goods is store sales, which is obtained by using commercial services to audit retail store inventory changes and/or shipments from wholesalers. For example, A. C. Nielsen monitors retail store inventories and SAMI monitors shipments from wholesalers. If there is not a long distribution pipeline or if such a pipeline is predictable, factory shipments provide a third useful measure.

The basic analytic approach is to observe test market share after 9 to 12 months and project it to a national level. The advantage of this approach is its simplicity and the low cost of analysis. One disadvantage is that many anomalies occur in test market which may make the projections inaccurate. Another disadvantage is that many opportunities are missed because of the lack of diagnostic information.

Experimentation

As the emphasis of test marketing shifts to profit improvement, experimentation on marketing variables has been emphasized (Hardin, 1966). Early approaches used different strategies in different cities, but this is expensive and allows few variations. Alternatively, one can use "controlled store testing" to vary marketing strategies within cities by sampling or coupon only sections of the market or by varying the price and promotion across retail stores. In industrial products, salesmen could use different selling strategies on designated sample groups of customers.

The advantage of experimentation is that the diagnostic information generates profit improvement opportunities. However, experimentation requires substantial analytic capability in the fields of experimental design, measurement, statistical estimation, and decision modeling. Experimentation also is limited to the relatively few alternative strategies possible in a reasonable experimental design and subject to the same projectability problems as a replication test market.

Behavioral Model Based Analysis

A third approach attempts to compensate for the potential problems inherent in national replication and experimentation. Behavioral model based analysis recognizes that a test market city is different from a national introduction and collects the information necessary to correct for these differences. True experimentation may or may not be done, but in either case, a detailed behavioral model of the consumer is used to analyze the measurements in such a way that forecasts can be made for the effect of modified marketing strategies.

Model based analysis enhances the accuracy of forecasts and provides a wide range of diagnostic information, but such analysis requires good management science capabilities and a willingness to invest considerable time and money. The level of analysis will vary by situation depending upon the resources of the organization and the product's current position relative to the decision frontier.

The remainder of the chapter provides the tools necessary to analyze a test market. We begin with the behavioral foundations and then turn to the forecasting models. No manager will use all of these techniques, of course, but to select the technique appropriate for a given situation each manager should understand the basic ideas underlying the techniques.

BEHAVIORAL FOUNDATIONS OF TEST MARKET ANALYSIS

Sales do not reach their long-run level immediately. Many complex phenomena occur during the early stages of a new product introduction. Early promotion is aimed at generating trial, but repeat purchases are necessary for many products. Awareness, opinions, and purchase intent propagates by word of mouth. To properly "read" a test market, we must understand and model these and other phenomena. This topic, called the diffusion of innovations, has received extensive study by sociologists, economists, and marketers. In this section, we summarize their key findings and interpret them in the light of improving forecasts and diagnostic information. (For a more comprehensive survey of the field see Midgley, 1977.)

Diffusion Process

One of the underlying concepts of the diffusion process is that different consumers adopt an innovation at different times after it becomes available. These innovators then exert a word-of-mouth influence on others. For purposes of analysis, consumers are classified as innovators, early adopters, early majority, late majority, and laggards, according to when they adopt

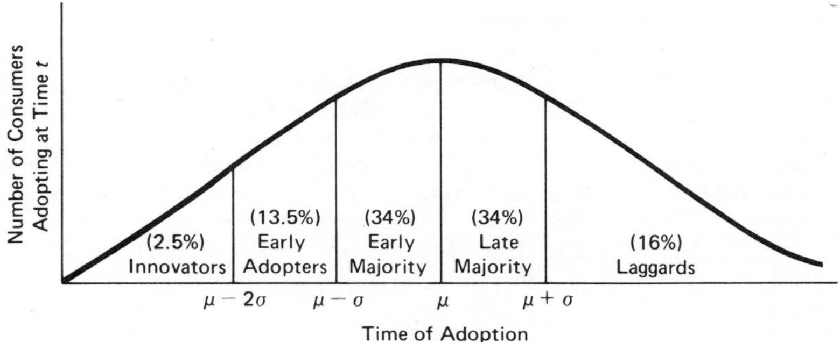

Figure 15.2. *Categorization of Consumers According to Time of Adoption*

(Rogers, 1962; Rogers and Shoemaker, 1971) or their propensity to adopt (Midgley, 1977). Figure 15.2 shows a categorization of consumers according to their adoption time. The demarcation lines depend upon the distance in standard deviations (σ) from the mean adoption time (μ).

There are several implications of this diffusion process. As the process moves from innovators to "others," there are several influences on trial rates. All else equal, the "others" are less likely to try, but if the innovators are opinion leaders, the word-of-mouth effect increases the likelihood of trial. However, as more people adopt, there are fewer left to adopt and trial rates decrease. These effects were the basis of Bass's diffusion model (Chapter 5). Bass's model and other more complex models can be used to measure and project these diffusion trends. Diffusion also affects repeat purchase. If the product is better suited to the innovators, their repeat sales may be higher than the others. Alternatively, innovators per se may be more vulnerable to competitive new products. In either case, the process should be monitored to understand and forecast adoption for industrial products and consumer durables and to forecast trial and repeat for frequently purchased products.

As the diffusion process occurs adopters should not be considered mere recipients of the innovation. There is an emerging body of research that indicates adopters are active in improving the product (Rogers, 1978). Re-inventions made by users can be substantial improvements in the product. Organizations should carefully monitor how consumers not only adopt, but adapt the innovation. This information may lead to new profit opportunities.

Characteristics of Innovators

Since innovators are crucial to the diffusion process, it is important to understand their characteristics. If innovators can be identified, then media

Table 15.1. Characteristics of Innovators[a]

Characteristics or Relationships	Number of Generalizations with Each Type of Relationship to Innovativeness (Percent) Positive	None	Negative	Conditional	Total	Total Number of Generalizations
Social Characteristics						
1. Education	75	16	5	4	100	193
2. Literacy	70	22	4	4	100	27
3. Income	80	11	6	3	100	112
4. Age	32	40	18	10	100	158
Attitudinal Characteristics						
1. Knowledgeability	79	17	1	3	100	66
2. General Attitude Toward Change	74	14	8	4	100	159
3. Achievement Motivation	65	23	0	12	100	17
4. Educational Aspirations	83	9	4	4	100	23
5. Business Orientation	60	20	20	0	100	5
6. Satisfaction with Life	29	28	43	0	100	7
7. Empathy	75	0	25	0	100	4
8. Mental Rigidity	21	25	50	4	100	24
Social Relationships						
1. Cosmopolitism	81	11	3	5	100	73
2. Mass Media Exposure	86	12	0	2	100	49
3. Contact with Change Agencies	92	7	0	1	100	136
4. Deviancy from Norms of the Social System	54	14	28	4	100	28
5. Group Participation	79	10	6	5	100	156
6. Interpersonal Communication Exposure	70	15	15	0	100	40
7. Opinion Leadership	64	22	7	7	100	14

[a]*Adapted from Rogers and Stanfield (1968), pp. 240–242*

and other promotions can be targeted better. To better understand the characteristics of innovators, Rogers and Stanfield (1968) have classified over 2,400 research studies. Some of the results of these studies are summarized in Table 15.1. In general, this table indicates that an innovator is likely to be educated and knowledgeable with a high income, to have a positive attitude toward change and high aspirations, and to be linked to external information sources of media and change agents.

In industrial markets, the important adopter is the innovative firm. These markets have not been as well studied as consumer markets, but some generalizations are beginning to emerge. For example, Webster (1969) argues firms who have a high level of aspiration and can tolerate the risk involved in adoption are likely to be the first to adopt. Mansfield et al. (1971) found, in a study of adoption of numerical control technology in the tool and die industry, that early adoptors tended to be larger firms whose presidents were highly educated. Non-users tended to have low levels of knowledge about numerical controls.

Communication Process and Opinion Leadership

Word-of-mouth is an integral component of the diffusion process (Robertson, 1971; Midgley, 1977). After adopting, innovators communicate their experience to others (Katz and Lazarsfeld, 1955). Later adopters look to the innovators and early adopters for opinion leadership that will encourage or discourage them from adopting.

A classic study was done on room air conditioners. Whyte (1954) observed that air conditioners were being concentrated in clusters of neighboring apartments rather than uniformly spread over the potential housing units. Word of mouth from an adopter to neighbors had resulted in additional adoptions in the neighborhood.

Subsequent research has shown the word-of-mouth communication and opinion leadership process to be complex. Figure 15.3 shows a typical influence pattern in a neighborhood group. The primary group (solid boxes) had various degrees of specialization in influence. Subsidiary friendships (broken-line boxes) reinforced and propagated the process.

This communication process in not limited to consumer products. Coleman, Katz, and Menzel (1957 and 1966) show that there is social interaction among doctors and that socially integrated doctors adopt new products earlier.

It has been difficult to identify the traits of opinion leaders. Although innovators tend to be opinion leaders, all opinion leaders are not innovators. Opinion leadership seems to be specialized by product area (Silk, 1966). Opinion leadership is relative—a leader has more information than a follower (Robertson, 1978). The concept of opinion leadership is important, but the complexity of this process, the lack of generalized opinion leaders,

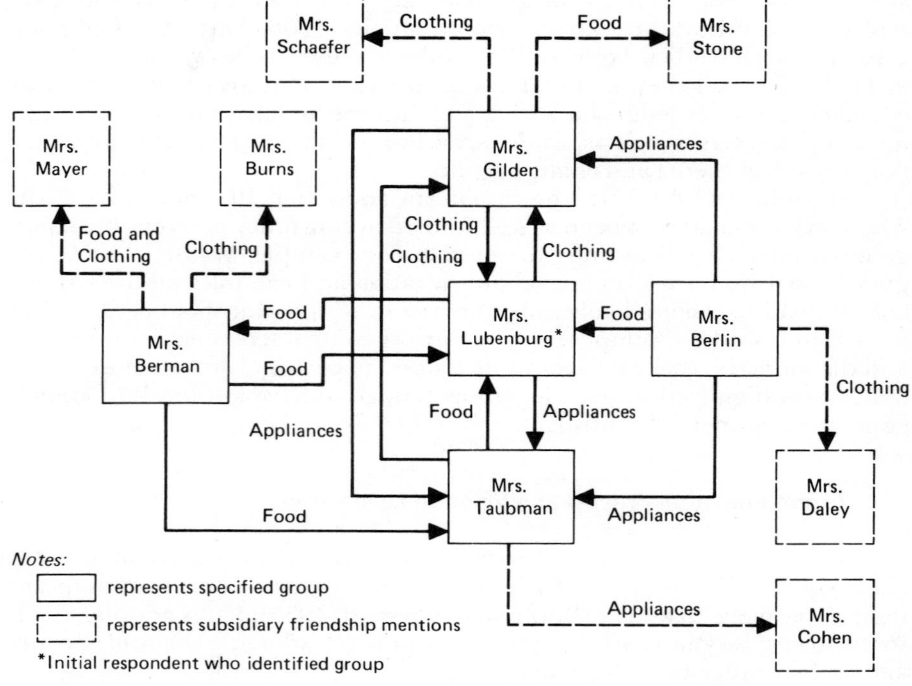

Figure 15.3. *Influence Sources for Food, Clothing, and Appliances within a Typical Informal Neighborhood Group (Robertson, 1978, p. 217)*

and importance of the specific personal interaction situation have made this a difficult area to model.

Although the web of personal relationships and communication are complex, they are real and should be recognized as phenomena underlying forecasting of a new product.

Hierarchy of Effects

The diffusion process is important among consumers. The dynamic process for each consumer is also important. In Chapter 11, we introduced the concepts of awareness, availability, trial, and repeat. We expanded this in Chapter 13 to include the details of advertising exposure. That process is exposure, attention, comprehension, acceptance, and purchase. The essence of both models is that the consumer receives and processes information through a hierarchy of steps, then acts on that information to search, find, try and perhaps repeat purchase of the product. Search will not occur without awareness and intent, trial will not occur unless the consumer can obtain the product, and repeat cannot occur without trial. There are alternative ways to model this process including the AIDA model (awareness, interest, desire, and action—Strong, 1925) and the hierarchy of effects model (unaware, aware, knowledge, liking, preference, conviction, and purchase—Lavidge and Steiner, 1961). Each works. The important point is to use some model to represent the dynamic process so that the conditions of purchase can be measured and understood. We have found the simple model in Figure 15.4 to be particularly effective for analyzing test markets.

Together, the diffusion of innovations among consumers and the dynamic model of information processing provide the behavioral foundations for test market analysis. We turn now to methods which use these ideas to analyze test markets.

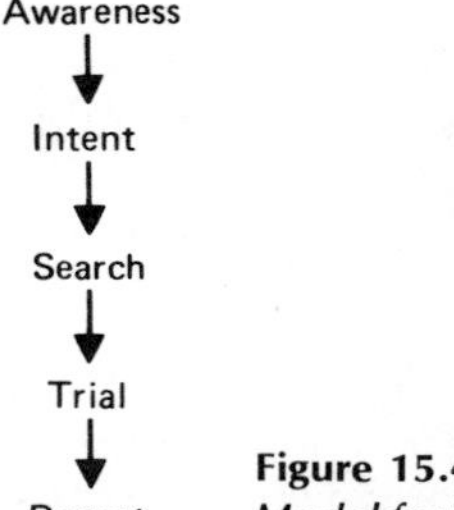

Figure 15.4. *A Useful Hierarchical Model for Test Market Analysis*

METHODS TO ANALYZE TEST MARKETS

If the test market strategy is to replicate national experience, then forecasting is done by simply projecting the test market share to the national level. If the test market strategy is experimentation, then the best marketing mix is the one yielding the greatest profit where the choice is made from the experimental design. If the test market strategy combines either of these approaches with behavioral model based analysis, then more accurate projections and better diagnostic information can be achieved. Whether the new product is a service, an industrial good, or a consumer product, any forecasting procedure must not only include a projection model, but also the capability to adjust the forecast for differences between test and national markets and to project the forecast to alternative marketing strategies.

The behavioral science phenomena identified in the previous section underlie detailed analyses. Some analytic methods explicitly recognize these phenomena, collect data to understand them, and make projections based on the diffusion of innovation process. Other methods implicitly reflect them in their models.

Data collection in test marketing can support complex models. Data sources include:

- store audits of sales,
- telephone surveys of awareness, attitude, trial and repeat purchase, and usage,
- warehouse sales withdrawals,
- consumer diary or panel records of purchasing,
- factory shipments of product, and
- salesmen's call reports.

A good analysis plan will include data from these various sources to allow accurate forecasting and useful diagnostics.

We begin by discussing panel data and stochastic models that consider trial and repeat phenomena for frequently purchased consumer products. Recursive and macro-flow models build from this base to include more behavioral phenomena and provide more managerial insight.

Panel Data Projection Models

The first step toward analyzing a test market is to decompose total sales into trial and repeat as shown in Figure 15.5. These measures are normally obtained from a consumer panel. Trial and repeat measures are used for early projection of results. For example, sales can be higher in months 3 or 4 than 8 or 9 due to high trial purchase. Failure to recognize this has led some firms to cut off the test and go national after a few months based on the "fantastic" sales. Trial and repeat measures are also diagnostic information because they indicate whether the success/failure of the product is due to good promotion (trial) or to a good product (repeat).

Forecasts are made by re-expressing the sales patterns shown in Figure

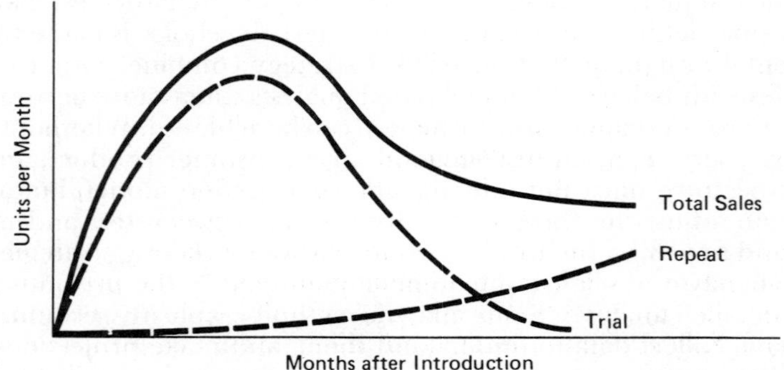

Figure 15.5. *Typical Sales Patterns for Trial, Repeat and Total Sales*

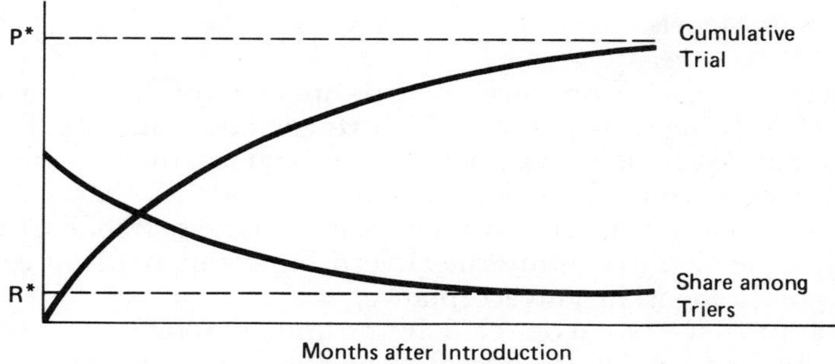

Figure 15.6. *Trial–Repeat Model (long-run share = P*·R*)*

15.5 as cumulative trial and repeat rates shown in Figure 15.6 (Parfitt and Collins, 1968). If no major changes are made in the product, the marketing strategy, or the market itself, then the cumulative trial (percent of the market who will ever try the new product) will reach some penetration level, P^*. The repeat rate (share of purchases among those who have tried the product) will level off to some equilibrium share, R^*. The long-run share is estimated by multiplying these two values. In cases of segmented markets, the values are calculated for each group and summed after weighting by a buying-level index. Early forecasts are made by projecting the curves from the panel data obtained for the first few months. Accuracy is reported to be good (Parfitt and Collins, 1968), but only limited diagnostic information is produced.

Often the test market is not totally representative of the national market. In this case, a number of correction factors must be multiplied times the forecast to account for differences in distribution, consumption rates, seasonality, attitudes, and consumer demographics (Ahl, 1970). Occasionally, judgmental correction factors are used to account for differences in marketing strategies between test and national.

The basic Parfitt-Collins model provides good estimates of equilibrium share, but stochastic models, such as described for pretest market analysis, provide better estimates of the path to this equilibrium (Eskin, 1973; Fourt and Woodlock, 1960). These "depth of repeat" models are used in the same way as the stochastic models described in Chapter 14 except that they use more direct observations to determine the proportion of people who repeat after one purchase, after two purchases, etc.

The most complex trial and repeat purchase models are stochastic models which structure individual purchases by the use of probability distributions to reflect heterogeneity in the population and non-stationarity in purchase (Massy, 1969; Herniter and Cook, 1978). Although analytically attractive, these stochastic models have not achieved wide acceptance due to their complexity, estimation difficulties and costs, and the inability to be effectively communicated to management.

Recursive Models

Trial and repeat are only two of the components of the dynamic model in Figure 15.4. If the manager wants to further understand the acceptance process so that he can better diagnose the test market, then measures must be taken for more components of the dynamic model. In particular, measures are needed with respect to awareness, intent, and search. Further measures may be needed to capture the advertising components of exposure, attention, comprehension, and acceptance.

One approach is based on recursive formulations (Charnes et al., 1966 and 1968). In these models, each level of the dynamic model is derived from a regression equation based on the next lower level. For example, the level of awareness depends on the level of advertising and on previous awareness, trial depends upon awareness and previous trial, etc. The test market is observed for a period of time long enough to estimate the parameters in the equation and then the regression equations are used to estimate long-run share.

The most recent and completely documented recursive model is called TRACKER (Blattberg and Golanty, 1978). In this model, the change in awareness (A_t) depends non-linearly upon the gross rating points (GRPs) and trial results from awareness. The trial in period t, T_t, is given by:

$$T_t = T_{t-1} + \alpha_2(A_t - A_{t-1}) + \beta_2(A_{t-1} - T_{t-1}) \quad \textbf{(15.1)}$$

where α_2 and β_2 are again estimated by regression. Note that the fraction of people who recently became aware, $A_t - A_{t-1}$, can have a different trial rate than those who were previously aware but have not tried, $A_{t-1} - T_{t-1}$. This trial is adjusted by a price elasticity and repeat purchases are represented as the percentage of new triers who are still users at a given point in time. Sales are the sum of the trial and repeat sales.

The advantage of such recursive models is that they are relatively low cost (for example, $15,000 for data collection and analysis, Blattberg and Golanty, 1978) and can accurately predict test market results from three months of test market. They are useful in understanding trial and repeat purchases, and the process by which awareness produces trial and repeat. They are most useful for product categories where data is available from prior new product introductions so parameters can be precisely estimated.

The disadvantage of recursive models is that they measure only part of the dynamic hierarchy. Detailed diagnostic information on the behavioral process is not available and the panel data is not well integrated into the model.

We feel that recursive models are most appropriate for managers who want low cost forecasts from actual test market data. If forecasts have already been obtained from pretest market analysis, the manager may wish more diagnostic information. If the test market investment is sufficiently large (say $1 to $1.5 million), the manager may wish to augment the early forecasts with more expensive ($50,000 to $100,000), but more detailed measurement.

Macro-Flow Models

Most managers need more behavioral diagnostic information than the recursive models can provide. A modeling technique to fill this need is macro-flow modeling (Urban, 1970). In this approach, the manager and analysts identify the behavioral phenomena that are important to the strategic managerial decisions. These phenomena and other key behavior necessary to interrelate the strategic phenomena are connected in a macro-flow diagram such as Figure 15.7 where the boxes indicate consumer "states" and the arrows represent "flows" from state to state. The number of states can vary from 10 in a very simple model to over 500 in a complex model. More states give greater diagnostic information, but cost more to measure. Panel, survey, and store audit data are used to measure the number of people in each state in each time period and to estimate flows between states. Once the flows are known, the macro-flow model is used to project test market results.

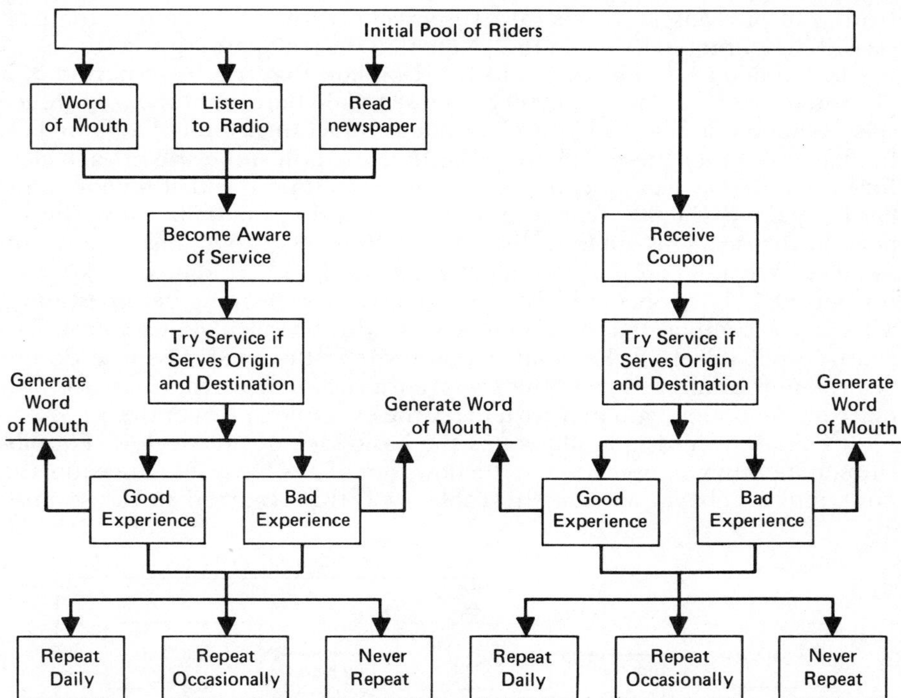

Figure 15.7. *Example of a Macro-flow Model for a New Transportation Service (Hauser and Koppelman, 1977, p. 649)*

Macro-flow analysis is best illustrated by example. Consider the simple three-state model in Figure 15.8 which is a macro-flow representation of a simple depth of repeat model. Suppose we have observed the test model for a few months and know:

- the target group is 10 million households and all are potential triers,
- the percent of families who try the new product is 10 percent in month 1, 5 percent in month 2, 1 percent in month 3, and zero thereafter,
- sixty percent of the preference class will buy our new product when they need to repeat,
- eighty percent of the loyal class will buy our new product when they need to repeat,
- every family needs to repeat once per month.

In practice, these flow rates are obtained from panel or survey data. (Appendix 15.2 shows how to calculate flows, if the number of people in each state is known.)

Now let us use this information to calculate the sales in several periods. In period 1, all states except trial class are empty, thus sales will occur only from trial purchases. In this case 10 percent of the 10 million in the trial class. This is entered in Table 15.2. Note that the trial class is now reduced in size to 9 million (10 million minus the 1 million that tried). In period 2, 5 percent of the trial class (450,000 families) and 60 percent of the preference class (600,000 families) will purchase the product for a total of 1.05 million families. We enter these sales in Table 15.2 and adjust the size of each class for period 3. (For example, the preference class started with 1 million families but gained 450,000 from trial purchases and lost 600,000 from the repeat purchase. This gives 1,000,000 + 450,000 − 600,000 = 850,000 families.) We now compute the sales from period 3 and update the class sizes for period 4. The concept is the same, but the bookkeeping becomes somewhat more complicated. You may wish to do the calculations yourself to reproduce Table 15.2. Remember that those in the loyal class who do not buy the new brand return to the preference class. For further practice with macro-flow models, you may wish to try the example in Appendix 15.3.

The above example illustrates the basic idea of macro-flow. Equilibrium, if it occurs, happens when the flows out of a behavioral state equal the flows into a behavioral state. In Table 15.2, this occurred in the seventh

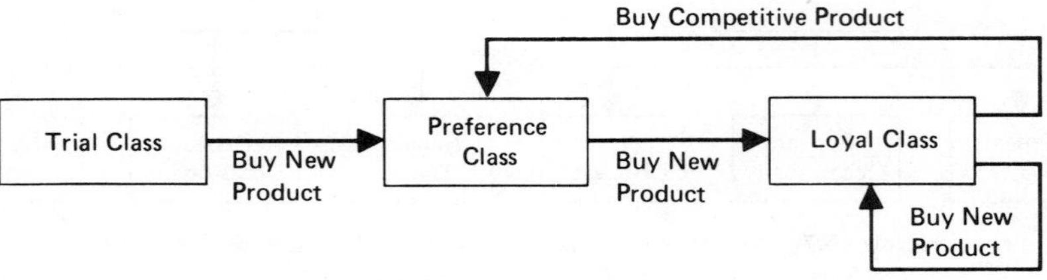

Figure 15.8. *Simple Macro-flow Model*

period. This equilibrium is determined by the flow rates and the initial states. As soon as the flow rates stabilize, forecasts can be made. Note that the states are defined such that they are mutually exclusive (no consumer is in more than one state simultaneously) and collectively exhaustive (each consumer is in at least one state). Thus, the number of people in each state always sums to the total number of consumers. Diagnostic information is obtained by examining the flow rates and the number of people in the behavioral states. In Table 15.2, we see that in equilibrium much of the sales come from loyal customers. This is evidence that the product has good performance that encourages repeat purchase, but for some unidentified reason, perhaps poor advertising, only 15.4 percent of the population has tried the product. One can make similar interpretations from the flow rates.

Table 15.2. Calculations for Simple Macro-flow Diagram (Numbers are in thousands)

Period	*Trial Class Size*	*Trial Sales*	*Preference Class Size*	*Preference Sales*	*Loyal Class Size*	*Loyal Sales*	*Total Sales*
1	10,000	1,000	0	0	0	0	1,000
2	9,000	450	1,000	600	0	0	1,050
3	8,550	86	850	510	600	480	1,076
4	8,464	0	546	328	990	792	1,120
5	8,464	0	416	250	1,120	896	1,146
6	8,464	0	390	234	1,146	917	1,151
7	8,464	0	385	231	1,151	920	1,151
8	8,464	0	385	231	1,151	920	1,151
Long run	8,464	0	385	231	1,151	920	1,151

Usually the manager has many more questions than the simple model in Figure 15.8 can answer. Thus, in practice, macro-flow models are more complex. States are included for advertising, couponing, distribution, etc. Flows can occur once per time period or more often for such advertising states as exposure, attention, comprehension, and acceptance. Flows can be added for word-of-mouth influence, forgetting of appeals, returns to trial states, etc. Not every consumer will have the same usage rate, thus parallel models may be used for light, medium, and heavy users with transitions among states taking place at different times in different models.

Macro-flow is an extremely flexible concept that enables each manager to customize the analysis model to his test market. If an effect, say coupon response, is important to the marketing strategy, it should be included in the model and measured. But the manager should resist modelling every effect he can think of since the cost of analysis and measurement and the difficulty of interpretation increase as the macro-flow diagram becomes more complex.

A number of test market models have been designed based on the macro-flow concept. Midgley (1977) uses innovation theory to derive one relatively complex model, while Assmus (1975) uses a simple eleven-state model, called NEWPROD, to forecast new product sales. Urban (1970) has designed an extremely flexible model called SPRINTER. In SPRINTER, the manager can select from three basic models (Mod I, Mod II, and Mod III) and customize the analysis within each model (Urban and Karash, 1971). Furthermore, this model provides flexible computer software to use the model for forecasting and to search for improvements in profit that can result from changes in the marketing mix. Some regression equations are built into the model to link the marketing mix changes to changes in flows between states. Each of the macro-flow models appear to be reasonably accurate and useful to managers. We further describe SPRINTER to illustrate the power and flexibility of this approach.

Mod I begins with the dynamic model of awareness, intent, search, trial, and repeat that was illustrated in Figure 15.4. Awareness change depends upon advertising spending, intent rates depend upon appeal effectiveness, and availability depends upon distribution. Measures include questionnaires to measure ad recall and predisposition to try.

Mod II extends Mod I by considering each stage of the dynamic model in more detail. Explicit equations are included linking advertising spending, price, and other marketing variables to the flows so that forecasts can be made for alternative marketing mixes. For example, overall awareness is segmented into four specific awareness states, defined as: (1) unaware, (2) aware of product only, (3) aware of ads and product, and (4) aware of specific advertising appeals. Advertising response functions are explicitly stated as the percentage made aware by a particular advertising expenditure. The rate of intent in each awareness class in Mod II is used to determine the total number of people with intent to try. Those with intent look for and find the product in proportion to its availability, after which some fraction will buy it. This fraction is a function of the price of the new product relative to that of the competitive products. Those with no intent also go through the in-store process, and some may buy the product because of point-of-purchase activity. Those who purchase move to the preference class; those who do not buy experience forgetting, they move from higher to lower awareness classes and remain in the trial class.

These and other behavioral phenomena included in search, trial, repeat, and experience classes make Mod II a useful intermediate model for many test markets. (Figure 15.9 gives the experience classes for Mod II.)

Even more capabilities are available in Mod III where phenomena like word-of-mouth communication can be included. For example, Figure 15.10 gives one possible set of behavioral states for the preference class in Mod III. Since no application will include all the possible detailed options, a typical application will use a moderate level of detail to capture the essential behavioral dynamics. The case reported later in this chapter demonstrates such an intermediate level model and how it can be used to plan a test market, track early sale results, and forecast national sales levels.

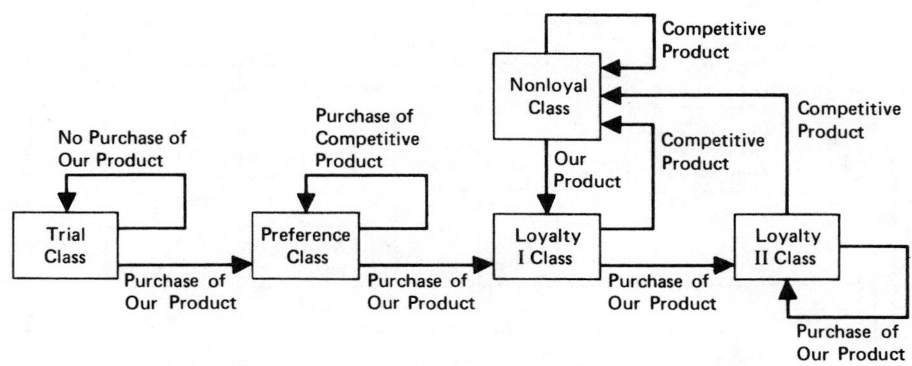

Figure 15.9. *SPRINTER Mod II Product Experience Classes*

Durable and Industrial Products and Services

Most of the previous models can be used for consumer services as well as consumer products. For example, a successful new transit service will be described by high trial ridership and a large volume of repeat usage. A new HMO is tried for one year, but in the long run, year-to-year repeat must be high enough to sustain a successful health plan. Financial and insurance services display similar patterns. In these cases, the macro-flow models outlined in the previous section can be utilized to interpret test market, pilot program, or demonstration project results.

The analysis and managerial strategy is somewhat different for consumer durables and many industrial products. For the first few years, sales will depend upon one-time purchases of high-priced items. Replacement sales may not occur for five or ten years. The sales pattern for new durable products tends to appear as an inverted "U" shape, such as that we saw in Figure 15.2. The time to reach the highest levels may be several years after introduction because the diffusion of innovation process takes place much more slowly.

In many industrial and durable products, test markets are not done and emphasis shifts to interpreting early national or regional sales data to forecast if the product will succeed. In analyzing these data, it is critical to understand the diffusion process through both primary selling, advertising, and word-of-mouth communication. Furthermore, the test must be run sufficiently long to allow projection of the sales peak. Detailed test market models are not yet available for durables and industrial products. At present, the best approach seems to be a macro-flow or diffusion model (such as Bass's model described in Chapter 5). In the absence of test markets and models, the innovating organization is forced to use more judgment and bear more risk.

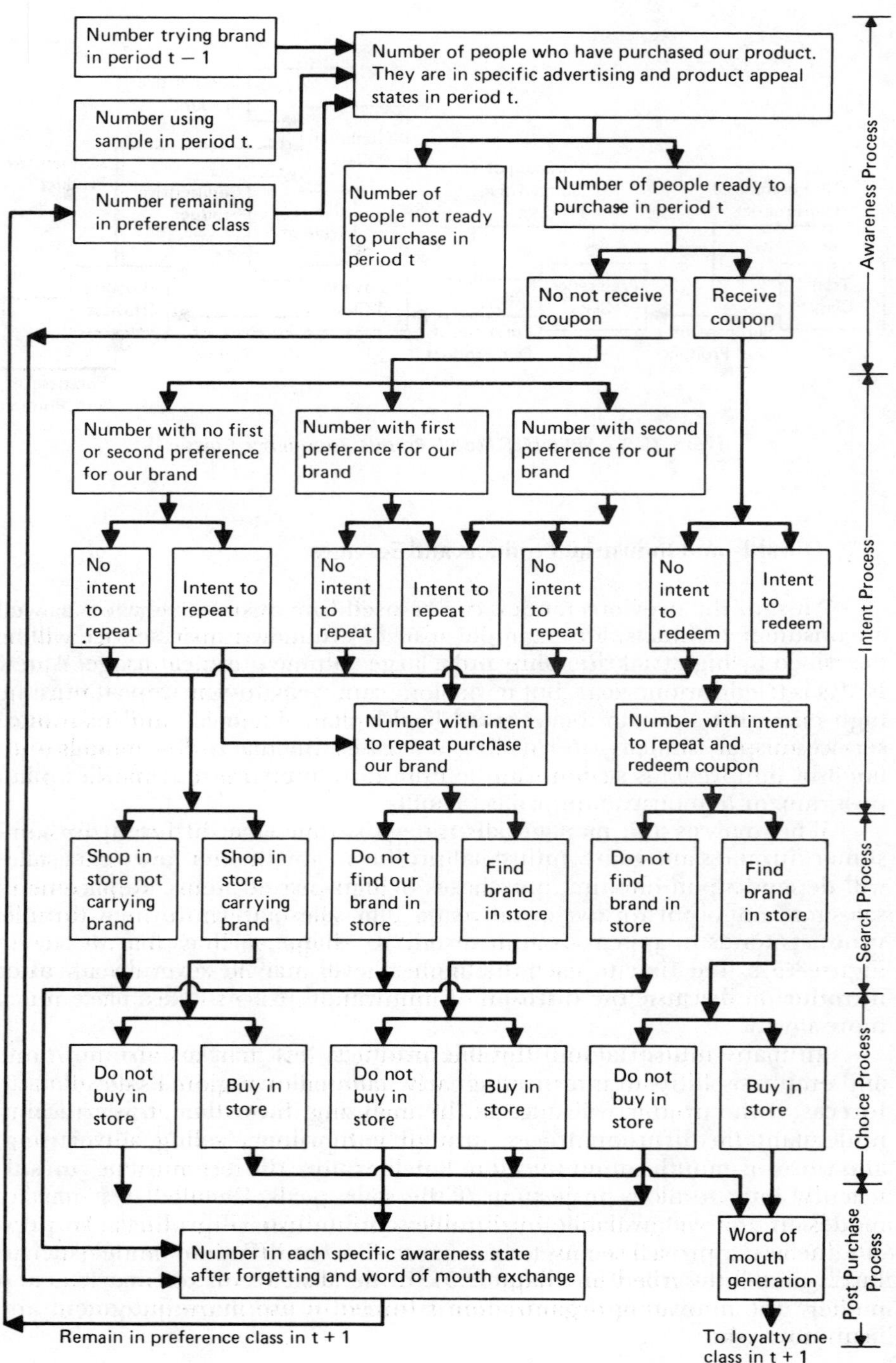

Figure 15.10. *SPRINTER Mod III Behavioral States for Preference Class (Urban, 1970, p. 820)*

Selecting an Analysis Method

In this section, we have presented a variety of analysis methods ranging from the simple trial/repeat panel projections to the more complete recursive models to the flexible macro-flow models. These models vary in complexity and cost as shown in Figure 15.11. All models report good forecasting accuracy; the difference in the models is the degree to which they can provide actionable managerial diagnostics. We suggest that the new product manager understand the basics of each analysis method and select the model most appropriate for the specific test market analysis.

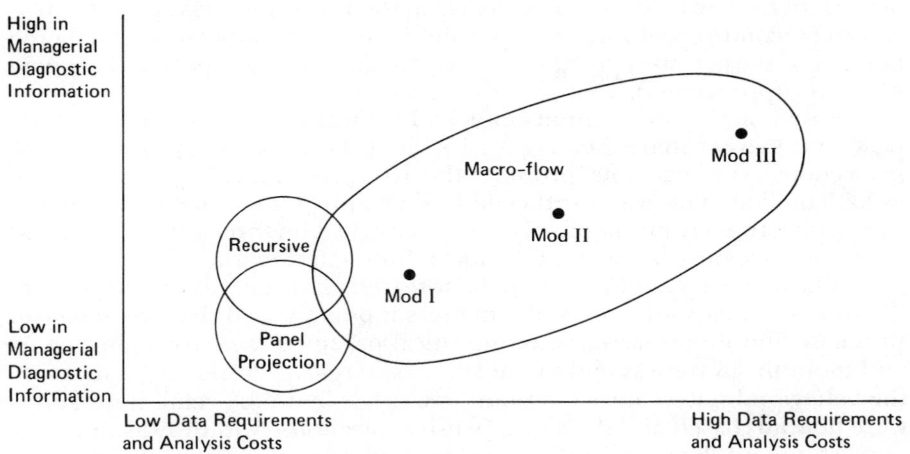

Figure 15.11. *Categorization of Test Market Analysis Methods*

CASE: HERITAGE PASTRY

To demonstrate the use of a test market model and measurement system, we present the case of a new pastry mix that was developed by a major food manufacturer. There were many cake mixes on the market, but no pastry mixes. The premium cake mix market was growing and analysis indicated a potential for pastry recipes. A line of cream puffs and torts called Heritage Pastry (name disguised) was developed with a core benefit proposition of "elegant and easy to prepare desserts." Advertising fulfilled the CBP by showing the pastry on a crystal dish on a formal dining table. Concept and home use tests were positive and management felt the product was ready for test market. A macro-flow model was used to plan and analyze the test.

Planning the Test Market

Before the product was taken to test market, initial forecasts were made to establish consumer response goals and to provide input to the decision of whether to proceed with the new product. Early analyses identified the target group as households with income over $15,000 where the housemaker bakes at least once a month. This was estimated to be 15 million households. Awareness, intent, and availability were projected from concept tests and planned spending levels. See Figure 15.12a, b, c. Panel data for premium cake mixes was used to estimate the frequency of purchase of this type of mix. See Figure 15.12d. The first repeat rate was estimated at 50 percent (50 percent of those who tried would repeat at their next purchase opportunity) and the second repeat rate at 88 percent based on product testing data (88 percent of those who repeat purchased one time would repeat at their next purchase opportunity).

Based on the above inputs, SPRINTER was used to estimate trial, repeat, and market share. See Figure 15.12e, f. Financial analysis used these projections to estimate total profit for the first three years of full-scale launch as $2.5 million. This was a sufficient business proposition for management to approve the test market for Heritage Pastry. The forecasts of share and consumer response became the standard for the test market.

The next step was to develop the research plan. Three cities were identified. A sensitivity analysis of the model's inputs showed that frequency of purchase and intent were the most critical parameters. A consumer panel and monthly awareness and intent surveys were commissioned to measure the behavioral states. Since distribution was less in doubt, retail store audits were not purchased and the less expensive warehouse withdrawal and company shipment data were utilized to measure sales. The research data cost $150,000 and model analysis cost $30,000. Advertising, product production, and distribution in the cities cost $1 million.

Tracking the Test Market

There was great expectation as the test market began. Would the two years of development be a success? The brand manager in particular felt it *had* to succeed if he was to move ahead rapidly in the organization. After two months, the first month share was available—3 percent! The goal (Figure 15.12f) was one-half of 1 percent. Congratulations all around. There was even talk of aborting the test market and going national. But why had the sales been so high? This question became more important as the second period share became available—1.2 percent versus the goal of 1.0 percent. The brand now exceeded the goal for two months and there was evidence that a competitor was ready to introduce a pastry mix. However, the brand manager began to be cautious while others pushed to go national. Months three and four shares were above goal and pressure increased to go national. The brand manager was now actively resisting these pressures. In month

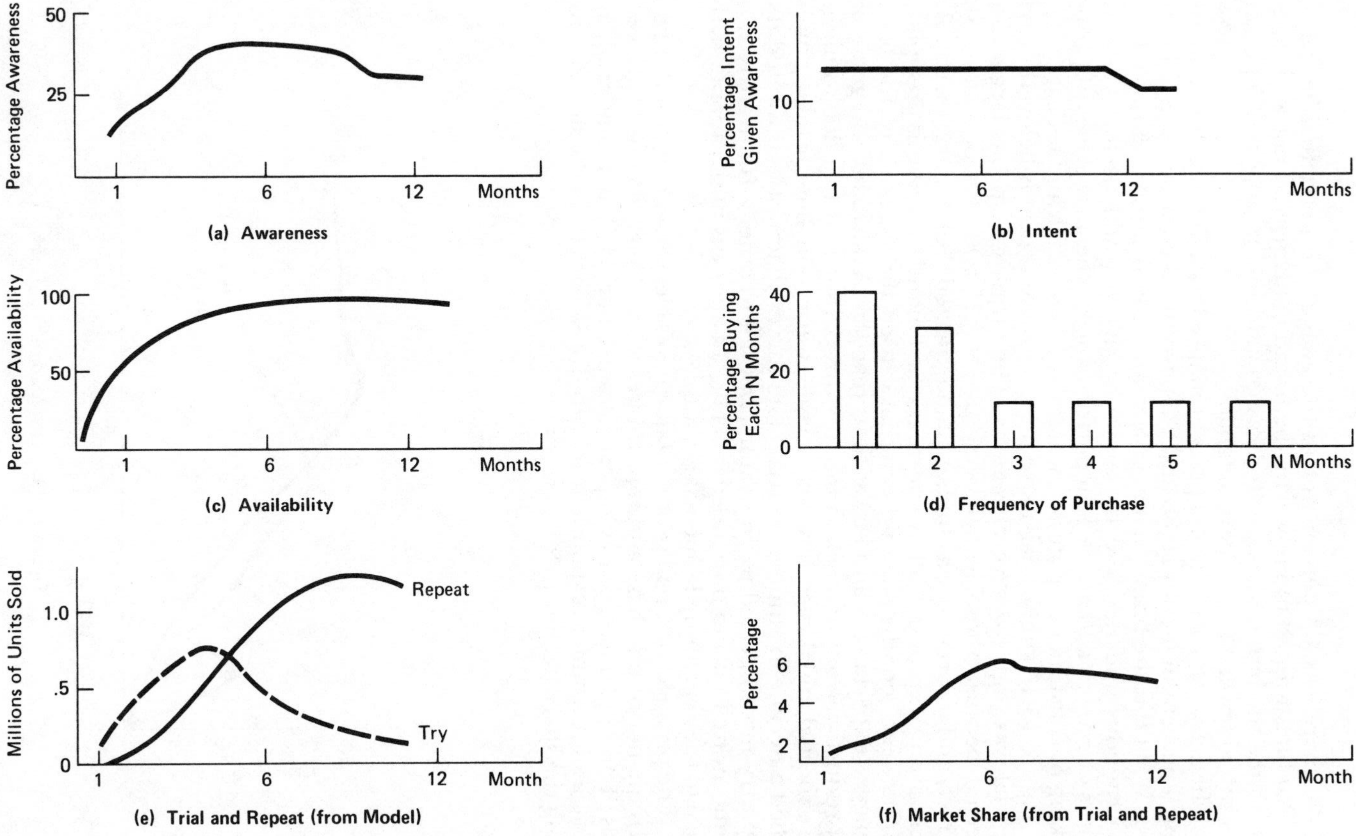

Figure 15.12. *Estimates of Growth Made Prior to Test Market*

five, the share was 4.3 percent versus the goal of 4.0 percent. The V.P. of New Products wanted to know why the brand manager did not want to go. At this time the brand manager forecasted a decline in sales to 2.5 percent by month nine versus the 5 percent goal.

The actual share fell to 4.0 percent in month six, 3.6 percent in month seven, 3 percent in month eight, and 2.5 percent in month nine. See Figure 15.13. What did the brand manager learn in month three that caused him to be cautious and how did he know the share would decline after five months of growth and above goal performance? The answer lies in proper tracking of consumer response—awareness, intent, trial, and repeat.

Figure 15.14 shows trial above the goal while repeat is below the goal. The trial was high for several reasons. The target group was larger than expected because those with income under $15,000 also purchased the product, and awareness was higher than expected. See Figure 15.14c. But intent given awareness was lower after the first month. It was this precipitous drop between months one and two that caused the sobering of the brand manager's expectations and his cautious reaction. Availability was good, but the product was out of stock in months three, four and five due to higher than expected initial sales. Overall trial sales were good, but the falling intent implied a decline in cumulative trial. The survey diagnostics indicated that the pastry mix appealed to innovators, but not to others. Innovative cooks became aware first and purchased quickly, but after the initial adoption, the appeal to the remaining homemakers was low.

The prospect of a real product failure was clearly seen in month five as the frequency of purchase and repeat rates were observed to be below standards. This explained the low repeat sales per period. The falling trial, low repeat, and consumer response by month five were used in the model's projecting the share decline to 2.5 percent.

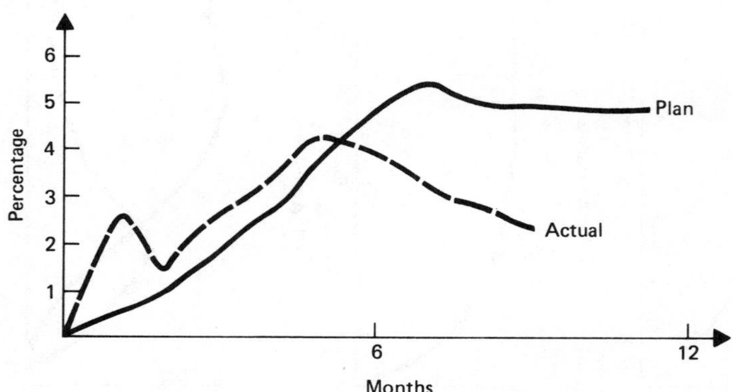

Figure 15.13. *Test Market Share*

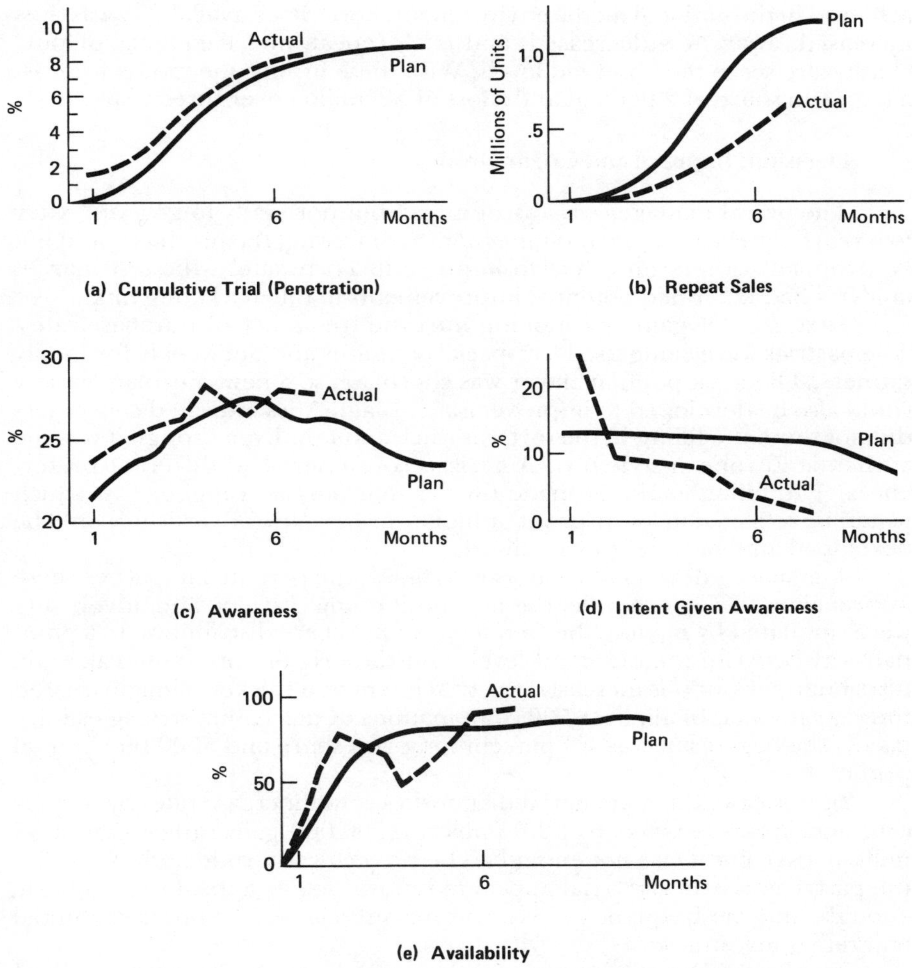

Figure 15.14. *Test Market Results Compared with Prior Forecasts*

Forecasting National Sales

A share of 2.5 percent was well below the forecast 5.0 percent necessary for the projected profit, but conditions may change in national introduction. Before declaring Heritage Pastries a failure, the brand manager considered changes that would occur between test and national. One change that would improve sales was the fact that the out-of-stock condition would not occur nationally. On the other hand, the same levels of awareness could not be achieved in the national media environment. Forecasted availability was thus increased, awareness decreased, and trial, repeat, and frequency of purchase were set to the observed levels. With these inputs, the model forecast a long-run share of 2 percent and a loss of $2 million over three years.

Decision: Improve and Go, or Drop

The brand manager was discouraged, but not ready to give up. After two years of development and nine months of testing, the product could not be dropped without an effort to improve it. Fortunately, the test market analyses had identified potential improvements in the marketing mix.

First, the "elegant" positioning hurt the frequency of purchase rates. The pastries were being used for special occasions and not weekly for family dinners. The most popular flavor was chocolate, so a new chocolate variety could also be developed to improve usage. Diagnostics indicated consumers did not think the filling in the tort was thick enough. Even though it was the authentic European-style tort, American consumers had different preferences. The filling could be made thicker, but only at a higher cost which would be reflected in lower profit or higher prices. New advertising could be developed and more coupons utilized.

The macro-flow model was used to assess the potential impact of these improvements. For example, the new positioning, variety, and advertising were simulated by moving the frequency of purchase distribution to a point half-way between the observed levels and the original premium cake mix distribution. Coupons increased the trial intent and thicker filling increased the repeat rates. In all, over 500 combinations of marketing strategies were tried. The best result was a 3 percent long-run share and $500,000 in total profit.

In this case, the managerial diagnostics could increase sales by 50 percent and improve profit by $2.5 million ($500,000 gain rather than a $2 million loss), but it was not enough. The project was terminated. Although the pastry mix had high trial appeal to innovators, its appeal was not wide enough, and the frequency of use was not sufficient to pay back the initial marketing investment.

The brand manager presented these results to the President, V.P. of Marketing, and V.P. of New Products. Although they were not pleased by the result, they felt that without the brand manager's careful work, the firm would have gone national and lost $10,000,000 and some good will with the trade. The brand manager was promoted and put in charge of the first major effort to develop soy-bean-based products for consumer and institution markets.

GO/NO GO ANALYSIS

Not all decisions are as clear-cut as the NO decision for Heritage Pastry. Many times profit is good, but not outstanding, and a balancing of risk versus return must be made relative to the goals of the firm. In this section, we review some analytic approaches to this problem and discuss the emotional reactions that may be involved in such decisions.

Decision Frontier Revisited

In Chapter 13 we introduced the concept of a decision frontier (Figure 13.1) Test market analysis provided profit forecasts which reduce risk and managerial diagnostics which increase expected benefit. After test market, we must face the managerial decision of whether to go to full-scale launch (GO), drop the product from further consideration (NO GO), or collect further data and attempt to improve the product (ON). We can represent these decisions on the same diagram as the decision frontier (see Figure 15.15). Note that if there is no uncertainty in the forecast, there is one level of minimum expected benefit (*M*) to divide the GO and NO areas. There is no ON alternative since the uncertainty is zero and more information cannot change it. As risk increases, a greater expected benefit is needed for a GO decision. This is the decision frontier. We have added a lower line which defines the boundary between the NO GO and ON decision areas. It is usually downward sloping as shown, but may be horizontal, or even upward sloping. As shown, this line indicates that as risk increases the organization becomes more cautious about dropping a product from further analysis.

If a test market was run, we can assume the product was in the ON region prior to test market. Figure 15.15 shows three possible outcomes. In outcome 1 the test market has produced better forecasts (less uncertainty) and has identified sufficient improvements in the marketing mix to move the product past the decision frontier. Outcome 1 would mean a decision to proceed to full-scale launch. In outcome 2, the test market has reduced uncertainty and improved the marketing mix, but there is still too much risk

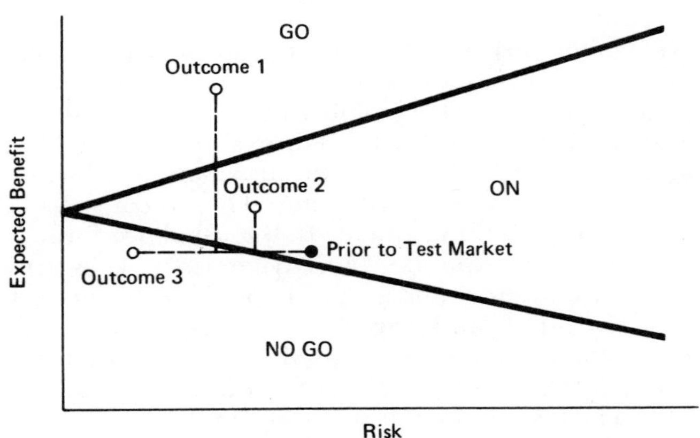

Figure 15.15. *Schematic Representation of GO/NO GO Analysis*

involved in launching the product. Outcome 2 would mean a consideration of further analysis, but not a GO decision. Finally, in outcome 3, once uncertainty is reduced it is clear the product should be dropped from further consideration.

Figure 15.15 gives the intuitive basis for the GO and NO GO decisions. We can quantify risk and expected benefit to further support the managerial decision. Some managers may wish to use these venture analysis techniques, others may compute expected benefit and judgmentally estimate risk. In either case the GO/ON/NO GO regions provide a useful decision tool.

Quantifying Risk and Expected Benefit

One way to represent expected benefit is by the expected level of profit discounted by the organization's target rate of return (Urban, 1968). Risk is measured by the standard deviation of the discounted profit. Straight line decision frontiers represent

- GO if the probability of achieving the target ROI is greater than, or equal to, some cutoff level, X, and
- NO GO if the probability of achieving the target ROI is less than, or equal to, another cutoff level, Y.

The slope of the decision frontiers are higher if the probabilities are greater. For example, the straight-line decision frontier may represent the decision criteria of, "We want a 90 percent chance of achieving a rate of 15 percent before we will GO national," or the decision criteria of, "We want a 75 percent probability of making at least $5 million in profit in the first three years." Mathematical formulae are given in Appendix 15.4. Other measures such as market share or sales which are used to quantify expected benefit produce similar results. Decision analysis with risk averse utility functions produces decision frontiers which may or may not be straight lines.

The most widely accepted method of measuring risk is venture analysis (Hertz, 1964). The underlying concept is to use the test market analyses to estimate probability distributions for each demand and cost factor. Then Monte Carlo simulation is used to calculate the total expected sales and profits and the distributions around them. For example, Figure 15.16 shows a simple example of the critical factors and distributions which describe the probability of alternative events occurring. The outcome is the expected sales, profit, and return on investment. In the usual new product case, the simulation is much more complex. The dynamics of sales growth can be described by a distribution around an "S" shaped growth curve. The costs of ingredients and the timing and magnitude of competitive response may be uncertain. Different distributions are used for different marketing strategies. For example, the expected value of market share may be five share points higher at a 50 percent higher spending level.

These complexities can be captured in computerized venture models. Pessemier (1977) has developed a venture model especially for new product

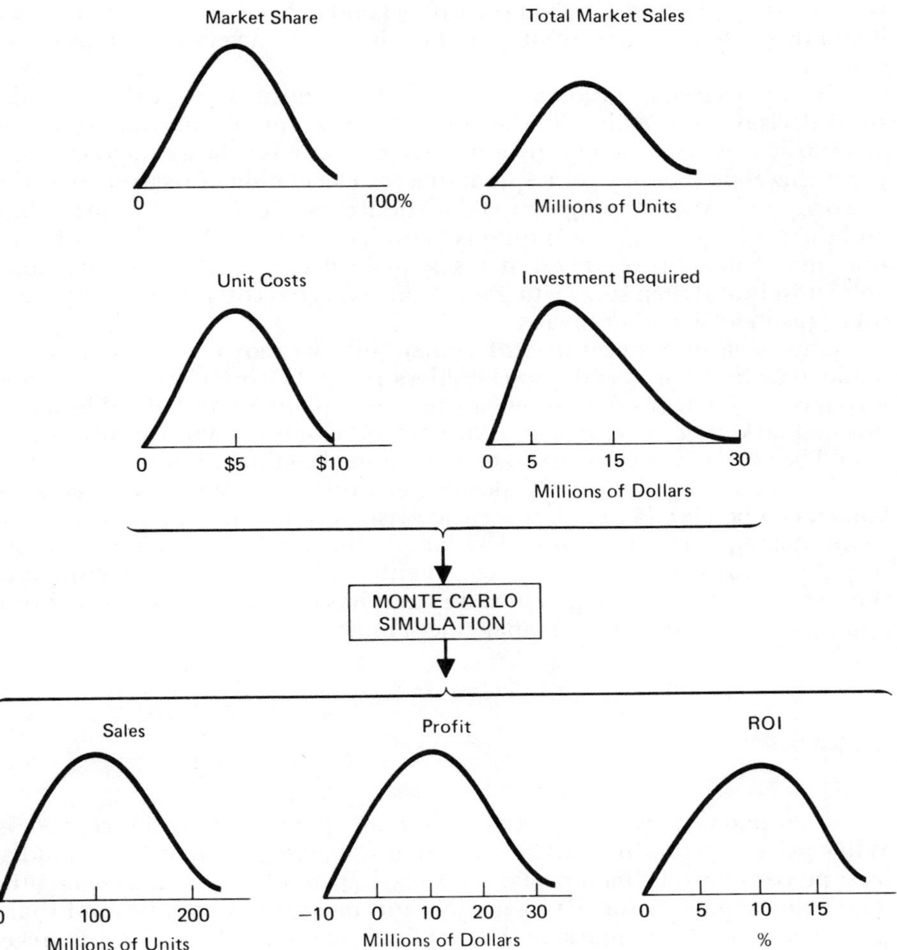

Figure 15.16. *Venture Analysis — Simple Example*

investment decisions which can be used to analyze the best price level and the GO/NO GO decision. In these cases a decision tree is used to define the alternatives (e.g., various prices) and a Monte Carlo simulation is used to calculate the probabilities of various profit outcomes for each strategy. (For other examples of venture analysis, see Beattie, 1969, and Urban, 1968.)

Managing Enthusiasm

While rational and analytical methods are available, the GO/NO decision environment is usually charged with emotion. Some people have invested much time and effort in the project and believe in it. Others may see it as a threat to their organizational domain or take a very conservative approach.

In the extreme, rational methods may be misused to achieve the desired decision. For example, the profit forecast for the product could be arbitrarily raised by the new product advocates. In one large chemical company, this abuse became so frequent that the calculation of risk/return ratios was dropped. Another abuse is to fill in the acceptable bottom line profit first and then make enough assumptions to justify it. For example, a transportation service made revenue estimates by first calculating a politically acceptable subsidy and then subtracting this from expected costs. This revenue was then "justified" by other means.

Enthusiasm is necessary, of course, but a rational forecast must be made in order to protect the stockholders and public interests. One method is to have the new product sponsor's forecast and analysis audited by a corporate marketing or financial service group. Then substantive differences would be resolved before top management made a final commitment.

With such resolution, the positive emotion necessary for a successful launch can be maintained, but accurate risks and returns can be presented to top management for a final decision. At this level, the risk/return input must be integrated with a final assessment of the product's contribution to the firm's long-run strategy objectives and its compatibility to the political and social environment top management must face.

SUMMARY

Test markets are an important component in the new product process. While not every product will be taken to test market, test markets should at least be considered. If a test market is used, it should be used intelligently. The proven state of the art of models and measures is well beyond share projections from test markets that replicate national plans. A test market should not only forecast sales, but also generate response data to maximize profits. Data support through experimentation and statistical analysis should be carefully designed to supply the response estimates necessary to track test market performance, adjust for differences between test market and national markets, and maximize profits. Both risk and return should be included in the GO/NO GO decision and procedures should be instituted to prevent biased estimates.

If the product is successfully tested by analysts and accepted by management, a GO decision will be made. The next chapter discusses the issues underlying the launch of a new product. If a NO GO decision is made, the firm will abort the product and allocate its attention to other projects in the design and testing phase, or initiate new projects.

REVIEW QUESTIONS

15.1 Should new products always be taken to test market? Why or why not?

15.2 Is it ever possible that the best decision will be to launch the new product despite a *bad* test market? If so, under what conditions?

15.3 When management wants to use a decision analysis for the decision as to whether or not to test market, they must assign subjective probabilities to the occurrence of different events. For example, they must quantify the probability the product will be successful. When forming subjective probabilities, the numbers generated must obey different rules. For example, they must all be greater than zero and less than one. Suppose that the probability of the product being successful (as evaluated before test market) is identical to the probability that the test market will be successful. Under this special situation what is the relationship between the following probabilities: (1) the probability of national success given a bad test market, (2) the probability of national success given a good test market, and (3) the before-test-market probability of national success?

15.4 Exotic Games Inc. is launching a new form of video game. To help study its adoption a diffusion of innovation approach was adopted. Innovators were found to be young, educated adults living in large cities. Given this information, what implications does it have for:

a. Media Selection
b. Retail Distribution
c. Packaging

15.5 A company is doing a decision analysis to determine whether to adopt a GO or NO GO decision. After quantifying test market information they find the probability of gaining $2,000,000 is 0.5 and the probability of losing $1,000,000 is also 0.5. The company decides to GO. Why could another company facing this same problem decide not to GO?

15.6 What diagnostic information should a manager expect from test market analysis? How can this information be used to improve the new product and marketing mix?

15.7 The following table shows consumer purchases for a consumer panel consisting of eight individuals. Time periods are shown from the first week after the launch of a new brand through the tenth week of the test market. C indicates purchases of a competitive brand; N indicates purchases of a new brand.

Assuming this sample panel is completely representative of the market, forecast the long term market share for the new brand.

Week After Launch	*1*	*2*	*3*	*4*	*5*	*6*	*7*	*8*	*9*	*10*
Consumer										
1	C	N	N	N	N	N	N	N	N	N
2	C	N	N	N	C	N	N	N	C	N
3	C	C	C	C	C	C	C	C	C	C
4	C	C	C	C	C	C	C	C	C	C
5	C	C	N	N	C	N	N	C	N	C
6	C	C	C	C	C	C	C	C	C	C
7	C	C	C	C	C	C	C	C	C	C
8	C	C	C	C	N	C	N	C	N	C

15.8 Mr. Labas, the vice president of Growth and Development at Tingus Industries, does not run test markets. When asked to explain this Mr. Labas said, "Why run a test market? Even if the results are bad there is so much enthusiasm after working so hard on the new product that we will launch the new product anyway." Comment on this strategy.

15.9 What are the advantages and disadvantages of panel data when monitoring test markets?

15.10 Modify the Blattberg and Golanty recursive model to include other marketing mix variables besides advertising.

Chapter 15 Appendices

15.1 Calculation of Probabilities for Decision Trees

From pretest market or from prior experience, we calculate the unconditional probabilities of success and failure, p(success) and p(failure). We often know the capabilities of a test market and can express this as the conditional probability of observing a good (or bad) result given the true state of the world. That is, we know p(good result | product is a success), p(bad result | product is a success), p(good result | product is a failure), and p(bad result | product is a failure). To use the decision tree, we must use these probabilities to compute the unconditional probabilities for test market outcomes, p(good result) and p(bad result), and the conditional probabilities for success (failure) given test market outcome, for example, p(product is a success | good test market result). To do these calculations, we use Bayes' theorem. We first switch to more general notation.

Let S_i = the state of the world, e.g., S_1 = success and S_2 = failure

ϕ_j = the outcome of the test, e.g., ϕ_1 = good outcome and ϕ_2 = bad outcome

$P(S_i)$ = prior probability the state of the world is S_i

$P(\phi_j \mid S_i)$ = conditional probability of outcome ϕ_j given the state of the world is S_i

$P(\phi_j)$ = unconditional probability of outcome ϕ_j

$P(S_i \mid \phi_j)$ = conditional probability that the state of the world is S_i given that the test outcome is ϕ_j

To compute $P(\phi_j)$, we simply use the rules of conditional probability:

$$\textbf{(15.1A-1)} \qquad P(\phi_j) = \sum_i P(\phi_j \mid S_i)P(S_i)$$

To compute $P(S_i \mid \phi_j)$, we use Bayes' theorem:

$$\textbf{(15.1A-2)} \qquad P(S_i \mid \phi_j) = \frac{P(\phi_j \mid S_i)\,P(S_i)}{P(\phi_j)}$$

The probabilities $P(\phi_j)$ and $P(S_i \mid \phi_j)$ are then used in the decision tree. Note that we can use this analysis for any number of outcomes, ϕ_j, and states of the world, S_i.

To compute the probabilities in Figure 15.1, we assumed that $P(S_i) = 0.5$, $P(\phi_1 \mid S_1) = 0.9$, and $P(\phi_2 \mid S_2) = 0.9$. $P(\phi_2 \mid S_1)$ and $P(\phi_1 \mid S_2)$ both equal 0.1 since $\sum_j P(\phi_j \mid S_i) = 1$ by definition. Similarly, $P(S_2) = 1 - P(S_1)$. Using these values we get:

$$\begin{aligned} P(\phi_1) &= P(\phi_1 \mid S_1)P(S_1) + P(\phi_1 \mid S_2)P(S_2) \\ &= (0.9)\,(0.5) + (0.1)\,(0.5) = 0.5 \\ P(S_1 \mid \phi_1) &= \frac{P(\phi_1 \mid S_1)P(S_1)}{P(\phi_1)} \\ &= \frac{(0.9)\,(0.5)}{(0.5)} = 0.9 \end{aligned}$$

In this example, the resulting probabilities are much like the input probabilities, but in general this is not the case. You may wish to try $P(S_1) = 0.8$ or $P(\phi_1 \mid S_1) \neq P(\phi_2 \mid S_2)$.

15.2 Calculation of Flows in Macro-Flow Diagrams

To understand the macro-flow model, it is best if you understand how to go forward. That is, you should be able to calculate the number of people in each state for the next T periods given the flow rates and the initial number of people in each state. In many test market questionnaires, you determine only the number of people in each state. You must use this information to calculate the flows.

Suppose there is an m-state model underlying the consumer response process. Call these states $S_1, S_2, \ldots, S_m$. Suppose we measure the number of people in each state for the first T periods. We want to calculate the flows, f_{ijt}, from state S_i to state S_j in the tth period.

First suppose that f_{ijt} does not depend upon t. Let $F = \{f_{ij}\}$ be the $(m \times m)$ matrix of flows. Let n_{it} be the number of people in state S_i at time t, and let N_t be the $(1 \times m)$ vector $\{n_{1t}, n_{2t}, \ldots, n_{mt}\}$. Let N_0 be the number of people initially in each state at time $t = 0$. Then the state transitions from $t - 1$ to t is given by:

(15.2A-1) $N_t = N_{t-1}F$

Or in general

15.2A-2 $N_t = N_0(F)^t$

Where $(F)^t$ means F multiplied by itself t times. To calculate F from $N_1, N_2, \ldots, N_t$, we simply solve the simultaneous equations $N_t = N_{t-1}F$ for $t = 1$ to T. Each set of equations gives us $m - 1$ non-redundant equations. Thus, in total we have $(m - 1)T$ equations. If all the elements of F were non-zero, then we would need $m(m - 1)$ equations since

$$f_{ii} \neq 1 - \sum_{j \neq i} f_{ij}$$

This means m time periods are needed. This is usually too long. Thus, it is necessary to use behavioral theory (e.g., awareness → intent → search → trial → repeat) to limit the number of potential non-zero flows to some reasonable number. For example, in

awareness → intent → search → trial → repeat

there are only six flows with five states versus a possible twenty flows.

If f_{ij} depends upon t, then we define $F_t = \{f_{ijt}\}$ and equation (15.2A-1) becomes $N_t = N_{t-1}F_t$. In this case, we must have less than m non-zero flows or limit the nonstationarity to a relatively few flows.

In either case, if there are more observations than there are non-zero flows, use linear regression to estimate F or F_t.

Those readers familiar with Markov theory will recognize equations (15.2A-1) and (15.2A-2) as Markov equations when F is a stochastic matrix. (That is, when $\sum_j f_{ij} = 1$.) In this case, all the results of Markov theory can be used and the equilibrium market is obtained by solving the following equation for the long-run number of people in each state, π_i. Let $\Pi = \{\pi_1, \pi_2, \ldots, \pi_m\}$ such that $\sum_i \pi_i =$ total people. Then:

(15.2A-3) $\Pi = \Pi F$

If the process is non-stationary, use projected equilibrium flows.

Table 15.2A-1 Number of People in Each State for a Three-State, Macro-Flow Process

Period	S_1	S_2	S_3
$t = 1$	10,000	0	0
$t = 2$	9,000	1,000	0
$t = 3$	8,100	1,300	600
$t = 4$	7,290	1,730	980

Consider the example in Table 15.2A-1. Suppose we know that the process is given by $S_1 \rightarrow S_2 \leftrightarrows S_3$, that is, only f_{12}, f_{23}, f_{32}, and f_{ii} are non-zero. Using equation (15.2A-1) we get:

$$\left.\begin{aligned} n_{12} &= 9{,}000 = 10{,}000(1 - f_{12}) \\ n_{22} &= 1{,}000 = 10{,}000 f_{12} + 0 \cdot (1 - f_{23}) + 0 \cdot f_{32} \\ n_{23} &= 0 = 0 \cdot f_{23} + 0 \cdot (1 - f_{32}) \end{aligned}\right\} \rightarrow f_{12} = 0.10$$

$$\left.\begin{aligned} n_{13} &= 8{,}100 = 9{,}000(1 - 0.1) \\ n_{32} &= 1{,}300 = 9{,}000(0.1) + 1{,}000(1 - f_{23}) + 0 \cdot f_{32} \\ n_{33} &= 600 = 1{,}000 \cdot f_{23} + 0 \cdot (1 - f_{32}) \end{aligned}\right\} \rightarrow f_{23} = 0.60$$

$$\left.\begin{aligned} n_{14} &= 7{,}290 = 8{,}100(1 - 0.1) \\ n_{24} &= 1{,}730 = 8{,}100(0.1) + 1{,}300(1 - 0.6) + 600 \cdot f_{32} \\ n_{34} &= 980 = 1{,}300(0.6) + 600(1 - f_{32}) \end{aligned}\right\} \rightarrow f_{32} = 0.67$$

Using equation (15.2A-3), we calculate equilibrium.

$$\begin{aligned} \pi_1 &= 0.9\pi_1 \\ \pi_2 &= 0.1\pi_1 + 0.4\pi_2 + 0.67\pi_3 \\ \pi_3 &= 0.6\pi_2 + 0.33\pi_3 \\ 10{,}000 &= \pi_1 + \pi_2 + \pi_3 \end{aligned}$$

which gives $\pi_1 = 0$, $\pi_2 = 5{,}263$, and $\pi_3 = 4{,}737$.

An alternative to the equilibrium Markov calculation is a direct computer simulation that is conducted until equilibrium is observed.

15.3 Macro-Flow Example

Grandma's Golden Goodies are being test marketed. They have been out now for five months and we have sufficient data to attempt to project sales for month ten. After studying the market, you have come up with a model that represents the behavioral process of Goodies' consumers. The model is as follows:

1. There are 100,000 Goodie consumers in the target market.
2. Sixty percent of the consumers buy Goodies every month.
3. The other 40 percent can resist temptation long enough so that they only buy in every other month. (Half of these consumers buy in January, March, May, July, September and November; the other half buy in February, April, June, August, October, and December.)
4. The advertising agency has promised that in each of the first 5 months, 20 percent of those unaware will become aware. This awareness rate will drop to 10 percent per month for months 6 to 10.
5. In each month, ⅔ of those aware will try the Golden Goodies.
6. After trying the Golden Goodies, consumers can be classified as either loyal to Grandma or disloyal. Grandma loyalists buy her Goodies 80 percent of the time and the disloyal buy 20 percent of the time. The probability of a consumer being loyal after the first try is 50–50. Once loyal, one does not become disloyal and vice versa.
7. In any given month, 10 percent who buy Grandma's Goodies become satiated and return to the unaware state. (They can later be reached by the advertising.)

Part I. First Set up Flow Diagram

1. Market size: There are 100,000 consumers in the category, not necessarily Grandma's customers.
2. Frequency categories:
   Heavy users: $0.6 \times 100{,}000 = 60{,}000$
   Light users: odd months $0.4(\frac{1}{2}) \times 100{,}000 = 20{,}000$
   even months $0.4(\frac{1}{2}) \times 100{,}000 = 20{,}000$
   The odd/even breakout must be explicit because advertising and hence flow to awareness depends upon month. Thus, there are three flow charts, one for each frequency of use category. We continue only for heavy users, but remember total flows and total usage come from treating all three categories independently—then summing total flows and usage.
3. Flow from [Start] to [Aware] depends on advertising.

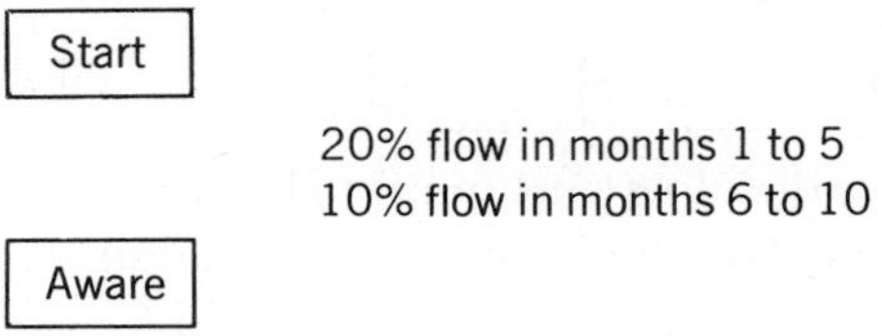

[Start]

20% flow in months 1 to 5
10% flow in months 6 to 10

[Aware]

4. Flow *out* of aware is ⅔. We do not yet know where they will go.
5. Since ⅔ of the aware state try, 10 percent of them become satiated; thus we know ⅔ (0.1) = 0.07 flow

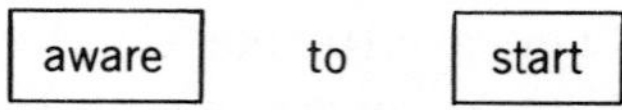

We must keep satiation effect in mind for later analysis.

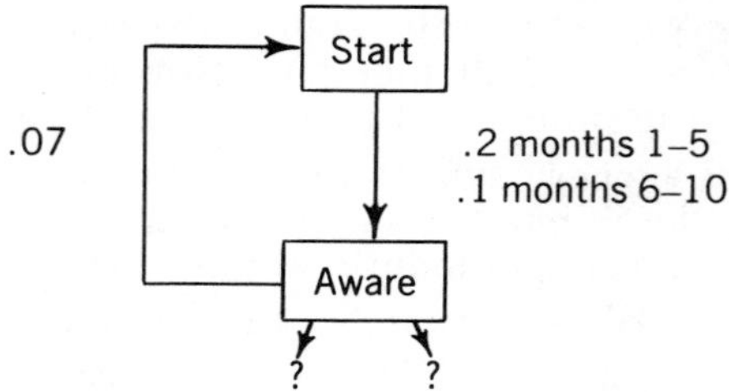

6. Those who try from the aware state flow either to loyal, disloyal, or start. Since 0.07 out of a total 0.67 flow to start, this leaves 0.6 to flow to either loyal or disloyal.

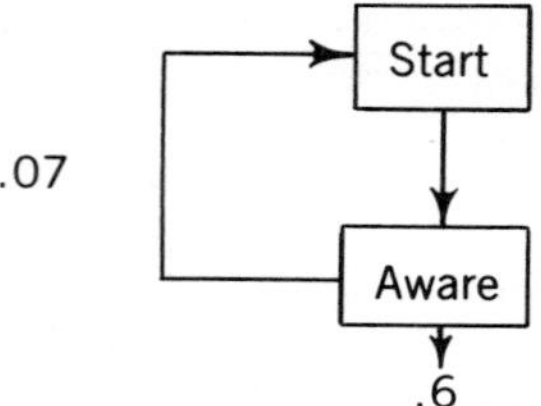

But we know these are split 50–50 between loyal and disloyal. Thus

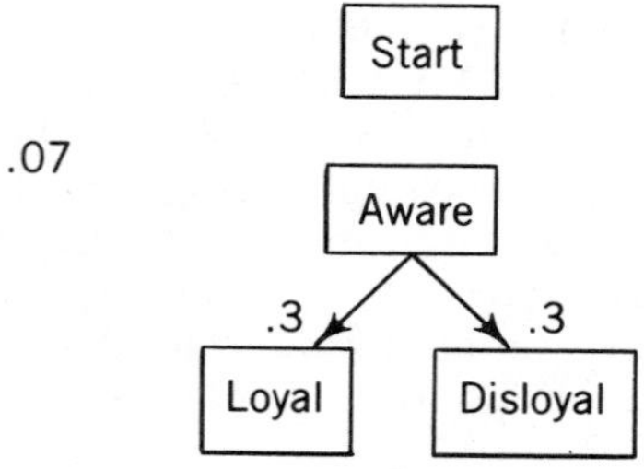

7. Once loyal . . . means no one flows

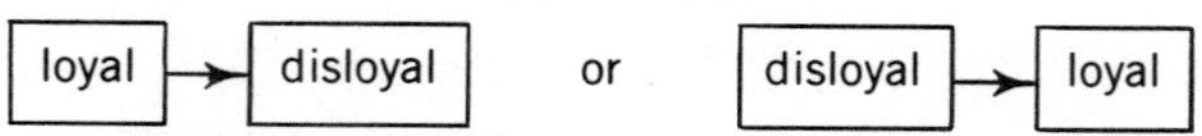

But we do allow consumers to become satiated.

8. We calculate the flows due to satiation. From loyal: Usage rate is 0.8,

and 0.1 of the users become satiated; thus in any period .08 become satiated and flow

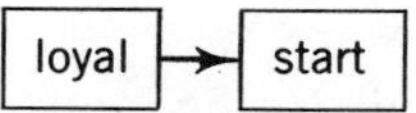

From disloyal: Usage rate is 0.2, and 0.1 become satiated; thus 0.02

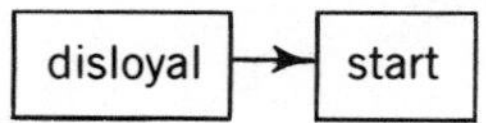

Thus

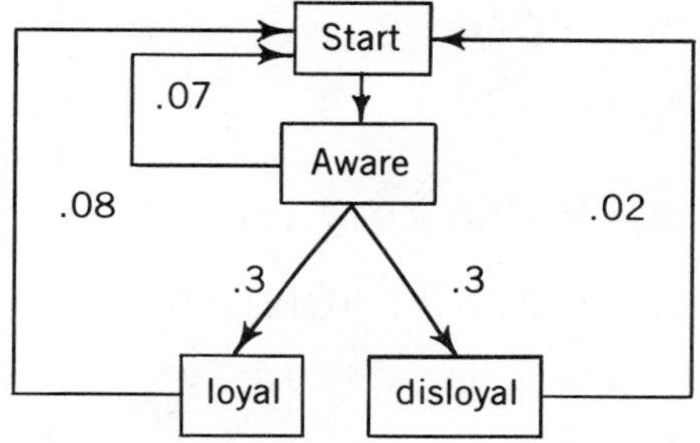

We want to explicitly model the usage, i.e., the flows

Thus loyal: 0.8 use the product and 0.9 do not become satiated. Since no one flows loyal to disloyal, this gives 0.8(0.9) = 0.72 flowing loyal to loyal.

Similarly (0.2) (0.9) = 0.18 flow disloyal to disloyal

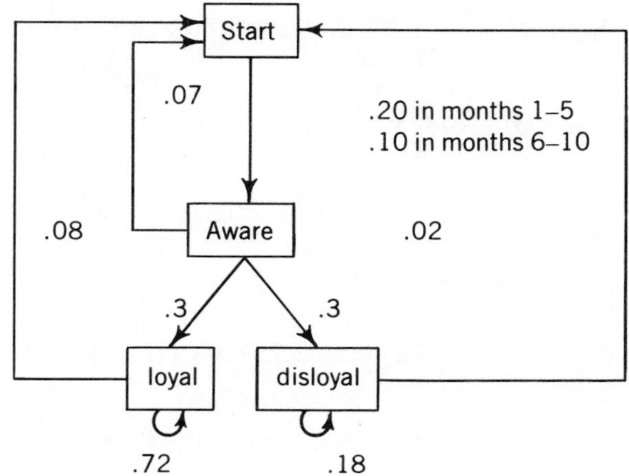

Part II. Calculate the Number of People in Each State at Each Time Period

We do this only for heavy users; it must be done for both heavy and light users. (U = unaware or starting state, A = aware, L = loyal, D = disloyal. Calculations are for thousands of consumers.)

Month 0: $U = 60$ in **Start**

Month 1: $U = 60 - .2(\overset{U}{60}) = 48$

$A = 0.2(\overset{U}{60}) = 12$

Month 2: $U = \overset{U}{48} - 0.2(\overset{U}{48}) + 0.07(\overset{A}{12}) = 39.2$

$A = \overset{A}{12} + 0.2(\overset{U}{48}) - 0.07(\overset{A}{12}) - 0.3(\overset{A}{12}) - 0.3(\overset{A}{12}) = 13.6$

flow rate and A → D (pointing to 0.3); *number of people in A at end of month 1* (pointing to 12)

$L = 0.3(\overset{A}{12}) = 3.6$

$D = 0.3(\overset{A}{12}) = 3.6$

flow rate in A → D (pointing to 0.3); *number of people in A at end of month 1* (pointing to 12)

Month 3: $U = \overset{U}{39.2} - 0.2(\overset{U}{39.2}) + 0.07(\overset{A}{13.6}) + 0.08(\overset{L}{3.6}) + 0.02(\overset{D}{3.6})$
$= 32.7$

$A = \overset{A}{13.6} + 0.2(\overset{U}{39.2}) - 0.07(\overset{A}{13.6}) - 0.3(\overset{A}{13.6}) - 0.3(\overset{A}{13.6})$
$= 12.3$

$L = \overset{L}{3.6} - 0.08(\overset{L}{3.6}) + 0.3(\overset{A}{13.6})$

$D = \overset{D}{3.6} - 0.02(\overset{D}{3.6}) + 0.3(\overset{A}{13.6})$

Month 4: $U = \overset{U}{32.7} - 0.2(\overset{U}{32.7}) + 0.07(\overset{A}{12.3}) + 0.08(\overset{L}{7.4}) + 0.02(\overset{D}{7.6})$
$= 27.8$

$A = \overset{A}{12.3} + 0.2(\overset{U}{32.7}) - 0.07(\overset{A}{12.3}) - 0.3(\overset{A}{12.3}) - 0.3(\overset{A}{12.3})$
$= 10.6$

$$\overset{L}{L} = \overset{L}{7.4} - 0.08(\overset{A}{7.4}) + 0.3(12.3) = 10.5$$

$$\overset{D}{D} = \overset{D}{7.6} - 0.02(\overset{A}{7.6}) + 0.3(12.3) = 11.1$$

etc.

Part III. Calculate Purchases

We do this for heavy users only; light users are left as excercise. Simply count the flows on all branches of the macro-flow diagram except the flow from start to aware.

Month	*Flows*		*Purchases (000's) by Heavy Users*	
Month 0	0			
Month 1	0			
Month 2	0.07(12) 0.3(12) 0.3(12)	$\overset{A}{0.67(12)}$	= 8 (Trial)	8.0
Month 3	0.07(13.6) 0.3(13.6) 0.3(13.6)	$\overset{A}{0.67(12)}$	= 8.08 (Trial)	
	0.08(3.6) 0.72(3.6)	$\overset{L}{0.8(3.6)}$	= 2.9 (Loyal Buys)	11.6
	0.02(3.6) 0.18(3.6)	$\overset{D}{0.2(3.6)}$	= 0.7 (Disloyal Buys)	
	purchase probability		*number of people in category*	
Month 4	0.0712.3) 0.3(12.3) 0.3(12.3)	$\overset{A}{0.67(12.3)}$	= 8.2 (Trial)	
	0.08(7.4) 0.72(7.4)	$\overset{L}{0.8(7.4)}$	= 6.0 (Loyal Buys)	15.7
	0.02(7.6) 0.18(7.6)	$\overset{D}{0.2(7.6)}$	= 1.5 (Disloyal Buys)	
etc.				

15.4 Calculation of Decision Boundaries

The decision frontier is a straight line if a GO decision is made when the probability of achieving a target, M is greater than, or equal to, some cutoff level, X. If outcomes are normally distributed with variance, σ^2, then the probability that a given plan exceeds (or equals) the target is given by the cumulative normal distribution. Let us suppose that a plan has an expected benefit, EB, and a risk (standard deviation), σ. Then the GO region is given by

$$\textbf{(15.4A-1)} \qquad \varphi\left(\frac{\text{EB} - M}{\sigma}\right) \geq X$$

where φ is the cumulative normal distribution. Thus equation (15.4A-1) translates to:

$$\textbf{(15.4A-2)} \qquad \frac{(\text{EB} - M)}{\sigma} \geq \varphi^{-1}(X)$$

where $\varphi^{-1}(X)$ is the value for which the cumulative normal distribution gives a probability of X. Solving algebraically yields

$$\textbf{(15.4A-3)} \qquad \text{EB} - \varphi^{-1}(X)\sigma \geq M$$

which gives a straight line in {EB, σ} space such as that shown in Figure 15.15.

Using similar analysis we can define the NO GO region as

$$\textbf{(15.4A-4)} \qquad \text{EB} + \varphi^{-1}(Y)\sigma \leq M$$

where Y is the cutoff probability of achieving the target. Similar analyses apply for other probability distributions.

Some useful values of $\varphi^{-1}(\cdot)$ are $\varphi^{-1}(.50) = 0.0$, $\varphi^{-1}(.60) = .25$, $\varphi^{-1}(0.70) = 0.53$, $\varphi^{-1}(0.80) = 0.84$, $\varphi^{-1}(0.90) = 1.28$, $\varphi^{-1}(0.95) = 1.65$, and $\varphi^{-1}(0.99) = 2.33$.

PART V

PRODUCT INTRODUCTION AND PROFIT MANAGEMENT

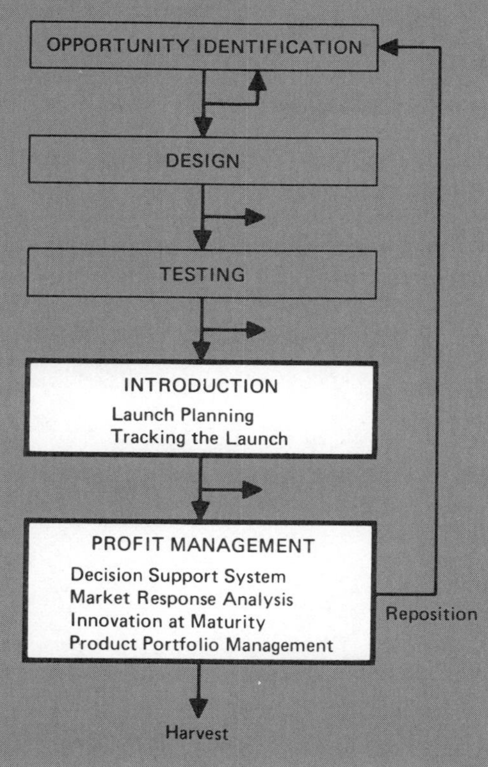

Chapter 16

Launching the Product

Once a GO decision is made, an organization must marshall its resources to plan and implement a full-scale introduction. At this point many of the strategic decisions will have been made and tested. Management has a physical product, advertising copy, a promotion plan, price, and a distribution plan. The key markets are known and the likelihood of competitive response has been gauged. The product will have a high probability of success since it has passed the testing phase criteria. The management task in full-scale introduction is to achieve the potential success through careful management of the launch.

Although successive testing and development has reduced the risks, the stakes are now higher. For example, in 1977 S. C. Johnson spent over $7 million on television and magazine advertising, along with a $7 million sampling campaign to introduce Agree Creme Rinse and Hair Conditioner. Then in 1978 they launched Agree shampoo with a $30 million campaign. During the same period Gillette planned to spend about $15 million to get buyers to try its Ultra Max shampoo; in the same year Gillette spent over $8 million to introduce Atra automatic tracking razors.

In durable goods, the costs of introduction include investment in production as well as in marketing. A new kitchen stove may require $10 million for production line set up, and a new model car may require substantially more. Industrial products do not require as much marketing expenditure, but may entail more in production. For example Boeing's new line of jets will require investment equal to more than the firm's net worth. While the magnitudes depend upon the product category, it is clear that by far the largest investment in new product development is in full-scale launch.

This means careful plans for the launch must be made and the introduction must be carefully controlled and tracked to assure that an adequate return on this investment will result.

LAUNCH PLANNING

In planning the launch an organization must coordinate the marketing mix and production. The required tasks can be outlined in a critical path analysis to support decisions ensuring that all components at the introduction are ready in time for the launch. This coordination and timing becomes particularly crucial if competitors also are developing a new product for the target category. In this section, we first discuss the coordination and timing necessary for launch and then indicate some of the organizational issues in the management of the new product launch.

Coordination of Marketing and Production

If the launch is to be successful, the product advertising, selling, promotion, and distribution strategy must be implemented effectively. The sales force is trained and motivated to allocate the planned effort. The channels of distribution are established and combined to stock, display, and service the product. Advertising media are purchased and ad copy manufactured. Samples and coupons are produced and mailed.

While marketing activities are significant, the major pre-launch effort is in production. Machines are purchased and set up in a mass production system. Materials are procured and inventoried. In some cases new plants must be built and staffed. Quality control standards are implemented and the production facility tuned to consistently produce the desired product specifications. If the quality is not assured, real risks of national failure will exist.

The marketing and production activities must be closely coordinated. One important aspect of this coordination is the timing of the manufacturing start-up. A start-up that is too early may create large, expensive inventories and result in product deterioration. If the start-up is too late, there may not be enough supply to meet the growing demand and large back orders. Insufficient supplies cause opportunities to be missed and cause good will to be lost with consumers and channel members. For example, Abernathy and Baloff (1972) report that before Douglas Aircraft Company merged with McDonnell Aircraft it encountered severe problems in delivering its DC-8 and DC-9 aircraft. Referring to this problem, *The Wall Street Journal* (1966) reported:

> The fact that the company has sailed into a financial squall . . . has roots in production problems that are causing serious delays in deliveries of jet liners. In one instance, at least, a big international carrier who . . . was alerted that the planes would be delayed, complains it had to go out and lease additional planes,

with the prospect of operating at a loss a service that was to have been highly profitable from the start.

In planning the start-up date, organizations must consider that full capacity will not be reached immediately. For example, Figure 16.1 illustrates what can happen if production and introduction are begun at the same time, but production grows slowly with experience. In this case, back orders will occur resulting in long waits and customer dissatisfaction. If production starts too early, large inventories will result.

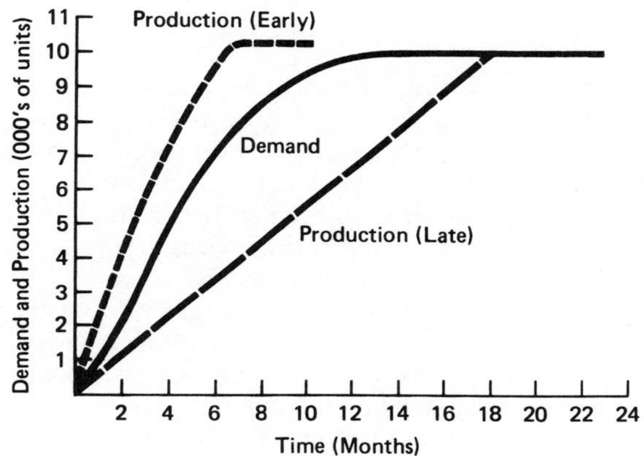

Figure 16.1. *Production Start-up and Demand Growth (adapted from Abernathy and Baloff, 1972, p. 31)*

To manage these problems a joint plan must be developed for marketing and production, which should include the relative timing of production, introduction, and the marketing mix. For example, marketing may delay advertising or promotion to avoid too much demand too soon, or manufacturing may produce a large inventory so that out-of-stock situations will not leave the company vulnerable to competitive products. This joint planning of the functions has the potential to result in greatly increased profits.

Joint planning can run into organizational snags. Marketing and production may each have "political muscle" which must be controlled to achieve an effective production plan. To overcome potential difficulties, Abernathy and Baloff (1972) suggest the plan should:

- focus on the performance profile over the introductory period, not merely offer a few target points that are to be met at the end or along the way;
- represent plans in units that are common to all functions involved (such as sales and net inventory levels), rather than in specialized functional measures;

- offer flexibility once the plan is implemented, by basing target performance on activity levels containing margins for adjustment and change;
- control performance with respect to the plan.

If these goals are achieved within the organizational environment, then management will have an effective programmed start-up over the introductory period.

Timing of Launch

Besides the relative timing of production and marketing, we must select the specific timing of the launch. Without competitive pressure, most organizations would proceed through introduction planning at a safe but steady pace. An overly cautious approach can delay entry and result in a missed opportunity. Usually pressures exist for a fast launch. Competitors may read your test market and begin their own crash program. You even may find they have just entered the market for your new product. In each case your organization may feel the need to begin its own crash program.

Such pressures for crash programs are difficult to resist, but should be resisted until they can be carefully considered. A crash program may be just the response the competitor wants if it forces you to make a key mistake that may compromise your position. On the other hand, you must take some risk to succeed. This is never an easy decision. The management task is to balance the expected gain and the risk to arrive at a timing for the launch.

The first, and most obvious, consideration is some quantification of the gains to be achieved through early introduction. Among the questions to be considered are: "How much harm can the competitor do if they enter soon after you?" "How firmly entrenched will they be if they enter before you?" To answer these questions, we suggest your organization evaluate the relative strength of the competitive product. If it is inferior, then you lose little by waiting. Usually there is an advantage of being first in a market, but this is not true if the first in does not have a good positioning. For example, General Foods was first with a freeze-dried coffee called Maxim, but was dominated by the second entry, Taster's Choice. Maxim required one-half a teaspoon per cup rather than the usual one teaspoon per cup. As a result many consumers brewed a bitter cup of coffee. Even after re-blending and moving to the usual one teaspoon per cup recipe, it has not matched Taster's Choice share. If your CBP is based on a superior product, do not rush to the market if it would compromise your positioning claims. A pretest market laboratory study of your competitor's product can be very useful in assessing the competitive threat.

While revenue may increase for an early launch that allows you to be first in the market, costs also increase. Crash programs usually result in inefficiencies which increase cost. Beyond a certain point costs increase rapidly for small gains in time.

Management also should consider the risks of early launch. One well-known food product manufacturer tried to beat a competitor with an early launch, but did so before lining up a variety of suppliers. The supplier of a key ingredient went on strike and the manufacturer had to find alternative suppliers at a high cost. Such risks are inherent in innovation, but should explicitly be considered before rushing the launch.

There are good reasons to launch early, but many organizations miss opportunities by launching too rapidly. The possible loss of revenue must be traded off against the increased costs of a crash program and greater risk of failure. Figure 16.2 shows a hypothetical set of time and cost curves. Costs, risks, and revenue increase, but the gain in revenue probably is not worth the risks and costs involved. Only if the revenue loss by not responding to a competitive threat with a crash effort was large, and the costs low, would it be worth the risk.

These tradeoffs are usually made judgmentally, but formal decision analysis (Keeney and Raiffa, 1976) could be applied. It is important to take a rational look at the tradeoffs rather than to panic and rush a premature launch.

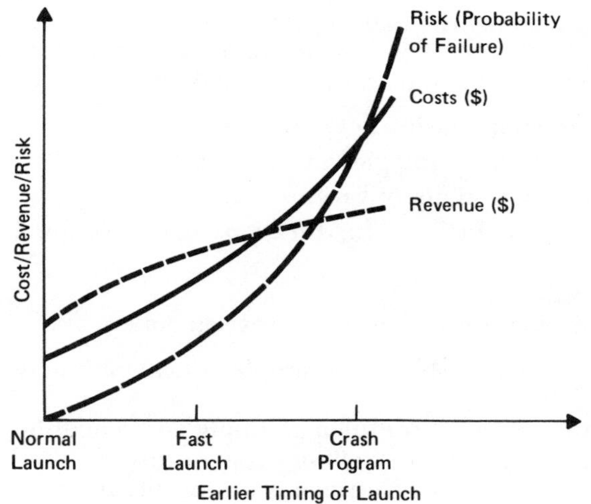

Figure 16.2. *Hypothetical Cost, Risk, and Revenue Curves of Launch Timing*

Critical Path Analysis

One method of managing the issues of launch timing and coordination is called critical path analysis. It structures the sequence of activities and identifies those that are most critical to the success of the timing; it also supplies standards to coordinate the various activities.

When a launch is planned, obviously a wide range of supporting activities must be carried out in the marketing and production areas. Some activities are independent and can proceed simultaneously. For example, the development of advertising copy does not depend upon the negotiations with raw material suppliers. But some activities are sequential. For example, an organization must develop the advertising copy before it begins introductory advertising, or it must begin to stock up on raw materials before it begins production.

To achieve the target dates for launch and to generate time-cost tradeoffs such as shown in Figure 16.2, management must enumerate and schedule all the details of the launch. This is best done by some form of critical path analysis (CPA) such as the program evaluation and review technique (PERT) or the critical path method (CPM).

While the details vary, the concept of CPA is simple. Related tasks are laid out in paths according to the order in which they are to be completed. Independent tasks are indicated by parallel paths. The time required to complete each task is determined; then these times are summed to give the total time for each path. The longest path gives the time required for launch and is called the "critical path." To launch the product early, resources must be directed at shortening the critical path.

According to Wong (1964) the advantages of critical path analysis for new product launch is that it provides information to:

- show the interrelationship between tasks,
- evaluate alternative strategies and approaches,
- pinpoint responsibilities of various task forces,
- check progress at intervening durations against original plans and objectives,
- forecast bottlenecks,
- replan and redesign a project with revised data.

For example, Figure 16.3 shows one simple critical path network for a new product launch.

Critical path analysis is useful in planning and scheduling, but it is also useful for monitoring and controlling progress toward the launch. In the PERT variation, management estimates optimistic, likely, and pessimistic times; decisions are based on both the average times and the variance in times. In the CPM variation, the optimal project duration is identified by trading off the cost of reducing the critical path versus the opportunity cost due to delays in product introduction. For more details, see Wong (1964).

Management of the Launch

As the product is moved toward launch, there may be a shift in management responsibility. In many cases, the design and testing phases will have been handled by a new product team consisting of specialists who can

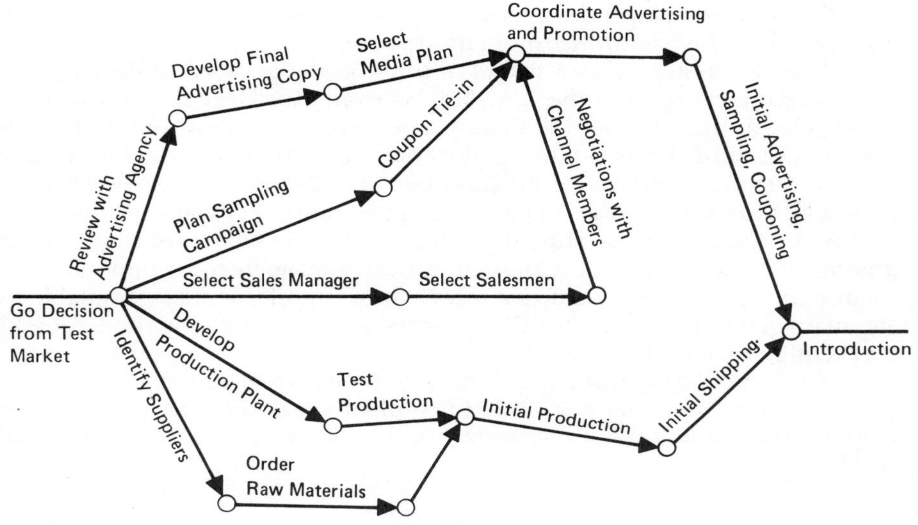

Figure 16.3. *Simple Critical Path Analysis Example*

manage and enhance creativity and who know how to gather and act upon consumer information. In launch planning, different skills are needed to plan the details of production and of the marketing mix. The launch demands a large amount of resources which could overwhelm projects in other phases of the new product program. Thus organizations face the difficult problem of how much control to shift from the new product developers to the product managers.

There are advantages of a shift to a person such as a brand manager. Brand managers are implementation specialists who can handle the management details of a launch. Such shifts free the new product team to devote their skills to other new product projects. But a shift is not without dangers. Without clear-cut responsibility, the product can fall into "limbo" with no one really committed to it. After all, the new product may not be viewed as the brand manager's project and he may lack commitment. Furthermore, the brand manager may not have the intimate knowledge of the market or of the CBP that is necessary to overcome the final hurdles of launch. Finally, the management of a product with millions of dollars in revenue may be viewed by the new product team as a reward for the earlier hard work and risks. They may seek this reward for themselves or may begrudge the reward being given to someone who has not had to bear the risks.

There is no one answer to the very difficult question of transition of management. In some organizations transition is most effective by granting control to the established product group. Others may wish to delay transition until the product has reached a mature state (1 to 2 years after launch).

In still others the new product venture team becomes the management team and stays with the product through its life cycle.

The best system would preserve the knowledge used to develop the product strategy, but add the new skills necessary for large scale implementation. One approach would be to have a new product person join the launch team under the direction of the product manager. Another would be to have the product manager and production personnel work on the new product team for the first 12 months of launch and then have them take over responsibility. In all cases, there should be close communication and cooperation among the new product development group and the introduction management team. Important diagnostic information should be made available to the management team so that they understand how and why the CBP and marketing mix have been developed.

The issues of organization of the new product will be considered more comprehensively in Chapter 18. Launch planning is only one aspect of the overall problem, but it is an important aspect of implementing a successful product.

TRACKING THE LAUNCH

Need to Monitor

Unexpected events will take place before and during the launch of your new product. Although the test market modeling and forecasting prepares you for the sales response and growth, events outside your control are likely to occur.

One set of such events are economic changes that may occur between test market and launch, or during the launch. For example, before the oil embargo of 1973 there was a desire for higher powered cars. Foreign and domestic manufacturers had developed cars based on the Wankel rotary engine. Based on the rising cost and decreasing availability of fuel, consumer interest shifted to better mileage. General Motors dropped its Wankel project after investing seven years and $100 million. The Monza body style which was designed for the Wankel was introduced with a conventional internal combustion engine. Toyo Kogyo (Mazda) did not recognize the impact, followed their initial forecast, achieved only a fraction of their forecast sales, and had thousands more production workers than they needed. Mazda responded with a redesign and in 1978 introduced a conventionally powered line based on a "racy yet moderately priced" appeal. Similar unforecasted shifts in the economy, in consumer tastes, or in foreign influence can cause dramatic differences between test and launch. It is important to monitor the environment between the GO decision and the national launch to be sure the environmental assumptions underlying the GO decision have not changed.

Consumer preferences may also change. For example, your CBP for a new analgesic may be "safe and effective" with an emphasis on "safe," but during the period prior to launch tastes have shifted in the direction of "effective." By recognizing and reacting to this change you can improve profit potential by repositioning the product to emphasize effectiveness and by generating new advertising copy. Perceived price barriers may rise. In one new product launch the initial price was fixed by what management perceived as a $1.00 barrier in that category. But inflation pushed competitive products past the barrier, opening up new opportunities for a premium priced product or for a larger size package.

Distribution channels may change. If a channel, such as a large retailer, is now carrying a product category formerly carried only by specialty stores, your display strategy may have to be changed.

During the launch, economic and distribution variables may or may not change, but you can be almost certain that competitors will respond to your introduction. These must be monitored and diagnosed so that appropriate actions can be taken. Competitive reactions are intense, but varied. It is not uncommon to see a competitor increase advertising expenditures substantially to counter a new product introduction. While it is doubtful that high spending levels can be maintained indefinitely, they can cause havoc in the launch plan. If the new product is a strong product, competitors may wage the battle by attempting to undermine trial. For example, to counter a strong sampling campaign by "Aim" toothpaste, Procter and Gamble undertook an advertising campaign which appealed to loyal "Crest" users and discouraged them from trying the unnamed sample brought by the mailman.

Other competitors may try to exercise power in the distribution channel by giving channel members incentives to emphasize their product rather than the new product. Such strategies are particularly effective if shelf space is limited in the retail outlet. All these occurrences emphasize the need to monitor the pre-launch and launch environment carefully. The lack of response to a change or an inappropriate response can undermine the profitability and perhaps the success of the product. Each change must be monitored and diagnosed. Then revisions must be made to maximize the product's profit.

In the next sections we outline concepts of a control system to support management of the launch and adaptive control procedures to assume an appropriate and timely response. We then describe a case of tracking the launch of a new health and beauty aid.

Control System

Market intelligence to identify profit opportunities and to formulate responses to unexpected changes is best obtained through an explicit control system. A basic control system is schematically illustrated in Figure 16.4. Plans are set and they determine forecasts which are compared to actual results to enable the diagnosis of problems, planning of response, and updating of forecasts.

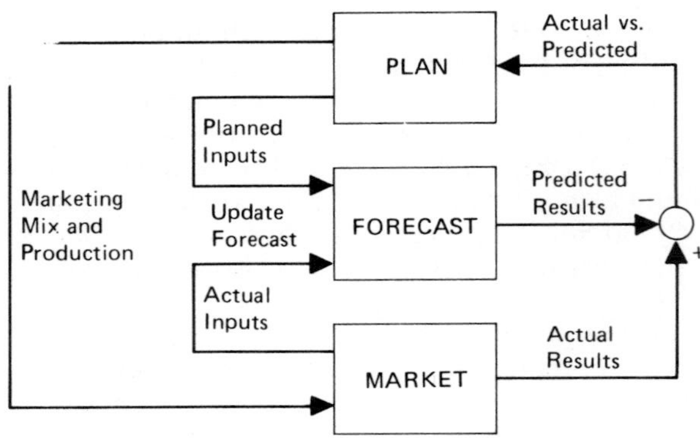

Figure 16.4. *Product Launch Information System*

The basic plan is set by the test or pretest experience. The model that was used to analyze the test (pretest) market provides forecasts of awareness, trial, repeat and, if possible, more detailed consumer response measures. The implementation of the product and marketing mix in the market influences the actual results. Since the marketing mix and production were planned based on this forecast, any deviation of consumer response from forecasts is a signal for analysis. The "actual" and "predicted" results are compared with respect to the behavioral process as well as overall sales. Based on this comparison new plans are developed and new forecasts based on planned changes. New market trends are produced for control of the remainder of the launch. This systematic comparison of actual and predicted results is called "tracking"—this activity continues throughout the launch with periodic updating of the plan and the forecast.

Consumer response data is important during launch as well as during test market. If a detailed analysis model was used during test market, it should continue to be used during launch. Suppose that a macro-flow model was used to forecast sales and suppose that a monthly awareness survey and store audit data are being collected during the launch to provide data for the model. By comparing actual vs. predicted flows through each and every state, management can quickly identify where and why forecast sales differ from actual. Actions can be taken to modify strategy. If a recursive model or if a statistical data projection model was used, each variable in the prediction equations should be remeasured and compared to its forecast value. In either case, the analyses and the strategy identification are performed using the same analytic tools introduced in Chapter 15.

It is also important to recognize that the plan is adaptive in the sense that it is continually updated by market response data. This type of updating makes the plan less susceptible to failures in execution. Any mistake is quickly identified through the feedback mechanism and rectified in the next period. The speed of updating is important, since actions must be taken quickly before the major funds to support the launch have been expended. Usually monthly control is exercised, but some firms track weekly results.

Figure 16.4 is a schematic that can readily be implemented through any of the test market models. At the end of this section we illustrate a macro-flow model of tracking with a case, but first we turn to a formal system for adaptive control through experimentation.

Adaptive Control and Experimentation

The system of control outlined in Figure 16.4 can be used based on market sales or market research measures. These are useful measures in monitoring and forecasting of the diffusion of innovation. However, it also can be implemented by additional response information gained from market experimentation during the launch, growth, or maturity phases of the life cycle. This information is used to improve the setting of price, advertising, and promotion levels.

The initial levels of the marketing mix were set based on an analysis of the tradeoffs in the cost of obtaining the sales versus the profit generated. For example, Figure 16.5 illustrates one market response function used to forecast sales from advertising expenditures. We use the general function $f_t(a)$ to describe response. If the response functions are accurate and if the environment has not changed dramatically, then actual sales should be close to predicted sales and the launch should proceed as planned. If, on the other hand, the response function shifts, the best level of expenditure will change. Thus it is important to monitor the market response functions throughout the product launch and life cycle.

Suppose we are interested in controlling the market response to advertising. One suggested way to measure this response might be to increase advertising in all markets by some amount, say 20 percent over forecast, and observe sales results. This may be undesirable for two reasons. First, the increase is a change from the forecast level which was previously identified

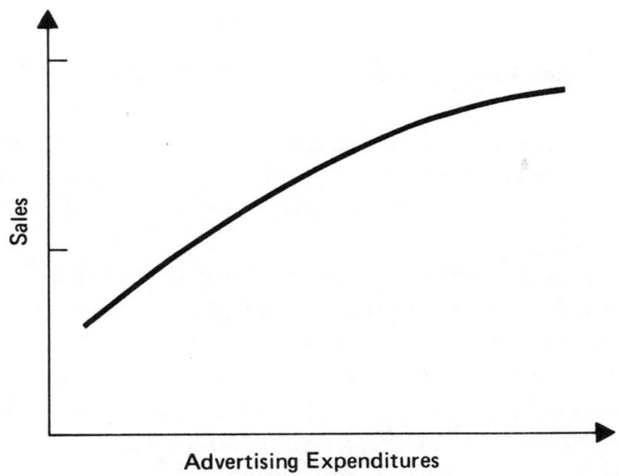

Figure 16.5. *Market Response Function: $f_t(a)$*

as the best strategy. It is likely that the advertising change will result in decreased profits, perhaps a larger decrease than the information is worth. Second, the market is not a controlled laboratory. Other marketing mix strategies may change, new competitive products may enter, or the basic environment may change. All these effects will influence sales and there may be no way to isolate the effect due to the change in advertising.

Both reasons argue that if experimentation is to be done, it should not be done in all submarkets at the same time. If advertising experimentation is done in only a fraction of the submarkets, then the non-experimental markets can act as control groups.

Little (1966) has developed a technique, called "adaptive control," to determine both the number of markets in which to experiment and the size (e.g., advertising change) of the experiment. His approach is illustrated in Figure 16.6. The advertising rate is varied by an amount (Δ). It is reduced from its initial level (a_0) to a lower level ($a_0 - \Delta/2$) in some markets (n) and is increased to a higher level ($a_0 + \Delta/2$) in others (n).

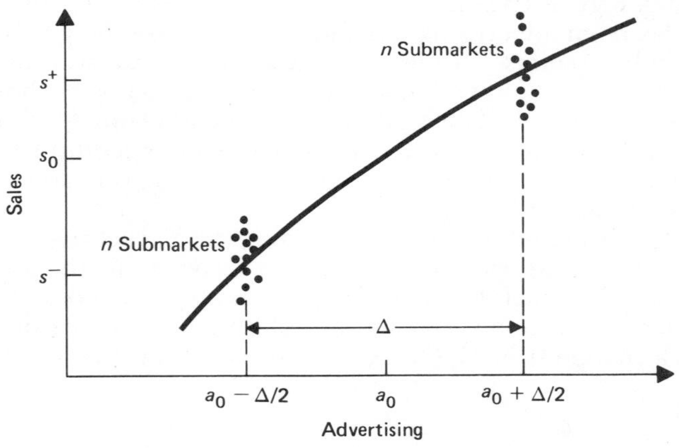

Figure 16.6. *Adaptive Control (Little, 1966, p. 1080)*

In any given time period, t, the best base level, $a_0(t)$, is a function of the current best estimate of market response function, $f_t(a)$, and the cost of advertising. That is, without experimentation the manager would select $a_0(t)$ to maximize profit.

Since advertising response, $f_t(a)$, is not known with certainty, the profit may not be optimal. There is some potential loss in profit due to this uncertainty. This "expected loss," is quantified and the value of sample information is computed. The manager then weighs the value of sample information against the expected loss in sales due to experimentation. Based on this comparison, the parameters of the experiment, n and Δ, are selected to maximize the expected profit in the next period, period $t + 1$.

The computations are straightforward, but tedious. (See Appendix 16.1 for the special case of $f_t(a)$ being a quadratic function.) After the parameters of the experiment (n and Δ) are calculated, the optimal advertising in period $t + 1$ is given by Little as:

$$\textbf{(16.1)} \qquad a_0(t+1) = a_0(t) + c(ms^+ - ms^- - \Delta)$$

where c is a constant (see Appendix 16.1) and m is the gross margin on sales; s^+ is the sales rate observed in the markets where advertising was increased, and s^- is the sales rate observed in the markets where advertising was decreased. The first term is the previous expenditure and the second term is the profit due to changing advertising by the amount Δ multiplied by a constant (c). This equation says advertising is revised proportionally to the change in profit. Thus, even without using the equations in Appendix 16.1, a manager can use adaptive control. One simply increases (or decreases) the advertising rate by an amount proportional to the observed net profit rate (or loss) due to the advertising change in the experimental markets.

Fortunately, the method of adaptive control is surprisingly robust and works well even if n and Δ are not chosen optimally. The important idea is that the periodic measurements serve to keep the manager in touch with the market and help the manager learn from experience in an organized way.

In many ways adaptive control works like a home thermostat, which "controls temperature under widely varying conditions of heat loss and does it without solving heat flow equations on a digital computer" (Little, 1966, p. 1094). Similarly, adaptive control measures potential profit gain (or loss) and responses by increasing (or decreasing) the advertising.

We have illustrated adaptive control for advertising response, but the same technique can be used for any element or combination of elements of the marketing mix. Similar equations can be developed for market variables other than advertising and as combinations in the marketing mix (Little, 1977). During the launch the models can be extended to a dynamic form (Pekelman and Tse, 1980), but the concept of using experimental information to improve the setting of market mix problems is the same.

Case: Health and Beauty Aid

In the previous sections we outlined the concepts of tracking and adaptive control based on experimentation. We now illustrate one method of using a control system to track a launch. In this case a macro-flow model (Sprinter MOD III; Urban, 1970), was used to analyze the test market and set norms for the product launch of a new beauty aid. The same model then provided the basis for tracking the launch. First, the series of unexpected events and actual management responses in the first nine months of the launch are reported. Then the launch is tracked with a model to demonstrate profit gain possible from careful control measures and analysis.

National Launch Tracking. After a 12-month test market, the product went national with a $5 million advertising and promotion campaign supported by missionary sales efforts. Within a few weeks of introduction, feedback from salesmen indicated that the product was "not moving." The causes of this problem were found by examining the results of the national awareness and usage questionnaires carried out four weeks after introduction. These surveys showed that the awareness rates were down 20 percent from the predicted value and that the trial rates for those who were aware were 10 percent below expectation. The reduction of the conditional trial rate was because the innovators nationally were not responding as rapidly as in the test cities. The 20 percent reduction in awareness was due in part to an error in translating the national advertising budget to the test-market cities. Too much advertising was inserted in the test market and the observed test levels were therefore artificially high. The remaining reduction in awareness seems to have been due to a low national response to the advertising. The firm responded to this information by doubling advertising.

At the beginning of the third month of national introduction, the major competitive firm unexpectedly introduced a brand to compete directly with the new product. They backed this introduction with a 50 percent increase in their advertising level. This new, competitive product advertising lowered trial rates and reduced the proportion of people who translated preference to intent to repurchase. These effects were monitored in the second national awareness survey. This survey was carried out ten weeks after introduction.

This three-city awareness survey also indicated some behavioral changes in addition to the effects of the competitor's new product. In particular, based on a comparison of the response levels in the cities, it was found that the awareness response function had shifted back to the level specified prior to introduction. The trial rates for the specific awareness classes also returned to their expected levels. This recovery was apparently due to the innovators being held out of the market by the initially low awareness levels and entering later than expected. The slow start of the product caused the innovators to spill over into the first five months rather than just the first three months, as had been observed in the test cities.

Six months after introduction, media audits showed the competitor had become very aggressive and had doubled his advertising relative to expectations. This new competitive rate was nearly equal to the total industry advertising in the previous year. The firm introducing the new product responded to this competitive activity with a continued high level of advertising in months 5, 6, and 7, but had to reduce spending in months 8, 9, and 10, since they had depleted the product's advertising budget. In periods 8, 9, and 10 the competitor also reduced his rates of advertising to his previous level. As a result of these changes the expected profit for the first three years was far below planned levels. A profit of $500,000 over two years was projected for the brand.

National Launch—Model Based Control

A macro-flow model had been built during the test market and early launch of the brand. It was parameterized for the brand based on test market data and used to track the national launch. Information collected during the market introduction was used to diagnose basic problems, update the model's parameters, and search for a best response to the new information and the diagnosed problems.

As indicated above, the first month's national awareness questionnaires indicated that the advertising response function was lower than expected and that the trial rates were below expectation. The model was utilized to examine alternate advertising and price responses. The best advertising strategy was to hold to the original plan. In contrast, the firm actually doubled advertising. The model indicated the firm's action would reduce profit by more than $200,000.

The second set of new information indicated that the competitor had introduced a new brand and increased advertising. At this same time, trial rates and the advertising response function had recovered to their expected levels. Since this was diagnosed as the late arrival of the innovators, the trial rates of periods 3, 4, 5 and 6 were raised 10 percent from their reference values to reflect the spill-over of innovators into later periods, a decision that was based on subjective managerial judgment. Alternatives were again searched and an increase of 20 percent in advertising and an additional 10 percent reduction in price were found to be the best response to the increased competitive activity and the basic behavioral response changes. The remainder of the adaptive testing was based on these changes having been implemented in period 4. The price change could have been implemented by a price-off deal.

As indicated earlier, in period 6 media audits indicated that the competitor had doubled his advertising expenditure. Since it was felt that this was a short-run strategy change, the model was updated by increasing the competitive expenditures only in periods 6, 7, and 8. The best response to this aggressive competitive action was to hold to the previously recommended advertising level (20 percent more than reference). In period 8 the media audits reflected the competitor's return to the previous level (50 percent greater than reference) and the best response by the firm was to reduce advertising 20 percent. This decrease was implemented in period 9.

The adaptive testing procedure for the first 10 periods and the projected results, based on the assumption that the period 9 strategy was used until period 36, generated a cash-flow profit of $2 million. As noted above, the company's actual strategy of higher prices and its nonoptimal adaptive strategy generated only $500,000, so the combination of the better introductory plan and the national adaptive strategy determination generated an estimated additional $1.5 million of cash-flow profit.

The model based tracking was done in months 5 and 6 so before they are accepted, the accuracy of the model in forecasting actual shares based on

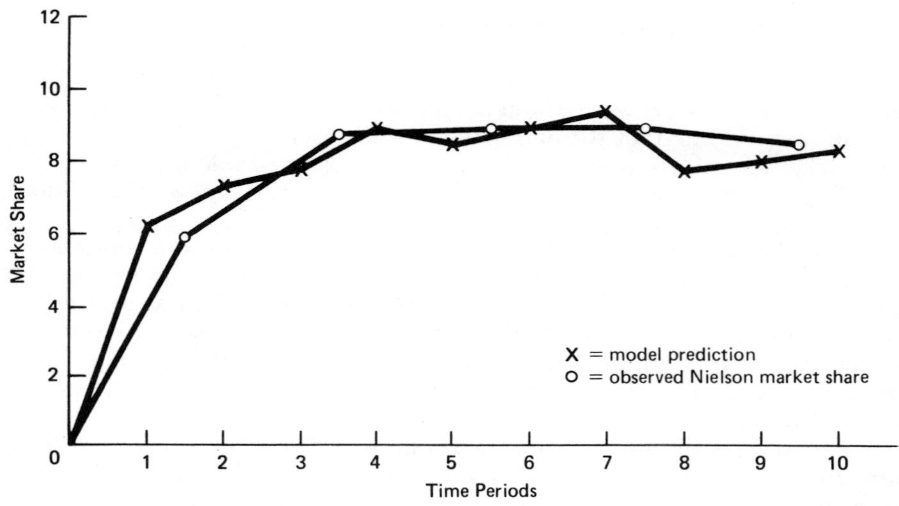

Figure 16.7. *National Tracking Case (Urban, 1970, p. 84)*

the actual managerial decisions should be investigated. The success of the model in duplicating the actual national introduction market shares is shown in Figure 16.7. The real market shares are based on Nielsen store audits and the model predictions are based on the prior test-market estimates updated for the changes in national environment described in the previous paragraphs. The model seems to be very accurate in its updated forecasts. These forecasts were made in the ninth month after introduction, but before the Nielsen market shares for months 8 and 9 were available. These forecasts do not reflect live forecasting tests for months 1 to 6, but month 8 and 9 tests are future forecasts based only on past data. The model predicted a downturn in the share for months 7, 8 and 9. Subsequently, the Nielsen market-share report showed this to be accurate not only to the extent of predicting the turn, but also the amount of the drop. Model testing also was carried out at the microlevel. For example, the growth of availability predicted by the model closely matched the Nielsen measurement of availability. The testing of the model on the national data indicated it to be valid in terms of management's standards for the accuracy required in a new-product decision model.

With the establishment of the accuracy of the forecasts of the model, management realized that many profit opportunities had been missed by not correctly tracking and controlling the launch. Therefore, they were willing to use the model to consider future action. One decision was whether to improve or drop the brand. The model indicated that the plan could be achieved if the price level was reduced by 10 percent and a 40 percent better advertising campaign developed. The firm decided to try to develop the new advertising campaign and reduce the price. Improvements in advertising resulted, but not of the required magnitude. The brand stayed on the market, but the share only reached 50 percent of the objective and profits were only minimally acceptable. Profits could have been much higher and the brand's mediocre performance raised to acceptable levels had the model been used throughout the launch to prevent the errors of overresponding to competition with advertising increases and not with price reductions that are utilized as a defensive tool.

While many firms do extensive test market analysis and forecasting, few use the same discipline in controlling the launch. This case indicates that tracking the launch is at least as important as test market analysis. We recommend firms build a capability to track their product launches to maximize profit. It appears that the use of the adaptive control and a model is potentially quite rewarding.

DURABLE AND INDUSTRIAL PRODUCTS

The full-scale launch of durable and industrial products is similar in many ways to the full-scale launch of frequently purchased products. The marketing mix and production must be coordinated; timing relative to competitors is crucial; critical path analysis is useful to achieve timing objectives; tracking is important; many of the same behavioral phenomena occur; and the analytic tools of adaptive control are the same. But there are key differences that must be faced when launching a consumer durable or an industrial product. In this section we consider some of the differences that are important in full-scale introduction.

In many durable and industrial products a test market will not have been run due to the expense of a pilot plant or because of the length of time necessary to get a reading on sales. Thus early tracking serves the evaluation and refinement functions that would have been carried out by test market of a frequently purchased product. Adaptive control can play an even greater role because of the rapid learning that can occur during the launch.

Sales in the first few years will result from one time purchases. Repeat purchase will occur later in the life cycle and be dominated by replacement decisions as the initial products wear out. Although product satisfaction is not directly reflected in repeat purchasing rates as in frequently purchased brands, it is important in the diffusion of innovation since positive word of mouth information from a buyer to prospective customer is critical to success. In durable and industrial products, service is of great importance. For example, copying machine manufacturers maintain large staffs to service copying machines that have been purchased or leased so that product satisfaction will be high and positive recommendations will be generated.

Another difference between frequently purchased and durable consumer or industrial products is the role of price. The price of frequently purchased brands is relatively stable, but the price of new consumer durable or industrial products can fall over time due to economies of scale, industry learning, and competition. These phenomena should be understood and forecast to make realistic plans to maximize profit.

In this section we concentrate on diffusion and price response and present a series of procedures to project industry sales and prices. In durable and industrial markets understanding these phenomena is necessary before using the techniques presented earlier in this chapter to plan and control the full-scale launch. In the next chapter we return to the industrial marketing issues related to sales effort and advertising for a mature product.

Early Projections of Sales

In Chapter 5 we introduced market growth models as a means to forecast growth for a market (e.g., Bass, 1969). In some durable and industrial products, the new product will establish a new market. In others the new product will enter a rapidly growing, but existing market. In either case a model is useful in forecasting the total market. In the case of an existing market, some model of share similar to those for consumer products will be needed to forecast sales for the innovating organization.

A growth model can be used in two ways during product launch. First, it can be used to establish norms before the launch. These approximate norms are input to the initial planning process. Then, when the product is launched and real-market information becomes available, the parameters of the model can be updated to provide information that is necessary to update managerial tactics.

Recall that in the growth model proposed by Bass (review Chapter 5), sales growth depends on the number of initial innovators, $p(0)$, the total number of potential buyers (m), and a parameter to reflect the rate of diffusion (q). In establishing launch norms, each of these three values needs to be estimated. Then they can be substituted into the model (equation 5.3) to produce a sales forecast.

To estimate $p(0)$ we must first identify the percentage of the population that are innovators and the likelihood that they will purchase the product; $p(0)$ is then obtained by multiplying these percentages. One could present concept descriptions to a representative sample of potential customers, measure their propensity to purchase through preference and intent scales, and measure situational and demographic characteristics likely to identify innovators. Regressions with preference and intent as dependent variables and the characteristics as explanatory variables provide a basis by which to isolate innovators. The likelihood that these innovators purchase is then obtained by a choice model (see the methods of Chapter 11).

In the telecommunication case discussed earlier (Chapters 9 to 12), innovators were most likely to be scientists and managers whose communication needs are for interactions requiring moderate interaction time (10–30 minutes) and who do not now use visuals, but would like to use visuals. This represented about 4.3 percent of the communication interactions, and this group had a 20.6 percent chance of trying the new device. The initial innovation probability then was estimated by multiplying 0.043 times 0.206 to give $p(0) = 0.009$. (Hauser, 1978). Another method of measuring $p(0)$ is by laboratory measurement (Chapter 14). Both these estimation methods are new and serve only as initial estimates of initial acceptance since their predictive accuracy has not been established. However they are likely to be superior to estimates based only on subjective judgments.

The next parameter is q, which represents the influence that those consumers who have purchased the product will exert on those who have not yet purchased the product. In Bass's model, q times the percentage who have purchased gives the probability of purchase for a customer who has not yet purchased. This is the most difficult parameter to obtain prior to launch, thus we suggest that a range be obtained for q rather than a single estimate. This range can then be used in sensitivity analyses. One way to obtain this range is to examine related product categories. For example, one might examine the diffusion rate for home fire alarms to estimate q for home burglar alarms. In the telecommunication study, the diffusion rates for television-related products were used to obtain an estimate of q in the range of 0.40 to 0.90.

The final parameter is the market size, m. Based on reaction to new telecommunication concepts, intent was translated (review Chapter 11) to estimate 11 percent to 15 percent as the ultimate penetration for the telecommunications device. Measures of preference and intent with judgment can establish a range for the initial estimates of the ultimate market size.

The parameters $p(0)$, q, and m were then used in the algebraic equations to obtain estimates such as shown in Figure 16.8. Note that we obtain a range of estimates rather than a single estimate. This realistically reflects the uncertainty prior to the launch of the new communication device and enables management to design a flexible plan that can handle the multiple contingencies.

Once the product is introduced, actual sales data becomes available and more accurate estimates of $p(0)$, q, and m, are obtained. The procedures to obtain these estimates are the same procedures (algebraic solution and/or regression of $S(t)$ vs. $Y(t)$) that were used in Chapter 5. As data from each new period comes in, these parameters can be reestimated or updated to provide better forecasts and to refine marketing and production tactics. In this way your organization can use up-to-date market intelligence with this forecasting capability to control the launch of an industrial product.

Price Dynamics

In many consumer durable and industrial products, we expect that the manufacturer will be able to cut costs due to learning that comes with production experience. Thus as more units are produced, the marginal cost of producing another unit will drop. But the price also depends on production cost, and therefore prices will drop as more customers purchase the new

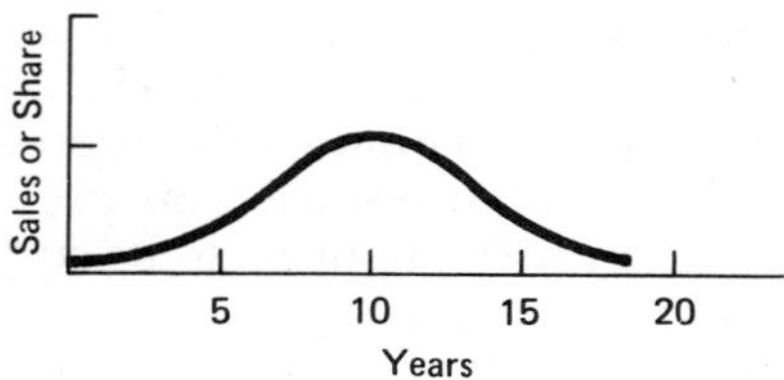

(a) Low Diffusion Rate, Smaller Market

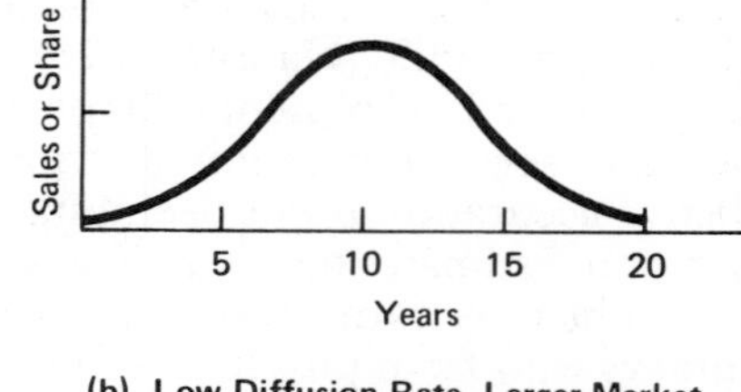

(b) Low Diffusion Rate, Larger Market

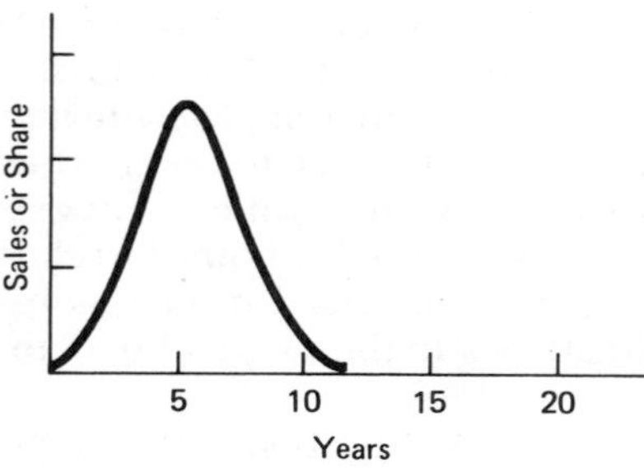

(c) High Diffusion Rate, Smaller Market

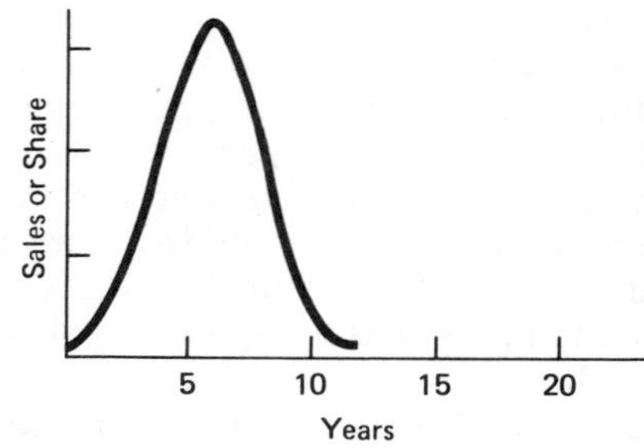

(d) High Diffusion Rate, Larger Market

Figure 16.8. *Forecasts to Establish Norms for a New Telecommunication Service*

product. The price drop will in turn accelerate purchases which may then lower costs further.

This experience effect complicates pricing strategies and opens up new managerial considerations. One pricing tactic is to enter with a high price and skim the market then drop the price as other organizations gain experience. A different tactic is to price low in anticipation of the learning effect to establish good word-of-mouth diffusion and accelerate sales so the low costs will occur and the firm will have a dominant share. In order to identify the appropriate tactic, management needs a forecast of sales and industry price that includes both diffusion phenomena and industry learning.

Integrating learning and diffusion is conceptually easy, but analytically quite difficult. Much research is being done in this area (Robinson and Lakhani, 1975; Bass, 1978; Dolan and Jeuland, 1979; and Abell and Hammond, 1979, pp. 115–116). To illustrate this type of model consider extending the forecasts for color television to include price forecasts (Bass, 1978). Figure 16.9 shows total industry sales, predicted prices, and actual prices. This illustrates the potential accuracy of forecasts that are attainable for price. Although no one best model has been identified, we can expect good models to be available soon to predict sales and prices of durable products.

The availability of specific models of diffusion and price dynamics for durable and industrial products makes tracking the launch more effective. Better forecasts can be made, more effective diagnoses can take place, and higher profits can be achieved.

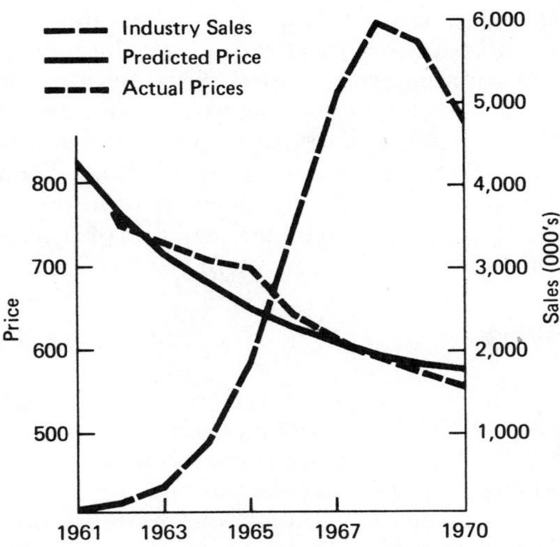

Figure 16.9. *Forecasts of Color TV Sales and Prices (Bass, 1978)*

SERVICES: FAMILY PLANNING CASE

The above sections emphasize the need for careful tracking and understanding of the dynamics of the launch for consumer and industrial products. This is also true in introducing new services. In this section we describe a case of the introduction of a new health delivery service. It serves as another example of the concept of model based tracking and control of the early growth phases of an innovation.

Background and Model

Population growth has become a national and worldwide issue as our growing population puts a renewed strain on our natural and economic resources. One way to address this problem is by providing people with the contraception, education, and medical services so they can plan their families. This case illustrates the launch of a system of family planning clinics in Atlanta, Georgia. Various clinics existed before this analysis, but the innovation was the deployment of an integrated service (see Urban, 1974, for details of this case).

The Atlanta Area Family Planning Council (AAFPC) developed a macro-flow model to track the growth of family planning in Atlanta. At that time there were three basic service granting agencies: (1) Grady Charity Hospital; (2) Planned Parenthood and World Population; and (3) the Fulton and DeKalb County Health Departments. The purpose of the analysis was to provide forecasts and to integrate plans and budgets for the system.

The analysis began with a simple macro-flow model of patients' clinic visits, acceptance, and continuance. Before long the model was found to be inadequate for many managerial requirements. Because of the flexibility of macro-flow, the model evolved to a basic process (Figure 16.10a) with more detailed descriptions of consumer flows in post partum (after delivery) hospital clinics and non-post partum community clinics. For example, Figure 16.10b shows the flow structure for agencies handling non-post partum patients in a community clinic and their choice of one of contraceptive methods (pills, I.U.D., diaphragm, sterilization, etc.).

Input and Fitting

The basic source of input was the Center for Disease Control and its client-record system. This data recorded each patient visit and births in the target population. This data supported trial, repeat, and timing estimation. Outreach workers called on prospective patients and data were collected on a sample basis to determine their impact on appointments and visits. Tabulations were made to find the response rates.

Initial data estimates were made based on the client-record and outreach data of months 1 to 12. After these flow-parameter estimates were put in the model, changes were made so that the model output of "active" and new patients fitted the actual over the first 18 months. "Active" patients were those who accepted contraceptives on their last visit and were not yet due for their next appointment. The fit for the total number of actives is shown in Figure 16.11. Fitting was also done to assure that the model replicated the real data for each method and for new patients at each agency.

Tracking Results

Although the model fitted past data encouragingly well, such fits were the result of considerable massaging of the data. Testing the model was

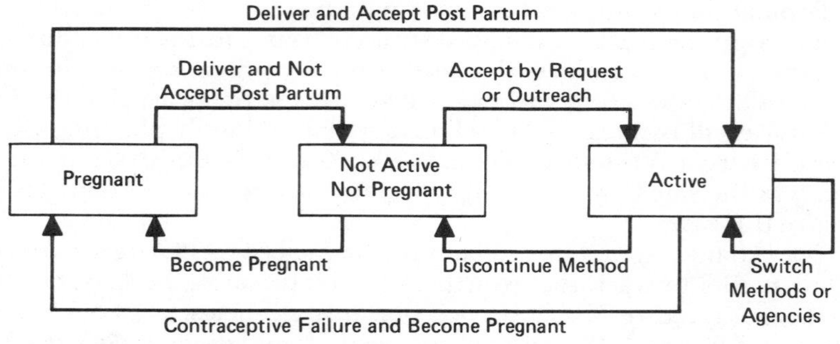

(a) Target Group Sections and Their Interactions

Figure 16.10. *Models for the Analysis of Family Planning Programs (Urban, 1974, pp. 207 and 209)*

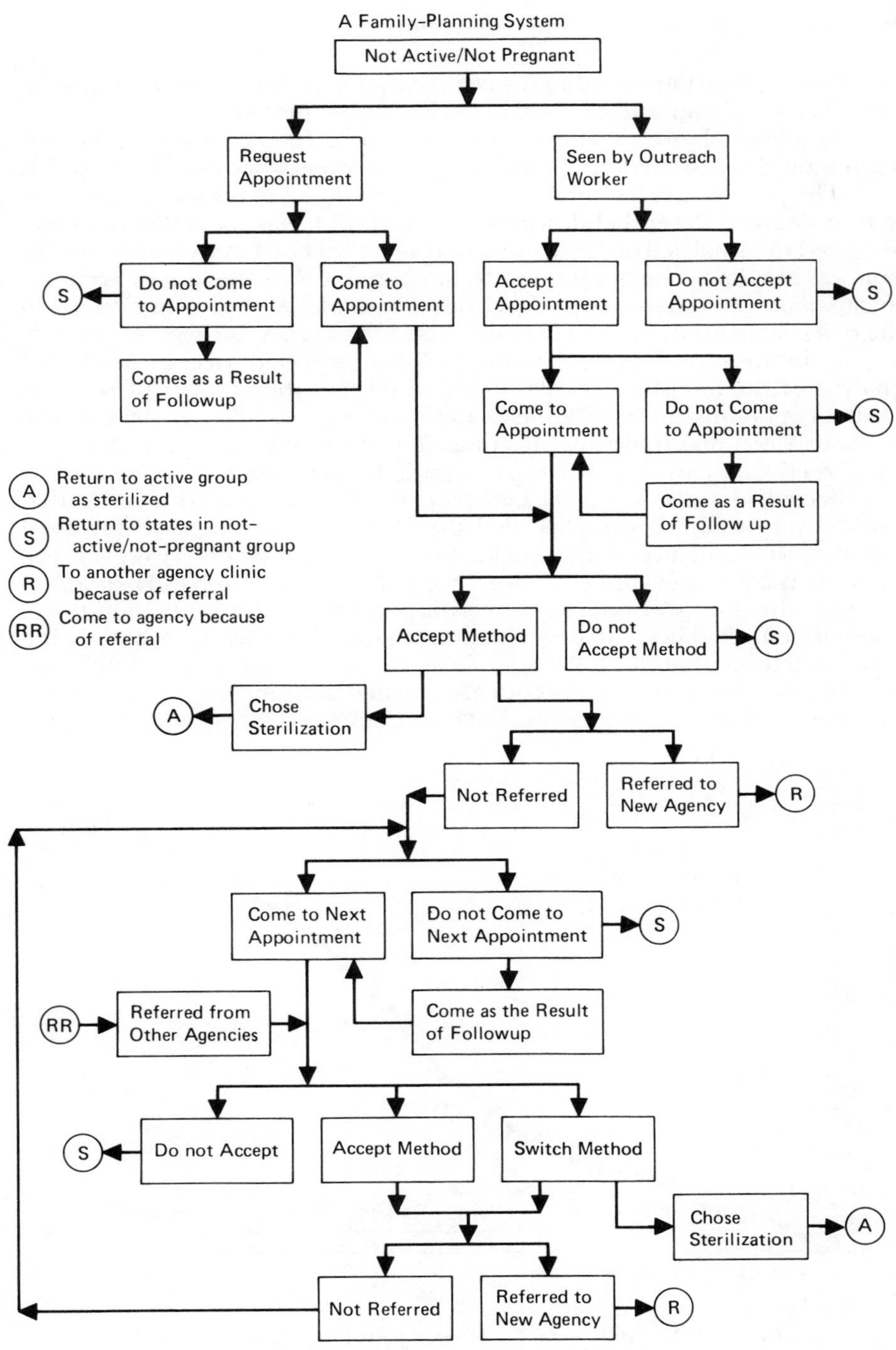

(b) The Nonpost-Partum Agency-Flow Structure

Figure 16.10 *(continued)*

based on comparing actual and predicted patient flows over a six-month period of saved data and over a real twelve-month period.

Conditional forecasts were made for months 19 to 24 based on the first 12 months data estimation. This initial prediction is shown in Figure 16.11.

The prediction was lower than the actual figure. This was particularly true at Planned Parenthood, where the prediction was for stable performance and the total active curve increased sharply. The Grady Clinic prediction was also low. The question to be answered was: is the lack of accuracy due to poor input, random error, inadequate model structure, or changes in the real system itself? Answering this question is an exercise in problem finding, or, in this case, finding the reasons for unexpected success. A detailed analysis of the months 19 to 24 data showed that the number of requests (walk-in appointments) at Planned Parenthood increased from 100 a month to about 250 a month during this time. The initial tracking prediction was based on the past average of 100 per month. Revising the input to reflect the unexpected increase in actual new-patient inflow produced a curve that tracked very well. This implied that the other inputs were probably good and that the structure of the model was reasonably sound. However, the rapid increase of new patients called for diagnosis. There had been an increase in the number of outreach recruitment calls, but very few additional appointments had been made with outreach workers. The hypothesis being investigated was that there was an indirect outreach effect in which calls did not only increase appointments but also created awareness and interest that was demonstrated by clinic visits. Data were collected for new-patient clinic

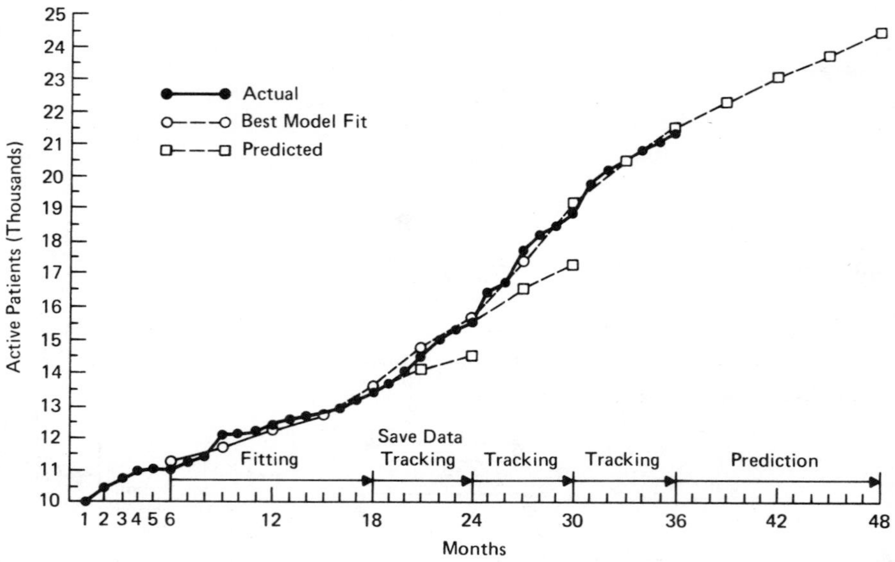

Figure 16.11. *Active Family-planning Patients Using Modern Methods at Major Service Agencies in Atlanta, Georgia*

arrivals to see if outreach calls correlated with voluntary requests for appointments.

The lack of correspondence between actual and predicted patient loads at Grady was found to be due to a new volunteer-run clinic opened to serve the youth community and the subsequent increase of about 50 new patients per month. After these adjustments for new patients at Planned Parenthood and Grady, the active tracking appeared good (see Figure 16.11). Tracking was also carried out at the specific method level. This tracking and a new analysis of the months 19 to 24 data indicated a shift in the composition of method selection toward the pill.

The tracking over the saved-data period (months 19 to 24) was encouraging, but additional tracking over the months 25 to 36 allowed for additional refinements of the model input and understanding of the system structure. The conditional prediction made in month 24 is shown in Figure 16.11. Again the prediction was low. The Planned Parenthood agency increased its new-patient rate to 450 a month, or 200 more than predicted. It was found that Grady referred 50 percent more people per month to the county clinics than expected. New non-postpartum clinic growth added 50 more patients per month. Finally, county-health-department outreach was more effective than anticipated.

The volatility of the system reflected in these changes emphasizes the need for tracking and an effective adaptive planning model. With the input updated, the model tracked well, but again the question of why the new-patient rate increased was asked. Data indicated the indirect outreach effects to be real. Four times as many people contacted by outreach workers came to a clinic without an appointment than came with an appointment. Inputs were revised and new predictions made for months 31 to 36. In month 37, examination of predicted and actual values for months 31 to 36 showed close correspondence. This was encouraging and reflected good predictive performance at the total system level. This was generally true at the agency level except at Planned Parenthood, where actual was less than predicted. Refitting indicated that a decay in continuance rates could explain this. Additional diagnosis showed that the decay was due to an increasing proportion of white college girls. This raised questions of priority between college girls and indigent mothers, since clinic capacity was limited. Grady actual was 3 percent higher than predicted, owing to less referral in December and more new patients in months 35 and 36. In summary, the fitting and tracking of the system growth helped diagnose problems, raised new issues, and provided new insights into the response dynamics.

Managerial Use of the Analysis

The model was used by the Director of AAFPC. He used it to develop an overall system plan and as a tool to aid agencies in their planning of growth of the service system. Special planning sessions were conducted with member agencies so that these managers could better understand their patient response, improve forecasts, and develop goals and plans. Although formal measurement of the impact is difficult, the managers have reacted positively and used the model to determine the effects of outreach workers and to predict the number of new patients and the change in the cost per year of contraceptive protection that would result from undertaking an outreach program. In another agency, the outreach data and model runs indicated weakness in the success of outreach workers in making appointments and led to new training sessions based on new communication appeals.

The managerial control process produced some new insights. In particular, the fitting and tracking exercise has been valuable since it required a detailed analysis of why predictions were not as good as desired. The indirect effect of outreach workers was one such new insight that resulted from the model use. Another resulted from tracking birth flows. It was found that 25 percent of the deliveries from the target group of indigent women were not done at the Grady Hospital, but at private hospitals. Originally, managers had believed that virtually all target-group deliveries were done at Grady on a charity basis. This insight has resulted in a new outreach program to the maternity wards of these private hospitals. The analytic approach fostered by the model led to this new insight and change in the system behavior.

After fitting the model for the months 25 to 30, a conditional forecast was made for months 31 to 48. This was the basis for the next tracking period, but also was needed for the budget request for the next year (months 37–48). HEW planning requirements specified the next year's budget was needed in month 34. The model proved valuable to managers in generating the forecast for the remainder of the current year so that past funds could be justified and next year's needs could be estimated. It should also be mentioned that the environment surrounding the budgeting was frantic due to the proposal deadline. The model allowed rapid simulation and predictions, so that an effective proposal could be formulated on time. The first budget forecast was based upon a budget sufficient to meet capacity requirements. The forecast showed 24,600 actives by month 48 and an average cost per year of contraceptive protection of $69.91 over the next three years. However, the funding agency in Washington had requested that last year's budget amount be held for the next year. In order to show the effects of this constrained budget, the model was run again with the new patient arrival rates decreased until existing capacity could serve adequately the active groups. This budget-constrained run indicated 11.7 percent fewer actives, 240 women per month being refused service, and an increase of 1.6 percent in the cost per year of contraceptive protection. These forecasts were included in the budget request and the explicit cost/benefit justification was cited as a contributing factor in the subsequent granting of the larger budget amount. Although the larger budget was obtained for months 36 to 48, the model forecast for the following year indicated doubt that needed funds could be obtained from existing sources. This led to more attention being given to generating new funding sources (welfare, Medicare), procedures for allocation between agencies, and methods of clinic screening for only the target-group members most in need.

In addition to an orderly forecasting procedure, various strategic alternatives were considered. First, an outreach program to postpartum non-Grady patients was simulated. With an estimate of the number of calls allocated to this new program and their effect, it was found actives (those who had received contraceptives on their last visit) increased 1 percent over three years and cost per year of protection decreased slightly. The second strategy was to increase the capacity to perform sterilizations. Requests had been twice the capacity. This strategy resulted in a small increase in actives in three years, but a 5 percent reduction in the birth rate. However, the cost per year of protection increased. Since sterilizations were priced at $300 each, they did not pay back in three years (recall an overall figure of about $70 per year of protection). In fact, at this price it would take five years to pay back. Sterilization in the short run was not very attractive managerially as a method with this cost and pricing. New technology, more efficient procedure, or negotiations to reduce the cost could make sterilization more attractive. Other strategies were tried, but the gains due to the new strategies, although significant, were small (less than 5 percent). It became clear to program managers that the target group was being saturated. This insight has led them to widen their program to include more of Georgia. The improvements due to strategic analysis were important, but an equal benefit of the analysis has been a better perception of the system dynamics. The discipline of tracking and controlling the launch led to better budgeting and planning for the growth of the family planning system.

SUMMARY

Full-scale launch of a new product is the phase of new product development that commands the largest commitment in time, money, and managerial resources. No matter how well the product is designed and tested, the launch presents risks to achievement of profit goals. Marketing and production must be coordinated, and the timing of the launch carefully planned, or profits will be jeopardized. Unexpected changes in the consumer, competitive, technological, and economic environments present risks. The need to monitor these events is critical. Appropriate revisions of the launch plan to reflect these changes can maintain the desired level of profit and present opportunities for improvement of plans. A key lesson of this chapter is the need to monitor consumer and competitive response to gain the market intelligence necessary to capitalize on these opportunities. In consumer, industrial, and service innovations, the control of the launch is equally as important as any phase in the new product development process and deserves equal management attention and analytic support.

After the product is launched successfully, sales, marketing, and production will continue to be improved until product maturity occurs. The next chapter discusses issues concerning long-term profit management and provides procedures for managing the mature product.

REVIEW QUESTIONS

16.1 Why should management spend the time and money to monitor new products during national introduction?

16.2 What are the advantages of performing a critical path analysis before product introduction?

16.3 You have spent two years developing a new pie mix, but six months prior to full scale launch your major competitor finds out about your new product concept and is rushing to beat you to market. The competitor does not have access to your technology. Should you begin a crash program to get to market in three, rather than six, months? What information do you need to make your decision and how would you go about obtaining the necessary information?

16.4 Why should you compare predicted results to actual results rather than just monitoring actual results?

16.5 Consider the mathematical implication of Little's control model expressed as follows:

$a_0(t + 1) = a_0(t) + c(ms^+ - ms^- - \Delta)$

The symbols are defined as indicated by the text. What is the recommended effect on advertising expenditures of an increase in m, s^+, s^-, or Δ? Why?

16.6 Explain the intuition behind adaptive control? How would adaptive control be used to set price?

16.7 What special problems does a new product manager face when monitoring durable and industrial products? Services?

16.8 Production costs and prices drop rapidly in markets for high technology products such as calculators and computers. Why? What implications do these decreases have for launching a new high technology product?

16.9 What diagnostic information should a new product manager look for when monitoring a new product launch? How is this information obtained?

16.10 Baldai Industries, a manufacturer of large appliances, has developed a new line of European-look dining room furniture called Stalas-Kėdė. Develop a hypothetical plan to help the product manager, Mr. Galva, launch this new line of furniture.

Chapter 16 Appendix

16.1 Adaptive Control*

This appendix gives the mathematical details of adaptive control. The basic idea as described in the text is (1) to use market experimentation to improve knowledge about the sales response function, and (2) to use the updated sales response function to select the optimal level for the marketing mix element under consideration. The basic experiment in each time period t is to select $2n$ markets and spend $a_t^0 - (\Delta/2)$ dollars per household in n markets, $a_t^0 + (\Delta/2)$ dollars per household in n markets, and a_t^0 dollars per household in the remaining markets where a_t^0 is the optimal spending (given past knowledge) and n, Δ are set by the experimenter.

Underlying Market Models. The assumed goal is to maximize profit. Let π_t be the profit rate for time period t, s_t be the sales rate in time period t, m be the gross margin per unit sales, a_t be the cost rate of the marketing mix element, and c_t be the rate for other costs. Then profit is given by:

(16.1A-1) $$\pi_t = ms_t - a_t - c_t$$

Assume that the sales response function is a quadratic function with unknown constants α, β, and γ to be determined by the adaptive control model. That is:

(16.1A-2) $$s_t = \alpha_t + \beta_t a_t - \gamma_t a_t^2$$

*This appendix is based on an article by Little (1966).

Using differential calculus, the optimal level of the market mix element, a_t^o, is given by:

$$\text{(16.1A-3)} \qquad a_t^o = \frac{m\beta_t - 1}{2m\gamma_t}$$

If the organization sets the marketing mix element at a_t rather than a_t^o, it experiences an effective loss, l_t, of:

$$\text{(16.1A-4)} \qquad l_t = \pi_t(a_t^o) - \pi_t(a_t) = m\gamma_t(a_t - a_t^o)^2$$

Little assumes that $\gamma_t = \gamma$ is constant and known, that α_t is independent of time period with a high variance, and that β_t has an autoregressive component, i.e.:

$$\text{(16.1A-5)} \qquad \beta_t = \varphi\beta_{t-1} + (1 - \varphi)\beta_o + \epsilon_\beta$$

where φ and β_o are constants and ϵ_β is a normal zero mean random variable with variance σ_β^2.

Sales Experiment. Sales rates vary by market. Assume that market specific effects can be modeled by random error which is normally distributed with zero mean and variance given by σ^2. Furthermore, assume the errors are uncorrelated across markets. Suppose the experiments are run in the $2n$ markets. Let s_t^+ be the observed mean sales rate in the markets with spending rates $a_t + (\Delta/2)$. Let s_t^- be the observed mean sales rate in the markets with spending rates $a_t - \Delta/2$. Then the experimental mean for β_t is:

$$\text{(16.1A-6)} \qquad \hat{\beta}_t = \frac{(s_t^+ - s_t^-)}{\Delta} + 2\gamma a_t^o$$

Furthermore, $\hat{\beta}_t$ is a normal random variable with mean β_t and variance $2\sigma^2/n\Delta^2$.

Updating the Model and Setting the Spending Rate. One method to update both β_t and a_t^o is to use exponential smoothing. That is, select some number k $(0 \leq k \leq 1)$ and update β_t by:

$$\text{(DR1)} \qquad \beta_{t+1} = k\beta_t + (1 - k)\hat{\beta}_t$$

Then set the optimal spending rate by the following decision rule:

$$\text{(DR2)} \qquad a_{t+1}^o = \frac{(m\beta_{t+1} - 1)}{2m\gamma}$$

This becomes clearer if we define $\hat{a}_t^o = (m\hat{\beta}_t - 1)/2m\gamma$, then DR1 and DR2 combine to:

(DR3) $$a_{t+1}^o = ka_t^o + (1 - k)\hat{a}_t^o$$

In other words, the optimal spending rate is simply a combination of the previous period's spending rate and the experimentally determined optimal spending rate. If the experiment is rough, the k is close to 1.0 and spending rates vary slowly. If the experiment is very accurate, then $1 - k$ is close to 1.0 and the spending rate depends strongly on the experimental outcome. Another way of writing this is:

(16.1A-7) $$a_{t+1}^o = a_t^o + \left[\frac{(1 - k)}{2m\gamma\Delta}\right][ms_t^+ - ms_t^- - \Delta]$$

which is equation (16.1) from the text with slightly altered notation and

$$c = \left[\frac{(1 - k)}{2m\gamma\Delta}\right]$$

The decision rules DR1, DR2, and DR3 are intuitively appealing. Little uses Bayesian analysis to justify these rules with optimal updating. With Bayesian analysis, Little shows the optimal updating, β_{t+1}^o, is given by:

(DR1a) $$\beta_{t+1}^o = \varphi k\beta_t^o + \varphi(1 - k)\hat{\beta}_t + (1 - \varphi)\beta_o$$

with $k = \nu/(\nu + \nu')$ where ν' the a priori variance for β_t and ν is the variance for $\hat{\beta}_t$. Little further shows that the a priori variance is given in steady state by:

(16.1A-8) $$\nu' = \tfrac{1}{2}[\sigma_\beta^2 - (1 - \varphi^2)\nu] + \tfrac{1}{2}\{[\sigma_\beta^2 - (1 - \varphi^2)\nu]^2 + 4\sigma_\beta^2\nu\}^{1/2}$$

If $\varphi \rightarrow 1$, then DR1a becomes DR1 and equation 16.1A-8 simplifies. Remember $\nu = 2\sigma^2/n\Delta^2$.

(16.1A-9) $$\nu' = \tfrac{1}{2}\sigma_\beta^2\left\{1 + \left[1 + \frac{8\sigma^2}{\sigma_\beta^2 n\Delta^2}\right]^{1/2}\right\}$$

thus the updating constant k is given by known parameters. Similar analyses show the exact version of DR3 to be:

(DR3a) $$a_{t+1}^o = \varphi ka_t^o + \varphi(1 - k)\hat{a}_t + (1 - \varphi)a_o$$

where $a_o = (m\beta_o - 1)/2m\gamma)$ and k is calculated as in DR1a.

Optimal Experiment. In the above analysis we assumed that the parameters of the experiment were known. But running the experiment has costs since presumably $a_t^o + (\Delta/2)$ and $a_t^o - (\Delta/2)$ represent deviations from optimal spending rates. If β_t^* is the (unknown) true value for β_t, then the optimal spending rate under perfect information is $a_t^* = (m\beta_t^* - 1)/2m\gamma$. Instead, we choose a_t^o given by DR3 or DR3a. The loss (equation 16.1A-4) is given by:

(16.1A-10) $$l_t = m\gamma(a_t^o - a_t^*)^2$$

Since the expected value of a_t^* is a_t^o, the expected loss is $m\gamma$ times the variance of a_t^* which is $(1/4)\nu'$. That is, the expected loss is $\bar{l}_t = m\nu'/4\gamma$.

But we deliberately set the spending levels at non-optimal rates. For example, in the n markets with spending rates $a_t^o - (\Delta/2)$, the loss rate is given by $l_t^- = m\gamma(a_t^o - (\Delta/2) - a_t^*)^2$. Substituting, it is easy to show that the expected loss in those markets is given by $\bar{l}_t^- = m(\nu' + 4\gamma^2\Delta^2)/16\gamma$.

If there are N markets and H households per market, then the total expected loss rate, T, is given by:

(16.1A-11) $$T = \frac{\text{NHm}\nu'}{(4\gamma + \frac{1}{2}Hm\gamma n\Delta^2)}$$

To select the optimal experiment, we must select n and Δ to minimize T. Note that these parameters appear as $n\Delta^2$. Thus, we can choose any n and Δ as long as the quality $n\Delta^2$ remains constant. This flexibility is practical and valuable. Little reports that finding the optimal $n\Delta^2$ is straightforward but tedious. In the special case of $\varphi = 1$, the minimum is found with respect to a dimensionless quality z, where $n\Delta^2 = 8\sigma^2/\sigma_\beta^2 z$. Substituting this definition in the above equations, Little shows the optimal z can be obtained by the following equation:

(16.1A-12) $$\frac{z}{(1 - z)^{1/4}} = \frac{8\gamma\sigma}{(\sigma_\beta\sqrt{N})}$$

which can be solved graphically. Therefore, given the system constants, z is determined from equation (16.1A-12), and the experimental constants are determined from $n\Delta^2 = 8\sigma^2/\sigma_\beta^2 z$.

For numerical examples, simulation results, sensitivity analysis, and more details, see Little (1966).

Chapter 17

Managing the Mature Product

Your organization has designed, tested, and introduced a new product. You have a product and marketing strategy that has been successful in test market and has been transformed into an effective full-scale introduction. But product management is not finished. There is a significant effort required to realize the profit and reap the rewards of innovation and risk taking. The marketing and production budgets must be controlled to maximize profits. Advertising, selling, distribution, price, and promotion strategies must be modified in response to changes in competitive situations, economic environment, and consumer tastes. These actions need to be coordinated with the other products in the firm's product line. The product portfolio must be managed by appropriately allocating resources across products, and by selectively rejuvenating or dropping other products in the line.

While profit management is a central task in managing the mature product, innovation is still necessary to defend the product against competition and to extend its life by adding product modifications or introducing slightly different brands to tap smaller market segments. These modified products, often called "flankers," can be very profitable. In addition, innovation may be needed to improve the efficiency of the production process, reduce costs, and increase profits.

Eventually, the product will reach the decline phase of its life cycle. Management then faces the decision of whether to "cash in" and drop the product or to rejuvenate the product through repositioning for new market needs. Major innovation can launch the product on a new life cycle for continued growth in profits and sales.

In this chapter we consider the mature and decline phases of a product's life. We use response analysis as a method to manage profit for the mature product. Decision support systems provide the necessary information, and models of profit response to marketing mix variables provide the managerial tools. We illustrate these concepts through a case application for a major food product. The final sections of the chapter consider strategies for product management late in the life cycle and the relationship of individual products to the overall product portfolio.

PROFIT MANAGEMENT: THE CONCEPT OF RESPONSE ANALYSIS

Profit management is a critical problem for executives. Major resources are being allocated in an uncertain and diverse world and they should be carefully committed. Consider setting the product and marketing variables. In Chapter 12 we discussed setting the marketing mix for the early phases of the life cycle. Conceptually, the task was to establish the levels of advertising (A), selling (S), distribution (D), price (p) and promotion (R) to control sales (P) and costs (C) to maximize profit (π) where:

(17.1) $$P = f(A,S,D,p,R)$$

(17.2) $$C = g(P)$$

(17.3) $$\pi = (p-C)P-A-D-S-R$$

We have already discussed some simple models and concepts for addressing this problem prior to the launch of a new product. As the product matures, a set of levels for the marketing variables evolve based on these analyses.

In managing the mature products, managerial decisions are considered as modifications of that strategy. In most cases we deal with fine-tuning the marketing strategies. For example we may consider raising the advertising budget by 20 percent, or reallocating the budget between advertising and trade promotions, or changing the price in response to competition.

Early in the development process the attention is on developing a basic strategy. As a result, much of the emphasis is on primary data collection and analysis. Once the product is in full-scale market, the emphasis shifts to tactical decisions, fire-fighting, and continuing maintenance of market share. The degree of primary market research data collection and analysis diminishes. The net return on additional information is less and the expenditures on massive new market research studies often cannot be justified when compared to alternative direct expenditures for advertising, promotion, or distribution.

Fortunately, as the product matures, management can tap the data accumulated from market experience with the product. Before the product was launched we had to forecast the response of consumers, competitors, and suppliers by using survey and test market measurements and models that approximated the "real world" as closely as possible. These methods are subject to many types of errors and are considerably expensive; therefore we must also use diagnostics and judgment to direct strategy. However, once the product is launched, we can observe consumer and competitor response and can experience the reactions of suppliers, distributors, and other channel members. All data collection and analyses become information capital which can be accumulated and integrated into the decision process.

Once experiential data is accumulated, more accurate response models can be developed. That is, management can develop tables and curves which represent how sales or profit will respond to changes in marketing mix variables. Making decisions based on those tables and curves is called "response analysis." For example, if the basic positioning is held constant for two years, and the only changes are modifications to the advertising budget, then a simple "sales response to advertising" model can be estimated to fit the two years of data. Such a model can be developed since the basic positioning (and other marketing mix elements) act as an initial strategy that causes "base level" sales. The advertising response is then represented as a deviation from these base level sales. This simplicity may produce greater accuracy, and often makes it possible to select an "optimal" budget as long as we consider budgets that are within the limits of past expenditures.

Despite the fact that more sales data is available, the management of a mature product is not simple; it presents many challenges. It is a rare product that manages itself. Consumer tastes evolve and change. Competitors enter with their new products or retaliate with changes in the marketing mix of existing products. Response analysis helps the manager forecast and monitor these changes, but it is ultimately the manager who must understand the new situation and modify his marketing and production strategies.

MARKETING DECISION SUPPORT SYSTEMS

The success of response analysis depends on the quality of market response information. A good decision support information system supplies the needed information. To better understand the relationship of data and models to managerial decisions, examine Figure 17.1.

Figure 17.1 emphasizes the importance of the manager and illustrates that the models, the data, the statistical analysis, and information summaries (display) are integrated to provide usable information to the manager so that he can make better decisions. The data is the result of special research studies and monitoring the response to past decisions. The statistical analysis summarizes that data and reduces it to a usable form. The management science models use the data, statistics, and managerial judgment to provide decision directives. The display interface allows the manager to effectively utilize the data, statistics, and models for decision making.

We begin our discussion of decision support systems by examining the managerial requirements for the management science models that reside within it.

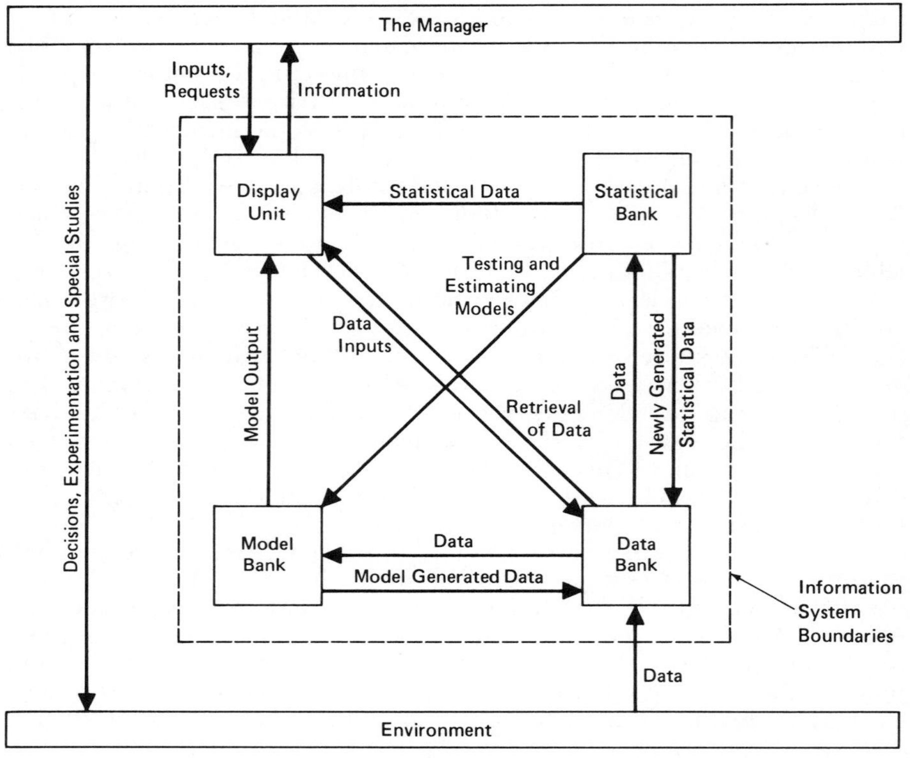

Figure 17.1. *Decision Information System Structure (Montgomery and Urban, 1969, p. 18)*

The Concept of a Decision Calculus

To be effective, a decision information system must be used by managers. Thus, the decision information system must be implemented with managerial problems in mind. Such decision oriented implementation has been widely studied. For example, see Churchman and Shainblatt (1965), Naert and Leeflang (1978), Urban (1974), and Lilien (1975).

One of the most significant approaches to implementation was developed by Little (1970). He identified a series of guidelines that he called "decision calculus." These guidelines serve both the model builder and the product manager by providing a basis of communication. For the model builder they indicate what is necessary before the model can be useful for management and for the manager these guidelines indicate some of the criteria by which to judge the models. Decision calculus models are characterized by both an understanding of how the model is to be used and an abstraction of the process that is to be analyzed and controlled. In our view, the models in the information system to support profit management of a mature product should meet these criteria.

In Little's (1970, pp. 469–470) words:

> From experience gained so far, it is suggested that a decision calculus should be:
>
> 1. *Simple.* Simplicity promotes ease of understanding. Important phenomena should be put in the model and unimportant ones left out. Strong pressure often builds up to put more and more detail into a model. This should be resisted, until the users demonstrate they are ready to assimilate it.
>
> 2. *Robust.* Here I mean that a user should find it difficult to make the model give bad answers. This can be done by a structure that inherently constrains answers to a meaningful range of values.
>
> 3. *Easy to control.* A user should be able to make the model behave the way he wants it to. For example, he should know how to set inputs to get almost any outputs. This seems to suggest that the user could have a preconceived set of answers and simply fudge the inputs until he gets them. That sounds bad. Should not the model represent objective truth?
>
> Wherever objective accuracy is attainable, I feel confident that the base majority of managers will seize it eagerly. Where it is not, which is most of the time, the view here is that the manager should be left in control. Thus, the goal of parameterization is to represent the operation as the manager sees it. I rather suspect that if the manager cannot control the model he will not use it for fear it will coerce him into actions he does not believe in. However, I do not expect the manager to abuse the capability because he is honestly looking for help.
>
> 4. *Adaptive.* The model should be capable of being updated as new information becomes available. This is especially true of the parameters, but to some extent of structure too.
>
> 5. *Complete on Important Issues.* Completeness is in conflict with simplicity. Structures must be found that can handle many phenomena without bogging down. An important aid to completeness is the incorporation of subjective judgments. People have a way of making better decisions than their data seem to warrant. It is clear that they are able to process a variety of inputs and come up with aggregate judgments about them. So, if you can't lick 'em, join 'em. I say this without taking away from the value of measurement. Many, if not most, of the big advances in scientific knowledge come from measurement. Nevertheless, at any given point in time, subjective estimates will be valuable for quantities that are currently difficult to measure or which cannot be measured in the time available before a decision must be made.
>
> One problem posed by the use of subjective inputs is that they personalize the model to the individual or group that makes the judgments. This makes the model, at least superficially, more fragile and less to be trusted by others than, say a totally empirical model. However, the model with subjective estimates may often be a good deal tougher because it is more complete and conforms more realistically to the world.
>
> 6. *Easy to Communicate With.* The manager should be able to change inputs easily and obtain outputs quickly. On-line, conversational Input/Output and time-shared computing make this possible.

> Every effort should be made to express input requests in operational terms. The internal parameterization of the model can be anything, but the requests to the user for data should be in his language. Thus, coefficients and constants without clear operational interpretation are to be discouraged. Let them be inferred by the computer from inputs that are easier for the user to work with. Expressing inputs and outputs as differences from reference values often helps.

It is important to recognize that decision calculus is not any specific model or type of model, but rather a philosophy of model building. Although many academically interesting models can and have been built, experience in the last ten years suggests that the general decision calculus criteria must be met by models in the decision information system if decision implementation is desired. This experience also has led to some refinement of the decision calculus criteria. A model must be easy to communicate with, but it is unlikely that a manager will use it in an on-line computer mode. Managers rarely sit at terminals. They have assistants who make the runs and report results. The on-line model is important in providing rapid response and managerial understanding, but the manager is usually buffered from the terminal by a decision support interface specialist who can efficiently run the model in response to the manager's requests.

Another refinement has been to develop evolutionary capabilities (Urban and Karash, 1971) so that a manager can begin with a simple model, but quickly expand his understanding to encompass a more complex model that is necessary to capture the completeness of the decision problem.

To illustrate the decision calculus approach consider a simple advertising budgeting model for a mature product (Little, 1970). Based on data and judgment, the manager or analyst develops a sales response model such as that shown in Figure 17.2. The manager controls advertising which affects sales. In particular there is some minimum market share even without advertising, but this share can be increased with advertising. The amount of the increase depends upon the level of advertising and can be represented as a curve like the curve in Figure 17.2.

Advertising is not a static phenomena. The effects decay over time.

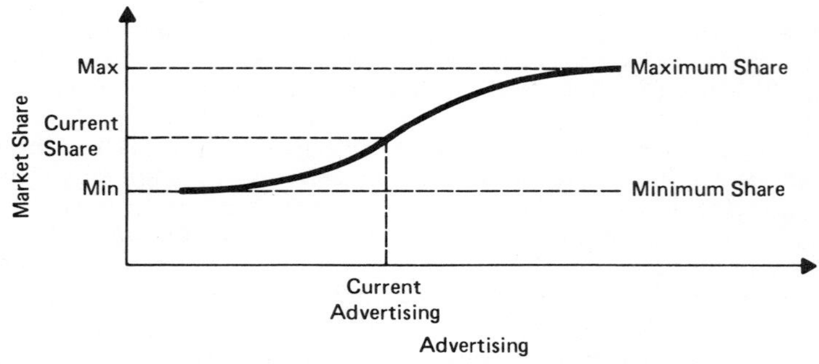

Figure 17.2. *Sales Response to One Period*

This decay is represented in Figure 17.3 as a decline of sales over time in the absence of advertising. In particular Figure 17.3 says that the effect of advertising will decline by a fixed percentage, α, for each time period. Mathematically, Little's model, ADBUDG, can be represented by the following equation.

$$S_t = \lambda + \alpha(S_{t-1} - \lambda) + \beta f(A_t) \tag{17.4}$$

where S_t = market share in time t,

A_t = advertising in time t,

λ = the long run minimum that would occur without advertising

α = the fixed percentage representing the decay in the effect of advertising,

$f(A_t)$ = the direct effect of advertising as given by the curve in Figure 17.2,

β = a constant representing the magnitude of the direct effect of advertising.

The first term is the minimum sales, the second is the carry-over effect of past sales, and the third is the effect of current period advertising on sales. Once the manager or data analysis provide the information to determine α, β, λ, and $f(A_t)$, the decision support system can be used to provide the sales implications of changes in advertising. Profit is then determined based on industry sales, contribution margin, and advertising costs.

ADBUDG was one of the first decision calculus models to be implemented. The approach also has been used to determine geographic allocation of advertising (Urban, 1975). Most recently it has been extended by Little (1975) to include more variables and phenomena. The case, presented later in this chapter, demonstrates these extensions.

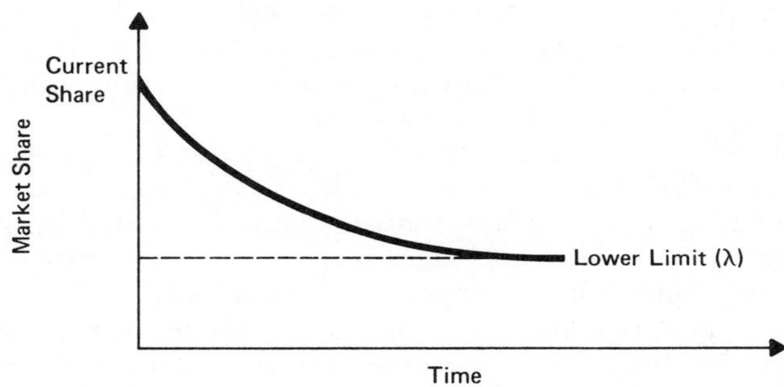

Figure 17.3. *Decay in Sales in No Advertising*

Data

Information can be obtained from many sources. If data is collected in the design and testing phases, it can and should be updated and used. Beyond that there are many additional sources of information each with its advantages and disadvantages. The best managerial strategy is to be open to and utilize multiple information sources depending upon their availability and quality. We briefly review three information sources that are extremely useful for response analysis: judgment, experimentation, and statistical analysis of market history. Other data sources such as archival information and test market tracking measures have been discussed in chapters 8 and 15.

Managerial Judgment. Managers make implicit judgments about marketing budgets, prices, advertising, etc. Therefore, at a minimum, we can use their judgment to obtain response functions. Usually, judgments can be improved if they are obtained in an organized way and from more than one person.

Little (1975) uses a simple, but effective process. He assembles a group of knowledgeable people such as the brand group. The response function is defined and each member of the group is provided with a table of control variables (e.g., advertising levels) and asked to provide his estimates of the response functions. These are displayed anonymously on the blackboard. Little reports: "People usually identify their own estimates and a lively discussion follows as to why certain values are picked. Sometimes misunderstandings about what was to be estimated are uncovered. People may introduce considerations that lead others to change their values. Finally, a consensus position—usually a median or something close to it—is proposed, perhaps modified, then adopted."

For example, Figure 17.4 gives an example of a judgmentally desired response curve. Clearly, opinions differ, but they are explicit and lead to better understanding. The group understands the final position which then provides a good starting point for later analysis.

Experimentation. Of course, judgments should not be depended upon if better data can be obtained. In our opinion, since judgments may not be accurate (see Chakravarti, Mitchell, and Staelin, 1979), they should be supplemented and improved by data wherever possible. One of the best sources of data is from market experiments.

Explicit experimentation provides greater detail on some aspects of market response, but experiments are difficult and expensive. Alternative levels of variables need to be tried, control groups identified, and competing hypotheses examined. Other variables such as environmental effects, competitive actions, changes in nonexperimental marketing mix variables, and population changes must all be identified and considered in the analysis.

In particular, the adaptive control techniques used in product launch should continue through the life cycle providing improved information and improved strategies as the product matures. For other discussions of the

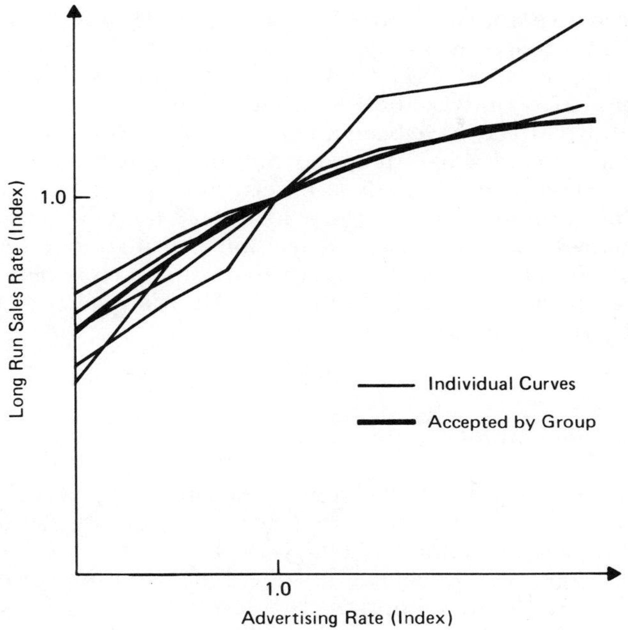

Figure 17.4. *Example of a Judgmentally Determined Curve of Sales Response to Advertising (Little, 1975, p. 660)*

techniques of experiments see Campbell and Stanley (1966) and Cochran and Cox (1957). For applications see Rao (1970) and Urban (1975).

Statistical Analysis of Market History. A good tracking system monitors sales and the levels of the marketing mix variables. The more observations there are over a wider range of strategies, the more useful this tracking data is in determining response functions. Statistical analysis and, in particular, econometrics is a well developed field that is valuable for estimating response functions. (See Parsons and Schultz, 1976, for a comprehensive review in marketing.)

In most econometric models, market share or sales is the dependent variable, and advertising and/or other marketing variables are explanatory variables. The simplest models link one period sales to advertising by nonlinear functions (e.g., Kotler, 1964; Urban, 1969; Balachandran and Gensch, 1974; and Rao and Miller, 1975).

Dynamics are commonly included by modeling lagged effects. One approach is to include the independent variable levels from previous periods. Another is by the use of previous sales as a variable in prediction of current sales. These distributed lag models have been widely used (e.g., Palda, 1964; Montgomery and Silk, 1972; and Clarke, 1973).

An alternative approach that has been gaining acceptance in recent years includes inertial effects in both the dependent and explanatory variables (Box and Jenkins, 1970; Helmer and Johansson, 1977; Glass, Willson, and Guttman, 1975; Pindyck and Rubinfeld, 1976).

Competitive effects have been analyzed by including variables for several firms in the equation affecting sales. Simultaneous equations have been used to include competitive and dynamic effects (e.g., see Bass, 1969).

Econometric modeling is a large field and has high potential to serve as a mechanism to estimate response in a decision information system designed to support profit management of a mature product. Econometrics can provide usable information that fits within the decision calculus framework (e.g., Lambin, 1972 and 1976).

Planning and Control

If the decision-information system shown in Figure 17.1 is effective, it will contain decision calculus models whose parameters are estimated by experimentation and econometric analysis. These inputs will be complemented by managerial judgment at the model/manager interface to provide a decision support tool for setting the best level of advertising, sales, distribution, promotion, and price variables. The outcome is a one year and a three-to-five year plan for the product. While the most immediate interest is in the first year's annual plan, one should look at the longer term effects before setting the first year budgets.

Once the plan is established, a control system must be instituted to assure it is met and to adjust for unexpected events. We introduced the notion of a simple control system for product launch in Chapter 16. Figure 17.5 elaborates on that planning and control system. It begins with plans generated by the models from the decision information system (box 1). Once the basic marketing plan is selected, it is implemented via operations and impacts on the market. The actual results are compared to the goals from planning in circle A and to the predictions based on tracking with management's model of the marketplace (from box 2) in circle B. If the actual results match the predicted results and management's goals, then the plan is continued. If they are not in agreement, a market diagnosis is performed (in box 3), the model is updated (in box 4), and modifications to the marketing plan are considered and evaluated (in box 5).

By quarterly tracking actual, predicted, and desired results, changes can be made to budgets in order to best achieve the plans, or the plan can be revised to best achieve the firm's goals given the changes that may have occurred in the market. Effective management of a mature product requires the development of a decision support system consisting of a decision information and control system. With such systems the full profit potential for the product can be earned.

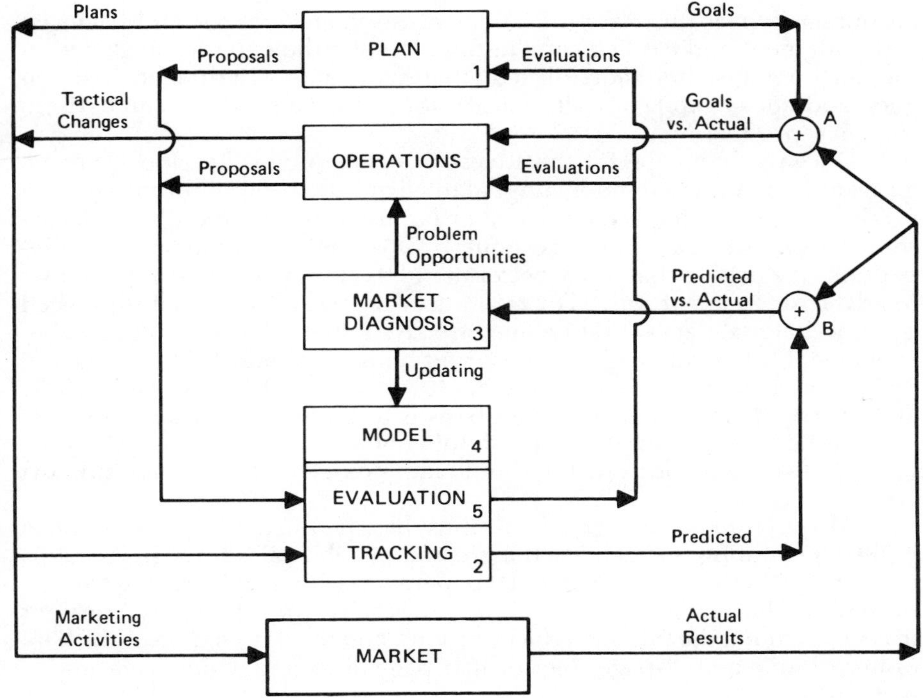

Figure 17.5. *The Market Control System (Little, 1975, p. 631)*

MARKETING RESPONSE ANALYSIS

Once management has designed and implemented the basic structure of a marketing decision support system, the system can be used for decisions with respect to each of the elements of the marketing mix including advertising, selling, distribution, prices, and promotion. In this section we review selected methods and models for each of the marketing mix elements. From this review you should be aware of what is available and how it is used for managerial decisions. For details see Montgomery and Urban (1969), Kotler (1971), and Fitzroy (1976).

Advertising

Advertising objectives and the types of advertising change as the product matures. In design, we concentrate on initial copy to portray the core benefit proposition (CBP) and achieve trial. In the mature product, the task is to portray the CBP to entice non-buyers to switch to our product and encourage repeat purchase. Creativity must be encouraged and balanced with judgment and explicit copy testing. Despite the experience gained in the category, it is often more difficult to judgmentally discriminate between good and not so good copy alternatives for a mature product than to do so for a new product.

Once the best copy for the advertising campaign is developed, managerial attention turns to budgeting. Many firms use rules of thumb such as: (1) the same spending level as last year (adjusted for inflation), (2) allocate some percent of forecast sales revenue, or (3) budget enough to match competition. We believe that much better methods exist based on sales response to advertising expenditure. For example, the ADBUDG model supported by experimental data would be one effective initial approach. More elaborate models of advertising, response and budgeting exist (Little, 1975 and 1979; Rao and Miller, 1975; Lambin, 1976; Sasieni, 1971; Wittink, 1977; and Clarke, 1978). Their common basis is to budget by sales response to advertising. The modern manager should not wonder what he is getting for his advertising expenditure, but should know based on his decision support system.

After setting the budget, media must be scheduled. The basic problem in media planning is to achieve the greatest possible sales for a given media budget. To do this the manager, in cooperation with an advertising agency, must select the media plan. That is, they must select the number of insertions in each medium, the timing of the insertions, and the designation of the kind of insertion (e.g., full-page color or half-page black and white magazine insertions).

A number of models have been developed to address this complex problem (e.g., Engel and Warshaw, 1964; Little and Lodish, 1969; and Ness and Sprague, 1972). Careful use of these models can improve the effectiveness of the media schedule. Since much media exposure data is now computerized, such models can utilize this data to find the schedule that maximizes response (usually reach and frequency), subject to the budget constraint. As media costs rise, firms are becoming more concerned with efficiency and should use models to maximize the impact of their media expenditure.

In summary, there are three advertising decisions: (1) what copy to use, (2) how much to advertise, and (3) where to allocate advertising. The first decision should be based on the CBP and copy testing results, the second and the third should be set by using budgeting and media models. We have introduced the key decisions and have indicated some of the criteria to make these decisions. For more detail, the reader should refer to Aaker and Myers (1976), Mandell (1974), and Wright, Warner, Winter, and Ziegler (1977).

Selling Effort

For industrial products there is more emphasis on personal selling than on advertising, but even for frequently purchased consumer products personal selling to wholesalers can play a crucial role. In Chapter 12 we discussed how the salesman can be used to communicate the core benefit proposition and how simple models can be used to set the initial sales level.

As in advertising, we gain more information on sales response as the product matures. With the more exact sales response information available, we become more interested in the sales force budget and the details of assigning salesmen to accounts and territories. The solution approach to these problems bears a close resemblance to the models for advertising. First, fix the overall budget and then decide how to allocate salesmen to accounts. (See Davis and Webster, 1968, for a review of sales management.)

The size of the sales force should depend upon the sales and profit response to changes in the sales effort. As in advertising, the decisions should be made on the basis of response. (See Semlow, 1959; and Waid, Clark, and Ackoff, 1956). Since response depends upon how salesmen are allocated to accounts and territories, these models have been extended to reflect the effects of changing the size of sales force on the allocation of effort and subsequent sales response. One approach is to have individual salesmen estimate the response for each potential customer in terms of the probability of initial or repeat sales at various levels of call effort (number of calls per month). CALLPLAN (Lodish, 1971) is one model that uses such input to specify the number of calls on each customer to maximize sales subject to a total constraint on the number of possible calls. By changing this constraint, the response to selling effort can be estimated and the effects of changes in the size of sales force examined.

A number of improved models have recently been developed (e.g., Zoltners, 1976; Beswick, 1977; Beswick and Cravens, 1977; Lodish, 1975; Lucas, Weinberg, and Clowes, 1975; Parasuranan and Day, 1977), behavioral research is being done (e.g., Bagozzi, 1978), and it appears that sales managers will have new tools to more effectively allocate sales effort and plan the overall sales effort. We predict these models will become important decision support tools in the next ten years and will facilitate response analysis for sales management.

Distribution

The choice of a channel of distribution is an important decision for a new product, but for the mature product the channel usually is given. The channel member still conducts many activities that affect the sales of the product. For example, in consumer products the retailer controls retail price, retail advertising, quality of shelf positions and facings, and in-store promotion displays. In established products, the manufacturer wants to impact upon these activities through sales effort, trade promotion, wholesale price, and package size assortment. There is also a feedback influence from sales levels because retailers tend to promote and display items that consumers demand most.

The ultimate function of distribution is to increase sales or profit. Although the distribution incentives vary, each of these control variables can be modeled analogously to advertising. For each variable or combination of variables a response curve should be estimated. One can use judgment or detail the effects on awareness, availability, trial, and repeat as was done in Chapters 11 and 12. Once the response curves are obtained, tradeoffs are made to select the best allocation of distribution incentives subject to budget constraints. The overall response curve is then estimated and used when allocating resources among distribution incentives and the rest of the marketing mix.

For a complete discussion of distribution management as well as total channel strategy, we refer the reader to Stern and El-Ansary (1977). One of the key lessons is to budget by sales response and use careful models and measures to understand and estimate the underlying components of response.

Price

Price is a component of the core benefit proposition. It is set in design and modified through pretest and test market. It remains a control variable for the mature product. Higher prices give more return per unit sold, but usually decrease the sales volume. Furthermore, competitors often enter or retaliate with lower prices. Thus, the mature product manager must recognize and attempt to quantify the sales response to price. Again, judgment, experimentation, statistical analysis can be used, supplemented by models based on the learning curve and price elasticity. Given the sales response, the manager can select the price to best fulfill the organization goals. For example, the goal of a price change may be to increase short-run profits, maximize long-run profits, encourage growth, stabilize the competitive price structure, enhance the "fair" image of the organization, discourage entrants, avoid government investigation, or maintain the loyalty of middlemen.

Determining the sales response to price curve is often difficult, but must be done if the organization is to be effective in using price as a marketing control variable. For more details, see Monroe and Bitta (1978), or Green (1963).

Promotion

Promotion is a key element of the new product introduction because it produces early trial and gets the product off the ground. In Chapter 12 we considered the use of sampling, couponing, and price-off, and how to compare the cost and benefits to select the best introductory strategy. As the product matures, promotion remains a vital marketing mix control variable. Some product categories such as food products are characterized by frequent competitive dealing in which over 50 percent of the volume is sold with a deal promotion. In industrial products it is not uncommon for the salesman to offer a special promotion, price, and service package.

The response to such promotion is usually characterized by a period of high sales followed by a period of depressed sales due to stocking up by channel members or consumers. The magnitude of these effects depends upon the magnitude of the promotion. A typical time pattern is shown in Figure 17.6a. Note that the total gain in sales is more than is lost during the period of depressed sales. If this were not the case, the promotion would be difficult to justify. To apply this pattern we also must know the sales response to the promotion curve as shown in Figure 17.6b. To estimate the total response to promotion, read the magnitude of the response from the sales response curve, for given promotional intensity. This becomes the magnitude of the maximum increase in the time pattern in Figure 17.6a. Next apply the time pattern indices to the base level of sales to get predicted sales. The increase in sales is compared to the cost of promotion to determine the

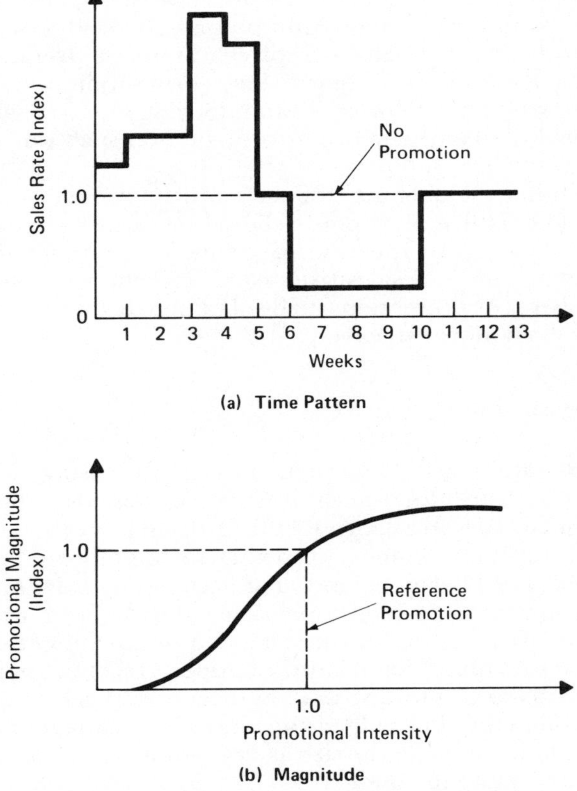

Figure 17.6. *Sales Response to Promotion (Little, 1975, p. 641)*

return on promotion expenditure. This is weighted against other marketing investments.

All consumers may not react to promotions in the same manner. In these cases the overall response may be disaggregated for each of several segments. For example, in frequently purchased products, higher income families may be more likely to have larger houses, more space to stock up on deals, and hence a higher response. On the other hand, their economic incentive may not be as great as low income families. Similarly, variables such as car ownership, age of children, or employment status can all impact on deal proneness. The product manager may want to consider the demographic mix of his trade areas and treat them differentially in modeling response. For more details on deal proneness, see Blattberg, Buesing, Peacock, and Sen (1978), and Montgomery (1971).

Although most promotions will produce sales increases as modeled in Figure 17.6, sometimes unexpected results may occur. If the promotion lowers the price too much, there could be an adverse effect on perceived quality which decreases sales. A consumer may not read a magazine that is received free, but would read it if some nominal price (say half-price) was required. Continued promotion of a frequently purchased product could give the product a low quality image. Savarin coffee was once a premium New York brand "served at the Waldorf," but with long-term promotion and little marketing support it became a "price" brand. (See Scott, 1976; and Sternthal, Craig, and Leavitt, 1980, for more information on such possible phenomena.)

Promotion is a valuable tool for the mature product as well as for the new product. The nature of response should be carefully modeled and experimentally validated. In the case of frequently purchased brands, the growing availability of daily sales from electronic check-out records means improved measures of promotion can be obtained, better models estimated, and profit managed more effectively.

Marketing Mix Interactions

Promotion cannot be set independently of advertising, nor advertising independently of distribution incentives, nor distribution incentives independently of price, sales force, product features or any other component of the marketing mix. The elements of the marketing mix are highly interdependent in terms of consumer and channel response. In addition to response interaction, another interdependence derives from the fact that all elements of the mix compete for the same marketing budget. For example, suppose that there is a $6 million annual budget. If $5 million is allocated to advertising and $900,000 to distribution incentive, this leaves little budget for other mix elements. The role of the marketing manager is to maximize sales, given the budget, using the size as given or achieve the greatest profit by setting and allocating the budget.

There is considerable judgment required in setting the levels of the variables in most managerial settings. With the underlying concepts outlined above integrated into a marketing mix model that models the interdependencies, profit can be more effectively managed. Several models of this type have been developed (e.g., Kotler, 1964; Lambin, 1972 and 1976; Pessemier, 1977; Little, 1975).

In the following case we discuss one model called BRANDAID and demonstrate how profit can be effectively managed by structuring response, utilizing judgments and statistical inputs, and tracking market performance.

DECISION SUPPORT CASE—BRANDAID

Model Structure

BRANDAID is an extension of the decision calculus concepts of ADBUDG to include additional variables and response phenomena. The market system considered by BRANDAID is shown in Figure 17.7. The participants include the manufacturer, retailer, consumer and competitor. Variables include price, promotion, sales effort, packaging, and advertising.

The model is based on sales response to each of these variables (Appendix 17.1 includes the mathematical equations). The sales level is computed by combining the individual sales response functions such as represented in the graphs in Figures 17.4 and 17.6, which show the proportionate change index for sales produced by a change in a marketing variable. The response index when multiplied times the base sales level provides an estimate of the sales effect of that change in the variable. Mathematically, sales is given by:

$$\textbf{(17.5)} \qquad s_t = s_0 f_1(x_{1t}) f_2(x_{2t}) \cdots f_m(x_{mt})$$

where s_0 = base sales

$f_i(x_{it})$ = the sales response function for the ith marketing mix element.

The use of this form makes it easy to add or delete details in computations since an element can be dropped from consideration by setting its index to 1.0. It also represents a multiplicative form of interdependencies between the variables. In the model, equation (17.5) is extended for competitive and dynamic effects at the industry and firm level (see Appendix 17.1). When necessary for managerial decisions, the computations are performed for each package size, market segment, or geographic area.

The model is supported by judgmental, historical, and experimental data analyzed by statistical procedures. It is integrated as part of the decision support system. Tracking is done based on the control system shown in Figure 17.5 so that annual and quarterly brand plans can be revised to best achieve the firm's goals.

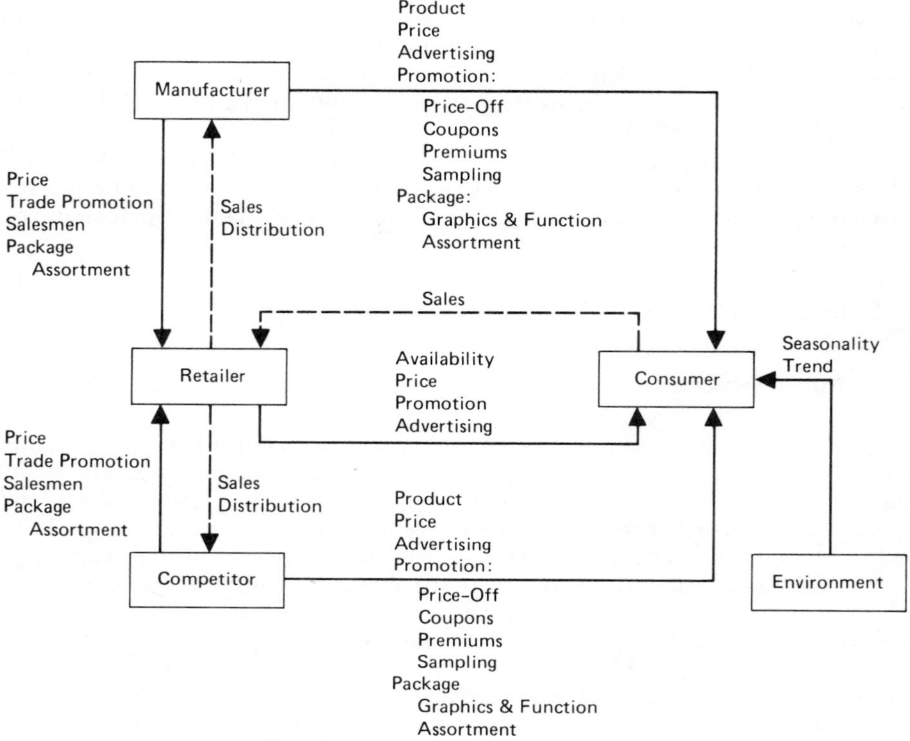

Figure 17.7. *Marketing System for One Consumer Packaged Good (Little, 1975, p. 632)*

Case History

We now illustrate the analyses with a case history. The product is a well-established brand of packaged goods sold through grocery stores. For confidentiality, the brand is called GROOVY. Figure 17.8 shows GROOVY sales (warehouse shipments) by month for months 1 to 36. Note the many fluctuations. One question asked was what caused these and how should they be managed?

In the initial application, the analysis was at the national level and the planning period was one month. Advertising, promotion, price, and seasonality were treated in the analysis. The management team, consisting of the brand manager, plus individuals with skills in marketing research, advertising, sales analysis, and management science, met about one-half day per week for a period of three months. The response curves were first estimated judgmentally, then checked econometrically. The peaks were the results of special promotions. Both advertising and promotion effects were estimated

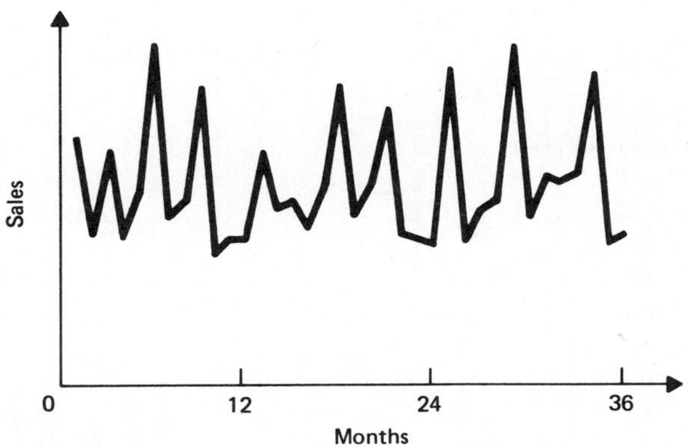

Figure 17.8. *GROOVY Sales (Little, 1975, p. 663)*

in the analysis, but management had more confidence in the promotion response function.

The first use of the model was for the annual brand plan. Table 17.1 gives (in coded form) five of the many plans generated. The brand manager presented these and others to upper management along with his recommendation for a plan with an additional promotion, increased advertising, and advertising reallocated to different periods. These changes implied a significant improvement in profit. Since there was some doubt on the advertising response function, the final decision was to approve the promotion, retain the existing advertising level, and begin a field measurement program in advertising.

Table 17.1. Alternative Annual Brand Plans[a]

Plan	*Advertising*	*Promotion*	*Relative Profit*
1	30% increase, previous allocation	Jan., June, Nov.	$1,210,000
2	6% increase, previous allocation	June, Nov.	980,000
3	6% increase, previous allocation	None	0
4	30% increase, previous allocation	Jan., June, Nov.	1,290,000
5	50% increase, new allocation	Jan., June, Nov.	1,390,000

[a]*Little, (1975), p. 664*

After the initial plan was developed, tracking began, and the precision of the response function improved. Comparison of actual to predicted sales uncovered a promotion missing in the historical data. Price data was obtained and the price response estimates improved. Figure 17.9 shows the indices developed from the response functions that were multiplied together to produce predicted sales (see equation 17.4). Figure 17.10 compares actual sales to the sales predicted by the model. The results are very good.

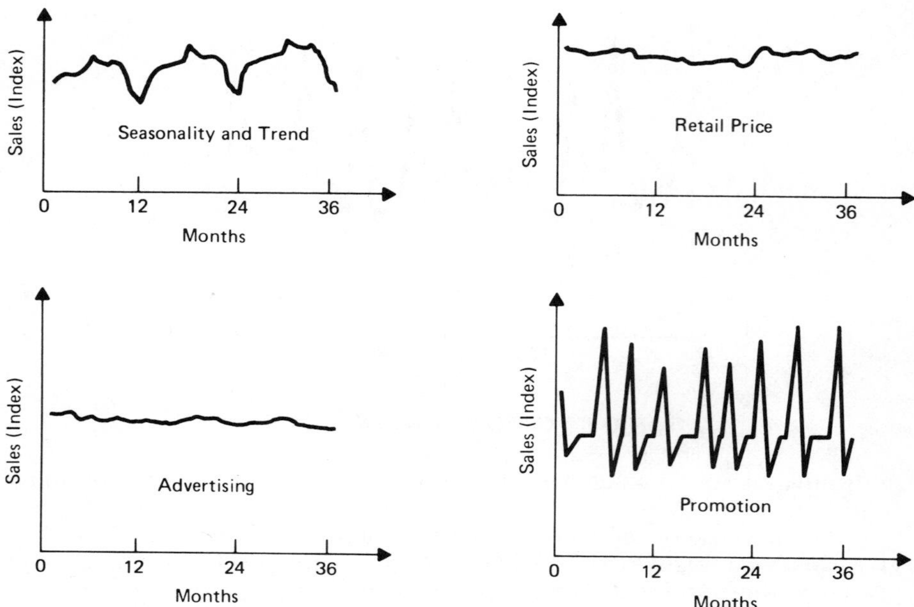

Figure 17.9. *The Results of Historical Company Actions as Modeled Through the Response Fuctions (Little, 1975, p. 666)*

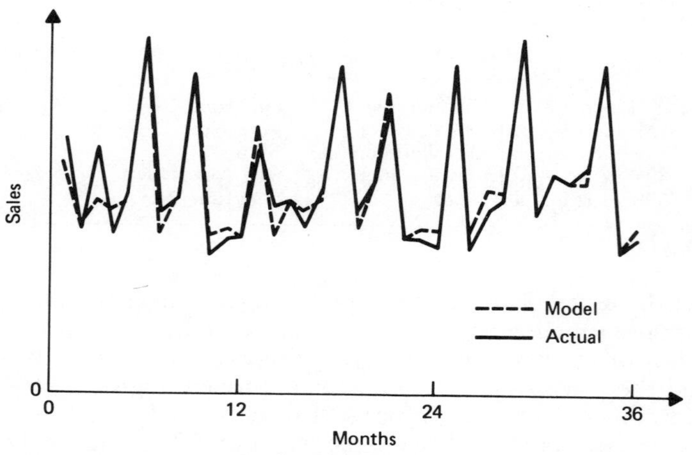

Figure 17.10. *Comparison of Predicted Sales with Actual Sales (Little, 1975, p. 667)*

The model worked well on past data, but how well did it predict the future? The answer is shown in Figure 17.11a. The model did well in the first months, but then deviated from actual sales. A closer inspection reveals that the model did well until a strike and a new package size disrupted sales. Market research results indicating the response effect of the new package size were entered and the strike effect was retrospectively fit to correspond to the actual results (Figure 17.11b). After adjusting to these unexpected results, the model tracked well. In the subsequent five years the model tracked well and has proved useful for price, promotion, and promotion planning.

An example of managerial decision support with the model occurred after the tracking periods in month 78. The year-to-date sales were substantially ahead of the previous year, but based on model forecasts and analyses the brand manager announced the brand was in trouble. An additional promotion had been run in month 72 that had not been implemented in month 60. In addition, price had been increased in March. Sales were depressed more than they would have been from a normal price effect, because the change pushed the retail price across a psychological barrier of 50¢. Its effect on sales was masked in part by a large TV special and coordinated promotion which featured the brand among others.

Tracking months 72 to 78 and model computations made it clear that, although year-to-date sales were good, much of the annual advertising budget had been spent, the price had been increased, and the sales picture for the rest of the year was bleak. The brand manager proposed an additional promotion. On the strength of his model supported case, management accepted the recommendation and profits for the year were increased. This is a good illustration of the importance of detailed monitoring of the marketplace. Without prior warning, by the time the losses were detected in actual sales, it would have been difficult to plan and execute the promotion.

Summary of the Case. Over a period of five years, analysis of Groovy improved annual planning and resulted in increased profits. Response based on statistical and experimental analyses has helped improve the model's ability to predict the sales effects of revision in the marketing mix. The model based control system has led to a quick and definitive diagnosis of environmental and consumer preference variations. It has provided a tool for effective reaction to those unexpected events so that profit can be effectively managed. BRANDAID is only one of many models that could be used to manage a mature product, but this case implies that a model based design support system can aid management of mature products.

INNOVATION IN THE MATURE PHASE OF THE LIFE CYCLE

New product development is an iterative process which can continue until the product is dropped. Creativity and good managerial sense are im-

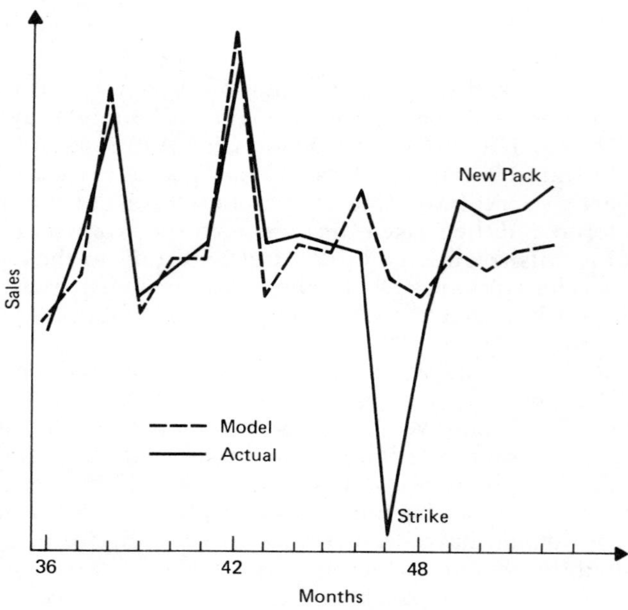

(a) Without New Phenomena

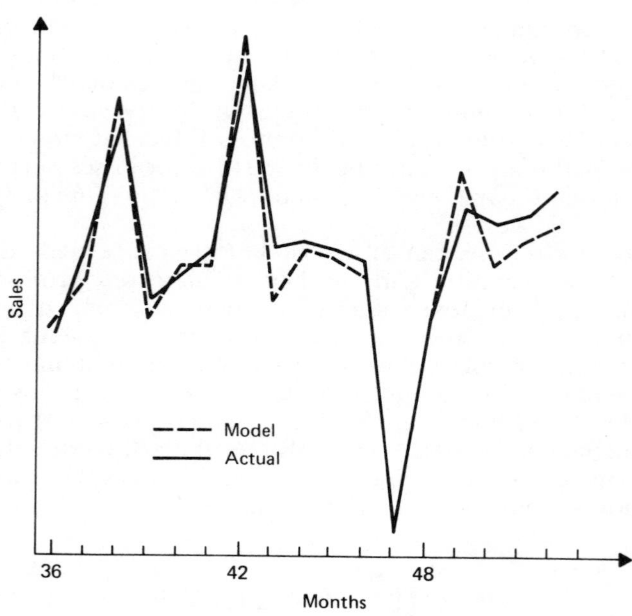

(b) With Strike and Package Effect

Figure 17.11. *Comparison of A Priori Predictions with Actual Sales (Little, 1975, p. 668 669)* portant to insure that profit rewards are fully realized from the product. Up to this point we have concentrated on the innovation as a whole and in particular the CBP. As the product matures, production innovation is possible to cut costs, and market expansion through "flanker" brands is possible to increase revenue.

Production Innovation

Production process innovation can reduce costs, improve product performance, and increase profits (Abernathy and Utterback 1978). Specialization increases at this time and economies of scale occur due to large volumes of production. In the mature phase of the life cycle, products tend to be less differentiated and price competition increases. In this phase production efficiencies can enable lower prices, higher market shares, and increased volumes of production. While product innovation is most important in the early phases of the life cycle, in many industries process innovation is most significant in the later phases where productivity increases and economies of scale lead to concentration in the number of manufacturers (Utterback and Abernathy, 1975). The lowering of costs described as the "experience curve" in large part is due to innovations in production and do not result automatically as cumulative volumes increase. Firms should direct attention to innovating the production process in the mature phase of the life cycle as a source of increasing and lengthening the flow of profit.

Flankers

Once the basic product is established in the market, the organization may be able to expand into new benefit segments with minor product innovation. These flankers expand the product line to tap specific subsegments and to defend the product line from competitive product elaboration. For example, the product variation may be a "new flavor" or new "extra strength" variety. Flankers can also extend the life cycle by adding more enthusiasm to the advertising and marketing, or by widening the product's appeal.

Flankers are often tested by pretest market procedures and introduced nationally. Most are not test marketed since the costs and risks are small. Creative efforts to generate product extensions plus a good low cost screening procedure can lead to lengthening the life cycle.

MANAGING THE DECLINE PHASE

Eventually the product reaches the decline phase of its life cycle. Whether the decline results from shifts in consumer tastes, rises in production costs and price, regulations, or pressure from a superior product, the decline phase should be managed. Management must decide whether to drop the product, milk the profit from the product as it dies, or enter a new life cycle (recycle) through a major repositioning.

Dropping the Product

Continued support of a product must be looked at as an investment. If the marginal revenue from a product drops below its marginal cost and if there are no countervailing considerations such as firm image or tie-ins with other products, the product should be dropped (Kotler, 1965; Hise and McGinnis, 1975; Hamelman and Mazze, 1972). Even if the marginal costs are less than the revenue, the product may be "pruned" from the line if there are more attractive investments in other products produced by the organization.

While eliminating a declining product can be emotionally difficult, it must be done if the organization is to thrive. The method of elimination can be done by termination or milking. In cases where the product still has some consumer loyalty and inertia generated through years of production and marketing, the organization can milk high profits by withdrawing marketing support as the product dies.

Repositioning and Rejuvenation

An alternative to the drop decision is major repositioning against a new market based on major product innovations (Levitt, 1965). Figure 17.12 shows this effort to rejuvenate as a recycling of the product back to the initial phases of the new product development process. The current definition of the market is reexamined to indicate the best opportunities. Ideas for CBP repositioning are generated and translated into concepts and products for evaluation and refinement. If they are successful, testing and introduction would take place.

The repositioning strategy would be most successful when the brand name of the old product is strong and would lend credibility to the new positioning. For example, Volkswagen repositioned its new line of cars by including the "Rabbit" and a racy Italian-styled model called "Scirocco." The Beetle was dropped as a major line item, but the image of reliability and economy of Volkswagen carried over to the new line. This rejuvenation of the life cycle was based on major product innovation and a good psychological positioning. With the repositioning, VW was launched into the growth phase of a new life cycle.

PRODUCT PORTFOLIO MANAGEMENT

The new product becomes part of a product line which is one component in the portfolio of products offered by the organization. Each product

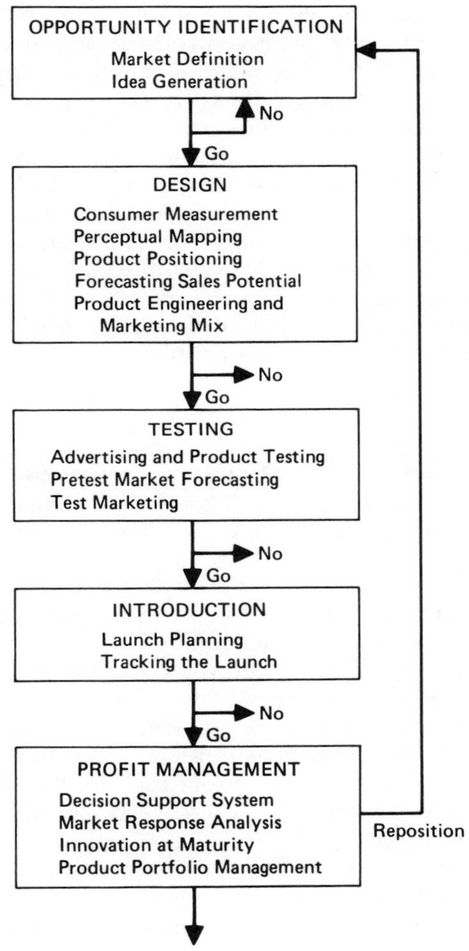

Figure 17.12. *Recycling the Declining Product*

line should be carefully positioned in this portfolio and resources carefully allocated between them. Considering this problem from a product line perspective, many of the techniques discussed in this book are applicable to the product line as a whole. This section reviews the concepts of product line management and suggests how the techniques of the book can be used to support portfolio decisions.

Positioning the Product Line

A product line is a set of related products. For example, Kellogg's cereals, Campbell soups, Christian Brothers wines, and General Motors' Chevrolets are all product lines. A product line is more than a collection of products. It is an entity with an image and composition. For example, the quality image of the line can be raised with a few prestige products or compromised with one or more "bargain" products. Since the line as a whole faces many of the problems faced by a single product, the techniques of this book can be used for the product line. For example, we can view a "line" as a "product" to exploit markets identified in opportunity identification, or we can use idea generation techniques for the line. The line has an image which can be analyzed by perceptual maps and positioning studies. When a single product is tested in a pretest or test market, the information on cannibalization and draw can be used for product line decisions. Product line positioning can be addressed through management science models introduced in this text. The success of this endeavor depends on creativity and persistence in translating product specific techniques to product line problems.

Composition of the Product Line

In addition to overall image, management must address the issue of selecting specific products to make up the product lines. A good product line will cover all the important market segments, but have little overlap between the products in the line.

In Chapter 5 we indicated some approaches to identifying markets. These methods are useful in managing the product line since they help identify which products overlap and where gaps exist. If each product in the line is a different branch of a hierarchically defined market, the coverage and duplication levels will be good. For example, in the case of coffee cited in Chapter 5 (Table 5.7), General Foods had good coverage of the market while Nestlé only covered the instant coffee market. In that example, the question was raised whether General Foods was suffering some duplication by Brim and Sanka instant competing in the same sub-market, and if Nestlé should widen its product line by adding a ground coffee.

Product lines often are the result of incremental growth in the past and do not represent a coordinated product line. It is important to examine the product line for coverage, but it is also important to avoid duplication. Often firms are reluctant to drop products. But products that are late in their life cycle and compete in market segments in which the firm has newer products should be considered for elimination in the light of product line considerations. Product proliferation prevents the concentration of resources on high growth and share products. A careful analysis of the loss of revenue due to dropping the old product and the gain in revenue due to redeployment of the released resources is necessary for successful product line management.

Allocating Resources

The planning of a portfolio of product lines should be addressed first from a global perspective. This portfolio perspective is to allocate the organization's scarce resources to achieve the desired balance or risk and expected benefit across the product lines that make up the product portfolio.

One approach to portfolio analysis is the concept of "stars," "dogs," "cows," and "questionables" that was outlined in Chapter 5 (Boston Consulting Group, 1970; Abell and Hammond, 1979, pp. 173–194; and Day, 1977). In this framework, product lines are placed in one of four classes dependent on their rate of growth and the firm's market share. "Stars" have high growth and high market share, "dogs" low growth and low share, "cows" low growth and high share, and "questionables" high growth and low share. The resources would be allocated toward "stars" and away from "dogs." If the market share response to resource allocation to "questionables" was profitable, it would also be undertaken. "Cows" require little additional support and serve to generate the cash necessary to support "stars" and to move "questionables" to "star" status.

This overall framework is useful in identifying the strategic notion of building share in growth product areas. After allocating resources to product lines, one also should consider allocation within lines and the specific interdependencies between the products in that line. Some products may be complementary and help each other, and others may to some degree substitute for each other. If complementary or substitution effects exist, these interdependencies should be reflected in the marketing mix of each product.

The interaction at the marketing level has not been widely studied. One approach to interdependency is by cross elasticity (Urban, 1969), but due to data and technical estimation problems, managers usually use judgment to assess the effects of one product's price, advertising, and selling strategy on others.

In addition to demand interactions, other interdependencies may exist due to a constraint on the total funds available to support the product line. Even if there are no market or production interdependencies, allocating the best individual budget for each product may not be feasible since the aggregate spending amount is limited. In these cases the marginal profit for each unit of allocation should be equated to maximize the total profit.

Sometimes environmental constraints force product line allocations to be made. For example, auto manufacturers must meet government efficiency (CAFE) standards for their total sales of cars. This means if a below average miles per gallon car is sold, an above average miles per gallon car must be sold to meet the overall fleet average. This interaction has led auto makers to increase prices of large cars with low efficiency and to direct resources toward small high efficiency cars. Although a "Pinto" does not compete directly with an "LTD," they are interdependent due to the gasoline consumption requirements on Ford's total sales. The task is to set the prices and marketing mix to maximize profit for the total line while meeting the government efficiency standards.

Risk

The final explicit consideration in managing the product portfolio is balancing risk against expected benefit. In testing (Chapter 13, Figure 13.1) we used the concept of risk-benefit tradeoffs to establish a decision frontier. A similar concept can be used in portfolio decisions.

In order to grow, an organization must take some risks, but it must also proceed with enough caution to avoid catastrophic failure. So it must balance risk and potential benefit. For example, a good product portfolio could include a few high risk, high potential benefit products balanced by some low risk, moderate potential benefit products.

A useful way to approach this problem is to quantify risk and expected benefit for each potential product in the portfolio, quantify the interdependencies in risk and benefit, and compute the risk and expected benefit for various realistic combinations of products. The alternative portfolios are then selected based on the organization's preferences with respect to risk and benefit. (For formal procedures to compare risk and benefit, see Keeney and Raiffa, 1976.)

SUMMARY

The profit earned in the mature phase of the life cycle is the reward for the creative innovation, investment, and risk taking in new product development processes. To assure that these rewards occur, management should develop an information and control system based on models, measurements, statistics, and managerial judgments. With a good control system, management can manage the product's advertising, selling, distribution, price, and promotion to adapt to changes in the consumer, competitive, economic, political, and technical environments.

As a product matures, its place in the product line and portfolio must be understood. Resources need to be allocated to maximize the total profit of the portfolio and not just the profit of the new product. To do this requires a detailed understanding of the interdependencies between products and the marginal response to resource allocations.

The consideration of composition of the product line may lead to dropping some products as the decline phase of the life cycle occurs. Those which overlap or are in the decline phases of their life cycle are candidates for elimination. However, before a product is eliminated, managerial action to improve profit should be considered, including production innovation to reduce costs and improve profit, and major market repositioning to rejuvenate the life cycle.

REVIEW QUESTIONS

17.1 What is a "flanker" and why can it be very profitable?

17.2 What intuitive interpretation can you give to the sales response curve found in Little's ADBUDG?

17.3 When employing a decision calculus approach, what are the various sources of information? What are the advantages and disadvantages of each source? Is there any advantage in gathering information from several sources?

17.4 What types of problems are encountered when setting the marketing mix for a mature product which are not present when introducing a new product?

17.5 What short term and long term effects does a deal or promotional effort have on a mature product? Would you expect sales immediately after a deal or promotion has ended to be higher or lower than before the promotion? Why?

17.6 What are the basic steps to determine the best price for a mature product? Would this price be higher or lower than the best price during the product's first introduction?

17.7 How is the management of a product line different than the management of a single product?

17.8 How does a "product portfolio" differ from a "product line"?

17.9 Why might a good product portfolio include some high risk, high potential products in addition to low risk, moderate potential products?

17.10 You are the product manager for a line of office equipment. How would you use the following concepts in managing this line?

a. Core Benefit Proposition
b. Adaptive Control
c. Opportunity Identification

Chapter 17 Appendix

17.1 Equations of BRANDAID*

BRANDAID is a marketing-mix model developed by Little (1975) to provide sales forecasting for combined marketing mix strategies. In the text we discussed the managerial issues, presented a numerical example, and an application. Here we summarize the mathematical equations. We would like to stress that BRANDAID is a flexible model. Applications should fit the model to the situation, not the situation to the model. Most applications will use only some of the equations, although some applications will need to expand and generalize the model. These generalizations will be especially important if there are complex interdependencies among various elements of the marketing mix.

Let π_t be the profit, let g_t be the gross contribution per unit sales, let s_t be the sales, let $c_t(i)$ be the cost of the ith marketing mix element, and let $c_t(o)$ be fixed cost—all subscripted by time t to indicate that there are estimates for time period t. The goal is to maximize profit given by:

(17.1A-1) $$\pi_t = g_t s_t - \sum_{i=0}^{M} c_t(i)$$

Sales is given by base sales, s_o, modified by the response functions, f_{it}, where f_{it} is a real valued function equal to 1.0 for base level spending and representing percentage increases (decreases) in sales response. The form of interdependency is multiplicative, i.e.:

(17.1A-2) $$s_t = s_o \prod_{i=1}^{M} f_{it}$$

*This appendix is based on an article by Little (1975).

Note that equation (17.1A-2) is equivalent in concept to equation (17.5) in the text with revised notation.

Advertising. Let x_{1t} be the advertising rate in time t. Since advertising is a complex quantity, we let it be composed of its component parts. That is, let h_{1t} be media efficiency (exposures/\$), k_{1t} be copy effectiveness (dimensionless), and let a_{1t} be the spending rate, all in time period t. Then the advertising rate is given by:

(17.1A-3) $$x_{1t} = \frac{h_{1t}k_{1t}a_{1t}}{h_{1o}k_{1o}a_{1o}}$$

where h_{1o}, k_{1o}, and a_{1o} represent base levels. Let f_{1t} be the advertising index in time period t.

Since advertising is likely to have a lagged effect, the advertising response is modeled with an autoregressive component plus a direct effect:

(17.1A-4) $$f_{1t} = \varphi_1 f_{1,t-1} + (1 - \varphi_1) r_1(x_{1t})$$

where $r_1(x_{1t})$ is the direct effect and φ_1 is a constant. The model can be customized in certain circumstances. For example, a memory effect can be included by substituting $\hat{x}_{1t}$ for x_{1t} in equation (17.1A-4) where $\hat{x}_{1t}$ is given by:

(17.1A-5) $$\hat{x}_{1t} = \beta_1 \hat{x}_{1,t-1} + (1 - \beta_1) x_{1t}$$

where β_1 is a memory constant. One can further customize the model by making total advertising a weighted sum of different types of advertising. (See Little, 1975, p. 639.)

Promotion. As explained in the text, we can expect a time pattern to be associated with promotion. Let $q(\tau)$ be the time pattern relative to base for the τth period after the promotion. Let x_{2t} be the promotion intensity and let $r_2(x_{2t})$ be a scale factor that depends upon the promotion intensity, x_{2t}. Usually only a portion of the product line is promoted. Let l be that portion. Furthermore, let b be the fraction of sales gain that comes from cannibalizing the rest of the product line. Since reference sales usually contain promotions, let f_2 be the reference sales if no promotions are made. Then for a promotion run at time, t_p, the sales index, f_{2t}, that results from that promotion is given by:

(17.1A-6) $$f_{2t} = f_2\{1 + l r_2(x_{2t_p}) [q(t - t_p) - 1] (1 - b)\}$$

If more than one promotion is run, then index each promotion in the schedule by p. The promotion index is then given by:

(17.1A-7) $$f_{2t} = f_2\{1 + \sum_p l_p r_{2p} (x_{2t_p}) [q_p(t - t_p) - 1] (1 - b_p)\}$$

As with advertising, one can customize the model to cover efficiency and consumer effectiveness. Simply formulate a promotion index in a manner analogous to equation (17.1A-3).

In a promotion, we must also model the variable costs. Let $c_t(2)$ be the cost of promotion at time t. Let $c_t(0, p)$ be the fixed cost of promotion p at time t. Let $[\tau_1, \tau_2]$ be the interval during which promotion allowance is paid on sales. Define $I(t) = \{p \mid (t - t_p)\epsilon[\tau_1, \tau_2]\}$. Then

(17.1A-8) $$c_t(2) = \sum_p c_t(0, p) + \sum_{p\epsilon I(t)} \cdot \left\{ P_p(t_p) l_p s_t \left(\frac{f_2}{f_{2t}}\right) [1 + (q_p(t - t_p) - 1) r_p(x_{2t_p})] \right\}$$

where $P_p(t_p)$ is the promotion offer at time t_p. $P_p(t_p)$ may or may not equal x_{2tp}, depending on whether x_{2t_p} is modified by cover efficiency and consumer effectiveness.

Price. There will be an overall effect due to price response, plus possibly an effect due to end-pricing. Let p_t be the manufacturer's price, let x_{3t} be the normalized price index, let $r_3(x_{3t})$ be the response to price, and let $\psi(p_t)$ be an additional effect due to retail price-ending. Then the price index, f_{3t}, is given by:

(17.1A-9) $$f_{3t} = r_3(x_{3t})\psi(p_t)$$

where $x_{3t} = p_t/p_0$ where p_0 is the reference price.

Sales Force. Let a_{4t} be the sales force effort rate ($/customer/yr), let h_{4t} be the cover efficiency (calls/$), and let k_{4t} be the effectiveness on store (effectiveness/call). Then the normalized sales force effort index, x_{4t}, is given by:

(17.1A-10) $$x_{4t} = \frac{h_{4t}k_{4t}a_{4t}}{h_{4o}k_{4o}a_{4o}}$$

where h_{4o}, k_{4o}, and a_{4o} are reference values. To complete the response, we include a memory effect and an autoregressive component, i.e.:

(17.1A-11) $$\hat{x}_{4t} = \beta_4 \hat{x}_{4,t-1} + (1 - \beta_4)x_{4t}$$

(17.1A-12) $$f_{4t} = \varphi_4 f_{4,t-1} + (1 - \varphi_4) r_4(\hat{x}_{4t})$$

where β_4 and φ_4 are constants.

Other Influences. Other influences such as seasonality, trend, pack assortment, and package changes can be handled by direct indices. Trend can also be treated by a growth rate. Let r_{5t} be the per period growth rate in period t, then the trend index, f_{5t}, is given by:

$$\textbf{(17.1A-13)} \qquad f_{5t} = f_5 \prod_{\tau=1}^{\tau=t} [1 + r_{5\tau}]$$

where f_5 is the base level index at $t = 0$.

Competition. Competition is handled analogously to the way we handled direct sales effects. Each effect (competitive advertising, competitive promotion, competitive price, etc.) can go into the model as a direct index or a submodel, depending upon the detail of information available about competition activity. These combine to produce "unadjusted" competitive sales, s'_{bt}, i.e.:

$$\textbf{(17.1A-14)} \qquad s'_{bt} = s_{bo} \prod_{i=1}^{M} f_{bit}$$

where equation (17.1A-14) is analogous to equation (17.1A-2), except that b indexes the brand and s_{bo} is the base level sales for brand b. Let γ_{bc} be the fraction of brand c's unadjusted incremental sales that comes from brand b, then the adjusted sales, s_{bt}, for brand b are given by:

$$\textbf{(17.1A-15)} \qquad s_{bt} = s'_{bt} - \sum_{c \neq b} \gamma_{bc}(s'_{ct} - s_{co})$$

A useful assumption is that a brand draws its incremental sales from competing brands proportional to their sales. Then

$$\textbf{(17.1A-16)} \qquad \gamma_{bc} = \frac{s_{bo}(1 - \gamma_{cc})}{\sum_{c \neq b} s_{co}}$$

where γ_{cc} is a constant equal to

$$1 - \sum_{c \neq b} \gamma_{bc}$$

Retail Distribution. The model can be extended to include retailer (and wholesaler) activities. Let $I_m = \{i_1, \ldots, i_m\}$ be the set of manufacturer activities (as described above). Let $I_R = \{i_1, \ldots, i_R\}$ be a set of retailer activities and let $I_E = \{i_1, \ldots, i_E\}$ be a set of environmental and other influences. Then

$$\textbf{(17.1A-17)} \qquad s_t = s_o \prod_{i \in I_m} f_{it} \cdot \prod_{i \in I_R} f_{it} \cdot \prod_{i \in I_E} f_{it}$$

Thus indices can be created for such activities as salesman effort, trade promotion, package size assortment, shelf facings, point of purchase effects, and end-aisle displays. Direct effects, memory components, and autoregressive components can all be handled in a manner analogous to equations (17.1A-3) to (17.1A-5).

The models and submodels that are available in BRANDAID are summarized in Table 17.1A-1.

Table 17.1A-1 Principal Sales Influences and Their Treatment in BRANDAID. Any Sales Influence Can Depend on Brand, Package Type, Time, Segment, or Combination of These[a]

Sales Influences	Model Options	
	Direct Index	Response Submodel
Manufacturer's Control Variables		
M1. Product Characteristics	√	
M2. Price	√	√
M3. Advertising	√	√
M4. Consumer promotion		
(a) Price-Off	√	√
(b) Sampling	√	√
(c) Coupons	√	√
(d) Premiums	√	√
(e) Other	√	
M5. Trade Promotion		
(a) Price-Off	√	√
(b) Other	√	
M6. Salesman Effort	√	√
M7. Package		
(a) Graphics and Function	√	
(b) Assortment	√	
M8. Production Capacity	√	√
M9. Other	√	
Environmental Influences		
E1. Seasonality	√	
E2. Trend	√	√
E3. Other	√	
Retailer Activities		
R1. Availability	√	√
R2. Price	√	
R3. Promotion	√	
R4. Advertising	√	
R5. Consumer Sales at Fixed Distribution		√
R6. Other	√	

[a]*Little, 1975, p. 653*

PART VI

IMPLEMENTING THE NEW PRODUCT DEVELOPMENT PROCESS

CHAPTER 18

Organizing the Development Effort

Decision making within the ordered sequential analysis of the development process is best made within an organizational structure that effectively mobilizes the diverse resources from R&D, engineering, marketing, sales, production, finance, and top management. To be most effective a formal assignment of responsibility and authority must be made, working relationships defined, goals and reward structures established, and staff recruited, trained, and motivated. Without an organizational entity that has the defined responsibility to produce new products, few innovations will result. Some group must make the emotional and resource commitment to develop new products.

A good formal organization is important, but some "well-structured" organizations fail, while some new products succeed despite the organization. With enough energy, a "champion" can make the process work in almost any organization. Three of the cases cited in this book succeeded to a certain stage in spite of an organizational weakness, but could go no further until that weakness was overcome. The squid study was funded by a government agency (National Oceanic and Atmospheric Administration). A market opportunity was identified and a new product was designed, but without an organization to gain the support of fishermen, distributors, and branded fish packaging companies, no product was brought to the market. The HMO study was done for a small HMO without a new products director who could shepherd the innovation through the internal implementation process. Many recommendations were put into effect through the energy of the medical director, the business manager, and others, but not as quickly or as effectively as if a new service director were hired. In the Sloan School of Management case, the new master's program was not established until a separate dean was appointed to oversee the implementation of the changes. Although innovation can occur within any structure, it is better to have an organizational structure that works for, rather than against you.

This chapter discusses some of the organizational issues of new product development. Alternative organizational structures are presented and critiqued. In this discussion we distinguish between formal and informal roles. Formal organizations (as defined by an organization chart) can help or hinder innovation, but below any formal organization lies an equally strong informal organization which really determines how well the process is implemented. Some formal organizations encourage good informal systems while others do not. The informal system of responsibility is defined by the roles various actors play in the development process. A good new products manager should recognize the strengths and weaknesses of the formal organization and should ensure that the informal roles are filled when the development process is put into effect. To give you practice in addressing organizational problems, we close the chapter with the identification of some of the common organizational problems that occur and outline some approaches to solving them.

FORMAL ORGANIZATION

There are many ways to organize. In any particular company the best organizational structure will depend upon the strengths and weaknesses of the firm, the orientation and skills of the officers, and previous organizational history. This section reviews and critiques many of the alternative organizational forms for new product development. Any particular firm may wish to combine two or more structures.

Research and Development

One home for new products, especially high technology products, is in the R&D department. The advantage of an R&D department is that the new product development effort is near the technological research and the product development capability. But the influence of the market, the consumer, and intermediate users may be under-represented if the department has primarily an R&D outlook. Another problem is that the time frame may be too long, jeopardizing management's rate of growth objectives.

If management weighs the pros and cons and places new product development within R&D, they must decide whether to organize around underlying scientific groups or as specific project groups with assigned new product responsibility. In a study of 19 labs (7 chemical, 5 drug, 4 petroleum, 3 electronic), Mansfield et al. (1971, pp. 40-42) found that the form or organization chosen depended upon the degree of basic research. Most firms separated basic research from development if over 24 percent of R&D funds went to basic research. In development efforts most labs used a product or project form of organization, while in basic research they were more likely to use the scientific discipline or functional form of organization.

In another study, Allen (1977, pp. 211–220) evaluates project and functional organizations. Based on 38 large, long-term projects (median size $4 million, average duration 3.4 years), Allen found the functional organizational form produced results that were technically rated as being more successful, but were somewhat more subject to cost overruns. Based on his analyses, he suggests that the choice between functional (i.e. scientific discipline based) and project organizational forms depends on the project duration and the rate of change of the technical knowledge (see Figure 18.1). For short duration projects or those based on technologies that are not changing rapidly, project organization is recommended. For assignments that involve rapidly changing knowledge bases or are of long duration, a functional organization is appropriate. Since most new product efforts are applied R&D projects of shorter duration, the project form of organization would be appropriate in most cases. For example, if the product were new aircraft instruments utilizing minicomputers, the project form would be most appropriate. Alternatively, a new product effort requiring basic R&D based on emerging technologies might find the functional form of organization best.

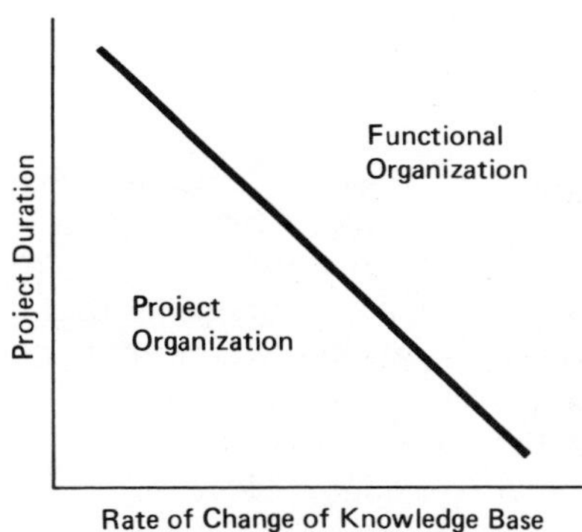

Figure 18.1. *Organizational Structure as a Function of Project Duration and Rate of Change of the State of Knowledge in the Fields Involved in a Project (adapted from Allen, 1977, p. 219)*

Marketing: Product Managers

Some firms assign the responsibility for new products to the marketing department, which has the advantage of placing a heavy emphasis on consumers and markets. However, there are dangers. The time frame of marketing may be too short, and new products may represent only minor improvements in existing products; also major technological opportunities may be neglected.

Since resources from departments besides marketing are necessary to develop new products, many firms have evolved the product manager system (Gemmil and Wilemon, 1972) to link the various functional areas that underlie a new product's success. The product manager has overall responsibility for a product and must work with the sales, finance, R&D, and production departments to coordinate an organized marketing of his brand. Although the manager does not have authority over these groups, he gains cooperation based on his expertise, personal influence, and ability to supply evaluative information to management about specific managers and departments.

Although the product manager form of organization is most commonly used for established products, some firms assign product managers to develop new products, and others encourage new product work by managers of established products. The advantage of this system is that the product manager is already familiar with the complexity of product management. The disadvantage is that the product manager tends to be accustomed to and rewarded for short-term results. If the new product is an extra responsibility, it may tend to be neglected. As a result, product manager systems may tend to produce too many product line extensions or minor modifications. Major product design and development projects may not fit well with the usual product manager's skills and orientation. Product manager systems are most appropriate for incremental growth through product improvement rather than "breakthrough" innovations.

New Products Department

Because of the long-term, high-risk nature of products, many firms establish a separate department to integrate and coordinate the company's capabilities and bear the responsibility for product innovation (see Lorsch and Lawrence, 1965). This organizational form has the advantages of making clear that innovation is a high priority activity, balancing R&D and marketing, bringing a diverse set of skills into its staff, and freeing itself from the short-run pressures of existing business. Such a group may be called a "growth and development department" to emphasize its responsibility to develop major new innovations that build new business areas for the company.

One disadvantage of a new product department is that instead of linking R&D and marketing, it may be viewed as an interloper by both groups. The R&D and marketing directors are power centers in many companies and they may resent a third force based on new products. A clear commitment from the chief executive officer of an organization is needed to make a new products department work. Another problem is that the new products department may be too structured to facilitate the identification and utilization of entrepreneurial talent. Although it can be an organizational base of major developmental efforts to enter strategic markets, it may not attract people with individual ideas and the ability and determination to pursue them.

Entrepreneurial Division

If a firm does not facilitate internal entrepreneurs, they are likely to leave. Roberts (1968) found that one electronics organization was the source of 39 independent companies which grew to a total sales volume of over twice the parent company. Some firms have attempted to capture this talent by forming an entrepreneurial division (Roberts and Frohman, 1972). Entrepreneurs are given security and the opportunity to create and develop their ideas in a nonbureaucratic setting. Funds are available to recruit a product team or buy support services from other parts of the company. The outcome can be a new product for the company which the entrepreneur may manage over its life cycle, or a spinoff company in which the parent may have a substantial share. If the idea does not develop, the manager can return to his original division. This is an important advantage for the firm. Instead of driving the entrepreneur outside the firm to try his idea and search for a new career if it fails, an internal method has been provided so that if he fails, the company retains the valuable employee.

One of the problems in an entrepreneurial division is that the risks and the expected costs of success are high. Since the benefits may be correspondingly high with the creation of a major innovation, the firm may find this activity a viable component in its growth and development strategy and a supplement to a new products department.

Corporate Structures

Often new product development is a high priority within the corporation's goals. Some firms use a high level corporate structure to represent the various points of view of management. There are at least three common organizational forms: (1) new product committee, (2) task force, and (3) small staff reporting to the chief executive officer (CEO).

The new product committee is usually made up of the vice presidents of the major departments and the president. This group may set priorities, screen ideas, and coordinate the implementation of product ideas (Johnson and Jones, 1957). The details of the development effort cannot be handled by the committee, so they must be accomplished in the component departments and coordinated by the committee. One advantage of this system is that it directs top management's attention toward innovation.

The task force form of organization gives more direct responsibility to the corporate group, but usually the task force is a special corporate structure of limited duration. The advantage of this system is that the firm's top talent is mustered to meet major challenges. For example, Motorola used a task force to successfully penetrate the market for electronic automobile ignition, combustion, and emission controls. The disadvantage is that the task force tends to work on a project rather than a program which may result in the sacrifice of some of the risk spreading and synergies associated with a program effort (review Chapter 3).

The third form of corporate structure is a small staff that reports directly to the president or the chairman of the board. This group reflects the strategic perspective of top management and attempts to initiate projects in the organization that will fill the long-run product/market posture the firm desires. This staff would do early market definition and idea generation work and, after developing specific projects for departments, monitor the progress of the new product. The advantage of this system is top level responsibility and continuing effort, but in order to make it work, the staff must be able to coordinate and encourage the follow-up efforts in the departments.

Overall corporate structures have the advantage of top level interest and corresponding power to get the job done, but they must face the challenge of making sure that the details of new product development are carried out effectively.

Matrix Organization and Venture Groups

A matrix form of organization has been used effectively by many firms in developing new products (Galbraith, 1971, 1973).In this form a person reports to two supervisors. For example, an R&D person in integrated circuits may report to the head of this functional area in R&D and to the director of a project to develop control circuits for microwave ovens. These dual reporting relationships can be used to integrate efforts in new products. A person may be assigned to a "venture group" that is given resources and freedom to meet the responsibility of developing a particular product (Hanan, 1969; Jones and Wilemon, 1971; Hill and Hlavacek, 1972). A person may be assigned full time or part-time to the team. The team leader and the original supervisor would both make performance evaluations and promotion recommendations. Since the venture group tends to stay together only for the duration of the project, the decisions on personnel advancement are influenced by the group's performance and the individual's other activities. If the duration of the effort is short and part-time, the original supervisor will be a stronger influence. If the duration is long and full time, the team leader's input will dominate.

The advantage of the matrix system is that it allows integration of diverse skills at the working level along with a clear definition of the priority of innovation. It does require flexibility due to the dual reporting relationship, but can be effectively handled by mature management. Its disadvantage is the complexity and the mixture of responsibility that can cause conflict if not managed correctly.

Multi-Divisional Groups

Most large firms have many divisions. One approach suggested by the principle of decentralization is to place the new product activity at the lowest level where the information and the development capabilities exist—the divisional level. But the existence of many divisional-level new product groups can produce inefficiencies and duplication of effort across divisions.

An alternative approach, which overcomes some of the duplication, is to have corporate groups such as R&D, marketing research, advertising, strategic planning, and operations research that would service the divisional new product departments. The divisional new product departments would have a small staff, but budget to buy services from the corporate groups. This ensures the availability of the required talent for the division, but allows the corporate group to pool projects so developmental efforts are not duplicated. The disadvantage is the need to establish effective communication to ensure that the corporate groups are receptive to the divisional departments and that the divisional departments understand the services that are available from the corporate groups.

In some cases, a corporate new products group might also be given the task of building businesses where no division now exists.

Outside Suppliers

Not all firms will have the resources to organize an internal new product development effort and will want to consider outside services. Even in multi-divisional structures it is healthy for corporate service groups to be ready to meet competitive services available outside the firm.

Although most services are reputable, some are not, and all vary in the degree of their service and capability. It is important that the purchasing firm understand the basic concepts of the new product development process so it can evaluate the alternative services and get the most for its expenditure.

Outside services are numerous. "Boutiques" will create, design, and test the product. Advertising agencies will often be available to develop product positioning, specify marketing plans, and run test markets as well as develop advertising copy. Various major consulting companies (e.g., Arthur D. Little; Booz, Allen & Hamilton, Inc.) offer research and management services for new product development. Market research firms can collect and analyze data, and many offer marketing consulting as well. Valuable support may be obtained from suppliers of raw materials. They may do R&D to create new end uses for their products. In the case of Teflon cookware, Du Pont not only developed the product, but advertised it to consumers to pull it through the manufacturing and distribution channels.

A wide range of outside services exists. They tend to be costly, but may represent a viable alternative for some firms, especially small firms, to acquire creative talent and services for specific projects. Sometimes the total cost of an internal staff is higher than using outside suppliers on an interim basis or as a new creative stimulus.

Top Management Involvement

One common element in all organizational structures is the need for top management involvement. Although involvement is needed, too much interaction could be undesirable if it becomes viewed as "meddling" by the staff. Figure 18.2 describes a good balance of involvement and defines the decisions required of top management in an organization that has a growth and development department.

First, a clear policy on innovation should be formulated to describe

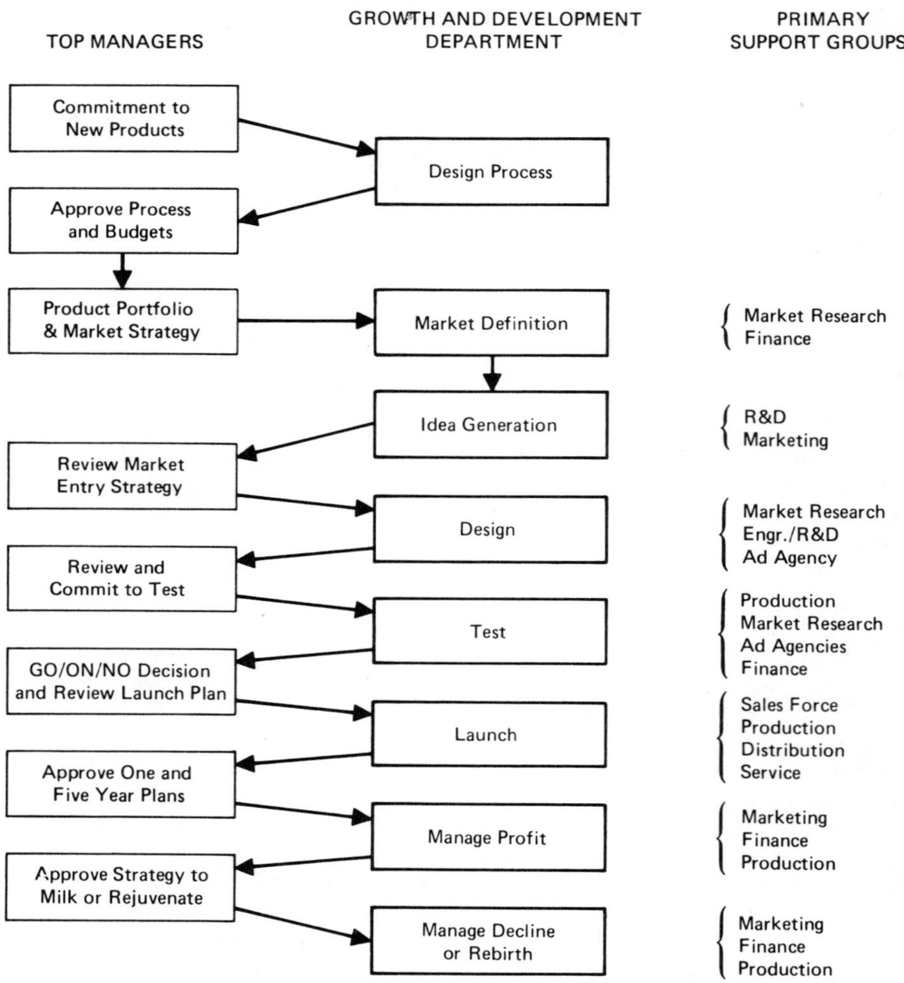

Figure 18.2. *Top Management Involvement in New Product Development Process*

growth goals, specify the organizational structure, enumerate constraints, and define the new product commitment. This is the basic input necessary for the growth and development department to design the sequence of steps that will comprise its development process and the time and resources necessary to meet the firm's growth goals. Top management then should review these resource requirements. This is important since, as indicated earlier, the expected time and cost of innovation is higher than many would suspect (review Chapter 3). Top management must have a realistic expectation of the times and costs involved. With this approval and a clear understanding of top management's corporate product and market strategy, opportunity identification and idea generation are undertaken.

The entry strategy should be reviewed for consistency with corporate strategy and budgets approved for the specific new product development opportunity. After a core benefit proposition (CBP) is specified, positioned, and fulfilled by a physical product, top management should review the progress and decide if the testing phase should begin.

After the testing (e.g., product and market tests) is complete, top management must make the GO/NO GO decision, or recycle the product for more testing and improvement with an ON decision.

If a launch is undertaken, top management should review the launch plan and the subsequent annual and five-year plans that will lead the product to maturity. The final involvement is in the decision to rejuvenate or milk the product when it reaches the decline phase of its life cycle.

Although more management involvement could be productive, Figure 18.2 describes a good level. In committing to a proactive growth and development strategy, top management should also commit to a reasonable level of involvement in the process, or success will be unlikely.

Emerging View

We have indicated a number of formal organizational forms for new product development. All require top management involvement, but they are different in many aspects. Each has its strengths and each has its weaknesses and no one form will serve all needs. The number of complex alternatives is large and there is little scientific evidence to indicate which form is best under which circumstances. Therefore, you must rely upon judgment, anecdotal experience, and your capabilities in selecting the right organizational form for your firm.

For the large firm, current thinking suggests a growth and development department with a small staff, supplemented by venture teams drawing members on a project basis from operating functional areas such as R&D and marketing. The department would also have funds to buy services from corporate or outside organizations. Major innovations using new technologies or tapping new markets would be addressed by this group, while product line extensions would be the responsibility of product managers in the marketing department.

A portfolio of approaches may exist. A corporate entrepreneurial division could exist to assure that entrepreneurial talent and ideas are utilized to create new businesses. For example, 3M has such a division in addition to new product efforts within divisions. The large firm could have a complex formal organization based on matrix management concepts with responsibility assigned to a new products department on a divisional basis, but supported by corporate services and development.

Small companies would probably rely on a task force or new product committee and outside suppliers for new product development. Development probably would be an intermittent activity and not warrant a separate new product department.

Although there is not a large body of knowledge on formal organizations, there is a real need for effective organization of the new product development effort if successful products are to be created. To select a good organizational form for your situation, you should understand the new product development process, recognize the strengths and weaknesses of your firm, and use your managerial common sense. In making this decision, do not evaluate only the formal organization; also evaluate how the formal organization encourages or hinders the informal organization, and vice versa.

INFORMAL ORGANIZATION

If the product is to succeed, the various development tasks must be done by individuals. These tasks will not be accomplished by simply placing names on an organizational chart. They must be assumed by people working together. Interpersonal relationships and personalities of the people in the organization will affect results. The new product manager and top management must recognize and organize these relationships and encourage their proper functioning (see Farris, 1971). Individuals will fill various informal organization roles within the formal structure. In this section we define some of the major informal roles that should be recognized and executed.

Champion

> And it is worth noting that nothing is harder to manage, more risky in the undertaking, or more doubtful of success than to set up as the introducer of a new order. Such an innovator has as enemies all the people who were doing well under the old order, and only halfhearted defenders in those who hope to profit from the new—*Machiavelli*.

A new product needs an advocate who will champion its cause within the organization. The idea has to be "sold" to many people. Objections must be overcome and energy and enthusiasm to proceed generated. Some of this may be done in formal presentations, but much is done by informal contacts at social or business functions. The champion must believe in the idea and be willing to take career risks to defend it and attract resources to further it.

Protector

An allied role is that of the protector. This is a senior person who can intercede to save the champion when the established power centers of the company fight back. The protector does not overtly advocate the idea, but strongly defends the champion's right to do so. The protector is also a "coach" to the champion. He tells him of friendship, power, and interpersonal workings of the organization and advises the champion on how and when to make the correct contacts and statements to further the project. The protector also provides legitimacy and maturity for the product development effort.

Auditor

While the champion pushes on, it is necessary that the organization not be misled by enthusiasm. The auditor balances the enthusiasm of the champion by assuring that the sales forecasts are realistic and accurate. Analytic capabilities are important in this function. The champion is critical and the skills to make this role effective are not common. The auditor helps the organization make sure that the champion is being directed toward the correct product project so his capabilities will be used effectively. The auditor function may be fulfilled by a member of the product team or by an independent person or group outside the development team.

Controller

Allied to the auditor's role is the controller of the new product effort. The controller must watch the schedules and budgets. Cost overruns should be minimized and efforts made to meet time objectives. Most R&D cost overruns occur on projects that are eventual failures (Meadows, 1968). There is apparently a reluctance to give up on an idea even after it becomes clear that it should be terminated. The controller should be willing to treat past inputs to a project as sunk costs and allocate funds based on future results.

Creator/Inventor

The new product effort needs creators and inventors to successfully design and market new products. Unfortunately, this creative talent can be vastly underutilized if the formal structure imposes too many constraints. Creativity does not come automatically by establishing a new products committee, group, or department. In fact, a rigid structure can stifle creativity. Creative scientists and marketing managers must be identified and involved in the new product process.

If formal creative procedures (Chapter 6) are used, the people who demonstrate the most originality should be identified and involved in the design and problem-solving process. In the technological area, creativity also depends on scientific information. Managers should recognize and manage the "gatekeeper" effect (Chapter 6) in which a few people become the "gate" through which much new information flows. The creative scientist and the gatekeeper should be nourished and supported. Tapping their skills will be based on a clear understanding of individual capabilities and the problem-solving process in the lab.

The creative talent of a firm may be difficult to pinpoint. Giving a wide variety of people the chance to participate in problem solving will help identify those who have the most creativity. Everyone is creative to some degree if put in the right circumstances. The good manager encourages those circumstances and rewards creativity. For example, a production man was recruited for a creative group as part of a cross-section of managers. When asked to evaluate his own creativity, he said, "I cannot do much with new concepts. I am not a creative type." After three days in a group session he came up with the single most innovative product concept identified during the meeting. In another situation a large midwestern bank found that they were losing their most creative people, partly because these people were not allowed to initiate their own projects.

Leader

New product development is a team effort which needs a leader. The leader must recruit a staff, mold the talent of this group into a powerful team, and develop effective plans. The leader must motivate, teach, and train members of the team.

In motivating the team, the leader should recognize that the formal reward of pay is not the most effective motivator. Group recognition, career advancement, and the granting of responsibility are more powerful (see Myers, 1964). The simple assignment of interesting tasks is a big motivator—particularly when they are successfully accomplished. Working relationships are also important. Friendly, cooperative functioning within the team that recognizes the needs of individuals on the team will build morale. The leader must maintain this enthusiasm through the periods when the product future looks bleak. The leader must show strength in pursuing the project through these difficult times and may have to demonstrate courage to kill the project should it clearly prove unwise to continue. Termination of projects that will fail is critical to maintaining the group's respect.

New product development requires special skills. The skills necessary for new product work are often increased on the job and the leader must manage the human resources of the group to develop the capabilities of each person to his maximum level (Schein, 1977).

Motivation and training are necessary to keep the team functioning. The leader also provides the interface with other parts of the organization by acting as a translator and an integrator. The new product idea must be rephrased in R&D terms, marketing jargon, or in the top management semantics. This translator function is one of facilitating communication and being sure the idea is correctly perceived by all parties. In the integrator function, the leader collects diverse information from numerous sources in the organization to support decision making and coordinate actions (Galbraith, 1973). He is important in developing a consensus and integrating different points of view (Souder, 1977).

The leader must fulfill many functions and, in most cases, is the person formally heading the new product effort. There have been cases where the head of new products was the "champion" and the team "leader" was an unrecognized member of the group who earned this position of respect from his peers.

Another level of leadership exists at the top of an organization. If the president or CEO does not demonstrate strong support for product innovation, the new product effort will probably fail. If the CEO shows real leadership in this area, it can be very successful. Mr. Land at Polaroid is a good example of a man who spearheaded the innovation effort from the top of the organization and introduced a number of successful new instant cameras.

Strategist

It is often easy to become involved with new products for their own sake. To avoid this trap and to keep the firm's strategic goals in the forefront, the new product development effort needs a strategist with a long-run managerial perspective. For example, up until 1976, Evanston, Illinois, had a dial-a-ride system that served the elderly residents. The system was innovative, but expensive, and provided poor service. Recognizing that the city's goal was better service for the elderly rather than simply innovation, the city government replaced the dial-a-ride system with a subsidized taxi system that provided better service and was less expensive for the city and for the elderly.

The role of the strategist is to set goals for growth in sales and profit, select market priorities, and allocate resources. It is not innovation per se that is important, but innovation as a means to achieve long-range goals. The strategist must select the overall development strategy (Chapter 3) and set the budget.

We suggest the strategist be a person at a high level in the company (see Figure 18.2). Some top management people do very little strategic thinking, so this function may fall on the director of new products. The director may undertake market definition studies (Chapter 5) and request a budget based on strategic analysis instead of waiting for a message from the top. After top management has reacted and revised the director's plan a working strategy can result.

Judge

As new product development proceeds, honest differences of opinion are certain to occur. People will disagree about whether the product should be introduced or not. Some will feel that the product is reducing their sphere of influence and power or that it will jeopardize some successful on-going ideas. Conflicts can result in these situations.

It is important to have a method of resolving or at least managing this conflict. One role is by a mediator who attempts to find a common ground of agreement and widen this to a working consensus. But in many cases a judicial role must be invoked. The opposing sides should be heard, and then a GO or NO GO decision made on the product.

It is healthy to involve top management in resolving major differences since the new product commitments are large and strategic. This should not be neglected by corporate management (see Gluck and Foster, 1975, for a revealing case history). Some top managers like an approve or disapprove proposal and therefore someone below the top level decides which alternatives will be given a GO/ON/NO GO review. Other top managers like to see a wide range of alternatives. Some want advocates of each to present them and then to exercise judicial authority in making final commitments. Many styles exist, but the judicial process must be clearly defined.

Relationship Between Formal and Informal Roles

Organization is not complete when a chart is drawn. The informal roles of champion, protector, auditor, controller, creator/inventor, leader, strategist, and judge must come into being within the formal organization. Some of these may correspond to the director of new products, but other roles will correspond to other people within the organization.

For example, Figure 18.3 is a map of communication in a R&D laboratory (Allen, 1977). It is clear that formal and informal communication is affected by the formal organization sections (A to D). However, the informal gatekeeping function often is present. A statistical analysis of two laboratories indicated that although the formal structure is an important determinant of communication, the informal organization makes its own independent contributions of nearly equal magnitude.

In considering formal and informal interaction, recall the major effects the geography of the office plays (review Figure 6.4 and Allen, 1977). The closer people are together, the more they communicate. A venture team with adjacent offices will have more interaction than one that is geographically separated and only is in close proximity during meetings. This is particularly important when integrating R&D and marketing. Although it is more expensive, it may be wise for the R&D people to have two offices—one in R&D and one in the new products department—so they can be close together on days allocated to the new product project. The use of office geography is important, but only beginning to be understood as a mechanism to improve the organizational functioning of the new product team.

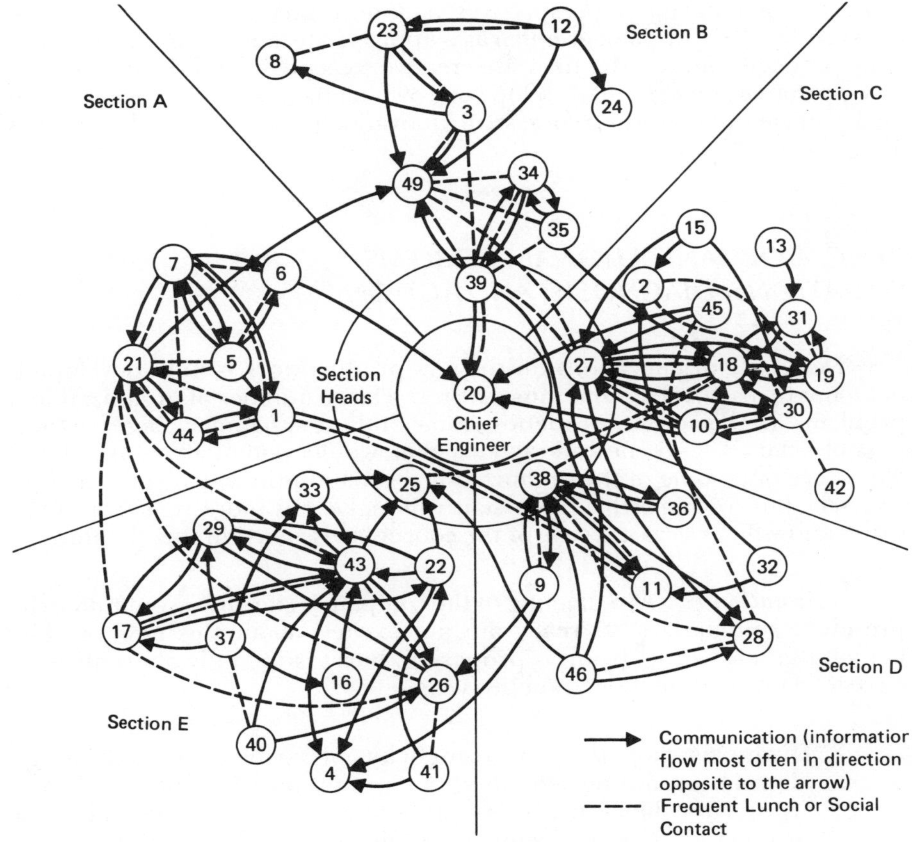

Figure 18.3. *Communication Network in a Typical R&D Laboratory Showing the Influence of Formal and Information Organization (Allen, 1977, p. 208)*

Implementing Informal Roles

Some concepts have been proposed, but you must exercise your own judgment in developing an organization. Experimentation is encouraged. The product manager and matrix forms of organization were developed by managers facing difficult problems and not by academics. Evaluation of these two organizational structures indicate they can be effective. New forms need to be created. In organizing your efforts, do not feel constrained to rigid charts and formal organizational thinking. Try to capture the concepts outlined in the discussion of the formal organization: link R&D and marketing, assign priority and responsibility, encourage entrepreneurship, integrate top management perspectives and necessary working level skills, decentralize development as much as is appropriate, and do not neglect new product talent outside the firm. Be creative, receptive, and flexible, recognize people problems, and develop your own managerial style. Encouraging an organization to produce successful innovation is difficult, but the rewards are great.

TYPICAL ORGANIZATIONAL PROBLEMS: SITUATIONS, DIAGNOSIS, AND ACTION

Designing organizations for the new product work is still a developing field, but firms now face and must address a large number of organizational problems. In this section we identify some of these situations, diagnose their possible causes, and indicate actions that would potentially solve them. These are only some of the organizational situations in which you may find yourself, but we hope this discussion will make you more sensitive to the issues and will serve as a review of the concepts discussed in this chapter.

Situation 1. At a meeting of the company's executive committee the president summarizes, "I really am not excited about this new product launch the marketing group is proposing, but it is the only alternative we have to get something to market this year."

> **Diagnosis:** Here the firm is caught by not having generated enough alternatives and developing them into pretest concepts and prototypes. Now they must accept more risk than they should. The new product work is apparently being done in marketing and the new product effort is not of large enough scale.
>
> **Action:** Set up a new products department and give it top management priority. Budget for a stream of ideas and involve management in the selection before test market. Develop an explicit decision process with GO/ON/NO GO elimination steps.

Situation 2. "We spend 10 million dollars per year on R&D and we have not had one major new product success in five years."

> **Diagnosis:** This firm apparently is relying on R&D to carry out all the new product work. R&D's perspective is probably too highly oriented towards technology.
>
> **Action:** Link R&D and Marketing in a new product venture group or department.

Situation 3. President laments, "Why are we always caught behind our competitors in product innovation? Why is it that when we are first in, we lose our competitive advantage within a year?"

> **Diagnosis:** The competitor apparently has an effective new product effort. The competitor can innovate and quickly copy and improve on "first-in" products.
>
> **Action:** Strengthen your new product development effort to be competitive: more people and funds. Stress major innovation. Get more R&D and creative talent working to identify breakthroughs. Do more strategic analysis earlier to define market opportunities. Preempt the competitive second-but-better entry by being sure the new product is in the best position and has the best physical features to back it up.

Situation 4. The Director of Growth and Development complains: "We have been trying to get into that category for the past five years and have not yet succeeded."

> **Diagnosis:** A number of reasons can explain the failure to "crack" a category, but one prime reason may be that the firm has artificially constrained itself to a low growth and highly competitive category. Alternatively, the category may be open to innovation, but the firm has not found the opening.
>
> **Action:** The Director of Growth and Development should question the strategic decision to enter that category by undertaking a market identification study. Perceptual mapping should be done and any gaps (or the lack thereof) should be identified.

Situation 5. R&D Director: "We send five breakthrough new products to marketing each year and they don't do anything with them!" Product Manager: "We want a product modified to expand our market and R&D cannot deliver or comes back with an answer too late." Marketing Director: "Ideas are cheap! What we need is an *action* man who can take an idea to market." Sales Director: "We get the craziest new products. They divert our effort from already profitable products."

> **Diagnosis:** Clearly there is little communication and coordination between the groups. R&D is not getting market input. The product manager is thinking of line extensions and the marketing director is portraying his bias toward fast results. The sales director probably already has problems meeting his sales goals and is not convinced that new products will help.

Action: Share developmental goals. Leadership and coordination are needed. Top management could use a task force or product committee to integrate at the upper management levels. A matrix form of development with a junior person from each area in a venture team could be considered as a supplement to top level coordination. A new products director who can be a team leader is needed, and a protector should be identified for him.

Situation 6. President: "We have had four new product directors in the last six years and none of them could deliver sales and profit results."

Diagnosis: A number of reasons could exist for this, but apparently the new product directors are being judged by results after one year. The expected time to get a product from idea to market is four to five years (see chapter 3 and 19).

Action: Set evaluation criteria based on progress through the process and successful completion of the design, pretest, test market, and launch phases. Give the new product director adequate tenure and budget to meet the sales goals. If you want major innovation, pushing for one-year results will prevent it by forcing consideration of minor modifications of products to get to the market fast.

Situation 7. Founder and Chairman of the Board: "Why can't we grow a business in the same way as I built this company? Start small, work hard, and produce a good product."

Diagnosis: The founder was an entrepreneur. Today's markets are much more complex and highly structured and require more sophistication, but it is still possible to find opportunities by entrepreneurship.

Action: Set up an entrepreneurship division to support people with ideas and determination. Look for entrepreneurs inside and outside the company. Start many projects (50–100) since only a few (2 or 3) will succeed. Major new opportunities may be found in this way that would otherwise have been overlooked. This activity can be used to enhance the efforts of a new products department.

Situation 8. President: "All our new products look like our existing products. Why can't we come up with some really new ideas?" Product Manager: "I try, but it is difficult to find breakthrough ideas when I must spend most of my time solving the problems of my existing product."

Diagnosis: The product manager is too involved with firefighting on the product he is now managing. The product manager does not have the time or is unwilling to take the risk involved in major innovation.

Action: Set up a new product group or department with responsibility for major innovations. Continue to reward product managers for "flankers," but set up a reward system for the new department to encourage careful analysis aiming toward breakthroughs. Give this department sufficient time and resources to undertake the risk of major innovation.

Situation 9. R&D Director: "Our lab work is good, but we don't do much with it, and either a competitor exploits it or one of our employees leaves and starts his own company."

Diagnosis: The firm's technological capabilities are not being utilized and entrepreneurial talent is being wasted.

Action: If the new technology does not match the priorities of the new products department, let the inventor take it to the entrepreneurial division and see if he can make it work. If it does, a new product is born. If not, the inventor's talent is retained in the firm. Be sure R&D is part of the venture team so that technological opportunities can be utilized in new products.

Situation 10. Chairman of New Product Committee: "This product has been worked on for five years, but if I understand this group, there is a feeling we need to do another pricing study and be sure our guarantee plan is financially stable."

Diagnosis: This is probably a product that should be killed. ON-ON-ON instead of a GO or NO GO decision is wasting resources.

Action: Pose the GO/NO GO decision and have one person (say the president) make it. The committee is probably trying to be nice to the sponsor when a hard business decision is called for. The new product effort should concentrate on major winners, not middle majority ideas that are likely to generate mediocre profits if they do get to market.

Situation 11. President: "Well, we really failed in the market! What went wrong?" Market Research Director: "We had many negative indicators all along—but no one would listen." Product Manager (champion): "Market research always tries to come back and say I told you so. If we followed them, we would never introduce a product! If firm X had not cut their prices to cost, we would have had a winner." Vice President of Growth and Development: "I have never understood those market research models and I have always felt new products is a place where managerial judgment is most critical. We missed, but new products is a risky business."

Diagnosis: Apparently the champion was very effective and marketing research very ineffective in communication. The champion probably could sell anything. The protector (Vice President of Growth and Development) supported him. The models and data were not understood by the Vice President and therefore not heard by top management.

Action: Market research personnel should be on the venture team. A special new products research group could be established in the growth and development department to ensure effective communication. Market research and the analytic people should be trained to understand and communicate with management. Management should be trained to understand market research and realize when their judgment is needed. Both the quantitative and qualitative skills must be utilized for the implementation of a successful new product development effort.

Situation 12. Product Manager: "Why should I spend part of my budget on market research? After all, our firm has survived in this category for almost ten years without such spending."

Diagnosis: The firm has survived, but has it thrived? Here the product manager is unwilling to invest money and effort on intangible research. This strategy causes opportunities to be missed and, in the long run, causes the firm to be caught with only products that are in decline stages of their life cycles.

Action: Implement a new product development process and reward innovation. Set a budget based on expected benefits and thus quantify the intangible so that it can be balanced against other investment opportunities.

Situation 13. ". . . . ." Fill in the blanks when you are on the job. This is the situation you will face!

Diagnosis: Be sensitive to organizational issues. Spend time trying to understand what is going on around you. Think of the informal as well as the formal organization. A position is filled by a person and is not merely a job description. People have personal as well as organizational needs. What would you do in their job?

Action: We hope this chapter has helped give you some perspectives and concepts, but you will have to experiment. We hope through this experimentation you will advance the state of the art in organizing the new product effort.

REVIEW QUESTIONS

18.1 What are the major decisions a company must make concerning the organization of the new product development effort?

18.2 What are the advantages and disadvantages of a product manager approach to organization?

18.3 In the matrix form of organization, how might channels of authority and responsibility be confused? What rewards would an individual have for cooperating with the innovative effort?

18.4 When choosing a "venture group," how might individuals be chosen? What criteria should be applied?

18.5 What are the advantages of categorizing the informal roles of people influencing the new product effort? What managerial conclusions can be made based on these classifications?

18.6 How do you judge outside suppliers?

18.7 Discuss the role of top management in new product development.

18.8 Discuss how you would encourage and reward people for filling the informal roles necessary for new product development.

18.9 Situation 14: Vice President of Growth and Development, "My staff is too technical, all I ever get is numbers and computer printouts." Analyze.

18.10 Situation 15: Chairman of the Board, "My new product group is motivated. There is always action and excitement. We have had five new products in the last two years and expect five more products this year. My problem is that while sales are growing, profits are declining." Analyze.

Chapter 19

Customizing the New Product Development Process

This book has presented many concepts and analytic tools to aid in the successful development of a new product. In many cases we have presented a portfolio of techniques so that each organization can select those analyses most appropriate to their situation. Part of this customization process was discussed directly or illustrated through the cases and examples. Many of these examples were frequently purchased products developed by large companies. This is not surprising, since such companies tend to have more experience and resources to use in the development of consumer-oriented products. Whenever possible we have cited the recent applications to services, consumer durables, and industrial products.

But many of you are not with (or planning to be with) large consumer goods companies, and are probably asking whether the techniques are applicable to your situation. If you are with a firm with gross annual revenues of $2 million, you may not wish to invest $100,000 in analysis. If your company sells computers worth $50,000 to $100,000, you may not be concerned with direct consumer advertising or with the trial-repeat behavior. Even if you are with a large consumer goods company, you may find that your personal style or organizational pressures force you to emphasize certain components of the development process. How then can this text help you?

Consider an electrical engineer designing a digital heart rate monitor. The engineer is probably aware of the properties of a portfolio of electrical components and of numerous ways to connect these components, but he would be a poor engineer if every product he designed used every one of those components. The skill and art of engineering is not knowing just one set of components and one way to put them together, but rather knowing the technical properties of each component and being able to decide how they relate to the complete product design. The good engineer uses those components in a creative way to do the job. Your task is similar. The concepts and tools of this book represent a portfolio of techniques. By knowing their strengths and weaknesses and by knowing what they do and at what cost, you can creatively customize the new product development process to fit the needs of your organization.

For example, Tom Hatch (the manager in Chapter 4) was with a large consumer goods company. But even in this case, he customized the process to match his managerial style and strategic emphasis. He relied heavily on advertising and on the use of analytical as well as qualitative methods. In his use of models, he has found it personally valuable to use a test market trial and repeat purchase model (SPRINTER) early in the opportunity identification phase to assess the sales potential and profit implications of his judgment. Although SPRINTER is usually applied at the testing phase, the use of it as a unifying structure for the progression from idea generation to launch is an appropriate customization of the modeling technology. By understanding the concept and tools, Tom Hatch was able to select a process that fit his organization. We hope you will do the same.

This chapter raises the issues involved in customization and outlines how we would customize the process to large, small, and entrepreneurial businesses and for frequently purchased, durable, industrial, and technological product businesses and services. The central theme of customization is to select and assemble the tools in a way that balances risk, cost, expected benefit, and the needs of the organization.

After an organization has tailored a set of decision steps, concepts, and models for its use, a budget should be set to enable its implementation. We indicate some of the issues in budgeting and close the chapter with an identification of some of the constraints that limit the new product development effort.

VARIATIONS BY SCALE OF THE ORGANIZATION

Opportunity identification, design, testing, and introduction are necessary to the development of a new product. All organizations should go through these phases in one way or another. Large organizations with the ability to commit sufficient resources can afford detailed analysis and would be well advised to spend resources up front to avoid failure and enhance success. Other organizations may not have the same resources (time, money, personnel), but should still use the key concepts of the development process. Table 19.1 outlines some of the variations in the process that result from customization based on scale of the organization.

The first column represents the typical large firm like General Electric, General Foods, or IBM. Substantial sums of money are available to do the appropriate research, usually drawing on over $100 million in sales. These firms can minimize their risk since they have capital available to do analytic studies in the market definition, design, and test market phases. Tom Hatch's description of the development process he used at Miles Laboratory would fall in column one. In the large firm emphasis is placed on understanding the market and targeting creative effort to build a new product in this area with R&D support. The development of the core benefit proposition (CBP) is the critical focus of the design phase. The successful product delivers perceived CBP benefits to the consumer, but these benefits must be substantiated by real product features. Analytic modeling and market research efforts are undertaken to thoroughly understand the linkages of physical product features to the positioning of the CBP versus competition and the implications for preference and choice. The large firm is also able to identify and exploit segments in the market by supplying benefits in the CBP that appeal differentially to some groups.

Table 19.1. Variations in New Product Development Activities by Size of Organization

	Size of Organization		
Phase of Development	*Large Firm or Division*	*Small Business*	*Entrepreneur*
Opportunity Identification	Analytic Market Definition & Segmentation Study Creative Group Sessions R&D	Focus Groups Creative Groups	Look to Consumers Generate Alternatives
Design	Development of CBP Models of Perception, Preference, and Choice Product Features to Fulfill CBP Benefit Segmentation R&D Marketing Link	CBP Orientation Concept Test Product Placement Engineering	CBP Concept Consumer Relations Engineering Financial Support
Testing	Formal Advertising Testing Lab and Consumer Product Testing Pretest Market Model Analysis Test Market	Product Testing Pretest Market Model Analysis Monitor Roll Out and Improve	Prototype and "In-use" Tests Sell Some Revise Product Sell More
Introduction	National Launch Adaptive Control	National Penetration	Build Business

After developing the advertising and promotion, a large firm would formally test the copy. Both lab and consumer product testing would be done to ensure that the CBP has been fulfilled and to develop the best product formulation. Pretest models would be used to minimize the risk of failure and a large test market would be considered for final validation and for improving the product's profit. The national launch would be tracked carefully and changes made adaptively to meet competitive actions and changes in consumer preference.

The second column indicates recommended steps in each phase for a smaller business with less than $100 million in sales (often substantially less, say $10–20 million or even $1–2 million). Because the smaller business does not have as large a revenue base to draw on for support, it typically cannot undertake all the analytic activity of column one nor will it be able to draw the same magnitude of resources from marketing, research and development, and production departments. This does not mean a small firm cannot produce major innovation. A creative breakthrough backed with a modest amount of analytic support can produce a winner. National Brewing Company successfully went national with Colt 45 malt liquor while retaining their traditional local beer franchise in Baltimore. Noxell Corporation had a modest sales volume based on Noxema, but was able to double sales by introducing Cover Girl makeup.

Although the small business may not be able to fund a large market definition study ($50–$100 thousand), it should talk to consumers through focus groups (about $1,000 per group session) to get a feel for consumer perceptions and preferences and to orient its development towards meeting consumers' needs. Internal creative groups are not expensive ($2,000 to $3,000) and will help assure that a range of creative alternatives has been considered. The small firm cannot afford to take the first idea that comes along and allocate its limited resources to it. Small firms need major creative ideas so they can eventually compete against the corporate giants in the market. A careful identification of a market segment and powerful CBP are needed.

Development of the CBP can be supported by consumer testing (concept and product). These are low cost studies and help assure that the production and engineering staff have the correct product. In conducting such studies, some information on consumer perception and preferences should be generated and some formal models may be used. Pretest market analysis should become a major tool since most small firms will not spend $1 million on test marketing. With a small amount of funds for pretest markets ($25,000 to $50,000), the small business can match the analytic sophistication of large firms. After a successful pretest market analysis, the small firm would probably roll out from market to market rather than go national. By monitoring each market roll-out, they can learn as they go, improving the product and marketing strategy. National penetration would be developed over a number of years and profits would be earned to pay for further expansion. Thus, although small firms may not have the resource base of large firms, they can compete successfully by focusing on a creative CBP and using the low cost technology of pretest market simulation to reduce risks.

Entrepreneurs have tewer resources than even small businesses and, as a result, must bear more risk. Entrepreneurs usually work alone using their personal skills and energy to get a product off the ground. Entrepreneurs can rarely invest in major market research, but they should not use this as an excuse to avoid a consumer orientation. The concepts presented in this book still apply and can greatly improve the entrepreneur's chances of success, but the entrepreneur may have to be a little more creative in getting the consumer information. The entrepreneur is typically a person who has an idea and desperately believes in it. Often the idea is generated through technical or production considerations. Faith in the idea is critical, but we strongly recommend that entrepreneurs expose their ideas to potential consumers early and spend a good deal of time talking to and observing these consumers. This understanding of the consumers' environment can substantially improve the idea. It is also important to generate alternative ideas. Developing commitment to one particular idea too early is dangerous. Certainly an entrepreneur cannot undertake formal market research studies to define markets, but early exposure to consumers is an important method of reducing risk and improving the product concept.

In designing the product, entrepreneurs should concentrate on a CBP just as big firms would. It is easy to be enthralled by the concepts underlying technology, but the product will be purchased by consumers only if it delivers perceived benefits. Informal concept reactions can be obtained from consumers at low cost. Although Chapter 8 discussed measurement briefly, it gave references for further study. With additional reading an entrepreneur could write a questionnaire and personally interview a small sample of consumers. This effort will pay back in an improved product design.

A clear CBP is needed to do the engineering necessary to develop a product. The CBP and a venture plan are necessary to generate funds from venture capital companies and from investors. Even simple consumer research can help gain acceptance of the plan and better communicate the product. After a prototype is built, laboratory and consumer input is critical to improving its performance and CBP. Testing usually takes the form of selling one or more units of the product. Each sale generates new information on perceived benefits and on "in-use" problems. Solving the problems and emphasizing the strength can improve the product. The business evolves and grows. There is usually no explicit launch, but a continued adaptation of the product and its marketing. The concepts of tracking (Chapters 15 and 16) are extremely useful in managing product evolution and growth.

As you may suspect, entrepreneurship is risky. Although Xerox and Polaroid did develop from entrepreneurial efforts, there are thousands of entrepreneurs who have failed. The chances for success are low. There is a high risk, but a high return is possible if a successful product is created. This effort can be undertaken only by those with tremendous energy, creativity, and persistence. We feel that the concepts and simple analytic tools presented in this book can reduce the risks substantially and improve the productivity of entrepreneurial effort.

VARIATIONS BY PRODUCT TYPE (DURABLE, INDUSTRIAL, SERVICE, HIGH TECHNOLOGICAL)

The basic process of new product development is generic in that it can be applied to innovations in a wide variety of product types. Only the emphasis shifts. Table 19.2 highlights some of the differences in implementing the process, depending upon whether the product type is a frequently purchased good, a consumer durable, an industrial product, a high technology product, or a service. Table 19.2 suggests that each of the four phases must be customized to best serve the innovating organization.

In the first phase, opportunity identification, the organization should

Table 19.2. Variations in New Product Development Activities by Product Type

	Product Type				
Phase of Development	*Consumer Frequently Purchased*	*Consumer Durable*	*Industrial*	*High Techno-logical*	*Service*
Opportunity Identification	Market Definition Creative Groups	Market Definition Engineering/ Marketing Creative Groups	User Identification Needs Analysis Technology/ Marketing R&D	R&D Technical Forecast Users	Needs Analysis Service Plans
Design	CBP Psychological Perception/ Preference/ Choice Features Advertising	CBP Psychological & Physical Perception/ Preference/ Choice Features Advertising Diffusion	CBP Physical & Psychological Buying Process Features Engineering Selling	CBP Physical Buying Process R&D Communication Selling Diffusion	CBP Benefits Delivery Perception/ Preference/ Choice Communication
Testing	Consumer Pretest Mkt. Test Market	Laboratories & Consumer Pretest Models	Laboratory Tests In-use Tests	Laboratory Tests In-use Tests	Demonstration or Pilot Programs
Introduction	Launch Adaptive Control	Launch Adaptive Control	Launch Adaptive Control	Launch Adaptive Control	Launch Adaptive Control

understand market needs and be creative for innovations in any of the product types. In consumer markets formal models are available to understand these needs and creative group procedures are well developed. In other areas less formal methods for market definition may be used. Services such as energy, conservation, and family-planning may be difficult since a market may not exist at all. Opportunities may have to be defined by creative structuring of the alternatives and the development of service plans that generate benefits which can be effectively communicated. For example, a health maintenance organization is an innovative concept. Contrast this with minor revisions in existing health services by cost control.

During opportunity identification, differences also exist in the role of R&D. In technological and industrial products R&D plays a large role and exerts considerable influence. Although we have cited much evidence that consumer need is highly associated with success, we must be careful not to eliminate basic R&D. When lasers were developed, there were no known uses. Today laser technology has paid off in such diverse fields as cutting steel, eye surgery, and geological surveying. It should be pointed out that even in the case of lasers, the weapon needs of the military were a prime consideration in the large amount of R&D funds allocated to basic development. Needs are important, but we must keep a balance between applied and basic R&D. While applied R&D has a faster payback, a component of speculative basic R&D is needed in a good technological innovation program. In other product types the role of R&D will be less pronounced. New consumer durables can begin from developments in R&D, but the emphasis often shifts to engineering to fulfill consumer needs. In consumer goods the emphasis is usually on matching the product to identified needs, and in services the emphasis varies from high technology (e.g., communications) to innovative uses of existing technology (e.g., dial-a-bus), to new ways of using professional skills (e.g., management education).

In the design stage, the core benefit proposition is an important underlying concept in all product types. In technological products, remember users do not buy the technology, per se, but what it will do for them. In consumer products the CBP tends to emphasize psychological components while industrial products tend to emphasize objective criteria. But even in industrial products, psychological attributes are important to understanding response. Each member of the decision-making unit may have different evaluation criteria. Usually there are a mix of objective engineering criteria and psychological criteria such as "reliability" and the "quality of technical backup service." A common problem for all product types is understanding the decision process and how the new product is perceived and evaluated versus other products. Much of the analysis capability has been developed for consumer products, but recent applications have shown that these techniques are directly applicable to services and many industrial products. A creative manager should be able to adapt these techniques to most industrial and high technological products.

Another common design concept is the need for product features and actual performance to fulfill the CBP. In consumer products such features may be engineered easily, while in technological products they may be developed only after substantial R&D effort.

All CBP's must be effectively communicated, but the method of communication varies. Consumer products emphasize advertising, while industrial and technological products emphasize personal selling. In public services, advertising, selling, and publicly supported media are used. Word-of-mouth is an important component in the diffusion of innovation for all product types.

Testing is undertaken in all areas. When the CBP is more psychological, consumer testing is more important. When the CBP is more objective, laboratory or controlled-use testing are important. Initial testing of services tends to be based on small demonstration sites. Although the methods vary, the purpose is the same: to see that the final product fulfills the CBP at the desired cost, and meets reliability and quality standards. Test marketing is used for frequently purchased brands, but because of high fixed start-up costs, consumer durable and industrial products usually are not test marketed. In these industries much more emphasis must be placed on pretest market models and evaluations. In this book we have described proven methods for branded products, but these approaches have also been used in durables. General Motors has spent over $1 million on laboratory simulations for new models on several occasions. While pretest market analysis has not been used extensively in industrial or technological goods, it is likely that feasible models and measures will be available to carry out these tasks in the future.

The final phase of development is the product launch. If the development process has been followed, the product should have good potential. But no matter how good the product appears, things can go wrong. Thus we suggest that the consumers, the market, the environment, and competitors be carefully monitored no matter what product type is being developed. Adaptive control (informal or formal) can lead to improvements in strategy that can make or break the success of the launch.

There are differences in the development processes across product types, but there are many common concepts. These concepts are:

- talking to potential users early,
- developing a CBP,
- supporting the CBP with features,
- careful design of communication,
- pre-market testing,
- adaptive control of the launch.

Your management task is to creatively utilize these concepts in a development process that matches your organization's scale and product type.

VARIATIONS BY STRATEGIC EMPHASIS

The previous two sections outlined how the priorities in the development process vary by scale and product type. But many organizations find themselves with scarce resources.

In most organizations you will find a history of common belief on how to proceed. A new manager in that organization (or an existing manager about to change the process) must understand the options throughout the development process to make a well-informed decision on overall strategy. Like any innovation, innovations in strategy can bring great rewards to the innovator, but like any innovation, these rewards are accompanied by risk. You should weigh the benefits against the risks and customize the process with your skills.

Search and Screen versus Up-Front Development

If sufficient resources are available for new product development, then (as was argued in Chapter 3) there are great potential benefits that result from investment in early studies to define markets, forecast technology, and understand perceptions, preferences, and choice. This managerial strategy can be termed "up-front" because it allocates substantial funds to formulating product/market strategies before concepts are created and tested. This is not the traditional mode of operation in many organizations. The alternative to an up-front strategy is "search and screen," that is, generate many ideas and test them in a low-cost screening procedure. Although search and screen is less efficient and runs the risk of not achieving a "best" positioning, there are situations where search and screen may be the more appropriate strategy (see Table 19.3).

Table 19.3. Up-Front Versus Search and Screen Strategy

Up-Front Strategy	*Search and Screen Strategy*
Major New Market	Substantial Knowledge of Existing Market
Breakthrough Innovation	"Flanker" or Minor Innovation
Strong Competition	Small Volume Products
Sufficient Resources	Low Advertising and Investment
Capability and Commitment	

If your organization is looking for major new business opportunities in a market with which you are unfamiliar or if you are trying to identify an innovation likely to open a new market, then up-front investment or high-risk entrepreneurial effort are needed. If competition is strong, then you must protect yourself from second-but-better strategies by obtaining the best positioning the first time out. An up-front strategy requires sufficient resources, capability to interpret the analysis, and a commitment by top management.

Not all innovations are breakthrough innovations. Some organizations have been moderately successful, maintaining share and even growing by working to better exploit existing markets. Under these conditions we can expect that the organization really "knows its market" and can develop potential modifications to products based on its experience. For example, one large consumer product firm has a strategy of creating many (5–10) concepts in its traditional market area, converting them into rough ads, and testing them in a pretest market model. If one is a success, the project goes on. If none are successful, the diagnostics collected in the laboratory research are used to recycle the ideas into a formal design effort. Although this strategy results in a high rate of failure (7 of 8 failed) at the pretest level, the organization thought that the cost of pretest analysis (less than $50,000 per test) was substantially below what up-front investment would be. In this particular case, the cumulative cost was higher because the organization found out by experience that it did not know its basic product category as well as it thought.

Another situation where search and screen may be best is the case of small volume products which cannot pay back the investment in a design study. If a product is not advertised heavily and does not require investment in production facilities, then it may be feasible to screen products by placing them in the store and obtaining diagnostics through small sample surveys. This will work if the firm has a relatively low profile that is not at risk if a number of successive products fail.

The more assured the firm is of the strategic desirability of being in a market and the more clearly it understands the behavioral and competitive response, the more a firm can emphasize appropriately the search and screening phase. Although there are situations where managerial style or the nature of the product prevent up-front work, we find it worthwhile in most situations to consider up-front studies.

Acquisition and Licensing

The perspective of this book is that of the innovating organization. The strategic emphasis is toward high utilization of marketing and R&D. Some firms find their strength lies in other areas and approach growth and development from a financial perspective. Products are acquired on attractive terms, a decentralized organization with a financial control is established, and the product divisions are asked to make the products grow to meet specified financial objectives. Some firms have used this strategy very successfully. Beatrice Foods has become a $6 billion company through such an acquisition strategy. Many others have failed and lost substantial sums of money.

For the organization with financial muscle, acquisition may be a viable strategy, but acquisition strategies can be greatly improved through marketing analysis and innovation. For example, the CBP, product features, and advertising for the acquired firm should be examined. Although an acquisition may be attractive from a purely financial point of view, higher rewards can be obtained if the products and marketing mix are improved. The acquired firm's position in its market should be examined and new products introduced if opportunities are uncovered. Any innovations should proceed through the development process of design, testing, and introduction. Firms that can produce marketing and financial improvements for an acquisition will be highly rewarded.

Organizations with financial power can pursue a strategy of acquisition. Organizations with little or no excess financial resources, but with a viable patent, may wish to consider licensing of that patented innovation to other organizations in exchange for royalties. This strategy is also attractive if an organization cannot utilize an innovation in its product markets or if substantial modifications are necessary to design and sell effective products in other markets. Such arrangements are attractive to the purchasing organization since they save R&D expense, lower risks, and gain access to technology. After a patent is granted, royalties can be collected for seventeen years. Organizations buying, selling, or licensing patent rights should recognize the legal dangers. The laser was patented and licensed in 1959 by Schawlow and Townes and royalties were collected. However, in 1977 a new patent was granted R. Gordon Gould for his work in 1957, which is now legally viewed as the basic patent. In another case it was after more than ten years of legal action that MIT received $10 million in royalties from IBM and other computer manufacturers for Jay Forrester's invention of the magnetic core memory.

Production

Another strategy for product development is based on innovation in production. Costs are reduced by production efficiency and the price for the "new" product is reduced. This is viable strategy when price is an important attribute for buyers. Production innovation capabilities can be leveraged by marketing innovation. Marketing (e.g., conjoint analysis) can guide the cost reduction effort by indicating what attributes consumers will be willing to give up to obtain the lower price. Occasionally, the production innovation gives the organization an opportunity to be aggressive by reducing costs and repositioning the CBP to meet new needs at a lower cost which results in higher total profit.

Summary

The emphasis of marketing, R&D, engineering, finance, and production in an organization's overall strategy will depend upon the strength of that organization. There are situations where search and screen, acquisition, licensing, and production strategies better match an organization's capabilities than a strategy or major innovation through up-front investment. However, in each of these cases, many of the analysis techniques and concepts developed by considering up-front strategies are applicable and can improve the overall innovation strategy.

GEOGRAPHIC EXPANSION AND INTERNATIONAL MARKETS

The most general innovation strategy is a new market and a new product, although many of the examples throughout this book have been new products in existing markets. Another way to innovate is by taking a proven product and introducing it into a new market. One form of this strategy is geographic marketing which includes regional and international expansion. For example, Procter & Gamble has gone East market by market with Folger's coffee. Mexican fast food is being accepted at different rates in various areas of the country. Many of Beatrice Foods' acquisitions were local or regional brands which it then took to national markets.

In regional expansion the emphasis in strategy is somewhat different. The firm has experience with the product which has a proven CBP and advertising strategy. Consumers in new regions are different but can be expected to have somewhat similar tastes. For example, many foods such as Coca-Cola, Thomas' English Muffins, and many brands of beer vary by area to meet local tastes even though the basic product is the same. Some design work and fine-tuning may be necessary, but the emphasis is on testing. The differences between regions can often be identified with a pretest market laboratory study. Once a product is introduced to a new region it is valuable to use a test market model to track the acceptance in each area. Each new market updates the model inputs and helps develop the capabilities to accurately forecast sales growth in the expansion area.

The largest view of geographic expansion is international marketing. In this case the geographic differences are greater. Language, culture, competitors, distribution channels, media, and regulations differ—to name just a few. It is often infeasible to test market, so a pretest market analysis in each country is appropriate to gauge success. Diagnostics and inputs for perceptual mapping and advertising response may be added so the product CBP can be customized for the local tastes. Often joint ventures or licensing are used to penetrate the market. In these cases local knowledge can be combined with a modest market research study to be sure the product position is appropriate and the risks of failure are low. Throughout the world people buy products that are perceived as having benefits over existing products. The importance of attributes and responses to advertising may vary, but the behavioral process of response is similar. The opportunities for international expansion can be tapped by applying the concepts of the new product development process with modification based on the constraints of the "new" country.

BUDGETING FOR GROWTH AND DEVELOPMENT

The budget for a new product development process will depend upon how the process is customized to meet the organization's needs. In setting this budget, top management should recognize the inherent risk involved in innovation. Sufficient resources must be allocated to cover failures in all phases of development. If a program is to be implemented, this budget should be on an ongoing rate so that the yearly investment is sufficient to achieve the yearly new product goals. For example, if an organization wants one new product every two years and if the expected cost to develop a new product is \$6.4 million, then that organization should budget at least \$3.2 million per year to its new product development effort. If less than the expected cost is budgeted, it is likely that the organization's growth and development goals will not be met. It is particularly important for the chief executive officer of a company to authorize an adequate budget and establish a reasonable time frame for results. If this is not done, the Vice President of Growth and Development may be fired because unrealistic time or expectations were set for results and the project was underbudgeted.

Large organizations will have the resources to budget for expected costs and will have the flexibility to cover variations from average. Other organizations may have to take more risk.

To calculate expected costs we need the cost per phase of the development process and the probability that an idea will pass that phase successfully. In Chapter 3 we provided estimates for these costs and probabilities for an average consumer good and an average industrial good based on documented evidence and on our personal experience. Variances are large, so you should plan on updating these estimates for your particular situation. For example, you may have an in-house procedure for test market or you may find that introduction costs are higher in your market. Experience can lower costs. If your organization has an ongoing process, then the costs and probabilities can be calculated from your experience. To establish future budgets it is helpful to keep careful records on the expenditures and rewards for all new product projects including successes *and* failures. The calculation for expected cost is:

$$EC = C(0) + \frac{C(1)}{P(1)*P(2)*P(3)*P(4)} + \frac{C(2)}{P(2)*P(3)*P(4)} + \frac{C(3)}{P(3)*P(4)} + \frac{C(4)}{P(4)} \tag{19.1}$$

where EC = expected investment for development of a winner

$C(i)$ = cost of completing one project in a phase where $i = 0 \Rightarrow$ opportunity identification, $i = 1 \Rightarrow$ design, $i = 2 \Rightarrow$ pretest, $i = 3 \Rightarrow$ test market, $i = 4 \Rightarrow$ introduction

$P(i)$ = probability of success for a project in phase i.

The rationale behind this formula is that each phase has to start with enough projects to cover failures in each successive phase. A similar calculation should be done for development time except that cost, $C(i)$, is replaced with time, $T(i)$, in equation (19.1).

Table 19.4 shows profiles of cost and time for large, small, and entrepreneurial ventures in typical consumer markets based on the subjective judgments of the authors. The large company represents major investment in each phase and is the case covered in Chapter 3. The actual costs of a successful product is lower for a small business and an entrepreneurial venture, but the probability of success also drops. Multiplying the probabilities at each phase gives the chance of direct success on one pass through the system. The value declines from 0.204 for a large firm to 0.050 for a small business and 0.009 for an entrepreneurial venture. The expected cost for a winner increases from $8.6 million to $12.1 million.

Budgeting for the small firm or entrepreneur is very different than for a large organization. Instead of budgeting a continuing development process, the new product work is often a one-time effort. Usually the funds are approved phase by phase for the program and the project receives support so long as it is viewed desirable on a risk-and-return basis. The usual spending would be above the actual cost for a success and, in many cases, the program would terminate in failure rather than continued search for a winner.

For the entrepreneur, budgeting is not relevant. The work is begun and goes on as long as financial support can be attracted or until the project fails. The probability of success is shown as 0.009, which would indicate that 99.1 out of 100 will fail if funds are not available to repeat phases in the

Table 19.4. Typical Budgets Varying by Scale of the Organization

	Large Company		*Small Business*		*Entrepreneur*	
Phase of Development	*Probability*	*Cost (000's)*	*Probability*	*Cost (000's)*	*Probability*	*Cost (000's)*
Opportunity Identification	—	$ 100	—	$ 20	—	$ 5
Design	0.50	200	0.30	100	0.10	20
Testing						
Pretest Market	0.60	50	0.40	50	0.30	50
Test Market	0.80	1,000	0.60	500[a]	0.50	300[b]
Introduction	0.85	5,000	0.70	5,000	0.60	5,000
Probability of Success	0.204		0.050		0.009	
Cost of a Product if It Is Successful	6,350,000		5,670,000		5,375,000	
Expected Cost to Develop One Successful Product	8,556,000		10,635,000		12,116,000	

[a] *Includes first roll-out market*
[b] *Includes first 6 months of building business*

process. Entrepreneurs can start again on a new project, but the risks remain high. For those who are willing to pursue entrepreneurial activities, the gains from a success could be high, but the chances of success are low. As a society we can feel justified in providing high rewards for individuals who take such risks.

CONSTRAINTS ON NEW PRODUCT DEVELOPMENT

In implementing a new product development effort, many constraints limit the organization's freedom. Some grow out of the need to use intermediaries to distribute products, others from public interest groups, and of course, government regulation. We highlight briefly some of the aspects of these constraints and consider strategies to deal with them.

Channels

In many product markets middlemen serve a physical distribution, inventory, selling, or servicing function. These wholesalers, brokers, distributors, or retailers are independent decision makers. For example, a retail food store decides whether it will stock a new product—not the manufacturer. By making the product attractive to the retailer through an adequate margin, special deals, high quality, and consumer advertising the firm can gain shelf space. Some firms directly influence the retail outlet by the use of missionary salesmen who stock the shelf and install special displays for their brand. In industrial markets, sales effort may be directed both at distributors and customers. In some cases the distributor is an order taker, while in others he carries out the complete selling function.

In all cases, the desires of the channel participant must be considered to assure adequate distribution and selling effort. Some researchers have attempted to understand how the channel responds by using explicit models (Montgomery, 1975; Stern and El-Ansary, 1977). It is also reasonable to conduct perception and preference studies with middlemen to see what new product design and promotion actions would result in the decision to accept the product and aggressively sell it.

In many situations channel constraints can be overcome by bargaining with exclusive rights, commissions, service quality, and other rewards. Each concession by one party comes at a cost and is traded for a benefit. In this book we do not discuss negotiation (see Nierenberg, 1968, for more information), but if you know what each concession costs each organization and how each channel action benefits each organization, then you are better equipped to negotiate. Many of the analyses in this book arm you with that information.

Public Interest Groups

Consumer advocacy groups have had an influence on new product development. Their attacks on issues such as TV violence have resulted in some manufacturers withdrawing ads from some programs. Concern over advertising to children may affect the marketing of new foods. These groups have highlighted product safety and the product liability of manufacturers. The cost of product liability insurance has risen dramatically as claims have increased in volume and the size of awards have grown.

One way to deal with public interest groups is to proactively design the new product to avoid intervention. If a product is free of design and manufacturing defects and consumers are given clear warning of dangers associated with a product's use, then your organization may face fewer liability claims. Building product safety into the CBP is often a good strategy. Attention to these matters will result in better products and lower liability cost.

The consumer watchdog groups can influence manufacturers indirectly through consumer feedback that can reduce sales or directly by sponsoring government regulation. These pressures are likely to increase in the future.

One way to deal with this constraint is to test products before regulators see them to be sure they exceed legal requirements and develop an effective service policy and consumer feedback service. This will not only lessen pressure from consumer advocacy groups, but could improve the product and the ultimate sales. Although consumer pressures cannot be eliminated, many can be avoided by careful design of major innovations and straightforward communication of the product's benefits.

Regulation

Government regulation is growing. The number of pages of federal regulation has grown by over 33 percent from 1970 to 1975 and the spending by major regulatory agencies has increased even more. Dow Chemical spent $147 million in 1975 to meet domestic regulations (*Business Week,* April 4, 1977). Procter and Gamble spent over 25 percent of their R&D budget to meet safety and environmental concerns of regulators (Harness, 1977). Regulation presents important constraints on new product development. For example, Professor William Abernathy of Harvard University suggests, "The overall effect of regulation on the auto industry has been to build an envelope around the internal combustion device and the whole car structure. 'Don't do anything really new, don't change.' That's what these regulations say" (*Business Week,* July 3, 1978). Dealing with regulation constraints requires not only legal resources, but a strategy that reflects on all aspects of growth and development.

One major area of regulation is health and safety. The Federal Drug

Administration, Consumer Product Safety Commission, and Office of Safety and Health Administration are major participants in writing regulations to implement legislative actions in health and safety. These regulations affect product design, labeling, and testing. For example, drug companies must now present rigorous evidence, sometimes requiring hundreds of volumes, to document the lack of dangerous side effects and the effectiveness of their products. Labeling on products represents a highly regulated area. The label must catch the attention of a reasonably prudent consumer, be comprehensible to the average user, and fairly indicate the potential dangers. A chemical company used a label approved by the Department of Agriculture, but it was found deficient in court since farm laborers with limited education and poor reading ability could not understand it. Regulation is intricate and filled with bureaucratic traps. Any reasonable size organization should have sufficient staff assigned to keep abreast of all the implications of health and safety regulations.

Another area that impacts heavily on new products is the regulation of marketing practices. The Federal Trade Commission directs its attention to unfair trade practices, attempts to maintain competition, and attempts to protect consumers. An example of government action is in the area of unfair and deceptive advertising. Listerine mouthwash was required to spend $10 million in corrective advertising to say that Listerine does not "lessen the severity of colds." The Antitrust division of the Justice Department takes action to prevent concentration of industry, and the Federal Communication Commission regulates broadcasting. These agencies interpret their enabling legislation, write regulations, and bring suits for marketing practices they view as illegal. In some cases firms can sign a consent decree and stop the disapproved practice, but in others they must go to court to fight the motion. These responses to a complaint are complex and require substantial legal input.

A recent case in the cereal industry has raised another issue. The government claims that by introducing products for every possible market segment, cereal companies have restricted competition and earned excess profits. Although it is not clear how this case will be resolved, it demonstrates the widening impact of regulation. If this case were sustained, firms could not proliferate brands in an existing category and they would have to license their brands to all manufacturers.

There are suits in process against Xerox and Kodak. In Kodak's case the district court ruled that Kodak should make some of its new product designs available to competitors before introduction. Although this decision was overturned in a federal appeals court, these suits indicate that more concern will have to be given to the impact of a large firm developing new products in a product market where it already has a very high market share. These legal threats also raise the danger that the rewards for new products will be so reduced that the role of innovation will decrease and society will suffer.

How should a firm deal with these regulatory constraints? One approach is to "fight it out." That is, do what will maximize profit and then use a large legal staff to defend actions to the letter of the law by long drawn out legal appeals with legal costs of $10 to $20 million or more. A second approach is to change the constraints by lobbying, public relations, and advertising to consumers. In the late 1970's, the oil industry advertised heavily to convince voters that they were searching and would find oil so that government regulation would not be needed. A third strategy is to avoid the constraints. For example, manufacturers of hair dye reformulated their products to avoid a potentially carcinogenic substance rather than fight a proposed warning label. The first firm to institute the change gained a competitive advantage. This type of strategy calls for organizations to develop product innovations so that advertising does not have to resort to questionable claims and marketing practices to attract customers.

The most mature strategy of dealing with restrictions is to preempt them. For example, instead of waiting for regulations on energy consumption, manufacturers could recognize the building political pressures and genuine social concern generated by high energy prices and allocate developmental efforts towards a CBP that includes low energy consumption in a bundle of benefits that consumers will find attractive. Another example is in export markets. Balance of payment difficulties may lead to regulation, but organizations might preempt this by direct developmental effort and marketing their products overseas. As a final example, organizations may preempt health regulations by developing new health service plans for their employees which control costs while providing high quality and personalized care. Reducing alcoholism can improve the health and productivity of workers. This may be a wise innovation that preempts subsequent regulation.

We have identified some of the constraints and responses, but we feel one of the most effective overall responses is through major innovation. Although regulations identify errors of commission, there is a greater error in the omission of innovation. Corporate responsibility should direct the organization to serve society's needs. One of the most beneficial ways of doing so is by creating new products to fill these needs. In return for innovation, society grants the organization profit. It is our view that organizations should concentrate on major innovations where benefits will be widely perceived rather than fighting out regulatory constraints on minor product changes. It should be noted the the FTC dropped its complaint against Quaker Oats in part because of Quaker's record of innovation in products such as "100% Natural Cereal."

Firms should not restrict innovation only to their traditional businesses. Entry into new categories will promote competition and result in new products which serve society better. We hope the process outlined in this book will help organizations create such major innovations and enter significant new markets. Although there will be times when specific regulatory actions will have to be met in the courts, a preemptive strategy based on solid innovation is highly desirable.

SUMMARY

The new product development process is quite general and can be adapted to a wide variety of situations, but it is not a "cookbook" procedure. The concepts are sound and the analytic tools are powerful, but they need an intelligent manager to use them. Part of this use is a customization of the process to fit the scale of the organization, product type, strategic emphasis, geographic location, budget, and constraints. This chapter has outlined some of the basic considerations in customization, but the final success depends on your ability to use creatively the ideas presented here in an actual new product development process.

REVIEW QUESTIONS

19.1 Discuss how the introduction of a new product may vary as the size of the organization varies. As the product type varies.

19.2 Suppose you are the manager of a small business and wish to test a possible new product. How would your strategy vary under the following conditions?
   a. You are entering a market dominated by many small businesses.
   b. You are entering a market dominated by one small business.
   c. You are entering a market dominated by several large firms.
   d. You are entering a market composed of large firms, small businesses and several entrepreneurs.
   e. You are creating your own market.
   f. Your product will compete in two different markets.

19.3 Consider a large firm and a small business each considering launching of exactly the same new product. Why might their approaches toward design, testing, and introduction differ? How might you expect them to differ?

19.4 How should a firm allocate its funds between basic research and applied research? How would this allocation depend on the industry, the number of firms in the industry, the size of the research staff, and the size of the company?

19.5 How would the testing of a new product be different when the CBP is more objective than psychological?

19.6 What are the advantages of an "up-front development effort" over a "search and screen development effort"? When would each strategy be recommended?

19.7 Suppose there are definite geographic differences in tastes for a new beverage line. How would this situation affect new product design and testing?

19.8 After extensive experience a company finds the following probabilities and costs to be fairly representative of their new product development process:

Phase in Process	*Cost of Phase (per product)*	*Probability a product will complete phase*
Design	15,000	0.3
Pretest	50,000	0.4
Test Market	500,000	0.3
Introduction	1,500,000	0.7

a. What is the expected investment for development of a successful new product?
b. Suppose the firm could improve up-front design or test market procedures so that the success probability for national launch was raised to 0.80. How would improving these testing procedures affect expected investment?
c. Suppose the firm could weed out more national failures but could do so only with an increase in the cost of test marketing. How much of a testing improvement would be required to justify a 1 percent increase in testing cost.

19.9 What constraints would you expect when launching a new line of 20-speed bicycles? How would you overcome these constraints?

19.10 How would you develop a new concept in fast foods for the following cultures (1) American, (2) European, (3) African, (4) Asian? For specificity you may wish to select one country in each category and develop a plan for that country. What considerations are unique to each culture?

Chapter 20

New Product Strategy Revisited

This book began with a managerial perspective on proactive and reactive new product strategies. We identified situations in which one strategy is most applicable and then suggested that there is a common set of concepts and techniques that are applicable across a wide range of strategies. Throughout the book these concepts and techniques were presented in the context of proactive strategies. We now return to the same strategic perspective armed with the lessons from Chapters 5 through 19. We indicate which concepts and techniques are applicable to reactive strategies, provide a checklist to help you implement your overall new product strategy, and review how to avoid new product failures. We close by examining the future of new product development.

PROACTIVE VERSUS REACTIVE STRATEGIES

An organization is proactive if it explicitly allocates its resources to search for consumer needs and technological opportunities, design new products, and preempt undesirable future events. A successful *proactive strategy* is based on an effective execution of each of the phases of the new product development process by the initiating firm. It is appropriate if the firm:

1. has an aggressive policy toward growth,
2. is willing to introduce new products and enter new markets,
3. can protect its innovation by patents or market position,

4. is targeting towards high volume or high margin markets,
5. has the financial resources, staff and time required, and
6. can prevent its innovation from being overwhelmed by competition.

If your organization faces a different situation, then the best strategy may be a reactive strategy in which you respond to competitive pressures. There are a range of reactive strategies depending upon when and how strongly your organization can respond to innovation by competitors and consumers. Figure 20.1 compares a proactive strategy to four reactive strategies—responsive, second-but-better, imitative, and defensive. As the strategies become more reactive your organization carries out fewer development functions because they are completed by other participants in the innovation process. Nonetheless, once you explicitly enter the process, the best strategy is to utilize market information and aggressively manage your development effort to achieve your managerial goals.

In a responsive strategy, the user develops a prototype to meet his needs and thereby carries out the idea generation and early design functions. The manufacturer responds by enhancing this design and widening its appeal to a broader market by defining an attractive core benefit proposition. The product testing and introduction are conducted by the producer.

In a second but better strategy another organization has created the product idea, designed, and tested it. The producer using the second-but-better strategy takes the competitive product and improves it if he wishes to be successful. This calls for design work to improve the physical product or its psychological positioning. The procedures outlined in Chapters 7–12 are useful in this effort. The second organization must know the underlying dimensions consumers use in evaluating the product and their importance so improvements can be made that generate sufficient market share to justify the entry costs. After design, product tests and pretest market models may be used. It is unlikely that a test market would be run, but some testing usually is necessary to minimize the risk of national failure. This strategy is most successful when the innovating organization has left opportunities for a second entry by careless positioning or production practices.

In an imitative strategy the producer carries out even fewer functions. In this case the product is not improved, but merely copied. The product should be tested by in-lab and consumer use situations (see Chapters 13 and 14), but design and test market analyses are not usually done. This strategy is based on the imitator lowering costs and getting a share of a rapidly growing market where the innovating organization has little market or patent protection on the product idea.

Defensive strategies involve changes in existing products to minimize the effects of competitors' new products. The competitor has done all the development. No counter entry is planned. Instead, the marketing mix of the existing products is adjusted by changes in price, promotion, or advertising to blunt the impact of competitors. These defensive actions may be combined with an imitative strategy in an attempt to further counter the impact of a competitive new product.

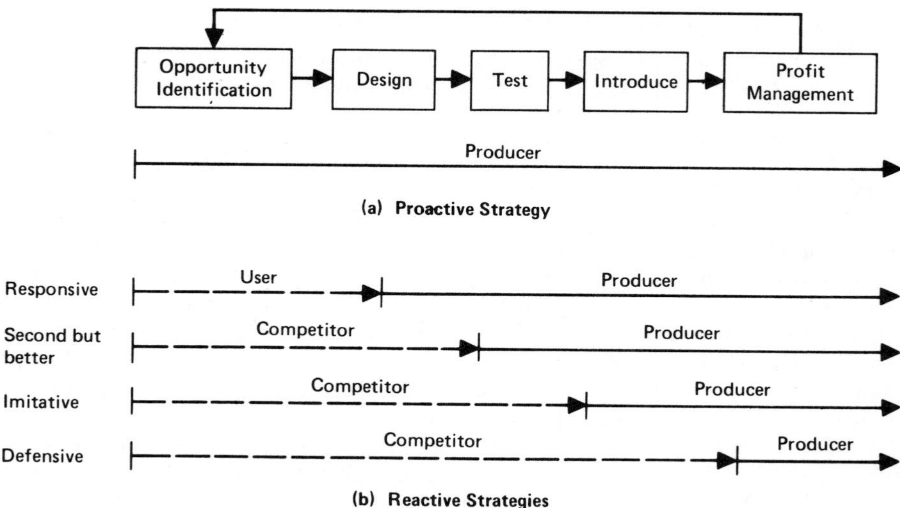

Figure 20.1. *Proactive vs. Reactive Strategies*

The selection of a specific *reactive strategy* depends on a number of relative factors, but in general reactive strategies may be best when:

1. the organization sees its strength in managing existing products,
2. markets are too small to recover development costs,
3. little protection is available for innovation in these markets, and
4. the organization has insufficient resources to conduct development.

The intelligent manager recognizes these forces and chooses the level of investment in new product development that is appropriate for his organization. To choose the best strategy we recommend that you understand the costs, risks, benefits, and capabilities of each phase in the development process.

Although we feel proactive strategies are usually most effective, you should understand and be able to use reactive strategies also. Even firms that usually are proactive may have to utilize a reactive strategy if a competitor introduces a new product first. Then they must defend their existing products and copy or improve upon the competitive innovation. We recommend that you be familiar with all phases of the development process at a level that will enable you to manage the analysis and use the concepts that are appropriate to reduce the risk and enhance the rewards of innovation.

CHECKLIST FOR NEW PRODUCT DEVELOPMENT

When developing new products it is important to maintain a wide perspective so you can allocate resources to each phase of the development process to best achieve your managerial goals. In this section we present a checklist to review the key lessons of the book. If you now can answer each of these questions for your organization, you have the basic management strategy to successfully design and market new products.

(1) Does Our Development Strategy Reflect the Best Long-Term Interest of Our Organization? It is easy to concentrate on short-term results from small product enhancements. The long term should be reflected in a proactive strategy based on user needs and R&D. Do not overlook major innovations in the effort to get short-term payback. For example, basic research as well as development uncovers the new technologies that will be the basis of future new businesses. Basic research in market structures and market needs will target development toward major opportunities. (Review Chapters 3, 5, and 6.)

(2) Are We Allocating Enough Money, Time, and Talent to the Up-Front Phases of New Product Development? Early efforts pay off in lowering expected risks and costs of successful innovation (review Chapter 3). Although up-front investment may appear to add cost and extra steps to the process, it enables the identification of major innovations, minimizes the risk of failure, and maximizes creativity. If resources are available, budget the development based on the expected costs of developing a successful product (review Chapter 19). Without an adequate budget it is unlikely that the growth and development goals will be met on schedule.

(3) Are We Targeting the Right Market? Be sure to consider your product/market strategy early in the process. A fantastic product idea will not be sufficient if it does not fit into the perspective of the corporate growth and development plan. Market definition studies identify markets that match your firm's capabilities, have potential growth, and are vulnerable to your entry (review Chapter 5). If a good market is found, careful analysis helps you create ideas and design products for it. Tests of these products or concepts and prototypes insure that the potential can be realized (review Chapter 12).

(4) Are R&D and Marketing Working Together? You need both of these functions. Successful innovations fulfill consumer needs and demands. Marketing identifies these needs and develops the psychological appeals. R&D develops the technologies and products to fulfill the appeal. (Review Chapters 2 and 6). Integration of these efforts requires that communication be established between these functions within the organizational structure (review Chapter 18).

(5) Are We Fully Utilizing Our Creative Skills? Your organization has great creative potential you may not be aware of. Many additional ideas and insights result if you conduct explicit creative idea generating activities (review Chapter 6). Successful utilization of creative skills requires cycling between analytic and creative modes of thinking. Positioning is creative.

Choice modeling is analytical. Developing copy is artistic, while testing is rational. Pulling together the test market strategy requires insight, forecasting demand requires quantitative skills, and diagnosis requires creative problem solving (review Chapters 6–17). Risks can be controlled but not eliminated.

A portfolio of activities taps both high risk, high return areas and low risk, moderate return areas. An organization can balance full-scale development activities with an entrepreneurial division to nurture and channel creativity *within* rather than outside of the organization.

(6) Have We Developed a Good Core Benefit Proposition for Our Products? The basic consumer decision is with respect to the benefits you are delivering to users rather than the production characteristics of the physical product. This requires that you define the target group carefully and position your benefits against competitors and consumer preferences in order to define major opportunities (review Chapters 7, 9, and 10).

(7) Do Our Products Fulfill Their Core Benefit Propositions? It is easier to identify a positioning opportunity than to build a product to fill it. Good products require both physical and psychological innovation. To fulfill the core benefit proposition your physical product should deliver its promised benefits with real physical performance features and advertising. Checking a product's positioning after consumers use it and comparing this to consumer perceptions of a concept description ensures that the benefits are being delivered (review Chapters 1, 12, 13).

(8) Should We Be Using Pretest Market Models? In frequently purchased consumer products it is clear that pretest market models significantly reduce risk and development costs (review Chapters 3 and 14). These models are being developed for consumer durables and industrial products. We expect that in the near future they will be available to reduce risks and improve profits for these industries.

(9) When We Test Market Are We Utilizing the Full Potential of the Information We Could Obtain? With the advent of pretest market models, test market failure will become less common. Test markets now can do more than merely replicate the national plan. Experiments and consumer response measures supply data for behavioral models that help diagnose improvements, maximize profit, and make conditional forecasts. Good test market models identify the best positioning, product, and marketing mix for full-scale launch (review Chapter 15).

(10) Have We Implemented a Control System for the Launch of the New Product? The advantages of developmental efforts do not stop when the product goes national. Because of environmental changes, forecasting errors, or competitive response, the national launch probably will not go as planned. With a control system you can identify the unexpected events, make a timely response, and improve performance. There may be as many profit opportunities in the revision of the launch by a control system as in revisions of the product plan after test market (review Chapter 16).

(11) Are We Maximizing the Profit Potential from our Maturing Products? Profit rewards for risk taking and creativity come as the product matures. A decision support system maximizes these rewards through response analysis. There are opportunities to revise the marketing mix, reduce cost by process innovation, add flankers, or reposition the product to increase profits (review Chapter 17).

(12) Do We Have the Best People Managing and Working on Our New Products? A structured process is useless without the people to make it real. Success comes from a new product management team (marketing, R&D, engineering, advertising agency, production, and finance) that believes in the discipline of the process, but can be creative within it. An effective team utilizes qualitative and quantitative methodologies within a customized process. Confidence and technical competence backed by energy, drive, commitment and an effective formal and informal organization structure are needed (review Chapters 18 and 19).

Evaluation and reward should reflect the time and risks of development. Promotion should not be based only on successful launches, but also on successful completion of each phase in the product development process.

(13) Is Our Process Up-to-Date with Respect to Qualitative and Quantitative Techniques? Many advanced, but proven techniques have been presented in this book and many more will be developed in the future. You can use these techniques intelligently if you know what they can and cannot do, and when they are appropriate. You should know their strengths and their pitfalls and make sure they reflect your management needs. A mature manager is neither enthralled with the elegance of a technique nor intimidated by complex analysis. An analysis method should be used only if it pays off in better decisions and actions. We have tried to give you the perspective to manage the technology. We believe this perspective will enhance your creativity, reduce the risks you face, and lead to a better long-run track record for successful new product innovation.

REASONS FOR FAILURE REVISITED

Careful consideration of the strategic issues reviewed by the checklist directs your overall strategy and helps you avoid errors of omission. As you implement your strategy you should also avoid errors of commission. There are a number of reasons why new products fail. Chapter 3 identified the major reasons and suggested how to avoid them. We now review these pitfalls and indicate how to use the concepts and techniques of Chapters 5 to 19 to overcome them.

Figure 20.2 reviews the reasons for failure and indicates where in the development process you should take action to resolve them. The one rea-

Reason For Failure	Opportunity Identification		Design			Testing			Introduction	Profit Management
	Market Definition	*Idea Generation*	*Positioning*	*Forecasting*	*Completing Design*	*Product and Advertising Testing*	*Pretest Market Forecasting*	*Test Marketing*	*Monitor Launch*	*Decision Support System*
Market too small	√			√			√	√		
Poor Match for Company	√									
Not New/Not Different		√	√			√				
No Real Benefit			√			√				
Poor Positioning			√							
Little Support from Channel	√				√			√		
Forecasting Error				√			√	√	√	
Competitive Response	√		√				√	√	√	√
Changes in Environment	√				√			√	√	√
Insufficient Return on Investment	√			√	√		√	√	√	√

Figure 20.2. *Reasons for Failure and How to Avoid Them*

son for failure not shown in this figure is "organizational problems." This was discussed in Chapter 18; we recommend that you direct top management's attention to this important issue.

Market Too Small

This error should be avoided as early as possible in the development process. Careful understanding of the competitive boundaries and the market's growth should be the first step in new product development (Chapter 5). In order to see if the product will attract substantial sales, test the concept and product to generate a forecast at the design phase (Chapter 11). The final checks on market size are by pretest models and test markets (Chapters 14 and 15), but in most cases this reason for failure should be resolved earlier in the process.

Poor Match for Company

All growing markets are not appropriate for all companies. A strategic analysis to determine if a market matches the unique competences of the company should be conducted early in the process (review Chapter 5).

Not New/Not Different

Parity products are not enough. Creative methods are needed to generate real advantages (Chapter 6) and uniquely position the product as different from competition (Chapter 9). This uniqueness should be in directions important to the consumer (Chapter 10) and evident in advertising and the products' performance (Chapter 13).

No Real Benefit

The basis for the positioning is delivery of benefits to users. The Core Benefit Proposition (Chapter 7) is a key concept in developing and testing the product to be sure it delivers real benefits (Chapters 10 and 13).

Poor Positioning

If you understand your target consumer's decision process and determine the dimensions of evaluation and their importance, you can position your product in an area of the perceptual space where there are few competitors, but high consumer preference (Chapter 10).

Little Support from the Channel of Distribution

Understanding channel structures and their likely response helps you identify target markets and channels for entry (Chapter 5). You should plan specific functions and rewards for channel members to encourage the desired actions (Chapter 12). In cases of very active channels, the product should be concept tested with middlemen to see if they would stock and support it (Chapter 11). The final check is in test market (Chapter 15).

Forecasting Error

Forecasting deserves attention throughout the process. GO/NO GO decision points are based on forecasts. As the product advances through the process, the forecasts become more accurate. Rough forecasts first occur at the design phase (Chapter 11), are refined by pretest market analysis (Chapter 14), finalized at test market (Chapter 15), and monitored during launch (Chapter 16). By carefully applying the forecasting tools described in this book you should be able to avoid the pitfall of poor forecasting.

Competitive Response

It is often in the best interests of your competitors to attack and undermine your innovations. You should understand the competitive structure and practices of the markets you want to enter. If you do not want to compete on this basis or are not willing to commit the resources to do so, do not enter the market (review Chapter 5).

You should position your product carefully so that you do not leave yourself vulnerable to competitive entry (Chapters 9 and 10). Careful positioning means developing a major product advantage and preemptively positioning it.

In pretest you may want to test your vulnerability by introducing likely competitive new products to consumers. You can use a mocked-up advertisement and package for the competitor to see how you would stand up to competition. If a competitor is market testing their counter entry use it in a pretest market analysis (Chapter 14).

In test market, you can measure competitive actions and use a model to make forecasts based on alternative competitive scenarios (Chapter 15). Your GO national plan should presume some competitive reaction. You need a margin for error in the forecasts if you face competitive pressure.

If you carefully control the launch you can diagnose and rapidly react to competitive actions (Chapter 16). The problems of competitive response are difficult, but careful planning can minimize the adverse effects of retaliation.

Changes in the Environment

You can determine how sensitive your market is to environmental fluctuations by carefully monitoring the environment through the design, test, launch, and mature phases of development by a control system (Chapters 11, 15, 16, and 17).

Insufficient Return on Investment

The "bottom line" is the final indicator of success. The product may be a sales success, but if it does not achieve its profit goals it has failed. Careful attention to profit is needed when selecting growing high margin markets, designing product innovations, setting premium margins, planning the marketing mix, analyzing changes in the testing phase, controlling the launch, and managing the mature product. With the decision process and procedures described in this book you should be able to design major innovations that have high profit potential, also you should be able to manage the innovations to realize the potential return.

FUTURE OF NEW PRODUCT DEVELOPMENT

Product development is exciting. It combines the thrill of creativity with the challenge of avoiding failure. To succeed one must be both imaginative and realistic. The challenges and rewards are great, but so is the risk.

The successful new product developer understands and uses a wide range of the managerial functions including marketing, R&D, engineering, finance, production, and administration. He keeps a long term managerial perspective, but balances it against the short term organizational needs for profit performance. Most importantly the new product developer channels creativity with a disciplined process and a set of effective qualitative and quantitative decision support techniques.

In the future, challenges will be even greater. As organizations become more sophisticated competition will increase. Environmental changes will continue at a rapid pace which will increase the need and opportunity for innovation. We cannot expect inflation, regulation, resource constraints, technology, or consumer tastes to remain stable.

For example, consider the increasing shortages of traditional energy sources. The need for more energy efficient products or products using alternative energy sources (e.g., solar energy) is clear. Opportunities exist to re-engineer many industrial and durable products based on new ratios of energy costs and capital costs. New energy conservation concepts and services will be needed. The design and marketing of this range of products is important if we are to avoid the adverse effects of energy shortage on inflation, balance of payments, employment, and social programs.

Change in energy availability is but one example of the many possible futures we face. The many changes in the status quo present problems for unchanging organization, but represent real opportunities for those organizations that adapt and evolve with new market offerings. The organizations that will not just survive, but thrive, will use an effective new product development process.

As organizations examine their role in society and our changing environment, one of the important rationales for their existence is based on innovation to fill societal and consumer needs. Profit is justified as the reward for risk taking and innovation. Major new products that increase the physical, economic, psychological, social, and aesthetic well-being of people are a major method for organizations to fulfill their social responsibility. We hope this book has helped equip you with a managerial perspective and set of tools to meet this responsibility by designing and marketing new products.

REVIEW QUESTIONS

20.1 Review opportunity identification, design, testing, launch, and profit management for each of the reactive strategies.

20.2 Discuss the checklist for new product development. How can you answer each question? What organizational issues are involved?

20.3 Discuss the future, as you see it, for new product development.

Bibliography

AAKER, D., and J. MYERS, *Advertising Management* (Englewood Cliffs, NJ: Prentice-Hall, 1975).

———., and C. B. WEINBERG, "Interactive Marketing Models," *Journal of Marketing,* 39 (October 1975), 16–23.

ABELL, D. F., and J. S. HAMMOND, *Strategic Market Planning* (Englewood Cliffs, NJ: Prentice-Hall, 1979).

ABERNATHY, W. J., and N. BALOFF, "Interfunctional Planning for New Product Introduction," *Sloan Management Review,* 14, No. 2 (Winter 1972–73), 25–44.

———, and K. WANE, "Limits of the Learning Curve," *Harvard Business Review,* 52, No. 5 (September-October 1974), 109–19.

———, and J. M. UTTERBACK, "Patterns of Industrial Innovation," *Technology Review,* 80, No. 7 (June-July 1978), 1–9.

ADLER, L., "Time Lag in New Product Development," *Journal of Marketing Research* (January 1966), 17–21.

AHL, D. H., "New Product Forecasting Using Consumer Panels," *Journal of Marketing Research,* 7, No. 2 (May 1970), 159–67.

ALBERS, S., and K. BROCKHOFF, "A Procedure for New Product Positioning in an Attribute Space," *European Journal of Operational Research,* 1 (1977), 230–38.

ALBRIGHT, R. L., S. R. LERMAN, and C. F. MANSKI, "Report on the Development of an Estimation Program for the Multinomial Probit Model," Cambridge Systematics, Inc. report prepared for the Federal Highway Administration, U.S.D.O.T. (October 1977).

ALLAIRE, Y., "A Model for the Evaluation of Risk and Additional Information in New Product Decisions," *INFOR,* 13, No. 1 (February 1975), 36–47.

———, "The Measurement of Heterogeneous Semantic Perceptual, and Preference Structures," unpublished Ph.D. dissertation, M.I.T. (1973).

ALLEN, T. J., "Communication Networks in R&D Laboratories," *R and D Management,* 1, No. 1 (1970), 14–21.

———, *Managing the Flow of Technology* (Cambridge, MA: M.I.T. Press, 1977).

———, "Studies of the Problem-Solving Process in Engineering Design," *IEEE Transactions on Engineering Management,* Vol. EM-13, No. 2 (June 1966), 72–83.

ALLISON, R. E., and K. P. UHL, "Influence of Beer Brand Identification on Taste Perception," *Journal of Marketing Research,* 1, No. 3 (August 1964), 36–39.

American Newspaper Publishers Association, Bureau of Advertising, "What Can One Newspaper Ad Do? An Experimental Study of Newspaper Advertising Communication and Results" (1969).

AMSTUTZ, A. E., *Computer Simulation of Competitive Market Response* (Cambridge, MA: M.I.T. Press, 1967).

ANGELUS, T. L., "Why Do Most New Products Fail?" *Advertising Age,* 40 (March 24, 1969), 85–86.

ANSOFF, H. I., "Strategies for Diversification," *Harvard Business Review* (September-October 1957), 113–24.

ARNOLD, J. E., "Useful Creative Techniques," in *Source Book for Creative Thinking,* eds. S. J. Parnes and H. F. Harding (New York: Charles Scribner's Sons, 1962).

ARROW, K. J., *Social Choice and Individual Values,* 2nd ed. (New York: John Wiley & Sons, Inc., 1963).

ARTHUR D. LITTLE, INC., *Patterns and Problems of Technical Innovation in American Industry,* Report C65344 to the National Science Foundation (Cambridge, MA: Arthur D. Little, Inc., 1959).

ASSMUS, G., "Newprod: The Design and Implementation of a New Product Model," *Journal of Marketing,* 39, No. 1 (January 1975), 16–23.

AXELROD, J. N., "Attitude Measures that Predict Purchase," *Journal of Advertising Research,* 8, No. 1 (March 1968), 3–18.

———, "14 Rules for Building an MIS," *Journal of Advertising Research,* 10, No. 3 (June 1970), 3–11.

AYRES, R. U., *Technological Forecasting and Long Range Planning,* (New York: McGraw-Hill Book Company, 1969).

BAGOZZI, R. P., "Salesforce Performance and Satisfaction as a Function of Individual Difference, Interpersonal, and Situational Effects," *Journal of Marketing Research,* 15, No. 4 (November 1978), 517–31.

BAKER, N. R., "R&D Project Selection Models: An Assessment," *IEEE Transactions on Engineering Management,* Vol. EM-21, No. 4 (November 1974), 165–71.

———, and W. POUND, "R and D Project Selection: Where We Stand," *IEEE Transactions on Engineering Management,* Vol. EM-11, No. 4 (December 1964), 124–34.

———, J. SIEGMAN, and A. H. RUBENSTEIN, "The Effects of Perceived Needs and Means on the Generation of Ideas for Industrial Research and Development Projects," *IEEE Transactions on Engineering Management,* Vol. EM-14, No. 14 (December 1967), 156–63.

———, W. E. SOUDER, D. R. SHUMWAY, P. M. MAHER, and A. H. RUBENSTEIN, "A Budget Allocation Model for Large Hierarchical R&D Organizations," *Management Science,* 23, No. 1 (September 1976), 59–70.

BALACHANDRAN, V., and D. H. GENSCH, "Solving the 'Marketing Mix' Problem Using

Geometric Programming," *Management Science,* 21, No. 2 (October 1974), 160–71.

BASS, F. M., "A New Product Growth Model for Consumer Durables," *Management Science,* 15, No. 5 (January 1969), 215–27.

______, "A Simultaneous Equation Regression Study of Advertising and Sales of Cigarettes," *Journal of Marketing Research,* 6, No. 3 (August 1969), 291–01.

______, "Fishbein and Brand Preference: A Reply," *Journal of Marketing Research,* 9 (November 1972), 461.

______, "Marketing Research Expenditures—A Decision Model," *Journal of Business,* 36, No. 1 (January 1963), 77–90.

______, "The Relationships Between Diffusion Rates, Experience Curves, and Demand Elasticities for Consumer Durable Technological Innovations," presented at *Interfaces Between Marketing and Economics* (Rochester, NY, University of Rochester, April 1978) and forthcoming, *Journal of Business* (1980).

______, "The Theory of Stochastic Preference and Brand Switching," *Journal of Marketing Research,* 11 (February 1974), 1–20.

______, and D. G. CLARKE, "Testing Distributed Lag Models of Advertising Effect," *Journal of Marketing Research,* 9 (August 1972), 298–301.

______, A. JEULAND, and G. P. WRIGHT, "Equilibrium Stochastic Choice and Market Penetration Theories: Derivations and Comparisons," *Management Science,* 22, No. 10 (June 1976), 1051–63.

______, and R. T. LONSDALE, "An Exploration of Linear Programming in Media Selection," *Journal of Marketing Research* (May 1966), 179–88.

______, E. A. PESSEMIER, and D. R. LEHMANN, "An Experimental Study of Relationships Between Attitudes, Brand Preference, and Choice," *Behavioral Science,* 17, No. 6 (November 1972), 532–41.

______, and W. W. TALARZYK, "An Attitude Model for the Study of Brand Preference," *Journal of Marketing Research,* 9 (February 1972), 93–96.

______, and W. L. WILKIE, "A Comparative Analysis of Attitudinal Predictions of Brand Preference," *Journal of Marketing Research,* 10 (August 1973), 262–69.

BASU, S. and R. G. SCHROEDER, "Incorporating Judgments in Sales Forecasts: Application of the Delphi Method at American Horst and Derrick," *Interfaces,* 7, No. 3 (May 1977), 18–27.

BEATTIE, D. W., "Marketing a New Product," *Operations Research Quarterly,* 20 (December 1969), 429–35.

BECKWITH, N. E., and U. V. KUBILIUS, "Empirical Evidence of Halo Effects in Store Image Research by Estimating True Locations," Working Paper, University of Pennsylvania, Philadelphia, PA (1978).

______, and D. R. LEHMANN, "Halo Effects in Multiattribute Attitude Models: An Appraisal of Some Unresolved Issues," *Journal of Marketing Research,* 13 (November 1976), 418–21.

______, and ______, "The Importance of Halo Effects in Multi-Attribute Attitude Models," *Journal of Marketing Research,* 12 (August 1975), 265–75.

BELK, R. W., "Situational Variables and Consumer Behavior," *Journal of Consumer Research,* 2 (December 1975), 157–64.

BEN-AKIVA, M. E., "Structure of Passenger Travel Demand Model," Ph.D. thesis, Department of Civil Engineering, M.I.T. (1973).

BERGER, A., "Factors Influencing the Locus of Innovation Activity Leading to Scientific Instrument and Plastics Innovations," unpublished S. M. thesis, Sloan School of Management, M.I.T. (June 1975).

BERKMAN, J., D. BROWNSTONE, G. M. DUNCAN, and D. MCFADDEN, "QUAIL User's Manual," Institute of Transportation Studies, University of California, Berkeley (1976).

BESWICK, C. A., "Allocating Selling Effort via Dynamic Programming," *Management Science,* 23, No. 7 (March 1977), 667–78.

______, and D. W. CRAVENS, "A Multistage Decision Model for Salesforce Management," *Journal of Marketing Research,* 14, No. 2 (May 1977), 135–44.

BETTMAN, J. R., "The Structure of Consumer Choice Processes," *Journal of Marketing Research,* 8 (November 1971), 465–71.

______, "Toward a Statistics for Consumer Decision Nets," *Journal of Consumer Research,* 1, No. 1 (June 1974), 51–62.

BIRD, M., and A. S. C. EHRENBERG, "Non-Awareness and Non-Usage," *Journal of Advertising Research,* 6 (December 1966), 4–8.

BLATTBERG, R., T. BUESING, P. PEACOCK, and S. SEN, "Identifying the Deal Prone Segment," *Journal of Marketing Research,* 15 (August 1978), 369–77.

______, and J. GOLANTY, "Tracker: An Early Test Market Forecasting and Diagnostic Model for New Product Planning," *Journal of Marketing Research,* 15, No. 2 (May 1978), 192–202.

BLIN, J. M., and J. A. DODSON, JR., "A Multiple Criteria Decision Model for Repeated Choice Situation," in *Multiple Criteria Problem Solving,* ed. S. Zionts (New York: Springer-Verlag, 1978).

BLOOM, D., A. JAY, and T. TWYMAN, "The Validity of Advertising Pretests," *Journal of Advertising Research,* 17, No. 2 (April 1977), 14.

BOOZ, ALLEN & HAMILTON, *Management of New Products,* (New York: Booz, Allen and Hamilton, Inc., 1971).

BOSTON CONSULTING GROUP, *Perspectives on Experience* (Boston, MA: Boston Consulting Group, 1970).

BOUCHARD, T. J., and M. HARE, "Size, Performance, and Potential in Brainstorming Groups," *Journal of Applied Psychology,* 54, No. 1 (January 1970), 51–55.

BOX, G.E.P., and G. M. JENKINS, *Times Series Analysis Forecasting and Control* (San Francisco, CA: Holden-Day, Inc., 1970).

BOYD, J. W., JR., R. WESTFALL, and S. F. STASCH, *Marketing Research: Text and Cases,* 4th ed. (Homewood, IL: Richard D. Irwin, Inc., 1977).

BOYDEN, J., "A Study of the Innovation Process in the Plastics Additives Industry," unpublished S.M. thesis, Sloan School of Management, M.I.T. (January 1976).

BRAUN, M. A., and V. SRINIVASAN, "Amount of Information as a Determinant of Consumer Behavior Towards New Products," Reprint Series Report No. 220, Graduate School of Business, Stanford University, Stanford, CA (1975).

BRIGHT, J. R., "Evaluating Signals of Technological Change," *Harvard Business Review,* 48, No. 1 (January-February 1970), 62–79.

BRISCOE, G., "Some Observations on New Industrial Product Failures," *Industrial Marketing Management,* 2 (February 1973), 151–62.

BRITT, S. H., "How Advertising Can Use Psychology's Rules of Learning," *Printer's Ink,* 252 (September 23, 1955), 74, 77, and 80.

BRUNER, J. S., J. J. GOODNAW, and G. R. AUSTIN, *A Study of Thinking* (New York: John Wiley & Sons, Inc., 1956).

BRUNSWIK, E., *The Conceptual Framework of Psychology* (Chicago: University of Chicago Press, 1952).

BULTEZ, A., C. DERBAIX, and A. J. SILK, "Developing Advertising Alternatives: Is the Magic Number One, or Could It Be Four?" Working Paper, Sloan School of Management, M.I.T. (1976).

BURGER, P., "COMP: A New Product Forecasting System," Working Paper #123–72, Graduate School of Management, Northwestern University (1972).

______, R. J. LAVIDGE, and H. N. GUNDEE, "COMP: A Comprehensive System for the Evaluation of New Products," Working Paper, Elrich and Lavidge, Inc., Chicago, IL (October 1978).

BURSK, E. C., and S. A. GREYSER, *Advanced Cases in Marketing Management,* (Englewood Cliffs, NJ: Prentice-Hall, Inc., 1968), 156–65.

BUTLER, D. H., "Development of Statistical Marketing Models," in *Speaking of Hendry* (Croton-on-Hudson, NY: Hendry Corporation, 1976), 125–45.

______, and B. F. BUTLER, *Hendrodynamics: Fundamental Laws of Consumer Dynamics* (Croton-on-Hudson, NY: Hendry Corporation, August 1966).

BUZZELL, R. D., "Predicting Short-Term Changes in Market Share as a Function of Advertising Strategy," *Journal of Marketing Research,* 1, No. 3 (August 1964), 27–31.

______, B. T. GALE, and R. C. M. SULTAN, "Market Share—A Key to Profitability," *Harvard Business Review,* 53, No.1 (January-February 1975), 97–106.

______, and R. E. M. NOURSE, *Product Innovation in Food Processing, 1954–1964* (Boston, Division of Research, Graduate School of Business Administration, Harvard University, 1967).

CADBURY, N. D., "When, Where, and How to Test Market," *Harvard Business Review,* 53, No. 3 (May-June 1975), 96–105.

CALDER, B. J., "Focus Groups and the Nature of Qualitative Marketing Research," *Journal of Marketing Research,* 14 (August 1977), 353–64.

CAMPBELL, B. M., "The Existence of Evoked Set and Determinants of Its Magnitude in Brand Choice Behavior," unpublished Ph.D. dissertation, Columbia University, 1969.

CAMPBELL, D. T., and D. W. FISKE, "Convergent and Discriminant Validation by the Multitrait-Multimethod Matrix," *Psychological Bulletin,* 56, No. 2 (March 1959).

______, and J. C. STANLEY, *Experimental and Quasi-experimental Designs for Research* (Chicago: Rand McNally, 1966).

CARMONE, F. J., P. E. GREEN, and A. K. JAIN, "Robustness of Conjoint Analysis: Some Monte Carlo Results," *Journal of Marketing Research,* XV (May 1978), 300–303.

CARROLL, J. D., "Individual Differences and Multidimensional Scaling," in *Multidimensional Scaling: Theory and Application in the Behavioral Sciences,* eds. R. N. SHEPARD, A. K. ROMNEY, and S. NERLOVE, Vol. 1 (New York: Seminar Press, Inc., 1972).

______, and J. J. CHANG, "A General Index of Nonlinear Correlation and Its Application to the Interpretation of Multidimensional Scaling Solutions," *American Psychologist,* 19 (1964), 540.

______, and ______, "An Alternate Solution to the Metric Unfolding Problem," *Psychometric Society,* St. Louis (March 1971).

———, and ———, "Analysis of Individual Differences in Multidimensional Scaling via an N-way Generalization of the Eckart-Young Decomposition," *Psychometrika,* 35 (1970), 283–319.

———, and ——— "Relating Preference Data to Multidimensional Scaling Solutions via a Generalization of Coombs' Unfolding Model," Bell Telephone Laboratories, Murray Hill, NJ (1967).

CARTER, C. F., and B. R. WILLIAMS, *Industry and Technical Progress: Factors Governing the Speed of Application of Science* (London: Oxford University Press, 1957).

CATTIN, P., and D. R. WITTINK, "A Monte Carlo Study of Metric and Nonmetric Estimation Methods for Multiattribute Models," research paper No. 341, Graduate School of Business, Stanford University (November 1976).

CETRON, M. J., *Technological Forecasting: A Practical Approach* (New York: Technology Forecasting Institute, 1969), 58.

CHAKRAVATI, D., A. MITCHELL, and R. STAELIN, "Judgment Based Marketing Decision Models: An Experimental Investigation of the Decision Calculus Approach," *Management Science,* 25, No. 3 (March 1979), 251–63.

CHARLES RIVER ASSOCIATES, INC. (CRA), "A Disaggregate Behavioral Model of Urban Travel Demand," Federal Highway Administration, U.S. Department of Transportation, Washington, D.C. (1972).

CHARLTON, P., A. S. C. EHRENBERG, and B. PYMONT, "Buyer Behavior under Mini-Test Conditions," *Journal of the Market Research Society,* 14, No. 3 (July 1972), 171–83.

———, and B. PYMONT, "Evaluating Marketing Alternatives," *Journal of the Market Research Society,* 17, No. 2 (April 1975), 90–103.

CHARNES, A., W. W. COOPER, J. K. DEVOE, and D. B. LEARNER, "DEMON: Decision Mapping via Optimal GO-ON Networks—A Model for Marketing New Products," *Management Science,* 12, No. 11 (1966), 865–87.

———, ———, ———, and ———, "DEMON: Mark II External Equations Approach to New Product Marketing," *Management Science,* 14, No. 9 (May 1968), 513–24.

———, ———, ———, and ———, "DEMON: Mark II External Solutions and Approximations," *Management Science,* 14, No. 11 (July 1968), 682–91.

CHOFFRAY, J. M., and G. L. LILIEN, *Market Planning for New Industrial Products* (New York: John Wiley & Sons, Inc., forthcoming 1980).

———, and ———, "The Market for Solar Cooling: Perceptions, Response, and Strategy Implications," *Studies in the Management Sciences,* 10 (1978), 209–26.

CHOW, G. C., "Tests of Equality Between Sets of Coefficients on Two Linear Regressions," *Econometrica,* 28, No. 3 (July 1960).

CHURCHILL, G. A., JR., *Marketing Research: Methodological Foundations* (Hinsdale, IL: The Dryden Press, 1976).

CHURCHMAN, L. W., and A. K. SHAINBLATT, "The Researcher and the Manager: A Dialectic of Implementation," *Management Science,* 11, No. 4 (February 1965), B69–B87.

CLANCY, K., and L. E. OSTLUND, "Commercial Effectiveness Measures," *Journal of Advertising Research,* 16, No. 1 (February 1976), 29–34.

CLARKE, D. G., "Sales-Advertising, Cross-Elasticities, and Advertising Competition," *Journal of Marketing Research,* 10, No. 3 (August 1973), 250–61.

______, "Strategic Advertising Planning," *Management Science,* 24, No. 16 (December 1978), 1687–99.

CLAYCAMP, H., and L. E. LIDDY, "Prediction of New Product Performance: An Analytical Approach," *Journal of Marketing Research,* 6, No. 3 (November 1969), 414–20.

COCHRAN, W. G., and G. M. COX, *Experimental Designs,* 2nd ed. (New York: John Wiley & Sons, Inc., 1957).

COLEMAN, J. C., E. KATZ, and H. MENZEL, *Medical Innovation: A Diffusion Study,* (Indianapolis: Bobbs-Merrill, 1966).

______, ______, and ______, "The Diffusion of an Innovation Among Physicians," *Sociometry,* 20, No. 4 (December 1957), 253–70.

COOLEY, W. W., and P. R. LOHNES, *Multivariate Data Analysis* (New York: John Wiley & Sons, Inc., 1971).

COOMBS, C. H., R. M. DAWES, and A. TVERSKY, *Mathematical Psychology: An Elementary Introduction* (Englewood Cliffs, NJ: Prentice-Hall, 1970).

COOPER, R. G., "Why New Industrial Products Fail," *Industrial Marketing Management,* 4, No. 2 (December 1975), 315–26.

COX, D. R., *Planning of Experiments* (New York: John Wiley & Sons, Inc., 1958).

COX, K. K., J. B. HIGGINBOTHAM, and J. BURTON, "Applications of Focus Group Interviewing in Marketing," *Journal of Marketing,* 40, No. 1 (January 1976), 77–80.

COX, W. E., "Product Life Cycles as Marketing Models," *Journal of Business,* (October 1967), 375–84.

CRAWFORD, C. M., "Marketing Research and the New Product Failure Rate," *Journal of Marketing,* 41 (April 1977), 51–61.

______, "Unsolicited Product Ideas—Handle with Care," *Research Management,* 18 (January 1975), 19–24.

DALAL, S. R., and V. SRINIVASAN, "Determining Sample Size for Pretesting Comparative Effectiveness of Advertising Copies," *Management Science,* 23, No. 12 (August 1977), 1284–94.

DALBY, J., "Practical Refinements to the Cross-Impact Matrix Technique of Technological Forecasting," in *Industrial Applications of Technological Forecasting: Its Utilization in R&D Management,* eds. J. Cetron and C. A. Ralph (New York: John Wiley & Sons, Inc., 1971).

DALKEY, N. C., "The Delphi Method: An Experimental Study of Group Opinion," Rand Report RM-5888-PR, The Rand Corporation, Santa Monica, CA (1969).

______, and O. HELMEN, "An Experimental Application of the Delphi Method to the Use of Experts," *Management Science,* 9 (April 1963), 458.

DAVIDSON, J. H., "Why Most New Consumer Brands Fail," *Harvard Business Review,* 54 (March-April 1976), 117–21.

DAVIS, K., and F. WEBSTER, *Sales Force Management* (New York: The Ronald Press, 1968).

DAY, G. S., "Diagnosing the Product Portfolio," *Journal of Marketing,* 41 (April 1977), 29–38.

______, A. D. SHOCKER, and R. K. SRIVASTAVA, "Consumer Oriented Approaches to Identifying Product Markets," *Journal of Marketing,* 43, No. 4, (Fall, 1979) 8–19.

DAY, R. L., "Linear Programming in Media Selection," *Journal of Advertising Research* (June 1963), 40–44.

DHALLA, N. K., and S. YUSPEH, "Forget the Product Life Cycle Concept!" *Harvard Business Review* (January-February 1976), 102–12.

DIXON, W. J., ed., *BMDP: Biomedical Computer Programs* (Berkeley, CA: University of California Press, 1975).

DODDS, W., "Application of the Bass Model in Long-Term New Product Forecasting," *Journal of Marketing Research,* 10 (August 1973), 308–11.

DODSON, J. A., and E. MULLER, "Models of New Product Diffusion Through Advertising and Word of Mouth," *Management Science,* 15 (November 1978), 1568–78.

DOLAN, R. J., and A. P. JEULAND, "The Experience Curve Concept: Implications for Optimal Pricing Strategies," working paper, Graduate School of Business, University of Chicago (revised February 1979).

DOLDE, W., R. STAELIN, T. YAO, "Estimating Response Rates for Different Market Segments from Questionnaire Data," working paper 57–78–79, Carnegie Mellon University, Pittsburgh, PA (revised April 1979).

DORFMAN, R., and P. O. STEINER, "Optimal Advertising and Optimal Quality," *American Economic Review* (December 1954), 826–36.

DOYLE, P., and I. FENWICK, "The Pitfalls of AID Analysis," *Journal of Marketing Research,* 12 (November 1975), 408–13.

DRUCKER, P., *Management Tasks, Responsibilities & Practices* (New York: Harper & Row, 1974).

EHRENBERG, A. S. C., "Predicting the Performance of New Brands," *Journal of Advertising Research,* 11 (December 1971), 3–10.

______, *Repeat-Buying* (Amsterdam: North-Holland Press, 1971).

______, and G. J. GOODHARDT, "Repeat Buying of a New Brand," *British Journal of Marketing,* 2 (Autumn 1968), 200–205.

EINHORN, H. J., and R. M. HOGARTH, "Unit Weighting Schemes for Decision Making," *Organizational Behavior and Human Performance,* 13 (1975), 171–92.

ENGEL, J. F., D. T. KOLLAT, and R. D. BLACKWELL, *Consumer Behavior* (New York: Holt, Rinehart & Winston, Inc., 1968).

______, and M. R. WARSHAW, "Allocating Advertising Dollars by Linear Programming," *Journal of Advertising Research,* 4, No. 3 (September 1964), 42–48.

ENOS, J. L., in *The Rate and Direction of Inventive Activity: Economic and Social Factors,* ed. R. R. Nelson, (Princeton, NJ: Princeton University Press, 1962), 299–322.

ESKIN, G. J., "Dynamic Forecasts of New Product Demand Using a Depth of Repeat Model," *Journal of Marketing Research,* 10, No. 2 (May 1973), 115–29.

______, and J. MALEC, "A Model for Estimating Sales Potential Prior to Test Market," *Proceedings of the American Marketing Association Educators' Conference* (Chicago, American Marketing Association, 1976), 230–33.

FARQUHAR, P. H., "A Fractional Hypercube Decomposition Theorem for Multi-attribute Utility Functions," *Operations Research,* 23, No. 5 (September-October 1975), 941–67.

______, "A Survey of Multi-attribute Utility Theory and Applications," *Studies in Management Sciences,* 6 (1977), 59–89.

FARRIS, G., "Organizing Your Informal Organization," *Innovation*, No. 25 (October 1971), 2–11.

FISHBEIN, M., "Attitude and the Prediction of Behavior," in *Readings in Attitude Theory and Measurement,* ed. M. Fishbein (New York: John Wiley & Sons, Inc., 1967).

———, "The Prediction of Behaviors from Attitudinal Variables," in *Advances in Communication Research,* eds. C. D. Mortensen and K. K. Sereno (New York: Harper & Row, 1972).

FISHBURN, P. C., "Additive Representations of Real-Valued Functions on Subsets of Product Sets," *Journal of Mathematical Psychology,* 8 (1971), 382–88.

———, "Lexicographic Orders, Utilities, and Decision Rules: A Survey," *Management Science,* 20, No. 11 (July 1974), 1442–71.

———, *Mathematics of Decision Theory,* UNESCO (1972).

———, *Utility Theory for Decision Making* (New York: John Wiley & Sons, Inc., 1970).

———, "Von Neumann-Morgenstern Utility Functions on Two Attributes," *Operations Research,* 22, No. 1 (January-February 1974), 35–45.

FISHER, F. M., "Tests of Equality Between Sets of Coefficients in Two Linear Regressions: An Expository Note," *Econometrica,* 38, No. 2 (March 1970), 361–66.

FITZROY, P. T., *Analytic Methods for Marketing Management* (New York: McGraw-Hill Book Company, 1976).

FOURT, L. A., and J. W. WOODLOCK, "Early Prediction of Market Success for Grocery Products," *Journal of Marketing,* 25, No. 2 (October 1960), 31–38.

FRANK, R. E., W. F. MASSY, and Y. WIND, *Market Segmentation* (Englewood Cliffs, NJ: Prentice-Hall, 1972).

FREIMER, M., and L. SIMON, "The Evaluation of Potential New Product Alternatives," *Management Science,* 13 (February 1967), 279–92.

FRIEDMAN, M., and L. J. SAVAGE, "The Expected-Utility Hypothesis and the Measurability of Utility," *The Journal of Political Economy,* 60 (1952), 463–74.

GALBRAITH, J. R., "Matrix Organizational Designs," *Business Horizons,* 12, No. 1 (February 1971), 29–40.

———, *Organization Design* (Reading, MA: Addison-Wesley, 1973).

GALLAGHER, R., *Information Theory and Reliable Communication* (New York: John Wiley & Sons, Inc., 1968).

GARISH, B., D. HORSKY, K. SRIKANTH, "Optimal Product Positioning of a New Product," working paper, University of Rochester, Rochester, NY (June 1979).

GEMMILL, G. R., and D. L. WILEMON, "The Product Manager as an Influence Agent," *Journal of Marketing,* 36, No. 1 (January 1972), 26–30.

GENSCH, D., "Computer Models in Advertising Media Selection," *Journal of Marketing Research* (November 1968), 414–24.

———, and W. W. RECKER, "The Multinomial, Multiattribute Logit Choice Model," *Journal of Marketing Research,* 16 (February 1979), 124–32.

GERTENFELD, A., *Effective Management of Research and Development* (Reading, MA: Addison-Wesley, 1970).

GILLETTE COMPANY, *A Word About Ideas* (Boston, MA: The Gillette Co., 1972).

GINTER, J. L., "An Experimental Investigation of Attitude Change and Choice of a New Brand," *Journal of Marketing Research,* 11 (February 1974), 30–40.

GLASS, G. U., V. L. WILLSON, and J. M. GUTTMAN, *Design and Analysis of Time-Series Experiments* (Boulder, CO: Colorado Associated University Press, 1975).

GLUCK, F., and R. N. FOSTER, "Managing Technological Change: A Box of Cigars for Brad," *Harvard Business Review,* 53, No. 5 (September-October 1975), 139–50.

GORDON, W. J. J., *Synectics: The Development of Creative Capacity* (New York: Harper & Row, 1961).

GREEN, P. E., *Analyzing Multivariate Data* (Hinsdale, IL: The Dryden Press, 1978).

______, "Bayesian Decision Theory in Pricing Strategy," *Journal of Marketing,* 27, No. 1 (January 1963), 5–14.

______, "On the Robustness of Multidimensional Scaling Techniques," *Journal of Marketing Research,* 12 (February 1975), 73–81.

______, and F. J. CARMONE, "A BIB/Logit Approach to Conjoint Analysis," working paper, Wharton School, University of Pennsylvania (April 1977).

______, and ______, *Multidimensional Scaling and Related Techniques in Marketing Analysis* (Boston, MA: Allyn and Bacon, 1970).

______, ______, and D. P. WACHSPRESS, "On the Analysis of Qualitative Data in Marketing Research," *Journal of Marketing Research,*" 14 (February 1977), 52–59.

______, and W. S. DESARBO, "Additive Decomposition of Perceptions Data via Conjoint Analysis," *Journal of Consumer Research,* 5, No. 1 (June 1978), 58–65.

______, and M. T. DEVITA, "An Interaction Model of Consumer Utility," *Journal of Consumer Research,* 2 (September 1975), 146–53.

______, and R. E. FRANK, "Bayesian Statistics and Marketing Research," *Applied Statistics,* 15, No. 3 (1966), 173–90.

______, A. MAHESHWARI, and V. R. RAO, "Dimensional Interpretation and Configuration Invariance in Multidimensional Scaling – Empirical Study," *Multivariate Behavioral Research,* 4 (April 1969), 159–80.

______, and V. R. RAO, *Applied Multidimensional Scaling* (New York: Holt, Rinehart & Winston, Inc., 1972).

______, ______, and W. S. DESARBO, "Incorporating Group-Level Similarity Judgments in Conjoint Analysis," *Journal of Consumer Research,* 5, No. 3 (December 1978), 187–93.

______, and V. SRINIVASAN, "Conjoint Analysis in Consumer Research: Issues and Outlook," *Journal of Consumer Research,* 5, No. 2 (September 1978), 103–23.

______, and D. S. TULL, *Research for Marketing Decisions* (Englewood Cliffs, NJ: Prentice-Hall, 1978).

______, and Y. WIND, *Multiattribute Decisions in Marketing* (Hinsdale, IL: The Dryden Press, 1973).

______, and ______, "New Way to Measure Consumer's Judgments," *Harvard Business Review* (July-August 1975), 107–17.

GREENSTREET, R. L., and R. J. CONNOR, "Power of Tests for Equality of Covariance Matrices," *Technometrics,* 16, No. 1 (February 1974), 27–30.

GROSS, I., "The Creative Aspects of Advertising," *Sloan Management Review,* 14, No. 1 (Fall 1972), 83–109.

GRUBER, A., "Purchase Intent and Purchase Probability," *Journal of Advertising Research,* 10 (1970), 23–28.

GUMBEL, E. J., *Statistics of Extremes* (New York: Columbia University Press, 1958).

GUTTMAN, L., "The Quantification of a Class of Attributes: A Theory and Method of

Scale Construction," in *The Predictions of Personal Adjustment,* ed. P. HORST et al. (New York: Social Science Research Council, 1941).

______, "The Determinacy of Factor Score Matrices with Implications for Five Other Basic Problems of Common-Factor Theory," *British Journal of Statistical Psychology,* 8, (1965), 65–81.

HADLEY, G., *Linear Algebra* (Reading, MA: Addison-Wesley, 1964).

HALEY, R. I., "Benefit Segmentation: A Decision Oriented Research Tool," *Journal of Marketing,* 32 (July 1968), 30–35.

______, "We Shot an Arrowhead into the Air," *Proceedings, 16th Annual Conference* (New York, Advertising Research Foundation, 1970), 25–30.

HAMBERG, D., "Invention in the Industrial Research Laboratory," *The Journal of Political Economy,* 71 (April 1963), 95.

HAMELMAN, P. W., and E. M. MAZZE, "Improving Product Abandonment Decisions," *Journal of Marketing,* 36, No. 2 (April 1972), 20–26.

HAMERMESH, R. G., M. J. ANDERSON, and J. E. HARRIS, "Strategies for Low Market Share Businesses," *Harvard Business Review,* 56, No. 3 (May-June, 1978), 95–102.

HANAN, M., "Corporate Growth Through Venture Management," *Harvard Business Review,* 47, No. 1 (January-February 1969).

HARDIN, D. K., "A New Approach to Test Marketing," *Journal of Marketing,* 30, No. 4 (October 1966), 28–31.

HARMON, H. H., *Modern Factor Analysis* (Chicago: University of Chicago Press, 1967).

HARNESS, E. G., "Views on Corporate Responsibility" (Cincinnati, OH: The Procter & Gamble Company, 1977).

HAUSER, J. R., "Consumer Preference Axioms: Behavioral Postulates for Describing and Predicting Stochastic Choice," *Management Science,* 24, No. 13 (September 1978), 1331–41.

______, "Forecasting and Influencing the Adaption of Technological Innovations Applications to Telecommunications Innovations," working paper, Transportation Center, Northwestern University, Evanston, IL (October, 1978).

______, "Testing and Accuracy, Usefulness, and Significance of Probabilistic Models: An Information Theoretic Approach," *Operations Research,* 26, No. 3 (May-June 1978), 406–21.

______, and F. S. KOPPELMAN, "Alternative Perceptual Mapping Techniques: Relative Accuracy and Usefulness," *Journal of Marketing Research* (November 1979).

______, and ______, "Designing Transportation Services: A Marketing Approach," *Transportation Research Forum,* 18, No. 1 (October 1977), 628–52.

______, and S. M. SHUGAN, "Extended Conjoint Analysis with Intensity Measures and Computer Assisted Interviews," *Advances in Consumer Research,* 5 (October 1977).

______, and ______, "Intensity Measures of Consumer Preferences," *Operations Research* 28, No. 2 (March–April, 1980).

______, and P. SIMMIE, "Profit Maximizing Perceptual Positioning: A Theory for the Selection of Physical Features and Price," working paper, Graduate School of Management, Northwestern University, Evanston, IL (revised February 1980).

______, and P. STOPHER, "Choosing an Objective Function Based on Modeling Consumer Perceptions and Preferences," *Proceedings of Systems, Man, and Cybernetics* (November 1976).

———, A. M. Tybout, and F. S. Koppelman, "Consumer-Oriented Transportation Service Planning: Consumer Analysis and Strategies," in *Applications of Management Science,* ed. R. Schultz (JAI Press, 1979).

———, and G. L. Urban, "A Normative Methodology for Modeling Consumer Response to Innovation," *Operations Research,* 25, No. 4 (July-August 1977), 579–619.

———, and ———, "Assessment of Attribute Importances and Consumer Utility Functions: Von Neumann-Morgenstern Theory Applied to Consumer Behavior," *Journal of Consumer Research,* 5 (March 1979), 251–62.

———, and K. Wisniewski, "Consumer Analysis for General Travel Destinations," Technical Report, Transportation Center, Northwestern University, Evanston, IL (March 1979).

Hayhurst, R., D. F. Midgley, and G. S. C. Wills, eds., *Creating and Marketing New Products,* (London: Crosby Lockwood Staples, 1973).

Heller, H. B., "The Ostrich and the Copy Researcher: A Comparative Analysis," paper presented at the December 1971 Meeting of the Advertising Effectiveness Research Group, New York Chapter, American Marketing Association.

Helmer, R., and J. K. Johansson, "An Exposition of Box-Jenkins Transfer Function Analysis with an Application to Advertising-Sales Relationship," *Journal of Marketing Research,* 14, No. 2 (May 1977), 227–39.

Herniter, J., "A Comparison of the Entropy and the Hendry Model," *Journal of Marketing Research,* 11 (February 1974), 21–29.

———, "An Entropy Model of Brand Purchase Behavior," *Journal of Marketing Research,* 10 (November 1973), 361–75.

———, and V. J. Cook, "A Multidimensional Stochastic Model of Consumer Purchase Behavior," in *Behavioral and Management Science in Marketing,* eds. A. J. Silk and H. J. Davis, (New York: John Wiley & Sons, Inc., 1978), 237–68.

Herstein, I. N., and J. Milner, "An Axiomatic Approach to Measurable Utility," *Econometrica,* 21 (1953), 291–97.

Hertz, D. B., *New Power for Management,* (New York: McGraw-Hill Book Company, 1969).

———, "Risk Analysis in Capital Investment," *Harvard Business Review,* 42, No. 1 (January-February 1964), 95–106.

Hill, R. M., and J. D. Hlavacek, "The Venture Team: A New Concept in Marketing Organization," *Journal of Marketing,* 36, No. 3 (July 1972), 44–50.

Hise, R. T., and M. A. McGinnis, "Product Elimination, Practice, Policies, and Ethics," *Business Horizons,* 18, No. 30 (June 1975), 25–32.

Hoffman, G. A., *Urban Underground Highways and Parking Facilities* (Santa Monica, CA : Rand Corporation, 1963).

Horsky, D., "An Empirical Analysis of Optimal Advertising Policy," *Management Science,* 23, No. 10 (June 1977), 1037–49.

———, "Market Share Response to Advertising: An Example of Theory Testing," *Journal of Marketing Research,* 14 (February 1977), 10–21.

Howard, J. A., and W. M. Morgenroth, "Information Processing Model of Executive Decisions," *Management Science,* 14 (March 1968), 416–28.

______, and J. N. SHETH, *The Theory of Buyer Behavior,* (New York: John Wiley & Sons, Inc., 1969), 83–114.

HOWARD, R. A., "System Analysis of Semi-Markov Processes," *IEEE Transactions in Military Electronics,* MIL-8, No. 2 (April 1964), 114–24.

HUBER, J., "Predicting Preferences on Experimental Bundles of Attributes: A Comparison of Models," *Journal of Marketing Research,* 12 (August 1975), 290–97.

______, and D. SHELUGA, "The Analysis of Graded Paired Comparisons in Marketing Research," working paper, Purdue University (May 1977).

HUMPHREYS, L. G., and D. R. ILGEN, "Note on a Criterion for the Number of Common Factors," *Educational and Psychological Measurement,* 29 (1969), 571–78.

HURTER, A., and A. RUBENSTEIN, "Market Penetration by New Innovations: The Technological Literature," *Technological Forecasting,* 2 (1978), 197–221.

ILLINOIS INSTITUTE of TECHNOLOGY RESEARCH INSTITUTE, *Technology in Retrospect and Critical Events in Science,* Report to the National Science Foundation (Chicago 1968).

ISENSON, R., "Project Hindsight: An Empirical Study of the Sources of Ideas Utilized in Operational Weapon Systems," in *Factors in the Transfer of Technology,* eds. W. GRUBER and D MARQUIS. (Cambridge, MA: M.I.T. Press, 1969), 157.

JACOBY, J., "Model of Multi-Brand Loyalty," *Journal of Advertising Research,* 11 (June 1971), 25–31.

JAIN, A. K., and C. PINSON, "The Effect of Order of Presentation of Similarity Judgments on Multidimensional Scaling Results: An Empirical Comparison," *Journal of Marketing Research,* 13 (November 1976), 435–39.

JENSEN, N. E., "An Introduction to Bernoullian Utility Theory. I. Utility Functions," *Swedish Journal of Economics,* 69 (1967), 163–83.

JEWKES, J., D. SAWERS, and R. STILLERMAN, *The Sources of Invention* (London: W. W. Norton, 1970).

JOHANSSON, J. K., D. L. MAC LACHLAN, and R. F. YALCH, "Halo Effects in Multi-attribute Attitude Models: Some Unresolved Issues," *Journal of Marketing Research,* 13 (November 1976), 414–17.

JOHNSON, R. M., "A Simple Method for Pairwise Monotone Regression," *Psychometrika,* 40 (June 1975), 163–68.

______, "Beyond Conjoint Analysis: A Method of Pairwise Trade-off Analysis," *Advances in Consumer Research* (October 1975).

______, "Market Segmentation: A Strategic Management Tool," *Journal of Marketing Research,* 8 (February 1971), 13–18.

______, "Multiple Discriminant Analysis Applications to Marketing Research," Market Facts, Inc. (January 1970).

______, "Tradeoff Analysis of Consumer Values," *Journal of Marketing Research,* 11 (May 1974), 121–27.

JOHNSON, S. C., and C. JONES, "How to Organize for New Products," *Harvard Business Review,* 37, No. 3 (May-June 1957), 49–62.

JOHNSTON, J., *Econometric Methods* (New York: McGraw-Hill Book Company, 1972).

JONES, K. A., and D. L. WILEMON, "Emerging Patterns in New Venture Management," *Research Management,* 15, No. 6 (November 1972), 14–27.

JUSTER, F. T., "Consumer Buying Intentions and Purchase Probability: An Experiment in Survey Design," *Journal of American Statistical Association,* 61 (1966), 658–96.

KAHN, F., and L. LIGHT, "Copytesting—Communication vs. Persuasion," in *Advances in Consumer Research,* Vol. 2, ed. M. J. SCHLINGER, *Proceedings* (November, 1974 Conference of the Association for Consumer Research), 595–605.

KALWANI, M. U., "The Entropy Concept and the Hendry Partitioning Approach," working paper, Sloan School of Management, M.I.T., Cambridge, MA (1979).

______, and D. MORRISON, "A Parsimonious Description of the Hendry System," *Management Science,* 23, No. 5 (January 1977), 467–77.

KANNO, M. "Effects on Communication Between Labs and Plants of the Transfer of R&D Personnel," unpublished M.S. thesis, Sloan School of Management, M.I.T., Cambridge, MA (May 1968).

KATZ, E., and P. F. LAZARSFELD, *Personal Influence* (New York: Free Press, 1955).

KEENEY, R. L., "A Decision Analysis with Multiple Objectives: The Mexico City Airport," *The Bell Journal of Economics and Management Science,* 4 (1973), 101–17.

______, "Multiplicative Utility Functions," *Operations Research,* 12, No. 1 (January 1974), 22–33.

______, "Utility Functions for Multiattributed Consequences," *Management Science,* 18 (1972), 276–87.

______, and H. RAIFFA, *Decision Analysis with Multiple Objectives* (New York: John Wiley & Sons, Inc., 1976).

KELLY, G. A., *The Psychology of Personal Constructs,* Vol. 1 (New York: W. W. NORTON, 1955).

KENDALL, M. G., *Rank Correlation Methods* (London: Griffin, 1955).

KLAHR, D., "A Monte Carlo Investigation of the Statistical Significance of Kruskal's Non-metric Scaling Procedure," *Psychometrika,* 34, No. 3 (September 1969), 319–30.

KLOMPMAKER, J. E., G. D. HUGHES, and R. I. HALEY, "Test Marketing in New Product Development," *Harvard Business Review,* 54, No. 3 (May-June 1976), 128–38.

KOPPELMAN, F. S., "Guidelines for Aggregate Travel Prediction Using Disaggregate Choice Models," *Transportation Research Record,* No. 610, Transportation Research Board, Washington, D.C. (1976).

______, and J. R. HAUSER, "Destination Choice Behavior for Non-grocery Shopping Trips," *Transportation Research Record,* No. 673, Transportation Research Board, Washington, D.C. (1979), pp. 157–65.

KOTLER, P., *Marketing Decision Making: A Model Building Approach* (New York: Holt, Rinehart & Winston, 1971).

______, *Marketing Management: Analysis, Planning, and Control,* 3rd ed. (Englewood Cliffs, NJ: Prentice-Hall, 1976).

______, "Marketing Mix Decisions for New Products," *Journal of Marketing Research,* 1, No. 1 (February 1964), 43–49.

______, "Phasing Out Weak Products," *Harvard Business Review,* 43, No. 2 (March-April 1965), 107–18.

KRANTZ, D. H., R. D. LUCE, P. SUPPES, and A. TVERSKY, *Foundations of Measurement* (New York: Academic Press, 1971).

KRISHNAN, K. S., "Incorporating Thresholds of Indifference in Probabilistic Choice Models," *Management Science,* 23, No. 11 (July 1977), 1224–33.

KRUSKAL, J. B., "Analysis of Factorial Experiments by Estimating Monotone Transformations of the Data," *Journal of Royal Statistical Society,* Series B, 27 (1965), 251–63.

———, "Multidimensional Scaling by Optimizing Goodness of Fit to a Non-metric Hypothesis," *Psychometrika,* 29 (1964), 1–27.

LAMBIN, J., "A Computer On-Line Marketing Mix Model," *Journal of Marketing Research,* 9, No. 2 (May 1972), 119–26.

———, *Advertising, Competition and Market Conduct in Oligopoly Over Time* (Amsterdam: North-Holland, 1976).

LANCASTER, K., *Consumer Demand: A New Approach* (New York: Columbia University Press, 1971).

LANGRISH, J., "Technology Transfer: Some British Data," *R&D Management,* Vol. 1, No. 133 (June 1971).

LANITIS, T., "How to Generate New Product Ideas," *Journal of Advertising Research,* 10, No. 3 (June 1970), 31–35.

LAVIDGE, R. J., and G. A. STEINER, "A Model for Predictive Measurements of Advertising Effectiveness," *Journal of Marketing,* 25, No. 6 (October 1961), 59–62.

LAWTON, S. B., and W. H. LAWTON, "An Autocatalytic Model for the Diffusion of Educational Innovations," *Educational Administration Quarterly,* 15, No. 1 (Winter 1979), 19–46.

LEHMANN, D. R., *Market Research and Analysis* (Homewood, IL: Richard D. Irwin, Inc., 1979).

LERMAN, S. R., C. F. MANSKI, and T. ATHERTON, "Alternative Sampling Procedure for Disaggregate Choice Model Estimation," *Transportation Research Record,* No. 592, Transportation Research Board, Washington, D.C. (1976).

LEVITT, T., "Exploit the Product Life Cycle," *Harvard Business Review,* 43, No. 6 (November-December 1965), 81–94.

LEWIS, A. C., T. L. SADOSKY, and T. CONNOLLY, "The Effectiveness of Group Brainstorming in Engineering Problem Solving," *IEEE Transactions on Engineering Management,* EM–22, No. 3 (August 1975), 119–24.

LIGHT, L., L. PRINGLE, and E. F. SNOW, "NEWS Report," working paper, No. 126, Graduate School of Industrial Administration, Carnegie-Mellon University (May 1968).

LIKERT, R., "A Technique for the Measurement of Attitudes," *Archives of Psychology,* No. 140 (1932).

LILIEN, G. L., "Model Relativism: A Situational Approach to Model Building," *Interfaces,* 5, No. 3 (May 1975), 11–18.

———, and J. D. C. LITTLE, "The ADVISOR Project: A Study of Industrial Marketing Budgets," *Sloan Management Review,* 17, No. 3 (Spring 1976), 17–31.

LINSTON, H., and M. TUROFF, eds., *The Delphi Method: Techniques and Applications* (Reading, MA: Addison-Wesley, 1975).

LITTLE, J. D. C., "A Model of Adaptive Control of Promotional Spending," *Operations Research,* 14, No. 6 (November-December 1966), 1075–98.

———, "Aggregate Advertising Models: The State of the Art," *Operations Research,* 27, No. 4 (July-August, 1979), 629–67.

———, "BRANDAID: A Marketing Mix Model, Structure, Implementation, Calibration, and Case Study," *Operations Research,* 23, No. 4 (July-August 1975), 628–73.

———, "Models and Managers: The Concept of a Decision Calculus," *Management Science,* 16, No. 8 (April 1970), 466–85.

———, "Optimal Adaptive Control: A Multivariate Model for Marketing Applications," *IEEE Transactions on Automatic Control,* Vol. AC-22, No. 2 (April 1977), 187–95.

______, and L. M. LODISH, "A Media Planning Calculus," *Operations Research,* 17, No. 1 (January-February 1969), 1–35.

______, and J. F. SHAPIRO, "A Theory for Pricing Non-features Products in Supermarkets," working paper 931–77, Sloan School of Management, M.I.T. (May 1977), forthcoming *Journal of Business.*

LODISH, L. M., "Callplan: An Interactive Saleman's Call Planning System," *Management Science,* 18, No. 4, Part II (December 1971), 25–40.

______, "Sales Territory Alignment to Maximize Profit," *Journal of Marketing Research,* 12, No. 1 (February 1975), 30–36.

LORSCH, J. W., and P. R. LAWRENCE, "Organizing for Product Innovation," *Harvard Business Review,* 43, No. 1 (January-February 1965), 109–22.

LUCAS, M. C., C. WEINBERG, and K. CLOWES, "Sales Response as a Function of Territory Potential and Sales Representative Workload," *Journal of Marketing Research,* 12, No. 3 (August 1975), 298–305.

LUCE, R. D., *Individual Choice Behavior* (New York: John Wiley & Sons, Inc., 1959).

______, "The Choice Axiom After Twenty Years," *Journal of Mathematical Psychology,* 15 (1977), 215–33.

______, and P. SUPPES, "Preference, Utility, and Subjective Probability," in *Handbook of Mathematical Psychology,* Vol. 3, eds. R. D. LUCE, R. R. BUSH, and E. GALANTER, (New York: John Wiley & Sons, Inc., 1965), 249–410.

MACHIAVELLI, N., *The Prince: A New Translation and Background Interpretation,* trans. and ed. by R. M. ADAMS (New York: W. W. Norton, 1977).

MAGEE, J. F., "Decision Trees for Decision Making," *Harvard Business Review,* 42, No. 4 (July-August 1964), 126–39.

MAHAJAN, V. and E. MULLER, "Innovation Diffusion and New Product Growth Models in Marketing," *Journal of Marketing,* 43, No. 4 (Fall 1979), 55–68.

MALONEY, J. L., "Curiosity versus Disbelief in Advertising," *Journal of Advertising Research,* 2, No. 2 (June 1962), 2–8.

MANDELL, M. L., *Advertising,* 2nd ed. (Englewood Cliffs, NJ: Prentice Hall, 1974).

MANSFIELD, E., *Industrial Research and Technological Innovation* (New York: W. W. NORTON, 1968).

______, *The Economics of Technological Change* (New York: W. W. Norton, 1968).

______, and J. RAPOPORT, "The Costs of Industrial Product Innovation," *Management Science,* 21, No. 12 (August 1975), 1380–86.

______, ______, J. SCHNEE, S. WAGNER, and M. HAMBERGER, *Research and Innovation in the Modern Corporation* (New York: W. W. Norton, 1971).

______, and S. WAGNER, "Organizational and Strategic Factors Associated with Probabilities of Success in Industrial R and D," *Journal of Business* (April 1975).

MANSKI, C. F., "Maximum Score Estimation of the Stochastic Utility Models of Choice," *Journal of Econometrics,* 3 (1975), 205–28.

______, and S. R. LERMAN, "The Estimation of Choice Probabilities from Choice Based Samples," *Econometrica,* 45, No. 8 (November 1977).

Market Facts, Inc., "High Efficiency Electric Motors: Focus Group Results," *Contract Job Report No. 9312.* (Washington, D.C.: Department of Energy, August 1978).

______, and PEAT, MARWICK, MITCHELL CO., "A Marketing Approach to Carpool Demand Analysis," (Washington, D.C.: Department of Energy, April 1976), *Contract Job No. C-04-50179-00.*

MARQUARDT, R., J. MAKENS, and H. LARZELERE, "Measuring the Utility Added by Branding and Grading," *Journal of Marketing Research,* 2, No. 1 (February 1965), 45–50.

MARQUIS, D. G., "The Anatomy of Successful Innovation," *Innovation,* 1, No. 7 (1969), 28–37.

MARSCHAK, J., "Rational Behavior, Uncertain Prospects, and Measurable Utility," *Econometrica,* 18 (1950), 111–41.

MARSCHAK, T., T. K. GLENNAN, and R. SUMMERS, *Strategy for R&D: Studies in Microeconomics of Development* (New York: Springer-Verlag, 1967).

MARSHALL, A. W., and W. H. MECKLING, "Predictability of Costs, Time, and Success of Development," in *The Rate and Direction of Inventive Activity: Economic and Social Factors,* ed. National Bureau of Economic Research, (Princeton, NJ: Princeton University Press, 1962) 461–76.

MASSY, W. F., "Forecasting Demand for New Convenience Products," *Journal of Marketing Research,* 6, No. 4 (November 1969), 405–12.

______, R. E. FRANK, and T. M. LODAHL, *Purchasing Behavior and Personal Attributes* (Philadelphia: University of Pennsylvania Press, 1968).

Materials Advisory Board, Division of Engineering, National Research Council, *Report of the Ad Hoc Committee on Principles of Research-Engineering Interaction,* Publication MAB-222-M (Washington, D.C.: National Academy of Sciences – National Research Council, July 1966), 15–16.

MCCONNELL, J. D., "The Price-Quality Relationship in an Experimental Setting," *Journal of Marketing Research,* 5, No. 3 (August 1968), 300–303.

MCFADDEN, D., "A Comment on Discriminant Analysis vs. Logit Analysis," *Annals of Economic and Social Measurement* (May 4, 1976), 511–23.

______, "Conditional Logit Analysis of Qualitative Choice Behavior," in *Frontiers in Econometrics,* ed. P. ZAREMBKA, pp. 105–42 (New York: Academic Press, 1970).

______, "Econometric Models for Probabilistic Choice Among Products," forthcoming *Journal of Business* (1980).

______, "Quantal Choice Analysis: A Survey," *Annals of Economic and Social Measurement* (May 4, 1976), 363–90.

______, "The Revealed Preferences at a Government Bureaucracy: Theory," *The Bell Journal of Economics and Management Sciences,* 6, No. 2 (Autumn 1975), 401–16.

______, and H. WILLS, "XLOGIT, a Program for Multinominal Logit Analysis," Travel Demand Forecasting Project, Institute for Transportation and Traffic Engineering, University of California, Berkeley (1975).

MEADOWS, D. L., "Estimate Accuracy and Project Selection Models in Industrial Research," *Industrial Management Review,* 9, No. 3 (Spring 1968), 105–19.

MIDGLEY, D. F., *Innovation and New Product Marketing* (London: Croom Helm, 1977).

______, and G. R. DOWLING, "Innovativeness: The Concept and Its Measurement," *Journal of Consumer Research,* 4, No. 4 (March 1978), 229–47.

MILLER, J. E., *Innovation, Organization, and Environment* (Sherbrooke, Quebec, Canada: University of Sherbrooke, 1971).

______, R. N. SHEPARD, and J. J. CHANG, "An Analytical Approach to the Interpretation of Multidimensional Scaling Solutions," *American Psychologist,* 19 (1964), 579–80.

MITTELSTAEDT, R. A., S. L. GROSSBART, W. W. CURTIS, and S. P. DEVERE, "Optimal

Stimulation Level and the Adoption Decision Process," *Journal of Consumer Research*, 3, No. 2 (June 1976), 84–94.

MONROE, K. B., and A. J. D. BITTA, "Models for Pricing Decisions," *Journal of Marketing Research,* 15, No. 3 (August 1978), 413–28.

MONTANELLI, R. G., JR., and L. G. HUMPHREYS, "Latent Roots of Random Data Correlation Matrices with Squared Multiple Correlations on the Diagonal: A Monte Carlo Study," *Psychometrika,* 41, No. 3 (September 1976).

MONTGOMERY, D. B., "Consumer Characteristics Associated with Dealing: An Empirical Example," *Journal of Marketing Research,* 8, No. 1 (February 1971), 118–20.

______, "New Product Distribution—An Analysis of Supermarket Buyer Decision," *Journal of Marketing Research,* 12, No. 3 (August 1975), 255–64.

______, and A. SILK, "Estimating Dynamic Effects of Market Communications Expenditures," *Management Science,* 18, No. 10 (June 1972) 485–501.

______, ______, and C. E. ZARAGOZA, "A Multiple-Product Sales Force Allocation Model," *Management Science,* 18, No. 4, Part II (December 1971), P-3 to P-24.

______, and G. L. URBAN, *Management Science in Marketing* (Englewood Cliffs, NJ: Prentice-Hall, 1969).

MOOD, A. M., and F. A. GRAYBILL, *Introduction to the Theory of Statistics* (New York: McGraw-Hill Book Company, 1963).

Moody's OTC Industrial Manual (New York: Moody's Investors Service, Inc., 1978), p. 689.

MORRISON, D. G., "Purchase Intentions and Purchase Behavior," *Journal of Marketing,* 43, No. 2 (Spring 1979), 65–74.

______, Reliability of Tests: A Technique Using the 'Regression to the Mean' Fallacy," *Journal of Marketing Research,* 10 (February 1973), 91–93.

MOSKOWITZ, H. R., "Subjective Ideals and Sensory Optimization in Evaluating Perceptual Dimensions in Food," *Journal of Applied Psychology,* 56, No. 1 (1972), 60–66.

MUELLER, W. F., in *The Rate and Direction of Inventive Activity: Economic and Social Factors,* ed. R. R. NELSON, (Princeton, NJ: Princeton University Press, 1962), 299–322.

MYERS, J. G., "An Interactive Computer Approach to Product Positioning," *Proceedings of the Attitude Research Conference,* Hilton Head, SC (February 11-15, 1976).

MYERS, J. H., "Benefit Structure Analysis: A New Tool for Product Planning," *Journal of Marketing,* 40, No. 4 (October 1976), 23–33.

MYERS, M. S., "Who Are Your Motivated Workers?" *Harvard Business Review,* 42, No. 1 (January-February 1964), 73–88.

MYERS, S., and D. G. MARQUIS, *Successful Industrial Innovation: A Study of Factors Underlying Innovation in Selected Firms,"* NSF 69–17 (Washington, D.C.:National Science Foundation, 1969).

NAERT, P. A., and P. S. M. LEEFLANG, *Building Implementable Models* (Boston, MA: Martinus Nijhoff/Leiden, 1978).

NARAYANA, C. L., and R. J. MARKIN, "Consumer Behavior and Product Performance: An Alternative Conceptualization," *Journal of Marketing,* 39 (October 1975), 1–6.

NELSON, C. R., *Applied Time Series Analysis for Managerial Forecasting* (San Francisco: Holden-Day, Inc., 1973).

NESLIN, S. A., "Linking Product Features to Perceptions: Applications and Evaluation of Graded Paired Comparisons," *Proceedings of AMA Educators' Conference*, Chicago, IL (August 1978).

______, "Analyzing Consumer Response to Health Innovations: The Concept of Preference Inertia," working paper, (Cambridge, MA: M.I.T., May 1976).

______, and G. L. URBAN, "Taste Testing Squid Chowder," working paper, Sloan School of Management (Cambridge, MA: M.I.T., 1977).

NESS, D., and C. R. SPRAGUE, "An Interactive Media Decision Support System," *Sloan Management Review*, 14, No. 1 (Fall 1972), 51–62.

NEVERS, J. V., "Extensions of a New Product Growth Model," *Sloan Management Review*, 13, No. 2 (Winter 1972), 77–90.

NICOSIA, F., *Consumer Decision Processes* (Englewood Cliffs, NJ: Prentice-Hall, 1966).

NIE, N. H., G. H. HULL, J. G. JENKINS, K. STEINBRENNER, and D. H. BENT, *SPSS: Statistical Package for the Social Sciences*, 2nd ed. (New York: McGraw-Hill Book Company, 1975).

NIELSEN MARKETING SERVICE, "New Brand or Superbrand?" *The Nielsen Researcher*, No. 5 (1971), 4–10.

______, "New Product Success Ratios," *The Nielsen Researcher* (1979), 2–9.

NIERENBERG, G. I., *The Art of Negotiating: Psychological Strategies for Gaining Advantageous Bargains* (New York: Hawthorn Books, 1968).

NORRIS, K. P., "The Accuracy of Project Cost and Duration Estimates in R&D," *R and D Management*, 2, No. 1 (October 1971), 25–36.

NORTH, H. Q., and D. L. PYKE, " 'Probes' of the Technological Future," *Harvard Business Review*, 47, No. 2 (May-June 1969), 68–82.

NOYCE, R. N., "Microelectronics," *Scientific American*, 237, No. 3 (September 1977), 63–69.

O'MEARA, J. T., "Selecting Profitable Products," *Harvard Business Review*, 39, No. 1 (January-February 1961), 83–89.

OPPENHEIM, A. N., *Questionnaire Design and Attitude Measurement* (New York: Basic Books, Inc., 1966).

OSBORN, A. F., *Applied Imagination*, (New York: Charles Scribner's Sons, 1963).

PALDA, K. S., *The Measurement of Cumulative Advertising Effects* (Englewood Cliffs, NJ: Prentice Hall, 1964).

PARASURANAN, A., and R. L. DAY, "A Management Oriented Model for Allocating Sales Efforts," *Journal of Marketing Research*, 14, No. 1 (February 1977), 22–23.

PARFITT, J. H., and B. J. K. COLLINS, "Use of Consumer Panels for Brand Share Prediction," *Journal of Marketing Research*, 5, No. 2 (May 1968), 131–46.

PARKER, B. R., and V. SRINIVASAN, "A Consumer Preference Approach to the Planning of Rural Primary Health-Care Facilities," *Operations Research*, 24, No. 5 (September-October 1976), 991–1025.

PARNES, S. J., and H. F. HARDING, eds., *A Source Book for Creative Thinking* (New York: Charles Scribner's Sons, 1962).

PARSONS, L. J., and R. L. SCHULTZ, *Marketing Models and Econometric Research* (New York: North Holland Publishing Co., 1976).

PAYNE, S. L., *The Art of Asking Questions* (Princeton, NJ: Princeton University Press, 1951).

Pekelman, D., and S. Sen, "Improving Prediction in Conjoint Measurement," *Journal of Marketing Research*, 16 (May 1979), 211–20.

———, and ———, "Measurement and Estimation of Conjoint Utility Functions," *Journal of Consumer Research*, 5, No. 4 (March 1979), 263–71.

———, and E. Tse, "Experimentation and Control in Advertising: An Adaptive Control Approach," *Operations Research*, 28, No. 2, (March–April 1980).

Peplow, M. E., "Design Acceptance," in *The Design Method*, ed. S. A. Gregory (London: Butterworth, 1960).

Pessemier, E. A., *New Product Decision: An Analytical Approach* (New York: McGraw-Hill Book Company, 1966).

———, *Product Management: Strategy and Organization* (New York: John Wiley & Sons, Inc./Hamilton, 1977).

———, P. Burger, R. Teach, and D. Tigert, "Using Laboratory Brand Preference Scales to Predict Consumer Brand Purchases," *Management Science*, 17 (February 1971), B–371 to B–385.

Pindyck, R. S., and D. L. Rubinfeld, *Econometric Models and Economic Forecasting* (New York: McGraw-Hill Book Company, 1976).

Pollack, R. A., "Additive Von Neumann-Morgenstern Utility Functions," *Econometrica*, 35 (1967), 485–595.

Polli, R., and V. Cook, "Validity of the Product Life Cycle," *Journal of Business*, 42, No. 4 (October 1969), 385–400.

Pratt, J. W., H. Raiffa, and R. Schlaifer, *Introduction to Statistical Decision Theory* (New York: McGraw-Hill Book Company, 1965).

Prince, G. M., *The Practice of Creativity* (New York: Collier Books, 1972).

Punj, G. N., and R. Staelin, "The Choice Process for Graduate Business Schools," *Journal of Marketing Research*, 15 (November 1978), 588–98.

Pymont, B. C., D. Reay, and P. G. M. Standen, "Towards the Elimination of Risk from Investment in New Products: Experience with Micro-Market Testing," paper presented at the 1976 ESOMAR Congress, Venice, Italy (September 1976).

Quinn, J. B., "Technology Forecasting," *Harvard Business Review*, 45, No. 2 (March-April 1967), 73–90.

Raiffa, H., "Assessments of Probabilities," unpublished manuscript (January 1969).

———, *Decision Analysis: Introductory Lectures on Choices Under Uncertainty* (Reading, MA: Addison-Wesley, 1968).

———, and R. Schlaifer, *Applied Statistical Decision Theory* (Boston, MA: Harvard University Press, 1961).

Rao, A. G., *Quantitative Theories in Advertising* (New York: John Wiley & Sons, Inc., 1970).

———, and P. B. Miller, "Advertising/Sales Response Functions," *Journal of Advertising Research*, 15, No. 2 (April 1975), 7–15.

Rao, V., "The Salience of Price in the Perception and Evaluation of Product Quality: A Multidimensional Measurement Model and Experimental Test," unpublished Ph.D. thesis, University of Pennsylvania (1970).

———, and J. E. Cox, Jr., "Sales Forecasting Methods: A Survey of Recent Developments," Marketing Science Institute Report No. 78–119, Cambridge, MA (December 1978).

RATHBONE, R. R., *Communicating Technical Information* (Reading, MA: Addison-Wesley, June 1967).

RHODES, R., *What ADTEL Has Learned* (ADTEL Publication, 1977).

RICHARD, S. F., "Multivariate Risk Aversion, Utility Independence, and Separable Utility Functions," *Management Science*, 22, No. 1 (1975), 12–21.

ROBERTS, E. B., "A Basic Study of Innovation: How to Keep and Capitalize on Their Talents," *Research Management*, 11, No. 4 (July 1968), 249–66.

______, "A Simple Model of R&D Project Dynamics," *R and D Management*, 5, No. 1 (October 1974), 1–15.

______, "Exploratory and Normative Technological Forecasting: A Critical Appraisal," *Technological Forecasting*, 1 (1969), 113–27.

______, and A. L. FROHMAN, "Internal Entrepreneurship: Strategy for Growth," *The Business Quarterly*, 37, No. 1 (Spring 1972), 71–78.

______, and H. A. WAINER, "New Enterprises on Route 128" *Science Journal*, 4, No. 12 (December 1968), 78–83.

ROBERTSON, A. B., B. ACHILLADELIS, and P. JERVIS, *Success and Failure in Industrial Innovation: Report on Project Sappho* (London: Centre for the Study of Industrial Innovation, 1972).

ROBERTSON, T. S., "Diffusion Theory and the Concept of Personal Influence," in *Behavioral and Management Science in Marketing*, eds. H. L. Davis and A. J. Silk, (New York: John Wiley & Sons, Inc., 1978), 214–36.

______, *Innovative Behavior and Communication* (New York: Holt, Rinehart, and Winston, 1971).

ROBINSON, B., and C. LAKHANI, "Dynamic Price Models for New-Product Planning," *Management Science*, 21, No. 10 (June 1975), 1113–22.

ROBINSON, P., C. FARRIS, and Y. WIND, *Industrial Buying and Creative Marketing* (Boston, MA: Allyn and Bacon, 1967).

ROGERS, E. M., "Re-inventing During the Innovation Process," working paper, Stanford University, Palo Alto, CA, Institute for Communication Research (1978).

______, *The Diffusion of Innovation* (New York: The Free Press, 1962).

______, and F. F. SHOEMAKER, *Communications of Innovations: A Cross-Cultural Approach* (New York: The Free Press, 1971).

______, and J. D. STANFIELD, "Adoption and Diffusion of New Products: Emerging Generalizations and Hypotheses," in *Applications of the Sciences in Marketing Management*, ed. F. Bass et al., pp. 227–50 (New York: John Wiley & Sons, Inc., 1968).

ROOT, H. P., "A Computer Simulation Model for the Financial Analysis of New Products," Bureau of Business Research, School of Business, University of Michigan (November 1970).

ROSENBURG, M. J., "Cognitive Structure and Attitude Effect," *Journal of Abnormal and Social Psychology*, 53 (1956), 367–72.

ROTHWELL, R., et al., "SAPPHO Updated—Project SAPPHO Phase II," *Research Policy*, 3 (1974), 258–91.

RUBENSTEIN, A. H., and H. SCHREDER, "Management Differences in Assessing Probabilities of Technical Success for R&D Projects," *Management Science*, 24, No. 2 (October 1977), 137–48.

RUBINSON, J. R., and F. M. BASS, "A Note on a Parsimonious Description of the

Hendry System," working paper, No. 658, Krannert Graduate School, Purdue University, West Lafayette, IN (March 1978).

RUMMEL, R. J., *Applied Factor Analysis* (Evanston, IL: Northwestern University Press, 1970).

RYAN, M. J., and E. H. BONFIELD, "The Fishbein Extended Model and Consumer Behavior," *Journal of Consumer Research*, 2, No. 2 (September 1975), 118–36.

RYANS, A. B., "Estimating Consumer Preferences for a New Durable Brand in an Established Product Class," *Journal of Marketing Research*, 11 (November 1974), 434–43.

SASIENI, M. W., "Optimal Advertising Expenditure," *Management Science*, 18 (1971), 64–72.

SCHEFFE, H., "An Analysis of Variance for Paired Comparisons," *Journal of the American Statistical Association*, 47 (September 1952), 381–400.

SCHEIN, E. H., "Increasing Organizational Effectiveness Through Better Human Resource Planning and Development," *Sloan Management Review*, 19, No. 1 (Fall 1977), 1–20.

SCHLAIFER, R., *Analysis of Decisions Under Uncertainty* (New York: McGraw-Hill Book Company, 1969).

SCHOEFFLER, S. R., R. D. BUZZELL, and D. F. HEANY, "Impact of Strategic Planning on Profit Performance," *Harvard Business Review*, 52, No. 2 (March-April 1974), 137–45.

SCHWERIN RESEARCH CORPORATION, "Classic Study," *Schwerin Research Bulletin*, 3 (December 1955), 2–4.

SCOTT, C. A., "The Effects of Trial and Incentives on Repeat Purchase Behavior," *Journal of Marketing Research*, 13, No. 3 (August 1976), 263–69.

SEMLOW, W. J., "How Many Salesmen Do You Need?" *Harvard Business Review*, 38, No. 3 (May-June 1959), 126–32.

SHAPLEY, L. S., "Cardinal Utility from Intensity Comparisons," RAND Report R-1683-PR (Santa Monica, CA: Rand Corporation, July 1975).

SHERWIN, C. W., and R. S. ISENSON, "Project Hindsight," *Science*, 156, No. 3782 (1967), 1571–77.

SHETH, J. N., "Reply to Comments on the Nature and Uses of Expectancy-Value Models in Consumer Attitude Research," *Journal of Marketing Research*, 9 (November 1972), 462–65.

———, and W. W. TALARZYK, "Perceived Instrumentality and Value Importance as Determinants of Attitudes," *Journal of Marketing Research*, 9 (February 1972), 6–9.

SHOCKER, A. D., and V. SRINIVASAN, "A Consumer-Based Methodology for the Identification of New Product Ideas," *Management Science*, 20, No. 6 (February 1974), 921–37.

———, and ———, "Multiattribute Approaches to Product Concept Evaluation and Generation: A Critical Review," *Journal of Marketing Research*, 16 (May 1979), 159–80.

SHUGAN, S. M., and J. R. HAUSER, "P.A.R.I.S.: An Interactive Market Research Information System," working paper, Department of Marketing, Graduate School of Management, Northwestern University, Evanston, IL (May 1977).

SICHERMAN, A., "An Interactive Computer Program for Assessing and Using Multiattribute Utility Functions," Operations Research Center, Technical Report No. 111 (Cambridge, MA: M. I. T., June 1975).

SIGFORD, J. V., and R. H. PARVIN, "Project PATTERN: A Normative Methodology for Determining Relevance in Complex Decision Making," *IEEE Transactions on Engineering Management*, Vol. EM-12, No. 1 (March 1965), 2–7.

SILK, A. J., "Overlap Among Self-Designated Opinion Leaders: A Study of Selected Dental Products and Services," *Journal of Marketing Research*, 3, No. 3 (August 1966), 255–59.

______, "Test-Retest Correlations and the Reliability of Copy Testing," *Journal of Marketing Research*, 14, No. 4 (November 1977), 476–86.

______, "The Influence of Advertising's Affective Qualities on Consumer Response," in *Buyer/Consumer Information Processing*, eds. G. D. Hughes and M. L. Ray, (Chapel Hill, NC: University of North Carolina Press, 1974), 157–86.

______, and M. KALWANI, "Structure of Repeat Buying for New Packaged Goods," Sloan School of Management, M. I. T., Cambridge, MA (1978).

______, and G. L. URBAN, "Pre-Test Market Evaluation of New Packaged Goods: A Model and Measurement Methodology," *Journal of Marketing Research*, 15, No. 2 (May 1978), 171–91.

SIMMIE, P., "Alternative Perceptual Models: Reproducibility, Validity, and Data Integrity," *Proceedings of American Marketing Association Educators Conference* (Chicago, IL, American Marketing Association, 1978), 12–16.

SINGSON, R. L., "Multidimensional Scaling Analysis of Store Image and Shopping Behavior," *Journal of Retailing*, 51, No. 2 (Summer 1975).

SONQUIST, J. A., E. L. BAKER, and J. N. MORGAN, *Searching for Structure*, Survey Research Center, Institute for Social Research, University of Michigan, Ann Arbor, MI (1973).

SOUDER, W. E., "Effectiveness of Nominal and Interacting Group Decision Processes for Integrating R&D and Marketing," *Management Science*, 23, No. 6 (February 1977), 595–605.

SRINIVASAN, V., "A General Procedure for Estimating Consumer Preference Distributions," *Journal of Marketing Research*, 12 (November 1975), 377–89.

______, "A Theoretical Comparison of the Predictive Power of the Multiple Regression and Unit Weighting Procedures," paper presented at the Joint National Meeting of the Operations Research Society of America/The Institute of Management Science, Atlanta, GA (November 7–9, 1977).

______, and A. D. SHOCKER, "Estimating the Weights for Multiple Attributes in a Composite Criterion Using Pairwise Judgments," *Psychometrika*, 38, No. 4 (December 1973), 473–93.

______, and ______, "Linear Programming Techniques for Multidimensional Analysis of Preferences," *Psychometrika*, 38, No. 3 (September 1973), 337–69.

Standard and Poor's Industry Surveys, "The Consumer Market: Healthy Growth Seen in 1977" (October 1977), c64.

______, Vol. 2 (January 1979), 05 and 038–045.

STANTON, F., "What Is Wrong with Test Marketing?," *Journal of Marketing*, 31, No. 2 (April 1967), 43–47.

STASCH, S. F., "Linear Programming and Space-Time Considerations," *Journal of Advertising Research*, 4 (December 1965), 40–47.

STEFFLRE, V., "Market Structure Studies: New Products for Old Markets and New Markets (Foreign) for Old Products," in *Applications of the Sciences in Marketing Management*, eds. F. Bass, C. W. King, and E. A. Pessemier, (New York: John Wiley & Sons, 1968), 251–68.

———, "Multidimensional Scaling as a Model for Individual and Aggregate Perception and Cognition," *Proceedings*, American Marketing Association Educators' Conference, Chicago, IL (August 1978).

———, "New Products: Organizational and Technical Problems and Opportunities," in *Analytic Approaches to Product and Marketing Planning*, ed. A. D. Shocker (Cambridge, MA: Marketing Science Institute, May 1979).

———, "Some Applications of Multidimensional Scaling to Social Science Problems," in *Multidimensional Scaling: Theory and Applications in the Behavioral Sciences*, Vol. 2, eds. A. K. Romney, R. N. Shepard, and S. B. Nerlove (New York: Seminar Press, 1972).

STERN, L. W., and A. I. EL-ANSARY, *Marketing Channels* (Englewood Cliffs, NJ: Prentice-Hall, 1977).

STERNTHAL, B., C. S. CRAIG, and C. LEAVITT, *Consumer Behavior Theory and Strategy* (Englewood Cliffs, NJ: Prentice-Hall, 1980).

STEVENS, S. S., "Ratio Scales of Opinion," in *Handbook of Measurement and Assessment in Behavioral Sciences*, ed. D. K. Whitla, (Reading, MA: Addison-Wesley, 1968), 171–99.

STRONG, E. K., *The Psychology of Selling* (New York: McGraw-Hill Book Company, 1925).

SUDMAN, S., "On the Accuracy of Recording of Consumer Panels: Part I and Part II," *Journal of Marketing Research* (May 1964), pp. 14–20 and (August 1964), pp. 69–83.

———, and R. FERBER, "A Comparison of Alternative Procedures for Collecting Consumer Expenditure Data for Frequently-Purchased Products," *Journal of Marketing Research*, 11 (May 1974), 128–35.

———, and ———, "Experiments in Obtaining Consumer Expenditures by Diary Methods," *Journal of the American Statistical Association*, 66 (December 1971), 725–35.

TANNENBAUM, M., et al., *Report of the Ad Hoc Committee on Principles of Research/Engineering Interaction*, report No. MAB 222-M, National Academy of Sciences-National Research Council Material Advisory Board, Washington, D.C. (1966).

TAUBER, E. M., "Forecasting Sales Prior to Test Market," *Journal of Marketing*, 41, No. 1 (January 1977), 80–84.

TORGERSON, W. S., *Theory and Method of Scaling* (New York: John Wiley & Sons, Inc., 1958).

TUCKER, W. T., "The Development of Brand Loyalty," *Journal of Marketing Research*, 1, No. 3 (August 1964), 32–35.

TVERSKY, A., "Elimination by Aspects: A Theory of Choice," *Psychological Review*, 79, No. 4 (1972), 281–99.

URBAN, G. L., "A Mathematical Modeling Approach to Product Line Decisions," *Journal of Marketing Research*, 6, No. 1 (February 1969), 40–47.

———, "A Model for Managing a Family-Planning System," *Operations Research*, 22, No. 2 (March-April 1974), 205–33.

———, "A New Product Analysis and Decision Model," *Management Science*, 14, No. 8 (April 1968), 490–517.

______, "Allocating Ad Budgets Geographically," *Journal of Advertising Research*, 15, No. 6 (December 1975), 7–18.

______, "Building Models for Decision Makers," *Interfaces*, 4, No. 3 (May 1974), 1–11.

______, "PERCEPTOR: A Model for Product Positioning," *Management Science*, 21, No. 8 (April 1975), 858–71.

______, "Product Planning in the Aerospace Industry," unpublished manuscript thesis, School of Business, University of Wisconsin, Madison, WI (1964).

______, "SPRINTER mod III: A Model for the Analysis of New Frequently Purchased Consumer Products," *Operations Research*, 18, No. 5 (September-October 1970), 805–53.

______, P. JOHNSON and R. BRUDNICK "Market Entry Strategy Formulation : A Hierarchical Model and Consumer Measurement Approach" working paper, Sloan School of Management, M. I. T., Cambridge, MA (1979).

______, and R. KARASH, "Evolutionary Model Building," *Journal of Marketing Research*, 8, No. 1 (February 1971), 62–66.

______, and S. NESLIN, "The Design and Marketing of New Educational Programs," working paper, Sloan School of Management, M. I. T., Cambridge, MA (1977).

U. S. DEPARTMENT OF TRANSPORTATION, "Feasibility Study of Shared-Ride Auto Transit," report IT-06-0144-77-1, Urban Mass Transportation Administration, Service and Methods Demonstration Program (September 1977).

UTTERBACK, J. M., "Innovation in Industry and the Diffusion of Technology," *Science*, 183 (February 15, 1974), 620–26.

______, "The Process of Innovation: A Study of the Origination and Development of Ideas for New Scientific Instruments," *IEEE Transactions on Engineering Management*, Vol. EM-18, No. 4 (November 1971), 124–131.

______, and W. J. ABERNATHY, "A Dynamic Model of Process and Product Innovation," *Omega*, 3, No. 6 (1975), 639–56.

______, and J. W. BROWN, "Monitoring Technological Opportunities," *Business Horizons*, 15 (October 1972), 5–15.

VINSON, W. D., and D. F. HEANY, "Is Quality Out of Control?," *Harvard Business Review*, 55, No. 6 (November-December 1977), 114–22.

VON HIPPEL, E., "Has a Customer Already Developed Your Next Product?," *Sloan Management Review*, 18, No. 2 (Winter 1977), 63–75.

______, "Successful Industrial Products from Consumers' Ideas," *Journal of Marketing*, 42, No. 1 (January 1978), 39–49.

______, "The Dominant Role of the User in Semiconductor and Electronic Subassembly Process Innovation," *IEEE Transactions on Engineering Management* (May 1977).

______, "The Dominant Role of Users in the Scientific Instrument Innovation Process," *Research Policy*, 5 (July 1976), 212–39.

______, "Transferring Process Equipment Innovations from User-Innovators to Equipment Manufacturing Firms," *R&D Management* (October 1977).

VON NEUMANN, J., and O. MORGENSTERN, *The Theory of Games and Economic Behavior*, 2nd ed. (Princeton, NJ: Princeton University Press, 1947).

WACHSLER, R. A., L. G. PRINGLE, and E. I. BRODY, "NEWS: A Systematic Method for Diagnosing New Product Marketing Plans and Developing Actionable Recommendations," Batten, Barton, Durstine, and Osburn, Inc., working paper, New York (February 1972).

WAID, C., D. F. CLARK, and R. L. ACKOFF, "Allocation of Sales Effort in the Lamp Division of General Electric Company," *Operations Research*, 4, No. 6 (December 1956), 629–47.

WAINER, H., "Estimating Coefficients in Linear Models: It Do Not Make No Nevermind," *Psychological Bulletin*, 83 (1976), 213–17.

The Wall Street Journal (October 24, 1966), 32.

WEBSTER, F. E., "New Product Adoption in Industrial Markets, A. Framework for Analysis," *Journal of Marketing*, 33, No. 3 (July 1969), 35–39.

WELLS, W. D., and G. GUBAR, "Life Cycle Concept in Marketing Research," *Journal of Marketing Research*, 3 (November 1966), 355–63.

WHYTE, W. H., "The Web of Word-of-Mouth," *Fortune*, 50, No. 5 (November 1954), 140–43.

WICKER, A. W., "An Examination of the 'Other Variables' Explanation of Attitude-Behavior Inconsistency," *Journal of Personality and Social Psychology*, 19 (1971), 18–30.

WIERENGA, B., *An Investigation of Brand Choice Processes*, (Rotterdam: University of Rotterdam Press, 1974).

WILKIE, W. L., and E. A. PESSEMIER, "Issues in Marketing's Use of Multiattribute Attitude Models," *Journal of Marketing Research*, 10 (November 1973), 428–41.

WILSON, C. L., "Use of Linear Programming to Optimize Media Schedules in Advertising," in *Proceedings of the 46th National Conference of the American Marketing Association*, ed. H. Gomer, pp. 178–91 (1963).

WILSON, N. M., R. W. WEISSBERG, and J. R. HAUSER, "Advanced Dial-a-Ride Algorithms Research Project-Final Report," M. I. T. Department of Civil Engineering, UMTA Grant MA-11-0024, Cambridge, MA (March 1976).

WIND, Y., and L. K. SPITZ, "Analytical Approach to Marketing Decisions in Health-Care Organizations," *Operations Research*, 24, No. 5 (September-October 1976), 973–90.

WITTINK, D. R., "Exploring Territorial Differences in the Relationship Between Marketing Variables," *Journal of Marketing Research*, 14 (May 1977), 145–55.

WOLFE, C. H., *Modern Radio Advertising* (New York: Funk & Wagnalls, 1949).

WONG, Y., "Critical Path Analysis for New Product Planning," *Journal of Marketing*, 28, No. 4 (October 1964), 53–59.

WRIGHT, J. S., D. S. WARNER, W. L. WINTER, JR., and S. K. ZIEGLER, *Advertising*, 4th ed. (New York: McGraw-Hill Book Company, 1977).

YANKELOVICH, SKELLY AND WHITE, *LTM Estimating Procedures* (New York: Yankelovich, Skelly and White, Inc., undated).

YOUNG, S., "Copy Testing Without Magic Numbers," *Journal of Advertising Research*, 12 , No. 1 (February 1972), 3–12.

ZOLTNERS, A. A., "Integer Programming Models for Sales Territory Alignment to Maximize Profit," *Journal of Marketing Research*, 13, No. 4 (November 1976), 426–30.

ZUFRYDEN, F. S., "A Composite Heterogeneous Model of Brand Choice and Purchase Timing Behavior," *Management Science*, 24, No. 2 (October 1977), 121–36.

———, "ZIPMAP—A Zero-One Integer Programming Model for Market Segmentation and Product Positioning," *Journal of the Operational Research Society*, 30, No. 1 (1979), 63–76.

Author Index

D

E

F

G

H

I

J

K

L

M

N

O

P

Q

R

S

T

U

V

W

Y

Z

Concept and Technique Index

A

B

C

D

E

F

G

H

I

L

M

S

T

U

V

W